# More Progresses
## In Analysis

# ore Progresses
# In Analysis

**Proceedings of the 5th International ISAAC Congress**

Catania, Italy     25 – 30 July 2005

Editors

## H. G. W. Begehr
Freie Universität Berlin, Germany

## F. Nicolosi
Università di Catania, Italy

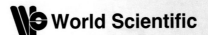 **World Scientific**

NEW JERSEY · LONDON · SINGAPORE · BEIJING · SHANGHAI · HONG KONG · TAIPEI · CHENNAI

*Published by*

World Scientific Publishing Co. Pte. Ltd.

5 Toh Tuck Link, Singapore 596224

*USA office:* 27 Warren Street, Suite 401-402, Hackensack, NJ 07601

*UK office:* 57 Shelton Street, Covent Garden, London WC2H 9HE

**British Library Cataloguing-in-Publication Data**
A catalogue record for this book is available from the British Library.

**MORE PROGRESSES IN ANALYSIS**
**Proceedings of the 5th International ISAAC Congress**

ISBN-13 978-981-283-562-8
ISBN-10 981-283-562-8

Printed in Singapore by B & JO Enterprise

# Preface

The 5th International ISAAC Congress took place from July 25 to 30, 2005 at the University of Catania in Sicily, Italy. There were about 380 participants. The congress was sponsored by
Azienda Provinciale del Turisma di Catania
Banca Monte dei Paschi di Siena
Dipartimento di Mathematica e Informatica dell'Universitá di Catania
Facoltá di Ingegneria dell'Universitá di Catania
Facoltá di Lettre e Filosofia dell'Universitá di Catania
Facoltá di Scienze Matematiche, Fisiche, Naturali dell'Universitá di Catania
ST Microelectronics
Universitá degli Studi di Catania

ISAAC is grateful for the support. One main problem of ISAAC is the lack of money. Most of its members are life members who just pay the life membership fee of - at the time being - € 200. Regular members paying annual fees of € 20, are rare. Therefore the congresses depend on the congress fees and on support from organizations. In time of economical weakness support from societies and companies are difficult to get.

At the opening ceremony two ISAAC awards were presented to
Dimitri V. Georgievskii, Professor at the Mechanical and Mathematical Department of Moscow State University for his contributions in continuum and fluid dynamics,
and to
Terence Tao, Professor at the Department of Mathematics of the University of California, Los Angeles, Fields Medallist from 2006, for his contributions in harmonic analysis and partial differential equations.

The price was equipped with € 500 for the participating first listed price winner. Unfortunately Terence Tao did not come.

At the congress Professor Dr. Oleg Besov from the Steklov Institute of Mathematics in Moscow was honoured as Honorary Member of ISAAC. For a citation see this volume at the beginning. Included are also citations of two eminent ISAAC members on the occasion of their 75th and 65th birthdays, respectively, R.P. Gilbert and V.I. Burenkov. R.P. Gilbert is the founder, Founding President, and now Honorary President, V. I. Burenkov a Vice President and long served board member of ISAAC.

During the congress the ISAAC board has elected Man Wah Wang from York University in Toronto as the new President. By the statutes re-election of President is only once possible. H. Begehr, FU Berlin, became Secretary and Treasurer, S. Zhang, University of Delaware, was confirmed as Secretary and Webmaster. After the congress ISAAC members elected the new Vice Presidents electronically. They are for the 2 years until 2007 E. Brüning, University KwaZulu-Natal, South Africa, V. Burenkov, Cardiff University, Great Britain, S. Saitoh, Gunma University, Japan. Also a new board was elected at the same time. The board members are listed at the ISAAC home page

www.mathisaac.org

Besides this home page there is still the second one

www.math.fu-berlin.de/rd/AG/ISAAC/

The ISAAC board has fixed the site of the next International ISAAC Congress. As suggested by Prof. A.O. Celebi it will be organized in August 2007 at the Middle East Technical University in Ankara.

Besides its congresses ISAAC is organizing, co-organizing and supporting workshops and conferences. In this connection the following events from recent years are to be mentioned:

Recent Trends in Applied Complex Analysis, June 1–5, 2004, Middle East Technical University, Ankara, Turkey (Proceedings are published in the Journal of Applied Functional Analysis 2, 1–3(2007) under the title Snapshot in Applied Complex Analysis, edited by H. Begehr, A.O. Celebi, and R.P. Gilbert),

Pseudo-Differential Operators and Related Topics, June 22–25, 2005, Växjö University, Sweden (Proceedings have appeared as Pseudo-differential operators and related topics. Operator Theory: Advances and Applications 164, Birkhäuser, Basel, 2006, edited by P. Boggiatto, L. Rodino, J. Toft, M.W. Wong).

Analytic Methods of Analysis and Differential Equations (AMADE-2006), September 13–19, 2006, Belarusian State University and Institute of Mathematics of the Belarusian National Academy of Sciences in co-operation with the Moscow State University, in Minsk, Belarus,

ISAAC Workshop on Pseudo-Differential Operators, Partial Differential Equations and Time-Frequency Analysis, December 11–15, 2006, Fields Institute, Toronto (Proceedings are published under title Pseudo-differential operators. Partial differential equations and time-frequency analysis as Fields Inst. Comm. 52(2007), edited by L. Rodino, B.-W. Schulze, M.W. Wong).

Inverse problems, Homogenization and Related Topics in Analysis on the occasion of Professor Robert Gilbert's 75th birthday, January 13 -15, 2007, Department of Mathematics, University of Central Florida, Orlando, Florida,

ISAAC Conference on Complex Analysis, Partial Differential Equations, and Mechanics of Continua in conjunction with the IUTAM Symposium on Relation of Shell, Plate, Beam, and 3D Models both dedicated to the centenary of Ilia Vekua, April 23–27, 2007, I. Vekua Institute of Applied Mathematics of Iv. Javakhishvili Tbilisi State University, Tbilisi, Georgia.

At the Catania congress two more Special ISAAC Interest Groups were established:

Integral Transforms chaired by S. Saitoh and A. Kilbas,

Reproducing Kernels chaired by S. Saitoh and A Berlinet.

Since the Toronto congress the proceedings of the ISAAC congresses are split into two parts. The Special Interest Group on Pseudo-Differential Operators are publishing the proceedings for their session in an own series. The first volume is: Advances in pseudo-differential operators. Operator Theory: Advances and Applications 155, Birkhäuser, Basel, 2004, edited by R. Achino, P. Bogiatto, M.W. Wong. For this Catania congress it is: Modern trends in pseudo-differential operators. Operator Theory: Advances and Applications 172, Birkhäuser, Basel, 2007, edited by J. Toft, M.W. Wong, H. Zhou)

Therefore this session is also not included in this volume.

The sessions of the congress are besides the plenary talks

I.1 Spaces of Differentiable Functions and Applications, organized by V. Burenkov

I.2 Variable Exponent Analysis and Applications, organized by St. Samko

I.3 Reproducing Kernels and Related Topics, organized by D. Alpay, A. Berlinet, S. Saitoh

I.4 Integral Transforms and Applications, organized by A. Kilbas, S. Saitoh, V. Tuan, A.I. Zayed

I.5 Toeplitz and Toeplitz-like Operators, organized by S. Grudski, N. Vasiliveski

I.6 Wavelets, organized by R. Hochmuth, M. Holschneider

I.7 Harmonic analysis and Applications, organized by A. Tabacco, A. Kumar

I.8 Pseudo-Differential Operators, organized by J. Toft, M.W. Wong

I.9 Stochastic Analysis, organized by N. Jacob, Y. Xiao

II.1 Quantitative Analysis of Partial Differential Equations, organized by M. Reissig, J. Wirth

II.2 Boundary Value Problems and Integral Equations, organized by P. Krutitskii

II.3 Elliptic and Parabolic Nonlinear Problems, organized by F. Nicolosi

II.4 Variational Methods for Nonlinear Equations, organized by B. Ricceri

II.5 Asymptotic Stability and Long Time Behaviour of Nonlinear PDE Dynamics, organized by G. Avalos, I. Lasiecka

II.6 Boundary Element Methods, organized by G.C. Hsiao

III.1 Complex Analysis and Potential Theory, organized by M. Lanza de Cristoforis, P. Tamrazov

III.2 Dirac operators in Analysis and Related Topics, organized by J. Ryan, I. Sabadini

III.3 Complex Analysis and Functional Equations, organized by V.V. Napalkov, V.Yu. Novokshenov

III.4 Complex and Functional Analytic Methods in Partial Differential Equations, organized by H. Begehr, D.-Q. Dai, A. Soldatov

III.5 Complex Analytic Methods in the Applied Sciences, organized by V.V. Mityushev, S.V. Rogosin

III.6 Value Distribution Theory and Related Topics, organized by P.C. Hu, P. Li, C.C. Yang

III.7 Geometric Theory of Real and Complex Functions, organized by G. Barsegian

IV.1 Multiscale Modelling and Homogenization, organized by A. Bourgeat, G. Panasenko

IV.2 Mathematical and Computational Aspects of Kinetic Models, organized by A. Majorana

IV.3 Numerical Methods for Kinetic Equations, organized by G. Russo

IV.4 Inverse Problems, Theory and Numerical Methods, organized by M. Klibanov, M. Yamamoto

IV.5 Mathematical Problems in Continuum Mechanics, organized by M. Eglit

IV.6 Mathematical Biology and Medicine, organized by R.P. Gilbert, A. Wirgin, Y. Xu

The session organizers were responsible for collecting the contributions and for getting them refereed. They also were asked to write some introduction to their sessions. Hence detailed prefaces are shifted to the single sessions.

The differently typed manuscripts were again unified in style by Barbara G. Wengel all by herself. This took much more time as expected because her duties at FU Berlin have unexpectedly quickly changed a lot after the editor from FU Berlin was retired in 2004. This is the main reason for the delay of the publication of this volume. The editors thank her for her great enthusiasm. Thanks are due also to the helpers of the organization of the congress.

# CITATIONS

# Professor O.V. Besov Honorary Member of ISAAC

Victor I. Burenkov

Professor O.V. Besov is an outstanding Russian mathematician whose name is well-known in the world of Analysis. It suffices to recall "Besov spaces" named after him which are widely used in real analysis (especially in the theory of function spaces, the interpolation theory and the approximation theory), partial differential equations (especially in the theory of boundary value problems), numerical methods and other mathematical dicsiplines.

Besov spaces were introduced in the sixties of the twentieth century and allowed O.V. Besov to give a complete description of the traces of functions in Sobolev spaces.

For further fourty years he, together with many other mathematicians throughout the world, continued investigation of these and other related spaces and obtained a number of seminal results, which were partially summarised in the famous book by the three authors O.V. Besov, V.P. Il'in, S.M. Nikol'skii "Integral representations of functions and embedding theorems".

O.V. Besov is a corresponding member of the Russian Academy of Sciences, works at the Steklov Mathematical Institute (Moscow) for more than fourty years and is currently the head of the department of the theory of functions of that institute, works part time as a professor of the department of mathematics of the Moscow Institute of Physics and Technology (also for more than fourty years), is a member of the editorial boards of several journals.

He has published more that 100 papers, was awarded a number of prizes, was an invited speaker at the International Congress of Mathematicians (Nice, 1970), a plenary speaker at many international conferences, in particular at the 5th Congress of the ISAAC.

# A TRIBUTE TO THE 65$^{th}$-BIRTHDAY OF PROF. VICTOR I. BURENKOV

Massimo Lanza de Cristoforis

Professor Victor Ivanovich Burenkov has completed his post-graduate studies in Moscow in 1967 under the guidance of S.M. Nikol'skii, a world leader in Function Space Theory, who had been in turn a student of A.N. Kolmogorov.

For about forty years Prof. Burenkov has taken part in the same Moscow seminar that eminent mathematicians such as O.V. Besov and S.L. Sobolev have attended.

Because of his international reputation, in the mid nineties the University of Cardiff has offered him a position.

Prof. Burenkov is a world leader in the field of function spaces, a subject of great importance in real analysis, and in functional analysis, and in particular in approximation theory, and in harmonic analysis, and in the theory of partial differential equations. A subject on which eminent mathematicians such as O.V. Besov, A.P. Calderon, J.L. Lions, V.G. Maz'ya, S.M. Nikol'skii, S.L. Sobolev, E.M. Stein, H. Triebel, A. Zygmund have worked.

Prof. Burenkov has authored or co-authored more than 100 scientific publications and has obtained seminal results.

He has developed the method of mollifiers with variable step, which has enabled to prove a series of significant results for the problem of approximating functions by smooth functions.

He has obtained deep results in the construction of optimal extension operators and in the two-sided estimation of the minimal norm of an extension operator, which have made him a world leader in the field.

Prof. Burenkov has also obtained important and sharp results in the field of integral inequalities for the derivatives of a function, and important results on Fourier multipliers in weighted Lebesgue spaces. His elegant and powerful results reveal a deep insight into mathematical knowledge.

An important feature of his research is its universality, and he has always been interested in effective applications of his work. He has found applications of function space theory to several fields such as the theory of hypoelliptic equations, and the theory of ill-posed problems for integral equations, and spectral theory of differential operators, and also to quantum mechanics.

Prof. Burenkov is well known not only for the level of his research, but also for his expository qualities and for his ability as a teacher. His lectures are characterized by an extreme care to make even technically difficult material easily accessible to a large audience. He is known for his ability to communicate the central ideas of mathematical issues. He has a considerable production of educational nature. His book 'Sobolev spaces on domains' is considered to be one of the best on the subject, and is suitable both to experts and to beginners. His ability as a teacher has been long recognized and he has been invited to deliver series of lectures in many institutions world wide. In particular, he has delivered the lecture course "The main ideas in the theory of Sobolev spaces" in more than 10 Universities throughout the world.

Prof. Burenkov has invested an immense effort in educating generations of young mathematicians, and about twenty students have obtained a PhD under his guidance, and many of those have been collaborating with him even after graduation.

Prof. Burenkov is an enthusiastic promoter of analysis all over the world. In particular, he has attended all five ISAAC Congresses, and has jointly organized with Prof. S. Samko the session 'Function spaces and applications', one of the most successful ones. Well deservedly has he been elected Vice-President of ISAAC, the International Society for Analysis, its Applications and Computation.

Prof. Burenkov has also served as a member of the editorial board of several journals and book series, and has devoted his energy to the academic life of the instituions he has belonged to.

Recently, Prof. Burenkov has been awarded a Honorary Professorship at the L.N. Gumilyov Eurasian National University of the Republic of Kazakhstan, a deserved honor for an outstanding mahematician of uncommon human qualities who has produced an enormous quantity of work and who has devoted his life to research and diffusion of mathematics.

# Robert Pertsch Gilbert: Citation for his 75th Birthday

Heinrich Begehr

Professor Gilbert is the leading analyst in the field of complex analytic methods for partial differential equations throughout the world. In a natural way these methods were applied long time ago in mathematical physics as e.g. for potential theory and fluid flows. In the 30th and 40th of the last century mainly in the Soviet Union and in the USA this area was intensively developed through e.g. I. Muskhelishvili, I.N. Vekua, L. Bers, W. Haack and others. On one hand the applications to elasticity theory and shell theory on the other the treatment of certain elliptic systems and equations in the plane via complex analysis had shown how powerful and elegant complex methods are.

As a young mathematician with a physical background R.P. Gilbert has started in the late 50s with his investigation of singularities of solutions to certain differential equations in higher dimensions in particular for GASP (generalized axially symmetric potentials). In connection with GASP-Theory he has studied Bergman integral operators and Riemann functions. This has led him to the so-called "method of ascent" and the Bergman-Gilbert operator. This latter has served R. Carroll as a motivation for his general "transmutation theory". After one decade of research he had become known to the Georgian school around I. Muskhelishvili and I.N. Vekua already. This contact was the beginning of his important world wide international cooperation.

In connection with the Ph.D. project of G.N. Hile he initiated the theory of generalized hyperanalytic functions, describing the theory of solutions to first order elliptic systems of 2n equations in the plane. Moreover, basic integral representation formulas in Clifford analysis have been given in this thesis. This theory was later further developed by the Gent school around R. Delanghe. And it was R.P. Gilbert who in the eighties made this Clifford analysis popular in mainland China. His 1983 jointly written monograph appeared at the beginning of the newly risen interest in Clifford analysis. It also has anticipated later interest in applying function theory of several complex variables to partial differential equations.

Besides having graduated in mathematics R.P. Gilbert has also graduated in physics. Hence, he always was and is interested in applied problems especially from mathematical physics. He has studied problems in fluid dynamics, underwater acoustics, nonlinear waves, Hele-Shaw flows, planar filtration, porous media, biological mechanics. The methods applied are from complex analysis, potential theory, inverse problems as e.g. inverse scattering, homogenization, approximation theory, numerical analysis.

Many of his publications are joint ones with a variety of co-authors from all over the world. He has built these international co-operations with visiting prestigious institutions often on the basis of highly ranked awards like the Alexander von Humboldt Senior Scientist Award and the British Science Council Research Award. Through his grants many of his coworkers have visited the University of Delaware. A second group originated from his former Ph.D.-students. At present he is leading a strong USA-French research group on underwater acoustics. This group is very productive and its publications culminate in various book projects.

More than many other scientists Dr. Gilbert takes care of his students in particular his Ph.D. students. Instead of just advising them he immediately starts to collaborate with them and this collaboration often continues beyond the day they receive their Ph.D. degrees. Many of his co-authors are thus former Ph.D. students of his. In the just mentioned USA-French research group two of his former Ph.D. students are involved.

R.P. Gilbert has a deep interest in teaching. Once in the eighties visiting the University of Delaware myself I was witness of his teaching on symbolic computation - at that time using Macsyma - applied to problems from differential equations. His enthusiasm in this subject led to several book publications and one of the co-authors, W. Koepf, before a pure function theorist, became a specialist in symbolic computation and only because of this has found a tenure position at a German university.

Besides his eminent, important and numerous scientific work including several monographs, course books and edited proceedings initiated and co-organized by him beginning as early as 1970, R.P. Gilbert has two more great contributions for the mathematical community. They both originated from his desire to promote the area of analysis in a time where discrete mathematics and numerical methods attract the young generation of mathematicians. One of these contributions is the foundation of two journals "Applicable Analysis" and "Complex Variables, Theory and Applications". Years before the majority of grants for mathematicians in the US were shifted from pure to applied mathematics he had the vision of the importance of applied mathematics in conjunction to pure mathematics. Both journals have shown to be perfect junction between pure and applied mathematics and proved to be very successful scientific journals (despite their extremely high prices). The same philosophy led R.P. Gilbert to found ISAAC, the International Society for Analysis, its Applications and Computation. Analysis with its applications and together with computational methods has to be considered as a unity. To promote this area and to attract young mathematicians to this field regularly international congresses are organized, thus opening a certain window on this huge field in mathematics. Moreover, at each congress, young talented researchers are awarded for their achievements in analysis, its applications and computation. Between these congresses workshops and special conferences are organized. How successful ISAAC is can be seen from their publications. In the ISAAC series with Kluwer 10 proceedings volumes have appeared since 1998, 5 more with other publishers.

R.P. Gilbert has become a central figure in mathematical analysis. He is the world leading expert in complex analysis devoted to partial differential equations. Since decades he is leading an international, very productive research team. He has attained some results in analysis with will forever be linked to his name. His contributions to the mathematical community in the form of his journals and the ISAAC distinguish him from all other mathematicians of our time. R.P. Gilbert is a remarkable analyst with great merits for the mathematical science.

# CONTENTS

Preface                                                                                    v

**Citations**                                                                              ix

Professor O.V. Besov Honorary Member of ISAAC                                              xi
  *Victor I. Burenkov*
A Tribute to the 65th Birthday of Prof. Victor I. Burenkov                                 xiii
  *Massimo Lanza de Cristoforis*
Robert Pertsch Gilbert: Citation for his 75th Birthday                                     xv
  *Heinrich Begehr*

**Plenary Lectures**                                                                       1

Carleman estimates and applications to uniqueness of the continuation
and inverse problems                                                                       3
  *V. Isakov*
From microelectronics to nanoelectronics: mathematical challenges                          13
  *A.M. Anile and V. Romano*
A characterization of the Calderón projector for the biharmonic equation                   23
  *G.C. Hsiao and W.L. Wendland*
Shock reflection by obstacle                                                               39
  *S. Chen*
Weighted function spaces with constant and variable smoothness                             55
  *O.V. Besov*

**I.1 Spaces of Differentiable Functions and Applications
(V.I. Burenkov)**                                                                          67

Regularization of the Cauchy problem for the system of elasticity
theory in $R^m$                                                                            69
  *O.I. Makhmudov and I.E. Niyozo*
First Lyapunov method for the abstract parabolic equation                                  83
  *V.A. Trenogin*

**I.2 Variable Exponent Analysis and Applications (St. Samko)**     91

Some results on variable exponent analysis     93
    *X.-L. Fan*

Further results on variable exponent trace spaces     101
    *L. Diening and P. Hästö*

Variable exponent spaces on metric measure spaces     107
    *T. Futamura, P. Harjulehto, P. Hästö, Y. Mizuta and T. Shimomura*

**I.3 Reproducing Kernels and Related Topics**
**(D. Alpay, A. Berlinet, S. Saitoh)**     123

Topics on the Bergman kernel for some balls     125
    *K. Fujita*

Applications of reproducing kernels to linear singular integral equations
through the Tikhonov regularization     135
    *H. Itou and S. Saitoh*

Reproducting kernel Hilbert spaces and random measures     143
    *C. Suquet*

Reproducing kernels in probability and statistics     153
    *A. Berlinet*

Splines with non positive kernels     163
    *S. Canu, C.S. Ong and X. Mary*

Algebraic, differential, integral and spectral properties of Mercer-like-
kernels     175
    *J. Buescu and A.C. Paixão*

**I.4 Integral Transforms and Applications**
**(A. Kilbas, S. Saitoh, V. Tuan, A.I. Zayed)**     189

Analytical and numerical real inversion formulas of the
Laplace transform     191
    *T. Matsuura and S. Saitoh*

Integral transform with the extended generalized Mittag-Leffler function     201
    *A.A. Kilbas and A.A. Koroleva*

Relationships between conditional Fourier-Feynman transform and
conditional convolution product of unbounded functions over Wiener
paths in abstract Wiener space     211
    *B.I. Seung and D.H. Cho*

Change of scale formulas for Wiener integrals and Fourier-Feynman
transforms     221
   *I. Yoo, T.S. Song, B.S. Kim and K.S. Chang*

Sobolev type spaces associated with the Kontorovich-Lebedev transform     231
   *S.B. Yakubovich*

Fourier type analysis and quantum mechanics     241
   *S. Watanabe*

The singular value decomposition for generalized transform of
randon type in $R^n$     251
   *J. Wang*

Eye direction by stereo image processing     259
   *K. Tsuji and M. Aoyagi*

Some integral equations with modified Bessel functions     269
   *J.M. Rappoport*

The zeta function for learning theory and resolution of singularities     279
   *M. Aoyagi and S. Watanabe*

The background and survey of recent results in the theory of functions
of $\omega$-bounded type in the half-plane     289
   *A.M. Jerbashian*

Fractional modeling and applications     301
   *S. Kempfle, K. Krüger and I. Schäfer*

Euler-type fractional differential equations     311
   *A.A. Kilbas, M. Rivero and J.J. Trujillo*

**I.5 Toeplitz and Toeplitz-like Operators (S. Grudski, N. Vasilevski)**     325

Local properties of the Segal-Bargmann projection and $\Psi_0$-algebras     327
   *W. Bauer*

$C^*$-algebras of Bergman type operators with piecewice continuous
coefficients on bounded domains     339
   *Yu.I. Karlovich and L.V. Pessoa*

A criterion for lateral invertibility of matrix Wiener-Hopf plus
Hankel operators with good Hausdorff sets     349
   *A.P. Nolasco and L.P. Castro*

Potential type operators on Carleson curves acting on weighted
Hölder spaces     359
   *V. Rabinovich*

Upper and lower indices of a certain class of monotonic functions in
connection with Fredholmness of singular integral operators 369
  *N. Samko*

The asymptotic behavior of the trace of generalized truncated integral
convolutions 377
  *O.N. Zabroda*

Multiplication and Toeplitz operators on the analytic Besov spaces 387
  *N. Zorboska*

## I.6 Wavelets (R. Hochmuth, M. Holschneider) 397

S-asymptotic and S-asymptotic expansion of distributional wavelet
transform 399
  *K. Saneva and A. Bučkovska*

A wavelet-based vectorial approach for an integral formulation of
antenna problems 407
  *F.P. Andriulli, A. Tabacco and G. Vecchi*

New reproducing subgroups of $Sp(2,\mathbb{R})$ 415
  *E. Cordero, F. De Mari and A. Tabacco*

## I.8 Pseudo-Differential Operators (J. Toft, M.W. Wong) 421

$L^2$ stability and boundedness of the Fourier integral operators applied
to the theory of the Feynman path integral 423
  *W. Ichinose*

Smooth functional derivatives in Feynman path integrals by time
slicing approximation 429
  *N. Kumano-Go and D. Fujiwara*

## I.9 Stochastic Analysis (N. Jacob, Y. Xiao) 439

Identification and series decomposition of anisotropic Gaussian fields 441
  *A. Ayache, A. Bonami and A. Estrade*

Subordination in fractional diffusion processes via continuous time
random walk 451
  *R. Gorenflo, F. Mainardi and A. Vivoli*

Functional spaces and operators connected with some Lévy white noises 467
  *E. Lytvynov*

**II.1 Quantitative Analysis of Partial Differential Equations**
**(M. Reissig, J. Wirth)** 481

Identification of linear dynamic systems 483
    *A. Kryvko and V.V. Kucherenko*

Strongly hyperbolic complex systems, reduced dimension and Hermitian
systems II 493
    *J. Vaillant*

Stability of stationary solutions of nonlinear hyperbolic systems with
multiple characteristics 511
    *A. Kryvko and V.V. Kucherenko*

The fundamental solution for one class of degenerate elliptic equations 521
    *M.S. Salakhitdinov and A. Hasanov*

On the spectrum of Schrödinger operators with oscillating long-range
potenials 533
    *K. Mochizuki*

Regular global solutions of semi-linear evolution equations with singular
pseudo-differential principal part 543
    *D. Gourdin, H. Kamoun and O.B. Khalifa*

Asymptotic behaviour for Kirchhoff equation 553
    *T. Matsuyama*

Some results on global existence and energy decay of solutions to the
Cauchy problem for a wave equation with a nonlinear dissipation 561
    *A. Benaissa*

Non-negative solutions of the Cauchy problem for semilinear wave
equations and non-existence of global non-negative solutions 571
    *H. Uesaka*

On the large time behavior of solutions to semilinear systems of the wave
equation 581
    *S. Katayama and H. Kubo*

Levi conditions for higher order operators with finite degeneracy 591
    *F. Colombini and G. Taglialatela*

A theory of diagonalized systems of nonlinear equations
and application to an extended Cauchy-Kowalevsky theorem 603
    *S. Miyatake*

## II.2 Boundary Value Problems and Integral Equations (P. Krutitskii) — 615

Influence of Signorini boundary condition on bifurcation in
reaction-diffussion systems — 617
  *M. Kučera*

Wave propagation in a 3-d optical waveguide II: Numerical examples — 627
  *G. Ciraolo and O. Alexandrov*

On a Dirichlet boundary value problem for coupled second order
differential equations — 637
  *M. Hihnala and S. Seikkala*

Existence of a classical solution and non-existence of a weak solution
to the harmonic Dirichlet problem in a planar domain with cracks — 647
  *P.A. Krutitskii*

## II.3 Elliptic and Parabolic Nonlinear Problems (F. Nicolosi) — 657

Harnack inequalities for energy forms on fractals sets — 659
  *M.A. Vivaldi*

Elliptic systems in divergence form with discontinuous coefficients — 669
  *A. Tarsia*

Directional localization of solutions to elliptic equations
with nonstandard anisotropic growth conditions — 681
  *S.N. Antontsev and S.I. Shmarev*

Bifurcation direction and exchange of stability for an elliptic unilateral BVP — 691
  *J. Eisner, M. Kučera and L. Recke*

Non local Harnack inequalities for a class of partial differential equaitons — 701
  *U. Boscain and S. Polidoro*

Γ-convergence for strongly local Dirichlet forms in open sets with holes — 711
  *M. Biroli and N.A. Tchou*

Mountain pass techniques for some classes of nonvariational problems — 721
  *M. Girardie, S. Mataloni and M. Matzeu*

On some Schrödinger type equations — 729
  *S. Polidoro and M.A. Ragusa*

Existence and uniqueness of classical solutions to certain nonlinear parabolic
integrodifferential equations and applications — 737
  *D.R. Akhmetov, M.M. Lavrentiev, Jr. and R. Spigler*

On existence and asymptotic behavior of the solutions of quasilinear
degenerate parabolic equations in unbounded domains    747
   *S. Bonafede and F. Nicolosi*

Multiple solitary waves for non-homogeneous Klein-Gordon-Maxwell
equations    753
   *A.M. Candela and A. Salvatore*

A quadratic Bolza-type problem in stationary space times with critical
growth    763
   *R. Bartolo, A.M. Candela and J.L. Fiores*

Regularity of minimizers of some degenerate integral functionals    771
   *V. Cataldo, S. D'Asero and F. Nicolosi*

On the optimal retraction problem for $\gamma$-Lipschitz mappings and applications    775
   *G. Trombetta*

Some remarks on Nirenberg-Gagliardo interpolation inequality in
anisotropic case    785
   *F. Nicolosi and P. Cianci*

Hausdorff dimension of singular sets of Sobolev functions and applications    793
   *D. Žubrinić*

**II.4 Variational Methods for Nonlinear Equations (B. Ricceri)**    803

Elliptic eigenvalue problems on unbounded domains involving sublinear terms    805
   *A. Kristály*

Elliptic boundary value problems involving oscillating nonlinearities    815
   *G. Anello*

Infinitely many solutions for the Dirichlet problem for the p-Laplacian    823
   *F. Cammaroto, A Chinnì and B. Di Bella*

Infinitely many solutions to Dirichlet and Neumann problems for
quasilinear elliptic systems    833
   *A.G. Di Falco*

Multiplicity results for a Neumann-type problem involving the p-Laplacian    843
   *F. Cammaroto, A. Chinnì and B. Di Bella*

Multiple solutions to a class of elliptic differential equations    853
   *G. Cordaro*

One can hear the shapes of some non-convex drums    863
   *W. Matsumoto*

Multiplicity results for two points boundary value problems    873
   *A. Chinnì, F. Cammaroto and B. Di Bella*

Multiple solutions for an ordinary second order system     881
    R. Livrea

Some multiplicity results for second order non-autonomous systems     889
    F. Faraci

A purely vectorial critical point theorem     897
    B. Ricceri

On minimization in infinite dimensional Banach spaces     901
    E. Brüning

Existence results for nonlinear hemivariational inequalities     921
    P. Candito

**III.1 Complex Analysis and Potential theory**
**(M. Lanza de Cristoforis, P. Tamrazov)**     929

Moser's conjecture on Grunsky inequalities and beyond     931
    S. Krushkal

Finely meromorphic functions in contour-solid problems     945
    T. Aliyev Azeroglu and P.M. Tamrazov

A singular domain perturbation problem for the Poisson equation     955
    M. Lanza de Cristoforis

Iteration dynamical system of discrete Laplacians and the evolution
of extinct animals     967
    K. Kosaka and O. Suzuki

Commutative algebras of hypercomplex monogenic functions and solutions
of elliptic type equations degenerating on an axis     977
    S. Plaksa

Quaternionic background of the periodicity of petal and sepal structures
in some fractals of the flower type     987
    J. Lawrynowicz, St. Marchiafava and M. Nowak-Kȩpczyk

**III.2 Dirac operators in Analysis and Related Topics**
**(J. Ryan, I. Sabadini)**     997

On some relations between real, complex and quaternionic linear spaces     999
    M.E. Luna-Elizarrarás and M. Shapiro

Holomorphic functions and regular quaternionic functions
on the hyperkähler space $\mathbb{H}$     1009
    A. Perotti

A new Dolbeault complex in quaternionic and Clifford analysis    1019
  *F. Colombo, I. Sabadini, A. Damiano and D.C. Struppa*

Some integral representations and its applications in Clifford analysis    1033
  *Z.X. Zhang*

Clifford algebra applied to the heat equation    1043
  *P. Cerejeiras and F. Sommen*

Cauchy-type integral formulas for *k*-hypermonogenic functions    1051
  *S.-L. Eriksson*

Deconstructing Dirac operators I: Quantitative Hartogs-Rosenthal theorems    1065
  *M. Martin*

Hypermonogenic and holomorphic Cliffordian functions    1075
  *E. Lehman*

A fractal renormalization theory of infinite dimensional Clifford algebra
and renormalized Dirac operator    1085
  *J. Lawrynowicz, K. Nôno and O. Suzuki*

Analytic functions in algebras    1095
  *Y. Krasnov*

## III.4 Complex and Functional Analytic Methods in Partial Differential Equations (H. Begehr, D.-Q. Dai, A. Soldatov)    1107

On distribution of zeros and asymptotics of some related quantities
for orthogonal polynomials on the unit circle    1109
  *Z.H. Du and J.Y. Du*

Representation of pseudoanalytic functions in the space    1119
  *P. Berglez*

Hilbert boundary vlaue problem for a class of metaanalytic
functions on the unit circumference    1127
  *Y.F. Wang*

Four boundary value problems for the Cauchy-Riemann
equation in a quarter plance    1137
  *S.A. Abdymanapov, H. Begehr, G. Harutyunyan and A.B. Tungatarov*

Mixed boundary value problem for inhomogeneous
poly-analytic-harmonic equation    1149
  *A. Kumar and R. Prakash*

Initial value problem for a system of equations of crystal optics    1163
  *N.A. Zhura*

On representation of solutions of second order elliptic systems
in the plane   1171
  *A.P. Soldatov*

Quantitative transfer of smallness for solutions of elliptic equations
with analytic coefficients and their gradients   1185
  *E. Malinnikova*

The solution of spectral problems for the curl and Stokes operators with
periodic boundary conditions and some classes of explicit solutions of
Navier-Stokes equations   1195
  *R.S. Saks*

About one class of linear first order overdetermined systems
with interior singular and super-singular manifolds   1207
  *N. Rajabov*

On the construction of the general solutions of the classes of
Abel's equations of the second kind   1219
  *D.E. Panayotounakos*

Qualitative theory of nonlinear ODE's in algebras   1229
  *Y. Krasnov and S. Zur*

**III.5 Complex Analytic Methods in the Applied Sciences
(V.V. Mityushev, S.V. Rogosin)**   1241

The Weierstrass $\mathcal{P}$-function is not topologically elementary   1243
  *V.V. Mityushev and S.V. Rogosin*

On application of the monotone operator method to solvability of nonlinear
singular integral equations   1247
  *S.V. Rogosin*

Analytical and numerical results for the effective conductivity of 2D
composite materials with random position of circular reinforcements   1259
  *E.V. Pesetskaya, T. Fiedler, A. Öchsner, J. Grácio and S.V. Rogosin*

Strains in tissue development: A vortex description   1271
  *R. Wojnar*

An analytic solution of the Ornstein-Zernike equation for a fluid and
its application   1283
  *M. Yasutomi*

**III.6 Value Distribution Theory and Related Topics
(P.C. Hu, P. Li, C.C. Yang)**                                   1293

Finite Fourier transforms and the zeros of the Riemann $\xi$-function II    1295
  G. Csordas and C.-C. Yang
Condition pour les zéros de la fonction holomorphe, bornée et second
problém de Cousin dans le bidisque-unité de $\mathbb{C}^2$              1303
  K. Katô
Criteria for biholomorphic convex mappings on $p$-ball in $C^n$         1311
  M.-Sh. Liu

**III.7 Geometric Theory of Real and Complex Functions
(G. Barsegian)**                                                1321

Radial cluster set of a bounded holomorphic map in the unit ball of $\mathbb{C}^n$   1323
  T. Matsushima

**IV.2 Mathematical and Computational Aspects of Kinetic Models
(A. Majorana )**                                                1327

Efficiency considerations for the Boltzmann-Poisson system for
semiconductors                                                  1329
  A. Domaingo and A. Majorana
Kinetic theory applications in evaporation/condensation flows of
polyatomic gases                                                1339
  A. Frezzotti
Maths against cancer                                            1351
  F. Pappalardo, S. Motta, P.-L. Lollini and E. Mastriani

**IV.4 Inverse Problems, Theory and Numerical Methods
(M. Klibanov, M. Yamamoto)**                                    1361

Lipschitz stability in an inverse hyperbolic problem by boundary
observations                                                    1363
  M. Bellassoued and M. Yamamoto
Numerical Cauchy problems for the Laplace equation              1375
  T. Matsuura, S. Saitoh and M. Yamamoto
Conditional stability in reconstruction of initial temperatures  1385
  M. Yamamoto and J. Zou

**IV.6 Mathematical Biology and Medicine
(R.P. Gilbert, A. Wirgin. Y. Xu)**                                    1391

Inverse problem for wave propagation in a perturbed layered
half-space with a bump                                               1393
   *R.P. Gilbert, N. Zhang, N. Zeev and Y. Xu*

A time domain method to model viscoelastic wave propagation
in long cortical bones                                               1407
   *J.-P. Groby, E. Ogam, A. Wirgin, Z.E.A. Fellah and C. Tsogka*

Eigenmode analysis of the cortical osseous tissue-marrow coupled system   1419
   *E. Ogam, Z.E.A. Fellah, J.-P. Groby and A. Wirgin*

An inverse problem for the free boundary model of ductal carcinoma
*in situ*                                                            1429
   *Y.S. Xu*

Comparing mathematical models of the human liver based on BSP test   1439
   *L. Čelechovská-Kozáková*

Implementation of adaptive randomizations for clinical trials        1449
   *E.R. Miller*

List of Session Organizers                                           1461

List of Authors                                                      1463

# PLENARY LECTURES

# CARLEMAN ESTIMATES AND APPLICATIONS TO UNIQUENESS OF THE CONTINUATION AND INVERSE PROBLEMS

VICTOR ISAKOV*

*Department of Mathematics and Statistics*
*Wichita State University*
*Wichita, KS 67260, USA*
*E-mail: victor.isakov@wichita.edu*

We give some recent results on Carleman type estimates for systems of partial differential equations with emphasis on applications to continuum mechanics. In particular we discuss isotropic elasticity with residual stress. We show how to derive stability estimates in the Cauchy problem and we study increased stability for the Helmholtz equation. We formulate uniqueness and stability of determining residual stress from few boundary data.

**Key words:** General existence and uniqueness theorems, inverse problems
**Mathematics Subject Classification:** Primary: 35A05; Secondary: 35R30.

## 1. Carleman estimates

Let $\mathbf{A}$ be the $m \times m$-matrix linear partial differential operator with $L_\infty(\Omega)$-coefficients and with the principal part

$$\begin{pmatrix} A(1) \ldots & 0 \\ & \ldots & \\ 0 & \ldots & A(m) \end{pmatrix}$$

where $A(1), ..., A(m)$ are scalar linear partial differential operators of second order with $C^1(\overline{\Omega})$ real-valued coefficients in a bounded Lipschitz domain $\Omega \subset \mathbf{R}^n$. Let $\mathbf{u} = (u_1, ..., u_m)$. We remind that $H_{(k)}(\Omega)$ is the Sobolev space with the norm $\| \cdot \|_{(k)}$, $\partial_j = \frac{\partial}{\partial x_j}$ and $\nu$ is the outer unit normal to $\partial\Omega$. By $C$ we denote constants which depend only on $\mathbf{A}, \Omega$, and $\Gamma$. Any additional dependence will be indicated.

Let $\psi \in C^2(\overline{\Omega})$, $\nabla\psi \neq 0$ in $\overline{\Omega}$. We will use

$$\varphi = e^{\sigma\psi} \tag{1}$$

**Theorem 1.1.** *Let $\psi$ be pseudo-convex and $\nabla\psi$ be noncharacteristic with respect to $A(1), ..., A(m)$ in $\overline{\Omega}$. Then there are $C, \sigma < C$ such that*

$$\int_\Omega (\tau^3|\mathbf{u}|^2 + \tau|\nabla\mathbf{u}|^2)e^{2\tau\varphi} \leq C(\int_\Omega |\mathbf{A}\mathbf{u}|^2 e^{2\tau\varphi} + \int_{\partial\Omega} (\tau^3|\mathbf{u}|^2 + \tau|\nabla\mathbf{u}|^2)e^{2\tau\varphi}) \tag{2}$$

*Work partially supported by grant DMS 04-05976 of the National Science Foundation.

*for all* $\mathbf{u} \in H_{(2)}(\Omega)$ *when* $C < \tau$.

The weighted ($L^1$-)estimates (2) were introduced in 1939 by T. Carleman to obtain first uniqueness of the continuation results for (systems of) partial differential equations with nonanalytic coefficients [2].

Under strong pseudo-convexity assumptions this result for general scalar partial differential operators was proven by Hörmander [4] in 1962 and the most general anisotropic case is considered by Isakov in 1980 [8] (on compactly supported $u$). General isotropic Carleman estimates with boundary terms were given by Tataru [14]. Relaxation of strong pseudo-convexity for second order operators can be found for example in [9]. For hyperbolic operators and particular weights (2) with (semi)explicit boundary terms is obtained in [13]. Some basic ideas of the proof are given in an outline of the proof of Theorem 2.2. In general, available proofs are technically involved since they use techniques of differential quadratic forms or a refined form of the Garding's inequality.

A significance of Theorem 1.1 is due to the fact that several classical systems of partial differential equations of mathematical physics (like isotropic Maxwell's and elasticity systems) can be transformed into principally diagonal systems. Hence one can derive from (2) important results about uniqueness and stability of the continuation and controllability of such systems [3], as well as uniqueness and stability of determination of source terms and coefficients from additional lateral boundary data.

As an interesting anisotropic example we consider a (principally non-diagonalizable) system of isotropic elasticity with residual stress. In Theorem 1.2 $n = 4, x_4 = t, x' = (x_1, x_2, x_3), \mathbf{u} = (u_1, u_3, u_3)$ is the displacement vector of elastic medium and $\nabla, div, \mathbf{e}_1, \mathbf{e}_2, \mathbf{e}_3$ are operators of vector analysis and the standard base in $\mathbf{R}^3$.

We introduce

$$\mathbf{A}_{e,R}\mathbf{u} = \rho\partial_t^2\mathbf{u} - \mu\Delta\mathbf{u} - \nabla(\lambda div\mathbf{u}) - \sum_{j=1}^{3} \nabla\mu \cdot (\nabla u_j + \partial_j\mathbf{u})\mathbf{e}_j + \mathbf{R}\mathbf{u} \qquad (3)$$

where positive $\lambda, \lambda + \mu, \rho \in C^4(\overline{\Omega})$, and the residual stress operator

$$\mathbf{R}\mathbf{u} = \sum_{j,k=1}^{3} r_{jk}\partial_j\partial_k\mathbf{u}, \ r_{jk} = r_{kj}.$$

The residual stress is assumed to be divergence free:

$$\nabla \cdot \mathbf{R} = 0 \quad \text{in} \quad \Omega. \qquad (4)$$

We will make the following physically motivated assumption

$$|r_{jk}|_1(\Omega) < \varepsilon_0. \qquad (5)$$

**Theorem 1.2.** *Let $\psi$ be pseudo-convex and $\nabla\psi$ be noncharacteristic with respect to the scalar operators $\rho\partial_t^2 - \mu\Delta, \rho\partial_t^2 - (\mu + 2\lambda)\Delta$ on $\overline{\Omega}$.*

*Then there are $\varepsilon_0(\Omega, \rho, lambda, \mu), C, \sigma < C$, such that*

$$\int_\Omega (\tau^2 |\mathbf{u}|^2 + |div\mathbf{u}|^2 + |curl\mathbf{u}|^2 + \tau^{-1}|\nabla \mathbf{u}|^2)e^{2\tau\varphi} \leq C \int_\Omega |\mathbf{A}_{e,R}\mathbf{u}|^2 e^{2\tau\varphi}, \mathbf{u} \in H^0_{(2)}(\Omega)$$

(6)

*when $C < \tau$.*

Proofs [7], [11], use reduction to the extended "upper diagonal" system for $(u_1, ..., u_7)$ where $u_4 = div\mathbf{u}$, $(u_5, u_6, u_7) = curl\mathbf{u}$ and Carleman estimates in Sobolev spaces of negative order.

## 2. Uniqueness and stability of the continuation

We consider the Cauchy problem

$$\mathbf{Au} = \mathbf{f} \text{ in } \Omega, \ \mathbf{u} = \mathbf{g}_0, \partial_\nu \mathbf{u} = \mathbf{g}_1 \text{ on } \Gamma \subset \partial\Omega.$$

(7)

Let $\Omega_\varepsilon = \Omega \cap \{\varepsilon < \psi\}$.

**Theorem 2.1.** *Let $\psi$ satisfy the conditions of Theorem 1.1 and*

$$\psi < 0 \text{ on } \partial\Omega \setminus \Gamma.$$

(8)

*Then there are $C, \kappa \in (0, 1)$ depending only on $\Omega, \Gamma, \mathbf{A}, \varepsilon$, such that*

$$\|\mathbf{u}\|_{(1)}(\Omega_\varepsilon) \leq C(F + \|\mathbf{u}\|_{(1)}^{1-\kappa}(\Omega)F^\kappa)$$

(9)

*for any solution $\mathbf{u}$ to (7).*

Here $F = \|\mathbf{f}\|_{(0)}(\Omega) + \|\mathbf{g}_0\|_{(1)}(\Gamma) + \|\mathbf{g}_1\|_{(0)}(\Gamma)$.

**Proof** Let $\chi$ be a $C^\infty$ function that is 1 on $\Omega_{\varepsilon/2}$ and 0 near $\partial\Omega \setminus \Gamma$, $0 \leq \chi \leq 1$. We can choose $\chi$ so that $|\partial^\beta \chi| \leq C\varepsilon^{-|\beta|}, |\beta| \leq 2$.

Using Leibniz' formula for the differentiation of the product we conclude that $\mathbf{A}(\chi\mathbf{u}) = \chi\mathbf{Au} + \mathbf{A}_1\mathbf{u}$, where the last operator involves only $\partial^\alpha \mathbf{u}$ with $|\alpha| \leq 1$ and has coefficients bounded by $C(\varepsilon)$. Observe that $\mathbf{A}_1 = 0$ on $\Omega_{\frac{\varepsilon}{2}}$. By applying to $\chi\mathbf{u}$ the estimate (2), shrinking the integration domain on the left side to $\Omega_\varepsilon$ and using that $\chi = 1$ on $\Omega_\varepsilon$, we get

$$\sum_{|\alpha| \leq 1} \|e^{\tau\varphi}\partial^\alpha \mathbf{u}\|_2^2(\Omega_\varepsilon) \leq$$

$$C(\|e^{\tau\varphi}\mathbf{f}\|_2^2(\Omega) + C(\varepsilon)\sum \|e^{\tau\varphi}\partial^\alpha \mathbf{u}\|_2^2(\Omega \setminus \Omega_{\frac{\varepsilon}{2}}) +$$

$$\tau^3(\|e^{\tau\varphi}\chi\mathbf{u}\|_{(0)}^2(\Gamma) + \|e^{\tau\varphi}\nabla(\chi\mathbf{u})\|_{(0)}^2(\Gamma))).$$

We have $\varepsilon_1 < \varphi$ on $\Omega_\varepsilon$ where $\varepsilon_1 = e^{\sigma\varepsilon}$, $\varphi < \Phi$ where $\Phi$ is sup $\varphi$ over $\Omega$, and $\varphi \leq e^{\frac{\sigma\varepsilon}{2}}$ on $\Omega \setminus \Omega_{\frac{\varepsilon}{2}}$. Replacing $\varphi$ by its minimum on the left side and by its maximum over closures of the integration domains on the right side we obtain

$$e^{2\tau\varepsilon_1} \sum_{|\alpha| \leq 1} \|\partial^\alpha \mathbf{u}\|_2^2(\Omega_\varepsilon)$$

$$\leq C(e^{2\tau\Phi}\|\mathbf{f}\|_2^2(\Omega) + e^{2\tau\varepsilon_2}C(\varepsilon)\sum_{|\alpha|\leq 1}\|\partial^\alpha\mathbf{u}\|_2^2(\Omega\setminus\Omega_{\varepsilon/2})) +$$

$$C(\varepsilon)\tau^3 e^{2\tau\Phi}(\|\mathbf{g}_0\|_{(1)}(\Gamma) + \|\mathbf{g}_1\|_{(0)}(\Gamma)))$$

where we used the Cauchy data (7). Taking square roots and dividing both sides by $e^{\tau\varepsilon_1}$, we yield

$$\|\partial^\alpha\mathbf{u}\|_2(\Omega_\varepsilon) \leq C(\varepsilon)(\tau^3 e^{\tau(\Phi-\varepsilon_1)}F + e^{-\tau(\varepsilon_1-\varepsilon_2)}\|\mathbf{u}\|_{(1)}(\Omega))$$

$$\leq C(\varepsilon)(e^{\tau(\Phi-\varepsilon_2)}F + e^{-\tau(\varepsilon_1-\varepsilon_2)}\|\mathbf{u}\|_{(1)}(\Omega))$$

because $\tau^3 e^{-(\varepsilon_1-\varepsilon_2)t} \leq C(\varepsilon)$. Let us choose

$$\tau = \max\{(\Phi + \varepsilon_1 - 2\varepsilon_2)^{-1}\ln(\frac{\|\mathbf{u}\|_{(1)}(\Omega)}{F}), C\},$$

where $C > 0$ is needed to ensure that $\tau > C$. Due to this choice, the second term on the right side does not exceed the first one. After substituting the above expression $\tau$, we obtain (9) with $\kappa = \frac{\varepsilon_1-\varepsilon_2}{\Phi+\varepsilon_1-2\varepsilon_2}$.

The proof is complete.

**Remark** Under the conditions of Theorem 1.2 the Hölder type bound (9) can be derived from (6) with indices $1.5, 0.5$ of Sobolev spaces (instead of $1, 0$).

Now we consider a particular domain $\Omega \subset \{0 < x_n\}$. We let $\Gamma = \partial\Omega \cap \{0 < x_n\}$ and $\Omega(d) = \Omega \cap \{0 < x_n\}$.

**Theorem 2.2.** *Let* $A = -\Delta - k^2 a_0^2$, $a_0 \in C^1(\overline{\Omega})$, $0 < a_0$ *on* $\overline{\Omega}$ *and*

$$0 < a_0 + \nabla a_0 \cdot x + \beta_n \partial_n a_0, \ 0 \leq \partial_n a_0 \ \text{on} \ \overline{\Omega} \tag{10}$$

*for some positive* $\beta_n$.

*Then for any* $\varepsilon$ *there are* $C, C(\varepsilon), \kappa(d) \in (0,1)$ *not depending on* $k$ *such that*

$$\|u\|_{(0)}(\Omega(d)) \leq C(F + \varepsilon\|u\|_{(1)}(\Omega) + C(\varepsilon)\frac{\|u\|_{(1)}^{1-\kappa}F(k,d)^\kappa}{d^2 k}) \tag{11}$$

*where* $F(k,d) = F + (kd^{-0.5} + d^{-1.5})\|u\|_{(0)}(\Gamma)$, *for all* $u$ *solving (7)*.

We outline a proof for constant $a_0 = 1$ [5].

By using extension theorems we can extend our problem into the strip $\{0 < x_n < h\}$. We split $u = u_1 + u_2$ into "low frequency" part $u_1$:

$$U_1(\xi, x_n) = U(\xi, x_n), \text{ if } |\xi| < \frac{k}{2},$$

where $U(\xi, x_n)$ is the Fourier transformation of $u(x', x_n)$ in $x'$. The equation (7) for $u$ is transformed into the ODE

$$\partial_n^2 U_1 + (k^2 - |\xi|^2)U_1 = F_1$$

which is $x_n$-hyperbolic, and hence one has the standard energy integral giving a "stable part" of (11).

To bound unstable "high frequency" part $u_2$ we observe that

$$\|u_2\|_{(0)} \leq \frac{1}{k}\|u_2\|_{(1)} \leq \frac{1}{k}\|u\|_{(1)}$$

The last term can be bounded by Theorem 2.1 if $C$ does not depend on $k$. Theorem 2.1 with $k$-independent $C$ follows from the Carleman estimate (2) with $C$ not depending on $k$. We sketch a derivation of this estimate with $\varphi(x) = |x + \beta|^2, \beta = (0, ..., 0, \beta_n)$.

To get (2) we use the substitution $u = e^{-\tau\varphi}v$ eliminating $e^{\tau\varphi}$ and replacing $A$ by $A(, \partial - \tau\nabla\varphi)$. For $v$ we need $L_2$-bounds by $A(, \partial - \tau\nabla\varphi)$. To obtain them one can use an idea from the theory of (higher order) hyperbolic equations:

$$|A(, \partial - \tau\nabla\varphi)v|^2 \geq |A(, \partial - \tau\nabla\varphi)v|^2 - |A(, \partial + \tau\nabla\varphi)v|^2 =$$

$$-16\tau\Delta v(x+\beta)\cdot\nabla v - 8\tau n v\Delta v - 16\tau(4\tau^2|x+\beta|^2+k^2)v(x+\beta)\cdot\nabla v - 8\tau(4\tau|x+\beta|^2+k^2)v^2$$

and integrate by parts the last (divergent) expression.

For variable $a_0$ one can use "freezing" $a_0$ in $x'$, partitioning of unity, and Carleman estimates for the wave operator $a_0^2\partial_t^2 - \Delta$ [10].

Fritz John [J] showed that stability of the continuation into outside of convex hull of $\Gamma$ is not increasing with growing $k$, but, on the contrary, it is deteriorating.

For $x_n$-hyperbolic equations and $\Omega = \Omega' \times (-T, T), \Gamma = \partial\Omega' \times (-T, T)$ there is the following Lipschitz (optimal) stability estimate.

**Theorem 2.3.** *Let $\psi$ satisfy the conditions of Theorem 1.1 and*

$$0 < \psi \text{ on } \Omega' \times \{0\}, \ \psi < 0 \text{ on } \overline{\Omega'} \times \{-T, T\}. \tag{12}$$

*Then there is $C$ such that*

$$\int_{\Omega'}(|\nabla\mathbf{u}|^2 + |\mathbf{u}|^2)(, t) \leq C(\int_{\Omega}|\mathbf{f}|^2 + \int_{\Gamma}(|\mathbf{u}|^2 + |\nabla\mathbf{u}|^2)) \tag{13}$$

*for any $\mathbf{u}$ solving (7).*

**Proofs** in particular cases were obtained in 1980-s by Lasiecka and Triggiani and Lop Fat Ho using multipliers methods of Friedrichs and Morawetz. The idea of combining Carleman estimates with energy inequalities is due to Klibanov and Tataru (around 1992), and we used it to derive the most general form of Lipschitz stability from (2).

**Remark.** $\Gamma'$ can be replaced by a "large part" of $\partial\Omega'$. Results for $\Gamma$ of arbitrary size are not known.

## 3. Uniqueness and stability in inverse problems

We first consider the inverse source problem

$$\mathbf{Au} = \mathbf{Bf}, \ \partial_t\mathbf{f} = 0 \text{ on } \Omega = \Omega' \times (-T, T) \tag{14}$$

where one looks for $(\mathbf{u}, \mathbf{f})$ from

$$\mathbf{u} = \mathbf{g}_0, \ \partial_\nu \mathbf{u} = \mathbf{g}_1 \text{ on } \Gamma = \Gamma' \times (-T, T) \tag{15}$$

where $\Gamma' \subset \partial\Omega'$, $x_n = t$.

**Theorem 3.1.** *Let $\boldsymbol{A}$ does not involve $\partial_t \partial_j, j = 1, ..., n - 1$ and its coefficients be $t$-independent. Let $\mathbf{B}, \partial_t \mathbf{B} \in C(\overline{\Omega})$ and*

$$\varepsilon_0 < |det\mathbf{B}|; \text{ on } \Omega' \times \{0\}, \ \psi < 0 \text{ on } \overline{\Omega'} \times \{-T, T\}. \tag{16}$$

*Let the Carleman type estimate hold:*

$$\int_\Omega \tau^2 |\mathbf{u}|^2 e^{2\tau\varphi} \leq C \int_\Omega |A\mathbf{u}|^2 e^{2\tau\varphi}$$

*for all $\mathbf{u} \in H^0_{(2)}(\Omega)$ when $C < \tau$, and*

$$\psi < 0 \text{ on } \partial\Omega \setminus \Gamma, \ \psi(\cdot, t) < \psi(\cdot, 0) \text{ when } 0 < |t| < T. \tag{17}$$

*Then there are $C, \kappa$ depending on $\varepsilon$ such that*

$$\|\mathbf{f}\|_{(0)}(\Omega_\varepsilon') \leq C(F_1 + M^{1-\kappa} F_1^\kappa) \tag{18}$$

*where*

$$F_1 = \sum_{j=0}^{3} (\|\partial_t^j \mathbf{g}_0\|_{(2.5)}(\Gamma) + \|\partial_t^j \mathbf{g}_1\|_{(1.5)}(\Gamma)),$$

$$M = \sum_{j=0}^{3} \|\partial_t^j \mathbf{u}\|_{(2)}(\Omega).$$

**Proof** combines methods of [1] and [7] extending the basic idea of Carleman to the inverse source problem.

**Example:** (identification of the residual stress $r_{jk}$ for constant $\rho, \lambda, \mu$ [11]).

Let $\mathbf{R}(1), \mathbf{R}(2)$ generate solutions $\mathbf{u}(1), \mathbf{u}(2)$ to the systems $\mathbf{A}_{e,R(j)}\mathbf{u}(j) = 0$ in $\Omega$ with the initial data $\mathbf{u}_0, \mathbf{u}_1$. It turns out that a single set of Cauchy data is sufficient to recover the residual stress. To guarantee the uniqueness, we impose some non-degeneracy condition on the initial data $(\mathbf{u}_0, \mathbf{u}_1)$. More precisely, we assume that

$$det\mathbf{M} = \begin{pmatrix} \partial_1^2 \mathbf{u}_0 \ 2\partial_1\partial_2 \mathbf{u}_0 ... \ \partial_3^2 \mathbf{u}_0 \\ \partial_1^2 \mathbf{u}_1 \ 2\partial_1\partial_2 \mathbf{u}_1 ... \ \partial_3^2 \mathbf{u}_1 \end{pmatrix} \neq 0 \quad \text{on } \overline{\Omega}. \tag{19}$$

Note that $\mathbf{M}(x)$ is a $6 \times 6$ matrix-valued function. For example, one can check that $\mathbf{u}_0(x) = (x_1^2, x_2^2, x_3^2)^\top$ and $\mathbf{u}_1(x) = (x_2 x_3, x_1 x_3, x_1 x_2)^\top$ satisfy (19).

Let $\mathbf{R}(1), \mathbf{R}(2)$ produce the same boundary data. The function $\psi(x, t) = |x - \beta|^2 - \theta^2 T^2 - d_1^2$ is pseudo-convex with respect to $\rho\partial_t^2 - \mu\Delta, \rho\partial_t^2 - (\mu + 2\lambda)\Delta$ on $\overline{\Omega}$.

Subtracting the equations $\mathbf{A}_{e,R(2)}\mathbf{u}(2) = 0$ and $\mathbf{A}_{e,R(1)}\mathbf{u}(1) = 0$ we get

$$\mathbf{A}_{e,R(2)}\mathbf{u} = \sum_{j=1}^{3} \partial_j \mathbf{u}(1) f_{jk}$$

where $\mathbf{u} = \mathbf{u}(2) - \mathbf{u}(1), f_{jk} = r_{jk}(1) - r_{jk}(2)$. Due to our assumptions on the initial data, the matrix $\mathbf{B}$ formed by $\partial_j \mathbf{u}(), \partial_j \partial_t \mathbf{u}(1)$ satisfies (16).

Now uniqueness (and Hölder stability) of $\mathbf{f}$ follows from Theorem 3.1. We give more precise results proven in [11].

For that purpose we introduce the norm of the differences of the data

$$F = \sum_{\beta=2}^{4} (\|\partial_t^\beta(\mathbf{u}(;2) - \mathbf{u}(;1))\|_{(\frac{5}{2})}(\Gamma \times (-T,T)) + \|\partial_t^\beta \sigma_\nu(\mathbf{u}(;2) - \mathbf{u}(;1))\|_{(\frac{3}{2})}(\Gamma \times (-T,T))).$$

Denote $d = \inf |x|$ and $D = \sup |x|$ over $x \in \Omega$. We assume that

$$0 < d. \tag{20}$$

The residual stress will satisfy the boundary condition

$$\mathbf{R}\nu = 0 \quad \text{on} \quad \partial\Omega, \tag{21}$$

Let $R(\varepsilon_0, M)$ be the class of residual stresses defined by

$$R(\varepsilon_0, E) = \{|\mathbf{R}|_6(\overline{\Omega}) < E : \mathbf{R} \text{ is symmetric and satisfies } (4), (21), \text{and} (5)\}.$$

Indeed, Due to our assumptions on $\rho, \lambda, \mu$ we can choose positive $\theta$ so that

$$\theta^2 < \frac{\mu}{\rho}, \quad \theta^4 < \frac{\mu}{\rho} \frac{d^2}{T^2}. \tag{22}$$

**Theorem 3.2.** *Assume that the domain $\Omega$ satisfies (20), $\theta$ satisfies (22), and*

$$|x|^2 - d_1^2 < 0 \text{ when } x \in (\partial\Omega \setminus \Gamma), \text{ and } D^2 - \theta^2 T^2 - d_1^2 < 0. \tag{23}$$

*Let the initial data $(\mathbf{u}_0, \mathbf{u}_1)$ satisfy (19).*

*Then there exist an $\varepsilon_0$ and constants $C, \gamma < 1$, depending on $\varepsilon$, such that for $\mathbf{R}(;1), \mathbf{R}(;2) \in R(\varepsilon_0, E)$ one has*

$$\|\mathbf{R}(;2) - \mathbf{R}(;1)\|_{(0)}(\Omega(\varepsilon)) \leq C F^\gamma. \tag{24}$$

The domain $\Omega(\varepsilon)$ is discussed in , section 3.4.

If $\Gamma$ is the whole lateral boundary and $T$ is sufficiently large, then a much stronger (and in a certain sense best possible) Lipschitz stability estimate holds.

**Theorem 3.3.** *Let $d_1 = d$. Assume that the domain $\Omega$ satisfies (20),*

$$D^2 < 2d^2, \tag{25}$$

*and*

$$\frac{D^2 - d^2}{\theta^2} < T^2. \tag{26}$$

*Let the initial data $(\mathbf{u}_0, \mathbf{u}_1)$ satisfy (19). Let $\Gamma = \partial\Omega$.*

*Then there exist an $\varepsilon_0$ and $C$ such that for $\mathbf{R}(;1), \mathbf{R}(;2) \in R(\varepsilon_0, E)$ satisfying the condition*

$$\mathbf{R}(;1) = \mathbf{R}(;2) \quad \text{on} \quad \Gamma \times (-T,T), \tag{27}$$

*one has*

$$\|\mathbf{R}(;2) - \mathbf{R}(;1)\|_{(0)}(\Omega) \leq C F \tag{28}$$

10

## 4. Open Problems

We list few outstanding questions.

**Problem 1.**

Derive Carleman estimates for some some **principally nondiagonalizable** (hyperbolic) systems.

Transversally isotropic linear elasicity system is of particular interest.

**Problem 2.** Derive Carleman estimates for second order equations with **explicit boundary terms**.

This probably can be done integrating by parts.

**Problem 3.** Prove uniqueness of $c = c(x)$ in the hyperbolic initial boundary value problem

$$\partial_t^2 u - \Delta u + cu = 0 \text{ in } \Omega \times (0, T), \ \Omega = \{0 < x_n\},$$

$$u = \partial_t u = 0 \text{ on } \Omega \times \{0\}, \ \partial_n u = 1 \text{ on } \partial\Omega \times (0, T)$$

from the additional data

$$u = g_0 \text{ on } \partial\Omega \times (0, T).$$

## References

1. A. Bukhgeim, M. Klibanov. Global uniqueness of a class of multidimensional inverse problems. *Soviet Math. Dokl.*, **24** (1981), 244-247.
2. T. Carleman. Sur un probleme d'unicite pour les systemes d'equations aux derivees partielles a deux variables independentes. *Ark. Mat. Astr. Fys.*, **26B** (1939), 1-9.
3. M. Eller, V. Isakov, G. Nakamura, D. Tataru. Uniqueness and stability in the Cauchy problem for Maxwell's and elasticity systems. *College France Seminar*, Vol. 16, Elsevier-Gauthier Villars "Series in Appl. Math.", **7** (2002), 329-349.
4. L. Hörmander. *Linear Partial Differential Operators*. Springer-Verlag, 1963.
5. T. Hrycak, V. Isakov. On increased stability in the Cauchy problem for the Helmholtz equation. *Inverse Problems*, **20** (2004), 697-712.
6. O. Imanuvilov. On Carleman estimates for hyperbolic operators. *Asympt. Anal.*, **32** (2002), 185-220.
7. O. Imanuvilov, V. Isakov, M. Yamamoto. An inverse problem for the dynamical Lame system with two sets of boundary data. *Comm. Pure Appl. Math.*, **56** (2003), 1366-1382.
8. V. Isakov. Carleman Type Estimates in an Anisotropic Case and Applications. *J. Diff. Equat.*, **105** (1993), 217-239.
9. V. Isakov. *Inverse Problems for PDE.*, Springer-Verlag, 1998; new enlarged edition, 2006.
10. V. Isakov. Increased stability in the continuation for the Helmholtz equation with variable coefficient. *AMS Contemp. Math.* (2006) (tp appear).
11. V. Isakov, J.-N. Wang, M. Yamamoto. Uniqueness and stability of determining the residual stress by one measurement. *Comm. Part. Diff. Equat.* (2007) (to appear) .
12. F. John. Continuous dependence on data for solutions of partial differential equations with a prescribed bound. *Comm. Pure Appl. Math.*, **13** (1960), 551-585.

13. I. Lasiecka, R. Triggiani, P.F Yao. Inverse/observability estimates for second-order hyperbolic equations with variable coefficients. *J. Math. Anal. Appl.*, **235** (2000), 13-57.

14. D. Tataru. Carleman estimates and unique continuation for solutions to boundary value problems. *J. Math. Pures Appl.*, **75** (1996), 367-408.

# FROM MICROELECTRONICS TO NANOELECTRONICS: MATHEMATICAL CHALLENGES *

A. M. ANILE and V. ROMANO

*Dipartimento di Matematica e Informatica, Università di Catania,
Viale A.Doria 6, I-95125 Catania, Italy
E-mail: anile@dmi.unict.it, romano@dmi.unict.it*

According to the 2001 international technology roadmap for semiconductors, MOSFETs with physical channel length less than 10nm will be mass produced in 2016. Such devices would have approximately 10 silicon atoms along the effective channel length. This poses challenging problems in the mathematical modeling of charge transport e.g. one has a major relevance of the discrete nature of the dopant distribution for intrinsic stochastic parameter variations. These topics will be discussed in this article along with a critical review of the already existing models.

**Key words:** Nanoelectronics, electromagnetic coupling, transport theory, continuum models

**Mathematics Subject Classification:** 82-08, 82D37, 65N06, 65N30

## 1. Introduction

In today semiconductor technology, the miniaturization of devices is more and more progressing. For example let us consider the Intel 4004 Micro Processor 1971 which was constituted by 10 mm details, 2300 components and 64 kHz speed and compare it with the Intel Pentium 4 Extream Edition micro processor 2005 which has about 50 millions components, 90 nm details and 3.8 Ghz.

According to the 2001 international technology roadmap for semiconductors, MOSFETs with physical channel length less than 10nm will be mass produced in 2016. Such devices would have approximately 10 silicon atoms along the effective channel length.

With shrinking dimensions, the simulation of submicron semiconductor devices requires advanced transport models. In particular one has to face with the following problems. Continuum approximations loose validity and it is necessary a kinetic description of carrier transport in devices; coupling effects become more important (electromagnetic, thermal and with circuits); the atomistic effects are no longer negligible and it is necessary to take into account the discrete structure of matter

---

*This work is supported by M.I.U.R. (PRIN 2004 *Problemi Matematici delle teorie cinetiche*), P.R.A. (ex 60 %), CNR (grant n. 00.00128.ST74) and the RTN Marie Curie project COMSON, grant n. 019417

in the process simulations.

The traditional approach to the research and development of new electron devices was based on the analysis of huge experimental data. Typically a new prototype was proposed and, after it was fabricated, the operational functionalities were tested. If the prototype achieved the desired performance and met some criteria related to budget problems then it was built on large scale. However this trial and error approach is becoming too expensive for ultra-small devices.

The modern trend is to substitute the experimental step with the tool of the computer aided design (TCAD) cheaper and more feasible, allowing for example the study of innovative ideas as quantum devices or the control of single physical effects.

For devices with characteristic lengths $L_g$ longer than 0.5 $\mu$m simple macroscopic models as the drift-diffusion (see [1-4] and reference therein) or the energy-transport (see [4,5] and references therein) are accurate enough. These models are implemented in the most part of commercial simulators and require acceptable computing time. When $0.1\mu m \leq L_g < 0.5\mu m$ it is necessary ro resort to more sophisticated macroscopic model, e.g. those proposed in [6], or to a kinetic approach, based on the semiclassical Boltzmann equation[7-9]. The electrons are considered as particles with an energy related to the conduction and /or valence bands and undergone collisions with themselves and the ions of the crystal inside which they move. The numerical solutions of the system of equations in the semiclassical case is performed with a Monte Carlo approach which leads to very time consuming simulations, often not suited for CAD purposes.

Below 0.1 $\mu$m the effects of discrete impurities must be also taken into account, the electron-electron interaction is no longer negligible and quantum effects play an important role.

## 2. The models

Here the several mathematical models are presented and their main features highlighted.

The electrons inside a semiconductors (see [10] for a review of the basic concepts of solid state physics) move under the action of an electromagnetic field whose spatial and temporal behavior is described by the Maxwell equations for the electric field $\mathbf{E}(\mathbf{x}, t)$ and the magnetic induction $\mathbf{B}(\mathbf{x}, t)$

$$\nabla \cdot (\varepsilon \mathbf{E}) = \rho, \quad \nabla \cdot \mu \mathbf{B} = 0, \tag{1}$$

$$\mathrm{curl}\, \mathbf{E} = -\mu \frac{\partial \mathbf{B}}{\partial t}, \quad \mathrm{curl}\, \mathbf{B} = \epsilon \mathbf{J} + \epsilon \frac{\partial \mathbf{E}}{\partial t}, \tag{2}$$

where $\rho$ is the density charge, $\epsilon$ the dielectric constant, $\mu$ the permeability, $\mathbf{J}$ the total density current vector. In the quasi-static approximation only the Poisson equation $(1)_1$ is considered and the magnetic field neglected. This approximation is rather justified for the present devices, but will loss validity for shorter ones.

## 2.1. *The kinetic description*

The charge transport in semiconductors is described at the kinetic level by the semiclassical Boltzmann equation for the charge carriers, coupled to the Poisson equation for the electric potential. The charge carriers can occupy several conduction bands (electrons) or valence bands (holes) according to the energy level described in quantum mechanics by the stationary Schrödinger equation with the hamiltonian of the perfect crystal lattice (see [10]).

Often analytical approximations of the energy bands are employed. By observing that the electrons occupy the neighborhoods of the local minima, the so-called valleys, around any local minimum the parabolic band approximation is used

$$E_A = \frac{\hbar^2 |\mathbf{k}|^2}{2m^*}, \quad \mathbf{k} \in \mathbf{R^3}, \tag{3}$$

with $m^*$ the effective electron mass, $\hbar \mathbf{k}$ the crystal momentum, $E_A$ energy in the A valley measured from the value corresponding to the bottom of the valley and $\hbar$ the Planck constant $h$ divided by $2\pi$.

A more accurate approximation is given by the Kane dispersion relation, which takes into account the non-parabolicity at high energy

$$E_A(k)\left[1 + \alpha E_A(k)\right] = \frac{\hbar^2 k^2}{2m^*}, \quad \mathbf{k} \in \mathbf{R^3}, \tag{4}$$

where $\alpha$ is the non parabolicity parameter.

The transport equation in the A-th valley reads

$$\frac{\partial f_A}{\partial t} + v_A^i(\mathbf{k})\frac{\partial f_A}{\partial x^i} - \frac{qE^i}{\hbar}\frac{\partial f_A}{\partial k^i} = C_A[f]. \tag{5}$$

$f_A(\mathbf{x}, t, \mathbf{k})$ is a one-particle distribution function, $q$ represents the absolute value of the electron charge, $\mathbf{E}$ is the electric field, obtained by solving the Poisson equation for the electric potential $\phi$

$$E_i = -\frac{\partial \phi}{\partial x_i}, \quad \epsilon \Delta \phi = -q(N_D - N_A - n), \tag{6}$$

$N_D$ and $N_A$ being the donor and acceptor density respectively and $n$ the total electron number density $n = \sum_A \int_{\mathbf{R^3}} f_A d^3\mathbf{k}$. Similar considerations apply to holes too.

$C_A[f]$ is the collision term which describe the scattering mechanisms like electron-phonon interaction, interaction with impurities, electron-electron scattering and interaction with stationary imperfections of the crystal as vacancies, external and internal crystal boundaries.

The equation (5) is of integral differential type in the independent variables of space, time and wave-vector. Due to the high dimensionality and the complexity of the collision operator the solution is usually obtained with a stochastic approach via Monte Carlo simulation [7–9,11], even if recently direct numerical methods have been designed [12] when analytical models for the energy bands are employed. In any case the main drawback is the excessive CPU time which limits the use of the semiclassical model in TCAD.

## 2.2. *The drift-diffusion model*

For devices with typical length greater than one microm from a suitable scaling it is possible [1-3] to recover a simplified macroscopic model, the drift-diffusion one

$$\frac{\partial n}{\partial t} + \frac{1}{q}\mathrm{div}\mathbf{J} = R, \qquad \frac{\partial p}{\partial t} - \frac{1}{q}\mathrm{div}\mathbf{J_p} = -R, \tag{7}$$

$$\mathbf{J_n} = q\left(-D_n\nabla n + \mu_n n\nabla\phi\right), \qquad \mathbf{J_p} = -q\left(-D_p\nabla p + \mu_p p\nabla\phi\right), \tag{8}$$

where $n$ and $p$ are the densities of electrons and holes, $D_n$ and $D_p$ the diffusion coefficients, $\mu_n$ and $\mu_p$ the field mobilities, $R$ the generation-ricombination term. Various models for the mobility and diffusion coefficient have been proposed on the base of MC data and the interested reader can see the book of Selberherr [2]. From a mathematical point of view the drift-diffusion model is a system of nonlinear parabolic equations coupled to the elliptic Poisson equation. It has received a thorough investigation: several results of existence and uniqueness have been obtained and very accurate numerical schemes both with finite differences and finite elements, as that of Scharfetter and Gummel, have been formulated. The interested reader is referred to [13] and reference therein.

## 2.3. *The energy-transport model*

Despite its widespread popularity, the drift-diffusion model does not contain the temperature as dynamical variable being based upon the hypothesis of isothermal flow. This makes the drift-diffusion model not able to describe the high field phenomena occurring in submicron devices. An improvement is represented by the energy-transport models where, in addition to the continuity equations, also a balance equation for the energy is included. In the unipolar case the general form of the models is constituted by the system

$$\frac{\partial n}{\partial t} + \mathrm{div}(n\mathbf{V}) = 0, \tag{9}$$

$$\frac{\partial(nW)}{\partial t} + \mathrm{div}(n\mathbf{S}) - ne\mathbf{V}\cdot\nabla\phi = nC_W, \tag{10}$$

$$\mathrm{div}\left(\epsilon\nabla\phi\right) = -e(N_D - N_A - n). \tag{11}$$

Where $W$, $\mathbf{V}$ and $\mathbf{S}$ are the electron energy, velocity and energy-flux. $C_W$ is the energy production term.

Usually the energy is related to the electron absolute temperature $T$ by the monatomic gas equation of state $W = \frac{3}{2}k_B T$. The differences between the several proposed energy-transport models are represented by the expressions of fluxes and energy relaxation time.

The Stratton model is characterized by the constitutive relations [14]

$$C_W = -\frac{\frac{3}{2}(k_B T - k_B T_L)}{\tau_W},$$ (12)

$$n\mathbf{V} = -\tilde{\mu}_0 \left( \nabla n - \frac{en}{k_B T} \nabla \phi \right),$$ (13)

$$n\mathbf{S} = -\frac{3}{2} \tilde{\mu}_0 \left[ \nabla (k_B n T) - en\nabla\phi \right],$$ (14)

where $\tilde{\mu}_0 = \frac{\mu_0 k_B T_L}{e}$, $\mu_0$ is the low-field mobility and $\tau_W$ the energy relaxation time, usually taken as a constant. Chen and al [15] used the Stratton model but with the Caughey-Thomas formulas for the mobility [2].

Another well known energy-transport model is that obtained by Degond et al [16,17] by a spherical harmonics expansion, and for this reason called SHE model, but also deduced by Lyumkis et al [18]

$$C_W = -\frac{2}{\sqrt{\pi}} \frac{k_B T - k_B T_L}{\tau_0 (T/T_L)^{1/2}},$$ (15)

$$n\mathbf{V} = -\frac{2\mu_0}{e\sqrt{\pi}} (k_B T_L)^{1/2} \left[ \nabla (n(k_B T)^{1/2}) - \frac{e\,n}{(k_B T)^{1/2}} \nabla\phi \right],$$ (16)

$$n\mathbf{S} = -\frac{4\mu_0}{e\sqrt{\pi}} (k_B T_L)^{1/2} \left[ \nabla (n(k_B T)^{3/2}) - e\,n(k_B T)^{1/2}\nabla\phi \right],$$ (17)

where $\tau_0$ is a constant.

A thermodynamic approach based on linear irreversible thermodynamics has been adopted in [19,20].

The model is constituted by a system of coupled nonlinear parabolic equations endowed with a convex entropy which allows a symmetrization in terms of the dual entropy variables. Results concerning existence and uniqueness, stability of equilibrium state have been obtained in[21−23]. Numerical scheme have been developed in [24−27].

Quantum corrections have been included in the drift-diffusion and energy transport models by adding corrections based on the Böhm potential [5]. A different approach has been followed in [28] where discrete energy quantum levels are calculated by solving the stationary Schrödinger equations.

## 2.4. *The hydrodynamical model*

The hydrodynamical models comprises also the balance equation for momentum and more in general for additional macroscopic variables.

Among the various hydrodynamical models, the one introduced by Blotekjaer [29] and subsequently thoroughly investigated by Baccarani and coworkers [30] (which we shall denote by BBW) is incorporated in several commercial simulation packages.

The BBW hydrodynamical model for electrons consists of the continuity equation, momentum equation, energy equation, Poisson's equation. In the unipolar case reads

$$\frac{\partial n}{\partial t} + \nabla \cdot (n\mathbf{V}) = 0 \tag{18}$$

$$\frac{\partial}{\partial t}(nV^i) + \frac{\partial}{\partial x^j}\left(nV^iV^j + \frac{nk_BT}{m^\star}\delta^{ij}\right) + \frac{neE^i}{m^\star} = -\frac{nV^i}{\tau_p} \tag{19}$$

$$\frac{\partial}{\partial t}\left(\frac{1}{2}nm^\star\mathbf{V}^2 + \frac{3}{2}nk_BT\right) + \nabla \cdot \left[(\frac{1}{2}nm^\star\mathbf{V}^2 + \frac{5}{2}nk_BT)\mathbf{V}\right.$$

$$\left. -\kappa\nabla T\right] + ne\mathbf{E} \cdot \mathbf{V} = -n\frac{W - W_0}{\tau_w} \tag{20}$$

$$\nabla \cdot (\epsilon\nabla\phi) = e(N_A - N_D + n). \tag{21}$$

These equations, were not for the collision terms, would be the same as the balance equations for a charged heat conducting fluid coupled to Poisson's equation.

Gardner [31,32] and Gardner, Jerome and Rose [33] numerically integrated the BBW model for the ballistic diode in the stationary case. In [31] the system of equations was discretized by using central differences (if the flow is everywhere subsonic) or the second upwind method (for transonic flow). The discretized system is then linearized by using Newton's method with a damping factor. In this way Gardner was able to show evidence for an electron shock wave in the diode. In [34] Gardner's results have been recovered by using a viscosity method.

Numerical solutions in the nonstationary case have been obtained in [35] by using an ENO scheme. The same results have been obtained in [36] by using a central scheme.

More recently the BBW model has been employed for simulating electron devices by considering the relaxation times, diffusion coefficient as function of the temperature to be determined by fitting of MC data. In this way non parabolic effects can be also included. Moreover the same equation of state is fitted by introducing an effective mass depending on the electron temperature as well.

However the BBW hydrodynamical model suffers from serious theoretical drawbacks: an inconsistency with the Onsager reciprocity relations, presents several free parameters which must be calibrated in each simulation, the comparison with MC simulations are not very accurate.

This has prompted the development of more sophisticated hydrodynamical models. In particular in [37-40] (for a complete review the interested reader is referred to [6]) a hydrodynamical model based on the maximum entropy principle (MEP) has been formulated. It is physically sound, can be set in the framework of the more recent thermodynamical theories of non equilibrium [41,42] and the simulations of some benchmark cases, as $n^+$- n-$n^+$ silicon diode, MESFETs and MOSFETs, shows that it is highly accurate and is computational suited for CAD purposes [43]. In figures (1), (2) we have reported the characteristic curves for a MOSFET by comparing the results obtained with Monte Carlo simulations and the various models known in the literature (for the details the interested reader is referred to [43]). It is evident the superiority of the MEP one.

The mathematical analysis of the MEP model is still at the beginning. A results of existence and uniqueness has been established in [44] for the Cauchy problem in the whole space while some results of stability have been obtained in [45,46] both for mono and bidimensional cases. An analysis of the boundary value problem is still missing and represents an interesting challenging problem.

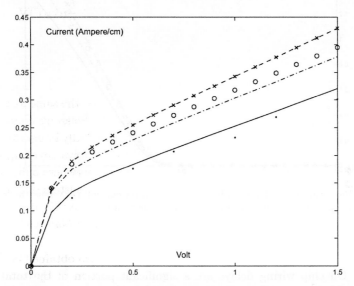

Fig. 1. Characteristic curve for the MOSFET. The drain current $I_D$ versus the drain-source applied voltage $V_D$ at $V_G = 0$. The dots are the MC solution, the continuous line is the MEP model, the xxx and dashed line are the Stratton and Chen model respectively, the ooo line is the SHE model, the dotted dashed line is the reduced hydrodynamical model.

## 2.5. *Future perspectives in the mathematical modeling in nanoelectronics*

On the microscale the continuum like equations (drift-diffusion, energy transport, moment equations, with increasing degree of accuracy, and with some quantum corrections added) still represent the only viable approach: the mathematical models and methods are those of classical PDE analysis. On the nanoscale a similar framework based on the kinetic/atomistic approach is still at an early stage. The mathematical models and methods probably would be a blend of stochastic PDE and quantum kinetic theory.

The dopant distribution should be described with an atomistic simulator by random function and the coupling to the device simulator (hydrodynamic or energy transport) will lead to stochastic PDEs.

On chip interconnects are a further limiting factor to the overall performance

20

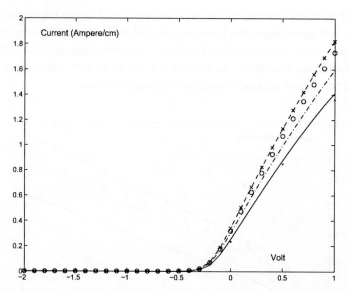

Fig. 2. Characteristic curve for the MOSFET. The drain current $I_D$ versus the gate applied voltage $V_G$) at $V_S = 0$. The notation is as in the previous figure.

of circuits. On chip wiring delays are a significant portion of the total chip delay time. Metal-insulator-semiconductor interconnects are of paramount interest. Electromagnetic field coupling with device charge carriers transport is required for accurate description. Since computationally very intensive, model order reduction will be required.

Performing the step from micro to nanoelectronics, the semiconductor industry is confronted with very high levels of integration introducing coupling effects that are not observed before. The complexity of this problem is beyond the possibilities of the software and design environments used within the microelectronics industry at present. Furthermore, it places additional requirements on the researchers of the future, since they must be able to understand all aspects of the problems faced by the industry.

To meet these new scientific and training challenges, the Marie Curie RTN **COMSON** project on *COupled Multiscale Simulation and Optimization in Nanoelectronics* merges the know-how of the three major European semiconductor industries (Infineon, Philips, ST-Microelectronics) with the combined expertise, in all fields of interest, of specialized university groups (Bucarest, Catania, Wuppertal) for developing adequate mathematical models and numerical schemes and realizing them in a common demonstrator platform. On the one hand, to test mathematical methods and approaches, so as to assess whether they are capable of addressing the industrys problems; on the other hand, to adequately educate young researchers by obtaining immediate hands-on experience and awareness of the state-of-the-art of the problems to be faced.

## References

1. P. Markowich, C. Ringhofer, and C. Schmeiser. *Semiconductor equations*. Springer, Wien, 1990.
2. S. Selberherr, *Analysis and simulation of semiconductor devices*, Wien - New York, Springer-Verlag, 1984.
3. A.Schenk. *Advanced Physical Models for Silicon Device Simulation*. Springer Wien New York.
4. J. W. Jerome, *Analysis of charge transport*, Springer (1996).
5. A. Jüngel, *Quasi-hydrodynamic semiconductor equations*, Basel, Birkhäuser (2001).
6. A.M. Anile, G. Mascali, V.Romano. Recent developments in hydrodynamical modeling of semiconductors (2003) 1:54 in Mathematical Problems in Semiconductor Physics, Lecture Notes in Mathematics 1832, Springer (2003).
7. C. Jacoboni and L. Reggiani, *Rev. Mod. Physics* **55**, 645 (1983).
8. C. Jacoboni C. and P. Lugli, *The Monte Carlo Method for Semiconductor Device Simulation*, Springer: Wien, New York, 1989
9. K. Tomizawa, *Numerical simulation of sub micron semiconductor devices*, Artech House, Boston, 1993.
10. N. C. Ashcroft and N. D. Mermin, ASHCROFT, *Solid State Physics*, Holt-Sounders, Philadelphia, 1976.
11. M. V. Fischetti and S. E. Laux, *J. Appl. Physics* **80**, 2234 (1996).
12. J. A. Carrillo, I. Gamba, A. Majorana,C-W. Shu, *J. Comp. Physics*, **184** 498 (2003).
13. F. Brezzi, L. D. Marini, S. Micheletti, P. Pietra, R. Sacco, S. Wang, *Discretization of semiconudctor device problemd (I)*, Handbook of Numerical Analysis Vol. XIII, Elsevier-North Holland (2005).
14. R. Stratton, *Phys. Rev.* **126**, 2002 (1962).
15. D. Chen, E. C. Kan, U. Ravaioli, C-W. Shu, R. Dutton, *IEEE on Electron Device Letters* **13**, 26 (1992).
16. N. Ben Abdallah, P. Degond and S. Genieys, *J. Stat. Phys.* **84**, 205 (1996).
17. N. Ben Abdallah and P. Degond, *J. Math. Phys.* **37**, 3306 (1996).
18. E. Lyumkis, B. Polsky, A. Shir and P. Visocky, *Compel* **11**, 311 (1992).
19. G. Wachutka, *Compel* **10**, 311 (1991).
20. and in H. Brand and S. Selberherr, *IEEE trans. On electron devices* **42**, 2137 (1995).
21. G. Albinus, H. Gajewski and R. Hünlich, *Nonlinearity* **16**, 367 (2002) .
22. P. Degond, S. Genieys and A. Jüngel, *J. Math. Pures Appl.* **76**, 991 (1997).
23. P. Degond, S. Genieys and A. Jüngel, *Math. Meth. Appl. Sci.* **21**, 1399 (1998).
24. A. Marrocco and P. Montarnal, *C. R. Acad. Sci. Paris, Sèr. I* **323**, 535 (1996).
25. A. Marrocco, P. Montarnal and B. Perthame, Simulation of the energy-transport and simplified hydrodynamical models for semiconductors devices using mixed fionite elements. In Proceedings ECCOMAS 96, Wiley (1996).
26. A. Jüngel and P. Pietra, *Math. Models Meth. Appl. Sci.* **7**, 935 (1997).
27. J. W. Jerome and C-W. Shu. Energy models for one-carrier transport in semiconductor devices, in *Semiconductors part II*, The IMA volumes in Mathematics and its Applications, N. M. Coughran, J. Cole, P. Lloyd, J. K. White editors (1994), pp. 185–207.
28. M. Ancona and G. Iafrate, *Phys. Rev. B* **39** 9536 (1989).
29. K. Blotekjaer, *IEEE Trans. on Electron Devices* **17**, 38 (1970).
30. G. Baccarani and M. R., *Solid-state Electronics* **29**, 970 (1982).
31. C. L. Gardner, *IEEE Transaction on Electron devices* **38**, 392 (1991).

22

32. C. L. Gardner, *IEEE Trans.Electron Devices* **40**, 455 (1993).
33. C. L. Gardner, J.W. Jerome and D.J. Rose, *IEEE Trans. CAD CAD* **8**, 501 (1989).
34. A.M. Anile, C.R. Maccora, M. Pidatella, *Compel*, 1 (1995).
35. E. Fatemi, J. Jerome e S. Osher, *IEEE Transaction on Computer-Aided Design* **10**, 232 (1991).
36. V. Romano and G. Russo, *Mathematical Models and Methods in Applied Sciences* **10**, 833 (2000).
37. A.M. Anile, V.Romano, *Continuum Mech. Thermodyn.* **11**, 307 (1999).
38. V.Romano, *Continuum Mech. Thermodyn.* **12**, 31 (2000).
39. V.Romano, *Math. Meth. Appl. Sci.* **24**, 439 (2001).
40. G. Mascali and V. Romano, *Physica A* **352**, 459(2005).
41. I. Müller and T. Ruggeri. *Rational Extended Thermodynamics*. Springer, New York, 1998.
42. D. Jou, J. Casas-Vazquez and G. Lebon, *Extended irreversible thermodynamics*, Springer-Verlag, Berlin, 1993.
43. A. M. Anile, A. Marrocco, V. Romano and J. M. Sellier, 2D Numerical Simulation of the MEP Energy-Transport Model with a Mixed Finite Elements Scheme, to appear in J. Comp. Electronics.
44. G. Alì and A. M. Anile, *Model. Simul. Sci. Eng. Technol.* 59 (2004).
45. A. M. Blokhin, R. S. Bushmanov, V. Romano, *Int. J. Engineering Science* **42** 915 (2004).
46. A. M. Blokhin, R. S. Bushmanov, V. Romano, *Global existence for the system of the macroscopic balance equations of charge transport in semiconductors*, to appear in J. Math. Anal. Appl.

# A CHARACTERIZATION OF THE CALDERÓN PROJECTOR FOR THE BIHARMONIC EQUATION

G. C. HSIAO

*Department of Mathematical Sciences, University of Delaware*
*Newark, DE 19716, USA*
*E-mail: hsiao@math.udel.edu*

W. L. WENDLAND

*Institut für Angewandte Analysis und Numerische Simulation, Universität Stuttgart*
*70569 Stuttgart, Germany*
*E-mail: wendland@mathematik.uni-stuttgart.de*

In this paper, we present some recent results on a new characterization of the Calderón projector for the biharmonic equation and its application of boundary integral equation methods to the fundamental problems of the biharmonic equation. Specifically, we will consider the solutions of the Dirichlet and Neumann problems by the integral equations of the first kind. We are particularly interested in the direct integral formulations of problems. As will be seen, a simplified and systematic notation may be introduced for the corresponding boundary integral operators in the associated Calderón projector so that their properties can be identified as easily as those in the case for the Laplacian. Emphasis will be placed upon the variational formulations and mathematical foundations of the corresponding boundary integral equation methods.

**Key words:** Biharmonic equation, mixed boundary value problems, boundary integral equations, variational formulations
**Mathematics Subject Classification:** 35J35, 45A05, 45F99, 35S05

## 1. Introduction

It is now well-known that in the direct formulation of the integral equation method based on the Green representation formula for the solution of the Laplace equation, one obtains a system of two boundary integral equations for the trace and the normal derivative of the solution on the boundary, and defines the so-called Calderón projector for the Cauchy data. The Calderón projector plays a fundamental role in the study of boundary value problems by the boundary element methods (see, e.g., Hsiao and Wendland [9]). The purpose of the present paper is to see the feasibility of extending the approach for the Laplace equation to the biharmonic equation. Clearly for the latter, it is much more involved; there is double the amount of Cauchy data and 16 boundary operators needed to be considered. However, as will be seen, the recent systematic characterization of the Calderón projector in Hsiao and Wendland [10] has simplified the approach in the same manner as in the case for the Laplace equation.

The paper is organized as follows: In Section 2, we formulate the boundary problems and present some preliminary results for the weak solutions of the interior Dirichlet and Neumann boundary value problems for the biharmonic equation. Section 3 contains the main result for the Calderón projector consisting of 16 basic boundary integral operators. Section 4 contains basic results for the exterior boundary value problems. We present the results concerning existence and uniqueness of the system of boundary integral equations of the first kind in Section 5. Finally in the last section, Section 6, we conclude the paper by presenting the generalized first Green formula for the biharmonic equation which are relevant to the Gårding inequality for the corresponding boundary integral operators in Section 5.

## 2. Boundary Value Problems

Through out the paper let $\Omega$ be a bounded domain in $\mathbb{R}^2$ with a smooth boundary $\Gamma$ and $\Omega^c$ be the exterior domain $\mathbb{R}^2 \setminus \bar{\Omega}$. To discuss boundary value problems for the biharmonic equation

$$(-\Delta)^2 u = 0 \tag{1}$$

in $\Omega$ (or in $\Omega^c$), it is best to begin with Green's formula for the Eq. (1) in $\Omega$. As is well known, for the fourth-order differential equations, Green's formula generally varies and depends on how to apply the integration by parts formulae. In order to include boundary conditions arising for the thin plate, we rewrite the biharmonic operator $\Delta^2$ in terms of the Poisson ratio $\nu$ in the form

$$\Delta^2 u = \frac{\partial^2}{\partial x_1^2}\left(\frac{\partial^2 u}{\partial x_1^2} + \nu\frac{\partial^2 u}{\partial x_2^2}\right) + 2(1-\nu)\frac{\partial^2}{\partial x_1 \partial x_2}$$
$$\left(\frac{\partial^2 u}{\partial x_1 \partial x_2}\right) + \frac{\partial^2}{\partial x_2^2}\left(\frac{\partial^2 u}{\partial x_2^2} + \nu\frac{\partial^2 u}{\partial x_1^2}\right).$$

Now integration by parts leads to the first Green formula in the form

$$\int_\Omega (\Delta^2 u)v\,dx = a(u,v) - \int_\Gamma \left\{(Mu)\frac{\partial v}{\partial n} + (Nu)v\right\}ds, \tag{2}$$

where $a(u,v)$ is the bilinear form

$$a(u,v) := \int_\Omega \left\{\nu\Delta u\Delta v + (1-\nu)\left(\frac{\partial^2 u}{\partial x_1^2}\frac{\partial^2 v}{\partial x_1^2}\right.\right.$$
$$\left.\left. + 2\frac{\partial^2 u}{\partial x_1 \partial x_2}\frac{\partial^2 v}{\partial x_1 \partial x_2} + \frac{\partial^2 u}{\partial x_2^2}\frac{\partial^2 v}{\partial x_2^2}\right)\right\}dx,$$

and $Mu$ and $Nu$ are the boundary integral operators defined by

$$Mu|_\Gamma := \nu\Delta u + (1-\nu)\left(\frac{\partial^2 u}{\partial x_1^2}n_1^2 + 2\frac{\partial^2 u}{\partial x_1 \partial x_2}n_1 n_2\right.$$

$$\left. + \frac{\partial^2 u}{\partial x_2^2}n_2^2\right)$$

$$Nu|_\Gamma := -\frac{\partial}{\partial n}\Delta u + (1-\nu)\frac{d}{ds}\left\{\left(\frac{\partial^2 u}{\partial x_1^2} - \frac{\partial^2 u}{\partial x_2^2}\right)n_1 n_2\right.$$

$$\left. - \frac{\partial^2 u}{\partial x_1 \partial x_2}(n_1^2 - n_2^2)\right\},$$

where $\mathbf{n} = (n_1, n_2)$ is the exterior unit normal to $\Omega$. The Green formula (2) suggests the *Dirichlet boundary conditions* :

$$u|_\Gamma = \varphi_1 \quad \text{and} \quad \frac{\partial u}{\partial n}|_\Gamma = \varphi_2 \quad \text{on} \quad \Gamma,$$

and the *Neumann boundary conditions* :

$$Mu|_\Gamma = \sigma_1 \quad \text{and} \quad Nu|_\Gamma = \sigma_2 \quad \text{on} \quad \Gamma,$$

where $\varphi_1, \varphi_2$ or $\sigma_1, \sigma_2$ are prescribed data satisfying proper regularity conditions to be specified. In thin plate theory, when $u$ stands for the deflection of the middle surface of the plate, the Dirichlet condition specifies the displacement and the angle of rotation of the plate at the boundary, whereas the Neumann condition provides the bending moment and shear force at the boundary. Clearly, various linear combinations will lead to other mixed boundary conditions (see, e.g., Cakoni, Hsiao and Wendland [2]). As usual, the given Neumann data $\sigma_1$, and $\sigma_2$ are required to satisfy the *compatibility condition*:

$$\int_\Gamma \{(\sigma_1)\frac{\partial v}{\partial n} + (\sigma_2)v\}ds = 0 \quad \forall\, v \in \mathcal{R}, \tag{3}$$

where $\mathcal{R} = \{v|v = \mathbf{a}\cdot x + b\ ,\ \forall \mathbf{a}\in\mathbb{R}^2, b\in\mathbb{R}\}$ is the eigenspace of the bilinear form $a(\cdot,\cdot)$. Looking at (2), one might tempt to replace the differential operators $Mu$ and $Nu$ by $\Delta u$ and $-\frac{\partial}{\partial n}\Delta u$ in the Neumann boundary conditions, which corresponds to the special case when the Poisson ratio $\nu = 1$. Then the compatibility condition (3) requires that it should hold for all harmonic functions $v$. However, the space of harmonic functions in $\Omega$ has infinite dimension, and it does not lead to a regular boundary value problem (see Agmon [1]). We remark that the compatibility condition (3) is equivalent to the following system of equations:

$$\int_\Gamma \sigma_2(y)ds_y = 0, \tag{4}$$

$$\int_\Gamma \{\sigma_1(y)n_y + \sigma_2(y)y\}ds_y = 0 \tag{5}$$

since $v = 1$ and $y = (y_1, y_2)$ form a basis of the eigenspace $\mathcal{R}$. As will be seen later, these conditions are closely related to the growth conditions of the solutions of the biharmonic equation in the exterior domain $\Omega^c$, which are needed in order to ensure the uniqueness of the solution in the exterior domain $\Omega^c$. In fact, for the exterior boundary value problems we need to append Eq. (1) with the radiation condition:

$$u(x) = \left( A_0 r + \frac{\mathbf{A} \cdot x}{|x|} \right) r \, \log r + O(r) \tag{6}$$

as $r = |x| \to \infty$ for given constant $A_0$ and constant vector $\mathbf{A}$. We are interested in the weak solutions of the Dirichlet and the Neumann problems for the biharmoinc equation (1) by the method of boundary integral equations. For this purpose, we need some basic results. We begin with the following.

**Lemma 1.** (S. Agmon [1]). *The bilinear form $a(\cdot, \cdot)$ is coercive over $H^2(\Omega)$ iff $-3 < \nu < 1$. Here the coerciveness means that $a(\cdot, \cdot)$ satisfies a Gårding inequality in the form:*

$$a(v, v) \geq c_0 ||v||^2_{H^2(\Omega)} - \lambda_0 ||v||^2_{H^0(\Omega)} \tag{7}$$

*for all $v \in H^2(\Omega)$, where $c_0 > 0$ and $\lambda_0 \geq 0$ are constants.*

It is well known that Gårding's inequality (7) implies the validity of the Fredholm's alternative. As a consequence we have the following results.

**Theorem 1.** *Let $(\varphi_1, \varphi_2) \in H^{3/2}(\Gamma) \times H^{1/2}(\Gamma)$ be given. Then there exists a unique solution $u \in H^2(\Omega)$ of the interior Dirichlet problem:*

$$\Delta^2 u = 0 \quad in \quad \Omega$$

$$u = \varphi_1 \quad and \quad \frac{\partial u}{\partial n} = \varphi_2 \quad on \quad \Gamma.$$

**Theorem 2.** *Given $(\sigma_1, \sigma_2) \in H^{-1/2}(\Gamma) \times H^{-3/2}(\Gamma)$, there exists a solution $u \in H^2(\Omega)$ of the interior Neumann problem*

$$\Delta^2 u = 0 \quad in \quad \Omega$$

$$Mu = \sigma_1 \quad and \quad Nu = \sigma_2 \quad on \quad \Gamma,$$

*provided $(\sigma_1, \sigma_2)$ satisfies the compatibility condition:*

$$\int_\Gamma (\sigma_1 \frac{\partial v}{\partial n} + \sigma_2 v) ds = 0 \quad for \quad v = 1, x_1, x_2.$$

*Moreover, the solution is unique up to a linear function $p(x) \in \mathcal{R}$.*

The proofs of these theorems are straightforward and are omitted. For the classical results, we refer to the books by Gakhov [6] and by Muskhelishvili [11]. We need similar results for the exterior Dirichlet and Neumann boundary value problems and we will return to this later.

## 3. Calderón Projector

For the interior boundary value problems for (1), the starting point is the representation formula

$$u(x) = \int_\Gamma \left\{ E(x,y)Nu(y) + \frac{\partial E(x,y)}{\partial n_y} Mu(y) \right\} ds_y$$

$$- \left\{ M_y E(x,y)\frac{\partial u}{\partial n}(y) + N_y E(x,y)u(y) \right\} ds_y$$

for $x \in \Omega$, where $E(x,y)$ is the fundamental solution for the biharmonic equation given by

$$E(x,y) = \frac{1}{8\pi}|x-y|^2 log|x-y|$$

which satisfies

$$\Delta_x^2 E(x,y) = \delta(x,y) \quad \text{in} \quad \mathbb{R}^2.$$

As in case of the Laplacian, we may rewrite $u$ in the form

$$u(x) = V(Mu, Nu) - W\left(u, \frac{\partial u}{\partial n}\right), \tag{8}$$

where

$$V(Mu, Nu) := \int_\Gamma \{E(x,y)Nu(y) + \frac{\partial E}{\partial n_y}(x,y)Mu(y)\} ds_y$$

$$= \frac{1}{8\pi}\int_\Gamma |x-y|^2 log|x-y|Nu(y)ds_y$$

$$+ \frac{1}{8\pi}\int_\Gamma n_y \cdot (y-x) \times$$

$$(2\log|x-y|+1)Mu(y)ds_y$$

$$W(u, \frac{\partial}{\partial n}u) := \int_\Gamma \{M_y E(x,y)\frac{\partial}{\partial n}u(y)$$

$$+ N_y E(x,y)u(y)\} ds_y$$

$$= \frac{1}{8\pi}\int_\Gamma \left\{(2(1+\nu)log|x-y| + (1+3\nu))\right.$$

$$+ 2(1-\nu)\frac{((y-x)\cdot n_y)^2}{|x-y|^2}\right\}\frac{\partial}{\partial n}u(y)ds_y$$

$$+ \frac{1}{2\pi}\int_\Gamma \left\{\frac{\partial}{\partial n_y}\log(\frac{1}{|x-y|}) - \frac{1}{2}(1-\nu) \times\right.$$

$$\frac{d}{ds_y}\left(\frac{(y-x)\cdot n_y^\perp(y-x)\cdot n_y}{|x-y|^2}\right)\right\}u(y)ds_y$$

are the *simple-* and *double layer- potentials*, respectively, and

$$(u|_\Gamma, \frac{\partial u}{\partial n}|_\Gamma) \in H^{3/2}(\Gamma) \times H^{1/2}(\Gamma)$$
$$(Mu|_\Gamma, Nu|_\Gamma) \in H^{-1/2}(\Gamma) \times H^{-3/2}(\Gamma)$$

are the (modified) Cauchy data. By applying the trace and boundary differential operators to the representation of u, this leads to the following four basic *boundary integral equations*:

$$u|_\Gamma = \int_\Gamma E(x,y)Nu(y)ds_y + \int_\Gamma \frac{\partial E(x,y)}{\partial n_y}Mu(y)ds_y$$
$$- \int_\Gamma M_y E(x,y)\frac{\partial u}{\partial n}ds_s$$
$$+ \frac{1}{2}u(x) - \int_\Gamma N_y E(x,y)u(y)ds_y$$

$$\frac{\partial u}{\partial n}|_\Gamma = \int_\Gamma \frac{\partial E(x,y)}{\partial n_x}Nu(y)ds_y + \int_\Gamma \frac{\partial^2 E(x,y)}{\partial n_x \partial n_y}Mu(y)ds_y$$
$$+ \frac{1}{2}\frac{\partial u}{\partial n} + (-\int_\Gamma \frac{\partial}{\partial n_x}M_y E(x,y)\frac{\partial u}{\partial n}ds_y)$$
$$- \int_\Gamma \frac{\partial}{\partial n_x}N_y E(x,y)u(y)ds_y$$

$$M_x u|_\Gamma = \int_\Gamma M_x E(x,y)Nu(y)ds_y$$
$$+ \frac{1}{2}Mu(x) - (-\int_\Gamma M_x \frac{\partial E(x,y)}{\partial n_y}Mu(y)ds_y)$$
$$- \int_\Gamma M_x M_y E(x,y)\frac{\partial u}{\partial n}ds_y$$
$$- \int_\Gamma M_x N_y E(x,y)u(y)ds_y$$

$$N_x u|_\Gamma = \frac{1}{2}Nu(x) + \int_\Gamma N_x E(x,y)Nu(y)ds_y$$
$$+ \int_\Gamma N_x \frac{\partial E(x,y)}{\partial n_y}Mu(y)ds_y$$
$$- \int_\Gamma N_x M_y E(x,y)\frac{\partial u}{\partial n}ds_y$$
$$- N_x \int_\Gamma N_y E(x,y)u(y)ds_y,$$

which contain 16 boundary integral operators. We may rewrite the above equations in the form:

$$
\begin{pmatrix} u \\ \frac{\partial}{\partial n}u \\ Mu \\ Nu \end{pmatrix}\Bigg|_{\Gamma} = \left(\frac{1}{2}\mathcal{I} + \mathcal{K}\right) \begin{pmatrix} u \\ \frac{\partial}{\partial n}u \\ Mu \\ Nu \end{pmatrix}\Bigg|_{\Gamma}. \tag{9}
$$

Here $\mathcal{I}$ is the identity matrix operator and $\mathcal{K}$ is the matrix of the 16 boundary integral operators

$$
\mathcal{K} := \begin{pmatrix} -K_{11} & V_{12} & V_{13} & V_{14} \\ D_{21} & K_{22} & V_{23} & V_{24} \\ D_{31} & D_{32} & -K_{33} & V_{34} \\ D_{41} & D_{42} & D_{43} & K_{44} \end{pmatrix},
$$

which are defined explicitly on the Cauchy data:

$$
(\varphi_1, \varphi_2, \sigma_1, \sigma_2)^{\top} = (u, \frac{\partial u}{\partial n}, Mu, Nu)^{\top}|_{\Gamma}.
$$

More precisely, we define

$$
K_{11}\varphi_1(x) := \lim_{\Omega \ni z \to x \in \Gamma} W(\varphi_1, 0)(z) + \frac{1}{2}\varphi_1(x)
$$
$$
V_{12}\varphi_2(x) := \lim_{\Omega \ni z \to x \in \Gamma} -W(0, \varphi_2)(z)
$$
$$
V_{13}\sigma_1(x) := \lim_{\Omega \ni z \to x \in \Gamma} V(\sigma_1, 0)(z)
$$
$$
V_{14}\sigma_2(x) := \lim_{\Omega \ni z \to x \in \Gamma} V(0, \sigma_2)(z)
$$

for the first row of $\mathcal{K}$; and for the second row,

$$
D_{21}\varphi_1(x) := \lim_{\Omega \ni z \to x \in \Gamma} -n_x \cdot \nabla_z W(\varphi_1, 0)(z)
$$
$$
K_{22}\varphi_2(x) := \lim_{\Omega \ni z \to x \in \Gamma} -n_x \cdot \nabla_z W(0, \varphi_2)(x) - \frac{1}{2}\varphi_2(x)
$$
$$
V_{23}\sigma(x) := \lim_{\Omega \ni z \to x \in \Gamma} n_x \cdot \nabla_z V(\sigma_1, 0)(z)
$$
$$
V_{24}\sigma_2(x) := \lim_{\Omega \ni z \to x \in \Gamma} n_x \cdot \nabla_z V(0, \sigma_2)(z),
$$

and the third row,

$$
D_{31}\varphi_1(x) := \lim_{\Omega \ni z \to x \in \Gamma} -M_z W(\varphi_1, 0)(z)
$$
$$
D_{32}(x) := \lim_{\Omega \ni z \to x \in \Gamma} -M_z W(0, \varphi_2)(z)
$$
$$
K_{33}\sigma_1(x) := \lim_{\Omega \ni z \to x \in \Gamma} -M_z V(\sigma_1, 0)(z) + \frac{1}{2}\sigma_1(x)
$$
$$
V_{34}\sigma_2(x) := \lim_{\Omega \ni z \to x \in \Gamma} M_z V(0, \sigma_2)(z).
$$

Finally for last row, we define

$$D_{41}\varphi_1(x) := \lim_{\Omega \ni z \to x \in \Gamma} -N_z W(\varphi_1, 0)(z)$$

$$D_{42}\varphi_2(x) := \lim_{\Omega \ni z \to x \in \Gamma} -N_z W(0, \varphi_2)(z)$$

$$D_{43}\sigma_1(x) := \lim_{\Omega \ni z \to x \in \Gamma} N_z V(\sigma_1, 0)(z)$$

$$K_{44}\sigma_2(x) := \lim_{\Omega \ni z \to x \in \Gamma} N_z V(0, \sigma_2)(z) - \frac{1}{2}\sigma_2(x).$$

The Calderón projector associated with the biharmonic equation in $\Omega$ is now defined by

$$C_\Omega := \frac{1}{2}\mathcal{I} + \mathcal{K} \tag{10}$$

$$= \begin{pmatrix} \frac{1}{2}I - K_{11} & V_{12} & V_{13} & V_{14} \\ D_{21} & \frac{1}{2}I + K_{22} & V_{23} & V_{24} \\ D_{31} & D_{32} & \frac{1}{2}I - K_{33} & V_{34} \\ D_{41} & D_{42} & D_{43} & \frac{1}{2}I + K_{44} \end{pmatrix}.$$

The principal symbols of $C_\Omega$ can be computed as in Hsiao and Wendland [10]:

$$\begin{pmatrix} \frac{1}{2L} & \frac{(1+\nu)}{4}|\xi|^{-1} & 0 & \frac{L^2}{4}|\xi|^{-3} \\ \frac{(1+\nu)}{4L^2}|\xi| & \frac{1}{2L} & \frac{1}{4}|\xi|^{-1} & 0 \\ 0 & \frac{(1-\nu)(1+3\nu)}{4L^2}|\xi| & \frac{1}{2L} & -\frac{(1+\nu)}{4}|\xi|^{-1} \\ \frac{(1-\nu)(1+3\nu)}{4L^4}|\xi|^3 & 0 & -\frac{(1+\nu)}{4L^2}|\xi| & \frac{1}{2L} \end{pmatrix},$$

where the parameter $L$ denotes the arc-length of the parametric representation of the boundary $\Gamma = \{y(t) : t \in [0, L]\}$. As a consequence, we now have the mapping property of $C_\Omega$:

**Theorem 3.** *Let* $\Gamma \in C^2$. *Then the Calderón projector* $C_\Omega := ((C_{ij}))$ *is a matrix of pseudo-differential operators* $C_{ij}$ *of order* $i - j - \delta_{2,|i-j|}$. *The mapping*

$$C_\Omega : \prod_{k=0}^{3} H^{3/2-k}(\Gamma) \to \prod_{k=0}^{3} H^{3/2-k}(\Gamma)$$

*is continuous. Moreover,*

$$C_\Omega^2 = C_\Omega. \tag{11}$$

As a consequence of (11), one finds in particular the following relations:

$$V_{12}D_{21} + V_{13}D_{31} + V_{14}D_{41} = \left(\frac{1}{4}I - K_{11}^2\right),$$
$$D_{21}V_{12} + V_{23}D_{32} + V_{24}D_{42} = \left(\frac{1}{4}I - K_{22}^2\right),$$
$$D_{31}V_{13} + D_{32}V_{23} + V_{34}D_{43} = \left(\frac{1}{4}I - K_{33}^2\right),$$
$$D_{41}V_{14} + D_{42}V_{24} + D_{43}V_{34} = \left(\frac{1}{4}I - K_{44}^2\right).$$

In the same manner as in case for the Laplacian, for any solution $u$ of (1) in $\Omega^c$, we may introduce the Calderón projector $C_{\Omega^c}$ in the exterior domain for the biharmonic equation. Then clearly, we have

$$C_{\Omega^c} = \mathcal{I} - C_\Omega, \tag{12}$$

where $\mathcal{I}$ denotes the identity matrix operator. This relation then provides the corresponding boundary integral equations for the exterior boundary value problems. In order to discuss the properties of the boundary integral equations consisting of the boundary operators in the Calderón projector, we now consider the exterior boundary value problems.

## 4. Exterior Boundary Value Problems

For the exterior problems, we begin with the representation of the solution $u$ of (1) in terms of the Cauchy data $(u, \frac{\partial}{\partial n} u, Mu, Nu)|_\Gamma$,

$$u(x) = W(u, \frac{\partial u}{\partial n})(x) - V(Mu, Nu)(x) + p(x), \quad x \in \Omega^c, \tag{13}$$

where

$$p(x) \in \mathcal{R} = \{v | v = \mathbf{a} \cdot x + b , \ \forall \mathbf{a} \in \mathbb{R}^2, b \in \mathbb{R}\}.$$

From (13), one can show that the solution $u$ admits the asymptotic development

$$u(x) = (A_0 r + \frac{\mathbf{A} \cdot x}{|x|}) r \log r + O(r)$$

as $r = |x| \to \infty$ with

$$A_0 = -\frac{1}{8\pi} \int_\Gamma Nu(y) ds$$

$$\mathbf{A} = \frac{1}{4\pi} \int_\Gamma (Mu(y) \, n_y + Nu(y) \, y) \, ds.$$

As will be seen, the constant $A_0$ and the vector $\mathbf{A}$ are needed to be specified for the uniqueness of the exterior problems. In fact we have the following.

**Theorem 4.** *Let $(\varphi_1, \varphi_2) \in H^{3/2}(\Gamma) \times H^{1/2}(\Gamma)$ be given. Then there exists a unique solution $u \in H^2_{loc}(\Omega^c)$ of the exterior Dirichlet problem:*

$$\Delta^2 u = 0 \quad in \quad \Omega^c,$$

$$u = \varphi_1 \quad and \quad \frac{\partial u}{\partial n} = \varphi_2 \quad on \quad \Gamma,$$

$$u(x) = \left( A_0 r + \frac{\mathbf{A} \cdot x}{|x|} \right) r \log r + O(r)$$

*as $r = |x| \to \infty$ for any given $(A_0, \mathbf{A}) \in \mathbb{R} \times \mathbb{R}^2$.*

In particular, we may set $(A_0, \mathbf{A}) = (0, 0)$, and require that

$$u = O(|x|), \quad \text{as} \quad |x| \to \infty.$$

Then Theorem 4 remains valid. On the other hand for the exterior Neumann problem, we have the following.

**Theorem 5.** *Given* $(\sigma_1, \sigma_2) \in H^{-1/2}(\Gamma) \times H^{-3/2}(\Gamma)$, *the exterior Neumann problem*

$$\Delta^2 u = 0 \quad \text{in} \quad \Omega^c,$$

$$Mu = \sigma_1 \quad \text{and} \quad Nu = \sigma_2 \quad \text{on} \quad \Gamma,$$

*has a solution* $u \in H^2_{loc}(\Omega^c)$ *satisfying*

$$u(x) - \left( A_0 r + \frac{\mathbf{A} \cdot x}{|x|} \right) r \, log \, r = O(r) \tag{14}$$

*as* $r = |x| \to \infty$ *with* $A_0$ *and* $\mathbf{A}$ *given by*

$$A_0 = -\frac{1}{8\pi} \int_\Gamma \sigma_2(y) ds$$

$$\mathbf{A} = \frac{1}{4\pi} \int_\Gamma (n\sigma_1(y) \, n_y + \sigma_2(y) \, y) \, ds.$$

*Moreover, if in addition to* $\sigma$, *the linear function* $p \in \mathcal{R}$ *is given, then the solution is unique with the behavior* (13) *and* (14) *at infinity.*

We remark that the condition $O(r)$ may be replaced by $o(r)$ in Eq. (14), in which case the uniqueness in Theorem 5 again holds. We close this section by including, as in the case of the Laplacian, the useful identity of the Gauss type:

$$W\left(p, \frac{\partial p}{\partial n}\right) = \begin{cases} -p & \text{for } x \in \Omega, \\ -\frac{1}{2}p & \text{for } x \in \Gamma, \\ 0 & \text{for } x \in \Omega^c, \end{cases}$$

for any $p \in \mathcal{R} = \{v | v = \mathbf{a} \cdot x + b \,, \,\, \forall \mathbf{a} \in \mathbb{R}^2, b \in \mathbb{R}\}$.

## 5. Boundary Integral Equations of the First Kind

We now present the reduction of the boundary value problems to boundary integral equations of the first kind.

### 5.1. Interior Dirichlet Problem

We begin with the interior Dirichlet boundary value problem. We seek a solution in the form of Eq. (8), namely

$$u(x) = V(\sigma_1, \sigma_2) - W(\varphi_1, \varphi_2) \quad \text{in} \quad \Omega, \tag{15}$$

where
$$(\sigma_1, \sigma_2) = (Mu, Nu) \in H^{-1/2}(\Gamma) \times H^{-3/2}(\Gamma)$$
are the unknown Cauchy data and
$$(\varphi_1, \varphi_2) = (u|_\Gamma, \frac{\partial u}{\partial n}|_\Gamma) \in H^{3/2}(\Gamma) \times H^{1/2}(\Gamma)$$
are the given Dirichlet data. For the integral equations of the first kind, we employ the second and the first row of Eq. (10) from Eq. (9) which leads to the following system of the first kind integral equations:

$$\mathbf{V}\sigma := \begin{pmatrix} V_{23} & V_{24} \\ V_{13} & V_{14} \end{pmatrix} \begin{pmatrix} \sigma_1 \\ \sigma_2 \end{pmatrix} = \mathbf{f}_i, \tag{16}$$

where $\mathbf{f}_i$ is given in term of the Dirichlet data $(\varphi_1, \varphi_2)$ :

$$\mathbf{f}_i := \begin{pmatrix} -D_{21} & \frac{1}{2}I - K_{22} \\ \frac{1}{2}I + K_{11} & -V_{12} \end{pmatrix} \begin{pmatrix} \varphi_1 \\ \varphi_2 \end{pmatrix}.$$

In order to discuss the existence and uniqueness for the solution of Eq. (16), we first consider for any $(\lambda_1, \lambda_2) \in H^{-1/2}(\Gamma) \times H^{-3/2}(\Gamma)$, the potential
$$v(x) = \mathbf{V}(\lambda_1, \lambda_2) \quad \text{for} \quad x \notin \Gamma.$$

Then $v$ is a solution of (1) in $\Omega \cup \Omega^c$, and we see that

$$\mathbf{V}\lambda|_\Gamma = \begin{pmatrix} \frac{\partial v}{\partial n}|_\Gamma \\ v|_\Gamma \end{pmatrix}, \quad \text{and} \quad \lambda = \begin{pmatrix} [Mv]|_\Gamma \\ [Nv]|_\Gamma \end{pmatrix},$$

where $[w]|_\Gamma$ denotes the jump of $w$ across the boundary $\Gamma$. Thus, we have the relation between the boundary- and the domain bilinear forms

$$\langle \mathbf{V}\lambda, \lambda \rangle = \int_\Gamma (\frac{\partial v}{\partial n}|_\Gamma \lambda_1 + v|_\Gamma \lambda_2)ds$$
$$= \int_\Gamma (\frac{\partial v}{\partial n}|_\Gamma [Mv] + v|_\Gamma [Nv])ds$$
$$= a_\Omega(v, v) + a_{\Omega^c}(v, v). \tag{17}$$

In view of (17), following Costabel and Wendland [5] we show that the boundary operator $\mathbf{V}$ satisfies a Gårding inequality on the product space $H^{-1/2}(\Gamma) \times H^{-3/2}(\Gamma)$ by introducing the cut-off function in the exterior domain $\Omega^c$. That is, $\mathbf{V}$ is a so-called strongly elliptic operator for which the classical Fredholm alternative holds. Also from Eq. (17), we see that the uniqueness of the solution of (16) follows from that of the interior and exterior Dirichlet problems. Theorem 1 and Theorem 4 suggest that we should solve the system (16) subject to the constraints (4), (5), i.e.,

$$\int_\Gamma \sigma_2(y)ds_y = 0, \tag{18}$$

$$\int_\Gamma \{\sigma_1(y)n_y + \sigma_2(y)y\}ds_y = 0. \tag{19}$$

Indeed, the system (16) together with (18) and (19) has a unique pair of solutions $(\sigma_1, \sigma_2)$. In fact, the system (16) together with (18) and (19) is equivalent to the following modified system as in the case of the Laplacian (see, e.g.Hsiao and MacCamy [7], and Hsiao and Wendland [8]), and we have the following result:

**Theorem 6.** *Given* $(\varphi_1, \varphi_2) \in H^{3/2}(\Gamma) \times H^{1/2}(\Gamma)$, *the modified system*

$$\mathbf{V}\sigma + \mathbf{M}(x)\,\omega = \mathbf{f}_i \tag{20}$$

*together with* (18) *and* (19) *has a unique solution*

$$(\omega, \sigma) \in \mathbb{R}^3 \times H^{-1/2}(\Gamma) \times H^{-3/2}(\Gamma),$$

*where* $\mathbf{M}$ *denotes the rigid motion matrix*

$$\mathbf{M}(x)\,\omega = \begin{pmatrix} 0 & n_1(x) & n_2(x) \\ 1 & x_1 & x_2 \end{pmatrix} \begin{pmatrix} \omega_0 \\ \omega_1 \\ \omega_2 \end{pmatrix}.$$

We remark that corresponding to the modified system (20), the representation for the solution of the interior Dirichlet problem then assumes the form

$$u(x) = V(\sigma_1, \sigma_2) - W(\varphi_1, \varphi_2) + p(x)$$

with $p(x) = \omega_0 + \omega_1 x_1 + \omega_2 x_2$, where $(\omega, \sigma)$ is the unique solution of the modified system in Theorem 6.

## 5.2. Exterior Dirichlet problem

For the exterior Dirichlet problem, by using $\mathcal{C}_{\Omega^c}$ and the representation (13), this leads to the following system for the unknowns $\sigma = (\sigma_1, \sigma_2)^\top$:

$$\mathbf{V}\sigma + \mathbf{M}(x)\,\omega = \mathbf{f}_e, \tag{21}$$

together with then side conditions:

$$\int_\Gamma \sigma_2(y)ds_y = -8\pi A_0, \tag{22}$$

$$\int_\Gamma \{\sigma_1(y)n_y + \sigma_2(y)y\}ds_y = 4\pi \mathbf{A}. \tag{23}$$

Here $p(x) = -\omega_0 - \omega_1 x_1 - \omega_2 x_2$, and $\mathbf{f}_e$ is given as

$$\mathbf{f}_e := \begin{pmatrix} D_{21} & \frac{1}{2}I + K_{22} \\ \frac{1}{2}I - K_{11} & V_{12} \end{pmatrix},$$

moreover $(A_0, \mathbf{A})$ are given so that the solution $u$ of the exterior Dirichelt problem has the asymptotic behavior (14) at infinity. A comparison of (21), (22), and (23) with the corresponding equations (20), (18) and (19) in Theorem 6 for the interior Dirichlet problem shows that they are the same equations with only different right hand sides. Thus, we conclude that the system (21), (22), and (23) has a unique solution

$$(\omega, \sigma) \in \mathbb{R}^3 \times H^{-1/2}(\Gamma) \times H^{-3/2}(\Gamma)$$

as expected.

### 5.3. *Interior Neumann Problem*

Now for the interior Neumann problem, we seek a solution of the form (8)

$$u(x) = -W(\varphi_1, \varphi_2) - V(\sigma_1, \sigma_2) \quad \text{in} \quad \Omega, \tag{24}$$

where

$$(\varphi_1, \varphi_2) = (u, \frac{\partial u}{\partial n})|_\Gamma \in H^{3/2}(\Gamma) \times H^{1/2}(\Gamma)$$

are the unknown Cauchy data and

$$(\sigma_1, \sigma_2) = (Mu, Nu)|_\Gamma \in H^{-1/2}(\Gamma) \times H^{-3/2}(\Gamma)$$

are the given Neumann data which are required to satisfy the compatibility conditions in Theorem 2. For the integral equations of the first kind, we employ the 3rd and 4th rows of $\mathcal{C}_\Omega$ in (10). Then from (9) we obtain the system

$$\mathbf{D}\varphi := \begin{pmatrix} D_{41} & D_{42} \\ D_{31} & D_{32} \end{pmatrix} \begin{pmatrix} \varphi_1 \\ \varphi_2 \end{pmatrix} = \mathbf{g}_i, \tag{25}$$

where $\mathbf{g}_i$ is given in terms of the Neumann data $(\sigma_1, \sigma_2)$ :

$$\mathbf{g}_i := \begin{pmatrix} -D_{43} & \frac{1}{2}I - K_{44} \\ \frac{1}{2}I + K_{33} & -V_{34} \end{pmatrix} \begin{pmatrix} \sigma_1 \\ \sigma_2 \end{pmatrix}.$$

Similar to the Dirichlet problem, let $(\phi_1, \phi_2) \in H^{3/2}(\Gamma) \times H^{1/2}(\Gamma)$ be any function and set

$$v(x) = -W(\phi_1, \phi_2) \quad \text{for} \quad x \notin \Gamma.$$

Then $v$ is a solution of (1) in $\Omega \cup \Omega^c$,

$$\mathbf{D}\phi|_\Gamma = \begin{pmatrix} Nv|_\Gamma \\ Mv|_\Gamma \end{pmatrix}, \quad \text{and} \quad \phi = \begin{pmatrix} [v]|_\Gamma \\ [\frac{\partial v}{\partial n}]|_\Gamma \end{pmatrix},$$

where $[w]|_\Gamma$ denotes the jump of $w$ across the boundary $\Gamma$. Similarly, we have the relation

$$\langle \mathbf{D}\phi, \phi \rangle = \int_\Gamma (Nv|_\Gamma \, [v] + Mv|_\Gamma \, [\frac{\partial u}{\partial n}]) ds$$

$$= a_\Omega(u, u) + a_{\Omega^c}(u, u).$$

The above relation implies that the boundary operator $\mathbf{D}$ now has the eigenspace $\{(p, \frac{\partial p}{\partial n})^\top \mid p \in \mathcal{R}\}$, although $\mathbf{D}$ satisfies a Gårding inequality on the product space $H^{3/2}(\Gamma) \times H^{1/2}(\Gamma)$. As a consequence, we have the following theorem.

**Theorem 7.** *Given* $(\sigma_1, \sigma_2) \in H^{-1/2}(\Gamma) \times H^{-3/2}(\Gamma)$
*satisfying the compatibility condition*

$$\int_\Gamma (\sigma_2 p + \sigma_1 \frac{\partial p}{\partial n}) ds = 0 \quad \forall p \in \mathcal{R},$$

*then the system of boundary integral equations of the first kind for the Neumann problem* (25) *has a unique solution* $(\varphi_1, \varphi_2)$ *in the quotient space*

$$[H^{3/2}(\Gamma) \times H^{1/2}(\Gamma)]/\mathcal{R}.$$

We remark that the right hand side of the system (25) satisfies the orthogonal conditions in order to ensure that the Fredholm alternative holds, since it is known that for the direct approach the right-hand always lie in the range of the operator under consideration.

For the computational purpose, by including additional normalization conditions we may augment the system (25) so that it is uniquely solvable. We will pursue this after we discuss the exterior Neumann problem.

### 5.4. *Exterior Neumann Problem*

For the exterior Neumann problem, we seek a solution in the form of (13):

$$u(x) = W(\varphi_1, \varphi_2) - V(\sigma_1, \sigma_2) + p(x), \tag{26}$$

where $\varphi = (\varphi_1, \varphi_2)^\top$ is the unknown, while $(\sigma_1, \sigma_2)$ as well as the linear function $p(x)$ are given. In the present case, it is worth mentioning that in contrast to the interior Neumann problem the given required Neumann data are not required to satisfy the compatibility conditions in Theorem 2. The system of boundary integral equations now reads

$$\mathbf{D}\varphi := \begin{pmatrix} D_{41} & D_{42} \\ D_{31} & D_{32} \end{pmatrix} \begin{pmatrix} \varphi_1 \\ \varphi_2 \end{pmatrix} = \mathbf{g}_e, \tag{27}$$

where $\mathbf{g}_e$ is given in terms of the Neumann data $(\sigma_1, \sigma_2)$ :

$$\mathbf{g}_e := -\begin{pmatrix} D_{43} & \frac{1}{2}I + K_{44} \\ \frac{1}{2}I - K_{33} & V_{34} \end{pmatrix} \begin{pmatrix} \sigma_1 \\ \sigma_2 \end{pmatrix}.$$

We note that the given linear function $p(x)$ doest not appear in the right hand side of (27), since both $Mp$ and $Np$ vanish. Clearly, Theorem 7 remains valid for the system (27) but without requiring the compatibility condition to hold.

We now return to the modified versions of (25) and (27). As in the case for the Dirichlet problems, we now modify the system (25) and (27) by adding the rigid motion term

$$\mathbf{M}(x)\omega = \begin{pmatrix} 0 & n_1(x) & n_2(x) \\ 1 & x_1 & x_2 \end{pmatrix} \begin{pmatrix} \omega_0 \\ \omega_1 \\ \omega_2 \end{pmatrix}$$

and consider the modified system:

$$\mathbf{D}\,\varphi + \mathbf{M}(x)\,\omega = \mathbf{g} \tag{28}$$

together the side conditions:

$$\int_\Gamma \varphi_1 ds = 0, \quad \int_\Gamma \varphi_1\, y + \varphi_2\, n_y ds = \mathbf{0}. \tag{29}$$

Here the right hand side of (28) $\mathbf{g}$ denotes either $\mathbf{g}_i$ or $\mathbf{g}_e$. The augmented system (28) and (29) is now uniquely solvable. We summary the result in the following.

**Theorem 8.** *Given* $(\sigma_1, \sigma_2) \in H^{-1/2}(\Gamma) \times H^{-3/2}(\Gamma)$
*(satisfying the compatibility condition*

$$\int_\Gamma (\sigma_2 p + \sigma_1 \frac{\partial p}{\partial n}) ds = 0 \quad \forall p \in \mathcal{R}$$

*in the case of interior Neumann problem), then the augmented system of boundary integral equations (28) and (29) has a unique solution* $(\varphi_1, \varphi_2)$ *in the product space* $H^{3/2}(\Gamma) \times H^{1/2}(\Gamma)$.

## 6. Concluding Remarks

We remark that one may also solve both interior and exterior boundary value problems by the boundary integral equations of the second kind and obtain their modified versions, employing the Calderón projectors and $\mathcal{C}_\Omega$. Details are available in Hsiao-Wendland [10]. Alternatively, one may solve the problems by employing boundary integral equations from the indirect approach. In this connection, we refer to the work of Hsiao-MacCamy [7] and Wendland et al [4].

We note that in order to establish Gårding's inequalities for the boundary bilinear forms for $\mathbf{V}$ and $\mathbf{D}$ in the same manner as in Costabel and Wendland [5], one needs the so-called generalized first Green formula in the Sobolev space

$$H^2(\Omega, \Delta^2) :=$$
$$\{w \in H^2(\Omega) : \Delta^2 w \in \widetilde{H}^{-2}(\Omega) = (H^2(\Omega))'\}$$

as well as in the space $H^2(\widetilde{\Omega^c}, \Delta^2)$ where $\widetilde{\Omega^c} = \Omega^c \cap B_R(0)$ is the truncated exterior domain in the neighborhood of $\Gamma$, and $B_R(0)$ is the disk with radius $R$ sufficiently large containing $\Omega$ in its interior. To conclude the paper, we now state the *Generalized first Green formula* for the biharmonic equation (1) below.

**Lemma 2.** *For fixed* $u \in H^2(\Omega, \Delta^2)$, *the mapping*

$$\gamma v = (v, \frac{\partial v}{\partial n}) \mapsto < \tau u, \gamma v >_\Gamma := a_\Omega(u, \mathcal{Z}\gamma v)$$

$$- \int_\Omega \Delta^2 u \mathcal{Z} \gamma v dx$$

*is a continuous linear functional $\tau u$ on*
$\gamma v \in H^{\frac{3}{2}}(\Gamma) \times H^{\frac{1}{2}}(\Gamma)$ *that for $u \in H^4(\Omega)$,*
$\tau u = (Nu, Mu)$. *The mapping*

$$\tau : H^2(\Omega, \Delta^2) \to H^{-\frac{3}{2}}(\Gamma) \times H^{-\frac{1}{2}}(\Gamma)$$

*with $u \mapsto \tau u$ is continuous. Here $\mathcal{Z}$ is a right inverse to the trace operator of $\gamma$.*

For a proof of Lemma 2, we refer to the monograph by Hsiao and Wendland [10].

## Acknowledgments

This research was supported in part by the German Research Foundation DFG under the Grant SFB 404 Multifield Problems in Continuum Mechanics.

## References

1. S. Agmon, *Lectures on Elliptic Boundary Value Problems* (D. Van Nostrand Co., Inc., 1965).
2. F. Cakoni, G.C. Hsiao and W.L. Wendland, *On the boundary integral equation method for a mixed boundary value problem of the biharmonic equation, Complex Variables* **50**, 681–696 (2005).
3. G. Chen and J. Zhou, *Boundary Element Methods* (Academic Press, 1992).
4. M. Costabel, E. Stephan and W.L. Wendland, *On boundary integral equations of the first kind for the bi-Laplacian in a polygonal plane domain, Ann. Scuola Norm. Sup. Pisa Cl. Sci.* **10**, 197–241 (1986).
5. M. Costabel and W. L. Wendland, *Strong ellipticity of boundary integral operators, J. Reine Angew. Math.* **372**, 34–63 (1986).
6. F.D. Gakhov, *Boundary Value Problems* (Pergamon Press, New York, 1966).
7. G.C. Hsiao and R.C. MacCamy, *Solution of boundary value problems by integral equations of the first kind, SIAM Review,* **15**, 687–705 (1973).
8. G. C. Hsiao and W.L. Wendland, *A finite element method for some integral equations of the first kind, J. Math. Anal. Appl.* **34**, 1-19 (1977).
9. G. C. Hsiao and W. L. Wendland, *Boundary element methods: foundation and error analysis* in *Encyclopedia of Computational Machanics* E.Stein, R.deBrost and T.J.R.Hughes ed. (John Wiley & Sons 2004, 339–373).
10. G.C. Hsiao and W.L. Wendland, *Boundary Integral Equations: Variational Methods* (Springer-Verlag: Heidelberg, to appear).
11. N.I. Muskhelishvili, Some basic Problems of the mathematical Theory of Elasticity (Noordhoff, Groningen, Groningen, 1953).

# SHOCK REFLECTION BY OBSTACLE

SHUXING CHEN*

*School of Mathematical Sciences,*
*Fudan University, Shanghai, 200433, China*
*E-mail: sxchen@public8.sta.net.cn*

The shock reflection problem is discussed, which is the basic phenomenon in fluid dynamics and related to quasilinear hyperbolic partial differential equations.

**Key words:** Shock reflection, Euler system, Elliptic-hyperbolic system, Compressible flow, Mach reflection

**Mathematics Subject Classification:** Primary 35L65; Secondary 35L67

## 1. Introduction

The lecture is mainly concerned with the shock reflection problem. Shock wave is a basic phenomenon in fluid dynamics. It is also most important in studying quasilinear hyperbolic equations, because singularities can be developed from a smooth solution to the nonlinear hyperbolic equations generally. When a shock meets an obstacle in its propagation process, it will be reflected by this obstacle. Depending on the shape of the shock front and the obstacle, as well as the parameters of the fluid on both sides of the shock, various different patterns of shock reflection occur. To describe the phenomena of shock reflection one has to treat complicated problems of partial differential equations, among them great many are still open. In this lecture we would like to give a brief description on the study of this subject.

## 2. Plane shock reflection

Let start with a simplest case: a plane shock hit a plane wall. First we assume that the shock moves with a speed $V$ perpendicular to the wall, as well as to the shock front itself. The parameters ahead of the shock are the pressure $p_1$, the density $\rho_1$ and the velocity $u_1 = 0$. Correspondingly, behind the shock front these parameters are $p_0, \rho_0$ and $u_0$ with $p_0 > p_1$. Then the Rankine-Hugoniot conditions on shock

---

*The paper is partially supported by National Natural Science Foundation of China, the Key Grant of National Ministry of Science and Technology of China and the Doctorial Foundation of National Educational Ministry

give the following equalities.

$$\rho_0(u_0 - V) = \rho_1(u_1 - V), \tag{1}$$

$$p_0 + \rho_0(u_0 - V)^2 = p_1 + \rho_1(u_1 - V)^2, \tag{2}$$

$$\frac{1}{2}(u_0 - V)^2 + i_0 = \frac{1}{2}(u_1 - V)^2 + i_1, \tag{3}$$

where $i$ is enthalpy, which equals $\dfrac{\gamma p}{(\gamma - 1)\rho}$ for polytropic gas with adiabatic exponent $\gamma$.

When the shock is reflected, then $(p_0, \rho_0, u_0)$ becomes the state ahead of the shock front. Meanwhile, the state behind the shock is described by the parameters $p_2, \rho_2$ and $u_2 = 0$, satisfying $p_2 > p_0$ and

$$\rho_0(u_0 - V') = \rho_2(u_2 - V'), \tag{4}$$

$$p_0 + \rho_0(u_0 - V')^2 = p_2 + \rho_1(u_2 - V')^2, \tag{5}$$

$$\frac{1}{2}(u_0 - V')^2 + i_0 = \frac{1}{2}(u_2 - V')^2 + i_2, \tag{6}$$

where $V'$ is the velocity of the reflected shock determined by the above Rankine-Hugoniot conditions.

Next we consider the case when a plane shock hits a plane wall with an angle $\beta$. If the velocity of the shock is $V$, then the intersection of the incident shock and the wall moves with a speed $V / \sin \beta$. By taking a coordinate system moving with the intersection both the incident shock and the reflected shock are stationary. Assume that $x$-axis is taken along the wall, then the parameters ahead the shock front and the parameters behind the shock front obey the relation as follows [14] :

$$\rho_0 u_{n0} = \rho_1 u_{n1}, \tag{7}$$

$$p_0 + \rho_0 u_{n0}^2 = p_1 + \rho_1 u_{n1}^2, \tag{8}$$

$$u_{t0} = u_{t1}, \tag{9}$$

$$\frac{1}{2}q_0^2 + i_0 = \frac{1}{2}q_1^2 + i_1, \tag{10}$$

where $u_{n0}, u_{t0}$ are the components of the velocity $\vec{u_0}$ in the normal direction and the tangential direction respectively, similar meaning is for $u_{n1}$ and $u_{t1}$.

The relation of the parameters ahead the shock front and the parameters behind the shock front can be described by shock tthe polar, which is a locus of all possible states behind the shock front connecting with a fixed state ahead of shock front. When the angle $\beta$ of the incident shock front with the wall is given, one can find a unique point on the shock polar, which corresponds to the state behind the shock front. To determine the location of the reflected shock as well as the state behind the reflected shock, one can again use the shock polar. However, for a given turning angle of velocity of the flow, the location of the possible shock is not unique. There are generally two possible locations for the reflected shocks, both are capable to connect with the state ahead of shock satisfying Rankine-Hugoniot conditions and entropy condition [14]. According to the strength of the reflected shock we distinguish

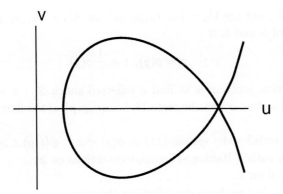

Fig. 1.   Shock polar

them as weaker shock reflection and the strong shock reflection. Generally, for the weaker reflection the flow behind shock is still relatively supersonic and is stable, while for the strong reflection the flow behind the shock is relatively subsonic and is only conditionally stable. Similar phenomena also occur in supersonic flow past a wedge ( see [6,10,15] ).

Another important phenomenon is that if the angle between the shock front and the wall is somehow large ( for instance, near to $\pi/2$ ), then the simple reflection of oblique shock is impossible. The phenomenon can also be observed from the shock polar. Deleting the superfluous part of the shock polar ( corresponding the case when the entropy condition is violated ) the shock polar is a closed curve apart from the origin: see Fig.1. Then a ray starting from the origin with sufficiently large slope will not intersect the shock polar, so that the simple reflection could not occur. In this case Mach reflection will occur, see the last section of this paper.

## 3.  Shock reflection by a smooth surface

When a shock front hits an obstacle with curved surface, the reflection of shock has to be described by partial differential equations. If the surface of the obstacle is smooth, then the local existence of the problem on shock reflection has been established. [9] For notational simplicity let us consider the case of space-dimension 2. The system of conservation laws for inviscid compressible flow takes the form

$$\frac{\partial}{\partial t}\begin{pmatrix} \rho \\ \rho u \\ \rho v \\ \rho(e + \frac{1}{2}q^2) \end{pmatrix} + \frac{\partial}{\partial x}\begin{pmatrix} \rho u \\ p + \rho u^2 \\ \rho uv \\ \rho u(i + \frac{1}{2}q^2) \end{pmatrix} + \frac{\partial}{\partial y}\begin{pmatrix} \rho v \\ \rho uv \\ p + \rho v^2 \\ \rho v(i + \frac{1}{2}q^2) \end{pmatrix} = 0, \quad (11)$$

where $(u, v)$ are the velocity components, $p, \rho, e, i$ stand for pressure, density, inner energy and enthalpy respectively, $q^2 = u^2 + v^2$.

Assume that the moving shock front is $S : x = Vt$, and the surface of the obstacle is $\Sigma : x = \phi(y)$ with $\phi(0) = \phi'(0) = 0$. For $t < 0$ the states on the both sides of the shock front are $\mathbf{U_0} = (u_0, v_0, p_0, \rho_0)$ and $\mathbf{U_1} = (u_1, v_1, p_1, \rho_1)$. For $t \geq 0$ the intersection of $S$ and $\Sigma$ is

$$\sigma: \quad x = \phi(y), \ t = \frac{1}{V}\phi(y). \tag{12}$$

The shock reflection problem is to find a reflected shock $S_1 : x = \psi(t, y)$ passing through the intersection $\sigma$ and the state $\mathbf{U} = (u, v, p, \rho)$ of the flow between $S_1$ and $\Sigma$, such that

(1) $\mathbf{U}(t, x, y)$ satisfies the system (11) in $\phi(y) < x < \psi(t, y), t > 0$.
(2) $\mathbf{U}$ and $\mathbf{U_1}$ satisfy Rankine-Hugoniot conditions on $S_1$.
(3) $u - \phi_y v = 0$ on $\Sigma$.

For such a problem we have the following theorem:

**Theorem 3.1.** *Assume that the states $\mathbf{U_0}$, $\mathbf{U_1}$ satisfy Rankine-Hugoniot conditions on $S$ and $p_0 > p_1$, $u_1 = v_1 = 0$. The function $\phi(y)$ is sufficiently smooth and satisfies $\phi(0) = \phi'(0) = 0$. Then there exists a function $x = \psi(t, y)$ defined in a neighborhood of the origin, satisfying $\psi(\phi(y)/V, y) = \phi(y)$, and a local solution $\mathbf{U}(t, x, y)$ defined in $\phi(y) < x < \psi(t, y)$ satisfying the above requirements (1) to (3). $(\psi(t, y), \mathbf{U}(t, x, y))$ forms a perturbation of the normal shock reflection locally.*

By using a coordinate transformation the problem can be reduced to a nonlinear Goursat problem in an angular domain, whose two lateral planes are the image of the unknown shock front $S_1$ and the surface of the obstacle. By introducing a weighted Sobolev space one can establish energy estimates for the solution of linearized problem in this norm. To look for the solution of nonlinear problem, we first construct an approximate solution near the edge of the angular domain, which is a kind of expansion of finite power series of the unknown function in fact. Then starting from it one can establish an iteration process to modify the approximate solution successively, and establish a convergent sequence of approximate solutions. The limit of the sequence solves the nonlinear Goursat problem. The details of the proof can be found in [9].

## 4. Shock reflection by a ramp

When the surface of the obstacle is not smooth, interaction of reflected shock occurs. The interaction will cause very complicated picture. Therefore, to get a clear understanding one has to reduce the problem to some typical cases. When the space-dimension is two, the simplest case is that the obstacle is a wedge, formed by two planes. When an incident shock hits the wedge, it will be reflected by two lateral planes of the wedge. In many cases, one can study these two reflective shock independently. Therefore, a typical problem is so-called *shock reflection by a ramp*. That is, an incident plane shock perpendicular to the horizontal line is moving forward and meets a plane ramp with angle of inclination $\tau$, then the shock front is reflected by the ramp. According to the angle of inclination of the ramp and the parameters of the shock various reflection pattern may occur. Basically, they are divided into two classes: regular reflection and Mach reflection, which will be discussed in this and the next sections.

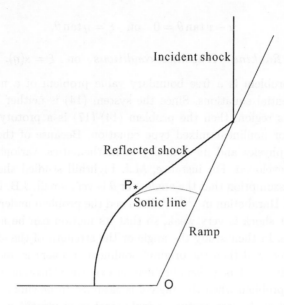

Fig. 2.  Regular reflection of shock by ramp

Suppose that we consider the problem on the half plane $y > 0$. The ramp locates at $y = x \tan \tau$. At the time $t = 0$ the shock front $x = const.$ with velocity $V$ arrives at the origin. For $t > 0$, the shock front continuously goes forward, then it will be reflected by the ramp. When the angle $\theta = \pi/2 - \tau$ is small, then the intersection $(Vt, Vt \tan \tau)$ is outside of the influence domain of the origin. In this case the reflected shock and the flow field near the intersection can be determined as an oblique reflection of plane shock [8,31]. Afterwards, the straight reflected shock bends down and forms an expanding bubble. Such a flow pattern is called *regular reflection*. In this case the problem is to determine the flow in the region bounded by

$$y = 0, \quad x = y \tan \theta, \quad \text{and the reflected shock } x = \zeta(t, y). \tag{13}$$

Since both the differential system (11) governing the flow and the boundary condition $u \cos \theta - v \sin \theta = 0$ are invariant under the transformation $t \to \alpha t$, $x \to \alpha x, y \to \alpha y$, then we may only consider the self-similar solution of the problem. Let $\xi = x/t, \eta = y/t, \zeta = ts(\eta)$, the system (11) is reduced to

$$(-\xi \frac{\partial}{\partial \xi} - \eta \frac{\partial}{\partial \eta}) \begin{pmatrix} \rho \\ \rho u \\ \rho v \\ \rho(e + \frac{1}{2}q^2) \end{pmatrix} + \frac{\partial}{\partial \xi} \begin{pmatrix} \rho u \\ p + \rho u^2 \\ \rho u v \\ \rho u(i + \frac{1}{2}q^2) \end{pmatrix} + \frac{\partial}{\partial \eta} \begin{pmatrix} \rho v \\ \rho u v \\ p + \rho v^2 \\ \rho v(i + \frac{1}{2}q^2) \end{pmatrix} = 0, \tag{14}$$

Meanwhile, the boundary conditions are reduced to

$$v = 0 \quad \text{on} \quad \eta = 0, \tag{15}$$

$$u - v \tan \theta = 0 \quad \text{on} \quad \xi = \eta \tan \theta, \tag{16}$$

$$Rankine - Hugoniot \quad conditions \quad \text{on} \quad \xi = s(\eta). \tag{17}$$

The above problem is a free boundary value problem of a nonlinear system of partial differential equations. Since the system (14) is neither pure hyperbolic nor elliptic in its region, then the problem (14)-(17) is a prototype of boundary value problem for nonlinear mixed type equation. Because of the importance of the problem in physics and its difficulty in mathematics, various approximation methods were developed. For instance, M.J. Lighthill studied shock reflection in [25,26] under the assumption that the angle $\tau$ or $\theta$ is very small, J.B. Keller, A. Black, J. Hunter and E. Harabetian in [19,22,24] discussed the problem under the assumption that the incident shock is very weak, so that its motion can be approximated by an acoustic wave. In their study the angle or the strength of the shock is taken as a small parameter, and then the original nonlinear problem is linearized by using this parameter. In [8] S. Chen also obtained an existence theorem for the linearized shock reflection problem when the angle $\theta$ is assumed to be small.

Since the end of the last century, people start to study the nonlinear problem directly. Due to the complexity of the full Euler system various new models to describe the compressible flow are employed. In [5-7] the authors took UTSD (Unsteady Transonic Small Disturbance) equation as a model to study shock reflection by a ramp systematically. The equation is a simplification of the full Euler system, and was introduced by M. Brio, J. Hunter and others (see [2,34]). The model has the following form:

$$u_t + u u_x + v_y = 0, \quad v_x - u_y = 0. \tag{18}$$

It can also be reduced to a second order equation as

$$\left( (u - \rho) u_\rho + \frac{u}{2} \right)_\rho + u_{\eta\eta} = 0 \tag{19}$$

This model offers a good approximation near the intersection of the shock and the ramp. An advantage of the model is that in the equation (19) the coefficients of the second derivatives only depend on the unknown function, but not on its first derivatives. This feature does not holds generally for a nonlinear second order equation derived from a nonlinear first order system. Noticing this advantage B.L.Keyfitz, S.Canic and E.H.Kim carefully established the approximate process of the free boundary value problem for a quasi-linear degenerate elliptic equation. The process finally leads to the existence of solutions to the shock reflection problem [6,7]. The method is also applied to deal with similar problem for nonlinear wave equation [3,23]

$$\rho_{tt} = \nabla(c^2(\rho)\nabla\rho). \tag{20}$$

Another good model as a simplification of the full Euler system is potential flow equation. Consider the irrotational flow, one can introduce a velocity potential $\Phi(x, y)$ satisfying $\nabla\Phi = (u, v)$. By using potential $\Phi$ the system (11) can be reduced to

$$(H(\nabla\Phi))_t + (\Phi_x H(\nabla\Phi))_x + (\Phi_y H(\nabla\Phi))_y = 0, \tag{21}$$

where $H(\nabla\Phi) = h^{-1}(-(\Phi_t + \frac{1}{2}|\nabla\Phi|^2))$ and $h$ is enthalpy, which equals $\dfrac{\gamma p}{(\gamma - 1)\rho}$ for polytropic gas. Similar to above discussion one can consider the self-similar solution of (21). By introducing $\xi = x/t, \eta = y/t, \psi(\xi, \eta) = \Phi(t, x, y)$ (20) is reduced to

$$(-\xi\frac{\partial}{\partial\xi} - \eta\frac{\partial}{\partial\eta})H + \frac{\partial}{\partial\xi}(\psi_\xi H) + \frac{\partial}{\partial\eta}(\psi_\eta H) = 0. \tag{22}$$

Correspondingly, the shock reflection problem is reduced to a free boundary value problem of (22). Since the coefficients of (22) depend on the derivatives of the unknown functions, the method in [7,23] does not work in this case. Therefore, more careful estimate and treatment are inevitable. In [13] G.Q.Chen and M.Feldman have shown their result for regular reflection by using the model of potential flow equation. Besides, we also noticed that Y.Zheng is attacking the more difficulty case, when the full Euler system model is considered, see [35].

## 5. Mach reflection

### 5.1. *Mach configuration*

When the angle $\theta$ exceeds a critical value, the simple shock reflection pattern near the intersection of the incident shock front and the ramp could not occur. Instead, the intersection of the incident shock and the reflected shock will stay away from the ramp, and connected by a so-called Mach stem to the ramp. The triple intersection causes a new shock structure first confirmed by von Neumann [30]. An important fact is that the structure solely containing three shock fronts separating the neighborhood of the triple intersection into three zones with different continuous states does not exist [14,32]. Many physical experiments indicate that there is a slip line issuing from the triple intersection. Such a local wave pattern is called *Mach configuration* or *Mach structure*. Correspondingly, the whole process of reflection is called *Mach reflection*. Due to the appearance of the slip line, which corresponds to a contact discontinuity in a given flow, one has to use the full Euler system to describe Mach reflection, because both nonlinear wave equation or potential flow equation do not contain linear degenerate characteristics. Unfortunately, people do not know much on Mach reflection mathematically so far. Particularly, the global existence of Mach reflection is a completely open problem.

### 5.2. *Flat Mach Configuration and its perturbation*

Mach configuration also appears in many other problems involving shock reflection or interaction. It is natural and necessary to start work on getting a clear under-

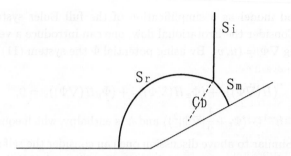

Fig. 3. Mach reflection of shock by ramp

standing of the local Mach configuration. Recently, we have obtained the stability of Mach configuration for stationary flow and pseudo-stationary flow [11,12]. Next we introduce the result on the stability of Mach configuration in pseudo-stationary compressible flow and the sketch of its proof.

As mentioned in the previous section for the shock reflection by a ramp we can restrict ourselves to consider the self-similar solution, which only depends on the variables $\xi = x/t$ and $\eta = y/t$. By introducing the relative velocity $U = u - \xi, V = v - \eta$, $(U, V)$ and the pressure $p$, the entropy $s$ satisfy the Euler system with the following form:

$$
\begin{pmatrix} \rho U & & 1 & \\ & \rho U & & \\ 1 & & a^{-2}\rho^{-1}U & \\ & & & U \end{pmatrix} \frac{\partial}{\partial \xi} \begin{pmatrix} U \\ V \\ p \\ s \end{pmatrix}
$$

$$
+ \begin{pmatrix} \rho V & & & \\ & \rho V & 1 & \\ & 1 & a^{-2}\rho^{-1}V & \\ & & & V \end{pmatrix} \frac{\partial}{\partial \eta} \begin{pmatrix} U \\ V \\ p \\ s \end{pmatrix} + \begin{pmatrix} \rho U \\ \rho V \\ 2 \\ 0 \end{pmatrix} = 0. \qquad (23)
$$

Meanwhile, on the shock front and the slip line the parameters of the flow satisfy Rankine-Hugoniot conditions

$$[\rho U]\psi_\xi - [\rho V] = 0, \qquad (24)$$

$$[p + \rho U^2]\psi_\xi - [\rho UV] = 0, \qquad (25)$$

$$[\rho UV]\psi_\xi - [p + \rho V^2] = 0, \qquad (26)$$

$$[\rho \tilde{E}U + pU]\tilde{\psi}_\xi - [\rho \tilde{E}V + pV] = 0, \qquad (27)$$

where $\tilde{E} = e + \frac{1}{2}(U^2 + V^2)$.

Mach configuration is composed of three shock fronts and one slip line in a flow field. All these curves bearing discontinuity of flow parameters intersect at one point. A special case is that all these curves are straight rays starting from the intersection, and the flow parameters are piecewise constant, i.e. all flow parameters in the sectors bounded by these rays are constant. Such a wave configuration is called *flat Mach configuration*. Obviously, the system (23) is satisfied in each sector, then the flow parameters and the slope of the rays separating these sectors have to satisfy Rankine-Hugoniot conditions (24)-(27). Among the four sectorial regions the flow in one sector is relatively supersonic flow, which is called upstream flow. Let the coordinate system moves with the intersection together, then the origin is relatively fixed ( as it in the self-similar coordinate system ) and the upstream flow is supersonic. The supersonic flow passes two shock fronts: one is supersonic-supersonic shock $S_1$ ( amounts to incident shock ), and the other is supersonic-subsonic shock $S_2$ ( amounts to Mach stem ). The supersonic flow behind $S_1$ will passes another shock $S_3$ ( amounts to reflected shock ) and then becomes to subsonic flow. Two subsonic flows in two different regions are adjacent and separated by a contact discontinuity $D$. When the upstream flow and the slope of the incident shock are given, then all flow in other three regions and the rays separating these regions can be determined by using shock polar ( see [12] ).

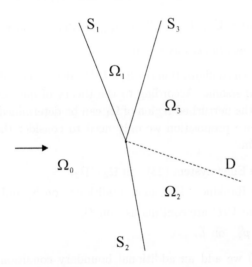

Fig. 4.   Flat Mach configuration

When the upstream flow is perturbed, then the whole configuration is also perturbed. Our aim is to prove that the configuration still holds, and all elements including the parameters in a bounded domain, as well as the location of all shock fronts and the slip line, are perturbed.

Assume that $\mathbf{U}_i^0, \mathbf{U}_i$ ($0 \le i \le 3$) are the unperturbed and the perturbed states in $\Omega_i$ respectively. $\eta = \psi_i^0 \xi$, $\eta = \psi_i(\xi)$, are the slope of the unperturbed and the perturbed curves bearing singularity of solution respectively. Also assume that $\alpha_0$

is a number in $(0,1)$,

$$N_\varepsilon = \{\mathbf{U_0}(\xi, \eta); \mathbf{U}_0(0,0) = \mathbf{U}_0^0, \|\mathbf{U}_0 - \mathbf{U}_0^0\|_{C^{2,\alpha_0}(\Omega_0^\varepsilon)} < \varepsilon\}, \qquad (28)$$

$$L_\varepsilon = \{\psi_1(\xi); \psi_0(0) = 0, \psi_0'(0) = \psi_1^0, \|\psi_1(\xi) - \psi_1^0\xi\|_{C^{2,\alpha_0}(-b,0)} < \varepsilon\}, \qquad (29)$$

then we have

**Theorem 5.1.** *Assume that* $(\mathbf{U}_0^0, S_1^0, \mathbf{U}_1^0, S_2^0, \mathbf{U}_2^0, D^0, \mathbf{U}_3^0, S_3^0)$ *forms a flat Mach configuration under suitable assumptions on the parameters. Assume that for given $R > 0, e > 0$ and small number $\varepsilon > 0$, the upstream flow $\mathbf{U_0} \in N_\varepsilon$ and $\psi_1(\xi) \in L_\varepsilon$, then there is $R_1 \in (0, R)$, $\ell < R_1$ and a Mach configuration $\{\mathbf{U}_0, S_1, \mathbf{U}_1, S_2, \mathbf{U}_2, D, \mathbf{U}_3, S_3\}$ in $C_{R_1}$, satisfying Euler system in corresponding domains, Rankine-Hugoniot conditions on the shock fronts and the slip line, and the pressure control on the cut-off boundary. Moreover,*

$$\psi_i(0) = 0, \psi_i'(0) = \psi_i^0, \ \|\psi_i(\xi) - \psi_i^0\xi\|_{C^{2,\alpha}(0,\ell)} < \varepsilon_1, \quad i = 2,3,4$$

$$\mathbf{U}_1(0,0) = \mathbf{U}_1^0, \ \|\mathbf{U}_1 - \mathbf{U}_1^0\|_{C^{2,\alpha}(\Omega_1 \cap C_{R_1})} < \varepsilon_1,$$

$$\mathbf{U}_i(0,0) = \mathbf{U}_i^0, \|\mathbf{U}_i - \mathbf{U}_i^0\|_{C^{1,\alpha}(\Omega_{iL})} < \varepsilon_1, \quad i = 2,3,$$

*where $\alpha \in (0, \alpha_0]$, $\varepsilon_1$ tends to zero as $\varepsilon \to 0$.*

Since for the flat Mach configuration the flow $\mathbf{U}_1^0$ is also supersonic, then its small perturbation is also supersonic. According to the theory of quasilinear hyperbolic system the flow $\mathbf{U}_1$ in the perturbed region of $\Omega_1$ can be determined independently. Then to prove the above proposition we only need to consider the following free boundary value problem:

$$(FB): \begin{cases} \text{The Euler system (23)} & \text{in } \Omega_2 \cup \Omega_3, \\ \text{The Rankine} - \text{Hugoniot conditions} & \text{on } S_2 \text{ and } S_3, \\ p \text{ and } V/U \text{ are continuous} & \text{on } D, \\ p = p_2^0 & \text{on } L. \end{cases} \qquad (30)$$

For the problem $(FB)$ we add an additional boundary condition $p = p_2^0$ on $L$. This is necessary because we only study the local structure in this stage and the system is not pure hyperbolic so that the state is not completely determined by its upstream part. Here $p = p_2^0$ is a simplified version of the condition given on $L$. The real physical condition should be assigned when a global existence is considered.

To solve the free boundary problem $(FB)$, one should overcome some crucial difficulties. First, the problem is a free boundary value problem, in which the shock fronts $S_2$ and $S_3$ are unknown, and should be determined together with the solution of the problem. Second, the slip line $D$ is also an unknown curve locating inside the flow field, so that the solution has discontinuity in the domain. Another difficulty is that the system (23) is neither hyperbolic nor elliptic, which is an elliptic-hyperbolic composed system. Finally, the corners of the domain also cause some trouble, because it decreases the smoothness of the precise solution and all approximate solutions for the elliptic equation near the corners. Next we will mainly explain the way to deal with the slip line bearing contact discontinuity and the decomposition technique to overcome the difficulty caused by the elliptic-hyperbolic composed system.

### 5.3. *Generalized Lagrange transformation*

Since the solution of the system (23) has discontinuity on the slip line, which is also a stream line, then it is convenient if one can introduce a transformation $T$ to straighten all stream lines. Such a transformation can be found by using generalized Lagrange transformation. Suppose that the solution $\mathbf{U} = (U, V, p, s)$ to the problem $(FB)$ is given, we define $\mu$ by

$$(U\frac{\partial}{\partial \xi} + V\frac{\partial}{\partial \eta})\mu - 2\mu = 0 \tag{31}$$

with the initial value $\mu = 1$ on the shock front, and then introduce variables $x, y$ by

$$\begin{cases} \dfrac{\partial x}{\partial \xi} = 1, \quad \dfrac{\partial x}{\partial \eta} = 0, \\ \dfrac{\partial y}{\partial \xi} = -\mu\rho V, \quad \dfrac{\partial y}{\partial \eta} = \mu\rho U, \\ x(\xi_0, \eta_0) = 0, \quad y(\xi_0, \eta_0) = 0. \end{cases} \tag{32}$$

Since $\mu$ satisfies the equation (31), then the function $x(\xi, \eta), y(\xi, \eta)$ are well defined. In fact, one can define

$$x = \xi, \quad y = \int_{(\xi_0, \eta_0)}^{(\xi, \eta)} -\mu\rho V d\xi + \mu\rho U d\eta, \tag{33}$$

The transform $T : (\xi, \eta) \mapsto (x, y)$ is called generalized Lagrange transformation. Under such a transformation the system of conservation laws becomes

$$\frac{\partial}{\partial x}(\frac{1}{\mu\rho U}) = \frac{\partial}{\partial y}(\frac{V}{U}), \tag{34}$$

$$\rho U \frac{\partial U}{\partial x} + \frac{\partial p}{\partial x} - \mu\rho V \frac{\partial p}{\partial y} + \rho U = 0, \tag{35}$$

$$\frac{1}{\mu}\frac{\partial V}{\partial x} + \frac{\partial p}{\partial y} + \frac{1}{\mu}\frac{V}{U} = 0, \tag{36}$$

$$\frac{\partial}{\partial x}(\tilde{E} + \frac{p}{\rho}) + \frac{\rho}{U}(U^2 + V^2) = 0. \tag{37}$$

Denote the image of $\eta = \psi(\xi)$ by $y = \chi(x)$, the Rankine-Hogoniot conditions are

$$
\begin{cases}
[\frac{1}{\rho U}]\frac{\chi'}{\mu} = -[\frac{V}{U}], \\
[\frac{1}{\rho U}(p + \rho U^2)]\frac{\chi'}{\mu} = -[\frac{pV}{U}], \\
[V]\frac{\chi'}{\mu} = [p], \\
[\tilde{E} + \frac{p}{\rho}] = 0.
\end{cases}
\tag{38}
$$

Then the problem $(FB)$ is reduced to a new free boundary value problem

$$
(FB)_1 : \begin{cases}
\text{The Euler system } (34) - (37) \text{ and } (31), \\
\text{Rankine} - \text{Hugoniot conditions } (38) \text{ and } \mu = 1 \quad \text{on } \Gamma_2 \text{ and } \Gamma_3, \\
p \text{ and } V/U \text{ are continuous} \quad \text{on } \Gamma_D, \\
p = p_0 \quad \text{on } \Gamma_L.
\end{cases}
\tag{39}
$$

where $\Gamma_2, \Gamma_3, \Gamma_D$ are the image of $S_2, S_3, D$ under the transformation $T$ respectively. In the problem $(FB)_1$ all stream lines including the slip line become lines parallel to $x$-axis.

### 5.4. Decomposition

Next we introduce the decomposition technique, which can also be applied in other cases involving Euler system.

Notice that the Euler system (23) is hyperbolic if $U^2 + V^2 > a^2$, and is neither pure hyperbolic nor elliptic if $U^2 + V^2 < a^2$. In the domain behind the reflected shock front and the Mach stem the pseudovelocity is subsonic, so (34)-(37) is an elliptic-hyperbolic composed system. To deal with such a composed system we reduce the system to a canonical form, which decouples the elliptic part and the hyperbolic part of the system at the level of the principal part. Write the system in the matrix form as

$$
A\frac{\partial}{\partial x}U + B\frac{\partial}{\partial y}U + D = 0,
\tag{40}
$$

where

$$
A = \begin{pmatrix} U & \frac{1}{\rho} & & \\ & U & & \\ \frac{1}{\rho} & \frac{U}{a^2\rho^2} & & \\ & & 1 & \\ & & & U \end{pmatrix}, \quad
B = \begin{pmatrix} & -\mu V & & \\ & \mu U & & \\ -\mu V & \mu U & 0 & \\ & & 0 & \\ & & & 0 \end{pmatrix}, \quad
D = \begin{pmatrix} U \\ V \\ \frac{2}{\rho} \\ 0 \\ -2 \end{pmatrix}
$$

The characteristic polynomial of this system is

$$D(\lambda) = det(\lambda A - B)$$
$$= \lambda^3 U(\frac{\lambda^2 U^2}{a^2 \rho^2} - (\mu V + \frac{\lambda}{\rho})^2 - \mu^2 U^2).$$

Then $D(\lambda) = 0$ has a root $\lambda = 0$ with multiplicity 3 and other two roots

$$\lambda_\pm = \frac{\mu a^2 \rho V \pm \mu a \rho U \sqrt{U^2 + V^2 - a^2}}{U^2 - a^2}.$$

Denote

$$\lambda_\pm = \lambda_R \pm i\lambda_I = \frac{\mu a^2 \rho V}{U^2 - a^2} \pm \frac{i\mu a \rho U \sqrt{a^2 - U^2 - V^2}}{U^2 - a^2}.$$

Multiplying the eigenvectors $\ell_\pm$ corresponding to $\lambda_\pm$ on (23), then separating the real part and imaginary part, we obtain

$$-V D_R U + U D_R V - \frac{\mu}{a\rho}\sqrt{a^2 - U^2 - V^2} D_I p + \frac{\lambda_R}{\rho} = 0, \qquad (41)$$

$$-V D_I U + U D_I V + \frac{\mu}{a\rho}\sqrt{a^2 - U^2 - V^2} D_R p + \frac{\lambda_I}{\rho} = 0, \qquad (42)$$

where $D_R = \frac{\partial}{\partial x} + \lambda_R \frac{\partial}{\partial y}, D_I = \lambda_I \frac{\partial}{\partial y}$. Therefore, the Euler system with (31) is finally decomposed to the form

$$\begin{cases} D_R W - e D_I p + \dfrac{\lambda_R}{\rho U^2} = 0, \\ D_I W + e D_R p + \dfrac{\lambda_I}{\rho U^2} = 0, \\ \dfrac{\partial U}{\partial x} + d_W \dfrac{\partial W}{\partial x} + d_p \dfrac{\partial p}{\partial x} + 1 = 0, \\ \dfrac{\partial p}{\partial x} - \dfrac{\gamma p}{\rho}\dfrac{\partial \rho}{\partial x} = 0, \\ U\dfrac{\partial \mu}{\partial x} - 2\mu = 0, \end{cases} \qquad (43)$$

where $e, d_W, d_p$ are functions of flow parameters, we omit their detailed expressions. In (43) the first two equations form an elliptic system, while the last three are hyperbolic. Such a decomposition will help us to derive necessary estimates.

### 5.5. *Estimates and existence*

The remaining work is to establish estimates for the linearized problem and by using a suitable approximation process to solve the nonlinear problem.

Since the domain where the system (40) or (43) is defined, has corners and contains a line bearing discontinuity of the unknown functions, then the regularity of the expected solution is restrictive. Applying the result in [17,18], we can establish a global $H^1$ estimate and a piecewise $H^2$ or Hölder estimate. These estimates lead us to the corresponding regularity of approximate solutions and the precise solution to the nonlinear problem.

In order to finally solve the free boundary value problem $(FB)_1$, we reduce the whole free boundary value problem to two problems: one is a regular boundary value problem for Euler system in a fixed domain, the other is an initial value problem of an ordinary differential equation updating the shock fronts. By seeking the solutions to these two problems we establish a map defined on a closed convex set in a suitable functional space, then the demonstration of the existence of the problem $(FB)_1$ amounts to prove the existence of a fixed point for such a given map. Such a framework was employed and developed in [5–7,9,10,28] etc. In [12] we combine all above treatment and prove the existence of solutions to free boundary value problem $(FB)_1$, and then obtain Theorem 5.1.

Let us end this lecture by showing some problems, which is quite open today.

1) To establish the global existence of solutions to Mach reflection.

2) When the incline angle $\tau$ of the ramp gradually decrease from $\pi/2$, then the regular reflection will change to Mach reflection. When and how a regular reflection transits to Mach reflection?

3) The Mach reflection as shown in Fig.4 is called *simple Mach reflection*. In some cases of shock reflection by a ramp more than one triple point will appear, then one will have other wave patterns like *double Mach reflection, complex Mach reflection*. To make these wave structures clear are also strongly expected.

# References

1. G.Ben-Dor, *Shock Waves Reflection Phenomena*. Springer-Verlag. New York, 1992.
2. M.Brio and J.K.Hunter, *Mach reflection for the two-dimensional Burgers equation*, Phys. D, **60** (1992), 148-207.
3. S.Canic and B.Kerfitz, *Riemann problems for the two-dimensional unsteady transonic small disturbance equation*, SIAM J. Appl. Math. **58** (1998), 636-665.
4. S.Canic and E.H.Kim, *A class of quasilinear degenerate elliptic problems* J. Differential Equations. **189** (2003), 71-98.
5. S.Canic, B.Kerfitz and G.Lieberman, *A proof of existence of perturbed steady transonic shocks via a free boundary problem*, Comm. Pure Appl. Math. **53** (2000),484-511.
6. S.Canic, B.Kerfitz and E.H.Kim, *A free boundary problems for unsteady transonic small disturbance equation: transonic regular reflection*, Methods and Appl. Anal. **7**(2000),313-336.
7. S.Canic, B.Kerfitz and E.H.Kim, *A free boundary problems for a quasilinear degenerate elliptic equation: transonic regular reflection*, Comm. Pure Appl. Math. **55** (2002), 71-92.
8. Shuxing Chen, *Linear approximation of shock reffection at a wedge with large angle*, Comm. P.D.E. **21** (1996), 1103-1114.
9. Shuxing Chen, *On reflection of multidimensional shock front*, J. Differential Equations **80** (1989), 199-236.
10. Shuxing Chen, *Existence of stationary supersonic flow past a pointed body*, Archive Rat. Mech. Anal. **156** (2001), 141-181.
11. Shuxing Chen, *Stability of Mach Configuration*, Comm. Pure Appl. Math. **59** (2006), 1-35.

12. Shuxing Chen, *Mach configuration in pseudo-stationary compressible flow* Jour. Amer. Math. Soc., **21**(2008), 23–61.

13. G.Q. Chen and M.Feldman, *Potential theory for shock reflection by a large-angle wedge*, Proc. Nat. Acad. Sci. U.S.A., **102** (2005), 15368-15372.

14. R. Courant and K.O. Friedrichs, *Supersonic Flow and Shock Waves*. Interscience Publishers Inc., New York, 1948.

15. Shuxing Chen and Beixiang Fang, *Stability of transonic shocks in supersonic flow past a wedge*, J. Differential Equations; **233** (2007), 103–135.

16. C.M. Dafermos, *Hyperbolic Conservation Laws in Continuum Physics*, Springer-Verlag, Berlin Heiderberg New York, 2002.

17. M.Dauge, *Elliptic Boundary Value Problems in Corner Domains – Smoothness and Asymptotics of Solutions*, Lecture Notes in Mathematics, **1341**, Springer-Verlag, Berlin, 1988.

18. P.Grisvard, *Elliptic Problems in nonsmooth domains*, Monographs and Studies in Mathematics **24**, Pitman, London 1987.

19. E.Harabetian, *Diffraction of a weak shock by a wedge*, Comm. Pure Appl. Math. **40** (1987), 849-863.

20. L.F.Henderson, *Region and Boundaries for Diffracting Shack Wave Systems*, Zeitshrift für Angewandte Mathematik und Mechanik, **67** (1987), 73-88.

21. J.Hunter and M.Brio, *Weak shock reflection*, J. Fluid Mech., **410** (2000), 235-261.

22. J.Hunter and J.B.Keller, *Weak shock diffraction*, Wave Motion, **6** (1984), 79-89.

23. K.Jegdic, B.L.Keyfitz and S.Canic, *Transonic regular reflection for the nonlinear wave system*, J. Hyperbolic Differential Equations **3** (2006), 443-474.

24. J.B.Keller and A.Blank, *Diffraction and reflection of pulses by wedges abd corners*, Comm. Pure Appl. Math., **4**, 75-84.

25. L.J.Lighthill, *The diffraction of a blast I*, Proc. Roy. Soc. London Ser., **198** (1949), 454-470.

26. L.J.Lighthill, *The diffraction of a blast II*, Proc. Roy. Soc. London Ser., **200** (1950), 554-565.

27. Y. Li and M.Vogelius, *Gradient estimates for solutions to divergence form elliptic equations with discontinuous coefficients*, Arch. Rat. Mech. Anal.,**153** (2000), 91-151.

28. A. Majda, *One perspective on open problems in multi-dimensional conservation laws*, IMA vol. Math. Appl. **29**(1991), 217-237.

29. C.S. Morawetz, *Potential theory for regular and Mach reflection of a shock at a wedge*, Comm. Pure Appl. Math. **47**(1994), 593-624.

30. John von Neumann, *Oblique Reflection of Shocks*. U.S. Dept. Comm. Off. of Tech. Serv., Washigton, D.C. PB-37079 (1943).

31. D.Serre, *Ecoulements de fluides parfaits en deux variables indépendantes de type espace. Réflexion d'un choc plan par un dièdre compressif*, Arch. Rat. Mech. Anal. **132** (1995), 15-36.

32. D.Serre, *Shock reflection in gas dynamics*, Handbook of Fluid Dynamics, **4**, Elsvier, North-Holland, 2005.

33. J.Smoller, *Shock waves and reaction-diffusion equations*, Springer-Verlag, New York, 1983.

34. Y.Zheng, *System of Conservation Laws, Two-Dimensional Riemann Problems*, Birkäuser Boston, 2001.

35. Y. Zheng, *Shock reflection for the Euler system* , Proceedings of the 10-th Inter. Conf. Hyperbolic Problems, Osaka, 2004.

# WEIGHTED FUNCTION SPACES WITH CONSTANT AND VARIABLE SMOOTHNESS

O. V. BESOV*

*Steklov Institute of Mathematics*
*Russian Academy of Sciences*
*Gubkina str. 8*
*119991 Moscow, Russia*
*E-mail: besov@mi.ras.ru*

The lecture consists of the following three parts:

I. *Weighted Sobolev spaces with constant smoothness $s \in \mathbb{N}$ defined on an irregular domain $G \subset \mathbb{R}^n$.* In this part, we will give some sufficient conditions for the embedding of weighted Sobolev spaces into weighted Lebesgue spaces and for compactness of such embedding.

II. *Weighted spaces $B_{p,q}^s(G)$ and $F_{p,q}^s(G)$ of functions that have variable smoothness $s = s(x)$ and are defined on a domain $G \subset \mathbb{R}^n$ with a locally Lipschitz boundary.* Here we will formulate embedding, interpolation, and extention theorems.

III. *Weighted spaces $B_{p,q}^s(\mathbb{R}^n)$ and $F_{p,q}^s(\mathbb{R}^n)$ of functions with variable smoothness.* Here we will discuss equivalent norms in terms of the Fourier transform of functions.

**Key words:** Sobolev spaces, Besov spaces, Triebel-Lizorkin spaces, variable smoothness irregular domains

**Mathematics Subject Classification:** 46E35

## I. Weighted Sobolev spaces with constant smoothness $s \in \mathbb{N}$ defined on an irregular domain

Below $G$ stands for a domain of the $n$-dimensional Euclidean space $\mathbb{R}^n$, $n \geq 2$; $B(x, R) = \{y\colon |y - x| < R\}$; $\rho(x) = \mathrm{dist}(x, \mathbb{R}^n \setminus G)$; the weight functions $v, w\colon G \to (0, \infty)$, $v, w \in L(G, \mathrm{loc})$; for $\delta > 0$, $G_\delta = \{x\colon x \in G, \ \rho(x) > \delta\}$; $G^{(R)} = G \cap B(0, R)$; $\chi$ is the characteristic function of the ball $B(0, 1)$ and the interval $(-1, 1)$; and $\frac{1}{p'} = 1 - \frac{1}{p}$.

For $E \subset G$, we set $|E|_w = \int_E w(x)\,dx$ and $|E| = |E|_1$.

Next, we assume that

$$s \in \mathbb{N}, \qquad 1 \leq p, r < q < \infty, \qquad s - \frac{n}{p} + \frac{n}{q} \geq 0. \qquad (\text{I.1})$$

Let

$$\|f|L_{p,v}(E)\| = \|fv^{-\frac{1}{p}}|L_p(E)\| = \left( \int_E |f|^p v\,dx \right)^{\frac{1}{p}},$$

Denote by $W_{p,v}^s(G)$ a Sobolev space of functions $f$ that have the Sobolev generalized derivatives $D^\alpha f$, $|\alpha| = s$, on $G$ and the finite norm

$$\|f|W_{p,v}^s(G)\| = \sum_{|\alpha|=s} \|D^\alpha f|L_{p,v}(G)\| + \|f|L_p(G_\delta)\|,$$

where $\delta > 0$ is sufficiently small, $W_p^s(G) = W_{p,1}^s(G)$.

It is well known that the Sobolev embeddings

$$W_p^s(G) \subset L_q(G), \qquad s \in \mathbb{N}, \quad 1 \leq p < q < \infty, \tag{I.2}$$

are true for a domain $G$ with regular boundary under conditions (I.1) and are compact if

$$s - \frac{n}{p} + \frac{n}{q} > 0. \tag{I.3}$$

In this paper, we find sufficient conditions for the compactness of embeddings of weighted and unweighted Sobolev spaces,

$$W_{p,v}^s(G) \subset L_{q,w}(G), \qquad s \in \mathbb{N}, \quad 1 \leq p < q < \infty, \tag{I.4}$$

and for the validity of the corresponding week-type estimates,

$$\sup_{\lambda>0} \lambda |\{x \in G : |f(x)| > \lambda\}|_w^{1/q} \leq C \|f|W_{p,v}^s(G)\|, \tag{I.5}$$

for domains $G$ with regular and irregular boundaries. These conditions are formulated in simple geometrical terms. We also present additional conditions for the compactness of embedding (I.4).

Suppose that $G \subset \mathbb{R}^n$, $\delta_0 \in (0,1)$, $R_0 > 0$, $C_0 \geq 1$, and every point $x \in G$ is assigned a piecewise smooth path $\gamma = \gamma_x \colon [0, t_x] \to G$ and a continuous piecewise smooth function $r_\gamma \colon [0, t_x] \to (0, \infty)$ possessing the properties

$$\gamma(0) = x, \qquad \gamma \subset B(x, R_0), \qquad \rho(\gamma(t_x)) \geq 2\delta_0, \qquad |\gamma'| \leq 1 \quad \text{a.e.},$$

$$0 < r_\gamma(t) \leq \delta_0 \rho_1(\gamma(t)), \qquad r_\gamma(t_x) \geq \delta_0^2, \qquad |r_\gamma'(t)| = \left|\frac{d}{dt} r_\gamma(t)\right| \leq C_0 \quad \text{for a.e. } t \in [0, t$$

$$\delta_0 r_\gamma(t') \leq r_\gamma(t'') \quad \text{if} \quad t', t'' \in [0, t_x], \quad B(\gamma(t'), \delta_0 r_\gamma(t')) \cap B(\gamma(t''), \delta_0 r_\gamma(t'')) \neq \emptyset.$$

Then, we will write $(G, \gamma, r_\gamma) \in \mathcal{G}(\delta_0, R_0, C_0)$.

For $h > 0$, $x \in G$, and $E \subset G$, we set

$$V_h(\gamma, r_\gamma, t) = r_\gamma(t)^{-\frac{n}{p'}} \left[ \int_{B(\gamma(t), r_\gamma(t)) \backslash B(x,h)} v(y)^{-\frac{p'}{p}} dy \right]^{\frac{1}{p'}}.$$

Let, for $a, b > 0$,

$$P_\alpha(a, b) = a^{\alpha-} b^{\alpha+} \quad \text{if} \quad \alpha \neq 0 \quad \text{and} \quad P_0(a, b) = \left| \ln \frac{a}{b} \right|,$$

and for $h > 0$,

$$N_1(h) = \sup_{x \in G} \left( \int_{B(x, \delta_0 \rho(x)) \setminus B(x,h)} P_{s-n}(|y - x|, r_\gamma(0)) v(y)^{-\frac{p'}{p}} dy \right)^{\frac{1}{p'}} |G \cap B(x,h)|_w^{\frac{1}{q}},$$

$$N_2(h) = \sup_{x \in G} \left( \int_0^{t_x} (t + r_\gamma(0))^{(s-1)p'} r_\gamma(t)^{\frac{1-n}{p}p'} V_h(\gamma, r_\gamma(t), t)^{p'} dt \right)^{\frac{1}{p'}} |G \cap B(x,h)|_w^{\frac{1}{q}},$$

and

$$N(h) = N_1(h) + N_2(h).$$

**Theorem I.1.** *Suppose that $(G, \gamma, r_\gamma) \in \mathcal{G}(\delta_0, R_0, C_0)$ and*

$$\sup_{h>0} N(h) < \infty.$$

*Then estimate (I.5) holds.*

*If, in addition, $s \geq 2$ and $p\tilde{q} < q$, then*

$$W_{p,v}(G) \subset L_{\tilde{q},w}(G),$$

*while if $s = 1$, then (I.4) holds.*

**Theorem I.2.** *Suppose that conditions of Theorem I.1 hold, $s = 1$, and the following conditions are satisfied:*

(i) *$N(h) \to 0$ as $h \to 0$;*
(ii) *for any $\varepsilon > 0$, there exist a representation $G = F^{(\varepsilon)} \cup G^{(\varepsilon)}$, where $|F^{(\varepsilon)}|_w < \varepsilon$ and $G^{(\varepsilon)}$ is an open set such that*

$$\inf\{v(x) \colon x \in G^{(\varepsilon)}\} > 0,$$

$$0 < \inf\{w(x) \colon x \in G^{(\varepsilon)}\} \leq \sup\{w(x) \colon x \in G^{(\varepsilon)}\} < \infty.$$

*Then the embedding $W_{p,v}^1(G) \subset L_{q,w}(G)$ is compact.*

Some theorems of this type can be found in [2].

As a consequence of these general results, we obtain sufficient conditions for the embedding of Sobolev spaces on $\sigma$-John domains $G$ for $v(x) = \rho(x)^a$ and $w(x) = \rho(x)^b$.

**Definition I.1.** For $\sigma \geq 1$, a bounded domain $G \subset \mathbb{R}^n$ is called a $\sigma$-John domain if, for certain $t^*, \kappa > 0$ and any $x \in G$, there exists a piecewise smooth path

$$\gamma \colon [0, t^*] \to G, \qquad \gamma(0) = x, \qquad \left| \frac{d}{dt} \gamma(t) \right| \leq 1 \quad \text{a.e.,}$$

such that $\rho(\gamma(t)) \geq \kappa t^\sigma$ for $0 < t \leq t^*$.

**Theorem I.3.** *Let* $G \subset \mathbb{R}^n$ *be a* $\sigma$-*John domain* $(\sigma \geq 1)$, $p > 1$ $(p \geq 1$ *if* $s = 1)$, $a \in (-\infty, \infty)$, $b \geq 0$, *and*

$$\text{(i)} \quad s - \frac{n}{p} + \frac{n}{q} > 0 \quad (\geq 0 \text{ if } s = 1),$$

$$\text{(ii)} \quad s - \frac{n}{p} + \frac{n}{q} - \frac{a}{p} + \frac{b}{q} \geq 0,$$

$$\text{(iii)} \quad s - \frac{\sigma(n - 1 + a) + 1}{p} + \frac{n + b}{q} \geq 0.$$

*Then, embedding* (I.4) *is valid.*

## II. Weighted spaces $B_{p,q}^s(G)$ and $F_{p,q}^s(G)$ of functions that have variable smoothness $s = s(x)$ and are defined on a domain $G \subset \mathbb{R}^n$ with a locally Lipschitz boundary

In this section, we study spaces $B_{p,q}^s(G)$ and $L_{p,q}^s(G) = F_{p,q}^s(G)$ of functions defined on a domain $G$ of the $n$-dimensional Euclidean space $\mathbb{R}^n$; here, either $G = \mathbb{R}^n$, or $G \subset \mathbb{R}^n$ has the form

$$G = \{x = (x', x_n) \in \mathbb{R}^n : x' \in \mathbb{R}^{n-1}, \ x_n > \phi(x')\}, \tag{II.1}$$

where $\phi$ is a Lipschitz function on $\mathbb{R}^{n-1}$:

$$\exists \Lambda > 0 : \ |\phi(x') - \phi(y')| \leq \Lambda |x' - y'| \quad \forall x', y' \in \mathbb{R}^{n-1}. \tag{II.2}$$

For $1 \leq p, q \leq \infty$,

$$B_{p,q}^s(G) = \{f : f \in L_p(G, \text{loc}), \ \|f|B_{p,q}^s(G)\| < \infty\}, \qquad \text{where}$$

$$\|f|B_{p,q}^s(G)\| = \|s_0 f|L_p(G)\| + \left[ \sum_{k=1}^{\infty} \sup_{|y| \leq 1} \left\| s_k \Delta^M (2^{-k} y, G) f|L_p(G) \right\|^q \right]^{1/q}; \tag{II.3}$$

for $1 < p, q < \infty$,

$$L_{p,q}^s(G) = \{f : f \in L_p(G, \text{loc}), \ \|f|L_{p,q}^s(G)\| < \infty\}, \qquad \text{where}$$

$$\|f|L_{p,q}^s(G)\| = \|s_0 f|L_p(G)\| + \left\| \left[ \sum_{k=1}^{\infty} s_k^q \left( \int_{|y| < 1} |\Delta^M (2^{-k} y, G) f| \, dy \right)^q \right]^{1/q} \right| L_p(G) \right\|. \tag{II.4}$$

Here $s = s(x) = \{s_k(x)\}_{k=0}^{\infty}$ is a variable smoothness that characterizes the behavior of functions in $B_{p,q}^s(G)$ and $L_{p,q}^s(G)$.

In the above formulas (II.3) and (II.4), the parameter $M$ is a positive integer and the difference $\Delta^M$ is defined as follows:

$$\Delta^M (y, G) f(x) = \Delta^M (y) f(x) = \sum_{k=0}^{M} (-1)^{M-k} \frac{M!}{k!(M-k)!} f(x + ky)$$

if $G = \mathbb{R}^n$ or $\text{dist}(x, \mathbb{R}^n \setminus G) > M|y|$, and

$$\Delta^M(y, G)f(x) = 0 \quad \text{if} \quad G \neq \mathbb{R}^n \quad \text{and} \quad \text{dist}(x, \mathbb{R}^n \setminus G) \leq M|y|.$$

Throughout this paper, we assume that some numbers $m'$, $m$, and $M$, $0 < m' \leq m < M$, are fixed.

**Definition II.1.** We say that $s \in S(G; m', m)$ if the smoothness $s = \{s_k(x)\}_0^\infty$ satisfies the following conditions:

$(1_s)$ $s_k : G \to (0, \infty)$ are continuous functions on $G$ for $k \in \mathbb{N}_0$;
$(2_s)$ there exists a constant $c_0 > 0$ such that

$$0 < s_k(x) \leq c_0 s_k(y) \quad \text{for} \quad x, y \in G, \quad |x - y| \leq 2^{-k};$$

$(3_s)$ there exist numbers $c'$, $c$, and $k_0$ such that

$$0 < c' \cdot 2^{(l-k)m'} \leq \frac{s_l(x)}{s_k(x)} \leq c \cdot 2^{(l-k)m} \quad \text{for} \quad x \in G, \quad k_0 \leq k < l.$$

Everywhere below, we will assume that $s \in S(G; m', m)$.

The spaces $B_{p,q}^s$ and $L_{p,q}^s$ generalize various well-known spaces for which the smoothness $s(x)$ has a certain specific form.

In the present section, we employ an integral representation of functions to prove a retraction theorem for $B_{p,q}^s(G)$ and $L_{p,q}^s(G)$. We then apply this theorem in order to obtain interpolation and embedding theorems for $B_{p,q}^s(G)$ and $L_{p,q}^s(G)$, as well as a theorem on the extension of functions from domains $G$ of type (II.1) to the whole $\mathbb{R}^n$, i.e., from $B_{p,q}^s(G)$ and $L_{p,q}^s(G)$ to $B_{p,q}^s(\mathbb{R}^n)$ and $L_{p,q}^s(\mathbb{R}^n)$, respectively.

Let $\Omega \in C_0^\infty(\mathbb{R}^n)$, $\int \omega(x)\,dx = 1$, be some special averaging kernel. Set

$$\Omega_k(y) = 2^{kn}\Omega(2^k y).$$

We have the following representation almost everywhere in $G$:

$$f = \Omega_0 * \Omega_0 * f + \sum_{k=1}^\infty [(\Omega_k * \Omega_k) - (\Omega_{k-1} * \Omega_{k-1})]f,$$

or, in other terms,

$$f = \Omega_0 * f_0 + \sum_{k=1}^\infty (\Omega_k + \Omega_{k-1}) * f_k = \sum_{k=0}^\infty \Omega_k^+ * f_k, \tag{II.5}$$

where

$$f_0 = \Omega_0 * f, \quad f_k = (\Omega_k - \Omega_{k-1}) * f, \quad k \geq 1. \tag{II.6}$$

Below we will need the following Banach spaces (with parameters $1 \leq p, q \leq \infty$) of real functions, function sequences, and sequence-valued functions:

$$L_p(G) = \left\{ f : \|f|L_p(G)\| = \left( \int |f(x)|^p \right)^{1/p} < \infty \right\},$$

$$l_q(L_p(G)) = \left\{ \{f_k\}_{k=0}^{\infty} : \|f_k|l_q(L_p(G))\| = \left( \sum_{k=0}^{\infty} \|f_k|L_p(G)\|^q \right)^{1/q} < \infty \right\},$$

$$L_p(G, l_q) = \left\{ \{f_k\}_{k=0}^{\infty} : \|f_k|L_p(G, l_q)\| = \left\| \left( \sum_{k=0}^{\infty} |f_k|^q \right)^{1/q} \, \Big| \, L_p(G) \right\| < \infty \right\},$$

$$L_{p,\rho}(G) = \{ f : \rho f \in L_p(G), \ \|f|L_{p,\rho}(G)\| = \|\rho f|L_p(G)\| < \infty \},$$

$$l_q(L_{p,s_k}(G)) = \{ \{f_k\}_{k=0}^{\infty} : \|f_k|l_q(L_{p,s_k}(G))\| = \|s_k f_k|l_q(L_p(G))\| < \infty \},$$

$$L_p(G, l_{q,s}) = \{ \{f_k\}_{k=0}^{\infty} : \|f_k|L_p(G, l_{q,s})\| = \|s_k f_k|L_p(G, l_q)\| < \infty \}.$$

**Definition II.2.** A Banach space $Y$ is called a *retract* of a Banach space $X$ if there exist linear continuous operators $\mathcal{R} : X \to Y$ (retraction) and $\mathcal{S} : Y \to X$ (coretraction) such that $I_Y = \mathcal{R}\mathcal{S} : Y \to Y$ is the identity operator.

**Theorem II.1.** *The space $B_{p,q}^s(G)$ with $1 \leq p, q \leq \infty$ and $s \in S(G; m', m)$ is a retract of the space $l_q(L_{p,s_k}(G))$.*

*The space $L_{p,q}^s(G)$ with $1 < p, q < \infty$ and $s \in S(G; m', m)$ is a retract of the space $L_p(G, l_{q,s})$.*

*The operators $\mathcal{S}$ (coretraction) and $\mathcal{R}$ (retraction) are given by*

$$(\mathcal{S}f)_k(x) = f_k(x), \qquad k \in \mathbb{N}_0, \quad x \in G, \tag{II.7}$$

$$\mathcal{R}h(x) = \sum_{k=0}^{\infty} (\Omega_k^+ * h_k)(x), \qquad x \in G. \tag{II.8}$$

**Lemma II.1.** *Let $s \in S(G; m', m)$, $1 \leq p \leq r \leq \infty$, $1 \leq q \leq \infty$, $N \in \mathbb{N}$, and*

$$\gamma := \min \left\{ m' - \frac{n}{p} + \frac{n}{r}, \ N - m + \frac{n}{p} - \frac{n}{r} \right\},$$

$$t(x) = \{t_k(x)\}_{k=0}^{\infty} = \left\{ s_k(x) \cdot 2^{-k(\frac{n}{p} - \frac{n}{r})} \right\}_{k=0}^{\infty}.$$

*Let $\mathcal{S}$ and $\mathcal{R}$ be defined by (II.7) and (II.8), respectively. Then*

$$\mathcal{S} : B_{p,q}^s(G) \to l_q(L_{r,t}(G)) \tag{II.9}$$

*and, for $\gamma > 0$,*

$$\mathcal{R} : l_q(L_{p,s_k}(G)) \to B_{r,q}^t(G), \tag{II.10}$$

*where the norm in $B_{r,q}^t(G)$ is defined in terms of $N$th-order differences.*

**Lemma II.2.** *Let $s \in S(G; m', m), 1 < p \le r \le \infty, 1 < q < \infty,$ and*

$$t(x) = \{t_k(x)\}_{k=0}^\infty = \left\{ s_k(x) \cdot 2^{-k(\frac{n}{p} - \frac{n}{r})} \right\}_{k=0}^\infty.$$

*Let $S$ and $\mathcal{R}$ be defined by (II.7) and (II.8).*
*Then*

$$S : L_{p,q}^s(G) \to L_r(G, l_{q,t}) \tag{II.11}$$

*and, for $m' - \frac{n}{p} + \frac{n}{r} > 0,$*

$$\mathcal{R} : L_p(G, l_{q,s}) \to L_{p,q}^s(G). \tag{II.12}$$

Theorem 2.1, combined with the general interpolation theorems for spaces of Banach-valued sequences and functions , allows one to obtain the following interpolation theorem for $B_{p,q}^s(G)$ and $L_{p,q}^s(G)$.

**Theorem II.2.** *Let $s^{(0)}, s^{(1)} \in S(G; m', m)$ and*

$$1 \le p_0, p_1 < \infty, \qquad 1 \le q_0 < \infty, \qquad 1 \le q_1 \le \infty, \qquad 0 < \theta < 1,$$

$$\frac{1}{p} = \frac{1-\theta}{p_0} + \frac{\theta}{p_1}, \qquad \frac{1}{q} = \frac{1-\theta}{q_0} + \frac{\theta}{q_1}, \qquad p = q,$$

$$s_k^{(\theta)}(x) = (s_k^{(0)}(x))^{1-\theta} (s_k^{(1)}(x))^\theta.$$

*Then*

$$\left[ B_{p_0,q_0}^{s^{(0)}}(G), B_{p_1,q_1}^{s^{(1)}}(G) \right]_\theta = B_{p,p}^{s^{(\theta)}}(G). \tag{II.13}$$

*In addition, if $1 \le q_0, q_1 < \infty$, then*

$$\left( B_{p_0,q_0}^{s^{(0)}}(G), B_{p_1,q_1}^{s^{(1)}}(G) \right)_{\theta,p} = B_{p,p}^{s^{(\theta)}}(G); \tag{II.14}$$

*if $1 < p_0, p_1 < \infty$ and $1 < q_0, q_1 < \infty$, then*

$$\left( L_{p_0,q_0}^{s^{(0)}}(G), L_{p_1,q_1}^{s^{(1)}}(G) \right)_{\theta,p} = \left[ L_{p_0,q_0}^{s^{(0)}}(G), L_{p_1,q_1}^{s^{(1)}}(G) \right]_\theta = L_{p,p}^{s^{(\theta)}}(G) = B_{p,p}^{s^{(\theta)}}(G).$$

As a corollary to Lemmas 1 and 2, we obtain the following embedding theorem.

**Theorem II.3.** *Let*

$$s \in S(G; m', m), \qquad 1 \le p < r \le \infty, \qquad 1 \le q \le \infty, \qquad N \in \mathbb{N},$$

$$\gamma := \min\left\{ m' - \frac{n}{p} + \frac{n}{r}, \; N - m + \frac{n}{p} - \frac{n}{r} \right\} > 0,$$

$$t(x) = \{t_k(x)\}_{k=0}^\infty = \left\{ s_k(x) \cdot 2^{-k(\frac{n}{p} - \frac{n}{r})} \right\}_{k=0}^\infty.$$

*Then*

$$B_{p,q}^s(G) \subset B_{r,q}^t(G). \tag{II.15}$$

*In addition, if* $1 < p \le r < \infty$ *and* $1 < q < \infty$, *then*

$$L_{p,q}^s(G) \subset L_{r,q}^t(G). \tag{II.16}$$

*Here, the norms in* $B_{r,q}^t(G)$ *and* $L_{r,q}^t(G)$ *are defined by means of* $N$*th-order differences.*

Now we pass to the extension of functions that belong to $B_{p,q}^s(G)$ and $L_{p,q}^s(G)$ from $G$ to $\mathbb{R}^n$ while preserving their smoothness. The domain $G$ has the form (II.1). Such a problem can naturally be called a *problem of extension* of the above-mentioned function spaces. In the case of variable smoothness, we should first define the function $s(x) = \{s_k(x)\}_{k=0}^\infty$, $x \in G$, on the set $\mathbb{R}^n \setminus G$. It will be clear from what follows that the values of $s$ on the complement can be chosen greater than those on $G$. Thus, we can say that we "improve the smoothness" when extending the function spaces.

**Theorem II.4.** *Suppose that* $G \subset \mathbb{R}^n$ *is a domain of the form* (II.1) *and the vector function* $s$ *satisfies conditions* $(1_s)$–$(3_s)$ *on* $G$. *Let* $\tilde{s}$ *be an extension of* $s$ *from* $G$ *to* $\mathbb{R}^n$ *that satisfies conditions* $(1_s)$–$(3_s)$ *on* $\mathbb{R}^n$.

*Then there exists a linear bounded extension operator* $\mathcal{E}$,

$$\mathcal{E} : B_{p,q}^s(G) \to B_{p,q}^{\tilde{s}}(\mathbb{R}^n),$$

$$\mathcal{E} : L_{p,q}^s(G) \to L_{p,q}^{\tilde{s}}(\mathbb{R}^n).$$

## III. Weighted spaces $B_{p,q}^s(\mathbb{R}^n)$ and $F_{p,q}^s(\mathbb{R}^n)$ of functions with variable smoothness

In this section, we study the Banach spaces $B_{p,q}^a$ and $F_{p,q}^a$, $1 < p, q < \infty$, of functions defined on the $n$-dimensional Euclidean space $\mathbb{R}^n$. The smoothness properties of functions are determined by the finiteness of the weighted integral norms of their differences considered as functions of a point $x \in \mathbb{R}^n$ and the difference step. Namely, for $1 < p, q < \infty$, we consider the spaces

$$B_{p,q}^a = \left\{ u : u \in L_p(\mathbb{R}^n, \mathrm{loc}), \ \|u|B_{p,q}^a\| = \left[ \sum_{k=1}^\infty \sup_{|h| \le 1} \|a_k \Delta^M(2^{-k}h)u|L_p(\mathbb{R}^n)\|^q \right]^{\frac{1}{q}} \right.$$

$$\left. + \|a_0 u|L_p(\mathbb{R}^n)\| < \infty \right\}, \tag{III.1}$$

$$F_{p,q}^a = \left\{ u : u \in L_p(\mathbb{R}^n, \mathrm{loc}), \ \|u|F_{p,q}^a\| = \left\| \left[ \sum_{k=1}^\infty a_k^q \left( \int_{|h| < 1} |\Delta^M(2^{-k}h)u| \, dh \right)^q \right]^{\frac{1}{q}} \middle| L_p(\mathbb{R}^n) \right\| \right.$$

$$\left. + \|a_0 u|L_p(\mathbb{R}^n)\| < \infty \right\} \tag{III.2}$$

for a sequence $a(x) = \{a_k(x)\}_0^\infty$ of functions $a_k(x)$ of a certain class (more precisely, for $a \in \overline{R}(\omega; m, m')$; see Definition III.1). The spaces $B_{p,q}^a$ and $F_{p,q}^a$ are generalizations of thoroughly studied spaces with the smoothness $a(x)$ of a specific form.

Here, we show that the norms of the spaces $B_{p,q}^a$ and $F_{p,q}^a$ given in (1) and (2) are equivalent to other norms expressed in terms of the Fourier transforms of functions. These latter norms are the $l_q(L_p)$- and $L_p(l_q)$-norms of the sequence $\{a_k F^{-1}\phi_k F u\}_{k=0}^\infty$, where $\{\phi_k(\xi)\}_0^\infty$ is a special smooth dyadic partition of unity on $\mathbb{R}^n$.

We will need notations and certain generalizations, related to the theory of Fourier multipliers in the weighted spaces $l_q(L_p(\rho, \mathbb{R}^n))$ and $L_p(\mathbb{R}^n, \rho, l_q)$, which are borrowed, in particular, from the theory of ultradistributions.

**Definition III.1.** (i) Denote by $\mathfrak{M}$ the family of all functions $\omega \colon \mathbb{R}^n \to [0, \infty)$ such that $\omega(x) = \sigma(|x|)$, where $\sigma \colon [0, \infty) \to [0, \infty)$ is an increasing continuous concave function such that

$$\sigma(0) = 0, \qquad \int_0^\infty \frac{\sigma(t)}{1+t^2}\, dt < \infty, \qquad \sigma(t) \geq c + d\ln(1+t) \quad \text{for } t \geq 0,$$

where $c$ is a real number and $d$ is a positive number.

(ii) We will write $\omega \in \mathfrak{M}_E$ if $\omega \in \mathfrak{M}$ and the corresponding function $\sigma$ satisfies the additional condition

$$e^{\sigma(t)} = \sum_{k=0}^\infty a_k t^k \qquad \text{for } t \geq 0,$$

where $a_k \geq 0$ and $(k+l)!\, a_{k+l} \leq k!\, a_k\, l!\, a_l$ for $k, l \in \mathbb{N}_0$.

It is clear that $\omega \in \mathfrak{M}$ is subadditive, i.e.,

$$\omega(x+y) \leq \omega(x) + \omega(y) \qquad \text{for } x, y \in \mathbb{R}^n. \tag{III.3}$$

Typical examples of the functions $\omega \in \mathfrak{M}_E \subset \mathfrak{M}$ are

$$\omega(x) = d\ln(1+|x|) \qquad \text{for } d > 0, \tag{III.4}$$

$$\omega(x) = |x|^\beta \qquad \text{for } 0 < \beta < 1.$$

**Definition III.2.** Let $\omega \in \mathfrak{M}$.

(i) The symbol $D_\omega$ denotes the family of all complex-valued functions $\phi \in C_0^\infty(\mathbb{R}^n)$ such that

$$\|\phi\|_{\lambda\omega} = \int |F\phi(x)| e^{\lambda\omega(x)}\, dx < \infty \qquad \forall \lambda > 0.$$

(ii) The symbol $S_\omega$ denotes the family of all complex-valued functions $\phi \in L_1(\mathbb{R}^n)$ such that $\phi$ and $F\phi$ are infinitely differentiable and

$$p_{\alpha,\lambda}(\phi) = \sup_{x \in \mathbb{R}^n} e^{\lambda\omega(x)} |D^\alpha \phi(x)| < \infty, \tag{III.5}$$

$$\pi_{\alpha,\lambda}(\phi) = \sup_{x \in \mathbb{R}^n} e^{\lambda\omega(x)} |D^\alpha (F\phi)(x)| < \infty \tag{III.6}$$

for every $\alpha \in \mathbb{N}_0^n$ and every $\lambda > 0$.

(iii) The symbol $\overline{D}_\omega$ denotes the family of all sequences $\phi(x) = \{\phi_j(x)\}_0^\infty$ of functions $\phi_j \in C_0^\infty(\mathbb{R}^n)$ such that

$$\|\phi\|_{\lambda\omega} := \sup_{j \geq 0} \|\phi_j(2^j \cdot)\|_{\lambda\omega} = \sup_{j \geq 0} \|\phi_j\|_{\lambda\omega(2^j \cdot)} < \infty \qquad \text{(III.7)}$$

for every $\lambda > 0$.

**Remark III.1.** If $\omega$ is defined by (1.2), then $S_\omega = S$ is the ordinary Schwartz space of test functions and $D_\omega = D$ is the ordinary space of infinitely differentiable finite functions.

**Remark III.2.** The space $S_\omega$ equipped with a topology generated by semi-norms (1.3) and (1.4) is a complete locally convex space, $S_\omega \subset S$. By $S_\omega'$, we denote the space dual to $S_\omega$; the elements of $S_\omega'$ are called tempered ultradistributions or $\omega$-tempered distributions. The Fourier transform $F$ and the inverse Fourier transform $F^{-1}$ are defined on $S_\omega$ as follows ($f \in S_\omega'$):

$$(Ff, \phi) = (f, F\phi), \qquad (F^{-1}f, \phi) = (f, F^{-1}\phi) \qquad \forall \phi \in S_\omega.$$

**Definition III.3.** Let $\omega \in \mathfrak{M}$ and $\phi = \{\phi_j(x)\}_{j=0}^\infty$. We write $\phi \in \Phi_\omega$ if

(i) $\phi \in \overline{D}_\omega$,
(ii) $\operatorname{supp} \phi_0 \subset \{x \colon |x| \leq 2\}$ and $\operatorname{supp} \phi_j \subset \{x \colon 2^{j-1} \leq |x| \leq 2^{j+1}\}$ for $j \in \mathbb{N}$,
(iii) $\sum_{j=0}^\infty \phi_j(x) = 1$ for every $x \in \mathbb{R}^n$.

Note that the class $\Phi_\omega$ is nonempty. In particular, we can assume that

$$\phi_j(x) = \phi_1(2^{-j+1}x) \qquad \text{for } j \in \mathbb{N}.$$

**Definition III.4.** Let $\omega \in \mathfrak{M}$. Denote by $\overline{R}(\omega)$ the family of sequences $\rho(x) = \{\rho_k(x)\}_{k=0}^\infty$ of continuous functions $\rho_k$ on $\mathbb{R}^n$ such that

$$0 < \rho_k(x) \leq c_\rho \rho_k(y) e^{\omega(2^k(x-y))} \qquad \forall x, y \in \mathbb{R}^n \qquad \text{(III.8)}$$

for a certain constant $c_\rho > 0$ independent of $k$.

An example of $\rho \in \overline{R}(\omega)$ is given by $\rho(x) = \{\gamma_k e^{\omega(2^k x)}\}_{k=0}^\infty$ for any $\gamma_k > 0$. In particular,

$$\rho(x) = \{\gamma_k(1 + 2^k|x|)^d\}_{k=0}^\infty \in \overline{R}(d \ln(1 + |x|)), \qquad d > 0, \quad \forall \gamma_k > 0;$$

$$\rho(x) = \{\gamma_k e^{2^{k\beta}|x|^\beta}\}_{k=0}^\infty \in \overline{R}(|x|^\beta), \qquad 0 < \beta < 1.$$

**Definition III.5.** Let $1 < p, q < \infty$.

(i) $L_p = L_p(\mathbb{R}^n) = \left\{ f \colon \|f|L_p\| = \left( \int_{\mathbb{R}^n} |f(x)|^p dx \right)^{\frac{1}{p}} < \infty \right\}$.

(ii) $l_q(L_p) = \left\{ \{f_k\}_{k=0}^\infty \colon \|f_k|l_q(L_p)\| = \left( \sum_{k=0}^\infty \|f_k|L_p\|^q \right)^{\frac{1}{q}} < \infty \right\}$.

(iii) $L_p(l_q) = \left\{\{f_k\}_{k=0}^\infty : \|f_k|L_p(l_q)\| = \left\|\left(\sum_{k=0}^\infty |f_k|^q\right)^{\frac{1}{q}}\Big|L_p\right\| < \infty\right\}.$

(iv) For $\omega \in \mathfrak{M}$ and $\rho \in R(\omega)$, we set $L_{p,\rho} = \{f : \rho f \in L_p, \|f|L_{p,\rho}\| = \|\rho f|L_p\| < \infty\}.$

(v) For $\omega \in \mathfrak{M}$ and $\rho \in \overline{R}(\omega)$, we set $l_q(L_{p,\rho}) = \{\{f_k\}_0^\infty : \|f_k|l_q(L_{p,\rho})\| = \|\rho_k f_k|l_q(L_p)\| < \infty\}$ and $L_p(l_{q,\rho}) = \{\{f_k\}_0^\infty : \|f_k|L_p(l_{q,\rho})\| = \|\rho_k f_k|L_p(l_q)\| < \infty\}.$

(vi) For $\omega \in \mathfrak{M}$, $\rho \in \overline{R}(\omega)$, and $n_0 \in \mathbb{N}_0$, we set $L_p^A(l_{q,\rho}) = \{\{f_k\}_0^\infty : \{f_k\}_0^\infty \in L_p(l_{q,\rho}), \text{ supp } Ff_k \subset \{x : |x| \le 2^{k+n_0}\} \text{ for } k \in \mathbb{N}_0, \|f_k|L_p^A(l_{q,\rho})\| = \|f_k|L_p(l_{q,\rho})\| < \infty\}.$

**Definition III.6.** Let $\omega \in \mathfrak{M}_E$ and $0 < m' \le m < \infty$. We write $a = \{a_k(x)\}_0^\infty \in \overline{R}(\omega; m, m')$ if $a \in \overline{R}(\omega)$ and, for some constants $c > 0$ and $c' > 0$,

$$c' \cdot 2^{(l-k)m'} \le \frac{a_l(x)}{a_k(x)} \le c \cdot 2^{(l-k)m} \qquad \text{for } l \ge k \text{ and } x \in \mathbb{R}^n. \tag{III.9}$$

The spaces $B_{p,q}^a = B_{p,q}^a(\mathbb{R}^n)$ and $F_{p,q}^a = F_{p,q}^a(\mathbb{R}^n)$ are defined in (III.1) and (III.2) and are characterized by the behavior of the differences of functions that belong to these spaces. Here, we present the definitions of $B_{p,q}^a$ and $F_{p,q}^a$ that are expressed in terms of Fourier transforms and are equivalent to definitions (III.1) and (III.2), respectively.

**Definition III.7.** Let $\omega \in \mathfrak{M}$, $0 < m' \le m < \infty$, $a \in \overline{R}(\omega; m, m')$, $\phi = \{\phi_j\}_0^\infty \in \Phi_\omega$, and $1 < p, q < \infty$. Set

$$\widetilde{B}_{p,q}^a = \{u : u \in L(\mathbb{R}^n, \text{loc}) \cap S_\omega', \|u|\widetilde{B}_{p,q}^a\| = \|a_k F^{-1}\phi_k Fu|l_q(L_p)\| < \infty\}, \tag{III.10}$$

$$\widetilde{F}_{p,q}^a = \{u : u \in L(\mathbb{R}^n, \text{loc}) \cap S_\omega', \|u|\widetilde{F}_{p,q}^a\| = \|a_k F^{-1}\phi_k Fu|L_p(l_q)\| < \infty\}. \tag{III.11}$$

**Theorem III.1.** Let $\omega \in \mathfrak{M}$, $0 < m' \le m < \infty$, $a \in \overline{R}(\omega; m, m')$, and $\psi = \{\psi_j\}_0^\infty \in \Phi_\omega$. Then, the norm of (III.10) is changed to an equivalent norm if $\phi$ is replaced by $\psi$. The same is valid for the norm of (III.11).

**Theorem III.2.** Let $\omega \in \mathfrak{M}_E$, $0 < m' \le m < \infty$, $a \in \overline{R}(\omega; m, m')$, $\phi = \{\phi_j\}_0^\infty \in \Phi_\omega$, and $1 < p, q < \infty$. Then,

(i) $B_{p,q}^a = \widetilde{B}_{p,q}^a$, and the norms of (III.1) and (III.10) are equivalent;

(ii) $F_{p,q}^a = \widetilde{F}_{p,q}^a$, and the norms of (III.2) and (III.11) are equivalent.

Some references to previous results and history of the problems under consideration can be found in [1-4].

### References

1. O. V. Besov, "Sobolev's embedding theorem for a domain with irregular boundary," Mat. Sb. **192** (3), 3–26 (2001) [Sb. Math. **192**, 323–346 (2001)].

2. O. V. Besov, "On the compactness of embeddings of weighted Sobolev spaces on a domain with irregular boundary," Tr. Mat. Inst. Steklova, Ross. Akad. Nauk **232**, 72–93 (2001) [Proc. Steklov Inst. Math. **232**, 66–87 (2001)].

3. O. V. Besov, "Interpolation, embedding, and extention of spaces of functions of variable smoothness," Tr. Mat. Inst. Steklova, Ross. Akad. Nauk **248**, 52–63 (2005) [Proc. Steklov Inst. Math. **248**, 47–58 (2001)].

4. O. V. Besov, "Equivalent normings of spaces of functions of variable smoothness," Tr. Mat. Inst. Steklova, Ross. Akad. Nauk **243**, 87–95 (2003) [Proc. Steklov Inst. Math. **243**, 80–88 (2003)].

## I.1 Spaces of Differentiable Functions and Applications

Organizer: V.I. Burenkov

# REGULARIZATION OF THE CAUCHY PROBLEM FOR THE SYSTEM OF ELASTICITY THEORY IN $R^m$

O.I. MAKHMUDOV and I.E. NIYOZOV

*Samarkand State University*
*Departament of Mechanics and Mathematics*
*University Boulevard 15*
*703004 Samarkand, Uzbekistan*
*E-mail: olimjan@yahoo.com, makhmudovo@rambler.ru,iqbol@samdu.uz, iqboln@rambler.ru*

In this paper we consider the regularization of the Cauchy problem for a system of second order differential equations with constant coefficients.

**Key words**: Cauchy problem, Lame system, elliptic system, ill-posed problem, Carleman matrix, regularization, Laplace equation.

**Mathematics Subject Classification:** 35J25, 74B05

## Introduction

As is well known, the Cauchy problem for elliptic equations is ill-posed, the solution of the problem is unique, but unstable (Hadamard's example). For ill-posed problems, one does not prove existence theorems, the existence is assumed a priori. Moreover, the solution is assumed to belong to a given subset of a function space, usually a compact one [4]. The uniqueness of the solution follows from the general Holmgren theorem [9].

Let $x = (x_1, , x_m)$ and $y = (y_1, , y_m)$ be points in $R^m$, $D_\rho$ be a bounded simply connected domain in $R^m$ whose boundary consists of a cone surface

$$\Sigma: \quad \alpha_1 = \tau y_m, \quad \alpha_1^2 = y_1^2 + \ldots + y_{m-1}^2, \quad \tau = tg\frac{\pi}{2\rho}, \quad y_m > 0, \quad \rho > 1$$

and a smooth surface $S$, lying in the cone.

In the domain $D_\rho$, consider the system of elasticity theory

$$\mu\Delta U(x) + (\lambda + \mu)grad \ divU(x) = 0;$$

here $U = (U_1, \ldots, U_m)$ is the displacement vector, $\Delta$ is the Laplace operator, $\lambda$ and $\mu$ are the Lame constants. For brevity, it is convenient to use matrix-valued notation. Let us introduce the matrix differential operator

$$A(\partial_x) = \|A_{ij}(\partial_x)\|_{m \times m},$$

where

$$A_{ij}(\partial_x) = \delta_{ij}\mu\Delta + (\lambda + \mu)\frac{\partial^2}{\partial x_i \partial x_j}.$$

Then the elliptic system can be written in matrix form

$$A(\partial_x)U(x) = 0. \tag{1}$$

**Statement of the problem.** Assume the Cauchy data of a solution $U$ are given on $S$,

$$U(y) = f(y), \quad y \in S,$$

$$T(\partial_y, n(y))U(y) = g(y), \quad y \in S, \tag{2}$$

where $f = (f_1, \ldots, f_m)$ and $g = (g_1, \ldots, g_m)$ are prescribed continuous vector functions on $S$, $T(\partial_y, n(y))$ is the strain operator, i.e.,

$$T(\partial_y, n(y)) = \|T_{ij}(\partial_y, n(y))\|_{m \times m} = \left\| \lambda n_i \frac{\partial}{\partial y_j} + \mu n_j \frac{\partial}{\partial y_i} + \mu \delta_{ij} \frac{\partial}{\partial n} \right\|_{m \times m}.$$

$\delta_{ij}$ is the Kronecker delta, and $n(y) = (n_1(y), \ldots, n_m(y))$ is the unit normal vector to the surface $S$ at a point $y$.

It is required to determine the function $U(y)$ in $D$, i.e., find an analytic continuation of the solution of the system of equations in a domain from the values of $f$ and $g$ on a smooth part $S$ of the boundary.

On establishing uniqueness in theoretical studies of ill - posed problems, one comes across important questions concerning the derivation of estimates of conditional stability and the construction of regularizing operators.

Suppose that, instead of $f(y)$ and $g(y)$, we are given their approximations $f_\delta(y)$ and $g_\delta(y)$ with accuracy $\delta$, $0 < \delta < 1$ (in the metric of $C$) which do not necessarily belong to the class of solutions. In this paper, we construct a family of functions $U(x, f_\delta, g_\delta) = U_{\sigma\delta}(x)$ depending on the parameter $\sigma$ and prove that under certain conditions and a special choice of the parameter $\sigma(\delta)$ the family $U_{\sigma\delta}(x)$ converges in the usual sense to the solution $U(x)$ of problem (1),(2), as $\delta \to 0$.

Following A.N. Tikhonov, $U_{\sigma\delta}(x)$ is called a regularized solution of the problem. A regularized solution determines a stable method of approximate solution of problem [13].

Using results of [4,14] concerning the Cauchy problem for the Laplace equation, we succeeded in constructing a Carleman matrix in explicit form and on in constructing a regularized solution of the Cauchy problem for the system (1). Since we refer to explicit formulas, it follows that the construction of the Carleman matrix in terms of elementary and special functions is of considerable interest. For $m = 2, 3$ the problem under consideration coincides with the Cauchy problem for the system of elasticity theory describing statics of an isotropic elastic medium. In these cases problem (1),(2) was studied for special classes of domains in [5-8].

Further, the Cauchy problem for systems describing steady-state elastic vibrations, for systems of thermoelasticity, and for systems of Navier-Stokes was studied in [15,2,10,3].

Earlier, is was proved in [12],[11] that the Carleman matrix exists in any Cauchy problem for solutions of elliptic systems whenever the Cauchy data are given on a boundary set of positive measure.

## 1. Construction of the matrix of fundamental solution for the system of elasticity of a special form

**Definition 1.** $\Gamma(y,x) = ||\Gamma_{ij}(y,x)||_{m \times m}$, is called the matrix of fundamental solutions of system (1), where

$$\Gamma_{ij}(y,x) = \frac{1}{2\mu(\lambda + 2\mu)}((\lambda+3\mu)\delta_{ij}q(y,x)-(\lambda+\mu)(y_j-x_j)\frac{\partial}{\partial x_i}q(y,x)), \quad i,j = 2,...,m,$$

$$q(y,x) = \begin{cases} \frac{1}{(2-m)\omega_m} \cdot \frac{1}{|y-x|^{m-2}}, & m > 2 \\ \frac{1}{2\pi}ln|y-x|, & m = 2, \end{cases}$$

and $\omega_m$ is the area of unit sphere in $R^m$.

The matrix $\Gamma(y,x)$ is symmetric and its columns and rows satisfy equation (1) at an arbitrary point $x \in R^m$, except $y = x$. Thus, we have

$$A(\partial_x)\Gamma(y,x) = 0, \quad y \neq x.$$

Developing Lavrent'ev's idea concerning the notion of Carleman function of the Cauchy problem for the Laplace equation [4], we introduce the following notion.

**Definition 2.** By a Carleman matrix of problem (1),(2) we mean an $(m \times m)$ matrix $\Pi(y,x,\sigma)$ satisfying the following two conditions:

$$1) \quad \Pi(y,x,\sigma) = \Gamma(y,x) + G(y,x,\sigma),$$

where $\sigma$ is a positive numerical parameter and, with respect to the variable $y$, the matrix $G(y,x,\sigma)$ satisfies system (1) everywhere in the domain $D$.

2) The relation holds

$$\int_{\partial D \backslash S} (|\Pi(y,x,\sigma)| + |T(\partial_y,n)\Pi(y,x,\sigma)|)ds_y \leq \varepsilon(\sigma),$$

where $\varepsilon(\sigma) \to 0$, as $\sigma \to \infty$, uniformly in $x$ on compact subsets of $D$ ; here and elsewhere $|\Pi|$ denotes the Euclidean norm of the matrix $\Pi = ||\Pi_{ij}||$, i.e., $|\Pi| = (\sum\limits_{i,j=1}^{m} \Pi_{ij}^2)^{\frac{1}{2}}$. In particular $|U| = (\sum\limits_{i=1}^{m} U_i^2)^{\frac{1}{2}}$ for a vector $U = (U_1,...,U_m)$.

**Definition 3.** A vector function $U(y) = (U_1(y),...,U_m(y))$ is said to be regular in $D$, if it is continuous together with its partial derivatives of second order in $D$ and those of first order in $\overline{D} = D\bigcup\partial D$.

In the theory of partial differential equations, an important role is played by representations of solutions of these equations as functions of potential type. As an example of such representations, we show the formula of Somilian-Bettis [11] below.

**Theorem 1.** Any regular solution $U(x)$ of equation (1) in the domain $D$ is represented by the formula

$$U(x) = \int_{\partial D} (\Gamma(y,x)\{T(\partial_y,n)U(y)\} - \{T(\partial_y,n)\Gamma(y,x)\}^*U(y))ds_y, \quad x \in D. \quad (3)$$

Here $A^*$ is conjugate to $A$.

Suppose that a Carleman matrix $\Pi(y,x,\sigma)$ of the problem (1),(2) exists. Then for the regular functions $v(y)$ and $u(y)$ the following holds

$$\int_{\partial D_\rho} [v(y)\{A(\partial_y)U(y)\} - \{A(\partial_y)v(y)\}^*U(y)]dy$$

$$= \int_{\partial D_\rho} [v(y)\{T(\partial_y,n)U(y)\} - \{T(\partial_y,n)v(y)\}U(y)]ds_y.$$

Substituting $v(y) = G(y,x,\sigma)$ and $u(y) = U(y)$, a regular solution system (1), into the above equality, we get

$$\int_{\partial D_\rho} [G(y,x,\sigma)\{A(\partial_y)U(y)\} - \{A(\partial_y)G(y,x,\sigma)\}^*U(y)]dy = 0. \quad (4)$$

Adding (3) and (4) gives the following theorem.

**Theorem 2.** Any regular solution $U(x)$ of equation (1) in the domain $D_\rho$ is represented by the formula

$$U(x) = \int_{\partial D_\rho} (\Pi(y,x,\sigma)\{T(\partial_y,n)U(y)\} - \{T(\partial_y,n)\Pi(y,x,\sigma)\}^*U(y))ds_y, \quad x \in D_\rho,$$

$$(5)$$

where $\Pi(y,x,\sigma)$ is a Carleman matrix.

Suppose that $K(\omega)$, $\omega = u + iv$ ($u$, $v$ are real), is an entire function taking real values on the real axis and satisfying the conditions

$$K(u) \neq 0, \quad \sup_{v \geq 1} |v^p K^{(p)}(\omega)| = M(p,u) < \infty, \quad p = 0, ..., m, \quad u \in R^1.$$

Let

$$s = \alpha^2 = (y_1 - x_1)^2 + ... + (y_{m-1} - x_{m-1})^2.$$

For $\alpha > 0$, we define the function $\Phi(y,x)$ by the following relations:
if $m = 2$, then

$$-2\pi K(x_2)\Phi(y,x) = \int_0^\infty Im[\frac{K(i\sqrt{u^2 + \alpha^2} + y_2)}{i\sqrt{u^2 + \alpha^2} + y_2 - x_2}]\frac{udu}{\sqrt{u^2 + \alpha^2}}, \quad (6)$$

if $m = 2n + 1$, $n \geq 1$, then

$$C_m K(x_m)\Phi(y,x) = \frac{\partial^{n-1}}{\partial s^{n-1}} \int_0^\infty Im[\frac{K(i\sqrt{u^2 + \alpha^2} + y_m)}{i\sqrt{u^2 + \alpha^2} + y_m - x_m}]\frac{du}{\sqrt{u^2 + \alpha^2}}; \quad (7)$$

where $C_m = (-1)^{n-1} \cdot 2^{-n}(m-2)\pi\omega_m(2n-1)!$,
if $m = 2n$, $n \geq 2$, then

$$C_m K(x_m)\Phi(y,x) = \frac{\partial^{n-2}}{\partial s^{n-2}} Im \frac{K(\alpha i + y_m)}{\alpha(\alpha + y_m - x_m)}, \tag{8}$$

where $C_m = (-1)^{n-1}(n-1)!(m-2)\omega_m$.

**Lemma 1.** The function $\Phi(y,x)$ can be expressed as

$$\Phi(y,x) = \frac{1}{2\pi}\ln\frac{1}{r} + g_2(y,x), \quad m = 2, \ r = |y-x|,$$

$$\Phi(y,x) = \frac{r^{2-m}}{\omega_m(m-2)} + g_m(y,x), \quad m \geq 3, \ r = |y-x|,$$

where $g_m(y,x)$, $m \geq 2$ is a function defined for all values of $y, x$ and harmonic in the variable $y$ in all of $R^m$.

With the help of function $\Phi(y,x)$ we construct a matrix:

$$\Pi(y,x) = \|\Pi_{ij}(y,x)\|_{m \times m} = \left\| \frac{\lambda+3\mu}{2\mu(\lambda+2\mu)}\delta_{ij}\Phi(y,x) - \frac{\lambda+\mu}{2\mu(\lambda+2\mu)}(y_j - x_j)\frac{\partial}{\partial y_i}\Phi(y,x) \right\|_{m \times m},$$

$$i,j = 1,2,...,m. \tag{9}$$

## 2. The solution of problems (1), (2) in domain $D_\rho$

I. Let $x_0 = (0,\dots,0,x_m) \in D_\rho$. We adopt the notation

$$\beta = \tau y_m - \alpha_0, \ \gamma = \tau x_m - \alpha_0, \ \alpha_0^2 = x_1^2 + \dots + x_{m-1}^2, \ r = |x-y|,$$

$$s = \alpha^2 = (y_1 - x_1)^2 + \dots + (y_{m-1} - x_{m-1})^2, \ w = i\tau\sqrt{u^2 + \alpha^2} + \beta, \ w_0 = i\tau\alpha + \beta.$$

We now construct Carleman's matrix for the problem (1), (2) for the domain $D_\rho$. The Carleman matrix is explicitly expressed by Mittag-Löffler's a entire function. It is defined by series [1]

$$E_\rho(w) = \sum_{n=0}^{\infty} \frac{w^n}{\Gamma\left(1 + \frac{n}{\rho}\right)}, \quad \rho > 0, \quad E_1(w) = \exp w,$$

where $\Gamma(\cdot)$ is the Euler function.

Denote by $\gamma = \gamma(1,\theta)$, $0 < \theta < \frac{\pi}{\rho}$, $\rho > 1$ the contour in the complex plane $w$, running in the direction nondecreasing $argw$ and consisting of the following part's.
1)ray $argw = -\theta$, $|w| \geq 1$,
2)arc $-\theta \leq argw \leq \theta$ of the circle $|w| = 1$,
3)ray $argw = \theta$, $|w| \geq 1$.

The contour $\gamma$ divides complex plane into two parts: $D^-$ and $D^+$ lying on the left and on the right from $\gamma$, respectively. Suppose that $\frac{\pi}{2\rho} < \theta < \frac{\pi}{\rho}$, $\rho > 1$. Then the formula holds

$$E_\rho(w) = \exp w^\rho + \Psi_\rho(w), \quad w \in D^+$$

$$E_\rho(w) = \Psi_\rho(w), \quad E'_\rho(w) = \Psi'_\rho(w), \quad w \in D^-, \tag{10}$$

where

$$\Psi_\rho(w) = \frac{\rho}{2\pi i} \int_\gamma \frac{\exp \zeta^\rho}{\zeta - w} d\zeta, \quad \Psi'_\rho(w) = \frac{\rho}{2\pi i} \int_\gamma \frac{\exp \zeta^\rho}{(\zeta - w)^2} d\zeta. \tag{11}$$

$$Re\Psi_\rho(w) = \frac{\Psi_\rho(w) + \Psi_\rho(\overline{w})}{2} = \frac{\rho}{2\pi i} \int_\gamma \frac{\exp \zeta^\rho (\zeta - Rew)}{(\zeta - w)(\zeta - \overline{w})} d\zeta,$$

$$Im\Psi_\rho(w) = \frac{\Psi_\rho(w) - \Psi_\rho(\overline{w})}{2i} = \frac{\rho Imw}{2\pi i} \int_\gamma \frac{\exp \zeta^\rho}{(\zeta - w)(\zeta - \overline{w})} d\zeta, \tag{12}$$

$$\frac{Im\Psi'_\rho(w)}{Imw} = \frac{\rho}{2\pi i} \int_\gamma \frac{2\exp \zeta^\rho (\zeta - Rew)}{(\zeta - w)^2 (\zeta - \overline{w})^2} d\zeta.$$

In what follows, we take $\theta = \frac{\pi}{2\rho} + \frac{\varepsilon_2}{2}$, $\rho > 1$, $\varepsilon_2 > 0$. It is clear that if $\frac{\pi}{2\rho} + \varepsilon_2 \le |argw| \le \pi$, then $w \in D^-$ and $E_\rho(w) = \Psi_\rho(w)$.

Set

$$E_{k,q}(w) = \frac{\rho}{2\pi i} \int_\gamma \frac{\zeta^q \exp \zeta^\rho}{(\zeta - w)^k (\zeta - \overline{w})^k} d\zeta, \quad k = 1, 2, \ldots, \quad q = 0, 1, 2, \ldots.$$

If $\frac{\pi}{2\rho} + \frac{\varepsilon_2}{2} \le |argw| \le \pi$, then the inequalities are valid

$$|E_\rho(w)| \le \frac{M_1}{1 + |w|}, \quad |E'_\rho(w)| \le \frac{M_2}{1 + |w|^2},$$

$$|E_{k,q}(w)| \le \frac{M_3}{1 + |w|^{2k}}, \quad k = 1, 2, \ldots, \tag{13}$$

where $M_1, M_2, M_3$ are constants.

Suppose that in formula (10) $\theta = \frac{\pi}{2\rho} + \frac{\varepsilon_2}{2} < \frac{\pi}{\rho}$, $\rho > 1$. Then $E_\rho(w) = \Psi_\rho(w)$, $\cos \rho\theta < 0$ and

$$\int_\gamma |\zeta|^q \exp(\cos \rho\theta |\zeta|^q)|d\zeta| < \infty, \quad q = 0, 1, 2, \ldots. \tag{14}$$

In this case for sufficiently large $|w|$ ($w \in D^+$, $\overline{w} \in D^-$), we have

$$\min_{\zeta \in \gamma} |\zeta - w| = |w| \sin \frac{\varepsilon_2}{2}, \quad \min_{\zeta \in \gamma} |\zeta - \overline{w}| = |w| \sin \frac{\varepsilon_2}{2}. \tag{15}$$

Now from (10) and

$$\frac{1}{\zeta - w} = -\frac{1}{w} + \frac{\zeta}{w(\zeta - \overline{w})},$$

$$\frac{1}{\zeta - \overline{w}} = -\frac{1}{\overline{w}} + \frac{\zeta}{\overline{w}(\zeta - \overline{w})}, \tag{16}$$

for large $|w|$ we obtain

$$\left| E_\rho(w) - \Gamma^{-1}\left(1 - \frac{1}{\rho}\right)\frac{1}{w} \right| \leq \frac{\rho}{2\pi \sin \frac{\varepsilon_2}{2}} \frac{1}{|w|^2}.$$

$$\int_\gamma |\zeta| \exp\left[\cos \rho\theta |\zeta|^\rho\right] |d\zeta| \leq \frac{const}{|w|^2},$$

$$\Gamma^{-1}\left(1 - \frac{1}{\rho}\right) = \frac{\rho}{2\pi i} \int_\gamma \exp\left(\zeta^\rho\right) d\zeta.$$

From this it follows that

$$|E_\rho(w)| \leq \frac{M_1}{1 + |w|}.$$

From (11),(15) and

$$\frac{1}{(\zeta - w)^2} = \frac{1}{w^2} - \frac{2\zeta}{w^2(\zeta - w)} + \frac{\zeta^2}{w^2(\zeta - w)^2}$$

for large $|w|$, we obtain

$$\left| E'_\rho(w) - \Gamma^{-1}\left(1 - \frac{1}{\rho}\right)\frac{1}{w^2} \right| \leq \frac{const}{|w|^3}$$

or

$$|E'_\rho(w)| = \frac{M_2}{1 + |w|^2}.$$

For $k = 1, 2, \ldots$ we have from (16)

$$\frac{1}{(\zeta - w)^k(\zeta - \overline{w})^k} = \left[ \frac{(-1)^k}{w^k} + \ldots + \frac{\zeta^k}{w^k(\zeta - w)^k} \right] \left[ \frac{(-1)^k}{\overline{w}^k} + \ldots + \frac{\zeta^k}{\overline{w}^k(\zeta - \overline{w})^k} \right]$$

$$= \frac{1}{|w|^{2k}} - \frac{k}{|w|^{2k+1}|\zeta - w|} + \ldots.$$

From this for large $|w|$, (14) and (15) we get

$$\left| E_{k,q}(w) - \Gamma^{-1}\left(1 - \frac{1}{\rho}\right)\frac{1}{|w|^{2k}} \right| \leq \frac{const}{|w|^{2k+1}}$$

or

$$|E'_{k,q}(w)| = \frac{M_3}{1 + |w|^{2k}}, \quad k = 1, 2, \ldots.$$

Therefore, since

$$(\zeta - w)(\zeta - \overline{w}) = \zeta^2 - 2\zeta(y_m - x_m) + u^2 + \alpha^2 + (y_m - x_m)^2, \quad \alpha^2 = s,$$

then

$$\frac{\partial^{n-1}}{\partial s^{n-1}} \frac{1}{(\zeta - w)(\zeta - \overline{w})} = \frac{(-1)^{n-1}(n-1)!}{(\zeta - w)^n(\zeta - \overline{w})^n}.$$

Now from (11) we obtain

$$\frac{d^{n-1}}{ds^{n-1}} Re E_\rho(w) = \frac{(-1)^{n-1}(n-1)!\rho}{2\pi i} \int_\gamma \frac{(\zeta - (y_m - x_m)) \exp \zeta^\rho}{(\zeta - w)^n(\zeta - \overline{w})^n} d\zeta,$$

$$\frac{d^{n-1}}{ds^{n-1}} \frac{Im E_\rho(w)}{\sqrt{u^2 + \alpha^2}} = \frac{(-1)^{n-1}(n-1)!\rho}{\pi i} \int_\gamma \frac{\exp \zeta^\rho}{(\zeta - w)^n(\zeta - \overline{w})^n} d\zeta,$$

Then from (12) we have

$$\left| \frac{d^{n-1}}{ds^{n-1}} Re E_\rho(w) \right| \leq \frac{const \cdot r}{1 + |w|^2}$$

$$\left| \frac{d^{n-1}}{ds^{n-1}} \frac{Im E_\rho(w)}{\sqrt{u^2 + \alpha^2}} \right| \leq \frac{const \cdot r}{1 + |w|^2}.$$

Now for $\sigma > 0$, we set in formulas (6)-(9)

$$K(w) = E_\rho(\sigma^{\frac{1}{\rho}} w), \quad K(x_m) = E_\rho(\sigma^{\frac{1}{\rho}} \gamma). \tag{17}$$

Then, for $\rho > 1$ we obtain

$$\Phi(y, x) = \Phi_\sigma(y, x) = \frac{\varphi_\sigma(y, x)}{c_m E_\rho(\sigma^{\frac{1}{\rho}} \gamma)}, \quad y \neq x,$$

where $\varphi_\sigma(y, x)$ is defined as follows:
if $m = 2$, then

$$\varphi_\sigma(y, x) = \int_0^\infty Im \frac{E_\rho(\sigma^{\frac{1}{\rho}} w)}{i\sqrt{u^2 + \alpha^2} + y_2 - x_2} \frac{u du}{\sqrt{u^2 + \alpha^2}};$$

if $m = 2n + 1$, $n \geq 1$, then

$$\varphi_\sigma(y, x) = \frac{d^{n-1}}{ds^{n-1}} \int_0^\infty Im \frac{E_\rho(\sigma^{\frac{1}{\rho}} w)}{i\sqrt{u^2 + \alpha^2} + y_m - x_m} \frac{u du}{\sqrt{u^2 + \alpha^2}}, y \neq x;$$

if $m = 2n$, $n \geq 2$, then

$$\varphi_\sigma(y, x) = \frac{d^{n-2}}{ds^{n-2}} Im \frac{E_\rho\left(\sigma^{\frac{1}{\rho}} w\right)}{\alpha(i\alpha + y_m - x_m)}, y \neq x.$$

We now define the matrix $\Pi(y, x, \sigma)$ by formula (9) for $\Phi(y, x) = \Phi_\sigma(y, x)$. In the work [14] there is proved

**Lemma 2.** The function $\Phi_\sigma(y,x)$ can be expressed as

$$\Phi_\sigma(y,x) = \frac{1}{2\pi} ln\frac{1}{r} + g_2(y,x,\sigma), \quad m = 2, \quad r = |y - x|,$$

$$\Phi_\sigma(y,x) = \frac{r^{2-m}}{\omega_m(m-2)} + g_m(y,x,\sigma), \quad m \geq 3, \quad r = |y - x|,$$

where $g_m(y,x,\sigma)$, $m \geq 2$ is a function defined for all $y,x$ and harmonic in the variable $y$ in all of $R^m$.

Using Lemma 2, we obtain.

**Theorem 3.** The matrix $\Pi(y,x,\sigma)$ given by (7)-(9) is a Carleman matrix for problem (1), (2).

We first consider some properties of function $\Phi_\sigma(y,x)$

I. Let $m = 2n+1$, $n \geq 1$, $x \in D_\rho$, $y \neq x, \sigma \geq \sigma_0 > 0$, then

1) for $\beta \leq \alpha$ the following inequality holds:

$$|\Phi_\sigma(y,x)| \leq C_1(\rho)\frac{\sigma^{m-2}}{r^{m-2}} \exp(-\sigma\gamma^\rho),$$

$$\left|\frac{\partial\Phi_\sigma}{\partial n}(y,x)\right| \leq C_2(\rho)\frac{\sigma^m}{r^{m-1}} \exp(-\sigma\gamma^\rho), \quad y \in \partial D_\rho,$$

$$\left|\frac{\partial}{\partial x_i}\frac{\partial\Phi_\sigma}{\partial n}(y,x)\right| \leq C_3(\rho)\frac{\sigma^{m+2}}{r^m} \exp(-\sigma\gamma^\rho), \quad i = 1, ..., m, \tag{18}$$

2) for $\beta > \alpha$ the following inequalities hold:

$$|\Phi_\sigma(y,x)| \leq C_4(\rho)\frac{\sigma^{m-2}}{r^{m-2}} \exp(-\sigma\gamma^\rho + \sigma Re\omega_0^\rho),$$

$$\left|\frac{\partial\Phi_\sigma}{\partial n}(y,x)\right| \leq C_5(\rho)\frac{\sigma^m}{r^{m-1}} \exp(-\sigma\gamma^\rho + \sigma Re\omega_0^\rho), \quad y \in \partial D_\rho,$$

$$\left|\frac{\partial}{\partial x_i}\frac{\partial\Phi_\sigma}{\partial n}(y,x)\right| \leq C_6(\rho)\frac{\sigma^{m+2}}{r^m} \exp(-\sigma\gamma^\rho + \sigma Re\omega_0^\rho), \quad i = 1, ..., m. \tag{19}$$

II. Let $m = 2n$, $n \geq 2$, $x \in D_\rho$, $x \neq y$, $\sigma \geq \sigma_0 > 0$, then

1) for $\beta \leq \alpha$ the following inequalities hold:

$$|\Phi_\sigma(y,x)| \leq \tilde{C}_1(\rho)\frac{\sigma^{m-3}}{r^{m-2}} \exp(-\sigma\gamma^\rho),$$

$$\left|\frac{\partial\Phi_\sigma}{\partial n}(y,x)\right| \leq \tilde{C}_2(\rho)\frac{\sigma^m}{r^{m-1}} \exp(-\sigma\gamma^\rho), \quad y \in \partial D_\rho,$$

$$\left|\frac{\partial}{\partial x_i}\frac{\partial\Phi_\sigma}{\partial n}(y,x)\right| \leq \tilde{C}_3(\rho)\frac{\sigma^{m+2}}{r^m} \exp(-\sigma\gamma^\rho), \quad y \in \partial D_\rho, \quad i = 1, ..., m, \tag{20}$$

2) for $\beta > \alpha$ the following inequalities hold:

$$|\Phi_\sigma(y,x)| \leq \tilde{C}_4(\rho)\frac{\sigma^{m-2}}{r^{m-2}}\exp(-\sigma\gamma^\rho + \sigma Re\omega_0^\rho),$$

$$\left|\frac{\partial\Phi_\sigma}{\partial n}(y,x)\right| \leq \tilde{C}_5(\rho)\frac{\sigma^m}{r^{m-1}}\exp(-\sigma\gamma^\rho + \sigma Re\omega_0^\rho), \quad y \in \partial D_\rho,$$

$$\left|\frac{\partial}{\partial x_i}\frac{\partial\Phi_\sigma}{\partial n}(y,x)\right| \leq \tilde{C}_6(\rho)\frac{\sigma^{m+2}}{r^m}\exp(-\sigma\gamma^\rho + \sigma Re\omega_0^\rho), \quad y \in \partial D_\rho, \quad i = 1,...,m. \quad (21)$$

III. Let $m = 2$, $x \in D_\rho$, $x \neq y$, $\sigma \geq \sigma_0 > 0$, then
1) if $\beta \leq \alpha$, then

$$|\Phi_\sigma(y,x)| \leq C_7(\rho)E^{-1}(\sigma^{\frac{1}{\rho}}\gamma)ln\frac{1+r^2}{r^2},$$

$$\left|\frac{\partial\Phi_\sigma}{\partial y_i}(y,x)\right| \leq C_8(\rho)\frac{E_\rho^{-1}(\sigma^{\frac{1}{\rho}}\gamma)}{r}, \quad (22)$$

2) if $\beta > \alpha$, then

$$|\Phi_\sigma(y,x)| \leq \tilde{C}_7(\rho)E^{-1}(\sigma^{\frac{1}{\rho}}\gamma)(ln\frac{1+r^2}{r^2})\exp(\sigma Re\omega_0^\rho),$$

$$\left|\frac{\partial\Phi_\sigma}{\partial y_i}(y,x)\right| \leq \tilde{C}_8(\rho)E_\rho^{-1}(\sigma^{\frac{1}{\rho}}\gamma)\frac{1}{2}\exp(\sigma Re\omega_0^\rho). \quad (23)$$

Here all coefficients $C_i(\rho)$ and $\tilde{C}_i(\rho)$, $i = 1,\ldots,8$, depend on $\rho$.

**Proof Theorem 3.** From the definition of $\Pi(y,x,\sigma)$ and Lemma 1, we have

$$\Pi(y,x,\sigma) = \Gamma(y,x) + G(y,x,\sigma),$$

where

$$G(y,x,\sigma) = ||G_{kj}(y,x,\sigma)||_{m\times m}$$

$$= \left\|\frac{\lambda+3\mu}{2\mu(\lambda+2\mu)}\delta_{kj}g_m(y,x,\sigma) - \frac{\lambda+\mu}{2\mu(\lambda+2\mu)}(y_j-x_j)\frac{\partial}{\partial y_i}g_m(y,x,\sigma)\right\|_{m\times m}.$$

Prove that $A(\partial_y)G(y,x,\sigma) = 0$. Indeed, since $\Delta_y g_m(y,x,\sigma) = 0$, $\Delta_y = \sum\limits_{k=1}^{m}\frac{\partial^2}{\partial y_k^2}$ and for the $j$th column $G^j(y,x,\sigma)$ :

$$divG^j(y,x,\sigma) = \frac{1}{2\mu(\lambda+2\mu)}\cdot\frac{\partial}{\partial y_j}g_m(y,x,\sigma),$$

then for the $k$th components of $A(\partial_y)G^j(y,x,\sigma)$ we obtain

$$\sum\limits_{i=1}^{m}A(\partial_y)_{ki}G_{ij}(y,x,\sigma) = \mu\Delta_y[\frac{\lambda+3\mu}{2\mu(\lambda+2\mu)}\cdot\delta_{kj}g_m(y,x,\sigma) - \frac{\lambda+\mu}{2\mu(\lambda+2\mu)}(y_j-x_j)\frac{\partial}{\partial y_k}g_m(y,x,\sigma)]$$

$$+(\lambda+\mu)\frac{\partial}{\partial y_k}divG^j(y,x,\sigma)=-\frac{\lambda+\mu}{2\mu(\lambda+2\mu)}\frac{\partial^2}{\partial y_j^2}g_m(y,x,\sigma)+\frac{\lambda+\mu}{2\mu(\lambda+2\mu)}\frac{\partial^2}{\partial y_j^2}g_m(y,x,\sigma)=0$$

Therefore, each column of the matrix $G(y,x,\sigma)$ satisfies to system (1) in the variable $y$ everywhere on $R^m$.

The second condition of Carleman's matrix follows from inequalities (18)-(23). The theorem proved.

For fixed $x \in D_\rho$ we denote by $S^*$ the part of $S$, where $\beta \ge \alpha$. It $x = x_0 = (0,\ldots,0,x_m) \in D_\rho$, then $S = S^*$. In the point $(0,\ldots,0) \in D_\rho$, suppose that

$$\frac{\partial U}{\partial n}(0) = \frac{\partial U}{\partial y_m}(0), \quad \frac{\partial \Phi_\sigma(0,x)}{\partial n} = \frac{\partial \Phi_\sigma(0,x)}{\partial y_m}.$$

Let

$$U_\sigma(y) = \int\limits_{S^*} [\Pi(y,x,\sigma)\{T(\partial_y,n)U(y)\} - \{T(\partial_y,n)\Pi(y,x,\sigma)\}^*U(y)]ds_y, \quad x \in D_\rho.$$

(24)

**Theorem 4.** Let $U(x)$ be a regular solution of system (1) in $D_\rho$, such that

$$|U(y)| + |T(\partial_y,n)U(y)| \le M, \quad y \in \Sigma. \tag{25}$$

Then,

1) if $m = 2n+1$, $n \ge 1$, and for the $x \in D_\rho$, $\sigma \ge \sigma_0 > 0$, the following estimate is valid:

$$|U(x) - U_\sigma(x)| \le MC_1(x)\sigma^{m+1}\exp(-\sigma\gamma^\rho).$$

2) In case $m = 2n$, $n \ge 1$, $x \in D_\rho$, $\sigma \ge \sigma_0 > 0$, the following estimate is valid

$$|U(x) - U_\sigma(x)| \le MC_2(x)\sigma^m\exp(-\sigma\gamma^\rho),$$

where

$$C_k(x) = C_k(\rho)\int\limits_{\partial D_\rho}\frac{ds_y}{r^m}, \quad k = 1,2,$$

$C_k(\rho)$ is a constant depending on $\rho$.

**Proof.** From formula (5)

$$U(x) = \int\limits_{S^*} [\Pi(y,x,\sigma)\{T(\partial_y,n)U(y)\} - \{T(\partial_y,n)\Pi(y,x,\sigma)\}^*U(y)]ds_y$$

$$+ \int\limits_{\partial D_\rho\backslash S^*} [\Pi(y,x,\sigma)\{T(\partial_y,n)U(y)\} - \{T(\partial_y,n)\Pi(y,x,\sigma)\}^*U(y)]ds_y, \quad x \in D_\rho,$$

therefore, (24) implies

$$|U(x)-U_\sigma(x)| \le \int\limits_{\partial D_\rho\backslash S^*} [\Pi(y,x,\sigma)\{T(\partial_y,n)U(y)\}-\{T(\partial_y,n)\Pi(y,x,\sigma)\}^*U(y)]ds_y$$

$$\leq \int\limits_{\partial D_\rho \backslash S^*} [|\Pi(y,x,\sigma)| + |T(\partial_y,n)\Pi(y,x,\sigma)|] [|T(\partial_y,n)\Pi(y,x,\sigma)| + |U(y)|] \, ds_y.$$

Therefore for $\beta \leq \alpha$ we obtain from inequalities (18)-(23), and condition (25) for $m = 2n+1, \ n \geq 1$

$$|U(x) - U_\sigma(x)| \leq MC_1(\rho)\sigma^{m+1} \exp(-\sigma\gamma^\rho) \int\limits_{\partial D_\rho} \frac{ds_y}{r^m},$$

and for $m = 2n, \ n \geq 1$ we obtain

$$|U(x) - U_\sigma(x)| \leq MC_2(\rho)\sigma^m \exp(-\sigma\gamma^\rho) \int\limits_{\partial D_\rho} \frac{ds_y}{r^m}.$$

The theorem proved.

Now we write out a result that allows us to calculate $U(x)$ approximately if, instead of $U(y)$ and $T(\partial_y,n)U(y)$ their continuous approximations $f_\delta(y)$ and $g_\delta(y)$ are given on the surface $S$:

$$\max_S |U(y) - f_\delta(y)| + \max_S |T(\partial_y,n)U(y) - g_\delta(y)| \leq \delta, \ 0 < \delta < 1. \quad (26)$$

We define a function $U_{\sigma\delta}(x)$ by setting

$$U_{\sigma\delta}(x) = \int\limits_{S^*} [\Pi(y,x,\sigma)g_\delta(y) - \{T(\partial_y,n)\Pi(y,x,\sigma)\}^* f_\delta(y)] ds_y, \ x \in D_\rho, \quad (27)$$

where $\sigma = \frac{1}{R^\rho} ln\frac{M}{\delta}, \ R^\rho = \max\limits_{y \in S} Re\omega_0^\rho.$

Then the following theorem holds.

**Theorem 5.** Let $U(x)$ be a regular solution of system (1) in $D_\rho$ such that

$$|U(y)| + |T(\partial_y,n)U(y)| \leq M, \ y \in \partial D_\rho.$$

Then,

1) if $m = 2n+1, \ n \geq 1$, the following estimate is valid:

$$|U(x) - U_{\sigma\delta}(x)| \leq C_1(x)\delta^{(\frac{\gamma}{R})^\rho} \left(ln\frac{M}{\delta}\right)^{m+1},$$

2) if $m = 2n, \ n \geq 1$, the following estimate is valid:

$$|U(x) - U_{\sigma\delta}(x)| \leq C_2(x)\delta^{(\frac{\gamma}{R})^\rho} \left(ln\frac{M}{\delta}\right)^m,$$

where

$$C_k(x) = C_k(\rho) \int\limits_{\partial D_\rho} \frac{ds_y}{r^m}, \ k = 1,2.$$

**Proof.** From formula (5) and (27) we have

$$U(x) - U_{\sigma\delta}(x) = \int\limits_{\partial D_\rho \backslash S^*} [\Pi(y,x,\sigma)\{T(\partial_y,n)U(y)\} - \{T(\partial_y,n)\Pi(y,x,\sigma)\}^* U(y)] ds_y$$

$$+ \int_{S^*} [\Pi(y,x,\sigma)\{T(\partial_y,n)U(y) - g_\delta(y)\} + \{T(\partial_y,n)\Pi(y,x,\sigma)\}^*(U(y) - f_\delta(y))] ds_y = I_1 + I_2.$$

By Theorem 4 for $m = 2n + 1$, $n \geq 1$,

$$|I_1| = MC_1(\rho)\sigma^{m+1}\exp(-\sigma\gamma^\rho)\int_{\partial D_\rho}\frac{ds_y}{r^m},$$

and for the $m = 2n$, $n \geq 1$

$$|I_1| = MC_2(\rho)\sigma^m\exp(-\sigma\gamma^\rho)\int_{\partial D_\rho}\frac{ds_y}{r^m}.$$

Now consider $|I_2|$ :

$$|I_2| = \int_{S^*} (|\Pi(y,x,\sigma)| + |T(\partial_y,n)\Pi(y,x,\sigma)|)\,(|T(\partial_y,n)U(y) - g_\delta(y)| + |U(y) - f_\delta(y)|)\,ds_y.$$

By Lemma 2 and condition (28) we obtain for $m = 2n + 1$, $n \geq 1$

$$|I_2| = \tilde{C}_1(\rho)\sigma^{m+1}\delta\exp(-\sigma\gamma^\rho + \sigma Re w_0^\rho)\int_{\partial D_\rho}\frac{ds_y}{r^m}$$

and for $m = 2n$, $n \geq 1$,

$$|I_2| = \tilde{C}_2(\rho)\sigma^m\delta\exp(-\sigma\gamma^\rho + \sigma Re w_0^\rho)\int_{\partial D_\rho}\frac{ds_y}{r^m}.$$

Therefore, from

$$\sigma = \frac{1}{R^\rho}\ln\frac{M}{\delta}, \quad R^\rho = \max_{y \in S} Re w_0^\rho.$$

we obtain the desired result.

**Corollary 1.** The limit relation

$$\lim_{\sigma \to \infty} U_\sigma(x) = U(x), \quad \lim_{\delta \to 0} U_{\sigma\delta}(x) = U(x)$$

hold uniformly on any compact set from $D_\rho$.

## References

1. M.M. Dzharbashyan. Integral Transformations and Representations of Functions in a Complex Domain. [in Russian], Nauka, Moscow 1966.
2. T.I. Ishankulov and I.E. Niyozov. *Regularization of solution of the Cauchy problem for Navier-Stokes.* Uzb.Math.J., No.1, 35-42 (1997).
3. V.D. Kupradze, T.V. Burchuladze, T.G. Gegeliya, ot.ab. *Three-Dimensional Problems of the Mathematical Theory of Elasticity, etc.* [in Russian], Nauka, Moscow, 1976.
4. M.M. Lavrent'ev. Some Ill-Posed Problems of Mathematical Physics [in Russian], Computer Center of the Siberian Division of the Russian Academy of Sciences, Novosibirck (1962) 92p.

82

5. O.I. Makhmudov. *The Cauchy problem for a system of equation of the spatial theory of elasticity in displacements.* Izv. Vyssh. Uchebn. Zaved. Math. [Russian Math. (Iz.VUZ)], **380**, No.1, 54-61 (1994).

6. O.I. Makhmudov and I.E. Niyozov. *Regularization of the solution of the Cauchy problem for a system of equations in the theory of elasticity in displacements.* Sibirsk. Math. zh. [Sibirian Math. j.], **39**, No.2, 369-376 (1998).

7. O.I. Makhmudov and I.E. Niyozov. *On a Cauchy problem for a system of equations of elasticity theory.* Differensialnye uravneniya [Differential equations], **36**, No.5, 674-678 (2000).

8. O.I. Makhmudov and I.E. Niyozov. *Regularization of the solution of the Cauchy problem for a system of elasticity theory in an infinite domain.* Math. zhametki. [Math.Notes], **68**, No.4, 548-553 (2000).

9. I.G. Petrovskii. Lectures on Partial Differential Equations [in Russians], Fizmatgiz, Moscow, (1961).

10. A.A. Shlapunov. *On the Cauchy problem for the Lame system preprint di Mathematica,* Scuola Normale Superiore 40 (1994)ZAMM **76** No. 4, 215-221 (1996).

11. N.N. Tarkhanov. *On the Carleman matrix for elliptic systems.* Dokl. Acad. Nauk SSSR [Soviet Math.Dokl.], **284**, No.2, 294-297 (1985).

12. N.N. Tarkhanov. The Cauchy Problem for Solutions of Elliptic Equations. Akademie-Verlag, Berlin (1985).

13. A.N. Tikhonov. *Solution of ill-posed problems and the regularization method.* Dokl. Acad. Nauk USSR [Soviet Math.Dokl.], **151**, 501-504 (1963).

14. Sh.Ya. Yarmukhamedov. *Cauchy problem for the Laplace equation.* Dokl. Acad. Nauk SSSR [Soviet Math.Dokl.], **235**, No.2, 281-283 (1977).

15. Sh.Ya. Yarmukhamedov, T.I. Ishankulov and O.I. Makhmudov. *Cauchy problem for the system of equations of elasticity theory space.* Sibirsk Math. zh. [Sibirian Math.J.], **33**, No.1, 186-190 (1992).

# FIRST LYAPOUNOV METHOD FOR THE ABSTRACT
# PARABOLIC EQUATION *

### V. A. TRENOGIN

*Department of Mathematics*
*Moscow State Steel and Alloys Institute,*
*119094 Leninsky prosp. 4*
*Moscow, Russia*
*E-mail: trenogin@km.ru*

In the Banach space for quasilinear differential equation with unbounded operator the existence and uniqueness theorems of generalized and classical solutions on the positive semiaxis tending to zero on infinity are proved. These results may be interpreted as generalizations of the known Lyapounov theorem about the asyptotical stability by linear approximation.

**Key words:** Banach space, diffential equation, Lyapunov theorem
**Mathematics Subject Classification:** 47H15

Let $X$ be real or complex Banach space. Consider the differential equation (DE)

$$\dot{x} = Ax + R(x). \tag{1}$$

Below the following condition are supposed to be fulfilled.

**I.** The closed linear operator $A$ with dense in X domain $D(A)$ is the generator of the semigroup $U(t) = exp(At)$ of $C_0$ class. U(t) is exponentially decreasing, i.e. there exist the constants $M > 0$ and $\alpha > 0$ such that for all $t \in \Re^+ = [0, +\infty)$ the inequality $||U(t)|| \leq Mexp(-\alpha t)$ is valid.

**II.** For the defined in the ball $S = \{x \in X : ||x|| < p\}$ nonlinear operator $R(x)$ there exist the constants $C > 0$ and $\beta > 0$ such that in $S$ the inequality $||R(x_1) - R(x_2)|| \leq Cmax^\beta(||x_1||, ||x_2||))||x_1 - x_2||$ takes place.

The abstract continuously differentiable on $\Re^+ = [0, \infty)$ function $x = x(t)$ belonging to $D(A)$ and satisfying (1) is named the classical solution.

**Corollary 1** From II the inequality $||R(x)|| \leq C||x||^{1+\beta}, \forall x \in S$ follows.

The condition $R(0) = 0$ implies that DE (1) has on $\Re^+$ the trivial classical solution $x(t) = 0$.

For DE (1) and its linearization

$$\dot{x} = Ax. \tag{2}$$

---

*The research has been suported by the grant 05.01.00422a of the Russian Found of Fundamental Research.

84

the initial data

$$x(0) = x_0. \tag{3}$$

are posed. The DE (1) can be interpreted as the abstract dynamical system, its solution as trajectory in the phase space $X$ and the point $x = 0$ as the state of equilibrium. Below we give the conditions when the known Lyapounov theorem about asymptotical stability of the trivial solution of DE (1) is the stability on its linear approximation DE (2). By using the semigroup theory and differential equations in Banach space theory one can transfer A.M. Lyapounov results on the abstract case.

Begin with the generalized solution of Cauchy problem (1),(3). Consider the integral equation

$$x(t) = U(t)x_0 + \int_0^t U(t-s)R(x(s))ds. \tag{4}$$

The continuous on $\Re^+$ solution of (4) will be called the generalized solution of (1),(3). From the conditions I-II it follows that for any continuous function $x(t)$, $t \in \Re^+$ with the values in $S$ the right-hand-side of (4) is also continuous function on $\Re^+$. Our nearest purpose is to investigate solutions of the integral equation (4) exponentially decreasing at $t \to +\infty$ solutions of the initial equation (4). Introduce the suitable family Banach spaces of abstract functions.

**Definition** Let $\gamma > 0$ hold. Denote by $C_\gamma$ the set of all abstract continuous on $\Re^+$ functions $x(t)$ with values in $X$ with the natural operations of addition and multiplication on the scalars for which the norm $|||x|||_\gamma = sup_{\Re^+}||x(t)||exp(\gamma t)$ is finite.

Note that $C_\gamma$ is the Banach space.

**Lemma 1** Define the linear operator by the formula $(Dx_0)(t) = U(t)x_0$. Then $D \in L(X, C_\alpha)$, $||D|| \leq M$ and $U(t)x_0$ is the generalized solution of Cauchy problem (2),(3).

**Proof** $||exp(\alpha t)U(t)x_0|| \leq M||x_0||$. Hence $|||Dx_0|||_\alpha \leq M||x_0||$.

Introduce in $C_\alpha$ the open ball $S_\alpha = \{x(t) \in C_\alpha : |||x|||_\alpha < p\}$.

**Lemma 2** Let $x(t) \in S_\alpha$. Define the nonlinear operator $F$ by the formula $F(x)(t) = \int_0^t U(t-s)R(x(s))ds$. Then $F(x)(t) \in C_\alpha$ and there exists the independent from $x$ constant $K > 0$ and such that

$$|||F(x)|||_\alpha \leq K|||x|||_\alpha^{1+\beta} \tag{5}$$

**Proof** Condition I and corollary 1 imply the following estimate

$||exp(\alpha t)F(x)(t)|| \leq exp(\alpha t) \int_0^t exp(-\alpha(t-s))exp(-\alpha(1+\beta)s)||x(s)exp(\alpha s)||^{1+\beta}ds$
$\leq MC|||x|||_\alpha^{1+\beta} \int_0^t exp(-\alpha\beta s)ds = MC|||x|||_\alpha^{1+\beta} \frac{1-exp(-\alpha\beta t)}{\alpha\beta} \leq K|||x|||_\alpha^{1+\beta}$, where $K = \frac{MC}{\alpha\beta}$.

**Lemma 3** Let $x_1(t), x_2(t) \in S_\alpha$. Then the following inequality is valid

$$|||F(x_1) - F(x_2)|||_\alpha \leq Kmax^\beta(|||x_1|||_\alpha, |||x_2|||_\alpha)|||x_1 - x_2|||_\alpha \tag{6}$$

**Proof** From I, II and the definition of the space $C_\alpha$ the inequality follows
$\|R(x_1) - R(x_2)\| \leq Cmax^\beta(\|x_1(t)exp(\alpha t)\|, \|x_2(t)exp(\alpha t)\|)\|x_1(t)exp(\alpha t) - x_2(t)exp(\alpha t)\|exp(-\alpha(\beta+1)t) \leq Cmax^\beta(\|x_1\|_\alpha, \|x_2\|_\alpha)\|x_1-x_2\|_\alpha exp(-\alpha(\beta+1)t)$.
Now the argument of the Lemma 2 proof can be repeated. The proof is finished similarly Lemma 2.

The Lemmas 1-3 are the base for the subsequent arguments. The integral equation (4) can be regarded as the operator equation

$$x = Dx_0 + F(x) \tag{7}$$

with unknown $x \in C_\alpha$ and parameter $x_0 \in X$. For its solving the implicit operator theorem can be used. Here the direct proof is given in order to obtain the maximal domain of stability with respect to $x_0$.

**Theorem 1** *Introduce the numbers $r_* = (K(\beta+1))^{-1/\beta}$ and $\rho_* = \frac{\beta r_*}{M(\beta+1)}$. Then for any $x_0$ such that $\|x_0\| \leq \rho_*$ the equation (7) has in the ball $\||x\||_\alpha \leq r_*$ the unique solution $x = x(x_0)$. This solution is continuous with respect to $x_0$ in the same ball $\|x_0\| \leq \rho_*$ with $x(0) = 0$.*

**Proof** Write the equation (7) in the form $x = \Phi(x, x_0)$, where $\Phi(x, x_0) = Dx_0 + F(x)$. According to Lemmas 1,2 the operator $\Phi$ maps the ball $D_r = \{x \in S_\alpha : \||x\||_\alpha\} \leq r$ to the space $C_\alpha$. Moreover according to Lemma 3 for all $x_1, x_2 \in D_r$ the inequality $\||\Phi(x_1, x_0)-\Phi(x_2, x_0)\||_\alpha \leq Kr^\beta\||x_1-x_2\||_\alpha$ is fulfilled. Let $r \in (0, K^{-\frac{1}{\beta}})$ be hold. Then the operator $\Phi$ is the contractive operator on $D_r$ with the contraction coefficient $q = q(r) = Kr^\beta$. If additionally the condition $\|\Phi(0, x_0)\| \leq (1-q)r$ takes place then $\Phi$ will be mapping of the ball $D_r$ into itself (see, for example, [3], no. 33.2). In our case this condition has the form $\|Dx_0\| \leq \phi(r)$, where $\phi(r) = r - Kr^{\beta+1}$. Now according to the corollary to contraction mapping principle (see [3], no. 33.2) for any $x_0$ satisfying the inequality $\|Dx_0\| \leq \phi(r)$ in the ball $D_r$ there exists the unique solution of the equation (7).

We will show that the radius $r$ may be selected in the best way in order to implicit operator $x(x_0)$ would have the most wide domain of definition as the function of $x_0$. The function $\phi(r)$ is positive on the interval $(0, K^{-\frac{1}{\beta}})$ and is equal to zero on its endpoints. At $r = r_* = (K(\beta+1))^{-\frac{1}{\beta}}$ the function $\phi(r)$ attains its maximum $\frac{\beta r_*}{\beta+1}$. Consequently, if $\|Dx_0\| \leq \frac{\beta r_*}{\beta+1}$, then the operator $\Phi$ in the ball $D_{r_*}$ is contractive with the relevant contraction coefficient $q_* = q(r_*) = \frac{1}{\beta+1}$. Moreover, $\Phi$ maps this ball into itself. The indicated bound is fulfilled if (Lemma 1) $\|x_0\| \leq \rho_* = \frac{\beta r_*}{M(\beta+1)}$.

**Corollary 2** Restore to integral equation (6). The existence and uniqueness of its solution $x = x(t, x_0) \in D_{r_*}$ defined for all $x_0$ $\|x_0\| \leq \rho_*$ is proved. The inequality $\|x(t, x_0)\| \leq r_* exp(-\alpha t)$ is valid. Further for brevity $x_*(t) = x_*(t, x_0)$ is designated the generalized solution of Cauchy problem (1),(3) in the ball $D_{r_*}$, defined at $\|x_0\| \leq \rho_*$.

**Corollary 3** The equation (2) is the linearization of (1), trivial solution of which is asymptotically stable by Lemma 1. Theorem 1 establishes the asymptotical stability of the generalized trivial solution of nonlinear equation (1) at the restriction II on its nonlinear part.

**Corollary 4** The continuity of $x(t, x_0)$ relatively to $x_0$ in the space $C_\alpha$ means the following statement. Fix $x_0$ with $||x_0|| \le \rho_*$. Then for any $\epsilon > 0$ there exists $\eta > 0$ such that for all $x' \in X$ with $||x'|| \le \rho_*$ such that $||x' - x_0|| < \eta$ the inequality $||x(t, x') - x(t, x_0)|| < \epsilon . exp(-\alpha t)$ is fulfilled.

**Corollary 5** The generalized solution $x_*(t) = x_*(t, x_0)$ can be calculated at the usage of traditional iterative procedure. Organize the sequence of the functions $x_1(t) = 0$, $x_{n+1}(t) = U(t)x_0 + \int_0^t U(t - s)R(s, x_n(s))ds$, $n = 1, 2, ...$. Then $\{x_n\} \subset C_\alpha$ and $|||x_n|||_\alpha \in D_{r_*}$. Besides, $x_n(t) \to x_*(t)$, $n \to \infty$ $C_\gamma$. The following estimate of the convergence rate is valid $|||x_n - x_*|||_\gamma \le \frac{M||x_0||}{\beta(\beta+1)^{n-2}}$.
Proposition about the iterational procedure convergence for our case is the part of the proof of contraction mapping principle.

Below we indicate the conditions where it is possible step by step to show that the obtained generalized solution $x_*(t)$ is the classical one.

**Corollary 6** Let $x(t) \in C_\alpha$. For any $h > 0$ take $m_h(x) = sup_{t\in\Re^+}||x(t + h) - x(t)|| \exp(\alpha t)$. If $x \in C_\alpha$ then $m_h(x) < +\infty$. Really $||x(t + h) - x(t)|| \exp(\alpha t) \le (||x(t+h)|| + ||x(t)||) \exp(\alpha t) \le ||x(t+h)|| \exp(\alpha(t+h) + ||x(t)|| \exp(\alpha t) \le 2|||x|||_\alpha$
Through $H_\alpha$ denote the subset of functions from $C_\alpha$ satisfying the Lipschitz condition $sup_{h>0}(m_h(x)h^{-1}) < \infty$.

**Lemma 4** If $x_0 \in D(A)$ then $U(t)x_0 \in H_\alpha$.

**Proof** This fact is widely known in the theory of differential equations in Banach spaces (see, for example [2]). Really, one has $exp(\alpha t)(U(t + h) - U(t))x_0 = exp(\alpha t) \int_0^1 U(t+\theta h)Ax_0 d\theta h$ whence $||exp(\alpha)(U(t+h) - U(t))x_0|| \le M||Ax_0||$. Note that $U(t)x_0$ is also differentiable on $\Re^+$.

**Lemma 5** If $y(t) = \int_0^t U(t-s)R(x_*(s))ds$ then the inequality $m_h(y) \le q_*m_h(x_*) + q_*r_*\alpha\beta h$ is valid.

**Proof** Firstly note that
$y(t) = \int_0^t U(\theta)R(x_*(t - \theta))d\theta$.
$y(t + h) = \int_0^t U(\theta)R(x_*(t + h - \theta))d\theta$.
Consequently $exp(\alpha t)(y(t + h) - y(t)) = u(t) + v(t)$, where
$u(t) = exp(\alpha t) \int_0^t U(\theta)(R(x_*(t + h - \theta)) - R(x_*(t - \theta))d\theta$
$v(t) = exp(\alpha t) \int_t^{t+h} U(\theta)R(x_*(t + h - \theta))d\theta$.
The proof will be finished if the inequalities $||u(t)|| \le q_*m_h(x_*)$, $||v(t)|| \le q_*r_*\alpha\beta h$ would obtained. As in the Lemma 3 proof one has the estimates $||R(x_*(t + h)) - R(x_*(t))|| \le Cmax^\beta(||x_*(t + h)exp(\alpha t)||, ||x_*(t)exp(\alpha t)||)||x_*(t + h)exp(\alpha t) - x_*(t)exp(\alpha t)||exp(-\alpha(\beta + 1)t) \le Cmax^\beta(||x_*(t + h)exp(\alpha(t+h)||, ||x_*(t)exp(\alpha t)||)||x_*(t+h)exp(\alpha t) - x(t)exp(\alpha t)||exp(-\alpha(\beta+1)t) \le C|||x_*|||_\alpha^{\beta+1}m_h(x_*)exp(-\alpha(\beta + 1)t) \le Cr_*^\beta m_h(x_*)exp(-\alpha(\beta + 1)t)$.
Hence $||u(t)|| \le \int_0^t Mexp(\alpha(t - \theta))||R(x_*(t + h - \theta)) - R(x_*(t - \theta))||d\theta \le MCr_*^\beta m_h(x_*) \int_0^t exp(-\alpha\beta(t - \theta))d\theta \le \frac{MC}{\alpha\beta}r_*^\beta m_h(x_*) = Kr_*^\beta = q_*m_h(x_*)$. Analogously one has $||v(t)|| \le exp(\alpha t) \int_t^{t+h} Mexp(\alpha\theta)||R(x_*(t + h - \theta))||d\theta \le MCr_*^{\beta+1} \int_t^{t+h} exp(\alpha - \theta t)exp(-\alpha(\beta + 1)(t + h - \theta)d\theta \le MCr_*^{\beta+1}h = q_*r_*\alpha\beta h$.

**Lemma 6** If $x_0 \in D(A)$ then $x_*(t) \in H_\alpha$.

**Proof** According to Lemmas 4, 5 one has $\|(x_*(t + h) - x_*(t))exp(\alpha t)\| \le M\|Ax_0\|h + q_*m_h(x_*) + q_*r_*\alpha\beta h$. Passing to supremum on the left side of this bound we obtain $m_h(x_*) \le M\|Ax_0\|h + q_*m_h(x_*) + q_*r_*\alpha\beta h$. Hence $m_h(x_*) \le (1 - q_*)(M\|Ax_0\| + q_*r_*\alpha\beta)h$.

**Lemma 7** $R(x_*(t)) \in H_\alpha$.

**Proof** $\|R(x_*(t + h)) - R(x_*(t))\| \le Cr_*^{\beta+1}\|x_*(t + h) - x_*(t)\| \le Cr_*^\beta m_h(x)$

**Theorem 2** *Let $x_0 \in D(A)$. Then $x_*(t) = x_*(t, x_0)$ is the classical solution of the Cauchy problem (1),(3).*

**Proof** By Lemma 5 $x_*(t) \in H_\alpha$. By Lemma 6 $R(x_*(t)) \in H_\alpha$. Accordingly to theory of differential equations in Banach space (see for example [3]) if $x_0 \in D(A)$ and $f(t)$ satisfies the Hölder condition then Cauchy problem $\dot{x} = Ax + f(t)$, $x(0) = x_0$ has the classical unique solution which is expressed by the formula $x(t) = U(t)x_0 + \int_0^t U(t - s)f(s)ds$. Let us take $f(t) = R(t, x_*(t))$. Then $x(t) = x_*(t)$ and the proof of the theorem is finished.

Our arguments are also valid for analytical case of the operator $R$. Instead of the condition II introduce the following one.

**III.** Nonlinear operator $R(x)$ is analytical at the point $x = 0$, i.e. it can be represented by the series

$$R(x) = \sum_{k=2}^{+\infty} R_k x^k, \tag{8}$$

where $R_k$ are $k$-linear bounded operators in $X$ and the majorizing series $\sum_{k=2}^{+\infty}\|R_k\|r^{k-2}$ is convergnet at $0 < r < r_0$, $r_0 > 0$ is the convergence radius.

**Corollary 5** Fix $r \in (0, r_0)$ . From III it follows that there exists the constant $C_1 > 0$ such that $\|R_k\| \le C_1 r^{-k+2}$ , $k = 2, 3, \ldots\ldots$

**Lemma 8** *Let the condition III is valid. Then the operator $F_k(x)$ defined by formula $F_k(x)(t) = \int_0^t U(t - s)R_k x^k(s)ds$ is $k$ -linear bounded operator from $L(C_\alpha)$, and $\||F_k\||_\alpha \le \frac{M.\|R_k\|}{\alpha}$.*

**Proof** $\|exp(\alpha t)F_k(x)(t)\| \le exp(\alpha t)\int_0^t Mexp(-\alpha(t - s))\|R_k\|.\|x(s)\|^k ds = M\|R_k\|\int_0^t exp(\alpha t)exp(-k\alpha.s)\|exp(\alpha s)x(s)\|^k ds \le M\|R_k\|.\||x\||_\alpha^k \int_0^t exp((-(k - 1)\alpha)s)ds \le \frac{M\|R_k\|}{\alpha}\||x\||_\alpha^k$

**Lemma 9** *Let the condition III is valid. Then the operator $F(x)$ acting in the space $C_\alpha$ according to formula $(F(x)(t) = \int_0^t U(t - s)R(x(s))ds$ is analytical one in this space at the point $x = 0$.*

**Proof** By Lemma 8 $(F(x)(t) = \sum_{k=2}^{+\infty} F_k(x)(t)$ and the series on the right-hand-side converges in $C_\alpha$ if $\||x\||_\alpha \le r$.

Consider the integral equation (4) as the operator equation

$$x = Dx_0 + \sum_{k=2}^{+\infty} F_k(x)(t) \tag{9}$$

with the unknown $x \in C_\alpha$ and parameter $x_0 \in X$. One can apply to equation (9) the theorem on implicit operator in analytical case and make the assertion that this equation gives in the neighborhood of the point $x = 0$ of the space $C_\alpha$ unique implicit operator $x = x(t, x_0)$ which is analytical one at the point $x_0 = 0$.

However we have the possibility to obtain the more precise results and in particular to indicate the lower bound for the convergence radius of the series representing the implicit operator. The solution of the equation (9) we will seek in the form of the series in Banach space $C_\alpha$ on the powers of the initial value $x_0 \in X$:

$$x = \sum_{l=1}^{+\infty} X_l x_0^l. \tag{10}$$

Substitute the series (10) in the equation (9), equate the components with the same in respect to $x_0$ powers operators and obtain

$$X_1 x_0 = D x_0, \quad X_2 x_0^2 = F_2(X_1 x_0)^2, \quad X_3 x_0^3 = 2F_2 X_1 x_0 . X_2 x_0^2 + F_3(X_1 x_0)^3, \ldots$$

This equations system is recurrent one, from which all members of the series (10) may be step by step computable. Set for brevity $C_1 = \frac{MC}{\alpha}$.

Majorizing Cauchy-Goursat equation formed for (9) has the following form

$$\xi = M\eta + \frac{C_1 \xi^2}{1 - \frac{\xi}{r}}. \tag{11}$$

This equation has the unique root $\xi = \xi(\eta)$ satisfying the condition $\xi(0) = 0$. It is equal $\xi = \frac{2M\eta}{1 + \frac{M\eta}{r} + \sqrt{(1 - \frac{M\eta}{r})^2 - 4MC_1\eta}}$ and defined for $\eta < \frac{1}{2MC_1 + M/r + 2M\sqrt{C_1^2 + C_1/r}} = \eta(r)$.

Thus the series (10) converges at $||x_0|| < \eta(r)$. Because the decreasing of $\eta(r)$ the series (10) is convergent for $||x_0|| < \eta(r_0)$. If $r_0 = +\infty$ then this series converges for $||x_0|| < \frac{1}{4MC_1} = \frac{\alpha}{4M^2C}$.

Note that the more is the domain of convergence of the series representing the operator $R(x)$ then the more will be the domain of the absolute convergence of the series representing the generalized solution.

Our considerations can be corrected if $R_2 x^2 = 0, \quad \ldots R_{m-1} x^{m-1} = 0, \quad R_m x^m \neq 0$.

Thus the following proposition is proved.

**Theorem 3** *Let the operator $R(x)$ is analytical at point $x = 0$ and their radius of absolute convergence is equal to $r_0 \in (0, +\infty]$. Then the Cauchy problem (1),(3) has in the sufficiently small neighborhood of the point $x = 0$ the unique generalized solution $x(x_0) = x(t, x_0)$. This solution is analytical operator in the space $C_\alpha$ with respect to $x_0$ at the point $x_0 = 0$. The absolute convergence radius of the series representing $x(x_0)$ has the following lower bound : the absolute convergence radius of the series (10) is not less then $\eta(r_0)$.*

Consider now the question: will be the generalized solution the classical one? Applying the Lagrange formula

$F(x_1) - F(x_2) = \int_0^1 F'(x_1(1 - \theta) + x_2\theta)(x_1 - x_2)d\theta$, where $F'(x) = \sum_2^{+\infty} kF_k x^{k-1}$ one has $||F(x_1) - F(x_2)|| \leq \int_0^1 \sum_{k=2}^{+\infty} k||F_k||||x_1(1 - \theta) + x_2\theta)d\theta||^{k-1}||x_1 - x_2|| \leq \sum_{k=2}^{+\infty} k||F_k||max^{k-1}(||x_1, ||x_2||)||x_1 - x_2||$.

Fix $\xi < r < r_0$, then for all $x_1, x_2$ from the ball $||x|| \leq \xi$ we obtain the condition II where $C = \sum_2^{+\infty} k||F_k||\xi^{k-2}$ and $\beta = 1$. newline The following proposition can be proved.

**Theorem 4** *Let $x_0 \in D(A)$. Then there exists the positive number $\rho_{**}$ such that in the ball $||x_0|| \leq \rho_{**}$ the generalized analytical solution from theorem 3 is classical one.*

The simplest examples to the proved theorems give the A.M. Lyapounov classical results for the systems of ordinary differential equations. Here the semigroup is the matrix exponential and the condition I is fulfilled if the real parts of all matrix $A$ eigenvalues are negative. In our more general case we havn't possibility to use Jordan normal form. The Lyapounov functions method is also inapplicable.

Isolated parts of this paper were discussed in [3,6]. These results are the essential parts of the Lyapounov first method. Note that the case where the operator $A$ depends from $t$ is technically complicated chapter of the differential equations in Banach space theory. We hope to investigate this situation in the future.

The obtained results have the applications to functional differential equations and to reaction-diffusion equation.

## References

1. Lyapounov A.M. Collection of works. vol 2, (in Russian) Moscow-Leningrad., 1956
2. Kartashov M.P., Rozhdestvensky B.L. Ordinary differential equations and introduction to varitional calculus. Moscow.-Nauka, 1986
3. Trenogin V.A. Functional analysis. (Russian) Moscow.- Phizmathlit, 2002 (Nauka-1980, 1993, French translation Moscow, Mir, 1980)
4. Trenogin V.A. First Lyapounov method in analytical case for dynamical system. (Russian) Proc. of 12-th Math. School of Moscow State Social University, Moscow, 2005
5. Trenogin V.A. First Lyapounov method for the abstract parabilic equations. Abstracts of ISAAC-2005, pp. 15-16, University of Catania.
6. Trenogin V.A. Abstract dynamical systems and the first Lyapounov method. (Russian) Abstracts of International Conference, deducated to 100-birthday of Academician S.M. Nikol'sky, Moscow, Russian Acad. Sci. Publ, 2005, p. 225

## I.2 Variable Exponent Analysis and Applications

Organizer: St. Samkov

# SOME RESULTS ON VARIABLE EXPONENT ANALYSIS *

XIAN-LING FAN

*Department of Mathematics, Lanzhou University,*
*Lanzhou 730000, China*
*E-mail: fanxl@lzu.edu.cn*

We present some recent results on variable exponent Lebesgue spaces, which include: A variant of the definition of the norm in the variable exponent Lebesgue space; The Amemiya norm equals the Orlicz norm in the variable exponent Lebesgue space; An exact inequality involving the Luxemburg norm and the conjugate-Orlicz norm in the variable exponent Lebesgue space. We also present some results and open problems on the solutions of the $p(x)$−Laplacian equations.

**Key words:** Variable exponent Lebesgue space, norm, $p(x)$-Laplacian equation, eigenvalue

**Mathematics Subject Classification:** 46E30, 35J70.

This paper is divided into two sections. In section 1 we present some recent results on the variable exponent Lebesgue spaces. In section 2 we present some results and open problems on the solutions of the $p(x)$-Laplacian equations.

## 1. Variable Exponent Lebesgue Spaces

### 1.1. *A variant of the definition of the norm in space $L^{p(x)}_{w(x)}(\Omega)$*

Let $\Omega$ be an open subset of $R^N$, $S(\Omega) = \{u : u : \Omega \to R \text{ is measurable}\}$, $p \in S(\Omega)$, $w \in S(\Omega)$, $p : \Omega \to [1, \infty]$ and $w : \Omega \to (0, \infty)$. Denote

$$\Omega_1 = \{x \in \Omega : p(x) = 1\}, \ \Omega_+ = \{x \in \Omega : p(x) \in (1, \infty)\}, \ \Omega_\infty = \{x \in \Omega : p(x) = \infty\}.$$

For the variable exponent (weighted) Lebesgue space $L^{p(x)}_{w(x)}(\Omega)$ see [1,2,11,13,15,16] and references therein. The general definition of the norm in space $L^{p(x)}_{w(x)}(\Omega)$ is as follows (see e.g. [2,13,16]). Define

$$\rho(u) = \int_{\Omega \setminus \Omega_\infty} w(x)|u(x)|^{p(x)} dx + \operatorname*{ess\,sup}_{\Omega_\infty} |u(x)|, \ \forall u \in S(\Omega), \tag{1}$$

$$L^{p(x)}_{w(x)}(\Omega) = \{u \in S(\Omega) : \exists \lambda > 0 \text{ such that } \rho(u/\lambda) < \infty\},$$

---

*This research is supported by the National Science Foundation of China (10371052)

$$\|u\|_\rho = \inf\{\lambda > 0 : \ \rho(u/\lambda) < 1\}.$$

Note that in (1) $\rho(u)$ consists of two parts, where one is an integral, but the other is not an integral.

For the case when $\Omega_\infty \neq \emptyset$, we give a variant of the definition of the norm in space $L^{p(x)}_{w(x)}(\Omega)$ as follows (see [3,7]). We define an uniform function $\Phi$ on whole $\Omega$ by

$$\Phi(x, t) = w(x)t^{p(x)}, \quad \forall t \geq 0, \ \forall x \in \Omega, \tag{2}$$

with the convention

$$t^\infty = \begin{cases} 0, & 0 \leq t \leq 1 ; \\ \infty, & t > 1 . \end{cases} \tag{3}$$

We define the $\Phi$-integral $I_\Phi$, the space $L^{p(x)}_{w(x)}(\Omega)$ and the norm $\|\cdot\|_\Phi$ by

$$I_\Phi(u) = \int_\Omega \Phi(x, |u(x)|)\,dx, \quad \forall u \in S(\Omega),$$

$$L^{p(x)}_{w(x)}(\Omega) = \{u \in S(\Omega) : \ \exists \lambda > 0 \text{ such that } I_\Phi(u/\lambda) < \infty\},$$

$$\|u\|_\Phi = \inf\{\lambda > 0 : \ I_\Phi(u/\lambda) < 1\}.$$

The Norm $\|\cdot\|_\rho$ and the norm $\|\cdot\|_\Phi$ are equivalent. Especially, when $\Omega_\infty = \emptyset$ they are the same.

By our definition, the function $\Phi(x, t)$ defined by (2) and (3) is an Orlicz function with a parameter $x$, which is called a Musielak-Orlicz function. The space $\left(L^{p(x)}_{w(x)}(\Omega), \|\cdot\|_\Phi\right)$ with the norm $\|\cdot\|_\Phi$ defined via the integral of the Musielak-Orlicz function $\Phi(x, t)$ is a space of Orlicz type, which is called a Musielak-Orlicz space. By the general definition, the space $\left(L^{p(x)}_{w(x)}(\Omega), \|\cdot\|_\rho\right)$ with the norm $\|\cdot\|_\rho$ defined via the modular $\rho$ is a Musielak-Nakano modular space (see [14]), when $\Omega_\infty \neq \emptyset$ it is not a Musielak-Orlicz space.

In general, to consider the space $L^{p(x)}_{w(x)}(\Omega)$ as a Musielak-Orlicz space is more convenient than consider it as a Musielak-Nakano modular space. It can be seen from the following section 1.2 and 1.3 that, in the case that $\Omega_\infty \neq \emptyset$, our definition is more suitable than the general definition at times.

### 1.2. *Amemiya norm and Orlicz norm*

Let $\Omega$, $p$, $w$, $\Phi$ and $\left(L^{p(x)}_{w(x)}(\Omega), \|\cdot\|_\Phi\right)$ be as above. Denote $L^\Phi(\Omega) = L^{p(x)}_{w(x)}(\Omega)$ and $I_\Phi(u) = \int_\Omega \Phi(x, |u(x)|)\,dx$. The norm $\|\cdot\|_\Phi$ is called the Luxemburg norm in $L^\Phi(\Omega)$. The Amemiya norm $\|\cdot\|_\Phi^A$ in $L^\Phi(\Omega)$ is defined by

$$\|u\|_\Phi^A = \inf\left\{\frac{1}{k}\left(I_\Phi(ku) + 1\right) : k > 0\right\}.$$

Denote by $\Phi^*(x, \cdot)$ the complementary to $\Phi(x, \cdot)$, that is

$$\Phi^*(x, s) = \sup\{ts - \Phi(x, t) : t > 0\}, \quad \forall x \in \Omega, \quad \forall s \in [0, \infty).$$

The Orlicz norm $\|\cdot\|_\Phi^*$ in $L^\Phi(\Omega)$ is defined by

$$\|u\|_\Phi^* = \sup\left\{\left|\int_\Omega u(x)v(x)dx\right| : v \in L^{\Phi^*}(\Omega), I_{\Phi^*}(v) \le 1\right\}.$$

We have proved the following

**Theorem 1.1.** [3]  $\|u\|_\Phi^A = \|u\|_\Phi^*$ *for* $u \in L^\Phi(\Omega)$.

In fact, in [4] we have proved that the statement of Theorem 1.1 is also true for general Musielak-Orlicz spaces. The corresponding result for the Orlicz spaces has been proved by Hudzik and Maligranda [12].

Musielak [14] and Samko [16] have proved that the Luxemburg norm, the Amemiya norm and the Orlicz norm are equivalent in the Musielak-Nakano modular space and in the variable exponent Lebesgue space with the modular (1) respectively.

### 1.3. *An exact inequality involving Luxemburg norm and conjugate-Orlicz norm*

Let $w = 1$ and $\Phi(x,t) = t^{p(x)}$. In this case we denote $L_{w(x)}^{p(x)}(\Omega) = L^{p(x)}(\Omega)$ and $\|u\|_\Phi = \|u\|_{p(x)}$. Denote by $p^o(x)$ the conjugate number to $p(x)$ , that is

$$p^o(x) = \begin{cases} \dfrac{p(x)}{p(x) - 1}, & x \in \Omega_+ ; \\ \infty, & x \in \Omega_1 ; \\ 1, & x \in \Omega_\infty . \end{cases}$$

The conjugate-Orlicz norm $\|u\|_{p(x)}^o$ in $L^{p(x)}(\Omega)$ is defined by

$$\|u\|_{p(x)}^o = \sup\left\{\left|\int_\Omega u(x)v(x)dx\right| : v \in L^{p^o(x)}(\Omega), \int_\Omega |v(x)|^{p^o(x)}\,dx \le 1\right\}.$$

Denote $p_- = \operatorname{ess\,inf}_\Omega p(x)$ and $p_+ = \operatorname{ess\,sup}_\Omega p(x)$. We define $d_{(p_-,\,p_+)}$, a positive constant depending only on $p_-$ and $p_+$, as follows.

When $1 < p_- < p_+ < \infty$,

$$d_{(p_-,\,p_+)} = \left(\frac{(p_- - 1)^{p_- - 1}}{p_-^{p_-}}\right)^{\frac{p_+ - 1}{p_+ - p_-}} \left(\frac{p_+^{p_+}}{(p_+ - 1)^{p_+ - 1}}\right)^{\frac{p_- - 1}{p_+ - p_-}}$$

$$+ \left(\frac{p_-^{p_-}}{(p_- - 1)^{p_- - 1}} \frac{(p_+ - 1)^{p_+ - 1}}{p_+^{p_+}}\right)^{\frac{1}{p_+ - p_-}}. \tag{4}$$

In other cases $d_{(p_-,\,p_+)}$ is the corresponding limit of (4), that is:

when $1 = p_- < p_+ < \infty$, $d_{(1,\,p_+)} = \lim_{r\downarrow 1} d_{(r,\,p_+)} = 1 + \left(\dfrac{(p_+ - 1)^{p_+ - 1}}{p_+^{p_+}}\right)^{\frac{1}{p_+ - 1}}$,

when $1 < p_- < p_+ = \infty$, $d_{(p_-,\,\infty)} = \lim_{q\uparrow\infty} d_{(p_-,\,q)} = \dfrac{(p_- - 1)^{p_- - 1}}{p_-^{p_-}} + 1$,

when $1 = p_-$ and $p_+ = \infty$, $d_{(1,\,\infty)} = \lim_{r\downarrow 1,\,q\uparrow\infty} d_{(r,\,q)} = 2$,

when $p_- = p_+$, $d_{(p_-, p_+)} = \lim_{q \to p_-} d_{(p_-, q)} = \lim_{r \to p_+} d_{(r, p_+)} = 1$.
We have proved the following

**Theorem 1.2.** [7]  *The inequality*

$$\|u\|_{p(x)} \le \|u\|_{p(x)}^{o} \le d_{(p_-, p_+)} \|u\|_{p(x)}, \quad \forall u \in L^{p(x)}(\Omega)$$

*holds and is exact, that is*

$$\sup_{u \in L^{p(x)}(\Omega) \setminus \{0\}} \frac{\|u\|_{p(x)}}{\|u\|_{p(x)}^{o}} = 1, \quad \sup_{u \in L^{p(x)}(\Omega) \setminus \{0\}} \frac{\|u\|_{p(x)}^{o}}{\|u\|_{p(x)}} = d_{(p_-, p_+)}.$$

Kováčik and Rákosník [13] have proved that, under the norm $\|\cdot\|_{\rho}$, there holds the inequality

$$\|u\|_{\rho}^{o} \le r_{p(x)} \|u\|_{\rho}, \quad \forall u \in L^{p(x)}(\Omega),$$

where

$$r_{p(x)} = \|\chi_{\Omega_1}\|_{\infty} + \|\chi_{\Omega_+}\|_{\infty} + \|\chi_{\Omega_\infty}\|_{\infty} + \frac{1}{p_-(\Omega_+)} - \frac{1}{p_+(\Omega_+)},$$

and $\chi$ is the characteristic function.

Let us compare the constant $d_{(p_-, p_+)}$ with the constant $r_{p(x)}$. The constant $r_{p(x)} \in [1, 4]$ is not best and is not continuous in $(p_-, p_+)$ but our constant $d_{(p_-, p_+)} \in [1, 2]$ is best and is continuous in $(p_-, p_+)$. In the case that $\Omega_+ \ne \emptyset$, we have the following conclusions:

If $\Omega_1 \cup \Omega_\infty \ne \emptyset$, then $r_{p(x)} > 2 \ge d_{(p_-, p_+)}$.

If $\Omega_1 = \Omega_\infty = \emptyset$, then $r_{p(x)} = 1 + \frac{1}{p_-} - \frac{1}{p_+} \ge d_{(p_-, p_+)}$, and in this case, $r_{p(x)} = d_{(p_-, p_+)}$ if and only if $p_- = p_+$ or $p_- = 1$ and $p_+ = \infty$.

## 2. $p(x)$−Laplacian Equations

### 2.1. *Results on existence and multiplicity of solutions of $p(x)$−Laplacian equations*

The $p(x)$−Laplacian equation is a generalization of the $p$−Laplacian equation. We have extended some basic results on existence and multiplicity of solutions of the $p$−Laplacian equations to the $p(x)$−Laplacian equations (see [5,6,8,10]). The $p(x)$−Laplacian equations have been studied also by other authors, e.g. Acerbi, Coscia, Marcellini, Mingione and Zhikov et al.. The relevant works were listed e.g. as references in [1].

Consider the $p(x)$−Laplacian Dirichlet problem

$$\begin{cases} -\operatorname{div}\left( |\nabla u|^{p(x)-2} \nabla u \right) = f(x, u) & \text{in } \Omega \\ u = 0 & \text{on } \partial\Omega, \end{cases} \tag{5}$$

where $\Omega$ is an open domain in $R^N$, $p \in C(\overline{\Omega})$ and $1 < p_- < p_+ < \infty$. We only consider the subcritical case, that is

$$|f(x,t)| \le c_1 + c_2 |t|^{q(x)-1}, \quad \forall (x,t) \in \Omega \times R,$$

$$q \in C(\overline{\Omega}), \quad q(x) < p^*(x), \quad \forall x \in \overline{\Omega},$$

where $p^*(x)$ is the Sobolev critical exponent of $p(x)$ :

$$p^*(x) := \begin{cases} \frac{Np(x)}{N-p(x)} & \text{if } p(x) < N \\ \infty & \text{if } p(x) \ge N. \end{cases}$$

A typical form of (5) is the following problem

$$\begin{cases} -\operatorname{div}\left(|\nabla u|^{p(x)-2}\nabla u\right) = \lambda |u|^{r(x)-2} u + \mu |u|^{q(x)-2} u & \text{in } \Omega \\ u = 0 & \text{on } \partial\Omega, \end{cases} \tag{6}$$

where $\Omega$ is bounded,

$$r_+ < p_-, \ p_+ < q_-, \ q(x) < p^*(x). \tag{7}$$

In [8] we have extended some basic results on $p-$Laplacian equations to the problem (6). In [5] we have considered the case with singular coefficients, a typical form of which is the following Hardy-Sobolev subcritical problem:

$$\begin{cases} -\operatorname{div}\left(|\nabla u|^{p(x)-2}\nabla u\right) = \lambda \dfrac{1}{|x|^{s_1(x)}} |u|^{r(x)-2} u + \mu \dfrac{1}{|x|^{s_2(x)}} |u|^{q(x)-2} u & \text{in } \Omega \\ u = 0 & \text{on } \partial\Omega, \end{cases} \tag{8}$$

where $\Omega$ is bounded,

$$r(x) < \frac{N - s_1(x)}{N} p^*(x), \quad q(x) < \frac{N - s_2(x)}{N} p^*(x). \tag{9}$$

Note that (6) is a special case of (8) with $s_1(x) \equiv s_2(x) \equiv 0$. In [6] we have considered the case that $\Omega = R^N$. Here we only state a result of [5] as follows.

**Theorem 2.1.** *[5] Suppose that the conditions (7) and (9) are satisfied. Then*

(1) *for every $\mu > 0$ and $\lambda \in (-\infty, +\infty)$, the problem (8) has a sequence of solutions $\{\pm u_k\}$ such that the energy $\varphi(\pm u_k) \to +\infty$ as $k \to \infty$.*
(2) *for every $\lambda > 0$ and $\mu \in (-\infty, +\infty)$, the problem (8) has a sequence of solutions $\{\pm v_k\}$ such that the energy $\varphi(\pm v_k) < 0$ and $\varphi(\pm v_k) \to 0$ as $k \to \infty$.*

Note that in Theorem 2.1 it is not necessary to suppose that $p$ satisfies the log-Hölder condition. In fact, in this theorem we suppose that $p \in C(\overline{\Omega})$, $q \in C(\overline{\Omega})$, $q(x) < p^*(x)$ for $x \in \overline{\Omega}$ and $\Omega$ is bounded, and we know that under these assumptions there is a compact embedding $W_0^{1,\,p(x)}(\Omega) \hookrightarrow L^{q(x)}(\Omega)$ (see e.g. [13]).

## 2.2. *The eigenvalues of $p(x)$-Laplacian*

Consider the eigenvalue problem

$$
\begin{cases}
-\operatorname{div}\left(|\nabla u|^{p(x)-2}\,\nabla u\right) = \lambda\,|u|^{p(x)-2}\,u & \text{in } \Omega \\
u = 0 & \text{on } \partial\Omega,
\end{cases}
\tag{10}
$$

where $\Omega$ is bounded and $1 < p_- < p_+ < \infty$. Denote

$$
\lambda_* = \inf\{\lambda : \lambda \text{ is a eigenvalue of } (10)\}.
\tag{11}
$$

It is well known that, in the constant exponent case, $\lambda_*$ is the first eigenvalue of the $p$-Laplacian and $\lambda_* > 0$. However, in the variable exponent case, by our result, in general $\lambda_* = 0$. Our result is stated as follows (see [9]).

**Theorem 2.2.** [9] *Consider the eigenvalue problem (10) and let $\lambda_*$ be as in (11). We have the following conclusions.*

*(1) When $N = 1$, $\lambda_* > 0$ if and only if the function $p(x)$ is monotone.*
*(2) When $N > 1$, if there is a vector $h \in R^N \backslash \{0\}$ such that for any $x \in \Omega$ the function $g(t) = p(x + th)$ is monotone, then $\lambda_* > 0$.*
*(3) If there is an open subset $U \subset \Omega$ and a point $x_0 \in U$ such that*

$$
p(x_0) < (\text{or } >)\, p(x), \quad \forall x \in \partial U,
$$

*then $\lambda_* = 0$.*

## 2.3. *Some open problems*

(1) Does there exist a nontrivial solution of (6) in the case that $r(x) < p(x)$ for each $x \in \Omega$ but $r_+ \geq p_-$, and $q(x) > p(x)$ for each $x \in \Omega$ but $q_- \leq p_+$ ?
(2) Consider the problem

$$
\begin{cases}
-\operatorname{div}\left(|\nabla u|^{p(x)-2}\,\nabla u\right) = \lambda\,|u|^{r(x)-2}\,u & \text{in } \Omega \\
u = 0 & \text{on } \partial\Omega,
\end{cases}
\tag{12}
$$

where $\Omega$ is bounded, $\lambda > 0$ and $r_+ < p_-$. It is well known that the problem (12) has at least one positive solution (see [8]). Is the positive solution of (12) unique?
(3) Let $\Omega$ be bounded. For any $t > 0$, define

$$
M_t = \left\{ u \in W_0^{1,\,p(x)}(\Omega) : \int_\Omega \frac{|u|^{p(x)}}{p(x)}\,dx = t \right\}.
$$

Consider the constrained minimization problem:

$$
\min_{u \in M_t} \int_\Omega \frac{|\nabla u|^{p(x)}}{p(x)}\,dx.
\tag{13}
$$

It is clear that, for each $t > 0$, the problem (13) has at least one positive minimizer. Is such positive minimizer unique?

We know that if $u$ is a positive minimizer of (13), then $(u, \lambda)$ is a solution of (12) with $\lambda = \int_\Omega |\nabla u|^{p(x)} dx / \int_\Omega |u|^{p(x)} dx$. We wonder whether the converse proposition of this statement is true, that is, if $(u, \lambda)$ is a solution of (12) and $u$ is positive, must $u$ then be a positive minimizer of (13) with $t = \int_\Omega \frac{|u|^{p(x)}}{p(x)} dx$?

## References

1. L. Diening, P. Hästö and A. Nekvinda, Open problems in variable exponent Lebesgue and Sobolev spaces, *FSDONA04 Proceedings* (Drábek and Rákosník (eds.)), Milovy, Czech Republic, 2004, 38-58.
2. D. E. Edmunds, J. Lang and A. Nekvinda, On $L^{p(x)}$ norms, *R. Soc. Lond. Proc. Ser. A Math. Phys. Eng. Sci.*, **455**(1999), no. 1981, 219-225.
3. X-L. Fan, Amemiya norm equals Orlicz norm in variable exponent Lebesgue spaces, *Chinese Sciencepaper Online*, (http://www.paper.edu.cn/process/download.jsp?file = 200405-86)
4. X-L. Fan, Amemiya norm equals Orlicz norm in Musielak-Orlicz spaces, to appear in *Acta Math. Sinica*.
5. X-L. Fan, Solutions for $p(x)$-Laplacian Dirichlet problems with singular coefficients, *J. Math. Anal. Appl.*, **312** (2005), 464-477.
6. X-L. Fan and X-Y. Han, Existence and multiplicity of solutions for $p(x)$-Laplacian equations in $R^N$, *Nonlinear Anal.*, **59**(2004), 173-188.
7. X-L. Fan and W-M. Liu, An exact inequality involving Luxemburg norm and conjugate-Orlicz norm in $L^{p(x)}(\Omega)$, *Chinese J. Contemp. Math.*, **27:2** (2006), 147-158.
8. X-L. Fan and Q-H. Zhang, Existence of solutions for $p(x)$-Laplacian Dirichlet problems, *Nonlinear Anal.*, **52**(2003), 1843-1852.
9. X-L. Fan, Q-H. Zhang and D. Zhao, Eigenvalues of $p(x)$-Laplacian Dirichlet problem, *J. Math. Anal. Appl.*, **302**(2005), 306-317.
10. X-L. Fan, Y-Z. Zhao and Q-H. Zhang, A strong maximum principle for $p(x)$-Laplace equations, *Chinese J. Comtemp. Math.*, **24:3**(2003),277-282.
11. P. Harjulehto and P. Hästö, An overview of variable exponent Lebesgue and Sobolev spaces, *Future Trends in Geometric Function Theory* (D. Herron (ed.), RNC Workshop), Jyväskylä, 2003, 85-93.
12. H. Hudzik and L. Maligranda, Amemiya norm equals Orlicz norm in general, *Indag. Math. N.S.*, **11:4**(2000), 573-585.
13. O. Kováčik and J. Rákosník, On spaces $L^{p(x)}(\Omega)$ and $W^{k,p(x)}(\Omega)$, *Czechoslovak Math. J.*, **41**(1991), 592-618.
14. J. Musielak, Orlicz Spaces and Modular Spaces, Lecture Notes in Math., Vol. 1034, Springer-Verlag, Berlin, 1983.
15. M. Růžička, Electrorheological fluids: modeling and mathematical theory, Lecture Notes in Math., Vol. 1748, Springer-Verlag, Berlin, 2000.
16. S. G. Samko, Convolution type operators in $L^{p(x)}$, *Integr. Transform. and Special Funct.*, **7**(1998), 123-144.

# FURTHER RESULTS ON VARIABLE EXPONENT TRACE SPACES

LARS DIENING*

*Section of Applied Mathematics, Eckerstr. 1,*
*Freiburg University, 79104 Freiburg/Breisgau, Germany*
*E-mail: diening@mathematik.uni-freiburg.de*

PETER HÄSTÖ†.

*Department of Mathematics and Statistics, P.O. Box 68,*
*FI-00014 University of Helsinki, Finland*
*E-mail: peter.hasto@helsinki.fi*

Recently, the trace space of Sobolev functions with variable exponents has been characterized by the authors [L. Diening and P. Hästö: Variable exponent trace spaces, Preprint (2005)]. In this note we relax the assumptions on the exponent need for some basic results on trace spaces, like a characterization of zero boundary value spaces in terms of traces.

**Key words:** Variable exponent, Lebesgue space, Sobolev space, trace, density of smooth functions
**Mathematics Subject Classification:** 46E35, 46E30

## 1. Introduction

From the point of boundary value problems it is very important to study the trace spaces of the natural energy space. Indeed, a partial differential equation is in many cases solvable if and only if the boundary values are in the corresponding trace space. In the case of electrorheological fluids[17] the energy space is a Sobolev space with variable exponent, namely $W^{1,p(\cdot)}$. These spaces are defined as follows: For an open set $\Omega \subset \mathbb{R}^n$ let $p \colon \Omega \to [1,\infty)$ be a measurable bounded function, called a variable exponent on $\Omega$ with $p^+ := \operatorname{ess\,sup} p(x) < \infty$. Further, let $p^- := \operatorname{ess\,inf} p(x)$. The *variable exponent Lebesgue space* $L^{p(\cdot)}(\Omega)$ consists of all measurable functions $f \colon \Omega \to \mathbb{R}$ for which the modular

$$\varrho_{L^{p(\cdot)}(\Omega)}(f) = \int_\Omega |f(x)|^{p(x)}\, dx$$

is finite. Then $\|f\|_{L^{p(\cdot)}(\Omega)} = \inf\big\{\lambda > 0 \colon \varrho_{L^{p(\cdot)}(\Omega)}(f/\lambda) \leq 1\big\}$ defines a norm on $L^{p(\cdot)}(\Omega)$. The space $W^{1,p(\cdot)}(\Omega)$ is the subspace of $L^{p(\cdot)}(\Omega)$ such that $|\nabla f| \in L^{p(\cdot)}(\Omega)$.

---

*Supported by Landesstiftung Baden-Württemberg
†Supported by the Academy of Finland

The norm $\|f\|_{W^{1,p(\cdot)}(\Omega)} = \|f\|_{L^{p(\cdot)}(\Omega)} + \|\nabla f\|_{L^{p(\cdot)}(\Omega)}$ makes $W^{1,p(\cdot)}(\Omega)$ a Banach space. For basic properties of $L^{p(\cdot)}$ and $W^{1,p(\cdot)}$ we refer to Kováčik and Rákosník[15] or Fan and Zhao[9].

We are interested in domains $\Omega$ with Lipschitz boundary but for the sake of simplicity we assume that $\Omega$ is just the halfspace $\mathbb{H} = \mathbb{R}^n \times (0, \infty)$. Corresponding results for Lipschitz domains can then be achieved via flattening of the boundary by local Bi-Lipschitz mappings. We write $\mathbb{R}^n$ instead of $\mathbb{R}^n \times \{0\}$ for the boundary of $\mathbb{H}$. The trace space of $W^{1,p(\cdot)}(\mathbb{H})$ is naturally defined to be the quotient space of the traces of functions from $W^{1,p(\cdot)}(\mathbb{H})$, i.e.

$$\|f\|_{\mathrm{Tr}\, W^{1,p(\cdot)}(\mathbb{H})} = \inf \left\{ \|F\|_{W^{1,p(\cdot)}(\mathbb{H})} : F \in W^{1,p(\cdot)}(\mathbb{H}) \text{ and } \mathrm{tr} F = f \right\}.$$

Note that the trace $\mathrm{Tr}\, F$ is well defined, since every $F \in W^{1,p(\cdot)}(\mathbb{H})$ is in $W^{1,1}_{\mathrm{loc}}(\overline{\mathbb{H}})$. Trace spaces of Sobolev spaces with variable exponents first appeared in the study of the Laplace equation $-\Delta u = f$ on the half space with $f \in L^{p(\cdot)}(\mathbb{H})$ and prescribed boundary values[7,8].

Although the definition above is the most natural one, it depends on the exponent $p$ in the interior of the domain. Nevertheless, it was found by the authors[6] that if $p$ is globally log-Hölder continuous, then the definition of the trace space depends only on the values of $p$ on the boundary (see Proposition 1.1 below) — we say that the exponent $p$ is *globally log-Hölder continuous* if there exist constants $c > 0$ and $p_\infty \in (1, \infty)$ such that for all points $|x - y| < \frac{1}{2}$ and all points $z$

$$|p(x) - p(y)| \leq \frac{c}{\log(1/|x-y|)} \quad \text{and} \quad |p(z) - p_\infty| \leq \frac{c}{\log(e + |z|)}$$

hold. Let us denote by $\mathcal{P}(\Omega)$ the class of globally log-Hölder continuous variable exponents $p$ on $\Omega \subset \mathbb{R}^m$ with $1 < p^- \leq p^+ < \infty$.

The log-Hölder condition appears quite naturally in the context of variable exponent spaces: For example, we know that the Hardy-Littlewood maximal operator $M$ is bounded on $L^{p(\cdot)}(\Omega)$ if $p \in \mathcal{P}(\Omega)$[3,4]. Global log-Hölder continuity is the best possible modulus of continuity to imply the boundedness of the maximal operator[3,16]. Note that the translation operators are not continuous on $L^{p(\cdot)}$, but if the maximal operator is bounded, it is at least possible to use the technique of mollifiers.

Let us summarize the results[6] on $\mathrm{Tr}\, W^{1,p(\cdot)}(\mathbb{R}^n)$ that we will need:

**Theorem 1.1.** *Let $p_1, p_2 \in \mathcal{P}(\mathbb{H})$ with $p_1|_{\mathbb{R}^n} = p_2|_{\mathbb{R}^n}$. Then with equivalence of norms we have $\mathrm{Tr}\, W^{1,p_1(\cdot)}(\mathbb{H}) = \mathrm{Tr}\, W^{1,p_2(\cdot)}(\mathbb{H})$.*

This proposition is proved by the following useful extension theorem:

**Proposition 1.1.** *Let $p \in \mathcal{P}(\mathbb{R}^{n+1})$. Then there exists a bounded, linear extension operator $\mathcal{E} : W^{1,p(\cdot)}(\mathbb{H}) \to W^{1,p(\cdot)}(\mathbb{R}^{n+1})$.*

**Proposition 1.2.** *Let $X \subset \mathbb{R}^n$. If $p \in \mathcal{P}(X)$, then there exists an extension $\tilde{p} \in \mathcal{P}(\mathbb{R}^n)$.*

**Remark 1.1.** Due to Propositions 1.1 and 1.2 it is possible to define in some cases a trace space just by the knowledge of the values of $p$ on the boundary $\mathbb{R}^n$. Indeed, if $p \in \mathcal{P}(\mathbb{R}^n)$ then we can extend $p$ by Proposition 1.2 to some $q \in \mathcal{P}(\mathbb{H})$. It is now possible to consider the trace space $T := \operatorname{Tr} W^{1,p(\cdot)}(\mathbb{H})$. Proposition 1.1 ensures that the definition of $T$ does not depend on the extension $q$ (up to isomorphism). Thus, it is possible to define the trace space $(\operatorname{Tr} W^{1,p(\cdot)})(\mathbb{R}^n) := \operatorname{Tr} W^{1,q(\cdot)}(\mathbb{H})$ for $p \in \mathcal{P}(\mathbb{R}^n)$, where $q$ is an arbitrary extension of $p$ with $q \in \mathcal{P}(\mathbb{H})$.

Although the definition of the trace space above is very natural it is not so useful for deciding if a function is a $W^{1,p(\cdot)}$-trace. For this purpose it is better to have an intrinsic norm, i.e. a norm only in terms of the values on the boundary. The following theorem[6] provides such an characterization for all globally log-Hölder continuous exponents:

**Theorem 1.2.** *Let $p \in \mathcal{P}(\mathbb{R}^n)$ and let $q \in \mathcal{P}(\overline{\mathbb{H}})$ be an arbitrary extension of $p$, i.e. $p(x) = q(x, 0)$ for all $x \in \mathbb{R}^n$. Then the function $f$ belongs to the trace space $(\operatorname{Tr} W^{1,p(\cdot)})(\mathbb{R}^n) \cong \operatorname{Tr} W^{1,q(\cdot)}(\mathbb{H})$ if and only if*

$$\int_{\mathbb{R}^n} |f(x)|^{p(x)} dx + \int_0^1 \int_{\mathbb{R}^n} \left( \tfrac{1}{r} M^\sharp_{B^n(x,r)} f \right)^{p(x)} dx \, dr < \infty,$$

*where $M^\sharp_{B^n(x,r)}$ denotes the sharp operator*

$$M^\sharp_{B^n(x,r)} f = \int_{B^n(x,r)} \left| f(y) - \int_{B^n(x,r)} f(z) \, dz \right| dy$$

*and $B^n(x,t)$ denotes the n-dimensional ball with center $x$ and radius $t$.*

In this article we extend the results by the authors[6] and consider spaces with more general exponents. The standing assumption of Diening and Hästö[6] was that the exponent is globally log-Hölder continuous. In this article we work with the considerably weaker assumption that the exponent is such that smooth functions are dense in our Sobolev space. Note that smooth functions are certainly dense if the maximal operator is bounded.

The main result is a characterization of the variable exponent Sobolev functions with zero boundary values – we show that these are just the functions which have trace zero.

## 2. Trace spaces when smooth functions are dense

In order to work with classical derivatives in our proofs, we need to prove the density of smooth functions in our function space. In the variable exponent case this question is far from trivial, as convolutions cannot be used in general, see the articles[10,13,14,18-20]. So in this section we will simply assume that smooth functions are dense in the ambient space.

104

Notice the difference between the spaces $C_0^\infty(\overline{\mathbb{H}})$ and $C_0^\infty(\mathbb{H})$: in the former space functions simply have bounded support, in the latter the support of the function is bounded and disjoint from the boundary $\mathbb{R}^n$ of $\mathbb{H}$.

**Theorem 2.1.** *Suppose that $C_0^\infty(\overline{\mathbb{H}})$ is dense in $W^{1,p(\cdot)}(\mathbb{H})$. Then $C_0^\infty(\mathbb{R}^n)$ is dense in* $\operatorname{Tr} W^{1,p(\cdot)}(\mathbb{H})$.

**Proof.** Let $f \in \operatorname{Tr} W^{1,p(\cdot)}(\mathbb{H})$, and let $F \in W^{1,p(\cdot)}(\mathbb{H})$ be such that $\operatorname{Tr} F = f$. Then if $\varphi_i \in C_0^\infty(\overline{\mathbb{H}})$ tend to $F$ in $W^{1,p(\cdot)}(\mathbb{H})$, we see that $\varphi_i|_{\mathbb{R}^n} \to f$ in $\operatorname{Tr} W^{1,p(\cdot)}(\mathbb{H})$. $\square$

Recall the definition of the Sobolev space of functions with zero boundary value: the space $W_0^{1,p(\cdot)}(\mathbb{H})$ is the completion of $C_0^\infty(\mathbb{H})$ in $W^{1,p(\cdot)}(\mathbb{H})$. (Other definitions are better, when smooth functions are not dense[11,12].) We next characterize $W_0^{1,p(\cdot)}(\mathbb{H})$ in terms of traces. For this we need to recall the definition of the Sobolev $p(\cdot)$-capacity: for $E \subset \mathbb{R}^n$ we define

$$\operatorname{cap}_{p(\cdot)}(E) = \inf_u \int_{\mathbb{R}^n} |u(x)|^{p(x)} + |\nabla u(x)|^{p(x)} \, dx,$$

where the infimum is taken over functions $u \in W^{1,p(\cdot)}(\mathbb{R}^n)$, $0 \le u \le 1$, which equal 1 in an open set containing $E$. Basic properties of this capacity, like monotony and sub-additivity, were derived by Harjulehto, Hästö, Koskenoja and Varonen[12].

The next result was previously proven by the authors under the stronger assumption that $p$ is log-Hölder continuous, see Diening and Hästö[6], Theorem 3.5. In that case the proof can be made much simpler using convolution.

**Theorem 2.2.** *Suppose that $p^+ < \infty$ and $C_0^\infty(\overline{\mathbb{H}})$ is dense in $W^{1,p(\cdot)}(\mathbb{H})$. Then $F \in W^{1,p(\cdot)}(\mathbb{H})$ belongs to $W_0^{1,p(\cdot)}(\mathbb{H})$ if and only if $\operatorname{Tr} F = 0$.*

**Proof.** Let $F \in W^{1,p(\cdot)}(\mathbb{H})$ with $\operatorname{Tr} F = 0$. Multiplying $F$ by a Lipschitz cut-off we see that it suffices to prove the claim for $F$ with support in $B := B^{n+1}(0, r)$. Below, we will prove the claim for non-negative functions $F$. But the general claim follows from this, since we can write $F = F_+ - F_-$, where $F_+, F_- \in W^{1,p(\cdot)}(\mathbb{H})$ are non-negative functions with trace zero. We know that the first order Sobolev space is a lattice, i.e. if $F, G \in W^{1,p(\cdot)}(\mathbb{H})$, then $\max\{F, G\}, \min\{F, G\} \in W^{1,p(\cdot)}(\mathbb{H})$. Furthermore, $\min\{F, a\} \to F$ in $W^{1,p(\cdot)}(\mathbb{H})$ for a constant $a$ tending to $\infty$. Thus we can also assume that $F$ is bounded.

So let $F$ be non-negative, bounded and with support in $B$, and fix $\varepsilon > 0$. By Harjulehto, Hästö, Koskenoja and Varonen[12] there exists a $p(\cdot)$-quasicontinuous function $F^* \in W^{1,p(\cdot)}(\mathbb{R}^{n+1})$ which equals $F$ almost everywhere in $\mathbb{H}$ and is identically zero in $\mathbb{R}^{n+1} \setminus \mathbb{H}$. Recall that quasicontinuity (by definition) means that we can choose an open $E$ such that $F^*|_{\mathbb{R}^{n+1}\setminus E}$ is continuous in $\mathbb{R}^{n+1} \setminus E$ and $\operatorname{cap}_{p(\cdot)}(E) < \varepsilon^{p^++1}$. Thus we can find a function $\varphi \in W^{1,p(\cdot)}(\mathbb{R}^{n+1})$, $0 \le \varphi \le 1$, which equals 1 on $E$ and has $p(\cdot)$-modular at most $\varepsilon^{p^++1}$. Let $\psi \colon \mathbb{R}^{n+1} \to [0, 1]$ be a

$2/\varepsilon$-Lipschitz function, which equals 1 for $x_{n+1} < \varepsilon$ and 0 for $x_{n+1} \geq 2\varepsilon$. We denote the support of $\psi$ by $V$.

Since $\overline{B} \backslash E$ is compact, we see (by continuity) that we can choose a neighborhood $U \subset \mathbb{R}^{n+1}$ of $\mathbb{R}^n \times \{0\}$ such that $F^* < \varepsilon$ in $U \setminus E$. Then we define

$$\tilde{F} = (1 - \varphi\psi) \max\{F^* - \varepsilon, 0\}.$$

Since $F^*$ is less than $\varepsilon$ in $(U \setminus E) \cup (-\mathbb{H})$, and since $1 - \varphi\psi = 0$ in

$$E \cap \{(x, t) \in \mathbb{R}^n \times \mathbb{R} : t < \varepsilon\},$$

we find that $\tilde{F} = 0$ in $U \cup (-\mathbb{H})$, which means that $\tilde{F}$ has compact support in $\mathbb{H}$. In the next estimate we think of $\tilde{F} - F$ as

$$\left( \max\{F^* - \varepsilon, 0\} - F \right) - \varphi\psi \max\{F^* - \varepsilon, 0\}.$$

Thus we find that

$$\varrho_{1,p(\cdot)}(\tilde{F} - F) = \int_B \left( |\tilde{F}(z) - F(z)|^{p(z)} + |\nabla\tilde{F}(z) - \nabla F(z)|^{p(z)} \right) dz$$

$$\leq 2^{p^+} |B| \varepsilon + 2^{p^+} \int_{\{F^* < \varepsilon\}} |\nabla F(z)|^{p(z)} dz$$

$$+ 2^{p^+} \int_B \left( |F(z)\,\varphi(z)\psi(z)|^{p(z)} + |\nabla(F(z)\,\varphi(z)\psi(z))|^{p(z)} \right) dz$$

The first two terms on the right-hand-side clearly tend to zero with $\varepsilon$ (since $\nabla F = 0$ almost everywhere in the set $\{F = 0\}$). To estimate the rightmost integral, we note that $\psi \equiv 0$ in $B \setminus V$, and further use that $|\varphi| \leq 1$ and $|\psi| \leq 1$. Thus we have

$$\varrho_{1,p(\cdot)}(F\,\varphi\,\psi) \leq 3^{p^+} \int_{B \cap V} \left( |F(z)|^{p(z)} + |\nabla F(z)|^{p(z)} \right) dz$$

$$+ 3^{p^+} \|F\|_\infty^{p^+} \int_B \left[ \left( \|\psi\|_\infty |\nabla\varphi(z)| \right)^{p(z)} + \left( \|\nabla\psi\|_\infty |\varphi(z)| \right)^{p(z)} \right] dz$$

$$\leq 3^{p^+} \varrho_{W^{1,p(\cdot)}(B \cap V \cap \mathbb{H})}(F) + \left( \frac{6\|F\|_\infty}{\varepsilon} \right)^{p^+} \varrho_{1,p(\cdot)}(\varphi).$$

Since the measure of $B \cap V \cap \mathbb{H}$ is at most a constant times $\varepsilon$, and since $\varrho_{1,p(\cdot)}(\varphi) \leq \varepsilon^{p^+ + 1}$, we see that this upper bound goes to zero with $\varepsilon$.

Since $\varepsilon$ was arbitrary, we have shown that $F$ can be approximated by Sobolev functions with compact support in $\mathbb{H}$. So it remains to show that Sobolev functions with compact support in $\mathbb{H}$ can be approximated by functions in $C_0^\infty(\mathbb{H})$. But this is easy to do using a cut-off function $\psi$ as before.

For the converse, if $F \in W_0^{1,p(\cdot)}(\mathbb{H})$, then, by definition, $F = \lim \varphi_i$ in $W^{1,p(\cdot)}(\mathbb{H})$, where $\varphi_i \in C_0^\infty(\mathbb{H})$. Since $\operatorname{Tr} \varphi_i = \varphi_i|_{\mathbb{R}^n} \equiv 0$, the claim follows by continuity of $\operatorname{Tr} : W^{1,p(\cdot)}(\mathbb{H}) \to \operatorname{Tr} W^{1,p(\cdot)}(\mathbb{H})$. (Notice that the proof of the converse does not require the density of smooth functions.) $\qquad\square$

## References

1. R. Adams: *Sobolev spaces*, Pure and Applied Mathematics, Vol. 65, Academic Press, New York–London, 1975.
2. C. Capone, D. Cruz-Uribe, and A. Fiorenza: The fractional maximal operator on variable $L^p$ spaces, preprint (2004).
3. D. Cruz-Uribe, A. Fiorenza and C. J. Neugebauer: The maximal function on variable $L^p$ spaces, *Ann. Acad. Sci. Fenn. Math.* **28** (2003), 223–238; **29** (2004), 247–249.
4. L. Diening: Maximal function on generalized Lebesgue spaces $L^{p(\cdot)}$, *Math. Inequal. Appl.* **7** (2004), no. 2, 245–254.
5. L. Diening: Riesz potential and Sobolev embeddings of generalized Lebesgue and Sobolev spaces $L^{p(\cdot)}$ and $W^{k,p(\cdot)}$, *Math. Nachr.* **263** (2004), no. 1, 31–43.
6. L. Diening and P. Hästö: Variable exponent trace spaces, Rev. Mat. Iberoamericana, to appear.
7. L. Diening and M. Růžička: Integral operators on the halfspace in generalized Lebesgue spaces $L^{p(\cdot)}$, part I, *J. Math. Anal. Appl.* **298** (2004), no. 2, 559–571.
8. L. Diening and M. Růžička: Integral operators on the halfspace in generalized Lebesgue spaces $L^{p(\cdot)}$, part II, *J. Math. Anal. Appl.* **298** (2004), no. 2, 572–588.
9. X.-L. Fan and D. Zhao: On the spaces spaces $L^{p(x)}(\Omega)$ and $W^{m,p(x)}(\Omega)$, *J. Math. Anal. Appl.* **263** (2001), 424–446.
10. X.-L. Fan, S. Wang and D. Zhao: Density of $C^\infty(\Omega)$ in $W^{1,p(x)}(\Omega)$ with discontinuous exponent $p(x)$. Math. Nachr. 279 (2006), no. 1-2, 142–149.
11. P. Harjulehto, Variable exponent Sobolev spaces with zero boundary values, preprint (2004). Available at
   http://mathstat.helsinki.fi/analysis/varsobgroup/.
12. P. Harjulehto, P. Hästö, M. Koskenoja and S. Varonen: Sobolev capacity on the space $W^{1,p(\cdot)}(\mathbb{R}^n)$, *J. Funct. Spaces Appl.* **1** (2003), no. 1, 17–33.
13. P. Hästö: Counter-examples of regularity in variable exponent Sobolev spaces, *The p-Harmonic Equation and Recent Advances in Analysis* (Manhattan, KS, 2004), Contemp. Math. 367, Amer. Math. Soc., Providence, RI, 2005.
14. P. Hästö: On the density of smooth functions in variable exponent Sobolev space, preprint (2004). Available at
   http://mathstat.helsinki.fi/analysis/varsobgroup/.
15. O. Kováčik and J. Rákosník: On spaces $L^{p(x)}$ and $W^{1,p(x)}$, *Czechoslovak Math. J.* **41(116)** (1991), 592–618.
16. L. Pick and M. Růžička: An example of a space $L^{p(x)}$ on which the Hardy-Littlewood maximal operator is not bounded, *Expo. Math.* **19** (2001), 369–371.
17. M. Růžička: *Electrorheological Fluids: Modeling and Mathematical Theory*, Springer-Verlag, Berlin, 2000.
18. S. Samko: Denseness of $C_0^\infty(\mathbb{R}^n)$ in the generalized Sobolev spaces $W^{m,p(x)}(\mathbb{R}^n)$, pp. 333–342 in *Direct and Inverse Problems of Mathematical Physics (Newark, DE, 1997)*, Int. Soc. Anal. Appl. Comput. 5, Kluwer Acad. Publ., Dordrecht, 2000.
19. V. V. Zhikov: Averaging of functionals of the calculus of variations and elasticity theory, *Math. USSR-Izv.* **29** (1987), no. 1, 33–66. [Translation of *Izv. Akad. Nauk SSSR Ser. Mat.* **50** (1986), no. 4, 675–710, 877.]
20. V.V. Zhikov: On the density of smooth functions in Sobolev-Orlicz spaces. (Russian) Zap. Nauchn. Sem. S.-Peterburg. Otdel. Mat. Inst. Steklov. (POMI) 310 (2004), Kraev. Zadachi Mat. Fiz. i Smezh. Vopr. Teor. Funkts. 34, 67–81, 226; translation in J. Math. Sci. (N. Y.) 132 (2006), no. 3, 285–294.

# VARIABLE EXPONENT SPACES ON METRIC MEASURE SPACES

TOSHIHIDE FUTAMURA

*Department of Mathematics, Daido Institute of Technology*
*Nagoya 457-8530, Japan*

PETTERI HARJULEHTO*, PETER HÄSTÖ*

*Department of Mathematics and Statistics, P.O. Box 68,*
*FI-00014 University of Helsinki, Finland*
*E-mails: petteri.harjulehto@helsinki.fi, peter.hasto@helsinki.fi*

YOSHIHIRO MIZUTA

*Division of Mathematical and Information Sciences*
*Faculty of Integrated Arts and Sciences*
*Hiroshima University, Higashi-Hiroshima 739-8521, Japan*

TETSU SHIMOMURA

*Department of Mathematics, Graduate School of Education, Hiroshima University*
*Higashi-Hiroshima 739-8524, Japan*

In this survey we summarize recent results from the variable exponent, metric measure space setting, and some other closely related material. We show that the variable exponent arises very naturally in the metric measure space setting. We also give an example which shows that the maximal operator can be bounded for piecewise constant, nonconstant, exponents.

**Key words:** Variable exponent, variable dimension, Hausdorff measure, Lebesgue space, Riesz potential, Hajłasz space, metric measure space
**Mathematics Subject Classification:** Primary 46E35; Secondary 28A78, 28A80, 42B20, 46E30

## 1. Introduction

The theory of Sobolev spaces on metric measure spaces has been developed by several researchers during the past ten or so years, see e.g. Heinonen[20]. For the existence of a viable theory, it turns out that one should assume that we are dealing with a metric measure space with doubling measure in which a certain Poincaré inequality holds (more on this later). There are many examples of metric measure spaces with a doubling measure supporting a Poincaré inequality. A. Björn[2], has collected the following examples: Unweighted and weighted (for example with Muckenhoupt-

---

*Supported by the Academy of Finland

type weights) Euclidean spaces including, Riemannian manifolds with nonnegative Ricci curvature, graphs, and the Heisenberg group with the Lebesgue measure and a certain metric.

The study of variable exponent spaces has likewise gained impetus only during the last five years, see e.g. Diening, Hästö and Nekvinda[6] for a review of results in the Euclidean setting. Variable exponent spaces have been proposed e.g. for use in the analysis of certain fluids with complicated behavior.

In this survey we summarize recent results from the variable exponent, metric measure space setting, and some other closely related results. We will show that the variable exponent arises very naturally in the metric measure space setting e.g. when deriving optimal Sobolev embeddings. In the appendix we study a space which shows that the maximal operator can be bounded for piece-wise constant (but non-constant) exponents.

## *Definitions*

By a *metric measure space* we mean a triple $(X, d, \mu)$, where $X$ is a set, $d$ is a metric on $X$ and $\mu$ is a non-negative Borel regular outer measure on $X$ which is finite in every bounded set. For simplicity, we often write $X$ instead of $(X, d, \mu)$. For $x \in X$ and $r \geq 0$ we denote by $B(x, r)$ the open ball centered at $x$ with radius $r$. A metric measure space $X$ or a measure $\mu$ is said to be *doubling* if there is a constant $C \geq 1$ such that

$$\mu(B(x, 2r)) \leq C\mu(B(x, r)) \tag{1}$$

for every open ball $B(x, r) \subset X$. The constant $C$ in (1) is called the *doubling constant* of $\mu$. The doubling property is equivalent to the following: there exist constants $Q$ and $C_Q$ such that if $B(y, R)$ is an open ball in $X$, $x \in B(y, R)$ and $0 < r \leq R < \infty$, then

$$\frac{\mu(B(x, r))}{\mu(B(y, R))} \geq C_Q \left(\frac{r}{R}\right)^Q. \tag{2}$$

For example, in $\mathbb{R}^n$ with the Lebesgue measure (2) holds with $Q$ equal to the dimension $n$.

We say that the measure $\mu$ is *lower Ahlfors Q–regular* if there exists a constant $C > 0$ such that $\mu(B) \geqslant C \operatorname{diam}(B)^Q$ for every ball $B \subset X$. The measure $\mu$ is *Ahlfors Q–regular* if $\mu(B) \approx \operatorname{diam}(B)^Q$ for every ball $B \subset X$. If $X$ is a bounded doubling metric measure space (so that $\mu(X) < \infty$ and $\operatorname{diam}(X) < \infty$), then it is lower Ahlfors $Q$-regular.

## 2. Lebesgue spaces

We call a measurable function $p \colon X \to [1, \infty)$ a *variable exponent*. For $A \subset X$ we define $p_A^+ = \operatorname{ess\,sup}_{x \in A} p(x)$ and $p_A^- = \operatorname{ess\,inf}_{x \in A} p(x)$; we use the abbreviations

$p^+ = p_X^+$ and $p^- = p_X^-$. For a $\mu$–measurable function $u\colon X \to \mathbb{R}$ we define the *modular*

$$\varrho_{p(\cdot)}(u) = \int_X |u(y)|^{p(y)} d\mu(y)$$

and the *norm*

$$\|u\|_{p(\cdot)} = \inf\{\lambda > 0\colon \varrho_{p(\cdot)}(u/\lambda) \le 1\}.$$

Sometimes we use the notation $\|u\|_{p(\cdot),X}$ when we also want to indicate in what metric space the norm is taken. The *variable exponent Lebesgue spaces on* $X$, $L^{p(\cdot)}(X, d, \mu)$, consists of those $\mu$–measurable functions $u\colon X \to \mathbb{R}$ for which there exists $\lambda > 0$ such that $\varrho_{p(\cdot)}(\lambda u) < \infty$.

The following facts are easily proven, see Kováčik and Rákosník[22] for the Euclidean case and Harjulehto, Hästö & Pere[18], Section 3, for the metric space case:

- $\|\cdot\|_{p(\cdot)}$ is a norm;
- if $p^+ < \infty$, then $\varrho_{p(\cdot)}(f_i) \to 0$ if and only if $\|f_i\|_{p(\cdot)} \to 0$;
- the Hölder inequality $\|fg\|_1 \le C\|f\|_{p(\cdot)}\|g\|_{p'(\cdot)}$ holds $(p > 1)$;
- the space $L^{p(\cdot)}(X)$ is a Banach space;
- if $X$ is a locally compact doubling space and $p^+ < \infty$, then continuous functions with compact support are dense in $L^{p(\cdot)}(X)$.

The following condition has emerged as the right one to guarantee regularity of variable exponent Lebesgue spaces in the Euclidean setting. We say that $p$ is log-*Hölder continuous* if

$$|p(x) - p(y)| \le \frac{C}{-\log d(x,y)},$$

when $d(x,y) \le 1/2$. Following Diening, Lemma 3.2[4], it was shown in Lemma 3.6[18] that if $p$ is log-Hölder continuous, and $\mu$ is lower Ahlfors $Q$-regular, then for all balls $B \subset X$ we have $\mu(B)^{p_B^- - p_B^+} \le C$.

It was shown by Diening[4] that the log-Hölder condition is sufficient for the local boundedness of the maximal operator. Recall that the maximal operator is defined by

$$Mu(x) = \sup_{r>0} \fint_{B(x,r)} |u(x)| \, d\mu(x).$$

Moreover, an example by Pick and Růžička[25] shows that this is the best possible modulus of continuity under which this claim holds. It turns out that log-Hölder continuity is still sufficient for local boundedness in the metric measure spaces setting, however, it is no longer necessary in the modulus of continuity sense:

**Theorem 2.1.** [Theorem 4.3 [18]] *Let $X$ be a bounded doubling space. Suppose that $p$ is* log-*Hölder continuous with $1 < p^- \le p^+ < \infty$. Then*

$$\|\mathcal{M}f\|_{p(\cdot)} \le C\|f\|_{p(\cdot)}.$$

To show that log-Hölder continuity is not necessary, the following example was given in Example 4.5[18]. Let $X_1 = \{(x,0) \in \mathbb{R}^2 : 0 \leq x < 1/4\}$ and $X_2 = \{(x,y) \in B(0,1/2) : x < 0\}$ and define $(X,\mu) = (X_1,m_1) \cup (X_2,m_2)$, where $m_i$ denotes the $i$-dimension Lebesgue measure. We set the exponent $p$ equal to $s$ on $X_1$ and to $t$ on $X_2$ $(s,t > 1)$.

In Theorem 4.7[18], it was shown that the maximal operator is bounded in this space for certain values of $s$ and $t$, but not for all. The situation is shown in Figure 1. Harjulehto, Hästö and Pere[18] were not able to settle the boundedness of the maximal operator for the critical case $s = t/(2-t)$ (for $t < 2$), the upper curve in the figure). In the appendix of this article we give a new simpler proof for the boundedness of the maximal operator, which also works in the critical case.

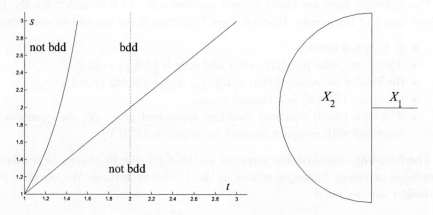

FIGURE 1. A sketch of the area where the maximal operator is bounded (left) and the space $X$ (right).

It was pointed out by Lars Diening that the measure in this example is not doubling.

Open Problem 1. Is there an example of a Lebesgue space with doubling measure and piecewise constant exponent on which the maximal operator is bounded?

Variable dimension spaces will be defined below in Section 4. The dimension seems to play a fundamental role in the example above. Thus we may ask

Open Problem 2. Is it possible to derive an optimal modulus of continuity for the exponent in terms of the variability of the dimension in the metric measure space setting?

## 3. Hajłasz and Newtonian spaces

There are many different ways to define Sobolev spaces, with a fixed exponent, on metric measure spaces. In this chapter we discuss how two commanly used spaces, namely Hajłasz and Newtonian spaces, can be generalized to the variable exponent case. We restrict our attention to exponents with $1 < p^- \leq p^+ < \infty$.

## Hajłasz spaces

We say that a $p(\cdot)$-integrable function $u$ belongs to *Hajłasz space* $M^{1,p(\cdot)}(X)$ if there exists a non-negative $g \in L^{p(\cdot)}(X)$ such that

$$|u(x) - u(y)| \leq d(x,y)\,(g(x) + g(y)) \tag{3}$$

for $\mu$-almost every $x,y \in X$. The function $g$ is called a *Hajłasz gradient* of $u$. In $\mathbb{R}^n$ we may use $\mathcal{M}(|\nabla u|)$ as a Hajłasz gradient of $u \in W^{1,q}(\mathbb{R}^n)$, where $q > 1$ is a constant. We equip $M^{1,p(\cdot)}(X)$ with the norm

$$\|u\|_{M^{1,p(\cdot)}(X)} = \|u\|_{p(\cdot)} + \inf \|g\|_{p(\cdot)},$$

where the infimum is taken over all Hajłasz gradients of $u$. Following the original arguments of P. Hajłasz[12] it is easy to prove that $M^{1,p(\cdot)}(X)$ is a Banach space and Lipschitz continuous functions are dense. Integrating both sides of (3) over $y$ in a bounded space $X$ of finite measure and using Hölder's inequality we find that

$$|u(x) - u_X| \leq \operatorname{diam}(X)\Big(g(x) + \frac{C\|1\|_{p'(\cdot)}}{\mu(X)}\|g\|_{p(\cdot)}\Big),$$

where $u_X$ denotes the integral average of $u$ over $X$. This point-wise estimate leads easily to the *Poincaré inequality*

$$\|u - u_X\|_{p(\cdot)} \leq C\big(p^-, p^+, \mu(X)\big)\operatorname{diam}(X)\|g\|_{p(\cdot)}.$$

If the measure is atomless, then $L^{p(\cdot)}$ is reflexive. Using this and Mazur's lemma we find that for every $u \in M^{1,p(\cdot)}(X)$ there exists a unique Hajłasz gradient of $u$ which minimizes the norm. By unique we mean that if $g$ and $g'$ are minimal Hajłasz gradients of $u$, then $\|g - g'\|_{L^{p(\cdot)}(X)} = 0$.

## Newtonian spaces

A curve $\gamma$ in $X$ is a non-constant continuous map $\gamma : I \to X$, where $I = [a,b]$ is a closed interval in $\mathbb{R}$.

Let $\Gamma$ be a family of rectifiable curves. We denote by $F(\Gamma)$ the set of all Borel measurable functions $\rho : X \to [0, \infty]$ such that

$$\int_\gamma \rho\, ds \geq 1$$

for every $\gamma \in \Gamma$, where $ds$ represents integration with respect to path length. We define the $p(\cdot)$-*modulus of* $\Gamma$ by

$$\mathrm{M}_{p(\cdot)}(\Gamma) = \inf_{\rho \in F(\Gamma)} \int_X \rho(x)^{p(x)}\, d\mu(x).$$

If $F(\Gamma) = \emptyset$, then we set $\mathrm{M}_{p(\cdot)}(\Gamma) = \infty$. The arguments from $\mathbb{R}^n$ imply that the $p(\cdot)$-modulus is an outer measure on the space of all curves of $X$, for a proof see Lemma 2.1[17].

A non-negative Borel measurable function $\rho$ on $X$ is a $p(\cdot)$-*weak upper gradient* of $u$ if there exists a family $\Gamma$ of rectifiable curves with $\mathrm{M}_{p(\cdot)}(\Gamma) = 0$ and

$$|u(x) - u(y)| \leqslant \int_\gamma \rho \, ds$$

for every rectifiable curve $\gamma \notin \Gamma$ with endpoints $x$ and $y$. In the Euclidean case we would use $|\nabla u|$ as the upper gradient.

The *Newtonian space* $N^{1,p(\cdot)}(X)$ is the collection of functions in $L^{p(\cdot)}(X)$ with a weak upper gradient in $L^{p(\cdot)}(X)$ equipped with the norm

$$\|u\|_{N^{1,p(\cdot)}(X)} = \|u\|_{p(\cdot)} + \inf \|\rho\|_{p(\cdot)},$$

where the infimum is taken over all weak upper gradients of $u$. The Newtonian space $N^{1,p(\cdot)}(X)$ is a Banach space, Theorem 3.4[19].

A metric measure space $X$ is said to *support a* $(1,q)$-*Poincaré inequality* if there exists a constant $C > 0$ such that for all open balls $B$ in $X$ and all pairs of functions $u$ and $\rho$ defined on $B$ the inequality

$$\fint_B |u - u_B| \, d\mu \leqslant C \operatorname{diam}(B) \left( \fint_B \rho^q \, d\mu \right)^{\frac{1}{q}}$$

holds whenever $\rho$ is an upper gradient of $u$ on $B$ and $u$ integrable on $B$.

If $q > 1$ is a constant and the space $X$ supports a $(1,q)$-Poincaré inequality, then the Newtonian space $N^{1,q}(X)$ is reflexive[3]. The same condition also implies that Lipschitz continuous functions are dense in Newtonian space. In Harjulehto, Hästö & Pere[19] it was shown that a $(1,1)$-Poincaré inequality implies the density of Lipschitz functions in Newtonian space.

**Open Problem 3.** Are Lipschitz continuous functions dense in $N^{1,p(\cdot)}(X)$ when the space $X$ supports a $(1, p(\cdot))$-Poincaré inequality?

**Open Problem 4.** Show that $N^{1,p(\cdot)}(X)$ is reflexive, assuming either a $(1,1)$- or a $(1, p(\cdot)$-Poincaré inequality.

We end this chapter by studying when Hajłasz, Newtonian and classical Sobolev spaces agree. All the proof can be found in Harjulehto, Hästö & Pere[19].

**Theorem 3.1.** *We have the following relations between Hajłasz, Newtonian and classical Sobolev spaces:*

*(1)* $M^{1,p(\cdot)}(\mathbb{R}^n) \subset W^{1,p(\cdot)}(\mathbb{R}^n)$;

*(2) If the maximal operator is bounded from $L^{p(\cdot)}(\mathbb{R}^n)$ to itself, then* $M^{1,p(\cdot)}(\mathbb{R}^n) = W^{1,p(\cdot)}(\mathbb{R}^n)$;

*(3) If $\Omega \subset \mathbb{R}^n$ is open, then* $N^{1,p(\cdot)}(\Omega) \subset W^{1,p(\cdot)}(\Omega)$;

*(4) If $\Omega \subset \mathbb{R}^n$ is open, $1 < p^- \leqslant p^+ < \infty$ and $C^1(\Omega)$ is dense in $W^{1,p(\cdot)}(\Omega)$, then $N^{1,p(\cdot)}(\Omega) = W^{1,p(\cdot)}(\Omega)$;*

*(5) $M^{1,p(\cdot)}(X) \subset N^{1,p(\cdot)}(X)$;*

*(6) If $X$ supports a $(1,1)$-Poincaré inequality and if the Hardy-Littlewood maximal operator is bounded from $L^{p(\cdot)}(X)$ to itself, then $M^{1,p(\cdot)}(X) = N^{1,p(\cdot)}(X)$.*

Recently P. Hajłasz[13] has generalized the definition of $M^{1,1}(\mathbb{R}^n)$ so that $M^{1,1}(\mathbb{R}^n) = W^{1,1}(\mathbb{R}^n)$. This raises the question whether the assumption of the boundedness of the maximal operator in (2), above, can be relaxed if this generalized definition is used.

## 4. Sobolev embeddings $p < Q$

Of the various methods for proving Sobolev inequalities in constant exponent spaces, the one based on the Riesz potential is best suited for the variable exponent case.

Let $\alpha > 0$ be fixed. We define the Riesz potential as

$$I_\alpha^\Omega |u|(x) = \int_\Omega \frac{|u(y)| \, d(x,y)^\alpha}{\mu\big(B(x,d(x,y))\big)} \, d\mu(y).$$

(For technical reasons we sometimes use a modified Riesz potential $J_\alpha^\Omega$ from Hajłasz and Koskela[14] with the property that $I_\alpha^\Omega |u|(x) \leq J_\alpha^\Omega u(x)$ for almost every $x \in X$ if the measure $\mu$ is doubling.) Riesz potentials in the variable exponent, Euclidean setting have to been studied for instance in [5,10,21,24].

It was shown recently in Harjulehto, Hästö and Latvala[16] that it is possible to study variable dimension, variable exponent spaces without any extra work due to the the variable dimension. In this context the exact definition of variable exponent is as follows. If $Q \colon X \to (0,\infty)$ is a BOUNDED function, then we say that $\mu$ is *Ahlfors $Q(\cdot)$-regular* if

$$\mu(B(x,r)) \approx r^{Q(x)}$$

for all $x \in X$ and $r \leq \operatorname{diam} X$. As an example of a variable exponent space, a variable parameter von Koch curve was constructed in Section 3[16], see Figure 1.

Various embedding theorems for the modified Riesz potential were proven in Harjulehto, Hästö & Latvala[16]. We quote here only one representative and easily understandable result:

**Theorem 4.1.** [Corollary 5.4[16]] *Let $\mu$ be lower Ahlfors $Q(\cdot)$-regular and doubling in a bounded metric space $X$. Let $p$ be log-Hölder continuous in $X$ and let $1 < p^- \leq p^+ < \infty$. If $1 < \inf \frac{Q(x)}{p(x)}$, then for each ball $B \subset X$ we have*

$$\|u - u_B\|_{L^{p^*(\cdot)}(B)} \leq C\|g\|_{L^{p(\cdot)}(5B)}, \quad p^*(x) = \frac{Q(x)p(x)}{Q(x) - p(x)},$$

*for every $u \in M^{1,p(\cdot)}(X)$.*

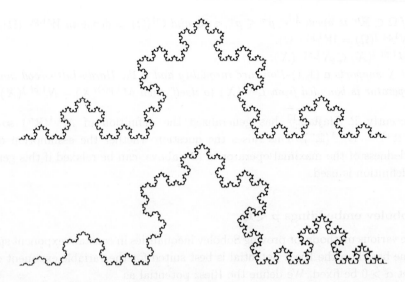

FIGURE 2. An ordinary Koch curve (upper) and a variable dimension Koch curve (lower).

The question of necessary condition for Sobolev embeddings remains largely open, both in the Euclidean and the metric space cases. The only known counter-example is due to Kováčik and Rákosník Example 3.2[22], which was slightly improved in Diening, Hästö & Nekvinda[6], where it was shown that there exists a continuous exponent $p$ on a regular domain $\Omega \subset \mathbb{R}^2$ such that

$$W^{1,p(\cdot)}(\Omega) \not\hookrightarrow L^{p^*(\cdot)}(\Omega).$$

The exponent in the this example was not uniformly continuous. Thus Diening, Hästö and Nekvinda[6] asked the following:

Open Problem 5. Are there counter-examples to the Sobolev embedding in regular domains for uniformly continuous exponents?

Open Problem 6. We saw before that the maximal operator features vastly different behavior in the metric space case as compared with the Euclidean one. Are similar examples relevant for the Sobolev embedding?

## 5. Exponential inequalities

First we consider a Trudinger-type inequality in the case $p(x) = Q(x)$. It turns out that the embeddings in this case are much the same as in the classical case. Then we consider the case when the exponent tends to the dimension at some point from above. Depending on the speed at which the exponent approaches the critical value, we get different embeddings. Section 5 deals with the case there $p(x) = Q$ at one point or on a sphere.

## Trudinger-type embeddings

The introduction of variable dimension spaces allowed generalizations of the Trudinger embedding (i.e. $p = n$) to the variable exponent case. Recall that in the classical case the Trudinger embedding states that $W^{1,n}$ embeds into $\exp L^{n'}$, where the exponent $n' = n/(n-1)$ is the best possible. This generalizes as follows:

**Theorem 5.1.** *Let $X$ be a bounded connected doubling space, and assume that $\mu$ is lower Ahlfors $Q(\cdot)$-regular, where $Q$ is log-Hölder continuous and $1 < Q^- \leq Q^+ < \infty$. Then there is a constant $C_1$, depending on $X$, such that*

$$\fint_B \exp\left(\frac{C_1|u(x) - u_B|}{\|g\|_{L^{Q(\cdot)}(5B)}}\right)^{Q'(x)} d\mu(x) \leq 2$$

*for every ball $B \subset X$ and for each $u \in M^{1,Q(\cdot)}(X)$ with Hajłasz gradient $g$.*

## The super-critical case

For fixed $x_0 \in X$, where $X$ is a doubling metric measure spaces and $Q$ is as in (2), let us consider an exponent $p(x)$ such that

$$p(x) \geq \frac{Q}{\alpha} \qquad \text{for } x \in B(x_0, r_0).$$

Set $B_0 = B(x_0, r_0)$ for simplicity. We further assume that

$$p(x) = \frac{Q}{\alpha} + \frac{a \log\left(\log(1/|x_0 - x|)\right)}{\log(1/|x_0 - x|)} + \frac{b}{\log(1/|x_0 - x|)} \tag{4}$$

for $x \in B_0$, where $0 < r_0 < 1/4$, $a \geq 0$ and $b$ is a real number. If $a > 0$, then we can take $r_0$ so small that

$$p(x) > \frac{Q}{\alpha} \qquad \text{when } x \in B_0 \setminus \{x_0\}.$$

Let $f$ be a nonnegative measurable function on $B_0$ with $\|f\|_{p(\cdot)} < \infty$. In connection with Sobolev's imbedding theorem, the borderline value for $a$ will be shown to be

$$a = \frac{Q - \alpha}{\alpha^2}.$$

**Theorem 5.2.** [Theorems 5.6 and 5.9 [11]] *If $0 \leq a < (Q - \alpha)/\alpha^2$, then there exist positive constants $c_1$ and $c_2$ such that*

$$\int_{B_0} \exp\left(c_1(I_\alpha f(x))^{Q/(Q-\alpha-a\alpha^2)}\right) d\mu(x) \leq c_2$$

*for all nonnegative measurable functions $f$ on $B_0$ with $\|f\|_{p(\cdot)} \leq 1$.*
*If $a = (Q - \alpha)/\alpha^2$, then there exist positive constants $c_1$ and $c_2$ such that*

$$\int_{B_0} \exp\left(\exp(c_1(I_\alpha f(x))^{Q/(Q-\alpha)})\right) d\mu(x) \leq c_2$$

*for all nonnegative measurable functions $f$ on $B_0$ with $\|f\|_{p(\cdot)} \leq 1$.*

Suppose next that $X = \mathbb{R}^n$ (and $Q = n$). Let $\varphi$ be a non-increasing function the interval on $(0, \infty)$ and consider $p(x)$ satisfying

$$p(x) = \frac{n}{\alpha} + \frac{\log \varphi(|x_0 - x|)}{\log(1/|x_0 - x|)} \qquad \text{for } x \in B_0.$$

**Theorem 5.3.** [Theorem7.2 [11]] *Let $f$ be a nonnegative measurable function on $B_0$ with $\|f\|_{p(\cdot)} \leq 1$. If*

$$\int_0^1 \varphi(t)^{-\alpha^2/(n-\alpha)} t^{-1} dt < \infty, \tag{5}$$

*then $I_\alpha f$ is continuous on $B_0$ and*

$$\lim_{x \to x_0} \Phi(|x - x_0|)^{-1} |I_\alpha f(x) - I_\alpha f(x_0)| = 0,$$

*where $\Phi(r) = \left( \int_0^r \varphi(t)^{-\alpha^2/(n-\alpha)} t^{-1} dt \right)^{(n-\alpha)/n}$.*

As a special case we consider $\varphi(r) = c(\log(1 + (1/r)))^a$ with $c > 0$ and $a \geq 0$. Then $\varphi$ satisfies (5) if and only if $a > (n - \alpha)/\alpha^2$. Suppose that $p(\cdot)$ satisfies (4) for $a > (n - \alpha)/\alpha^2$ and let $f$ be a nonnegative measurable function on $B_0$ with $\|f\|_{p(\cdot)} < \infty$. Then $I_\alpha f$ is continuous on $B_0$ and

$$\lim_{x \to x_0} (\log(1/|x - x_0|))^A |I_\alpha f(x) - I_\alpha f(x_0)| = 0,$$

where $A = (a\alpha^2 - (n - \alpha))/n$.

Futamura, Mizuta and Shimomura[11] also considered the case of a variable exponent $p$ satisfying

$$p(x) = \frac{n}{\alpha} + \frac{a \log(\log(1/(1 - |x|)))}{\log(1/(1 - |x|))} + \frac{b}{\log(1/(1 - |x|))}$$

in the unit ball of $\mathbb{R}^n$, where $a \geq 0$ and $b$ is a real number. Assuming that $p(x) > \frac{n}{\alpha}$ when $x_n \neq 0$ they derived analogous results on exponential and double exponential integrability of Riesz potentials.

*Generalizations*

Let $p(r)$ be a continuous function on $[0, \infty)$ such that

$$p(r) = p_0 + \frac{a \log(\log(1/r))}{\log(1/r)} + \frac{b}{\log(1/r)}$$

for $0 < r < r_0 < 1/4$; set $p(0) = p_0$ and $p(r) = p(r_0)$ for $r > r_0$. Let $K$ be a compact set in $\mathbb{R}^n$. Set

$$p(x) = p(\delta_K(x)),$$

where $\delta_K(x)$ denotes the distance of $x$ from $K$.

(i) If $K = \{x_0\}$, then $p(\cdot)$ is nothing but the first case.

(ii) If $K = \partial B(0,1)$, then $p(\cdot)$ is nothing but the second case.

**Open Problem 7.** Under what conditions on the set $K$ above do Sobolev inequalities still hold in the variable exponent spaces?

We can investigate the behavior of our spaces further by taking a closer look at the critical case, and defining the exponent $p(\cdot)$ by

$$p(x) = \frac{n}{\alpha} + \frac{b}{\log(1/|x_0 - x|)} + \frac{n - \alpha}{\alpha^2} \frac{\log(\log(1/|x_0 - x|))}{\log(1/|x_0 - x|)}$$
$$+ \frac{a \log(\log(\log(1/|x_0 - x|)))}{\log(1/|x_0 - x|)}.$$

Even more generally, we can consider an exponent with continuity modulus given by some function $\varphi$.

**Open Problem 8.** With $p$ as before, is it possible to derive exponential integrability results similar to those in Edmunds, Gurka and Opic[7,8].

## Appendix A.

## 6. The boundedness of the maximal operator for a discontinuous exponent

In this appendix we consider the boundedness of the maximal operator on a certain metric measure space. The result is an improvement of a result by Harjulehto, Hästö and Pere[18].

Take positive integers $\ell$ and $m$, and set $n = \ell + m$. For a point $x \in \mathbb{R}^n$, we write $x = (x', x'')$, where $x' = (x_1, x_2, \ldots, x_\ell) \in \mathbb{R}^\ell$ and $x'' = (x_{\ell+1}, x_{\ell+2}, \ldots, x_n) \in \mathbb{R}^m$. Let

$$X_1 = \left\{ x = (x', 0) \in \mathbb{R}^n : |x| < 1/4, x_1 > \tfrac{\sqrt{3}}{2}|x| \right\}$$

and

$$X_2 = \left\{ x = (x', x'') \in \mathbb{R}^n : |x| < 1/2, x_1 < 0 \right\}.$$

Consider the metric measure space $(X, \mu)$ given by

$$(X, \mu) = (X_1, H_\ell) \cup (X_2, H_n),$$

where $H_i$ denotes the $i$-dimensional Lebesgue measure. For $s > 1$ and $t > 1$, set

$$p(x) = \begin{cases} s & \text{if } x \in X_1, \\ t & \text{if } x \in X_2. \end{cases}$$

**Theorem A.1.** *Consider the maximal operator on the space $L^{p(\cdot)}(X)$, where $X$ and $p$ are defined above.*

*(i) If $t > s$, then $\mathcal{M}$ is not bounded.*
*(ii) If $s \geq t \geq n/m$, then $\mathcal{M}$ is bounded.*

*(iii)* If $t < n/m$ and $t \leq s \leq \ell t/(n - mt)$, then $\mathcal{M}$ is bounded.

*(iv)* If $t < n/m$ and $s > \ell t/(n - mt)$, then $\mathcal{M}$ is not bounded.

To prove the theorem, we use the following result.

**Lemma A.1.** *In case $s \geq t$, $\mathcal{M}$ is bounded if and only if*

$$\|\mathcal{M}f_2\|_{s,X_1} \leq C\|f_2\|_{t,X_2} \tag{A.1}$$

*for every $f \in L^{p(\cdot)}(X)$, where $f_2 = f\chi_{X_2}$.*

**Proof.** By the classical boundedness of maximal functions in metric spaces and Hölder's inequality, we see that

$$\begin{aligned}
\|\mathcal{M}f\|_{p(\cdot)} &\leq \|\mathcal{M}f_1\|_{s,X_1} + \|\mathcal{M}f_2\|_{s,X_1} + \|\mathcal{M}f\|_{t,X_2} \\
&\leq C\|f_1\|_{s,X} + \|\mathcal{M}f_2\|_{s,X_1} + C\|f\|_{t,X} \\
&\leq C(\|f_1\|_{p(\cdot)} + \|f\|_{p(\cdot)}) + \|\mathcal{M}f_2\|_{s,X_1},
\end{aligned}$$

where $f_1 = f\chi_{X_1}$ and $f_2 = f\chi_{X_2}$. Hence (A.1) implies that $\mathcal{M}$ is bounded.

Conversely suppose $\mathcal{M}$ is bounded on $L^{p(\cdot)}(X)$. Then we have

$$\|\mathcal{M}f_2\|_{s,X_1} \leq \|\mathcal{M}f_2\|_{p(\cdot)} \leq C\|f_2\|_{p(\cdot)} = C\|f_2\|_{t,X_2},$$

which implies (A.1). $\square$

Now we move on to the proof of the theorem.

**Proof.** To prove (i), consider the function

$$f(y) = \begin{cases} |y'|^{-2\ell/(s+t)} & \text{when } y \in X_1, \\ 0 & \text{otherwise.} \end{cases}$$

Then we see that

$$\int_X |f(y)|^{p(y)} d\mu(y) = \int_{X_1} |y'|^{-2\ell s/(s+t)} dH_\ell(y) < \infty$$

since $-2\ell s/(s+t) + \ell > 0$.

Let $\tilde{X}_2 = \{x \in X_2 : -x_1 > \frac{\sqrt{3}}{2}|x|\}$. If $x \in \tilde{X}_2$, then, letting $r = |x| + |x|^{n/\ell}$ and noting that $\mu(B(x,r)) \approx |x|^n$, we find that

$$\begin{aligned}
\mathcal{M}f(x) &\geq \frac{1}{\mu(B(x,r))} \int_{B(x,r) \cap X_1} |f(y)| \, dH_\ell(y) \\
&\geq C|x|^{-n} \int_{B(0,|x|^{n/\ell}) \cap X_1} |f(y)| \, dH_\ell(y) \\
&\geq C|x|^{-2n/(s+t)},
\end{aligned}$$

so that

$$\int_X |\mathcal{M}f(x)|^{p(x)} d\mu(x) \geq C \int_{\tilde{X}_2} |x|^{-2nt/(s+t)} dH_n(x) = \infty$$

since $-2nt/(s+t)+n < 0$.

Next we prove (ii) and (iii). Take a ball $B(x,r)$ such that $x \in X_1$ and $B(x,r) \cap X_2 \neq \emptyset$. For $x \in X_1$, we have

$$\frac{1}{\mu(B(x,r))} \int_{B(x,r)} |f_2(y)| \, d\mu(y)$$

$$= \frac{1}{H_\ell(B(x,r) \cap X_1) + H_n(B(x,r) \cap X_2)} \int_{B(x,r) \cap X_2} |f_2(y)| \, dH_n(y)$$

$$\leq \frac{C}{\min\{r^\ell, 1\}} \int_{B(x,r) \cap X_2} |f_2(y)| \, dH_n(y)$$

$$\leq C \int_{B(x,r) \cap X_2} |x-y|^{-\ell} |f_2(y)| \, dH_n(y),$$

which implies that

$$Mf_2(x) \leq C \int_{X_2} |x-y|^{m-n} |f_2(y)| \, dH_n(y).$$

If $s \geq t$ when $t \geq n/m$, or $\ell t/(n - mt) \geq s \geq t$ when $t < n/m$, then we have by the Sobolev imbedding theorem (see Adams and Fournier[1])

$$\|Mf_2\|_{s,X_1} \leq C \left( \int_{X_1} \left( \int_{X_2} |x-y|^{m-n} |f_2(y)| \, dH_n(y) \right)^s dH_\ell(x) \right)^{1/s} \leq C\|f_2\|_{t,X_2},$$

which proves (ii) and (iii) by the aid of Lemma A.1.

To prove (iv), consider the function

$$g(x) = |x|^{-n/t} \left( \log \left( 1/|x| \right) \right)^{-2/t} \chi_{X_2}.$$

Then

$$\int_X g(y)^{p(y)} \, d\mu(y) = \int_{X_2} |y|^{-n} \left( \log \left( 1/|y| \right) \right)^{-2} dH_n(y)$$

$$\leq C \int_0^{1/2} r^{-1} \left( \log \left( 1/r \right) \right)^{-2} dr < \infty,$$

so $g \in L^{p(\cdot)}(X)$. Furthermore, we have for $x \in X_1$

$$Mg(x) \geq \frac{1}{\mu(B(x,2|x|))} \int_{B(x,2|x|)} g(y) \, d\mu(y)$$

$$\geq C|x|^{-\ell} \int_{B(0,|x|) \cap X_2} |y|^{-n/t} \left( \log \left( 1/|y| \right) \right)^{-2/t} dH_n(y)$$

$$\geq C|x|^{-\ell} \int_0^{|x|} r^{-n/t+n-1} \left( \log \left( 1/r \right) \right)^{-2/t} dr$$

$$\geq C|x|^{-n/t+n-\ell} \left( \log \left( 1/|x| \right) \right)^{-2/t},$$

120

so that

$$\int_{X_1} (\mathcal{M}g(x))^s \, dH_\ell(x) \geq C \int_{X_1} |x|^{-sn/t+sm} \left(\log\left(1/|x|\right)\right)^{-2s/t} dH_\ell(x)$$

$$\geq C \int_0^{1/4} r^{-sn/t+sm+\ell-1} \left(\log\left(1/r\right)\right)^{-2s/t} dr = \infty$$

since $-sn/t + sm + \ell < 0$. Hence $\mathcal{M}g \notin L^{p(\cdot)}(X)$. $\qquad\square$

## References

1. R. A. Adams and J. F. Fournier, Sobolev spaces, Second Edition, Academic Press, 2003.
2. A. Björn: Characterizations of $p$-superharmonic functions on metric spaces, *Studia Math.* **169** (2005), no. 1, 45–62.
3. J. Cheeger: Differentiability of Lipschitz functions on measure spaces, *Geom. Funct. Anal.* **9** (1999), 428–517.
4. L. Diening: Maximal function on generalized Lebesgue spaces $L^{p(\cdot)}$, *Math. Inequal. Appl.* **7** (2004), no. 2, 245–254.
5. L. Diening: Riesz potential and Sobolev embeddings of generalized Lebesgue and Sobolev spaces $L^{p(\cdot)}$ and $W^{k,p(\cdot)}$, *Math. Nachr.* **263** (2004), no. 1, 31–43.
6. L. Diening, P. Hästö and A. Nekvinda: Open problems in variable exponent Lebesgue and Sobolev spaces, *FSDONA04 Proceedings* (Drabek and Rakosnik (eds.); Milovy, Czech Republic, 2004), 38–58.
7. D. E. Edmunds, P. Gurka and B. Opic: Double exponential integrability, Bessel potentials and embedding theorems, *Studia Math.* **115** (1995), 151–181.
8. D. E. Edmunds, P. Gurka and B. Opic: Sharpness of embeddings in logarithmic Bessel-potential spaces, *Proc. Royal Soc. Edinburgh.* **126** (1996), 995–1009.
9. T. Futamura and Y. Mizuta: Continuity properties of Riesz potentials for functions in $L^{p(\cdot)}$ of variable exponent, *Math. Inequal. Appl.* **8** (2005), no. 4, 619–631.
10. T. Futamura, Y. Mizuta and T. Shimomura: Sobolev embeddings for Riesz potential space of variable exponent, *Math. Nachr.* **279** (2006), 1463–1473.
11. T. Futamura, Y. Mizuta and T. Shimomura: Sobolev embeddings for variable exponent Riesz potentials on metric spaces, *Ann. Acad. Sci. Fenn. Math.* **31** (2006), no. 2, 495–522.
12. P. Hajłasz: Sobolev spaces on arbitrary metric spaces, *Potential Anal.* **5** (1996), 403–415.
13. P. Hajłasz: A new characterization of the Sobolev space, *Studia Math.* **159** (2003), no. 2, 263–275.
14. P. Hajłasz and P. Koskela: Sobolev met Poincaré, *Mem. Amer. Math. Soc.* **145** (2000), no. 688.
15. P. Harjulehto and P. Hästö: A capacity approach to the Poincaré inequality and Sobolev imbeddings in variable exponent Sobolev spaces, *Rev. Mat. Complut.* **17** (2004), 129–146.
16. P. Harjulehto, P. Hästö and V. Latvala: Sobolev embeddings in metric measure spaces with variable dimension, *Math. Z.* **254** (2006), 591–609.
17. P. Harjulehto, P. Hästö and O. Martio: Fuglede's theorem in variable exponent Sobolev space, *Collect. Math.* **55** (2004), no. 3, 315–324.
18. P. Harjulehto, P. Hästö and M. Pere: Variable exponent Lebesgue spaces on metric spaces: the Hardy-Littlezood maximal operator, *Real Anal. Exchange* **30** (2004/2005), 87–104.

19. P. Harjulehto, P. Hästö and M. Pere: Variable exponent Sobolev spaces on metric measure spaces, *Funct. Approx. Comment. Math.*, to appear.
20. J. Heinonen: *Lectures on Analysis on Metric Spaces*, Springer-Verlag, New York, 2001.
21. V. Kokilashvili and S. Samko: On Sobolev theorem for Riesz type potentials in the Lebesgue spaces with variable exponent, *Z. Anal. Anwendungen* **22** (2003), no. 4, 899–910.
22. O. Kováčik and J. Rákosník: On spaces $L^{p(x)}$ and $W^{1,p(x)}$, *Czechoslovak Math. J.* **41(116)** (1991), 592–618.
23. Y. Mizuta and T. Shimomura: Continuity of Sobolev functions of variable exponent on metric spaces, *Proc. Japan Acad. Ser. A Math. Sci.* **80** (2004), 96–99.
24. Y. Mizuta and T. Shimomura: Sobolev's inequality for Riesz potentials with variable exponent satisfying a log-Hölder condition at infinity, *J. Math. Anal. Appl.* **311** (2005), no. 1, 268–288.
25. L. Pick and M. Růžička: An example of a space $L^{p(x)}$ on which the Hardy–Littlewood maximal operator is not bounded, *Expo. Math.* **19** (2001), 369–371.

## I.3 Reproducing Kernels and Related Topics

Organizers: Daniel Alpay, Alain Berlilnet, Saburou Saitoh

The session on "Reproducing Kernels And Related Topics" held in Catania in 2005 as part of the 5th ISAAC Congress gave rise to six papers.

The first one by Alain Berlinet analyzes the developments in Probability and Statistics relevant to the theory of reproducing kernels. In the same fields a paper by Stéphane Canu, Chen Soon Ong and Xavier Mary deals with splines defined from non positive kernels and Charles Suquet presents limit theorems for random measures and stochastic processes obtained through embeddings in reproducing kernel Hilbert spaces.

Jorge Buescu and A. Paixão explore properties of kernels related to Mercer theorem. The paper by Keiko Fujita is devoted to Bergman kernels.

Finally, the paper by Hiromichi Itou and Saburou Saitoh gives solutions of singular integral equations by combining Tikhonov regularization and reproducing kernel theory.

We take advantage of the present opportunity to announce the creation of a special interest group in the ISAAC in order to systematize the scientific activity related to reproducing kernels consistently and conservatively. A website is under construction to create a community of researchers involved in the domain. It will be a tool for research, communication, discussion and references on kernels. A link will soon appear on the ISAAC home page.

# TOPICS ON THE BERGMAN KERNEL FOR SOME BALLS

K.FUJITA

*Faculty of Culture and Education*
*Saga University,*
*Saga, 840-8502, JAPAN*
*E-mail: keiko@cc.saga-u.ac.jp*

In our previous papers ([1] to [6]), we studied the Bergman kernels on the "$N_p$-balls". We represented their Bergman kernels by means of the double series expansion using spherical harmonics. In this note, first we will review our results on the Bergman kernels focused on the two dimensional $N_p$-balls which are equivalent to the $L_p$-balls defined by the $L_p$-norm, and then we will consider an analytic continuation of complex harmonic functions on the $N_p$-balls through the integral representations with their "harmonic" Bergman kernels.

**Key words:** Bergman kernel
**Mathematics Subject Classification:** 32A25, 30E20

## 1. Introduction

For a bounded domain $G$ of $\mathbf{C}^n$, we denote by $\mathcal{O}(G)$ the space of holomorphic functions on $G$ equipped with the topology of uniform convergence on compact sets. Put

$$HO(G) = \left\{ f \in \mathcal{O}(G); \int_G |f(w)|^2 dV_G(w) < \infty \right\},$$

where $dV_G(w)$ is the normalized Lebesgue measure on $G$. We know the existence of the Bergman kernel $B_G(z,w)$ for $HO(G)$, and $f \in HO(G)$ can be represented by

$$f(w) = \int_G f(z)\overline{B_G(z,w)}dV_G(z), \quad w \in G.$$

However we know its explicit form only for a few domains such as the complex Euclidean ball $\tilde{B}_2(r) = \{z \in \mathbf{C}^n; \|z\|^2 < r^2\}$:

$$B_{\tilde{B}_2(r)}(z,w) = \frac{r^{2n+2}}{(r^2 - z \cdot \overline{w})^{n+1}}. \tag{1}$$

For the Lie ball $\tilde{B}(r) = \{z \in \mathbf{C}^n; \|z\|^2 + \sqrt{\|z\|^4 - |z^2|^2} < r^2\}$, L.K.Hua gave the explicit form in [8]:

$$B_{\tilde{B}(r)}(z,w) = \frac{r^{4n}}{(r^4 - 2r^2 z \cdot \overline{w} + z^2\overline{w}^2)^n}. \tag{2}$$

Further, for the dual Lie ball $\tilde{B}_1(r) = \{z \in \mathbf{C}^n; \|z\|^2 + |z^2| < 2r^2\}$, K.Oeljeklaus, P.Pflug, E.H.Youssfi gave the explicit form in [10]:

$$B_{\tilde{B}_1(r)}(z,w) = \frac{\displaystyle\sum_{j=0}^{[n/2]} \frac{(n-1)! X^{n-1-2j} Y^j}{(2j+1)!(n-2j)!}(2nX - (n-2j)(X^2 - Y))}{(X^2 - Y)^{n+1}}, \qquad (3)$$

where $X = 1 - \left(\frac{z}{\sqrt{2r}}\right) \cdot \left(\frac{\overline{w}}{\sqrt{2r}}\right)$, $Y = \frac{z^2}{2r^2} \times \frac{\overline{w}^2}{2r^2}$.

Generalizing these balls, we define the $N_p$-balls in Section 2. For the $N_p$-balls, the Bergman kernel is not known explicitly except for the above three balls. However it can be expressed by the double series expansion by means of the homogenous harmonic polynomials (see Section 3). In Section 4, we will consider an analytic continuation of harmonic functions by using the integral representation with the harmonic Bergman kernel. In section 5, we review the Bergman kernels and the harmonic Bergman kernels in explicit forms and study the analytic continuation through their explicit forms. Lastly, we translate our results for two dimensional $N_p$-balls into $L_p$-balls. For the $L_4$-balls, Jong-Do Park, in [11], gave an explicit form of the Bergman kernel. He also treated Szegö kernels and consider the reproducing kernels for the complex ellipsoids in [11].

The author would like to thank Professor Mitsuo Morimoto for useful discussions.

## 2. $N_p$-norm and $N_p$-ball

To introduce the $N_p$-balls, first we review the $N_p$-norms. We denote the Lie norm by $L(z)$:

$$L(z) = \sqrt{\|z\|^2 + \sqrt{\|z\|^4 - |z^2|^2}}, \quad z \in \mathbf{C}^n.$$

Put

$$L(z,w) = z \cdot w + \sqrt{(z \cdot w)^2 - z^2 w^2}, \quad M(z,w) = z \cdot w - \sqrt{(z \cdot w)^2 - z^2 w^2}.$$

In the following, we will use the notation $X^n(z,w)$ for $(X(z,w))^n$, etc. Consider the functions

$$N_p(z,w) = \left(\frac{L^{p/2}(z,w) + M^{p/2}(z,w)}{2}\right)^{1/p},$$

and $N_p(z) = N_p(z,\overline{z})$:

$$N_p(z) = \left(\frac{\left(\|z\|^2 + \sqrt{\|z\|^4 - |z^2|^2}\right)^{p/2} + \left(\|z\|^2 - \sqrt{\|z\|^4 - |z^2|^2}\right)^{p/2}}{2}\right)^{1/p}$$

$$= \left(\frac{L^p(z) + (|z^2|/L(z))^p}{2}\right)^{1/p}. \qquad (4)$$

Then $\lim_{p\to\infty} N_p(z) = L(z)$. Further we proved that $N_p(z)$ is a norm if $p \geq 1$ in [9, Corollary 14] and that $N_p(z)$ is a monotone increasing function in $p$ in [6, Lemma 1.1]. Note that $N_2(z) = \|z\|$ and $N_1(z) = \sqrt{(\|z\|^2 + |z^2|)/2} = L^*(z) = \sup\{z; |z \cdot w|; L(w) \leq 1\}$ is the dual Lie norm.

Now, for $p \geq 1$, we define the $N_p$-ball $\tilde{B}_p(r)$ of radius $r$ by

$$\tilde{B}_p(r) = \{z \in \mathbf{C}^n; N_p(z) < r\}, \quad \tilde{B}(r) \equiv \tilde{B}_\infty(r) = \{z \in \mathbf{C}^n; L(z) < r\}.$$

We call $\tilde{B}_1(r)$ the dual Lie ball, $\tilde{B}_2(r)$ the complex Euclidean ball and $\tilde{B}(r)$ the Lie ball. By the definition, $\tilde{B}_p(r)$ is an open convex subset and balanced in $\mathbf{C}^n$. In particular, $\tilde{B}_p(r)$ is a domain of holomorphy in $\mathbf{C}^n$.

By (4), we have $2^{-1/p}L(z) \leq N_p(z) \leq L(z)$. Therefore

$$\tilde{B}(r) \subset \tilde{B}_p(r) \subset \tilde{B}(2^{1/p}r).$$

Since

$$L(z,w) + M(z,w) = 2z \cdot w, \quad L(z,w)M(z,w) = z^2 w^2,$$
$$M(z,w) = -L(iz, iw), \quad L(z,w) = -M(iz, iw),$$

the denominators of the Bergman kernels (3), (1) and (2) of $\tilde{B}_1(r)$, $\tilde{B}_2(r)$ and $\tilde{B}(r)$, respectively, are written by

$$r^{4(n+1)}\left(1 - \left(\tfrac{z}{\sqrt{2}r}\right)\cdot\left(\tfrac{\overline{w}}{\sqrt{2}r}\right) - \tfrac{z^2}{2r^2} \times \tfrac{\overline{w}^2}{2r^2}\right)^{n+1} = (r^2 - N_1^2(z,\overline{w}))^{n+1}(r^2 + N_1^2(iz, i\overline{w}))^{n+1},$$
$$(r^2 - z \cdot \overline{w})^{n+1} = (r^2 - N_2^2(z,\overline{w}))^{n+1},$$
$$(r^4 - 2r^2 z \cdot \overline{w} + z^2\overline{w}^2)^n = (r^2 - L(z,\overline{w}))^n(r^2 - M(z,\overline{w}))^n.$$

Because $N_p(z,\overline{z}) = N_p(z)$ and $N_\infty(z,\overline{z}) = L(z)$, these representations of the kernels indicate us that all the points $z = w$ on the boundary of $\tilde{B}_p(r)$ are singular points of their kernels.

In the following, we consider the Bergman kernels on the two dimensional $N_p$-balls. For $z = (z_1, z_2) \in \mathbf{C}^2$,

$$N_p(z) = \left(\frac{|z_1 + iz_2|^p + |z_1 - iz_2|^p}{2}\right)^{1/p}, \quad N_\infty(z) = \max|z_1 \pm iz_2|.$$

Thus for $p$ with $1 \leq p < \infty$, the $N_p$-norm in $\mathbf{C}^2$ is equivalent to the $L_p$-norm defined by

$$L_p(\zeta) = (|\zeta_1|^p + |\zeta_2|^p)^{1/p}, \quad \zeta = (\zeta_1, \zeta_2) \in \mathbf{C}^2$$

and the Lie norm is equivalent to the supremum norm:

$$L_\infty(\eta) = \sup\{|\eta_1|, |\eta_2|\}, \quad \zeta = (\zeta_1, \zeta_2) \in \mathbf{C}^2.$$

Because $\{L(z,w), M(z,w)\} = \{z_1w_1 + z_2w_2 \pm i(z_1w_2 - z_2w_1)\}$ in $\mathbf{C}^2$, we may take $L(z,w) = z_1w_1 + z_2w_2 + i(z_1w_2 - z_2w_1)$, $M(z,w) = z_1w_1 + z_2w_2 - i(z_1w_2 - z_2w_1)$.

## 3. Double series expansion in $\mathbf{C}^2$

Holomorphic functions on the $N_p$-ball in $\mathbf{C}^2$ can be expanded into the double series by using the homogeneous harmonic extended Tchebycheff polynomials. (For general dimension, see [7], for example.)

For $-1 \leq x \leq 1$, the Tchebycheff polynomial of degree $k$ is defined by

$$T_k(x) = \frac{1}{2}\left((x + i\sqrt{1-x^2})^k + (x - i\sqrt{1-x^2})^k\right).$$

We define the homogeneous extended Tchebycheff polynomial of degree $k$ by

$$\tilde{T}_k(z,w) = (\sqrt{z^2})^k(\sqrt{w^2})^k T_k\left(\frac{z}{\sqrt{z^2}} \cdot \frac{w}{\sqrt{w^2}}\right)$$

$$= \frac{\left(z \cdot w + \sqrt{(z \cdot w)^2 - z^2 w^2}\right)^k + \left(z \cdot w - \sqrt{(z \cdot w)^2 - z^2 w^2}\right)^k}{2}$$

$$= \frac{L^k(z,w) + M^k(z,w)}{2}.$$

For $z, w \in \mathbf{C}^2$, $\tilde{T}_k(z,w)$ is a complex harmonic function in $z$ or in $w$; that is,

$$\Delta_z \tilde{T}_k(z,w) \equiv \left(\frac{\partial^2}{\partial z_1^2} + \frac{\partial^2}{\partial z_2^2}\right)\tilde{T}_k(z,w) = 0. \tag{5}$$

By the well-known formula, the Bergman kernel for the Lie ball (2) is expanded into the double series as follows:

$$\frac{r^8}{(r^4 - 2r^2 z \cdot w + z^2 w^2)^2} = \sum_{k=0}^{\infty}\sum_{l=0}^{[k/2]} \frac{(k-l+1)(l+1)N(k-2l)}{r^{2k}} \left(z^2\right)^l \left(w^2\right)^l \tilde{T}_{k-2l}(z,w), \tag{6}$$

where $N(k)$ is the dimension of the space of homogeneous harmonic polynomials of degree $k$ in $\mathbf{C}^2$; $N(0) = 1$, $N(k) = 2$, $k = 1, 2, \cdots$. For the $N_p$-ball, we could represent the Bergman kernel by the double series expansion as in the right-hand side of (6). Besides the Bergman kernel for $H\mathcal{O}(\tilde{B}_p(r))$, we will also consider the Bergman kernel for the subspace

$$H\mathcal{O}_\Delta(\tilde{B}_p(r)) = \left\{f \in H\mathcal{O}(\tilde{B}_p(r)); \Delta_z f(z) = 0\right\}.$$

We denote by $B_{p,r}^\Delta(z,w)$ the Bergman kernel for $H\mathcal{O}_\Delta(\tilde{B}_p(r))$ and call it harmonic Bergamn kernel.

**Theorem 3.1.** ([3, Theorem 2.3], [2, Theorem 3.2]) In $\mathbf{C}^2$, the Bergman kernel $B_{p,r}(z,w)$ for $H\mathcal{O}(\tilde{B}_p(r))$ and the harmonic Bergman kernel $B_{p,r}^\Delta(z,w)$ for

$H\mathcal{O}_\Delta(\tilde{B}_p(r))$ *are given as follows*:

$$B_{p,r}(z,w) = \sum_{k=0}^{\infty}\sum_{l=0}^{[k/2]} \frac{N(k-2l)\Gamma(\frac{2}{p})^2\Gamma(\frac{2k+4}{p}+1)}{\Gamma(\frac{4}{p}+1)\Gamma(\frac{2k-2l+2}{p})\Gamma(\frac{2l+2}{p})2^{\frac{2k}{p}}r^{2k}}(z^2)^l(\overline{w}^2)^l\tilde{T}_{k-2l}(z,\overline{w})$$

$$= \sum_{k=0}^{\infty}\sum_{l=0}^{\infty} \frac{\Gamma(\frac{2}{p})^2\Gamma(\frac{2k+2l+4}{p}+1)}{\Gamma(\frac{4}{p}+1)\Gamma(\frac{2k+2}{p})\Gamma(\frac{2l+2}{p})2^{\frac{2k+2l}{p}}r^{2k+2l}}(L(z,\overline{w}))^k(M(z,\overline{w}))^l.$$

$$B_{p,r}^\Delta(z,w) = \sum_{k=0}^{\infty} \frac{N(k)\Gamma(\frac{2}{p})\Gamma(\frac{2k+4}{p}+1)}{\Gamma(\frac{4}{p}+1)\Gamma(\frac{2k+2}{p})2^{\frac{2k}{p}}r^{2k}}\tilde{T}_k(z,\overline{w})$$

$$= \sum_{k=0}^{\infty} \frac{\Gamma(\frac{2}{p})\Gamma(\frac{2k+4}{p}+1)}{\Gamma(\frac{4}{p}+1)\Gamma(\frac{2k+2}{p})2^{\frac{2k}{p}}r^{2k}}\left(L^k(z,\overline{w})+M^k(z,\overline{w})\right)-1.$$

## 4. Analytic continuation

For the Bergman kernel $B_{p,r}(z,w)$, $p = 1, 2, \infty$, we checked all the points $z = w$ on the boundary of $\tilde{B}_p(r)$ are singular points. However the harmonic Bergman kernel $B_{p,r}^\Delta(z,w)$ can be continued out side of the domain $\tilde{B}_p(r) \times \tilde{B}_p(r)$ except for $p = \infty$; that is, we have the following propositions:

**Proposition 4.1.** $B_{p,r}^\Delta(z,\overline{w})$ *is a holomoprhic function in*

$$D = \left\{(z,w) \in \mathbf{C}^2 \times \mathbf{C}^2; \; L(z)L(w) < 2^{2/p}r^2\right\}.$$

*Moreover, for* $(z,w) \in D$ *we have*

$$\Delta_z B_{p,r}^\Delta(z,w) = 0, \quad \Delta_w \overline{B_{p,r}^\Delta(z,w)} = 0. \tag{7}$$

**Proof.** Since we know $|\tilde{T}_k(z,w)| \le L(z)L(w)$ and

$$\limsup_{k\to\infty} \left|\frac{\Gamma(\frac{2}{p})\Gamma(\frac{2k+4}{p}+1)}{\Gamma(\frac{4}{p}+1)\Gamma(\frac{2k+2}{p})2^{\frac{2k}{p}}r^{2k}}\right|^{1/k} = \frac{1}{2^{2/p}r^2},$$

$B_{p,r}^\Delta(z,w)$ is holomorphic for $(z,w)$ with $L(z)L(w) < 2^{2/p}r^2$ and we can see (7) by (5). $\qquad\square$

**Proposition 4.2.** *Let* $1 \le p < \infty$ *and* $f \in H\mathcal{O}_\Delta(\tilde{B}_p(r))$. *For* $w \in \tilde{B}(2^{1/p}r)$, *define*

$$F(w) = \int_{\tilde{B}_p(r)} f(z)\overline{B_{p,r}^\Delta(z,w)}dV_{\tilde{B}_p(r)}. \tag{8}$$

*Then* $F \in \mathcal{O}(\tilde{B}(2^{1/p}r))$, $\Delta_w F(w) = 0$ *and* $F(w) = f(w)$ *for* $w \in \tilde{B}_p(r)$.

**Proof.** By Proposition 4.1, $B_{p,r}^\Delta(z,\overline{w}) \in \mathcal{O}\left(\tilde{B}[2^{1/p}r] \times \tilde{B}(2^{1/p}r)\right)$, where we denote $\tilde{B}[r] = \{z \in \mathbf{C}^2; L(z) \le r\}$. Thus the right-hand side of (8) is well-defined for $w \in \mathcal{O}(\tilde{B}(2^{1/p}r))$ and $\Delta F(w) = 0$ by (7). $F(w) = f(w)$ on $\tilde{B}_p(r)$ is clear. $\qquad\square$

## 5. Bergman kernels and harmonic Bergman kernels

We can sum up the infinite sums in Theorem 3.1 in explicit forms for $p = 1, 2, 4, \infty$. The explicit forms are usuful to consider the analytic continuation.

### 5.1. *Bergman kernels in explicit forms*

By (3),

$$
B_{1,r}(z,w) = \frac{\left((1 - \frac{z}{\sqrt{2}r} \cdot \frac{\overline{w}}{\sqrt{2}r})^2 - \frac{z^2}{(\sqrt{2}r)^2} \frac{\overline{w}^2}{(\sqrt{2}r)^2}\right)(1 + \frac{z}{\sqrt{2}r} \cdot \frac{\overline{w}}{\sqrt{2}r}) + \frac{8}{3} \frac{z^2}{(\sqrt{2}r)^2} \frac{\overline{w}^2}{(\sqrt{2}r)^2}}{\left((1 - \frac{z}{\sqrt{2}r} \cdot \frac{\overline{w}}{\sqrt{2}r})^2 - \frac{z^2}{(\sqrt{2}r)^2} \frac{\overline{w}^2}{(\sqrt{2}r)^2}\right)^3}
$$

$$
= \frac{r^{12}\left((1 - \frac{z}{\sqrt{2}r} \cdot \frac{\overline{w}}{\sqrt{2}r})^2 - \frac{z^2}{(\sqrt{2}r)^2} \frac{\overline{w}^2}{(\sqrt{2}r)^2}\right)(1 + \frac{z}{\sqrt{2}r} \cdot \frac{\overline{w}}{\sqrt{2}r}) + \frac{8}{3}r^8 \frac{z^2}{(\sqrt{2}r)^2} \frac{\overline{w}^2}{(\sqrt{2}r)^2}}{(r^2 - N_1^2(z,\overline{w}))^3 (r^2 + N_1^2(iz, i\overline{w}))^3}. \tag{9}
$$

By (1),

$$
B_{2,r}(z,w) = \frac{r^6}{(r^2 - z \cdot \overline{w})^3} = \frac{r^6}{(r^2 - N_2^2(z,\overline{w}))^3}. \tag{10}
$$

In [11], Jong-Do Park represented $B_{4,r}(z,w)$ explicitly as follows: Put

$$
f(a,b) = 3 - 6a + 6b + 3a^2 - 6ab - b^2, \quad g(a,b) = (2 - a - b - a^2 + 2ab - b^2)(1 - a - b),
$$

$$
F(Y_1, Y_2) = \frac{Y_1(\pi + 2\arcsin Y_1)f(Y_1^2, Y_2^2)}{4(1 - Y_1^2)^{3/2}} + \frac{Y_2(\pi + 2\arcsin Y_2)f(Y_2^2, Y_1^2)}{4(1 - Y_2^2)^{3/2}}
$$
$$
+ \frac{g(Y_1^2, Y_2^2)}{2(1 - Y_1^2)(1 - Y_2^2)} + 2\pi Y_1 Y_2, \tag{11}
$$

and $F_0(z,w) = F\left(L(z,\overline{w})/(2^{1/2}r^2), M(z,\overline{w})/(2^{1/2}r^2)\right)$. Then we have

$$
B_{4,r}(z,w) = \frac{r^{12}F_0(z,w)}{(r^4 - 2(z \cdot \overline{w})^2 + z^2\overline{w}^2)^3} = \frac{r^{12}F_0(z,w)}{(r^4 - N_4^4(z,\overline{w}))^3}. \tag{12}
$$

By (2),

$$
B_{\infty,r}(z,w) = \frac{r^8}{(r^4 - 2r^2 z \cdot \overline{w} + z^2\overline{w}^2)^2} = \frac{r^8}{(r^2 - L(z,\overline{w}))^2(r^2 - M(z,\overline{w}))^2}. \tag{13}
$$

Looking at the denominator in the above Bergman kernels (9), (10), (12) and (13), we see that all the points $z = w$ on the boundary of $\tilde{B}_p(r)$, $p = 1, 2, 4, \infty$, are singular points.

### 5.2. *Harmonic Bergman kernels in explicit forms*

Next we review the harmonic Bergman kernels in explicit forms and review the analytic continuation of harmonic fucntions through their explicit forms.

For $p = 1$,

$$
B_{1,r,\Delta}(z,w) = \frac{P(\frac{z}{2r} \cdot \frac{\overline{w}}{2r}, \frac{z^2}{2r^2} \frac{\overline{w}^2}{2r^2})}{\left(1 - 2\frac{z}{2r} \cdot \frac{\overline{w}}{2r} + \frac{z^2\overline{w}^2}{(2r)^4}\right)^4}, \tag{14}
$$

where $P(\frac{z}{2r} \cdot \frac{\overline{w}}{2r}, \frac{z^2}{2r^2}\frac{\overline{w}^2}{2r^2})$ is the polynomial in $s = \frac{z}{2r} \cdot \frac{\overline{w}}{2r} \in \mathbf{C}$ and $t = \frac{z^2}{4r^2}\frac{\overline{w}^2}{4r^2} \in \mathbf{C}$ determined by

$$P(s,t) = 1 + 2s - 24t + 60st + 4t^2$$
$$+ 18st^2 - 80s^2t - 4t^3 + 8st^3 - 24s^2t^2 + 40s^3t - t^4. \tag{15}$$

(The coefficient of $st^3$ was written 48 in $^2$, 8 is correct.)

The denominator of (14) is factorized by

$$1 - 2\frac{z}{2r} \cdot \frac{\overline{w}}{2r} + \frac{z^2\overline{w}^2}{(2r)^4} = \left(1 - \frac{L(z,\overline{w})}{(2r)^2}\right)\left(1 - \frac{M(z,\overline{w})}{(2r)^2}\right).$$

This representation indicates us that Propositions 4.1 and 4.2 hold for $p = 1$.

For $p = 2$,

$$B_{2,r,\Delta}(z,w) = \frac{r^{12}\left(2 - 3\frac{z\cdot\overline{w}}{r^2} + 3(\frac{z}{r} \cdot \frac{\overline{w}}{r})^2 - \frac{3}{2}\frac{z^2\overline{w}^2}{r^4} - (\frac{z}{r} \cdot \frac{\overline{w}}{r})^3 + \frac{3z\cdot\overline{w}z^2\overline{w}^2}{4r^6}\right)}{(r^4 - r^2 z \cdot \overline{w} + \frac{z^2\overline{w}^2}{4})^3} - 1.$$

Note that

$$r^4 - r^2 z \cdot \overline{w} + z^2\overline{w}^2/4 = \left(r^2 - L(z,\overline{w})/2\right)\left(r^2 - M(z,\overline{w})/2\right).$$

This representation indicates us that Propositions 4.1 and 4.2 hold for $p = 2$.

For $p = 4$, put

$$G(X) = \sum_{k=0}^{\infty} \frac{\sqrt{\pi}\Gamma(\frac{k}{2} + 2)}{\Gamma(\frac{k+1}{2})} X^k = \frac{3X(\pi + 2\arcsin X)}{4(1 - X^2)^{5/2}} + \frac{X^2 + 2}{2(1 - X^2)^2}. \tag{16}$$

$$B_{4,r,\Delta}(z,w) = G\left(\frac{L(z,\overline{w})}{(2^{1/4}r)^2}\right) + G\left(\frac{M(z,\overline{w})}{(2^{1/4}r)^2}\right) - 1.$$

This representation indicates us that Propositions 4.1 and 4.2 hold for $p = 4$.

For $p = \infty$, we have

$$B_{r,\Delta}(z,w) = \frac{r^8 - z^2\overline{w}^2(4r^4 - 4r^2 z \cdot \overline{w} + z^2\overline{w}^2)}{(r^4 - 2r^2 z \cdot \overline{w} + z^2\overline{w}^2)^2}$$
$$= \frac{r^8 - z^2\overline{w}^2(4r^4 - 4r^2 z \cdot \overline{w} + z^2\overline{w}^2)}{(r^2 - L(z,\overline{w}))^2(r^2 - M(z,\overline{w}))^2}.$$

This representation implies that all the points $z = w$ on the boundary of $\tilde{B}(r)$ are singular points of $B_{r,\Delta}(z,w)$.

## 6. Bergman kernels for 2-dimensional $L_p$-balls

For $\zeta = (\zeta_1, \zeta_2)$, $\eta = (\eta_1, \eta_2) \in \mathbf{C}^2$, consider the function

$$L_p(\zeta,\eta) = \left((\zeta_1 \cdot \eta_1)^{p/2} + (\zeta_2 \cdot \eta_2)^{p/2}\right)^{1/p}.$$

Then $L_p(\zeta) = L_p(\zeta, \overline{\zeta})$. For $p \geq 1$, we denote the 2-dimensional $L_p$-ball by $\tilde{B}_{L_p}(r)$:

$$\tilde{B}_{L_p}(r) = \{\zeta \in \mathbf{C}^2; L_p(\zeta) < r\}, \quad 1 \leq p \leq \infty.$$

Put

$$z_1 + iz_2 = \sqrt{2}\zeta_1, \quad z_1 - iz_2 = \sqrt{2}\zeta_2; \quad w_1 + iw_2 = \sqrt{2}\eta_1, \quad w_1 - iw_2 = \sqrt{2}\eta_2.$$

By using the matrix

$$A = \begin{pmatrix} 1/\sqrt{2} & i/\sqrt{2} \\ 1/\sqrt{2} & -i/\sqrt{2} \end{pmatrix},$$

we can write

$$A \begin{pmatrix} z_1 \\ z_2 \end{pmatrix} = \begin{pmatrix} \zeta_1 \\ \zeta_2 \end{pmatrix}.$$

Under these correspondence we have

$$z \cdot w = \zeta_1 \eta_2 + \zeta_2 \eta_1, \quad z \cdot \overline{w} = \zeta \cdot \overline{\eta}, \quad z^2 = 2\zeta_1\zeta_2, \quad \|z\|^2 = \|\zeta\|^2.$$

$$N_p(z) = 2^{1/2-1/p} L_p(\zeta), \quad L(z) = \sqrt{2} L_\infty(\zeta), \quad \tilde{T}_k(z,w) = \frac{(2\zeta_1\eta_1)^k + (2\zeta_2\eta_2)^k}{2}.$$

Note that $L_\infty(z) \le L_p(z) \le 2^{1/p} L_\infty(z)$ and

$$\tilde{B}_{L_\infty}(2^{1/p}r) \subset \tilde{B}_{L_p}(r) \subset \tilde{B}_{L_\infty}(r).$$

Considering the fact that $\Delta_z = \frac{\partial^2}{\partial z_1^2} + \frac{\partial^2}{\partial z_2^2} = 2\frac{\partial^2}{\partial\zeta_1\partial\zeta_2}$, we set

$$HO_{\Delta'}(\tilde{B}_{L_p}(r)) = \left\{ f \in HO(\tilde{B}_{L_p}(r)); \frac{\partial^2}{\partial\zeta_1\partial\zeta_2} f(\zeta) = 0 \right\}.$$

Since $\{z \in \mathbf{C}^2; N_p(z) < 2^{1/2-1/p}r\} = \{\zeta \in \mathbf{C}^2; L_p(\zeta) < r\}$ and $|\det A| = 1$, as a corollary to Theorem 3.1, we have

**Corollary 6.1.** *In* $\mathbf{C}^2$, *the Bergman kernel* $B_{Lp,r}(\zeta,\eta)$ *for* $HO(\tilde{B}_{L_p}(r))$ *and the Bergman kernel* $B_{Lp,r}^{\Delta'}(\zeta,\eta)$ *for* $HO_{\Delta'}(\tilde{B}_{L_p}(r))$ *are given as follows:*

$$B_{Lp,r}(\zeta,\eta) = \sum_{k=0}^{\infty} \sum_{l=0}^{[k/2]} \frac{N(k-2l)\Gamma(\frac{2}{p})^2\Gamma(\frac{2k+4}{p}+1)(\zeta_1\zeta_2\overline{\eta}_1\overline{\eta}_2)^{k/2}\tilde{T}_{k-2l}\left(\frac{\zeta\cdot\overline{\eta}}{2\sqrt{\zeta_1\zeta_2\overline{\eta}_1\overline{\eta}_2}}\right)}{\Gamma(\frac{4}{p}+1)\Gamma(\frac{2k-2l+2}{p})\Gamma(\frac{2l+2}{p})r^{2k}}$$

$$= \sum_{k=0}^{\infty} \sum_{l=0}^{\infty} \frac{\Gamma(\frac{2}{p})^2\Gamma(\frac{2k+2l+4}{p}+1)}{\Gamma(\frac{4}{p}+1)\Gamma(\frac{2k+2}{p})\Gamma(\frac{2l+2}{p})r^{2k+2l}}(\zeta_1\overline{\eta}_1)^k(\zeta_2\overline{\eta}_2)^l.$$

$$B_{Lp,r}^{\Delta'}(\zeta,\eta) = \sum_{k=0}^{\infty} \frac{N(k)\Gamma(\frac{2}{p})\Gamma(\frac{2k+4}{p}+1)}{\Gamma(\frac{4}{p}+1)\Gamma(\frac{2k+2}{p})r^{2k}}(\zeta_1\zeta_2\overline{\eta}_1\overline{\eta}_2)^{k/2}\tilde{T}_k\left(\frac{\zeta\cdot\overline{\eta}}{2\sqrt{\zeta_1\zeta_2\overline{\eta}_1\overline{\eta}_2}}\right)$$

$$= \sum_{k=0}^{\infty} \frac{\Gamma(\frac{2}{p})\Gamma(\frac{2k+4}{p}+1)}{\Gamma(\frac{4}{p}+1)\Gamma(\frac{2k+2}{p})r^{2k}}\left((\zeta_1\overline{\eta}_1)^k + (\zeta_2\overline{\eta}_2)^k\right) - 1.$$

The last equality in Corollary 6.1 implies

**Corollary 6.2.** *For* $\eta_0$ *with* $L_p(\eta_0) = r$, $B_{Lp,r}^{\Delta'}(\zeta,\eta_0) \in \mathcal{O}(\tilde{B}_{L_\infty}(r))$.

Moreover we have

**Corollary 6.3.** *For $f \in HO_{\Delta'}(\tilde{B}_{L_p}(r))$ and $\eta \in \tilde{B}_{L_p}(r)$, define*

$$F(\eta) = \int_{\tilde{B}_{L_p}(r)} f(\zeta)\overline{B_{L_p,r}^{\Delta'}(\zeta,\eta)}dV_{\tilde{B}_{L_p}(r)}.$$

*Then $F \in \mathcal{O}(\tilde{B}_{L_\infty}(r))$ and $F$ is an analytic continuation of $f$.*

For $p = 1, 2, 4, \infty$, we have the following explicit Bergman kernels;

$$B_{L1,r}(\zeta,\eta) = \frac{3r^6\left((r^2 - \zeta \cdot \overline{\eta})^2 - 4\zeta_1\zeta_2\overline{\eta}_1\overline{\eta}_2\right)(r^2 + \zeta \cdot \overline{\eta}) + 32r^8\zeta_1\zeta_2\overline{\eta}_1\overline{\eta}_2}{3\left((r^2 - \zeta \cdot \overline{\eta})^2 - 4\zeta_1\zeta_2\overline{\eta}_1\overline{\eta}_2\right)^3}.$$

$$= \frac{3r^6\left((r^2 - \zeta \cdot \overline{\eta})^2 - 4\zeta_1\zeta_2\overline{\eta}_1\overline{\eta}_2\right)(r^2 + \zeta \cdot \overline{\eta}) + 32r^8\zeta_1\zeta_2\overline{\eta}_1\overline{\eta}_2}{3(r^2 - L_1^2(\zeta,\overline{\eta}))^3(r^2 + L_1^2(i\zeta,i\overline{\eta}))}.$$

$$B_{L2,r}(\zeta,\eta) = \frac{r^6}{(r^2 - \zeta \cdot \overline{\eta})^3} = \frac{r^6}{(r^2 - L_2^2(\zeta,\overline{\eta}))^3}.$$

$$B_{L4,r}(\zeta,\eta) = \frac{r^{12}F\left(2\zeta_1\eta_1/r^2, 2\zeta_2\eta_2/r^2\right)}{(r^4 - (\zeta_1\overline{\eta}_1)^2 - (\zeta_2\overline{\eta}_2)^2)^3} = \frac{r^{12}F\left(2\zeta_1\eta_1/r^2, 2\zeta_2\eta_2/r^2\right)}{(r^4 - L_4^4(\zeta,\overline{\eta}))^3},$$

where $F$ is defined by (11)

$$B_{\infty,r}(\zeta,\eta) = \frac{r^8}{(r^2 - \zeta_1\overline{\eta}_1)^2(r^2 - \zeta_2\overline{\eta}_2)^2}.$$

$$B_{1,r,\Delta'}(\zeta,\eta) = \frac{r^{16}P\left(\frac{\zeta}{\sqrt{2}r} \cdot \frac{\overline{\eta}}{\sqrt{2}r}, \frac{\zeta_1\zeta_2}{2r^2}\frac{\overline{\eta}_1\overline{\eta}_2}{2r^2}\right)}{(r^2 - \zeta_1\overline{\eta}_1)^4(r^2 - \zeta_2\overline{\eta}_2)^4}, \quad \text{where } P(s,t) \text{ is defined by (15).}$$

$$B_{2,r,\Delta'}(\zeta,\eta) = \frac{r^{12}(2 - 3\frac{\zeta \cdot \overline{\eta}}{r^2} + 3(\frac{\zeta}{r} \cdot \frac{\overline{\eta}}{r})^2 - \frac{3}{2}\frac{\zeta_1\zeta_2\overline{\eta}_1\overline{\eta}_2}{r^4} - (\frac{\zeta}{r} \cdot \frac{\overline{\eta}}{r})^3 + \frac{3\zeta \cdot \overline{\eta}\zeta_1\zeta_2\overline{\eta}_1\overline{\eta}_2}{4r^6})}{(r^2 - \zeta_1\overline{\eta}_1)^3(r^2 - \zeta_2\overline{\eta}_2)^3} - 1.$$

$$B_{4,r,\Delta'}(\zeta,\eta) = G\left(\frac{\zeta_1\overline{\eta}_1}{r^2}\right) + G\left(\frac{\zeta_2\overline{\eta}_2}{r^2}\right) - 1, \quad \text{where } G(X) \text{ is defined by (16).}$$

$$B_{r,\Delta'}(\zeta,\eta) = \frac{r^8 - \zeta_1\zeta_2\overline{\eta}_1\overline{\eta}_2(4r^4 - 2r^2\zeta \cdot \overline{\eta} + \zeta_1\zeta_2\overline{\eta}_1\overline{\eta}_2)}{(r^2 - \zeta_1\overline{\eta}_1)^2(r^2 - \zeta_2\overline{\eta}_2)^2}$$

Note that the denominators of the last five equalities imply that all the points $\zeta = \eta$ on the boundary of $\tilde{B}_{L_\infty}(r)$ are singular points.

## References

1. K.Fujita, Bergman transformation for analytic functionals on some balls, Microlocal Analysis and Complex Fourier Analysis, World Scientific publisher, 2002, 81-98.
2. K.Fujita, Harmonic Bergman kernel for some balls, Universitatis Iagellonicae Acta Mathematica, **41**(2003), 225–234.

134

3. K.Fujita, Bergman kernel for the two-dimensional balls, Complex Variables Theory and Application, Vol. 49, No. 3 (2004), 215–225.
4. K. Fujita, Bergman kernel for complex harmonic functions on some balls, Advances in analysis, Proceedings of the 4th International ISAAC Congress, World Scientific publishers, 2005, 429–437.
5. K.Fujita, Some remark on the Bergman kernel for the dual Lie ball, Proceedings of The 12th International Conference on Finite or Infinite Dimensional Complex Analysis and Applications, Kyushu University Press, 2005, 59–66.
6. K.Fujita and M.Morimoto, Holomorphic functions on the dual Lie ball and related topics, Proceedings of Eighth International Colloquium on Finite or Infinite Dimensional Complex Analysis, Shandong Science and Technology Press, 2000, 33-37.
7. K.Fujita and M.Morimoto, On the double series expansion of holomorphic functions, J. Math. Anal. Appl. 272(2002), 335-348.
8. L.K.Hua, Harmonic Analysis of Functions of Several Complex Variables in Classical Domain, Moskow 1959, (in Russian); Translations of Math. Monographs vol. 6, Amer. Math. Soc., Providence, Rhode Island, 1979.
9. M.Morimoto and K.Fujita, Between Lie norm and dual Lie norm, Tokyo J. Math., 24(2001), 499–507.
10. K.Oeljklaus, P.Pflug and E.H.Youssfi, The Bergman kernels of the minimal ball and applications, Ann.Inst. Fourier, 47, 3 (1997), 915–928.
11. Jong-Do Park, Explicit computations of the Bergman kernel and Toeplitz products on the Bergman spaces, doctoral thesis of Seoul National University, 2004.

# APPLICATIONS OF REPRODUCING KERNELS TO LINEAR SINGULAR INTEGRAL EQUATIONS THROUGH THE TIKHONOV REGULARIZATION

H. ITOU*

*Department of Mathematics,*
*Faculty of Engineering,*
*Gunma University,*
*Kiryu 376-8515, Japan,*
*E-mail:h-itou@math.sci.gunma-u.ac.jp*

S. SAITOH†

*Department of Mathematics,*
*Faculty of Engineering,*
*Gunma University,*
*Kiryu 376-8515, Japan,*
*E-mail:ssaitoh@math.sci.gunma-u.ac.jp*

By a new concept and method we shall give practical and numerical solutions of linear singular integral equations by combining the two theories of the Tikhonov regularization and reproducing kernels.

**Key words:** Singular integral equation, reproducing kernel, Tikhonov regularization, sampling theory, Carleman's equation
**Mathematics Subject Classification:** 45E01, 30C40

## 1. Introduction

Singular integral equations are presently encountered in a wide range of mathematical models, for instance in acoustics, fluid dynamics, elasticity and fracture mechanics. See, for example, Refs. [4], [6,7]. As a typical singular integral equation of them, we shall consider the Carleman's equation over a real interval, for any $L_2(-1,1)(:= L_2)$ function $g$ and for complex valued $L_2$ (or bounded integrable) functions $a, b$

---
*Work was supported by the Gunma University Foundation for the Promotion of Science and Engineering.
†Work was partially supported by the Grant-in-Aid for the Scientific Research (C)(2) (No. 16540137) from the Japan Society for the Promotion Science and by the Mitsubishi Foundation, the 36th, Natural Sciences, No. 20 (2005-2006).

$$(Ly)(t) = a(t)y(t) + \frac{b(t)}{\pi i}\text{p.v.}\int_{-1}^{+1}\frac{y(\zeta)}{\zeta - t}\,\mathrm{d}\zeta = g(t) \quad \text{on} \quad -1 < t < 1. \qquad (1.1)$$

According to Ref. [7], the operator $L$, satisfying a condition $a^2(t) - b^2(t) \neq 0$ for $-1 < t < 1$, is called a regular type operator. It is well known that the equation (1.1) always has an explicit solution for a regular type one. See also Ref. [6]. However, when $a^2(t) - b^2(t) = 0$, there exist solutions if and only if $g$ satisfies a special condition (see, Section 5, and also, cf. Ref. [4]). The analysis of this case is important for the kinked crack problem. Accordingly, we shall introduce a new method which gives simple and natural approximate solutions for linear singular integral equations including the case where the condition of a regular type operator is violated. We can deal with a general linear singular integral equation, however, for simplicity, we shall state the results for this most typical case.

Indeed, here we shall introduce a new approach for some general linear singular integral equations with bounded linear integral operators by transforming the integral equations to integral equations of Fredholm of the second type with sufficiently smooth coefficients and by using the two theories of the Tikhonov regularization and reproducing kernels. See Section 4.

## 2. Paley-Wiener space and reproducing kernels

We shall consider the integral transform, for $L_2(-\pi/h, +\pi/h), (h > 0)$ functions $g$

$$f(z) = \frac{1}{2\pi}\int_{-\pi/h}^{\pi/h} g(t)e^{-izt}\,\mathrm{d}t. \qquad (2.2)$$

In order to identify the image space following the theory of reproducing kernels (Ref. [8]), we form the reproducing kernel

$$K_h(z, \overline{u}) = \frac{1}{2\pi}\int_{-\pi/h}^{\pi/h} e^{-izt}\overline{e^{-iut}}\,\mathrm{d}t$$

$$= \frac{1}{\pi(z - \overline{u})}\sin\frac{\pi}{h}(z - \overline{u}). \qquad (2.3)$$

The image space of (2.2) is called the Paley-Wiener space $W\left(\frac{\pi}{h}\right)(:= W_h)$ comprised of all analytic functions of exponential type satisfying, for some constant $C$ and as $z \to \infty$

$$|f(z)| \leq C\exp\left(\frac{\pi|z|}{h}\right)$$

and

$$\int_{\mathbf{R}} |f(x)|^2\,\mathrm{d}x < \infty.$$

From the identity

$$K_h(jh, j'h) = \frac{1}{h}\delta(j, j')$$

(the Kronecker's $\delta$), since $\delta(j, j')$ is the reproducing kernel for the Hilbert space $\ell^2$, from the general theory of integral transforms and the Parseval's identity we have the isometric identities in (2.2)

$$\frac{1}{2\pi}\int_{-\pi/h}^{\pi/h} |g(t)|^2 \, dt = h\sum_j |f(jh)|^2$$

$$= \int_{\mathbf{R}} |f(x)|^2 \, dx.$$

That is, the reproducing kernel Hilbert space $H_{K_h}$ with $K_h(z, \overline{u})$ is characterized as a space comprising the Paley-Wiener space $W_h$ with the norm squares defined above. Here we used the well-known result that $\{jh\}_j$ is a uniqueness set for the Paley-Wiener space $W_h$; that is, $f(jh) = 0$ for all $j$ implies $f \equiv 0$. Then, the reproducing property of $K_h(z, \overline{u})$ states that

$$f(x) = (f(\cdot), K_h(\cdot, x))_{H_{K_h}} = h\sum_j f(jh)K_h(jh, x)$$

$$= \int_{\mathbf{R}} f(\xi)K_h(\xi, x) \, d\xi,$$

in particular, for $x \in \mathbf{R}$. This representation is the sampling theorem which represents the whole data $f(x)$ in terms of the discrete data $\{f(jh)\}_j$. Furthermore, for a general theory for the sampling theory and error estimates for some finite points $\{hj\}_j$, see Ref. [8].

## 3. Reproducing kernels and the Tikhonov regularization

The good application of reproducing kernels to the Tikhonov regularization is given by the following propositions:

**Proposition 3.1.** [2,9] *Let $H_K$ be a Hilbert space admitting the reproducing kernel $K(p, q)$ on a set $E$. Let $L : H_K \to \mathcal{H}$ be a bounded linear operator on $H_K$ into a Hilbert space $\mathcal{H}$. For $\lambda > 0$ introduce the inner product in $H_K$ and call it $H_{K_\lambda}$ as*

$$(f_1, f_2)_{H_{K_\lambda}} = \lambda(f_1, f_2)_{H_K} + (Lf_1, Lf_2)_{\mathcal{H}}, \tag{3.4}$$

*then $H_{K_\lambda}$ is the Hilbert space with the reproducing kernel $K_\lambda(p, q)$ on $E$ and satisfying the equation*

$$K(\cdot, q) = (\lambda I + L^*L)K_\lambda(\cdot, q), \tag{3.5}$$

*where $L^*$ is the adjoint of $L : H_K \to \mathcal{H}$.*

**Proposition 3.2.** [9] *Let* $H_K$, $L$, $\mathcal{H}$, $E$ *and* $K_\lambda$ *be as in* PROPOSITION 3.1. *Then, for any* $\lambda > 0$ *and for any* $g \in \mathcal{H}$, *the extremal function in*

$$\inf_{f \in H_K} \left( \lambda \|f\|_{H_K}^2 + \|Lf - g\|_{\mathcal{H}}^2 \right) \tag{3.6}$$

*exists uniquely and the extremal function is represented by*

$$f_{\lambda,g}^*(p) = (g, LK_\lambda(.,p))_{\mathcal{H}} \tag{3.7}$$

*which is the member of* $H_K$ *attaining the infimum in* (3.6).

In (3.7), when $g$ contains errors or noises, we need its error estimation. For this, we can obtain the general result:

**Theorem 3.1.** *In* (3.7), *we obtain the estimate*

$$|f_{\lambda,g}^*(p)| \leq \frac{1}{\sqrt{\lambda}} \sqrt{K(p,p)} \|g\|_{\mathcal{H}}.$$

For the properties and error estimates for the limit

$$\lim_{\lambda \to 0} f_{\lambda,g}^*(p),$$

see Refs. [11,12]. In particular, when there exists the Moore-Penrose generalized solution for the operator equation

$$Lf = g,$$

the limit converges uniformly to the Moore-Penrose generalized solution on any subset of $E$ such that $K(p,p)$ is bounded.

For many concrete applications of these general theorems, see, for example, Refs. [1,5], [9,10].

## 4. Construction of approximate solutions by solving Fredholm's integral equation

Following the idea and method in Section 3, we shall consider the extremal problem:

$$\inf_{f \in W_h} \left\{ \lambda \|f\|_{W_h}^2 + \|Lf - g\|_{L_2}^2 \right\}. \tag{4.8}$$

Note that $L$ is a bounded linear operator from $W_h$ into $L_2$, as we see from the Cauchy-Schwarz inequality and a boundedness

$$\int_{-1}^{1} \left| \frac{1}{\pi} \text{p.v.} \int_{-1}^{1} \frac{F(\xi)}{\xi - \eta} \, d\xi \right|^2 d\eta \leq \int_{-1}^{1} |F(\xi)|^2 \, dx$$

of the finite Hilbert transform (cf. Ref. [6,13]). We wish to construct the reproducing kernel for the Hilbert space with the norm square

$$\lambda \|f\|_{W_h}^2 + \|Lf\|_{L_2}^2. \tag{4.9}$$

From Proposition 3.1 this reproducing kernel $K_\lambda(t, t')$ is calculated by solving an integral equation of Fredholm of the second kind:

$$\frac{1}{\lambda} K_h(t, t') = K_\lambda(t, t') + \frac{1}{\lambda}(LK_\lambda(\cdot, t')(p), (LK_h(\cdot, t))(p))_{L_2}. \tag{4.10}$$

Then, the extremal function in $f_\lambda^*$ in (4.8) is given by

$$f_\lambda^*(t) = (g, LK_\lambda(\cdot, t))_{L_2}. \tag{4.11}$$

By applying the operator $L$ to (4.10) with respect to functions of $t$, we have

$$\frac{1}{\lambda} L_t K_h(t, t') = L_t K_\lambda(t, t') + \frac{1}{\lambda} L_t(LK_\lambda(\cdot, t')(p), (LK_h(\cdot, t))(p))_{L_2}. \tag{4.12}$$

Therefore,

$$LK_\lambda(\cdot, t)$$

is given as the solution of the integral equation of Fredholm of the second kind for fixed $t$. Note that the functions in (4.12)

$$L_t K_h(t, t')$$

and

$$L_t \overline{LK_h(\cdot, t)}$$

are calculated by using Fourier's integral and the formula: For the Fourier transform $\mathcal{F}$,

$$\mathcal{F}[f](\xi) = \int_{-\infty}^{\infty} f(t)e^{-i\xi t}\, dt, \tag{4.13}$$

and for the Hilbert transform $\mathcal{H}$,

$$[\mathcal{H}y](t) = \frac{1}{\pi}\, \text{p.v.} \int_{-\infty}^{\infty} \frac{y(\zeta)}{\zeta - t}\, d\zeta,$$

$$\mathcal{H}F = \mathcal{F}^{-1}\left(-i\text{sgn}\xi(\mathcal{F}F)(\xi)\right). \tag{4.14}$$

## 5. Carleman's equation for the case of the whole line and with complex constant coefficients

As a typical example of applying our method, we shall consider approximate solutions for Carleman's equation for the case of the whole line and with complex constant coefficients $a$, $b$:

$$(\tilde{L}y)(t) = ay(t) + \frac{b}{\pi i} \text{ p.v.} \int_{-\infty}^{\infty} \frac{y(\zeta)}{\zeta - t} \, d\zeta = g(t) \quad \text{on} \quad -\infty < t < \infty. \quad (5.15)$$

In the same way as (4.12), we obtain

$$\frac{1}{\lambda} \tilde{L}_t K_h(t, t') = \tilde{L}_t K_\lambda(t, t') + \frac{1}{\lambda} \tilde{L}_t (\tilde{L} K_\lambda(\cdot, t'))(p), (\tilde{L} K_h(\cdot, t))(p))_{L_2}. \quad (5.16)$$

By using the Fourier transforms we can find the solution of (5.16)

$$\tilde{L}_t K_\lambda(t, t') = \frac{a - b}{\lambda + |a - b|^2} \frac{1}{2\pi i(t - t')} \cdot \left(1 - e^{-\frac{i\pi}{h}(t - t')}\right)$$

$$+ \frac{a + b}{\lambda + |a + b|^2} \frac{1}{2\pi i(t - t')} \cdot \left(e^{\frac{i\pi}{h}(t - t')} - 1\right). \quad (5.17)$$

**Theorem 5.1.** *In (5.15), for any function $g \in L_2$, the best approximate solution $f_{\lambda,h,g}^*$ is represented by*

$$f_{\lambda,h,g}^*(t) = \frac{1}{2\pi} \int_{-\infty}^{\infty} g(\xi) \left[ \frac{\overline{(a + b)}}{\lambda + |a + b|^2} \int_{-\pi/h}^{0} e^{i\eta(\xi - t)} \, d\eta \right.$$

$$\left. + \frac{\overline{(a - b)}}{\lambda + |a - b|^2} \int_{0}^{\pi/h} e^{i\eta(\xi - t)} \, d\eta \right] d\xi.$$

*For a function $f \in W_h$ if we take $g$ as $\tilde{L}f = g$, then we obtain the result*

$$\lim_{\lambda \to 0} f_{\lambda,h,g}^*(t) = f(t),$$

*uniformly.*

Note that the regular type case, $a^2 - b^2 \neq 0$, we can take $\lambda = 0$; that is, we do not need the Tikhonov regularization in our problem. It is a trivial case.

In Theorem 5.1, we see directly

$$\tilde{L}_t f_{\lambda,h,g}^*(t) = \frac{1}{2\pi} \int_{-\infty}^{\infty} g(\xi) \left[ \frac{|a + b|^2}{\lambda + |a + b|^2} \int_{-\pi/h}^{0} e^{i\eta(\xi - t)} \, d\eta \right.$$

$$\left. + \frac{|a - b|^2}{\lambda + |a - b|^2} \int_{0}^{\pi/h} e^{i\eta(\xi - t)} \, d\eta \right] d\xi.$$

Therefore, for the cases $\lambda = 0$ and $a^2 - b^2 \neq 0$

$$\lim_{h \to 0} \tilde{L}_t f_{0,h,g}^*(t) = g(t)$$

at the points $t$ where $g$ is continuous.

In particular, when $a = 0$ and $b = i$; that is, for the Hilbert transform case, we obtain for $\lambda = 0$

**Corollary 5.1.** *For the extremal problem*

$$\inf_{f \in W_h} \left\{ \|\mathcal{H}f - g\|_{L_2}^2 \right\}, \tag{5.18}$$

*the extremal function $f^*_{H,h,g}$ attaining the infimum exists uniquely and it is given by*

$$f^*_{H,h,g}(t) = \frac{1}{\pi} \int_{-\infty}^{\infty} g(\xi) \frac{1}{\xi - t} \left[ \cos \frac{\pi}{h}(\xi - t) - 1 \right] d\xi$$

*and then,*

$$\left[ \mathcal{H}f^*_{H,h,g} \right](t) = (g, K_h(\cdot, t))_{L_2};$$

*that is, $\left[ \mathcal{H}f^*_{H,h,g} \right]$ is the orthogonal projection of $g$ onto the Paley-Wiener space $W_h$.*

Corollary 5.1 also means to give an approximate Hilbert transform for any $L_2$ function $g$ by an ordinary integral; that is,

$$\lim_{h \to 0} \frac{1}{\pi} \int_{-\infty}^{\infty} g(\xi) \frac{1}{\xi - t} \left[ 1 - \cos \frac{\pi}{h}(\xi - t) \right] d\xi = [\mathcal{H}g](t)$$

at the points $t$ where $g$ is continuous.

In particular, note that for the singular cases, $a^2 - b^2 = 0$, the integral equations have the solutions only for very special functions $g$.

For example, if $a = \pm b$, since $\mathcal{H}(\mathcal{H}y) = -y$, from (5.15),

$$a\mathcal{H}y \mp \frac{a}{i} y = \mathcal{H}g$$

and so, we see that $g$ must satisfy the relation

$$\mathcal{H}g = \pm ig,$$

(see, for the details, Ref. [4], pp. 270). The following corollary gives the solutions for the singular cases for general $L_2$ functions $g$, and coincides with the results for the very special functions $g$ in Ref. [4].

**Corollary 5.2.** *If $a = b = 1$, then we obtain*

$$\lim_{\lambda \to 0} f^*_{\lambda,h,g}(t) := f^*_{+0,h,g}(t)$$

$$= \frac{1}{4} \left[ i f^*_{H,h,g}(t) + (g, K_h(\cdot, t))_{L_2} \right].$$

*Furthermore,*

$$\lim_{h \to 0} \tilde{L}_t f^*_{+0,h,g}(t) = \frac{i}{2} f^*_{H,h,g}(t) + \frac{1}{2}(g, K_h(\cdot, t))_{L_2}.$$

*If the condition $\mathcal{H}g = ig$ for existence of the solutions is satisfied, then*

$$\lim_{h \to 0} f^*_{+0,h,g}(t) = \frac{1}{2} g(t)$$

*and*

$$\lim_{h \to 0} \tilde{L}_t f^*_{+0,h,g}(t) = g(t)$$

*at the points $t$ where $g(t)$ is continuous.*

If $a = 1, b = -1$, then we can obtain the corresponding results.

Therefore, in Theorem 5.1, we obtain the explicit representations of the approximate solutions including the singular cases. Surprisingly enough, we can obtain the explicit representations of the "solutions" for any $L_2$ function $g$.

## References

1. M. Asaduzzaman, T. Matsuura and S. Saitoh, *Constructions of approximate solutions for linear differential equations by reproducing kernels and inverse problems*, Advances in Analysis, Proceedings of the 4th International ISAAC Congress, World Scientific (2005), **30**, 335–344.
2. D-W, Byun and S. Saitoh, *Best approximation in reproducing kernel Hilbert spaces*, Proc. of the 2nd International Colloquium on Numerical Analysis, VSP-Holland, (1994), 55–61.
3. H. W. Engl, M. Hanke and A. Neubauer, *Regularization of Inverse Problems*, Mathematics and Its Applications **376**, Kluwer Academic Publishers, 2000.
4. R. Estrada and R. P. Kanwal, *Singular Integral Equations*, Birkhäuser, Boston, 2000.
5. T. Matsuura and S. Saitoh, *Analytical and numerical solutions of linear ordinary differential equations with constant coefficients*, Journal of Analysis and Applications, **3** (2005), 1-17.
6. S. G. Mikhlin and S. Prössdorf, *Singular Integral Operators*, Springer-Verlag, Berlin, 1986.
7. N. I. Muskhelishvili, *Singular Integral Equations*, Noordhoff, Groningen, 1972.
8. S. Saitoh, *Integral Transforms, Reproducing Kernels and their Applications*, Pitman Res. Notes in Math. Series **369**, Addison Wesley Longman Ltd, UK, 1997.
9. S. Saitoh, *Approximate Real Inversion Formulas of the Gaussian Convolution*, Applicable Analysis, **83** (2004), 727-733.
10. S. Saitoh, *Applications of Reproducing Kernels to Best Approximations, Tikhonov Regularizations and Inverse Problems*, Advances in Analysis, Proceedings of the 4th International ISAAC Congress, World Scientific (2005), **39**, 439–446.
11. S. Saitoh, *Best approximation, Tikhonov regularization and reproducing kernels*, Kodai. Math. J. **28**(2005), 359–367.
12. S. Saitoh, *Tikhonov regularization and the theory of reproducing kernels*, Proceedings of the 12th ICFIDCAA (to appear).
13. F. Stenger, *Numerical Methods Based on Sinc and Analytic Functions*, Springer Series in Computational Mathematics **20**, Springer-Verlag, New York, 1993.

# REPRODUCING KERNEL HILBERT SPACES AND RANDOM MEASURES

CHARLES SUQUET

*Laboratoire P. Painlevé, UMR CNRS 8524,*
*Bât M2, Cité Scientifique, Université Lille I*
*F59655 Villeneuve d'Ascq Cedex, France*

We show how to use Guilbart's embedding of signed measures into a R.K.H.S. to study some limit theorems for random measures and stochastic processes.

**Key words:** Epidemic change, functional central limit theorem, Hölderian invariance principle, mean measure, random measure, reproducing kernel

**Mathematics Subject Classification:** 60G57, 60F17, 62G10

## 1. R.K.H.S. and metrics on signed measures

In the late seventies, C. Guilbart [4, 5] introduced an embedding into a reproducing kernel Hilbert space (R.K.H.S.) $\mathcal{H}$ of the space $\mathcal{M}$ of signed measures on some topological space $\mathfrak{X}$. He characterized the inner products on $\mathcal{M}$ inducing the weak topology on the subspace $\mathcal{M}^+$ of bounded positive measures and established in this setting a Glivenko-Cantelli theorem with applications to estimation and hypothesis testing. In this contribution we present a constructive approach of Guilbart's embedding following [20]. This embedding provides a Hilbertian framework for signed random measures. We shall discuss some applications of this construction to limit theorems for random measures and partial sums processes.

Let $\mathfrak{X}$ be a metric space and let $\mathcal{M}$ denote the space of *signed measures* on the Borel $\sigma$-field of $\mathfrak{X}$. A signed measure $\mu$ is the difference of two positive bounded measures. We denote by $(\mu^+, \mu^-)$ its Hahn-Jordan decomposition and by $|\mu| = \mu^+ + \mu^-$ its total variation measure. We consider the class of reproducing kernels having the following representation

$$K(x,y) = \int_{\mathbb{U}} r(x,u)\overline{r(y,u)}\rho(du), \quad x,y \in \mathfrak{X}, \tag{1}$$

where $\rho$ is a positive measure on some measurable space $(\mathbb{U}, \mathcal{U})$ and the function $r : \mathfrak{X} \times \mathbb{U} \to \mathbb{C}$ satisfies

$$\sup_{x \in \mathfrak{X}} \|r(x,\,.\,)\|_{L^2(\rho)} < \infty. \tag{2}$$

We denote by $\mathcal{H}$ the reproducing kernel Hilbert space associated with $K$. It is easily checked (Prop.2 in [20]) that under (2), $r(.,u)$ is $\mu$-integrable over $\mathfrak{X}$ for $\rho$-almost

$u \in \mathbb{U}$. We assume moreover that

$$\text{if } \mu \in \mathcal{M} \text{ and } \int_{\mathcal{X}} r(x,u)\mu(\mathrm{d}x) = 0 \text{ for } \rho\text{-almost } u, \text{ then } \mu = 0. \qquad (3)$$

The essential facts about the embeddings of $\mathcal{M}$ into $\mathcal{H}$ and $L^2(\rho)$ are gathered in the following theorem which is proved in [20].

**Theorem 1.1.** *Under (1), (2) and (3), the following properties hold.*

a) *Let $E$ be the closed subspace of $L^2(\rho)$ spanned by $\{r(x,.), \ x \in \mathcal{X}\}$. A function $h : \mathcal{X} \to \mathbb{C}$ belongs to $\mathcal{H}$ if and only if there is a unique $g \in L^2(\rho)$ such that*

$$h(x) = \int_{\mathbb{U}} g(u)\overline{r(x,u)}\rho(\mathrm{d}u), \quad x \in \mathcal{X}. \qquad (4)$$

*The representation (4) defines an isometry of Hilbert spaces $\Psi : \mathcal{H} \to E, \ h \mapsto g$.*
b) *$K$ induces an inner product on $\mathcal{M}$ by the formula*

$$\langle \mu, \nu \rangle_K := \int_{\mathcal{X}^2} K(x,y)\mu \otimes \nu(\mathrm{d}x, \mathrm{d}y), \quad \mu, \nu \in \mathcal{M}. \qquad (5)$$

c) *$(\mathcal{M}, \langle ., . \rangle_K)$ is isometric to a dense subspace of $\mathcal{H}$ by*

$$\mathfrak{I} : \mathcal{M} \to \mathcal{H}, \quad \mu \longmapsto \mathfrak{I}_\mu := \int_{\mathcal{X}} K(x,.)\mu(\mathrm{d}x). \qquad (6)$$

*Moreover we have*

$$\langle h, \mathfrak{I}_\mu \rangle = \int_{\mathcal{X}} h \, \mathrm{d}\mu, \quad \langle \mathfrak{I}_\mu, h \rangle = \int_{\mathcal{X}} \overline{h} \, \mathrm{d}\mu, \quad h \in \mathcal{H}, \mu \in \mathcal{M}. \qquad (7)$$

d) *The isometric embedding $\zeta = \Psi \circ \mathfrak{I} : \mu \mapsto \zeta_\mu$ of $\mathcal{M}$ into $L^2(\rho)$ satisfies*

$$\zeta_\mu(u) = \int_{\mathcal{X}} r(x,u)\mu(\mathrm{d}x), \quad u \in \mathbb{U}. \qquad (8)$$

Let us examine some examples where Theorem 1.1 applies.

**Example 1.1.** Take for $\rho$ the counting measure on $\mathbb{U} = \mathbb{N}$ and define $r$ by $r(x,i) := f_i(x), \ x \in \mathcal{X}, \ i \in \mathbb{N}$, where the sequence of functions $f_i : \mathcal{X} \to \mathbb{R}$ separates the measures, i.e. the only $\mu \in \mathcal{M}$ such that $\int_{\mathcal{X}} f_i \, \mathrm{d}\mu = 0$ for all $i \in \mathbb{N}$ is the null measure. To have a bounded kernel we also assume that $\sum_{i \in \mathbb{N}} \|f_i\|_\infty^2 < \infty$. Then

$$K(x,y) = \sum_{i \in \mathbb{N}} f_i(x)f_i(y), \quad x, y \in \mathcal{X}^2.$$

$\mu$ is represented in $\ell^2(\mathbb{N})$ by $\zeta_\mu = \left(\int_{\mathcal{X}} f_i \, \mathrm{d}\mu\right)_{i \in \mathbb{N}}$ and in $\mathcal{H}$ by $\mathfrak{I}_\mu = \sum_{i \in \mathbb{N}} \left(\int_{\mathcal{X}} f_i \, \mathrm{d}\mu\right) f_i$. It easily follows from (4) that every $f_i$ belongs to $\mathcal{H}$.

**Example 1.2.** Take $\mathcal{X} = \mathbb{U} = \mathbb{R}^d$, with $r(x,u) := \exp(i\langle x, u \rangle), \ x, u \in \mathbb{R}^d$ and choose $\rho$ as a bounded positive measure on $\mathbb{R}^d$. This gives the continuous *stationary* kernels

$$K(x,y) = \int_{\mathbb{R}^d} \exp(i\langle x - y, u \rangle)\rho(\mathrm{d}u), \quad x, y \in \mathbb{R}^d.$$

Here $\zeta_\mu(u) = \int_{\mathbb{R}^d} \exp(i\langle x, u\rangle)\mu(dx) =: \hat{\mu}(u)$, is the characteristic function of $\mu$ and $\mathfrak{I}_\mu(x) = \int_{\mathbb{R}^d} \exp(-i\langle x, u\rangle)\hat{\mu}(u)\rho(du)$. These kernels are used in [20] to study the convergence rate in the CLT.

**Example 1.3.** Take $\mathfrak{X} = \mathbb{U} = [0, 1]$, $\rho = \lambda + \delta_1$, where $\lambda$ is the Lebesgue measure and $\delta_1$ the Dirac mass at the point 1. With $r(x, u) := \mathbf{1}_{[x,1]}(u)$, we obtain $K(x, y) = 2 - \max(x, y)$ and $\zeta_\mu(u) = \mu([0, u])$.

**Remark 1.1.** The usual topologies on $\mathcal{M}$ are generated by functionals $f \mapsto \int_\mathfrak{X} f \, d\mu$, $f \in F$, where $F$ is some family of continuous functions defined on $\mathfrak{X}$. When $\mathfrak{X}$ is locally compact, $F = C(\mathfrak{X})$, the space of all bounded continuous functions on $\mathfrak{X}$ gives the weak topology while restricting to $F = C_0(\mathfrak{X})$ the space of continuous function converging to zero at infinity gives the vague topology. By convergence to zero at infinity we mean that for every positive $\varepsilon$ there is a compact subset $A$ of $\mathfrak{X}$ such that $|f(x)| < \varepsilon$ for every $x \in \mathfrak{X} \setminus A$. In the special case where $\mathfrak{X}$ is compact, $C(\mathfrak{X}) = C_0(\mathfrak{X})$. Endowed with the supremum norm, $C_0(\mathfrak{X})$ is a Banach space with topological dual $\mathcal{M}$ (Riesz's theorem). Now if we choose in Example 1.1 the $f_i$'s in $C_0(\mathfrak{X})$, a simple Hahn-Banach argument gives the density of $\mathcal{H}$ in $C_0(\mathfrak{X})$. In this setting, let $(\mu_n)_{n\geq 1}$ be a sequence in $\mathcal{M}$ such that $\sup_{n\geq 1} |\mu_n|(\mathfrak{X}) < \infty$. Then weak and strong convergence in $\mathcal{H}$ of $\mathfrak{I}_{\mu_n}$ to $\mathfrak{I}_\mu$ are equivalent to the weak convergence in $\mathcal{M}$ of $\mu_n$ to $\mu$.

## 2. Some limit theorems for random measures

### 2.1. *Random measures*

A random measure $\mu^\bullet$ is a random element in a set $\mathfrak{M}$ of measures equipped with some $\sigma$-field $\mathcal{G}$, i.e. a measurable mapping

$$\mu^\bullet : (\Omega, \mathcal{F}, P) \longrightarrow (\mathfrak{M}, \mathcal{G}), \quad \omega \mapsto \mu^\omega.$$

Here $(\Omega, \mathcal{F}, P)$ is a probability space and the law or distribution of $\mu^\bullet$ (under $P$) is the image measure $P \circ (\mu^\bullet)^{-1}$ on $\mathcal{G}$. Among the well known examples of random measures let us mention the empirical process $\mu_n^\bullet = n^{-1}\sum_{i=1}^n \delta_{X_i}$, where the $X_i$'s are random elements in the space $\mathfrak{X}$ and the point processes $\sum_{i=1}^N \delta_{Y_i}$, where $N$ and the $Y_i$'s are random. In the classical theory, e.g. Kallenberg [7], $\mathfrak{X}$ is locally compact with a countable basis of neighborhoods, $\mathfrak{M}$ is the set of *positive* Radon measures on the Borel $\sigma$-field of $\mathfrak{X}$ and $\mathfrak{M}$ is endowed with the Borel $\sigma$-field $\mathcal{G}$ of the vague topology. This framework of positive measures is sufficient to the classical study of point processes and positive random measures. But the above setting does not cover the case of signed measures. Still random signed measures appear naturally by centering of positive ones [6]. Guilbart's embedding of $\mathcal{M}$ in an R.K.H.S. $\mathcal{H}$ provides the background for a Hilbertian theory of *signed* random measures. This way we can exploit the nice probabilistic properties of Hilbert spaces and obtain useful limit theorems like CLT or FCLT.

From now on, we assume for simplicity that $\mathfrak{X}$ is metric locally compact and that $K$ is as in Example 1.1 with the $f_i$'s in $C_0(\mathfrak{X})$. Identifying $\mathcal{H}$ with a completion of $\mathcal{M}$, we call random measure a random element $\mu^\bullet$ in $\mathcal{H}$ such that $P(\mu^\bullet \in \mathcal{M}) = 1$. The *observations* of such a random measure are the random variables $\langle h, \mu^\bullet \rangle_K = \int_{\mathfrak{X}} h \, d\mu^\bullet$, $h \in \mathcal{H}$, accounting (7). Some natural measurability questions raised by our definition of random measures are positively answered in [19]: $\mathcal{M}$ is a Borel subset of $\mathcal{H}$, $|\mu^\bullet|$ is also a random measure, the $\int_{\mathfrak{X}} f \, d\mu^\bullet$'s, $f \in C_0(\mathfrak{X})$, and $|\mu^\bullet|(\mathfrak{X})$ are random variables.

## 2.2. *Strong law of large numbers*

If $\mathbf{E}\|\mu^\bullet\|_K$ is finite, the random measure $\mu^\bullet$ is Bochner integrable and $\mathbf{E}\mu^\bullet$ is defined as a deterministic element of $\mathcal{H}$. Then $\mu^\bullet$ is also Pettis integrable, when

$$\mathbf{E}\langle h, \mu^\bullet \rangle_K = \langle h, \mathbf{E}\mu^\bullet \rangle_K, \quad h \in \mathcal{H}. \tag{9}$$

The following theorem is an immediate application of the strong law of large numbers in separable Banach spaces, see e.g. [9].

**Theorem 2.1.** *Let* $\mu_1^\bullet, \ldots, \mu_n^\bullet, \ldots$ *be independent identically distributed copies of* $\mu^\bullet$. *If* $\mathbf{E}\|\mu^\bullet\|_K$ *is finite, then*

$$\nu_n^\bullet := \frac{1}{n}\sum_{i=1}^{n} \mu_i^\bullet \xrightarrow[\text{a.s.}]{\mathcal{H}} \mathbf{E}\mu^\bullet. \tag{10}$$

*Conversely, if* $\nu_n^\bullet$ *converges almost surely in* $\mathcal{H}$ *to some limit* $\ell$, *this limit is deterministic,* $\mathbf{E}\|\mu^\bullet\|_K$ *is finite and* $\ell = \mathbf{E}\mu^\bullet$.

Although $\nu_n^\bullet$ is obviously a random measure, it is not clear that the same holds true for its a.s. limit $\mathbf{E}\mu^\bullet$. When $\mathbf{E}\mu^\bullet$ belongs to $\mathcal{M}$, we call it the *mean measure of* $\mu^\bullet$. In this case, (9) can be recast as

$$\mathbf{E}\langle h, \mu^\bullet \rangle_K = \int_{\mathfrak{X}} h \, d(\mathbf{E}\mu^\bullet), \quad h \in \mathcal{H}. \tag{11}$$

Here is a simple sufficient condition for the existence of the mean measure.

**Proposition 2.1.** *The membership of* $\mathbf{E}\mu^\bullet$ *in* $\mathcal{M}$ *follows from the finiteness of* $\mathbf{E}|\mu^\bullet|(\mathfrak{X})$ *if* $\mathfrak{X}$ *is locally compact,* $K$ *is continuous on* $\mathfrak{X}^2$ *and* $K(x,.) \in C_0(\mathfrak{X})$ *for every* $x \in \mathfrak{X}$.

The proof (cf. Prop. XI.1.2 in [17]) relies on the characterization of measures in $\mathcal{H}$ by

$$g \in \mathcal{I}(\mathcal{M}) \quad \text{iff} \quad \sup_{f \in \mathcal{H}, \|f\|_\infty \leq 1} |\langle f, g \rangle| < \infty, \tag{12}$$

using the fact that when finite, the supremum in (12) equals $|\mu|(\mathfrak{X})$, where $\mu := \mathcal{I}^{-1}(g)$, together with the elementary estimate

$$\|\mu\|_K \leq \big(\sup_{\mathfrak{X}^2} K\big)^{1/2} |\mu|(\mathfrak{X}), \quad \mu \in \mathcal{M}. \tag{13}$$

**Corollary 2.1.** *If* $\mathbf{E}|\mu^\bullet|(\mathfrak{X}) < \infty$, *let* $\mu$ *be the mean measure of* $\mu^\bullet$. *Then the a.s. convergence of* $\nu_n^\bullet$ *to* $\mu$ *holds both in* $\mathcal{H}$ *and in the weak topology on* $\mathcal{M}$.

The a.s. convergence in $\mathcal{H}$ obviously follows from Theorem 2.1 by applying (13) to $\mu^\bullet$. By Remark 1.1, (10) implies the a.s. weak convergence in $\mathcal{M}$ of $\nu_n^\bullet$ to $\mu$ provided that $\sup_{n \geq 1} |\nu_n^\bullet|(\mathfrak{X}) < \infty$. This uniform boundedness follows from the estimate $|\nu_n^\bullet|(\mathfrak{X}) \leq n^{-1} \sum_{i=1}^n |\mu_i^\bullet|(\mathfrak{X})$ and of the a.s. convergence of this upper bound to $\mathbf{E}|\mu^\bullet|(\mathfrak{X})$ by the strong law of large numbers applied to the i.i.d. random variables $|\mu_i^\bullet|(\mathfrak{X})$.

### 2.3. *Central limit theorem for i.i.d. summands*

In any separable Hilbert space $H$, the central limit theorem for a sum of i.i.d. random elements is equivalent to the square integrability of the summands. This nice property does not extend to general Banach spaces, because the CLT is deeply connected to the geometry of the space [9]. A square integrable random element $X$ in $H$ is always *pregaussian*, i.e. there is a Gaussian random element in $H$ with the same covariance structure as $X$.

**Theorem 2.2.** *Let* $\mu_1^\bullet, \ldots, \mu_n^\bullet, \ldots$ *be i.i.d. copies of* $\mu^\bullet$. *If* $\mathbf{E}\|\mu^\bullet\|_K^2 < \infty$, *then*

$$S_n^* := \frac{1}{\sqrt{n}} \sum_{i=1}^n (\mu_i^\bullet - \mathbf{E}\mu^\bullet) \xrightarrow[in \ law]{\mathcal{H}} \gamma^\bullet, \tag{14}$$

*where* $\gamma^\bullet$ *is a Gaussian random element in* $\mathcal{H}$ *with* $\mathbf{E}\gamma^\bullet = 0$ *and covariance given by*

$$\mathrm{Cov}(\gamma^\bullet)(f, g) = \mathbf{E}\left(\int_{\mathfrak{X}} f \, d\mu^\bullet \int_{\mathfrak{X}} g \, d\mu^\bullet\right) - \left(\mathbf{E}\int_{\mathfrak{X}} f \, d\mu^\bullet\right)\left(\mathbf{E}\int_{\mathfrak{X}} g \, d\mu^\bullet\right), \tag{15}$$

*for every* $f, g \in \mathcal{H}$.

*Conversely, if* $S_n^*$ *converges in law in* $\mathcal{H}$, *its limit is Gaussian and* $\mathbf{E}\|\mu^\bullet\|_K^2 < \infty$.

**Corollary 2.2.** *If* $\mathfrak{X}$ *is locally compact and* $\mathbf{E}|\mu^\bullet|(\mathfrak{X})^2 < \infty$, *then both* $\mu^\bullet$ *and* $\mu^\bullet \otimes \mu^\bullet$ *have mean measures, say* $\mu$ *and* $\nu$ *and (14) holds. In this case, (15) can be recast as*

$$\mathrm{Cov}(\gamma^\bullet)(f, g) = \int_{\mathfrak{X}^2} f \otimes g \, d\nu - \left(\int_{\mathfrak{X}} f \, d\mu\right)\left(\int_{\mathfrak{X}} g \, d\mu\right).$$

**Example 2.1.** (CLT for empirical measure) Let $X$ be a random element $(\Omega, \mathcal{F}, P) \to (\mathfrak{X}, \mathcal{B}_{\mathfrak{X}})$ with unknown distribution $\mu = P \circ X^{-1}$. Denote by $X_1, \ldots, X_n$, i.i.d. copies of $X$ and put $\mu_i^\bullet := \delta_{X_i}$, $i = 1, \ldots, n$. Then $n^{-1} \sum_{i=1}^n \delta_{X_i}$ is the *empirical measure* associated with the sample $X_1, \ldots, X_n$. The CLT in $\mathcal{H}$ for the empirical measure was obtained by Berlinet [2] by a direct approach. It can also be seen as a special case of Corollary 2.2. Indeed here $\mu^\bullet = \delta_X$, so $|\mu^\bullet|(\mathfrak{X}) = 1$, $\mathbf{E}\mu^\bullet = \mu = P \circ X^{-1}$ and $\mathbf{E}(\mu^\bullet \otimes \mu^\bullet) =: \nu$ is the image measure of $P \circ X^{-1}$ by the mapping $x \mapsto (x, x)$. Hence

$$\sqrt{n}\left(\frac{1}{n}\sum_{i=1}^n \delta_{X_i} - \mu\right) \xrightarrow[in \ law]{\mathcal{H}} \gamma^\bullet,$$

where the covariance of the Gaussian centered random element $\gamma^{\bullet}$ is given by

$$\mathrm{Cov}(\gamma^{\bullet})(f,g) = \int_{\mathfrak{X}} fg \, \mathrm{d}\mu - \left( \int_{\mathfrak{X}} f \, \mathrm{d}\mu \right) \left( \int_{\mathfrak{X}} g \, \mathrm{d}\mu \right).$$

## 2.4. CLT for Donsker random measure and FCLT in $L^2[0,1]$

It is also possible to obtain central limit theorems for sums of non i.i.d. random measures, like the Donsker random measure

$$\nu_n^{\bullet} := \frac{1}{s_n} \sum_{i=1}^{n} X_i \delta_{\frac{i}{n}}, \quad n \geq 1, \tag{16}$$

where the $X_i$'s are mean zero random variables, possibly dependent, with $s_n^2 := \mathbf{E}S_n^2$ and $S_n = \sum_{i=1}^{n} X_i$. An application of such CLT is a functional central limit theorem (FCLT) in $L^2[0,1]$ for the partial sums processes

$$W_n(t) := s_n^{-1} S_{[nt]}, \quad t \in [0,1]. \tag{17}$$

This application was suggested by P. Jacob to P.E. Oliveira and the author. The weak convergence of $W_n$ is classically studied in the Skorohod space $D(0,1)$ which is continuously embedded in $L^2[0,1]$. As many test statistics are functionals continuous in $L^2[0,1]$ sense of $W_n$ or of the empirical process, see [12] and [10], the weaker topological framework of $L^2[0,1]$ has its own interest. This way we can hope to relax the assumptions on the dependence structure of the underlying variables $X_i$'s. Here we just sketch the method and refer to [11, 12] for more precise results.

Let us choose $\mathfrak{X} = [0,1]$ with the kernel of Example 1.3. Then

$$\zeta_{\nu_n^{\bullet}}(t) = \nu_n^{\bullet}([0,t]) = s_n^{-1} S_{[nt]} = W_n(t), \quad t \in [0,1]. \tag{18}$$

Hence by the isometry between the Hilbert spaces $\mathcal{H}$ and $L^2[0,1]$,

$$\nu_n^{\bullet} \xrightarrow[\text{in law}]{\mathcal{H}} \gamma^{\bullet} \iff W_n \xrightarrow[\text{in law}]{L^2[0,1]} W, \tag{19}$$

where under mild assumptions, the limiting process $W$ is identified as a Brownian motion by a simple covariance computation. Now the relevant CLT for $\nu_n^{\bullet}$ may be established by checking the following conditions.

a) The inner products $\langle h, \nu_n^{\bullet} \rangle_K$ converge in law to $\langle h, \gamma^{\bullet} \rangle_K$ for any fixed $h \in \mathcal{H}$.
b) The sequence $(\nu_n^{\bullet})_{n \geq 1}$ is tight in $\mathcal{H}$, i.e. for any positive $\varepsilon$, there is a compact subset $C_\varepsilon$ of $\mathcal{H}$ such that $\inf_{n \geq 1} P(\nu_n^{\bullet} \in C_\varepsilon) \geq 1 - \varepsilon$.

The first condition reduces to a CLT in $\mathbb{R}$ for triangular arrays because

$$\langle h, \nu_n^{\bullet} \rangle_K = \frac{1}{s_n} \sum_{i=1}^{n} X_i \langle h, \delta_{\frac{i}{n}} \rangle_K = \frac{1}{s_n} \sum_{i=1}^{n} h\left(\frac{i}{n}\right) X_i. \tag{20}$$

By an adaptation of a classical Prohorov's result (Th.1.13 in [14]), sufficient conditions for the tightness of $(\nu_n^\bullet)_{n\geq 1}$ are

$$\sup_{n\geq 1} \mathbf{E}\|\nu_n^\bullet\|_K^2 < \infty, \tag{21}$$

$$\lim_{n\to\infty} \sup_{n\geq 1} \mathbf{E} \sum_{i\geq N} |\langle f_i, \nu_n^\bullet\rangle_K|^2 = 0, \tag{22}$$

for some Hilbertian basis $(f_i)_{i\in\mathbb{N}}$ of $\mathcal{H}$. Concerning (21) which does not come from Th.1.13 in [14], see the remark after Theorem 5 in [21].

Now the heart of the matter is in the following elementary estimate.

$$\mathbf{E} \sum_{i\geq N} |\langle f_i, \nu_n^\bullet\rangle_K|^2 = \sum_{i\geq N} \mathbf{E} \left( \int f_i \, d\nu_n^\bullet \right)^2$$

$$= \sum_{i\geq N} \frac{1}{s_n^2} \sum_{j,k=1}^n \mathbf{E}(X_j X_k) f_i\left(\frac{j}{n}\right) f_i\left(\frac{k}{n}\right)$$

$$\leq \left( \frac{1}{s_n^2} \sum_{j,k=1}^n |\mathbf{E}(X_j X_k)| \right) \sup_{x\in[0,1]} \sum_{i\geq N} f_i(x)^2. \tag{23}$$

The first factor in (23) may be bounded uniformly in $n$, subject to good covariance estimates for the $X_j$'s. The second factor goes to zero due to Dini's theorem (the $f_i$'s being continous like any element of $\mathcal{H}$). Moreover (21) obviously follows from (23) with $N = 0$ in the same setting.

To sum up, the FCLT in $L^2[0,1]$ for the partial sums process $W_n$ based on some dependent sequence $(X_j)_{j\geq 1}$ is obtained under the estimate $\sum_{j,k=1}^n |\mathbf{E}(X_j X_k)| = O(s_n^2)$ and a one-dimensional CLT for the triangular arrays (20).

## 2.5. *Functional central limit theorems*

We discuss now the extension to random measures of the classical FCLT for random variables. First note that polygonal lines in $\mathcal{M}$ make sense, due to $\mathcal{M}$'s vector space structure. Let $\mu^\bullet$ be a signed random measure and the $\mu_i^\bullet$'s be i.i.d. copies of $\mu^\bullet$. We denote by $\xi_n^\bullet$ the $\mathcal{M}$-valued stochastic process indexed by $[0,1]$, whose paths are polygonal lines with vertices $(k/n, n^{-1/2} S_k)$, $k = 0, 1, \ldots, n$, $S_k := \mu_1^\bullet + \cdots + \mu_k^\bullet$.

Combining Theorem 2.2 with Kuelbs FCLT [8], we immediately obtain the FCLT for $\xi_n^\bullet$ in the space $\mathcal{C}([0,1], \mathcal{H})$ of continuous functions $[0,1] \to \mathcal{H}$.

**Theorem 2.3.** *The following statements are equivalent.*

a) $\mathbf{E}\|\mu^\bullet\|_K^2 < \infty$ and $\mathbf{E}\mu^\bullet = 0$,

b) $\xi_n^\bullet$ *converges in law in* $\mathcal{C}([0,1], \mathcal{H})$ *to some* $\mathcal{H}$-valued Brownian motion $W$, i.e. *a Gaussian process with independent increments such that* $W(t) - W(s)$ *has the same distribution as* $|t - s|^{1/2} \gamma^\bullet$, *where* $\gamma^\bullet$ *is a Gaussian random element in* $\mathcal{H}$ *with null expectation and same covariance structure as* $\mu^\bullet$.

As the paths of $\xi_n^\bullet$ are Lipschitz $\mathcal{H}$-valued functions, it is natural to look for a stronger topological framework than $\mathcal{C}([0,1],\mathcal{H})$ for the FCLT. A clear limitation in this quest comes from the modulus of uniform continuity of the limiting process, $\omega(W,u) := \sup_{0 \le t-s \le u} \|W(t) - W(s)\|_{\mathcal{H}}$. Indeed by a simple projection argument and Lévy's well known result, $\omega(W,u)$ cannot be better than $u^{1/2}\ln(1/u)$. This forbids any weak convergence of $\xi_n^\bullet$ in some Hölder topology based on a weight function stronger than $u^{1/2}\ln(1/u)$. Introduce the separable Hölder spaces $\mathrm{H}_\rho^o([0,1],\mathcal{H})$ of functions $f : [0,1] \to \mathcal{H}$, such that

$$\|f\|_\rho := \|f(0)\|_{\mathcal{H}} + \omega_\rho(f,1) < \infty \quad \text{and} \quad \lim_{u \to 0} \omega_\rho(f,u) = 0,$$

where

$$\omega_\rho(f,u) := \sup_{0 < t-s \le u} \frac{\|f(t) - f(s)\|_{\mathcal{H}}}{\rho(t-s)}.$$

We assume moreover that the weight functions $\rho$ are of the form $\rho(u) = u^\alpha L(1/u)$, $0 < \alpha \le 1/2$, where $L$ is continuous normalized slowly varying at infinity. The $\mathrm{H}_\rho^o([0,1],\mathcal{H})$ weak convergence of $\xi_n^\bullet$ to $W$ requires stronger integrability of $\mu^\bullet$ than Condition a) in Theorem 2.3. Combining Theorem 2.2 with the Hölderian FCLT in [15], leads to the FCLT for $\xi_n^\bullet$ in the space $\mathrm{H}_\rho^o([0,1],\mathcal{H})$.

**Theorem 2.4.** *Assume that there is a $\beta > 1/2$ such that*

$$t^{1/2}\rho(1/t)\ln^{-\beta}(t) \text{ is non decreasing on some } [a,\infty). \tag{24}$$

*Then the following statements are equivalent.*

a) $\mathbf{E}\mu^\bullet = 0$ *and*

$$\text{for every } A > 0, \quad \lim_{t \to \infty} t\, P\big(\|\mu^\bullet\|_K \ge At^{1/2}\rho(1/t)\big) = 0. \tag{25}$$

b) $\xi_n^\bullet$ *converges in law in $\mathrm{H}_\rho^o([0,1],\mathcal{H})$ to the $\mathcal{H}$-valued Brownian motion $W$ of Th. 2.3.*

When $\alpha < 1/2$, Condition (24) is automatically satisfied and it is enough to take $A = 1$ in (25). To clarify Condition (25), let us consider two important special cases. When $\rho(t) = t^\alpha$ for some $0 < \alpha < 1/2$, (25) reduces to $P\big(\|\mu^\bullet\|_K \ge t\big) = o(t^{-p(\alpha)})$, with $p(\alpha) := (1/2 - \alpha)^{-1}$ and this is slightly weaker than $\mathbf{E}\|\mu^\bullet\|_K^{p(\alpha)} < \infty$. When $\rho(t) = t^{1/2}\ln^\beta(c/t)$ for some $\beta > 1/2$, then (25) is equivalent to the finiteness of $\mathbf{E}\exp(d\|\mu^\bullet\|_K^{1/\beta})$ for each $d > 0$.

Following [16], we present briefly a statistical application of Theorem 2.4 to the detection of epidemic change in the expectation of a random measure. In what follows, $\mu_k^\bullet$, $k = 1, \ldots, n$ are always i.i.d. copies of the *mean zero* random measure $\mu^\bullet$. Based on the observation of the random measures $\nu_1^\bullet, \ldots, \nu_n^\bullet$, we want to test the null hypothesis

$$(H_0): \nu_k^\bullet = \mu_k^\bullet, \ k = 1, \ldots, n,$$

against the so called epidemic alternative

$$(H_A) \qquad \nu_k^\bullet = \begin{cases} \mu_c + \mu_k^\bullet & \text{if } k \in \mathbb{I}_n := \{k^* + 1, \dots, m^*\} \\ \mu_k^\bullet & \text{if } k \in \mathbb{I}_n^c := \{1, \dots, n\} \setminus \mathbb{I}_n \end{cases}$$

where $\mu_c \neq 0$ is some deterministic signed measure which may depend on $n$. To achieve this goal, we use some weighted dyadic increments statistics which behave like continuous functionals of $\xi_n^\bullet$ in Hölder topology. Consider partial sums

$$S_n(a, b) = \sum_{na < k \leq nb} \nu_k^\bullet, \quad 0 \leq a < b \leq 1.$$

Let us denote by $\mathbb{D}_j$ the set of dyadic numbers in $[0, 1]$ of level $j$, i.e. $\mathbb{D}_0 = \{0, 1\}$, and $\mathbb{D}_j = \{(2l-1)2^{-j}; 1 \leq l \leq 2^{j-1}\}, j \geq 1$. Write for $r \in \mathbb{D}_j, j \geq 0, r^- := r - 2^{-j}$ and $r^+ := r + 2^{-j}$. Then define the dyadic increments statistics $\mathrm{DI}(n, \rho)$ by

$$\mathrm{DI}(n, \rho) := \frac{1}{2} \max_{1 \leq j \leq \log n} \frac{1}{\rho(2^{-j})} \max_{r \in \mathbb{D}_j} \left\| S_n(r^-, r) - S_n(r, r^+) \right\|_K. \qquad (26)$$

Here "log" stand for the logarithm with basis 2 $(\log(2^j) = j)$ while "ln" denotes the natural logarithm $(\ln(e^t) = t)$.

**Theorem 2.5.** *Assume that the weight function $\rho$ satisfies (24) and that the mean zero random measure $\mu^\bullet$ satisfies (25). Then under $(H_0)$, $n^{-1/2}\mathrm{DI}(n, \rho)$ converges in law to a non negative random variable $Z$ with distribution function*

$$P(Z \leq z) = \prod_{j=1}^{\infty} \left( P(\|\gamma^\bullet\|_K \leq 2^{(j+1)/2}\rho(2^{-j})z) \right)^{2^{j-1}}, \quad z \geq 0, \qquad (27)$$

*where $\gamma^\bullet$ is a mean zero Gaussian random element in $\mathcal{H}$ with the same covariance as $\mu^\bullet$. The convergence of the product (27) is uniform on any interval $[\varepsilon, \infty), \varepsilon > 0$.*

Theorem 2.5 is easily obtained from Theorem 2.2 and from [16] Th. 2 and Prop. 3. For general estimates on the convergence rate in (27), see Prop. 4 in [16]. The consistency of the sequence of test statistics $n^{-1/2}\mathrm{DI}(n, \rho)$ follows from the next result which is an easy adaptation of Th. 5 in [16].

**Theorem 2.6.** *Let $\rho$ satisfying (24). Under $(H_A)$, write $l^* := m^* - k^*$ for the length of epidemics and assume that*

$$\lim_{n \to \infty} n^{1/2} \frac{u_n \|\mu_c\|_K}{\rho(u_n)} = \infty, \quad \text{where} \quad u_n := \min\left\{ \frac{l^*}{n}; 1 - \frac{l^*}{n} \right\}. \qquad (28)$$

*Then*

$$n^{-1/2}\mathrm{DI}(n, \rho) \xrightarrow[n \to \infty]{\mathrm{pr}} \infty.$$

To discuss Condition (28), assume for simplicity that $\mu_c$ does not depend on $n$. When $\rho(t) = t^\alpha$, (28) allows us to detect *short epidemics* such that $l^* = o(n)$ and $l^* n^{-\delta} \to \infty$, where $\delta = (1 - 2\alpha)(2 - 2\alpha)^{-1}$. When $\rho(t) = t^{1/2} \ln^\beta(c/t)$ with $\beta > 1/2$, (28) is satisfied provided that $u_n = n^{-1} \ln^\gamma n$, with $\gamma > 2\beta$. This leads to detection of short epidemics such that $l^* = o(n)$ and $l^* \ln^{-\gamma} n \to \infty$. In both cases one can detect symmetrically *long epidemics* such that $n - l^* = o(n)$ .

152

## References

1. A. Berlinet, *Espaces autoreproduisants et mesure empirique. Méthodes splines en estimation fonctionnelle*, Thesis, University of Lille 1, France (1980).
2. A. Berlinet, Variables aléatoires à valeurs dans les espaces à noyau reproduisant, *C.R.A.S.* **290**, série A, 973–975 (1980).
3. A. Berlinet and Ch. Thomas-Agnan, *Reproducing kernel Hilbert spaces in probability and statistics*, Kluwer Academic Publishers, Boston, Dordrecht, London (2004).
4. C. Guilbart, *Étude des produits scalaires sur l'espace des mesures. Estimation par projection. Tests à noyaux.* Thèse d'Etat, Lille 1, France (1978).
5. C. Guilbart, Produits scalaires sur l'espace des mesures, *Annales de l'Institut Henri Poincaré*, Section B, **15**, 333–354 (1979).
6. P. Jacob, Convergence uniforme à distance finie des mesures signées, *Annales de l'Institut Henri Poincaré*, Section B, **15**, 355–373 (1979).
7. O. Kallenberg, *Random measures*, Academic Press (1983).
8. J. Kuelbs, The invariance principle for Banach space valued random variables, *J. Multivariate Anal.* **3**, 161–172 (1973).
9. M. Ledoux and M. Talagrand, *Probability in Banach Spaces*, Springer-Verlag, Berlin, Heidelberg (1991).
10. B. Morel and Ch. Suquet, Hilbertian invariance principles for the empirical process under association, *Mathematical Methods of Statistics* **11**, No 2, 203–220 (2002).
11. P.E. Oliveira and Ch. Suquet, An invariance principle in $L^2(0,1)$ for non stationary $\varphi$-mixing sequences, *Comm. Math. Univ. Carolinae* **36**, 2, 293–302 (1995).
12. P.E. Oliveira and Ch. Suquet, An $L^2(0,1)$ invariance principle for LPQD random variables, *Portugaliae Mathematica* **53**, 367–379 (1995).
13. P.E. Oliveira and Ch. Suquet, $L^2(0,1)$ weak convergence of the empirical process for dependent variables, *Lecture Notes in Statistics* **103**, A. Antoniadis and G. Oppenheim (Eds), Wavelets and Statistics, 331–344 (1995).
14. Y.V. Prohorov, Convergence of random processes and limit theorems in probability theory, *Theor. Prob. Appl.* **1**, 157–214 (1956).
15. A. Račkauskas and Ch. Suquet, Necessary and sufficient condition for the Hölderian functional central limit theorem, *J. of Theoretical Probab.* **17**, No 1, 221–243 (2004).
16. A. Račkauskas and Ch. Suquet, Testing epidemic changes of infinite dimensional parameters, *Statistical Inference for Stochastic Processes* **9**, 111–134 (2006).
17. Ch. Suquet, *Espaces autoreproduisants et mesures aléatoires*, Thesis, University of Lille 1, France (1986).
18. Ch. Suquet, Une topologie pré-hilbertienne sur l'espace des mesures à signe bornées, *Pub. Inst. Stat. Univ. Paris* **35**, 51–77 (1990).
19. Ch. Suquet, Convergences stochastiques de suites de mesures aléatoires à signe considérées comme variables aléatoires hilbertiennes, *Pub. Inst. Stat. Univ. Paris* **37**, 1-2, 71–99 (1993).
20. Ch. Suquet, Distances euclidiennes sur les mesures signées et application à des théorèmes de Berry-Esséen, *Bull. Belg. Math. Soc.* **2**, 161–181 (1995).
21. Ch. Suquet, Tightness in Schauder decomposable Banach spaces, *Amer. Math. Soc. Transl.* (2) Vol. **193**, 201–224 (1999).

# REPRODUCING KERNELS IN PROBABILITY AND STATISTICS

ALAIN BERLINET

*I3M, UMR CNRS 5149,*
*Université de Montpellier,*
*place Bataillon, 34 095 Montpellier cedex, France*
*E-mail: berlinet@stat.math.univ-montp2.fr*

Since the first works laying its foundations as a subfield of Complex Analysis, the theory of reproducing kernels has proved to be a powerful tool in many fields of Pure and Applied Mathematics. The aim of this paper is to give some idea of how and why this theory interacts with Probability and Statistics.

**Key words:** Reproducing kernels, positive type functions, stochastic processes, nonparametric estimation, random measures, law of iterated logarithm

**Mathematics Subject Classification:** 46E22, 60B11, 62G05

## 1. Introduction

In this paper we present a few selected topics in Probability and Statistics where reproducing kernels play an important role. For basic definitions, details and the treatment of other subjects in the same fields the reader is referred to Berlinet and Thomas-Agnan ([1]).

The next section deals with applications in the theory of Stochastic Processes. The early role of reproducing kernels in this field is explained by the fact that covariance functions are reproducing kernels and that the converse is true. Section 3 deals with nonparametric estimation. Here two basic properties are implying reproducing kernels. The first one is the fact that unbiasedness in density estimation can be read as a reproducing property in a $L^2$ space. The second one is that higher order kernels, widely used in functional estimation, can be written as product of reproducing kernels with densities. In Section 4 we will see how the embedding method can be successfully applied to the study of measures and random measures. The role of reproducing kernels in the Law of the Iterated Logarithm is shortly reported in Section 5. Finally, in Section 6 we will briefly mention some other subfields of Probability and Statistics where reproducing kernels play a well established or a promising role.

## 2. Stochastic Processes

Historically the theory of Stochastic Processes is the first subject in Probability and Statistics where reproducing kernels were applied, mainly by Parzen in the early 1950's. It is a point in the theory of reproducing kernels that they can be equivalently defined as functions of two variables with the reproducing property in some Hilbert space or with the positiveness property. The famous Moore-Aronszajn theorem establishes the equivalence between these two properties. Both aspects are met in Probability and Statistics.

Expectations define integral operators

$$E(X_t) = \int X_t(\omega) \, dP(\omega)$$

and covariances define positive type functions

$$0 \leq E \left| \sum_{i=1}^{n} a_i X_{t_i} \right|^2 = \sum_{i=1}^{n} \sum_{j=1}^{n} a_i \overline{a}_j E(X_{t_i} \overline{X}_{t_j}).$$

To be more precise let $L^2(\Omega, \mathcal{A}, P)$ be the Hilbert space of second order random variables on some probability space $(\Omega, \mathcal{A}, P)$. The inner product in this space is defined by

$$< X, Y > = E(X\overline{Y}) = \int X\overline{Y} \, dP.$$

Let $X_t$, $t$ ranging in some set $T$, be a second order stochastic process defined on $(\Omega, \mathcal{A}, P)$ with values in $\mathbb{R}$ or $\mathbb{C}$. Assuming their existence we will denote by $m$ the mean function of the process

$$m(t) = E(X_t)$$

by $R$ the second moment function

$$R(t, s) = E(X_t \overline{X}_s)$$

and by $K$ the covariance function

$$K(t, s) = R(t, s) - m(t)\overline{m(s)}.$$

For instance the Brownian motion has covariance function $\min(t, s)$ and a Ornstein-Uhlenbeck process has a covariance function of the form $\exp(-\beta|t - s|)$.

Properties of the process like continuity or differentiability in mean square are equivalent to relevant properties of its covariance function $K$.

A central result due to Loève is the following.

**Theorem 2.1.** *R is a second moment function of a second order stochastic process indexed by T if and only if R is a function of positive type on $T \times T$.*

To prove that a function $R$ of positive type is a covariance function one can build a gaussian process with $R$ as covariance function (see Loève ($^6$) or Neveu ($^7$)). A striking fact is that the Hilbert subspace of $L^2(\Omega, \mathcal{A}, P)$ generated by the process is isomorphic to the RKHS with kernel the covariance function $K$.

The link established by Loève has been used to translate some problems related to stochastic processes into functional ones. Such equivalence results are interesting to use functional methods for solving stochastic problems but also to use stochastic methods for improving functional algorithms, as in filtering and spline problems. They belong to the large field of interactions between approximation theory and statistics. Bayesian numerical analysis for example has been surveyed by Diaconis ([4,5]). In the 1960's, Parzen popularized the use of Mercer and Karhunen representation theorems to write formal solutions to best linear prediction problems for stochastic processes. This lead Wahba in the 1970's to reveal the spline nature of the solution of some filtering problems.

The treatment of stochastic processes is made easier when representations as stochastic integrals or infinite series are available. Here is one of these representations obtained as a consequence of Mercer theorem.

**Theorem 2.2.** *(Karhunen-Loève expansion). If $K$ is continuous, there exists a sequence of random variables $\zeta_n$ such that $E(\zeta_n \zeta_m) = \lambda_n \delta_{m,n}$ and*

$$X_t = \sum_{n=1}^{\infty} \zeta_n \phi_n(t)$$

*where $(\phi_n, \lambda_n), n \in \mathbb{N}$ are eigenelements of the kernel $K$.*

The advantage of this kind of representation of the process is that it isolates the manner in which the random function $X_t(\omega)$ depends upon $t$ and upon $\omega$.

Many applications can be found in stochastic filtering, Kriging models and spline functions, ridge regression (or Tikhonov regularization), extraction and detection problems.

## 3. Nonparametric estimation

Consider the problem of estimating **unbiasedly** the value $f(x)$ of an unknown continuous density $f$ from observations $X_1, X_2, \ldots, X_n$ having density $f$ with respect to the Lebesgue measure $\lambda$. For this we have to find a measurable function $\varphi(X_1, X_2, \ldots, X_n, x)$ such that

$$E_f[\varphi(X_1, X_2, \ldots, X_n, x)] = f(x).$$

We know from the Bickel-Lehmann theorem ([2]) that if it were possible to estimate unbiasedly $f(x)$ then an unbiased estimate based on one observation would exist. This means that there would exist a function $K(., x)$ such that

$$E[K(X, x)] = \int K(y, x) f(y) d\lambda(y) = f(x),$$

where $X$ has density $f$. In other words the function $K$ would have a reproducing property in the set of possible densities.

More precisely we have the following result ([2]).

**Theorem 3.1.** *Suppose that the vector space $\mathcal{H}$ spanned by the set $\mathcal{D}$ of possible densities with respect to the measure $\nu$ is included in $L^2(\nu)$. In order that there exists an estimate $K(X, x)$ of $f(x)$ satisfying*

$$\forall x \in \mathbb{R}, \qquad K(.,x) \in \mathcal{H} \quad and \quad E(K(X,x)) = f(x),$$

*where $X$ has density $f$, it is necessary and sufficient that $\mathcal{H}$, endowed with the inner product of $L^2(\nu)$, be a pre-Hilbert space with reproducing kernel $K$.*

Without strong assumptions on the underlying density we are not able to construct unbiased density estimates. This is a common situation in infinite dimensional models. Let us try to reduce the bias.

Consider a sequence $(X_i)_{i \in \mathbb{N}}$ of real-valued random variables with common unknown density $f$ and the standard Akaike-Parzen-Rosenblatt kernel estimate ([2])

$$f_n(x) = \frac{1}{nh_n} \sum_{i=1}^{n} K\left(\frac{x - X_i}{h_n}\right)$$

where $(h_n)_{n \in \mathbb{N}}$ is a sequence of positive real numbers tending to zero and $K$ is a bounded measurable function integrating to one.

In this estimation setting each observed Dirac measure $\delta_{X_i}$ is replaced with a distribution around $X_i$ defined by $K$ and $h_n$ and the final estimate $f_n$ mixes the $n$ distributions with equal weight $1/n$.

The expectation of $f_n(x)$ is

$$E f_n(x) = \frac{1}{h_n} \int_{\mathbb{R}} K\left(\frac{x - v}{h_n}\right) f(v) \, d\lambda(v).$$

Hence, by a change of variable and the fact that $K$ integrates to one, one gets the bias

$$E f_n(x) - f(x) = \int_{\mathbb{R}} [f(x - h_n u) - f(x)] K(u) d\lambda(u).$$

If the $p^{th}$ order derivative of $f$ $(p \geq 2)$ exists and if $K$ has finite moments

$$\mu_j(K) = \int_{\mathbb{R}} u^j K(u) \, d\lambda(u), \qquad 1 \leq j \leq p,$$

a simple Taylor series expansion gives

$$E f_n(x) - f(x) = \sum_{k=1}^{p-1} h_n^k \frac{(-1)^k}{k!} f^{(k)}(x) \mu_k(K) + O(h_n^p).$$

Therefore the asymptotic bias is reduced whenever the first moments of $K$ vanish. This motivates the following definition.

**Definition 3.1.** (Higher order kernels). Let $p \geq 2$. A bounded measurable function $K$ integrating to one is said to be a kernel of order $p$ if and only if $\mu_p(K)$ is finite and non null and

$$\mu_j(K) = 0, \qquad 1 \leq j \leq (p - 1).$$

A key property is that most kernels of order $(r+1), r \geq 1$, can be written as products

$$\mathcal{K}_r(x,0)K_0(x)$$

where $K_0$ is a probability density function (with finite moments up to order $r$) and $\mathcal{K}_r(.,.)$ is the reproducing kernel of the subspace $\mathbb{P}_r$ of $L^2(K_0\lambda)$ of polynomials of degree at most $r$.

These kernels are widely used in nonparametric estimation. By solving optimization problems with this kind of kernel one can find good $K_0$ to get higher order kernels satisfying asymptotic optimality criteria and computationally efficient.

## 4. Embedding method for measures

When a problem involves elements of some abstract set $\mathcal{S}$ the first attempt to shift it in a hilbertian framework consists in associating an element of a space $\ell^2(X)$ to any element of the originally given set $\mathcal{S}$. If any element $s$ of $\mathcal{S}$ is characterized by a family $\{s_\alpha, \alpha \in X\}$ of complex numbers satisfying

$$\sum_{\alpha \in X} |s_\alpha|^2 < \infty,$$

the mapping

$$\mathcal{S} \longrightarrow \ell^2(X)$$
$$s \longmapsto \{s_\alpha, \alpha \in X\}$$

defines the natural embedding of $\mathcal{S}$ into $\ell^2(X)$.

Dirac measures on a set $E$ can be considered as elements of any RKHS $H_K$ of functions on $E$:

both $\delta_x$ and $K(.,x)$ represent the evaluation functional at $x$.

Now, how can the general embedding methodology be applied to more general sets of measures?

A measure $\mu$ on $(E, \mathcal{T})$ is characterized by the set of its values

$$\{\mu(A) : A \in \mathcal{T}\} = \left\{ \int \mathbf{1}_A \, d\mu : A \in \mathcal{T} \right\}$$

or more generally by a set of integrals

$$\left\{ \int f \, d\mu : f \in \mathcal{F} \right\}$$

where $\mathcal{F}$ is some family of functions.

For instance a probability measure $P$ on $\mathbb{R}^d$ is uniquely determined by its characteristic function

$$\phi_P(t) = \int_{\mathbb{R}^d} e^{i<t,x>} \, dP(x), \qquad t \in \mathbb{R}^d,$$

or equivalently by the set of integrals of the family

$$\mathcal{F} = \left\{ e^{i<t,.>} : t \in \mathbb{R}^d \right\}.$$

Sets of power functions, of continuous bounded functions and many other families $\mathcal{F}$ can be considered to study measures and their closeness or convergence.

Suppose that to deal with some problem related to a set $\mathcal{M}$ of signed measures on a measurable space $(E, \mathcal{T})$ we can consider a set $\mathcal{F}$ of complex functions on $E$ and the families of integrals

$$I_\mu = \left\{ \int f \, d\mu : f \in \mathcal{F} \right\}$$

where $\mu$ belongs to $\mathcal{M}$. If, for any $\mu$ in $\mathcal{M}$, we have

$$\sum_{f \in \mathcal{F}} \left| \int f \, d\mu \right|^2 < \infty,$$

we can work in the Hilbert space $\ell^2(\mathcal{F})$.

The inner product of $I_\mu$ and $I_\nu$ in this space is given by

$$< I_\mu, I_\nu >_{\ell^2(\mathcal{F})} = \sum_{f \in \mathcal{F}} \left( \int f \, d\mu \right) \left( \int \overline{f} \, d\nu \right).$$

Assuming that we may apply the Fubini theorem and exchange sum and integral one gets

$$< I_\mu, I_\nu >_{\ell^2(\mathcal{F})} = \sum_{f \in \mathcal{F}} \left( \int f \otimes \overline{f} \, d(\mu \otimes \nu) \right)$$

$$= \int \left( \sum_{f \in \mathcal{F}} f \otimes \overline{f} \right) d(\mu \otimes \nu).$$

Here $I_\mu$ and $I_\nu$ are not functions on $E$. They are sequences of complex numbers indexed by the class $\mathcal{F}$ or equivalently they can be considered as functions on $\mathcal{F}$. Setting formally

$$K = \sum_{f \in \mathcal{F}} f \otimes \overline{f} \tag{1}$$

we get through the general approach the following expression

$$< I_\mu, I_\nu >_{\ell^2(\mathcal{F})} = \int K \, d(\mu \otimes \nu).$$

Formula (1) holds true whenever $\mathcal{F}$ can be chosen as a complete orthonormal system in some separable RKHS $\mathcal{H}$ with reproducing kernel $K$.

Let us present a very simple example: the example of moments.

Let $E = [0, 0.5]$, $\mathcal{T}$ be its Borel $\sigma$-algebra and $\mathcal{M}$ be the set of signed measures on $(E, \mathcal{T})$. Any element $\mu$ of the set $\mathcal{M}$ is characterized by the sequence $\mathcal{I}_\mu = \{\mu_i : i \in \mathbb{N}\}$ where

$$\mu_i = \int_E x^i d\mu(x)$$

is the moment of order $i$ of $\mu$. Here the class $\mathcal{F}$ is equal to the set of monomials $\{x^i : i \in \mathbb{N}\}$.
As we have

$$\forall i \in \mathbb{N}, \ 0 \le \mu_i \le 2^{-i}\mu(E),$$

the sequence $\mathcal{I}_\mu$ is in $\ell^2(\mathbb{N})$. Identifying $\mu$ and $\mathcal{I}_\mu$ we get, by using Fubini theorem and exchanging sum and integral (the integrated functions are nonnegative),

$$< \mu, \nu >_\mathcal{M} = < \mathcal{I}_\mu, \mathcal{I}_\nu >_{\ell^2(\mathbb{N})} = \sum_{i \in \mathbb{N}} \mu_i \nu_i = \int_{E \times E} \frac{1}{1 - xy} d(\mu \otimes \nu)(x, y).$$

For $a$ in $E$ and $\nu = \delta_a$ we have

$$\forall i \in \mathbb{N}, \ \nu_i = a^i$$

and

$$< \mu, \delta_a > = \int_E \frac{1}{1 - ax} d\mu(x) = \sum_{i \in \mathbb{N}} \mu_i a^i.$$

As the sequence of moments $\mathcal{I} = \{\mu_i, i \in \mathbb{N}\}$, the entire function

$$\varphi_\mu : \quad E \longrightarrow \mathbb{R}$$

$$x \longmapsto \varphi_\mu(x) = \sum_{i \in \mathbb{N}} \mu_i x^i$$

characterizes the measure $\mu$. It follows that the set of functions $\Phi = \{\varphi_\mu, \mu \in \mathcal{M}\}$ endowed with the inner product

$$< \varphi_\mu, \varphi_\nu >_\Phi = \sum_{i \in \mathbb{N}} \mu_i \nu_i$$

induced by the inner product of $\ell^2(\mathbb{N})$ is a prehilbertian subspace with reproducing kernel

$$K(x, y) = \frac{1}{1 - xy} = < \varphi_{\delta_x}, \varphi_{\delta_y} >_\Phi = < \mathcal{I}_{\delta_x}, \mathcal{I}_{\delta_y} >_{\ell^2(\mathbb{N})} = < \delta_x, \delta_y >_\mathcal{M}.$$

In the present context the distance of two signed measures on $E$ is equal to the $\ell^2$-distance of their sequences of moments.
We have seen how a set of signed measures on a measurable set $(E, \mathcal{T})$ can be embedded in a RKHS $\mathcal{H}$ of functions on $E$ with reproducing kernel $K$. Under suitable assumptions we have the following formula

$$< \mathcal{I}_\mu, \mathcal{I}_\nu >_\mathcal{H} = \int K \, d(\mu \otimes \nu)$$

a particular case of which is

$$< I_{\delta_x}, I_{\delta_y} >_{\mathcal{H}} = K(x,y), \qquad (x,y) \in E \times E.$$

However the embedding of measures in RKHS gives rise to a list of problems to be analyzed more precisely in a general setting (see the paper by Suquet ([8]) in the present volume).

1) Under what conditions can an inner product $< .,. >_{\mathcal{M}}$ be defined on a set $\mathcal{M}$ of signed measures?

2) How does the inner product depend on the reproducing kernel?

3) Is any inner product on a set of measures of the kind defined above?

4) What can be the limit of a sequence of measures converging in the sense of the inner product?

5) What are the relationships between the topology induced on $\mathcal{M}$ by the inner product and other topologies on $\mathcal{M}$ such as the weak topology?

6) What kind of results can be obtained through RKHS methods?

## 5. Law of the Iterated Logarithm

In this section we briefly mention the role of reproducing kernels in a nice piece of Probability theory: the Law of the Iterated Logarithm.

Consider a sequence $(X_i)_{i \geq 1}$ of independent real random variables defined on a probability space $(\Omega, \mathcal{A}, P)$ with mean 0 and variance 1. Let, for $n \geq 1$, $S_n = \sum_{i=1}^{n} X_i$. Of what order of magnitude is $S_n$ as $n$ increases to infinity? In other words can we find a deterministic sequence $(a_n)$ such that almost surely

$$\limsup_{n \longrightarrow \infty} \frac{S_n}{a_n} = -\liminf_{n \longrightarrow \infty} \frac{S_n}{a_n} = 1?$$

The answer is positive and the solution is given by

$$a_n = (2n \log \log n)^{1/2}, n \geq 3.$$

Any result of this kind is called a Law of the Iterated Logarithm (LIL) after the *iterated logarithm* appearing in the expression of $a_n$.

Khinchin (1923, 1924) discovered the LIL for binomial variables. In 1929 Kolmogorov established the LIL for bounded, independent not necessarily identically distributed random variables. Many papers treated this kind of problem for random variables or stochastic processes under various hypotheses.

The LIL given above was proved by Hartman and Wintner (1941) in the case of identically distributed random variables. Strassen extended their result and proved a converse (1964, 1966). More precisely the Strassen's LIL states that for independent identically distributed random variables, with the same distribution as $X$, we have

$$EX = 0 \text{ and } \left[ E(X^2) \right]^{1/2} = \sigma < \infty$$

if and only if, almost surely,

$$\lim_{n \to \infty} d\left(\frac{S_n}{a_n}, [-\sigma, \sigma]\right) = 0$$

and the set of limit points of the sequence $(S_n/a_n)$ is equal to the interval $[-\sigma, \sigma]$. Now suppose that

$$EX = 0 \text{ and } \left[E(X^2)\right]^{1/2} = 1$$

and let $\ell_n$ be the continuous function on $(0, 1)$ obtained by linearly interpolating $S_i/a_n$ at $i/n$, $0 \leq i \leq n$. Let $H_0^1(0, 1)$ be the subspace of elements of the Sobolev space $H^1(0, 1)$ vanishing at 0. $H^1(0, 1)$ is endowed with the norm defined by

$$\|u\|^2 = u(0)^2 + \int_0^1 \left(u'(x)\right)^2 \, d\lambda(x).$$

We have the following fundamental result.

**Theorem 5.1.** *(Strassen theorem). If $(X_i)_{i \geq 1}$ is a sequence of independent real random variables with mean 0 and variance 1 the set of limit points of the sequence $(\ell_n)_{n \geq 3}$ with respect to the uniform topology is, with probability one, equal to the closed unit ball of $H_0^1(0, 1)$.*

The space $H_0^1(0, 1)$ is the RKHS of the Brownian motion and its unit ball named the Strassen set.

This kind of fundamental result had a lot of applications in last decades in Probability and Statistics. Many of them take advantage of the following corollary on functions of $\ell_n$.

**Corollary 5.1.** *Let $\varphi$ be a continuous map defined on the space of continuous functions on $[0, 1]$ endowed with the uniform norm and taking its values in some Haussdorf space. Under the assumptions of Strassen's theorem, with probability one, the sequence $(\varphi(\ell_n))_{n \geq 3}$ is relatively compact and the set of its limit points is the transformed by $\varphi$ of the Strassen set.*

Now let us turn to the case of independent identically random variables $(X_i)$ with values in some separable Banach space $B$ with dual denoted by $B'$. We suppose that for any $u$ in $B'$ we have

$$E[u(X)] = 0 \qquad \text{and } E[u^2(X)] < \infty.$$

In this context a RKHS appears in the definition of the set of limit points of the sequence $(S_n/a_n)$ and we have the following result.

**Theorem 5.2.** *(Kuelbs theorem). If the sequence $(S_n/a_n)$ is almost surely relatively compact in $B$ then, almost surely,*

$$\lim_{n \to \infty} d\left(\frac{S_n}{a_n}, B(0, 1)\right) = 0$$

162

*where $B(0,1)$ is the unit ball of the RKHS associated with the covariance structure of $X$ and the set of limit points of the sequence $(S_n/a_n)$ is equal to $B(0,1)$ which is a compact set.*

We end this review paper by mentioning some topics were reproducing kernels have proved to be powerful tools.

## 6. Miscellaneous applications

Space is missing to deal with many other fields were reproducing kernels are successfully applied:

- Learning and Decision Theory, where extensions to non positive kernels are developed (see the paper by Canu *et al* [3] in the present volume).
- Analysis of Variance in Function Spaces
- Strong Approximation
- Generalized Method of Moments
- Computational Aspects

The last thirty years have seen a continuous rise of the use of reproducing kernels in Probability and Statistics. The recent burst of the field of Learning Theory (Functional Classification and Support Vector Machines) attests that the scope of their applications is far from being exhausted.

## References

1. A. Berlinet and C. Thomas-Agnan, *Reproducing Kernel Hilbert Spaces in Probability and Statistics*, Kluwer Academic Publishers (2004).
2. D. Bosq and J. P. Lecoutre, *Théorie de l'Estimation Fonctionnelle*. Economica, Paris (1987).
3. S. Canu, C. Soon Ong and X. Mary, *Splines with Non Positive Kernels*, Proceedings of the 5th ISAAC Congress, Catania 2005, World Scientific.
4. P. Diaconis, Bayesian Numerical Analysis. *Statistical decision theory and related topics IV*, J. Berger and S. Gupta eds, pp. 163-176, (1988).
5. P. Diaconis and S. Evans, A different construction of Gaussian fields from Markov chains: Dirichlet covariances. *Ann. Inst. Henri Poincare*, B 38, pp. 863-878, (2002).
6. M. Loeve, *Probability Theory*. Springer, New York, (1978).
7. J. Neveu, *Processus aléatoires gaussiens*. Séminaire Math. Sup., Les presses de l'Université de Montréal, (1968).
8. C. Suquet, *Reproducing Kernel Hilbert Spaces and Random Measures*, Proceedings of the 5th ISAAC Congress, Catania 2005, World Scientific.

# SPLINES WITH NON POSITIVE KERNELS

STÉPHANE CANU[1], CHENG SOON ONG[2] and XAVIER MARY[3]

*1- PSI - FRE CNRS 2645 - INSA de Rouen*
*76801 St Etienne du Rouvray, France*
*Stephane.Canu@insa-rouen.fr*

*2- Max Planck Institute for Biological Cybernetics*
*Spemannstrasse 38, 72076 Tuebingen, Germany*
*chengsoon.ong@tuebingen.mpg.de*

*3- ENSAE-CREST-LS*
*3 avenue Pierre Larousse, 92240 Malakoff, France*
*xavier.mary@ensae.fr*

Non parametric regression methods can be presented in two main clusters. The one of smoothing splines methods requiring positive kernels and the other one known as Nonparametric Kernel Regression allowing the use of non positive kernels such as the Epanechnikov kernel. We propose a generalization of the smoothing spline method to include kernels which are still symmetric but not positive semi definite (they are called indefinite). The general relationship between smoothing splines, Reproducing Kernel Hilbert Spaces (RKHS) and positive kernels no longer exists with indefinite kernels. Instead the splines are associated with functional spaces called Reproducing Kernel Krein Spaces (RKKS) endowed with an indefinite inner product and thus not directly associated with a norm. Smoothing splines in RKKS have many of the interesting properties of splines in RKHS, such as orthogonality, projection and representer theorem. We show that smoothing splines can be defined in RKKS as the regularized solution of the interpolation problem. Since no norm is available in an RKKS, Tikhonov regularization cannot be defined. Instead, we propose the use of conjugate gradient type iterative methods, with early stopping as a regularization mechanism. Several iterative algorithms are collected which can be used to solve the optimization problems associated with learning in indefinite spaces. Some preliminary experiments with indefinite kernels for spline smoothing reveal the computational efficiency of this approach.

**Key words:** Spline approximation, Krein spaces, spaces with indefinite metric, non-parametric regression

**Mathematics Subject Classification:** 41A15, 46C20, 47B50, 62G08

## 1. Inroduction

Spline functions are a widely used tool in non parametric curve estimation. The underlying theory is one of positive kernels and Reproducing Kernel Hilbert Spaces (RKHS). Thus, positivity of the kernel is both a requirement and a limitation to the use of this method. Recently, positive kernels have been popularized in the statistical learning community with Vapnik's support vector machine and more generally with kernel machines. An advantage of kernel methods pointed out in these works, is their ability to handle large data sets (typically millions of data points). In this framework, the starting point is the kernel. From the kernel, the associated RKHS is built and the problem is stated in this functional space. But the solution of this problem only requires knowledge about the kernel. No explicit formulation of the RKHS is needed. Its existence is sufficient to justify the proposed algorithm.

However, in many practical problems with large data sets and a lack of models, the natural application dependent kernels are non positive ones. These problems arise in fields such as of text mining, biostatistics and astronomy. Practitioners also report good results when using indefinite kernels although, up to now, no theory was available. This paper aims at presenting such theory, a theory of non positive kernels, associated functional spaces and the equivalent splines in this framework.

This paper is organized as follows. In the first part, the functional framework associated with non positive kernels is described. It is shown that indefinite kernels are associated with Reproducing Kernel Krein Spaces (RKKS) endowed with an indefinite inner product and thus not directly associated with a norm. Apart from the norm, all the interesting properties required for learning in RKHS are preserved in RKKS. The second part deals with splines and their definition without using a norm. This definition is straightforward for interpolation splines, but care is required to define a generalization of approximation splines. Based on these definitions, the third part discusses different implementations and proposes an iterative conjugate gradient type algorithm adapted for indefinite kernel matrices.

## 2. Reproducing Kernel Krein Spaces

Krein spaces are indefinite inner product spaces endowed with a Hilbert topology. They can also be seen as a kind of generalized Hilbert space in the sense that their inner product is no longer positive. Before we delve into the definitions and some basic properties of Krein spaces, we give an example.

**Example 2.1.** *4 dimensional space-time.* Indefinite spaces were first introduced into the solution of physical problems via the 4-dimensional Minkowski space of special relativity. There we have 3 negative and one positive dimensions (see Chapter 2 of the reference[6] for the same example with 2 positive and 1 negative dimensions), and the inner product is given by

$$\langle (x_1, y_1, z_1, t_1), (x_2, y_2, z_2, t_2) \rangle = -x_1 x_2 - y_1 y_2 - z_1 z_2 + t_1 t_2$$

and it is no longer necessarily positive. In this space the vector $v = (1, 1, 1, \sqrt{3})$ is a neutral vector (such that $\langle v, v \rangle = 0$). More generally all vectors belonging to the cone $x^2 + y^2 + z^2 - t^2 = 0$ are neutral vectors (the so called light cone in relativity).

### 2.1. *Krein spaces*

As can be seen from Example 2.1, there are several differences between a Krein space and a Hilbert space. The main difference lies in the fact that we allow a more general inner product (detailed expositions can be found in books[4,2]).

**Definition 2.1.** *Inner product.* Let $\mathcal{K}$ be a vector space on the scalar field $\mathbb{R}^{\mathrm{a}}$. An inner product $\langle .,.\rangle_{\mathcal{K}}$ on $\mathcal{K}$ is a bilinear form verifying:

- $\forall f, g \in \mathcal{K}^2, \quad \langle f, g\rangle_{\mathcal{K}} = \langle g, f\rangle_{\mathcal{K}}$                     (symmetric)
- $\forall f, g, h \in \mathcal{K}^3 \quad \langle \alpha f + g, h\rangle_{\mathcal{K}} = \alpha\langle f, h\rangle_{\mathcal{K}} + \langle g, h\rangle_{\mathcal{K}}$   (linear)
- $\{\forall g \in \mathcal{K}, \quad \langle f, g\rangle_{\mathcal{K}} = 0\} \Rightarrow f = 0$         (nondegenerate)

An inner product is said to be: positive if $\forall f \in \mathcal{K}, \quad \langle f, f\rangle_{\mathcal{K}} \geq 0$, negative if $\forall f \in \mathcal{K}, \quad \langle f, f\rangle_{\mathcal{K}} \leq 0$, otherwise it is indefinite.

A vector space $\mathcal{K}$ endowed with the inner product $\langle .,.\rangle_{\mathcal{K}}$ is called an *inner product space*. Two vectors $f, g$ of an inner product space are said to be *orthogonal* if $\langle f, g\rangle_{\mathcal{K}} = 0$.

**Definition 2.2.** *Krein space.* An inner product space $(\mathcal{K}, \langle .,.\rangle_{\mathcal{K}})$ is a Krein space if there exist two Hilbert spaces $(\mathcal{H}_+, \langle .,.\rangle_{\mathcal{H}_+})$ and $(\mathcal{H}_-, \langle .,.\rangle_{\mathcal{H}_-})$ included in $\mathcal{K}$ where:

- $\forall f \in \mathcal{K}, \exists! f_+ \in \mathcal{H}_+$ and $f_- \in \mathcal{H}_-$ such that $f = f_+ + f_-$
- $\forall f, g \in \mathcal{K}^2, \langle f, g\rangle_{\mathcal{K}} = \langle f_+, g_+\rangle_{\mathcal{H}_+} - \langle f_-, g_-\rangle_{\mathcal{H}_-}$

These spaces are then orthogonal with respect to the inner product and the Krein space may be seen as the direct difference of the two Hilbert spaces

$$\mathcal{K} = \mathcal{H}_+ \ominus \mathcal{H}_-$$

Note that such a decomposition is not unique in general.

**Definition 2.3.** *Direct sum and strong topology.* Let $\mathcal{K}$ be a Krein space and $\mathcal{K} = \mathcal{H}_+ \ominus \mathcal{H}_-$ a decomposition. The *direct sum* of $\mathcal{H}_+$ and $\mathcal{H}_-$ defines a Hilbert space, denoted $|\mathcal{K}| = \mathcal{H}_+ \oplus \mathcal{H}_-$, endowed with the (positive) inner product

$$\langle f, g\rangle_{|\mathcal{K}|} = \langle f_+, g_+\rangle_{\mathcal{H}_+} + \langle f_-, g_-\rangle_{\mathcal{H}_-}$$

with the induced norm $\|f\|_{|\mathcal{K}|}^2 := \langle f, f\rangle_{|\mathcal{K}|}$. $|\mathcal{K}|$ is the smallest Hilbert upper bound of the Krein space $\mathcal{K}$ and one defines the strong topology on $\mathcal{K}$ as the Hilbert topology of $|\mathcal{K}|$. The topology does not depend on the chosen decomposition.

Note that for all $f \in \mathcal{K} : |\langle f, f\rangle_{\mathcal{K}}| \leqslant \|f\|_{|\mathcal{K}|}^2$. $\mathcal{K}$ is said to be *Pontryagin* if it admits a decomposition with a negative part of finite dimension, that is $dim(\mathcal{H}_-) < \infty$, and *Minkowski* if it admits a decomposition with both positive and negative part of finite dimension, that is $dim(\mathcal{H}_+) < \infty$ and $dim(\mathcal{H}_-) < \infty$.

Let $\Omega \subseteq \mathbb{R}^d$ be the domain. The set of genuine functions defined on the domain $\Omega$ whose value lie in $\mathbb{R}$ is defined to be $\mathbb{R}^{\Omega}$. The evaluation functional tells us the value of a function at a certain point, and an RKKS is a subset of $\mathbb{R}^{\Omega}$ where all the evaluation functionals are continuous.

---

[a]Like Hilbert spaces, Krein spaces can be defined on $\mathbf{R}$ or $\mathbb{C}$. We use $\mathbf{R}$ in this paper.

**Definition 2.4.** *Evaluation functional.* For all $x \in \Omega$, the evaluation functional $T$ at point $x$ is defined as,

$$T_x : \mathcal{K} \to \mathbb{R}$$
$$f \mapsto T_x(f) = f(x)$$

**Definition 2.5.** *Reproducing Kernel Krein Space (RKKS).* A Krein space $(\mathcal{K}, \langle ., . \rangle_\mathcal{K})$ is a reproducing kernel Krein space (see chapter 7 in the reference[1]) if:

- $\mathcal{K} \subset \mathbb{R}^\Omega$
- $\forall x \in \Omega$, $T_x$ is continuous on $\mathcal{K}$ endowed with its strong topology.

### 2.2. From Krein spaces to kernels

We prove an analog to the Moore-Aronszajn theorem[9], which tells us that for every kernel there is an associated Krein space, and for every Krein space, there is a unique kernel.

**Proposition 2.1.** Reproducing kernel of an RKKS. *Let $(\mathcal{K}, \langle ., . \rangle_\mathcal{K})$ be a Reproducing Kernel Krein Space and $\mathcal{K} = \mathcal{H}_+ \ominus \mathcal{H}_-$ a decomposition.*
*Then*

- *$\mathcal{H}_+$ and $\mathcal{H}_-$ are RKHS (with kernel functions $k_+$ and $k_-$)*
- *there exists a unique symmetric function $k(x, y)$ belonging to $\mathcal{K}$ as a function of a single variable, such that: $\forall f \in \mathcal{K}, \langle f(.), k(., x) \rangle_\mathcal{K} = f(x)$*
- *$k = k_+ - k_-$*

**Proof.** Since $(\mathcal{K}, \langle ., . \rangle_\mathcal{K})$ is an RKKS, the evaluation functional is continuous with respect to the strong topology hence for the Hilbert topology of $(|\mathcal{K}|, \langle ., . \rangle_{|\mathcal{K}|})$. It follows that $\mathcal{H}_+$ and $\mathcal{H}_-$, as Hilbert subspaces of an RKHS, are RKHS. Then let $f = f_+ + f_-$. The evaluation functional can be expressed as

$$T_x(f) = T_x(f_+) + T_x(f_-)$$
$$= \langle f_+, k_+(x) \rangle_{\mathcal{H}_+} + \langle f_-, k_-(x) \rangle_{\mathcal{H}_-}$$
$$= \langle f_+, k_+(x) \rangle_{\mathcal{H}_+} - \langle f_-, -k_-(x) \rangle_{\mathcal{H}_-}$$
$$= \langle f, \underbrace{k_+(x) - k_-(x)}_{k(.,x)} \rangle_\mathcal{K}.$$

It has to be symmetric since the inner product is symmetric, but it is not necessarily positive. It is unique since the inner product is non-degenerate. □

### 2.3. From kernels to Reproduicing Kernel Krein Spaces

Let $k(x, y)$ be a symmetric real valued function, $k : \Omega \times \Omega \longrightarrow \mathbb{R}$.

**Proposition 2.2.**
*(Theorem 8.9[8] or Theorem 2.28 p.86[7]) . The following propositions are equivalent:*

- *there exist (at least) one RKKS with kernel $k$*
- *$k$ admits a positive decomposition, that is there exists two positive kernels $k_+$ and $k_-$ such that $k = k_+ - k_-$*
- *$k$ is dominated by some positive kernel $p$ that is, $p - k$ and $p + k$ are positive.*

The two last conditions are equivalent by choosing $p = k_+ + k_-$. Note that with such a kernel decomposition, the Krein space $\mathcal{K} = H_+ \ominus H_-$ is a natural choice. There is *no bijection but an onto mapping* between the set of RKKS and the set of generalized kernels defined in the vector space generated out of the cone of positive kernels. This is not a major problem since in the forthcoming section, our derivation only requires the existence of an RKKS.

## 3. Functional estimation in an RKKS

### 3.1. *Notations*

Assume we have a set of observations $\mathbf{x}_i, y_i, i = 1, n$ with $\mathbf{x}_i \in \Omega$ and $y_i \in \mathbb{R}$. Let $\mathbf{y} = (y_1, ..., y_n)^\top \in \mathbb{R}^n$. Let $\mathcal{H}$ be a reproducing kernel Hilbert or Krein space with kernel $k$. The evaluation operator $T$ and its adjoint are defined by:

$$T : \mathcal{H} \longrightarrow \mathbb{R}^n \qquad\qquad T^\star : \mathbb{R}^n \longrightarrow \mathcal{H}$$
$$f \longmapsto (f(\mathbf{x}_1), ..., f(\mathbf{x}_n))^\top \qquad \alpha \longmapsto T^\star \alpha = \sum_{i=1}^n \alpha_i k(\mathbf{x}_i, x)$$

The image of operator $T^\star$ is $\text{Im}(T^\star) = \{f \in \mathcal{H} \mid \exists \alpha \in \mathbb{R}^n \text{ such that } f(x) = \sum_{i=1}^n \alpha_i k(\mathbf{x}_i, x)\}$. The null space of operator $T$ is $\text{Null}(T) = \{f \in \mathcal{H} \mid f(\mathbf{x}_i) = 0, \ i = 1, n\}$. Based on the evaluation functional and its adjoint the Gram matrix $K = TT^\star$ is defined such that $K_{ij} = k(\mathbf{x}_i, \mathbf{x}_j)$.

**Definition 3.1.** *Interpolation set.* For a given set of observations $\mathbf{x}_i, y_i, i = 1, n$, the interpolation set in $\mathcal{H}$ is defined by $\mathcal{S} = \{f \in \mathcal{H} \mid Tf = y\}$.

Note that all these definitions hold for both reproducing kernel Hilbert and Krein spaces.

### 3.2. *Interpolation in an RKKS*

In this framework the minimal norm interpolating problem is the following:

**Definition 3.2.** *Interpolation Spline in an RKHS.* For a given set of observations $\mathbf{x}_i, y_i, i = 1, n$, the interpolation spline in a RKHS $\mathcal{H}$ is the solution of the following minimization problem:

$$\min_{f \in \mathcal{H}} \|f\|_{\mathcal{H}}^2 \quad \text{such that: } Tf = \mathbf{y}$$

**Theorem 3.1.** Equivalent definition for splines (Theorem 58[3]). *If the interpolation set associated with the interpolation spline problem is not empty then the interpolation spline in an RKHS, denoted by $\tilde{f}$, is the unique orthogonal projection of the interpolation set on the image of $T^\star$.*

**Proof.** $\mathcal{H}$ can be decomposed in two orthogonal subspaces such that $\mathcal{H} = \mathrm{Im}(T^\star) \oplus$ Null($T$) and $\forall f \in \mathcal{H}\ \exists f^\star \in \mathrm{Im}(T^\star)$ and $f_r \in \mathrm{Null}(T)$ such that $f = f^\star + f_r$. It follows that $Tf = Tf^\star$ and because of the minimal norm principle, the interpolation spline verifies $f_r = 0$. The interpolation spline $\widetilde{f}$ is the unique solution of the following problem:

$$\text{find } f \in \mathrm{Im}(T^\star) \text{ such that } Tf = y$$

It verifies: $\widetilde{f} = T^\star \alpha$ where vector $\alpha$ is the solution of the following linear system: $K\alpha = \mathbf{y}$.

Thus, $\forall f \in \mathcal{S}$ and $\forall \mathbf{v} \in \mathbb{R}^n$, $\mathbf{v}^\top T f - \mathbf{v}^\top T \widetilde{f} = 0$
$$\mathbf{v}^\top T f - \mathbf{v}^\top T \widetilde{f} = 0$$
$$\langle T^\star \mathbf{v}, f - \widetilde{f} \rangle_{\mathcal{H}} = 0$$

since $\widetilde{f}$ belongs to $\mathrm{Im}(T^\star)$ it is also the orthogonal projection of $\mathcal{S}$ on $\mathrm{Im}(T^\star)$. It is unique since $\mathrm{Im}(T^\star)$ is closed and convex. $\qquad\square$

This theorem gives an alternative way to define the interpolation spline without using a norm. While no norm is directly available in a Krein space, the second definition still holds and we have:

**Definition 3.3.** *Interpolation Spline in an RKKS.* The interpolation spline in an RKKS is the orthogonal projection of the interpolating functions on the set spanned by the kernel.

It can be computed in the same way as in the Hilbert case: first solve $K\alpha = \mathbf{y}$ and then the interpolation spline is the function $\widetilde{f}(x) = \sum_{i=1}^{n} \alpha_i k(\mathbf{x}_i, x)$.

### 3.3. *Smoothing splines in an RKKS*

Smoothing splines deal with the case where the target values are known up to a certain amount of error. In this case we have $Tf = y + \varepsilon$ where $\varepsilon$ denotes some error. The principle of smoothing splines suggests minimizing the norm of the error vector together with some regularity constraints, leading to:

**Definition 3.4.** *Smoothing splines in an RKHS (1).* For a given $\lambda$, a smoothing spline is the solution of the following minimization problem:

$$\mathcal{P}_\lambda : \quad \min_{f \in \mathcal{H}} \sum_{i=1}^{n} (f(\mathbf{x}_i) - y_i)^2 + \lambda \|f\|_{\mathcal{H}}^2$$

It is worth noticing that this minimization problem may be seen as a minimal decomposition in complementary spaces. Precisely, instead of decomposing $\mathcal{H} = \mathrm{Im}(T^\star) \oplus \mathrm{Im}(T^\star)^\perp$, change the inner product on $\mathrm{Im}(T^\star)$ to a new one $\langle .,. \rangle_\lambda$ such that $\|f\|_{\mathcal{H}} \leq \|f\|_\lambda$, $\forall f \in \mathrm{Im}(T^\star)$. Then there exists a unique complementary space $Q_\lambda$ (contractively included in $\mathcal{H}$) and a unique minimal decomposition which actually solves the problem.

As a matter of fact, depending on the value of parameter $\lambda$, the smoothing spline is not a single solution but rather a sequence of solutions. The sequence may be defined as a decreasing sequence of Hilbert norms on $\text{Im}(T^*)$ *i.e.*

$$\lambda < \lambda' \Rightarrow \|f\|_{\lambda'} \leq \|f\|_\lambda, \; \forall f \in \text{Im}(T^*)$$

This sequence can also be defined by using a sequence of embedded sets $\mathcal{H}_\lambda$:

**Definition 3.5.** *Smoothing splines in an RKHS (2).* Let $c_\lambda$ be a sequence of increasing positive real numbers. Then for $\mathcal{H}_\lambda = \{f \in \mathcal{H} \mid \|f\|_\mathcal{H}^2 \leq c_\lambda\}$ a smoothing spline is the sequence of solutions of the following minimization problem:

$$\mathcal{P}_\lambda: \quad \min_{f \in \mathcal{H}_\lambda} \sum_{i=1}^n (f(\mathbf{x}_i) - y_i)^2$$

Note that for $c_\lambda$ large enough, the solution to the problem $\mathcal{P}_\lambda$ is also the interpolation spline $\tilde{f}$. Thus the sequence $\tilde{f}_\lambda$ of solutions of problems $\mathcal{P}_\lambda$ converges towards $\tilde{f}$. The definition of this sequence depends only on the definition of the sequence of embedded spaces. This suggest the following analogous definition:

**Definition 3.6.** *Smoothing splines in an RKKS.* Let $\mathcal{K}_\mu$ be a sequence of embedded reproducing kernel Krein subspaces converging towards $\text{Im}(T^*)$ when $\mu$ increases towards $\infty$ (we have $\mu_1 < \mu_2 \Rightarrow \mathcal{K}_{\mu_1} \subseteq \mathcal{K}_{\mu_2}$). Then a smoothing spline is the sequence of solutions of the following minimization problem:

$$\mathcal{P}_\mu: \quad \min_{f \in \mathcal{K}_\mu} \sum_{i=1}^n (f(\mathbf{x}_i) - y_i)^2$$

The remaining problem is to define the sequence of embedded spaces. In this framework the smoothing effect is due to an early stopping in the regularization path.

## 4. Implementation of smoothing splines in an RKKS

### 4.1. *Three different regularization strategies*

There exists three different ways to build a sequence of embedded spaces. These subspaces can be defined by limiting their size (it is the penalization approach), by explicitly using more and more generating functions (such as basis functions) or through an iterative process building implicitly at each iteration a more complex solution. We will show that, for smoothing splines in an RKKS, the latter is preferable.

The first regularization principle is the classical penalization approach also known as Tikhonov regularization. But since no norm is available in a Krein space a different variational principle has to be used. Instead of looking for a minimum the problem can be restated using the stabilization principle. The stabilizer of some functional is the point where the gradient vanishes.

$$\mathcal{P}_\lambda: \underset{f \in \mathcal{K}}{\text{stabilize }} \|Tf - \mathbf{y}\|^2 + \lambda \langle f, f \rangle_\mathcal{K}$$

This problem is the same as the one of the Hilbertian case but also presents a major drawback. Just like the RKHS spline the solution of this problem is given by the solution of: $(K + \lambda I)\alpha = \mathbf{y}$. But in the RKKS case, the matrix $K$ is no longer positive definite and this strategy, because it increases all eigenvalues by a factor $\lambda$, can lead to a singular system. An interesting way to bypass this problem would be the one of complementary spaces. However the building of a decreasing sequence of indefinite inner products on $\mathrm{Im}(T^\star)$ is far more difficult than in the positive case. The second regularizing strategy consists in building explicit subspaces. Since the interpolating solution lies in $\mathrm{Im}(T^\star)$ the truncated spectral factorization approach suggests using the eigenvectors of the Gram matrix $K$. At each step the following problem has to be solved

$$\mathcal{P}_\ell : \min_{\alpha \in \mathcal{K}_\ell} \|K\alpha - \mathbf{y}\|^2$$

where $\mathcal{K}_\ell = \mathrm{span}\{\phi_1, ..., \phi_\ell\}$, $\phi_\ell$ being the $\ell^{\mathrm{th}}$ eigenvector of $K$. But this technique is not computationally suitable since it requires the computation of the spectrum of $K$. The third regularization strategy consists in implicitly building the solution subspace by using an iterative approach. There are two well known iterative approaches: gradient iterations and Krylov subspace techniques. Gradient iterations (also known as Landweber-Friedman iterations) propose for a given stepsize $\rho$:

$$\alpha_\ell = \alpha_{\ell-1} - \rho \nabla_\alpha (\|K\alpha_{\ell-1} - \mathbf{y}\|^2).$$

Krylov subspace (also known as conjugate gradient type) methods propose to build a sequence of iteration polynomials $q_\ell$ to get $\alpha_\ell = q_{\ell-1}(K)\mathbf{y}$. When possible, the later is preferable because it converges faster for the same computational complexity as the fixed stepsize gradient. Furthermore, there exists Krylov subspace approaches dealing with non positive Gram matrices such as the Minimal Residual II (MR II) algorithm.

### 4.2. *MR II: a Krylov subspace algorithm for indefinite matrix*

The MR II algorithm[5] is an iterative procedure computing the following sequence:

$$\alpha_\ell = \alpha_{\ell-1} - \rho \mathbf{d}_\ell$$

where $\rho$ is the stepsize and $\mathbf{d}_\ell$ the descent direction. The sequence converges towards the solution of the linear interpolating system $K\alpha = \mathbf{y}$. The MR II algorithm is also a minimal residual algorithm for symmetric indefinite linear systems with starting vector $K\mathbf{y}$ instead of $\mathbf{y}$. By minimal residual we mean that it minimizes at each step the squared residual error:

$$\min_{\alpha \in \mathbf{R}^n} \|K\alpha - \mathbf{y}\|^2.$$

At each step, coefficient $\rho$ is chosen to minimize this residual, *i.e.* such that the differential of the cost function vanishes. Descent directions $\mathbf{d}_\ell$ are built to be $K^2$

conjugates to decouple the cost function. Thus we have:

$$\alpha = \sum_{i=1}^{n} a_i d_i \quad \text{with} \quad d_i^\top KK d_j = 0, i \neq j$$

These directions $d_\ell$ are determined based on a three term recurrence formula (a standard Lanczos process). This kind of algorithm belongs to the family of Krylov subspace algorithms since it admits a polynomial interpretation:

$$\alpha_\ell = q_{\ell-1}(K)\mathbf{y} \qquad r_\ell = p_\ell(K)\mathbf{y}$$
$$r_\ell = \mathbf{y} - K\alpha_\ell \Longrightarrow p_\ell(K) = I - Kq_{\ell-1}(K)$$

where $q_{\ell-1}$ and $p_\ell$ are two families of orthogonal polynomials is some sense. For regularization purposes[b] the Krylov space over which the residual is minimized is chosen as follows:

$$\alpha_\ell \in \text{span}\{K\mathbf{y}, K^2\mathbf{y}, ..., K^{\ell-1}\mathbf{y}\}$$

Algorithm 4.1 summarizes MR II. Note that for practical reasons it is preferable to compute another vector sequence storing $K d_\ell$ omitted here for brevity and clarity.

---

**Algorithm 4.1** : *Kernel Spline MR II*

---

1. $(\alpha_1, r_1, d_0, d_{-1}, c, \ell) \leftarrow$ initialize $(0, \mathbf{y}, K\mathbf{y}/c, 0, (\mathbf{y}^\top K^4 \mathbf{y})^{\frac{1}{2}}, 1)$
**while** SolutionReached ==FALSE
    2. Compute next solution
        2.1 Compute the stepsize      $\rho \leftarrow r_\ell^\top K d_{\ell-1}$
        2.2 Compute the next solution    $\alpha_{\ell+1} \leftarrow \alpha_\ell + \rho d_{\ell-1}$
    3. Update the residual          $r_{\ell+1} \leftarrow \mathbf{y} - K\alpha_{\ell+1}$
    4. Compute a new descent direction
        4.1 Compute the stepsize      $b \leftarrow d_{\ell-1}^\top K^3 d_{\ell-1}$
        4.2 Update the direction      $d_\ell \leftarrow K d_{\ell-1} - b d_{\ell-1} - c d_{\ell-2}$
        4.3 Compute the stepsize      $c \leftarrow (d_\ell^\top K^2 d_\ell)^{\frac{1}{2}}$
        4.4 Normalize the direction    $d_\ell \leftarrow \frac{d_\ell}{c}$
    5. Counting the number of iterations    $\ell \leftarrow \ell + 1$
    **if** $\|\mathbf{r}\| < \varepsilon$ or $\ell = n$
        6. SolutionReached $\leftarrow$ TRUE
    **end if**
**end while**

---

MRII algorithm is not *the only* solution of the approximation spline in indefinite reproducing kernel space but one among several others. It presents the advantage of being very fast and regularizing. Its main drawback is that it is not flexible

---

[b]Regularizing properties of this method are detailed in chapter 6[5].

regarding the choice of the regularization parameter. Because the regularization process is controlled through an early stopping mechanism, it may happen that the "truth" is lying between two iterations.

But since the RKKS framework is fruitful, many other algorithms remain to be found.

## 5. Conclusion

In this paper we have shown how to do splines using non positive kernels. A functional framework has been designed together with new definitions of what splines are. These definitions are coherent with usual splines in reproducing kernel Hilbert spaces (RKHS) and can be applied to non positive kernels.

Regarding the functional framework, we show that reproducing kernel spaces are not necessarily Hilbertian and reproducing kernel Krein spaces (RKKS) can be built from an indefinite kernel in almost the same way as RKHS is built from positive kernels. In this framework, the representation of the evaluation functional through kernels still holds. The main difference between RKHS and RKKS is that no norm is induced by the inner product in the latter structure. So usual splines were revisited to be defined without norms. Interpolating splines are seen as an orthogonal projection while approximation splines are defined as a regularized family of solutions to the interpolation problem. A solution to this problem has been proposed through the use of MR II, an efficient iterative conjugate gradient technique, suitable for indefinite matrices. Note that large problems can be solved using this algorithm since its complexity is $\mathcal{O}(kn^2)$ where $k$ is the number of iteration needed.

We have reasons to believe that this complexity can be practically reduced. The strategy to achieve such a result would be the same as in SVM, to consider a regularization mechanism involving a sparse representation (a lot of components of the solution vector set to 0). Note also that this framework could be extended to non symmetric kernels[7] and to the pattern recognition case.

## References

1. D. Alpay The Schur algorithm, reproducing kernel spaces and system theory, volume 5, SMF/AMS Texts and Monographs of the Société mathématique de France (2001)
2. T.Ya. Azizov and I.S. Iokhvidov, Linear operators in spaces with an indefinite metric, Wiley (1989)
3. A. Berlinet and C. Thomas-Agnan, Reproducing Kernel Hilbert Spaces in Probability and Statistics, Kluwer Academic Pub (2004)
4. J. Bognar, Indefinite Inner Product Spaces, Springer-Verlag (1974)
5. M. Hanke Conjugate Gradient Type Methods for Ill-Posed Problems, Pitman Research Notes in Mathematics 327, (1995)
6. B. Hassibi, A. H. Sayed, and T. Kailath, Indefinite-Quadratic Estimation and Control: A Unified Approach to $H^2$ and $H^\infty$ Theories, SIAM, Philadelphia (1999)
7. X. Mary, Hilbert subspaces, subdualities and applications, PhD thesis, Laboratory PSI, INSA Rouen (2003)

8. J. Rovnyak, Methods of Krein space operator theory, Toeplitz lectures, Interpolation theory, systems theory, and related topics, Oper. Theory Adv. Appl., vol. 134, Birkhauser, Basel, 2002, pp. 31–66, (1999)
9. G. Wahba, Spline Models for Observational Data, SIAM, CBMS-NSF Regional Conference Series in Applied Mathematics, volume 59 (1990)

# ALGEBRAIC, DIFFERENTIAL, INTEGRAL AND SPECTRAL PROPERTIES OF MERCER-LIKE-KERNELS

JORGE BUESCU*

*Dep. Matemática*
*Instituto Superior Técnico*
*1049-001 Lisboa, Portugal*
*E-mail: jbuescu@math.ist.utl.pt*


A. C. PAIXÃO

*Dep. Mecânica*
*ISEL*
*Lisboa, Portugal*

We present a survey of recent results by the authors which show that reproducing kernels enjoy a closely knit interplay of algebraic, differential, integral and spectral properties (whenever all are defined). The matrix characterization of reproducing kernels implies a family of differential diagonal dominance inequalities, which in turn acts as the appropriate integrability condition when these are kernels of $L^2$ integral operators. This fact allows complete and optimal determination of eigenvalue asymptotics.

**Key words:** Reproducing kernels, positive integral operators, eigenvalues
**Mathematics Subject Classification:** 45C05, 45P05

## 1. Introduction

Recent developments have shown a renewed interest in the study of 'positive definite matrices' in the sense of Moore or, equivalently, reproducing kernels[28,29]. The large number of applications of this subject, ranging from Complex Analysis to Operator Theory, Statistical Learning Theory or Integral Transforms extends to problems in areas such as signal processing, from which we can trace the early questions motivating our early work[3]. Indeed, many physical phenomena are modelled by random processes; for second order processes, reconstruction of the signal by sampling requires consideration of the autocorrelation function both in the time and frequency domains. This function is by construction a reproducing kernel. It thus becomes a problem of interest for applications to study this class of two variable functions from the point of view of Fourier transforms and taking into account the structural role played by the diagonal[3,4].

This study is naturally carried out in the context of $L^2(\mathbb{R}^2)$ functions; it has

---

*Work partially supported by CAMGSD through FCT/POCTI/FEDER.

been shown[5] that many of the relevant properties carry through the Fourier trans-formation. This context, on the other hand, suggests the convenience of clarifying the links between reproducing kernels and the kernels of positive integral operators, allowing the use of well known properties from the respective general theory.

We address this issue in §3, showing that under the assumptions of continuity and summability along the diagonal, positivity implies that a reproducing kernel $k(x, y)$ is the kernel of a positive integral operator $K$ in $L^2(\mathbb{R})$ and, moreover, that this operator is Hilbert-Schmidt and thus necessarily compact since $k$ is in $L^2(\mathbb{R}^2)$. While emphasizing the importance of the diagonal, this fact implies, according to standard spectral theory, that $k$ is expressed by an $L^2$-convergent bilinear series constructed from the eigenfunctions of $K$ associated with the countable sequence of positive eigenvalues of the operator.

The spectrum and bilinear eigenfunction series expansion thus become useful concepts on which to base the present study. Specific properties that may be asserted with respect to these notions may then serve as a basis for characterizing the kernel $k$ and deriving consequences. In this context, our attention is naturally driven to the contents of the classical theorem of Mercer, which establishes that, if $k$ is continuous positive definite kernel defined on a compact interval $I$, then the eigenfunction series is absolutely and uniformly convergent and the associated operator is trace class. Mercer's theorem does not apply in the case where $I = \mathbb{R}$. However, in §3 we show that the assumptions of continuity and summability along the diagonal together with the further requirement that $k(x, x) \to 0$ as $|x| \to \infty$ provide the necessary and sufficient conditions under which the contents of Mercer's theorem still hold for kernels defined on unbounded domains. This fact suggests the use of the term *Mercer-like* to specify any kernel verifying the above hypothesis. Mercer-like kernels can equivalently be defined as the continuous positive definite kernels which are integrable and uniformly continuous along the diagonal, thus coinciding, if the domain is compact, with the continuous positive definite kernels to which Mercer's theorem applies.

The issue of eigenvalue distribution of an operator associated with a Mercer-like kernel $k$ arises naturally in this context. Compactness of these operators assures the spectral characteristics upon which the study of eigenvalues can be undertaken without the restraints and adaptations needed for other kinds of operators. The case where the domain of $k$ is compact is very well studied, and a number of results have been established which describe sharply how the rate of decay of eigenvalues depends on positivity and the regularity of the kernel[13,14,23–26]. Most of these results rely on the use of techniques based on the best approximation, in the trace norm, of the operator $K$ by finite rank operators. In §4 we show that these techniques can be appropriately extended to the case of operators associated with Mercer-like kernels defined on unbounded domains. Our results show that in this case the eigenvalue distribution depends not only on the regularity – differentiability and Lipschitz continuity – of the kernel, but also on its rate of decay along the diagonal, yielding the above referred results for the compact domain case when the support of $k$ is compact. Very different approaches to related issues have been followed in some works[2,18,21,22], setting the discussion in the context of general integral operators and providing for more or less complex solutions to deal with the general lack of compactness of the operator. However, and contrarily to our main concern, no account has been taken for the particular effect of positivity in the works addressing the issue from this general point of view.

Along this study, a natural use is made of certain classes of differentiable Mercer-like kernels. On the other hand, a number of properties relative to differentiation are known to hold in the case of continuous positive definite kernels defined on compact domains[17]. These facts suggest the question of whether these properties may be extended to Mercer-like kernels defined on unbounded domains. In the second part of §3 we give a positive answer to this question by showing that appropriately differentiable Mercer-like kernels have correspondingly differentiable eigenfunctions and admit an eigenfunction series expansion which may be termwise differentiated while maintaining strong convergence properties. The partial mixed symmetric derivatives of $k$ play a central role in this study: they are shown to define reproducing kernels on their own and, as a consequence of the aforementioned properties, to have a bilinear series expansion irrespective of being associated with a Hilbert-Schmidt operator. Under further hypothesis on their behavior along the diagonal they are shown to be Mercer-like kernels, in which case sharp estimates for norm bounds of eigenfunctions can be established.

Integrability conditions necessary for the proof of these results are derived from properties established in §2, a fact which finally leads our discussion back to reproducing kernels. These properties are seen to have its most fundamental expression in the context of the general definition of positive matrices in the sense of Moore and generalize the basic diagonal dominance inequality $|k(x,y)|^2 \leq k(x,x)k(y,y)$ (which holds for any Moore matrix $k$ defined on an arbitrary set $E$). However, the resulting class of two parameter inequalities has an analytical character which can only be expressed if, as in the case of $\mathbb{R}$, the domain of the kernel is endowed with a convenient differential structure and $k$ is appropriately differentiable. Proofs are constructed directly from the definition of Moore matrices using algebraic arguments associated with separation in matrix blocks and the study of order two submatrices of positive semidefinite matrices as well as the possibility of expressing derivatives as limits of finite differences. In particular, they seem to provide an answer as to the exact methods by which the proof of the classical diagonal dominance inequality can be extended to account for the consequences of the rich algebraic information contained in the definition of a Moore matrix.

Since the method is adaptable to other instances of the set $E$, we include an application to the important case where $E = \Omega \subset \mathbb{C}^n$ and the reproducing kernel $k$ satisfies an appropriate holomorphy condition. Application to other contexts is naturally conceivable.

As a final remark, we observe that both the results in §2 and their applications in §3 may be interpreted within the abstract Krein[19] and reproducing kernel[28,29] approaches, based on the construction of the reproducing kernel Hilbert space $\mathcal{H}_k$. This interpretation, however, implies the clarification of some nontrivial difficulties which are avoided by the direct algebraic approach that we have chosen.

## 2. Differentiable reproducing kernels: an algebraic approach

### 2.1. *Moore matrices and reproducing kernels*

A *positive definite matrix in the sense of Moore*, or Moore matrix for short, is a function $k : E \times E \to \mathbb{C}$ such that

$$\sum_{i,j=1}^{n} k(x_i, x_j)\, \overline{\xi_i}\, \xi_j \geq 0 \tag{1}$$

for all $n \in \mathbb{N}$, $(x_1, \cdots, x_n) \in E^n$ and $(\xi_1, \cdots, \xi_n) \in \mathbb{C}^n$. Moore matrices enjoy the following basic and well-known properties:

$$\forall x, y \in E \quad k(x, y) = \overline{k(y, x)}$$
$$\forall x \in E \quad k(x, x) \geq 0, \tag{2}$$
$$\forall x, y \in E \quad |k(x, y)|^2 \leq k(x, x)\, k(y, y).$$

These properties are immediate consequences of the inequalities obtained in (1) for $n = 1$ and $n = 2$ and will be referred to generically as *diagonal dominance*. This section will show that (1) implies, by consideration of the cases $n \geq 2$, a generalization of (2) which, in contrast with the classical case, has a specific analytical nature which is only apparent when $E$ admits a differentiable structure and $k$ is appropriately differentiable. The construction is made in two contexts: that of functions of two real variables, having in mind applications to integral operators, and that of holomorphic functions of several complex variables. Our approach is through the linear algebraic structure provided by (1). The Moore-Aronszajn theorem[1] [20] [28] [29], which characterizes positive definite matrices in the sense of Moore as reproducing kernels, is not used except for terminology.

### 2.2. *Two results on positive semidefinite matrices*

We mention two algebraic results relative to block matrix decomposition of positive semidefinite matrices crucial for the proof of the main results in this section [6] [7] [8].

Let $m$ and $r$ be positive integers and $A$ be a square matrix of order $r(m + 1)$. In Proposition 2.1 below we denote by $A^{pq}$ the order $r$ square submatrices of $A$, with $p, q = 0, \ldots, m$, resulting from the partition of $A$ into the $m+1$ square blocks defined by $[A^{pq}]_{ij} \equiv \left[a_{ij}^{pq}\right] = [a_{st}]$ for $s = i + pr$, $t = j + qr$ and $i, j = 1, \ldots, r$.

**Proposition 2.1.** *Let $A$ be an $r(m + 1)$ square matrix. For each $X = (x_0, \ldots, x_m) \in \mathbb{C}^{m+1}$, define the $r \times r$ matrix*

$$\mathcal{A}(X) = \sum_{p,q=0}^{m} A^{pq}\, x_p \overline{x_q}.$$

If $A$ is positive semidefinite, then for every $X \in \mathbb{C}^{m+1}$ the matrix $\mathcal{A}(X)$ is positive semidefinite.

**Proposition 2.2.** *Let $T$ be a square matrix of order $r_1 + r_2$ partitioned in the block form*

$$T = \left[ \begin{array}{c|c} A & B \\ \hline D & C \end{array} \right]$$

*where $A = [a_{ij}]$, $B = [b_{iq}]$, $C = [c_{pq}]$, $D = [d_{pj}]$ with $i, j = 1, \ldots, r_1$ and $p, q = 1, \ldots, r_2$. Define $\alpha_{ij}^{pq} = a_{ij}c_{pq} - b_{iq}d_{pj}$ and let $X = (x_1, \ldots, x_{r_1}) \in \mathbb{C}^{r_1}$, $Y = (y_1, \ldots, y_{r_2}) \in \mathbb{C}^{r_2}$. Then, if $T$ is positive semidefinite, we have*

$$\sum_{i,j=1}^{r_1} \sum_{p,q=1}^{r_2} \alpha_{ij}^{pq} x_i \overline{x_j}\, y_p \overline{y_q} \geq 0.$$

## 2.3. Differential inequalities for reproducing kernels

### 2.3.1. Inequalities for differentiable reproducing kernels in $\mathbb{R}^2$

We explore some of the consequences of the differentiability hypothesis for a reproducing kernel (or Moore matrix) $k$ defined on the interval $I \subset \mathbb{R}^6$. Continuity or differentiability at boundary points should be understood in the usual sense of appropriate one-sided limits.

**Definition 2.1.** Let $I \subset \mathbb{R}$ be an interval. $k : I^2 \to \mathbb{C}$ is said to be *in class* $\mathcal{S}_n(I)$ if for every $m_1 = 0, 1, \ldots n$ and $m_2 = 0, 1, \ldots n$, the partial derivatives $\dfrac{\partial^{m_1+m_2}}{\partial y^{m_2} \partial x^{m_1}} k(x, y)$ are continuous in $I^2$.

It is possible to show that the above derivatives may be represented as limits of adequate finite differences of $k$. Using proposition 2.1 we obtain

**Theorem 2.2.** *Let $I \subseteq \mathbb{R}$ be an interval and $k(x, y)$ be a reproducing kernel in class $\mathcal{S}_n(I)$. Then, for every $0 \leq m \leq n$,*

$$k_m(x, y) \equiv \frac{\partial^{2m}}{\partial y^m \partial x^m} k(x, y)$$

*is a reproducing kernel in class $\mathcal{S}_{n-m}(I)$.*

Conjugate symmetry of $k$ and a particular version of proposition 2.2 imply the following generalization of the diagonal dominance inequalities (2):

**Theorem 2.3.** *Let $I \subseteq \mathbb{R}$ be an interval and $k(x, y)$ a reproducing kernel in class $\mathcal{S}_n(I)$. Then, for every $m_1, m_2 = 0, 1, \ldots, n$ and $x, y \in I$ we have*

$$\left| \frac{\partial^{m_1+m_2}}{\partial y^{m_2} \partial x^{m_1}} k(x, y) \right|^2 \leq \frac{\partial^{2m_1}}{\partial y^{m_1} \partial x^{m_1}} k(x, x) \frac{\partial^{2m_2}}{\partial y^{m_2} \partial x^{m_2}} k(y, y). \tag{3}$$

180

### 2.3.2. *Inequalities for holomorphic reproducing kernels in $\mathbb{C}^n$*

In this section we treat the case where $E = \Omega$ is a domain in $\mathbb{C}^{n}$[7,8]. We say that $k(Z, U)$ is a *holomorphic reproducing kernel* if it is a reproducing kernel and is sesquiholomorphic (that is, holomorphic in the first variable and anti-holomorphic in the second) as a function in $\mathbb{C}^{2n}$. This class is the appropriate starting point for our study. It is easily shown that biholomorphic reproducing kernels are necessarily constant.

We shall use the following notation: a point in $\mathbb{C}^n$ is denoted by un uppercase variable, e.g. $Z = (z_1, \ldots, z_n)$, with $z_j \in \mathbb{C}$. A set of $n$ integers $M \in \mathbb{N}^n$ will be denoted by $M = (m_1, \ldots, m_n)$ with $m_j \in \mathbb{N}$ and referred to as a multi-index. We shall also denote $|M| = \sum_{j=1}^{n} m_j$.

For a sesquiholomorphic function and, in particular, for a holomorphic reproducing kernel all derivatives of the form

$$\frac{\partial^{|M_1+M_2|}}{\partial \overline{U}^{M_2} \partial Z^{M_1}} k(Z, U) \tag{4}$$

exist, are continuous and are independent of the order of differentiation.

Having established the possibility of representing the above derivatives by limits of appropriate finite differences of $k$ and using proposition 2.1 we obtain

**Theorem 2.4.** *Let $\Omega \subset \mathbb{C}^n$ be a domain and $k : \Omega^2 \to \mathbb{C}$ be a holomorphic reproducing kernel in $\Omega$. Then, for every multi-index $M \in \mathbb{N}^n$,*

$$D_M k(Z, U) \equiv k_M(Z, U) \equiv \frac{\partial^{2|M|}}{\partial \overline{U}^{M} \partial Z^{M}} k(Z, U)$$

*is a holomorphic reproducing kernel in $\Omega$.*

Using proposition 2.2 it is now possible to show the following generalization of the diagonal dominance inequalities (2):

**Theorem 2.5.** *Let $\Omega \subset \mathbb{C}^n$ and $k : \Omega^2 \to \mathbb{C}$ a holomorphic reproducing kernel. For every $M_1, M_2 \in \mathbb{N}^n$ and all $Z, U \in \Omega$ we have*

$$\left| \frac{\partial^{|M_1+M_2|}}{\partial \overline{U}^{M_2} \partial Z^{M_1}} k(Z, U) \right|^2 \leq \frac{\partial^{2|M_1|}}{\partial \overline{U}^{M_1} \partial Z^{M_1}} k(Z, Z) \, \frac{\partial^{2|M_2|}}{\partial \overline{U}^{M_2} \partial Z^{M_2}} k(U, U). \tag{5}$$

### 2.4. *The RKHS approach*

Theorems 2.2 and 2.4 and the corresponding families of inequalities (3) and (5) may be interpreted in the context of the general theory of reproducing kernels (see e.g. Krein[19], Saitoh[28,29]). Indeed, it is possible to show that inequalities (2.3) and (5) may be derived, within the abstract RKHS, as a form of the Cauchy-Schwarz inequality. On the other hand, the proofs we offer of theorems 2.2, 2.3, 2.4 and 2.5 bypass the difficulties associated to this general construction and show that the corresponding results are a direct consequence of the algebraic-analytical properties implied by the definition of reproducing kernel itself.

## 3. Positive integral operators and reproducing kernels

### 3.1. *Positive definite kernels in bounded and unbounded domains*

Given an interval $I \subseteq \mathbb{R}$, a linear operator $K : L^2(I) \to L^2(I)$ is integral if there exists a measurable function $k$ (called the kernel of $K$) such that, for all $\phi \in L^2(I)$,

$$K(\phi) = \int_I k(x,y)\,\phi(y)\,dy$$

almost everywhere. If, for all $\phi \in L^2(I)$, $K$ satisfies

$$\int_I \int_I k(x,y)\,\overline{\phi(x)}\,\phi(y)\,dx\,dy \geq 0 \tag{6}$$

then it is a positive operator and the corresponding kernel $k(x,y)$ is a *positive definite kernel* on $I$. We focus in the case where $k$ is a positive definite kernel and $k(x,y) \in L^2(I^2)$. In this case, $K$ is a Hilbert-Schmidt operator, thus compact. Its spectrum consists of a finite or countable sequence of non-negative eigenvalues of finite multiplicity accumulating at most at 0. Standard operator theory yields the expansion

$$k(x,y) = \sum_{i \geq 1} \lambda_i\,\phi_i(x)\,\overline{\phi_i(y)}, \tag{7}$$

where the $\{\phi_i\}_{i \geq 1}$ are an $L^2(I)$-orthonormal system spanning the range of $K$, $\{\lambda_i\}_{i \geq 1}$ are the nonzero eigenvalues of $K$ and the series (7) is $L^2$-convergent. In the case where $I$ is compact and $k$ is continuous, the classical theorem of Mercer states that the eigenfunctions $\phi_i(x)$ in (7) are continuous, the corresponding series converges absolutely and uniformly and the operator $K$ is trace class with

$$\mathrm{tr}\,K = \int_I k(x,x)\,dx = \sum_{i \geq 1} \lambda_i. \tag{8}$$

It also also possible to show[17] that, for kernels satisfying appropriate differentiability assumptions, series (7) may be termwise differentiated and the resulting series converge absolutely and uniformly on $I$ to the corresponding derivatives of $k$.

The next definition will be instrumental in establishing the conditions under which the above results may be extended to the case of unbounded domains.

**Definition 3.1.** A positive definite kernel $k(x,y)$ on an interval $I \subseteq \mathbb{R}$ (possibly unbounded) is called a *Mercer-like kernel* if

(1) $k(x,y)$ is continuous in $I^2$,
(2) $k(x,x) \in L^1(I)$,
(3) $k(x,x)$ is uniformly continuous on $I$.

Notice that if $I$ is compact, the first condition trivially implies the other two, thus resulting in the classical setting of Mercer's theorem. Also notice that a Mercer-like kernel is always, in particular, an $L^2$-positive definite kernel.

In § 3.3 the class of Mercer-like kernels will be seen to specify necessary and sufficient conditions for the non-compact analogue of Mercer's theorem to hold[4,5]. In §3.4 we explore consequences of the condition $k \in \mathcal{S}_n(I)$, including the possibility of termwise differentiation of the series expansion (7) for a Mercer-like[11] kernel $k$.

In the next section we establish the precise relation between positive definite kernels and reproducing kernels. The general theory of reproducing kernels is directly used.

## 3.2. *Positive definite kernels as reproducing kernels*

Based on what happens for continuous positive definite kernels on compact intervals $I$, the following result, which will allow the use of the inequalities established for reproducing kernels in the study of positive definite kernels, may be established.

**Corollary 3.2.** *Let* $k : I^2 \to \mathbb{C}$ *be a continuous function such that* $k(x,x) \in L^1(I)$. *Then* $k$ *is a positive definite kernel if and only if it is a reproducing kernel. In these conditions,* $k \in L^2(I^2)$ *and*

$$k(x,y) = \sum_{i \geq 0} \mu_i \, \phi_i(x) \, \overline{\phi_i(y)}, \tag{9}$$

*where* $\{\phi_i\}_{i \geq 0}$ *are the* $L^2$-*orthonormal eigenfunctions of the operator* $K$ *which span its range,* $\mu_i$ *are the corresponding eigenvalues, which form a positive sequence accumulating at most at* 0, *and the series* (9) *is* $L^2$-*convergent.*

## 3.3. *Mercer-like kernels*

We next discuss the conditions under which Mercer's theorem extends to kernels on unbounded intervals. We use $\mathbb{R}$ as a model of such a domain, bearing in mind that the results generalize in the obvious way to other types of intervals. The following result establishes the crucial properties of Mercer-like kernels:

**Theorem 3.3.** *Let* $k : \mathbb{R}^2 \to \mathbb{C}$ *be a Mercer-like kernel. Then*

(i) *Eigenfunctions* $\phi_i$ *of the associated operator* $K$ *corresponding to nonzero eigenvalues are uniformly continuous and vanish at infinity.*

(ii) *The bilinear series* (9) *converges absolutely and uniformly to* $k$.

(iii) *The operator* $K$ *is trace class with* $\operatorname{tr} K = \int_{-\infty}^{+\infty} k(x,x)\,dx = \sum_{i \geq 0} \mu_i$.

Uniform continuity of $k(x,x)$ is essential for the validity of the results in Theorem 3.3; counterexamples are easily provided.

### 3.4. *Mercer-like kernels in classes $\mathcal{S}_n(\mathbb{R})$ and $\mathcal{A}_n(\mathbb{R})$*

To explore the consequences of differentiability hypotheses on Mercer-like kernels, we begin by using the results described in § 2.3 and § 3.2 to show the following lemma.

**Lemma 3.4.** *Let $k(x,y)$ be a Mercer-like kernel in class $\mathcal{S}_n(\mathbb{R})$, $n \geq 0$. Then:*

*(1) if $\phi_i$ is an eigenfunction associated to an eigenvalue $\lambda_i \neq 0$, then $\phi_i$ is in $C^n(\mathbb{R})$;*
*(2) each $k_m$ is a reproducing kernel in class $\mathcal{S}_{n-m}(\mathbb{R})$ and*

$$k_m(x,y) = \frac{\partial^{2m} k}{\partial y^m \partial x^m}(x,y) = \sum_{i \geq 1} \lambda_i \phi_i^{(m)}(x) \overline{\phi_i^{(m)}(y)} \tag{10}$$

*absolutely, pointwise in $\mathbb{R}^2$ and uniformly on compact sets of $\mathbb{R}^2$ for every $m = 0, \ldots, n$.*

**Corollary 3.5.** *Let $k(x,y)$ be a Mercer-like kernel in class $\mathcal{S}_n(\mathbb{R})$, $n \geq 0$. Then, in addition to the statements in lemma 3.4,*

$$\frac{\partial^{m_1+m_2}}{\partial y^{m_2} \partial x^{m_1}} k(x,y) = \sum_{i=1}^{\infty} \lambda_i \phi_i^{(m_1)}(x) \overline{\phi_i^{(m_2)}(y)}$$

*absolutely, pointwise in $\mathbb{R}^2$ and uniformly on compact sets of $\mathbb{R}^2$ for every $m_1$, $m_2 = 0, 1, \ldots, n$.*

Consider the following definition:

**Definition 3.6.** A function $k : \mathbb{R}^2 \to \mathbb{C}$ is said to be in class $\mathcal{A}_0(\mathbb{R})$ if:

(1) $k(x,y)$ is continuous in $\mathbb{R}^2$;
(2) $k(x,x) \in L^1(\mathbb{R})$;
(3) $k(x,x)$ is uniformly continuous on $\mathbb{R}$.

If $n \geq 1$ is an integer, $k$ is said to be in class $\mathcal{A}_n(\mathbb{R})$ if $k \in \mathcal{S}_n(\mathbb{R})$ and

$$k(x,y), \ \frac{\partial^2}{\partial y \partial x} k(x,y), \ \ldots, \ \frac{\partial^{2n}}{\partial y^n \partial x^n} k(x,y)$$

are in class $\mathcal{A}_0(\mathbb{R})$.

Observe that Mercer-like kernels are precisely the positive definite kernels in class $\mathcal{A}_0(\mathbb{R})$. In the sequence we write $\mathcal{K}_m \equiv \int_{-\infty}^{\infty} k_m(x,x)\, dx$ for every $k$ in class $\mathcal{A}_n(\mathbb{R})$ and $m = 0, \ldots, n$, and denote by $H^n(\mathbb{R})$ the Sobolev space $W^{n,2}(\mathbb{R})$. Note that for $m > 0$ the expansion (10) in lemma 3.4 is not, in general, the Schmidt series of an integral operator. Let $k(x,y)$ be a Mercer-like kernel in class $\mathcal{A}_n(\mathbb{R})$, $n \geq 0$. If $m \leq n$, then $\mathcal{K}_m$ is a Mercer-like kernel. Denoting

$$k_l(x,y) = \sum_{i \geq 1} \lambda_i^{[l]} \phi_i^{[l]}(x) \overline{\phi_i^{[l]}(y)},$$

where $\lambda_i^{[l]} \neq 0$ are the nonzero eigenvalues of the operator $K_l$ with kernel $k_l$ and $\{\phi_i^{[l]}\}_{i \geq 1}$ are the corresponding normalized eigenfunctions, the following result may be derived from lemma 3.4.

**Theorem 3.7.** *Let $k(x,y)$ be a Mercer-like kernel in class $\mathcal{A}_n(\mathbb{R})$, $n \geq 0$. Then*

*(1) If $\phi_i^{[l]}$ is an eigenfunction associated with an eigenvalue $\lambda_i^{[l]} \neq 0$, then $\phi_i^{[l]}$ is in $C^{n-l}(\mathbb{R}) \cap H^{n-l}(\mathbb{R})$ and*

$$\|\phi_i^{[l](m-l)}\|_{L^2(\mathbb{R})} \leq \left( \frac{\mathcal{K}_m}{\lambda_i^{[l]}} \right)^{1/2} \tag{11}$$

*for all $0 \leq l \leq m \leq n$.*

*(2) Each $k_m$ is a positive definite kernel in class $\mathcal{A}_{n-m}(\mathbb{R})$ and*

$$k_m(x,y) = \sum_{i \geq 1} \lambda_i^{[l]} \phi_i^{[l](m-l)}(x) \overline{\phi_i^{[l](m-l)}(y)} \tag{12}$$

*uniformly and absolutely in $\mathbb{R}^2$ for $0 \leq l \leq m \leq n$.*

*(3) The integral operator $K_m$ with kernel $k_m$ is trace class with*

$$tr(K_m) = \mathcal{K}_m = \int_{-\infty}^{\infty} k_m(x,x)\,dx = \sum_{i=1}^{\infty} \lambda_i^{[l]} \|\phi_i^{[l](m-l)}\|_{L^2(\mathbb{R})}^2. \tag{13}$$

Sharp bounds for $L^2$ and Sobolev norms of eigenfunctions may be obtained from this result.

## 4. Eigenvalue distribution for positive operators

### 4.1. *Introduction and preparatory results*

In this section we study the eigenvalue distribution of positive integral operators associated with the kernels defined in §3, namely Mercer-like kernels with domain $I^2$, where $I$ is an unbounded real interval[9,10]. The case where $I$ is compact is widely studied and the results well-known. Reade[23-27], Cochran and Lukas[14], Chang and Ha[13], for instance, establish upper bounds for the decay of eigenvalues of the form $\lambda_n = o(\frac{1}{n^\gamma})$ or $\lambda_n = O(\frac{1}{n^\gamma})$, where $\gamma$ depends on the smoothness hypotheses on the kernel, namely its differentiability and Lipschitz continuity classes. More general results are known for operators with compact or noncompact domain (see, e. g. Birman and Solomyak[2], Gohberg e Krein[15], König[18], Pietsch[21]), but these general results fail to take into account the specific consequences of the positivity of the operator.

We base our study on the class of differentiable Mercer-like kernels. The symmetric derivatives $k_m$ of these kernels play a vital role, in particular as a consequence of the properties known in the case where $I$ is compact.

In order to derive our main results, we use methods introduced by Ha[16] and Reade[23] in the context of kernels defined on a compact interval (here fixed as $[0, L]$)

as well as a relevant corollary. The discussion of the best approximation in the trace norm by symmetric kernels in $L^2([0, L]^2)$ with rank $\leq N$ is relevant. Also crucial are the relevant properties of the square root of a positive operator and the definition of an appropriate class of finite rank operators to be used in the approximation of the positive operator $K$ associated with the continuous positive definite kernel $k$ defined in $[0, L]^2$. Finally, based on a result of Ha[16] we prove and use an estimate for the study of eigenvalue distribution of the operator with kernel $k_m$, $m > 0$, where $k$ is a positive definite kernel in class $\mathcal{S}_m([0, L])$. This estimate plays a key role in the study of asymptotics of eigenvalues of $K$. Complete proofs of the results below may be found elsewhere[9,10].

## 4.2. Asymptotics of eigenvalue distribution

We now state our main results on eigenvalue distribution, taking $I = [0, +\infty[$ as a model unbounded interval. Let $k$ be a Mercer-like kernel in $[0, +\infty[$, and denote by $k^L$ the restriction of $k$ to $[0, L]^2$. $k$ and $k^L$ are both positive definite kernels associated with trace class operators. Defining $T_k(N) = \sum_{n=N+1}^{\infty} \lambda_n(k)$ one may show:

**Lemma 4.1.** *Let $k \in \mathcal{A}_0([0, +\infty[)$ and $k^L$ be the restriction of $k$ to $[0, L]^2$. Then, for sufficiently large $N$, $T_k(N) \leq T_{k^L}(N) + \int_L^{\infty} k(x, x)\, dx$.*

The following definitions are necessary for the statement of theorem 4.4.

**Definition 4.2.** *Let $k : [0, +\infty[ \to \mathbb{C}$. We say that $k(x, y)$ is uniformly continuous with respect to $y$ on the diagonal if for every $\epsilon > 0$ there is $\delta > 0$ such that $|k(x, x) - k(x, y)| < \epsilon$ whenever $|x - y| < \delta$.*

**Definition 4.3.** *Let $k : [0, +\infty[ \to \mathbb{C}$ and $\alpha \in ]0, 1]$. We say that $k(x, y)$ is $\alpha$-Lipschitz (written $Lip^{\alpha}$) with respect to $y$ on the diagonal if there is a positive constant $A$ such that $|k(x, x) - k(x, y)| \leq A|x - y|$ for every $(x, y) \in [0, +\infty[^2$.*

The main result in this section is obtained by means of the evaluation of $T_k(N)$ in terms of $T_{k^L}(N)$ which results from lemma 4.1 together with the application of the results mentioned in § 4.1 regarding best approximations to operators with kernel $k^L$ as well as its derivatives. Two technical results, details of which may be found elsewhere[10], are used; the first is essential in the proof of theorem 4.4, while the second shows that its results cannot be improved.

**Theorem 4.4.** *Let $m \geq 0$ and suppose $k(x, y)$ is a Mercer-like kernel in class $\mathcal{A}_0([0, +\infty[) \cap \mathcal{S}_m([0, +\infty[)$. Let $\{\lambda_n\}_{n \in \mathbb{N}}$ be the sequence of eigenvalues of the integral operator with kernel $k$. Then the following statements hold.*

*1.1 Suppose $k_m(x, y)$ is uniformly continuous with respect to $y$ on the diagonal. Then:*

i) If $\beta > 1$, $\int_L^\infty k(x,x)\,dx = O\left(1/L^{\beta-1}\right)$ (resp. $\int_L^\infty k(x,x)\,dx = o\left(1/L^{\beta-1}\right)$)
as $L \to +\infty$ and $\gamma = \dfrac{(2m+1)\beta}{2m+\beta}$, then $\lambda_n = O\left(1/n^\gamma\right)$ (resp. $\lambda_n = o(1/n^\gamma)$).

ii) If $\int_L^\infty k(x,x)\,dx = O\left(1/L^{\beta-1}\right)$ as $L \to +\infty$ for all $\beta > 1$, then $\lambda_n = o\left(1/n^\gamma\right)$
for all $\gamma < 2m+1$.

iii) If $k(x,x)$ has compact support, then $\lambda_n = o\left(1/n^{2m+1}\right)$.

1.2 Suppose $k_m(x,y)$ is $Lip^\alpha$ with respect to $y$ on the diagonal. Then:

i) If $\beta > 1$, $\int_L^\infty k(x,x)\,dx = O\left(1/L^{\beta-1}\right)$ as $L \to +\infty$ and $\gamma = \dfrac{(2m+\alpha+1)\beta}{2m+\alpha+\beta}$,
then $\lambda_n = O\left(1/n^\gamma\right)$.

ii) If $\int_L^\infty k(x,x)\,dx = O\left(1/L^{\beta-1}\right)$ as $L \to +\infty$ for all $\beta > 1$, then $\lambda_n = o\left(1/n^\gamma\right)$
for all $\gamma < 2m+\alpha+1$.

iii) If $k(x,x)$ has compact support, then $\lambda_n = O\left(1/n^{2m+\alpha+1}\right)$.

2.1 Suppose $k_m(x,y)$ is continuously differentiable with respect to $x$ and that $\dfrac{\partial k_m}{\partial x}$
is uniformly continuous with respect to $y$ on the diagonal. Then:

i) If $\beta > 1$, $\int_L^\infty k(x,x)\,dx = O\left(1/L^{\beta-1}\right)$ (resp. $\int_L^\infty k(x,x)\,dx = o\left(1/L^{\beta-1}\right)$) as
$L \to +\infty$ and $\gamma = \dfrac{(2m+2)\beta}{2m+1+\beta}$, then $\lambda_n = O\left(1/n^\gamma\right)$ (resp. $\lambda_n = o(1/n^\gamma)$).

ii) If $\int_L^\infty k(x,x)\,dx = O\left(1/L^{\beta-1}\right)$ as $L \to +\infty$ for all $\beta > 1$, then $\lambda_n = o\left(1/n^\gamma\right)$
for all $\gamma < 2m+2$.

iii) If $k(x,x)$ has compact support, then $\lambda_n = o\left(1/n^{2m+2}\right)$.

2.2 Suppose $k_m(x,y)$ is continuously differentiable with respect to $x$ and that $\dfrac{\partial k_m}{\partial x}$
is $Lip^\alpha$ with respect to $y$ on the diagonal. Then:

i) If $\beta > 1$, $\int_L^\infty k(x,x)\,dx = O\left(1/L^{\beta-1}\right)$ as $L \to +\infty$ and $\gamma = \dfrac{(2m+\alpha+2)\beta}{2m+\alpha+1+\beta}$, then $\lambda_n = O\left(1/n^\gamma\right)$.

ii) If $\int_L^\infty k(x,x)\,dx = O\left(1/L^{\beta-1}\right)$ as $L \to +\infty$ for all $\beta > 1$, then $\lambda_n = o\left(1/n^\gamma\right)$
for all $\gamma < 2m+\alpha+2$.

iii) If $k(x,x)$ has compact support, then $\lambda_n = O\left(1/n^{2m+\alpha+2}\right)$.

Theorem 4.4 determines the rate of decay of eigenvalues of $K$ as a function of the smoothness of the kernel $k$ and of the decay rate of $k(x,x)$ at infinity. In the case of compactly supported kernels these results coincide, as should be expected, with the known results for operators defined on compact intervals. The known optimality of those results and the content of the technical propositions strongly suggest that the results in theorem 4.4 are optimal.

The essential features of theorem 4.4 may be expressed in the following corollaries.

**Corollary 4.5.** Let $k(x,y)$ be a Mercer-like kernel defined on $[0,+\infty[^2$. Suppose $k$ is of class $C^p$ in $[0,+\infty[^2$ and that the partial derivatives up to order $p$ are uniformly continuous with respect to $y$ on the diagonal, and let $\{\lambda_n\}_{n\in\mathbb{N}}$ be the sequence of *eigenvalues of the integral operator with kernel $k$. If $\beta > 1$ and $\int_L^\infty k(x,x)\,dx = O(1/L^{\beta-1})$ (resp. $\int_L^\infty k(x,x)\,dx = o(1/L^{\beta-1})$) as $L \to +\infty$, then $\lambda_n = O(1/n^\gamma)$ (resp. $\lambda_n = o(1/n^\gamma)$ ) for $\gamma = \dfrac{(p+1)\beta}{p+\beta}$.*

**Corollary 4.6.** *Let $k(x,y)$ be a Mercer-like kernel defined on $[0, +\infty[^2$. Suppose $k$ is of class $C^p$ in $[0, +\infty[^2$ and that the partial derivatives up to order $p$ are satisfy an $\alpha$-Lipschitz condition with respect to $y$ on the diagonal and let $\{\lambda_n\}_{n\in\mathbb{N}}$ be the sequence of eigenvalues of the integral operator with kernel $k$. If $\beta > 1$ and $\int_L^\infty k(x,x)\,dx = O(1/L^{\beta-1})$ as $L \to +\infty$, then $\lambda_n = O(1/n^\gamma)$ for $\gamma = \dfrac{(p+1+\alpha)\beta}{p+\alpha+\beta}$.*

## References

1. N. Aronszajn, Theory of reproducing kernels. Trans. Amer. Math. Soc. **68** (1950), 337–404.
2. M. Birman, M. Solomyak, Estimates of singular numbers of integral operators. Russ. Math. Surv. **32** (1977), 15–89.
3. J. Buescu, F. Garcia, I. Lourtie, $L^2(\mathbb{R})$ nonstationary processes and the sampling theorem. IEEE Sign. Proc. Lett. **8**, 4 (2001), 117–119.
4. J. Buescu, Positive integral operators in unbounded domains. J. Math. Anal. Appl. **296** (2004), 244–255.
5. J. Buescu, F. Garcia, I. Lourtie, A. C. Paixão, Positive definiteness, integral equations and Fourier transforms. Jour. Int. Eq. Appl. **16**, 1 (2004), 33–52.
6. J. Buescu, A. C. Paixão, Positive definite matrices and reproducing kernel inequalities. Jour. Math. Anal. Appl., to appear.
7. J. Buescu, A. C. Paixão, Inequalities for holomorphic reproducing kernels. Submitted.
8. J. Buescu, A. C. Paixão, A linear algebraic aproach to holomorphic reproducing kernels in $\mathbb{C}^n$. Lin. Alg. Appl., to appear.
9. J. Buescu, A. C. Paixão, Eigenvalues of positive integral operators on unbounded domains. Submitted.
10. J. Buescu, A. C. Paixão, Eigenvalue distribution of positive definite kernels on unbounded domains. Submitted.
11. J. Buescu, A. C. Paixão, Positive definite matrices and integral equations on unbounded domains. Diff. Int. Eq., to appear.
12. J. Buescu, A. C. Paixão, Inequalities for differentiable reproducing kernels and an application to positive operators. Submitted.
13. C. Chang, C. Ha, On eigenvalues of differentiable positive definite kernels. Integr. Equ. Oper. Theory **33** (1999), 1–7.
14. J. Cochran, M. Lukas, Differentiable positive definite kernels and Lipschitz continuity. Math. Proc. Camb. Phil. Soc. **104** (1988), 361–369.
15. I. Gohberg, M. Krein, *Introduction to the theory of linear nonselfadjoint operators in Hilbert space.* A.M.S., Providence, 1969.
16. C. Ha, Eigenvalues of differentiable positive definite kernels. SIAM J. Math. Anal. **17** (1986), 2, 415–419.
17. T. Kadota, Term-by-term differentiability of Mercer's expansion. Proc. Amer. Math. Soc. **18** (1967), 69–72.
18. H. König, Eigenvalue distribution of compact operators. Operator Theory: Advances and Applications, vol. 16. Birkhäuser, Basel, 1986.

188

19. M. G. Krein, Hermitian positive kernels on homogeneous spaces I. Amer. Math. Soc. Transl. (2), **34** (1963), 69–108.
20. E. H. Moore, *General Analysis*. Memoirs of Amer. Philos. Soc. Part I (1935), Part II (1939).
21. A. Pietsch, Zur Fredholmschen Theorie in lokalconvexe Räumen. Stud. Math. **28** (1966/67), 161–179.
22. A. Pietsch, Eigenvalues of integral operators II. Math. Ann. **262** (1983), 343–376.
23. J. Reade, Eigenvalues of positive definite kernels. SIAM J. Math. Anal. **14** (1983), 1, 152–157.
24. J. Reade, Eigenvalues of Lipschitz kernels. Math. Proc. Camb. Phil. Soc. **93** (1983), 1, 135–140.
25. J. Reade, Eigenvalues of positive definite kernels II. SIAM J. Math. Anal. **15** (1984), 1, 137–142.
26. J. Reade, Positive definite $C^p$ kernels. SIAM J. Math. Anal. **17** (1986), 2, 420–421.
27. J. Reade, Eigenvalues of smooth positive definite kernels. Proc. Edimburgh Math. Soc. **35** (1990), 41–45.
28. S. Saitoh, *Theory of reproducing kernels and its applications*. Pitman Research Notes in Mathematics Series, **189**, Longman, 1988.
29. S. Saitoh, *Integral transforms, reproducing kernels and their applications*. Pitman Research Notes in Mathematics Series, **369**, Longman, 1997.

## I.4   Integral Transforms and Applications

Organizers: A. Kilbas, S. Saitoh, V. Tuan, A.I. Zayed

# ANALYTICAL AND NUMERICAL REAL INVERSION FORMULAS OF THE LAPLACE TRANSFORM

T. MATSUURA

*Department of Mechanical Engineering, Faculty of Engineering,*
*Gunma University,Kiryu, Gunma 376-8515, Japan*
*E-mail: matsuura@me.gunma-u.ac.jp*

S. SAITOH

*Department of Mathematics, Faculty of Engineering,*
*Gunma University,Kiryu, Gunma 376-8515, Japan*
*E-mail: ssaitoh@math.sci.gunma-u.ac.jp*

We shall give very natural, analytical, numerical and approximate real inversion formulas of the Laplace transform for natural reproducing kernel Hilbert spaces by using the ideas of best approximations, generalized inverses and the theory of reproducing kernels having a good connection with the Tikhonov regularization. These approximate real inversion formulas may be expected to be practical to calculate the inverses of the Laplace transform by computers when the real data contain noises or errors. We shall illustrate examples, by using computers.

**Key words:** Laplace transform, numerical inversion formula, Tikhonov regularization, reproducing kernel

**Mathematic Subject Classifications:** 44A10, 65R10, 30C40

## 1. Introduction

We shall give very natural, analytical, numerical and approximate real inversion formulas of the Laplace transform

$$(\mathcal{L}F)(p) = \int_0^\infty e^{-pt} F(t) dt, \quad p > 0 \qquad (1.1)$$

for functions $F$ of some natural function space. This integral transform is, of course, very fundamental in mathematical science. The inversion formula for the Laplace transform is, in general, given by a complex form, however, we are interested in and are requested to obtain its real inversion formulas in many practical problems. However, its real inversion formulas will be very involved and one might think that its real inversion formulas will be essentially involved, because we must catch "analyticity" from the real data or discrete data. Note that the image functions of the Laplace transform are analytic on some half complex plane. See [1,2,3,6,11,12] and the recent related article [4] for real inversion formulas of the Lapace transform. In this paper, we shall give new type and very natural approximate real inversion formulas from the viewpoints of best approximations, generalized inverses and the Tikhonov regularization by combining these fundamental ideas and methods by means of the theory of reproducing kernels. We may think that these approximate real inversion formulas are practical and natural. We can give good error estimates in our inversion formulas. Furthermore, we shall illustrate examples, by using computers.

## 2. Background Theorems

We shall use basically the following two general theorems.

**Theorem 1.** ([3,6]) Let $H_K$ be a Hilbert space admitting the reproducing kernel $K(p,q)$ on a set $E$. Let $L:H_K \to \mathcal{H}$ be a bounded linear operator on $H_K$ into $\mathcal{H}$. For $\lambda > 0$ introduce the inner product in $H_K$ and call it $H_{K_\lambda}$ as

$$\langle f_1, f_2 \rangle_{H_{K_\lambda}} = \lambda \langle f_1, f_2 \rangle_{H_K} + \langle Lf_1, Lf_2 \rangle_{\mathcal{H}}, \tag{2.2}$$

then $H_{K_\lambda}$ is the Hilbert space with the reproducing kernel $K_\lambda(p,q)$ on $E$ and satisfying the equation

$$(\lambda I + L^*L)K_\lambda(\cdot, q) = K(\cdot, q) \tag{2.3}$$

where $L^*$ is the adjoint of $L : H_K \to \mathcal{H}$.

**Theorem 2.** ([8,11]) Let $H_K$, $L$, $\mathcal{H}$, $E$ and $K_\lambda$ be as in Theorem 1. Then, for any $\lambda > 0$ and for any $g \in \mathcal{H}$, the extremal function in

$$\inf_{f \in H_K} \left( \lambda \|f\|_{H_K}^2 + \|Lf - g\|_{\mathcal{H}}^2 \right) \tag{2.4}$$

exists uniquely and the extremal function is represented by

$$f_{\lambda,g}^*(p) = \langle g, LK_\lambda(\cdot, p) \rangle_{\mathcal{H}} \tag{2.5}$$

which is the member of $H_K$ which attains the infimum in (2.4).

For the properties and error estimates for the limit

$$\lim_{\lambda \to 0} f_{\lambda,g}^*(p) \tag{2.6}$$

see [9,10]. In particular, when there exists the Moore-Penrose generalized solution for the operator equation

$$Lf = g$$

in (2.4), the limit (2.6) converges uniformly to the Moore-Penrose generalized solution on any subset of $E$ such that $K(p,p)$ is bounded.

## 3. A Natural Situation for Real Inversion Formulas

In order to apply the general theory in Section 2 to the real inversion formula of the Lapace transform, we shall recall the "natural situation" based on [7].

We shall introduce the simple reproducing kernel Hilbert space (RKHS) $H_K$ comprised of absolutely continuous functions $F$ on the positive real line $\mathbf{R}^+$ with finite norms

$$\left\{ \int_0^\infty |F'(t)|^2 \frac{1}{t} e^t dt \right\}^{1/2}$$

and satisfying $F(0) = 0$. This Hilbert space admits the reproducing kernel $K(t, t')$

$$K(t, t') = \int_0^{\min(t,t')} \xi e^{-\xi} d\xi \tag{3.7}$$

(see [6], pages 55-56). Then we see that

$$\int_0^\infty |(\mathcal{L}F)(p)p|^2 dp \leq \frac{1}{2} \|F\|_{H_K}^2; \tag{3.8}$$

that is, the linear operator on $H_K$

$$(\mathcal{L}F)(p)p$$

into $L_2(\mathbf{R}^+, dp) = L_2(\mathbf{R}^+)$ is bounded ([7]). For the reproducing kernel Hilbert spaces $H_K$ satisfying (3.8), we can find some general spaces ([7]). Therefore, from the general theory in Section 2, we obtain

**Theorem 3.** ([7]). For any $g \in L_2(\mathbf{R}^+)$ and for any $\lambda > 0$, the best approximation $F_{\lambda,g}^*$ in the sense

$$\inf_{F \in H_K} \left\{ \lambda \int_0^\infty |F'(t)|^2 \frac{1}{t} e^t dt + \|(\mathcal{L}F)(p)p - g\|_{L_2(\mathbf{R}^+)}^2 \right\}$$

$$= \lambda \int_0^\infty |F_{\lambda,g}^{*\prime}(t)|^2 \frac{1}{t} e^t dt + \|(\mathcal{L}F_{\lambda,g}^*)(p)p - g\|_{L_2(\mathbf{R}^+)}^2 \tag{3.9}$$

exists uniquely and we obtain the respresentation

$$F_{\lambda,g}^*(t) = \int_0^\infty g(\xi) \left( \mathcal{L}K_\lambda(\cdot, t) \right)(\xi) \xi d\xi. \tag{3.10}$$

Here, $K_\lambda(\cdot, t)$ is determined by the functional equation

$$K_\lambda(t, t') = \frac{1}{\lambda} K(t, t') - \frac{1}{\lambda} ((\mathcal{L}K_{\lambda,t'})(p)p, (\mathcal{L}K_t)(p)p)_{L_2(\mathbf{R}^+)} \tag{3.11}$$

for

$$K_{\lambda,t'} = K_\lambda(\cdot, t')$$

and

$$K_t = K(\cdot, t)$$

## 4. New Algorithm

In this paper, we shall propose a new algorithm to solve numerically the equation (3.11) which is, in general, an integral equation of Fredholm of the second kind. Our algorithm will give a new type discretization whose effectivity will be proved by examples, since to solve the equation (3.11) is decisively important to obtain the concrete representation (3.10).

We take a complete orthonormal system $\{\varphi_j\}_{j=1}^{\infty}$ of the Hilbert space $L_2(\mathbf{R}^+)$, for example, we can take

$$\varphi_j(p) = e^{-\frac{p}{2}} L_j(p);$$

for

$$L_j(p) = \sum_{m=1}^{j} (-1)^{j-m} \frac{j! p^{j-m}}{m![(j-m)!]^2}$$

$$= \sum_{k=0}^{j} (-1)^k \frac{1}{k!} {}_jC_k p^k.$$

For a sufficiently large $N$, and for fixed $\{\lambda_j\}_{j=1}^{N} (\lambda_j > 0)$, we consider the extremal problem for (3.9)

$$\inf_{F \in H_K} \left\{ \lambda \|F\|_{H_K}^2 + \sum_{j=1}^{N} \lambda_j |((\mathcal{L}F)(p)p - g(p), \varphi_j(p))_{L_2(\mathbf{R}^+)}|^2 \right\}. \tag{4.12}$$

That is,

$$\|(\mathcal{L}F)(p)p - g(p)\|_{L_2(\mathbf{R}^+)}^2$$

is replaced by

$$\sum_{j=1}^{N} \lambda_j |((\mathcal{L}F)(p)p - g(p), \varphi_j(p))_{L_2(\mathbf{R}^+)}|^2.$$

Then, we shall give an algorithm constructing the reproducing kernel $K_{\lambda,\lambda_j}^{(N)}(t, t')$ of the Hilbert space $H_{K_{\lambda,\lambda_j}^{(N)}}$ with the norm square

$$\lambda \|F\|_{H_K}^2 + \sum_{j=1}^{N} \lambda_j |((\mathcal{L}F)(p)p, \varphi_j(p))_{L_2(\mathbf{R}^+)}|^2. \tag{4.13}$$

We shall start with the first step. The reproducing kernel $K^{(1)}(t,t')$ of the Hilbert space with the norm square

$$\lambda\|F\|_{H_K}^2 + \sum_{j=1}^{1}\lambda_j|((\mathcal{L}F)(p)p, \varphi_j(p))_{L_2(\mathbf{R}^+)}|^2 \tag{4.14}$$

is given by

$$K^{(1)}(t,t') = K^{(0)}(t,t')$$

$$-\frac{\lambda_1(\varphi_1(p),(\mathcal{L}K^{(0)}(\cdot,t))(p)p)_{L_2(\mathbf{R}^+)}((\mathcal{L}K^{(0)}(\cdot,t'))(q)q, \varphi_1(q))_{L_2(\mathbf{R}^+)}}{1 + \lambda_1(\mathcal{L}(\varphi_1(p),(\mathcal{L}K^{(0)}(\cdot,t'))(p)p)_{L_2(\mathbf{R}^+)}(q)q, \varphi_1(q))_{L_2(\mathbf{R}^+)}}, \tag{4.15}$$

for

$$K^{(0)}(t,t') = \frac{1}{\lambda}K(t,t').$$

For the second step, the reproducing kernel $K^{(2)}(t,t')$ of the Hilbert space with the norm square

$$\lambda\|F\|_{H_K}^2 + \sum_{j=1}^{2}\lambda_j|((\mathcal{L}F)(p)p, \varphi_j(p))_{L_2(\mathbf{R}^+)}|^2 \tag{4.16}$$

is given by

$$K^{(2)}(t,t') = K^{(1)}(t,t')$$

$$-\frac{\lambda_2(\varphi_2(p),(\mathcal{L}K^{(1)}(\cdot,t))(p)p)_{L_2(\mathbf{R}^+)}((\mathcal{L}K^{(1)}(\cdot,t'))(q)q, \varphi_2(q))_{L_2(\mathbf{R}^+)}}{1 + \lambda_2(\mathcal{L}(\varphi_2(p),(\mathcal{L}K^{(1)}(\cdot,t'))(p)p)_{L_2(\mathbf{R}^+)}(q)q, \varphi_2(q))_{L_2(\mathbf{R}^+)}}, \tag{4.17}$$

by using the reproducing kernel $K^{(1)}(t,t')$. In this way, we can obtain the desired representation of $K_{\lambda,\lambda_j}^{(N)}(t,t')$. Then, we obtain

Theorem 4. For any $g \in L_2(\mathbf{R}^+)$, the extremal function $f_{\lambda,\lambda_j}^{(N)}$ in the extremal problem (4.12) is given by

$$f_{\lambda,\lambda_j}^{(N)}(t) = \sum_{j=1}^{N}\lambda_j(g,\varphi_j)_{L_2(\mathbf{R}^+)}(\varphi_j(p),(\mathcal{L}K_{\lambda,\lambda_j}^{(N)}(\cdot,t))(p)p)_{L_2(\mathbf{R}^+)}. \tag{4.18}$$

We consider a general extremal problem in (4.12) by considering a general weight $\{\lambda_j\}$. This means that for a larger $\lambda_{j_0}$, the speed of the convergence

$$((\mathcal{L}F)(p)p, \varphi_{j_0}(p))_{L_2(\mathbf{R}^+)} \to (g(p), \varphi_{j_0}(p))_{L_2(\mathbf{R}^+)}$$

is higher. This technique is a very important for practical applications. For examples, see [5].

## 5. Error Estimates

In Theorem 4, when the data $g$ contain errors or noises, we need the estimation of our solutions $f_{\lambda,\lambda_j}^{(N)}(t)$ in terms of $g$. For this, we can obtain a good estimation in the form:

**Theorem 5.** In Theorem 4, we obtain the estimate

$$|f_{\lambda,\lambda_j}^{(N)}(t)| \leq \frac{1}{\sqrt{\lambda}} \left( \sum_{j=1}^{N} \lambda_j \right) \left( \|g\|_{L_2(\mathbf{R}^+)} \right). \tag{5.19}$$

## 6. Inverses for More General Functions

By a suitable transform, our inversion formula in Theorem 3 is applicable for more general functions as follows:

We assume that $F$ satisfies the properties (P):

$$F \in C^1[0,\infty),$$

$$F'(t) = o(e^{\alpha t}), \quad 0 < \alpha < k - \frac{1}{2},$$

and

$$F(t) = o(e^{\beta t}), \quad 0 < \beta < k - \frac{1}{2}.$$

Then, the function

$$G(t) = \{F(t) - F(0) - tF'(0)\}e^{-kt} \tag{6.20}$$

belongs to $H_K$. Then,

$$(\mathcal{L}G)(p) = f(p+k) - \frac{F(0)}{p+k} - \frac{F'(0)}{(p+k)^2}. \tag{6.21}$$

Therefore, if we know $F(0)$ and $F'(0)$, then from

$$g(p) = (\mathcal{L}G)(p)$$

by Theorem 3, we obtain $G(t)$ and so, from the identity

$$F(t) = G(t)e^{kt} + F(0) + tF'(0) \tag{6.22}$$

we have the inverse $F(t)$ from the data $f(p), F(0)$ and $F'(0)$ through the above procedures.

## 7. Numerical Experiments

At first, note that

$$K(t,t') = \begin{cases} -te^{-t} - e^{-t} + 1 & \text{for} \quad t \le t' \\ -t'e^{-t'} - e^{-t'} + 1 & \text{for} \quad t \ge t'. \end{cases}$$

$$(\mathcal{L}K(\cdot,t'))(p)$$

$$= e^{-t'p}e^{-t'} \left[ \frac{-t'}{p(p+1)} + \frac{-1}{p(p+1)^2} \right] + \frac{1}{p(p+1)^2}. \tag{7.23}$$

$$\int_0^\infty e^{-qt'} (\mathcal{L}K(\cdot,t'))(p)dt' = \frac{1}{pq(p+q+1)^2}. \tag{7.24}$$

We shall give a numerical experiment for the typical example

$$F_0(t) = \begin{cases} -te^{-t} - e^{-t} + 1 & \text{for} \quad 0 \le t \le 1 \\ 1 - 2e^{-1} & \text{for} \quad 1 \le t, \end{cases}$$

whose Laplace transform is

$$(\mathcal{L}F_0)(p) = \frac{1}{p(p+1)^2} \left[ 1 - (p+2)e^{-(p+1)} \right]. \tag{7.25}$$

In Figure 1, we calculate $K^{(N)}(t,t')$ for $\lambda = \lambda_j = 1$.

Figure 2, (a) is the graph of the function $y = F_0(t)$. In Figure 2, (b),(c), and (d), we calculate (4.18) for $N = 1, 3, 5$ and for $\lambda = 0.05$ and $\lambda_j = 1$.

We can not say that our algorithm for the real inversion formula is sufficiently good at this moment. However, for a good computer we expect that our algorithm becomes sufficiently effective.

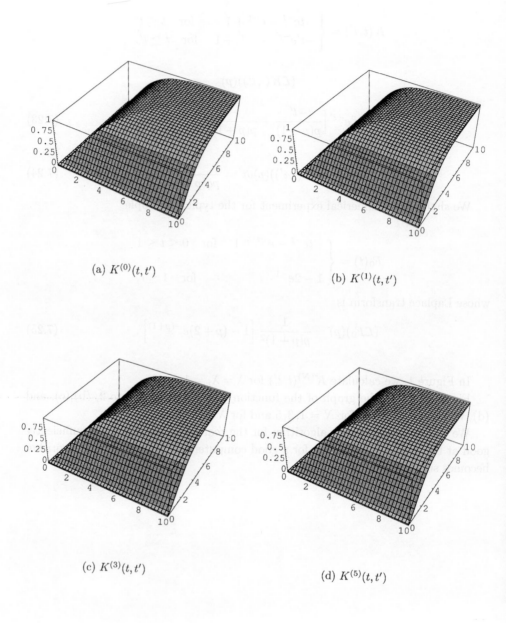

(a) $K^{(0)}(t,t')$

(b) $K^{(1)}(t,t')$

(c) $K^{(3)}(t,t')$

(d) $K^{(5)}(t,t')$

Fig. 1.   $K^{(N)}(t,t')$ for $N = 1, 3, 5$ and for $\lambda = 0.05$ and $\lambda_j = 1$.

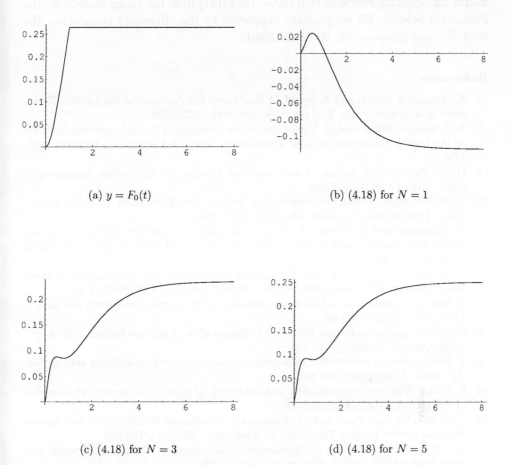

(a) $y = F_0(t)$

(b) (4.18) for $N = 1$

(c) (4.18) for $N = 3$

(d) (4.18) for $N = 5$

Fig. 2.   (4.18) for $\lambda = 0.05$ and $\lambda_j = 1$

## Acknowledgements

T. Matsuura is supported in part by the Gunma University Foundation for the Promotion of Science and Engineering. S. Saitoh is supported in part by the Grant-in-Aid for Scientific Research (C)(2)(No. 16540137) from the Japan Society for the Promotion Science. We are partially supported by the Mitsubishi Foundation, the 36th, Natural Sciences, No. 20 (2005-2006).

## References

1. K. Amano, S. Saitoh and A. Syarif, *A Real Inversion Formula for the Laplace Transform in a Sobolev Space*, Z. Anal. Anw. **18**(1999), 1031-1038.

2. K. Amano, S. Saitoh and M. Yamamoto, *Error estimates of the real inversion formulas of the Laplace transform*, Integral Transforms and Special Functions, **10**(2000), 165-178.

3. D.-W. Byun and S. Saitoh, *A real inversion formula for the Laplace transform*, Z. Anal. Anw. **12**(1993), 597-603.

4. V. V. Kryzhniy, *Regularized Inversion of Integral Transformations of Mellin Convolution Type*, Inverse Problems, **19**(2003), 1227-1240.

5. T. Matsuura and S. Saitoh, *Analytical and numerical solutions of linear ordinary differential equations with constant coefficients*, Journal of Analysis and Applications, **3**(2005), pp. 1-17.

6. S. Saitoh, *Integral Transforms, Reproducing Kernels and their Applications*, Pitman Res. Notes in Math. Series **369**, Addison Wesley Longman Ltd (1997), UK.

7. S. Saitoh, *Approximate real inversion formulas of the Laplace transform*, Far East J. Math. Sci. **11**(2003), 53-64.

8. S. Saitoh, *Approximate Real Inversion Formulas of the Gaussian Convolution*, Applicable Analysis, **83**(2004), 727-733.

9. S. Saitoh, *Best approximation, Tikhonov regularization and reproducing kernels*, Kodai. Math. J. **28**(2005), 359–367.

10. S. Saitoh, *Tikhonov regularization and the theory of reproducing kernels*, Proceedings of the 12th ICFIDCAA (to appear).

11. S. Saitoh, Vu Kim Tuan and M. Yamamoto, *Conditional Stability of a Real Inverse Formula for the Laplace Transform*, Z. Anal. Anw. **20**(2001), 193-202.

12. S. Saitoh, N. Hayashi and M. Yamamoto (eds.), *Analytic Extension Formulas and their Applications*, (2001), Kluwer Academic Publishers.

# INTEGRAL TRANSFORM WITH THE EXTENDED GENERALIZED MITTAG-LEFFLER FUNCTION *

A.A. KILBAS and A.A. KOROLEVA

*Belarusian State University,*
*Independence Avenue 4,*
*220050 Minsk, Belarus*
*E-mail: kilbas@bsu.by; akoroleva@tur.by*

Integral transform involving special function generalizing classical Mittag-Leffler function in the kernel, is considered. Conditions for the existence and series expansions for such a special function are given. The representation of the considered integral transform in the form of the **H**-transform with the *H*-function in the kernel is established. Mapping properties such as the boundedness, the representation, the range and the inversion of this transforms in weighted spaces of summable functions are characterized.

**Keywords:** Integral transforms with special functions as kernels, generalized Mittag-Leffler function, H-function
**Mathematic Subject Classifications:** 44A15, 33E12, 33C60

## 1. Introduction

Classical Mittag–Leffler functions $E_\alpha(z)$ and $E_{\alpha,\beta}(z)$ are defined by the following power series

$$E_\alpha(z) = \sum_{k=0}^{\infty} \frac{z^k}{\Gamma(\alpha k + 1)} \quad (\alpha > 0; \ z \in \mathbb{C}), \tag{1}$$

$$E_{\alpha,\beta}(z) = \sum_{k=0}^{\infty} \frac{z^k}{\Gamma(\alpha k + \beta)} \quad (\alpha > 0; \ \beta \in \mathbb{R}; \ z \in \mathbb{C}) \tag{2}$$

thus $E_\alpha(z) = E_{\alpha,1}(z)$. The main properties of functions (1) and (2) and their application may be found in Section 18.1 of the notebook by Erdelyi, etc. [6] and in books by Dzrbashyan [2], [3].

In particular, the function $E_{\alpha,\beta}(z)$ is an entire function of complex $z \in \mathbb{C}$ of order $\rho = 1/\alpha$ and type $\sigma = 1$. The exponential function and trigonometric and hyperbolic cosine and sine functions are expressed through functions (1) and (2):

$$E_1(z) = e^z, \quad E_2(-z^2) = \cos z, \quad E_2(z^2) = \cosh z,$$

*This work is supported by Belarusian Fundamental Research Fund, Project F05MC-050

$$zE_{2,2}(-z^2) = \sin z, \quad zE_{2,2}(z^2) = \sinh z.$$

Recently interest to investigation of various properties of functions (1.1) and (1.2) has considerably increased thanking to their connections with the theory of fractional calculus; in particular, with the solution of the so-called integral and differential equations of fractional order.

Dzhrbashyan [1] introduced the generalized Mittag–Leffler function

$$E_{\alpha_1,\beta_1;\alpha_2,\beta_2}(z) = \sum_{k=0}^{\infty} \frac{z^k}{\Gamma(\alpha_1 k + \beta_1)\Gamma(\alpha_2 k + \beta_2)} \tag{3}$$

with positive $\alpha_1 > 0$, $\alpha_2 > 0$, real $\beta_1$, $\beta_2 \in \mathbb{R}$ and complex $z \in \mathbb{C}$. He investigated some properties of function (3) and proved, that the function $E_{\alpha_1,\beta_1;\alpha_2,\beta_2}(z)$ is the entire function of $z$ of order

$$\rho = (\alpha_1 + \alpha_2)^{-1}$$

and type

$$\sigma = \left(\frac{\alpha_1 + \alpha_2}{\alpha_1}\right)^{\frac{\alpha_1}{\alpha_1+\alpha_2}} \left(\frac{\alpha_1 + \alpha_2}{\alpha_2}\right)^{\frac{\alpha_2}{\alpha_1+\alpha_2}}.$$

Let's note, that some special Bessel type functions [5], [10] are expressed in terms of function (1.5): the Bessel function of the first kind

$$J_\nu(z) = \left(\frac{z}{2}\right)^\nu E_{1,\nu+1;1,1}\left(-\frac{z^2}{4}\right); \tag{4}$$

the Struve function

$$H_\nu(z) = \left(\frac{z}{2}\right)^{\nu+1} E_{1,\nu+3/2;1,3/2}\left(-\frac{z^2}{4}\right); \tag{5}$$

the Lommel function

$$S_{\mu,\nu}(z) = \frac{z^{\mu+1}}{4} \Gamma\left(\frac{\mu-\nu+1}{2}\right)\Gamma\left(\frac{\mu+\nu+1}{2}\right) E_{1,\frac{\mu-\nu+1}{2};1,\frac{\mu+\nu+1}{2}}\left(-\frac{z^2}{4}\right); \tag{6}$$

the Bessel–Maitland function

$$J_\nu^\mu(z) = E_{\mu,\nu+1;1,1}(-z); \tag{7}$$

the generalized Bessel–Maitland function

$$J_{\nu,\lambda}^\mu(z) = \left(\frac{z}{2}\right)^{\nu+2\lambda} E_{\mu,\nu+\lambda+1;1,\lambda+1}\left(-\frac{z^2}{4}\right). \tag{8}$$

It is directly verified, that the generalized Mittag–Leffler function $E_{\alpha_1,\beta_1;\alpha_2,\beta_2}(z)$ can be expressed in terms of the so-called Mellin-Barnes integral

$$E_{\alpha_1,\beta_1;\alpha_2,\beta_2}(z) = \mathbb{E}_{\alpha_1,\beta_1;\alpha_2,\beta_2}(z),$$

where

$$\mathbb{E}_{\alpha_1,\beta_1;\alpha_2,\beta_2}(z) = \frac{1}{2\pi i} \int_\gamma \frac{\Gamma(s)\Gamma(1-s)}{\Gamma(\beta_1 - \alpha_1 s)\Gamma(\beta_2 - \alpha_2 s)}(-z)^{-s}ds \quad (z \neq 0). \tag{9}$$

Here $\gamma$ is a specially chosen infinite contour, which separates all poles of the gamma function $\Gamma(s)$ to the left, and all poles of the gamma function $\Gamma(1-s)$ to the right.

## 2. Extended generalized Mittag-Leffler function

For $z \in \mathbb{C}$ ($z \neq 0$) and complex parameters $\alpha_1, \alpha_2 \in \mathbb{C}$ ($\mathrm{Re}(\alpha_1 + \alpha_2) \neq 0$) and $\beta_1, \beta_2 \in \mathbb{C}$ we define the function $\mathbb{E}_{\alpha_1,\beta_1;\alpha_2,\beta_2}(z)$ by relation (9), where $\gamma$ is one of the following contours:

(a) $\gamma = \gamma_{-\infty}$ is a left loop located in a horizontal strip starting at the point $-\infty + i\varphi_1$ and terminating at the point $-\infty + i\varphi_2$, $\quad -\infty < \varphi_1 < \varphi_2 < \infty$.

(b) $\gamma = \gamma_{+\infty}$ is a right loop which located in the horizontal strip starting at the point $+\infty + i\varphi_1$ and terminating at the point $+\infty + i\varphi_2$, $\quad -\infty < \varphi_1 < \varphi_2 < \infty$

Conditions for the existence of $\mathbb{E}_{\alpha_1,\beta_1;\alpha_2,\beta_2}(z)$ depend on conditions of the convergence of integral (9), which depend on a choice of a contour of integration $\gamma$ and parameters $\alpha_1$, $\alpha_2$.

**Theorem 1.** *Let* $z \in \mathbb{C}$ *($z \neq 0$), and let* $\alpha_1, \alpha_2, \beta_1, \beta_2 \in \mathbb{C}$ *be such numbers that* $\mathrm{Re}(\alpha_1 + \alpha_2) \neq 0$. *The extended generalized Mittag–Leffler function* $\mathbb{E}_{\alpha_1,\beta_1;\alpha_2,\beta_2}(z)$ *exists in the following cases:*

$$(a) \quad \gamma = \gamma_{-\infty} \quad and \quad \mathrm{Re}(\alpha_1 + \alpha_2) > 0; \tag{10}$$

$$(b) \quad \gamma = \gamma_{+\infty} \quad and \quad \mathrm{Re}(\alpha_1 + \alpha_2) < 0. \tag{11}$$

**Proof.** Theorem 1 is proved on the basis of definition (9) with using the asymptotic expansion of the gamma function $\Gamma(z)$, az $z \to \infty$; see relation 1.18(2) in 4.

Using representation (9) and the residue theory, from Theorem 1 we deduce expansions of $\mathbb{E}_{\alpha_1,\beta_1;\alpha_2,\beta_2}(z)$ in power series with respect to $z$ and $1/z$ in respective cases (10) and (11).

**Theorem 2.** *Let* $z \in \mathbb{C}$ *($z \neq 0$), and let* $\alpha_1, \alpha_2, \beta_1, \beta_2 \in \mathbb{C}$, $\mathrm{Re}(\alpha_1 + \alpha_2) \neq 0$.

*(a) If the conditions in (10) are satisfied, then* $\mathbb{E}_{\alpha_1,\beta_1;\alpha_2,\beta_2}(z)$ *has the following representation:*

$$\mathbb{E}_{\alpha_1,\beta_1;\alpha_2,\beta_2}(z) = \sum_{k=0}^{\infty} c_k z^k, \quad c_k = \frac{1}{\Gamma(\alpha_1 k + \beta_1)\Gamma(\alpha_2 k + \beta_2)}. \tag{12}$$

*(b) If the conditions in (11) are valid, then*

$$\mathbb{E}_{\alpha_1,\beta_1;\alpha_2,\beta_2}(z) = \sum_{k=0}^{\infty} \frac{d_k}{z^{k+1}}, \quad d_k = \frac{1}{\Gamma(-\alpha_1 k - \alpha_1 + \beta_1)\Gamma(-\alpha_2 k - \alpha_2 + \beta_2)}.$$

The power expansion (12) for $\mathbb{E}_{\alpha_1,\beta_1;\alpha_2,\beta_2}(z)$ with $\mathbb{L} = \mathbb{L}_{-\infty}$ and $\mathrm{Re}(\alpha_1 + \alpha_2) > 0$ coincides with the power series (3). Therefore the relation (9) can be considered as the extension of function (3) from real to any complex values of parameters $\alpha_1, \alpha_2, \beta_1, \beta_2$. In this connection we call $\mathbb{E}_{\alpha_1,\beta_1;\alpha_2,\beta_2}(z)$ the *extended generalized Mittag–Leffler function*.

## 3. $\mathbb{E}_{\alpha_1,\beta_1;\alpha_2,\beta_2}$-Transform as the H-Transform

We consider the integral transform

$$(\mathbb{E}_{\alpha_1,\beta_1,\,\alpha_2,\beta_2} f)(x) = \int_0^\infty \mathbb{E}_{\alpha_1,\beta_1;\alpha_2,\beta_2}(-xt) f(t) dt \quad (x > 0), \tag{13}$$

with real $\alpha_1, \beta_1, \alpha_2, \beta_2 \in \mathbb{R}$, involving the extended generalized Mittag-Leffler function (9) in the kernel. We shall study the mapping properties such as the boundedness, the range, the representation and the inversion of transform (13) on the space $L_{\nu,r}$ of Lebesgue mesuarable functions on $\mathbb{R}_+ = (0, +\infty)$ such that

$$\int_0^\infty |t^\nu f(t)| \frac{dt}{t} < \infty \quad (1 \leq r \leq \infty, \ \nu \in \mathbb{R}).$$

Our investigations are based on the representation of this transform in the form of more general integral transform with the so-called $H$-function in the kernel. Such a function $H_{p,q}^{m,n}(z)$ is defined for integers $m, n, p, q$ ($0 \leq m \leq q, 0 \leq n \leq p$), for complex $a_i, b_j \in \mathbb{C}$ and positive $\alpha_i > 0$, $\beta_j > 0$ ($1 \leq i \leq p;\ 1 \leq j \leq q$) by the following equation:

$$H_{p,q}^{m,n}(z) = H_{p,q}^{m,n}\left[ z \left| \begin{matrix} (a_i, \alpha_i)_{1,p} \\ (b_j, \beta_j)_{1,q} \end{matrix} \right. \right] = \frac{1}{2\pi i} \int_\gamma \mathbb{H}_{p,q}^{m,n}(s) z^{-s} ds,$$

with

$$\mathbb{H}_{p,q}^{m,n}(s) := \frac{\prod_{j=1}^m \Gamma(b_j + \beta_j s) \prod_{l=1}^n \Gamma(1 - a_l - \alpha_l s)}{\prod_{l=n+1}^r \Gamma(a_l + \alpha_l s) \prod_{j=m+1}^q \Gamma(1 - b_j - \beta_j s)}.$$

Here $\gamma$ is a specially chosen contour, and an empty product, if it occurs, being taken to be one. One may find the theory of this function in the books by Mathai and Saxena [9] (Chapter 2), by Srivastava, Gupta and Goyal [11] (Chapter 1), by Prudnikov, Brychkov and Marichev [10] (Section 8.3) and by Kilbas and Saigo [8] (Chapters 1 and 2).

By (9), the extended generalized Mittag-Leffler function $\mathbb{E}_{\alpha_1,\beta_1;\alpha_2,\beta_2}(z)$ is represented in the form of the **H**-function, and such a representation is different in dependence on values of real parameters $\alpha_1$ and $\alpha_2$:

$$\mathbb{E}_{\alpha_1,\beta_1;\alpha_2,\beta_2}(z) = H_{1,3}^{1,1}\left[ z \left| \begin{matrix} (0,1) \\ (0,1), (1 - \beta_1, \alpha_1), (1 - \beta_2, \alpha_2) \end{matrix} \right. \right] \quad (\alpha_1 > 0,\ \alpha_2 > 0)$$

$$\tag{14}$$

205

$$\mathbb{E}_{\alpha_1,\beta_1;\alpha_2,\beta_2}(z) = H_{2,2}^{1,1}\left[z\left|\begin{array}{l}(0,1),(\beta_2,-\alpha_2)\\(0,1),(1-\beta_1,\alpha_1)\end{array}\right.\right] \quad (\alpha_1>0,\ \alpha_2<0), \quad (15)$$

$$\mathbb{E}_{\alpha_1,\beta_1;\alpha_2,\beta_2}(z) = H_{2,2}^{1,1}\left[z\left|\begin{array}{l}(0,1),(\beta_1,-\alpha_1)\\(0,1),(1-\beta_2,\alpha_2)\end{array}\right.\right], \quad (16)$$

$$\mathbb{E}_{\alpha_1,\beta_1;\alpha_2,\beta_2}(z) = H_{3,1}^{1,1}\left[z\left|\begin{array}{l}(0,1),(\beta_1,-\alpha_1),(\beta_2,-\alpha_2)\\(0,1)\end{array}\right.\right] \quad (\alpha_1<0,\alpha_2<0). \quad (17)$$

According to the above, the transform $\mathbb{E}_{\alpha_1,\beta_1;\alpha_2,\beta_2}f$ is a special case of more general $\mathbf{H}$-transform of the form

$$(\mathbf{H}f)(x) = \int_0^\infty H_{p,q}^{m,n}\left[xt\left|\begin{array}{l}(a_i,\alpha_i)_{1,p}\\(b_j,\beta_j)_{1,q}\end{array}\right.\right]f(t)dt. \quad (18)$$

In terms of the Mellin transform

$$(\mathbf{M}f)(s) = \int_0^\infty f(x)x^{s-1}dx, \quad (19)$$

the $\mathbf{H}$-transform has the property

$$(\mathbf{MH}f)(s) = \mathbb{H}_{p,q}^{m,n}(s)(\mathbf{M}f)(1-s).$$

By (14)-(18), the transform $\mathbb{E}_{\alpha_1,\beta_1;\alpha_2,\beta_2}f$ has the following representations:

$$(\mathbb{E}_{\alpha_1,\beta_1;\alpha_2,\beta_2}f)(x) = \int_0^\infty H_{1,3}^{1,1}\left[xt\left|\begin{array}{l}(0,1)\\(0.1),(1-\beta_1,\alpha_1),(1-\beta_2,\alpha_2)\end{array}\right.\right]f(t)dt, \quad (20)$$

$$(\mathbb{E}_{\alpha_1,\beta_1;\alpha_2,\beta_2}f)(x) = \int_0^\infty H_{2,2}^{1,1}\left[xt\left|\begin{array}{l}(0,1),(\beta_2,-\alpha_2)\\(0.1),(1-\beta_1,\alpha_1)\end{array}\right.\right]f(t)dt, \quad (21)$$

$$(\mathbb{E}_{\alpha_1,\beta_1;\alpha_2,\beta_2}f)(x) = \int_0^\infty H_{2,2}^{1,1}\left[z\left|\begin{array}{l}(0,1),(\beta_1,-\alpha_1)\\(0.1),(1-\beta_2,\alpha_2)\end{array}\right.\right]f(t)dt, \quad (22)$$

and

$$(\mathbb{E}_{\alpha_1,\beta_1;\alpha_2,\beta_2}f)(x) = \int_0^\infty H_{3,1}^{1,1}\left[xt\left|\begin{array}{l}(0,1),(\beta_1,-\alpha_1),(\beta_2,-\alpha_2)\\(0.1)\end{array}\right.\right]f(t)dt, \quad (23)$$

in the cases (i) $\alpha_1>0$, $\alpha_2>0$, (ii) $\alpha_1>0$, $\alpha_2<0$, (iii) $\alpha_1<0$, $\alpha_2>0$ and (iv) $\alpha_1<0$, $\alpha_2<0$, respectively.

The mapping properties, such as the boundedness, the range, the representation and the invertibility of the **H**-transform on the $L_{\nu,r}$-spaces with any $\nu \in \mathbb{R}$ and $1 \le r \le \infty$ were given in the above book by Kilbas and Saigo [8] (Theorems 3.6-3.7, Theorems 4.1-4.10 and Theorems 4.11-4.14).

We note that for $f \in L_{\nu,r}$ the Mellin transform $\mathbb{M}$ is defined by

$$(\mathbb{M}f)(s) = \int_{-\infty}^{\infty} e^{\tau s} f(e^{\tau}) d\tau \quad (s = \sigma + it; \ \sigma, t \in \mathbb{R}),$$

and if $f \in L_{\nu,2} \bigcap L_{\nu,1}$ and $\operatorname{Re}(s) = \nu$, then this transform coincides with the usual Mellin transform (19). We also note that the range $\mathbf{H}(L_{\nu,r})$ of the **H**-transform is different in nine cases and it is characterized in terms of the following transforms: Erdelyi–Kober type fractional integration operators $I_{0+;\sigma,\eta}^{\alpha} f$ and $I_{-;\sigma,\eta}^{\alpha} f$ defined for $\alpha \in \mathbb{C}$ $(\operatorname{Re}(\alpha) > 0)$, $\sigma > 0$ and $\eta \in \mathbb{C}$ by

$$\left(I_{0+;\sigma,\eta}^{\alpha} f\right)(x) = \frac{\sigma x^{-\sigma(\alpha+\eta)}}{\Gamma(\alpha)} \int_{0}^{x} (x^{\sigma} - t^{\sigma})^{\alpha-1} t^{\sigma\eta+\sigma-1} f(t) dt \quad (x > 0), \qquad (24)$$

$$\left(I_{-;\sigma,\eta}^{\alpha} f\right)(x) = \frac{\sigma x^{\sigma\eta}}{\Gamma(\alpha)} \int_{0}^{x} (t^{\sigma} - x^{\sigma})^{\alpha-1} t^{\sigma(1-\alpha-\eta)-1} f(t) dt \quad (x > 0); \qquad (25)$$

the modified Laplace transform $\mathbb{L}_{k,\alpha} f$:

$$(\mathbb{L}_{k,\alpha} f)(x) = \int_{0}^{\infty} (xt)^{-\alpha} e^{-|k|(xt)^{1/k}} f(t) dt \quad (x > 0), \qquad (26)$$

with $k \in \mathbb{R}$ $(k \ne 0)$ and $\alpha \in \mathbb{C}$; the modified Hankel transform $H_{k,\eta} f$:

$$(H_{k,\eta} f)(x) = \int_{0}^{\infty} (xt)^{1/k-1/2} J_{\eta}\left(|k|(xt)^{1/k}\right) f(t) dt \quad (x > 0),$$

with $k \in \mathbb{R}$ $(k \ne 0)$ and $\eta \in \mathbb{C}$ $(\operatorname{Re}(\eta) > -3/2)$; and the elementary transform

$$(M_{\zeta} f)(x) = x^{\zeta} f(x) \quad (\zeta \in \mathbb{C}).$$

Note that when $\sigma = 1$, (24) and (25) coincide with the so-called Erdelyi-Kober operators [7][Section 18.1]:

$$\left(I_{\eta,\alpha}^{+} f\right)(x) \equiv \left(I_{0+;1,\eta}^{\alpha} f\right)(x) = \frac{x^{-\alpha-\eta}}{\Gamma(\alpha)} \int_{0}^{x} (x - t)^{\alpha-1} t^{\eta} f(t) dt \quad (x > 0),$$

$$\left(K_{\eta,\alpha}^{-} f\right)(x) \equiv \left(I_{-;1,\eta}^{\alpha} f\right)(x) = \frac{x^{\eta}}{\Gamma(\alpha)} \int_{x}^{\infty} (t - x)^{\alpha-1} t^{-\alpha-\eta} f(t) dt \quad (x > 0),$$

while for $k = 1$ and $\alpha = 0$ (26) yields the classical Laplace transform

$$(\mathbb{L}f)(x) \equiv (\mathbb{L}_{1,0} f)(x) = \int_{0}^{\infty} e^{-xt} f(t) dt \quad (x > 0).$$

For $1 \le r < \infty$ we shall use the notation $r'$ and $\gamma(r)$ as follows:

$$\frac{1}{r} + \frac{1}{r'} = 1, \quad \gamma(r) = \max\left[\frac{1}{r}, \frac{1}{r'}\right].$$

## 4. $L_{\nu,r}$-theory of $\mathbb{E}_{\alpha_1,\beta_1;\alpha_2,\beta_2}$-transform when $\alpha_1 > 0$ and $\alpha_2 > 0$

Here we present $L_{\nu,r}$- theory of transform (13) in the case (i) when $\alpha_1 > 0$ and $\alpha_2 > 0$. By (20) and (18), using the $L_{\nu,r}$-theory of the **H**-transform, we deduce the boundedness, the range and the representation of this transform. From [8] (Theorem 4.3) we deduce the first result.

**Theorem 3.** *Let $\alpha_1 > 0$ and $\alpha_2 > 0$ be such that $\alpha_1 + \alpha_2 = 2$, and let $0 < \nu < 1$ and $1 < r < \infty$ be such that $2\nu + \Re(\beta_1 + \beta_2) - 5/2 \geq \gamma(r)$.*

*(a) The transform $\mathbb{E}_{\alpha_1,\beta_1;\alpha_2,\beta_2}$ defined on $L_{\nu,2}$ can be extended to $L_{\nu,r}$ as an element of $[L_{\nu,r}, L_{1-\nu,s}]$ for all $s$ with $r \leq s < \infty$ such that $s' \geq [2\nu + \mathrm{Re}(\beta_1 + \beta_2) - 5/2]^{-1}$ with $1/s + 1/s' = 1$.*

*(b) If $1 < r \leq 2$, then the transform $\mathbb{E}_{\alpha_1,\beta_1;\alpha_2,\beta_2}$ is one-to-one on $L_{\nu,r}$, and there holds the equality:*

$$(\mathbb{M}\mathbb{E}_{\alpha_1,\beta_1;\alpha_2,\beta_2}f)\,(s) = \frac{\Gamma(s)\Gamma(1-s)}{\Gamma(\beta_1 - \alpha_1 s)\Gamma(\beta_2 - \alpha_2 s)}(\mathbb{M}f)(1-s) \quad (\mathrm{Re}(s) = 1 - \nu).$$

*(c) If the following condition holds*

$$s \neq \frac{\beta_1 + k}{\alpha_1}, \quad s \neq \frac{\beta_2 + l}{\alpha_2} \quad (k, l = 0, 1, 2, \cdots) \text{ for } \Re(s) = 1 - \nu, \tag{27}$$

*then the transform $\mathbb{E}_{\alpha_1,\beta_1;\alpha_2,\beta_2}$ is one-to-one on $L_{\nu,r}$. Moreover, $\Re(\beta_1 + \beta_2) > -1$ and there holds*

$$\mathbb{E}_{\alpha_1,\beta_1;\alpha_2,\beta_2}(L_{\nu,r}) = \left(M_{(1-(\beta_1+\beta_2)/2}H_{2,\beta_1+\beta_2}\right)\left(L_{\nu-1+\mathrm{Re}(\beta_1+\beta_2)/2,r}\right). \tag{28}$$

*When the condition in (27) is not satisfied, then $\mathbb{E}_{\alpha_1,\beta_1;\alpha_2,\beta_2}(L_{\nu,r})$ is a subset of the right-hand side of (28).*

*(d) If $f \in L_{\nu,r}$ and $g \in L_{\nu,s}$, $1 < r < \infty$, $1 < s < \infty$ and $1/s + 1/r \geq 1$ and $2\nu + \mathrm{Re}(\beta_1 + \beta_2) - 5/2 \geq \max[\gamma(r), \gamma(s)]$, then there holds the relation*

$$\int_0^\infty f(x)\,(\mathbb{E}_{\alpha_1,\beta_1;\alpha_2,\beta_2}g)\,(x)dx = \int_0^\infty (\mathbb{E}_{\alpha_1,\beta_1;\alpha_2,\beta_2}f)\,(x)g(x)dx. \tag{29}$$

*(e) If $f \in L_{\nu,r}$, $\lambda \in \mathbb{C}$, $h > 0$ and $2\nu + \mathrm{Re}(\beta_1 + \beta_2) - 5/2 \geq -\gamma(r)$, then $\mathbb{E}_{\alpha_1,\beta_1;\alpha_2,\beta_2}f$ is given by*

$$(\mathbb{E}_{\alpha_1,\beta_1;\alpha_2,\beta_2}f)\,(x) = hx^{1-(\lambda+1)/h}\frac{d}{dx}x^{(\lambda+1)/h} \times$$

$$\times \int_0^\infty H_{2,4}^{1,2}\left[xt \,\middle|\, \begin{array}{l} (-\lambda, h), (0, 1) \\ (0,1), (1 - \beta_1, \alpha_1), (1 - \beta_2, \alpha_2), (-\lambda - 1, h) \end{array}\right] f(t)dt$$

*for $\mathrm{Re}(\lambda) > (1 - \nu)h - 1$, while*

$$(\mathbb{E}_{\alpha_1,\beta_1;\alpha_2,\beta_2}f)\,(x) = -hx^{1-(\lambda+1)/h}\frac{d}{dx}x^{(\lambda+1)/h} \times$$

$$\times \int_0^\infty H_{2,4}^{2,1}\left[xt\left|\begin{array}{c}(0,1),(-\lambda,h)\\(-\lambda-1,h),(0,1),(1-\beta_1,\alpha_1),(1-\beta_2,\alpha_2)\end{array}\right.\right]f(t)dt$$

*for* $\mathrm{Re}(\lambda)<(1-\nu)h-1$.

*If* $2\nu+\mathrm{Re}(\beta_1+\beta_2)>3$, *then* $\mathbb{E}_{\alpha_1,\beta_1;\alpha_2,\beta_2}f$ *is given by (13) and (20).*

Next statement follows from [8] (Theorem 4.5).

**Theorem 4.** *Let* $0<\nu<1$, $1\le r\le s<\infty$, *and let* $\alpha_1>0$ *and* $\alpha_2>0$ *be such that* $\alpha_1+\alpha_1<2$.

*(a) The transform* $\mathbb{E}_{\alpha_1,\beta_1;\alpha_2,\beta_2}$ *defined on* $L_{\nu,2}$ *can be extended to* $L_{\nu,r}$ *as an element of* $[L_{\nu,r},L_{1-\nu,s}]$.

*When* $1\le r\le 2$, *then* $\mathbb{E}_{\alpha_1,\beta_1;\alpha_2,\beta_2}$ *is a one-to-one transform from* $L_{\nu,r}$ *onto* $L_{1-\nu,s}$.

*(b) If* $f\in L_{\nu,r}$ *and* $g\in L_{\nu,s'}$ *with* $1/s+1/s'=1$, *then the relation (29) holds.*

Further three statements, following from [8] ( Theorems 4.6, 4.7 and 4.9), characterize the boundedness and the range of the $\mathbb{E}_{\alpha_1,\beta_1;\alpha_2,\beta_2}$-transform in the cases $\alpha_1+\alpha_2<1$, $\alpha_1+\alpha_2=1$ and $1<\alpha_1+\alpha_2<2$, respectively.

**Theorem 5.** *Let* $\alpha_1>0$ *and* $\alpha_2>0$ *be such that* $\alpha_1+\alpha_2<1$, *and let* $0<\nu<1$, $\omega=\alpha_1-\beta_1+\alpha_2-\beta_2+1$, $a_2^*=1-\alpha_1-\alpha_2$ *and* $1<r<\infty$ .

*(a) If the condition in (27) is satisfied, or if* $1<r\le 2$ *then the transform* $\mathbb{E}_{\alpha_1,\beta_1;\alpha_2,\beta_2}$ *is one-to-one on* $L_{\nu,r}$.

*(b) If* $\mathrm{Re}(\omega)\ge 0$ *and the condition in (27) is valid, then*

$$\mathbb{E}_{\alpha_1,\beta_1;\alpha_2,\beta_2}(L_{\nu,r})=\left(\mathbb{L}\mathbb{L}_{a_2^*,-\omega/a_2^*}\right)(L_{1-\nu,r}). \tag{30}$$

*When the condition in (27) is not valid,* $\mathbb{E}_{\alpha_1,\beta_1;\alpha_2,\beta_2}(L_{\nu,r})$ *is the subset of the right-hand side of (30).*

*(c) If* $\mathrm{Re}(\omega)<0$ *and the condition in (27) is satisfied, then*

$$\mathbb{E}_{\alpha_1,\beta_1;\alpha_2,\beta_2}(L_{\nu,r})=\left(K_{0,-\omega}^{-}\mathbb{L}\mathbb{L}_{a_2^*,0}\right)(L_{1-\nu,r}). \tag{31}$$

*When the condition in (27) is not satisfied,* $\mathbb{E}_{\alpha_1,\beta_1;\alpha_2,\beta_2}(L_{\nu,r})$ *is the subset of the right-hand side of (31).*

**Theorem 6.** *Let Let* $\alpha_1>0$ *and* $\alpha_2>0$ *be such that* $\alpha_1+\alpha_2=1$, *let* $0<\nu<1$, $\omega=3/2-\beta_1-\beta_2$, *and let* $1<r<\infty$ .

*(a) If the condition in (27) is satisfied, or if* $1<r\le 2$, *then transform* $\mathbb{E}_{\alpha_1,\beta_1;\alpha_2,\beta_2}$ *is one-to-one on* $L_{\nu,r}$.

*(b) If* $\mathrm{Re}(\omega)\ge 0$ *and the condition in (27) is valid, then*

$$\mathbb{E}_{\alpha_1,\beta_1;\alpha_2,\beta_2}(L_{\nu,r})=\mathbb{L}_{1,-\omega}(L_{\nu,r}). \tag{32}$$

*When the condition in (27) is not valid,* $\mathbb{E}_{\alpha_1,\beta_1;\alpha_2,\beta_2}(L_{\nu,r})$ *is the subset of the right-hand side of (32).*

*(c) If* $\mathrm{Re}(\omega)<0$ *and the condition in (27) is satisfied, then*

$$\mathbb{E}_{\alpha_1,\beta_1;\alpha_2,\beta_2}(L_{\nu,r})=\left(K_{0,-\omega}^{-}\mathbb{L}\right)(L_{\nu,r}). \tag{33}$$

When the condition in (27) is not satisfied, $\mathbb{E}_{\alpha_1,\beta_1;\alpha_2,\beta_2}(L_{\nu,r})$ is the subset of the right-hand side of (33).

**Theorem 7.** Let $\alpha_1 > 0$ and $\alpha_2 > 0$ be such that $1 < \alpha_1 + \alpha_2 < 2$, $a_2^* = 1 - \alpha_1 - \alpha_2$, and let $0 < \nu < 1$ and $1 < r < \infty$ .

(a) If the condition in (27) is satisfied,, or if $1 < r \leq 2$ then transform $\mathbb{E}_{\alpha_1,\beta_1;\alpha_2,\beta_2}$ is one-to-one on $L_{\nu,r}$.

(b) Let $\omega, \eta, \zeta \in \mathbb{C}$ be chosen as

$$\omega = \eta + \beta_1 + \beta_2 - \frac{3}{2}; \quad \mathrm{Re}(\eta) \geq \gamma(r) + 2a_2^*(\nu - 1) + 1 - \mathrm{Re}(\beta_1 + \beta_2);$$

$$\mathrm{Re}(\eta) > \nu - 1; \quad \mathrm{Re}(\zeta) < 1 - \nu.$$

If the condition in (27) is satisfied, then

$$\mathbb{E}_{\alpha_1,\beta_1;\alpha_2,\beta_2}(L_{\nu,r}) =$$

$$= \left( M_{\frac{1}{2}+\frac{\omega}{2a_2^*}} H_{-2a_2^*,2a_2^*\zeta+\omega-1} \mathbb{L}_{\alpha_1+\alpha_2-2,\frac{1}{2}+\eta-\frac{\omega}{2a_2^*}} \right) \left( L_{\frac{3}{2}-\nu+\frac{\mathrm{Re}(\omega)}{2a_2^*},r} \right). \tag{34}$$

When the condition in (27) is not satisfied, $\mathbb{E}_{\alpha_1,\beta_1;\alpha_2,\beta_2}(L_{\nu,r})$ is the subset of the right-hand side of (34).

Using [8] (Theorem 4.13) we deduce the inversion of the transform (13).

**Theorem 8.** Let $\alpha_1 > 0$ and $\alpha_2 > 0$ be such that $\alpha_1 + \alpha_2 = 2$, and let $0 < \nu < 1$ and $1 < r < \infty$ be such that

$$1 - \min\left[\frac{\mathrm{Re}(\beta_1)}{\alpha_1}, \frac{\mathrm{Re}(\beta_2)}{\alpha_2}\right] < \nu < \frac{7}{4} - \frac{\mathrm{Re}(\beta_1+\beta_2)}{2}, \quad \nu \geq \frac{5}{4} + \frac{\mathrm{Re}(\beta_1+\beta_2)}{2} + \frac{\gamma(r)}{2},$$

and let $\lambda \in \mathbb{C}$ and $h \in \mathbb{R}$.

If $f \in L_{\nu,r}$, then the inversion of $\mathbb{E}_{\alpha_1,\beta_1;\alpha_2,\beta_2}$-transform (3.1) is given by

$$f(x) = h x^{1-(\lambda+1)/h} \frac{d}{dx} x^{(\lambda+1)/h} \times$$

$$\times \int_0^\infty H_{2,4}^{2,1} \left[ xt \, \middle| \, \begin{matrix} (-\lambda, h), (0, 1) \\ (\beta_1 - \alpha_1, \alpha_1), (\beta_2 - \alpha_2, \alpha_2), (0, 1), (-\lambda - 1, h) \end{matrix} \right] \left( \mathbb{E}_{\alpha_1,\beta_1;\alpha_2,\beta_2} f \right)(t) dt$$

when $\mathrm{Re}(\lambda) > \nu h - 1$, while for $\mathrm{Re}(\lambda) < \nu h - 1$,

$$f(x) = -h x^{1-(\lambda+1)/h} \frac{d}{dx} x^{(\lambda+1)/h} \times$$

$$\times \int_0^\infty H_{2,4}^{3,0} \left[ xt \, \middle| \, \begin{matrix} (0, 1), (-\lambda, h) \\ (-\lambda - 1, h), (\beta_1 - \alpha_1, \alpha_1), (\beta_2 - \alpha_2, \alpha_2) \end{matrix} \right] \left( \mathbb{E}_{\alpha_1,\beta_1;\alpha_2,\beta_2} f \right)(t) dt.$$

**Remark 1.** The $L_{\nu,r}$-theory of the transform (21), (22) and (23) in respective cases (ii) $\alpha_1 > 0$, $\alpha_2 < 0$ (iii) $\alpha_1 < 0$, $\alpha_2 > 0$ and (iv) $\alpha_1 < 0$, $\alpha_2 < 0$ can be also deduced from the corresponding $L_{\nu,r}$-theory of the **H**-transform (18) presented in [8] (Chapter 4).

**Remark 2.** The results obtained for the transform $\mathbb{E}_{\alpha_1,\beta_1;\alpha_2,\beta_2}f$ can be applied to construct $L_{\nu,r}$-theory of integral transforms of the form (13) in which $\mathbb{E}_{\alpha_1,\beta_1;\alpha_2,\beta_2}(z)$ is replaced by the generalized Mittag-Leffler function (3) and by the Bessel functions (4)-(8).

## References

1. M. M. Dzhrbashyan, On integral transforms generated by generalized Mittag-Leffler function, (Russian), *Izv. Akad. Nauk Armyan. SSR, Ser. Fiz.-Mat.*, **13**, no. *3*, 21-63 (1960).
2. M. M. Dzhrbashyan, *Integral Transforms and Representations of Functions in the Complex Domain.* (Russian), Moscow, Nauka (1968).
3. M. M. Dzhrbashyan, *Harmonic Analysis and Boundary Value Problems in the Complex Domain*, Operator theory: Advances and Applications, 65, Birkhäuser Verlag, Basel (1993).
4. A. Erdelyi, W. Magnus, F. Oberhettinger and F. G. Tricomi F.G., *Higher Transcendental Functions*, Vol. 1, McGraw-Hill Book. Corp., New York (1953); Reprinted Krieger, Melbourne, Florida (1981).
5. A. Erdelyi, W. Magnus, F. Oberhettinger and F. G. Tricomi F.G., *Higher Transcendental Functions*, Vol. 2, McGraw-Hill Book. Corp., New York (1953); Reprinted Krieger, Melbourne, Florida (1981).
6. A. Erdelyi, W. Magnus, F. Oberhettinger and F. G. Tricomi F.G., *Higher Transcendental Functions*, Vol. 3, McGraw-Hill Book. Corp., New York (1954); Reprinted Krieger, Melbourne, Florida (1981).
7. S.G.Samko, A.A.Kilbas and O.I.Marichev, *Fractional Integrals and Derivatives. Theory and Applications.* Gordon and Breach, Yverdon (2003).
8. A. A. Kilbas and M. Saigo, *H-Transform. Theory and Aplpications.* Chapman and Hall/CRC, Boca Raton-London- New York-Washington, D.C (2004).
9. A. M. Mathai and R. K. Saxena R.K. *The H-Function with Applications in Statistics and Other Disciplines.* Halsted Press [John Wiley and Sons], New York-London-Sydney (1978).
10. A. P. Prudnikov, Yu. A. Brychkov and O. I. Marichev, *Integrals and Series*, Vol. 3, *Special Functions*, Gordon and Breach, New York, etc. (1989).
11. H. M. Srivastava H.M., K. C. Gupta and S. L. Goyal S.L. *The H-function of One and Two Variables with Applications.* South Asian Publishers, New-Delhi-Madras (1982).

# RELATIONSHIPS BETWEEN CONDITIONAL FOURIER-FEYNMAN TRANSFORM AND CONDITIONAL CONVOLUTION PRODUCT OF UNBOUNDED FUNCTIONS OVER WIENER PATHS IN ABSTRACT WIENER SPACE

B. I. SEUNG

*Department of Mathematics, Kyonggi University, Kyonggido Suwon 443-760, Korea*
*E-mail: biseung@kyonggi.ac.kr*

D. H. CHO*

*Department of Mathematics, Kyonggi University, Kyonggido Suwon 443-760, Korea*
*E-mail: j94385@kyonggi.ac.kr*

In this paper, using a simple formula for conditional Wiener integrals over Wiener paths in abstract Wiener space, we evaluate the conditional Fourier-Feynman transform and conditional convolution product for the product of cylinder type functions and functions in a Banach algebra which is equivalent to the Fresnel class. And then, we show that the conditional Fourier-Feynman transform of the conditional convolution product can be expressed as a product of the conditional Fourier-Feynman transform of each function over Wiener paths in abstract Wiener space.

**Key words:** Conditional convolution product, conditional Fourier-Feynman transform, conditional Wiener integral, cylinder type function, Fresnel class

**Mathematics subject Classification:** 28C20

## 1. Introduction and preliminaries

Let $C_0[0,T]$ denote the classical Wiener space, that is, the space of real-valued continuous functions $x$ on $[0,T]$ with $x(0) = 0$. As mentioned in [2], a concept of conditional Wiener integrals on the space was introduced by Yeh[16,15] and using a simple formula for conditional Wiener integrals on $C_0[0,T]$, the analytic conditional Feynman integrals and analytic conditional Fourier-Feynman transforms with conditional convolution products of various functions were evaluated[1,4,8,13,12,16,15].

In [10], $C_0(\mathbb{B})$, the space of abstract Wiener space-valued continuous functions $x$ on $[0,T]$ with $x(0) = 0$, was introduced and Ryu[14] developed several properties on the space. In [2], the class $\mathcal{A}_{r,u}^{(1)}$ of cylinder type functions on $C_0(\mathbb{B})$ was introduced and the authors in [2] defined the analytic conditional Fourier-Feynman transform and the conditional convolution product on the space $C_0(\mathbb{B})$. And then, using a simple formula[3] for conditional Wiener integrals on the space $C_0(\mathbb{B})$, they

---
*This work was supported by Kyonggi University Research Grant.

showed that the analytic conditional Fourier-Feynman transform of the conditional convolution product of functions in $\mathcal{A}_{r,u}^{(1)}$, is a product of the analytic conditional Fourier-Feynman transform of each function.

In this paper, we evaluate the analytic conditional Fourier-Feynman transform and conditional convolution product for the product of the functions in $\mathcal{A}_{r,u}^{(1)}$ and the functions in the Banach algebra $\mathcal{F}(C_0(\mathbb{B}); u)^5$ which is equivalent to the Fresnel class. And then, we show that the analytic conditional Fourier-Feynman transform of the conditional convolution product can be expressed as a product of the analytic conditional Fourier-Feynman transform of each function over Wiener paths in abstract Wiener space. Note that the results are a generalization of those in citechoch.

Let $(\mathcal{H}, \mathbb{B}, m)$ be an abstract Wiener space[11] and for each $h \in \mathcal{H}$ and $x_1 \in \mathbb{B}$, let $(h, x_1)^\sim$ be the stochastic inner product[9] of $h$ and $x_1$. Note that for each $h(\neq 0)$ in $\mathcal{H}$, $(h, \cdot)^\sim$ is a Gaussian random variable on $\mathbb{B}$ with mean zero and variance $|h|^2$ and it is well-known that if $\{h_1, h_2, \cdots, h_n\}$ is an orthogonal set in $\mathcal{H}$, then the random variables $(h_j, \cdot)^\sim$'s are independent. Moreover, if both $h$ and $x_1$ are in $\mathcal{H}$, then $(h, x_1)^\sim = \langle h, x_1 \rangle$ where $\langle \cdot, \cdot \rangle$ denotes the inner product on $\mathcal{H}$.

Let $C_0(\mathbb{B})$ denote the space of all continuous paths $x : [0, T] \to \mathbb{B}$ with $x(0) = 0$. Then $C_0(\mathbb{B})$ is a real separable Banach space with the norm $\|x\|_{C_0(\mathbb{B})} \equiv \sup_{0 \leq t \leq T} \|x(t)\|_\mathbb{B}$ and the Brownian motion in $\mathbb{B}$ induces a probability measure $m_\mathbb{B}$ on $(C_0(\mathbb{B}), \mathcal{B}(C_0(\mathbb{B})))$ which is mean zero Gaussian. A complex-valued measurable function on $C_0(\mathbb{B})$ is said to be Wiener integrable if it is integrable with respect to $m_\mathbb{B}$.

Now, we introduce the Wiener integration theorem[14].

**Theorem 1.1.** *Let* $0 = t_0 \leq t_1 \leq \cdots \leq t_k \leq T$ *and let* $f : \mathbb{B}^k \to \mathbb{C}$ *be a Borel measurable function. Then, we have*

$$\int_{C_0(\mathbb{B})} f(x(t_1), \cdots, x(t_k)) \, dm_\mathbb{B}(x)$$

$$\overset{*}{=} \int_{\mathbb{B}^k} f\left(\sqrt{t_1 - t_0}\, x_1, \cdots, \sum_{j=1}^{k} \sqrt{t_j - t_{j-1}}\, x_j\right) d\left(\prod_{j=1}^{k} m\right)(x_1, \cdots, x_k),$$

*where by* $\overset{*}{=}$ *we mean that if either side exists, then both sides exist and they are equal.*

**Definition 1.1.** *Let* $F : C_0(\mathbb{B}) \to \mathbb{C}$ *be Wiener integrable and* $X : C_0(\mathbb{B}) \to B$ *be a random variable, where $B$ is a real normed linear space. The conditional expectation* $E[F|X]$ *of $F$ given $X$ on $B$ is called the conditional Wiener integral[2] of $F$ given $X$.*

Let

$$\tau : 0 = t_0 < t_1 < \cdots < t_k = T \tag{1}$$

be a partition of $[0,T]$ and let $x$ be in $C_0(\mathbb{B})$. Define the polygonal function $[x]$ of $x$ on $[0,T]$ by

$$[x](t) = \sum_{j=1}^{k} \chi_{(t_{j-1},t_j]}(t) \left[ x(t_{j-1}) + \frac{t - t_{j-1}}{t_j - t_{j-1}}(x(t_j) - x(t_{j-1})) \right] \qquad (2)$$

for $t \in [0,T]$. For $\vec{\xi} = (\xi_1, \cdots, \xi_k) \in \mathbb{B}^k$, let $[\vec{\xi}]$ be the polygonal function of $\vec{\xi}$ on $[0,T]$ given by (2) with replacing $x(t_j)$ by $\xi_j$ for $j = 0, 1, \cdots, k$ ($\xi_0 = 0$).

The following lemma[3] is useful to define and evaluate the analytic conditional Wiener and Feynman integrals of functions on $C_0(\mathbb{B})$.

**Lemma 1.1.** *Let $F$ be defined and Wiener integrable on $C_0(\mathbb{B})$. Let $X_\tau : C_0(\mathbb{B}) \to \mathbb{B}^k$ be a random variable given by $X_\tau(x) = (x(t_1), \cdots, x(t_k))$. Then for every Borel measurable subset $B$ of $\mathbb{B}^k$, we have*

$$\int_{X_\tau^{-1}(B)} F(x) \, dm_{\mathbb{B}}(x) = \int_B E[F(x - [x] + [\vec{\xi}])] \, dP_{X_\tau}(\vec{\xi}) \qquad (3)$$

*where $P_{X_\tau}$ is the probability distribution of $X_\tau$ on $(\mathbb{B}^k, \mathcal{B}(\mathbb{B}^k))$.*

A subset $E$ of $C_0(\mathbb{B})$ is called a scale-invariant null set if $m_{\mathbb{B}}(\lambda E) = 0$ for every $\lambda > 0$ and a property on $C_0(\mathbb{B})$ is said to hold scale-invariant almost everywhere (in abbreviation, s-a.e.) if it holds except for a scale-invariant null set. For a function $F : C_0(\mathbb{B}) \to \mathbb{C}$ and $\lambda > 0$, let $F^\lambda(x) = F(\lambda^{-\frac{1}{2}}x)$ and $X_\tau^\lambda(x) = X_\tau(\lambda^{-\frac{1}{2}}x)$. Suppose that $E[F^\lambda]$ exists for each $\lambda > 0$. By the definition of the conditional Wiener integral (Definition 1.1) and (3), we have

$$E[F^\lambda | X_\tau^\lambda](\vec{\xi}) = E[F(\lambda^{-\frac{1}{2}}(x - [x]) + [\vec{\xi}])]$$

for $P_{X_\tau^\lambda}$-a.e. $\vec{\xi} \in \mathbb{B}^k$, where $P_{X_\tau^\lambda}$ is the probability distribution of $X_\tau^\lambda$ on $(\mathbb{B}^k, \mathcal{B}(\mathbb{B}^k))$. If $E[F(\lambda^{-\frac{1}{2}}(x - [x]) + [\vec{\xi}])]$ has the analytic extension $J_\lambda^*(F)(\vec{\xi})$ on $\mathbb{C}_+ \equiv \{z \in \mathbb{C} : \text{Re } z > 0\}$ as a function of $\lambda$, then it is called the analytic conditional Wiener integral of $F$ given $X_\tau$ over $C_0(\mathbb{B})$ with parameter $\lambda$ and denoted by

$$E^{anw_\lambda}[F | X_\tau](\vec{\xi}) = J_\lambda^*(F)(\vec{\xi})$$

for $\vec{\xi} \in \mathbb{B}^k$. Moreover, if for a non-zero real $q$, $E^{anw_\lambda}[F | X_\tau](\vec{\xi})$ has a limit as $\lambda$ approaches to $-iq$ through $\mathbb{C}_+$, then it is called the analytic conditional Feynman integral of $F$ given $X_\tau$ over $C_0(\mathbb{B})$ with parameter $q$ and denoted by

$$E^{anf_q}[F | X_\tau](\vec{\xi}) = \lim_{\lambda \to -iq} E^{anw_\lambda}[F | X_\tau](\vec{\xi}).$$

Now, we introduce a useful integral formula which is applied in the proofs of several results.

**Lemma 1.2.** *Let $a \in \mathbb{C}_+$ and $b$ be a real number. Then, we have*

$$\int_{\mathbb{R}} \exp\{-au^2 + ibu\} du = \left(\frac{\pi}{a}\right)^{\frac{1}{2}} \exp\left\{-\frac{b^2}{4a}\right\}.$$

## 2. Conditional analytic Fourier-Feynman transforms

Throughout the remainder of this paper, let $0 < u \leq T$ be fixed, but arbitrary. Let $\mathcal{H}$ be a real separable infinite dimensional Hilbert space, let $r$ be a positive integer and let $\{e_1, \cdots, e_r\}$ be an orthonormal set in $\mathcal{H}$. Let

$$X_r(x_1) = ((e_1, x_1)^\sim, \cdots, (e_r, x_1)^\sim)$$

for $x_1$ in $\mathbb{B}$. For $1 \leq p < \infty$, let $\mathcal{A}_{r,u}^{(p)}$ be the space of all cylinder type functions $F_r$ of the form

$$F_r(x) = f(X_r(x(u))) \tag{4}$$

for s-a.e. $x$ in $C_0(\mathbb{B})$ where $f \in L_p(\mathbb{R}^r)$. Let $\mathcal{A}_{r,u}^{(\infty)}$ be the space of all functions of the form (4) with $f \in L_\infty(\mathbb{R}^r)$, the space of essentially bounded functions on $\mathbb{R}^r$. Note that we can take $f$ to be Borel measurable without loss of generality.

Let $\mathcal{M}(\mathcal{H})$ be the class of all $\mathbb{C}$-valued Borel measures on $\mathcal{H}$ with bounded variation and let $\mathcal{F}(C_0(\mathbb{B}); u)$ be the space of all s-equivalence classes of functions $F$ which for $\sigma \in \mathcal{M}(\mathcal{H})$, have the form

$$F(x) = \int_{\mathcal{H}} \exp\{i(h, x(u))^\sim\} d\sigma(h) \tag{5}$$

for $x \in C_0(\mathbb{B})$. Note that $\mathcal{F}(C_0(\mathbb{B}); u)$ is a Banach algebra which is equivalent to Fresnel class with the norm $\|F\| = \|\sigma\|$, the total variation of $\sigma$ in $\mathcal{M}(\mathcal{H})^5$.

For convenience, we introduce a useful notations using Gram-Schmidt orthonormalization process. Given $h \in \mathcal{H}$, we obtain the orthonormal set $\{e_1, \cdots, e_r, e_{r+1}\}$ as follows;

$$c_j(h) = \begin{cases} \langle h, e_j \rangle & \text{for } j = 1, \cdots, r; \\ \sqrt{|h|^2 - \sum_{l=1}^{r} [c_l(h)]^2} & \text{for } j = r+1 \end{cases} \tag{6}$$

and

$$e_{r+1} = \frac{1}{c_{r+1}(h)} \left( h - \sum_{j=1}^{r} c_j(h) e_j \right)$$

if $c_{r+1}(h) \neq 0$. Then we have

$$h = \sum_{j=1}^{r+1} c_j(h) e_j \quad \text{and} \quad |h|^2 = \sum_{j=1}^{r+1} [c_j(h)]^2. \tag{7}$$

Note that (7) holds trivially for the case $c_{r+1}(h) = 0$.

For a given real number $p$ with $1 < p \leq \infty$, suppose $p$ and $p'$ are related by $\frac{1}{p} + \frac{1}{p'} = 1$ (possibly $p' = 1$ if $p = \infty$). Let $G_n$ and $G$ be measurable functions such that, for each $\gamma > 0$,

$$\lim_{n \to \infty} \int_{C_0(\mathbb{B})} |G_n(\gamma x) - G(\gamma x)|^{p'} dm_{\mathbb{B}}(x) = 0.$$

Then we write

$$\operatorname*{l.i.m.}_{n \to \infty} (w_s^{p'})(G_n) \approx G$$

and similar definition is understood when $n$ is replaced by a continuously varying parameter.

**Definition 2.1.** Let $F$ be defined on $C_0(\mathbb{B})$ and let $X_\tau$ be given as in Lemma 1.1. For $\lambda \in \mathbb{C}_+$ and for a.e. $\vec{\xi} \in \mathbb{B}^k$, let

$$T_\lambda[F|X_\tau](y, \vec{\xi}) = E^{anw_\lambda}[F(\cdot + y)|X_\tau](\vec{\xi})$$

for s-a.e. $y \in C_0(\mathbb{B})$ if it exists. For a non-zero real $q$ and for a.e. $\vec{\xi} \in \mathbb{B}^k$, we define the $L_1$ analytic conditional Fourier-Feynman transform $T_q^{(1)}[F|X_\tau]$ of $F$ given $X_\tau$ by the formula

$$T_q^{(1)}[F|X_\tau](y, \vec{\xi}) = \lim_{\lambda \to -iq} T_\lambda[F|X_\tau](y, \vec{\xi})$$

if it exists for s-a.e. $y \in C_0(\mathbb{B})$ and for $1 < p \le \infty$ we define the $L_p$ analytic conditional Fourier-Feynman transform $T_q^{(p)}[F|X_\tau]$ of $F$ given $X_\tau$ by the formula

$$T_q^{(p)}[F|X_\tau](\cdot, \vec{\xi}) \approx \operatorname*{l.i.m.}_{\lambda \to -iq} (w_s^{p'})(T_\lambda[F|X_\tau](\cdot, \vec{\xi})),$$

where $\lambda$ approaches to $-iq$ through $\mathbb{C}_+$.

Using Theorem 1.1 and Lemma 1.2, the following lemma is obtained by the fact that the stochastic inner product is mean-zero Gaussian. For detailed proof, see [6].

**Lemma 2.1.** *Let $a > 0$, $h \in \mathcal{H}$ and a partition of $[0, T]$ be given by (1). Further, let $F_r$ be given by (4) and suppose that $t_{p^*-1} < u < t_{p^*}$ for some $p^* \in \{1, \cdots, k\}$. Then we have*

$$E[\exp\{ia(h, x(u) - [x](u))^\sim\} F_r(a(x - [x]))]$$

$$\overset{*}{=} \left(\frac{\Gamma}{2\pi a^2}\right)^{\frac{r}{2}} \int_{\mathbb{R}^r} f(\vec{u}_r) \exp\left\{\frac{a^2}{2\Gamma} \left[\sum_{j=1}^{r} \left[\frac{\Gamma}{a^2} iu_j + c_j(h)\right]^2 - |h|^2\right]\right\} d\vec{u}_r$$

*with $\Gamma = \frac{t_{p^*} - t_{p^*-1}}{(t_{p^*} - u)(u - t_{p^*-1})}$ and $\vec{u}_r = (u_1, \cdots, u_r)$, where by $\overset{*}{=}$ we mean that if either side exists, then both sides exist and they are equal, and $c_j(h)$ is given by (6).*

Now, we have the following theorem from Lemma 2.1, the change of variable theorem and Morera's theorem.

**Theorem 2.1.** *Let $G_r$ be given by*

$$G_r(x) = F(x)F_r(x) \tag{8}$$

*for s-a.e. $x$ in $C_0(\mathbb{B})$, where $F_r \in \mathcal{A}_{r,u}^{(p)}$ $(1 \le p \le \infty)$ and $F \in \mathcal{F}(C_0(\mathbb{B}); u)$ are given by (4) and (5), respectively. Further, let $X_\tau$ be given as in Lemma 1.1. Then, for $\lambda \in \mathbb{C}_+$ and a.e. $\vec{\xi} \in \mathbb{B}^k$, $T_\lambda[G_r|X_\tau](y, \vec{\xi})$ exists for s-a.e. $y \in C_0(\mathbb{B})$. Moreover,* *when $t_{p^*-1} < u < t_{p^*}$ for some $p^* \in \{1, \cdots, k\}$, we have*

$$T_\lambda[G_r|X_\tau](y, \vec{\xi}) \tag{9}$$

$$= \left(\frac{\lambda\Gamma}{2\pi}\right)^{\frac{r}{2}} \int_\mathcal{H} \exp\{i(h, y(u) + [\vec{\xi}](u))^\sim\} \int_{\mathbb{R}^r} f(\vec{u}_r) \exp\left\{\frac{1}{2\lambda\Gamma}\left[\sum_{j=1}^r [\lambda\Gamma i(u_j\right.\right.$$

$$\left.\left. - (e_j, y(u) + [\vec{\xi}](u))^\sim) + c_j(h)]^2 - |h|^2\right]\right\} d\vec{u}_r d\sigma(h)$$

*where $\Gamma$, $\vec{u}_r$ and $c_j(h)$ are given as in Lemma 2.1. When $u = t_{p^*}$ for some $p^* \in \{1, \cdots, k\}$, we have*

$$T_\lambda[G_r|X_\tau](y, \vec{\xi}) = G_r(y + [\vec{\xi}]). \tag{10}$$

**Remark 2.1.** When $u = t_{p^*}$ for some $p^* \in \{1, \cdots, k\}$, $T_q^{(p)}[G_r|X_\tau](y, \vec{\xi})(1 \le p \le \infty, q \in \mathbb{R} - \{0\})$ exists and it is given by the right hand side of (10) for any function $G_r$ on $C_0(\mathbb{B})$. Hence, it suffices to consider the case $t_{p^*-1} < u < t_{p^*}$ for some $p^* \in \{1, \cdots, k\}$.

By Theorem 2.1 and the dominated convergence theorem, we have the following theorem.

**Theorem 2.2.** *Let the assumptions and notations be given as in Theorem 2.1 with one exception $p = 1$. Further, let $t_{p^*-1} < u < t_{p^*}$ for some $p^* \in \{1, \cdots, k\}$. Then, for a non-zero real $q$ and a.e. $\vec{\xi} \in \mathbb{B}^k$, $T_q^{(1)}[G_r|X_\tau](y, \vec{\xi})$ exists for s-a.e. $y \in C_0(\mathbb{B})$ and it is given by (9) with $\lambda = -iq$, that is,*

$$T_q^{(1)}[G_r|X_\tau](y, \vec{\xi})$$

$$= \left(\frac{q\Gamma}{2\pi i}\right)^{\frac{r}{2}} \int_\mathcal{H} \exp\{i(h, y(u) + [\vec{\xi}](u))^\sim\} \int_{\mathbb{R}^r} f(\vec{u}_r) \exp\left\{\frac{i}{2q\Gamma}\left[\sum_{j=1}^r [q\Gamma(u_j\right.\right.$$

$$\left.\left. - (e_j, y(u) + [\vec{\xi}](u))^\sim) + c_j(h)]^2 - |h|^2\right]\right\} d\vec{u}_r d\sigma(h).$$

Let $\hat{M}(\mathbb{R}^r)$ be the set of all functions $\phi$ on $\mathbb{R}^r$ given by

$$\phi(\vec{u}_r) = \int_{\mathbb{R}^r} \exp\{i\langle \vec{u}_r, \vec{v}_r \rangle\} d\rho(\vec{v}_r), \tag{11}$$

where $\vec{u}_r \in \mathbb{R}^r$ and $\rho$ is a complex Borel measure of bounded variation over $\mathbb{R}^r$.

Now, with the function $\phi$, we have the following theorem which describes the analytic conditional Fourier-Feynman transform. The proof follows from Lemma 1.2, Theorem 2.1, Morera's theorem and the dominated convergence theorem.

**Theorem 2.3.** *Let $\phi$ be given by (11), let $K_r$ be given by the right-hand side of (4) with replacing $f$ by $\phi$ and let $G_r = FK_r$ where $F$ is given by (5). Further, let $1 \le p \le \infty$ and let $X_\tau$ be given as in Lemma 1.1. Then, for $\lambda \in \mathbb{C}_+$ and a.e.*

$\vec{\xi} \in \mathbb{B}^k$, $T_\lambda[G_r|X_\tau](y, \vec{\xi})$ exists for s-a.e. $y \in C_0(\mathbb{B})$ and when $t_{p^*-1} < u < t_{p^*}$ for some $p^* \in \{1, \cdots, k\}$, it is given by

$$T_\lambda[G_r|X_\tau](y, \vec{\xi})$$

$$= \int_{\mathcal{H}} \exp\{i(h, y(u) + [\vec{\xi}](u))^\sim\} \int_{\mathbb{R}^r} \exp\left\{i \sum_{j=1}^r v_j(e_j, y(u) + [\vec{\xi}](u))^\sim\right.$$

$$\left. - \frac{1}{2\lambda\Gamma}\left[|h|^2 + 2\sum_{j=1}^r v_j c_j(h) + \sum_{j=1}^r v_j^2\right]\right\} d\rho(\vec{v}_r) d\sigma(h)$$

with $\vec{v}_r = (v_1, \cdots, v_r)$, where $\Gamma$ is given as in Lemma 2.1 and $c_j(h)$ is given by (6). Moreover, for a non-zero real $q$ and a.e. $\vec{\xi} \in \mathbb{B}^k$, $T_q^{(p)}[G_r|X_\tau](y, \vec{\xi})$ exists for s-a.e. $y \in C_0(\mathbb{B})$ and when $t_{p^*-1} < u < t_{p^*}$ for some $p^* \in \{1, \cdots, k\}$, it is given by

$$T_q^{(p)}[G_r|X_\tau](y, \vec{\xi})$$

$$= \int_{\mathcal{H}} \exp\{i(h, y(u) + [\vec{\xi}](u))^\sim\} \int_{\mathbb{R}^r} \exp\left\{i \sum_{j=1}^r v_j(e_j, y(u) + [\vec{\xi}](u))^\sim\right.$$

$$\left. + \frac{1}{2qi\Gamma}\left[|h|^2 + 2\sum_{j=1}^r v_j c_j(h) + \sum_{j=1}^r v_j^2\right]\right\} d\rho(\vec{v}_r) d\sigma(h).$$

## 3. Conditional convolution products and relationships with analytic conditional Fourier-Feynman transforms

In this section, we evaluate conditional convolution products and investigate the relationships between the analytic conditional Fourier-Feynman transforms and conditional convolution products.

Now, let $X_\tau$ be given as in Lemma 1.1 and let $F, G$ be defined on $C_0(\mathbb{B})$. We define the conditional convolution product $[(F * G)_\lambda|X_\tau]$ of $F$ and $G$ given $X_\tau$ by the formula, for a.e. $\vec{\xi} \in \mathbb{B}^k$

$$[(F * G)_\lambda|X_\tau](y, \vec{\xi})$$

$$= \begin{cases} E^{anw_\lambda}\left[F\left(\dfrac{y + \cdot}{2^{\frac{1}{2}}}\right)G\left(\dfrac{y - \cdot}{2^{\frac{1}{2}}}\right)\bigg|X_\tau\right](\vec{\xi}), & \lambda \in \mathbb{C}_+; \\[3mm] E^{anf_q}\left[F\left(\dfrac{y + \cdot}{2^{\frac{1}{2}}}\right)G\left(\dfrac{y - \cdot}{2^{\frac{1}{2}}}\right)\bigg|X_\tau\right](\vec{\xi}), & \lambda = -iq, \quad q \in \mathbb{R} - \{0\} \end{cases}$$

if they exist for s-a.e. $y \in C_0(\mathbb{B})$. If $\lambda = -iq$, then we denote $[(F * G)_\lambda|X_\tau](y, \vec{\xi})$ by $[(F * G)_q|X_\tau](y, \vec{\xi})$.

Now, we begin with the following theorem which evaluates conditional convolution products. The proof follows from Theorem 1.1, Lemma 1.2, Fubini's theorem, the change of variable theorem, Morera's theorem and the fact that the stochastic inner product is mean-zero Gaussian.

**Theorem 3.1.** For $i = 1, 2$, let $F_{ri}$, $F_i$ be given by (4), (5) with replacing $f, \sigma$ by $f_i, \sigma_i$, respectively. Further, let $G_{ri} = F_i F_{ri}(F_{ri} \in \mathcal{A}_{r,u}^{(p)}, 1 \leq p \leq \infty)$ and $X_\tau$ be given as in Lemma 1.1. Then, for $\lambda \in \mathbb{C}_+$ and a.e. $\vec{\xi} \in \mathbb{B}^k$, $[(G_{r1} * G_{r2})_\lambda | X_\tau](y, \vec{\xi})$ exists for s-a.e. $y \in C_0(\mathbb{B})$. Moreover, when $t_{p^*-1} < u < t_{p^*}$ for some $p^* \in \{1, \cdots, k\}$, we have

$$
[(G_{r1} * G_{r2})_\lambda | X_\tau](y, \vec{\xi})
$$
$$
= \left(\frac{\lambda\Gamma}{2\pi}\right)^{\frac{r}{2}} \int_{\mathcal{H}} \int_{\mathcal{H}} \exp\left\{\frac{i}{\sqrt{2}}[(h_1, y(u) + [\vec{\xi}](u))^\sim + (h_2, y(u) - [\vec{\xi}](u))^\sim]\right\}
$$
$$
\times \int_{\mathbb{R}^r} f_1\left(\frac{1}{\sqrt{2}}[X_r(y(u) + [\vec{\xi}](u)) + \vec{u}_r]\right) f_2\left(\frac{1}{\sqrt{2}}[X_r(y(u) - [\vec{\xi}](u)) - \vec{u}_r]\right)
$$
$$
\times \exp\left\{\frac{1}{2\lambda\Gamma}\left[\sum_{j=1}^{r}\left[\lambda\Gamma i u_j + \frac{1}{\sqrt{2}}c_j(h_1 - h_2)\right]^2 - \frac{1}{2}|h_1 - h_2|^2\right]\right\} d\vec{u}_r d\sigma_1(h_1) d\sigma_2(h_2)
$$

where $\Gamma$, $\vec{u}_r$ and $c_j(h)$ are given as in Lemma 2.1 with replacing $h$ by $h_1 - h_2$, and when $u = t_{p^*}$ for some $p^* \in \{1, \cdots, k\}$, we have

$$
[(G_{r1} * G_{r2})_\lambda | X_\tau](y, \vec{\xi}) = G_{r1}\left(\frac{1}{\sqrt{2}}(y + [\vec{\xi}])\right) G_{r2}\left(\frac{1}{\sqrt{2}}(y - [\vec{\xi}])\right).
$$

If $f_i$ in Theorem 3.1 is the Fourier transform of a complex Borel measure of bounded variation over $\mathbb{R}^r$, then we have the following theorem by Lemma 1.2 and the dominated convergence theorem.

**Theorem 3.2.** For $i = 1, 2$, let $\phi_i$ be given by (11) with replacing $\rho$ by $\rho_i$ and let $t_{p^*-1} < u < t_{p^*}$ for some $p^* \in \{1, \cdots, k\}$. Then, under the assumptions and notations given as in Theorem 3.1 with replacing $f_i$ by $\phi_i$, we have, for $\lambda \in \mathbb{C}_+$ and a.e. $\vec{\xi} \in \mathbb{B}^k$,

$$
[(G_{r1} * G_{r2})_\lambda | X_\tau](y, \vec{\xi})
$$
$$
= \int_{\mathcal{H}} \int_{\mathcal{H}} \int_{\mathbb{R}^r} \int_{\mathbb{R}^r} \exp\left\{\frac{i}{\sqrt{2}}\left[\left(h_1 + \sum_{j=1}^{r} z_j e_j, y(u) + [\vec{\xi}](u)\right)^\sim + \left(h_2 + \sum_{j=1}^{r} w_j e_j,\right.\right.\right.
$$
$$
\left.\left.\left. y(u) - [\vec{\xi}](u)\right)^\sim\right] - \frac{1}{4\lambda\Gamma}\left[|h_1 - h_2|^2 + 2\sum_{j=1}^{r}(z_j - w_j)c_j(h_1 - h_2) + \sum_{j=1}^{r}(z_j\right.\right.
$$
$$
\left.\left. - w_j)^2\right]\right\} d\rho_1(\vec{z}_r) d\rho_2(\vec{w}_r) d\sigma_1(h_1) d\sigma_2(h_2)
$$

for s-a.e. $y \in C_0(\mathbb{B})$ where $\vec{z}_r = (z_1, \cdots, z_r)$ and $\vec{w}_r = (w_1, \cdots, w_r)$. Moreover, for a non-zero real $q$, $[(G_{r1} * G_{r2})_q | X_\tau](y, \vec{\xi})$ exists and it is given by the right hand side of the equation with replacing $\lambda$ by $-iq$.

For $p = 1$ in Theorem 3.1, using the same method given as in the proof of Theorem 11 in [7], we have the following convolution product by the dominated convergence theorem.

**Theorem 3.3.** Let the assumptions and notations be given as in Theorem 3.1 with one exception $p = 1$. Then, for a non-zero real $q$ and a.e. $\vec{\xi} \in \mathbb{B}^k$,

$[(G_{r1} * G_{r2})_q | X_\tau](y, \vec{\xi})$ exists for s-a.e. $y \in C_0(\mathbb{B})$. Further, when $t_{p^*-1} < u < t_{p^*}$ for some $p^* \in \{1, \cdots, k\}$, it is given by the right hand side of the first equation in Theorem 3.1 with $\lambda = -iq$, that is,

$$[(G_{r1} * G_{r2})_q | X_\tau](y, \vec{\xi})$$

$$= \left(\frac{q\Gamma}{2\pi i}\right)^{\frac{r}{2}} \int_{\mathcal{H}} \int_{\mathcal{H}} \exp\left\{\frac{i}{\sqrt{2}}[(h_1, y(u) + [\vec{\xi}](u))^\sim + (h_2, y(u) - [\vec{\xi}](u))^\sim]\right\}$$

$$\times \int_{\mathbb{R}^r} f_1\left(\frac{1}{\sqrt{2}}[X_r(y(u) + [\vec{\xi}](u)) + \vec{u}_r]\right) f_2\left(\frac{1}{\sqrt{2}}[X_r(y(u) - [\vec{\xi}](u)) - \vec{u}_r]\right)$$

$$\times \exp\left\{\frac{i}{2q\Gamma}\left[\sum_{j=1}^r \left[q\Gamma u_j + \frac{1}{\sqrt{2}}c_j(h_1 - h_2)\right]^2 - \frac{1}{2}|h_1 - h_2|^2\right]\right\} d\vec{u}_r d\sigma_1(h_1) d\sigma_2(h_2),$$

and when $u = t_{p^*}$ for some $p^* \in \{1, \cdots, k\}$, it is given by

$$[(G_{r1} * G_{r2})_q | X_\tau](y, \vec{\xi}) = G_{r1}\left(\frac{1}{\sqrt{2}}(y + [\vec{\xi}])\right) G_{r2}\left(\frac{1}{\sqrt{2}}(y - [\vec{\xi}])\right).$$

By Theorems 2.1, 3.1 and the change of variable theorem, we have the following theorem which evaluates the Fourier-Wiener transform of the convolution product.

**Theorem 3.4.** *Under the assumptions given as in Theorem 3.1, for $\lambda \in \mathbb{C}_+$ and for a.e. $\vec{\xi}_1, \vec{\xi}_2 \in \mathbb{B}^k$, we have*

$$T_\lambda[[(G_{r1} * G_{r2})_\lambda | X_\tau](\cdot, \vec{\xi}_1) | X_\tau](y, \vec{\xi}_2)$$

$$= \left[T_\lambda[G_{r1}|X_\tau]\left(\frac{1}{\sqrt{2}}y, \frac{1}{\sqrt{2}}(\vec{\xi}_2 + \vec{\xi}_1)\right)\right]\left[T_\lambda[G_{r2}|X_\tau]\left(\frac{1}{\sqrt{2}}y, \frac{1}{\sqrt{2}}(\vec{\xi}_2 - \vec{\xi}_1)\right)\right]$$

*for s-a.e. $y \in C_0(\mathbb{B})$.*

By Theorems 2.3, 3.2, 3.4 or Theorems 2.2, 3.3, 3.4, we have the final result;

**Theorem 3.5.** *Under the assumptions given as in Theorem 3.2 or 3.3, for a non-zero real $q$ and for a.e. $\vec{\xi}_1, \vec{\xi}_2 \in \mathbb{B}^k$, we have*

$$T_q^{(1)}[[(G_{r1} * G_{r2})_q | X_\tau](\cdot, \vec{\xi}_1) | X_\tau](y, \vec{\xi}_2)$$

$$= \left[T_q^{(1)}[G_{r1}|X_\tau]\left(\frac{1}{\sqrt{2}}y, \frac{1}{\sqrt{2}}(\vec{\xi}_2 + \vec{\xi}_1)\right)\right]\left[T_q^{(1)}[G_{r2}|X_\tau]\left(\frac{1}{\sqrt{2}}y, \frac{1}{\sqrt{2}}(\vec{\xi}_2 - \vec{\xi}_1)\right)\right]$$

*for s-a.e. $y \in C_0(\mathbb{B})$.*

**Remark 3.1.** In (5), let $\sigma \in \mathcal{M}(\mathcal{H})$ be the Dirac measure concentrated at 0. Then it is not difficult to show that $1 \in \mathcal{F}(C_0(\mathbb{B}); u)$ and hence we can regard Theorem 4.8 in [2] as a special case of Theorem 3.5.

## References

1. K. S. Chang and J. S. Chang, *Evaluation of some conditional Wiener integrals*, Bull. Korean Math. Soc. **21** (1984), 99-106.

2. K. S. Chang, D. H. Cho, B. S. Kim, T. S. Song, and I. Yoo, *Conditional Fourier-Feynman transform and convolution product over Wiener paths in abstract Wiener space*, Integral Transforms Spec. Funct. **14** (2003), no. 3, 217-235.

3. K. S. Chang, D. H. Cho and I. Yoo, *A conditional analytic Feynman integral over Wiener paths in abstract Wiener space*, Int. Math. J. **2** (2002), no 9, 855-870.

4. S. J. Chang and D. L. Skoug, *The effect of drift on conditional Fourier-Feynman transforms and conditional convolution products*, Int. J. Appl. Math. **2** (2000), no 4, 505-527.

5. D. H. Cho, *Conditional analytic Feynman integral over product space of Wiener paths in abstract Wiener space*, Rocky Mountain J. Math. (2005), to appear.

6. D. H. Cho, *Change of scale formulas for conditional Wiener integrals as integral transforms over Wiener paths in abstract Wiener space*, Integral Transforms Spec. Funct. (2005), submitted.

7. D. H. Cho, *Conditional Fourier-Feynman transform and convolution product over Wiener paths in abstract Wiener space : an $L_p$ theory*, J. Korean Math. Soc. **41** (2004), no. 2, 265-294.

8. D. M. Chung and D. L. Skoug, *Conditional analytic Feynman integrals and a related Schrödinger integral equation*, SIAM J. Math. Anal. **20** (1989), 950-965.

9. G. Kallianpur and C. Bromley, *Generalized Feynman integrals using analytic continuation in several complex variables*, Stochastic analysis and applications, 217-267, Dekker, 1984.

10. J. Kuelbs and R. LePage, *The law of the iterated logarithm for Brownian motion in a Banach space*, Trans. Amer. Math. Soc. **185** (1973), 253-264.

11. H. H. Kuo, *Gaussian measures in Banach spaces*, Lecture Notes in Math. **463**, Springer-Verlag, 1975.

12. C. Park and D. L. Skoug, *Conditional Yeh-Wiener integrals with vector-valued conditioning functions*, Proc. Amer. Math. Soc. **105** (1989), 450-461.

13. C. Park and D. L. Skoug, *A simple formula for conditional Wiener integrals with applications*, Pacific J. Math. **135** (1988), 381-394.

14. K. S. Ryu, *The Wiener integral over paths in abstract Wiener space*, J. Korean Math. Soc. **29** (1992), no. 2, 317-331.

15. J. Yeh, *Inversion of conditional Wiener integrals*, Pacific J. Math. **59** (1975), 623-638.

16. J. Yeh, *Inversion of conditional expectations*, Pacific J. Math. **52** (1974), 631-640.

# CHANGE OF SCALE FORMULAS FOR WIENER INTEGRALS
# AND FOURIER-FEYNMAN TRANSFORMS*

IL YOO

*Department of Mathematics, Yonsei University,*
*Wonju 220-710, Korea*
*E-mail: iyoo@yonsei.ac.kr*

TEUK SEOB SONG

*Department of Computer Engineering, Mokwon University, Daejon 302-729, Korea*
*E-mail: teukseob@gmail.com*

BYOUNG SOO KIM

*School of Liberal Arts, Seoul National University of Technology,*
*Seoul 139-743, Korea*
*E-mail: mathkbs@snut.ac.kr*

KUN SOO CHANG

*Department of Mathematics, Yonsei University,*
*Seoul 120-749, Korea*
*E-mail: kunchang@yonsei.ac.kr*

Cameron and Storvick introduced change of scale formulas for Wiener integrals of bounded functions in the Banach algebra $S$ of analytic Feynman integrable functions on classical Wiener space. In this article, we survey change of scale formulas for Wiener integrals and introduce the recent results about change of scale formulas for Wiener integrals obtained by Fourier-Feynman transforms.

**Key words:** Wiener integral, Feynman integral, change of scale formula, Fourier-Feynman transform
**Mathematics Subject Classification:** 28C20

## 1. Introduction

It has long been known that Wiener measure and Wiener measurability behave badly under the change of scale transformation[3] and under translations[2]. However Cameron and Storvick[5] expressed the analytic Feynman integral for a rather large class of functionals as a limit of Wiener integrals. In doing so, they[6] discovered nice change of scale formulas for Wiener integrals on classical Wiener space $C_0[0, T]$.

---
*Research Supported by the Basic Science Research Institute Program, Korea Research Foundation under Grant KRF 2003-005-C00011.

In Refs. [16]-[19], Yoo, Skoug, Chang, Kim, Song and Yoon extended these results to classical Yeh-Wiener space and an abstract Wiener space $(H, B, \nu)$. In particular, Yoo and Skoug[16] established a change of scale formula for Wiener integrals of functions in the Fresnel class $\mathscr{F}(B)$ on abstract Wiener space, and then they[17] developed this formula for a more generalized Fresnel class $\mathscr{F}_{A_1, A_2}$ than the Fresnel class. Recently in Ref. [18], the authors established change of scale formulas for Wiener integrals of functionals not necessarily bounded or continuous.

In this article, we survey change of scale formulas for Wiener integrals on classical Wiener space and introduce the recent results about change of scale formulas for Wiener integrals obtained by Fourier-Feynman transforms.

## 2. A Change of Scale Formula for Functions in $\mathcal{S}$

In this section, we introduce the change of scale formula for Wiener integrals of functions in a Banach algebra on classical Wiener space introduced by Cameron and Storvick [4].

Let $C_0[0, T]$ denote Wiener space, that is, the space of real-valued continuous functions $x$ on $[0, T]$ such that $x(0) = 0$. Let $M$ be the class of all Wiener measurable subsets of $C_0[0, T]$ and let $m$ denote Wiener measure.

Let $\mathbb{C}$ and $\mathbb{C}_+$ denote the complex numbers and the complex numbers with positive real part, respectively.

A subset $E$ of $C_0[0, T]$ is said to be scale-invariant measurable provided $\rho E$ is measurable for each $\rho > 0$, and a scale-invariant measurable set $N$ is said to be scale-invariant null provided $m(\rho N) = 0$ for each $\rho > 0$. A property that holds except on a scale-invariant null set is said to hold scale-invariant almost everywhere (s-a.e.). If two functionals $F$ and $G$ are equal s-a.e., then we write $F \approx G$.

**Definition 2.1.** Let $F$ be a $\mathbb{C}$-valued scale-invariant measurable functional on $C_0[0, T]$ such that the Wiener integral

$$J_F(\lambda) = \int_{C_0[0,T]} F(\lambda^{-1/2} x) \, dm(x)$$

exists as a finite number for all $\lambda > 0$. If there exists an analytic function $J_F^*(\lambda)$ in $\mathbb{C}_+$ such that $J_F^*(\lambda) = J_F(\lambda)$ for all $\lambda > 0$, then we define $J_F^*(\lambda)$ to be the analytic Wiener integral of $F$ over $C_0[0, T]$ with parameter $\lambda$, and for $\lambda \in \mathbb{C}_+$ we write

$$\int_{C_0[0,T]}^{\mathrm{anw}_\lambda} F(x) \, dm(x) = J_F^*(\lambda). \tag{1}$$

Let $q$ be a non-zero real number and let $F$ be a functional such that $\int_{C_0[0,T]}^{\mathrm{anw}_\lambda} F(x) \, dm(x)$ exists for all $\lambda \in \mathbb{C}_+$. If the following limit (2) exists, then we call it the analytic Feynman integral of $F$ over $C_0[0, T]$ with parameter $q$ and we write

$$\int_{C_0[0,T]}^{\mathrm{anf}_q} F(x) \, dm(x) = \lim_{\lambda \to -iq} \int_{C_0[0,T]}^{\mathrm{anw}_\lambda} F(x) \, dm(x) \tag{2}$$

where $\lambda \to -iq$ through $\mathbb{C}_+$.

The Banach algebra $\mathcal{S}$ consists of functionals expressible in the form

$$F(x) = \int_{L_2[0,T]} \exp\{i\langle v, x\rangle\}\, d\mu(v) \tag{3}$$

for s-a.e. $x$ in $C_0[0,T]$ where $\mu$ is an element of $M(L_2[0,T])$, the space of $\mathbb{C}$-valued countably additive Borel measures on $L_2[0,T]$, and $\langle v, x\rangle$ denotes the Paley-Wiener-Zygmund stochastic integral $\int_0^T v(s)\, d\tilde{x}(s)$.

Cameron and Storvick showed that every element in $\mathcal{S}$ is analytic Feynman integrable, and they expressed its analytic Feynman integral as a limit of Wiener integrals. And also, they obtained the following change of scale formula for Wiener integrals on classical Wiener space $C_0[0,T]$.

**Theorem 2.2.** *Let $F \in \mathcal{S}$ be given by (3). Then $F$ is analytic Feynman integrable and*

$$\int_{C_0[0,T]}^{\mathrm{anf}_q} F(x)\, dm(x) = \int_{L_2[0,T]} \exp\left\{-\frac{i}{2q}\|v\|_2^2\right\} d\mu(v) \tag{4}$$

*for each non-zero real $q$.*

**Theorem 2.3.** *Let $F \in \mathcal{S}$ and let $\{\alpha_n\}$ be a complete orthonormal sequence in $L_2[0,T]$. Let $\{\lambda_n\}$ be a sequence of complex numbers with $\mathrm{Re}\,\lambda_n > 0$ such that $\lambda_n \to -iq$. Then*

$$\int_{C_0[0,T]}^{\mathrm{anf}_q} F(x)\, dm(x) = \lim_{n\to\infty} \lambda_n^{n/2} \int_{C_0[0,T]} \exp\left\{\frac{1-\lambda_n}{2}\sum_{k=1}^{n}\langle \alpha_k, x\rangle^2\right\} F(x)\, dm(x) \tag{5}$$

*for each non-zero real $q$.*

**Theorem 2.4.** *Let $F$ and $\{\alpha_n\}$ be given as in Theorem 2.3. Then for each $\rho > 0$*

$$\int_{C_0[0,T]} F(\rho x)\, dm(x) = \lim_{n\to\infty} \rho^{-n} \int_{C_0[0,T]} \exp\left\{\frac{\rho^2-1}{2\rho^2}\sum_{k=1}^{n}\langle \alpha_k, x\rangle^2\right\} F(x)\, dm(x). \tag{6}$$

The space $\mathcal{S}$ is not closed with respect to pointwise or even uniform convergence[12] and thus it's closure $\bar{\mathcal{S}}^u$ under uniform convergence is a larger space than $\mathcal{S}$. The above change of scale formula can be extended to the functionals in $\bar{\mathcal{S}}^u$.

**Theorem 2.5.** *Let $\rho$ and $\{\alpha_n\}$ be given as in Theorem 2.4. Then the change of scale formula (6) holds for each $F \in \bar{\mathcal{S}}^u$.*

## 3.   A Change of Scale Formula for Cylinder Type Functions

In this section, we establish the change of scale formula for Wiener integrals of cylinder type functions on classical Wiener space.

Let $\{\alpha_1, \cdots, \alpha_l\}$ be an orthonormal set of functions in $L_2[0,T]$. For $1 \leq p < \infty$, let $\mathscr{A}_l^{(p)}$ be the class of cylinder type functions $F$ on $C_0[0,T]$ of the form

$$F(x) = \Psi(\langle \alpha_1, x \rangle, \cdots, \langle \alpha_l, x \rangle) \tag{7}$$

s-a.e. where $\Psi$ is in $L_p(\mathbb{R}^l)$. Let $\mathscr{A}_l^{(\infty)}$ be the space of functions of the form (7) with $\Psi \in C_0(\mathbb{R}^l)$, the space of bounded continuous functions which vanish at infinity.

The next is the existence theorem of the analytic Wiener and analytic Feynman integral for cylinder type functions in $F \in \mathscr{A}_l^{(p)}$.

**Theorem 3.1.** *Let $F \in \mathscr{A}_l^{(p)}$ be given by (7) with $1 \leq p < \infty$. Then for $\lambda \in \mathbb{C}_+$, $F$ is analytic Wiener integrable and*

$$\int_{C_0[0,T]}^{\mathrm{anw}_\lambda} F(x)\, dm(x) = \left(\frac{\lambda}{2\pi}\right)^{l/2} \int_{\mathbb{R}^l} \Psi(\vec{u}) \exp\left\{-\frac{\lambda}{2} \|\vec{u}\|^2\right\} d\vec{u}. \tag{8}$$

*In particular, if $p = 1$, $F$ is analytic Feynman integrable and*

$$\int_{C_0[0,T]}^{\mathrm{anf}_q} F(x)\, dm(x) = \left(\frac{-iq}{2\pi}\right)^{l/2} \int_{\mathbb{R}^l} \Psi(\vec{u}) \exp\left\{\frac{iq}{2} \|\vec{u}\|^2\right\} d\vec{u} \tag{9}$$

*for each non-zero real $q$.*

In the following theorems, we express the analytic Feynman integral for cylinder type functions $F \in \mathscr{A}_l^{(p)}$ as a limit of Wiener integrals. And then, we obtain the change of scale formula for Wiener integrals of cylinder type functions on classical Wiener space $C_0[0,T]$.

**Theorem 3.2.** *Let $F \in \mathscr{A}_l^{(p)}$ be given by (7) with $1 \leq p < \infty$. Then for $\lambda \in \mathbb{C}_+$,*

$$\int_{C_0[0,T]}^{\mathrm{anw}_\lambda} F(x)\, dm(x) = \lambda^{l/2} \int_{C_0[0,T]} \exp\left\{\frac{1-\lambda}{2} \sum_{k=1}^{l} \langle \alpha_k, x \rangle^2\right\} F(x)\, dm(x). \tag{10}$$

**Corollary 3.3.** *Let $F \in \mathscr{A}_l^{(1)}$ be given by (7) and let $q$ be a non-zero real number. Let $\{\lambda_n\}$ be a sequence of complex numbers with $\mathrm{Re}\lambda_n > 0$ such that $\lambda_n \to -iq$. Then*

$$\int_{C_0[0,T]}^{\mathrm{anf}_q} F(x)\, dm(x) = \lim_{n \to \infty} \lambda_n^{l/2} \int_{C_0[0,T]} \exp\left\{\frac{1-\lambda_n}{2} \sum_{k=1}^{l} \langle \alpha_k, x \rangle^2\right\} F(x)\, dm(x). \tag{11}$$

**Theorem 3.4.** *Let $F \in \mathscr{A}_l^{(p)}$ be given by (7) with $1 \leq p < \infty$. Then for every $\rho > 0$,*

$$\int_{C_0[0,T]} F(\rho x)\, dm(x) = \rho^{-l} \int_{C_0[0,T]} \exp\left\{\frac{\rho^2 - 1}{2\rho^2} \sum_{k=1}^{l} \langle \alpha_k, x \rangle^2\right\} F(x)\, dm(x). \tag{12}$$

## 4. A Change of Scale Formula for Unbounded Functions

In this section, we develope the change of scale formula for Wiener integrals of some unbounded functions on classical Wiener space. In particular, we consider the functions having the form

$$F(x) = G(x)\Psi(\langle \alpha_1, x \rangle, \cdots, \langle \alpha_l, x \rangle) \tag{13}$$

for $G$ is given by (3) and $\Psi = \psi + \phi$ where $\psi \in L_p(\mathbb{R}^l)$, $1 \le p < \infty$ and $\phi \in \hat{M}(\mathbb{R}^l)$, the class of Fourier transform of a measure of bounded variation over $\mathbb{R}^l$, namely,

$$\phi(\vec{s}) = \int_{\mathbb{R}^l} \exp\Big\{ \sum_{k=1}^{l} s_k t_k \Big\} d\eta(\vec{t}) \tag{14}$$

where $\eta$ is a complex Borel measure of bounded variation on $\mathbb{R}^l$.

The next is the existence theorem of the analytic Wiener and analytic Feynman integral for functions having the form (13).

**Theorem 4.1.** *Let $F$ be given by (13) with $1 \le p < \infty$. Then for $\lambda \in \mathbb{C}_+$, $F$ is analytic Wiener integrable and*

$$\int_{C_0[0,T]}^{\text{anw}_\lambda} F(x)\, dm(x)$$

$$= \Big( \frac{\lambda}{2\pi} \Big)^{l/2} \int_{L_2[0,T]} \int_{\mathbb{R}^l} \exp\Big\{ \frac{1}{2\lambda} \Big[ \sum_{k=1}^{l} (i\lambda u_k + \langle \alpha_k, v \rangle)^2 - \|v\|_2^2 \Big] \Big\} \psi(\vec{u})\, d\vec{u}\, d\mu(v) \tag{15}$$

$$+ \int_{L_2[0,T]} \int_{\mathbb{R}^l} \exp\Big\{ -\frac{1}{2\lambda} \Big[ \|v\|_2^2 + \sum_{k=1}^{l} 2t_k \langle \alpha_k, v \rangle + \|\vec{t}\|^2 \Big] \Big\} d\eta(\vec{t})\, d\mu(v).$$

*In particular, if $p = 1$, $F$ is analytic Feynman integrable and*

$$\int_{C_0[0,T]}^{\text{anf}_q} F(x)\, dm(x)$$

$$= \Big( \frac{-iq}{2\pi} \Big)^{l/2} \int_{L_2[0,T]} \int_{\mathbb{R}^l} \exp\Big\{ \frac{i}{2q} \Big[ \sum_{k=1}^{l} (qu_k + \langle \alpha_k, v \rangle)^2 - \|v\|_2^2 \Big] \Big\} \psi(\vec{u})\, d\vec{u}\, d\mu(v) \tag{16}$$

$$+ \int_{L_2[0,T]} \int_{\mathbb{R}^l} \exp\Big\{ -\frac{i}{2q} \Big[ \|v\|_2^2 + \sum_{k=1}^{l} 2t_k \langle \alpha_k, v \rangle + \|\vec{t}\|^2 \Big] \Big\} d\eta(\vec{t})\, d\mu(v)$$

*for each non-zero real $q$.*

In the following theorems, we express the analytic Feynman integral for functions of the form (13) as a limit of Wiener integrals. And then, we have the change of scale formula for Wiener integrals of functions of the form (13).

**Theorem 4.2.** *Let $F$ be given by (13) with $1 \le p < \infty$ and let $\{\alpha_n\}$ be a complete orthonormal sequence in $L_2[0,T]$. Then for $\lambda \in \mathbb{C}_+$,*

$$\int_{C_0[0,T]}^{\text{anw}_\lambda} F(x)\, dm(x) = \lim_{n \to \infty} \lambda^{n/2} \int_{C_0[0,T]} \exp\Big\{ \frac{1-\lambda}{2} \sum_{k=1}^{n} \langle \alpha_k, x \rangle^2 \Big\} F(x)\, dm(x). \tag{17}$$

**Corollary 4.3.** *Let $F$ be given by (13) with $p = 1$ and let $\{\alpha_n\}$ be a given as in Theorem 4.2. Let $q$ be a non-zero real number and $\{\lambda_n\}$ be a sequence of complex numbers with $\mathrm{Re}\lambda_n > 0$ such that $\lambda_n \to -iq$. Then*

$$\int_{C_0[0,T]}^{\mathrm{anf}_q} F(x)\, dm(x) = \lim_{n\to\infty} \lambda_n^{n/2} \int_{C_0[0,T]} \exp\Big\{\frac{1-\lambda_n}{2}\sum_{k=1}^{n}\langle \alpha_k, x\rangle^2\Big\} F(x)\, dm(x).$$

$$(18)$$

**Theorem 4.4.** *Let $F$ and $\{\alpha_n\}$ be given as in Theorem 4.2. Then for every $\rho > 0$,*

$$\int_{C_0[0,T]} F(\rho x)\, dm(x) = \lim_{n\to\infty} \rho^{-n} \int_{C_0[0,T]} \exp\Big\{\frac{\rho^2-1}{2\rho^2}\sum_{k=1}^{n}\langle \alpha_k, x\rangle^2\Big\} F(x)\, dm(x).$$

$$(19)$$

## 5. Change of Scale Formulas for Wiener Integrals and Fourier-Feynman Transforms

In this section, we introduce the concept of Fourier-Feynman transforms of functions discussed in Sections 2 and 3, and establish the change of scale formulas for Wiener integrals obtained by Fourier-Feynman transforms.

We now begin with introducing the Fourier-Feynman transform using the analytic Wiener and Feynman integrals given by (1) and (2). For $\lambda \in \mathbb{C}_+$ and $y \in C_0[0,T]$, we let

$$T_\lambda(F)(y) = \int_{C_0[0,T]}^{\mathrm{anw}_\lambda} F(x+y)\, dm(x). \qquad (20)$$

In the standard Fourier transform theory, the integrals involved are often interpreted in the mean; a similar concept is useful in the Fourier-Feynman transform theory. Let $1 < p < \infty$ and let $H_n$ and $H$ be scale-invariant measurable functionals such that, for each $\rho > 0$,

$$\lim_{n\to\infty} \int_{C_0[0,T]} |H_n(\rho x) - H(\rho x)|^{p'}\, dm(x) = 0 \qquad (21)$$

where $p$ and $p'$ are related by $\frac{1}{p} + \frac{1}{p'} = 1$. Then we write

$$\mathrm{l.\,i.\,m.}_{n\to\infty}(w_s^{p'})(H_n) \approx H \qquad (22)$$

and we call $H$ the scale-invariant limit in the mean of order $p'$ of $H_n$. A similar definition is understood when $n$ is replaced by a continuously varying parameter.

**Definition 5.1.** Let $q$ be a non-zero real number. For $1 < p < \infty$ the $L_p$ analytic Fourier-Feynman transform $T_q^{(p)}(F)$ of $F$ is defined by the formula

$$T_q^{(p)}(F)(y) = \mathrm{l.\,i.\,m.}_{\lambda\to-iq}(w_s^{p'})T_\lambda(F)(y) \qquad (23)$$

whenever this limit exists where $\lambda$ approaches $-iq$ through $\mathbb{C}_+$.

We also define the $L_1$ analytic Fourier-Feynman transform $T_q^{(1)}(F)$ of $F$ by the formula

$$T_q^{(1)}(F)(y) = \lim_{\lambda \to -iq} T_\lambda(F)(y) \qquad (24)$$

for s-a.e. $y$ where $\lambda \to -iq$ through $\mathbb{C}_+$.

**Theorem 5.2.** *Let $F \in \mathcal{S}$ be given by (3) and let $1 \leq p < \infty$. Then for non-zero real $q$,*

$$T_q^{(p)}(F)(y) = \int_{L_2[0,T]} \exp\left\{i\langle v, y \rangle - \frac{i}{2q}\|v\|_2^2\right\} d\mu(v) \qquad (25)$$

*for s-a.e. $y \in C_0[0,T]$.*

In the next theorems, we express the Fourier-Feynman transform for functions in the Banach algebra $\mathcal{S}$ as a limit of Wiener integrals. And then, we have the change of scale formulas for Wiener integrals of functions in $\mathcal{S}$ on classical Wiener space $C_0[0,T]$.

**Theorem 5.3.** *Let $F \in \mathcal{S}$ be given by (3). Let $q$ be a non-zero real number and let $\{\alpha_n\}$ be a complete orthonormal sequence in $L_2[0,T]$. Let $\{\lambda_n\}$ be a sequence of complex numbers with $\mathrm{Re}\lambda_n > 0$ such that $\lambda_n \to -iq$. Then*

$$T_q^{(1)}(F)(y) = \lim_{n\to\infty} \lambda_n^{n/2} \int_{C_0[0,T]} \exp\left\{\frac{1-\lambda_n}{2}\sum_{k=1}^n \langle \alpha_k, x\rangle^2\right\} F(x+y)\, dm(x), \qquad (26)$$

*and for $1 < p < \infty$*

$$T_q^{(p)}(F)(y) = \mathrm{l.\,i.\,m.}_{n\to\infty}(w_s^{p'})\lambda_n^{n/2} \int_{C_0[0,T]} \exp\left\{\frac{1-\lambda_n}{2}\sum_{k=1}^n \langle \alpha_k, x\rangle^2\right\} F(x+y)\, dm(x) \qquad (27)$$

*for s-a.e. $y \in C_0[0,T]$.*

**Corollary 5.4.** *Let $F$ and $\{\alpha_n\}$ be given as in Theorem 5.3. Then for every $\rho > 0$,*

$$\int_{C_0[0,T]} F(\rho x + y)\, dm(x)$$
$$= \lim_{n\to\infty} \rho^{-n} \int_{C_0[0,T]} \exp\left\{\frac{\rho^2-1}{2\rho^2}\sum_{k=1}^n \langle \alpha_k, x\rangle^2\right\} F(x+y)\, dm(x) \qquad (28)$$

*for s-a.e. $y \in C_0[0,T]$.*

**Theorem 5.5.** *Let $F \in \mathscr{A}_l^{(1)}$ be given by (7). Then $T_q^{(1)}(F)$ exists for all non-zero real $q$ and*

$$T_q^{(1)}(F)(y) = g(-iq; \langle \alpha_1, y\rangle, \cdots, \langle \alpha_l, y\rangle) \in \mathscr{A}_l^{(\infty)} \qquad (29)$$

*for s-a.e. $y$ in $C_0[0,T]$, where*

$$g(\lambda; \vec{w}) = \left(\frac{\lambda}{2\pi}\right)^{l/2} \int_{\mathbb{R}^l} f(\vec{u}) \exp\left\{-\frac{\lambda}{2}\|\vec{u} - \vec{w}\|^2\right\} d\vec{u}. \qquad (30)$$

**Theorem 5.6.** *Let $F \in \mathscr{A}_l^{(p)}$ be given by (7) with $1 < p \leq 2$. Then $T_q^{(p)}(F)$ exists for all non-zero real $q$ and is given by the formula*

$$T_q^{(p)}(F)(y) = g(-iq; \langle \alpha_1, y \rangle, \cdots, \langle \alpha_l, y \rangle) \in \mathscr{A}_l^{(p')} \tag{31}$$

*for s-a.e. $y$ in $C_0[0,T]$ where $g$ is given by (30).*

In the following theorems, we express the Fourier-Feynman transform for cylinder type functions $F \in \mathscr{A}_l^{(p)}$ as a limit of Wiener integrals. And then, we obtain the change of scale formulas for Wiener integrals of cylinder type functions on classical Wiener space $C_0[0,T]$.

**Theorem 5.7.** *Let $F \in \mathscr{A}_l^{(1)}$ be given by (7). Let $q$ and let $\{\lambda_n\}$ be given as in Theorem 5.3. Then*

$$T_q^{(1)}(F)(y) = \lim_{n \to \infty} \lambda_n^{l/2} \int_{C_0[0,T]} \exp\left\{\frac{1 - \lambda_n}{2} \sum_{k=1}^{l} \langle \alpha_k, x \rangle^2\right\} F(x + y) \, dm(x) \tag{32}$$

*for s-a.e. $y \in C_0[0,T]$.*

**Theorem 5.8.** *Let $F \in \mathscr{A}_l^{(p)}$ be given by (7) with $1 < p \leq 2$. Let $q$ and $\{\lambda_n\}$ be given as in Theorem 5.3. Then*

$$T_q^{(p)}(F)(y) = \operatorname*{l.i.m.}_{n \to \infty}(w_s^{p'}) \lambda_n^{l/2} \int_{C_0[0,T]} \exp\left\{\frac{1 - \lambda_n}{2} \sum_{k=1}^{l} \langle \alpha_k, x \rangle^2\right\} F(x + y) \, dm(x)$$

$$\tag{33}$$

*for s-a.e. $y \in C_0[0,T]$.*

**Corollary 5.9.** *Let $F \in \mathscr{A}_l^{(p)}$ be given by (7) with $1 \leq p \leq 2$. Then for every $\rho > 0$,*

$$\int_{C_0[0,T]} F(\rho x + y) \, dm(x) = \rho^{-l} \int_{C_0[0,T]} \exp\left\{\frac{\rho^2 - 1}{2\rho^2} \sum_{k=1}^{l} \langle \alpha_k, x \rangle^2\right\} F(x + y) \, dm(x)$$

$$\tag{34}$$

*for s-a.e. $y \in C_0[0,T]$.*

### References

1. J.M. Ahn, G.W. Johnson and D.L. Skoug, Functions in the Fresnel Class of an Abstract Wiener Space, *J. Korean Math. Soc.*, **28** (1991), 245–265.
2. R.H. Cameron, The Translation Pathology of Wiener Space, Duke Math. J., **21** (1954), 623–628.
3. R.H. Cameron and W.T. Martin, The Behavior of Measure and Measurability under Change of Scale in Wiener Space, *Bull. Amer. Math. Soc.*, **53** (1947), 130–137.
4. R.H. Cameron and D.A. Storvick, Some Banach Algebras of Analytic Feynman Integrable Functionals, *in Analytic Functions, Kozubnik, Lecture Notes in Math.*, **798** (1980), 18–67.
5. R.H. Cameron and D.A. Storvick, Relationships between the Wiener Integral and the Analytic Feynman Integral, *Supplemento ai Rendiconti del Circolo Matematico di Palermo, Serie II-Numero*, **17** (1987), 117–133.

6. R.H. Cameron and D.A. Storvick, Change of Scale Formulas for Wiener Integrals, *Supplemento ai Rendiconti del Circolo Matematico di Palermo, Serie II-Numero*, **17** (1987), 105–115.

7. R.H. Cameron and D.A. Storvick, New Existence Theorems and Evaluation Formulas for Analytic Feynman Integrals, *Deformations of Mathematics Structures*, 297–308, Kluwer, Dordrecht, 1989.

8. K.S. Chang, Scale-Invariant Measurability in Yeh-Wiener Space, *J. Korean Math. Soc.*, **21** (1982), 61–67.

9. D.M. Chung, Scale-Invariant Measurability in Abstract Wiener Space, *Pacific J. Math.*, **130** (1987), 27–40.

10. L. Gross, Abstract Wiener Space, *Proc. 5th Berkeley Sympos. Math. Stat. Prob.*, **2** (1965), 31–42.

11. G.W. Johnson and D.L. Skoug, Scale-Invariant Measurability in Wiener Space, *Pacific J. Math.*, **83** (1979), 157–176.

12. G.W. Johnson and D.L. Skoug, Stability Theorems for the Feynman Integral, *Supplemento ai Rendiconti del Circolo Matematico di Palermo, Serie II-Numero*, **8** (1985), 361–377.

13. H.H. Kuo, Gaussian Measures in Banach Spaces, *Lecture Notes in Math.*, **463** (1975), Springer-Verlag, Berlin.

14. I. Yoo and K.S. Chang, Notes on Analytic Feynman Integrable Functions, *Rocky Mountain J. Math.*, **23** (1993), 1133–1142.

15. I. Yoo, K.S. Chang, D.H. Cho, B.S. Kim and T.S. Song, Fourier-Feynman Transforms on Wiener Spaces, *Stochastic Analysis and Mathematical Physics*, 183–201, World Scientific, 2004.

16. I. Yoo and D.L. Skoug, A Change of Scale Formula for Wiener Integrals on Abstract Wiener Spaces, *Intern. J. Math. Math. Sci.*, **17** (1994), 239–248.

17. I. Yoo and D.L. Skoug, A Change of Scale Formula for Wiener Integrals on Abstract Wiener Spaces II, *J. Korean Math. Soc.*, **31** (1994), 115–129.

18. I. Yoo, T.S. Song, B.S. Kim and K.S. Chang, A Change of Scale Formula for Wiener Integrals of Unbounded Functions, *Rocky Mountain J. Math.*, **34** (2004), 371-389.

19. I. Yoo and G.J. Yoon, Change of Scale Formulas for Yeh-Wiener Integrals, *Commun. Korean Math. Soc.*, **6** (1991), 19–26.

# SOBOLEV TYPE SPACES ASSOCIATED WITH THE KONTOROVICH-LEBEDEV TRANSFORM

SEMYON B. YAKUBOVICH*

*Department of Pure Mathematics,*
*Faculty of Sciences,*
*University of Porto,*
*Campo Alegre st., 687*
*4169-007 Porto*
*Portugal*
*E-mail: syakubov@fc.up.pt*

We construct the Sobolev type space $S_p^{N,\alpha}(\mathbf{R}_+)$ with the finite norm

$$||u||_{S_p^{N,\alpha}(\mathbf{R}_+)} = \left( \sum_{k=0}^{N} \int_0^\infty |A_x^k u|^p x^{\alpha_k p - 1} dx \right)^{1/p} < \infty,$$

where $\alpha = (\alpha_0, \alpha_1, \ldots, \alpha_N), \alpha_k \in \mathbf{R}, k = 0, \ldots, N$, and $A_x$ is the differential operator of the form

$$A_x u = x^2 u(x) - x \frac{d}{dx} \left[ x \frac{du}{dx} \right],$$

and $A_x^k$ means $k$-th iterate of $A_x$, $A_x^0 u = u$. It is shown that the Kontorovich-Lebedev transformation

$$(KLf)(x) = \int_0^\infty K_{i\tau}(x) f(\tau) d\tau, \ x \in \mathbf{R}_+$$

maps the weighted space $L_p(\mathbf{R}_+; \omega(\tau) d\tau), \ 2 \le p \le \infty$ into $S_p^{N,\alpha}(\mathbf{R}_+)$. Elementary properties for the space $S_p^{N,\alpha}(\mathbf{R}_+)$ are exhibited. Boundedness and inversion properties for the Kontorovich-Lebedev transform are studied. In the Hilbert case ($p = 2$) the isomorphism between these spaces is established for the special type of weights and Plancherel's type theorem is formulated.

**Key words:** Sobolev spaces, Kontorovich-Lebedev transform, modified Bessel function, Hardy inequality, Plancherel theorem, imbedding theorem
**Mathematics Subject Classification:** 44A15, 46E35, 26D10

## 1. Introduction

In this paper we extend the theory of the Kontorovich-Lebedev transformation [8,10]

$$(KLf)(x) = \int_0^\infty K_{i\tau}(x) f(\tau) d\tau, \tag{1.1}$$

---

*Work was supported, in part, by the "Centro de Matemática" of the University of Porto.

on the following Sobolev type space $S_p^{N,\alpha}(\mathbb{R}_+), 1 \le p < \infty$ with the finite norm

$$\|u\|_{S_p^{N,\alpha}(\mathbb{R}_+)} = \left( \sum_{k=0}^{N} \int_0^\infty |A_x^k u|^p x^{\alpha_k p - 1} dx \right)^{1/p} < \infty. \tag{1.2}$$

Here $\alpha = (\alpha_0, \alpha_1, \ldots, \alpha_N), \alpha_k \in \mathbb{R}, k = 0, \ldots, N$, and $A_x$ is the differential operator, which has eigenfunction the modified Bessel function $K_\nu(x)$ with eigenvalue $-\nu^2$ and can be written in the form

$$A_x u = x^2 u(x) - x \frac{d}{dx}\left[ x \frac{du}{dx} \right], \quad A_x K_\nu = -\nu^2 K_\nu(x). \tag{1.3}$$

In particular, the case of the pure imaginary subscript (an index) $\nu = i\tau$ corresponds to the kernel of the Kontorovich-Lebedev transform (1.1). As usual we denote by $A_x^k$ the $k$-th iterate of $A_x$, $A_x^0 u = u$. The differential operator (1.3) was used for instance in [4,17] in order to construct the spaces of testing functions to consider the Kontorovich-Lebedev transform on distributions (see also in [10]). Recently (see [15]) it is involved to investigate the corresponding class of the Kontorovich-Lebedev convolution integral equations.

In the following, $(x,\tau) \in \mathbb{R}_+ \times \mathbb{R}_+$, $K_{i\tau}(x)$ is real-valued and smooth with respect to $x$ and $\tau$. It is called also the Macdonald function (cf. [1,8] p. 355). The modified Bessel function has the asymptotic behaviour (cf. [1], relations (9.6.8), (9.6.9), (9.7.2))

$$K_\nu(z) = \left( \frac{\pi}{2z} \right)^{1/2} e^{-z}[1 + O(1/z)], \quad z \to \infty, \tag{1.4}$$

and near the origin

$$K_\nu(z) = O\left( z^{-|\mathrm{Re}\,\nu|} \right), \quad z \to 0, \tag{1.5}$$

$$K_0(z) = O(\log z), \quad z \to 0. \tag{1.6}$$

Meanwhile, when $x$ is restricted to any compact subset of $\mathbb{R}_+$ and $\tau$ tends to infinity we have the following asymptotic ([11], p. 20)

$$K_{i\tau}(x) = \left( \frac{2\pi}{\tau} \right)^{1/2} e^{-\pi\tau/2} \sin\left( \frac{\pi}{4} + \tau \log \frac{2\tau}{x} - \tau \right) [1 + O(1/\tau)], \quad \tau \to +\infty. \tag{1.7}$$

The modified Bessel function can be represented by the integrals of the Fourier and Mellin types [1,8,11]

$$K_\nu(x) = \int_0^\infty e^{-x \cosh u} \cosh \nu u \, du, \tag{1.8}$$

$$K_\nu(x) = \frac{1}{2} \left( \frac{x}{2} \right)^\nu \int_0^\infty e^{-t - \frac{x^2}{4t}} t^{-\nu-1} dt. \tag{1.9}$$

We also note that the product of the Macdonald functions of different arguments can be represented by the Macdonald formula [1,6,11]

$$K_{i\tau}(x) K_{i\tau}(y) = \frac{1}{2} \int_0^\infty e^{-\frac{1}{2}\left( u \frac{x^2+y^2}{xy} + \frac{xy}{u} \right)} K_{i\tau}(u) \frac{du}{u}. \tag{1.10}$$

In this paper we deal with the Lebesgue weighted $L_p(\mathbb{R}_+; \omega(x)dx)$ spaces with respect to the measure $\omega(x)dx$ with the norm

$$||f||_p = \left( \int_0^\infty |f(x)|^p \omega(x)dx \right)^{1/p}, \ 1 \le p < \infty, \tag{1.11}$$

$$||f||_\infty = \text{ess sup}|f(x)|. \tag{1.12}$$

In particular, we will use the spaces $L_{\nu,p} \equiv L_p(\mathbb{R}_+; x^{\nu p-1}dx), \ 1 \le p \le \infty, \nu \in \mathbb{R}$, which are related to the Mellin transforms pair [7-9]

$$f^*(s) = \int_0^\infty f(x)x^{s-1}dx, \tag{1.13}$$

$$f(x) = \frac{1}{2\pi i} \int_{\nu-i\infty}^{\nu+i\infty} f^*(s)x^{-s}ds, \ s = \nu + it, \ x > 0. \tag{1.14}$$

The integrals (1.13)- (1.14) are convergent, in particular, in mean with respect to the norm of the spaces $L_2(\nu - i\infty, \nu + i\infty; ds)$ and $L_2(\mathbb{R}_+; x^{2\nu-1}dx)$, respectively. In addition, the Parseval equality of the form

$$\int_0^\infty |f(x)|^2 x^{2\nu-1}dx = \frac{1}{2\pi} \int_{-\infty}^\infty |f^*(\nu + it)|^2 dt \tag{1.15}$$

holds true.

As it is proved in [12,13], the Kontorovich-Lebedev operator (1.1) is an isomorphism between the spaces $L_2(\mathbb{R}_+; [\tau \sinh \pi\tau]^{-1}d\tau)$ and $L_2(\mathbb{R}_+; x^{-1}dx)$ with the identity for the square of norms

$$\int_0^\infty |(KLf)(x)|^2 \frac{dx}{x} = \frac{\pi^2}{2} \int_0^\infty |f(\tau)|^2 \frac{d\tau}{\tau \sinh \pi\tau}, \tag{1.16}$$

and the Plancherel equality of type

$$\int_0^\infty (KLf)(x)\overline{(KLg(x)} \frac{dx}{x} = \frac{\pi^2}{2} \int_0^\infty f(\tau)\overline{g(\tau)} \frac{d\tau}{\tau \sinh \pi\tau}, \tag{1.17}$$

where $f, g \in L_2(\mathbb{R}_+; [\tau \sinh \pi\tau]^{-1}d\tau)$. We note that the convergence of the integral (1.1) in this case is with respect to the norm (1.11) for the space $L_2(\mathbb{R}_+; x^{-1}dx)$.

However, our goal is to study the Kontorovich-Lebedev transformation in the space (1.2). First in the sequel we will exhibit imbedding properties for the spaces $S_p^{N,\alpha}(\mathbb{R}_+)$ and we will find integral representations for the functions from $S_p^{N,\alpha}(\mathbb{R}_+)$. Finally we will study the boundedness and inversion properties for the Kontorovich-Lebedev transformation as an operator from the weighted $L_p$-space $L_p(\mathbb{R}_+; \omega(x)dx)$ into the space $S_p^{N,\alpha}(\mathbb{R}_+)$. When $p = 2, \alpha = 0$ it corresponds to the Plancherel type theorem for the Kontorovich-Lebedev transformation. Moreover, it forms an isomorphism for the special type of weights between these spaces. Concerning detail proofs of the results see the forthcoming paper [16].

## 2. Elementary properties for the space $S_p^{N,\alpha}(\mathbb{R}_+)$

From the norm definition (1.2) and elementary inequalities it follows that there are positive constants $C_1, C_2$ such that

$$C_1 \sum_{k=0}^{N} \left( \int_0^\infty |A_x^k u|^p x^{\alpha_k p-1} dx \right)^{1/p} \leq \left( \sum_{k=0}^{N} \int_0^\infty |A_x^k u|^p x^{\alpha_k p-1} dx \right)^{1/p}$$

$$\leq C_2 \sum_{k=0}^{N} \left( \int_0^\infty |A_x^k u|^p x^{\alpha_k p-1} dx \right)^{1/p}. \tag{2.1}$$

Hence by (1.11) we have the equivalence of norms

$$C_1 \sum_{k=0}^{N} \|A_\cdot^k u\|_{L_p(\mathbb{R}_+; x^{\alpha_k p-1} dx)} \leq \|u\|_{S_p^{N,\alpha}(\mathbb{R}_+)} \leq C_2 \sum_{k=0}^{N} \|A_\cdot^k u\|_{L_p(\mathbb{R}_+; x^{\alpha_k p-1} dx)}.$$

$$\tag{2.2}$$

One can show by standard methods that $S_p^{N,\alpha}(\mathbb{R}_+), 1 \leq p < \infty$ is a Banach space. Considering the space $S_p^{1,\alpha}(\mathbb{R}_+)$ and applying the the classical Hardy's inequality [2]

$$\int_0^\infty x^{-r} \left| \int_0^x f(t)dt \right|^p dx \leq \text{const.} \int_0^\infty x^{p-r} |f(x)|^p dx,$$

one can prove an imbedding theorem into Sobolev's weighted space ${}_0W_p^1(\mathbb{R}_+; x^{\gamma p-1} dx)$ with the norm

$$\|u\|_{{}_0W_p^1(\mathbb{R}_+; x^{\gamma p-1} dx)} = \left( \int_0^\infty |u'(x)|^p x^{\gamma p-1} dx \right)^{1/p}.$$

Indeed, we have the following result.

**Theorem 1.** Let $1 < p < \infty$, $\alpha = (2 - \beta, -\beta), \beta > 0$. The imbedding

$$S_p^{1,\alpha}(\mathbb{R}_+) \subset {}_0W_p^1(\mathbb{R}_+; x^{(1-\beta)p-1} dx)$$

is true.

One can derive integral representations for functions from the space $S_p^{N,\alpha}(\mathbb{R}_+)$ by using for any $u(x) \in L_{\nu,p}$, $\nu \in \mathbb{R}$ and $\varepsilon \in (0, \pi)$ the following regularization operator

$$u_\varepsilon(x) = \frac{x \sin \varepsilon}{\pi} \int_0^\infty \frac{K_1((x^2 + y^2 - 2xy \cos \varepsilon)^{1/2})}{(x^2 + y^2 - 2xy \cos \varepsilon)^{1/2}} u(y) dy, \quad x > 0. \tag{2.3}$$

Hence one can prove the Bochner type representation theorem.

**Theorem 2.** Let $u(x) \in L_{\nu,p}$, $0 < \nu < 1$, $1 \leq p < \infty$. Then

$$u(x) = \lim_{\varepsilon \to 0} u_\varepsilon(x),$$

with respect to the norm in $L_{\nu,p}$. Besides, for $1 < p < \infty$ this limit exists for almost all $x > 0$.

Appealing to this theorem we will approximate functions from $S_p^{N,\alpha}(\mathbb{R}_+)$ by regularization operator (2.3). Indeed we will prove

**Theorem 3.** *Operator (2.3) is well defined on functions from $S_p^{N,\alpha}(\mathbf{R}_+)$ with $\alpha = (\alpha_0, \alpha_1, \ldots, \alpha_N)$, where $0 < \alpha_k < 1$, $k = 0, 1, \ldots, N$ and $1 \le p < \infty$. Besides*

$$u(x) = \lim_{\varepsilon \to 0} u_\varepsilon(x),$$

*with respect to the norm in $S_p^{N,\alpha}(\mathbf{R}_+)$.*

**Proof.** Indeed, taking some function $u \in S_p^{N,\alpha}(\mathbf{R}_+)$ we then choose a sequence $\{\varphi_n\} \in C_0^\infty(\mathbf{R}_+)$, which converges to $u$. This immediately implies that $A_x^k \varphi_n \to A_x^k u$, $n \to \infty$ with respect to the norm in $L_{\alpha_k, p}$, $k = 0, 1, \ldots, N$, respectively.

Defining by

$$\varphi_{\varepsilon,n}(x) = \frac{x \sin \varepsilon}{\pi} \int_0^\infty \frac{K_1((x^2 + y^2 - 2xy \cos \varepsilon)^{1/2})}{(x^2 + y^2 - 2xy \cos \varepsilon)^{1/2}} \varphi_n(y) dy, \quad x > 0, \qquad (2.4)$$

we employ the relation (2.16.51.8) in [6]

$$\int_0^\infty \tau \sinh((\pi - \varepsilon)\tau) K_{i\tau}(x) K_{i\tau}(y) d\tau$$

$$= \frac{\pi}{2} xy \sin \varepsilon \frac{K_1((x^2 + y^2 - 2xy \cos \varepsilon)^{1/2})}{(x^2 + y^2 - 2xy \cos \varepsilon)^{1/2}}, \quad x, y > 0, \ 0 < \varepsilon \le \pi$$

and we substitute it in (2.4). Changing the order of integration by the Fubini theorem we find

$$\varphi_{\varepsilon,n}(x) = \frac{2}{\pi^2} \int_0^\infty \tau \sinh((\pi - \varepsilon)\tau) K_{i\tau}(x) \int_0^\infty K_{i\tau}(y) \varphi_n(y) \frac{dy}{y} d\tau.$$

Meantime, we apply the operator $A_x^k$, $k = 0, 1 \ldots, N$ (1.3) through both sides of the latter integral. Then via its uniform convergence with respect to $x \in (x_0, X_0) \subset \mathbf{R}_+$ and by using the equality $A_x^k K_{i\tau}(x) = \tau^{2k} K_{i\tau}(x)$, we integrate by parts and come out with

$$A_x^k \varphi_{\varepsilon,n} = \frac{2}{\pi^2} \int_0^\infty \tau \sinh((\pi - \varepsilon)\tau) K_{i\tau}(x) \int_0^\infty \tau^{2k} K_{i\tau}(y) \varphi_n(y) \frac{dy}{y} d\tau$$

$$= \frac{2}{\pi^2} \int_0^\infty \tau \sinh((\pi - \varepsilon)\tau) K_{i\tau}(x) \int_0^\infty K_{i\tau}(y) A_y^k \varphi_n \frac{dy}{y} d\tau.$$

This is equivalent to

$$A_x^k \varphi_{\varepsilon,n} = \frac{x \sin \varepsilon}{\pi} \int_0^\infty \frac{K_1((x^2 + y^2 - 2xy \cos \varepsilon)^{1/2})}{(x^2 + y^2 - 2xy \cos \varepsilon)^{1/2}} A_y^k \varphi_n dy.$$

Hence

$$A_x^k \varphi_{\varepsilon,n} - (A_x^k u)_\varepsilon = \frac{x \sin \varepsilon}{\pi} \int_0^\infty \frac{K_1((x^2 + y^2 - 2xy \cos \varepsilon)^{1/2})}{(x^2 + y^2 - 2xy \cos \varepsilon)^{1/2}} \left[ A_y^k \varphi_n - A_y^k u \right] dy$$

and due to Theorem 2 we have that $\lim_{n \to \infty} A_x^k \varphi_{\varepsilon,n} = (A_x^k u)_\varepsilon$ with respect to the norm in $L_{\alpha_k, p}$ for each $\varepsilon \in (0, \pi)$. Furthermore by Theorem 2 we derive that

$$\left\| (A_\cdot^k u)_\varepsilon - A_\cdot^k u \right\|_{L_{\alpha_k, p}} \to 0, \ \varepsilon \to 0, \ k = 0, 1, \ldots, N.$$

236

If we show that almost for all $x > 0$ $(A_x^k u)_\varepsilon = A_x^k u_\varepsilon$, $k = 0, 1, 2, \ldots, N$ then via (2.2) we complete the proof of Theorem 3. When $k = 0$ it is defined by (2.3). At the same time according to Du Bois-Reymond lemma it is sufficient to show that for any $\psi \in C_0^\infty(\mathbf{R}_+)$

$$\int_0^\infty \left[ (A_x^k u)_\varepsilon - A_x^k u_\varepsilon \right] \frac{\psi(x)}{x} dx = 0. \tag{2.5}$$

We have

$$\int_0^\infty \left[ (A_x^k u)_\varepsilon - A_x^k u_\varepsilon \right] \frac{\psi(x)}{x} dx = \int_0^\infty \left[ (A_x^k u)_\varepsilon - A_x^k \varphi_{\varepsilon,n} \right] \frac{\psi(x)}{x} dx$$

$$+ \int_0^\infty \left[ A_x^k \varphi_{\varepsilon,n} - A_x^k u_\varepsilon \right] \frac{\psi(x)}{x} dx = \int_0^\infty \left[ (A_x^k u)_\varepsilon - A_x^k \varphi_{\varepsilon,n} \right] \frac{\psi(x)}{x} dx$$

$$+ \int_0^\infty \left[ \varphi_{\varepsilon,n} - u_\varepsilon \right] \frac{A_x^k \psi}{x} dx.$$

Now as it is easily seen the right-hand side of the last equality is less than an arbitrary $\delta > 0$ when $n \to \infty$. Thus we prove (2.5) and we complete the proof of Theorem 3.

## 3. The Kontorovich - Lebedev transformation in $S_p^{N,\alpha}(\mathbf{R}_+)$

Finally in this section we announce the main results concerning the boundedness of the Kontorovich-Lebedev transformation (1.1) in the Sobolev type space $S_p^{N,\alpha}(\mathbf{R}_+)$. First we consider the Hilbert case $p = 2$. Namely, let operator (1.1) be as $KL : L_2(\mathbf{R}_+; \omega_\alpha(\tau)d\tau) \to S_2^{N,\alpha}(\mathbf{R}_+)$, where the weight $\omega_\alpha(\tau)$ is defined by

$$\omega_\alpha(\tau) = \pi^{3/2} \sum_{k=0}^N \frac{2^{-2\alpha_k-1} \tau^{4k} |\Gamma(2\alpha_k + i\tau)|^2}{\Gamma(2\alpha_k + 1/2)}, \quad \alpha_k > 0, \ k = 0, 1, \ldots, N.$$

In the limit case $\alpha_k = 0$, $k = 0, 1, \ldots, N$ one gets

$$\omega_0(\tau) = \frac{\pi^2}{2} \frac{1 - \tau^{4(N+1)}}{(1 - \tau^4)\tau \sinh \pi\tau}.$$

The following Plancherel type theorem takes place.

**Theorem 4.** *Let* $f \in L_2(\mathbf{R}_+; \omega_0(\tau)d\tau)$. *Then the integral* (1.1) *for the Kontorovich-Lebedev transform converges in the mean square sense with respect to the norm in the space* $S_2^{N,0}(\mathbf{R}_+)$ *and the sequence*

$$f_n(\tau) = \frac{2}{\pi^2} \tau \sinh \pi\tau \int_{1/n}^n K_{i\tau}(x)(KLf)(x)\frac{dx}{x}$$

*converges in the mean to* $f(\tau)$ *with respect to the norm in* $L_2(\mathbf{R}_+; \omega_0(\tau)d\tau)$. *Moreover, the following Plancherel identity is true*

$$\sum_{k=0}^N \int_0^\infty A_x^k KLf \, \overline{A_x^k KLh} \frac{dx}{x} = \frac{\pi^2}{2} \int_0^\infty f(\tau)\overline{h(\tau)} \frac{1 - \tau^{4(N+1)}}{1 - \tau^4} \frac{d\tau}{\tau \sinh \pi\tau},$$

*where $f, h \in L_2(\mathbf{R}_+; \omega_0(\tau)d\tau)$. In particular,*

$$||KLf||^2_{S_2^{N,0}(\mathbf{R}_+)} = ||f||^2_{L_2(\mathbf{R}_+;\omega_0(\tau)d\tau)}$$

*that is*

$$\sum_{k=0}^{N} \int_0^{\infty} |A_x^k KLf|^2 \frac{dx}{x} = \frac{\pi^2}{2} \int_0^{\infty} |f(\tau)|^2 \frac{1 - \tau^{4(N+1)}}{1 - \tau^4} \frac{d\tau}{\tau \sinh \pi\tau}.$$

*Finally, for almost all $\tau$ and $x$ from $\mathbf{R}_+$ the reciprocal formulas take place*

$$(KLf)(x) = g(x) = \frac{d}{dx} \int_0^{\infty} \int_0^x K_{i\tau}(y) f(\tau) dy d\tau,$$

$$f(\tau) = \frac{2}{\pi^2} \frac{(1 - \tau^4) \sinh \pi\tau}{1 - \tau^{4(N+1)}} \frac{d}{d\tau} \int_0^{\infty} \int_0^{\tau} y K_{iy}(x) \frac{1 - y^{4(N+1)}}{1 - y^4} (KLf)(x) \frac{dy dx}{x}.$$

**Remark 1.** When $N = 0$ we immediately obtain Plancherel identities (1.16), (1.17). The latter relations become then reciprocal formulas for the Kontorovich-Lebedev transformation in $L_2$- space with respect to the weight $\frac{\pi^2}{2}[\tau \sinh \pi\tau]^{-1}$ (see [11,12]).

In order to interpolate the norm of the Kontorovich-Lebedev operator on the $L_p$-case, where $2 \le p \le \infty$ let us consider $KL : L_p(\mathbf{R}_+; \rho_{p,\alpha}(\tau)d\tau) \to S_p^{N,\alpha}(\mathbf{R}_+)$, where the weighted function $\rho_{p,\alpha}(\tau)$ will be indicated below. In the case $p = \infty$ we understand the norm in the space $S_{\infty}^{N,\alpha}(\mathbf{R}_+)$ as (see (1.2))

$$||u||_{S_{\infty}^{N,\alpha}(\mathbf{R}_+)} = \lim_{p \to \infty} \left( \sum_{k=0}^{N} \int_0^{\infty} |A_x^k u|^p x^{\alpha_k p - 1} dx \right)^{1/p}.$$

From the equivalence of norms (2.2) we immediately derive that

$$C_1 \sum_{k=0}^{N} ||A_x^k u||_{L_{\alpha_k,\infty}} \le ||u||_{S_{\infty}^{N,\alpha}(\mathbf{R}_+)} \le C_2 \sum_{k=0}^{N} ||A_x^k u||_{L_{\alpha_k,\infty}},$$

where the norm in $L_{\nu,\infty}$ is defined by (see (1.11), (1.12))

$$||f||_{L_{\nu,\infty}} = \operatorname{ess\,sup} |x^{\nu} f(x)| = \lim_{p \to \infty} \left( \int_0^{\infty} |f(x)|^p x^{\nu p - 1} dx \right)^{1/p}.$$

We begin to derive an inequality for the modulus of the modified Bessel function $|K_{i\tau}(x)|$. We will apply it below to estimate the $L_{\nu,\infty}$-norm for the $(KLf)(x)$. Indeed, taking the Macdonald formula (1.10), we employ the Schwarz inequality and invoke (1.9) with relation (2.16.33.2) from [6] to obtain

$$K_{i\tau}^2(x) = \frac{1}{2} \int_0^{\infty} e^{-u - \frac{x^2}{2u}} K_{i\tau}(u) \frac{du}{u} \le \frac{1}{2} \left( \int_0^{\infty} e^{-2u - \frac{x^2}{u}} u^{-2\nu - 1} du \right)^{1/2}$$

$$\times \left( \int_0^{\infty} K_{i\tau}^2(u) u^{2\nu - 1} du \right)^{1/2} = \pi^{1/4} 2^{(\nu-3)/2} x^{-\nu} K_{2\nu}^{1/2} \left( 2\sqrt{2}x \right)$$

$$\times \left(\frac{\Gamma(\nu)}{\Gamma(\nu+1/2)}\right)^{1/2} |\Gamma(\nu+i\tau)|, \quad \nu > 0.$$

Hence we get

$$|K_{i\tau}(x)| \le \pi^{1/8} 2^{(\nu-3)/4} \left(\frac{\Gamma(\nu)}{\Gamma(\nu+1/2)}\right)^{1/4} |\Gamma(\nu+i\tau)|^{1/2} x^{-\nu/2} K_{2\nu}^{1/4}\left(2\sqrt{2x}\right).$$

Invoking inequality $x^\beta K_\beta(x) \le 2^{\beta-1}\Gamma(\beta), \beta > 0$ (see (1.9)) we derive the inequality

$$x^\nu |K_{i\tau}(x)| \le 2^{(2\nu-5)/4}\Gamma^{1/2}(\nu)|\Gamma(\nu+i\tau)|^{1/2}, \quad x, \nu > 0.$$

Thus we find that

$$x^\nu |(KLf)(x)| \le \|f\|_\infty x^\nu \int_0^\infty |K_{i\tau}(x)|d\tau \le 2^{(2\nu-5)/4}\Gamma^{1/2}(\nu)\|f\|_\infty \int_0^\infty |\Gamma(\nu+i\tau)|^{1/2}d\tau$$

$$= C_\nu \|f\|_\infty,$$

where $C_\nu > 0$ is the constant

$$C_\nu = 2^{(2\nu-5)/4}\Gamma^{1/2}(\nu) \int_0^\infty |\Gamma(\nu+i\tau)|^{1/2}d\tau, \quad \nu > 0.$$

Therefore we obtain that the Kontorovich-Lebedev operator $KL : L_\infty(\mathbf{R}_+; d\tau) \to L_{\nu,\infty}$ is bounded. It is of type $(\infty, \infty)$ and

$$\|KLf\|_{L_{\nu,\infty}} \le C_\nu \|f\|_\infty.$$

But using the norm inequality, which is proved in [13] we derive that this operator is of type $(2, 2)$ too. Consequently, by the Riesz-Thorin convexity theorem [3] the Kontorovich-Lebedev transformation is of type $(p, p)$, where $2 \le p \le \infty$ i.e. maps the space $L_p(\mathbf{R}_+; |\Gamma(2\nu+i\tau)|^2 d\tau)$ into $L_{\nu,p}$. Moreover for $2 \le p < \infty$ we arrive at the inequality

$$\int_0^\infty |(KLf)(x)|^p x^{\nu p-1}dx \le B_{p,\nu} \int_0^\infty |f(\tau)|^p |\Gamma(2\nu+i\tau)|^2 d\tau, \quad \nu > 0,$$

where we denoted by $B_{p,\nu}$ the constant

$$B_{p,\nu} = \pi^{3/2} 2^{-(3-p/2)\nu-5p/4+3/2} \frac{\Gamma^{p/2-1}(\nu)}{\Gamma(2\nu+1/2)} \left(\int_0^\infty |\Gamma(\nu+i\mu)|^{1/2}d\mu\right)^{p-2}.$$

Thus we obtain

$$\|KLf\|_{S_p^{N,\alpha}(\mathbf{R}_+)} \le \|f\|_{L_p(\mathbf{R}_+;\rho_{p,\alpha}(\tau)d\tau)},$$

where

$$\rho_{p,\alpha}(\tau) = \sum_{k=0}^N B_{p,\alpha_k} \tau^{2kp} |\Gamma(2\alpha_k+i\tau)|^2, \quad \alpha_k > 0, k = 0, 1, \dots, N.$$

In particular, we have $\rho_{2,\alpha}(\tau) = \omega_\alpha(\tau)$. So the boundedness of the Kontorovich-Lebedev transformation (1.1) is proved. Finally we show that for all $x > 0$ it exists as a Lebesgue integral for any $f \in L_p(\mathbf{R}_+; \rho_{p,\alpha}(\tau)d\tau)$, $p > 2$. Indeed, it will immediately follow from the inequality

$$\int_0^\infty |K_{i\tau}(x)f(\tau)|\, d\tau \le \|f\|_{L_p(\mathbf{R}_+;|\Gamma(2\nu+i\tau)|^2 d\tau)}$$

$$\times \left( \int_0^\infty |K_{i\tau}(x)|^q |\Gamma(2\nu+i\tau)|^{-2q/p} d\tau \right)^{1/q}, \quad q = \frac{p}{p-1},$$

and from the convergence of the latter integral with respect to $\tau$. This is easily seen from (1.7) and the Stirling asymptotic formula for gamma-functions [1] since the integrand behaves as $O\left( e^{\pi\tau q(\frac{1}{p}-\frac{1}{2})}\tau^{\frac{q}{p}(1-4\nu)-\frac{q}{2}} \right)$, $\tau \to +\infty$.

### References

1. M. Abramowitz, I.A. Stegun, *Handbook of Mathematical Functions*, Dover, New York, 1972.
2. E.M. Dynkin, B.P. Osilenker, Weighted estimates for singular integrals and their applications, *Itogi Nauki i Tekhn.*, **21** (1983), 42- 129 (in Russian).
3. R.E. Edwards, *Fourier Series II*, Holt, Rinehart and Winston, New York, 1967.
4. B. Lisena, On the generalized Kontorovich-Lebedev transform, *Rend. di Matematica, Ser. YII*, **9** (1989), 87- 101.
5. Prudnikov, A.P., Brychkov, Yu.A. and Marichev, O.I. *Integrals and Series: Elementary Functions.* Gordon and Breach, New York, 1986.
6. Prudnikov, A.P., Brychkov, Yu.A. and Marichev, O.I. *Integrals and Series: Special Functions.* Gordon and Breach, New York, 1986.
7. Prudnikov, A.P., Brychkov, Yu.A. and Marichev, O.I. *Integrals and Series: More Special Functions.* Gordon and Breach, New York, 1989.
8. I.N. Sneddon, *The Use of Integral Transforms*, McGraw-Hill, New York, 1972.
9. E.C. Titchmarsh, *An Introduction to the Theory of Fourier Integrals*, Clarendon Press, Oxford, 1937.
10. S.B. Yakubovich and B. Fisher, On the theory of the Kontorovich-Lebedev transformation on distributions, *Proc. of the Amer. Math. Soc.*, **122** (1994), N 3, 773-777.
11. S.B. Yakubovich, *Index Transforms*, World Scientific Publishing Company, Singapore, New Jersey, London and Hong Kong, 1996.
12. S.B. Yakubovich and J. de Graaf, On Parseval equalities and boundedness properties for Kontorovich-Lebedev type operators, *Novi Sad J. Math.*, **29** (1999), N 1, 185-205.
13. S.B. Yakubovich, On the integral transfomation associated with the product of gamma-functions, *Portugaliae Mathematica*, **60** (2003), N 3, 337-351.
14. S.B. Yakubovich, On the Kontorovich-Lebedev transformation, *J. of Integral Equations and Appl.* **15** (2003), N 1, 95-112.
15. S.B. Yakubovich, Integral transforms of the Kontorovich-Lebedev convolution type, *Collect. Math.* **54** (2003), N 2, 99-110.
16. S.B. Yakubovich, The Kontorovich-Lebedev transformation on Sobolev type spaces, *Sarajevo J. Math.* **1** (**14**) (2005), N 2 (to appear).
17. A.H. Zemanian, The Kontorovich-Lebedev transformation on distributions of compact support and its inversion, *Math. Proc. Cambridge Philos. Soc.* **77** (1975), 139- 143.

# FOURIER TYPE ANALYSIS AND QUANTUM MECHANICS

SHUJI WATANABE*

*Department of Mathematics, Faculty of Engineering,*
*Gunma University*
*4-2 Aramaki-machi, Maebashi 371-8510, Japan*
*Email: watanabe@fs.aramaki.gunma-u.ac.jp*

We discuss Fourier type analysis originating from quantum mechanics. The usual Fourier transform is an example of our Fourier type analysis. For simplicity we deal with the 1-dimensional space. The Fourier transform is suitable for differential operators in $\mathbb{R}$ with constant coefficients. On the other hand, our Fourier type analysis is suitable for differential operators *in bounded or unbounded open intervals with variable coefficients. Here some variable coefficients are singular.* We construct an integral transform that transforms a certain differential operator with a singular variable coefficient into the multiplication by $iy$ ($i = \sqrt{-1}$, $y \in \mathbb{R}$). We find that our transform is a generalized Fourier transform. We then define spaces of Sobolev type using our transform, and show an embedding theorem for each space. We apply both our transform and our embedding theorem to partial differential equations in bounded or unbounded open intervals with singular variable coefficients so as to discuss some properties of the solutions.

**Key words:** Fourier analysis, quantum mechanics, ordinary differential operators with singular coefficients, Sobolev spaces, embedding theorem, partial differential equations
**Mathematics Subject Classification:** 42A38, 81-99

## 1. Introduction

This is based on a joint work [11] with Y. Ohnuki. Let $-\infty \le a < b \le \infty$. Let $f$ be a diffeomorphism of $(a, b)$ onto $\mathbb{R}$:

$$\xi = f(x), \qquad \xi \in \mathbb{R}, \qquad x \in (a, b),$$

and let

$$f(c) = 0, \qquad a < c < b.$$

Set

$$\widetilde{x} = f^{-1}(-\xi).$$

We deal with the following operator in $L^2(a, b)$:

$$\mathcal{D} = \frac{1}{f'}\frac{\partial}{\partial x} - \frac{f''}{2(f')^2} - q\frac{\sqrt{|f'|}}{f}R\frac{1}{\sqrt{|f'|}}. \tag{1}$$

---

*The author is supported in part by the Gunma University Foundation for the Promotion of Science and Engineering.

Here, $q > -1/2$ and $R$ denotes the reflection operator given by

$$Rv(x) = Rv\left(f^{-1}(\xi)\right) = v\left(f^{-1}(-\xi)\right) = v(\tilde{x}).$$

The expression for our operator therefore becomes

$$\mathcal{D}u(x) = \frac{1}{f'(x)}\frac{\partial u}{\partial x}(x) - \frac{f''(x)}{2f'(x)^2}u(x) - q\frac{\sqrt{|f'(x)|}}{f(x)}\frac{u(\tilde{x})}{\sqrt{|f'(\tilde{x})|}}.$$

**Remark 1.1.** Our operator is a linear differential operator *in a bounded or unbounded open interval* $(a, b)$. Moreover, its coefficients are *variable coefficients, and one of them is singular* since $f(x) = 0$ at $x = c$.

We denote by $f$ the multiplication by $f$ and regard it as an operator in $L^2(a, b)$. We also denote by $y$ the multiplication by $y$ and regard it as an operator in $L^2(\mathbb{R})$. Here, $y \in \mathbb{R}$. We construct an integral transform $U$ that transforms $\mathcal{D}$ into the multiplication by $iy$ ($i = \sqrt{-1}$). Our transform is associated both with $\mathcal{D}$ and with $f$, and they satisfy Wigner's commutation relations [18] in quantum mechanics:

$$\{\mathcal{D}, [f, \mathcal{D}]\} = -2\mathcal{D}, \qquad \{f, [f, \mathcal{D}]\} = -2f,$$

where $\{A, B\} = AB + BA$. So our Fourier type analysis originates from quantum mechanics. Using our transform $U$ we define spaces of Sobolev type, and show an embedding theorem for each space. Our embedding theorem is a generalization of the Sobolev embedding theorem. We apply both our transform and our embedding theorem to partial differential equations *in bounded or unbounded open intervals with singular variable coefficients* so as to discuss properties of the solutions.

We now give some examples of our operator (1).

**Example 1.** Let $a = -\infty$, $b = \infty$, $f(x) = x$ and let $q = 0$. Then, by (1),

$$\mathcal{D} = \frac{\partial}{\partial x}, \qquad x \in \mathbb{R},$$

and hence our transform $U$ reduces to the Fourier transform in this case (see Remark 2.2 below). Therefore, our transform can be regarded as a generalized Fourier transform, and the Fourier transform is an example of our Fourier type analysis.

Our operator appears in many quantum-mechanical systems, as is shown just below.

**Example 2.** Let $a = -\infty$, $b = \infty$ and let $q = 0$. In this case, each function $f$ corresponds to a point transformation in quauntum mechanics. We [9] first define and discuss a point transformation as a canonical transformation in quauntum mechanics from the view point of functional analysis. Our operator $-i\mathcal{D}$ then corresponds to the new momentum operator given by the point transformation.

**Example 3.** Let $a = 0$, $b = \infty$, $f(x) = \ln x$ and let $q = 0$. In this case, our operator $-i\mathcal{D}$ corresponds to the dilatation operator of quantum mechanics. We [10] studied the essential selfadjointmess of $-i\mathcal{D}$ and showed that a Mellin transform transforms $-i\mathcal{D}$ into the multiplication by $y$ ($y \in \mathbb{R}$).

**Example 4.** Let $a = 0$, $b = \pi$, $f(x) = -\ln\tan(x/2)$ and let $q = 0$. Our operator $-iD$ in this case corresponds to the momentum operator appearing in quantum mechanics on $S^1$ based on Dirac Formalism [1,8]. We [15–17] discussed the selfadjointmess of $-iD$ and constructed an integral transform that transforms $-iD$ into the multiplication by $y$ $(y \in \mathbb{R})$. See also Soltani [12] for related material.

**Example 5.** Let $a = -\infty$, $b = \infty$ and $f(x) = x$. Our operator $-iD$ then corresponds to the momentum operator of a bose-like oscillator governed by Wigner's commutation relations. See Yang [19], and Ohnuki and Kamefuchi [6,7].

For $n = 0, 1, 2, 3, \ldots$, let (cf. (4.31) of Ref. 6 and (23.80) of Ref. 7)

$$\begin{cases} u_{2n}(x) = K_n^{q+\frac{1}{2}} \sqrt{|f'(x)|}\, |f(x)|^q\, L_n^{q-\frac{1}{2}}(f(x)^2) \exp\left(-\frac{f(x)^2}{2}\right), \\ u_{2n+1}(x) = K_n^{q+\frac{3}{2}} \sqrt{|f'(x)|}\, f(x)\, |f(x)|^q\, L_n^{q+\frac{1}{2}}(f(x)^2) \exp\left(-\frac{f(x)^2}{2}\right). \end{cases}$$

Here $K_n^\nu = (-1)^n \sqrt{n!/\Gamma(n+\nu)}$ with $\Gamma$ the gamma function, and $L_n^\nu$ is a generalized Laguerre polynomial. Note that $u_n \in L^2(a,\, b)$.

**Remark 1.2.** Ohnuki and Kamefuchi [6,7] obtained the functions $u_n$, when $a = -\infty$, $b = \infty$ and $f(x) = x$.

Let $V$ be the set of finite linear combinations of $u_n$'s. A straightforward calculation gives the following.

**Lemma 1.1.** *The set $\{u_n\}_{n=0}^\infty$ is a complete orthonormal set of $L^2(a,\, b)$. Consequently, $V$ is dense in $L^2(a,\, b)$.*

Using Nelson's analytic vector theorem [5] we can show the following.

**Proposition 1.1.** *The operator $(-iD) \restriction V$ is essentially selfadjoint, and so is the multiplication operator $f \restriction V$.*

We denote by $(\cdot,\, \cdot)_{L^2(a,\, b)}$ the inner product of $L^2(a,\, b)$:

$$(u_1,\, u_2)_{L^2(a,\, b)} = \int_a^b u_1(x)\, \overline{u_2(x)}\, dx, \quad u_1,\, u_2 \in L^2(a,\, b)$$

and by $\|\cdot\|_{L^2(a,\, b)}$ the norm $\|\cdot\|_{L^2(a,\, b)} = \sqrt{(\cdot,\, \cdot)_{L^2(a,\, b)}}$. We also denote by $(\cdot,\, \cdot)_{L^2(\mathbb{R})}$ the inner product of $L^2(\mathbb{R})$, and by $\|\cdot\|_{L^2(\mathbb{R})}$ the norm $\|\cdot\|_{L^2(\mathbb{R})} = \sqrt{(\cdot,\, \cdot)_{L^2(\mathbb{R})}}$.

## 2. An Integral Transform

Set

$$\varphi(y,\, x) = \frac{\sqrt{|yf(x)f'(x)|}}{2}\left\{ J_{q-1/2}(|yf(x)|) + i\,\mathrm{sgn}(yf(x))J_{q+1/2}(|yf(x)|) \right\},$$

where $x \in (a,\, b)$, $y \in \mathbb{R}$ and $J_\nu$ denotes the Bessel function of the first kind.

**Remark 2.1.** Ohnuki and Kamefuchi [6,7] obtained the function $\varphi(y, x)$, when $a = -\infty$, $b = \infty$ and $f(x) = x$.

We consider the following integral transform:

$$Uu(y) = \int_a^b \overline{\varphi(y, x)} u(x)\, dx, \qquad u \in V,$$

where $y \in \mathbb{R}$. Note that $Uu \in L^2(\mathbb{R})$. The operator $U$ satisfies

$$(Uu_1,\, Uu_2)_{L^2(\mathbb{R})} = (u_1,\, u_2)_{L^2(a,\, b)}, \qquad u_1,\, u_2 \in V.$$

Combining this fact with Lemma 1.1 gives the following.

**Theorem 2.1.** *The transform $U$ becomes a unitary operator from $L^2(a, b)$ to $L^2(\mathbb{R})$.*

A straightforward calculation gives that our transform $U$ transforms our operator $-i\mathcal{D}$ into the multiplication operator $y$:

**Proposition 2.1.**

$$U\left(-i\mathcal{D}\right) U^* = y.$$

This proposition immediately implies the following.

**Corollary 2.1.** *Let $-i\mathcal{D}$ be the selfadjoint operator in $L^2(a, b)$ given above. Then the operator $\mathcal{D}^2$ generates an analytic semigroup $\{\exp(t\mathcal{D}^2) : t > 0\}$ on $L^2(a, b)$.*

**Remark 2.2.** If $a = -\infty$, $b = \infty$, $f(x) = x$ and $q = 0$, then

$$\mathcal{D} = \frac{\partial}{\partial x}, \quad \varphi(y, x) = \frac{1}{\sqrt{2\pi}} \exp\left[iyx\right].$$

Here, $(x, y) \in \mathbb{R} \times \mathbb{R}$. Our transform $U$ reduces to the Fourier transform in this case, and hence can be regarded as a generalized Fourier transform.

**Remark 2.3.** We constructed our transform on the basis of the study of the Hankel transform. Kilbas and Borovco [4] considered a more general integral transform including the Hankel transform.

## 3. Spaces of Sobolev Type

**Definition 3.1.** For $\nu \geq 0$, we define spaces of Sobolev type:

$$\mathcal{H}^\nu(a, b) = \left\{ u \in L^2(a, b) : \int_{\mathbb{R}} \left(1 + |y|^2\right)^\nu |Uu(y)|^2\, dy < \infty \right\}.$$

A straightforward calculation gives that each $\mathcal{H}^\nu(a, b)$ is a Hilbert space with inner product

$$(u_1,\, u_2)_{\mathcal{H}^\nu(a,\, b)} = \int_{\mathbb{R}} \left(1 + |y|^2\right)^\nu Uu_1(y)\, \overline{Uu_2(y)}\, dy, \qquad u_1,\, u_2 \in \mathcal{H}^\nu(a, b)$$

and norm $|u|_{\mathcal{H}^\nu(a,\,b)} = \sqrt{(u,\,u)_{\mathcal{H}^\nu(a,\,b)}}$ .

**Remark 3.1.** If $a = -\infty$, $b = \infty$, $f(x) = x$ and $q = 0$, then our transform $U$ reduces to the Fourier transform as is stated above, and hence $\mathcal{H}^\nu(a,\,b)$ to the usual Sobolev space $H^\nu(\mathbb{R})$ in this case.

Definition 3.1 together with Proposition 2.1 immediately implies the following.

**Corollary 3.1.** *(A)* $\quad \mathcal{H}^0(a,\,b) = L^2(a,\,b)$.
*(B)* $\quad \mathcal{H}^{\nu'}(a,\,b) \subset \mathcal{H}^\nu(a,\,b)$, $\qquad \nu' \geq \nu$.
*(C)* $\quad |u|_{\mathcal{H}^\nu(a,\,b)} \leq |u|_{\mathcal{H}^{\nu'}(a,\,b)}$, $\qquad u \in \mathcal{H}^{\nu'}(a,\,b)$, $\qquad \nu' \geq \nu$.
*(D)* $\quad$ *Let* $|y|^\nu$ *be the selfadjoint multiplication operator and* $D(|y|^\nu)$ *its domain. Then* $U\mathcal{H}^\nu(a,\,b) = D(|y|^\nu)$.

We need the following to prove our embedding theorem in the next section.

**Lemma 3.1.** *Let* $\nu > \frac{1}{2}$ *and let* $0 \leq k < \nu - \frac{1}{2}$. *Then*
$$|y|^k U u \in L^1(\mathbb{R}), \qquad u \in \mathcal{H}^\nu(a,\,b).$$

## 4. An Embedding Theorem of Sobolev Type

As is well known, the usual Sobolev embedding theorem tells us only about smoothness of each element. On the other hand, our embedding theorem tells us both about smoothness of $u$ $(u \in \mathcal{H}^\nu(a,\,b))$ and about continuity of $u/f^n$, as is shown below. So our embedding theorem is a generalization of the Sobolev embedding theorem.

**Definition 4.1.** Let $f$ be as above. For $n = 0, 1, 2, \ldots$, we define
$$S_f^n(a,\,b) = \left\{ u(x) : u, \; \frac{u}{f^n} \in C(a,\,b) \right\}.$$

**Remark 4.1.** If $u \in S_f^n(a,\,b)$, then $u/f^n$ is continuous on $(a,\,b)$.

The following is our embedding theorem.

**Theorem 4.1.** *Let* $q \geq 0$. *Suppose* $\nu > \dfrac{1}{2}$ *and* $\nu \neq m + \dfrac{1}{2}$ $(m = 1, 2, 3, \ldots)$. *Then*
$$\mathcal{H}^\nu(a,\,b) \subset C^\alpha(a,\,b) \cap S_f^\beta(a,\,b),$$

*where* $\quad (k = 0, 1, 2, \ldots)$

$$\alpha = \begin{cases} \left[\nu - \frac{1}{2}\right] & (q = 2k), \\ \min\left(\left[\nu - \frac{1}{2}\right], q - 1\right) & (q = 2k + 1), \\ \min\left(\left[\nu - \frac{1}{2}\right], [q]\right) & (otherwise) \end{cases}$$

*and*

$$\beta = \begin{cases} \min\left(\left[\nu - \frac{1}{2}\right], q\right) & (q = 2k), \\ \min\left(\left[\nu - \frac{1}{2}\right], q - 1\right) & (q = 2k + 1), \\ \min\left(\left[\nu - \frac{1}{2}\right], [q]\right) & (otherwise). \end{cases}$$

**Remark 4.2.** If $a = -\infty$, $b = \infty$, $f(x) = x$ and $q = 0$, then our transform $U$ and our space $\mathcal{H}^\nu(a, b)$ reduce to the Fourier transform and to the Sobolev space $H^\nu(\mathbb{R})$, respectively. Moreover, $\alpha = [\nu - \frac{1}{2}]$ and $\beta = 0$ in this case. Our embedding theorem thus reduces to the usual Sobolev embedding theorem:

$$H^\nu(\mathbb{R}) \subset C^{[\nu - 1/2]}(\mathbb{R}).$$

So our embedding theorem is a generalization of the Sobolev embedding theorem.

## 5. Applications to Partial Differential Equations

Our Fourier type analysis can treat partial differential equations *in bounded or unbounded open intervals with singular variable coefficients*, where

$$\sum_{k=0}^{m} a_k \mathcal{D}^k \qquad \text{in} \quad (a, b)$$

appears. Here, $a_k$ are constants. In what follows, for simplicity, we confine ourselves to treating such partial differential equations that the operator

$$a_2 \mathcal{D}^2 + a_0$$

appears. We assume that $q \geq 0$.

Note that (see (1))

$$\mathcal{D}^2 u(x) = \frac{1}{f'(x)^2} \frac{\partial^2 u}{\partial x^2}(x) - \frac{2f''(x)}{f'(x)^3} \frac{\partial u}{\partial x}(x) + \frac{5f''(x)^2 - 2f'(x)f'''(x)}{4f'(x)^4} u(x)$$

$$-q\frac{\sqrt{f'(x)}}{f(x)} \left\{ \frac{1}{f'(\tilde{x})^{3/2}} \frac{\partial u}{\partial x}(\tilde{x}) + \frac{1}{f'(x)} \frac{\partial}{\partial x}\left( \frac{u(\tilde{x})}{\sqrt{f'(\tilde{x})}} \right) \right\}$$

$$+q\frac{\sqrt{f'(x)}}{f(x)} \left\{ \frac{1}{f(x)} + \frac{f''(\tilde{x})}{2f'(\tilde{x})^2} \right\} \frac{u(\tilde{x})}{\sqrt{f'(\tilde{x})}} + q^2 \frac{u(x)}{f(x)f(\tilde{x})},$$

where $\tilde{x} = f^{-1}(-\xi)$ and $x = f^{-1}(\xi)$. Note also that

$$\mathcal{D}^2 u(x) = \frac{\partial^2 u}{\partial x^2}(x),$$

when $a = -\infty$, $b = \infty$, $f(x) = x$ and $q = 0$.

**Definition 5.1.** For $|a|, |b| < \infty$, let $c = (a + b)/2$. Here, $c = f^{-1}(0)$. Set $(x_R + x)/2 = c$, where $a < x_R < b$. A function $u$ defined on $(a, b)$ is said to be an even function with respect to $x = c$ if $u(x_R) = u(x)$. A function $u$ defined on $(a, b)$ is said to be an odd function with respect to $x = c$ if $u(x_R) = -u(x)$.

**Example 5.1.** Let $a = 0$, $b = \pi$. The function: $x \mapsto \ln \tan(x/2)$, $0 < x < \pi$ is an odd function with respect to $x = \pi/2$.

Suppose that $f$ is an odd function with respect to $x = c$. Then $x_R = \tilde{x}$, and hence $f(\tilde{x}) = -f(x)$ and $f'(\tilde{x}) = f'(x)$. The expressions for $\mathcal{D}u$ and $\mathcal{D}^2u$ therefore become somewhat simpler:

$$\mathcal{D}u(x) = \frac{1}{f'(x)}\frac{\partial u}{\partial x}(x) - \frac{f''(x)}{2f'(x)^2}u(x) - q\frac{u(\tilde{x})}{f(x)},$$

$$\mathcal{D}^2u(x) = \frac{1}{f'(x)^2}\frac{\partial^2 u}{\partial x^2}(x) - \frac{2f''(x)}{f'(x)^3}\frac{\partial u}{\partial x}(x) + \frac{5f''(x)^2 - 2f'(x)f'''(x)}{4f'(x)^4}u(x)$$
$$+q\frac{u(\tilde{x})}{f(x)^2} - q^2\frac{u(x)}{f(x)^2}.$$

Moreover, if $u$ is an even function with respect to $x = c$, then $u(\tilde{x}) = u(x)$. Hence

$$\mathcal{D}u(x) = \frac{1}{f'(x)}\frac{\partial u}{\partial x}(x) - \frac{f''(x)}{2f'(x)^2}u(x) - q\frac{u(x)}{f(x)},$$

$$\mathcal{D}^2u(x) = \frac{1}{f'(x)^2}\frac{\partial^2 u}{\partial x^2}(x) - \frac{2f''(x)}{f'(x)^3}\frac{\partial u}{\partial x}(x) + \frac{5f''(x)^2 - 2f'(x)f'''(x)}{4f'(x)^4}u(x)$$
$$+q\frac{u(x)}{f(x)^2} - q^2\frac{u(x)}{f(x)^2}.$$

Note that the variable $\tilde{x}$ disappears in the expressions for $\mathcal{D}u(x)$ and $\mathcal{D}^2u(x)$. A similar argument holds for an odd function $u$.

We first deal with the following problem in $L^2(a, b)$ with singular variable coefficients:

$$\begin{cases} \dfrac{\partial u}{\partial t}(t, x) = \mathcal{D}^2u(t, x), & t > 0, \quad x \in (a, b), \\[2mm] u(0, x) = u_0(x), & x \in (a, b), \end{cases} \tag{2}$$

where $u_0 \in L^2(a, b)$. When $a = -\infty$, $b = \infty$, $f(x) = x$ and $q = 0$, this problem reduces to the initial value problem for the usual heat equation. See Watanabe and Watanabe [13], and Watanabe [14] for related material. For applications of transform methods to partial differential equations, see e.g. Duffy [2].

**Remark 5.1.** If $f$ is an odd function with respect to $x = c$ and if $u_0$ is decomposed into the sum of the even function with respect to $x = c$ and the odd one, then the variable $\tilde{x}$ disappears in (2).

Let us look for $u(t, \cdot) \in \mathcal{H}^2(a, b)$ satisfying the problem (2). By Corollary 2.1, the operator $\mathcal{D}^2$ generates an analytic semigroup $\{\exp(t\mathcal{D}^2) : t > 0\}$ on $L^2(a, b)$. Combining Corollary 3.1 with Theorem 4.1 thus yields the following.

**Corollary 5.1.** Let $m = 1, 2, 3, \ldots$. For $u_0 \in L^2(a, b)$, there is a unique solution $u \in C\left([0, \infty); L^2(a, b)\right) \cap C^1\left((0, \infty); \mathcal{H}^{2m}(a, b)\right)$ of the problem (2) satisfying

$$u(t, \cdot) = \exp(t\mathcal{D}^2)u_0 \in C^\alpha(a, b) \cap S_f^\beta(a, b),$$

*where* $(k = 0, 1, 2, \dots)$

$$\alpha = \begin{cases} \infty & (q = 2k), \\ q-1 & (q = 2k+1), \\ [q] & (otherwise), \end{cases} \qquad \beta = \begin{cases} q & (q = 2k), \\ q-1 & (q = 2k+1), \\ [q] & (otherwise). \end{cases}$$

Suppose $u_0 \in C_0^\infty(a, b)$. We now try to write the solution in an explicit form. By Proposition 2.1, our transform $U$ turns (2) into

$$\begin{cases} \dfrac{dv}{dt} = -y^2 v, \quad t > 0, \\[2mm] v(0) = U u_0, \end{cases}$$

where $v = Uu$. Therefore, $v(t) = e^{-ty^2} U u_0$. A straightforward calculation gives that

$$
\begin{aligned}
u(t, x) &= \int_{\mathbb{R}} \varphi(y, x) e^{-ty^2} U u_0(y)\, dy \\
&= \frac{\sqrt{|f(x)|\, f'(x)}}{4t} e^{-\frac{f(x)^2}{4t}} \int_a^b e^{-\frac{f(\xi)^2}{4t}} \sqrt{|f(\xi)|\, f'(\xi)}\; u_0(\xi) \times \\
&\quad \times \left\{ I_{q-\frac{1}{2}}\left(\frac{|f(x)f(\xi)|}{2t}\right) + \operatorname{sgn}\left(f(x)f(\xi)\right) I_{q+\frac{1}{2}}\left(\frac{|f(x)f(\xi)|}{2t}\right) \right\} d\xi.
\end{aligned}
\tag{3}
$$

Here, $I_\nu$ is a modified Bessel function and the following formula (see (23), p.51 of Ref. 3) is used:

$$\int_0^\infty e^{-a^2 y^2}\, y\, J_\nu(py)\, J_\nu(qy)\, dy = \frac{1}{2\,a^2} \exp\left(-\frac{p^2 + q^2}{4\,a^2}\right) I_\nu\left(\frac{pq}{2\,a^2}\right),$$

where $\Re(\nu) > -1$ and $|\arg a| < \pi/4$.

**Corollary 5.2.** *Suppose* $u_0 \in C_0^\infty(a, b)$. *Let $u$ be the solution of the problem* (2) *given by Corollary 5.1. Then the solution is explicitly given by* (3).

We second deal with the following problem in $L^2(a, b)$ with singular variable coefficients:

$$\begin{cases} \dfrac{\partial^2 u}{\partial t^2}(t, x) = \mathcal{D}^2 u(t, x), \quad t \in \mathbb{R}, \quad x \in (a, b), \\[3mm] u(0, x) = u_0(x), \quad \dfrac{\partial u}{\partial t}(0, x) = u_1(x), \quad x \in (a, b), \end{cases} \tag{4}$$

where $u_0 \in \mathcal{H}^2(a, b)$, $u_1 \in \mathcal{H}^1(a, b)$. When $a = -\infty$, $b = \infty$, $f(x) = x$ and $q = 0$, this problem reduces to the initial value problem for the usual wave equation.

**Remark 5.2.** If $f$ is an odd function with respect to $x = c$ and if each of $u_0$ and $u_1$ is decomposed into the sum of the even function with respect to $x = c$ and the odd one, then the variable $\tilde{x}$ disappears in (4).

Since $-\mathcal{D}^2$ with domain $\mathcal{H}^2(a, b)$ and $\sqrt{-\mathcal{D}^2}$ with domain $\mathcal{H}^1(a, b)$ are both nonnegative selfadjoint operators, the problem (4) is well-posed. Therefore, $u(t, \cdot) \in \mathcal{H}^2(a, b)$. The following is an immediate consequence of Theorem 4.1.

**Corollary 5.3.** *Suppose $u_0 \in \mathcal{H}^2(a, b)$ and $u_1 \in \mathcal{H}^1(a, b)$. Then the problem (4) is well-posed, and the solution satisfies*

$$u(t, \cdot) \in C^\alpha(a, b) \cap S_f^\beta(a, b),$$

*where*   $(k = 0, 1, 2, \dots)$

$$\alpha = \begin{cases} 1 & (q = 2k), \\ \min(1, q-1) & (q = 2k+1), \\ \min(1, [q]) & (otherwise), \end{cases} \quad \beta = \begin{cases} \min(1, q) & (q = 2k), \\ \min(1, q-1) & (q = 2k+1), \\ \min(1, [q]) & (otherwise). \end{cases}$$

We finally deal with the following problem in $L^2(a, b)$ with singular variable coefficients. Given a $g \in L^2(a, b)$ we look for a solution $u \in \mathcal{H}^2(a, b)$ satisfying

$$-\mathcal{D}^2 u(x) + \lambda^2 u(x) = g(x) \quad \text{in} \quad (a, b), \tag{5}$$

where $\lambda > 0$.

**Remark 5.3.** If $f$ is an odd function with respect to $x = c$ and if $g$ is decomposed into the sum of the even function with respect to $x = c$ and the odd one, then the variable $\tilde{x}$ disappears in (5).

**Corollary 5.4.** *For $g \in L^2(a, b)$, there is a unique solution $u \in \mathcal{H}^2(a, b)$ of the problem (5), and the estimate*

$$|u|_{\mathcal{H}^2(a, b)} \le C \|g\|_{L^2(a, b)}$$

*holds for some constant $C > 0$, independent of the solution $u$. Consequently, the solution $u$ continuously depends on the data $g$. Moreover,*

$$u \in C^\alpha(a, b) \cap S_f^\beta(a, b),$$

*where $\alpha$ and $\beta$ are those in Corollary 5.3.*

# References

1. P. A. M. Dirac, *Lectures on quantum mechanics*, Belfer Graduate School of Science, Yeshiva Univ., New York, 1964.
2. D. G. Duffy, *Transform methods for solving partial differential equations*, 2nd ed., CRC Press, Boca Raton, FL, 2004.
3. A. Erdélyi (ed.), *Tables of integral transforms*, vol. II, McGraw-Hill, New York, 1954.
4. A. A. Kilbas and A. N. Borovco, *Hardy-Titchmarsh and Hankel type transforms in $L_{\nu,\tau}$-spaces*, Integ. Transf. Spec. Funct. **10** (2000), 239–266.
5. E. Nelson, *Analytic vectors*, Ann. Math. **70** (1959), 572–615.
6. Y. Ohnuki and S. Kamefuchi, *On the wave-mechanical representation of a Bose-like oscillator*, J. Math. Phys. **19** (1978), 67–78.

250

7. Y. Ohnuki and S. Kamefuchi, *Quantum field theory and parastatistics*, Univ. of Tokyo Press, Tokyo, 1982/Springer-Verlag, Berlin, Heidelberg and New York, 1982.

8. Y. Ohnuki and S. Kitakado, *Fundamental algebra for quantum mechanics on $S^D$ and gauge potentials*, J. Math. Phys. **34** (1993), 2827–2851.

9. Y. Ohnuki and S. Watanabe, *Characterization of point transformations in quantum mechanics*, J. Anal. Appl. **1** (2003), 193–205.

10. Y. Ohnuki and S. Watanabe, *The dilatation operator in quantum mechanics and its applications*, Far East J. Math. Sci. **15** (2004), 353–367.

11. Y. Ohnuki and S. Watanabe, *Fourier type analysis originating from quantum mechanics*, submitted.

12. F. Soltani, *Practical inversion formulas in a quantum mechanical system*, Applicable Analysis **84** (2005), 759–767.

13. M. Watanabe and S. Watanabe, *The explicit solution of a diffusion equation with singularity*, Proc. Amer. Math. Soc. **126** (1998), 383–389.

14. S. Watanabe, *The explicit solutions to the time-dependent Schrödinger equations with the singular potentials $k/(2x^2)$ and $k/(2x^2) + \omega^2 x^2/2$*, Commun. Partial Differential Equations **26** (2001), 571–593.

15. S. Watanabe, *Quantum mechanics on $S^1$ based on Dirac formalism*, Applicable Analysis **82** (2003), 25–34.

16. S. Watanabe, *An integral transform and its applications*, Integ. Transf. Spec. Funct. **14** (2003), 537–549.

17. S. Watanabe, *An embedding theorem of Sobolev type*, Integ. Transf. Spec. Funct. **15** (2004), 369–374.

18. E. P. Wigner, *Do the equations of motion determine the quantum mechanical commutation relations ?*, Phys. Rev. **77** (1950), 711–712.

19. L. M. Yang, *A note on the quantum rule of the harmonic oscillator*, Phys. Rev. **84** (1951), 788–790.

# THE SINGULAR VALUE DECOMPOSITION FOR GENERALIZED TRANSFORM OF RADON TYPE IN $R^n$ *

JINPING WANG

*Department of Mathematics, Ningbo University, Ningbo, 315211, P.R. China*
*E-mail: wangjp66cn@yahoo.com.cn*

The singular value decomposition for the generalized transform of Radon type is derived when the generalized transform of Radon type is restricted to functions which are square integrable on $R^n$ with respect to the weight $W_n$. Furthermore, an approximation inversion formula about the measured data is also obtained.

**Key words:** Generalized transform of Radon type, singular value decompositions (SVD), spherical harmonics

**Mathematics Subject Classification:** 44A15, 65R10

## 1. Introduction

The singular value decomposition(SVD) for Radon transform has a long history in computed tomograph and many results, see Davison[1], Louis[3], Maass[4]. Quite different techniques have been used in these papers. From this one immediately may obtain an inversion formula, range characterizations and some results on the ill-posedness of the inverse problem [3]. The generalized transform of Radon type is linear operator $\mathcal{R}$ that acts on functions on Euclidean space $R^n$ by integration over hyperplanes with respect to some measure. Specially, if $\omega$ is a point on the unit $S^{n-1}$, $s \in R$, and $\mu$ is a constant, then the generalized transform of Radon type $\mathcal{R}$ is defined by

$$\mathcal{R}f(\omega, s) = \int_{R^n} f(x) \exp\{\mu \langle x, \omega \rangle\} \delta(s - \langle x, \omega \rangle) \mathrm{d}x, \qquad (1.1)$$

where $<,>$ denotes the standard inner product on $R^n$. When $\mu \equiv 0$, $\mathcal{R}$ is the classical Radon transform (in this case we shall still denote by $\mathcal{R}$). The problem of inverting the generalized transform of Radon type arises in many applications. For example, in medical radiology. Inverting the Radon transform arises in computed tomography[3]. Another special case of interest in radiology is the exponential

---
*Supported by K.C.Wong Education Foundation, Hong Kong and the Natural Science Foundation of Zhejiang Province(Y606093);Scientific Research Fund of Zhejiang Provincial Education Department(20051760); the NNSF(10471069); Professor(Doctoral) Foundation of Ningbo University.

Radon transform which occurs in two-dimensional single photon emission computed tomography(SPECT)[2].

A standard method of constructing the singular value decomposition is described below. Suppose that $\mathcal{R}$ is a continuous linear map between the Hilbert spaces $H$ and $K$, and that $\mathcal{R}\mathcal{R}^* : K \longrightarrow K$ has a complete eigenfunctions $g_i, i = 1, 2, \ldots$, of $\mathcal{R}\mathcal{R}^*$ whose associated eigenvalues $\lambda_i$ are positive, and set

$$f_i = (1/\sqrt{\lambda_i})\mathcal{R}^* g_i, \ i = 1, 2, \ldots \tag{1.2}$$

then the $\{g_i\}$ form an orthonormal basis for the range of $\mathcal{R}$ and the $\{f_i\}$ form an orthonormal basis for the orthogonal complement of the kernel of $\mathcal{R}$. Furthermore given an $f \in H$,

$$\mathcal{R}f = \sum_{i=1}^{\infty} \lambda_i^{1/2} \langle f, f_i \rangle_H \, g_i, \tag{1.3}$$

where $\langle , \rangle_H$ denotes the inner product in $H$. Also, for given $g \in K$, which might represent measured data, the $\tilde{f}$ of the smallest norm which minimizes $\|\mathcal{R}\tilde{f} - g\|$ is given by

$$\tilde{f} = \sum_{i=1}^{\infty} \frac{\langle g, g_i \rangle_K}{\sqrt{\lambda_i}} f_i. \tag{1.4}$$

This paper is devoted to carrying out this construction, where $H = L^2(R^n, W_n)$, $W_n(x) = \pi^{n/2} \exp(|x|^2)$, $K = L^2(S^{n-1} \times R, W_1)$, $W_1 = \sqrt{\pi} \exp(s^2 - 2\mu s)$.

## 2. Results

We can now consider the generalized transform of Radon type $\mathcal{R}$ as operator from $H$ into $K$, we have the following results:

**Lemma 1** *For $\omega \in S^{n-1}, s \in R$. Since it is often convenient to view $\mathcal{R}f(\omega, s)$ as a family of functions of $s$ parameterized by $\omega$, throughout this paper,we use the abbreviated notation $\mathcal{R}_\omega f(s) = \mathcal{R}f(\omega, s)$ and take $H$ and $K$ as above. Then it is true that, $1/W_n$ and $\exp(-\mu^2)/W_1$ are normalized, $\mathcal{R}_\omega(1/W_n)(s) = \exp(-\mu s)/W_1(s)$.*
**Proof** Here the proof is fulfilled by simple calculation and Davison[1].

**Lemma 2** *Let operators $\mathcal{R}_\omega$ and $H, K$ be as above, then for all $\omega \in S^{n-1}, \mathcal{R}_\omega$ is a continuous linear map from $H$ into $L^2(R, W_1), \|\mathcal{R}_\omega\| = 1$ and*

$$[\mathcal{R}_\omega^* g](x) = (1/W_n) \cdot B_\omega(gW_1), \tag{2.1}$$

*where $B_\omega$ is the back-projection operator,*

$$(B_\omega g)(x) = g(\langle x, \omega \rangle)\exp(\mu \langle x, \omega \rangle). \tag{2.2}$$

**Proof** We prove that $B\omega$ is a norm preserving operator :

$$L^2(R, 1/W_1) \longrightarrow L^2(R^n, 1/W_n),$$

where $B_\omega^* f = W_1 \cdot \mathcal{R}_\omega(f/W_n)$.

And hence we have operators

$$\mathcal{R}_\omega : L^2(R^n, W_n) \longrightarrow L^2(R^1, W_1),$$

$$I_{W_1} : L^2(R^1, W_1) \longrightarrow L^2(R^n, 1/W_1),$$

$$B_\omega : L^2(R^n, 1/W_1) \longrightarrow L^2(R^n, 1/W_n),$$

$$I_{1/W_n} : L^2(R^n, 1/W_n) \longrightarrow L^2(R^n, W_n).$$

Here $I_{W_n}$ denotes pointwise multiplication by $W_n$. " * " denotes adjoint operator. It is easy to verify that $I_{W_n}$ and $I_{W_1}$ are Hilbert space isomorphism. Hence from the above operators, then

$$\mathcal{R}_\omega^* = I_{1/W_n} \cdot B_\omega \cdot I_{W_1},$$

and

$$\|\mathcal{R}_\omega\| = \|\mathcal{R}_\omega^*\| = \|B_\omega\| = 1.$$

Furthermore $B_\omega^* = I_{W_1} \cdot \mathcal{R}_\omega \cdot I_{1/W_n}$.

**Lemma 3** *Operators $\mathcal{R}_\omega$, and $H, K$ as above, then $\mathcal{R}$ is a continuous linear operator from $H$ into $K$ with $\|\mathcal{R}\| = \omega_n^{1/2}$ ($\omega_n$ is the surface area of $S^{n-1}$). If $G \in K$, then*

$$(\mathcal{R}^*G)(x) = \int_{S^{n-1}} (\mathcal{R}_\omega^* G_\omega)(x) \mathrm{d}\Omega(\omega), \tag{2.3}$$

*where $G_\omega(s) = G(\omega, s), s \in R, \omega \in S^{n-1}, x \in R^n$.*

**Proof** Let $f \in L^2(R^n, W_n)$, then

$$\|\mathcal{R}\|^2 = \int_{S^{n-1}} \int_{-\infty}^{+\infty} |\mathcal{R}_\omega f|^2 W_1 \mathrm{d}s \mathrm{d}\Omega(\omega)$$

$$= \int_{S^{n-1}} \|\mathcal{R}_\omega f\|^2 \mathrm{d}\Omega(\omega) \le \omega_n \|f\|^2.$$

Hence $\|\mathcal{R}\| \le \omega_n^{1/2}$. To prove equality take $f = 1/W_n$. Then $\|f\|_{W_n} = 1$ and, since $\mathcal{R}(1/W_n) = 1/W_1$, further $\|\mathcal{R}\| = \omega_n^{1/2}$. By definition of adjoint operator, the proof of (2.3) is easy, here we omit.

**Lemma 4** *The operators $\mathcal{R}_\omega$, and $H, K$ as above, let*

$$\Phi_m(s) = \exp(-\mu s)(1/W_1) \cdot H_m(s),$$

*for $m = 0, 1, \ldots$ . Here $H_m$ denotes the mth Hermite polynomial, then $\{\Phi_m\}$ form a complete orthogonal basis for $L^2(R, W_1)$ and for any pair of unit vectors $\xi, \omega \in S^{n-1}$*

$$\mathcal{R}_\xi \mathcal{R}_\omega^* \Phi = \langle \xi, \omega \rangle^m \Phi_m(s). \tag{2.4}$$

**Proof** The completeness and orthogonality of the $\{\Phi_m\}$ follows immediately from the definition of the weight $W_1$ and corresponding properties of the $H_m$.

By rotating coordinates we may assume the case of $\xi = (1, 0, \cdots, 0)$ as well as $\omega = (\cos\theta, \sin\theta, 0, \cdots, 0)$, using(2.1) and (2.2),

$$(\mathcal{R}_\omega^* \Phi_m)(x) = \frac{B_\omega(W_1 \Phi_m)(x)}{W_n} = \frac{H_m(\langle x, \omega \rangle)}{W_n}.$$

so

$$\mathcal{R}_\xi(\mathcal{R}_\omega^* \Phi_m) = \mathcal{R}(\mathcal{R}_\omega^* \Phi_m)(\xi, s)$$

$$= \frac{e^{-s^2 + \mu s}}{\pi} \int_\infty^{+\infty} H_m(s\cos\theta + t\sin\theta)e^{-t^2} dt$$

$$= \langle \xi, \omega \rangle^m \Phi_m(s)$$

where by use of the equality in[6]

$$\int_{-\infty}^{+\infty} H_m(s\cos\theta + t\sin\theta)e^{-t^2} dt = \sqrt{\pi}\cos^m\theta \cdot H_m(s).$$

**Lemma 5** *For $m = 0, 1, \ldots$, let $V_m$ be the subspace of $K$ consisting of functions of the form $G(\omega)\Phi_m(s)$ where $G \in L^2(S^{n-1})$ Then*
   *(1) The $\{V_m\}$ are orthogonal and span $K$,*
   *(2) The $\{V_m\}$ are invariant under $\mathcal{R}\mathcal{R}^*$ and*
   *(3) $(\mathcal{R}\mathcal{R}^* G\Phi_m)(\omega, s) = \Phi_m(s) \int_{S^{n-1}} \langle \omega, \xi \rangle^m G(\xi)d\Omega(\xi).$* (2.5)

**Proof** (1) The orthogonality of the $\{V_m\}$ follows from the orthogonality of the functions $\Phi_m$. Similarly the completeness of the $\Phi_m$ ensures that the $V_m$ span $K$.

We prove (2) and (3) together. Let $G(\omega)\Phi_m(s) \in V_m$, then

$$\mathcal{R}\mathcal{R}^* (G\Phi_m)(\omega, s) = \mathcal{R}_\omega[\mathcal{R}^*(G\Phi_m)](s)$$

$$= \Phi_m(s) \int_{S^{n-1}} G(\xi) \cdot \langle \omega, \xi \rangle^m d\Omega(\xi),$$

where the last two equalities follow from (2.3) and (2.4), respectively.

From Lemma 1-5 the eigenproblem for $\mathcal{R}\mathcal{R}^*$ is thus reduced to solving the eigenproblem for each of the integral operators which take $G \in L^2(S^{n-1})$ into $\int_{S^{n-1}} \langle \omega, \xi \rangle^m G(\xi) d\Omega(\xi)$. It is apparent that, the operators are invariant under arbitrary rotations, which implies that an eigenspace of the operator must be a direct sum of subspaces of spherical harmonics. Applying the Funk-Hecke theorem allows us to find the eigenvalues explicitly, which yields the following theorem.

**Theorem 1** *Let $SH^l$ denote the subspace of spherical harmonics of order $l$. Then the eigenfunctions of $\mathcal{R}\mathcal{R}^*$ are all of the form $Y_l(\omega)\Phi_m(s)$, where $Y_l \in SH^l$, with* *corresponding eigenvalues $\lambda_{lm}$ given by*

$$\lambda_{lm} = \frac{\omega_{n-1}}{C_l^\nu(1)} \int\limits_{-1}^{+1} t^m C_l^\nu(t) \left(1 - t^2\right)^{\nu-1/2} dt, \qquad (2.6)$$

*where $\nu = n/2 - 1$, and $l = 0, 1, 2, \cdots$.*
**Proof** By Funk-Hecke theorem[6] and (2.5)

$$\mathcal{R}\mathcal{R}^* \left(Y_l\Phi_m\right)(\omega, s) = \Phi_m(s)Y_l(\omega)\frac{\omega_{n-1}}{C_l^\nu(1)} \int\limits_{-1}^{+1} t^m C_l^\nu(t) \left(1 - t^2\right)^{\nu-1/2} dt$$

$$= \lambda_{lm} Y_l(\omega)\Phi_m(s).$$

The theorem is proved.

By[6] and simple calculation, we obtain

$$\int\limits_{-1}^{+1} s^m C_l^\nu(s) \left(1 - s^2\right)^{\nu-1/2} ds = \frac{(2\nu)_l (2\rho + 1)_l \Gamma(\gamma + 1/2)\Gamma(\rho + 1/2)}{2^l l! \Gamma(l + \nu + \rho + 1)} \qquad (2.7)$$

where $(a)_l = \Gamma(a + l)/\Gamma(a)$, if $m = l + 2\rho$, with $\rho, a$ non negative integer, otherwise the integral, and $\lambda_{lm}$, is zero.

To complete the construction of the singular value decomposition of $\mathcal{R}$, we must calculate $\mathcal{R}^*$ of the eigenfunctions of $\mathcal{R}\mathcal{R}^*$. The next theorem gives the result.

**Theorem 2** *Let $Y_l(\omega)\Phi_m(s)$ be an eigenfunction of $\mathcal{R}\mathcal{R}^*$. Then for $\xi \in S^{n-1}$ and $r\xi = x \in R^n$,*

$$\mathcal{R}^* \left(Y_l\Phi_m\right)(r\xi) = Y_l(\xi)\frac{\omega_{n-1}}{C_l^\nu(1)} \cdot \frac{P_{lm}(r)}{W_n(r)},$$

*where $P_{lm}(r) = r^l Q_{lm}(r^2), Q_{lm}(r)$ is a polynomial of degree $(m - l)/2$ given by*

$$Q_{lm}(r^2) = (-1)^{(m-l)/2} \left(\tfrac{m-l}{2}\right)! 2^m \int\limits_{-1}^{1} t^m C_l^\nu(t) \left(1 - t^2\right)^{\nu-1/2} dt$$

$$\times L_{(m-l)/2}^{(l+\nu)} \left(r^2\right).$$

*Here $L_k^{(\alpha)}$ denotes the kth generalized Laguerre polynomial[6].*
**Proof** By Lemma 3

$$\mathcal{R}^* \left(Y_l\Phi_m\right)(r\xi) = \int\limits_{S^{n-1}} \left(\mathcal{R}_\omega^* \left(Y_l\Phi_m\right)_\omega\right) d\Omega(\omega)$$

$$= \int\limits_{S^{n-1}} \frac{B_\omega \left(W_1 \Phi_m Y_l(\omega)\right)(r\xi)}{W_n(r\xi)} d\Omega(\omega)$$

$$= \int\limits_{S^{n-1}} \frac{Y_l(\omega) H_m \left(r \langle \omega, \xi \rangle\right)}{W_n(r)} d\Omega(\omega).$$

Applying the Funk-Hecke theorem the integral becomes

$$\frac{\omega_{n-1}Y_l(\xi)}{C_l^\nu(1)W_n(r)} \int\limits_{-1}^{+1} H_m(rt)C_l^\nu(t)\left(1-t^2\right)^{\nu-1/2} dt,$$

where $\nu = n/2 - 1$. Here we are reduced to calculating

$$P_{lm}(r) = \int\limits_{-1}^{+1} H_m(rt)C_l^\nu(t)\left(1-t^2\right)^{\nu-1/2} dt.$$

If we write

$$H_m(rt) = \sum_{\substack{j=0 \\ m-j \, even}}^{m} a_{mj}r^j t^j$$

and

$$t^j = \sum_{\substack{k=0 \\ j-k \, even}}^{j} c_{jk}C_k^\nu(t)$$

and apply the orthogonality of the $C_k^\nu$, we see that

$$\int\limits_{-1}^{+1} H_m(rt)C_l^\nu(t)\left(1-t^2\right)^{\nu-1/2} dt$$

$$= r^l \left[\int\limits_{-1}^{+1} (C_l^\nu(t))^2 \left(1-t^2\right)^{\nu-1/2} dt\right] \tag{2.8}$$

$$\times \sum_{\substack{j=l \\ m-j \, even}}^{m} a_{mj}c_{jl}r^{j-l}$$

where $(m-l)/2$ is a non-negative integer, and that the integral is zero otherwise. One can verify that $a_{mm}c_{ml}$, the coefficient of the largest power of $r$ in (2.8), is nonzero, hence $P_{ml}(r)$ has the form $r^l Q_{lm}(r^2)$ where $Q_{lm}$ is a polynomial of degree $(m-l)/2$ in $r^2$.

For a fixed $l$, as $m$ ranges over $l, l+2, l+4, \ldots$, the polynomials $Q_{lm}(u)$ form a triangular family, and are determined up to a constant by their orthogonality relationship, which we now determine.

For $m \neq m'$ we have that

$$\int\limits_{R^n} \mathcal{R}^*(Y_l\Phi_m)\mathcal{R}^*(Y_l\Phi_{m'})W_n(x)dx = 0,$$

since the $Y_l\Phi_m$ ($m = l, l+2, \cdots$) are orthogonal eigenfunctions of $\mathcal{R}\mathcal{R}^*$. Rewrite this relationship in polar coordinates using the form that we have established for

$\mathcal{R}^*(Y_l\Phi_m)$, then it yields that

$$\int_0^{+\infty} Q_{lm}(u)Q_{lm'}(u)e^{-u}u^{l+n/2-1}du = 0.$$

From the uniqueness of orthogonal triangular families and the definition of $W_n$ we conclude that $Q_{lm}(u)$ is proportional to $L_{(m-l)/2}^{(l+\nu)}$ and the coefficient of the highest power of $u$ is

$$(-1)^{\frac{m-l}{2}}(\frac{m-l}{2})!2^m \int_{-1}^{+1} t^m C_l^\nu(t)\left(1-t^2\right)^{\nu-1/2} dt.$$

and the theorem is proved.

**Theorem 3** *Let $H, K$ be defined as Introduction, then $\mathcal{R} : H \longrightarrow K$ is injective.*
**Proof** For $m = l, l+2, l+4, \ldots$, set

$$f_{lmk}(\omega, s) = Y_{lk}(\omega)\Phi_m(s), \omega \in S^{n-1}, s \in R$$

and

$$F_{lmk}(x) = \mathcal{R}^*(f_{lmk})(x), x \in R^n.$$

Theorem 2 shows how to write $F_{lmk}$ explicitly. Since $f_{lmk}$ is an eigenfunction of $\mathcal{R}\mathcal{R}^*$, we have

$$\mathcal{R}F_{lmk} = \lambda_{lm}f_{lmk},$$

where $\lambda_{lm}$ is given by (2.6).

It follows that the functions $(1/W_n(r))r^l Q_{lm}(r^2)$ are complete in the $H$. Since given $f \in L^2(R^n, W_n)$ we may write

$$f(r\omega) = \sum_{l=0}^{+\infty}\sum_{k=1}^{N(l)} b_{lk}Y_{lk}(\omega)g_{lk}(r)$$

where $g_{lk} \in L^2((0, +\infty), r^{n-1}W_n(r))$.

We can then write

$$g_{lk}(r) = \sum_{\substack{m=l \\ m-l\, even}}^{+\infty} c_{lmk}\frac{r^l Q_{ml}(r^2)}{W_n(r)},$$

so

$$f(r\omega) = \sum_{l=0}^{+\infty}\sum_{k=0}^{N(l)}\sum_{\substack{m=l \\ m-l\, even}}^{+\infty} b_{lk}c_{lmk}F_{lmk}(r\omega).$$

We thus have proved that $\mathcal{R} : H \to K$ is injective, since the $F_{lmk}$ span the perpendicular complement of the kernel of $\mathcal{R}$.

Note that the functions $f_{lmk}$ and $F_{lmk}$ must be normalized for us to obtain the singular value decomposition outlined in the Introduction.

258

## Acknowledgments

The author is very grateful to the support of K.C. Wang Education Foundation, Hong Kong, and Prof. Jinyuan Du for his assistance and very instructive discussion.

## References

1. Davison M E. A singular value decomposition for the Radon transform in $n$-dimensional Euclidean space. *Numer. Funct. Anal. Optimiz.* 1981, 3: 321-340.
2. Hertle A. On the injectivity of the attenuated Radon transform. *Amer. Math. Soc.* 1984, 92: 201-205.
3. Louis A K. Medical imaging: state of the art and future development. *Inverse Problems.* 1992, 8: 709-738.
4. Maass P. The interior Radon transform. *SIAM J. Appl. Math,* 1992, 52(3):710-724.
5. Katsevich A. New range theorems for the dual Radon transform. *Trans. Amer. Math. Soc.* 2000, 353: 1089-1102.
6. Abramowitz M, Stegum I A eds, *Handbook of mathematical Functions with formulas, graphs and mathamatical tables.* Dover,1965.

# EYE DIRECTION BY STEREO IMAGE PROCESSING

KUMIKO TSUJI

*Fukuoka Medical technology, Teikyo University,*
*4-3-124 Shin-Katsutachi-Machi,*
*Ohmuta, 836 ,Japan*
*E-mail c74206g@wisdom.cc.kyushu-u.ac.jp*

MIKI AOYAGI

*Department of Mathematics, Sophia University,*
*7-1 Kioi-cho,Chiyoda-ku, Tokyo, Japan, 102-8554,*
*E-mail miki-a@sophia.ac.jp*

The measurement system of a movement of an eye direction is important in ophthalmology. A human inspector usually examines the movement. The purpose of this paper is to analyze two eyes of a human inspector and to show the possibility of inspection by an image processing in stereo cameras with computer programming. In order to have the same functions performed by human eyes, two cameras are used, whose distance and direction to the center of a pupil circle are the same ones of human eyes.

**Key words:** Eye direction, stereo image processing, pattern recognition
**Mathematics Subject Classification:** 68T10, 92C50

## 1. Introduction

Types of disability in the vision are now detected by classifying manners of nystagmus in ophthalmology [5]. A relationship between smooth pursuit of an eye movement and visual perception has been studied [4], by observing disabled children. In the paper [4], they used an electrooculogram and a video at the same time. The electrooculogram measures electric potential differences between a nose and an ear, by applying voltage to the nose and the ear. The video records an eye movement. By this method, only a movement is known but a detailed eye direction is not yet measured. In addition, there are more information when an inspector sees children eyes by his own eyes. So, it is important to develop a machine to determine an eye direction, which plays a role of inspector's eyes. The purpose of this paper is to understand the functions of inspection with two eyes and to show the possibility of inspection by an image processing in stereo cameras with computer programming. To understand the role of inspection with two eyes, movements of stereo cameras are controlled by angle sensors as follows: (1) the distance between two cameras is 60mm as the average distance between human eyes; (2) an attitude of two cameras is controlled to direct to the center of a pupil circle in the initial procedure. An eye direction algorithm for an eye tracking has been developed using many frames of observed eye sphere images by CCD cameras [3]. The camera projection has been considered as affine. A model of an image pupil has been constructed by assuming that a contour image is an ellipse by an affine projection [3]. However these affine projections do not give sufficient accuracy in the case when the camera and the eye sphere are near each other. In this paper, we introduce the method to calculate an eye direction, using a perspective projection. The problem in general is that the center of a pupil does not map to the center of an image ellipse in the perspective projection. In order to settle this problem, we propose an algorithm to control attitudes of cameras. If the attitudes of the two cameras are controlled to direct toward the center of the pupil, then the eye direction is obtained, since the image of the center of the pupil attains the origin in each camera coordinate. Furthermore we propose the procedure, called an 'epipolar radius procedure', by which the attitudes of the two cameras are correctly obtained, even when the two cameras do not direct correctly to the center of the pupil.

## 2. Coordinate systems for stereo cameras and object eye

In this section, the eye sphere, the pupil circle, and the locations and the angles of the cameras are defined, for coordinate systems of stereo cameras and an object eye.

### 2.1. World coordinate system

An object point in 'a world right hand coordinate system' is denoted as $\mathbf{X} = \begin{pmatrix} X \\ Y \\ Z \end{pmatrix}$.

Its origin lies at the center of an eye sphere. $X$ axis is the horizontal axis, $Y$ axis is the vertical axis and $Z$ axis directs to the midpoint of the two camera locations.

### 2.2. Eye sphere

Let $r_0$ be the radius of an eye sphere and $a_0$ be the radius of a pupil circle. The average values in human eyes are $r_0 = 12$mm and $a_0 = 4$mm.

A half sphere point $\mathbf{q}(\beta, \gamma, r_0)$ in $Z \geq 0$ is expressed as

$$\mathbf{q}(\beta, \gamma, r_0) = r_0 \mathbf{u}_3 = r_0 \, {}^t(\cos\gamma \sin\beta, \sin\gamma \sin\beta, \cos\beta),$$

where $r_0$ is the radius of the sphere. $\mathbf{u}_3$ is the vector given by rotating the point ${}^t(0, 0, r_0)$ on $Z$-axis around $Y$ axis by the angle $\beta$ in $0 \leq \beta < \frac{\pi}{2}$ and next rotating around $Z$ axis by the angle $\gamma$ in $0 \leq \gamma < 2\pi$. That is, $\mathbf{u}_3$ is the third column vector in the rotation matrix $T$, which is defined as

$$T = \begin{pmatrix} \cos\gamma\cos\beta & -\sin\gamma & \cos\gamma\sin\beta \\ \sin\gamma\cos\beta & \cos\gamma & \sin\gamma\sin\beta \\ -\sin\beta & 0 & \cos\beta \end{pmatrix} = (\mathbf{u}_1, \mathbf{u}_2, \mathbf{u}_3).$$

An eye direction $(\theta, \phi)$ is defined as $\theta = \arctan(\cos\gamma\tan\beta)$ and $\phi = \arctan(\sin\gamma\tan\beta)$ in [3]. In this paper, the eye direction is defined by $\mathbf{u}_3$, since $\mathbf{u}_3$ is obtained if and only if $\theta$ and $\phi$ are obtained:

$$\tan\gamma = \frac{\tan\phi}{\tan\theta}, \qquad \tan^2\beta = \tan^2\theta + \tan^2\phi.$$

### 2.3. Pupil circle

The pupil circle with the center $\mathbf{p} = \sqrt{r_0^2 - a_0^2}\,\mathbf{u}_3$ is defined as

$$a_0\cos\psi\mathbf{u}_1 + a_0\sin\psi\mathbf{u}_2 + \sqrt{r_0^2 - a_0^2}\,\mathbf{u}_3,$$

for an arbitrary $\psi$. Note that the pupil circle lies on the eye sphere and the center of the pupil disc lies on the same direction with $\mathbf{q}$. Its distance $\sqrt{r_0^2 - a_0^2}$ is shorter than $r_0$.

### 2.4. Rotation matrices for camera angle sensors

In this section, the rotation matrices between the two camera coordinates and the world coordinate are defined, as cameras are controlled to direct to the center of the pupil. Let $\mathbf{O}_r = \begin{pmatrix} d \\ 0 \\ O_Z \end{pmatrix}$ and $\mathbf{O}_l = \begin{pmatrix} -d \\ 0 \\ O_Z \end{pmatrix}$ be the locations of the right and left cameras, respectively, where $O_Z$ is the distance from the center of the sphere to the midpoint of the segment with the two camera locations, and $2d$ is the distance of two cameras.

Let the right and left coordinates be attached to the lenses of the cameras as $\mathbf{X}_r = \begin{pmatrix} X_r \\ Y_r \\ Z_r \end{pmatrix}$ and $\mathbf{X}_l = \begin{pmatrix} X_l \\ Y_l \\ Z_l \end{pmatrix}$ with the origins at the centers of the right and left cameras. Here $Z_r$ and $Z_l$ direct to the center of the pupil. $Y_r$ and $Y_l$ are axes perpendicular to the space constructed from the vectors $\mathbf{p} - \mathbf{O}_r$ and $\mathbf{p} - \mathbf{O}_l$. That is, $Y = Y_r = Y_l$. $X_r$ is an axis perpendicular to $(Z_r, Y_r)$ plane and $X_l$ an axis perpendicular to $(Z_l, Y_l)$ plane.

Then an object point $\mathbf{X}$ satisfies

$$\mathbf{X} = R_r\mathbf{X}_r + \mathbf{O}_r, \qquad \mathbf{X} = R_l\mathbf{X}_l + \mathbf{O}_l.$$

Here $R_r$ and $R_l$ are rotation matrices such that $R_r = [\mathbf{r}_1\mathbf{r}_2\mathbf{r}_3]$ and $R_l = [\mathbf{l}_1\mathbf{l}_2\mathbf{l}_3]$, where

$$\mathbf{r}_1 = \mathbf{r}_2 \times \mathbf{r}_3 = \frac{\mathbf{l}_3 \times \mathbf{r}_3}{\|\mathbf{l}_3 \times \mathbf{r}_3\|} \times \mathbf{r}_3, \quad \mathbf{r}_2 = \mathbf{l}_2 = \frac{\mathbf{l}_3 \times \mathbf{r}_3}{\|\mathbf{l}_3 \times \mathbf{r}_3\|}, \quad \mathbf{r}_3 = \frac{\mathbf{p} - \mathbf{O}_r}{\|\mathbf{p} - \mathbf{O}_r\|},$$

$$\mathbf{l}_1 = \frac{\mathbf{l}_3 \times \mathbf{r}_3}{\|\mathbf{l}_3 \times \mathbf{r}_3\|} \times \mathbf{l}_3, \quad \mathbf{l}_3 = \frac{\mathbf{p} - \mathbf{O}_l}{\|\mathbf{p} - \mathbf{O}_l\|}.$$

The rotation matrices are determined by information of the camera angle sensors. Then the attitudes of the camera angles are controlled by the rotation matrices.

### 3. Calculation of an object point X and relation between right and left image points

'A perspective projection' is defined by the two equations

$$\mathbf{x}_r = \frac{1}{Z_r}\begin{pmatrix} X_r \\ Y_r \end{pmatrix}, \qquad \mathbf{x}_l = \frac{1}{Z_l}\begin{pmatrix} X_l \\ Y_l \end{pmatrix},$$

where $\mathbf{x}_r$ and $\mathbf{x}_l$ are the right and left image points, respectively.

The above equations with an unknown $\mathbf{X}$ are rewritten as follows. The first equation is the vector formula of the projection for the right image:

$$Z_r \mathbf{x}_r = {}^t\tilde{R}_r(\mathbf{X} - \mathbf{O}_r), \tag{1}$$

where $Z_r = {}^t\mathbf{r}_3(\mathbf{X} - \mathbf{O}_r)$, ${}^t\tilde{R}_r$ is the first and second rows in ${}^tR_r$ and ${}^t\mathbf{r}_3$ is the third row in ${}^tR_r$, i.e., $\begin{pmatrix} {}^t\tilde{R}_r \\ {}^t\mathbf{r}_3 \end{pmatrix} = {}^t R_r$. The second equation is the projection for the left image:

$$Z_l x_l = {}^t\mathbf{l}_1(\mathbf{X} - \mathbf{O}_l), \tag{2}$$
$$Z_l y_l = {}^t\mathbf{l}_2(\mathbf{X} - \mathbf{O}_l), \tag{3}$$

where $Z_l = {}^t\mathbf{l}_3(\mathbf{X} - \mathbf{O}_l)$. Here ${}^t\mathbf{l}_1$, ${}^t\mathbf{l}_2$ and ${}^t\mathbf{l}_3$ be three row vectors in ${}^tR_l$, i.e., $\begin{pmatrix} {}^t\mathbf{l}_1 \\ {}^t\mathbf{l}_2 \\ {}^t\mathbf{l}_3 \end{pmatrix} = {}^tR_l$. Then we have the following theorem, which is about (a) the formula determining the object point $\mathbf{X}$ from the image pairs and (b) the Longuet-Higgin's relation.

**Theorem 3.1.**

*Denote the determinant of a matrix $M$ by $\mid M \mid$ and the cofactor matrix by $M^c$.*

(a) The object point $\mathbf{X}$ is determined by $\mathbf{X} = B^{-1}\mathbf{d}$ where $B = \begin{pmatrix} \mathbf{x}_r {}^t\mathbf{r}_3 - {}^t\tilde{R}_r \\ {}^t\mathbf{l}_1 - {}^t\mathbf{l}_3 x_l \end{pmatrix}$ and $\mathbf{d} = \begin{pmatrix} -{}^t\tilde{R}_r\mathbf{O}_r + {}^t\mathbf{r}_3\mathbf{O}_r\mathbf{x}_r \\ ({}^t\mathbf{l}_1 - {}^t\mathbf{l}_3 x_l)\mathbf{O}_l \end{pmatrix}$, by using (1) and (2).

(b) The Longuet-Higgin's relation between the image pairs $\mathbf{x}_l$ and $\mathbf{x}_r$, is obtained from (3):

$$y_l = \frac{a_0(\mathbf{x}_r)x_l + a_1(\mathbf{x}_r)}{a_2(\mathbf{x}_r)x_l + a_3(\mathbf{x}_r)}.$$

Coefficients $a_i(\mathbf{x}_r), (i = 0, 1, 2, 3)$ are

$$a_0(\mathbf{x}_r) = {}^t\mathbf{l}_2(\mathbf{b}_2 - \mid B_2 \mid \mathbf{O}_l),$$
$$a_1(\mathbf{x}_r) = {}^t\mathbf{l}_2(\mathbf{b}_1 - \mid B_1 \mid \mathbf{O}_l),$$
$$a_2(\mathbf{x}_r) = {}^t\mathbf{l}_3(\mathbf{b}_2 - \mid B_2 \mid \mathbf{O}_l),$$
$$a_3(\mathbf{x}_r) = {}^t\mathbf{l}_3(\mathbf{b}_1 - \mid B_1 \mid \mathbf{O}_l),$$

*where*

$$b_1 = B_1^c(d_1 + d_2), \qquad b_2 = B_1^c d_3 + B_2^c d_1,$$

$$B_1 = \begin{pmatrix} x_r{}^t r_3 - {}^t \tilde{R}_r \\ {}^t l_1 \end{pmatrix}, \qquad B_2 = \begin{pmatrix} x_r{}^t r_3 - {}^t \tilde{R}_r \\ -{}^t l_3 \end{pmatrix},$$

$$d_1 = \begin{pmatrix} -{}^t \tilde{R}_r O_r + {}^t r_3 O_r x_r \\ 0 \end{pmatrix}, \qquad d_2 = \begin{pmatrix} 0 \\ {}^t l_1 O_l \end{pmatrix}, \qquad d_3 = -\begin{pmatrix} 0 \\ {}^t l_3 O_l \end{pmatrix}.$$

**Proof.** (a) Substituting $Z_r = {}^t r_3(X - O_r)$ into (1), we have ${}^t r_3(X - O_r)x_r = {}^t \tilde{R}_r(X - O_r)$. Again by substituting $Z_l = {}^t l_3(X - O_l)$ into (2), we obtain ${}^t l_3(X - O_l)x_l = {}^t l_1(X - O_l)$.

(b) From Eq.(3), we have ${}^t l_3(X - O_l)y_l = {}^t l_2(X - O_l)$. Therefore, we have the relation $({}^t l_2 - y_l{}^t l_3)B^{-1}d = -{}^t l_3 O_l y_l + {}^t l_2 O_l$. That is,

$$({}^t l_2 - y_l{}^t l_3)B^c d + \mid B \mid ({}^t l_3 O_l y_l - {}^t l_2 O_l) = 0.$$

Since $d = d_1 + d_2 + d_3 x_l$, we have $B^c d = b_1 + b_2 x_l$.

The fact that $\mid B \mid = \mid B_1 \mid + \mid B_2 \mid x_l$ and $B^c = B_1^c + B_2^c \begin{pmatrix} 1 & 0 & 0 \\ 0 & 1 & 0 \\ 0 & 0 & 0 \end{pmatrix} x_l$, completes

the proof.

q.e.d.

**Remark** In [1], Theorem 3.1 is shown in case that one of cameras is translated, that is, ${}^t O_r = (0, 0, 0)$, and $R_r = R_l = E$, where $E$ is the unit matrix and $O_l$ is any vector. So Theorem 3.1 is an extended result of [1] since we assume any camera direction and any camera location.

## 4. Initial procedure

The images of the pupil centers are given by $x_r = 0$ and $x_l = 0$. By putting $x_r = 0$ and $x_l = 0$ in Theorem 3.1 (a), the eye direction $u_3$ is obtained by the formula depending on camera angles: $u_3 = \dfrac{B_0^{-1} d_0}{\| B_0^{-1} d_0 \|}$, where $B_0 = \begin{pmatrix} -{}^t \tilde{R}_r \\ {}^t l_1 \end{pmatrix}$, $d_0 = \begin{pmatrix} -{}^t \tilde{R}_r O_r \\ {}^t l_1 O_l \end{pmatrix}$. This is proved from the fact that we have

$$ {}^t \tilde{R}_r(p - O_r) = 0, \qquad {}^t l_1(p - O_l) = 0,$$

if $x_r = 0$ and $x_l = 0$, and that $p = \sqrt{r_0^2 - a_0^2}\, u_3$ is the unique solution of the above equations.

The initial procedure is to calculate the eye direction $u_3$ by $\dfrac{B_0^{-1} d_0}{\| B_0^{-1} d_0 \|}$.

If the camera directs to the center of the pupil exactly, the correct eye direction is calculated by the initial procedure. However, since the cameras are controlled to direct to the center of the pupil by human hands, not knowing the exact location of the pupil center, the camera only directs in the neighborhood of the center of the pupil. So the outward normal $u_3$ at the wrong point of $p$ is calculated by the initial procedure. In the next procedure, the camera angle sensor is adjusted to direct to the correct $p$, by putting $p = \sqrt{r_0^2 - a_0^2}\, u_3$ and by calculating correct $\beta$ and $\gamma$, using the epipolar radius procedure in the following sections.

## 5. Image of the pupil circle

Substituting $\mathbf{X} = (\mathbf{u}_1, \mathbf{u}_2)\mathbf{v} + \mathbf{p}$ into $\mathbf{X}_r = {}^t R_r(\mathbf{X} - \mathbf{O}_r)$, we have

$$\mathbf{X}_r = \begin{pmatrix} {}^t\mathbf{r}_1(\mathbf{u}_1, \mathbf{u}_2)\mathbf{v} \\ {}^t\mathbf{r}_2(\mathbf{u}_1, \mathbf{u}_2)\mathbf{v} \\ {}^t\mathbf{r}_3(\mathbf{u}_1, \mathbf{u}_2)\mathbf{v} + {}^t\mathbf{r}_3(\mathbf{p} - \mathbf{O}_r) \end{pmatrix} = \begin{pmatrix} A_r\mathbf{v} \\ {}^t\mathbf{a}_r\mathbf{v} + \zeta_r \end{pmatrix}.$$

The transformation from the pupil circle to the right image is expressed as $\mathbf{x}_r = \frac{A_r\mathbf{v}}{{}^t\mathbf{a}_r\mathbf{v}+\zeta_r}$, where $\mathbf{v} = a_0\begin{pmatrix}\cos\psi \\ \sin\psi\end{pmatrix}$, $A_r = \begin{pmatrix}{}^t\mathbf{r}_1(\mathbf{u}_1\mathbf{u}_2) \\ {}^t\mathbf{r}_2(\mathbf{u}_1\mathbf{u}_2)\end{pmatrix}$, ${}^t\mathbf{a}_r = {}^t\mathbf{r}_3(\mathbf{u}_1\mathbf{u}_2)$, $\zeta_r = {}^t\mathbf{r}_3(\mathbf{p} - \mathbf{O}_r) = \|\mathbf{p} - \mathbf{O}_r\|$. Similarly, the transformation to the left image is expressed as $\mathbf{x}_l = \frac{A_l\mathbf{v}}{{}^t\mathbf{a}_l\mathbf{v}+\zeta_l}$, where $A_l = \begin{pmatrix}{}^t\mathbf{l}_1(\mathbf{u}_1\mathbf{u}_2) \\ {}^t\mathbf{l}_2(\mathbf{u}_1\mathbf{u}_2)\end{pmatrix}$, ${}^t\mathbf{a}_l = {}^t\mathbf{r}_3(\mathbf{u}_1\mathbf{u}_2)$, $\zeta_l = \|\mathbf{p} - \mathbf{O}_l\|$.

### 5.1. Center of image ellipse

The center in the image ellipse of the pupil circle is obtained in the following theorem for the perspective projection. By the perspective projection, the image of the center $\sqrt{r_0^2 - a_0^2}\mathbf{u}_3$ is not the center of the image ellipse as shown in the following theorem.

**Theorem 5.1.** Let $\Phi = \frac{\zeta_r^2}{a_0^2}{}^t A_r^{-1} A_r^{-1} - {}^t A_r^{-1}\mathbf{a}_r\, {}^t\mathbf{a}_r A_r^{-1}$. Assume that $\mid \Phi \mid > 0$ and $\phi_{11}(1+c) > 0$, where $\phi_{11}$ is the $(1,1)$ element in $\Phi$ and $c = \frac{a_0^2}{\zeta^2 - a_0^2}$. Then the image of the circle is the ellipse:

$$ {}^t(\mathbf{x} - \Psi)\Phi(\mathbf{x} - \Psi) = 1 + c. $$

The ellipse center is $\Psi = \frac{-A_r\mathbf{a}_r}{(\zeta_r/a_0)^2 - 1}$. We have $\Psi \neq \mathbf{0}$, except that both $\begin{pmatrix}{}^t\mathbf{u}_1 \\ {}^t\mathbf{u}_2\end{pmatrix}\mathbf{r}_1$ and $\begin{pmatrix}{}^t\mathbf{u}_1 \\ {}^t\mathbf{u}_2\end{pmatrix}\mathbf{r}_2$ are perpendicular to $\begin{pmatrix}{}^t\mathbf{u}_1 \\ {}^t\mathbf{u}_2\end{pmatrix}\mathbf{r}_3$.

**Proof.** The inverse mapping of $\frac{A_r\mathbf{v}}{{}^t\mathbf{a}_r\mathbf{v}+\zeta_r}$ is $\mathbf{v} = \frac{\zeta_r A_r^{-1}\mathbf{x}}{1 - {}^t\mathbf{a}_r A_r^{-1}\mathbf{x}}$, since $\mathbf{x}({}^t\mathbf{a}_r\mathbf{v} + \zeta_r) - A_r\mathbf{v} = 0$ and $(\mathbf{x}^t\mathbf{a}_r A_r^{-1} - E)A_r\mathbf{v} = -\zeta_r\mathbf{x}$. Put ${}^t\mathbf{a}_A = {}^t\mathbf{a}_r A_r^{-1}$. Then

$$ (\mathbf{x}\mathbf{a}_A - E)^{-1}\mathbf{x} = -\mathbf{x}/(1 - {}^t\mathbf{a}_A\mathbf{x}), $$

since $\mid \mathbf{x}^t\mathbf{a}_A - E \mid = 1 - {}^t\mathbf{a}_A\mathbf{x}$ and $(\mathbf{x}^t\mathbf{a}_A - E)^c\mathbf{x} = -\mathbf{x}$. Here $E$ is the $2 \times 2$ unit matrix.

q.e.d.

## 6. Epipolar plane and epipolar radius

Epipolar geometry is defined as follows in [2].

**Definition 6.1.** [2] A plane generated from the three points which consist of the two camera locations and the object point $\mathbf{X}$, is called an 'epipolar plane'. An intersection line of the epipolar plane and the image plane by the camera is called an 'epipolar line'. The epipolar lines for several objects intersect at one point. This point is called an 'epipole'. Epipolar lines and epipoles constitute a fan-shaped figure. This paticular geometry such as the fan-shaped figure is called an 'epipolar geometry'.

The epipole in the image by the right (left) camera is the image of the left (right) camera location. This is denoted as $\mathbf{e}_r$ (or $\mathbf{e}_l$) [2].

Applying this definition to the object point $\mathbf{X}$, which is the center of the pupil circle, a special epipolar radius is defined as follows, which lies on the special epipolar line passing through the image of the center.

Consider an epipolar plane, which is constructed from the three points, i.e., two camera locations $\mathbf{O}_r$, $\mathbf{O}_l$ and the center $\mathbf{p}$ of the pupil circle: a point in the epipolar plane is expressed as $\mathbf{Q} = \alpha_1(\mathbf{O}_l - \mathbf{p}) + \alpha_2(\mathbf{O}_r - \mathbf{p}) + \mathbf{p}$, where $\alpha_1$ and $\alpha_2$ take real scalar values.

**Definition 6.2.** 'An epipolar radius' is defined as a segment $[\mathbf{f}_r, \mathbf{p}_r]$ (or $[\mathbf{f}_l, \mathbf{p}_l]$) projected from the specified radius $[\mathbf{f}, \mathbf{p}]$ on the intersection of the epipolar plane and the pupil disc.

Then the epipolar radius lies on the epipolar line which connects the image of the center $\mathbf{p}_r$ (or $\mathbf{p}_l$) of the pupil circle and the epipole $\mathbf{e}_r$ (or $\mathbf{e}_l$).

If a point on the object pupil circl is not on the epipole radius and lies on the epipolar plane, then the image of the pupil circle concentrates into the epipole line and the 'epipolar radius procedure' does not work.

The epipolar $\mathbf{e}_r$ (or $\mathbf{e}_l$) is obtained in the following lemma. The intersection line of the epipolar plane and the plane of the pupil disc is also given in the following lemma.

Proof is performed by *Cramér's* rule and is omitted here.

**Lemma 6.1.**

(1) *The epipoles are determined by* $\mathbf{e}_r = \binom{r_1^1}{r_2^1}/r_3^1$, $\mathbf{e}_l = \binom{l_1^1}{l_2^1}/l_3^1$, *where* $r_i^1$ *is a first element in* $\mathbf{r}_i$ *and* $l_i^1$ *is a first element in* $\mathbf{l}_i$ *for* $i = 1, 2, 3$. *These definitions depend only the camera angles.*

(2) *Let* $\mathbf{Q}$ *be an epipolar plane* $\mathbf{Q} = \alpha(\mathbf{O}_l - \mathbf{p}) + \beta(\mathbf{O}_r - \mathbf{p}) + \mathbf{p}$, *where* $\mathbf{p}$ *is the center of the pupil circle. Let* $v_1\mathbf{u}_1 + v_2\mathbf{u}_2 + \mathbf{p}$ *be the pupil circle on the eye sphere at* $\mathbf{p}$, *where* $\mathbf{p} = \sqrt{r_0^2 - a_0^2}\mathbf{u}_3$. *Here* $\mathbf{u}_1$ *and* $\mathbf{u}_2$ *are tangent vectors at* $r_0\mathbf{u}_3$ *on the sphere,* $v_1 = a_0 \cos\psi$ *and* $v_2 = a_0 \sin\psi$. *Then the specified radius* $[\mathbf{f}, \mathbf{p}]$ *on the intersection of the epipolar plane and the pupil disc is given by the angle* $\psi$ *of* $\mathbf{v} =^t (v_1, v_2)$ *on the pupil circle, i.e.,*

$$\psi = \arctan\left(-\frac{\sin\gamma(\sqrt{r_0^2 - a_0^2} - \cos\beta O_Z)}{\cos\gamma(\sqrt{r_0^2 - a_0^2}\cos\beta - O_Z)}\right).$$

## 7. Three pairs of image points $\mathbf{x}_r$ and $\mathbf{x}_l$

### 7.1. *Independent selection of object points* $\mathbf{X}_i$ *on pupil circle*

In the following theorem, it is shown that the eye direction is obtained from the three points on the pupil circle, if these points on the pupil circle are independent. The three points are calculated from three pairs of image points $\mathbf{x}_r$ and $\mathbf{x}_l$.

**Theorem 7.1.** *The following conditions are assumed.*

(1) *The three pairs of image points* $(\mathbf{x}_r, \mathbf{x}_l)_i$, $(i = 0, 1, 2)$ *are known, where each pair is constituted of the right image point* $\mathbf{x}_r$ *and the left image point* $\mathbf{x}_l$.

(2) *The angles and the locations of the cameras are known.*

(3) *Define a point* $\mathbf{X}_i$ *on the pupil circle by substituting* $(\mathbf{x}_r, \mathbf{x}_l)_i$ *into* $\mathbf{X}_i = B^{-1}\mathbf{d}$. $\mathbf{X}_i$ *are arranged in counterclockwise for* $i = 0, 1, 2$ *and* $\mathbf{X}_i$ *are independent.*

*Then the exterior product is the same direction with the eye direction* $\mathbf{u}_3$: $\mathbf{u}_3 = \frac{\mathbf{X}_2 - \mathbf{X}_1}{\|\mathbf{X}_2 - \mathbf{X}_1\|} \times \frac{\mathbf{X}_0 - \mathbf{X}_1}{\|\mathbf{X}_0 - \mathbf{X}_1\|}$.

**Proof.** Put $\mathbf{v}_i = a_0 \binom{\cos \psi_i}{\sin \psi_i}$ and put $\mathbf{X}_i = (\mathbf{u}_1, \mathbf{u}_2)\mathbf{v}_i + \mathbf{p}$. Since $\mathbf{u}_3 = \mathbf{u}_1 \times \mathbf{u}_2$ and $\mathbf{X}_i$ are arranged in counterclockwise for $i = 0, 1, 2$, it holds that $\psi_i$ increases and that $(\mathbf{X}_2 - \mathbf{X}_1) \times (\mathbf{X}_0 - \mathbf{X}_1) = 4a_0^2 S \mathbf{u}_3$, where $S = \sin \frac{\psi_2 - \psi_1}{2} \sin \frac{\psi_1 - \psi_0}{2} \sin \frac{\psi_2 - \psi_0}{2}$.

q.e.d.

## 7.2. *Optimal selection of* $\mathbf{X}_i$

The pattern matching method to recognize the planer curve with characteristic points has been introduced in [2]. A characteristic point is a bitangent point, a inflection point, or so on. A bitangent point is a point where its tangent line is tangent to the curve at more than two points. That is, if two points on the curve has the same tangent line, then we call each point a bitangent point. A point where its planer curve's curvature is 0, is called a inflection point. If there exist more than five characteristic points on the planer curve, then the projective invariant value of the cross ratio is calculated and the pattern of the planer curve is recognized by the projective invariant value. However there exit no characteristic points on the pupil circle. Therefore the idea of the cross ratio might not be applicable.

In this paper, instead of these characteristic points, we take independent three points $\mathbf{X}^{(i)}, i = 0, 1, 2$ on the object pupil circle. In the proposed algorithm, we call three points $\mathbf{X}^{(i)}$ optimal if their magnitude of the exterior product $\|(\mathbf{X}^{(2)} - \mathbf{X}^{(1)}) \times (\mathbf{X}^{(0)} - \mathbf{X}^{(1)})\|$ takes the maximum value. The maximum value is determined by only the radius of the pupil circle.

The fact that the maximum value is attained means that the calculated center of the pupil is correct. So the cameras direct to the center of the pupil and the eye direction is obtained accurately.

**Corollary 7.1.** *The norm* $\|(\mathbf{X}_2 - \mathbf{X}_1) \times (\mathbf{X}_0 - \mathbf{X}_1)\|$ *takes the maximum value* $\frac{3\sqrt{3}}{2}a_0^2$ *where the three points are corresponding to* $\psi_i = \psi_0 + \frac{2\pi i}{3}$.

**Proof.** Consider the angles such that $\frac{\partial S}{\partial \psi_1} = 0$ and $\frac{\partial S}{\partial \psi_2} = 0$. Then $3(\psi_1 - \psi_0) = 2\pi$ by using $\sin \frac{\psi_2 - 2\psi_1 + \psi_0}{2} = 0$ and $\sin \frac{2\psi_2 - \psi_1 - \psi_0}{2} = 0$.

q.e.d.

## 8. Epipolar radius procedure

In this section, we propose the epipolar radius procedure when the images of the cameras are pixel. Assume that the distance $2d$ of two cameras, the distance $O_z$ between the midpoint of two camera locations and the center of the eye sphere, the radius $r_0$ of the eye sphere and the radius $a_0$ of the pupil circle are known. Let the right and left pixel image points (say 100 discrete points in each ) be given on the right and the left contour image of the pupil circle. Let $\mathbf{y}_r$ denote the variable of the right pixel image points and $\mathbf{y}_l$ the variable of the left pixel image points.

The eye direction $\mathbf{u}_3$ is calculated by the following epipolar radius procedure for the pixel image points after calculating $(\beta,\gamma)$ by the initial procedure.

(1) *Select the optimal points $\mathbf{f}_{ri}$ and $\mathbf{f}_{li}$.* — The angle of the right epipolar radius is given by $\psi_0 = \arctan(\frac{\sin\gamma(\sqrt{r_0^2-a_0^2}-\cos\beta O_z)}{-\cos\gamma(\sqrt{r_0^2-a_0^2}\cos\beta-O_z)})$ for $\beta$ and $\gamma$. Calculate the right and the left image points $\mathbf{f}_r^{(i)} = \frac{A_r\mathbf{v}_i}{{}^t\mathbf{a}_r\mathbf{v}_i+\zeta_r}$ and $\mathbf{f}_l^{(i)} = \frac{A_l\mathbf{v}_i}{{}^t\mathbf{a}_l\mathbf{v}_i+\zeta_l}$, by using the points $\mathbf{v}_i = a_0\binom{\cos(\psi_0+\frac{2\pi i}{3})}{\sin(\psi_0+\frac{2\pi i}{3})}$ for $i = 0,1,2$ .

(2) *Select the practical image points $\mathbf{y}_{ri}$ nearest to the images of the optimal points $\mathbf{f}_r^{(i)}$.* — Let $\mathbf{y}_{ri}$ be the point to minimize $|\,\mathbf{y}_r - \mathbf{f}_r^{(i)}\,|$ in the contour of the image ellipse for $i = 0,1,2$. Here $|\,\mathbf{x}\,|$ denotes the sum of the absolute values of the elements in $\mathbf{x} = \binom{x}{y}$: $|\,\mathbf{x}\,| = |\,x\,| + |\,y\,|$.

(3) *Select the practical left image points $\mathbf{y}_{li}$.* — If the camera directs not to the center $\mathbf{p}$, then the epipolar radius $\mathbf{v}_0$ has the error depending on $(\beta,\gamma)$. The left image points are assigned by the Longuet Higgin's relation, which has no error depending on $(\beta,\gamma)$. This relation for each left image point may give two solutions. In order to select one of these, one point is selected, which is near to each left image point $\mathbf{f}_l^i$. Calculate the point $\mathbf{y}_{(li)}$ to minimize $|\,y_l - \frac{a_0(\mathbf{y}_{ri})x_l+a_1(\mathbf{y}_{ri})}{a_2(\mathbf{y}_{ri})x_l+a_3(\mathbf{y}_{ri})}\,| + \|\mathbf{y}_l - \mathbf{f}_l^{(i)}\|$ for $i = 0,1,2$ for the right image points $\mathbf{y}_{ri}$.

(4) *Three points $\mathbf{X}_i$ on pupil circle.* — The three points $\mathbf{X}_i$ on the object pupil circle are calculated from the three pairs of the image points $(\mathbf{y}_{ri},\mathbf{y}_{li})$ by substituting $(\mathbf{x}_r,\mathbf{x}_l) = (\mathbf{y}_{ri},\mathbf{y}_{li})$ into $\mathbf{X}_i = B^{-1}\mathbf{d}$, where $B = \binom{\mathbf{x}_r\,{}^t\mathbf{r}_3-{}^t\tilde{R}_r}{{}^t l_1-{}^t l_3 x_l}$, $\mathbf{d} = \binom{-{}^t\tilde{R}_r O_r + {}^t\mathbf{r}_3 O_r \mathbf{x}_r}{({}^t l_1-{}^t l_3 x_l)O_l}$.

(5) *Eye direction $\mathbf{u}_3$.* — Calculate the exterior product $\mathbf{X} = (\mathbf{X}_2-\mathbf{X}_1)\times(\mathbf{X}_0-\mathbf{X}_1)$. If $|\,\frac{3\sqrt{3}}{2}a_0^2 - \|\mathbf{X}\|\,| < \epsilon$ then stop the procedure and if $|\,\frac{3\sqrt{3}}{2}a_0^2 - \|\mathbf{X}\|\,| > \epsilon$ then calculate $\mathbf{u}_3 = \frac{\mathbf{X}_2-\mathbf{X}_1}{\|\mathbf{X}_2-\mathbf{X}_1\|} \times \frac{\mathbf{X}_0-\mathbf{X}_1}{\|\mathbf{X}_0-\mathbf{X}_1\|}$. Repeat the initial procedure in the section 4 and the epipolar radius procedure in this section, by putting $\mathbf{p} = \sqrt{r_0^2 - a_0^2}\mathbf{X}$. Here $\epsilon$ is a small value depending on the accuracy of the precision of the computer programming, the pixel accuracy in the camera image and the accuracy of the camera angles measured by the camera angle sensors.

## References

1. T. Moons, L. Van Gool, M.Van Diest and E.J. Pauwels, *In J.L. Mundy, A. Zisserman, and D.A. Forsyth, editors, Applications of Invariance in Computer Vision, Springer-*

268

*Verlag,* **LNCS 825**, 297-316, (1994).

2. J. Sato, *Computervision-Geometry of Vision, corona publishing,* 172 pages (1999) (Japanese).

3. T. Takegami, T. Gotoh and G. Ohyama, *IEICE D-II* **Vol.J84-D-II No. 8** , 1580, (2001) (Japanese).

4. A. Sera and Y. Sengoku, *J. of Japanese association of occupational therapists,* **VOL 21**, 307 (2002) (Japanese).

5. K. Akiyama, T. Gotoh and H. Eguchi, *OPHTHALMOLOGY, kaibashobo co.,* 302pages (2004) (Japanese).

# SOME INTEGRAL EQUATIONS WITH MODIFIED BESSEL FUNCTIONS

## FUNCTIONS

JURI M. RAPPOPORT*

*Department of Mathematical Sciences,*
*Russian Academy of Sciences,*
*Vlasov street, Building 27, Apt.8,*
*Moscow 117335, Russia,*
*E-mail:jmrap@landau.ac.ru*

The properties of the modified KONTOROVITCH-LEBEDEV transforms and their kernels are considered. The solutions of some inhomogeneous integral equations are derived by means of the use of these transforms. The applications for boundary value problems of mathematical physics in the wedge domains are given. The numerical solution is conducted.

**Key words:** Kontorovitch-Lebedev intgral transforms, inversion formulas, MacDonald function, inhomogeneous integral equation

**Mathematics Subject Classification:** 44A15, 33C10, 45E10

## 1. Some properties of the functions $ReK_{\frac{1}{2}+i\beta}(x)$ and $ImK_{\frac{1}{2}+i\beta}(x)$

In this section a new properties of the kernels of modified KONTOROVITCH–LEBEDEV integral transforms are deduced and some of their known properties are collected, which are necessary later on.

It is possible to write the kernels of these transforms in the form $ReK_{\frac{1}{2}+i\beta}(x) = \frac{K_{\frac{1}{2}+i\beta}(x)+K_{\frac{1}{2}-i\beta}(x)}{2}$ and $ImK_{\frac{1}{2}+i\beta}(x) = \frac{K_{\frac{1}{2}+i\beta}(x)-K_{\frac{1}{2}-i\beta}(x)}{2i}$, where $K_\nu(x)$ is the modified BESSEL function of the second kind (also called MACDONALD function).

The functions $ReK_{\frac{1}{2}+i\beta}(x)$ and $ImK_{\frac{1}{2}+i\beta}(x)$ have integral representations [1]

$$ReK_{\frac{1}{2}+i\beta}(x) = \int_0^\infty e^{-x\cosh t} \cosh\frac{t}{2}cos(\beta t)dt, \tag{1}$$

$$ImK_{\frac{1}{2}+i\beta}(x) = \int_0^\infty e^{-x\cosh t} \sinh\frac{t}{2}sin(\beta t)dt. \tag{2}$$

The vector-function $(y_1(x), y_2(x))$ with the components $y_1(x) = ReK_{\frac{1}{2}+i\beta}(x)$, $y_2(x) = ImK_{\frac{1}{2}+i\beta}(x)$ is the solution of the system of differential equation[1,2]

---

*Work was partially supported by the U. S. Civilian Research and Development Foundation, Grant RM1-361 and by the Program of the Fundamental Research "Mathematical Modeling" from the Presidium of the Russian Academy of Sciences, Grant 1035.

$$\frac{d^2 y_1}{dx^2} + \frac{1}{x}\frac{dy_1}{dx} - (1 + \frac{\frac{1}{4} - \beta^2}{x^2})y_1 + \frac{\beta}{x^2}y_2 = 0,$$

$$\frac{d^2 y_2}{dx^2} + \frac{1}{x}\frac{dy_2}{dx} - \frac{\beta}{x^2}y_1 - (1 + \frac{\frac{1}{4} - \beta^2}{x^2})y_2 = 0. \tag{3}$$

The functions $ReK_{\frac{1}{2}+i\beta}(x)$ and $ImK_{\frac{1}{2}+i\beta}(x)$ are even and odd functions, respectively of the variable $\beta$,

$$ReK_{\frac{1}{2}+i\beta}(x) = ReK_{\frac{1}{2}-i\beta}(x),$$

$$ImK_{\frac{1}{2}+i\beta}(x) = -ImK_{\frac{1}{2}-i\beta}(x).$$

It follows from (1)-(2) that it is possible to write $ReK_{\frac{1}{2}+i\beta}(x)$ in the form of the FOURIER cosinus-transform

$$ReK_{\frac{1}{2}+i\beta}(x) = \left(\frac{\pi}{2}\right)^{\frac{1}{2}} F_C[e^{-x\cosh t}\cosh\frac{t}{2}; t \to \beta], \tag{4}$$

and $ImK_{\frac{1}{2}+i\beta}(x)$ in the form of the FOURIER sinus-transform

$$ImK_{\frac{1}{2}+i\beta}(x) = (\frac{\pi}{2})^{\frac{1}{2}} F_S[e^{-x\cosh t}\sinh\frac{t}{2}; t \to \beta]. \tag{5}$$

The inversion formulas have the respective forms

$$F_C[ReK_{\frac{1}{2}+i\beta}(x); \beta \to t] = (\frac{\pi}{2})^{\frac{1}{2}} e^{-x\cosh t}\cosh\frac{t}{2},$$

$$F_S[ImK_{\frac{1}{2}+i\beta}(x); \beta \to t] = (\frac{\pi}{2})^{\frac{1}{2}} e^{-x\cosh t}\sinh\frac{t}{2} \tag{6}$$

or, in integral form,

$$\int_0^\infty ReK_{\frac{1}{2}+i\beta}(x)\cos(t\beta)d\beta = \frac{\pi}{2} e^{-x\cosh t}\cosh\frac{t}{2}, \tag{7}$$

$$\int_0^\infty ImK_{\frac{1}{2}+i\beta}(x)\sin(t\beta)d\beta = \frac{\pi}{2} e^{-x\cosh t}\sinh\frac{t}{2}. \tag{8}$$

Differentiating equations (7) and (8) with respect to $t$, we obtain

$$\int_0^\infty \beta ReK_{\frac{1}{2}+i\beta}(x)\sin(t\beta)d\beta = \frac{\pi}{2}(x\sinh t\cosh\frac{t}{2} - \sinh\frac{t}{2})e^{-x\cosh t},$$

$$\int_0^\infty \beta ImK_{\frac{1}{2}+i\beta}(x)\cos(t\beta)d\beta = \frac{\pi}{2}(\cosh\frac{t}{2} - x\sinh t\sinh\frac{t}{2})e^{-x\cosh t}. \tag{9}$$

It follows from (7) that

$$\int_0^\infty ReK_{\frac{1}{2}+i\beta}(x)d\beta = \frac{\pi}{2}e^{-x},$$

and from (9) that

$$\int_0^\infty \beta ImK_{\frac{1}{2}+i\beta}(x)d\beta = \frac{\pi}{2}e^{-x}.$$

Differentiating (7) and (8) $2n$ times with respect to $t$, we obtain

$$\int_0^\infty \beta^{2n} Re K_{\frac{1}{2}+i\beta}(x)\cos(t\beta)d\beta = \frac{\pi}{2}(-1)^n D_t^{2n}(e^{-x\cosh t}\cosh\frac{t}{2}),$$

$$\int_0^\infty \beta^{2n} Im K_{\frac{1}{2}+i\beta}(x)\sin(t\beta)d\beta = \frac{\pi}{2}(-1)^n D_t^{2n}(e^{-x\cosh t}\sinh\frac{t}{2}),$$

from which there follows, for $t=0$,

$$\int_0^\infty \beta^{2n} Re K_{\frac{1}{2}+i\beta}(x)d\beta = \frac{\pi}{2}(-1)^n D_t^{2n}(e^{-x\cosh t}\cosh\frac{t}{2})_{t=0}.$$

Differentiating (7) and (8) $2n+1$ times with respect to $t$, we obtain

$$\int_0^\infty \beta^{2n+1} Re K_{1/2+i\beta}(x)\sin(t\beta)d\beta = \frac{\pi}{2}(-1)^{n+1} D_t^{2n+1}(e^{-x\cosh t}\cosh\frac{t}{2}),$$

$$\int_0^\infty \beta^{2n+1} Im K_{1/2+i\beta}(x)\cos(t\beta)d\beta = \frac{\pi}{2}(-1)^n D_t^{2n+1}(e^{-x\cosh t}\sinh\frac{t}{2}).$$

whence, for $t=0$,

$$\int_0^\infty \beta^{2n+1} Im K_{1/2+i\beta}(x)d\beta = \frac{\pi}{2}(-1)^n D_t^{2n+1}(e^{-x\cosh t}\sinh\frac{t}{2})_{t=0}.$$

For the computation of certain integrals of the functions $Re K_{\frac{1}{2}+i\beta}(x)$ and $Im K_{\frac{1}{2}+i\beta}(x)$ integral identities are useful which reduce this problem to the computation of some other integrals over elementary functions.

**Proposition 1.1.**

If $f$ is absolutely integrable on $[0,\infty)$, then the following identities hold,

$$\int_0^\infty Re K_{\frac{1}{2}+i\beta}(x)f(\beta)d\beta = (\frac{\pi}{2})^{\frac{1}{2}} \int_0^\infty e^{-x\cosh t}\cosh\frac{t}{2} F_C(t)dt, \tag{10}$$

$$\int_0^\infty Im K_{\frac{1}{2}+i\beta}(x)f(\beta)d\beta = (\frac{\pi}{2})^{\frac{1}{2}} \int_0^\infty e^{-x\cosh t}\sinh\frac{t}{2} F_S(t)dt, \tag{11}$$

where $F_C(t)$ is the FOURIER cosinus-transform of $f(\beta)$, and $F_S(t)$ the FOURIER sinus-transform of $f(\beta)$.

Proof:

Multiplying both sides of the equalities (4) and (5) by $f(\beta)$, integrating with respect to $\beta$ from 0 to $\infty$, and applying FUBINI'S theorem for singular integrals with parameter, we obtain (10) and (11).

**Proposition 1.2.**

If $f$ is absolutely integrable on $[0,\infty)$, then the following identities hold

$$\int_0^\infty Re K_{\frac{1}{2}+i\beta}(x)F_C(\beta)d\beta = (\frac{\pi}{2})^{\frac{1}{2}} \int_0^\infty e^{-x\cosh t}\cosh\frac{t}{2} f(t)dt, \tag{12}$$

$$\int_0^\infty Im K_{\frac{1}{2}+i\beta}(x)F_S(\beta)d\beta = (\frac{\pi}{2})^{\frac{1}{2}} \int_0^\infty e^{-x\cosh t}\sinh\frac{t}{2} f(t)dt. \tag{13}$$

Proof:

This follows from (7)-(8) and from FUBINI'S theorem.

The equations (10)-(13) are useful for the simplification and the calculation of different integrals containing $ReK_{1/2+i\beta}(x)$ and $ImK_{1/2+i\beta}(x)$.

For example, let $f(\beta) = e^{-\alpha\beta}$, then $F_C(t) = \sqrt{\frac{2}{\pi}}\frac{\alpha}{\alpha^2+t^2}$, $F_S(t) = \sqrt{\frac{2}{\pi}}\frac{t}{\alpha^2+t^2}$ and

$$\int_0^\infty ReK_{\frac{1}{2}+i\beta}(x)e^{-\alpha\beta}d\beta = \alpha \int_0^\infty (\alpha^2+t^2)^{-1}e^{-x\cosh t}\cosh\frac{t}{2}dt,$$

$$\int_0^\infty ReK_{\frac{1}{2}+i\beta}(x)\frac{1}{\alpha^2+\beta^2}d\beta = \frac{\pi}{2\alpha}\int_0^\infty e^{-\alpha t - x\cosh t}\cosh\frac{t}{2}dt,$$

$$\int_0^\infty ImK_{\frac{1}{2}+i\beta}(x)e^{-\alpha\beta}d\beta = \int_0^\infty t(\alpha^2+t^2)^{-1}e^{-x\cosh t}\sinh\frac{t}{2}dt,$$

$$\int_0^\infty ImK_{\frac{1}{2}+i\beta}(x)\frac{\beta}{\alpha^2+\beta^2}d\beta = \frac{\pi}{2}\int_0^\infty e^{-\alpha t - x\cosh t}\sinh\frac{t}{2}dt.$$

We use the representation (1) for the evaluation of the LAPLACE transformation of $ReK_{\frac{1}{2}+i\beta}(x)$. We have

$$L[ReK_{\frac{1}{2}+i\beta}(x);\beta] = \int_0^\infty \cos(\beta t)\cosh\frac{t}{2}\int_0^\infty e^{-(p+\cosh t)x}dxdt$$

$$= \int_0^\infty \frac{\cos(\beta t)\cosh\frac{t}{2}}{\cosh t + \cosh\alpha}dt \quad (p = \cosh\alpha)$$

$$= \sqrt{\frac{\pi}{2}}F_C(\frac{\cosh\frac{t}{2}}{\cosh t + \cosh\alpha}) = \frac{\pi}{2}\frac{\cos(\alpha\beta)}{\cosh\frac{\alpha}{2}\cosh(\pi\beta)}.$$

Equivalently, we can write

$$L^{-1}[\frac{\cos(\beta\cosh^{-1}p)}{\sqrt{\frac{p+1}{2}}}] = (\frac{\pi}{2})^{-1}\cosh(\pi\beta)ReK_{\frac{1}{2}+i\beta}(x). \tag{14}$$

For the evaluation of the LAPLACE transform of $ImK_{\frac{1}{2}+i\beta}(x)$ we utilize the representation (2). We have

$$L[ImK_{\frac{1}{2}+i\beta}(x);p] = \sqrt{\frac{\pi}{2}}F_S(\frac{\sinh\frac{t}{2}}{\cosh t + \cosh\alpha}) = \frac{\pi}{2}\frac{\sin(\alpha\beta)}{\cosh(\pi\beta)\sinh\frac{\alpha}{2}},$$

or, equivalently,

$$L^{-1}[\frac{\sin(\beta\cosh^{-1}p)}{\sqrt{\frac{p-1}{2}}}] = \sqrt{\frac{\pi}{2}}\cosh(\pi\beta)ImK_{\frac{1}{2}+i\beta}(x).$$

We note that these equations can also be obtained directly from the formula for the LAPLACE transforms [3] of $K_\nu(x)$ by separating real and imaginary parts.

It follows from (1) that for all $\beta \in [0,\infty)$

$$|ReK_{\frac{1}{2}+i\beta}(x)| \le K_{\frac{1}{2}}(x) = (\frac{\pi}{2x})^{\frac{1}{2}}e^{-x}, \tag{15}$$

and it follows from (2) that for all $\beta \in [0, \infty)$

$$|ImK_{\frac{1}{2}+i\beta}(x)| \leq \int_0^\infty e^{-x\cosh t}\sinh\frac{t}{2}dt = (\frac{\pi}{2x})^{\frac{1}{2}}e^x[1 - \phi((2x)^{\frac{1}{2}})] \leq B\frac{e^{-x}}{x}, \quad (16)$$

where $B$ is some positive constant.

For future use, an analysis of the behavior of the modified BESSEL function $K_{\frac{1}{2}+i\beta}(x)$ and inequalities are necessary [4-6]:

**Lemma 1.1.** *The following inequalities hold for $x > 0$*

$$|ReK_{\frac{1}{2}+i\beta}(x)| \leq c|\beta|e^{-\frac{\pi|\beta|}{2}}x^{-\frac{3}{4}} + (\frac{2\pi}{x})^{\frac{1}{2}}e^{-x}e^{-\pi|\beta|}, \quad (17)$$

$$|ImK_{\frac{1}{2}+i\beta}(x)| \leq c_0|\beta|e^{-\frac{\pi|\beta|}{2}}x^{-\frac{3}{4}}, \quad (18)$$

*where $c_0$ and $c$ are some positive constants.*

## 2. The integral equations and Parseval equalities for the modified Kontorovitch–Lebedev integral transforms.

Integral transforms containing integration with respect to the index of the BESSEL function play an important role for the solution of some classes of the problems in mathematical physics [3]. In particular, for the solution of mixed boundary value problems for the HELMHOLTZ equation in wedge-shaped and conic domains, the modified KONTOROVITCH–LEBEDEV transforms [1,2] are used,

$$F_+(\beta) = \int_0^\infty f(x)\frac{K_{\frac{1}{2}+i\beta}(x) + K_{\frac{1}{2}-i\beta}(x)}{2}dx, \quad 0 \leq \beta < \infty, \quad (19)$$

$$F_-(\beta) = \int_0^\infty f(x)\frac{K_{\frac{1}{2}+i\beta}(x) - K_{\frac{1}{2}-i\beta}(x)}{2i}dx, \quad 0 \leq \beta < \infty. \quad (20)$$

The inversion formulas have the form (21) and (22), respectively,

$$f(x) = \frac{4}{\pi^2}\int_0^\infty \cosh(\pi\beta)F_+(\beta)\frac{K_{\frac{1}{2}+i\beta}(x) + K_{\frac{1}{2}-i\beta}(x)}{2}d\beta, \quad 0 < x < \infty, \quad (21)$$

$$f(x) = \frac{4}{\pi^2}\int_0^\infty \cosh(\pi\beta)F_-(\beta)\frac{K_{\frac{1}{2}+i\beta}(x) - K_{\frac{1}{2}-i\beta}(x)}{2i}d\beta, \quad 0 < x < \infty. \quad (22)$$

The integral transforms (19)-(22) are met in the solution of some classes of dual integral equations with kernels containing the MACDONALD function [2], for example,

$$\int_0^\infty M(\beta)\beta\tanh(\gamma\beta)K_{i\beta}(x)d\beta = 0, \quad x < a,$$

$$\int_0^\infty M(\beta)K_{i\beta}(x)d\beta = f(x), \quad x > a,$$

or

$$\int_0^\infty M(\beta)K_{i\beta}(x) = 0, \quad x < a,$$

$$\int_0^\infty M(\beta)\beta\tanh(\gamma\beta)K_{i\beta}(x)d\beta = xf(x), \quad x > a.$$

The functions $ReK_{\frac{1}{2}+i\beta}(x)$ and $ImK_{\frac{1}{2}+i\beta}(x)$ are solutions of the homogeneous integral equations of the second kind [1]

$$\varphi(x) = \lambda \int_0^\infty K(x,y)\varphi(y)dy, 0 < x < \infty,$$

with kernels $K(x,y) = K_1(x+y) \pm K_0(x+y)$, and respective eigenvalue $\lambda = \frac{\cosh(\pi\beta)}{\pi}$, $0 < \beta < \infty$.

The solutions of corresponding inhomogeneous equations

$$\varphi(x) = f(x) + \lambda \int_0^\infty K(x,y)\varphi(y)dy, \ 0 < x < \infty,$$

for values of the parameter $\lambda < 1/\pi$ are derived by means of the use of the transforms (19)-(22).

Thus, the modified KONTOROVITCH–LEBEDEV integral transforms $F_+(\beta)$ and $F_-(\beta)$ of the function $f(x)$ defined on the positive real semiaxis are defined by the formulas

$$REK[f(x);\beta] = \int_0^\infty f(x)ReK_{\frac{1}{2}+i\beta}(x)dx, \tag{23}$$

$$IMK[f(x);\beta] = \int_0^\infty f(x)ImK_{\frac{1}{2}+i\beta}(x)dx, \tag{24}$$

and are written in the form

$$F_+(\beta) = REK[f(x);\beta], \ F_-(\beta) = IMK[f(x);\beta].$$

It follows from the inequalities (15)-(16) that for all $\beta \in (0,\infty)$

$$|F_+(\beta)| \le \int_0^\infty |f(x)|K_{\frac{1}{2}}(x)dx \tag{25}$$

and

$$|F_-(\beta)| \le B \int_0^\infty |f(x)|\frac{e^{-x}}{x}dx. \tag{26}$$

## Proposition 2.1.

*If the following conditions for function $f(x)$ are valid*
1. $\frac{f(x)}{x} \in L(0,\frac{1}{2})$,
2. $f(x)e^{-x}x^{-\frac{1}{2}} \in L(\frac{1}{2},\infty)$,

*then the modified KONTOROVITCH–LEBEDEV integral transforms $F_+(\beta)$ and $F_-(\beta)$ are well defined, the integrals (23)-(24) converge uniformly in $\beta$ and determine a continuous functions of $\beta$ bounded for $\beta \in [0,\infty)$.*

Proof:

The proposition follows from the inequalities (15)-(16) and from criteria of uniform convergence of integrals with a parameter.

The integrals (25)-(26) are thus well defined and the integrals (23)-(24) converge uniformly in $\beta$ and determine a continuous function of $\beta$.

The inversion formulas for the transforms (23)-(24) have respectively the following form [1]

$$\frac{f(x+0)+f(x-0)}{2} = \frac{4}{\pi^2}\int_0^\infty \cosh(\pi\beta)F_+(\beta)ReK_{\frac{1}{2}+i\beta}(x)d\beta, \ 0 < x < \infty, \quad (27)$$

and

$$\frac{f(x+0)+f(x-0)}{2} = \frac{4}{\pi^2}\int_0^\infty \cosh(\pi\beta)F_-(\beta)ImK_{\frac{1}{2}+i\beta}(x)d\beta, \ 0 < x < \infty. \quad (28)$$

**Proposition 2.2.**
  *If the following conditions are fulfilled*
  *1.* $F_+(\beta)e^{\frac{\pi|\beta|}{2}} \in L(0,\infty)$,
  *2.* $F_-(\beta)e^{\frac{\pi|\beta|}{2}} \in L(0,\infty)$,
  *then the inverse modified* KONTOROVITCH–LEBEDEV *integral transforms (27)-(28) exist.*

Proof:
The proposition is obtained by using the inequalities from Lemma 1.1.
The derived estimations of BESSEL functions and Lemma 1.1 give the possibility to prove the following inversion theorems of the modified KONTOROVITCH–LEBEDEV integral transforms [4-6].

**Theorem 2.1.** *If the following conditions on the function* $f(x)$ *and the transform (19),* $F_+(\beta)$, *are fulfilled.*
  *1.* $f(x) \in L(1,\infty)$,
  *2.* $f(x)x^{-\frac{3}{4}} \in L(0,1)$,
  *3.* $F_+(\beta)e^{\frac{\pi\beta}{2}} \in L(0,\infty)$,
  *then the inversion formula (21) holds for almost all* $x$.

**Theorem 2.2.** *If the following conditions on the function* $f(x)$ *defined on the interval* $(0,\infty)$ *are satisfied:*
  *1. The variation of the function* $f(x)$ *is bounded on every finite interval* $(a,b)$, $0 < a < b$,
  *2.* $f(x)x^{-\frac{1}{2}} \in L(0,\frac{1}{2})$,
  *3.* $f(x)xe^x \in L(\frac{1}{2},\infty)$,
  *then the inversion formulas (27)-(28) hold.*

Example. Let $f(x) = e^{-x\cosh\alpha}, \alpha > 0$. Then $F_+(\beta) = \frac{\pi}{2}\frac{\cos(\alpha\beta)}{\cosh\frac{\alpha}{2}\cosh(\pi\beta)}$.
It is easily seen that $f(x) = e^{-x\cosh\alpha}$ satisfies the conditions of Theorem 2.2.

**Theorem 2.3.** *If* $f(x)$ *is a finite function of bounded variation satisfying the condition*
  *1.* $f(x)x^{-\frac{1}{2}} \in L(0,\frac{1}{2})$,
  *then the inversion formulas (27)-(28) hold.*

Theorem 2.3 is a corollary of Theorem 2.2.

The justification the solution of the singular integral equations connected with the modified KONTOROVITCH–LEBEDEV integral transformations and the proof of the PARSEVAL equations for these transforms are given.

The following relation [1] is valid

$$\frac{\cosh(\pi\beta)}{\pi} \int_0^\infty (K_1(x+y) + K_0(x+y)) ReK_{\frac{1}{2}+i\beta}(y)dy = ReK_{\frac{1}{2}+i\beta}(x), \quad 0 < x < \infty, \tag{29}$$

Using the transformation (19) and relation (29) it is possible to obtain the solution of the inhomogeneous integral equation

$$\varphi_+(x) = f_+(x) + \lambda \int_0^\infty (K_1(x+y) + K_0(x+y))\varphi_+(y)dy, \ 0 < x < \infty, \tag{30}$$

where $f_+(x)$ is a defined function, $\lambda$ is a parameter satisfying the condition $\lambda < \frac{1}{\pi}$.

Multiplying the right and the left sides of (30) with $ReK_{\frac{1}{2}+i\beta}(x)$ and integrating with respect to $x$ from 0 to $\infty$, we obtain $\phi_+(\beta) = \frac{F_+(\beta)}{1-\frac{\lambda\pi}{\cosh(\pi\beta)}}$, where

$$\phi_+(\beta) = \int_0^\infty \varphi_+(x)ReK_{\frac{1}{2}+i\beta}(x)dx, \ F_+(\beta) = \int_0^\infty f_+(x)ReK_{\frac{1}{2}+i\beta}(x)dx.$$

Applying the inverse modified KONTOROVITCH–LEBEDEV integral transformation, we find the formal solution

$$\varphi_+(x) = \frac{4}{\pi^2} \int_0^\infty \cosh(\pi\beta)\phi_+(\beta)ReK_{\frac{1}{2}+i\beta}(x)d\beta$$

$$= \frac{4}{\pi^2} \int_0^\infty \frac{F_+(\beta)\cosh^2(\pi\beta)}{\cosh(\pi\beta) - \lambda\pi} ReK_{\frac{1}{2}+i\beta}(x)d\beta. \tag{31}$$

**Theorem 2.4.** *If the following conditions are satisfied:*
*1)* $f_+(x) \in C(0, \infty)$,
*2)* $f_+(x) \in L(1, \infty)$,
*3)* $f_+(x)x^{-\frac{3}{4}} \in L(0, 1)$,
*4)* $F_+(\beta)\beta e^{\frac{\pi\beta}{2}} \in L(0, \infty)$,
*then the integral (31) exists and determines a solution of the equation (30) continuous on* $(0, \infty)$.

The absolute convergence (31) and the continuity of $\varphi_+(x)$ for $x \in (0, \infty)$ and $\lambda < \frac{1}{\pi}$ follows from Lemma 1.1 and condition 4 of the theorem.

We now substitute $\varphi_+(x)$, determined by the formula (31), into the integral (30) and note that the following sequence of the equalities is valid:

$$\lambda \int_0^\infty (K_1(x+y) + K_0(x+y))\varphi_+(y)$$

$$= \lambda \int_0^\infty (K_1(x+y) + K_0(x+y))\frac{4}{\pi^2}\int_0^\infty \frac{F_+(\beta)\cosh^2(\pi\beta)}{\cosh(\pi\beta) - \lambda\pi} ReK_{\frac{1}{2}+i\beta}(y)d\beta dy$$

$$= \lambda\frac{4}{\pi^2}\int_0^\infty \frac{F_+(\beta)\cosh^2(\pi\beta)}{\cosh(\pi\beta) - \lambda\pi}\int_0^\infty (K_1(x+y) + K_0(x+y))ReK_{\frac{1}{2}+i\beta}(y)dyd\beta$$

$$= \lambda\frac{4}{\pi^2}\int_0^\infty \frac{F_+(\beta)\cosh^2(\pi\beta)}{\cosh(\pi\beta) - \lambda\pi}\frac{\pi}{\cosh(\pi\beta)}ReK_{\frac{1}{2}+i\beta}(x)d\beta.$$

Here we made use of the equation (29) and *Fubini's* theorem on interchanging the limits in an improper integral depending on a parameter. Its conditions fulfilled in view of condition 4 of the theorem and inequality (17).

It follows from the conditions 1)-4) of the theorem that one can use Theorem 2.1 on the inversion of the modified KONTOROVITCH–LEBEDEV integral transforms. We have

$$\lambda\pi\frac{4}{\pi^2}\int_0^\infty \frac{F_+(\beta)\cosh(\pi\beta)}{\cosh(\pi\beta) - \lambda\pi}ReK_{\frac{1}{2}+i\beta}(x)d\beta$$

$$= \frac{4}{\pi^2}\int_0^\infty \frac{F_+(\beta)\cosh^2(\pi\beta)}{\cosh(\pi\beta) - \lambda\pi}ReK_{\frac{1}{2}+i\beta}(x)d\beta$$

$$- \frac{4}{\pi^2}\int_0^\infty F_+(\beta)\cosh(\pi\beta)ReK_{\frac{1}{2}+i\beta}(x)d\beta = \varphi_+(x) - f_+(x).$$

The theorem is proved.

Example. Consider the equation

$$\varphi(x) = e^{-x\cos\alpha} + \lambda\int_0^\infty (K_1(x+y) + K_0(x+y))\varphi(y)dy, \quad \lambda < \frac{1}{\pi}.$$

Then on the basis of Theorem 2.4 the solution is found in the following way:

$$\varphi(x) = \frac{\pi}{2\cos\frac{\alpha}{2}}\int_0^\infty \frac{\cosh(\alpha\beta)\cosh^2(\pi\beta)}{\cosh(\pi\beta) - \lambda\pi}ReK_{\frac{1}{2}+i\beta}(x)d\beta.$$

Let $f(x)$ and $g(x)$ be arbitrary real functions on $[0, \infty)$. Then we have

**Theorem 2.5.** *If the following conditions are satisfied:*
1) $f(x) \in L_2(0, \infty)$, $g(x) \in L_2(0, \infty)$,
2) $f(x) \in L(1, \infty)$, $g(x) \in L(1, \infty)$,
3) $f(x)x^{-\frac{3}{4}} \in L(0, 1)$, $g(x)x^{-\frac{3}{4}} \in L(0, 1)$,
4) $F_+(\beta)\beta e^{\frac{\pi\beta}{2}} \in L(0, \infty)$, $G_+(\beta)\beta e^{\frac{\pi\beta}{2}} \in L(0, \infty)$,
*then the* PARSEVAL *equation holds*

$$\frac{4}{\pi^2}\int_0^\infty \cosh(\pi\beta)F_+(\beta)G_+(\beta)d\beta = \int_0^\infty f(x)g(x)dx. \tag{32}$$

The proof of the theorem follows from Theorem 2.1 on the inversion of the modified KONTOROVITCH–LEBEDEV integral transforms, FUBINI'S theorem, and inequality (17).

Remark. It is possible by a more detailed analysis to see that Theorem 2.5 is valid if the conditions 2)-4) are fulfilled only for one function, $f(x)$, and for $g(x)$ the condition $g(x)x^{-\frac{3}{4}}$ is satisfied.

**Corollary 2.1.** *If the conditions 1)-4) of Theorem 2.5 for the function $f(x)$ are satisfied, then the following relation is valid,*

$$\frac{4}{\pi^2} \int_0^\infty \cosh(\pi\beta) F_+^2(\beta) d\beta = \int_0^\infty f^2(x) dx. \tag{33}$$

The formulas (32), (33) permit to simplify the evaluation of many integrals containing the MACDONALD function and other special functions.

The applications for boundary value problems in wedge domains and numerical solution are described in [7-10].

### References

1. N. N. Lebedev and I. P. Skalskaya, *Some integral transforms related to* KONTOROVITCH– LEBEDEV *transforms*, The questions of the mathematical physics, Leningrad, Nauka, (1976), 68–79 [in Russian].

2. N. N. Lebedev and I. P. Skalskaya, *The dual integral equations connected with the* KONTOROVITCH–LEBEDEV *transform*, Prikl. Matem. and Mechan., **38** (1974), N 6, 1090–1097 [in Russian].

3. S. Saitoh, *Integral Transforms, Reproducing Kernels and their Applications*, Pitman Res. Notes in Math. Series (369), Addison Wesley Longman Ltd., UK, 1997.

4. V. B. Poruchikov and J.M. Rappoport, *Inversion formulas for modified* KONTOROVITCH–LEBEDEV *transforms*, Diff. Uravn., **20** (1984), N 3, 542–546 [in Russian].

5. J. M. Rapppoport, *Some properties of modified* KONTOROVITCH-LEBEDEV *integral transforms*, Diff. Uravn., **21** (1985), N 4, 724–727 [in Russian].

6. J. M. Rappoport, *Some results for modified* KONTOROVITCH–LEBEDEV *integral transforms*, Proceedings of the 7th International Colloquium on Finite or Infinite Dimensional Complex Analysis, Marcel Dekker Inc., (2000), 473–477.

7. J. M. Rappoport, *The canonical vector-polynomials at computation of the* BESSEL *functions of the complex order*, Comput. Math. Appl., **41** (2001), N 3/4, 399–406.

8. J. M. Rappoport, *Some numerical quadrature algorithms for the computation of the* MACDONALD *function*, Proceedings of the Third ISAAC Congress. Progress in Analysis. Volume 2. World Scientific Publishing, (2003), 1223–1230.

9. B. R. Fabijonas, D. L. Lozier and J. M. Rappoport, *Algorithms and codes for the* MACDONALD *function: Recent progress and comparisons*, Journ. Comput. Appl. Math., **161** (2003), N 1, 179–192.

10. J. M. Rappoport, *Dual integral equations for some mixed boundary value problems*, Proceedings of the 4th ISAAC Congress. Advances in Analysis. World Scientific Publishing, (2005), 167–176.

# THE ZETA FUNCTION FOR LEARNING THEORY AND RESOLUTION OF SINGULARITIES *

MIKI AOYAGI

*Department of Mathematics, Sophia University,*
*7-1 Kioi-cho, Chiyoda-ku, Tokyo, Japan, 102-8554,*
*E-mail*

SUMIO WATANABE

*Precision and Intelligence Laboratory, Tokyo Institute of Technology,*
*4259 Nagatsuda, Midori-ku, Yokohama, 226-8503,*
*E-mail swatanab@pi.titech.ac.jp*

Recently, the problem of obtaining the maximum pole of the zeta function is posed by the learning theory. The zeta function is defined by the integral of the Kullback function and an a priori probability density function. The poles of the zeta function give the asymptotic form of the generalization error of any hierarchical learning model [5,6]. However, for several examples, upper bounds of the main term in the asymptotic form were obtained but not the exact values, because of their computational complexities. In this paper, we obtain the explicit value of the main term for a three layered neural network, which is one of the hierarchical learning models.

**Key words:** Zeta function, resolution of singularities, maximum pole, non-regular learning machines, Kullback function

**Mathematic subject Classifications:** 14J17, 93E35

## 1. Introduction

The purpose of the learning system is image or speech recognition, artificial intelligence, control of a robot, genetic analysis, data mining, time series prediction, or so on. These data are very complicated, not generated by simple normal distributions, since they are influenced by many factors. Learning models to analyze such data have to have complicated structure, too.

Hierarchical learning models such as a layered neural network, reduced rank regression, a normal mixture model and a Boltzmann machine are known as effective learning models. However, they can not be analyzed by using the classic theory of regular statistical model. A few mathematical theories for such learning models are known. So it is necessary and crucial to construct fundamental mathematical theories.

---
*This work is supported by the Ministry of Education, Science, Sports and Culture in Japan, Grant-in-Aid for Scientific Research 16700218.

The learning system is formulated as follows. Let $x$ be an input with a probability density function $q(x)$ and $y$ an output. We assume that $n$ training samples $\{x_i\}_{i=1}^n$ are randomly selected from $q(x)$ and that $\{y_i\}_{i=1}^n$ are obtained by the conditional probability density function $q(y|x)$. $(x^n, y^n) := \{(x_i, y_i)\}_{i=1}^n$ is a set of learning data. If the purpose of the learning system is handwriting recognition, the input $x$ is a script and $y$ is its category. Since observation data usually have noise, probability density functions are used.

The aim of the learning system is to estimate the function $q(y|x)$.

Let us consider a learning model $p(y|x, w)$ which infers a probabilistic output $y$ from a given input $x$, where $w$ is a parameter.

By using training samples $(x^n, y^n)$, the learning model $p(y|x, w)$ changes the parameter $w$ and tries to approximate $q(y|x)$. Assume that $p(y|x, (x^n, y^n))$ is the result density function after learning, by using a proper algorithm.

Here we define the difference between two probability density functions $q(y|x)$ and $p(y|x)$, namely Kullback function (relative entropy):

$$K(q\|p) = \int q(y|x) \log \frac{q(y|x)}{p(y|x)} q(x) \mathrm{d}x \mathrm{d}y.$$

This function is always positive and satisfies $K(q\|p) = 0$ if and only if $q = p$.

The expectation value of the Kullback function over training samples between the true density function $q(y|x)$ and the predictive density function $p(y|x, (x^n, y^n))$ is called the generalization error. It clarifies how precisely $p(y|x, (x^n, y^n))$ can approximate $q(y|x)$.

There are usually considered to be two different types of direct and inverse problems. The direct problem would be to solve the generalization error with a known true density function $q(y|x)$. The inverse problem is to find proper learning models and learning algorithms to minimize the generalization error under the condition of the unknown true density function $q(y|x)$. The inverse problem is important for practical usage, but in order to tackle the inverse problem, first the direct problem has to be solved.

Recently, the progress has been achieved in analyzing and evaluating the direct problem in Bayesian estimation [5,6].

Consider an arbitrary fixed a priori probability density function $\psi(w)$ on the parameter set $W$. Then, the a posteriori probability density function $p(w|(x^n, y^n))$ is written by

$$p(w|(x^n, y^n)) = \frac{1}{Z_n} \psi(w) \prod_{i=1}^n p(y_i|x_i, w),$$

where

$$Z_n = \int_W \psi(w) \prod_{i=1}^n p(y_i|x_i, w) \mathrm{d}w.$$

So the average inference $p(y|x, (x^n, y^n))$ of the Bayesian density function is given by

$$p(y|x, (x^n, y^n)) = \int p(y|x, w)p(w|(x^n, y^n))dw.$$

Assume that the true probability density function is expressed by $p(y|x, w^*)$, where $w^*$ is fixed.

Let $G(n)$ be the generalization error (the learning efficiency):

$$G(n) = E_n\{\int p(y|x, w^*) \log \frac{p(y|x, w^*)}{p(y|x, (x^n, y^n))} q(x)dxdy\},$$

where $E_n\{\cdot\}$ is the expectation value over all sets of $n$ training samples. Our purpose is to know the asymptotic form of this generalization error.

Here we define the average stochastic complexity (the free energy):

$$F(n) = -E_n\{\log \int \exp(-nK_n(w))\psi(w)dw\},$$

where

$$K_n(w) = \frac{1}{n}\sum_{i=1}^n \log \frac{p(X_i|w^*)}{p(X_i|w)}.$$

This is also important function since it satisfies

$$G(n) = F(n+1) - F(n).$$

Define the zeta function $J(z)$ of the learning model by

$$J(z) = \int K(w)^z \psi(w)dw,$$

where $K(w)$ is the Kullback information:

$$K(w) = \int p(y|x, w^*) \log \frac{p(y|x, w^*)}{p(y|x, w)} q(x)dxdy.$$

Then, for the maximum pole $-\lambda$ of $J(z)$ and its order $\theta$, we have

$$F(n) = \lambda \log n - (\theta - 1) \log\log n + O(1), \tag{1}$$

and

$$G(n) \cong \lambda/n - (\theta - 1)/(n \log n), \tag{2}$$

where $O(1)$ is a bounded function of $n$. Therefore, our aim is to obtain $\lambda$ and $\theta$.

The values $\lambda$ and $\theta$ can be calculated by using the blowing-up process.

In spite of these mathematical foundations, for several examples, only upper bounds of $\lambda$ were obtained but not the exact values, by two main reasons as follows.

(1) By Hironaka's Theorem [4], it is known that the desingularization of an arbitrary polynomial can be obtained by using the blowing-up process. However the desingularization of any polynomial in general, although it is known as a finite process, is very difficult. Furthermore, there are other problems in obtaining the desingularization such that

- most of the Kullback functions are degenerate (over $\mathbb{R}$) with respect to their Newton polyhedrons,
- singularities of the Kullback functions are not isolated,
- the Kullback functions are not simple polynomials, i.e., they have parameters, for example, $p$ of $\sum_{n=1}^{p}(\sum_{m=1}^{p} a_m b_m^{2n-1})^2$, which is one of the three layered neural networks.

We note that there are many classical results to calculate the maximum poles of the zeta functions using the desingularization of plane curves in the dimension two. Also there have been many investigations for the case of the prehomogeneous spaces, which corresponds to a special case. The Kullback functions do not occur in the prehomogeneous spaces. Therefore, to obtain the desingularization of the Kullback functions is a new problem even in mathematics, since most of these singularities have not been investigated.

(2) Since the main purpose is to obtain the maximum pole, to get the desingularization is not enough. We need some techniques to compare poles as real numbers. However, as far as we know, no some theorems for comparing poles have been developed, so far.

Recently, in the paper [1], we have clarified the values $\lambda$ and $\theta$ in case of the reduced rank regression which is the three layered neural network with linear hidden units. In this paper, we use a recursive blowing-up, an inductive comparing method and a toric resolution to obtain these values in case of the three layered neural network.

## 2. Generalization error of a three-layer neural network

Consider a three-layer neural perceptron with one input unit, $p$ hidden units, and one output unit which is trained to approximate the true density function represented by the model with $\tilde{r}$ ($\tilde{r} < p$) hidden units.

Denote the input value by $x \in \mathbb{R}$ with a probability density function $q(x)$. Then the output value $y \in \mathbb{R}$ of the three layered neural network is given by $y = f(x, w) + (\text{noise})$, where $f(x, w) = \sum_{m=1}^{p} a_m^{(w)} \tanh(b_m^{(w)} x)$.

Consider the statistical model

$$p(y|x, w) = \frac{1}{\sqrt{2\pi}} \exp(-\frac{1}{2}(y - f(x, w))^2).$$

Assume that the true density function which is included in the learning model is $p(y|x, w^*)$, where $w^* = (a_1^*, \ldots, a_p^*, b_1^*, \ldots, b_p^*)$. Let $W^*$ be the true parameter set:

$$W^* = \{\tilde{w} \in W \mid f(x, \tilde{w}) = f(x, w^*) \text{ for any } x\}.$$

Suppose that the a priori probability density function $\psi(w)$ is a $C^\infty -$ function with compact support $W$ where $\psi(w^*) > 0$.

Let $K(w)$ be the Kullback function: $K(w) = \int p(y|x, w^*) \log \frac{p(y|x,w^*)}{p(y|x,w)} q(x) dx dy$.

Then the maximum pole of $\int_W K(w)^z \psi(w) dw$ is equal to that of

$$J(z) = \int_W \{\sum_{n=1}^{P}(\sum_{m=1}^{p} a_m^{(w)} b_m^{(w)}{}^{2n-1} - \sum_{m=1}^{p} a_m^* b_m^*{}^{2n-1})^2\}^z dw,$$

where $P$ is a sufficient large integer by Lemma 5 [5].

Let $-\lambda$ be the maximum pole of $J(z)$ and $\theta$ its order.

## 3. Main Theorems

The result in Main Theorem 2 is related to the three layered neural network. It is obtained from Main Theorem 1, which contains more general cases.

Let $w = (a_1^{(w)}, a_2^{(w)}, \cdots, a_p^{(w)}, b_1^{(w)}, b_2^{(w)}, \cdots, b_p^{(w)})$ be a parameter, $w^* = (a_1^*, a_2^*, \cdots, a_p^*, b_1^*, b_2^*, \cdots, b_p^*)$ constant values and $U^*$ a sufficiently small neighborhood of $w^*$.

Let $Q$ be an arbitrary natural number. Set

$$J_{Q,w^*}(z) = \int_{U^*} \{\sum_{n=1}^{P}(\sum_{m=1}^{p} a_m^{(w)} b_m^{(w)}{}^{Q(n-1)+1} - \sum_{m=1}^{p} a^*{}_m b_m^*{}^{Q(n-1)+1})^2\}^z \prod_{m=1}^{p} da_m^{(w)} db_m^{(w)}.$$

In case of the three layered neural network, $Q$ is equal to two.

Let $b_1^{**Q}, \ldots, b_r^{**Q}$ be different real numbers in $\{b_i^{*Q} \mid b_i^{*Q} \neq 0\}$ from each other: $\{b_1^{**Q}, \ldots, b_r^{**Q} \mid b_i^{**Q} \neq b_j^{**Q}, 1 \leq i \neq j \leq r\} = \{b_i^{*Q} \mid b_i^{*Q} \neq 0\}$, and let

$$a^{**}{}_i = -(\sum_{\{m \mid b^*{}_m^Q = b^{**}{}_i^Q, 1 \leq m \leq p\}} a^*{}_m b^*{}_m)/b^{**}{}_i.$$

Also set $a_0^{**} = 0$, $b_0^{**} = 0$.

Put $\begin{cases} B_\tau^{(w)} = \{i \mid b_i^{*Q} = b_\tau^{**Q}\}, \, s_\tau = \#B_\tau^{(w)}, \, 1 \leq \tau \leq r, \\ B_0^{(w)} = \{i \mid b_i^* = 0\}, \qquad s_0 = \#B_0^{(w)}, \end{cases}$

where $\#$ implies the number of elements.

Let $\tilde{r}$ ($\tilde{r} \leq r$) be the total number of $a_\tau^{**} \neq 0$. We may assume that $a_1^{**} \neq 0, \cdots, a_{\tilde{r}}^{**} \neq 0, a_{\tilde{r}+1}^{**} = 0, \cdots, a_r^{**} = 0$. Then we have

$$\sum_{m=1}^{p} a_m^* b_m^*{}^{Q(n-1)+1} = -\sum_{m=1}^{\tilde{r}} a_m^{**} b_m^{**Q(n-1)+1},$$

for any $n \in \mathbb{N}$.

Set $H_\tau = \sum_{b_m \in B_\tau^{(w)}} a_m^{(w)} b_m^{(w)}{}^{Q(n-1)+1} + a_\tau^{**} b_\tau^{**Q(n-1)+1}.$

By the definitions, we have

$$J_{Q,w^*}(z) = \int_{U^*} \{\sum_{n=1}^{P}(\sum_{\tau=1}^{r} H_\tau)^2\}^z \prod_{m=1}^{p} da_m^{(w)} db_m^{(w)}.$$

Let

$$J_\tau(z) = \int_{U^* \cap \mathbb{R}^{s_\tau}} \{\sum_{n=1}^{P} H_\tau^2\}^z \prod_{m \in B_\tau^{(w)}} da_m^{(w)} db_m^{(w)},$$

and $-\lambda_{Q,w^*}$, $-\lambda_\tau$ be the maximum poles of $J_{Q,w^*}(z)$ and $J_\tau(z)$, respectively.

**Main Theorem 1** *(1) We have $\lambda_{Q,w^*} = \sum_{\tau=0}^{r} \lambda_\tau$.*
*(2)*

$$\lambda_0 = \frac{Q(\tilde{n}_0^2 + \tilde{n}_0) + 2s_0}{4Q\tilde{n}_0 + 4},$$

$$\lambda_{\tau_1} = \frac{n_{\tau_1} + n_{\tau_1}^2 + 2s_{\tau_1}}{4n_{\tau_1}}, \text{ for } 1 \leq \tau_1 \leq \tilde{r},$$

$$\lambda_{\tau_2} = \frac{n_{\tau_2} + n_{\tau_2}^2 + 2(s_{\tau_2} - 1)}{4n_{\tau_2}}, \text{ for } \tilde{r} < \tau_2 \leq r,$$

*where* $\begin{cases} \tilde{n}_0 = \max\{i \in \mathbb{Z} \mid Q(i^2 - i) + 2i \leq 2s_0\}, \\ n_{\tau_1} - 1 = \max\{i \in \mathbb{Z} \mid i^2 + i \leq 2s_{\tau_1}\}, \quad \text{for } 1 \leq \tau_1 \leq \tilde{r}, \\ n_{\tau_2} - 1 = \max\{i \in \mathbb{Z} \mid i^2 + i \leq 2(s_{\tau_2} - 1)\}, \text{ for } \tilde{r} < \tau_2 \leq r. \end{cases}$

*(3) Set*

$$\Theta = \{\tau_0, \tau_1, \tau_2 \mid Q(\tilde{n}_0^2 - \tilde{n}_0) + 2\tilde{n}_0 = 2s_0, s_0 \geq 1,$$
$$(n_{\tau_1} - 1)^2 + n_{\tau_1} - 1 = 2s_{\tau_1}, s_{\tau_1} > 1, 1 \leq \tau_1 \leq \tilde{r},$$
$$(n_{\tau_2} - 1)^2 + n_{\tau_2} - 1 = 2(s_{\tau_2} - 1), s_{\tau_2} > 1, \tilde{r} < \tau_2 \leq r\}.$$

*Then $\theta = \#\Theta + 1$.*

**Main Theorem 2** *The maximum pole $-\lambda$ and its order $\theta$ in (1) and (2) is obtained by setting $Q = 2$ in Main Theorem 1. More precisely, we have $\lambda = \max_{\tilde{w}^* \in W^*} \lambda_{Q,w^*}$ with its order $\theta = \#\Theta + 1$.*

## 4. Proof of Main Theorem 1

To prove Main Theorem 1, we use three steps. The first step is the desingularization and the preparation to compare poles. The second one is to obtain the explicit value of the maximum pole and the third one is about the order of the maximum pole. Since the whole proof takes about 60 pages in the preprint [2], here we show only the key points.

Define

$$\Psi = \{\sum_{n=1}^{P}(\sum_{m=1}^{p} a_m^{(w)} b_m^{(w)Q(n-1)+1} - \sum_{m=1}^{p} a_m^* b_m^{*Q(n-1)+1})^2\}^z \prod_{m=1}^{p} da_m^{(w)} db_m^{(w)}.$$

Put the auxiliary function $f_{n,l}$ by

$$f_{n,l}(x_1, \cdots, x_l) = \begin{cases} \sum_{j_1 + \cdots + j_l = n-l} x_1^{Qj_1} \cdots x_l^{Qj_l}, \\ \qquad \text{if } n - l \geq 0, \\ 0, \text{ if } n - l < 0. \end{cases}$$

Let $n, \ell \in \mathbb{N}$. Set $C_i' = \sum_{m=i}^{\ell} a_m' b_m' (b_m'^Q - b_1'^Q) \cdots (b_m'^Q - b_{i-1}'^Q)$ for $i = 1, \cdots, \ell$. Then

$$\sum_{m=1}^{\ell} a_m' b_m'^{Q(n-1)+1} = f_{n,1}(b_1') C_1' + f_{n,2}(b_1', b_2') C_2' + \cdots + f_{n,\ell}(b_1', \cdots, b_\ell') C_\ell'.$$

Let

$$C_i = \sum_{m=i}^{p} a_m^{(w)} b_m^{(w)} (b_m^{(w)^Q} - b_1^{(w)^Q}) \cdots (b_m^{(w)^Q} - b_{i-1}^{(w)^Q})$$

$$+ \sum_{m=1}^{\tilde{r}} a_m^{**} b^{**}{}_m (b_m^{**^Q} - b_1^{(w)^Q}) \cdots (b_m^{**^Q} - b_{i-1}^{(w)^Q})$$

for $i \leq p$, and

$$C_i = \sum_{m=i-p}^{\tilde{r}} a_m^{**} b_m^{**} (b_m^{**^Q} - b_1^{(w)^Q}) \cdots (b_m^{**^Q} - b_p^{(w)^Q})(b_m^{**^Q} - b_1^{**^Q}) \cdots (b_m^{**^Q} - b_{i-p-1}^{**}{}^Q)$$

for $p < i \leq p + \tilde{r}$.

We may assume that $b_1^{*Q} = b_1^{**Q}, \cdots, b_r^{*Q} = b_r^{**Q}$ and that if $b_m^{*Q} \neq b_{m'}^{*Q}$ then $b_m^{(w)^Q} \neq b_{m'}^{(w)^Q}$ on $U^*$. Then since $b_i^* \neq 0$, $b_i^{*Q} - b_j^{*Q} \neq 0$, for $1 \leq i < j \leq r$, we can change the variables from $a_i$ to $d_i$ by $d_i = C_i$ for $1 \leq i \leq r$.

Next consider the case $i > r$. Let

$$a_i = \begin{cases} a_i^{(w)} & i = r+1, \ldots, p, \\ a_{i-p}^{**} & i = p+1, \ldots, p+\tilde{r}, \end{cases}$$

and

$$b_i = \begin{cases} b_i^{(w)}, & \text{if } b_i^* = 0, \\ b_i^{(w)^Q} - b_i^{*Q}, & \text{if } b_i^{*Q} = b_\tau^{**Q} \neq 0, a_\tau^{**} \neq 0, \\ b_i^{(w)^Q} - b_\tau^{(w)^Q}, & \text{if } b_i^{*Q} = b_\tau^{**Q} \neq 0, a_\tau^{**} = 0, \\ & \qquad \text{for } 1 \leq i \leq p, \\ 0 & \text{for } i = p+1, \cdots, p+\tilde{r}. \end{cases}$$

When we distinguish $a^*$, $b^*$ from $a^{(w)}$, $b^{(w)}$, we call $a$, $b$ constants. Let $J_i^{(1)} = b_i^{*Q}$, $i = 1, \ldots, \tilde{r}, r+1, \ldots, p+\tilde{r}$. Put $p' = p + \tilde{r}$. Then we have

$$C_i = \sum_{\substack{i \leq m \leq p', \\ J_m^{(1)} = 0}} g(i,m) a_m b_m \prod_{\substack{1 \leq i' \leq i-1, \\ J_{i'}^{(1)} = J_m^{(1)}}} (b_m{}^Q - b_{i'}{}^Q)$$

$$+ \sum_{\substack{i \leq m \leq p', \\ J_m^{(1)} \neq 0, a_m^* = 0}} g(i,m) a_m b_m \prod_{\substack{1 \leq i' \leq i-1, \\ J_{i'}^{(1)} = J_m^{(1)}}} (b_m - b_{i'})$$

$$+ \sum_{\substack{i \leq m \leq p', \\ J_m^{(1)} \neq 0, a_m^* \neq 0}} g(i,m) a_m \prod_{\substack{1 \leq i' \leq i-1, \\ J_{i'}^{(1)} = J_m^{(1)}}} (b_m - b_{i'}), \tag{3}$$

for $r < i \leq p'$, where $g(i,m) \neq 0$ on $U^*$.

The maximum pole of $\int_W \Psi$ is equal to that of $\int_W \Psi'$, where

$$\Psi' = \{d_1^2 + \cdots + d_r^2 + C_{r+1}^2 + \cdots + C_{\tilde{r}+p}^2\}^z \tag{4}$$

$$\prod_{m=1}^{r} \mathrm{dd}_m \prod_{m=r+1}^{p} \mathrm{da}_m \prod_{m=\tilde{r}+1}^{r} \mathrm{db}_m^{(w)} \prod_{m=1}^{\tilde{r}} \mathrm{db}_m \prod_{m=r+1}^{p} \mathrm{db}_m,$$

by using Lemmas 2 and 3 in the paper [1].

Let $J^{(\alpha)}$ be elements in $\mathbb{R}^\alpha$. Denote $J^{(\alpha)} = (J^{(\alpha')}, *)$ by $J^{(\alpha)} > J^{(\alpha')} (\alpha > \alpha')$ and $J^{(\alpha)} = (0, \cdots, 0)$ by $J^{(\alpha)} = 0^{(\alpha)}$ or $J^{(\alpha)} = 0$. Set $\mathbb{Z}_+ = \mathbb{N} \cup \{0\}$. We need to calculate poles of the following function by using the blowing-up process together with an inductive method of $k, K, \alpha$.

**Inductive statement**

Set $E = \{m \mid e_m \text{ is non-constant }\}$,
$s(J^{(\alpha)}) = \#\{m \mid m \geq k+1, J_m^{(\alpha)} = J^{(\alpha)}, m \in E\}$,
$s(i, J^{(\alpha)}) = \#\{m \mid k+1 \leq m \leq i-1, J_m^{(\alpha)} = J^{(\alpha)}, m \in E\}$, for $J^{(\alpha)} \in \mathbb{R}^\alpha$, and
$E_\tau = \{m \mid J_m^{(\alpha)} = (b_\tau^{**Q}, *)\}$.

(a) $k = k_0 + \cdots + k_r$, $K = K_0 + \cdots + K_r$, where $k_0, \ldots, k_r \in \mathbb{Z}_+$, $K_0, \ldots, K_r \in \mathbb{Z}_+$, $K_0 \geq k_0$, $K_\tau \geq k_\tau + 1$ for $1 \leq \tau \leq r$. Set $k_i^{(\alpha)} = k_i$.

(b)

$$\Psi' = \{ \prod_{\tau=0}^{r} (v_{1\tau}^{t_{1\tau}} v_{2\tau}^{t_{2\tau}} \cdots v_{k_\tau \tau}^{t_{k_\tau \tau}}) \big(d_1^2 + d_2^2 + \cdots + d_K^2\big) + \sum_{i=K+1}^{p'} C_i^2 \}^z$$

$$\prod_{\tau=0}^{r} \prod_{l=1}^{k_\tau} v_{1\tau}^{q_{l\tau}} \prod_{m=1}^{K} \mathrm{dd}_m \prod_{\tau=0}^{r} \prod_{l=1}^{k_\tau} \mathrm{dv}_{l\tau} \prod_{m=K+1}^{p} \mathrm{da}_m \prod_{\substack{k+1 \leq m \leq p \\ m \in E}} \mathrm{de}_m \prod_{m=\tilde{r}+1}^{r} \mathrm{db}_m^{(w)}.$$

Here, $t_{l\tau}, q_{l\tau} \in \mathbb{Z}_+$. Also, there exist $RJ^{(\alpha)} \subset \mathbb{R}^\alpha$, $t(i, J, (l, \tau)) \in \mathbb{Z}_+$ and functions $g(i, m) \neq 0$, such that

$$C_i = \prod_{\tau=0}^{r} (v_{1\tau}^{t(i,0,(1,\tau))} v_{2\tau}^{t(i,0,(2,\tau))} \cdots v_{k_\tau \tau}^{t(i,0,(k_\tau,\tau))}) \sum_{\substack{i \leq m \leq p' \\ J_m^{(\alpha)} = 0}} g(i, m) a_m e_m \prod_{\substack{k+1 \leq i' < i \\ J_{i'}^{(\alpha)} = 0}} (e_m^Q - e_{i'}^Q)$$

$$+ \sum_{J \in RJ^{(\alpha)}} \prod_{\tau=0}^{r} (v_{1\tau}^{t(i,J,(1,\tau))} v_{2\tau}^{t(i,J,(2,\tau))} \cdots v_{k_\tau \tau}^{t(i,J,(k_\tau,\tau))}) \sum_{\substack{i \leq m \leq p' \\ J_m^{(\alpha)} = J}} g(i, m) a_m e_m \prod_{\substack{k+1 \leq i' < i \\ J_{i'}^{(\alpha)} = J}} (e_m - e_{i'})$$

$$+ \sum_{J \notin RJ^{(\alpha)}, J \neq 0} \prod_{\tau=0}^{r} (v_{1\tau}^{t(i,J,(1,\tau))} v_{2\tau}^{t(i,J,(2,\tau))} \cdots v_{k_\tau \tau}^{t(i,J,(k_\tau,\tau))}) \sum_{\substack{i \leq m \leq p' \\ J_m^{(\alpha)} = J}} g(i, m) a_m \prod_{\substack{k+1 \leq i' < i \\ J_{i'}^{(\alpha)} = J}} (e_m - e_{i'}).$$

(c) $J_{i'}^{(\alpha)} \neq J_i^{(\alpha)}$ for $k < i' < i \leq K$ and $J_i^{(\alpha)} \notin RJ^{(\alpha)} \cup \{0\}$ for $k < i \leq K$.

(d) $t(i, J_m^{(\alpha)}, (l, \tau)) \geq t_{l\tau}/2$ for all $J_m^{(\alpha)}$, $i \leq m \leq p'$ and there exist non-negative integers $D_{J^{(\mu)},(l,\tau)}$ such that

$$t(i, J_m^{(\alpha)}, (l,\tau)) = \sum_{J_m^{(\alpha)} \geq 0^{(\mu)}} D_{0^{(\mu)},(l,\tau)}(Qs(i, 0^{(\mu)}) + 1)$$

$$+ \sum_{\substack{J_m^{(\alpha)} \geq J^{(\mu)} \\ J^{(\mu)} \in RJ^{(\mu)}}} D_{J^{(\mu)},(l,\tau)}(s(i, J^{(\mu)}) + 1) + \sum_{\substack{J_m^{(\alpha)} \geq J^{(\mu)} \\ J^{(\mu)} \notin RJ^{(\mu)}, J^{(\mu)} \neq 0}} D_{J^{(\mu)},(l,\tau)} s(i, J^{(\mu)}).$$

(e) There exist $g_{(l,\tau),\tau'} \in \mathbb{Z}_+$, $\eta^{(\xi)}_{\ell,(l,\tau),\tau'} \in \mathbb{Z}_+$ such that

$$\frac{t_{l\tau}}{2} = \sum_{\xi=1}^{g_{(l,\tau),\tau'}} (1 + \eta^{(\xi)}_{1,(l,\tau),\tau'} + \cdots + \eta^{(\xi)}_{K_\tau,(l,\tau),\tau'}) \text{ for all } \tau',$$

and $g_{(l,\tau),\tau'} \leq \sum_{J_m^{(\alpha)} \geq J^{(\mu)}} D_{J^{(\mu)},(l,\tau)}$ for $m \in E_{\tau'}$.

$$0 \leq \eta^{(\xi)}_{1,(l,\tau),0} \leq Q, 0 \leq \eta^{(\xi)}_{1,(l,\tau),0} + \eta^{(\xi)}_{2,(l,\tau),0} \leq 2Q,$$

$$\vdots$$

$$0 \leq \eta^{(\xi)}_{1,(l,\tau),0} + \eta^{(\xi)}_{2,(l,\tau),0} + \cdots + \eta^{(\xi)}_{K_0,(l,\tau),0} \leq QK_0,$$

$$\eta^{(\xi)}_{1,(l,\tau),\tau'} = 0, 0 \leq \eta^{(\xi)}_{2,(l,\tau),\tau'} \leq 1, 0 \leq \eta^{(\xi)}_{2,(l,\tau),\tau'} + \eta^{(\xi)}_{3,(l,\tau),\tau'} \leq 2,$$

$$\vdots$$

$$0 \leq \eta^{(\xi)}_{2,(l,\tau),\tau'} + \eta^{(\xi)}_{3,(l,\tau),\tau'} + \cdots + \eta^{(\xi)}_{K_\tau,(l,\tau),\tau'} \leq K_\tau - 1,$$

for $\tau' \geq 1$.

(f) Let $\varphi^{(\xi)}_{(l,\tau),\tau'} :=$
$$\begin{cases} s_0 + \eta^{(\xi)}_{1,(l,\tau),0} + 2\eta^{(\xi)}_{2,(l,\tau),0} + \cdots + K_0\eta^{(\xi)}_{K_0,(l,\tau),0}, \\ \qquad \text{if } \tau' = 0, \\ s_{\tau'} + 1 + \eta^{(\xi)}_{1,(l,\tau),\tau'} + 2\eta^{(\xi)}_{2,(l,\tau),\tau'} + \cdots + K_{\tau'}\eta^{(\xi)}_{K_{\tau'},(l,\tau),\tau'}, \\ \qquad \text{if } 1 \leq \tau' \leq \tilde{r}, \\ s_{\tau'} + \eta^{(\xi)}_{1,(l,\tau),\tau'} + 2\eta^{(\xi)}_{2,(l,\tau),\tau'} + \cdots + K_{\tau'}\eta^{(\xi)}_{K_{\tau'},(l,\tau),\tau'}, \\ \qquad \text{if } \tau' > \tilde{r}, \end{cases}$$

There exist $\phi_{(l,\tau),\tau'} \in \mathbb{Z}_+$ such that

$$q_{(l,\tau),\tau'} = \sum_{\xi=1}^{g_{(l,\tau),\tau'}} \varphi^{(\xi)}_{(l,\tau),\tau'} + \phi_{(l,\tau),\tau'} + \sum_{m=k+1, m \in E_\tau}^{p'} (-g_{(l,\tau),\tau'} + \sum_{J_m^{(\alpha)} \geq J^{(\mu)}} D_{J^{(\mu)},(l,\tau)}),$$

and $q_{l\tau} + 1 = \sum_{\tau'=0}^{r} q_{(l,\tau),\tau'}$.

**The end of inductive statement**

Statements (d), (e) and (f) are needed to compare poles. The proof of the induction statement will appear in the preprint [2].

The second and third steps is to prove the following theorem by using a toric resolution.

**Theorem 1**     *Fix $\tilde{\tau}$. We call the case satisfying the following three conditions Special Case. (1) $J_m^{(\alpha)}$ is the same for any $m \in E_{\tilde{\tau}}$, (2) $s(J_m^{(\alpha)}) =$*
$$\begin{cases} s_{\tilde{\tau}} - k_{\tilde{\tau}}, & 0 \le \tilde{\tau} \le \tilde{r}, \\ s_{\tilde{\tau}} - 1 - k_{\tilde{\tau}}, & \tilde{r} < \tilde{\tau} \le r, \end{cases} \text{ and (3) } s(K+1, J^{(\mu)}) \le \begin{cases} K_{\tilde{\tau}} - k_{\tilde{\tau}}, & \text{if } 0 \le \tilde{\tau} \le \tilde{r}, \\ K_{\tilde{\tau}} - k_{\tilde{\tau}} - 1, & \text{if } \tilde{r} < \tilde{\tau} \le r, \end{cases}$$
*for $\mu \le \alpha$. Assume that all $B_{\tilde{\tau}}^{(w)}$ satisfies special conditions.*

*Then there exists the pole $\lambda_{Q,w}$ whose order $\theta$ in Main Theorem 1.*

The detail will appear in the preprint [2].

**Remark** (1) Main Theorem 1 in case of $Q = 1$ can be applied to the asymptotic expansion of the generalization error of the normal mixture for the special case [7].
(2) $\tilde{r}$ represents the number of hidden units for the true density function.

**Example of Main Theorem 2**

Assume that $W^*$ includes all $\tilde{w}^* \in \mathbb{R}^{2p}$ satisfying $f(x, \tilde{w}^*) = f(x, w^*)$. Let $w_0^*$ be the constant such that $\lambda_{2,w_0^*} = \lambda := \max_{\tilde{w}^* \in W^*} \lambda_{2,w^*}$.

By some computations in [2], we have

$$\lambda_{2,w_0^*} = \tilde{r} + \frac{i^2 + i + p - \tilde{r}}{4i + 2}, \theta = \begin{cases} 1, & \text{if } i^2 < p - \tilde{r}, \\ 2, & \text{if } i^2 = p - \tilde{r}, \end{cases} \text{ for } p - \tilde{r} + 1 \ge 10,$$

where $i = \max\{\ell \in \mathbb{Z} \mid \ell^2 \le p - \tilde{r}\}$, and $w_0^*$ satisfies $s_0 = p - \tilde{r}$, $s_1 = 1, \ldots, s_{\tilde{r}} = 1$. Also we have

$$\lambda_{2,w_0^*} = \tilde{r} - 1 + \frac{j + j^2 + 2(p - \tilde{r} + 1)}{4j}, \theta = \begin{cases} 1, & \text{if } (j-1)^2 + j - 1 < 2(p - \tilde{r} + 1), \\ 2, & \text{if } (j-1)^2 + j - 1 = 2(p - \tilde{r} + 1), \end{cases}$$

for $p - \tilde{r} + 1 < 10$, where $j - 1 = \max\{\ell \in \mathbb{Z} \mid \ell^2 + \ell \le 2(p - \tilde{r} + 1)\}$ and $w_0^*$ satisfies $s_0 = 0$, $\{s_1, \ldots, s_{\tilde{r}}\} = \{1, \ldots, 1, p - \tilde{r} + 1\}$.

For instance, if $\tilde{r} = 1$, $p = 4$ then $\lambda = 5/3$, $\theta = 1$. If $\tilde{r} = 1$, $p = 10$ then $\lambda = 5/2$, $\theta = 2$.

## References

1. M. Aoyagi and S. Watanabe, Stochastic Complexities of Reduced Rank Regression in Bayesian Estimation, Neural Networks, **18**, 924-933, 2005.
2. M. Aoyagi and S. Watanabe, Stochastic Complexities of Three-Layer Neural Perceptron in Bayesian Estimation, preprint.
3. W. Fulton, Introduction to toric varieties, Annals of Mathematics Studies, vol. 131, Princeton University Press, 1993.
4. H. Hironaka, Resolution of Singularities of an algebraic variety over a field of characteristic zero. Math. Ann. **79** 109-326, 1964.
5. S. Watanabe, Algebraic analysis for nonidentifiable learning machines. Neural Computation. **13** (4) 899-933, 2001.
6. S. Watanabe, Algebraic geometrical methods for hierarchical learning machines. Neural Networks. **14** (8) 1049-1060, 2001.
7. S. Watanabe, K. Yamazaki and M. Aoyagi, Kullback Information of Normal Mixture is not an Analytic Function, Technical report of IEICE, NC2004, 41-46, 2004.

# THE BACKGROUND AND SURVEY OF RECENT RESULTS IN THE THEORY OF FUNCTIONS OF $\omega$-BOUNDED TYPE IN THE HALF-PLANE

ARMEN M. JERBASHIAN

*Institute of Mathematics*
*National Academy of Sciences of Armenia*
*24-b Marshal Baghramian Avenue*
*375019 Yerevan, Armenia*
*E-mail: armen_ jerbashian@yahoo.com*

After highlighting the historical background of the considered problems, the survey gives the recently established basic statements of the general theory of functions of $\omega$-bounded type in the upper half-plane. The starting point is the canonical representation of some Banach spaces $A^p_{\omega,\gamma}$ of holomorphic functions. For $p = 2$ (i.e. in the case of Hilbert spaces) there is a theorem on the orthogonal projection from the corresponding $L^2_\omega$ to $A^2_\omega$, a Paley-Wiener type theorem and a theorem on a natural isometry between $A^2_{\omega,0}$ and the Hardy space $H^2$, which is an integral operator along with its inversion. A theorem on projection from $L^p_{\omega,0}$ to $A^p_{\omega,0}$ is given and it is proved that $(A^p_{\omega,0})^* = A^q_{\omega,0}$ $(1/p + 1/q = 1)$ under several conditions on $\omega$. Then the canonical representations of Nevanlinna–Djrbashian type classes of $\delta$–subharmonic functions are given. The considered functions can have arbitrary growth near the finite points of the real axis.

**Key words:** Weighted spaces of analytic functions, tempered spaces of distributions
**Mathematics Subject Classifications:** 46E15, 30H05, 42B35

## 1. History and background

The concept of meromorphic functions of $\alpha$-bounded type in the unit disc $|z| < 1$ came from the result of R.Nevanlinna on the density of zeros and poles of functions which may be not of bounded type in $|z| < 1$ but are such that

$$\int_0^1 (1-r)^\alpha T(r,f)dr = \sup_{0<r<1} \int_0^1 (1-t)^\alpha T(rt,f)dt < +\infty \quad (\alpha > -1), \quad (1.1)$$

i.e. the Riemann–Liouville $\alpha+1$-th fractional primitive of Nevanlinna's characteristic is bounded. In Nevanlinna's book of 1936 ([1], Section 216), one can see his statement that zeros $\{a_k\}$ and poles $\{b_n\}$ of such functions satisfy the conditions

$$\sum_{|a_k|<1} (1-|a_k|)^{2+\alpha} < +\infty \quad \text{and} \quad \sum_{|b_n|<1} (1-|b_n|)^{2+\alpha} < +\infty. \quad (1.2)$$

Consideration of the classes (1.1) of meromorphic functions was natural particularly due to many results on the Banach spaces $A^p_\alpha$ $(\alpha > -1, p \geq 1)$ of holomorphic functions

$$\iint_{|\zeta|<1} (1-|\zeta|)^\alpha |f(\zeta)|^p d\sigma(\zeta) = \sup_{0<r<1} \int_0^1 (1-t)^\alpha t dt \int_0^{2\pi} |f(tre^{i\vartheta})|^p d\vartheta < +\infty \quad (1.3)$$

(where $\sigma(\zeta)$ is Lebesgue's area measure), which were widely known. One can see that $A_\alpha^p$ are defined by boundedness of the $\alpha + 1$-th fractional primitive of the $p$-th integral means. So, Nevanlinna just replaced the $p$-th integral means by his characteristic becoming the integral means of $\log^+ |f|$ for a holomorphic function.

As to the spaces $A_\alpha^p$, the origins of investigations related to the simplest, unweighted $A_0^2$ can be found in a work of L.Biberbach [2] (1914) and many other publications related to approximations by rational functions: *for $A_0^2$, see T.Carleman* [3] *(1922), S.Bergman* [4] *(1929), W.Wirtinger* [5] *(1932). For $A_\alpha^p$ (still without representation formula) see Hardy–Littlewood* [6] *(1932), and M.V.Keldysch* [7] *(1941). A good reference list of the old publications on $A_0^2$ can be found in the book of J.L.Walsh* [8] *of 1956 on interpolation and approximation by rational functions.*

The misunderstandings in the historical background of $A_\alpha^p$ spaces and even Nevanlinna's class (1.1) can be easily seen by comparing the monograph of A.E.Djrbashian–F.A.Shamoian [9] of 1988 (see, for instance, p. 8) with a similar book published by H.Hedenmalm–B.Korenblum–K.Zhu [10] in 2000, and observed in the lists of references of these books and [11-17]].

It should be noted that *the representation formula of the weighted spaces $A_\alpha^p$ was first found by M.M.Djrbashian* [18] *in 1945* (see also its detailed and complemented version [19a] of 1948). But the work of 1945 was mainly aimed at investigating a more complicated problem: it improved Nevanlinna's result on the density of zeros and poles of the meromorphic class (1.1) by giving its canonical factorization as some Blaschke type products and a surface integral of $\log |f|$ in the exponent.

The results of M.M.Djrbashian of 1945–1948 were widely known in the former USSR since they were included in his Doctor of Physical–Mathematical Sciences Thesis "Metrical theorems on completeness and representability of analytic functions" defended in 1949 in Moscow State University with references of M.V.Keldysch, A.O.Gelfond and A.I.Markushevich.

Later, application of fractional integro-differentiation and a more general operator directly to the considered functions led M.M.Djrbashian (see Chapter IX in [20-24] and [25]) to the factorization theory of his Nevanlinna type $N\{\omega\}$ classes, the sum of which coincides with the whole set of functions meromorphic in $|z| < 1$. The new theory in a sense was more preferable, particularly, since the integrals in the factorizations of $N\{\omega\}$ were over the unit circle while those of Nevanlinna's class (1.1) (and of the representation of $A_\alpha^p$) are over the surface of the unit disc.

---

[a]The PDF files of the English translations of these works can be downloaded from the author's homepage at: http://math.sci.am/People/ArmenJerbashian.html

Nevertheless, the results of 1945–1948, where the fractional integration is applied to some quantities characterizing the growth of considered functions, remain in considerable interest as they find development and application in numerous contemporary investigations connected with some particular questions related to the general Nevanlinna's weighted class (1.1) and $A_\omega^p$. *This was the background for the author's recent work* [26,27] *aimed at constructing the main analytic apparatus of these classes or spaces through improving the methods developed by M.M.Djrbashian for the construction of his $N\{\omega\}$–factorization theory.*

After that, the author started an investigation of

## 2. The spaces $A_{\omega,\gamma}^p$ in the half-plane

**2.1.** The construction of a theory of such spaces and of a theory in the half-plane similar to Nevanlinna–Djrbashian factorization theory in the disc was an unsolved problem for a long time. It was posed by M.M.Djrbashian from 1970's many times after he constructed the factorization theory of his $N\{\omega\}$ classes [23] in the disc. The following procedure for finding the representing $\omega$-kernel was obvious: *if in* $z \in G^+ = \{z : Im\ z > 0\}$

$$C_\omega = \int_0^{+\infty} e^{itz} \frac{dt}{I_\omega(t)}, \quad \text{where} \quad I_\omega(t) = \int_0^{+\infty} e^{-tx} d\omega(x),$$

*then the Liouville type general operator $L_\omega f(z) \equiv \int_0^{+\infty} f(z+it)d\omega(t)$ transforms it to the ordinary Cauchy kernel: $L_\omega C_\omega(z) = -(iz)^{-1}$, $z \in G^+$, and particularly*

$$I_\omega(t) = \Gamma(2+\alpha)t^{-(1+\alpha)} \quad \text{and} \quad C_\omega = (-iz)^{-(2+\alpha)} \quad \text{for} \quad \omega(t) = t^{1+\alpha}.$$

Nevertheless, the theory could not be constructed since there was no approach permitting to operate with the Fourier–Laplace transform in a way similar to the one used for the Fourier–Taylor series in the $N\{\omega\}$ theory in the disc. An approach of such type was given by S.M.Gindikin [28]. But this approach was not sufficient for the construction of the theory, it led only to some partial results which were obtained by A.H.Karapetyan [35–38].

Fortunately, the necessary argument with the Fourier–Laplace transform was found and it led to a new $\omega$-theory in the half-plane. The initial results are recently published in [29], more complete results are published in [30], and some of results included in this survey are fresh and still unpublished.

**2.2.** Define $A_{\omega,\gamma}^p$ ($0 < p < +\infty$, $-\infty < \gamma \le 2$) as the set of those functions $f(z)$ holomorphic in the upper half-plane $G^+ = \{z : Im\ z > 0\}$, which for sufficiently small $\rho > 0$ satisfy the Nevanlinna condition

$$\liminf_{R \to +\infty} \frac{1}{R} \int_\beta^{\pi-\beta} \log^+ |f(Re^{i\vartheta})| \left( \sin \frac{\pi(\vartheta-\beta)}{\pi-2\beta} \right)^{\pi/\kappa-1} d\vartheta = 0, \qquad (2.1)$$

where $\beta = \arcsin \frac{\rho}{R} = \frac{\pi}{2} - \kappa$ and, simultaneously,

$$\|f\|_{p,\omega,\gamma}^p \equiv \iint_{G^+} |f(z)|^p \frac{d\mu_\omega(z)}{(1+|z|)^\gamma} < +\infty, \qquad (2.2)$$

where $d\mu_\omega(x+iy) = dxd\omega(2y)$ and it is supposed that $\omega(t) \in \Omega_\alpha$ $(-1 \le \alpha < +\infty)$, i.e. $\omega(t)$ is given in $[0, +\infty)$ and such that

(1) $\omega(t) \nearrow$ (is non-decreasing) in $(0, +\infty)$, $\omega(0) = \omega(+0)$ and there exists a sequence $\delta_k \downarrow 0$ such that $\omega(\delta_k) \downarrow$ (is strictly decreasing);
(2) $\omega(t) \asymp t^{1+\alpha}$ for $\Delta_0 \le t < +\infty$ and some $\Delta_0 \ge 0$

$(f(t) \asymp g(t)$ means that $m_1 f(t) \le g(t) \le m_2 f(t)$ for some constants $m_{1,2} > 0)$. $L_{\omega,\gamma}^p$ is assumed to be the Lebesgue space defined solely by (2.2).

**Remark 2.1.** $A_{\omega,\gamma}^p = (i+z)^{\gamma/p} A_{\omega,0}^p$.

**Remark 2.2.** For $\omega(t) = t^{1+\alpha}$ $(\alpha > -1)$, $\gamma = 0$ and $p \ge 1$ the spaces $A_{\omega,\gamma}^p$ coincide with the well-known $A_\alpha^p$ in the half-plane (see [31-33]). In this case (2.2) implies (2.1), and this implication is true even in more general cases but *the condition (1.1) cannot be derived from (1.2) in the general case.*

By the results of [34], under (2.2) the condition (2.1) is equivalent to

$$\liminf_{R \to +\infty} \frac{1}{R} \int_\beta^{\pi-\beta} |f(Re^{i\vartheta})|^p \left( \sin \frac{\pi(\vartheta - \beta)}{\pi - 2\beta} \right)^{\pi/\kappa-1} d\vartheta < +\infty \qquad (2.1')$$

for any $\rho > 0$, and also to any of the following conditions:

$$\liminf_{R \to +\infty} \frac{1}{R} \int_0^\pi \log^+ |f(Re^{i\vartheta} + i\rho)| \sin \vartheta d\vartheta = 0,$$

$$\liminf_{R \to +\infty} \frac{1}{R} \int_0^\pi |f(Re^{i\vartheta} + i\rho)|^p \sin \vartheta d\vartheta < +\infty.$$

**Proposition 2.1.** $A_{\omega,\gamma}^p$ $(1 \le p < +\infty, -\infty < \gamma < 1, \omega \in \Omega_\alpha, \alpha \ge -1)$ *is a Banach space with the norm (2.2).*

## 3. Representation over strips

The following theorem gives the canonical representations of $A_{\omega,\gamma}^p$ spaces when the measure $\omega$ has bounded support, i.e. $\omega(t) = const$ for $0 < \Delta < t < +\infty$.

**Theorem 3.1.** *Let* $f(z) \in A_{\omega,\gamma}^p(G^+)$ *for some* $1 \le p < +\infty$, $-\infty < \gamma < 1$ *and* $\omega(x)$ *satisfying the condition (i) and such that* $\omega(t) = \omega(\Delta) < +\infty$ $(\Delta < t < +\infty)$ *for some* $\Delta > 0$. *Then*

$$f(z) = \frac{1}{2\pi} \iint_{G^+} f(w) C_\omega(z - \overline{w}) d\mu_\omega(w), \quad z \in G^+, \qquad (3.1)$$

$$f(z) = \frac{1}{\pi} \iint_{G^+} \{Re f(w)\} C_\omega(z - \overline{w}) d\mu_\omega(w), \quad z \in G^+, \qquad (3.2)$$

*where both integrals are absolutely and uniformly convergent inside* $G^+$.

## 4. Representation over the whole half-plane

**4.1.** The estimate of the following lemma is one of the main tools for letting $\Delta \to +\infty$ in formulas (3.1) and (3.2) and obtaining a representation of $A^p_{\omega,\gamma}$ by some integrals which can be over the whole half-plane.

**Lemma 4.1.** *Let $\omega(x) \in \Omega_\alpha$ for some $\alpha \geq -1$. Then for any non-integer $\beta \in ([\alpha] - 1, \alpha)$ and any $\rho > 0$ there exists a constant $M \equiv M_{\rho,\beta} > 0$ such that*

$$|C_\omega(z)| \leq M|z|^{-(2+\beta)}, \quad z \in G^+_\rho.$$

Due to this estimate, the passage $\Delta \to +\infty$ in formulas (3.1) and (3.2) leads to

**Theorem 4.1.** *Let $f(z) \in A^p_{\omega,\gamma}$ for some $1 \leq p < +\infty$ and $-\infty < \gamma < 1 - (1 + \alpha)(p - 1)$. Then*

$$f(z) = \frac{1}{2\pi} \iint_{G^+} f(w)C_\omega(z - \overline{w})d\mu_\omega(w), \quad z \in G^+, \tag{4.1}$$

$$f(z) = \frac{1}{\pi} \iint_{G^+} \{Re\, f(w)\}C_\omega(z - \overline{w})d\mu_\omega(w), \quad z \in G^+, \tag{4.2}$$

*where both integrals are absolutely and uniformly convergent inside $G^+$.*

**Remark 4.1.** The representation (4.1) for $1 \leq p \leq 2$ and absolute continuous measure $d\omega$ has been established by A.H.Karapetyan [35–38] in some general mixed-norm weighted spaces in radial tube domains from $\mathbb{C}^n$.

## 5. Orthogonal projection and isometry

**Theorem 5.1.** *The orthogonal projection of $L^2_{\omega,0}$ onto $A^2_{\omega,0}$ ($\omega \in \Omega_\alpha$, $-1 \leq \alpha < 0$, $\omega(0) = 0$) is written in the form*

$$P_\omega f(z) = \frac{1}{2\pi} \iint_{G^+} f(w)C_\omega(z - \overline{w})d\mu_\omega(w), \quad f \in L^2_{\omega,0}.$$

The next assertion is the similarity of the Paley–Wiener theorem.

**Theorem 5.2.** *The class $A^2_{\omega,0}$ ($\omega \in \Omega_\alpha$, $-1 \leq \alpha < 0$, $\omega(0) = 0$) coincides with the set of functions representable in the form*

$$f(z) = \frac{1}{\sqrt{2\pi}} \int_0^{+\infty} e^{itz} \frac{\Phi(t)}{\sqrt{I_\omega(t)}}dt, \quad z \in G^+, \quad \Phi(t) \in L^2(0, +\infty). \tag{5.1}$$

*If this representation is true, then $\|f\|_{A^2_{\omega,0}} = \|\Phi\|_{L^2(0,+\infty)}$ and*

$$\Phi(t) = \frac{1}{\sqrt{I_\omega(t)}} \int_0^{+\infty} e^{-tv}\widehat{f_v}(t)d\omega(2v), \tag{5.2}$$

*where $\widehat{f_v}(t) = \underset{R \to +\infty}{\text{l.i.m.}} \frac{1}{\sqrt{2\pi}} \int_{-R}^R e^{-itu}f(u + iv)du.$*

**Remark 5.1.** *Let $S$ be the set of those $\omega(x)$, which belong to $\Omega_\alpha$ for some $-1 \leq \alpha < 0$ and $\omega(0) = 0$. Then the set $\cup_{\omega \in S} A^2_{\omega,0}$ coincides with the set of all those functions, which are representable in the form*

$$f(z) = \int_0^{+\infty} e^{itz} \Psi(t)dt, \quad z \in G^+,$$

*where $e^{-\varepsilon t}\Psi(t) \in L^2(0, +\infty)$ for every $\varepsilon > 0$.*

**5.2.** The next theorem particularly gives more information about the function $\Phi$ in (5.1)–(5.2).

**Theorem 5.3.** *Let $\omega(t) \in \Omega_\alpha$ $(-1 \leq \alpha < 0$, $\omega(0) = 0)$ and let $\widetilde{\omega}(x)$ be its Volterra square*

$$\widetilde{\omega}(x) = \int_0^x \omega(x - t)d\omega(t), \quad 0 < x < +\infty$$

*(which belongs to $\Omega_{1+2\alpha}$ and $\widetilde{\omega}(0) = 0$). Then $I^2_\omega(x) = I_{\widetilde{\omega}}(x)$, $0 < x < +\infty$, and $A^2_{\widetilde{\omega},0}$ coincides with the set of functions representable in the form*

$$f(z) = \frac{1}{2\pi} \int_{-\infty}^{+\infty} \varphi(t)C_\omega(z - t)dt, \quad z \in G^+, \quad \varphi \in L^2(-\infty, +\infty). \tag{5.3}$$

*For any $f \in A^2_{\widetilde{\omega},0}$, $L_\omega f = \varphi_0$ is the unique function of Hardy's space $H^2 \equiv H^2_0$, such that (5.3) is true with $\varphi = \varphi_0$. Besides, $\|\varphi_0\|_{H^2} = \|f\|_{A^2_{\widetilde{\omega},0}}$ and $\varphi - \varphi_0 \perp H^2$ for any $\varphi \in L^2(-\infty, +\infty)$ with which (5.3) is true. The operator*

$$L_\omega f(z) \equiv \int_0^{+\infty} f(z + i\sigma)d\omega(\sigma)$$

*is an isometry $A^2_{\widetilde{\omega},0} \longrightarrow H^2$, and the integral of (5.3) defines $(L_\omega)^{-1}$ in $H^2$.*

**Remark 5.2.** Under the conditions of Theorem 5.3, $2\pi I_\omega(t)\Phi(t)$ is the Fourier transform of $L_\omega f(t)$ in the representation (5.1) of $A^2_{\widetilde{\omega},0}$.

**Remark 5.3.** The explicitely written isometry (5.3) permits to convert any additive result known in the Hardy space $H^2$ to its similarity in $A^2_{\widetilde{\omega}}$. Particularly, this is true for the well-known approximations by rational functions in $H^2$, the images of which are the kernels $C_\omega$ in $A^2_{\widetilde{\omega}}$.

# 6. The projection $L^p_{\omega,0} \to A^p_{\omega,0}$ and the conjugate space of $A^p_{\omega,0}$

Using the recently found sharp estimates of M.M.Djrbashian kernels [39], the following assertions related to the projection $L^p_{\omega,0} \to A^p_{\omega,0}$ and the conjugate space of $A^p_{\omega,0}$ are recently proved by V.A.Jerbashian.

**Theorem 6.1.** *Let $1 < q < +\infty$, $q \neq 2$, and let the functions $\omega_1 \in \Omega_{\kappa_1}$ and $\omega_2 \in \Omega_{\kappa_2}$ $(\kappa_{1,2} > -1)$ be continuously differentiable in $[0, +\infty)$ and satisfy one of the following conditions (A) and (B):*

**(A)** $t^{-1}w_2'(t) \nearrow$ or, alternatively, $t^{-1}w_2'(t) \searrow$ but $t^{-\delta}w_2'(t) \nearrow$ for some $\delta \in (0,1)$.

Besides, $t^{-\alpha_{1,2}}w_2'(t) \searrow$ and $t^{-\beta_{1,2}}w_2'(t) \nearrow$ in $(0,+\infty)$ (6.1)

for some $\alpha_{1,2} > 0$ and $\beta_{1,2} > 0$ such that

$$\alpha_1 < 1 + \beta_1 + \beta_2, \quad 1 + \alpha_1 < q(1+\beta_2), \quad q(\alpha_2 - \beta_2) < 2 + \beta_1. \quad (6.2)$$

**(B)** $t^{-1}w_2(t) \searrow$ and $t^{-\delta}w_2(t) \nearrow$ for some $\delta \in (0,1)$. Besides, $w_2'(t) \searrow$ and (6.1) is true for some $\alpha_{1,2} \in (-1,0)$ and $\beta_{1,2} \in (-1,0)$ satisfying (6.2)

Then

$$P_{w_2}f(z) = \frac{1}{2\pi} \iint_{G^+} f(w)C_{w_2}(z-\overline{w})d\mu_{w_2}(w)$$

is a bounded operator acting from $L^q_{w_1,0}$ to $A^q_{w_1,0}$.

**Remark 6.1.** The requiremets of Theorem 6.2 permit the equality $w_1(t) \equiv w_2(t)$ $(0 < t < +\infty)$. Denoting $w \equiv w_{1,2}$, one can see that Theorem 6.2 states that the representation formula (4.1) gives a bouded projection $L^p_{w,0} \longrightarrow A^p_{w,0}$ under the requirements of Theorem 4.1 and Theorem 6.2 (with $q$ replaced by $p$).

**Theorem 6.2.** Let $1 < p < +\infty$, $p \neq 2$, let $1/p + 1/q = 1$ and let a function $w \in \Omega_\kappa$ $(\kappa > -1, (1+\kappa)(p-1) > 1)$ satisfy the conditions of Theorem 6.1 for $w_1 \equiv w_2$. Then the set of bounded linear functionals over $A^p_{w,0}$ is described by the formula

$$\Phi(f) = \frac{1}{2\pi} \iint_{G^+} f(z)\overline{g(z)}d\mu_w(z), \quad g \in A^q_{w,0} \quad \left(\frac{1}{p}+\frac{1}{q}=1\right),$$

and $\left(A^p_{w,0}\right)^* = A^q_{w,0}$ in the sense of isomorphism.

**Remark 6.2.** Obviously, $A^p_{w,0} = A^p_{\tilde{w},0}$ for any continuously differentiable functions $w$ and $\tilde{w}$ such that $w'(x) \asymp \tilde{w}'(x)$ $(x \geq 0)$. Besides, the requirements of Theorem 4.3 are satisfied for $w(x) = \int_0^x t^\alpha \log^\lambda \left(1+\frac{a}{t}\right) dt$ $(x \geq 0)$, where $\alpha > -1$, $\lambda \geq 0$ and $a > 0$ are any numbers. Hence, the assertion of Theorem 4.3 is true for the functions $w \in \Omega_\alpha$ $(\alpha > -1)$ such that $w'(x) \asymp x^\alpha \log^\lambda \left(1+\frac{a}{x}\right)$ $(x \geq 0)$ for some $\alpha > -1$, $\lambda \geq 0$ and $a > 0$.

## 7. Weighted classes of harmonic functions

**Theorem 7.1.** Let $U(z)$ be a harmonic function in $G^+$, such that for $\rho > 0$ small enough and some $\gamma \in (-\infty, 2]$

$$\liminf_{R \to +\infty} \frac{1}{R} \int_\beta^{\pi-\beta} |U(Re^{i\vartheta})| \left(\sin \frac{\pi(\vartheta - \beta)}{\pi - 2\beta}\right)^{\pi/\kappa - 1} d\vartheta = 0,$$

$$\iint_{G^+} |U(z)| \frac{d\mu_w(z)}{(1+|z|)^\gamma} < +\infty,$$

where $\beta = \arcsin \frac{\rho}{R} = \frac{\pi}{2} - \kappa$ and $d\mu_w(x+iy) = dxdw(2y)$. If additionally

1°. $\omega(t)$ satisfies the condition (i) (in Section 2) and $\omega(t) = \omega(\Delta)$ ($\Delta < t < +\infty$) for some $\Delta > 0$ or, alternatively, 2°. $\omega(t) \in \Omega_\alpha$ for some $\alpha \geq -1$ and $\gamma < 1$, then

$$U(z) = \frac{1}{\pi} \iint_{G^+} U(w) \operatorname{Re} \{C_\omega(z - \overline{w})\} d\mu_\omega(w), \quad z \in G^+,$$

where the integral is absolutely and uniformly convergent inside $G^+$.

## 8. Nevanlinna-Djrbashian type classes in the half-plane

**8.1.** Henceforth, it will be assumed that $U(z)$ is a $\delta$-subharmonic function in $G^+$ and $\nu$ is its associated measure, i.e. $U(z) = U_1(z) - U_2(z)$, where $U_{1,2}(z)$ are subharmonic in $G^+$, posses Riesz associated measures $\nu_{1,2}$, and $\nu = \nu_1 - \nu_2$. Besides, we shall assume that the measure $\nu$ is minimally decomposed in the Jordan sense: $\nu = \nu_+ - \nu_-$, where (supp $\nu_+$) $\cap$ (supp $\nu_-$) $= \emptyset$ and $\nu_\pm$ are the positive and negative variations of $\nu$. We shall say that two $\delta$-subharmonic functions are equal in $G^+$, i.e. $U(z) = V(z)$, where $V(z) = V_1(z) - V_2(z)$ (and $V_{1,2}(z)$ are subharmonic in $G^+$), if $U_1(z) + V_2(z) = U_2(z) + V_1(z)$ everywhere in $G^+$.

Consider Tsuji's characteristic of the form

$$\mathcal{L}(y, U) \equiv \frac{1}{2\pi} \int_{-\infty}^{+\infty} U^+(x + iy) dx + \int_y^{+\infty} n_+(t) dt, \quad 0 < y < +\infty,$$

where

$$n_+(t) = \iint_{G_t^+} d\nu_-(\zeta), \quad G_t^+ = \{\zeta : \operatorname{Im} \zeta > t\}.$$

Assuming that $\mathcal{L}(y, -U)$ is defined similarly, by means of $U^-$ and $\nu_+$, one has to note that generally $\mathcal{L}(y, U)$, $\mathcal{L}(y, -U)$ or both these quantities can be infinite. Nevertheless, there are some conditions (see [40], Chapter 5) under which $\mathcal{L}(y, \pm U)$ ($0 < y < +\infty$) are finite and connected by a special form B.Ya.Levin's formula:

$$\mathcal{L}(y, U) = \mathcal{L}(y, -U), \quad 0 < y < +\infty,$$

which is a natural similarity of the well known equilibrium relation for Nevanlinna's characteristics.

**Definition 8.1.** $\widetilde{\Omega}_\alpha$ ($-1 < \alpha < +\infty$) is the set of those functions $\omega(x)$, $\omega(0) = 0$, which are continuous and strictly increasing in $[0, +\infty)$, continuously differentiable in $(0, +\infty)$ and such that $\omega'(x) \asymp x^\alpha$ ($\Delta < x < +\infty$) for any $\Delta > 0$.

The below definition of the half-plane similarity of Nevanlinna's weighted class is natural since B.Ya.Levin's equilibrium is not true for any $\delta$-subharmonic functions.

**Definition 8.2.** $\mathcal{N}_\omega$ ($\omega(x) \in \widetilde{\Omega}_\alpha$, $-1 < \alpha < +\infty$) is the set of those functions, which are $\delta$-subharmonic in $G^+$ and such that

$$\int_0^{+\infty} [\mathcal{L}(y, U) + \mathcal{L}(y, -U)] \, d\omega(2y) < +\infty.$$

**8.2.** Before stating the canonical representation of $\delta$-subharmonic functions from $\mathcal{N}_\omega$, note that the following assertion is true.

**Theorem 8.1.** *Let* $\omega(x) \in \widetilde{\Omega}_\alpha$ $(-1 < \alpha < +\infty)$. *Then the function*

$$b_\omega(z, \zeta) = \exp\left\{ -\int_0^{2Im\,\zeta} C_\omega(z - \zeta + it)\omega(t)dt \right\}, \quad Im\,z > Im\,\zeta,$$

*has holomorphic continuation to the whole* $G^+$, *where it has unique, simple zero at the point* $z = \zeta$. *If* $\{z_k\} \subset G^+$ *is a sequence satisfying the density condition*

$$\sum_k \int_0^{2Im\,z_k} \omega(t)dt < +\infty, \tag{8.1}$$

*then the following Blaschke type product is uniformly convergent inside* $G^+$:

$$B_\omega(z, \{z_k\}) = \prod_k b_\omega(z, z_k)$$

**Remark 8.1.** For $\omega(t) \equiv const$ the above mentioned product and its convergence condition become

$$B_\omega(z, \{z_k\}) = \prod_k \frac{z - z_k}{z - \overline{z_k}}, \quad \sum_k Im\,z_k < +\infty.$$

**Theorem 8.2.** *Let* $U(z) \in \mathcal{N}_\omega$ *for some* $\omega(x) \in \widetilde{\Omega}_\alpha$ $(-1 < \alpha < +\infty)$. *Then*

$$\iint_{G^+} |U(z)|d\mu_\omega(z) < +\infty \quad and \quad \iint_{G^+} \left( \int_0^{2Im\,\zeta} \omega(t)dt \right) d\nu_\pm(\zeta) < +\infty,$$

*and for any* $\rho > 0$

$$\iint_{G_\rho^+} Im\,\zeta d\nu_\pm(\zeta) < +\infty.$$

*Besides, the Green type potentials* $P_\omega^{(\pm)}(z) \equiv \iint_{G^+} \log|b_\omega(z, \zeta)|d\nu_\pm(\zeta)$ *by the positive and negative variations of the associated measure of* $U(z)$ *are convergent, and the following representation is valid in* $G^+$:

$$U(z) = \iint_{G^+} \log|b_\omega(z, \zeta)|d\nu(\zeta) + \frac{1}{\pi} \iint_{G^+} U(w)Re\left\{C_\omega(z - \overline{w})\right\}d\mu_\omega(w). \tag{8.2}$$

**Remark 8.2.** If a sequence $\{z_k\} \subset G^+$ satisfies the density condition (8.1), then in the particular case $U(z) = \log|f(z)|$, where $f(z)$ is a function meromorphic in $G^+$, and $U(z)$ belongs to $\mathcal{N}_\omega$ (and $\omega(x)$ is as required in Theorem 8.2) the representation (8.2) becomes the following factorization:

$$f(z) = \frac{B_\omega(z, \{a_n\})}{B_\omega(z, \{b_m\})} \exp\left\{ \frac{1}{\pi} \iint_{G^+} \log|f(w)|C_\omega(z - \overline{w})d\mu_\omega(w) + iC \right\}, \quad z \in G^+,$$

*where* $C$ *is a real number and* $\{a_n\}$, $\{b_m\} \subset G^+$ *are the zeros and poles of* $f(z)$.

## Acknowledgments

The author expresses deep gratitude to Department of Armenian Communities of Calouste Gulbenkian Foundation for sponsoring the author's scientific activities, which particularly led to this publication.

## References

1. R. Nevanlinna, *Springer Verlag* (1936).
2. L. Biberbach, *Palermo Rendiconti*, **38**, 98 (1914).
3. T. Carleman, *Arkhiv för Mat., Astr. och Fys.*, **17** (1922).
4. S. Bergman, *Math. Z.*, **29**, 641 (1929).
5. W. Wirtinger, *Motatshefte für Math. und Phys.*, **39**, 377 (1935).
6. Gh. Hardy, J. E. Littlewood, *Mathematische Zeitschrift*, **34**, 403 (1932).
7. M. V. Keldysch, *Coptes Rendus, Academie des Sciences USSR*, **30**, 778 (1941).
8. J. L. Walsh, *Amer. Math. Soc. Coll. Publ.*, **XX** (1956).
9. A. E. Djrbashian, F. A. Shamoian, *Teubner-Texte zur Math.*, **b105** (1988).
10. H. Hedenmalm, B. Korenblum, K. Zhu, *Springer-Verlag*, (2000).
11. G. M. Gubreev, A. M. Jerbashian, *J. of Operator Theory*, **26**, 155 (1991).
12. F. A. Shamoian, *Math. Notes*, **52**, 727 (1993).
13. F. A. Shamoian, *Sib. Math. J.*, **40**, 1420 (1999).
14. F. A. Shamoian, E. N. Shubabko, *Operator Theory; Advances and Applications, Birkhauser Verlag*, **113**, 332 (2000).
15. F. A. Shamoian, E. N. Shubabko, *Investigations on linear operators and function theory, POMI, St. Petersburg*, **29**, 242 (2001).
16. F. A. Shamoian, *Sib. Math. J.*, **31**, 197 (1990).
17. K. L. Avetisyan, *Analysis Math.*, **26**, 161 (2000).
18. M. M. Djrbashian, *Dokl. Akad. Nauk. Arm. SSR*, **3**, 1 (1945).
19. M. M. Djrbashian, *Soobsch. Inst. Matem. i Mekh. Akad. Nauk Arm. SSR*, **2**, 3 (1948).
20. M. M. Djrbashian, *Nauka*, (1966).
21. M. M. Djrbashian, *Math. USSR Izv.*, **2**, 1027 (1968).
22. M. M. Djrbashian, *Math. USSR Sbornik*, **8**, 493 (1969).
23. M. M. Djrbashian, *Proceedings of the ICM, Vancouver, 1974*, **2**, 197 (1975).
24. M. M. Djrbashian, V. S. Zakarian, *Nauka*, (1993).
25. A. M. Jerbashian, *Izv. Akad. Nauk Arm. SSR, Matematika*, **30**, 39 (1995).
26. A. M. Jerbashian, *Preprint 2002-01, Institute of Mathematics, National Ac. of Sci. of Armenia* (2002).
27. A. M. Jerbashian, *Complex Variables*, **50**, 155 (2005).
28. S. M. Gindikin, *Uspehi Mat. Nauk*, **19**, 3 (1964).
29. A. M. Jerbashian, *Operator Theory: Advances and Applications, Birkhauser Verlag*, **158**, 141 (2005).
30. A. M. Jerbashian, *National Academy of Sciences of Armenia Reports*, **2** (2005).
31. R. R. Coifman, R. Rochberg, *Astérisque*, **77**, 12 (1980).
32. F. Ricci, M. Taibleson, *Annali Scuola Normale Superiore - Pisa, Classe di Scienze, Ser. IV*, **X**, 1 (1983).
33. M. M. Djrbashian, A. E. Djrbashian, *Dokl. Akad. Nauk USSR*, **285**, 547 (1985).
34. A. M. Jerbashian, *J. of Contemp. Math. Analysis*, **28**, 42 (1993).
35. A. H. Karapetyan, *Reference by the Author to a Candidate Thesis* (1988).
36. A. H. Karapetyan, *J. of Contemp. Math. Analysis*, **25** (1990).
37. A. H. Karapetyan, *J. of Contemp. Math. Analysis*, **26**, (1991).
38. A. H. Karapetyan, *J. of Contemp. Math. Analysis*, **27** (1992).

39. A. M. Jerbashian, *Archives of Inequalities and Applications*, **1**, 399 (2003).
40. A. M. Jerbashian, *Advances in Complex Analysis and Applications, Springer* (2005).

# FRACTIONAL MODELING AND APPLICATIONS

S. KEMPFLE and K. KRÜGER

*Helmut–Schmidt–University,*
*Holstenhofweg 85,*
*D–22043 Hamburg, Germany*
*E-mail: Siegmar.Kempfle@hsu-hh.de, Klaus.Krueger@hsu-hh.de*

I. SCHÄFER

*Federal Armed Forces Underwater Acoustics and*
*Marine Geophysics Research Institute*
*Klausdorfer Weg 2–24,*
*D–24148 Kiel, Germany*
*E-mail: Ingo.Schaefer@hsu-hh.de*

In this paper we use fractional pseudo–differential operators in distributional spaces. They are defined via a functional calculus based on Fourier transforms. This procedure is briefly sketched. It turned out that this approach provides very good few–parameter models, particularly for processes which are governed by memory effects. In previous papers we have shown this along measurements and calculations on viscoelastic media. Here we present results on electrodynamic environments.

**Key words:** Fractional Differential Operators, Functional Calculus, Electrodynamic Coils

**Mathematics Subject Classification:** 26A33, 34G10, 47A60, 78A25

## 1. Fractional pseudo-differential operators

### 1.1. *The* $\mathbf{L}_2$*-approach*

We consider the formal linear differential expression

$$\mathcal{A} := D^{\nu_n} + a_{n-1}D^{\nu_{n-1}} + \cdots + a_1 D^{\nu_1} + a_0 \,,$$

$$(a_k \in \mathbb{R}, \quad 0 < \nu_1 < \cdots < \nu_n =: \deg \mathcal{A}) \tag{1}$$

We call it "fractional", if at least one $\nu_k$ is non integer, otherwise we call it "integer". To establish $\mathcal{A}$ as a pseudo–differential operator ($\Psi$DO) in $\mathbf{L}_2$ we use the unitary FOURIER transform of some $\varphi \in \mathbf{L}_2 := \mathbf{L}_2(\mathbb{R})$ which is defined as

$$\widehat{\varphi}(\omega) := \mathcal{F}\varphi(\omega) = \frac{1}{\sqrt{2\pi}}\, p.v. \int\limits_{-\infty}^{\infty} e^{-i\omega t}\varphi(t)\,dt \,, \quad \omega \in \mathbb{R}$$

Here "*p.v.*" denotes the "principal value". The definition works in two steps

302

**Definition 1.1.**

(1) We associat to $\mathcal{A}$ a so–called symbol:

$$a(\omega) := (i\omega)^{\nu_n} + a_{n-1}(i\omega)^{\nu_{n-1}} + \cdots + a_1(i\omega)^{\nu_1} + a_0. \tag{2}$$

(2) Now we set for some $x(t) \in \mathbf{L}_2$

$$\mathcal{A}x(t) := \mathcal{F}^{-1}\left\{a(\omega)\widehat{x}(\omega)\right\}. \tag{3}$$

It is well known (see e.g. [1]) that the set of all $x(t)$ such that $a(\omega)\widehat{x}(t) \in \mathbf{L}_2$ is dense in $\mathbf{L}_2$. Hence $\mathcal{A}$ acts as a densely defined, linear and closed $\Psi$DO in $\mathbf{L}_2$.

**Remark 1.1.** From a mathematical point of view one is nearly free in the choice of possible measurable definitions of the fractional powers in the symbol (2). But to get physically consistent solutions of associated differential equations it is necessary to define all noninteger powers $(i\omega)^\nu$ as principle branches on the logarithmic RIEMANNian surface with branchcut $\mathbb{R}_0^-$ (see e.g. [2]). In accordance with computer systems like *Mathematica, Maple, Mathlab*, etc., this is precisely

$$(i\omega)^{\nu_k} := |\omega|^\nu \exp\left(i\,\mathrm{sign}(\omega)\,\nu\pi/2\right). \tag{4}$$

**Remark 1.2.** Consequently a fractional derivative of order $\nu$ is given via

$$D^\nu x(t) := \mathcal{F}^{-1}\left\{(i\omega)^\nu \widehat{x}(\omega)\right\}.$$

But the point that we start from an entire operator (3) is the feature of the so–called *functional calculus*, namely, that the analysis of fractional operators and solutions of associated equations can be done as analysis of the symbol (2).

## 1.2. $\Psi DO$ in distributional spaces

The above done definition can be embedded in all function spaces in which the FOURIER transformation can be established. Easily this can be done into the space of tempered distributions $\mathcal{S}'$ (see e.g. [3]), because the FOURIER transformation acts as an isomorphism on the test space $\mathcal{S}$. But it turns out in practice that $\mathcal{S}'$ is not sufficient, e.g., for modeling non stable processes, where exponential growth of the solutions may occur. An appropriate function space is the space of distributions $\mathcal{D}'$ where the test space $\mathcal{D}$ consists of all $C^\infty$– functions with compact support.

**Remark 1.3.** For applications the following notes are helpful:

(1) The space $\mathbf{L}_{\mathrm{loc}}$ of locally integrable functions $\mathbb{R} \to \mathbb{C}$ can be identified with the space of *regular distributions* in $\mathcal{D}'$:

$$\text{If } f \in \mathbf{L}_{\mathrm{loc}}, \quad \text{then} \quad f \stackrel{\mathrm{id.}}{:=} \langle f, \varphi \rangle := \int_{-\infty}^{\infty} f(t)\,\varphi(t)\,\mathrm{d}t \quad \forall \varphi \in \mathcal{D}.$$

(2) Any element $f \in \mathcal{D}'$ is $\mathcal{D}'$–limit of $\mathbf{L}_{\mathrm{loc}}$–sequences.

(3) Every $\mathcal{D}'$–element has unique derivatives of arbitrary order which fulfil the common differentiation rules. Vice versa every $\mathcal{D}'$–element has primitives which only differ by constants.

(4) In general, the approach to products in $\mathcal{D}'$ is very expensive (see e.g. [4]). But, sufficient for most applications, one can use properties (1), (2) to embed point-wise $\mathbf{L}_{\mathrm{loc}}$–products and $\mathcal{D}'$–limits of according sequences into $\mathcal{D}'$.

(5) In physics and engineering it has become common to use consequently the function notation instead of functionals. E.g., the DIRAC–impact is briefly written $\delta(t)$.

It is not topic of this paper to describe in detail the more sophisticated procedure to define FOURIER transforms on $\mathcal{D}'$. Here we refer to the literature (e.g. [5-7]). Roughly, it works as follows:

Firstly, the FOURIER transform $\mathcal{F}$ is substituted by the so–called FOURIER–LAPLACE transform $\mathcal{F}_{\mathbb{C}}$ by changing the variable $\omega \in \mathbb{R}$ into $\zeta \in \mathbb{C}$. Now, $\mathcal{F}_{\mathbb{C}}$ provides a one– to–one mapping on a very good natured complex function space, usually denoted $\mathcal{Z}$ and often called "space of *hyperfunctions*". All its elements are entire functions which are rapidly decreasing with respect to the real part of $\zeta$ and "exponentially" bounded with respect to the imaginary part (see e.g. [8]). The one–to–one property ensures the existence of an inverse transform $\mathcal{F}_{\mathbb{C}}^{-1}$.

Obviously, $\mathcal{F}$ maps $\mathcal{D}$ on $\mathcal{Z}|_{\mathbb{R}}$. Every $\varphi(z) \in \mathcal{Z}$ can be reconstructed as unique analytic continuation of its restriction $\varphi|_{\mathbb{R}}(z)$. This way the inverse FOURIER transform is also well defined on $\mathcal{F}(\mathcal{D}) = \mathcal{Z}|_{\mathbb{R}}$. Now, FOURIER transforms togeteher with their inverses can easily be established on the dual spaces (spaces of bounded linear functionals) $\mathcal{D}', \mathcal{Z}'$ via

**Definition 1.2.**

(1) For $f \in \mathcal{D}'$ define: $\mathcal{F}\{f\} := \left\langle \widehat{f}, \phi \right\rangle := \left\langle f, \widehat{\phi} \right\rangle \quad \forall \phi \in \mathcal{Z}|_{\mathbb{R}}$,

(2) For $g \in \mathcal{Z}'$ define: $\mathcal{F}\{g\} := \langle \widehat{g}, \varphi \rangle := \langle f, \widehat{\varphi} \rangle \quad \forall \varphi \in \mathcal{D}$,

**Remark 1.4.** Of course this definition is not very helpful to calculate explicitely some $\widehat{f}, \widehat{g}$. But it is very important that this definition preserves by nature all rules of FOURIER transforms. Thus, one can e.g. start from any known FOURIER transform and generate chains of $\mathcal{D}'$–correspondencies via the differentiation laws.

We are now ready to extend Definition 1.1 literally from $\mathbf{L}_2$ into $\mathcal{D}'$.

## 1.3. *Properties*

(1) In contrast to other fractional calculi (see e.g. [9,10]) integer ordered operators are included ex definition.

(2) Moreover, our operators inherit the algebraic properties from the class of symbols. In fact, they form a commutative $\mathbb{R}$-algebra. Particularly, we get from

the associativity of the operator product for fractional operators $\mathcal{A}_1, \mathcal{A}_2$ with symbols $a_1, a_2$ the so–called semigroup property:

$$\mathcal{A}_{12} := \mathcal{A}_1 \mathcal{A}_2 = \mathcal{A}_2, \mathcal{A}_1 \quad \text{with symbol} \quad a_{12} = a_1 a_2$$

(3) Physical laws have to be translation invariant with respect to time. It is well known that approaches via one sided LAPLACE–transforms do not have this property in contrast to our approach. Precisely,

Let $x_\tau(t) := x(t - \tau)$ denote a shifted function (fixed $\tau$) and $\mathcal{A}$ from definition 1.1. Then

$$\mathcal{A} x_\tau(t) = \mathcal{A} x(t - \tau) \quad \text{for all} \quad t \in \mathbb{R}$$

(4) Let again $\mathcal{A}$ be a fractional $\Psi$DO via Definition 1.1. It is easily seen that $1/a \in \mathcal{D}'$. This way $1/a$ acts as symbol for the inverse operator $\mathcal{A}^{-1}$

$$\mathcal{A}^{-1} \mathcal{A} = \mathcal{A} \mathcal{A}^{-1} = \mathcal{F}^{-1} a \, \mathcal{F} \mathcal{F}^{-1} \frac{1}{a} \mathcal{F} = .. = id$$

(5) In accordance with item 1 the integer ordered differentiation rule for convolutions can be extended to our fractional derivatives (as far as the convolutions exist). We state briefly (for a proof see e.g. [11])

$$D^\nu(f * g) = (D^\nu f) * g = f * (D^\nu g)$$

(6) From this and the semigroup property one can easily verify (see also [11]) that our derivatives coincide partially with the RIEMANN–LIOUVILLE– and the CAPUTO–definition with lower bound $-\infty$:

Let $n - 1 < q < n$, $n \in \mathbb{N}$. Then it holds, if the accordant integral exists

$$D^q f(t) = {}_{-\infty} D_t^q \, f(t) = \frac{1}{\Gamma(n-q)} \frac{d^n}{dt^n} \int_{-\infty}^{t} \frac{f(\tau) \, d\tau}{(t-\tau)^{q-n+1}}$$

$$= {}_{-\infty}^{C} D_t^q \, f(t) = \frac{1}{\Gamma(n-q)} \int_{-\infty}^{t} \frac{f^{(n)}(\tau) \, d\tau}{(t-\tau)^{q-n+1}}$$

## 1.4. Scope

The crucial point is the product $(i\omega)^\nu \hat{x}(\omega)$. For $\nu \notin \mathbb{N}$ the symbol is of course no $\mathcal{D}'$– or $\mathcal{Z}'$– multiplier. But from $\nu > 0$, it follows that the pointwise product of $(i\omega)^\nu$ with $L_{\text{loc}}$–functions is again locally integrabel. Concerning singular distributions, we conclude from $(i\omega)^\nu \in C^\infty(\mathbb{R} \backslash \{0\})$ that difficulties can only occur, if $\hat{x}(\omega)$ is singular in the origin. Thus the scope of our approach is sufficiently large for nearly almost applications. This is demonstrated by the following list of fractional $\mathcal{D}'$–derivatives. The reader may check, that they include indeed the integer derivatives.

**Examples 1.1.** The unit step function is denoted by $\theta(t)$.

(1) $\quad D^\nu \delta(t) = \dfrac{\theta(t)}{\Gamma(-\nu)} t^{-\nu-1} \quad (\nu \notin \mathbb{N})$

(2) $\quad D^\nu c = 0 \qquad (\nu \in \mathbb{R}^+)$

(3) $\quad D^\nu \{\theta(t)\, t^p\} = \dfrac{\theta(t)\,\Gamma(p+1)}{\Gamma(p+1-\nu)}\, t^{p-\nu} \qquad p, \nu > 0$

(4) $\quad D^\nu \sin(at+b) = a^\nu \sin\left(at+b+\dfrac{a\pi}{2}\right) \quad \text{for} \quad a > 0$

(5) $\quad D^\nu e^{at+b} = a^\nu e^{at+b}$

(6) $\quad D^\nu \{t\, e^{at+b}\} = e^{at+b}\left(a^\nu t + \nu a^{\nu-1}\right) \quad (a \neq 0).$

(7) $\quad D^\nu \{t2\, e^{at+b}\} = e^{at+b}\left(a^\nu t2 + 2\nu a^{\nu-1} t + \nu(\nu-1) a^{\nu-2}\right)$

(8) $\quad D^\nu \{t^m\, e^{at+b}\} = e^{at+b}\left(a^\nu t^m + \binom{m}{1}\nu a^{\nu-1} t^{m-1} + \ldots\right)$

(9) $\quad D^\nu \left\{e^{at} \sin(\sigma t)\right\} = e^{at} r^\nu \sin(\sigma t + \nu \vartheta) \quad (\sigma \in \mathbb{R}^+)$

$\quad$ where $r = \sqrt{a2 + \sigma^2}; \ \tan(\vartheta) = |a|/\sigma.$

(10) $\quad D^\nu t^r = (-1)^\nu \dfrac{\Gamma(\nu - r)}{\Gamma(-r)} t^{r-\nu} \qquad (r, r - \nu \notin \mathbb{N}_0)$

(11) $\quad D^n t^r = \dfrac{\Gamma(r+1)}{\Gamma(r+1-n)} t^{r-n} \qquad (r \in \mathbb{R}; \ n \in \mathbb{N}_0)$

## 2. Applications

In previous papers (e.g. [12-14]) we have intensively described the handling of our approach. Moreover we discussed the behaviour of solutions (impulse responses) including stability and causality. Thus we omit in this paper a repetition and concentrate now in presenting concrete results to demonstrate the capability of this approach to applications. We emphasize that this approach is predestinated for modeling memory effects as well as steady state solutions, simply, because the "past" is ex definition included via FOURIER transforms.

### 2.1. *Viscoelastic rods*

We resume some results from previous papers (e.g. [14-16]). On viscoelastic rods of different materials impulse responses and frequency responses were measured and compared with calculations via some classical and a simple fractional model. It turned out that only the fractional model provided good results over a broad range

306

of frequencies (up to 12000 Hz) as well as over a broad range of masses. Even effects like hysteresis and dispersion were modeled, in opposite to the classical models.

## 2.2. Electrodynamic coils with iron cores

To be able to assess the suitability of the conventional and the fractional model, different real magnetic core coils are examined. Measured were transfer functions as well as step responses.

### 2.2.1. Transfer functions

The transfer function of some system is defined as the quotient of output and input in frequency domain. The setup for measurement on an iron core coil is depicted in Fig. 1. An ohmic resistor $R_M$ is connected in series with the respective coil consisting of an ohmic Resistor $R_{Cu}$ and an inductivity $L_\alpha$.

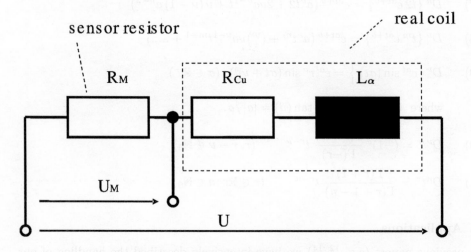

Fig. 1. Set up for measurement of transfer function.

From KIRCHHOFF rules one gets easily the mathematical model

$$U = U_M + \frac{R_{Cu}}{R_M}U_M + \frac{L_\alpha}{R_M}D^\alpha U_M$$

and via FOURIER transforms the transfer function

$$\widehat{G} := \frac{\widehat{U}}{\widehat{U}_M} = \frac{R_M + R_{Cu} + (i\omega)^\alpha L_\alpha}{R_M}.$$

For $\alpha = 1$ this is the conventional model whereas the fractional one is given for $0 < \alpha < 1$. It turns out that in case of ferrite core coils the transfer functions of both models are not significant different. But in case of soft iron cores, due to the high selfinductivity, the classical model is in opposite to the fractional one not able to model the system sufficiently good. This can be impressively seen in Fig. 2. We remark that soft iron core coils are in use for high fidelity loadspeakers. In our model the adequate fractional derivative was of degree about $\alpha = 0.6$.

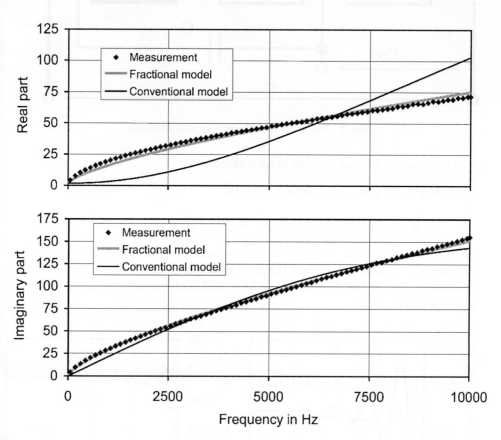

Fig. 2.   Transfer functions of a soft iron core coil.

### 2.2.2. Step responses

To measure step responses, i.e., the input is given by $U_M(t) = \theta(t)$, the setup has to be modified by an additional capacity parallel to the coil, as depicted in Fig. 3.

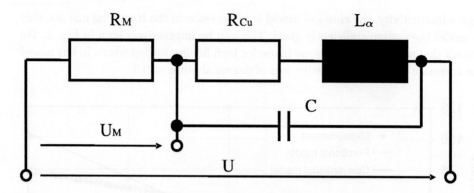

Fig. 3. Set up for measurement of step responses.

The equation for this network in time domain is

$$R_M U + R_M R_{C_u} C\dot{U} + R_M C L_\alpha D^{\alpha+1} U$$

$$= (R_M + R_{C_u})U_M + L_\alpha D^\alpha U_M + R_M R_{C_u} C\dot{U}_M + R_M C L_\alpha D^{\alpha+1} U_M .$$

Again, $\alpha = 1$ is the conventional model and again, in case of soft iron core coils, the conventional equation models not the real behaviour of the system. This shows the left hand picture of Fig. 4, whereas the fractional model is in very good accordance with the measurement.

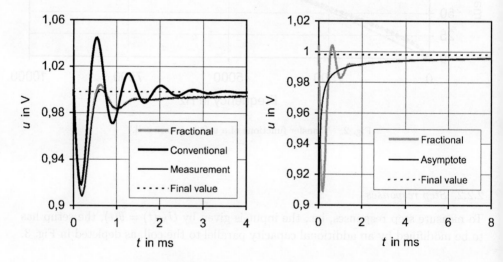

Fig. 4. Step responses of an soft iron core coil.

Typically for fractional models is the rapid damping of oscillations which decay into a totally monotonic creeping function of power law with negative exponent (see e.g. [12,14]). This is demonstrated in the right hand picture of Fig. 4.

# References

1. G. Grubb, Pseudodifferential boundary problems and applications. *DMV Jahresbericht* **99**, No 3 (1997), 110-121.
2. S. Kempfle and I. Schäfer, Physical consistency of fractional ΨDO, *Mathematica Balkanica* **18** (3–4), (2004), 345–353.
3. E. M. Stein and G. Weiss, *Introduction to Fourier Analysis on Euclidean Spaces.* University Press, Princeton (1971).
4. M. Reed and B. Simon, *Methods of Modern Mathematical Physics.* Vol. 1, Academic Press, New York (1980).
5. A.H.Zemanian, *Distribution Theory and Transform Analysis.* Reprint, Dover Publications, New York (1987).
6. I. M. Gelfand and G. E. Shilov, *Generalized Functions.* vol. 2, Academic Press, New York (1968).
7. W. Walter, *Einführung in die Theorie der Distributionen.*, 3.Aufl. Bibliographisches Institut Mannheim (1994).
8. W. Rudin, *Functional Analysis*, McGraw–Hill, New York (1973).
9. S. G. Samko, A. A. Kilbas and O. I. Marichev, *Fractional Integrals and Derivatives.* Gordon and Breach Science Publishers, London (1993).
10. I. Podlubny, *Fractional Differential Equations.* Academic Press, San Diego (1999).
11. S. Kempfle, I. Schäfer, Functional Calculus and a Link to Fractional Calculus. *Fractional Calculus & Applied Analysis*, **5** (4) (2002), 411-426.
12. S. Kempfle and H. Beyer, Global and causal solutions of fractional differential equations *Transform Methods & Special Funtions, Varna '96, Proc. 2nd Intern. Workshop*, (Eds. P.Rusev, I. Dimovski, V. Kiryakova), IMI-BAS, Sofia (1998), 210-226.
13. S. Kempfle, Causality criteria for solutions of linear fractional differential operators. *Fractional Calculus & Applied Analysis*, **1** (4) (1998), 351-364.
14. S. Kempfle, I. Schäfer and H.Beyer, Fractional Calculus via Functional Calculus: Theory and Applications, *Nonlinear Dynamics* **29** (2002), 99-127.
15. I. Schäfer, Fraktionale Zeitableitungen zur Beschreibung viskoelastischen Materialverhaltens, *Dissertation (Thesis)*, Hamburg (2001).
16. I. Schaefer, H.–J. Seifert, Description of the impulse response in rods by fractional derivatives. *ZAMM* **82**, (2002), 423-427.

# EULER-TYPE FRACTIONAL DIFFERENTIAL EQUATIONS *

A.A. KILBAS

*Belarusian State University,*
*Independence Avenue 4,*
*220050 Minsk, Belarus*
*E-mail: kilbas@bsu.by; akoroleva@tur.by*

M. RIVERO and J.J. TRUJILLO

*Faculdad de Matematicas,*
*Universidad de La Laguna,*
*38271 La Laguna-Tenerife, Spain*
*E-mail: jtruill@ull.es*

Two linear non-homogeneous ordinary differential equations with the Liouville derivatives of fractional order are considered. Using the direct and inverse Mellin integral transforms, a general approach is given to deduce particular solutions of the considered equations. In the special case particular solutions of such equations are expressed in terms of the generalized Wright function, of the Gauss hypergeometric function and of the Euler psi function.

**Key words:** Linear non-homogeneous differential equations with Liouville fractional derivatives, Mellin transforms
**Mathematic Subject Classifications:** 34A05, 26A33, 44A99

## 1. Introduction

Let $D_{0+}^{\alpha}y$ and $D_{-}^{\alpha}y$ be the left- and the right-hand sided Liouville fractional derivatives of order $\alpha > 0$ defined by [6, (5.8)]

$$(D_{0+}^{\alpha}y)(x) = \left(\frac{d}{dx}\right)^n \frac{1}{\Gamma(n-\alpha)} \int_0^x \frac{y(t)dt}{(x-t)^{\alpha-n+1}} \quad (x>0;\; n=-[-\alpha]), \quad (1.1)$$

$$(D_{-}^{\alpha}y)(x) = \frac{1}{\Gamma(n-\alpha)} \left(-\frac{d}{dx}\right)^n \int_x^\infty \frac{y(t)dt}{(t-x)^{\alpha-n+1}} \quad (n=-[-\alpha];\; x>0). \quad (1.2)$$

Consider the linear non-homogeneous differential equations

$$\sum_{k=0}^{m} A_k x^{\alpha+k} \left(D_{0+}^{\alpha+k}y\right)(x) = f(x) \quad (x>0;\; \alpha>0), \quad (1.3)$$

---

*This work is supported by Belarusian Fundamental Research Fund, Project F05MC-050

312

$$\sum_{k=0}^{m} B_k x^{\alpha+k} \left( D_-^{\alpha+k} y \right)(x) = f(x) \quad (x > 0; \ \alpha > 0),\tag{1.4}$$

with real constants $A_k, \ B_k \in \mathbb{R} \ (k = 0, \cdots, m)$.

When $\alpha = n \in \mathbb{N}$, then, in accordance with (1.1) and (1.2), $(D_{0+}^{\alpha} y)(x) = y^{(n)}(x)$ and $(D_-^{\alpha} y)(x) = (-1)^n y^{(n)}(x)$. Therefore when $\alpha = 0$, equations (1.3) and (1.4) are reduced to the linear ordinary differential equations

$$\sum_{k=0}^{m} A_k x^k y^{(k)}(x) = f(x), \quad (x > 0; \ n \in \mathbb{N})\tag{1.5}$$

and

$$\sum_{k=0}^{m} (-1)^k A_k x^k y^{(k)}(x) = f(x) \quad (x > 0; \ n \in \mathbb{N}),\tag{1.6}$$

respectively. These equations are known as Euler equations; and their solutions are reduced to solutions of ordinary linear differential equations by using the change $t = \log|x|$; for example, see [11, equation 5.2.27]. Therefore we call (1.3) and (1.4) *Euler-type fractional differential equations*.

We present a general approach to solve the linear non-homogeneous equations (1.3) and (1.4) by using the one-dimensional direct and inverse Mellin integral transforms $M\varphi$ and $M^{-1}g$. The Mellin transform of a function $\varphi(x)$ of a real variable $x \in \mathbb{R}_+ = (0, \infty)$ is defined by

$$(M\varphi)(s) := \int_0^\infty t^{s-1} \varphi(t) dt \quad (s \in \mathbb{C}),\tag{1.7}$$

and the inverse Mellin transform is given for $x \in \mathbb{R}_+$ by the formula

$$(M^{-1}g)(x) := \frac{1}{2\pi i} \int_{\gamma-i\infty}^{\gamma+i\infty} x^{-s} g(s) ds \quad (\gamma = \mathrm{Re}(s)).\tag{1.8}$$

One may find properties of these Mellin transforms in the books by Sneddon [7] and Ditkin and Prudnikov [1].

We use such an approach to establish the explicit solutions of equations (1.3) and (1.4) in the case $m = 1$ in terms of the special cases of the generalized Wright function $_p\Psi_q(z)$. Such a function is defined for complex $z \in \mathbb{C}$, $a_i, b_j \in \mathbb{C}$, and real $\alpha_i, \beta_j \in \mathbb{R} \ (i = 1, 2, \cdots p; \ j = 1, 2, \cdots, q)$ by the series [2, Section 4.1]:

$$_p\Psi_q(z) \equiv {}_p\Psi_q \left[ \begin{matrix} (a_i, \alpha_i)_{1,p} \\ (b_i, \beta_i)_{1,q} \end{matrix} \ \Big| \ z \right] = \sum_{k=0}^{\infty} \frac{\prod_{i=1}^{p} \Gamma(a_i + \alpha_i k)}{\prod_{j=1}^{q} \Gamma(b_j + \beta_j k)} \frac{z^k}{k!}.\tag{1.9}$$

This function was introduced by Wright [8] who in [8–10] established main terms of its asymptotic expansions at infinity under the condition

$$\sum_{j=1}^{q} \beta_j - \sum_{i=1}^{p} \alpha_i > -1.\tag{1.10}$$

Properties of (1.9) were investigated in [3], where, in particular it was proved that $_p\Psi_q(z)$ is an entire function provided that (1.10) is valid.

Note that Podlubny [5, Section 6.1) indicated that the Mellin transform (1.5) can be applied to solve the Cauchy problem for the simplest fractional differential equation (1.3) with $m = 1$ and $0 < \alpha < 1$:

$$x^{\alpha+1}\left(D_{0+}^{\alpha+1}y\right)(x) + x^{\alpha}\left(D_{0+}^{\alpha}y\right)(x) = f(x) \quad (x > 0). \tag{1.11}$$

## 2. General approach

The Mellin transform method for solving equations (1.3) and (1.4) is based on the following relations:

$$\left(\mathbb{M}x^{\alpha+k}D_{0+}^{\alpha+k}y\right)(s) = \frac{\Gamma(1-s)}{\Gamma(1-s-\alpha-k)}(\mathbb{M}y)(s) \tag{2.1}$$

and

$$\left(\mathbb{M}x^{\alpha+k}D_{-}^{\alpha+k}y\right)(s) = \frac{\Gamma(s+\alpha+k)}{\Gamma(s)}(\mathbb{M}y)(s), \tag{2.2}$$

for the left- and right-hand sided Liouvile fractional derivatives (1.1) and (1.2), respectively. These formulas are valid for suitable functions $y(x)$; see [6, §7.3].

Applying the Mellin transform to (1.3), (1.4) and using (2.1), (2.2), we have

$$\left[\sum_{k=0}^{m} A_k \frac{\Gamma(1-s)}{\Gamma(1-s-\alpha-k)}\right](\mathbb{M}y)(s) = (\mathbb{M}f)(s) \tag{2.3}$$

and

$$\left[\sum_{k=0}^{m} B_k \frac{\Gamma(s+\alpha+k)}{\Gamma(s)}\right](\mathbb{M}y)(s) = (\mathbb{M}f)(s), \tag{2.4}$$

respectively. Applying the inverse Mellin transform to (2.3) and (2.4), we derive solutions of equations (1.3) and (1.4) in the respective forms:

$$y(x) = \left(\mathbb{M}^{-1}\left[\frac{1}{P_{\alpha}^1(1-s)}(\mathbb{M}f)(s)\right]\right)(x), \quad P_{\alpha}^1(s) = \sum_{k=0}^{m} A_k \frac{\Gamma(s)}{\Gamma(s-\alpha-k)} \tag{2.5}$$

and

$$y(x) = \left(\mathbb{M}^{-1}\left[\frac{1}{P_{\alpha}^2(s)}(\mathbb{M}f)(s)\right]\right)(x), \quad P_{\alpha}^2(s) = \sum_{k=0}^{m} B_k \frac{\Gamma(s+\alpha+k)}{\Gamma(s)}. \tag{2.6}$$

Now we introduce the functions:

$$G_{\alpha}^1(x) = \left(\mathbb{M}^{-1}\left[\frac{1}{P_{\alpha}^1(1-s)}\right]\right)(x), \quad P_{\alpha}^1(s) = \sum_{k=0}^{m} A_k \frac{\Gamma(s)}{\Gamma(s-\alpha-k)}, \tag{2.7}$$

$$G_{\alpha}^2(x) = \left(\mathbb{M}^{-1}\left[\frac{1}{P_{\alpha}^2(s)}\right]\right)(x), \quad P_{\alpha}^2(s) = \sum_{k=0}^{m} B_k \frac{\Gamma(s+\alpha+k)}{\Gamma(s)}. \tag{2.8}$$

Applying the well known Mellin convolution property:

$$\left( M \left( \int_0^\infty k \left( \frac{x}{t} \right) f(t) \frac{dt}{t} \right) \right) (s) = (Mk)(s)(Mf)(s), \tag{2.9}$$

we present solution (2.5) in the form:

$$y(x) = \int_0^\infty G_\alpha^1(t) f(xt) dt, \tag{2.10}$$

while solution (2.6) in the form of the Mellin convolution of $G_\alpha^2(x)$ and $f(x)$:

$$y(x) = \int_0^\infty G_\alpha^2 \left( \frac{x}{t} \right) f(t) \frac{dt}{t}. \tag{2.11}$$

By an analogy with the ordinary differential equations, functions $G_\alpha^1(x)$ and $G_\alpha^2(x)$ in (2.7) and (2.8) can be called as the *Mellin fractional analogues* of the Green function. Further we deduce their explicit representations for equations (1.3) and (1.4) in the case $m = 1$.

## 3. Equations with Left-Sided Fractional Derivatives

We consider equation (1.3) with $m = 1$:

$$x^{\alpha+1} \left( D_{0+}^{\alpha+1} y \right) (x) + \lambda x^\alpha \left( D_{0+}^\alpha y \right) (x) = f(x) \quad (x > 0; \ \alpha > 0; \ \lambda \in \mathbb{R}). \tag{3.1}$$

Particular solutions of this equation are different in the cases $\lambda \neq n+1$ and $\lambda = n+1$ ($n \in \mathbb{N}_0$). In the first case a particular solution of (3.1) is expressed in terms of the generalized Wright function (1.9) with $p = 1$ and $q = 2$ of the form:

$$_1\Psi_2 \left[ \begin{array}{c} (a,1) \\ \\ (b,-1),(c,1) \end{array} \middle| \ z \right] \quad (a,b,c,z \in \mathbf{C}), \tag{3.2}$$

and of the Gauss hypergeometric function $_2F_1(a,b;c;z)$. Such a function with $a,b,c \in \mathbb{C}$ ($c \neq 0, -1, -2, \cdots$) and $z \in \mathbb{C}$ is defined by the hypergeometric series:

$$_2F_1(a,b;c;z) = \sum_{k=0}^\infty \frac{(a)_k (b)_k}{(c)_k} \frac{z^k}{k!} \quad (|z| < 1), \tag{3.3}$$

where $(a)_k$ is the Pochhammer symbol:

$$(z)_0 = 1 \quad \text{and} \quad (z)_n = z(z+1) \cdots (z+n-1) \quad (z \in \mathbb{C}, \ n \in \mathbb{N}). \tag{3.4}$$

Note that the series in (3.3) converges absolutely for $|z| < 1$ and $|z| = 1$, $\mathrm{Re}(c - a - b) > 0$ [2, Section 2.1].

There hold the following assertions.

**Lemma 1.** *The generalized Wright function (3.2) is defined for complex $z \in \mathbb{C}$ such that $|z| < 1$ and $|z| = 1$, $\mathrm{Re}(c - a + b - 1) > 0$.*

**Lemma 2.** *Let $a, b, c, z \in \mathbb{C}$ $(a \neq 0, -1, -2, \cdots)$ and let $z \in \mathbb{C}$ be such that $|z| < 1$ or $|z| = 1$, $\mathrm{Re}(c - a + b - 1) > 0$. Then*

$$
{}_1\Psi_2 \left[ \begin{matrix} (a, 1) \\ (b, -1), (c, 1) \end{matrix} \, \middle| \, z \right] = \frac{\Gamma(a)}{\Gamma(b)\Gamma(c)} \, {}_2F_1(a, 1 - b; c; -z) \quad (a, b, c, z \in \mathbb{C}). \tag{3.5}
$$

**Proof.** Lemma 1 follows from $[^3$, Theorem 1.3], while Lemma 2 from (1.9), (3.3) and the following property of the gamma function:

$$
\Gamma(a - k) = (-1)^k \frac{\Gamma(a)}{(1 - a)_k} \quad (a \neq 0, -1, -2, \cdots). \tag{3.6}
$$

**Theorem 1.** *Let $\alpha > 0$ and $\lambda \in \mathbb{R}$ $(\lambda \neq n + 1; \; n \in \mathbb{N}_0)$. Then the fractional differential equation (3.1) is solvable, and its particular solution is given by*

$$
y(x) = \int_0^1 G^1_{\alpha, \lambda}(t) f(xt) dt, \tag{3.7}
$$

*where*

$$
G^1_{\alpha, \lambda}(x) = x^{-\alpha} \left\{ \frac{\Gamma(1 - \lambda)}{\Gamma(\alpha + 1 - \lambda)} x^{\lambda - 1} - {}_1\Psi_2 \left[ \begin{matrix} (1 - \lambda, 1) \\ (\alpha, -1), (2 - \lambda, 1) \end{matrix} \, \middle| \, -x \right] \right\}
$$

$$
= x^{-\alpha} \left\{ \frac{\Gamma(1 - \lambda)}{\Gamma(\alpha + 1 - \lambda)} x^{\lambda - 1} - \frac{1}{(1 - \lambda)\Gamma(\alpha)} \, {}_2F_1(1 - \lambda, 1 - \alpha; 2 - \lambda; x) \right\}. \tag{3.8}
$$

**Proof.** (3.1) is equation (1.3) with $m = 1$, $A_1 = 1$ and $A_0 = \lambda$, for which (2.3) takes the form:

$$
\frac{(\lambda - s - \alpha)\Gamma(1 - s)}{\Gamma(1 - s - \alpha)} \, (\mathfrak{M}y)(s) = (\mathfrak{M}f)(s). \tag{3.9}
$$

Hence the Mellin fractional analogue of the Green function in (2.7) is given by

$$
G^1_\alpha(x) = \left( \mathfrak{M}^{-1} \left[ \frac{\Gamma(s - \alpha)}{(s - \alpha - 1 + \lambda)\Gamma(s)} \right] \right)(x) =: G^1_{\alpha, \lambda}(x), \tag{3.10}
$$

and, in accordance with (2.10), a particular solution of (3.1) has the form:

$$
y(x) = \int_0^\infty G^1_{\alpha, \lambda}(t) f(xt) dt. \tag{3.11}
$$

First we show that $G^1_{\alpha, \lambda}(x) = 0$ for $x > 1$. By (3.9), we have

$$
\left( \mathfrak{M} G^1_{\alpha, \lambda} \right)(s) = (\mathfrak{M}G_1)(s) \, (\mathfrak{M}G_2)(s), \tag{3.12}
$$

where

$$
(\mathfrak{M}G_1)(s) = \frac{\Gamma(s - \alpha)}{\Gamma(s)}, \quad (\mathfrak{M}G_2)(s) = \frac{1}{s - \alpha - 1 + \lambda}. \tag{3.13}
$$

The direct application of the Mellin transform leads to the following relations:

$$G_1(x) = \begin{cases} \dfrac{x^{-\alpha}(1-x)^{\alpha-1}}{\Gamma(\alpha)}, & 0 < x < 1, \\ 0, & x > 1; \end{cases} \quad G_2(x) = \begin{cases} x^{-\alpha-1+\lambda}, & 0 < x < 1, \\ 0, & x > 1. \end{cases} \quad (3.14)$$

Thus, according to (2.9), we have

$$G_{\alpha,\lambda}^1(x) = \int_0^\infty G_1\left(\frac{x}{t}\right) G_2(t) \frac{dt}{t}, \qquad (3.15)$$

and it follows from (3.14) that $G_{\alpha,\lambda}^1(x) = 0$ for $x > 1$. Therefore, (3.11) yields (3.7).

Now we show that $G_{\alpha,\lambda}^1(x)$ is given by (3.8). By (3.10) and (1.8), we have

$$G_{\alpha,\lambda}^1(x) = \frac{1}{2\pi i} \int_{\gamma-i\infty}^{\gamma+i\infty} \frac{\Gamma(s-\alpha)}{(s-\alpha-1+\lambda)\Gamma(s)} x^{-s} ds. \qquad (3.16)$$

Since $\lambda \neq n+1$ $(n \in \mathbb{N}_0)$, then the pole $s = \alpha+1-\lambda$ of the integrand in (3.16) does not coincide with any pole $s_k = \alpha - k$ $(k \in \mathbb{N}_0)$ of $\Gamma(s-\alpha)$. If we choose $\gamma > \max[\alpha+1-\operatorname{Re}(\lambda), \alpha]$, then evaluation of the residues at the above poles yields

$$G_{\alpha,\lambda}^1(x) = \frac{\Gamma(1-\lambda)}{\Gamma(\alpha+1-\lambda)} x^{-\alpha+\lambda-1} - \sum_{k=0}^{\infty} \frac{(-1)^k}{k!} \frac{\Gamma(k+1-\lambda)}{\Gamma(k+2-\lambda)\Gamma(\alpha-k)} x^{-\alpha+k}.$$

Hence, by (3.2), (1.9) and (3.5), $G_{\alpha,\lambda}^1(x)$ is given by (3.8), and Theorem 1 is proved.

**Example 1.** Equation (3.1) with $\alpha = 1/2$ and $\lambda \neq n+1$ $(n \in \mathbb{N}_0 := \{0,1,2,\cdots\})$:

$$x^{3/2}\left(D_{0+}^{3/2} y\right)(x) + \lambda x^{1/2}\left(D_{0+}^{1/2} y\right)(x) = f(x) \quad (x > 0;\ \alpha > 0;\ \lambda \in \mathbb{R}) \quad (3.17)$$

has a particular solution

$$y(x) = \int_0^1 G_{1/2,\lambda}^1(t) f(xt) dt, \qquad (3.18)$$

where $G_{1/2,\lambda}^1(x)$ is given by (3.8) with $\alpha = 1/2$.

**Example 2.** The function

$$y(x) = \frac{1}{1-\lambda} \int_0^1 \left(t^{\lambda-2} - \frac{1}{t}\right) f(xt) dt. \qquad (3.19)$$

is a particular solution of the differential equation of second order:

$$x^2 y''(x) + \lambda x y'(x) = f(x) \quad (x > 0;\ \lambda \neq 1;\ n \in \mathbb{N}_0). \qquad (3.20)$$

Note that, for $\lambda \neq n+1$ $(n \in \mathbb{N}_0)$, (3.7) with $\alpha = 1$ yields the solution (3.19), but it is directly verified that (3.19) satisfies the equation (3.20) with any $\lambda \neq 1$.

Next we give a particular solution to equation (3.1) with $\lambda = n+1$ $(n \in \mathbb{N}_0)$:

$$x^{\alpha+1}\left(D_{0+}^{\alpha+1} y\right)(x) + (n+1)x^\alpha\left(D_{0+}^\alpha y\right)(x) = f(x) \quad (x > 0;\ \alpha > 0;\ n \in \mathbb{N}_0), \quad (3.21)$$

in terms of the Euler psi function

$$\psi(z) = \frac{d}{dz} \log \Gamma(z) = \frac{\Gamma'(z)}{\Gamma(z)} \quad (z \in \mathbb{C}) \qquad (3.22)$$

and of the Euler constant

$$\gamma = -\psi(1) = -\Gamma'(1); \tag{3.23}$$

see [2, Section 1.7]. The results are different in the cases $n \in \mathbb{N}$ and $n = 0$.

**Theorem 2.** Let $\alpha > 0$ and $n \in \mathbb{N}_0$ be such that $\alpha \neq 1, \cdots, n$ for $n > 0$. Then the fractional differential equation (3.21) is solvable, and its particular solution has the form:

$$y(x) = \int_0^1 G^1_{\alpha,n+1}(t)f(xt)dt, \tag{3.24}$$

where

$$G^1_{\alpha,n+1}(x) = \frac{(-1)^{n-1}x^{n-\alpha}}{n!\Gamma(\alpha-n)}\left[\log(x) + \sum_{j=0}^{n-1}\frac{1}{j-n} + \psi(\alpha-n) + \gamma\right]$$

$$+ \sum_{k=0 \ (k\neq n)}^{\infty}\frac{(-1)^kx^{k-\alpha}}{k!(n-k)\Gamma(\alpha-k)}. \tag{3.25}$$

**Proof.** Let $n > 0$. By the proof of Theorem 1 the solution of equation (3.21) has the form (3.7) with $\lambda = n+1$, which yields (3.24). To prove the explicit representations for $G^1_{\alpha,n+1}(x)$, we use relation (3.16) with $\lambda = n+1$:

$$G^1_{\alpha,n+1}(x) = \frac{1}{2\pi i}\int_{\gamma-i\infty}^{\gamma+i\infty}\frac{\Gamma(s-\alpha)}{(s-\alpha+n)\Gamma(s)}x^{-s}ds. \tag{3.26}$$

The integrand in (3.26) has a pole of the second order at $s_n = \alpha - n$ and simple poles at $s_k = \alpha - k$ for $k \in \mathbb{N}_0$, $(k \neq n)$. If we choose $\gamma > \alpha$, then evaluation of the residues at the above poles and using the asymptotic relation for the Gamma function:

$$\Gamma(z) = \frac{\Gamma(z+n)}{(z)_n} \quad (\mathrm{Re}(z) > -n; \ n \in \mathbb{N}; \ z \notin \mathbb{Z}_0^- := \{0, -1, -2, \cdots\}), \tag{3.27}$$

we have

$$G^1_{\alpha,n+1}(x) = \mathrm{Res}_{s=\alpha-n}\left[\frac{\Gamma(s-\alpha)}{\Gamma(s)(s-\alpha+n)}x^{-s}\right]$$

$$+ \sum_{k=0 \ (k\neq n)}^{\infty}\mathrm{Res}_{s=\alpha-k}\left[\frac{\Gamma(s-\alpha)}{(s-\alpha+n)\Gamma(s)}x^{-s}\right]$$

$$= x^{n-\alpha}\frac{\Gamma'(1) - \Gamma(1)\log(x) - \Gamma(1)\left[\sum_{j=0}^{n-1}\frac{1}{j-n} + \frac{\Gamma'(\alpha-n)}{\Gamma(\alpha-n)}\right]}{(-n)\cdots(-1)\Gamma(\alpha-n)}$$

$$+ \sum_{k=0 \ (k\neq n)}^{\infty}\frac{(-1)^kx^{k-\alpha}}{k!(n-k)\Gamma(\alpha-k)}.$$

From here, in accordance with (3.22) and (3.23), $G^1_{\alpha,n+1}(x)$ is given by (3.25), which completes the proof of the theorem for $n > 0$ The case $n = 0$ is proved similarly.

**Example 3.** Equation (3.21) with $\alpha = 1/2$:

$$x^{3/2}\left(D_{0+}^{3/2}y\right)(x)+(n+1)x^{1/2}\left(D_{0+}^{1/2}y\right)(x) = f(x) \quad (x > 0; \ \alpha > 0; \ n \in \mathbb{N}_0) \quad (3.28)$$

has a particular solution

$$y(x) = \int_0^1 G^1_{1/2,n+1}(t)f(xt)dt, \tag{3.29}$$

where $G^1_{1/2,n+1}(x)$ is given by (3.25) with $\alpha = 1/2$.

**Example 4.** The following ordinary differential equation of order $n + m + 1$:

$$x^{n+m+1}y^{(n+m+1)}(x) + (n+1)x^{n+m}y^{(n+m)}(x) = f(x) \quad (x > 0; \ \alpha > 0; \ n,m \in \mathbb{N}) \tag{3.30}$$

has a particular solution given by

$$y(x) = \int_0^1 G^1_{n+m,n+1}(t)f(xt)dt, \tag{3.31}$$

where

$$G^1_{n+m,n+1}(x) = \frac{(-1)^{n-1}x^{-m}}{n!(m-1)!}\left[\log(x) + \sum_{j=0}^{n-1}\frac{1}{j-n} + \psi(m) + \gamma\right]$$

$$+ \sum_{k=0,\ (k\neq n)}^{n+m-1}\frac{(-1)^k x^{k-n-m}}{k!(n-k)(n+m-k-1)!}. \tag{3.32}$$

**Example 5.** The following ordinary differential equation of order $m + 1$:

$$x^{m+1}y^{(m+1)}(x) + x^m y^{(m)}(x) = f(x) \quad (x > 0; \ m \in \mathbb{N}) \tag{3.33}$$

has a particular solution of the form

$$y(x) = \int_0^1 G^1_{m,1}(t)f(xt)dt, \tag{3.34}$$

where

$$G^1_{m,1}(x) = -\frac{x^{-m}}{(m-1)!}\left[\log(x) + \psi(m) + \gamma\right] - \sum_{k=1}^{m-1}\frac{(-1)^k x^{k-m}}{k!k(m-k-1)!}. \tag{3.35}$$

In particular,

$$y(x) = -\int_0^1 \frac{\log(t)}{t}f(xt)dt \tag{3.36}$$

is a particular solution of the equation:

$$x^2 y''(x) + xy'(x) = f(x) \quad (x > 0). \tag{3.37}$$

**Remark 1.** In Theorem 2 we obtained a particular solution of equation (3.21) for $\alpha > 0$ and $n \in \mathbb{N}_0$, except for the case when $n \in \mathbb{N}$ and $\alpha \neq 1, \cdots, n$. This is due to the fact that the function $G^1_{\alpha, n+1}(x)$, given by (3.25), is not defined for such $\alpha = k$ $(k = 1, \cdots, n)$. This means that the Laplace transform method can not be used to obtain a particular solution to the following ordinary differential equation of order $k + 1$ $(k = 1, \cdots n)$:

$$x^{k+1} y^{(k+1)}(x) + (n+1)x^k y^{(k)}(x) = f(x) \quad (x > 0; \quad n \in \mathbb{N}; \quad k = 1, \cdots, n). \quad (3.38)$$

**Remark 2.** The solution of equation (3.21) with $n = 0$ in the form (3.24) was obtained by Podlubny [5, Example 6.1], but his formula (6.10) for $G^1_{\alpha,1}(x)$ has an error. In fact, $\log(x) + \psi(\alpha)$ should be replaced by $-\log(x) - \psi(\alpha)$.

## 4. Equations with Right-Sided Fractional Derivatives

We consider equation (1.4) with $m = 1$:

$$x^{\alpha+1} \left(D_-^{\alpha+1} y\right)(x) + \lambda x^\alpha \left(D_-^\alpha y\right)(x) = f(x) \quad (x > 0; \quad \alpha > 0; \quad \lambda \in \mathbb{R}). \quad (4.1)$$

As in the case of equation (3.1), its particular solutions will be different for $\alpha + \lambda \neq n$ and $\alpha + \lambda = n$ $(n \in \mathbb{N}_0)$. In the first case, a particular solution is expressed in terms of the generalized Wright function (3.2) and of the Gauss hypergeometric function (3.3).

**Theorem 3.** *Let $\alpha > 0$ and $\lambda \in \mathbb{C}$ are such that $\alpha + \lambda \neq n$ $(n \in \mathbb{N}_0)$. Then the fractional differential equation (4.1) is solvable, and its particular solution has the form:*

$$y(x) = \int_x^\infty G^2_{\alpha,\lambda}\left(\frac{x}{t}\right) f(t) \frac{dt}{t}, \quad (4.2)$$

*where*

$$G^2_{\alpha,\lambda}(x) = \frac{\Gamma(-\lambda - \alpha)}{\Gamma(-\lambda)} x^{\alpha+\lambda} - {}_1\Psi_2 \left[ \begin{array}{c} (-\alpha - \lambda, 1) \\ (\alpha, -1), (1 - \alpha - \lambda, 1) \end{array} \middle| -x \right]$$

$$= \frac{\Gamma(-\lambda - \alpha)}{\Gamma(-\lambda)} x^{\alpha+\lambda} + \frac{1}{(\alpha+\lambda)\Gamma(\alpha)} \, {}_2F_1(-\alpha - \lambda, 1 - \alpha; 1 - \alpha - \lambda; x). \quad (4.3)$$

**Proof.** (4.1) is equation (1.4) with $m = 1$, $B_1 = 1$, $B_0 = \lambda$, and relation (2.4) takes the form:

$$\frac{(s + \alpha + \lambda)\Gamma(s + \alpha)}{\Gamma(s)} (\mathbb{M}y)(s) = (\mathbb{M}f)(s). \quad (4.4)$$

Hence the Mellin fractional analogue of the Green function in (2.8) is given by

$$G^2_\alpha(x) = \left( \mathbb{M}^{-1} \left[ \frac{\Gamma(s)}{(s + \alpha + \lambda)\Gamma(s + \alpha)} \right] \right)(x) =: G^2_{\alpha,\lambda}(x), \quad (4.5)$$

and, according to (2.11), a particular solution of equation (4.1) is given by

$$y(x) = \int_0^\infty G_{\alpha,\lambda}^2\left(\frac{x}{t}\right) f(t) \frac{dt}{t}.$$ (4.6)

We show that $G_{\alpha,\lambda}^2(x) = 0$ for $x < 1$. By (4.5),

$$\left(\mathrm{M}G_{\alpha,\lambda}^2\right)(s) = (\mathrm{M}G_3)(s)(\mathrm{M}G_4)(s),$$ (4.7)

where

$$(\mathrm{M}G_s)(s) = \frac{\Gamma(s)}{\Gamma(s+\alpha)}, \quad (\mathrm{M}G_4)(s) = \frac{1}{s+\alpha+\lambda}.$$ (4.8)

The direct application of the Mellin transform leads to the following relations:

$$G_3(x) = \begin{cases} \frac{x^{\alpha-1}}{\Gamma(\alpha)}, & 0 < x < 1, \\ 0, & x > 1; \end{cases} \quad G_4(x) = \begin{cases} x^{\alpha+\lambda}, & 0 < x < 1, \\ 0, & x > 1. \end{cases}$$ (4.9)

Thus, according to (2.9), we have

$$G_{\alpha,\lambda}^2(x) = \int_0^\infty G_3\left(\frac{x}{t}\right) G_4(t) \frac{dt}{t},$$ (4.10)

and it follows from (4.9) that $G_{\alpha,\lambda}^2(x) = 0$ for $x > 1$. Therefore, (4.6) yields (4.2).
Now we show that $G_{\alpha,\lambda}^2(x)$ is given by (4.3). By (4.5) and (2.2), we get

$$G_{\alpha,\lambda}^2(x) = \frac{1}{2\pi i} \int_{\gamma-i\infty}^{\gamma+i\infty} \frac{\Gamma(s)}{(s+\alpha+\lambda)\Gamma(s+\alpha)} x^{-s} ds =: G_{\alpha,\lambda}^2(x).$$ (4.11)

Since $\alpha + \lambda \neq n$ $(n \in \mathbb{N}_0)$, then the pole $s = -\lambda - \alpha$ in the integrand of (4.11) does not coincide with any pole $s_k = -k$ $(k \in \mathbb{N}_0)$ of $\Gamma(s)$. If we choose $\gamma > \max[-\alpha - \mathrm{Re}(\lambda), 0]$, then evaluation of the residues at the above poles yields

$$G_{\alpha,\lambda}^2(x) = \mathrm{Res}_{s=-\lambda-\alpha}\left[\frac{\Gamma(s)x^{-s}}{(s+\alpha+\lambda)\Gamma(s+\alpha)}\right] + \sum_{k=0}^\infty \mathrm{Res}_{s=-k}\left[\frac{\Gamma(s)x^{-s}}{(s+\alpha+\lambda)\Gamma(s+\alpha)}\right]$$

$$= \frac{\Gamma(-\alpha-\lambda)}{\Gamma(-\lambda)} x^{\alpha+\lambda} - \sum_{k=0}^\infty \frac{(-1)^k}{k!} \frac{\Gamma(-\lambda-\alpha+k)}{\Gamma(1-\lambda-\alpha+k)\Gamma(\alpha-k)} x^k.$$

Hence, by (3.2), (1.9) and (3.5), $G_{\alpha,\lambda}^2(x)$ is given by (4.3), and thus Theorem 3 is proved.

**Example 6.** The equation (4.1) with $\alpha = 1/2$ and $\lambda \neq n - 1/2$ $(n \in \mathbb{N}_0)$:

$$x^{3/2}\left(D_-^{3/2}y\right)(x) + \lambda x^{1/2}\left(D_-^{1/2}y\right)(x) = f(x) \quad (x > 0; \ \alpha > 0; \ \lambda \in \mathbb{R})$$ (4.12)

has a particular solution of the form

$$y(x) = \int_x^\infty G_{1/2,\lambda}^2\left(\frac{x}{t}\right) f(t) \frac{dt}{t},$$ (4.13)

where $G_{1/2,\lambda}^2(x)$ is given by (4.3) with $\alpha = 1/2$,

**Example 7.** The function

$$y(x) = \frac{1}{1+\lambda} \int_x^\infty \left[1 - \left(\frac{x}{t}\right)^{\lambda+1}\right] f(t)dt \qquad (4.14)$$

is a particular solution of the equation:

$$x^2 y''(x) - \lambda x y'(x) = f(x) \quad (x > 0; \ \lambda \neq -1). \qquad (4.15)$$

Note that, for $\lambda \neq n - 1$ ($n \in \mathbb{N}_0$), (4.2) with $\alpha = 1$ yields the solution (4.14), but it is directly verified that (4.14) satisfies equation (4.15) with any $\lambda \neq -1$.

Now we give a particular solution to equation (4.1) with $\lambda = n - \alpha$ ($n \in \mathbb{N}_0$):

$$x^{\alpha+1} \left(D_-^{\alpha+1}y\right)(x) + (n-\alpha)x^\alpha \left(D_-^\alpha y\right)(x) = f(x) \quad (x > 0; \ \alpha > 0; \ n \in \mathbb{N}_0). \quad (4.16)$$

As in the case of the equation (3.17) with the Liouville left-sided derivatives, the solution will be expressed in terms of the psi function $\psi(z)$ and the Euler constant $\gamma$, and the results will be different in the cases $n \in \mathbb{N}$ and $n = 0$.

**Theorem 4.** *Let $\alpha > 0$ and $n \in \mathbb{N}_0$ be such that $\alpha \neq 1, \cdots, n$ for $n > 0$. Then the fractional differential equation (4.16) is solvable, and its particular solution has the form:*

$$y(x) = \int_x^\infty G_{\alpha,n-\alpha}^2 \left(\frac{x}{t}\right) f(t) \frac{dt}{t}, \qquad (4.17)$$

*where*

$$G_{\alpha,n-\alpha}^2(x) = \frac{(-1)^{n-1}x^n}{n!\Gamma(\alpha-n)} \left[\log(x) + \sum_{j=0}^{n-1} \frac{1}{j-n} + \psi(\alpha-n) + \gamma\right]$$

$$+ \sum_{\substack{k=0 \ (k\neq n)}}^\infty \frac{(-1)^k x^k}{k!(n-k)\Gamma(\alpha-k)}. \qquad (4.18)$$

**Proof.** Let $n > 0$. By the proof of Theorem 3, the solution of equation (4.16) has the form (4.2) with $\lambda = n - \alpha$, which yields (4.17)). To prove the explicit representation for $G_{\alpha,n-\alpha}^2(x)$, we use the relation (4.11) with $\lambda = n - \alpha$:

$$G_{\alpha,n-\alpha}^2(x) = \frac{1}{2\pi i} \int_{\gamma-i\infty}^{\gamma+i\infty} \frac{\Gamma(s)}{(s+n)\Gamma(s+\alpha)} x^{-s} ds. \qquad (4.19)$$

The integrand in (4.19) has a pole of the second order at $s_n = -n$ and simple poles at $s_k = -k$ for $k \in \mathbb{N}_0$ ($k \neq n$). If we choose $\gamma > 0$, then evaluation of the residues at the above poles, with using the relation (3.27), yields

$$G_{\alpha,n-\alpha}^2(x) = x^n \frac{\Gamma'(1) - \Gamma(1)\log(x) - \Gamma(1)\left[\sum_{j=0}^{n-1} \frac{1}{j-n} + \frac{\Gamma'(\alpha-n)}{\Gamma(\alpha-n)}\right]}{(-n)\cdots(-1)\Gamma(\alpha-n)}$$

$$+ \sum_{\substack{k=0 \ (k\neq n)}}^\infty \frac{(-1)^k x^{k-\alpha}}{k!(n-k)\Gamma(\alpha-k)}.$$

From here, in accordance with (3.22) and (3.23), $G_{\alpha,n-\alpha}(x)$ is given by (4.18), which completes the proof of the theorem for $n > 0$ The case $n = 0$ is proved similarly.

**Example 8.** Equation (4.16) with $\alpha = 1/2$:

$$x^{3/2}\left(D_-^{3/2}y\right)(x) + \left(n - \frac{1}{2}\right)x^{1/2}\left(D_-^{1/2}y\right)(x) = f(x) \quad (x > 0; \quad \alpha > 0; \quad n \in \mathbb{N}_0)$$

(4.20)

has a particular solution in the form

$$y(x) = \int_x^\infty G_{1/2,n-1/2}^2\left(\frac{x}{t}\right)f(t)\frac{dt}{t},$$

(4.21)

where $G_{1/2,n-1/2}^2(x)$ is given by (4.18) with $\alpha = 1/2$.

**Example 9.** The following ordinary differential equation of order $n + m + 1$:

$$x^{n+m+1}y^{(n+m+1)}(x) + mx^{n+m}y^{(n+m)}(x) = f(x) \quad (x > 0; \quad n, m \in \mathbb{N})$$

(4.22)

has a particular solution given by

$$y(x) = (-1)^{n+m+1}\int_x^\infty G_{n+m,-m}^2\left(\frac{x}{t}\right)f(t)\frac{dt}{t},$$

(4.23)

where

$$G_{n+m,-m}^2(x) = \frac{(-1)^{n-1}x^n}{n!(m-1)!}\left[\log(x) + \sum_{j=0}^{n-1}\frac{1}{j-n} + \psi(m) + \gamma\right]$$

$$+ \sum_{k=0,\ (k\neq n)}^{n+m-1}\frac{(-1)^kx^k}{k!(n-k)(n+m-k-1)!}.$$

(4.24)

**Example 10.** The following ordinary differential equation of order $m + 1$:

$$x^{m+1}y^{(m+1)}(x) + mx^my^{(m)}(x) = f(x) \quad (x > 0; \quad m \in \mathbb{N})$$

(4.25)

has a particular solution given by

$$y(x) = (-1)^{m+1}\int_x^\infty G_{m,-m}^2\left(\frac{x}{t}\right)f(t)\frac{dt}{t},$$

(4.26)

$$G_{m,-m}^2(x) = -\frac{1}{(m-1)!}\left[\log(x) + \psi(m) + \gamma\right] - \sum_{k=1}^{m-1}\frac{(-1)^kx^k}{k!k(m-k-1)!}.$$

(4.27)

Specifically,

$$y(x) = -\int_x^\infty \log\left(\frac{x}{t}\right)f(t)\frac{dt}{t}$$

(4.28)

is a particular solution of the equation (3.37). It is clear that this solution coincides with the solution given by (3.36).

**Remark 3.** In Theorem 4 we obtained a particular solution to equation (4.1) for all $\alpha > 0$ and $n \in \mathbb{N}_0$, except for the case when $n \in \mathbb{N}$ and $\alpha \neq 1, \cdots, n$. This is due to the fact that the function $G^2_{\alpha,n-\alpha}(x)$, given by (4.3), is not defined for such $\alpha = k \ (k = 1, \cdots, n)$. This means that the Laplace transform method can not be used to obtain a particular solution to the following ordinary differential equation of order $k \ (k = 1, \cdots, n)$:

$$x^{k+1} y^{(k+1)}(x) + (n-k) x^k y^{(k)}(x) = f(x) \quad (x > 0; \ n \in \mathbb{N}; \ k = 1, \cdots, n). \quad (4.27)$$

**Remark 4.** In Sections 3 and 4 we have applied a general approach, developed in Section 2, to establish particular solutions to equations (1.3) and (1.4) with $m = 1$. Such an approach can be also applied to obtaining particular solutions of equations (1.3) and (1.4) with $m \geq 2$.

**Acknowledgements:** This work was supported, in part, by Belarusian Fundamental Research Fund (Project F06R-106) and University of La Laguna.

### References

1. V.A. Ditkin, and A.P. Prudnikov, *Integral Transforms and Operational Calculus*, Pergamon Press, Oxford, 1965.
2. A. Erdelyi, W. Magnus, F. Oberhettinger and F.G. Tricomi, *Higher Transcendental functions*, Vol.I, McGraw-Hill, New York-Toronto -London, 1953.
3. A.A. Kilbas, M. Saigo and J.J. Trujillo, On the generalized Wright function. *Frac. Calc. Appl. Anal.*, **5** (2002), *no. 4*, 437-460.
4. A.A. Kilbas and J.J. Trujillo, Differential equations of fractional order: methods, results and problems - I. *Appl. Anal.* **78** (2001), 153-192.
5. I. Podlubny, *Fractional Differential Equations*, Mathematics in Sciences and Engineering. 198, Academic Press, San-Diego, 1999.
6. S.G. Samko, A.A. Kilbas and O.I. Marichev, *Fractional Integrals and Derivatives. Theory and Applications.* Gordon and Breach, Yverdon *et alibi*, 1993. Extension of Russian edition of 1987.
7. I.N. Snedon, *Fourier Transforms.* Dover, New York, 1995. The first edition in McGraw Hill-Book, New York, etc., 1951.
8. E.M. Wright, The asymptotic expansion of the generalized hypergeometric function, *J. London Math. Soc.*, **10** (1935), 287-293.
9. E.M. Wright, The asymptotic expansion of integral functions defined by Taylor series, *Philos. Trans. Roy. Soc. London, Ser. A*, **238** (1940), 423-451.
10. E.M. Wright, The asymptotic expansion of the generalized hypergeometric funciton, *Proc. London Math. Soc. (2)*, **46** (1940), 389-408.
11. V.I. Zaitsev and A.D. Polianin, *Handbook on Ordinary Differential Equations* (Russian). Factorial, M., 1997.

## I.5   Toeplitz and Toeplitz-like Operators

Organizers: S. Grudski, N. Vasilevski

The theory of Toeplitz operators has been developing intensively for the last 50 to 60 years. Being a common ground for operator theory, complex analysis, and function spaces, the theory of Toeplitz operators not only uses the ideas and results of these fields, but is a source of new problems and new methods in all of them, as well as being a useful tool for many applications.

The main goal of Session I.5 "Toeplitz and Toeplitz-like operators" was to bring together experts actively working on this subject and in various related areas where Toeplitz operators play an essential role, such as asymptotic linear algebra, quantization, approximation, singular integral and convolution type operators, etc.

The results presented in the Session, together with many fruitful informal discussions, reflect current tendencies in the theory of Toeplitz operators and related areas.

The papers published in this volume are quite representative of the contributions of the Session. Besides providing the reader with an overview of contemporary results, we hope they may suggest directions for future investigations as well.

# LOCAL PROPERTIES OF THE SEGAL-BARGMANN PROJECTION AND $\Psi_0$-ALGEBRAS

W. BAUER

*Johannes Gutenberg-Universität Mainz,*
*Fachbereich Physik, Mathematik und Informatik,*
*Staudinger Weg 9,*
*55128 Mainz, Germany*
*Email: BauerWolfram@web.de*

As a contribution to the theory of Toeplitz operators and Fréchet operator algebras we study $\Psi_0$- and $\Psi^*$-algebras defined by smooth vector fields and containing the Segal-Bargmann projection. From these one obtains corresponding algebras in $\mathcal{L}(H_2)$ where $H_2$ denotes the Segal-Bargmann space. Subsequent to the results in [6], [7] many consequences for such objects arise and here we only mention the stability of $\Psi^*$-algebras under holomorphic functional calculus following results by Waelbroeck (1954) [13]. In general, by passing from a ring of operators to its closure (a Banach- or $C^*$-algebra) one looses local $C^\infty$-properties such as pseudo- or micro-locality. The finer structure of $\Psi_0$-algebras allows to preserve local behavior of the single operators within the algebra. Motivated by the Hörmander classes $\Psi_{\rho,\delta}^0$ where $0 \leq \delta \leq \rho \leq 1$ and $\delta < 1$ several methods for constructing sub-multiplicative Fréchet operator algebras with (locally) spectral invariance have been developed in e.g. [7], [9], [10] and [11]. In the present paper we use commutator methods with finite sets of closed operators to obtain algebras which in our sense are localized in complex cones in $\mathbb{C}^n$.

**Key words:** $\Psi_0$-algebras, Segal-Bargmann space, commutator methods, spectral invariance, cone localization
**Mathematics Subject Classification:** 47B35, 47B47, 46H35, 46E22

## 1. Fréchet operator algebras by commutator methods

We remind of some notions which were introduced by B. Gramsch in [6]:

**Definition 1.1** Let $\mathcal{B}$ be a Banach-algebra with unit $e$ and let $\mathcal{F}$ be a continuously embedded Fréchet algebra in $\mathcal{B}$ with $e \in \mathcal{F}$, then $\mathcal{F}$ is called $\Psi_0$-*algebra* if it is *locally spectral invariant* in $\mathcal{B}$, i.e. there is $\varepsilon > 0$ with

$$\{ a \in \mathcal{F} : \| e - a \|_{\mathcal{B}} < \varepsilon \} \subset \mathcal{F}^{-1}.$$

If $\mathcal{B}$ is a $C^*$-algebra and $\mathcal{F}$ is a symmetric $\Psi_0$-algebra in $\mathcal{B}$, then $\mathcal{F}$ is called $\Psi^*$-*algebra*. In this case $\mathcal{F}$ is *spectral invariant*, i.e. $\mathcal{F} \cap \mathcal{B}^{-1} = \mathcal{F}^{-1}$.

Moreover, $\mathcal{F}$ is called *sub-multiplicative* or *locally m-convex* (E. Michael, 1952) $\Psi_0$- resp. $\Psi^*$-algebra if its topology is generated by a countable system $[\, q_j : j \in \mathbb{N} \,]$ of sub-multiplicative semi-norms with $q_j(e) = 1$ for all $j \in \mathbb{N}$.

Let us recall a general method for generating $\Psi_0$- resp. $\Psi^*$-algebras which can be found in [7]. Let $H$ be a Hilbert space and $\{\mathcal{F}, (q_j)\} \subset \mathcal{L}(H)$ be a sub-multiplicative $\Psi^*$-algebra. Without loss of generality we assume that $q_0 := \| \cdot \|_{\mathcal{L}(H)}$. With any finite system $\mathcal{V}$ of closed and densely defined operators on $H$ and following [7] we define with $A \in \mathcal{V}$:

- $\mathcal{I}(A) := \{ a \in \mathcal{F} : a(\mathcal{D}(A)) \subset \mathcal{D}(A) \}$,
- $\mathcal{B}(A) := \{ a \in \mathcal{I}(A) : [A, a] := Aa - aA \text{ extends to an element } \delta_A(a) \in \mathcal{F} \}$.

Inductively we define a decreasing scale $\{ \Psi_j^{\mathcal{V}} : j = 0, 1, \cdots \}$ of algebras in $\mathcal{F}$ by:

- $\Psi_0^{\mathcal{V}} := \mathcal{F}$ with semi-norms $q_{0,j} := q_j$ for $j \in \mathbb{N}$,
- $\Psi_1^{\mathcal{V}} := \bigcap_{A \in \mathcal{V}} \mathcal{B}(A)$,
- $\Psi_k^{\mathcal{V}} := \{ a \in \Psi_{k-1}^{\mathcal{V}} : \delta_A a \in \Psi_{k-1}^{\mathcal{V}} \text{ for all } A \in \mathcal{V} \}$ where $k \geq 2$,
- $\Psi_\infty^{\mathcal{V}} := \bigcap_{k \in \mathbb{N}_0} \Psi_k^{\mathcal{V}}$.

Further, we inductively endow the algebra $\Psi_n^{\mathcal{V}}$ for $n \geq 1$ with the semi-norms:

$$q_{n,j}(a) = q_{n-1,j}(a) + \sum_{A \in \mathcal{V}} q_{n-1,j}(\delta_A a) \qquad \text{for } a \in \Psi_n^{\mathcal{V}} \text{ and } j \in \mathbb{N}$$

and $\Psi_\infty^{\mathcal{V}}$ with the system of norms $(q_{n,j})_{n,j \in \mathbb{N}}$. Then by results in [7] it is known that the space $\Psi_\infty^{\mathcal{V}}$ is a sub-multiplicative $\Psi_0$-algebra in $\mathcal{L}(H)$. If each $A \in \mathcal{V}$ is symmetric, then the algebras $\Psi_n^{\mathcal{V}}$ are symmetric and $\Psi_\infty^{\mathcal{V}}$ is a $\Psi^*$-algebra in $\mathcal{L}(H)$. Moreover, there is a notion of *generalized Sobolev spaces* corresponding to the decreasing scale $\{ \Psi_j^{\mathcal{V}} : j \in \mathbb{N}_0 \}$ and of *abstract regularity*.

The *(locally) spectral invariance* is preserved under projections in the following sense. Let $\mathcal{F} \subset \mathcal{B}$ be (locally) spectral invariant in $\mathcal{B}$ and $p = p^2 \in \mathcal{F}$, then consider the projected algebra $\mathcal{F}_p := p \mathcal{F} p$. It is easy to see that $\mathcal{F}_p$ again is locally spectral invariant in the space $\mathcal{B}_p := p \mathcal{B} p$. If in addition $\mathcal{B}$ is a $C^*$-algebra, $\mathcal{F}$ is symmetric in $\mathcal{B}$ and $p = p^2 = p^*$, then $\mathcal{F}_p$ is symmetric and spectral invariant in $\mathcal{B}_p$.

Let $M \subset H$ be a linear subspace. For each system $\mathcal{S}_m := [X_1, \cdots, X_m]$ where $X_j \in L(M)$ and $B \in L(M)$ we call $m$ the *length* of $\mathcal{S}_m$. We inductively define the *iterated commutators* $\text{ad}[\emptyset](B) := B$ and:

- $\text{ad}[\mathcal{S}_1](B) := [X_1, B] = X_1 B - B X_1$,
- $\text{ad}[\mathcal{S}_{m+1}](B) := \text{ad}[X_{m+1}](\text{ad}[\mathcal{S}_m](B))$.

With this notion it follows for finite systems $\mathcal{S}_j$ and $\mathcal{S}_k$ in $L(M)$:

$$\text{ad}[\mathcal{S}_j](\text{ad}[\mathcal{S}_k](B)) = \text{ad}[\mathcal{S}_k, \mathcal{S}_j](B). \qquad (1.1)$$

From now on assume that for all sets $\mathcal{V}$ of closed operators $A : H \supset \mathcal{D}(A) \to H$, there is a dense subspace $D \subset H$ such that each domain of definition $\mathcal{D}(A)$ is the closure of $D$ with respect to the graph norm $\| \cdot \|_{\text{gr}} := \| \cdot \| + \|A \cdot \|$, i.e. $A$ is the minimal closed extension from $D$. We give a necessary condition for $a \in \mathcal{F}$ to belong to the algebra $\Psi_k^{\mathcal{V}}$ generated by commutators with $\mathcal{V}$.

**Proposition 1.1** *Let $k \in \mathbb{N} \cup \{\infty\}$ and $a \in \mathcal{F}$. With $D \subset H$ as above we assume that the following property $(E_k)$ holds:*

$(E_k)$: *Let $D$ be invariant under all the operators $A \in \mathcal{V}$ as well as under $a \in \mathcal{F}$ such that the commutators $\mathrm{ad}[\,A\,](a) : H \supset D \to H$ are well-defined for any system $\mathcal{A}$ in*

$$\mathcal{S}_k(\mathcal{V}) := \Big\{ [\,A_1, \cdots, A_j\,] : \text{ where } A_l \in \mathcal{V} \text{ and } 1 \leq j \leq k \Big\}.$$

*Assume that all iterated commutators $\mathrm{ad}[\,\mathcal{A}\,](a)$ where $\mathcal{A} \in \mathcal{S}_k(\mathcal{V})$ admit continuous extensions to operators $C(\mathcal{A}, a) \in \mathcal{F}$.*

*Then $a \in \Psi_k^{\mathcal{V}}$ and for any system $\mathcal{A} \in \mathcal{S}_k(\mathcal{V})$ the operator $C(\mathcal{A}, a)$ is a continuous extension of $\mathrm{ad}[\,\mathcal{A}\,](a) : H \subset \mathcal{D}(A) \to H$ for any $A \in \mathcal{V}$.*

The proof of Proposition 1.1 is a consequence of the following easy observation:

**Lemma 1.1** *Let $a \in \mathcal{L}(H)$ and let $A$ be a closed operator on $H$ such that the inclusion $a(D) \subset \mathcal{D}(A)$ holds. Assume that the commutator $\mathrm{ad}[A](a) : H \supset D \to H$ admits an extension $\delta_A a \in \mathcal{L}(H)$. Then it follows that*

(a) $a(\mathcal{D}(A)) \subset \mathcal{D}(A)$.
(b) *The operator $\delta_A a$ is an extension of $\mathrm{ad}[\,A\,](a) : H \supset \mathcal{D}(A) \to H$.*

**Proof:** Let $x \in \mathcal{D}(A)$ and choose a sequence $(y_k)_k \subset D$ such that $\| x - y_k \|_{\mathrm{gr}} \to 0$ as $k \to \infty$. It follows that $\lim_{k \to \infty} a y_k = a x$ and because $\mathrm{ad}[\,A\,](a)$ admits a bounded extension $\delta_A a$ to $H$ the sequence

$$A a y_k = \mathrm{ad}[\,A\,](a) y_k + a A y_k = \delta_A a \, y_k + a A y_k$$

is convergent in $H$. By assumption $A$ is closed and so it follows that $a x \in \mathcal{D}(A)$ which implies $(a)$. We obtain $[\,A, a\,](x) = \mathrm{ad}[\,A\,](a) x = \delta_A a (x)$ and from this we conclude $(b)$. $\qquad\square$

**Remark 1.1** In general, by applying the closed graph theorem for an operator $a \in \mathcal{I}(A)$ we can show that the commutator $\mathrm{ad}[\,A\,](a) : (\,\mathcal{D}(A), \| \cdot \|_{\mathrm{gr}}\,) \to H$ is well-defined and continuous.

**Proof of Proposition 1.1:** Let $k = 1$, then with our notations above we have:

$$\Psi_1^{\mathcal{V}} = \bigcap_{A \in \mathcal{V}} \mathcal{B}(A) \subset \mathcal{L}(H).$$

We have to check the inclusion $a(\,\mathcal{D}(A)\,) \subset \mathcal{D}(A)$ for all $A \in \mathcal{V}$. This in fact follows from assumption $(E_1)$ in Proposition 1.1 together with the dense inclusion $D \subset (\,\mathcal{D}(A), \| \cdot \|_{\mathrm{gr}}\,)$ for all $A \in \mathcal{V}$ and with Lemma 1.1 (a). Moreover, by Lemma 1.1, (b) the operator $C(\,[\,A\,], a\,) \in \mathcal{F}$ is a continuous extension of the commutator $\mathrm{ad}[\,A\,](a) : H \supset \mathcal{D}(A) \to H$ for $A \in \mathcal{V}$ and so it follows that $a \in \Psi_1^{\mathcal{V}}$.

Now, for $k > 1$ and by induction we assume that $a \in \Psi_j^V$ where $1 \le j < k$. Let $A \in V$ and consider the commutator $\mathrm{ad}[\, A \,](a) : H \supset \mathcal{D}(A) \to H$ which has an extension $\delta_A a \in \mathcal{F}$. Again by (1.1) and assumption $(E_k)$ the commutator:

$$\mathrm{ad}[\, \mathcal{A}_j \,]\,(\delta_A a) = \mathrm{ad}[\, A, \mathcal{A}_j \,](a) : H \supset D \to D \subset H$$

has a continuous extension to $C([\, A, \mathcal{A}_j\,], a) \in \mathcal{F}$ for all systems $\mathcal{A}_j \in \mathcal{S}_j(V)$. Hence by induction we conclude that $\delta_A a \in \Psi_j^V$ and it follows that $a \in \Psi_{j+1}^V$. $\qquad\square$

## 2. Localization of the Segal-Bargmann projection

Let us specialize the constructions above to operators on the *Segal-Bargmann space* of Gaussian square integrable entire functions on $\mathbb{C}^n$. In general, for defining new $\Psi_0$-algebras (resp. $\Psi^*$-algebras) $\mathcal{F}_p := p\,\mathcal{F}\,p$ from a given one $\mathcal{F}$ by using projections $p = p^2 = p^*$ it is of interest whether or not $p$ belongs to $\mathcal{F}$. Here we want to study the case of the Toeplitz projection and we start with some definitions. For any $c > 0$ let $\mu_c$ be the normed Gaussian measure on $\mathbb{C}^n$ with respect to the density

$$d\,\mu_c = c^n \pi^{-n} \exp\left( -c\,\| \cdot \|^2 \right)\, dv$$

where $v$ is the usual Lebesgue measure and $\| \cdot \|$ denotes the Euclidean norm on $\mathbb{C}^n$. We also will write $\mu := \mu_1$ and with the space $\mathcal{H}(\mathbb{C}^n)$ of entire functions on $\mathbb{C}^n$ we consider the Hilbert spaces:

$$H_1 := L^2(\mathbb{C}^n, \mu), \quad \text{and} \quad H_2 := H_1 \cap \mathcal{H}(\mathbb{C}^n)$$

with innerproduct $\langle \cdot, \cdot \rangle_2$ and norm $\| \cdot \|_2$.

By classical results $H_2$ is a reproducing kernel Hilbert space which is canonically isomorphic to the *Boson Fock space*. It often is referred to as *Segal-Bargmann space* ( c.f. [1]). The reproducing kernel $K$ is given by

$$K(\, z, w\,) := \exp\left( z \cdot \bar{w} \right), \qquad z, w \in \mathbb{C}^n \qquad\qquad (2.1)$$

and by $P \in \mathcal{L}(\, H_1\,)$ we denote the orthogonal projection from $H_1$ onto the closed subspace $H_2$. For any symbol $f \in L^\infty(\mathbb{C}^n)$ the Toeplitz operator $T_f \in \mathcal{L}(\, H_2\,)$ is defined by $T_f g = P(\, fg\,)$. We set $\mathcal{F} := \mathcal{L}(\, H_1\,)$ and we want to determine classes $\mathcal{X}$ of vector fields on $\mathbb{C}^n$ sufficiently large such that it holds $P \in \Psi_\infty^V$ for all finite systems $V \subset \mathcal{X}$. According to Proposition 1.1 we have to construct a suitable space $\mathcal{D} \subset H_1$. Moreover, we must find norm estimates on all iterated commutators of vector fields on $\mathbb{C}^n$ with the Toeplitz projection $P$ considered as operators on $\mathcal{D}$.

We start by defining a dense subspace $\mathcal{D} \subset \mathcal{C}^\infty(\mathbb{C}^n) \cap H_1$ which is invariant under all partial derivatives, multiplications by smooth functions with derivatives having only polynomial growth and by the Toeplitz projection $P$. Consider

$$L_{\exp}(\mathbb{C}^n) := \left\{ f \in \mathcal{C}^\infty(\mathbb{C}^n) : \exists\, 0 < c < 2^{-1},\ \exists\, d > 0 \text{ with } |f(z)| \le d \exp\left( c\|z\|^2 \right) \right\}.$$

Then it holds $L_{\exp}(\mathbb{C}^n) \subset C^\infty(\mathbb{C}^n) \cap H_1$ and we can define $\mathcal{D} \subset L_{\exp}(\mathbb{C}^n)$ by:

$$\mathcal{D} := \{\, f \in L_{\exp}(\mathbb{C}^n) : \text{ all partial derivatives of } f \text{ belong to } L_{\exp}(\mathbb{C}^n)\,\}. \qquad (2.2)$$

Clearly $\mathcal{D}$ is invariant under partial derivatives and multiplications by functions with derivatives having only polynomial growth. In particular, $\mathcal{D}$ contains all complex polynomials in $z$ and $\bar{z}$ and so it forms a dense subspace of $H_1$. Let us show that it also is invariant under $P$. We inductively define a sequence $(a_n)_{n \in \mathbb{N}}$ of real numbers by:

$$a_1 := \frac{1}{4} \qquad \text{and} \qquad a_{n+1} := [\,4 \cdot (1 - a_n)\,]^{-1}$$

for $n \geq 1$. It is an easy computation that $(a_n)_{n \in \mathbb{N}}$ has the following properties:

(a) $a_n < \frac{1}{2}$, $\quad \forall\, n \in \mathbb{N}$,

(b) $(a_n)_{n \in \mathbb{N}}$ is strictly increasing,

(c) $\lim_{n \to \infty} a_n = \frac{1}{2}$.

**Proposition 2.1** *The space $\mathcal{D} \subset C^\infty(\mathbb{C}^n) \cap H_1$ is invariant under $P$.*

**Proof:** First we show that $L_{\exp}(\mathbb{C}^n)$ is invariant under $P$. Let $g \in L_{\exp}(\mathbb{C}^n)$ and choose numbers $0 < c < \frac{1}{2}$ and $D_1 > 0$ such that

$$|\,g(z)\,| \leq D_1 \exp\left(c\,\|\,z\,\|^2\right) \qquad \text{for all } z \in \mathbb{C}^n.$$

By (a), (b) and (c) above there is $n_0 \in \mathbb{N}$ such that $c < a_{n_0} < \frac{1}{2}$. Using the transformation formula for the integral together with the reproducing property of $K$ we obtain:

$$
\begin{aligned}
|\,Pg(z)\,| &\leq \int_{\mathbb{C}^n} |\,g\,\exp\left(\langle z, \cdot\rangle\right)|\, d\mu \\
&\leq D_1\, \pi^{-n} \int_{\mathbb{C}^n} \exp\left(\operatorname{Re}\langle z, \cdot\rangle - \left[\,1 - a_{n_0}\,\right]\|\cdot\|^2\right) dv \\
&= D_1\left(1 - a_{n_0}\right)^{-n} \int_{\mathbb{C}^n} \exp\left(2\operatorname{Re}\left\langle 2^{-1}\left(1 - a_{n_0}\right)^{-\frac{1}{2}} z, \cdot\right\rangle\right) d\mu \\
&= D_2 \exp\left(a_{n_0+1}\,\|\,z\,\|^2\right),
\end{aligned}
$$

where $D_2 := D_1 \cdot (1 - a_{n_0})^{-n}$. From (a) above we conclude that $Pg \in L_{\exp}(\mathbb{C}^n)$. In order to prove the invariance of $\mathcal{D}$ under $P$ note that by a straightforward calculation it holds

$$\frac{\partial^\alpha}{\partial z^\alpha} Pf = P\{\,\bar{z}^\alpha f\,\}, \qquad \text{for all } f \in \mathcal{D} \text{ and } \alpha \in \mathbb{N}_0^n$$

which together with the invariance of $L_{\exp}(\mathbb{C}^n)$ by multiplication with polynomials and under the Toeplitz projection implies the desired result. $\qquad \square$

Let us consider $P$, $\partial_i := \frac{\partial}{\partial z_i}$ and $\bar{\partial}_i := \frac{\partial}{\partial \bar{z}_i}$ as operators on the dense subspace $\mathcal{D}$ of $H_1$. As usual for any multi-index $\alpha \in \mathbb{N}_0^n$ we write $\partial^\alpha := \partial_n^{\alpha_n} \cdots \partial_1^{\alpha_1}$ and $\bar{\partial}^\alpha := \bar{\partial}_n^{\alpha_n} \cdots \bar{\partial}_1^{\alpha_1}$. For $f, g \in \mathcal{D}$ and the inner-product $\langle \cdot, \cdot \rangle_2$ of $H_1$ we have:

$$\langle \bar{\partial}_k f, g \rangle_2 = \langle f, [\bar{z}_k - \partial_k] g \rangle_2 \qquad \text{and} \qquad \langle \partial_k f, g \rangle_2 = \langle f, [z_k - \bar{\partial}_k] g \rangle_2.$$

Hence we set $\partial_k^* := z_k - \bar{\partial}_k$ and $(\bar{\partial}_k)^* = \overline{\partial_k^*} = \bar{z}_k - \partial_k$ for $k = 1, \cdots, n$. The following result follows by an application of the *Schur test* and it will be useful to show boundedness of integral operators on $H_1$:

**Proposition 2.2** *Let $L : \mathbb{C}^n \times \mathbb{C}^n \to \mathbb{C}$ be a measurable function such that:*

$$|L(x, y)| \leq |F(x - y)| \exp\left( \operatorname{Re} \langle x, y \rangle \right)$$

*where $F \in L^1(\mathbb{C}^n, \mu_{\frac{1}{2}})$. Then the integral operator $A$ on $H_1$ with kernel $L$ is bounded on $H_1$. Moreover, its operator norm can be estimated by $\| A \| \leq 2^n \| F \|_{L^1(\mathbb{C}^n, \mu_{\frac{1}{2}})}$.*

**Proof:** Consider the positive function $p = q = \exp(\frac{1}{2} \| \cdot \|^2)$ on $\mathbb{C}^n$. Then it follows that:

$$\int_{\mathbb{C}^n} |L(\cdot, y)| \, p \, d\mu \leq \frac{1}{\pi^n} \int_{\mathbb{C}^n} |F(\cdot - y)| \, \exp\left( \operatorname{Re} \langle \cdot, y \rangle - \frac{1}{2} \| \cdot \|^2 \right) dv$$

$$= 2^n \, p(y) \, \| F \|_{L^1(\mathbb{C}^n, \mu_{\frac{1}{2}})}.$$

Similar we get $\int |L(x, \cdot)| \, p \, d\mu \leq 2^n \, p(x) \, \| F \|_{L^1(\mathbb{C}^n, \mu_{\frac{1}{2}})}$. Now, applying the *Schur test* (c.f. [12]), we obtain the desired result. $\qquad \square$

**Lemma 2.1** *With $0 \neq \alpha \in \mathbb{N}_0^n$ and as operators on $\mathcal{D}$ we obtain:*

(a) $[P, \partial^\alpha] := P \partial^\alpha - \partial^\alpha P = 0.$

(b) *The commutator $[P, \bar{\partial}^\alpha] = P\bar{\partial}^\alpha$ has a bounded extensions to $H_1$.*

**Proof:** The commutator relation (a) follows by a straightforward calculation using integration by parts. We omit the details and we only prove (b):

$$P\bar{\partial}^\alpha f(x) = \langle \bar{\partial}^\alpha f, K(\cdot, x) \rangle_2 = \langle f, [\bar{\cdot} - \partial]^\alpha K(\cdot, x) \rangle_2 = \langle f, [\bar{\cdot} - \bar{x}]^\alpha K(\cdot, x) \rangle_2.$$

Because $\bar{\partial}^\alpha P = 0$ on $\mathcal{D}$ for any $\alpha \neq 0$ it follows that $[P, \bar{\partial}^\alpha] = P \bar{\partial}^\alpha$. Now, we conclude from Proposition 2.2 with $F := \| \cdot \|^{|\alpha|} \in L^1(\mathbb{C}^n, \mu_{\frac{1}{2}})$ that the commutators $[P, \bar{\partial}^\alpha]$ can be extended to continuous operators on $H_1$. $\qquad \square$

Consider the space $\mathcal{X}_{\mathrm{lin}}(\mathbb{C}^n) := \operatorname{span}\{ \partial_j, \bar{\partial}_j : j = 1, \cdots, n \}$ of all vector fields with constant coefficients. Let $\mathcal{Z}$ be a finite system in $\mathcal{X}_{\mathrm{lin}}(\mathbb{C}^n)$ of length $r$. Then by induction and applying Lemma 2.1 it can be shown that:

$$\operatorname{ad}[\mathcal{Z}](P) \in \operatorname{span}\left\{ P\bar{\partial}^\alpha : \alpha \in \mathbb{N}_0^n \text{ and } |\alpha| = r \right\} \subset \mathcal{L}(\mathcal{D}).$$

In particular, from Lemma 2.1 (b) it immediately follows that all the iterated commutators $\mathrm{ad}[\,\mathcal{Z}\,](P)$ admit bounded and linear extensions to $H_1$. Let us consider vector fields $\mathcal{Y}$ with coefficients in the space of bounded smooth functions on $\mathbb{C}^n$. We give $X \in \mathcal{Y}$ with rapidly oscillating coefficients such that $[\,X,P\,]$ not even is bounded on $H_2$.

**Example 2.1** In case of dimension $n = 1$ we define $f(z) := \exp(\,i|\,z\,|^2\,)$ for $z \in \mathbb{C}$. Let us consider the smooth vector field $X := f\partial$, then we show that the commutator

$$[\,X,P\,] = M_f\,[\,\partial,P\,] + [\,M_f,P\,]\partial = [\,M_f,P\,]\partial$$

does not admit a bounded extension from the dense subspace $\mathcal{D} \cap H_2$ to $H_2$. With the operator $T_f = P\,M_f$ and the orthonormal basis $[\,e_j := (j!)^{-\frac{1}{2}} z^j : j \in \mathbb{N}_0\,] \subset \mathcal{D}$ of $H_2$ it is a straightforward computation that $T_f$ is diagonal of the following form $T_f e_j = [\,1 - i\,]^{-(j+1)}\,e_j$ for $j \in \mathbb{N}_0$. Let us compute the norm of $[\,M_f,P\,]\partial e_j$. It follows with $\partial e_j = \sqrt{j}\,e_{j-1}$ that:

$$[\,M_f,P\,]\partial e_j = \sqrt{j}\,\{\,M_f - T_f\,\}\,e_{j-1} = \sqrt{j}\,\{\,f - [\,1-i\,]^{-j}\,\}\,e_{j-1},$$

$$\|\,[\,M_f,P\,]\partial e_j\,\|_2 \geq \sqrt{j}\,\{\,\|\,fe_{j-1}\,\|_2 - 2^{-\frac{j}{2}}\,\|\,e_{j-1}\,\|_2\,\} = \sqrt{j}\,\left[\,1 - 2^{-\frac{j}{2}}\,\right].$$

Hence the numerical sequence $(\,\|\,[\,M_f,P\,]\partial e_j\,\|_2\,)_j$ is unbounded as $j$ tends to infinity and so the operator $[\,X,P\,]$ does not admit a bounded extension to $H_2$.

Motivated by our calculations above we will pose certain conditions on the oscillation of the coefficients $Xz_j$ and $X\bar{z}_j$ $(j = 1,\cdots,n)$ at infinity for vector fields $X \in \mathcal{X}$. Then $P$ will be contained in each $\Psi_0$-algebra generated by finite systems $\mathcal{V} \in \mathcal{X}$. These conditions naturally will arise for radial extensions of smooth functions on the complex sphere.

Let us denote by $\mathcal{C}^\infty_{\mathrm{pol}}(\mathbb{C}^n)$ the space of all functions in $\mathcal{C}^\infty(\mathbb{C}^n)$ such that $f$ and all its partial derivatives only have polynomial growth. We consider the following space $\mathcal{L} \subset L(\mathcal{D})$ of operators on the invariant space $\mathcal{D}$ [c.f. (2.2)] containing all vector fields with coefficients in $\mathcal{C}^\infty_{\mathrm{pol}}(\mathbb{C}^n)$:

$$\mathcal{L} := \mathrm{span}\{\,a\partial_j, b\bar{\partial}_j, \partial_j c, \bar{\partial}_j d, e : \text{ with } a,b,c,d,e \in \mathcal{C}^\infty_{\mathrm{pol}}(\mathbb{C}^n) \text{ and } j = 1,\cdots,n\,\}.$$

For any operator $Z = \sum_{j=1}^n\{\,a_j\partial_j + b_j\bar{\partial}_j + \partial_j c_j + \bar{\partial}_j d_j\,\} + e$ we define its *adjoint operator* $Z^*$ and its *conjugation* $\bar{Z}$ by:

$$Z^* := \sum_{j=1}^n\{\,\partial_j^*\bar{a}_j + \bar{\partial}_j^*\bar{b}_j + \bar{c}_j\partial_j^* + \bar{d}_j\bar{\partial}_j^*\,\} + \bar{e}, \quad \bar{Z} := \sum_{j=1}^n\{\,\bar{a}_j\bar{\partial}_j + \bar{b}_j\partial_j + \bar{\partial}_j\bar{c}_j + \partial_j\bar{d}_j\,\} + \bar{e}.$$

Because $\mathcal{C}^\infty_{\mathrm{pol}}(\mathbb{C}^n)$ is invariant under multiplication with complex polynomials and with the definition of $\partial_j^*$ and $\bar{\partial}_j^*$ it follows that $\bar{Z}$ and $Z^*$ again are contained in $\mathcal{L}$. Moreover, it clearly holds $\overline{Z^*} = (\bar{Z})^*$. For any $g \in \mathcal{D}$ and $z \in \mathbb{C}^n$ one has:

$$\{\,[\,Z,P\,]g\,\}(z) = Z_z\langle\,g, K(\cdot,z)\,\rangle_2 - \langle\,Z_u\,g, K(\cdot,z)\,\rangle_2 = \langle\,g, [\,\bar{Z}_z - Z_u^*\,]K(\cdot,z)\,\rangle_2.$$

Here the subscript in $Z_z$ (resp. $Z_u$) indicates that we consider $Z$ as an operator w.r.t. the variable $z$ (resp. $u$). Hence $[Z, P]$ has the integral kernel $L_Z$ given by

$$\overline{L_Z(u, z)} = (\bar{Z}_z - Z_u^*)\, K(u, z), \qquad \text{where} \qquad u, z \in \mathbb{C}^n. \tag{2.3}$$

**Proposition 2.3** *Let $r \in \mathbb{N}$ and $\mathcal{A} := [Z_1, \cdots, Z_r]$ be a finite system in $\mathcal{L}$ and consider:*

$$\overline{L_{\mathcal{A}}(u, z)} = (\bar{Z}_{r,z} - Z_{r,u}^*) \cdots (\bar{Z}_{1,z} - Z_{1,u}^*) K(u, z). \tag{2.4}$$

*Then the commutator $\mathrm{ad}[\mathcal{A}](P)$ is an integral operator on $\mathcal{D}$ with respect to the Gaussian measure $\mu$ and the integral kernel $L_{\mathcal{A}}$.*

Proposition 2.3 follows by induction from (2.3). In order to prove the boundedness of the commutators $\mathrm{ad}[\mathcal{A}](P)$ for certain systems $\mathcal{A} \subset \mathcal{L}$ of vector fields we estimate the kernels (2.4) and then apply Proposition 2.2. For a vector field

$$Y = \sum_{j=1}^{n} \{a_j \partial_j + b_j \bar{\partial}_j\} \in \mathcal{L}$$

we decompose the adjoint operator $Y^* \in \mathcal{L}$ into a derivation part and a multiplication part. A straightforward calculation shows $Y^* = -\bar{Y} - \overline{f_{Y,2}} + f_{Y,1}$ where

$$f_{Y,1}(z) := \sum_{j=1}^{n} \{z_j \bar{a}_j(z) + \bar{z}_j \bar{b}_j(z)\} \quad \text{and} \quad f_{Y,2}(z) := \sum_{j=1}^{n} \{[\partial_j a_j](z) + [\bar{\partial}_j b_j](z)\}.$$

Hence with the vector field $\delta_Y := \bar{Y}_z + \bar{Y}_u$ on $\mathcal{C}^\infty(\mathbb{C}^{2n})$ we have the following decomposition of the operators appearing in (2.4):

$$\bar{Y}_z - Y_u^* = \delta_Y - [f_{Y,1} - \overline{f_{Y,2}}](u).$$

In the following we will consider the case where all the coefficients $a_j$ and $b_j$ have bounded derivatives of all orders. Then the function $f_{Y,2}$ above will be bounded on $\mathbb{C}^n$. With the reproducing kernel $K$ in (2.1) we define $T_Y \in \mathcal{C}^\infty(\mathbb{C}^{2n})$ by:

$$T_Y(u, z) := \{K^{-1} \cdot [\delta_Y - f_{Y,1}(u)] K\}(u, z). \tag{2.5}$$

Using these notations we can calculate the integral kernels in (2.4) more explicitly. Given a finite system $\mathcal{A} = [Z_1, \cdots, Z_r]$ of vector fields $Z_j \in \mathcal{L}$, we define:

$$\delta_{\mathcal{A}}^{(1)} := \mathrm{id} \qquad \text{and} \qquad \delta_{\mathcal{A}}^{(k)} := \delta_{Z_k} \cdots \delta_{Z_2}, \quad \text{for } 2 \leq k \leq r.$$

Let $\mathbf{A}$ be the set consisting of all systems which are permutations of $\mathcal{A}$. Then with (2.5) we consider the algebra $M^{\mathcal{A}}(\mathbb{C}^{2n}) \subset \mathcal{C}^\infty(\mathbb{C}^{2n})$ generated by:

$$Q(\mathcal{A}) := \{1, \delta_{\mathcal{C}}^{(k)} T_{C_1} : \text{with } 1 \leq k \leq r-1 \text{ and } [\mathcal{C}, C_1] \in \mathbf{A}\}. \tag{2.6}$$

In the following we denote by $\mathcal{C}_b^\infty(\mathbb{C}^n)$ the space of all bounded smooth functions on $\mathbb{C}^n$ having bounded derivatives of all orders.

**Proposition 2.4** *Let all the vector fields $Z_j$ in $\mathcal{A}$ have coefficients in $\mathcal{C}_b^\infty(\mathbb{C}^n)$. Then the integral kernel $L_{\mathcal{A}}$ in (2.4) has the form of a finite sum:*

$$\overline{L_{\mathcal{A}}(u,z)} = \sum_{j \in J} \{\, B_j \cdot T_j \cdot K \,\}(u,z) \qquad (2.7)$$

*where $T_j \in M^{\mathcal{A}}(\mathbb{C}^{2n})$ and with $B_j \in \mathcal{C}_b^\infty(\mathbb{C}^{2n})$ for all $j$ in the finite set $J$.*

**Proof:** Let $r = 1$, i.e. $\mathcal{A} = [\,Z_1\,]$, then by Proposition 2.3 and with definition (2.5) we have:

$$\overline{L_{\mathcal{A}}(u,z)} = (\,\bar{Z}_{1,z} - Z_{1,u}^*\,)K(u,z) = \{\, T_{Z_1}(u,z) + \overline{f_{Z_1,2}}(u)\,\} K(u,z). \qquad (2.8)$$

Because $Z_1$ has coefficients in $\mathcal{C}_b^\infty(\mathbb{C}^n)$ it is clear that $\overline{f_{Z_1,2}} \in \mathcal{C}_b^\infty(\mathbb{C}^{2n})$. Moreover, we have $T_{Z_1} \in M^{\mathcal{A}}(\mathbb{C}^{2n})$ and so (2.8) has the form (2.7).

Now, let $r > 1$ and set $\mathcal{A}_{r-1} := [\,Z_1, \cdots, Z_{r-1}\,]$. Then there are functions $h_1 \in \mathcal{C}^\infty(\mathbb{C}^{2n})$ and $h_2 \in \mathcal{C}_b^\infty(\mathbb{C}^{2n})$ such that $\bar{Z}_{r,z} - Z_{r,u}^* = \delta_{Z_r} + h_1(u) + h_2(u)$. It follows that:

$$\overline{L_{\mathcal{A}}(u,z)} = [\,\delta_{Z_r} + h_1(u)\,]\,\overline{L_{\mathcal{A}_{r-1}}(u,z)} + h_2(u) \cdot \overline{L_{\mathcal{A}_{r-1}}(u,z)}. \qquad (2.9)$$

By induction the second term in (2.9) already has the desired form (2.7). In order to treat the first term we compute:

$$H_j(u,z) := [\,\delta_{Z_r} + h_1(u)\,]\,\{\, B_j \cdot T_j \cdot K \,\}(u,z) \quad \text{with} \quad T_j \in M^{\mathcal{A}_{r-1}}(\mathbb{C}^{2n}).$$

The operator $\delta_{Z_r}$ is a derivation and using $[\,\delta_{Z_r} + h_1(u)\,]K = T_{Z_r} \cdot K$ we find:

$$H_j(u,z) = [\,\{\,(\,\delta_{Z_r} B_j\,)T_j K\,\} + \{\, B_j(\,\delta_{Z_r} T_j\,)K\,\} + \{\, B_j T_j T_{Z_r} K\,\}\,](u,z)$$

for all $j$. We have $\delta_{Z_r} B_j \in \mathcal{C}_b^\infty(\mathbb{C}^{2n})$ and because $\delta_{Z_r}$ is a derivation it is easy to see that the functions $\delta_{Z_r} T_j$ and $T_{Z_r}$ are contained in $M^{\mathcal{A}}(\mathbb{C}^{2n})$. □

It will be our next aim to estimate the functions in $Q(\mathcal{A})$ defined in (2.6). With any vector field $Y \in \mathcal{L}$ as above we find by a straightforward calculation that the function $T_Y$ in (2.5) can be decomposed into $T_Y = F_Y + G_Y$ where:

$$F_Y(u,z) := \sum_{j=1}^{n} u_j\,[\,\bar{a}_j(z) - \bar{a}_j(u)\,] \quad \text{and} \quad G_Y(u,z) := \sum_{j=1}^{n} \bar{b}_j(u)\,[\,\bar{z}_j - \bar{u}_j\,].$$

Moreover, for a vector field $Z \in \mathcal{L}$ and with $\delta_Z = \bar{Z}_z + \bar{Z}_u$ we obtain:

$$\delta_Z F_Y = \sum_{j=1}^{n} \{\, u_j\,[\,(\,\bar{Z}\bar{a}_j\,)(z) - (\,\bar{Z}\bar{a}_j\,)(u)\,] + (\,\bar{Z}_u u_j\,)(u)\,[\,\bar{a}_j(z) - \bar{a}_j(u)\,]\,\},$$

$$\delta_Z G_Y = \sum_{j=1}^{n} \{\, \bar{b}_j(u)\,[\,(\,\bar{Z}_z \bar{z}_j\,)(z) - (\,\bar{Z}_u \bar{u}_j\,)(u)\,] + (\,\bar{Z}b_j\,)(u)\,[\,\bar{z}_j - \bar{u}_j\,]\,\}. \qquad (2.10)$$

We define an algebra of functions having *linear vanishing oscillation at infinity*:

$$\mathcal{VO}(\mathbb{C}^n) := \{\, g \in \mathcal{C}_b^\infty(\mathbb{C}^n) : \exists\, C > 0 \text{ with } |f(z) - f(u)| \cdot \|u\| \leq C \cdot (1 + \|z - u\|)\,\}.$$

One can check that the space $\mathcal{VO}(\mathbb{C}^n)$ contains the $\mathcal{C}_b^\infty(\mathbb{C}^n)$-module $\mathcal{O}_{-1}(\mathbb{C}^n)$ of all smooth functions vanishing at infinity to the order $-1$. Define:

$$\mathcal{VO}^\infty(\mathbb{C}^n) := \{ \, f \in \mathcal{VO}(\mathbb{C}^n) : \text{ where } \bar\partial^\beta \partial^\alpha f \in \mathcal{VO}(\mathbb{C}^n) \text{ for all } \alpha, \beta \in \mathbb{N}_0^n \, \}.$$

The function algebra $\mathcal{VO}^\infty(\mathbb{C}^n)$ contains $\mathcal{O}_{-1}^\infty(\mathbb{C}^n)$, the space of all $f \in \mathcal{VO}(\mathbb{C}^n)$ such that all partial derivatives $\bar\partial^\beta \partial^\alpha f$ with $|\alpha + \beta| \neq 0$ are elements of $\mathcal{O}_{-1}(\mathbb{C}^n)$. Let us consider the following spaces of smooth vector fields in $\mathcal{L}$ which both will lead to bounded iterated commutators on $\mathcal{D}$ with the Toeplitz projection $P$:

- $\mathcal{X}_1(\mathbb{C}^n) := \mathrm{span}\{ \, a_j \partial_j, b_j \bar\partial_j : \text{ where } a_j, b_j \in \mathcal{VO}^\infty(\mathbb{C}^n) \text{ for } j = 1, \cdots, n \, \}$,
- $\mathcal{X}_2(\mathbb{C}^n) := \mathrm{span}\{ \, a_j \partial_j : a_j \in \mathcal{O}_{-1}^\infty(\mathbb{C}^n) \, \} \oplus \mathrm{span}\{ \, b_j \bar\partial_j : b_j \in \mathcal{C}_b^\infty(\mathbb{C}^n) \, \}$.

Clearly, $\mathcal{O}_{-1}^\infty(\mathbb{C}^n)$ is invariant under $Z, \bar{Z} \in \mathcal{X}_2(\mathbb{C}^n)$. Moreover, we observe that the algebra $\mathcal{VO}^\infty(\mathbb{C}^n)$ is invariant under $Z, \bar{Z} \in \mathcal{X}_1(\mathbb{C}^n)$. For $j = 1, 2$ let us define the following subspaces of $\mathcal{C}^\infty(\mathbb{C}^{2n})$:

$$\mathcal{F}_j(\mathbb{C}^{2n}) := \{ \, F_Z : Z \in \mathcal{X}_j(\mathbb{C}^n) \, \},$$
$$\mathcal{G}(\mathbb{C}^{2n}) := \mathrm{span}\{ \, b_l(u) \cdot (\bar{z}_l - \bar{u}_l) : b_l \in \mathcal{C}_b^\infty(\mathbb{C}^n) \, \}.$$

The next result is a direct consequence from our definitions of $F_Y$ and $\mathcal{G}(\mathbb{C}^{2n})$:

**Lemma 2.2** *For $j \in \{1, 2\}$ fix functions $f \in \mathcal{F}_j(\mathbb{C}^{2n})$ and $g \in \mathcal{G}(\mathbb{C}^{2n})$. Then it holds*

$$(i): \quad |f(u, z)| \leq c_f \left(1 + \| u - z \|\right) \qquad (ii): \quad |g(u, z)| \leq c_g \| u - z \|.$$

*where $c_f$ and $c_g$ are suitable positive numbers.*

From the equations (2.10) it immediately follows that for $j = 1, 2$, the spaces

$$\mathcal{F}_j(\mathbb{C}^{2n}) + \mathcal{C}_b^\infty(\mathbb{C}^{2n}) \qquad \text{and} \qquad \mathcal{G}(\mathbb{C}^{2n}) + \mathcal{C}_b^\infty(\mathbb{C}^{2n})$$

are invariant under the derivations $\delta_Z$ for all $Z \in \mathcal{X}_j(\mathbb{C}^n)$. Hence for any finite system $\mathcal{A}_j \subset \mathcal{X}_j(\mathbb{C}^n)$ and with the decomposition $T_Y = F_Y + G_Y$ we obtain that an estimate as in Lemma 2.2, (i) holds for all $f \in Q(\mathcal{A}_j)$ in (2.6).

**Theorem 2.1** *Let $\mathcal{A}_j$ be a finite system of operators in $\mathcal{X}_j(\mathbb{C}^n)$ for $j = 1, 2$. Then the commutator $\mathrm{ad}[\,\mathcal{A}_j\,](P)$ admits a bounded extension from $\mathcal{D}$ to $H_1$.*

**Proof:** From our remark above it is easy to see that all functions $f \in M^{\mathcal{A}_j}(\mathbb{C}^{2n})$ fulfill:

$$|f(u, z)| \leq c \cdot \left(1 + \| u - z \|^m\right)$$

where $c > 0$ and $m \in \mathbb{N}$ are suitable numbers. Applying Proposition 2.4 we conclude that the commutator $\mathrm{ad}[\,\mathcal{A}_j\,](P)$ is an integral operator with integral kernel $L_{\mathcal{A}}$ and

$$| L_{\mathcal{A}_j}(u, z)| \leq \tilde{c} \cdot \left(1 + \| u - z \|^m\right) \cdot \exp\left(\mathrm{Re}\langle u, z \rangle\right).$$

where $\tilde{c} > 0$. Because of $F := (1 + \| \cdot \|^m) \in L^1(\mathbb{C}^n, \mu_{\frac{1}{2}})$ the assertion follows from Proposition 2.2. $\qquad \square$

**Remark 2.1** As a linear subspace $\mathcal{O}_{-1}^{\infty}(\mathbb{C}^n)$ contains all $f \in \mathcal{C}^{\infty}(\mathbb{C}^n)$ which outside a ball of radius $R > 0$ coincide with the radial extension of a smooth function on the *complex unit sphere* $S^{2n-1} \subset \mathbb{C}^n$. Hence for any open subset $U \subset S^{2n-1}$ there are non-zero vector fields $X \in \mathcal{X}_2(\mathbb{C}^n)$ which are supported in the complex cone $\mathcal{C}_U := \{ x \in \mathbb{C}^n : x \cdot \| x \|^{-1} \in U \}$. This observation together with Proposition 1.1 and Theorem 2.1 will lead to a *cone-localization* of the Toeplitz projection and corresponding $\Psi_0$-algebras.

## Acknowledgments

The author thanks Prof. L. Coburn for his invitation to the State University of New York (SUNY) at Buffalo in 2003. He is greatly indebted to Prof. B. Gramsch who proposed an application of the general theory in [6] and [7] to algebras of Toeplitz operators for many hints.

## References

1. V. BARGMANN, *On a Hilbert space of analytic functions and an associated integral transform,* Comm. Pure Appl. Math. 14, pp. 187-214, (1961).
2. W. BAUER, *Toeplitz operators on finite and infinite dimensional spaces with associated $\Psi^*$-Fréchet algebras,* Ph. D. thesis, submitted to J. Gutenberg-University of Mainz, (June 2005).
3. W. BAUER, *Hilbert-Schmidt Hankel operators on the Segal-Bargmann space,* Proc. Amer. Math. Soc. 132, pp. 2989-2998, (2004).
4. W. BAUER, *Mean oscillation and Hankel operators on the Segal-Bargmann space,* Int. Equ. Oper. Theory 52, pp. 1-15, (2005).
5. C.A. BERGER, L.A. COBURN, *Toeplitz operators on the Segal Bargmann space,* Trans. Amer. Math. Soc. 301, pp. 813-829, (1994).
6. B. GRAMSCH, *Relative Inversion in der Störungstheorie von Operatoren und $\Psi$-Algebren,* Math. Ann. 269, pp. 27-71, (1984).
7. B. GRAMSCH, J. UEBERBERG, K. WAGNER, *Spectral invariance and submultiplicativity for Fréchet algebras with applications to pseudo-differential operators and $\Psi^*$-quantization,* in: Operator Theory: Advances and Applications , vol. 57, pp. 71-98. Birkhäuser, Basel, (1992).
8. B. GRAMSCH , K.G. KALB, *Pseudo-locality and hypoellipticity in operator algebras,* Semesterbericht Funktionalanalysis, Universität Tübingen, pp. 51-61, Sommersemester (1985).
9. R. LAUTER, *Pseudodifferential analysis on conformally compact spaces,* Memoirs of the AMS 163 No. 777, (2003).
10. R. LAUTER, R. MONTHUBERT, V. NISTOR, *Spectral invariance for certain algebras of pseudodifferential operators,* Jour. of the Inst. of Math. Jussieu 4, No. 3, pp. 405-442, (2005).
11. E. SCHROHE, *Fréchet algebra techniques for boundary value problems: Fredholm criteria and functional calculus via spectral invariance,* Math. Nachr. 199, pp. 145-185, (1999).
12. K. STROETHOFF, *Hankel and Toeplitz operators on the Fock space,* Michigan Math. J. 39, pp. 3-16, (1992).
13. L. WAELBROECK, *Le calcul symbolique dans les algèbres commutatives,* J. Math. Pures Appl., 33 No. 9, pp. 147-186, (1954).

# C*-ALGEBRAS OF BERGMAN TYPE OPERATORS WITH PIECEWISE CONTINUOUS COEFFICIENTS ON BOUNDED DOMAINS*

YU. I. KARLOVICH

*Facultad de Ciencias,*
*Universidad Autónoma del Estado de Morelos,*
*Av. Universidad 1001, Col. Chamilpa,*
*C.P. 62209, Cuernavaca, Morelos, México*
*E-mail: karlovich@buzon.uaem.mx*

L. V. PESSOA

*Departamento de Matemática,*
*Instituto Superior Técnico,*
*Av. Rovisco Pais,*
*1049 - 001, Lisboa, Portugal*
*E-mail: lpessoa@math.ist.utl.pt*

The $C^*$-algebra $\mathfrak{A}_{n,m}(U)$ generated by the $n$ poly-Bergman and $m$ anti-poly-Bergman projections and by operators of multiplication by piecewise continuous functions on the Lebesgue space $L^2(U)$ over the bounded multiply connected domain $U \subset \mathbb{C}$ is studied. A symbol calculus for the $C^*$-algebra $\mathfrak{A}_{n,m}(U)$ is constructed and a Fredholm criterion for the operators $A \in \mathfrak{A}_{n,m}(U)$ is obtained.

**Keywords:** Poly-Bergman and anti-poly-Bergman projections, local principle, isomorphism, $C^*$-algebra, symbol calculus, Fredholm criterion.
**Mathematics Subject classification:** 47L15, 47A53, 47G10

## 1. Introduction

Given an arbitrary domain $U \subset \mathbb{C}$, let $\mathcal{B} := \mathcal{B}(L^2(U))$ be the $C^*$-algebra of all bounded linear operators on the Hilbert space $L^2(U) := L^2(U, dA)$ where $dA(z) = dxdy$ is the Lebesgue area measure, and let $\mathcal{K} := \mathcal{K}(L^2(U))$ be the closed two-sided ideal of all compact operators in $\mathcal{B}$. We denote by $B_{U,n}$ and $\widetilde{B}_{U,n}$ the orthogonal projections of $L^2(U)$ onto the poly-Bergman spaces $\mathcal{A}_n^2(U)$ and anti-poly-Bergman spaces $\widetilde{\mathcal{A}}_n^2(U)$, respectively, where $\mathcal{A}_n^2(U)$ and $\widetilde{\mathcal{A}}_n^2(U)$ are the Hilbert subspaces of $L^2(U)$ that consist of n-differentiable functions such that, respectively, $(\partial/\partial\overline{z})^n f = 0$ and $(\partial/\partial z)^n f = 0$ (see, e.g. [1,5,17]).

*Both authors were partially supported by FCT project POCTI/MAT/59972/2004 (Portugal). The first author was also supported by PROMEP (México). The second author was also supported by the FCT (Portugal) and FSE.

Let now $U$ be a bounded (connected) multiply connected domain in $\mathbb{C}$ with infinitely smooth boundary $\partial U$ oriented such that $U$ is on the left of $\partial U$, and let $\overline{U}$ be the closure of $U$ in $\mathbb{C}$. Given a finite union $\mathfrak{L}$ of Lyapunov curves in $\overline{U}$ with a finite set $\mathfrak{L} \cap \partial U$, we denote by $PC(\mathfrak{L})$ the $C^*$-subalgebra of $L^\infty(U)$ consisting of all continuous functions on $\overline{U} \setminus \mathfrak{L}$ which have one-sided limits at the points of $\mathfrak{L}$. We assume that $\mathfrak{L}$ satisfies the conditions:

($\mathfrak{L}1$) for each $z \in U$ there exist numbers $r_z > 0$ and $n_z \in \mathbb{N}$ such that every disk $D(z, r)$ of radius $r \in (0, r_z)$ centered at $z$ is divided by $\mathfrak{L}$ into $n_z$ domains with $z$ as a common limit point;

($\mathfrak{L}2$) for each $z \in \partial U \cap \mathfrak{L}$ there exists a neighborhood $V_z$ of $z$ such that $V_z \cap \mathfrak{L}$ consists of an $n_z - 1$ Lyapunov arcs having only the point $z$ in common and forming at this point pairwise distinct angles $\theta_k = \theta_k(z)$ with $\partial U$ where $0 < \theta_1 < \ldots < \theta_{n_z-1} < \pi$.

Thus, for a sufficiently small neighborhood $V_z$ of a $z \in \overline{U}$, the set $V_z \cap (U \setminus \mathfrak{L})$ consists of $n_z$ connected components $\Omega_k = \Omega_k(z)$ whose closures contain $z$.

According to [5, Lemmas 3.5.2 and 2.9.1], the poly-Bergman and anti-poly-Bergman projections $B_{U,n}$ and $\widetilde{B}_{U,n}$ for a bounded multiply connected domain $U$ with a sufficiently smooth boundary are represented via the two-dimensional singular integral operators $S_U, S_U^* \in \mathcal{B}(L^2(U))$ by the following formulas:

$$B_{U,n} = I - \left(S_U\right)^n \left(S_U^*\right)^n + K_n, \quad \widetilde{B}_{U,n} = I - \left(S_U^*\right)^n \left(S_U\right)^n + \widetilde{K}_n \quad (n \in \mathbb{N}), \quad (1)$$

where $K_n, \widetilde{K}_n \in \mathcal{K}(L^2(U))$, $K_n = \widetilde{K}_n = 0$ for the unit disk $\mathbb{D} = \{z \in \mathbb{C} : |z| < 1\}$ and the upper half-plane $\Pi = \{z \in \mathbb{C} : \operatorname{Im} z > 0\}$ (see [5] and [14]), and

$$(S_U f)(z) = -\frac{1}{\pi} \int_U \frac{f(w)}{(w - z)^2} \, dA(w), \quad (S_U^* f)(z) = -\frac{1}{\pi} \int_U \frac{f(w)}{(\overline{w} - \overline{z})^2} \, dA(w).$$

Given numbers $n, m \in \mathbb{N}$, we study the $C^*$-algebra

$$\mathfrak{A}_{n,m}(U) := \operatorname{alg} \left\{ B_{U,1}, \ldots, B_{U,n}, \widetilde{B}_{U,1}, \ldots, \widetilde{B}_{U,m}; \mathfrak{L} \right\} \subset \mathcal{B}(L^2(U)) \quad (2)$$

generated by the poly-Bergman projections $B_{U,1}, \ldots, B_{U,n}$, anti-poly-Bergman projections $\widetilde{B}_{U,1}, \ldots, \widetilde{B}_{U,m}$, and by the multiplication operators $aI$ ($a \in PC(\mathfrak{L})$).

The $C^*$-algebra generated by the Bergman projection of a bounded multiply connected domain $U$ with a smooth boundary $\partial U$ and by multiplications by piecewise continuous functions having one-sided limits at points of a finite union of curves intersecting $\partial U$ at distinct points was investigated in [16]. Generalization of this work to the case of piecewise continuous coefficients admitting more than two one-sided limits at the points of $\partial U$ was obtained in [9] in the case of unit disk $\mathbb{D}$. The $C^*$-algebra generated by the harmonic Bergman projection being a one-dimensional perturbation of the operator $B_{\mathbb{D},1} + \widetilde{B}_{\mathbb{D},1}$, by operators of multiplication by piecewise continuous functions, and by all compact operators was studied in [10]. More general $C^*$-algebras $\mathfrak{A}_{n,m}(\Pi)$ generated by $n$ poly-Bergman and $m$ anti-poly-Bergman projections of the upper half-plane $\Pi$ and by multiplication operators $aI$ ($a \in PC(\mathfrak{L})$)

were studied in [6,7]. In [6] a *-isomorphism of the $C^*$-algebras $\mathfrak{A}_{1,1}(\Pi)$ and $\mathfrak{A}_{1,1}(\mathbb{D})$ is also established.

Making use of the Allan-Douglas local principle (see, e.g., [4, Theorem 1.34]), limit operators techniques (see, e.g., [3]), orthogonal decomposition of $L^2(\Pi)$ [17], formulas (1) for $U = \Pi$ [14], and the Plamenevsky results [12] (also see [8]) on two-dimensional singular integral operators with coefficients admitting homogeneous discontinuities, we reduce studying the $C^*$-algebras $\mathfrak{A}_{n,m}(\Pi)$ in [6,7] to simpler $C^*$-algebras associated with the points $z \in \dot{\overline{\Pi}}$ and pairs $(z, \lambda) \in (\dot{\mathbb{R}} \cap \mathcal{L}) \times \mathbb{R}$ where $\dot{\overline{\Pi}} := \overline{\Pi} \cup \dot{\mathbb{R}}$ and $\dot{\mathbb{R}} := \mathbb{R} \cup \{\infty\}$. In contrast to the papers [16], [9] and [10], whose methods do not allow to study the $C^*$-algebras $\mathfrak{A}_{n,m}(\Pi)$ in case $n + m \geq 2$, the study of local algebras associated with the points $z \in \dot{\mathbb{R}} \cap \mathcal{L}$ is based in [6,7] on a new symbol calculus elaborated in [6] for unital $C^*$-algebras generated by $N$ orthogonal projections sum of which equals the unit and by $M = n + m$ one-dimensional orthogonal projections possessing some additional properties, and also on the relations for the Gauss hypergeometric function.

In the present paper we extend the results of [6,7] to arbitrary bounded multiply connected domains $U$ with infinitely smooth boundaries. In Section 2 we construct isomorphisms of local algebras $\mathfrak{A}_{n,m,z}^{\pi}(U)$ defined by the Allan-Douglas local principle onto corresponding local algebras $\mathfrak{A}_{n,m,0}^{z,\pi}(\Pi)$. In Section 3 applying the results of [7] we construct a symbol calculus for the $C^*$-algebra $\mathfrak{A}_{n,m}(U)$ and deduce a Fredholm criterion for the operators $A \in \mathfrak{A}_{n,m}(U)$.

## 2. Isomorphisms of local algebras

Fix $n, m \in \mathbb{N}$ and consider the $C^*$-algebra $\mathfrak{A}_{n,m}(U)$ given by (2). By [6, Lemma 2.6] that remains valid for arbitrary domains $U \subset \mathbb{C}$, the $C^*$-algebra $\mathfrak{A}_{n,m}(U)$ contains the ideal $\mathcal{K} = \mathcal{K}(L^2(U))$ of all compact operators in $\mathcal{B} = \mathcal{B}(L^2(U))$. To obtain a Fredholm criterion for the operators $A \in \mathfrak{A}_{n,m}(U)$ we need to study the invertibility of the cosets $A^{\pi} := A + \mathcal{K}$ in the quotient $C^*$-algebra $\mathfrak{A}_{n,m}^{\pi}(U) := \mathfrak{A}_{n,m}(U)/\mathcal{K}$. To this end we will apply the Allan-Douglas local principle to the algebra $\mathfrak{A}_{n,m}^{\pi}(U)$ over some its central $C^*$-subalgebra. If two $C^*$-algebras $\mathcal{A}_1$ and $\mathcal{A}_2$ are (isometrically) *-isomorphic, we will write $\mathcal{A}_1 \cong \mathcal{A}_2$ and say that $\mathcal{A}_1$ and $\mathcal{A}_2$ are isomorphic.

According to [15] (also see [2, Section 8.2]), an operator $A \in \mathcal{B}(L^2(U))$ is called an *operator of local type* if the commutators $cA - AcI$ are compact for all $c \in C(\overline{U})$. From [11, Chapter X, Theorem 7.1] it follows that the singular integral operators $S_U$ and $S_U^*$ are of local type. Hence, (1) implies the following.

**Lemma 2.1.** *For every $n \in \mathbb{N}$ and every function $a \in C(\overline{U})$, the commutators $aB_{U,n} - B_{U,n}aI$ and $a\widetilde{B}_{U,n} - \widetilde{B}_{U,n}aI$ are compact on the space $L^2(U)$.*

By Lemma 2.1, all the operators in the $C^*$-algebra $\mathfrak{A}_{n,m}$ are of local type, and $\mathcal{Z}^{\pi} := \{cI + \mathcal{K} : c \in C(\overline{U})\}$ is a central subalgebra of the $C^*$-algebra $\mathfrak{A}_{n,m}^{\pi}(U)$. Obviously, $\mathcal{Z}^{\pi} \cong C(\overline{U})$, and therefore the maximal ideal space of $\mathcal{Z}^{\pi}$ can be identified with $\overline{U}$. For every point $z \in \overline{U}$, let $J_{U,z}^{\pi}$ denote the closed two-sided ideal of the $C^*$-algebra $\mathfrak{A}_{n,m}^\pi(U)$ generated by the maximal ideal

$$I_{U,z}^\pi := \{cI + \mathcal{K} : c \in C(\overline{U}),\ c(z) = 0\} \subset \mathcal{Z}^\pi.$$

According to [2, Proposition 8.6], the ideal $J_{U,z}^\pi$ has the form

$$J_{U,z}^\pi = \{(cA)^\pi : c \in C(\overline{U}),\ c(z) = 0,\ A \in \mathfrak{A}_{n,m}(U)\}.$$

With every $z \in \overline{U}$ we associate the local $C^*$-algebra $\mathfrak{A}_{n,m,z}^\pi(U) := \mathfrak{A}_{n,m}^\pi(U)/J_{U,z}^\pi$.

By the Allan-Douglas local principle, we obtain the following.

**Theorem 2.1.** *An operator $A \in \mathfrak{A}_{n,m}(U)$ is Fredholm on the space $L^2(U)$ if and only if for every $z \in \overline{U}$ the coset $A_z^\pi := A^\pi + J_{U,z}^\pi$ is invertible in the local $C^*$-algebra $\mathfrak{A}_{n,m,z}^\pi(U)$, and $\|A^\pi\| = \max\{\|A_z^\pi\| : z \in \overline{U}\}$.*

The cosets $A^\pi, B^\pi \in \mathfrak{A}_{n,m}^\pi(U)$ are called locally equivalent at a point $z \in \overline{U}$ if $A^\pi - B^\pi \in J_{U,z}^\pi$, and in that case we write $A^\pi \overset{z}{\sim} B^\pi$.

Analogously to [7, Lemma 3.4] we obtain the following.

**Lemma 2.2.** *For every $n \in \mathbb{N}$, the cosets $B_{U,n}^\pi$ and $\widetilde{B}_{U,n}^\pi$ are locally equivalent to zero at every point $z \in U$.*

Given domains $U, G \subset \mathbb{C}$ and a conformal mapping $\varphi : U \to G$, we define the unitary shift operator $W_\varphi : L^2(G) \to L^2(U)$, $W_\varphi f = (f \circ \varphi)\varphi'$. From [6, Lemma 6.1] that remains valid for conformal mappings $\varphi : \mathbb{D} \to G$ we deduce the following.

**Lemma 2.3.** *If $\varphi$ is a conformal diffeomorphism of the closed unit disk $\overline{\mathbb{D}}$ onto $\overline{G}$ and the derivative $\varphi'$ satisfies the Hölder condition in $\overline{\mathbb{D}}$, then the operators $W_\varphi S_G W_\varphi^{-1} - (\overline{\varphi'}/\varphi')S_\mathbb{D}$, $W_\varphi S_G^* W_\varphi^{-1} - (\varphi'/\overline{\varphi'})S_\mathbb{D}^*$ are compact on the space $L^2(\mathbb{D})$.*

Since $S_\mathbb{D}$ and $S_\mathbb{D}^*$ are operators of local type, Lemma 2.3 and the relation (1) imply the following corollary.

**Corollary 2.1.** *If $\varphi$ is a conformal diffeomorphism of $\overline{\mathbb{D}}$ onto $\overline{G}$ and the derivative $\varphi'$ satisfies the Hölder condition in $\overline{\mathbb{D}}$, then for every $n \in \mathbb{N}$,*

$$\left(W_\varphi B_{G,n} W_\varphi^{-1}\right)^\pi = B_{\mathbb{D},n}^\pi, \quad \left(W_\varphi \widetilde{B}_{G,n} W_\varphi^{-1}\right)^\pi = \widetilde{B}_{\mathbb{D},n}^\pi.$$

Let $\partial U = \bigcup_{k=0}^r \Gamma_k$ where the Jordan curves $\Gamma_k$ are infinitely smooth and the bounded domain $G$ with the boundary $\Gamma_0$ contains all other Jordan curves $\Gamma_k$.

If $z \in \Gamma_0$ then, by the Kellogg-Warschawski theorem (see, e.g., [13, Theorem 3.6]), there exists a conformal mapping $\varphi_z$ of the open unit disk $\mathbb{D}$ onto $G$ such that $\varphi_z(-1) = z$, $\varphi_z^{(k)}$ are homeomorphisms of $\overline{\mathbb{D}}$ onto $\overline{G}$ for all $k = 0, 1, \ldots$, and $\varphi_z$ preserves the angles between smooth curves in $\mathbb{D}$ outgoing from the point $-1$ and between their images in $G$ outgoing from the point $z$ (see [13, Proposition 4.10]). We also consider the conformal mapping $\psi : \Pi \to \mathbb{D}$, $\zeta \mapsto (\zeta - i)/(\zeta + i)$ where $\Pi = \{\zeta \in \mathbb{C} : \operatorname{Im}\zeta > 0\}$. Hence $\psi_z = \varphi_z \circ \psi$ is a conformal mapping of $\Pi$ onto $\mathfrak{D}_0 := G$, $\psi_z(0) = z$, and again $\psi$ preserves the angles between smooth curves in $\Pi$ outgoing from the point $0$ and between their images in $\mathbb{D}$ at the point $-1$.

If $z \in \Gamma_k$ $(k = 1, 2, \ldots, r)$ then, choosing a point $z_k$ in a bounded domain $U_k$ with the boundary $\Gamma_k$, we infer that $\beta_k : \zeta \mapsto 1/(\zeta - z_k)$ is a conformal mapping of $U$ onto the bounded multiply connected domain $\beta_k(U)$ such that the domain $G_k = \mathbb{C} \setminus \overline{\beta_k(U_k)}$ bounded by the Jordan curve $\beta_k(\Gamma_k)$ contains all other curves $\beta_k(\Gamma_s)$ $(s = 0, 1, \ldots, r;\ s \neq k)$. Let $\varphi_{k,z}$ be a conformal mapping of $\mathbb{D}$ onto $G_k$ such that $\varphi_{k,z}(-1) = \beta_k(z)$. Then $\psi_z := \beta_k^{-1} \circ \varphi_{k,z} \circ \psi$ is a conformal mapping of $\Pi$ onto $\mathfrak{D}_k := \dot{\mathbb{C}} \setminus \overline{U}_k$ where $\dot{\mathbb{C}} = \mathbb{C} \cup \{\infty\}$, and $\psi_z(0) = z$.

For $z \in \Gamma_k$ $(k = 0, 1, \ldots, r)$, along with the $C^*$-algebras $\mathfrak{A}_{n,m}(U)$, $\mathfrak{A}_{n,m}^\pi(U)$ and $\mathfrak{A}_{n,m,z}^\pi(U)$ we define the $C^*$-algebras

$$\mathfrak{A}_{n,m}^z(\Pi) := \mathrm{alg}\left\{B_{\Pi,1}, \ldots, B_{\Pi,n}, \widetilde{B}_{\Pi,1}, \ldots, \widetilde{B}_{\Pi,m}; \psi_z^{-1}(\mathfrak{L} \cup \partial U \setminus \Gamma_k)\right\} \subset \mathcal{B}(L^2(\Pi)),$$

$$\mathfrak{A}_{n,m}^{z,\pi}(\Pi) = \mathfrak{A}_{n,m}^z(\Pi)/\mathcal{K}(L^2(\Pi)), \quad \mathfrak{A}_{n,m,0}^{z,\pi}(\Pi) := \left\{A^\pi + J_{\Pi,0}^\pi : A \in \mathfrak{A}_{n,m}^z(\Pi)\right\},$$

where the closed two-sided ideal $J_{\Pi,0}^\pi$ of the $C^*$-algebra $\mathfrak{A}_{n,m}^{z,\pi}(\Pi)$ is given by

$$J_{\Pi,0}^\pi := \left\{(cA)^\pi : c \in C(\overline{\Pi}),\ c(0) = 0,\ A \in \mathfrak{A}_{n,m}^z(\Pi)\right\}.$$

Let $\Lambda(U)$ be the $C^*$-algebra of all operators of local type in $\mathcal{B}(L^2(U))$. To every point $z \in \overline{U}$ we assign the closed two-sided ideal

$$\widehat{J}_{U,z}^\pi := \left\{(cA)^\pi : c \in C(\overline{U}),\ c(z) = 0,\ A \in \Lambda(U)\right\}, \tag{3}$$

of the $C^*$-algebra $\Lambda^\pi(U) := \Lambda(U)/\mathcal{K}(L^2(U))$.

**Lemma 2.4.** *For every $z \in \Gamma_k$ $(k = 0, 1, \ldots, r)$, the local $C^*$-algebras $\mathfrak{A}_{n,m,z}^\pi(U)$ and $\mathfrak{A}_{n,m,0}^{z,\pi}(\Pi)$ are isomorphic, and the isomorphism is given by*

$$\begin{aligned}
B_{U,i}^\pi + J_{U,z}^\pi &\mapsto B_{\Pi,i}^\pi + J_{\Pi,0}^\pi &\quad &(i = 1, 2, \ldots, n), \\
\widetilde{B}_{U,j}^\pi + J_{U,z}^\pi &\mapsto \widetilde{B}_{\Pi,j}^\pi + J_{\Pi,0}^\pi &\quad &(j = 1, 2, \ldots, m), \\
(aI)^\pi + J_{U,z}^\pi &\mapsto ((\widetilde{a} \circ \psi_z)I)^\pi + J_{\Pi,0}^\pi &\quad &(a \in PC(\mathfrak{L})),
\end{aligned} \tag{4}$$

*where $\widetilde{a} \in PC(\mathfrak{L} \cup \partial U \setminus \Gamma_k)$, $\widetilde{a}(\zeta) = a(\zeta)$ for $\zeta \in U$ and $\widetilde{a}(\zeta) = 0$ for $\zeta \in \mathfrak{D}_k \setminus \overline{U}$.*

**Proof.** Let $z \in \Gamma_0$ and consider the local $C^*$-algebra $\mathfrak{A}_{n,m,z}^\pi(U)$. By analogy with [7, Lemma 4.3], the equality $\widehat{J}_{U,z}^\pi \cap \mathfrak{A}_{n,m,z}^\pi(U) = J_{U,z}^\pi$ implies that the $C^*$-algebras

$$\mathfrak{A}_{n,m,z}^\pi(U) = \{A^\pi + J_{U,z}^\pi : A \in \mathfrak{A}_{n,m}(U)\}, \quad \widehat{\mathfrak{A}}_{n,m,z}^\pi(U) := \{A^\pi + \widehat{J}_{U,z}^\pi : A \in \mathfrak{A}_{n,m}(U)\}$$

are isomorphic, and the isomorphism is given by $\nu_z : A^\pi + J_{U,z}^\pi \mapsto A^\pi + \widehat{J}_{U,z}^\pi$. Let $\chi_U$ be the characteristic function of $U$,

$$\mathfrak{A}_{n,m}(G) := \mathrm{alg}\left\{B_{G,1}, \ldots, B_{G,n}, \widetilde{B}_{G,1}, \ldots, \widetilde{B}_{G,m}; \mathfrak{L} \cup \partial U \setminus \Gamma_0\right\} \subset \mathcal{B}(L^2(G)),$$

and let $\widehat{J}_{G,z}^\pi$ be given by (3) with $U$ replaced by $G$. Identifying the cosets

$$\begin{aligned}
B_{U,i}^\pi + \widehat{J}_{U,z}^\pi &\text{ with } \left(\chi_U B_{G,i} \chi_U I\right)^\pi + (\chi_U I)^\pi \widehat{J}_{G,z}^\pi (\chi_U I)^\pi &\quad &(i = 1, 2, \ldots, n), \\
\widetilde{B}_{U,j}^\pi + \widehat{J}_{U,z}^\pi &\text{ with } \left(\chi_U \widetilde{B}_{G,j} \chi_U I\right)^\pi + (\chi_U I)^\pi \widehat{J}_{G,z}^\pi (\chi_U I)^\pi &\quad &(j = 1, 2, \ldots, m), \\
(aI)^\pi + \widehat{J}_{U,z}^\pi &\text{ with } (\chi_U \widetilde{a} I)^\pi + (\chi_U I)^\pi \widehat{J}_{G,z}^\pi (\chi_U I)^\pi &\quad &(a \in PC(\mathfrak{L})),
\end{aligned}$$

and taking into account the relations

$$A^\pi + \widehat{J}^\pi_{G,z} = \left(\chi_U A \chi_U I\right)^\pi + \widehat{J}^\pi_{G,z} \quad (A \in \mathfrak{A}_{n,m}(G)),$$

we conclude that the $C^*$-algebras $\widehat{\mathfrak{A}}^\pi_{n,m,z}(U)$ and $\widehat{\mathfrak{A}}^\pi_{n,m,z}(G)$ are isomorphic, and the isomorphism $\mu_z : \widehat{\mathfrak{A}}^\pi_{n,m,z}(U) \to \widehat{\mathfrak{A}}^\pi_{n,m,z}(G)$ is given by

$$
\begin{aligned}
B^\pi_{U,i} + \widehat{J}^\pi_{U,z} &\mapsto B^\pi_{G,i} + \widehat{J}^\pi_{G,z} \quad (i = 1, 2, \ldots, n), \\
\widetilde{B}^\pi_{U,j} + \widehat{J}^\pi_{U,z} &\mapsto \widetilde{B}^\pi_{G,j} + \widehat{J}^\pi_{G,z} \quad (j = 1, 2, \ldots, m), \\
(aI)^\pi + \widehat{J}^\pi_{U,z} &\mapsto (\widetilde{a}I)^\pi + \widehat{J}^\pi_{G,z} \, (a \in PC(\mathfrak{L})).
\end{aligned}
$$

Corollary 2.1 and the relations

$$W_{\varphi_z}(\widetilde{a}I)W^*_{\varphi_z} = (\widetilde{a} \circ \varphi_z)I \ \ (\widetilde{a} \in PC(\mathfrak{L} \cup \partial U \setminus \Gamma_0)), \quad (W_{\varphi_z})^\pi \widehat{J}^\pi_{G,z}(W^*_{\varphi_z})^\pi = \widehat{J}^\pi_{\mathbb{D},-1}$$

imply that $\widehat{\mathfrak{A}}^\pi_{n,m,z}(G) \cong \widehat{\mathfrak{A}}^{z,\pi}_{n,m,-1}(\mathbb{D}) := \left\{ A^\pi + \widehat{J}^\pi_{\mathbb{D},-1} : A \in \mathfrak{A}_{n,m}(\mathbb{D}) \right\}$ where

$$\mathfrak{A}^z_{n,m}(\mathbb{D}) := \mathrm{alg}\left\{ B_{\mathbb{D},1}, \ldots, B_{\mathbb{D},n}, \widetilde{B}_{\mathbb{D},1}, \ldots, \widetilde{B}_{\mathbb{D},m}; \varphi_z^{-1}(\mathfrak{L} \cup \partial U \setminus \Gamma_0) \right\},$$

and the isomorphism

$$\lambda_z : \widehat{\mathfrak{A}}^\pi_{n,m,z}(G) \to \widehat{\mathfrak{A}}^{z,\pi}_{n,m,-1}(\mathbb{D}), \quad A^\pi + \widehat{J}^\pi_{G,z} \mapsto (W_{\varphi_z})^\pi A^\pi (W^*_{\varphi_z})^\pi + \widehat{J}^\pi_{\mathbb{D},-1}$$

is given on the generators of the $C^*$-algebra $\widehat{\mathfrak{A}}^\pi_{n,m,z}(G)$ by

$$
\begin{aligned}
B^\pi_{G,i} + \widehat{J}^\pi_{G,z} &\mapsto B^\pi_{\mathbb{D},i} + \widehat{J}^\pi_{\mathbb{D},-1} \quad (i = 1, 2, \ldots, n), \\
\widetilde{B}^\pi_{G,j} + \widehat{J}^\pi_{G,z} &\mapsto \widetilde{B}^\pi_{\mathbb{D},j} + \widehat{J}^\pi_{\mathbb{D},-1} \quad (j = 1, 2, \ldots, m), \\
(\widetilde{a}I)^\pi + \widehat{J}^\pi_{G,z} &\mapsto ((\widetilde{a} \circ \varphi_z)I)^\pi + \widehat{J}^\pi_{\mathbb{D},-1} \, (\widetilde{a} \in PC(\mathfrak{L} \cup \partial U \setminus \Gamma_0)).
\end{aligned}
$$

If the conformal mapping $\psi : \Pi \to \mathbb{D}$ is given by $\zeta \mapsto (\zeta - i)/(\zeta + i)$, then applying the unitary operator $W_\psi : L^2(\mathbb{D}) \to L^2(\Pi)$, $f \mapsto (f \circ \psi)\psi'$, we obtain

$$W_\psi S_\mathbb{D} W^*_\psi = S_\Pi uI, \quad W_\psi S^*_\mathbb{D} W^*_\psi = \overline{u} S^*_\Pi \tag{5}$$

where $u(\zeta) = -(\zeta + i)^2/(\overline{\zeta} - i)^2$ for $\zeta \in \Pi$. Hence, (1) and (5) imply that

$$\left(W_\psi B_{\mathbb{D},k} W^*_\psi\right)^\pi = \left(I - (S_\Pi uI)^k(\overline{u} S^*_\Pi)^k\right)^\pi, \quad \left(W_\psi \widetilde{B}_{\mathbb{D},k} W^*_\psi\right)^\pi = \left(I - (\overline{u} S^*_\Pi)^k(S_\Pi uI)^k\right)^\pi,$$

and therefore for every $k \in \mathbb{N}$,

$$\left(W_\psi B_{\mathbb{D},k} W^*_\psi\right)^\pi + \widehat{J}^\pi_{\Pi,0} = (B_{\Pi,k})^\pi + \widehat{J}^\pi_{\Pi,0}, \quad \left(W_\psi \widetilde{B}_{\mathbb{D},k} W^*_\psi\right)^\pi + \widehat{J}^\pi_{\Pi,0} = (\widetilde{B}_{\Pi,k})^\pi + \widehat{J}^\pi_{\Pi,0}.$$

Since also $(W_\psi)^\pi \widehat{J}^\pi_{\mathbb{D},-1}(W^*_\psi)^\pi = \widehat{J}^\pi_{\Pi,0}$, we conclude that the map

$$\delta_z : \widehat{\mathfrak{A}}^{z,\pi}_{n,m,-1}(\mathbb{D}) \to \widehat{\mathfrak{A}}^{z,\pi}_{n,m,0}(\Pi), \quad A^\pi + \widehat{J}^\pi_{\mathbb{D},-1} \mapsto (W_\psi)^\pi A^\pi (W^*_\psi)^\pi + \widehat{J}^\pi_{\Pi,0}$$

given on the generators of the $C^*$-algebra $\widehat{\mathfrak{A}}^{z,\pi}_{n,m,-1}(\mathbb{D})$ by

$$
\begin{aligned}
B^\pi_{\mathbb{D},i} + \widehat{J}^\pi_{\mathbb{D},-1} &\mapsto B^\pi_{\Pi,i} + \widehat{J}^\pi_{\Pi,0} \quad (i = 1, 2, \ldots, n), \\
\widetilde{B}^\pi_{\mathbb{D},j} + \widehat{J}^\pi_{\mathbb{D},-1} &\mapsto \widetilde{B}^\pi_{\Pi,j} + \widehat{J}^\pi_{\Pi,0} \quad (j = 1, 2, \ldots, m), \\
((\widetilde{a} \circ \varphi_z)I)^\pi + \widehat{J}^\pi_{\mathbb{D},-1} &\mapsto ((\widetilde{a} \circ \psi_z)I)^\pi + \widehat{J}^\pi_{\Pi,0} \, (\widetilde{a} \in PC(\mathfrak{L} \cup \partial U \setminus \Gamma_0)).
\end{aligned}
$$

is an isomorphism of $\widehat{\mathfrak{A}}^{z,\pi}_{n,m,-1}(\mathbb{D})$ onto the $C^*$-algebra $\widehat{\mathfrak{A}}^{z,\pi}_{n,m,0}(\Pi) = \{A^\pi + \widehat{J}^\pi_{\Pi,0} :$
$A \in \mathfrak{A}^z_{n,m}(\Pi)\}$. Finally, by analogy with $\mathfrak{A}^\pi_{n,m,z}(U) \cong \widehat{\mathfrak{A}}^\pi_{n,m,z}(U)$, we infer that
$\widehat{\mathfrak{A}}^{z,\pi}_{n,m,0}(\Pi) \cong \mathfrak{A}^{z,\pi}_{n,m,0}(\Pi)$ where the isomorphism $\eta_z : \widehat{\mathfrak{A}}^{z,\pi}_{n,m,0}(\Pi) \to \mathfrak{A}^{z,\pi}_{n,m,0}(\Pi)$ is
given by $A^\pi + \widehat{J}^\pi_{\Pi,0} \mapsto A^\pi + J^\pi_{\Pi,0}$ for every $A \in \mathfrak{A}^z_{n,m}(\Pi)$.

Thus, if $z \in \Gamma_0$, then the map $\theta_z := \eta_z \circ \delta_z \circ \lambda_z \circ \mu_z \circ \nu_z$ given by (4) is an
isomorphism of the $C^*$-algebra $\mathfrak{A}^\pi_{n,m,z}(U)$ onto the $C^*$-algebra $\mathfrak{A}^{z,\pi}_{n,m,0}(\Pi)$.

The case of $z \in \Gamma_k$ $(k = 1, 2, \ldots, r)$ is reduced to the previous one by applying
the isomorphism

$$\gamma_z : \mathfrak{A}^\pi_{n,m,z}(U) \to \mathfrak{A}^\pi_{n,m,\beta_k(z)}(\beta_k(U)), \quad A^\pi + J^\pi_{U,z} \mapsto (W^*_{\beta_k})^\pi A^\pi (W_{\beta_k})^\pi + J^\pi_{\beta_k(U),\beta_k(z)},$$

where $J^\pi_{\beta_k(U),\beta_k(z)}$ is the closed two-sided ideal of the $C^*$-algebra $\mathfrak{A}^\pi_{n,m}(\beta_k(U))$, and

$$\mathfrak{A}_{n,m}(\beta_k(U)) := \mathrm{alg}\left\{ B_{\beta_k(U),1}, \ldots, B_{\beta_k(U),n}, \widetilde{B}_{\beta_k(U),1}, \ldots, \widetilde{B}_{\beta_k(U),m}; \beta_k(\mathfrak{L}) \right\}.$$

Indeed, applying locally the Kellogg-Warschawski theorem and Corollary 2.1, we
infer from the Allan-Douglas local principle that for $a \in PC(\mathfrak{L})$ and all $i, j$,

$$(W^*_{\beta_k})^\pi B^\pi_{U,i}(W_{\beta_k})^\pi = B^\pi_{\beta_k(U),i}, \qquad (W^*_{\beta_k})^\pi \widetilde{B}^\pi_{U,j}(W_{\beta_k})^\pi = \widetilde{B}^\pi_{\beta_k(U),j},$$

$$(W^*_{\beta_k})^\pi (aI)^\pi (W_{\beta_k})^\pi = ((a \circ \beta_k^{-1})I)^\pi, \quad (W^*_{\beta_k})^\pi J^\pi_{U,z}(W_{\beta_k})^\pi = J^\pi_{\beta_k(U),\beta_k(z)}.$$

Since $\beta_k(z) \in \beta_k(\Gamma_k)$ and the domain $G_k$ bounded by the Jordan curve $\beta_k(\Gamma_k)$
contains all other Jordan curves $\beta_k(\Gamma_s)$, it remains to apply the isomorphism

$$\theta_{\beta_k(z)} : \mathfrak{A}^\pi_{n,m,\beta_k(z)}(\beta_k(U)) \to \mathfrak{A}^{z,\pi}_{n,m,0}(\Pi),$$

$$B^\pi_{\beta_k(U),i} + J^\pi_{\beta_k(U),\beta_k(z)} \mapsto B^\pi_{\Pi,i} + J^\pi_{\Pi,0} \qquad (i = 1, 2, \ldots, n),$$

$$\widetilde{B}^\pi_{\beta_k(U),j} + J^\pi_{\beta_k(U),\beta_k(z)} \mapsto \widetilde{B}^\pi_{\Pi,j} + J^\pi_{\Pi,0} \qquad (j = 1, 2, \ldots, m),$$

$$((a \circ \beta_k^{-1})I)^\pi + J^\pi_{\beta_k(U),\beta_k(z)} \mapsto ((\widetilde{a} \circ \psi_z)I)^\pi + J^\pi_{\Pi,0} \qquad (a \in PC(\mathfrak{L})).$$

Finally, the map $\theta_{\beta_k(z)} \circ \gamma_z$ is the isomorphism (4) of the $C^*$-algebra $\mathfrak{A}^\pi_{n,m,z}(U)$
onto the $C^*$-algebra $\mathfrak{A}^{z,\pi}_{n,m,0}(\Pi)$. $\qquad\square$

## 3. Main results

Let $\delta_{j,k}$ be the Kronecker symbol and, for every $a \in PC(\mathfrak{L})$ and every $z \in \overline{U}$,

$$a_k(z) := \lim_{\zeta \to z, \, \zeta \in \Omega_k(z)} a(\zeta) \qquad (k = 1, 2, \ldots, n_z), \tag{6}$$

where $\Omega_k(z)$ are connected components of the set $V_z \cap (U \setminus \mathfrak{L})$. Analogously to [7,
Lemma 4.2], applying Lemma 2.2 and Lemma 2.3 one can prove the following.

**Lemma 3.1.** *For the $C^*$-algebra $\mathfrak{A}_{n,m}(U)$ given by (2) the following holds:*

(i) *if $z \in U$, then $\mathfrak{A}^\pi_{n,m,z}(U) \cong \mathbb{C}^{n_z}$, where $n_z \in \mathbb{N}$ is given by condition $(\mathfrak{L}1)$,
and the corresponding isomorphism of $\mathfrak{A}^\pi_{n,m,z}$ onto $\mathbb{C}^{n_z}$ is given by*

$$(B_{U,i})^\pi_z \mapsto (0, \ldots, 0) \quad (i = 1, \ldots, n), \quad (\widetilde{B}_{U,j})^\pi_z \mapsto (0, \ldots, 0) \quad (j = 1, \ldots, m),$$

$$(aI)^\pi_z \mapsto (a_1(z), a_2(z) \ldots, a_{n_z}(z)) \quad (a \in PC(\mathfrak{L}));$$

(ii) *if $z \in \partial U \setminus \mathfrak{L}$, then $\mathfrak{A}^{\pi}_{n,m,z} \cong \mathbb{C}^{n+m+1}$ and the corresponding isomorphism is defined on the generators of the $C^*$-algebra $\mathfrak{A}^{\pi}_{n,m,z}(U)$ by*

$$(B_{U,i})^{\pi}_z \mapsto \left(\sum_{\nu=1}^{i} \delta_{\nu,s}\right)^{n+m+1}_{s=1} \quad (i = 1, 2, \ldots, n),$$

$$(\widetilde{B}_{U,j})^{\pi}_z \mapsto \left(\sum_{\nu=1}^{j} \delta_{n+\nu,s}\right)^{n+m+1}_{s=1} \quad (j = 1, 2, \ldots, m),$$

$$(aI)^{\pi}_z \mapsto \left(a(z)\delta_{s,s}\right)^{n+m+1}_{s=1} \quad (a \in PC(\mathfrak{L})).$$

**Proof.** Part (i) follows from Lemma 2.2 and from the local equivalence $(aI)^{\pi} \overset{z}{\sim} \left(\sum_{k=1}^{n_z} a_k(z)\chi_{\Omega_k(z)}I\right)^{\pi}$ where $\chi_{\Omega_k(z)}$ are characteristic functions of $\Omega_k(z)$.

If $z \in \partial U \setminus \mathfrak{L}$, we infer from Lemma 2.4 that $\mathfrak{A}^{\pi}_{n,m,z} \cong \mathfrak{A}^{z,\pi}_{n,m,0}(\Pi)$. Then part (ii) follows from [7, Lemma 4.2]. □

Fix $z \in \partial U \cap \mathfrak{L}$ and put $N = n_z$. Let $\theta_k = \theta_k(z)$ $(k = 1, 2, \ldots, N-1)$ be the angles defined in ($\mathfrak{L}2$). Hence $0 \le \theta_{k-1} < \theta_k \le \pi$ for $k = 1, 2, \ldots, N$, where $\theta_0 = 0$ and $\theta_N = \pi$. According to [7], with every $\lambda \in \mathbb{R}$ we associate the inner products

$$\alpha^k_{j,r}(\lambda) := \langle \chi_k v_{\lambda,j}, \chi_k v_{\lambda,r}\rangle \quad (k = 1, 2, \ldots, N; \ j, r = 1, 2, \ldots, M := n+m) \quad (7)$$

in the Hilbert space $L^2(\mathbb{T}_+)$ over the upper semi-circle $\mathbb{T}_+ = \{e^{i\theta} : \theta \in [0, \pi]\}$, where $\chi_k$ is the characteristic function of the arc $\{e^{i\theta} : \theta_{k-1} \le \theta \le \theta_k\}$,

$$v_{\lambda,k}(t) = (-1)^{k-1} G(\lambda) \, t^{i\lambda-1} F(1-k, 1-i\lambda; 1; 1-t^{-2}) \quad (k = 1, 2, \ldots, n),$$

$$v_{\lambda,n+k}(t) = (-1)^{k-1} G(-\lambda) \, t^{-i\lambda+1} F(1-k, 1-i\lambda; 1; 1-t^2) \quad (k = 1, 2, \ldots, m)$$

for $t \in \mathbb{T}_+$, and $G(\lambda) = \left(2\lambda/(1 - e^{-2\pi\lambda})\right)^{1/2}$ for $\lambda \in \mathbb{R}$. Here $F(-m, b; c; z)$ is the Gauss $(2,1)$-hypergeometric function given by

$$F(-m, b; c; z) = \sum_{n=0}^{m} \frac{(-m)_n (b)_n}{(c)_n n!} \, z^n \quad (m = 0, 1, 2, \ldots; \ b, z \in \mathbb{C}; \ c \in \mathbb{C} \setminus Y_m),$$

where $Y_m = \{0, -1, -2, \ldots, -m+1\}$ and $(x)_n = x(x+1)\ldots(x+n-1)$ for $x \in \mathbb{C}$. Following [7, Section 6], we introduce the matrix functions

$$\mathcal{M}_j(\lambda) = \left(B^{k,r}_j(\lambda)\right)^N_{k,r=1} \quad (j = 1, 2, \ldots, M) \quad (8)$$

where the $M \times M$ blocks $B^{k,r}_j(\lambda)$ $(j = 1, 2, \ldots, M)$ are given by

$$B^{k,r}_j(\lambda) = \begin{bmatrix} \overline{\langle e^k_{\lambda,1}, v_{\lambda,j}\rangle}\langle e^r_{\lambda,1}, v_{\lambda,j}\rangle & \cdots & \overline{\langle e^k_{\lambda,1}, v_{\lambda,j}\rangle}\langle e^r_{\lambda,j}, v_{\lambda,j}\rangle & 0 & \cdots & 0 \\ \vdots & \ddots & \vdots & & \ddots & \vdots \\ \overline{\langle e^k_{\lambda,j}, v_{\lambda,j}\rangle}\langle e^r_{\lambda,1}, v_{\lambda,j}\rangle & \cdots & \overline{\langle e^k_{\lambda,j}, v_{\lambda,j}\rangle}\langle e^r_{\lambda,j}, v_{\lambda,j}\rangle & 0 & \cdots & 0 \\ 0 & \cdots & 0 & 0 & \cdots & 0 \\ \vdots & \ddots & \vdots & & \ddots & \vdots \\ 0 & \cdots & 0 & 0 & \cdots & 0 \end{bmatrix},$$

$$\langle e^k_{\lambda,s}, v_{\lambda,j}\rangle = \begin{cases} 0 & \text{if } j = 1, 2, \ldots, s-1, \\ \dfrac{D^k_{s,j}(\lambda)}{\sqrt{D^k_{s-1}(\lambda)D^k_s(\lambda)}} & \text{if } j = s, s+1, \ldots, M, \end{cases}$$

$$D_s^k(\lambda) = D_{s,s}^k(\lambda) \neq 0, \quad D_{s,j}^k(\lambda) = \begin{vmatrix} \alpha_{1,1}^k(\lambda) & \alpha_{2,1}^k(\lambda) & \dots & \alpha_{s,1}^k(\lambda) \\ \vdots & \vdots & \ddots & \vdots \\ \alpha_{1,s-1}^k(\lambda) & \alpha_{2,s-1}^k(\lambda) & \dots & \alpha_{s,s-1}^k(\lambda) \\ \alpha_{1,j}^k(\lambda) & \alpha_{2,j}^k(\lambda) & \dots & \alpha_{s,j}^k(\lambda) \end{vmatrix}$$

for $s, j = 1, 2, \dots, M$, and $\alpha_{j,r}^k(\lambda)$ given by (7) are calculated in [7].

**Theorem 3.1.** [7, Theorem 6.6] *For every* $j = 1, 2, \dots, M$, *the matrix function* $\mathcal{M}_j(\cdot)$ *given by* (8) *belongs to the space* $C(\overline{\mathbb{R}}, \mathbb{C}^{MN \times MN})$, *and*

$$\lim_{\lambda \to +\infty} \mathcal{M}_j(\lambda) = \operatorname{diag}\{\delta_{j,s}\}_{s=1}^{MN}, \quad \lim_{\lambda \to -\infty} \mathcal{M}_j(\lambda) = \operatorname{diag}\{\delta_{M(N-1)+j,s}\}_{s=1}^{MN}. \quad (9)$$

Lemma 2.4 and [7, Theorem 7.5] imply the following.

**Theorem 3.2.** *If* $z \in \partial U \cap \mathfrak{L}$ *is a common endpoint of* $n_z - 1$ *arcs of* $\mathfrak{L}$, *then the local* $C^*$-*algebra* $\mathfrak{A}_{n,m,z}^\pi(U)$ *is isomorphic to a* $C^*$-*subalgebra* $\mathbb{C}^N \oplus \mathfrak{C}_z$ *of* $\mathbb{C}^N \oplus C(\overline{\mathbb{R}}, \mathbb{C}^{MN \times MN})$, *where* $N = n_z$, $M = n + m$. *The isomorphism is given by*

$$\left(B_{U,i}\right)_z^\pi \mapsto (0, \dots, 0) \oplus \left(\lambda \mapsto \sum_{\nu=1}^i \mathcal{M}_\nu(\lambda)\right) \qquad (i = 1, 2, \dots, n),$$

$$\left(\widetilde{B}_{U,j}\right)_z^\pi \mapsto (0, \dots, 0) \oplus \left(\lambda \mapsto \sum_{\nu=1}^j \mathcal{M}_{n+\nu}(\lambda)\right) \qquad (j = 1, 2, \dots, m),$$

$$\left(aI\right)_z^\pi \mapsto (a_1(z), \dots, a_N(z)) \oplus \left(\lambda \mapsto \operatorname{diag}\{a_k(z)I_M\}_{k=1}^N\right) \quad (a \in PC(\mathfrak{L})),$$

*where* $a_k(z)$ $(k = 1, 2, \dots, N)$ *are given by* (6) *and the matrix functions* $\mathcal{M}_\nu(\cdot) \in C(\overline{\mathbb{R}}, \mathbb{C}^{MN \times MN})$ *for* $\nu = 1, 2, \dots, M$ *are defined by* (8) *and* (9).

Now we can construct a symbol calculus for the $C^*$-algebra $\mathfrak{A}_{n,m}(U)$ and establish a Fredholm criterion for the operators $A \in \mathfrak{A}_{n,m}^\pi(U)$. Applying Theorem 2.1, Lemma 3.1 and Theorem 3.2 we obtain the main result of the paper.

**Theorem 3.3.** *The* $C^*$-*algebra* $\mathfrak{A}_{n,m}^\pi(U) = \mathfrak{A}_{n,m}(U)/\mathcal{K}(L^2(U))$ *is isomorphic to the* $C^*$-*subalgebra* $\Phi(\mathfrak{A}_{n,m}^\pi(U))$ *of the* $C^*$-*algebra*

$$\Psi_{n,m,\mathfrak{L}} := \left(\bigoplus_{z \in \overline{U}} \mathbb{C}^{n_z}\right) \oplus \left(\bigoplus_{z \in \partial U \setminus \mathfrak{L}} \mathbb{C}^M\right) \oplus \left(\bigoplus_{z \in \partial U \cap \mathfrak{L}} \mathfrak{C}_z\right),$$

*and the isomorphism* $\Phi : \mathfrak{A}_{n,m}^\pi(U) \to \Phi(\mathfrak{A}_{n,m}^\pi(U))$ *is given by*

$$\Phi(B_{U,i}^\pi) := \left(\bigoplus_{z \in \overline{U}} (0, \dots, 0)\right) \oplus \left(\bigoplus_{z \in \partial U \setminus \mathfrak{L}} \widetilde{e}_i\right) \oplus \left(\bigoplus_{z \in \partial U \cap \mathfrak{L}} \left(\lambda \mapsto \sum_{\nu=1}^i \mathcal{M}_\nu(\lambda)\right)\right),$$

$$\Phi(\widetilde{B}_{U,j}^\pi) := \left(\bigoplus_{z \in \overline{U}} (0, \dots, 0)\right) \oplus \left(\bigoplus_{z \in \partial U \setminus \mathfrak{L}} \widetilde{e}_{n+j}\right) \oplus \left(\bigoplus_{z \in \partial U \cap \mathfrak{L}} \left(\lambda \mapsto \sum_{\nu=1}^j \mathcal{M}_{n+\nu}(\lambda)\right)\right),$$

$$\Phi((aI)^\pi) := \left(\bigoplus_{z \in \overline{U}} (a_1(z), \dots, a_{n_z}(z))\right) \oplus \left(\bigoplus_{z \in \partial U \setminus \mathfrak{L}} (a(z), \dots, a(z))\right)$$

$$\oplus \left(\bigoplus_{z \in \partial U \cap \mathfrak{L}} \left(\lambda \mapsto \operatorname{diag}\{a_k(z)I_M\}_{k=1}^{n_z}\right)\right),$$

where $i = 1, 2, \ldots, n$, $j = 1, 2, \ldots, m$, $a \in PC(\mathfrak{L})$, $n_z$ is the number of connected components $\Omega_k(z)$ of the set $V_z \cap (U \setminus \mathfrak{L})$ for a small neighborhood $V_z$ of $z \in \overline{U}$, $M = n + m$, $\mathfrak{C}_z$ is the $C^*$-subalgebra of $C(\overline{\mathbb{R}}, \mathbb{C}^{Mn_z \times Mn_z})$ determined in Theorem 3.2, $\widetilde{e}_j = (\delta_{j,s})_{s=1}^{M}$, the matrix functions $\mathcal{M}_\nu(\cdot) \in C(\overline{\mathbb{R}}, \mathbb{C}^{Mn_z \times Mn_z})$ for $\nu = 1, 2, \ldots, M$ are defined by (8) and (9), and $a_k(z)$ $(k = 1, 2, \ldots, n_z)$ are given by (6).

An operator $A \in \mathfrak{A}_{n,m}(U)$ is Fredholm on the space $L^2(U)$ if and only if its symbol $\Phi(A^\pi)$ is invertible in the $C^*$-algebra $\Psi_{n,m,\mathfrak{L}}$, that is, if

$$\big([\Phi(A^\pi)](z)\big)_k \neq 0 \text{ for all } z \in \overline{U} \text{ and all } k = 1, 2, \ldots, n_z;$$

$$\big([\Phi(A^\pi)](z)\big)_j \neq 0 \text{ for all } z \in \partial U \setminus \mathfrak{L} \text{ and all } j = 1, 2, \ldots, M;$$

$$\det\big([\Phi(A^\pi)](z)\big)(\lambda) \neq 0 \text{ for all } z \in \partial U \cap \mathfrak{L} \text{ and all } \lambda \in \overline{\mathbb{R}},$$

where $\big([\Phi(A^\pi)](z)\big)_k$ for $z \in \overline{U}$ are the $k$-entries of the vector $[\Phi(A^\pi)](z) \in \mathbb{C}^{n_z}$ and $\big([\Phi(A^\pi)](z)\big)_j$ for $z \in \partial U \setminus \mathfrak{L}$ are the $j$-entries of the vector $[\Phi(A^\pi)](z) \in \mathbb{C}^M$.

## References

1. M. B. Balk, *Polyanalytic Functions*. Akademie Verlag, Berlin, 1991.
2. A. Böttcher and Yu. I. Karlovich, *Carleson Curves, Muckenhoupt Weights, and Toeplitz Operators*. Birkhäuser, Basel 1997.
3. A. Böttcher, Yu. I. Karlovich and V. S. Rabinovich, *J. Operator Theory* **43**, 171 (2000).
4. A. Böttcher and B. Silbermann, *Analysis of Toeplitz Operators*. Akademie-Verlag, Berlin 1989 and Springer-Verlag, Berlin 1990.
5. A. Dzhuraev, *Methods of Singular Integral Equations*. Longman Scientific & Technical, 1992.
6. Yu. I. Karlovich and L. Pessoa, *Integr. Eq. Oper. Theory* **52**, 219 (2005).
7. Yu. I. Karlovich and L. Pessoa, *C*-Algebras of Bergman type operators with piecewise continuous coefficients*, Integr. Eq. Oper. Theory, to appear.
8. A. N. Karapetyants, V. S. Rabinovich and N. L. Vasilevski, *Integr. Eq. Oper. Theory* **40**, 278 (2001).
9. M. Loaiza, *Integr. Eq. Oper. Theory* **46**, 215 (2003).
10. M. Loaiza, *Bol. Soc. Mat. Mexicana (3)* **10**, No. 2, 179 (2004).
11. S. G. Mikhlin and S. Prössdorf, *Singular Integral Operators*. Springer, Berlin 1986.
12. B. A. Plamenevsky, *Algebras of Pseudodifferential Operators*. Kluwer, Dordrecht 1989.
13. Ch. Pommerenke, *Boundary Behaviour of Conformal Maps*. Springer, Berlin 1992.
14. J. Ramírez and I. M. Spitkovsky, *Factorization, Singular Operators and Related Problems*, Proc. of the Conf. in Honour of Professor Georgii Litvinchuk. Kluwer, Dordrecht, 273 (2003).
15. I. B. Simonenko and Chin Ngok Min, *Local Method in the Theory of One-Dimensional Singular Integral Equations with Piecewise Continuous Coefficients. Noetherity.* University Press, Rostov on Don 1986 (Russian).
16. N. L. Vasilevski, *Soviet Math. (Izv. VUZ)* **30**, No. 2, 14 (1986).
17. N. L. Vasilevski, *Integr. Eq. Oper. Theory* **33**, 471 (1999).

# A CRITERION FOR LATERAL INVERTIBILITY OF MATRIX WIENER-HOPF PLUS HANKEL OPERATORS WITH GOOD HAUSDORFF SETS

A. P. NOLASCO* and L. P. CASTRO

*Department of Mathematics,*
*University of Aveiro,*
*3810-145 Aveiro, PORTUGAL*
*E-mails: {ANolasco, LCastro}@mat.ua.pt*

We will present a criterion for left, right and both-sided invertibility of matrix Wiener-Hopf plus Hankel operators with the same Fourier symbol in the Wiener subclass of the almost periodic algebra. The criterion is based on the value of a certain mean motion constructed from a particular Hausdorff set which is bounded away from zero.

**Key words:** Wiener-Hopf plus Hankel operator, invertibility, almost periodic function, mean motion, Hausdorff set

**Mathematics Subject Classification:** 47B35, 47A68, 47A05, 42A75

## 1. Introduction

We will consider matrix Wiener-Hopf plus Hankel operators of the form

$$WH_\Phi = W_\Phi + H_\Phi : [L^2_+(\mathbb{R})]^n \to [L^2(\mathbb{R}_+)]^n ,$$

with $W_\Phi$ and $H_\Phi$ being matrix Wiener-Hopf and Hankel operators defined by

$$W_\Phi = r_+ \mathcal{F}^{-1} \Phi \cdot \mathcal{F} : [L^2_+(\mathbb{R})]^n \to [L^2(\mathbb{R}_+)]^n$$

$$H_\Phi = r_+ \mathcal{F}^{-1} \Phi \cdot \mathcal{F}J : [L^2_+(\mathbb{R})]^n \to [L^2(\mathbb{R}_+)]^n ,$$

respectively. The operators $WH_\Phi$ are a special kind of the Wiener-Hopf plus Hankel operators $W_\Phi + H_\Psi$ with possible different Fourier symbols, $\Phi \neq \Psi$, belonging to $[L^\infty(\mathbb{R})]^{n \times n}$. These two types of Wiener-Hopf plus Hankel operators were object of recent fundamental research within the framework of some specific Fourier symbol classes, as well as object of concrete use in particular applied problems (e.g., in problems arising from Mathematical-Physics). In view of some examples in these directions, we like to refer the reader to the works of Basor and Ehrhardt[3], Castro and Speck [7], Castro, Speck and Teixeira[8–10], Ehrhardt[11], Lebre, Meister and Teixeira[15], Meister, Speck and Teixeira[16], Roch and Silbermann[18], and Teixeira[20,21], for instance.

---
*A. P. Nolasco is sponsored by *Fundação para a Ciência e a Tecnologia* (Portugal), under the grant number SFRH/BD/11090/2002.

As about the notations, here and in what follows, we use $[L^2_+(\mathbb{R})]^n$ to denote the subspace of $[L^2(\mathbb{R})]^n$ formed by all the matrix functions supported on the closure of $\mathbb{R}_+ = (0, +\infty)$, $r_+$ represents the operator of restriction from $[L^2_+(\mathbb{R})]^n$ into $[L^2(\mathbb{R}_+)]^n$, $\mathcal{F}$ denotes the Fourier transformation, $J$ is the reflection operator given by the rule $J\varphi(x) = \widetilde{\varphi}(x) = \varphi(-x)$, $x \in \mathbb{R}$, and $\Phi$ is a $n \times n$ matrix function with elements belonging to the so-called $APW$ algebra.

For defining the $APW$ functions, let us first consider the algebra of almost periodic functions, usually denoted by $AP$, i.e., the smallest closed subalgebra of $L^\infty(\mathbb{R})$ that contains all the functions $e_\lambda$ ($\lambda \in \mathbb{R}$) where

$$e_\lambda(x) = e^{i\lambda x} , \qquad x \in \mathbb{R} .$$

Since every function in $AP$ may be represented by a series, but not every function in $AP$ may be represented by an absolutely convergent series, $APW$ is precisely the subclass of all functions $\varphi \in AP$ which can be written in the form of an absolutely convergent series, i.e.,

$$APW = \left\{ \varphi = \sum_j \varphi_j e_{\lambda_j} : \varphi_j \in \mathbb{C}, \lambda_j \in \mathbb{R}, \sum_j |\varphi_j| < \infty \right\}.$$

In what follows, we will use the notation $\mathcal{G}B$ for the group of all invertible elements of a Banach algebra $B$. By a similar result of *Bohr's Theorem for AP functions*, it holds that for each $\phi \in \mathcal{G}APW$ there exists a real number $\kappa(\phi)$ and a function $\psi \in APW$ such that

$$\phi(x) = e^{i\kappa(\phi)x} e^{\psi(x)} , \tag{1}$$

for all $x \in \mathbb{R}$ (cf. Theorem 8.11 in Böttcher, Karlovich and Spitkovsky[4]). Since $\kappa(\phi)$ is uniquely determined, $\kappa(\phi)$ is usually called the *mean motion* of $\phi$.

## 2. Operator relations

In this section, following the spirit of the *Gohberg-Krupnik-Litvinchuk identity* (see Karapetiants and Samko[12], Kravchenko and Litvinchuk[13], Krupnik[14] or Roch and Silbermann[18], for instance) we will present a relation between Wiener-Hopf plus Hankel operators and Wiener-Hopf operators (upon the use of an additional operator). Such relation will have a significant role in the proof of the invertibility criterion for Wiener-Hopf plus Hankel operators presented ahead. For such a purpose, we will first recall some different types of relations between bounded linear operators.

Let us consider two bounded linear operators $T : X_1 \to X_2$ and $S : Y_1 \to Y_2$, acting between Banach spaces. The operators $T$ and $S$ are said to be *equivalent*, and we will denote this by $T \sim S$, if there are two boundedly invertible linear operators, $E : Y_2 \to X_2$ and $F : X_1 \to Y_1$, such that

$$T = E\,S\,F. \tag{2}$$

It directly follows from (2) that if two operators are equivalent, then they belong to the same *regularity class* [5,6,19]. More precisely, one of these operators is invertible, one-sided invertible, Fredholm, one-sided regularizable, generalized invertible or only normally solvable, if and only if the other operator enjoys the same property.

The so-called Δ–*relation after extension* was introduced by Castro and Speck[6] for bounded linear operators acting between Banach spaces, e.g. $T : X_1 \to X_2$ and $S : Y_1 \to Y_2$. We say that $T$ is Δ–related after extension to $S$ if there is a bounded linear operator acting between Banach spaces $T_\Delta : X_{1\Delta} \to X_{2\Delta}$ and invertible bounded linear operators $E$ and $F$ such that

$$\begin{bmatrix} T & 0 \\ 0 & T_\Delta \end{bmatrix} = E \begin{bmatrix} S & 0 \\ 0 & I_Z \end{bmatrix} F, \tag{3}$$

where $Z$ is an additional Banach space and $I_Z$ represents the identity operator in $Z$. In the particular case when $T_\Delta : X_{1\Delta} \to X_{2\Delta} = X_{1\Delta}$ is the identity operator, we say the $T$ and $S$ are *equivalent after extension operators*.

It follows from (3) that if we have $T$ being Δ–related after extension to $S$, then the transfer of regularity properties can only be guaranteed in one direction, that is, from operator $S$ to operator $T$, as stated in Theorem 2.1 of Castro and Speck[6] work. It is clear that this restriction occurs only here and in contrast to what happens with the transfer of regularity properties between two equivalent (after extension) operators, where the transfer can be done in both directions.

**Lemma 2.1.** *Let* $\Phi \in \mathcal{G}[L^\infty(\mathbb{R})]^{n \times n}$. *Then the matrix Wiener-Hopf plus Hankel operator* $WH_\Phi : [L_+^2(\mathbb{R})]^n \to [L^2(\mathbb{R}_+)]^n$ *is* Δ–*related after extension to the Wiener-Hopf operator* $W_{\Phi\widetilde{\Phi^{-1}}} : [L_+^2(\mathbb{R})]^n \to [L^2(\mathbb{R}_+)]^n$ *(with the same matrix size as the original one).*

**Proof.** We start by extending $WH_\Phi$ on the left by the zero extension operator $\ell_0 : [L^2(\mathbb{R}_+)]^n \to [L_+^2(\mathbb{R})]^n$, and therefore obtaining

$$WH_\Phi \sim \ell_0 WH_\Phi : [L_+^2(\mathbb{R})]^n \to [L_+^2(\mathbb{R})]^n.$$

Choosing the notation $P_+ = \ell_0 r_+$ and $P_- = I_{[L^2(\mathbb{R})]^n} - P_+$, we will now extend

$$\ell_0 WH_\Phi = P_+ \mathcal{F}^{-1}(\Phi \cdot + \Phi \cdot J)\mathcal{F}_{|P_+[L^2(\mathbb{R})]^n}$$

to the full $[L^2(\mathbb{R})]^n$ space by using the identity in $[L_-^2(\mathbb{R})]^n$. Next we will extend the obtained operator to $[L^2(\mathbb{R})]^{2n}$ with the help of the auxiliar paired operator

$$\mathcal{T} = \mathcal{F}^{-1}(\Phi \cdot - \Phi \cdot J)\mathcal{F}P_+ + P_- : [L^2(\mathbb{R})]^n \to [L^2(\mathbb{R})]^n.$$

Altogether, we have

$$\left[\begin{array}{cc|c} \ell_0 WH_\Phi & 0 & 0 \\ 0 & I_{P_-[L^2(\mathbb{R})]^n} & 0 \\ \hline 0 & 0 & \mathcal{T} \end{array}\right] = E_1 \mathcal{W}_1 F_1$$

with

$$E_1 = \frac{1}{2}\begin{bmatrix} I_{[L^2(\mathbb{R})]^n} & J \\ I_{[L^2(\mathbb{R})]^n} & -J \end{bmatrix},$$

$$F_1 = \begin{bmatrix} I_{[L^2(\mathbb{R})]^n} & I_{[L^2(\mathbb{R})]^n} \\ J & -J \end{bmatrix}\begin{bmatrix} I_{[L^2(\mathbb{R})]^n} - P_-\mathcal{F}^{-1}(\Phi\cdot - \Phi\cdot J)\mathcal{F}P_+ & 0 \\ 0 & I_{[L^2(\mathbb{R})]^n} \end{bmatrix},$$

$$\begin{aligned}
\mathcal{W}_1 &= \begin{bmatrix} \mathcal{F}^{-1}\Phi\cdot\mathcal{F} & 0 \\ \mathcal{F}^{-1}\widetilde{\Phi}\cdot\mathcal{F} & 1 \end{bmatrix}P_+ + \begin{bmatrix} 1 & \mathcal{F}^{-1}\Phi\cdot\mathcal{F} \\ 0 & \mathcal{F}^{-1}\widetilde{\Phi}\cdot\mathcal{F} \end{bmatrix}P_- \\
&= \begin{bmatrix} 1 & \mathcal{F}^{-1}\Phi\cdot\mathcal{F} \\ 0 & \mathcal{F}^{-1}\widetilde{\Phi}\cdot\mathcal{F} \end{bmatrix}(\mathcal{F}^{-1}\Psi\cdot\mathcal{F}P_+ + P_-) \\
&= \begin{bmatrix} 1 & \mathcal{F}^{-1}\Phi\cdot\mathcal{F} \\ 0 & \mathcal{F}^{-1}\widetilde{\Phi}\cdot\mathcal{F} \end{bmatrix}(P_+\mathcal{F}^{-1}\Psi\cdot\mathcal{F}P_+ + P_-)(I_{[L^2(\mathbb{R})]^{2n}} + P_-\mathcal{F}^{-1}\Psi\cdot\mathcal{F}P_+),
\end{aligned}$$

where in the last definition of operator $\mathcal{W}_1$ we are using $P_\pm$ defined in $[L^2(\mathbb{R})]^{2n}$ and

$$\Psi = \begin{bmatrix} 0 & -\Phi\widetilde{\Phi^{-1}} \\ 1 & \widetilde{\Phi^{-1}} \end{bmatrix}.$$

We point out that the paired operator

$$I_{[L^2(\mathbb{R})]^{2n}} + P_-\mathcal{F}^{-1}\Psi\cdot\mathcal{F}P_+ : [L^2(\mathbb{R})]^{2n} \to [L^2(\mathbb{R})]^{2n}$$

used above is an invertible operator with inverse given by

$$I_{[L^2(\mathbb{R})]^{2n}} - P_-\mathcal{F}^{-1}\Psi\cdot\mathcal{F}P_+ : [L^2(\mathbb{R})]^{2n} \to [L^2(\mathbb{R})]^{2n}.$$

Therefore, we have just explicitly demonstrated that $WH_\Phi$ is $\Delta$–related after extension to

$$\mathcal{W}_\Psi = r_+\mathcal{F}^{-1}\Psi\cdot\mathcal{F} : [L^2_+(\mathbb{R})]^{2n} \to [L^2(\mathbb{R}_+)]^{2n}.$$

Further, we have

$$\begin{bmatrix} \mathcal{W}_{\Phi\widetilde{\Phi^{-1}}}\ell_0 & 0 \\ 0 & I_{[L^2(\mathbb{R}_+)]^n} \end{bmatrix} = \mathcal{W}_\Psi\ell_0 r_+\mathcal{F}^{-1}\begin{bmatrix} \widetilde{\Phi^{-1}} & 1 \\ -1 & 0 \end{bmatrix}\mathcal{F}\ell_0 : [L^2(\mathbb{R}_+)]^{2n} \to [L^2(\mathbb{R}_+)]^{2n}$$

which shows an explicit equivalence after extension relation between $\mathcal{W}_{\Phi\widetilde{\Phi^{-1}}}$ and $\mathcal{W}_\Psi$. This together with the $\Delta$–relation after extension between $WH_\Phi$ and $\mathcal{W}_\Psi$ concludes the proof. $\square$

**Corollary 2.1.** *Let* $\Phi \in \mathcal{G}[L^\infty(\mathbb{R})]^{n\times n}$. *The Wiener-Hopf plus Hankel operator* $WH_\Phi : [L^2_+(\mathbb{R})]^n \to [L^2(\mathbb{R}_+)]^n$ *has the same regularity properties as the Wiener-Hopf operator* $W_{\Phi\widetilde{\Phi^{-1}}} : [L^2_+(\mathbb{R})]^n \to [L^2(\mathbb{R}_+)]^n$.

**Proof.** The assertion is a direct consequence of the $\Delta$–relation after extension between the two operators presented in Lemma 2.1, which ensures that:

(i)
$$\operatorname{im} \begin{bmatrix} WH_\Phi & 0 \\ 0 & T \end{bmatrix} \text{ is closed if and only if } \operatorname{im} W_{\widetilde{\Phi\Phi^{-1}}} \text{ is closed;}$$

(ii)
$$\left( [L^2(\mathbb{R}_+)]^n \times [L^2(\mathbb{R})]^n \right) \setminus \operatorname{im} \begin{bmatrix} WH_\Phi & 0 \\ 0 & T \end{bmatrix} \simeq [L^2(\mathbb{R}_+)]^n \setminus \overline{\operatorname{im} W_{\widetilde{\Phi\Phi^{-1}}}} \, ;$$

(iii)
$$\ker \begin{bmatrix} WH_\Phi & 0 \\ 0 & T \end{bmatrix} \simeq \ker W_{\widetilde{\Phi\Phi^{-1}}} \, . \qquad\qquad \square$$

## 3. An invertibility criterion based on a mean motion depending on a Hausdorff set

Let us recall that the *Hausdorff set* (or *numerical range*) of a complex matrix $\Theta \in \mathbb{C}^{n \times n}$ is defined as

$$\mathcal{H}(\Theta) = \{ (\Theta\eta, \eta) : \eta \in \mathbb{C}^n, \|\eta\| = 1 \}.$$

If $\Phi \in [APW]^{n \times n}$, then (due to the definition of $APW$) we have that $\mathcal{H}(\Phi(x))$ is well-defined for all $x \in \mathbb{R}$. In this way, the Hausdorff set of $\Phi$ is said to be *bounded away from zero* (or "good") if

$$\inf_{x \in \mathbb{R}} \operatorname{dist}\Big( \mathcal{H}\big(\Phi(x)\big), \, 0 \Big) > 0$$

or, equivalently, if there is an $\varepsilon > 0$ such that

$$|(\Phi(x)\eta, \eta)| \geq \varepsilon \|\eta\|^2 \text{ for all } x \in \mathbb{R} \text{ and all } \eta \in \mathbb{C}^n.$$

Consider $\Phi \in [APW]^{n \times n}$ and $\eta \in \mathbb{C}^n \setminus \{0\}$. If the Hausdorff set of $\Phi$ is bounded away from zero, then the function $(\Phi\eta, \eta)$ given by

$$(\Phi\eta, \eta)(x) = (\Phi(x)\eta, \eta), \qquad x \in \mathbb{R}$$

is invertible in $APW$. Therefore, the mean motion of $(\Phi\eta, \eta)$, denoted by $\kappa\big((\Phi\eta, \eta)\big)$, is well-defined for all $\eta \in \mathbb{C}^n \setminus \{0\}$. Moreover, by a theorem due to Babadzhanyan and Rabinovich (see Theorem 9.9 in Böttcher, Karlovich and Spitkovsky[4], and cf. also the works of Babadzhanyan and Rabinovich[1,2]), we have that $\kappa\big((\Phi\eta, \eta)\big)$ is independent of $\eta \in \mathbb{C}^n \setminus \{0\}$.

**Theorem 3.1.** *Let us consider $\Phi \in \mathcal{G}[APW]^{n \times n}$ such that the Hausdorff set of $\widetilde{\Phi\Phi^{-1}}$ is bounded away from zero.*

(a) *If $\kappa\big((\widetilde{\Phi\Phi^{-1}}\eta, \eta)\big) = 0$, then $WH_\Phi$ is invertible.*

(b) *If $\kappa\big((\widetilde{\Phi\Phi^{-1}}\eta, \eta)\big) > 0$, then $WH_\Phi$ is left-invertible.*

(c) *If $\kappa\big((\widetilde{\Phi\Phi^{-1}}\eta, \eta)\big) < 0$, then $WH_\Phi$ is right-invertible.*

**Proof.** The assertion is now a consequence of the $\Delta$–relation after extension presented in the last section, and of the corresponding result for Wiener-Hopf operators (cf. Corollary 9.10 in Böttcher, Karlovich ans Spitkovsky[4]). In fact, first, the hypothesis in (a), (b), and (c) give us the invertibility, left-invertibility, and right-invertibility of $W_{\Phi\widetilde{\Phi^{-1}}}$, respectively. Secondly, by using Corollary 2.1, these three cases lead us to the final conclusion about the Wiener-Hopf plus Hankel operator. $\qquad\square$

At this point, it is interesting to remark that although we are using the $\Delta$–relation after extension (and therefore the help of an additional operator $T$), the last result comes out only for our main object (i.e., $WH_{\Phi}$) and in a description of all the universe of possible real numbers of the corresponding mean motion. We would also like to point out that the condition of the Hausdorff set of $\Phi\widetilde{\Phi^{-1}}$ be bounded away from zero is obviously a fundamental condition in our main result, and it is clear that not every matrix function in $\mathcal{G}[APW]^{n\times n}$ yields such property. For instance,

$$\Phi(x) = \begin{bmatrix} 2e^{e^{ix}} & e^{e^{ix}} \\ e^{-e^{ix}} & e^{-e^{ix}} \end{bmatrix}, \quad x \in \mathbb{R}$$

is invertible in $[APW]^{n\times n}$ but produces a matrix function $\Phi\widetilde{\Phi^{-1}}$ which does not have a Hausdorff set bounded away from zero.

## 4. Example

To illustrate the previous theorem, we will present in this last section a concrete case of an invertible matrix Wiener-Hopf plus Hankel operator $WH_{\Phi_p}$ with an $[APW]^{2\times 2}$ Fourier symbol $\Phi_p$.

Let us then consider the particular matrix-valued function

$$\Phi_p(x) = \begin{bmatrix} 2e^{e^{ix}} & e^{e^{ix}} - 1 \\ e^{e^{-i3x}} + 1 & e^{e^{-i3x}} \end{bmatrix}, \quad x \in \mathbb{R}.$$

From the similar result of *Bohr's Theorem for AP functions* (cf. (1)), it follows that $\Phi_p \in [APW]^{2\times 2}$. Moreover, $\Phi_p$ is invertible and we have

$$\Phi_p^{-1}(x) = \begin{bmatrix} e^{-e^{ix}} & -e^{-e^{-i3x}} - 1 \\ -e^{-e^{ix}} + 1 & 2e^{-e^{-i3x}} \end{bmatrix}, \quad x \in \mathbb{R}.$$

It can be easily seen that $\Phi_p^{-1} \in [APW]^{2\times 2}$, which yields that $\Phi_p \in \mathcal{G}[APW]^{2\times 2}$. As about the form of $\Phi_p\widetilde{\Phi_p^{-1}}$, we have in this case

$$(\Phi_p\widetilde{\Phi_p^{-1}})(x) = \begin{bmatrix} e^{e^{ix} - e^{-ix}} & 0 \\ 0 & e^{e^{-i3x} - e^{i3x}} \end{bmatrix}, \quad x \in \mathbb{R}.$$

Let us now show that the Hausdorff set of $\Phi_p \widetilde{\Phi_p^{-1}}$ is bounded away from zero. Considering $\eta = (\eta_1, \eta_2)^{\mathsf{T}} \in \mathbb{C}^2$, such that $\|\eta\| = 1$, it follows that

$$\left( \left( \Phi_p \widetilde{\Phi_p^{-1}} \right)(x)\, \eta, \eta \right) = \left( \begin{bmatrix} \eta_1 \, e\, e^{ix} - e^{-ix} \\ \eta_2 \, e\, e^{-i3x} - e^{i3x} \end{bmatrix}, \begin{bmatrix} \eta_1 \\ \eta_2 \end{bmatrix} \right)$$

$$= |\eta_1|^2 e\, e^{ix} - e^{-ix} + |\eta_2|^2 e\, e^{-i3x} - e^{i3x},$$

for all $x \in \mathbb{R}$. Therefore,

$$\mathcal{H}\left( \left( \Phi_p \widetilde{\Phi_p^{-1}} \right)(x) \right)$$

$$= \left\{ |\eta_1|^2 e\, e^{ix} - e^{-ix} + |\eta_2|^2 e\, e^{-i3x} - e^{i3x} \; : \; \eta = \begin{bmatrix} \eta_1 \\ \eta_2 \end{bmatrix} \in \mathbb{C}^2, \|\eta\| = 1 \right\}, \quad (4)$$

for $x \in \mathbb{R}$. Then, we have that

$$\mathrm{dist}\left( \mathcal{H}\left( \left( \Phi_p \widetilde{\Phi_p^{-1}} \right)(x) \right), 0 \right) = \left| |\eta_1|^2 e\, e^{ix} - e^{-ix} + |\eta_2|^2 e\, e^{-i3x} - e^{i3x} \right|, \quad (5)$$

with $\eta = (\eta_1, \eta_2)^{\mathsf{T}} \in \mathbb{C}^2$ such that $\|\eta\| = 1$.

Let us first analyze the particular case when $|\eta_1| = 1$ (and $\eta_2 = 0$). In such a case, it is clear that $\mathrm{dist}\left( \mathcal{H}\left( \left( \Phi_p \widetilde{\Phi_p^{-1}} \right)(x) \right), 0 \right) = 1$, for all $x \in \mathbb{R}$.

Assume now that $|\eta_1| \neq 1$. From (5), we have in this case

$$\mathrm{dist}\left( \mathcal{H}\left( \left( \Phi_p \widetilde{\Phi_p^{-1}} \right)(x) \right), 0 \right) = \left| |\eta_1|^2 e^{-i2\sin(3x)} \left( e^{i2(\sin(x) + \sin(3x))} + c \right) \right|$$

$$= |\eta_1|^2 \left| e^{i2(\sin(x) + \sin(3x))} + c \right|,$$

where

$$c = \frac{1 - |\eta_1|^2}{|\eta_1|^2} > 0. \quad (6)$$

We will now verify that

$$\left| e^{i2(\sin(x) + \sin(3x))} + c \right| \neq 0,$$

for all $x \in \mathbb{R}$ and $c > 0$. Considering the notation $\theta = 2(\sin(x) + \sin(3x))$, we have

$$\left| e^{i2(\sin(x) + \sin(3x))} + c \right|^2 = [\cos(\theta) + c]^2 + \sin^2(\theta) = 1 + 2c\cos(\theta) + c^2$$

and notice that $c^2 + 2c\cos(\theta) + 1 = 0$ if and only if $c = -\cos(\theta) \pm i|\sin(\theta)|$. Therefore, on one hand, if $\theta = 0$ we obtain $c = -1$, and on the other hand, if $\theta \neq 0$ then $c$ is a complex (non-real) number. Both cases are impossible due to (6).

Altogether, we obtain

$$\inf_{x\in\mathbb{R}} \text{dist}\left(\mathcal{H}\left(\left(\Phi_p\widetilde{\Phi_p^{-1}}\right)(x)\right),0\right) > 0, \tag{7}$$

i.e., the Hausdorff set of $\Phi_p\widetilde{\Phi_p^{-1}}$ is bounded away from zero (or "good").

Now for computing the corresponding mean motion, we start by considering $\eta = (1,0)^{\top}$. From (4), it follows that

$$\left(\left(\Phi_p\widetilde{\Phi_p^{-1}}\right)(x)\,\eta,\eta\right) = e^{e^{ix} - e^{-ix}}, \qquad x \in \mathbb{R}.$$

Since $e^{ix} - e^{-ix} \in APW$, from the analogue of *Bohr's Theorem for AP functions*, we have that

$$\kappa\left(\left(\left(\Phi_p\widetilde{\Phi_p^{-1}}\right)\eta,\eta\right)\right) = 0 \tag{8}$$

for $\eta = (1,0)^{\top}$. Due to (7) and (8), and according to the *Babadzhanyan and Rabinovich Theorem* (mentioned above), we conclude that $\kappa\big(((\Phi_p\widetilde{\Phi_p^{-1}})\eta,\eta)\big) = 0$ for all $\eta \in \mathbb{C}^n\backslash\{0\}$. Finally, applying Theorem 3.1, we derive that $WH_{\Phi_p}$ is in fact an invertible operator.

## Acknowledgment

This work was supported in part by *Unidade de Investigação Matemática e Aplicações* of Universidade de Aveiro through the Portuguese Science Foundation (*FCT–Fundação para a Ciência e a Tecnologia*).

## References

1. R. G. Babadzhanyan and V. S. Rabinovich, *A system of integral-difference equations on the half line* (Russian). Akad. Nauk Armyan. SSR Dokl. 81 (1985), no. 3, 107–111.
2. R. G. Babadzhanyan and V. S. Rabinovich, *Factorization of almost periodic operator functions* (Russian). Differential and Integral Equations and Complex Analysis (Russian), 13–22, Kalmytsk. Gos. Univ., Elista, 1986.
3. E. L. Basor and T. Ehrhardt, *Factorization theory for a class of Toeplitz + Hankel operators*, J. Operator Theory 51 (2004), no. 2, 411–433.
4. A. Böttcher, Yu. I. Karlovich and I. M. Spitkovsky, *Convolution Operators and Factorization of Almost Periodic Matrix Functions*, Birkhäuser, Basel, 2002.
5. L. P. Castro, *Regularity of convolution type operators with PC symbols in Bessel potential spaces over two finite intervals*, Math. Nachr. 261/262 (2003), 23–36.
6. L. P. Castro and F.-O. Speck, *Regularity properties and generelized inverses of delta-related operators*, Z. Anal. Anwend. 17 (1998), 577–598.
7. L. P. Castro and F.-O. Speck, *Inversion of matrix convolution type operators with symmetry*, Port. Math. (N.S.) 62 (2005), no. 2, 193–216.
8. L. P. Castro, F.-O. Speck and F. S. Teixeira, *Explicit solution of a Dirichlet-Neumann wedge diffraction problem with a strip*, J. Integral Equations Appl. 15 (2003), 359–383.

9. L. P. Castro, F.-O. Speck and F. S. Teixeira, *On a class of wedge diffraction problems posted by Erhard Meister*, Oper. Theory Adv. Appl. 147 (2004), 211–238.

10. L. P. Castro, F.-O. Speck and F. S. Teixeira, *A direct approach to convolution type operators with symmetry*, Math. Nachr. 269-270 (2004), 73–85.

11. T. Ehrhardt, *Invertibility theory for Toeplitz plus Hankel operators and singular integral operators with flip*, J. Funct. Anal. 208 (2004), 64–106.

12. N. Karapetiants and S. Samko, *Equations with Involutive Operators*, Birkhäuser, Boston, 2001.

13. V. G. Kravchenko and G. S. Litvinchuk, *Introduction to the Theory of Singular Integral Operators with Shift*, Kluwer Academic Publishers Group, Dordrecht, 1994.

14. N. Ya. Krupnik, *Banach Algebras with Symbol and Singular Integral Operators*, Birkhäuser Verlag, Basel, 1987.

15. A. B. Lebre, E. Meister and F. S. Teixeira, *Some results on the invertibility of Wiener-Hopf-Hankel Operators*, Z. Anal. Anwend. 11 (1992), 57–76.

16. E. Meister, F.-O. Speck, F. S. Teixeira, *Wiener-Hopf-Hankel operators for some wedge diffraction problems with mixed boundary conditions*, J. Integral Equations Appl. 4 (1992), 229–255.

17. A. P. Nolasco and L. P. Castro, *Factorization of Wiener-Hopf plus Hankel operators with APW Fourier symbols*, Int. J. Pure Appl. Math. 14 (2004), 537–550.

18. S. Roch and B. Silbermann, *Algebras of Convolution Operators and their Image in the Calkin Algebra*, Report MATH (90-05), Akademie der Wissenschaften der DDR, Karl-Weierstrass-Institut für Mathematik, Berlin, 1990.

19. F.-O. Speck, *General Wiener-Hopf Factorization Methods*, Pitman, London, 1985.

20. F. S. Teixeira, *On a class of Hankel operators: Fredholm properties and invertibility*, Integral Equations Operator Theory 12 (1989), 592–613.

21. F. S. Teixeira, *Diffraction by a rectangular wedge: Wiener-Hopf-Hankel formulation*, Integral Equations Operator Theory 14 (1991), 436–454.

# POTENTIAL TYPE OPERATORS ON CARLESON CURVES ACTING ON WEIGHTED HÖLDER SPACES

V. RABINOVICH*

*Dep. of Telecommunication,*
*ESIME-Zacatenco, National Politechnic Institut,*
*Ed.1, Av.IPN, Mexico 07738, D.F.*
*MEXICO*
*e-mail: vladimir.rabinovich@gmail.com*

We study potential type operators on certain Carleson curves acting on the weighted Hölder spaces. The curves under consideration are locally Lyapunov except for a finite set $F$ of singular points. The normal vector $\nu(y)$ to the curve $\Gamma$ does not have a limit at the singular points and, moreover, $\nu(y)$ may be an oscillating and rotating vector-function in neighborhoods of the singular points. We establish a Fredholm theory of the potential type operators in the Hölder spaces $H_{s,w}(\Gamma, \mathbf{C}^n)$ where $w$ is a weight satisfying an analog of the Mackenhoupt condition.

**Keywords:** Potential operators, Fredholmness, essential spectrum
**Mathematics Subject Classification:** 31A10

## 1. Introduction

We consider the operators

$$Bu(x) = a(x)u(x) + \int_\Gamma b(x,y)\frac{(\nu(y), x-y)}{|x-y|^2}u(y)dl_y, \quad x \in \Gamma \qquad (1)$$

on a curve $\Gamma$ which is locally Lyapunov except for a finite sets $F$ of singular points. We refer to the points in $F$ as the singular points of $\Gamma$. In formula (1) $dl_y$ is the oriented Lebesgue measure on $\Gamma$, $a, b$ are bounded matrix-functions on $\Gamma$ continuous except for singular points where these functions have slowly oscillating discontinuities, $\nu(y)$ is the normal vector to the curve $\Gamma$ at the point $y \in \Gamma\backslash F$. We assume that vector $\nu(y)$ does not approach a limit at singular points. Moreover, we allow $\nu(y)$ to be an oscillated and rotated vector function in a neighborhood of each singular point. Singular points of the latter kind will be called *vorticity points*. We require that the curves under consideration satisfy the well-known Carleson condition ( see for instance [1], [2]). The class of such operators includes the operators of harmonic potentials, wave potentials and many others important in the mathematical physics operators on a class of Carleson curves. The results obtained in the paper allow to consider interior and exterior boundary value problems in planar domains whose boundary has vorticity points.

We construct the Fredholm theory of the operators (1) on the Hölder spaces $\Lambda_{s,w}(\Gamma, \mathbb{C}^N)$, where $s \in (0,1]$ and a weight $w$ has singularities at the point in $F$, and satisfy a certain conditions connected with the order $s$ of the Hölder space.

The potential operators (1) in the case if $\Gamma$ is a so-called Radon curve without peaks, and $a$ and $b$ are piece-wise continuous functions are considered by many authors, starting with the classic Radon's paper [14]. Later these operators acting on the spaces $L_{p,w}(\Gamma)$, where $p \in (1,\infty)$, $w$ is a power weight were investigated by Y.B. Lopatinskiy [6], I. Daniluk [3], V. Shelepov [17] and other authors (see, the surveys [8], [7]). Note that the class of curves studied here is not contained in the classes of curves treated in the aforementioned papers.

In the paper [12], see also the paper [5] the potential type operators on the curves with vorticity points have been considered on the spaces $L_{p,w}(\Gamma)$, $(p \in (1,\infty), w$ is a Mackenhoupt weight ) where the Mellin pseudodifferential operators technique developed earlier in the papers [9], [16], [11], [10], [4], [5] have been applied to the potential type operators.

For investigation of Fredholm property of potential type operators acting on weighted Hölder spaces we applied the Mellin pseudodifferential operators acting on the weighted Hölder spaces following to the paper [13], where singular integral operators on a class of Carleson curves acting on weighted Hölder spaces have been studied.

## 2. Mellin pseudodifferential operators on weighted Hölder spaces

In this Chapter we give an auxiliary material with respect to the local Fredholmness of Mellin pseudodifferential operators acting on multiplicative Hölder spaces on $\mathbb{R}_+$ following to the paper [13].

### 2.1. Multiplicative Hölder spaces on $\mathbb{R}_+$

We consider here the Hölder spaces on the semiaxis $\mathbb{R}_+ = \{r \in \mathbb{R} : r > 0\}$ with respect to the multiplicative structure of the group $\mathbb{R}_+$.

**Definition 2.1.** By $\tilde{\Lambda}^s(\mathbb{R}_+)$, where $s \in (0,1)$ we denote the class of bounded functions $u$ such that

$$\|u\|_{\tilde{\Lambda}^s(\mathbb{R}_+)} = \|u\|_{L^\infty(\mathbb{R}_+)} + \sup_{t \in \mathbb{R}_+, \lambda \in \mathbb{R}_+ \setminus \{1\}} \frac{|u(\lambda t) - u(t)|}{|\log \lambda|^s}$$

Note that the mapping $\eta : \mathbb{R}_+ \to \mathbb{R}, \eta(r) = -\log r$ generates the isomorphisms $\eta^* : \Lambda^s(\mathbb{R}) \to \tilde{\Lambda}^s(\mathbb{R}_+)$, where $\Lambda^s(\mathbb{R})$ is the standard Hölder space on $\mathbb{R}$ with the norm

$$\|u\|_{\Lambda^s(\mathbb{R})} = \|u\|_{L^\infty(\mathbb{R})} + \sup_{t \in \mathbb{R}, y \neq 0} \frac{|u(x+y) - u(x)|}{|y|^\lambda}.$$

We set $\Lambda_0^s(\mathbb{R}_+) = \{u \in \Lambda^s(\mathbb{R}_+) : \lim_{x \to +0} u(x) = 0\}$. The following proposition gives a relation between the multiplicative and the standard Hölder spaces on the semiaxis $\mathbb{R}_+$.

**Proposition 2.1.** *The operator* $u \to x^s u$ *is an isomorphism from* $\tilde{\Lambda}^s(\mathbb{R}_+)$ *on* $\Lambda_0^s(\mathbb{R}_+)$.

## 2.2. Local Fredholmness of Mellin pseudodifferential operators on Hölder classes

We say that a complex-valued function $a$ defined on $\mathbb{R}_+ \times \mathbb{R}$ *belongs to the class* $\mathcal{S}_{1,0}^0$ if $a \in C^\infty(\mathbb{R}_+ \times \mathbb{R})$ and satisfies the estimates

$$|a|_{r,t} = \sum_{\alpha \le r, \beta \le t} \sup_{\mathbb{R}_+ \times \mathbb{R}} \left|(r\partial_r)^\beta \partial_\lambda^\alpha a(r,\lambda)\right| \langle \lambda \rangle^\alpha < \infty$$

for all $\alpha, \beta \in \mathbb{N}_0 = \{0,1,2,3....\}$, where $\langle \lambda \rangle = (1 + |\lambda|^2)^{1/2}$.

**Definition 2.2.** Let $a \in \mathcal{S}_{1,0}^0$. The operator

$$(Op_M(a)u)(r) = (2\pi)^{-1} \int_\mathbb{R} d\lambda \int_{\mathbb{R}_+} a(r,\lambda) \left(\frac{r}{\rho}\right)^{i\lambda} u(\rho) \frac{d\rho}{\rho} \qquad (2)$$

where $u \in C_0^\infty(\mathbb{R}_+)$, is called the Mellin pseudodifferential operator with symbol $a$.

The class of operators of the form (2) with $a \in \mathcal{S}_{1,0}^0$ is denoted by $OPS_{1,0}^0$.

The Mellin pseudodifferential operators in the class $OPS_{1,0}^0$ are the transplantation on $\mathbb{R}_+$ of pseudodifferential operators in the class $OPS_{1,0}^m$, by means of the mapping $\phi : \mathbb{R}_+ \to \mathbb{R}, \phi(r) = -\log r$.

By $\mathbb{S}(\mathbb{R}_+)$ we denote the class of functions $\varphi$ on $\mathbb{R}_+$ such that $\varphi(\exp x) \in S(\mathbb{R})$. From the boundedness of usual pseudodifferential operators in $OPS_{1,0}^0$ on $S(\mathbb{R})$ it follows that an operator $A \in OPS_{1,0}^0$ is a bounded operator on $\mathbb{S}(\mathbb{R}_+)$. An operator $A^t$ is called formally adjoint to the operator $A$ if

$$\int_{\mathbb{R}_+} (Au)(r)\bar{v}(x) \frac{dr}{r} = \int_{\mathbb{R}_+} u(r)\overline{(A^tv)(r)} \frac{dr}{r} \qquad (3)$$

for arbitrary functions $u, v \in \mathbb{S}(\mathbb{R}_+)$. Let $A = Op_M(a) \in OPS_{1,0}^0$. Then the formally adjoint operator $A^t \in OPS_{1,0}^0$. Thus formula (3) allow us to consider pseudodifferential operators on the space of distributions $\mathbb{S}'(\mathbb{R}_+)$, and consequently on the space of Hölder functions.

We say that a symbol $a(\in \mathcal{S}_{1,0}^0)$ is *slowly oscillating* at the point 0, if

$$\left|(r\partial_r)^\beta \partial_\lambda^\alpha a(r,\lambda)\right| \le C_{\alpha\beta}(r) \langle \lambda \rangle^{-\alpha},$$

where

$$\lim_{r \to 0} C_{\alpha\beta}(r) = 0,$$

for all $\alpha \in \mathbb{N}_0$ and $\beta \in \mathbb{N}$.

We denote by $SO_0$ the class of slowly oscillating at the point 0 symbols, and by $\mathcal{OPSO}_0$ the corresponding class of pseudodifferential operators.

**Proposition 2.2.** Let $A = Op_M(a) \in OPS_{1,0}^0$. Then $A$ is a bounded operator on $\tilde{\Lambda}^s(\mathbb{R}_+)$, and

$$\|Op_M(a)\|_{\tilde{\Lambda}^s(\mathbb{R}_+) \to \tilde{\Lambda}^s(\mathbb{R}_+)} \leq C \, |a|_{l_1,l_2}, l_1 > 2, l_2 > 1.$$

Let $\chi$ be a $C^\infty-$function such that $\chi(r) = 1$ if $r > 1$ and $\chi(r) = 0$ if $r < 0$, and $\chi_R(r) = \chi(r/R)$. We set $\tilde{\chi}_R(r) = \chi(\frac{-\log r}{R})$. This function belongs $\mathbb{S}(\mathbb{R}_+)$ and it is equal 1 if $r < e^{-R}$ and equal 0 if $r > 1$.

**Definition 2.3.** An operator $A : \tilde{\Lambda}^s(\mathbb{R}_+) \to \tilde{\Lambda}^s(\mathbb{R}_+)$ is called a locally Fredholm operator at the point 0 if there exist operators $\mathcal{L}_R, \mathcal{R}_R \in \mathcal{B}\left(\tilde{\Lambda}^s(\mathbb{R}_+)\right)$ such that

$$\mathcal{L}_R \tilde{\psi}_R A \tilde{\chi}_R I = \tilde{\chi}_R I + T'_R, \tilde{\chi}_R A \tilde{\psi}_R \mathcal{R}_R = \tilde{\chi}_R I + T''_R,$$

where $T'_R, T''_R$ are compact operators on $\tilde{\Lambda}^s(\mathbb{R}_+)$.

We denote by $\widetilde{\mathbb{R}}$ the two-point compactification of $\mathbb{R}$ homeomorphic to the segment $[-1, 1]$. We denote by $\mathcal{S}_{1,0}^0(\widetilde{\mathbb{R}})$ the class of symbols $a \in \mathcal{S}_{1,0}^0$ such that $a$ is extended to a continuous function on $\mathbb{R}_+ \times \widetilde{\mathbb{R}}$, and the corresponding class of pseudodifferential operators is denoted by $\mathcal{OPS}_{1,0}^0(\widetilde{\mathbb{R}})$. We set

$$\mathcal{OPSO}_0(\widetilde{\mathbb{R}}) = \mathcal{OPS}_{1,0}^0(\widetilde{\mathbb{R}}) \cap \mathcal{OPSO}_0.$$

We will make use of Mellin pseudodifferential operators in Hölder spaces of vector-valued functions. By $\tilde{\Lambda}^s(\mathbb{R}_+, \mathbb{C}^N)$ we denote the space of vector-valued functions $u = (u_1, ..., u_N)$, where $u_j \in \tilde{\Lambda}_-^s(\mathbb{R}_+)$ with the norm

$$\|u\|_{\Lambda^s(\mathbb{R}_+, \mathbb{C}^N)} = \max_{1 \leq j \leq N} \|u_j\|_{\Lambda^s(\mathbb{R}_+)}, \tag{4}$$

and by $\mathcal{OPSO}_0(\widetilde{\mathbb{R}}, \mathbb{C}^{N \times N})$ the class of matrix-valued pseudodifferential operators $Op_M(a(r, \lambda)) = Op_M((a_{ij}(r, \lambda))_{i,j=1}^N)$, where $a_{ij}(r, \lambda) \in \mathcal{OPSO}_0(\widetilde{\mathbb{R}})$.

**Theorem 2.1.** Let $A = Op_M(a) \in \mathcal{OPSO}_0(\widetilde{\mathbb{R}}, \mathbb{C}^{N \times N})$. Then $A : \tilde{\Lambda}^s(\mathbb{R}_+, \mathbb{C}^N) \to \tilde{\Lambda}^s(\mathbb{R}_+, \mathbb{C}^N)$ is a locally Fredholm operator at the point 0 if and only if

$$\lim_{\varrho \to +0} \inf_{0 < r < \varrho, \lambda \in \mathbb{R}} |\det a(r, \lambda)| > 0. \tag{5}$$

Let $w \in C^\infty(\mathbf{R}_+)$. We say that a vector-function $u \in \tilde{\Lambda}_w^s(\mathbb{R}_+, \mathbb{C}^N)$ if

$$\|u\|_{\tilde{\Lambda}_w^s(\mathbb{R}_+, \mathbb{C}^N))} = \|wu\|_{\tilde{\Lambda}^s(\mathbb{R}_+, \mathbb{C}^N)} < \infty.$$

In what follows we consider weights $w = \exp v$, where the function $v$ satisfies the conditions: there is an interval $(0, s), s > 0$ such that

$$\sup_{r \in (0,s)} \left| \left( r \frac{d}{dr} \right)^k v(r) \right| < \infty \tag{6}$$

for all $k \in \mathbb{N}$, and there is an interval $(c,d) \ni 0$ such that

$$\varkappa_w(r) = rv'(r) \in (c,d) \qquad (7)$$

if $r \in (0,s)$. We say that a weight satisfying conditions (6), (7) is slowly oscillating at the point 0 if the conditions (6) and (7) hold and

$$\lim_{r \to 0} r\varkappa'_w(r) = 0.$$

We will denote by $\mathcal{R}(c,d)$ the class of slowly varying weights.

**Definition 2.4.** We say that the a matrix-function $a(r,\lambda) \in S\mathcal{O}_0(\widetilde{\mathbb{R}} \times (a,b), \mathbb{C}^{N \times N})$ if $a(r,\lambda)$ has an analytically continuation with respect to $\lambda$ in the strip : $\Pi = \{\lambda \in \mathbb{C}: \operatorname{Im}\lambda \in (c,d)\}$, and

$$\sup_{\mathbb{R}_+ \times \Pi} |(r\partial_r)^\beta \partial^\alpha a_{ij}(r,\lambda)| \langle \lambda \rangle^\alpha < \infty.$$

We denote by $\mathcal{OPSO}_0(\widetilde{\mathbb{R}} \times (a,b), \mathbb{C}^{N \times N})$ the corresponding class of Mellin pseudodifferential operators with analytical symbols.

It has been proved in [13] that operators in $\mathcal{OPSO}_0(\widetilde{\mathbb{R}} \times (a,b), \mathbb{C}^{N \times N})$ are bounded on the space $\tilde{\Lambda}^s_w(\mathbb{R}_+, \mathbb{C}^N))$ if $w \in \mathcal{R}(c,d)$.

**Theorem 2.2.** Let $A = Op_M(a) \in \mathcal{OPSO}_0(\widetilde{\mathbb{R}}, \mathbb{C}^{N \times N})$, $w \in \mathcal{R}(c,d)$. Then $A : \tilde{\Lambda}^s_w(\mathbb{R}_+, \mathbb{C}^N) \to \tilde{\Lambda}^s_w(\mathbb{R}_+, \mathbb{C}^N)$ is a locally Fredholm operator at the point 0 if and only if

$$\lim_{\varrho \to +0} \inf_{0 < r < \varrho, \lambda \in \mathbb{R}} |\det a(r, \lambda + i\varkappa_w(r))| > 0. \qquad (8)$$

## 3. Operators of potential type on Jordan curves with vorticity points

We will consider the potential type operators on oriented Jordan closed curves $\Gamma$ in the complex plane $\mathbb{C}$. We say that that the function $a$ belongs to $\mathcal{E}^\infty(0,r)$ if $a \in C^\infty(0,r)$ and satisfies the estimates

$$\sup_{t \in (0,r)} \left| \left( t\frac{d}{dt} \right)^k a(t) \right| < \infty, k = 0, 1, \ldots\ldots$$

and we say that $a (\in \mathcal{E}^\infty(0,r))$ belongs to $\mathcal{E}^\infty_{sl}(0,r)$ if

$$\lim_{t \to +0} t\frac{da(t)}{dt} = 0.$$

Let $\Gamma$ be a curve in the complex plane $\mathbb{C}$. By definition, $z \in \Gamma$ is a vorticity point of $\Gamma$ if there is a neighborhood $U^z$ of the point $z$ such that

$$\Gamma \cap U^z_\pm = \{x = z + t \exp i w^z_\pm(t), \ t \in [0,r)\} \qquad (9)$$

where $U_+^z$, $(U_-^z)$ is a right (left) half-neighborhood of the point $z$, respectively. We suppose that

$$\omega_\pm^z(t) = \theta_0^z(t) + \theta_\pm^z(t), \ t \in [0, r)$$

where the functions $\delta_z(t) = t\frac{d\theta_0^z(t)}{dt} \in \mathcal{E}_{sl}^\infty(0, r), \theta_\pm^z(t) \in \mathcal{E}_{sl}^\infty(0, r)$. Moreover, we suppose that

$$0 \le \theta_-^z(t) \le M_- < m_+ \le \theta_+^z(t) \le M_+ < 2\pi, \ z \in F. \tag{10}$$

Note that the function $\delta_z(t)$ is responsible for the rotation of the curve $\Gamma$ at the point $z$, and we call $\delta_z(t)$ the twisting coefficient of $\Gamma$ at the point $z$, while the functions $\theta_\pm^z(t)$ describe the oscillations of the curve $\Gamma$ at the point $z$. The typical example of the function $\theta_0^z$ is $\theta_0^z(t) = \delta \log t$ with the constant twisting coefficient $\delta \in \mathbb{R}$. Note that the variable twisiting coefficient $\delta(t) = \sin \log^\alpha(1 + \log^2 t)^{1/2}) \in \mathcal{E}_{sl}^\infty(0, r)$ if $0 \le \alpha < 1$.

Let $\Gamma$ be an oriented closed Jordan curve on the complex plane $\mathbb{C}$ with a finite set $F$ of the vorticity points and such that $\Gamma \setminus F$ is a locally Lyapunov curve. Let

$$A_\Gamma u(x) = \frac{1}{\pi} \int_\Gamma g(x, y) \frac{(\nu(y), x - y)}{|x - y|^2} u(y) dl_y, \quad x \in \Gamma$$

be a potential type operator on the curve $\Gamma$, where $g \in L^\infty(\Gamma \times \Gamma, B(\mathbb{C}^N))$ is a bounded matrix-function continuous on $(\Gamma \times \Gamma) \setminus (F \times F)$ such that in a neigborhood of the point $(z, z) \in F \times F$ the functions:

$$g_{++}^z(t, \tau) = g(z + t \exp i\omega_+^z(t), z + \tau \exp i\omega_+^z(\tau)),$$
$$g_{+-}^z(t, \tau) = g(z + t \exp i\omega_+^z(t), z + \tau \exp i\omega_-^z(\tau)),$$
$$g_{-+}^z(t, \tau) = g(z + t \exp i\omega_-^z(t), z + \tau \exp i\omega_+^z(\tau)),$$
$$g_{--}^z(t, \tau) = g(z + t \exp i\omega_-^z(t), z + \tau \exp i\omega_-^z(\tau)),$$

defined in a neighborhood of the point $(0, 0)$ have entries $g_{ij}(t, \tau)$ satisfying the conditions

$$\left| (t\partial_t)^\alpha (\tau\partial_\tau)^\beta g_{ij}(t, \tau) \right| \le C_{\alpha, \beta}^{ij}, (t, \tau) \in (0, r) \times (0, r)$$

for enough small $r > 0$, and

$$\lim_{t \to +0} \sup_{\tau \in (0, r)} |t\partial_t g_{ij}(t, \tau)| = 0, \lim_{\tau \to +0} \sup_{t \in (0, r)} |t\partial_t g_{ij}(t, \tau)| = 0.$$

We say that $u \in \Lambda^s(\Gamma, \mathbb{C}^N), 0 < s < 1$ if there exists of neighborhood $F_\varepsilon$ of $F$ such that $u \in \Lambda^s(\Gamma \setminus F_\varepsilon, \mathbb{C}^N)$, that is $u$ is a continuous function on $\Gamma \setminus F_\varepsilon$, and

$$\|u\|_{\Lambda^s(\Gamma \setminus F_\varepsilon)} = \|u\|_{L^\infty(\Gamma \setminus F_\varepsilon, \mathbb{C}^N)} + \sup_{t, \tau \in \Gamma \setminus F_\varepsilon, |t - \tau| > 0} \frac{\|u(t) - u(\tau)\|_{\mathbb{C}^N}}{|t - \tau|^s} < \infty,$$

and $u(z + te^{i(\theta_0^z(t) + \theta_\pm^z(t))}) \in \Lambda_0^s((0, r), \mathbb{C}^N)$. A norm in $\Lambda^s(\Gamma)$ is introduced by the evident way.

The weighted Hölder space $\Lambda_w^s(\Gamma, \mathbb{C}^N); 0 < s < 1$ is the space of the vector-valued functions $u$ such that $wu \in \Lambda^s(\Gamma, \mathbb{C}^N)$ where $w$ is a positive continue function on $\Gamma \backslash F$. In a neighborhood of a singular point $z$ the weight $w$ has the following representation $w(x) = \exp v^z(|x - z|)$, $|x - z| < r$, where the function $v^z(t)$ is such that $\beta^z(t) = t \frac{dv^z(t)}{dt} \in \mathcal{E}_{sl}^\infty(0, r)$. Moreover we suppose that

$$\beta_w^z(t) = t \frac{dv^z(t)}{dt} \in (s, 1+s),\ t \in (0, r) \tag{11}$$

Let $U^z$ be a small neighborhood of the point $z$. We set

$$\Phi_z f(t) = (t^{-s} e^{v^z(t)} f(z + t \exp i\omega_+^z(t)), t^{-s} e^{v^z(t)} f(z + t \exp i\omega_-^z(t)), t \in (0, r)$$

where $w$ satisfies condition (11). Applying Proposition 2.1 one can prove that there exists $r > 0$ such that

$$\Phi_z : \Lambda_w^s(\Gamma \cap U^z, \mathbb{C}^N) \to \tilde{\Lambda}^s((0, r), \mathbb{C}^{2N})$$

is the Banach spaces isomorphism.

The following proposition is the corner-stone of the local study of the potential type operators at vorticity points of the curve $\Gamma$.

**Proposition 3.1.** *Let $\chi_z$ be a cut-off function of the point $z$ that is $\chi_z \in C^\infty(\Gamma), \chi_z(x) = 1$ in a neighborhood of the point $z$ and $\mathrm{supp}\chi_z(x) \subset U_z$. Then, $\Phi_z \chi_z A \chi_z \Phi_z^{-1}$ is a Mellin pseudodifferential operator in the class $OPSO_0(\tilde{\mathbb{R}}, \mathbb{C}^{N\times N})$. Its symbol $\sigma^z(A)(t, \lambda)$ within to a symbol of a Mellin pseudodifferential operator with zero local norm is given by the formula*

$$\sigma^z(A_\Gamma)(t, \lambda) = \begin{pmatrix} a_{++}^z(t, \lambda) & a_{-+}^z(t, \lambda) \\ a_{+-}^z(t, \lambda) & a_{--}^z(t, \lambda) \end{pmatrix}$$

*where*

$$a_{++}^z(t, \lambda) = \frac{i}{2} \frac{\sin \frac{2\pi\delta^z(t)(\lambda+i(-s+\beta_w^z(t)))}{1+\delta^z(t)^2}}{\sinh \frac{\pi(\lambda+i(-s+\beta_w^z(t)))}{1+i\delta^z(t)} \sinh \frac{\pi(\lambda+i(-s+\beta_w^z(t)))}{1+i\delta^z(t)}} g_{++}(t, t)$$

$$a_{--}^z(t, \lambda) = -\frac{i}{2} \frac{\sin \frac{2\pi\delta^z(t)(\lambda+i(-s+\beta_w^z(t)))}{1+\delta^z(t)^2}}{\sinh \frac{\pi(\lambda+i(-s+\beta_w^z(t)))}{1+i\delta^z(t)} \sinh \frac{\pi(\lambda+i(-s+\beta_w^z(t)))}{1+i\delta^z(t)}} g_{--}(t, t)$$

$$a_{-+}^z(t, \lambda) = \frac{1}{2} \left[ \frac{\exp(\theta^z(t) - \pi)\frac{\lambda+i(-s+\beta_w^z(t))}{1+i\delta^z(t)}}{\sinh \frac{\pi(\lambda+i(-s+\beta_w^z(t)))}{1+i\delta^z(t)}} \right.$$

$$\left. - \frac{\exp -(\theta^z(t) - \pi)\frac{\lambda+i(-s+\beta_w^z(t))}{1-i\delta^z(t)}}{\sinh \frac{\pi(\lambda+i(-s+\beta_w^z(t)))}{1-i\delta^z(t)}} \right] g_{-+}(t, t)$$

$$a^z_{+-}(t, \lambda) = -\frac{1}{2} \left[ \frac{\exp - (\theta^z(t) - \pi)\frac{\lambda + i(-s + \beta^z_w(t))}{1 + i\delta^z(t)}}{\sinh \frac{\pi(\lambda + i(-s + \beta^z_w(t)))}{1 + i\delta^z(t)}} \right.$$

$$\left. - \frac{\exp(\theta^z(t) - \pi)\frac{\lambda + i(-s + \beta^z_w(t))}{1 - i\delta^z(t)}}{\sinh \frac{\pi(\lambda + i(-s + \beta^z_w(t)))}{1 - i\delta^z(t)}} \right] g_{+-}(t, t),$$

where $\theta^z(t) = \theta^z_+(t) - \theta^z_-(t)$.

The proof of this proposition is similar to the proof of Proposition 15 [12]. The appearence of $-s$ in the symbol $\sigma^z(A_\Gamma)(t, \lambda)$ is explained by the connection between the usual Hölder space $\Lambda^s_0\left((0, r), \mathbb{C}^N\right)$ and the Hölder space $\tilde{\Lambda}^s\left((0, r), \mathbb{C}^N\right)$ given in Proposition 2.1.

**Theorem 3.1.** *Let for every point* $z \in F$ *condition (11) hold. Then* $A_\Gamma :$ $\Lambda^s_w(\Gamma, \mathbb{C}^N) \to \Lambda^s_w(\Gamma, \mathbb{C}^N), 0 < s < 1$ *is a bounded operator.*

The proof is similar to the proof of boundedness of singular integral operators on curves with vorticity points from [13]. This proof uses an admissible partition of unity on $\Gamma$, Proposition 3.1, and Proposition 2.2.

Let

$$B_\Gamma u(x) = a(x)u(x) + A_\Gamma u(x)$$

where the matrix-function $a$ has the components in the space $C(\Gamma \setminus F)$, and in a neighborhood of every point $z \in F$

$$a^z_\pm(t) = a(z + t \exp i\omega_\pm(t)), t \in (0, r)$$

are matrix-functions with components in $\mathcal{E}^\infty_{sl}(0, r)$.

We set

$$a_z(t) = \begin{pmatrix} a^z_+(t) & \\ & a^z_+(t) \end{pmatrix},$$

$$\sigma^z(B_\Gamma)(t, \lambda) = a_z(t) + \sigma^z(A_\Gamma)(t, \lambda).$$

The main result of the paper is the following theorem.

**Theorem 3.2.** *The potential operator*

$$B_\Gamma : \Lambda^s_w(\Gamma, \mathbb{C}^N) \to \Lambda^s_w(\Gamma, \mathbb{C}^N), 0 < s < 1$$

*is a Fredholm operator if and only if :*

$$1) \inf_{x \in \Gamma} |\det a(x)| > 0,$$

$$2) \underline{\lim}_{t \to 0} \inf_{\lambda \in \mathbb{R}} |\det \sigma^z(B_\Gamma)(t, \lambda)| > 0$$

*for every singular point $z \in F$. If the conditions 1) and 2) are fulfilled, then*

$$Ind\, B_\Gamma = -\sum_{z \in F} \frac{1}{2\pi} \lim_{t \to 0} \left[\arg \det(I + a^z(t)^{-1}\sigma^z(A_\Gamma)(t,\lambda)\right]_{\lambda=-\infty}^{\infty}. \qquad (12)$$

The proof is similar to the proof of Theorem 38 in [13] and it is based on an admissible partition of unity, the construction of local regularizators applying Proposition 3.1 and Theorem 2.1.

# References

1. D. David: Opérateurs integraux singuliers sur certaines courbes du plan complexe. Ann. Sci. École Norm. Sup. 17 (1984), 157–189.
2. E.M.Dynkin and B.P. Osilenker: Weighted norm estimates for singular integrals and their applications. J. Soviet Math. 30 (1985), 2094–2154.
3. I.I. Daniluk, Nonregular Boundary Value Problems on the Plane. M. Nauka,1975 [Russian]
4. A. Böttcher, Yu.I.Karlovich, V.S. Rabinovich. Mellin pseudodifferential operators with slowly varying symbols and singular integrals on Carleson curves with Muckenhoupt weights. Manuscripta math., 95, 1998, 363-376
5. A.Bottcher, Yu.I.Karlovich, V.S.Rabinovich. Singular integral operators with complex conjugation from the viewpoint of pseudodifferential operators. Operator Theory: Advances and Applications, vol.121, 2001, Birkhäuser Verlag Basel/ Switzerland, p. 36-59.
6. Lopatinskiy, Ya.B.: On a type of singular integral equations: Theor. and Appl. Math. Lvov State Univ. 1963, No.2, p.53-57.
7. V.G. Maz'ya, Boundary integral equations. Itogi Nauki i Techniki. Sovremennie Problemi Matematiki. Fundamentalnie Napravlenia V.27, p.131-237 [Russian]
8. Z. Presdorf, Linear integral equations. Itogi Nauki i Techniki. Sovremennie Problemi Matematiki. Fundamentalnie Napravlenia.V.27, p.5-130 [Russian]
9. V.S. Rabinovich: Singular integral operators on a composed contour with oscillating tangent and pseudodifferential Mellin operators. Soviet Math. Dokl. 44 (1992), 791–796.
10. V.S.Rabinovich. Algebras of singular integral operators on complicated contours with nodes being of logarithmic whirl points. Izvestia AN Rossii, ser. mathem., v.60, No.6, 1996,169-200.(In Russian) English translation: Izvestia: Mathematics 60:6, 1996, 1261-1292
11. V.S.Rabinovich. Mellin pseudodifferential operators technique in the theory of singular integral operators on some Carleson curves. Operator Theory: Advances and Applications, Vol.102, 1998, Birkhauser Verlag Basel / Switzerland, p.201-218
12. V. S. Rabinovich, Potential type operators on curves with vorticity points, Zeitschrift für Analysis und ihre Anwendungen, 18(1999) 4, 1065-1081
13. V. S. Rabinovich, N. Samko, S. Samko, Local Fredholm spectrums and Fredholm properties of singular integral operators on Carleson curves acting on weighted Hölder spaces, Accepted to the journal "Integral Equations and Operator Theory".
14. I. Radon. On boundary value problems for logarithmic potential. Uspehi Math. Nauk,1946, v.1, No.3-4. p.96-124 [Russian]. [Ukrainian].
15. V.S. Rabinovich, Singular integral operators on a composed contour with oscillating tangent and pseudodifferential Mellin operators. Soviet Math. Dokl. 44 (1992), 791–796.

16. V.S. Rabinovich, Singular integral operators on composed contours and pseudodifferential operators. Matem. Zametki 58 (1995), 65–85 [Russian].

17. V.Yu. Shelepov, On index and spectrum of integral operators of potential type along the Radon curves. Mathem. sbornik, 181, No.6, p.751-778 [Russian]

# UPPER AND LOWER INDICES OF A CERTAIN CLASS OF MONOTONIC FUNCTIONS IN CONNECTION WITH FREDHOLMNESS OF SINGULAR INTEGRAL OPERATORS

NATASHA SAMKO*

*Faculdade de Ciências e Tecnologia*
*Universidade do Algarve*
*Campus de Gambelas, Faro 8005, Portugal*
*e-mail: nsamko@ualg.pt*

In connection with Fredholmness of singular integral operators in weighted generalized Hölder spaces $H_0^\omega(\Gamma, \rho)$, there is studied a certain subclass of monotonic functions $\omega(x)$ of the class of Zygmund-Bary-Stechkin which may oscillate between two power functions. In particular, there are derived explicit formulas for the best lower and upper exponents of the power-type estimation of $\omega(x)$, these exponents in general not coinciding with Boyd-type indices of $\omega(x)$.

**Key words:** Indices of monotonic functions, continuity modulus, Zygmund conditions, Bary-Stechkin class, Boyd-type indices, Fredholm operators
**Mathematics Subject Classification:** 26A48, 54C35, 26A16

## 1. Introduction

The Fredholm nature of the singular integral operators with piecewise continuous coefficients is well known in various function spaces of integrable functions, for example in weighted Lebesgue or Orlich spaces, or in general in Banach function spaces, we refer to the books [4,5,2], Subsection 9.6. In particular, the effect of the massiveness of the essential spectra of singular integral operators caused by the generality of weights or that of curves, is well known.

The corresponding theory of singular integral operators in weighted subspaces of continuous functions is much less developed. In Hölder spaces $H^\lambda(\Gamma, \rho)$ with power weights, the results on Fredholmness of singular integral operators were known from the papers of R.Duduchava [3], see also their presentation in [5].

In [8,9] we considered the generalized Hölder spaces $H_0^\omega(\Gamma, \rho)$ with characteristics $\omega$ and weights $\rho$ more general than power ones and obtained the statement on Fredholmness, together with a formula for the index, for singular integral operators on Lyapunov type curves in such spaces in the case when both characteristic and weight do not oscillate (*equilibrated* characteristic in terminology of the papers [8,9]). The Boyd-type indices $m_\omega$ and $M_\omega$ of the characteristic $\omega(h)$ coincide in this case:

*Supported by CEMA T (Centro de Matemática e Aplicações), IST, Lisbon

$m_\omega = M_\omega$.

In this case no massive spectra may appear and the result on Fredholmness is in a certain sense similar to Gohberg-Krupnik's result on Fredholmness of singular integral operators in the Lebesgue spaces with power weight.

A possibility to get results with the massive spectra in Hölder type spaces may appear when one admits either oscillating characteristic $\omega$ or oscillating weight $\rho$ (*non-equilibrated* characteristics and weights with non-coinciding Boyd-type indices ). Such a new situation was considered in [11], where the appearance of the "lunes" generating the massivity of the spectra is due to the presence of oscillation of $\omega$ so that the massive spectra appears even in non-weighted case and on nice curves, the situation similar to Orlicz spaces (see [2], Subsection 10.5 on Fredholmness in Orlicz spaces). The case when both the characteristic and weight are non-equilibrated will appear in [7].

In this connection, for the goals of investigation of the singular integral operators in generalized Hölder spaces, it becomes important to study the subclass of non-equilibrated functions in the Zygmund-Bary-Stechkin class $\Phi$ of continuity moduli. In [10] we showed that this subclass is rich enough by studying a certain family of non-equilibrated functions of monotonic functions with non-coinciding lower and upper indices $m_\omega$ and $M_\omega$ of the type of the Boyd indices (more precisely, of the type of the Matuzsewska-Orlicz indices).

Monotonic functions in this family, called $(\alpha, \beta)$-functions, originally oscillate between $c_1 x^\alpha$ and $c_2 x^\beta$, $0 < \alpha < \beta < 1$, $0 \le x \le \ell$. In general, the lower and upper indices $m_\omega$ and $M_\omega$ of an $(\alpha, \beta)$-function $\omega$ do not necessarily coincide with $\alpha$ and $\beta$, but may be arbitrary numbers such that $\alpha \le m_\omega \le M_\omega \le \beta$.

We recall (see [8,10]) that the indices $m_\omega$ and $M_\omega$ of a function $\omega(x) \in \Phi$ lie in the interval $0 < m_\omega \le M_\omega < 1$ and $\omega$ satisfies the estimates

$$c_3 x^{M_\omega + \varepsilon} \le \omega(x) \le c_4 x^{m_\omega - \varepsilon} \tag{1}$$

with an arbitrarily small $\varepsilon > 0$ (constants $c_3$ and $c_3$ in general being dependent on $\varepsilon$).

In this paper we continue the study of the family of $(\alpha, \beta)$-functions and in particular show that the real power-type bounds for a function in $\Phi$ may be different from those defined by the indices $m_\omega$ and $M_\omega$. Namely, it may happen that

$$c_5 x^{\Lambda_\omega + \varepsilon} \le \omega(x) \le c_6 x^{\lambda_\omega - \varepsilon} \tag{2}$$

where $\Lambda_\omega < M_\omega$ and $\lambda_\omega > m_\omega$, so that bounds (2) are certainly more exact than (1). We give in particular concrete examples of monotonic $(\alpha, \beta)$-functions for which

$$\alpha \le m_\omega < \lambda_\omega \le \Lambda_\omega < M_\omega \le \beta$$

and the values of $\lambda_\omega$ and $\Lambda_\omega$ best possible for (2) may be arbitrary numbers in the interval $(m_\omega, M_\omega)$. Thus, the indices $m_\omega$ and $M_\omega$ may be different from the best exponents of power-type bounds of $\omega(x)$ and therefore, the indices $m_\omega$ and $M_\omega$ prove to be non-necessarily responsible for the best power-type bounds for a monotonic function, they are rather responsible for the nature of oscillation of $\omega(x)$.

For the family of $(\alpha, \beta)$-functions we prove explicit formulas for calculating the best bounds $\lambda_\omega$ and $\Lambda_\omega$, see Theorem 3.1.

Notation:

a.d.= almost decreasing,

a.i.= almost increasing,

$C, C_1, \dots$ stand for absolute positive constants in estimations;

$W = \{\omega \in \mathbb{C}([0, \ell]) : \omega(0) = 0, \ \omega(x) > 0 \ \text{for} \ x > 0, \ \omega(x) \ \text{is a.i.}\}, \quad 0 < \ell < \infty.$

## 2. Preliminaries

### 2.1. The Zygmund-Bary-Stechkin class $\Phi$.

**Definition 2.1.** A function $\omega \in W$ is said to belong to the Zygmund-Bary-Stechkin class $\Phi$ if

$$\int_0^h \frac{\omega(x)}{x} \, dx \le C\omega(h) \quad \text{and} \quad \int_h^\ell \frac{\omega(x)}{x^2} \, dx \le C\frac{\omega(h)}{h}, \quad 0 < h \le \ell. \quad (3)$$

Let

$$\overline{\Omega}(x) = \limsup_{h \to 0} \frac{\omega(hx)}{\omega(h)} \quad \text{and} \quad \underline{\Omega}(x) = \liminf_{h \to 0} \frac{\omega(hx)}{\omega(h)}.$$

The numbers

$$m_\omega = \sup_{x>1} \frac{\ln \underline{\Omega}(x)}{\ln x} = \sup_{0<x<1} \frac{\ln \overline{\Omega}(x)}{\ln x} = \lim_{x \to 0} \frac{\ln \overline{\Omega}(x)}{\ln x}$$

and

$$M_\omega = \sup_{x>1} \frac{\ln \overline{\Omega}(x)}{\ln x} = \lim_{x \to \infty} \frac{\ln \overline{\Omega}(x)}{\ln x}$$

(see [8,10,6] and references therein), are known as the lower and upper indices of the function $\omega(x)$. It is known that $0 \le m_\omega \le M_\omega \le \infty$ for $\omega \in W$ and

$$0 < m_\omega \le M_\omega < 1 \quad \text{for} \quad \omega \in \Phi,$$

the conditions $m_\omega > 0$ and $M_\omega < 1$ being equivalent to the first and second inequalities in (3), respectively. The indices $m_\omega$ and $M_\omega$ may be calculated by the formulas

$$m_\omega = \sup \left\{ \lambda_1 \in (0,1) : \ \frac{\omega(x)}{x^{\lambda_1}} \ \text{is almost increasing} \right\}, \quad (4)$$

$$M_\omega = \inf \left\{ \lambda_2 \in (0,1) : \ \frac{\omega(x)}{x^{\lambda_2}} \ \text{is almost decreasing} \right\}. \quad (5)$$

In the sequel by $\lambda_\omega$ and $\Lambda_\omega$ we denote the best possible constants in (4), that is,

$$\lambda_\omega = \sup\left\{\lambda_1 \in (0,1): \quad \sup_{x \in (0,\ell]} \frac{\omega(x)}{x^{\lambda_1}} < \infty\right\}, \tag{6}$$

$$\Lambda_\omega = \inf\left\{\lambda_2 \in (0,1): \quad \inf_{x \in (0,\ell]} \frac{\omega(x)}{x^{\lambda_2}} > 0\right\}, \qquad \lambda_\omega \le \Lambda_\omega. \tag{7}$$

Note that these are the numbers $m_\omega$ and $M_\omega$ which are responsible for the Fredholm characterization of singular operators in the generalized holder spaces.

## 2.2. $(\alpha, \beta)$-functions.

We recall that $(\alpha, \beta)$-functions introduced in [10] were defined as continuous piecewise power functions in the following way. Let $\ell = 1$ and let $I_{2n+1} = [a_{2n+2}, a_{2n+1}]$ and $I_{2n} = [a_{2n+1}, a_{2n}]$, where $\{..., a_n, a_{n-1}, ..., a_1, a_0\}$ is an arbitrary partition of $[0,1]$ by a sequence of points such that $\cdots < a_n < a_{n-1} < \cdots < a_1 < a_0 = 1$ and $\lim_{n\to\infty} a_n = 0$. Then by an $(\alpha, \beta)$-function we call a function of the form

$$\omega(x) = \begin{cases} c_{2n+1} x^\beta, & \text{if } x \in I_{2n+1}, \\ c_{2n}\, x^\alpha, & \text{if } x \in I_{2n}, \quad n = 0, 1, 2, ... \end{cases} \tag{8}$$

where we take $c_0 = 1$ and afterwards the coefficients $c_1, c_2, c_3, ...$ are chosen, step by step, in such a way that the resulting function $\omega(x)$ is continuous. When $\beta \ne \alpha$, we have $c_{2n} < 1$ and $c_{2n} \downarrow$, while $c_{2n+1} > 1$ and $c_{2n+1} \uparrow$, see details in [10]. We recall also that the continuity of "gluing" in (8) involves the following relation between the coefficients $c_n$ and the partition points $a_n$:

$$a_{2n} = \left(\frac{c_{2n}}{c_{2n-1}}\right)^{\frac{1}{\beta-\alpha}}, \qquad a_{2n+1} = \left(\frac{c_{2n}}{c_{2n+1}}\right)^{\frac{1}{\beta-\alpha}} \tag{9}$$

where $\beta \ne \alpha$, the case $\beta = \alpha$ being trivial. Thus, conversely, given $c_{2n} \downarrow$ and $c_{2n+1} \uparrow$, we can calculate the corresponding partition points $a_n$. Any $(\alpha, \beta)$-function belongs to $\Phi$ ([10], Lemma 5.3).

We recall that

$$a_1^{\alpha-\beta} x^\beta \le \omega(x) \le x^\alpha, \tag{10}$$

see [10], formula (5.8).

## 3. The formula for the indices

**Lemma 3.1.** Let $0 < \alpha < \beta < 1$, let $\omega$ be an $(\alpha, \beta)$-function and $\lambda, \mu \in (\alpha, \beta)$. The following equivalencies hold

$$\sup_{x \in (0,1]} \frac{\omega(x)}{x^\lambda} = A \quad \Longleftrightarrow \quad \sup_n c_{2n+1}^{\lambda-\alpha} c_{2n}^{\beta-\lambda} = A^{\beta-\alpha} \tag{11}$$

*and*

$$\inf_{x \in (0,1]} \frac{w(x)}{x^\mu} = B \quad \Longleftrightarrow \quad \inf_n c_{2n-1}^{\mu-\alpha} c_{2n}^{\beta-\mu} = B^{\beta-\alpha}. \tag{12}$$

**Proof.** Let us consider the inequality $w(x) \leq Ax^\lambda$ first on the intervals $I_{2n+1} = [a_{2n+2}, a_{2n+1}]$. For $x \in I_{2n+1}$ we have, taking into account that $\beta - \lambda > 0$,

$$w(x) \leq Ax^\lambda \quad \Longleftrightarrow \quad c_{2n+1}x^\beta \leq Ax^\lambda \quad \Longleftrightarrow \quad c_{2n+1}x^{\beta-\lambda} \leq A \quad \Longleftrightarrow \quad c_{2n+1}a_{2n+1}^{\beta-\lambda} \leq A. \tag{13}$$

In view of (9) we then have

$$w(x) \leq Ax^\lambda \quad \text{for} \quad x \in \bigcup_n I_{2n+1} \quad \Longleftrightarrow \quad c_{2n+1}^{\lambda-\alpha} c_{2n}^{\beta-\lambda} \leq A^{\beta-\alpha} \quad \text{for all} \quad n. \tag{14}$$

Similarly for $x \in I_{2n} = [a_{2n+1}, a_{2n}]$ we have

$$w(x) \leq Ax^\lambda \quad \Longleftrightarrow \quad c_{2n}x^\alpha \leq Ax^\lambda \quad \Longleftrightarrow \quad c_{2n}x^{\alpha-\lambda} \leq A \quad \Longleftrightarrow \quad c_{2n}a_{2n+1}^{\alpha-\lambda} \leq A \tag{15}$$

where we used the fact that $\alpha - \lambda < 0$, and then by (9) we arrive at the inequality $c_{2n+1}^{\lambda-\alpha} c_{2n}^{\beta-\lambda} \leq C_2$, that is, we have exactly the same equivalence (14) on $\bigcup_n I_{2n}$ as well, which yields (11), if we take also into account that the function $c_{2n+1}x^{\beta-\lambda}$ in (13) is increasing on $I_{2n+1}$ and attains its maximum at the point $a_{2n+1}$.

Similarly, formula (12) is proved. $\qquad\square$

**Theorem 3.1.** *Let $0 < \alpha < \beta < 1$. The bounds $\lambda_w$ and $\Lambda_w$ of an $(\alpha, \beta)$-function $w(x)$ are calculated in terms of the coefficients $c_{2n+1}$ and $c_{2n}$ by means of the formulas*

$$\lambda_w = \alpha + \frac{\beta - \alpha}{1 + \limsup_{n \to \infty} \frac{\ln c_{2n+1}}{\ln \frac{1}{c_{2n}}}} = \beta - \frac{\beta - \alpha}{1 + \liminf_{n \to \infty} \frac{\ln \frac{1}{c_{2n}}}{\ln c_{2n+1}}} \tag{16}$$

*and*

$$\Lambda_w = \alpha + \frac{\beta - \alpha}{1 + \liminf_{n \to \infty} \frac{\ln c_{2n-1}}{\ln \frac{1}{c_{2n}}}} = \beta - \frac{\beta - \alpha}{1 + \limsup_{n \to \infty} \frac{\ln \frac{1}{c_{2n}}}{\ln c_{2n-1}}}. \tag{17}$$

**Proof.** Let us prove formula (16). By (10) it suffices to treat only the values $\lambda \in (\alpha, \beta)$. Then relation (11) is applicable. Passing to logarithms in (11), after easy transformations we obtain that all the possible values of $\lambda \in (\alpha, \beta)$ for which $w(x) \leq Ax^\lambda$ with $A = A(\lambda)$, are given by

$$\lambda \leq \alpha + \frac{\beta - \alpha}{1 + \frac{\ln c_{2n+1}}{\ln \frac{1}{c_{2n}}}} + \frac{A^{\beta-\alpha}}{\ln \frac{c_{2n+1}}{c_{2n}}}. \tag{18}$$

Observe that $\frac{c_{2n+1}}{c_{2n}} = \frac{1}{a_{2n}^{\beta-\alpha}} \to 0$ as $n \to \infty$, so that the last term in (18) tends to zero as $n \to \infty$ and we get

$$\sup \lambda = \alpha + \liminf_{n \to \infty} \frac{\beta - \alpha}{1 + \frac{\ln c_{2n+1}}{\ln \frac{1}{c_{2n}}}}.$$

By the property

$$\liminf_{n\to\infty} \frac{1}{f(n)} = \frac{1}{\limsup\limits_{n\to\infty} f(n)} \tag{19}$$

we get the first relation in (16). The second relation is obtained from the first one directly with property (19) taken into account.

Formula (17) is similarly proved: we have

$$\omega(x) \geq Bx^\mu \quad \Longleftrightarrow \quad c_{2n-1}^{\mu-\alpha} c_{2n}^{\beta-\mu} \geq B > 0; \tag{20}$$

then we pass to logarithms and after easy transformations similar to those in (18) arrive at

$$\mu \geq \alpha + \frac{\beta - \alpha}{1 + \frac{\ln c_{2n-1}}{\ln \frac{1}{c_{2n}}}} + \frac{B^{\beta-\alpha}}{\ln \frac{c_{2n-1}}{c_{2n}}}$$

which yields formula (17). □

**Corollary 3.1.** *An $(\alpha,\beta)$-function has coinciding bounds $\lambda_\omega = \Lambda_\omega$, if and only if*

$$\limsup_{n\to\infty} \frac{\ln c_{2n+1}}{\ln \frac{1}{c_{2n}}} = \liminf_{n\to\infty} \frac{\ln c_{2n-1}}{\ln \frac{1}{c_{2n}}}. \tag{21}$$

*In particular, this is the case if $c_{2n} = e^{-an^N + P_{N-1}(n)}$ and $c_{2n+1} = e^{an^N + Q_{N-1}(n)}$, where $N \geq 1$ is an integer, $a > 0$ and $b > 0$ are constants and $P_{N-1}(n)$ and $Q_{N-1}(n)$ are arbitrary polynomials of order $N - 1$.*

**Corollary 3.2.** *Let $a > 0, b > 0, 0 < c < 1, A > 1, T > 1$. The $(\alpha,\beta)$-functions with $0 < \alpha < \beta < 1$ given in the following examples*
1) $c_{2n} = \frac{1}{(n+1)^a}, \quad c_{2n+1} = (n+1)^b$,
2) $c_{2n} = 2^{-an}, \quad c_{2n+1} = 2^{bn}$,
3) $c_{2n} = A^{-aT^n}, \quad c_{2n+1} = A^{bT^n}$,
4) $c_{2n} = e^{-A^{(n-c)^2}}, \quad c_{2n+1} = e^{A^{n^2}}$,
*where $A > 1, T > 1$, have the following best exponents $\lambda_\omega$ and $\Lambda_\omega$:*
1) $\lambda_\omega = \Lambda_\omega = \frac{b\alpha + a\beta}{a+b}$,
2) $\lambda_\omega = \Lambda_\omega = \frac{b\alpha + a\beta}{a+b}$,
3) $\lambda_\omega = \frac{b\alpha + a\beta}{a+b}, \quad \Lambda_\omega = \frac{b\alpha + aT\beta}{aT+b} > m_\omega$,
4) $\lambda_\omega = m_\omega = \alpha, \quad \Lambda_\omega = M_\omega = \beta$,
*respectively. Thus, functions (8) corresponding to the examples in 1), 2) and 4) have coinciding bounds $\lambda_\omega$ and $\Lambda_\omega$.*

Observe that the example in 4) may be generalized with the same result $m_\omega = \alpha, \quad M_\omega = \beta$ as
4') $c_{2n} = e^{-A^{u_n}}, \quad c_{2n+1} = e^{A^{v_n}}$,
where $u_n \to \infty, v_n \to \infty$ and $\lim\limits_{n\to\infty} (v_n - u_n) = \lim\limits_{n\to\infty} (u_n - v_{n+1}) = +\infty$.

In the paper [1] there was considered an example of the Young function defining an Orlicz space in connection with calculation of their Boyd indices. That example corresponds to the above example 1) and provides different Boyd indices but coinciding bounds $\lambda_\omega$ and $\Lambda_\omega$.

**Remark 3.1.** If

$$\inf_n c_{2n}c_{2n-1} = C > 0, \tag{22}$$

then from (12) it follows that

$$\sqrt{C}x^{\frac{\alpha+\beta}{2}} \le \omega(x) \le x^\alpha.$$

Similarly, if

$$\sup_n c_{2n}c_{2n+1} = C < \infty, \tag{23}$$

then according to (11)

$$a_1^{\alpha-\beta}x^\beta \le \omega(x) \le \sqrt{C}x^{\frac{\alpha+\beta}{2}}.$$

Observe that the example of a non-equilibrated function $\omega$ considered in [1], corresponds to the choice

$$c_{2n} = e^{-\frac{\beta-\alpha}{2}n(n+1)} \quad \text{and} \quad c_{2n+1} = e^{\frac{\beta-\alpha}{2}(n+1)(n+2)}$$

which satisfies (22). In that example

$$m_\omega = \alpha, M_\omega = \beta, \quad \text{but} \quad x^{\frac{\alpha+\beta}{2}} \le \omega(x) \le Cx^{\frac{\alpha+\beta}{2}-\varepsilon}$$

for an arbitrarily small $\varepsilon > 0, C = C(\varepsilon)$.

### References

1. V. D. Aslanov and Yu. I. Karlovich. One-sided invertibility of functional operators in reflexive Orlicz spaces. *Akad. Nauk Azerbaidzhan. SSR Dokl.*, 45(11-12):3–7, 1989.
2. A. Böttcher and Yu. Karlovich. *Carleson Curves, Muckenhoupt Weights, and Toeplitz Operators*. Basel, Boston, Berlin: Birkhäuser Verlag, 1997. 397 pages.
3. R.V. Duduchava. Singular integral equations in weighted Hölder spaces (in Russian). *Matem. Issledov.*, 5(3):58–82, 1970.
4. I. Gohberg and N. Krupnik. *One-Dimensional Linear Singular Integral equations, Vol. I. Introduction*. Operator theory: Advances and Applications, **53**. Basel-Boston: Birkhauser Verlag, 1992. 266 pages.
5. I. Gohberg and N. Krupnik. *One-Dimensional Linear Singular Integral equations, Vol. II. General Theory and Applications*. Operator theory: Advances and Applications, **54**. Basel-Boston: Birkhauser Verlag, 1992. 232 pages.
6. N.K. Karapetiants and N.G. Samko. Weighted theorems on fractional integrals in the generalized Hölder spaces $H_0^\omega(\rho)$ via the indices $m_\omega$ and $M_\omega$. *Fract. Calc. Appl. Anal.*, 7(4), 2004.
7. N.G. Samko. Singular Integral Operators in Weighted Spaces of Continuous Functions with an Oscillating Continuity Moduli and Oscillating Weights. *To appear*.

8. N.G. Samko. Singular integral operators in weighted spaces with generalized Hölder condition. *Proc. A. Razmadze Math. Inst*, 120:107–134, 1999.
9. N.G. Samko. Criterion of Fredholmness of singular operators with piece-wise continuous coefficients in the generalized Hölder spaces with weight. In *Proceedings of IWOTA 2000, Setembro 12-15, Faro, Portugal*, pages 363–376. Birkhäuser, In: "Operator Theory: Advances and Applications", v. 142, 2002.
10. N.G. Samko. On non-equilibrated almost monotonic functions of the Zygmund-Bary-Stechkin class. *Real Anal. Exch.*, 30(2), 2005.
11. N.G. Samko. Singular integral operators in weighted spaces of continuous functions with non-equilibrated continuity modulus. *Preprint 2, Departamento de Matematica, Instituto Superior Tecnico, Lisbon*, (2/2005):1–23, January, 2005.

# THE ASYMPTOTIC BEHAVIOR OF THE TRACE OF GENERALIZED TRUNCATED INTEGRAL CONVOLUTIONS

O. N. ZABRODA

*Technical University Chemnitz,*
*Department of mathematic,*
*Reichenhainer Str. 41,*
*09126, Chemnitz, Germany*
*E-mail: olga.zabroda@s2003.tu-chemnitz.de*

The paper deals with generalized truncated integral convolutions. We study the spectral properties of the operator family $\{\mathcal{A}_\tau(a)\}_{\tau>0}$ with the symbol $a(x, y, \xi) = c_a + (\mathcal{F}\widehat{a})(x, y, \xi)$ defined on $[0, 1] \times [0, 1] \times \mathbb{R}$. The operator $\mathcal{A}_\tau(a) \in \mathrm{End}(L_2([0, \tau]))$ for $\tau > 0$ is defined as

$$(\mathcal{A}_\tau(a)f)(t) = c_a f(t) + \int_0^\tau \widehat{a}\left(\frac{t}{\tau}, \frac{s}{\tau}, t - s\right) f(s)ds, \quad t \in [0, \tau], \ f \in L_2([0, \tau]).$$

We investigate the asymptotic invertibility of $\mathcal{A}_\tau(a)$ for $\tau$ tending to $+\infty$ and construct an almost inverse operator. This makes it possible to describe a complex set in a neighbourhood of which the spectrum asymptotically concentrates. Finally, we prove a Szegö type theorem.

**Key words:** Szegö limit theorem, convolution operator, eigenvalues, spectral theory
**Mathematics Subject Classification:** Primary 47B35, Secondary 15A18

## 1. Introduction

The theory of regular discrete and integral convolutions and their applications is at present well advanced. The first results on the asymptotic behavior of the spectrum of discrete convolutions were published by Szegö in [1]. His paper was the start of investigations on this topic. However, most investigations concentrated on regular convolution operators.

Meanwhile, interest in generalized convolutions has also developed. We mention in this connection the papers [2-8,10], where generalized discrete convolutions of various types have been considered, and also the paper of H. Widom [11], where the generalized integral convolution operator has been studied.

The subject of this paper is the truncated generalized integral convolution operator $\mathcal{A}_\tau(a)$ ($\tau > 0$) acting on the space $L_2([0, \tau])$ as follows:

$$(\mathcal{A}_\tau(a)f)(t) = c_a f(t) + \int_0^\tau \widehat{a}\left(\frac{t}{\tau}, \frac{s}{\tau}, t - s\right) f(s)ds, \quad t \in [0, \tau], \ f \in L_2([0, \tau]).$$

The author investigates the asymptotic behavior of the spectrum of this operator for $\tau \to \infty$ by methods which were first applied in the joint paper [8] for studying the discrete analogue of the operator $\mathcal{A}_\tau(a)$.

## 2. Main results

Let us introduce the following notation:

$\mathbb{N}$, $\mathbb{Z}$, $\mathbb{R}$, $\mathbb{C}$ are the sets of natural, integer, real and complex numbers, respectively, $\mathbb{R}^+ = [0, +\infty)$, $\mathbb{R}^- = (-\infty, 0)$;

$\Sigma$ is the set of all measurable subsets of $\mathbb{R}$;

$L_p(U)$ for $p \geqslant 1$, $U \in \Sigma$ is the space of complex-valued functions $f$ defined and absolute integrable with power $p$ on $U$, with the norm

$$\|f\|_{L_p(U)} = \left( \int_U |f(t)|^p dt \right)^{1/p}.$$

$L_p = L_p(\mathbb{R})$;

$(\mathcal{F}f)(\xi)$ for $f(t) \in L_1 \cup L_2$ is the Fourier-transformation:

$$(\mathcal{F}f)(\xi) = \int_{\mathbb{R}} e^{i\xi t} f(t) dt, \quad \xi \in \mathbb{R};$$

$\mathrm{Hom}(K_1, K_2)$ is the Banach space of bounded linear operators from a Banach space $K_1$ to a Banach space $K_2$, $\mathrm{End}(K_1) = \mathrm{Hom}(K_1, K_1)$.

**Definition 2.1.** By $\mathrm{Kr}_{x,y}$ we denote the normed space of functions $a(x, y, \xi)$ defined on $[0, 1] \times [0, 1] \times \mathbb{R}$ and having the form $a(x, y, \xi) = c_a + (\mathcal{F}\hat{a})(x, y, \xi)$, where $c_a \in \mathbb{C}$, $\hat{a}(x, y, t)$ is a function defined on $[0, 1] \times [0, 1] \times \mathbb{R}$ satisfying the following conditions: $\sup_{x,y} |\hat{a}(x, y, t)|$ belongs to the space $L_1$ and

$$\int_{\mathbb{R}} |t| \sup_{x,y} |\hat{a}(x, y, t)|^2 dt < +\infty.$$

The norm in $\mathrm{Kr}_{x,y}$ is defined by

$$\|a\|_{\mathrm{Kr}_{x,y}} = |c_a| + \int_{\mathbb{R}} \sup_{x,y} |\hat{a}(x, y, t)| dt + \left( \int_{\mathbb{R}} |t| \sup_{x,y} |\hat{a}(x, y, t)|^2 dt \right)^{1/2}.$$

One can show that $\mathrm{Kr}_{x,y}$ is a Banach algebra. It can be regarded as a generalization of the Krein algebra $\mathrm{Kr}$, which will be separately considered below in Section 4.

By $\mathcal{A}_\tau(a) \in \mathrm{End}(L_2([0, \tau]))$, $\tau > 0$, for $a \in \mathrm{Kr}_{x,y}$, $a(x, y, \xi) = c_a + (\mathcal{F}\hat{a})(x, y, \xi)$ we denote the operator acting as follows:

$$(\mathcal{A}_\tau(a)f)(t) = c_a f(t) + \int_0^\tau \hat{a}\left(\frac{t}{\tau}, \frac{s}{\tau}, t - s\right) f(s) ds, \quad t \in [0, \tau], \ f \in L_2([0, \tau]).$$

The function $a(x, y, \xi)$ is called the symbol of the operator $\mathcal{A}_\tau(a)$, the function $\alpha_\tau(t, s) = \hat{a}\left(\frac{t}{\tau}, \frac{s}{\tau}, t - s\right)$ $(t, s \in [0, \tau])$ is its kernel.

$R(a)$ is the set of all $\lambda \in \mathbb{C}$ for which the following conditions are fulfilled:
1) $a(x, x, \xi) - \lambda \neq 0$ for all $x \in [0, 1]$, $\xi \in \mathbb{R}$,
2) the function $a(0, 0, \xi) - \lambda$ has winding number zero on $\mathbb{R}$.
We define $S(a) = \mathbb{C} \backslash R(a)$.
The following statement holds:

**Theorem 2.1.** *Let $a(x, y, \xi) \in \mathrm{Kr}_{x,y}$ have the form $a(x, y, \xi) = (\mathcal{F}\hat{a})(x, y, \xi)$ $(c_a = 0)$, where $\hat{a}(x, y, t)$ is continuous in the variables $x$, $y$ uniformly in $t$; let the function $a(x, x, \xi)$ belong to the space $L_1$ for every fixed $x \in [0, 1]$.*

*Suppose also that $\mathcal{A}_\tau(a) = \mathcal{A}_\tau(\mathcal{F}\hat{a})$ is a kernel operator for every $\tau > 0$.*

*Let a function $\varphi$ be analytic in some neighbourhood $U$ of the set $S(a)$ and $\varphi(0) = 0$.*

*Then there exists $\tau_0 > 0$ such that the spectrum of the operator $\mathcal{A}_\tau(a)$ for $\tau > \tau_0$ is contained in $U$. Furthermore, $\varphi(\mathcal{A}_\tau(a))$ is a kernel operator and the following asymptotic holds:*

$$\lim_{\tau \to \infty} \frac{1}{\tau} \operatorname{tr} \varphi(\mathcal{A}_\tau(a)) = \frac{1}{2\pi} \int_0^1 \left\{ \int_{\mathbb{R}} \varphi(a(x, x, \xi)) d\xi \right\} dx.$$

## 3. Some information on the operator ideals

The theory of operator ideals $\mathfrak{S}_p(H)$ $(p \in [1, +\infty))$, in particular, of the classes of kernel operators $\mathfrak{S}_1(H)$ and Hilbert-Schmidt operators $\mathfrak{S}_2(H)$, is described in detail in [12]. We cite here only the definitions and important facts.

Let $H$ be a separable Hilbert space. $\mathfrak{S}_\infty(H)$ is the set of all linear compact operators on $H$.

Let $\mathcal{A} \in \mathfrak{S}_\infty(H)$. The eigenvalues of the operator $(\mathcal{A}^*\mathcal{A})^{1/2}$ are called $s$-numbers of the operator $\mathcal{A}$. Let us denote by $\{s_j(\mathcal{A})\}_{j \in \mathbb{N}}$ the non-increasing sequence of the $s$-numbers of $\mathcal{A}$ (if the dimension of the image of $(\mathcal{A}^*\mathcal{A})^{1/2}$ is finite and equal to $n$ then we put $s_j(\mathcal{A}) = 0$ for $j > n$).

$\mathfrak{S}_p(H)$ for $p \in [1, +\infty)$ is the set of all compact operators $\mathcal{A}$ for which

$$\sum_{j=1}^{\infty} s_j^p(\mathcal{A}) < \infty.$$

The norm in $\mathfrak{S}_p(H)$ is defined as follows:

$$\|\mathcal{A}\|_{\mathfrak{S}_p(H)} = \left( \sum_{j=1}^{\infty} s_j^p(\mathcal{A}) \right)^{1/p}.$$

It should be mentioned that $\|\mathcal{A}\| \leqslant \|\mathcal{A}\|_{\mathfrak{S}_p(H)}$ for every $p \in [1, +\infty)$ (the operator norm is denoted by $\| \cdot \|$ as usually).

It is known that $\mathfrak{S}_p(H)$ is an ideal in $\mathrm{End}(H)$ possessing the following property:

**Proposition 3.1.** *If $\mathcal{A} \in \mathfrak{S}_p(H)$ and $\mathcal{B} \in \mathrm{End}(H)$ then $\mathcal{AB} \in \mathfrak{S}_p(H)$ and $\|\mathcal{AB}\|_{\mathfrak{S}_p(H)} \leqslant \|\mathcal{A}\|_{\mathfrak{S}_p(H)} \|\mathcal{B}\|$.*

380

We consider important facts about the class of kernel operators $\mathfrak{S}_1(H)$.

It is well known that the sum of all eigenvalues (taking multiplicities into account) of the operator $\mathcal{A} \in \mathfrak{S}_1(H)$ is finite. This sum is denoted by $\mathrm{tr}(\mathcal{A})$ (the trace of the operator).

The following properties hold:

1. $|\mathrm{tr}(\mathcal{A})| \leqslant ||\mathcal{A}||_{\mathfrak{S}_1(H)}$.

2. If $\mathcal{A} \in \mathfrak{S}_\infty(H)$, $\mathcal{B} \in \mathrm{End}(H)$ and, moreover, $\mathcal{AB} \in \mathfrak{S}_1(H)$, $\mathcal{BA} \in \mathfrak{S}_1(H)$, then $\mathrm{tr}(\mathcal{AB}) = \mathrm{tr}(\mathcal{BA})$.

3. If $\mathcal{A} \in \mathfrak{S}_1(H)$, $P_1, P_2 \in \mathrm{End}(H)$ and $P_1 P_2 = 0$, then $\mathrm{tr}(P_1 \mathcal{A} P_2) = 0$.

4. If $\mathcal{A} \in \mathfrak{S}_1(H)$, $P$ is orthogonal projector on $H$, then $\mathrm{tr}(P\mathcal{A}) = \mathrm{tr}(P\mathcal{A}P)$.

5. If $\mathcal{A} \in \mathfrak{S}_1(H)$ is a non-negative operator, then $\mathrm{tr}\,\mathcal{A} = ||\mathcal{A}||_{\mathfrak{S}_1(H)}$.

## 4. The Krein algebra

By $\mathfrak{K}$ we denote the set of functions $f(t)$ defined on $\mathbb{R}$ and square integrable with weight $|t|$. The norm in $\mathfrak{K}$ is defined as follows:

$$||f||_{\mathfrak{K}} = \left( \int_{\mathbb{R}} |t||f(t)|^2 dt \right)^{1/2}.$$

By Kr we denote the set of functions on $\mathbb{R}$ of the form $b(\xi) = c + (\mathcal{F}\hat{b})(\xi)$, $\xi \in \mathbb{R}$, where $\hat{b} \in L_1 \cap \mathfrak{K}$ and $c \in \mathbb{C}$ (by properties of the Fourier transformation, $c = b(\infty)$).

The set Kr with naturally defined algebraic operations and the norm

$$||b||_{\mathrm{Kr}} = |c| + ||\hat{b}||_{L_1} + ||\hat{b}||_{\mathfrak{K}}$$

(where $b(\xi) = c + (\mathcal{F}\hat{b})(\xi)$) is a Banach algebra with unity.

The algebra Kr was considered first by Krein (see [13], it is denoted there by $W_\cap$). It is stated there that Kr is the natural class of symbols for Szegö type theorems. The following statement is also proved in [13]:

**Theorem 4.1.** *The function $b \in$ Kr is invertible in the algebra Kr if and only if it satisfies the condition $b(\xi) \neq 0$ for every $\xi \in \mathbb{R}$.*

The set of all invertible elements in Kr is denoted by $G(\mathrm{Kr})$.

By $G_\pm(\mathrm{Kr})$ we denote the set of all elements $b \in$ Kr satisfying the following two conditions:

1) $b \in G(\mathrm{Kr})$,

2) the function $b(\xi)$ has winding number zero on $\mathbb{R}$.

Let $b \in$ Kr have the form $b(\xi) = b(\infty) + (\mathcal{F}\hat{b})(\xi)$ $(\hat{b} \in \mathfrak{K})$. $\mathcal{L}(b) \in \mathrm{End}(L_2)$ is the convolution operator defined by

$$(\mathcal{L}(b)f)(t) = b(\infty)f(t) + \int_{\mathbb{R}} \hat{b}(t-s)\,f(s)ds, \quad t \in \mathbb{R},\ f \in L_2.$$

The operator $\mathcal{L}(b)$ is invertible for $b \in G(\mathrm{Kr})$, since in this case $\mathcal{L}^{-1}(b) = \mathcal{L}(b^{-1})$.

Let us denote also by $P_{U,V} \in \mathrm{Hom}(L_2(U), L_2(V))$, $U, V \in \Sigma$, $V \subset U$, the projector on $L_2(V)$; $J_{V,U} \in \mathrm{Hom}(L_2(V), L_2(U))$, $U, V \in \Sigma$, $V \subset U$, is the operator of continuation by zero.

Furthermore let $\mathcal{L}_U(b) = P_{\mathbb{R},U}\mathcal{L}(b)J_{U,\mathbb{R}}$ for $U \in \Sigma$.

As it is known (see for example [14]), the operators $\mathcal{L}_{\mathbb{R}^-}(b)$ and $\mathcal{L}_{\mathbb{R}^+}(b)$ are invertible for $b \in G_\pm(\mathrm{Kr})$.

We note also that $\mathrm{Kr} \in \mathrm{Kr}_{x,y}$. Thus, the algebra $\mathrm{Kr}_{x,y}$ can be regarded as an expansion of the Krein algebra. It is the natural class of symbols for the generalized convolutions.

## 5. The invertibility of the operator $\mathcal{A}_\tau(a)$. The estimation of the trace of the inverse operator.

The important question of this paper is the invertibility of the operator $\mathcal{A}_\tau(a)$. For the solution of this problem we use a so-called almost inverse operator. The principle of its construction is analogous to that in the discrete case (see [8]). The proof of the corresponding proposition for the almost inverse operator likewise does not differ in the main from the discrete situation. However, the author was guided by results from [15] for substantiating the details of the proof.

Let $a(x, y, \xi) \in \mathrm{Kr}_{x,y}$.

Let us denote by $a^\gamma$, $\gamma \in [0,1]$, the function defined on $\mathbb{R}$ as follows:

$$a^\gamma(\xi) = a(\gamma, \gamma, \xi), \quad \xi \in \mathbb{R}.$$

Suppose also that $a^\gamma(\xi) \in G(\mathrm{Kr})$ for every $\gamma \in [0,1]$ and $a^0(\xi) \in G_\pm(\mathrm{Kr})$. Obviously, in this case $a^1(\xi) \in G_\pm(\mathrm{Kr})$.

By $\mathcal{B}_{\tau,n}(a)$, $n \in \mathbb{N}$, we denote the operator acting on the space $L_2([0,\tau])$ of the following form:

$$\mathcal{B}_{\tau,n}(a) = J_{[0,\tau/n],[0,\tau]}P_{\mathbb{R}^+,[0,\tau/n]}\mathcal{L}_{\mathbb{R}^+}^{-1}\left(a^0\right)J_{[0,2\tau/n],\mathbb{R}^+}P_{[0,\tau],[0,2\tau/n]}$$

$$+ \sum_{k=2}^{n-1} J_{[(k-1)\tau/n,k\tau/n],[0,\tau]}P_{\mathbb{R},[(k-1)\tau/n,k\tau/n]}\mathcal{L}^{-1}\left(a^{(k-1)/n}\right)$$

$$\times J_{[(k-2)\tau/n,(k+1)\tau/n],\mathbb{R}}P_{[0,\tau],[(k-2)\tau/n,(k+1)\tau/n]}$$

$$+ J_{[(n-1)\tau/n,\tau],[0,\tau]}P_{(-\infty,\tau],[(n-1)\tau/n,\tau]}\mathcal{L}_{(-\infty,\tau]}^{-1}\left(a^1\right)$$

$$\times J_{[(n-2)\tau/n,\tau],(-\infty,\tau]}P_{[0,\tau],[(n-2)\tau/n,\tau]}.$$

**Proposition 5.1.** *Let $\mathcal{U}$ be a compact set in $\mathrm{Kr}_{x,y}$ consisting of elements $a(x,y,\xi) = c_a + (\mathcal{F}\hat{a})(x,y,\xi)$, where $\hat{a}(x,y,t)$ is continuous in the variables $x$, $y$ uniformly in $t$, which satisfy the following conditions:*

*1) $a^\gamma(\xi) \in G(\mathrm{Kr})$ for every fixed $\gamma \in [0,1]$,*

*2) $a^0(\xi) \in G_\pm(\mathrm{Kr})$.*

*Then for every $\varepsilon > 0$ there exists $n_0(\varepsilon) \in \mathbb{N}$ such that for every $n > n_0(\varepsilon)$ there* *is* $\tau(n) > 0$ *satisfying the following condition:*

$$\sup_{a \in \mathcal{U}} \sup_{\tau > \tau(n)} \|\mathcal{B}_{\tau,n}(a)\mathcal{A}_\tau(a) - E_\tau\| < \varepsilon,$$

*where* $E_\tau$ *is the identity operator on* $L_2([0,\tau])$.

It can be shown that if the conditions of Proposition 5.1 are fulfilled, the family of operators $\{\mathcal{B}_{\tau,n}(a)\}_{a \in \mathcal{U}, \tau > 0, n \in \mathbb{N}}$ is bounded. Taking this fact into account, we obtain the following result as a corollary from Proposition 5.1:

**Corollary 5.1.** *Let the conditions of Proposition 5.1 be fulfilled. Then:*

*1) there exists* $\tau_0 > 0$ *such that the operator* $\mathcal{A}_\tau(a)$ *is invertible for all* $\tau > \tau_0$, $a \in \mathcal{U}$;

*2) for every* $\varepsilon > 0$ *there exists* $n_0(\varepsilon) \in \mathbb{N}$ *such that for every* $n > n_0(\varepsilon)$ *there is a corresponding* $\tau(n) > 0$ *satisfying the following condition:*

$$\sup_{a \in \mathcal{U}} \sup_{\tau > \tau(n)} \|\mathcal{A}_\tau^{-1}(a) - \mathcal{B}_{\tau,n}(a)\| < \varepsilon .$$

Let us denote by $\widetilde{\mathcal{B}}_{\tau,n}(a)$ the operators which can be obtained from $\mathcal{B}_{\tau,n}(a)$ when one replaces $\mathcal{L}_{\mathbb{R}^+}^{-1}(a^0)$ and $\mathcal{L}_{(-\infty,\tau]}^{-1}(a^1)$ by $\mathcal{L}^{-1}(a^0)$, $\mathcal{L}^{-1}(a^1)$, respectively. Using the properties of kernel operators (see [12]) and the results of the paper [15], one can show that

$$\sup_{a \in \mathcal{U}} \frac{1}{\tau} \left\| \widetilde{\mathcal{B}}_{\tau,n}(a) - \mathcal{B}_{\tau,n}(a) \right\|_{\mathfrak{S}_1(L_2[0,\tau])} \xrightarrow[\tau \to \infty]{} 0.$$

Furthermore, the following important statement on the trace of the operator $\mathcal{A}_\tau^{-1}(a)$ can be proved (the proof is laborious and will be omitted).

**Proposition 5.2.** *Let* $\mathcal{U}$ *be a compact in* $\mathrm{Kr}_{x,y}$ *consisting of functions of the form* $a(x,y,\xi) = c_a + (\mathcal{F}\hat{a})(x,y,\xi)$ *which satisfy the conditions of Proposition 5.1. Suppose also that the function* $(\mathcal{F}\hat{a})(x,x,\xi)$ *belongs to the space* $L_1$ *for any fixed* $x \in [0,1]$ *and* $\mathcal{A}_\tau(\mathcal{F}\hat{a})$ *is a kernel operator for every* $\tau > 0$.

*Then for every* $\varepsilon > 0$ *there exists* $n_0(\varepsilon) \in \mathbb{N}$ *such that for every* $n > n_0(\varepsilon)$ *there is a corresponding* $\tau(n) > 0$ *satisfying the following condition:*

$$\sup_{a \in \mathcal{U}} \sup_{\tau > \tau(n)} \left| \frac{1}{\tau} \mathrm{tr}\left( \mathcal{A}_\tau^{-1}(a) - \widetilde{\mathcal{B}}_{\tau,n}(a) \right) \right| < \varepsilon.$$

Let the conditions of Proposition 5.2 be fulfilled. Then for $n \in \mathbb{N}$ and all $\tau > 0$ for which the operator $\mathcal{A}_\tau(a)$ is invertible, we denote

$$\Delta_{\tau,n}(a) := \mathcal{A}_\tau^{-1}(a) - \widetilde{\mathcal{B}}_{\tau,n}(a)$$

and for other $\tau > 0$ we put $\Delta_{\tau,n}(a) = 0$.

## 6. An asymptotic representation of the operator $\varphi(\mathcal{A}_\tau(a))$

Let $a(x, y, \xi)$ be an element of $\mathrm{Kr}_{x,y}$ satisfying the conditions of Theorem 2.1.

It is clear that if $\Gamma$ is compact in $R(a)$, then the family of functions $\{a - \lambda\}_{\lambda \in \Gamma}$ is compact and satisfies the conditions of Proposition 5.2.

We present first an auxiliary result.

**Lemma 6.1.** *Let $a(x, y, \xi)$ be an element from $\mathrm{Kr}_{x,y}$ satisfying the conditions of Theorem 2.1; let $\Gamma$ be compact in $R(a)$. Then the value $\|\Delta_{\tau, n}(a - \lambda)\|_{\mathfrak{S}_1(L_2([0,\tau]))}$ is continuous in $\lambda \in \Gamma$ for all fixed $\tau > 0$, $n \in \mathbb{N}$.*

By $l_1(\mathbb{N})$ we denote the standard space of summable sequences $x = \{x_k\}_{k \in \mathbb{N}}$ with the norm

$$\|x\|_{l_1(\mathbb{N})} = \sum_{k \in \mathbb{N}} |x_k|.$$

Let $\{\psi_k\}_{k \in \mathbb{N}}$ be an orthogonal normalized basis in $L_2([0, \tau])$. Since

$$\sum_{k=1}^{\infty} |(\Delta_{\tau, n}(a - \lambda)\psi_k, \psi_k)| \leqslant \|\Delta_{\tau, n}(a - \lambda)\|_{\mathfrak{S}_1(L_2([0,\tau]))},$$

then (according to Lemma 6.1) the sequences $\{(\Delta_{\tau, n}(a - \lambda)\psi_k, \psi_k)\}_{k \in \mathbb{N}}$ for $\lambda \in \Gamma$ generate a compact set in $l_1(\mathbb{N})$ ($\tau > 0$ and $n \in \mathbb{N}$ are fixed here).

Therefore the series

$$\mathrm{tr}\, \Delta_{\tau, n}(a - \lambda) = \sum_{k=1}^{\infty} (\Delta_{\tau, n}(a - \lambda)\psi_k, \psi_k) \tag{1}$$

converges absolutely and uniformly in $\lambda \in \Gamma$.

**Proposition 6.1.** *Let $a \in \mathrm{Kr}_{x,y}$ satisfy the conditions of Theorem 2.1; let the function $\varphi(z)$ be analytic in a domain $U$ containing the set $S(a)$. Then:*

*1) the spectrum of the operator $A_\tau(a)$ is contained in $U$ for $\tau > 0$ large enough;*

*2) for every $\varepsilon > 0$ there exists $n_0(\varepsilon) \in \mathbb{N}$ such that for every $n > n_0(\varepsilon)$ there is $\tau(n) > 0$ such that the following asymptotic representation holds for $\tau > \tau(n)$:*

$$\varphi(\mathcal{A}_\tau(a)) = J_{[0, \tau/n], [0, \tau]} P_{\mathbb{R}^+, [0, \tau/n]} \mathcal{L}\left(\varphi(a^0)\right) J_{[0, 2\tau/n], \mathbb{R}^+} P_{[0, \tau], [0, 2\tau/n]}$$

$$+ \sum_{k=2}^{n-1} J_{[(k-1)\tau/n, k\tau/n], [0, \tau]} P_{\mathbb{R}, [(k-1)\tau/n, k\tau/n]} \mathcal{L}\left(\varphi\left(a^{(k-1)/n}\right)\right)$$

$$\tag{2}$$

$$\times J_{[(k-2)\tau/n, (k+1)\tau/n], \mathbb{R}} P_{[0, \tau], [(k-2)\tau/n, (k+1)\tau/n]}$$

$$+ J_{[(n-1)\tau/n, \tau], [0, \tau]} P_{(-\infty, \tau], [(n-1)\tau/n, \tau]} \mathcal{L}\left(\varphi(a^1)\right)$$

$$\times J_{[(n-2)\tau/n, \tau], (-\infty, \tau]} P_{[0, \tau], [(n-2)\tau/n, \tau]} + \Delta_{\varphi, \tau, n}(a),$$

*where $\Delta_{\varphi, \tau, n}(a) \in \mathrm{End}(L_2([0, \tau]))$ and $\frac{1}{\tau} |\mathrm{tr}\, \Delta_{\varphi, \tau, n}(a)| < \varepsilon$ (for $\tau > \tau(n)$).*

**Proof:** The first statement follows directly from Theorem 5.1. We prove the second one.

Let $\varepsilon > 0$. Let $U_1$ be a set satisfying the condition $S(a) \subset U_1 \subset \overline{U_1} \subset U$ ($\overline{U_1}$ denotes the conclusion of $U_1$). Let also $\Gamma$ be a Jordan curve in $U \backslash U_1$.

Let us consider the family $\mathcal{U}$ of functions having the form $a - \lambda$, $\lambda \in \Gamma$. It is compact in $\mathrm{Kr}_{x,y}$. Let $\tau_0' > 0$ be such that the operator $\mathcal{A}_\tau(a - \lambda)$ is invertible for all $\tau > \tau_0'$, $\lambda \in \Gamma$.

Let us consider the representation

$$\varphi(z) = \int_\Gamma \frac{\varphi(\lambda)}{\lambda - z} d\lambda, \quad z \in S(a).$$

For $\tau > \tau_0'$ we obtain

$$\varphi(\mathcal{A}_\tau(a)) = -\int_\Gamma \varphi(\lambda)(\mathcal{A}_\tau(a) - \lambda E_\tau)^{-1} d\lambda = -\int_\Gamma \varphi(\lambda)\mathcal{A}_\tau^{-1}(a - \lambda) d\lambda.$$

In accordance with Proposition 5.2 there exists $n(\varepsilon) \in \mathbb{N}$ such that for every $n > n(\varepsilon)$ there is a corresponding $\tau(n) > 0$ such that for every $\tau > \tau(n)$ the operator $\mathcal{A}_\tau(a - \lambda)$ can be represented in the form

$$\mathcal{A}_\tau^{-1}(a - \lambda) = \widetilde{\mathcal{B}}_{\tau, n}(a - \lambda) + \Delta_{\tau, n}(a - \lambda),$$

where $\displaystyle \sup_{\lambda \in \Gamma} \frac{1}{\tau} |\mathrm{tr}\, \Delta_{\tau, n}(a - \lambda)| < \frac{\varepsilon}{\mathrm{mes}(\Gamma) \max_{\lambda \in \Gamma} \varphi(\lambda)}$ and $\mathrm{mes}(\Gamma)$ is the length of the curve $\Gamma$.

It should be noted, that for every $b \in \mathrm{Kr}$

$$\varphi(\mathcal{L}(b)) = -\int_\Gamma \varphi(\lambda)\mathcal{L}^{-1}(b - \lambda) d\lambda = \mathcal{L}(\varphi(b)).$$

Then

$$\int_\Gamma \varphi(\lambda)\widetilde{\mathcal{B}}_{\tau, n}(a - \lambda) d\lambda =$$

$$= J_{[0, \tau/n], [0, \tau]} P_{\mathbb{R}^+, [0, \tau/n]} \mathcal{L}\left(\varphi(a^0)\right) J_{[0, 2\tau/n], \mathbb{R}^+} P_{[0, \tau], [0, 2\tau/n]}$$

$$+ \sum_{k=2}^{n-1} J_{[(k-1)\tau/n, k\tau/n], [0, \tau]} P_{\mathbb{R}, [(k-1)\tau/n, k\tau/n]} \mathcal{L}\left(\varphi\left(a^{(k-1)/n}\right)\right)$$

$$\times J_{[(k-2)\tau/n, (k+1)\tau/n], \mathbb{R}} P_{[0, \tau], [(k-2)\tau/n, (k+1)\tau/n]}$$

$$+ J_{[(n-1)\tau/n, \tau], [0, \tau]} P_{(-\infty, \tau], [(n-1)\tau/n, \tau]} \mathcal{L}\left(\varphi(a^1)\right)$$

$$\times J_{[(n-2)\tau/n, \tau], (-\infty, \tau]} P_{[0, \tau], [(n-2)\tau/n, \tau]}.$$

Let us denote

$$\Delta_{\varphi, \tau, n}(a) = -\int_\Gamma \varphi(\lambda)\Delta_{\tau, n}(a - \lambda) d\lambda.$$

Since the series (1) converges absolutely and uniformly in $\lambda \in \Gamma$, the following estimates hold:

$$\frac{1}{\tau}|\operatorname{tr}\Delta_{\varphi,\tau,n}(a)| \leqslant \frac{1}{\tau}\int_{\Gamma}|\varphi(\lambda)||\operatorname{tr}\Delta_{\tau,n}(a-\lambda)|d\lambda < \varepsilon.$$

Thus, the statement is proved. ∎

## 7. The proof of Theorem 2.1

In this section we prove the main result of this paper. First, we mention the well known fact that $\mathcal{L}_{[\alpha,\beta]}(b)$ is the kernel operator for $b \in \operatorname{Kr} \cap L_1$ and $0 \leqslant \alpha < \beta$ and

$$\operatorname{tr}\mathcal{L}_{[\alpha,\beta]}(b) = (\beta - \alpha)\frac{1}{2\pi}\int_{\mathbb{R}}b(\xi)d\xi.$$

Let us start with the proof.

**Proof of Theorem 2.1:** It is clear that $0 \in S(a)$. Since $\varphi(0) = 0$, the function $\varphi(z)$ can be represented in the form $\varphi(z) = z\,\varphi_1(z)$, where $\varphi_1(z)$ is analytic in $U$.

Then $\varphi(A_\tau(a)) = A_\tau(a)\,\varphi_1(A_\tau(a)) \in \mathfrak{S}_1(L_2([0,\tau]))$, because $A_\tau(a) \in \mathfrak{S}_1(L_2([0,\tau]))$.

Let us calculate the trace of the operator $\varphi(A_\tau(a))$ applying the representation (2). For every $\varepsilon > 0$ there exists $n_0(\varepsilon) \in \mathbb{N}$ and for every $n > n_0(\varepsilon)$ there is a corresponding $\tau(n) > 0$ such that the following representation holds for $\tau > \tau(n)$:

$$\operatorname{tr}\varphi(A_\tau(a)) = \tau\frac{1}{n}\frac{1}{2\pi}\int_{\mathbb{R}}\varphi\left(a^0(\xi)\right)d\xi + \tau\sum_{k=2}^{n-1}\frac{1}{n}\frac{1}{2\pi}\int_{\mathbb{R}}\varphi\left(a^{(k-1)/n}(\xi)\right)d\xi$$

$$+ \tau\frac{1}{n}\frac{1}{2\pi}\int_{\mathbb{R}}\varphi\left(a^1(\xi)\right)d\xi + \operatorname{tr}\Delta_{\varphi,\tau,n}(a),$$

where $\Delta_{\varphi,\tau,n}(a) \in \operatorname{End}(L_2([0,\tau]))$ and $\frac{1}{\tau}|\operatorname{tr}\Delta_{\varphi,\tau,n}(a)| < \frac{\varepsilon}{2}$ (for $\tau > \tau(n)$).

Let us fix $n' > n_0(\varepsilon)$ such that

$$\left|\frac{1}{2\pi}\int_0^1\left\{\int_{\mathbb{R}}\varphi(a(x,x,\xi))d\xi\right\}dx - \left(\frac{1}{n}\frac{1}{2\pi}\int_{\mathbb{R}}\varphi\left(a^0(\xi)\right)d\xi\right.\right.$$

$$\left.\left.+\sum_{k=2}^{n-1}\frac{1}{n}\frac{1}{2\pi}\int_{\mathbb{R}}\varphi\left(a^{(k-1)/n}(\xi)\right)d\xi + \frac{1}{n}\frac{1}{2\pi}\int_{\mathbb{R}}\varphi\left(a^1(\xi)\right)\right)d\xi\right| < \frac{\varepsilon}{2},$$

Let $\tau_0 = \tau(n')$. It is clear that for every $\tau > \tau_0$ the following inequality holds:

$$\left|\frac{1}{\tau}\operatorname{tr}\varphi(A_\tau(a)) - \frac{1}{2\pi}\int_0^1\left\{\int_{\mathbb{R}}\varphi(a(x,x,\xi))d\xi\right\}dx\right| < \varepsilon.$$

Thus, for every $\varepsilon > 0$ there exists $\tau_0 > 0$ such that for every $\tau > \tau_0$ hold

$$\left|\frac{1}{\tau}\operatorname{tr}\varphi(A_\tau(a)) - \frac{1}{2\pi}\int_0^1\left\{\int_{\mathbb{R}}\varphi(a(x,x,\xi))d\xi\right\}dx\right| < \varepsilon.$$

This proves the statement. ∎

## References

1. G. Szegö, *On certain hermitian forms associated with the Fourier series of positive functions.* Festkrift Marcel Riesz, Lund, 228-238 (1952).
2. T. Ehrhardt and B. Shao, *Asymptotic behavior of variable-coefficient Toeplitz determinants.* J. Math. Anal. Appl., Volume 7, Issue 1, 71-92 (2001).
3. D. Fasino and S. Serra Capizzano, *From Toeplitz matrix sequences to zero distribution of orthogonal polynomials.* Contemporary Math., vol. 323, pp. 329-339 (2003).
4. A. Kuijlaars and S. Serra Capizzano *Asymptotic zero distribution of orthogonal polynomials with discontinuously varying recurrence coefficiens.* J. Approx. Theory, vol. 113, pp. 142-155 (2001).
5. S. Serra Capizzano, *Generalized locally Toeplitz sequences: spectral analysis and applications to discretized partial differential equations.* Linear Algebra Appl., vol. 366, pp. 371-402 (2003).
6. P. Tilli, *Locally Toeplitz sequences: spectral properties and applications.* Linear Algebra Appl., vol. 278, pp. 91-120 (1998).
7. I. Simonenko, *Szegö-type limit theorems for determinants of truncated generalized multidimensional discrete convolutions.* Dokl. Rus. Akad. Nauk, vol. 373 (5), pp. 588-589 (2000).
8. O. Zabroda and I. Simonenko, *Asymptotic invertibility of truncated operators of one-dimensional generalized discrete convolution and the Szegö-type limit theorem.* Preprint 1230. Moscow: VINITI. Dep. 02.07.02, 21 p (2002).
9. O. Zabroda and I. Simonenko, *Asymptotic invertibility of truncated operators of one-dimensional generalized discrete convolution and the Szegö-type limit theorem. III.* Preprint 1557. Moscow: VINITI. Dep. 08.08.03, 24 p (2003).
10. H. Widom, *Szegö's theorem and a complete symbolic calculus for pseudo-differential operators.* Princeton University Press, Seminar on Singularities of Solutions (1978).
11. I. Gohberg and M. Krein, *Introduction to the theory of linear nonselfadjoint operators.* Nauka, Moscow (1965).
12. M. Krein, *On some new Banach algebras and Wiener-Levy theorems for Fourier series and integrals.* Math. research., vol. 1, pp. 82-109 (1966).
13. I. Gohberg and M. Krein, *Systems of integral equations on a half line with kernels depending on the difference of arguments.* Uspekhi Mat. Nauk., vol. XIII, issue 2 (80), pp. 3-72 (1958).
14. E. Maksimenko, *Asymptotic of the generalized trace of the truncated matrix convolution operators on widening segments.* Preprint 2228. Moscow: VINITI. Dep. 20.12.02, 24 p (2002).
15. I. Gohberg and I. Feldman, *Convolution equations and projection methods for their solution.* Nauka, Moscow (1971).
16. A. Kozak and I. Simonenko, *Projection methods for the solution of multidimensional discrete convolution equations.* Sibir. Mat. Zh., vol 21 (2), pp. 119-127 (1980).

# MULTIPLICATION AND TOEPLITZ OPERATORS ON THE ANALYTIC BESOV SPACES

N. ZORBOSKA*

*University of Manitoba,*
*Department of Mathematics,*
*Winnipeg, MB R3T 2N2, Canada*
*E-mail: zorbosk@cc.umanitoba.ca*

We give an integral description of the boundedness of multiplication and Toeplitz operators on the analytic Besov spaces.

**Key words:** Toeplitz operators, multipliers, Besov spaces
**Mathematics Subject Classification:** 47B35, 32A37

## 1. Introduction and Preliminaries

Toeplitz operators have provided one of the most explored examples of concrete operators on the spaces of analytic functions. Most of the existing results consider the classical Hardy space, or the classical and weighted Bergman spaces. In this paper we will investigate the basic problem of boundedness of the Toeplitz operators on the Besov spaces.

Let $\mathbb{D}$ be the unit disk in the complex plane and let $H(\mathbb{D})$ denote the space of functions analytic on $\mathbb{D}$. For $p > 1$, the analytic Besov space $B_p$ is defined by

$$B_p = \{g \in H(\mathbb{D}) : \|g\|_{B_p}^p = \int_{\mathbb{D}} (1 - |z|^2)^{p-2} |g'(z)|^p dA(z) < \infty\},$$

where $dA(z) = \frac{1}{\pi} dx dy$ is the normalized Lebesgue measure on $\mathbb{D}$.

Note that if $d\lambda = \frac{dA(z)}{(1-|z|^2)^2}$ denotes the Mobious invariant measure on $\mathbb{D}$, then g belongs to $B_p$ if and only if $(1 - |z|^2)|g'(z)|$ belongs to $L^p(d\lambda)$.

Let $P$ denote the classical Bergman projection operator

$$P(g)(z) = \int_{\mathbb{D}} \frac{g(z)}{(1 - \bar{w}z)^2} dA(w),$$

for $g \in L^1(dA)$. The proofs of the following few facts about Besov spaces can be found in [9].

---

*Work partially supported by an NSERC grant.

(i) $g \in B_p$ if and only if $g \in PL^p(d\lambda)$.

(ii) $B_2$ is the Dirichlet space $D$.

(iii) $B_\infty$ is the Bloch space $B$.

(iv) Each $B_p$ is a Banach space with norm $||| \ \ |||_p$ defined by $|||g|||_p = \|g\|_{B_p} + |g(0)|$

(v) For $1 < p < \infty, (B_p)^* = B_q$ where $\dfrac{1}{p} + \dfrac{1}{q} = 1$, and $B = (L_a^1(dA))^*$.

(vi) For $1 < p < q \leq \infty$, we have that $B_p \subset B_q$.

Let $f$ be in $L^1(dA)$. The Toeplitz operator induced by $f$ is defined by

$$T_f(g) = P(fg),$$

for every g in $B_p$. By the closed graph theorem, the Toeplitz operator $T_f$ is bounded on $B_p$ if and only if $P(fg)$ belongs to $B_p$, for every $g$ in $B_p$. In the case when $f$ is analytic, the Toeplitz operator $T_f$ is usually referred to as a multiplication operator $M_f$. So, as above, $M_f$ is bounded on $B_p$ if and only if $f$ belongs to the multiplier space $\mathcal{M}(B_p)$. Thus, determining when $M_f$ is bounded on $B_p$ is equivalent to determining the multiplier space for $B_p$.

The boundedness of Toeplitz operators on the Bloch space has already been addressed before. Arazy determined the bounded multiplication operators in [1].

**Theorem A[1]** *The multiplication operator $M_f$ is bounded on the Bloch space $B$ if and only if $f \in H^\infty$ and*

$$\sup_{z \in \mathbb{D}}(1 - |z|^2)|f'(z)| \log \frac{2}{1 - |z|^2} < \infty.$$

For a non-analytic, bounded $f$, the boundedness of $T_f$ on $B$ has been determined by Wu, Zhao and the author in [7].

**Theorem B[7]** *For $f \in L^\infty$, the Toeplitz operator $T_f$ is bounded on the Bloch space $B$ if and only if*

$$\sup_{z \in \mathbb{D}}(1 - |z|^2)|(Pf)'(z)| \log \frac{2}{1 - |z|^2} < \infty.$$

The result is a natural generalization of the analytic case. The same paper also contains a result on the boundedness of $T_f$ for a larger class of unbounded inducing functions $f$, on more general Bloch-type spaces.

For $1 < p < \infty$, there have been a number of papers on multiplier spaces of $B_p$. The first one is Stegenga's paper [5], which considers the case $p = 2$. It determines the multipliers of of the Dirichlet space in terms of Carleson measures and capacities. Other papers that use Carleson measure type of characterization of the multiplier space of $B_p$ are, for example, [3] and [6].

In this paper we will address the boundedness of the Toeplitz operator on Besov spaces with a description that is closer to the Bloch space approach in Theorems A and B from above.

Throughout the paper we will use a generic letter $c$ to denote a constant that might change its value from one line to another. We will use the notation $A \sim B$, whenever there exist two positive constants $c_1$ and $c_2$ such that $c_1 A \leq B \leq c_2 A$.

For $z, w$ in $\mathbb{D}$, let $\psi_w(z) = \frac{w-z}{1-\bar{w}z}$ be a *Möbius* transformation of $\mathbb{D}$. Let $\beta(z, w)$ denote the hyperbolic distance in $\mathbb{D}$ defined by $\beta(z, w) = \frac{1}{2} \log \frac{1+|\psi_z(w)|}{1-|\psi_z(w)|}$. For $r > 0$, let $D(w, r) = \{z \in \mathbb{D} : \beta(z, w) < r\}$ be a hyperbolic disk with radius $r$. We will use the fact that for $z \in D(w, r)$, we have that $|1 - \bar{w}z| \sim 1 - |z|^2 \sim 1 - |w|^2$. For a proof, and more on the hyperbolic metric see, for example, [9].

## 2. General Inducing Function

In this section we will give some sufficient and some neccessary conditions for the boundedness of the Toeplitz operator $T_f$ on $B_p$. If $f$ is bounded and if $p = \infty$, the boundedness of $T_f$ is determined by Theorem B. Note that in general, the inducing function $f$ does not have to be bounded, as already shown in [7]. The first theorem in this section shows that if, on the other hand, $f$ belongs to a class of functions with bounded oscilation, then $T_f$ bounded implies that f has to be bounded.

**Theorem 2.1** Let $BMO^1 = \{f \in L^1(D) : \sup_{w \in D} \|f \circ \psi_w - f(w)\|_{L^1} < \infty\}$, and let $f \in BMO^1$. If $T_f$ is bounded on $B_p$, then $f$ has to be bounded.

**Proof.** Let $h_w(z) = \frac{1-|w|^2}{1-\bar{w}z} \psi_w(z)$, for $w \in \mathbb{D}$. Note that $|h_w'(z)| \leq 3 \frac{1-|w|^2}{|1-\bar{w}z|^2}$ and, since $\int_{\mathbb{D}} \frac{(1-|z|^2)^{p-2}}{|1-\bar{w}z|^{2p}} dA(z) \leq \frac{c}{(1-|w|^2)^p}$, we have that $\|h_w\|_{B_p} \leq c$. Thus, if $T_f$ is bounded, $\|T_f h_w\|_{B_p} \leq c$. Also, $h_w'(w) = \frac{1}{|w|^2-1}$ and so

$$|f(w)| = \left(1 - |w|^2\right) |f(w) h_w'(w)|$$

$$\leq (1 - |w|^2) |(P(f h_w))'(w)| + (1 - |w|^2) |(P(f h_w))'(w) - f(w) h_w'(w)|$$

$$\leq c \left( \int_{D(w,r)} (1 - |z|^2)^{p-2} |P(f h_w)'(z)|^p dA(z) \right)^{\frac{1}{p}}$$

$$+ (1 - |w|^2) \left| \int_{\mathbb{D}} \frac{(f(u) - f(w)) h_w(u) \bar{u}}{(1 - \bar{u}w)^3} dA(u) \right|$$

$$\leq c \|T_f h_w\|_{B_p} + (1 - |w|^2)^2 \int_{\mathbb{D}} \frac{|f(u) - f(w)|}{|1 - \bar{u}w|^4} |\psi_w(u)| dA(u)$$

$$\leq c + \int_{\mathbb{D}} |f(u) - f(w)| \, |\psi_w'(u)|^2 dA(u)$$

$$\leq c + \int_{\mathbb{D}} |f \circ \psi_w(z) - f(w)| dA(u)$$

$$\leq c + \sup_{w \in \mathbb{D}} \|f \circ \psi_w(z) - f(w)\|_{L^1} < \infty.$$

$\square$

Next we will provide sufficient and nesseccary conditions for $T_f$ to be bounded, whenever $f$ is bounded and $2 < p < \infty$. We start first with a lemma.

**Lemma 2.2** *Let* $2 < p < \infty$ *and let* $g \in B_p$. *Then*

$$\int_{\mathbb{D}} (1 - |z|^2)^{p-2} \left( \int_{\mathbb{D}} \frac{|g(u) - g(z)|}{|1 - \bar{u}z|^3} dA(u) \right) dA(z) \le c \|g\|_{B_p}^p.$$

**Proof.** For $g \in B_p$, it has been proven in [9] that

$$\int_{\mathbb{D}} \int_{\mathbb{D}} \frac{|g(u) - g(z)|^p}{|1 - \bar{u}z|^4} dA(u) \, dA(z) \le c \|g\|_{B_p}^p.$$

Using Holder's inequality and Forelli-Rudin estimates (see [9]), we have that

$$\int_{\mathbb{D}} (1 - |z|^2)^{p-2} \left( \int_{\mathbb{D}} \frac{|g(u) - g(z)|}{|1 - \bar{u}z|^3} dA(u) \right) dA(z)$$

$$\le \int_{\mathbb{D}} (1 - |z|^2)^{p-2} \left( \int_{\mathbb{D}} \frac{dA(u)}{|1 - \bar{u}z|^{3q - 4\frac{q}{p}}} \right)^{\frac{p}{q}} \left( \int_{\mathbb{D}} \frac{|g(u) - g(z)|^p}{|1 - \bar{u}z|^4} dA(u) \right) dA(z)$$

$$\le c \int_{\mathbb{D}} (1 - |z|^2)^{p-2} \left( \frac{1}{(1 - |z|^2)^{2-q}} \right)^{\frac{p}{q}} \left( \int_{\mathbb{D}} \frac{|g(u) - g(z)|^p}{|1 - \bar{u}z|^4} dA(u) \right) dA(z)$$

$$\le c \int_{\mathbb{D}} \int_{\mathbb{D}} \frac{|g(u) - g(z)|^p}{|1 - \bar{u}z|^4} dA(u) dA(z) \le c \|g\|_{B_p}^p.$$

Note that we have used the fact that $3q - 4\frac{q}{p} - 2 = 2 - q > 0$, and that $(2 - q)\frac{p}{q} = p - 2$. □

**Theorem 2.3** *Let* $f \in L^\infty$ *and let* $2 < p < \infty$.

(i) *If* $T_f$ *is bounded on* $B_p$, *then* $f$ *is such that*

$$\sup_{w \in \mathbb{D}} \int_{D(w,r)} (1 - |z|^2)^{p-2} |(Pf)'(z)|^p \left( \log \frac{2}{1 - |z|^2} \right)^{p-1} dA(z) < \infty.$$

(ii) *If*

$$\int_{\mathbb{D}} (1 - |z|^2)^{p-2} |(Pf)'(z)|^p \left( \log \frac{2}{1 - |z|^2} \right)^{p-1} dA(z) < \infty,$$

*then* $T_f$ *is bounded on* $B_p$.

**Proof.**

(i) For $w$ in $\mathbb{D}$, the function $K_w(z) = \log \frac{2}{1 - \bar{w}z}$ belongs to $B_p$ and $\|K_w\|_{B_p} = \left( \log \frac{2}{1 - |w|^2} \right)^{\frac{1}{p}}$. Since $T_f$ is bounded on $B_p$ we have that

$$\|T_f K_w\|_{B_p}^p \le c \|K_w\|_{B_p}^p \le c \log \frac{2}{1 - |w|^2}.$$

On the other hand,

$$(T_f K_w)'(z) = (P(f K_w))'(z) = \int_{\mathbb{D}} \frac{f(u) K_w(u) \bar{u}}{(1 - \bar{u}z)^3} dA(u)$$

$$= \int_{\mathbb{D}} \frac{f(u)(K_w(u) - K_w(z)) \bar{u}}{(1 - \bar{u}z)^3} dA(u) + K_w(z) \int_{\mathbb{D}} \frac{f(u) \bar{u}}{(1 - \bar{u}z)^3} dA(u).$$

Thus

$$\int_{\mathbb{D}} (1 - |z|^2)^{p-2} |(Pf)'(z)|^p \left| \log \frac{2}{1 - \bar{w}z} \right|^p dA(z)$$

$$= \int_{\mathbb{D}} (1 - |z|^2)^{p-2} \left| \int_{\mathbb{D}} \frac{f(u) \bar{u}}{(1 - \bar{u}z)^3} dA(u) \right|^p |K_w(z)|^p dA(z)$$

$$\leq \int_{\mathbb{D}} (1 - |z|^2)^{p-2} |P(f K_w)'(z)|^p dA(z)$$

$$+ \int_{\mathbb{D}} (1 - |z|^2)^{p-2} \left| \int_{\mathbb{D}} \frac{f(u)(K_w(u) - K_w(z)) \bar{u}}{|1 - \bar{u}z|^3} dA(u) \right|^p dA(z)$$

$$\leq \|T_f K_w\|_{B_p}^p + \|f\|_{\infty}^p \int_{\mathbb{D}} (1 - |z|^2)^{p-2} \left( \int_{\mathbb{D}} \frac{|K_w(u) - K_w(z)| \bar{u}}{|1 - \bar{u}z|^3} dA(u) \right)^p dA(z)$$

$$\leq c \|K_w\|_{B_p}^p + I_1.$$

The integral in $I_1$ represents the integral in Lemma 2.3, with the function $g$ replaced by $K_w$. Hence, using Lemma 2.2, we have that $I_1 \leq c \|f\|_{\infty}^p \|K_w\|_{B_p}^p$, and so we get that

$$\int_{\mathbb{D}} (1 - |z|^2)^{p-2} |(Pf)'(z)|^p \left| \log \frac{2}{1 - \bar{w}z} \right|^p dA(z) \leq c \log \frac{2}{1 - |w|^2}.$$

Since for $r > 0$ and $z$ in $D(w, r)$, we have that $|1 - \bar{w}z| \sim 1 - |z|^2 \sim 1 - |w|^2$, we have that

$$\sup_{w \in \mathbb{D}} \int_{D(w,r)} (1 - |z|^2)^{p-2} |(Pf)'(z)|^p \left( \log \frac{2}{1 - |z|^2} \right)^{p-1} dA(z)$$

$$\leq \sup_{w \in \mathbb{D}} \left( \log \frac{2}{1 - |w|^2} \right)^{-1} \int_{\mathbb{D}} (1 - |z|^2)^{p-2} |(Pf)'(z)|^p \left| \log \frac{2}{1 - \bar{w}z} \right|^p dA(z) \leq c.$$

(ii) We will use that $P(fg)'(z) = \int_{\mathbb{D}} \frac{f(u)(g(u) - g(z)) \bar{u}}{(1 - \bar{u}z)^3} dA(u) + g(z)(Pf)'(z)$, and that for every $g$ in $B_p$,

$$|g(z)| \leq c \|g\|_{B_p} (\log \frac{2}{1 - |z|^2})^{\frac{1}{q}}.$$

(For the proof on the growth of $B_p$ functions see, for example [9].)

392

Thus, we have that

$$\|T_f g\|^p_{B_p} = \int_{\mathbb{D}} (1-|z|^2)^{p-2} \, |P(fg)'(z)| \, P dA(z)$$

$$\leq \|f\|_\infty \int_{\mathbb{D}} (1-|z|^2)^{p-2} \left( \int_{\mathbb{D}} \frac{|g(u)-g(z)|}{|1-\bar{u}z|^3} dA(u) \right) dA(z)$$

$$+\|g\|^p_{B_p} \int_{\mathbb{D}} (1-|z|^2)^{p-1} \left( \log \frac{p}{1-|z|^2} \right)^{p-1} |(Pf)'(z)|^p dA(z) = I_1 + I_2.$$

By Lemma 2.2, we have that $I_1 \leq c\|f\|_\infty \|g\|^p_{B_p}$. The condition under (ii) implies that $I_2 \leq c\|g\|^p_{B_p}$, and so $\|T_f g\|^p_{B_p} \leq c\|g\|^p_{B_p}$. $\qquad\square$

**Question 1.** Is it true that for $f$ bounded and $p > 2$ we have that $T_f$ is bounded on $B_p$ if and only if $\int_{\mathbb{D}} (1-|z|^2)^{p-2} |(Pf)'(z)|^p \left( \log \frac{2}{1-|z|^2} \right)^{p-1} dA(z) < \infty$? Is the same true for $1 < p \leq 2$?

Next we consider some special non-analytic cases for the inducing function of the Toeplitz operator.

**Theorem 2.4** Let $f$ be such that $\bar{f}$ is analytic. Then $T_f$ is bounded on $B_p$ if and only if $\bar{f} \in H^\infty$.

**Proof.** For $f \in L^1(dA)$, $f$ analytic and $g_w(z) = \frac{1}{(1-\bar{w}z)^2}$, we have that

$$P(fg_w)(z) = \int_{\mathbb{D}} \frac{f(u)g_w(u)}{(1-\bar{u}z)^2} dA(u)$$

$$= \int_{\mathbb{D}} \frac{f(u)}{(1-u\bar{z})^2} \frac{1}{(1-\bar{u}w)^2} dA(u) = \frac{f(w)}{(1-\bar{z}w)^2} = f(w)g_w(z).$$

Thus $T_f(g_w) = f(w)g_w$.

Let $T_f$ be bounded on $B_p$. Since for every $w \in \mathbb{D}$, $g_w(z)$ belongs to $B_p$, we get from above that $|f(w)| \leq \|T_f\|$, i.e. that $\bar{f} \in H^\infty$.

On the other hand, if we assume that $\bar{f} \in H^\infty$, we want to prove that for every $g \in B_p$, we have that $P(fg) \in B_p$, i.e. that $(1-|z|^2)|(P(fg))'(z)| \in L^p(d\lambda)$.

Using the integral representation from Lemma 4.2.8 in[9] for the analytic function g, we have that

$$(1-|z|^2)|P(fg)'(z)| = (1-|z|^2) \left| \int_{\mathbb{D}} \frac{f(u)g(u)\bar{u}}{(1-\bar{u}z)^3} dA(u) \right|$$

$$= (1-|z|^2) \left| \int_{\mathbb{D}} \frac{f(u)\bar{u}}{(1-\bar{u}z)^3} \left( \int_{\mathbb{D}} \frac{(1-|w|^2)g'(w)}{\bar{w}(1-\bar{w}u)^2} dA(w) \right) dA(u) \right|$$

$$= (1-|z|^2) \left| \int_{\mathbb{D}} \frac{(1-|w|^2)g'(w)}{\bar{w}} \left( \int_{\mathbb{D}} \frac{\overline{f(u)}u}{(1-\bar{z}u)^3} \frac{1}{(1-\bar{u}w)^2} dA(u) \right) dA(w) \right|$$

$$= (1-|z|^2) \left| \int_{\mathbb{D}} \frac{(1-|w|^2)g'(w)f(w)}{(1-\bar{w}z)^3} dA(w) \right|.$$

Thus, by Theorem 1.9 from [4], we have that $(1 - |z|^2)P(fg)'(z) \in L^p(d\lambda)$ whenever $(1 - |w|^2)g'(w)f(w) \in L^p(d\lambda)$. But this is certainly satisfied whenever $\bar{f} \in H^\infty$ and $g \in B^p$, since then

$$\int_{\mathbb{D}} (1 - |w|^2)^{p-2}|g'(w)|^p|f(w)|^p dA(w) \le \|\bar{f}\|_\infty^p \|g\|_{B^p}^p. \qquad \square$$

Note that for the case when $p = \infty$, Theorem 2.4 can be seen to be true directly by using that $(L_a^1)^* = B$, that $(T_{\bar{f}})^* = T_f$ and that the multiplier space of $L_a^1$ is $H^\infty$. The duality for $1 < p < \infty$ though is achieved by a more complicated pairing, and the adjoint of $T_f$ does not have such a simple form, as when $p = \infty$.

**Corollary 2.5** *Let $f$ be harmonic.*

(i) *If $T_f$ and $T_{\bar{f}}$ are both bounded on $B_p$, then $f$ has to be bounded.*

(ii) *If $Pf$ and $P\bar{f}$ are both in $\mathcal{M}(B_p)$, then $T_f$ and $T_{\bar{f}}$ are both bounded on $B_p$.*

(iii) *If $f$ is real harmonic, then $T_f$ is bounded if and only if $Pf \in \mathcal{M}(B_p)$.*

**Proof.** Since $f$ is harmonic, $f = h + \bar{g}$, where $h$ and $g$ are analytic and $g(0) = 0$, and so $Pf = h$ and $P\bar{f} = g$.

(i) If $T_f$ and $T_{\bar{f}}$ are both bounded on $B_p$, then $h$ and $g$ belong to $B_p$. But $BMO^1 \cap H(\mathbb{D}) = B$ and so $h$ and $\bar{g}$ belong to $BMO^1$. Thus, $f$ belongs to $BMO^1$, and by Theorem 2.1, $f$ has to be bounded.

(ii) We have that $M_h$ and $M_g$ are bounded on $B_p$. Since also $\bar{h}$ and $\bar{g}$ have to be bounded, by Theorem 2.4, $T_{\bar{h}}$ and $T_{\bar{g}}$ are also bounded. Thus, $T_f$ and $T_{\bar{f}}$ are both bounded.

(iii) Follows from (ii) and the fact that $f = h + \bar{h}$, with $h$ analytic. $\qquad \square$

Note that when $p = \infty$, part (iii) of Corollary 2.5 has been proven in [2].

## 3. Analytic inducing function

Throughout this section we will assume that the inducing function for the Toeplitz operator is always analytic. Recall that in that case, the operator $M_f$ is bounded on $B_p$ if and only if $f$ belongs to the multiplier space of $B_p$. Since Theorem A determines the multiplier space for $B_p$ when $p = \infty$, let us also assume for the rest of this section that $1 < p < \infty$.

For analytic $f$, we have that $Pf = f$ and thus, replacing $(Pf)'$ with $f'$ in Theorem 2.3, we get a necessary and a sufficient condition for $f$ to belong to the multiplier space of $B_p$, for $p > 2$. Note that if $f$ is in the multiplier space of $B_p$, then $f \in B_p \subset B = BMO^1 \cap H(\mathbb{D})$, and so, by Theorem 2.1, $f$ has to be bounded.

394

The boundedness of the multiplier functions can also be proven by a purely Banach space theory methods. See, for example, [9].

In this section, by using the analyticity of $f$, we will extend Theorem 2.3 to the remaining case $p \leq 2$.

**Theorem 3.1** *Let $f$ be analytic and bounded, and let $1 < p < \infty$.*

*(i) If $M_f$ is bounded on $B_p$, then $f$ is such that*

$$\sup_{w \in \mathbb{D}} \int_{D(w,r)} (1 - |z|^2)^{p-2} |f'(z)|^p \left( \log \frac{2}{1 - |z|^2} \right)^{p-1} dA(z) < \infty.$$

*(ii) If*

$$\int_{\mathbb{D}} (1 - |z|^2)^{p-2} |f'(z)|^p \left( \log \frac{2}{1 - |z|^2} \right)^{p-1} dA(z) < \infty,$$

*then $M_f$ is bounded on $B_p$.*

**Proof.** For f and g in $B_p$, we have that $P(fg)'(z) = (fg)'(z) = f'(z)g(z) + f(z)g'(z)$. Also, as before, $|g(z)| \leq c\|g\|_{B_p} (\log \frac{2}{1-|z|^2})^{\frac{1}{q}}$.

(i) Let $M_f$ be bounded on $B_p$ and let $K_w(z) = \log \frac{2}{1-\bar{w}z}$. Recall that $\|K_w\|_{B_p} = c \left( \log \frac{2}{1-|w|^2} \right)^{\frac{1}{p}}$ and so

$$\int_{\mathbb{D}} (1 - |z|^2)^{p-2} |f'(z)|^p \left| \log \frac{2}{1 - \bar{w}z} \right|^p dA(z)$$

$$\leq \int_{\mathbb{D}} (1 - |z|^2)^{p-2} |(fK_w)'(z)|^p dA(z) + \int_{\mathbb{D}} (1 - |z|^2)^{p-2} |f(z)|^p |K'_w(z)|^p dA(z)$$

$$\leq \|M_f K_w\|_{B_p}^p + \|f\|_\infty^p \|K_w\|_{B_p}^p \leq c \log \frac{2}{1 - |w|^2},$$

where c does not depend on w. Thus, same as in the proof of Theorem 2.3, we have that

$$\sup_{w \in \mathbb{D}} \int_{D(w,r)} (1 - |z|^2)^{p-2} |f'(z)|^p \left( \log \frac{2}{1 - |z|^2} \right)^{p-1} dA(z)$$

$$\leq \sup_{w \in \mathbb{D}} \left( \log \frac{2}{1 - |w|^2} \right)^{-1} \int_{\mathbb{D}} (1 - |z|^2)^{p-2} |f'(z)|^p \left| \log \frac{2}{1 - \bar{w}z} \right|^p dA(z) < \infty.$$

(ii) Let g be a function in $B_p$. Then

$$\|M_f g\|_{B_p}^p = \int_{\mathbb{D}} (1 - |z|^2)^{p-2} |(fg)'(z)|^p dA(z)$$

$$\leq c \int_{\mathbb{D}} (1 - |z|^2)^{p-2} |f'(z)g(z)|^p dA(z) + c\|f\|_\infty^p \int_{\mathbb{D}} (1 - |z|^2)^{p-2} |g'(z)|^p dA(z)$$

$$\leq c\|g\|_{B_p}^p \int_{\mathbb{D}} (1 - |z|^2)^{p-2} |f'(z)|^p \left( \log \frac{2}{1 - |z|^2} \right)^{\frac{p}{q}} dA(z) + c\|f\|_\infty^p \|g\|_{B_p}^p$$

$$\leq c\|g\|_{B_p}^p,$$

where we have used that $\frac{p}{q} = p - 1$. Thus $M_f$ is bounded on $B_p$. $\qquad\square$

**Corollary 3.2** *If $M_f$ is bounded on $B_p$, then $f \in H^\infty$ and*

$$\sup_{z \in \mathbb{D}}(1 - |z|^2)|f'(z)| \left( \log \frac{2}{1 - |z|^2} \right)^{1 - \frac{1}{p}} < \infty.$$

**Proof.** The proof follows from part (i) of Theorem 3.1. and the fact that, since $f'(z)$ and $\log \frac{2}{1 - \bar{w}z}$ are analytic functions, we have that

$$(1 - |z|^2)^p|f'(z)|^p \left( \log \frac{2}{1 - |z|^2} \right)^{p-1}$$

$$\leq \frac{1}{(1 - |z|^2)^2} \int_{D(z,r)} (1 - |w|^2)^p|f'(w)|^p \left| \log \frac{2}{1 - \bar{w}z} \right|^{p-1} dA(w)$$

$$\leq \left( \log \frac{2}{1 - |z|^2} \right)^{-1} \int_{D(z,r)} (1 - |w|^2)^{p-2}|f'(w)|^p \left| \log \frac{2}{1 - \bar{w}z} \right|^p dA(w) < \infty. \qquad\square$$

**Remark 1.** Let $F_{f,p}(z) = (1 - |z|^2)|f'(z)| \left( \log \frac{2}{1 - |z|^2} \right)^{1 - \frac{1}{p}}$. Note that, since $\lambda$ is not a finite measure on $\mathbb{D}$, the fact that $F_{f,p} \in L^p(d\lambda)$ implies that $F_{f,p} \in L^\infty$. The converse is not always true; i.e., when $p < \infty$, there are bounded functions that do not belong to $L^p(d\lambda)$. For the limiting case $p = \infty$, the two spaces coincide and we get (after replacing $\frac{1}{p}$ by 0) that the condition in part (ii) of Theorem 3.1 and the condition in Corollary 3.2 are the same, and are equal to the Arazy's condition in Theorem A.

**Remark 2.** Note that from the proof of Theorem 3.1, and from the fact that $(1 - |w|^2) \sim |1 - \bar{w}z| \sim (1 - |w|^2)$, for every $z$ in $D(w,r)$, we also get that if $f \in \mathcal{M}(B_p)$, then, for $r > 0$ we have that

$$\sup_{w \in \mathbb{D}} \left( \log \frac{1}{1 - |w|^2} \right)^{p-1} \frac{1}{(1 - |w|^2)^2} \int_{D(w,r)} (1 - |z|^2)^p|f'(z)|^p dA(z) < \infty.$$

Let $\mu_{f,p}$ denote the measure defined by $d\mu_{f,p}(z) = (1 - |z|^2)^p|f'(z)|^p$. Then the above condition is a Carleson measure condition. More precisely, the above condition is equivalent to saying that the measure $\mu_{f,p}$ is a 2-Carleson, $p-1$ logarithmic measure, as defined in [8].

The same result in a slightly different form also appears in [6]. Namely, if $M_f$ is bounded on $B_p$, Lemma 3.1 from [6] essentially shows that the measure $\mu_f$ is a 2-Carleson, $p-1$ logarithmic measure. The same paper has a further characterization of the multipiers of $B_p$ in terms of capacities, which is a generalization of the result in [5] for the case $p = 2$. The latter paper provides an example of a bounded analytic function $f$ such that $\mu_f$ is a 2-Carleson, 1-logarithmic measure, but $M_f$ is not bounded on $B_2$. Thus, Carleson measure conditions are not the right way of describing the multiplier space of a Besov space, at least not when $p = 2$.

396

**Question 2.** For $f$ analytic and $1 < p < \infty$, is it true that $M_f$ is bounded on $B_p$ if and only if $f \in H^\infty$ and $(1 - |z|^2)|f'(z)| \left( \log \frac{2}{1-|z|^2} \right)^{1-\frac{1}{p}} \in L^p(d\lambda)$?

## References

1. J. Arazy, *Multipliers of Bloch functions,* University of Haifa Mathem. Public. Series 54,1982.
2. K.R.M. Attele, *Toeplitz and Hankel operators on Bergman one space,* Hokkaido Math. J. 21 (1992), 279-293.
3. N. Arcozzi, R. Rochberg and E. Sawyer,*Carleson measures for analytic Besov spaces,* Rev. Mat. Iberoamericana 18 (2002), 443-510.
4. H. Hedenmalm, B. Korenblum and K. Zhu, Theory of Bergman spaces, Springer, New York, 2000.
5. D.A. Stegenga, *Multipiliers of the Dirichlet spaces,* Illinois J. of Math. 24 (1980), 113-139.
6. Z. Wu, *Carleson measures and multipliers for Dirichlet spaces,* J. of Func. Anal. 169 (1999), 148-163.
7. Z. Wu, R. Zhao and N. Zorboska *Toeplitz operators on the Bloch-type spaces,* to appear in the Proc Amer. Math. Soc.
8. R. Zhao, *On logarithmic Carleson measures,* Acta Sci. Math. (Szeged) 69 (2003), 605-618.
9. K. Zhu, Operator Theory on Function Spaces, Marcel Dekker, New York, 1990.

## I.6 Wavelets

Organizers: R. Hochmuth, M. Holschneider

The aim of Section I.6 about wavelets was to bring together leading scientists and young researchers. The invited Keynote-Speaker was A. Tabacco (Turin). The considered topics covered in particular actual research on the continuous wavelet transform, which was directed to analytical questions, and on the discrete wavelet transform, which allows to establish efficient and fast numerical algorithms for treating problems from applications. The following three papers give a good impression about the variety of themes discussed in Catania.

The paper by F.P. Andriulli, A. Tabacco and G. Vecchi presents a multiresolution approach for solving an integral equation, which has applications in electromagnetics: it allows to calculate efficiently approximations of the electric field radiated by an antenna. In the second paper E. Cordero, F. De Mari and A. Tabacco prove, that the restriction of the metaplectic representation to a class of three dimensional subgroups of the symplectic group $Sp(2, \mathbb{R})$ gives rise to a reproducing formula for $L^2(\mathbb{R}^2)$ functions. Metaplectic representations are relevant in the context of harmonic analysis in phase space or time-frequency analysis. Finally, K. Saneva and A. Buckovska introduce the notion of S-asymptotic and investigate the asymptotic behaviour of the distributional wavelet transform at infinity.

# S-ASYMPTOTIC AND S-ASYMPTOTIC EXPANSION OF DISTRIBUTIONAL WAVELET TRANSFORM

K. SANEVA* and A. BUČKOVSKA

*Faculty of Electrical Engineering, University 'Ss. Cyril and Methodius',
1000 Skopje, Macedonia
E-mail: saneva@etf.ukim.edu.mk*

In this paper we investigated the asymptotic behaviour at infinity of the distributional wavelet transform. Using the notion of S-asymptotic at infinity of a distribution from $\mathcal{S}'(\mathbb{R})$ and $\mathcal{K}'_1(\mathbb{R})$ we obtained results for the ordinary asymptotic behaviour at infinity of its wavelet transform. We also analyzed the asymptotic expansion at infinity of the wavelet transform of distributions from $\mathcal{S}'(\mathbb{R})$ and $\mathcal{K}'_1(\mathbb{R})$ with appropriate S-asymptotic expansion.

**Key words:** Wavelet transform, S-asymptotic behaviour, S-asymptotic expansion
**Mathematics Subject Classification:** 46F12, 42C40

## 1. Introduction

The asymptotic behaviour of solutions of mathematical models, classical or generalized, using Abelian and Tauberian type theorems, has found applications in various fields of pure and applied mathematics, physics, and engineering. In the last three decades many definitions of asymptotic behaviour of distributions are elaborated and applied to analysis of the asymptotic behaviour of some integral transforms such as Laplace transform, Fourier transform, Stieltjes transform, Weierstrass transform. (See Refs. [1] and [2] for more details.)

In Ref. [3] we determined the ordinary asymptotic behaviour at 0 of the wavelet transform $\mathcal{W}_g f(b, a)$ with respect to the both variables $a$ and $b$ assuming that $f \in \mathcal{S}'(\mathbb{R})$ has quasiasymptotic behaviour at 0. The definition of quasiasymptotic behaviour at $0^+$ is extended in Ref. [4], and quasiasymptotic behaviour at $x_0^+$, $x_0 > 0$ of a distribution from $\mathcal{S}'_{x_0^+}(\mathbb{R})$ is defined. We applied this definition to analysis of the asymptotic behaviour of the distributional wavelet transform, and obtained result for the ordinary asymptotic behaviour at $0^+$ of the wavelet transform $\mathcal{W}_g f(b, a)$ with respect to the variable $a$.[4] We also analyzed[4] the ordinary asymptotic expansion at $0^+$ of the wavelet transform of distributions from $\mathcal{S}'_+(\mathbb{R})$ $(\mathcal{S}'_{b^+}(\mathbb{R}))$ with respect to the both variables $a$ and $b$ (with respect to the variable $a$) with appropriate quasiasymptotic expansion at $0^+$ (at $b^+$).

The asymptotic behaviour at infinity of the wavelet transform of an exponential distribution from $\mathcal{K}'_1(\mathbb{R})$ is analyzed in Ref. [5], and an Abelian type result for the

S-asymptotic behaviour of the wavelet transform of a distribution from $\mathcal{K}'_1(\mathbb{R})$ with appropriate S-asymptotic is proved.

In this paper we analyzed the asymptotic behaviour and the asymptotic expansion of distributional wavelet transform at infinity. The quasiasymptotic and quasiasymptotic expansion at $0^+$ and at $x_0^+$, $x_0 > 0$ which we used in the previous investigations are local properties of distributions, but at infinity they are of global nature.[1] Therefore we used the notion of S-asymptotic and S-asymptotic expansion which are local properties of distributions at infinity.[1,6,7] By assuming that $f \in \mathcal{S}'(\mathbb{R})$ or $f \in \mathcal{K}'_1(\mathbb{R})$ has S-asymptotic (S-asymptotic expansion) at infinity we determined the ordinary asymptotic behaviour (expansion) at infinity of its wavelet transform $\mathcal{W}_g f(b, a)$ with respect to the variable $b$.

## 2. Definitions and known results

By $\mathcal{S}(\mathbb{R})$ is denoted the space of rapidly decreasing smooth functions defined on the real line, supplied with the usual topology. Its dual is well known space of tempered distributions $\mathcal{S}'(\mathbb{R})$. We refer to Ref. [8] for the properties of Schwartz space $\mathcal{S}'(\mathbb{R})$. We also apply the definition of the spaces of a highly localized functions over the real line $\mathbb{R}$ and half-plain $H = \{(b, a) : b \in \mathbb{R}, a > 0\}$. These spaces are denoted by $S_+(\mathbb{R})$, $S_-(\mathbb{R})$, $S_0(\mathbb{R})$, $S(H)$, and they are described in Ref. [9]. It is proved[9] that $S_+(\mathbb{R})$, $S_-(\mathbb{R})$, and $S_0(\mathbb{R})$ are closed subspaces of $\mathcal{S}(\mathbb{R})$. $S_+(\mathbb{R})$ $(S_-(\mathbb{R}))$ consists of those functions in $\mathcal{S}(\mathbb{R})$ whose Fourier transforms are supported by the positive (negative) frequencies only. And, the space $S_0(\mathbb{R})$ consists of functions from $\mathcal{S}(\mathbb{R})$ for which all the moments vanish.

$\mathcal{K}_1(\mathbb{R})$ is the space of smooth functions $\varphi$ on $\mathbb{R}$ for which all the norms

$$\nu_k(\varphi) := \sup_{i \le k, x \in \mathbb{R}} \{ e^{k|x|} |\varphi^{(i)}(x)| \}, \quad k \in \mathbb{N}_0$$

are finite. Its dual space is denoted by $\mathcal{K}'_1(\mathbb{R})$, and its topological properties are the same as for $\mathcal{S}'(\mathbb{R})$. From the fact that any distribution from $\mathcal{K}'_1(\mathbb{R})$ is a finite sum of distributional derivatives of continuous functions which do not tend to infinity faster than a function of the form $e^{kx}$ for some $k \in \mathbb{R}$, we often used the name 'exponential distribution' for an element of $\mathcal{K}'_1(\mathbb{R})$.

We use the definition of wavelet transform of tempered and exponential distribution from Refs. [9] and [5], respectively. The basic functions of wavelet analysis are obtained from the wavelet $g$ by dilation and translation:

$$g_{b,a}(t) = T_b D_a g(t) = \frac{1}{a} g\left(\frac{t - b}{a}\right), t \in \mathbb{R}, b \in \mathbb{R}, a > 0.$$

*The wavelet transform* with respect to any wavelet $g \in S_0(\mathbb{R})$ $(g \in \mathcal{K}_1(\mathbb{R}))$ of any distribution $f \in \mathcal{S}'(\mathbb{R})$ $(f \in \mathcal{K}'_1(\mathbb{R}))$ is given by

$$\mathcal{W}_g f(b, a) = \langle f(t), \bar{g}_{b,a}(t) \rangle, \quad (b, a) \in H, t \in \mathbb{R}.$$

We will also use the following notation:

$$\mathcal{W}_g f(b, a) = \mathcal{W}_{g_a} f(b) = \langle f(t), \bar{g}_a(t - b) \rangle, \quad \text{where} \quad g_a(\cdot) = \frac{1}{a} g\left(\frac{\cdot}{a}\right).$$

According to Ref. [10] we will give the definition of S-asymptotic at infinity of distributions from $\mathcal{S}'(\mathbb{R})$ and $\mathcal{K}'_1(\mathbb{R})$. Many properties of the S-asymptotic can be found in Refs. [1], [6], [10], and [11].

We always denote in this paper by $L$ a function which is a *slowly varying* at infinity, i.e. which is a positive continuous function defined on $(a, +\infty)$, $a \in \mathbb{R}$ such that $\lim_{h \to +\infty} L(hx)/L(h) = 1$ for every $x > 0$.

Let $f \in \mathcal{S}'(\mathbb{R})$ ($f \in \mathcal{K}'_1(\mathbb{R})$), and $c(h)$, $h > h_0$ be a continuous positive function. It is said that $f$ has *S-asymptotic* at infinity related to the function $c(h)$ if there exists $u \in \mathcal{S}'(\mathbb{R})$ ($u \in \mathcal{K}'_1(\mathbb{R})$), $u \neq 0$ such that

$$\lim_{h \to +\infty} \left\langle \frac{f(x+h)}{c(h)}, \varphi(x) \right\rangle = \langle u(x), \varphi(x) \rangle, \quad \forall \varphi \in \mathcal{S}(\mathbb{R}) \qquad (\forall \varphi \in \mathcal{K}_1(\mathbb{R})).$$

Then, we write $f \sim^s u$ related to $c(h)$ in $\mathcal{S}'(\mathbb{R})$ ($\mathcal{K}'_1(\mathbb{R})$).

It is known[10,11] that if $f$ has S-asymptotic in $\mathcal{S}'(\mathbb{R})$ ($\mathcal{K}'_1(\mathbb{R})$) then

$$u(x) = C \ \text{ and } \ c(h) = L(e^h) \ \left( u(x) = Ce^{\alpha x} \text{ and } c(h) = e^{\alpha h}L(e^h) \right), \ h > h_0$$

for some constant $C \neq 0$, $\alpha \in \mathbb{R}$, and some slowly varying function $L$ at infinity.

According to Ref. [7] we will give the definition of S-asymptotic expansion of first kind at infinity of distributions from $\mathcal{S}'(\mathbb{R})$ ($\mathcal{K}'_1(\mathbb{R})$). We will call it here just S-asymptotic expansion at infinity . We refer to Refs. [1] and [7] for the properties of the S-asymptotic expansion.

Let $\{d_m\}$, $m \in \mathbb{N}$ be a sequence of continuous positive functions different from zero in $(a_m, +\infty)$, $a_m > 0$ such that $d_{m+1}(h) = o(d_m(h))$, $h \to +\infty$. Let $\{u_m\}$, $m \in \mathbb{N}$ be a sequence in $\mathcal{S}'(\mathbb{R})$ ($\mathcal{K}'_1(\mathbb{R})$) such that $u_m \neq 0$ for $m = 1, \ldots, p < \infty$, and $u_m = 0$ for $m > p$, or $u_m \neq 0$ for $m \in \mathbb{N}$.

Denote by $\sum$ the set of pairs of sequences $(d_m(h), u_m)$, and by $\sum_1$ the subset of $\sum$ consisting of elements $(d_m(h), u_m)$ for which

$$u_m \sim^s v_m \quad \text{related to } d_m(h) \quad \text{in} \quad \mathcal{S}'(\mathbb{R}) \quad (\mathcal{K}'_1(\mathbb{R})),$$

where $v_m \in \mathcal{S}'(\mathbb{R})$ ($v_m \in \mathcal{K}'_1(\mathbb{R})$), and $v_m \neq 0$ for $m = 1, \ldots, p < \infty$, or $m \in \mathbb{N}$.
It follows from Ref. [7] that

$$d_m(h) = L_m(e^h) \ \left( d_m(h) = e^{\alpha_m h}L_m(e^h) \right), \quad h > a_m > 0,$$

and

$$v_m(x) = C_m \ \left( v_m(x) = C_m e^{\alpha_m x} \right),$$

where $C_m \neq 0$, $\alpha_m \in \mathbb{R}$, and $L_m$ is a slowly varying function at infinity for $m = 1, \ldots, p < \infty$, or $m \in \mathbb{N}$.

Let $f \in \mathcal{S}'(\mathbb{R})$ ($f \in \mathcal{K}'_1(\mathbb{R})$), and $(d_m(h), u_m) \in \sum_1$. If

$$\lim_{h \to +\infty} \left\langle \frac{f(x+h) - \sum_{i=1}^m u_i(x+h)}{d_m(h)}, \varphi(x) \right\rangle = 0, \quad \forall \varphi \in \mathcal{S}(\mathbb{R}) \quad (\forall \varphi \in \mathcal{K}_1(\mathbb{R}))$$

for $m = 1, \ldots, p < \infty$, or $m \in \mathbb{N}$, then we say that $f$ has *S-asymptotic expansion* at infinity related to $(d_m(h), u_m)$, and we write

$$f \overset{s.e.}{\sim} \overset{p\,(\infty)}{\underset{i=1}{\sum}} u_i \;\Big|\; \{d_m(h)\} \quad \text{in} \quad \mathcal{S}'(\mathbb{R}) \quad (\mathcal{K}'_1(\mathbb{R})).$$

## 3. Asymptotic behaviour of distributional wavelet transform at infinity

In the next theorems we consider the behaviour of $\mathcal{W}_g f(b + h, a)$ when $h \to +\infty$, assuming that $f \in \mathcal{S}'(\mathbb{R})$ (Theorem 3.1), or $f \in \mathcal{K}'_1(\mathbb{R})$ (Theorem 3.2) has S-asymptotic at infinity.

**Theorem 3.1.** *Let $f \in \mathcal{S}'(\mathbb{R})$ has S-asymptotic at infinity with a limit $C \neq 0$ related to the continuous and positive function $c(h)$, $h > h_0$. Then for its wavelet transform with respect to any wavelet $g \in S_0(\mathbb{R})$ we have*

$$\mathcal{W}_{g_a} f(b + h) = o(c(h)) \quad \text{when} \quad h \to +\infty.$$

**Proof.** From the definition of the distributional wavelet transform we have

$$\lim_{h \to +\infty} \frac{\mathcal{W}_{g_a} f(b + h)}{c(h)} = \lim_{h \to +\infty} \left\langle \frac{f(t)}{c(h)}, \, \bar{g}_a(t - b - h) \right\rangle.$$

If we put $t - h = x$ we obtain

$$\lim_{h \to +\infty} \frac{\mathcal{W}_{g_a} f(b + h)}{c(h)} = \lim_{h \to +\infty} \left\langle \frac{f(x + h)}{c(h)}, \, \bar{g}_a(x - b) \right\rangle.$$

Since $g \in S_0(\mathbb{R})$ it follows that

$$\bar{g}_a(\cdot - b) = \frac{1}{a} \bar{g}\left( \frac{\cdot - b}{a} \right) \in \mathcal{S}(\mathbb{R}).$$

So, from the assumption $f \sim^s C$, $C \neq 0$ related to $c(h)$ it follows that

$$\lim_{h \to +\infty} \left\langle \frac{f(x + h)}{c(h)}, \, \bar{g}_a(x - b) \right\rangle = \left\langle C, \, \bar{g}_a(x - b) \right\rangle = \mathcal{W}_{g_a} C(b).$$

From the fact that $g$ is a wavelet and by change of the variable in the integral, $t = (x - b)/a$ we get

$$\mathcal{W}_{g_a} C(b) = C \int_{-\infty}^{+\infty} \frac{1}{a} \bar{g}\left( \frac{x - b}{a} \right) dx = C \int_{-\infty}^{+\infty} \bar{g}(t)dt = 0.$$

So, $\displaystyle\lim_{h \to +\infty} \mathcal{W}_{g_a} f(b + h)/c(h) = 0$. $\qquad\qquad\square$

**Remark:** Let $f \in \mathcal{S}'(\mathbb{R})$ and $c(h)$ satisfy the assumptions of Theorem 3.1. If we put $b = 0$ we obtain result for the behaviour at infinity of $\mathcal{W}_g f(b, a)$ with respect to the variable $b$:

$$\mathcal{W}_{g_a} f(h) = o(c(h)) \quad \text{when} \quad h \to +\infty.$$

In the case when $f \in \mathcal{K}'_1(\mathbb{R})$ we have the following theorem:

**Theorem 3.2.** Let $f \in \mathcal{K}'_1(\mathbb{R})$ has S-asymptotic at infinity with a limit $Ce^{\alpha x}$, $C \neq 0$, $\alpha \in \mathbb{R}$ related to the continuous and positive function $c(h)$, $h > h_0$. Then for its wavelet transform with respect to any wavelet $g \in \mathcal{K}_1(\mathbb{R})$ we have

$$W_{g_a} f(b+h) \sim M_{\alpha,a,b} c(h) \quad \text{when} \quad h \to +\infty,$$

where $M_{\alpha,a,b} = C e^{\alpha b} \int_{-\infty}^{+\infty} e^{\alpha t} \bar{g}_a(t) dt.$

**Proof.** The proof is similar to the proof of Theorem 3.1.

$$\lim_{h \to +\infty} \frac{W_{g_a} f(b+h)}{c(h)} = \lim_{h \to +\infty} \left\langle \frac{f(x+h)}{c(h)}, \bar{g}_a(x-b) \right\rangle.$$

Since $g \in \mathcal{K}_1(\mathbb{R})$ it follows that

$$\bar{g}_a(\cdot - b) = \frac{1}{a} \bar{g}\left(\frac{\cdot - b}{a}\right) \in \mathcal{K}_1(\mathbb{R}).$$

So, from the assumption $f \sim^s Ce^{\alpha x}$, $C \neq 0$, $\alpha \in \mathbb{R}$ related to $c(h)$ we have

$$\lim_{h \to +\infty} \left\langle \frac{f(x+h)}{c(h)}, \bar{g}_a(x-b) \right\rangle = \left\langle Ce^{\alpha x}, \bar{g}_a(x-b) \right\rangle = CW_{g_a} e^{\alpha x}(b).$$

By change of the variable $t = x-b$ in the integral $W_{g_a} e^{\alpha x}(b) = \int_{-\infty}^{+\infty} e^{\alpha x} \bar{g}_a(x-b) dx$ we get

$$W_{g_a} e^{\alpha x}(b) = e^{\alpha b} \int_{-\infty}^{+\infty} e^{\alpha t} \bar{g}_a(t) dt.$$

So, $\lim_{h \to +\infty} W_{g_a} f(b+h)/c(h) = M_{\alpha,a,b}$, where $M_{\alpha,a,b} = C e^{\alpha b} \int_{-\infty}^{+\infty} e^{\alpha t} \bar{g}_a(t) dt$. (The last integral converges since $g \in \mathcal{K}_1(\mathbb{R})$). $\square$

**Remark 1:** Let $f \in \mathcal{K}'_1(\mathbb{R})$ and $c(h)$ satisfy the assumptions of Theorem 3.2. If we put $b = 0$ we obtain result for the behaviour at infinity of $W_g f(b,a)$ with respect to the variable $b$:

$$W_{g_a} f(h) \sim M_{\alpha,a} c(h) \quad \text{when} \quad h \to +\infty,$$

where $M_{\alpha,a} = C \int_{-\infty}^{+\infty} e^{\alpha t} \bar{g}_a(t) dt.$

Using the Parseval formula we may write the wavelet transform of $u(x) = Ce^{-\alpha|x|}$, $C \neq 0$, $\alpha > 0$ in the Fourier space

$$W_{g_a} u(0) = \left\langle u(x), \bar{g}_a(x) \right\rangle = \frac{1}{2\pi} \left\langle \hat{u}(x), \hat{\bar{g}}_a(x) \right\rangle,$$

where by $\hat{u}$ is denoted the Fourier transform of $u$. Since $\hat{u}(\omega) = 2C\alpha/(\alpha^2 + \omega^2)$ we get

$$W_{g_a} u(0) = \frac{C\alpha}{\pi} \int_{-\infty}^{+\infty} \frac{\bar{\hat{g}}(a\omega)}{\alpha^2 + \omega^2} \, d\omega. \tag{1}$$

By Remark 1 and Eq. (1) we obtain the next result:

**Remark 2:** If $f \in \mathcal{K}_1'(\mathbb{R})$ has S-asymptotic with a limit $Ce^{-\alpha|x|}$, $C \neq 0$, $\alpha > 0$ related to the positive and continuous function $c(h)$, $h > h_0$, then for any wavelet $g \in \mathcal{K}_1(\mathbb{R})$ we have $W_{g_a} f(h) \sim M_{\alpha,a} c(h)$ when $h \to +\infty$, where

$$M_{\alpha, a} = \frac{C\alpha}{\pi} \int_{-\infty}^{+\infty} \frac{\bar{\hat{g}}(a\omega)}{\alpha^2 + \omega^2} \, d\omega.$$

**Example 1.** Let $f \in \mathcal{K}_1'(\mathbb{R})$ has S-asymptotic with a limit $Ce^{\alpha x}$, $C \neq 0$, $\alpha \in \mathbb{R}$ related to the continuous and positive function $c(h)$, $h > h_0$. Then for its wavelet transform with respect to the wavelet

$$g(x) = \frac{1}{2\sqrt{\pi\beta}} e^{-\frac{x^2}{4\beta}} \in \mathcal{K}_1(\mathbb{R}), \; \beta > 0$$

we have $W_{g_a} f(h) \sim M_{\alpha,a} c(h)$ when $h \to +\infty$.
By Remark 1 we get

$$M_{\alpha,a} = C \int_{-\infty}^{+\infty} e^{\alpha t} \bar{g}_a(t) dt$$

$$= \frac{C}{2a\sqrt{\pi\beta}} \int_{-\infty}^{+\infty} e^{-\frac{(t-a)^2}{4\beta} + \alpha t} dt$$

$$= \frac{C}{a\sqrt{\pi}} e^{\alpha(a+\alpha\beta)} \int_{-\infty}^{+\infty} e^{-x^2} dx$$

$$= \frac{C}{a} e^{\alpha(a+\alpha\beta)}.$$

**Example 2.** The function $\Theta(x)Chx = \Theta(x)(e^x + e^{-x})/2$ defines an element from $\mathcal{K}_1'(\mathbb{R})$. It is known[6] that $\Theta(x)Chx$ has S-asymptotic with a limit $e^x/2$ related to the function $c(h) = e^h$. Then for the behaviour at infinity of its wavelet transform with respect to any wavelet $g \in \mathcal{K}_1(\mathbb{R})$ we obtain $W_{g_a} f(h) \sim M_a e^h$ when $h \to +\infty$, where

$$M_a = \frac{1}{2} \int_{-\infty}^{+\infty} e^t \bar{g}_a(t) \, dt.$$

## 4. Asymptotic expansion of distributional wavelet transform at infinity

In the next theorem we consider the asymptotic expansion of the wavelet transform $W_g f(b + h, a)$, $f \in \mathcal{S}'(\mathbb{R})$ ($f \in \mathcal{K}_1'(\mathbb{R})$) when $h \to +\infty$.

**Theorem 4.1.** *Let* $f \in \mathcal{S}'(\mathbb{R})$ ($f \in \mathcal{K}_1'(\mathbb{R})$) *has S-asymptotic expansion at infinity with respect to* $(L_m(e^h), u_m)$ $\left((e^{\alpha_m h} L_m(e^h), u_m)\right)$, $h > a_m > 0$, $\alpha_m \in \mathbb{R}$, *and* $L_m$

*is a slowly varying function at infinity for $m = 1, \ldots, p < \infty$, or $m \in \mathbb{N}$. Then, for any wavelet $g \in S_0(\mathbb{R})$ $(g \in \mathcal{K}_1(\mathbb{R}))$,*

$$\lim_{h \to +\infty} \frac{\mathcal{W}_{g_a} f(b+h) - \sum_{i=1}^{m} \mathcal{W}_{g_a} u_i(b+h)}{L_m(e^h)} = 0$$

$$\left( \lim_{h \to +\infty} \frac{\mathcal{W}_{g_a} f(b+h) - \sum_{i=1}^{m} \mathcal{W}_{g_a} u_i(b+h)}{e^{\alpha_m h} L_m(e^h)} = 0 \right),$$

$m = 1, \ldots, p < \infty$, or $m \in \mathbb{N}$.

The proof is similar to the proof of Theorem 4.1. in Ref. [4].

**Remark:** Let the assumptions of Theorem 4.1. are satisfied. If we put $b = 0$ then we obtain

$$\mathcal{W}_{g_a} f(h) \sim \sum_{i=1}^{p(\infty)} \mathcal{W}_{g_a} u_i(h), \quad h \to +\infty$$

related to $L_m(e^h)$ $\left( e^{\alpha_m h} L_m(e^h) \right)$, $m = 1, \ldots, p < \infty$, or $m \in \mathbb{N}$.

# References

1. S. Pilipović, B. Stanković, A. Takači, *Asymptotic Behaviour and Stieltjes Transformation of Distribution*, Taubner-Texte zur Mathematik, band **116** (1990).
2. V. S. Vladimirov, Ju. Drožinov, B. I. Zavialov. *Tauberian Theorems for Generalized Functions*, Nauka, Moscow (1986) (in Russian).
3. K. Saneva, A. Bučkovska, *Asymptotic Behavior of the Distributional Wavelet Transform at 0*, Math. Balk., New Series Vol. **17**, Fasc. 1-2, 437-441 (2003).
4. K. Saneva, A. Bučkovska, *Asymptotic Expansion of Distributional Wavelet Transform*, Integral Transforms and Special Functions, Vol. **17**, Numbers 2-3, 85-91 (2006).
5. A. Takači, N. Teofanov, *A note of wavelets and S-asymptotics*, Matematički vesnik, **49**, 215-220 (1997).
6. S. Pilipović, B. Stanković, *S-asymptotic of a distribution*, Studia Mat. Bulg., **10**, 147-156 (1989).
7. S. Pilipović, *Asymptotic Expansions of Schwartz's Distributions*, Publ. Inst. Math. Belgrade, **45**, 119-127 (1989).
8. L. Schwartz, *Théory des Distributions I*. 2nd ed. Hermann, Paris (1957).
9. M. Holschneider, *Wavelets, an analysis tool*, Clarendon press, Oxford (1995).
10. S. Pilipović, *S-asymptotic of tempered and $\mathcal{K}_1'$ distributions, Part 1, 2, 4*, Univ. u Novom Sadu, Zb. Rad. Prir. Mat. Fak., 15 (**1**), 47-58, 59-67 (1985); 18 (**2**), 191-195 (1988).
11. S. Pilipović, *Quasiasymptotics and S-asymptotic in $\mathcal{S}'$ and $\mathcal{D}'$*, Publ. de l'Inst. Math., 58 (**72**), 13-20 (1995).

# A WAVELET-BASED VECTORIAL APPROACH
# FOR AN INTEGRAL FORMULATION OF ANTENNA PROBLEMS

F.P. ANDRIULLI

*The Radiation Laboratory*
*University of Michigan*
*1301 Beal Avenue*
*48109 Ann Arbor, Michigan E-mail: fandri@umich.edu*


A. TABACCO

*Dipartimento di Matematica*
*Politecnico di Torino*
*Corso Duca degli Abruzzi, 24*
*10129 Torino, Italy E-mail: anita.tabacco@polito.it*


G. VECCHI

*Dipartimento di Elettronica*
*Politecnico di Torino*
*Corso Duca degli Abruzzi, 24*
*10129 Torino, Italy E-mail: giuseppe.vecchi@polito.it*

In this paper we present an overview of a multiresolution approach for solving an integral equation arising in an applied electromagnetics problem. The problem under consideration is to find the electric field radiated by an antenna, given the expression for the excitation. The structures analyzed are assumed to be made of a Perfect Electric Conductor, i.e. they must satisfy Dirichlet boundary conditions. The problem is considered in the integral formulation, numerically discretized with boundary elements. After defining a set of encapsulated spaces for the discretized solution, a multiresolution treatment is addressed. Each space is first separated into a solenoidal and not solenoidal component and then a vector wavelet basis is obtained for each of them, starting from a scalar decomposition, via a proper scalar to vector mapping. Preliminary results are presented in order to show the actual sparsity of the arising linear system and the fast behaviour of iterative solvers.

**Key words:** Multiresolution analysis, wavelets, integral equations, method of moments, triangular meshes, antennas

**Mathematics Subject Classification:** 65N30, 35B25

## 1. Introduction

The problem of finding the electromagnetic field radiated by an antenna, and its network parameters (impedance, scattering matrix) is of key importance in applied electromagnetics, and the related numerical simulation tools are part of all practical antenna design. The typical antenna structure is constituted of metal conductors (typically assumed ideal), and of dielectric bodies that are piecewise homogeneous; the related analysis thus amounts to the solution of Maxwell equations with boundary conditions to be enforced on the surface of material discontinuity; it is to be noted that a fully three-dimensional electromagnetic problem is intrinsically vector-valued, and in general cannot be transformed in a straightforward manner to multiple scalar problems. Various numerical methods, including also wavelet techniques, have been devised and applied to antenna analysis over the past decades, either based on the differential formulation of the boundary-value problem, or on its integral formulations via boundary elements approaches; for a review, one can e.g. refer to[2,5] and specifically for wavelet approaches to[3,9,6,1].

## 2. Integral Equation and Multiresolution Analysis

In our setting, we will consider antennas that can be modelled by a two dimensional surface $\Gamma$, that need not be close. The equation to be solved is the so called Electric Field Integral Equation (EFIE) that arises from a boundary integral formulation of Maxwell's equations as outlined in[2]

$$
\begin{aligned}
-i\omega\mu \iint_{\Gamma} \frac{e^{ik|r-r'|}}{4\pi|r-r'|} J(r')\, dS(r') \\
+ \frac{1}{\epsilon} \nabla_s \iint_{\Gamma} \frac{e^{ik|r-r'|}}{4\pi|r-r'|} \nabla_s \cdot J(r')\, dS(r') - E_{in}^{tan}(r) = 0
\end{aligned}
\tag{1}
$$

where a time dependence $\exp(-i\omega t)$ has been assumed and suppressed; $\epsilon$ and $\mu$ denote the permeability and the permittivity of the free space and $k = \omega\sqrt{\epsilon\mu}$ is the wave-number; $\nabla_s$ indicates the surface divergence. $E_{in}^{tan}$ is the forcing term, a function of the antenna feeding and $J$, the equation unknown, represents the surface current on the antenna surface. From the knowledge of $J$, all the antenna parameters of practical interest (like the radiated field or the antenna impedance) can be obtained.

The numerical solution of the problem is customarily obtained via finite-element approach with a Galerkin choice, where $J$ is projected onto the finite dimensional space $X$

$$
J = J_X + e, \qquad J_X = \sum_i I_i x_i
\tag{2}
$$

This gives rise to the finite-dimensional linear system

$$
\left[ Z^A + Z^\phi \right] [I] = [V]
\tag{3}
$$

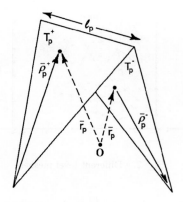

Fig. 1.   Rao-Wilton-Glisson function

where

$$\left[Z^A\right]_{p,q} = i\omega\mu \int_\Gamma \boldsymbol{x}_p(\boldsymbol{r}) \cdot \int_\Gamma \frac{e^{ik|\boldsymbol{r}-\boldsymbol{r}'|}}{4\pi |\boldsymbol{r}-\boldsymbol{r}'|} \boldsymbol{x}_q(\boldsymbol{r}') \, \mathrm{d}S(\boldsymbol{r}') \, \mathrm{d}S(\boldsymbol{r}) \tag{4}$$

$$\left[Z^\phi\right]_{p,q} = -\frac{i}{\omega\epsilon} \int_\Gamma \nabla_s \cdot \boldsymbol{x}_p(\boldsymbol{r}) \cdot \int_\Gamma \frac{e^{ik|\boldsymbol{r}-\boldsymbol{r}'|}}{4\pi |\boldsymbol{r}-\boldsymbol{r}'|} \nabla_s \cdot \boldsymbol{x}_q(\boldsymbol{r}') \, \mathrm{d}S(\boldsymbol{r}') \, \mathrm{d}S(\boldsymbol{r}) \tag{5}$$

$$[V]_i = -\int_\Gamma \boldsymbol{x}_i(\boldsymbol{r}) \cdot E_i^{tan}(\boldsymbol{r}) \, \mathrm{d}S(\boldsymbol{r}) \tag{6}$$

$$[I]_i = \quad I_i \tag{7}$$

We start defining projection spaces $X_j$ in which the solution can be projected. In doing this we need a triangular mesh on $\Gamma$. For sake of simplicity we will assume that $\Gamma$ is a faceted structure i.e. the triangularization can be obtained using planar triangles (examples can be found in Figures 3 and 4). If this is not verified, a continuous bijection is needed in addition to the following construction[10]. In antenna problems, a widely used finite element space is the one introduced in[4] and applied to computational electromagnetics in[7], spanned by the so called *Rao-Wilton-Glisson* (*RWG*) functions. These are defined on the inner edges of the mesh and, referring to Figure 1, their expression in local coordinates is

$$\boldsymbol{f}_p = \begin{cases} \dfrac{l_n}{2A_p^+}\rho_p^+ & \boldsymbol{r} \in T_p^+ \\[2mm] \dfrac{l_n}{2A_p^-}\rho_p^- & \boldsymbol{r} \in T_p^- \\[2mm] 0 & \text{otherwise} \end{cases} \tag{8}$$

Let us define

$$X_j^{RWG} = \mathrm{span}\left\{ \boldsymbol{f}_1, \dots \boldsymbol{f}_{n_{ed}^j} \right\}$$

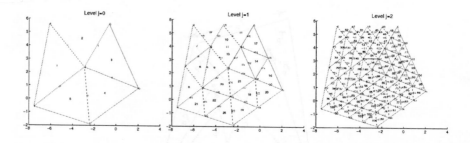

Fig. 2.   Different level meshes

where $j$ labels a particular mesh and $n_{ed}^j$ is the number of inner edges relative to the mesh $j$.

Considering the antenna surface $\Gamma$ and a mesh at level $j$ defined on it, we can obtain a finer mesh bisecting each edge (refer to Figure 2). The obtained mesh will be labelled $j+1$. Thus on each mesh one can define a RWG space, obtaining spaces $X_j^{RWG}$, $X_{j+1}^{RWG}$, $X_{j+2}^{RWG}$, ... which satisfy the inclusion relationship

$$X_j^{RWG} \subset X_{j+1}^{RWG} \qquad (9)$$

as proved in[1].

Relationship (9) makes spaces $X_j^{RWG}$ eligible for a wavelet construction

$$X_{j+1}^{RWG} = X_j^{RWG} \oplus W_j^{RWG}$$

According to[1] the wavelet detail $W_j^{RWG}$ is further decomposed in a direct summation of three spaces

$$W_j^{RWG} = W_j^{TE} \oplus W_j^{qTM} \oplus W_j^{B}.$$

where $W_j^{TE}$ is defined so that

$$(\nabla_s \cdot) W_j^{TE} = 0$$

and the basis chosen is constituted by functions of the type

$$l_n^j(\boldsymbol{r}) = \nabla_s \times \hat{n} \Lambda_n^j(\boldsymbol{r}) \quad n = 1, \ldots, n_{nodes} \qquad (10)$$

where $\hat{n}$ is the direction normal to the antenna surface, $\Lambda_n^j$ is the scalar wavelet basis introduced in[10] build using the scalar linear Lagrange nodal interpolating functions defined hierarchically on inner nodes (equal to one at one inner node and linearly going to zero on all neighboring nodes). The functions $l_n^j(\boldsymbol{r})$ itself are a wavelet basis as shown in[1]. The space $W_j^{qTM}$ is chosen in order to be a complement (not orthogonal) to the space $W_j^{TE}$ in each triangular cell of level $j$ and $W_j^{B}$ takes properly into account the functions necessary for the completeness as outlined in[1].

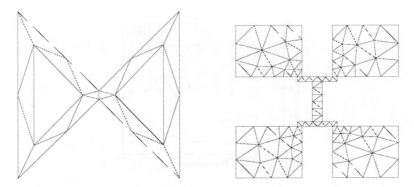

Fig. 3. Bowtie (left) and Patch Array (right) antennas

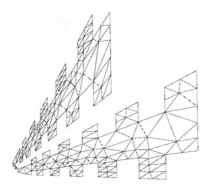

Fig. 4. Log-Periodic antenna

## 3. Numerical Results

The surfaces $\Gamma$ that will be considered represent some antenna structures actually usable in applications. The first structure is a wide band dipole known as Bowtie antenna. In Figure 3, left, the structure is presented with a mesh that is relative to the coarsest level $j = 0$. Two levels (j=0, 1, 2) of the wavelets presented in this work have been used for this structure, so each edge of the coarsest level presented in Figure 3 has been bisected twice. The second structure is a sub-array of four printed patches (Figure 3), right, a frequent configuration in the microstrip technology. Also for this structure the figure represents the coarsest level mesh, and two level of wavelets have been used. The last structure is a so called Log-Periodic antenna (Figure 4), which is a well known wide band antenna; for this structure one level of wavelets has been used.

The performances of the new basis will be tested considering the number of iterations needed by the conjugate gradient type solver[8] to obtain a fixed precision $(10^{-10})$ on the solution. In Figure 5, we present the results for the Bowtie antenna; the number of iterations of the modified conjugate gradient needed to solve the system obtained with the usual Rao-Wilton-Glisson basis is compared with the

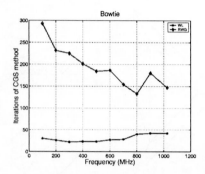

Fig. 5.    Bowtie: number of CG iterations of RWG basis vs Wavelet basis

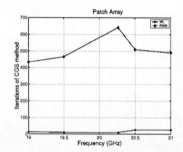
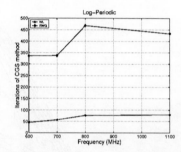

Fig. 6.    Patch Array (left) and Log-Periodic (right): number of CG iterations of RWG basis vs Wavelet basis

number of iterations needed with the basis presented in this work. This comparison is made for several frequencies $f = \frac{\omega}{2\pi}$, where $\omega$ is the constant which appears in (1). The same analysis is performed for the other two structures and the results are presented in Figure 6.

## References

1. F. P. Andriulli, A. Tabacco, and G. Vecchi. A Multiresolution Approach to the Electric Field Integral Equation in Antenna Problems, Siam J. Scient. Comp., 29:1–21, (2007)
2. D. Colton and R. Kress. *Integral Equation Methods in Scattering Theory*. Wiley, 1983.
3. J. Goswami, A. Chan, and C. Chui. On solving first-kind integral equations using wavelets on a bounded interval. *IEEE Trans. Antennas Propagation*, 43(6):614–622, June 1995.
4. J. C. Nedelec. A new family of mixed finite element. *Numerische Mathematik*, 50:57–81, 1986.
5. A. F. Peterson, S. L. Ray, and R. Mittra. *Computational Methods for Electromagnetics*. Wiley-IEEE Press, 1997.
6. P. Pirinoli, G. Vecchi, and L. Matekovits. Multiresolution analysis of printed antennas and circuits: a dual-scalar approach. *IEEE Trans. Antennas Propagation*, pages 858–874, Sept. 1998.
7. S. M. Rao, D. R. Wilton, and A. W. Glisson. Elettromagnetic scattering by surfaces of arbitrary shape. *IEEE Trans. Antennas Propagation*, AP-30(3):409–418, May 1982.

8. P. Sonneveld. Cgs: A fast Lanczos-type solver for nonsymmetric linear systems. *SIAM J. Sci. Stat. Comput.*, 10(1):36–52, Jan. 1989.

9. F. Vipiana, P.Pirinoli, and G.Vecchi. A multi-resolution method of moments for triangular meshes. *IEEE Trans. Antennas Propagation*, 53:2247–2258, 2005.

10. T. von Petersdorff and C. Schwab. Fully discrete multiscale Galerkin BEM. In *Multiscale Wavelet Methods for PDEs*. Academic Press, 1997.

# NEW REPRODUCING SUBGROUPS OF $Sp(2, \mathbb{R})$*

E. CORDERO

*Dipartimento di Matematica*
*Politecnico di Torino*
*Corso Duca degli Abruzzi, 24*
*10129 Torino, Italy E-mail: cordero@dm.unito.it*

F. DE MARI

*DIPTEM*
*Piazzale J. F. Kennedy, Pad. D.*
*16129 Genova, Italy E-mail: demari@dima.unige.it*

A. TABACCO

*Dipartimento di Matematica*
*Politecnico di Torino*
*Corso Duca degli Abruzzi, 24*
*10129 Torino, Italy E-mail: anita.tabacco@polito.it*

We prove that the restriction of the metaplectic representation to a class of three dimensional subgroups of the symplectic group $Sp(2, \mathbb{R})$ gives rise to a reproducing formula for $L^2(\mathbb{R}^2)$ functions.

**Key words:** Metaplectic representation, symplectic group, wavelets, Wigner distribution

**Mathematics Subject Classification:** 42C15, 17B45

## 1. Introduction

Let $G$ denote the symplectic group $Sp(d, \mathbb{R})$ and $\mu$ the metaplectic representation of (the double cover of) $G$. It is well-known that for some subgroups $H$ of $G$ there exist one or more "window" functions $\phi \in L^2(\mathbb{R}^d)$ such that the formula

$$f = \int_H \langle f, \mu(h)\phi \rangle \mu(h)\phi \, dh \qquad (1)$$

holds weakly for all $f \in L^2(\mathbb{R}^d)$, where $dh$ is left Haar measure on $H$. For other subgroups no such windows exist. Understanding why some groups satisfy (1), and are then called *reproducing*, and why others do not, finding invariants for the reproducing ones and possibly classifying them, are all very natural problems, which

---

*This work is partially supported by the GNAMPA Project "Laplaciani generalizzati, analisi tempo-frequenza e teoria delle rappresentazioni", coord. S. Meda.

fall within the scopes of various areas, from coherent states in physics [1] to group representations [5] and to wavelet and Gabor analysis [7]. For the relevance of the metaplectic representation in the context of harmonic analysis in phase space or time-frequency analysis, the reader is referred to [6] and [7], and the references therein. It is to be noticed that one could even widen the problem and allow more general *subsets* H of G, not necessarily subgroups. This broader perspective, however, is not considered here.

This paper is part of an ongoing project that addresses several questions related to (1). A complete classification of reproducing subgroups in the case $d = 1$ is given in [4] and many interesting facts have been proved in [8] in a somewhat different setting. Some new results in higher dimensions are in [2] and [3], although most interesting examples are given for $d = 2$. In [3] we have proved that a class of semidirect products $G_\beta = H_\beta \rtimes SO(2) \subset Sp(2, \mathbb{R})$ satisfy (1) with suitable windows (see below for the explicit definition). This class includes the group $SIM(2)$ of similitude transformations of the plane, i.e. the group generated by translations, rotations and dilations. From the structural point of view, semidirect products with a compact factor seem to play an important role in the general picture of reproducing groups. Here we consider the following issue related to them: it is almost trivial to show that the semidirect product of a reproducing group with a compact group is again reproducing; is the converse statement true? We provide an example of a class of reproducing semidirect products for which the non-compact factor is also reproducing. In other words, we show that the groups $H_\beta$ are themselves reproducing.

## 2. Notation and preliminaries

Since the result of this paper holds for $d = 2$, we specialize to this case. The symplectic group is as usual

$$Sp(2, \mathbb{R}) = \left\{ g \in GL(4, \mathbb{R}) : {}^t g J g = J \right\},$$

where $J = \begin{bmatrix} 0 & I_2 \\ -I_2 & 0 \end{bmatrix}$ defines the standard symplectic form

$$\omega(x, y) = {}^t x J y, \qquad x, y \in \mathbb{R}^4.$$

The metaplectic representation $\mu$ of (the two-sheeted cover of) the symplectic group arises as intertwining operator between the standard Schrödinger representation of the Heisenberg group and the representation that is obtained from it by composing it with the action of the symplectic group by automorphisms on the Heisenberg group (see e.g. [6]). For elements $Sp(2, \mathbb{R})$ in special form, the metaplectic representation can be computed explicitly in a simple way. For $f \in L^2(\mathbb{R}^2)$ we have

$$\mu \left( \begin{bmatrix} A & 0 \\ 0 & {}^t A^{-1} \end{bmatrix} \right) f(x) = (\det A)^{-1/2} f(A^{-1} x) \tag{2}$$

$$\mu \left( \begin{bmatrix} I & 0 \\ C & I \end{bmatrix} \right) f(x) = \pm e^{-i\pi \langle Cx, x \rangle} f(x) \tag{3}$$

$$\mu(J) = i \mathcal{F}^{-1},$$

where $\mathcal{F}$ denotes the Fourier transform

$$\mathcal{F}f(\xi) = \int_{\mathbb{R}^2} f(x)e^{-2\pi i\langle x,\xi\rangle}\, dx, \qquad f \in L^1(\mathbb{R}^2) \cap L^2(\mathbb{R}^2).$$

## 3. The groups $H_\beta$

In [3] we have introduced the 3-dimensional groups $H_\beta$ defined by:

$$H_\beta = \left\{ \begin{bmatrix} e^{-t/2}R_{\beta t/2} & 0 \\ \Sigma_y e^{-t/2}R_{\beta t/2} & e^{t/2}R_{\beta t/2} \end{bmatrix} : t \in \mathbb{R},\ y \in \mathbb{R}^2 \right\},$$

where $R_\theta$ is the rotation

$$R_\theta = \begin{bmatrix} \cos\theta & \sin\theta \\ -\sin\theta & \cos\theta \end{bmatrix}$$

and $\Sigma_y$ is the $2\times 2$ symmetric and traceless matrix parametrized by $y = (y_1, y_2) \in \mathbb{R}^2$ as follows:

$$\Sigma_y = \begin{bmatrix} y_1 & y_2 \\ y_2 & -y_1 \end{bmatrix}.$$

The parameter $\beta$ is a real number and groups corresponding to different values of $\beta$ are mutually non-conjugate within $Sp(2,\mathbb{R})$. For a discussion of the relevance of conjugation in this circle of ideas, see [4]. By setting $\tau = e^t$, we can reparametrize as folllows

$$H_\beta = \left\{ \begin{bmatrix} \tau^{-1/2}R_{\beta/2\log\tau} & 0 \\ \Sigma_y \tau^{-1/2}R_{\beta/2\log\tau} & \tau^{1/2}R_{\beta/2\log\tau} \end{bmatrix} : \tau > 0,\ y \in \mathbb{R}^2 \right\},$$

with corresponding group law

$$h(\tau, y) \cdot h(\sigma, z) = h(\tau\sigma, y + \tau R_{\beta\log\tau}z).$$

An easy calculation shows that a left Haar measure on $H_\beta$ is

$$d\lambda(\tau, y) = \frac{d\tau\, dy}{\tau^2}.$$

**Theorem 3.1.** *The subgroups $H_\beta$ are reproducing subgroups.*

**Proof.** We prove the result in two steps. First, we show that $H_\beta$ is isomorphic to the semidirect product $\mathbb{R}^2 \rtimes D_\beta$, where $D_\beta$ is the dilation subgroup of $GL(2,\mathbb{R})$ given by

$$D_\beta = \{\tau R_{\beta\log\tau}, \tau > 0\}.$$

Clearly, $\beta$ plays a trivial role: all the groups $D_\beta$ are mutually isomorphic; what counts is just the parametrization. Secondly, we show that the metaplectic representation restricted to $H_\beta$ is equivalent to the wavelet representation of the semidirect product $\mathbb{R}^2 \rtimes D_\beta$, for which the result is known.

For the isomorphism, if $A_\tau = \tau R_{\beta \log \tau}$ and $A_\sigma = \sigma R_{\beta \log \sigma}$ in $D_\beta$, then

$$A_\tau A_\sigma = \tau \sigma R_{\beta \log \tau} R_{\beta \log \sigma} = \tau \sigma R_{\beta(\log \tau + \log \sigma)} = \tau \sigma R_{\beta \log(\tau \sigma)} = A_{\tau \sigma}.$$

Hence the product in $\mathbb{R}^2 \rtimes D_\beta$ is

$$(y, A_\tau)(z, A_\sigma) = (y + A_\tau z, A_{\tau \sigma}) = (y + \tau R_{\beta \log \tau} z, A_{\tau \sigma}),$$

in full agreement with the product in $H_\beta$. In other words we shall make the identification $(y, A_\tau) \in \mathbb{R}^2 \rtimes D_\beta \leftrightarrow h(\tau, y) \in H_\beta$.

The natural wavelet representation of $\mathbb{R}^2 \rtimes D_\beta$ is given by

$$\nu(y, A)f(x) = |\det A|^{-1/2} f(A^{-1}(x - y)), \quad (y, A) \in \mathbb{R}^2 \rtimes D_\beta.$$

It is shown in [8] (Theorem 1.7) that a semidirect product $H = \mathbb{R}^2 \rtimes D$ satisfies (1) if and only if

$$\Delta_\phi(\xi) = \int_D |\widehat{\phi}(A^{-1}\xi)|^2 \, dA = 1 \quad \text{for a.e. } \xi \in \mathbb{R}^2,$$

where $dA$ is the left Haar measure on $D$. This is indeed the case at hand, because it is not hard to find $\psi$ such that

$$\Delta_\psi(\xi) = \int_{D_\beta} |\widehat{\psi}(A^{-1}\xi)|^2 \, dA$$

$$= \int_0^{+\infty} |\widehat{\psi}(\tau R_{\beta \log \tau} \xi)|^2 \, \frac{d\tau}{\tau^3} = 1 \quad \text{for a.e. } \xi \in \mathbb{R}^2.$$

For instance, we can take a band-limited function $\psi \in L^2(\mathbb{R}^2)$ such that supp $\widehat{\psi} \subset \{z : c < |z| < C\}$ with $0 < c < C < \infty$.

At this point we appeal to Theorem 12 in [3], where it is shown that the wavelet representation $\nu$ is equivalent to the restriction of the metaplectic representation $\mu$ to $H_\beta$. For the reader's convenience we briefly review the steps through which one shows the equivalence. A proof of all the statements that follow are to be found in [3] or can be routinely verified via direct computation. First, one computes the explicit form of $\mu$ restricted to $H_\beta$ using the formulae (2) and (3). Secondly, one computes the Fourier transform version of $\nu$, that is, $\pi = \mathcal{F} \circ \nu$. Then, one considers the mapping

$$\Phi : \dot{\mathbb{R}} \times \mathbb{R}_+ \to \mathbb{R}^2 \setminus \{(0,0)\}, \qquad x \mapsto \left(\frac{x_1^2 - x_2^2}{2}, x_1 x_2\right)$$

and shows that it is a diffeomorphism. Even functions on $\mathbb{R}^2$ can be parametrized up to null-sets by functions on $\dot{\mathbb{R}} \times \mathbb{R}_+$, by even extension. Under this identification, the mapping

$$\mathcal{U}f(u) = \|u\|^{-1/2} f(\Phi^{-1}(u)), \qquad u \in \dot{\mathbb{R}} \times \mathbb{R}_+$$

defines an isometry of $L^2_{\text{even}}(\mathbb{R}^2)$ onto $L^2(\mathbb{R}^2)$. Finally, using the explicit expressions of $\pi = \mathcal{F} \circ \nu$ and $\mu$ one shows that $\mathcal{U}$ intertwines $\pi = \mathcal{F} \circ \nu$ and $\mu$ on $H_\beta$, i.e., $\pi(h) \circ \mathcal{U} = \mathcal{U} \circ \mu(h)$ for every $h \in H_\beta$. $\qquad \square$

# References

1. S. T. Ali, J. P . Antoine, J. P .Gazeau, *Coherent states, wavelets and their generalizations*, Springer-Verlag, New York (2000).
2. E. Cordero, F. De Mari, K. Nowak and A. Tabacco, Reproducing groups for the metaplectic representation, Operator Theory: Advances and Applications, **164**, 227 (2006).
3. E. Cordero, F. De Mari, K. Nowak and A. Tabacco, Analytic features of reproducing groups for the metaplectic representation, Journal of Fourier Analysis and Applications, **12** (2006), 157–180.
4. F. De Mari and K. Nowak, Analysis of the affine transformations of the time-frequency plane, *Bull. Austral. Math. Soc.* **63**, 195 (2001).
5. J. Dixmier, *Les C*-Algèbres et leurs représentations*, Gauthier-Villars Éditeur, Paris (1969).
6. G. B. Folland, *Harmonic Analysis in Phase Space*, Princeton University Press (1989).
7. K. Gröchenig, *Foundations of Time-Frequency Analysis*, Birkhäuser, Boston (2001).
8. R. S. Laugesen, N. Weaver, G. L. Weiss, and E. N. Wilson, A characterization of the higher dimensional groups associated with continuous wavelets. *J. Geom. Anal.*, **12**, 89 (2002).

## I.8  Pseudo-Differential Operators

Organizers: J. Toft, M.W. Wong

1.8. Pseudo-Differential Operators

Organized J. Toft, M.W. Wong

# $L^2$ STABILITY AND BOUNDEDNESS OF THE FOURIER INTEGRAL OPERATORS APPLIED TO THE THEORY OF THE FEYNMAN PATH INTEGRAL

WATARU ICHINOSE*

*Department of Mathematical Science*
*Shinshu University*
*Matsumoto 390-8621, Japan*
*e-mail:ichinose@math.shinshu-u.ac.jp*

The Fourier integral operators or oscillatory integral operators are studied, which are essentially used for the mathematical theory of the Feynman path integrals. The stability in $L^2$ and $L^2$-boundedness for such operators are proved.

**Key words:** Fourier integral operator, Feynman path integral
**Mathematics Subject Classification:** 35S30, 81Q05

## 1. Main theorem

Let $x \in R^n, \dot{x} \in R^n$ and $t \in [0, T]$, where $T > 0$ is arbitrary. We consider the Lagrangian function

$$\mathcal{L}(t, x, \dot{x}) = \frac{1}{2}|\dot{x}|^2 + \dot{x} \cdot A(t, x) - V(t, x),$$

where $V \in R$ and $A = (A_1, \ldots, A_n) \in R^n$ are the electromagnetic potentials and are assumed to be continuous in $[0, T] \times R^n$. For a path $q : [s, t] \to R^n$ let

$$S(t, s; q) := \int_s^t \mathcal{L}\left(\theta, q(\theta), \frac{dq}{d\theta}(\theta)\right) d\theta$$

be the classical action.

Let $\mathbf{A} = (-V, A) \in R^{n+1}$ and $\mathbf{x} = (t, x) \in R^{n+1}$. Then we have from the definitions of the potentials

$$d(\mathbf{A} \cdot d\mathbf{x}) = -\sum_{j=1}^{n} E_j(t, x) dt \wedge dx_j$$

$$+ \sum_{1 \leq j < k \leq n} B_{jk}(t, x) dx_j \wedge dx_k,$$

* Research partially supported by Grant-in-Aid for Scientific No.13640161, Ministry of Education, Science, and Culture, Japanese Government.

where $E = (E_1, \cdots, E_n) \in R^n$ denotes the electric strength and $(B_{jk})_{1 \leq j < k \leq n} \in R^{n(n-1)/2}$ the magnetic strength tensor.

Let

$$q_{x,y}^{t,s}(\theta) = y + \frac{\theta - s}{t - s}(x - y), \ \theta \in [s, t]$$

denote the straight line connecting $y$ at $\theta = s$ and $x$ at $\theta = t$. Let $p(x, w)$ be a function in $R^{2n}$. In this paper we study the Fourier integral operator

$$(P(t,s)f)(x) = \sqrt{\frac{1}{2\pi i(t-s)}}^{n}$$
$$\times \int \left( \exp iS(t, s; q_{x,y}^{t,s}) \right) p\left( x, \frac{x-y}{\sqrt{t-s}} \right) f(y) dy$$

for $s < t$.

Let $C^\infty = C^\infty(R^n)$ be the space of all infinitely differentiable functions in $R^n$ and $C_0^\infty = C_0^\infty(R^n)$ the space of all functions in $C^\infty(R^n)$ with compact support. We denote the space of all square integrable functions in $R^n$ with inner product $(\cdot, \cdot)$ and norm $\| \cdot \|$ by $L^2 = L^2(R^n)$. For an $x = (x_1, \ldots, x_n) \in R^n$ and a multi-index $\alpha = (\alpha_1, \ldots, \alpha_n)$ we write $|\alpha| = \sum_{j=1}^n \alpha_j$, $\partial_x^\alpha = (\partial/\partial x_1)^{\alpha_1} \cdots (\partial/\partial x_n)^{\alpha_n}$ and $<x> = \sqrt{1 + |x|^2}$.

**Theorem.** Assume that $\partial_x^\alpha E_j(t, x), \partial_x^\alpha B_{jk}(t, x)$ and $\partial_x^\alpha A_j(t, x)$ are continuous in $[0, T] \times R^n$ for all $\alpha$ and that

$$|\partial_x^\alpha E_j(t, x)| \leq C_\alpha, \ |\alpha| \geq 1,$$
$$|\partial_x^\alpha B_{jk}(t, x)| \leq C_\alpha(1 + |x|)^{-(1+\delta)}, \ |\alpha| \geq 1,$$
$$|\partial_x^\alpha A_j(t, x)| \leq C_\alpha, \ |\alpha| \geq 1$$

where constants $\delta = \delta_\alpha > 0$ may depend on $\alpha$. Then there exists a constant $\rho^* > 0$ such that we have for $0 < t - s \leq \rho^*$:

(1) Let $p(x, w) \in C^\infty(R^{2n})$ whose all derivatives are bounded. Then $P(t, s)$ is a bounded operator on $L^2(R^n)$.

(2) (stability) Let $p(x, w) = 1$. Then there exists a constant $K \geq 0$ such that

$$\|P(t, s)f\| \leq e^{K(t-s)}\|f\|$$

for all $f \in L^2$.

Theorem is applied to the proof of the convergence of the Feynman path integral defined by the time slicing method through broken line paths (cf. [2,3,10-12]).

**Remark.** We know the Asada-Fujiwara result [1] in 1978. Assume

$$(t - s)\left| \det \left( \partial_{x_i} \partial_{y_j} \phi(t, s; x, y) \right) \right| \geq \text{Const.} > 0,$$
$$(t - s)\left| \partial_x^\alpha \partial_y^\beta \phi(t, s; x, y) \right| \leq C_{\alpha,\beta} < \infty, \ |\alpha| \geq 1, |\beta| \geq 1.$$

Then we have

$$\left\| \sqrt{\frac{1}{2\pi i(t-s)}}^{\,n} \int (\exp i\phi(t,s;\cdot,y))\, p(\cdot,y)f(y)dy \right\|$$
$$\le \text{Const.}\|f\|.$$

This result was applied to the proof of the convergence of the Feynman path integral by the time-slicing method through piecewise classical paths (cf. [4,5,18]). We note that the amplitude function $p(x,(x-y)/\sqrt{t-s})$ in our Theorem is more general than the amplitude function $p(x,y)$ in the Asada-Fujiwara result. Now set $\rho = t - s$. The classical action is written as

$$S(t,s;q_{x,y}^{t,s}) = \frac{|x-y|^2}{2(t-s)} + (x-y)\cdot \int_0^1 A(t-\theta\rho,$$
$$x - \theta(x-y))d\theta - \rho \int_0^1 V(t-\theta\rho, x-\theta(x-y))d\theta.$$

Hence, even if $p(x,w) = 1$, we can't apply the Asada-Fujiwara result to our Theorem.

## 2. The outline of the proof

We write

$$S(q_{x,y}^{t,s}) = S(t,s;q_{x,y}^{t,s})$$

and set

$$\mathbf{q}_{x,y}^{t,s}(\theta) = (\theta, q_{x,y}^{t,s}(\theta)) \in R^{n+1} \ (\theta \in [s,t]).$$

Then we have

$$S(q_{x,y}^{t,s}) = \frac{|x-y|^2}{2(t-s)} + \int_{\mathbf{q}_{x,y}^{t,s}} \mathbf{A}\cdot d\mathbf{x}.$$

So,

$$S(q_{z,y}^{t,s}) - S(q_{z,x}^{t,s}) = (x-y)\cdot\frac{1}{t-s}\left(z - \frac{x+y}{2}\right)$$
$$+ \left(\int_{\mathbf{q}_{z,y}^{t,s}} - \int_{\mathbf{q}_{z,x}^{t,s}}\right)\mathbf{A}\cdot d\mathbf{x}.$$

Apply the Stokes theorem. Then we have

$$S(q_{z,y}^{t,s}) - S(q_{z,x}^{t,s}) = (x-y) \cdot \frac{1}{t-s} \left( z - \frac{x+y}{2} \right)$$

$$+ \int_{\mathbf{q}_{x,y}^{s,s}} \mathbf{A} \cdot d\mathbf{x} + \iint_\Delta d(\mathbf{A} \cdot d\mathbf{x})$$

$$= (x-y) \cdot \frac{1}{t-s} \left( z - \frac{x+y}{2} \right)$$

$$+ (x-y) \cdot \int_0^1 A(s, y + \theta(x-y)) d\theta$$

$$+ \iint_\Delta \left( -\sum_{j=1}^n E_j dt \wedge dx_j + \sum_{1 \le j < k \le n} B_{jk} dx_j \wedge dx_k \right),$$

where $\Delta = \Delta(t, s, x, y, z)$ is the 2-dimensional plane with oriented boundary consisting of $-\mathbf{q}_{x,y}^{s,s}, \mathbf{q}_{z,y}^{t,s}$ and $-\mathbf{q}_{z,x}^{t,s}$. Introduce the local coordinates $0 \le \sigma_1, \sigma_2 \le 1$ in $\Delta$ by

$$(\tau(\sigma), \zeta(\sigma)) := (t - \sigma_1(t-s), z + \sigma_1(x-z) + \sigma_1 \sigma_2(y-x)) \in \Delta.$$

Then we have

$$S(q_{z,y}^{t,s}) - S(q_{z,x}^{t,s}) = (x-y) \cdot \frac{\Phi(t, s; x, y, z)}{t-s},$$

where $\Phi = (\phi_1, \ldots, \phi_n) \in R^n$ and

$$\phi_j = z_j - \frac{x_j + y_j}{2} + (t-s) \int_0^1 A_j(s, x + \theta(y-x)) d\theta$$

$$- (t-s) \sum_{k=1}^n (z_k - x_k) \int_0^1 \int_0^1 B_{jk}(\tau(\sigma), \zeta(\sigma)) d\sigma_1 d\sigma_2$$

$$- (t-s)^2 \int_0^1 \int_0^1 \sigma_1 E_j(\tau(\sigma), \zeta(\sigma)) d\sigma_1 d\sigma_2.$$

So,

$$\frac{\partial \phi_i}{\partial z_j} = \delta_{ij} + (t-s)\phi_{ij}'(t, s; x, y, z) \ (i, j = 1, 2, \ldots, n),$$

where $\partial_x^\alpha \partial_y^\beta \partial_z^\gamma \phi_{ij}'(t, s; x, y, z)$ are bounded in $0 \le s \le t \le T$ and $(x, y, z) \in R^{3n}$ for all $\alpha, \beta$ and $\gamma$. Hence, there exists a constant $\rho^* > 0$ such that the mapping

$$R^n \ni z \to \xi = \Phi(t, s; x, y, z) \in R^n$$

is homeomorphic for each fixed $0 \le t - s \le \rho^*, x$ and $y \in R^n$ and we have $\det \partial\Phi/\partial z \ge 1/2$, where $\partial\Phi/\partial z$ denotes the matrix whose (i,j) component is $\partial\phi_i/\partial z_j$. We write the inverse mapping of $R^n \ni z \to \xi = \Phi(t, s; x, y, z) \in R^n$ as $R^n \ni \xi \to z = z(t, s; x, \xi, y) \in R^n$.

For the sake of simplicity let $p(x, w) = 1$. Let $0 \leq \chi \in C_0^\infty(R^n)$ such that $\chi(0) = 1$. We denote the formally self-adjoint operator of $P(t,s)$ in $L^2$ by $P(t,s)^*$. Let $f \in C_0^\infty(R^n)$. We can easily have for $0 < t - s \leq \rho^*$

$$P(t,s)^* \chi(\epsilon \cdot)^2 P(t,s)f = \left(\frac{1}{2\pi(t-s)}\right)^n \int f(y)dy$$

$$\times \int \chi(\epsilon z)^2 \left\{\exp i(S(q_{z,y}^{t,s}) - S(q_{z,x}^{t,s}))\right\} dz$$

$$= \left(\frac{1}{2\pi(t-s)}\right)^n \int f(y)dy \int \chi(\epsilon z)^2$$

$$\times \left\{\exp i(x - y) \cdot \frac{\Phi(t, s; x, y, z)}{t - s}\right\} dz$$

$$= \left(\frac{1}{2\pi(t-s)}\right)^n \int f(y)dy \int \chi\left(\epsilon z(t, s; x, \xi, y)\right)^2$$

$$\times e^{i(x-y)\cdot\xi/(t-s)} \left|\det \frac{\partial\Phi}{\partial z}(t, s; x, y, z)\right|^{-1} d\xi$$

$$= \left(\frac{1}{2\pi(t-s)}\right)^n \int f(y)dy \int \chi\left(\epsilon z(t, s; x, \xi, y)\right)^2$$

$$\times e^{i(x-y)\cdot\xi/(t-s)}(1 + (t - s)h(t, s; x, \xi, y))d\xi,$$

where $\partial_\xi^\alpha \partial_x^\beta \partial_y^{\beta'} h(t, s; x, \xi, y)$ are bounded in $0 \leq t - s \leq \rho^*$ and $(x, \xi, y) \in R^{3n}$ for all $\alpha, \beta$ and $\beta'$. So setting $\eta = \xi/(t - s)$, then we have

$$\lim_{\epsilon \to 0} P(t,s)^* \chi(\epsilon\cdot)^2 P(t,s)f$$

$$= \lim_{\epsilon \to 0} \left(\frac{1}{2\pi}\right)^n \int f(y)dy \int \chi\left(\epsilon z(t, s; x, \xi, y)\right)^2$$

$$\times e^{i(x-y)\cdot\eta}(1 + (t - s)h(t, s; x, \xi, y))d\eta$$

$$= f + (t - s)H(t, s; X, (t - s)D_X, X')f,$$

where $H(t, s; X, (t - s)D_X, X')f$ is the pseudo-differential operator with double symbol. Hence, applying the Calderón-Vaillancourt theorem[14] , we have a constant $K \geq 0$ such that

$$\lim_{\epsilon \to 0} \|\chi(\epsilon\cdot)P(t,s)f\|^2 = \lim_{\epsilon \to 0} \left(P(t,s)^* \chi(\epsilon\cdot)^2 P(t,s)f, f\right)$$

$$= (f + (t - s)Hf, f)$$

$$\leq (1 + 2K(t - s))\|f\|^2.$$

By the Fatou lemma we can see $P(t,s)f \in L^2$ for all $f \in C_0^\infty(R^n)$ and

$$\|P(t,s)f\| \leq e^{K(t-s)}\|f\|.$$

Thus we could prove the stability in Theorem. See [10] for the proof of (1) of Theorem and the detailed proof.

428

The following is an important open problem.

Can we prove the $L^2$ boundedness of $P(t,s)$, and the stability of $P(t,s)$ when $p(x,w) = 1$, for general potentials $A(t,x)$ and $V(t,x) \geq -\text{Const.}|x|^2 - \text{Const.}$?

## References

1. K. Asada and D. Fujiwara, On some oscillatory integral transformations in $L^2(R^n)$. Japan. J. Math., **4** (1978), 299-361.

2. R. P. Feynman, Space-time approach to non-relativistic quantum mechanics, Rev. Mod. Phys., **20** (1948), 367-387.

3. R. P. Feynman and A. R. Hibbs, Quantum Mechanics and Path Integrals, McGraw-Hill, New York, 1965.

4. D. Fujiwara, A construction of the fundamental solution for the Schrödinger equation, J. D'Analyse Math., **35** (1979), 41-96.

5. D. Fujiwara and T. Tsuchida, The time slicing approximation of the fundamental solution for the Schrödinger equation with electromagnetic fields, J. Math. Soc. Japan, **49** (1997), 299-327.

6. C. Groche and F. Steiner, Handbook of Feynman Path integrals, Springer, Berlin-Heidelberg-New York, 1998.

7. K. Huang, Quantum Field Theory: From Operators to Path Integrals, John Wiley and Sons, New York, 1998.

8. W. Ichinose, A note on the existence and $\hbar$-dependency of the solution of equations in quantum mechanics, Osaka J. Math., **32** (1995), 327-345.

9. W. Ichinose, On the formulation of the Feynman path integral through broken line paths, Commun. Math. Phys., **189** (1997), 17-33.

10. W. Ichinose, On convergence of the Feynman path integral formulated through broken line paths, Rev. Math. Phys., **11** (1999), 1001-1025.

11. W. Ichinose, The phase space Feynman path integral with gauge invariance and its convergence, Rev. Math. Phys., **12** (2000), 1451-1463.

12. W. Ichinose, Convergence of the Feynman path integral in the weighted Sobolev spaces and the representation of correlation functions, J. Math. Soc. Japan, **55** (2003).

13. G. W. Johnson and M. L. Lapidus, The Feynman Integral and Feynman's Operational Calculus, Oxford Univ. Press, Oxford, 2000.

14. H. Kumano-go, Pseudo-Differential Operators, MIT Press, Cambridge, 1981.

15. L. H. Ryder, Quantum Field Theory, Cambridge University Press, Cambridge, 1985.

16. L. S. Schulman, Techniques and Applications of Path Integration, John Wiley and Sons, New York, 1981.

17. A. Truman, The polygonal path formulation of the Feynman path integral, Lecture Notes in Phys., **106**, Springer, 1979, pp. 73-102.

18. K. Yajima, Schrödinger evolution equations with magnetic fields, J. D'Analyse Math., **56** (1991), 29-76.

# SMOOTH FUNCTIONAL DERIVATIVES IN FEYNMAN PATH INTEGRALS BY TIME SLICING APPROXIMATION

NAOTO KUMANO-GO[*][†]

*Mathematics, Kogakuin University,*
*1-24-2 Nishishinjuku, Shinjuku-ku, Tokyo 163-8677, JAPAN*
*E-mail:ft24343@ns.kogakuin.ac.jp, and*
*GFM, Lisbon University,*
*Av. Prof. Gama Pinto 2, P-1649-003 Lisboa, PORTUGAL*

DAISUKE FUJIWARA[‡]

*Department of Mathematics, Gakushuin University,*
*1-5-1 Mejiro Toshima-ku Tokyo 171-8588, JAPAN*
*E-mail:daisuke.fujiwara@gakushuin.ac.jp*

This note is an exposition of our recent papers [22,9]. We give a fairly general class of functionals on a path space so that Feynman path integral has a mathematically rigorous meaning. More precisely, for any functional belonging to our class, the time slicing approximation of Feynman path integral converges uniformly on compact subsets of the configuration space. Our class of functionals is closed under addition, multiplication, translation, real linear transformation and functional differentiation. The invariance under translation and orthogonal transformation, the integration by parts with respect to functional differentiation, the interchange of the order with Riemann-Stieltjes integrals, the interchange of the order with a limit, the semiclassical approximation and the fundamental theorem of calculus hold for Feynman path integral.

**Key words:** Path integrals, Fourier integral operators, stoachastic analysis
**Mathematics Subject Classification:** 81S40, 35S30, 60H07

## 1. Introduction

In 1948, R. P. Feynman [3] expressed the integral kernel of the fundamental solution for the Schrödinger equation, using the path integral as follows:

$$\int e^{\frac{i}{\hbar}S[\gamma]}\mathcal{D}[\gamma]\,. \tag{1}$$

---

[*]Work partially supported by MEXT.KAKENHI 15740094, 18740077 of Japan.
[†]Work partially supported by POCTI/MAT/34924 of Portugal.
[‡]Work partially supported by JSPS.KAKENHI (C)15540184, (C) 17540170

Here $0 < \hbar < 1$ is Planck's parameter, $\gamma : [0, T] \to \mathbf{R}^d$ is a path with $\gamma(0) = x_0$ and $\gamma(T) = x$, and $S[\gamma]$ is the action along the path $\gamma$ defined by

$$S[\gamma] = \int_0^T \frac{1}{2} \left| \frac{d\gamma}{dt} \right|^2 - V(t, \gamma(t)) dt. \tag{2}$$

The path integral is a new sum of $e^{\frac{i}{\hbar} S[\gamma]}$ over all the paths. Feynman explained his new integral as a limit of the finite dimensional integral, which is now called the time slicing approximation. Furthermore, Feynman suggested a new analysis on a path space with the functional integration

$$\int F[\gamma] e^{\frac{i}{\hbar} S[\gamma]} \mathcal{D}[\gamma],$$

and the functional differentiation $(DF)[\gamma][\eta]$. (cf. Feynman-Hibbs[4], L. S. Schulman [26]) However, in 1960, R. H. Cameron [2] proved that the completely additive measure $e^{\frac{i}{\hbar} S[\gamma]} \mathcal{D}[\gamma]$ does not exist.

Therefore, in this note, using the time slicing approximation, we prove the existence of Feynman path integrals

$$\int e^{\frac{i}{\hbar} S[\gamma]} F[\gamma] \mathcal{D}[\gamma], \tag{3}$$

with smooth functional derivatives $(DF)[\gamma][\eta]$. More precisely, we give a fairly general class $\mathcal{F}^\infty$ of functionals $F[\gamma]$ on the path space $C([0, T] \to \mathbf{R}^d)$ so that for any $F[\gamma] \in \mathcal{F}^\infty$, the time slicing approximation of (3) converges uniformly on any compact set of the configuration space $\mathbf{R}^{2d}$ with respect to the endpoints $(x, x_0)$.

There were some mathematical works which proved the time slicing approximation of (1) converges uniformly on any compact subset of $\mathbf{R}^{2d}$. See D. Fujiwara [5,7,8], H. Kitada and H. Kumano-go [17], K. Yajima [28], N. Kumano-go [20], D. Fujiwara and T. Tsuchida [13], and W. Ichinose [14]. However all these works treated (1), that is the particular case of (3) with $F[\gamma] \equiv 1$.

Many people tried to give a mathematically rigorous meaning to Feynman path integral. E. Nelson [24] succeeded in connecting Feynman path integral to Wiener measure by analytic continuation with respect to a parameter. K. Itô [16] succeeded in defining Feynman path integrals as an improper oscillatory integral over a Hilbert manifold of paths. Albeverio and Høegh Krohn [1] and J. Rezende [25] applied Itô's idea and discussed many problems.

## 2. Main Results

Let $\Delta_{T,0}$ be an arbitrary division of the interval $[0, T]$ into subintervals, i.e.,

$$\Delta_{T,0} : T = T_{J+1} > T_J > \cdots > T_1 > T_0 = 0. \tag{4}$$

Set $x_{J+1} = x$. Let $x_j$, $j = 1, 2, \ldots, J$ be arbitrary points of $\mathbf{R}^d$. Let

$$\gamma_{\Delta_{T,0}} = \gamma_{\Delta_{T,0}}(t, x_{J+1}, x_J, \ldots, x_1, x_0), \tag{5}$$

be the broken line path which connects $(T_j, x_j)$ and $(T_{j-1}, x_{j-1})$ by a line segment for any $j = 1, 2, \ldots, J, J+1$. Set $t_j = T_j - T_{j-1}$ and $|\Delta_{T,0}| = \max_{1 \leq j \leq J+1} t_j$.

As Feynman [3] had first defined by the time slicing approximation, we define the Feynman path integrals (3) by

$$\int e^{\frac{i}{\hbar} S[\gamma]} F[\gamma] \mathcal{D}[\gamma] = \lim_{|\Delta_{T,0}| \to 0} \prod_{j=1}^{J+1} \left( \frac{1}{2\pi i \hbar t_j} \right)^{d/2} \int_{\mathbf{R}^{dJ}} e^{\frac{i}{\hbar} S[\gamma_{\Delta_{T,0}}]} F[\gamma_{\Delta_{T,0}}] \prod_{j=1}^{J} dx_j \,, \quad (6)$$

whenever the limit exists.

**Remark.** $S[\gamma_{\Delta_{T,0}}]$ and $F[\gamma_{\Delta_{T,0}}]$ are functions of a finite number of variables $x_{J+1}$, $x_J, \ldots, x_1, x_0$, i.e.,

$$S[\gamma_{\Delta_{T,0}}] = S_{\Delta_{T,0}}(x_{J+1}, x_J, \ldots, x_1, x_0) \,,$$
$$F[\gamma_{\Delta_{T,0}}] = F_{\Delta_{T,0}}(x_{J+1}, x_J, \ldots, x_1, x_0) \,. \quad (7)$$

Therefore Feynman[3] omitted the first step $S[\gamma_{\Delta_{T,0}}]$, $F[\gamma_{\Delta_{T,0}}]$ and wrote the form of functions of the right hand side of (7). Furthermore, many books about Feynman path integrals abandon the first step $S[\gamma_{\Delta_{T,0}}]$ in order to use the Trotter formula, i.e.,

$$S[\gamma_{\Delta_{T,0}}] = \sum_{j=1}^{J+1} \frac{(x_j - x_{j-1})^2}{2t_j} - \sum_{j=1}^{J+1} \int_{T_{j-1}}^{T_j} V\left(t, \frac{t - T_{j-1}}{T_j - T_{j-1}} x_j + \frac{T_j - t}{T_j - T_{j-1}} x_{j-1}\right) dt$$

$$\neq \sum_{j=1}^{J+1} \frac{(x_j - x_{j-1})^2}{2t_j} - \sum_{j=1}^{J+1} V(T_{j-1}, x_{j-1}) \,.$$

However, we keep the first step $S[\gamma_{\Delta_{T,0}}]$, $F[\gamma_{\Delta_{T,0}}]$.

**Remark.** Even when $F[\gamma] \equiv 1$, the integrals of the right hand side of (6) does not converge absolutely. Therefore, we treat integrals of this type as oscillatory integrals. (cf. H. Kumano-go [18], H. Kumano-go and K. Taniguchi [19], D. Fujiwara, N. Kumano-go and K. Taniguchi [12], N. Kumano-go [21,22])

**Remark.** If $|\Delta_{T,0}| \to 0$, the number $J$ of the integrals of the right hand side of (6) tends to $\infty$. Therefore, we use the first step $S[\gamma_{\Delta_{T,0}}]$, $F[\gamma_{\Delta_{T,0}}]$.

Our assumption of the potential $V(t, x)$ of (2) is the following:

**Assumption 2.1.** $V(t, x)$ *is a real-valued function of* $(t, x) \in \mathbf{R} \times \mathbf{R}^d$, *and, for any multi-index* $\alpha$, $\partial_x^\alpha V(t, x)$ *is continuous in* $\mathbf{R} \times \mathbf{R}^d$. *For any integer* $k \geq 2$, *there exists a positive constant* $A_k$ *such that for any multi-index* $\alpha$ *with* $|\alpha| = k$,

$$|\partial_x^\alpha V(t, x)| \leq A_k \,.$$

In order to state the definition of the class $\mathcal{F}^\infty$ of functionals $F[\gamma]$, we explain the functional derivatives in this note.

**Definition 2.1.** (Functional derivatives). For any division $\Delta_{T,0}$ of (4), we assume

$$F_{\Delta_{T,0}}(x_{J+1}, x_J, \ldots, x_1, x_0) \in C^{\infty}(\mathbf{R}^{d(J+2)}).$$

For any broken line paths $\gamma : [0, T] \to \mathbf{R}^d$ and $\eta_l : [0, T] \to \mathbf{R}^d$, $l = 1, 2, \ldots, L$, we define the functional derivative $(D^L F)[\gamma] \prod_{l=1}^{L} [\eta_l]$ by

$$(D^L F)[\gamma] \prod_{l=1}^{L} [\eta_l] = \left( \prod_{l=1}^{L} \frac{\partial}{\partial \theta_l} \right) F\left[ \gamma + \sum_{l=1}^{L} \theta_l \eta_l \right] \Bigg|_{\theta_1 = \theta_2 = \cdots = \theta_L = 0}.$$

When $L = 0$, we also write $(D^L F)[\gamma] \prod_{l=1}^{L} [\eta_l] = F[\gamma]$.

**Remark.** Let $\Delta_{T,0}$ of (4) contain all time where the broken line path $\gamma$ or the broken line path $\eta$ breaks. Set $\gamma(T_j) = x_j$ and $\eta(T_j) = y_j$, $j = 0, 1, \ldots, J, J+1$. Then, for any $\theta \in \mathbf{R}$, $\gamma + \theta \eta$ is the broken line path which connects $(T_j, x_j + \theta y_j)$ and $(T_{j-1}, x_{j-1} + \theta y_{j-1})$ by a line segment for $j = 1, 2, \ldots, J, J+1$. Hence we have

$$F[\gamma + \theta \eta] = F_{\Delta_{T,0}}(x_{J+1} + \theta y_{J+1}, x_J + \theta y_J, \ldots, x_1 + \theta y_1, x_0 + \theta y_0).$$

Therefore, we can write $(DF)[\gamma][\eta]$ as a finite sum as follows:

$$(DF)[\gamma][\eta] = \frac{d}{d\theta} F[\gamma + \theta \eta] \Bigg|_{\theta=0} = \sum_{j=0}^{J+1} (\partial_{x_j} F_{\Delta_{T,0}})(x_{J+1}, x_J, \ldots, x_1, x_0) \cdot y_j.$$

**Definition 2.2.** (The class $\mathcal{F}$ of functionals $F[\gamma]$). Let $F[\gamma]$ be a functional on the path space $C([0, T] \to \mathbf{R}^d)$ such that the domain of $F[\gamma]$ contains all of broken line paths at least. We say that $F[\gamma]$ belongs to the class $\mathcal{F}^{\infty}$ if $F[\gamma]$ satisfies Assumption 2.2. For simplicity, we write $F[\gamma] \in \mathcal{F}^{\infty}$.

**Assumption 2.2.** *Let $m$ be a non-negative integer and $\rho(t)$ be a function of bounded variation on $[0, T]$. For any non-negative integer $M$, there exists a positive constant $C_M$ such that*

$$\left| \left( D^{\sum_{j=0}^{J+1} L_j} F \right) [\gamma] \prod_{j=0}^{J+1} \prod_{l_j=1}^{L_j} [\eta_{j,l_j}] \right| \leq (C_M)^{J+2} (1 + \|\gamma\|)^m \prod_{j=0}^{J+1} \prod_{l_j=1}^{L_j} \|\eta_{j,l_j}\|,$$

$$\left| \left( D^{1 + \sum_{j=0}^{J+1} L_j} F \right) [\gamma][\eta] \prod_{j=0}^{J+1} \prod_{l_j=1}^{L_j} [\eta_{j,l_j}] \right|$$

$$\leq (C_M)^{J+2} (1 + \|\gamma\|)^m \int_0^T |\eta(t)| d|\rho|(t) \prod_{j=0}^{J+1} \prod_{l_j=1}^{L_j} \|\eta_{j,l_j}\|,$$

*for any division $\Delta_{T,0}$ defined by (4), any $L_j = 0, 1, \ldots, M$, any broken line path $\gamma : [0, T] \to \mathbf{R}^d$, any broken line path $\eta : [0, T] \to \mathbf{R}^d$, and any broken line paths $\eta_{j,l_j} : [0, T] \to \mathbf{R}^d$, $l_j = 1, 2, \ldots, L_j$ whose supports exist in $[T_{j-1}, T_{j+1}]$. Here $0 = T_{-1} = T_0$, $T_{J+1} = T_{J+2} = T$, $\|\gamma\| = \max_{0 \leq t \leq T} |\gamma(t)|$ and $|\rho|(t)$ is the total variation of $\rho(t)$.*

**Remark.** Note that the support of the broken line path $\eta_{j,l_j}$ exists in $[T_{j-1}, T_{j+1}]$ for any $j = 0, 1, \ldots, J, J+1$. Roughly speaking, the broken line paths $\eta_{j,l_j}$, $j = 0, 1, \ldots, J, J+1$ slice the time interval $[0, T]$.

**Theorem 2.1.** *(Existence of Feynman path integral). Let $T$ be sufficiently small. Then, for any $F[\gamma] \in \mathcal{F}^\infty$, the right hand side of (6) converges uniformly on any compact set of the configuration space $(x, x_0) \in \mathbf{R}^{2d}$, together with all its derivatives in $x$ and $x_0$.*

**Remark.** The size of $T$ depends only on $d$ and $A_k$ of Assumption 2.1.

**Theorem 2.2.** *(Smooth algebra). For any $F[\gamma]$, $G[\gamma] \in \mathcal{F}^\infty$, any broken line path $\zeta : [0, T] \to \mathbf{R}^d$ and any real $d \times d$ matrix $P$, we have the following.*

(1) $F[\gamma] + G[\gamma] \in \mathcal{F}^\infty$, $\quad F[\gamma]G[\gamma] \in \mathcal{F}^\infty$.
(2) $F[\gamma + \zeta] \in \mathcal{F}^\infty$, $\quad F[P\gamma] \in \mathcal{F}^\infty$.
(3) $(DF)[\gamma][\zeta] \in \mathcal{F}^\infty$.

**Remark.** In other words, $\mathcal{F}^\infty$ is closed under addition, multiplication, translation, real linear transformation and functional differentiation. Applying Theorem 2.2 to the examples of Theorems 2.3 (1)(2), Theorem 2.4 (1) and Theorem 2.6, we can produce many functionals $F[\gamma] \in \mathcal{F}^\infty$.

**Assumption 2.3.** *Let $m$ be a non-negative integer. $B(t, x)$ is a function of $(t, x) \in \mathbf{R} \times \mathbf{R}^d$. For any multi-index $\alpha$, $\partial_x^\alpha B(t, x)$ is continuous on $\mathbf{R} \times \mathbf{R}^d$, and there exists a positive constant $C_\alpha$ such that*

$$|\partial_x^\alpha B(t, x)| \leq C_\alpha (1 + |x|)^m .$$

**Theorem 2.3.** *(Interchange with Riemann-Stieltjes integrals). Let $0 \leq T' \leq T'' \leq T$ and $0 \leq t \leq T$. Let $\rho(t)$ be a function of bounded variation on $[T', T'']$. Suppose $B(t, x)$ satisfy Assumption 2.3. Then we have the following.*

(1) *The value at a fixed time $t$*

$$F[\gamma] = B(t, \gamma(t)) \in \mathcal{F}^\infty .$$

(2) *The Riemann-Stieltjes integral*

$$F[\gamma] = \int_{T'}^{T''} B(t, \gamma(t)) d\rho(t) \in \mathcal{F}^\infty .$$

(3) *Let $T$ be sufficiently small. Then we have*

$$\int_{T'}^{T''} \left( \int e^{\frac{i}{\hbar} S[\gamma]} B(t, \gamma(t)) \mathcal{D}[\gamma] \right) d\rho(t)$$

$$= \int e^{\frac{i}{\hbar} S[\gamma]} \left( \int_{T'}^{T''} B(t, \gamma(t)) d\rho(t) \right) \mathcal{D}[\gamma] .$$

**Remark.** We explain the key of the proof of Theorem 2.3 (3) roughly. In order to use the Trotter formula, many books about Feynman path integrals approximate the position of the particle at time $t$ by the endpoint $x_j$ or $x_{j-1}$. On the other hand, using the number $j$ so that $T_{j-1} \le t \le T_j$, we keep the position of the particle at time $t$, i.e.,

$$\gamma_{\Delta_{T,0}}(t) = \frac{t - T_{j-1}}{T_j - T_{j-1}} x_j + \frac{T_j - t}{T_j - T_{j-1}} x_{j-1},$$

inside the finite dimensional oscillatory integral of (6). Therefore, we can use the continuity of the broken line path $\gamma_{\Delta_{T,0}}(t)$ with respect to $t$.

**Proof of Theorem 2.3(3).** Note that $B(t, \gamma_{\Delta_{T,0}}(t))$ is a continuous function of $t$ on $[T', T'']$, together with all its derivatives in $x_j$, $j = 0, 1, \ldots, J, J+1$. By Lebesgue's dominated convergence theorem after integrating by parts by $x_j$, $j = 1, 2, \ldots, J$ (Oscillatory integrals), for any division $\Delta_{T,0}$,

$$\prod_{j=1}^{J+1} \left( \frac{1}{2\pi i \hbar t_j} \right)^{d/2} \int_{\mathbf{R}^{dJ}} e^{\frac{i}{\hbar} S[\gamma_{\Delta_{T,0}}]} B(t, \gamma_{\Delta_{T,0}}(t)) \prod_{j=1}^{J} dx_j$$

is also a continuous function of $t$ on $[T', T'']$. By Theorem 2.1, the convergence of

$$\int e^{\frac{i}{\hbar} S[\gamma]} B(t, \gamma(t)) \mathcal{D}[\gamma]$$

$$= \lim_{|\Delta_{T,0}| \to 0} \prod_{j=1}^{J+1} \left( \frac{1}{2\pi i \hbar t_j} \right)^{d/2} \int_{\mathbf{R}^{dJ}} e^{\frac{i}{\hbar} S[\gamma_{\Delta_{T,0}}]} B(t, \gamma_{\Delta_{T,0}}(t)) \prod_{j=1}^{J} dx_j,$$

is uniform with respect to $t$ on $[T', T'']$. Therefore,

$$\int e^{\frac{i}{\hbar} S[\gamma]} B(t, \gamma(t)) \mathcal{D}[\gamma]$$

is also a continuous function of $t$ on $[T', T'']$ and Riemann-Stieltjes integrable. Furthermore, by the uniform convergence, we can interchange the order of $\int_{T'}^{T''} \cdots dt$ and $\lim_{|\Delta_{T,0}| \to 0}$.

$$\int_{T'}^{T''} \left( \int e^{\frac{i}{\hbar} S[\gamma]} B(t, \gamma(t)) \mathcal{D}[\gamma] \right) dt$$

$$= \int_{T'}^{T''} \lim_{|\Delta_{T,0}| \to 0} \prod_{j=1}^{J+1} \left( \frac{1}{2\pi i \hbar t_j} \right)^{d/2} \int_{\mathbf{R}^{dJ}} e^{\frac{i}{\hbar} S[\gamma_{\Delta_{T,0}}]} B(t, \gamma_{\Delta_{T,0}}(t)) \prod_{j=1}^{J} dx_j dt$$

$$= \lim_{|\Delta_{T,0}| \to 0} \int_{T'}^{T''} \prod_{j=1}^{J+1} \left( \frac{1}{2\pi i \hbar t_j} \right)^{d/2} \int_{\mathbf{R}^{dJ}} e^{\frac{i}{\hbar} S[\gamma_{\Delta_{T,0}}]} B(t, \gamma_{\Delta_{T,0}}(t)) \prod_{j=1}^{J} dx_j dt.$$

By Fubini's theorem after integrating by parts by $x_j$, $j = 1, 2, \ldots, J$ (Oscillatory integrals), we have

$$
= \lim_{|\Delta_{T,0}| \to 0} \prod_{j=1}^{J+1} \left( \frac{1}{2\pi i \hbar t_j} \right)^{d/2} \int_{\mathbf{R}^{dJ}} e^{\frac{i}{\hbar} S[\gamma_{\Delta_{T,0}}]} \int_{T'}^{T''} B(t, \gamma_{\Delta_{T,0}}(t)) dt \prod_{j=1}^{J} dx_j
$$

$$
= \int e^{\frac{i}{\hbar} S[\gamma]} \left( \int_{T'}^{T''} B(t, \gamma(t)) dt \right) \mathcal{D}[\gamma]. \quad \square
$$

**Assumption 2.4.** $f(b)$ *is an analytic function of* $b \in \mathbf{C}$ *on a neighborhood of zero, i.e., there exist positive constants* $\mu > 0$, $A > 0$ *such that*

$$
\|f\|_{\mu, A} = \sup_{n, |b| \leq \mu} \frac{|\partial_b^n f(b)|}{A^n n!} < \infty.
$$

**Theorem 2.4.** *(Interchange with a limit). Let* $0 \leq T' \leq T'' \leq T$. *Let* $\rho(t)$ *be a function of bounded variation on* $[T', T'']$. *Suppose* $B(t, x)$ *satisfy Assumption 2.3 with* $m = 0$. *Let* $f(b)$ *and* $f_k(b)$, $k = 1, 2, 3, \ldots$ *be analytic functions such that* $\lim_{k \to \infty} \|f_k - f\|_{\mu, A} = 0$. *Then we have the following.*

(1) $F[\gamma] = f \left( \displaystyle\int_{T'}^{T''} B(t, \gamma(t)) d\rho(t) \right) \in \mathcal{F}^\infty$.

(2) *Let* $T$ *be sufficiently small. Then we have*

$$
\lim_{k \to \infty} \int e^{\frac{i}{\hbar} S[\gamma]} f_k \left( \int_{T'}^{T''} B(t, \gamma(t)) d\rho(t) \right) \mathcal{D}[\gamma]
$$

$$
= \int e^{\frac{i}{\hbar} S[\gamma]} f \left( \int_{T'}^{T''} B(t, \gamma(t)) d\rho(t) \right) \mathcal{D}[\gamma].
$$

**Corollary.** (Perturbation expansion formula). *Let* $T$ *be sufficiently small. Let* $\rho(t)$ *and* $B(t, x)$ *be the same as in Theorem 2.4. Then we have*

$$
\int e^{\frac{i}{\hbar} S[\gamma] + \frac{i}{\hbar} \int_{T'}^{T''} B(\tau, \gamma(\tau)) d\rho(\tau)} \mathcal{D}[\gamma]
$$

$$
= \sum_{n=1}^{\infty} \left( \frac{i}{\hbar} \right)^n \int_{T'}^{T''} d\rho(\tau_n) \int_{T'}^{\tau_n} d\rho(\tau_{n-1}) \cdots \int_{T'}^{\tau_2} d\rho(\tau_1)
$$

$$
\times \int e^{\frac{i}{\hbar} S[\gamma]} B(\tau_n, \gamma(\tau_n)) B(\tau_{n-1}, \gamma(\tau_{n-1})) \cdots B(\tau_1, \gamma(\tau_1)) \mathcal{D}[\gamma].
$$

**Theorem 2.5.** *(Semiclassical approximation). Let* $T$ *be sufficiently small. Let* $F[\gamma] \in \mathcal{F}^\infty$ *and the domain of* $F[\gamma]$ *be continuously extended to* $C([0, T] \to \mathbf{R}^d)$ *with respect to the norm* $\|\gamma\| = \max_{0 \leq t \leq T} |\gamma(t)|$. *Let* $\gamma^{cl}$ *be the classical path with* $\gamma^{cl}(0) = x_0$ *and* $\gamma^{cl}(T) = x$, *and* $D(T, x, x_0)$ *be the Morette-Van Vleck determinant. Define* $\Upsilon(\hbar, T, x, x_0)$ *by*

$$
\int e^{\frac{i}{\hbar} S[\gamma]} F[\gamma] \mathcal{D}[\gamma] = \left( \frac{1}{2\pi i \hbar T} \right)^{d/2} e^{\frac{i}{\hbar} S[\gamma^{cl}]} \left( D(T, x, x_0)^{-1/2} F[\gamma^{cl}] + \hbar \Upsilon(\hbar, T, x, x_0) \right).
$$

*Then, for any multi-indices $\alpha$, $\beta$, there exists a positive constant $C_{\alpha,\beta}$ such that*

$$|\partial_x^\alpha \partial_{x_0}^\beta \Upsilon(\hbar, T, x, x_0)| \leq C_{\alpha,\beta}(1 + |x| + |x_0|)^m.$$

**Remark.** If $\hbar \to 0$, the remainder term $\hbar \Upsilon(\hbar, T, x, x_0) \to 0$.

**Theorem 2.6.** *(New curvilinear integrals along path on path space). Let $0 \leq T' \leq T'' \leq T$. Let $m$ be non-negative integer. Let $Z(t, x)$ be a vector-valued function of $(t, x) \in \mathbf{R} \times \mathbf{R}^d$ into $\mathbf{R}^d$ such that, for any multi-index $\alpha$, $\partial_x^\alpha Z(t, x)$ and $\partial_x^\alpha \partial_t Z(t, x)$ are continuous on $[0, T] \times \mathbf{R}^d$, and there exists a positive constant $C_\alpha$ such that*

$$|\partial_x^\alpha Z(t, x)| + |\partial_x^\alpha \partial_t Z(t, x)| \leq C_\alpha(1 + |x|)^m,$$

*and $\partial_x Z(t, x)$ is a symmetric matrix, i.e., ${}^t(\partial_x Z) = \partial_x Z$. Then the curvilinear integrals along paths of Feynman path integral*

$$F[\gamma] = \int_{T'}^{T''} Z(t, \gamma(t)) \cdot d\gamma(t) \in \mathcal{F}^\infty.$$

*Here $Z \cdot d\gamma$ is the inner product of $Z$ and $d\gamma$ in $\mathbf{R}^d$.*

**Remark.** In order to explain the difference with known curvilinear integrals on a path space, please forgive very rough sketch. As examples of curvilinear integrals on a path space, Itô integral [15] and Stratonovich integral [27] for the Brownian motion $\mathbf{B}(t)$ are successful in Malliavin analysis [23]. If we can set $\mathbf{B}(T_j) = x_j$, Itô integral is approximated by initial points, i.e.,

$$\int_{T'}^{T''} Z(t, \mathbf{B}(t)) \cdot d\mathbf{B}(t) \approx \sum_j Z(T_{j-1}, x_{j-1}) \cdot (x_j - x_{j-1}).$$

and Stratonovich integral is approximated by middle points, i.e.,

$$\int_{T'}^{T''} Z(t, \mathbf{B}(t)) \circ d\mathbf{B}(t) \approx \sum_j Z\left(\frac{T_j + T_{j-1}}{2}, \frac{x_j + x_{j-1}}{2}\right) \cdot (x_j - x_{j-1}).$$

And many books about Feynman path integrals use endpoints or middle points. On the other hand, if $\gamma = \gamma_{\Delta_{T,0}}$, our new curvilinear integrals is the classical curvilinear integrals itself along the broken line path $\gamma_{\Delta_{T,0}}$, i.e.,

$$\int_{T'}^{T''} Z(t, \gamma_{\Delta_{T,0}}(t)) \cdot d\gamma_{\Delta_{T,0}}(t). \tag{8}$$

In other words, Itô integral and Stratonovich integral are limits of Riemannian sum. On the other hand, our new integral is a limit of curvilinear integral.

**Theorem 2.7.** *(Fundamental theorem of calculus). Let $m$ be non-negative integer and $0 \leq T' \leq T'' \leq T$. $g(t, x)$ is a function of $(t, x) \in \mathbf{R} \times \mathbf{R}^d$ such that $g(t, x)$ and $\partial_t g(t, x)$ satisfy Assumption 2.3. Let $T$ be sufficiently small. Then we have*

$$\int e^{\frac{i}{\hbar} S[\gamma]} \Big( g(T'', \gamma(T'')) - g(T', \gamma(T')) \Big) \mathcal{D}[\gamma]$$

$$= \int e^{\frac{i}{\hbar} S[\gamma]} \left( \int_{T'}^{T''} (\partial_x g)(t, \gamma(t)) \cdot d\gamma(t) + \int_{T'}^{T''} (\partial_t g)(t, \gamma(t)) dt \right) \mathcal{D}[\gamma].$$

**Remark.** (8) is the key of the proof of Theorem 2.7.

**Proof of Theorem 2.7.** By Theorem 2.3(1) and Theorem 2.2(1), we have

$$G_1[\gamma] = g\left(T'', \gamma(T'')\right) - g\left(T', \gamma(T')\right) \in \mathcal{F}^\infty.$$

We note that $\,^t(\partial_x^2 g) = (\partial_x^2 g)$. By Theorem 2.6, Theorem 2.3(2) and Theorem 2.2(1), we have

$$G_2[\gamma] = \int_{T'}^{T''} (\partial_x g)(t, \gamma(t)) \cdot d\gamma(t) + \int_{T'}^{T''} (\partial_t g)(t, \gamma(t)) dt \in \mathcal{F}^\infty.$$

By the fundamental theorem of calculus for any broken line path $\gamma_{\Delta_{T,0}}$, we have $G_1[\gamma_{\Delta_{T,0}}] = G_2[\gamma_{\Delta_{T,0}}]$. By Theorem 2.1, we get

$$\int e^{\frac{i}{\hbar}S[\gamma]} G_1[\gamma] \mathcal{D}[\gamma]$$

$$= \lim_{|\Delta_{T,0}| \to 0} \prod_{j=1}^{J+1} \left(\frac{1}{2\pi i \hbar t_j}\right)^{d/2} \int_{\mathbf{R}^{dJ}} e^{\frac{i}{\hbar}S[\gamma_{\Delta_{T,0}}]} G_1[\gamma_{\Delta_{T,0}}] \prod_{j=1}^{J} dx_j$$

$$= \lim_{|\Delta_{T,0}| \to 0} \prod_{j=1}^{J+1} \left(\frac{1}{2\pi i \hbar t_j}\right)^{d/2} \int_{\mathbf{R}^{dJ}} e^{\frac{i}{\hbar}S[\gamma_{\Delta_{T,0}}]} G_2[\gamma_{\Delta_{T,0}}] \prod_{j=1}^{J} dx_j$$

$$= \int e^{\frac{i}{\hbar}S[\gamma]} G_2[\gamma] \mathcal{D}[\gamma]. \quad \square$$

**Theorem 2.8.** *(Translation). Let $T$ be sufficiently small. For any $F[\gamma] \in \mathcal{F}^\infty$ and any broken line path $\eta : [0, T] \to \mathbf{R}^d$,*

$$\int_{\gamma(0)=x_0, \gamma(T)=x} e^{\frac{i}{\hbar}S[\gamma+\eta]} F[\gamma + \eta] \mathcal{D}[\gamma] = \int_{\gamma(0)=x_0+\eta(0), \gamma(T)=x+\eta(T)} e^{\frac{i}{\hbar}S[\gamma]} F[\gamma] \mathcal{D}[\gamma].$$

**Corollary.** (Invariance under translation). *Let $T$ be sufficiently small. For any $F[\gamma] \in \mathcal{F}^\infty$ and any broken line path $\eta : [0, T] \to \mathbf{R}^d$ with $\eta(0) = \eta(T) = 0$,*

$$\int_{\gamma(0)=x_0, \gamma(T)=x} e^{\frac{i}{\hbar}S[\gamma+\eta]} F[\gamma + \eta] \mathcal{D}[\gamma] = \int_{\gamma(0)=x_0, \gamma(T)=x} e^{\frac{i}{\hbar}S[\gamma]} F[\gamma] \mathcal{D}[\gamma].$$

**Theorem 2.9.** *(Integration by parts). Let $T$ be sufficiently small. For any $F[\gamma] \in \mathcal{F}^\infty$ and any broken line path $\eta : [0, T] \to \mathbf{R}^d$ with $\eta(0) = \eta(T) = 0$,*

$$\int e^{\frac{i}{\hbar}S[\gamma]} (DF)[\gamma][\eta] \mathcal{D}[\gamma] = -\frac{i}{\hbar} \int e^{\frac{i}{\hbar}S[\gamma]} (DS)[\gamma][\eta] F[\gamma] \mathcal{D}[\gamma].$$

**Theorem 2.10.** *(Orthogonal transformation). Let $T$ be sufficiently small. For any $F[\gamma] \in \mathcal{F}^\infty$ and any $d \times d$ orthogonal matrix $Q$,*

$$\int_{\gamma(0)=x_0, \gamma(T)=x} e^{\frac{i}{\hbar}S[Q\gamma]} F[Q\gamma] \mathcal{D}[\gamma] = \int_{\gamma(0)=Qx_0, \gamma(T)=Qx} e^{\frac{i}{\hbar}S[\gamma]} F[\gamma] \mathcal{D}[\gamma].$$

**Corollary.**(Invariance under orthogonal transformation). *Let $T$ be sufficiently small. For any $F[\gamma] \in \mathcal{F}^{\infty}$, any $d \times d$ orthogonal matrix $Q$ and any broken line path $\eta : [0, T] \to \mathbf{R}^d$,*

$$\int_{\gamma(0)=0,\gamma(T)=0} e^{\frac{i}{\hbar} S[Q\gamma+\eta]} F[Q\gamma + \eta] \mathcal{D}[\gamma] = \int_{\gamma(0)=\eta(0),\gamma(T)=\eta(T)} e^{\frac{i}{\hbar} S[\gamma]} F[\gamma] \mathcal{D}[\gamma].$$

As the other application of our formulation, recently, D. Fujiwara wrote down the second term of the semi-classical approximation of Feynman path integrals. If $F[\gamma] \equiv 1$, this second term coincides with the one given by G. D. Birkhoff. For the details, see D. Fujiwara and N. Kumano-go [11,10].

## References

1. S. Albeverio and Høegh -Krohn. *Mathematical theory of Feynman path integrals.* Lecture notes of Mathematics, 523, Springer, Berlin (1976).
2. R. H. Cameron, *J. Math. and Phys.* **39**, 126 (1960).
3. R. P. Feynman, *Rev. Modern Phys.* **20**, 367 (1948).
4. R. P. Feynman and A. R. Hibbs, *Quantum Mechanics and Path Integrals,* McGraw-Hill College (1965).
5. D. Fujiwara, *Duke Math. J.* **47**, 559 (1980).
6. D. Fujiwara, *Nagoya Math. J.* **124**, 61 (1991).
7. D. Fujiwara, *Lecture Notes in Math. Springer* **1540**, 39 (1993).
8. D. Fujiwara, *Mathematical methods for Feynman path integrals, Springer-Verlag Tokyo* (1999).
9. D. Fujiwara and N. Kumano-go, *Bull. Sci. math.* **129**, 57 (2005).
10. D. Fujiwara and N. Kumano-go, *Funkcial. Ekvac.*, to appear.
11. D. Fujiwara and N. Kumano-go, *J. Math. Soc. Japan*, to appear.
12. D. Fujiwara, N. Kumano-go and K. Taniguchi, *Funkcial. Ekvac.* **40**, 459 (1997).
13. D. Fujiwara and T. Tsuchida, *J. Math. Soc. Japan* **49**, 299 (1997).
14. W. Ichinose, *Comm. Math. Phys.* **189** 17 (1997).
15. K. Itô, *Proc. Imp. Acad. Tokyo* **20** 519 (1944).
16. K. Itô, *Proc. 5th Berkeley Sympos. Math. Statist. and Prob.* **2**, 145, University of California press, Berkeley (1967).
17. H. Kitada and H. Kumano-go. *Osaka J. Math.* **18** 291 (1981).
18. H. Kumano-go. *Pseudo-Differential Operators.* The MIT press, Cambridge, Massachusetts, and London, England (1981).
19. H. Kumano-go and K. Taniguchi. *Funkcial. Ekvac.* **22**, 161 (1979).
20. N. Kumano-go. *J. Math. Sci. Univ. Tokyo* **2**, 441 (1995).
21. N. Kumano-go, *Osaka J. Math.* **35**, 357 (1998).
22. N. Kumano-go, *Bull. Sci. math.* **128**, 197 (2004).
23. P. Malliavin, *Stochastic analysis.* Springer, Berlin, Heidelberg, New York (1997).
24. E. Nelson, *J. Math. Phys.* **5**, 332 (1964).
25. J. Rezende, *Rev. Math. Phys.* **8**, 1161 (1996).
26. L. S. Schulman, *Techniques and Applications of Path Integration.* John Wiley & Sons Inc (1981)
27. R. L. Stratonovich, *J. SIAM Control* **4**, 362 (1966).
28. K. Yajima, *J. Analyse Math.* **56**, 29 (1991).

## I.9   Stochastic Analysis

Organizers: N. Jacob, Y. Xiao

Already for the ISAAC Congress in Toronto an attempt was made to run a session called "Stochastic Analysis" (understood in a wide sense). With the ISAAC Congress in Catania this session not only continued to run but for the first time invited lectures of this session contributed to the Proceedings ISAAC05-Catania. May be it is worth adding a few remarks on what is the intention of including "Stochastic Analysis" to the regular programme of ISAAC congresses.

By Stochastic Analysis (in a wide sense) we understand most of all those parts of stochastics, especially the theory of stochastic processes, which heavily rely on tools from analysis, or, especially when thinking on Stochastic Analysis in a more narrow sense (as usually probabilists do), those parts which need to develop new analytic tools - for example It's integral or analysis in infinite dimensions.

The first rigorous models of stochastic processes rely much on analysis, especially the theory of partial differential equations and Fourier analysis. Later it was realized that stochastic processes are important tools for studying partial differential equations - the most known example seems to be the Feynman-Kac formula. Modern developments include It's calculus and stochastic differential equations, Malliavin calculus, white noise analysis, fractional calculus, pseudodifferential operator theory as well as analysis on metric measure spaces or stochastic partial differential equations - just to mention a few topics.

It is obvious that research in these directions heavily depends on analysis, but one must note that many results from stochastic analysis give new insights in problems of analysis. Introducing a session on "Stochastic Analysis" into the ISAAC programme is a further attempt of breaking barriers, aiming to cross boundaries.

This holds the more when including problems from mathematical modelling. The three articles to follow cover quite different directions: Analysis of Random Fields, Modelling using Fractional Calculus, and Infinite Dimensional Analysis. We are convinced that they may suite the purpose stated above: to help bridging between different subject areas.

# IDENTIFICATION AND SERIES DECOMPOSITION OF ANISOTROPIC GAUSSIAN FIELDS

A. AYACHE

UMR CNRS 8524 Laboratoire Paul Painlevé, Université Lille 1, 59655 Villeneuve d'Ascq Cedex, France
and
UMR CNRS 8179 LEM, IAE de Lille (Université Lille 1), 104, Avenue du peuple Belge 59043 Lille Cedex, France
E-mail: Antoine.Ayache@math.univ-lille1.fr


A. BONAMI

MAPMO-UMR 6628, Université d'Orléans, 45067 Orléans Cédex 2, France
E-mail: Aline.Bonami@univ-orleans.fr


A. ESTRADE

MAP5-UMR 8145, Université René Descartes, 45, rue des Saints-Pères, 75270 Paris Cédex 06, France
E-mail: anne.estrade@univ-paris5.fr

Anisotropy of a Gaussian field with stationary increments is related with the anisotropy of its spectral density. Such Gaussian fields can be used for modelling anisotropic homogeneous media, which leads to try to simulate these ones and identify parameters. This paper is a first attempt in these two directions. We first show how such Gaussian fields can be written as a random series, which is a first step for simulation purposes. There is anisotropy of the Gaussian field when its spectral density has a different power law in each direction. The exponent in a given direction can be obtained from an integration of the field on the orthogonal hyperplanes, as proved by the two last authors in [4]. We then show how this property can be used to propose an estimator of this exponent.

**Keywords:** Anisotropy, identification, estimator, Gaussian field
**Mathematics Subject Classification:** 60G15, 60G17, 60G18, 62F10

## 1. Introduction

We consider a general real-valued Gaussian field with zero mean and stationary increments, which we assume as given by its spectral representation

$$X(t) := \int_{I\!\!R^d} (e^{it.\xi} - 1) g(\xi) \, dW(\xi) \, , \, t \in I\!\!R^d \tag{1}$$

where $\{W(\xi); \xi \in I\!\!R^d\}$ is the complex-valued Lévy Brownian field obtained by Fourier transformation of the real-valued Lévy Brownian field. We assume that the function $|g|^2$, which is known as *the spectral density of the Gaussian field $X$*, is an even function that satisfies the integrability assumption

$$\int_{I\!R^d} (1 \wedge |\xi|^2) |g(\xi)|^2 d\xi < \infty. \tag{2}$$

We assume that the function $g$ itself is positive.

In fact, we shall make use of a sharper integrability condition, called **Assumption** $D(m)$, where $m$ is some real number, $m \in (0,1)$. This last one is given by

$$D(m) \qquad \int_{I\!R^d} (|\xi|^{2m} \wedge |\xi|^2) |g(\xi)|^2 d\xi < \infty. \tag{3}$$

Under this assumption, it is elementary to see that, for $t, t'$ in a compact set of $I\!R^d$, there exists a constant $C$ such that

$$\mathbf{E}(|X(t) - X(t')|^2) \leq C|t - t'|^{2m}. \tag{4}$$

It then follows from the equivalence of Gaussian moments and from Kolmogorov's method (see for example [1]) that $X$, as a function of $t$, is almost surely Hölder continuous with parameter $r$, for every $r < m$. More precisely, a.s. for any compact set $K$ of $I\!R^d$, there exists a positive finite random variable $A$ such that,

$$|X(t) - X(t')| \leq A|t - t'|^r \quad t, t' \in K. \tag{5}$$

We refer to [4] and [5] for details and references.

We first prove that the field $X$ can also be represented as a sum of $\sum s_n(t)\epsilon_n$, with $\epsilon_n$ a sequence of independent standard Gaussian variables. Roughly speaking, the coefficients $s_n$ are obtained from an orthonormal basis through the action of $g$ as a Fourier multiplier. In particular, for the Fractional Brownian Motion with Hurst parameter $H$, then $g(\xi) = |\xi|^{-H-d/2}$ and the operator involved is a fractional primitive. Explicit formulas have been developed in this case in [10] to obtain a practical method of simulation, using wavelets and Multi Resolution Analysis. Here we content ourselves to give a general formula, which is not contained in [10] in the case of the Fractional Brownian Motion. More work would be necessary to make it operational for numerical purposes.

Next, in order to reflect anisotropy, we assume that $g$ has a different behavior at infinity depending on the direction, that is, decreases like $|\xi|^{-\beta(\kappa)-d/2}$ in the direction $\kappa$ (we will give precise definitions later on). For each direction $\kappa$, we are interested in the identification of the asymptotical exponent $\beta(\kappa)$. For obvious symmetry reasons, we can assume that $\kappa$ is the vertical direction, which we will do from now on.

We rely on the main result of [4], which is the fact that taking integrals on the horizontal hyperplanes gives a Gaussian process with stationary increments and spectral density that decreases also like $|\xi|^{-\beta(\kappa)-d/2}$. We then adapt Istas-Lang estimators [6] to our context. Moreover, we show that the integrals on the horizontal hyperplanes can be replaced by finite sums, which correspond to discretization, under the condition that the discretization lag be sufficiently small compared to the lag in the vertical direction.

Section 2 is devoted to random series, Section 3 to estimators. We postpone to the Appendix an identification result for Gaussian processes with stationary increments that we shall use.

Finally, we recall that such Gaussian fields have been introduced in [4] to model the density of bones or the grey level of bones radiographs. The identification method that we present here has been tested on simulated pictures [7] and on trabecular bones [8].

## 2. Random series expansions of anisotropic models

As we said in the introduction, we assume that the spectral density $|g|^2$ satisfies Assumption $D(m)$ and write $X$, given by (1), as the sum of an almost surely uniformly convergent random series. To this end we introduce an arbitrary orthonormal basis of the Hilbert space $L^2(\mathbb{R}^d)$, which we note $\{h_n\}_{n\in\mathbb{N}}$. Then $\{s_n\}_{n\in\mathbb{N}}$ is the sequence of functions defined as

$$s_n(t) = \int_{\mathbb{R}^d} (e^{it\cdot\xi} - 1)g(\xi)h_n(\xi)\, d\xi, \tag{6}$$

for every $n \in \mathbb{N}$ and $t \in \mathbb{R}^d$. These functions are well-defined and continuous, since $|(e^{it\cdot\xi} - 1)g(\xi)h_n(\xi)|$ is bounded by an integrable function of $\xi$ for $t$ varying in a compact set. Indeed, in such a compact set, it is bounded by $C(|\xi| \wedge 1)|g(\xi)h_n(\xi)|$, and one concludes using Cauchy-Schwarz Inequality and (2). In fact, under the assumption $D(m)$, one sees easily that $s_n$ satisfies a Hölder condition of order $m$.

Let $\{\epsilon_n\}_{n\in\mathbb{N}}$ be the sequence of independent standard Gaussian random variables defined for every $n \in \mathbb{N}$ as

$$\epsilon_n = \int_{\mathbb{R}^d} \overline{h_n(\xi)}\, dW(\xi). \tag{7}$$

We are now in position to state the main result of this section.

**Theorem 2.1.** *When the spectral density $|g|^2$ satisfies assumption $D(m)$ for some real number $m \in (0,1)$, then the corresponding field $X$, given by (1), can be represented as*

$$X(t) = \sum_{n=0}^{+\infty} s_n(t)\epsilon_n, \tag{8}$$

*where, almost surely, the series is uniformly convergent in $t$ on each compact subset of $\mathbb{R}^d$.*

**Proof.** The proof of Theorem 2.1 is inspired from the proof of Proposition 3 in [2]. First let us suppose that $t \in \mathbb{R}^d$ is arbitrary and fixed. By expanding the kernel $\xi \mapsto (e^{it\cdot\xi} - 1)g(\xi)$ in the orthonormal basis $\{\overline{h}_n\}_{n\in\mathbb{N}}$ and by using the isometry property of the stochastic integral $\int_{\mathbb{R}^d}(\cdot)\, dW$, it follows from (2) that the series $\sum_{n=0}^{+\infty} s_n(t)\epsilon_n$ is convergent to $X(t)$ in the $L^2(\Omega)$-norm (where $\Omega$ is the underlying probability space). To conclude, it is sufficient to prove that the convergence of this series also holds almost surely, uniformly in $t$, on each compact subset $K$ of $\mathbb{R}^d$. For simplicity we will assume that $K = [-R, R]^d$. For $N \in \mathbb{N}$ and $t \in \mathbb{R}^d$, we set

$$X_N(t) := \sum_{n=0}^{N} s_n(t)\epsilon_n. \tag{9}$$

Since the functions $s_n$ are continuous and the random variables $\epsilon_n$ are symmetric and independent, Itô-Nisio Theorem (see Theorem 2.1.1 in [9]) implies that for proving the uniform convergence of the series (8) on $[-R, R]^d$ it is sufficient to show that the sequence $(X_N)_{N \in \mathbb{N}}$ is weakly relatively compact in $\mathcal{C}([-R, R]^d)$, the space of continuous functions on $[-R, R]^d$ equipped with the usual topology of uniform convergence. Using the fact that the $\epsilon_n$'s are independent standard Gaussian random variables and the fact that for every $t$, the series $\sum_{n=0}^{+\infty} s_n(t)\epsilon_n$ is convergent to $X(t)$ in the $L^2(\Omega)$-norm, one obtains that

$$\mathbf{E}(|X_N(t) - X_N(t')|^2) \leq \mathbf{E}(|X(t) - X(t')|^2) \leq C|t - t'|^{2m}, \tag{10}$$

for every $N \in \mathbb{N}$, $t, t' \in [-R, R]^d$. We have used Inequality (4) for the right hand inequality. Observe that the constant $C$ does not depend on $N$, $t$ and $t'$. Finally, using (10), the equivalence of Gaussian moments and Theorem 12.3 in [3], it follows that the sequence $(X_N)_{N \in \mathbb{N}}$ is weakly relatively compact in $\mathcal{C}([-R, R]^d)$.  □

One can obtain as well the convergence of the series in Hölder classes of order $r$, for $r < m$.

## 3. Identification of anisotropic asymptotical exponents

As we said in the introduction, we can fix the direction $\kappa$ to be the vertical direction, generated by the unit vector $\kappa := (0, \cdots, 1)$. We assume that $g$ has a power law in the direction $\kappa$ and want an estimator of the corresponding exponent $\beta := \beta(\kappa)$.

Let us first give definitions, instead of the vague assertion that $g$ decreases like $|\xi|^{-\beta-d/2}$ in the vertical direction. We assume that $g$ satisfies the assumption $C(\beta)$, which is defined now.

We use the notation $\xi = (\xi', \xi_d)$, with $\xi' \in \mathbb{R}^{d-1}$. We introduce the following assumptions on $g$.

**Assumption $C^-(\beta)$:** for $\alpha < \beta$, there exists constants $A, c > 0$ such that

$$|g(\xi)| \leq |\xi|^{-\alpha - \frac{d}{2}} \quad \text{for } |\xi| > A \text{ and } |\xi'| \leq c|\xi_d| .$$

**Assumption $C^+(\beta)$:** for $\alpha > \beta$, there exists constants $A, c > 0$ such that

$$|g(\xi)| \geq |\xi|^{-\alpha - \frac{d}{2}} \quad \text{for } |\xi| > A \text{ and } |\xi'| \leq c|\xi_d| .$$

**Assumption** $C(\beta)$: both $C^-(\beta)$ and $C^+(\beta)$ are satisfied.

In one dimension, if $g$ satisfies $C(\beta)$, then the process $X$ has $\beta$ as critical Hölder exponent [4],[5]. In higher dimension, one reduces to one dimension through an averaging process that we describe now.

We fix a window function $\varphi$, which is smooth and compactly supported on $\mathbb{R}^{d-1}$. It is not necessary for $\varphi$ to be in the class $\mathcal{C}^\infty$, but we assume this for simplification. We then define

$$\langle X \rangle(t) := \int_{\mathbb{R}^{d-1}} X(t',t)\varphi(t') \, dt' \, , \quad t \in \mathbb{R} \, . \tag{11}$$

We know [4] that $\langle X \rangle$ is itself a Gaussian process with zero mean and stationary increments. Moreover, one has the following lemma [4].

**Lemma 3.1.** *The spectral density of* $\langle X \rangle$ *is equal to*

$$|\langle g \rangle|^2(\eta) := \int_{\mathbb{R}^{d-1}} |g(\xi',\eta)|^2 |\widehat{\varphi}(\xi')|^2 \, d\xi' \, , \quad \eta \in \mathbb{R} \, . \tag{12}$$

*Moreover, if $g$ is bounded outside a compact set and satisfies Assumption $C(\beta)$, then $\langle g \rangle$ satisfies $C(\beta + \frac{d-1}{2})$.*

So, in order to identify the parameter $\beta$, we can rely on a one dimensional result that we state now. The following proposition is adapted from Istas-Lang [6].

**Proposition 3.1.** *Let $Z = \{Z(t); t \in \mathbb{R}\}$ be a Gaussian process with zero mean, stationary increments and spectral density $f$. Assume that $f$ has a derivative outside 0, that $f$ satisfies $C(\beta)$ and $f'$ satisfies $C^-(\beta + \frac{1}{2})$.*

*For $K \geq \beta + \frac{1}{2}$ a fixed integer, we denote by $V_N(Z)$ the generalized quadratic variation of $Z$, given, for $N > K$ by*

$$V_N(Z) := \sum_{p=0}^{N-K} \left( \sum_{k=0}^{K} (-1)^k C_K^k Z \left( \frac{p+k}{N} \right) \right)^2 . \tag{13}$$

*Then, with probability one,*

$$\lim_{N \to \infty} \frac{\log V_N(Z)}{\log N} = -2\beta + 1 \, .$$

We postpone the proof to the appendix, and go on with the identification problem for $\beta$, the vertical exponent of $g$. Using the previous proposition, we know that

$$\lim_{N \to \infty} \frac{\log V_N(\langle X \rangle)}{\log N} = -2\beta - d + 2$$

whenever the process $\langle X \rangle$ satisfies the required assumption with $\beta + \frac{d-1}{2}$ in place of $\beta$. It is the case when the following conditions are satisfied:

$(C1)$ outside a compact set, the spectral density $|g|^2$ is bounded and possesses a bounded partial derivative with respect to the vertical coordinate;

(C2) $|g|^2$ satisfies the condition $C(\beta)$;

(C3) $\partial_d(|g|^2)$ satisfies the condition $C^-(\beta + \frac{1}{2})$.

Indeed, under these assumptions, it is clear that the spectral density of $\langle X \rangle$, that is $|\langle g \rangle|^2$, satisfies the conditions of Proposition 3.1 if we choose $K \geq \beta + \frac{d}{2}$. It is a direct consequence of Formula (12), derivation of the integral in the $\eta$ variable and Lemma 3.1.

While the estimator $\frac{\log V_N(\langle X \rangle)}{\log N}$ has a theoretical interest to recover the value $\beta$, it requires to compute horizontal integrals, which is not realistic. We will see that we have also an asymptotic when replacing $\langle X \rangle$ by approximations of it. These last ones are given by finite sums of values of $X$ on a lattice, which may be computed for real data.

For $N \geq 2$ an integer, we define $I_N$ by

$$I_N(s) := N^{-\nu(d-1)} \times \sum X(N^{-\nu}\mathbf{j}, s)\varphi(N^{-\nu}\mathbf{j}) , \quad s \in \mathbb{R} . \tag{14}$$

The sum is taken for $\mathbf{j} \in \mathbb{Z}^{d-1}$. Remark that it is a finite sum since $\varphi$ has compact support. The exponent $\nu$ will be fixed later on and has to be large enough in order to get a good approximation of the integral.

We then compute the generalized quadratic variation of this approximation. It produces, with the same notations as in (13),

$$T_N(X) := \sum_{p=0}^{N-K} \left( \sum_{k=0}^{K} (-1)^k C_K^k I_N\left(\frac{p+k}{N}\right) \right)^2 . \tag{15}$$

We can now state the theorem.

**Theorem 3.1.** *Let $X$ be Gaussian field with zero mean, stationary increments and spectral density $|g|^2$, which satisfies $D(m)$ for some real number $m \in (0,1)$. Assume that $|g|^2$ satisfies conditions (C1), (C2) and (C3). If $K \geq \beta + \frac{d}{2}$ and $\nu m > \beta + \frac{d-1}{2}$ then, with probability one,*

$$\lim_{N \to \infty} \frac{\log T_N(X)}{\log N} = -2\beta - d + 2 .$$

**Proof.** By triangle inequality,

$$|(T_N)^{\frac{1}{2}} - (V_N)^{\frac{1}{2}}| \leq \left[ \sum_{p=0}^{N-K} \left( \sum_{k=0}^{K} (-1)^k C_K^k \left( I_N\left(\frac{p+k}{N}\right) - \langle X \rangle \left(\frac{p+k}{N}\right) \right) \right)^2 \right]^{\frac{1}{2}} .$$

Since $\varphi$ is Lipschitz and $X$ is Hölder regular of order $r$, with $r < m$ (see (5)), there exists a positive finite random variable $C$ such that, a.s. and for all $t \in [0,1]$,

$$|I_N(t) - \langle X \rangle(t)| \leq CN^{-\nu r} .$$

Hence, putting together the last two inequalities, we get

$$|(T_N)^{\frac{1}{2}} - (V_N)^{\frac{1}{2}}| \le CN^{-\nu r + \frac{1}{2}} .$$

On the other hand, for $\varepsilon > 0$, using Proposition 3.1, we know that there exists a positive finite random variable $C'$ such that a.s.

$$(V_N)^{-\frac{1}{2}} \le C' N^{\varepsilon + \frac{1}{2} + \beta + \frac{d}{2} - 1} .$$

From these two estimates and since $\nu > \frac{2\beta + d - 1}{2m}$, we conclude that

$$\lim_{N \to +\infty} |(T_N)^{\frac{1}{2}} - (V_N)^{\frac{1}{2}}|(V_N)^{-\frac{1}{2}} = 0 \; a.s. \tag{16}$$

Finally, we write

$$\frac{\log T_N}{\log N} = \frac{\log V_N}{\log N} + \frac{\log\left(1 + ((T_N)^{\frac{1}{2}} - (V_N)^{\frac{1}{2}})(V_N)^{-\frac{1}{2}}\right)}{\log N^{\frac{1}{2}}}$$

and use (16) as well as Proposition 3.1 to conclude. □

We have not given the rate of convergence. Let us mention that the fact that the estimator has asymptotically a Gaussian law has been obtained by Biermé (work in progress and [5]). Her assumptions are slightly different from ours.

Let us also mention one classical difficulty: the lattice that we use is related to the direction that we consider. On real data, which is given by values on a lattice (corresponding, for instance, with pixels when images are concerned), computing the discretized integrals on hyperplanes orthogonal to a given direction will only be possible for particular directions, such as the axes, the diagonals, etc.. Otherwise, not enough values will be available.

Finally, on real data, one may be tempted to use the same lag, both in the vertical direction and on horizontal hyperplanes. Then the rate of convergence of the estimator $-\frac{\log T_N(X)}{2 \log N} - \frac{d}{2} + 1$ towards $\beta$ is slowed.

## Appendix: Identification of the exponent for a 1D-process

We now prove Proposition 3.1. Let us quote that, although the same asymptotic as in [6] is used, the assumptions are different: those we are requiring rely directly on the spectral density of $Z$, and not on the variance of the increments of $Z$. Moreover, our assumptions are weaker and only imply the existence of a critical exponent for the increments variance at the origin, whereas those of [6] are sufficient for the existence of an asymptotic development.

**Proof of Prop.3.1.** The sketch of the proof is classical. Let us first use the fact that $Z$ is a centered Gaussian process with stationary increments to get the following statements.

For $p = 0 \ldots, N - K$ define

$$Z_N(p) := \sum_{k=0}^{K} d_k Z\left(\frac{p+k}{N}\right)$$

where $d_k = (-1)^k C_K^k$. Note that $(Z_N(p))_{p=0,\ldots,N-K}$ is a Gaussian centered stationary sequence with

$$\mathbf{E}(V_N(Z)) = (N - K + 1)\mathrm{Var}(Z_N(0)) \qquad (17)$$

and

$$\mathrm{Var}(V_N(Z)) = 2 \sum_{p,p'=0}^{N-K} \mathrm{Cov}(Z_N(p), Z_N(p'))^2 . \qquad (18)$$

In the next lemma the covariance of $Z_N$ is computed and estimated in relation with the asymptotic behavior of the spectral density $f$.

**Lemma 3.2.** *Define* $\Gamma_N(p) := Cov(Z_N(p), Z_N(0))$ *for* $p = 0, \ldots, N - K$. *Then*

$$\Gamma_N(p) = 2^{2(K-1)} \int_{\mathbb{R}} \cos\left(\frac{p\xi}{N}\right) \sin^{2K}\left(\frac{\xi}{2N}\right) f(\xi) \, d\xi .$$

*Moreover, assume that* $f$ *satisfies Assumption* $C^+(\beta)$ *for some positive* $\beta$. *Then, for* $\alpha > \beta$, *there exists* $c > 0$ *such that, for* $N \geq 2$,

$$\Gamma_{N,K}(0) \geq c \, N^{-2\alpha} .$$

*Assume that* $K \geq \beta + \frac{1}{2}$, *that* $f$ *satisfies Assumption* $C^-(\beta)$, *and admits a derivative outside a compact set that satisfies* $C^-(\beta + \frac{1}{2})$. *Then, for* $\alpha < \beta$, *there exists a positive constant* $C$ *such that for* $N \geq 2$ *and* $p = 0 \ldots, N - K$,

$$|\Gamma_N(p)| \leq \frac{C}{1+p} \, N^{-2\alpha} .$$

**Proof.** For $p = 0 \ldots, N - K$,

$$\Gamma_N(p) = \sum_{1 \leq k,k' \leq K} d_k d_{k'} \mathrm{Cov}\left(Z\left(\frac{p+k}{N}\right) - Z(0), Z\left(\frac{k'}{N}\right) - Z(0)\right)$$

$$= \frac{1}{2} \int_{\mathbb{R}} h_K\left(p, \frac{\xi}{2N}\right) f(\xi) \, d\xi$$

where

$$h_K(p, \xi) = - \sum_{1 \leq k,k' \leq K} d_k d_{k'} \sin^2((p + k - k')\xi) = 2^{2K-1} \cos(2p\xi) \sin^{2K}(\xi) .$$

Using Assumption $C^+(\beta)$, for $\alpha > \beta$, there exists a constant $A$ such that

$$\Gamma_N(0) \geq c \int_{|\xi|>A} h_K(0, \frac{\xi}{2N})|\xi|^{-(2\alpha+1)} \, d\xi$$

$$\geq N^{-2\alpha} \int_{|\xi|>\frac{A}{2N}} h_K(0, \xi)|\xi|^{-(2\alpha+1)} \, d\xi$$

which gives the first inequality.

For the second one, we take $\alpha < \beta$ and $A > 0$ from Assumption $C^-(\beta)$ satisfied by $f$ and $C^-(\beta + \frac{1}{2})$ satisfied by $f'$. Remark that we directly get that the integral on $[-A, +A]$ is bounded by $CN^{-2K}$. The required inequality for the corresponding term follows from the fact that $2K \geq 2\beta + 1$. For the remaining integral, when $p \neq 0$, we use an integration by part. The only term for which the proof is not direct is

$$\int_{|\xi| > A} \frac{N}{p} \sin\left(\frac{p\xi}{N}\right) \frac{d}{d\xi}\left(\sin^{2K}\left(\frac{\xi}{2N}\right) f(\xi)\right) \, d\xi,$$

which is bounded by

$$\frac{C}{1+p} N \int_{|\xi| > A} |g'_{N,K}(\xi)| \, d\xi,$$

where $g_{N,K}(\xi) = \sin^{2K}\left(\frac{\xi}{2N}\right) f(\xi)$. Doing the change of variable $\xi \mapsto \frac{\xi}{2N}$ and estimating the derivative leads to the required conclusion.

For $p = 0$, the same upper-bound can be obtained without any integration by part. This proves the second inequality and concludes the proof of Lemma 3.2. $\square$

Let us proceed now to the proof of Proposition 3.1. Starting from (17) and (18) and using the previous lemma, we get the following asymptotics: for all $\varepsilon > 0$, there exists positive constants $c$ and $C$ such that for all integer $N \geq 2$,

$$c\, N^{-2\beta+1-\varepsilon} \leq \mathbf{E}(V_N(Z)) \leq C\, N^{-2\beta+1+\varepsilon} \qquad (19)$$
$$c\, N^{-4\beta+1-\varepsilon} \leq \mathrm{Var}(V_N(Z)) \leq C\, N^{-4\beta+1+\varepsilon}. \qquad (20)$$

Let $\eta$ be a positive real number and denote by $A_N$ the event

$$A_N = \left\{ \left| \frac{V_N(Z)}{\mathbf{E}(V_N(Z))} - 1 \right| > \eta \right\}.$$

By the Markov inequality,

$$\mathbb{P}(A_N) \leq \eta^{-4} |\mathbf{E}(V_N(Z))|^{-4} \mathbf{E}\left( |V_N(Z) - \mathbf{E}(V_N(Z))|^4 \right).$$

A simple but tedious computation shows that there exists a non-negative constant $c$ such that

$$\mathbf{E}\left( |V_N(Z) - \mathbf{E}(V_N(Z))|^4 \right) \leq c\, \mathrm{Var}(V_N(Z))^2.$$

Taking into account (19) and (20), we see that there exists a non-negative constant $C$ such that, for $N \in \mathbb{N}$,

$$\mathbb{P}(A_N) \leq C\eta^{-4} N^{-2+6\varepsilon}.$$

Therefore the series $\sum_N \mathbb{P}(A_N)$ is convergent. Applying Borel-Cantelli lemma gives, with probability one,

$$\left| \frac{V_N(Z)}{\mathbf{E}(V_N(Z))} - 1 \right| \leq \eta \text{ for all } N \text{ large enough.}$$

This yields

$$\lim_{N\to\infty} \frac{V_N(Z)}{\mathbf{E}(V_N(Z))} = 1 \quad a.s.$$

and the estimate (19) allows to conclude. □

## References

1. A. Ayache & J.Lévy Véhel. The Generalized Multifractional Brownian Motion. *Statistical Inference for Stochastic Processes* 3, 1-2, 7-18 (2000).
2. A. Ayache & Y. Xiao. Asymptotic Growth Properties and Hausdorff dimensions of Fractional Brownian Sheets. *Jour. Fourier Anal. Appl.* 11, Issue 4, (2005).
3. P. Billingsley, Convergence of probability measure, John Wiley & Sons, New York (1968).
4. A. Bonami & A. Estrade, Anisotropic analysis of some Gaussian models. *Jour. Fourier Anal. Appl.* 9 (2003) 215-236.
5. H. Biermé, Champs aléatoires: autosimilarité, anisotropie et analyse directionnelle. *PhD report, Orléans* 2005.
6. J. Istas & G. Lang, Quadratic variations and estimation of the local Hölder index of a Gaussian process. *Ann. I.H.P.* 33 (1997) 407-436.
7. R. Jennane, R. Harba, E. Perrin, A. Bonami & A. Estrade Analyse de champs browniens fractionnaires anisotropes. *GRETSI 2001* (2001) 99-102.
8. G. Lemineur, R. Harba, S. Bretteil, R. Jennane, A. Estrade, A. Bonami & C. L. Benhamou, Relation entre la régularité de fractals 3D et celle de leurs projections 2D. Application à l'os trabéculaire. *GRETSI 2003*, Vol. II (2003) 40-43.
9. S. Kwapień & N. A. Woyczyński, Random Series and Stochastic Integrals: Single and Multiple, Birkhäuser, Boston (1992).
10. Y. Meyer, F. Sellan & M.S. Taqqu, Wavelets, generalized white noise and fractional integration : the synthesis of fractional brownian motion. *Jour. Fourier Anal. Appl.* 5 (1999) 465-494.

# SUBORDINATION IN FRACTIONAL DIFFUSION PROCESSES VIA CONTINUOUS TIME RANDOM WALK *

R. GORENFLO

*Department of Mathematics and Informatics, Free University of Berlin,*
*Arnimallee 3, D-14195 Berlin, Germany*
*E-mail: gorenflo@mi.fu-berlin.de*


F. MAINARDI[†] and A. VIVOLI

*Department of Physics, University of Bologna and INFN,*
*Via Irnerio 46, I-40126 Bologna, Italy*
*E-mail: mainardi@bo.infn.it    vivoli@bo.infn.it*

The well-scaled transition to the diffusion limit in the framework of the theory of continuous-time random walk (CTRW) is presented starting from its representation as an infinite series that points out the subordinated character of the CTRW itself. This formula allows us to treat the CTRW as a discrete-space discrete-time random walk that in the continuum limit tends towards a generalized diffusion process governed by a space-time fractional diffusion equation. The essential assumption is that the probabilities for waiting times and jump-widths behave asymptotically like powers with negative exponents related to the orders of the fractional derivatives. Plots of simulations for some case-studies are given in order to display the sample paths for the fractional diffusion processes, generally non Markovian, that are obtained by the composition of two Markovian processes.

**Key words:** Random walks, anomalous diffusion, fractional calculus, stochastic processes, renewal theory, subordination, asymptotic power laws, stable probability densities

**Mathematical Subject Classification:** 26A33, 33C60, 44A10, 45K05, 47G30, 60G18, 60G50, 60G51, 60G55, 60J60, 60J70

## 1. Introduction

Surveying the literature of the past fifteen years we can observe an ever increasing interest in modelling *anomalous diffusion* processes, namely in diffusion processes deviating essentially from Gaussian behaviour which is characterized by evolution of the second centered moment like the first power of time. The reader interested to these processes is referred to several educational/review papers and books, including 27,28,32,36,41,44,46,48,49.

---

*This work has been carried out in the framework of a joint research project for *Fractional Calculus Modelling* (www.fracalmo.org)
†Corresponding author

In Section 2 we recall the simplest models for anomalous diffusion based on fractional calculus. They are obtained by replacing in the classical diffusion equation the partial derivatives with respect to space and/or time by derivatives of non-integer order, in such a way that the resulting Green function can still be interpreted as a probability density evolving in time differently from the Gaussian type.

A more general approach to anomalous diffusion is provided by the so-called *continuous time random walk* (CTRW) introduced in Statistical Mechanics by Montroll and Weiss [30], see also [29,45], which differs from the usual models in that the steps of the walker occur at random times generated by a renewal process. The sojourn probability density of this process is known to be governed by an integral equation and expressed in terms of a relevant series expansion, as it will be recalled in Section 3. The concept of CTRW, can be understood by considering a random walk subordinated to a *renewal process*, see *e.g.* [8], as pointed out by a number of authors, see *e.g.* [1,13,19,24,25,38].

It is well known that the space-time fractional diffusion (STFD) equation and its variants, including the *fractional Fokker-Planck equation*, can be derived from the CTRW integral equation, see *e.g.* [2-4,16,26-28,32,38-40], and references therein. More rigorously the passage from CTRW to STFD can be carried out via a properly scaled transition to the diffusion limit (under appropriate assumptions on waiting times and jumps), as shown in [43] and in a number of papers of our research group, see *e.g.* [11,13,14,35].

It is the purpose of this paper to offer another scheme of well-scaled transition to the diffusion limit, namely a scheme based on subordination, starting from the series expansion of the sojourn probability of the CTRW: this will be described in Section 4.

Then we shall consider the problem how to construct the sample paths for the STFD based on the above diffusion limit of the CTRW. In Section 5 we shall provide the analytical approach to this problem whereas in Section 6 we shall describe the numerical schemes and provide the sample paths for four case studies of fractional diffusion. Finally, the main conclusions are drawn in Section 7.

## 2. The space-time fractional diffusion

We begin by considering the Cauchy problem for the (spatially one-dimensional) space-time fractional diffusion equation

$$_tD_*^\beta\, u(x,t) \;=\; {_x}D_\theta^\alpha\, u(x,t)\,, \quad u(x,0) = \delta(x)\,, \tag{2.1}$$

where $-\infty < x < +\infty$, $t \geq 0$, $\alpha$, $\theta$, $\beta$ are real parameters restricted to the ranges

$$0 < \alpha \leq 2\,, \quad |\theta| \leq \min\{\alpha, 2-\alpha\}\,, \quad 0 < \beta \leq 1\,. \tag{2.2}$$

Here $_tD_*^\beta$ denotes the *Caputo fractional derivative* of order $\beta$, acting on the time variable $t$, and $_xD_\theta^\alpha$ denotes the *Riesz-Feller fractional derivative* of order $\alpha$ and skewness $\theta$, acting on the space variable $x$.

Let us note that the solution $u(x,t)$ of the Cauchy problem (2.1), known as the fundamental solution of the space-time fractional diffusion equation, is a probability density in the spatial variable $x$, evolving in time $t$. In the case $\alpha = 2$ and $\beta = 1$ we recover the standard diffusion equation for which the fundamental solution is the Gaussian density with variance $\sigma^2 = 2t$.

Writing the transforms of Laplace and Fourier as

$$\mathcal{L}\{f(t); s\} = \widetilde{f}(s) := \int_0^\infty e^{-st} f(t)\, dt\,, \ \mathcal{F}\{g(x); \kappa\} = \widehat{g}(\kappa) := \int_{-\infty}^{+\infty} e^{+i\kappa x}\, g(x)\, dx\,,$$

we have the corresponding transforms of $_tD_*^\beta f(t)$ and $_xD_\theta^\alpha g(x)$ as

$$\mathcal{L}\{\,_tD_*^\beta f(t)\} = s^\beta\, \widetilde{f}(s) - s^{\beta-1}\, f(0)\,, \tag{2.3}$$

$$\mathcal{F}\{\,_xD_\theta^\alpha g(x)\} = -|\kappa|^\alpha\, i^{\theta\,\mathrm{sign}\,\kappa}\, \widehat{g}(\kappa)\,. \tag{2.4}$$

Notice that $i^{\theta\,\mathrm{sign}\,\kappa} = \exp[i\,(\mathrm{sign}\,\kappa)\,\theta\,\pi/2]$. For the mathematical details the interested reader is referred to [12,31] on the Caputo derivative, and to [33] on the Feller potentials. For the general theory of pseudo-differential operators and related Markov processes the interested reader is referred to the excellent volumes by Jacob [17]. For our purposes let us here confine ourselves to recall the representation in the Laplace-Fourier domain of the (fundamental) solution of (2.1) as it results from the application of the transforms of Laplace and Fourier. Using $\widehat{\delta}(\kappa) \equiv 1$ we have from (2.1)

$$s^\beta\, \widehat{\widetilde{u}}(\kappa, s) - s^{\beta-1} = -|\kappa|^\alpha\, i^{\theta\,\mathrm{sign}\,\kappa}\, \widehat{\widetilde{u}}(\kappa, s)\,,$$

hence

$$\widehat{\widetilde{u}}(\kappa, s) = \frac{s^{\beta-1}}{s^\beta + |\kappa|^\alpha\, i^{\theta\,\mathrm{sign}\,\kappa}}\,. \tag{2.5}$$

For explicit expressions and plots of the fundamental solution of (2.1) in the space-time domain we refer the reader to [20]. There, starting from the fact that the Fourier transform $\widehat{u}(\kappa, t)$ can be written as a Mittag-Leffler function with complex argument, the authors have derived a Mellin-Barnes integral representation of $u(x,t)$ with which they have proved the non-negativity of the solution for values of the parameters $\{\alpha,\ \theta,\ \beta\}$ in the range (2.2) and analyzed the evolution in time of its moments. In particular for $\{0 < \alpha < 2, \beta = 1\}$ we obtain the stable densities of order $\alpha$ and skewness $\theta$. The representation of $u(x,t)$ in terms of Fox $H$-functions can be found in [22].

## 3. The continuous-time random walk

A CTRW is generated by a sequence of independent identically distributed (*iid*) positive random waiting times $T_1, T_2, T_3, \ldots$, each having the same probability density function $\phi(t)$, $t > 0$, and a sequence of *iid* random jumps $X_1, X_2, X_3, \ldots$, in

**R**, each having the same probability density $w(x)$, $x \in \mathbf{R}$. Let us remark that, for ease of language, we use the word density also for generalized functions in the sense of Gel'fand & Shilov [10], that can be interpreted as probability measures. Usually the *probability density functions* are abbreviated by *pdf*. We recall that $\phi(t) \geq 0$ with $\int_0^\infty \phi(t)\,dt = 1$ and $w(x) \geq 0$ with $\int_{-\infty}^{+\infty} w(x)\,dx = 1$.

Setting $t_0 = 0$, $t_n = T_1 + T_2 + \ldots T_n$ for $n \in \mathbf{N}$, the wandering particle makes a jump of length $X_n$ in instant $t_n$, so that its position is $x_0 = 0$ for $0 \leq t < T_1 = t_1$, and $x_n = X_1 + X_2 + \ldots X_n$, for $t_n \leq t < t_{n+1}$. We require the distribution of the waiting times and that of the jumps to be independent of each other. So, we have a compound renewal process (a renewal process with reward), compare [8].

By natural probabilistic arguments we arrive at the integral equation for the probability density $p(x,t)$ (a density with respect to the variable $x$) of the particle being in point $x$ at instant $t$, see *e.g.* [13,15,23,34,35],

$$p(x,t) = \delta(x)\,\Psi(t) + \int_0^t \phi(t - t') \left[ \int_{-\infty}^{+\infty} w(x - x')\,p(x',t')\,dx' \right] dt', \qquad (3.1)$$

in which the *survival function*

$$\Psi(t) = \int_t^\infty \phi(t')\,dt' \qquad (3.2)$$

denotes the probability that at instant $t$ the particle is still sitting in its starting position $x = 0$. Clearly, (3.1) satisfies the initial condition $p(x,0) = \delta(x)$.

In the Laplace-Fourier domain Eq. (3.1) reads as

$$\widehat{\widetilde{p}}(\kappa,s) = \widetilde{\Psi}(s) + \widehat{w}(\kappa)\,\widetilde{\phi}(s)\widehat{\widetilde{p}}(\kappa,s)\,,$$

and using $\widetilde{\Psi}(s) = (1 - \widetilde{\phi}(s))/s$, explicitly

$$\widehat{\widetilde{p}}(\kappa,s) = \frac{1 - \widetilde{\phi}(s)}{s}\,\frac{1}{1 - \widehat{w}(\kappa)\,\widetilde{\phi}(s)}\,. \qquad (3.3)$$

This Laplace-Fourier representation is known in physics as the the *Montroll-Weiss equation*, so named after the authors, see [30], who derive it in 1965 as the basic equation for the CTRW. By inverting the transforms one can find the evolution $p(x,t)$ of the sojourn density for time $t$ running from zero to infinity. In fact, recalling that $|\widehat{w}(\kappa)| < 1$ and $|\widetilde{\phi}(s)| < 1$, if $\kappa \neq 0$ and $s \neq 0$, Eq. (3.3) becomes

$$\widehat{\widetilde{p}}(\kappa,s) = \widetilde{\Psi}(s) \sum_{n=0}^\infty [\widetilde{\phi}(s)\,\widehat{w}(\kappa)]^n = \sum_{n=0}^\infty \widetilde{v}_n(s)\,\widehat{w}_n(\kappa)\,, \qquad (3.4)$$

and we promptly obtain the *series representation of the continuous time random walk*, see *e.g.* [8] (Ch. 8, Eq. (4)) or [45] (Eq.(2.101)),

$$p(x,t) = \sum_{n=0}^\infty v_n(t)\,w_n(x) = \Psi(t)\,\delta(x) + \sum_{n=1}^\infty v_n(t)\,w_n(x)\,, \qquad (3.5)$$

where the functions $v_n(t)$ and $w_n(x)$ are obtained by repeated convolutions in time and in space, $v_n(t) = (\Psi * \phi^{*n})(t)$, and $w_n(x) = (w^{*n})(x)$, respectively. In particular,

$v_0(t) = (\Psi * \delta)(t) = \Psi(t)$, $v_1(t) = (\Psi * \phi)(t)$, $w_0(x) = \delta(x)$, $w_1(x) = w(x)$. In the R.H.S of Eq (3.5) we have isolated the first singular term related to the initial condition $p(x, 0) = \Psi(0)\delta(x) = \delta(x)$. The representation (3.5) can be found without detour over (3.1) by direct probabilistic reasoning and transparently exhibits the CTRW as a subordination of a random walk to a renewal process: it can be used as starting point to derive the Montroll-Weiss equation, as it was originally recognized by Montroll and Weiss [30]. Though (3.5), while being an attractive general formula, is unlikely to lead to explicit answers to rather simple problems, we consider it as a basic and useful formula for our analysis, as it will be shown later on.

A special case of the integral equation (3.1) is obtained for the *compound Poisson process* where $\phi(t) = me^{-mt}$ (with some positive constant $m$). Then, the corresponding master equation reduces after some manipulations, that best are carried out in the Laplace-Fourier domain, to the *Kolmogorov-Feller equation*:

$$\frac{\partial}{\partial t} p(x, t) = -m\, p(x, t) + m \int_{-\infty}^{+\infty} w(x - x')\, p(x', t)\, dx'. \tag{3.6}$$

Then, the solution obtained via the series representation reads

$$p(x, t) = \sum_{k=0}^{\infty} \frac{(mt)^k}{k!} e^{-mt}\, w_k(x). \tag{3.7}$$

We note that only in this case the corresponding stochastic process is *Markovian*.

## 4. Subordination in stochastic processes

In recent years a number of papers have appeared where explicitly or implicitly subordinated stochastic processes have been treated in view of their relevance in physical and financial applications, see e.g. [1,3,5,24,27,35,37,42,44,47] and references therein. Historically, the notion of subordination was originated by Bochner, see [6,7]. In brief, according to Feller [9] (using our notation) a *subordinated process* $x = X(t) = Y(T_*(t))$ is obtained by randomizing the time clock of a stochastic process $Y(t_*)$ using a new clock $t = T(t_*)$, the non-decreasing right-continuous random functions $t = T(t_*)$ and $t_* = T_*(t)$ being inverse to each other (inverse in the appropriate sense). The resulting process $X(t)$ is said to be subordinated to $Y(t_*)$, called the *parent process*. The process $T(t_*)$ is called the *directing process*, and $t_*$ is often referred to as the operational time. In particular, assuming $Y(t_*)$ to be a Markov process with a spatial probability density function (*pdf*) of $x$, evolving in time $t_*$, $q_{t_*}(x) \equiv q(x, t_*)$, and $T_*(t)$ to be a process with non-negative independent increments with *pdf* of $t_*$ depending on a parameter $t$, $r_t(t_*) \equiv r(t_*, t)$, then the subordinated process $X(t) = Y(T_*(t))$ is governed by the spatial *pdf* of $x$ evolving with $t$, $p_t(x) \equiv p(x, t)$, given by the *integral formula of subordination* (compare with Eq (7.1), Ch. X in [9] and with Eq. (3.1) in [21])

$$p_t(x) = \int_0^{\infty} q_{t_*}(x)\, r_t(t_*)\, dt_*. \tag{4.1}$$

If the parent process $Y(t_*)$ is *self-similar* of the kind that its *pdf* $q_{t_*}(x)$ is such that, with a probability density $q(x)$ and a positive number $\gamma$,

$$q_{t_*}(x) \equiv q(x, t_*) = t_*^{-\gamma} q\left(\frac{x}{t_*^\gamma}\right), \tag{4.2}$$

then Eq. (4.1) reads

$$p_t(x) = \int_0^\infty q\left(\frac{x}{t_*^\gamma}\right) r_t(t_*) \frac{dt_*}{t_*^\gamma}. \tag{4.3}$$

In the series representation (3.5) for the CTRW the running index $n$ corresponds to the so-called "operational time" $t_*$ in the subordination formula for a continuous (stable) process. We will pass in (3.5) to the diffusion limit under the "power law" assumptions

$$1 - \widetilde{\phi}(s) \sim \lambda s^\beta, \quad \lambda > 0, \quad s \to 0^+, \tag{4.4}$$

$$1 - \widehat{w}(\kappa) \sim \mu|\kappa|^\alpha \, i^{\,\theta \mathrm{sign}\,\kappa}, \quad \mu > 0, \quad \kappa \to 0, \tag{4.5}$$

where $\beta$, $\alpha$ and $\theta$ are restricted as in (2.2). By (4.4) and (4.5) the asymptotic power law behaviour of the functions $\phi(t)$ and $w(x)$ at infinity is determined by some Tauber type lemmata as laid out e.g. in [14].

The idea is to treat the series expansion (starting from $n = 0$) in (3.5) as an approximation to an improper Riemann integral. Being interested on behaviour in "large time" and "wide space" we change the units of measurement in order to make large time intervals and space distances appear numerically of moderate size, moderate time intervals and space distances of small size. To this aim we replace waiting times $T$ by $\tau T$, jumps $X$ by $hX$, and then send the positive *scaling factors* $\tau$ and $h$ to zero, observing a *scaling relation* that will become mandatory in our calculations.

For the CTRW this means replacing $\phi(t)$ by $\phi_\tau(t) = \phi(t/\tau)/\tau$, $w(x)$ by $w_h(x) = w(x/h)/h$, correspondingly $\widetilde{\phi}(s)$ by $\widetilde{\phi}_\tau(s) = \widetilde{\phi}(\tau s)$, $\widehat{w}(\kappa)$ by $\widehat{w}_h(\kappa) = \widehat{w}(h\kappa)$. Decorating (3.5) by indices $h$ and $\tau$ gives

$$p_{h,\tau}(x, t) = \sum_{n=0}^\infty v_{\tau,n}(t)\, w_{h,n}(x), \tag{4.6}$$

yielding in the Fourier-Laplace domain

$$\widetilde{\widehat{p}}_{h,\tau}(\kappa, s) = \sum_{n=0}^\infty \frac{1 - \widetilde{\phi}(\tau s)}{s} \left(\widetilde{\phi}(\tau s)\right)^n \left(\widehat{w}(h\kappa)\right)^n. \tag{4.7}$$

Separately we treat the powers $\left(\widetilde{\phi}(\tau s)\right)^n$ and $(\widehat{w}(h\kappa))^n$, so avoiding the problematic simultaneous inversion of the diffusion limit from the Fourier-Laplace domain into the physical domain. Observing from (4.4)

$$\left(\widetilde{\phi}(\tau s)\right)^n \sim \left(1 - \lambda(\tau s)^\beta\right)^n, \tag{4.8}$$

we relate the running index $n$ to the presumed operational time $t_*$ by

$$n \sim \frac{t_*}{\lambda \tau^\beta}, \qquad (4.9)$$

and for <u>fixed</u> $s$ (as required by the continuity theorem of probability theory), by sending $\tau \to 0$ we get

$$\left(\widetilde{\phi}(\tau s)\right)^n \sim \left(1 - \lambda \tau^\beta s^\beta\right)^{t_*/(\lambda \tau^\beta)} \to \exp\left(-t_* s^\beta\right). \qquad (4.10)$$

Here $s$ corresponds to physical time $t$, and in Laplace inversion we must treat $t_*$ as a parameter. Hence, in physical time $\exp(-t_* s^\beta)$ corresponds to

$$\bar{g}_\beta(t, t_*) = t_*^{-1/\beta} \, \bar{g}_\beta(t_*^{-1/\beta} t), \qquad (4.11)$$

with $\widetilde{\bar{g}}_\beta(s) = \exp(-s^\beta)$. Here $\bar{g}_\beta(t, t_*)$ is the totally positively skewed stable density (with respect to the variable $t$) evolving in operational time $t_*$ according to the "space"- fractional equation

$$\frac{\partial}{\partial t_*} \bar{g}_\beta(t, t_*) = {}_t D_{-\beta}^\beta \, \bar{g}_\beta(t, t_*), \quad \bar{g}_\beta(t, 0) = \delta(t), \qquad (4.12)$$

where $t$ is playing the role of the spatial variable.

Analogously, observing from (4.5)

$$(\widehat{w}(h\kappa))^n \sim \left(1 - \mu(h|\kappa|)^\alpha \, i^{\theta \mathrm{sign}\, \kappa}\right)^n, \qquad (4.13)$$

and with the aim of obtaining a meaningful limit we now set

$$n \sim \frac{t_*}{\mu h^\alpha}, \qquad (4.14)$$

and find, by sending $h \to 0^+$, the relation

$$(\widehat{w}(h\kappa))^n \sim \left(1 - \mu(h|\kappa|)^\alpha \, i^{\theta \mathrm{sign}\, \kappa}\right)^{t_*/(\mu h^\alpha)} \to \exp\left(-t_* |\kappa|^\alpha \, i^{\theta \mathrm{sign}\, \kappa}\right), \qquad (4.15)$$

the Fourier transform of a $\theta$-skewed $\alpha$-stable density $f_{\alpha,\theta}(x, t_*)$ evolving in operational time $t_*$. This density is the solution of the space-fractional equation

$$\frac{\partial}{\partial t_*} f_{\alpha,\theta}(x, t_*) = {}_x D_\theta^\alpha \, f_{\alpha,\theta}(x, t_*), \quad f_{\alpha,\theta}(x, 0) = \delta(x). \qquad (4.16)$$

The two relations (4.9) and (4.14) between the running index $n$ and the presumed operational time $t_*$ require the (asymptotic) *scaling relation*

$$\lambda \tau^\beta \sim \mu h^\alpha, \qquad (4.17)$$

that for purpose of computation we simplify to $\lambda \tau^\beta = \mu h^\alpha$. Replacing $t_*$ by $t_{*,n} = n \lambda \tau^\beta$, using the asymptotic results (4.10) and (4.15) obtained for the powers $\left(\widetilde{\phi}(\tau s)\right)^n$ and $(\widehat{w}(h\kappa))^n$, furthermore noting $[1 - \widetilde{\phi}(\tau s)]/s \sim s^{\beta-1} \lambda \tau^\beta$, we finally obtain from (4.7) the Riemann sum (with increment $\lambda \tau^\beta$)

$$\widetilde{\widetilde{p}}_{h,\tau}(\kappa, s) \sim s^{\beta-1} \sum_{n=0}^{\infty} \exp\left[-n\lambda\tau^\beta \left(s^\beta + |\kappa|^\alpha \, i^{\theta \mathrm{sign}\, \kappa}\right)\right] \lambda \tau^\beta, \qquad (4.18)$$

and hence the integral

$$\widehat{\widetilde{p}}_{h,\tau}(\kappa, s) \sim s^{\beta-1} \int_0^\infty \exp\left[-t_*\left(s^\beta + |\kappa|^\alpha i^{\theta \operatorname{sign}\kappa}\right)\right] dt_*. \tag{4.19}$$

For the *limiting process* $u_\beta(x,t)$ this means

$$\widehat{\widetilde{u}}_\beta(\kappa, s) = \int_0^\infty s^{\beta-1} \exp\left[-t_*\left(s^\beta + |\kappa|^\alpha i^{\theta \operatorname{sign}\kappa}\right)\right] dt_*. \tag{4.20}$$

Observe that the RHS of this equation is just another way of writing the RHS of equation (2.5) which is the Laplace-Fourier solution of the STFD equation (2.1). By inverting the transforms we get after some manipulations (compare [24]) in physical space-time the *integral formula of subordination*

$$u_\beta(x,t) = \int_0^\infty f_{\alpha,\theta}(x, t_*)\, g_\beta(t_*, t)\, dt_* \tag{4.21}$$

with

$$g_\beta(t_*, t) = \frac{t}{\beta}\, \bar{g}_\beta\left(t\, t_*^{-1/\beta}\right) t_*^{-1/\beta-1} \tag{4.22}$$

standing for the density $r_t(t_*)$ in equation (4.1).

There are two processes involved. One is the unidirectional motion along the $t_*$ axis representing the operational time. This motion happens in physical time $t$ and the *pdf* for the operational time having value $t_*$ is (as density in $t_*$, evolving in physical time $t$) given by (4.22). In fact, by substituting $y = t\, t_*^{-1/\beta}$ we find

$$\int_0^\infty g_\beta(t_*, t)\, dt_* \equiv \int_0^\infty \bar{g}_\beta(t, t_*)\, dt = 1, \quad \forall\, t > 0. \tag{4.23}$$

The operational time $t_*$ stands in analogy to the counting index $n$ in Eqs. (3.5) and (4.7). The other process is the process described by Eq. (4.16), a spatial probability density for sojourn of the particle in point $x$ evolving in operational time $t_*$,

$$\bar{u}_\beta(x, t_*) = f_{\alpha,\theta}(x, t_*). \tag{4.24}$$

To find the *pdf* $u(x,t) = u_\beta(x,t)$ for sojourn in point $x$, evolving in physical time $t$, we must *average* $\bar{u}_\beta(x, t_*)$ with the *weight function* $g_\beta(t_*, t)$ over the interval $0 < t_* < \infty$ according to (4.22).

Remark: Our derivation of the subordination formulas (4.21)-(4.22) from the CTRW model clearly is deficient in rigour as we have ignored the problems of interchanging some passages to the limit. Our intention is to provide intuitive insight into the structure of the processes. Purists are invited to fill in the missing details. For a strictly analytical derivation we refer to [21], see there Eqs (4.23) and (7.1)-(7.2).

## 5. Sample path for space-time fractional diffusion

In the series representation (3.5) of the CTRW the running index $n$ (the number of jumps having occurred up to physical time $t$) is a *discrete operational time*, proceeding in unit steps. To this index $n$ corresponds the physical time $t = t_n$, the sum of the first $n$ waiting times, and in physical space the position $x = x_n$, the sum of the first $n$ jumps, see Section 3.

Rescaling space and physical time by factors $h$ and $\tau$ respectively, obeying the *scaling relation* (4.17), we introduce, by sending $\{h \to 0, \tau \to 0\}$, the *continuous operational time*

$$t_* \sim n\lambda\tau^\beta \sim n\mu h^\alpha. \tag{5.1}$$

Then, in the series representation (3.5) we have <u>two</u> discrete <u>Markov</u> processes (discrete in operation time $n$), namely a random walk in the space variable $x$, and another random walk (only in positive direction) of the physical time $t$, making a forward jump at every instant $n$.

In the diffusion limit the spatial process becomes an $\alpha$-stable process for the position $x = \bar{x} = \bar{x}(t_*)$, whereas the unilateral time process becomes a unilateral (positively directed) $\beta$-stable process for the physical time $t = \bar{t} = \bar{t}(t_*)$. A sample path of a diffusing particle in physical coordinates can be produced by combining in the $(t, x)$ plane the two random functions

$$\begin{cases} x = \bar{x} = \bar{x}(t_*), \\ t = \bar{t} = \bar{t}(t_*), \end{cases} \tag{5.2}$$

both evolving in operational time $t_*$, both being Markovian and obeying stochastic differential equations

$$\begin{cases} d\bar{x} = d(\text{Lévy noise of order } \alpha \text{ and skewness } \theta), \\ d\bar{t} = d(\text{one sided Lévy noise of order } \beta). \end{cases} \tag{5.3}$$

This gives us in the $(t, x)$ plane the $t_*$- parametrized particle path, and by elimination of $t_*$ we get it as $x = x(t)$.

<u>Concerning notation</u>: It is good to make a conceptual distinction between the position $\bar{x}$ of an individual particle and the variable $x$, likewise between the physical time position $\bar{t}$ and the physical time variable $t$. When there are many particles we have overall densities for them and for these densities fractional diffusion equations. The *pdf* for the particle being in point $\bar{x} = x$ at operational time $t_*$, that we denote by $\bar{u}_\beta(x, t_*) = f_{\alpha,\theta}(x, t_*)$, satisfies the evolution equation (Eq. (4.16) re-written with $\bar{u}_\beta$)

$$\frac{\partial}{\partial t_*} \bar{u}_\beta(x, t_*) = {}_x D_\theta^\alpha \bar{u}_\beta(x, t_*), \quad \bar{u}(x, 0) = \delta(x). \tag{5.4}$$

The *pdf* for the physical time being in $\bar{t} = t$ at operational time $t_*$, that we denote by $\bar{v}(t, t_*) = \bar{g}_\beta(t, t_*)$, obeys the skewed fractional equation

$$\frac{\partial}{\partial t_*} \bar{v}(t, t_*) = {}_t D_{-\beta}^\beta \bar{v}(t, t_*), \quad \bar{v}(t, 0) = \delta(t). \tag{5.5}$$

<u>Remind</u>: In operational time two Markovian random functions $\bar{x}(t_*)$, $\bar{t}(t_*)$ occur, as random processes, individually for each particle. In physical coordinates we have the $t_*$-parametrized random path described by (5.2).

<u>Remark</u>: It is instructive to look what happens for the limiting value $\beta = 1$. In this case the Laplace transform of $\bar{g}_\beta(t, t_*) = \bar{g}_1(t, t_*)$ is $\exp(-t_* s)$, implying $\bar{g}_1(t, t_*) = \delta(t - t_*)$, the delta density concentrated on $t = t_*$. So, in this case, $t = t_*$, operational time and physical time coincide.

## 6. Numerical results

In this Section, after describing the numerical schemes adopted, we shall show the sample paths for two case studies of symmetric ($\theta = 0$) fractional diffusion processes: $\{\alpha = 2,\ \beta = 0.80\}$, $\{\alpha = 1.5,\ \beta = 0.90\}$, As explained in the previous sections, for each case we need to construct the sample paths for three distinct processes, the parent process $x = Y(t_*)$, the directing process $t = T(t_*)$ (both in the operational time) and, finally, the subordinated process $x = X(t)$, corresponding to the required fractional diffusion process. We shall depict the above sample paths in Fig. 1, Fig. 2 and Fig. 3, respectively, devoting the left and the right plates to the different case studies. For this purpose we proceed as follow.

First, let operational time $t_*$ assume only integer values, say $t_{*,n} = n$, $n = 0, 1, \ldots, 10000$. Then produce 10000 independent identically distributed random deviates, say $X_1, X_2, \ldots, X_{10000}$ having a symmetric stable probability distribution of order $\alpha$, see the book by Janicki [18] for a useful and efficient method to do that. Now, with the points

$$x_0 = 0, \quad x_n = \sum_{k=1}^{n} X_k, \quad n \geq 1, \tag{6.1}$$

the couples $(t_{*,n}, x_n)$, plotted in the $(t_*, x)$ plane (operational time, physical space) can be considered as points of a sample path $\{x(t_*) : 0 \leq t_* \leq 10000\}$ of a symmetric Lévy motion with order $\alpha$ corresponding to the integer values of operational time $t_* = t_{*,n}$. In this identification of $t_*$ with $n$ we use the fact that our stable laws for waiting times and jumps imply $\lambda = \mu = 1$ in the asymptotics (4.4) and (4.5) and $\tau = h = 1$ as initial scaling factors in (4.6) and (4.17).

In order to complete the sample path we agree to connect every two successive points $(t_{*,n}, x_n)$ and $(t_{*,n+1}, x_{n+1})$ by a horizontal line from $(t_{*,n}, x_n)$ to $(t_{*,n+1}, x_n)$, and a vertical line from $(t_{*,n+1}, x_n)$ to $(t_{*,n+1}, x_{n+1})$. Obviously, that is not the 'true' Lévy motion from point $(t_{*,n}, x_n)$ to point $(t_{*,n+1}, x_{n+1})$, but from the theory of CTRW we know this kind of sample path to converge to Lévy motion paths in the diffusion limit. However, as the successive values of $t_{*,n}$ and $x_n$ are generated by successively adding the relevant standardized stable random deviates, the obtained sets of points in the three coordinate planes: $(t_*, t)$, $(t_*, x)$, $(t, x)$ can, in view of infinite divisibility and self-similarity of the stable probability distributions, be considered as snapshots of the corresponding true random processes occurring in

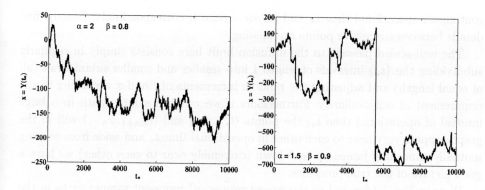

Fig. 1. A sample path for the parent process $x = Y(t_*)$.
LEFT: $\{\alpha = 2, \ \beta = 0.80\}$, RIGHT: $\{\alpha = 1.5, \ \beta = 0.90\}$.

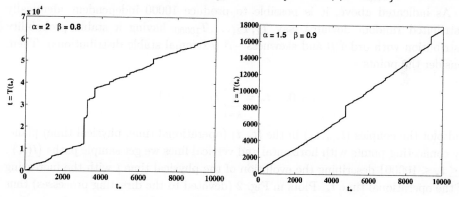

Fig. 2. A sample path for the directing process $t = T(t_*)$.
LEFT: $\{\alpha = 2, \ \beta = 0.80\}$, RIGHT: $\{\alpha = 1.5, \ \beta = 0.90\}$.

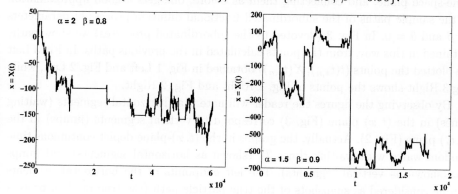

Fig. 3. A sample path for the subordinated process $x = X(t)$.
LEFT: $\{\alpha = 2, \ \beta = 0.80\}$, RIGHT: $\{\alpha = 1.5, \ \beta = 0.90\}$.

continuous operational time $t_*$ and physical time $t$, correspondingly. Clearly, fine details between successive points are missing.

The well-scaled passage to the diffusion limit here consists simply in regularly subdividing the $\{t_*\}$ intervals of length 1 into smaller and smaller subintervals (all of equal length) and adjusting the random increments of $t$ and $x$ according to the requirement of self-similarity. Furthermore if we watch a sample path in a large interval of operational time $t_*$, the points $(t_{*,n}, x_n)$ and $(t_{*,n+1}, x_{n+1})$ will in the graphs appear very near to each other in operational time $t_*$ and aside from missing mutually cancelling jumps up and down (extremely near to each other) we have a good picture of the true processes.

Plots in Fig. 1 (devoted to the parent processes) represent sample paths in the $(t_*, x)$ plane, produced in the way explained above, for Lévy motions of order $\alpha$ and skewness $\theta = 0$ (symmetric stable distributions).

As indicated above, it is possible to produce 10000 independent identically distributed random deviates, say $T_1, T_2, \ldots, T_{10000}$ having a stable probability distribution with order $\beta$ and skewness $-\beta$ (extremal stable distributions). Then, consider the points

$$t_0 = 0, \quad t_n = \sum_{k=1}^{n} T_k, \quad n \geq 1, \tag{6.2}$$

and plot the couples $(t_{*,n}, t_n)$ in the $(t_*, t)$ (operational time, physical time) plane. By connecting points with horizontal and vertical lines we get sample paths $\{t(t_*) : 0 \leq t_* \leq 10000\}$ describing the evolution of the physical time $t$ with the increasing of the operational time $t_*$. Plots in Fig. 2 (devoted to the directing processes) thus represent sample paths in the $(t_*, t)$ plane of unilateral Lévy motions of order $\beta$.

Now, plotting points $(t(t_{*,n}), x(t_{*,n}))$ in the $(t, x)$ plane, namely the physical time-space plane, and connecting them as before, one gets a good approximation of the sample paths of the subordinated fractional diffusion process of parameters $\alpha$, $\beta$ and $\theta = 0$. In Fig. 3 (devoted to the subordinated processes) we show paths obtained in this way, from the points calculated in the previous paths. In Fig.3 Left we plotted the points $(t(t_{*,n}), X(t_{*,n}))$ obtained in Fig. 1 Left and Fig. 2 Left, while Fig.3 Right shows the points of Fig. 1 Right and Fig. 2 Right.

By observing the figures the reader will note that horizontal segments (waiting times) in the $(t, x)$ plane (Fig. 3) correspond to vertical segments (jumps) in the $(t, t_*)$ plane (Fig. 2). Actually, the graphs in the $(t, x)$-plane depict continuous time random walks with waiting times $T_k$ (shown as horizontal segments) and jumps $X_k$ (shown as vertical segments). The left endpoints of the horizontal segments can be considered as snapshots of the true particle path (the true random process to be simulated), the segments being segments of our ignorance. In the interval $t_n < t \leq t_{n+1}$ the true process (namely the spatial variable $x = X(t)$) may jump up and down (infinitely) often, the sum (or integral) of all these ups (counted positive) and downs (counted negative) amounting to the vertical jump $X_{n+1}$.

Finer details will become visible by choosing in the operational time $t_*$ the step length $\tau << 1$ (instead of length 1 as we have done) and correspondingly the waiting times and spatial jumps as $\tau^{1/\beta}$ multiplied by a standard extreme $\beta$-stable deviate, $\tau^{1/\alpha}$ multiplied by a standard (in our special case: symmetric) $\alpha$-stable deviate, respectively, as required by the self-similarity properties of the stable probability distributions. In a forthcoming paper we will describe in detail what happens for finer and finer discretization of the operational time $t_*$.

## 7. Conclusions

Starting from the series representation (3.5) of the CTRW, by considering there the running index of summation as discrete operational time and passing to the diffusion limit in a well-scaled way, we have shown how to arrive at the integral formula of subordination in fractional diffusion. Furthermore, we have explained how, in analogy to the construction of particle paths in CTRW, particle paths in space-time fractional diffusion can be obtained by composition of two stable (hence Markovian) processes (one for the physical time, the other for the position in space, both processes running in operational time). By this composition we get in physical space-time the particle path, parametrized by the operational time. The essential games are played in operational time, for construction of a particle path we avoid to explicitly run the tricky random process generating from physical time the operational time.

## Acknowledgements

The Authors are grateful to E. Barkai and M.M. Meerschaert for inspiring e-mail exchange of opinions, and to A.V. Chechkin and I.M. Sokolov for illuminating discussions.

## References

1. B. Baeumer, M.M. Meerschaert: Stochastic solutions for fractional Cauchy problems, *Fractional Calculus and Appl. Analysis* **4**, 481-500 (2001).
2. E. Barkai, Fractional Fokker-Planck equation, solution, and application, *Phys. Rev. E* **63**, 046118-1/18 (2001).
3. E. Barkai, CTRW pathways to the fractional diffusion equation, *Chem. Phys.* **284**, 13–27 (2002).
4. E. Barkai, R. Metzler, J. Klafter, From continuous time random walk to fractional Fokker-Planck equation, *Phys. Rev. E* **61**, 132-138 (2000).
5. O.E. Barndorff-Nielsen, T. Mikosch, S.I. Resnick (Editors), *Lévy Processes: Theory and Applications*, Birkhäuser, Boston (2001).
6. S. Bochner, *Harmonic Analysis and the Theory of Probability*, University of California Press, Berkeley (1955).
7. S. Bochner, Subordination of non-Gaussian stochastic processes, *Proc. Nat. Acad. Sciences, USA* **48**, 19–22 (1962).
8. D.R. Cox, *Renewal Theory*, Methuen, London (1967).
9. W. Feller, *An Introduction to Probability Theory and its Applications*, Vol. 2, Wiley, New York (1971).

10. I. M. Gel`fand and G. E. Shilov, *Generalized Functions*, Volume I. Academic Press, New York and London (1964).

11. R. Gorenflo and E. Abdel-Rehim, ¿From power laws to fractional diffusion: the direct way, *Vietnam Journal of Mathematics* **32** SI, 65-75 (2004).

12. R. Gorenflo and F. Mainardi, Fractional calculus: integral and differential equations of fractional order, in: A. Carpinteri and F. Mainardi (Editors), *Fractals and Fractional Calculus in Continuum Mechanics*, Springer Verlag, Wien (1997), pp. 223-276.

13. R. Gorenflo and F. Mainardi, Fractional diffusion processes: probability distributions and continuous time random walk, in: G. Rangarajan and M. Ding (Editors), *Processes with Long Range Correlations*, Springer-Verlag, Berlin (2003), pp. 148-166. [Lecture Notes in Physics, No. 621]

14. R Gorenflo and F. Mainardi, Simply and multiply scaled diffusion limits for continuous time random walks, in: S. Benkadda, X. Leoncini and G. Zaslavsky (Editors), Proceedings of the International Workshop on Chaotic Transport and Complexity in Fluids and Plasmas Carry Le Rouet (France) 20-25 June 2004, *IOP (Institute of Physics) Journal of Physics: Conference Series* **7**, 1-16 (2005).

15. R. Gorenflo, F. Mainardi, E. Scalas and M. Raberto Fractional calculus and continuous-time finance III: the diffusion limit, in: M. Kohlmann and S. Tang (Editors), *Mathematical Finance*, Birkhäuser Verlag, Basel (2001), pp. 171-180.

16. R. Hilfer, On fractional diffusion and continuous time random walks, *Physica A* **329**, 35-39 (2003).

17. N. Jacob, *Pseudodifferential Operators - Markov Processes*, Vol I: Fourier Analysis and Semigroups Vol II: Generators and Their Potential Theory, Vol. III: Markov Processes and Applications, Imperial College Press, London (2001), (2002), (2005).

18. A. Janicki, *Numerical and Statistical Approximation of Stochastic Differential Equations with Non-Gaussian Measures*, Monograph No 1, H. Steinhaus Center, Technical University, Wroclaw, Poland (1996).

19. M. Kotulski, Asymptotic distributions of continuous-time random walks: a probabilistic approach, *J. Stat. Phys.* **81**, 777–792 (1995)..

20. F. Mainardi, Yu. Luchko and G. Pagnini, The fundamental solution of the space-time fractional diffusion equation, *Fractional Calculus and Appl. Analysis* **4**, 153-192 (2001). [Reprinted in NEWS 010401, see http://www.fracalmo.org]

21. F. Mainardi, G. Pagnini and R. Gorenflo, Mellin transform and subordination laws in fractional diffusion processes, *Fractional Calculus and Appl. Analysis* **6**, 441-459 (2003).

22. F. Mainardi, G. Pagnini and R.K. Saxena, Fox *H* functions in fractional diffusion, *J. Computational and Appl. Mathematics* **178**, 321-331 (2005).

23. F. Mainardi, M. Raberto, R. Gorenflo and E. Scalas, Fractional calculus and continuous-time finance II: the waiting-time distribution, *Physica A* **287**, 468–481 (2000).

24. M.M. Meerschaert, D.A. Benson, H.P Scheffler and B. Baeumer, Stochastic solutions of space-fractional diffusion equation, *Phys. Rev. E* **65**, 041103-1/4 (2002).

25. M.M. Meerschaert, D.A. Benson, H.P Scheffler and P. Becker-Kern, Governing equations and solutions of anomalous random walk limits, *Phys. Rev. E* **66**, 060102-1/4 (2002).

26. R. Metzler, J. Klafter, I.M. Sokolov, Anomalous transport in external fields: Continuous time random walks and fractional diffusion equations extended, *Phys. Rev, E* **58**, 1621-1633 (1998).

27. R. Metzler and J. Klafter, The random walk's guide to anomalous diffusion: a fractional dynamics approach, *Phys. Reports* **339**, 1-77 (2000).

28. R. Metzler and J. Klafter, The restaurant at the end of the random walk: Recent developments in the description of anomalous transport by fractional dynamics, *J. Phys. A. Math. Gen.* **37**, R161-R208 (2004).

29. E.W. Montroll and H. Scher, Random walks on lattices, IV: Continuous-time walks and influence of absorbing boundaries, *J. Stat. Phys.* **9**, 101-135 (1973).

30. E.W. Montroll and G.H. Weiss, Random walks on lattices, II, *J. Math. Phys.* **6**, 167-181 (1965).

31. I. Podlubny, *Fractional Differential Equations*, Academic Press, San Diego (1999).

32. A. Piryatinska, A.I. Saichev and W.A. Woyczynski, Models of anomalous diffusion: the subdiffusive case, *Physica A* **349**, 375-420 (2005).

33. S.G. Samko, A.A. Kilbas and O.I. Marichev, *Fractional Integrals and Derivatives: Theory and Applications*, Gordon and Breach, New York (1993).

34. E. Scalas, R. Gorenflo and F. Mainardi, Fractional calculus and continuous-time finance, Physica A **284**, 376-384 (2000).

35. E. Scalas, R. Gorenflo and F. Mainardi, Uncoupled continuous-time random walks: Solution and limiting behavior of the master equation, *Phys. Rev. E* **69**, 011107-1/8 (2004).

36. M.F. Shlesinger, G.M. Zaslavsky, and J. Klafter, Strange kinetics, *Nature* **363**, 31-37 (1993).

37. I.M. Sokolov, Thermodynamics and fractional Fokker-Planck equation, *Phys. Rev. E* **63**, 056111-1/8 (2001).

38. I.M. Sokolov, Lévy flights from a continuous-time process, *Phys. Rev. E* **63** 011104-1/10 (2001).

39. I.M. Sokolov, J. Klafter, A. Blumen, Linear response in complex systems: CTRW and the fractional Fokker-Planck equations, *Physica A* **302**, 268-278 (2001).

40. I.M. Sokolov and J. Klafter, ¿From diffusion to anomalous diffusion: a century after Einstein's Brownian motion, *Chaos* **15**, 026103-026109 (2005).

41. I.M. Sokolov, J. Klafter and A. Blumen, Fractional Kinetics, *Physics Today* **55**, 48-54 (2002).

42. A.A. Stanislavski, Black-Scholes model under subordination. *Physica A* **318** (2003), 469-474.

43. V.V. Uchaikin and V.V. Saenko, Stochastic solution of partial differential equations of fractional orders, *Siberian Journal of Numerical Mathematics* **6**, 197-203 (2003).

44. V.V. Uchaikin and V.M. Zolotarev, *Chance and Stability. Stable Distributions and their Applications*, VSP, Utrecht (1999).

45. G.H. Weiss, *Aspects and Applications of Random Walks*, North-Holland, Amsterdam (1994).

46. B.J. West, M. Bologna and P. Grigolini. *Physics of Fractal Operators*, Springer Verlag, New York (2003).

47. M.M. Wyss and W. Wyss, Evolution, its fractional extension and generalization, *Fractional Calculus and Appl. Analysis* **4**, 273-284 (2001).

48. G.M. Zaslavsky, Chaos, fractional kinetics and anomalous transport, *Phys. Reports* **371** (2002) 461-580.

49. G.M. Zaslavsky. *Hamiltonian Chaos and Fractional Dynamics*, Oxford University Press, Oxford (2005).

# FUNCTIONAL SPACES AND OPERATORS CONNECTED WITH SOME LÉVY WHITE NOISES

E. LYTVYNOV

*Department of Mathematics*
*University of Wales Swansea*
*Singleton Park*
*Swansea SA2 8PP*
*U.K.*
*E-mail: e.lytvynov@swansea.ac.uk*

We review some recent developments in white noise analysis and quantum probability. We pay a special attention to spaces of test and generalized functionals of some Lévy white noises, as well as to the structure of quantum white noise on these spaces.

**Key words:** Extended Fock spac, Gaussian white noise, Lévy white noise, Poisson white noise, square of white noise

**Mathematics Subject Classification:** 60H40, 60G51

## 1. Gaussian white noise and Fock space

Since the work of Hida[10] of 1975, Gaussian white noise analysis has become an established theory of test and generalized functions of infinitely many variables, see e.g. Refs. 11, 6 and the references therein.

Let us shortly recall some basic results of Gaussian analysis. In the space $L^2(\mathbb{R}) := L^2(\mathbb{R}, dx)$, consider the harmonic oscillator

$$(Hf)(t) := -f''(t) + (t^2 + 1)f(t), \qquad f \in C_0^\infty(\mathbb{R}).$$

This operator is self-adjoint and we preserve the notation $H$ for its closure. For each $p \in \mathbb{R}$, define a scalar product

$$(f, g)_p := (H^p f, g)_{L^2(\mathbb{R})}, \qquad f, g \in C_0^\infty(\mathbb{R}).$$

Let $\mathcal{S}_p$ denote the Hilbert space obtained as the closure of $C_0^\infty(\mathbb{R})$ in the norm $\|\cdot\|_p$ generated by the scalar product $(\cdot, \cdot)_p$. Then, for any $p > q$, the space $\mathcal{S}_p$ is densely and continuously embedded into $\mathcal{S}_q$, and if $p - q > 1/2$, then this embedding is of Hilbert–Schmidt type. Furthermore, for each $p > 0$, $\mathcal{S}_{-p}$ is the dual space of $\mathcal{S}_p$ with respect to zero space $L^2(\mathbb{R})$, i.e., the dual paring $\langle f, \varphi \rangle$ between any $f \in \mathcal{S}_{-p}$ and any $\varphi \in \mathcal{S}_p$ is obtained as the extension of the scalar product in $L^2(\mathbb{R})$. The above conclusions are, in fact, corollaries of the fact that the sequence of Hermite functions on $\mathbb{R}$,

$$e_j = e_j(t) = (\sqrt{\pi} 2^j j!)^{-1/2} (-1)^j e^{t^2/2} (d/dt)^j e^{-t^2}, \qquad j \in \mathbb{Z}_+ := \{0, 1, 2, \dots\},$$

forms an orthonormal basis of $L^2(\mathbb{R})$ such that each $e_j$ is an eigenvector of $H$ with eigenvalue $(2j+2)^2$.

Then

$$S := \operatorname*{proj\,lim}_{p \to \infty} S_p$$

is the Schwartz space of infinitely differentiable, rapidly decreasing functions on $\mathbb{R}$, and its dual

$$S' = \operatorname*{ind\,lim}_{p \to \infty} S_{-p}$$

is the Schwartz space of tempered distributions.

We denote by $C(S')$ the $\sigma$-algebra on $S'$ which is generated by cylinder sets in $S'$, i.e., by the sets of the form

$$\{\omega \in S' : (\langle \omega, \varphi_1 \rangle, \dots, \langle \omega, \varphi_N \rangle) \in A\},$$

where $\varphi_1, \dots, \varphi_N \in S$, $N \in \mathbb{N}$, and $A \in \mathcal{B}(\mathbb{R}^N)$.

By the Minlos theorem, there exists a unique probability measure $\mu_G$ on $(S', C(S'))$ whose Fourier transform is given by

$$\int_{S'} e^{i\langle \omega, \varphi \rangle} d\mu_G(\omega) = \exp\left[-(1/2)\|\varphi\|^2_{L^2(\mathbb{R})}\right], \qquad \varphi \in S. \tag{1}$$

The measure $\mu_G$ is called the (Gaussian) white noise measure. Indeed, using formula (1), it is easy to see that, for each $\varphi \in S$,

$$\int_{S'} \langle \omega, \varphi \rangle^2 d\mu_G(\omega) = \|\varphi\|^2_{L^2(\mathbb{R})}.$$

Hence, extending the mapping

$$L^2(\mathbb{R}) \supset S \ni \varphi \mapsto \langle \cdot, \varphi \rangle \in L^2(S', \mu_G)$$

by continuity, we obtain a random variable $\langle \cdot, f \rangle \in L^2(S', \mu_G)$ for each $f \in L^2(\mathbb{R})$. Then, for each $t \in \mathbb{R}$, we define

$$X_t := \begin{cases} \langle \cdot, \mathbf{1}_{[0,t]} \rangle, & t \geq 0, \\ -\langle \cdot, \mathbf{1}_{[t,0]} \rangle, & t < 0. \end{cases} \tag{2}$$

It is easily seen that $(X_t)_{t \in \mathbb{R}}$ is a version of Brownian motion, i.e., finite-dimensional distributions of the stochastic process $(X_t)_{t \in \mathbb{R}}$ coincide with those of Brownian motion. We now informally have, for all $t \in \mathbb{R}$, $X_t(\omega) = \int_0^t \omega(t)\,dt$, so that $X_t'(\omega) = \omega(t)$. Thus, elements $\omega \in S'$ can be thought of as paths of the derivative of Brownian motion, i.e., Gaussian white noise.

Let us recall that the symmetric Fock space over a real separable Hilbert space $\mathcal{H}$ is defined as

$$\mathcal{F}(\mathcal{H}) := \bigoplus_{n=0}^{\infty} \mathcal{F}^{(n)}(\mathcal{H})n!.$$

Here $\mathcal{F}^{(n)}(\mathcal{H}) := \mathcal{H}_{\mathbb{C}}^{\odot n}$, where $\odot$ stands for symmetric tensor product and the lower index $\mathbb{C}$ denotes complexification of a real space. Thus, for each $(f^{(n)})_{n=0}^{\infty} \in \mathcal{F}(\mathcal{H})$,

$$\|(f^{(n)})_{n=0}^{\infty}\|_{\mathcal{F}(\mathcal{H})}^2 = \sum_{n=0}^{\infty} \|f^{(n)}\|_{\mathcal{F}^{(n)}(\mathcal{H})}^2 \, n! \,.$$

The central technical point of the construction of spaces of test and generalized functionals of Gaussian white noise is the Wiener–Itô–Segal isomorphism $I_{\mathrm{G}}$ between the Fock space $\mathcal{F}(L^2(\mathbb{R}))$ and the complex space $L^2(\mathcal{S}' \to \mathbb{C}, \mu)$, which, for simplicity of notations, we will denote by $L^2(\mathcal{S}', \mu)$.

There are different ways of construction of the isomorphism $I_{\mathrm{G}}$, e.g., using multiple stochastic integrals with respect to Gaussian random measure. For us, it will be convenient to follow the approach which uses the procedure of orthogonalization of polynomials, see e.g. Ref. 6 for details.

A function $F(\omega) = \sum_{i=0}^{n} \langle \omega^{\otimes i}, f^{(i)} \rangle$, where $\omega \in \mathcal{S}'$, $n \in \mathbb{Z}_+$, and each $f^{(i)} \in \mathcal{S}_{\mathbb{C}}^{\odot i}$, is called a continuous polynomial on $\mathcal{S}'$, and $n$ is called the order of the polynomial $F$. The set $\mathcal{P}$ of all continuous polynomials on $\mathcal{S}'$ is dense in $L^2(\mathcal{S}', \mu_{\mathrm{G}})$. For $n \in \mathbb{Z}_+$, let $\mathcal{P}^{(n)}$ denote the set of all continuous polynomials on $\mathcal{S}'$ of order $\leq n$, and let $\mathcal{P}_{\mathrm{G}}^{(n)}$ be the closure of $\mathcal{P}^{(n)}$ in $L^2(\mathcal{S}', \mu_{\mathrm{G}})$. Let $\mathfrak{P}_{\mathrm{G}}^{(n)}$ stand for the orthogonal difference $\mathcal{P}_{\mathrm{G}}^{(n)} \ominus \mathcal{P}_{\mathrm{G}}^{(n-1)}$ in $L^2(\mathcal{S}', \mu_{\mathrm{G}})$. Then we easily get the orthogonal decomposition

$$L^2(\mathcal{S}', \mu_{\mathrm{G}}) = \bigoplus_{n=0}^{\infty} \mathfrak{P}_{\mathrm{G}}^{(n)} \,.$$

Next, for any $f^{(n)} \in \mathcal{S}_{\mathbb{C}}^{\otimes n}$, we define $:\langle \omega^{\otimes n}, f^{(n)} \rangle:_{\mathrm{G}}$ as the orthogonal projection of $\langle \omega^{\otimes n}, f^{(n)} \rangle$ onto $\mathfrak{P}_{\mathrm{G}}^{(n)}$. The set of such projections is dense in $\mathfrak{P}_{\mathrm{G}}^{(n)}$. Furthermore, for any $f^{(n)}, g^{(n)} \in \mathcal{S}_{\mathbb{C}}^{\odot n}$, we have:

$$\int_{\mathcal{S}'} \overline{:\langle \omega^{\otimes n}, f^{(n)} \rangle:_{\mathrm{G}}} \times :\langle \omega^{\otimes n}, g^{(n)} \rangle:_{\mathrm{G}} \, d\mu_{\mathrm{G}}(\omega) = (f^{(n)}, g^{(n)})_{\mathcal{F}^{(n)}(L^2(\mathbb{R}))} n! \,. \qquad (3)$$

Let $\mathcal{F}_{\mathrm{fin}}(\mathcal{S})$ denote the set of all sequences $(f^{(n)})_{n=0}^{\infty}$ such that each $f^{(i)} \in \mathcal{S}_{\mathbb{C}}^{\odot i}$ and for some $N \in \mathbb{N}$ $f^{(n)} = 0$ for all $n \geq N$. The $\mathcal{F}_{\mathrm{fin}}(\mathcal{S})$ is a dense subset of $\mathcal{F}(L^2(\mathbb{R}))$. For any $f = (f^{(n)})_{n=0}^{\infty} \in \mathcal{F}_{\mathrm{fin}}(\mathcal{S})$, we set

$$(I_{\mathrm{G}}f)(\omega) = \sum_{n=0}^{\infty} :\langle \omega^{\otimes n}, f^{(n)} \rangle:_{\mathrm{G}} \in L^2(\mathcal{S}', \mu_{\mathrm{G}}). \qquad (4)$$

By (3), we can extend $I_{\mathrm{G}}$ by continuity to get a unitary operator

$$I_{\mathrm{G}} : \mathcal{F}(L^2(\mathbb{R})) \to L^2(\mathcal{S}', \mu_{\mathrm{G}}).$$

For any function $f^{(n)} \in \mathcal{F}^{(n)}(L^2(\mathbb{R}))$, we will use the evident notation $:\langle \omega^{\otimes n}, f^{(n)} \rangle:_{\mathrm{G}}$. Then, for each $f = (f^{(n)})_{n=0}^{\infty} \in \mathcal{F}(L^2(\mathbb{R}))$, $I_{\mathrm{G}}f$ is given by formula (4).

For any $\varphi \in \mathcal{S}$, let $\langle \cdot, \varphi \rangle \cdot$ denote the operator of multiplication by $\langle \cdot, \varphi \rangle$ in $L^2(\mathcal{S}', \mu_{\mathrm{G}})$. We set

$$A_{\mathrm{G}}(\varphi) := I_{\mathrm{G}}^{-1} \langle \cdot, \varphi \rangle \cdot I_{\mathrm{G}}.$$

Then, $\mathcal{F}_{\text{fin}}(\mathcal{S}) \subset \text{Dom}(A_G(\varphi))$, $A_G(\varphi)\mathcal{F}_{\text{fin}}(\mathcal{S}) \subset \mathcal{F}_{\text{fin}}(\mathcal{S})$, and $A_G(\varphi)$ is essentially self-adjoint on $\mathcal{F}_{\text{fin}}(\mathcal{S})$. Furthermore, $A_G(\varphi)$ has the following representation on $\mathcal{F}_{\text{fin}}(\mathcal{S})$:

$$A_G(\varphi) = A^+(\varphi) + A^-(\varphi). \tag{5}$$

Here $A^+(\varphi)$ is the creation operator: for $f^{(n)} \in \mathcal{S}_{\mathbb{C}}^{\odot n}$

$$A^+(\varphi)f^{(n)} = \varphi \odot f^{(n)} \in \mathcal{S}_{\mathbb{C}}^{\odot(n+1)},$$

and $A^-(\varphi)$ is the annihilation operator: $A^-(\varphi)f^{(n)}$ belongs to $\mathcal{S}_{\mathbb{C}}^{\odot(n-1)}$ and is given by

$$(A^-(\varphi)f^{(n)})(t_1, \ldots, t_{n-1}) = \int_{\mathbb{R}} \varphi(t)f^{(n)}(t, t_1, \ldots, t_{n-1})\, dt.$$

For each $\varkappa \in [-1, 1]$ and $p \in \mathbb{R}$, we denote

$$\mathcal{F}_{\varkappa}(\mathcal{S}_p) := \bigoplus_{n=0}^{\infty} \mathcal{F}^{(n)}(\mathcal{S}_p)(n!)^{1+\varkappa},$$

and for each $\varkappa \in [0, 1]$ we set

$$\mathcal{F}_{\varkappa}(\mathcal{S}) := \text{proj} \lim_{p \to \infty} \mathcal{F}_{\varkappa}(\mathcal{S}_p).$$

It is easy to see that each $\mathcal{F}_{\varkappa}(\mathcal{S})$ is a nuclear space. Furthermore, the dual space of $\mathcal{F}_{\varkappa}(\mathcal{S})$ with respect to zero space $\mathcal{F}(L^2(\mathbb{R}))$ is

$$\mathcal{F}_{-\varkappa}(\mathcal{S}') := \text{ind} \lim_{p \to \infty} \mathcal{F}_{-\varkappa}(\mathcal{S}_{-p}).$$

Thus, we get a standard triple

$$\mathcal{F}_{\varkappa}(\mathcal{S}) \subset \mathcal{F}(L^2(\mathbb{R})) \subset \mathcal{F}_{-\varkappa}(\mathcal{S}').$$

The test space $\mathcal{F}_1(\mathcal{S})$ is evidently the smallest one between the above spaces, whereas its dual space $\mathcal{F}_{-1}(\mathcal{S}')$ is the biggest one.

For each $F = (F^{(n)}) \in \mathcal{F}_{-1}(\mathcal{S}')$, the $S$-transform of $F$ is defined by

$$(SF)(\xi) := \sum_{n=0}^{\infty} \langle F^{(n)}, \xi^{\otimes n} \rangle, \qquad \xi \in \mathcal{S}_{\mathbb{C}}, \tag{6}$$

provided the series on the right hand side of (6) converges absolutely.

The $S$-transform uniquely characterizes an element of $\mathcal{F}_{-1}(\mathcal{S}')$. More exactly, let $\text{Hol}_0(\mathcal{S}_{\mathbb{C}})$ denotes the set of all (germs) of functions which are holomorphic in a neighborhood of zero in $\mathcal{S}_{\mathbb{C}}$. The following theorem was proved in Ref. 16.

**Theorem 1.1.** *The $S$-transform is a one-to-one map between $\mathcal{F}_{-1}(\mathcal{S}')$ and $\text{Hol}_0(\mathcal{S}_{\mathbb{C}})$.*

Note that the choice of $\varkappa > 1$ would imply that the $S$-transform is not well-defined on $\mathcal{F}_{-\varkappa}(\mathcal{S}')$. On the other hand, all the spaces $\mathcal{F}_{-\varkappa}(\mathcal{S}')$ and $\mathcal{F}_\varkappa(\mathcal{S})$ with $\varkappa \in [0, 1]$ admit a complete characterization in terms of their $S$-transform, see e.g. Refs. 6, 19, 16.

Taking into account that the product of two elements of $\mathrm{Hol}_0(\mathcal{S}_\mathbb{C})$ remains in this set, one defines a Wick product $F_1 \diamond F_2$ of $F_1, F_2 \in \mathcal{F}_{-1}(\mathcal{S}')$ through the formula

$$S(F_1 \diamond F_2) = S(F_1)S(F_2). \tag{7}$$

Furthermore, if $F \in \mathcal{F}_{-1}(\mathcal{S}')$ and $f$ is a holomorphic function in a neighborhood of $(SF)(0)$ in $\mathbb{C}$, then one defines $f^\circ(F) \in \mathcal{F}_{-1}(\mathcal{S}')$ through

$$S(f^\circ(F)) = f(SF). \tag{8}$$

Using the unitary operator $I_\mathrm{G}$, all the above definitions and results can be reformulated in terms of test and generalized functions on $\mathcal{S}'$ whose dual paring is generated by the scalar product in $L^2(\mathcal{S}', \mu_\mathrm{G})$. In particular, one defines spaces of test functions $(\mathcal{S})_\mathrm{G}^\varkappa := I_\mathrm{G}\mathcal{F}_\varkappa(\mathcal{S})$ and their dual spaces $(\mathcal{S})_\mathrm{G}^{-\varkappa}$, $\varkappa \in [0, 1]$. For $\varkappa = 0$, these are the Hida test space and the space of Hida distrubutions, respectively (e.g. Refs 11, 6). For $\varkappa \in (0, 1)$, these spaces were introduced and studied by Kondratiev and Streit [19], and for $\varkappa = 1$, by Kondratiev, Leukert and Streit[16]. Note also that, in the case of a Gaussian product measure, such spaces for all $\varkappa \in [0, 1]$ were studied by Kondratiev[15].

The Wick calculus of generalized Gaussian functionals based on the definitions (7), (8) and the unitary operator $I_\mathrm{G}$ has found numerous applications, in particular, in fluid mechanics and financial mathematics, see e.g. Refs. 12, 9.

Additionally to the description of the above test spaces $(\mathcal{S})_\mathrm{G}^\varkappa$ in terms of their $S$-transform, one can also give their inner description, e.g. Refs. 19, 16. Let $\mathcal{E}(\mathcal{S}_\mathbb{C}')$ denote the space of all entire functions on $\mathcal{S}_\mathbb{C}'$. For each $\beta \in [1, 2]$, we denote by $\mathcal{E}_{\min}^\beta(\mathcal{S}_\mathbb{C}')$ the subset of $\mathcal{E}(\mathcal{S}_\mathbb{C}')$ consisting of all entire functions of the $\beta$-th order of growth and minimal type. That is, for any $\Phi \in \mathcal{E}_{\min}^\beta(\mathcal{S}_\mathbb{C}')$, $p \geq 0$, and $\varepsilon > 0$, there exists $C > 0$ such that

$$|\Phi(z)| \leq C \exp(\varepsilon |z|_{-p}^\beta), \qquad z \in \mathcal{S}_{-p,\mathbb{C}}.$$

Next, we denote by $\mathcal{E}_{\min}^\beta(\mathcal{S}')$ the set of all functions on $\mathcal{S}'$, which are obtained by restricting functions from $\mathcal{E}_{\min}^\beta(\mathcal{S}_\mathbb{C}')$ to $\mathcal{S}'$. The following theorem unifies the results of Refs. 19, 16, see also Refs. 6, 11,

**Theorem 1.2.** *For each $\varkappa \in [0, 1]$, we have*

$$(\mathcal{S})_\mathrm{G}^\varkappa = \mathcal{E}_{\min}^{2/(1+\varkappa)}(\mathcal{S}'). \tag{9}$$

*The equality (9) is understood in the sense that, for each $f = (f^{(n)}) \in \mathcal{F}_\varkappa(\mathcal{S})$, the following realization was chosen for $\Phi := I_\mathrm{G}f$:*

$$\Phi(\omega) = \sum_{n=0}^\infty \langle :\omega^{\otimes n}:_\mathrm{G}, f^{(n)} \rangle, \qquad \omega \in \mathcal{S}',$$

*where $:\omega^{\otimes n}:_\mathrm{G} \in \mathcal{S}'^{\odot n}$ is defined by the recurrence relation*

$$:\omega^{\otimes 0}:_\mathrm{G} = 1, \qquad :\omega^{\otimes 1}:_\mathrm{G} = \omega,$$

$$:\omega^{\otimes(n+1)}:_\mathrm{G}(t_1, \ldots, t_{n+1}) = \left(:\omega^{\otimes n}:_\mathrm{G}(t_1, \ldots, t_n)\omega(t_{n+1})\right)^{\sim}$$

$$- n\left(:\omega^{\otimes(n-1)}:_\mathrm{G}(t_1, \ldots, t_{n-1})1(t_n)\delta(t_{n+1} - t_n)\right)^{\sim}, \qquad n \in \mathbb{N}.$$

*Here $\delta(\cdot)$ denotes the delta function at zero and $(\cdot)^{\sim}$ denotes symmetrization.*

For each $t \in \mathbb{R}$, we define an annihilation operator at $t$, denoted by $\partial_t$, and a creation operator at $t$, denoted by $\partial_t^{\dagger}$, by

$$(\partial_t f^{(n)})(t_1, \ldots, t_{n-1}) := n f^{(n)}(t_1, \ldots, t_{n-1}, t), \qquad f^{(n)} \in \mathcal{S}_{\mathbb{C}}^{\odot n},$$

$$\partial_t^{\dagger} F^{(n)} := \delta_t \odot F^{(n)}, \qquad F^{(n)} \in \mathcal{S}_{\mathbb{C}}'^{\odot n},$$

where $\delta_t$ denotes the delta function at $t$. The operators $\partial_t$ and $\partial_t^{\dagger}$ can then be extended to linear continuous operators on $\mathcal{F}_{\varkappa}(\mathcal{S})$ and $\mathcal{F}_{-\varkappa}(\mathcal{S}')$, respectively, and $\partial_t^{\dagger}$ becomes the dual operator of $\partial_t$. Then,

$$A^{-}(\varphi) = \int_{\mathbb{R}} \varphi(t)\partial_t \, dt,$$

$$A^{+}(\varphi) = \int_{\mathbb{R}} \varphi(t)\partial_t^{\dagger} \, dt,$$

and so by (5),

$$A_\mathrm{G}(\varphi) = \int_{\mathbb{R}} \varphi(t)(\partial_t + \partial_t^{\dagger}) \, dt,$$

the above integrals being understood in the Bochner sense. The operators

$$W_\mathrm{G}(t) := \partial_t + \partial_t^{\dagger}, \qquad t \in \mathbb{R}, \tag{10}$$

are called quantum Gaussian white noise. Realized on $(\mathcal{S})_\mathrm{G}^1$, the operator $\partial_t$ becomes the operator of Gâteaux differentiation in direction $\delta_t$:

$$(\partial_t F)(\omega) = \lim_{\varepsilon \to 0}(F(\omega + \varepsilon\delta_t) - F(\omega))/\varepsilon, \qquad F \in (\mathcal{S})_\mathrm{G}^1, \ t \in \mathbb{R}, \ \omega \in \mathcal{S}',$$

see e.g. Ref. 11.

## 2. Poisson white noise

The measure $\mu_\mathrm{P}$ of (centered) Poisson white noise is defined on $(\mathcal{S}', \mathcal{C}(\mathcal{S}'))$ by

$$\int_{\mathcal{S}'} e^{i\langle\omega,\varphi\rangle} \, d\mu_\mathrm{P}(\omega) = \exp\left[\int_{\mathbb{R}}(e^{i\varphi(t)} - 1 - i\varphi(t)) \, dt\right], \qquad \varphi \in \mathcal{S}.$$

Under $\mu_\mathrm{P}$, the stochastic process $(X_t)_{t\in\mathbb{R}}$, defined by (2), is a centered Poisson process, which is why $\omega \in \mathcal{S}'$ can now be thought of as a path of Poisson white noise (in fact, $\mu_\mathrm{P}$ is concentrated on infinite sums of delta functions shifted by $-1$).

The procedure of orthogonalization of continuous polynomials in $L^2(\mathcal{S}', \mu_P)$ leads to a unitary isomorphism $I_P$ between the Fock space $\mathcal{F}(L^2(\mathbb{R}))$ and $L^2(\mathcal{S}', \mu_P)$. The counterpart of formula (5) now looks as follows:

$$A_P(\varphi) = A^+(\varphi) + A^0(\varphi) + A^-(\varphi),$$

where $A^0(\varphi)$ is the neutral operator:

$$(A^0(\varphi)f^{(n)})(t_1, \ldots, t_n) := \left( \sum_{i=1}^{n} \varphi(t_i) \right) f^{(n)}(t_1, \ldots, t_n), \qquad f^{(n)} \in \mathcal{S}_{\mathbb{C}}^{\odot n}.$$

In terms of the operators $\partial_t$ and $\partial_t^\dagger$, the neutral operator has the following representation:

$$A^0(\varphi) = \int_{\mathbb{R}} \varphi(t)\partial_t^\dagger \partial_t \, dt.$$

Thus, the quantum (centered) Poisson white noise is given by

$$W_P(t) = \partial_t + \partial_t^\dagger \partial_t + \partial_t^\dagger \tag{11}$$

(see e.g. Refs. 13, 21 ).

Next, using the unitary operator $I_P$, we obtain a scale of spaces of test functions $(\mathcal{S})_P^\varkappa$ and generalized functions $(\mathcal{S}')_P^{-\varkappa}$. For any $f^{(n)} \in \mathcal{S}_{\mathbb{C}}^{\odot n}$, the orthogonal projection $:\langle \omega^{\otimes n}, f^{(n)} \rangle:_P$ has a $\mu_P$-version $\langle :\omega^{\otimes n}:_P, f^{(n)} \rangle$, where $:\omega^{\otimes n}:_P \in \mathcal{S}'^{\odot n}$ are given by the following recurrence relation (see Ref. 22):

$$:\omega^{\otimes 0}:_P = 1, \quad :\omega^{\otimes 1}:_P = \omega,$$

$$:\omega^{\otimes (n+1)}:_P(t_1, \ldots, t_{n+1}) = \left( :\omega^{\otimes n}:_P(t_1, \ldots, t_n)\omega(t_{n+1}) \right)^{\widetilde{\phantom{x}}}$$
$$- n\left( :\omega^{\otimes (n-1)}:_P(t_1, \ldots, t_{n-1})1(t_n)\delta(t_{n+1} - t_n) \right)^{\widetilde{\phantom{x}}}$$
$$- n\left( :\omega^{\otimes n}:_P(t_1, \ldots, t_n)\delta(t_{n+1} - t_n) \right)^{\widetilde{\phantom{x}}}, \qquad n \in \mathbb{N}.$$

However, in the Poisson case, the following statement holds[22].

**Theorem 2.1.** *For each* $w \in \mathcal{S}'$, *denote* $D(w) := (:\omega^{\otimes n}:_P)_{n=0}^{\infty}$. *Then* $D(w) \in \mathcal{F}_{-1}(\mathcal{S}')$, *and if* $w \neq 0$, *then* $D(w) \notin \mathcal{F}_{-\varkappa}(\mathcal{S}')$ *for all* $\varkappa \in [0, 1)$.

It is straightforward to see that the $D(w)$ in the Poisson realization is just the delta function at $w$, denoted by $\delta_w$. Thus, Theorem 2.1 implies:

**Corollary 2.1.** *For each* $w \in \mathcal{S}'$, $\delta_w \in (\mathcal{S}')_P^{-1}$, *and if* $w \neq 0$, *then* $\delta_w \notin (\mathcal{S}')_P^{-\varkappa}$ *for all* $\varkappa \in [0, 1)$.

The above corollary shows that the test spaces $(\mathcal{S})_P^\varkappa$ with $\varkappa \in [0, 1)$ do not posses nice inner properties, and therefore they are not appropiate for applications. On the other hand, we have[20]:

**Theorem 2.2.** *We have:*

$$(\mathcal{S})_P^1 = \mathcal{E}_{\min}^1(\mathcal{S}').$$

Thus, $\mathcal{E}^1_{\min}(\mathcal{S}')$ appears to be a universal space for both Gaussian and Poisson white noise analysis.

The annihilation operator $\partial_t$ realized on $(\mathcal{S})^1_P$ becomes a difference operator[22]:

$$(\partial_t F)(\omega) = F(\omega + \delta_t) - F(\omega), \qquad F \in (\mathcal{S})^1_P, \ t \in \mathbb{R}, \ \omega \in \mathcal{S}'.$$

We note that one can also study a more general white noise measure of Poisson type for which the corresponding quantum white noise is given by

$$\partial_t + \lambda \partial_t^\dagger \partial_t + \partial_t^\dagger, \qquad (12)$$

where $\lambda \in \mathbb{R}_+$, $\lambda \neq 0$, see e.g. Ref. 26.

## 3. Lévy white noise and extended Fock space

We will now discuss the case of a Lévy white noise without Gaussian part. Let $\nu$ be a probability measure on $(\mathbb{R}, \mathcal{B}(\mathbb{R}))$ such that $\nu(\{0\}) = 0$. We will also assume that there exists $\varepsilon > 0$ such that $\int_{\mathbb{R}} \exp(\varepsilon|s|) \, \nu(ds) < \infty$. The latter condition implies that $\nu$ has all moments finite, and moreover, the set of all polynomials on $\mathbb{R}$ is dense in $L^2(\mathbb{R}, \nu)$.

We define a centered Lévy white noise as a probability measure $\mu_\nu$ on $(\mathcal{S}', \mathcal{C}(\mathcal{S}'))$ with Fourier transform

$$\int_{\mathcal{S}'} e^{i\langle \omega, \varphi \rangle} \, \mu_\nu(d\omega) = \exp\left[ \varkappa \int_{\mathbb{R}} \int_{\mathbb{R}} (e^{is\varphi(t)} - 1 - is\varphi(t)) \frac{1}{s^2} \nu(ds) \, dt \right], \qquad \varphi \in \mathcal{S},$$

where $\varkappa > 0$. For notational simplicity, we will assume that $\varkappa = 1$, which is a very weak restriction.

Under the measure $\mu_\nu$, the stochastic process $(X_t)_{t \in \mathbb{R}}$, defined by (2), is a centered Lévy process with Lévy measure $\frac{1}{s^2} \nu(ds)$:

$$\int_{\mathcal{S}'} e^{iuX_t(\omega)} \, \mu_\nu(d\omega) = \exp\left[ |t| \int_{\mathbb{R}} (e^{isu \, \mathrm{sign}(t)} - 1 - isu \, \mathrm{sign}(t)) \frac{1}{s^2} \nu(ds) \right], \qquad u \in \mathbb{R}.$$

Hence, $\omega \in \mathcal{S}'$ can be thought of as a path of Lévy white noise.

By the above assumptions, the set $\mathcal{P}$ of all continuous polynomials on $\mathcal{S}'$ is dense in $L^2(\mathcal{S}', \mu_\nu)$. Therefore, through the procedure of orthogonalization of polynomials, one gets an orthogonal decomposition

$$L^2(\mathcal{S}', \mu_\nu) = \bigoplus_{n=0}^{\infty} \mathfrak{P}_\nu^{(n)},$$

and the set of all orthogonal projections $:\langle \omega^{\otimes n}, f^{(n)} \rangle:_\nu$ of $\langle \omega^{\otimes n}, f^{(n)} \rangle$ onto $\mathfrak{P}_\nu^{(n)}$ is dense in $\mathfrak{P}_\nu^{(n)}$.

However, in contrast to the Gaussian and Poisson cases, the scalar product of any $:\langle \omega^{\otimes n}, f^{(n)} \rangle:_\nu$ and $:\langle \omega^{\otimes n}, g^{(n)} \rangle:_\nu$ in $L^2(\mathcal{S}', \mu_\nu)$ is not given by the scalar product of $f^{(n)}$ and $g^{(n)}$ in the Fock space, but by a much more complex expression, see Refs. 24, 7 for an explicit formula in the general case, and formulas (14), (15) below in a special case. Still it is possible to construct a unitary isomorphism $I_\nu$

between the so-called extended Fock space $F_\nu(L^2(\mathbb{R})) = \bigoplus_{n=0}^\infty F_\nu^{(n)}(L^2(\mathbb{R}))n!$ and $L^2(\mathcal{S}', \mu_\nu)$. In the above construction, $F_\nu^{(n)}(L^2(\mathbb{R}))$ is a Hilbert space that is obtained as the closure of $\mathcal{S}_{\mathbb{C}}^{\odot n}$ in the norm generated by the scalar product

$$(f^{(n)}, g^{(n)})_{F_\nu^{(n)}(L^2(\mathbb{R}))} := \frac{1}{n!} \int_{\mathcal{S}'} \overline{:\langle \omega^{\otimes n}, f^{(n)} \rangle:_\nu} \times :\langle \omega^{\otimes n}, g^{(n)} \rangle:_\nu \mu_\nu(d\omega).$$

We next set

$$A_\nu(\varphi) := I_\nu^{-1}\langle \cdot, \varphi \rangle \cdot I_\nu, \qquad \varphi \in \mathcal{S}.$$

Our next aim is to derive an explicit form of the action of these operators. This can be done[24,7], however, the property that $\mathcal{F}_{\text{fin}}(\mathcal{S})$ is invariant under the action of $A_\nu(\varphi)$, generally speaking, does not hold. The following theorem[7,24,23] identifies all the Lévy noises for which this property is preserved.

**Theorem 3.1.** *Under the above assumptions, the property*

$$A_\nu(\varphi)\mathcal{F}_{\text{fin}}(\mathcal{S}) \subset \mathcal{F}_{\text{fin}}(\mathcal{S}), \qquad \varphi \in \mathcal{S},$$

*holds if and only if $\nu$ is the measure of orthogonality of a system of polynomials $(p_n(t))_{n=0}^\infty$ on $\mathbb{R}$ which satisfy the following recurrence relation:*

$$tp_n(t) = \sqrt{(n+1)(n+2)}p_{n+1}(t) + \lambda(n+1)p_n(t) + \sqrt{n(n+1)}p_{n-1}(t),$$
$$n \in \mathbb{Z}_+, \ p_0(t) = 1, \ p_{-1}(t) = 0, \tag{13}$$

*for some $\lambda \in \mathbb{R}$.*

Let us consider the situation described in Theorem 3.1 in more detail. We will denote by $\nu_\lambda$ the measure $\nu$ which corresponds to the parameter $\lambda \in \mathbb{R}$ through (13). We first mention that the condition of orthogonality of the polynomials satisfying (13) uniquely determines the measure $\nu_\lambda$. It is easy to show that, for $\lambda > 0$, the measure $\nu_{-\lambda}$ is the image of the measure $\nu_\lambda$ under the mapping $\mathbb{R} \ni t \mapsto -t \in \mathbb{R}$, which is why we will only consider the case $\lambda \geq 0$. In fact, we have (see e.g. Ref. 8): for $\lambda \in [0, 2)$,

$$\nu_\lambda(ds) = \frac{\sqrt{4 - \lambda^2}}{2\pi} \left| \Gamma\left(1 + i(4 - \lambda^2)^{-1/2}s\right) \right|^2$$
$$\times \exp\left[ -s2(4 - \lambda^2)^{-1/2} \arctan\left(\lambda(4 - \lambda^2)^{-1/2}\right)\right] ds$$

($\nu_\lambda$ is a Meixner distribution), for $\lambda = 2$

$$\nu_2(ds) = \chi_{(0,\infty)}(s)e^{-s}s\, ds$$

($\nu_2$ is a gamma distribution), and for $\lambda > 2$

$$\nu_\lambda(ds) = (\lambda^2 - 4) \sum_{k=1}^\infty \left(\frac{\lambda - \sqrt{\lambda^2 - 4}}{\lambda + \sqrt{\lambda^2 - 4}}\right)^k k\, \delta_{\sqrt{\lambda^2 - 4}\,k}$$

($\nu_\lambda$ is now a Pascal distribution).

In what follows, we will use the lower index $\lambda$ instead of $\nu_\lambda$.

The stochastic process $(X_t)_{t\in\mathbb{R}}$ under $\mu_\lambda$ is a Meixner process for $|\lambda| < 2$, a gamma process for $|\lambda| = 2$ and a Pascal process for $|\lambda| > 2$. In other words, for each $t \in \mathbb{R}$, $t \neq 0$, the distribution of the random variable $X_t$ under $\mu_\lambda$ is of the same class of distributions as the measure $\nu_\lambda$.

Next, for each $\lambda$, the scalar product in the space $F^{(n)}(L^2(\mathbb{R})) = F^{(n)}_\lambda(L^2(\mathbb{R}))$ is given as follows[23,24]. For each $\alpha \in \mathbb{Z}_+$, $1\alpha_1 + 2\alpha_2 + \cdots = n$, $n \in \mathbb{N}$, and for any function $f^{(n)} : \mathbb{R}^n \to \mathbb{R}$ we define a function $D_\alpha f^{(n)} : \mathbb{R}^{|\alpha|} \to \mathbb{R}$ by setting

$$(D_\alpha f^{(n)})(t_1,\ldots,t_{|\alpha|}):=f^{(n)}(t_1,\ldots,t_{\alpha_1},\underbrace{t_{\alpha_1+1},t_{\alpha_1+1}}_{2\text{ times}},\underbrace{t_{\alpha_1+2},t_{\alpha_1+2}}_{2\text{ times}},\ldots,\underbrace{t_{\alpha_1+\alpha_2},t_{\alpha_1+\alpha_2}}_{2\text{ times}},$$
$$\underbrace{t_{\alpha_1+\alpha_2+1},t_{\alpha_1+\alpha_2+1},t_{\alpha_1+\alpha_2+1}}_{3\text{ times}},\ldots). \tag{14}$$

Here $|\alpha| := \alpha_1 + \alpha_2 + \cdots$. Then, for any $f^{(n)}, g^{(n)} \in \mathcal{S}_\mathbb{C}^{\odot n}$,

$$(f^{(n)}, g^{(n)})_{F^{(n)}(L^2(\mathbb{R}))} = n! \sum_{\alpha\in\mathbb{Z}_+:\, 1\alpha_1+2\alpha_2+\cdots=n} \frac{n!}{\alpha_1!\, 1^{\alpha_1}\alpha_2!\, 2^{\alpha_2}\cdots}$$
$$\times \int_{\mathbb{R}^{|\alpha|}} \overline{(D_\alpha f^{(n)})(t_1,\ldots,t_{|\alpha|})} \times (D_\alpha g^{(n)})(t_1,\ldots,t_{|\alpha|})\, dt_1\cdots dt_{|\alpha|}. \tag{15}$$

The following theorem[23,24] describes the action of $A_\lambda(\varphi)$ on $\mathcal{F}_{\text{fin}}(\mathcal{S})$.

**Theorem 3.2.** *For each $\lambda \in \mathbb{R}$ and $\varphi \in \mathcal{S}$, we have on $\mathcal{F}_{\text{fin}}(\mathcal{S})$:*

$$A_\lambda(\varphi) = A^+(\varphi) + \lambda A^0(\varphi) + \mathfrak{A}^-(\varphi).$$

*Here $\mathfrak{A}^-(\varphi)$ is the restriction to $\mathcal{F}_{\text{fin}}(\mathcal{S})$ of the adjoint operator of $A^+(\varphi)$ in $F(L^2(\mathbb{R})) = \bigoplus_{n=0}^\infty F^{(n)}(L^2(\mathbb{R}))n!$, and*

$$\mathfrak{A}^-(\varphi) = A^-(\varphi) + A_1^-(\varphi),$$

*where*

$$(A_1^-(\varphi)f^{(n)})(t_1,\ldots,t_{n-1}) = n(n-1)\big(\varphi(t_1)f^{(n)}(t_1,t_1,t_2,t_3,\ldots,t_n)\big)^\sim. \tag{16}$$

*Furthermore, each $A_\lambda(\varphi)$ is essentially self-adjoint on $\mathcal{F}_{\text{fin}}(\mathcal{S})$.*

We see that $A_\lambda(\varphi)$ has creation, neutral, and annihilation parts. Therefore, the family of self-adjoint commuting operators $(A_\lambda(\varphi))_{\varphi\in\mathcal{S}}$ is a Jacobi field in $F(L^2(\mathbb{R}))$ (compare with Ref. 5 and the references therein).

From Theorem 3.2, one concludes that, for any $f^{(n)} \in \mathcal{S}_\mathbb{C}^{\odot n}$, the orthogonal projection $:\langle \omega^{\otimes n}, f^{(n)}\rangle:_\lambda$ has a $\mu_\lambda$-version $\langle :\omega^{\otimes n}:_\lambda, f^{(n)}\rangle$, where $:\omega^{\otimes n}:_\lambda \in \mathcal{S}'^{\otimes n}$ are given by the following recurrence relation:

$$:\omega^{\otimes 0}:_\lambda = 1, \quad :\omega^{\otimes 1}:_\lambda = \omega,$$
$$:\omega^{\otimes(n+1)}:_\lambda(t_1,\ldots,t_{n+1}) = \big(:\omega^{\otimes n}:_\lambda(t_1,\ldots,t_n)\omega(t_{n+1})\big)^\sim$$
$$- n\big(:\omega^{\otimes(n-1)}:_\lambda(t_1,\ldots,t_{n-1})1(t_n)\delta(t_{n+1}-t_n)\big)^\sim$$
$$- \lambda n\big(:\omega^{\otimes n}:_\lambda(t_1,\ldots,t_n)\delta(t_{n+1}-t_n)\big)^\sim$$
$$- n(n-1)\big(:\omega^{\otimes(n-1)}:_\lambda(t_1,\ldots,t_{n-1})\delta(t_n-t_{n-1})\delta(t_{n+1}-t_n)\big)^\sim, \qquad n \in \mathbb{N}.$$

We will now construct a space of test functions. It is possible to show[17] that the Hilbert space $\mathcal{F}_1(\mathcal{S}_1)$ is densely and continuously embedded into $F(L^2(\mathbb{R}))$. This embedding is understood in the sense that $\mathcal{F}_1(\mathcal{S}_1)$ is considered as the closure of $\mathcal{F}_{\text{fin}}(\mathcal{S})$ in the respective norm. Therefore, the nuclear space $\mathcal{F}_1(\mathcal{S})$ is densely and continuously embedded into $F(L^2(\mathbb{R}))$. We will denote by $(\mathcal{S})^1_\lambda$ the image of $\mathcal{F}_1(\mathcal{S})$ under $I_\lambda$.

Using a result from the theory of test and generalized functions connected with a generalized Appell system of polynomials[18,14], one proves the following theorem.

**Theorem 3.3.** *For each $\lambda \in \mathbb{R}$,*

$$(\mathcal{S})^1_\lambda = \mathcal{E}^1_{\min}(\mathcal{S}').$$

Thus, $\mathcal{E}^1_{\min}(\mathcal{S}')$ appears to be a universal space for our purposes.

Taking to notice that $(\mathcal{S})^1_\lambda$ is the image of $\mathcal{F}_1(\mathcal{S})$ under $I_\lambda$, we will identify the dual space $(\mathcal{S}')^{-1}_\lambda$ of $(\mathcal{S})^1_\lambda$ with $\mathcal{F}_{-1}(\mathcal{S}')$. Notice, however, that now the dual pairing between elements of $(\mathcal{S}')^{-1}_\lambda$ and $(\mathcal{S})^1_\lambda$ is obtained *not* through the scalar product in $L^2(\mathcal{S}', \mu_\lambda)$, or equivalently in $F(L^2(\mathbb{R}))$, but through the scalar product in the usual Fock space $\mathcal{F}(L^2(\mathbb{R}))$. In particular, such a realization of the dual space $(\mathcal{S}')^{-1}_\lambda$ is convenient for developing Wick calculus on it.

By (16), the operator $A^-_1(\varphi)$ has the following representation through the operators $\partial_t$ and $\partial^\dagger_t$:

$$A^-_1(\varphi) = \int_{\mathbb{R}} \varphi(t) \partial^\dagger_t \partial_t \partial_t \, dt.$$

Therefore, by Theorem 3.2, we get:

$$A_\lambda(\varphi) = \int_{\mathbb{R}} \varphi(t)(\partial^\dagger_t + \lambda \partial^\dagger_t \partial_t + \partial_t + \partial^\dagger_t \partial_t \partial_t) \, dt.$$

Hence, the corresponding quantum white noise, denoted by $W_\lambda(t)$, is given by

$$W_\lambda(t) = \partial^\dagger_t + \lambda \partial^\dagger_t \partial_t + \partial_t + \partial^\dagger_t \partial_t \partial_t. \tag{17}$$

Realized on the space $(\mathcal{S})^1_\lambda$, the operator $\partial_t$ acts as follows[23]:

$$(\partial_t F)(\omega) = \int_{\mathbb{R}} \frac{F(\omega + s\delta_t) - F(\omega)}{s} \nu_\lambda(ds), \qquad F \in (\mathcal{S})^1_\lambda, \ t \in \mathbb{R}, \ \omega \in \mathcal{S}'.$$

## 4. The square of white noise algebra

As we have seen in Sections 1 and 2, the Gaussian white noise is just the sum of the annihilation operator $\partial_t$ and the creation operator $\partial^\dagger_t$, whereas the Poisson white noise is obtained by adding to the Gaussian white noise a constant times the product of $\partial^\dagger_t$ and $\partial_t$. Let us also recall that the operators $\partial_t$, $\partial^\dagger_t$, $t \in \mathbb{R}$, satisfy the canonical commutation relations:

$$[\partial_t, \partial_s] = [\partial^\dagger_t, \partial^\dagger_s] = 0,$$
$$[\partial_t, \partial^\dagger_s] = \delta(t - s), \tag{18}$$

where $[A, B]:=AB - BA$.

It was proposed in Ref. 2 to develop a stochastic calculus for higher powers of white noise, in other words, for higher powers of the operators $\partial_t$, $\partial_t^\dagger$. This problem was, in fact, influenced by the old dream of T. Hida that the operators $\partial_t$, $\partial_t^\dagger$ should play a fundamental role in infinite-dimensional analysis.

We will now deal with the squares of $\partial_t$, $\partial_t^\dagger$. The idea is to introduce operators $B_t$ and $B_t^\dagger$ which will be interpreted as $\partial_t^2$ and $(\partial_t^\dagger)^2$, to derive from (18) the commutation relations satisfied by $B_t$, $B_t^\dagger$, and $N_t := B_t^\dagger B_t$ and then to consider the quantum white noise $B_t + B_t^\dagger + \lambda N_t$, where $\lambda \in \mathbb{R}$ (compare with (10), (11), and (12)). However, when doing this, one arrives at the expression $\delta(\cdot)^2$ — the square of the delta function. It was proposed in Refs. 3, 4 to carry out a renormalization procedure by employing the following equality, which may be justified in the framework of distribution theory:

$$\delta(\cdot)^2 = c\delta(\cdot).$$

Here $c \in \mathbb{C}$ is arbitrary. This way one gets the following commutation relations:

$$[B_t, B_s^\dagger] = 2c\delta(t - s) + 4\delta(t - s)N_s,$$
$$[N_t, B_s^\dagger] = 2\delta(t - s)B_s^\dagger,$$
$$[N_t, B_s] = -2\delta(t - s)B_s,$$
$$[N_t, N_s] = [B_t, B_s] = [B_t^\dagger, B_s^\dagger] = 0. \tag{19}$$

As usual in mathematical physics, the rigorous meaning of the commutation relations (19) is that they should be understood in the smeared form. Thus, we introduce the smeared operators

$$B(\varphi) := \int_{\mathbb{R}} \varphi(t)B_t\, dt, \ \ B^\dagger(\varphi) := \int_{\mathbb{R}^d} \varphi(t)B_t^\dagger\, dt, \ \ \ N(\varphi) := \int_{\mathbb{R}} \varphi(t)N_t\, dt,$$

where $\varphi \in S$, and then the commutation relations between these operators take the form

$$[B(\varphi), B^\dagger(\psi)] = 2c\langle \varphi, \psi \rangle + 4N(\varphi\psi),$$
$$[N(\varphi), B^\dagger(\psi)] = 2B^\dagger(\varphi\psi),$$
$$[N(\varphi), B(\psi)] = -2B(\varphi\psi),$$
$$[N(\varphi), N(\psi)] = [B(\varphi), B(\psi)] = [B^\dagger(\varphi), B^\dagger(\psi)] = 0, \qquad \phi, \psi \in S. \tag{20}$$

The operator algebra with generators $B(\varphi), B^\dagger(\varphi), N(\varphi)$, $\varphi \in S$, and a central element 1 with relations (20) is called the square of white noise (SWN) algebra.

In Ref. 3, it was shown that a Fock representation of the SWN algebra exists if and only if the constant $c$ is real and strictly positive. In what follows, it will be convenient for us to choose the constant $c$ to be 2, though this choice is not essential.

Using the notations of Section 3, we define

$$B_t = 2(\partial_t + \partial_t^\dagger \partial_t \partial_t), \ \ B_t^\dagger = 2\partial_t^\dagger, \ \ N_t = 2\partial_t^\dagger \partial_t.$$

Then it is straightforward to show that the corresponding smeared operators

$$B(\varphi) = 2\mathfrak{A}(\varphi), \ B^\dagger = 2A^+(\varphi), \ N(\varphi) = 2A^0(\varphi) \tag{21}$$

form a representation of a SWN algebra in $F(L^2(\mathbb{R}))$. We also refer to Ref. 1 for a unitarily equivalent representation, see also Ref. 25.

Thus, the quantum white noises $W_\lambda(t)$, $\lambda \in \mathbb{R}$, (see (17)) can be thought of as a class of (commuting) quantum processes obtained from the SWN algebra (21).

## References

1. L. Accardi, U. Franz, and M. Skeide, Renormalized squares of white noise and other non-Gaussian noises as Lévy processes on real Lie algebras, *Comm. Math. Phys.* **228** (2002), 123–150.
2. L. Accardi, Y. G. Lu, and I. Volovich, Nonlinear extensions of classical and quantum stochastic calculus and essentially infinite dimensional analysis, *in* "Probability Towards 2000", eds. L. Accadri and C. Heyde, Lecture Notes in Statistics, Vol. 128, Springer-Verlag, New York, 1998, pp. 1–33.
3. L. Accardi, Y. G. Lu, and I. Volovich, White noise approach to classical and quantum stochastic calculi, Volterra Preprint, 1999, to appear in "Lecture Notes of the Volterra International School, Trento, Italy, 1999".
4. L. Accardi, Y. G. Lu, and I. Volovich, A white-noise approach to stochastic calculus, *Acta Appl. Math.* **63** (2000), 3–25.
5. Yu. M. Berezansky, Commutative Jacobi fields in Fock space, *Integral Equations Operator Theory* **30** (1998), 163–190.
6. Yu. M. Berezansky and Yu. G. Kondratiev, "Spectral Methods in Infinite Dimensional Analysis", Kluwer Acad. Publ., Dordrecht, 1995.
7. Yu. M. Berezansky, E. Lytvynov, and D. A. Mierzejewski, The Jacobi field of a Lévy process, *Ukrainian Math. J.* **55** (2003), 853–858.
8. T. S. Chihara, "An Introduction to Orthogonal Polynomials", Gordon and Breach Sci. Pbl., New York/London/Paris, 1978.
9. R. J. Elliott and J. van der Hoek, A general fractional white noise theory and applications to finance, *Math. Finance* **13** (2003), 301–330.
10. T. Hida, "Analysis of Brownian Functionals", Carlton Mathematical Lecture Notes, Vol. 13, Carlton University, Ottawa, 1975.
11. T. Hida, H.-H. Kuo, J. Potthoff, and L. Streit, "White Noise: An Infinite Dimensional Calculus", Acad. Publ., Dordrecht, 1993.
12. H. Holden, B. Øksendal, J. Ubøe, T. Zhang, "Stochastic Partial Differential Equations. A Modeling, White Noise Functional Approach," Birkhäuser, Boston, 1996.
13. Y. Ito and I. Kubo, Calculus on Gaussian and Poisson white noises, *Nagoya Math. J.* **111** (1988), 41–84.
14. N.A. Kachanovsky and S. V. Koshkin, Minimality of Appell-like systems and embeddings of test function spaces in a generalization of white noise analysis, *Methods Funct. Anal. Topology* **5** (1999), no. 3, 13–25.
15. Yu. G. Kondratiev, Generalized functions in problems of infinite dimensional analysis, PhD thesis, Kiev University, 1978.
16. Yu. G. Kondratiev, P. Leukert, and L. Streit, Wick calculus in Gaussian analysis, *Acta Appl. Math.* **44** (1996), 269–294.
17. Y. Kondrtatiev and E. Lytvynov, Operators of gamma white noise calculus, *Infin. Dimen. Anal. Quant. Prob. Rel. Top.* **3** (2000), 303–335.

18. Yu. G. Kondratiev, J.L. Silva, and L. Streit, Generalized Appell systems, *Methods Funct. Anal. Topology* **3** (1997), no. 3, 28–61.

19. Yu. G. Kondratiev and L. Streit, Spaces of white noise distributions: Constructions, descriptions, applications. I, *Rep. Math. Phys.* **33** (1993), 341–366.

20. Yu. G. Kondratiev, L. Streit, W. Westerkamp, and J. Yan, Generalized functions in infinite dimensional analysis, *Hiroshima Math. J.* **28** (1998), 213–260.

21. E. Lytvynov, Multiple Wiener integrals and non-Gaussian white noises, *Methods Funct. Anal. Topology* **1** (1995), no. 1, 61–85.

22. E. Lytvynov, A note on spaces of test and generalized functions of Poisson white noise, *Hiroshima Math. J.* **28** (1998), 463–480.

23. E. Lytvynov, Polynomials of Meixner's type in infinite dimensions—Jacobi fields and orthogonality measures, *J. Funct. Anal.* **200** (2003), 118–149.

24. E. Lytvynov, Orthogonal decompositions for Lévy processes with an application to the Gamma, Pascal, and Meixner processes, *Infin. Dimens. Anal. Quantum Probab. Relat. Top.* **6** (2003), 73–102.

25. E. Lytvynov, The square of white noise as a Jacobi field, *Infin. Dimens. Anal. Quantum Probab. Relat. Top.* **7** (2004), 619–629.

26. E. Lytvynov, A. L. Rebenko, and G.V. Shchepan'uk, Wich theorems in non-Gaussian white noise calculus, *Rep. Math. Phys.* **39** (1997), 217–232.

## II.1 Quantitative Analysis of Partial Differential Equations

Organizers: M. Reissig, Jens Wirth

After Berlin (2001) and Toronto (2003) it was the third time that Michael Reissig from University of Freiberg organized a session devoted to the theory of partial differential equations. The title of this session was chosen to give colleagues from different research topics the opportunity to present their recent results. 26 lectures were given in this session. 12 among them have been invited lectures with a length of 45 minutes, 14 have been 30 minutes lectures. The lectures covered the research topics "theory of hyperbolic equations", "theory of Schrödinger equations" and "Cauchy-Kowalewsky theory".

The lectures devoted to the "theory of hyperbolic equations" discussed questions as energy decay or $L^p - L^q$ decay estimates for its solutions connected with the influence of mass or dissipation terms. The question for blow-up of solutions or the existence of global attractors was considered. Several lectures presented new results for the well-posedness of weakly hyperbolic systems and the related question for sharp Levi conditions.

The lectures devoted to the "theory of Schrödinger equations" discussed among other things the spectrum and the smoothing property for Schrödinger operators with oscillating long range or magnetic potential.

One plenary speaker, Prof. Chen Shuxing (Fudan,Shanghai) who was proposed by the session organizers, has given a plenary talk entitled "Shock reflection by obstacles".

The following colleagues presented lectures in Session II.1:

Benaissa (Sidi bel Abbes), Cicognani (Bologna), Georgiev (Pisa), Jachmann (Freiberg), Tacksun Jung (Kunsan), Kagawa (Tokyo), Kajitani (Tokai), Karp (Karmiel), Kinoshita (Tsukuba), Kubo (Osaka), Kucherenko (Mexico), Matsuyama (Pisa,Tokai), van der Mee (Cagliari), Miyatake (Nara), Mochizuki (Tokyo), Nakao (Kyushu), Nishitani (Osaka), Park (Seoul), Reissig (Freiberg), Ruzhansky (London), Sugimoto (Osaka), Taglialatela (Bari), Uesaka (Tokyo), Vaillant (Paris), Wirth (Freiberg), Yamaguchi (Tokai)

# IDENTIFICATION OF LINEAR DYNAMIC SYSTEMS *

A. KRYVKO and V.V. KUCHERENKO

*Instituto Politecnico Nacional - ESFM,*
*Av. IPN, U.P.A.L.M., Edif. 9, Mexico, DF, Mexico, cp 07738*
*E-mail: valeri@esfm.ipn.mx*

A new approach for the inverse problem for the linear dynamic system is considered. The matrix of stable dynamic system and the amplitudes of harmonic forces are determined through the stationary response at harmonic input.

**Key words:** Inverse problem, frequency domain, algorithm
**Mathematics Subject Classification:** 34A55, 65L09

## 1. Problem definition

Let us consider the vibrations of a stable linear dynamic system in $\mathbb{C}^n$ under a harmonic force in some fixed frame

$$\frac{dy}{dt} = Ay + F(\omega)e^{i\omega t}. \tag{1}$$

Here $y(t) \in \mathbb{C}^n$, $A$ is a constant $(n \times n)$-matrix, and the amplitude $F(\omega)$ is a polynomial in $\omega$: $F(\omega) = \sum_{j=0}^{\nu} f_j \omega^j$, $f_j \in \mathbb{C}^n$. For example, the polynomial amplitudes $F(\omega)$ occur in vibration problems under disbalances[1,2]. The general solution of system (1) is[3]

$$y(t) = \widetilde{y}(\omega)e^{i\omega t} + \sum_{j=1}^{q} e^{\lambda_j t} Q_j(t); \tag{2}$$

$$\widetilde{y}(\omega) := (i\omega - A)^{-1} F(\omega),$$

where $q$ is the number of different eigenvalues $\lambda_j, j = 1, ..., q$, of the matrix $A$, and the degree of the vector polynomial $Q_j(t)$ is greater then or equal to the algebraic multiplicity $\nu_j$ of the eigenvalue $\lambda_j$. For the stable systems $Re\lambda_j < 0$ and the solution of system (2) tends to the stationary solution $\overline{y}(t) := \widetilde{y}(\omega)e^{i\omega t}$ regardless of the initial data as $t \to +\infty$.

For many engineering and scientific models the initial data and coefficients $f_j$ are uncertain and we can only measure the amplitude $\widetilde{y}(\omega)$ of the stationary solution[4] at

*This research was partially supported by EDI scholarship of IPN, SNI scholarship of Mexico, and CONACYT projects 35009E, 4297-F.

484

different frequencies $\omega$. Furthermore, many real dynamic systems are nonlinear and can be described by linear model (1) only after some transient process[1,2]. Thus for such systems it is natural to identify the matrix $A$ from the amplitudes of stationary response $\tilde{y}(\omega_j)$ measured at different frequencies $\omega_j$, $j = 1, ..., l$, $l \geq n^4$.

As a rule in experiments due to some uncontrollable stochastic processes the solution $y(t)$ is a stochastic process. If the spectral density of stationary stochastic process is a uniformly bounded function then with probability 1 it holds that $\tilde{y}(\omega) = \lim_{T \to +\infty} T^{-1} \int_0^T y(t)e^{-i\omega t - \gamma t}dt$. When the spectral density of stationary process is constant (white noise) then $T^{-1} \int_0^T y(t)e^{-i\omega t - \gamma t}dt$ is a normally distributed stochastic quantity with variance $\sigma$, such that $\sigma \to 0$ as $T \to +\infty^5$.

The key problem is to determine the conditions over the coefficients $f_j$, $j = 0, ..., \nu$, and matrix $A$, such that this conditions will provide the uniqueness of identification of the matrix $A$ and coefficients $f_j$, $j = 0, ..., \nu$, for almost all deterministic data $\tilde{y}(\omega_j)$, $j = 1, ..., l$; $l \geq n$.

To extend the class of retrievable matrices $A$ it is suggested to consider several problems (1) with the same matrix $A$ but different amplitudes $F^\alpha(\omega) = \sum_{j=0}^\nu f_j^\alpha \omega^j$ of the harmonic force. Amplitudes $\tilde{y}_\alpha(\omega_j)$ of the stationary response are measured at frequencies $\omega_1, ..., \omega_l$, for each set of coefficients $\{f_0^\alpha, ..., f_\nu^\alpha\}$. Let $\lambda_j$, $j = 1, ..., q$, be the eigenvalues of the matrix $A$. Assume that there exist $b_j$ Jordan blocks of dimension more than zero and $a_j$ Jordan blocks of dimension equal to zero, belonging to the eigenvalue $\lambda_j$ (by definition the order of Jordan block $J$ is equal to the order of the matrix $J$ minus one). We define $\varkappa_A$ as

$$\varkappa_A := \max_{1 \leq j \leq q} \{(b_j + a_j)\}. \tag{3}$$

Denote by $P_A$ the minimal polynomial of the matrix $A$ and by $\deg P_A$ its degree.

For the uniqueness of the identification we introduce a **resolving condition**.

**Definition 1.1.** The set of vector-coefficients $\{f_\mu^\alpha \in \mathbb{C}^n : \alpha = 1, ..., \varkappa; \mu = 0, ..., \nu\}$ and the $(n \times n)$–matrix $A$ satisfy **the resolving condition** if

$$\text{rank} \left[ F(-iA) \vdots AF(-iA) \vdots \cdots \vdots A^{n-1}F(-iA) \right] = n. \tag{4}$$

Here the columns of the matrix $F(-iA)$ are the vectors $\sum_{\mu=0}^\nu (-i)^\mu A^\mu f_\mu^\alpha$.

If $F^\alpha(\omega) = F_0$, then the resolving condition is the condition of controllability[6].

*Suppose the matrix $A$ is fixed and the matrices $F^j$, $j = 0, ..., \nu$ are randomly chosen. In this case we prove that if $\varkappa \geq \varkappa_A$ then with probability equal to one the matrices $F^j$, $j = 0, ..., \nu$, satisfy the resolving condition. Thus under this random choice the inverse problem has a unique solution with probability equal to one.*

## 2. Algorithm of identification

First let us consider the inverse problem with one amplitude $F(\omega)$ and describe the identification algorithm for the matrix $A$ and the set of vector coefficients $f_j$,

$j = 0, ..., \nu$, through the stationary response amplitudes $\tilde{y}(\omega_k)$, $k = 1, .., (n + \nu + 1)$ (evidently, $(n + \nu + 1) \geq \deg P_A + \nu + 1$).

Since $\tilde{y}(\omega) = [i\omega - A]^{-1} F(\omega)$, we obtain the following linear system with respect to the elements of matrix $A$ and vector coefficients $f_j$

$$A\tilde{y}(\omega_k) + F(\omega_k) = i\omega_k \tilde{y}(\omega_k), \quad k = 1, \ldots, n + \nu + 1, \tag{5}$$

where the values $\tilde{y}(\omega_k)$ are known.

Denote by $y_{km}$, $m = 1, ..., n$, $k = 1, ..., (n + \nu + 1)$, the components of the vectors $\tilde{y}(\omega_k)$ and let $\|y_{km}\|$ be the matrix such that the vector $\tilde{y}(\omega_k)$ is its $k$-th row. Then system (5) can be reduced to the linear system with respect to the rows $a_{lm}, l = 1, ..., n$, of the matrix $A$ and components $f_{jm}, m = 1, ..., n$, of the vector $f_j$

$$\sum_{m=1}^{n} y_{km} a_{lm} + \sum_{j=0}^{\nu} \omega_k^j f_{jl} = i\omega_k y_{kl}, \tag{6}$$

where $k = 1, ..., (n + \nu + 1)$. For a fixed index $l$ and variable indexes $k, j, m$ system (6) is a linear system with respect to the $(n+\nu+1)$ unknowns $a_{l1}, ..., a_{ln}, f_{0l}, ..., f_{\nu l}$. We can determine the matrix coefficients $a_{lj}$ and the vectors $f_j$ varying the index $l$ from 1 to $n$ in system (6). Note that for exact values of $\tilde{y}(\omega_j)$ system (6) always has a solution (the coefficients $f_j$ and the parent matrix $A$). The basic problem is to determine the conditions for the uniqueness of the solution.

Let us remark that for an arbitrary matrix $A$ it is impossible to obtain the uniqueness of solution of system (5) by putting some restrictions on the coefficients $f_0, ..., f_\nu$. With this objection in mind it is necessary to consider $\varkappa$ systems (5) with the same matrix and different input amplitudes $F^\alpha(\omega)$, $\alpha = 1, ..., \varkappa$, i.e.

$$A\tilde{y}_\alpha(\omega_k) + F^\alpha(\omega_k) = i\omega_k \tilde{y}_\alpha(\omega_k), \tag{7}$$

where $k = 1, ..., (n + \nu + 1)$; $1 \leq \alpha \leq \varkappa$. The number $\varkappa$ is such that $\varkappa \geq \varkappa_A$, where the number $\varkappa_A$ is defined in (3). Denote by $y_{km}^\alpha$, $m = 1, ..., n$; $k = 1, ..., n+\nu+1$, the components of the vectors $\tilde{y}_\alpha(\omega_k)$ and let $\|y_{km}^\alpha\|$ be the matrix with rows $\tilde{y}_\alpha(\omega_k)$. Now we can reduce system (7) to the linear system $\sum_{m=1}^{n} y_{km}^\alpha a_{lm} + \sum_{j=0}^{\nu} \omega_k^j f_{jl}^\alpha = i\omega_k y_{kl}^\alpha$, with respect to the rows of the matrix $A$ and the components the vectors $f_j^\alpha$. For fixed indexes $l, \alpha$ and variable indexes $k, j, m$ this system is a linear system with respect to $a_{l1}, ..., a_{ln}, f_{0l}^\alpha, ..., f_{\nu l}^\alpha$, with corresponding $(n \times n)$ −matrix

$$\begin{pmatrix} y_{11}^\alpha & \cdots & y_{1n}^\alpha & 1 & \omega_1 & \cdots & \omega_1^\nu \\ \vdots & \ddots & \vdots & \vdots & \vdots & \ddots & \vdots \\ y_{(n+\nu+1)n}^\alpha & \cdots & y_{(n+\nu+1)n}^\alpha & 1 & \omega_{n+\nu+1} & \cdots & \omega_{n+\nu+1}^\nu \end{pmatrix}. \tag{8}$$

## 3. Main theorem

Let $\mathbb{C}^{n\varkappa(\nu+1)}$ be the space of coefficients $f_j^\alpha, j = 0, ..., \nu; \alpha = 1, ..., \varkappa$, and $\Phi^\varkappa \in \mathbb{C}^{n\varkappa(\nu+1)}$ be the unit vector defined as

$$\Phi^\varkappa = \left( \frac{f_0^1}{\Psi^\varkappa}, ... \frac{f_0^\varkappa}{\Psi^\varkappa}; ...; \frac{f_\nu^1}{\Psi^\varkappa}, ..., \frac{f_\nu^\varkappa}{\Psi^\varkappa} \right), \quad \Psi^\varkappa = \sqrt{\sum_{\alpha=1}^\varkappa \sum_{j=0}^\nu (f_j^\alpha, f_j^\alpha)}, \quad f_k^\alpha \in \mathbb{C}^n. \quad (9)$$

Evidently, $\Phi^\varkappa \in S^{2n\varkappa(\nu+1)-1}$, where $S^{2n\varkappa(\nu+1)-1}$ is the unit sphere of real dimension $2n\varkappa(\nu+1) - 1$. Let *mess* denote the normalized to unity standard measure on the unit sphere $S^{2n\varkappa(\nu+1)-1}$.

The determinant of system (7) is an analytic function with respect to $a_{ij}$, $f_j$, $\omega_k$. Moreover, this determinant is an homogeneous first-order function with respect to the coefficients $f_j^\alpha, j = 0, ..., \nu; \alpha = 1, ..., \varkappa$. Thus to determine the conditions such that the determinant of system (7) is not equal to zero, it is enough to consider the components of the unit vector $\Phi^\varkappa$ in the space of coefficients $\mathbb{C}^{n\varkappa(\nu+1)}$.

Let $Nr(A)$ be the set of vectors $\Phi^\varkappa \in S^{2n\varkappa(\nu+1)-1}$, such that system (7) does not have a unique solution for $\Phi^\varkappa \in Nr(A)$.

**Theorem 3.1.** *I) Assume that for some matrix A and coefficients $f_\mu^\alpha$, the values $\widetilde{y}_\alpha (\omega_j)$, at $(\deg P_A + \nu + 1)$ or more frequencies $\omega_j$, are given by $\widetilde{y}_\alpha (\omega_j) = [i\omega_j - A]^{-1} f^\alpha (\omega_j)$, where $\alpha = 1, ..., \varkappa; \varkappa \geq \varkappa_A; \mu = 0, ..., \nu$. If the dimension $n \geq 2$, then:*

*1. System (7) has a unique solution $A, f_\mu^\alpha$ if and only if the vector $\Phi^\varkappa \notin Nr(A)$;*
*2. The set $Nr(A)$ is a closed one and $messNr(A) = 0$;*
*3. The kernel of system (7) does not depend on the choice of frequencies $\omega_j$;*

*II) If for some set $\{\widetilde{y}^\alpha (\omega_j) \mid j = 1, ..., d; \ d \geq (\deg P_A + \nu + 1)\}$, the matrix A and coefficients $f_k^\alpha, \alpha = 1, ..., \varkappa; \ k = 0, ..., \nu$, satisfy equation (7) and the solution is unique, then the matrix A and coefficients $f_k^\alpha$ satisfy the resolving condition.*

*III) If the dimension $n$ is equal to 1, then:*
*1. $A \in \mathbb{C}, f_\beta \in \mathbb{C}$, and the coefficients $f_0^*, ..., f_\nu^*$, and complex number A can be determined uniquely if and only if $F(-iA) \neq 0$.*
*2. In the case when the rank of matrix (8) is equal to $(\nu + 2)$ there exist a unique complex number A and coefficients $f_0^*, ..., f_\nu^*$, such that $(i\omega - A)^{-1} F(\omega)|_{\omega=\omega_j} = y_j$ and $F(A) \neq 0$.*

## 4. Numerical experiments

### 4.1. *First-order system*

In order to obtain the numerical solution of system (6) we describe the detailed algorithm of identification of the matrix $A$ and vectors $f_j^\alpha$. Suppose that the largest Jordan block of $(n \times n)$-matrix $A$ in system (1) has dimension $r$. In this case for the unique identification of the coefficients of system (1), it is necessary to consider $\varkappa = r + 1$ systems (this follows from Theorem 3.1):

$$A\widetilde{y}_\alpha (\omega_k) + F^\alpha (\omega_k) = i\omega_k \widetilde{y}_\alpha (\omega_k), \quad k = 1, ..., n + \nu + 1; \alpha = 1, ..., \varkappa. \quad (10)$$

As before, we denote by $y_{km}^\alpha$, $m = 1, ..., n$; $k = 1, ..., n + \nu + 1$, the components of the vectors $\widetilde{y}_\alpha(\omega_k)$ and let $\|y_{km}^\alpha\|$ be the matrix with rows $\widetilde{y}_\alpha(\omega_k)$. Thus we can reduce system (10) to the linear system with respect to the rows $a_{lm}$, $l = 1, \ldots, n$, of the matrix $A$ and components $f_{jm}^\alpha$, $m = 1, \ldots, n$, of the vectors $f_j^\alpha$ :

$$\sum_{m=1}^n y_{km}^\alpha a_{lm} + \sum_{j=0}^\nu \omega_k^j f_{jl}^\alpha = i\omega_k y_{kl}^\alpha, \tag{11}$$

where $k = 1, \ldots, n + \nu + 1; \alpha = 1, ..., \varkappa; l = 1, ..., n$. For fixed indexes $l, \alpha$ and variable indexes $k, j, m$ system (11) is a linear system with respect to the unknowns $a_{l1}, ..., a_{ln}, f_{0l}^\alpha, ..., f_{\nu l}^\alpha$. Now, we define the following matrices

$$M_1 := \begin{pmatrix} 1 & \omega_1 & \cdots & \omega_1^\nu \\ \vdots & \vdots & \ddots & \vdots \\ 1 & \omega_{\nu+1} & \cdots & \omega_{\nu+1}^\nu \end{pmatrix}, M_2 := \begin{pmatrix} 1 & \omega_{\nu+2} & \cdots & \omega_{\nu+2}^\nu \\ \vdots & \vdots & \ddots & \vdots \\ 1 & \omega_{n+\nu+1} & \cdots & \omega_{n+\nu+1}^\nu \end{pmatrix},$$

$$D_1^\alpha := \begin{pmatrix} y_{11}^\alpha & \cdots & y_{1n}^\alpha \\ \vdots & \ddots & \vdots \\ y_{(\nu+1)1}^\alpha & \cdots & y_{(\nu+1)n}^\alpha \end{pmatrix}, D_2^\alpha := \begin{pmatrix} y_{(\nu+2)1}^\alpha & \cdots & y_{(\nu+2)n}^\alpha \\ \vdots & \ddots & \vdots \\ y_{(n+\nu+1)1}^\alpha & \cdots & y_{(n+\nu+1)n}^\alpha \end{pmatrix},$$

$$g_{1l}^\alpha := \left( i\omega_1 y_{1l}^\alpha, ..., i\omega_{\nu+1} y_{(\nu+1)l}^\alpha \right)^T, g_{2l}^\alpha := \left( i\omega_{(\nu+2)} y_{(\nu+2)l}^\alpha, ..., i\omega_{(n+\nu+1)} y_{(n+\nu+1)l}^\alpha \right)^T.$$

Consequently we obtain the linear system $D^\alpha A_l = g_l^\alpha$, with respect to the unknowns $a_{l1}, ..., a_{ln}$. Here $A_l = (a_{l1}, ..., a_{ln})^T$ and the matrices $D^\alpha, g_l^\alpha$ are defined as $D^\alpha := D_2^\alpha - M_2 M_1^{-1} D_1^\alpha$, $g_l^\alpha := g_{2l}^\alpha - M_2 M_1^{-1} g_{1l}^\alpha$.

**Proposition 4.1.** *Let* $D^\alpha$ *be a* $(k \times m) - matrix$, $m \geq k, \alpha = 1, ..., \varkappa$, *then:*
*1) The kernel of the system*

$$D^\alpha x = g_l^\alpha, \ g_l^\alpha \in \mathbb{C}^m, x \in \mathbb{C}^k, \alpha = 1, ..., \varkappa, \tag{12}$$

*is equal to the kernel of the system* $\left[ \sum_{\alpha=1}^\varkappa (D^\alpha)^* D^\alpha \right] x = 0$;
*2) Suppose that system (12) is solvable and*

$$\left[ \sum_{\alpha=1}^\varkappa (D^\alpha)^* D^\alpha \right] x = \sum_{\alpha=1}^\varkappa (D^\alpha)^* g_l^\alpha \tag{13}$$

*is a nondegenerate system. In this case equation (13) uniquely determines the solution of system (12).*

Note that system (12) can become unsolvable due to the experimental errors in $\widetilde{y}_\alpha(\omega_k)$. However it is possible to perform the computations using more simple system (13). Namely Theorem 3.1 implies that with the probability equal to 1 it holds that $Ker\left( \sum_{\alpha=1}^\varkappa (D^\alpha)^* D^\alpha \right) = 0$ for $\varkappa \geq r + 1$, and randomly chosen unit vectors $\{f_0^\alpha, ..., f_\nu^\alpha | \alpha = 1, ..., \varkappa\}$.

In what follows we apply our method of identification to the matrix $A$ of order 7 with one simple Jordan block of order 2 and vector-polynomial $F(\omega)$ of degree 2. In this section we consider system (1) with the matrix $A$ of the form

$$A = T^{-1}LT. \tag{14}$$

Here $T = ||t_{ij}||$ is $(7 \times 7)$ −matrix, $T^{-1}$ is the inverse matrix of $T$, and

$$L := \begin{pmatrix} \lambda_1 & 1 & 0 & 0 & 0 & 0 & 0 \\ 0 & \lambda_1 & 1 & 0 & 0 & 0 & 0 \\ 0 & 0 & \lambda_1 & 0 & 0 & 0 & 0 \\ 0 & 0 & 0 & \lambda_2 & 0 & 0 & 0 \\ 0 & 0 & 0 & 0 & \lambda_2 & 0 & 0 \\ 0 & 0 & 0 & 0 & 0 & \lambda_3 & 0 \\ 0 & 0 & 0 & 0 & 0 & 0 & \lambda_4 \end{pmatrix}.$$

The values $t_{ij}, i,j = 1,...,7$, the frequencies $\omega_k, k = 1,...,10$, and the eigenvalues $\lambda_l, l = 1,...,4$, where randomly generated with the uniform distribution on the intervals $[-50, 50]$, $[0, 50]$ and $[-50, -1]$, respectively. The vectors $f_j^\alpha \in \mathbb{R}^7, j = 0,1,2; \alpha = 1,...,4$, were also randomly generated with the uniform distribution in such a way that the norm of the vectors $\Phi^\varkappa = \left(f_0^1/\Psi^\varkappa, ... f_0^\varkappa/\Psi^\varkappa; ...; f_2^1/\Psi^\varkappa, ..., f_2^\varkappa/\Psi^\varkappa\right)$ is equal to 1, where $\Psi^\varkappa := \sqrt{\sum_{\alpha=1}^\varkappa \sum_{j=0}^2 \left(f_j^\alpha, f_j^\alpha\right)}$.

Using the program "Mathematica 4" we performed the following experiments:

1) Randomly generated matrix $A$ of form (14) and vectors $f_j^\alpha, j = 0,1,2; \alpha = 1,...,\varkappa$, were identified from system (13) for $\varkappa = 3$. This calculations were performed for one fixed matrix $A$ and 500 randomly generated sets $\left\{F^\alpha(\omega) : F^\alpha(\omega) = \sum_{j=0}^2 f_j^\alpha \omega^j, \alpha = 1,2,3\right\}$. Graphs of distribution of calculation accuracy for the matrix $A$ (Fig. 1) and vectors $f_j^\alpha$ (Fig. 2):

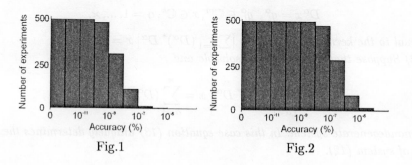

Fig.1                    Fig.2

show the number of times when the matrix $A$ and vectors $f_j^\alpha, j = 0,1,2; \alpha = 1,2,3$, were not identified with the corresponding accuracy. Here the accuracy of calculation for the matrix $A = ||a_{ij}||$, was calculated as $100 \cdot \max_{1 \le i,j \le 7} \{a_{ij}/\tilde{a}_{ij}\}$, and for the vectors $f_j^\alpha = \left(f_{j1}^\alpha, ..., f_{j7}^\alpha\right), j = 0,1,2; \alpha = 1,2,3$, as $100 \cdot$

$\max_{0\leq j\leq 2,1\leq i\leq 7,1\leq\alpha\leq 3}\left\{f_{ji}^{\alpha}/\widetilde{f_{ji}^{\alpha}}\right\}$, where $\tilde{a}_{ij}$ and $\tilde{f}_{ji}^{\alpha}$ are the components of the calculated matrix $\tilde{A}$ and calculated vectors $\tilde{f}_{j}^{\alpha}$, respectively.

When the vector-polynomials $F^{\alpha}(\omega)$ are known system (11) became more simple. The accuracy distribution calculated as above according to the random choice of vector-polynomials $F^{\alpha}(\omega), \alpha = 1, ..., \varkappa$, is shown on Fig. 3.

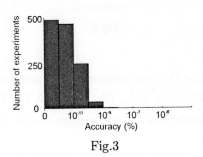

Fig.3

2) Experimental results show that it is impossible to determine the dimension of the kernel of $D^{\alpha}$ and the number $\varkappa$ in (13) computing the eigenvalues of $D^{\alpha}$. To determine the number $\varkappa$ that provides the uniqueness of solution of system (13) we propose to perform successive computations of the matrix $A$ with $\varkappa$ equal to $1, 2, 3, ..., k, ..$, and compare the accuracy of calculations on the steps $k$ and $(k-1)$, until the required accuracy is reached. Figures 4,5,6 show the accuracy of calculations of one fixed randomly generated matrix $A$ of form (14) for 500 randomly generated sets of vector-polynomials $\left\{F^{\alpha}(\omega) : F^{\alpha}(\omega) = \sum_{j=0}^{2} f_{j}^{\alpha}\omega^{j}, \alpha = 1, ..., \varkappa\right\}$ with $\varkappa = 3, 4, 5$, respectively (for $\varkappa = 1, 2$, it is impossible to compute the matrix $A$):

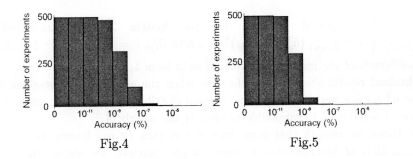

Fig.4                           Fig.5

3) We have performed identification of the matrix $A$ in the case when the experimental errors arise. Namely the amplitudes of stationary response $y(\omega_{k})$ were perturbed by 400 normal distributed noises with mean 0 and variance $\sigma$. We considered four cases $\sigma = 1, 0.1, 0.001, 0.0001$. For each value of variance $\sigma$ we calculated

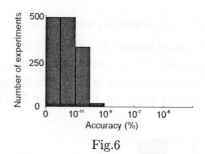

Fig.6

the average matrix $\widehat{A}$ after the computation of 400 matrices $\widetilde{A}$ as in Item 1 for a set of vectors $\left\{F^{\alpha}\left(\omega\right) | F^{\alpha}\left(\omega\right) = \sum_{j=0}^{2} f_{j}^{\alpha}\omega^{j}, \alpha = 1, 2, 3\right\}$. Table 1 shows the maximal relative error of calculations for the average matrix $\widehat{A}$ obtained for different variances $\sigma$ of the noise.

|  | $\sigma = 1$ | $\sigma = 0.01$ | $\sigma = 0.001$ | $\sigma = 0.0001$ |
|---|---|---|---|---|
| Relative error | 140.299 | 310.641 | 23.9098 | 0.211148 |

Table 1

Here the maximal relative error of calculations for the average matrix $\widehat{A} = \|\widehat{a}_{ij}\|$ was calculated as $100 \cdot \max_{1 \leq i,j \leq 7} \left\{a_{ij}/\widehat{a}_{ij}\right\}$.

Table 2 shows the maximal variance for the matrix $\widetilde{A}$ computed for different variances $\sigma$ of the noise.

|  | $\sigma = 1$ | $\sigma = 0.01$ | $\sigma = 0.001$ | $\sigma = 0.0001$ |
|---|---|---|---|---|
| Maximal variance | 39.1205 | 596.943 | 25.2041 | 0.261192 |

Table 2

Here the maximal variance for the matrix $\widetilde{A}$ was calculated as $\max_{1 \leq i,j \leq 7} \left\{N^{-1} \sum_{\gamma=1}^{N} \left(\widetilde{a}_{ij}^{\gamma}\right)^{2} - \left(\widehat{a}_{ij}\right)^{2}\right\}$, where $\widetilde{a}_{ij}^{\gamma}$, $i = 1, ..., 7$; $j = 1, ..., 7$, are the coefficients of the matrix $\widetilde{A}$ calculated as in Item 1, $N = 400$.

Obtained results show that in the case when the amplitudes of stationary response $y\left(\omega_{k}\right)$ are perturbed by some normal distributed noises with mean 0 and variance $\sigma \sim 0.001$, it is impossible to identify the matrix $A$.

4) Above we remark that sometimes the vectors $f_{j}^{\alpha}$ are known. In this case the algorithm of identification is more simple. Namely, we obtain the system $\sum_{m=1}^{n} y_{km}^{\alpha} a_{lm} = i\omega_{k} y_{kl}^{\alpha} - \sum_{j=0}^{\nu} \omega_{k}^{j} f_{jl}^{\alpha}$, with respect to the unknowns $a_{l1}, ..., a_{ln}$, where $k = 1, ..., n+\nu+1; \alpha = 1, ..., \varkappa; l = 1, ..., n$. The average matrix $\widehat{A}$ was calculated after the computation of 900 matrices $\widetilde{A}$. Table 3 shows the maximal relative error of calculations and maximal variance for the average matrix $\widehat{A}$ obtained for one set of vectors $\left\{F^{\alpha}\left(\omega\right) : F^{\alpha}\left(\omega\right) = \sum_{j=0}^{2} f_{j}^{\alpha}\omega^{j}, \alpha = 1, 2, 3\right\}$ in the case when the

vectors $y(\omega_k)$ were perturbed by a noise with mean 0 and variance $\sigma=0.0001$:

| | $f_j^\alpha$ are known | $f_j^\alpha$ are unknown |
|---|---|---|
| Accuracy | 0.0320115 | 1.0903 |
| Variance | 0.0000314207 | 0.00577234 |

Table 3

Here the maximal relative error of calculations and maximal variance for the average matrix $\widehat{A}$ were computed as above.

## 5. Systems of second order

Let us consider the system of second order

$$M\frac{d^2y}{dt^2} + C\frac{dy}{dt} + Ky = F(\omega)e^{i\omega t}, \quad F(\omega) = \sum_{\mu=0}^{\nu} f_\mu \omega^\mu, \tag{15}$$

where $y \in \mathbb{C}^n$; $f_\mu \in \mathbb{C}^n$; $C, K$ are $(n \times n)$-matrices, and $M$ is the identity $(n \times n)$ −matrix.

It is possible to reduce (15) to a first-order system:

$$\begin{cases} \frac{dy}{dt} = z \\ \frac{dz}{dt} = -Cz - Ky + F(\omega)e^{i\omega t} \end{cases}, \quad A = \begin{pmatrix} 0 & I \\ -K & -C \end{pmatrix}. \tag{16}$$

Obviously, if the vectors $\widetilde{F}(-iA) := \sum_{j=0}^{\nu}(-iA)^j \widetilde{f}_j$, where $\widetilde{f}_j = (0, f_j)^T$; $\alpha = 1, ..., \varkappa$; $j = 1, ..., q$, satisfy the resolving condition (4) and Theorem 3.1 is valid for system (16) at $(2n + \nu + 1)$ frequencies $\omega_j$.

In this case we have a natural restriction on the vector coefficients $\widetilde{f}_j^\alpha$. Namely, their $n$ first components are equal to zero; and instead of the sphere $S^{4n\varkappa(\nu+1)-1}$ it is necessary to consider the resolving condition over the sphere $S^{2n\varkappa(\nu+1)-1}$.

*Here we considered the case when matrix $A$ has $2n$ different eigenvalues* $\lambda_1, ..., \lambda_{2n}$. Therefore the matrix $A^*$ (i.e. the adjoint matrix of $A$) also has $2n$ different eigenvalues $\overline{\lambda}_1, ..., \overline{\lambda}_{2n}$. Suppose $Ae_k^* = \lambda_k e_k^*$, for $k = 1, ..., 2n$, and let $\widetilde{N}_r(A)$ be the set defined as the union of $2n$ planes $\pi_k$, defined as $\pi_k : (\sum_{j=0}^{\nu} \widetilde{f}_j(-i\lambda_k)^j, e_k^*) = 0$, i.e. $\widetilde{N}_r(A) = \cup_{k=1}^{2n}\pi_k$.

Consider the vector $\Phi = \left(\frac{f_0}{\Psi}, ..., \frac{f_\nu}{\Psi}\right)$, with $\Psi^2 = \sum_{j=1}^{\nu}(f_j, f_j)$, over the unite sphere $S^{2n(\nu+1)-1}$ of the real dimension $2n(\nu + 1) - 1$ in the space of coefficients $f_0, ..., f_\nu$. By definition $\widehat{N}_r(A) = \widetilde{N}_r(A) \cap S^{2n(\nu+1)-1}$.

**Lemma 5.1.** *Let $A$ be the matrix of form (16) and assume that the matrix $A$ has $2n$ different eigenvalues. In this case it holds that:*

*1) On the vectors $F(\omega) = \sum_{\mu=0}^{\nu} \widetilde{f}_\mu \omega^\mu$, $\widetilde{f}_\mu = (0, f_\mu)^T$, $f_\mu \in \mathbb{C}^n$, resolving conditions (4) are equivalent to the condition that the vector $\Phi$ does not belong to the set $\widehat{N}_r(A)$;*

*2) The set $\widehat{N}_r(A)$ is a closed one, and over the sphere $S^{2n(\nu+1)-1}$ it holds true that $mess\widehat{N}_r(A) = 0$.*

492

## References

1. Rao J.S. *Rotor dynamics* 1996 (New Delhi: New Age International Publisher)
2. Vance J.M. *Rotordynamics of Turbomachinery* 1998 (Jon Wiley and Sons, Inc.)
3. Earl A.Coddington, Norman Levinson *Theory of ordinary differential equations* 1984 (Malabar, Fla.: Krieger Publishing Company)
4. A. Antonio-Garcia, J. Gomez-Mancilla, V.V. Kucherenko *Identification of rotordynamics parameters for the Jefcott rotor-bearing model* 2002 (Honolulu, Hawaii, USA: The 9th International Symposium on Transport Phenomena and Dynamics of Rotating Machinery (ISROMAC-9))
5. Yu. V. Prohorov, Yu. A. Rozanov *Probability Theory* 1969 (Berlin, Heidelbery,New-York: Springer-Verlag)
6. Frank M. Callier and Charles A. Desoer *Linear System Theory* 1994 (New-York, Berlin, Heidelbery London: Springer-Verlag (Springer Texts in Electrical Engineering))
7. John T. Moore *Elements of linear algebra and matrix theory* 1968 (New York: McGraw-Hill)
8. Shlomo Sternberg *Lectures on differential geometry* 1964 (Englewood Cliffs, NJ: Prentice-Hall)

# STRONGLY HYPERBOLIC COMPLEX SYSTEMS, REDUCED DIMENSION AND HERMITIAN SYSTEMS II

JEAN VAILLANT

*Université Pierre et Marie Curie*
*Mathématiques, BC 202*
*4, Place Jussieu 75252 Paris, France*
*E-mail: vlnt@dm.jussieu.fr*

We consider the first order system $a(D) = ID_0 + \sum_{1}^{n} a_k D_k$, where $I$ is the identity matrix and $a_k$ is a complex-valued $m \times m$ matrix. We define the *reduced dimension* $d(a)$ of $a$. Then we prove, that if $m \geq 4$, $d(a) \geq m^2 - 3$ and if $a$ is strongly hyperbolic, then $a$ is prehermitian, that means, there exists a suitable $T$ such that $T^{-1}a(\xi)T$ is hermitian for all $\xi$.

**Key words:** Complex hyperbolic, systems of partial differential equations, reduced dimension

**Mathematics Subject Classification:** 35L40

## 1. Introduction

We consider a first order system:

$$a(D) = ID_0 + \sum_{1}^{n} a_k D_k,$$

where $I$ is the identity matrix and $a_k$ is a complex valued $m \times m$ matrix.

Let $a(\xi)$ be the principal symbol of $a(D)$:

$$a(\xi) = I\xi_0 + \sum_{1}^{n} a_k \xi_k.$$

We define the *reduced dimension* of $a$:

$d(a)$ = dimension of the real vector space of $M(\mathbb{C})$ generated by:
$(I, \ldots, \operatorname{Re} a_k, \ldots, i \operatorname{Im} a_k \ldots)$. $1 \leq k \leq n$ .

We have stated [10-12] that $d(a) = d(T^{-1}aT)$, where $T$ is invertible $\in M(\mathbb{C})$; $d({}^t a) = d(\bar{a}) = d(a)$. Also, if $a$ is diagonalisable, $d(a) = d(\det a)$.

We recall the theorem by Kasahara, Yamaguti:

*The matrix-valued operator $a$ is strongly hyperbolic if, and only if, $a$ is uniformly diagonalizable; that means:*

*i)* $\forall \xi$ *the zeros in* $\tau$ *of* $\det(\tau I + a(\xi)) = 0$ *are real; in other words if we denote:*

$$p(\xi') = \sum_1^n a_k \xi_k, \ \forall \xi' \ \text{the eigenvalues of } p(\xi') \text{ are real;}$$

*ii)* $p(\xi')$ *is diagonalizable, if* $\tau$ *is an eigenvalue of multiplicity* $\mu$*, the dimension of the corresponding vector space of eigenvectors is* $\mu$*; so there exists* $\Delta(\xi')$ *such that:* $\Delta^{-1}(\xi')p(\xi')\Delta(\xi')$ *is diagonal* $\forall \xi'$*;*

*iii)* *there exists* $\Delta(\xi')$*, such that (uniformly):*

$$\|\Delta(\xi')\| \leq M, \quad |\det \Delta(\xi')| \geq \varepsilon > 0.$$

**Lemma 1.1.** *We denote by* $\phi^i_j$ *the entries of the matrix* $a$*. We assume* $a$ *is diagonalizable and we denote:*

$$V = \text{vector space spanned by the real linear forms: } \operatorname{Re} \phi^i_j, \operatorname{Im} \phi^i_j, \ i > j,$$

*then:*

*i)* $1 \leq i \leq m$*:* $\phi^i_i(\xi) = \xi_0 + \chi_i(\xi') + i\lambda_i(\xi')$*,* $\chi_i, \lambda_i$ *real valued,* $\lambda_i \in V$*,*

*ii)* $p < q$*,* $\operatorname{Re} \phi^p_q \in V$*,* $\operatorname{Im} \phi^p_q \in V$*.*

*Proof.* [1,6]. $\qquad\qquad\qquad\qquad\qquad\qquad\qquad\qquad\qquad\qquad\qquad\qquad\qquad\qquad\square$

**Consequence.** *The reduced dimension* $d(a) \leq 2m(m-1)/2 + m = m^2$*.*

The proof of the theorem is divided in 4 parts, according to the dimension of the vector space $V$. In this paper we study two cases. In §2 the dimension of $V$ is $m^2 - m - 1$; in §3 it is $m^2 - m$. In another paper [13] we study the cases, where the dimension of $V$ is $m^2 - m - 3$ or $m^2 - m - 2$.

We state the following result:

**Theorem 1.** *If* $m \geq 4$*,* $d(a) \geq m^2 - 3$*, and if* $a$ *is strongly hyperbolic, then* $a$ *is prehermitian, that means:*

$$\exists T \in M(\mathbb{C}), \ \text{such that: } T^{-1}a(\xi)T \ \text{is hermitian} \ \forall \xi.$$

*Remarks.* 1. If $m = 4$, the theorem was stated in [12].

2. If we replace the assumption of uniform diagonalizability by the weaker one of diagonalizability and the assumption of the reduced dimension by the stronger one: $d(a) \geq m^2 - 2$ $(m \geq 3)$, due to [10,11] the theorem holds; so here we study only the case $m^2 - 3$.

The result in the case $m = 2$ was obtained by [5] without any assumption to the dimension.

3. The real case and the extension of the results to the case of variable coefficients were considered in [6,1,7-9,2-4].

**Lemma 1.2.** *Denote by* $E_j$ *the* $m \times m$ *matrix, all the non diagonal elements of which are 0, all the diagonal elements are equal to 1 except the* $(j,j)$ *element which* *is equal to* $-i$ *($i^2 = -1$). Then:*

$$E_j^{-1}(\phi)E_j = \begin{pmatrix} \phi_1^1 & & & & \\ & & -i\phi_k^j & & \\ & i\phi_j^k & \phi_k^k & & \\ & & & & \phi_m^m \end{pmatrix}.$$

**Lemma 1.3.** *If b is prehermitian, there exists a hermitian positive definite matrix H such that:*

$$bH = H\,{}^t\bar{b}. \tag{1}$$

*Proof.* There exists $T$ such that:

$$T^{-1}bT = {}^t\overline{(T^{-1}bT)},$$

and we denote by

$$H = T^t\overline{T}. \qquad \square$$

*Notations* 1.4. We denote by $h_{ij}$ the non diagonal elements of $H$, $h_{ij} = \overline{h_{ji}}$; we denote by $h_i > 0$ the diagonal elements.

We assume now $m \geq 5$.

We need effective symmetrization of submatrices of $a(\xi)$ in many cases; the efficiency of symmetrization is also of self contained interest. In some special cases, using the particularity of the cases, the proof of the following lemmas can be simplified.

*Notations* 1.5. Denote by $z$ a basis of $V$; we choose the elements of $z$ as coordinates; if $\text{Re}\,\phi_j^i$, $i > j$, belongs to $z$, we denote it by $\xi_j^i$; if $\text{Im}\,\phi_j^i$ $i > j$, belongs to $z$, we denote it by $\eta_j^i$; we denote: $z_j^i = \xi_j^i + i\eta_j^i$.

If $\dim V = m^2 - m + k - 3$, $k = 0, 1, 2, 3$; we can choose $m - k - 1$ real diagonal forms $\chi_j$, and $\xi_0$ as coordinates; the other real parts of diagonal elements have the form: $\xi_0 + \sum_j c_{\ell j}\chi_j + \psi_\ell(z)$.

**Lemma 1.6.1.** *Assume*

*1) a is uniformly diagonalizable,*

*2) $d(a) = m^2 - 3$,*

*3) the $(m-1)$ first elements of the last line of $a(\xi)$ have their real and imaginary parts belonging to $z$: $\{\xi_1^m, \eta_1^m, \ldots, \xi_k^m, \eta_k^m, \ldots, \xi_{m-1}^m, \eta_{m-1}^m\}$; denote by $z^m$ this set; denote by $z' = C_z z^m$.*

*Assume, moreover,*

4) $a(\xi) = \begin{vmatrix} \delta_1 & & & & \phi_m^1 \\ & & & \phi_q^p(z), p < q & \\ \phi_i^l(z), i < l & \delta_j & & & \phi_m^j \\ & & & & \\ & & \delta_{m-1} & \phi_m^{m-1} \\ z_1^m & z_j^m & z_{m-1}^m & \delta_m \end{vmatrix}.$

We have denoted: $\delta_j = \xi_0 + \chi_j + i\lambda_j(z)$ *to simplify the typography;* $\delta_k = \xi_0 + i\lambda_k(z)$.

*Assume* $\chi_l \neq \chi_{l'}$ *if* $l \neq l'$, $\chi_l \neq 0$, *if* $l \neq k$.

*Then by change of basis in* $\mathbb{C}^m$ *and also in* $\mathbb{R}^{n+1}$ *with* $(1, 0, \ldots, 0)$ *unchanged, we obtain the new form of* $a(\xi)$ *(we do not change the notations for the new a and its elements):*

$$a(\xi) = \begin{vmatrix} \delta_1(z^m) & & & \phi_m^1 \\ & & k_l^i \overline{\phi}_i^l(z') + \phi_l^i(z^m) & \phi_m^i \\ & \delta_j(z^m) & & \\ \phi_i^l(z), i < l & & & \phi_m^{m-1} \\ z_1^m & z_j^m & & \delta_m(z^m) \end{vmatrix},$$

$\delta_j(z^m) = \xi_0 + \chi_j + i\lambda_j(z^m)$, $\delta_k(z^m) = \xi_0 + i\lambda_k(z^m)$, $k_l^i > 0$, $k_j^i = k_l^i k_j^l$, $1 \leq i < l < j \leq m-1$.

*Proof.* Let $z_1^m = \cdots = z_{m-1}^m = 0$; consider the $(m-1) \times (m-1)$ matrix $b$ obtained by cancelling the last line and column of $a(\xi)$ the elements of which are the restricted $\phi_j^i(z')$.

By induction $b$ is prehermitian and it exists (Lemma 1.3) $H$ such that (1.1) is satisfied.

Considering the element in the $j^{\text{th}}$ line, $k^{\text{th}}$ column, in this identity, $m-1 \geq j > k \geq 1$, and especially the terms in $\chi$, we obtain.

We obtain: $(\chi_j - \chi_k) h_{kj} \equiv 0$ , so $H$ is diagonal; we deduce easily, $\phi_i^i(z') = k_\ell^i \overline{\phi}_i^l(z')$, $\lambda_j(z') = 0$, and also the relations between the $k_l^i$. $\qquad \square$

**Lemma 1.6.2.** *Assume as before 1), 2), 3) and now 4) is:*

$$a(\xi) = \begin{pmatrix} \delta_1 + \psi(z) & & & \phi_m^1 \\ & \xi_0 + \chi_1 + i\lambda_2(z) & & \\ & & \phi_q^p(z), p < q & \\ & & \delta_j & \\ & \phi_i^l(z), i < l & & \\ z_1^m & z_2^m & z_j^m & \delta_m \end{pmatrix},$$

$\psi(z)$ *real;* $\chi_l \neq \chi_{l'}$ *if* $l \neq l'$.

*Then, by a change of basis in* $\mathbb{C}^m$ *we obtain for* $a(\xi)$ *the form,*

$$a(\xi) = \begin{pmatrix} \delta_1(z^m) + \psi(z^m) & k_2^1 \overline{\phi_1^2}(z') + \phi_2^1(z^m) & k_j^1 \overline{\phi_1^j}(z') + \phi_j^1(z^m) & \phi_m'^1(z) \\ \phi_1^2(z') + \phi_1^2(z^m) & \xi_0 + \chi_1 + i\lambda_2(z^m) & k_j^2 \overline{\phi_2^j}(z') + \phi_j^2(z^m) & \phi_m^2(z) \\ \phi_1^j(z') + \phi_1^j(z^m) & \phi_2^j(z) & \delta_j(z^m) & \phi_m^j(z) \\ z_1^m & z_2^m & z_j^m & \delta_m(z^m) \end{pmatrix},$$

$$k_j^i > 0, \quad k_j^i = k_l^i k_j^l, \quad 1 \leq i < l < j \leq m - 1.$$

More details are contained in the proof.

*Proof.* We have $\lambda_m(z') = 0$.

Let $z_1^m = z_2^m = \cdots = z_{m-1}^m = 0$ in $a(\xi)$, consider the $(m-1) \times (m-1)$ matrix $b(\xi)$ obtained by the restricted elements $\phi_j^i(\xi)$, $1 \leq i, j \leq m-1$; $b(\xi)$ is prehermitian by induction and it exists a matrix $H$ (Lemma 1.3) such that (1.1) is satisfied.

We explicit this identity. We obtain easily by considering the terms in $\chi$ that: $h_{ij} = 0$, $\forall i, j$, $i \neq j$ except $h_{12}$.

We have: $\lambda_j(z') = 0$, $m - 1 \geq j \geq 3$. So $\lambda_1(z') + \lambda_2(z') = 0$. We explicit the element in the second line and second column in (1.1):

$$\phi_1^2(z')h_{12} + 2ih_2\lambda_2(z') = \overline{h_{12}\phi_1^2(z')}.$$

We deduce

$$h_2\lambda_2(z') + \text{Im}\left(h_{12}\phi_1^2(z')\right) = 0, \tag{2}$$

$$\phi_1^2(z')\left(\frac{h_{12}}{h_2}\right)^2 + 2i\frac{h_{12}}{h_2}\lambda_2(z') = \frac{|h_{12}|^2}{h_2^2}\overline{\phi_1^2}(z'). \tag{3}$$

Explicit the element in the $j^{\text{th}}$ line, $k^{\text{th}}$ column; $j > k > 2$, $\phi_k^j(z')h_k = h_j\overline{\phi_j^k}(z')$.

So, $\phi_j^k(z) = \frac{h_k}{h_j}\overline{\phi_k^j}(z') + \phi_j^k(z^m)$.

Explicit the element in the $j^{\text{th}}$ line, $j > 2$, and $2^{\text{nd}}$ column: $\phi_1^j(z')h_{12} + \phi_2^j(z')h_2 = h_j\overline{\phi_2^j}(z')$, so: $\phi_j^2(z') = \dfrac{h_2}{h_i}\overline{\phi_2^j}(z') + \dfrac{h_{12}}{h_j}\overline{\phi_1^j}(z')$.

Explicit the element in the $j^{\text{th}}$ line, $1^{\text{st}}$ column:

$$\phi_j^1(z') = \frac{h_1}{h_j}\overline{\phi_1^j}(z') + \frac{h_{12}}{h_j}\overline{\phi_2^j}(z').$$

Explicit the element in the $2^{\text{nd}}$ line, $1^{\text{st}}$ column:

$$\phi_1^2(z')h_1 + \overline{h_{12}}\left(\chi_1 + i\lambda_2(z')\right) = \overline{h_{12}}\left(\chi_1 + \psi(z') - i\lambda_1(z') + h_2\overline{\phi_2^1}(z')\right);$$

then $\phi_2^1(z') = \dfrac{h_1}{h_2}\overline{\phi_1^2}(z') - \dfrac{h_{12}}{h_2}\psi(z')$.

We denote by $E_{12}$ the $m \times m$ matrix, all the elements of which are zero, except the diagonal elements which are equal to $= 1$ and the element in the $1^{\text{st}}$ line, $2^{\text{nd}}$ column which is equal to $h_{12}$. We transform the preceding matrix $a(\xi)$ by $E_{12}$ and we obtain $E_{12}^{-1}a(\xi)E_{12}$.

We state this matrix is the announced matrix in the lemma. The element in the $1^{\text{st}}$ line, $1^{\text{st}}$ column is:

$$\chi_1 + \psi_1'(z) - i\lambda_2(z') - \frac{h_{12}}{h_2}\phi_1^2(z') + i\lambda_1(z^m),$$

we use (2) to obtain the result.

The element in the $1^{\text{st}}$ line, $2^{\text{nd}}$ column is:

$$-2i\frac{h_{12}}{h_2}\lambda_2(z') - \left(\frac{h_{12}}{h_2}\right)^2\phi_1^2(z') + \frac{h_1}{h_2}\overline{\phi_1^2}(z') + \text{terms in } (z^m);$$

the formula (3) implies: this term is equal to: $\dfrac{h_1h_2 - |h_{12}|^2}{(h_2)^2}\overline{\phi_1^2}(z') + \text{terms in } (z^m);$

let: $k_2^1 = \dfrac{h_1h_2 - |h_{12}|^2}{(h_2)^2} > 0$; we obtain the result.

In the same manner, the element in the $1^{\text{st}}$ line $j^{\text{th}}$ column $(j < m)$ is: $k_j^1\overline{\phi_1^j}(z') + \phi_j^1(z^m)$ with: $k_j^1 = \dfrac{h_1h_2 - |h_{12}|^2}{h_2h_j} > 0$.

The element in the $1^{\text{st}}$ line, $m^{\text{th}}$ column is: $\phi_m^1(z) - \dfrac{h_{12}}{h_2}\phi_m^2(z) = \phi_m'^1(z)$.

The element in the $j^{\text{th}}$ line, $2^{\text{nd}}$ column is: $\phi_j^2(z) + \dfrac{h_{12}}{h_2}\phi_1^1(z) = \phi_2'^j(z)$, and in the $2^{\text{nd}}$ line, $j^{\text{th}}$ column is: $\phi_j^2(z) = k_j^2\overline{\phi_2'^j}(z') + \phi_j^2(z^m)$, $\qquad k_j^2 = \dfrac{h_2}{h_j}$.

In the $j^{\text{th}}$ line, $k^{\text{th}}$ column, $j > k > 2$, we have: $\phi_k^j(z)$ and in the $k^{\text{th}}$ line, $j^{\text{th}}$ column we have:

$$k_j^k\overline{\phi_k^j}(z') + \phi_j^k(z^m); \qquad k_j^k = \frac{h_k}{h_j}.$$

We verify easily the properties of the $k_q^p$; and we obtain the lemma. $\qquad\square$

**Lemma 1.6.3.** *Assume as before* 1), 2), 3) *and now* 4) $a(\xi)$ *is:*

$$a(\xi) = \begin{pmatrix} \delta_1 + \psi_1(z) & & & & \\ & \xi_0 + \chi_1 + i\lambda_2(z) & & \phi_q^p(z) & \\ & & \xi_0 + \psi_3 + i\lambda_3(z) & & \\ & & & \xi_0 + i\lambda_4(z) & \\ & \phi_j^i(z) & & & \\ z_1^m & z_2^m & z_3^m & z_4^m & \delta_m \end{pmatrix},$$

$\psi_1(z)$, $\psi_3(z)$ *real;* $\chi_l \neq \chi_{l'}$ *if* $l \neq l'$, $\chi_l \neq 0$.

*Then by a change of basis in* $\mathbb{C}^m$, *we obtain for* $a(\xi)$ *the form:*

$$a(\xi) = \begin{pmatrix} \delta_1(z^m) + \psi_1(z) & & & & \phi_m^1 \\ & \xi_0 + \chi_1 + i\lambda_2(z^m) & & k_q^p \overline{\phi_p^q}(z) + \phi_q^p(z^m) & \\ & & \xi_0 + \psi_3(z) + i\lambda_3(z^m) & & \phi_m'^3 \\ & & & \xi_0 + i\lambda_4(z^m) & \\ & \phi_j^i(z), i > j & & & \\ z_1^m & z_2^m & z_3^m & z_4^m & \delta_m(z^m) \end{pmatrix},$$

$k_q^p > 0$, $1 \leq p < q \leq m-1$, $k_q^p = k_r^p k_q^r$.

Let

$$z_1^m = z_2^m = \cdots z_{m-1}^m = 0, \quad \text{then} \quad \lambda_m(z') = 0$$

as before (1.1) is satisfied.

We explicit this identity. We obtain easily considering the term in $\chi$ that: $h_{ij} = 0$, $\forall i \neq j$ except $h_{12}, h_{34}$.

Explicit the element of (1.1) in the $j^{\text{th}}$ line, $j^{\text{th}}$ column, $m - 1 \geq j \geq 5$, we obtain: $\lambda_j(z') = 0$.

Explicit the element in the $4^{\text{th}}$ line, $4^{\text{th}}$ column, we obtain:

$$\phi_3^4(z')h_{34} + 2i\lambda_4(z')h_4 = \overline{h}_{34}\overline{\phi}_3^4(z'), \tag{4}$$

hence $\phi_3^4(z')\,(h_{34})^2 + 2i\lambda_4(z')h_4 h_{34} - |h_{34}|^2 \overline{\phi}_3^4(z') = 0$.

Explicit the element of the $3^{\text{rd}}$ line, $3^{\text{rd}}$ column, we obtain: $2i\lambda_3(z')h_3 + \phi_4^3(z')\overline{h}_{34}(z') = h_{34}\overline{\phi}_4^3(z')$.

In the $2^{\text{nd}}$ line, $2^{\text{nd}}$ column, we have

$$\phi_1^2(z')h_{12} + 2i\lambda_2(z')h_2 = \overline{h}_{12}\overline{\phi}_1^2(z'), \tag{5}$$

$$\phi_1^2(z')\,(h_{12})^2 + 2i\lambda_2(z')h_2 h_{12} - |h_{12}|^2\overline{\phi}_1^2(z') = 0. \tag{6}$$

In the $1^{\text{st}}$ line, $1^{\text{st}}$ column we have

$$2i\lambda_1(z')h_1 + \overline{h}_{12}\phi_2^1(z') = h_{12}\overline{\phi}_2^1(z'). \tag{7}$$

In the $j^{\text{th}}$ line, $\text{k}^{\text{th}}$ column, $\phi_k^j(z')h_{jk} = h_j\overline{\phi}_j^k(z')$, $5 \leq k < j \leq m - 1$, and $\phi_j^k(z') = \dfrac{h_k}{h_j}\overline{\phi}_k^j(z')$.

In the $4^{\text{th}}$ line, $3^{\text{rd}}$ column we have

$$\phi_3^4(z')h_3 + i\lambda_4(z')\overline{h}_{34} = \overline{h}_{34}\left[\phi_3(z') - i\lambda_3(z')\right] + h_4\overline{\phi}_4^3(z'),$$

$$\phi_4^3(z') = \frac{h_3}{h_4}\overline{\phi}_3^4(z') + \frac{h_{34}}{h_4}\left[-i\left(\lambda_4 + \lambda_3\right)(z') - \psi_3(z')\right].$$

In the $j^{\text{th}}$ line, $4^{\text{th}}$ column we have $\phi_j^4 = \dfrac{h_4}{h_j}\overline{\phi}_4^j + \dfrac{\overline{h_{34}}}{h_j}\overline{\phi}_3^j$, $5 \leq j \leq m - 1$.

In the $j^{\text{th}}$ line, $3^{\text{rd}}$ column we have $\phi_j^3 = \dfrac{h_3}{h_j}\overline{\phi}_3^j + \dfrac{h_{34}}{h_j}\overline{\phi}_4^j$, $5 \leq j \leq m - 1$.

In the $j^{\text{th}}$ line, $2^{\text{nd}}$ column we have $\phi_j^2 = \dfrac{h_2}{h_j}\overline{\phi}_2^j + \dfrac{h_{12}}{h_j}\overline{\phi}_1^j$, $5 \leq j \leq m - 1$.

In the $j^{\text{th}}$ line, $1^{\text{st}}$ column we have $\phi_j^1 = \dfrac{h_1}{h_j}\overline{\phi}_1^j + \dfrac{h_{12}}{h_j}\overline{\phi}_2^j$, $5 \leq j \leq m - 1$.

In the $4^{\text{th}}$ line, $2^{\text{nd}}$ column we have

$$\phi_4^2(z') = \frac{h_2}{h_4}\overline{\phi}_2^4(z') + \frac{\overline{h_{12}}}{h_4}\overline{\phi}_1^4(z') - \frac{h_{34}}{h_4}\phi_3^2(z'). \tag{8}$$

In the $4^{\text{th}}$ line, $1^{\text{st}}$ column we have

$$\phi_4^1(z') = \frac{h_1}{h_4}\overline{\phi}_1^4(z') + \frac{h_{12}}{h_4}\overline{\phi}_2^4(z') - \frac{h_{34}}{h_4}\phi_3^1(z'). \tag{9}$$

In the $3^{\text{rd}}$ line, $2^{\text{nd}}$ column we have

$$\phi_3^2(z') = \frac{h_2}{h_3}\overline{\phi}_2^3(z') + \frac{\overline{h_{12}}}{h_4}\overline{\phi}_1^3(z') - \frac{h_{34}}{h_3}\phi_4^2(z'). \tag{10}$$

Let $H_1 = h_3h_4 - |h_{34}|^2 > 0$. Then (8) and (9) imply:

$$\phi_3^2(z') = \frac{h_2h_4}{H_1}\overline{\phi}_2^3(z') + \frac{\overline{h_{12}}h_4}{H_1}\overline{\phi}_1^3(z') - \frac{\overline{h_{34}}h_2}{H_1}\overline{\phi}_2^4(z') - \frac{\overline{h_{34}}\overline{h_{12}}}{H_1}\overline{\phi}_1^4(z'),$$

$$\phi_4^2(z') = \frac{h_2h_3}{H_1}\overline{\phi}_2^4(z') + \frac{\overline{h_{12}}h_3}{H_1}\overline{\phi}_1^4(z') - \frac{h_{34}h_2}{H_1}\overline{\phi}_2^3(z') - \frac{h_{34}\overline{h_{12}}}{H_1}\overline{\phi}_1^3(z').$$

The $3^{\text{rd}}$ line, $1^{\text{st}}$ column gives

$$\phi_3^1(z') = \frac{h_1}{h_3}\overline{\phi}_1^3(z') + \frac{h_{12}}{h_3}\overline{\phi}_2^3(z') - \frac{\overline{h_{34}}}{h_3}\phi_4^1(z'), \tag{11}$$

and (9) et (11) imply

$$\phi_3^1(z') = \frac{h_1h_4}{H_1}\overline{\phi}_1^3(z') + \frac{h_{12}h_4}{H_1}\overline{\phi}_2^3(z') - \frac{\overline{h_{34}}h_1}{H_1}\overline{\phi}_1^4(z') - \frac{h_{12}\overline{h_{34}}}{H_1}\overline{\phi}_2^4(z'),$$

$$\phi_4^1(z') = \frac{h_1h_3}{H_1}\overline{\phi}_1^4(z') + \frac{h_{12}h_3}{H_1}\overline{\phi}_2^4(z') - \frac{h_{34}h_1}{H_1}\overline{\phi}_1^3(z') - \frac{h_{34}h_{12}}{H_1}\overline{\phi}_2^3(z').$$

In the 2<sup>nd</sup> line, 1<sup>st</sup> column we have

$$\phi_2^1(z') = \frac{h_1}{h_2}\overline{\phi}_1^2(z') - \frac{h_{12}}{h_2}\psi_1(z') \ .$$

We replace $\phi_2^1(z)$ in (7) and we obtain:

$$2i\lambda_1(z')h_1 + \overline{h}_{12}\left[\frac{h_1}{h_2}\overline{\phi}_1^2(z') - \frac{h_{12}}{h_2}\psi_1(z')\right] = h_{12}\left[\frac{h_1}{h_2}\phi_1^2(z') - \frac{\overline{h_{12}}}{h_2}\psi_1(z')\right] \ ,$$

hence,

$$2i\lambda_1(z')h_1 + \frac{\overline{h}_{12}h_1}{h_2}\overline{\phi}_1^2(z') - \frac{h_{12}h_1}{h_2}\phi_1^2(z') = 0.$$

We compare with (5) and obtain

$$\lambda_1(z') + \lambda_2(z') = 0 \ .$$

We deduce $\lambda_3(z') + \lambda_4(z') = 0$, and also

$$-2i\lambda_1(z')h_2 + h_{12}\phi_1^2(z') - \overline{h}_{12}\overline{\phi}_1^2(z') = 0,$$

and with (4)

$$2i\lambda_3(z')h_4 = h_{34}\phi_3^4(z') - \overline{h}_{34}\overline{\phi}_3^4(z') = 0.$$

We denote by $E_2$ the $m \times m$ matrix, the diagonal elements of which are equal to 1; the other elements are 0, except in the first line, second column where we have $\frac{h_{12}}{h_2}$ and in the third line, fourth column where we have $\frac{h_{34}}{h_4}$. We transform $a(\xi)$ by $E_2$ and obtain,

$$E_2^{-1}a(\xi)E_2 \ .$$

We replace the elements $\phi_q^p(z')$, $p < q$ thanks to the preceding formulas. We calculate the elements of the new matrix "$a(\xi)$": Let

$$\xi_0' = \xi_0 + \mathrm{Re}\,\frac{h_{34}}{h_4}\phi_3^4,$$

$$\chi_j' = \chi_j - \mathrm{Re}\,\frac{h_{34}}{h_4}\phi_3^4, \quad j \geq 5,$$

$$\chi_1' = \chi_1 - \mathrm{Re}\left(\frac{h_{12}}{h_2}\phi_1^2 - \frac{h_{34}}{h_4}\phi_3^4\right) \ .$$

We obtain the diagonal of the new $a(\xi)$. We get also

$$\phi_1^2 = \phi_1'^2, \quad \phi_2^1(z') = k_2'^1\overline{\phi}_1^2(z') + \phi_2^1(z^m) \ ,$$

with

$$k_2'^1 = \frac{h_1h_2 - |h_{12}|^2}{(h_2)^2}, \quad \phi_1'^3 = \phi_1^3 - \frac{h_{34}}{h_4}\phi_1^4, \quad \phi_3'^1(z) = k_3'^1\overline{\phi_1'^3}(z') + \phi_3'^1(z^m) \ ,$$

with

$$k_3'^1 = \frac{(h_1 h_2 - |h_{12}|^2) h_4}{H_1 h_2}, \qquad \phi_1'^j = \phi_1^j, \quad j \geq 4, \qquad \phi_4'^1(z) = k_4'^1 \overline{\phi}_1^4(z') + \phi_4^1(z^m),$$

$$k_4'^1 = \frac{h_1 h_2 - |h_{12}|^2}{h_2 h_4}, \qquad \phi_j'^1(z) = k_j'^1 \overline{\phi}'_1^j, \quad 5 \leq j \leq m-1, \qquad k_j'^1 = \frac{h_1 h_2 - |h_{12}|^2}{h_2 h_j},$$

$$\phi_m'^1 = \phi_m^1 - \frac{h_{12}}{h_2}\phi_m^2, \qquad \phi_2'^3 = \phi_2^3 - \frac{h_{34}}{h_4}\phi_2^4 + \frac{h_{12}}{h_2}\phi_1^3 - \frac{h_{12}h_{34}}{h_2 h_4}\phi_1^4,$$

$$\phi_3'^2(z) = k_3'^2 \overline{\phi}_2^3(z') + \phi_3^2(z^m), \qquad k_3'^2 = \frac{h_2 h_4}{H_1} \qquad \phi_2'^4 = \phi_2^4 + \frac{h_{12}}{h_2}\phi_1^4,$$

$$\phi_4'^2(z) = k_4'^2 \overline{\phi}_2^4(z') + \phi_4^2(z^m), \qquad k_4'^2 = \frac{h_2}{h_4}, \qquad \phi_2'^j = \phi_2^j + \frac{h_{12}}{h_2}\phi_1^j, \quad 5 \leq j \leq m-1,$$

$$\phi_j^2(z) = \phi_j'^2(z) = k_j'^2 \overline{\phi}'_2^j(z') + \phi_j'^2(z^m), \qquad k_j'^2 = \frac{h_2}{h_j}, \qquad \phi_m'^2 = \phi_m^2,$$

$$\phi_3'^4 = \phi_3^4, \qquad \phi_4'^3(z) = k_4'^3 \overline{\phi}'_3^4(z') + \phi_4'^3(z^m), \qquad k_4'^3 = \frac{H_1}{h_4^2},$$

$$\phi_3'^j = \phi_3^j, \quad m \geq j \geq 5, \qquad \phi_j^3(z) = k_j'^3 \overline{\phi}'_3^j(z') + \phi_j'^3(z^m), \quad 5 \leq j \leq m-1,$$

$$k_j'^3 = \frac{h_3}{h_j}, \qquad \phi_m'^3 = \phi_m^3 - \frac{h_{34}}{h_4}\phi_m^4, \qquad \phi_4'^j = \phi_4^j + \frac{h_{34}}{h_4}\phi_3^j, \quad 5 \leq j \leq m,$$

$$\phi_j'^4(z) = k_j'^4 \overline{\phi}'_4^j(z') + \phi_j'^4(z^m), \quad 5 \leq j \leq m-1, \qquad k_j'^4 = \frac{h_4}{h_j},$$

$$\phi_m'^j = \phi_m^j, \quad 4 \leq j \leq m, \qquad \phi_k^j(z) = \frac{h_j}{h_k}\overline{\phi}_j^k(z') + \phi_k^j(z^m), \quad 5 \leq k \leq j \leq m-1.$$

We obtain the relation between the $k_q'^p$.

**Lemma 1.6.4.** *Assume 1) 2) 3) and 4); $a(\xi)$ has the following structure:*

$$a(\xi) = \begin{pmatrix} \delta_1 + \psi_1(z) & & & \\ & \xi_0 + \chi_1 + \psi_2(z) + i\lambda_2(z) & & \phi_q^p(z) \\ & & \xi_0 + \chi_1 + i\lambda_3(z) & \\ & & & \delta_4 \\ & \phi_j^i(z) & & \\ z_1^m & z_2^m & z_3^m & z_4^m \quad \delta_m \end{pmatrix}.$$

*Then by a change of basis in $\mathbb{C}^m$, we obtain for $a(\xi)$ the form:*

$$a(\xi) = \begin{pmatrix} \delta_1(z^m)+\psi_1(z) & & & & \\ & \xi_0+\chi_1+\psi_2(z)+i\lambda_2(z^m) & & k_q^p\overline{\phi_p^q}(z')+\phi q^p(z^m) & \\ & & \xi_0+\chi_1+i\lambda_3(z^m) & & \\ & & & \delta_4(z^m) & \\ & & & & \\ & \phi_j^i(z),i>j & & & \\ & & & & \\ z_1^m & z_2^m & z_3^m & z_4^m & \delta_m(z^m) \end{pmatrix}.$$

*Proof.* We give only the sketch of the proof.

Let $z_1^m = \cdots = z_{m-1}^m = 0$; as before we obtain a matrix $H$; all the non diagonal elements of $H$ are 0 except $h_{12}, h_{13}, h_{23}$.

We explicit the formula (1.1). We denote by

$$E_3 = \begin{pmatrix} 1 & \dfrac{\overline{h_{23}}h_{13} - h_{12}h_3}{h_2 h_3 - |h_{23}|^2} & \dfrac{h_{12}h_{23} - h_{13}h_2}{h_2 h_3 - |h_{23}|^2} & \\ & 1 & -\dfrac{h_{23}}{h_3} & 0 \\ & & 1 & \\ & & & \ddots \\ & 0 & & 1 \end{pmatrix}.$$

We transform $a(\xi)$ by $E_3$ and obtain: $E_3^{-1}a(\xi)E_3$. Thus we obtain the new form of $a(\xi)$. $\qquad\square$

## 2. Dimension of the space $V = m^2 - m - 1$

*Notations.* Among the forms $\operatorname{Re}\phi_j^i$ and $\operatorname{Im}\phi_j^i$, $m(m-1)-1$ are chosen as coordinates $(\xi_j^i, \eta_j^i)$; $z_j^i = \xi_j^i + i\eta_j^i$; $z = \{\xi_j^i, \eta_j^i\}$, $m-3$ diagonal forms $\chi_j$ and $\xi_0$ are the last coordinates.

$$a(\xi) = \begin{pmatrix} \varepsilon_1 + \psi_1(z) & & & & \\ & \delta_2 & & \phi_q^p(z) \; p < q & \\ \phi_j^i(z) \; i > j & & & & \\ & & \delta_{m-2} & & \\ & & & \xi_0 + i\lambda_{m-1}(z) & \\ & & & & \varepsilon_m + \psi_m(z) \end{pmatrix}.$$

Let $\varepsilon_i = \displaystyle\sum_{2 \le j \le m-2} c_{ij}\chi_j + i\lambda_i(z)$, $i = 1, m$.

We remark that we can assume that the last line (except the last element) is formed by linearly independent elements (otherwise we consider the first column); the two cases are equivalent by symmetry.

One form $\operatorname{Re}\phi^i_j$ or $\operatorname{Im}\phi^i_j$, $i > j$, is dependent; all the cases can be reduced to the following two cases:

case A: $\operatorname{Re}\phi^3_2$ is the dependent form,

case B: $\operatorname{Re}\phi^2_1$ is the dependent form.

First we study the case A.

Let $z^m_1 = \cdots = z^m_{m-1} = 0$; by the Lemmas 1.6.1, 1.6.2, we obtain

$$a(\xi)=\begin{pmatrix} \varepsilon_1(z^m)+\psi_1(z)\, k^1_2\bar{z}^2_1+\phi^1_2(z^m) & k^1_3\bar{z}^3_1+\phi^1_3(z^m) & & * \\ z^2_1 & \delta_2(z^m) & k^2_3\big(\operatorname{Re}\phi^3_2(z')+i\eta^3_2\big)+\phi^2_3(z^m) & & * \\ z^3_1 & \phi^3_2 & \delta_3(z^m) & & * \\ & & & k^p_q\psi^q_p(z')+\phi^p_q(z^m) & * \\ & & \phi^i_j & & * \\ & & & \xi_0+i\lambda_{m-1}(z) & * \\ z^m_1 & z^m_2 & z^m_3 & & * \end{pmatrix}$$

and the relations between the $k^p_q$, $1 \le p < q < m - 1$. Let

$$\varepsilon_i(z^m) = \xi_0 + \sum c_{ij}\chi_j + i\lambda_i(z^m), \quad i = 1, m.$$

Let $z^2_1 = \cdots = z^m_1 = 0$, we obtain in the same manner

$$a(\xi)=\begin{pmatrix} \varepsilon_1(z^m_1)+\psi_1(z)\, k^1_2\bar{z}^2_1+\phi^1_2(z^m) & k^1_3\bar{z}^3_1+\phi^1_3(z^m) & & * \\ z^2_1 & \delta_2(z^m_1) & k^2_3\big[\operatorname{Re}\phi^3_2(z'')+i\eta^3_2\big]+\phi^2_3(z^m_1) & & \\ z^3_1 & \phi^3_1(z) & \delta_3(z^m_1) & & \\ & & & k^p_q\phi^q_p(z'')+\phi^p_q(z^m) & * \\ & & \phi^i_j(z) & & * \\ & & & \xi_0+i\lambda_{m-1}(z) & * \\ z^m_1 & z^m_2 & & & * \end{pmatrix},$$

the last column is $\big(\phi^1_m, \quad , \, k^p_m\overline{z^m_p} + \phi^p_m(z_1), \quad , \, \varepsilon_m(z^m_1) + \psi_m(z)\big)$.

Here $z_1 = (z^2_1,\ldots,z^m_1)$, $z'' = \{$independent $\xi^i_j, \eta^i_j$ except $\xi^1_1, \eta^m_1\}$; we have relations between the $k^p_q$, $2 \le p < q \le m$.

Let $z^2_1 = \cdots = z^{m-1}_1 = 0$, $k^2_m\overline{z^m_2} + \phi^2_m(z^m_1) = 0, \ldots, k^{m-1}_m\overline{z^m_{m-1}} + \phi^{m-1}_m(z^m_1) = 0$, that is to say $\xi^m_2, \eta^m_2,\ldots,\xi^m_{m-1},\eta^m_{m-1}$ are expressed by $\xi^m_1, \eta^m_1$; the following $(m - 2) \times (m - 2)$ matrix is obtained:

$$\begin{pmatrix} \delta_2(z^m_1)\, k^2_3\big[\operatorname{Re}\phi^3_2() + i\eta^3_2\big] + \phi^2_3(z^m_1) & k^2_{m-1}\overline{z^{m-1}_2} + \phi^2_{m-1}(z^m_1) \\ \phi^3_2() & \delta_3(z^m_1) & k^3_{m-1}\overline{z^{m-1}_3} + \phi^3_{m-1}(z^m_1) \\ & & \\ & & \\ & & \\ z^{m-1}_1 & z^{m-1}_3 & \delta_{m-1}(z^m_1) \end{pmatrix},$$

it is prehermitian by induction, and we obtain immediatly:

$$\lambda_2(z) = \cdots = \lambda_{m-1}(z) = 0, \quad \lambda_1(z) + \lambda_m(z) = 0,$$

$$\phi_q^p(z) = k_q^p \overline{\phi_p^q}(z), \quad 2 \le p < q \le m-1.$$

Let $z_1^{m-1} = \cdots = z_{m-2}^{m-1} = 0$; $z_1^m = \cdots = z_{m-2}^m$; and let us consider the $2 \times 2$ matrix which is obtained. We get $\phi_m^{m-1}\left(z_1^2, \ldots, z_1^{m-2}\right) = 0$.

We get also $\phi_j^1(z_{m-1}^m) = 0$, $2 \le j \le m-2$.

Let $z_1^{m-1} = \cdots = z_{m-2}^{m-1} = 0$; $k_m^{m-1} \overline{z}_{m-1}^m + \phi_m^{m-1}(z_1^m) = 0$, the last equation means that we express $\xi_{m-1}^m$ and $\eta_{m-1}^m$ as linear function of $\xi^m$ and $\eta^m$.

We consider the $(m-1) \times (m-1)$ matrix which is obtained after cancelling the $(m-1)^{\text{th}}$ line and column; by induction it is prehermitian and by adaptation of the Lemmas 1.6.1, 1.6.2, 1.6.3 we obtain that: $p(\xi')$ has the following new form:

$$p(\xi') = \begin{pmatrix} \sum c_{1j}\chi_j + \psi_1(z) & k_2^1 \overline{z}_1^2 & k_3^1 \overline{z}_1^3 & k_{m-1}^1 \overline{z}_1^{m-1} + \phi_{m-1}^1(z^m) & * \\ z_1^2 & \chi_2 & k_3^2 \overline{\phi}_2^3 & & * \\ z_1^3 & & \phi_2^3 & & * \\ & & & & * \\ & & & & * \\ z_1^{m-1} & z_2^{m-1} & & k_m^{m-1} \overline{z}_{m-1}^m + \phi_m^{m-1}(z_1^{m-1}, z_1^m) & * \\ z_1^m & z_2^m & & z_{m-1}^m & * \end{pmatrix},$$

the last column is
$$(\phi_m^1(z_1^{m-1}, \ldots, z_{m-2}^{m-1}, z_{m-1}^m, z_1^m), k_m^2 z_i^m + \phi_m^2(z_1^{m-1}), \ldots, \sum c_{mj}\chi_j + \psi_m(z)).$$

Let $z_2^{m-1} = \cdots = z_{m-2}^{m-1} = 0$; $z_2^m = \cdots = z_{m-2}^m = 0$; $z_1^2 = \cdots = z_1^{m-2} = 0$; the $3 \times 3$ matrix which is obtained is

$$\begin{pmatrix} \sum c_{1j}\chi_j + \psi_1(z) & k_{m-1}^1 \overline{z}_1^{m-1} + \phi_{m-1}^1(z_1^m, z_{m-1}^m) & \phi_m^1\left(z_1^{m-1}, z_{m-1}^m, z_1^m\right) \\ z_1^{m-1} & 0 & k_m^{m-1} \overline{z}_{m-1}^m + \phi_m^{m-1}\left(z_1^{m-1}, z_1^m\right) \\ z_1^m & z_{m-1}^m & \sum c_{mj}\chi_j + \psi_m(z) \end{pmatrix},$$

and it is prehermitian

$$\phi_{m-1}^1(z_1^m, z_{m-1}^m) = 0; \qquad \phi_m^1\left(z_1^{m-1}, z_{m-1}^m, z_1^m\right) = k_m^1 \overline{z}_1^m;$$
$$\phi_m^{m-1}(z_1^{m-1}, z_1^m) = 0; \qquad k_m^1 = k_{m-1}^1 k_m^{m-1}.$$

Let $z_1^{m-1} = z_2^{m-1} = \cdots = z_{m-2}^{m-1} = 0$ and let us develop $\det p(\xi')$ along the $(m-1)^{\text{th}}$ line we obtain

$$\overline{z}_{m-1}^m \phi_{m-1}^1\left(z_2^m, \ldots, z_{m-2}^m\right) \chi_2 \cdots \chi_{m-2} z_1^m$$

it is real for any independent variables and

$$\phi_{m-1}^1\left(z_2^m, \ldots, z_{m-2}^m\right) = 0.$$

Let $z_1^{m-1} = 0$ and develop $\det p(\xi')$ along the last column; the coefficient of $\prod_{2 \le k \le m-2} \chi_k$ is real and $\phi_m^1(z) = k_m^1 \overline{z}_1^m$.

Let $z_2^m = \cdots = z_{m-1}^m = 0$; develop $p(\xi')$ along the last line; it is real; we obtain easily

$$\phi_m^2(z_1^{m-1}) = 0; \quad \phi_m^{m-2}\left(z_1^{m-1}\right) = 0,$$

and the desired result.

Now let us devote to the case B, $\operatorname{Re}\phi_1^2(z)$ is the dependent form.

We summarize the proof.

Let $z_1^m = \cdots = z_{m-1}^m = 0$; as before we symmetrize the obtained matrix $b$ by using the Lemma 1.6.

Then let $z_1^{m-1} = \cdots = z_{m-2}^{m-1} = 0$; $z_1^m = \cdots = z_{m-2}^m = 0$.

It appears the prehermitian matrix

$$\begin{pmatrix} i\lambda_{m-1}\left(z_{m-1}^m\right) & \phi_m^{m-1}\,() \\ z_{m-1}^m & \sum c_{1j}\chi_j + \psi^m\,() + i\lambda_m\left(z_{m-1}^m\right) \end{pmatrix}$$

and a $(m-2) \times (m-2)$ prehermitian matrix with the left upper corner. We deduce

$$\lambda_j\left(z_{m-1}^m\right) = 0, \quad \forall j;$$

$$\forall i,j,\ 1 \leq i < j \leq m-2,\ (i,j) \neq (1,2),\quad \phi_j^i(z) = k_j^i \bar{z}_i^j + \phi_j^2\left(z_1^m, \ldots, z_{m-2}^m\right),$$

$$z'' = \{\text{independent } \xi_l^k,\ \eta_l^k,\ \text{except } z_1^m, \ldots, z_{m-2}^m\},$$

$$\phi_2^1(z) = k_2^1 \overline{\phi_1^2}(z'') + \phi_2^1\left(z_1^m, \ldots, z_{m-2}^m\right),$$

$$\phi_m^{m-1}(z) = k_{m-1}^m \bar{z}_{m-1}^m + \phi_m^{m-1}\left(z_1^{m-1}, \ldots, z_{m-2}^{m-1}, z_1^m, \ldots, z_{m-2}^m\right).$$

Then let $z_1^{m-1} \equiv \cdots z_{m-2}^{m-1} = 0$; $k_{m-1}^m \bar{z}_{m-1}^m + \phi_m^{m-1}\left(z_1^{m-1}, z_{m-2}^{m-1}\right) = 0$.

We obtain $\lambda_j(z) = 0, \quad \forall j$:

$$\forall i,j,\ 1 \leq i < j \leq m-2,$$

$$\phi_j^i(z) = k_j^i \bar{z}_i^j,\ 1 \leq j \leq m-2,$$

$\phi_m^j$ depends only on $z_1^{m-1}, \ldots, z_{m-2}^{m-1}, z_1^m, \ldots, z_{m-1}^m$.

Let $z_1^3 = z_2^3 = z_2^4 = \cdots = z_3^m = 0$, we obtain

$$\phi_{m-1}^j(z) = k_{m-1}^j \bar{z}_1^{m-1} + \phi_{m-1}^j(z_3^m), \quad 1 \leq j \leq m-2, \quad j \neq 3,$$

$$\phi_m^j(z) = k_m^j \bar{z}_j^m + \phi_m^j(z_3^{m-1}, z_3^m), \quad 1 \leq j \leq m-2, \quad j \neq 3,$$

$$\phi_m^{m-1}(z) = k_m^{m-1} \bar{z}_{m-1}^m + \phi_m^{m-1}(z_3^{m-1}, z_3^m).$$

Let $z_1^{m-1} = \cdots = z_{m-2}^{m-1} = 0$, $k_{m-1}^m \bar{z}_{m-1}^m + \phi_m^{m-1}(z_3^m) = 0$, we obtain

$$\phi_m^j(z) = k_m^j \bar{z}_1^m + \phi_m^j(z_3^{m-1}), \quad 1 \leq j \leq m-2, \quad j \neq 3,$$

$$\phi_m^3(z) = k_m^3 \bar{z}_3^m + \phi_m^3(z_1^{m-1}, z_{m-2}^{m-1}, z_3^m, \ldots, z_{m-1}^m).$$

Let $z_1^3 = z_2^3 = z_1^4 = z_4^4 = \cdots = z_1^m = z_2^m = 0$, we obtain

$$\phi_{m-1}^3(z) = k_{m-1}^3 \bar{z}_3^{m-1} + \phi_{m-1}^3(z_1^m, z_2^m), \quad \phi_{m-1}^j(z) = k_{m-1}^j \bar{z}_j^{m-1},\ 4 \leq j \leq m-2,$$

$$\phi_m^3(z) = k_m^3 \bar{z}_3^m + \phi_m^3(z_1^{m-1}, z_2^{m-1}), \quad \phi_m^j(z) = k_m^j \bar{z}_j^m,\ 4 \leq j \leq m-1.$$

The coefficient of $(\eta_1^2)^2$ in $\det p(\xi')$ is real; we deduce

$$\phi_{m-1}^3(z) = k_{m-1}^3 \bar{z}_3^{m-1}.$$

Let $z_3^m = 0$ and let us consider the real part of $\det p(\xi')$, we get

$$\phi_m^j(z) = k_m^j \bar{z}_j^m, \quad j = 1, 2, 3.$$

Then, we obtain by considering the same determinant,

$$\phi_{m-1}^j(z) = k_{m-1}^j \bar{z}_j^{m-1}, \quad j = 1, 2.$$

We have the needed relations between the $k_q^p$.

## 3. Dimension of the space $V = m^2 - m$

*Notations.* The forms $\operatorname{Re} \phi_j^i$, $\operatorname{Im} \phi_j^i$, $i > j$, are chosen as coordinates $\xi_j^i$, $\eta_j^i$; $m - 4$ forms $\chi_j$ and $\xi_0$ are the other coordinates,

$$a(\xi) = \begin{pmatrix} \varepsilon_1 + \psi_1(z) & & & & \phi_q^p(z) \;\; p < q \\ & \varepsilon_2 + \psi_2(z) & & & \\ & & z_j^i & & \\ & & & \delta_{m-2} & \\ & & & & \xi_0 + i\lambda_{m-1}(z) \\ & & & & & \varepsilon_m + \psi_m(z) \end{pmatrix}.$$

Let $\varepsilon_i = \xi_0 + \sum\limits_{3 \le j \le m-2} c_{ij}\chi_j + i\lambda_i(z)$, $i = 1, 2, m$.

Let $z_1^m = \cdots = z_{m-1}^m = 0$ and using the Lemmas 1.6.1, 1.6.2, 1.6.3, 1.6.4 we obtain results as before for the obtained $(m-1) \times (m-1)$ matrix and we transform again $a(\xi)$.

Let $z_1^2 = \cdots = z_1^{m-1} = 0$, we obtain results as before for the obtained $(m-1) \times (m-1)$ matrix and we transform again $a(\xi)$.

Let $z_1^2 = \cdots = z_1^m = 0$, $k_m^2 \bar{z}_2^m + \phi_m^2(z_1^m) = 0, \ldots, k_m^{m-1} \bar{z}_{m-1}^m + \phi_m^{m-1}(z_1^m) = 0$, we consider the obtained $(m-2) \times (m-2)$ matrix and we deduce the following new form of $p(\xi')$:

$$p(\xi') = \begin{pmatrix} \tilde{\varepsilon}_1(z^m) + \psi_1(z) & & k_j^1 \bar{z}_1^j + \phi_j^1(z^m) & \phi_m^1(z) \\ z_1^2 & \sum\limits_{1 \le j \le m-2} c_{2j}\chi_j + \psi_2(z) & & \\ & & k_j^k \bar{z}_k^j & k_m^j \bar{z}_j^m + \phi_m^j(z_1) \\ z_1^j & & \chi_j & \\ & & 0 & \\ z_1^m & z_2^m & & \tilde{\varepsilon}_m(z_1^m) + \psi_m(z) \end{pmatrix}.$$

Let $\tilde{\varepsilon}_i(z^m) = \sum\limits_{3 \le j \le m-2} c_{ij}\chi_j + i\lambda_i(z^m)$.

Let $z_1^{m-1} = \cdots = z_{m-2}^{m-1} = 0$; $z_1^m = \cdots = z_{m-2}^m = 0$; we obtain

$$\phi_m^{m-1}(z) = k_m^{m-1} \bar{z}_{m-1}^m + \phi_m^{m-1}\left(z_1^{m-1}, z_1^m\right),$$

$$\phi_j^1(z) = k_j^1 \bar{z}_1^j + \phi_j^1(z_1^m, \ldots, z_{m-2}^m), \quad 2 \le j \le m - 2.$$

Let $z_1^{m-1} = \cdots = z_{m-2}^{m-1} = 0$; $k_m^{m-1} \bar{z}_{m-1}^m + \phi_m^{m-1}(z_1^m) = 0$.

After cancelling the $(m-1)^{\text{th}}$ lines and column of $a(\xi)$ we obtain the abridged form of the matrix $b(\xi)$

$$\begin{pmatrix} \varepsilon_1(z^m)+\psi_1(z) & & & \\ & \sum c_{2j}\chi_j+\psi_2(z) & \phi_q^p(z) & \\ z_1^j & z_2^j & \chi_j & \\ & & & \varepsilon_m(z^m)+\psi_m(z) \end{pmatrix}.$$

By Lemma 1.3, there exists a matrix $H$ such that (1.1) is satisfied.

Consider the elements in the $(m-1)^{\text{th}}$ lines and $(m-1)^{\text{th}}$ columns of this equation, we obtain

$$\sum_{1\leq k\leq m-2} z_k^m h_{km-1} - \sum_{i\leq k\leq m-2} \overline{h}_{km-1}\overline{z}_k^m + 2i\lambda_m(z_1^m)h_{m-1} = 0.$$

We deduce $h_{km-1}=0$, $2\leq k\leq m-2$.

Afterwards, we consider the element in the $(m-2)^{\text{th}}$ lines, $(m-2)^{\text{th}}$ column, in the $j^{\text{th}}$ line, $j^{\text{th}}$ column, $j\geq 2$, and we obtain $h_{kj}=0$, $\forall k\neq j$ except $h_{1m-1}$.

Then, we use the Lemma 1.6.1 and an easy adaptation of the Lemma 1.6.2. Now,

$$p(\xi')=\begin{pmatrix} \sum_{1\leq j\leq m-2}c_{1j}\chi_j+\psi_1(z) & k_k^1\overline{z}_1^k\; k_{m-1}^1\overline{z}_1^{m-1}+\phi_{m-1}^1(z^m) & \phi_m^1(z^{m-1},z_1^m) \\ z_1^k & k_k^j\overline{z}_j^k & k_m^k\overline{z}_1^m+\phi_m^k(z^{m-1}) \\ z_1^j & z_k^j & \\ & & k_m^{m-1}\overline{z}_{m-1}^m+\phi_m^{m-1}(z_1^{m-1},z_1^m) \\ z_1^m & z_k^m & z_{m-1}^m & \sum c_{mj}\chi_j+\psi_m(z) \end{pmatrix},$$

here, $z^{m-1}=\left(z_1^{m-1},\ldots,z_{m-2}^{m-1},z_{m-1}^m\right)$.

Let $z_1^3=z_2^3=z_3^4=\cdots=z_3^m=0$, we consider the $(m-1)\times(m-1)$ matrix obtained by cancelling the $3^{\text{rd}}$ column of $a(\xi)$ and we deduce

$$p(\xi')=\begin{pmatrix} \sum_{3\leq j\leq m-2}c_{1j}\chi_j+\psi_1(z)\; k_j^1\overline{z}_1^j\; k_{m-1}^1\overline{z}_1^{m-1}+\phi_{m-1}^1(z_3^m) & k_m^1\overline{z}_1^m+\phi_m^1(z_3^m) \\ z_1^2 & & \\ z_1^3 & k_j^3\overline{z}_3^j & k_m^3\overline{z}_3^m+\phi_m^3(z_1^{m-1}) \\ z_1^k & & k_m^k\overline{z}_k^m \\ z_1^{m-1} & & k_m^{m-1}\overline{z}_{m-1}^m \\ z_1^m & z_{m-1}^m & \sum_{3\leq j\leq m-2}c_{mj}\chi_j+\psi_m(z) \end{pmatrix}.$$

At last, let $z_1^2=z_2^3=\cdots=z_2^m=0$, we obtain

$$\phi_m^1(z)=k_m^1\overline{z}_1^m, \quad \phi_m^3(z)=k_m^3\overline{z}_3^m, \quad \phi_{m-1}^1(z)=k_{m-1}^1\overline{z}_1^{m-1}.$$

Summarizing the relations between the $k_q^p$ are verified and the result is proved.

# References

1. T. Nishitani. *Symmetrization of a class of hyperbolic systems with real constant coefficients.* Ann. Scuola Norm. Sup. Pisa, Cl. sc. **21**, (1994) 97–130.
2. T. Nishitani, J. Vaillant. *Smoothly symmetrizable systems and the reduced dimensions.* Tsukuba J. Math. **25**, No. 1, 2001, 165–177.
3. T. Nishitani, J. Vaillant. *Smoothly symmetrizable systems and the reduced dimensions II.* Tsukuba J. Math. **27**, No. 2, 2003, 389–403.
4. T Nishitani, J. Vaillant. *Smoothly symmetrizable complex systems and the real reduced dimension.* to appear in Tsukuba J. Math.
5. G. Strang. *On strong hyperbolicity.* J. Math. Kyoto Univ. **6**, (1967), 397–417.
6. J. Vaillant. *Symmétrisabilité des matrices localisées d'une matrice fortement hyperbolique en un point multiple.* Ann. Scuola Norm. Sup. Pisa Cl. Sci. **5**, (1978), 405–427.
7. J. Vaillant. *Systèmes fortement hyperboliques $4 \times 4$, dimension reduite et symetrie.* Ann. Scuola Norm. Sup. Pisa Ser. IV, col. XXIX, Fasc. 4 (2000), 839–890.
8. J. Vaillant. *Systèmes uniformément diagonisables, dimension réduite et symétrie.* I Bulletin de la Société Royale des Sciences de Liège, **70**, 4–5–6 (2001), 407–433.
9. J. Vaillant. *Systèmes uniformément diagonisables, dimension réduite et symétrie II.* In *Partial Diff. Equations and Math. Physics in memory of Jean LERAY* (K. Kajitani and J. Vaillant, Editors) Birkhäuser 2002, 195–224.
10. J. Vaillant. *Diagonalizable complex systems, reduced dimension and hermitian system I.* In *Hyperbolic Problems and Related Topics.* (F. Colombini, T. Nishitani Editors) International Press, (2003), 409–422.
11. J. Vaillant. *Diagonalizable complex systems, reduced dimension and hermitian system II.* Pliska Sud. Math. Bulgar. **15** (2002), 131–148.
12. J. Vaillant. *Complex strongly hyperbolic $4 \times 4$ systems, reduced dimension and hermitian systems.* Bulletin des Sciences Mathématiques **129** (2005) 415–456
13. J. Vaillant. *Strongly hyperbolic complex systems, reduced dimension and hermitian systems I.* Eds. M. Padula and L. Zanghirati, Hyperbolic problems and regularity questions, Birkhäuser, 2006, 217–231.

# STABILITY OF STATIONARY SOLUTIONS OF NONLINEAR HYPERBOLIC SYSTEMS WITH MULTIPLE CHARACTERISTICS*

A. KRYVKO and V.V. KUCHERENKO

*Instituto Politecnico Nacional - ESFM,*
*Av. IPN, U.P.A.L.M., Edif. 9, Mexico, DF, Mexico, cp 07738*
*E-mail: valeri@esfm.ipn.mx*

A linearized system of hydrodynamics for ideal compressible fluids is considered. It is shown that in the case when the sound speed in the medium is equal to zero the characteristics change their multiplicity. For this case the asymptotic solution of the Cauchy problem with high frequency initial data is constructed. We prove that in this case the linearized system is unstable with respect to the high frequency perturbations.

**Key words:** Multiplicity, linearization, stability
**Mathematics Subject Classification:** 39A11

Linearization of nonlinear equations is used to investigate the waves in physical mediums.

The law of energy conservation is satisfied for the physical systems of partial differential equations. As a rule, the solutions of linearized systems have their own energy estimates. These estimates determine the wave modes that occur in such systems.

Let us discuss the relation between the energy estimates and the wave modes that occur in linearized systems. Suppose that the linearized system is a real hyperbolic first-order system $L$ of the form

$$Lu = -i\frac{\partial u}{\partial t} + (-i)\sum_{r=1}^{n} A_r(x)\frac{\partial u}{\partial x_r} + (-i)B(x)u = f(x,t), \qquad (1)$$

here $B; A_r, r = 1, ..., n$, are $C^\infty(\mathbb{R}^n)$ real matrix functions, $u \in \mathbb{R}^n$. We say that the system (1) has an energy estimate if there exists a nonnegatively definite $(n \times n)$ −matrix $E(x) : (E(x)\eta, \eta) \geq 0$, $E(x) \in C^\infty(\mathbb{R}^n)$, such that for each function $u(x,t) \in C^\infty(\mathbb{R}^n \times \mathbb{R}^1_+)$ with compact union of supports $\cup_{t=0}^{T} Supp\, u(x,t)$ it

*This research was partially supported by EDI scholarship of IPN, SNI scholarship of Mexico, and CONACYT projects 35009E, 4297-F.

holds that

$$\left. \int_{\mathbb{R}^n} \left( E\left(x\right) u\left(x,\tau\right), u\left(x,\tau\right)\right) dx \right|_0^t \leq C \left\{ \int_0^t \int_{\mathbb{R}^n} \left(u\left(x,\tau\right), u\left(x,\tau\right)\right) dx d\tau + \right.$$

$$\left. \int_0^t \int_{\mathbb{R}^n} \left| Lu\left(x,\tau\right)\right|^2 dx d\tau \right\}. \tag{2}$$

Suppose that there exist constants $m, M \in (0, \infty)$ such that $M \|\eta\|^2 \geq (E\left(x\right)\eta, \eta) \geq m \|\eta\|^2$ and the energy estimate (2) holds true. In this case in[1] it was proved that the symbol $\sum_{r=1}^n A_r\left(x\right)\xi_r$ is diagonalizable for all $(x, \xi) \in \mathbb{R}^{2n}$ and the diagonalizing matrices $S\left(x, \xi\right), S^{-1}\left(x, \xi\right)$ :

$$S\left(x, \xi\right) \left\| \sum_{r=1}^n A_r\left(x\right)\xi_r \right\| S^{-1}\left(x, \xi\right) = \left\| \delta_{ij}\lambda_j\left(x, \xi\right)\right\|,$$

are uniformly bounded in $K \times (\mathbb{R}^n \backslash 0)$ for any compact set $K \subset \mathbb{R}^n$.

Therefore the energy estimate (2) with a nonnegative definite matrix $E$ implies that the symbol $\sum_{r=1}^n A_r\left(x\right)\xi_r$ does not have Jordan blocks.

This fact facilitates the investigation of complex physical systems. Consider the system of magnetohydrodynamics (abbreviation MHD)[5]

$$div\ B = 0\,;\ B_t - curl\left[u \times B\right] = 0;\ \rho_t + div\left(\rho u\right) = 0, \\ \rho u_t + \rho\left(u, \nabla\right) u + c^2\left(\rho\right) \nabla \rho - \tfrac{1}{\mu}\left(curl\ B\right) \times B = 0, \tag{3}$$

for $x \in \mathbb{R}^3, t > 0$. Let us linearize system (3) at the stationary solution $u_0, B_0, \rho_0$ and denote the obtained system by LMHD. Then for the $C^\infty$−solution $u, B, \rho$ of LMHD the following energy estimate holds:

$$\left. \int_{\mathbb{R}^3} \left( \frac{1}{2\mu} B^2 + \frac{\rho_0}{2} u^2 + \frac{c^2\left(\rho_0\right)}{\rho_0} \rho^2 \right) dx \right|_0^t \leq C \int_0^t \int_{\mathbb{R}^3} \left( B^2 + u^2 + \rho^2 \right) dx d\tau. \tag{4}$$

Linearizing the system of hydrodynamics for compressible ideal fluids with external forces $f(x, t)$

$$\rho u_t + \rho\left(u, \nabla\right) u = -c^2\left(\rho\right) \nabla \rho + f(x, t) \\ \rho_t + div\left(\rho u\right) = 0, \qquad \text{here}\quad c^2\left(\rho\right) := \partial p(\rho)/\partial \rho \tag{5}$$

at its stationary solution $u_0, \rho_0$ we obtain the following energy estimate for the solution of the linearized system of hydrodynamics (abbreviation LSHD)

$$\left. \int_{\mathbb{R}^3} \left( \frac{\rho_0}{2} u^2 + \frac{c^2\left(\rho_0\right)}{2\rho_0} \rho^2 \right) dx \right|_0^t \leq C \int_0^t \int_{\mathbb{R}^3} \left( u^2 + \rho^2 \right) dx d\tau + \int_0^t \int_{\mathbb{R}^3} uf dx d\tau. \tag{6}$$

Thus we see that if it holds that $M \geq c^2\left(\rho_0\right)/\rho_0 \geq m > 0$, for some $M, m \in (0, \infty)$, then the quadratic forms in the left-hand side of energy estimates (4), (6) are nondegenerate. Consequently we obtain that the principal symbols of the systems LMHD and LSHD are diagonalizable. Evidently, these diagonalizable symbols for LSHD and LMHD can have eigenvalues of variable multiplicity[7,8,9]. Below we show that if energy form (6) is degenerated (i.e. $c^2(\rho_0)/\rho_0 = 0$) then the main symbol of system (5) has a Jordan block.

The adiabatic process of wave propagation in an ideal compressible fluid is considered. In such a case the nonlinear system (5) has the energy integral[10]:

$$\int_{\mathbb{R}^3} \left\{\{\rho u^2/2 + \rho\varepsilon(\rho)\} \, dx\right\}\Big|_0^t = \int_0^t \int_{\mathbb{R}^3} (uf) \, dxdt, \tag{7}$$

where $\varepsilon(\rho)$ is the internal energy

$$\varepsilon(\rho) = \varepsilon(\rho_0) + \int_{\rho_0}^{\rho} \frac{p(\rho)}{\rho^2} \, d\rho.$$

In the present paper we investigate the phenomenon of generation of large amplitude waves in the linearized systems when the high-frequency wave (the wave front of solution) passes through the points of multiplicity. Note that even in the case when the eigenvalues of the symbol $\sum_{r=1}^{n} A_r(x)\xi_r$ have constant multiplicity the beam focusing and local amplitude growth occur. However, the integral of squared absolute value of the field remains bounded. For example the Airy function $(2\pi h)^{-1/2} \int \exp\left(ih^{-1}\left[xp - p^3/3\right]\right) dp$ at the point $x = 0$ has the order $h^{-1/6}$, but the integral of its squared absolute value over a compact set is uniformly bounded for $h \in (0,1]$ in virtue of Parseval's equality. This means that focusing leads to redistributions of energy but not to its growth. So it turns out that when the waves pass the points of multiplicity a significant growth of the integral of the square of solution's absolute value (energy growth) occurs only in the case when Jordan blocks arise in the principal symbol at the points of multiplicity.

Results obtained in[1] imply that this can happen only when the energy form of the linearized system in the left-hand side of energy estimate (2) degenerates.

Linearizing system (5) at some stationary solution $u_0, \rho_0 \in C^\infty(\mathbb{R}^n)$ we obtain

$$\rho_0 \frac{\partial u}{\partial t} + \rho_0 (u_0, \nabla) u = -\frac{\partial p(\rho_0)}{\partial \rho_0} \nabla \rho + B(u, \rho), \frac{\partial \rho}{\partial t} + \rho_0 div\, u + (u_0, \nabla) \rho = a(u, \rho). \tag{8}$$

Here $B(u, \rho), a(u, \rho)$ denote the corresponding zero-order terms, i.e. the terms that do not contain the derivatives of $u$ and $\rho$. In what follows system (8) is considered.

The dependence of the pressure $p$ on the density $\rho$ is determined for the adiabatic process of wave propagation. As a rule $dp/d\rho \geq 0$ and the sound speed $c(\rho)$ in the fluid is equal to $\sqrt{dp/d\rho}$ [6].

The principal symbol of system (8) has the form

$$D := \begin{bmatrix} \rho_0(\eta + (u_0, \xi)) & 0 & 0 & c^2\xi_1 \\ 0 & \rho_0(\eta + (u_0, \xi)) & 0 & c^2\xi_2 \\ 0 & 0 & \rho_0(\eta + (u_0, \xi)) & c^2\xi_3 \\ \rho_0\xi_1 & \rho_0\xi_2 & \rho_0\xi_3 & \rho_0(\eta + (u_0, \xi)) \end{bmatrix}.$$

Thus characteristic roots $\eta = \lambda_j(x, \xi)$ satisfy the equation

$$\det D = (\eta + (u_0, \xi))^2 (\eta + (u_0, \xi) - c|\xi|)(\eta + (u_0, \xi) + c|\xi|) = 0.$$

Evidently at the points $x$ such that $c(\rho_0(x)) = 0$ the roots $\eta_\pm = -(u_0, \xi) \pm c|\xi|$ coincide with the root $\eta = -(u_0, \xi)$. Simple calculations show that at the points $c(\rho_0(x)) = 0$ there are three eigenvectors and one adjoint vector, i.e. a Jordan block arises. This occurs because the quadratic form in the energy estimate (6) is degenerated at the point such that $c(\rho_0(x)) = 0$.

In what follows we suppose that over the set of multiplicity $\Sigma$ it holds that $\{\eta + \lambda_j, \eta + \lambda_k\}|_\Sigma \neq 0$, for the coincident at $\Sigma$ roots. Then for the case under consideration we obtain

$$\Sigma = \{x, \xi : c(\rho_0(x)) = 0\}, \{(u_0, \xi), c|\xi|\}|_\Sigma = (u_0, \nabla)c|\xi| \, |_\Sigma \neq 0. \tag{9}$$

So if $u_0|_\Sigma \neq 0$, then formula (9) implies that $\nabla c|_\Sigma \neq 0$. Thus we obtain that $\Sigma$ is a smooth $(2n - 1)$−surface in $\mathbb{R}^{2n}_{x,\xi}$.

For the LSHD symbol the root $\eta = -(u_0, \xi)$ has two $C^\infty$−eigenvectors in $\mathbb{R}^{2n}_{x,\xi}$: $\{u_m = k_m(\xi), \rho_m = 0\}$, here $k_m(\xi)$, $m = 1, 2$, are two linearly independent vectors orthogonal to $\xi$. Furthermore, the eigenvectors $e_\pm = \{u_\pm = \pm c\xi, \rho = \rho_0\}$, such that $e_\pm \in C^\infty(\mathbb{R}^n_x \times (\mathbb{R}^n \backslash 0))$ correspond to the roots $\eta_\pm = -(u_0, \xi) \pm c|\xi|$, respectively. Over the set $\Sigma$ the $C^\infty$−vector $(e_+ - e_-)/(\eta_+ - \eta_-)$ transforms in the adjoint vector $\{u = \xi, \rho = 0\}$.

In the case when $dp/d\rho < 0$ the type of the equation changes and under a small highly oscillating perturbation the solution will have exponential growth with respect to the frequency. This means that the Cauchy problem for system (8) becomes ill-posed. Such cases (i.e. when $dp/d\rho < 0$) occur in the Wander-Vaals gas[6]. The physicists change the dependence of $p$ on $\rho$ for the Wander-Vaals gas in such a way that it holds $dp/d\rho \geq 0$. This was explained by some additional physical effects of phase transition[6].

In what follows we consider the case $d\rho/dp \geq 0$ and $d\rho/dp(\rho_0) = 0$ at some point $\rho_0$. Then equation (5) has the stationary solution $u = u_0$, $\rho = \rho_0 = Const \neq 0$, $u_0 = Const \neq 0$, (this also occurs in the Wander-Vaals gas[6]). Evidently, in this case the minor terms $B(u, \rho), a(u, \rho)$ in (8) are equal to zero.

The solution of the Cauchy problem $u|_{t=0} = k \exp\{i(x, k) h^{-1}\}$, $\rho|_{t=0} = 0, |k| = 1$ for equation (8) has the form

$$u = k \exp\{i[(x, k) - (u_0, k)t] h^{-1}\};$$

$$\rho = -\frac{i}{h} t \rho_0 \exp\{i[(x, k) - (u_0, k)t] h^{-1}\}.$$

Therefore the density $\rho$ of the high-frequency excitation rapidly increases as $h \longrightarrow 0$ at any point $t > 0$. Evidently, during a small period of time $h^\gamma, 0 \leq \gamma \leq 1$, the density $\rho$ increases by the factor $h^{-1+\gamma}$.

Now we consider the case when the stationary solution $u_0, \rho_0$ is not constant and satisfies condition (9). Below we show that if the minor terms of LSHD satisfy some conditions then the integral of the square of the absolute value of the LSHD solution with high frequency initial data

$$\{\phi(x), \psi(x)\}^T \exp\left(\frac{i}{h} S_0(x)\right); h \in (0, 1] \tag{10}$$

increases by factor $h^{-\mu}$ during a short period of time $h^\delta$ for some $\mu, \delta > 0$. This phenomenon we call the instability of the LSHD solution with respect to high frequency excitations for any finite time interval $[0, t]$, $t > 0$ (abbreviation HFinstability). Here the values $\mu, \delta$ depend on the terms of lower order. Multiplying the function (10) by a small parameter we obtain the elementary excitation[11]. Due to the energy integral (7) elementary excitation of the initial data for the nonlinear equation (5) will not imply a large growth of energy (7). That is the HFinstability of the linearized equation does not imply the HFinstability of the nonlinear problem. Therefore in case of multiplicity and HFinstability the linear approximation is not an adequate approximation of the nonlinear problem. In the case of multiplicity the structure of the nonlinear problems will be rather complicated.

One can prove the uniqueness of LSHD solutions with loss of smoothness, i.e. if $u(x, 0), \rho(x, 0) \in H_{s+N_0}$ (for some $N_0$ and all $s \geq 0$) then LSHD solutions exist and $\partial_t^j \{u(x, t), \rho(x, t)\} \in H_{s-j}$, with $s - j > 0$.

To estimate the amplitude growth of linearized systems (that is its HFinstability) it is sufficient to consider the formally asymptotic solutions of the Cauchy problem with data (10) up to $O(h^N)$ ($N < \infty$). Thus in what follows we present the results obtained by the asymptotic methods in the case when the change of multiplicity leads to the generation of Jordan blocks. Obtained asymptotic solutions are called FAS (formally asymptotic solutions). In order to apply the results to more general systems (for example to the equation of LMHD), we generalize the properties of the eigenvalues and eigenvectors of the LSHD.

We perform the ulterior constructions in some conical with respect to $\xi$ neighborhood $\Omega_{x,\xi}$ of the point $(x, \xi) \in \Sigma$, i.e. microlocally. Suppose that $\Sigma \cap \Omega_{x,\xi}$ is a $C^\infty$−manifold outside the points $\xi = 0$. Let $\lambda_1(x, \xi)$ be the eigenvalue of multiplicity $r$ of the matrix symbol $A(x, \xi) = \sum_{j=1}^n A_j(x) \xi_j$ of system (1), and suppose that there exist $r$ eigenvectors $e_j(x, \xi), j = 1, ..., r$, corresponding to $\lambda_1(x, \xi)$ in $\Omega_{x,\xi} \backslash \Sigma$. Suppose that $\lambda_1 \in S^1(\Omega_{x,\xi})$ and $e_j \in S^0(\Omega_{x,\xi})$ for $j = 1, ..., r$. Now, let $\lambda_2, \lambda_3$ be the eigenvalues of the symbol, such that

$$\lambda_1|_\Sigma = \lambda_2|_\Sigma = \lambda_3|_\Sigma; \ \lambda_1 \neq \lambda_2, \ \lambda_1 \neq \lambda_3, \ \lambda_2 \neq \lambda_3 \text{ in } \Omega_{x,\xi} \backslash \Sigma;$$

and

$$\lambda_j \in S^1(\Omega_{x,\xi}) \text{ for } j = 1, 2, 3. \tag{11}$$

We assume that the symbol $A(x, \xi)$ does not have any other eigenvalues coinciding with $\lambda_1$ and different from $\lambda_2, \lambda_3$ on the set $\Sigma$.

Let $e_{r+1}, e_{r+2}$ be the unique eigenvectors corresponding to the eigenvalues $\lambda_2, \lambda_3$ in $\Omega_{x,\xi} \backslash \Sigma$. Suppose that

$$e_{r+1}, e_{r+2} \in S^0(\Omega_{x,\xi}); \ |\xi| \frac{e_{r+1} - e_{r+2}}{\lambda_2 - \lambda_3} \in S^0(\Omega_{x,\xi}); \ |\xi| \left. \frac{e_{r+1} - e_{r+2}}{\lambda_2 - \lambda_3} \right|_\Sigma \neq 0, \tag{12}$$

and

$$\{\lambda_1, \lambda_2\}|_\Sigma \neq 0, \{\lambda_1, \lambda_3\}|_\Sigma \neq 0, \{\lambda_2, \lambda_3\}|_\Sigma \neq 0. \tag{13}$$

The case of mulitplicity of second order with smooth roots was considered in the paper [12].

Using the approach of [2] we obtain the following lemma.

**Lemma 0.1.** *Suppose that the conditions (11),(12),(13) hold true, then there exist matrix-symbols* $P_r(x,\xi) \in S^{-r}(\Omega_{x,\xi}), 0 \le r < \infty,$ *such that:*

*1) The matrix-valued symbol* $P_0(x,\xi)$ *is an invertible matrix.*

*2) Substituting* $u = Pv$, *where* $P$ *is a pseudodifferential operator (P.D.O.) corresponding to the symbol* $P = \sum_{r=0}^{N} P_r$, *reduces system (1) to the system*

$$\frac{\partial v}{\partial t} + b\left(x, -i\frac{\partial}{\partial x}\right)v = 0, \ 0 \le t \le T, \tag{14}$$

*microlocally in* $\Omega_{x,\xi}$. *Here* $b\left(x, -i\frac{\partial}{\partial x}\right)$ *is a P.D.O. with a block-diagonal symbol* $b(x,\xi)$ *of the form*

$$b = \begin{bmatrix} b_{11} & 0 \\ 0 & b_{22} \end{bmatrix}, b_{11} = \sum_{q=-N}^{1} b_{11}^q, b_{22} = \sum_{q=-N}^{1} b_{22}^q.$$

*Here* $b_{jj}^q \in S^{-q}(\Omega_{x,\xi}), j = 1, 2,$ *and*

$$b_{11}^1 = \begin{bmatrix} \lambda_1 I & 0 & 0 \\ 0 & \lambda_2 & d \\ 0 & 0 & \lambda_3 \end{bmatrix}, \tag{15}$$

*where* $I$ *is the identity* $(r \times r) - matrix$ *and* $d = |\xi|$. *Moreover, the spectrums of the operators* $b_{11}^1$ *and* $b_{22}^1$ *do not intersect.*

*3) If the wave front of solution* $v(x,t)$ *of equation (14) belongs to* $\Omega_{x,\xi}$ *for all* $t \in [0,T]$ *then*

$$\frac{\partial}{\partial t}(Pv) + \sum_{r=1}^{n} A_r(x)\frac{\partial}{\partial x_r}(Pv) = R_{N+1}v,$$

*where* $R_{N+1}$ *is a P.D.O. with symbol* $R_{N+1}(x,\xi) \in S^{-(N+1)}(\Omega_{x,\xi})$.

*Note that changing the basis we can take the element* $d$ *in (15) to an arbitrary* $S^1(\Omega_{x,\xi}) - function$ *not equal to zero at* $\Sigma$.

We reduce the construction of FAS to the construction of two FAS for the equations (14) with symbols $b_{11}, b_{22}$. Let us construct FAS for the Cauchy problem $v|_{t=0} = \phi(x)\exp\left(\frac{i}{h}S_0(x)\right), \phi \in C_0^\infty(\mathbb{R}^n), S_0 \in C^\infty(\mathbb{R}^n);$

$$(-i)\frac{\partial v}{\partial t} + B_{11}\left(x, -i\frac{\partial}{\partial x}\right)v = 0, \ 0 \le t \le T, \tag{16}$$

where the P.D.O. $B_{11}$ has the symbol $b_{11}$. Let $\Lambda_0$ be a Lagrangian manifold in $\Omega_{x,\xi}$ defined as $\Lambda_0 = \{x = x_0, \xi = \nabla S_0(x_0) : x_0 \in Supp\ \phi\}$.

We construct FAS of problem (16), in p-representation with respect to $t$, by analogy with the integral representation for Weber functions[3].

Below we use so-called $h^{-1}-$P.D.O.[4]

$$A\left(x, -ih\frac{\partial}{\partial x}\right)u(x) := \frac{1}{(2\pi h)^n}\int \exp\left(\frac{i}{h}\xi[x-y]\right)a(x,\xi)u(y)\,dy\,d\xi =$$

$$\frac{1}{(2\pi)^n}\int \exp(i\eta[x-y])a(x,h\eta)u(y)\,dy\,d\eta.$$

Multiplying system (16) by $h$ we obtain

$$(-ih)\frac{\partial v}{\partial t} + B^1_{11}\left(x, -ih\frac{\partial}{\partial x}\right)v + \sum_{q=-N}^{0}h^{1+|q|}B^q_{11}\left(x, -ih\frac{\partial}{\partial x}\right)v = 0, \qquad (17)$$

with the same initial data as for equation (16).

Let $U(t)$ be the group of the unitary operator generated by the Friedrichs extension of the symmetric P.D.O. $\lambda(x, -ih\partial/\partial x) + \lambda^*(x, -ih\partial/\partial x)$ (here $*$ denotes the formally conjugate operator). We search the solution in the form $v = Uw$. In the domain $|\lambda_2 - \lambda_3| \le h^{1/2-\delta}$, $|t| \le h^{1/2-\delta}$, $0 < \delta \le 1/8$ the function $w$ is searched in the $p$-representation with respect to $t$ :

$$w(t,x,h) = \int_C \exp\{i/h^{-1}[tp + S(p,x)]\}\widehat{w}(p,x,h)\,dp, \qquad (18)$$

here the contour $C$ bypasses from the below the point $p = 0$ by the circle $r = h^{1/2+2\delta}$; and over the contour $C$ it holds that $|p\|_C \le h^{1/2-\delta}$. In the domain $|\lambda_2 - \lambda_3| \ge h^{1/2-2\delta}$ the function $w$ can be constructed as WKB solution. These two solutions can be patched together using the partition of unity.

Define the symbols $\widetilde{\lambda}_j(t,x,\xi) = (\lambda_j - \lambda_2) \circ g^t_2(x,\xi)$, $j = 1,3$, $\widetilde{d} := d \circ g^t_2(x,\xi)$, where $g^t_2(x,\xi)$ is the Hamiltonian flow generated by the Hamiltonian $\lambda_2$. The components of the vector function $y := \exp\{i/h^{-1}[S(p,x)]\}\widehat{w}(p,x,h)$ satisfy the equations

$$(p + \widehat{\widetilde{\lambda}}_1)y_j = -h\sum_{k=1}^{r+2}\widehat{\widetilde{B}}_{j,k}y_k; \quad py_{r+1} + \widehat{\widetilde{d}}y_{r+1} = -h\sum_{k=1}^{r+2}\widehat{\widetilde{B}}_{r+1,k}y_k;$$

$$(p + \widehat{\widetilde{\lambda}}_{r+2})y_{r+2} = -h\sum_{k=1}^{r+2}\widehat{\widetilde{B}}_{r+2,k}y_k; \quad j = 1, ..., r, \qquad (19)$$

where $\widehat{\widetilde{B}}_{j,k}$ are some $h^{-1}$P.D.O., and $\widehat{\widetilde{\lambda}}_j = \widetilde{\lambda}_j(ih\partial/\partial p, x, -ih\partial/\partial x)$. We search the solution in the class of functions $\exp\{i/h^{-1}[S(p,x)]\}\widehat{w}(p,x,h)$, with $C^\infty$ phase function $S$ and singular amplitudes of the form

$$p^\sigma\{\sum_{k,l,j=0}^{M}(h/p)^j h^l(\ln p)^l p^k c_{k,l,j}(x,h)\}, \quad M < \infty, \qquad (20)$$

where $c_{k,l,j}(x,h)$ are polynomials in $h$ with $C^\infty(R^n)$ coefficients, and the function $\sigma(x) \in C^\infty$. If the functions $y_k$, $k \ne r+1$, belong to this class then we can determine the function $y_{r+1}$ up to some precision $h^{\gamma N_1}$ for some $\gamma > 0$:

$$y_{r+1} = -p^{-1}\widehat{\widetilde{d}y}_{r+2} + \sum_{j=1}^{N_1}(-1)^j p^{-1}\left(hp^{-1}\widehat{\widetilde{B}}_{r+1,r+1}\right)^j \left(-\widehat{\widetilde{d}y}_{r+2} - h\sum_{k\neq r+1}\widehat{\widetilde{B}}_{r+1,k}y_k\right).$$

$$(21)$$

Substituting expression (21) in equations (19) we obtain that the phase function $S(p,x)$ must satisfy one of the equations $p+\widetilde{\lambda}_j(-\partial S/\partial p, x, \partial S/\partial x), j = 1,3$, and the amplitudes will satisfy some transport equation with singular coefficients. Solutions of this transport equation will belong to the class (20).

Let $b^0_{(11)i,j}(x,\xi)$ be the elements of the matrix symbol $b^0_{11}(x,\xi)$, corresponding to the operator $B^0_{11}$ of equation (17). Explicit calculations give

$$\sigma = -ib^0_{(11)r+1,r+2}d/\{\lambda_2,\lambda_3\}(x,\nabla_x S(p,x)).$$

Let $\Lambda_j(t), j = 1,2,3$, be the manifolds $g^t_j(\Lambda_0), j = 1,2,3$, where $g^t_j$ are the displacements along the trajectories of the Hamiltonian systems with Hamiltonians $\lambda_j, j = 1,2,3$, respectively. Suppose that at some point $(x_0,\xi_0) \in \Lambda_0$ of the phase space $\mathbb{R}^{2n}_{x,\xi}$ the trajectory $g^t_j(x_0,\xi_0)$ crosses the manifold $\Sigma$ at the moment of time $t^j_0 = t(x_0,\xi_0)$. At this moment of time we consider the trajectories $g^{t-t^j_0}_k(x_j,\xi_j)$ proceeding from the point $(x_j,\xi_j) = g^{t^j_0}_j(x_0,\xi_0)$, for $k \neq j$. These new trajectories $g^{t-t^j_0}_k(x_j,\xi_j)$ are called generated trajectories. The family of generated trajectories

$$\left\{g^{t-t^j_0}_k(x_j,\xi_j) : (x_j,\xi_j) = g^{t^j_0}_j(x_0,\xi_0), (x_0,\xi_0) \in \Lambda_0, k \neq j\right\}$$

forms new Lagrangian manifolds $\Lambda_{kj} \subset \mathbb{R}^{2n}_{x,\xi}$.

Suppose that at the initial time the initial data vector has the form

$$\phi = \{0,...,0,1,(\lambda_3-\lambda_2)/d\}^T\Big|_{\Lambda_0}\phi_0(x)\exp\{ih^{-1}S_0\}.$$

It turns out that the wave front of the principal FAS term will be concentrated on the manifolds $\Lambda_3(t), \Lambda_{23}$ and $\Lambda_{13}$. Assume that

$$\Lambda_3(t), \Lambda_{23}(t), \Lambda_{13}(t) \subset \Omega_{x,\xi} \text{ for } 0 \leq t \leq T$$

and

$$\Lambda_0 \subset \Omega_{x,\xi}, \ \Lambda_0 \cap \Sigma = \varnothing.$$

Now we suppose that at the moment of time $\bar{t}$ it holds

$$\Lambda_3(\bar{t}) \cap \Sigma = \varnothing, \Lambda_{23}(\bar{t}) \cap \Sigma = \varnothing, \Lambda_{13}(\bar{t}) \cap \Sigma = \varnothing,$$

and the manifold $\Lambda_3(t)$ intersects with $\Sigma$ for $t$ such that $0 < \bar{t}_1 < t < \bar{t}$. To simplify our reasoning we can assume that the manifolds $\Lambda_3(t), \Lambda_{23}(t), \Lambda_{13}(t)$ have diffeomorphic projections $\pi_x$ of $\mathbb{R}^{2n}_{x,\xi}$ onto the plain $\xi = 0$, for all $t \in [0,\bar{t}]$. This signifies that there are no focal points. The general case can be investigated using the canonical Maslov operator.

**Theorem 0.1.** *There exists a FAS solution of problem (17) and it has the following form:*

*1) For $t \in [0, \overline{t_1})$ the FAS solution is the WKB asymptotic*

$$FAS_N(x, t, h) = C(x, t) \{0, ..., 0, 1, (\lambda_3 - \lambda_2)/d\}\big|_{\Lambda_3(t)}^T \times$$

$$\exp\left(\frac{i}{h} S_3(x, t)\right)(1 + O(h)); \quad \partial_t^\alpha C(x, t) \in C_0^\infty(\mathbb{R}^n), 0 \le \alpha \le \infty, \quad (22)$$

*where $S_3$ is the phase function on the manifold $\Lambda_3(t) = g_3^t \circ \Lambda_0$.*

*2) Suppose that $\mathrm{Re}\sigma(x, \xi)\big|_{\Omega_{x,\xi}} - 1/2 < 0$. At the moments of time $t \ge \overline{t}$ FAS has the form*

$$FAS_N(x, t, h) = h^{-1/2 + \sigma \circ g_2^{t - t_0^3(x_1, \xi_1)}} \widetilde{C}(x, t) \{0, ..., 0, 1, 0\}^T$$
$$\times \exp\left(\frac{i}{h} S_{23}(x, t)\right)(1 + o(1)), \quad (23)$$

*where $S_{23}$ is the phase function on the manifold $\Lambda_{23}$, $g^{t - t_0^3(x_1, \xi_1)}(x_1, \xi_1) = (x, \partial S_{23}/\partial x)$, and $o(1) \longrightarrow 0$ as $h \longrightarrow 0$.*

*There exists a function $\phi_0(x) \in C_0^\infty(\pi_x(\Omega_{x,\xi}))$ such that*

$$\int_{\mathbb{R}^n} |FAS_N(x, t, h)|^2 \, dx \ge h^{-\delta}$$

*for $t \ge \overline{t}$ uniformly with respect to $h \in (0, 1]$, however*

$$\beta > \int_{\mathbb{R}^n} |FAS_N(x, 0, h)|^2 \, dx \ge \alpha,$$

*for $0 < \alpha, \beta < \infty$.*

*3) The residual terms in formulas (22),(23) are $C_0^\infty$-functions that satisfy the estimates*

$$\left|\partial_t^\beta \partial_x^\alpha O(h^l)\right| \le C_{l,\beta,\alpha} h^{l - |\beta| - |\alpha|};$$

*4) $FAS_N$ exists for all $N < \infty$ and satisfy equation (17) with an accuracy up to $O(h^N)$ $(N < \infty)$.*

## References

1. V. Ivrii and V. Petkov, *The necessary conditions of correctness of Cauchy problem for nonstrictly hyperbolic equations*, Uspehi Matematicheskih Nauk (**179**) v.XXIX, N5, pp 3-70, 1974
2. W. Wazov, *Linear Turning Point Theory*, John Wiley and Sons, New York, 1985
3. V.V. Kucherenko and Yu. Osipov, *The Cauchy problem for nonstrictly hyperbolic equations*, Math USSR Sbornik, v. 48, N1, pp 81-109, 1984
4. V. Maslov and M. Fedoruk, *The quasiclassical approach to equations of quantum mechanics*, Nauka Moscow, 1976
5. R.Courant and D.Hilbert, *Method of Mathematical Physics VII*, John Wiley and Sons, New York, 1989
6. Kerson Huang, *Statistical Mechanics*, John Wiley and Sons, New York, 1987

520

7. V.L.Ginzburg,*The propagation of electromagnetic waves in plasma*, Nauka Moscow 1967

8. Dobrokhotov, S.Yu. and Shafarevich, A.I.,*Parametrix and asymptotic behavior of localized solutions of the Navier-Stokes equations in* $\mathbb{R}^3$ *that are linearized on a smooth flow*, Math. Notes, Vol. **51**, N.1-2, pp 47-54, 1992

9. Dobrokhotov S. Yu. and Shafarevich A. I.,*Asymptotic solutions of linearized navier-stokes equations*, Math. Notes, Vol. **53**, N. 1, pp 19-26, 1993

10. L.D.Landau and E.M.Lifshitz,*Hydrodynamics*, Nauka Moscow 1988

11. M.A.Leontovich,*Introduction in thermodynamics. Statistical physics.* Nauka Moscow 1983.

12. Karen Yagdjian, *The Cauchy problem for hyperbolic operator*, Akademie Verlag, Berlin, 1997.

# THE FUNDAMENTAL SOLUTION FOR ONE CLASS OF DEGENERATE ELLIPTIC EQUATIONS

M.S. SALAKHITDINOV and A. HASANOV

*Institute of Mathematics, Uzbek Academy of Sciences,*
*Tashkent-700125, F.Hodjaev str. 29, Uzbekistan*
*Email: mathinst@uzsci.net; anvarhasanov@yahoo.com*

Fundamental solutions are constructed for the equation

$$Lu = y^m z^k u_{xx} + x^n z^k u_{yy} + x^n y^m u_{zz} = 0, \ (m, n, k = \text{const} > 0),$$

in the first octant $x > 0$, $y > 0$, $z > 0$, by the aid of hypergeometric functions of Lauricella $F_A$. Further, elementary transformations lead to a decomposition of hypergeometric functions of Lauricella $F_A$ as a product of simple hypergeometric Gauss functions. On the base of the obtained expansions it is proved that the constructed fundamental solutions have an asymptotic behavior of order $\frac{1}{r}$ as $r \to 0$.

**Key words:** Fundamental solution, degenerate elliptic equation, hypergeometric system of equations, Appell functions, Lauricella functions, doubly hypergeometric functions, integral representations

**Mathematics Subject Classification:** Primary 35M10; Secondary 33C20, 33C65

## 1. Introduction

A lot of problems connected with applied questions lead to the study of boundary value problems for degenerate partial differential equations of second order; for example, the problem of adiabatic plane parallel without whirlwind gas flows; the problem of flow of supersonic string from the vessel with plane walls; the problem of supersonic pail of torrent into the wedge when the zone between the main wave and sonic speed is being formed until the sonic speed; and finally problems near sonic gas flows [1].

Basic researches for degenerate elliptic equations with one line of degeneration, that is, for the model equation

$$y^m u_{xx} + u_{yy} = 0, \ m = \text{const} > 0, \tag{1.1}$$

can be found in works of F. Tricomi [2], E. Holmgren [3], S. Gellerstedt [4,5] P. Germain, R. Bader [6]. For the equation (1.1) the following fundamental solutions were constructed:

$$q_1(x, y; x_0, y_0) = k_1 \left(r_1^2\right)^{-\beta} F(\beta, \beta; 2\beta; 1 - \sigma), \tag{1.2}$$

$$q_2\left(x, y; x_0, y_0\right) = k_2 \left(r_1^2\right)^{-\beta} (1 - \sigma)^{1-2\beta} F\left(1 - \beta, 1 - \beta; 2 - 2\beta; 1 - \sigma\right), \qquad (1.3)$$

where

$$\left.\begin{array}{c} r^2 \\ r_1^2 \end{array}\right\} = (x - x_0)^2 + \frac{4}{(m+2)^2} \left(y^{\frac{m+2}{2}} \begin{array}{c} - \\ + \end{array} y_0^{\frac{m+2}{2}}\right)^2, \quad \sigma = \frac{r^2}{r_1^2}, \quad \beta = \frac{m}{2(m+2)}, \qquad (1.4)$$

$F\left(a, b; c; z\right)$ – is the hypergeometric function of Gauss [9,10], that is,

$$F\left(a, b; c; z\right) = \sum_{n=0}^{\infty} \frac{(a)_n\, (b)_n}{(c)_n\, n!} z^n,$$

$$F\left(a, b; c; z\right) = \frac{\Gamma\left(c\right)}{\Gamma\left(b\right)\Gamma\left(c - b\right)} \int_0^1 t^{b-1} (1 - t)^{c-b-1} (1 - tz)^{-a}\, dt,$$

respectively, for $\Re c > \Re b > 0$.

In the paper [7] the elliptic equation with two lines of degeneration of the form

$$y^m u_{xx} + x^n u_{yy} = 0, \; m, n = \text{const} > 0 \qquad (1.5)$$

was investigated and the following fundamental solutions were constructed:

$$q_1\left(x, y; x_0, y_0\right) = k_1 \left(r^2\right)^{-\alpha-\beta} F_2\left(\alpha + \beta; \alpha, \beta; 2\alpha, 2\beta; \sigma_1, \sigma_2\right), \qquad (1.6)$$

$$q_2\left(x, y; x_0, y_0\right) = k_2 \left(r^2\right)^{-\alpha-\beta} \sigma_1^{1-2\alpha} F_2\left(1 - \alpha + \beta; 1 - \alpha, \beta; 2 - 2\alpha, 2\beta; \sigma_1, \sigma_2\right), \qquad (1.7)$$

$$q_3\left(x, y; x_0, y_0\right) = k_3 \left(r^2\right)^{-\alpha-\beta} \sigma_2^{1-2\beta} F_2\left(1 + \alpha - \beta; \alpha, 1 - \beta; 2\alpha, 2 - 2\beta; \sigma_1, \sigma_2\right), \qquad (1.8)$$

$$q_4\left(x, y; x_0, y_0\right) = k_4 \left(r^2\right)^{-\alpha-\beta} \sigma_1^{1-2\alpha} \sigma_2^{1-2\beta}$$
$$\times F_2\left(2 - \alpha - \beta; 1 - \alpha, 1 - \beta; 2 - 2\alpha, 2 - 2\beta; \sigma_1, \sigma_2\right), \qquad (1.9)$$

where

$$\left.\begin{array}{c} r^2 \\ r_1^2 \\ r_2^2 \end{array}\right\} = \left(\frac{1}{q}x^q + \frac{1}{q}x_0^q\right)^2 \begin{array}{c} \\ - \\ \end{array} + \left(\frac{1}{p}y^p + \frac{1}{p}y_0^p\right)^2 \begin{array}{c} \\ - \\ \end{array}, \quad \sigma_1 = \frac{r^2 - r_1^2}{r^2}, \quad \sigma_2 = \frac{r^2 - r_2^2}{r^2}, \qquad (1.10)$$

$$\alpha = \frac{n}{2(n+2)}, \quad \beta = \frac{m}{2(m+2)}, \quad q = \frac{n+2}{2}, p = \frac{m+2}{2}. \qquad (1.11)$$

Here $F_2(a; b, b'; c, c'; x, y)$ is the hypergeometric function of Appell [9,10], that is,

$$F_2(a; b_1, b_2; c_1, c_2; x, y) = \sum_{n,m=0}^{\infty} \frac{(a)_{m+n} (b_1)_m (b_2)_n}{(c_1)_m (c_2)_n \, m! n!} x^m y^n, \qquad (1.12)$$

which has the integral representation

$$F_2(a; b_1, b_2; c_1, c_2; x, y) = \frac{\Gamma(c_1) \Gamma(c_2)}{\Gamma(b_1) \Gamma(b_2) \Gamma(c_1 - b_1) \Gamma(c_2 - b_2)}$$

$$\times \int_0^1 \int_0^1 \xi^{b_1 - 1} \eta^{b_2 - 1} (1 - \xi)^{c_1 - b_1 - 1} (1 - \eta)^{c_2 - b_2 - 1} (1 - x\xi - y\eta)^{-a} \, d\xi d\eta, \qquad (1.13)$$

for $\Re b_1 > 0$, $\Re b_2 > 0$, $\Re(c_1 - b_1) > 0$, $\Re(c_2 - b_2) > 0$.

With the aid of the constructed fundamental solutions some boundary value problems for the equation (1.5) were solved.

## 2. Fundamental solutions

Let us consider the equation

$$Lu = y^m z^k u_{xx} + x^n z^k u_{yy} + x^n y^m u_{zz} = 0, \quad (m, n, k = const > 0), \qquad (2.1)$$

in the first octant $x > 0, y > 0, z > 0$.

We search solutions of the equation (2.1) as

$$u = (r^2)^{-\alpha - \beta - \gamma - \frac{1}{2}} \omega(\xi, \eta, \zeta), \qquad (2.2)$$

where

$$\left. \begin{array}{c} r^2 \\ r_1^2 \\ r_2^2 \\ r_3^2 \end{array} \right\} = \left( \frac{1}{q} x^q \begin{array}{c} - \\ + \\ - \\ - \end{array} \frac{1}{q} x_0^q \right)^2 + \left( \frac{1}{p} y^p \begin{array}{c} - \\ - \\ + \\ - \end{array} \frac{1}{p} y_0^p \right)^2 + \left( \frac{1}{l} z^l \begin{array}{c} - \\ - \\ - \\ + \end{array} \frac{1}{l} z_0^l \right)^2, \qquad (2.3)$$

$$\xi = \frac{r^2 - r_1^2}{r^2}, \, \eta = \frac{r^2 - r_2^2}{r^2}, \, \zeta = \frac{r^2 - r_3^2}{r^2}, \qquad (2.4)$$

$$\alpha = \frac{n}{2(n+2)}, \, \beta = \frac{m}{2(m+2)}, \, \gamma = \frac{k}{2(k+2)}, \qquad (2.5)$$

$$q = \frac{n+2}{2}, \, p = \frac{m+2}{2}, \, l = \frac{k+2}{2}. \qquad (2.6)$$

Substituting (2.2) into the equation (2.1) we have

$$
\begin{cases}
\xi(1-\xi)\omega_{\xi\xi} - \xi\eta\omega_{\xi\eta} - \xi\zeta\omega_{\xi\zeta} + \left[2\alpha - \left(2\alpha+\beta+\gamma+\tfrac{3}{2}\right)\xi\right]\omega_\xi - \alpha\eta\omega_\eta - \alpha\zeta\omega_\zeta \\
\qquad\qquad\qquad\qquad\qquad\qquad\qquad\qquad\qquad - \left(\alpha+\beta+\gamma+\tfrac{1}{2}\right)\alpha\omega = 0, \\
\eta\left(1-\eta\right)\omega_{\eta\eta} - \xi\eta\omega_{\xi\eta} - \eta\zeta\omega_{\eta\zeta} + \left[2\beta - \left(\alpha+2\beta+\gamma+\tfrac{3}{2}\right)\eta\right]\omega_\eta - \beta\xi\omega_\xi - \beta\zeta\omega_\zeta \\
\qquad\qquad\qquad\qquad\qquad\qquad\qquad\qquad\qquad - \left(\alpha+\beta+\gamma+\tfrac{1}{2}\right)\beta\omega = 0, \\
\zeta\left(1-\zeta\right)\omega_{\zeta\zeta} - \xi\zeta\omega_{\xi\zeta} - \eta\zeta\omega_{\eta\zeta} + \left[2\gamma - \left(\alpha+\beta+2\gamma+\tfrac{3}{2}\right)\zeta\right]\omega_\zeta - \gamma\xi\omega_\xi - \gamma\eta\omega_\eta \\
\qquad\qquad\qquad\qquad\qquad\qquad\qquad\qquad\qquad - \left(\alpha+\beta+\gamma+\tfrac{1}{2}\right)\gamma\omega = 0.
\end{cases}
$$

$$(2.7)$$

The system of equations (2.7) has the following solutions [9]:

$$
\omega_1\left(\xi,\eta,\zeta\right) = F_A\left(\alpha+\beta+\gamma+\frac{1}{2};\alpha,\beta,\gamma;2\alpha,2\beta,2\gamma;\xi,\eta,\zeta\right), \qquad (2.8)
$$

$$
\omega_2\left(\xi,\eta,\zeta\right) = F_A\left(-\alpha+\beta+\gamma+\frac{3}{2};1-\alpha,\beta,\gamma;2-2\alpha,2\beta,2\gamma;\xi,\eta,\zeta\right)\xi^{1-2\alpha}, \quad (2.9)
$$

$$
\omega_3\left(\xi,\eta,\zeta\right) = F_A\left(\alpha-\beta+\gamma+\frac{3}{2};\alpha,1-\beta,\gamma;2\alpha,2-2\beta,2\gamma;\xi,\eta,\zeta\right)\eta^{1-2\beta}, \quad (2.10)
$$

$$
\omega_4\left(\xi,\eta,\zeta\right) = F_A\left(\alpha+\beta-\gamma+\frac{3}{2};\alpha,\beta,1-\gamma;2\alpha,2\beta,2-2\gamma;\xi,\eta,\zeta\right)\zeta^{1-2\gamma}, \quad (2.11)
$$

$$
\omega_5\left(\xi,\eta,\zeta\right) = F_A\left(-\alpha-\beta+\gamma+\frac{5}{2};1-\alpha,1-\beta,\gamma;2-2\alpha,2-2\beta,2\gamma;\xi,\eta,\zeta\right)
$$
$$
\times\ \xi^{1-2\alpha}\eta^{1-2\beta}, \qquad (2.12)
$$

$$
\omega_6\left(\xi,\eta,\zeta\right) = F_A\left(-\alpha+\beta-\gamma+\frac{5}{2};1-\alpha,\beta,1-\gamma;2-2\alpha,2\beta,2-2\gamma;\xi,\eta,\zeta\right)
$$
$$
\times\ \xi^{1-2\alpha}\zeta^{1-2\gamma}, \qquad (2.13)
$$

$$
\omega_7\left(\xi,\eta,\zeta\right) = F_A\left(\alpha-\beta-\gamma+\frac{5}{2};\alpha,1-\beta,1-\gamma;2\alpha,2-2\beta,2-2\gamma;\xi,\eta,\zeta\right)
$$
$$
\times\ \eta^{1-2\beta}\zeta^{1-2\gamma}, \qquad (2.14)
$$

$$
\omega_8\left(\xi,\eta,\zeta\right) = F_A\left(-\alpha-\beta-\gamma+\frac{7}{2};1-\alpha,1-\beta,1-\gamma;2-2\alpha,2-2\beta,2-2\gamma;\xi,\eta,\zeta\right)
$$
$$
\times\ \xi^{1-2\alpha}\eta^{1-2\beta}\zeta^{1-2\lambda}. \qquad (2.15)
$$

Taking account of the hypergeometric functions (2.8) to (2.15) we get the following fundamental solutions for (2.2):

$$
q_1\left(x,y,z;x_0,y_0,z_0\right)
$$
$$
= k_1\left(r^2\right)^{-\alpha-\beta-\gamma-\frac{1}{2}}F_A\left(\alpha+\beta+\gamma+\tfrac{1}{2};\alpha,\beta,\gamma;2\alpha,2\beta,2\gamma;\xi,\eta,\zeta\right), \qquad (2.16)
$$

525

$$q_2\left(x, y, z; x_0, y_0, z_0\right)$$
$$= k_2 \xi^{1-2\alpha} \left(r^2\right)^{-\alpha-\beta-\gamma-\frac{1}{2}} F_A\left(-\alpha+\beta+\gamma+\tfrac{3}{2}; 1-\alpha, \beta, \gamma; 2-2\alpha, 2\beta, 2\gamma; \xi, \eta, \zeta\right),$$
$$(2.17)$$

$$q_3\left(x, y, z; x_0, y_0, z_0\right)$$
$$= k_3 \eta^{1-2\beta} \left(r^2\right)^{-\alpha-\beta-\gamma-\frac{1}{2}} F_A\left(\alpha-\beta+\gamma+\tfrac{3}{2}; \alpha, 1-\beta, \gamma; 2\alpha, 2-2\beta, 2\gamma; \xi, \eta, \zeta\right),$$
$$(2.18)$$

$$q_4\left(x, y, z; x_0, y_0, z_0\right)$$
$$= k_4 \zeta^{1-2\gamma} \left(r^2\right)^{-\alpha-\beta-\gamma-\frac{1}{2}} F_A\left(\alpha+\beta-\gamma+\tfrac{3}{2}; \alpha, \beta, 1-\gamma; 2\alpha, 2\beta, 2-2\gamma; \xi, \eta, \zeta\right),$$
$$(2.19)$$

$$q_5\left(x, y, z; x_0, y_0, z_0\right) = k_5 \xi^{1-2\alpha} \eta^{1-2\beta} \left(r^2\right)^{-\alpha-\beta-\gamma-\frac{1}{2}}$$
$$\times F_A\left(-\alpha-\beta+\gamma+\tfrac{5}{2}; 1-\alpha, 1-\beta, \gamma; 2-2\alpha, 2-2\beta, 2\gamma; \xi, \eta, \zeta\right),$$
$$(2.20)$$

$$q_6\left(x, y, z; x_0, y_0, z_0\right) = k_6 \xi^{1-2\alpha} \zeta^{1-2\gamma} \left(r^2\right)^{-\alpha-\beta-\gamma-\frac{1}{2}}$$
$$\times F_A\left(-\alpha+\beta-\gamma+\tfrac{5}{2}; 1-\alpha, \beta, 1-\gamma; 2-2\alpha, 2\beta, 2-2\gamma; \xi, \eta, \zeta\right),$$
$$(2.21)$$

$$q_7\left(x, y, z; x_0, y_0, z_0\right) = k_7 \eta^{1-2\beta} \zeta^{1-2\gamma} \left(r^2\right)^{-\alpha-\beta-\gamma-\frac{1}{2}}$$
$$\times F_A\left(\alpha-\beta-\gamma+\tfrac{5}{2}; \alpha, 1-\beta, 1-\gamma; 2\alpha, 2-2\beta, 2-2\gamma; \xi, \eta, \zeta\right),$$
$$(2.22)$$

$$q_8\left(x, y, z; x_0, y_0, z_0\right) = k_8 \xi^{1-2\alpha} \eta^{1-2\beta} \zeta^{1-2\gamma} \left(r^2\right)^{-\alpha-\beta-\gamma-\frac{1}{2}}$$
$$\times F_A\left(-\alpha-\beta-\gamma+\tfrac{7}{2}; 1-\alpha, 1-\beta, 1-\gamma; 2-2\alpha, 2-2\beta, 2-2\gamma; \xi, \eta, \zeta\right),$$
$$(2.23)$$

where the constants $k_i$, $i = 1, \cdots, 8$, will be determined after solving appropriate boundary problems. Here $F_A$ is the hypergeometric function of Lauricella [9], that is,

$$F_A\left(\alpha; \beta_1, \beta_2, \beta_3; \gamma_1, \gamma_2, \gamma_3; x, y, z\right) = \sum_{m,n,p=0}^{\infty} \frac{(\alpha)_{m+n+p} (\beta_1)_m (\beta_2)_n (\beta_3)_p}{(\gamma_1)_m (\gamma_2)_n (\gamma_3)_p \, m!n!p!} x^m y^n z^p.$$

The hypergeometric function of Lauricella has the integral representation

$$F_A\left(\alpha; \beta_1, \beta_2, \beta_3; \gamma_1, \gamma_2, \gamma_3; x, y, z\right)$$
$$= \frac{\Gamma(\gamma_1)\Gamma(\gamma_2)\Gamma(\gamma_3)}{\Gamma(\beta_1)\Gamma(\beta_2)\Gamma(\beta_3)\Gamma(\gamma_1-\beta_1)\Gamma(\gamma_1-\beta_1)\Gamma(\gamma_2-\beta_2)\Gamma(\gamma_3-\beta_3)}$$
$$\times \int_0^1\int_0^1\int_0^1 \xi^{\beta_1-1}\eta^{\beta_2-1}\zeta^{\beta_3-1}(1-\xi)^{\gamma_1-\beta_1-1}(1-\eta)^{\gamma_2-\beta_2-1}(1-\zeta)^{\gamma_3-\beta_3-1}$$
$$\times (1-x\xi-y\eta-z\zeta)^{-\alpha}\, d\xi d\eta d\zeta,$$
$$(2.24)$$

$\Re\beta_i > 0$, $\Re(\gamma_i - \beta_i) > 0$, $i = 1, 2, 3$.

The constructed fundamental solutions (2.16) to (2.23) have the following properties:

$$\frac{\partial}{\partial x}q_1|_{x=0} = \frac{\partial}{\partial y}q_1|_{y=0} = \frac{\partial}{\partial z}q_1|_{z=0} = 0, \qquad q_2|_{x=0} = \frac{\partial}{\partial y}q_2|_{y=0} = \frac{\partial}{\partial z}q_2|_{z=0} = 0,$$

$$\frac{\partial}{\partial x}q_3|_{x=0} = q_3|_{y=0} = \frac{\partial}{\partial z}q_3|_{z=0} = 0, \qquad \frac{\partial}{\partial x}q_4|_{x=0} = \frac{\partial}{\partial y}q_4|_{y=0} = q_4|_{z=0} = 0,$$

$$q_5|_{x=0} = q_5|_{y=0} = \frac{\partial}{\partial z}q_5|_{z=0} = 0, \qquad q_6|_{x=0} = \frac{\partial}{\partial y}q_6|_{y=0} = q_6|_{z=0} = 0,$$

$$\frac{\partial}{\partial x}q_7|_{x=0} = q_7|_{y=0} = q_7|_{z=0} = 0, \qquad q_8|_{x=0} = q_8|_{y=0} = q_8|_{z=0} = 0.$$

These properties are used during the study of boundary problems of Dirichlet or Neumann type for the equation (2.1) in bounded or in infinite domains.

## 3. Decomposition of hypergeometric functions of Lauricella $F_A$

We show, that the hypergeometric function of Lauricella has the following decomposition:

$$F_A\left(\alpha; \beta_1, \beta_2, \beta_3; \gamma_1, \gamma_2, \gamma_3; x, y, z\right)$$
$$= \sum_{i,j,k=0}^{\infty} \frac{(\alpha)_{i+j+k}\,(\beta_1)_{i+j}\,(\beta_2)_{i+k}\,(\beta_3)_{j+k}}{(\gamma_1)_{i+j}\,(\gamma_2)_{i+k}\,(\gamma_3)_{j+k}\,i!j!k!} x^{i+j}y^{i+k}z^{j+k}$$
$$\times F\left(\alpha+i+j, \beta_1+i+j; \gamma_1+i+j; x\right) F\left(\alpha+i+j+k, \beta_2+i+k; \gamma_2+i+k; y\right)$$
$$\times F\left(\alpha+i+j+k, \beta_3+j+k; \gamma_3+j+k; z\right).$$

$$(3.1)$$

**Proof.** Using the following equality

$$(1 - x\xi - y\eta - z\zeta)^{-\alpha}$$

$$= (1 - x\xi + x\xi y\eta - y\eta - z\zeta)^{-\alpha}\left[1 - \frac{xy\xi\eta}{1 - x\xi + x\xi y\eta - y\eta - z\zeta}\right]^{-\alpha},$$

$$\left[1 - \frac{xy\xi\eta}{1 - x\xi + x\xi y\eta - y\eta - z\zeta}\right]^{-\alpha} = \sum_{i=0}^{\infty} \frac{(\alpha)_i}{i!}\left[\frac{xy\xi\eta}{1 - x\xi + x\xi y\eta - y\eta - z\zeta}\right]^i,$$

we obtain

$$(1 - x\xi - y\eta - z\zeta)^{-\alpha} = \sum_{i=0}^{\infty} \frac{(\alpha)_i}{i!}(xy\xi\eta)^i (1 - x\xi + x\xi y\eta - y\eta - z\zeta)^{-\alpha-i}. \qquad (3.2)$$

Similarly,

$$(1 - x\xi + x\xi y\eta - y\eta - z\zeta)^{-\alpha-i}$$

$$= (1 - x\xi)^{-\alpha-i} (1 - y\eta - z\zeta)^{-\alpha-i} \left[ 1 - \frac{xz\xi\zeta}{(1 - x\xi)(1 - y\eta - z\zeta)} \right]^{-\alpha-i},$$

$$\left[ 1 - \frac{xz\xi\zeta}{(1 - x\xi)(1 - y\eta - z\zeta)} \right]^{-\alpha-i} = \sum_{j=0}^{\infty} \frac{(\alpha+i)_j}{j!} \left[ \frac{xz\xi\zeta}{(1 - x\xi)(1 - y\eta - z\zeta)} \right]^{j}$$

we find

$$(1 - x\xi + x\xi y\eta - y\eta - z\zeta)^{-\alpha-i}$$

$$= \sum_{j=0}^{\infty} \frac{(\alpha+i)_j}{j!} (xz\xi\zeta)^j (1 - x\xi)^{-\alpha-i-j} (1 - y\eta - z\zeta)^{-\alpha-i-j}. \qquad (3.3)$$

Substituting (3.3) into (3.2) we have

$$(1 - x\xi - y\eta - z\zeta)^{-\alpha}$$

$$= \sum_{i,j=0}^{\infty} \frac{(\alpha)_{i+j}}{i!j!} (xy\xi\eta)^i (xz\xi\zeta)^j (1 - x\xi)^{-\alpha-i-j} (1 - y\eta - z\zeta)^{-\alpha-i-j}. \qquad (3.4)$$

It is not difficult to see that the following identities are valid:

$$(1 - y\eta - z\zeta)^{-\alpha-i-j}$$

$$= (1 - y\eta)^{-\alpha-i-j} (1 - z\zeta)^{-\alpha-i-j} \left[ 1 - \frac{yz\eta\zeta}{(1 - y\eta)(1 - z\zeta)} \right]^{-\alpha-i-j},$$

$$\left[ 1 - \frac{yz\eta\zeta}{(1 - y\eta)(1 - z\zeta)} \right]^{-\alpha-i-j} = \sum_{k=0}^{\infty} \frac{(\alpha+i+j)_k}{k!} \left[ \frac{yz\eta\zeta}{(1 - y\eta)(1 - z\zeta)} \right]^{k},$$

$$(1 - y\eta - z\zeta)^{-\alpha-i-j}$$

$$= \sum_{k=0}^{\infty} \frac{(\alpha+i+j)_k}{k!} (yz\eta\zeta)^k (1 - y\eta)^{-\alpha-i-j-k} (1 - z\zeta)^{-\alpha-i-j-k}. \qquad (3.5)$$

Thus, from equalities (3.4) and (3.5) we can deduce

$$(1 - x\xi - y\eta - z\zeta)^{-\alpha} = \sum_{i,j,k=0}^{\infty} \frac{(\alpha)_{i+j+k}}{i!j!k!} x^{i+j} y^{i+k} z^{j+k} \xi^{i+j} \eta^{i+k} \zeta^{j+k}$$

$$\times (1 - x\xi)^{-\alpha-i-j} (1 - y\eta)^{-\alpha-i-j-k} (1 - z\zeta)^{-\alpha-i-j-k}. \qquad (3.6)$$

Substituting the identity (3.6) into the integral representation (2.24) and changing the order of summation and integration, we find

$$
F_A\left(\alpha; \beta_1, \beta_2, \beta_3; \gamma_1, \gamma_2, \gamma_3; x, y, z\right)
$$
$$
= \sum_{i,j,k=0}^{\infty} \frac{(\alpha)_{i+j+k}}{i!j!k!} x^{i+j} y^{i+k} z^{j+k}
$$
$$
\times \frac{\Gamma(\gamma_1)}{\Gamma(\beta_1)\Gamma(\gamma_1-\beta_1)} \int_0^1 \xi^{\beta_1-1+i+j}\left(1-\xi\right)^{\gamma_1-\beta_1-1}\left(1-x\xi\right)^{-\alpha-i-j} d\xi \qquad (3.7)
$$
$$
\times \frac{\Gamma(\gamma_2)}{\Gamma(\beta_2)\Gamma(\gamma_2-\beta_2)} \int_0^1 \eta^{\beta_2-1+i+k}\left(1-\eta\right)^{\gamma_2-\beta_2-1}\left(1-y\eta\right)^{-\alpha-i-j-k} d\eta
$$
$$
\times \frac{\Gamma(\gamma_3)}{\Gamma(\beta_3)\Gamma(\gamma_3-\beta_3)} \int_0^1 \zeta^{\beta_3-1+j+k}\left(1-\zeta\right)^{\gamma_3-\beta_3-1}\left(1-z\zeta\right)^{-\alpha-i-j-k} d\zeta.
$$

Taking into account the integral representation for the hypergeometric function of Gauss we get

$$
\frac{\Gamma(\gamma_1)}{\Gamma(\beta_1)\Gamma(\gamma_1-\beta_1)} \int_0^1 \xi^{\beta_1-1+i+j}\left(1-\xi\right)^{\gamma_1-\beta_1-1}\left(1-x\xi\right)^{-\alpha-i-j} d\xi
$$
$$
= \frac{(\beta_1)_{i+j}}{(\gamma_1)_{i+j}} F\left(\alpha+i+j, \beta_1+i+j; \gamma_1+i+j; x\right), \qquad (3.8)
$$

$$
\frac{\Gamma(\gamma_2)}{\Gamma(\beta_2)\Gamma(\gamma_2-\beta_2)} \int_0^1 \eta^{\beta_2-1+i+k}\left(1-\eta\right)^{\gamma_2-\beta_2-1}\left(1-y\eta\right)^{-\alpha-i-j-k} d\eta
$$
$$
= \frac{(\beta_2)_{i+k}}{(\gamma_2)_{i+k}} F\left(\alpha+i+j+k, \beta_2+i+k; \gamma_2+i+k; y\right), \qquad (3.9)
$$

$$
\frac{\Gamma(\gamma_3)}{\Gamma(\beta_3)\Gamma(\gamma_3-\beta_3)} \int_0^1 \zeta^{\beta_3-1+j+k}\left(1-\zeta\right)^{\gamma_3-\beta_3-1}\left(1-z\zeta\right)^{-\alpha-i-j-k} d\zeta
$$
$$
= \frac{(\beta_3)_{j+k}}{(\gamma_3)_{j+k}} F\left(\alpha+i+j+k, \beta_3+j+k; \gamma_3+j+k; z\right). \qquad (3.10)
$$

Substituting the identities (3.8) to (3.10) into (3.7), we finally get the decomposition (3.1). $\qquad \square$

We note that it is possible to prove the decomposition (3.1) by the method of comparison of coefficients in identity (3.1). Moreover, at $z = 0$ it is possible to get the known decomposition of the Appell function $F_2$, consisting of a product of hypergeometric functions of Gauss in one variable [11], that is,

$$
F_2\left(\alpha; \beta_1, \beta_2; \gamma_1, \gamma_2; x, y\right)
$$
$$
= \sum_{i=0}^{\infty} \frac{(\alpha)_i (\beta_1)_i (\beta_2)_i}{(\gamma_1)_i (\gamma_2)_i i!} x^i y^i F\left(\alpha+i, \beta_1+i; \gamma_1+i; x\right) F\left(\alpha+i, \beta_2+i; \gamma_2+i; y\right). \qquad (3.11)
$$

## 4. Properties of fundamental solutions

We shall prove that the obtained fundamental solutions (2.16) to (2.23) have the asymptotic behavior $\frac{1}{r}$ as $r \to 0$. We shall consider the fundamental solution

$q_1(x, y, z; x_0, y_0, z_0)$. Using (3.8) we can write $q_1(x, y, z; x_0, y_0, z_0)$ as

$$
q_1(x, y, z; x_0, y_0, z_0) = k_1(r^2)^{-\alpha-\beta-\gamma-\frac{1}{2}}
$$
$$
\times \sum_{i,j,k=0}^{\infty} \frac{(\alpha+\beta+\gamma+\frac{1}{2})_{i+j+k}\,(\alpha)_{i+j}\,(\beta)_{i+k}\,(\gamma)_{j+k}}{(2\alpha)_{i+j}\,(2\beta)_{i+k}\,(2\gamma)_{j+k}\,i!j!k!}\,\xi^{i+j}\eta^{i+k}\zeta^{j+k}
$$
$$
\times F\left(\alpha+\beta+\gamma+\tfrac{1}{2}+i+j, \alpha+i+j; 2\alpha+i+j; \xi\right) \tag{4.1}
$$
$$
\times F\left(\alpha+\beta+\gamma+\tfrac{1}{2}+i+j+k, \beta+i+k; 2\beta+i+k; \eta\right)
$$
$$
\times F\left(\alpha+\beta+\gamma+\tfrac{1}{2}+i+j+k, \gamma+j+k; 2\gamma+j+k; \zeta\right).
$$

Taking into account the formula of analytic continuation for hypergeometric functions of Gauss [10] we have

$$
F(a, b; c, x) = \frac{\Gamma(c)\Gamma(b-a)}{\Gamma(b)\Gamma(c-a)}(-x)^{-a} F\left(a, 1+a-c; 1+a-b; \tfrac{1}{x}\right)
$$
$$
+ \frac{\Gamma(c)\Gamma(a-b)}{\Gamma(a)\Gamma(c-b)}(-x)^{-b} F\left(b, 1+b-c; 1+b-a; \tfrac{1}{x}\right). \tag{4.2}
$$

At the large values of the argument the following equality is true:

$$
F\left(\alpha+\beta+\gamma+\tfrac{1}{2}+i+j, \alpha+i+j; 2\alpha+i+j; \tfrac{r^2-r_1^2}{r^2}\right)
$$
$$
= \lambda_1\left(-\tfrac{r^2-r_1^2}{r^2}\right)^{-\alpha-\beta-\gamma-\frac{1}{2}-i-j} + \lambda_2\left(-\tfrac{r^2-r_1^2}{r^2}\right)^{-\alpha-i-j} \tag{4.3}
$$
$$
+ O\left(\left(\tfrac{r^2-r_1^2}{r^2}\right)^{-\alpha-\beta-\gamma-\frac{3}{2}-i-j}\right) + O\left(\left(\tfrac{r^2-r_1^2}{r^2}\right)^{-\alpha-i-j-1}\right),
$$

$$
F\left(\alpha+\beta+\gamma+\tfrac{1}{2}+i+j+k, \beta+i+k; 2\beta+i+k; \tfrac{r^2-r_2^2}{r^2}\right)
$$
$$
= \lambda_3\left(-\tfrac{r^2-r_2^2}{r^2}\right)^{-\alpha-\beta-\gamma-\frac{1}{2}-i-j-k} + \lambda_4\left(-\tfrac{r^2-r_2^2}{r^2}\right)^{-\beta-i-k} \tag{4.4}
$$
$$
+ O\left(\left(\tfrac{r^2-r_2^2}{r^2}\right)^{-\alpha-\beta-\gamma-\frac{3}{2}-i-j-k}\right) + O\left(\left(\tfrac{r^2-r_2^2}{r^2}\right)^{-\beta-i-k-1}\right),
$$

$$
F\left(\alpha+\beta+\gamma+\tfrac{1}{2}+i+j+k, \gamma+j+k; 2\gamma+j+k; \tfrac{r^2-r_3^2}{r^2}\right)
$$
$$
= \lambda_5\left(-\tfrac{r^2-r_3^2}{r^2}\right)^{-\alpha-\beta-\gamma-\frac{1}{2}-i-j} + \lambda_6\left(-\tfrac{r^2-r_3^2}{r^2}\right)^{-\gamma-j-k} \tag{4.5}
$$
$$
+ O\left(\left(\tfrac{r^2-r_3^2}{r^2}\right)^{-\alpha-\beta-\gamma-\frac{3}{2}-i-j-k}\right) + O\left(\left(\tfrac{r^2-r_3^2}{r^2}\right)^{-\gamma-j-k-1}\right),
$$

where $\lambda_\tau, \tau = 1, \cdots, 6$ are known constants.

Substituting (4.3) to (4.5) into the equality (4.1), it is easy to show, that the fundamental solution of the equation (2.16) has the feature $\frac{1}{r}$ as $r \to 0$. By similar arguments one can prove, that the other fundamental solutions of equation (3.1) have the same asymptotic behavior $\frac{1}{r}$ as $r \to 0$.

## 5. Conclusion

In the conclusion we note, that it is possible to find fundamental solutions for the equation

$$y^m u_{xx} + x^n u_{yy} - \lambda^2 x^n y^m u = 0, \ m, n = \text{const} > 0, \tag{5.1}$$

where $\lambda$ is a real number. In this case fundamental solutions will be searched as

$$u = \left(r^2\right)^{-\alpha-\beta} \omega\left(\xi, \eta, \zeta\right),$$

where

$$\xi = \frac{r^2 - r_1^2}{r^2}, \ \eta = \frac{r^2 - r_2^2}{r^2}, \ \zeta = -\frac{\lambda^2 r^2}{4}, \ \alpha = \frac{n}{2(n+2)}, \ \beta = \frac{m}{2(m+2)}.$$

The functions $r^2$, $r_1^2$, $r_2^2$ are defined by equality (1.10). Similarly we get a system of equations concerning the unknown functions $\omega\left(\xi, \eta, \zeta\right)$.

This method is also applicable for elliptic differential equations of the form

$$L(u) \equiv \left[ \sum_{i=1}^{n} \prod_{\substack{j=1 \\ j \neq i}}^{n} x_j^{m_j} \frac{\partial^2}{\partial x_i^2} - \lambda^2 \prod_{i=1}^{n} x_i^{m_i} \right] u$$

$$\equiv x_2^{m_2} x_3^{m_3} \cdots x_n^{m_n} u_{x_1 x_1} + x_1^{m_1} x_3^{m_3} \cdots x_n^{m_n} u_{x_2 x_2} + \cdots + x_1^{m_1} x_2^{m_2} \cdots x_{n-1}^{m_{n-1}} u_{x_n x_n}$$

$$- \lambda^2 x_1^{m_1} x_2^{m_2} \cdots x_n^{m_n} u = 0. \tag{5.2}$$

Fundamental solutions of the equation (5.2) are searched as

$$u = \left(r^2\right)^{\frac{2-n}{2}-l} \omega\left(\xi_1, \xi_2, ..., \xi_n, \zeta\right),$$

where

$$\xi_i = \frac{r^2 - r_i^2}{r^2}, \ \zeta = -\frac{\lambda^2 r^2}{4}, \ l = \sum_{i=1}^{n} \beta_i, \ \beta_i = \frac{m_i}{2(m_i+2)}, \ i = 1, 2, \cdots, n$$

$$r^2 = \sum_{i=1}^{n} \left( \frac{2}{m_i+2} x_i^{\frac{m_i+2}{2}} - \frac{2}{m_i+2} x_{0i}^{\frac{m_i+2}{2}} \right)^2,$$

$$r_i^2 = \left( \frac{2}{m_i+2} x_i^{\frac{m_i+2}{2}} + \frac{2}{m_i+2} x_{0i}^{\frac{m_i+2}{2}} \right)^2 + \sum_{\substack{j=1 \\ j \neq i}}^{n} \left( \frac{2}{m_j+2} x_j^{\frac{m_j+2}{2}} - \frac{2}{m_j+2} x_{0j}^{\frac{m_j+2}{2}} \right)^2.$$

### Acknowledgment

We are grateful to Professor M. Reissig for his attention and suggestions during the preparation of this note.

# References

1. Frankel F.I.*Selected works on gas dynamics*. Moscow, Nauka,1973.

2. Tricomi F. *Ancora sull'equazione* $yz_{xx} + z_{yy} = 0$. Rendiconti della Reale Accademia Nazionale dei Lincei VI, 1927.

3. Holmgren E. *Sur un probleme aux limates pour l'equation* $y^m z_{xx} + z_{yy} = 0$. Arkiv Mat. Astr., och. Fysik, 1926, 19B, 14.

4. Gellerstedt S. *Sur un probleme aux limites pour une equation lineaire aux derivees partielles du second orde de tihe mixte*. Thesis, Uppsala, 1935.

5. Gellerstedt S. *Sur un probleme aus limites pour l'equation* $y^{2s} z_{xx} + z_{yy} = 0$. Arkiv Mat., Astr., och Fysik, 1935, 25A, 10.

6. Germain P., Bader R. *Sur quelques problems relatifs a l'equation du type mixte de Tricomi*. Publ. ONERA. 1952, 56.

7. Hasanov A. *About a mixed problem for the equation* $\operatorname{sign} y \, |y|^m \, u_{xx} + x^n u_{yy} = 0$. Izv. AN UzSSR, ser. Fiz.-mat.nauk.1982, 2, 28-32.

8. Horn J. *Über die Konvergenz der hypergeometrischen Reihen zweier und dreier Veränderlicher*. Math. Ann. 34, 1889, 544-600.

9. Appell Payl, Kampe de Feriet M.J. *Functions hypergeometriques et hyperspheriques. Polynomes d'Hermite*. Gauthier - Villars. Paris, 1926.

10. Erdelyi, A. (Ed) *Higher transcendental functions*. vol. 1. New York: McGraw Hill Book. Co. 1953.

11. Burchnall J.L., Chaundy T.W. *Expansions of Appell's double hypergeometric functions*. 1940: Quart. J. Math. Oxford Ser. 11, 249-270

# ON THE SPECTRUM OF SCHRÖDINGER OPERTORS WITH OSCILLATING LONG-RANGE POTENTIALS

KIYOSHI MOCHIZUKI

*Department of Mathematics, Chuo University*
*Kasuga, Bunkyo-ku, Tokyo 112-8551, Japan*

In this paper we consider the Schrödinger operators with oscillating long-range potentials. We make an improvement of results of Jäger and Rejto on the growth estimates of generalized eigenfunctions and apply it to show the principle of limiting absorption.

**Key words:** Oscillating long-range potentials, growth estimats of eigenfunctions, principle of limiting absorption

**Mathematics Subject Classification:** 35J10, 35P10

## 1. Introduction

In [3] Jäger and Rejto studied the Schrödinger operators with short-range perturbations of von Neumann-Wigner potentials:

$$L = -\Delta + \frac{c \sin b|x|}{|x|} + V_3(x), \tag{1}$$

where $V_3(x)$, $x \in \mathbf{R}^3$, is a real valued function behaving like $O(|x|^{-1-\delta})$ ($\delta > 0$) as $|x| \to \infty$. They illustrated how to establish the principle of limiting absorption for this potential and for a compact subinterval $J$ of the positive, real axis $\mathbf{R}_+$ such that

$$\text{dist}\left\{J, \frac{b^2}{4}\right\} > \frac{1}{2}|bc|. \tag{2}$$

This work is closely related to the earlier one of Mochizuki-Uchiyama [6]. The principle of limiting absorption is proved in [6] for more general oscillating long-range potentials, and if the operator is restricted to (1), the condition on the interval $J$ becomes

$$\inf J > \frac{b^2}{4} + \frac{1}{\min\{2, 4\delta\}}|bc|. \tag{3}$$

Note that in [3] is not proved the principle of limiting absorption itself, but a most important ingredient of its proof is given. It is the so called uniqueness or growth estimate of the generalized eigenfunctions. As in the previous results (see eg., Kato [4], Eidus [1], Mochizuki-Uchiyama [7]), the growth estimates are proved by formulating a differential inequality for a functional of solutions. A new point of the proof of [3] (see also [2]) is that they adopted a functional which includes an approximate phase of the operator (1). Their approximate phase is the same one as is introduced in [6], where a similar functional identity is formulated to establish the principle of limiting absorption. But it is not used there to show the growth estimate of generalized eigenfunctions.

The purpose of this paper is to make a slight improvement of the result of [3], and apply it to show the principle of limiting absorption. Our theorems are stated under semi-abstract conditions on the potentials. If we adapt them for the operator (1), the growth estimate is obtained for the interval $J$ satisfying (2) and the principle of limiting absorption is established for $J$ satisfying

$$\text{dist}\left\{J, \frac{b^2}{4}\right\} > \frac{1}{\min\{2, 4\delta\}}|bc|. \tag{4}$$

In this paper we do not restrict ourselves to the 3 dimensional case.

Our theorems are also applicable to the operators

$$L = -\Delta + \frac{c\sin(\log|x|)}{\log|x|} + V_3(x), \tag{5}$$

where $V_3(x) = O(|x|^{-1}\{\log|x|\}^{-1-\delta})$ as $|x| \to \infty$. In this case the growth estimate holds for each $J \subset \mathbf{R}_+$, and the principle is obtained for $J$ satisfying

$$\inf J > \frac{1}{4\delta}|c|. \tag{6}$$

We should mention that condition (2) or (6) does not cover all the compact subintervals of $\mathbf{R}_+$, and the principle still remains unsolved for some intervals. As is already shown in [6], Example I-2, condition (6) can be replaced by "each $J \subset \mathbf{R}_+$" if $V_3(x) = O(|x|^{-1-\delta})$ in (5).

We formulate a functional identity of solutions to the Schrödinger equation (Proposition 1) by use of the weighted energy method. It slightly modifies the one used in [6], and the growth estimate of generalized eigenfunctions (Theorem 1) and the principle of limiting absorption (Theorem 2) are both proved based on this identity. The functional identity used in [3] is obtained by considering the equation as an operator-valued ordinary differential equation. So, the expression is apparently much different from ours.

## 2. Asymptotic phase and related functional identity

The Schrödinger operator $L$ on the function $u$ is given by

$$Lu = -\Delta u + V(x)u, \quad x \in \mathbf{R}^n,$$

where $-\Delta$ is the $n$-dimensional Laplacian and $V = V(x)$ is the potential. We require that $V(x)$ is a real-valued, locally $L^2$-function which goes to 0 as $r = |x| \to \infty$. In case $n \geq 4$ we restrict more the order of singularities of $V$. Then $L$ becomes a selfadjoint operator in $L^2(\mathbf{R}^n)$ with domain $D(L) = H^2(\mathbf{R}^n)$. In the following, we restrict ourselves to the potential

$$V(x) = V_1(r) + V_3(x).$$

We require

$(A1)$ $\qquad\qquad V_1(r) = O(1), \ V_1'(r) = \dfrac{dV_1}{dr}(r) = O(r^{-1}),$

$$|V_1''(r) + aV_1(r)| + |V_3(x)| \leq C\mu(r) \ \text{ for some } a \geq 0,$$

where $\mu$ is a positive smooth function of $r > 0$ such that

$$\mu(r) = o(r^{-1}) \text{ as } r \to \infty \text{ and } \mu(r) \in L^1((0,\infty)). \tag{7}$$

Without loss of generality we can assume $\mu(r) \geq C(1+r)^{-2}$.

**Remark** A similar treatment is possible to the potential

$$V(x) = V_1(r) + V_2(x) + V_3(x)$$

with long-range part $V_2(x)$ satisfying

$$r^{-1}|V_2(x)| + |\partial_r V_2(x)| \leq C\mu(r).$$

The following lemma is already proved in [6] (Lemmas 8.1, 8.2 and 8.3).

**Lemma 2.1.** *We put for $0 < \gamma \leq 2$,*

$$E^{\pm}(\gamma) = \limsup_{r \to \infty}\left[\pm\frac{1}{\gamma}\{rV_1'(r) + \gamma V_1(r)\}\right]. \tag{8}$$

*Then $E^{\pm}(\gamma)$ is nonincreasing and continuous in $\gamma$ and we have*

$$\limsup_{r \to \infty}\{\pm V_1(r)\} \leq E^{\pm}(2). \tag{9}$$

*Moreover, if $a > 0$ in (A1), then*

$$V_1(r) = O(r^{-1}) \text{ as } r \to \infty. \tag{10}$$

Now we choose $J = [\lambda_0, \lambda_1] \subset \mathbf{R}_+$ to satisfy

$$\lambda_0 > \frac{a}{4} + E^{+}(2) \text{ or } \lambda_1 < \frac{a}{4} - E^{-}(2). \tag{11}$$

Let $\zeta = \lambda \pm i\epsilon$ ($\lambda \in J$, $0 < \epsilon < \epsilon_0$) and $f \in L^2$, and let $u$ be a solution to the equation

$$-\Delta u + V(x)u - \zeta u = f(x), \ x \in \mathbf{R}^n. \tag{12}$$

For a complex-valued function $p = p(r, \zeta)$, we put $v = e^p u$. Then $v$ solves

$$-\Delta v + 2p'\tilde{x} \cdot \nabla v + \left(V(x) - \zeta + p'' + \frac{n-1}{r}p' - (p')^2\right)v = e^p f(x), \tag{13}$$

where $\tilde{x} = x/r$. We choose

$$p(r, \zeta) = \rho(r, \zeta) + \sigma(r),$$

where $\sigma(r)$ is a nonnegative smooth function and $\rho(x, \zeta)$ is an improper integral of approximate phase $\rho'$ of solution $u$, As is constructed in [6], approximate phase is characterized by the equation

$$q_1(r, \zeta) \equiv V_1(r) - \zeta + \rho'' + \frac{n-1}{r}\rho' - \rho'^2 = O(r^{-2}) \quad as \quad r \to \infty,$$

and is realized for large $r$ by the function

$$\rho'(r, \zeta) = -i\sqrt{\zeta - \eta V_1(r)} + \frac{n-1}{2r} - \frac{\eta V_1'(r)}{4(\zeta - \eta V_1(r))}, \tag{14}$$

where $\eta = \eta(\zeta) = 4\zeta/(4\zeta - a)$ and $\sqrt{z}$ denotes the branch of the square roots of $z \in \mathbf{C} \setminus [0, \infty)$ with $\mathrm{Im}\sqrt{z} > 0$. As for the details, see Appendix of[6].

Let $\psi = \psi(r)$ be a positive smooth function of $r > 0$. We multiply by $2\psi e^{-2\mathrm{Re}\rho}\tilde{x} \cdot \nabla\bar{v}$ on the both sides of (13), take the real part and integrate by parts over $B(R, t) = \{x; R < |x| < t\}$. Then we obtain

**Proposition 2.1.** *For each solution of (12) the following identity holds.*

$$-\left\{\int_{S(t)} - \int_{S(R)}\right\}\psi\left\{2|\tilde{x} \cdot \theta_\sigma|^2 - |\theta_\sigma|^2 + \sigma'^2|u_\sigma|^2\right\} dS + \int_{B(R,t)}$$

$$\times\psi\left\{\left(2\mathrm{Re}\rho' + \frac{\psi'}{\psi} - \frac{n-1}{r}\right)|\tilde{x} \cdot \theta_\sigma|^2 + \left(2\mathrm{Re}\rho' - \frac{\psi'}{\psi} - \frac{n-3}{r}\right)(|\theta_\sigma|^2 - |\tilde{x} \cdot \theta_\sigma|^2)\right.$$

$$+4\sigma'|\tilde{x} \cdot \theta_\sigma|^2 + 2\mathrm{Re}\left[\left(q_1 + V_3 + \sigma'' + \frac{n-1}{r}\sigma' - 2\rho'\sigma'\right)u_\sigma\tilde{x} \cdot \bar{\theta}_\sigma\right] + \left[2\sigma'\sigma''\right.$$

$$\left.+ \left(\frac{n-1}{r} + \frac{\psi'}{\psi} - 2\mathrm{Re}\rho'\right)\sigma'^2\right]|u_\sigma|^2\right\} dx = \int_{B(R,t)} 2\psi\mathrm{Re}\left[f_\sigma\tilde{x} \cdot \bar{\theta}_\sigma\right] dx,$$

*where $0 < R < t$, $S(r) = \{x; |x| = r\}$ and we have put $u_\sigma = e^\sigma u$, $f_\sigma = e^\sigma f$ and $\theta_\sigma = \nabla u_\sigma + \tilde{x}\rho' u_\sigma$.*

In the following let $K_\pm = \{\zeta = \lambda \pm i\epsilon; \lambda \in J, 0 < \epsilon \le \epsilon_0\}$, where $\epsilon_0 > 0$ is chosen sufficiently small, and let $\bar{K}_\pm$ be the closure of $K_\pm$. We write $\rho_\pm = \rho(r, \zeta)$ for each $\zeta \in \bar{K}_\pm$. Especially, we distinguish the two approximate phase $\rho'_+(r, \lambda)$ and $\rho'_-(r, \lambda)$ to be

$$\rho'_\pm(r, \lambda) = \rho'(r, \lambda \pm i0) = \lim_{\epsilon\downarrow 0} \rho'(r, \lambda \pm i\epsilon).$$

## 3. Growth properties of generalized eigenfunctions

In this section we consider the solution $u$ of the homogeneous equation

$$-\Delta u + V(x)u - \lambda u = 0, \quad \lambda \in J. \tag{15}$$

We shall show that the growth estimate of solutions given by Jäger and Rejto [3] is extended to the problem with conditions $(A1)$ and $J$ satisfying $(11)$.

Let us denote $\theta_{\sigma\pm} = \nabla u_\sigma + \tilde{x}\rho_\pm u_\sigma$. Then Proposition 2.1 also holds for this $u$ with $\rho = \rho_\pm$, $\theta_\sigma = \theta_{\sigma\pm}$ and $f_\sigma \equiv 0$.

By Lemma 2.1 and relation $(14)$ there exists $R_0 > 0$ such that for each $\lambda \in J$ and $r > R_0$,

$$\mathrm{Re}\rho'_\pm = \frac{n-1}{2r} - \frac{\eta V'_1}{4(\lambda - \eta V_1)}, \quad \mathrm{Im}\rho'_\pm = \mp\sqrt{\lambda - \eta V_1(r)},$$

where $\eta = 4\lambda/(4\lambda - a)$.

**Lemma 3.1.** *There exists $R_1 \geq R_0$ such that for $r \geq R_1$,*

$$\mathrm{Im}\left[\int_{S(r)} (\partial_r u_\sigma)\bar{u}_\sigma dS\right] = 0, \tag{16}$$

$$\int_{S(r)} |u_\sigma|^2 dS \leq C \int_{S(r)} |\tilde{x} \cdot \theta_{\sigma\pm}|^2 dS, \tag{17}$$

$$2\mathrm{Re}\rho'_\pm - \frac{n-2}{r} \geq \frac{C}{r} > 0. \tag{18}$$

**Theorem 3.1.** *Under $(A1)$ and $(11)$ let $u$ satisfy equation $(15)$. If $u$ does not have compact support, then*

$$\liminf_{r\to\infty} \int_{S(r)} |\partial_r u + \rho'_\pm u|^2 dS \neq 0. \tag{19}$$

We shall only show $(19)$ with $\rho_+$ writing $\rho = \rho_+$ and $\theta_\sigma = \theta_{\sigma+}$.
We use two functionals of solution $u$ to $(15)$:

$$F(r) = \int_{S(r)} (2|\tilde{x} \cdot \theta|^2 - |\theta|^2)dS,$$

where $\theta = \nabla u + \tilde{x}\rho' u$, and

$$F_{\sigma,\tau}(r) = \int_{S(r)} \left\{2|\tilde{x} \cdot \theta_\sigma|^2 - |\theta_\sigma|^2 + (\sigma'^2 - \tau)|u_\sigma|^2\right\} dS,$$

where $\tau = \tau(r)$ is another positive function.

*Proof of Theorem 3.1 under an additional assumption* In this part we prove Theorem 1 requiring an additional assumption that there exists a sequence $r_k \to \infty$ such that $F(r_k) > 0$.

Put $\sigma \equiv 0$, $\psi \equiv 1$ and $R \geq R_1$ in Proposition 2.1 with $f \equiv 0$. Then it follows that

$$F(t) - F(R) = \int_{B(R,t)} \left\{ \left( 2\mathrm{Re}\rho' - \frac{n-1}{r} \right) |\tilde{x} \cdot \theta|^2 \right.$$

$$\left. + \left( 2\mathrm{Re}\rho' - \frac{n-3}{r} \right) (|\theta|^2 - |\tilde{x} \cdot \theta|^2) + 2\mathrm{Re}\left[ (q_1 + V_3) u\tilde{x} \cdot \bar{\theta} \right] \right\} dx.$$

Differentiating both sides in $t$ and noting (17), (18) and the fact $|q_1(r, \lambda) + V_3(x)| \leq C\mu(r)$, we obtain

$$\frac{d}{dt} F(t) \geq \int_{S(t)} \left\{ \left( 2\mathrm{Re}\rho' - \frac{n-1}{r} - 2C\mu \right) (2|\tilde{x} \cdot \theta|^2 - |\theta|^2) \right. \tag{20}$$

$$+ 2 \left( 2\mathrm{Re}\rho' - \frac{n-2}{r} - C\mu \right) (|\theta|^2 - |\tilde{x} \cdot \theta|^2) \Bigg\} dS$$

$$\geq \left( 2\mathrm{Re}\rho' - \frac{n-1}{t} - 2C\mu \right) F(t)$$

in $t \geq R_2 (\geq R_1)$. Note that

$$2\mathrm{Re}\rho' - \frac{n-1}{r} = \frac{1}{2} \frac{d}{dr} \log(\lambda - \eta V_1).$$

Then since $\mu \in L^1([R_1, \infty))$, choosing $r_k \geq R_1$ and integrating (20) over $(r_k, t)$, we have

$$\frac{F(t)}{F(r_k)} \geq \left\{ \frac{\lambda - \eta V_1(t)}{\lambda - \eta V_1(r_k)} \right\}^{1/2} \exp \left\{ -2C \int_{r_k}^{\infty} \mu \, dr \right\},$$

which proves the uniform positivity near infinity of $F(t)$. $\qquad\square$

*Proof of Theorem 3.1 under the complementary assumption* In this part we prove Theorem 1 requiring the complementary assumption of the preceding proof. Namely we assume that $F(r) \leq 0$ for $r > R_0$ and does not have compact support.

We choose $\psi = r^2 \sqrt{\lambda - \eta V_1}$ in Proposition 1 with $f \equiv 0$. Then we have

$$\frac{\psi'}{\psi} = \frac{-\eta V_1'}{2(\lambda - \eta V_1)} + \frac{2}{r} = 2\mathrm{Re}\rho' - \frac{n-3}{r},$$

and it follows that

$$\left\{ \int_{S(t)} - \int_{S(R)} \right\} \psi \left\{ 2|\tilde{x} \cdot \theta_\sigma|^2 - |\theta_\sigma|^2 + \sigma'^2 |u_\sigma|^2 \right\} dS = \int_{B(R,t)} \tag{21}$$

$$\times \psi \left\{ 4 \left( \mathrm{Re}\rho' - \frac{n-2}{2r} + \sigma' \right) |\tilde{x} \cdot \theta_\sigma|^2 + 2\mathrm{Re}\left[ \left( q_1 + V_3 + \sigma'' + \frac{n-1}{r}\sigma' \right) \right. \right.$$

$$\left. \left. - 2\rho'\sigma' \right) u_\sigma \tilde{x} \cdot \bar{\theta}_\sigma \right] + \left( 2\sigma'\sigma'' + \frac{2}{r}\sigma'^2 \right) |u_\sigma|^2 \right\} dx.$$

Subtract the identity

$$\left\{ \int_{S(t)} - \int_{S(R)} \right\} \psi \tau |u_\sigma|^2 dS = \int_{B(R,t)} \psi \left\{ 2\mathrm{Re}\left[ \tau u_\sigma \tilde{x} \cdot \bar{\theta}_\sigma \right] + \left( \frac{2}{r}\tau + \tau' \right) |u_\sigma|^2 \right\} dx$$

from (21) and differentiate both sides in $t$. Then we have

$$\frac{d}{dt}\left[ \psi F_{\sigma,\tau}(t) \right] = \int_{S(t)} \psi \left\{ 4\left( \mathrm{Re}\rho' - \frac{n-2}{2r} \right) |\tilde{x} \cdot \theta_\sigma|^2 + 2\mathrm{Re}\left[ (q_1 + V_3)u_\sigma \tilde{x} \cdot \bar{\theta}_\sigma \right] \right.$$

$$+ 4\sigma'|\partial_r u_\sigma + \mathrm{Re}\rho' u_\sigma|^2 + 2\left( \sigma'' + \frac{n-1}{r}\sigma' - 2\sigma'\mathrm{Re}\rho' - \tau \right)$$

$$\left. \times \mathrm{Re}[u_\sigma(\partial_r \bar{u}_\sigma + \mathrm{Re}\rho' \bar{u}_\sigma)] + \left( 2\sigma'\sigma'' + \frac{2}{r}\sigma'^2 - \frac{2}{r}\tau - \tau' \right) |u_\sigma|^2 \right\} dS.$$

Here

$$\int_{S(t)} 2\mathrm{Re}[(q_1 + V_3)u_\sigma \tilde{x} \cdot \bar{\theta}_\sigma] dS \leq \int_{S(t)} C\mu |\tilde{x} \cdot \theta_\sigma|^2 dS.$$

Moreover, note the inequalities

$$2\left( \sigma'' + \frac{n-1}{r}\sigma' - 2\sigma'\mathrm{Re}\rho' \right) \mathrm{Re}[u_\sigma(\partial_r \bar{u}_\sigma + \mathrm{Re}\rho' \bar{u}_\sigma)]$$

$$\leq 2\sigma'|\partial_r u_\sigma + \mathrm{Re}\rho' u_\sigma|^2 + \frac{1}{2}\sigma'\left( \frac{\sigma''}{\sigma'} + \frac{n-1}{r} - 2\mathrm{Re}\rho' \right)^2 |u_\sigma|^2,$$

$$-2\tau \mathrm{Re}[u_\sigma(\partial_r \bar{u}_\sigma + \mathrm{Re}\rho' \bar{u}_\sigma)] \leq 2\sigma'|\partial_r u_\sigma + \mathrm{Re}\rho' u_\sigma|^2 + \frac{1}{2}\sigma'^{-1}\tau^2 |u_\sigma|^2.$$

Then

$$\frac{d}{dt}\left[ \psi F_{\sigma,\tau}(t) \right] \geq \int_{S(t)} \psi \left\{ 2\left( 2\mathrm{Re}\rho' - \frac{n-2}{r} - C\mu \right) |\tilde{x} \cdot \theta_\sigma|^2 \right.$$

$$- \left( \frac{1}{2}\sigma'^{-1}\tau^2 + \tau' + \frac{2}{r}\tau \right) |u_\sigma|^2$$

$$\left. - \frac{1}{2}\sigma'\left( \frac{\sigma''}{\sigma'} + \frac{n-1}{r} - 2\mathrm{Re}\rho' \right)^2 |u_\sigma|^2 + \left( 2\sigma'\sigma'' + \frac{2}{r}\sigma'^2 \right) |u_\sigma|^2 \right\} dS.$$

Now we choose $\sigma$ and $\tau$ with $m \geq 1$ and $1/3 < \epsilon < 1$ as follows:

$$\sigma(r) = \frac{m}{1-\epsilon} r^{1-\epsilon}, \quad \tau(r) = r^{-2\epsilon} \log r. \tag{22}$$

Then as $r \to \infty$,

$$\frac{1}{2}\sigma'^{-1}\tau^2 + \tau' + \frac{2}{r}\tau \leq \frac{1}{2m} r^{-3\epsilon}(\log r)^2 + 2r^{-1-2\epsilon}(\log r + 1) = o(r^{-1}),$$

$$\frac{1}{2}\sigma'\left(\frac{\sigma''}{\sigma'} + \frac{n-1}{r} - 2\mathrm{Re}\rho'\right)^2 = mO(r^{-2-\epsilon}),$$

$$2\sigma'\sigma'' + \frac{2}{r}\sigma'^2 = 2(1-\epsilon)m^2 r^{-1-2\epsilon} > 0.$$

Thus, using again (17), we see that

$$\frac{d}{dt}[\psi F_{\sigma,\tau}(t)] \geq \int_{S(t)} 2\psi\left(2\mathrm{Re}\rho' - \frac{n-2}{r} - o(r^{-1})\right)|\tilde{x}\cdot\theta_\sigma|^2 dS.$$

It then follows from (18) that $d[\psi F_{\sigma,\tau}]/dt \geq 0$ for any $m \geq 1$ in $t \geq R_3$ if $R_3 \geq R_2$ is chosen sufficiently large.

By assumption $\int_{S(R_4)}|u_\sigma|^2 dS > 0$. So, there exists $R_4 > R_3$ such that $u_\sigma(R_4) \neq 0$, and hence $F_{\sigma,\tau}(R_4) \to \infty$ as $m \to \infty$. We choose a large $m$ satisfying $F_{\sigma,\tau}(R_4) > 0$, and fix it. Then it is concluded that $F_{\sigma,\tau}(t) > 0$ for $t > R_4$.

Finally, we note that $F_{\sigma,\tau}(r)$ is expressed as

$$F_{\sigma,\tau}(r) = e^{2\sigma}\left\{F(r) + \sigma'\frac{d}{dr}\int_{S(r)}|u|^2 dS + (2\sigma'^2 - \tau + 2\sigma'\mathrm{Re}\rho')\int_{S(r)}|u|^2 dS\right\}.$$

Here, by assumption $F(r) \leq 0$ near infinity. Moreover, the third term of right becomes nonpositive when $r$ goes large. Thus, it follows that

$$\frac{d}{dr}\int_{S(r)}|u|^2 dS > 0$$

for $r$ large enough, and the desired conclusion holds. □

## 4. The Principle of Limiting Absorption

In this section we require the additional condition to show the principle of limiting absorption.

(A2)   There exist $\mu_i = \mu_i(r)$, $i = 1,2$, verifying (7) such that

$$\mu(r)^2 \leq \mu_1(r)\mu_2(r), \quad \mu_1(r) \leq \mu_2(r). \tag{23}$$

Moreover, if we put

$$\varphi_i(r) = \left\{\int_r^\infty \mu_i(s)ds\right\}^{-1},$$

then there exists $R_5 \geq R_0$ such that

$$\varphi_1'(r) \geq \varphi_2'(r), \quad \frac{\varphi_1'(r)}{\varphi_1(r)} \leq \frac{1}{r} \quad \text{for } r > R_5 \tag{24}$$

and

$$E_{\varphi_1}^\pm \equiv \limsup_{r\to\infty}\left[\pm\left\{\frac{\varphi_1(r)}{2\varphi_1'(r)}V_1'(r) + V_1(r)\right\}\right] < \infty. \tag{25}$$

(A3)   The unique continuation property holds for $-\Delta + V(x)$.

Under (A1) and (A2) assume $J = [\lambda_0, \lambda_1]$ to satisfy

$$\lambda_0 > \frac{a}{4} + \max\{E^+(2), E^+_{\varphi_1}\} \text{ or } \lambda_1 < \frac{a}{4} - \max\{E^-(2), E^-_{\varphi_1}\}. \tag{26}$$

**Lemma 4.1.** *(i)  We have for any $R > 0$,*

$$\varphi'_i(r) = \mu_i(r)\varphi_i(r)^2 \notin L^1([R, \infty)).$$

*(ii)  There exists $R_6 \geq R_5$ such that for each $\zeta = \lambda \pm i\epsilon \in K_\pm$ and $r > R_6$,*

$$2\mathrm{Re}\rho'_\pm + \frac{\varphi'_1(r)}{\varphi_1(r)} - \frac{n-1}{r} \geq C\frac{\varphi'_1(r)}{\varphi_1(r)} > 0.$$

For function $\xi = \xi(r) > 0$ let $L^2_\xi = L^2_\xi(\mathbf{R}^n)$ be the weighted $L^2$-space with norm

$$\|f\|_\xi = \left\{ \int_{\mathbf{R}^n} \xi(r)|f(x)|^2 dx \right\}^{1/2} < \infty.$$

**Theorem 4.1.** *Assume (A1), (A2) and choose $J$ to satisfy (26). Let $R(\zeta) = (L - \zeta)^{-1}$, $\zeta \in K_\pm$, be the resolvent of the Schrödinger operator $L = -\Delta + V(x)$. Then $R(\zeta)$ is continuously extended to $\bar{K}_\pm$ as an operator from $L^2_{\mu_1^{-1}}$ to $L^2_{\mu_2}$. More specifically,*

$$\sup_{\zeta \in K_\pm} \|R(\zeta)f\|_{\mu_2} \leq C\|f\|_{\mu_1^{-1}}. \tag{27}$$

The proof of this theorem is based on Theorem 3.1 and the following two lemmas.

**Lemma 4.2.** *Let $\zeta \in K_\pm$ and $f \in L^2_{\mu_1^{-1}}$. Then $u = R(\zeta)f$ satisfies for $R \geq R_6$,*

$$\|\theta_\pm\|^2_{\varphi'_1, B(R, \infty)} \leq C\left\{ \|u\|^2_{\mu_2} + \|f\|^2_{\mu_1^{-1}} \right\}.$$

**Lemma 4.3.** *Let $\zeta \in K_\pm$ and $f \in L^2_{\mu_1^{-1}}$. Then there exists $R_7 \geq R_6$ such that $u = R(\zeta)f$ satisfies for $R \geq R_7$,*

$$\|u\|^2_{\mu_2, B(R, \infty)} \leq \varphi_2(R)^{-1}\left\{ \|\tilde{x} \cdot \theta_\pm\|^2_{\varphi'_2, B(R, \infty)} + \|u\|^2_{\mu_2} + \|f\|^2_{\mu_1^{-1}} \right\}$$

*Proof of Theorem 4.1* Let $\{\zeta_k, f_k\} \subset K_\pm \times L^2_{\mu_1^{-1}}$ converges to $\{\zeta_0, f_0\}$ as $k \to \infty$. Since the other case is easier, we assume that $\zeta_0 = \lambda \pm i0 \in J$. Let $u_k = R(\zeta_k)f_k$. Note the inequalities

$$\varphi'_1(r) \geq \varphi'_2(r), \quad \mu_1(r)^{-1} \geq \mu_2(r)^{-1}$$

hold for $r$ large. Then since $\varphi_2(R)^{-1} \to 0$ as $R \to \infty$, the Rellich compactness criterion, Lemmas 4.2 and 4.3 show that $\{u_k\}$ is compact in $L^2_{\mu_2}$ if it is bounded in the same space. Moreover, Lemma 4.2 shows that every accumulation point $u_0 \in L^2_{\mu_2}$ satisfies the inequality

$$\|\partial_r u_0 + \rho'_\pm u_0\|_{\varphi'_1} < \infty.$$

542

The boundedness (27) of $\{u_k\}$ is proved by contradiction. In fact, assume that there exists a subsequence, which we also write $\{u_k\}$, such that $\|u_k\|_{\mu_2} \to \infty$ as $k \to \infty$. Put $v_k = u_k/\|u_k\|_{\mu_2}$. Then as is explained above, $\{\zeta_k, v_k\}$ has a convergent subsequence, and if we denote the limit by $\{\zeta_0, v_0\}$, then it satisfies the homogeneous equation (15) with $\zeta = \zeta_0$ and also

$$\|v_0\|_{\mu_2} = 1, \quad \|\partial_r v_0 + \rho'_\pm v_0\|_{\varphi'_1} < \infty, \tag{28}$$

where $\rho'_\pm = \rho'(r, \lambda \pm i0)$. The second inequality implies

$$\liminf_{r \to \infty} \int_{S(r)} |\partial_r v_0 + \rho'_\pm v_0|^2 dS = 0$$

since $\varphi'_1(r) \notin L^1([R, \infty))$ for any $R > 0$ by Lemma 3 (i). Comparing this with (19) of Theorem 3.1, we see that $v_0$ has a compact support in $x \in \mathbf{R}^n$. Hence, $v_0 \equiv 0$ by the unique continuation property for solutions to (15). But this contradicts to the first equation of (28).

We have shown that the sequence $\{u_k\}$ is precompact in $L^2_{\mu_2}$ and satisfies inequality (27). But if we apply Theorem 3.1 once more, then $\{u_k\}$ itself is shown to converge. $\qquad \square$

## References

1. D. M. Eidus, *The principle of limiting amplitude*, Uspekhi Math. Nauk **24** (1969), 91-156 (Russian Math. Surveys, **24** (1969), 97-167).
2. W. Jäger and P. Rejto, *Limiting absorption principle for some Schrödinger operators with exploding potentials. II*, J. Math. Anal. Appl. **95** (1983), 169-194.
3. W. Jäger and P. Rejto, *On a theorem of Mochizuki and Uchiyama about oscillating long range potentials*, Operator Theory and its Applications (Winnipeg, MB,1998), 305-329, Fields Inst. Commun. 25, Amer. Math. Soc., Providence, RI, 2000.
4. T. Kato, *Growth properties of solutions of the reduced wave equation with a variable coefficient*, Comm. Pure Appl. Math. **12** (1959), 403-425.
5. K. Mochizuki, *Spectral and scattering theory for second order elliptic differential operators in an exterior domain*, Lecture Notes Univ. Utah, Winter and Spring 1972.
6. K. Mochizuki and J. Uchiyama, *Radiation conditions and spectral theory for 2-body Schrödinger operators with "oscillating" long range potentials I*, J. Math. Kyoto Univ. **18** (1978), 377-408.
7. K. Mochizuki and J. Uchiyama, *On eigenvalues in the continuum of 2-body or many-body Schrödinger operators*, Nagoya Math. J. **70** (1978), 125-141.

# REGULAR GLOBAL SOLUTIONS OF SEMI-LINEAR EVOLUTION EQUATIONS WITH SINGULAR PSEUDO-DIFFERENTIAL PRINCIPAL PART

D. GOURDIN, H. KAMOUN, O. BEN KHALIFA

*University of Paris 6*
*Département de Mathématiques*
*Université P. et M.Curie Paris VI*
*175 rue du Chevaleret*
*75013 Paris, France*
*E-mail: gourdin@math.jussieu.fr*

We give here examples of equations of type

$$\partial_{tt}^2 y - p(t, D_x)y = f(Dy), \tag{1}$$

where p is a singular pseudo-differential operator and f is a small $\mathcal{C}^\infty$ non-linear perturbation $(f(0) = 0)$ with regular global solutions when the Cauchy data are regular and small.

**Key words:** Global Cauchy problem, Sobolev spaces, semi-linear evolution equations, singular operators
**Mathematics Subject Classification:** 35C20, 35L70, 35S10

## 1. Introduction

In his paper "Nonstrictly Hyperbolic Nonlinear Systems" W. Craig studied $N \times N$ second order systems with a nonlinear part being a composition of a nonlinear function with a pseudo-differential operator applied to the unknown under weak hyperbolicity assumptions, and a preliminary example from mechanics is

$$\partial_{tt}^2 y - F(by, ay, c\partial_t y, t, x) = 0, \tag{2}$$

$$y(0, x) = f(x), \ \partial_t y(0, x) = g(x), \tag{3}$$

$b(t, x, \partial_t, D_x)$ is a pseudo-differential operator of second order, $a(t, x, D_x)$, $c(t, x, D_x)$ are pseudo-differential operators of order $0 \leq d < 2$ $(x \in \mathbb{R})$. One gets solutions, regular, but local in time t. Here we solve the equation (1) from the abstract with $p(t, D_x)$ chosen such that (following lacunas of Gårding)

$$y = \frac{t}{w_n} \int_{C^{n-1}} g(x + tz)dz, \quad y = \frac{t}{w_n} \int_{D^{n-1}} \frac{g(x + tz)}{\sqrt{1 - |z|^2}}dz,$$

respectively, is a solution of the following Cauchy problem:

$$\partial_{tt}^2 y - p(t, D_x)y = 0, \ y(0, x) = 0, \ y_t(0, x) = g(x). \tag{4}$$

## 2. Pseudo-differential operators $p(t, D_x)$ of singular type

We want to check which pseudo-differential operators $p(t, D_x)$ are available so that the Cauchy problem

$$\partial_{tt}^2 y - p(t, D_x)\, y = 0, \tag{5}$$

$$y(t = 0) = 0, \quad y_t(t = 0) = g \tag{6}$$

admits the solution

$$y = \frac{t}{w_n} \int_C g(x + tz)\, dz, \tag{7}$$

where $C \in \mathbb{R}^n$ is a regular, compact, orientable hypersurface with area $w_n$ and parametrization in polar co-ordinates

$$\begin{cases} x_1 = r \cos\theta_1 \, \cos\theta_2 \cdots \cos\theta_{n-2} \, \cos\theta_{n-1} \\ x_2 = r \cos\theta_1 \, \cos\theta_2 \cdots \cos\theta_{n-2} \, \sin\theta_{n-1} \\ \qquad \cdots\cdots\cdots \\ x_{n-1} = r \cos\theta_1 \sin\theta_2 \\ x_n = r \sin\theta_1 \\ \theta_1, \cdots, \theta_{n-2} \in [-\frac{\pi}{2}, \frac{\pi}{2}], \quad \theta_{n-1} \in [0, 2\pi] \\ r = \rho(\theta_1, \cdots, \theta_{n-1}) \quad \rho : [-\frac{\pi}{2}, \frac{\pi}{2}]^{n-2} \times [0, 2\pi] \to \mathbb{R}^+; \end{cases} \tag{8}$$

$\rho$ is a smooth, periodic, positive function with period $\pi$ with respect to $\theta_1, \cdots, \theta_{n-2}$ and period $2\pi$ with respect to $\theta_{n-1}$;
$dx = dx_1 \cdots dx_n = r^{n-1} \cos^{n-2}\theta_1 \cos^{n-3}\theta_2 \cdots \cos\theta_{n-2} dr d\theta_1 \cdots d\theta_{n-1}$ $(n \geq 2)$. By Fourier transformation $\mathcal{F}_{x \to \xi}$ we get:

**Theorem 2.1.** *When $C$ is fixed, $p(t, D_x)$ is determined by its singular symbol:*

$$p(t, \xi) = \frac{\int_C e^{it(z \cdot \xi)}(2i(z \cdot \xi) - t(z \cdot \xi)^2)dz}{t \int_C e^{it(z \cdot \xi)}dz}$$

*and* $\qquad \lim_{t \to 0^+} p(t, \xi)$ *exists*

$$\Leftrightarrow \forall \xi, \ \int_C (z \cdot \xi)\, dz = \xi \cdot \int_C z\, dz = 0 \Leftrightarrow \int_C z dz = 0.$$

*Moreover,* $\qquad \lim_{t \to 0} p(t, \xi) = -3 \dfrac{\int_C (z \cdot \xi)^2\, dz}{w_n = \int_C dz}.$

## 3. Condition of independence of $p(t, \xi)$ with respect to $t$

**Lemma 3.1.** *From Theorem 2.1 we get:*

$$p(t, \xi) = \frac{1}{t^2} p(1, t\xi), \quad \forall\, t > 0, \ \forall\, \xi \in \mathbb{R}^n.$$

**Lemma 3.2.** *The symbol $p(t, \xi)$ is independent of $t$ if and only if $p(t, \xi)$ is positively homogeneous with degree $2$ with respect to $\xi$ if and only if $p(1, \xi)$ is positively homogeneous with degree $2$ with respect to $\xi$. Of course, the independence of $p(t, \xi)$ with respect to $t$ is a matter of choice of the hypersurface $C$ imbedded into $\mathbb{R}^n$.*

**Proposition 3.1.** *The symbol $p(t, \xi)$ is independent of t if and only if the hypersurface $C$ verifies for all $\xi \in \mathbb{R}^n$:*

$$\left\{ \int_C e^{iz \cdot \xi} [-3(z \cdot \xi)^2 - i(z \cdot \xi)^3] dz \right\} \left\{ \int_C e^{iz \cdot \xi} dz \right\} \tag{9}$$

$$=$$

$$\left\{ \int_C e^{iz \cdot \xi} [1 + i(z \cdot \xi)] dz \right\} \left\{ \int_C e^{iz \cdot \xi} [2i(z \cdot \xi) - (z \cdot \xi)^2] dz \right\}.$$

Denoting by $m_j = m_j(\xi)$ each side of (9) ($j = 1, 2$), these functions are analytic (product of Fourier transform of compact distributions with support C). We get

$$(9) \Leftrightarrow \left( \frac{\partial^{|l|}}{\partial \xi_1^{l_1} \cdots \partial \xi_n^{l_n}} m_1 \right) (\xi = 0) = \left( \frac{\partial^{|l|}}{\partial \xi_1^{l_1} \cdots \partial \xi_n^{l_n}} m_2 \right) (\xi = 0) \quad (\forall \, l \in \mathbb{N}^n)$$

and the following statements:

**Proposition 3.2.** *The symbol $p(t, \xi)$ is independent of t if and only if the hypersurface $C$ verifies the following equalities (i) and (ii):*

$$i) \quad \int_C z^l \, dz = 0 \quad \forall \, l \in \mathbb{N}^n \text{ with } |l| \text{ odd};$$

$$ii) \quad 3 \left\{ \sum_{\substack{p+q=l \\ |q| \geq 3}} \left( \sum_{\substack{|r| = 2 \\ r \leq q}} 1 + \sum_{\substack{|r| = 3 \\ r \leq q}} 1 \right) + \sum_{\substack{p+q=l \\ |q| = 2}} \left( \int_C z^p dz \right) \left( \int_C z^q dz \right) \right\}$$

$$=$$

$$\left\{ \sum_{\substack{p+q=l \\ |q| \geq 2}} \left( \sum_{\substack{|r| = 1 \\ r \leq q}} 1 + \sum_{\substack{|r| = 2 \\ r \leq q}} 1 \right) \left( 1 + \sum_{|s| = 1} 1 \right) \left( \int_C z^p dz \right) \left( \int_C z^q dz \right) \right\}$$

*for all $l \in \mathbb{N}^n$, $|l| \geq 4$ is even.*

These equalities (i), (ii) allow us to calculate by induction the even moments of $C$ which are defined by $m_l = \int_C z^l dz$ with $|l|$ even ($|l| \geq 4$); they are polynomial functions of volume $\int_C dz = w_n$ and of moments of order 2 of C: $\int_C z_j z_k dz = m_{ij}$ for all $1 \leq i, j \leq n$.

This is a characterization of hypersurfaces $C$ such that $p(t, \xi)$ is independent of $t$.

If these equalities are fulfilled for $C$ we get

$$\begin{cases} p(t, \xi) = \lim_{t \to 0} p(t, \xi) = \dfrac{-3 \sum_{1 \leq i,j \leq n} \int z_i z_j (\xi_i \xi_j) \, dz}{w_n} \\ \quad = -\dfrac{3}{w_n} \sum_{1 \leq i,j \leq n} m_{ij} \xi_i \xi_j. \end{cases} \tag{10}$$

Of course, we know that for $n = 3$ there is such a hypersurface $C$, it is $C = S^2(0, 1)$ and $m_{ij} = \delta_{ij} \, m = \delta_{ij} \frac{w_3}{3}$.

From now on, we suppose $p(t, \xi)$ depends on t and we calculate $p(t, \xi)$ when $C = S^{n-1}(0, 1)$.

## 4. Calculation of $p(t, \xi)$ when n is odd and $C = S^{n-1}(0,1)$

With Theorem 2.1 we get

**Proposition 4.1.** *When n is odd, it is possible to have an exact calculation of $p(t, \xi)$ thanks to Theorem 2.1; using the change of variables $u = \sin\theta_1$ in the integrals we have for $C = S^{n-1}(0,1)$:*

$$\text{when } n = 1, \quad p(t,\xi) = -|\xi|^2 - 2\frac{t|\xi|}{t^2}\frac{\sin t|\xi|}{\cos t|\xi|},$$

$$\text{when } n = 3, \quad p(t,\xi) = -|\xi|^2,$$

$$\text{when } n = 5, \quad p(t,\xi) = -|\xi|^2 + \frac{6}{t^2} + 2\frac{|\xi|}{t}\frac{t|\xi|}{t\,|\xi|}\frac{\sin t|\xi|}{\cos t|\xi| - \sin t|\xi|}.$$

For a general odd number $n = 2m + 1$ we have for $m \geq 1$ that

$$P_{2m+1} = p(t,\xi)$$

$$= -|\xi|^2 + \frac{\int_0^1 \cos[t|\xi|u]\,(1-u^2)^m\,du}{\int_0^1 \cos[t|\xi|u]\,(1-u^2)^{m-1}\,du}|\xi|^2 - \frac{2}{t}\frac{\int_0^1 \sin[t|\xi|u]\,u\,(1-u^2)^{m-1}\,du}{\int_0^1 \cos[t|\xi|u]\,(1-u^2)^{m-1}\,du}|\xi|$$

which can be calculated by induction on m.

## 5. Calculation of $p(t, \xi)$ when n is even and $C = S^{n-1}(0,1)$

Denote $n = 2m$ we get

$$p(t,\xi) = -|\xi|^2 + \frac{\int_0^{\frac{\pi}{2}} \cos[t|\xi|\sin\theta_1]\,\cos^{2m}\theta_1\,d\theta_1}{\int_0^{\frac{\pi}{2}} \cos[t|\xi|\sin\theta_1]\,\cos^{2(m-1)}\theta_1\,d\theta_1}|\xi|^2$$

$$- \frac{2}{t}\frac{\int_0^{\frac{\pi}{2}} \sin[t|\xi|\sin\theta_1]\,\sin\theta_1\cos^{2(m-1)}\theta_1\,d\theta_1}{\int_0^{\frac{\pi}{2}} \cos[t|\xi|\sin\theta_1]\,\cos^{2(m-1)}\theta_1\,d\theta_1}|\xi|.$$

But here we cannot get an exact calculation by the change of variables $u = \sin\theta_1$ and induction (in connection with the theory of lacunas of Gårding). With the stationary phase method we have the following behaviour of $p(t, \xi)$:

$$p(t,\xi) \sim -|\xi|^2 - \frac{|\xi|}{t}\tan\left(t|\xi| - \frac{\pi}{4}\right) \quad (n = 2) \text{ if } |\xi| \to \infty.$$

**Remark 5.1.** By the method of descent we can, for $n = 2m$, keep the expression of $p(t, \xi)$ obtained with $n = 2m + 1$ and change (7) by

$$y = \frac{2t}{w_{2m+1}}\int_{D_{2m}=\{z \in \mathbb{R}^{2m}, |z| \leq 1\}} \frac{g(x+tz)}{\sqrt{1-|z|^2}}\,dz.$$

We remark also that, keeping (7) for general n, we have

$$p(t,\xi) \to -\frac{3}{n}|\xi|^2 \text{ if } t \to 0,$$

$$p(t,\xi) \to -|\xi|^2 \text{ if } t \to \infty.$$

## 6. Interpretation of the operator $L = \frac{\partial^2}{\partial t^2} - p(t, D_x)$ with singular $p(t, \xi)$ when $n = 5$

We must give a meaning to the operator $L$:

$$Lu = \Box u - Nu, \tag{11}$$

$$Nu = \frac{6}{t^2} + Mu, \tag{12}$$

where

$$Mu = 2\frac{\wedge}{t} (2\pi)^{-\frac{5}{2}} \int_{\mathbb{R}^{n=5}} e^{ix \cdot \xi} \frac{t|\xi| \sin t|\xi|}{t|\xi| \cos t|\xi| - \sin t|\xi|} \hat{u} \, d\xi \tag{13}$$

and $\wedge$ is the pseudo-differential operator with symbol $|\xi|$ .
We give the definition of the operator M by

$$Mu = (2\pi)^{-\frac{5}{2}} \lim_{\epsilon \to 0^+} \int_{C_{\mathbb{R}^5} \cup_{k=1}^{\infty} \{\xi, \, \rho_k - \epsilon < |\xi| < \rho_k + \epsilon\}} e^{ix \cdot \xi} \left(\frac{2|\xi|}{t}\right) \frac{t|\xi| \sin t|\xi|}{t|\xi| \cos t|\xi| - \sin t|\xi|} \hat{u}(t, \xi) d\xi,$$

where $\rho_k = \frac{\lambda_k}{t}$ and $\lambda_k$ verify

$$\begin{cases} \lambda \cos \lambda - \sin \lambda = 0, \\ -\frac{\pi}{2} + k\pi < \lambda_k < \frac{\pi}{2} + k\pi, \\ k \in \mathbb{N}^*, \end{cases}$$

and we prove that $L \left(\frac{t}{w_n} \int_{S^{n-1}} g(x + tz) \, dz\right) = 0$, $n = 5$, $g \in \mathcal{S}(\mathbb{R}^5)$.
With polar co-ordinates in $\mathbb{R}^5_\xi$: $\xi = \rho w, \rho = |\xi|, w \in S^4_\xi(0, 1)$ using $\frac{-t^2 \rho \sin t\rho}{t\rho \cos t\rho - \sin t\rho} = \frac{d}{d\rho} [log(t\rho \cos t\rho - \sin t\rho)]$ and after integration by parts we obtain for the new function H that

$$H = \int_0^{+\infty} [\, log(t\rho \cos t\rho - \sin t\rho)\,]$$

$$\times \frac{d}{d\rho} \left[\frac{2\rho^2}{t^2(1+\rho)^5} \int_{\mathbb{R}^{n=5}_y} e^{i||x-y||\rho \sin \theta_1} (1 + \wedge)^5 \Delta^{2-\frac{1}{2}} g(t, y) \, dy\right] d\rho$$

is an absolutely convergent integral when $u \in \mathcal{S}$.
We get, defining $w \in S^4(0, 1)$ by angles $\theta_1, \theta_2, \theta_3 \in [-\frac{\pi}{2}, \frac{\pi}{2}], \theta_4 \in [0, 2\pi]$, the following proposition:

**Proposition 6.1.**
*The operator L is defined for $t > 0$ by*

$$Lu = \Box u - Nu = \Box u - \left(\frac{6}{t^2} u + Mu\right)$$

$$= \Box u - \left(\frac{6}{t^2} u + (2\pi)^{-5} \int_{S_w^4} dw \int_0^{+\infty} [\log|t\rho \cos t\rho - \sin t\rho|] \right.$$

$$\left. \times \frac{d}{d\rho} \left[ \frac{2\rho^2}{t^2(1+\rho)^5} \int_{\mathbb{R}_y^{n=5}} e^{i\|x-y\|\rho \sin\theta_1} (1+\wedge)^5 \Delta^{2-\frac{1}{2}} g(t,y) \, dy \right] d\rho \right)$$

for all $u \in C^2(\mathbb{R}^+, \mathcal{S}(\mathbb{R}^5))$ and satisfies $L(\frac{t}{w_n} \int_{S^4(0,1)} g(x + tz)dz) = 0$, for all $g \in \mathcal{S}(\mathbb{R}^5)$. Moreover, $L(0, D_x) = \frac{\partial^2}{\partial t^2} - \frac{3}{5}\Delta$.

## 7. Extension of Strichartz inequalities

We have

$$Lu = \Box u - Nu, \qquad (14)$$

$$Ng = \frac{1}{16\pi^3} \int_{\mathbb{R}_y^5} (1+\wedge)^5 \Delta^{2-\frac{1}{2}} g(t,y) \, dy \left[ \int_0^{\frac{\pi}{2}} \cos^3\theta_1 \, d\theta_1 + \int_{-\frac{\pi}{2}}^0 \cos^3\theta_1 \, d\theta_1 \right]$$

$$\times \underbrace{\left\{ \int_0^{+\infty} e^{i\|x-y\|\rho \sin\theta_1} \left( \frac{6}{t^2} + \frac{2\rho t}{t^2} \frac{t\rho \sin t\rho}{t\rho \cos t\rho - \sin t\rho} \right) \frac{\rho}{(1+\rho)^5} \, d\rho \right\}}_{I}.$$

We compute I by applying the residual theorem with different pathes $\Gamma$ in $\mathbb{C}$ when $\theta_1 \in [0, \frac{\pi}{2}]$ or $\theta_1 \in [-\frac{\pi}{2}, 0]$.
If $\epsilon \to 0$ and $R \to +\infty$ we get the following statement:

**Proposition 7.1.** *For* $Lu = \Box u - Nu$ *and* $u \in C^2(\mathbb{R}^+, \mathcal{S}(\mathbb{R}^5))$ *the operator* $N$ *is described by the following convolution:*

$$Ng = [(1+\wedge)^5 \Delta^{2-\frac{1}{2}} u] *$$

$$\left\{ \left\{ \frac{-20\pi^2}{t} \sum_{k=1}^{+\infty} \frac{t^5}{(\lambda_k + t)^5} \right\} \left[ \frac{1}{\|x\|} - \frac{2\sin\|x\|\rho_k}{\|x\|^2 \rho_k} - \frac{2(\cos\|x\|\rho_k - 1)}{\|x\|^3 \rho_k^2} \right] \right.$$

$$-2 \int_0^{+\infty} \left( \frac{6}{t^2} - 2\sigma \frac{\sigma \sinh \sigma t}{\sigma t \cosh \sigma t - \sinh \sigma t} \right) \frac{(5\sigma^4 - 10\sigma^2 + 1)}{(\sigma^2 + 1)^5}$$

$$\left. \times \left[ \frac{1}{\|x\|} + \frac{2e^{-\|x\|\sigma}}{\|x\|^2 \sigma} + \frac{2e^{-\|x\|\sigma} - 1}{\|x\|^3 \sigma^2} \right] d\sigma \right\}.$$

*Moreover, there is a linear continuous extension of* $N : W^{8,1}(\mathbb{R}^5) \to L^r(\mathbb{R}^5)$ *for all* $r > n = 5$ *which is an integro pseudo-differential operator of order 8.*

The proof of extension is obtained with precise Strichartz computation on the quadratic hypersurface $S^4(0,1)$ thanks to polar co-ordinates and Young's inequality of convolution.

## 8. $L^p - L^q$ inequality for $Ly = (\Box - N)y = f$, $y|_{t=0} = 0$, $y_t|_{t=0} = g$

**Proposition 8.1.** *There is unique solution $y$ given by*

$$y(t, x) = \frac{t}{w_n} \int_{S^4} g(x + tz) \, dz = \Omega(t) \, g \tag{15}$$

*for the Cauchy problem*

$$Ly = 0, \; y|_{t=0} = 0, \; y_t|_{t=0} = g, \tag{16}$$

*when $g \in S(\mathbb{R}^5)$.*
*The uniqueness of $y$ is obtained with Holmgren's method and the existence result with the former construction of $L$.*

### 8.1. $L^2 - L^2$ inequality for (8.2)

Considering the more general Cauchy problem

$$Ly = (\Box - N)y = f, \; y|_{t=0} = 0, \; y_t|_{t=0} = g, \tag{17}$$

with Duhamel's principle we have

$$y = \Omega(t)g(x) + \int_0^t \Omega(t - r)f(r)dr, \tag{18}$$

and by Fourier transformation we get

$$||Dy||_{L^2} = ||\nabla_x y||_{L^2} + ||\partial_t y||_{L^2} \leq ||g||_{L^2} + \int_0^t ||f(r,.)||_{L^2} dr \; \forall \, t \geq 0. \tag{19}$$

### 8.2. $L^1 - L^\infty$ inequality for (8.2)

With the same arguments as in John-Klainerman's theory and as in the book of Racke [7] we get

$$||D\Omega(t)g||_{L^\infty} \leq c \, (1 + t)^{-3} \, ||g||_{W^{6,1}}. \tag{20}$$

So, applying the complex interpolation method we have

**Theorem 8.1.** $(n = 5)$
*Let $2 \leq q \leq +\infty$, $\frac{1}{p} + \frac{1}{q} = 1$ for all $N_p \geq (n+1)(1 - \frac{2}{q})$ there exists a constant $c = c(q, n = 5)$ such that for all $g \in W^{N_p, p}$, and for all $t \geq 0$ we have*

$$||D\Omega(t)g||_{L^q} \leq c \, (1 + t)^{-3(1 - \frac{2}{q})} \, ||g||_{W^{N_p, p}}. \tag{21}$$

## 9. Local existence and uniqueness of the solution of the semi-linear Cauchy problem $Ly = f(Dy)$, $y|_{t=0} = 0$, $y_t|_{t=0} = g$

We consider the sequence $(y_n)_n$ defined by

$$\begin{cases} Ly_{n+1} = f(Dy_n), \\ y_{n+1}|_{t=0} = 0, \ y_{t,n+1}|_{t=0} = g, \end{cases}$$

and we use the inequality (19) to get

$$* \quad ||\nabla^r Dy_{n+1}||_2 \leq ||\nabla^r g||_2 + \int_0^t ||\nabla^r (f \circ Dg_n)||_2 \, (\theta) \, d\theta,$$

$$** \quad ||\nabla^r (Dy_k - Dy_l)||_2 \leq \int_0^t ||\nabla^r (f \circ Dy_{k-1} - f \circ Dy_{l-1})||_2 \, (\theta) \, d\theta.$$

We prove the existence of positive constants $R$ and $T_*$ such that for all $n \in \mathbb{N}$

$$|Dy_n|_{s,T_*} = \sup_{0 \leq t \leq T_*} ||y_n||_{W^{s,2}(\mathbb{R}^5)} \leq R \quad \left(s > \frac{n}{2}\right). \tag{22}$$

So we get the following result:

**Theorem 9.1.** *Let* $g \in W^{s,2}$, $s \in \mathbb{N}$, $s > \frac{n}{2} + 1$ *and* $c_1 = K_s \, ||g||_{s,2}$,
$c_2 > c_1$ *fixed.*
*There exist* $T > 0$ *and a unique classical solution* $y \in C_b^2([0,T] \times \mathbb{R}^5)$ *of the Cauchy problem*

$$\begin{cases} Ly = f(Dy), \\ y|_{t=0} = 0, \ y_t|_{t=0} = g, \end{cases}$$

*when* $f \in C^\infty$, $f(0) = 0$, $df(0) = 0$ *with* $\sup\{|Dy(t,x)|; \ (t,x) \in [0,T] \times \mathbb{R}^5\} \leq c_2$
*and* $Dy \in C^0([0,T], W^{s,2}(\mathbb{R}^5)) \cap C^1([0,T], W^{s-1,2}(\mathbb{R}^5))$.

## 10. Global existence

For the local solution of the previous section we derive the global estimate

$$||Dy(t)||_{s,2} \leq c \, ||g||_{s,2} \, \exp\left\{c \int_0^t ||\nabla u||_\infty(r) \, dr\right\}$$

with $t \in [0,T]$ and $c$ independent of T. We get with the same argument as in [7]:

**Theorem 10.1.** *Considering the Cauchy problem*

$$\begin{cases} Ly = \Box y - N(t, D_x) \, y = f(Dy), \\ y|_{t=0} = 0, \ y_t|_{t=0} = g, \end{cases}$$

*where* $f \in C^\infty$ *and* $f(0) = 0$, $df(0) = 0$, $N(t, D_x)$ *is the operator defined in* §5, *that is, a singular pseudo-differential operator with symbol* $N(t,\xi) = \frac{6}{t^2} + 2\frac{|\xi|}{t} \frac{t|\xi| \sin t|\xi|}{t|\xi| \cos t|\xi| - \sin t|\xi|}$, *there exists an integer* $s_0 > \frac{n}{2} + 1$ *and a real number* $\delta > 0$ *such that we have the following property:*

*if* $g \in W^{s,2}(\mathbb{R}^5) \cap W^{s,p}(\mathbb{R}^5)$ *with* $p = \frac{4}{3}$ *and* $\|g\|_{s,2} + \|g\|_{s,p} < \delta$, *the semi-linear Cauchy problem has a unique global solution* $y$.
*Moreover,*

$$Dy = (y_t, \nabla_x y) \in \mathcal{C}^0([0,\infty[, W^{s,2}(\mathbb{R}^5)) \cap \mathcal{C}^1([0,\infty[, W^{s-1,2}(\mathbb{R}^5)),$$

$$\|Dy(t)\|_\infty + \|Dy(t)\|_4 \leq c\, t^{-\frac{1}{2}},$$

*and* $\|Dy(t)\|_{s,2}$ *is bounded on* $[0, +\infty[$.

**Remark 10.1.** We can get the same theorem for $n = 4$ and the same $N(t,\xi) = \frac{6}{t^2} + 2\frac{|\xi|}{t}\frac{t|\xi| \sin t|\xi|}{t|\xi| \cos t|\xi| - \sin t|\xi|}$ with $\xi = (\xi_1, \xi_2, \xi_3, \xi_4)$ thanks to the method of descent (for linear arguments ) and some nonlinear perturbation $f(Dy)$ of the right-hand side of $\Box y - Ny = f(Dy)$.
We conjecture that there is possibility of generalization to any $n \geq 3$.

# References

1. Robert A. Adams, *Sobolev Spaces*, Pure and Applied Mathematics, **65**, Academic Press, 1975.
2. O. Ben Khalifa, D. Gourdin and H. Kamoun, *Problème de Cauchy global régulier pour des équations semi-linéaires à partie principale pseudo différentielle singulière*, Institut de Maths de Jussieu **U M R 7586** Université Paris 6 et Paris 7 (CNRS) Prépublication 391 (Mai 2005 ); à paraître. Séminaire EDP linéaires et non lineaires et physique math (B. Gaveau , D. Gourdin , J. Vallant) Exposé du 8 juin 2005.
3. Walter Craig, *Nonstrictly Hyperbolic Nonlinear Systems* , Math. Ann. **277** (1987), 213-232.
4. R. Courant and D. Hilbert, *Methods of mathematical physics. Volume II: Partial differential equations. Transl. and rev. from the German Original. Reprint of the 1st Engl. ed. 1962.*, Wiley Classics Edition, John Wiley & Sons, New York 1989.
5. Jean Dieudonné, *Calcul infinitesimal. 2e ed., rev. et corr. (French)*, Collection Methodes, Hermann, Paris 1980
6. J. Ginibre and G. Velo, *Inégalités de Strichartz généralisées de l'équation des ondes* , Séminaire sur les Eqations aux Dérivées partielles. Ecole Polytecnique Palaiseau, Exp $n^0 17$ (1995).
7. Reinhardt Racke, *Lectures on Nonlinear Evolution Equations. Initial Value Problems*, Aspects of Mathematics, Fried. Vieweg & Sohn, Braunschweig 1992.
8. Robert S. Strichartz, *Restrictions of Fourier transforms to quadratic surfaces and decay of solution of wave equations*, Duke Mathematical Journal. Vol **44** $n^0$ **3** (1977), 705-714.
9. Michael E.Taylor,*Partial differential equations III. Non linear Equations*, Applied Mathematical sciences, **117**, Springer-Verlag, Berlin Heidelberg New York 1996.
10. M. Berger, D. Gauduchon and E. Mazet, *Le spectre d'une Variété Riemanienne* , Lecture Notes in Mathematics **194**, Springer-Verlag, Berlin Heidelberg New York 1971.

# ASYMPTOTIC BEHAVIOUR FOR KIRCHHOFF EQUATION

TOKIO MATSUYAMA

*Department of Mathematics*
*Tokai University*
*Hiratsuka*
*Kanagawa, 259-1292*
*Japan*
*E-mail: matsu@sm.u-tokai.ac.jp*

In this article we shall investigate asymptotic profiles for Kirchhoff equation. More precisely, it will be shown that the solution consists of a free wave, non-free wave and a remainder term. These asymptotic behaviours are based on global-in-time existence theorems of two types; one is the existence of $H^1$-solution with small $H^1$ data belonging to suitable Sobolev spaces of $L^p$ type, and another is an existence theorem obtained by Yamazaki [8].

**Key words:** Kirchhoff equation, asymptotic profiles, scattering states
**Mathematics Subject Classification:** 35L05, 35L10

## 1. Introduction

This article is the résúme of recent results [7]. Let us consider the following Cauchy problem of Kirchhoff equation:

$$
\text{(K)} \quad
\begin{cases}
\partial_t^2 u - \left( 1 + \displaystyle\int_{\mathbb{R}^n} |\nabla u|^2 \, dx \right) \Delta u = 0, & (x,t) \in \mathbb{R}^n \times (0,\infty), \\
u(x,0) = u_0(x), \quad \partial_t u(x,0) = u_1(x), & x \in \mathbb{R}^n.
\end{cases}
$$

We shall describe the asymptotic behaviour of solutions to problem (K), i.e., we want to find asymptotic profiles of solutions for the Kirchhoff equation. Yamazaki [8] introduced a certain set $Y_k$, $k > 1$, of initial data to obtain a global-in-time existence theorem with small data of low regularity, and then, the scattering theory was developped in the case when $k > 2$. This generalized the results [4,5]. *A motivation in this article arises from a problem whether the scattering states exist or not in the case when $1 < k \leq 2$.* Roughly speaking, it depends on the decay rate of data; if the data behave like $L^1$-functions, then solutions are asymptotically free (see Theorem 3.3), while, if data do not behave like $L^1$-functions, then such a phenomenon does not occur (see Theorem 3.2). For deriving these asymptotics, we need a delicate analysis of an oscillatory integral associated with Kirchhoff equation. Such an oscillatory integral was introduced by [5] (see also [2,3,8]), and we will develop an asymptotic expansion of the oscillatory integral by using the asymptotic behaviours of Bessel potentials ([1]).

Throughout this article, we fix the notation as follows: $\dot{H}^{s,p} = \dot{H}^{s,p}(\mathbb{R}^n)$, $H^{s,p} = H^{s,p}(\mathbb{R}^n)$, $\dot{H}^s = \dot{H}^{s,2}(\mathbb{R}^n)$, $H^s = H^{s,2}(\mathbb{R}^n)$. We say that $u \in X^{s_0}(0,\infty)$ ($\dot{X}^{s_0}(0,\infty)$ resp.) if

$$\left.\begin{array}{c} u \in C\left([0,\infty); H^{s_0}\right) \quad \left(C([0,\infty); \dot{H}^{s_0}) \text{ resp.}\right) \\[2mm] (\nabla, \partial_t)\, u \in \mathcal{B}^0\left([0,\infty); H^{s_0-1}\right) \cap \mathcal{B}^1\left([0,\infty); H^{s_0-2}\right) \end{array}\right\} \quad \text{for } s_0 \geq 1,$$

where $\mathcal{B}^j\left([0,\infty); X\right)$ is the space of all functions whose derivatives up to the $j$-th order are all bounded and continuous on $[0,\infty)$ with values in a Banach space $X$.

## 2. Global-in-time existence theorems

In this section we shall introduce global-in-time existence theorems to develop the asymptotic behaviour. The first one is the folllowing:

**Theorem 2.1.** Let $n \geq 4$ and $s_0 \geq 1$. Assume that the data $u_0$, $u_1$ satisfy $u_0 \in H^{s_0} \cap H^{n\left(\frac{1}{p'}-\frac{1}{p}\right)+1,p'}$, $u_1 \in H^{s_0-1} \cap H^{n\left(\frac{1}{p'}-\frac{1}{p}\right),p'}$, and

$$\|u_0\|_{H^{n\left(\frac{1}{p'}-\frac{1}{p}\right)+1,p'}} + \|u_1\|_{H^{n\left(\frac{1}{p'}-\frac{1}{p}\right),p'}} \ll 1,$$

where $\frac{2(n-1)}{n-3} < p \leq \infty$ and $\frac{1}{p} + \frac{1}{p'} = 1$. Then the problem (K) has a unique solution $u(x,t)$ in the class $X^{s_0}(0,\infty)$ so that $1+\|\nabla u(\cdot,t)\|_{L^2}^2$ converges to a uniquely determined constant $c_\infty^2 \equiv c_\infty^2(u_0,u_1)$ as $t$ goes to infinity in the following manner:

$$1 + \|\nabla u(\cdot,t)\|_{L^2}^2 = c_\infty^2 + O(t^{-k(n)+1}) \quad \text{as } t \to \infty,$$

where $k(n) = \frac{n-1}{2}\left(\frac{1}{p'}-\frac{1}{p}\right) (>1)$.

We shall give a remark on Theorem 2.1. By the Sobelev imbedding theorem we see that

$$H^{n\left(\frac{1}{p'}-\frac{1}{p}\right),p'} \subset H^{\frac{n}{n-1}} \subset H^1.$$

Hence the data in Theorem 2.1 with $s_0 = 1$ satisfy

$$u_0 \in H^{\frac{n}{n-1}+1}, \quad u_1 \in H^{\frac{n}{n-1}}.$$

This class is contained in $H^2 \times H^1$, but not contained in the class of [9]. Hence Theorem 2.1 generalizes [9], and the solution $u$ belongs to $X^{\frac{2n-1}{n-1}}(0,\infty)(\subset X^2(0,\infty))$.

As to the global-in-time existence theorem including low dimensional cases, we cite the next theorem (see [8]) in order to find asymptotic profiles for Kirchhoff equation. For this purpose, let us introduce a set $Y_k$ ($k > 1$), which was introduced by Yamazaki as follows:

$$Y_k := \left\{ \{\phi,\psi\} \in \dot{H}^{\frac{3}{2}} \times H^{\frac{1}{2}}\,;\, |\{\phi,\psi\}|_{Y_k} < \infty \right\},$$

where

$$|\{\phi,\psi\}|_{Y_k} := \sup_{\tau\in\mathbb{R}}(1+|\tau|)^k \left| \int_{\mathbb{R}^n} e^{i\tau|\xi|}|\xi|^3|\hat\phi(\xi)|^2\, d\xi \right|$$

$$+ \sup_{\tau\in\mathbb{R}}(1+|\tau|)^k \left| \int_{\mathbb{R}^n} e^{i\tau|\xi|}|\xi||\hat\psi(\xi)|^2\, d\xi \right|$$

$$+ \sup_{\tau\in\mathbb{R}}(1+|\tau|)^k \left| \int_{\mathbb{R}^n} e^{i\tau|\xi|}|\xi|^2 \Re\left(\hat\phi(\xi)\overline{\hat\psi(\xi)}\right)\, d\xi \right|.$$

Then we have the following:

**Theorem A** [8] *Let* $n \geq 1$ *and* $s_0 \geq \frac{3}{2}$. *If the data* $u_0, u_1$ *satisfy* $u_0 \in \dot{H}^{s_0}$, $u_1 \in H^{s_0-1}$, *and*

$$\|\nabla u_0\|_{L^2} + \|u_1\|_{L^2} + |\{u_0, u_1\}|_{Y_k} \ll 1 \quad \text{for some } k > 1,$$

*then the problem (K) has a unique solution* $u(x,t) \in \dot{X}^{s_0}(0,\infty)$ *having the following property: there exists a constant* $c_\infty \equiv c_\infty(u_0, u_1) > 0$ *so that*

$$1 + \|\nabla u(\cdot,t)\|_{L^2}^2 = c_\infty^2 + O(t^{-k+1}) \quad \text{as } t \to \infty.$$

**Remark.** It was proved in [8] that the constant $c_\infty$ appearing in Theorem 2.1 and Theorem A for $k > 2$ is uniquely determined by the following equation:

$$c_\infty = \sqrt{1 + \frac{1}{2}\left(\|\nabla u_0\|_{L^2}^2 + \frac{1}{c_\infty^2}\|u_1\|_{L^2}^2\right)}.$$

## 3. Asymptotic profiles

Based on Theorem 2.1 and Theorem A, we can consider asymptotic profiles of solutions for problem (K). Since the Kirchhoff equation inherits various properties from linear equations by applying results for linear problems (see [6]), the structure of solutions for the Kirchhoff equation can be determined. Namely, the solution $u$ consists of a **free wave, non-free wave** and a remainder term. Here we give definitions of free and non-free waves according to [6].

**Definition.** *(i) The function* $v = v(x,t)$ *is a **free wave** if* $v$ *satisfies the equation*

$$(\partial_t^2 - c_\infty^2\Delta)\,v = 0.$$

*(ii) Let* $\sigma \in \mathbb{R}$. *Then we say that* $w = w(x,t)$ *is a **non-free wave** in* $\dot{H}^\sigma \times \dot{H}^{\sigma-1}$ *if* $w$ *satisfies*

$$(\partial_t^2 - c_\infty^2\Delta)\,w \neq 0,$$

*and* $\{w, \partial_t w\}$ *never decays in* $\dot{H}^\sigma \times \dot{H}^{\sigma-1}$, *and is never asymptotically free in* $\dot{H}^\sigma \times \dot{H}^{\sigma-1}$ *as* $t$ *goes to infinity, i.e.,* $\{w, \partial_t w\}$ *is never asymptotic to any free* *wave in $\dot{H}^\sigma \times \dot{H}^{\sigma-1}$ as t goes to infinity.*

From now on, we set $\kappa(n) = k(n) = \frac{n-1}{2}\left(\frac{1}{p'} - \frac{1}{p}\right)$, $n \geq 4$, if we treat solutions from Theorem 2.1, or $\kappa(n) = k$, $n \geq 1$, if we treat solutions from Theorem A. Then we have the following statements:

**Theorem 3.1.** *(i) Let $u(x,t)$ be the solution in Theorem 2.1 ($n \geq 4$). Then $u(x,t)$ has the following structure: There exist a free wave $v_\infty^{(1)}(x,t)$, a function $w_\infty(x,t)$ belonging to $X^{s_0}(0,\infty)$ and satisfying $(\partial_t^2 - c_\infty^2 \Delta)w_\infty(x,t) = f(x,t)$ with a certain nontrivial function $f(x,t)$ such that, for every $\sigma \in [1, s_0]$,*

$$u(x,t) = v_\infty^{(1)}(x,t) + w_\infty(x,t) + O(t^{-\kappa(n)+1}) \quad \text{in } \dot{H}^\sigma \text{ as } t \to \infty, \tag{1}$$

*where $v_\infty^{(1)}(x,t)$, $w_\infty(x,t)$ and $f(x,t)$ depend on $\sqrt{1 + \|\nabla u(\cdot,t)\|_{L^2}^2}$.*

*(ii) Let $u(x,t)$ be the solution in Theorem A ($n \geq 1$). Then (1) holds as long as $u_0 \in \dot{H}^\sigma$ and $u_1 \in \dot{H}^{\sigma-1}$ for some $\sigma \leq s_0$.*

We can deduce from Theorem 4.4 in [6] that it is sufficient for the non-free property of $w_\infty(x,t)$ in Theorem 3.1 that $\sqrt{1 + \|\nabla u(\cdot,t)\|_{L^2}^2} - c_\infty$ is not integrable over $(0,\infty)$. The next theorem gives the non-free property of $w_\infty(x,t)$.

**Theorem 3.2.** *Let $1 < k \leq 2$, $\eta \in (\frac{k-1}{2}, \frac{k+1}{2})$ and $0 \leq \varepsilon < \frac{k-1}{2}$. Assume that $n > \max(k-1, 3-k)$. Let us take the data so that*

$$|D|u_0(x) = u_1(x) = \delta|D|^{-\varepsilon}\langle x \rangle^{-\frac{n+k-1}{2}} \quad \text{for some } 0 < \delta \ll 1.$$

*Then $\{u_0(x), u_1(x)\} \in (\dot{H}^s \times \dot{H}^{s-1}) \cap Y_{k-2\varepsilon}$ for every $s \geq 2\varepsilon + (1-\eta)(1-\varepsilon)$. Furthermore, let $u(x,t)$ be the solution in Theorem A with these data. Then $w_\infty(x,t)$ in Theorem 3.1 is a non-free wave in $\dot{H}^{2\varepsilon+(1-\eta)(1-\varepsilon)} \times \dot{H}^{2\varepsilon+(1-\eta)(1-\varepsilon)-1}$.*

**Remark.** If $\varepsilon = \frac{\eta}{\eta+1}$, then $2\varepsilon + (1-\eta)(1-\varepsilon) = 1$, and hence, $w_\infty(x,t)$ is a non-free wave in the energy space $\dot{H}^1 \times L^2$.

If $\sqrt{1 + \|\nabla u(\cdot,t)\|_{L^2}^2} - c_\infty$ is integrable over $(0,\infty)$, $w_\infty(x,t)$ can be decomposed into a free wave part and a decaying one in Theorem 3.1. Then the next theorem holds.

**Theorem 3.3.** *Let $u(x,t)$ be the solution in Theorem 2.1. Then $u(x,t)$ is, in general, asymptotically free and $\sqrt{1 + \|\nabla u(\cdot,t)\|_{L^2}^2} - c_\infty$ belongs to $L^1(0,\infty)$. Namely, $u(x,t)$ is asymptotic to a free wave $v_\infty^{(2)}(x,t) \in X^{s_0}(0,\infty)$ so that, for every $\sigma \in [1, s_0]$,*

$$u(x,t) = v_\infty^{(2)}(x,t) + o(1) \quad \text{in } \dot{H}^\sigma \text{ as } t \to \infty. \tag{2}$$

*More precisely we have the following: (i) In the case when $p' \neq 1$, let us take the data so that*

$$u_0(x) = \delta\langle x \rangle^{-\ell_0}, \quad u_1(x) = \delta|D|\langle x \rangle^{-\ell_1}, \quad 0 < \delta \ll 1 \tag{3}$$

for some $\ell_0$, $\ell_1 \in \left(\frac{n}{p'}, n\right]$. Then (2) holds and further, for every $\sigma \in [0, s_0 - 1]$,

$$\nabla u(x,t) = \nabla v_\infty^{(2)}(x,t) + O(t^{-[2\ell-n+1]}) \quad in\ H^{\sigma-1}\ as\ t \to \infty, \tag{4}$$

where $\ell = \min(\ell_0, \ell_1)$ and $[2\ell - n + 1] > \frac{2}{n-1} + 3$.

(ii) In the case when $p' = 1$, let us take the data as in (3) for $\ell_0$, $\ell_1 > n$. Then (2) holds and for every $\sigma \in [0, s_0 - 1]$,

$$\nabla u(x,t) = \nabla v_\infty^{(2)}(x,t) + O(t^{-(n+2)}) \quad in\ H^{\sigma-1}\ as\ t \to \infty. \tag{5}$$

**Remark.** (i) Let $u(x,t)$ be the solution in Theorem 2.1 with the assumption $k(n) > 2$ ($n \geq 6$). In this case, since the data belong to $Y_{k(n)}$, it follows from Theorem 1.2 of [8] that the solution $u(x,t)$ is asymptotically free for every data satisfying the assumption in Theorem 2.1, while the asymptotic rate in (5) is replaced by $t^{-k(n)+1}$. (ii) In the case when $k > 2$ ($n \geq 1$) in Theorem A, Yamazaki already proved the asymptotically free property for every small data in $Y_k$ (see [8]).

## 4. Implicit representations of free and non-free waves

In this section we shall give implicit representations of $v_\infty^{(1)}(x,t)$, $w_\infty(x,t)$ and $v_\infty^{(2)}(x,t)$. Let us define

$$c(t) = \sqrt{1 + \|\nabla u(\cdot, t)\|_{L^2}^2} \quad and \quad \vartheta(t) = \int_0^t c(\tau)\, d\tau.$$

Then we consider the system of fundamental solutions $\{V_0(\xi, t), V_1(\xi, t)\}$ such that

$$\begin{cases} V_0'' + c^2(t)|\xi|^2 V_0 = 0, \\ V_0(\xi, 0) = 1, \quad V_0'(\xi, 0) = 0, \end{cases} \qquad \begin{cases} V_1'' + c^2(t)|\xi|^2 V_1 = 0, \\ V_1(\xi, 0) = 0, \quad V_1'(\xi, 0) = 1. \end{cases}$$

It was proved in [6] that there exists

$$\lim_{t \to \infty} \begin{pmatrix} \cos(\vartheta(t)|\xi|) & -\dfrac{\sin(\vartheta(t)|\xi|)}{c(t)|\xi|} \\ c(0)|\xi|\sin(\vartheta(t)|\xi|) & \dfrac{c(0)\cos(\vartheta(t)|\xi|)}{c(t)} \end{pmatrix} \begin{pmatrix} V_\ell(\xi, t) \\ V_\ell'(\xi, t) \end{pmatrix}, \quad \ell = 0, 1,$$

and $\{\alpha_\ell(\xi), \beta_\ell(\xi)\}$ can be represented by

$$\begin{pmatrix} \alpha_\ell(\xi) \\ \beta_\ell(\xi) \end{pmatrix} = \lim_{t \to \infty} \begin{pmatrix} \cos(\vartheta(t)|\xi|) & -\dfrac{\sin(\vartheta(t)|\xi|)}{c(t)|\xi|} \\ c(0)|\xi|\sin(\vartheta(t)|\xi|) & \dfrac{c(0)\cos(\vartheta(t)|\xi|)}{c(t)} \end{pmatrix} \begin{pmatrix} V_\ell(\xi, t) \\ V_\ell'(\xi, t) \end{pmatrix}.$$

Furthermore, $\{\alpha_\ell(\xi), \beta_\ell(\xi)\}$ have the following growth order:

$$\begin{cases} |\alpha_0(\xi)| \lesssim 1, & |\beta_0(\xi)| \lesssim |\xi|, \\ |\alpha_1(\xi)| \lesssim |\xi|^{-1}, & |\beta_1(\xi)| \lesssim 1. \end{cases}$$

Then the free wave $v_\infty^{(1)}(x,t)$ and the function $w_\infty(x,t)$ given in Theorem 3.1 can be represented by

$$v_\infty^{(1)}(x,t) = \sum_{\ell=0,1} \mathcal{F}^{-1}\left[\left(\alpha_\ell(\xi)\cos(c_\infty|\xi|t) + \beta_\ell(\xi)\frac{\sin(c_\infty|\xi|t)}{\tilde{c}(0)|\xi|}\right)\hat{u}_\ell(\xi)\right](x),$$

$$w_\infty(x,t) = \sum_{\ell=0,1} \mathcal{F}^{-1}\left[\left(\alpha_\ell(\xi)\varphi_c(|\xi|,t) + \beta_\ell(\xi)\frac{\varphi_s(|\xi|,t)}{\tilde{c}(0)|\xi|}\right)\hat{u}_\ell(\xi)\right](x),$$

respectively, where

$$\varphi_c(|\xi|,t) = 2\sin\left(\frac{c_\infty|\xi|t + \vartheta(t)|\xi|}{2}\right)\sin\left(\frac{c_\infty|\xi|t - \vartheta(t)|\xi|}{2}\right),$$

$$\varphi_s(|\xi|,t) = -2\cos\left(\frac{c_\infty|\xi|t + \vartheta(t)|\xi|}{2}\right)\sin\left(\frac{c_\infty|\xi|t - \vartheta(t)|\xi|}{2}\right),$$

and $w_\infty(x,t)$ satisfies the equation

$$\left(\partial_t^2 - c_\infty^2\Delta\right)w_\infty(x,t) = \frac{c^2(t) - c_\infty^2}{c(t)}\partial_t f_1(x,t) + c'(t)f_1(x,t).$$

The function $f_1(x,t)$ is given by

$$f_1(x,t) = \sum_{\ell=0,1} \mathcal{F}^{-1}\left[\left(-\alpha_\ell(\xi)|\xi|\sin(\vartheta(t)|\xi|) + \frac{\beta_\ell(\xi)}{c(0)}\cos(\vartheta(t)|\xi|)\right)\hat{u}_\ell(\xi)\right](x).$$

In particular, if we take the data as in Theorem 3.2, then it follows from [7] that $c(t) - c_\infty$ is non-integrable, and hence, $w_\infty(x,t)$ is a non-free wave in $\dot{H}^{2\varepsilon+(1-\eta)(1-\varepsilon)} \times \dot{H}^{2\varepsilon+(1-\eta)(1-\varepsilon)-1}$. Finally, the free wave $v_\infty^{(2)}(x,t)$ given in Theorem 3.3 can be represented by

$$v_\infty^{(2)}(x,t) = \sum_{\ell=0,1} \mathcal{F}^{-1}\left[\left(\alpha_\ell(\xi)\cos(c_\infty|\xi|t+\psi|\xi|)+\beta_\ell(\xi)\frac{\sin(c_\infty|\xi|t+\psi|\xi|)}{c(0)|\xi|}\right)\hat{u}_\ell(\xi)\right](x),$$

where

$$\psi := \int_0^\infty \left(\sqrt{1 + \|\nabla u(\cdot,t)\|_{L^2}^2} - c_\infty\right)dt.$$

## Acknowledgments

This research was done during the author stayed in Dipartimento di Matematica, Università di Pisa. The author would like to thank Professor Sergio Spagnolo and the staff of Università di Pisa for their kind hospitality.

# References

1. N. Aronszajn and K. T. Smith, *Theory of Bessel potentials. Part I.*, Ann. Inst. Fourier, Grenoble **11** (1964), 385 – 475.
2. P. D'Ancona and S. Spagnolo, *A class of nonlinear hyperbolic problems with global solutions*, Arch. Rational Mech. Anal. **124** (1993), 201 – 219.
3. P. D'Ancona and S. Spagnolo, *Nonlinear perturbations of the Kirichhoff equation,* Comm. Pure Appl. Math. **47** (1994), 1005 – 1029.
4. M. Ghisi, *Asymptotic behavior of Kirchhoff equation*, Ann. Mat. Pura Appl. (4) **171** (1996), 293 – 312.
5. J. M. Greenberg and S. C. Hu, *The initial-value problem for a stretched string*, Quart. Appl. Math. **38** (1980), 289 – 311.
6. T. Matsuyama, *Asymptotic profiles for wave equations with time-dependent coefficients*, submitted (2005).
7. T. Matsuyama, *Asymptotic analysis for Kirchhoff equation*, preprint (2005).
8. T. Yamazaki, *Scattering for a quasilinear hyperbolic equation of Kirchhoff type*, J. Differential Equations **143** (1998), 1 – 59.
9. T. Yamazaki, *Global solvability for the Kirchhoff equations in exterior domains of dimension larger than three*, Math. Methods Appl. Sci. **27** (2004), 1893 – 1916.

# SOME RESULTS ON GLOBAL EXISTENCE AND ENERGY DECAY OF SOLUTIONS TO THE CAUCHY PROBLEM FOR A WAVE EQUATION WITH A NONLINEAR DISSIPATION

ABBES BENAISSA

*Université Djillali Liabès*
*Faculté des Sciences*
*Département de mathématiques*
*B. P. 89, Sidi Bel Abbès 22000, ALGERIA.*
*E-mail: benaissa_ abbes@yahoo.com*

In this article we give some results on the existence of global decaying solutions to the Cauchy problem for a wave equation with a nonlinear dissipative term in the Sobolev spaces $H^1(\mathbb{R}^n)$ and $H^2(\mathbb{R}^n)$.

**Key words:** Nonlinear wave equation, golbal existence, decay rate
**Mathematics Subject Classification:** 35L70, 35B40

## 1. Introduction and main results

We consider the Cauchy problem for the nonlinear wave equation with a nonlinear dissipation and source terms of the type

$$\text{(P)} \quad \begin{cases} u'' - \Delta_x u + \lambda^2(x)u + \sigma(t)g(u') = |u|^{p-1}u \text{ in } \mathbb{R}^n \times [0, +\infty[, \\ u(x,0) = u_0(x), \quad u'(x,0) = u_1(x) \text{ in } \mathbb{R}^n, \end{cases}$$

where $g : \mathbb{R} \to \mathbb{R}$ is a continuous non-decreasing function and $\lambda$ and $\sigma$ are positive functions.

For the Cauchy problem $(P)$ with $\lambda \equiv 1$ and $\sigma \equiv 1$, and when $g(x) = \delta|x|^{m-1}x$ $(m \geq 1)$ Todorova [12] (see also [11] and [9]) proved that the energy decay rate is $E(t) \leq (1+t)^{-\frac{2-n(m-1)}{(m-1)}}$ for $t \geq 0$. She used a general method introduced by Nakao [8] with the condition that the data have compact support. Unfortunately, this method does not seem to be applicable in the case of more general functions $\lambda$ and $\sigma$.

Our purpose in this paper is to give a global solvability result in the class $H^1$ and energy decay estimates of the solutions to the Cauchy problem $(P)$ for a weak linear perturbation and a nonlinear dissipation when $p \leq \dfrac{n}{n-2}$. To treat the case when $p > \dfrac{n}{n-2}$, we need further regularity of $(u_0, u_1)$. The last result concerns the case of bounded dissipation.

We use a new method recently introduced by Martinez [7] to study the decay rate of solutions to the wave equation $u'' - \Delta_x u + g(u') = 0$ in $\Omega \times \mathbb{R}^+$, where $\Omega$ is a bounded domain of $\mathbb{R}^n$. This method is based on a new integral inequality that generalizes a result of Haraux [6]. So we proceed the argument combining the method in [7] with the concept of a modified stable set on $H^1(\mathbb{R}^n)$. Here the modified stable set is the extended $\mathbb{R}^n$ version of Sattinger's stable set.

The functions $\lambda(x), \sigma(t)$ and $g$ satisfy the following hypotheses:

- $\lambda(x)$ is a locally bounded measurable function defined on $\mathbb{R}^n$ and satisfies

$$\lambda(x) \geq d(|x|), \tag{1}$$

where $d$ is a decreasing function such that $\lim_{y \to \infty} d(y) = 0$,

- $\sigma : \mathbb{R}_+ \to \mathbb{R}_+$ is a non increasing function of class $C^1$ on $\mathbb{R}_+$,
- $g : \mathbb{R} \to \mathbb{R}$ is a nonincreasing $C^0$ function such that

$$g(v)v > 0 \quad \text{for all } v \neq 0,$$

moreover, we suppose that there exist $C_i > 0$; $i = 1, 2, 3, 4$ such that

$$C_3|v|^m \leq |g(v)| \leq C_4|v|^{\frac{1}{m}} \text{ if } |v| \leq 1, \tag{2}$$

$$C_1|v| \leq |g(v)| \leq C_2|v|^r \text{ for all } |v| \geq 1, \tag{3}$$

where $m \geq 1$ and $1 \leq r \leq \dfrac{n+2}{(n-2)^+}$.

Before stating the global existence theorem and the decay property to the problem $(P)$, we shall introduce the notion of the modified stable set. Let

$$\begin{aligned} K(u) &= \|\nabla_x u\|_2^2 + \|u\|_2^2 - \|u\|_{p+1}^{p+1} && \text{if } \lambda \equiv \text{const} \\ I(u) &= \|\nabla_x u\|_2^2 - \|u\|_{p+1}^{p+1} && \text{if } \lambda \not\equiv \text{const} \end{aligned}$$

for $u \in H^1(\mathbb{R}^n)$. Then we define the modified stable sets $\widetilde{W}^*$ and $\widetilde{W}^{**}$ by

$$\widetilde{W}^* \equiv \{u \in H^1(\mathbb{R}^n) \backslash K(u) > 0\} \cup \{0\} \text{ if } \lambda \equiv \text{const}$$

and

$$\widetilde{W}^{**} \equiv \{u \in H^1(\mathbb{R}^n) \backslash I(u) > 0\} \cup \{0\} \text{ if } \lambda \not\equiv \text{const}.$$

Next, let $J(u)$ and $E(t)$ be the potential and the energy associated with the problem $(P)$, respectively:

$$J(u) = \frac{1}{2}\|\nabla_x u\|_2^2 + \frac{1}{2}\|\lambda(x)u\|_2^2 - \frac{1}{p+1}\|u\|_{p+1}^{p+1} \quad \text{for } u \in H^1(\mathbb{R}^n),$$

$$E(t) = \frac{1}{2}\|u'\|_2^2 + J(u).$$

We first state a lemma that will be needed later (see [7]).

**Lemma 1.1.** *Let $E : \mathbb{R}_+ \to \mathbb{R}_+$ be a non increasing function and $\phi : \mathbb{R}_+ \to \mathbb{R}_+$ an increasing $C^2$ function such that*

$$\phi(0) = 0 \quad and \quad \phi(t) \to +\infty \quad as \quad t \to +\infty.$$

*Assume that there exist $p \geq 1$ and $A > 0$ such that*

$$\int_S^{+\infty} E(t)^{\frac{p+1}{2}}(t)\phi'(t)\, dt \leq AE(S), \quad 0 \leq S < +\infty.$$

*Then we have*

$$E(t) \leq cE(0)(1 + \phi(t))^{\frac{-2}{p-1}} \ \forall t \geq 0, \quad if \ \ p > 1$$

*and*

$$E(t) \leq cE(0)e^{-\omega\phi(t)} \ \forall t \geq 0, \quad if \ \ p = 1,$$

*where $c$ and $\omega$ are positive constants independent of the initial energy $E(0)$.*

The following local existence theorems are proved in the standard manner (e.g. see [10] and [11])

**Theorem 1.1.** *Let $1 < p \leq \dfrac{n+2}{n-2}$ $(1 < p < \infty$ if $n = 1,2)$ and assume that $(u_0, u_1) \in H^1(\mathbb{R}^n) \times L^2(\mathbb{R}^n)$ and $u_0$ belongs to the modified stable set $\widetilde{W}^*$. Then there exists $T > 0$ such that the Cauchy problem $(P)$ has a unique solution $u(t)$ on $\mathbb{R}^n \times [0, T)$ in the class*

$$u(t, x) \in C([0, T); H^1(\mathbb{R}^n)) \cap C^1([0, T); L^2(\mathbb{R}^n)),$$

*satisfying*

$$u(t) \in \widetilde{W}^*.$$

*This solution can be continued in time as long as $u(t) \in \widetilde{W}^*$.*

**Theorem 1.2.** *Let $(u_0, u_1) \in H^2 \times H^1$. Suppose that*

$$1 \leq p \leq \frac{n}{n-4} \quad (1 \leq p \leq \infty \text{ if } n \leq 4)$$

*Then under the hypothesis (1), (2) and (3), the problem $(P)$ admits a unique local solution $u(t)$ on some interval $[0, T[$, $T \equiv T(u_0, u_1) > 0$, in the class $W^{2,\infty}([0, T[; L^2) \cap W^{1,\infty}([0, T[; H^1) \cap L^\infty([0, T[; H^2)$, satisfying the finite propagation speed property.*

We denote the life span of the solution $u(t, x)$ of the Cauchy problem $(P)$ by $T_{\max}$. First we consider the case $\lambda(x) \equiv const$ $(\lambda(x) \equiv 1$ without loss of generality). We construct a stable set in $H^1(\mathbb{R}^n)$.

Setting

$$C_0 \equiv K \left\{ \frac{2(p+1)}{(p-1)} \right\}^{\frac{(p-1)}{2}} \tag{4}$$

we assume

$$\int_0^\infty \sigma(\tau)\, d\tau = +\infty \quad \text{if } m = 1, \tag{5}$$

$$\int_0^\infty (1+\tau)^{-\frac{n(m-1)}{2}} \sigma(\tau)\, d\tau = +\infty \quad \text{if } m > 1. \tag{6}$$

Our result reads as follows (see [1]):

**Theorem 1.3.** *Let $u(t,x)$ be a local solution of the problem $(P)$ on $[0, T_{\max})$ with initial data $u_0 \in \widetilde{\mathcal{W}}^*$, $u_1 \in L^2(\mathbb{R}^n)$ with sufficiently small initial energy $E(0)$ so that*

$$C_0 E(0)^{\frac{(p-1)}{2}} < 1.$$

*Then $T_{\max} = \infty$. Furthermore, the global solution of the Cauchy problem $(P)$ has the following energy decay property:*
*Under (2), (3) and (5) there exists a positive constant $\omega$ such that:*

$$E(t) \leq E(0) \exp\left( 1 - \omega \int_0^t \sigma(\tau)\, d\tau \right) \quad \forall t > 0. \tag{7}$$

*Under (2), (3) and (6) there exists a positive constant $C(E(0))$ depending on $E(0)$ in a continuous way such that:*

$$E(t) \leq \left( \frac{C(E(0))}{\displaystyle\int_0^t (1+\tau)^{-\frac{n(m-1)}{2}} \sigma(\tau)\, d\tau} \right)^{\frac{2}{(m-1)}} \quad \forall t > 0. \tag{8}$$

Secondly, we consider the case $\lambda(x) \not\equiv const$ and we assume that

$$\frac{n+4}{n} \leq p \leq \frac{n}{n-2}.$$

**Case 1)** Let $\sigma(t) = \mathcal{O}(\tilde{d}(t))$, where $\tilde{d}(t) = d(L+t)$.
  If $m = 1$, we suppose that

$$\int_0^\infty \sigma(\tau)\, d\tau = +\infty \tag{9}$$

with

$$\begin{cases} (\tilde{d}(t))^{-\frac{4-(n-2)(p-1)}{2}} \exp\left( 1 - \omega \int_0^t \sigma(\tau)\, d\tau \right)^{\frac{p-1}{2}} < \infty \\ \text{and} \\ (\tilde{d}(t))^{-1} \exp\left( \frac{1}{2} - \frac{\omega}{2} \int_0^t \sigma(\tau)\, d\tau \right) < \infty. \end{cases} \tag{10}$$

If $m > 1$, we suppose that

$$\int_0^\infty (1+\tau)^{-\frac{n(m-1)}{2}} \sigma(\tau)\, d\tau = \infty \tag{11}$$

with

$$\frac{(\tilde{d}(t))^{-\frac{4-(n-2)(p-1)}{2}}}{\left(\int_0^t (1+\tau)^{-\frac{n(m-1)}{2}}\sigma(\tau)\,d\tau\right)^{\frac{p-1}{m-1}}} < \infty \quad \text{and} \quad \frac{(\tilde{d}(t))^{-1}}{\left(\int_0^t (1+\tau)^{-\frac{n(m-1)}{2}}\sigma(\tau)\,d\tau\right)^{\frac{1}{m-1}}} < \infty.$$

(12)

**Case 2)** Let $\tilde{d}(t) = \mathcal{O}(\sigma(t))$.
If $m = 1$, we suppose that for some $0 \le \alpha < 1$

$$\int_0^\infty \frac{\tilde{d}^2(\tau)}{\sigma^\alpha(\tau)}\,d\tau = +\infty$$

(13)

with

$$\begin{cases} (\tilde{d}(t))^{-\frac{4-(n-2)(p-1)}{2}} \exp\left(1 - \omega \int_0^t \frac{\tilde{d}^2(\tau)}{\sigma^\alpha(\tau)}\,d\tau\right)^{\frac{p-1}{2}} < \infty \\ \text{and} \\ (\tilde{d}(t))^{-1} \exp\left(\frac{1}{2} - \frac{\omega}{2}\int_0^t \frac{\tilde{d}^2(\tau)}{\sigma^\alpha(\tau)}\,d\tau\right) < \infty. \end{cases}$$

(14)

If $m > 1$, we suppose that for some $0 \le \alpha < 1$

$$\int_0^\infty (1+\tau)^{-\frac{n(m-1)}{2}}\sigma^{-\frac{(1+\alpha)(1+m)-2}{2}}(\tau)\tilde{d}^{m+1}(\tau)\,d\tau = \infty$$

(15)

with

$$\begin{cases} \dfrac{(\tilde{d}(t))^{-\frac{4-(n-2)(p-1)}{2}}}{\left(\int_0^t (1+\tau)^{-\frac{n(m-1)}{2}}\sigma^{-\frac{(1+\alpha)(1+m)-2}{2}}(\tau)\tilde{d}^{m+1}(\tau)\,d\tau\right)^{\frac{p-1}{m-1}}} < \infty \\ \text{and} \\ \dfrac{(\tilde{d}(t))^{-1}}{\left(\int_0^t (1+\tau)^{-\frac{n(m-1)}{2}}\sigma^{-\frac{(1+\alpha)(1+m)-2}{2}}(\tau)\tilde{d}^{m+1}(\tau)\,d\tau\right)^{\frac{1}{m-1}}} < \infty. \end{cases}$$

(16)

Then we have the following theorem.

**Theorem 1.4.** *Let $(u_0, u_1) \in H^1 \times L^2$, $u_0 \in \widetilde{\mathcal{W}}^{**}$, and let the initial energy $E(0)$ be sufficiently small. We have the following cases:*

- $\sigma(t) = \mathcal{O}(\tilde{d}(t))$:
  *Suppose (2), (3), (9) and (10) or (2), (3), (11) and (12). Then the problem (P) admits a unique global solution $u(t) \in C([0,\infty); H^1) \cap C^1([0,\infty); L^2)$ and we obtain the same decay property as in Theorem 1.3.*
- $\tilde{d}(t) = \mathcal{O}(\sigma(t))$:
  *Suppose (2), (3), (13) and (14) or (2), (3), (15) and (16). Then the problem (P) admits a unique global solution $u(t) \in C([0,\infty); H^1) \cap C^1([0,\infty); L^2)$.*

*Furthermore, the global solution of the Cauchy problem (P) has the following energy decay property*

$$E(t) \leq E(0) exp \left( 1 - \omega \int_0^t \frac{\tilde{d}^2(\tau)}{\sigma^\alpha(\tau)} \, d\tau \right) \quad \forall t > 0 \qquad (17)$$

*with a positive constant $\omega$ if $m = 1$ and*

$$E(t) \leq \left( \frac{C(E(0))}{\int_0^t (1+\tau)^{-\frac{n(m-1)}{2}} \sigma^{-\frac{(1+\alpha)(1+m)-2}{2}}(\tau) \tilde{d}^{m+1}(\tau) \, d\tau} \right)^{\frac{2}{(m-1)}} \quad \forall t > 0$$

$$(18)$$

*with a positive constant $C(E(0))$ depending on $E(0)$ in a continuous way if $m > 1$.*

## 1.1. *The case $p > \dfrac{n}{n-2}$*

Consider now the problem $(P)$ with $\sigma(t) \equiv 1$ and $g : \mathbb{R} \to \mathbb{R}$ is an increasing $C^1$ function and we suppose that there exist $C_i > 0$; $i = 1, 2, 3, 4$ such that

$$C_3|v|^m \leq |g(v)| \leq C_4|v| \text{ if } |v| \leq 1, \qquad (19)$$

$$C_1|v| \leq |g(v)| \leq C_2|v|^r \text{ for all } |v| \geq 1, \qquad (20)$$

where $m \geq 1$ and $1 \leq r \leq \dfrac{n+2}{(n-2)_+}$. We set

$$E_1(t) = \|u''(t)\|_2^2 + \|\nabla_x u_t(t)\|_2^2 + \|\lambda(x)u'(t)\|_2^2.$$

If $m = 1$, we suppose that

$$\int_0^\infty \tilde{d}^2(\tau) \, d\tau = +\infty \qquad (21)$$

and that the conditions (13)and (14) are satisfied for $\sigma = 1$.
If $m > 1$, we suppose that

$$\int_0^\infty (1+\tau)^{-\frac{n(m-1)}{2}} \tilde{d}^{m+1}(\tau) \, d\tau = \infty \qquad (22)$$

and that the conditions (15)and (16) are satisfied for $\sigma = 1$.
We assume further that

the initial energy $E(0)$ is sufficiently small. $\qquad (23)$

Our result reads as follows (see [2]):

**Theorem 1.5.** *Let $(u_0, u_1) \in H^2 \times H^1$, suppose that*

$$\max\left\{ \frac{4}{n}, \frac{2}{(n-2)} \right\} \leq p - 1 < \frac{2}{(n-4)_+}, \qquad (24)$$

$$\int_0^\infty \left( \int_0^t (1+\tau)^{-\frac{n(m-1)}{2}} \tilde{d}^{m+1}(\tau) \, d\tau \right)^{-\frac{2k}{m-1}} dt < +\infty, \tag{25}$$

where $k = \dfrac{2 - (n-4)(p-1)}{4}$.

*(i)* When $p \leq \dfrac{n+2}{n-2}$, there exists a positive constant $\varepsilon$ independent of $(u_0, u_1)$ and $L$ such that if, in addition to (23), the condition

$$C(E(0))(E(0) + E_1(0))^\tau \leq \varepsilon \tag{26}$$

with $\tau = \dfrac{((n-2)(p-1)-2)}{4}$, is satisfied, the problem (P) admits a unique global solution in the class

$$u(t) \in W^{2,\infty}([0,\infty); L^2) \cap W^{1,\infty}([0,\infty); H^1) \cap L^\infty([0,\infty); H^2).$$

*Furthermore, the global solution of the Cauchy problem (P) has the following energy decay property:*

$$E(t) \leq E(0) exp\left(1 - \omega \int_0^t \tilde{d}^2(\tau) \, d\tau \right) \quad \forall t > 0 \;\; if \; m = 1, \tag{27}$$

$$E(t) \leq \left( \frac{C(E(0))}{\displaystyle\int_0^t (1+\tau)^{-\frac{n(m-1)}{2}} \tilde{d}^{m+1}(\tau) \, d\tau} \right)^{\frac{2}{(m-1)}} \quad \forall t > 0 \; if \; m > 1, \tag{28}$$

$$E_1(t) \leq C \quad for \; t \geq 0. \tag{29}$$

*(ii)* When $p > \dfrac{n+2}{n-2}$, we assume further, in addition to (26)

$$E_1^\delta(0)(E(0) + E_1(0)) \leq \varepsilon_1 \tag{30}$$

for a certain small $\varepsilon_1$ independent of $(u_0, u_1)$ and $L$, where

$$\delta = \frac{4 - (n-4)(p-1)}{(n-2)(p-1) - 4}.$$

*Then, the same conclusion as the part (i) holds.*

## 1.2. The case of bounded dissipation

Consider now the problem (P) with $\sigma(t) \equiv 1$ and $g : \mathbb{R} \to \mathbb{R}$ is an increasing $C^1$ function and we suppose that there exist $C_i > 0$; $i = 1, 2, 3, 4$ such that

$$C_3|v|^m \leq |g(v)| \leq C_4|v| \; if \; |v| \leq 1, \tag{31}$$

$$C_1 \leq |g(v)| \leq C_2|v|^r \; for \; all \; |v| \geq 1, \tag{32}$$

where $m \geq 1$ and $1 \leq r \leq \dfrac{n+2}{(n-2)_+}$.

If $m = 1$, we suppose that

$$\int_0^\infty \tilde{d}^2(\tau)\, d\tau = +\infty, \tag{33}$$

and if $m > 1$, we suppose that

$$\int_0^\infty (1+\tau)^{-\frac{n(m-1)}{2}} \tilde{d}^{m+1}(\tau)\, d\tau = \infty. \tag{34}$$

Our result reads as follows (see [3]).

**Theorem 1.6.** *Assume that $(u_0, u_1) \in H^2 \times H^1$ with compact support, the problem $(P)$ admits a global solution which satisfies the following decay estimate: If $m = 1$ and $n = 1, 2$, then*

$$E(t) \le C(E(0)) exp\left(1 - \omega \int_0^t \tilde{d}^2(\tau)\, d\tau\right) \quad \forall t > 0, \tag{35}$$

*where $C$ and $\omega$ are positive constants.*
*When $n \ge 3$, then*

$$E(t) \le \frac{C(E(0))}{\left(\displaystyle\int_0^t \tilde{d}^2(\tau)\, d\tau\right)^{\frac{4}{n-2}}} \quad \forall t > 0. \tag{36}$$

*If $m > 1$, then*

$$E(t) \le \left(\frac{C(E(0))}{\displaystyle\int_0^t (1+\tau)^{-\frac{n(m-1)}{2}} \tilde{d}^{m+1}(\tau)\, d\tau}\right)^{\frac{2}{\left(\max\left\{m, \frac{n}{2}\right\} - 1\right)}} \quad \forall t > 0, \tag{37}$$

*where $C$ is a positive constant.*

## References

1. A Benaissa and S. Mokeddem, *Global existence of solutions to the Cauchy problem for a wave equation with a weak nonlinear damping and sources terms*, Abstract and Applied Analysis **11** (2004), 935-955.
2. A. Benaissa and N. Amroun, *Some remarks on Global existence to the Cauchy problem of the wave equation with nonlinear dissipation*, Mathematische Nachrichten, accepted.
3. A Benaissa and S. Mokeddem, *Decay estimates of solutions to the Cauchy problem of the wave equation with bounded nonlinear dissipation*, ZAA (accepted).
4. J. Dieudonné, Calcul infinitésimal, Collection Methodes, Herman, Paris, 1968.
5. V. Georgiev and G. Todorova, *Existence of a solution of the wave equation with nonlinear damping and source terms*, J. Diff. Equat. **109** (1994), 295-308.
6. A. Haraux, *Two remarks on dissipative hyperbolic problems*, in: Research Notes in Mathematics, Pitman, 1985, p. 161-179.
7. P. Martinez, *A new method to decay rate estimates for dissipative systems*, ESAIM Control Optim. Calc. Var. **4** (1999), 419-444.

8. M. Nakao, *A difference inequality and its applications to nonlinear evolution equations,* J. Math. Soc. Japan **30** (1978), 747-762.

9. M. Nakao, *Energy decay of the wave equation with a nonlinear dissipative term ,* Funkcialaj Ekvacioj **26** (1983), 237-250.

10. M. Nakao and K. Ono, *Existence of global solutions to the Cauchy problem for semilinear dissipative wave equations,* Math. Z. **214** (1993), 325-342.

11. M. Nakao and K. Ono, *Global existence to the Cauchy problem of the semilinear wave equation with a nonlinear dissipation,* Funkcialaj Ekvacioj **38** (1995), 417-431.

12. G. Todorova, *Stable and unstable sets for the Cauchy problem for a nonlinear wave equation with nonlinear damping and source terms,* J. Math. Anal. Appl. **239** (1999), 213-226.

# NON-NEGATIVE SOLUTIONS OF THE CAUCHY PROBLEM FOR SEMILINEAR WAVE EQUATIONS AND NON-EXISTENCE OF GLOBAL NON-NEGATIVE SOLUTIONS

HIROSHI UESAKA

*Department of Mathematics*
*College of Science and Technology*
*Nihon University*
*E-mail: uesaka@math.cst.nihon-u.ac.jp*

In this note we consider the Cauchy problem for a class of semilinear wave equations in two space dimension $\mathbf{R}^2$. We first briefly report on results stating that non-negative local or global-in-time solutions exist for suitable non-compactly initial data. Then we employ these results to show that global solutions do not exist for initial data satisfying additional conditions on the order of vanishing at infinity.

**Key words:** Systems of semi-linear wave equations, non-negative solutions, non-existence of global solutions

**Mathematics Subject Classification:** 35L70, 35B05, 35L55

## 1. Introduction

We shall consider our subject only in $\mathbf{R}^2$ and report the results, but we can treat the same one in $\mathbf{R}^1$ and in $\mathbf{R}^3$, too.

We shall consider the following semilinear wave operator,

$$Pu := \partial_t^2 u - \triangle u - F(u), \tag{1}$$

where $F$ is of the form

$$F(u) := au + f(u), \ a = a(x,t) \geq 0 \text{ for } (x,t) \in \mathbf{R}^2 \times [0,T).$$

**Example.** A typical case of interest would be $F(u) = au + u^p$ with the exponent $p > 1$.

We shall consider the Cauchy problem for the semilinear wave operator $P$,

$$\mathbf{P} : \begin{cases} Pu = 0 \text{ in } \mathbf{R}^2 \times [0,T), \\ u(x,0) = \phi(x), \ \partial_t u(x,0) = \psi(x) \text{ in } \mathbf{R}^2, \end{cases} \tag{2}$$

where $0 < T \leq \infty$.

We will suppose in this note that $\mathbf{P}$ has a sufficiently smooth unique solution in $\mathbf{R}^2 \times [0,T)$. The solution may exist in $\mathbf{R}^2 \times [0,T)$ only for a finite $T < \infty$. If it exists up to $T = \infty$, it is called a global solution.

We shall show that the solution of **P** is non-negative if suitable initial data are given. Moreover, we shall give a proof of non-existence of non-negative global solutions if the initial data satisfy additional conditions. We shall employ the results on non-negativity of solutions to show the non-existence of global solutions.

Concerning the non-negativity of solutions we shall only give an indication of proof for the relevant results. However, for the non-existence of global solutions complete proofs will be given.

Only few results on non-negativity of solutions for wave equations seem to be published. Protter and Weinberger have proved non-negativity for linear equations (see [9]). Caffarelli and Friedman [3] have proved such results for semilinear equations. Antonini and Merle [1] state Caffarelli and Friedman's result on non-negativity of solutions and they give a simpler proof. Non-negativity of solutions for the same Cauchy problem as (2) has been proved in [13], but the obtained results were incomplete.

Our approach to show non-negativity of solutions in this note is based on Protter-Weinberger's idea for treating the linear case and is different from Caffarelli and Friedman's one.

We remark that solutions of the Cauchy problem for another class of semilinear wave equations than (1), i.e. $\partial_t^2 u - \Delta u + F(u) = 0$, where $F(u) = a(x,t)u + f(u)$ with $a(x,t) > 0$, which conserves the energy, changes their signs in some sense as time goes to $\infty$ for any non zero initial data (refer to [5] and [12], [13]).

Finally we shall make another remark. We can consider non-negativity of solutions of the Cauchy problems not only for a single equation but also for a system of semilinear wave equations.

We consider the system of semilinear wave equations in $\mathbf{R}^2 \times [0, T)$,

$$\begin{cases} \partial_t^2 u_1 - c_1^2 \Delta u_1 = F_1(u_1, u_2), \\ \partial_t^2 u_2 - c_2^2 \Delta u_2 = F_2(u_1, u_2), \end{cases} \tag{3}$$

where the propagation velocities $c_1, c_2$ are assumed to fulfil $c_1 \geq c_2 > 0$.

We prescribe initial conditions in $\mathbf{R}^2$ to the solutions of the above system,

$$\begin{cases} u_1(x,0) = \phi_1(x), \ \partial_t u_1(x,0) = \psi_1(x), \\ u_2(x,0) = \phi_2(x), \ \partial_t u_2(x,0) = \psi_2(x). \end{cases} \tag{4}$$

The semilinear term $F_k(u_1, u_2)$ is given by

$$F_k(u_1, u_2) = a_1 u_1 + a_2 u_2 + f_k(u_1, u_2), \ a_k = a_k(t, x) \geq 0.$$

*Examples.* The typical cases of interest for $F_k(u_1, u_2)$, $k = 1, 2$ are:

1.  $F_1(u_1, u_2) = \alpha_1 u_2^{p_1}$, $F_2(u_1, u_2) = \alpha_2 u_1^{p_2}$, where $\alpha_k > 0$, $p_k > 1$,

2.  $F_k(u_1, u_2) = (\alpha_k u_1 + \beta_k u_2)^{p_k}$, where $\alpha_k, \beta_k > 0$, $p_k > 1$ and $\begin{vmatrix} \alpha_1 & \alpha_2 \\ \beta_1 & \beta_2 \end{vmatrix} \neq 0$.

Examples as $F_k(u_1, u_2) = \alpha_k u_1^{p_k} u_2^{q_k}$ are excluded from our study.

We can prove that the Cauchy problem (3) and (4) has non-negative solutions as well as the Cauchy problem (2).

## 2. Non-Negative Solutions

We shall consider mainly single equations and state the results and the outline of the proof. At the end of this section we shall state some facts on the non-negativity of solutions for systems.

Using the well-known representation formula for solutions the Cauchy problem (2), we can transform it into the following integral equation:

$$u(x,t) = \frac{\partial}{\partial t} \frac{1}{2\pi} \int\int_{D_t} \frac{\phi(x+\xi)}{\sqrt{t^2 - |\xi|^2}} d\xi_1 d\xi_2 + \frac{1}{2\pi} \int\int_{D_t} \frac{\psi(x+\xi)}{\sqrt{t^2 - |\xi|^2}} d\xi_1 d\xi_2$$
$$+ \frac{1}{2\pi} \int_0^t d\tau \int\int_{D_{t-\tau}} \frac{F(u(x+\xi,\tau))}{\sqrt{(t-\tau)^2 - |\xi|^2}} d\xi_1 d\xi_2, \tag{5}$$

where $x = (x_1, x_2) \in \mathbf{R}^2, \xi = (\xi_1, \xi_2) \in \mathbf{R}^2, |x| = \sqrt{x_1^2 + x_2^2}$ and $D_t = \{\xi \in \mathbf{R}^2 | |\xi| < t\}$.

Several particular types of non-linearities will be treated in this section by transforming the Cauchy problem to the corresponding semilinear wave equation into integral form according to (5). We remark here that this idea can also be employed for diagonal wave equations systems with suitable semilinear coupling terms.

We impose the following assumptions on the non-linearity $F$ and the initial data $\phi$ and $\psi$ from **P**.

## Assumption 2.1

(1) The function $F = F(u)$ is $F(u) := au + f(u)$, where $a$ is a sufficiently smooth non-negative function $a : \mathbf{R}^2 \times [0, \infty) \to \mathbf{R}$ and $f : \mathbf{R} \to \mathbf{R}$ is a $C^1$-function,

(2) there is a sufficiently smooth non-negative function $a_0 : \mathbf{R}^2 \to \mathbf{R}$ such that $a(x,t) \geq a_0(x)$ for all $x \in \mathbf{R}^2$ and $t \in [0, \infty)$,

(3) we require for $f$ that $f'(s) \geq -\min\{a_0(x)| x \in \mathbf{R}^2\}$ for all $s \in \mathbf{R}$,

(4) $\phi \in C^4(\mathbf{R}^2)$ and $\phi \geq 0$,

(5) $\psi \in C^3(\mathbf{R}^2)$ and $\psi \geq 0$,

(6) the function $\phi$ satisfies $\triangle\phi + a_0\phi + f(\phi) \geq 0$.

Under these assumptions the following statement holds:

**Theorem 2.1.** *Suppose that **Assumption** 2.1 holds. Let $u$ be a $C^3$-solution of **P**. If $u$ exists for $(x,t)$ with $x \in \mathbf{R}^2$ and $0 < t < \infty$, then*

$$u(x,t) \geq u(x,0) = \phi(x) \geq 0.$$

In order to prove the main theorem we need the following auxiliary Cauchy problem with an initial velocity $\psi(x) \equiv 0$:

$$\mathbf{P_0} : \begin{cases} (\partial_t^2 - \triangle)v = F_0(v) \text{ in } \mathbf{R}^2 \times [0, T), \\ v(x,0) = \phi(x), \ \partial_t v(x,0) = 0 \text{ in } \mathbf{R}^2. \end{cases} \tag{6}$$

We impose the following assumptions on the non-linearity $F_0$ and data $\phi$ from $\mathbf{P_0}$:

**Assumption 2.2**

(1) The functions $f, \phi$ and $a_0$ are the same ones as in **Assumption 2.1**,
(2) $F_0(v) := a_0 v + f(v)$,
(3) the function $\chi$ which is defined by $\chi := \Delta\phi + F_0(\phi)$ is non-negative in $\mathbf{R}^2$.

Then we can show that the solution $v$ of $\mathbf{P_0}$ is non-negative.

**Theorem 2.2.** *Suppose that **Assumption 2.2** holds. Let $v$ be a $C^3$-solution of $\mathbf{P_0}$. If $v$ exists for $(x,t)$ with $x \in \mathbf{R}^2$ and $0 < t < \infty$, then*

$$v(x,t) \geq v(x,0) = \phi(x) \geq 0.$$

Let $u$ and $v$ be the solutions from **Theorem 2.1** and **Theorem 2.2** respectively. A comparison relation between $u$ and $v$ holds. Let us subtract $(\partial_t^2 - \Delta)v = F_0(v)$ from $(\partial_t^2 - \Delta)u = F(u)$. Then we can get a relation for $u - v$ and represent it by an integral equation. From this integral equation we can show

$$u(x,t) \geq v(x,t). \tag{7}$$

The above comparison relation is proved by a contradiction argument due to Keller, which will be also used in the proof of the following result.

*We shall give the outline of the proof of **Theorem 2.2**.*

Let $w(x,t) = \partial_t v(x,t)$. Differentiating (6) with respect to $t$, we can show that $w(x,t)$ satisfies the relation

$$\mathbf{P_1} : \begin{cases} (\partial_t^2 - \Delta)w = F_0'(v)w \text{ in } \mathbf{R}^2 \times [0,T), \\ w(x,0) = 0, \; \partial_t w(x,0) = \chi(x) \text{ in } \mathbf{R}^2. \end{cases} \tag{8}$$

We shall show that $w(x,t) \geq 0$, then we can get $v(x,t) \geq 0$ because $v$ is increasing in $t$ and $v(x,0) = \phi(x) \geq 0$.

By the same way as getting the relation (5), we have from (8)

$$w(x,t) = I_1(x,t) + I_2(x,t), \tag{9}$$

where

$$I_1(x,t) = \frac{1}{2\pi} \iint_{D_t} \frac{\chi(x+\xi)}{\sqrt{t^2 - |\xi|^2}} d\xi_1 d\xi_2,$$

$$I_2(x,t) = \frac{1}{2\pi} \int_0^t d\tau \iint_{D_{t-\tau}} \frac{F_0'(v(x+\xi,\tau))w(x+\xi,\tau)}{\sqrt{(t-\tau)^2 - |\xi|^2}} d\xi_1 d\xi_2.$$

Obviously, $I_1(x,t) > 0$ holds because of $\chi(x) \geq 0$. The proof of $I_2 \geq 0$ is not so straightforward. We use Keller's idea (see [7]). Let $(x_0, t_0)$ be any point where $w$ exists in the half space $t \geq 0$. We denote the backward characteristic cone with a vertex in $(x_0, t_0)$ by

$$C(x_0,t_0) = \{(x,t) \in \mathbf{R}^2 \times [0,\infty) | |x - x_0| \leq (t_0 - t)\}.$$

We assume that there exists a point $(x, t) \in C(x_0, t_0)$ such that $w(x, t) < 0$. Then we can show that the assumption leads to a contradiction, and consequently, we have shown $w \geq 0$ in $C(x_0, t_0)$. Hence we can conclude that $v \geq 0$ and $u \geq 0$ hold in $\mathbf{R}^2 \times [0, T)$.

We consider the Cauchy problem for the system (3) with the initial condition (4). We assume for the Cauchy problem almost the same assumptions as Assumption 2.1, the principle conditions are for $k = 1, 2$

$$\phi_k \geq 0, \quad \psi_k \geq 0 \quad \text{and} \quad c_k^2 \Delta \phi_k + F_k(\phi_1, \phi_2) \geq 0.$$

We remark here that the same ideas for treating a single equation can be used, hence the non-negativity of solutions can be proved.

We are going to publish a paper with complete proofs on non-negativity of solutions of the Cauchy problems for single semilinear wave equations or for diagonal wave equations system with suitable semilinear coupling terms somewhere.

## 3. Non-Existence of Global Solutions

As an application of the results about the existence of non-negative solutions from the previous section, we shall show that no global solution exists for the Cauchy problem of a semilinear wave equation if the initial data vanish at infinity to the order less than a certain positive number $k + 1$.

We shall treat the problem $\mathbf{P}$ with $F(u) = Au^p$ $(A > 0, p > 1)$. The critical value $p_0(2) = \frac{3+\sqrt{17}}{2}$ for the exponent $p$ is well-known. If $1 < p \leq p_0(2)$, there is no global solution of (2) for any initial data. If $p > p_0(2)$, there is a global solution for sufficiently small initial data with compact support. For non-compact initial data, provided that $p > p_0(2)$, a different situation, that is, the non-existence of global solutions arises if the initial data satisfies some condition on the order at infinity.

Let us consider the problem with non-compact initial data in this note. Several results on existence and non-existence of global solutions in $\mathbf{R}^2$ have been obtained (see [8], [10] and [11]).

We shall confine our subject to the non-existence of global solutions for $p > p_0(2)$.

Let $u$ be the solution of (2). It seems to be that results on non-existence of global solutions have been obtained only for the case of the initial position $u(x, 0) \equiv 0$. We shall consider here the case $u(x, 0) \geq 0$.

We impose the following assumptions on $\mathbf{P}$ and $\mathbf{P_0}$ .

### Assumption 3.1

(1) The functions $F(u) = F_0(u) = Au^p$, where the exponent $p > p_0(2)$ and with a positive constant $A$,

(2) the data $\phi$ and $\psi$ satisfies the same conditions as those of **Assumption 2.1**,

(3) the function $\chi$ satisfies $\chi(x) = \Delta\phi(x) + A\phi(x)^p \geq \frac{C_0}{(1+|x|)^{1+k}}$ for $(x, t) \in \mathbf{R}^2$, where $C_0$ is a positive constant and $0 < k < \frac{2}{p-1} + 1$.

This assumption is contained in **Assumption 2.1-2.2** from Section 2. Thus the solution $u$ of **P** and the solution v of **P₀** are non-negative.

The established result on non-existence of global solutions is as follows:

Let $u$ be the solution of **P** with $F(u) = Au^p$. If $k$ satisfies $0 < k < \frac{2}{p-1}$ with $p > p_0(2)$ and the initial data satisfies $u(x,0) = \phi(x) \equiv 0$ and $\partial_t u(x,0) = \psi(x) \geq \frac{C_0}{(1+|x|)^{1+k}}$, then $u$ does not exist globally in time (see [10], [11]).

We shall show the non-existence of global solutions even if $\phi(x) \geq 0$.

**Theorem 3.1.** *Suppose that **Assumption 3.1** holds. Let $u$ be a $C^3$-solution of (2). If $k$ satisfies $0 < k < \frac{p+1}{p-1} = \frac{2}{p-1} + 1$, $u$ does not exist globally in time.*

Let $v$ be the solution of (6). Using the comparison relation $u \geq v$ from (7) in Section 2, it is sufficient to show that $v$ takes $\infty$ in finite time.

The way of the proof is almost the same argument as those by Asakura [2] and Tsutaya [10].

First we consider the linear problem,

$$\begin{cases} \Box w_0 = 0, \\ w_0(x,0) = 0, \quad \partial_t w_0(x,0) = \chi(x). \end{cases} \tag{10}$$

**Lemma 3.1.** *The sufficiently smooth solution $w_0$ of (9) exists uniquely and satisfies*

$$w_0(x,t) \geq \frac{C_0 t}{(1+t+|x|)^{1+k}}.$$

□

For the proof we refer to Proposition 5.1 in [10].

*We shall begin to prove the theorem.*

Let $v$ be the non-negative solution of (6) and hence $w = v_t$ is the solution of (8).

We shall show that $v$ takes $\infty$ in finite time by using a contradiction argument. Accordingly we assume that there exists a global $C^3$-solution. Then this assumption will lead to a contradiction.

The function $w$ becomes a global $C^2$-solution of (8) because of this supposition.

Noting that $w \geq 0$ and so $F'(v)w \geq 0$, by the representation formula (5) we have

$$w(x,t) \geq w_0(x,t) \geq \frac{C_0 t}{(1+t+|x|)^{1+k}}. \tag{11}$$

First we prove that $w = v_t$ takes $\infty$ in finite time if the solution keeps its regularity. We prove it by a recurrence argument.

We assume:

$$w(x,t) \geq \frac{C t^a}{(1+t+|x|)^b},$$

where $C > 0$, $a \geq 1$ and $b > 1$. Then

$$v(x,t) = \int_0^t w(x,s)ds + \phi(x) \geq C \int_0^t \frac{s}{(1+s+|x|)^b}ds$$

$$\geq \frac{C}{(1+t+|x|)^b} \int_0^t s^a ds = \frac{C}{1+a}\frac{t^{a+1}}{(1+t+|x|)^b}. \tag{12}$$

Then inserting (11) into the right-hand of the first line in the following equation, we have

$$w(x,t) \geq \frac{1}{2\pi}\int_0^t d\tau \iint_{D_{t-\tau}} \frac{F_0'(v(x+\xi,\tau))w(x+\xi,\tau)}{\sqrt{(t-\tau)^2-|\xi|^2}}d\xi_1 d\xi_2$$

$$\geq \frac{1}{2\pi}\int_0^t ds \int_0^{t-s} \frac{r}{\sqrt{(t-s)^2-r^2}}dr \int_{|\omega|=1} \frac{pAC^p s^{p(a+1)-1}}{(1+a)^{p-1}(1+s+|x|)^{bp}}d\omega$$

$$\geq \frac{AC^p}{(1+a)^p}\frac{1}{(a+1)p+1}\frac{t^{(a+1)p+1}}{(1+t+|x|)^{bp}}. \tag{13}$$

We follow the argument by Tsutaya[10].

We define the sequences $\{a_n\}, \{b_n\}, \{C_n\}$ by

$$a_n = (a_{n-1}+1)p+1, \ b_n = pb_{n-1}, \ C_n = \frac{AC_{n-1}^p}{(1+a_{n-1})^p}\frac{1}{(a_{n-1}+1)p+1},$$

$$a_0 = 1, \ b_0 = 1+k.$$

Then we have

$$a_n = \frac{2p^{n+1}-(p+1)}{p-1}, \ b_n = (1+k)p^n, \ (n=1,2,\dots). \tag{14}$$

Moreover, we have

$$C_n = \frac{A(p-1)^{p+1}}{2^p}\frac{C_{n-1}^p}{(p^n-1)^p\{2p^{n+1}-(p+1)\}}. \tag{15}$$

Setting $B = \frac{A}{p}\left(\frac{p-1}{2}\right)^{p+1}$ we conclude

$$C_n > B \cdot 2p\frac{C_{n-1}^p}{(p^n)^p 2p^{n+1}} > B\frac{1}{p^{(p+1)n}}C_{n-1}^p$$

$$> B^{1+p}\frac{C_{n-2}^{p^2}}{p^{p(p+1)n+p(p+1)(n-1)}} > \cdots$$

$$> B^{\frac{p^{n-1}-1}{p-1}}\frac{C_0^{p^n}}{p^{(p+1)S_{n-1}}}, \tag{16}$$

where $S_{n-1} = \sum_{k=0}^{n-1}(n-k)p^k = p^n\sum_{j=1}^n \frac{j}{p^j}$. We remark that $\sum_{j=1}^\infty \frac{j}{p^j} < \infty$. Thus we have

$$w(x,t) \geq B^{\frac{-1}{p-1}}t^{-\frac{p+1}{p-1}}\exp\left(p^n K\right), \tag{17}$$

where

$$K = \ln(BC_0) - (p+1)\sum_{j=1}^{\infty} \frac{j}{p^j} \ln p + \frac{2p}{p-1}\ln t - (1+k)\ln(1+t+|x|). \quad (18)$$

The inequality $0 < 1 + k < \frac{2p}{p-1}$ holds from the hypothesis of the theorem. By choosing $t$ large enough, we can find a positive number $\delta$ depending on $x$ such that

$$K \geq \delta > 0. \quad (19)$$

Then we have

$$w(x,t) \geq B^{\frac{-1}{p-1}} t^{-\frac{p+1}{p-1}} \exp(p^n \delta), \quad (n = 1, 2, \cdots). \quad (20)$$

Let $n \to \infty$ in (20). Then we get $w(x,t) = \infty$ for a sufficiently large $t$. Thus $v(x,t) = \infty$ for sufficiently large $t$, too. This is a contradiction. Thus $v$ attains $\infty$ in finite time, and then $u$ also attains $\infty$ in finite time if the supposed regularity of $u$ holds.

□

Finally, we make a remark on the range of $k$. We have treated initial data related to a rational function $\frac{C_0}{(1+|x|)^{k+1}}$ to show the non-existence of global solutions. In the established result global solutions exist if $\frac{2}{p-1} \leq k$. Our result says that global solutions do not exist for $0 < k < \frac{2}{p-1} + 1$. These two results contradict seemingly each other for $\frac{2}{p-1} \leq k < \frac{2}{p-1} + 1$. Of course there is no discrepancy between them. The assumption of the established result is as follows:

$$\phi(x) = 0, \ \psi(x) \geq \frac{C_0}{(1+|x|)^{k+1}}.$$

On the other hand the assumption on initial data in our theorem is:

$$\chi(x) = \Delta\phi(x) + A\phi(x)^p \geq \frac{C_0}{(1+|x|)^{k+1}} \text{ and } \psi(x) \geq 0.$$

## References

1. C.Antonini and F. Merle, *Optimal bounds on positive blow-up solutions for a semilinear wave equation*, International Mathematical Research Notices, No.21, (2001), 1141-1167.
2. F. Asakura, *Existence of a global solution to a semi-linear wave equation with slowly decreasing initial data in three space dimensions*, Comm. PDEs., 11 (1986), 1459-1487.
3. L.A.Caffarelli and A. Friedman, *The blow-up boundary for nonlinear wave equations*, Transaction of A.M.S.,**297** (1986), 223-241.
4. R. Glassey, *Existence in the large for $\Box u = F(u)$ in two dimensions*, Math. Z., **178** (1981), 233-261.
5. A. Haraux, *Semi-linear hyperbolic problems in bounded domains*, Math. Report, vol.3 part 1, harwood, 1987.
6. F. John, *Blow-up solutions of nonlinear wave equations in three space dimensions*, Manuscripta Math., **28** (1979), 235-268.

7. J. B. Keller, *On solutions of nonlinear wave equations,* Comm. Pure. Appl. Math., **38** (1957), 523-530.

8. K. Kubota, *Existence of a global solution to a semi-linear wave equation with initial data of non-compact support in low space dimensions,* Hokkaido Math. J. , **22** (1993), 123-180

9. M. Protter and H. Weinberger, *Maximum principles in partial differential equations,* Springer, 1984.

10. K. Tsutaya, *A global existence theorem from semilinear wave equations with data of noncompact support in two space dimensions,* Comm.PDEs, **17** (1992), 1925-1954.

11. K. Tsutaya, *Global existence theorem for semilinear wave equations with noncompact data in two space dimensions,* J. Diff. Eqs., **104** (1993), 332-360.

12. H. Uesaka, *A pointwise oscillation property of semilinear wave equations with time-dependent coefficients,* N. Anal.TMA., **54** (2003), 1271-1283.

13. H. Uesaka, *Oscillation or nonoscillation property for semilinear wave equations,* J. Comput. Appl. Math., **164-165**(2004), 723-730.

# ON THE LARGE TIME BEHAVIOR OF SOLUTIONS TO SEMILINEAR SYSTEMS OF THE WAVE EQUATION

S. KATAYAMA

*Department of Mathematics, Wakayama University,*
*930 Sakaedani, Wakayama 640-8510, Japan*

H. KUBO *

*Department of Mathematics, Graduate School of Science,*
*Osaka University, Toyonaka, Osaka 560-0043, Japan*
*E-mail: kubo@math.sci.osaka-u.ac.jp*

We consider the asymptotic behavior of solutions to the Cauchy problem for semilinear systems of wave equations in three space dimensions. If the nonlinearity satisfies the *null condition*, then there exists a global small amplitude solution whose asymptotic behavior is characterized by the free solution. In this note we shall show that there exist small amplitude solutions to certain semilinear systems of wave equations whose large time behavior actually differs from that of any free solutions.

**Key words:** Hyperbolic system, asymptotic behavior, energy increase
**Mathematics Subject Classification:** 35L70

## 1. Introduction

In this note we consider the asymptotic behavior of solutions to the Cauchy problem for semilinear systems of wave equations:

$$\partial_t^2 u_i - \Delta u_i = F_i(\partial u) \quad \text{in} \quad \mathbb{R}^3 \times (0, \infty), \tag{1}$$

where $i = 1, \cdots, N$, $\Delta = \sum_{j=1}^3 \partial_j^2$, $\partial = (\partial_0, \partial_1, \partial_2, \partial_3)$, $\partial_j = \partial/\partial x_j$, $\partial_0 = \partial_t = \partial/\partial t$ and $u(x,t) = (u_1(x,t), \cdots, u_N(x,t))$ is a real-valued unknown function. Besides, $F_i \in C^1(\mathbb{R}^{4N})$ is a given function satisfying

$$F_i(0) = \nabla F_i(0) = 0.$$

Our purpose here is to show that there are examples of nonlinearities $F$ such that the corresponding equation (1) cannot be regarded as a perturbation from the system of homogeneous wave equations:

$$\partial_t^2 v_i - \Delta v_i = 0 \quad \text{in} \quad \mathbb{R}^3 \times (0, \infty), \tag{2}$$

---

*This work is supported by Grant-in-Aid for Science Research (No.17540157), JSPS.

even if we restrict our attention to small amplitude solutions. It is well-known that $v_i(t, x)$ behaves like $O(t^{-1})$ as $t \to \infty$ and is compactly supported near the light cone $t = |x|$ if the support of the initial data are bounded. In addition, its energy is preserved.

Observe that the quadratic nonlinearity is of critical order for the Cauchy problem to (1) concerning the *small data global existence* and *blowup* (see e.g. John[5]). When $F(\partial u)$ satisfies the so-called *null condition*, namely the quadratic part of it can be written as a linear combination of the following null forms:

$$Q_0(u_j, u_k) = (\partial_t u_j)(\partial_t u_k) - (\nabla u_j) \cdot (\nabla u_k), \tag{3}$$

$$Q_{ab}(u_j, u_k) = (\partial_a u_j)(\partial_b u_k) - (\partial_b u_j)(\partial_a u_k) \quad (0 \le a < b \le 3), \tag{4}$$

the problem for (1) admits a unique global solution for small initial data (see e.g. Christodoulou[3], Klainerman[6]). Moreover, it is known that the global solution approaches to some free solution (see Kubo–Ohta[8]). This means that the semilinear system (1) satisfying the *null condition* can be regarded as a perturbation from the free system (2).

Therefore, it is natural to ask what will happen when (1) does not satisfy the *null condition*. If we consider the scalar case, i.e., $N = 1$, the *null condition* seems necessary to ensure *small data global existence* (see e.g. Hanouzet–Joly[4]). However, when $N \ge 2$, in Alinhac[2] the global existence result for (1) with small initial data

$$u_j(0, x) = \varepsilon f_j(x), \quad (\partial_t u_j)(0, x) = \varepsilon g_j(x) \quad \text{for} \quad x \in \mathbb{R}^3, \tag{5}$$

is recently proved under ceratin weaker assumption on the nonlinearity. One of the typical example satisfying the condition is

$$\begin{cases} \partial_t^2 u_1 - \Delta u_1 = (\partial_1 u_1)(\partial_1 u_2 - \partial_2 u_1) \text{ in } \mathbb{R}^3 \times (0, \infty), \\ \partial_t^2 u_2 - \Delta u_2 = (\partial_2 u_1)(\partial_1 u_2 - \partial_2 u_1) \text{ in } \mathbb{R}^3 \times (0, \infty). \end{cases} \tag{6}$$

Notice that the nonlinearities in (6) do not satisfy the *null condition*. Our main goal of this paper is to show that this semilinear system cannot be regarded as a perturbation from the free system (2), by proving that the energy of the global solution possibly grows up as $t \to \infty$. In other words, for these nonlinearities, the effect of perturbation possibly remains so strong even in sufficiently large time that the perturbed solution cannot approach to any free solutions. To our knowledge, there are only few results around this for nonlinear wave equations (see e.g. Alinhac[1-2], Kubo–Kubota–Sunagawa[7], Lindblad–Rodnianski[9-10]).

The key of the proof of the existence result given in [2] is to introduce an auxiliary function $w = \partial_1 u_2 - \partial_2 u_1$. Then we have

$$\partial_t^2 w - \Delta w = Q_{12}(w, u_1) \qquad \text{in} \quad \mathbb{R}^3 \times (0, \infty), \tag{7}$$

$$w(x, 0) = \varepsilon f_3(x), \quad (\partial_t w)(x, 0) = \varepsilon g_3(x) \quad \text{for} \quad x \in \mathbb{R}^3, \tag{8}$$

where $Q_{12}(w, u_1) = (\partial_1 w)(\partial_2 u_1) - (\partial_2 w)(\partial_1 u_1)$ and

$$f_3 = \partial_1 f_2 - \partial_2 f_1, \quad g_3 = \partial_1 g_2 - \partial_2 g_1. \tag{9}$$

Since the right-hand side of (7) is written in term of the null form, there exists a unique solution $v_3$ of (2) for $j = 3$ satisfying

$$\lim_{t \to \infty} \|w(t) - v_3(t)\|_E = 0 \tag{10}$$

and $\|v_3(0)\|_E < \infty$ (see the Subsection 2.4 below for details). Here we have defined the energy

$$\|v(t)\|_E := \left( \int_{\mathbb{R}^3} \left( |\partial_t v(t,x)|^2 + |\nabla_x v(t,x)|^2 \right) dx \right)^{\frac{1}{2}} \tag{11}$$

for a smooth function $v = v(t,x)$.

Now, we state a result which shows that the asymptotic profile of the perturbed solution $(u_1, u_2)$ is actually different from any free solution. In fact, (12) means that $(u_1, u_2)$ *never* approaches them as $t$ tends to infinity.

**Theorem 1.1.** *There exist initial data $f_1$, $f_2$, $g_1$ and $g_2 \in C_0^\infty(\mathbb{R}^3)$ such that for the global solution $(u_1, u_2)$ to the Cauchy problem (6) and (5)*

$$\lim_{t \to \infty} \|u_j(t)\|_E = \infty \quad (j = 1, 2) \tag{12}$$

*holds, provided that $\varepsilon$ is sufficiently small.*

Next, we consider the asymptotic profile of the purterbed solution $(u_1, u_2)$ to the problem (6) and (5). To this end, we can rewrite the system (6) as

$$\partial_t \vec{u}_i(t) = \sum_{j=1}^{3} A_j \partial_j \vec{u}_i(t) + B_i(t) \vec{u}_i(t) + F_i(t) \quad \text{for} \quad t \geq 0, \ i = 1, 2, \tag{13}$$

where $\vec{u}_i(t) = {}^t(\nabla_x u_i, \partial_t u_i)$, $F_i(t) = {}^t(0,0,0,(w - v_3)(\partial_i u_1))$ and

$$A_j = \begin{pmatrix} 0 & 0 & 0 & \delta_{1j} \\ 0 & 0 & 0 & \delta_{2j} \\ 0 & 0 & 0 & \delta_{3j} \\ \delta_{1j} & \delta_{2j} & \delta_{3j} & 0 \end{pmatrix} \quad (j = 1, 2, 3),$$

$$B_1(t) = \begin{pmatrix} 0 & 0 & 0 & 0 \\ 0 & 0 & 0 & 0 \\ 0 & 0 & 0 & 0 \\ v_3(t) & 0 & 0 & 0 \end{pmatrix}, \quad B_2(t) = \begin{pmatrix} 0 & 0 & 0 & 0 \\ 0 & 0 & 0 & 0 \\ 0 & 0 & 0 & 0 \\ 0 & v_3(t) & 0 & 0 \end{pmatrix}.$$

Here $\delta_{ij}$ stands for the Kronecker delta. We see that there exists a fundamental solution $E_i(t,s)$ which allows us to express the solution of (13) as

$$\vec{u}_i(t) = E_i(t,0)\vec{u}_i(0) + \int_0^t E_i(t,s)F_i(s)ds \quad \text{for} \quad t \geq 0, \ i = 1, 2, \tag{14}$$

where $\vec{u}_i(0) = {}^t(\nabla f_i, g_i)$ (for instance, we refer to Yosida[11]). Then we have the following result.

**Theorem 1.2.** *Let* $(\vec{u}_1, \vec{u}_2)$ *be the solution to (14). Then for any initial data* $f_1$, $f_2$, $g_1$ *and* $g_2 \in C_0^\infty(\mathbb{R}^3)$, *there exists* $(u_1^+, u_2^+) \in \{L^2(\mathbb{R}^3)\}^4 \times \{L^2(\mathbb{R}^3)\}^4$ *such that*

$$\lim_{t\to\infty} \|\vec{u}_j(t) - E_j(t,0)u_j^+\|_E = 0 \quad (j = 1, 2). \tag{15}$$

REMARK. We can extend Theorems 1 and 2 to the case where the null forms are added in the right-hand sides of the system (6). Moreover, the size of the system is not restrictive. Indeed, similar results can be obtained for the following system:

$$\begin{cases} \partial_t^2 u_1 - \Delta u_1 = -(\partial_t u_1)(\partial_1 u_3) - (\partial_t u_2)(\partial_t u_3) & \text{in} \quad \mathbb{R}^3 \times (0, \infty), \\ \partial_t^2 u_2 - \Delta u_2 = (\partial_1 u_3)(\partial_1 u_1 + \partial_t u_2) & \text{in} \quad \mathbb{R}^3 \times (0, \infty), \\ \partial_t^2 u_3 - \Delta u_3 = (\partial_2 u_1)(\partial_1 u_3) + (\partial_t u_2)(\partial_2 u_3) & \text{in} \quad \mathbb{R}^3 \times (0, \infty). \end{cases} \tag{16}$$

Analyzing the system, $u_3$ and $\widetilde{w} := \partial_1 u_1 + \partial_t u_2$ play the same role as $u_1$ and $w$ for (6), since the right-hand sides of the first and third equations in (16) can be expressed as $-\widetilde{w}(\partial_t u_3) - Q_{01}(u_1, u_3)$ and $\widetilde{w}(\partial_2 u_3) - Q_{12}(u_1, u_3)$, respectively.

## 2. Preliminaries

### 2.1. *Notations and Basic Identities*

As usual, we denote $r = |x|$ and $\partial_r = \sum_{j=1}^{3} \frac{x_j}{r} \partial_j$. For $1 \leq j, k \leq 3$, we set $\Omega_{jk} = x_j \partial_k - x_k \partial_j$. We also write $\omega_j = \frac{x_j}{|x|}$ for $j = 1, 2, 3$. Then we have

$$\partial_i = \omega_i \partial_r - \sum_{j\neq i} \frac{\omega_j}{r} \Omega_{ij} \quad \text{for} \quad i = 1, 2, 3, \tag{17}$$

$$\Box = \partial_r^2 + \frac{2}{r}\partial_r + \frac{1}{r^2} \sum_{1\leq j<k\leq 3} \Omega_{jk}^2. \tag{18}$$

We introduce

$$S = t\partial_t + \sum_{j=1}^{3} x_j \partial_j = t\partial_t + r\partial_r, \quad L_j = x_j\partial_t + t\partial_j \quad \text{for} \quad j = 1, 2, 3,$$

and also

$$L_r = \sum_{j=1}^{3} \omega_j L_j = r\partial_t + t\partial_r.$$

If we define $D_+$ and $D_-$ by $D_+ = \partial_t + \partial_r$ and $D_- = \partial_t - \partial_r$, then we have

$$D_+ = \frac{1}{t+r}(S + L_r). \tag{19}$$

Set $\Gamma = \{\Gamma_0, \Gamma_1, \ldots, \Gamma_{10}\} = \{S, (L_j)_{1\leq j\leq 3}, (\Omega_{ij})_{1\leq i<j\leq 3}, (\partial_a)_{0\leq a\leq 3}\}$. Using a multi-index $\alpha = (\alpha_0, \ldots, \alpha_{10})$, we write $\Gamma^\alpha$ for $\Gamma_0^{\alpha_0} \cdots \Gamma_{10}^{\alpha_{10}}$. Similarly, $\Omega^\beta$ denotes $\Omega_{12}^{\beta_1}\Omega_{13}^{\beta_2}\Omega_{23}^{\beta_3}$ for $\beta = (\beta_1, \beta_2, \beta_3)$.

Next we introduce the basic identities (20), (21) and (22) below. We set $V(t, r, \omega) = r v(r\omega, t)$ for $(t, r, \omega) \in \mathbb{R}_+ \times \mathbb{R}_+ \times S^2$, where $\mathbb{R}_+ = (0, \infty)$ and $S^2$ is the two dimensional sphere. Then we have

$$r \Box v = D_+ D_- V - \frac{1}{r} \sum_{1 \leq j < k \leq 3} \Omega_{jk}^2 v, \tag{20}$$

$$r \partial_i v = -\frac{\omega_i}{2}(D_- V) + r \left( \frac{\omega_i}{2(t+r)}(Sv + L_r v) + \frac{\omega_i}{2r} v - \sum_{j \neq i} \frac{\omega_j}{r} \Omega_{ij} v \right), \tag{21}$$

and

$$2 r^2 \left( |\partial_t v|^2 + |\nabla_x v|^2 \right) \tag{22}$$

$$= |D_+ V|^2 + |D_- V|^2 + 2 \sum_{i=1}^{3} \left( \sum_{j \neq i} \omega_j \Omega_{ij} v \right)^2 - 2 r^2 \left( 2 \frac{v}{r}(\partial_r v) + \left( \frac{v}{r} \right)^2 \right).$$

### 2.2. Lower bounds for free solutions

For $R > 0$ and $x \in \mathbb{R}^3$, we define

$$B_R(x) = \left\{ y \in \mathbb{R}^3; |y - x| \leq R \right\}, \text{ and } S_R(x) = \left\{ y \in \mathbb{R}^3; |y - x| = R \right\}.$$

We consider the homogeneous wave equation:

$$\begin{cases} \Box w_0 = 0 & \text{in } \mathbb{R}^3 \times \mathbb{R}_+, \\ w_0(x, 0) = 0, \ (\partial_t w_0)(x, 0) = g_3(x) & \text{for } x \in \mathbb{R}^3. \end{cases} \tag{23}$$

It is well known that $w_0$ can be represented as

$$w_0(x, t) = \frac{1}{4\pi t} \int_{S_t(x)} g_3(y) dS_y, \tag{24}$$

where $dS_y$ is the surface element of the sphere $S_t(x)$. Employing this expression, we can show the following.

**Lemma 2.1.** *Let $\Lambda$ be a small conic neighborhood of $(-1, -1, 0)$, and*

$$\Omega = \left\{ y \in \mathbb{R}^3; \text{ there exists } x \in \Lambda \text{ such that } x \cdot y \geq 0 \right\},$$

*where $x \cdot y$ denotes the inner product of $\mathbb{R}^3$. Assume that*

$$g_3 \geq 0 \text{ in } \Omega; \ g_3(x) \geq c_1 > 0 \text{ for any } x \in \Omega \text{ with } |x| \leq 4\delta \tag{25}$$

*for some $\delta > 0$. Then there exists a positive constant $C_1$ such that*

$$w_0(x, t) \geq C_1 (1 + t)^{-1} \tag{26}$$

*holds for any $(x, t) \in \Lambda \times [3\delta, \infty)$ satisfying $t + \delta \leq |x| \leq t + 2\delta$.*

## 2.3. Basic estimates for the perturbed solutions

Let $(u_1, u_2)$ and $w$ be the global solution for the problem (6) with (5) and (7) with (8), respectively. First we note that, for sufficiently small $\varepsilon$, the estimates

$$\sum_{|\alpha| \leq 2} |\Gamma^\alpha u_i(x,t)| \leq C\varepsilon(1+t)^{-1+C\varepsilon}, \tag{27}$$

$$\sum_{|\alpha| \leq 2} |\Gamma^\alpha \partial u_i(x,t)| \leq C\varepsilon(1+t)^{-1+C\varepsilon}(1+|r-t|)^{-1}, \tag{28}$$

$$\sum_{|\alpha| \leq 2} |\Gamma^\alpha w(x,t)| \leq C\varepsilon(1+t)^{-1}(1+|r-t|)^{-2/3} \tag{29}$$

hold for all $(x,t) \in \mathbb{R}^3 \times \mathbb{R}_+$, $i = 1, 2$, together with

$$\sum_{|\alpha| \leq 2} \|\Gamma^\alpha(\partial u_i)(t)\|_{L^2(\mathbb{R}^3)} \leq C\varepsilon(1+t)^{C\varepsilon} \quad \text{for } t > 0 \text{ and } i = 1, 2, \tag{30}$$

due to Alinhac[2], where $C$ is a constant independent of $\varepsilon$. Investigating its proof, we also have

$$|w(x,t) - \varepsilon w_0(x,t)| \leq C\varepsilon^2(1+t)^{-1} \quad \text{for all } (x,t) \in \mathbb{R}^3 \times \mathbb{R}_+, \tag{31}$$

where $w_0(x,t)$ is the solution to (23).

Based on these estimates, one can derive bounds of $L^2$ norms of $u_1, u_2$ and $w$, by making use of the following classical estimate:

$$\|v(t)\|_{L^2(\mathbb{R}^3)} \leq C\left(\|v(0)\|_{L^2(\mathbb{R}^3)} + \|\partial_t v(0)\|_{L^{\frac{6}{5}}(\mathbb{R}^3)}\right) + C\int_0^t \|\Box v(\tau)\|_{L^{\frac{6}{5}}(\mathbb{R}^3)} d\tau.$$

**Lemma 2.2.** *Let $f_1$, $f_2$, $g_1$ and $g_2 \in C_0^\infty(\mathbb{R}^3)$. Then we have*

$$\sum_{|\alpha| \leq 2} \|\Gamma^\alpha w(t)\|_{L^2(\mathbb{R}^3)} \leq C\varepsilon, \tag{32}$$

$$\sum_{|\alpha| \leq 2} \|\Gamma^\alpha u_i(t)\|_{L^2(\mathbb{R}^3)} \leq C\varepsilon + C\varepsilon^2(1+t)^{\frac{2}{3}+C\varepsilon} \tag{33}$$

*for $t > 0$ and $i = 1, 2$. If we further assume that $f_1$ and $g_1$ are radially symmetric, then we have*

$$\sum_{|\beta|=2} \|\Omega^\beta u_1(t)\|_{L^2} \leq C\varepsilon^2(1+t)^{\frac{2}{3}+C\varepsilon}. \tag{34}$$

## 2.4. Asymptotic behavior of the auxiliary function

Let $u_1, u_2$ and $w$ be as in the previous subsection. We define

$$R[F](x,t) = -\frac{1}{4\pi t}\int_t^\infty \int_{S_t(x)} F(y, t-s)\, dS_y\, ds \quad \text{in } \mathbb{R}^3 \times [0, \infty) \tag{35}$$

and

$$v_3(x,t) = w(x,t) - R[Q_{12}(w, u_1)](x,t) \quad \text{in } \mathbb{R}^3 \times [0, \infty). \tag{36}$$

We claim that

$$|v_3(x,t) - w(x,t)| \leq C\varepsilon^2 \langle |x| + t \rangle^{-2+C\varepsilon} \quad \text{in} \quad \mathbb{R}^3 \times [0,\infty) \tag{37}$$

and

$$\|w(t) - v_3(t)\|_E \leq C\varepsilon^2 \langle t \rangle^{-2+C\varepsilon} \quad \text{for} \quad t > 0. \tag{38}$$

It is well-known that $v_3$ satisfies (2) with $j = 3$ and

$$\|w(t) - v_3(t)\|_E \leq C \int_t^\infty \|Q_{12}(w, u_1)(\tau)\|_{L^2(\mathbb{R}^3)} \, d\tau \tag{39}$$

holds (see e.g. Proposition 2.6 in [8]). We recall the following basic estimate for $R[F](x,t)$. For the proof, see e.g. Proposition 2.6 in [8] with $a = c = 1$ and $n = 3$ (i.e. $m = 1$).

**Lemma 2.3.** *Let $\mu > 2$, $\kappa > 1$ and $F \in C(\mathbb{R}^3 \times [0,\infty))$. Then there exists a constant $C_1 = C_1(\mu, \kappa) > 0$ such that*

$$\sup_{(x,t)\in\mathbb{R}^3\times[0,\infty)} \langle t + |x| \rangle^{\mu-1} |R[F](x,t)| \tag{40}$$

$$\leq C_1 \sup_{(x,t)\in\mathbb{R}^3\times[0,\infty)} \langle t + |x| \rangle^\mu \langle t - |x| \rangle^\kappa |F(x,t)|.$$

Now, applying (40) as $F = Q_{12}(w, u_1)$, one can see from (36) and the estimates introduced in the previous subsection that (37) follows. While, (39) yields (38).

## 3. Outline of the proof of the theorems

Since the proof of Theorem 1.2 is rather standard, we shall concentrate on the proof of Theorem 1.1. We choose the initial data in such a way that $f_1 = f_2 = 0$ and that $g_1$, $g_2 \in C_0^\infty(\mathbb{R}^3)$ are radially symmetric functions satisfying (25) and $\|g_1\|_{L^2(D_\delta)} > 0$, where $D_\delta = \{x \in \Lambda \, ; \, \delta \leq |x| \leq 2\delta\}$.

First we prove (12) for $j = 1$. Set $U_1(t, r, \omega) = ru_1(r\omega, t)$, and choose $\widetilde{\Lambda} = S^2 \cap \Lambda$. If we set $\Theta := (t + \delta, t + 2\delta) \times \widetilde{\Lambda}$ and

$$E[U_1](t) := \left( \iint_\Theta (D_-U_1)^2(t, r, \omega) dS_\omega dr \right)^{\frac{1}{2}}, \tag{41}$$

then we see from (22), (30) and (33) that

$$\|u_1(t)\|_E^2 \geq \frac{1}{2} E[U_1](t)^2 - C\varepsilon^2 \tag{42}$$

for $t > 0$, provided that $\varepsilon$ is chosen to be small enough. Therefore, it suffices to estimate $E[U_1](t)$ from below. It follows from (6), (20) and (21) with $v = u_1$ and $i = 1$ that

$$D_+D_-U_1 + \frac{\omega_1}{2} wD_-U_1 = \mathcal{R}, \tag{43}$$

where

$$\mathcal{R}(t, r\omega)$$

$$= \frac{1}{r} \sum_{1 \le j < k \le 3} \Omega_{jk}^2 u_1 + rw \left( \frac{\omega_1}{2(t+r)}(Su_1 + L_r u_1) + \frac{\omega_1}{2r} u_1 - \sum_{j \ne 1} \frac{\omega_j}{r} \Omega_{1j} u_3 \right).$$

Integrating (43) multiplied by $D_- U_1$ over $\Theta$, we obtain

$$\frac{d}{dt}\left(E[U_1](t)^2\right) + \iint_\Theta \omega_1 w(r\omega, t)(D_- U_1)^2(t, r, \omega) dS_\omega dr$$

$$= 2 \iint_\Theta \mathcal{R}(t, r\omega)(D_- U_1)(t, r, \omega) dS_\omega dr. \tag{44}$$

Recalling the assumptions $f_1 = 0$ and $\|g_1\|_{L^2(D_\delta)} > 0$ and using (29), (30), (33) and (34), we can deduce

$$E[U_1](3\delta) \ge C_0 \varepsilon \tag{45}$$

with some positive constant $C_0$, provided that $\varepsilon$ is small enough. While, we can derive a lower bound of $w(x, t)$ for $(x, t) \in \Lambda \times [3\delta, \infty)$ with $t + \delta \le |x| \le t + 2\delta$. In fact, for such $(x, t)$, we find from (26) and (31) that

$$w(x, t) \ge (C_1 \varepsilon - C\varepsilon^2)(1+t)^{-1} \ge (C_1 \varepsilon/2)(1+t)^{-1}, \tag{46}$$

provided that $\varepsilon$ is sufficiently small.

Now, from (44) and (46) we obtain

$$\frac{d}{dt} E[U_1](t) \ge C_2 \varepsilon (1+t)^{-1} E[U_1](t) - C_3 \varepsilon^2 (1+t)^{-\frac{4}{3}+\varepsilon}$$

for $t \ge 3\delta$, since we may assume $\omega_1 \le -1/2$ in $\Lambda$. Here $C_2 = C_1/4$ and $C_3$ is a positive constant. Hence this estimate, together with (45), leads to

$$E[U_1](t) \ge C_4 \varepsilon (1+t)^{C_2 \varepsilon} \tag{47}$$

for $t \ge 3\delta$ with some positive constant $C_4$, provided that $\varepsilon$ is small enough. Thus we find (12) for $j = 1$.

Next we prove (12) for $j = 2$. Since $w = \partial_1 u_2 - \partial_2 u_1$, it follows that

$$\|u_2(t)\|_E \ge \|\partial_2 u_1(t)\|_{L^2(\mathbb{R}^3)} - \|w(t)\|_{L^2(\mathbb{R}^3)}.$$

By (32) it is enough to consider $\|\partial_2 u_1(t)\|_{L^2(\mathbb{R}^3)}$. Proceeding as before, we obtain

$$\|\partial_2 u_1(t)\|_{L^2(\mathbb{R}^3)} \ge \frac{1}{4} E[U_1](t) - C\varepsilon.$$

From (47), we find the desired estimate. This completes the proof. $\qquad \square$

# References

1. S. Alinhac, *An example of blowup at infinity for a quasilinear wave equation*, Astérisque, Autour de l'analyse microlocale **284** (2003), 1–91.
2. S. Alinhac, *Semilinear hyperbolic systems with blowup at infinity*, Indiana Unvi. Math. J. **55** (2006), 1209–1232.
3. D. Christodoulou, *Global solutions of nonlinear hyperbolic equations for small initial data*, Comm. Pure Appl. Math. **39** (1986) 267–282.
4. B. Hanouzet and J. L. Joly, *Explosion pour des problémes hyperboliques semi-linéaires avec second membre non compatible*, C. R. Acad. Sci. Paris **301** (1985), 581–584.
5. F. John, *Blow-up of solutions for quasi-linear wave equations in three space dimensions*, Comm. Pure Appl. Math. **34** (1981), 29–51.
6. S. Klainerman, *The null condition and global existence to nonlinear wave equations*, Lectures in Applied Math. **23** (1986), 293 – 326.
7. H. Kubo, K. Kubota and H. Sunagawa, *Large time behavior of solutions to semilinear systems of wave equations*, Math. Ann. **335** (2006), 435–478.
8. H. Kubo and M. Ohta, *On the global behaviour of classical solutions to coupled systems of semilinear wave equations*, "New trends in the theory of hyperbolic eqiations" (M. Reissig and B.-W. Schulze ed.), Birkhäuser, 2005.
9. H. Lindblad and I. Rodnainski, *The weak null condition for Einstein's equations*, C. R. Acad. Sci. Paris **336** (2003), 901–906.
10. H. Lindblad and I. Rodnianski, *Global existence for the Einstein vacuum equations in wave coordinates*, to appear in Comm. Math. Phys.
11. K. Yosida, *"Functional Analysis"*, Springer-Verlag, 1987.

# LEVI CONDITIONS FOR HIGHER ORDER OPERATORS WITH FINITE DEGENERACY

FERRUCCIO COLOMBINI, GIOVANNI TAGLIALATELA

*Dipartimento di Matematica*
*Università di Pisa*
*Largo Bruno Pontecorvo 5*
*56127 Pisa, Italy*
*E-mail: colombini@dm.unipi.it*

*Dipartimento di Scienze Economiche*
*Area Matematica – IV Piano*
*Università di Bari*
*via Camillo Rosalba 53*
*70124 BARI, Italy*
*E-mail: taglia@dse.uniba.it*

We consider the Cauchy problem for a weakly hyperbolic equation assuming that the characteristic roots vanish with finite speed. We give some sufficient conditions in order the Cauchy problem to be well posed in Gevrey spaces and $\mathcal{C}^\infty$.

**Key words:** Hyperbolic equations with finite degeneracy, Levi conditions
**Mathematics Subject Classification:** 35L30

## 1. Introduction

In this note we deal with the Cauchy Problem

$$
\begin{cases}
L(t, \partial_t, \partial_x)u = \displaystyle\sum_{j=0}^{m-1} M_j(t, \partial_t, \partial_x)u, & (t, x) \in [-T, T] \times \mathbb{R}^n, \\
\partial_t^j u(t_0, x) = u_j(x), & x \in \mathbb{R}^n, \quad j = 0, \ldots, m-1,
\end{cases}
\tag{CP}
$$

where

$$
L(t, \partial_t, \partial_x) = \partial_t^m + \sum_{\substack{\alpha_0 + |\alpha| = m \\ \alpha_0 < m}} a_{\alpha_0,\alpha}(t)\partial_t^{\alpha_0}\partial_x^\alpha, \qquad a_{\alpha_0,\alpha} \in \mathcal{C}^m\big([-T,T]\big),
$$

$$
M_j(t, \partial_t, \partial_x) = \sum_{\alpha_0 + |\alpha| = j} b_{\alpha_0,\alpha}(t)\partial_t^{\alpha_0}\partial_x^\alpha, \qquad b_{\alpha_0,\alpha} \in \mathcal{C}^0\big([-T,T]\big),
$$

and we will give sufficient conditions so that the Cauchy problem (CP) is well-posed in $\mathcal{C}^\infty$ or in the Gevrey classes. We recall that a function $f \in \mathcal{C}^\infty(\mathbb{R}^n)$ belongs to the Gevrey space $\gamma^d = \gamma^d(\mathbb{R}^n)$ if for any compact $K$ there exists a constant $C_K$

such that

$$\sup_{x \in K} |\partial_x^\alpha f(x)| \le C_K \, |\alpha|!^d \quad \text{for any } \alpha \in \mathbb{N}^n.$$

We know that the Cauchy problem (CP) is well-posed in $C^\infty$ or in the Gevrey classes, only if $L$ is hyperbolic, that is the solutions $\tau_j = \tau_j(t, \xi)$, $j = 1, \ldots, m$, of the equation $L(t, \tau, \xi) = 0$ are real for any $t \in [-T, T]$ and $\xi \in \mathbb{R}^n$ (see [10–12]). Moreover, if the roots are simple, the Cauchy problem (CP) is well-posed in $C^\infty$ and in all Gevrey classes.

On the other hand, when the roots may coincide the problem is more difficult to handle, since several difficulties may appear. We describe two of them by two examples.

In [14] (see also [5]) it is shown that the Cauchy problem for the operator

$$\partial_t^2 - \exp(-2/t^\alpha) b^2 (1/t) \partial_x^2,$$

where $\alpha > 0$ and $b$ is a positive, smooth, non-constant periodic function, is well-posed in $C^\infty$ if, and only if, $\alpha \ge \dfrac{1}{2}$. This shows that oscillations of the characteristic roots near multiple points may destroy the well-posedness.

In [8] (see also [13]) it is proved that the Cauchy problem for the operator

$$\partial_t^2 - t^{2\kappa} \partial_x^2 + t^\ell \partial_x$$

is well-posed in $C^\infty$ if, and only if, $\ell \ge \kappa - 1$, which shows that the lower order terms should be dominated by the principal symbol. These conditions are usually called *Levi conditions*.

In this note, we assume that the operator $L$ has *finite degeneracy*, that is, its discriminant

$$\Delta(t, \xi) = \prod_{j \ne k} (\tau_j(t, \xi) - \tau_k(t, \xi))$$

vanishes only at finite order. More precisely, we assume:

**Assumption 1.** There exist $\kappa_1 \le \kappa_2 \le \cdots \le \kappa_m$, such that

$$|\tau_j(t, \xi) - \tau_k(t, \xi)| \approx |t|^{\kappa_j} |\xi|, \quad \text{if } j < k.$$

To control the oscillations near multiple points, we assume the following:

**Assumption 2.** There exists $\Lambda_m \ge 0$ such that:

$$|t|^{1+\Lambda_m} |\tau_j'(t, \xi)| \lesssim |\tau_j(t, \xi) - \tau_k(t, \xi)| \quad \text{if } j \ne k.$$

Here and in the following $A(t, \xi) \approx B(t, \xi)$ (resp. $A(t, \xi) \lesssim B(t, \xi)$) means that there exist $c_1, c_2 > 0$ (resp. there exists $c > 0$) such that

$$c_1 \, A(t, \xi) \le B(t, \xi) \le c_2 \, A(t, \xi) \quad (\text{resp. } A(t, \xi) \le c \, B(t, \xi)),$$

for any $t \in [-T, T]$ and $\xi \in \mathbb{R}^n$.

For the lower order terms we assume the following Levi condition:

**Assumption 3.** There exist $\Lambda_j \geq 0$, $j = 1, \ldots, m-1$, such that:

$$|t|^{\Lambda_j} \left| M_j\big(t, \tau_k(t,\xi), \xi\big) \right| \lesssim \left| \partial_\tau^{m-j} L\big(t, \tau_k(t,\xi), \xi\big) \right| \quad {}^\forall t \in [-T, T], \, {}^\forall \xi \in \mathbb{R}^n,$$

for $j = 1, \ldots, m-1$ and $k = 1, 2, \ldots, j+1$.

Hence we have:

**Theorem 1.** *Let $L$ be a hyperbolic operator, satisfying the Assumptions 1, 2, and 3.*
Let $d_m = \max\left(1 + \dfrac{\kappa_1 + 1}{\Lambda_m}, 2 + \dfrac{2}{\Lambda_m}\right)$,

$$d_{j,k} = \begin{cases} +\infty & \text{if } \min(s_{j-k}, \Lambda_j) \leq m - j, \\ \left(1 - \dfrac{(m-j)(\kappa_h + 1)}{s_h + (m - k - h)\kappa_h - [s_{j-k} - \Lambda_j]^+}\right)^{-1} & \text{if } \min(s_{j-k}, \Lambda_j) > m - j, \end{cases}$$

where $h = h(j,k) = \min\left\{ l \in \mathbb{N} \setminus \{0\} \mid s_l + l \geq m - k + [s_{j-k} - \Lambda_j]^+ \right\}$, and $s_j = \kappa_1 + \kappa_2 + \cdots + \kappa_j$. Finally, let $d_0 = \min\left(d_m, \min\limits_{0 \leq k \leq j \leq m-1} d_{j,k}\right)$.

Then (CP) is $\gamma^d$ well-posed for all $1 < d < d_0$.
If $d_0 = +\infty$, then (CP) is also $C^\infty$ well-posed.

In particular, we have:

**Theorem 2.** *If $\Lambda_j \leq m - j$, for $j = 1, 2, \ldots, m$, then (CP) is well-posed in all Gevrey classes and in $C^\infty$.*

**Example.** ([8], [13]) Let $m = 2$ and

$$L(t, \partial_t, \partial_x) = \partial_t^2 - a(t)\partial_x^2, \qquad M_1(t, \partial_t, \partial_x) = b_0(t)\partial_t + b_1(t)\partial_x.$$

We have the following equivalences:

| | | |
|---|---|---|
| Hyperbolicity | $\Longleftrightarrow$ | $a(t) \geq 0$ |
| Assumption 1 | $\Longleftrightarrow$ | $a(t) \approx |t|^{2\kappa} |\xi|^2$ |
| Assumption 2 | $\Longleftrightarrow$ | $|t|^{1+\Lambda_2} |a'(t)| \lesssim \sqrt{a(t)}$ |
| Assumption 3 | $\Longleftrightarrow$ | $|t|^{\Lambda_1} |b_1(t)| \lesssim \sqrt{a(t)}$, |

hence: $d_0 = \min\left(1 + \dfrac{\kappa+1}{\Lambda_2}, \dfrac{\Lambda_1 + \kappa}{\Lambda_1 - 1}\right)$.

In particular, if $a(t) = t^{2\kappa}$ and $b_1(t) = t^\ell$, we have $\Lambda_2 = 0$ and $\Lambda_1 = \kappa - \ell$, so that $d_0 = \dfrac{2\kappa - \ell}{\kappa - \ell - 1}$, if $\ell < \kappa - 1$ and $d_0 = +\infty$, if $\ell \geq \kappa - 1$.

**Remark.** Assumption 3 is equivalent to

$$|t|^{\Lambda_j} \left| M_j\big(t, \tau_k(t,\xi), \xi\big) \right| \lesssim |t|^{s_k + (j-k)\kappa_k} |\xi|^j,$$

for $j = 1, \ldots, m-1$ and $k = 1, 2, \ldots, j+1$. In particular

$$\left| M_j\big(t, \tau_{j+1}(t,\xi), \xi\big) \right| \lesssim |t|^{[s_j - \Lambda_j]^+} |\xi|^j, \tag{1}$$

where $[a]^+ = \max(a, 0)$. Indeed, since $L(\tau) = (\tau - \tau_1) \cdots (\tau - \tau_m)$, we have

$$\partial_\tau^{m-j} L(\tau_k) = \sum_{\substack{1 \le l_1 < l_2 < \cdots < l_j \le m \\ k \notin \{l_1, \ldots, l_j\}}} (\tau_k - \tau_{l_1}) \cdots (\tau_k - \tau_{l_j}),$$

and, by Assumption 1,

$$\left| \partial_\tau^{m-j} L(\tau_k) \right| \approx \prod_{l \in \{1, \ldots, j+1\} \setminus \{k\}} |\tau_k - \tau_l|.$$

## Some known results

Many papers are devoted to the Cauchy problem for higher order equations. We report here only few results which are near to ours.

In [4], Colombini and Orrú considered higher order homogeneous equations with finite degeneracy and they proved that the Assumption 2 (with $\Lambda_m = 0$) is necessary and sufficient in order for the Cauchy problem to be well-posed in $C^\infty$.

Colombini and Ishida [3] (see also [7]) proved that Assumption 2, is sufficient in order for the Cauchy problem to be well-posed in $\gamma^d$, if $1 < d < 1 + \dfrac{\kappa_1 + 1}{\Lambda_m}$. (We should remark that their result is applicable also to some classes of operators having infinite order degeneracy.)

Kajitani, Wakabayashi and Yagdjian [9] considered Levi conditions for $C^\infty$ well-posedness for higher order equations admitting also infinite order degeneracy and $x$-dependence.

In all the previous results, the multiple points (the zeros of the discriminant) are supposed to be isolated. In [6], d'Ancona and Kinoshita considered higher order equations whose principal symbol may have multiple points also on a dense set.

## Strong hyperbolicity in Gevrey classes

Theorem 1, also allows us to state a result on the strong hyperbolicity in Gevrey classes.

**Theorem 3.** *If Assumption 1 is satisfied, then the Cauchy problem is well posed in $\gamma^d$, for any $d < d^*$, with*

$$d^* = \left( 1 - \frac{\kappa_h + 1}{s_h + (m-h)\kappa_h} \right)^{-1} = \frac{s_h + (m-h)\kappa_h}{s_h + (m-h-1)\kappa_h - 1},$$

*where $h = \min\{ l \mid s_l + l \ge m \}$.*

Indeed, if Assumption 1 holds true, then Assumption 2 holds true with $\Lambda_m = \kappa_{m-1} - 1$, and Assumption 3 holds true with $\Lambda_j = s_j$. By simple calculations we have

$$d^* = \min_{j,k} d_{j,k} = d_{m-1,0} \le d_m,$$

hence Theorem 3 is a consequence of Theorem 1.

*Example.* If $m = 2$, then

$$d^* = \begin{cases} +\infty & \text{if } \kappa_1 = 1, \\ \dfrac{2\kappa_1}{\kappa_1 - 1} & \text{if } \kappa_1 \geq 2. \end{cases}$$

*Example.* ([2]) If $m = 3$, then

$$d^* = \min\left(\frac{3\kappa_1}{2\kappa_1 - 1}, \frac{\kappa_1 + 2\kappa_2}{\kappa_1 + \kappa_2 - 1}\right) = \begin{cases} 2 + \dfrac{1}{\kappa_2} & \text{if } \kappa_1 = 1, \\ \dfrac{3\kappa_1}{2\kappa_1 - 1} & \text{if } \kappa_1 \geq 2. \end{cases}$$

*Example.* If $m = 4$, then

$$d^* = \min\left(\frac{4\kappa_1}{3\kappa_1 - 1}, \frac{\kappa_1 + 3\kappa_2}{\kappa_1 + 2\kappa_2 - 1}\right) = \begin{cases} \dfrac{3}{2} + \dfrac{1}{2\kappa_2} & \text{if } \kappa_1 = 1, \\ \dfrac{3\kappa_2 + 2}{2\kappa_2 + 1} & \text{if } \kappa_1 = 2, \\ \dfrac{4\kappa_1}{3\kappa_1 - 1} & \text{if } \kappa_1 \geq 3. \end{cases}$$

*Example.* If $\kappa_j = 1$, $j = 1, \ldots, \left[\frac{m+1}{2}\right]$, then $d^* = \dfrac{m}{m-2}$.

*Example.* If $\kappa_1 \geq m - 1$, then $d^* = \dfrac{m\kappa_1}{(m-1)\kappa_1 - 1} = \dfrac{m}{m - 1 - 1/\kappa_1}$.

Note that we always have $d^* > \dfrac{m}{m-1}$ (cf. [1]).

## Operators not verifying Assumption 1

The technique we use to prove Theorem 1 can be used also to study well-posedness for operators whose characteristic roots cannot be ordered as in Assumption 1. However it seems to be more difficult to state general results without Assumption 1. For the sake of simplicity we consider only a fourth order operator, whose characteristic roots verify the following Assumption instead of Assumption 1.

**Assumption 1'.** There exist $\kappa_1 \leq \ell_1 \leq \ell_2$, such that:

$$\tau_1 - \tau_2 \approx t^\kappa, \quad \tau_3 - \tau_4 \approx t^\kappa, \quad \tau_1 - \tau_3 \approx t^{\ell_1}, \quad \tau_2 - \tau_4 \approx t^{\ell_2}.$$

We define $d_m$ as before and

$$\tilde{d}_{j,k} = \begin{cases} +\infty & \text{if } \min(s_{j-k}(k), \Lambda_j) \leq m - j, \\ \left(1 - \dfrac{(m-j)(\kappa_h + 1)}{s_h(k) + (m - k - h)\kappa_h - [s_{j-k}(k) - \Lambda_j]^+}\right)^{-1} & \text{if } \min(s_{j-k}(k), \Lambda_j) > m - j, \end{cases}$$

where

$$s_h(0) = \begin{cases} \kappa & \text{if } h = 1 \\ 2\kappa & \text{if } h = 2 \\ 2\kappa + \ell_2 & \text{if } h = 3 \end{cases}, \qquad s_h(k) = \begin{cases} \kappa & \text{if } h = 1 \\ \kappa + \ell_1 & \text{if } h = 2 \end{cases}, \quad \text{if } k > 0.$$

596

Note that $\widetilde{d}_{j,k}$ is defined as $d_{j,k}$ in Theorem 1, but in $\widetilde{d}_{j,k}$ the $s_h$ depend on $k$. We have:

**Theorem 4.** *Let* $\widetilde{d}_0 = \min\left( d_3 \,,\, \min\limits_{1\leq k\leq j\leq 3} \widetilde{d}_{j,k} \right).$

*If $L$ verifies Assumptions 1', 2, and 3, then the Cauchy problem (CP) is $\gamma^d$ well-posed for any $1 < d < \widetilde{d}_0$.*

*If, moreover, $\widetilde{d}_0 = +\infty$, then the Cauchy problem (CP) is also $C^\infty$ well-posed.*

Note that if $\kappa = \ell_1$, then Assumption 1' is equivalent to Assumption 1 with $\kappa_1 = \kappa_2$.

## 2. Idea of the Proof of Theorem 1

The proof of Theorem 1 is based on an energy estimate. Since the basic ideas are similar to those in [4] (see also [3,9]), we sketch only the essential differences, the complete proof will be published elsewhere.

Let $v(t,\xi)$ be the Fourier transform of $u(t,x)$ with respect to the $x$-variables, $v$ satisfies the ordinary differential equation depending on the parameter $\xi$:

$$L(t,\partial_t,i\xi)\,v = \sum_{j=0}^{m-1} M_j(t,\partial_t,i\xi)\,v \qquad (v = \widehat{u}).$$

As usual, we split the strip $[0,T] \times \mathbb{R}^n$ into two *zones*. Let $t_1(\xi) := |\xi|^{-(1-1/d_0)}$, where $d_0$ is defined in Theorem 1, we define the *Pseudodifferential* and *Hyperbolic Zone* by:

$$Z_{pd} = \left\{ (t,\xi) \in \mathbb{R}^n \,\middle|\, |t| \leq t_1(\xi), |\xi| \geq 1 \right\},$$
$$Z_{hyp} = \left\{ (t,\xi) \in \mathbb{R}^n \,\middle|\, |t| \geq t_1(\xi), |\xi| \geq 1 \right\}.$$

In each zone we will use a different energy.

### 2.1. *Estimation in the Pseudodifferential Zone*

For $(t,\xi) \in Z_{pd}$ we define the Sobolev-type energy

$$E_1(t,\xi) = \sum_{j=0}^{m-1} |\xi|^{2(m-j-1)} |v^{(j)}|^2.$$

By standard calculations we have

$$E_1'(t,\xi) \lesssim |\xi|\, E_1(t,\xi),$$

hence, by Gronwall's Lemma:

$$E_1(t'',\xi) \leq \exp\!\left( C\,|\xi|^{1/d_0} \right) E_1(t',\xi), \tag{2}$$

for any $(t',\xi), (t'',\xi) \in Z_{pd}$.

## 2.2. *Estimation in the Hyperbolic Zone*

Let

$$[v]_0^2 = |v|^2, \qquad [v]_1^2 = \sum_{j=1,\dots,m} |L_j\, v|^2, \qquad [v]_2^2 = \sum_{\substack{j,k=1,\dots,m \\ j\neq k}} |L_{jk}\, v|^2, \qquad \cdots$$

$$\cdots \qquad [v]_j^2 = \sum_{1\leq k_1 < k_2 < \cdots < k_j \leq m} \left| L_{k_1,\dots,k_j}(t,\partial_t, i\xi)v \right|^2 \qquad (1 \leq j \leq m),$$

where

$$L_k\, v = v' - i\tau_k v, \qquad\qquad k = 1,2,\dots,m,$$
$$L_{jk}\, v = v'' - i(\tau_j + \tau_k)v' - \tau_j\tau_k v, \qquad j,k \in \{1,2,\dots,m\}, \ j \neq k,$$
$$\vdots$$

and, in general, $L_{k_1,\dots,k_j}(t,\partial_t, i\xi)$ is the operator with symbol:

$$L_{k_1,\dots,k_j}(t,\tau,i\xi) = \prod_{l=1,\dots,j} \left(\tau - i\tau_{k_l}(t,\xi)\right).$$

For $(t,\xi) \in Z_{hyp}$, we define the energy:

$$E_2(t,\xi) = \sum_{j=0}^{m-1} \left(\frac{|\xi|^{1/d_0}}{t}\right)^{2(m-1-j)} [v]_j^2.$$

For $(t,\xi) \in Z_{hyp}$, $E_2$ is equivalent to $E_1$, in the sense that there exist $M_1, M_2$, such that:

$$|\xi|^{M_1}\, E_1(t,\xi) \lesssim E_2(t,\xi) \lesssim |\xi|^{M_2}\, E_1(t,\xi), \qquad \text{for all } (t,\xi) \in Z_{hyp}. \qquad (3)$$

We have

$$E_2'(t,\xi) \lesssim 2\sum_{j=0}^{m-1} \left(\frac{|\xi|^{1/d_0}}{t}\right)^{2(m-1-j)} \partial_t\, [v]_j^2.$$

In order to estimate $\partial_t\, [v]_j^2$ we need some preliminaries.

By definition, we have

$$L_1\, v - L_2\, v = -i(\tau_1 - \tau_2)v,$$

hence,

$$|v| \leq \frac{|L_1\, v - L_2\, v|}{|\tau_1 - \tau_2|} \lesssim \frac{[v]_1}{|\tau_1 - \tau_2|}. \qquad (4)$$

Similarly,

$$L_{12}\, v - L_{13}\, v = -i(\tau_2 - \tau_3)\, L_1\, v,$$

hence,

$$|L_1\, v| \leq \frac{|L_{12}\, v - L_{13}\, v|}{|\tau_2 - \tau_3|} \lesssim \frac{[v]_2}{|\tau_2 - \tau_3|}. \qquad (5)$$

Now, combining (4) and (5), we get

$$|v| \lesssim \frac{[v]_2}{|\tau_1 - \tau_2| \, |\tau_2 - \tau_3|} \, ,$$

since $|\tau_2 - \tau_3| \lesssim |\tau_1 - \tau_3|$. In general, we have:

**Lemma 2.1.** *Let* $l_1, l_2, \ldots, l_p \in \{1, 2, \ldots, m\}$, $p \leq m-2$, *with* $l_i \neq l_j$ *if* $i \neq j$. *Let* $k_1, k_2, \ldots, k_q \in \{1, 2, \ldots, m\} \setminus \{l_1, l_2, \ldots, l_p\}$ *with* $k_1 < k_2 < \cdots < k_q$, $q \geq 2$ *and* $p + q \leq m$. *Then we have*

$$\left| L_{l_1, \ldots, l_p} \, v \right| \lesssim \frac{[v]_{p+q-1}}{|\tau_{k_1} - \tau_{k_2}| \, |\tau_{k_2} - \tau_{k_3}| \cdots |\tau_{k_{q-1}} - \tau_{k_q}|} \, . \tag{6}$$

In order to estimate $\partial_t \, [v]_j^2$ we must estimate $\partial_t \left| L_{k_1, \ldots, k_j} \, v \right|^2$. We consider here only $\partial_t \, [v]_2$, but the general case can be treated in a similar way. By definition we have

$$\partial_t \, L_{12} \, v = L_{123} \, v + i\tau_3 \, L_{12} \, v - i\tau_1' \, L_2 \, v - i\tau_2' \, L_1 \, v \, ,$$

hence,

$$\begin{aligned}
\partial_t \left| L_{12} \, v \right|^2 &= 2 \operatorname{Re}(L_{123} \, v, L_{12} \, v) + 2 \operatorname{Re}(i\tau_3 \, L_{12} \, v, L_{12} \, v) \\
&\quad - 2 \operatorname{Re}(i\tau_1' \, L_2 \, v, L_{12} \, v) - 2 \operatorname{Re}(i\tau_2' \, L_1 \, v, L_{12} \, v) \\
&\leq 2 \left| L_{123} \, v \right| \, \left| L_{12} \, v \right| + 2 \left| \tau_1' \right| \left| L_2 \, v \right| \left| L_{12} \, v \right| + 2 \left| \tau_2' \right| \left| L_1 \, v \right| \left| L_{12} \, v \right| \, .
\end{aligned}$$

The terms containing the derivatives of the characteristic roots are estimated using Assumption 2. For instance, if $\Lambda_m = 0$, i.e. $|t| \left| \dfrac{\tau_j'}{\tau_j - \tau_k} \right| \leq C$, recalling (5), we can see that

$$\partial_t \, [v]_2^2 \lesssim [v]_3 \, [v]_2 + \frac{1}{t} \, [v]_2^2 \, .$$

In general, we have

$$\partial_t \, [v]_j^2 \lesssim [v]_{j+1} \, [v]_j + \frac{|\xi|^{1/d_0}}{t} \, [v]_j^2 \, , \qquad j = 0, 1, \ldots, m-1 \, ,$$

which gives

$$E_2'(t, \xi) \leq |Mv| \, \sqrt{E_2(t, \xi)} + \frac{|\xi|^{1/d_0}}{t} \, E_2(t, \xi)$$

since $[v]_m = |Lv| = |Mv|$.

We claim that

$$|t| \, |Mv| \lesssim |\xi|^{1/d_0} \, \sqrt{E_2(t, \xi)} \, , \quad \text{for any } (t, \xi) \in Z_{hyp} \, . \tag{7}$$

Indeed, if we prove (7), we get:

$$E_2'(t, \xi) \leq \frac{|\xi|^{1/d_0}}{t} \, E_2(t, \xi) \, ,$$

hence, by Gronwall's Lemma, we get:

$$E_2(t'',\xi) \lesssim |\xi|^{C_1} \exp(C_2\,|\xi|^{1/d_0})\,E_2(t',\xi) \tag{8}$$

for any $(t',\xi), (t'',\xi) \in Z_{hyp}$. Combining (2), (3) and (8), we conclude the proof of Theorem 1.

To prove (7) we need some preliminaries.

Let $M(X)$ be a polynomial of degree $j$ in one variable, we define

$$\Delta_0[M](X_0) = M(X_0),$$

$$\Delta_1[M](X_0,X_1) = \frac{\Delta_0[M](X_0) - \Delta_0[M](X_1)}{X_0 - X_1} = \frac{M(X_0) - M(X_1)}{X_0 - X_1},$$

$$\vdots$$

$$\Delta_k[M](X_0,\dots,X_{k-1},X_k) = \frac{\Delta_k[M](X_0,\dots,X_{k-1}) - \Delta_k[M](X_0,\dots,X_k)}{X_{k-1} - X_k}.$$

It's easy to see that if $M(X)$ is a polynomial of degree $j$, then $\Delta_k[M](X_0,X_1,\dots,X_k)$ is a symmetric polynomial of degree $j-k$ in $X_0, X_1, \dots, X_k$. Indeed, let us consider, for example, $M(X) = b_0 X^2 + b_1 X + b_2$, we have

$$\Delta_1[M](X_0,X_1) = \frac{b_0 X_0^2 + b_1 X_0 + b_2 - b_0 X_1^2 - b_1 X_1 - b_2}{X_0 - X_1} = b_0(X_0 + X_1) + b_1,$$

$$\Delta_2[M](X_0,X_1,X_2) = \frac{b_0(X_0 + X_1) + b_1 - b_0(X_0 + X_2) - b_1}{X_1 - X_2} = b_0.$$

If $M_j(t,\partial_t,i\xi)$ is a differential operator in $\tau$, whose coefficients are polynomial in $\xi$, we may define in an obvious way $\Delta_k[M_j]$ as $\Delta_k[M_j(t,\partial_t,i\cdot)]$. The $\Delta_k[M_j]$ inherit some vanishing conditions from the Levi conditions on $M_j$ (cf. (1)).

**Lemma 2.2.** *If $M_j$ verifies Assumption 3 then we have*

$$|\Delta_k[M_j](\tau_{j+1-k},\dots,\tau_{j+1})| \lesssim |t|^{[s_{j-k}-\Lambda_j]^+}\,|\xi|^{j-k}\,,$$

*where $s_l = \kappa_1 + \cdots + \kappa_l$.*

We show here only the case $m=3$, $j=2$, $k=1$. Assumption 3 implies that

$$|t|^{\Lambda_2}\,|M_2(\tau_2)| \lesssim |\tau_2 - \tau_1|\,|\tau_2 - \tau_3|\,, \qquad |t|^{\Lambda_2}\,|M_2(\tau_3)| \lesssim |\tau_3 - \tau_1|\,|\tau_3 - \tau_2|\,,$$

hence,

$$\frac{|\Delta_1[M_2](\tau_2,\tau_3)|}{|\tau_2 - \tau_1|} = \frac{|M_2(\tau_2)|}{|\tau_2 - \tau_1|\,|\tau_2 - \tau_3|} + \frac{|\tau_3 - \tau_1|}{|\tau_2 - \tau_1|}\,\frac{|M_2(\tau_3)|}{|\tau_3 - \tau_2|\,|\tau_3 - \tau_1|} \le t^{-\Lambda_2}\,|\xi|\,,$$

since, by Assumption 1, $|\tau_3 - \tau_1| \lesssim |\tau_2 - \tau_1|$.

**Proposition 2.3.** *Let $M_j(t,\partial_t,i\xi)$ be a differential operator of order $j$, then we have the following decomposition:*

$$M_j = \sum_{k=0}^{j} i^{j-k}\Delta_k[M](\tau_{j+1-k},\dots,\tau_{j+1})\mathcal{L}_k\,,$$

*where* $\mathcal{L}_0 = 1$, $\mathcal{L}_1 = L_m$, $\mathcal{L}_2 = L_{m-1,m}$, $\ldots$, $\mathcal{L}_j = L_{m-j+1,\ldots,m}$, $\ldots$

We show only the case $m = 3$ and $j = 2$. The proof in the general case is similar. Since

$$v' = \mathcal{L}_1 v + i\tau_3 v, \qquad v'' = \mathcal{L}_2 v + i(\tau_2 + \tau_3)v' + \tau_2\tau_3 v,$$

we have

$$b_0 v'' + ib_1 v' - b_2 v = b_0 \mathcal{L}_2 v + i\big(b_0(\tau_2 + \tau_3) + b_1 \mathcal{L}_1 v - \big(b_0\tau_3^2 + b_1\tau_3 + b_2\big)v$$
$$= \Delta_2[M_2](\tau_1, \tau_2, \tau_3)\,\mathcal{L}_2 v + i\Delta_1[M_2](\tau_2, \tau_3)\,\mathcal{L}_1 v - \Delta_0[M_2](\tau_3)\,v\,.$$

Now we return to the proof of (7). Thanks to Proposition 2.3, it's enough to prove, for $0 \le k \le j \le m - 1$:

$$|t|\,|\Delta_k[M](\tau_{j+1-k}, \ldots, \tau_{j+1})\mathcal{L}_k| \lesssim |\xi|^{1/d_0}\,\sqrt{E_2(t,\xi)}\,. \tag{9}$$

Combining Lemmas 2.1 and 2.2, we get, for $h = 1, \ldots, j - k$:

$$|\Delta_k[M_j](\tau_{j+1-k}, \ldots, \tau_{j+1})|\,|\mathcal{L}_k v| \lesssim \frac{|t|^{[s_{j-k}-\Lambda_j]^+ - s_h + m - k - h}}{t_1^{m-1-k-h}\,|\xi|^{m-1-j}}\,\frac{1}{|t|}\,\sqrt{E_2(t,\xi)}\,.$$

Now, choosing $h = h(j,k)$ as in Theorem 1, we get (9). For example, if $\min(s_{j-k}, \Lambda_j) \le m - j$, choosing $h = j - k$, we have

$$|\Delta_k[M_j](\tau_{j+1-k}, \ldots, \tau_{j+1})|\,|\mathcal{L}_k v| \lesssim \frac{1}{|t|}\,\sqrt{E_2(t,\xi)}\,,$$

since $[s_{j-k} - \Lambda_j]^+ - s_{j-k} + m - j \ge 0$ and $t_1 \ge |\xi|^{-\frac{1}{d_0}} \ge |\xi|^{-1}$.

## References

1. M. D. Bronšteĭn, *The Cauchy problem for hyperbolic operators with characteristics of variable multiplicity*, Trudy Moskov. Mat. Obshch. **41** (1980), 83–99.
2. F. Colombini, *Quelques remarques sur le problème de Cauchy pour des équations faiblement hyperboliques*, Journées "Équations aux Dérivées Partielles" (Saint-Jean-de-Monts, 1992), École Polytech., Palaiseau, 1992, pp. Exp. No. XIII, 6.
3. F. Colombini and H. Ishida, *Well-posedness of the Cauchy problem in Gevrey classes for some weakly hyperbolic equations of higher order*, J. Anal. Math. **90** (2003), 13–25.
4. F. Colombini and N. Orrú, *Well-posedness in $C^\infty$ for some weakly hyperbolic equations*, J. Math. Kyoto Univ. **39** (1999), no. 3, 399–420.
5. F. Colombini and S. Spagnolo, *An example of a weakly hyperbolic Cauchy problem not well posed in $C^\infty$*, Acta Math. **148** (1982), 243–253.
6. P. d'Ancona and T. Kinoshita, *On the wellposedness of the Cauchy problem for weakly hyperbolic equations of higher order*, Math. Nach. **278** (2005), n. 10, 1147–1162.
7. H. Ishida and K. Yagdjian, *On a sharp Levi condition in Gevrey classes for some infinitely degenerate hyperbolic equations and its necessity*, Publ. Res. Inst. Math. Sci. **38** (2002), no. 2, 265–287.
8. V.Ja. Ivrii, *Cauchy problem conditions for hyperbolic operators with characteristics of variable multiplicity for Gevrey classes*, Sib. Math. J. **17** (1977), 921–931.

9. K. Kajitani, S. Wakabayashi, and K. Yagdjian, *The hyperbolic operators with the characteristics vanishing with the different speeds*, Osaka J. Math. **39** (2002), no. 2, 447–485.

10. P. D. Lax, *Asymptotic solutions of oscillatory initial value problems*, Duke Math. J. **24** (1957), 627–646.

11. S. Mizohata, *Some remarks on the Cauchy problem*, J. Math. Kyoto Univ. **1** (1961), 109–127.

12. T. Nishitani, *On the Lax-Mizohata theorem in the analytic and Gevrey classes*, J. Math. Kyoto Univ. **18** (1978), no. 3, 509–521.

13. K. Shinkai and K. Taniguchi, *Fundamental solution for a degenerate hyperbolic operator in Gevrey classes*, Publ. Res. Inst. Math. Sci. **28** (1992), no. 2, 169–205.

14. Tarama, S., *On the second order hyperbolic equations degenerating in the infinite order – Example*, Math. Japon. **42** (1995), no. 3, 523–533.

# A THEORY OF DIAGONALIZED SYSTEMS OF NONLINEAR EQUATIONS AND APPLICATION TO AN EXTENDED CAUCHY-KOWALEVSKY THEOREM

SADAO MIYATAKE

*Department of Mathematics*
*Nara Women's University*
*Kitauoya-Nishimachi, Nara 630-8506, Japan*

We study the Cauchy problem for the diagonalized system of nonlinear hyperbolic equations of first order. This problem is equivalent to the Cauchy problem for Hamilton system, which is the system of generalized ordinary differential equations of composite type. The solution of evolution type in the space of bounded continuous functions in $R^n$ up to second order derivatives, is constructed by the limit of approximate solutions to the Hamilton system. The problem for quasi-linear diagonalized system introduced by B.Riemann is solved in the above space as a typical example. Our theorem can be applied to extend the local theorem of the Cauchy-Kowalevsky type into the global theory. Namely the solution can be considered as wide as possible until the natural boundary.

**Key words:** Cauchy problem, nonlinear systems of first order, Cauchy-Kovalevsky theorem, Hamiltonian system

**Mathematics Subject Classification:** 35A10, 35F25

## 1. Introduction

In 1860, B. Riemann [3] considered the equations of gas dynamics

$$(1) \quad \begin{cases} \dfrac{\partial u}{\partial t} + u\dfrac{\partial u}{\partial x} = -\varphi(\rho)\dfrac{1}{\rho}\dfrac{\partial \rho}{\partial x}, \\ \dfrac{\partial \rho}{\partial t} + u\dfrac{\partial \rho}{\partial x} = -\rho\dfrac{\partial u}{\partial x}, \end{cases}$$

where $\varphi(\rho)$ is assumed to be a monotone increasing function, for example, $\varphi(\rho) = \rho^k$, $k > 1$. He transformed the system (1) to the following quasi-linear diagonalized system

$$(2) \quad \begin{cases} \dfrac{\partial r}{\partial t} + c_1(r,s)\dfrac{\partial r}{\partial x} = 0, \\ \dfrac{\partial s}{\partial t} + c_2(r,s)\dfrac{\partial s}{\partial x} = 0, \end{cases}$$

introducing so-called Riemann invariants: $r = \frac{f(\rho)+u}{2}$, $s = \frac{f(\rho)-u}{2}$, $f(\rho) = \int_1^\rho \sqrt{\varphi'(\sigma)}\frac{d\sigma}{\sigma}$. Using the solution of the system (2), he discussed the propagation of the solution and the appearance of shock waves.

In this paper we consider as a generalization of the system from (2) the solution to the following Cauchy problem:

$$(3) \qquad \begin{cases} \dfrac{\partial u^{(i)}}{\partial t} + H^{(i)}\left(t, x, u^{(1)}, u^{(2)}, \cdots, u^{(m)}, \dfrac{\partial u^{(i)}}{\partial x}\right) = 0, \\ u^{(i)}(0, x) = u_0^{(i)}(x), \quad x \in \mathrm{R}^n, \ i = 1, 2, \ldots, m, \end{cases}$$

in the framework of the classical function space $\mathcal{B}^2(\mathrm{R}^n)$. We assume that each $H^{(i)}(t, x, u, p)$ belongs to $\mathcal{B}^2((0, N) \times G(N))$ for any natural number $N$, where $G(N) = \mathrm{R}^n \times (-N, N)^m \times (-N, N)^n$. The solution satisfying

$$u(t, \cdot) \in \bigcap_{j=0}^{1} C^j([0, T); \mathcal{B}^{2-j}(\mathrm{R}^n)).$$

is unique.

In order to construct the solution $u(t, x)$, we exhibit that the problem (3) is equivalent to the following problem (3'):

$$(3') \quad \begin{cases} \dfrac{dX_k^{(i)}}{dt} = H_{p_k}^{(i)}(t, X^{(i)}, U(t; Y(t; X^{(i)})), P^{(i)}), \\ \dfrac{dP_k^{(i)}}{dt} = -H_{x_k}^{(i)}(t, X^{(i)}, U(t; Y(t; X^{(i)})), P^{(i)}) \\ \qquad\quad - \sum_{j \ne i} H_{u^{(j)}}^{(i)}(t, X^{(i)}, U(t; Y(t; X^{(i)})), P^{(i)}) P^{(j)}(t, Y^{(j)}(t; X^{(i)})) \\ \qquad\quad - H_{u^{(i)}}^{(i)}(t, X^{(i)}, u(t, U(t; Y(t; X^{(i)})), P^{(i)}) P_k^{(i)}, \\ \dfrac{dU^{(i)}}{dt} = -H^{(i)}(t, X^{(i)}, U(t; Y(t; X^{(i)})), P^{(i)}) \\ \qquad\quad + \sum_{j=1}^{n} H_{p_j}^{(i)}(t, X(i), U(t; Y(t; X^{(i)})), P^{(i)}) P_j^{(i)}, \quad t \in (0, T), \\ X_k^{(i)}(0) = y_k, \quad P_k^{(i)}(0) = \dfrac{\partial u_0^{(i)}}{\partial x_k}(y), \quad U^{(i)}(0) = u_0^{(i)}(y), \\ k = 1, 2, \ldots, n, \ i = 1, 2, \ldots, m, \end{cases}$$

where $U(t, Y(t; X^{(i)})) = (U^{(1)}(t, Y^{(1)}(t; X^{(i)})), \cdots, U^{(m)}(t; Y^{(m)}(t; X^{(i)})))$. The solutions of the two problems have the relations

$$\begin{cases} u^{(i)}(t, X^{(i)}(t; y)) = U^{(i)}(t; y), \quad u^{(i)}(t, x) = U^{(i)}(t; Y^{(i)}(t; x)), \\ P_k^{(i)}(t; y) = \dfrac{\partial u^{(i)}}{\partial x_k}(t, X^{(i)}(t; y)), \end{cases}$$

where $y = Y^{(i)}(t; x)$ is the inverse of $x = X^{(i)}(t; y)$. Since the above system (3') involves the composite terms $U^{(i)}(t, Y^{(1)}(t; X^{(i)}))$ and $P^{(j)}(t, Y^{(j)}(t; X^{(i)}))$, we call it *composite system*. However, as we see later, it is possible for us to apply the usual successive approximation method, in order to construct the solution. Since (3') has the unique solution in the corresponding function spaces, we can denote the lifespan of the solution by

$$\mathcal{LOC}(u_0) = \sup\{T \geq 0; (X - y, P, U) \in C^0([0, T]; \mathcal{B}^1(\mathbb{R}^n))\},$$

where

$$\begin{cases} X - y = (X^{(1)} - y, \cdots, X^{(m)} - y) \\ \quad = (X_1^{(1)} - y_1, \cdots, X_n^{(i)} - y_n, X_1^{(2)} - y_1, \cdots, X_n^{(m)} - y_n), \\ P = (P^{(1)}, \cdots, P^{(m)}) = (P_1^{(1)}, \cdots, P_n^{(1)}, P_1^{(2)} \cdots, P_n^{(m)}), \\ U = (U^{(1)}, \cdots, U^{(m)}). \end{cases}$$

**Theorem 1.1.** *Assume that the real-valued functions $H^{(i)}(t, x, u, p)$ and their derivatives* $\dfrac{\partial H^{(i)}}{\partial x_j}$, $\dfrac{\partial H^{(i)}}{\partial u_k}$ *and* $\dfrac{\partial H^{(i)}}{\partial p_j}$, $i = 1, 2, \ldots, m$, $j = 1, 2, \ldots, n$, $k = 1, 2, \ldots, m$, *belong to $\mathcal{B}^1((0, N) \times G(N))$ for any natural number $N$. Suppose $u_0 \in \mathcal{B}^2(\mathbb{R}^n)$, which means $u_0^{(i)} \in \mathcal{B}^2(\mathbb{R}^n)$ for $i = 1, 2, \ldots, m$. Then there exists a positive constant $T$, so that the Cauchy problem (3) has the unique solution $u(t, x) = (u^{(1)}, u^{(2)}, \cdots, u^{(m)})$ satisfying $u(t, \cdot) \in \bigcap\limits_{j=0}^{2} C^j([0, T]; \mathcal{B}^{2-j}(\mathbb{R}^n))$. Namely, the lifespan of the solution $\mathcal{LP}(u_0) = \sup\{T \geq 0; u(t, \cdot) \in C^0([0, T]; \mathcal{B}^2(\mathbb{R}^n))\}$ is verified to be positive for any initial data $u_0 \in \mathcal{B}^2(\mathbb{R}^n)$. We can construct the solution in the form*

$$u(t, x) = U(t; Y(t; x)) = (U^{(1)}(t; Y^{(1)}(t; x)), \cdots, U^{(m)}(t; Y^{(m)}(t; x)))$$

*by using the solution to the problem (3′). $\mathcal{LP}(u_0)$ is equal to $\mathcal{LOC}(u_0)$.*

If necessary the solution $u(t, x)$ is denoted also by $u(t, x; u_0)$.

**Corollary 1.1.** *We assume the same conditions as in Theorem 1.1 for $H^{(i)}(t, x, u, p)$, $i = 1, 2, \ldots, m$. Suppose that the initial data $\hat{u}_0 \in \mathcal{B}^2(\mathbb{R}^n)$ satisfies $\mathcal{LP}(\hat{u}_0) > t_0 > 0$. Then there exists a positive number $\epsilon_0$ so that, for any initial data $u_0$ in the $\delta_0$ neighbourhood of $\hat{u}_0$ $\{u_0 \in \mathcal{B}^2(\mathbb{R}^n); \|u_0 - \hat{u}_0\|_{\mathcal{B}^2(\mathbb{R}^n)} < \delta_0\}$, the solution $u = u(t, x; u_0)$ to the problem (3) exists and satisfies $u(t, \cdot; u_0) \in C^0([0, t_0]; \mathcal{B}^2(\mathbb{R}^n))$, where $\|u_0 - \hat{u}_0\|_{\mathcal{B}^2(\mathbb{R}^n)}$ means $\sum\limits_{i=1}^{m} \|u_0^{(i)} - \hat{u}_0^{(i)}\|_{\mathcal{B}^2(\mathbb{R}^n)}$. Moreover, as $\|u_0 - \hat{u}_0\|_{\mathcal{B}^2(\mathbb{R}^n)}$ tends to zero, $u(t, \cdot; u_0)$ converges to $u(t, \cdot; \hat{u}_0)$ in $\mathcal{B}^2(\mathbb{R}^n)$ uniformly in $t \in [0, t_0]$.*

## 2. Method of the proof

We use the successive approximation method to solve the problem (3). First we put

$$(3.0) \qquad u^{(i,0)}(t, x) = u_0^{(i)}(x), \quad i = 1, 2, \ldots, n, \quad t \geq 0.$$

Then we define $u^{(i,j)}(t, x)$, $j \in \mathbb{N}$, by the solution to the problem

$$(3.j) \qquad \begin{cases} \dfrac{\partial u^{(i,j)}}{\partial t} + H^{(i)}\left(t, x, u^{(1,j-1)}, u^{(2,j-1)}, \cdots, u^{(m,j-1)}, \dfrac{\partial u^{(i,j)}}{\partial x}\right) = 0, \\ u^{(i,j)}(0, x) = u_0^{(i)}(x), \quad x \in \mathbb{R}^n, \quad i = 1, 2, \ldots, n. \end{cases}$$

Note that $(3.j)$ is the problem for the single equation, since $u^{(i,j-1)}$ are regarded as known functions. If we denote $H^{(i)}(t,x,u^{(1,j-1)}(t,x),\cdots,u^{(m,j-1)}(t,x),p)$ by $H^{(i,j)}(t,x,p)$, $(3.j)$ for fixed $j$ is the problem $(*)$ below with $H(t,x,u,p) = H^{(i,j)}(t,x,p)$. Here we use the following results for a single equation. Namely, concerning the problem

$(*)$
$$\begin{cases} \dfrac{\partial u}{\partial t} + H\left(t,x,u,\dfrac{\partial u}{\partial x}\right) = 0, & (t,x) \in (0,T) \times \mathrm{R}^n, \\ u(0,y) = u_0(y), & y \in \mathrm{R}^n, \end{cases}$$

and the corresponding problem for the Hamilton system

$(**)$
$$\begin{cases} \dfrac{dX_k}{dt} = H_{p_k}(t,X(t),U(t),P(t)), \\ \dfrac{dP_k}{dt} = -H_{x_k}(t,X(t),U(t),P(t)) - H_u(t,X(t),U(t),P(t))\,P_k(t), \\ \dfrac{dU}{dt} = -H(t,X(t),U(t),P(t)) + \sum_{j=1}^{n} H_{p_j}(t,X(t),U(t),P(t))\,P_j(t), \\ X_k(0) = y_k,\ P_k(0) = \dfrac{\partial u_0}{\partial x_k}(y),\ U(0) = u_0(y),\ y \in \mathrm{R}^n,\ k = 1,2,\ldots,n \end{cases}$$

for $t \in (0,T)$, we have the following Theorems 2.1 and 2.2:

**Theorem 2.1.** *Let $H(t,x,u,p)$ satisfy the same conditions as $H^{(i)}(t,x,u,p)$ in Theorem 1.1. Assume that $u(t,x) \in \mathcal{B}^2((0,T)\times\mathrm{R}^n)$ is the solution to the Cauchy problem $(*)$. Now, let $X(t) = X(t;y)$ be the unique solution to the following initial value problem:*

(4)
$$\begin{cases} \dfrac{dX_k(t)}{dt} = \dfrac{\partial H}{\partial p_k}\left(t,X(t),u(t,X(t)),\dfrac{\partial u}{\partial x}(t,X(t))\right), & t \in (0,T), \\ X_k(0) = y_k,\ y_k \in \mathrm{R},\ k = 1,2,\ldots,n. \end{cases}$$

*Put* $P(t;y) = \dfrac{\partial u}{\partial x}(t,X(t;y))$ *and* $U(t;y) = u(t,X(t;y))$. *Then $(X(t;y),P(t;y),U(t;y))$ becomes the unique solution to the problem $(**)$ with $u_0(y) = u(0,y)$ satisfying*

$$(X(t;y) - y, P(t;y), U(t;y)) \in C^0([0,T); \mathcal{B}^0(\mathrm{R}^n)).$$

*The solution $u$ to the problem $(*)$ is verified to be unique in the space $\mathcal{B}^2((0,T)\times\mathrm{R}^n)$. $(X(t;y) - y, P(t;y), U(t;y))$ belongs to $C^0([0,T); \mathcal{B}^1(\mathrm{R}^n))$, if $u_0 = u(0,\cdot)$ belongs to $\mathcal{B}^2(\mathrm{R}^n)$.*

**Theorem 2.2.** *Suppose that $H(t,x,u,p)$ satisfies the same conditions as in Theorem 2.1. Assume that the initial data $u_0$ belongs to $\mathcal{B}^2(\mathrm{R}^n)$. Then there exists a positive constant $T$, so that we have the unique solution $u$ to the problem $(*)$ satisfying $u(t,\cdot) \in \bigcap_{j=0}^{2} C^j([0,T); \mathcal{B}^{2-j}(\mathrm{R}^n))$. We can find $T$ depending on $\|u_0\|_{\mathcal{B}^2}$ and the norm of $H = (H^{(1)}, H^{(2)}, \cdots, H^{(m)})$ in the space stated above. More precisely, we* can construct the solution $u(t,x)$ by the formula $u(t,x) = U(t; Y(t;x))$, using the solution to the problem $(**)$, where $Y(t;x)$ means the inverse of $X(t;y)$. Incidentally we have the fundamental relation

$$\frac{\partial u}{\partial x_k}(t,x) = P_k(t; Y(t;x)), \quad k = 1, 2, \cdots, n.$$

As for the estimates, therefore we can use $\|u(t,\cdot)\|_{\mathcal{B}^0} = \|U(t;\cdot)\|_{\mathcal{B}^0}$, and $\left\|\frac{\partial u}{\partial x}(t,\cdot)\right\|_{\mathcal{B}^0} = \|P(t;\cdot)\|_{\mathcal{B}^0}$. Hereafter we note $\mathcal{B}^k = \mathcal{B}^k(\mathbb{R}^n)$, $k = 1, 2, \cdots, n$.

## 3. The construction of the solution

The problem $(3.j)$ is related to the following Hamilton system for $i = 1, 2, \ldots, m$ and $j \in \mathbb{N}$:

(5)
$$
\begin{cases}
\dfrac{dX_k^{(i,j)}}{dt} = H_{p_k}^{(i)}(t, X^{(i,j)}, u^{[j-1]}(t, X^{(i,j)}), P^{(i,j)}), \\[2mm]
\dfrac{dP_k^{(i,j)}}{dt} = -H_{x_k}^{(i)}(t, X^{(i,j)}, u^{[j-1]}(t, X^{(i,j)}), P^{(i,j)}) \\[2mm]
\qquad - \displaystyle\sum_{l=1}^{m} H_{u^{(l)}}^{(i)}(t, X^{(i,j)}, u^{[j-1]}(t, X^{(i,j)}), P^{(i,j)}) \dfrac{\partial u^{(l,j-1)}}{\partial x_k}(t, X^{(i,j)}), \\[3mm]
\dfrac{dU^{(i,j)}}{dt} = -H^{(i)}(t, X^{(i,j)}, u^{[j-1]}(t, X^{(i,j)}), P^{(i,j)}) \\[2mm]
\qquad + \displaystyle\sum_{k=1}^{n} H_{p_k}^{(i)}(t, X^{(i,j)}, u^{[j-1]}(t, X^{(i,j)}), P^{(i,j)}) P_k^{(i,j)}, \quad t \in (0,T), \\[3mm]
X_k^{(i,j)}(0) = y_k, \quad P_k^{(i,j)}(0) = \dfrac{\partial u_0^{(i)}}{\partial x_k}(y), \quad U^{(i,j)}(0) = u_0^{(i)}(y), \quad k = 1, 2, \ldots, n, \quad y \in \mathbb{R}^n,
\end{cases}
$$

where $u^{[j-1]}$ stands for the vector $(u^{(1,j-1)}, u^{(2,j-1)}, \cdots, u^{(m,j-1)})$ for $j \geq 2$ and each $u^{(i,j-1)}$ is given by $U^{(i,j-1)}(t; Y^{(i,j-1)}(t;x))$, which we regard successively as a known function of $t$ and $x$. Here $y = Y^{(i,j-1)}(t;x)$ is the inverse of $x = X^{(i,j-1)}(t;y)$. Note that $u^{[0]}$ is a vector-valued function $(u_0^{(1)}, u_0^{(2)}, \cdots, u_0^{(m)})$, which is constant in $t$. Now put

$$M = \max_i \left\{ \|u_0^{(i)}\|_{\mathcal{B}^0}, \; \left\|\frac{\partial u_0^{(i)}}{\partial x}\right\|_{\mathcal{B}^0} \right\}$$

and represent the bounds of the left hand sides in (5) as follows:

$$
\begin{cases}
K_p(t) = \max_{i,k} \; \sup_{(x,u,p) \in G(2M), \, q \in [-2M, 2M]^n} \left| -H_{x_k}^{(i)}(t, x, u, p) - \displaystyle\sum_{j=1}^{m} H_{u^{(j)}}^{(i)}(t, x, u, p) \, q_j \right|, \\[3mm]
K_u(t) = \max_i \; \sup_{(x,u,p) \in G(2M)} \left| -H^{(i)}(t, x, u, p) - \displaystyle\sum_{k=1}^{n} H_{p_k}^{(i)}(t, x, u, p) \, p_k \right|.
\end{cases}
$$

Denote $\max\limits_{t\in[0,1]} \max\{K_p(t), K_u(t)\}$ simply by $K$. Put $T_0 = \min\left\{\dfrac{M}{K}, 1\right\}$, then from the integration form of (5) we have the estimate

$$\inf_{t\in[0,T_0]} \max_{i,j} \max\{\|P^{(i,j)}(t;\cdot)\|_{\mathcal{B}^0}, \|U^{(i,j)}(t;\cdot)\|_{\mathcal{B}^0}\} \leq 2M,$$

for $t \in [0, T_0]$. Since we see that $X^{(i,j)}(t;y) - y$ has the same properties from (5), $X^{(i,j)}(t;y) - y$, $P^{(i,j)}(t;y)$ and $U^{(i,j)}(t;y)$ are uniformly bounded in $\mathbb{R}^n$ for any $i$, $j \in \mathbb{N}$, $y \in \mathbb{R}^n$ and $t \in [0, T_0]$. Moreover, in next section we show the following regularity estimate:

*There exist positive constants $c_1$ and $c_2$, so that, for any $i$ and $j$, we have*

$$V^{(i,j)}(t) + W^{(i,j)}(t) + Z^{(i,j)}(t) \leq c_1(e^{c_2 t} - 1), \quad t \in [0, T_0],$$

*where*

$$\begin{cases} V^{(i,j)}(t) = \max\limits_{k,l}\left\|\dfrac{\partial X_k^{(i,j)}}{\partial y_l}(t;\cdot) - \dfrac{\partial X_k}{\partial y_l}(0;\cdot)\right\|_{\mathcal{B}^0}, \\[2ex] W^{(i,j)}(t) = \max\limits_{k,l}\left\|\dfrac{\partial P_k^{(i,j)}}{\partial y_l}(t;\cdot) - \dfrac{\partial P_k}{\partial y_l}(0;\cdot)\right\|_{\mathcal{B}^0}, \\[2ex] Z^{(i,j)}(t) = \max\limits_{l}\left\|\dfrac{\partial U^{(i,j)}}{\partial y_l}(t;\cdot) - \dfrac{\partial U}{\partial y_l}(0;\cdot)\right\|_{\mathcal{B}^0}. \end{cases}$$

Using this fact, in the next section, we show that for any fixed $i$ the sequences $\{X_k^{(i,j)}(t;\cdot)\}$, $\{P_k^{(i,j)}(t;\cdot)\}$ and $\{U^{(i,j)}(t;\cdot)\}$ are Cauchy sequences respectively in the space $\mathcal{B}^0(\mathbb{R}^n)$. Let us denote the limits by $X_k^{(i)}(t;\cdot)$, $P_k^{(i)}(t;\cdot)$ and $U^{(i)}(t;\cdot)$. We can show that $X^{(i)}(t;y)$ becomes a flow in the space $\mathbb{R}^n$ for $t \in [0, T_0]$. Now denote the inverse of $X^{(i)}(t;y) = x$ by $Y^{(i)}(t;x) = y$ and put $U^{(i)}(t;Y^{(i)}(t;x)) = u^{(i)}(t,x)$, then we can verify that $\{u^{(i)}(t,x)\}_{i=1}^m$ is the solution to the problem (3).

## 4. Regularity of the solution and the proof of Theorem 1.1

First we recall the case for a single equation. Taking the difference $\{(**)$ with $y$ replaced by $y + he_j\} - \{(**)\}$, we have the linear equations for the unknown functions

$$\left(\frac{X(t;y+he_j) - X(t;y)}{h}, \frac{P(t;y+he_j) - P(t;y)}{h}, \frac{U(t;y+he_j) - U(t;y)}{h}\right),$$

where the coefficients involve in $\{X, P, U\}$ but stay uniformly bounded in $(t, y) \in (0, T_0] \times \mathbb{R}^n$. We can verify the following estimate:

*There exist constants $c_1$ and $c_2$, so that we have*

$$v(t, t', h) + w(t, t', h) + z(t, t', h) \leq c_1(e^{c_2(t-t')} - 1), \quad 0 \leq t' < t < T_0,$$

*where*

$$\begin{cases} v(t,t',h) = \max_{k,j} \operatorname{ess.sup} \left| \dfrac{X_k(t;y+he_j)-X_k(t;y)}{h} - \dfrac{X_k(t';y+he_j)-X_k(t';y)}{h} \right|, \\ w(t,t',h) = \max_{k,j} \operatorname{ess.sup} \left| \dfrac{P_k(t;y+he_j)-P_k(t;y)}{h} - \dfrac{P_k(t';y+he_j)-P_k(t';y)}{h} \right|, \\ z(t,t',h) = \max_{j} \operatorname{ess.sup} \left| \dfrac{U(t;y+he_j)-U(t;y)}{h} - \dfrac{U(t';y+he_j)-U(t';y)}{h} \right|. \end{cases}$$

Therefore $X_k$, $P_k$, $k = 1, \ldots, n$, and $U$ are uniformly Lipschitz continuous in $y$. Hence $\dfrac{\partial P_k}{\partial y_j}$ and $\dfrac{\partial U}{\partial y_j}$, $k = 1, \ldots, n$, $j = 1, \ldots, n$, exist and we have

$$V(t,t') + W(t,t') + Z(t,t') \le c_1(e^{c_2\,(t-t')} - 1),$$

where

$$\begin{cases} V(t,t') = \max_{k,j} \left\| \dfrac{\partial X_k}{\partial y_j}(t;\cdot) - \dfrac{\partial X_k}{\partial y_j}(t';\cdot) \right\|_{\mathcal{B}^0}, \\ W(t,t') = \max_{k,j} \left\| \dfrac{\partial P_k}{\partial y_j}(t;\cdot) - \dfrac{\partial P_k}{\partial y_j}(t';\cdot) \right\|_{\mathcal{B}^0}, \\ Z(t,t') = \max_{j} \left\| \dfrac{\partial U}{\partial y_j}(t;\cdot) - \dfrac{\partial U}{\partial y_j}(t';\cdot) \right\|_{\mathcal{B}^0}, \end{cases}$$

for any fixed $t' \in [0,T_0)$. Therefore, $x = X(t;y)$ becomes a flow satisfying the following conditions for $t \in [0,T_0]$:
1) For fixed $i$, the mapping from $y$ to $x = X(t;y)$ is homeomorphic in $\mathrm{R}^n$ and the Jacobian is bounded and continuous in $\mathcal{B}^0(\mathrm{R}^n)$.
2) The derivatives $\dfrac{\partial^{1+|\alpha|} X_k}{\partial t \partial y^\alpha}(t;y) \in \mathcal{B}^0((0,T_0) \times \mathrm{R}^n)$ for $|\alpha| \le 1$.
3) The inverse $y = Y(t;y)$ has the same properties as $x = X(t;y)$ replaced $y$ by $x$. Therefore by the property of composite functions, we see that the continuity of $Z(t,t')$ yields $\dfrac{\partial u}{\partial x_j}(t,\cdot) \in C^0([0,T_0);\mathcal{B}^0)$. Similarly, the continuity of $W(t,t')$ gives us $\dfrac{\partial^2 u}{\partial x_k \partial x_j}(t,\cdot) \in C^0([0,T_0);\mathcal{B}^0)$.

Now step by step we consider the Hamilton system for the single equation (5) and take the difference $\{(5)$ with $y$ replaced by $y+he_j\} - \{(5)\}$. Then we have linear equations for

$$\begin{cases} \dfrac{X^{(i,l)}(t;y+he_j)-X^{(i,l)}(t;y)}{h}, \\ \dfrac{P^{(i,l)}(t;y+he_j)-P^{(i,l)}(t;y)}{h}, \\ \dfrac{U^{(i,l)}(t;y+he_j)-U^{(i,l)}(t;y)}{h}, \end{cases}$$

where the coefficients are uniformly bounded for all $i$, $l$, $j$ and $(t,y) \in [0,t_0] \times \mathrm{R}^n$.

Then we see that there exist constants $c_1$ and $c_2$, so that

$$
\begin{cases}
v(t,t',h,l) = \displaystyle\sup_{i,k,j,y} \left| \dfrac{X_k^{(i,l)}(t;y+he_j) - X_k^{(i,l)}(t;y)}{h} - \dfrac{X_k^{(i,l)}(t';y+he_j) - X_k^{(i,l)}(t';y)}{h} \right. \\[4mm]
w(t,t',h,l) = \displaystyle\sup_{i,k,j,y} \left| \dfrac{P_k^{(i,l)}(t;y+he_j) - P_k^{(i,l)}(t;y)}{h} - \dfrac{P_k^{(i,l)}(t';y+he_j) - P_k^{(i,l)}(t';y)}{h} \right| \\[4mm]
z(t,t',h,l) = \displaystyle\sup_{i,j,y} \left| \dfrac{U^{(i,l)}(t;y+he_j) - U^{(i,l)}(t;y)}{h} - \dfrac{U^{(i,l)}(t';y+he_j) - U^{(i,l)}(t';y)}{h} \right|.
\end{cases}
$$

satisfy

$$
v(t,t',h,l) + w(t,t',h,l) + z(t,t',h,l) \le c_1(e^{c_2\,(t-t')} - 1),
$$

uniformly for $h \ne 0$ and $l \in \mathbb{N}$. On the other hand, we can prove that the sequence

$$
\{(X^{(i,l)}(t;y), P^{(i,l)}(t;y), U^{(i,l)}(t;y))\}_{l=1}^{\infty}
$$

converges to the limit $(X^{(i)}(t;y), P^{(i)}(t;y), U^{(i)}(t;y))$ for any $i$ uniformly in $(t,y) \in [0,T_0] \times \mathbb{R}^n$. To show this fact, we take the difference $\{(5)\} - \{(5)$ with $l$ replaced by $l-1\}$. Then we have successively the following estimates. There exists a positive constant $C$ so that it holds

(6)
$$
\begin{cases}
\|X^{(i,l)}(t;\cdot) - X^{(i,l-1)}(t;\cdot)\|_{W^{0,\infty}(\mathbb{R}^n)} \le \dfrac{(Ct)^l}{l!}, \\[3mm]
\|P^{(i,l)}(t;\cdot) - P^{(i,l-1)}(t;\cdot)\|_{W^{0,\infty}(\mathbb{R}^n)} \le \dfrac{(Ct)^l}{l!}, \\[3mm]
\|U^{(i,l)}(t;\cdot) - U^{(i,l-1)}(t;\cdot)\|_{W^{0,\infty}(\mathbb{R}^n)} \le \dfrac{(Ct)^l}{l!},
\end{cases}
$$

for $t \in [0,T_0]$. Therefore the convergence $X_k^{(i)} = \lim X_k^{(i,l)}$, $P_k^{(i)} = \lim P_k^{(i,l)}$ and $U^{(i)} = \lim U^{(i,l)}$ are uniform in $(t,y) \in [0,T_0] \times \mathbb{R}^n$. Here tending $l$ to $\infty$ in the above estimate for any fixed $h \ne 0$ it follows

$$
v(t,t',h) + w(t,t',h) + z(t,t',h) \le c_1(e^{c_2\,(t-t')} - 1),
$$

where

$$
\begin{cases}
v(t,t',h) = \displaystyle\sup_{i,k,j,y} \left| \dfrac{X_k^{(i)}(t;y+he_j) - X_k^{(i)}(t;y)}{h} - \dfrac{X_k^{(i)}(t';y+he_j) - X_k^{(i)}(t';y)}{h} \right|, \\[4mm]
w(t,t',h) = \displaystyle\sup_{i,k,j,y} \left| \dfrac{P_k^{(i)}(t;y+he_j) - P_k^{(i)}(t;y)}{h} - \dfrac{P_k^{(i)}(t';y+he_j) - P_k^{(i)}(t';y)}{h} \right|, \\[4mm]
z(t,t',h) = \displaystyle\sup_{i,j,y} \left| \dfrac{U^{(i)}(t;y+he_j) - U^{(i)}(t;y)}{h} - \dfrac{U^{(i)}(t';y+he_j) - U^{(i)}(t';y)}{h} \right|.
\end{cases}
$$

Thus $X_k^{(i)}(t;y)$, $P_k^{(i)}(t;y)$ and $U^{(i)}(t;y)$ are uniformly Lipschitz continuous in $y$. Hence, $h$ tends to zero implies that they belong to $C^0([0,T_0];\mathcal{B}^1(\mathbb{R}^n))$. Here we define $u^{(i)}(t,x) = U^{(i)}(t;Y^{(i)}(t;x))$. Then we can see that $\{U^{(i,l)}(t;y) =$

$u^{(i,l)}(t, X^{(i,l)}(t; y))\}$ converges uniformly to $U^{(i)}(t; y) = u^{(i)}(t, X^{(i)}(t; y))$. Moreover, the integration of (5) gives

$$(7) \quad \begin{cases} \dfrac{dX_k^{(i)}}{dt} = H_{p_k}^{(i)}(t, X^{(i)}, u(t, X^{(i)}), P^{(i)}), \\[2mm] \dfrac{dP_k^{(i)}}{dt} = -H_{x_k}^{(i)}(t, X^{(i)}, u(t, X^{(i)}), P^{(i)}) \\[1mm] \qquad\quad -\displaystyle\sum_{j=1}^{m} H_{u^{(j)}}^{(i)}(t, X^{(i)}, u(t, X^{(i)}), P^{(i)}) \dfrac{\partial u^{(j)}}{\partial x_k}(t, X^{(i)}), \\[3mm] \dfrac{dU^{(i)}}{dt} = -H^{(i)}(t, X^{(i)}, u(t, X^{(i)}), P^{(i)}) \\[1mm] \qquad\quad +\displaystyle\sum_{j=1}^{n} H_{p_j}^{(i)}(t, X^{(i)}, u(t, X^{(i)}), P^{(i)}) P_j^{(i)}, \quad t \in (0, T), \\[3mm] X_k^{(i)}(0) = y_k, \quad P_k^{(i)}(0) = \dfrac{\partial u_0^{(i)}}{\partial x_k}(y), \quad U^{(i)}(0) = u_0^{(i,\cdot)}(y), \\[2mm] k = 1, 2, \ldots, n, \quad i = 1, 2, \ldots, m, \end{cases}$$

where $u(t, X^{(i)}) = (u^{(1)}(t, X^{(i)}), u^{(2)}(t, X^{(i)}), \cdots, u^{(m)}(t, X^{(i)}))$.

The construction of the solution to the problem (3) depends on the fundamental relations in the limit form:

$$(8) \quad \sum_k P_k^{(i)}(t; y) \frac{\partial X_k^{(i)}}{\partial y_j}(t; y) = \frac{\partial U^{(i)}}{\partial y_j}(t; y), \quad i = 1, 2, \ldots, m, \ j = 1, 2, \ldots, n.$$

We can verify (8) in the sense of distribution as in [9]. The relation (8) can be described also by

$$(9) \quad P_k^{(i)}(t; Y^{(i)}(t; x)) = \sum_j \frac{\partial U^{(i)}}{\partial y_j}(t; y) \frac{\partial Y_j^{(i)}}{\partial x_k}(t; X^{(i)}(t; y))$$

for $i = 1, 2, \ldots, m$ and $k = 1, 2, \ldots, n$.

Each component of $u(t, x) = (u^{(1)}(t, x), u^{(2)}(t, x), \ldots, u^{(m)}(t, x))$ is given also by $u^{(i)}(t, x) = \lim_{l \to \infty} U^{(i,l)}(t, Y^{(i,l)}(t; x))$. Now we can verify that $u(t, x)$ is the unique solution to the problem (3) as in [10]. Here we can see that the lifespan satisfies $\mathcal{LP}(u_0) \geq T_0 > 0$. $\mathcal{LP}(u_0)$ of the solution to the problem (3) is equal to the lifespan $\mathcal{LOC}(u_0)$ to the problem (3′).

Incidentally we give a comment. It is also important that the above construction of the solution to the problem (3) gives also numerical information on this solution, since this solution can be approximated, step by step, by the solutions of ordinary differential equations with parameters $y \in \mathbb{R}^n$.

## 5. Application

As an application, we can construct the solution to the Cauchy problem for the analytic partial differential equation

(10)
$$\begin{cases} \frac{du}{dz}(z_1, z) + H(z_1, z, u, \frac{du}{dz_1}) = 0, \\ u(z_1, 0) = u_0(z_1), \end{cases}$$

where $u_0(z_1)$ is an analytic function. The Cauchy-Kowalevsky theorem says that the problem has a local solution constructed by Taylor expansion, if $H(z_1, z, u, p)$ is an analytic function with respect to the four complex variables. We can reduce (10) to problems for a system of equations with real unknown functions of real variables. Using these solutions we can construct the solution globally as wide as possible.

Let us put

$$\begin{cases} z_1 = x_1 + iy_1, \ z = t + is, \ u = u_1 + iu_2, \ p = p_1 + ip_2, \\ H = H_1(x_1, y_1, t, s, u_1, u_2, p_1, p_2) + i\,H_1(x_1, y_1, t, s, u_1, u_2, p_1, p_2). \end{cases}$$

Then $H_1$ and $H_2$ are supposed to satisfy the Cauchy-Riemann relations

(11)
$$\begin{cases} H_{1,x_1} = H_{2,x_2}, \ H_{1,x_2} = -H_{2,x_1}, \ H_{1,t} = H_{2,s}, \ H_{1,s} = -H_{2,t}, \\ H_{1,u_1} = H_{2,u_2}, \ H_{1,u_2} = -H_{2,u_1}, \ H_{1,p_1} = H_{2,p_2}, \ H_{1,p_2} = -H_{2,p_1}. \end{cases}$$

We note

$$u = u_1(x_1, y_1, t, s) + i\,u_2(x_1, y_1, t, s).$$

Then, the Cauchy-Riemann relations show that the given equation (10) is equivalent to the following two relations:

(12 − 1)
$$\begin{cases} \frac{du_1}{dt} = -H_1(x_1, y_1, t, s, u_1, u_2, \frac{\partial u_1}{\partial x_1}, -\frac{\partial u_1}{\partial y_1}), \\ \frac{du_2}{dt} = -H_2(x_1, y_1, t, s, u_1, u_2, \frac{\partial u_2}{\partial y_1}, -\frac{\partial u_2}{\partial x_1}), \\ u_1(x_1, y_1, 0, 0) = u_{0,1}(x_1, y_1), \ u_2(x_1, y_1, 0, 0) = u_{0,2}(x_1, y_1), \end{cases}$$

(12 − 2)
$$\begin{cases} \frac{du_1}{ds} = -H_1(x_1, y_1, t, s, u_1, u_2, \frac{\partial u_1}{\partial x_1}, -\frac{\partial u_1}{\partial y_1}), \\ \frac{du_2}{ds} = -H_2(x_1, y_1, t, s, u_1, u_2, \frac{\partial u_2}{\partial y_1}, -\frac{\partial u_2}{\partial x_1}), \\ u_1(x_1, y_1, 0, 0) = u_1(x_1, y_1), \ u_2(x_1, y_1, 0, 0) = u_{0,2}(x_1, y_1). \end{cases}$$

Put $u_k(x_1, y_1, t, 0) = v_k(x_1, y_1, t), \ k = 1, 2.$ From (12 − 1) with $s = 0$, we have

(13 − 1)
$$\begin{cases} \frac{dv_1}{dt} = -H_1(x_1, y_1, t, 0, u_1, u_2, \frac{\partial u_1}{\partial x_1}, -\frac{\partial u_1}{\partial y_1}), \\ \frac{dv_2}{dt} = -H_2(x_1, y_1, t, 0, u_1, u_2, \frac{\partial u_2}{\partial y_1}, -\frac{\partial u_2}{\partial x_1}), \\ v_1(x_1, y_1, 0) = u_{0,1}(x_1, y_1), \ v_2(x_1, y_1, 0) = u_{0,2}(x_1, y_1), \end{cases}$$

$(12-2)$ and $(13-1)$ give us

$$(13-2) \quad \begin{cases} \frac{du_1}{ds} = -H_1(x_1, y_1, t, s, u_1, u_2, \frac{\partial u_1}{\partial x_1}, -\frac{\partial u_1}{\partial y_1}), \\ \frac{du_2}{ds} = -H_2(x_1, y_1, t, s, u_1, u_2, \frac{\partial u_2}{\partial y_1}, -\frac{\partial u_2}{\partial x_1}), \\ u_1(x_1, y_1, t, 0) = v_1(x_1, y_1, t), \quad u_2(x_1, y_1, t, 0) = v_{0,2}(x_1, y_1, t). \end{cases}$$

Namely, we have diagonalized systems by virtue of the Cauchy-Riemann relation of $u$. First we solve $(13-1)$ as an application of Theorem 1.1. Then using the initial data $v_1(x_1, y_1, t)$ and $v_1(x_1, y_1, t)$ we construct the solution $u_1(x_1, y_1, t, s)$ and $u_2(x_1, y_1, t, s)$ to the problem $(13-2)$ in the same way, using the corresponding Hamilton system. However we need to prove that $u_1$ and $u_2$ satisfy the Cauchy-Riemann relations with respect to $u = u_1 + i u_2$, $z_1 = x_1 + i y_1$ and $z = t + is$. Since $x_1$ and $y_1$ are parameters in the problems $(13-1)$ and $(13-2)$, the Cauchy-Riemann relations with respect to $x_1$ and $y_1$ follow directly from $(11)$ and the Cauchy-Riemann relations of the initial data. Then we can verify the Cauchy-Riemann relations with respect to $t$ and $s$, showing that all unknowns in the Hamilton systems satisfy the Cauchy-Riemann relations. Then we are able to see that $u = u_1(x_1, y_1, t, s) + i u_2(x_1, y_1, t, s)$ is the solution of the problem $(10)$. Remark that all the solutions to the Hamilton systems are locally continuous in the initial data. Moreover, we can see that the solution of the problem $(10)$ exists as long as the lifespan of the solution to the Hamilton system, which yields natural boundary from viewpoints of the analyticity of complex functions of many variables.

## References

1. B.RIEMANN, *La propagation d'ondes aeriennes planes*, Memoires de l'academie royale des science de Gettingen 8 (1760).
2. M.TSUJI - LI TA-SIEN, *Global classical solutions for nonlinear equations of first order*, Comm.Part.Diff. Eq. **10** (1985).
3. T.WAZEWSKI, *Sur l'unicité et limitation des intégrales de certains systemes d'équations aux dérivées partielles du premier ordre*, Annali di Matematica Pure ed Applicata **15** (1936-37).
4. R.COURANT - D.HILBERT, *Methods of mathematical physics*, Interscience Publishers, (1962)
5. C.CARATHÉODORY, *Calculus of variations and partial differential equations of the first order*, Holden-Day, (1965).
6. V.I.ARNOLD, *Mathematical methods of classical mechanics*, Springer-Verlag, (1983).
7. S.MIYATAKE, *Microlocal orders of singularities for distributions and an application to Fourieer integral operators*, J. Math. Kyoto Univ. **29** (1989).
8. S.MIYATAKE, *Hamilton flow and nonlinear hyperbolic equations of first order*, .Proceedings of the 3rd International ISAAC congres, vol.2, (2001).
9. S.MIYATAKE, *Cauchy problems for diagonalized system of nonlinear hyperbolic equations of first order*, Journal of Mathematical Fluid Dynamics **7** (2005).
10. MATHEMATICAL REVIEWS,1993-1999, *Nonliner hyperbolic system*, A.M.S., (2000).

## II.2 Boundary Value Problems and Integral Equations

Organizer: P. Krutitskii

# INFLUENCE OF SIGNORINI BOUNDARY CONDITION ON BIFURCATION IN REACTION-DIFFUSSION SYSTEMS[*]

MILAN KUČERA

*Mathematical Institute, Academy of Sciences of the Czech Republic*
*Žitná 25, 115 67 Prague 1; Czech Republic*
*Centre of Applied Mathematics, University of West Bohemia*
*Pilsen, Czech Republic*
*E-mail: kucera@math.cas.cz*

A reaction-diffusion system of activator-inhibitor type is studied with a bifurcation parameter which can describe e.g. the size of the domain. The influence of Signorini boundary conditions to a bifurcation of spatially nonhomogeneous stationary solutions (spatial patterns) is described. In contrast to all previous results of this type, in the case of unilateral conditions only for the activator, no Dirichlet boundary conditions on a part of the boundary are considered.

**Key words:** Reaction-diffusion systems, activator-inhibitor type, bifucation, Signosi boundary conditions
**Mathematics Subject Classification:** 35K57

## 1. Introduction

We will consider the reaction-diffusion system

$$u_t = d_1 \Delta u + b_{11} u + b_{12} v + n_1(u,v), \quad v_t = d_2 \Delta v + b_{21} u + b_{22} v + n_2(u,v) \quad (1)$$

in a bounded domain $\Omega$ in $\mathbb{R}^k$, $k \geq 2$, with a Lipschitzian boundary $\partial\Omega$. We assume that $b_{ij} \in \mathbb{R}$ $(i, j = 1, 2)$, $n_1, n_2$ are differentiable functions on $\mathbb{R}^2$, $n_j(0,0) = 0$, $\frac{\partial n_j}{\partial u}(0,0) = \frac{\partial n_j}{\partial v}(0,0) = 0$, $j = 1, 2$, $d_1, d_2$ are positive parameters (diffusion coefficients). Let $\Gamma_U$ be an open subset of $\partial\Omega$, meas $\Gamma_U > 0$. The boundary conditions

$$\begin{cases} u \geq 0, \ \frac{\partial u}{\partial n} \geq 0, \ \frac{\partial u}{\partial n} \cdot u = 0 \text{ on } \Gamma_U, \\ \frac{\partial u}{\partial n} = 0 \text{ on } \partial\Omega \setminus \Gamma_U, \ \frac{\partial v}{\partial n} = 0 \text{ on } \partial\Omega \end{cases} \quad (2)$$

will be of our main interest but also Signorini boundary conditions for $v$ will be discussed. Neumann boundary conditions

$$\frac{\partial u}{\partial n} = \frac{\partial v}{\partial n} = 0 \text{ on } \partial\Omega \quad (3)$$

---

[*]The research has been supported by the Grant 201/03/0671 of the Grant Agency of the Czech Republic.

or mixed Dirichlet-Neumann boundary conditions will give a certain background for our considerations. It will be always assumed that

$$b_{11} > 0, \ b_{12} < 0, \ b_{21} > 0, \ b_{22} < 0, \ b_{11} + b_{22} < 0, \ \det b_{ij} > 0. \tag{4}$$

If our system is related to a chemical reaction, $u$, $v$ describe the concentrations of reactants, then this assumption means that $u$ and $v$ is an activator and inhibitor, respectively. Further, (4) ensures that the trivial solution is stable as the solution of the system without any diffusion (i.e. of ODE's obtained for $d_1 = d_2 = 0$), but as a solution of (1) with the Neumann conditions (3) it is unstable for $d_1$, $d_2$ from some subdomain $D_U$ of $R_+^2$ and stable only for $d_1, d_2 \in D_S = R_+^2 \setminus \overline{D_U}$ (see Remark 2.1). Nontrivial stationary solutions to (1), (3) bifurcate from the trivial solutions when the parameters cross the border between $D_S$ and $D_U$. Under some assumptions, the bifurcation in the case of boundary conditions (2) can occur only in the interior of $D_U$ (a certain stabilizing effect of Signorini conditions) and in the case of unilateral conditions for $v$ it occurs also in $D_S$ (a destabilizing effect of Signorini conditions). The results of this type for various types of unilateral boundary conditions were proved in a series of papers (e.g. [1,7,13], for a certain survey see [3]). However, in all these papers, the Dirichlet boundary conditions for both $u$ and $v$ were prescribed on some part of $\partial\Omega$ which simplifies the situation. Theorem 2.1 and Corollary 2.1 below describe a stabilizing influence of Signorini conditions for $u$ in the case without Dirichlet conditions. Theorem 3.1 deals with a destabilizing influence of Signorini conditions for $v$. The old assumption about Dirichlet boundary conditions is again considered but a beter result than before is obtained. See Remark 3.1 for a comparison.

## 2. Stabilizing influence of Signorini conditions for activator

We will deal with the stationary solutions to (1), i.e. with the problem

$$\begin{aligned} d_1\Delta u + b_{11}u + b_{12}v + n_1(u, v) = 0, \\ d_2\Delta v + b_{21}u + b_{22}v + n_2(u, v) = 0 \end{aligned} \quad \text{in } \Omega \tag{5}$$

and its linearization

$$d_1\Delta u + b_{11}u + b_{12}v = 0, \ d_2\Delta v + b_{21}u + b_{22}v = 0 \quad \text{in } \Omega \tag{6}$$

with (2) which is in fact nonlinear again. By solutions we will always mean weak solutions introduced below. Set $H = W_2^1(\Omega)$, $\tilde{H} = H \times H$,

$$K = \{\varphi \in H; \ \varphi \geq 0 \text{ on } \Gamma_U \text{ in the sense of traces}\}.$$

Clearly, $K$ is a closed convex cone with its vertex at the origin in $H$. We introduce a weak solution to (5), (2) as $U = [u, v]$ satisfying the variational inequality

$$u \in K, \ v \in H,$$
$$\int_\Omega d_1\nabla u\nabla(\varphi - u) - (b_{11}u + b_{12}v + n_1(u, v))(\varphi - u)\,dx \geq 0 \text{ for all } \varphi \in K, \tag{7}$$
$$\int_\Omega d_2\nabla v\nabla\psi - (b_{21}u + b_{22}v + n_2(u, v))\psi\,dx = 0 \text{ for all } \psi \in H.$$

Similarly for all other boundary value problems considered. We will assume automatically the standard growth conditions on $n_j$ so that all integrals are well defined.

**Notation** $\kappa_j$, $j = 0, 1, 2, \ldots$, – eigenvalues of $\Delta u + \kappa u = 0$ with Neumann b. c.,

$e_j$, $j = 0, 1, 2, \ldots$ – the system of eigenfunctions of $\Delta u + \kappa u = 0$ with Neumann c.,

$$C_j := \left\{ d = [d_1, d_2] \in \mathbb{R}_+^2; \ d_2 = \frac{b_{12} b_{21}/\kappa_j^2}{d_1 - b_{11}/\kappa_j} + \frac{b_{22}}{\kappa_j} \right\}, \ j = 1, 2, \ldots \text{ (see Fig.1)},$$
$$D_S = \{ d \in \mathbb{R}_+^2; \ d \text{ is to the right from all } C_j, \ j = 1, 2, \ldots \},$$
$$D_U = \{ d \in \mathbb{R}_+^2; \ d \text{ is to the left from at least one } C_j \},$$
$$C_E = \overline{D_S} \cap \overline{D_U} - \text{the envelope of } C_j, \ j = 1, 2, \ldots,$$
$$C_r^R = \{ d = [d_1, d_2] \in C_E; \ r \le d_2 \le R \} \text{ (for } 0 < r < R),$$
$$C_r^R(\varepsilon) = \{ d = [d_1, d_2] \in \overline{D_U}; \ r \le d_2 \le R \ dist(d, C_E) < \varepsilon \} \text{ (for } \varepsilon > 0),$$
$$E(d) = \{ U = [u, v] \in \tilde{H}; \ [u, v] \text{ is a (weak) solution of (6), (3)} \},$$
$$E_I(d) = \{ U = [u, v] \in \tilde{H}; \ [u, v] \text{ is a (weak) solution of (6), (2)} \} \text{ (for } d \in \mathbb{R}_+^2).$$

By a critical point of (6), (3) or (6), (2) we mean $d = [d_1, d_2] \in \mathbb{R}_+^2$ such that $E(d) \neq \{0\}$ or $E_I(d) \neq \{0\}$, respectively.

**Remark 2.1.** (see e.g. [11]) The set $\cup_{j=1}^{+\infty} C_j$ is the set of all critical points of (6), (3). For $d \in D_S$, the trivial solution of (1), (3) is stable (all eigenvalues of the corresponding problem deciding about the stability have negative real parts). For $d \in D_U$, the trivial solution of (1), (3) is unstable (there is at least one positive eigenvalue of the corresponding eigenvalue problem). If $d \in C_m$ for $m = j, \ldots, j+k-1$ (i.e. either $k$ is the multiplicity of the eigenvalue $\kappa_j$, $C_j = \ldots = C_{j+k-1}$, or $d$ is the intersection point of two different $C_j$, $C_l$ and $k$ is the sum of the multiplicities of $\kappa_j$, $\kappa_l$) then $E(d) = Lin\{[\frac{d_2 \kappa_m - b_{22}}{b_{21}} e_m, e_m]\}_{m=j}^{j+k-1}$.

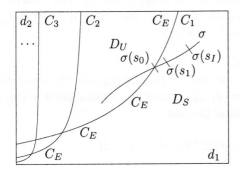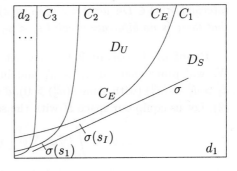

Fig. 1. The system of hyperbolas $C_j$, their envelope $C_E$, domain of stability $D_S$ (to the right from $C_E$) and instability $D_U$ (to the left from $C_E$), possible curves $\sigma$.

We will consider the following assumption:

$$\begin{aligned}&\text{for any } d \in C_r^R \text{ and any nontrivial solution } [u, v] \text{ of (6), (3),}\\&\text{the first component } u \text{ changes its sign on } \Gamma_U.\end{aligned} \tag{8}$$

620

**Remark 2.2.** If $C_r^R \subset C_p$, $C_r^R \cap C_j = \emptyset$ for all $C_j \neq C_p$ then the assumption (8) means that any eigenfunction of the Laplacian with Neumann b. c. corresponding to the $p$-th eigenvalue changes its sign on $\Gamma_U$. This follows from the form of solutions of (6), (3) (see Remark 2.1).

**Theorem 2.1.** Let (4) hold. Then there is no critical point of (6), (2) in $D_S$. If, moreover, $0 < r < R$ are such that (8) holds then there exists $\varepsilon > 0$ such that there is no critical point of (6), (2) in $C_r^R(\varepsilon)$.

Let us consider a smooth curve $\sigma : (0,\infty) \to \mathbb{R}_+^2$ and the problem

$$\begin{aligned} \sigma_1(s)\Delta u + b_{11}u + b_{12}v + n_1(u,v) = 0 \\ \sigma_1(s)\Delta v + b_{21}u + b_{22}v + n_1(u,v) = 0 \end{aligned} \quad \text{in } \Omega \tag{9}$$

with the boundary conditions (2) and with the bifurcation parameter $s$.

**Remark 2.3.** Let $[d_1, d_2] \in \mathbb{R}_+^2$ be given and set $\sigma_1(s) = d_1/s^2$, $\sigma_2(s) = d_2/s^2$ for $s \in (0,+\infty)$. Hence, our curve is a ray intersecting $[d_1, d_2]$ going to infinity for $s \to 0+$ and to the origin for $s \to +\infty$. Using the substitution $x = sx'$, we see that the problem (9) in $\Omega$ is equivalent to the problem (5) in the domain $\Omega_s$ of the same shape but with the size given by $s$, the volume of $\Omega_s$ is proportional to $s^k$.

By a bifurcation point of (9), (2) we mean $s_I > 0$ such that in any neighbourhood of $[s_0, 0]$ in $\tilde{H}$ there is a nontrivial solution $[s, U]$. If $s_0$ is a bifurcation point of (9), (2) then $\sigma(s_0)$ is a critical point of (6), (2) (cf. e.g. [6]).

**Corollary 2.1.** Let (4) hold. Then there is no bifurcation point $s_I$ of (9), (2) with $\sigma(s_I) \in D_S$. If (8) holds then there is $\varepsilon > 0$ such that there is no bifurcation point $s_I$ of (9), (2) with $\sigma(s_I) \in C_r^R(\varepsilon)$. In particular, if $\sigma(s_0) \in \text{int } C_r^R$ (the relative interior), $\sigma(s) \in D_S$ for $0 < s < s_0$, $\sigma(s) \in D_U$ for $s > s_0$ then there is $\delta > 0$ such that there is no bifurcation point $s_I < s_0 + \delta$ of (9), (2).

**Proof of Theorem 2.1:** First, let $d^0 = [d_1^0, d_2^0] \in C_E$ be fixed, $d^0 \neq [0,0]$. We will prove that for $d_1 > d_1^0$ and in the case of the assumption (8) also for $d_1 > d_1^0 - \varepsilon(d_2^0)$ (with some $\varepsilon(d_2^0) > 0$), $d = (d_1, d_2^0)$ cannot be a critical point of (6), (2). Let us equip our space $H$ with the scalar product

$$\langle \varphi, \psi \rangle = \int_\Omega (\nabla\varphi \cdot \nabla\psi + \eta\varphi\psi)\,dx \quad \text{for all } \varphi, \psi \in H$$

where $\eta \in (0, -b_{22}/d_2^0)$ is fixed. It is well-known that the corresponding norm $\|\cdot\|$ is equivalent to the usual Sobolev norm. Introduce an operator $A : H \to H$ by

$$\langle Au, \varphi \rangle = \int_\Omega u\varphi\,dx \quad \text{for all } u, \varphi \in H.$$

Clearly $A$ is linear, symmetric, positive and compact due to the compactness of the embedding $H \subset L_2$. We have $\langle \varphi, \psi \rangle = \int_\Omega \nabla\varphi \cdot \nabla\psi\,dx + \eta\langle A\varphi, \psi \rangle$. Weak formulation of (6), (2) with $d_2 = d_2^0$ is the variational inequality

$$u \in K, \ \langle d_1 u - d_1 \eta A u - b_{11} A u - b_{12} A v, \varphi - u \rangle \geq 0 \text{ for all } \varphi \in K,$$
$$v \in H, \ d_2^0 v - d_2^0 \eta A v - b_{21} A u - b_{22} A v = 0. \tag{10}$$

For $\eta$ considered, the operator $(d_2^0 I - (b_{22} + d_2^0 \eta) A)$ is invertible. Expressing $v$ from the equation in (10) and substituting it into the inequality we obtain

$$v = (d_2^0 I - (b_{22} + d_2^0 \eta) A)^{-1} b_{21} A u, \tag{11}$$
$$u \in K, \ \langle d_1 u - d_1 \eta A u - S u, \varphi - u \rangle \geq 0 \text{ for all } \varphi \in K, \tag{12}$$

where

$$S u = b_{11} A u + b_{12} b_{21} A (d_2^0 I - (b_{22} + d_2^0 \eta) A)^{-1} A u,$$

$S : H \to H$ is a linear, symmetric and compact operator. For any $\lambda \in \mathbb{R}$, set

$$M(\lambda) = max_{u \in H, \ \|u\|=1} \langle S u + \eta \lambda A u, u \rangle, \ M_I(\lambda) = max_{u \in K, \ \|u\|=1} \langle S u + \eta \lambda A u, u \rangle.$$

Clearly

$$M_I(\lambda) \leq M(\lambda) \text{ for all } \lambda \geq 0, \ M(\lambda) \leq M(\mu), \ M_I(\lambda) \leq M_I(\mu) \text{ for } \lambda \leq \mu. \tag{13}$$

Since the operators $S + \eta \lambda A$ are symmetric and compact, it is known (see e.g. [15], $\S 43.4$) that

$$\begin{aligned} &M(\lambda) \text{ is the largest eigenvalue of the operator } S + \eta \lambda A \\ &\text{and the maximum in the definition of } M(\lambda) \text{ is attained exactly} \\ &\text{in all corresponding normalized eigenvectors.} \end{aligned} \tag{14}$$

Let us prove that

$$M(d_1^0) = d_1^0, \ M(\lambda) < \lambda \text{ for all } \lambda > d_1^0. \tag{15}$$

The weak formulation of (6), (3) can be written in the form

$$d_1 u - d_1 \eta A u - b_{11} A u - b_{12} A v = 0, \ d_2 v - d_2 \eta A v - b_{21} A u - b_{22} A v = 0. \tag{16}$$

Due to Remark 2.1, there is a nontrivial $[u, v]$ satisfying (16) with $d = d^0$. Repeating for (16) the considerations performed above for (10), we get (11) and

$$d_1^0 u = S u + \eta d_1^0 A u. \tag{17}$$

It follows that $M(d_1^0) \geq d_1^0 > 0$. Assume that $M(d_1^0) > d_1^0$. Set

$$\lambda_0 = d_1^0, \ \lambda_n = M(\lambda_{n-1}) \text{ for } n = 1, 2, \dots$$

Due to (13) we obtain successively $\lambda_n \geq \lambda_{n-1}$, $n = 1, 2, \dots$, and therefore $\lambda_n \to \lambda_\infty$ for some $\lambda_\infty \in (\lambda_0, +\infty]$. Let us assume that $\lambda_\infty = +\infty$. According to (14), we can choose $u_n$ such that $\|u_n\| = 1$,

$$\lambda_n u_n = S u_n + \lambda_{n-1} \eta A u_n, \tag{18}$$

that means

$$\int_\Omega \lambda_n \nabla u_n \cdot \nabla \varphi + \eta (\lambda_n - \lambda_{n-1}) u_n \varphi \, dx = \langle S u_n, \varphi \rangle \text{ for all } \varphi \in H. \tag{19}$$

We can assume that $u_n \rightharpoonup u_\infty$ (weakly). Setting $v_n = (d_2^0 I - (b_{22} + d_2^0 \eta)A)^{-1} b_{21} A u_n$ as in (11) and using the compactness of $A$, we obtain $v_n \to v_\infty$ with $v_\infty = (d_2^0 I - (b_{22} + d_2^0 \eta)A)^{-1} b_{21} A u_\infty$ and $d_2^0 v_n - (b_{22} + d_2^0 \eta)A v_n - b_{21} A u_n = 0$, i.e.

$$\int_\Omega d_2^0 \nabla v_n \cdot \nabla \varphi - (b_{21} u_n + b_{22} v_n)\varphi \, \mathrm{d}x = 0 \text{ for all } \varphi \in H. \tag{20}$$

Setting $\varphi = u_n$ in (19) we obtain (by using the definition of $A$, $S$ and $v_n$)

$$\int_\Omega \lambda_n |\nabla u_n|^2 + \eta(\lambda_n - \lambda_{n-1})|u_n|^2 \, \mathrm{d}x = \int_\Omega b_{11} u_n^2 + b_{12} u_n v_n \, \mathrm{d}x.$$

Dividing by $\lambda_n$ and realizing that both integrands on the left hand side are nonnegative, we get $\int_\Omega |\nabla u_n|^2 \, \mathrm{d}x \to 0$. Consequently $\eta \int_\Omega u_n^2 \, \mathrm{d}x \to 1$ because of $\|u_n\| = 1$. Since $u_n \to u_\infty$ in $L_2(\Omega)$ due to the compactness of the embedding, we have $\eta \int_\Omega u_\infty^2 \, \mathrm{d}x = 1$. Hence, $\|u_\infty\| \geq \lim \|u_n\|$, therefore $u_n \to u_\infty$. In particular,

$$\int_\Omega |\nabla u_\infty|^2 \, \mathrm{d}x = 0, \text{ i.e. } u_\infty \text{ is constant.}$$

We can assume without loss of generality that $u_\infty > 0$ because we could replace $u_n$ by $-u_n$ in the above considerations. Then also $\int_\Omega u_n \, \mathrm{d}x > 0$ for $n$ large. Taking $\varphi = 1$ in (19), (20), we obtain

$$0 \leq \int_\Omega \eta(\lambda_n - \lambda_{n-1})u_n \, \mathrm{d}x = \int_\Omega b_{11} u_n + b_{12} v_n \, \mathrm{d}x, \quad \int_\Omega (b_{21} u_n + b_{22} v_n) \, \mathrm{d}x = 0.$$

Multiplying the last inequality and the last equation by $b_{22}$ and $b_{12}$, respectively, realizing that $b_{22} < 0$ and subtracting, we get by the limiting process

$$b_{11} b_{22} u_\infty - b_{12} b_{21} u_\infty \leq 0. \tag{21}$$

This implies $\det b_{ij} \leq 0$, which contradicts our assumption (4). Hence, $\lambda_\infty$ must be finite and the limiting process in (18) and in the definition of $v_n$ gives

$$\lambda_\infty u_\infty - \lambda_\infty \eta A u_\infty = b_{11} A u_\infty + b_{12} A v_\infty, \quad d_2^0 v_\infty - d_2^0 \eta A v_\infty = b_{21} A u_\infty + b_{22} A v_\infty.$$

Hence, $[\lambda_\infty, d_2^0]$ should lie on some $C_j$ by Remark 2.1 and $\lambda_\infty > d_1^0$ contradicts the assumption $d_0 \in C_E$ and the shape of $C_j$ (Fig. 1). Thus, $M(d_1^0) = d_1^0$ is proved. The assumption $M(\lambda) \geq \lambda$ for some $\lambda > d_1^0$ leads to a contradiction in the same way and (15) is proved.

Now, let $d^I = [d_1^I, d_2^0]$ be a critical point of (6), (2). Then there is a nontrivial solution $[u, v]$ of (10) with $d_1 = d_1^I$ and $u$ satisfies (12), $u \neq 0$. Choosing $\varphi = 2u$, $\varphi = 0$ in (12) we obtain

$$d_1^I \langle u, u \rangle = \langle Su + d_1^I \eta A u, u \rangle$$

and therefore $d_1^I \leq M_I(d_1^I)$. If it were $d_1^I > d_1^0$ then (13), (15) would imply $d_1^I \leq M_I(d_1^I) \leq M(d_1^I) < d_1^I$ and this contradiction shows that $d_1^I \leq d_1^0$. Since $d^0 \in C_E$ was arbitrary, the first assertion of Theorem 2.1 follows.

Now, let $d^0 \in C_r^R$ and let (8) be fulfilled. Assume that $M_I(d_1^0) = M(d_1^0)$. Then there is $u \in K$ with $\|u\| = 1$ such that

$$\langle Su + d_1^0 \eta A u, u \rangle = M_I(d_1^0) = M(d_1^0).$$

It follows from (14) that $u$ is an eigenvector of the operator $S + d_1^0 \eta A$. If we define $v$ by (11) then $[u,v]$ satisfies (16), i.e. $[u,v] \in E(d^0)$. This contradicts (8) which ensures that $u \notin K$ for nontrivial $[u,v] \in E(d^0)$. Hence, $M_I(d_1^0) < M(d_1^0)$. Let $d^I = [d_1^I, d_2^0]$ be a critical point of (6), (2) again. Using the results of the previous step and (13), (15), we get

$$d_1^I \leq M_I(d_1^I) \leq M_I(d_1^0) < M(d_1^0) = d_1^0. \tag{22}$$

Finally, let us assume that the second assertion of Theorem 2.1 holds with no $\varepsilon > 0$. Then there are critical points $[d_1^n, d_2^n]$ of (6), (2) such that $r \leq d_2^n \leq R$, $[d_1^n, d_2^n] \to [d_1^0, d_2^0]$ with some $[d_1^0, d_2^0] \in C_r^R$. It is easy to see that $[d_1^0, d_2^0]$ must be also a critical point of (6), (2) (cf. [13], Lemma 3). This is a contradiction because (22) should be true for $d_1^I = d_1^0$ by the previous step.

**Remark 2.4.** If the zero Dirichlet conditions are prescribed on a part of $\partial\Omega \setminus \Gamma_U$ as in Section 3 then we can take $\eta = 0$ and the proof is very simple (cf. [7,8]).

## 3. Destabilizing influence of Signorini conditions for inhibitor

Now, let $\Gamma_D$ be an additional open subset $\Gamma_D$ of $\partial\Omega$ such that $\Gamma_D \cap \Gamma_U = \emptyset$, $\operatorname{meas} \Gamma_D > 0$. We will consider boundary conditions

$$u = v = 0 \text{ on } \Gamma_D, \ \frac{\partial u}{\partial n} = 0 \text{ on } \partial\Omega \setminus \Gamma_D, \ \frac{\partial v}{\partial n} = 0 \text{ on } \partial\Omega \setminus (\Gamma_U \cup \Gamma_D),$$
$$v \geq 0, \ \frac{\partial v}{\partial n} \geq 0, \ \frac{\partial v}{\partial n} \cdot v = 0 \text{ on } \Gamma_U \tag{23}$$

and also classical boundary conditions

$$u = v = 0 \text{ on } \Gamma_D, \quad \frac{\partial u}{\partial n} = \frac{\partial v}{\partial n} = 0 \text{ on } \partial\Omega \setminus \Gamma_D. \tag{24}$$

We will use the notation from the previous section but $\kappa_j$ and $e_j$, $j = 1, 2, ...$ will be now eigenvalues and eigenfunctions of $\Delta u + \kappa u = 0$, $u = 0$ on $\Gamma_D$, $\frac{\partial u}{\partial n} = 0$ on $\partial\Omega \setminus \Gamma_D$. All assertions of Remark 2.1 remain valid if we replace (3) by (24). (Let us note that in the case of Neumann conditions, the hyperbolas $C_j$ are defined only for positive eigenvalues $\kappa_j$, $j = 1, 2, ...$, no hyperbola corresponds to $\kappa_0 = 0$).

Further, we will consider curves $\sigma$ satisfying

$$\lim_{s \to 0+} \sigma_1(s) = +\infty, \ \liminf_{s \to 0+} \sigma_2(s) > 0 \tag{25}$$

and intersecting a neighbourhood of some hyperbola $C_p$ with $p$ such that

there is an eigenfunction $e_0$ of the problem
$$\Delta u + \kappa u = 0, \ u = 0 \text{ on } \Gamma_D, \ \frac{\partial u}{\partial n} = 0 \text{ on } \partial\Omega \setminus \Gamma_D \tag{26}$$
corresponding to $\kappa_p$ and satisfying $e_0 > 0$ on $\Gamma_U$.

**Theorem 3.1.** *Let (4) and (26) be fulfilled, let $\Gamma_U \subset \partial\Omega$ be a $(k-1)$-dimensional smooth manifold in $\mathbb{R}^k$ with a smooth boundary. Then there is an open set $\mathcal{U}$ containing $C_p \cap C_E$ such that for any curve $\sigma$ satisfying (25) and $\sigma(s_1) \in \mathcal{U}$ for some $s_1 > 0$, there exists a bifurcation point $s_I > s_1$ with $\sigma_1(s_I) \leq \frac{b_{11}}{\kappa_1}$ of (9), (23). Moreover, this bifurcation is global in the sense of P. H. Rabinowitz [14].*

**Remark 3.1.** Let us describe the differences of Theorem 3.1 from the previous results. Some of them were formulated in a more abstract setting but in terms of our example, it was always necessary to assume $e_0 \geq \varepsilon$ (with some $\varepsilon > 0$) instead of $e_0 > 0$ in (26). Further, only curves $\sigma$ intersecting $C_p \cap C_E$ were always considered while in the present formulation, our curve $\sigma$ must only intersect a certain neighbourhood of $C_p \cap C_E$. This has the following consequence for the particular situation when $s$ corresponds to the growth of the domain, i.e. $\sigma_1(s) = d_1/s^2$, $\sigma_2(s) = d_2/s^2$ for $s \in (0,+\infty)$, with given $d_1$, $d_2 \in \mathbb{R}^2_+$ (see Remark 2.3). Elementary calculation shows that the hyperbolas $C_j$ have a joint tangent $T$. If $[d_1, d_2] \in \mathbb{R}^2_+$ lies above $T$ then $\sigma(s_1) \in C_1 \cap C_E$, $\sigma(s_2) = b_{11}/\kappa_1$ for some $s_1 > s_2 > 0$. Theorem 2.1 in $^3$ guarantees the existence of a bifurcation point $s_I \in (s_2, s_1)$ if (26) is fulfilled for $p = 1$ with $e_0 > 0$ replaced by $e_0 > \varepsilon$. If $[d_1, d_2]$ lies below $T$ then the whole curve $\sigma$ is in $D_S$ and no of the previous results gives any information. However, our Theorem 3.1 guarantees the existence of a bifurcation point even in this case if (26) is fulfilled for some $p$ and if the ray $\sigma$ is sufficiently close to $T$. Hence, the previous results stated that for couples $[d_1, d_2]$ above $T$, a bifurcation in case of the boundary conditions (23) occurs for smaller domains than for (24). Theorem 3.1 states in addition that in the case of Signorini conditions (23), the bifurcation occurs even for some diffusion coefficients for which it is excluded in case of classical boundary conditions (24).

**Proof of Theorem 3.1** is based on the same ideas as that of Theorem in $^{13}$ (where the particular case $\sigma_1(s) = s$, $\sigma_2(s) = 1$ is considered) or Theorem 1.1 in $^9$. The detailed proof will be contained in the forthcoming paper by J. Baltaev and the author. Here we will only emphasize the main ideas and differences from $^{13,9}$. Introduce the Hilbert space $H = \{\varphi \in W_2^1(\Omega); \; \varphi = 0 \text{ on } \Gamma_D\}$ with the scalar product

$$\langle \varphi, \psi \rangle = \int_\Omega \nabla\varphi \cdot \nabla\psi \, dx \quad \text{for all } \varphi, \; \psi \in H.$$

Under our assumption meas $\Gamma_D > 0$, the corresponding norm $\| \cdot \|$ is equivalent to the usual Sobolev norm on $H$. We can write a weak formulation of (9), (23) as

$$U \in \tilde{K}, \; \langle D(\sigma(s))U - BAU - N(U), \Phi - U \rangle \geq 0 \text{ for all } \Phi \in \tilde{K}$$

where $D(\sigma) = \begin{pmatrix} \sigma_1 & 0 \\ 0 & \sigma_2 \end{pmatrix}$, $B = \begin{pmatrix} b_{11} & b_{12} \\ b_{21} & b_{22} \end{pmatrix}$, $U = [u, v] \in \tilde{H} = H \times H$, $N(U) = [N_1(u,v), N_2(u,v)]$, $\tilde{K} = H \times K$. This is equivalent to the equation

$$U - (D(\sigma))^{-1}P_{\tilde{K}}[BAU + N(U)]) = 0 \tag{27}$$

where $P_{\tilde{K}}$ is the projection onto $\tilde{K}$ (see e.g. $^5$, cf. also $^{13}$). The main idea is to prove that the Leray-Schauder degree of the mapping from (27) with respect to the origin and the ball $B_\rho$ with the radius $\rho$ centered at the origin satisfies

$$\begin{aligned}
&\deg(I - D(\sigma(s))^{-1}P_{\tilde{K}}[BA + N], B_\rho, 0) = 1 \text{ for } s \in (0, \varepsilon), \\
&\deg(I - D(\sigma(s_1))^{-1}P_{\tilde{K}}[BA + N], B_\rho, 0) = 0 \text{ for } s_1 \text{ with } \sigma(s_1) \in \mathcal{U} \cap D_S
\end{aligned} \tag{28}$$

for sufficiently small $\varepsilon > 0$ and a small neighbourhood $\mathcal{U}$ of $C_p \cup C_E$. Then standard global bifurcation results of the Rabinowitz type imply that there is a global bifurcation for (27) between $s_1$ and $\varepsilon$. It follows from [2] that there is no bifurcation point $s_I$ with $\sigma_1(s_I) > \frac{b_{11}}{\kappa_1}$, and the assertion of Theorem 3.1 follows.

For the proof of the first assertion in (28), it is sufficient to show that our mapping is homotopically invariant to the identity for $s$ small, i.e. $\sigma_1(s)$, $\sigma_2(s)$ large. For the proof of the second assertion, the following basic condition is essential:

$$E_I(d) = E(d) \cap (\tilde{K}) \text{ for all } d \in C_p \tag{29}$$

where $E_I(d)$ and $E(d)$ are as in Section 2 but with (2) and (3) replaced by (23) and (24), respectively. In all previous papers, this condition was always ensured by the assumption that there is $[u, v] \in E(d)$ with $v$ lying in a certain pseudointerior of $K$ (for $d \in C_p$), which was a consequence of various modifications of [6], Lemma 1.1. Now it shows that the pseudointerior assumption is not necessary, only the condition (29) is sufficient. This condition is now guaranteed by the assumption (26) due to Lemma 3.1 below. Let us note that (26) does not imply the pseudointerior assumption which would require to replace $e_0 > 0$ by $e_0 \geq \varepsilon > 0$ in (26).

**Lemma 3.1.** *Assume that $\Gamma_U \subset \partial\Omega$ is a smooth $(k-1)$-dimensional manifold with a smooth boundary. Let $p$ be such that (26) is fulfilled. Then (29) holds.*

Proof: First, let us note that the weak formulation of (6), (23) is

$$u \in H, \ \int_\Omega d_1 \nabla u \cdot \nabla\varphi - (b_{11}u + b_{12}v)\varphi \, dx = 0 \text{ for all } \varphi \in H,$$
$$v \in K, \ \int_\Omega d_2 \nabla v \cdot \nabla(\psi - v) - (b_{21}u + b_{22}v)(\psi - v) \, dx \geq 0 \text{ for all } \psi \in K. \tag{30}$$

We assume $d \in C_p$ and due to Remark 2.1 there is a weak solution $[u_0, v_0]$ of (6), (24) with $v_0 = e_0$, i.e. we have

$$u_0 \in H, \ \int_\Omega d_1 \nabla u_0 \cdot \nabla\varphi - (b_{11}u_0 + b_{12}v_0)\varphi \, dx = 0 \text{ for all } \varphi \in H,$$
$$v_0 \in H, \ \int_\Omega d_2 \nabla v_0 \cdot \nabla\psi - (b_{21}u_0 + b_{22}v_0)\psi \, dx = 0 \text{ for all } \psi \in H. \tag{31}$$

Let us recall that if $w \in H$, $\Delta w \in L_2(\Omega)$ then $\frac{\partial w}{\partial n}$ can be introduced as a functional on $H$ defined by $[\frac{\partial w}{\partial n}, \varphi] = \int_\Omega (\Delta w \varphi + \nabla w \cdot \nabla\varphi) \, dx$ for all $\varphi \in H$ where $[\cdot, \cdot]$ denotes the dual pairing (see [4], Remark 5.2 for details). It is easy to show that (30) or (31) is valid if and only if (6) holds, $\Delta u \in L_2(\Omega)$, $\Delta v \in L_2(\Omega)$ and (23) or (24) hold in the sense of the functional mentioned. Clearly $E(d) \cap (\tilde{K}) \subset E_I(d)$ and it is sufficient to prove the opposite inclusion. Let $[u, v] \in E_I(d)$. Due to the above assertion, it is sufficiet to prove that $\frac{\partial v}{\partial n} = 0$ on $\Gamma_U$, i.e. $[\frac{\partial v}{\partial n}, \zeta] = 0$ for any $\zeta \in H$ such that $\zeta = 0$ on $\partial\Omega \setminus \Gamma_U$ (in the sense of traces). Let us consider such an arbitrary fixed $\zeta$. Under our assumption that $\Gamma_U$ is a smooth manifold with a smooth boundary, we have $W_2^{\frac{1}{2}}(\Gamma_U) = \overline{\mathcal{D}(\Gamma_U)}$ (see [10], Theorem 11.1). For $\varphi \in H$, let us denote by $T\varphi$ its trace. The traces of functions from $H$ lie in $W_2^{\frac{1}{2}}(\partial\Omega)$ and therefore there exist $w_n \in \mathcal{D}(\Gamma_U)$, $w_n \to T\zeta$ in $W_2^{\frac{1}{2}}(\Gamma_U)$. We can extend $w_n$ by zero to the remainder of $\partial\Omega$ and denoting these extended functions again by $w_n$ we have $w_n \in W_2^{\frac{1}{2}}(\partial\Omega)$.

626

There exists a linear continuous extension mapping $R : W_2^{\frac{1}{2}}(\partial\Omega) \rightarrow W_2^1(\Omega)$ (see e.g. [12], Chapter 2, Theorem 5.7) such that $TR\varphi = \varphi$ on $\partial\Omega$ for all $\varphi \in W_2^{\frac{1}{2}}(\partial\Omega)$. Setting $\zeta_n = R\omega_n + \zeta - RT\zeta$, we get $\zeta_n \in H$, $\zeta_n \to \zeta$. Since $e_0 > 0$ on $\Gamma_U$ by (26), $e_0$ is smooth and $T\zeta_n = \omega_n$ on $\Gamma_U$, $\omega_n \in \mathcal{D}(\Gamma_U)$, for any $n$ there is $\varepsilon_n > 0$ such that $e_0 \pm \varepsilon_n\zeta_n \in K$. Let us set $\varphi = u_0$, $\psi = v + v_0 \pm \varepsilon_n\zeta_n$ in (30) and $\varphi = u$, $\psi = v$ in (31). Subtracting the first expressions obtained in this way from (30) and (31), then the second expressions, and putting all information together, we get $\int_\Omega d_2\nabla v \cdot \nabla \zeta_n - (b_{21}u + b_{22}v)\zeta_n \, dx = 0$. Using the Green Theorem and (6), we obtain $[\frac{\partial v}{\partial n}, \zeta_n] = 0$. The limiting process gives $[\frac{\partial v}{\partial n}, \zeta] = 0$. Since $\zeta \in H$ was arbitrary such that $\zeta = 0$ on $\partial\Omega \setminus \Gamma_U$, we have $\frac{\partial v}{\partial n} = 0$ on $\Gamma_U$. Hence, $[u, v]$ is a weak solution of (6), (24), i.e. $[u, v] \in E(d)$.

## References

1. P. Drábek, M. Kučera and M. Míková: Bifurcation points of reaction-diffusion systems with unilateral conditions, Czechoslovak Math. J. 35 (1985), 639–660.
2. J. Eisner: Critical and bifurcation points of reaction-diffusion systems with conditions given by inclusions, Nonlinear Anal. 46 (2001), 69–90.
3. J. Eisner, M. Kučera: Spatial patterning in reaction-diffusion systems with nonstandard boundary conditions, Fields Institute Communications 25 (2000), 239–256.
4. J. Eisner, M. Kučera, L. Recke: Smooth continuation of solutions and eigenvalues for variational inequalities based on the implicit function theorem, J. Math. Anal. Appl. 274 (2002), 159–180.
5. D. Kinderlehrer and G. Stampacchia: An Introduction to Variational Inequalities and their Applications. Academic Press, New York, 1980.
6. M. Kučera: Bifurcation points of variational inequalities, Czechoslovak Math. J. 32 (107), (1982) 208–226.
7. M. Kučera: Stability and bifurcation problems for reaction-diffusion systems with unilateral conditions, Equadiff 6 (Ed.: Vosmanský, J. - Zlámal, M.), Brno, University J. E. Purkyně, (1986) 227–234 .
8. M. Kučera: Reaction-diffusion Systems: Bifurcation and stabilizing effect of conditions given by inclusions, Nonlinear Anal. 27 (1996), 249–260.
9. M. Kučera, M. Bosák: Bifurcation for quasi-variational inequalities of reaction-diffusion type, SAACM 3 (1993), 111–127.
10. J.-L. Lions, E. Magenes: Non-Homogeneous Boundary Value Problems and Applications. Springer-Verlag 1972.
11. M. Mimura, Y. Nishiura and M. Yamaguti: Some diffusive prey and predator systems and their bifurcation problems, Ann. N.Y. Acad. Sci. 316 (1979), 490–521.
12. J. Nečas: Les méthodes directes en théorie des équatios elliptiques. Academia, Prague 1967.
13. P. Quittner: Bifurcation points and eigenvalues of inequalities of reaction-diffusion type, J. Reine Angew. Math. 380 (1987), 1–13.
14. P. H. Rabinowitz: Some global results for non-linear eigenvalue problems, J. Functional Analysis 7 (1971), 487–513.
15. E. Zeidler: Nonlinear Functional Analysis and Applications III. Variational Methods and Optimization. Springer-Verlag 1984.

# WAVE PROPAGATION IN A 3-D OPTICAL WAVEGUIDE II
## NUMERICAL EXAMPLES

G. CIRAOLO

*Dipartimento di matematica U. Dini*
*Università di Firenze*
*Viale Morgagni 67/A*
*50134 Firenze, Italy*
*E-mail: ciraolo@math.unifi.it*


O. ALEXANDROV

*Department of Mathematics*
*University of California, Los Angeles*
*Los Angeles, CA 90095-1555*
*E-mail: aoleg@math.ucla.edu*

In this paper we apply the results we obtained in [2] to deduce explicit analytic formulas for the Green's function of the electromagnetic field propagating inside an optical fiber for the particular case of a step-index fiber. We use the obtained expressions to calculate numerically the electromagnetic field and represent it graphically.

**Key words:** Wave propagation, optical waveguide, Helmholtz equation, Green's function
**Mathematics Subject Classification:** 78A40, 78A50, 35J05

## 1. Introduction

In [2] we defined a transform which enabled us to study the wave propagation in a cylindrical optical fiber. We considered the whole $\mathbb{R}^3$ space and, as model equation, we used the *Helmholtz equation*

$$\Delta u + k^2 n(\sqrt{x^2 + z^2})^2 u = f. \tag{1}$$

Here, the positive number $k$ is called the *wavenumber*, and the function $f$ represents a source of energy. It is assumed that the optical fiber has a core, outside which the index of refraction $n(r)$ is constant, where $r := \sqrt{x^2 + z^2}$ is the radial coordinate in a cylindrical coordinate system $(r, \vartheta, z)$, in which $z$ runs along the fiber axis. Thus, for some $R > 0$ and $n_{cl} \geq 1$, $n(r) = n_{cl}$ for $r \geq R$. For $0 \leq r < R$, $n(r)$ is required to be a positive, bounded, and integrable function.

The main result of [2] was the construction of a representation formula (see (17)) for a solution $u$ of (1) satisfying suitable radiation conditions. Our results generalized a similar formula obtained by Magnanini and Santosa [5] in the two-dimensional case. In fact, both formulas clearly describe the division of the energy of the electromagnetic field: part of the energy coming from the source propagates inside the waveguide as a finite number of distinct *guiding modes*, while the remaining energy either decays exponentially along the fiber or is radiated outside. On the other hand, the 3-D case reveals a new feature: for special choices of the parameters, a new kind of guided modes appears which, rather than decaying exponentially outside the fiber, it vanishes as a power of the distance from the fiber's axis.

For the reader's convenience, in Section 2 we summarize the formulas obtained in [2] which are used in our calculations.

In this paper we study a particular case of (1) in which the Green's function of (1) can be calculated explicitly[a]. We consider a *step-index fiber*, which is characterized by the following index of refraction

$$n(r) = \begin{cases} n_{co}, & 0 < r < R, \\ n_{cl}, & r \geq R, \end{cases} \tag{2}$$

with $n_{co} > n_{cl}$.

In this case, it is relatively easy to derive from (17) explicit formulas. We present these calculations in Sections 3. We are then able to calculate numerically the obtained field and represent it graphically.

## 2. A summary of the results

In cylindrical coordinates (1) becomes

$$\frac{\partial^2 u}{\partial z^2} + \frac{1}{r}\frac{\partial}{\partial r}\left(r\frac{\partial u}{\partial r}\right) + \frac{1}{r^2}\frac{\partial^2 u}{\partial \vartheta^2} + k^2 n(r)^2 u = f. \tag{3}$$

Consider the homogeneous version of (3), that is, with $f = 0$ on the right-hand side. We look for a solution of this equation in separated variables

$$u(r, \vartheta, z) = e^{i\beta k z} e^{im\vartheta} \sqrt{r} w(r),$$

with $\beta \in \mathbb{C}$ and $m \in \mathbb{Z}$. Then $w(r)$ must satisfy the differential equation

$$w'' + \left\{l - q(r) - \frac{m^2 - 1/4}{r^2}\right\} w = 0, \quad r \in (0, \infty), \tag{4}$$

where

$$d^2 = k^2(n_0^2 - n_{cl}^2), \quad l = k^2(n_0^2 - \beta^2), \quad q(r) = k^2[n_0^2 - n(r)^2], \tag{5}$$

with $n_0$ being the maximum of $n(r)$. The function $q(r)$ is non-negative, with $q(r) = d^2$ for $r \geq R$. We will view (4) as an eigenvalue problem in $l \in \mathbb{C}$ and will call it the *associated eigenvalue problem* to (3).

The next lemma gives us a solution of (4) which is "well behaved" as $r \to 0$.

---

[a]We have done analogous computations in the case in which $n$ has a different form, for example when it models a coaxial cable. These results are available on the authors' web page in a more detailed version of this paper. See *www.math.ucla.edu/~aoleg* or *www.math.unifi.it/~ciraolo*.

**Lemma 2.1.** *There exists a solution $j_m(r,l)$ $(r > 0, l \in \mathbb{C}, m \in \mathbb{Z})$ of (4) such that*

$$\lim_{r \to 0} \frac{j_m(r,l)}{r^{|m|+1/2}} = 1, \quad \lim_{r \to 0} \frac{j'_m(r,l)}{(|m| + 1/2)\, r^{|m|-1/2}} = 1. \tag{6}$$

We will need some definitions and notation to state the next theorems.

Let $m \in \mathbb{Z}$ and $J_m$, $Y_m$ be the Bessel's functions of the first and second kind respectively. We use the following notations:

$$a_m(r,\lambda) = \sqrt{r} J_m\left(\sqrt{\lambda - d^2}\, r\right), \quad b_m(r,\lambda) = \sqrt{r} Y_m\left(\sqrt{\lambda - d^2}\, r\right), \quad \lambda > d^2. \tag{7}$$

It can be shown that

$$j_m(r,\lambda) = c_m(\lambda) a_m(r,\lambda) + d_m(\lambda) b_m(r,\lambda), \quad r \geq R, \tag{8}$$

where

$$c_m(\lambda) = \frac{\pi}{2}\{b'_m(R,\lambda) j_m(R,\lambda) - j'_m(R,\lambda) b_m(R,\lambda)\}, \tag{9a}$$

$$d_m(\lambda) = -\frac{\pi}{2}\{a'_m(R,\lambda) j_m(R,\lambda) - j'_m(R,\lambda) a_m(R,\lambda)\}. \tag{9b}$$

Furthermore, we define

$$k_m(r,\lambda) = \sqrt{r} K_m\left(\sqrt{d^2 - \lambda}\, r\right), \quad \lambda < d^2, r \geq R, \tag{10}$$

where $K_m$ is the modified Bessel function of the second kind. This function will decay exponentially as $r \to \infty$.

Next theorem gives us a transform in terms of the eigenfunctions $j_m(r,l)$ of the eigenvalue problem (4).

**Theorem 2.1.** *For every $m \in \mathbb{Z}$ it exists a non-decreasing function $\chi_m$ such that*

$$\langle d\chi_m, \eta \rangle = \frac{1}{\pi} \sum_{k=1}^{P_m} r_k^m \eta(\lambda_k^m) + \frac{1}{2} \int_{d^2}^{+\infty} \frac{\eta(\lambda)}{c_m(\lambda)^2 + d_m(\lambda)^2}\, d\lambda, \tag{11}$$

*for every $\eta \in C_0^\infty([0, +\infty))$, where*

$$r_k^m = \pi \left\{ \int_0^\infty j_m(r, \lambda_k^m)^2\, dr \right\}^{-1}, \quad k = 1, \ldots, P_m, \tag{12}$$

*and where $0 < \lambda_k^m \leq d^2$, $k = 1, \ldots, P_m$, $m \in \mathbb{Z}$, are finite in number and satisfy*

$$\frac{j'_m(R,\lambda)}{j_m(R,\lambda)} = \frac{k'_m(R,\lambda)}{k_m(R,\lambda)}, \quad \text{if } \lambda < d^2, \tag{13a}$$

$$\frac{j'_m(R,\lambda)}{j_m(R,\lambda)} = -\frac{|m| - 1/2}{R}, \quad \text{if } \lambda = d^2, |m| \geq 2. \tag{13b}$$

*Furthermore, we have that the inversion transform formula*

$$g(r) = \frac{1}{\pi} \int_{-\infty}^\infty j_m(r,\lambda) G_m(\lambda)\, d\chi_m(\lambda),$$

630

*and Parseval identity*

$$\int\limits_0^\infty |g(r)|^2\, dr = \frac{1}{\pi} \int\limits_{-\infty}^\infty |G_m(\lambda)|^2\, d\chi_m(\lambda),$$

*hold for any $g \in L^2(0,\infty)$, where we set*

$$G_m(\lambda) = \int\limits_0^\infty j_m(r,\lambda) g(r)\, dr. \tag{14}$$

We notice that, for $\lambda = \lambda_k^m < d^2$, (13a) implies

$$j_m(r,\lambda) = \frac{j_m(R,\lambda)}{k_m(R,\lambda)} k_m(r,\lambda), \quad r \geq R, \tag{15}$$

while for $\lambda = d^2$, $|m| \geq 2$, (13b) gives

$$j_m(r,\lambda) = \frac{j_m(R,\lambda)}{R^{1/2-|m|}} r^{1/2-|m|}, \quad r \geq R. \tag{16}$$

In the next theorem we provide a resolution formula for (1). We will not discuss about the uniqueness of such a solution; this topic will be studied elsewhere.

**Theorem 2.2.** *With the above assumptions, the solution of (3) can be represented as*

$$u(r,\vartheta,z) = \int\limits_{-\infty}^\infty \int\limits_0^\infty \int\limits_0^{2\pi} G(r,\rho;\vartheta,t;z,\zeta) f(\rho,t,\zeta)\, \rho\, dt\, d\rho\, d\zeta,$$

*where*

$$G(r,\rho;\vartheta,t;z,\zeta)$$
$$= \frac{1}{2\pi^2} \frac{1}{\sqrt{r\rho}} \sum_{m\in\mathbb{Z}} \int\limits_{-\infty}^{+\infty} \frac{e^{i|z-\zeta|\sqrt{k^2 n_0^2 - \lambda}}}{2i\sqrt{k^2 n_0^2 - \lambda}} e^{im(\vartheta-t)}\, j_m(\rho,\lambda)\, j_m(r,\lambda)\, d\chi_m(\lambda),$$

$$0 < r,\rho;\ 0 \leq \vartheta, t \leq 2\pi;\ z,\zeta \in \mathbb{R}, \tag{17}$$

*and $\chi_m$ is the non-decreasing function defined in Theorem 2.1.*

In the next two sections we will calculate explicit analytic formulas for the Green's function (17).

## 3. Step-index fibers

For a *step-index fiber*, the index of refraction satisfies (2). Then the function $q(r)$, defined by (5), is 0 if $0 \leq r < R$, and $d^2$ if $r \geq R$.

Let us find a formula for the function $j_m(r, \lambda)$ defined by Lemma 2.1. For $0 \leq r < R$, (4) takes the form

$$w'' + \left\{ \lambda - \frac{m^2 - 1/4}{r^2} \right\} w = 0.$$

Two linearly independent solutions of this equation are $\sqrt{r} J_m(r\sqrt{\lambda})$ and $\sqrt{r} Y_m(r\sqrt{\lambda})$. From (6) we deduce that $j_m(r, \lambda)$ must be a multiple of $\sqrt{r} J_m(r\sqrt{\lambda})$. By using the formulas (see identities (9.1.5) and (9.1.7) from [1]) $J_{-m}(z) = (-1)^m J_m(z)$, $m \in \mathbb{Z}$, $z \in \mathbb{C}$ and $J_m(z) \sim z^m (2m!)^{-m}$, $m \geq 0$, $z \in \mathbb{C}$, as $z \to 0$, we find that

$$j_m(r, \lambda) = \alpha_m(\lambda) \sqrt{r} J_m(r\sqrt{\lambda}), \quad 0 \leqslant r < R, \tag{18}$$

with $\alpha_m(\lambda) = (-1)^{(|m|-m)/2} 2^{|m|} |m|! \, \lambda^{-|m|/2}$, $m \in \mathbb{Z}$.

Let us calculate $j_m(r, \lambda)$, for $r > R$, and the measure $d\chi_m(\lambda)$. It will be convenient to represent $j_m(r, \lambda)$ as $j_m(r, \lambda) = \alpha_m(\lambda) \tilde{j}_m(r, \lambda)$. The measure $d\chi_m$ has a continuous part, for $\lambda > d^2$, and a discrete part, for $\lambda \leq d^2$. We will treat these cases separately.

Let $\lambda > d^2$. Denote $Q = \sqrt{\lambda - d^2}$. Define the operator

$$\mathcal{V}_x[f, g](\lambda) = x[\sqrt{\lambda} f(Qx) g'(x\sqrt{\lambda}) - Q f'(Qx) g(x\sqrt{\lambda})].$$

By using (7) - (9) we deduce

$$\tilde{j}_m(r, \lambda) = \begin{cases} \sqrt{r} \, J_m(r\sqrt{\lambda}), & 0 \leqslant r < R, \\ \frac{\pi}{2} \sqrt{r} \, [\beta_m(\lambda) J_m(Qr) + \gamma_m(\lambda) Y_m(Qr)], & r \geq R, \end{cases} \tag{19}$$

with

$$\beta_m(\lambda) = -\mathcal{V}_R[Y_m, J_m](\lambda), \quad \gamma_m(\lambda) = \mathcal{V}_R[J_m, J_m](\lambda). \tag{20}$$

From (8) and (11) we get

$$d\chi_m(\lambda) = 2\pi^{-1} \alpha_m(\lambda)^{-2} [\beta_m(\lambda)^2 + \gamma_m(\lambda)^2]^{-1} d\lambda, \quad \lambda \geqslant d^2.$$

Let now consider the case $\lambda \leq d^2$. Let $\lambda = \lambda_k^m$ be a discontinuity point of $\chi_m$ and $r_k^m$ be the jump of $\chi_m$ at $\lambda_k^m$. It will be convenient to represent $r_k^m$ as $r_k^m = \pi \alpha_m(\lambda)^{-2} \tilde{r}_k^m$. Also, in this case we will denote $\tilde{j}_m^g(r, \lambda) = \tilde{j}_m(r, \lambda)$. The motivation for this is to emphasize that for $\lambda \leq d^2$ we are dealing with *guided modes*, as opposed to *radiation modes* for $\lambda > d^2$. We need to consider separately the sub-cases $\lambda < d^2$ and $\lambda = d^2$.

Let $\lambda < d^2$. Denote $Q_0 = \sqrt{d^2 - \lambda}$. The condition for discontinuity of $\chi_m$ at $\lambda$ becomes in view of (10), (13a) and (18) is

$$\frac{\sqrt{\lambda} J_m'(\sqrt{\lambda} R)}{J_m(\sqrt{\lambda} R)} = \frac{Q_0 K_m'(Q_0 R)}{K_m(Q_0 R)}. \tag{21}$$

632

By using (15) we obtain

$$
\tilde{j}_m^g(r,\lambda) = \begin{cases} \sqrt{r}\, J_m(r\sqrt{\lambda}), & 0 \leqslant r < R, \\[3mm] \dfrac{J_m(\sqrt{\lambda}R)}{K_m(Q_0 R)} \sqrt{r}\, K_m(Q_0 r), & r \geq R. \end{cases} \tag{22}
$$

Now we will calculate the jump of $\chi_m$ at $\lambda_k^m$ by using (12). According to [6], see pp. 87-88, the following indefinite integral formulas hold:

$$
\int x Z_m(\alpha x)^2\, dx = \frac{x^2}{2}\left\{\left(1 - \frac{m^2}{\alpha^2 x^2}\right) Z_m(\alpha x)^2 + Z_m'(\alpha x)^2\right\},
$$

where $Z_m(x)$ is a solution of the Bessel equation, and

$$
\int x W_m(\alpha x)^2\, dx = \frac{x^2}{2}\left\{\left(1 + \frac{m^2}{\alpha^2 x^2}\right) W_m(\alpha x)^2 - W_m'(\alpha x)^2\right\},
$$

with $W_m(x)$ a solution of the modified Bessel equation. Then we can use (21) and (22) to deduce

$$
\tilde{r}_k^m = \frac{2\lambda_k^m (d^2 - \lambda_k^m)}{d^2} \cdot \frac{1}{\lambda_k^m R^2 J_m'(\sqrt{\lambda_k^m}\, R)^2 - m^2 J_m(\sqrt{\lambda_k^m}\, R)^2}. \tag{23}
$$

Lastly, let $\lambda = d^2$. From Theorem 2.1 and formulas (10) and (18) we obtain that the condition for discontinuity of $\chi_m$ at $\lambda$ is

$$
\frac{dJ_m'(Rd)}{J_m(Rd)} = -\frac{|m|}{R}. \tag{24}
$$

From (16) we get

$$
\tilde{j}_m^g(r,\lambda) = \begin{cases} \sqrt{r}\, J_m(rd), & 0 \leqslant r < R, \\[2mm] R^{|m|} J_m(dR)\, r^{1/2-|m|}, & r \geq R. \end{cases} \tag{25}
$$

In a similar fashion to the case $\lambda < d^2$ we find that, if $\chi_m(\lambda)$ is discontinuous at $\lambda = d^2$, then $|m| \geq 2$ and

$$
\tilde{r}_k^m = \frac{2(|m|-1)}{|m| R^2 J_m(dR)^2}. \tag{26}
$$

Now, we can define a transform in terms of $\tilde{j}_m(\rho,\lambda)$,

$$
\tilde{G}_m(\lambda) = \int_0^\infty \tilde{j}_m(\rho,\lambda)\, g(\rho)\, d\rho. \tag{27}
$$

For $\lambda = \lambda_k^m \leq d^2$ a discontinuity point for $\chi_m$ we will denote $\tilde{G}_m(\lambda)$ by $\tilde{G}_m^g(\lambda_k^m)$, again with the purpose of emphasizing that we are in the guided mode case.

We have the following corollary of Theorem 2.2.

**Corollary 3.1.** *Let* $f : \mathbb{R}^3 \to \mathbb{R}$ *be a continuous function with compact support and let* $\widetilde{\chi}_m : \mathbb{R} \to \mathbb{R}$ *be a function such that*

$$\langle d\widetilde{\chi}_m, \eta \rangle = \sum_{k=1}^{P_m(R,d)} \widetilde{r}_k^m \, \eta(\lambda_k^m) + \frac{2}{\pi^2} \int_{d^2}^{+\infty} \frac{\eta(\lambda)}{\beta_m(\lambda)^2 + \gamma_m(\lambda)^2} \, d\lambda,$$

*for all* $\eta \in C_0^\infty(\mathbb{R})$, *where* $\lambda_k^m \in (0, d^2]$, $k = 1 \ldots P_m(R, d)$, $m \in \mathbb{Z}$, *are the solutions of* (21) *or* (24), *the jumps* $\widetilde{r}_k^m$ *are defined by* (23) *or* (26), *and* $\beta_m(\lambda)$ *and* $\gamma_m(\lambda)$ *are given by* (20). *Hence, a Green's function for the step-index fiber is* $G = G(r, \vartheta, z; \rho, \eta, \zeta)$, *given by*

$$G = \frac{1}{2\pi^2} \frac{1}{\sqrt{r\rho}} \sum_{m \in \mathbb{Z}} \int_{-\infty}^{\infty} \frac{e^{i|z-\zeta|\sqrt{k^2 n_{co}^2 - \lambda}}}{2i\sqrt{k^2 n_{co}^2 - \lambda}} \, \widetilde{j}_m(\rho, \lambda) \, \widetilde{j}_m(r, \lambda) e^{im(\vartheta - \eta)} \, d\widetilde{\chi}_m(\lambda), \quad (28)$$

*with* $r, \rho > 0$, $0 \le \vartheta, t \le 2\pi$ *and* $z, \zeta \in \mathbb{R}$.

### 3.1. *Numerical examples.*

Using the explicit formulas deduced for the step-index fiber we are able to show some numerical examples.

Fig.1-5 show the real part of Green's function (28). In these examples, the wavenumber is $k = 10$, the indexes of refraction of the core and cladding are respectively $n_{co} = 2$, $n_{cl} = 1$, and the fiber radius is $R = 0.2$.

Under these assumptions, (21) has three solutions in $\lambda$, with $m$ being $-1, 0, 1$. We have $\lambda = 84.87$ for $m = 0$ and $\lambda = 205.65$ for $m = -1, 1$. These form the discrete spectrum, and correspond to the guided modes, represented by the finite sum in the Green's function.

The continuous spectrum, $\lambda > d^2$, is divided into two parts. The interval $d^2 < \lambda < k^2 n_{co}^2$ corresponds to $G^r$, the radiating part of Green's function, while for $\lambda > k^2 n_{co}^2$ we have $G^e$, the evanescent part of Green's function. To simplify the computations it has been useful to rewrite $G^r$ and $G^e$. By substituting in (28) $\beta_r = \sqrt{k^2 n_{co}^2 - \lambda}/k$ in $G^r$ and $\beta_e = \sqrt{\lambda - k^2 n_{co}^2}/k$ in $G^e$, we find:

$$G^r = -\frac{ik}{2\pi\sqrt{r\rho}} \sum_{m \in \mathbb{Z}} e^{im(\vartheta - \eta)} \int_0^{n_{cl}} e^{i|z-\zeta|k\beta_r} \left[ \widetilde{j}_m(\rho, \lambda) \widetilde{j}_m(r, \lambda) \widetilde{\chi}_m(\lambda) \right]_{\lambda = k^2 (n_{co}^2 - \beta_r^2)} d\beta_r, \quad (29)$$

$$G^e = \frac{k}{2\pi\sqrt{r\rho}} \sum_{m \in \mathbb{Z}} e^{im(\vartheta - \eta)} \int_0^{\infty} e^{-|z-\zeta|k\beta_e} \left[ \widetilde{j}_m(\rho, \lambda) \widetilde{j}_m(r, \lambda) \widetilde{\chi}_m(\lambda) \right]_{\lambda = k^2 (n_{co}^2 + \beta_e^2)} d\beta_e \quad (30)$$

In the sum over $\mathbb{Z}$ only the terms with $|m| \le 10$ have been retained. To compute the integrals we use the trapezoidal rule, with the sampling rates $\Delta\beta_r = 0.025$ and $\Delta\beta_e = 0.05$. The integral giving the evanescent part has been truncated at $\beta_e = 15$. Figures 1 and 2 are volume visualizations of the real part of Green's function in which is possible to see how the wave propagates inside and outside the fiber.

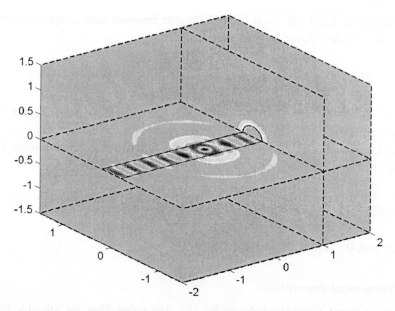

Fig. 1. Real part of Green's function with the source at the origin. The wave is cylindrically symmetric and most of the energy propagates inside the fiber. The continuous lines represent the border of the guide.

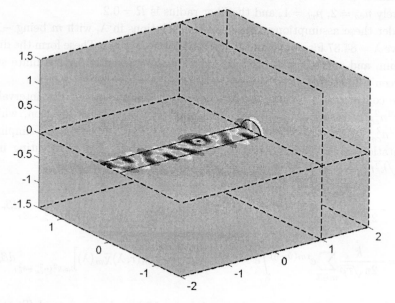

Fig. 2. Real part of Green's function with the source in $\rho = 0.1$, $\eta = 0$, $\zeta = 0$. All the guided modes are excited.

Because the source in positioned inside the waveguide, most of the energy remains in the fiber in both cases. In Fig. 1 the source is on the waveguide axis, exciting only

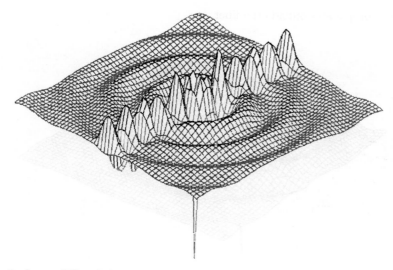

Fig. 3. Real part of Green's function in the $x - z$ plane, with the source at the origin. Only the $m = 0$ guided mode is present.

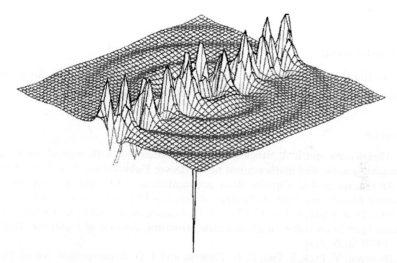

Fig. 4. Real part of Green's function in the $x - z$ plane with the source at $\rho = 0.1$, $\eta = 0$, $\zeta = 0$. Three guided modes are excited.

the guided mode corresponding to $m = 0$ and generating a cylindrically symmetric wave. In Fig. 2 the source is half the radius off axis, exciting all three guided modes.

Figures 3, 4 and 5 emphasize the different behavior of the wave amplitude inside and outside the fiber. These figures are a section of the global result, that is to say, they represent the real part of Green's function in the Cartesian plane $y = 0$. Fig. 3 and 4 are the equivalent of Fig. 1 and 2 respectively. In Fig. 5 the source is outside the fiber. Note that in this case the guided modes are weakly excited and most of

636

the energy propagates outside the fiber.

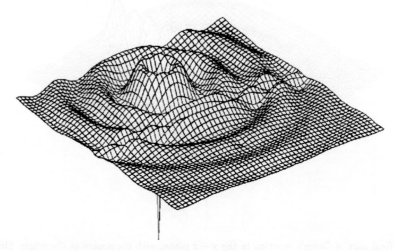

Fig. 5. Real part of Green's function in the $x - z$ plane with the source at $\rho = 1$, $\eta = 0$, $\zeta = 0$. The guided modes are weakly excited.

## Acknowledgement

We wish to thank our advisers, Fadil Santosa and Rolando Magnanini, for proposing this topic to us and for continual feedback while writing this paper.

## References

1. M. Abramowitz and I. A. Stegun, editors. *Handbook of mathematical functions with formulas, graphs, and mathematical tables* (Dover Publications, New York, 1992).
2. O. Alexandrov and G. Ciraolo. Wave propagation in a 3-D optical waveguide, *Mathematical Models and Methods in Applied Sciences* 14 (2004), no. 6, 819–852.
3. Y. Fink, D. J. Ripin, S. Fan, C. Chen, J. D. Joannopoulos, and E. L. Thomas. Guiding optical light in air using an all-dielectric structure, *Journal of Lightwave Technology* **17** (1999) 2039–2041.
4. M. Ibanescu, Y. Fink, S. Fan, E. L. Thomas, and J. D. Joannopoulos. An all-dielectric coaxial waveguide, *Science* **289** (2000) 415–419.
5. R. Magnanini and F. Santosa. Wave propagation in a 2-D optical waveguide, *SIAM J. Appl. Math.* **61** (2000/01) 1237–1252 (electronic).
6. W. Magnus, F. Oberhettinger, and R. P. Soni. *Formulas and theorems for the special functions of mathematical physics*. Third enlarged edition. Die Grundlehren der mathematischen Wissenschaften, Band 52 (Springer-Verlag, New York, 1966).

# ON A DIRICHLET BOUNDARY VALUE PROBLEM FOR COUPLED SECOND ORDER DIFFERENTIAL EQUATIONS

M. HIHNALA and S. SEIKKALA

*Division of Mathematics*
*Department of Electrical and Information Engineering*
*University of Oulu, 90570 Oulu, Finland*
*E-mail: Seppo.Seikkala@ee.oulu.fi*

Homogeneous Dirichlet data are studied for a coupled system of quasilinear ordinary differential equations of second order.

**Key words:** Dirichlet boundary value problem, coupled second order ordinary differential equation

**Mathematics Subject Classification:** 34A30, 34B05

## 1. Introduction

We shall consider the boundary value problem (BVP)

$$\begin{cases} \begin{pmatrix} x_1{}'' \\ x_2{}'' \end{pmatrix} + A \begin{pmatrix} x_1 \\ x_2 \end{pmatrix} = \begin{pmatrix} f_1(x_1, x_2) \\ f_2(x_1, x_2) \end{pmatrix} - \begin{pmatrix} b_1(t) \\ b_2(t) \end{pmatrix} \\ \\ x_1(0) = x_2(0) = x_1(\pi) = x_2(\pi) = 0, \end{cases} \tag{1}$$

where $f_1, f_2 : \mathbb{R}^2 \to \mathbb{R}$ and $b_1, b_2 : [0, \pi] \to \mathbb{R}$ are continuous functions. The matrix $A$ is diagonalizable and such that the corresponding homogeneous problem, with the given Dirichlet boundary conditions, has nontrivial solutions, i.e. we have a BVP at resonance. Problems of this type arise for example from mechanics (coupled oscillators) or from coupled circuits theory [1]. For example, for the spring-mass system of Figure 1 ($m_1$ and $m_2$ are the masses and $s_1$, $s_2$ and $S$ the stiffnesses)

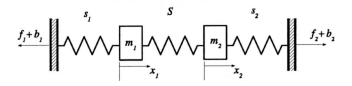

Fig. 1.   A coupled spring-mass system.

we have

$$A = \begin{pmatrix} a_1 & -\alpha \\ -\beta & a_2 \end{pmatrix}, \tag{2}$$

where $a_1 = \frac{s_1}{m_1} + \alpha$, $a_2 = \frac{s_2}{m_2} + \beta$, $\alpha = \frac{S}{m_1}$ and $\beta = \frac{S}{m_2}$. Special cases of (1) and (2), leading to a one dimensional null space of the differential operator on the left of (1) with the given Dirichlet boundary conditions, were studied in [2] and [3]. Here we shall continue the study of [3] and extend the study to the case of a two dimensional null space. Using both analytic and numerical techniques we will obtain multiple solutions for problem (1).

By change of variables $z = T^{-1}x$, where $T$ is a matrix diagonalizing $A$, having the eigenvectors of $A$ as columns, problem (1) is transformed into a system

$$\begin{cases} \begin{pmatrix} z_1'' + d_1 z_1 \\ z_2'' + d_2 z_2 \end{pmatrix} = T^{-1} \begin{pmatrix} f_1(x_1, x_2) \\ f_2(x_1, x_2) \end{pmatrix} - T^{-1} \begin{pmatrix} b_1(t) \\ b_2(t) \end{pmatrix} \\ z_1(0) = z_2(0) = z_1(\pi) = z_2(\pi) = 0, \end{cases} \tag{3}$$

where $x = \begin{pmatrix} x_1 \\ x_2 \end{pmatrix} = Tz$ and $d_1, d_2$ are the eigenvalues of $A$. For the matrix $A$ in (2) we may choose

$$T = \begin{pmatrix} \alpha & \gamma \\ \gamma & -\beta \end{pmatrix}, \quad T^{-1} = \frac{1}{\alpha\beta + \gamma^2} \begin{pmatrix} \beta & \gamma \\ \gamma & -\alpha \end{pmatrix}, \tag{4}$$

where either $\gamma = \frac{a_1 - a_2}{2} + \frac{\sqrt{(a_1-a_2)^2 + 4\alpha\beta}}{2}$ or $\gamma = \frac{a_1 - a_2}{2} - \frac{\sqrt{(a_1-a_2)^2 + 4\alpha\beta}}{2}$.

If either $d_1 = k^2$ or $d_2 = l^2$, where k and l are positive integers, then we have a resonance case and the null space of the differential operator $E$ in the coupled system (1),

$$Ex = \begin{pmatrix} x_1'' \\ x_2'' \end{pmatrix} + A \begin{pmatrix} x_1 \\ x_2 \end{pmatrix},$$

with the given Dirichlet boundary conditions, is spanned by either

$$\Phi(t) = T \begin{pmatrix} 1 \\ 0 \end{pmatrix} \sqrt{\frac{2}{\pi}} \sin kt$$

or by

$$\eta(t) = T \begin{pmatrix} 0 \\ 1 \end{pmatrix} \sqrt{\frac{2}{\pi}} \sin lt.$$

If both $d_1 = k^2$ and $d_2 = l^2$, then the null space of $E$ is two dimensional and spanned by the functions $\Phi$ and $\eta$.

## 2. One dimensional null space

Suppose that the eigenvalues of $A$ are $d_1 = k^2$, $d_2 \neq l^2$ where $k$ and $l$ are positive integers and that $T$ is given by (4). We will study the existence and multiplicity of solutions of (1) depending on the parameter $\bar{b}$ in the decomposition

$$b(t) = \begin{pmatrix} b_1(t) \\ b_2(t) \end{pmatrix} = \bar{b}\Psi(t) + \tilde{b}(t), \tag{5}$$

where $\tilde{b}$ is orthogonal to $\Psi(t) = \sqrt{\frac{2}{\pi}}\sqrt{\frac{1}{\beta^2+\gamma^2}} \begin{pmatrix} \beta \\ \gamma \end{pmatrix} \sin kt$ and $\bar{b} = (b|\Psi)$, the inner product $(x|y)$ in $X = C[0,\pi] \times C[0,\pi]$ being defined by

$$(x|y) = \int_0^\pi x(t)y(t)dt = \int_0^\pi [x_1(t)y_1(t) + x_2(t)y_2(t)]dt.$$

The motivation for decomposition (5) is that the BVP

$$Ex = b, \quad x(0) = x(\pi) = 0$$

has a solution if and only if $b$ is orthogonal to $\Psi$.

For a fixed $\lambda \in \mathbb{R}$ consider a 'regularized' BVP (1),

$$\begin{cases} \begin{pmatrix} z_1'' + d_1 z_1 \\ z_2'' + d_2 z_2 \end{pmatrix} = T^{-1}\begin{pmatrix} f_1(x_1, x_2) \\ f_2(x_1, x_2) \end{pmatrix} - T^{-1}\begin{pmatrix} b_1(t) \\ b_2(t) \end{pmatrix} - \bar{\delta}\psi \\ z_1(0) = z_2(0) = z_1(\pi) = z_2(\pi) = 0, \\ (z|\psi) = \lambda \end{cases} \tag{6}$$

where $x = Tz$, $\psi(t) = \sqrt{\frac{2}{\pi}}\begin{pmatrix} 1 \\ 0 \end{pmatrix}\sin kt$ and $\bar{\delta} = \bar{\delta}(\lambda)$ is defined by (8) below. The system (6) is equivalent to the integral equation system

$$z = \lambda\psi + KNz, \tag{7}$$

where

$$Nz = N\begin{pmatrix} z_1 \\ z_2 \end{pmatrix} = T^{-1}\begin{pmatrix} f_1(x_1, x_2) - b_1 \\ f_2(x_1, x_2) - b_2 \end{pmatrix},$$

the linear operator $K : X \to X$ is defined by

$$Kz = \begin{pmatrix} \int_0^\pi k_1(t, s)z_1(s)ds \\ \int_0^\pi k_2(t, s)z_2(s)ds \end{pmatrix},$$

with $k_1$ defined by

$$\begin{cases} Lk_1(t, s) = \delta(t - s) - \psi(t)\psi(s) \\ k_1(0, s) = k_1(\pi, s) = 0 \\ \int_0^\pi k_1(t, s)\sin t\, dt = 0, \end{cases}$$

$Lu = u'' + d_1 u$, and with $k_2$ being the ordinary Green's function for problem $z_2'' + d_2 z_2 = 0$, $z_2(0) = z_2(\pi) = 0$.

Henceforth we assume that for a fixed $\lambda \in \mathbb{R}$ the system (7) has at least one solution. This holds, for example, if $f_1$ and $f_2$ are bounded or satisfy a suitable Lipschitz condition. Any such solution we denote by $z^\lambda$ and furthermore $x^\lambda = Tz^\lambda$,

$$\bar{\delta}(\lambda) = \int_0^\pi T^{-1}[f(x^\lambda(t)) - b(t)]\psi(t)dt. \tag{8}$$

By defining $\delta(\lambda) = \frac{\alpha\beta+\gamma^2}{\sqrt{\beta^2+\gamma^2}}\tilde{\delta}(\lambda)$ and

$$\tilde{\delta}(\lambda) = \frac{\alpha\beta+\gamma^2}{\sqrt{\beta^2+\gamma^2}} \int_0^\pi T^{-1}[f(x^\lambda(t))]\psi(t)dt$$

we have $\delta(\lambda) = \tilde{\delta}(\lambda) - \bar{b}$.

We easily deduce that the BVP (3) has a solution if and only if the system

$$\begin{cases} z = \lambda\psi + KNz \\ \delta(\lambda) = 0. \end{cases}$$

has a solution. Note that $\tilde{\delta}(\lambda)$ depends on $\bar{b}$, however $\tilde{\delta}(\lambda)$ is independent of $\bar{b}$ because $x^\lambda$ is independent of $\bar{b}$. Denote

$$a = \inf\{\tilde{\delta}(\lambda) : \lambda \in \mathbb{R}, \ z^\lambda \text{ is a solution of (7)}\},$$
$$b = \sup\{\tilde{\delta}(\lambda) : \lambda \in \mathbb{R}, \ z^\lambda \text{ is a solution of (7)}\}.$$

The proof of the following theorem is, in principle, the same as that of Theorem 1 of [3].

**Theorem 1.** *If $f_1$ and $f_2$ are continuous and bounded, then the BVP (1) has (i) at least one solution if $\bar{b} \in (a,b)$ and (ii) no solution if $\bar{b} \notin [a,b]$. If $c \in (a,b)$ is a limit point of both $\{\tilde{\delta}(\lambda) : \lambda \in (-\infty, d]\}$ and $\{\tilde{\delta}(\lambda) : \lambda \in [d, \infty)\}$ for a $d \in \mathbb{R}$ and if $\bar{b} \in (a,b)\backslash\{c\}$, then problem (1) has at least two solutions.*

**Example 1.** For the problem

$$\begin{cases} x_1'' + 4x_1 - \frac{5}{2}x_2 = \tan^{-1}(x_1 + 2x_2) - b_1(t) \\ x_2'' - \frac{18}{5}x_1 + 4x_2 = \tan^{-1}(2x_1 - x_2) - b_2(t) \\ x_1(0) = x_2(0) = x_1(\pi) = x_2(\pi) = 0 \end{cases} \tag{9}$$

with $\tilde{b} \equiv 0$ we have $\alpha = \frac{5}{2}, \beta = \frac{18}{5}, \gamma = 3, d_1 = 1, d_2 = 7$ and

$$\tilde{\delta}(\lambda) = \eta \int_0^\pi \{\frac{18}{5}\tan^{-1}[x_1^\lambda(t) + 2x_2^\lambda(t)] +$$

$$3\tan^{-1}[2x_1^\lambda(t) - x_2^\lambda(t)]\} \sin t \ dt$$

where $\eta = \sqrt{\frac{2}{\pi}}\sqrt{\frac{1}{(\frac{18}{5})^2+9}}$. Since $\tan^{-1}(\pm\infty) = \pm\pi/2$ we deduce that $\lim\limits_{\lambda\to\infty}\tilde{\delta}(\lambda) = \eta\frac{33\pi}{5}$ and $\lim\limits_{\lambda\to-\infty}\tilde{\delta}(\lambda) = -\eta\frac{33\pi}{5}$. Hence for problem (9) we have $(-\eta\frac{33\pi}{5}, \eta\frac{33\pi}{5}) \subseteq (a,b)$. Thus, problem (9) has at least one solution if $\bar{b} \in (-\eta\frac{33\pi}{5}, \eta\frac{33\pi}{5}) \approx (-3.53, 3.53)$. Numerically we have found the curve $\tilde{\delta}(\lambda)$ shown in Figure 2 when $\bar{b} \equiv 0$. It indicates that problem (9), in case $\bar{b} \equiv 0$, has a unique solution when $\bar{b} \in (-3.53, 3.53)$ and no solution otherwise.

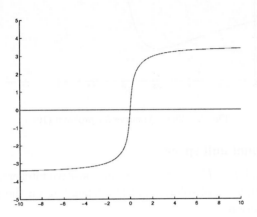

Fig. 2. The $\tilde{\delta}(\lambda)$ curve for problem (9).

**Example 2.** It can be shown that the problem

$$\begin{cases} x_1'' + 4x_1 - \frac{5}{2}x_2 = \frac{2}{3}\tan^{-1}(x_1 + 2x_2) - b_1(t) \\ x_2'' - \frac{18}{5}x_1 + 4x_2 = -\frac{4}{5}\tan^{-1}(2x_1 - x_2) - b_2(t) \\ x_1(0) = x_2(0) = x_1(\pi) = x_2(\pi) = 0 \end{cases} \tag{10}$$

with $\tilde{b} \equiv 0$, for small $|\bar{b}|$ has a solution. Also, $\lim\limits_{\lambda\to\infty}\tilde{\delta}(\lambda) = \lim\limits_{\lambda\to-\infty}\tilde{\delta}(\lambda) = 0$. Hence, by Theorem 1, Problem (10) has at least two solutions for small $|\bar{b}|$. Numerically we have found the curve $\tilde{\delta}(\lambda)$ given in Figure 3 below. It indicates that $(a,b) \approx (-0.348, 0.348)$.

**Theorem 2.** *Suppose $\gamma > 0$, $k$ is odd, $f_i(-\infty, -\infty) < f_i(\infty, \infty), i = 1, 2$ and*

$$\frac{2}{k}[\beta f_1(-\infty, -\infty) + \gamma f_2(-\infty, -\infty)] < \int_0^\pi [\beta b_1(t) + \gamma b_2(t)] \sin kt \, dt$$

$$< \frac{2}{k}[\beta f_1(\infty, \infty) + \gamma f_2(\infty, \infty)],$$

*then problem (3) has at least one solution.*

642

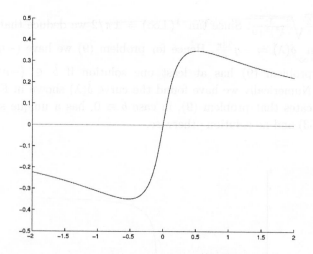

Fig. 3.   The $\bar{\delta}(\lambda)$ curve for problem (10).

## 3. Two dimensional null space

Let now $d_1 = k^2$ and $d_2 = l^2$, where k and l are positive integers. We will study the existence and multiplicity of solutions of (1) depending on the parameters $\bar{b}_1$ and $\bar{b}_2$ in the decomposition

$$b(t) = \begin{pmatrix} b_1(t) \\ b_2(t) \end{pmatrix} = \bar{b}_1 \Phi(t) + \bar{b}_2 \eta(t) + T\begin{pmatrix} \tilde{b}_1(t) \\ \tilde{b}_2(t) \end{pmatrix}, \tag{11}$$

where $\tilde{b}_1$ is orthogonal to $\psi_1(t) = \sqrt{\frac{2}{\pi}} \sin kt$, $\tilde{b}_2$ is orthogonal to $\psi_2(t) = \sqrt{\frac{2}{\pi}} \sin lt$, $\bar{b}_i = ((T^{-1}b)_i|\psi_i), i = 1, 2$, the inner product being the inner product of $L_2$. The motivation for decomposition (11) is that the BVP

$$Ex = b, \quad x(0) = x(\pi) = 0$$

has a solution if and only if $\bar{b}_1 = \bar{b}_2 = 0$. The problem (1) is transformed, as in Section 2, to problem (3),

$$\begin{cases} \begin{pmatrix} z_1'' + d_1 z_1 \\ z_2'' + d_2 z_2 \end{pmatrix} = T^{-1}\begin{pmatrix} f_1(x_1, x_2) \\ f_2(x_1, x_2) \end{pmatrix} - T^{-1}\begin{pmatrix} b_1(t) \\ b_2(t) \end{pmatrix} \\ z_1(0) = z_2(0) = z_1(\pi) = z_2(\pi) = 0, \end{cases}$$

Let $X$ be the set of all continuous functions $x : [0, \pi] \to \mathbb{R}^2$ and assume that $f_1$ and $f_2$ are continuous and bounded. For a fixed $\lambda = (\lambda_1, \lambda_2) \in \mathbb{R}^2$ consider the integral equation system

$$z = \lambda \psi + KNz, \tag{12}$$

where

$$Nz = N\begin{pmatrix} z_1 \\ z_2 \end{pmatrix} = T^{-1}\begin{pmatrix} f_1(x_1, x_2) - b_1 \\ f_2(x_1, x_2) - b_2 \end{pmatrix},$$

$x = Tz$, $\psi = (\psi_1, \psi_2)^T$, $\lambda\psi = (\lambda_1\psi_1, \lambda_2\psi_2)^T$, the linear operator $K : X \to X$ is defined by

$$Kz = \begin{pmatrix} \int\limits_0^\pi k_1(t,s)z_1(s)ds \\ \int\limits_0^\pi k_2(t,s)z_2(s)ds \end{pmatrix},$$

with $k_i$ defined by

$$\begin{cases} L_i k_i(t,s) = \delta(t-s) - \psi_i(t)\psi_i(s) \\ k_i(0,s) = k_i(\pi,s) = 0 \\ \int\limits_0^\pi k_i(t,s)\psi_i(t)dt = 0, \ i = 1,2, \end{cases}$$

$L_1 u = u'' + k^2 u$ and $L_2 u = u'' + l^2 u$.

Since $f_1$ and $f_2$ are continuous and bounded, then for a fixed $\lambda \in \mathbb{R}^2$ the system (7) has at least one solution. For any such solution $z^\lambda$ we denote $x^\lambda = Tz^\lambda$, $\delta(\lambda) = (\delta_1(\lambda), \delta_2(\lambda))$,

$$\delta_i(\lambda) = \int\limits_0^\pi (Nz)_i(t)\psi_i(t)dt$$

and

$$\tilde{\delta}_i(\lambda) = \delta_i(\lambda) + \bar{b}_i, \ i = 1,2.$$

Again we deduce that the BVP (3) has a solution if and only if the system

$$\begin{cases} z = \lambda\psi + KNz \\ \delta(\lambda) = 0. \end{cases}$$

has a solution.

**Example 3.** Consider the problem

$$\begin{cases} \begin{pmatrix} x_1'' \\ x_2'' \end{pmatrix} + \begin{pmatrix} 5 & -\frac{9}{2} \\ -\frac{32}{9} & 5 \end{pmatrix} \begin{pmatrix} x_1 \\ x_2 \end{pmatrix} = \\ \begin{pmatrix} \frac{9}{2}\sin(\frac{25}{36}x_1 + \frac{7}{32}x_2) + 4\sin(\frac{13}{36}x_1 - \frac{5}{16}x_2) \\ 4\sin(\frac{25}{36}x_1 + \frac{7}{32}x_2) - \frac{32}{9}\sin(\frac{13}{36}x_1 - \frac{5}{16}x_2) \end{pmatrix} - b(t), \\ x_1(0) = x_2(0) = x_1(\pi) = x_2(\pi) = 0, \end{cases} \tag{13}$$

where

$$b(t) = \begin{pmatrix} b_1(t) \\ b_2(t) \end{pmatrix} = \bar{b}_1\Phi(t) + \bar{b}_2\eta(t) + \begin{pmatrix} \frac{9}{2} & 4 \\ 4 & -\frac{32}{9} \end{pmatrix} \begin{pmatrix} \sin(3t) \\ \sin t \end{pmatrix}.$$

644

Problem (13) is then reduced to

$$\begin{cases} \begin{pmatrix} z_1'' + z_1 \\ z_2'' + 9z_2 \end{pmatrix} = \begin{pmatrix} \sin(4z_1 + 2z_2) \\ \sin(2z_1 + 4z_2) \end{pmatrix} - \begin{pmatrix} \bar{b}_1 \sin t + \sin(3t) \\ \bar{b}_2 \sin(3t) + \sin t \end{pmatrix} \\ z_1(0) = z_2(0) = z_1(\pi) = z_2(\pi) = 0. \end{cases} \tag{14}$$

It can be shown that with small $\bar{b}_1$ and $\bar{b}_2$ problem (13) has at least one solution. The zero level curves of $\delta_i(\lambda_1, \lambda_2), i = 1, 2,$ in Figure 4 indicate that for $b_1 = b_2 = 0$ there may exist infinitely many solutions.

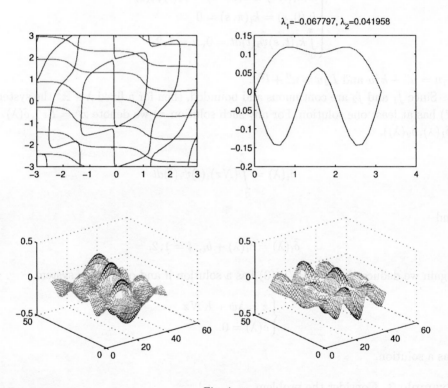

Fig. 4.

But if we have $\bar{b}_1 = \bar{b}_2 = 0.2$ on the right hand side of (14), then we get Figure 5. Assume now that $f_1(x_1, x_2) = f_1(ax_1 + bx_2), f_2(x_1, x_2) = f_2(cx_1 + dx_2), k$ and $l$ are odd and that $f_i(-\infty) < f_i(\infty), i = 1, 2.$ Denote $D_1 = \frac{2}{k}$ and $D_2 = \frac{2}{l}.$

**Theorem 3.** *Assume that $ad - bc \neq 0$ and that for $i = 1, 2,$*

$$D_i f_i(-\infty) < \int_0^\pi b_i(t)\psi_i(t)dt < D_i f_i(\infty) \tag{15}$$

*Then the BVP (1) has at least one solution.*

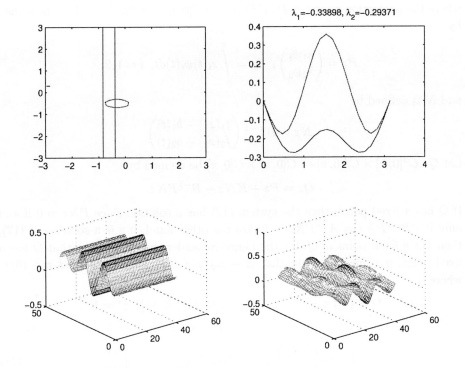

Fig. 5.

**Proof.** By the change of variables

$$\begin{pmatrix} z_1 \\ z_2 \end{pmatrix} = BT^{-1} \begin{pmatrix} x_1 \\ x_2 \end{pmatrix}$$

where $B$ is the matrix

$$B = \begin{pmatrix} a & b \\ c & d \end{pmatrix},$$

problem (1) is transformed into an equivalent problem

$$\begin{cases} \begin{pmatrix} z_1'' + d_1 z_1 \\ z_2'' + d_2 z_2 \end{pmatrix} = BT^{-1} \begin{pmatrix} f_1(z_1) \\ f_2(z_2) \end{pmatrix} - BT^{-1} \begin{pmatrix} b_1(t) \\ b_2(t) \end{pmatrix} \\ z_1(0) = z_2(0) = z_1(\pi) = z_2(\pi) = 0, \end{cases} \quad (16)$$

This, in turn, is equivalent to the system of equations

$$\begin{cases} z = Pz + KNz \\ PNz = 0, \end{cases} \quad (17)$$

where the projection operator $P : C^2[0, \pi] \times C^2[0, \pi] \to C[0, \pi] \times C[0, \pi]$ is defined by

$$Pz = \begin{pmatrix} \lambda_1 \psi_1 \\ \lambda_2 \psi_2 \end{pmatrix}, \quad \lambda_i = \int_0^\pi z_i(t) \psi_i(t) dt, \ i = 1, 2,$$

and $N$ is defined by

$$Nz = BT^{-1} \begin{pmatrix} f_1(z_1) - b_1(t) \\ f_2(z_2) - b_2(t) \end{pmatrix}$$

Let $Q : C^2[0, \pi] \times C^2[0, \pi] \to C[0, \pi] \times C[0, \pi]$ be defined by

$$Qz = Pz + KNz - B^{-1}PNz.$$

If Q has a fixed point, then the system (17) has a solution since $PNz = 0$ if and only if $PB^{-1}PNz = A^{-1}PNz = 0$. On the other hand, if z is a solution of (17), then z is a fixed point of $Q$. By the Leray–Schauder theory, to show that $Q$ has a fixed point, it suffices to show that $(I - \tau Q)z \neq 0$ for all $z \in \partial \Omega$ and $\tau \in (0, 1)$ where

$$\Omega = \{z \mid z = z^* + \tilde{z}, \ ||z^*|| < R, \ ||\tilde{z}|| < M\}, \ z^* = Pz,$$

for some chosen $R > 0$ and $M > 0$. Now, $(I - \tau Q)z \neq 0$, if $P(I - \tau Q)z \neq 0$ i.e. if

$$(1 - \tau) \begin{pmatrix} \lambda_1 \psi_1 \\ \lambda_2 \psi_2 \end{pmatrix} + \tau B^{-1} BT^{-1} \begin{pmatrix} \delta_1 \psi_1 \\ \delta_2 \psi_2 \end{pmatrix} \neq \begin{pmatrix} 0 \\ 0 \end{pmatrix}, \tag{18}$$

where

$$\delta_i = \int_0^\pi f_i[\lambda_i \psi_i(t) + \tilde{z}_i(t)] \psi_i(t) dt - \int_0^\pi b_i(t) \psi_i(t) dt, \ i = 1, 2.$$

Let $z \in \partial \Omega$. Then we have either 1) $||z^*|| = R$ and $||\tilde{z}|| \leq M$ or 2) $||z^*|| \leq R$ and $||\tilde{z}|| = M$. In case of 1) we have $\max\{|\lambda_1|, |\lambda_2|\} = R$. By condition (15) we choose $R$ so that $\delta_1 > 0$ if $\lambda_1 = R$, $\delta_1 < 0$ if $\lambda_1 = -R$, $\delta_2 > 0$ if $\lambda_2 = R$ and $\delta_2 < 0$ if $\lambda_2 = R$. Hence, in any of these cases, condition (18) holds. Hence $(I - \tau Q)z \neq 0$.

In case of 2) we choose $M > ||K|| \ ||Nz||$. Then $(I - \tau Q)z = 0$ is equivalent to $z = \tau(Pz - B^{-1}PNz) + \tau KNz$ which implies that $\tilde{z} = \tau KNz$ and hence $||\tilde{z}|| < M$, a contradiction. Thus $(I - \tau Q)z \neq 0$ and the proof is complete. □

## References

1. I.G. Main, *Vibrations and Waves in Physics*, Cambridge Univ. Press, Cambridge, UK, 1993.
2. A.Canada, Nonlinear ordinary boundary value problems under a combined effect of periodic and attractive nonlinearities, *J. Math. Anal. Appl.* **243 no. 1** (2001), 174–189.
3. S.Seikkala and M.Hihnala, A Resonance Problem for a Second Order Vector Differential Equation, in *Integral Methods in Science and Engineering*, Birkhäuser, Boston, (2004), 233–238.

# EXISTENCE OF A CLASSICAL SOLUTION AND NON–EXISTENCE OF A WEAK SOLUTION TO THE HARMONIC DIRICHLET PROBLEM IN A PLANAR DOMAIN WITH CRACKS

P.A. KRUTITSKII

*KIAM, Dept. 25, Miusskaya Sq. 4, Moscow 125047, Russia.*

The harmonic Dirichlet problem in a planar domain with smooth cracks of an arbitrary shape is considered in case, when the solution is not continuous at the ends the cracks. The well–posed formulation of the problem is given, theorems on existence and uniqueness of a classical solution are proved, the integral representation for a solution is obtained. With the help of the integral representation, the properties of the solution are studied. It is proved that a weak solution of the Dirichlet problem in question does not exist typically, though the classical solution exists.

**Key words:** Integral representation, Dirichlet problem, cracks, asymptotics, classical and weak solutions.

**Mathematics Subject Classification:** 35J05, 35J25, 31A25, 45E05

## 1. Introduction

Boundary value problems in planar domains with cracks are widely used in physics and in mechanics, and not in mechanics of solids only, but in fluid mechanics as well, where cracks (or cuts) model wings or screens in fluids. Integral representation of a classical solution to the harmonic Dirichlet problem in a planar domain with cracks of an arbitrary shape has been obtained by the method of integral equations in [1-5] in case when the solution is assumed to be continuous at the ends of the cracks. In the present paper this problem is considered in case when the solution is not continuous at the ends of the cracks. The well–posed formulation of the boundary value problem is given, theorems on existence and uniqueness of a classical solution are proved, the integral representation for a classical solution is obtained. Moreover, properties of the solution is studied with the help of this integral representation. It appears that the classical solution to the Dirichlet problem considered in the present paper exists, while the weak solution typically does not exist, though both the cracks and the functions specified in the boundary conditions are smooth enough. This result follows from the fact that the square of the gradient of a classical solution basically is not itegrable near the ends of the cracks, since singularities of the gradient are rather strong there. This result is very important for numerical analysis, it shows that finite elements and finite difference methods cannot be applied to numerical treatment of the Dirichlet problem in question directly, since all these methods imply existence of a weak solution. To use difference methods for numerical analysis one has to localize all strong singularities first and next to use difference method in a domain excluding the neighbourhoods of the singularities.

## 2. Formulation of the problem

By an open curve we mean a simple smooth non-closed arc of finite length without self-intersections [6].

In a plane in Cartesian coordinates $x = (x_1, x_2) \in R^2$ we consider a multiply connected domain bounded by simple open curves $\Gamma_1^1, \ldots, \Gamma_{N_1}^1 \in C^{2,\lambda}$ and simple closed curves $\Gamma_1^2, \ldots, \Gamma_{N_2}^2 \in C^{2,\lambda}$, $\lambda \in (0, 1]$, in such a way that all curves do not have common points, in particular, endpoints. We will consider both the case of an exterior domain and the case of an interior domain when the curve $\Gamma_1^2$ encloses all others. Set

$$\Gamma^1 = \bigcup_{n=1}^{N_1} \Gamma_n^1, \qquad \Gamma^2 = \bigcup_{n=1}^{N_2} \Gamma_n^2, \qquad \Gamma = \Gamma^1 \cup \Gamma^2.$$

The connected domain bounded by $\Gamma^2$ and containing curves $\Gamma^1$ will be called $\mathcal{D}$, so that $\partial \mathcal{D} = \Gamma^2$, $\Gamma^1 \subset \mathcal{D}$. We assume that each curve $\Gamma_n^j$ is parametrized by the arc length $s$:

$$\Gamma_n^j = \left\{ x : \; x = x(s) = \Big( x_1(s), x_2(s) \Big), s \in \big[ a_n^j, b_n^j \big] \right\}, n = 1, \ldots, N_j, \; j = 1, 2,$$

so that

$$a_1^1 < b_1^1 < \cdots < a_{N_1}^1 < b_{N_1}^1 < a_1^2 < b_1^2 < \cdots < a_{N_2}^2 < b_{N_2}^2$$

and the domain $\mathcal{D}$ is placed to the right when the parameter $s$ increases on $\Gamma_n^2$. The points $x \in \Gamma$ and values of the parameter $s$ are in one-to-one correspondence except the points $a_n^2$, $b_n^2$, which correspond to the same point $x$ for $n = 1, \ldots, N_2$. Further on, the set of the intervals

$$\bigcup_{n=1}^{N_1} [a_n^1, b_n^1], \quad \bigcup_{n=1}^{N_2} [a_n^2, b_n^2], \quad \bigcup_{j=1}^{2} \bigcup_{n=1}^{N_j} [a_n^j, b_n^j]$$

on the $Os$-axis will be denoted by $\Gamma^1$, $\Gamma^2$ and $\Gamma$ also.

Set $C^{j,r}(\Gamma_n^2) = \left\{ \mathcal{F}(s) : \; \mathcal{F}(s) \in C^{j,r}[a_n^2, b_n^2], \; \mathcal{F}^{(m)}(a_n^2) = \mathcal{F}^{(m)}(b_n^2), \right.$
$\left. m = 0, \ldots, j \right\}$, $j = 0, 1$, $r \in [0, 1]$ and $C^{j,r}(\Gamma^2) = \bigcap_{n=1}^{N_2} C^{j,r}(\Gamma_n^2)$. The tangent vector to $\Gamma$ in the point $x(s)$, in the direction of the increment of $s$, will be denoted by $\tau_x = (\cos \alpha(s), \sin \alpha(s))$, while the normal vector coinciding with $\tau_x$ after rotation through an angle of $\pi/2$ in the counterclockwise direction will be denoted by $\mathbf{n}_x = (\sin \alpha(s), -\cos \alpha(s))$. According to the chosen parametrization

$\cos \alpha(s) = x_1'(s)$, $\sin \alpha(s) = x_2'(s)$. Thus, $\mathbf{n}_x$ is an interior normal to $\mathcal{D}$ on $\Gamma^2$. By $X$ we denote the point set consisting of the endpoints of $\Gamma^1$:

$$X = \bigcup_{n=1}^{N_1} \left( x\left(a_n^1\right) \cup x\left(b_n^1\right) \right).$$

Let the plane be cut along $\Gamma^1$. We consider $\Gamma^1$ as a set of cracks (or cuts). The side of the crack $\Gamma^1$, which is situated on the left when the parameter $s$ increases, will be denoted by $(\Gamma^1)^+$, while the opposite side will be denoted by $(\Gamma^1)^-$.

We say that the function $u(x)$ belongs to the smoothness class $\mathbf{K}_1$, if

(1) $u \in C^0 \left( \overline{\mathcal{D} \setminus \Gamma^1} \setminus X \right) \cap C^2 \left( \mathcal{D} \setminus \Gamma^1 \right)$, $\quad \nabla u \in C^0 \left( \overline{\mathcal{D} \setminus \Gamma^1} \setminus \Gamma^2 \setminus X \right)$,

(2) in the neighbourhood of any point $x(d) \in X$ the equality

$$(1) \qquad \lim_{r \to +0} \int_{\partial S(d,r)} u(x) \frac{\partial u(x)}{\partial \mathbf{n}_x} dl = 0,$$

holds, where the curvilinear integral of the first kind is taken over a circumference $\partial S(d,r)$ of a radius $r$ with the center in the point $x(d)$, $\mathbf{n}_x$ is a normal in the point $x \in \partial S(d,r)$, directed to the center of the circumference and $d = a_n^1$ or $d = b_n^1$, $n = 1, \ldots, N_1$.

**Remark** By $C^0 \left( \overline{\mathcal{D} \setminus \Gamma^1} \setminus X \right)$ we denote the class of continuous in $\overline{\mathcal{D}} \setminus \Gamma^1$ functions, which are continuously extensible to the sides of the cracks $\Gamma^1 \setminus X$ from the left and from the right, but their limiting values on $\Gamma^1 \setminus X$ can be different from the left and from the right, so that these functions may have a jump on $\Gamma^1 \setminus X$. To obtain the definition of the class $C^0 \left( \overline{\mathcal{D} \setminus \Gamma^1} \setminus \Gamma^2 \setminus X \right)$ we have to replace $C^0 \left( \overline{\mathcal{D} \setminus \Gamma^1} \setminus X \right)$ by $C^0 \left( \overline{\mathcal{D} \setminus \Gamma^1} \setminus \Gamma^2 \setminus X \right)$ and $\overline{\mathcal{D}} \setminus \Gamma^1$ by $\mathcal{D} \setminus \Gamma^1$ in the previous sentence.

Problem $\mathbf{D}_1$. Find a function $u(x)$ from the class $\mathbf{K}_1$, so that $u(x)$ obeys the Laplace equation

$$(2a) \qquad u_{x_1 x_1}(x) + u_{x_2 x_2}(x) = 0,$$

in $\mathcal{D} \setminus \Gamma^1$ and satisfies the boundary conditions

$$(2b) \quad u(x)|_{x(s) \in (\Gamma^1)^+} = F^+(s), \quad u(x)|_{x(s) \in (\Gamma^1)^-} = F^-(s), \quad u(x)|_{x(s) \in \Gamma^2} = F(s).$$

If $\mathcal{D}$ is an exterior domain, then we add the following condition at infinity:

$$(2c) \qquad |u(x)| \leq \text{const}, \qquad |x| = \sqrt{x_1^2 + x_2^2} \to \infty.$$

All conditions of the problem $\mathbf{D}_1$ must be satisfied in a classical sense. The boundary conditions (2b) on $\Gamma^1$ must be satisfied in the interior points of $\Gamma^1$, their validity at the ends of $\Gamma^1$ are not required.

**Theorem 1** *If* $\Gamma \in C^{2,\lambda}$, $\lambda \in (0,1]$, *then there is no more than one solution to the problem* $\mathbf{D}_1$.

It is enough to prove that the homogeneous problem $\mathbf{D}_1$ admits the trivial solution only. The proof will be given for an interior domain $\mathcal{D}$. Let $u^0(x)$ be a solution to the homogeneous problem $\mathbf{D}_1$ with $F^+(s) \equiv F^-(s) \equiv 0$, $F(s) \equiv 0$. Let $S(d, \epsilon)$ be a disc of a small enough radius $\epsilon$ with the center in the point $x(d)$ $(d = a_n^1$ or $d = b_n^1$, $n = 1, ..., N_1)$. Let $\Gamma_{n,\epsilon}^1$ be a set consisting of such points of the curve $\Gamma_n^1$ which do not belong to discs $S(a_n^1, \epsilon)$ and $S(b_n^1, \epsilon)$. We choose a number $\epsilon_0$ so small that the following conditions are satisfied:

1) for any $0 < \epsilon \le \epsilon_0$ the set of points $\Gamma_{n,\epsilon}^1$ is a unique non-closed arc for each $n = 1, ..., N_1$,

2) the points belonging to $\Gamma \setminus \Gamma_n^1$ are placed outside the discs $S(a_n^1, \epsilon_0)$, $S(b_n^1, \epsilon_0)$ for any $n = 1, ..., N_1$,

3) discs of radius $\epsilon_0$ with centers in different ends of $\Gamma^1$ do not intersect.

Set $\Gamma^{1,\epsilon} = \cup_{n=1}^{N_1}\Gamma_{n,\epsilon}^1$, $S_\epsilon = \left(\cup_{n=1}^{N_1}[S(a_n^1, \epsilon) \cup S(b_n^1, \epsilon)]\right)$, $\mathcal{D}_\epsilon = \mathcal{D} \setminus \Gamma^{1,\epsilon} \setminus S_\epsilon$. Since $\Gamma^2 \in C^{2,\lambda}$, $u^0(x) \in C^0(\overline{\mathcal{D}} \setminus \Gamma^1)$ (remind that $u^0(x) \in \mathbf{K}_1$), and since $u^0|_{\Gamma^2} = 0 \in C^{2,\lambda}(\Gamma^2)$, and owing to the theorem on regularity of solutions of elliptic equations near the boundary [7], we obtain : $u^0(x) \in C^1(\overline{\mathcal{D}} \setminus \Gamma^1)$. Since $u^0(x) \in \mathbf{K}_1$, we observe that $u^0(x) \in C^1(\overline{\mathcal{D}_\epsilon})$ for any $\epsilon \in (0, \epsilon_0]$. By $C^1(\overline{\mathcal{D}_\epsilon})$ we mean $C^1(\mathcal{D}_\epsilon \cup \Gamma^2 \cup (\Gamma^{1,\epsilon})^+ \cup (\Gamma^{1,\epsilon})^- \cup \partial S_\epsilon)$. Since the boundary of a domain $\mathcal{D}_\epsilon$ is piecewise smooth, we write out Green's formula [8], p.328, for the function $u^0(x)$:

$$\|\nabla u^0\|_{L_2(\mathcal{D}_\epsilon)}^2$$

$$= \int_{\Gamma^{1,\epsilon}} (u^0)^+ \left(\frac{\partial u^0}{\partial \mathbf{n}_x}\right)^+ ds - \int_{\Gamma^{1,\epsilon}} (u^0)^- \left(\frac{\partial u^0}{\partial \mathbf{n}_x}\right)^- ds - \int_{\Gamma^2} u^0 \frac{\partial u^0}{\partial \mathbf{n}_x} ds + \int_{\partial S_\epsilon} u^0 \frac{\partial u^0}{\partial \mathbf{n}_x} dl,$$

By $\mathbf{n}_x$ the exterior (with respect to $\mathcal{D}_\epsilon$) normal on $\partial S_\epsilon$ at the point $x \in \partial S_\epsilon$ is denoted. By the superscripts $+$ and $-$ we denote the limiting values of functions on $(\Gamma^1)^+$ and on $(\Gamma^1)^-$ respectively. Since $u^0(x)$ satisfies the homogeneous boundary condition (2b) on $\Gamma$, we observe that $u^0|_{\Gamma^2} = 0$ and $(u^0)^\pm|_{\Gamma^{1,\epsilon}} = 0$ for any $\epsilon \in (0, \epsilon_0]$. Therefore

$$\|\nabla u^0\|_{L_2(\mathcal{D}_\epsilon)}^2 = \int_{\partial S_\epsilon} u^0 \frac{\partial u^0}{\partial \mathbf{n}_x} dl, \qquad \epsilon \in (0, \epsilon_0].$$

Setting $\epsilon \to +0$, taking into account that $u^0(x) \in \mathbf{K}_1$ and using the relationship (1), we obtain: $\|\nabla u^0\|_{L_2(\mathcal{D} \setminus \Gamma^1)}^2 = \lim_{\epsilon \to +0} \|\nabla u^0\|_{L_2(\mathcal{D}_\epsilon)}^2 = 0$. From the homogeneous boundary conditions (2b) we conclude that $u^0(x) \equiv 0$ in $\mathcal{D} \setminus \Gamma^1$, where $\mathcal{D}$ is an interior domain. If $\mathcal{D}$ is an exterior domain, then the proof is analogous, but we have to use the condition (2c) and the theorem on behaviour of a gradient of a harmonic function at infinity [8], p.373. The maximum principle cannot be used for the proof of the theorem even in the case of an interior domain $\mathcal{D}$, since the solution to the problem may not satisfy the boundary condition (2b) at the ends of the cracks, and it may be not continuous at the ends of the cracks.

## 3. Existence of a classical solution

Let us turn to the solution to the problem $\mathbf{D}_1$. Consider the double layer harmonic potential with the density $\mu(s)$ specified at the open arcs $\Gamma^1$:

(3)
$$w[\mu](x) = -\frac{1}{2\pi} \int_{\Gamma^1} \mu(\sigma) \frac{\partial}{\partial n_y} \ln|x - y(\sigma)| d\sigma.$$

**Theorem 2** *Let $\Gamma^1 \in C^{1,\lambda}$, $\lambda \in (0,1]$. Let $S(d,\epsilon)$ be a disc of a small enough radius $\epsilon$ with the center in the point $x(d)$ $(d = a_n^1$ or $d = b_n^1$, $n = 1, ..., N_1)$.*
*I. If $\mu(s) \in C^{0,\lambda}(\Gamma^1)$, then $w[\mu](x) \in C^0(R^2 \setminus \Gamma^1 \setminus X)$ and for any $x \in S(d,\epsilon)$, such that $x \notin \Gamma^1$, the inequality holds: $|w[\mu](x)| \leq$ const.*
*II. If $\mu(s) \in C^{1,\lambda}(\Gamma^1)$, then    1) $\nabla w[\mu](x) \in C^0(R^2 \setminus \Gamma^1 \setminus X)$;*
*2) for any $x \in S(d,\epsilon)$, such that $x \notin \Gamma^1$, the formulae hold*

$$\frac{\partial w[\mu](x)}{\partial x_1} = \frac{1}{2\pi} \frac{\mp\mu(d)}{|x - x(d)|} \sin\psi(x, x(d)) + \Omega_1(x), \quad \sin\psi(x, x(d)) = \frac{x_2 - x_2(d)}{|x - x(d)|},$$

$$\frac{\partial w[\mu](x)}{\partial x_2} = \frac{1}{2\pi} \frac{\pm\mu(d)}{|x - x(d)|} \cos\psi(x, x(d)) + \Omega_2(x), \quad \cos\psi(x, x(d)) = \frac{x_1 - x_1(d)}{|x - x(d)|},$$

$$|\Omega_j(x)| \leq \text{const} \cdot \ln\frac{1}{|x - x(d)|}, \quad j = 1, 2,$$

*the upper sign in the formulae is taken if $d = a_n^1$, while lower sign is taken if $d = b_n^1$;*
*3) for $w[\mu](x)$ the relationship holds*

$$\lim_{\epsilon \to +0} \int_{\partial S(d,\epsilon)} w[\mu](x) \frac{\partial w[\mu](x)}{\partial n_x} dl = 0,$$

*where the curvilinear integral of the first kind is taken over a circumference $\partial S(d,\epsilon)$, in addition, $n_x = (-\cos\psi(x, x(d)), -\sin\psi(x, x(d)))$ is a normal at the point $x \in \partial S(d,\epsilon)$, directed to the center of the circumference;*
*4) $|\nabla w[\mu](x)|$ belongs to $L_2(S(d,\epsilon))$ for any small $\epsilon > 0$ if and only if $\mu(d) = 0$.*
  Class $C^0(R^2 \setminus \Gamma^1 \setminus X)$ is defined in the remark to the definition of the class $\mathbf{K}_1$, if we set $\mathcal{D} = R^2$. The proof of the theorem is based on the representation of a double layer potential in the form of the real part of the Cauchy integral with the real density $\mu(\sigma)$:

$$w[\mu](x) = -\operatorname{Re}\Phi(z), \qquad \Phi(z) = \frac{1}{2\pi i} \int_{\Gamma^1} \mu(\sigma) \frac{dt}{t - z}, \qquad z = x_1 + ix_2,$$

where $t = t(\sigma) = (y_1(\sigma) + iy_2(\sigma)) \in \Gamma^1$. If $\mu(\sigma) \in C^{1,\lambda}(\Gamma^1)$, then for $z \notin \Gamma^1$:

$$\frac{d\Phi(z)}{dz} = -w'_{x_1} + iw'_{x_2} = -\frac{1}{2\pi i}\left(\sum_{n=1}^{N_1}\left\{\frac{\mu(b_n^1)}{t(b_n^1) - z} - \frac{\mu(a_n^1)}{t(a_n^1) - z}\right\} - \int_{\Gamma^1}\frac{e^{-i\alpha(\sigma)}\mu'(\sigma)}{t - z}dt\right).$$

Points I, II.1) and II.2) of Theorem 2 follow from these formulae and from the properties of Cauchy integrals, presented in $^6$. Points II.3) and II.4) can be proved by direct verification by using points I, II.1) and II.2).

We will construct the solution to the problem $\mathbf{D}_1$ in assumption that $F^+(s), F^-(s) \in C^{1,\lambda}(\Gamma^1)$, $\lambda \in (0,1]$, $F(s) \in C^0(\Gamma^2)$. We will look for a solution to the problem $\mathbf{D}_1$ in the form

$$(4) \qquad u(x) = -w[F^+ - F^-](x) + v(x),$$

where $w[F^+ - F^-](x)$ is the double layer potential (3), in which $\mu(\sigma) = F^+(\sigma) - F^-(\sigma)$. The potential $w[F^+ - F^-](x)$ satisfies the Laplace equation (2a) in $\mathcal{D} \setminus \Gamma^1$ and belongs to the class $\mathbf{K}_1$ according to Theorem 2. Limiting values of the potential $w[F^+ - F^-](x)$ on $(\Gamma^1)^\pm$ are given by the formula

$$w[F^+ - F^-](x)|_{x(s) \in (\Gamma^1)^\pm} = \mp(F^+(s) - F^-(s))/2 + w[F^+ - F^-](x(s)),$$

where $w[F^+ - F^-](x(s))$ is the direct value of the potential on $\Gamma^1$.

The function $v(x)$ in (4) must be a solution to the following problem.

**Problem D.** Find a function $v(x) \in C^0(\overline{\mathcal{D}}) \cap C^2(\mathcal{D} \setminus \Gamma^1)$, which obeys the Laplace equation (2a) in the domain $\mathcal{D} \setminus \Gamma^1$ and satisfies the boundary conditions
$$v(x)|_{x(s) \in \Gamma^1} = (F^+(s) + F^-(s))/2 + w[F^+ - F^-](x(s)) = f(s),$$
$$v(x)|_{x(s) \in \Gamma^2} = F(s) + w[F^+ - F^-](x(s)) = f(s).$$
(If $x(s) \in \Gamma^1$, then $w[F^+ - F^-](x(s))$ is the direct value of the potential on $\Gamma^1$).
If $\mathcal{D}$ is an exterior domain, then we add the following condition at infinity:

$$|v(x)| \leq \text{const}, \qquad |x| = \sqrt{x_1^2 + x_2^2} \to \infty.$$

All conditions of the problem $\mathcal{D}$ have to be satisfied in a classical sense. Obviously, $w[F^+ - F^-](x(s)) \in C^0(\Gamma^2)$. It follows from lemma 4(1) in [9], that $w[F^+ - F^-](x(s)) \in C^{1,\lambda/4}(\Gamma^1)$ (here by $w[F^+ - F^-](x(s))$ we mean the direct value of the potential on $\Gamma^1$). So, $f(s) \in C^{1,\lambda/4}(\Gamma^1)$ and $f(s) \in C^0(\Gamma^2)$.

We will look for the function $v(x)$ in the smoothness class $\mathbf{K}$.

We say that the function $v(x)$ belongs to the smoothness class $\mathbf{K}$ if

(1) $v(x) \in C^0(\overline{\mathcal{D}}) \cap C^2(\mathcal{D} \setminus \Gamma^1)$, $\nabla v \in C^0\left(\overline{\mathcal{D} \setminus \Gamma^1} \setminus \Gamma^2 \setminus X\right)$, where $X$ is a pointset consisting of the endpoints of $\Gamma^1$;

(2) in a neighbourhood of any point $x(d) \in X$ for some constants $C > 0$, $\delta > -1$ the inequality holds $|\nabla v| \leq C|x - x(d)|^\delta$, where $x \to x(d)$ and $d = a_n^1$ or $d = b_n^1$, $n = 1, \ldots, N_1$.

The definition of the functional class $C^0\left(\overline{\mathcal{D} \setminus \Gamma^1} \setminus \Gamma^2 \setminus X\right)$ is given in the remark to the to the definition of the smoothness class $\mathbf{K}_1$. Clearly, $\mathbf{K} \subset \mathbf{K}_1$, i.e. if $v(x) \in \mathbf{K}$, then $v(x) \in \mathbf{K}_1$.

It can be verified directly that if $v(x)$ is a solution to the problem $\mathbf{D}$ in the class $\mathbf{K}$, then the function (4) is a solution to the problem $\mathbf{D}_1$.

**Theorem 3.** *Let* $\Gamma \in C^{2,\lambda/4}$, $f(s) \in C^{1,\lambda/4}(\Gamma^1)$, $\lambda \in (0,1]$, $f(s) \in C^0(\Gamma^2)$. *Then the solution to the problem* $\mathbf{D}$ *in the smoothness class* $\mathbf{K}$ *exists and is unique.*

Theorem 3 has been proved in the following papers:
1) in [1,2], if $\mathcal{D}$ is an interior domain;

2) in [3], if $\mathcal{D}$ is an exterior domain and $\Gamma^2 \neq \emptyset$;

3) in [4,5], if $\Gamma^2 = \emptyset$ and so $\mathcal{D} = R^2$ is an exterior domain.

In all these papers, the integral representations for the solution to the problem $\mathbf{D}$ in the class $\mathbf{K}$ are obtained in the form of potentials, densities in which are defined from the uniquely solvable Fredholm integro–algebraic equations of the second kind and index zero. Uniqueness of a solution to the problem $\mathbf{D}$ is proved either by the maximum principle or by the method of energy (integral) identities. In the latter case we take into account that a solution to the problem belongs to the class $\mathbf{K}$. Note that the problem $\mathbf{D}$ is a particular case of more general boundary value problems studied in [2–5].

Note that Theorem 3 holds if $\Gamma \in C^{2,\lambda}$, $F^+(s), F^-(s) \in C^{1,\lambda}(\Gamma^1)$, $\lambda \in (0,1]$, $F(s) \in C^0(\Gamma^2)$. From Theorems 2, 3 we obtain the solvability of the problem $\mathbf{D}_1$.

**Theorem 4** *Let $\Gamma \in C^{2,\lambda}$, $F^+(s), F^-(s) \in C^{1,\lambda}(\Gamma^1)$, $\lambda \in (0,1]$, $F(s) \in C^0(\Gamma^2)$. Then a solution to the problem $\mathbf{D}_1$ exists and is given by the formula (4), where $v(x)$ is a unique solution to the problem $\mathbf{D}$ in the class $\mathbf{K}$, ensured by Theorem 3.*

**Remark** Let us check that the solution to the problem $\mathbf{D}_1$ given by formula (4) satisfies condition (1). Let $d = a_n^1$ or $d = b_n^1$ $(n = 1, ..., N_1)$ and $r$ is small enough, then substituting (4) in the integral in (1) we obtain

$$\int_{\partial S(d,r)} u(x) \frac{\partial u(x)}{\partial \mathbf{n}_x} dl$$

$$= \int_{\partial S(d,r)} w(x) \frac{\partial w(x)}{\partial \mathbf{n}_x} dl - \int_{\partial S(d,r)} w(x) \frac{\partial v(x)}{\partial \mathbf{n}_x} dl - \int_{\partial S(d,r)} v(x) \frac{\partial w(x)}{\partial \mathbf{n}_x} dl + \int_{\partial S(d,r)} v(x) \frac{\partial v(x)}{\partial \mathbf{n}_x} dl.$$

If $r \to 0$, then the first term tends to zero by Theorem 2(II.3). As metioned above, $v(x) \in \mathbf{K} \subset \mathbf{K}_1$, therefore the condition (1) holds for the function $v(x)$, so the fourth term tends to zero as $r \to 0$. The second term tends to zero as $r \to 0$, since $w(x)$ is bounded at the ends of $\Gamma^1$ according to Theorem 2(I), and since $v(x)$ satisfies condition 2) in the definition of the class $\mathbf{K}$. Noting that $v(x)$ is continuous at the ends of $\Gamma^1$ owing to the definition of the class $\mathbf{K}$, and using Theorem 2(II.2) for calculation of $\dfrac{\partial w(x)}{\partial \mathbf{n}_x}$ in the third term, we deduce that the third term tends to zero when $r \to 0$ as well. Consequently, the equality (1) holds for the solution to the problem $\mathbf{D}_1$ constructed in Theorem 4.

Uniqueness of a solution to the problem $\mathbf{D}_1$ follows from Theorem 1. The solution to the problem $\mathbf{D}_1$ found in Theorem 4 is, in fact, a classical solution. Let us discuss, under which conditions this solution to the problem $\mathbf{D}_1$ is not a weak solution.

## 4. Non–existence of a weak solution

Let $u(x)$ be a solution to the problem $\mathbf{D}_1$ defined in Theorem 4 by the formula (4). Consider a disc $S(d, \epsilon)$ with the center in the point $x(d) \in X$ and of radius $\epsilon > 0$ $(d = a_n^1$ or $d = b_n^1$, $n = 1, ..., N_1)$. In doing so, $\epsilon$ is a fixed positive number, which can be taken small enough. Since

$v(x) \in \mathbf{K}$, we have $v(x) \in L_2(S(d, \epsilon))$ and $|\nabla v(x)| \in L_2(S(d, \epsilon))$ (this follows from the definition of the smoothness class $\mathbf{K}$). Let $x \in S(d, \epsilon)$ and $x \notin \Gamma^1$. It follows from (4) that $|\nabla w[\mu](x)| \leq |\nabla u(x)| + |\nabla v(x)|$, whence $|\nabla w[\mu](x)|^2 \leq |\nabla u(x)|^2 + |\nabla v(x)|^2 + 2|\nabla u(x)| \cdot |\nabla v(x)| \leq 2(|\nabla u(x)|^2 + |\nabla v(x)|^2)$. Assume that $|\nabla u(x)|$ belongs to $L_2(S(d, \epsilon))$, then, integrating this inequality over $S(d, \epsilon)$, we obtain $\|\nabla w\|^2|_{L_2(S(d,\epsilon))} \leq 2(\|\nabla u\|^2|_{L_2(S(d,\epsilon))} + \|\nabla v\|^2|_{L_2(S(d,\epsilon))})$. Consequently, if $|\nabla u(x)| \in L_2(S(d, \epsilon))$, then $|\nabla w| \in L_2(S(d, \epsilon))$. However, according to Theorem 2, if $F^+(d) - F^-(d) \neq 0$, then $|\nabla w|$ does not belong to $L_2(S(d, \epsilon))$. Therefore, if $F^+(d) \neq F^-(d)$, then our assumption that $|\nabla u| \in L_2(S(d, \epsilon))$ does not hold, i.e. $|\nabla u| \notin L_2(S(d, \epsilon))$. Thus, if among numbers $a_1^1, ..., a_{N_1}^1, b_1^1, ..., b_{N_1}^1$ there exists such a number $d$ that $F^+(d) \neq F^-(d)$, then for some $\epsilon > 0$ we have $|\nabla u| \notin L_2(S(d, \epsilon)) = L_2(S(d, \epsilon) \backslash \Gamma^1)$, so $u \notin W_2^1(S(d, \epsilon) \backslash \Gamma^1)$, where $W_2^1$ is a Sobolev space of functions from $L_2$, which have generalized derivatives from $L_2$. We have proved

**Theorem 5** *Let conditions of Theorem 4 holds and among numbers $a_1^1, .., a_{N_1}^1, b_1^1, ..., b_{N_1}^1$ there exists such a number $d$, that $F^+(d) \neq F^-(d)$. Then the solution to the problem $\mathbf{D}_1$, ensured by Theorem 4, does not belong to $W_2^1(S(d, \epsilon) \backslash \Gamma^1)$ for some $\epsilon > 0$, whence it follows that it does not belong to $W_2^1(\mathcal{D} \backslash \Gamma^1)$. Here $S(d, \epsilon)$ is a disc of a radius $\epsilon$ with the center in the point $x(d) \in X$.*

If conditions of Theorem 5 hold, then the unique solution to the problem $\mathbf{D}_1$, constructed in Theorem 4, does not belong to $W_2^1(\mathcal{D} \backslash \Gamma^1)$, and so it is not a weak solution. We arrive to

**Corollary** *Let conditions of Theorem 5 hold, then a weak solution to the problem $\mathbf{D}_1$ in the space $W_2^1(\mathcal{D} \backslash \Gamma^1)$ does not exist.*

**Remark** Even if the number $d$, mentioned in Theorem 5, does not exist, then the solution $u(x)$ to the problem $\mathbf{D}_1$, ensured by Theorem 4, may be not a weak solution to the problem $\mathbf{D}_1$. This happens, for instance, in case if $\mathcal{D}$ is an exterior domain and if $\lim_{|x| \to \infty} u(x) \neq 0$, whence it follows that $u \notin L_2(\mathcal{D} \backslash \Gamma^1)$, and so $u \notin W_2^1(\mathcal{D} \backslash \Gamma^1)$. In this case a weak solution to the problem $\mathbf{D}_1$ in $W_2^1(\mathcal{D} \backslash \Gamma^1)$ does not exist as well.

Clearly, $L_2(\mathcal{D} \backslash \Gamma^1) = L_2(\mathcal{D})$, since $\Gamma^1$ is a set of zero measure. The research has been partly supported by the RFBR grant 08-01-00082.

## References

1. Krutitskii P.A. *The integral representation for a solution of the 2-D Dirichlet problem with boundary data on closed and open curves.* Mathematika (London), 2000, v.47, 339–354.
2. Krutitskii P.A. *The Dirichlet problem for the 2-D Laplace equation in a multiply connected domain with cuts.* Proc. Edinburgh Math. Soc., 2000, v.43, 325–341.
3. Krutitskii P.A. *The 2-Dimensional Dirichlet Problem in an External Domain with Cuts.* Zeitschr. Analysis u. Anwend., 1998, v.17, No.2, 361–378.
4. Krutitskii P.A. *A mixed problem for the Laplace equation outside cuts on the plane.* Differential Equations, 1997, v.33, No.9, 1184–1193.
5. Krutitskii P.A. *The mixed problem in an exterior cracked domain with Dirichlet condition on cracks.* Comp. and Math. with Appl., 2005, v.50, 769–782.

6. Muskhelishvili N.I. *Singular integral equations.* Nauka, Moscow, 1968. (In Russian; English translation: Noordhoff, Groningen, 1972.)

7. Gilbarg D., Trudinger N.S. *Elliptic partial differential equations of second order.* Springer-Verlag, Berlin-New York, 1977.

8. Vladimirov V.S. *Equations of mathematical physics.* Nauka, Moscow, 1981. (In Russian, English translation: Mir Publishers, Moscow, 1984.)

9. Krutitskii P.A., Kolybasova V.V. *On the mixed problem for the dissipative Helmholtz equation in a 2D cracked domain with the Neumann condition on cracks.* Complex Variables and Elliptic Equations, 2006, v.51, 1065–1089.

## II.3  Elliptic and Parabolic Nonlinear Problems

Organizer: F. Nicolosi

# HARNACK INEQUALITIES FOR ENERGY FORMS ON FRACTALS SETS*

MARIA AGOSTINA VIVALDI

*Dipartimento di Metodi e Modelli Matematici per le Scienze Applicate*
*Università degli Studi di Roma "La Sapienza"*
*Via A. Scarpa 16, 00161 Roma, Italy*
*E-mail: vivaldi@dmmm.uniroma1.it*

There is a huge literature about Harnack inequalities so here I restrict my talk to the case of weak solutions of elliptic equations in divergence form mentioning only fundamental contributions obtained by means of purely analytic tools and giving few historical references. I focus my attention on Harnack inequalities for either Dirichlet forms or p-homogeneous energy forms on fractal sets.

**Key words:** Harnack inequalities, Dirichlet forms, fractal sets

**Mathematics Subject Classification:** 35 B 05,35 B 65,35 J 20

J. Moser proved Harnack inequalities for weak solutions for second order uniformly, elliptic equations in divergence form (see [23]).

More precisely J. Moser proved that there exists a constant $C$ such that:

$$\sup_{B(x,R)} u \le C \inf_{B(x,R)} u, \quad C = C(n,\lambda,\Lambda), \tag{1}$$

wherever

$$\begin{cases} u \in H^1(\Omega), u \ge 0 \quad \text{in} \quad \Omega \\ a(u,v) = 0 \quad \forall v \in H_0^1(\Omega). \end{cases} \tag{2}$$

Where

$$a(u,v) = \int_\Omega \sum a_{ij}(x) u_{x_i}(x) v_{x_j}(x) \, dx, \tag{3}$$

$a_{ij} = a_{ji}$ are $\mathcal{L}^n - measurable$, such that $\lambda |\xi|^2 \le \sum a_{ij}(x) \xi_i \xi_j \le \Lambda |\xi|^2, 0 \le \lambda \le \Lambda$, $0 < R \le R_0$, $B(x, 4R_0)(x) \subset \Omega$, and $\Omega$ is an open connected set in $\mathbb{R}^n$.

According to the fundamental book of M. Fukushima [14], we recall the definition of Dirichlet forms. Let $(X, d)$ be a locally compact, separable, metric space and $m$ a positive Radon measure on $X$, with *supp* $m = X$. By $L^2(X, m)$ we denote the Hilbert space of the square integrable functions with respect to the measure $m$.

---

*Talk given by the author on July 26, 2005 at the ISAAC Congress, July 25-30, 2005 at Catania (Italy).

**Definition 0.1.** A Dirichlet form $a(\cdot,\cdot)$ with domain $D[a]$ is a real-valued, non-negative definite, symmetric, bilinear form closed and Markovian, where $D[a]$ is linear dense subspace of $L^2(X,m)$.

The word closed means that the domain $D[a]$ is complete under the intrinsic inner product $(u,v)_a := a(u,v) + (u,v)$, $(u,v) = \int_X u(x)v(x)m(dx)$. The word Markovian means that the domain enjoys the lattice property: if $u \in D[a]$ and $v = 0 \vee u \wedge 1$ then $v \in D[a]$ and $a(v,v) \le a(u,u)$. As a consequence of these properties it follows that $D[a] \cap L^\infty(X,m)$ is an algebra and we have:

$$a(uv, uv) \le 2\left(\|u\|^2_{L^\infty(X,m)} a(v,v) + \|v\|^2_{L^\infty(X,m)} a(u,u)\right)$$

(for the proof see Theorem 1.4.2 of [14]).

Furthermore we will consider Dirichlet forms that enjoy the two important properties of being regular and of diffusion type according to the following definitions.

**Definition 0.2.** A Dirichlet form $a(\cdot,\cdot)$ is of diffusion type if $a(u,v) = 0$, $\forall u, v \in D[a]$, $v$ constant $m - a.e.$ in a neighborhood of *supp u*.

**Definition 0.3.** A Dirichlet form $a(\cdot,\cdot)$ is regular in $L^2(X,m)$ if we have:

$$\overline{D[a] \cap C_0(X)}^{\|\cdot\|_\infty} = C_0(X) \quad and \quad \overline{D[a] \cap C_0(X)}^{\|\cdot\|_a} = D[a].$$

Here by $\|u\|_a$ we denote the intrinsic norm i.e. $(a(u,u) + (u,u))^{\frac{1}{2}}$ and $C_0(X) = \{f : X \to \mathbb{R} \quad continuous, \ supp\ f\ compact\ in\ X\}$.

**Remark 1.** The strong locality is a more restrictive assumption than the locality that requires the form $a(u,v)$ to vanish wherever the functions $u, v$ in the domain have disjoint supports. The regularity assumption implies that the domain is not trivial i.e. it does not reduce to constants but it is sufficiently rich.

According to the fundamental theory of A. Beurling & J. Deny [2] and the improvements due to M. L. Silverstein [27,28], Y. Lejean [17] and M. Fukushima [14]: there exists a unique Radon-measure-valued, non-negative definite, bilinear form $\mu$ on $D[a]$ such that

$$a(u,v) = \int_X \mu(u,v)\,dx \quad \forall u, v \in D[a] \tag{4}$$

$\mu$ is called the *energy measure* of $a$.

Hence any regular Dirichlet form of diffusion type can be expressed on its domain by an integral representation formula where $\mu(\cdot,\cdot)$ is a Radon-measure-valued, symmetric, bilinear form defined on the domain.

The measure $\mu(u,v)$ does not change sets of "intrinsic" capacity zero in $X$ and has a local character in $X$, that is the restriction of the measure $\mu(u,v)$ to any open subset $A$ of $X$ depends only on the restrictions of $u$ and $v$ to $A$:

$$1_A \mu(u_1, u_1) = 1_A \mu(u_2, u_2) \ if\ u_1 = u_2 \ m - a.e.\ on\ A,$$

where $1_A$ denotes the characteristic function of $A$ in $X$.

(For the proof see Proposition 1.5.2 of [17] and Lemma 5.4.6 of [14])

From Lemma 5.4.3 of [14] we derive the Schwarz rule: if $f \in L^2(X, \mu(u,u))$ and $g \in L^2(X, \mu(v,v))$ then we have: $2 |fg| |\mu(u,v)| \le f^2 \mu(u,u) + g^2 \mu(v,v)$.

Moreover $\mu(\cdot, \cdot)$ enjoys the following properties:

- *Leibnitz rule*: if $u, v \in D[a] \cap L^\infty(X, m)$ and $w \in D[a]$ then we have
  $\mu(uv, w) = u\mu(v, w) + v\mu(u, w)$.
- *Chain rule*: if $u \in D[a] \cap L^\infty(X, m), \eta \in C^1(\mathbb{R})$ and $v \in D[a]$ then we have
  $\eta(u) \in D[a] \cap L^\infty(X, m)$ and $\mu(\eta(u), v) = \eta'(u)\mu(u, v)$.

The Chain rule can be extended to the case in which $u \in D[a]$ if $\eta'$ is bounded in $\mathbb{R}$. (See [17] 2.1(a) and Theorem 5.4.3 of [14]).

Finally the *truncation Lemma* holds:

$$\mu(u^+, v) = 1_{\{u>0\}}\mu(u, v) \quad \forall u \in D[a], \ v \in D[a] \cap L^\infty(X, m)$$

and

$$\forall k \in \mathbb{R} \quad \mu\left((u-k)^+, v\right) = 1_{\{u>k\}}\mu(u, v).$$

**Remark 2.** If the set all continuous functions of the domain, whose energy measures have a bounded density with respect to the measure $m$, is rich enough to separate the points of $X$ then we can introduce the basic notion, that is the heart of this theory, of the *"intrinsic distance"*.

More precisely if $\forall x_1, x_2 \in X, x_1 \ne x_2 \ \exists \varphi \in D[a] \cap C(X)$ such that $\varphi(x_1) \ne \varphi(x_2)$ and $\mu(\varphi, \varphi)$ is absolutely continuous with respect to $m$, i.e.

$$\mu(\varphi, \varphi) \le m \quad \text{on } X$$

where $C(X) = \{f : X \to \mathbb{R} \ \text{continuous}\}$ then we can give the following:

**Definition 0.4.** $d^{(a)} : X \to [0, +\infty]$

$$d^{(a)} = \sup\{\varphi(x_1) - \varphi(x_2), \ \varphi \in D[a] \cap C(X) \ \mu(\varphi, \varphi) \le m \ \text{on } X\}.$$

It is easy to verify that $d^{(a)}$ satisfies the usual properties: $d^{(a)}(x_1, x_2) = 0 \iff x_1 = x_2, d^{(a)}(x_1, x_2) = d^{(a)}(x_2, x_1)$ and $d^{(a)}(x_1, x_2) \le d^{(a)}(x_1, x_3) + d^{(a)}(x_3, x_2)$.

From now on we denote by $B^{(a)}(\cdot, R)$ the intrinsic balls in $(X, d^{(a)})$ i.e.:

$$B^{(a)}(x, R) := \{t \in X : d^{(a)}(x, t) < R\}, R > 0.$$

Let me introduce now an abstract result, due to M. Biroli & U. Mosco (cfr [3]) that, to my opinion is beautiful and powerful. This approach provided an unified framework for many deep contributions and allowed the authors to establish new interesting results and to open the way to many further studies and researches. I will present some recent results in this direction.

We say that

**Definition 0.5.** $u$ is "harmonic" with respect to $a(\cdot, \cdot)$ if

$$\begin{cases} u \in D_{loc}[a, O], \\ a(u, v) = 0 \quad \forall v \in D_0(O) \end{cases} \tag{5}$$

where $O$ is a connected, relatively compact, open subset of $X$, $D_0(O) = \overline{D[a] \cap C_0(O)}^{\|\cdot\|_a}$ and $D_{loc}[a,O] = \{u$ measurable $: \forall \mathcal{A}$ relatively compact open set in $O$ $\exists w \in D[a]$ such that $u = w$ $m$-a.e. in $\mathcal{A}\}$.

The basic tool, that is the heart of this theory, is the notion of "*intrinsic distance*" actually a "suitable intrinsic distance" that is "intrinsically" related with the Dirichlet form $a(\cdot,\cdot)$ and at the same time induce on $X$ a metric topology equivalent to the initial given topology in $X$; our first assumption is:

$$\text{there exists an "intrinsic" distance } d^{(a)} \text{such that } (X, d) \backsim (X, d^{(a)}). \tag{6}$$

The second assumption requires that the measure $m$ is "doubling" with respect to the "intrinsic balls" i.e. there exists $C_0$ such that:

$$0 < m\left(B^{(a)}(x, 2R)\right) \leqslant C_0 m\left(B^{(a)}(x, R)\right) < +\infty \tag{7}$$

$$\forall x \in X, \quad 0 < R \leqslant R_0, \quad B^{(a)}(x, R) = \left\{t \in X : d^{(a)}(x, t) < R\right\}.$$

**Remark 3.** Let me recall that by assumption (7) the metric space $(X, d^{(a)})$ is an "homogeneous space" with dimension $\nu$ according to R.R. Coifman & G. Weiss [7] i.e.

$$m\left(B^{(a)}(x, R)\right) \leqslant 2m\left(B^{(a)}(x, S)\right)\left(\frac{R}{S}\right)^{\nu} \quad \nu = \frac{\lg C_0}{\lg 2}, \; 0 < S < R \leqslant \frac{R_0}{2}.$$

The third assumption is that "scaled" Poincaré inequalities with respect to the intrinsic balls hold, i.e.: there exist $C_1 > 0$ and $k_0 \geq 1$ integer such that

$$\int_{B^{(a)}\left(x, \frac{R}{k_0}\right)} |u - \bar{u}_R|^2 m\,(dx) \leq C_1 R^2 a\left(u|_{B^{(a)}(x,R)}, u|_{B^{(a)}(x,R)}\right) \quad u \in D_{loc}[a, O] \tag{8}$$

where $\bar{u}_R = \dfrac{\int_{B^{(a)}\left(x, \frac{R}{k_0}\right)} um(dx)}{m\left(B^{(a)}\left(x, \frac{R}{k_0}\right)\right)}$ and $O$ is a connected, relatively compact, open subset of $X$.

**Remark 4.** According to the theory of A. Beurling, J. Deny, (see [2]), (see also equality (4)) the right hand side in the inequality (8) has a rigorous meaning i.e.

$$\int_{B^{(a)}(x,R)} \mu(u, u)\,dx.$$

We are now in position to state:

**Theorem 0.1.** *Let $u$ be a non-negative harmonic function according to Definition 5. In the assumptions (6), (7) and (8) we have:*

$$\sup_{B^{(a)}(x,R)} u \leq C_H \inf_{B^{(a)}(x,R)} u \tag{9}$$

*with $C_H = C_H(C_0, C_1, k_0)$, $R \leq R_0$, $B^{(a)}(x, kR) \subset O, k \geq 1$ depending on $k_0$. Here $C_0$ is the constant appearing in the "doubling" condition (7) and $C_1$ and $k_0$ are the constants appearing in the Poincaré condition (8).*

The proof of the Harnack inequality also in the classical setting of J. Moser, is long and technical, the proof of Theorem 1 is longer and more difficult and hence I not explain it now. I refer to [3] and I simply discuss Theorem 1, in two classical, well known cases and in a few more recent examples.

The first example comes from the weighted uniform elliptic operators; the Dirichlet form being the associated bilinear form: the domain is the Sobolev weighted space $H_W^1$. The coefficients $a_{ij}$ are uniformly elliptic with respect to a weight $w$ that belongs to the Muckenhoupt class $A_2$ or is a weight associated with a quasi conformal transformation $F$ in $\mathbb{R}^n$.

We can choose as metric space $X$ the euclidean space $\mathbb{R}^n$ and as intrinsic distance the euclidean distance. Hence the intrinsic balls are actually the euclidean balls. More precisely

$$a(u,v) = \int_{\mathbb{R}^n} \sum a_{ij}(x)\, u_{x_i}(x)\, v_{x_j}(x)\, dx$$

$$\begin{cases} \lambda|\xi|^2 w(x) \leqslant \sum a_{ij}(x)\,\xi_i\xi_j \leqslant \Lambda|\xi|^2 w(x) & x \in \mathbb{R}^n \\ 0 < \lambda \leqslant \Lambda, \quad a_{ij} = a_{ji} \\ w \in A_2 \text{ or } w = |detF'|^{1-\frac{2}{n}} \end{cases} \tag{10}$$

$X = \mathbb{R}^n,\ d^{(a)}(\cdot,\cdot) = |\cdot|,\ B^{(a)}(\cdot,R) = B(\cdot,R),\ D[a] = H_W^1(\mathbb{R}^n),\ m(dx) = w(x)\,dx.$

The measure $m$ is the weighted Lebesgue measure in $\mathbb{R}^n$ and the doubling condition (7) is a property of the weight; the scaled Poincaré inequalities (8) are proved by E. Fabes, D. Jerison, C. Kenig and R. Serapioni (see [8] and [9]).

The second example comes from subelliptic operators and in particular from Hörmander squares. The Dirichlet form associated to a subelliptic operator satisfies a subellipticity condition in a fractional Sobolev space $H^\varepsilon$ for some positive $\varepsilon$. More precisely:

$$c\|u\|_{H^\varepsilon(\mathbb{R}^n)}^2 \leq a(u,u) + (u,u) \quad \varepsilon > 0. \tag{11}$$

Different (equivalent) choices of the "intrinsic distance" are proposed by C.L. Fefferman & D.Phong [10], C.L. Fefferman & A. Sanchez-Calle [11], condition (7) with respect to the Lebesgue measure: $m(dx) = \mathcal{L}^n(dx)$ is proved by A. Nagel, E. Stein & S. Weinger [24].

The scaled Poincaré inequalities (Condition (8)) are proved by D. Jerison [15] and D. Jerison & A. Sanchez-Calle [16].

These authors also provide examples that show the Poincaré inequalities may not hold on regions that do not coincide with intrinsic balls. Related results are obtained by B. Franchi & E. Lanconelli [12] and by B. Franchi & R. Serapioni [13].

Recent examples of Dirichlet forms came from the theory of quadratic energy forms associated with Brownian motions on fractals. In this interesting, new setting the question whether the Harnack inequality holds for harmonic functions is tricky: actually we can show example of fractals for which the answer is affirmative and fractals for which the answer is negative. There are also some open problems.

Let us consider the easy case of Koch type curves.

By J.H. Huctchinson theory (1981) the Koch curve $K\gamma$ can be defined as the unique closed set selfsimilar with respect to the given contractive similitudes $\psi_1^{(\gamma)}$ and $\psi_2^{(\gamma)}$ with contraction factor $|\alpha| = \sqrt{\frac{1}{4} + \gamma^2}$. Let us recall that $K_\gamma$ can be also obtained as the closure (in the Hausdorff distance) of sets $F_n^{(\gamma)}$ constructed by iteration of the similitudes starting from the initial set $F_0$ of the two points $z = 0$ and $z = 1$. More precisely:

$$\psi_1^{(\gamma)}(z) = \alpha\bar{z}, \quad \psi_2^{(\gamma)}(z) = \bar{\alpha}(\bar{z} - 1) + 1, \quad \alpha = \frac{1}{2} + i\gamma, \quad \gamma \in \left[0, \frac{1}{2}\right), \quad z \in \mathbb{C}.$$

$$K(\gamma) = \psi_1^{(\gamma)}(K(\gamma)) \bigcup \psi_2^{(\gamma)}(K(\gamma)).$$

Moreover

$$K(\gamma) = \overline{\bigcup_{n=1}^{+\infty} F_n^{(\gamma)}}, \quad F_0^{(\gamma)} = \{0, 1\}, \quad F_n^{(\gamma)} = \bigcup_{i=1}^{2} \psi_i^{(\gamma)}\left(F_{n-1}^{(\gamma)}\right), \quad n \geq 1.$$

Let us consider the Koch-roof that is the cartesian product of the "equilateral" Koch curve in $\mathbb{R}^2$ (here $\gamma = \frac{\sqrt{3}}{6}$) and the unit interval in $\mathbb{R}$. The heart of this theory is a suitable choice of the intrinsic quasi-distance that has to take into account the geometry of the fractal and in particular the Hausdorff dimension of the Koch curve and the scaling factor of the energy. More precisely any point $x$ in $X$ is a couple $(z, y)$ where $z$ belongs to the Koch curve $K$ in $\mathbb{R}^2$ and $y$ to the unit interval in $\mathbb{R}$. The energy form $a$ on the fractal layer $X = K \times I$ is defined by

$$a(u, v) = \int_I \int_K \mathcal{L}_z(u, v)(dz)dy + \int_K \int_I D_y u D_y v dy \mathcal{H}^\delta(dz). \tag{12}$$

Here, $\mathcal{L}_z(\cdot, \cdot)(dz)$ denotes the Lagrangean of the energy form $E_K$ associated with the Brownian motion on $K$ with domain $D_0(K)$, now acting on $u(z, y)$ and $v(z, y)$ as functions of $z \in K$ for a.e. $y \in I$; $\mathcal{H}^\delta(dx)$ is the Hausdorff measure acting on each section $K$ of $X$ for a.e. $y \in I$ (see [22] for details).

Hence a good choice of intrinsic quasi-distance is:

$$d^{(a)}(x, \tilde{x}) = |z - \tilde{z}|^\delta \vee |y - \tilde{y}|$$

where $x = (z, y)$ and $\tilde{x} = (\tilde{z}, \tilde{y})$.

A "natural" suitable choice of the measure $m$ in $X$ is then the product of the $\delta - dimensional$ Hausdorff measure in $K$ and the 1-dimensional Lebesgue measure on $I$: here $\delta$ depends on the contraction factor in the similitudes that (itself) depends on the number $\gamma$. In this example $\delta$ is equal to $\frac{\lg 4}{\lg 3}$.

In this situation we prove for the harmonic functions with respect to the Dirichlet form (12) the Harnack inequality with respect to the intrinsic balls. As the topology induced by the quasi-distance $d^{(a)}$ is equivalent to the topology induced by the euclidean distance in $\mathbb{R}^3$ we deduce the Hölder continuity of the harmonic functions with respect to the euclidean distance. More precisely we have:

**Theorem 0.2.** $X = K \times I$, $K \subset \mathbb{R}^2$, $K = K(\gamma)$, $\gamma = \frac{\sqrt{3}}{6}$, $I = [0,1] \subset \mathbb{R}$. *Let* $u \geq 0$ *be "harmonic" according to Definition 5 then there exist* $C > 0$ *and* $k \geq 1$ *such that*

$$\sup_{B^{(a)}(x,R)} u \leq C \inf_{B^{(a)}(x,R)} u, \quad C = C(C_0, C_1, k_0), \quad R \leq R_0, \quad B^{(a)}(x, kR_0) \subset X \quad (13)$$

*and*

$$\underset{B(x,R)}{osc\ u} \ \to 0 \ as \ R \to 0. \tag{14}$$

(See [6] for the proof).

Let us remark, incidentally, that here the homogeneous dimension $\nu$ (see Remark 3) is two and hence the elements of the domain $D[a]$ are not necessarily Hölder continuous. We take into account the "structure" of the fractal in constructing the Dirichlet form, and in establishing the scaled Poincaré inequalities. Finally the fact that the Dirichlet form $a(\cdot, \cdot)$ admits a density with respect to the measure $m$ plays an important role in the construction of useful "test functions" and hence in the proof of the Harnack inequality.

Analogous results hold for harmonic functions with respect to the Dirichlet forms on cartesian products of Koch curves.

We consider two Koch curves, possibly with different contraction factors, say $\gamma_1$ and $\gamma_2$. The points of $X$ are now the couples $(z_1, z_2)$ where $z_1$ belongs to the Koch curve $K^{(\gamma_1)}$ and $z_2$ to the Koch curve $K^{(\gamma_2)}$. We define the energy form in the spirit of the previous example (see (12)), by making some "natural" changes due to the "geometry" of our fractal

$$a(u,v) = \int_{K^{(\gamma_2)}} \int_{K^{(\gamma_1)}} \mathcal{L}_{\gamma_1}(u,v)(dz_1)\mathcal{H}^{\delta_2}(dz_2) + \int_{K^{(\gamma_1)}} \int_{K^{(\gamma_2)}} \mathcal{L}_{\gamma_2}(u,v)(dz_2)\mathcal{H}^{\delta_1}(dz_1).$$

Given two points $x = (z_1, z_2)$ and $\tilde{x} = (\tilde{z}_1, \tilde{z}_2)$ in $X$ our choice of intrinsic quasi-distance $d^{(a)}$ is:

$$d^{(a)}(x, \tilde{x}) = |z_1 - \tilde{z}_1|^{\delta_1} \vee |z_2 - \tilde{z}_2|^{\delta_2}.$$

Consequently "a doubling measure" $m$ on $X$ is the product measure of the Hausdorff measures $\mathcal{H}^{\delta_1}$ and $\mathcal{H}^{\delta_2}$, $\delta_i = \frac{\lg 4}{\lg \frac{4}{1+4\gamma_i^2}}, i = 1, 2$.

**Theorem 0.3.** $X = K(\gamma_1) \times K(\gamma_2)$, $\gamma_1, \gamma_2 \in \left(0, \frac{1}{2}\right)$. *Let* $u \geq 0$ *be "harmonic" according to Definition 5 then there exist* $C > 0$ *and* $k \geq 1$ *such that*

$$\sup_{B^{(a)}(x,R)} u \leq C \inf_{B^{(a)}(x,R)} u, \quad C = C(C_0, C_1, k_0), \quad R \leq R_0, \quad B^{(a)}(x, kR_0) \subset X \quad (15)$$

*and*

$$\underset{B(x,R)}{osc\ u} \ \to 0 \quad as \ R \to 0. \tag{16}$$

(See [6] for the proof).

The procedure for proving Theorem 3 is similar to the proof of Theorem 2 but needs some more technicalities.

Harmonic functions on certain random scale-irregular fractals exhibit an interesting analytic behavior; they are continuous functions, however uniform Harnack inequalities on decreasing balls do not hold for them: in order to be global energy minimizers, they are forced by the complicated fine geometry of the body to develop locally very sharp oscillations at every point.

More precisely environment dependent fractals are generated by families of euclidean similarities operating in random way that mimics the influence of the environment. The limit mixture and all relevant analytic estimates depend on the asymptotic frequency of the occurrence of each family. Let me discuss here briefly the simple case of the mixture of two Koch type curves. (see [18,21] and [20] for the Sierpinski gaskets).

Let $\xi$ be the sequence $\xi = (\xi_1, \xi_2, \xi_3, \dots), \xi_i \in \{\gamma_1, \gamma_2\}$

$$K^{(\xi)} = \bigcup_{n=1}^{+\infty} F_n^{(\xi)}, \quad F_0^{(\xi)} = \{0,1\}, \quad F_n^{(\xi)} = \bigcup_{i=1}^{2} \psi_i^{(\xi_n)}\left(F_{n-1}^{(\xi)}\right).$$

We denote by $h_j^{(\xi)}(n)$ the frequency of the occurrence of the family $j$ at the step $n$: $h_j^{(\xi)}(n) = \frac{1}{n} \sum_{i=1}^{n} 1_{\xi_i = \gamma_j}, \ j = 1, 2.$

We suppose that there exists an asymptotic frequency of occurrence $p_j$ such that: $h_j^{(\xi)}(n) \to p_j, \ (0 \leq p_j \leq 1) \ n \to +\infty, \ |h_j^{(\xi)}(n) - p_j| \leq \frac{g(n)}{n}, \ j = 1, 2, \ (n \geq 1)$ where $g$ is not decreasing, regular, $g(t) \leq g_0 t^{1-\varepsilon_0}$ as $t$ goes to infinity and $g \in \mathbb{R}^+, \ 0 < \varepsilon_0 \leq 1$. We set $\delta = \frac{\lg 4}{\sum_{j=1}^{2} p_j \lg \frac{4}{1+4\gamma_j^2}}$ and we choose as intrinsic quasi-distance: $d^{(a)}(x, \tilde{x}) = |x - \tilde{x}|^\delta$. The energy form is the limit of the "discrete" energy as $n \to +\infty$. (see [6,21] and also [18]).

In the setting of the Koch mixture any function $v$ in the domain $D[a]$ of the Dirichlet form is continuous (actually Hölder continuous with Hölder exponent $\beta < \frac{\lg 2}{\lg 3}$). On the other hand in the Harnack inequality for positive "harmonic" functions the constant $C$ depends on $R$. More precisely the following Theorem holds (see [21] and [6]):

**Theorem 0.4.** Let $X = K^{(\xi)}, \xi = (\xi_1, \xi_2, \xi_3, \dots), \xi_i \in \{\gamma_1, \gamma_2\}$. Then there exists $C > 0$ such that

$$\underset{B(x,R)}{osc}\ v \ \leq Ca\left(v_{|B(x,R)}, v_{|B(x,R)}\right)^{\frac{1}{2}} \cdot f(R), \quad \forall v \in D_{loc}[a] \tag{17}$$

where $f(R) \to 0$ as $R \to 0$. Moreover let $u$ be "harmonic" according to Definition 5 then:

$$\sup_{B^{(a)}(x,R)} u \leq C(R) \inf_{B^{(a)}(x,R)} u. \tag{18}$$

When the function $g(t)$ is bounded (as $t \to +\infty$) then the constant $C(R)$ in Harnack inequalities (18) is bounded (as $R \to 0^+$): hence a uniform Harnack inequality holds. If instead $g(t) \to +\infty$ (as $t \to +\infty$) (e.g. $g(t) \simeq t^{1-\varepsilon_0}$, $0 < \varepsilon_0 < 1$ as $t \to +\infty$) then the constant $C(R)$ in Harnack inequalities goes to infinity (as $R \to 0^+$).

**Remark 5.** Estimate (18) in Theorem 4 seems to be sharp as a consequence of some probabilistic results due to M.T. Barlow, B.M. Hambly (see [1]). Unfortunately, until now, we are not able to construct a pure analytic counter-example. Sharpness of estimate (18) is an open problem, a "tricky" open problem because our model is simple: namely in our mixture we have choosed at any step two contractive similitudes, the "discrete energies" have the same scaling factor ($\rho = 2$), the prefractal mixtures have the same homogeneous dimensions ($\nu = 1$) and the same spectral dimensions ($d_S = 1$); only the contraction coefficients in the similitudes are different $\left(|\alpha_i| = \sqrt{\frac{1}{4} + \gamma_i^2}\right)$ and hence Hausdorff dimensions and intrinsic distances are different $\left(\delta_i = \frac{\lg 4}{\lg \frac{4}{1+4\gamma_i^2}}\right)$.

As to the nonlinear cases of p-homogeneous energy forms on fractals, R. Capitanelli (see [5]) established Harnack inequalities for harmonic functions with respect to p-homogeneous energy on Koch type curves and in the more general setting of p-homogeneous Lagrangeans on metric fractals with homogeneous dimension $\nu < p$. In [5] are also proved some intermediate, interesting results as volume inequalities, scaled Poincaré inequalities and capacity inequalities. Abstract results, in the spirit of the previous Theorem 1 are stated by M. Biroli e P. Vernole in [4] assuming that the energy form is "Riemannian". More precisely the authors assume the absolute continuity of the p-homogeneous energy form with respect to the measure $m$, the topological equivalence between the initial topology and the new topology induced by the intrinsic distance constructed in terms of suitable test functions (see Remark 2), the doubling property of the measure $m$ with respect to the intrinsic balls and the Poincaré scaled inequalities.

# References

1. M. T. Barlow, B. M. Hambly, *Tansition density estimates for Brownian motion on scale irregular Sierpinski gaskets*, Ann. Inst. H. Poincaré, **33** (1997), Vol. 5 531–557.
2. A. Beurling, J. Deny, *Dirichlet spaces*, Proc. Nat. Acad. Sc. U.S.A., **45** (1959), 208–215.
3. M. Biroli, U. Mosco, *A Saint-Venant type principle for Dirichlet forms on discontinuous media*, Ann. Mat. Pure Appl., IV Serie CLXIX (1995), 125–181.
4. M. Biroli, P. Vernole, *Harnack inequality for harmonic functions relative to a non linear p-homogeneous Riemannian Dirichlet form*, Nonlinear Anal. T M A, to appear.
5. R. Capitanelli, *Harnack inequality for p-laplacians associated to homogeneous p-lagrangeans*, to appear.
6. R. Capitanelli, U. Mosco, M. A. Vivaldi, *Harnack inequality for fractal sets*, Preprint.
7. R. R. Coifman, G. Weiss, *Analyse harmonique non commutative sur certains espaces homogènes*, Lecture Notes in Math., **242**, Springer-Verlag, Berlin-Heidelberg-New York (1971).

8. E. Fabes, D. Jerison, C. Kenig, *The Wiener test for degenerate elliptic equations*, Ann. Inst. Fourier, **3** (1982), 151–183.

9. E. Fabes, C. Kenig, R. Serapioni, *The local regularity of solutions of degenerate elliptic equations*, Comm. Part. Diff. Eq., **7** (1982), 77–116.

10. C. L. Fefferman, D. Phong, *Subelliptic eigenvalue problems, Conference on Harmonic analysis*, Chicago, edited by W. Becker et al., Wadsworth (1981), 590–606.

11. C. L. Fefferman, A. Sanchez Calle, *Fundamental solution for second order subelliptic operators*, Ann. Math., **124** (1986), 247–272.

12. B. Franchi, E. Lanconelli, *An embedding theorem for Sobolev spaces related to non-smooth vector fields and Harnack inequality*, Comm. Part. Diff. Eq., **9** (1984), 1237–1264.

13. B. Franchi, R. Serapioni, *Pointwise estimates for a class of degenerate elliptic operators: a geometrical approach*, Ann. Sc. Norm. sup. Pisa, **14**, 4 (1987), 527–569.

14. M. Fukushima, *Dirichlet Forms and Markov Processes*, North-Holland Math., **23**, North-Holland and Kodansha, Amsterdam (1980).

15. D. Jerison, *The Poincaré inequality for vector fiels satisfying an Hörmander's condition*, Duke Math. J., **53** 2 (1986), 503–523.

16. D. Jerison, A. Sanchez Calle, *Subelliptic second order differential operators*, Lecture Notes in Math., **1277**, Springer-Verlag, Berlin-Heidelberg-New York (1987), 46–77.

17. Y. Lejean, *Measure associée à une forme de Dirichlet: Applications*, Bull. Soc. Math. France, **106**, (1978), 61–112.

18. U. Mosco, *Harnack Inequalities on Scale Irregular Sierpinski Gaskets*, In: Nonlinear Problems in Mathematical Physics and Related Topics II. Kluw Academic/Plenum Publishers, New York (2002), 305–328.

19. U. Mosco, *Harnack Inequalities on Recurrent Metric Fractals*, Proceedings of the Steklov Institute of Mathematics Vol. 236 (2002), 490–495.

20. U. Mosco, *Irregular similarity and Quasi-Metric scaling*, In Proceedings of the Conference: "Stochastic and Potential Theory" Saint-Priest de Gimel, (2002) to appear on Potential Analysis.

21. U. Mosco, *Gauged Sobolev Inequalities*, Preprint.

22. U. Mosco, M. A. Vivaldi, *Variational Problems with Fractal Layers*, Rend. Acc. N. delle Scienze detta dei XL, Memorie di Matematica e Applicazioni 121° (2003) Vol. XXVII, 237–251.

23. J. Moser, *On Harnack's inequality for elliptic differential equations*, Comm. Pure Appl. Math., **14**, (1961), 377–591.

24. A. Nagel, E. Stein, S. Weinger, *Balls and metrics defined by vector fields I: Basic properties*, Acta Math., **155**, (1985), 103–147.

25. L. Saloff Coste, *A note on Poincaré, Sobolev and Harnack inequalities*, Duke Math. J., Int. Math. Res. Notices, **2**, (1992), 27–38.

26. A. Sanchez Calle, *Fundamental solution and geometry of square of vector fields*, Inv. Math., **78** (1984), 143–160.

27. M. L. Silverstein, *Symmetric Markov Processes*, Lecture Notes in Math., **426**, Springer-Verlag, Berlin-Heidelberg-New York (1974).

28. M. L. Silverstein, *Boundary Theory for Symmetric Markov Processes*, Lecture Notes in Math., **516**, Springer-Verlag, Berlin-Heidelberg-New York (1976).

# ELLIPTIC SYSTEMS IN DIVERGENCE FORM WITH DISCONTINUOUS COEFFICIENTS

ANTONIO TARSIA

*Dipartimento di Matematica "L. Tonelli",*
*Università di Pisa*
*Largo B. Pontecorvo, 5. I-56127 Pisa, Italy*
*Email: tarsia@dm.unipi.it*

TheDirichlet problem is studied for elliptic systems in devergence form and discontinuous coefficents

**Key words:** Elliptic systems, divergence form, discontinuous coefficents, Dirichlet problem

**Mathematics Subject Classification:** 35J55

## 1. Introduction

We consider a family of $N \times N$ matrices, $A_{\alpha\beta}(x) = \{A_{\alpha\beta}^{hk}(x)\}_{h,k=1,\cdots,N}$, defined on an open bounded set $\Omega \subset \mathbb{R}^n$, $n > 1$, such that ([a])

$$A_{\alpha\beta} \in L^\infty(\Omega, \mathbb{R}^{N^2}), \ \forall \alpha, \beta, \tag{1}$$

$$\exists \nu > 0 : \sum_{|\alpha|=|\beta|=m} \lambda^{\alpha+\beta}(A_{\alpha\beta}(x)\eta|\eta) \geq \nu\|\lambda\|^{2m}\|\eta\|^2, \ \forall \lambda \in \mathbb{R}^n, \ \forall \eta \in \mathbb{R}^N \tag{2}$$

Moreover we set

$$Au = \sum_{|\alpha|=|\beta|=m} (-1)^m D^\alpha \left(A_{\alpha\beta}(x)D^\beta u\right). \tag{3}$$

We consider the Dirichlet problem:

$$\begin{cases} Au = f, & f \in H^{-m}(\Omega, \mathbb{R}^N) \\ u \in H_0^m(\Omega, \mathbb{R}^N). \end{cases} \tag{4}$$

---

[a] $\alpha, \beta$ are multi-indexes. That is $\alpha = (\alpha_1, \cdots, \alpha_n)$, $\beta = (\beta_1, \cdots, \beta_n)$, $\alpha_i, \beta_i \in \mathbb{N}$. $|\alpha| = \alpha_1 + \cdots + \alpha_n$, $|\beta| = \beta_1 + \cdots + \beta_n$, and if $\lambda \in \mathbb{R}^n$, then $\lambda^\alpha = \lambda_1^{\alpha_1} \cdots \lambda_n^{\alpha_n}$. Moreover $(\cdot|\cdot)_N$ and $\|\cdot\|_N$ are respectively the scalar product and the norm in $\mathbb{R}^N$. We omit the index if it there is not danger of ambiguity.
[b] Legendre-Hadamard condition.

Existence and uniqueness theorems are not known, about this problem, with the only hypotheses (1), (2), as it happens on the contrary if $A_{\alpha\beta} \in C^0(\overline{\Omega}, \mathbb{R}^{N^2})$ (see for example S. Campanato [4]).It is also possible to give a counterexample (see §3). In this paper we will show that problem (4) is well posed with more general hypotheses on the matrices $A_{\alpha\beta}$ than continuity: namely, the functions $A_{\alpha\beta}$ have to be piecewise VMO on $\Omega$ (see §5). In §3 we will display some auxiliary results. In §4 we will consider the Dirichlet problem with matrices having piecewise constant coefficients.

In order to lighten the reading of the paper and to emphasize the importance of the techniques used in the proofs we just consider the case where $\Omega$ is a n-dimensional cube (see Theorem (6.2) below). Quite similar results can be obtained for more general open sets by standard but tedious arguments.

## 2. Notations

If $\alpha = (\alpha_1, \cdots, \alpha_n)$ is a multi-index then we denote derivatives by the symbols

$$D^\alpha = D_1^{\alpha_1} \cdots D_n^{\alpha_n}, \quad D_i = \frac{\partial}{\partial x_i}.$$

$H^m(\Omega, \mathbb{R}^N)$ is the usual Sobolev space of functions defined on $\Omega$ and $\mathbb{R}^N$-valued, normed with

$$\|u\|_{H^m(\Omega, \mathbb{R}^N)} = \int_\Omega \sum_{|\alpha| \leq m} \|D^\alpha u\|^2 \, dx. \tag{5}$$

$H_0^m(\Omega, \mathbb{R}^N)$ is the closure of $C_0^\infty(\Omega, \mathbb{R}^N)$ with respect to the norm (5).
$H^{-m}(\Omega, \mathbb{R}^N)$ is the dual space of $H_0^m(\Omega, \mathbb{R}^N)$.
If $F \in H^{-m}(\Omega, \mathbb{R}^N)$ and $\varphi \in H_0^m(\Omega, \mathbb{R}^N)$ we denote by $\langle F, \varphi \rangle$ the corresponding duality, and we set

$$\|F\|_{H^{-m}(\Omega, \mathbb{R}^N)} = \sup \{ \langle F, \varphi \rangle : \varphi \in H^m(\Omega, \mathbb{R}^N), \|\varphi\| = 1 \}.$$

Moreover we denote by

$$Q(x_0, \rho) = \{ x = (x_1, \ldots, x_n) \in \mathbb{R}^n : \|x - x_0\|_C = \max\{|x_i - x_{0,i}| : i = 1, \ldots, n\} < \rho \}.$$

any non degenerate $n$-dimensional cube with edges parallel to the coordinate axes. We moreover set

$$\Omega(x, \rho) = Q(x, \rho) \cap \Omega, \quad |\Omega(x, \rho)| = \text{meas } \Omega(x, \rho),$$

$$v_{x,\rho} = \frac{1}{|\Omega(x, \rho)|} \int_{\Omega(x,\rho)} v(z) \, dz = \fint_{\Omega(x,\rho)} v(z) \, dz,$$

$$\gamma(v, \rho) = \sup_{0 < \sigma \leq \rho, \, x \in \Omega} \fint_{\Omega(x,\sigma)} |v(y) - v_{x,\sigma}| \, dy,$$

BMO($\Omega$): is the subspace of $L^1(\Omega)$–functions such that $\gamma(v,\rho) < +\infty$ for all $\rho$ such that $0 < \rho \le diam\ \Omega = \sup\{\|x - y\| : x, y \in \Omega\}$,

VMO($\Omega$): is the subspace of BMO($\Omega$)–functions such that $\lim_{\rho\to 0+}\ \gamma(v,\rho) = 0$,

$\mathcal{L}(v,\Omega)$: is the Lebesgue set of $v$ on $\Omega$.

## 3. A counterexample

In this section we show that the problem (4) is not well posed with the only hypotheses (1), (2). Indeed if it happens the operator $A$ is bjiective and continuous between the spaces $H_0^m(\Omega, \mathbb{R}^N)$ and $H_0^{-m}(\Omega, \mathbb{R}^N)$. Then, by *Banach open mapping theorem*, we can write

$$c\,\|u\|_{H_0^m(\Omega,\mathbb{R}^N)} \le \|Au\|_{H_0^{-m}(\Omega,\mathbb{R}^N)}, \quad \forall u \in H_0^m(\Omega, \mathbb{R}^N). \tag{6}$$

So we can take the usual cut-off function, that is $\theta \in C_0^\infty(\mathbb{R})$ such that

$$\theta(x) = \begin{cases} 1 \text{ se } x \in B(x_0,\rho) \\ 0 \text{ se } x \notin B(x_0,r), \end{cases} 0 < \rho < r, \quad |\theta| \le 1, \tag{7}$$

and multiplying the equation in problem (4) by $u\,\theta$, following the usual way (see for example S. Campanato [4]) and using inequality (6) instead of *ellipticity condition*, we can obtain Caccioppoli's inequality:

$$\|u\|_{H^m(B(x_0,\rho),\mathbb{R}^N)} \le C(\|Au\|_{H^{-m}(B(x_0,r),\mathbb{R}^N)}, +\|u\|_{H_0^{m-1}(B(x_0,r),\mathbb{R}^N)}), \tag{8}$$

for all $u \in H_0^m(\Omega, \mathbb{R}^N)$, where $x_0 \in \Omega$ and $dist(x_0, \partial\Omega) < r$.
But this contradicts the following counterexample of Giaquinta and Souček (see M. Giaquinta and J. Souček [5]).

*Example.* Let $B(0, e^{-2}) \subset \mathbb{R}^3$ and $N = 3$. Set

$$A_{ij}^{rs} = \delta_{rs}\delta_{ij} + \left(\frac{1}{\log^2\|x\|} - \frac{9}{8}\right) \varepsilon_{rst}\varepsilon_{ijk}\frac{x_t x_k}{\|x\|}, \quad x = (x_1, x_2, x_3), \tag{9}$$

where $\delta_{rs}$ is Kronecker's symbol and $\varepsilon_{rst}$ is the Levi Civita tensor (i.e. alternating tensor or isotropic tensor of rank 3). We have

$$\sum_{i,j=1}^{3}\sum_{r,s=1}^{3} A_{ij}^{rs}(x)\xi_r\xi_s\lambda_i\lambda_j = \|\xi\|^2\,\|\lambda\|^2, \quad \forall\lambda \in \mathbb{R}^3, \forall\xi \in \mathbb{R}^3. \tag{10}$$

Moreover

$$A_{ij}^{rs} \in L^\infty(B(0, e^{-2})) \quad \text{but} \quad A_{ij}^{rs} \notin \text{VMO}(B(0, e^{-2})).$$

Then the vector valued function

$$u(x) = \left(\|x\|^{\frac{1}{2}}\log\|x\|\right)^{-1}\frac{x}{\|x\|}, \quad x \in B(0, e^{-2}),$$

belongs to $H^1(B(0, e^{-2}), \mathbb{R}^3)$ and is a weak solution to

$$\sum_{i=1}^{3} \sum_{r,s=1}^{3} D_s[A_{ij}^{rs}(x) D_r u^i] = 0,$$

while does not verify Caccioppoli's inequality.

We can ask us whether the converse is true. It is easy to show that from Caccioppoli's inequality it follows:

$$\|u\|_{H^m(\Omega, \mathbb{R}^N)} \leq C(\|Au\|_{H^{-m}(\Omega, \mathbb{R}^N)}, + \|u\|_{H_0^{m-1}(\Omega, \mathbb{R}^N)}), \tag{11}$$

for all $u \in H_0^m(\Omega, \mathbb{R}^N)$, from this, by Lemma of Peetre (see J.L. Lions and E. Magenes, [7] cap 2,) we can write $dim\ kerA < +\infty$, but not in general $dim\ kerA = 0$. For example $Au = \Delta u + \lambda u$ (where $\lambda$ is an eigenvalue of $\Delta$) satisfies (11) but $dim\ kerA > 0$.

## 4. Near and almost near operators

The concept of *nearness* between operators was introduced by S. Campanato (see S. Campanato [1-3]) to study existence and uniqueness in nonvariational elliptic problems([b]). It is now formulated as follows.

**Definition 4.1.** Let $\mathcal{B}$ be a Banach space with norm $\|\cdot\|$, $\mathcal{X}$ a set, $A$, $B$ operators from $\mathcal{X}$ to $\mathcal{B}$. We say that $A$ is *near* $B$ if there exist two real positive constants $\alpha$, $k$, with $0 < k < 1$, such that for all $x_1, x_2 \in \mathcal{X}$:

$$\|B(x_1) - B(x_2) - \alpha[A(x_1) - A(x_2)]\| \leq k\|B(x_1) - B(x_2)\|. \tag{12}$$

The main result of this paper holds assuming a generalization of *nearness* among operators, that was introduced in A. Tarsia [10] as follows.

**Definition 4.2.** Let $\mathcal{B}$ be a Banach space with norm $\|\cdot\|$, $\mathcal{X}$ a set, $A$, $B$ operators from $\mathcal{X}$ to $\mathcal{B}$. We say that $A$ is *almost near* $B$ if there are:
• a set of operators $\{A_i\}_{i=1,\ldots,l+1}$, $l \geq 1$, from $\mathcal{X}$ to $\mathcal{B}$,
• a family $\{\mathcal{X}_i\}_{i=1,\ldots,l}$ of subsets of $\mathcal{X}$,
• two sets of real positive constants $\{\alpha_i\}_{i=1,\ldots,l}$, $\{k_i\}_{i=1,\ldots,l}$, $0 < k_i < 1$, such that:

$$A_1 = B \tag{13}$$

$$A_{l+1} = A \tag{14}$$

$$\|A_i(u) - A_i(v) - \alpha_i[A_{i+1}(u) - A_{i+1}(v)]\| \leq k_i\|A_i(u) - A_i(v)\| \tag{15}$$

$\forall u, v \in \mathcal{X}_i, i = 1, \ldots, l$.

---

[b]Other developments and applications are in A. Tarsia [8,9]

It is obvious that if $l = 1$ and $\mathcal{X}_1 = \mathcal{X}$ we obtain:

**Theorem 4.1.** *If $A$ is near $B$ then $A$ is almost near $B$.*

The following theorem is proved in A. Tarsia [10].

**Theorem 4.2.** *Let $A$ be almost near $B$ from $\mathcal{X}$ to $\mathcal{B}$, let $\{\mathcal{X}_i\}_{i=1,\ldots,l}$ be a family of subsets of $\mathcal{X}$, let $\{A_i\}_{i=1,\ldots,l+1}$, $l \geq 1$, be a set of operators from $\mathcal{X}$ to $\mathcal{B}$ that satisfy Definition 4.2.*
*Moreover assume that:*
*i) $\mathcal{X}$ is a Banach space,*
*ii) $\mathcal{X}_i$ are open sets in $\mathcal{X}$, $i = 1,\ldots,l$,*
*iii) $A_i : \mathcal{X} \to \mathcal{B}$ are linear operators, $i = 1,\cdots,l+1$.*
*If $B$ is continuous and bijective between $\mathcal{X}$ and $\mathcal{B}$, then $A$ is also continuous and bijective between $\mathcal{X}$ and $\mathcal{B}$.*

## 5. Matrices with piecewise constants coefficients

Let $\Omega$ be a nondegenerate $n$–dimensional cube with edges parallel to the axes. We denote by $\{Q_h\}_{h\in\mathbb{N}}$ a family of nondegenerate $n$–dimensional cubes with edges parallel to the axes contained in $\Omega$. Assume that

$$Q_h \cap Q_r = \emptyset, \quad h,\, r \in \mathbb{N},\ h \neq r, \tag{16}$$

$$\text{meas}(\Omega \setminus \cup_{h=1}^{\infty} Q_h) = 0 \tag{17}$$

Let $A_{\alpha\beta}$, $B_{\alpha\beta}$ be matrices such that

$$A_{\alpha\beta},\, B_{\alpha\beta} \in L^{\infty}(\Omega, \mathbb{R}^{N^2}),\ |\alpha| = |\beta| = m. \tag{18}$$

$$A_{\alpha\beta}(x) = A_{\alpha\beta}^h,\quad B_{\alpha\beta}(x) = B_{\alpha\beta}^h, \quad \forall x \in Q_h,\ h \in \mathbb{N}. \tag{19}$$

$$\exists \nu_a,\, \nu_b > 0 \text{ such that}$$

$$\sum_{|\alpha|=|\beta|=m} \lambda^{\alpha+\beta}\, (A_{\alpha\beta}(x)\eta|\eta) \geq \nu_a \|\lambda\|^{2m} \|\eta\|^2, \tag{20}$$

$$\sum_{|\alpha|=|\beta|=m} \lambda^{\alpha+\beta}\, (B_{\alpha\beta}(x)\eta|\eta) \geq \nu_b \|\lambda\|^{2m} \|\eta\|^2,\ \forall \lambda \in \mathbb{R}^n,\ \forall \eta \in \mathbb{R}^N. \tag{21}$$

**Theorem 5.1.**
*Under hypotheses (18)-(19)-(20)-(21), if $A,\, B : H_0^m(\Omega, \mathbb{R}^N) \longrightarrow H^{-m}(\Omega, \mathbb{R}^N)$ are defined by*

$$Au = \sum_{|\alpha|=|\beta|=m} (-1)^m\, D^\alpha \left(A_{\alpha\beta}(x)D^\beta u\right), \tag{22}$$

$$Bu = \sum_{|\alpha|=|\beta|=m} (-1)^m\, D^\alpha \left(B_{\alpha\beta}(x)D^\beta u\right),$$

*then $A$ is almost near $B$.*

674

**Proof.**

We define

$$\mathcal{A}_t u = (1-t)Bu + tAu, \quad t \in [0,1]. \tag{23}$$

It is known ($^c$) that, setting $\nu = \min(\nu_a, \nu_b)$, we have:

$$c\,\nu \int_\Omega \sum_{|\alpha|=m} \|D^\alpha u\|^2 \, dx \tag{24}$$

$$\leq \int_\Omega \sum_{|\alpha|=|\beta|=m} ([(1-t)B_{\alpha\beta} + tA_{\alpha\beta}]D^\beta u \mid D^\beta u) \, dx, \quad \forall u \in H_0^m(Q_h, \mathbb{R}^N), \ h \in \mathbb{N},$$

from which, by Lax-Milgram theorem, it follows:

$$c\nu \int_\Omega \sum_{|\alpha|=m} \|D^\alpha u\|^2 \, dx \leq \|\mathcal{A}_t u\|_{H^{-m}(\Omega,\mathbb{R}^N)}^2, \quad \forall u \in H_0^m(Q_h, \mathbb{R}^N), \ h \in \mathbb{N}. \tag{25}$$

Assume

$$\mathcal{P}_t(u) = \|\mathcal{A}_t u\|_{H^{-m}(\Omega,\mathbb{R}^N)}^2 - \frac{c\nu}{2} \int_\Omega \sum_{|\alpha|=m} \|D^\alpha u\|^2 \, dx.$$

Moreover we consider

$$W = H_0^m(\Omega, \mathbb{R}^N) \cap \left( \cap_{h=1}^{+\infty} H_0^m(Q_h, \mathbb{R}^N) \right).$$

On the whole vector space $W$, by (25), we have $\mathcal{P}_t(u) \geq 0$.

We need to claim that $W$ is strongly dense in $H_0^m(\Omega, \mathbb{R}^N)$.

Firstly we state that $W$ is weakly dense in $H_0^m(\Omega, \mathbb{R}^N)$.

Indeed, we consider functions $\theta_k \in C_0^\infty(\Omega) \cap (\cap_{h=1}^\infty C_0^\infty(Q_h))$, $k \in \mathbb{N}$, such that $0 \leq \theta_k \leq 1$ in $\Omega$ and $\theta_k(x) \to 1$, for almost all $x \in \Omega$. For all $u \in H_0^m(\Omega, \mathbb{R}^n)$ we set $u_k = u\,\theta_k$. Then $\{u_k\}_{k\in\mathbb{N}}$ is a sequence of functions in $H_0^m(\Omega, \mathbb{R}^N)$ such that $u_k \to u$ in $L^2(\Omega, \mathbb{R}^N)$ and $u_k \in H_0^m(\Omega, \mathbb{R}^N) \cap (\cap_{h=1}^{+\infty} H_0^m(\Omega, \mathbb{R}^N))$. We observe that $u_k \to u$ weakly in $H_0^m(\Omega, \mathbb{R}^N)$ because we have

$$\sum_{|\alpha|=m} \int_\Omega (D^\alpha u_k, D^\alpha \varphi) \, dx$$

$$= (-1)^m \sum_{|\alpha|=m} \int_\Omega (u_k, D^{2\alpha}\varphi) \, dx \longrightarrow (-1)^m \sum_{|\alpha|=m} \int_\Omega (u_k, D^{2\alpha}\varphi) \, dx$$

$$= \sum_{|\alpha|=m} \int_\Omega (D^\alpha u, D^\alpha \varphi) \, dx$$

The weak closure of $W$ in $H_0^m(\Omega, \mathbb{R}^N)$ is $H_0^m(\Omega, \mathbb{R}^N)$. But the strong closure of the vector space $W$ coincides with its weak closure. So $W$ is dense in $H_0^m(\Omega, \mathbb{R}^N)$. By this argument, we obtain

---

$^c$See for example S. Campanato $^4$ cap II.

$$\mathcal{P}_t(u-v) \geq 0, \quad \forall u,v \in H_0^m(\Omega, \mathbb{R}^N) \tag{26}$$

Then it follows, for all $u,v \in H_0^m(\Omega, \mathbb{R}^N)$ and $s,t \in [0,1]$ :

$$\|\mathcal{A}_t(u-v) - \mathcal{A}_s(u-v)\|_{H^{-m}(\Omega,\mathbb{R}^N)}$$

$$= \sup_{\varphi \in H_0^m(\Omega,\mathbb{R}^N), \|\varphi\|=1} |\langle \mathcal{A}_t(u-v) - \mathcal{A}_s(u-v), \varphi \rangle|$$

$$= \sup_{\varphi \in H_0^m(\Omega,\mathbb{R}^N), \|\varphi\|=1} \left| \int_\Omega \sum_{|\alpha|=|\beta|=m} ((t-s)(B_{\alpha\beta} - A_{\alpha\beta})D^\alpha(u-v) \,|\, D^\beta\varphi) \; dx \right|$$

$$\leq |t-s|M \left( \int_\Omega \sum_{|\alpha|=m} \|D^\alpha(u-v)\|^2 dx \right)^{\frac{1}{2}} \qquad \text{(by (26))}$$

$$\leq \frac{\sqrt{2}M|t-s|}{\sqrt{c\nu}} \|\mathcal{A}_t(u-v)\|_{H^{-m}(\Omega,\mathbb{R}^N)},$$

where $M$ depends on $\|A_{\alpha,\beta}\|_{L^\infty(\Omega,\mathbb{R}^{N^2})}$ and $\|B_{\alpha,\beta}\|_{L^\infty(\Omega,\mathbb{R}^{N^2})}$.

Setting $k_{s,t} = |t-s| \dfrac{\sqrt{2}M}{\sqrt{c\nu}}$, we can find a partition of the interval $[0,1]$, $0 = t_1 < \cdots < t_{l+1} = 1$, such that

$$k_i = k_{t_i,t_{i+1}} < 1, \quad i = 1, \cdots, l.$$

The proof is complete because we have found a family of linear operators $\{\mathcal{A}_{t_i}\}_{i=1,\cdots,l+1}$, acting between $H_0^m(\Omega, \mathbb{R}^N)$ and $H^{-m}(\Omega, \mathbb{R}^N)$, a family $\{\mathcal{X}_{t_i}\}_{i=1,\cdots,l}$, each of them coinciding with $H_0^m(\Omega, \mathbb{R}^N)$, and constants $0 < k_i < 1$, $i = 1, \cdots, l$ that verify Definition 4.2 with $\alpha_1 = \cdots = \alpha_l = 1$.

**Corollary 5.1.** *Let $\Omega$ a non degenerate n-dimensional cube with edges parallel to the coordinate axes. Every elliptic operator $A$ that satisfies hypotheses (18), (19), (20), (21) is an isomorphism between $H_0^m(\Omega, \mathbb{R}^N)$ and $H^{-m}(\Omega, \mathbb{R}^N)$.*

**Proof.**
Assume

$$B_{\alpha\beta} = \begin{pmatrix} 1 & \cdots\cdots & 0 \\ \cdots & 1 & \cdots\cdots \\ \cdots\cdots\cdots\cdots \\ 0 & \cdots\cdots & 1 \end{pmatrix}, \tag{27}$$

if $\alpha = (0, \cdots, \alpha_j, \cdots, 0)$, $\beta = (0, \cdots, \beta_j, \cdots, 0)$, $\alpha_j = \beta_j = m$, $j = 1, \cdots, n$, and

$$B_{\alpha\beta} = \begin{pmatrix} 0 & \cdots\cdots & 0 \\ \cdots & 0 & \cdots\cdots \\ \cdots\cdots\cdots\cdots \\ 0 & \cdots\cdots & 0 \end{pmatrix}, \tag{28}$$

otherwise.

The operator $B$ is an isomorphism between $H_0^m(\Omega, \mathbb{R}^N)$ and $H^{-m}(\Omega, \mathbb{R}^N)$, (see P. Grisvard, [6] and S. Campanato [4]). From this, remarking that the subsets $\mathcal{X}_{t_i}$ introduced in the proof of Theorem 5.1, are open sets, we obtain the thesis by Theorem 4.2 .

**Corollary 5.2.** *Under the assumptions of Theorem 5.1, there exist a real positive constant $c(\nu_a) > 0$ (depending only on the coercivity constant) and an open set $\mathcal{X} \subset H_0^m(\Omega, \mathbb{R}^N)$, such that for all $u, v \in \mathcal{X}$*

$$\left( c(\nu_a) \int_\Omega \sum_{|\alpha|=m} \|D^\alpha(u-v)\|^2 dx \right)^{\frac{1}{2}} \leq \|A(u-v)\|_{H^{-m}(\Omega, \mathbb{R}^N)}. \qquad (29)$$

**Proof.**

We follow the proof of Theorem 5.1 arriving to (26). From this we obtain that there is an open subset $\mathcal{X}_1 = H_0^m(\Omega, \mathbb{R}^N)$ such that $\mathcal{P}_1(u-v) > 0$, $\forall u, v \in \mathcal{X}_1$, and we complete the proof.

## 6. Matrices with piecewise VMO coefficients.

In order to obtain our main theorem (Theorem (6.2)), we need the following results proved in A. Tarsia [10].

**Lemma 6.1.** *Let $\Omega \subset \mathbb{R}^n$ be a bounded open convex set. Let $v_1, \cdots, v_k \in$ VMO$\cap L^\infty(\Omega)$. Then $\forall \varepsilon > 0$ and $\forall x_0 \in \cap_{i=1}^k \mathcal{L}(v_i, \Omega)$ a real positive constant $r(x_0, \varepsilon)$ exists such that $\forall r \in (0, r(x_0, \varepsilon))$ and $\forall x \in \cap_{i=1}^k \mathcal{L}(v_i, \Omega) \cap Q(x_0, r)$ we have*

$$|v_i(x) - \{v_i\}_{\Omega(x_0, r)}| \leq \varepsilon, \ i = 1, \cdots, k.$$

Let $\mathcal{Q}$ be the family of all the cubes of the kind $Q(x_0, r)$, for $x_0 \in \cap_{i=1}^k \mathcal{L}(v_i, \Omega)$ and $r \leq r(x_0, \varepsilon)$, with $r(x_0, \varepsilon)$ as in Lemma 6.1. Then $\mathcal{Q}$ covers $\cap_{i=1}^k \mathcal{L}(v_i, \Omega)$ in the *Vitali sense* (see R.L. Wheeden and, A. Zygmund, [11] ).

Then, by *Vitali Covering Lemma*, we get the following Theorem:

**Theorem 6.1.** *Let $\Omega$ be a bounded open convex set of $\mathbb{R}^n$ and $v_1, \ldots, v_k \in$ VMO$\cap L^\infty(\Omega)$. Then $\forall \varepsilon > 0$ a sequence of disjoint cubes $\{Q(x_j)\}_{j \in \mathbb{N}}$ in $\mathcal{Q}$ with $x_j \in \cap_{i=1}^k \mathcal{L}(v_i, \Omega)$, exists such that*

*(1) meas $\left( \cap_{i=1}^k \mathcal{L}(v_i, \Omega) \setminus \cup_{j=1}^\infty Q(x_j) \right) = 0$,*

*(2) $\sum_{j=1}^\infty |Q(x_j)| < +\infty$,*

*(3) $|v_i(x) - \{v_i\}_{\Omega(x_j)}| < \varepsilon$, $i = 1, \ldots, n$, $j \in \mathbb{N}$, $\forall x \in \cap_{i=1}^k \mathcal{L}(v_i, \Omega(x_j))$, where $\Omega(x_j) = Q(x_j) \cap \Omega$.*

Let $\Omega$ be a nondegenerate $n$–dimensional cube with edges parallel to the axes. We denote by $\{Q_h\}_{h\in\mathbb{N}}$ a family of nondegenerate $n$–dimensional cubes with edges parallel to the axes contained in $\Omega$. Assume that

$$Q_h \cap Q_r = \emptyset, \quad h,\, r \in \mathbb{N}, \; h \neq r, \tag{30}$$

$$\text{meas}(\Omega \setminus \cup_{h=1}^{\infty} Q_h) = 0. \tag{31}$$

Let $A_{\alpha\beta}$ be matrices such that

$$A_{\alpha\beta} \in L^{\infty}(\Omega, \mathbb{R}^{N^2}), \; |\alpha| = |\beta| = m. \tag{32}$$

$$\exists \nu_a > 0 : \sum_{|\alpha|=|\beta|=m} \lambda^{\alpha+\beta}(A_{\alpha\beta}(x)\eta|\eta) \geq \nu_a \|\lambda\|^{2m}\|\eta\|^2, \; \forall \lambda \in \mathbb{R}^n, \; \forall \eta \in \mathbb{R}^N, \tag{33}$$

$$A_{\alpha\beta} \in \mathsf{VMO}(Q_h, \mathbb{R}^{N^2}), \; |\alpha| = |\beta| = m, \quad h \in \mathbb{N}. \tag{34}$$

Under these hypotheses we prove the following theorem.

**Theorem 6.2.** *The operator*

$$Au = \sum_{|\alpha|=|\beta|=m} (-1)^m D^{\alpha}\left(A_{\alpha\beta}(x)D^{\beta}u\right) \tag{35}$$

*is an isomorphism between* $H_0^m(\Omega, \mathbb{R}^N)$ *and* $H^{-m}(\Omega, \mathbb{R}^N)$.

**Proof.**
We use Theorem 6.1 and consider the set of functions $A_{\alpha\beta}$, with $|\alpha| = |\beta| = m$. For all $\varepsilon > 0$, in every $Q_h$, $h \in \mathbb{N}$, we can find a countable family $\{Q(x_{h,s})\}_{s\in\mathbb{N}}$ of disjoint cubes with centers $x_{h,s}$ belonging to $\mathcal{L}(A_{\alpha\beta}, Q_h)$, $\forall \alpha, \beta, |\alpha| = |\beta| = m$, such that

$$\text{meas}\left(\cap_{|\alpha|=|\beta|=m}\mathcal{L}(A_{\alpha\beta}, Q_h) \setminus \cup_{s=1}^{+\infty}Q(x_{h,s})\right) = 0, \tag{36}$$

$$|A_{\alpha\beta}(x) - \{A_{\alpha\beta}\}_{\Omega(x_{h,s})}| < \varepsilon, \quad x \in \cap_{|\alpha|=|\beta|=m}\mathcal{L}(A_{\alpha\beta}, \Omega(x_{h,s})), \; s \in \mathbb{N}, \tag{37}$$

$$\sum_{s=1}^{\infty} |Q(x_{h,s})| < +\infty, \; \forall h \in \mathbb{N}. \tag{38}$$

We set

$$A_{\alpha\beta}^{\varepsilon}(x) = \begin{cases} A_{\alpha\beta}(x), & x \in \Omega \setminus \cup_{h,s=1}^{+\infty}Q(x_{h,s}) \\ \{A_{\alpha\beta}\}_{\Omega(x_{h,s})}, & x \in \Omega(x_{h,s}), \; h, s \in \mathbb{N}. \end{cases} \tag{39}$$

$$A^{\varepsilon}u = \sum_{|\alpha|=|\beta|=m} (-1)^m D^{\alpha}\left(A_{\alpha\beta}^{\varepsilon}(x)D^{\beta}u\right) \tag{40}$$

The operator $A^{\varepsilon}$ has piecewise constants coefficients and verifies the hypotheses of Corollary 5.1, therefore it is an isomorphism between $H_0^m(\Omega, \mathbb{R}^N)$ and $H^{-m}(\Omega, \mathbb{R}^N)$. Moreover we remark that if $\varepsilon$ is sufficiently small then $A$ is *almost near* $A^{\varepsilon}$. Indeed, for all $u, v \in H_0^m(\Omega, \mathbb{R}^N)$ we have:

$$\|\mathcal{A}^\varepsilon(u-v) - \mathcal{A}(u-v)\|_{H^{-m}(\Omega,\mathbb{R}^N)}$$

$$= \sup_{\varphi \in H_0^m(\Omega,\mathbb{R}^N), \|\varphi\|=1} |\langle \mathcal{A}^\varepsilon(u-v) - \mathcal{A}(u-v), \varphi \rangle|$$

$$= \sup_{\varphi \in H_0^m(\Omega,\mathbb{R}^N), \|\varphi\|=1} \left| \int_\Omega \sum_{|\alpha|=|\beta|=m} ((A_{\alpha\beta}^\varepsilon - A_{\alpha\beta})D^\alpha(u-v) \mid D^\beta\varphi) \; dx \right|$$

$$\leq \left( \int_\Omega \sum_{|\alpha|=|\beta|=m} \|(A_{\alpha\beta}^\varepsilon - A_{\alpha\beta})D^\alpha(u-v)\|^2 \; dx \right)^{\frac{1}{2}}$$

$$= \left( \sum_{h,s\in\mathbb{N}} \int_{Q(x_{h,s})} \sum_{|\alpha|=|\beta|=m} \|(A_{\alpha\beta}^\varepsilon - A_{\alpha\beta})D^\alpha(u-v)\|^2 \; dx \right)^{\frac{1}{2}}$$

$$\leq c\varepsilon \left( \sum_{h,s\in\mathbb{N}} \int_{Q(x_{h,s})} \sum_{|\alpha|=m} \|D^\alpha(u-v)\|^2 dx \right)^{\frac{1}{2}}$$

$$= c\varepsilon \left( \int_\Omega \sum_{|\alpha|=m} \|D^\alpha(u-v)\|^2 dx \right)^{\frac{1}{2}}.$$

On the other hand by Corollary 5.2 we know that for all real positive $\varepsilon$ an open set $\mathcal{X}_\varepsilon = H_0^m(\Omega,\mathbb{R}^N)$ exists such that for all $u, v \in \mathcal{X}_\varepsilon$

$$c(\nu_a) \left( \int_\Omega \sum_{|\alpha|=m} \|D^\alpha(u-v)\|^2 dx \right)^{\frac{1}{2}} \leq \|\mathcal{A}^\varepsilon(u-v)\|_{H^{-m}(\Omega,\mathbb{R}^N)}.$$

Let $\varepsilon > 0$ be such that

$$k = \frac{c\varepsilon}{c(\nu_a)} < 1.$$

We can find, in connection with it, an open set $\mathcal{X}_\varepsilon \subset H_0^m(\Omega,\mathbb{R}^N)$ such that for all $u, v \in \mathcal{X}_\varepsilon$

$$\|\mathcal{A}^\varepsilon(u-v) - \mathcal{A}(u-v)\|_{H^{-m}(\Omega,\mathbb{R}^N)} \leq k \|\mathcal{A}^\varepsilon(u-v)\|_{H^{-m}(\Omega,\mathbb{R}^N)}.$$

By Theorem 4.2 and Corollary 5.1 we obtain the thesis.

*Acknowledgements* - I wish to thank Prof. P. Acquistapace for his useful comments and advices.

**References**

1. S. Campanato, *Equazioni ellittiche non variazionali a coefficienti continui*, Ann. Mat. Pura Appl., 86 , pp. 125-154 (1970).

2. S. Campanato, *A Cordes type condition for nonlinear variational systems*, Rend. Accad. Naz. Sci. XL, Mem. Mat. (5), 13, No 1, pp. 307-321 (1989).
3. S. Campanato, *On the condition of nearness between operators*, Ann. Mat. Pura Appl. 167, pp. 243-256 (1994).
4. S. Campanato, *Sistemi ellittici in forma di divergenza. Regolarità all'interno*, Quaderni della Scuola Normale Superiore, Pisa (1980).
5. M. Giaquinta and J. Souček, *Caccioppoli's inequality and Legendre-Hadamard condition* Math. Ann. 270, pp.105-107 (1985).
6. P. Grisvard, *Elliptic problems in nonsmooth domains*, Pitman, London (1985).
7. J.L. Lions and E. Magenes, *Problèmes aux limites non homogènes et applications*, *vol.1,* Dunod, Paris (1968).
8. A. Tarsia, *Recent developments of the Campanato theory of near operators*, Le Matematiche, vol. LV, Supplemento n.2, pp. 197-208 (2000).
9. A.Tarsia, *Some topological properties preserved by nearness among operators and applications to PDE*, Czechoslovak Math. J. 46, pp. 115-133 (1996).
10. A. Tarsia, *On Invertible nonvariational elliptic operators*, to appear.
11. R.L. Wheeden and, A. Zygmund, *Measure and Integrals: An Introduction to Real Analysis,* Monographs and Textbook in Pure and Appl. Math, Marcel Dekker Inc. New York (1977).

# DIRECTIONAL LOCALIZATION OF SOLUTIONS TO ELLIPTIC EQUATIONS WITH NONSTANDARD ANISOTROPIC GROWTH CONDITIONS

S. N. ANTONTSEV*

*Departamento de Matemática*
*Universidade da Beira Interior, Portugal*
*E-mail: anton@ubi.pt*

S. I. SHMAREV†

*Departamento de Matemáticas*
*Universidad de Oviedo, Spain*
*E-mail: shmarev@orion.ciencias.uniovi.es*

We study the Dirichlet problem for a class of nonlinear elliptic equations with variable exponents of nonlinearity. We prove existence of solutions in a generalized Sobolev–Orlicz space and study the localization (vanishing) properties of the solutions. It is shown that unlike the equations with isotropic nonlinearity the anisotropy may cause localization of solutions in a separate direction. We use this property to establish sufficient conditions of solvability of the problems posed on unbounded domains without conditions at infinity. These condition relate the "asymptotic size" of the domain at infinity with the character of nonlinearity of the equation.

**Key words:** Dirichlet problem, elliptic equations with variable exponents of nonlinearity, Sobolev-Orliez space, localization properties, anisotropic nonlinearity
**Mathematics Subject Classification:** 35J60

## 1. Introduction

We study the Dirichet problem for the elliptic equation with inhomogeneous and anisotropic nonlinearity

$$\begin{cases} -\sum_i D_i \left( a_i(x,u)|D_i u|^{p_i(x)-2} D_i u \right) + c(x,u)|u|^{\sigma(x)-2} u = f & \text{in } \Omega, \\ u = 0 \text{ on } \Gamma. \end{cases} \quad (1)$$

The coefficients $a_i(x,r)$, and $c(x,r)$ are Carathéodory functions (measurable in $x$ for all $r \in \mathbb{R}$ and continuous in $r$ for almost all $x \in \Omega$). Unless explicitly stated, we always assume that $a_i$ and $c$ satisfy the conditions

*Work partially supported by the Project DECONT, FCT , (Portugal), UBI
†Work partially supported by grant MTM-2004-05417 of the Ministry of Science and Technology, (Spain) and Research Project HPRN–CT–2002–00274, (EC)

$$\forall x \in \overline{\Omega}, r \in \mathbb{R} \quad 0 < a_0 \le a_i(x, r) < \infty, \quad 0 \le c_0 \le c(x, r) < \infty \qquad (2)$$

with some constants $a_0$, $c_0$. The functions $p_i(x)$ and $\sigma(x)$ are bounded in $\Omega$: it is assumed that there exist constants $p^- > 1$, $p^+ < \infty$, $\sigma^+ < \infty$, $\sigma^- > 1$ such that for all $x \in \overline{\Omega}$

$$p_i(x) \in (p^-, p^+], \quad \sigma(x), \in (\sigma^-, \sigma^+], \quad \inf_{\Omega} p_i(x) > p^- \in (1, n). \qquad (3)$$

The functions $p_i(x)$ and $\sigma(x)$ are log–continuous in $\Omega$: for all $x, y \in \Omega$ such that $|x - y| \le 1$ the inequality

$$|\sigma(x) - \sigma(y)| + \sum_i |p_i(x) - p_i(y)| \le \omega(|x - y|) \qquad (4)$$

holds with a function $\omega(\tau)$ satisfying the condition $\omega(\tau) \ln \frac{1}{\tau} \le M$ for $\tau \in [0, 1]$, $M = const > 0$. We assume that the domain $\Omega$ satisfies the following conditions:

$$\begin{cases} \forall s \in (0, L) \text{ the set } \omega(x_1) = \Omega \cap \{x_1 = s\} \text{ is a simple-connected} \\ \text{domain in } \mathbb{R}^{n-1}, \ \partial\omega(t) \text{ is Lipschitz–continous} \\ \exists \kappa \in (0, 1): \ \kappa\lambda(s) \le \operatorname{diam}\omega(s) \le \lambda(s), \quad \lambda(0) \ge 0, \quad \lambda(L) \ge 0, \\ \text{where } \lambda(t) \text{ is a given continuous function.} \end{cases} \qquad (5)$$

Unless specially indicated, we will assume that $L < \infty$. The case $L = \infty$ corresponding to the unbounded domains will be specially discussed in the last section.

In Section 2 we introduce the generalized Sobolev–Orlicz spaces the solutions of problem (1) belong to. The detailed information on these spaces can be found in [5,7], an extended list of references is given in [4]. In Section 3 we prove that under assumptions (2)– (5) on the coefficients and the exponents of nonlinearity problem (1) has at least one a.e. bounded weak solution. The study of the localization properties is performed via the local energy method [3]. In Section 4 we define the local energy functions, derive the energy relation satisfies by every weak solution and formulate the ordinary nonlinear differential inequality for the energy function. The energy relation and the corresponding O.D.I. depend on the character of anisotropy of the equation. The conclusions about the localization properties of weak solutions are byproducts of the properties of the functions satisfying these O.D.I. The main result is the possibility of the *directional localization* of solutions of problem (1) with anisotropic nonlinearity. It is known that for the solutions of elliptic equations of the type "diffusion/absorption" with isotropic nonlinearity the following alternative holds: let $\Omega$ be a bounded domain in $\mathbb{R}^n$ and $u$ be a nonnegative weak solution of the exterior problem

$$-\Delta_p u \equiv -\operatorname{div}\left(|\nabla u|^{p-2}\nabla u\right) + c\, u^{\sigma-1} = 0 \quad \text{in } \Omega, \quad u \to 0 \text{ as } |x| \to \infty,$$

with the parameters $p > 1$, $\sigma > 1$, $c > 0$; then

$$p > \sigma \Longleftrightarrow \text{the strong maximum principle holds,}$$
$$p < \sigma \Longleftrightarrow \text{supp } u \text{ is compact}$$

(see [9,11,10], the detailed discussion of this issue is given in [8]). Thus, the localization of supp $u(x)$ is caused by a suitable diffusion/absorption balance and is impossible if $c = 0$. It turns out that the solutions of equations with anisotropic diffusion operator may be localized even in the absence of the absorption term. For another classes of diffusion–absorption equations this effect was studied in [3], [2].

## 2. Anisotropic generalized Sobolev–Orlicz spaces

Let $\Omega$ satisfy condition (5) and $p(x)$ satisfy conditions (2)–(4). By $L^{p(x)}(\Omega)$ we denote the space of measurable functions $f(x)$ on $\Omega$ such that

$$A_{p(\cdot)}(f) = \int_{\Omega} |f(x)|^{p(x)} \, dx < \infty.$$

The space $L^{p(x)}(\Omega)$ equipped with the norm

$$\|f\|_{p(\cdot)} \equiv \|f\|_{L^{p(x)}(\Omega)} = \inf \left\{ \lambda > 0 : A_{p(\cdot)}(f/\lambda) \leq 1 \right\}$$

becomes a Banach space. The Banach space $W^{1,p(x)}(\Omega)$ with $p(x) \in [p^-, p^+] \subset (1, \infty)$ is defined by

$$W^{1,p(x)}(\Omega) = \left\{ f \in L^{p(x)}(\Omega) : |\nabla f| \in L^{p(x)}(\Omega) \right\},$$
$$\|u\|_{W^{1,p(x)}(\Omega)} = \sum_i \|D_i u\|_{p(\cdot)} + \|u\|_{p(\cdot)}. \tag{6}$$

If condition (4) is fulfilled, then $C_0^\infty(\Omega)$ is dense in $W_0^{1,\,p(x)}(\Omega)$ and $W_0^{1,\,p(x)}(\Omega)$ can be defined as the closure of $C_0^\infty(\Omega)$ with respect to the norm (6). The equivalent norm of $W_0^{1,p(x)}(\Omega)$ is defined by $\sum_i \|D_i u\|_{p(\cdot)}$. If $p(x) \in C^0(\overline{\Omega})$, then $W^{1,p(x)}(\Omega)$ is separable and reflexive. If $p(x)$, $q(x) \in C^0(\overline{\Omega})$,

$$1 < q(x) \leq \sup_{\Omega} q(x) < \inf_{\Omega} p_*(x) \quad \text{with} \quad p_*(x) = \begin{cases} \dfrac{p(x)n}{n - p(x)} & \text{if } p(x) < n, \\ \infty & \text{if } p(x) > n, \end{cases}$$

then the embedding $W_0^{1,p(x)}(\Omega) \hookrightarrow L^{q(x)}(\Omega)$ is continuous and compact. The following inequalities hold:

(1)

$$\min \left( \|f\|_{p(\cdot)}^{p^-}, \|f\|_{p(\cdot)}^{p^+} \right) \leq A_{p(\cdot)}(f) \leq \max \left( \|f\|_{p(\cdot)}^{p^-}, \|f\|_{p(\cdot)}^{p^+} \right); \tag{7}$$

(2) Hölder's inequality: for $f \in L^{p(x)}(\Omega)$, $g \in L^{q(x)}(\Omega)$ with

$$\frac{1}{p(x)} + \frac{1}{q(x)} = 1, \quad 1 < p^- \le p(x) \le p^+ < \infty, \quad 1 < q^- \le q(x) \le q^+ < \infty$$

$$\int_\Omega |f\,g|\,dx \le 2\,\|f\|_{p(\cdot)}\,\|g\|_{q(\cdot)}. \tag{8}$$

(3) for every $1 \le q = const < p^-$ we have $\|f\|_q \le C\,\|f\|_{p(\cdot)}$ with the constant $C = 2\,\|1\|_{\frac{p(\cdot)}{p(\cdot)-q}}$.

Let $p_i(x)$ and $\sigma(x)$ satisfy conditions (2)– (4). We introduce the Banach space

$$\mathbf{V}(\Omega) = \left\{ u \,\middle|\, u \in L^{\sigma(x)}(\Omega) \cap W_0^{1,1}(\Omega), \quad D_i u \in L^{p_i(x)}(\Omega) \right\},$$

$$\|u\|_{\mathbf{V}} = \|u\|_{\sigma(\cdot),\Omega} + \sum_{i=1}^{n} \|D_i u\|_{p_i(\cdot),\Omega}$$

and its dual $\mathbf{V}'(\Omega)$. $\mathbf{V}(\Omega) \subset \mathbf{X} = W_0^{1,p^-}(\Omega) \cap L^{\sigma^-}(\Omega)$, so that $\mathbf{V}(\Omega)$ is reflexive and separable as a closed subspace of $\mathbf{X}$. We will denote

$$A_{\mathbf{p}(\cdot),\Omega}(\nabla u) = \sum_{i=1}^{n} \int_\Omega |D_i u|^{p_i(x)}\,dx.$$

## 3. Existence of bounded solutions

**Definition 3.1.** A function $u \in \mathbf{V}(\Omega)$ is said to be weak solution of problem (1) if (1) $u = 0$ on $\Gamma$ in the sense of traces, (2) for every test-function $\eta \in \mathbf{V}(\Omega)$

$$\sum_i \int_\Omega a_i |D_i u|^{p_i(x)-2} D_i u\, D_i \eta\, dx + \int_\Omega c|u|^{\sigma(x)-2} u\, \eta\, dx = \int_\Omega f\, \eta\, dx. \tag{9}$$

**Theorem 3.1.** *(Th.3.1[1]) Let the coefficients $a_i(x)$, $c(x)$ satisfy conditions (2), and the exponents $p_i(x)$ and $\sigma(x)$ satisfy conditions (3)–(4). Then for every $f \in L^2(\Omega)$ ($f \in L^2(\Omega) \cap L^{\sigma'(x)}(\Omega)$ is $c_0 > 0$) problem (1) has at least one weak solution which satisfies the estimate*

$$a_0 \sum_i \int_\Omega |D_i u|^{p_i(x)}\,dx + c_0 \int_\Omega |u|^{\sigma(x)}\,dx$$

$$\le C \begin{cases} \|f\|_{2,\Omega}^2 & \text{if } c_0 = 0 \text{ and } p^- > \dfrac{2n}{n+2}, \\[2mm] \|f\|_{2,\Omega}^2 + \displaystyle\int_\Omega |f|^{\sigma'(x)}\,dx & \text{if } c_0 > 0. \end{cases} \tag{10}$$

The theorem is proved via Galerkin's method with the use of monotonicity of the differential operator of problem (1). Condition (4) of log–continuity of $p_i(x)$ and $\sigma(x)$ allows one to approximate every element of $L^{p(x)}(\Omega)$ with a sequence of smooth functions.

**Theorem 3.2.** (a) *Let* $\|f\|_{\infty,\Omega} = K < \infty$ *and* $c_0 > 0$. *Then the weak solution of problem* (1) *satisfies the estimate* $\|u\|_{\infty,\Omega} \leq \max\left\{1;\, (K/c_0)^{1/(\sigma^- - 1)}\right\}$ *with the constant* $c_0$ *from condition* (2).

(b) *Let in the conditions of Theorem 3.1* $f(x) \in L^{(q-1)/q}(\Omega)$ *with some* $q \in (1, np^-/(n-p^-))$. *Then the solutions of problem* (1) *satisfy the estimate* $\|u\|_{\infty,\Omega} \leq K$ *with a constant* $M$ *depending on* $p^\pm$, $\|f\|$, $1/a_0$ *and* $n$.

**Proof.** Item (a) is proved in [Lemma 3.7][1]. Let us prove (b). Fix an arbitrary $k \in \mathbb{N}$ and take the function $\zeta(x) = \max\{0, u - k\}$ for the test–function in Definition 3.1. After some obvious transformations we have that

$$I_1 \equiv a_0 \sum_i \int_{\Omega_k} |D_i u|^{p_i(x)}\, dx \leq \|u - k\|_{q,\Omega_k} \|f\|_{q/(q-1),\Omega} \equiv I_2.$$

Not loosing generality we may assume that $|\Omega| \equiv \operatorname{meas}\Omega < 1$ and $A_{\mathbf{p}(x)}(\nabla u) < 1$. Using (7), (8) we estimate $I_1$ as follows:

$$I_1 \geq a_0 \sum_i \|D_i u\|_{p_i(\cdot),\Omega_k}^{p^+} \geq a_0 (2|\Omega_k|)^{-\frac{p^+}{p^-}} \|\nabla u\|_{p^-,\Omega_k}^{p^+}$$

On the other hand, applying the embedding theorems we have that

$$I_2 \leq c|\Omega_k|^{\frac{1}{q} - \frac{1}{p^-} + \frac{1}{n}} \|\nabla u\|_{p^-,\Omega_k} \|f\|_{q/(q-1),\Omega}.$$

Gathering these estimates we obtain the inequality

$$\|\nabla u\|_{p^-,\Omega_k}^{p^-} \leq \left(\frac{c}{a_0}\right)^{\frac{p^-}{p^+-1}} |\Omega_k|^{1 + \left(\frac{1}{q} + \frac{1}{n}\right)\frac{p^-}{p^+-1}} \|f\|_{q/(q-1),\Omega}^{\frac{p^-}{p^+-1}}.$$

The conclusion follows now from [Ch.2, Lemma 5.3][6]. $\qquad\qquad\qquad\square$

## 4. The energy relations

### 4.1. *The energy identity*

Set

$$\phi_k(x,s) = \begin{cases} 1 & \text{for } x_1 > s + \frac{1}{k}, \\ k(x_1 - s) & \text{for } x \in \left[s, s + \frac{1}{k}\right], \\ 0 & \text{for } x_1 < s, \qquad k \in \mathbb{N}, \end{cases}$$

686

and choose the function $u(x)\phi_k(x,s)$ for the test-function in the integral identity (9). The resulting identity has the form

$$\sum_{j=1}^{4} I_j(k,s) \equiv \sum_i \int_{\Omega \cap \{x_1 > s + 1/k\}} a_i |D_i u|^{p_i} \phi_k \, dx$$

$$+ k \int_{\Omega \cap \{s < x_1 < s + 1/k\}} a_1 \, u \, |D_1 u|^{p_1 - 2} D_1 u \, dx dt \tag{11}$$

$$+ \int_{\Omega \cap \{x_1 > s\}} c(x) |u|^{\sigma(x)} \phi_k \, dx - \int_{\Omega \cap \{x_1 > s\}} f \, u \, \phi_k \, dx = 0.$$

The inclusion $u \in \mathbf{V}(\Omega)$ yields the inclusions $a_i |D_i u|^{p_i} \phi_k$, $c(x) |u|^{\sigma(x)} \phi_k$, $f \, u \, \phi_k \in L^1(\Omega)$, which allows one to pass to the limit when $k \to \infty$ in $I_1$, $I_3$ and $I_4$:

$$\lim_{k \to \infty} I_1 = \sum_i \int_{\Omega \cap \{x_1 > s\}} a_i |D_i u|^{p_i} \, dx dt,$$

$$\lim_{k \to \infty} I_3 = \int_{\Omega \cap \{x_1 > s\}} c(x) |u|^{\sigma(x)} \, dx, \quad \lim_{k \to \infty} I_4 = -\int_{\Omega \cap \{x_1 > s\}} f \, u \, dx.$$

Now notice that by virtue of (11) $I_2$ is bounded uniformly with respect to $k$ provided that so are the integrals $I_1$, $I_2$, $I_3$. Writing $I_2$ in the form

$$I_2 = k \int_s^{s+1/k} dx_1 \int_{\omega(x_1)} a_i \, u \, |D_i u|^{p_i - 2} D_i u \, dx'$$

and applying the Lebesgue theorem we conclude that there exists

$$\lim_{k \to \infty} I_2(k,s) = \int_{\omega(s)} a_i \, u \, |D_i u|^{p_i - 2} D_i u \, dx'.$$

Let us assume that

$$f(x) = 0 \quad \text{a.e. in } \Omega \cap \{x : x_1 > l\} \text{ with some } l \in (0, L). \tag{12}$$

Then the energy relation takes on the form

$$\forall s > l \quad \sum_i \int_s^L dx_1 \int_{\omega(x_1)} a_i |D_i u|^{p_i} \, dx' + \int_s^L \int_{\omega(s)} c |u|^{\sigma(x)} \, dx'$$

$$= -\int_{\omega(s)} a_1 \, u \, |D_1 u|^{p_1 - 2} D_1 u \, dx'. \tag{13}$$

### 4.2. *The ordinary differential inequality.*

Let us introduce the function

$$J \equiv A_1 \int_{\omega(s)} |D_1 u|^{p_1(x)-1} |u| \, dx' \geq \left| \int_{\omega(s)} a_1 u |D_1 u|^{p_1-2} D_1 u \, dx' \right|.$$

By Hölder's inequality

$$J \leq A_1 \left\| |D_1 u|^{p_1(x)-1} \right\|_{\frac{p_1(\cdot)}{p_1(\cdot)-1},\omega(s)} \|u\|_{p_1(\cdot),\omega(s)},$$

According to (7)

$$\left\| |D_1 u|^{p_1(x)-1} \right\|_{\frac{p_1(\cdot)}{p_1(\cdot)-1},\omega(s)} \leq \max \left\{ A_{p_1(\cdot),\omega(s)}^{\frac{p_1^- -1}{p_1^-}}(D_1 u),\, A_{p_1(\cdot),\omega(s)}^{\frac{p_1^+ -1}{p_1^+}}(D_1 u) \right\} \tag{14}$$

$$= A_{p_1(\cdot),\omega(s)}^{\gamma_1(s)}(D_1 u)$$

with the exponent

$$\gamma_1(s) = \begin{cases} \dfrac{p_1^- - 1}{p_1^-} & \text{if } A_{p_1(\cdot),\omega(s)}(D_1 u) < 1, \\ \dfrac{p_1^+ - 1}{p_1^+} & \text{otherwise.} \end{cases}$$

The second factor can be estimated by virtue of the embedding theorem

$$\|u\|_{p_1(\cdot),\omega(s)} \leq C \max \left\{ \lambda^{1+\frac{n-1}{\beta}-\frac{n-1}{p_1^-}}(s),\, \lambda^{1+\frac{n-1}{\beta}-\frac{n-1}{p_1^+}}(s) \right\} \|\tilde{\nabla} u\|_{\beta,\omega(s)} \tag{15}$$

with $\beta = const > 0$ such that $\dfrac{1}{p_1(x)} > \dfrac{1}{\beta} - \dfrac{1}{n-1}$ in $\Omega$ if $\beta < n-1$. Let us claim that $\beta = \min\limits_{j\neq 1} p_j^-$. Then

$$\|\tilde{\nabla} u\|_{\beta,\omega(s)} \leq 2 \sum_{j\neq 1} \|1\|^{\frac{1}{\beta}}_{\frac{p_j(\cdot)}{p_j(\cdot)-\beta},\omega(s)} \||D_j u|^\beta\|^{\frac{1}{\beta}}_{\frac{p_j(\cdot)}{\beta},\omega(s)}$$

$$\leq 2 \sum_{j\neq 1} \max \left\{ \lambda(s)^{\frac{n-1}{\beta}-\frac{n-1}{p_j^-}},\, \lambda(s)^{\frac{n-1}{\beta}-\frac{n-1}{p_j^+}} \right\}$$

$$\times \max \left\{ A_{p_j(\cdot),\omega(s)}^{\frac{1}{p_j^-}}(D_j u),\, A_{p_j(\cdot),\omega(s)}^{\frac{1}{p_j^+}}(D_j u) \right\}.$$

Notice that $\forall j = 2,\ldots,n$

$$\max\left\{\lambda(s)^{\frac{n-1}{\beta}-\frac{n-1}{p_j^-}},\ \lambda(s)^{\frac{n-1}{\beta}-\frac{n-1}{p_j^+}}\right\}\leq\max\left\{\lambda(s)^{\frac{n-1}{\beta}-\frac{n-1}{p^-}},\ \lambda(s)^{\frac{n-1}{\beta}-\frac{n-1}{p^-}}\right\}.$$

Fix an arbitrary $s\in(l,L)$ and choose $k\in\{2,\ldots,n\}$ such that $A_{p_k(\cdot),\omega(s)}(D_k u)\geq A_{p_j(\cdot),\omega(s)}(D_j u)$ for all $j\geq2$. There are two possibilities: either $A_{p_k(\cdot),\omega(s)}(D_k u)<1$, or $A_{p_k(\cdot),\omega(s)}(D_k u)\geq1$. In the former case the inequality

$$\sum_{j\neq2}A_{p_j(\cdot),\omega(s)}^{\frac{1}{p_j^-}}(D_j u)\leq n A_{p_k(\cdot),\omega(s)}^{\frac{1}{p_k^+}}(D_k u)\leq n\left(\sum_{j\neq2}A_{p_j(\cdot),\omega(s)}(D_j u)\right)^{\frac{1}{q^+}}$$

holds with $q^+=\max_{j\neq1}p_j^+$, otherwise

$$\sum_{j\neq2}A_{p_k(\cdot),\omega(s)}^{\frac{1}{p^-}}(D_k u)\leq n A_{p_k(\cdot),\omega(s)}^{\frac{1}{p^-}}(D_k u)\leq n\left(\sum_{j\neq2}A_{p_j(\cdot),\omega(s)}(D_j u)\right)^{\frac{1}{q^-}}$$

with $q^-=\max_{j\neq1}p_j^-$. Thus, $J\leq K\,\rho(s)\,n\,A_{\mathbf{p}(\cdot),\omega(s)}^{\theta(s)}(\nabla u)$ with the exponent and the coefficient

$$\theta(s)=\gamma_1(s)+\tau_j(s),\qquad\tau_j(s)=\begin{cases}\dfrac{1}{q^+}&\text{if }\max_{j\geq2}A_{p_j(\cdot),\omega(s)}(D_j u)<1,\\[2mm]\dfrac{1}{q^-}&\text{otherwise,}\end{cases}$$

$$\rho(s)=\max\left\{\lambda(s)^{1+2\frac{n-1}{\beta}-2\frac{n-1}{p^-}},\ \lambda(s)^{1+2\frac{n-1}{\beta}-2\frac{n-1}{p^+}}\right\},$$

and with an absolute constant $K$ independent of $\lambda(s)$ and $u(x)$. Let us introduce the energy function

$$\Phi(s)\equiv\int_s^L dx_1\int_{\omega(s)}A_{\mathbf{p}(\cdot),\omega(s)}(\nabla u)\,dx'.$$

Equality (13) transforms then into the following inequality:

$$\begin{cases}\forall s>l\quad\Phi^{\frac{1}{\theta(s)}}(s)+\psi(s)\Phi'(s)\leq0,\quad s\in(l,L),\\[1mm]\Phi(L)=0,\quad\Phi'(s)\leq0,\quad\phi(s)=((K\,n\,\rho(s))/a_0)^{\frac{1}{\theta(s)}}.\end{cases}\tag{16}$$

**Lemma 4.1.** *Let the conditions of Theorem 3.1 be fulfilled, the exponents $p_i(x)$ satisfy the oscillation conditions*

$$\frac{1}{p_1(x)}>\frac{1}{\min_{j\neq1}p_j^-}-\frac{1}{n-1}\quad\text{if }\min_{j\neq1}p_j^-<n-1,\tag{17}$$

and the function $f(x)$ satisfies condition (12). Then for every weak bounded solution $u(x)$ the corresponding energy function $\Phi(s)$ is a solution of the ordinary differential inequality (16).

### 4.2.1. Analysis of the ordinary differential inequality

Let the energy function $\Phi(s)$ be uniformly bounded in the interval $(l, L)$ by a finite constant $M$. Then inequality (16) can be written in the form

$$\nu\,\Phi^\mu + \phi(s)\Phi'(s) \le 0, \quad \mu = \inf_{(l,L)} \frac{1}{\theta(s)}, \quad \nu = \inf_{(l,L)} M^{\frac{1}{\theta(s)} - \mu}. \tag{18}$$

Let us claim that $\mu < 1$. This is true if $\theta(s) > 1$, which is guaranteed by the condition $p_1^- > \max_{j \neq 1} p_j^+$.

**Lemma 4.2.** *Let the conditions of Lemma 4.1 be fulfilled, and the exponents* $p_i(x)$ *satisfy the conditions*

$$\begin{cases} 0 < \dfrac{1}{p_1^+} \le \dfrac{1}{p_1^-} < \dfrac{1}{p_2^+} & \text{if } n = 2, \\[2mm] \dfrac{1}{\min\limits_{j \neq 1} p_j^-} - \dfrac{1}{n-1} < \dfrac{1}{p_1^+} \le \dfrac{1}{p_1^-} < \displaystyle\sum_{j \neq 1} \dfrac{1}{\max_{j \neq 1} p_j^+} & \text{if } n \ge 3. \end{cases}$$

*Then one may indicate* $\epsilon_*$ *such that every solution* $\Psi(s)$ *of (1) satisfying the inequality* $\Psi(s) \le \epsilon_*$ *vanishes on an interval* $(s_0, L) \subset (l, L)$.

**Proof.** Let us integrate inequality (18) in the limits $(l, s)$,

$$\Phi^{1-\mu}(s) \le \epsilon_*^{1-\mu} - \nu(1-\mu) \int_l^s \frac{dt}{M^{1/\theta(t)}\psi(t)}, \tag{19}$$

and then take $\epsilon_* < \left( \nu(1-\mu) \displaystyle\int_l^L \dfrac{dt}{M^{1/\theta(t)}\psi(t)} \right)^{1/(1-\mu)}$. The nonnegative and monotone decreasing function $\Phi(s)$ vanishes at a point $s_0 \in (l, L)$, and $\Phi(s) \equiv 0$ for all $s \in [s_0, L]$. $\qquad\square$

**Theorem 4.1.** *Let under the conditions of Theorem 3.1 the data of problem (1) satisfy the conditions of Lemma 4.2. Then there exists* $\epsilon_*$ *such that every bounded weak solution of problem (1) satisfying* $\Phi(l) \le \epsilon_*$ *is localized in the direction* $x_1$: $u(x) = 0$ *a.e. in* $\Omega \cap \{x_1 > s_0\}$ *with some* $s_0 \in (l, L)$.

**Proof.** Since the energy function $\Phi(s)$ satisfies the ordinary differential inequality (18), it follows from Lemma 4.2 that $\Phi(s) = 0$ for all $s > s_0$, provided that $\Phi(s)$ is sufficiently small $\qquad\square$

**Corollary 4.1.** *The value of $\epsilon_*$ is arbitrary if the integral $\int_l^L \dfrac{dt}{\psi(t)}$ is divergent. If this is the case, every solution with finite total energy is localized in the direction $x_1$. The conditions of divergence read as conditions on the rate of vanishing of the function $\psi(t)$ as $t \to L$ i.e. on the shape of the problem domain near the point $t = L$.*

**Theorem 4.2.** *If in the conditions of Theorem 3.1 $L = \infty$, then there exists $\epsilon^* > 0$ such that problem (1) has a solution localized in the direction $x_1$ for every right–hand side $f$ satisfying the conditions (1) $f(x) \equiv 0$ for $x_1 > l$, (2) $\|f\| < \epsilon_*$.*

The proof follows the proof of Theorem 6.1 [2]. We consider the sequence $\{u_k\}$ of solution of problems (1) posed on bounded domains $\Omega_k = \Omega \cap \{x_1 < k\}$, $k \in \mathbb{N}$. For every finite $k$ this problem has a solution and for all $k$, from some $k_0$ on, the solution $u_k$ is localized in the direction $x_1$: $\operatorname{supp} u_k \subset \Omega_{k_0}$. Passing to the limit as $k \to \infty$ in the integral identity (9) for $u_k$, we obtain a localized weak solution of the problem in the unbounded domain. Notice that the limit value of the total energy $\epsilon_*$ can be arbitrary if $\int_l^\infty \dfrac{dt}{\psi(t)} = \infty$, which is the condition of the "asymptotic size" of the domain $\Omega$ at infinity.

### References

1. S. ANTONTSEV AND S. SHMAREV, *Elliptic equations and systems with nonstandard growth conditions: existence, uniqueness and localization properties of solutions* , Nonlinear Analysis Serie A: Theory and Methods (to appear).
2. ———, *On localization of solutions of elliptic equations with nonhomogeneous anisotropic degeneracy*, Siberian Math. Journal, 46(5) (2005), pp. 963–984.
3. S. N. ANTONTSEV, J. I. DÍAZ, AND S. SHMAREV, *Energy Methods for Free Boundary Problems:Applications to Non-linear PDEs and Fluid Mechanics*, Bikhäuser, Boston, 2002. Progress in Nonlinear Differential Equations and Their Applications, Vol. 48.
4. P. HARJULEHTO AND P. HÄSTÖ, *An overview of variable exponent Lebesgue and Sobolev spaces*, in Future trends in geometric function theory, vol. 92 of Rep. Univ. Jyväskylä Dep. Math. Stat., Univ. Jyväskylä, Jyväskylä, 2003, pp. 85–93.
5. O. KOVÁČIK AND J. RÁKOSNÍK, *On spaces $L^{p(x)}$ and $W^{k,p(x)}$*, Czechoslovak Math. J., 41 (1991), pp. 592–618.
6. O. A. LADYŽENSKAJA AND N. N. URAL'TSEVA, *Linear and quasilinear equations of elliptic equations*, Academic Press, New York, 1968. Translated from the Russian.
7. J. MUSIELAK, *Orlicz spaces and modular spaces*, vol. 1034 of Lecture Notes in Mathematics, Springer-Verlag, Berlin, 1983.
8. P. PUCCI AND J. SERRIN, *The strong maximum principle revisited*, J. Differential Equations, 196(1) (2004), pp. 1–66.
9. J. L. VÁZQUEZ *A strong maximum principle for some quasilinear elliptic equations* Appl. Math. Optim., 12, (1984) p.p. 191–202
10. P. PUCCI AND J. SERRIN, *A note on the strong maximum principle for elliptic differential inequalities*, J. Math. Pures Appl. (9), 79(1) (2000), pp. 57–71.
11. J. I. DÍAZ, *Nonlinear partial differential equations and free boundaries. Vol. I, Elliptic Equations* Pitman Research Notes in Mathematics 106, Boston, MA, (1985)

# BIFURCATION DIRECTION AND EXCHANGE OF STABILITY FOR AN ELLIPTIC UNILATERAL BVP*

JAN EISNER and MILAN KUČERA

*Mathematical Institute of the Academy of Sciences of the Czech Republic,*
*Žitná 25, 115 67 Prague 1, Czech Republic*
*E-mail: eisner@math.cas.cz, kucera@math.cas.cz*

LUTZ RECKE

*Institute of Mathematics of the Humboldt University of Berlin,*
*Unter den Linden 6, 10099 Berlin, Germany*
*E-mail: recke@mathematik.hu-berlin.de*

The direction of bifurcation of nontrivial solutions to the elliptic boundary value problem involving unilateral nonlocal boundary conditions is shown in a neighbourhood of bifurcation points of a certain type. Moreover, the stability and instability of bifurcating solutions as well as of the trivial solution is described in the sense of minima of the potential. In particular, an exchange of stability is observed.

**Key words:** Semilinear elliptic PDE, unilateral nonlocal boundary conditions, Crandall-Rabinowitz-type bifurcation, variational inequality on a nonconvex set, Lagrange multipliers

**Mathematics Subject Classification:** 35J85, 47J15, 49J40

## 1. Introduction

Let $\Omega$ be a bounded domain in $\mathbb{R}^N$ with a Lipschitzian boundary $\partial\Omega$, $1 < N < 5$, let $\Gamma_D$ and $\Gamma_j$, $j = 1, \ldots, n$, be pairwise disjoint open (in $\partial\Omega$) subsets of this boundary, meas $\Gamma_D > 0$. We consider a Crandall-Rabinowitz type bifurcation (a bifurcation from a trivial solution at a simple eigenvalue with exchange of stability, see [4]) for a

*The research has been supported by the grant IAA100190506 of the Grant Agency of the Academy of Sciences of the Czech Republic, by the Institutional Research Plan No. AVOZ10190503 of the Academy of Sciences of the Czech Republic, and by the DFG Research Center MATHEON "Mathematics for key technologies".

semilinear elliptic PDE with unilateral nonlocal boundary conditions:

$$\Delta u + pu + au^2 = 0 \quad \text{in } \Omega, \tag{1.1}$$

$$u = 0 \text{ on } \Gamma_D, \quad \frac{\partial u}{\partial \nu} = 0 \text{ on } \partial\Omega \setminus \left( \Gamma_D \cup \bigcup_{j=1}^{n} \Gamma_j \right), \tag{1.2}$$

$$\int_{\Gamma_j} \psi(u) \, d\Gamma \leq 0, \frac{\partial u}{\partial \nu} = c_j \psi'(u) \leq 0 \text{ on } \Gamma_j \text{ with some constant } c_j, \\ c_j \int_{\Gamma_j} \psi(u) \, d\Gamma = 0, \; j = 1, \ldots, n, \tag{1.3}$$

where $p \in \mathbb{R}$ is the bifurcation parameter, $a \in \mathbb{R}$ is a number, $\psi : \mathbb{R} \to \mathbb{R}$ is a $C^3$-smooth function and $\frac{\partial u}{\partial \nu}$ is the outer normal derivative. The second condition in (1.3) implies, in particular, that for any $j$, $\psi'(u)$ does not change sign on $\Gamma_j$ (i.e. $\psi'(u(x))$ is for all $x \in \Gamma_j$ either non-negative or non-positive). The whole condition (1.3) means that a certain $\psi$-average over any $\Gamma_j$ cannot exceed the zero value, the flux through any $x \in \Gamma_j$ is proportional to $\psi'(u(x))$ and can go only outwards from the domain $\Omega$. If the $\psi$-average over $\Gamma_j$ is strictly negative then there is no flux through $\Gamma_j$. Also the case $a = 0$, i.e. the linearized equation

$$\Delta u + pu = 0 \quad \text{in } \Omega \tag{1.4}$$

with (1.2), (1.3) and also with "linearized" (homogenized) boundary conditions

$$\int_{\Gamma_j} u \, d\Gamma \leq 0, \frac{\partial u}{\partial \nu} = c_j^0 \text{ on } \Gamma_j \text{ with some } c_j^0 \leq 0, \; c_j^0 \int_{\Gamma_j} u \, d\Gamma = 0, j = 1, \ldots, n, \tag{1.5}$$

will be of interest for us. The condition (1.5) means that the average over any $\Gamma_j$ cannot exceed the zero value, the flux through any $\Gamma_j$ is constant and can go only outwards from the domain $\Omega$. If the average over $\Gamma_j$ is strictly negative then there is no flux through $\Gamma_j$.

By solutions of all boundary value problems mentioned we mean weak solutions, that means solutions of variational inequalities introduced below. As usual in the case of variational inequalities, the linearized problem (1.4), (1.5) is (due to the inequalities in the boundary conditions) nonlinear again but is positively homogeneous. However, let us emphasize that (1.4) with the original boundary conditions (1.2), (1.3) is not even positively homogeneous if $\psi'$ is not constant. In [6] we have proved the existence of a smooth branch of solutions to (1.1)–(1.3) bifurcating from zero at a simple eigenvalue $p_0$ of (1.4), (1.5). Of course, the $u$-component of this branch emanates in the direction of the corresponding eigenfunction $u_0$. Now, our goal is to show that only $\psi''(0)$ together with $a$ and $u_0$ decide about the $p$-direction and stability of the bifurcation branch. Let us remark that also in the case $a = 0$ there is generically a direction of bifurcation.

We will show that there are at least three essential differences from the case of classical boundary conditions (see Fig. 2 of Section 3): First, only one half-branch of nontrivial solutions (not two, as for the classical boundary conditions) bifurcates from the branch of trivial solutions. Second, the bifurcating nontrivial solutions can be stable even if the trivial solution is unstable on both sides from the bifurcation point. And third, the bifurcating nontrivial solutions can be unstable even if the trivial solution loses stability at the bifurcation point, i.e. is stable on one side of the bifurcation point. Hence, there is not always exchange of stability if the bifurcation parameter crosses the first bifurcation point (even if there is a loss of stability of the trivial solution). Let us remark that the nonlinear term is quadratic, therefore there is a transcritical bifurcation as well as exchange of stability for the problem with classical boundary conditions (i.e. with $\Gamma_j = \emptyset$ for all $j = 1, \ldots, n$).

An analogous result for the particular case of constant $\psi'$ was described already in [5] but to consider the direction of the bifurcation branch it was necessary to have a nontrivial nonlinear perturbation in the equation because (1.3) becomes positively homogeneous in this case. Let us note that in [5], an abstract theory covering the case $\psi(\xi) = \xi$ was studied. The generalization of these abstract results including our present example is the subject of a forthcoming paper [7] based on an equivalence of the variational inequality with the Lagrange equation. The results can be understood as a certain modification of the well-known results for equations (see e.g. [11], Chapter 8.7) to variational inequalities. The method of Lagrange multipliers was used also in [6] to prove the existence of smooth bifurcation families of nontrivial solutions for a class of variational inequalities covering our present BVP.

Let us note that an abstract criterion of stability for variational inequalities without any relation to bifurcation was given in [10]. A loss of stability at the turning points for variational inequalities of a certain type was numerically proved in [3]. Stability and continuation for solutions to obstacle problems were studied in [8].

The main results are formulated in Theorems 2.2 and 2.3. At the end of the text (Section 3) we show possible bifurcation diagrams in the case of one obstacle (obtained with help of numerical computations).

## 2. Weak Formulation, Main Results

We will assume that the function $\psi$ satisfies

$$\psi(0) = 0, \quad \psi'(0) > 0 \tag{2.1}$$

and the growth conditions

$$|\psi(\xi)| \le c(1 + |\xi|^q), \quad \psi'(\xi) \le c(1 + |\xi|^{q-1}),$$
$$|\psi''(\xi)| \le c(1 + |\xi|^{q-2}), \quad |\psi'''(\xi)| \le c(1 + |\xi|^{q-3}) \quad \text{for all } \xi \in \mathbb{R} \tag{2.2}$$

with some $c > 0$ and $q \ge 3$ for $N = 2$, $q = 2\frac{N-1}{N-2}$ for $N = 3$ or $N = 4$. Let us remark that the first assumption in (2.1) ensures that $u = 0$ is a solution for any $p \in \mathbb{R}$ and the second implies that the condition (1.5) is a homogenization of (1.3).

In order to introduce a weak formulation, we consider the Hilbert space

$$H := \{u \in W^{1,2}(\Omega) : u = 0 \text{ on } \Gamma_D \text{ in the sense of traces}\}$$

with the inner product $\langle u, v \rangle := \int_\Omega \nabla u \cdot \nabla v \, dx$ for any $u, v \in H$. The corresponding norm $\| \cdot \|$ is equivalent to the usual Sobolev norm on $H$ under our assumptions.

Let us denote $\mathcal{A} := \{1, \ldots, n\}$ and define for any $u \in H$ a set

$$\mathcal{A}(u) := \left\{ \alpha \in \mathcal{A} : \int_{\Gamma_\alpha} \psi(u) \, d\Gamma = 0 \right\}.$$

Furthermore, let us introduce a closed set $K$ and a closed convex cone $K_0$ by

$$K := \left\{ u \in H : \int_{\Gamma_\alpha} \psi(u) \, d\Gamma \leq 0, \ \alpha \in \mathcal{A} \right\},$$
$$K_0 := \left\{ u \in H : \int_{\Gamma_\alpha} u \, d\Gamma \leq 0, \ \alpha \in \mathcal{A} \right\}$$

and functionals $g_\alpha : H \to \mathbb{R}$ by

$$g_\alpha(u) = \int_{\Gamma_\alpha} \psi(u) \, d\Gamma.$$

Then $\mathcal{A}(u) = \{\alpha \in \mathcal{A} : g_\alpha(u) = 0\}$ and $v_\alpha := \nabla g_\alpha(0)$, $\alpha \in \mathcal{A}$, satisfy $\langle v_\alpha, u \rangle = \psi'(0) \int_{\Gamma_\alpha} u \, d\Gamma$ for any $u \in H$. Under the assumption (2.2), standard considerations about Nemyckii operators and the continuity of the embedding of $W^{1,2}(\Omega)$ into $L^q(\partial\Omega)$ (see e.g. Theorem 4.2 in [9]) imply that the functionals $g_\alpha$ are $C^3$-smooth. Moreover, under the assumption (2.1) the elements $\nabla g_\alpha(u)$ are linearly independent for all $u \in H$. Realizing this and the definition of the local contingent cone to $K$ at $u$

$$T(K, u) := \{z \in H : \text{ there exist } w_n \in K, t_n > 0, w_n \to u, t_n(w_n - u) \to z\}$$

(see e.g. [1]) we have $K_0 = T(K, 0)$ and

$$T(K, u) = \left\{ v \in H : \int_{\Gamma_\alpha} \psi'(u) v \, d\Gamma \leq 0, \ \alpha \in \mathcal{A}(u) \right\} \text{ for all } u \in K.$$

We define a weak solution to (1.1)–(1.3) as a solution to the problem

$$p \in \mathbb{R}, u \in K : \int_\Omega \nabla u \cdot \nabla \varphi - (pu + au^2)\varphi \, dx \geq 0 \text{ for all } \varphi \in T(K, u) \quad (2.3)$$

and a weak solution to (1.4), (1.2), (1.5) as a solution to

$$p \in \mathbb{R}, u \in K_0 : \int_\Omega \nabla u \cdot \nabla(\varphi - u) - pu(\varphi - u) \, dx \geq 0 \text{ for all } \varphi \in K_0. \quad (2.4)$$

Finally, we define a functional $\Phi : \mathbb{R} \times H \to \mathbb{R}$ by

$$\Phi(p, u) := \int_\Omega \left( \frac{|\nabla u|^2 - pu^2}{2} - \frac{au^3}{3} \right) dx.$$

**Remark 2.1.** Of course, $K_0$, (2.3) and (2.4) is of the form

$$K_0 = \{u \in H : \langle v_\alpha, u \rangle \leq 0 \text{ for all } \alpha \in \mathcal{A}(0)\}, \quad (2.5)$$

$$p \in \mathbb{R}, \ u \in K : \langle F(p, u), \psi \rangle \leq 0 \text{ for all } \psi \in T(K, u), \quad (2.6)$$

$$p \in \mathbb{R}, \ u \in K_0 : \left\langle \frac{\partial F}{\partial u}(p, 0) u, \varphi - u \right\rangle \leq 0 \text{ for all } \varphi \in K_0, \quad (2.7)$$

respectively, with $F : \mathbb{R} \times H \to H$ defined by

$$\langle F(p, u), \varphi \rangle = \int_\Omega -\nabla u \cdot \nabla \varphi + (pu + au^2)\varphi \, dx \quad \text{for all } p \in \mathbb{R}, u, \varphi \in H.$$

For $N$ under consideration, the embedding theorems imply that $F$ is well defined and $C^\infty$-smooth. Hence, our problem fits into the abstract framework studied in [7]. Moreover, it is easy to see that the functional $\Phi$ satisfies the condition

$$\frac{\partial \Phi}{\partial u}(p, u)v = \langle -F(p, u), v \rangle \text{ for all } p \in \mathbb{R} \text{ and } u, v \in H, \tag{2.8}$$

i.e., $F(p, \cdot)$ is a potential operator for any $p \in \mathbb{R}$.

**Remark 2.2.** If $u \in H$ is such that $\Delta u \in L^2(\Omega)$, then the normal derivative $\frac{\partial u}{\partial \nu}$ can be defined as a linear bounded functional on the space $H$ by $\left[\frac{\partial u}{\partial \nu}, \varphi\right] = \int_\Omega (\Delta u \cdot \varphi + \nabla u \cdot \nabla \varphi) \, dx$ for all $\varphi \in H$, where $[\cdot, \cdot]$ is the dual pairing. The non-positivity of $\frac{\partial u}{\partial \nu}$ on $\Gamma_\alpha$ will be understood in the sense of such functional, i.e. $\left[\frac{\partial u}{\partial \nu}, \varphi\right] \le 0$ for all $\varphi \in H$, $\varphi \ge 0$ on $\Gamma_\alpha$, $\varphi = 0$ on $\partial\Omega \setminus \Gamma_\alpha$. The condition $\frac{\partial u}{\partial \nu} = c_\alpha \psi'(u)$ on $\Gamma_\alpha$ will mean that the functional $\frac{\partial u}{\partial \nu}$ can be represented on $\Gamma_\alpha$ by the function $c_\alpha \psi'(u) \in L^{q*}(\partial\Omega)$ (with $\frac{1}{q^*} + \frac{1}{q} = 1$, $q$ from (2.2)), i.e.

$$\left[\frac{\partial u}{\partial \nu}, \varphi\right] = c_\alpha \int_{\Gamma_\alpha} \psi'(u)\varphi \, d\Gamma \text{ for all } \varphi \in H, \varphi = 0 \text{ on } \partial\Omega \setminus \Gamma_\alpha. \tag{2.9}$$

**Observation 2.1.** A couple $(p, u) \in \mathbb{R} \times H$ satisfies (2.3) (i.e. it is a weak solution to (1.1)– (1.3)) if and only if $u$ is smooth in $\Omega$, $\Delta u \in L^2(\Omega)$, (1.1) is satisfied in the classical sense and (1.2), (1.3) are satisfied where $u$ on $\partial\Omega$ is understood in the sense of traces and $\frac{\partial u}{\partial \nu}$ is understood in the sense of the functional from Remark 2.2.

**Proof** is the same as that of Observation 5.3 in [6]. ∎

Let us fix a subset $\mathcal{A}_0 := \{\alpha_1, \ldots, \alpha_m\}$ of $\mathcal{A}$. Moreover, let $H_0$ be a subspace of $H$ given by $H_0 := \left\{ u \in H : \int_{\Gamma_\alpha} u \, d\Gamma = 0, \alpha \in \mathcal{A}_0 \right\}$. In the main assertions formulated below we will need the boundary value problem (1.4), (1.2),

$$\int_{\Gamma_\alpha} u \, d\Gamma = 0 \text{ for } \alpha \in \mathcal{A}_0, \tag{2.10}$$

$$\frac{\partial u}{\partial \nu} = c_\alpha \text{ on } \Gamma_\alpha \text{ with some } c_\alpha \in \mathbb{R} \text{ for any } \alpha \in \mathcal{A}_0, \tag{2.11}$$

$$\frac{\partial u}{\partial \nu} = 0 \text{ on } \Gamma_\alpha \text{ for } \alpha \in \mathcal{A} \setminus \mathcal{A}_0, \tag{2.12}$$

having a weak formulation

$$p \in \mathbb{R}, u \in H_0 : \int_\Omega \nabla u \cdot \nabla \varphi - pu\varphi \, dx = 0 \text{ for all } \varphi \in H_0. \tag{2.13}$$

Let us remark that the problem (2.13) can be written in the abstract framework of the paper [7] as $p \in \mathbb{R}, u \in H_0, P\frac{\partial F}{\partial u}(p, 0)u = 0$, where $P$ is the orthogonal projection of $H$ onto $H_0$.

The following theorem gives us the existence of a smooth branch bifurcating from the trivial solutions. The description of the direction and stability of this branch (which is our main goal) will be given in Theorems 2.2 and 2.3 below.

**Theorem 2.1.** *Let* $(p_0, u_0)$ *be a weak solution to (1.4), (1.2), (1.5) satisfying (2.10),*

$$\int_{\Gamma_\alpha} u_0 \, d\Gamma < 0 \text{ for } \alpha \in \mathcal{A} \setminus \mathcal{A}_0, \tag{2.14}$$

$$\frac{\partial u_0}{\partial \nu} = c_\alpha^0 \text{ on } \Gamma_\alpha \text{ with some } c_\alpha^0 < 0 \text{ for any } \alpha \in \mathcal{A}_0. \tag{2.15}$$

*Let us assume the following simplicity conditions:*

*If* $(p_0, v_0)$ *is a weak solution to (1.4), (1.2), (2.10)–(2.12) then* $v_0 = c u_0$, $c \in \mathbb{R}$.
*If* $(p_0, v_0)$ *is a weak solution to (1.4), (1.2), (1.5) then* $v_0 = c u_0$, $c \geq 0$.

*Then there exist* $\varepsilon > 0$, $s_0 > 0$ *and* $C^1$-*maps* $\hat{p} : [0, s_0) \to \mathbb{R}$ *and* $\hat{v} : [0, s_0) \to H_0$ *with* $\hat{p}(0) = p_0$, $\hat{v}(0) = 0$ *such that the following holds. The couple* $(p, u)$ *with* $|p - p_0| < \varepsilon$, $\|u\| < \varepsilon$ *and* $\|u\| \neq 0$ *is a weak solution to (1.1)–(1.3) if and only if* $p = \hat{p}(s)$, $u = \hat{u}(s) := s(u_0 + \hat{v}(s))$ *for a certain* $s \in (0, s_0)$. *In this case, moreover,* $u = \hat{u}(s)$ *satisfies (2.12) and*

$$\begin{array}{l} \int_{\Gamma_\alpha} \psi(u) \, d\Gamma = 0 \text{ for } \alpha \in \mathcal{A}_0, \\ \frac{\partial u}{\partial \nu} = c_\alpha \psi'(u) < 0 \text{ on } \Gamma_\alpha \text{ with some } c_\alpha \in \mathbb{R} \text{ for any } \alpha \in \mathcal{A}_0, \\ \int_{\Gamma_\alpha} \psi(u) \, d\Gamma < 0 \text{ for } \alpha \in \mathcal{A} \setminus \mathcal{A}_0. \end{array}$$

**Proof** *is the same as that of Theorem 5.4 in* [6]. ∎

**Remark 2.3.** *If* $N = 3$ *and a suitable growth estimate for the fourth derivative of* $\psi$ *is added to (2.2) then the functionals* $g_\alpha$ *are* $C^4$-*smooth and the mappings* $\hat{p}$, $\hat{v}$, $\hat{u}$ *in Theorem 2.1 are in fact* $C^2$-*smooth. If* $N = 2$ *and* $k > 2$ *is an arbitrary positive integer then under suitable growth estimates for derivatives of* $\psi$ *up to the order* $k + 1$, *the functionals* $g_\alpha$ *are* $C^{k+1}$-*smooth and the mappings* $\hat{p}$, $\hat{v}$, $\hat{u}$ *in Theorem 2.1 are in fact* $C^{k-1}$-*smooth.*

**Theorem 2.2.** *Let* $(p_0, u_0)$ *satisfy the assumptions of Theorem 2.1, let* $(\hat{p}(s), \hat{u}(s))$, $s \in [0, s_0)$, *be the bifurcation branch from Theorem 2.1. If*

$$a \int_\Omega u_0^3 \, dx + 3\psi''(0) \sum_{\alpha \in \mathcal{A}_0} c_\alpha^0 \int_{\Gamma_\alpha} u_0^2 \, d\Gamma < 0 \tag{2.16}$$

*then* $\hat{p}(s) > p_0$ *for all* $s \in (0, s_0)$ *and if, moreover,*

$$p_0 \text{ is the smallest eigenvalue of (1.4), (1.2), (2.10)–(2.12)} \tag{2.17}$$

*then* $\Phi(\hat{p}(s), \cdot)$ *attains a strong local minimum on* $K$ *in* $\hat{u}(s)$ *for all* $s \in (0, s_0)$. *If*

$$a \int_\Omega u_0^3 \, dx + 3\psi''(0) \sum_{\alpha \in \mathcal{A}_0} c_\alpha^0 \int_{\Gamma_\alpha} u_0^2 \, d\Gamma > 0 \tag{2.18}$$

*then* $\hat{p}(s) < p_0$ *and* $\Phi(\hat{p}(s), \cdot)$ *has no local minimum on* $K$ *in* $\hat{u}(s)$ *for all* $s \in (0, s_0)$.

**Proof.** The direction and stability of bifurcation branches was described in [5], Theorem 9 for a class of variational inequalities in the case when $K$ is a cone with its vertex at the origin in a Hilbert space. The problem (2.3) is included if $\psi(\xi) = \xi$, i.e. $K$ is a cone, and this particular case is described by [5], Theorem 14. These results can be generalized to an abstract class of variational inequalities where $K$ need not be a cone, i.e. inequalitites containing (2.3) with a general $\psi$. See [7], Theorem 5.5, which is proved by using local equivalence of a variational inequality to an equation with Lagrange multipliers. Theorem 2.2 follows from [7], Theorem 5.5 in the same way as [5], Theorem 14 from [5], Theorem 9. ∎

**Theorem 2.3.** *Let $p_0$ be the smallest eigenvalue of (1.4), (1.2), (1.5). Then $\Phi(p,\cdot)$ attains a strict local minimum on $K$ at $u = 0$ for any $p \in (0, p_0)$ and $\Phi(p,\cdot)$ has no local minimum on $K$ at $u = 0$ for any $p > p_0$.*

**Proof.** The assertion follows from [7], Theorem 5.6 in the same way as that of [5], Theorem 15 follows from [5], Theorem 10. ∎

**Remark 2.4.** Contrary to the fact that only one $p_0 \in \mathbb{R}$ can be the smallest eigenvalue of (2.4), the condition (2.17) can be fulfilled for more different values of $p_0$ corresponding to different subsets $\mathcal{A}_0$ of $\mathcal{A}$, i.e. to different subspaces $H_0$ of $H$ (maximally $2^n$). Hence, Theorem 2.2 can enable us to determine the stability of bifurcating branches in a neighbourhood of several parameters $p_0$.

Nevertheless, the complete exchange of stability (including the stability of the trivial solution) is ensured only for the unique positive $p_0$.

## 3. Example

Let $\Omega \subset \mathbb{R}^2$ be a rectangle $\{x \in (0,1),\ y \in (0,\ell)\}$, $\ell < 1$, $\Gamma_D := \{(x,0);\ x \in (0,1)\} \cup \{(0,y);\ y \in (0,\ell)\}$, $\Gamma_1 := \{(x,\ell);\ x \in (0,1)\}$, $n = 1$, $\mathcal{A} = \{1\}$. If $(p,u)$ satisfies (2.4), i.e. it is a weak solution to (1.4) with

$$u = 0 \text{ on } \Gamma_D, \quad \frac{\partial u}{\partial \nu} = 0 \text{ on } \partial\Omega \setminus (\Gamma_D \cup \Gamma_1), \tag{3.1}$$

$$\int_{\Gamma_1} u\,d\Gamma \le 0, \quad \frac{\partial u}{\partial \nu} \le 0, \quad \int_{\Gamma_1} u\,d\Gamma \cdot \frac{\partial u}{\partial \nu} = 0 \text{ on } \Gamma_1, \tag{3.2}$$

$$\frac{\partial u}{\partial \nu} = c_1^0 \text{ on } \Gamma_1 \text{ with some } c_1^0, \tag{3.3}$$

then one of the following conditions is fulfilled:

$$\int_{\Gamma_1} u\,d\Gamma < 0, \tag{3.4}$$

$$\int_{\Gamma_1} u\,d\Gamma = 0. \tag{3.5}$$

If (3.4) holds then we have

$$\frac{\partial u}{\partial \nu} = 0 \text{ on } \Gamma_1, \tag{3.6}$$

that means $p$ is an eigenvalue and $u$ is the corresponding eigenfunction with the proper sign of the classical mixed Dirichlet–Neumann boundary value problem (1.4), (3.1), (3.6).

The linearized problem in [5], Example, coincides with the present linearized problem (2.4), only our variational inequality (2.3) is more complicated because of the generality of $\psi$. In [5] we have proved the following facts about (2.4) (i.e. about (1.4) with (3.1)–(3.3)).

There exist sequences of couples $(p_{m,n}^N, u_{m,n}^N)$ and $(p_m^P, u_m^P)$, $m, n = 1, 2, \dots$, satisfying (1.4) with (3.1)–(3.3), (3.4) (hence also (3.6)) and (1.4) with (3.1)–(3.3), (3.5), respectively, and $0 < p_{1,1}^N < p_1^P < p_{2,1}^N \leq \cdots$. Moreover,

$$p_{m,n}^N = \left( \left( \tfrac{2m-1}{2} \right)^2 + \left( \tfrac{2n-1}{2\ell} \right)^2 \right) \pi^2,$$
$$u_{m,n}^N(x,y) = (-1)^m \sin \tfrac{(2m-1)\pi}{2} x \cdot \sin \tfrac{(2n-1)\pi}{2\ell} y, \quad m, n = 1, 2, \dots,$$

and $u_{m,n}^N$ satisfy (3.4) for all $m, n$ (i.e. they do not satisfy (3.5) for any $m, n$). In particular,

$$\mathcal{A}(u_{m,n}^N) = \emptyset \text{ for all } m, n \text{ (hence, } \mathcal{A}_0 = \emptyset \text{ in such cases).} \tag{3.7}$$

On the other hand, $u_m^P$ satisfy (3.5) and $\mathcal{A}(u_m^P) = \{1\}$ for all $m$. Furthermore,

$$\int_\Omega (u_{1,1}^N)^3 \, dx < 0, \quad \int_\Omega (u_1^P)^3 \, dx > 0 \tag{3.8}$$

(the second integral was computed numerically). Finally, the assumptions of Theorems 2.1, 2.2 and 2.3 are verified for the couples $(p_{1,1}^N, u_{1,1}^N)$ and $(p_1^P, u_1^P)$ for the particular case $\psi(\xi) = \xi$ (i.e. for the simplified conditions (2.16) and (2.18)).

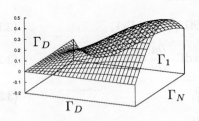 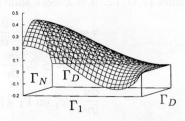

Fig. 1.   Two different viewpoints on the function $u_1^P$ corresponding to the parameter $p_1^P = 16.01$ for $\ell = 0.8$.

In the present paper the further discussion is more complex because of the presence of $\psi''(0)$. Because of (3.7) the terms in (2.16) and (2.18) with $\psi''(0)$ are trivially zero and due to (3.8), Theorem 2.2 gives that if $a > 0$ then the first bifurcating branch (emanating at $p_{1,1}^N$) goes to the right and is stable and if $a < 0$

then it goes to the left and is unstable. If $a = 0$ then we cannot decide about the direction of the first branch.

Due to (3.8), Theorem 2.2 gives that if $a > 0$ and $\psi''(0) < 0$ then the second bifurcating branch (starting at $p_0 = p_1^P > p_{1,1}^N$) goes to the left and is unstable and if $a < 0$ and $\psi''(0) > 0$ then it goes to the right and is stable.

Without loss of generality we can renorm the eigenfunction $u_0 = u_1^P$ so that the corresponding $c_1^0 = \frac{\partial u_0}{\partial \nu}\big|_{\Gamma_1} = -1$. Then the numerical computations e.g. for $\ell = 0.8$ give $\int_\Omega (u_1^P)^3\,dx = 0.0086287704$ and $\int_{\Gamma_1}(u_1^P)^2\,d\Gamma = 0.0172444952$. The left hand side in (2.16) and (2.18) is approximately equal to $3(0.167a - \psi''(0))$ and Theorem 2.2 implies that the second bifurcation branch starting at $p_0 = p_1^P$ goes to the right and is stable or goes to the left and is unstable if and only if $\psi''(0) > 0.167a$ or $\psi''(0) < 0.167a$, respectively. In particular, in the case $a > 0$, the second branch goes to the left and is unstable even in cases when $\psi$ is slightly convex at zero.

We get the following bifurcation diagram in a neigborhood of the two bifurcation points discussed above.

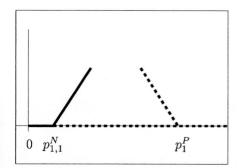 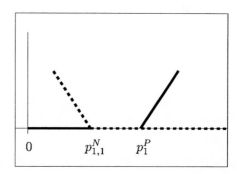

Fig. 2. Possible bifurcation diagrams. Left panel $a > 0$ and $\psi''(0) < 0$, right panel $a < 0$ and $\psi''(0) > 0$.

# References

1. J. P. Aubin and A. Celina, Differential inclusions. A Series of Comprehensive Studies in Mathematics 264 Springer-Verlag. Berlin, Heidelberg, New York, Tokyo, 1984.
2. S.-N. Chow and J. K. Hale, Methods of Bifurcation Theory. Springer-Verlag, New York, 1982.
3. F. Conrad, F. Issard-Roch, Loss of stability at turning points in nonlinear variational inequlities. *Nonlinear Anal., Theory Methods Appl.* **14** No. 4 (1990) 329–356.
4. M. G. Crandall, P. H. Rabinowitz, Bifurcation from simple eigenvalues. *J. Funct. Anal.* **8** (1971), 321–340.
5. J. Eisner, M. Kučera, L. Recke, Direction and Stability of Bifurcation Branches for Variational Inequalities. *J. Math. Anal. Appl.* **301**, No. 2 (2005), 276–294.
6. J. Eisner, M. Kučera, L. Recke, Smooth bifurcation for variational inequalities based on Lagrange multipliers. *Diff. Int. Equ.* **19**, No. 9 (2006), 981–1000.

7. J. Eisner, M. Kučera, L. Recke, Bifurcation direction and exchange of stability for variational inequalities based on Lagrange multipliers. *Nonlinear Anal.*, to appear.

8. E. Miersemann, H. Mittelmann, Stability and continuation of solutions to obstacle problems, *J. Comput. Appl. Math.* **35** (1991) 5–31.

9. J. Nečas, Les Méthodes Directes en Théorie des Equations Elliptiques, Academia, Praha (1967).

10. P. Quittner, On the principle of linearized stability for variational inequalities, *Math. Ann.* **283** (1986) 257–270.

11. E. Zeidler, Nonlinear Functional Analysis and its Applications. I: Fixed Point Theorems. Springer-Verlag, New York, 1986.

# NON LOCAL HARNACK INEQUALITIES FOR A CLASS OF PARTIAL DIFFERENTIAL EQUATIONS

U. BOSCAIN

*SISSA-ISAS*
*Via Beirut 2-4*
*34014 Trieste, Italy*
*E-mail: boscain@sissa.it*


S. POLIDORO

*Dipartimento di Matematica*
*Porta S. Donato 5*
*40126 Bologna, Italy*
*E-mail: polidoro@dm.unibo.it*

We are concerned with Gaussian lower bounds for positive solutions of a family of hypoelliptic partial differential equations on homogeneous Lie groups. We describe a method that relies on the repeated use of an invariant Harnack inequality and on a suitable optimal control problem. We also obtain an accurate Gaussian lower bound for the fundamental solution of Kolmogorov type operators on non-homogeneous groups.

**Key words:** Harnak inequality, hypoelliptic pde, homogeneous Lie group, Kolmogorov operator

**Mathematics Subject Classification:** 35H10

## 1. Introduction

We consider a class of linear second order operators in $\mathbb{R}^{N+1}$ of the form

$$L = \sum_{k=1}^{m} X_k^2 + X_0 - \partial_t. \tag{1}$$

In (1) the $X_k$'s are smooth vector fields on $\mathbb{R}^N$, i.e.

$$X_k(x) = \sum_{j=1}^{N} a_j^k(x)\partial_{x_j}, \qquad k = 0, \ldots, m,$$

where $z = (x, t)$ denotes the point in $\mathbb{R}^{N+1}$ and any $a_j^k$ is a $C^\infty$ function. For our purposes, in the sequel we also consider the $X_k$'s as vector fields in $\mathbb{R}^{N+1}$ and we denote

$$Y = X_0 - \partial_t. \tag{2}$$

We next state our main assumptions:

[H.1] there exists a homogeneous Lie group $\mathbb{G} = \left(\mathbb{R}^{N+1}, \circ, \delta_\lambda\right)$ such that

(i) $X_1, \ldots, X_m, Y$ are left translation invariant on $\mathbb{G}$;

(ii) $X_1, \ldots, X_m$ are $\delta_\lambda$-homogeneous of degree one and $Y$ is $\delta_\lambda$-homogeneous of degree two;

[H.2] (controllability) for every $(x, t), (y, s) \in \mathbb{R}^{N+1}$ with $t > s$, there exists an absolutely continuous path $\gamma : [0, t - s] \to \mathbb{R}^N$ such that

$$
\begin{cases}
\dot{\gamma}(\tau) = \sum_{k=1}^{m} \omega_k(\tau) X_k(\gamma(\tau)) + X_0(\gamma(\tau)) \\
\gamma(0) = x, \quad \gamma(t - s) = y.
\end{cases}
\tag{3}
$$

where $\omega_1, \ldots \omega_m$ belong to $L^2([0, t - s])$.

Operators $L$ of the form (1), verifying assumptions [H.1]-[H.2] have been considered by Kogoj and Lanconelli[1,2]. It is known that they satisfy the Hörmander condition:

$$
\text{rank Lie}\{X_1, \ldots, X_m, Y\}(z) = N + 1, \qquad \forall z \in \mathbb{R}^{N+1};
\tag{4}
$$

as a consequence, every operator $L$ of this kind is hypoelliptic (i.e. every distributional solution to $Lu = 0$ is smooth; see, for instance, Proposition 10.1 in Ref. [1]). Besides, $L$ has a fundamental solution $\Gamma$ which is smooth out of the pole and $\delta_\lambda$-homogeneous of degree $2 - Q$:

$$
\Gamma\left(\delta_\lambda z\right) = \lambda^{2-Q} \Gamma(z), \qquad \lambda > 0,
\tag{5}
$$

(here $Q$ denotes the homogeneous dimension of $\mathbb{G}$). Hence we are concerned with operators belonging to the general class of hypoelliptic operators on homogeneous groups first studied by Folland[3], Rothschild and Stein[4].

Our main goal is a Gaussian lower bounds for positive solutions to $Lu = 0$. We first recall the method introduced by Aronson[5], and used by Aronson and Serrin[6] in the study of uniformly parabolic equations. They show that, if $u$ is a positive solution of a parabolic equation, defined in the domain $\mathbb{R}^N \times ]T_0, T_1[$, then

$$
u(y, s) \le M^{1 + \frac{|x-y|^2}{t-s}} u(x, t),
$$

for every $(x, t), (y, s)$ such that $\frac{T_0 + T_1}{2} < s < t < T_1$. The result is proved by applying repeatedly an invariant Harnack inequality to a finite set of points lying on the segment of line connecting $(x, t)$ to $(y, s)$. In order to adapt the above method to our class of operators we will rely on the invariant (with respect to translations and to the dilations of the group $\mathbb{G}$) Harnack inequality proved by Kogoj and Lanconelli[1].

We recall that the Aronson's method has been used by Jerison and Sánchez-Calle[7], Kusuoka and Stroock[8], Varopoulos, Saloff-Coste and Coulhon[9] to the non-Euclidean setting of *heat kernels on Lie groups*. The ideas introduced in the above papers have been developed by many authors. The probabilistic approach due to

Kusuoka and Stroock has been recently employed by Bally and Kohatsu-Higa in the study of more general diffusion precesses[a].

Concerning the approach due to Varopoulos, Saloff-Coste and Coulhon, we quote the recent paper by by Alexopoulos[10], where operators (1) are considered, under the assumption that $X_0 \in \mathrm{Lie}\{X_1, \ldots, X_m\} = \mathbb{R}^N$. We finally recall that Polidoro[11] adapt the method to a family of homogeneous Kolmogorov operators (see Example 1.2 below), and that Pascucci and Polidoro[12] consider a more general class of operators satisfying assumptions [H.1]-[H.2].

Here we further develop that idea and we set the problem in the theory of optimal control with quadratic cost (we refer to the books by Jurdjevic[13] and Agrachev and Sachkov[14]). More specifically, we consider the functions $\omega_1, \ldots \omega_m$ as the *controls* of the path $\gamma$ in (3) and the following cost function:

$$\Phi(\omega) = \int_0^{t-s} |\omega(s)|^2 ds. \tag{6}$$

Our first result is the following estimate:

**Proposition 1.1.** *Let $u : \mathbb{R}^N \times ]T_0, T_1[$ be a positive solution to $Lu = 0$. There exist three positive constants $\theta \in ]0, 1[, h$ and $M > 1$, only depending on the operator $L$, such that*

$$u(y, s) \leq M^{1 + \frac{\Phi(\omega)}{h}} u(x, t).$$

*for every $(x, t), (y, s) \in \mathbb{R}^N \times ]T_0, T_1[$ such that $T_1 - \theta^2(T_1 - T_0) \leq s < t < T_1$. Here $\gamma : [0, t - s] \to \mathbb{R}^N$ is a solution to (3), and $\Phi(\omega)$ is the corresponding cost.*

Among the paths $\gamma$ satisfying (3), we look for one minimizing the cost:

$$\Psi(x, t, y, s) = \inf \Phi(\omega), \tag{7}$$

where the infimum is taken in the set of the paths satisfying (3). As a corollary of Proposition 1.1 we find:

**Proposition 1.2.** *Let $u : \mathbb{R}^N \times ]T_0, T_1[$ be a positive solution to $Lu = 0$. There exist three positive constants $\theta \in ]0, 1[, h$ and $M > 1$, only depending on the operator $L$, such that*

$$u(y, s) \leq M^{1 + \frac{\Psi(x,t,y,s)}{h}} u(x, t),$$

*for every $(x, t), (y, s) \in \mathbb{R}^N \times ]T_0, T_1[$ such that $T_1 - \theta^2(T_1 - T_0) \leq s < t < T_1$.*

A further direct consequence is the following lower bound for $\Gamma$:

---

[a]Talk "Lower bounds for the density of locally elliptic Ito processes", by Bally, and talk "Lower bounds for density estimates by Malliavin Calculus" by Kohatsu-Higa, held in the Congress: "Harnack Inequalities and Positivity for Solutions of Partial Differential Equations" Cortona (Italy), June 12-18, 2005

**Proposition 1.3.** *Let $L$ be the operator in (1) and let $\Gamma$ be its fundamental solution. There exists a positive constant $C$ such that*

$$\Gamma(x,t,y,s) \geq \frac{C}{(t-s)^{\frac{Q-2}{2}}} e^{-C\Psi(x,t,y,s)}, \qquad \forall (x,t),(y,s) \in \mathbb{R}^{N+1} \quad \text{with } t > s. \quad (8)$$

We next give some examples of operators satisfying assumptions [H.1]-[H.2].

**Example 1.1.** (HEAT OPERATORS ON CARNOT-CARATHEODORY GROUPS) Consider the operator $L$ in (1) under assumptions [H.1] and

$$\text{rank Lie}\{X_1, \ldots, X_m\}(x) = N, \qquad \forall x \in \mathbb{R}^N. \quad (9)$$

In this case $\mathbb{G} = (\mathbb{R}^N, \circ, \delta_\lambda)$ is a *Carnot-Caratheodory (or stratified) group* (see, for instance, Gromov[15], Bellaiche[16] and also Refs. 3 and 9).

When $X_0 \equiv 0$, the couple of conditions [H.1]-[H.2] and [H.1]-(9) are equivalent. In this case the operator $L$ in (1) is

$$L = \sum_{j=1}^m X_j^2 - \partial_t. \quad (10)$$

We recall the well-known Gaussian upper and lower bounds for heat kernels due to Jerison and Sánchez-Calle[7], Kusuoka and Stroock[8], Varopoulos, Saloff-Coste and Coulhon[9]

$$\frac{1}{C\sqrt{V(\mathcal{B}_{t-s}(x))}} \exp\left(-\frac{C\,d^2(x,y)}{t-s}\right) \leq \Gamma(x,t,y,s)$$

$$\leq \frac{C}{\sqrt{V(\mathcal{B}_{t-s}(x))}} \exp\left(-\frac{d^2(x,y)}{C\,(t-s)}\right), \quad (11)$$

for every $(x,t),(y,s) \in \mathbb{R}^N \times ]T_0, T_1[$ with $t > s$, where $d(x,y)$ denotes the Carnot-Carateodory distance and $V(\mathcal{B}_r(x))$ is the volume of the metric ball with center at $x$ and radius $r$. Note that this lower bound agrees with the one stated in Proposition 1.2, since

$$\Psi(x,t,y,s) = \frac{d^2(x,y)}{t-s} \qquad \text{when } X_0 = 0.$$

Next example is concerned with a family of operators that is not included in the class of Heat kernels on Carnot-Caratheodory groups considered above. Our method gives a lower bound of the fundamental solution that does not agree with (11); in fact it is known that the fundamental solutions of Kolmogorov operators does not satisfy the lower bound in (11) (see Remark 2.1 below).

**Example 1.2.** (KOLMOGOROV TYPE OPERATORS)

$$Ku \equiv \sum_{i,j=1}^{p_0} a_{i,j}\partial_{x_i x_j}u + \sum_{i,j=1}^{N} b_{i,j}x_i\partial_{x_j}u - \partial_t u,$$

where $A_0 = (a_{ij})_{i,j=1,\ldots,p_0}$ and $B = (b_{ij})_{i,j=1,\ldots,N}$ are real constant matrices, with $p_0 \leq N$, $A_0$ is symmetric and positive. We assume that $K$ satisfies condition [H.2].

The above example will be discussed in detail in Section 2. Here we briefly describe the main ideas underlying the proof of Propositions 1.1, 1.2 and 1.3, we refer to Ref. 17 for the details.

As said before, Aronson obtain his Gaussian lower bound by repeatedly applying an invariant Harnack inequality to a finite set of points lying on the segment of line connecting $(x,t)$ to $(y,s)$. The main difficulty in the framework of the Lie groups is to find a criterium to select an analogous finite set. Our idea is to relate that finite subset to the image of the path (3), according to a method introduced by Li and Yau[18].

The paper Ref. 18 provides a lower bound for positive solutions to parabolic equations in Riemannian manifolds. The lower bound is obtained by a pointwise estimate of the gradient of the solution $u$ which, in the case of the heat equation in $\mathbb{R}^{N+1}$, the estimate of the gradient reads

$$\partial_t u + \frac{N}{2t} u \geq \frac{|Du|^2}{u}. \tag{12}$$

The same approach was used by Cao and Yau[19] for a family of heat kernels on Lie group contained in Example 1.1, then by Pascucci and Polidoro[20] for a family of Kolmogorov operators on a homogeneous group. In Section 3 we use the optimal cost related to the Kolmogorov operator and we improve the result given in Ref. 20.

## 2. Kolmogorov operators

In this section we consider Kolmogorov operators

$$Ku \equiv \sum_{i,j=1}^{p_0} a_{i,j} \partial_{x_i x_j} u + \sum_{i,j=1}^{N} b_{i,j} x_i \partial_{x_j} u - \partial_t u, \tag{13}$$

where $A_0 = (a_{ij})_{i,j=1,\ldots,p_0}$ and $B = (b_{ij})_{i,j=1,\ldots,N}$ are real constant matrices, $A_0$ is symmetric and positive, and we assume that $K$ satisfies condition [H.2]. We first recall some known results, then we state and solve the optimal control problem (3) related to $K$.

Denote by $A_0^{\frac{1}{2}} = (\bar{a}_{ij})_{i,j=1,\ldots,p_0}$ the unique positive $p_0 \times p_0$ matrix such that $A_0^{\frac{1}{2}} \cdot A_0^{\frac{1}{2}} = A_0$, and by $A$ and $A^{\frac{1}{2}}$ the $N \times N$ matrices

$$A = \begin{pmatrix} A_0 & 0 \\ 0 & 0 \end{pmatrix}, \qquad A^{\frac{1}{2}} = \begin{pmatrix} A_0^{\frac{1}{2}} & 0 \\ 0 & 0 \end{pmatrix}. \tag{14}$$

Note that the operator $K$ can be written in the form (1) by choosing

$$X_i = \sum_{j=1}^{p_0} \bar{a}_{ij} \partial_{x_j}, \ i = 1, \ldots, p_0, \qquad X_0 = \langle x, B\nabla \rangle \tag{15}$$

where $\nabla = (\partial_{x_1}, \ldots, \partial_{x_N})$ and $\langle \cdot, \cdot \rangle$ are, respectively, the gradient and the inner product in $\mathbb{R}^N$.

Clearly, $K$ is hypoelliptic if, and only if, it satisfies the Hörmarder condition (4) (where $X_0, \ldots, X_{p_0}$ are defined in (15) and $Y = X_0 - \partial_t$). The hypoellipticity of $K$ is equivalent to any of the following conditions:

$H_1$  $\mathrm{Ker}(A^{\frac{1}{2}})$ does not contain non-trivial subspaces which are invariant for $B$;

$H_2$  there exists a basis of $\mathbb{R}^N$ such that $B$ has the form

$$
\begin{pmatrix}
* & B_1 & 0 & \ldots & 0 \\
* & * & B_2 & \ldots & 0 \\
\vdots & \vdots & \vdots & \ddots & \vdots \\
* & * & * & \ldots & B_n \\
* & * & * & \ldots & *
\end{pmatrix}
\tag{16}
$$

where $B_j$ is a matrix $p_{j-1} \times p_j$ of rank $p_j$, with

$$
p_0 \geq p_1 \geq \ldots \geq p_n \geq 1, \qquad p_0 + p_1 + \ldots + p_n = N,
$$

while $*$ are constant and arbitrary blocks; $H_3$  if we set

$$
E(s) = \exp(-sB^T), \qquad C(t) = \int_0^t E(s)AE^T(s)ds, \tag{17}
$$

then $C(t)$ is positive, for every $t > 0$.

For the equivalence of the above conditions we refer to Ref. 21. We explicitly remark that $H_1$ is equivalent to the well known Kalman condition: *the $N \times N^2$ matrix*

$$
\left[ A^{\frac{1}{2}}, B^T A^{\frac{1}{2}}, \ldots, (B^T)^{N-1} A^{\frac{1}{2}} \right] \tag{18}
$$

*has rank $N$*, which in turns is equivalent to the controllability condition [H.2] (see Ref. 22, Theorem 5, p. 81).

We recall that the group law related to the Kolmogorov operator is

$$
(x,t) \circ (\xi, \tau) = (\xi + E(\tau)x, t + \tau), \qquad (x,t), (\xi, \tau) \in \mathbb{R}^{N+1}. \tag{19}
$$

The fundamental solution of $K$, with singularity at any point $(y,s)$ is explicitly known:

$$
\Gamma(x,t,y,s) = \frac{(4\pi)^{-\frac{N}{2}}}{\sqrt{\det C(t-s)}} \exp\left( -\frac{1}{4} \langle C^{-1}(t-s)(x - E(t-s)y), x - E(t-s)y \rangle \right),
\tag{20}
$$

for $t > s$, and $\Gamma(x,t,y,s) = 0$ for $t \leq s$.

The problem (3) can be stated in the equivalent form

$$
\begin{cases}
\dot{\gamma}(\tau) = B^T \gamma(\tau) + A^{\frac{1}{2}} \omega(\tau) \\
\gamma(0) = x, \quad \gamma(t-s) = y,
\end{cases}
\tag{21}
$$

where $\omega(\tau) = (\omega_1(\tau), \ldots, \omega_{p_0}(\tau), 0, \ldots, 0)^T \in \mathbb{R}^N$. The following lemma states the remarkable fact that the optimal cost of the problem (21) has the same expression appearing in the fundamental solution of $K$ (20).

**Lemma 2.1.** *The solution* $\gamma : [0, t-s] \to \mathbb{R}^N$ *of* (21), *that minimizes the cost* (7), *is the solution related to the control*

$$\omega(\tau) = \left(A^{\frac{1}{2}}\right)^T E(\tau)^T C^{-1}(t-s)(x - E(t-s)y),$$

*the optimal cost is*

$$\Psi(\omega) = \int_0^{t-s} |\omega(\tau)|^2 d\tau = \langle C^{-1}(t-s)(x - E(t-s)y), x - E(t-s)y \rangle.$$

PROOF. Consider the Hamiltonian function $\mathcal{H}$ related to the control problem (21) in the interval $[0, t-s]$:

$$\mathcal{H}(x, p) = p\, B^T x, \qquad p = (p_1, \dots, p_N), \quad x = (x_1, \dots, x_N)^T. \qquad (22)$$

From the classical theorems of autonomous linear systems it follows that the optimal control is

$$\omega(\tau) = \left(A^{\frac{1}{2}}\right)^T p(\tau)^T, \qquad \text{for some solution to} \quad \dot{p} = -p\, B^T \qquad (23)$$

(see Ref. 22, Theorem 3, p. 180). We use the above identity in (21), and we compute the explicit solution:

$$\gamma(\tau) = E(-\tau)\left(x + C(\tau)p(0)^T\right)$$

for some constant vector $p(0)$ that is determined by the condition $\gamma(t-s) = y$. We find

$$p(0)^T = C^{-1}(t-s)(E(t-s)y - x).$$

This concludes the proof. $\qquad\qquad\qquad\qquad\qquad\qquad\qquad\qquad\qquad\qquad\square$

**Remark 2.1.** It is known that the Kolmogorov operator $K$ is invariant with respect to a dilations

$$\delta_\lambda = \text{diag}(\lambda\, I_{p_0}, \lambda^3 I_{p_1}, \dots, \lambda^{2n+1} I_{p_n}, \lambda^2), \qquad \lambda > 0,$$

($I_{p_k}$ is the $p_k \times p_k$ identity matrix) if, and only if, every $*$ block in (16) is null. In that case it is possible to define a norm on $\mathbb{R}^N$ which is homogeneous with respect to $\delta_\lambda$. However, unlike in the case the heat kernels, $\Psi(x, t, \xi, \tau)$ is not equivalent to $\frac{|(x,\xi)|_K^2}{t-\tau}$. For instance, in the simplest case of Kolmogorov operators in $K = \partial_x^2 + x\partial_y - \partial_t$, with $(x, y, t) \in \mathbb{R}^3$, the optimal cost is

$$\Psi(x, y, t, 0, 0, 0) = \frac{x^2}{t} + 3\frac{xy}{t^2} + 3\frac{y^2}{t^3}.$$

708

## 3. Gaussian estimates for Kolmogorov operators

We prove a sharp version of the Harnack inequality for Kolmogorov operators $K$ in (13) by using the argument due to Li and Yau[18]. We first prove the following gradient estimate for the positive solution ti $Ku = 0$, which is the analogous of (12).

**Proposition 3.1.** *Let $u$ be a positive solution of $Ku = 0$ in a strip $\mathbb{R}^N \times ]0, T[$, $T > 0$. Then*

$$-Yu + \varphi'(t)u \geq \frac{\langle ADu, Du \rangle}{u} \qquad \varphi(t) = \log\left(\sqrt{\det \mathcal{C}(t)}\right) \qquad (24)$$

*in $\mathbb{R}^N \times ]0, T[$.*

A non local Harnack inequality can be directly derived from Proposition 3.1.

**Theorem 3.1.** *Let $u$ be positive solution to $Ku = 0$ in $\mathbb{R}^N \times ]0, T[$. Let $(x,t), (y,s) \in \mathbb{R}^{N+1}$ be such that $0 < s < t < T$. Then*

$$u(x,t)\frac{\sqrt{\det \mathcal{C}(t)}}{\sqrt{\det \mathcal{C}(s)}} \geq u(y,s)\exp\left(-\frac{1}{4}\langle \mathcal{C}^{-1}(t-s)(x - E(t-s)y), x - E(t-s)y\rangle\right).$$

PROOF. Let $\gamma$ be a solution of the problem (21) and let $\omega = (\omega_1, \ldots, \omega_{p_0}, 0, \ldots, 0)$ be its control. Add the quantity $\frac{1}{4}u(\gamma(\tau), t-\tau)|\omega(\tau)|^2 - \langle A^{\frac{1}{2}}Du(\gamma(\tau), t-\tau), \omega(\tau)\rangle$ to both side of (24) evaluated at the point $(\gamma(\tau), t-\tau)$. We find

$$Yu(\gamma(\tau), t-\tau) \leq \varphi'(t-\tau)u(\gamma(\tau), t-\tau)-$$
$$\langle A^{\frac{1}{2}}Du(\gamma(\tau), t-\tau), \omega(\tau)\rangle + \frac{1}{4}u(\gamma(\tau), t-\tau)|\omega(\tau)|^2,$$

for every $\tau \in [0, t-s]$. Using the fact that $\gamma$ is a solution to (21), we then get

$$\frac{d}{d\tau}u(\gamma(\tau), t-\tau) \leq \varphi'(t-\tau)u(\gamma(\tau), t-\tau) + \frac{1}{4}u(\gamma(\tau), t-\tau)|\omega(\tau)|^2, \quad \text{for } \tau \in [0, t-s].$$

Dividing by $u(\gamma)$ and integrating in the variable $\tau$ over the interval $[0, t-s]$, we finally prove that

$$u(x,t)\sqrt{\det \mathcal{C}(t)} \geq u(y,s)\sqrt{\det \mathcal{C}(s)}\exp\left(-\frac{1}{4}\Phi(\omega)\right),$$

where as usual $\Phi(\omega)$ is the cost of the control $\omega$ (recall the definition of the function $\varphi$). We now consider the optimal path $\gamma$ and the thesis follows from Lemma 2.1. $\square$

PROOF OF PROPOSITION 3.1. We first show that the fundamental solution $\Gamma$ of $K$ verifies the equation

$$Y \log\left(\Gamma(x,t)\right) + \langle AD \log\left(\Gamma(x,t)\right), D \log\left(\Gamma(x,t)\right)\rangle = \varphi'(t) \qquad (25)$$

in $\mathbb{R}^N \times ]0, +\infty[$. Then we prove the gradient estimate (24) by means of a representation formula for positive solutions (see Ref. 23).

From (20) we immediately get

$$\log\left(\Gamma(x,t)\right) = -\frac{N}{2}\log(4\pi) - \varphi(t) - \frac{1}{4}\langle C^{-1}(t)x, x\rangle,$$

for any $(x,t) \in \mathbb{R}^N \times \mathbb{R}^+$. Then

$$D\log\left(\Gamma(x,t)\right) = -\frac{1}{2}C^{-1}(t)x,$$

$$-Y\log\left(\Gamma(x,t)\right) = \frac{1}{2}\langle x, BC^{-1}(t)x\rangle - \frac{1}{4}\left\langle \frac{d}{dt}C^{-1}(t)x, x\right\rangle - \varphi'(t).$$

(26)

On the other hand, since $\Gamma$ is a solution to $Ku = 0$, then $\log(\Gamma)$ satisfies

$$\text{div}\left(AD\log(\Gamma), D\log(\Gamma)\right) + Y\log(\Gamma) + \langle AD\log(\Gamma), D\log(\Gamma)\rangle = 0,$$

(27)

in $\mathbb{R}^N \times \mathbb{R}^+$. Therefore, if we set

$$C^{-1}(t) = (\tilde{c}_{ij}(t)),$$

and we use the first equation in (26), we find

$$-Y\log\left(\Gamma(x,t)\right) = \langle AD\log(\Gamma), D\log(\Gamma)\rangle + \sum_{i,j=1}^{N} a_{ij}\tilde{c}_{ij}(t).$$

Evaluating the above expression at $x = 0$, and the second one in (26), we finally find

$$\varphi'(t) = -\sum_{i,j=1}^{N} a_{ij}\tilde{c}_{ij}(t).$$

This proves (25). In order to conclude the proof, we shall use the usual representation formula: for any given $t_0 \in ]0,T[$, we have

$$u(x,t) = \int_{\mathbb{R}^N} \Gamma(x,t,y,t_0)u(y,t_0)dy, \qquad (x,t) \in \mathbb{R}^N \times ]t_0, T[.$$

(28)

We now rewrite (25) in the following equivalent form

$$-Y\Gamma + \varphi'(t)\Gamma = \frac{\langle AD\Gamma, D\Gamma\rangle}{\Gamma}.$$

Thus

$$-Yu + \varphi'(t)u = \int_{\mathbb{R}^N} \left(-Y\Gamma(\cdot,y,t_0) + \varphi'(t)\Gamma(\cdot,y,t_0)\right) u(y,t_0)dy$$

$$= \int_{\mathbb{R}^N} \left(\frac{\langle AD\Gamma(\cdot,y,t_0), D\Gamma(\cdot,y,t_0)\rangle}{\Gamma(\cdot,y,t_0)}\right) u(y,t_0)dy \geq$$

(by the Hölder inequality, since $u$ is positive)

$$\geq \langle A \int_{\mathbb{R}^N} D\Gamma(\cdot,y,t_0)u(y,t_0)dy, \int_{\mathbb{R}^N} D\Gamma(\cdot,y,t_0)u(y,t_0)dy\rangle \left(\int_{\mathbb{R}^N} \Gamma(\cdot,y,t_0)u(y,t_0)dy\right)^{-1}$$

and the proof is accomplished. $\qquad\square$

710

## References

1. A. E. KOGOJ AND E. LANCONELLI, *An invariant Harnack inequality for a class of hypoelliptic ultraparabolic equations*, Mediterr. J. Math., 1 (2004), pp. 51–80.

2. ———, *One-side Liouville theorems for a class of hypoelliptic ultraparabolic equations*, "Geometric Analysis of PDE and Several Complex Variables" Contemporary Mathematics Proceedings, (2004).

3. G. B. FOLLAND, *Subelliptic estimates and function spaces on nilpotent Lie groups*, Ark. Mat., 13 (1975), pp. 161–207.

4. L. P. ROTHSCHILD AND E. M. STEIN, *Hypoelliptic differential operators and nilpotent groups*, Acta Math., 137 (1976), pp. 247–320.

5. D. G. ARONSON, *Bounds for the fundamental solution of a parabolic equation*, Bull. Amer. Math. Soc., 73 (1967), pp. 890–896.

6. D. G. ARONSON AND J. SERRIN, *Local behavior of solutions of quasilinear parabolic equations*, Arch. Rational Mech. Anal., 25 (1967), pp. 81–122.

7. D. S. JERISON AND A. SÁNCHEZ-CALLE, *Estimates for the heat kernel for a sum of squares of vector fields*, Indiana Univ. Math. J., 35 (1986), pp. 835–854.

8. S. KUSUOKA AND D. STROOCK, *Applications of the Malliavin calculus. III*, J. Fac. Sci. Univ. Tokyo Sect. IA Math., 34 (1987), pp. 391–442.

9. N. T. VAROPOULOS, L. SALOFF-COSTE, AND T. COULHON, *Analysis and geometry on groups*, vol. 100 of Cambridge Tracts in Mathematics, Cambridge Univ. Press, 1992.

10. G. K. ALEXOPOULOS, *Sub-Laplacians with drift on Lie groups of polynomial volume growth.*, Mem. Am. Math. Soc., 739 (2002), p. 101 p.

11. S. POLIDORO, *A global lower bound for the fundamental solution of Kolmogorov-Fokker-Planck equations*, Arch. Rational Mech. Anal., 137 (1997), pp. 321–340.

12. ———, *Harnack inequalities and Gaussian estimates for a class of hypoelliptic operators*, to appear in Trans. Amer. Math. Soc., (2005).

13. V. JURDJEVIC, *Geometric control theory*, Cambridge Studies in Advanced Mathematics, 52, Cambridge Univ. Press, 1997.

14. A. A. AGRACHEV, YU. L. SACHKOV, *Control Theory from the Geometric Viewpoint*, (to appear), Encyclopedia of Mathematical Sciences, Vol. 87, Springer, 2004.

15. M. GROMOV, *Carnot-Caratheodory spaces seen from within*, Sub-Riemannian geometry, 79–323, Progr. Math., 144, Birkhäuser, Basel, 1996.

16. A. BELLAICHE, *The tangent space in sub-Riemannian geometry*, Sub-Riemannian geometry, 1–78, Progr. Math., 144, Birkhäuser, Basel, 1996.

17. U. BOSCAIN, AND S. POLIDORO, *Harnack inequalities and Gaussian estimates for a class of hypoelliptic operators*, (preprint).

18. P. LI AND S. T. YAU, *On the parabolic kernel of the Schrödinger operator*, Acta Math., 156 (1986), pp. 153–201.

19. H. D. CAO AND S.-T. YAU, *Gradient estimates, Harnack inequalities and estimates for heat kernels of the sum of squares of vector fields*, Math. Z., 211 (1992), pp. 485–504.

20. A. PASCUCCI AND S. POLIDORO, *On the Harnack inequality for a class of hypoelliptic evolution equations*, Trans. Amer. Math. Soc., 356 (2004), pp. 4383–4394.

21. E. LANCONELLI AND S. POLIDORO, *On a class of hypoelliptic evolution operators*, Rend. Sem. Mat. Univ. Politec. Torino, 52 (1994), pp. 29–63. Partial differential equations, II (Turin, 1993).

22. E. B. LEE AND L. MARKUS, *Foundations of optimal control theory*, John Wiley & Sons Inc., New York, 1967.

23. M. DI FRANCESCO AND S. POLIDORO, *Dirichlet problem and harnack inequality for Kolmogorov type operators in non-divergence form*, preprint, (2005).

# Γ-CONVERGENCE FOR STRONGLY LOCAL DIRICHLET FORMS IN OPEN SETS WITH HOLES

M. BIROLI

*Politecnico di Milano, Department of Mathematics*
*Piazza Leonardo da Vinci 32,*
*20133 Milano, Italy*
*E-mail: marco.biroli@polimi.it*

N.A. TCHOU

*Université Rennes 1, IRMAR*
*Campus de Beaulieu,*
*35042 Rennes, France*
*E-mail: Nicoletta.Tchou@univ-rennes1.fr*

We consider the homogenization problems with holes for strongly local Dirichlet forms in the cases in which we have Neumann homogeneous conditions on the boundaries of the holes. In this case the main difficulties arise from the absence of a group structure on the underlying space and then from the non-periodic distribution of the holes. The complete proofs of the results will appear in the paper [4].

**Key words:** Dirichlet forms, homogenization problems, variational convergences
**Mathematics Subject Classification:** 35J20, 40A30, 31C25, 43A85, 35H20

## 1. Introduction

In this paper we will give some results on the Γ-convergence of problems with holes relative to a Riemannian Dirichlet form. We consider the case of homogeneous Neumann boundary conditions on the boundary of holes. In this case one of the main difficulties is that we operate on a locally compact metrizable Hausdorff space, then no structure of group is available so there is no notion of periodicity on our space. In the case of Neumann boundary condition the first paper in which for a particular Euclidean elliptic problem with periodically distributed holes the convergence to a limit problem is rigorously proved [9], more general Euclidean cases are considered in [14]. L. Tartar make the remark that all the proofs depends on the existence of suitable extension operators (in the holes); the problem in Euclidean setting without periodicity but with the assumption of the existence of suitable extension operators (in the holes) has been solved in [8]. We remark that the question of existence of extension operators with the required properties is solved in Euclidean setting with some assumption on the regularity of the boundary of the holes but becomes very delicate in the subelliptic setting so it is important to investigate methods, which can operates in absence of classical extension theory; for paper in this direction see [14,1,7,11,10,20–23] for the Euclidean case and [5] for the periodic case in the Heisenberg (subelliptic) group.

We will now introduce the notion of Poincaré type or Riemannian Dirichlet forms.

Let $X$ be a metrizable locally compact Hausdorff space with a positive Radon measure $m$ such that $supp[m] = X$; we assume that we are given a *strongly local of diffusion type, regular Dirichlet form* in the Hilbert space $L^2(X, m)$, in the sense of M. Fukushima, [12], whose domain is denoted by $D[a]$. Such a form $a$ admits the following integral representation $a(u, v) = \int_X \alpha(u, v)(dx)$ for every $u, v \in D[a]$ where $\alpha(u, v)$ is a signed Radon measure on $X$, uniquely associated with the functions $u, v$. Moreover for any open subset $\Omega$ of $X$ the restriction of $\alpha(u, v)$ to $\Omega$ depends only on the restrictions of $u$ and $v$ to $\Omega$ and $\alpha(u, v) = 0$ whenever $u = cst$ on a neighbourhood of the support of $v$. Let $\Omega$ be an open set; by $D_0[a, \Omega]$ we denote the closure of $C_0(\Omega) \cap D[a]$ in $D[a]$. By $D_{loc}[a, \Omega]$ we denote the space of all $m$-measurable functions $u, v$ in $X$, that coincide $m$-a.e. on every compact subset of $\Omega$ with some function of $D[a]$ . The measure $\alpha(u, v)$ is defined unambigously in $\Omega$ for all $u, v \in D_{loc}[a, \Omega]$. We refer to [3,12,17,18] for the properties of $\alpha(u, v)$ with respect to Leibnitz, chain and troncature rules. We assume now that the space $X$ is endowed with a pseudo distance $d$ and is complete with respect to $d$ (we assume also that $d$ define a topology on $X$ equivalent to the initial one). We denote $B(x, r) = \{y; d(x, y) < r\}$, $B(r)$ will be balls $B(x, r)$ with a fixed center $x$. We say that the Dirichlet form is of *Poincaré type* if the following two assumptions hold:
($H_1$) There exists constants $0 < R_0 < +\infty$, $\nu > 0$ and $c_0 > 0$, such that

$$0 < c_0 \left(\frac{r}{R}\right)^{\nu} m(B(x, R)) \leq m(B(x, r))$$

for every $x \in X$ and every $0 < r < R_0$.
($H_2$) For every ball $B(x, r)$ such that $B(x, kr) \subseteq\subseteq \Omega$ and every $v \in D_{loc}[\Omega]$ the following *scaled Poincaré inequality* holds

$$\int_{B(x,r)} |v - v_{x,r}|^2 m(dx) \leq c_1 r^2 \int_{B(x,kr)} \alpha(v, v)(dx)$$

where $c_1, k \geq 1$, are constants independent of $x, r$.

**Remark 1.1** We observe that ($H_1$) is verifyied if a *duplication property* holds for the balls $B(x, r)$, $0 < r < R_0$, that is

$$m(B(x, 2r)) \leq c_0^* \, m(B(x, r)),$$

where $c_0^*$ is a positive constant independent of $x, r$. In this case we have $\nu \geq lg_2 c_0^*$.

**Remark 1.2** We recall,[12], that there a Choquet capacity associated to a given Dirichlet form of Poincarè type.
If $d$ is a distance on $X$ and $d \in D_{loc}[a]$ with $\alpha(d, d) \leq m$ we say that our Dirichlet form is a *Riemannian* Dirichlet form

**Remark 1.3** We observe that if

$$d_a(x,y) = sup\{\phi(x) - \phi(y); \forall \phi \ \ with \ \ \alpha(\phi, \phi) \le m\}$$

and $d_a(x,y)$ is finite and separating (i.e. if $x \ne y$ then $d_a(x,y) \ne 0$) then $d_a$ is a distance and $d_a \in D_{loc}[a]$ with $\alpha(d_a, d_a) \le m$.

**Remark 1.4** If $a$ is Riemannian there exists cut-off functions between balls (i.e. given two balls $B(x,r)$ and $B(x,R)$, $r < R$, there exists a function $\phi$ with $\phi = 1$ on $B(x,r)$, $\phi = 0$ out of $B(x,R)$ and $\alpha(\phi,\phi) \le \frac{1}{(R-r)^2}$

Finally we observe that our assumptions hold for Dirichlet forms associated to wide classes of
(a) weighted uniformly elliptic operators
(b) weighted degenerate elliptic operators generated by vector fields satisfying an Hörmander condition
(c) subelliptic operators
see [2,3], [15] for more details. Moreover also the Laplacian on p.c.f. fractals define a Poincarè type Dirichlet form, [19].

## 2. Γ-convergence in the case of homogeneous Neumann boundary condition on the boundary of holes

In this section we outline the proofs of the results, [4], since [4] is not yet available as publication. We consider a denumerable covering $\mathcal{B}_\epsilon$ of the space obtained by balls of radius $(1 + \delta)\epsilon$ ($\epsilon > 0$, $\delta > 0$ fixed), such that the homotetic balls of radius $\epsilon$ do not intersect. We assume also that $\mathcal{B}_\epsilon$ has a property of finite intersection uniform with respect to $\epsilon$ (i.e. every point of $X$ belongs to at most $Q$ balls in the covering, where $Q$ does not depend on $\epsilon$).

We denote by $B_{i,(1+\delta)\epsilon}$, $i = 1, 2, ..$, the balls in $\mathcal{B}_\epsilon$ and by $B_{i,\epsilon}$, $i = 1, 2, ..$, the homotetic balls of radius $\epsilon$. Consider now a relatively compact open set $\Omega$ with boundary $\Gamma$ and denote by $B_{j,\epsilon}$, $j = 1, 2, .., q$, a subfamily of the balls $B_{i,\epsilon}$, such that $B_{i,\epsilon(1+\delta)}$ is contained in $\Omega$ (the number of the balls $B_{j,\epsilon}$ is finite due to the homogeneous structure of $X$). In every ball $B_{j,\epsilon(1-\delta)}$ we consider a compact set $T_{j,\epsilon}$. We denote $\Omega_\epsilon = \Omega - \cup_j T_{j,\epsilon}$. Let $\theta_\epsilon$ be the characteristic function of $\Omega_\epsilon$ We assume that $\theta_\epsilon$ converges in the weak$^\star$ topology of $L^\infty(\Omega, m)$ to a function $\theta$ this property always hold at least after extraction of subsequences) with $0 < \sigma < \theta \le 1$. We consider on $X$ a Riemannian type form $a(u,v) = \int \alpha(u,v)(dx)$ with domain $D[a]$; we denote by $D_0[a, \Omega]$ the domain of the restriction of the form to $\Omega$ and we denote by $V_\epsilon(\Omega)$ the closure of the space of the functions $v$ in $C_0(\Omega) \cap D_{loc}[a, \Omega]$ for the norm

$$||v||_\epsilon = [\int_{\Omega_c} \alpha(v,v)(dx) + \int_{\Omega_c} |v|^2 m(dx)]^{\frac{1}{2}}$$

We assume that the following scaled Poincaré inequality hold

$(P_1)$ $$\int_{B_{j,\epsilon}-T_{j,\epsilon}} |v - \bar{v}_{j,\epsilon}|^2 m(dx) \leq C_2 \epsilon^2 \int_{B_{j,\epsilon(1+\delta)}-T_{j,\epsilon}} \alpha(v,v)(dx)$$

for every $v \in V_\epsilon(\Omega)$ where $C_2$ is a constant independent of $j$, $\epsilon$ and $\bar{v}_{j,\epsilon}$ is the average of $v$ for the measure $m(dx)$ on $B_{j,\epsilon} - B_{j,\epsilon(1-\delta)}$. From $(P_1)$ we can prove the following coercivity inequality

**Proposition 2.1** *There exists $\epsilon_0 > 0$ such that for $\epsilon \leq \epsilon_0$ we have*

$$\int_{\Omega_\epsilon} \alpha(v,v)(dx) \geq \lambda ||v||^2_{L^2(\Omega_\epsilon)}$$

*where $v \in V_\epsilon(\Omega)$ and $\lambda$ is a positive constant.*

Let now $\eta_{j,\epsilon}$ be the cut-off function of $B_{j,\epsilon(1-\delta)}$ with respect to $B_{j,\epsilon}$ and define the linear operator $P^\epsilon : V_\epsilon \to D_0[a, \Omega]$ as

$$P^\epsilon v = (1 - \sum_j \eta_{j,\epsilon})v + \sum_j \eta_{j,\epsilon}\bar{v}_{j,\epsilon}, \text{ on } \Omega_\epsilon$$

$$P^\epsilon v = \sum_j \eta_{j,\epsilon}\bar{v}_{j,\epsilon}, \text{ on } \Omega - \Omega_\epsilon$$

We have

**Proposition 2.2** *The linear operator $P^\epsilon$ is bounded. Moreover denoted by $\tilde{v}_\epsilon$ the extension of $v$ by $\bar{v}_{j,\epsilon}$ to $T_{j,\epsilon}$ ($\tilde{v}$ is so defined on $\Omega$) we have that $\lim_{\epsilon \to 0}(P^\epsilon v - \tilde{v}_\epsilon) = 0$ in $L^2(\Omega, m)$. Consider now a sequence $v_\epsilon$ such that $||v_\epsilon||_\epsilon$ is bounded; then there exists a subsequence, denoted again by $v_\epsilon$, such that $\lim_{\epsilon \to 0} P^\epsilon v_\epsilon = w$ weakly in $D_0[a, \Omega]$, strongly in $L^2(\Omega, m)$ and $\lim_{\epsilon \to 0}(P^\epsilon v_\epsilon - \widetilde{(v_\epsilon)}_\epsilon) = 0$ in $L^2(\Omega, m)$ (equivalently $||P^\epsilon v_\epsilon - v_\epsilon||_{L^2(\Omega_\epsilon, m)}$ converges to 0), more exactly*

$$||(P^\epsilon v_\epsilon - \widetilde{(v_\epsilon)}_\epsilon)||^2_{L^2(\Omega)} \leq C\epsilon^2 \int_{\Omega_\epsilon} \alpha(v_\epsilon, v_\epsilon)(dx)$$

$$||(P^\epsilon v_\epsilon - v_\epsilon)||^2_{L^2(\Omega_\epsilon)} \leq C\epsilon^2 \int_{\Omega_\epsilon} \alpha(v_\epsilon, v_\epsilon)(dx)$$

*$(\tilde{v}_\epsilon)_\epsilon$) converges to $w$ in $L^2(\Omega)$ and $||v_\epsilon - w||_{L^2(\Omega_\epsilon, m)}$ converges to 0. Moreover if $v$ is a function in $D[a, \Omega]$ then $(P^\epsilon v - v)$ converges to 0 in $L^2(\Omega, m)$.*

The proof is based on the Poincarè inequality $(P_1)$ and the compact embedding of $D[a, \Omega]$ into $L^2(X, m)$ [6].

In the following we will use $P^\epsilon$ as an approximate extension operator.

We consider now the following problems

$(2.1_\epsilon)$ $$\int_{\Omega_\epsilon} \alpha(u_\epsilon, v)(dx) = < f, P^\epsilon v >_{(D_0[a,\Omega])', D_0[a,\Omega]}$$

$$u_\epsilon \in V_\epsilon(\Omega) \quad \forall v \in V_\epsilon(\Omega)$$

The above problems have a unique solution $u_\epsilon$. Choosing $v = u_\epsilon$ we obtain that

$$\int_{\Omega_c} \alpha(u_\epsilon, u_\epsilon)(dx) \leq C$$

where $C$ is a constant independent of $\epsilon$. We define the linear operator $B_\epsilon$ : $(D_0[a, \Omega])' \to D_0[a, \Omega]$ as

$$B_\epsilon f = P^\epsilon u_\epsilon$$

It is easy to prove that the operators $B_\epsilon$ are uniformly bounded then, at least after extraction of a subsequence denoted again by $B_\epsilon$, weakly converge at every $f \in (D_0[a, \Omega])'$ to a linear operator $B_0$ : $(D_0[a, \Omega])' \to D_0[a, \Omega]$ (i.e. $lim_{\epsilon \to 0} B_\epsilon u = B_0 f$ weakly in $D_0[a, \Omega]$ for every fixed $f \in (D_0[a, \Omega])'$) From Proposition 2.1 we have that $P^\epsilon u_\epsilon$ is bounded in $D[a, \Omega]$ then (taking into account that $B_\epsilon$ weakly converges to $B_0$) we have that

$$lim_{\epsilon \to 0} P^\epsilon u_\epsilon = u_0$$

weakly in $D[a, \Omega]$ and strongly in $L^2(\Omega, m)$. Then the operator $B_0$ : $(D_0[a, \Omega])' \to D_0[a, \Omega]$ is defined as

$$B_0 f = u_0$$

At first we observe that for $f, g \in (D_0[a, \Omega])'$

$$< g, B_0 f >_{(D_0[a,\Omega])', D_0[a,\Omega]} = < f, B_0 g >_{(D_0[a,\Omega])', D_0[a,\Omega]}$$

$$< f, B_0 f >_{(D_0[a,\Omega])', D_0[a,\Omega]} = < f, u_0 >_{(D_0[a,\Omega])', D_0[a,\Omega]}$$

$$= lim_{\epsilon \to 0} < f, P^\epsilon u_\epsilon >_{(D_0[a,\Omega])', D_0[a,\Omega]} == lim_{\epsilon \to 0} \int_{\Omega_c} \alpha(u_\epsilon, u_\epsilon)(dx)$$

$$\geq C \, limsup_{\epsilon \to 0} ||P^\epsilon u_\epsilon||^2_{D_0[a,\Omega]} \geq C ||u_0||^2_{D_0[a,\Omega]} \geq C ||B_0 f||^2_{D_0[a,\Omega]}$$

We prove now that $B_0$ is coercive

(2.2) $$< f, B_0 f >_{(D_0[a,\Omega])', D_0[a,\Omega]} = < f, u_0 >_{(D_0[a,\Omega])', D_0[a,\Omega]}$$

$$= lim_{\epsilon \to 0} < f, P^\epsilon u_\epsilon >_{(D_0[a,\Omega])', D_0[a,\Omega]} = lim_{\epsilon \to 0} \int_{\Omega_\epsilon} \alpha(u_\epsilon, u_\epsilon)(dx)$$

We have also

$$\int_{\Omega_\epsilon} \alpha(u_\epsilon, v)(dx) = \int_{\Omega} \alpha(A^{-1} f, P^\epsilon v)(dx)$$

where $A$ is the operator defined by the form $a$. Choose now $v$ as the restriction to $\Omega_\epsilon$ of $A^{-1} f$, we obtain

$$\int_{\Omega_c} \alpha(u_\epsilon, A^{-1} f)(dx) = \int_{\Omega} \alpha(A^{-1} f, P^\epsilon A^{-1} f)(dx)$$

then

$$\frac{1}{2} \int_{\Omega_c} \alpha(u_\epsilon, u_\epsilon)(dx) + \frac{1}{2} \int_{\Omega_c} \alpha(A^{-1}f, A^{-1}f)(dx)$$

$$\geq \int_{\Omega} \alpha(A^{-1}f, P^\epsilon A^{-1}f)(dx)$$

We obtain in the limit as $\epsilon \to 0$

$$lim\epsilon \to 0 \int_{\Omega_\epsilon} \alpha(u_\epsilon, u_\epsilon)(dx) \geq \int_{\Omega} \alpha(A^{-1}f, A^{-1}f)(dx) = ||f||_{(D_0[a,\Omega])'}$$

then using (2.2)

$$< f, B_0 f >_{(D_0[a,\Omega])', D_0[a,\Omega]} \geq ||f||_{(D_0[a,\Omega])'}$$

Taking into account the previous properties we have that the operator $B_0$ is invertible and we denote by $A_0 : D_0[a, \Omega] \to (D_0[a, \Omega])'$ the operator $B_0^{-1}$. We observe that we can prove that $A_0$ is defined on all $D_0[a, \Omega]$ and is bounded and coercive from $D_0[a, \Omega]$ into $(D_0[a, \Omega])'$ such that

$$\lambda \quad < Au, u >_{(D_0[a,\Omega])', D_0[a,\Omega]}$$

$$\leq \quad < A_0 u, u >_{(D_0[a,\Omega])', D_0[a,\Omega]} \quad \leq \Lambda \quad < Au, u >_{(D_0[a,\Omega])', D_0[a,\Omega]}$$

so $A_0$ defines a bilinear form $a_0(u, v)$ on $D_0[a, \Omega] \times D_0[a, \Omega]$, which is positive and such that

$$\lambda a(u, u) \leq a_0(u, u) \leq \Lambda a(u, u)$$

for every $u \in D_0[a, \Omega]$. The above domination inequality implies that the bilinear form $a_0$ is regular $(D_0[a_0, \Omega] = D_0[a, \Omega]$ and the norms of the graph are equivalent in the two cases). Finally we observe that $u_0 = B_0 f$ is the unique solution of the problem

$$a_0(u_0, v) = < f, v >_{(D_0[a,\Omega])', D_0[a,\Omega]}, \quad \text{for every } v \in D_0[a, \Omega] = D_0[a_0, \Omega]$$

We will now prove that $a_0$ is a Dirichlet form. To prove that $a_0$ is a Dirichlet form it is enough to prove that for $a_0$ a Markov type property holds. The proof of Markov property uses essentially the following semicontinuity property:

**Proposition 2.3** *Let $v_\epsilon \in V_\epsilon$ be a sequence such that*

$$\int_{\Omega_\epsilon} \alpha(v_\epsilon, v_\epsilon)(dx) \leq C$$

*and $P^\epsilon v_\epsilon$ converges to $v_0$ as $\epsilon \to 0$ weakly in $D_0[a, \Omega]$, then*

$$liminf_{\epsilon \to 0} \int_{\Omega_\epsilon} \alpha(v_\epsilon, v_\epsilon)(dx) \geq a_0(v_0, v_0)$$

Now we assume that $a$ is such that $\alpha(u,u)(dx) = \alpha(u,u)(x)m(dx)$ where $\alpha(u,u)(.) \in L^1(\Omega, m)$ for all $u \in D_0[a, \Omega]$.

As in [18] we can prove also that $a_0$ is strongly local and we denote by $\alpha_0(u,u)(dx)$ its energy density. As a consequence of the boundness and coercivity of $a_0$ with respect to $a$ and we obtain that

$$\lambda\alpha(u,u)(dx) \leq \alpha_0(u,u)(dx) \leq \Lambda\alpha(u,u)(dx)$$

[18], then $\alpha_0(u,u)(dx) = \alpha_0(u,u)(x)m(dx)$ where $\alpha_0(u,u)(.) \in L^1(\Omega, m)$ for all $u \in D_0[a, \Omega]$. Moreover if $a$ is of a Riemannian form also $a_0$ is a Riemannian form with the same domain as $a$ with respect to a distance which is of the form $\Lambda^{-1}d$ where $d$ is the distance associated with $a$. We have so the following result:

**Theorem 2.4** *Let $u_\epsilon$ be the solution of the problems $(2.1_\epsilon)$. There exists a subsequence of $\epsilon$ denoted again by $\epsilon$ (that does not depend on $f$) such that (for every $f$) $P^\epsilon u_\epsilon$ strongly converges in $L^2(\Omega, m)$ to $u_0$; where $u_0$ is the solution of the problem*

$$(2.1_0) \qquad \int_\Omega \alpha_0(u_0, v)(x)m(dx) = < f, u_0 >_{(D_0[a,\Omega])', D_0[a,\Omega]}$$

$$u_0 \in D[a, \Omega], \quad \forall v \in D[a, \Omega]$$

*where $a_0(u,v) = \int_\Omega \alpha(u,v)m(dx)$ is a Riemannian type Dirichlet form with an energy density $\alpha_0(v,v) \in L^1(\Omega, m)$ and*

$$\lambda\alpha(u,u)(dx) \leq \alpha_0(u,u)(dx) \leq \Lambda\alpha(u,u)(dx)$$

We observe that the convergence result in Theorem 2.4 holds again if in the problems $(2.1_\epsilon)$ we replace $f$ by a sequence $f_\epsilon$ converging to $f$ in $(D[a,\Omega])'$; then we obtain:

**Theorem 2.5** *Let $u_\epsilon$ be solutions of the problems*

$$(2.1'_\epsilon) \qquad \int_{\Omega_c} \alpha(u_\epsilon, v)\, m(dx) = \int_{\Omega_c} f_\epsilon v\, m(dx)$$

$$u_\epsilon \in V_\epsilon(\Omega) \quad \forall v \in V_\epsilon(\Omega)$$

*where $f_\epsilon, f \in L^2(\Omega, m)$ and $f_\epsilon \to f$ in $L^2(\Omega, m)$. There exists a subsequence of $\epsilon$ denoted again by $\epsilon$ (that does not depend on $f$) such that (for every $f$) $P^\epsilon u_\epsilon$ strongly converges in $L^2(\Omega, m)$ to $u_0$; where $u_0$ is the solution of the problem*

$$(2.1'_0) \qquad \int_\Omega \alpha_0(u, v)\, m(dx) = \int_\Omega \theta f u_0\, m(dx)$$

$$u_0 \in D[a, \Omega] \quad \forall v \in D[a, \Omega]$$

*where $\alpha_0$ is as in Theorem 2.4.*

An easy consequence of Theorem 2.5 is the convergence of the global energy of the solution $u_\epsilon$ of (2.1'$_\epsilon$) (with respect to $\alpha$) in $\Omega_\epsilon$ to the global energy of the solution $u$ of (2.1'$_0$)(with respect to $\alpha_0$) in $\Omega$ i.e.

$$(2.3) \qquad lim_{\epsilon \to 0} \int_{\Omega_\epsilon} \alpha(u_\epsilon, u_\epsilon) m(dx) = \int_\Omega \alpha_0(u_0, u_0) m(dx)$$

We can also prove a local result of convergence of energies.

We first observe that denoting by $u_{\epsilon,\lambda}$ the solution of the problem

$$(2.1'_{\epsilon,\lambda}) \qquad \int_{\Omega_\epsilon} \alpha(u_{\epsilon,\lambda}, v) \ m(dx) + \lambda \int_{\Omega_\epsilon} u_{\epsilon,\lambda} v \ m(dx) = \int_{\Omega_\epsilon} fv \ m(dx)$$

$$u_\epsilon \in V_\epsilon(\Omega) \quad \forall v \in V_\epsilon(\Omega)$$

(where $\lambda > 0$). Assume that $P^\epsilon u_{\epsilon,\lambda}$ strongly converges in $L^2(\Omega, m)$ to $u_{0,\lambda}$; then $u_0$ is the solution of the problem

$$(2.1'_{0,\lambda}) \qquad \int_\Omega \alpha_0(u, v) \ m(dx) + \lambda \int_{\Omega_\epsilon} \theta u_0 v \ m(dx = \int_\Omega \theta f u_0 \ m(dx)$$

$$u_0 \in D[a, \Omega] \quad \forall v \in D[a, \Omega]$$

We also observe that if $f$ is bounded all the $u_{\epsilon,\lambda}$ are uniformly (with respect to $\epsilon$) bounded on $\Omega_\epsilon$ and that (2.3) hold again for $u_{\epsilon,\lambda}$ and $u_{0,\lambda}$.

Let now $f$ be bounded and $\phi \geq 0$ be in $D[a, \Omega]$ such that $A_0\phi \in L^2(\Omega)$ where $A_0$ is the operator corresponding to the form $A_0$. Let $\phi_\epsilon$ be the solution of the problem (2.1'$_\epsilon$) with $f = \theta^{-1}A_0\phi$ (we recall that $\theta \geq \sigma > 0$); we have that $\phi_\epsilon$ converges to $\phi$ in $L^2(\Omega)$ and $\int_{\Omega_\epsilon} \alpha(\phi_\epsilon, \phi_\epsilon) \ m(dx) \leq C$; moreover

$$lim_{\epsilon \to 0} \int_{\Omega_\epsilon} \alpha(\phi_\epsilon^-, \phi_\epsilon^-) \ m(dx) = 0$$

Using $u_{\epsilon,\lambda}\phi_\epsilon$ in (2.1'$_{\epsilon,\lambda}$) we obtain that

$$(2.4_\lambda) \qquad lim_{\epsilon \to 0} \int_{\Omega_\epsilon} \alpha(u_{\epsilon,\lambda}, u_{\epsilon,\lambda})\phi_\epsilon m(dx) \geq \int_\Omega \alpha_0(u_{0,\lambda}, u_{0,\lambda})\phi m(dx)$$

Using now as test function $u_{\epsilon,\lambda}\phi$ we can prove that

$$(2.5_\lambda) \qquad lim_{\epsilon \to 0}\theta_\epsilon \alpha(u_{\epsilon,\lambda}, u_{\epsilon,\lambda}) = \chi_\lambda \in L^1(\Omega)$$

weakly* in the space of measures. From (2.5$_\lambda$) and (2,4$_\lambda$) we obtain that

$$\chi_\lambda \geq \alpha_0(u_{0,\lambda}, u_{0,\lambda})$$

so the global convergence of energies give us the equality

$$\chi_\lambda = \alpha_0(u_{0,\lambda}, u_{0,\lambda})$$

Then

$$(2.6_\lambda) \qquad lim_{\epsilon \to 0}\theta_\epsilon \alpha(u_{\epsilon,\lambda}, u_{\epsilon,\lambda}) = \alpha_0(u_{0,\lambda}, u_{0,\lambda})$$

weakly$^*$ in the space of measures. The result can be generalized to the case $\lambda > 0$, $f \in L^2(\Omega)$ by approximation on $f$, so it is easily to deduce that the result holds also for $\lambda = 0$. We have so proved:

**Theorem 2.6** *Let $u_\epsilon$, $u_0$ be as in Theorem 2.5, then*

$$(2.6_0) \qquad lim_{\epsilon \to 0} \theta_\epsilon \alpha(u_\epsilon, u_\epsilon) = \alpha_0(u_{0,\lambda}, u_{0,\lambda})$$

*weakly$^*$ in the space of measures.*

Taking into account Theorem 2.2 we obtain:

**Theorem 2.7** *Assume $f \in L^2(\Omega, m)$. Let $u_\epsilon$ be solution of the problem $(2.1_\epsilon)$, $w_\epsilon$ be solution of the problem $(2.1'_\epsilon)$ with $f$ replaced by $\frac{f}{\theta}$, then $u_0$ be as in Theorem 2.5. There exists a subsequence of $\epsilon$ denoted again by $\epsilon$ (that does not depend on $f$) such that $P^\epsilon u_\epsilon$ and $P^\epsilon w_\epsilon$ converge to $u_0$ weakly in $D[a, \Omega]$ and strongly in $L^2(\Omega, m)$, where $u_0$ is the solution of the problem $(2.1_0)$; moreover we have that $\|u_\epsilon - w_\epsilon\|_\epsilon \to 0$ as $\epsilon \to 0$.*

From Theorem 2.6 and 2.7 we obtain:

**Theorem 2.8** *Assume $f \in (D[a, \Omega])'$ and let $u_\epsilon$ be the solution of problem $(2.1_\epsilon)$. We denote again by $u_\epsilon$ the subsequence considered in Theorem 2.3 converging to the solution $u_0$ of the problem $(2.1_0)$ then*

$$lim_{\epsilon \to 0} \theta_\epsilon \alpha(u_\epsilon, u_\epsilon) = \alpha_0(u_{0,\lambda}, u_{0,\lambda})$$

*weakly$^*$ in the space of measures.*

## References

1. G. Allaire, F. Murat: 1993, 'Homogenization of the Neumann problem with non-isolated holes' *Asymp. An.* **7**, pp. 81-95.
2. M. Biroli, U. Mosco: 1991, 'Formes de Dirichlet et estimations structurelles dans les milieux discontinus', *C.R.A.S. Paris* **315**, Sér I, pp. 193–198.
3. M. Biroli and U. Mosco: 1995, 'A Saint-Venant principle for Dirichlet forms on discontinuous media', *Ann. Mat. Pura Appl.*, (iv) **169**, pp. 125-181.
4. M. Biroli, N.A. Tchou: 2005 in preparation,
5. M. Biroli, N.A. Tchou, V.V. Zhikov: 1999, ' Homogenization for Heisenberg operator with Neumann boundary conditions', *Ricerche di Matematica*, **48** Supplemento, pp. 45-59.
6. M. Biroli, S. Tersian: 1997, 'On the existence of nontrivial solutions to a semilinear equation relative to a Dirichlet form', *Rend. Istituto Lombardo Acc. di Scienze e Lettere, A* **131**, pp. 151-168.
7. M. Briane: 1987, 'Poincarè-Writinger inequality for homogenization in perforated domains' *Boll. UMI* **11-B**, pp. 53-82.
8. M. Briane, A. Damlamian, P. Donato: 1998, 'H-convergence in perforated domains', *Nonlinear Partial Differential Equations and Applications, Collège de France Seminar* vol. XIII. Research Notes in Math. Pitman London pp. 62-100.

720

9.  D. Cioranescu, J. Saint Jean Paulin: 1979, 'Homogenization in open sets with holes',
    *J. Math. An. Appl.* **71**, pp. 590-607.
10. A. Damlamian, P. Donato: 2002, 'Which sequence of holes are admissible for periodic
    homogenization with Neumann boundary condition?', *ESAIM: Control, Optimizatio-
    nand Calculus of Variations*, **8**, pp. 555-585.
11. A. Damlamian, P. Donato: 2003, 'Periodic homogenization in perforated domains with
    Neumann conditions' *Homogenization 2001*, Gakuto international Series in Mathemat-
    ical Sciences **18**, 169-182.
12. M. Fukushima: 1980, *Dirichlet Forms and Markov Processes*, North Holland Math.
    Lib., North Holland Amsterdam.
13. E.Ja. Hruslov: 1972, 'The method of hortogonal projections and the Dirichlet problem
    in domains with a fine-grained boundary', *Math. USSR Sb.* **17**, pp. 37-59.
14. E.Ja. Hruslov: 1979, 'The asymptotic behavior of solutions of the second boundary
    value problems under fragmentation of the boundary of the domain', *Math. USSR Sb.*
    **35**, pp. 266-282.
15. G. Lu, 'On Harnack's inequality for a class of degenerate Schrödinger operators formed
    by vector fields', *preprint*.
16. V.A. Marčenko, E.Ja. Hruslov: 1974, *Boundary value problems in domains with fine-
    grained boundaries*, Nauk. Dumka, Kiev.
17. U. Mosco: 1992, 'Compact families of Dirichlet forms', Cours 3ème cycle, Université
    Paris VI.
18. U. Mosco: 1994, 'Composite media and asymptotic Dirichlet forms', *J. Func. An.* **123**,
    pp. 368–421.
19. U. Mosco: 1998, 'Dirichlet forms and self-similarity', *New Directions in Dirichlet
    forms*, AMS/IP Studies in Advanced Mathematics 8, International Press, Cambridge
    (MA), pp. 117-155.
20. G. Nguetseng: 2004, 'Homogenization in perforated domains beyond the periodic set-
    ting', *J. Math. Analysis Appl.*, **289**, pp. 608-628.
21. V.V. Zhikov: 1993, 'Averaging in perforated random domains of general type', *Math.
    Notes* **53**, pp. 30-42
22. V.V. Zhikov: 1994, 'On the homogenization of nonlinear variational problems in per-
    forated domains', *Russian J. of Math. Phys.* **2**, pp. 393-408
23. V.V. Zhikov: 1996, 'Connecttedness and homogenization. Examples of fractal conduc-
    tivity' *Math. USSR Sb.* **187**, pp. 1109-1147.

# MOUNTAIN PASS TECHNIQUES FOR SOME CLASSES OF NONVARIATIONAL PROBLEMS

M. GIRARDIE and S. MATALONI

*Dipartimento di Matematica*
*Università degli Studi di Roma Tre*
*Largo San Murialdo, 00146 Roma, Italia*

M. MATZEU

*Dipartimento di Matematica*
*Università degli Studi di Roma "Tor Vergata"*
*Via della Ricerca Scientifica,*
*00133 Roma, Italia*
*E-mail: matzeu@mat.uniroma2.it*

Existence results are presented for classical solutions to some nonvariational problems through a suitable approximation method of Mountain Pass critical points.

**Key words:** Mountain pass, nonvariational problems, quasilinear elliptic equation, classical solutions

**Mathematics Subject Classification:** 35B38, 35J60

In this paper we present some results about the existence of classical solutions to certain classes of nonvariational problems through a suitable approximation method of Mountain Pass critical points (see [2-4]).

What is the basic idea? Let us consider for example a quasilinear elliptic equation of the form

$$(\mathbf{P}) \begin{cases} -\Delta u(x) = f(x, u(x)\nabla u(x)), \ x \in \Omega, \\ u(x) = 0 \qquad x \in \partial\Omega \end{cases}$$

where $\Omega$ is an open bounded subset of $\mathbb{R}^N$ with a sufficiently smooth boundary $\partial\Omega$. Well, if one "freezes" the gradient variable, that is fix $\nabla u = \nabla w$, with $w$ arbitrary in the space $H_0^1(\Omega)$, then one gets a semilinear problem, so a variational one:

$$(\mathbf{P_w}) \begin{cases} -\Delta u_w(x) = f(x, u_w(x)\nabla w(x)), \ x \in \Omega, \\ u_w(x) = 0 \qquad x \in \partial\Omega \end{cases}$$

At this point one finds a solution $u_w$ of Mountain Pass type and discover that some estimates from above and from below hold for the $H_0^1$-norm of $u_w$ which are independent of $w$. This suggests to write an iterative scheme of the type

$$\begin{cases} -\Delta u_n(x) = f(x, u_n(x), \nabla u_{n-1}(x)), & x \in \Omega, \\ u_n(x) = 0 & x \in \partial\Omega \end{cases}$$

where one starts by fixing $w = u_0$, then considers a Mountain Pass solution $u_1$ associated with $w = u_0$ and so on for any natural number $n$. Then the estimate from above yields a sequence $\{u_n\}$ which is bounded in $H_0^1(\Omega)$, as well as in $C^1(\Omega)$, due to some standard boot-strap arguments based on Schauder's and Agmon-Douglas-Nirenberg's theorems.

At this point a suitable assumption of smallness of the local Lipschitz continuity constants of $f$ with respect to the first eigenvalue of $-\Delta$ in $\Omega$ enables to prove that actually $\{u_n\}$ converges to some $u$ which is indeed a classical solution of the initial problem. This solution $u$ is not trivial due to the estimate from below. For details see [2,3].

We have also considered other two types of nonvariational problems: integro-differential elliptic equations with non-symmetric kernels and fully non linear equations.

As for the first problem (see [4]) one considers the equation

$$(\mathcal{K}) \quad \begin{cases} -\Delta u = f(x, u) + \int_\Omega k(x, y) u(y)\, dy \text{ in } \Omega \\ u = 0 \text{ on } \partial\Omega \end{cases}$$

with

$$k(x, y) \neq k(y, x), \quad k \leq 0, \quad k \in L^2(\Omega \times \Omega)$$

$$\left( \int_\Omega k(x, y)^{(2^*)'} dy \right) \in L^{\frac{2^*}{p(2^*)'}}(\Omega)$$

$$\|k\|_{L^2(\Omega \times \Omega)} \leq h(\lambda_1, p, a_2, a_3, |\Omega|, \theta)$$

Assuming the following further conditions

$$\lambda_1^{-1}(L_1 + \|k\|_{L^2(\Omega \times \Omega)}) < 1$$

(where $L_1$ is the local Lipschitz continuity coefficient of $f$ w.r. to $u$), one can prove the existence of a positive solution $u$ to $(\mathcal{K})$.

Let us write now is a precise way the result abount fully nonlinear equations that is

$$(\mathcal{P}) \quad \begin{cases} -\Delta u = f(x, u(x), \nabla u(x), D^2 u(x)) \text{ in } \Omega \\ u = 0 \text{ on } \partial\Omega \end{cases}$$

One can state the following

**Theorem** Let $N \geq 3$, let $S^N$ the space of the $N \times N$ matrices and let $f$ be a function satisfying the following assumptions:

($f0$) $f : \overline{\Omega} \times \mathbb{R} \times \mathbb{R}^N \times \mathcal{S}^N$ is locally Lipschitz continuous w.r.t.
the first three variables and continuous w.r.t the last one.

($f1$) $\lim_{t \to 0} \frac{f(x,t,\xi,X)}{t} = 0$ uniformly for $x \in \overline{\Omega}, \xi \in \mathbb{R}^N, X \in \mathcal{S}^N$

($f2$) $\exists a_1 > 0, \ p \in \left(1, \frac{N+2}{N-2}\right)$ and $r, s \in (0,1)$ with $r + s < 1$:
$|f(x,t,\xi,X)| \le a_1(1 + |t|^p)(1 + |\xi|^r)(1 + ||X||^s),$
$\forall x \in \overline{\Omega}, \quad t \in \mathbb{R}, \xi \in \mathbb{R}^N, X \in \mathcal{S}^N$

($f3$) $\exists \theta > 2 : \ 0 < \theta F(x,t,\xi,X) \le tf(x,t,\xi,X)$
$\forall x \in \overline{\Omega}, \ t \in \mathbb{R} \setminus \{0\}, \xi \in \mathbb{R}^N, X \in \mathcal{S}^N$
where $F(x,t,\xi,X) = \int_0^t f(x,\tau,\xi,X) \, d\tau.$

($f4$) $\exists a_2, a_3 > 0 : \ F(x,t,\xi,X) \ge a_2|t|^\theta - a_3$
$\forall x \in \overline{\Omega}, \quad t \in \mathbb{R}, \xi \in \mathbb{R}^N, X \in \mathcal{S}^N$ .

($f5$) there exists a positive number $\rho$ depending explicitly on
$p, r, s, \theta, a_1, a_2, a_3, N, |\Omega|$ given in the previous hyptotheses, such that
$f(x,t,\xi,X) = g(x,t,\xi) \qquad \forall(x,t,\xi,X) \in B$ where
$B = \{(x,t,\xi,X) : \ x \in \overline{\Omega}, |t|, |\xi|, ||X|| \in [0,\rho]\},$
and $g$ is a $C^{1,\alpha}$ − function on $\overline{\Omega} \times \mathbb{R} \times \mathbb{R}^N$ such that ($f1$)-($f4$) hold
with $f$ replaced by $g$ .

Then there exist a positive and a negative solution $u$ of problem ($\mathcal{P}$) provided
the following relation holds:

$$(*)\lambda_1^{-\frac{1}{2}} L''_\rho + \lambda_1^{-1} L'_\rho < 1$$

where $\lambda_1$ is the first eigenvalue of the Laplacian with Dirichlet boundary conditions,
and $L'_\rho, L''_\rho$ are defined as follows

($f6$) $L'_\rho := \sup \left\{ \frac{|g(x,t',\xi) - g(x,t'',\xi)|}{|t'-t''|}, \quad x \in \overline{\Omega}, |t'|, |t''|, |\xi| \in [0,\rho] \right\},$

($f7$) $L''_\rho := \sup \left\{ \frac{|g(x,t,\xi') - g(x,t,\xi'')|}{|\xi'-\xi''|}, \quad x \in \overline{\Omega}, |t|, |\xi'|, |\xi''| \in [0,\rho] \right\},$

where $\rho$ is given in ($f5$).

An example of nonlinear term $f$ satisfying the hypothesis of the theorem is given
by

$$f(x,t,\xi,X) = \begin{cases} \tilde{a}_1 t|t|^{p-1}(1 + |\xi|^\beta) & \text{for } |t|, \ |\xi|, ||X|| \text{ small} \\ t|t|^{p-1}|\xi|^r||X||^s & \text{for } |t|, \ |\xi|, ||X|| \text{ large} \end{cases}$$

where $\beta > 1$ and $\tilde{a}_1 > 0$ sufficiently small.

Now let us give a sketch of the proof of Theorem. Firstly let us fix an arbitrary
$R > 0$ and consider the set

$$C_R = \{v \in \ H_0^1(\Omega) \cap C^{2,\alpha} : ||u||_{C^{2,\alpha}} \le R\},$$

Fix $w \in C_R$ and consider the functional

$$I_w(v) : \frac{1}{2}\int_\Omega |\nabla v|^2 - \int_\Omega F(x,v) \, dx, \ \forall v \in H_0^1(\Omega)$$

whose critical points $u_w$ are the solutions of the problem

$$(\mathcal{P}_w) \qquad \begin{cases} -\Delta u = f(x, u(x), \nabla w(x), D^2 w(x)) \text{ in } \Omega \\ u = 0 \text{ on } \partial\Omega \end{cases}$$

One can check that the following conditions are verified

$(AR1)$ $I_w(0) = 0$;

$(AR2)$ $\exists \rho_R > 0$, $\alpha_R > 0 : I_w(v) \geq \alpha_R \ \forall v \in H_0^1(\Omega) : ||v|| = \rho_R$;

$(AR3)$ $\exists \overline{v} \in H_0^1(\Omega)$ independent of $R$ such that $I_w(\overline{v}) < 0$, $||\overline{v}|| \geq \rho > \rho_R$, where $\rho$ is a constant independent of $R$;

$(AR4)$ the Palais - Smale condition is satisfied by $I_w$.

As for $(AR3)$, one fixes an arbitrary $v_0 \in H_0^1(\Omega)\backslash\{0\}$ and note that, due to $(f_4)$

$$I_w(tv_0) \leq t^2 \int |\nabla v_0|^2 - a_2 t^\vartheta \int |v_0|^\vartheta + a_3 |\Omega| \ \to -\infty \ \text{ as } t \to +\infty$$

$$\Longrightarrow \forall t \geq \overline{t} \ I(tv_0) < 0$$

in particular for $\overline{v} = \overline{t}v_0 \ I_w(\overline{v}) < 0$.

Also let us point out that

$$\rho_R < \rho = \overline{t}||v_0|| = ||\overline{v}|| \quad (\overline{v} \text{ is independent of } w \text{ and } R)$$

$(\rho_R \to 0 \text{ as } R \to +\infty)$.

Therefore, the Ambrosetti-Rabinowitz's theorem (see [1]) yields the following

**Proposition 1** *There exists a Mountain Pass critical point* $u_w$ *of* $I_w$, *that is*

$$I'_w(u_w) = 0, \ I_w(u_w) = \inf_{\gamma \in \Gamma} \max_{t \in [0,1]} I_w(\gamma(t))$$

*where*

$$\Gamma = \left\{ \gamma \in C^0([0,1], H_0^1) : \ \gamma(0) = 0, \gamma(1) = \overline{v} \right\}$$

$$I_w(u_w) \geq \alpha_R > 0 \quad (\Rightarrow \ u_w \not\equiv 0)$$

Actually one can establish also some estimates for $||u_w||$. Indeed the following proposition holds:

**Proposition 2** *There exist two positive constant numbers* $c_1(R)$, *depending on* $R$, *and* $c_2$ *independent of* $R$ *such that:*

$$c_1(R) \leq ||u_w|| \leq c_2 \quad \forall w \in C_R, \forall R > 0$$

The scheme of the proof of Proposition 2 is given as follows:

## Energy Estimate

$$I_w(u_w) \leq \max_{v \in \overline{0v}} I_w(v) \, .$$

$\overline{0v}$ = segment line joining 0 and $\overline{v}$ in $H_0^1(\Omega)$

$$\Rightarrow I_w(u_w) \leq \max_{t \in [0,1]} \left\{ \frac{t^2}{2} \int_\Omega |\nabla \overline{v}|^2 - a_2 |t|^\theta \int_\Omega |\overline{v}|^\theta - a_3 |\Omega| \right\} =$$

constant independent of $w$ and $R$.

$$\Rightarrow \|u_w\| \leq c_1 \, , \quad \text{constant independent of } w \text{ and } R \, .$$

## Norm Estimate from below

$$\forall R > 0 \, \forall w \in C_r, \, \exists c(R) > 0 : \, \|u_w\| \geq c(R)$$

This is true independently of the Mountain Pass nature of $u_w$, but only by the superlinear growth of $f$ w.r. to the $u-$ variable.

At this point one states a $C^{2,\alpha}$-estimate for $u_w$. The passage from the $H_0^1$-estimate to this one is due, in a standard way, to the Agmon-Douglis-Nirenberg's, to the Shauder's theorems and to the Sobolev embeddings. At any case it relies on condition $(f_2)$, then it arises some constant depending, in this case, on the $R$-variable, more precisely

$$\|u_w\|_{C^{c,\alpha}} \leq \mu (I + R)^{r+s} \quad \text{for some } \mu > 0 \, .$$

Since one would have that $C_R$ is stable under the transformation $w \to u_w$, it is natural to require that there exists some $\overline{R} > 0$ such that

$$\mu (I + \overline{R})^{r+s} \leq \overline{R} \, .$$

This condition is surely satisfied for any $\overline{R}$ sufficiently large if one assumes, as in $(f_2)$, $r + s < 1$. Therefore there exists some $\overline{R} > 0$ such that

$$w \in C_{\overline{R}} \Rightarrow u_w \in C_{\overline{R}} \, .$$

Now we are able to construct a sequence $\{u_n\}$ which converges to a classical solution of the fully nonlinear equation.

Actually we fix an element $u_0$ in $C_{\overline{R}}$ and consider $u_n$ defined iteratively as follows

$$\begin{cases} -\Delta u_n(x) = f(x, u_n(x) \nabla u_{n-1}(x), D^2 u_{n-1}(x)), \, x \in \Omega, \\ u_n(x) = 0 \quad x \in \partial \Omega \end{cases}$$

Now if we choose $\rho = \overline{R}$ in the assumption $(f_5)$ and use the function $g(x, u, \nabla u)$ in place of $f(x, u, \nabla u, D^2 u)$ as indicated in $(f_5)$, it is possible to check that the condition (*) allows to state that $\{u_n\}$ is a Cauchy sequence in $H_0^1$.

Indeed, following the iterative scheme, one has, for any $n \in \mathbb{N}$,

$$\int_\Omega \nabla u_{n+1}(x)(\nabla u_{n+1}(x) - \nabla u_n(x))\, dx =$$
$$\int_\Omega f(x, u_{n+1}(x), \nabla u_n, D^2 u_n)(u_{n+1}(x) - u_n(x))\, dx$$

$$\int_\Omega \nabla u_n(x)(\nabla u_{n+1}(x) - \nabla u_n(x))\, dx =$$
$$\int_\Omega f(x, u_n(x), \nabla u_{n-1}, D^2 u_{n-1})(u_{n+1}(x) - u_n(x))\, dx$$

$\Rightarrow$

$$\|u_{n+1} - u_n\|^2 \leq \int_\Omega |f(x, u_{n+1}, \nabla u_n, D^2 u_n) - f(x, u_n, \nabla u_n, D^2 u_{n-1})|\, |u_{n+1} - u_n|\, dx$$
$$+ \int_\Omega |f(x, u_n, \nabla u_n, D^2 u_{n-1}) - f(x, u_n, \nabla u_{n-1}, D^2 u_{n-1})|\, |u_{n+1} - u_n|\, dx$$

But

$$(x, u_{n+1}(x), \nabla u_n(x), D^2 u_n(x)),\ (x, u_n(x), \nabla u_n(x), D^2 u_{n-1}(x)),$$
$$(x, u_n(x), \nabla u_{n-1}(x), D^2 u_{n-1}(x))\text{all belong to } B\ .$$

$\Rightarrow f = g$ on these points. Then, by Poincaré inequality

$$\|u_{n+1} - u_n\|^2 \leq L'_\rho \lambda_1^{-1}\|u_{n+1} - u_n\|^2 + L''_\rho \lambda_1^{-\frac{1}{2}}\|u_n - u_{n-1}\| \cdot \|u_{n+1} - u_n\|$$

Therefore

$$\|u_{n+1} - u_n\| \leq \frac{L''_\rho \lambda_1^{-\frac{1}{2}}}{1 - L'_\rho \lambda_1^{-1}}\|u_n - u_{n-1}\|$$

and $\gamma = \frac{L''_\rho \lambda_1^{-\frac{1}{2}}}{1 - L'_\rho \lambda_1^{-1}}$ is less than 1 by (*)

$\Rightarrow \{u_n\}$ is a Cauchy sequence in $H_0^1(\Omega)$.

Therefore, due to the $C^{e,\alpha}$-estimate and the Ascoli-Arzela's theorem, one deduces that $\{u_n\}$ converges to some $u$ in $C^1$.

Let us describe now, by a scheme, the passage to the limit in the sequence of itarated problems.

**(A).** $u_n \to u$ in $C^2(\overline{\Omega})$
(in fact ($\clubsuit$) $|D^3(u_n(x))| \leq const$)
since $g \in C^{1,\alpha}$ and $\|u_n\|_{C^{1,\alpha}} \leq const$, so ($\clubsuit$) follows from Schauder Theorem.

**(B).** $u_n \to u$ in $C^2(\overline{\Omega}) \Rightarrow$

$$g(x, u_n(x), \nabla u_{n-1}(x)) = f(x, u_n(x), \nabla u_{n-1}(x), D^2 u_{n-1}(x))$$
$$\to g(x, u(x), \nabla u(x)) = f(x, u(x), \nabla u(x), D^2 u(x))$$
$$((x, u(x), \nabla u(x), D^2 u(x)) \in B!!)$$

**(C).** $\Delta u_n \to \Delta u$ uniformly in $\overline{\Omega}$.

Therefore $u$ is a solution of $\mathcal{P}$.

As for the existence of a strictly positive and a strictly negative solution of $(P)$, one argues as follows. For any $n \in \mathbb{N}$, some standard techniques allow to state that one can exhibit a strictly positive solution, say $u_n^+$, and a strictly negative solution, say $u_n^-$, to the iterated problems. The estimate from below in Proposition 2 allows to claim that their limits, respectively $u^+$ and $u^-$ are not identically zero. Finally the strict positivity of $u^+$ and the strict negativity of $u^-$ are standard consequences of the strong maximum principle.

## References

1. A. Ambrosetti, P.H. Rabinowitz, Dual variational methods in critical point theory and applications, *J. Funct. Anal.* **14** (1973), 349–381.
2. D. De Figuereido, M. Girardi, M. Matzeu, Semilinear elliptic equations with dependence on the gradient via Mountain–Pass techniques, *Differential and Integral Equations* **17**, 1–2 (2004), 119–126.
3. M. Girardi, M. Matzeu, Positive and negative solutions of a quasi–linear equation by a Mountain–Pass method and troncature techniques, *Nonlinear Analysis* **59** (2004), 199–210.
4. S. Mataloni, M. Matzeu, Semilinear integrodifferential problems with non–symmetric kernels via Mountain–Pass techniques, *Advanced Nonlinear Studies*, to appear, **5** (2005), 23–31.

# ON SOME SCHRÖDINGER TYPE EQUATIONS

SERGIO POLIDORO

*Dipartimento di Matematica, Università di Bologna*
*Piazza di Porta S. Donato, 5*
*40127 Bologna, Italia*
*E-mail address: polidoro@dm.unibo.it*

MARIA ALESSANDRA RAGUSA

*Dipartimento di Matematica, Università di Catania*
*Viale A. Doria, 6, 95125 Catania, Italia*
*E-mail address: maragusa@dmi.unict.it*

We present some results concerning a Harnack inequality for solutions of the ultra-parabolic equation

$$L_0\, u + Vu = 0,$$

where $L_0$ is the Kolmogorov operator in $\mathbb{R}^{n+1}$ and $V$ belongs to a Stummel-Kato class defined via the fundamental solution of $L_0$.

**Keywords:** Harnack inequality, ultraparabolic equation
**AMS Subject Classifications:** 35K70, 35J10, 35K20, 32A37, 35B65.

## 1. Introduction

Harnack inequality constitute an important part of the study of differential operators with wide fields of applications, mainly to the mathematical phisics. Now we present some results concerning a Harnack inequality for positive solutions of the hypoelliptic equation

$$L_0\, u + V\, u = 0, \tag{1.1}$$

where $L_0$ is the Kolmogorov operator in $\mathbb{R}^{n+1}$:

$$L_0 u \equiv \operatorname{div}\,(ADu) + \langle x, BDu \rangle - \partial_t u \tag{1.2}$$

$V$ belongs to a class of functions of Stummel-Kato type we will define later. In (1.2) $A = (a_{i,j})$, $B = (b_{i,j})$ are two $n \times n$ matrices with real constant entries; $z = (x,t) = (x_1, ..., x_n, t) \in \mathbb{R}^{n+1}$, $D = (\partial_{x_1}, ...., \partial_{x_n})$. We assume that the matrices have the following form in the form

$$
A = \begin{pmatrix} A_0 & 0 \\ 0 & 0 \end{pmatrix} \qquad
B = \begin{pmatrix}
0 & B_1 & 0 & \dots & 0 \\
0 & 0 & B_2 & \dots & 0 \\
\vdots & \vdots & \vdots & \ddots & \vdots \\
0 & 0 & 0 & \dots & B_r \\
0 & 0 & 0 & \dots & 0
\end{pmatrix} \tag{1.3}
$$

where $A_0$ is the identity $m_0 \times m_0$ matrix and every $B_j$ is a $m_{j-1} \times m_j$ block matrix with rank $m_j$, where $j = 1, 2, ..., r$, $m_0 \geq m_1 \geq \cdots \geq m_r \geq 1$ and $m_0 + m_1 + \cdots + m_r = n$. To motivate our interest in this kind of operators, we recall that arise in stochastic theory and in the theory of diffusion processes. We point out our attention in a Harnack inequality, invariant for positive solutions of (1.1). This problem has been considered by many authors in the last years and for a significant range of classes of operators $L_0$. For istance we recall the study made by Stampacchia in [23] who considers $L_0$ uniformly elliptic and show that if $V \in L^p, p > \frac{n}{2} \Rightarrow$ the solutions of $L_0 u + V u = 0$ are hölder continuous and it is true a uniform Harnack inequality.

Ladyženskaja and Ural'ceva in [15] prove that this is the best possible condition in the $L^p$ spaces in fact the authors show that the above result cannot be extended to $L_{loc}^{\frac{n}{2}}$, but it can be weakened in [1] by Aizeman and Simon, they prove a Harnack inequality in the case that $L_0$ is the Laplace operator and $V \in SK$, the class of locally summable functions in an open set $\Omega \subset \mathbb{R}^n$ such that

$$
\lim_{r \to 0} \left( \sup_{x \in \Omega} \int_{|x-y|<r} \frac{|V(y)|}{|x-y|^{n-2}} dy \right) = 0. \tag{1.4}
$$

We explicity observe that the Aizeman and Simon's result improves Stampacchia's result because

$$
L^p(\Omega) \subset SK(\Omega), \forall p > \frac{n}{2}.
$$

The proof by Aizeman and Simon is based on probabilistic techniques, but in the note [5] Chiarenza, E. Fabes, N. Garofalo extend the same inequality to the family of uniformly elliptic operators using analytic techniques.

Similar results are later obtained by Hinz and Kalf in [12], Simader in [22].

Results and techniques by Chiarenza, Fabes and Garofalo are semplified and extended by many authors and realized in different directions.

Gutierrez in [11] studies elliptic operators with degeneration of weight type.

Dal Maso and Mosco in [7] consider a class of relaxed Dirichlet problems having solutions that can be characterized as $\gamma$−limits of classic Dirichlet problems.

In this family of problems figure a Schrödinger type equation where the potential $V$ is replaced with a Borel measure $\mu$.

More recently the regularity problem for solutions of $L_0 u + V u = 0$ has been studied for a class of uniformly parabolic operators, we recall for istance the note [24] where Sturm suppose that the coefficients of the second order derivatives are continuous or the paper [25] by Zhang where are considered only measurable coefficients. In both cases is defined a space of parabolic Stummel-Kato class individualized by the fundamental solution of the heat equation $\Gamma_0$, as follows:

$$\sup_{(x,t)\,\in\,\mathbb{R}^{n+1}} \int_{(y,s)\in\Omega,\,t-r^2<s<t} \Gamma_0(x,t,y,s)|V(y,s)|dyds, \qquad (1.5)$$

$$\sup_{(y,s)\,\in\,\mathbb{R}^{n+1}} \int_{(x,t)\in\Omega,\,s<t<s+r^2} \Gamma_0(x,t,y,s)|V(x,t)|dxdt, \qquad (1.6)$$

tend to 0, as $r \to 0$.

In 1967 Hörmander studied in the celebrated note [13] the remarkable class of operators:

$$L_0 \equiv \sum_{j=1}^{m} X_j^2, \qquad (1.7)$$

where $X_0, X_1, \ldots, X_m$ are vector fields defined in an open set $\Omega \subset \mathbb{R}^n$ of the form

$$X_j = \sum_{i=1}^{n} a_{ij}(x)\partial_{x_i}, \quad j = 0, 1, \ldots m \qquad (1.8)$$

where $a_{ij} \in C^\infty(\Omega)$ verify hypoellipticity Hörmander condition. After the paper by Hörmander begin a systematic study to establish the principle properties of operators of type (1.7). The most important results of general character are:

- existence of the fundamental solution
- local estimate of the fundamental solution.

Let us recall, for istance, the study made by Folland in [10], Nagel, Stein and Wainger in [18], Sáncez-Calle in [21] and more recently by Jerison and Sáncez-Calle in [14].

Based on these instruments has been built a regularity theory for solutions of the equation

$$L_0 u = f$$

in Sobolev spaces and in spaces of hölder functions, similar to that one related to equations of elliptic and parabolic type (see [20] by Rotshild and Stein).

Regularity problem for solutions of Schrödinger equation $L_0 u + V u = 0$ has been studied, for example, by Citti, Garofalo, Lanconelli in [6]. This paper is different from the previous cited because here the potential $V$ of (1.1) is supposed only measurable. This note has been generalized in some directions:

- Lu in [17] considers the same kind of equation but with a degeneration of type $A_2$ weight;

- Biroli study the same problem for subelliptic p-laplace operators:

$$\sum_{j=1}^{m} X_j^* \left(|Xu|^{p-2}X_j u\right) + V|u|^{p-2}u = 0, \qquad (1.9)$$

where $Xu$ denotes the vector $(X_0u, ..., X_mu)$ and the fields $X_j$ verify Hörmander condition, see [2] and [3]. These two papers can be considered as a part of a systematic study made by Capogna, Danielli, Garofalo (see [4] and references therein).

We explicity observe that the above operator (1.9) is nonlinear then it is not possible to define a Kato class using the fundamental solution. Then Biroli in [2] use the following definition

$$\lim_{h \to 0} \left( \sup_{x \in \Omega} \int_0^r \left( \frac{1}{mis(B(x,s))} \left( |V(y)|dy \right) s^p \right) \frac{ds}{s} \right) = 0$$

where $B(x,s)$ is a sphere centered at $x$ with radius $s$ and $mis$ is the Lebesgue measure.

In the framework of Hörmander type operators, we consider not only the form of sum of square studied by Citti, Garofalo and Lanconelli in [6] but the operators belong to the more general class having the following form

$$L_0 = \sum_{j=1}^m X_j^2 + Y \tag{1.10}$$

where it is a sum of square and also a field of first order $Y$.

The operator $L_0$ in (1.2) can be express as in (1.10) if we set

$$Y = \langle x, BD \rangle - \partial_t, \quad X_j = \partial_{x_j}, \quad j = 1, \ldots, m_0.$$

In the note by Citti, Garofalo and Lanconelli the techniques used are that in the note [16] by Lanconelli and Polidoro.

We use some ideas by Zhang contained in [26], paper inspired by the note [8] and [9] by Fabes and Strook related to the homogeneous parabolic equation

$$L_0 u = 0.$$

## 2. Definitions and Results

Let us now give a definition of Stummel-Kato class more general of that one used in the above mentioned papers.

Let $\Omega$ be an open bounded open set in $\mathbb{R}^{n+1}$, $\Gamma_0$ the fundamental solution related to the Kolmogorov operator

$$L_0 u \equiv div\,(ADu) + \langle x, BDu \rangle - \partial_t u.$$

If $V \in L^1(\Omega)$ we set $\eta_V(h)$ and $\eta_V^*(h)$ respectively

$$\sup_{(x,t) \in \Omega} \int_{(y,s) \in \Omega, t-h^2 < s < t} \Gamma_0(x,t,y,s)|V(y,s)|dy\,ds,$$

$$\tag{2.1}$$

$$\sup_{(y,s) \in \Omega} \int_{(x,t) \in \Omega, s < t < s+h^2} \Gamma_0(x,t,y,s)|V(x,t)|dx\,dt$$

and we say that $V$ belongs to $SK(\Omega)$ related to $L_0$ if

$$\lim_{h\to 0} \eta_V(h) = 0, \quad \lim_{h\to 0} \eta_V^*(h) = 0.$$

If $A = I_n$, $B = 0$ the above operator $L_0$ is the heat operator and (2.1) is the same definition made by Zhang.

## 3. Main results

The main result is a Harnack inequality invariant for positive solutions of

$$L_0 u + V u = 0.$$

At first let us introduce some notations. Let $\Omega \subset \mathbb{R}^{n+1}$ a bounded open set, let us define

$$L = L_0 + V,$$

where $L_0$ is the Kolmogorov operator and $V \in SK(\Omega)$.

We say that $u$ is a weak solution of

$$Lu = 0, \quad if$$

1) $\exists\, p > 1 : u, \partial_1 u, ..., \partial_{m_0} u \in L^p_{loc}(\Omega)$,
2) $Vu \in L^1_{loc}(\Omega)$,
3) $-\int_\Omega \langle ADu, D\varphi \rangle + \int_\Omega uY^*\varphi + \int_\Omega uV\varphi = 0, \forall \varphi \in C_0^\infty(\Omega)$

where $Y = \langle x, BD \rangle - \partial_t$ and $Y^* = -Y$.

The operator $L_0$ is invariant respect to the homogeneous Lie group structure that we can explicity write:

- multiplication group:

$$(x,t) \cdot (\xi,\tau) = (\xi + E(\tau)x, t + \tau)$$

dilations group, defined by :

$$\left(D(\lambda), \lambda^2\right)_{\lambda > 0},$$

where $E(t) = \exp(-tB^T)$, $D(\lambda) = diag\left(\lambda I_{m_0}, \lambda^3 I_{m_1}, \ldots, \lambda^{2r+1} I_{m_r}\right)$, and $B$ is the matrix introduced above.

The integer $N = m_0 + 3m_1 + \cdots + (2r+1)m_r + 2$ is the homogeneous dimension of the group. We also define

$$Q_R(\xi,\tau,h) = \left\{(x,t) \in \mathbb{R}^{n+1} : \tau < t < \tau + h; \left|D(\tfrac{1}{R})(E(-t)x - E(-\tau)\xi)\right| < 1\right\},$$

$$S_R(\xi,\tau) = \left\{(x,t) \in \mathbb{R}^{n+1} : t = \tau; \left|D(\tfrac{1}{R})E(-\tau)(x - \xi)\right| < 1\right\},$$

$$S_R(\xi,\tau,h) = \left\{(x,t) \in \mathbb{R}^{n+1} : t = \tau + h; \left|D(\tfrac{1}{R})(E(-t)x - E(-\tau)\xi)\right| < 1\right\},$$

734

$$M_R(\xi,\tau,h) = \Big\{(x,t) \in \mathbb{R}^{n+1} : \tau < t < \tau + h; \Big|D\big(\tfrac{1}{R}\big)(E(-t)x - E(-\tau)\xi)\Big| = 1\Big\},$$

that denote respectively, the open set, its basis, its section at the height $t = \tau + h$ and its "lateral boundary".

We say that $u$ is solution of the Cauchy-Dirichlet problem

$$\begin{cases} Lu = 0 \text{ in } Q_R(\xi,\tau,h) \\ u = 0 \quad \text{in } M_R(\xi,\tau,h) \\ u = f \quad \text{in } S_R(\xi,\tau) \end{cases}$$

with $f \in C_0(S_R(\xi,\tau))$, if

(A-1)   $u$ is weak solution of $Lu = 0$ in $Q_R(\xi,\tau,h)$,

(A-2)   $\lim\limits_{(x,t)\to(y,s)} u(x,t) = 0, \qquad \forall(y,s) \in M_R(\xi,\tau,h)$,

(A-3)   $\lim\limits_{(x,t)\to(y,\tau)} u(x,t) = f(y,\tau), \quad \forall(y,\tau) \in S_R(\xi,\tau)$.

Moreover we say that $G$ is the Green function for $Q_R$ if, for every $f \in C_0(S_R)$, the function

$$u(x,t) = \int_{S_R} G(x,t,\xi,\tau)f(\xi)d\xi$$

is solution of the Cauchy-Dirichlet problem. Before we state the result contained in [19] let us consider the cylinder $Q_R(\xi,\tau,R^2)$ and, for every $\alpha,\beta,\gamma,\delta \in (0,1) : \alpha < \beta < \gamma$, let us set

$$Q^- = \{(x,t) \in Q_{\delta R}(\xi,\tau,R^2) : \tau + \alpha R^2 \le t \le \tau + \beta R^2\},$$

$$Q^+ = \{(x,t) \in Q_{\delta R}(\xi,\tau,R^2) : \tau + \gamma R^2 \le t\}.$$

**Theorem** Let $u \ge 0$ be a weak solution of $Lu = 0$ in $Q_R(\xi,\tau,R^2)$ and $V \in SK(Q_R(\xi,\tau,R^2))$.

Then there exists $M > 0$, dependent on $V$ and on the constants $\alpha,\beta,\gamma,\delta$, such that

$$\max_{Q^-} u \le M \min_{Q^+} u.$$

**Proof** At first we obtain the estimate of the modulus of continuity of the weak solutions of $Lu = 0$ making the additional assumption that the solution $u$ is bounded:

$$|u(z) - u(z_0)| \le \big(Cd(z,z_0)^{1/2} + 2\eta_V(5d(z,z_0)^{1/2})\big) \sup_{B_{4r}(z_0)} |u|$$

for every $z_0 \in \Omega$, $r \in (0,1)$ such that $B_{4r}(z_0) \subset \Omega$, for every $z \in B_{r^2}(z_0)$ being $C = C(L_0)$ a positive constant.

We built a Green function for suitable bounded open sets using the parametrix method and show that the set of bounded functions is a class of functions such that the Cauchy-Dirichlet problem is well-posed in the sense that exists and is unique the solution.

Then we use a suitable density argument that allows us to remove the extra assumption of boundedness of $u$ and prove the invariant Harnack inequality.

## References

1. M. Aizenman, B. Simon, *Brownian motion and Harnack's inequality for Schrödinger equations*, Comm. Pure Appl. Math. **35** (1982), 209–271.
2. M. Biroli, *Nonlinear Kato measures and nonlinear Schrödinger problems*, Rend. Acc. Naz. Sc. detta dei XL, Memorie di Matematica e Appl. **21** (1997), 235–252.
3. M. Biroli, *Schrödinger type and relaxed Dirichlet problems for the subelliptic p-Laplacian* Potential Anal. **15,1-2** (2001), 1–16.
4. L. Capogna, D. Danielli, N. Garofalo, *An embedding theorem and the Harnack inequality for nonlinear subelliptic equations*, Comm. in P.D.E. **18** (1993), 1765–1794.
5. F. Chiarenza, E. Fabes, N. Garofalo, *Harnack's inequality for Schrödinger operators and the continuity of solutions*, Proc. Amer. Math. Soc. **307** (1986), 415–425.
6. G. Citti, N. Garofalo, E. Lanconelli, *Harnack's inequality for sum of squares of vector fields*, Amer. J. of Math. **115** (1993), 699–734.
7. G. Dal Maso, U. Mosco, *Wiener criteria and energy decay for relaxed Dirichlet problems*, Arch. Rat. Mech. An. **95** (1986), 345–387.
8. E. B. Fabes, D. W. Stroock, *The $L^p$ integrability of Green's functions and fundamental solutions for elliptic and parabolic equations*, Duke Math. J. **51** (1984), 997–1016.
9. E. B. Fabes, D. W. Stroock, *A new proof of Moser's parabolic Harnack inequality using the old idea of Nash*, Arch. Rat. Mech. An. **96** (1986), 327–338.
10. G. B. Folland, *A fundamental solution for a subelliptic operator*, B.A.M.S. **79** (1973), 373–376.
11. C. E. Gutierrez, *Harnack's inequality for degenerate Schrödinger operators*, Trans Amer. Math. Soc. **312** (1989), 403–419.
12. A. Hinz, H. Kalf, *Subsolutions estimates and Harnack's inequality for Schrödinger operator*, J. Reine Angew. Math. **404** (1990), 118–134.
13. L. Hörmander, *Hypoelliptic second order differential equations*, Acta Math. **119** (1967), 147–171.
14. D. Jerison, A. Sánchez-Calle, *Estimates for the heat kernel for a sum of squares of vector fields* , Indiana. Univ. Math. **35** (4) (1986), 835 –854.
15. O. A. Ladyženskaja, N. N. Ural'ceva, *Linear and quasilinear elliptic equations*, Academic Press, New York, 1968.
16. E. Lanconelli, S. Polidoro, *On a class of hypoelliptic evolution operators*, Rend. Sem. Mat. Pol. Torino **51** (1993), 137–171.
17. G. Lu, *On Harnack's inequality for a class of strongly degenerate Schrödinger operators formed by vector fields*, Diff. Integral Eqs. **7** (1994), 73–100.
18. A. Nagel, E. M. Stein, S. Wainger, *Balls and metrics defined by vector fields I: basic properties*, Acta Math. **155** (1985), 103–147.
19. S. Polidoro, M. A. Ragusa, *A Green function and regularity results for an ultraparabolic equation with a singular potential*, Adv. in Diff. Eq. **7** (2002), 1281-1314.
20. L. P. Rothschild, E. M. Stein, *Hypoelliptic differential operators on nilpotent groups*, Acta Math. **137** (1977), 247–320.
21. A. Sánchez-Calle, *Fundamental solutions and geometry of the sum of squares of vector fields*, Invent. Math. **78** (1) (1984), 143–160.

736

22. C. G. Simader, *An elementary proof of Harnack's inequality for Schrödinger operators and related topics*, Math. Z. **203** (1990), 129–152.

23. G. Stampacchia, *Le problème de Dirichlet pour les équations elliptiques au second ordre à coefficients discontinus*, Ann. Inst. Fourier (Grenoble), **15** (1965), 189–256.

24. K. T. Sturm, *Harnack's inequality for parabolic operators with singular low order terms*, Math. Z. **216** (1994), 593–612.

25. Q. Zhang, *A Harnack inequality for Kolmogorov equations*, J. Math. Anal. Appl. **190** (1995), 402–418.

26. Q. Zhang, *On a parabolic equation with a singular lower order term*, Trans Amer. Math. Soc. **348** (1996), 2811–2844.

# EXISTENCE AND UNIQUENESS OF CLASSICAL SOLUTIONS TO CERTAIN NONLINEAR PARABOLIC INTEGRODIFFERENTIAL EQUATIONS AND APPLICATIONS

D. R. AKHMETOV

*Sobolev Institute for Mathematics, SB RAS*
*4, Acad. Koptyug prosp.*
*630090 Novosibirsk, Russia*
*E-mail: denis_r_akhmetov@yahoo.com*

M. M. LAVRENTIEV, Jr.

*Sobolev Institute for Mathematics, SB RAS*
*4, Acad. Koptyug prosp., and*
*Institute of Automation and Electrometry, SB RAS*
*1, Acad. Koptyug prosp.*
*630090 Novosibirsk, Russia*
*E-mail: mmlavr@nsu.ru*

R. SPIGLER

*Dipartimento di Matematica, Università "Roma Tre"*
*1, Largo S. L. Murialdo,*
*00146 Rome, Italy*
*E-mail: spigler@mat.uniroma3.it*

Recent results about existence, uniqueness, and some properties of the classical solutions to certain nonlinear partial differential equations of the parabolic type, affected by space-integrals, are reported. These are nonlinear Fokker-Planck type equations. The related problem concerns the dynamical statistical behavior of populations of infinitely many nonlinear random oscillators, a subject having numerous applications in Physics, Biology, Neural Networks, and Social Sciences. The corresponding numerical treatment is also outlined. These investigations have suggested some others, for instance, singularly perturbed partial differential equations whose analysis, rather surprisingly, does not require any boundary layer.

**Keywords:** Nonlinear parabolic integro differential equations, classical solutions, Fokker-Planck equation, nonlinear random oscillators
**AMS Mathematics Subject Classifiction:** 45K05, 82C31, 45G10

## 1. Introduction

In this paper, we are concerned with certain problems involving integrodifferential parabolic equations of the Fokker-Planck type. They are nonlinear through a space-integral term entering the drift coefficient, and describe the statistical time evolution of populations of infinitely many nonlinear random oscillators, subject to global (mean-field) coupling. One of the main features of these models is that they describe a rather ubiquitous phenomenon, that is the transition from incoherence to a synchronized state. It is noteworthy that, in practice, similar populations including only finitely many members are described very well by the limiting case of infinitely many oscillators. Numerical experiments show that even populations consisting only of few hundreds oscillators are described rather satisfactorily. In the recent review paper [5], the focus was on the physical aspects of these phenomena. In this paper we intend to concentrate on the mathematical analysis aspects.

Nowadays, we know that the dynamical behavior of such populations can model a rich variety of phenomena pertaining to Biology, Medicine, Physics, Engineering, and even Social Sciences. The original motivation, however, can be reconducted to some observations made by a British botanist concerning groups of fireflies. In the 1930s, a botanist observed that groups of fireflies, each virtually laying on a leave of a certain tree, many of such trees being located along a river for a couple of miles, were flashing simultaneously in the dark. A few fireflies collected by him and set on a wall of his hotel's room, in the dark room, started flashing first incoherently, but they pretty soon synchronized, and again the flashes occurred simultaneously.

A. T. Winfree was the first to propose, in 1967, to model phenomena like this in terms of an ensemble of simple phase oscillators, globally coupled to each other [23], see also [24]. Later, in 1975, Y. Kuramoto, in a famous paper [12] set up a model in terms of nonlinear globally coupled equations. Noise may affect such model, leading to a system of stochastic coupled differential equations, and a certain nonlinear integro-differential equation could be derived to describe systems of infinitely many oscillators, see also [18,20].

In [13,14], the Kuramoto (also called there Kuramoto-Sakaguchi) parabolic equation was first studied from the point of view of mathematical analysis. The Kuramoto equation is a nonlinear partial differential equation (PDE) of the parabolic type, indeed a Fokker-Planck equation where the space variable is an angle, the aforementioned phase, and the drift is quadratically nonlinear through an integral which accounts for the action of the infinitely many oscillators on any given one of them. An additional integration over all frequencies appears, which makes the model even more nonstandard from the mathematical viewpoint.

The Kuramoto model-equation can be derived by a system of $N$ nonlinearly coupled first-order stochastic differential equations, in the limit for $N \to \infty$. This was first done formally [9], and later rigorously [10].

On the other hand, some improvement of the original model was suggested in [11] and in [21,22] in terms of a system of possibly stochastic second-order differential equations. Proceeding as for the Kuramoto model, such a generalized model could then be described in the limit of infinitely many elements by a new partial differential equation [1,2]. The latter model was referred to as the "adaptive" or the "inertial" model. In fact, looking at the (stochastic) differential equations as describing mechanical or electrical oscillators, an inertial term such as mass or electrical capacitance appears. The ensuing equation turned out to be again of the Fokker-Planck type, indeed more similar to the classical (full) Fokker-Planck equation, since both "space" (an angle, as above) and "velocity" (an angular velocity) appear now in the PDE. Again, a quadratic nonlinearity enters the drift term of the PDE, through a certain integral term. Analysis of this equation is reported in [15,16,6].

A numerical treatment of both, the Kuramoto and the adaptive (or inertial) PDEs, has been reported in several papers, resorting to a variety of techniques. We recall [19] for a (rather formal) spectral approach to the Kuramoto equation, and [4] for a more rigorous spectral treatment. The adaptive equation has been solved numerically in a number of papers, see [1,2,4].

The investigations conducted especially in [15,16,6], have established that certain singular perturbation PDE problems could be handled without introducing any boundary-layer. This fact, rather surprising, suggested further investigations, in order to clarify this phenomenon, which turned out to be even more important since it could be found even within a number of *linear* singularly perturbed PDE problems [7,8].

Though the PDE problems considered here were originated by certain applications, and hence the nonlinear drift term has a special form, there is no doubt that they can be generalized to include some more general cases.

Here is the plan of the paper. In section 2, we formulate the problem concerning the Kuramoto equation, and present the main results about existence, uniqueness, and some properties of the classical solutions. In section 3, we do the same for the adaptive equation, for which all is much more technical. In section 4 we give a sketch of the numerical treatment for the Kuramoto equation. In section 5, finally, we report shortly on some related issues.

## 2. The Kuramoto equation

The Kuramoto equation is given by

$$\frac{\partial \rho}{\partial t} = D \frac{\partial^2 \rho}{\partial \theta^2} - \frac{\partial}{\partial \theta}(v\,\rho) \tag{1}$$

on the domain $\{(\theta, t) \in [0, 2\pi] \times [0, T]\}$, for some arbitrary $T > 0$, where the drift velocity, $v$, is

$$v \equiv v(\theta, t, \omega) := \omega + K\,r\,\sin(\psi - \theta), \tag{2}$$

$r \equiv r(t)$ and $\psi \equiv \psi(t)$ being defined by the complex relation

$$r\,e^{i\psi} = \int_0^{2\pi} \int_{-\infty}^{+\infty} e^{i\varphi}\,\rho(\varphi, t, \omega)\,g(\omega)\,d\omega d\varphi. \tag{3}$$

Here $D > 0$ represents the diffusion coefficient, $K > 0$ the coupling strength, and $\rho$ the transition probability density of the amplitude distribution of the oscillators. The variable $\omega$ should be picked up from the support of the "frequency distribution" $g(\omega)$.

To obtain the time evolution of the distribution function $\rho$, some additional data should be specified. Due to the nature of the space variable $\theta$, $2\pi$-periodicity with

$$\rho(\theta, t, \omega)|_{\theta=0} = \rho(\theta, t, \omega)|_{\theta=2\pi}, \qquad \left.\frac{\partial \rho}{\partial \theta}\right|_{\theta=0} = \left.\frac{\partial \rho}{\partial \theta}\right|_{\theta=2\pi} \tag{4}$$

is prescribed, along with initial data

$$\rho(\theta, t, \omega)|_{t=0} = \rho_0(\theta, \omega), \tag{5}$$

and normalization of the initial data

$$\int_0^{2\pi} \rho_0(\theta, \omega)\, d\theta = 1 \tag{6}$$

for all $\omega$.

In [13,14], the following theorem was proved:

**Theorem 1.** (Lavrentiev and Spigler, 2000). Assuming that the initial value $\rho_0(\theta, \omega)$ is $2\pi$-periodic in $\theta$, smooth, positive, and normalized as in (6), and that the frequency distribution $g(\omega)$ is nonnegative, compactly supported on the interval $[-G, G]$ and belongs to $L^1[-G, G]$, there exists $T > 0$ such that a classical solution, $\rho(\theta, t, \omega)$ to equation (1) satisfying the conditions (4) and (5) exists for $t \in [0, T]$. Such a solution is unique, positive, with $L^1$-norm with respect to $\theta$ equal to 1, and bounded, along with all its space derivatives, $\partial^m \rho/\partial \theta^m$, with respect to $G$.

The technique used to establish such results is a typical one, used for nonlinear parabolic equations, and is based on certain combinations of energy-like estimates and the Gronwall's lemma. However, some properties of the solution such as to be positive and normalized (assumed for the initial data and then proven to hold for the solution), were exploited. The results obtained were expected as "reasonable", but they had to be established anew, since it was impossible to refer to, or to rest on, existing analysis, due to the peculiarity of the present model.

## 3. The "adaptive" equation

The so-called "adaptive" equation, formally derived in [1], reads

$$\frac{\partial \rho}{\partial t} = \frac{D}{m^2} \frac{\partial^2 \rho}{\partial \omega^2} + \frac{1}{m} \frac{\partial}{\partial \omega} \left[ (\omega - \Omega - \mathcal{K}_\rho(\theta, t))\, \rho \right] - \omega \frac{\partial \rho}{\partial \theta}, \tag{7}$$

where we set

$$\mathcal{K}_\rho(\theta, t) := K \int_{-G}^{G} \int_{-\infty}^{+\infty} \int_0^{2\pi} g(\Omega') \sin(\theta' - \theta)\, \rho(\theta', \omega', t, \Omega')\, d\theta'\, d\omega'\, d\Omega'. \tag{8}$$

Hereafter, we set $D = 1$ and $m = 1$, $D$ denoting a diffusion coefficient (in the velocity space) and $m$ the mass of a representative particle. The function $g(\Omega) \in L^1[-G, G]$ has also here the meaning of "frequency distribution", and the constant $K > 0$ is the "coupling strength".

This type of equations is called sometimes "ultraparabolic", since it is parabolic in some space-like variable (the variable $\omega$), but it is not in another space-like variable (that is $\theta$), and hence it can be thought of as being degenerate with respect to the latter. As for the Kuramoto equation, some data should be prescribed, namely initial values and $2\pi$-periodicity in the angle $\theta$.

The "pathologies" of the present problem should be noted. In fact:

1) The governing equation (7) is of the *second order* with respect to $w$, but only of the *first order* with respect to $\theta$. Therefore, disregarding the integral term, this equation is neither of the parabolic nor of the hyperbolic type.

2) Equation (7) should be considered on the *slab* $(\theta, w, t) \in [0, 2\pi] \times \mathbf{R} \times [0, T]$, that is on a domain *unbounded* in $w$, which plays the role of a coefficient in the equation. This fact gives rise to *singularity* phenomena typical of equations with *unbounded coefficients*.

3) Equation (7) contains an integral term, where the integral is taken over an *unbounded* domain.

4) There is an additional variable, the natural frequency of the oscillators, $\Omega$, with respect to which no derivative appears, but which plays the role of an integration variable.

5) We are interested *only* in solutions *periodic* in $\theta$, while the governing equation contains only the first (i.e., a time-like) derivative with respect to $\theta$ (as in the case of "time-periodic solutions to parabolic equations").

6) The coefficient $w$, which multiplies the time-like derivative $\rho_\theta$, *changes its sign* in the domain that we are considering.

Therefore, the results available in the literature concerning nonlinear parabolic, or even integroparabolic equations, cannot be applied to the present case.

In [15,16], the existence of strong solutions was established to equation (7), which are nonnegative, $2\pi$-periodic in $\theta$, and normalized according to the relation

$$\int_0^{2\pi} \int_{-\infty}^{+\infty} \rho(\theta, w, t, \Omega)\, dw d\theta = 1 \tag{9}$$

for every $t \in [0, T]$ and every $\Omega \in [-G, G]$. In [6], existence and uniqueness of classical solutions to such problem were established. Existence was established at a price of requiring some more smoothness of the initial data. More precisely, the initial data are required to belong to a Hölder space that is included in the space $C^4$ rather than in $C^2$, as in [15,16]. The regularization technique involved an auxiliary problem, that is considering the fully parabolic equation

$$\frac{\partial \rho^{\varepsilon,N}}{\partial t} = \frac{\partial^2 \rho^{\varepsilon,N}}{\partial w^2} + \varepsilon \frac{\partial^2 \rho^{\varepsilon,N}}{\partial \theta^2} + \frac{\partial}{\partial w}\left(F_N \rho^{\varepsilon,N}\right) - \left(\Omega + \mathcal{K}_{\rho^{\varepsilon,N}}\right)\frac{\partial \rho^{\varepsilon,N}}{\partial w} - F_N \frac{\partial \rho^{\varepsilon,N}}{\partial \theta}, \tag{10}$$

for $\rho^{\varepsilon,N} \equiv \rho^{\varepsilon,N}(\theta, w, t, \Omega)$, where $F_N = F_N(w)$ is a suitable bounding function, subject to

$$\rho^{\varepsilon,N}(\theta, w, t, \Omega)|_{t=0} = \rho_0(\theta, w, \Omega), \tag{11}$$

and

$$\rho^{\varepsilon,N}|_{\theta=0} = \rho^{\varepsilon,N}|_{\theta=2\pi}, \quad \rho_\theta^{\varepsilon,N}|_{\theta=0} = \rho_\theta^{\varepsilon,N}|_{\theta=2\pi}. \tag{12}$$

Concerning such parabolically *regularized* problem, the following existence result was proved in [15,16]:

**Theorem 2.** Assume that the prescribed initial value, $\rho(\theta, \omega, t, \Omega)|_{t=0} = \rho_0(\theta, \omega, \Omega)$, belongs to the Hölder space $C^{2+\alpha_0}(Q)$, with $\alpha_0 \in (0,1)$, $Q := \{(\theta, \omega, \Omega) \in \mathbf{R} \times \mathbf{R} \times [-G, G]\}$, is $2\pi$-periodic in $\theta$, nonnegative in $Q$, normalized for every $\Omega \in [-G, G]$, according to

$$\int_0^{2\pi} \int_{-\infty}^{+\infty} \rho_0(\theta, \omega, \Omega) \, d\omega d\theta = 1,$$

and exponentially decaying in $\omega$ at infinity (along with some of its partial derivatives), according to the estimate

$$\sup_{\theta \in \mathbf{R}, \Omega \in [-G,G]} \left| D_{\theta,\omega,\Omega}^{l_1,l_2,l_3} \rho_0(\theta, \omega, \Omega) \right| \leq C_0 e^{-M_0 \omega^2}$$

for $\omega \in \mathbf{R}$ and integers $l_k \geq 0$ with $l_1 + l_2 + l_3 \leq 2$, $C_0$ and $M_0$ being positive constants.

Then, for every $\varepsilon > 0$ and for every $N > 0$, there exists a *classical* solution, $\rho^{\varepsilon,N}(\theta, \omega, t, \Omega)$, to problem (10)–(12) in $Q_T := \{(\theta, \omega, t, \Omega) \in [0, 2\pi] \times \mathbf{R} \times [0, T] \times [-G, G]\}$.

Moreover, such solution

1) is nonnegative in $Q_T$ and normalized as

$$\int_0^{2\pi} \int_{-\infty}^{+\infty} \rho^{\varepsilon,N}(\theta, \omega, t, \Omega) \, d\omega d\theta = 1,$$

for all $t \in [0, T]$ and $\Omega \in [-G, G]$;

2) decays exponentially at infinity in $\omega$, according to

$$\sup_{\theta \in [0,2\pi], t \in [0,T], \Omega \in [-G,G]} \left| D_{\theta,\omega,t,\Omega}^{l_1,l_2,l_3,l_4} \rho^{\varepsilon,N}(\theta, \omega, t, \Omega) \right| \leq C_\varepsilon e^{-M_\varepsilon \omega^2}$$

for $\omega \in \mathbf{R}$ and integers $l_k \geq 0$ with $l_1 + l_2 + 2l_3 + l_4 \leq 2$, the positive constants $C_\varepsilon$ and $M_\varepsilon$ depending on $\varepsilon$, $N$, $G$, $T$, $K \|g\|_{L^1[-G,G]}$, $C_0$, and $M_0$.

After that, in [6,16] passage to the limit with respect to regularization parameters introduced above, has been justified, and some qualitative properties of the solution to the original problem have been studied. It is worth noting that almost all tools of modern qualitative theory of parabolic PDEs have been used to obtain our results. We just enlist decay estimates of fundamental solutions to linear parabolic equations, the embedding theory for anisotropic Sobolev spaces, successive approximations, energy-like estimates uniform with respect to the parameters, compactness arguments on unbounded domains, and passage to the limit in nonlinear terms.

In order to keep this paper to within a moderate length, we omit to formulate long theorems, which are rather technical and possess a typical structure, similar to that of Theorem 2. For their precise statements the reader is referred to the literature, see [6,16]. Here we would like to stress an interesting and rather surprising phenomenon. All estimates obtained for the solutions to the regularized problem in (10)–(12) were shown to be uniform with respect to $\varepsilon$. This happens even for the derivatives of the solutions. This implies that no boundary layer is needed for the asymptotic description of the solutions to the so-called "singularly perturbed" problems. The latter are characterized by the fact that solutions obtained solving the formally derived limiting equations (obtained merely setting $\varepsilon = 0$ in the equation), cannot satisfy the full set of the originally prescribed boundary data. Apparently, this phenomenon was never reported in the existing literature, even though it could be expected when solutions enjoy sufficient regularity properties. We have analyzed such phenomenon in some detail, and in [7,8] some classes of singular perturbation problems for *linear* PDEs were considered where no boundary layer was needed for their asymptotics.

## 4. Numerical treatment

Simulations of various kind have been conducted, both on the system of SDEs and on the PDE, for both cases, of the Kuramoto and the "adaptive" equation. They are buried in [1-3,19], as well as in other papers mostly pertaining to the physical litera-ture. A more rigorous approach is that in [4], where a spectral numerical treatment for the Kuramoto equation was worked out.

Prior to describing briefly the numerical tests, we would like to comment on the spectral decomposition, which has been accomplished in [4]. Analytical estimates on decay of the Fourier coefficients of the solution to the Kuramoto equation (1) were derived and then compared with the results of direct numerical experiments. This comparison provided a measure of the quality and efficiency of the numerical approach. A similar method was also applied to the equation in (7), though not fully analyzed from a theoretical standpoint.

The spectral method for the Kuramoto equation consists in Fourier expanding the solution with respect to $\theta$,

$$\rho(\theta, t, \omega) = \sum_{n=-\infty}^{+\infty} \rho_n(t, \omega)\, e^{in\theta},$$

being

$$\rho_n(t, \omega) = \frac{1}{2\pi} \int_0^{2\pi} \rho(\theta, t, \omega)\, e^{-in\theta}\, d\theta,$$

for $n = 0, \pm 1, \pm 2, \ldots$. Substituting in (1) yields the coupled system of ODEs

$$\rho_n' = -n^2 D\rho_n - in\omega\rho_n - n\frac{K}{2}\left[\rho_{n+1}\int_{-\infty}^{+\infty}\rho_{-1}\,g\,d\omega - \rho_{n-1}\int_{-\infty}^{+\infty}\rho_1\,g\,d\omega\right],$$

for $n = 0, \pm 1, \pm 2, \ldots$; $' = \partial/\partial t$.

Now, for every function $\rho$, smooth in $\theta$, say $\rho \in C^k$, its Fourier coefficients, $\rho_n$, decay according to $|\rho_n| \leq \frac{M}{|n|^k}$, with $M$ independent of $n$. To obtain bounds for such an $M$, correspondingly to the solution $\rho$, use Parseval,

$$\left\| \frac{\partial^p \rho}{\partial \theta^p} \right\|_{L^2}^2 = 4\pi \sum_{-\infty}^{+\infty} n^{2p} |\rho_n|^2.$$

Hence, from every estimate like

$$\int_0^{2\pi} \left( \frac{\partial^p \rho}{\partial \theta^p} \right)^2 d\theta \leq C_p, \quad p \in \mathbf{N}^+,$$

the decay estimate

$$|\rho_n(t,\omega)| \leq \frac{M_p}{|n|^p}, \quad M_p := \frac{\sqrt{C_p}}{2\sqrt{\pi}},$$

follows for every $|n| \geq 1$, with $C_p$ independent of $n$. Therefore, with $N$ harmonics, $\rho$ can be approximated with a global error $\varepsilon_N$ with

$$\|\varepsilon_N(\theta, t, \omega)\|_{L^2(0,2\pi)} \leq \frac{\sqrt{C_p}}{(N+1)^p},$$

valid for every $p \geq 1$ (uniformly in $t, \omega$). Bounds for the constants $C_p$ can be obtained by "energy-like" estimates from the PDE. Moreover, the error bound can be improved by an "a posteriori error analysis". A pointwise (or even uniform) error estimate of the form $|\varepsilon_N(\theta, t, \omega)| \leq f(N,p)/(N+1)^{p-1}$ can also be derived.

## 5. Applications: Singular perturbations without boundary-layers

We consider as "applications" of the previous results some findings that were obtained as a consequence of the analysis conducted in [15,16,6]. In fact, investigating the "adaptive PDE" above revealed the existence of certain *singularly* perturbed PDE problems *without* boundary layers (from parabolic to ultraparabolic).

This phenomenon, observed in the regularization process of the "adaptive PDE", can be found even in *linear* singularly perturbed PDE problems. The key-explanation can be found in some special compatibility conditions on the data. In the case of the "adaptive" model, such conditions are represented by the periodicity properties of the initial data and of all coefficients as functions of $\theta$. The reader is referred to [7] for details.

As an example, very simple but enlightening, consider

$$u_t = \varepsilon u_{yy} + u_y,$$

on the half-strip $\Pi := \{(y,t) \in [0,1] \times [0,+\infty)\}$, with $u(y,0) = \varphi(y)$ and $(u, u_y)|_{y=0} = (u, u_y)|_{y=1}$.

We proved that, if $\varphi(y) \in C^\infty(\mathbf{R})$ and is periodic with period 1, then for every $\varepsilon > 0$ there is a unique classical solution, $u^\varepsilon(y,t)$, which is $C^\infty(\Pi)$ and such that, for every $k = 1, 2, \ldots$, there exists a constant $M_k$, independent of $\varepsilon$, such that

$$\|u^\varepsilon(y,t)\|_{C^k(\Pi)} \leq M_k, \quad \varepsilon \in (0,1).$$

A rather general result concerns the problem:

$$u_t + k(x, y, t)u_y = \varepsilon u_{yy} + a(x, y, t)u_{xx} + b(x, y, t)u_x + c(x, y, t)u + f(x, y, t),$$

on the set $\{(x, y, t) \in \mathbf{R} \times [0, 1] \times [0, T]\}$, with

$$(u, u_y)|_{y=0} = (u, u_y)|_{y=1}, \quad u(x, y, 0) = \varphi(x, y),$$

under the assumptions that the differential operator above, $L^\varepsilon$, be uniformly parabolic and with smooth coefficients, the source term, $f$, be smooth and with a suitable decay rate, the initial profile, $\varphi$, as well, and all coefficients of $L^\varepsilon$, $f$, and $\varphi$ be periodic in $y$, with period 1. Then, for every $\varepsilon > 0$ there is a unique bounded classical solution, periodic in $y$. Moreover, there exists a sequence $\varepsilon_n$, with $\lim_{n\to\infty} \varepsilon_n = 0$, such that the sequence of functions $u^{\varepsilon_n}$ converges as $n \to \infty$ to a strong solution of the formally obtained limiting problem. We refer the interested reader to [7,8].

## Acknowledgments

This work was sponsored by the GNFM of the Italian INdAM; D. R. Akhmetov was supported by the INTAS Program, YSF 03–55–778.

## References

1. J. A. Acebrón and R. Spigler, Adaptive frequency model for phase-frequency synchronization in large populations of globally coupled nonlinear oscillators, *Phys. Rev. Lett.* **81** (1998), 2229–2232.
2. J. A. Acebrón, L. L. Bonilla and R. Spigler, Synchronization in populations of globally coupled oscillators with inertial effects, *Phys. Rev.* **E 62** (2000), 3437–3454.
3. J. A. Acebrón, A. Perales and R. Spigler, Bifurcations and global stability of synchronized stationary states in the Kuramoto model for oscillator populations, *Phys. Rev.* **E 64** (2001), 016218-1/016218-5.
4. J. A. Acebrón, M. M. Lavrentiev Jr. and R. Spigler, Spectral analysis and computation for the Kuramoto-Sakaguchi integroparabolic equation, *IMA J. Numer. Anal.* **21** (2001), 239–263.
5. J. A. Acebrón, L. L. Bonilla, C. J. Pérez Vicente, F. Ritort and R. Spigler, The Kuramoto model: a simple paradigm for synchronization phenomena, *Rev. Modern Phys.* **77** (2005), 137–185.
6. D. R. Akhmetov, M. M. Lavrentiev Jr. and R. Spigler, Existence and uniqueness of classical solutions to certain nonlinear integro-differential Fokker-Planck type equations, *Electron. J. Differential Equations* **2002** (2002), No. 24, 17 pp. (electronic).
7. D. R. Akhmetov, M. M. Lavrentiev Jr. and R. Spigler, Singular perturbations for certain partial differential equations without boundary-layers, *Asymptot. Anal.* **35** (2003), 65–89.
8. D. R. Akhmetov, M. M. Lavrentiev Jr. and R. Spigler, Singular perturbations for parabolic equations with unbounded coefficients leading to ultraparabolic equations, *Differential Integral Equations* **17** (2004), 99–118.
9. L. L. Bonilla, Stable probability densities and phase transitions for mean-field models in the thermodynamic limit, *J. Statist. Phys.* **46** (1987), 659–678.

746

10. P. Dai Pra and F. den Hollander, McKean-Vlasov limit for interacting random processes in random media, *J. Statist. Phys.* **84** (1996), 735–772.
11. B. Ermentrout, An adaptive model for synchrony in the firefly Pteroptix malaccae, *J. Math. Biol.* **29** (1991), 571–585.
12. Y. Kuramoto, Self-entrainment of a population of coupled nonlinear oscillators, in: International Symposium on Mathematical Problems in Theoretical Physics, H. Araki ed., Lecture Notes in Physics **39**, Springer, New York, 1975, 420–422.
13. M. M. Lavrentiev Jr. and R. Spigler, Existence and uniqueness of solutions to the Kuramoto-Sakaguchi nonlinear parabolic integrodifferential equation, *Differential Integral Equations* **13** (2000), 649–667.
14. M. M. Lavrentiev Jr. and R. Spigler, The Kuramoto-Sakaguchi nonlinear parabolic integrodifferential equation, Partial differential equations (Praha, 1998), 248–253, Chapman & Hall/CRC Res. Notes Math., 406, Chapman & Hall/CRC, Boca Raton, FL, 2000.
15. M. M. Lavrentiev Jr., R. Spigler and D. R. Akhmetov, Nonlinear integroparabolic equations on unbounded domains: Existence of classical solutions with special properties (Russian), *Sibirsk. Mat. Zh.* **42** (2001), 585–609; translation in *Siberian Math. J.* **42** (2001), 495–516.
16. M. M. Lavrentiev Jr., R. Spigler and D. R. Akhmetov, Regularizing a nonlinear integroparabolic Fokker-Planck equation with space-periodic solutions: Existence of strong solutions (Russian), *Sibirsk. Mat. Zh.* **42** (2001), 825–848; translation in *Siberian Math. J.* **42** (2001), 693–714.
17. M. M. Lavrentiev Jr. and R. Spigler, Time-independent estimates and a comparison theorem for a nonlinear integroparabolic equation of the Fokker-Planck type, *Differential Integral Equations* **17** (2004), 549–570.
18. Sakaguchi, H., Cooperative phenomena in coupled oscillator systems under external fields, *Prog. Theor. Phys.* **79** (1988), 39–46
19. F. Sartoretto, R. Spigler and C. J. Pérez Vicente, Numerical solution of the Kuramoto-Sakaguchi equation governing populations of coupled oscillators, *Math. Models Methods Appl. Sci.* **8** (1998), 1023–1038.
20. S. H. Strogatz and R. E. Mirollo, Stability of incoherence in a population of coupled oscillators, *J. Statist. Phys.* **63** (1991), 613–635.
21. H. A. Tanaka, A. J. Lichtenberg and S. Oishi, First order phase transition resulting from finite inertia in coupled oscillator systems, *Phys. Rev. Lett.* **78** (1997), 2104–2107.
22. H. A. Tanaka, A. J. Lichtenberg and S. Oishi, Self-synchronization of coupled oscillators with hysteretic responses, *Phys. D* **100** (1997), 279–300.
23. A. T. Winfree, Biological rhythms and the behavior of populations of coupled oscillators, *J. Theoret. Biol.* **16** (1967), 15–42.
24. A. T. Winfree, The geometry of biological time, Springer, New York, 1980.

# ON EXISTENCE AND ASYMPTOTIC BEHAVIOR OF THE SOLUTIONS OF QUASILINEAR DEGENERATE PARABOLIC EQUATIONS IN UNBOUNDED DOMAINS

S. BONAFEDE *

*Dipartimento di Economia dei Sistemi Agro-Forestali*
*University of Palermo*
*Viale delle Scienze - 90128 Palermo, Italy*
*E-mail: bonafedes@unipa.it*

F. NICOLOSI

*Department of Mathematics*
*University of Catania*
*Viale A.Doria 6 - 95125 Catania, Italy*
*E-mail: fnicolosi@dmi.unict.it*

We prove the existence of bounded solutions of quasilinear degenerate parabolic equation of second order in unbounded domains and we study their asymptotic behavior near infinity.

**Key words:** Degenerate parabolic equation, unbounded domain, asymptotic behavior
**Mathematics Subject Classification:** 35K65

## 1. Introduction

We present existence and asymptotic behavior of bounded solutions for special class of degenerate nonlinear second order parabolic equations of divergence form

$$-\sum_{i=1}^{m} \frac{\partial}{\partial x_i} a_i(x,t,u,\nabla u) - c_0 u - f(x,t,u,\nabla u) + \frac{\partial u}{\partial t} = 0, \qquad (1)$$

$(x,t) \in Q = \Omega \times (0,+\infty)$ where $\Omega$ is an unbounded open set of $\mathbb{R}^m$, $c_0$ is a positive constant, $\nabla u$ is the gradient of unknown function $u$ and $f$ is nonlinear function which has quadratic growth with respect to gradient $\nabla u$.

## 2. Preliminaries

We assume that $f(x,t,u,p)$, $a_i(x,t,u,p)$ $(i=1,2,...,m)$ are Caratheodory's functions in $Q \times \mathbb{R} \times \mathbb{R}^m$, i.e. measurable with respect to $(x,t)$ for any $(u,p) \in \mathbb{R} \times \mathbb{R}^m$, continuous with respect to $(u,p)$ for a.e. $(x,t)$ in $Q$.

*Work partially supported by grant 60% of Italy.

We suppose the following inequalities

$$\lambda(|u|) \sum_{i=1}^{m} a_i(x,t,u,p)p_i \geq \nu(x)\psi(t)|p|^2, \tag{2}$$

$$uf(x,t,u,p) + c_1^2 + \lambda(|u|)\nu(x)\psi(t)|p|^2 + f_0(x,t) \geq 0, \tag{3}$$

$$|f(x,t,u,p)| \leq \lambda(|u|) \left[ f^*(x,t) + \nu(x)\psi(t)|p|^2 \right], \tag{4}$$

$$\frac{|a_i(x,t,u,p)|}{\sqrt{\nu\psi}} \leq \lambda(|u|) \left[ a^*(x,t) + \sqrt{\nu\psi}|p| \right] \tag{5}$$

hold, where $\nu : \Omega \to \mathbb{R}$, $\psi : ]0, +\infty[ \to \mathbb{R}$, $\lambda : [0, +\infty[ \to [1, +\infty[$ are functions with properties precised later and $c_1$ is a nonnegative real number such that $c_1 < c_0$. Moreover,

$$f_0(x,t) \in L^1(Q) \cap L^\infty(Q), \quad f^*(x,t) \in L^1(Q), \quad a^*(x,t) \in L^2(Q).$$

The conditions (2)-(5) determine the special structure of equation (1).

As model equation of this class we can consider the next equation

$$\frac{\partial u}{\partial t} = \sum_{i=1}^{m} \frac{\partial}{\partial x_i} \left( |x|^\alpha t^\beta \frac{\partial u}{\partial x_i} \right) - \gamma(x,t)u|u|^{p-2},$$

where $0 \leq \alpha < 2$, $\beta > 0$, $\gamma(x,t) \in L^1(Q)^+$ and $p \geq 2$.

Further we suppose that $\psi(t)$ and $\lambda(u)$ are measurable monotone nondecreasing function ,

$$\nu(x) \in L^1_{loc}(\Omega) , \quad \nu^{-1}(x) \in L^1_{loc}(\Omega).$$

For details concerning the above condition and definitions of our weighted Sobolev spaces see e.g. [2,5,6,12] and [13]. This paper is organized as follows. In section 3 we present the solvability of equation (1) by boundary and initial conditions

$$u(x,t) = 0, \quad (x,t) \in \partial\Omega \times (0, +\infty) \tag{6}$$

$$u(x,0) = 0, x \in \Omega. \tag{7}$$

In Section 4 we study the asymptotic behavior near infinity of a solution of equation (1) with condition (6) when $\Omega = \mathbb{R}^m$.

## 3. Existence

For any $n \in \mathbb{N}$ we denote

$$\Omega_n = \{x \in \Omega : |x| < n\} , \quad Q_n = \Omega_n \times ]0, n[.$$

In this section we assume the condition:

($\star$) for any $n \in \mathbb{N}$, there exist real positive numbers $g_n$ ($g_n > \frac{m}{2}$), $\tilde{g}_n$ such that $\nu^{-1}(x) \in L^{g_n}(\Omega_n)$ and $\dfrac{1}{\psi(t)} \in L^{\tilde{g}_n}(0,n)$. Furthermore the corresponding Sobolev weighted inequality

$$\left(\int_{\Omega_n} |u|^{\alpha_n} \, dx\right)^{\frac{1}{\alpha_n}} \leq \beta_n \left(\int_{\Omega_n} |u|^2 + \nu |\nabla u|^2 \, dx\right)^{\frac{1}{2}}$$

hold for an arbitrary function $u \in H_0^1(\nu, \Omega_n)$, $\beta_n > 0$, $\alpha_n > 2$ (see [11]).

Main result of this section is next theorem.

**Theorem 3.1.** *Let inequalities (2)-(5), condition ($\star$) be satisfied. Then the initial value problem (1),(6),(7) has at least one solution in $V^{1,0}(\nu\psi, Q) \cap L^\infty(Q)$.*

The main idea of the proof is based on approximation by bounded domains $Q_n$, where we obtain some suitable a priori estimates, and theory of monotone operators.

In fact, our assumptions, for any $n \in \mathbb{N}$, give us a function $u_n(x,t)$ such that

$$\int_{Q_n} \left\{ \sum_{i=1}^m a_i(x,t,u_n,\nabla u_n)\frac{\partial w}{\partial x_i} + c_0 u_n w + f(x,t,u_n,\nabla u_n)w - u_n \frac{\partial w}{\partial t} \right\} dx\,dt = 0$$

for any $w \in V^{1,1}(\nu\psi, Q_n) \cap L^\infty(Q_n)$ (see [7]);
moreover, for such function $u_n(x,t)$, the following inequalities

$$\operatorname{ess\,sup}_{Q_n} |u_n| \leq \left(\frac{\|f_0\|_\infty}{c_0 - c_1}\right)^{\frac{1}{2}} = K,$$

$$\left(\int_{Q_n} \left(|u_n|^2 + \sum_{i=1}^m \nu\psi \left|\frac{\partial u_n}{\partial x_i}\right|^2\right) dx\,dt\right)^{\frac{1}{2}} \leq \left\{\frac{K\lambda(K)\left[\|f^\star\|_1 + \|f_0\|_1^{\frac{1}{x}}\right]}{\min\left(\frac{1}{\lambda(K)}, c_0\right)}\right\}^{\frac{x}{2}} + 1$$

hold, whit $\chi \geq 1$ such that $\dfrac{\chi+1}{\lambda(K)} - \lambda(K) > 1$.

Hence it is possible to find a subsequence (denoted again by $\{u_n\}$) which converges weakly in $V^{1,0}(\nu\psi, Q)$ and weakly$^\star$ in $L^\infty(Q)$ to an element $u \in V^{1,0}(\nu\psi, Q) \cap L^\infty(Q)$.

To prove that $u(x,t)$ is a solution of problem (1),(6),(7) it will be sufficient pass to limit as $n$ goes to $+\infty$; to this aim we use the following

**Lemma 3.1.** *Let $\Omega_0$ be an open bounded subset of $\Omega$ and $b$ be a real positive number; $Q_0 = \Omega_0 \times ]0, b[$. Let $u(x,t) \in H^{1,0}(\nu\psi, Q_0)$ and $\{u_n\}$ be a sequence in $H^{1,0}(\nu\psi, Q_0)$ such that there exists a constant $\mu > 0$ for which $\int_{Q_0} |u_n|^2 + \nu\psi|\nabla u_n|^2 dx\,dt \leq \mu$ and $\lambda(|u_n(x,t)|) \leq \mu$ for almost $(x,t) \in Q_0$ and for any $n = 1,2,....$*

*Moreover, let us suppose*

$$\lim_{n\to+\infty} \int_{Q_0} |u_n(x,t) - u(x,t)|^2 dx\,dt = 0,$$

$$\lim_{n \to +\infty} \int_{Q_0} \sum_{i=1}^m [a_i(x, t, u_n(x,t), \nabla u_n(x,t)) -$$

$$-a_i(x, t, u_n(x,t), \nabla u(x,t))] \frac{\partial(u_n - u)}{\partial x_i} \, dx dt = 0.$$

*Then*

$$\lim_{n \to +\infty} \int_{Q_0} \nu \psi \sum_{i=1}^m \left| \frac{\partial(u_n - u)}{\partial x_i} \right|^2 \, dx dt = 0.$$

For more details about the proof of Theorem 3.1 see [3].

## Remark

It is possible to verify that $(\star)$ holds, for instance, if we take

$$\Omega = \{x \in \mathbb{R}^m : |x| > 1\} \text{ and}$$

$$\nu(x) = (|x| - 1)^\rho, \; 0 < \rho < \frac{2}{m}.$$

## 4. Asymptotic behaviour of solutions of nonlinear equation (1)

We take now $\Omega = \mathbb{R}^m$ and assume that it holds the following condition

$(\star\star)$ $\quad \nu(x) \in L^\infty_{loc}(\mathbb{R}^m) \, , \, \nu^{-1}(x) \in L^g_{loc}(\mathbb{R}^m) \; (g > \frac{m}{2}),$

$$\psi^{-1}(t) \in L^{\tilde{g}}(0, T) \text{ for any } T > 0 \; (\tilde{g} > 0).$$

The following theorem (see [4]) states the asymptotic behavior of the solutions near infinity.

**Theorem 4.1.** *Let inequalities (2)-(5), condition $(\star\star)$ be satisfied and let $R_0$ be a positive real number such that*

$$\text{supp } a^\star(x,t), \text{ supp } f_0(x,t), \text{ supp } f^\star(x,t) \subseteq \{x \in \mathbb{R}^m : |x| \le R_0\} \times [0, +\infty[.$$

*Take a function $u(x,t) \in W^{1,0}(\nu\psi, Q) \cap L^\infty(Q)$ weak solution of equation (1) with condition (7). Then for any $T > 0$ there exist two positive constants $\beta$ and $\tilde{\gamma}$, depending on $L = \text{ess sup}_Q |u(x,t)|$, such that*

$$H_R(T) \le \beta \left\{ \|f_0\|_{L^1(Q_T)} + \|f^\star\|_{L^1(Q_T)} \right\} e^{-\frac{\tilde{\gamma}(R - R_0)^2}{\eta(R)T\psi(T)}} \qquad \forall R > R_0,$$

*where*

$$H_R(T) = \int_{|x| > R} u^2(x, T) dx + \int_0^T \int_{|x| > R} \nu\psi |\nabla u|^2 \, dx dt,$$

*and*

$$\eta(R) = \sup_{R < |x| < 2R} \nu(x).$$

The results obtained in this note may be regarded as a continuation and completion of these contained in the paper [7] and extend to parabolic case the study on asymptotic behavior of weak solution of quasilinear degenerate elliptic equation stated in [9].

# References

1. R.A. Adams, *Sobolev Spaces*, Academic Press, New York, (1975).
2. S. Bonafede, *Existence results for a class of degenerate parabolic equations*, Rend.-Mat.-Appl. (7) **7** (1987), no. 2, 207–214.
3. S. Bonafede and F. Nicolosi, *Quasilinear degenerate parabolic equations in unbounded domains*, Comm. Appl. Anal. **8** (2004), 109–124.
4. S. Bonafede and F. Nicolosi, *On some properties of solutions of quasilinear degenerate parabolic equations in $I\!R^m \times (0,+\infty)$*, Math. Bohemica (2) **129** (2004), 113-123.
5. F. Guglielmino and F. Nicolosi, *W-solutions of boundary value problems for degenerate elliptic operators*, Ricerche di Matematica Suppl. Vol. **XXXVI** (1987), 59–72.
6. F. Guglielmino and F. Nicolosi, *Existence theorems for boundary value problems associated with quasilinear elliptic equations*, Ricerche di Matematica Vol. **XXXVII** , fasc. 1 (1988), 157–176.
7. F. Guglielmino and F. Nicolosi, *Existence results for boundary value problems associated with a class of quasilinear parabolic equations*, Current problems of analysis and mathematical physics (Italian) (Taormina, 1992), 95–117.
8. E. Hille and R.S. Phillips, *Functional analysis and semi-groups*, American Mathematical Society, Colloquium Pubblications, **31** (1957).
9. V. Kondratiev and F. Nicolosi, *On some properties of the solutions of quasilinear degenerate elliptic equations*, Math. Nachr. **182** (1996), 243–260.
10. O.A. Ladyzenskaja , V.A. Solonnikov and N.N. Ural'tseva, *Linear and quasi-linear equations of parabolic type*, Translation of mathematical monographs vol. **23**, A.M.S. Providence (1968).
11. M.K.V. Murthy and G. Stampacchia, *Boundary value problems for some degenerate-elliptic operators*, Ann. Mat. Pura Appl. (4) **80** (1968), 1–122.
12. F. Nicolosi, *Weak solutions of boundary value problems for parabolic operators that may degenerate*, Annali di Matematica (4) **125** (1980), 135–155.
13. F. Nicolosi, *Weak solutions of boundary value problems for degenerate parabolic operators in unbounded open sets*, Boll.Un.Mat.Ital.C (6) **4** (1985), no. 1, 269–278.

# MULTIPLE SOLITARY WAVES FOR NON-HOMOGENEOUS KLEIN-GORDON-MAXWELL EQUATIONS *

A.M. CANDELA[†] and A. SALVATORE[‡]

*Dipartimento di Matematica*
*Università degli Studi di Bari*
*Via E. Orabona 4*
*70125 Bari, Italy*
E-mails: [†] *candela@dm.uniba.it;* [‡] *salvator@dm.uniba.it*

By means of variational tools, in this paper we want to investigate the existence of multiple radial standing waves for a non–homogeneous Klein–Gordon equation coupled with Maxwell's equations.

**Key words:** Klein-Gordon-Maxwell equation, multiple solitray wave
**Mathematics Subject Classification:** 35Q40, 35Q51

## 1. Introduction

Let us consider the Klein–Gordon type equation

$$\frac{\partial^2 \psi}{\partial t^2} - \Delta\psi + m^2\psi - |\psi|^{p-2}\psi = g(x)\, e^{i\omega t}, \qquad x \in \mathbb{R}^3, \ t \in \mathbb{R}, \tag{1}$$

with $\psi : \mathbb{R}^3 \times \mathbb{R} \to \mathbb{C}$, $g : \mathbb{R}^3 \to \mathbb{R}$, $m > 0$, $p > 2$, $\omega \in \mathbb{R}$.

Recall that looking for *standing waves* which are solutions of (1) means finding $u : \mathbb{R}^3 \to \mathbb{R}$ such that

$$\psi(x,t) = u(x)\, e^{i\omega t} \tag{2}$$

solves the given equation. Clearly, if $g \equiv 0$ the constant $\omega \in \mathbb{R}$ can be arbitrarily choosen while, if $g \not\equiv 0$, $\omega$ is the constant appearing in the coefficient of $g$, i.e., the standing wave has the same pulsation $\omega$ of the source.

Thus, it is easy to check that the study of standing waves of the nonlinear Klein–Gordon equation is reduced to the study of solutions of a semilinear elliptic equation of type

$$-\Delta u + k_1 u = k_2|u|^{p-2}u + g, \qquad x \in \mathbb{R}^3, \tag{3}$$

with $k_1 = m^2 - \omega^2$ and $k_2 = 1$.

*This work was supported by M.I.U.R. (research funds ex 40 % and 60%).

In order to study equation (3), let us point out that its variational structure and standard arguments imply it is the Euler–Lagrange equation related to the functional

$$f(u) = \frac{1}{2}\int_{\mathbb{R}^3}|\nabla u|^2 dx + \frac{k_1}{2}\int_{\mathbb{R}^3}u^2 dx - \frac{k_2}{p}\int_{\mathbb{R}^3}|u|^p dx - \int_{\mathbb{R}^3}gu\,dx \qquad \text{on } H^1(\mathbb{R}^3).$$

If $g \equiv 0$ the existence of infinitely many radial solutions of (3) in $H^1(\mathbb{R}^3)$ has been proved in [12] if $k_1$, $k_2 > 0$ and $2 < p < 6$ as, here, the critical Sobolev exponent is 6 (see also [4] for a different proof based on the equivariant version of Mountain Pass Theorem stated in [1]).

If, on the contrary, it is $g \not\equiv 0$, this problem loses its symmetry, so, in general, very few existence and multiplicity results are known in the whole Euclidean space $\mathbb{R}^3$ (see, for example, [11]).

However, nonlinear problems of this type have also been much studied for the case of a bounded domain $\Omega$ with Dirichlet boundary conditions (see [8] and references therein). In this case infinitely many solutions of equation (3) in $H_0^1(\Omega)$ have been found for any $k_1 \in \mathbb{R}$, $k_2 > 0$ and $2 < p < 4$ (see Theorem 1.3 in [8] with $\mu = p$, $N = 1$). This result can be improved if $\Omega = B$ is a ball centered at 0 in $\mathbb{R}^3$ and $g$ has a radial symmetry, i.e., $g(x) = g(|x|)$: the existence of infinitely many radial solutions can be stated for each subcritical exponent $2 < p < 6$ (see Theorem 1.2 in [7]).

On the other hand, nonlinear Klein–Gordon equation (1) has been studied also when it interacts with an unknown electromagnetic field $(\mathbf{E},\mathbf{H})$. In this case, some existence results have been stated if $g \equiv 0$ (see [3,10]).

Encouraged by these results, here we want to look for radial solutions of a non–homogeneous Klein–Gordon equation in a ball $B$ with homogeneous boundary conditions when there is the interaction with an unknown electromagnetic field $(\mathbf{E},\mathbf{H})$. More precisely, since both $\mathbf{E}$ and $\mathbf{H}$ are not assigned, we have to study a system of equations in which are unknown either the wave function $\psi = \psi(x,t)$ and the gauge potentials

$$\mathbf{A} : B \times \mathbb{R} \to \mathbb{R}^3 \quad \text{and} \quad \Phi : B \times \mathbb{R} \to \mathbb{R},$$

which are related to $\mathbf{E}$, $\mathbf{H}$ by Maxwell's equations

$$\mathbf{E} = -\nabla\Phi - \frac{\partial\mathbf{A}}{\partial t}, \qquad \mathbf{H} = \nabla \times \mathbf{A}. \tag{4}$$

Thus, the aim of this paper is to investigate the existence of radial standing waves $\psi = \psi(x,t)$ of type (2) in the electrostatic case, or better when $\mathbf{A} \equiv 0$ and $\Phi$ is independent of time $t$.

Now, consider equation (1) stated in a ball $B$. Reasoning as in [3], this problem can be reduced to the system of equations

$$\begin{cases} -\Delta u + \left(m^2 - (\omega + e\Phi)^2\right)u - -|u|^{p-2}u = g, & x \in B, \\ -\Delta\Phi + u^2\Phi = -e\,\omega\,u^2, & x \in B, \\ u = \Phi = 0, & x \in \partial B, \end{cases} \tag{5}$$

where $e$ denotes the electric charge with $e^2 = 1$ (for more details, see Section 2).

If $g \equiv 0$, $e = 1$ and $B = \mathbb{R}^3$, the existence of infinitely many radially symmetric solutions $(u, \Phi)$, $u \in H^1(\mathbb{R}^3)$ and $\Phi \in L^6(\mathbb{R}^3)$ with $\nabla\Phi \in L^2(\mathbb{R}^3)$, has been proved in [3] under the assumptions $4 < p < 6$ and $|\omega| < m$. Later on, in [10] this result has been extended to the case $4 \leq p < 6$ and $0 < \omega < m$ or $2 < p < 4$ and $m\sqrt{p-2} > \sqrt{2}\omega > 0$.

Here, in a bounded ball $B$ centered in 0, we state the following result:

**Theorem 1.1.** *Let $m > 0$, $\omega \in \mathbb{R}^*$, $2 < p < 6$ and $g \in L^2(B)$ with $g(x) = g(|x|)$. Then, system (5) has infinitely many radially symmetric solutions $(u, \Phi)$ in $H_0^1(B) \times H_0^1(B)$ with $u \not\equiv 0$ and $\Phi \not\equiv 0$.*

**Remark 1.1.** If $\omega = 0$, system (5) reduces to an elliptic equation of type (3) so the existence of infinitely many radially symmetric solutions follows directly from Theorem 1.2 in [7].

**Remark 1.2.** Theorem 1.1 can be proved also if nonlinear term $|u|^{p-2}u$ is replaced with a more general superlinear odd function while, if function $g$ is not radially symmetric, it still holds but only if $2 < p < 4$ (for more details, see Remark 4.1).

This paper is organized as follows. In Section 2 we deduce system (5) describing equation (1) coupled with Maxwell's equations. In Section 3 we introduce the variational tools: a variational principle, stated in [3], which allows us to reduce the previous systems to a semilinear elliptic equation in the only variable $u$, and a perturbation method, introduced in [5], useful in order to state our multiplicity results. Finally, in Section 4 we prove our main theorem.

## 2. Coupled equations

In this section we show that system (5) arises in the study of solitary waves for Klein–Gordon equation (1) coupled with Maxwell's equations (4). To this aim, we adapt the arguments used in [3] to the non–homogeneous case with $g \neq 0$.

Firstly, let us consider the non–homogeneous Klein–Gordon equation (1) stated on a ball $B$ (centered in 0) with boundary conditions

$$\psi(x) = 0, \qquad x \in \partial B.$$

It is well known that its Lagrangian density is given by

$$L_{KG}(\psi)(x,t) = \frac{1}{2}\left(\left|\frac{\partial\psi}{\partial t}\right|^2 - |\nabla\psi|^2 - m^2|\psi|^2\right) + \frac{1}{p}|\psi|^p + Re\left(g(x)\,e^{i\omega t}\bar\psi\right). \quad (6)$$

If we assume that $\psi$ is a charged field with electric charge $e$, being $e^2 = 1$, then it causes an electromagnetic field $(\mathbf{E}, \mathbf{H})$, defined by Maxwell's equations (4), whose interaction with $\psi$ itself is described by the minimal coupling rule, i.e., making use in (6) of the formal substitution

$$\frac{\partial}{\partial t} \to \frac{\partial}{\partial t} + ie\Phi, \qquad \nabla \to \nabla - ie\mathbf{A}.$$

Thus, the corresponding Lagrangian density becomes

$$L_{KGM}(\psi, \Phi, \mathbf{A})(x, t) = \frac{1}{2} \left( \left| \frac{\partial \psi}{\partial t} + ie\Phi\psi \right|^2 - |\nabla\psi - ie\mathbf{A}\psi|^2 - m^2|\psi|^2 \right)$$
$$+ \frac{1}{p} |\psi|^p + Re\left( g(x) \, e^{i\omega t}\bar{\psi} \right).$$

If we write $\psi$ in polar form, i.e.,

$$psi(x, t) = u(x, t) \, e^{iS(x,t)}, \qquad \text{with } u, S : B \times \mathbb{R} \to \mathbb{R},$$

then the Lagrangian density becomes

$$L_{KGM}(u, S, \Phi, \mathbf{A})(x, t) = \frac{1}{2} \left( \left( \frac{\partial u}{\partial t} \right)^2 - |\nabla u|^2 - m^2 u^2 \right)$$
$$- \frac{u^2}{2} \left( |\nabla S - e\mathbf{A}|^2 - \left( \frac{\partial S}{\partial t} + e\Phi \right)^2 \right) + \frac{1}{p} |u|^p + gu \, \cos(\omega t - S).$$

On the other hand, the Lagrangian density of electromagnetic field $(\mathbf{E}, \mathbf{H})$ is

$$L_0(\Phi, \mathbf{A})(x, t) = \frac{|\mathbf{E}|^2 - |\mathbf{H}|^2}{2} = \frac{1}{2} \left( \left| \nabla\Phi + \frac{\partial \mathbf{A}}{\partial t} \right|^2 - |\nabla \times \mathbf{A}|^2 \right);$$

hence, the total action of the system "particle–electromagnetic field" is given by

$$L(u, S, \Phi, \mathbf{A}) = \iint (L_{KGM} + L_0) \, dx dt.$$

By making the first variation of $L$ with respect to $u$, $S$, $\Phi$ and $\mathbf{A}$, respectively, we obtain the following system of equations:

$$\frac{\partial^2 u}{\partial t^2} - \Delta u + \left( |\nabla S - e\mathbf{A}|^2 - \left( \frac{\partial S}{\partial t} + e\Phi \right)^2 + m^2 \right) u - |u|^{p-2}u \tag{7}$$
$$= g\cos(\omega t - S),$$

$$\frac{\partial}{\partial t} \left( \left( \frac{\partial S}{\partial t} + e\Phi \right) u^2 \right) - div\left( (\nabla S - e\mathbf{A}) u^2 \right) = gu\sin(\omega t - S), \tag{8}$$

$$div\left( \nabla\Phi + \frac{\partial \mathbf{A}}{\partial t} \right) = e\left( \frac{\partial S}{\partial t} + e\Phi \right) u^2, \tag{9}$$

$$\frac{\partial}{\partial t} \left( \nabla\Phi + \frac{\partial \mathbf{A}}{\partial t} \right) + \nabla \times (\nabla \times \mathbf{A}) = e\left( \nabla S - e\mathbf{A} \right) u^2. \tag{10}$$

At last, looking for standing waves with the same frequency $\omega$ of the source in the electrostatic case, or better considering the particular setting

$$u(x, t) = u(x), \qquad S(x, t) = \omega \, t, \qquad \Phi(x, t) = \Phi(x), \qquad \mathbf{A}(x, t) \equiv 0,$$

we have that equations (8) and (10) are identically satisfied while (7) and (9) are exactly the equations in system (5).

## 3. Variational tools

Now, our aim is solving system (5) by means of variational tools. Thus, firstly we introduce a suitable variational principle.

Defined functional $F_g : H_0^1(B) \times H_0^1(B) \to \mathbb{R}$ as

$$F_g(u, \Phi) = \frac{1}{2} \int_B (|\nabla u|^2 - |\nabla \Phi|^2 + (m^2 - (\omega + \Phi e)^2) u^2) dx - \frac{1}{p} \int_B |u|^p dx - \int_B g u \, dx,$$

standard arguments allow us to prove that $F_g$ is a $C^1$ map and its critical points are weak solutions of system (5). However, since $F_g$ is neither bounded from below nor from above, reasoning as in [3] we can introduce a new functional of the only variable $u$ so that its critical points are related to those ones of $F_g$.

To this purpose, we need the following result.

**Lemma 3.1.** *There exists a map* $\Phi : H_0^1(B) \to H_0^1(B)$ *such that for any* $u \in H_0^1(B)$ *function* $\Phi(u) \in H_0^1(B)$ *is the unique solution of equation*

$$-\Delta\Phi + u^2\Phi = -e \, \omega \, u^2. \tag{11}$$

*Moreover,* $\Phi(u)$ *satisfies the following properties:*

  (i) $e\omega\Phi(u) \leq 0$;
  (ii) $e\omega\Phi(u) \geq -\omega^2$ *in the set* $B_u = \{x \in B : u(x) \neq 0\}$;
  (iii) *if* $u$ *is radially symmetric, then* $\Phi(u)$ *is radial, too.*

**Proof.** The proof is essentially in [3] and [10] but here, for sake of completeness, we outline its main tools. Fixed $u \in H_0^1(B)$, it is $u^2 \in L^3(B)$ (by Sobolev Imbedding Theorem $H_0^1(B) \hookrightarrow L^6(B)$, hence $L^{\frac{6}{5}}(B) \hookrightarrow H^{-1}(B)$) and also $u^2 \in L^{\frac{3}{2}}(B)$, $u^2 \in H^{-1}(B)$ (by boundedness of $B$, we have continuous imbeddings between Lebesgue spaces: $L^3(B) \hookrightarrow L^{\frac{3}{2}}(B) \hookrightarrow L^{\frac{6}{5}}(B)$). Thus, considered the symmetric bilinear form

$$L_{u^2} : (v_1, v_2) \in H_0^1(B) \times H_0^1(B) \longmapsto \int_B \nabla v_1 \cdot \nabla v_2 \, dx + \int_B u^2 \, v_1 v_2 \, dx \in \mathbb{R},$$

it is easy to check that

$$L_{u^2}(v_1, v_2) \leq |u^2|_{\frac{3}{2}} \, |v_1|_6 \, |v_2|_6 \leq c \, |u|_3^2 \, \|v_1\| \, \|v_2\|$$

for all $(v_1, v_2) \in H_0^1(B) \times H_0^1(B)$ (i.e., $L_{u^2}$ is continuous) and

$$L_{u^2}(v, v) \geq \|v\|^2 \qquad \text{for all } v \in H_0^1(B)$$

(i.e., $L_{u^2}$ is coercive), where $c$ is a suitable constant, $|\cdot|_q$ is the $L^q$–norm, $\|\cdot\|$ is the standard $H_0^1(B)$–norm. Whence, by Lax–Milgram Theorem a unique function $\Phi(u) \in H_0^1(B)$ exists such that

$$L_{u^2}(\Phi(u), v) = -e \, \omega \int_B u^2 v \, dx \qquad \text{for all } v \in H_0^1(B);$$

hence, $\Phi(u)$ solves (11). Moreover, $\Phi(u)$ is the unique minimum point of functional

$$f_{u^2} : v \in H_0^1(B) \longmapsto \frac{1}{2} L_{u^2}(v, v) + e \, \omega \int_B u^2 v \, dx \in \mathbb{R}.$$

In order to verify *(i)–(ii)*, it is necessary to distinguish two different cases: $e\omega > 0$ and $e\omega < 0$. If $e\omega > 0$ then $-|\Phi(u)|$ is a minimum point of $f_{u^2}$ so, by uniqueness of $\Phi(u)$, it has to be $\Phi(u) = -|\Phi(u)| \leq 0$. Moreover, $(e\omega + \Phi(u))^- = \max\{0, -(e\omega + \Phi(u))\}$ is in $H_0^1(B)$, thus, by (11) it follows

$$\int_{B^-} |\nabla\Phi(u)|^2 dx + \int_{B^-} u^2(e\omega + \Phi(u))^2 dx = 0 \tag{12}$$

with $B^- = \{x \in B : \Phi(u)(x) < -e\omega\}$ which implies $\Phi(u) \geq -\omega$ in $B_u$. Hence, *(ii)* follows from multiplying by $e\omega > 0$ as $e^2 = 1$.

On the other hand, if $e\omega < 0$, then $|\Phi(u)|$ gives the minimum of $f_{u^2}$ so it is $\Phi(u) = |\Phi(u)| \geq 0$. Furthermore, $(e\omega + \Phi(u))^+ = \max\{0, e\omega + \Phi(u)\} \in H_0^1(B)$, so (12) still holds but replacing $B^-$ with $B^+ = \{x \in B : \Phi(u)(x) > -e\omega\}$. Thus, it has to be $\Phi(u) \leq -\omega$ in $B_u$ and *(ii)* holds.

Finally, adapting the arguments developed in the proof of Lemma 4.2 in [3] to a bounded ball, it follows that if $u$ is radially symmetric then $\Phi(u)$ is radially symmetric, too. ∎

**Remark 3.1.** Fixed $u \in H_0^1(B)$ by (11) it follows that

$$-\int_B |\nabla\Phi(u)|^2 dx = \int_B u^2(\Phi(u))^2 dx + e\omega \int_B u^2\Phi(u)dx. \tag{13}$$

**Proposition 3.1.** *There exists a $C^1$ functional $J_g : H_0^1(B) \to \mathbb{R}$ such that the following propositions are equivalent:*

*(i) $(u, \Phi) \in H_0^1(B) \times H_0^1(B)$ is a critical point of $F_g$;*
*(ii) $u$ is a critical point of $J_g$ and $\Phi = \Phi(u)$.*

**Proof.** Standard arguments imply that map $\Phi$, defined in Lemma 3.1, is $C^1$ on $H_0^1(B)$; furthermore, by definition its graph $G_\Phi$ is given by

$$G_\Phi = \left\{(u, \Phi) \in H_0^1(B) \times H_0^1(B) : \frac{\partial F_g}{\partial \Phi}(u, \Phi) = 0\right\}.$$

Hence, if we define $J_g(u) = F_g(u, \Phi(u))$, the previous arguments and (13) imply that $J_g$ is a $C^1$ functional such that

$$J_g(u) = \frac{1}{2}\int_B |\nabla u|^2 + \frac{m^2 - \omega^2}{2}\int_B u^2 dx - \frac{e\omega}{2}\int_B u^2\Phi(u)dx - \frac{1}{p}\int_B |u|^p dx - \int_B gudx,$$

$$J_g'(u) = \frac{\partial F_g}{\partial u}(u, \Phi(u)) + \frac{\partial F_g}{\partial \Phi}(u, \Phi(u))\Phi'(u) = \frac{\partial F_g}{\partial u}(u, \Phi(u)), \tag{14}$$

which implies the equivalence between *(i)* and *(ii)*. ∎

Whence, our problem is reduced to look for critical points of $J_g$ on $H_0^1(B)$. To this aim, it is enough to use Symmetric Mountain Pass Theorem in [1] if $g \equiv 0$ while, if the symmetry is broken by $g \not\equiv 0$, we need a method introduced by Bolle in [5] (see also [6]) but in the "weaker" version proved in [9].

The idea is considering a continuous path of functionals "linking" a symmetric functional $J_0$ to $J_g$ so that some min–max critical levels of $J_0$ "induce" critical levels of $J_g$.

Let $H$ be a Hilbert space equipped with the norm $\|\cdot\|$. Assume that $H = H_- \oplus H_+$, where $\dim(H_-) < +\infty$, and let $(e_k)_{k\geq 1}$ be an orthonormal base of $H_+$. Consider

$$H_0 = H_-, \quad H_{k+1} = H_k \oplus \mathbb{R}e_{k+1} \text{ if } k \in \mathbb{N};$$

so $(H_k)_k$ is an increasing sequence of finite dimensional subspaces of $H$.

Let $J : [0,1] \times H \to \mathbb{R}$ be a $C^1$–functional and, taken any $\theta \in [0,1]$, set $J_\theta = J(\theta, \cdot) : H \to \mathbb{R}$ and $J_\theta'(v) = \frac{\partial J}{\partial v}(\theta, v)$.

Let us set

$$c_k = \inf_{\gamma \in \Gamma} \sup_{v \in H_k} J_0(\gamma(v)), \tag{15}$$

with $\Gamma = \{\gamma \in C(H,H) : \gamma \text{ is odd and } \exists R > 0 \text{ s.t. } \gamma(v) = v \text{ if } \|v\| \geq R\}$.

Assume that

$(A_1)$ $J$ satisfies a weaker form of the classical Palais–Smale condition: any $((\theta_n, v_n))_n \subset [0,1] \times H$ such that

$$(J(\theta_n, v_n))_n \text{ is bounded} \quad \text{and} \quad \lim_{n\to+\infty} J_{\theta_n}'(v_n) = 0 \tag{16}$$

converges up to subsequences;

$(A_2)$ for any $b > 0$ there exists $C_b > 0$ such that if $(\theta, v) \in [0,1] \times H$ then

$$|J_\theta(v)| \leq b \implies \left|\frac{\partial J}{\partial \theta}(\theta, v)\right| \leq C_b \left(\|J_\theta'(v)\| + 1\right)(\|v\| + 1);$$

$(A_3)$ there exist two continuous maps $\eta_1, \eta_2 : [0,1] \times \mathbb{R} \to \mathbb{R}$, Lipschitz continuous with respect to the second variable, such that $\eta_1(\theta, \cdot) \leq \eta_2(\theta, \cdot)$ and if $(\theta, v) \in [0,1] \times H$ then

$$J_\theta'(v) = 0 \implies \eta_1(\theta, J_\theta(v)) \leq \frac{\partial J}{\partial \theta}(\theta, v) \leq \eta_2(\theta, J_\theta(v)); \tag{17}$$

$(A_4)$ $J_0$ is even and for each finite dimensional subspace $W$ of $H$ it results

$$\lim_{\substack{v \in W \\ \|v\| \to +\infty}} \sup_{\theta \in [0,1]} J(\theta, v) = -\infty .$$

For $i \in \{1, 2\}$, let $\psi_i : [0,1] \times \mathbb{R} \to \mathbb{R}$ be the flow associated to $\eta_i$, i.e., the solution of problem

$$\begin{cases} \frac{\partial \psi_i}{\partial \theta}(\theta, s) = \eta_i(\theta, \psi_i(\theta, s)) \\ \psi_i(0, s) = s. \end{cases}$$

Note that $\psi_i(\theta, \cdot)$ is continuous, non–decreasing on $\mathbb{R}$ and $\psi_1(\theta, \cdot) \leq \psi_2(\theta, \cdot)$. Set

$$\bar{\eta}_1(s) = \sup_{\theta \in [0,1]} |\eta_1(\theta, s)|, \qquad \bar{\eta}_2(s) = \sup_{\theta \in [0,1]} |\eta_2(\theta, s)| .$$

In this framework, the following abstract result can be proved (for more details and the proof, see Theorem 3 in [5] and Theorem 2.2 in [6]).

**Theorem 3.1.** *A constant $C \in \mathbb{R}$ exists such that if $k \in \mathbb{N}$ then*

    (a) *either $J_1$ has a critical level $\widetilde{c}_k$ with $\psi_2(1, c_k) < \psi_1(1, c_{k+1}) \leq \widetilde{c}_k$,*

    (b) *or $c_{k+1} - c_k \leq C\left(\bar{\eta}_1(c_{k+1}) + \bar{\eta}_2(c_k) + 1\right)$.*

## 4. Proof of Theorem 1.1

In this section we want to apply Bolle's method to functional $J_g$ on $H_0^1(B)$.

Firstly, let us point out that as $g$ is a radial function, we can reduce to study the critical points of $J_g$ restricted to the subspace of the radial functions

$$H_r = \left\{ u \in H_0^1(B) : u(x) = u(|x|) \right\}.$$

Indeed, by virtue of Lemma 3.1*(iii)* it follows easily that $H_r$ is a natural constraint for $J_g$, i.e. any critical point of $J_{g|H_r}$ is a critical point of $J_g$, too.

Now, consider the family of functionals

$$J : (\theta, u) \in [0, 1] \times H_r \longmapsto J_0(u) - \theta \int_B gu dx \in \mathbb{R},$$

where

$$J_0(u) = \frac{1}{2} \int_B |\nabla u|^2 + \frac{m^2 - \omega^2}{2} \int_B u^2 dx - \frac{e\omega}{2} \int_B u^2 \Phi(u) dx - \frac{1}{p} \int_B |u|^p dx$$

is even as, by definition, it is $\Phi(-u) = \Phi(u)$. Clearly, $J$ is a $C^1$ functional such that $J(0, \cdot) = J_0$, $J(1, \cdot) = J_g$. By simple computations and (14) it follows

$$\frac{\partial J}{\partial \theta}(\theta, u) = -\int_B gu dx,$$

$$J'_\theta(u)[v] = J'_g(u)[v] + (1 - \theta) \int_B gv dx$$

$$= \int_B \nabla u \cdot \nabla v dx + (m^2 - \omega^2) \int_B uv dx - 2e\omega \int_B uv\Phi(u) dx$$

$$- \int_B uv(\Phi(u))^2 dx - \int_B |u|^{p-2} uv dx - \theta \int_B gv dx$$

for all $\theta \in [0, 1]$, $u, v \in H_r$ (here, $J_\theta = J(\theta, \cdot)$).

The following lemma allows us to prove that functional $J$ verifies the assumptions of Bolle's abstract theorem (here and in the following, $a_i$ will denote some suitable positive constants).

**Lemma 4.1.** *Taken any $\delta \in \left(\frac{1}{p}, \frac{1}{2}\right)$ two constants $\beta_1(\delta)$, $\beta_2(\delta) > 0$ exist such that for any $(\theta, u) \in [0, 1] \times H_r$ it is*

$$\|u\|^2 + |u|_p^p \leq \beta_1(\delta)\left(J_\theta(u) - \delta J'_\theta(u)[u]\right) + \beta_2(\delta). \tag{18}$$

**Proof.** Fixed $\frac{1}{p} < \delta < \frac{1}{2}$, by the definition of $J$ and the expression of $J'_\theta$ it follows that

$$J_\theta(u) - \delta J'_\theta(u)[u] = \left(\frac{1}{2} - \delta\right)\left(\|u\|^2 + (m^2 - \omega^2)|u|_2^2\right) + \left(2\delta - \frac{1}{2}\right) e\omega \int_B u^2 \Phi(u) dx$$

$$+ \delta \int_B u^2 (\Phi(u))^2 dx + \left(\delta - \frac{1}{p}\right)|u|_p^p - (1 - \delta)\theta \int_B gu\, dx.$$

Now, by Lemma 3.1(i)–(ii) we have $\left(2\delta - \frac{1}{2}\right) e\omega \int_B u^2 \Phi(u) dx \geq -a_1 |u|_2^2$; hence,

$$J_\theta(u) - \delta J'_\theta(u)[u] \geq \left(\frac{1}{2} - \delta\right)\|u\|^2 + \left(\left(\frac{1}{2} - \delta\right)(m^2 - \omega^2) - a_1\right)|u|_2^2$$

$$+ \left(\delta - \frac{1}{p}\right)|u|_p^p - (1 - \delta)\theta|g|_2\,|u|_2$$

and the conclusion follows from $p > 2$. ∎

**Proof of Theorem 1.1** Firstly, let us prove that $J$ satisfies $(A_1)$; so let us consider a sequence $((\theta_n, u_n))_n \subset [0, 1] \times H_r$ such that (16) holds.

Clearly, (16) and (18) imply $(\|u_n\|)_n$ is bounded and, from (13), sequence $(\|\Phi(u_n)\|)_n$ is bounded, too. Moreover, by the expression of $J'_\theta$, (14), the first equation in (5) and (16) it follows

$$-\Delta u_n + \left(m^2 - (\omega + e\Phi(u_n))^2\right) u_n - |u_n|^{p-2} u_n - \theta_n g = \epsilon_n,$$

where $\epsilon_n \to 0$ and $-\Delta$ is an isomorphism from $H_r$ to its dual $H_r^{-1}$. Thus, since $H_r \hookrightarrow\hookrightarrow L^3(B)$ implies $L^{\frac{3}{2}}(B) \hookrightarrow\hookrightarrow H_r^{-1}$ (compact imbeddings) and both $(u_n\Phi(u_n))_n$ and $(u_n(\Phi(u_n))^2)_n$ are bounded in $L^{\frac{3}{2}}(B)$, the proof follows from standard Palais–Smale arguments and Rellich–Kondrachov Theorem.

By the expression of $\frac{\partial J}{\partial \theta}$ it is simple to prove that $J$ verifies $(A_2)$ since

$$\left|\frac{\partial J}{\partial \theta}(\theta, v)\right| \leq a_2 \|v\| \qquad \text{for all } (\theta, u) \in [0, 1] \times H_r.$$

On the other hand, by (18) we obtain

$$(\theta, v) \in [0, 1] \times H_r, \ J'_\theta(v) = 0 \implies \left|\frac{\partial J}{\partial \theta}(\theta, v)\right| \leq a_3 \left(J_\theta^2(v) + 1\right)^{\frac{1}{2p}};$$

thus, $J$ verifies $(A_3)$ with

$$\eta_2(\theta, s) = -\eta_1(\theta, s) = a_3 \left(s^2 + 1\right)^{\frac{1}{2p}}; \quad \text{hence, } \bar{\eta}_i(s) \equiv \eta_i(\theta, s) \text{ if } i \in \{1, 2\}. \quad (19)$$

Moreover, by Lemma 3.1(i)–(ii) it results $|e\omega\Phi(u)| = -e\omega\Phi(u) \leq \omega^2$ in $B_u$, so

$$J_\theta(u) \leq \frac{1}{2}\|u\|^2 + \frac{m^2 - \omega^2}{2}|u|_2^2 + \frac{\omega^2}{2}|u|_2^2 - \frac{1}{p}|u|_p^p + |g|_2|u|_2$$

$$\leq \frac{1}{2}\|u\|^2 + a_4|u|_2^2 - \frac{1}{p}|u|_p^p + a_5.$$

whence, $J$ verifies assumption $(A_4)$ as all the norms in a finite dimensional space are equivalent and $p > 2$.

Now, we are able to apply Theorem 3.1 and, arguing by contradiction, let us assume situation $(b)$ occurs for all $k$ large enough, i.e., critical levels $c_k$ of $J_0$, defined as in (15), are such that $c_{k+1} - c_k \leq a_6(c_{k+1}^{\frac{1}{p}} + c_k^{\frac{1}{p}} + 1)$ (by (19)). Whence, by Lemma 3.5 in [2] it is $c_k \leq a_7 k^{\frac{p}{p-1}}$ for all $k$ large enough.

On the other hand, by Lemma 3.1$(i)$ we obtain $J_0(u) \geq \frac{1}{2}\|u\|^2 - a_8|u|_p^p - a_9$; so, by applying the arguments developed in [7], the radial symmetry of the problem implies $c_k \geq a_{10} k^{\frac{p}{p-2}}$ in contradiction with the previous inequality. ∎

**Remark 4.1.** All the previous arguments can be used also if nonlinear term $|u|^{p-2}u$ is replaced by a more general function $h(x, u)$ which is odd in $u$, radial in $x$ and satisfies suitable growth conditions (see assumptions $(G_1)$ and $(G_2)$ in [7]). Furthermore, if function $g$ is not radially symmetric, then the contradiction at the end of the proof of Theorem 1.1 follows from the weaker growth estimate $c_k \geq a_{10} k^{\frac{2p}{3(p-2)}}$ only if $2 < p < 4$.

## References

1. A. Ambrosetti and P.H. Rabinowitz, Dual variational methods in critical point theory and applications, *J. Funct. Anal.* **14**, 349 (1973).

2. A. Bahri and H. Berestycki, A perturbation method in critical point theory and applications, *Trans. Amer. Math. Soc.* **267**, 1 (1981).

3. V. Benci and D. Fortunato, Solitary waves of the nonlinear Klein–Gordon equation coupled with the Maxwell equations, *Rev. Math. Phys.* **14**, 409 (2002).

4. H. Berestycki and P.L. Lions, Nonlinear scalar field equations I and II, *Arch. Rat. Mech. Anal.* **82**, 347 (1983).

5. P. Bolle, On the Bolza Problem, *J. Differential Equations* **152**, 274 (1999).

6. P. Bolle, N. Ghoussoub and H. Tehrani, The multiplicity of solutions in non–homogeneous boundary value problems, *Manuscripta Math.* **101**, 325 (2000).

7. A.M. Candela, G. Palmieri and A. Salvatore, Radial solutions of semilinear elliptic equations with broken symmetry. To appear on *Topol. Methods Nonlinear Anal.*

8. A.M. Candela, A. Salvatore and M. Squassina, Semilinear elliptic systems with lack of symmetry, *Dynam. Contin. Discrete Impuls. Systems* A **10**, 181 (2003).

9. M. Clapp, Y. Ding and S. Hernández–Linares, Strongly indefinite functionals with perturbed symmetries and multiple solutions of non symmetric elliptic systems. Preprint.

10. T. D'Aprile and D. Mugnai, Solitary waves for nonlinear Klein–Gordon–Maxwell and Schrödinger–Maxwell equations, *Proc. Roy. Soc. Edinburgh* **134A**, 893 (2004).

11. A. Salvatore, Multiple radial solutions for a superlinear elliptic problem in $\mathbb{R}^N$, in: Proc. Dynamical Systems and Applications (G.S. Ladde, N.G. Medhin & M. Sambandham Eds), *Dynam. Systems Appl.* **4**, 472 (2004).

12. W. Strauss, Existence of solitary waves in higher dimensions, *Comm. Math. Phys.* **55**, 149 (1977).

# A QUADRATIC BOLZA-TYPE PROBLEM IN STATIONARY SPACE TIMES WITH CRITICAL GROWTH

R. BARTOLO *

*Dipartimento di Matematica, Politecnico di Bari*
*Via G. Amendola 126/B, 70126 Bari, Italy*
*E-mail: rossella@poliba.it*

A.M. CANDELA*

*Dipartimento di Matematica, Università degli Studi di Bari*
*Via E. Orabona 4, 70125 Bari, Italy*
*E-mail: candela@dm.uniba.it*

J.L. FlORES†

*Departamento de Álgebra, Geometría y Topología*
*Facultad de Ciencias, Universidad de Málaga,*
*Campus Teatinos, 29071 Málaga, Spain*
*E-mail: floresj@agt.cie.uma.es*

In this note we prove the existence of connecting trajectories under the action of an external field in a stationary spacetime under optimal growth assumptions for both the potential and the coefficients of the metric.

**Keywords:** Bolza problem, standard stationary spacetimes, Lorentzian manifold
**AMS Mathematics Subject Classification:** 49J30

## 1. Introduction

The aim of this paper is to investigate the existence of solutions of a Bolza problem but settled in a class of Lorentzian manifolds called *standard stationary spacetimes*.

**Definition 1.1.** A Lorentzian manifold $(M, \langle \cdot, \cdot \rangle_L)$ given by a global splitting $M = M_0 \times \mathbb{R}$ is *(standard) stationary* if $(M_0, \langle \cdot, \cdot \rangle)$ is a finite dimensional connected Riemannian manifold and the metric is

$$\langle \zeta, \zeta' \rangle_L = \langle \xi, \xi' \rangle + \langle \delta(x), \xi \rangle \tau' + \langle \delta(x), \xi' \rangle \tau - \beta(x) \tau \tau' \tag{1}$$

for any $z = (x, t) \in M$ and $\zeta = (\xi, \tau)$, $\zeta' = (\xi', \tau') \in T_z M = T_x M_0 \times \mathbb{R}$, where $\delta$ and $\beta$ are respectively a smooth vector field and a smooth strictly positive scalar field

---

*Supported by M.I.U.R. (research funds ex 40% and 60%).
†Supported by a MEC Grant RyC-2004-382.

on $M_0$. In the particular case of $\delta \equiv 0$ the Lorentzian manifold is named *(standard) static*.

Given a function $V \in C^1(M \times \mathbb{R}, \mathbb{R})$, two points $z_0, z_1 \in M$ and an *arrival time* $T > 0$, our aim is to apply variational tools to find accurate conditions ensuring that $z_0$ and $z_1$ can be connected by means of trajectories under the action of potential $V$ in time $T$, thus extending to the stationary case the results obtained in [7] on Riemannian manifolds and in [1] on static spacetimes.

More precisely, we look for smooth solutions of problem

$$(P_T) \qquad \begin{cases} D_s^L \dot{z} + \nabla_L V(z, s) = 0 & \text{for all } s \in [0, T], \\ z(0) = z_0, \ z(T) = z_1, \end{cases}$$

where $D_s^L$ denotes the covariant derivative along $z$ induced by the Levi–Civita connection of metric $\langle \cdot, \cdot \rangle_L$, and $\nabla_L V(z, s)$ is the gradient of $V$ with respect to $z$.

Notice that problem $(P_T)$ in stationary spacetimes is interesting not only from a mathematical point of view but also from a physical one. In fact, these spacetimes represent time–independent gravitational fields as, for example, Kerr spacetime (for more details see [13]) and $(P_T)$ is a model problem involved in the study of relativistic particles under the action of electromagnetic fields (see Section 4.4a in [15]).

Our main theorem can be stated as follows.

**Theorem 1.1.** *Let* $(M, \langle \cdot, \cdot \rangle_L)$ *be a (standard) stationary spacetime with* $M = M_0 \times \mathbb{R}$ *and* $\langle \cdot, \cdot \rangle_L$ *as in (1). Moreover, let it be* $V \in C^1(M \times [0, \delta], \mathbb{R})$, $\delta > 0$ *and denote by* $d(\cdot, \cdot)$ *the distance canonically associated to Riemannian metric* $\langle \cdot, \cdot \rangle$ *on* $M_0$. *Suppose that:*

$(H_1)$ *Riemannian manifold* $(M_0, \langle \cdot, \cdot \rangle)$ *is complete and smooth (at least* $C^3$*);*
$(H_2)$ *there exist* $\mu_1, \mu_2 \geq 0$, $k_1, k_2 \in \mathbb{R}$ *and a point* $\bar{x} \in M_0$ *such that*

$$0 < \beta(x) \leq \mu_1 d^2(x, \bar{x}) + k_1 \quad \text{for all } x \in M_0,$$

$$\sqrt{\langle \delta(x), \delta(x) \rangle} \leq \mu_2 d(x, \bar{x}) + k_2 \quad \text{for all } x \in M_0;$$

$(H_3)$ *there exist* $\lambda \geq 0$, $k \in \mathbb{R}$ *such that*

$$V(z, s) \equiv V(x, s) \quad \text{for all } z = (x, t) \in M = M_0 \times \mathbb{R}, \ s \in [0, \delta], \qquad (2)$$

*and*

$$V(x, s) \leq \lambda d^2(x, \bar{x}) + k \quad \text{for all } x \in M_0, \ s \in [0, \delta]. \qquad (3)$$

*Then, each pair of points* $z_0, z_1 \in M$ *can be joined by at least one trajectory, solution of problem* $(P_T)$, *if arrival time* $T \in \ ]0, \delta]$ *is such that*

$$\lambda T^2 < \frac{\pi^2}{2}. \qquad (4)$$

**Remark 1.1.** If, in addition to the assumptions of Theorem 1.1, $M_0$ is non–contractible in itself, then a direct application of Ljusternik–Schnirelman Theory implies that any two points in $M$ are joined by infinitely many trajectories (see, e.g., [1]).

From a variational viewpoint, problem $(P_T)$ is equivalent to find critical points of the action functional

$$f_V(z) = \frac{1}{2}\int_0^T \langle \dot{z}, \dot{z}\rangle_L \, ds - \int_0^T V(z,s) \, ds \tag{5}$$

in a suitable set of curves $z : [0,T] \to M$ which satisfy boundary conditions $z(0) = z_0$, $z(T) = z_1$.

This critical point problem has been widely studied in the literature. In fact, in the particular case of $V \equiv 0$, i.e., when the problem is reduced to the study of geodesic connectedness on a stationary spacetime, the first pioneering result comes from [11] (and from [5] in the static case). There, exploiting the fact that the coefficients of the metric do not depend on $t$, the authors reduce this study to look for critical points of Riemannian functional

$$J(x) = \int_0^T \langle \dot{x}, \dot{x}\rangle \, ds + \int_0^T \frac{\langle \delta(x), \dot{x}\rangle^2}{\beta(x)} \, ds - K_t^2(x) \int_0^T \frac{1}{\beta(x)} \, ds,$$

with

$$K_t(x) = \left(\Delta_t - \int_0^T \frac{\langle \delta(x), \dot{x}\rangle}{\beta(x)} \, ds\right)\left(\int_0^T \frac{1}{\beta(x)} \, ds\right)^{-1}, \tag{6}$$

where $z_0 = (x_0, t_0)$, $z_1 = (x_1, t_1)$, $\Delta_t = t_1 - t_0$, and the domain of $J$ is a suitable space of curves $\Omega^T(x_0, x_1)$ joining $x_0$ to $x_1$ in a time $T$ (see Section 2 for more details).

Although in the last years multiple partial answers to this question have been provided (see [14,12,8]), the optimal result, which ensures the existence of critical points of $J$ just under hypotheses $(H_1)$ and $(H_2)$, has been stated only very recently in [2]. The key point of this result is the following proposition (see Lemma 2.6 in [2]):

**Proposition 1.1.** *Functional $J$ is bounded from below and coercive if conditions $(H_1)$ and $(H_2)$ hold.*

In this note we want to overcome the more general case of $V$ being non–trivial and independent of time coordinate $t$ (see (2)). As before, $(P_T)$ admits a variational formulation entirely based on the Riemannian part of the spacetime. More precisely, problem $(P_T)$ reduces to find critical points of functional

$$J_V(x) = \frac{1}{2}J(x) - \int_0^T V(x,s) \, ds \qquad \text{on } \Omega^T(x_0, x_1) \tag{7}$$

(see Proposition 2.1 for more details).

We remark that $(P_T)$, when set on a Riemannian manifold $(M_0, \langle \cdot, \cdot \rangle)$, turn into Riemannian Bolza problem

$$\begin{cases} D_s \dot{x} + \nabla_x V(x,s) = 0 & \text{for all } s \in [0,T], \\ x(0) = x_0, \; x(T) = x_1, \end{cases} \tag{8}$$

where $D_s$ denotes the covariant derivative along $x$ induced by the Levi–Civita connection of $\langle \cdot, \cdot \rangle$, $\nabla_x V(x,s)$ is the gradient of $V$ with respect to $x$ and $x_0, x_1 \in M_0$.

In particular, when $(M_0, \langle \cdot, \cdot \rangle)$ is an Euclidean space, (8) has been widely studied (e.g., see [6,9,10] and references therein).

Recently, in [7] it has been obtained a very accurate result stating the existence of a solution for problem (8) under hypotheses $(H_1)$, (3) and (4). The key point of this result is the following proposition (see Lemma 3.4 in [7]):

**Proposition 1.2.** *Functional*

$$J_T(x) = \frac{1}{2} \int_0^T \langle \dot{x}, \dot{x} \rangle \, ds \; - \; \int_0^T V(x,s) \, ds \qquad \text{on } \Omega^T(x_0, x_1)$$

*is bounded from below and coercive if conditions $(H_1)$, (3) and (4) hold.*

## 2. Variational setting and abstract tools

By the product structure of $M$, the infinite dimensional manifold $H^1([0,T],M)$ (first Sobolev space of curves on $M$) is diffeomorphic to product manifold $H^1([0,T],M_0) \times H^1([0,T],\mathbb{R})$ and can be equipped with a structure of Riemannian manifold by setting

$$\langle \zeta, \zeta \rangle_1 = \int_0^T \langle \xi, \xi \rangle \, ds \; + \; \int_0^T \langle D_s \xi, D_s \xi \rangle \, ds \; + \; \int_0^T \tau^2 ds \; + \; \int_0^T \dot{\tau}^2 ds,$$

for any $z = (x,t) \in H^1([0,T],M)$ and $\zeta = (\xi, \tau) \in T_z H^1([0,T],M)$, with

$$T_z H^1([0,T],M) \equiv T_x H^1([0,T],M_0) \times H^1([0,T],\mathbb{R}).$$

By Nash Embedding Theorem, as $M_0$ is at least $C^3$ we can assume that it is a submanifold of an Euclidean space $\mathbb{R}^N$, $\langle \cdot, \cdot \rangle$ is the restriction to $M_0$ of the Euclidean metric on $\mathbb{R}^N$ and $d(\cdot, \cdot)$ is the corresponding distance, i.e.,

$$d(\bar{x}_1, \bar{x}_2) = \inf \left\{ \int_a^b \sqrt{\langle \dot{\gamma}, \dot{\gamma} \rangle} ds : \; \gamma \in A_{\bar{x}_1, \bar{x}_2} \right\}$$

with $\bar{x}_1, \bar{x}_2 \in M_0$ and $\gamma \in A_{\bar{x}_1, \bar{x}_2}$ if $\gamma : [a,b] \to M_0$ is a piecewise smooth curve joining $\bar{x}_1$ to $\bar{x}_2$. Hence, it can be proved that $H^1([0,T],M_0)$ can be identified with the set of absolutely continuous curves $x : [0,T] \to \mathbb{R}^N$ with square summable derivative such that $x([0,T]) \subset M_0$.

Furthermore, as $(M_0, \langle \cdot, \cdot \rangle)$ is a complete Riemannian manifold, $H^1([0,T],M)$ is also a complete Riemannian manifold with respect to $\langle \cdot, \cdot \rangle_1$.

Let $Z$ be the smooth manifold of all $H^1([0,T],M)$–curves joining $z_0$ to $z_1$ and let $\Omega^T(x_0, x_1)$ denote the smooth submanifold of $H^1([0,T],M_0)$ formed by all

$H^1([0,T], M_0)$–curves joining $x_0$ to $x_1$ in $M_0$. Since $H^1([0,T], M)$ is diffeomorphic to a product manifold it follows

$$Z \equiv \Omega^T(x_0, x_1) \times W_T(t_0, t_1),$$

where

$$W_T(t_0, t_1) = \{t \in H^1([0,T], \mathbb{R}) : t(0) = t_0, \ t(T) = t_1\} = H_0^1([0,T], \mathbb{R}) + j^*$$

with

$$H_0^1([0,T], \mathbb{R}) = \{\tau \in H^1([0,T], \mathbb{R}) : \tau(0) = 0 = \tau(T)\},$$

$$j^* : s \in [0,T] \mapsto t_0 + \frac{s}{T}\Delta_t \in \mathbb{R}, \quad \Delta_t = t_1 - t_0.$$

Whence, $W_T(t_0, t_1)$ is a closed affine submanifold of $H^1([0,T], \mathbb{R})$ with tangent space $T_tW_T = H_0^1([0,T], \mathbb{R})$ for all $t \in W_T(t_0, t_1)$. Moreover, for all $x \in \Omega^T(x_0, x_1)$, it is

$$T_x\Omega^T(x_0, x_1) = \{\xi \in T_x H^1([0,T], M_0) : \xi(0) = 0 = \xi(T)\}.$$

Thus, taken any curve $z = (x,t) \in Z$ it is

$$T_zZ \equiv T_x\Omega^T(x_0, x_1) \times H_0^1([0,T], \mathbb{R}),$$

and $Z$ can be equipped with the following equivalent Riemannian structure:

$$\langle \zeta, \zeta \rangle_H = \langle (\xi, \tau), (\xi, \tau) \rangle_H = \int_0^T \langle D_s\xi, D_s\xi \rangle ds + \int_0^T \dot{\tau}^2 ds$$

for any $z = (x,t) \in Z$ and $\zeta = (\xi, \tau) \in T_zZ$.

As already remarked in Section 1, solving problem $(P_T)$ is equivalent to find critical points of functional $f_V$ in (5) on the manifold of curves $Z$ just defined. But, in general, $f_V$ is unbounded both from above and from below, so its critical levels cannot be directly investigated by means of classical topological methods. Nevertheless, when potential $V$ satisfies condition (2), the presence of a Killing vector field in any stationary spacetime allows one to introduce a new functional with better properties.

**Proposition 2.1.** *Assume that potential $V$ satisfies condition (2) and consider $z^* = (x^*, t^*) \in Z$. The following statements are equivalent:*

    *(i) $z^*$ is a critical point of action functional $f_V$ defined in (5);*

    *(ii) $x^*$ is a critical point of functional $J_V : \Omega^T(x_0, x_1) \to \mathbb{R}$ defined in (7) and $t^* = \Psi(x^*)$, with $\Psi : \Omega^T(x_0, x_1) \to W_T(t_0, t_1)$ such that*

$$\Psi(x)(s) = t_0 + \int_0^s \frac{\langle \delta(x(\sigma)), \dot{x}(\sigma) \rangle}{\beta(x(\sigma))} d\sigma + K_t(x) \int_0^s \frac{1}{\beta(x(\sigma))} d\sigma$$

*and $K_t(x)$ defined as in (6).*

*Moreover, it is $f_V(z^*) = J_V(x^*)$.*

**Proof.** As $f_V$ is a $C^1$ functional on $Z$ and $V$ is independent of time variable $t$, fixed $z = (x, t) \in Z$ and $\zeta = (\xi, \tau) \in T_z Z$ we have

$$f_V'(z)[(\xi, 0)] = \int_0^T \langle \dot{x}, D_s \xi \rangle \, ds + \int_0^T \langle \delta'(x)[\xi], \dot{x} \rangle \dot{t} \, ds$$
$$+ \int_0^T \langle \delta(x), \dot{\xi} \rangle \dot{t} \, ds - \frac{1}{2} \int_0^T \beta'(x)[\xi] \, \dot{t}^2 \, ds - \int_0^T \langle \nabla_x V(x, t), \xi \rangle \, ds,$$

$$f_V'(z)[(0, \tau)] = f'(z)[(0, \tau)] = \int_0^T \langle \delta(x), \dot{x} \rangle \dot{\tau} \, ds - \int_0^T \beta(x) \dot{t} \dot{\tau} \, ds,$$

where $\delta'$, $\beta'$ denotes the derivative of $\delta$, respectively $\beta$, with respect to the Riemannian structure on $M_0$ and

$$f(z) = \int_0^T \langle \dot{z}, \dot{z} \rangle_L \, ds$$

is the action functional corresponding to the geodesic equation in stationary spacetime $M$. Thus, it is enough reasoning as in the proof of Theorem 2.1 in [5]. ∎

Now, our problem is reduced to look for critical points of Riemannian functional $J_V$ on $\Omega^T(x_0, x_1)$. To this aim, we can apply the following classical abstract minimum theorem.

**Theorem 2.1.** *Let $\Omega$ be a complete Riemannian manifold and $F$ a $C^1$ functional on $\Omega$ which satisfies the Palais–Smale condition, i.e., any $(x_k)_k \subset \Omega$ such that*

$$(F(x_k))_k \text{ is bounded} \quad \text{and} \quad \lim_{k \to +\infty} F'(x_k) = 0$$

*converges in $\Omega$ up to subsequences. Then, if $F$ is bounded from below, it attains its infimum.*

## 3. Proof of Theorem 1.1

Our strategy is to apply abstract Theorem 2.1 to functional $J_V$ in (7) defined on the manifold of curves $\Omega^T(x_0, x_1)$. As already remarked in Section 2, functional $J_V$ is $C^1$ on $\Omega^T(x_0, x_1)$, which is complete (under assumption $(H_1)$). Hence, we just need to prove that $J_V$ is bounded from below and satisfies Palais–Smale condition. Even better, it is enough to prove that $J_V$ is bounded from below and coercive in $\Omega^T(x_0, x_1)$, i.e., $J_V(x_k) \to +\infty$ if $\|\dot{x}_k\| \to +\infty$ (here, $\|\cdot\|$ is the $L^2$–norm). In fact, if $J_V$ is coercive in $\Omega^T(x_0, x_1)$, then a sequence $(x_k)_k$ has to be bounded if $(J_V(x_k))_k$ is bounded, and Palais–Smale condition follows by the following lemma.

**Lemma 3.1.** *Let $(M_0, \langle \cdot, \cdot \rangle)$ be a Riemannian manifold which satisfies condition $(H_1)$ and let potential $V = V(x, s)$ be $C^1$ on $M_0 \times [0, T]$. If $(x_k)_k$ is a bounded sequence in $\Omega^T(x_0, x_1)$ such that $J_V'(x_k) \to 0$ then it converges in $\Omega^T(x_0, x_1)$ up to subsequences.*

**Proof.** If $(x_k)_k \subset \Omega^T(x_0, x_1)$ is a bounded sequence such that $J'_V(x_k) \to 0$ and we define $t_k = \Psi(x_k)$ (with $\Psi$ as in Proposition 2.1(ii)), then $(\|\dot{t}_k\|_\infty)_n$ has to be bounded and $x \in H^1(I, \mathbb{R}^N)$ exists such that, up to subsequences, it is $x_k \rightharpoonup x$ weakly in $H^1(I, \mathbb{R}^N)$. Then, reasoning as in the proof of Proposition 4.3 in [3], by completeness of $M_0$ it has to be $x \in \Omega^T(x_0, x_1)$ and $x_k \to x$ strongly in $\Omega^T(x_0, x_1)$ by means of Lemma 2.1 in [4]. ∎

**Lemma 3.2.** *If $(H_1)$, $(H_2)$ and $(H_3)$ hold, and $T \in ]0, \delta]$ satisfies condition (4), then $J_V$ is bounded from below and coercive in $\Omega^T(x_0, x_1)$.*

**Proof.** For any $\epsilon \in ]0, 1[$, we can write

$$J_V(x) = \frac{\epsilon}{2} J^\epsilon(x) + (1 - \epsilon) J_T^\epsilon(x), \tag{9}$$

where

$$J^\epsilon(x) = \int_0^T \langle \dot{x}, \dot{x} \rangle ds + \int_0^T \frac{\langle \delta(x), \dot{x} \rangle^2}{\overline{\beta}(x)} ds - \overline{K}_t^2(x) \int_0^T \frac{1}{\overline{\beta}(x)} ds,$$

$$\overline{K}_t(x) = \left( \overline{\Delta}_t - \int_0^T \frac{\langle \delta(x), \dot{x} \rangle}{\overline{\beta}(x)} ds \right) \left( \int_0^T \frac{1}{\overline{\beta}(x)} ds \right)^{-1},$$

with $\overline{\beta}(x) = \epsilon \beta(x)$ and $\overline{\Delta}_t = \Delta_t / \epsilon$ (clearly, $\overline{K}_t(x) = K_t(x)$), and

$$J_T^\epsilon(x) = \frac{1}{2} \int_0^T \langle \dot{x}, \dot{x} \rangle ds - \int_0^T \overline{V}(x, s) ds, \qquad \text{with } \overline{V}(x, s) = \frac{V(x, s)}{1 - \epsilon}.$$

From Proposition 1.1, $J^\epsilon$ is bounded from below and coercive. On the other hand, taken $\epsilon$ small enough so that

$$\frac{\lambda}{1 - \epsilon} T^2 < \frac{\pi^2}{2}$$

(which is possible by (4)), Proposition 1.2 ensures that $J_T^\epsilon$ is also bounded from below and coercive. Therefore, from (9) $J_V$ is bounded from below and coercive. ∎

**Proof of Theorem 1.1.** Obviously, from Lemmas 3.1 and 3.2 it follows that functional $J_V$ satisfies Palais–Smale condition; hence, Theorem 2.1 applies. So, $J_V$ attains its infimum, and thus, a curve solving $(P_T)$ must exist. ∎

**Remark 3.1.** Theorem 1.1 is optimal in the following sense:

(i) even with $\beta \equiv 1$ and $\delta \equiv 0$, there are counterexamples when hypothesis $(H_3)$ with (4) fails (see Example 3.6 in [7]);

(ii) even with $V \equiv 0$, there are counterexamples when $(H_2)$ fails (see Section 7 in [3]; see also Example 2.7 in [2] and previous discussion).

### Acknowledgment

The authors wish to thank Professor Miguel Sánchez for his useful remarks.

770

**References**

1. R. Bartolo and A.M. Candela, Quadratic Bolza problems in static spacetimes with critical asymptotic behavior. Preprint (2004).
2. R. Bartolo, A.M. Candela and J.L. Flores, Geodesic connectedness of stationary space-times with optimal growth. Preprint (2005).
3. R. Bartolo, A.M. Candela, J.L. Flores and M. Sánchez, Geodesics in static Lorentzian manifolds with critical quadratic behavior, *Adv. Nonlinear Stud.* **3**, 471 (2003).
4. V. Benci and D. Fortunato, On the existence of infinitely many geodesics on space-time manifolds, *Adv. Math.* **105**, 1 (1994).
5. V. Benci, D. Fortunato and F. Giannoni, On the existence of multiple geodesics in static space-times, *Ann. Inst. H. Poincaré Anal. Non Linéaire* **8**, 79 (1991).
6. P. Bolle, On the Bolza problem, *J. Differential Equations* **152**, 274 (1999).
7. A.M. Candela, J.L. Flores and M. Sánchez, A quadratic Bolza-type problem in a Riemannian manifold, *J. Differential Equations* **193**, 196 (2003).
8. A.M. Candela and A. Salvatore, Normal geodesics in stationary Lorentzian manifolds with unbounded coefficients, *J. Geom. Phys.* **44**, 171 (2002).
9. F.H. Clarke and I. Ekeland, Nonlinear oscillations and boundary value problems for Hamiltonian systems, *Arch. Rational Mech. Anal.* **78**, 315 (1982).
10. I. Ekeland, N. Ghoussoub and H. Tehrani, Multiple solutions for a classical problem in the calculus of variations, *J. Differential Equations* **131**, 229 (1996).
11. F. Giannoni and A. Masiello, On the existence of geodesics on stationary Lorentz manifolds with convex boundary, *J. Funct. Anal.* **101**, 340 (1991).
12. F. Giannoni and P. Piccione, An intrinsic approach to the geodesical connectedness of stationary Lorentzian manifolds, *Comm. Anal. Geom.* **7**, 157 (1999).
13. S.W. Hawking and G.F.R. Ellis, The Large Scale Structure of Space–Time, Cambridge University Press, London, 1973.
14. L. Pisani, Existence of geodesics for stationary Lorentz manifolds, *Boll. Unione Mat. Ital. A* **7**, 507 (1991).
15. R.S. Wald, General Relativity, University of Chicago Press, 1984.

# REGULARITY OF MINIMIZERS OF SOME DEGENERATE INTEGRAL FUNCTIONALS

V. CATALDO, S. D'ASERO and F. NICOLOSI

*Dipartimento di Matematica e Informatica,*
*Università di Catania*
*I-95125, Italy*
*E-mail: fnicolosi@dmi.unict.it*

In this paper we study qualitative properties of minimizers for a class of integral functionals, defined in a weighted space. In particular we obtain boundedness and Hölder regularity for the minimizers by using a modified Moser method with special test function.

**Keywords:** Integral functionals, minimizers, weighted spaces, Moser method
**AMS Mathematics Subject Classification:** 49J20

## 1. Introduction

We shall study some qualitative properties of minimizers for a class of functionals of high order:

$$I(u) = \int_\Omega \left\{ A(x, \nabla_2 u) + A_0(x, u) \right\} dx \qquad (1)$$

defined in a weighted space: $\mathring{W}^{1,q}_{2,p}(\nu, \mu, \Omega)$. Here $\nabla_2 u = \{D^\alpha u : |\alpha| = 1, 2\}$.

In particular we deal with the boundedness and the Hölder continuity of the minimizing solutions of the functional (3) under some growth condition concerning the coefficients of the functional and under the following degenerate elliptic condition:

$$A(x, \xi) \geq c_1 \left\{ \sum_{|\alpha|=1} \nu(x)|\xi_\alpha|^q + \sum_{|\alpha|=2} \mu(x)|\xi_\alpha|^p \right\} - f(x) \qquad (2)$$

We observe that the class of operators verifying an elliptic condition of such type, has been studied by Skrypnik in [4], where he proved the boundedness and Hölder's continuity of the solutions without any assumption on the relation between the dimension $n$, $p$ and the order of weak derivatives $m$.

## 2. Hypotheses and statement of the main result

Let $\Omega$ be an open, bounded set of $\mathbb{R}^n$, $(n \geq 2)$. Let $q$, $p$ be real numbers such that $p \geq 2$, $2p < q < n$. $\mathbb{R}^{n,2}$ is the space of all sets $\xi = \{\xi_\alpha \in \mathbb{R} : |\alpha| = 1, 2\}$ of real numbers; $x \in \Omega$ e $\xi \in \mathbb{R}^{n,2}$. Let $\nu(x)$ be a positive function definite in $\Omega$ such that $\nu \in L^1_{loc}(\Omega)$, $\left(\frac{1}{\nu}\right)^{\frac{1}{q-1}} \in L^1_{loc}(\Omega)$.

$W^{1,q}(\nu, \Omega)$ is the space of all functions $u \in L^q(\Omega)$ such that their derivatives in the sense of distribution $D^\alpha u$, $|\alpha| = 1$, are functions for which the following properties hold: $\nu^{\frac{1}{q}} D^\alpha u \in L^q(\Omega)$ se $|\alpha| = 1$; $W^{1,q}(\nu, \Omega)$ is a Banach space respect to the norm

$$\|u\|_{1,q,\nu} = \left( \int_\Omega |u|^q dx + \sum_{|\alpha|=1} \int_\Omega \nu |D^\alpha u|^q dx \right)^{\frac{1}{q}}.$$

$\mathring{W}^{1,q}(\nu, \Omega)$ is the closure of $C_0^\infty(\Omega)$ in $W^{1,q}(\nu, \Omega)$.

Let be $\mu(x)$ a positive functions definite in $\Omega$ such that : $\mu \in L^1_{loc}(\Omega)$, $\left(\frac{1}{\mu}\right)^{\frac{1}{p-1}} \in L^1_{loc}(\Omega)$. $W^{1,q}_{2,p}(\nu, \mu, \Omega)$ is the space of all functions $u \in W^{1,q}(\nu, \Omega)$, such that their derivatives in the sense of distribution $D^\alpha u$, $|\alpha| = 2$ are functions for which the following properties hold: $\mu^{\frac{1}{p}} D^\alpha u \in L^p(\Omega)$, $|\alpha| = 2$. $W^{1,q}_{2,p}(\nu, \mu, \Omega)$ is a Banach space respect to the norm: $\|u\| = \|u\|_{1,q,\nu} + \left( \sum_{|\alpha|=2} \int_\Omega \mu |D^\alpha u|^p dx \right)^{\frac{1}{p}}$. $\mathring{W}^{1,q}_{2,p}(\nu, \mu, \Omega)$ is the closure of $C_0^\infty(\Omega)$ in $W^{1,q}_{2,p}(\nu, \mu, \Omega)$.

More details concerning the previous weights can be found in [6], [7], [8].

We assume, moreover, that the function $\frac{1}{\nu} \in L^t(\Omega)$, with $t > \frac{n}{q}$.

Putting $\tilde{q} = \frac{nqt}{n(1+t)-qt}$, we can easily prove that a constant $\tilde{c} > 0$ exists such that if $u \in \mathring{W}^{1,q}(\nu, \Omega)$, the following inequality holds:

$$\int_\Omega |u|^q dx \le c_0 \sum_{|\alpha|=1} \int_\Omega \nu |D^\alpha u|^q dx \tag{3}$$

The principal part $A : \Omega \times \mathbb{R}^{n,2} \to \mathbb{R}$ of our functional is a Carathodory function and satisfies the degenerate condition (2) and the following growth condition

$$A(x, \xi) \le c_2 \left\{ \sum_{|\alpha|=1} \nu(x)|\xi_\alpha|^q + \sum_{|\alpha|=2} \mu(x)|\xi_\alpha|^p \right\} + f(x), \tag{4}$$

for almost all $x \in \Omega$ and for all $\xi = \{\xi_\alpha : |\alpha| = 1, 2\}$, where $c_1, c_2$ are positive constants and $f(x)$ is a nonnegative function, $f \in L^{t*}(\Omega)$, with $t* > \frac{nt}{qt-n}$ and such that $f_1 = \mu^{\frac{q}{q-2p}} \left(\frac{1}{\nu}\right)^{\frac{2p}{q-2p}} \in L^{t*}(\Omega)$.

Now we give hypotheses concerning the secondary term.

Let $A_0 : \Omega \times \mathbb{R} \to \mathbb{R}$ be a function such that for all $\eta \in \mathbb{R}$ the function $A_0(\cdot, \eta)$ is measurable in $\Omega$ e $A_0(x, \cdot)$ is convex in $\mathbb{R}$ for almost all $x \in \Omega$.

Moreover there exist $c_3$, $c_4$ positive constants and $f_0 \in L^{t*}(\Omega)$, nonnegative, such that almost everywhere in $\Omega$ and for all $\eta \in \mathbb{R}$ the following inequality holds:

$$-c_4|\eta|^q - f_0(x) \le A_0(x, \eta) \le c_3|\eta|^q + f_0(x) \tag{5}$$

Finally, we consider a nonempty, closed and convex set $V$ in $\mathring{W}^{1,q}_{2,p}(\nu,\mu,\Omega)$, which satisfies the following property:

If $v \in V, \varphi : \Omega \to \mathbb{R}, 0 \leq \varphi \leq 1$ in $\Omega$ and $\varphi v \in \mathring{W}^{1,q}_{2,p}(\nu,\mu,\Omega)$, then $v - \varphi v \in V$.

From the theory of convex and coercive functionals, there exists a minimizing solution of the functional (1).

At this moment we are able to give the first our result concerning the boundedness of the solutions.

**Theorem 2.1.** *Assume that all previous hypotheses are satisfied. Let $u$ be a minimizer of the functional $I$ in $V$.*

*Then $u \in L^\infty(\Omega)$.*

In order to obtain regularity result for the solution $u(x)$, we consider as closed and convex set $V$ the set $\mathring{W}^{1,q}_{2,p}(\nu,\mu,\Omega)$ and we need in addition the following hypothesis involving the weights: first we need to consider the weight $\nu(x)$ with bigger summability, this is, $\nu \in L^{t*}(\Omega)$ and then we suppose that there exists a positive constant $c_5$, such that for every ball of center $y \in \Omega$ and radius $\rho > 0$, $B(y,\rho)$, with $\overline{B(y,\rho)} \subset \Omega$, the following inequality:

$$\left\{ R^{-n} \int_{B(y,\rho)} \left(\frac{1}{\nu}\right)^t dx \right\}^{\frac{1}{t}} \left\{ R^{-n} \int_{B(y,\rho)} \nu^{t*} dx \right\}^{\frac{1}{t*}} \leq c_5$$

holds.

Now we are able to give the main result of this paper:

**Theorem 2.2.** *We assume that all previous hypotheses are satisfied. Then for every sub-domain $\Omega' \subset \Omega$ with $d(\Omega',\partial\Omega) > 0$ and for every $x, y \in \Omega'$*

$$|u(x) - u(y)| \leq c_6 [d(\Omega',\partial\Omega)]^\alpha |x - y|^\alpha$$

*where $c_6$ and $\alpha$ depending only on $n$, $q$, $p$, $c_1$-$c_5$, $\tilde{c}$, $t$, $t*$, $\|f_1\|$ and ess sup $|u|$.*

# References

1. M.Giaquinta, E.Giusti "On the regularity of the minima of variational integrals". *Acta Math.* 148 (1982), 31-46.
2. D.Giachetti, M.M.Porzio "Local regularity results for minima of functionals of the Calculus of variations". *Nonlinear Analysis* 39, 463–482 (2000)
3. A. Kovalevsky, F.Nicolosi "Boundedness of solutions of variational inequalities with nonlinear degenerated elliptic operators of high order". *Applicable Analysis* 65, 225–249 (1997)
4. Skrypnik I.V. "High order quasilinear elliptic equations with continuous generalized solutions". *Differential equations* 14 (1978), no.6, 786–795.
5. A. Kovalevsky, F.Nicolosi "On Holder continuity of solutions of equations and variational inequalities with degenerate nonlinear elliptic high order operators". *(Taormina, 1998)*, 205–220, Aracne, Rome, 2000.

6. S.Bonafede, S. D'Asero "Hölder continuity of solutions for a class of nonlinear elliptic variational inequalities of high order". *Nonlinear Analysis* 44, 657–667 (2001)

7. F.Guglielmino, F.Nicolosi "Existence theorems for boundary value problems associated with quasilinear elliptic equations". *Ricerche Mat.* 37 (1988) 157–176

8. P.Drabek, F.Nicolosi "Solvability of degenerate elliptic problems of higher order via Leray-Lions theorem". *Hiroshima Math.* 26 79–90.

9. O.Ladyzhenskaya, N.Ural'tseva "Linear and quasilinear elliptic equations". *Academic Press New York and London* 1968.

# ON THE OPTIMAL RETRACTION PROBLEM FOR γ-LIPSCHITZ MAPPINGS AND APPLICATIONS

GIULIO TROMBETTA

*Department of Mathematics,*
*University of Calabria,*
*87036 Arcavacata di Rende (CS), Italy*
*E-mail: trombetta@unical.it*

Let X be an infinite-dimensional Banach space with unit closed ball $B(X)$ and unit sphere $S(X)$. It is well known that there is a retraction of $B(X)$ onto $S(X)$, that is, a continuous mapping $R : B(X) \to S(X)$ with $Rx = x$ for all $x \in S(X)$, and moreover such a retraction can be chosen among Lipschitz mappings. Many authors have looked for the smallest possible Lipschitz constant for special classes of Banach spaces. A similar and related problem can be considered by replacing the Lipschitz constant by the γ-Lipschitz constant when γ is the Hausdorff measure of noncompactness.

Aim of this talk is to present recent contributions related to the above problem, and give some applications to fixed point and eigenvalue theory of nonlinear operators.

**Key words:** γ-Lipschitz mapping, optimal retraction problem, Hausdorff measure, nonlinear operators, fixed point, eigenvalue, Banach space

**Mathematics Subject Classification:** 54C15, 46B20

## 1. Introduction

Throughout, let $X$ be a Banach space, and let $B(X) = \{x \in X : ||x|| \leq 1\}$ and $S(X) = \{x \in X : ||x|| = 1\}$. A *retraction* $R : B(X) \to S(X)$ is a continuous mapping such that $x = Rx$, for all $x \in S(X)$. If $R : B(X) \to S(X)$ is a retraction, $-R$ is a continuous fixed-point free self-mapping of the unit ball. Therefore, by the Brouwer's fixed point theorem, when X is finite-dimensional there is no retraction from B(X) onto S(X).

The Scottish Book [23] contains the following question raised around 1935 by S. Ulam: "*Is there a retraction of the closed unit ball of an infinite-dimensional Hilbert space onto its boundary?*" In 1943, S. Kakutani [15] gave a positive answer to this question. Later J. Dugundjii [10] (1951) and V. Klee [16] (1953 ) gave a positive answer to Ulam's question in the more general setting of infinite-dimensional Banach spaces. It is due to Nowak [18], Benyamini and Sternfeld [4] the proof of the existence of a Lipschitz retraction in any infinite-dimensional Banach space.

Assume that $X$ is an infinite-dimensional Banach space. In the literature the optimal retraction problem for Lipschitz mappings is the problem of evaluating the infimum $k_0(X)$ of all numbers $k$ for which there exists a Lipschitz retraction, with

Lipschitz constant $k$. It is known that $k_0(X) \geq 3$ (see [12]), in any Banach space $X$. On the other hand all known constructions of Lipschitz retractions do not give a precise estimate of $k_0(X)$. For a survey on the subject we refer to [12] and the bibliography therein.

We are interested in a related problem: the optimal retraction problem for $\gamma$-Lipschitz mappings, where $\gamma$ denotes the Hausdorff measure of noncompactness. However we state the problem in a more general setting.

**Definition 1.1.** For a bounded set $A$ in $X$, the *Hausdorff measure of noncompactness* $\gamma(A)$ is the infimum of all $\epsilon > 0$ such that $A$ admits a finite $\epsilon$-net in $X$; the *Kuratowski measure of noncompactness* $\alpha(A)$ is the infimum of all $\epsilon > 0$ such that $A$ admits a finite covering by sets of diameter at most $\epsilon$; the *lattice measure of noncompactness* $\beta(A)$ is the supremum of all $\epsilon > 0$ such that $A$ contains a sequence $\{x_n\}$ with $\|x_n - x_k\| \geq \epsilon$, for $n \neq k$.

Throughout this paper $\psi$ will stand for either $\gamma$, $\alpha$ or $\beta$.
Let $T: dom(T) \subseteq X \to X$ be a continuous mapping.

**Definition 1.2.** We say that $T$ is $\psi$-*Lipschitz* with constant $k$ if for any bounded subset $A \subseteq dom(T)$ $\psi(TA) \leq k\psi(A)$; we say that $T$ is $\psi$-*condensing* if $\psi(TA) < \psi(A)$ for each bounded $A \subseteq dom(T)$ which is not relatively compact.

In the literature a $\psi$-Lipschitz mapping with constant $k$ is also called $k$-$\psi$-*contractive*, in particular *strict-$\psi$-contractive* if $k < 1$. Moreover a $\gamma$-Lipschitz mapping with constant $k$ is also called $k$-*ball contractive*.
We consider the following quantitative characteristic:

$$k_\psi(X) = \inf\{k \geq 1 : \text{there is a } k\text{-}\psi\text{-contractive retraction } R : B(X) \to S(X)\}.$$

In [26] Wośko has proved that in the space $X = C[0,1]$ for any $\epsilon > 0$ there exists a $(1 + \epsilon)$-ball contractive retraction of $B(X)$ onto $S(X)$, so that $k_\gamma(C[0,1]) = 1$. Moreover he has posed the question of estimating $k_\gamma(X)$ for special classical Banach spaces, and also the question whether or not there is a Banach space in which $k_\gamma(X)$ is a minimum. Actually, the estimate of $k_\psi$ is of interest in nonlinear fixed point and eigenvalue theory (see [2]). In [25] it was proved that $k_\psi(X) \leq 6$ for any infinite dimensional Banach space $X$, and $k_\psi(X) \leq 4$ for separable or reflexive Banach spaces. It has also been proved that $k_\psi(X) \leq 3$ whenever $X$ contains an isometric copy of $l^p$ with $p \leq (2 - \log 3/\log 2)^{-1}$. Results in other Banach spaces can be found in [7], [17], [21] and [22]. Recently, in [2] it has been proved that $k_\psi(X) = 1$ for the Banach spaces $X$ whose norm is monotone with respect to some basis.
In Section 2 we present the main results obtained in [8], we show that $k_\gamma(X) = 1$ in some classical Banach spaces of real valued measurable functions defined on $[0,1]$. We also consider the following quantitative characteristic:

$$\omega(T) = \sup\{k \geq 0 : \gamma(TA) \geq k\gamma(A) \text{ for every bounded } A \subseteq dom(T)\},$$

which is called the *Hausdorff lower measure of noncompactness* of $T$. Whenever $\omega(T) > 0$, $T$ is a proper mapping, that is, $T^{-1}K$ is compact for each compact subset $K$ of $X$. The retractions we construct have positive lower measure of noncompactness.

Moreover we give a positive answer to the question posed by Wośko, indeed in the Orlicz spaces we show that the value $k_\gamma(X) = 1$ is actually a minimum. As an application, in Section 3 we present the generalization, obtained in [9], of a theorem of Guo ([13]) to strict-$\psi$ contractions and $\psi$-condensing mappings, under a condition which depends on $k_\psi(X)$. The extension of the theorem being optimal in the Banach spaces $X$ in which $k_\psi(X) = 1$.

## 2. Proper $k$-ball contractive retractions

Let $\Sigma$ be the $\sigma$-algebra of all Lebesgue measurable subsets of $[0,1]$ equipped with the Lebesgue measure $\mu$, and write *a.e.* for $\mu$-almost everywhere. Let $\mathcal{M}_0 := \mathcal{M}_0([0,1], \Sigma, \mu)$ denote the space of all classes of Lebesgue measurable functions $f : [0,1] \to R$. Let $X$ be a Banach space of functions of $\mathcal{M}_0$, and $S$ denote the set of all simple functions of X.

We recall that a function $f \in X$ is said to have *absolutely continuous norm* if for every $\varepsilon > 0$ there is $\delta > 0$ such that $\|f\chi_D\| < \varepsilon$ for every $D \in \Sigma$ with $\mu(D) < \delta$. For $f \in X$ and $a \in [1,2]$, we set

$$f_a(t) = \begin{cases} f(at), & \text{if } t \in \left[0, \frac{1}{a}\right] \\ 0, & \text{if } t \in \left(\frac{1}{a}, 1\right]. \end{cases}$$

Throughout this section we assume that $(X, \|\cdot\|)$ is a *regular* Banach function space, that is, all functions of $X$ are of absolutely continuous norm, and the following *property of ideality* holds: $|g| \le |f|$ a.e. for some $f \in X$ and $g \in \mathcal{M}_0$ implies that $g \in X$ with $\|g\| \le \|f\|$ (see, for example, [3] and [24]).

Moreover, we assume that the space $X$ satisfies the following properties:

(P1) $X = \overline{S}^{\|\cdot\|}$, where $\overline{S}^{\|\cdot\|}$ denotes the closure of $S$ with respect to the norm $\|\cdot\|$;

(P2) there is a continuous decreasing function $\alpha : [1,2] \to R$ with $\alpha(1) = 1$ and $\alpha(2) > 0$ such that

$$\alpha(a)\|f\| \le \|f_a\| \le \|f\|, \tag{1}$$

for every $f \in X$ and $a \in [1,2]$.

Then it is easy to check that $f_a \in X$.

Let us begin to consider the mapping $Q : B(X) \to B(X)$ defined by

$$Qf(t) = f_{\frac{2}{1+\|f\|}}(t) = \begin{cases} f\left(\frac{2}{1+\|f\|}t\right), & \text{if } t \in \left[0, \frac{1+\|f\|}{2}\right] \\ 0, & \text{if } t \in \left(\frac{1+\|f\|}{2}, 1\right]. \end{cases}$$

778

Clearly we have $Qf = f$ for all $f \in S(X)$. Moreover the mapping $Q$ is continuous and, for any subset $A$ of $B(X)$, the following estimates for the Hausdorff measure of noncompactness of $QA$ hold.

**Proposition 2.1.** *Let* $A \subseteq B(X)$ *we have*

$$\alpha(2)\gamma(A) \leq \gamma(QA) \leq \gamma(A).$$

In order to prove that $k_\gamma(X) = 1$ we consider, for any $c > 0$, the compact mapping $P_c : B(X) \to X$ defined by setting

$$P_c f = \begin{cases} c(1 - \|Qf\|)\chi_{(\frac{1+\|f\|}{2},1]}, & \text{if } f \in B(X) \setminus S(X) \\ 0, & \text{if } f \in S(X). \end{cases}$$

Then the following theorem holds.

**Theorem 2.1.** *Let* $\varepsilon > 0$. *Then there exists* $c > 0$ *such that the mapping* $R : B(X) \to S(X)$ *defined by*

$$Rf = \frac{Qf + P_c f}{\|Qf + P_c f\|},$$

*is a* $(1 + \varepsilon)$-*ball contractive retraction. Moreover* $\omega(R) > 0$.

**Proof.** It is easy to verify that the mapping $R$ is a retraction. Since $P_c$ is compact, Proposition 2.1 implies, for any $A \subseteq B(X)$,

$$\alpha(2)\gamma(A) \leq \gamma((Q + P_c)A) \leq \gamma(A). \tag{2}$$

Now by the definition of $Q$ and $P$ and the property of ideality we have $\|Qf + P_c f\| \geq \max\{\|Qf\|, \|P_c f\|\}$. Using property (P2) we find

$$\|Qf\| \geq \alpha\Big(\frac{2}{1 + \|f\|}\Big)\|f\|,$$

and

$$\|P_c f\| \geq c(1 - \|f\|)\|\chi_{(\frac{1+\|f\|}{2},1]}\|.$$

So that we obtain

$$\|Qf + P_c f\| \geq \max\Big\{\alpha\Big(\frac{2}{1 + \|f\|}\Big)\|f\|, \ c(1 - \|f\|)\|\chi_{(\frac{1+\|f\|}{2},1]}\|\Big\}.$$

Let $0 < \delta < 1$ be given. Choose sufficiently big $c$ such that

$$\max\Big\{\alpha\Big(\frac{2}{1 + \|f\|}\Big)\|f\|, \ c(1 - \|f\|)\|\chi_{(\frac{1+\|f\|}{2},1]}\|\Big\} \geq 1 - \delta.$$

Therefore for such a number $c$ we find

$$\|Qf + P_c f\| \geq 1 - \delta,$$

which, together with the definition of $R$, implies

$$RA \subseteq [0, \frac{1}{1 - \delta}] \cdot (Q + P_c)A.$$

Consequently by the absorbing invariance property of $\gamma$ and the right-hand side of (2), we obtain

$$\gamma(RA) \leq \frac{1}{1-\delta}\gamma(A).$$

On the other hand,

$$\|Qf + P_c f\| \leq 1 + c\, \|\chi_{(\frac{1+\|f\|}{2},1]}\|,$$

hence setting $l_c = 1 + c\,\|\chi_{(\frac{1}{2},1]}\|$, we have

$$(Q + P_c)A \subseteq [0, l_c] \cdot RA.$$

The latter together with the left-hand side of (2) implies

$$\frac{\alpha(2)}{l_c}\gamma(A) \leq \gamma(RA).$$

We have proved that $R$ is a $(1/(1-\delta))$-ball contractive retraction with $w(R) \geq \alpha(2)/l_c$. Given $\epsilon > 0$, the theorem follows by the arbitrariness of $\delta$. $\qquad\square$

We now give some examples of regular Banach function spaces $X$ in which the properties (P1) and (P2) are satisfied. Therefore in all the spaces of the following examples for any $\epsilon > 0$ there is a retraction $R : B(X) \to S(X)$ which is $(1+\epsilon)$-ball contractive with $w(R) > 0$.

**Example 2.1.** Let $f*$ denote the decreasing rearrangement of a function $f \in \mathcal{M}_0$. The Lorentz space $L^{p,q} := L^{p,q}([0,1])(1 \leq q \leq p < \infty)$ consists of all $f \in \mathcal{M}_0$ for which the quantity

$$\|f\|_{p,q} = \left(\frac{q}{p}\int_{[0,1]}(t^{1/p}f*(t))^q \frac{dt}{t}\right)^{\frac{1}{q}}$$

is finite. Then

(1) $X = \overline{S}^{\|\cdot\|^{\Phi}} = L^{p,q}$, is a regular Banach function space;
(2) let $f \in L^{p,q}$ and $a \in [1,2]$, then

$$\left(\frac{1}{a}\right)^{\frac{1}{p}}\|f\|_{p,q} = \|f_a\|_{p,q}.$$

**Example 2.2.** Let $1 < p < \infty$. The grand Lebesgue space, denoted by $L^{p)} := L^{p)}([0,1])$, introduced in [14], is defined as the space of all functions $f \in \mathcal{M}_0$ such that

$$\|f\|_{p)} = \sup_{0<\epsilon<p-1}\left(\epsilon\int_{[0,1]}|f(t)|^{p-\epsilon}dt\right)^{1/(p-\epsilon)} < \infty.$$

The following properties hold:

(1) $X = \overline{S}^{\|\cdot\|_{p)}}$ is a regular Banach function space properly contained in $L^{p)}$;

780

(2) let $f \in X$ and $a \in [1,2]$, then

$$\frac{1}{a}\|f\|_{p)} \le \|f_a\|_{p)} \le \|f\|_{p)}.$$

**Example 2.3.** Let $L^{p)'} := L^{p)'}([0,1])$ $(1 < p < \infty)$ be the small Lebesgue space introduced in [11], in which the norm is defined as

$$\|f\|_{p)'} = \sup_{g \in L^{p)}} \frac{\int_{[0,1]} f(t)g(t)dt}{\|g\|_{p)}}.$$

We recall that the spaces $L^{p)}$ are characterized as dual spaces of $L^{p)'}$ (see [6]). The following properties hold:

(1) $X = \overline{S}^{\|\cdot\|_{p)'}} = L^{p)'}$ is a regular Banach function space;
(2) let $f \in L^{p)'}$ and $a \in [1,2]$, then

$$\frac{1}{a}\|f\|_{p)'} \le \|f_a\|_{p)'} \le \|f\|_{p)'}.$$

**Example 2.4.** Let $0 \le \beta < 1$. The Marcinkiewicz space $M_\beta := M_\beta([0,1])$ consists of all $f \in M_0$ for which

$$\|f\|_\beta = \sup \frac{1}{\mu(E)^\beta} \int_E |f(t)|dt < \infty.$$

where the supremum is taken over all $E \in \Sigma$ with $\mu(E) > 0$. The following properties hold:

(1) $X = \overline{S}^{\|\cdot\|_\beta}$ is a regular Banach function space properly contained in $M_\beta$;
(2) let $f \in M_\beta$ and $a \in [1,2]$, then

$$\left(\frac{1}{a}\right)^{1-\beta}\|f\|_\beta \le \|f_a\|_\beta \le \|f\|_\beta.$$

The following theorem states a condition under which, always under the hypothesis that $X$ is a regular Banach function space in which the properties (P1) and (P2) are satisfied, the best value $k_\gamma(X) = 1$ is achieved.

**Theorem 2.2.** *Assume that there is a compact mapping* $P : B(X) \to X$ *with* $Pf = 0$ *for all* $f \in S(X)$ *such that* $\|Qf + Pf\| = 1$ *for all* $f \in B(X)$. *Then the retraction*

$$Rf = Qf + Pf, \tag{3}$$

*is 1-ball contractive and* $\omega(R) \ge \alpha(2)$.

Hereafter we give the explicit formula of a 1-ball contractive retraction in Orlicz spaces.

**Example 2.5.** Let $\Phi : [0,\infty) \to [0,\infty)$ be a continuous strictly increasing Young function. Assume that $\Phi$ satisfies the $\Delta_2$-condition, that is, there is $c \in [0,\infty)$ such that $\Phi(2x) \le c\Phi(x)$ $(x \ge 0)$. For $f \in \mathcal{M}_0$ set

$$M^\Phi(f) = \int_{[0,1]} \Phi(|f(t)|)dt.$$

Let $L_\Phi := L_\Phi([0,1])$ be the Orlicz space generated by the Young function $\Phi$ endowed with the Luxemburg norm

$$\|f\|_\Phi = \inf\{u > 0 : M^\Phi\left(\frac{f}{u}\right) \le 1\}.$$

Then

(1) $X = \overline{S}^{\|\cdot\|_\Phi} = L_\Phi$, that is, $L_\Phi$ is a regular Banach function space;
(2) let $f \in L_\Phi$ and $a \in [1,2]$,

$$\frac{1}{a}\|f\|_\Phi \le \|f_a\|_\Phi \le \|f\|_\Phi.$$

We define the mapping $P_\Phi : B(L_\Phi) \to L_\Phi$ by setting

$$P_\Phi f = \begin{cases} \Phi^{-1}\left(\frac{2}{1-\|f\|_\Phi}(1 - M^\Phi(Qf))\right)\chi_{\left(\frac{1+\|f\|_\Phi}{2},1\right]}, & \text{if } f \in B(L_\Phi) \setminus S(L_\Phi) \\ 0, & \text{if } f \in S(L_\Phi), \end{cases}$$

then $\|Qf + P_\Phi f\|_\Phi = 1$. It follows that the mapping $R : B(L_\Phi) \to S(L_\Phi)$ defined by

$$Rf = Qf + P_\Phi f$$

is a 1-ball contractive retraction and $\omega(R) \ge \frac{1}{2}$.

## 3. Applications: an extension of Guo's theorem

Let $X$ be an infinite-dimensional Banach space, and let $\Omega$ be a bounded open subset of $X$. Denote by $\overline{\Omega}$ and $\partial\Omega$ the closure and the boundary of $\Omega$, respectively. A well known theorem of D. Guo [13] states that if a completely continuous mapping $F : \overline{\Omega} \to X$ satisfies: (i) $\inf_{x\in\partial\Omega}\|Fx\| > 0$ and (ii) $Fx \ne \lambda x$ for $x \in \partial\Omega$ and $0 < \lambda \le 1$, then the Leray-Schauder degree $\deg(I - F, \Omega, 0) = 0$.

When $F : \overline{\Omega} \to X$ is $\psi$-condensing and has no fixed points on $\partial\Omega$, there is defined in [1] an integer $\mathrm{ind}(F, \Omega)$, called the *fixed point index* of $A$, which coincides with the Nussbaum degree $\deg_N(I - F, \Omega, 0)$. As pointed out in [5], the restatement of Guo's theorem for a strict-$\alpha$-contraction under the hypotheses (i) and (ii) is false. We generalize Guo's theorem to strict-$\psi$-contractions and condensing mappings, under a condition which depends on $k_\psi(X)$. Assume the set $\Omega$ contains the origin. The main result is the following theorem.

**Theorem 3.1.** *Let* $F : \overline{\Omega} \to X$ *be a* $k$-$\psi$-*contraction* $(k < 1)$, *satisfying*

$$\inf_{x\in\partial\Omega}\|Fx\| > kk_\psi \sup_{x\in\partial\Omega}\|x\|.$$

*Assume that one of the following conditions holds:*
*(a) $kk_\psi < 1$ and $Fx \neq \lambda x$ for $x \in \partial\Omega$ and $kk_\psi < \lambda \leq 1$;*
*(b) $kk_\psi \geq 1$.*
*Then* $\text{ind}(F, \Omega) = 0$.

Whenever $k_\psi(X) = 1$, our extension of Guo's theorem is optimal. Moreover any generalization of Guo's theorem to noncompact mappings allows to obtain a degree theoretic proof of the Birkhoff Kellogg theorem and theorems on nonzero fixed points of noncompact mappings (see for example [19] and [20]). The following corollary generalizes the Birkhoff-Kellogg theorem to $k$-$\psi$-contractions.

**Corollary 3.1.** *Let $F : \overline{\Omega} \to X$ be a $k$-$\psi$-contraction (for any $k \geq 0$). Suppose that*

$$\inf_{x \in \partial\Omega} ||Fx|| > kk_\psi \sup_{x \in \partial\Omega} ||x||.$$

*Then there exist $\lambda > kk_\psi$ and $x_\lambda \in \partial\Omega$ such that $\lambda x_\lambda = Fx_\lambda$, and also there exist $\mu < -k_\psi k$ and $x_\mu \in \partial\Omega$ such that $\mu x_\mu = Fx_\mu$.*

The following corollary extends Guo's domain compression and expansion fixed point theorem to strict-$\psi$-contractions.

**Corollary 3.2.** *Let $\Omega_1$ and $\Omega_2$ bounded open sets in $X$, such that $0 \in \Omega_1$ and $\overline{\Omega}_1 \subseteq \Omega_2$. Let $F : \overline{\Omega}_2 \to X$ be a strict-$\psi$-contraction, with constant $k < 1/k_\psi$. Suppose that one of the following conditions holds*

$$\begin{cases} \inf_{x \in \partial\Omega_1} ||x|| > kk_\psi \sup_{x \in \partial\Omega_1} ||x|| \\ ||Fx|| \geq ||x|| \quad x \in \partial\Omega_1 \\ ||Fx|| \leq ||x|| \quad x \in \partial\Omega_2 \end{cases}$$

*or*

$$\begin{cases} \inf_{x \in \partial\Omega_2} ||x|| > kk_\psi \sup_{x \in \partial\Omega_2} ||x|| \\ ||Fx|| \geq ||x|| \quad x \in \partial\Omega_2 \\ ||Fx|| \leq ||x|| \quad x \in \partial\Omega_1 \end{cases}$$

*Then $F$ has at least a fixed point on $\overline{\Omega}_2 \setminus \Omega_1$.*

We now state the result which extends Guo's theorem to $\psi$-condensing mappings.

**Theorem 3.2.** *Let $F : \overline{\Omega} \to X$ be a $\psi$-condensing mapping, suppose that*

$$\inf_{x \in \partial\Omega} ||Fx|| > k_\psi \sup_{x \in \partial\Omega} ||x||.$$

*Then* $\text{ind}(F, \Omega) = 0$.

We conclude with the following example on the existence of eigenvalues of nonlinear integral operators.

**Example 3.1.** Let $I = [0,1]$ and $L_p := L_p[0,1]$ $(p > 1)$ and $g(t,s)$ be a continuous function defined on $I \times I$. Define the operator $G : L_p \to L_p$ by

$$Gx(t) = \int_I g(t,s)|x(s)|^p d s \quad (t \in I).$$

Let $\Omega$ be a bounded open subset of $L_p$ containing the origin and $H : \overline{\Omega} \to L_p$ be a $k$-ball contraction. We consider the operator $F : \overline{\Omega} \to L_p$ defined by

$$Fx = u\, Gx + Hx,$$

whenever $u \in (0,\infty)$.
Set $m = \inf_{x \in \partial\Omega} \|x\|_p$, $d = \sup_{x \in \partial\Omega} \|x\|_p$ and $c = \sup_{x \in \partial\Omega} \|Hx\|_p$ (clearly, we have $c < \infty$).

**Theorem 3.3.** *Suppose that*
*(i)* $a = \min\{\int_I g(t,s)dt,\ s \in I\} > 0$;
*(ii) the following inequality holds*

$$au > \frac{c+kd}{m^p}.$$

*Then $F$ has at least a positive eigenvalue $\lambda$ and a negative eigenvalue $\mu$ with corresponding eigenvectors on $\partial\Omega$.*

## References

1. R.R. Akhmerov, M.I. Kamenskii, A.S. Potapov and B.N. Sadovskii, *Condensing operators*, J. Sov. Math.(1982), 551-578 (translated from Mathematical Analysis, **18** (1980), 185-250 (Russian)).
2. J. Appell, N. A. Erkazova, S. Falcon Santana and M. Väth, *On some Banach space constants arising in nonlinear fixed point and eigenvalue theory*, Fixed Point Theory Appl. **4** (2004), 317-336.
3. C. Bennett and R. Sharpley, *Interpolation of operators*, Boston Academic Press, 1988.
4. Y. Benyamini and Y. Sternfeld, *Spheres in infinite-dimensional normed spaces are Lipschitz contractible*, Proc. Amer. Math. Soc. **88** (3) (1983), 439-445.
5. F. E. Browder , *Addendum. Remarks on the paper "Boundary conditions for condensing mappings"*, Nonlinear Analysis **8** (9) (1984), 1113.
6. C. Capone, A. Fiorenza, *On small Lebesgue spaces*, J. Funct. Spaces Appl. **3** (1) (2005), 73-89.
7. D. Caponetti and G. Trombetta, *On proper k-ball contractive retractions in the Banach space BC([0,∞))*, Nonlinear Func. Anal. Appl. (to appear).
8. D. Caponetti, A. Trombetta and G. Trombetta, *Proper 1-ball contractive retractions in the Banach spaces of measurable functions* , Bull. Austr. Math. Soc. (to appear).
9. D. Caponetti, A. Trombetta and G. Trombetta, *An extension of Guo's theorem via k-ψ-contractive retractions*, Nonlinear Anal. (to appear).
10. J. Dugundji, *An extension of Tietze's theorem*, Pacific J. Math. **1** (1951), 353-367.
11. A. Fiorenza, *Duality and reflexivity in grand Lebesgue spaces*, Collect. Math. **51** (2000), 131-148.

784

12. K. Goebel and W. A. Kirk, *Topics in metric fixed point theory*, Cambridge University Press, Cambridge, 1990.

13. D. J. Guo, *Eigenvalues and eigenvectors of nonlinear operators*, Chinese Ann. Math. **2** (1981), 65-80 [English].

14. T. Iwaniec and C. Sbordone, *On the integrability of the Jacobian under minimal hypotheses*, Arch. Rational Mech. Anal. **119** (2) (1992), 129-143.

15. S. Kakutani, *Topological properties of the unit sphere of a Hilbert space*, Proc. Imp. Acad. Tokio **19** (1943), 269-271.

16. V. L. Klee, *Convex bodies and periodic homeomorphisms in Hilbert spaces*, Trans. Amer. Math. Soc. **74** (1953), 10-43.

17. G. Lewicki and G. Trombetta, Almost contractive retractions in Orlicz spaces, Bull. Austr. Math. Soc. **68** (2003), 353-369.

18. B. Nowak, *On the Lipschitz retraction of the unit ball in infinite dimensional Banach spaces onto its boundary*, Bull. Acad. Polon. Sci., **27** (1979), 861-864.

19. J. X. Sun, *A generalization of Guo's theorem and applications*, J. Math. Anal. Appl. **126** (2) (1987), 566-573.

20. Y. Sun, *An extension of Guo's Theorem on domain compression and expansion*, Numer. Funct. Anal. Optimiz., **10** (586) (1989), 607-617.

21. G. Trombetta, *k-set contractive retractions in spaces of continuous functions*, Sci. Math. Jpn. **59** (1) (2004), 121-128.

22. A. Trombetta and G. Trombetta, *On the existence of $(\gamma_p)$ k-set contractive retractions in $L_p[0,1]$ spaces*, $1 \leq p < \infty$, Sci. Math. Jpn. **56** (2) (2002), 327-335.

23. S. Ulam, *Problem 36,* The Scottish Book, ed. R.D. Mauldin, Birkäuser, Basel and Boston, Mass., 1981.

24. M. Väth, *Ideal spaces*, Lect. Notes Math. 1664 (1997).

25. M. Väth, *On the minimal displacement problem of $\gamma$-Lipschitz maps and $\gamma$-Lipschitz retractions onto the sphere,* Z. Anal. Anwendungen **21** (4) (2002), 901-914.

26. J. Wośko, *An example related to the retraction problem,* Ann. Univ. Mariae Curie-Skłodowska **45** (1991), 127-130.

27. Y.Q. Yu and F.E. Browder, *Boundary conditions for condensing mappings*, Nonlinear Analysis **8** (3) (1984), 209-220.

# SOME REMARKS ON NIRENBERG-GAGLIARDO INTERPOLATION INEQUALITY IN ANISOTROPIC CASE

FRANCESCO NICOLOSI

Department of Mathematics, University of Catania
v.le A. Doria, 6
Catania, Italy.
E-mail: fnicolosi@dmi.unict.it

PAOLO CIANCI

Department of Mathematics, University of Catania,
v.le A. Doria, 6
Catania, Italy.
E-mail: cianci@dmi.unict.it

We establish an inequality of Nirenberg-Gagliardo kind for function that belonging to some anisotropic degenerate Sobolev space.

**Keywords:** Interpolation inequalities, anisotropic degenerate Sobolov space
**AMS Mathematics Subject Classification:** 26D10

## 1. Introduction

In this paper we shall find the following result:
If $u \in C_0^\infty(\Omega)$ then $\forall k \in \{2, \ldots, m-1\}$, $m \in \mathbb{N}$, $m \geq 3$,

$$
\sum_{|\alpha|=k} \left( \int_\Omega \nu_\alpha |D^\alpha u|^{p_\alpha} dx \right)^{\frac{1}{p_\alpha}} \leq c \sum_{|\alpha|=1} \left( \int_\Omega \nu_\alpha |D^\alpha u|^{p_\alpha} dx \right)^{\frac{1}{p_\alpha}} +
$$

$$
+ c \left( \sum_{|\alpha|=m} \left( \int_\Omega \nu_\alpha |D^\alpha u|^{p_\alpha} dx \right)^{\frac{1}{p_\alpha}} \right)^{\frac{k-1}{m-1}} \left\{ \sum_{|\alpha|=1} \left( \int_\Omega \nu_\alpha |D^\alpha u|^{p_\alpha} dx \right)^{\frac{1}{p_\alpha}} \right\}^{\frac{m-k}{m-1}} , (1)
$$

where the exponents $p_\alpha$ and the functions $\nu_\alpha$ will be precised later. This inequality can be extended by approximation to the functions $u \in \mathring{W}_{2,p}^{1,q}(\nu, \mu, \Omega)$, that is the closure of $C_0^\infty(\Omega)$ in $W_{m,p}^{1,q}(\nu, \mu, \Omega)$ with respect the norm

$$
\|u\| = \left( \int_\Omega |u|^{q_-} dx \right)^{\frac{1}{q_-}} + \sum_{|\alpha|=1} \left( \int_\Omega \nu_\alpha |D^\alpha u|^{q_\alpha} dx \right)^{\frac{1}{q_\alpha}} + \sum_{|\alpha|=m} \left( \int_\Omega \mu_\alpha |D^\alpha u|^{p_\alpha} dx \right)^{\frac{1}{p_\alpha}} .
$$

$$(2)$$

A result of this kind was obtained, relatively to isotropic case in [8], in that case the authors studied the problem

$$< Au, v - u >\, \geq 0 \qquad (3)$$

where A is the operator from $\mathring{W}_{2,p}^{1,q}(\nu, \mu, \Omega)$ into its dual defined by

$$< Au, v >= \int_{\Omega} \left\{ \sum_{|\alpha| \leq m} A_{\alpha}(x, \delta_m u) D^{\alpha} v \right\} dx \quad \forall u, v \in \mathring{W}_{2,p}^{1,q}(\nu, \mu, \Omega) \qquad (4)$$

where $\delta_m u = \{D^{\alpha}u\}_{|\alpha| \leq m}$. Studying this problem they used more times the inequality of Nirenberg-Gagliardo kind that they found, for instance proving that the problem have sense, or during the prove of a priori estimate for a solution of the problem, estimating the term containing intermediate derivatives, with terms containing first and last derivatives. So this can be seen as an application of the inequality of Nirenberg-Gagliardo kind. For other papers in which weighted functions are considered you can see [1-10].

## 2. Preliminary

We shall suppose the following: Let $n \in \mathbb{N}$, $n > 2$, $\Omega$ an open set of $\mathbb{R}^n$, $m \in \mathbb{N}$, $m \geq 3$. $\forall$ n-dimensional multi-index $\alpha$, $1 \leq |\alpha| \leq m$, let $q_{\alpha}$ be real numbers such that $q_{\alpha} \geq 2$, and let $\nu_{\alpha}$ be positive functions in $\Omega$ such that $\nu_{\alpha} \in L_{loc}^1(\Omega)$.

**Hypothesis 1** $\forall \alpha$, $1 < |\alpha| < m$, $\exists \nabla \nu_{\alpha}$ in weak sense.

In the sequel, if $\alpha$, $\beta$ are n-dimensional multi-indexes, $\alpha \leq \beta$ it means $\alpha_i \leq \beta_i$ $\forall i \in \{1, \ldots, n\}$.

**Definition 1.2.** $\forall \alpha$, $1 < |\alpha| < m$ and $\gamma$, with $|\gamma| = 1$, $\gamma \leq \alpha$, let $t_{\alpha\gamma} \in (-\infty, +\infty]$ and $r_{\alpha\gamma} \in (-\infty, +\infty]$ such that

$$\frac{1}{t_{\alpha\gamma}} = \frac{1}{q_{\alpha}} - \frac{1}{q_{\alpha-\gamma}}$$

$$\frac{1}{r_{\alpha\gamma}} = \frac{2}{q_{\alpha}} - \frac{1}{q_{\alpha-\gamma}} - \frac{1}{q_{\alpha+\gamma}}$$

and let $\sigma_{\alpha\gamma}$, $\mu_{\alpha\gamma}$ defined by

$$\sigma_{\alpha\gamma} = |\nabla \nu_{\alpha}| \left(\frac{1}{\nu_{\alpha}}\right)^{\frac{q_{\alpha}-1}{q_{\alpha}}} \left(\frac{1}{\nu_{\alpha-\gamma}}\right)^{\frac{1}{q_{\alpha-\gamma}}}$$

$$\mu_{\alpha\gamma} = \nu_{\alpha}^{\frac{2}{q_{\alpha}}} \left(\frac{1}{\nu_{\alpha-\gamma}}\right)^{\frac{1}{q_{\alpha-\gamma}}} \left(\frac{1}{\nu_{\alpha+\gamma}}\right)^{\frac{1}{q_{\alpha+\gamma}}}.$$

**Hypothesis 2** $\forall \alpha$, $1 < |\alpha| < m$, $\exists \gamma$, $|\gamma| = 1$ and $\gamma \leq \alpha$ such that

$$t_{\alpha\gamma} \in (1, +\infty] \qquad and \qquad \sigma_{\alpha\gamma} \in L^{t_{\alpha\gamma}}(\Omega)$$

$$r_{\alpha\gamma} \in (1, +\infty] \qquad and \qquad \mu_{\alpha\gamma} \in L^{r_{\alpha\gamma}}(\Omega).$$

## 3. Main Result

**Theorem 3.1.** *Let Hypotheses 1. and 2. be satisfied and let* $u \in C_0^\infty(\Omega)$, *then* $\forall k \in \{2, \ldots, m\}$

$$\sum_{|\alpha|=k} \left( \int_\Omega \nu_\alpha |D^\alpha u|^{p_\alpha} dx \right)^{\frac{1}{p_\alpha}} \le c \sum_{|\alpha|=1} \left( \int_\Omega \nu_\alpha |D^\alpha u|^{p_\alpha} dx \right)^{\frac{1}{p_\alpha}}$$

$$+ c \left( \sum_{|\alpha|=m} \left( \int_\Omega \nu_\alpha |D^\alpha u|^{p_\alpha} dx \right)^{\frac{1}{p_\alpha}} \right)^{\frac{k-1}{m-1}} \left\{ \sum_{|\alpha|=1} \left( \int_\Omega \nu_\alpha |D^\alpha u|^{p_\alpha} dx \right)^{\frac{1}{p_\alpha}} \right\}^{\frac{m-k}{m-1}} \quad (5)$$

*where the positive constant c depend only on* $m$, $\{q_\alpha : 1 \le |\alpha| \le m\}$ *and* $\{\nu_\alpha : 1 \le |\alpha| \le m\}$.

**Proof.** $\forall \alpha,\ 1 \le |\alpha| \le m$ let

$$I_\alpha = \left( \int_\Omega \nu_\alpha |D^\alpha u|^{q_\alpha} dx \right)^{\frac{1}{q_\alpha}}$$

and $\forall k \in \{1, \ldots, m\}$ let

$$I_k = \sum_{|\alpha|=k} I_\alpha.$$

Let us fix $k \in \{2, \ldots, m-1\}$ and let $\alpha$ be an arbitrary multi-index, $|\alpha| = k$. Due to Hypothesis 1.3. $\exists \gamma,\ |\gamma| = 1$ :

$$t_{\alpha\gamma} \in (1, +\infty] \quad and \quad \sigma_{\alpha\gamma} \in L^{t_{\alpha\gamma}}(\Omega)$$

$$r_{\alpha\gamma} \in (1, +\infty] \quad and \quad \mu_{\alpha\gamma} \in L^{r_{\alpha\gamma}}(\Omega).$$

For semplicity, let us suppose $t_{\alpha\gamma} \ne +\infty$, $r_{\alpha\gamma} \ne +\infty$. Using integration by part we get

$$\int_\Omega \nu_\alpha |D^\alpha u|^{q_\alpha} dx$$

$$= \int_\Omega \nu_\alpha |D^\alpha u|^{q_\alpha - 2} D^\alpha u D^\gamma (D^{\alpha-\gamma} u) dx$$

$$\le \int_\Omega |\nabla \nu_\alpha| |D^\alpha u|^{q_\alpha - 1} |D^{\alpha-\gamma} u| dx$$

$$+ q_\alpha \int_\Omega \nu_\alpha |D^\alpha u|^{q_\alpha - 2} |D^{\alpha-\gamma} u| |D^{\alpha+\gamma} u| dx. \quad (6)$$

788

Let us estimate the last 2 integrals, by Holder's inequality

$$\int_\Omega |\nabla\nu_\alpha||D^\alpha u|^{q_\alpha-1}|D^{\alpha-\gamma}u|dx \leq$$

$$\|\sigma_{\alpha\gamma}\|_{t_{\alpha\gamma}}\left(\int_\Omega \nu_\alpha|D^\alpha u|^{q_\alpha}dx\right)^{\frac{q_\alpha-1}{q_\alpha}}$$

$$\left(\int_\Omega \nu_{\alpha-\gamma}|D^{\alpha-\gamma}u|^{q_\alpha-\gamma}dx\right)^{\frac{1}{q_\alpha-\gamma}}$$

$$= \|\sigma_{\alpha\gamma}\|_{t_{\alpha\gamma}}I_\alpha^{q_\alpha-1}I_{\alpha-\gamma}. \tag{7}$$

and

$$\int_\Omega \nu_\alpha|D^\alpha u|^{q_\alpha-2}|D^{\alpha\gamma}u||D^{\alpha+\gamma}dx$$

$$\leq \|\mu_{\alpha\gamma}\|_{r_{\alpha\gamma}}\left(\int_\Omega \nu_\alpha|D^\alpha u|^{q_\alpha}dx\right)^{\frac{q_\alpha-2}{q_\alpha}}\left(\int_\Omega \nu_{\alpha-\gamma}|D^{\alpha-\gamma}u|^{q_\alpha-\gamma}dx\right)^{\frac{1}{q_\alpha-\gamma}}$$

$$\left(\int_\Omega \nu_{\alpha+\gamma}|D^{\alpha+\gamma}u|^{q_\alpha+\gamma}dx\right)^{\frac{1}{q_\alpha+\gamma}} = \|\mu_{\alpha\gamma}\|_{r_{\alpha\gamma}}I_\alpha^{q_\alpha-2}I_{\alpha-\gamma}I_{\alpha+\gamma}. \tag{8}$$

From (6)-(8) it follows

$$I_\alpha^{q_\alpha} \leq \|\sigma_{\alpha\gamma}\|_{t_{\alpha\gamma}}I_\alpha^{q_\alpha-1}I_{\alpha-\gamma} + q_\alpha\|\mu_{\alpha\gamma}\|_{r_{\alpha\gamma}}I_\alpha^{q_\alpha-2}I_{\alpha-\gamma}I_{\alpha+\gamma}$$

$$\leq c_1\left(I_\alpha^{q_\alpha-1}I_{\alpha-\gamma} + I_\alpha^{q_\alpha-2}I_{\alpha-\gamma}I_{\alpha+\gamma}\right) \tag{9}$$

where $c_1 > 0$ depend only on $\{q_\alpha : 1 \leq |\alpha| \leq m\}$ and $\{\nu_\alpha : 1 \leq |\alpha| \leq m\}$. Inequality (9) imply

$$I_\alpha^2 \leq c_1\left(I_\alpha I_{\alpha-\gamma} + I_{\alpha-\gamma}I_{\alpha+\gamma}\right) \leq$$

$$\frac{1}{2}I_\alpha^2 + \frac{1}{2}I_{\alpha-\gamma}^2 + c_1 I_{\alpha-\gamma}I_{\alpha+\gamma} \tag{10}$$

and so

$$I_\alpha \leq c_2\left(I_{\alpha-\gamma} + I_{\alpha-\gamma}^{\frac{1}{2}}I_{\alpha+\gamma}^{\frac{1}{2}}\right). \tag{11}$$

Because

$$I_{\alpha-\gamma} \leq I_{k-1} \quad and \quad I_{\alpha+\gamma} \leq I_{k+1}$$

from (11) we get

$$I_\alpha \leq c_2\left(I_{k-1} + I_{k-1}^{\frac{1}{2}}I_{k+1}^{\frac{1}{2}}\right) \qquad \forall k \in \{2,\ldots,m-1\} \tag{12}$$

Due to (12), using iteration tecnique, it follows

$$I_k \leq c\left(I_1 + I_m^{\frac{k-1}{m-1}}I_1^{\frac{m-k}{m-1}}\right) \qquad \forall k \in \{2,\ldots,m-1\} \tag{13}$$

and so the theorem is proved. $\qquad\qquad\square$

## 4. Example

Let us consider the case $m = 3$, ($n > 2$ arbitrary). $\forall \alpha$, $|\alpha| = 1$ and $|\alpha| = 3$, let $q_\alpha > 2$ such that

$$\max_{|\alpha|=3} q_\alpha < \min_{|\alpha|=1} q_\alpha \tag{14}$$

$\forall \alpha$, $|\alpha| = 2$, let us fix $\alpha'$, $|\alpha'| = 1$, $\alpha' \leq \alpha$, and let

$$\rho_\alpha \in (q_{\alpha+\alpha'}, q_{\alpha-\alpha'}) \tag{15}$$

$\forall \alpha$, $|\alpha| = 2$, let us define $q_\alpha$ in the following way

$$\frac{1}{q_\alpha} = \frac{1}{2q_{\alpha+\alpha'}} + \frac{1}{2\rho_\alpha}. \tag{16}$$

$\forall \alpha$, $|\alpha| = 1$ and $|\alpha| = 3$, let $\lambda_\alpha$ be positive numbers such that

$$\max_{|\alpha|=1} \lambda_\alpha < \min_{|\alpha|=3} \lambda_\alpha \tag{17}$$

and $\forall \alpha, |\alpha| = 2$, $\lambda_\alpha$ be numbers such that

$$\lambda_\alpha > q_\alpha \left( \frac{\lambda_{\alpha+\alpha'}}{q_{\alpha+\alpha'}} + 1 \right). \tag{18}$$

Let $\Omega = \{x \in \mathbb{R}^n : |x| < 2\}$ and $\forall \alpha, 1 \leq |\alpha| \leq 3$, let $\nu_\alpha$ be positive functions in $\Omega$ defined by

$$\nu_\alpha(x) = |x|^{\lambda_\alpha} \qquad \forall x \in \Omega - \{0\}. \tag{19}$$

Because $\forall \alpha, 1 \leq |\alpha| \leq 3$, the numbers $\lambda_\alpha$ are positive, Hypothesis 1.2. is true. Let us show that also Hp 1.3 holds.

Let $\alpha$ be an arbitrary multi-index, $|\alpha| = 2$. From $|\alpha'| = 1$ and $\alpha' \leq \alpha$, we have

$$\frac{1}{t_{\alpha\alpha'}} = \frac{1}{q_\alpha} - \frac{1}{q_{\alpha-\alpha'}} \tag{20}$$

$$\frac{1}{r_{\alpha\alpha'}} = \frac{2}{q_\alpha} - \frac{1}{q_{\alpha-\alpha'}} - \frac{1}{q_{\alpha+\alpha'}} \tag{21}$$

$$\sigma_{\alpha\alpha'} = |\nabla \nu_\alpha| \left( \frac{1}{\nu_\alpha} \right)^{\frac{q_\alpha-1}{q_\alpha}} \left( \frac{1}{\nu_{\alpha-\alpha'}} \right)^{\frac{1}{q_{\alpha-\alpha'}}} \tag{22}$$

$$\mu_{\alpha\alpha'} = \nu_\alpha^{\frac{2}{q_\alpha}} \left( \frac{1}{\nu_{\alpha-\alpha'}} \right)^{\frac{1}{q_{\alpha-\alpha'}}} \left( \frac{1}{\nu_{\alpha+\alpha'}} \right)^{\frac{1}{q_{\alpha+\alpha'}}}. \tag{23}$$

Thanks to (16) and (15)

$$\frac{1}{q_\alpha} - \frac{1}{q_{\alpha-\alpha'}} = \frac{1}{2q_{\alpha+\alpha'}} + \frac{1}{2\rho_\alpha} - \frac{1}{q_{\alpha-\alpha'}}$$
$$> \frac{1}{2q_{\alpha+\alpha'}} + \frac{1}{2q_{\alpha-\alpha'}} - \frac{1}{q_{\alpha-\alpha'}} = \frac{1}{2}\left( \frac{1}{q_{\alpha+\alpha'}} - \frac{1}{q_{\alpha-\alpha'}} \right), \tag{24}$$

and so, from (14)

$$\frac{1}{q_\alpha} - \frac{1}{q_{\alpha-\alpha'}} > 0.$$

Analogously

$$\frac{1}{q_\alpha} - \frac{1}{q_{\alpha+\alpha'}} < 0$$

hence

$$q_{\alpha+\alpha'} < q_\alpha < q_{\alpha-\alpha'}.$$

From this and (20) it follows

$$t_{\alpha\alpha'} \in (1, +\infty).$$

Let us prove that

$$\sigma_{\alpha\alpha'} \in L^{t_{\alpha\alpha'}}(\Omega). \qquad (25)$$

We have

$$|\nabla\nu_\alpha| = \lambda_\alpha |x|^{\lambda_\alpha - 1} \qquad a.e. \quad in \quad \Omega,$$

so due to (22)

$$\sigma_{\alpha\alpha'}(x) = \lambda_\alpha |x|^{-1 + \frac{\lambda_\alpha}{q_\alpha} - \frac{\lambda_{\alpha-\alpha'}}{q_{\alpha-\alpha'}}} \qquad a.e. \quad in \quad \Omega. \qquad (26)$$

Let us note that

$$-1 + \frac{\lambda_\alpha}{q_\alpha} - \frac{\lambda_{\alpha-\alpha'}}{q_{\alpha-\alpha'}} > 0 \qquad (27)$$

because from (18), (17) and (14)

$$\frac{\lambda_\alpha}{q_\alpha} > 1 + \frac{\lambda_{\alpha+\alpha'}}{q_{\alpha+\alpha'}} > 1 + \frac{\lambda_{\alpha-\alpha'}}{q_{\alpha-\alpha'}}.$$

Due to (26) and (27) we get (25). In the same way we can prove that

$$r_{\alpha\alpha'} \in (1, +\infty) \qquad and \qquad \mu_{\alpha\alpha'} \in L^{r_{\alpha\alpha'}}(\Omega).$$

## References

1. S. Bonafede, A weak maximum principle and estimates of Ess sup u for nonlinear degenerate elliptic equations, Czechoslovak Math. J., 46 (121)(1996),n. 2, 259-269.
2. P. Cianci, Boundedness of solutions of Dirichlet problem for a class of nonlinear elliptic equations with weights, Appl. Anal. 82 (2003) n. 5, 457-472.
3. P. Cianci, On the boundedness of functions from an anisotropic weighted space satisfying some integral inequalities, Acoustics, mechanics, and the related topics of mathematical analysis, 100–107, (2002).
4. G. R. Cirmi, Some regularity results for nonlinear degenerate elliptic equations, Differential Integral Equations 8 (1995), no. 1, 131–140.
5. S. D'Asero, Integral estimate for the gradients of solutions of local nonlinear variational inequalities with degeneration. Nonlinear Stud. 5 (1998), n. 1, 95-113.

6. F. Guglielmino, F. Nicolosi, Sulle W-soluzioni dei problemi al contorno per operatori ellittici degeneri, Ricerche di Matematica, 36 (1987), Supplemento.

7. F. Guglielmino, F. Nicolosi, Teoremi di esistenza per i problemi al contorno relativi alle equazioni ellittiche quasilineari, Ricerche di Matematica, 37 (1) 157-176 (1988).

8. Kovalevski A., Nicolosi F., Boundedness of solutions of variational inequalities with nonlinear degenerated elliptic operators of high order, Appl. Anal. 65 (1997), n. 3-4, 225-249.

9. Kovalevski A., Nicolosi F, Solvability of Dirichlet problem for a class of degenerate anisotropic equation with $L^1$ right hand side, Nonlinear Anal. 59 (2004), n. 3, 347-370.

10. S. Leonardi, On embedding theorems and Nemitskii's operator in weighted Lebesgue spaces, Ricerche di Mat. vol. XLII, I (1994).

# HAUSDORFF DIMENSION OF SINGULAR SETS OF SOBOLEV FUNCTIONS AND APPLICATIONS

DARKO ŽUBRINIĆ

*Department of Applied Mathematics*
*Faculty of Electrical Engineering & Computing*
*Unska 3, 10000 Zagreb, Croatia*
*E-mail: darko.zubrinic@fer.hr*

We describe a recent progress in finding optimal bounds of Hausdorff dimension of singular sets of functions in Lebesgue spaces, Sobolev spaces, Bessel potential spaces, Besov spaces, Lizorkin-Triebel spaces, and Hardy spaces. We are also interested in the question of existence of maximally singular functions in a given space of functions, that is, functions such that the Hausdorff dimension of their singular sets is maximal possible.

**Keywords:** Sobolev functions, Hausdorff dimenstion, singular sets, Bessel potential, Besov space, Lizockin-Triebel space, Hardy space, maximal singular functions
**AMS Mathematics Subject Classification:** 46E35, 28A78

## 1. Singular dimension and maximally singular functions

Our aim is to study Hausdorff dimension of singular sets of functions in various spaces of functions in a systematic way. Singularities of Sobolev functions and of solutions of quasilinear elliptic PDEs and systems have been considered by many authors[21], often with different meanings of the notion of singularity.

Assume that $u : \mathbb{R}^N \to \mathbb{R}$ is a Lebesgue measurable function (or $u : \Omega \to \mathbb{R}$, where $\Omega \subset \mathbb{R}^N$ is an open set). We say that $a \in \mathbb{R}^N$ is a *singular point* of $u$ if there exist positive constants $C$ and $\gamma$ such that

$$u(x) \geq C|x - a|^{-\gamma} \qquad (1)$$

a.e. in a neighbourhood of $a$. It is natural to consider the *singular set* of $u$ defined by:

$$\operatorname{Sing} u := \{a \in \mathbb{R}^N : a \text{ is a singular point of } u\}$$

The aim is to answer the question how large can be a singular set in the sense of Hausdorff dimension[2], that is, to find $\dim_H(\operatorname{Sing} u)$. More generally, given a space (or just a nonempty set) $X$ of measurable functions $u : \mathbb{R}^N \to \mathbb{R}$, we would like to find **singular dimension of** $X$, defined by:

$$\text{s-dim} X := \sup\{\dim_H(\operatorname{Sing} u) : u \in X\}. \qquad (2)$$

The notion of singular dimension has been introduced in Žubrinić[21]. It is clear that s-dim $X \leq N$. A function $v \in X$ is said to be **maximally singular**[23] in $X$ if the supremum in (2) is achieved, that is, there exists $v \in X$ such that

$$\text{s-dim}\, X = \dim_H(\text{Sing}\, v).$$

Basic problems we are interested in are the following: (a) given a space (or a nonempty set) $X$ of measurable functions from $\mathbb{R}^N$ to $\mathbb{R}$, find s-dim $X$; (b) find maximally singular functions in $X$, if any. We believe that it is natural to ask these questions for any given space of functions $X$.

## 2. Singular dimension of some function spaces

The problem of finding singular dimension of standard spaces of functions $X$ has been studied in the following cases: Lebesgue spaces $L^p(\mathbb{R}^N)$, Sobolev spaces $W^{k,p}(\mathbb{R}^N)$, Bessel potential spaces $L^{\alpha,p}(\mathbb{R}^N)$, Besov spaces $B_\alpha^{p,q}(\mathbb{R})$, Lizorkin-Triebel spaces $F_\alpha^{p,q}(\mathbb{R})$, and Hardy spaces. Of course, there are many other function spaces for which the problem of determining the singular dimension deserves to be studied.

Let us show a simple and elementary construction of maximally singular functions in Lebesgue spaces. The following result has been proved in Žubrinić[23]. Given a bounded set $A \subset \mathbb{R}^N$ and $\varepsilon > 0$ by $A_\varepsilon$ we denote the open $\varepsilon$-neighbourhood of $A$ (the so called Minkowski saussage around $A$ of radius $\varepsilon$, a term coined by B. Mandelbrot). By $d(x, A)$ we denote Euclidean distance from $x$ to $A$.

**Theorem 2.1.** *If* $1 \leq p < \infty$ *then* s-dim $L^p(\mathbb{R}^N) = N$, *and there exist maximally singular functions in* $L^p(\mathbb{R}^N)$:

$$v(x) := \sum_{k=1}^{\infty} 2^{-k} \frac{f_k(x)}{\|f_k\|_{L_p}}, \qquad (3)$$

*where*

$$f_k(x) := \begin{cases} d(x, A_k)^{-\gamma_k} & \text{for } x \in (A_k)_{\varepsilon_k}, \\ 0 & \text{for } x \in \mathbb{R}^N \setminus (A_k)_{\varepsilon_k}, \end{cases} \qquad (4)$$

$\varepsilon_k > 0$, $\overline{\dim}_B A_k < N$, $\dim_H A_k \to N$ *as* $k \to \infty$, *and*

$$0 < \gamma_k < \frac{N - \overline{\dim}_B A_k}{p}. \qquad (5)$$

The fact that $v$ is indeed in $L^p(\mathbb{R}^N)$ follows from the the right-hand side inequality in (5) and a result due to Harvey and Polking, see Theorem 4.1(a). The sets $A_k$ satisfying conditions of Theorem 2.1 can be easily constructed since for any given $s_k \in (0, N)$ there exists $A_k \subset \mathbb{R}^N$ such that $\dim_B A_k = \dim_H A_k = s_k$. Indeed, consider for example sets of the form of Cantor grill $A_k = [0,1]^{n_k} \times C^{(a_k)}$. Here $C^{(a_k)} \subset [0,1]$ is the generalized uniform Cantor set with suitably chosen parameter

$a_k \in (0, 1/2)$, see Falconer[2], and $n_k \in \mathbb{N} \cup \{0\}$ (when $n_k = 0$ we define $A_k = C^{(a_k)}$). Hence, by the product formula[2], $\dim_H A_k = \dim_B A_k = n_k + \log_{1/a_k} 2$. Clearaly, $\operatorname{Sing} u \supseteq \cup_k A_k$, so that by countable stability[2] of Hausdorff dimension we immediately obtain $\dim_H(\operatorname{Sing} u) = N$, since $\dim_H A_k \to N$.

In the case of $X = L^1(0, 1)$ an explicit construction of a maximally singular function can be given by (3), $p = 1$, defining $f_k \in L^1(0, 1)$ by $f_k(x) = d(x, C^{(a_k)})^{-\gamma_k}$, where we assume that $0 < \gamma_k < 1 - \log_{1/a_k} 2$, and $a_k \in (0, 1/2)$ are chosen so that $a_k \to 1/2$. It is interesting to note that the function $h : (0, 1) \to \mathbb{R}$, $h(x) = \int_0^x v(s) \, ds$, is absolutely continuous, and its derivative is maximally singular, that is, $\dim_H(\operatorname{Sing} h') = 1$. See Ref. 23.

The following result, obtained in Žubrinić[21] and in Horvat and Žubrinić[7], concerns singular dimension of Sobolev spaces and maximally singular functions in these spaces. Note that the condition $kp < N$ in Theorem 2.2 is natural, since by the Sobolev imbedding theorem for $kp > N$ all Sobolev functions are continuous, while for $kp = N$ Sobolev functions cannnot have singularities in the sense of (1), but may possess singularities of weaker (say logarithmic) type. Such singularities will be treated in Sec. 6.

**Theorem 2.2.** *If $k \geq 0$, $1 < p < \infty$, $kp < N$, then $s$-dim $W^{k,p}(\mathbb{R}^N) = N - kp$, and there exist maximally singular functions in $W^{k,p}(\mathbb{R}^N)$.*

We do not know if the result holds for $p = 1$ as well. Let us formulate a more general result, proved in Refs. 21 and 7, involving Bessel potential spaces $L^{\alpha,p}(\mathbb{R}^N)$. For their definition see Adams and Hedberg[1].

**Theorem 2.3.** *If $\alpha > 0$, $1 < p < \infty$, $\alpha p < N$, then $s$-dim $L^{\alpha,p}(\mathbb{R}^N) = N - \alpha p$, and there exist maximally singular functions in $L^{\alpha,p}(\mathbb{R}^N)$.*

The construction of maximally singular functions in Bessel potential spaces will be described in Sec. 5.

For Besov spaces $B_\alpha^{p,q}(\mathbb{R})$ and Lizorkin-Triebel spaces $F_\alpha^{p,q}(\mathbb{R})$ (see e.g. Hedberg and Adams[1] for their definition) we have the following result[25].

**Theorem 2.4.** *If $\alpha > 0$, $p, q \in (1, \infty)$, $\alpha p < N$, then*

$$s\text{-dim } B_\alpha^{p,q}(\mathbb{R}^N) = s\text{-dim } F_\alpha^{p,q}(\mathbb{R}^N) = N - \alpha p, \tag{6}$$

*and there exist maximally singular functions in $B_\alpha^{p,q}(\mathbb{R})$ and in $F_\alpha^{p,q}(\mathbb{R})$.*

We do not know the values of singular dimension of these spaces when $p = 1$ or $q = \infty$. For Hardy spaces (see e.g. Wojtaszczyk[19] for their definition and basic properties) we have the following result25.

**Theorem 2.5.** *For the Hardy space $H^1(\mathbb{R}^N)$ we have $s$-dim $H^1(\mathbb{R}^N) = N$, and there exist maximally singular functions.*

It would be interesting to know if there exists a Banach space of functions $X$ without maximally singular functions, such that s-dim $X > 0$.

## 3. Minkowski content and box dimension

It seems that the notion of Minkowski content is of fundamental importance for understanding Lebesgue integrability of functions with large singular sets, see Sec. 4. Let us recall the definition of this notion, see e.g. Mattila[13]. *The upper $s$-dimensional Minkowski content* of a bounded set $A \subset \mathbb{R}^N$, $s \geq 0$, is defined by

$$\mathcal{M}^{*s}(A) = \limsup_{\varepsilon \to 0} \frac{|A_\varepsilon|}{\varepsilon^{N-s}} \in [0, \infty],$$

where $|A_\varepsilon|$ is $N$-dimensional Lebesgue measure of $A_\varepsilon$. The upper box dimension of $A$ is defined by:

$$\overline{\dim}_B A := \inf\{s \geq 0 : \mathcal{M}^{*s}(A) = 0\}.$$

Analogously we can define lower $s$-dimensional Minkowski content $\mathcal{M}_*^s(A)$ and lower box-dimension $\underline{\dim}_B A$. An easy fact is the following (see Mattila[13] or Falconer[2] for this and many other results): $\dim_H A \leq \underline{\dim}_B A \leq \overline{\dim}_B A \leq N$. There are sets for which we may have strict inequalities[13]. Furthermore, it is possible to construct a class of fractal sets $A \subset \mathbb{R}^N$ for which their box dimensions are maximally separated, that is, $\underline{\dim}_B A = 0$ and $\overline{\dim}_B A = N$, see Žubrinić[27].

We say that $A$ is *Minkowski nondegenerate* if there exists $d \geq 0$ such that the corresponding Minkowski contents are different from 0 and $\infty$. If moreover $\mathcal{M}^{*d}(A) = \mathcal{M}_*^d(A) =: \mathcal{M}^d(A) \in (0, \infty)$, than $A$ is said to be *Minkowski measurable*. The notion of Minkowski nondegeneracy has been suggested in Ref. 27, see also Refs. 28 and 29.

As an example, the standard Cantor's triadic set $A$ is Minkowski nondegenerate, but not Minkowski measurable, as proved by Lapidus and Pomerance[10]:

$$\mathcal{M}_*^d(A) = (\log_{3/2} 9)(\log_4 3/2)^{\log_3 2}, \quad \mathcal{M}^{*d}(A) = 2^{2-\log_3 2}, \tag{7}$$

where $d := \dim_B A = \dim_H A = \log_3 2$. A similar result holds for generalized uniform Cantor sets, see Ref. 22.

Minkowski nondegenerate sets have been almost completely characterized by Lapidus and Pomerance[10] in the case of subsets of the real line. Their work is devoted to the study of generalized Weyl-Berry conjecture for fractal strings. Generalized Minkowski contents have been introduced and studied in He and Lapdius[6], see also Refs. 22 and 27. Important extension of the notion of fractal dimension to complex numbers has been introduced and studied by Lapidus and Van Frankenhuysen[11]. Relations of Minkowski content to Whitney cubes and $p$-capacity have been studied by Martio and Vuorninen[12].

## 4. Singular integrals generated by fractal sets

In this section we describe an easy construction of Lebesgue integrable functions which are singular on a prescribed set $A$. We start with a simple but important result, established essentially in Refs. 22 and 23, that we present here in slightly improved form. Part (a) is due to Harvey and Polking[5] (mentioned without proof on p. 42).

**Theorem 4.1.** *Let $A$ be a bounded set in $\mathbb{R}^N$.*

*(a) The following implication holds (Harvey and Polking):*

$$\gamma < N - \overline{\dim}_B A \quad \Longrightarrow \quad \int_{A_\varepsilon} d(x, A)^{-\gamma} dx < \infty. \tag{8}$$

*(b) If $\mathcal{M}_*^d(A) > 0$, where $d := \underline{\dim}_B A$, then*

$$\int_{A_\varepsilon} d(x, A)^{-\gamma} dx < \infty \quad \Longrightarrow \quad \gamma < N - \underline{\dim}_B A. \tag{9}$$

*(c) If $\underline{\dim}_B A = \overline{\dim}_B A =: \dim_B A$ and $\mathcal{M}_*^d(A) > 0$, then*

$$\int_{A_\varepsilon} d(x, A)^{-\gamma} dx < \infty \quad \Longleftrightarrow \quad \gamma < N - \dim_B A. \tag{10}$$

*(d) If $A$ is Minkowski nondegenerate, that is, $\mathcal{M}_*^d(A), \mathcal{M}^{*d}(A) \in (0, \infty)$, and $\gamma < N - \dim_B A$, then we have the following asymptotics:*

$$\int_{A_\varepsilon} d(x, A)^{-\gamma} dx \simeq \varepsilon^{N-d-\gamma} \quad \text{as } \varepsilon \to 0,$$

*i.e. the quotient of both sides is contained in a compact interval $[a, b] \subset (0, \infty)$ for small $\varepsilon$. For Minkowski measurable $A$ we have a more precise result:*

$$\int_{A_\varepsilon} d(x, A)^{-\gamma} dx \sim \frac{N-d}{N-d-\gamma} \mathcal{M}^d(A) \cdot \varepsilon^{N-d-\gamma},$$

*as $\varepsilon \to 0$, i.e. the quotient of both sides tends to 1.*

This and more general results can be seen in Refs. 22, 23, and 27. Equivalence (10) can be viewed as a generalization of the following simple fact: $|x|^{-\gamma}$ is integrable on the ball $B_\varepsilon(0)$ if and only if $\gamma < N$ (here $A = \{0\}$, so that $d(x, A) = |x|$ and $\dim_B A = 0$).

The proof of Theorem 4.1 rests on the following useful identity which enables to connect a singular integral with the Minkowski content of $A$. It has been obtained in Ref. 22, see also Ref. 23 for a more general formulation. For the sake of completeness we provide the proof.

**Lemma 4.1.** *For any bounded set $A \subset \mathbb{R}^N$, $\gamma > 0$, and $\varepsilon > 0$ we have*

$$\int_{A_\varepsilon} d(x, A)^{-\gamma} dx = \varepsilon^{-\gamma} |A_\varepsilon| + \gamma \int_0^\varepsilon s^{-\gamma-1} |A_s| \, ds. \tag{11}$$

**Proof.** We exploit a well known fact that for any measurable function $f : \mathbb{R}^N \to \mathbb{R}$, $f \geq 0$, and $\gamma > 0$, there holds

$$\int_{\mathbb{R}^N} f(x)^\gamma dx = \gamma \int_0^\infty t^{\gamma-1} |\{x \in \mathbb{R}^N : f(x) > t\}| \, dt, \tag{12}$$

see e.g. Folland[3] on p. 197. First we define $f(x) = d(x, A)^{-1}$ for $x \in A_\varepsilon$ and $f(x) = 0$ for $x \in \mathbb{R}^n \setminus A_\varepsilon$. Noting that the set $\{x : f(x) > t\}$ is equal to $A_{t^{-1}}$ for $t > \varepsilon^{-1}$, and to a constant set $A_\varepsilon$ for all $t < \varepsilon^{-1}$, we obtain

$$\int_{A_\varepsilon} d(x, A)^{-\gamma} dx = \gamma |A_\varepsilon| \int_0^{\varepsilon^{-1}} t^{\gamma-1} dt + \gamma \int_{\varepsilon^{-1}}^\infty t^{\gamma-1} |A_{t^{-1}}| \, dt.$$

The claim follows using the change of variable $s = t^{-1}$ in the last integral. $\qquad\square$

To prove Theorem 4.1(a), note that $\gamma < N - \overline{\dim}_B A$ implies that there exists $\sigma > \overline{\dim}_B A$ such that $\gamma < N - \sigma$. Since $\mathcal{M}^{*\sigma}(A) = 0$, there exists $C = C(\varepsilon, \sigma)$ such that $|A_s| \leq C \cdot s^{N-\sigma}$ for all $s \in (0, \varepsilon)$. From identity (11) it follows that

$$\int_{A_\varepsilon} d(x, A)^{-\gamma} dx \leq \frac{N - \sigma}{N - \sigma - \gamma} C(\varepsilon, \sigma) \, \varepsilon^{N-\sigma-\gamma} < \infty.$$

To prove Theorem 4.1(b), from $\mathcal{M}_*^d(A) > 0$ we conclude that there exists a positive constant $C = C(\varepsilon, d)$ such that $|A_s| \geq C \cdot s^{N-d}$ for all $s \in (0, \varepsilon)$. Hence, using (11) we obtain

$$\infty > \int_{A_\varepsilon} d(x, A)^{-\gamma} dx \geq \gamma \int_0^\varepsilon s^{-\gamma-1} |A_s| \, ds \geq \gamma C \int_0^\varepsilon s^{N-d-\gamma-1} ds,$$

which implies $\gamma < N - d$.

The remaining part of Theorem 4.1 is obtained similarly.

Condition $\mathcal{M}_*^d(A) > 0$ is essential for equivalence in (10) to hold. Namely, it is possible to construct a class of Minkowski degenerate fractal sets $A \subset \mathbb{A}^N$ such that $\mathcal{M}_*^d(A) = 0$, $d = \dim_B A$, for which we have the following surprising equivalence: $\int_{A_\varepsilon} d(x, A)^{-\gamma} dx < \infty \Leftrightarrow \gamma \leq N - \dim_B A$, see Theorem 4.2 in Žubrinić[27].

As an illustration of (10), take $A \subset [0, 1]$ to be the standard triadic Cantor set. Recalling that it is Minkowski nondegenerate, see (7), we have that

$$\int_0^1 d(x, A)^{-\gamma} dx < \infty \quad \Longleftrightarrow \quad \gamma < 1 - \log_3 2.$$

This equivalence can be obtained also directly, by elementary computation, involving only summation of gemetric series, see Žubrinić[23].

As another example we take the spiral $\Gamma$ defined by $r = \varphi^{-\alpha}$, $\varphi \geq 1$, in polar coordinates, and viewed as a subset of the unit disk $B_1(0)$. For fixed $\alpha \in (0, 1)$ the spiral $\Gamma$ has box dimension equal to $\frac{2}{1+\alpha}$, see Tricot[18]. It can be shown that it is Minkowski measurable in this case, see Žubrinić and Županović[28]. Hence, by (10) the function $d(x, \Gamma)^{-\gamma}$ is Lebesgue integrable on $B_1(0)$ if and only if $\gamma < 2 - \frac{2}{1+\alpha}$. The graph of this function is a smooth, connected surface which is singular along points corresponding to $\Gamma$. The rôle of fractal dimensions in dynamics is described in Županović and Žubrinić[29].

## 5. Maximally singular Sobolev functions

The rich repertoire of Lebesgue integrable functions with large sets of singularities, described in the preceding section, will enable us to construct a class maximally singular Sobolev functions. For this we shall need the Bessel potential space[1]

$$L^{\alpha,p}(\mathbb{R}^N) := \{G_\alpha * f : f \in L^p(\mathbb{R}^N)\},$$

where $G_\alpha$ is the Bessel potential kernel, $\alpha > 0$. Let us recall Calderón's theorem[1], which states that if $k \in \mathbb{N}$ and $p \in (1,\infty)$, then

$$L^{k,p}(\mathbb{R}^N) = W^{k,p}(\mathbb{R}^N).$$

In what follows we assume that $\alpha p < N$. We can describe a class of maximally singular functions in $L^{\alpha,p}(\mathbb{R}^N)$ explicitely as follows[23]:

$$v(x) := \sum_{k=1}^\infty 2^{-k} \frac{(G_\alpha * f_k)(x)}{\|f_k\|_{L_p}} = \left(G_\alpha * \sum_{k=1}^\infty 2^{-k} \frac{f_k}{\|f_k\|_{L_p}}\right)(x),$$

with $f_k$ as in (4), assuming that $\varepsilon_k > 0$, $\overline{\dim}_B A_k < N - \alpha p$, $\dim_H A_k \to N - \alpha p$ as $k \to \infty$, and

$$\alpha < \gamma_k < \frac{N - \overline{\dim}_B A_k}{p}. \tag{13}$$

The right-hand side inequality in (13) implies that $f_k \in L^p(\mathbb{R}^N)$, see Theorem 4.1(a), while the left-hand side inequality implies[21] that $A_k \subseteq \mathrm{Sing}\,(G_\alpha * f_k)$, hence $\cup_k A_k \subseteq \mathrm{Sing}\,v$. Exploiting countable stability of Hausdorff dimension we obtain $N - \alpha p \le \dim_H(\mathrm{Sing}\,v)$, which proves that s-dim $L^{\alpha,p}(\mathbb{R}^N) \ge N - \alpha p$.

The converse inequality follows combining the following two results concerning nonnegative $f \in L^p(\mathbb{R}^N)$, $\alpha p < N$, $1 < p < \infty$: (a) Reshetnyak[17] shoed that $\mathrm{Cap}_{\alpha,p}\{x : (G_\alpha * f)(x) = +\infty\} = 0$, hence (also by Reshetnyak),

$$\dim_H\{x : (G_\alpha * f)(x) = +\infty\} \le N - \alpha p; \tag{14}$$

(b) if $a$ is a point of singularity of $G_\alpha * f$, then $(G_\alpha * f)(a) = \infty$, that is,

$$\mathrm{Sing}\,(G_\alpha * f) \subseteq \{x : (G_\alpha * f)(x) = +\infty\},$$

see Refs. 21 and 24. Hence, $\dim_H \mathrm{Sing}\,(G_\alpha * f) \le N - \alpha p$, and therefore we have $\dim_H(\mathrm{Sing}\,v) =$ s-dim $L^{\alpha,p}(\mathbb{R}^N) = N - \alpha p$.

Using Calderón's theorem, for the Sobolev space $W^{k,p}(\mathbb{R}^N)$ we obtain s-dim $W^{k,p}(\mathbb{R}^N) = N - kp$, provided $kp < N$, $1 < p < \infty$. Furthermore, there exist maximally singular functions. Of course, the same result holds also for $W^{k,p}(\Omega)$, where $\Omega \subseteq \mathbb{R}^N$ is an open subset.

## 6. Extended singular set; case $\alpha p = N$

For $\alpha p = N$ no function from $L^{\alpha,p}(\mathbb{R}^N)$ can have singularities in the sense of (1). Therefore, given a measurable function $u : \mathbb{R}^N \to \mathbb{R}$ it is reasonable to consider the set e-Sing $u$ containing not only Sing $u$, but also weaker (e.g. logarithmic) singularities. Precisely, we define *extended singular set*[21] of $u$ as follows:

$$\text{e-Sing}\, u := \{a \in \mathbb{R}^N : \limsup_{r \to 0} \frac{1}{r^N} \int_{B_r(a)} u(y)\, dy = \infty\}. \tag{15}$$

We now introduce the **upper singular dimension**[21] of a set or space of functions $X$ by

$$\text{s-}\overline{\dim}\, X := \sup\{\dim_H(\text{e-Sing}\, u) : u \in X\}.$$

The following result has been proved in Žubrinić[21], see also Ref. 24.

**Theorem 6.1.** *If $1 < p < \infty$ and $\alpha p \leq N$, then $\text{s-}\overline{\dim}\, L^{\alpha,p}(\mathbb{R}^N) = N - \alpha p$. In particular, for $\alpha p = N$ we have $\dim_H(\text{e-Sing}\, u) = 0$ for any $u \in L^{\alpha,p}(\mathbb{R}^N)$.*

It can be shown that for $\alpha p \leq N$ and $f \in L^p(\mathbb{R}^N)$, $f \geq 0$, we have

$$\text{e-Sing}\,(G_\alpha * f) \subseteq \{x : (G_\alpha * f)(x) = +\infty\},$$

see Refs. 21 and 24, so that Theorem 6.1 follows from (14).

## 7. Problems for $p$-Laplace equations

It is well known that solutions of various PDEs may have singularities. As a simple model, let us consider the following $p$-Laplace equation in a bounded domain $\Omega \subseteq \mathbb{R}^N$, $1 < p < \infty$:

$$-\Delta_p u = f(x) \quad \text{in } \Omega \tag{16}$$
$$u = 0 \quad \text{on } \partial\Omega$$

where $f(x)$ is from a given set or space of functions, say $f \in L^{p'}(\Omega)$, where $p' = p/(p-1)$. For any such $f$ the above problem possess a unique weak solution in the Sobolev space $W_0^{1,p}(\Omega)$. It is natural to consider the set $X$ of solutions of (16) generated by all possible right-hand sides $f \in L^{p'}(\mathbb{R}^N)$, that is,

$$X := \{u \in W_0^{1,p}(\Omega) : \exists f \in L^{p'}(\Omega) \ -\Delta_p u = f\},$$

where the equation is to be understood in the sense of distributions[4]. The question of generating singularities of solutions of such equations in a single point has been studied in Ref. 20. It is natural to ask how large can be singular sets of weak solutions of (16), that is, to find s-dim $X$. Here is a partial answer to this question[26].

**Theorem 7.1.** *If $pp' < N$ then $\text{s-dim}\, X \geq N - pp'$. For $p = 2$, that is for the classical Laplace equation, we have the precise result:*

$$\text{s-dim}\, X = \begin{cases} N - 4 & \text{for } N \geq 4 \\ 0 & \text{for } N \leq 3. \end{cases}$$

*Moreover, in the case of $p = 2$ and $N \geq 5$ there exist maximally singular solutions.*

When $p = 2$ and $N \geq 5$ it is possible to construct a class of right-hand sides $f \in L^2(\Omega)$ for which the corresponding solutions are maximally singular, see Ref. 26. We do not know the value of s-dim $X$ for $p \neq 2$.

Weak solutions $u \in W_0^{1,p}(\Omega)$ of equation $-\Delta_p u = f(x)$, with $\Omega \subset \mathbb{R}^N$ bounded, are studied comparing them with solutions of $-\Delta_p u = d(x, A)^{-\gamma}$, using the Tolksdorf comparison principle. Here is an *a-priori* estimate[26] which enables us to derive Theorem 7.1.

**Theorem 7.2.** *If a nonnegative Sobolev function $u \in W^{1,p}(\Omega)$ is a supersolution of equation $-\Delta_p u = d(x, A)^{-\gamma}$, and if $p < \gamma < \min\{1 + \frac{N}{p'}, N - \overline{\dim}_B A\}$, then there exist positive constants $C$ and $D$ such that for a.e. $x \in \Omega$,*

$$u(x) \geq C \cdot d(x, A)^{-\frac{\gamma - p}{p-1}} - D.$$

*In particular, $A \subseteq \operatorname{Sing} u$.*

Generating singularities of solutions of general quasilinear elliptic equations of Leray-Lions type in a single point has been considered by Korkut, Pašić and Žubrinić[8], using the idea of control of solutions introduced by the second author. It has sense to study singular sets of solutions of such problems, as well as of variational inequalities, see Korkut, Pašić and Žubrinić[9]. Fractal properties of the graph of solutions of $p$-Laplace equations have been studied by Pašić[14-15], and by Pašić and Županović[16].

## Acknowledgements

I express my gratitude to Professor Francesco Nicolosi for his kind invitation to the 5th ISAAC Conference, 2005, held at the University of Catania, Italy.

## References

1. D.R. Adams, L.I. Hedberg, *Function Spaces and Potential Theory*, Springer Verlag, 1996.
2. K.J. Falconer, *Fractal Geometry*, John Wiley and Sons, 1990.
3. G.B. Folland, *Real Analysis, Modern Techniques and Their Applications*, 2nd ed., John Wiley & Sons, 1999.
4. D. Gilbarg, N.S. Trudinger, *Elliptic Partial Differential Equations of Second Order*, Springer–Verlag, 1977, 1983.
5. R. Harvey, J. Polking, Removable singularities of solutions of linear partial differential equations, *Acta Math.*, **125** (1970), 39–56.
6. C.Q. He, M. Lapidus, Generalized Minkowski content, spectrum of fractal drums, fractal strings and the Riemann zeta-function, *Mem. Amer. Math. Soc.*, **127** (1997),
7. L. Horvat, D. Žubrinić, Maximally singular Sobolev functions, *J. Math. Anal. Appl.*, **304** (2005), no. 2, 531–541.
8. L. Korkut, M. Pašić, D. Žubrinić, Some qualitative properties of solutions of quasilinear elliptic equtions and applications, *J. Differential Equations*, **170** (2001), no. 2, 247–280.

802

9. L. Korkut, M. Pašić, D. Žubrinić, A class of nonlinear elliptic variational inequalities: qualitative properties and existence of solutions, *Electron. J. Differential Equations*, 2002, No. 14, 14 pp.

10. M. Lapidus, C. Pomerance, The Riemann zeta-function and the one-dimensional Weyl-Berry conjecture for fractal drums, Proc. London Math. Soc. (3) **66** (1993), no. 1, 41–69.

11. M. Lapidus, M. van Frankenhuysen, *Fractal geometry and number theory. Complex dimensions of fractal strings and zeros of zeta functions*. Birkhäuser Boston, Inc., Boston, MA, 2000.

12. O. Martio, M. Vuorinen, Whitney cubes, $p$-capacity, and Minkowski content, *Expo. Math.* **5** (1987), 17–40.

13. P. Mattila, *Geometry of Sets and Measures in Euclidean Spaces. Fractals and Rectifiability*, Cambridge, 1995.

14. M. Pašić, Minkowski–Bouligand dimension of solutions of the one–dimensional $p$-Laplacian, *J. Differential Equations* **190** (2003) 268–305.

15. M. Pašić, Rectifiability of solutions of the one-dimensional $p$-Laplacian, *Electron. J. Differential Equations* 2005, No. 46, 8 pp.

16. M. Pašić, V. Županović, Some metric-singular properties of the graph of solutions of the one-dimensional $p$-Laplacian, *Electronic J. of Differential Equations*, **60** 2004(2004), 1–25.

17. Yu.G. Reshetnyak, On the concept of capacity in the theory of functions with generalized derivatives, *Sibirskij mat. žurnal*, **X**, No 5, (1969), 1108–1138 (Russian); *Siberian Math. J.*, **13** (1969), 818–842.

18. C. Tricot, *Curves and Fractal Dimension*, Springer–Verlag, 1995.

19. P. Wojtaszczyk, *A Mathematical Introduction to Wavelets*, Cambridge University Press, 1997.

20. D. Žubrinić, Generating singularities of solutions of quasilinear elliptic equations, *J. Math. Anal. Appl.*, **244** (2000), 10–16.

21. D. Žubrinić, Singular sets of Sobolev functions, *C. R. Acad. Sci., Analyse mathématique*, Paris, Série I, **334** (2002), 539–544.

22. D. Žubrinić, Minkowski content and singular integrals, *Chaos, Solitons and Fractals*, **17/1** (2003), 169–177.

23. D. Žubrinić, Singular sets of Lebesgue integrable functions, *Chaos, Solitons and Fractals*, **21** (2004) 1281–1287.

24. D. Žubrinić, Extended singular set of potentials, *Math. Inequal. Appl.*, **8** (2005), no. 2, 173–177.

25. D. Žubrinić, Maximally singular functions in Besov spaces, *Arch. Math. (Basel)*, to appear.

26. D. Žubrinić, Generating singularities of solutions of $p$-Laplace equations on fractal sets, 2005, submitted.

27. D. Žubrinić, Analysis of Minkowski contents of fractal sets and applications, submitted.

28. D. Žubrinić, V. Županović, Fractal analysis of spiral trajectories of some planar vector fields, *Bull. Sci. Math.*, **129/6** (2005), 457–485.

29. V. Županović, D. Žubrinić, Fractal dimensions in dynamics, in *Encyclopedia of Mathematical Physics*, Jean-Pierre Françoise, Greg Naber, Sheung Tsun Tsou (editors), Elsevier, 2006, to appear.

## II.4 Variational Methods for Nonlinear Equations

### Organizer: B. Ricceri

Section II.4 was devoted to the variational methods for nonlinear equations.

The contributions that follow entirely reflect the theme of the section.

Minimization of functionals, eigenvalues, boundary value problems for elliptic equations and systems, periodic solutions for second-order Hamiltonian systems, hemivariational inequalities are typical topics that one finds in these contributions.

Most of them share a common circle of ideas that allow the authors to avoid some assumptions that, on the contrary, one encounters almost always in the relevant literature.

# ELLIPTIC EIGENVALUE PROBLEMS ON UNBOUNDED DOMAINS INVOLVING SUBLINEAR TERMS*

A. KRISTÁLY

*University of Babeş-Bolyai*
*Department of Economics*
*400591 Cluj-Napoca, Romania*
*E-mail: alexandrukristaly@yahoo.com*

We present some multiplicity results concerning Schrödinger type equations which involve nonlinearity with sublinear growth at infinity. The results are based on some recent critical point theorems of B. Ricceri.

**Key words:** critical points, multiple solutions, sublinear nonlinearity, Schrödinger equation

**Mathematics Subject Classification:** 35J60, 35J20.

## 1. Historical background and Motivation

Let $f : \mathbb{R}^N \times \mathbb{R} \to \mathbb{R}$ be a continuous function, $V : \mathbb{R}^N \to \mathbb{R}$ a positive potential, and consider the problem

$$-\Delta u + V(x)u = f(x, u), \ x \in \mathbb{R}^N, \ u \in H^1(\mathbb{R}^N). \tag{P}$$

The study of (P) is motivated by mathematical physics; indeed, it is well-known that certain kinds of solitary waves in nonlinear Klein-Gordon or Schrödinger equations are solutions of (P), see Rabinowitz[12], Strauss[16]. Due to its importance, many papers are concerned with the existence and multiplicity of solutions of (P); without seek of completeness, we refer the reader to the works of Bartsch-Wang[3], Bartsch-Willem[4], Strauss[16], Willem[18]; for non-smooth approaches, where $f$ is allowed to be discontinuous, see Gazzola-Rădulescu[7] and Kristály[9]. The aforementioned papers have two common features: the nonlinearity $s \mapsto f(x, s)$ is *subcritical* and *superlinear at infinity*. To be more explicit, most of the authors use the well-known Ambrosetti-Rabinowitz type condition:

$(AR)$ There is $\eta > 2$ such that

$$0 < \eta F(x, s) \le f(x, s)s \ \text{ for each } x \in \mathbb{R}^N, \ s \in \mathbb{R} \setminus \{0\},$$

where $F(x, s) = \int_0^s f(x, t)dt$.

---

*This work was supported by the Istituto Nazionale di Alta Matematica.

And as we know, condition $(AR)$ implies the *superlinearity* of the function $s \mapsto f(x,s)$ *at infinity*, i.e., there exist some numbers $C > 0, s_0 > 0$ such that

$$|f(x,s)| \geq C|s|^{\eta-1} \quad \text{for each } x \in \mathbb{R}^N, \ |s| \geq s_0.$$

The main objective of this paper is to consider (P) when $s \mapsto f(x,s)$ has a *sublinear* growth *at infinity*. In order to avoid technicalities, we assume in the sequel that $f$ has the form $f(x,s) = W(x)f(s)$. With this choice, our basic hypotheses are

$(f1)$ There exist $c > 0$ and $q \in (0,1)$ such that

$$|f(s)| \leq c|s|^q \quad \text{for every } s \in \mathbb{R}, \text{ and}$$

$(W1)$ $W \in L^1(\mathbb{R}^N) \cap L^\infty(\mathbb{R}^N)$.

In such a case the energy functional associated to the studied problem is bounded below and coercive; thus the existence of at least *one* solution is always expected. However, one may happen that (P) has only the trivial solution, even if the nonlinearity fulfills $(f1)$. Indeed, if we consider for instance $V(x) \equiv 1$, $f(x,s) = \lambda W(x)\sin^2 s$ with $W : \mathbb{R}^N \to \mathbb{R}$ as above, and $0 < \lambda < (2\|W\|_{L^\infty}\kappa_2^2)^{-1}$ ($\kappa_2$ being the best Sobolev embedding constant of $H^1(\mathbb{R}^N) \hookrightarrow L^2(\mathbb{R}^N)$), then (P) has only the trivial solution. Thus, this fact motivates the study of an *eigenvalue problem* rather than problem (P). On account of this statement, we shall investigate the following eigenvalue problem

$$\begin{cases} -\Delta u + V(x)u = \lambda W(x)f(u), & x \in \mathbb{R}^N, \\ u \in H^1(\mathbb{R}^N), \end{cases} \tag{$P_\lambda$}$$

where the potential $W : \mathbb{R}^N \to \mathbb{R}$ and the continuous function $f : \mathbb{R} \to \mathbb{R}$ fulfil $(W1)$ and $(f1)$, respectively, while $\lambda \in \mathbb{R}$ is a parameter.

Note that the potential $V : \mathbb{R}^N \to \mathbb{R}$ also has an important role concerning the existence and behaviour of solutions of $(P_\lambda)$. When $V(x) = \text{const.} > 0$, or $V$ is radially symmetric, it is natural to look for radially symmetric solutions of $(P_\lambda)$, see e.g., Bartsch-Willem[4], Kristály[9], Kristály-Varga[10], Strauss[16], Willem[18]. Apart from Kristály-Varga[10], in the aforementioned papers the nonlinearity $f$ fulfills $(AR)$. Motivated by the work of Rabinowitz[12] (where $V \in C(\mathbb{R}^N, \mathbb{R})$, $\inf_{\mathbb{R}^N} V > 0$, and $V(x) \to +\infty$ as $|x| \to +\infty$), Bartsch-Wang[3] considered more general potentials:

$(BW)$ $V \in C(\mathbb{R}^N, \mathbb{R})$ satisfies $\inf_{\mathbb{R}^N} V > 0$, and for any $M > 0$

$$\mu(\{x \in \mathbb{R}^N : V(x) \leq M\}) < +\infty,$$

where $\mu$ denotes the Lebesgue measure in $\mathbb{R}^N$.

Under $(BW)$, Bartsch-Wang[3] proved the existence of infinitely many solutions of $(P_\lambda)$ (for any fixed $\lambda > 0$) when $f$ is subcritical, odd and verifies $(AR)$. Furtado-Maia-Silva[6] studied $(P_\lambda)$ in the case when $F$ (defined in $(AR)$) has some sort of resonance with a local nonquadraticity condition at infinity, while the potential $V$

verifies $(BW)$. Gazzola-Rădulescu[7] studied $(P_\lambda)$ when $V$ verifies $(BW)$, $f$ is not necessarily continuous and satisfies an appropriate non-smooth $(AR)$ condition.

Through this paper, we will assume on the potential $V$ that

$(V1)$ $V \in L^\infty_{loc}(\mathbb{R}^N)$, essinf$_{\mathbb{R}^N} V > 0$; and

$(V2)$ One of the following conditions is satisfied:

$(V2_a)$ $N \geq 2$ and $V : \mathbb{R}^N \to \mathbb{R}$ is radially symmetric;
$(V2_b)$ For any $M > 0$ and any $r > 0$ there holds:

$$\mu(\{x \in B(y,r) : V(x) \leq M\}) \to 0 \text{ as } |y| \to +\infty,$$

where $B(y,r)$ denotes the open ball in $\mathbb{R}^N$ with center $y$ and radius $r > 0$.

Note that hypotheses $(V1) - (V2_b)$ are weaker conditions than $(BW)$, see Bartsch-Pankov-Wang[2]. Requiring $(V1) - (V2_b)$, Bartsch-Liu-Weth[1] proved recently the existence of three solutions of $(P_\lambda)$ for any fixed $\lambda > 0$, $f$ subcritical, verifing $(AR)$.

In the next section we give two multiplicity results concerning problem $(P_\lambda)$ when $f$ verifies $(f1)$. Two different cases will be considered: (i) $f$ is 'superlinear' at the origin; (ii) $f$ does not satisfy any asymptotical property at the origin.

## 2. Main results

In view of $(V1)$, we introduce the Hilbert space

$$E = \left\{ u \in H^1(\mathbb{R}^N) : \int_{\mathbb{R}^N} V(x)u^2 < +\infty \right\}$$

which will be endowed with the inner product

$$(u,v)_E = \int_{\mathbb{R}^N} (\nabla u \nabla v + V(x)uv) \text{ for each } u,v \in E,$$

and with the induced norm $\|\cdot\|_E$. Note that solutions of $(P_\lambda)$ are being sought in $E$ which can be continuously embedded into $L^p(\mathbb{R}^N)$ whenever $2 \leq p < 2^*$. Here, $2^*$ denotes the critical Sobolev exponent, i.e., $2^* = 2N/(N-2)$ for $N \geq 3$ and $2^* = +\infty$ for $N = 1,2$.

Beside of $(W1)$ we require the following assumption on $W : \mathbb{R}^N \to \mathbb{R}$:

$(W2)$ $W \geq 0$, and $\sup_{R>0}$ essinf$_{|x| \leq R} W(x) > 0$.

### 2.1. The function $f$ is superlinear at the origin

On the function $f : \mathbb{R} \to \mathbb{R}$ we assume

$(f2)$ $f(s) = o(|s|)$ as $s \to 0$.
$(f3)$ $\sup_{s \in \mathbb{R}} F(s) > 0$, where $F(s) = \int_0^s f(t)dt$.

**Theorem 2.1.** *Let $f : \mathbb{R} \to \mathbb{R}$ be a continuous function such that $(f1) - (f3)$ are fulfilled. Assume that $V$, $W : \mathbb{R}^N \to \mathbb{R}$ satisfy $(V1) - (V2)$ and $(W1) - (W2)$, respectively, and $W$ is radially symmetric whenever $(V2_a)$ holds.*

*Then there exist an open interval $\Lambda \subseteq (0, \infty)$ and a number $\nu > 0$ such that for every $\lambda \in \Lambda$ problem $(P_\lambda)$ has at least two distinct nonzero weak solutions $u_\lambda^i$ $(i \in \{1, 2\})$ such that $\|u_\lambda^i\|_E \leq \nu$ for every $\lambda \in \Lambda$ and $i \in \{1, 2\}$. Moreover, these solutions are radially symmetric whenever $(V2_a)$ holds.*

### 2.2. *The function f is not superlinear at the origin*

Not only in Theorem 2.1 but also in the aforementioned papers (Bartsch-Liu-Weth[1], Bartsch-Wang[3], Bartsch-Willem[4], Gazzola-Rădulescu[7], Strauss[16], Willem[18]), the superlinearity of $f$ at the origin (i.e. hypothesis $(f2)$) is an indispensable fact. The aim of this subsection is to handle the situation when we drop $(f2)$. As it is expected, this step will be penalized: instead of standard weak solutions of $(P_\lambda)$ we will be able only to obtain multiple solutions for a closely related (perturbed) problem to $(P_\lambda)$. To state this result precisely, let us define the functional $\mathcal{F} : E \to \mathbb{R}$ for each $u \in E$ by

$$\mathcal{F}(u) = \int_{\mathbb{R}^N} W(x) F(u(x)) dx. \tag{1}$$

**Theorem 2.2.** *Let $f : \mathbb{R} \to \mathbb{R}$ be a nonzero, non-decreasing continuous function such that $(f1))$ is fulfilled. Assume that $V$, $W : \mathbb{R}^N \to \mathbb{R}$ satisfy $(V1) - (V2)$ and $(W1) - (W2)$, respectively, and $W$ is radially symmetric whenever $(V2_a)$ holds.*

*Then for each number $\zeta > 0$ there exist a number $\lambda > 0$ and $w \in \mathcal{F}^{-1}([0, \zeta[) \cap C_0^\infty(\mathbb{R}^N)$ such that the problem*

$$-\triangle u + V(x)u = \lambda W(x) f(u + w), \quad x \in \mathbb{R}^N \tag{$P_\lambda^w$}$$

*has at least three distinct weak solutions.*

Note that the above results can be proved on various unbounded domains with enough symmetry (for instance, we can consider strip-like domains). In the next section we will prove Theorems 2.1 and 2.2, respectively, applying two recent critical point results of Ricceri[13,15].

### 3. Proofs

A standard argument, which is based on the facts that $W \in L^1(\mathbb{R}^N) \cap L^\infty(\mathbb{R}^N)$, $f$ satisfies $(f1)$, and the embedding $E \hookrightarrow L^p(\mathbb{R}^N)$ is continuous $(2 \leq p < 2^*)$, shows that the functional $\mathcal{F} : E \to \mathbb{R}$, introduced in (1), is well defined, is of class $\mathcal{C}^1$, and satisfies

$$\mathcal{F}'(u)(v) = \int_{\mathbb{R}^N} W(x) f(u(x)) v(x) \quad \text{for each } u, v \in E. \tag{2}$$

Define now the functional $\mathcal{E} : E\times]0, +\infty[\to \mathbb{R}$ by

$$\mathcal{E}(u, \lambda) = \frac{1}{2}\|u\|_E^2 - \lambda\mathcal{F}(u) \quad \text{for each } (u, \lambda) \in E\times]0, +\infty[.$$

In view of (2), weak solutions of the problem $(P_\lambda)$ are precisely the critical points of $\mathcal{E}(\cdot, \lambda)$. Therefore, our attention will be restricted to find critical points of $\mathcal{E}(\cdot, \lambda)$.

We notify that when $(V1)-(V2_b)$ hold, the embedding $E \hookrightarrow L^p(\mathbb{R}^N)$ is compact when $2 \le p < 2^*$, cf. Bartsch-Pankov-Wang[2]. On the other hand, when $(V1)-(V2_a)$ hold, in general, the space $E$ cannot be compactly embedded into $L^p(\mathbb{R}^N)$. However, introducing the subspace of radially symmetric functions of $E$, i.e.

$$E_r = \{u \in E : u(gx) = u(x) \text{ for each } g \in O(N), \text{ a.e. } x \in \mathbb{R}^N\},$$

the embedding $E_r \hookrightarrow L^p(\mathbb{R}^N)$ is compact whenever $N \ge 2$ and $2 < p < 2^*$, cf. Strauss[16]. Taking into account that in this case $W$ is radially symmetric (see the hypotheses), the functional $\mathcal{E}(\cdot, \lambda)$ is $O(N)$-invariant, thus the critical points of the restricted functional $\mathcal{E}(\cdot, \lambda)$ to the space $E_r$, i.e. $\mathcal{E}_r = \mathcal{E}|_{E_r}$, are critical points of $\mathcal{E}(\cdot, \lambda)$, cf. Palais[11]. In order to consider simultaneously the two cases, we introduce some new notations. Let

$$X = \begin{cases} E_r, & \text{if } (V2_a) \text{ holds,} \\ E, & \text{if } (V2_b) \text{ holds,} \end{cases} \quad \text{and} \quad \mathcal{H}(\cdot, \lambda) = \begin{cases} \mathcal{E}_r(\cdot, \lambda), & \text{if } (V2_a) \text{ holds,} \\ \mathcal{E}(\cdot, \lambda), & \text{if } (V2_b) \text{ holds.} \end{cases}$$

On account of these notations, it is enough to find critical points of $\mathcal{H}(\cdot, \lambda)$ on $X$. We further denote by $\|\cdot\|_X$ and $\mathcal{F}_X$ the restriction of $\|\cdot\|_E$ and $\mathcal{F}$ to the space $X$, respectively.

### 3.1. Proof of Theorem 2.1

A simple calculation which is based on the hypotheses $(f1)$ and $(f2)$ shows that

**Lemma 3.1.** $\lim_{\rho\to0+} \frac{\sup\{\mathcal{F}_X(u): \|u\|_X < \sqrt{2\rho}\}}{\rho} = 0.$

Since $X \hookrightarrow L^p(\mathbb{R}^N)$ is compact $(2 < p < 2^*)$ one can show that $\mathcal{F}'_X$ is a compact operator. (For a similar argument, see Gonçalves-Miyagaki[8].) Thus, using Corollary 41.9 and Example 38.25 from Zeidler[19], as well as $(f1)$, one has

**Lemma 3.2.** *For any $\lambda > 0$, the functional $\mathcal{H}(\cdot, \lambda)$ is sequentially weakly lower semicontinuous and satisfies the Palais-Smale condition.*

The main ingredient in the proof of Theorem 2.1 is a Ricceri-type critical point theorem, see Ricceri[14,15]. Here, we recall a refinement of this result, established in Bonanno[5].

**Theorem 3.1.** *Let $Y$ be a separable and reflexive real Banach space, and let $\Phi, J : Y \to \mathbb{R}$ be two continuously Gâteaux differentiable functionals. Assume that there exists $x_0 \in Y$ such that $\Phi(x_0) = J(x_0) = 0$ and $\Phi(x) \ge 0$ for every $x \in Y$ and that there exists $x_1 \in Y$, $\rho > 0$ such that* *(i)* $\rho < \Phi(x_1)$ *and* $\sup_{\Phi(x)<\rho} J(x) < \rho\frac{J(x_1)}{\Phi(x_1)}$.

*Further, put*

$$\bar{a} = \frac{\zeta\rho}{\rho\frac{J(x_1)}{\Phi(x_1)} - \sup_{\Phi(x)<\rho} J(x)},$$

*with* $\zeta > 1$, *assume that the functional* $\Phi - \lambda J$ *is sequentially weakly lower semicontinuous, satisfies the Palais-Smale condition and*

*(ii)* $\lim_{\|x\|\to+\infty}(\Phi(x) - \lambda J(x)) = +\infty$,

*for every* $\lambda \in [0, \bar{a}]$.

*Then there is an open interval* $\Lambda \subset [0, \bar{a}]$ *and a number* $\nu > 0$ *such that for each* $\lambda \in \Lambda$, *the equation* $\Phi'(x) - \lambda J'(x) = 0$ *admits at least three distinct solutions in* $Y$ *having norm less than* $\nu$.

**Proof of Theorem 2.1.** Apply Theorem 3.1 with $Y = X$, $\Phi = \frac{1}{2}\|\cdot\|_X^2$, $J = \mathcal{F}_X$. Due to (W2) and (f3), there exist $R_0 > 0$ and $s_0 \in \mathbb{R}$ such that $W_{R_0} = \text{essinf}_{|x|\leq R_0} W(x) > 0$ and $F(s_0) > 0$, respectively. Choose further a number $0 < \varepsilon < 1$ such that

$$W_{R_0}F(s_0)\varepsilon^N - \|W\|_{L^\infty} \max_{[-|s_0|,|s_0|]} F(1 - \varepsilon^N) > 0. \qquad (3)$$

Moreover, let $u_\varepsilon \in X$ be such that $u_\varepsilon(x) = s_0$ for any $x \in B(0, \varepsilon R_0)$, $u_\varepsilon(x) = 0$ for any $x \in \mathbb{R}^N \setminus B(0, R_0)$, and $\|u_\varepsilon\|_{L^\infty} \leq |s_0|$. Denoting by $\omega_N$ the volume of the unit ball in $\mathbb{R}^N$, by means of (3) one has

$$\mathcal{F}_X(u_\varepsilon) \geq W_{R_0}F(s_0)\varepsilon^N R_0^N \omega_N - \|W\|_{L^\infty} \max_{[-|s_0|,|s_0|]} F(1 - \varepsilon^N)R_0^N \omega_N > 0.$$

Due to Lemma 3.1, one can fix a small number $\rho = \rho(\varepsilon) > 0$ such that $\sqrt{2\rho} < \|u_\varepsilon\|_X$ and

$$\frac{\sup\{\mathcal{F}_X(u) : \|u\|_X < \sqrt{2\rho}\}}{\rho} < \frac{2\mathcal{F}_X(u_\varepsilon)}{\|u_\varepsilon\|_X^2}.$$

Therefore, choosing $x_0 = 0$, $x_1 = u_\varepsilon$, $\zeta = 1 + \varepsilon$, and taking into account Lemma 3.2, the hypotheses of Theorem 3.1 are verified with

$$\bar{a} = \frac{1+\varepsilon}{2\mathcal{F}_X(u_\varepsilon)\|u_\varepsilon\|_X^{-2} - \sup\{\mathcal{F}_X(u) : \|u\|_X < \sqrt{2\rho}\}\rho^{-1}}.$$

Then there is an open interval $\Lambda \subset [0, \bar{a}]$ and a number $\nu > 0$ such that for any $\lambda \in \Lambda$, the functional $\mathcal{H}(\cdot, \lambda)$ admits at least three distinct critical points $u_\lambda^i \in X$ ($i \in \{1, 2, 3\}$), having norm less than $\nu$, concluding the proof of Theorem 2.1.

## 3.2. *Proof of Theorem 2.2*

A simple calculation, which is based on $(W1) - (W2)$ and on the fact that $f$ is non-decreasing and nonzero, shows that

**Lemma 3.3.** $\inf_X \mathcal{F}_X = 0$; $\sup_X \mathcal{F}_X = +\infty$.

The next lemma is crucial in the the proof of Theorem 2.2.

**Lemma 3.4.** *Let $\zeta > 0$ and $\gamma \in \mathbb{R}$ be two fixed numbers. Then there exist $u = u(\zeta, \gamma) \in X$ and $r = r(\zeta, \gamma) > 0$ such that $\mathcal{F}_X(u) = \zeta$, and $u(x) = \gamma$ for any $x \in B(0, r)$.*

**Proof.** Fix $\bar{\zeta} \in ]\zeta, +\infty[$, and $R_0 > 0$ such that $W_{R_0} = \text{essinf}_{|x| \le R_0} W(x) > 0$ (cf. $(W2)$). Fix $r > 0$ so small that

$$2r < R_0; \tag{4}$$

$$\|W\|_{L^\infty} \max_{[-|\gamma|, |\gamma|]} F\omega_N (2r)^N < \zeta; \tag{5}$$

$$\bar{\zeta}(1 - (2r/R_0)^N) > \zeta. \tag{6}$$

Let $u_0 \in X$ be such that $u_0(x) = \gamma$ for any $x \in B(0, r)$, $u_0(x) = 0$ for any $x \notin B(0, 2r)$, and $\|u_0\|_{L^\infty} \le |\gamma|$. Then, due to (5) one has

$$
\begin{aligned}
\mathcal{F}_X(u_0) &= \int_{\mathbb{R}^N} W(x) F(u_0(x)) dx \\
&\le \|W\|_{L^\infty} F(\gamma) \omega_N r^N + \|W\|_{L^\infty} \max_{[-|\gamma|, |\gamma|]} F\omega_N (2^N - 1) r^N \\
&< \zeta.
\end{aligned}
$$

On the other hand, since $\inf_{\mathbb{R}} F = 0$ and $\sup_{\mathbb{R}} F = +\infty$, there exists $\bar{\xi} \in \mathbb{R}$ such that

$$F(\bar{\xi}) = \bar{\zeta}(W_{R_0} \omega_N R_0^N)^{-1}. \tag{7}$$

According to (4), we may define $u_1 \in X$ such that $u_1(x) = \gamma$ for any $x \in B(0, r)$, and $u_1(x) = \bar{\xi}$ for any $x \in B(0, R_0) \setminus B(0, 2r)$. Since the functions $W$ and $F$ are non-negative, by (6) and (7) we have

$$\mathcal{F}_X(u_1) \ge F(\bar{\xi}) \text{essinf}_{2r \le |x| \le R_0} W(x) \omega_N (R_0^N - (2r)^N) > \zeta.$$

Define the set

$$S_r^\gamma = \{u \in X : u(x) = \gamma \text{ for each } x \in B(0, r)\}.$$

Taking into account the above constructions, we have two elements $u_0, u_1 \in S_r^\gamma$ such that $\mathcal{F}_X(u_0) < \zeta < \mathcal{F}_X(u_1)$. Since the function $X \ni u \mapsto \mathcal{F}_X(u)$ is continuous and the set $S_r^\gamma$ is connected (because it is convex), then there exists $u \in S_r^\gamma$ such that $\mathcal{F}_X(u) = \zeta$. □

**Lemma 3.5.** *Let $\zeta > 0$ be a fixed number. Then the set $\mathcal{F}_X^{-1}([\zeta, +\infty[)$ is not convex.*

*Proof.* Since $f$ is non-decreasing, then $\mathcal{F}_X'$ is monotone, cf. (2) and (W2). Therefore, $\mathcal{F}_X$ is a convex function and the level set $\mathcal{F}_X^{-1}(]-\infty, \zeta])$ is convex. If we assume that $\mathcal{F}_X^{-1}([\zeta, +\infty[)$ is also convex, then $\mathcal{F}_X^{-1}(\zeta) = \mathcal{F}_X^{-1}(]-\infty, \zeta]) \cap \mathcal{F}_X^{-1}([\zeta, +\infty[)$ will be convex as well. Thus, in order to get the conclusion of the lemma, it is enough to prove that $\mathcal{F}_X^{-1}(\zeta)$ is not convex.

To this end, recall that $f(0) = 0$ and $f$ is a nonzero function, i.e., there exists $\gamma \in \mathbb{R}$ such that $f(\gamma) \neq f(0) = 0$. By Lemma 3.4, there exist $u_0, u_1 \in X$ and $r > 0$ such that $u_0(x) = 0$ and $u_1(x) = \gamma$ for any $x \in B(0, r)$, and $u_0, u_1 \in \mathcal{F}_X^{-1}(\zeta)$. Arguing by contradiction, suppose that

$$\mathcal{F}_X(tu_0 + (1-t)u_1) = \zeta \quad \text{for each } t \in [0, 1]. \tag{8}$$

After a differentiation in (8) in rapport of $t$ and by using (2) one has

$$\int_{\mathbb{R}^N} W(x) f(tu_0(x) + (1-t)u_1(x))(u_0(x) - u_1(x)) dx = 0$$

for any $t \in [0, 1]$. Choosing in particular $t = 0$ and $t = 1$ in the above relation, one has

$$\int_{\mathbb{R}^N} W(x)[f(u_0(x)) - f(u_1(x))](u_0(x) - u_1(x)) dx = 0.$$

Since the potential $W$ is non-negative and $f$ is non-decreasing, we obtain

$$W(x)[f(u_0(x)) - f(u_1(x))](u_0(x) - u_1(x)) = 0 \quad \text{for a.e. } x \in \mathbb{R}^N.$$

On the other hand, hypothesis (W2) asserts the existence of a number $R_0 > 0$ such that $W_{R_0} = \operatorname{essinf}_{|x| \leq R_0} W(x) > 0$. Now, applying the last relation on the ball $B(0, \min\{R_0, r\})$ and exploring the choice of $u_0$ and $u_1$, respectively, we are led to $f(\gamma)\gamma = 0$, which contradicts the choice of the number $\gamma$. Thus, (8) is false, i.e., the set $\mathcal{F}_X^{-1}(\zeta)$ is not convex. □

Now, we are in the position to prove Theorem 2.2. Before to do this, we recall another recent critical point theorem of Ricceri which is derived by an ingenious application of a recent result of Tsar'kov[17], and it was applied to solve a two point boundary value problem for ordinary differential equations, see Ricceri[13].

**Theorem 3.2.** *Let $Y$ be a real Hilbert space and $J : Y \to \mathbb{R}$ a continuous Gâteaux differentiable, nonconstant functional, with compact derivative, such that*

$$\limsup_{\|x\| \to +\infty} \frac{J(x)}{\|x\|^2} \leq 0. \tag{9}$$

*Then, for each $r \in ]\inf_Y J, \sup_Y J[$ for which the set $J^{-1}([r, +\infty[)$ is not convex and for every set $S \subseteq Y$ dense in $Y$, there exist $x_0 \in J^{-1}(]-\infty, r[) \cap S$ and $\lambda > 0$ such that the equation*

$$x = \lambda J'(x) + x_0$$

*has at least three distinct solutions.*

**Proof of Theorem 2.2.** We apply Theorem 3.2 by choosing $Y = X$, $J = \mathcal{F}_X$, and $S = C_0^\infty(\mathbb{R}^N) \cap X$. As in the previous section, we have that $\mathcal{F}_X$ is of class $\mathcal{C}^1$, and $\mathcal{F}'_X$ is compact. Indeed, here we used only $(f1)$ and the fact that $W \in L^1(\mathbb{R}^N) \cap L^\infty(\mathbb{R}^N)$. Lemma 3.4 implies in particular that $\mathcal{F}_X$ is not a constant functional. By using $(f1)$, one has

$$\mathcal{F}_X(u) = \int_{\mathbb{R}^N} W(x)F(u(x))dx \leq c\kappa_2^{q+1}\|W\|_{L^{2/(1-q)}}\|u\|_X^{q+1}.$$

Thus, the inequality (9) is clearly verified, since $q < 1$. Taking into account Lemmas 3.3 and 3.5, for every $\zeta \in ]0, +\infty[$ there exist $w \in \mathcal{F}_X^{-1}(]-\infty, \zeta[) \cap C_0^\infty(\mathbb{R}^N) = \mathcal{F}_X^{-1}([0, \zeta[) \cap C_0^\infty(\mathbb{R}^N)$ and $\lambda > 0$ such that the equation

$$v = \lambda\mathcal{F}'_X(v) + w, \tag{10}$$

has three distinct solutions, say $v_i \in X$, $i \in \{1, 2, 3\}$. Due to (10) and (2), the elements $v_i$ are weak solutions of the equation

$$-\triangle v + V(x)v = \lambda W(x)f(v) - \triangle w + V(x)w, \quad x \in \mathbb{R}^N.$$

Therefore, the elements $u_i = v_i - w$ are weak solutions of $(P_\lambda^w)$. This concludes the proof.

## Acknowledgments

The author would like to thank Professor Biagio Ricceri for the kind invitation to deliver this talk at The 5th ISAAC Congress, July 25-30, 2005, University of Cantania, Italy.

## References

1. T. Bartsch, Z. Liu and T. Weth, Sign changing solutions of superlinear Schrödinger equations, *Comm. Partial Differential Equations*, **29**(2004), 25-42.
2. T. Bartsch, A. Pankov and Z.-Q. Wang, Nonlinear Schrödinger equations with steep potential well, *Comm. Contemp. Math.* **4** (2001), 549-569.
3. T. Bartsch and Z.-Q. Wang, Existence and multiplicity results for some superlinear elliptic problems on $\mathbb{R}^N$, *Comm. Partial Differential Equations*, **20** (1995), 1725-1741.
4. T. Bartsch and M. Willem, Infinitely many non-radial solutions of an Euclidean scalar field equation, *J. Func. Anal.* **117** (1995), 447-460.
5. G. Bonanno, Some remarks on a three critical points theorem, *Nonlinear Analysis TMA*, **54** (2003), 651-665.
6. M. F. Furtado, L. A. Maia and E. A. B. Silva, On a double resonant problem in $\mathbb{R}^N$, *Differential Integral Equations*, **15** (2002), 1335-1344.
7. F. Gazzola and V. Rădulescu, A nonsmooth critical point theory approach to some nonlinear elliptic equations in $\mathbb{R}^N$, *Differential Integral Equations*, **13** (2000), 47-60.
8. J. V. Gonçalves and O. H. Miyagaki, Multiple positive solutions for semilinear elliptic equations in $\mathbb{R}^N$ involving subcritical exponents, *Nonlinear Analysis TMA*, **32** (1998), 41-51.
9. A. Kristály, Infinitely many radial and non-radial solutions for a class of hemivariational inequalities, *Rocky Mountain J. Math.* **35**(2005), 1173-1190.

10. A. Kristály and Cs. Varga, On a class of quasilinear eigenvalue problems in $\mathbb{R}^N$, *Math. Nachr.*, 275 (2005), 1756–1765.

11. R. S. Palais, The principle of symmetric criticality, *Comm. Math. Phys.* **69** (1979) 19-30.

12. P. H. Rabinowitz, On a class of nonlinear Schrödinger equations, *Z. Angew. Math. Phys.* **43** (1992), 270-291.

13. B. Ricceri, A general multiplicity theorem for certain nonlinear equations in Hilbert spaces, *Proc. Amer. Math. Soc.* **133** (2005), 3255-3261.

14. B. Ricceri, On a three critical points theorem, *Arch. Math. (Basel)*, **75** (2002), 220-226.

15. B. Ricceri, Existence of three solutions for a class of elliptic eigenvalue problems, *Math. Comput. Modelling*, **32** (2000), 1485-1494.

16. W. A. Strauss, Existence of solitary waves in higher dimensions, *Comm. Math. Phys.* **55** (1977), 149-162.

17. I. G. Tsar'kov, Nonunique solvability of certain differential equations and their connection with geometric approximation theory, *Math. Notes*, **75** (2004), 259-271.

18. M. Willem, Minimax Theorems, Birkhäuser, Boston, 1995.

19. E. Zeidler, Nonlinear Functional Analysis and its Applications, vol. III, Springer-Verlag, 1984.

# ELLIPTIC BOUNDARY VALUE PROBLEMS INVOLVING OSCILLATING NONLINEARITIES

GIOVANNI ANELLO

*Department of Mathematics, University of Messina*
*98166 S.Agata, Messina*
*E-mail anello@dipmat.unime.it*

The aim of this paper is to present some multiplicity results for elliptic boundary value problems involving oscillating nonlinearities. These results are obtained by using a variational theorem of Ricceri and some developments of this latter.

**Key words:** Neumann Problem, oscillating nonlinearities, varational methods, critical points, weak solutions, strong solutions, multiple solutions
**Mathematics Subject Classification:** Primary 35J20, 35J25; Secondary 47H30

## 1. Introduction

We deal with a Neumann problem of the type

$$\begin{cases} -\Delta_p u = \mu f(x,u) + h(x,u) & \text{in} \quad \Omega \\ \\ \frac{\partial u}{\partial \nu} = 0 & \text{on} \quad \partial\Omega \end{cases} \qquad (P_\mu)$$

where:
$\Omega \subseteq \mathbb{R}^N$ is a bounded connected open set with smooth boundary $\partial\Omega$;
$\Delta_p u = \text{div}(|\nabla u|^{p-2}\nabla u)$ is the $p$-Laplacian operator with $p > 1$;
$\nu$ is the outer unit normal to $\partial\Omega$;
$f, h : \Omega \times \mathbb{R} \to \mathbb{R}$ are two Carathéodory functions;
$\mu \in \mathbb{R}$ is a parameter.
  We recall that:
a *weak solution of* $(P_\mu)$ is any $u \in W^{1,p}(\Omega)$ satisfying the equation

$$\int_\Omega (|\nabla u(x)|^{p-2}\nabla u(x)\nabla v(x))dx - \int_\Omega (\mu f(x,u(x)) + h(x,u(x)))v(x)dx = 0.$$

for all $v \in W^{1,p}(\Omega)$;
a *strong solution of* $(P_\mu)$ is any $u \in W^{2,p}(\Omega) \cap C^1(\overline{\Omega})$ satisfying the equation

$$-\Delta_p u = \mu f(x,u) + h(x,u)$$

almost everywhere in $\Omega$ and the boundary condition pointwise.

As it is well known, the weak solutions of problem $(P_\mu)$ turn out exactly the critical points of the energy functional

$$u \in W^{1,p}(\Omega) \longrightarrow \int_\Omega \left( \frac{1}{p} |\nabla u(x)|^p - \int_0^{u(x)} (\mu f(x,t) + g(x,t)) dt \right) dx.$$

Our aim is to present some results concerning the existence of multiple weak solutions and multiple strong solutions for problem $(P_\mu)$ under the basic assumption that at least one of the nonlinearities $f, g$ has an oscillating behavior with respect to the second variable.

A very suitable tool to study the multiplicity of weak solutions for elliptic problems is the following result

**Theorem 1.1.** (Theorem2.5 of Ref.[10]) *Let $X$ be a reflexive real Banach, and let $\Phi, \Psi : X \to \mathbb{R}$ be two sequentially weakly lower semicontinuous and Gâteaux differentiable functionals. Assume also that $\Psi$ is (strongly) continuous, and that it satisfies* $\lim_{\|u\| \to +\infty} \Psi(u) = +\infty$. *For each $\rho > \inf_X \Psi$, put*

$$\varphi(\rho) = \inf_{x \in \Psi^{-1}(]-\infty, \rho[)} \frac{\Phi(x) - \inf_{\overline{\Psi^{-1}(]-\infty,\rho[)}_w} \Phi}{\rho - \Psi(x)},$$

*where $\overline{\Psi^{-1}(]-\infty, \rho[)}_w$ is the closure of $\Psi^{-1}(]-\infty, \rho[)$ in the weak topology. Furthermore, set*

$$\gamma = \liminf_{\rho \to +\infty} \varphi(\rho)$$

*and*

$$\delta = \liminf_{\rho \to (\inf_X \Psi)^+} \varphi(\rho).$$

*Then, the following conclusions hold:*
*(a) If $\gamma < +\infty$, then, for each $\mu > \gamma$, the following alternative holds: either $\Phi + \mu\Psi$ has a global minimum, or there exists a sequence $x_n$ of critical points of $\Phi + \mu\Psi$ such that $\lim_{n \to \infty} \Psi(x_n) = +\infty$.*
*(b) If $\delta < +\infty$, then, for each $\mu > \delta$, the following alternative holds: either there exists a global minimum of $\Psi$ which is a local minimum of $\Phi + \mu\Psi$ or there exists a sequence $x_n$ of pairwise distinct critical points of $\Phi + \mu\Psi$, with $\lim_{n \to \infty} \Psi(x_n) = \inf_X \Psi$, which weakly converges to a global minimum of $\Psi$.*

In particular, the above result assures the existence of a sequence of pairwise distinct critical points for Gâteaux differentiable functionals under assumptions that, when we consider energy functionals, are satisfied just assuming an oscillating behavior on the nonlinearities. We refer the reader to Refs. [4-9,11] where he can find several applications of Theorem 1.1.

A common feature in these applications is the assumption $p > N$. In fact, the main question in applying Theorem 1.1 is to find suitable conditions in order that the number $\delta$ or the number $\gamma$ is finite. This circumstance can be obtained, as we have just said, assuming an oscillating behavior on the nonlinearities but, besides this, the embedding $C^0(\overline{\Omega}) \hookrightarrow W^{1,p}(\Omega)$ turns out necessary there.

So, a natural question is how to obtain multiplicity results in the framework of the oscillating nonlinearities removing the assumption $p > N$. In the next Section we give some answers to this question.

## 2. Main results

The proof of Theorem 1.1 is based on a general existence result of global minima (see Theorem 2.1 of Ref. [10]). Using this latter it is possible to obtain the following further multiplicity result of critical points which does not involve any estimation of the numbers $\delta, \gamma$.

**Theorem 2.1.** (Theorem 2.1 of Ref.[1]) *Let $E$ be a reflexive Banach space with norm $\| \cdot \|_E$ and let $\Phi, \Psi : E \to \mathbb{R}$ be two sequentially weakly lower semicontinuous and Gâteaux differentiable functionals. Suppose $\Psi$ (strongly) continuous. Moreover assume that there exist a sequence $\{u_n\}$ in $E$ with $\lim_{n\to\infty} \|u_n\|_E = +\infty$ and a sequence $\{r_n\}$ in $]0, +\infty[$ with $\lim_{n\to\infty}(\|u_n\|_E - r_n) = +\infty$ such that*

$$\inf_{\|v\|_E = r_n} \Psi(u_n + v) - \Psi(u_n) > 0$$

*for all $n \in \mathbb{N}$, and*

$$\limsup_{n\to\infty} \frac{\Phi(u_n) - \inf_{\|v\|_E \le r_n} \Phi(u_n + v)}{\inf_{\|v\|_E = r_n} \Psi(u_n + v) - \Psi(u_n)} < +\infty.$$

*Then, there exists $\rho^* > 0$ such that for all $\rho > \rho^*$ the functional $\rho\Psi + \Phi$ admits a sequence $\{w_n\}_{n\in\mathbb{N}}$ of critical points such that $\|u_n - w_n\|_E < r_n$ for all $n \in \mathbb{N}$. Hence, in particular, the sequence $\{w_n\}_{n\in\mathbb{N}}$ turns out unbounded.*

(**Remark:** in the original version of the above theorem one assumes also that the real function

$$r \to \inf_{\|v\|_E = r} \Psi(u_n + v)$$

is continuous in $]0, +\infty[$ for all $n \in \mathbb{N}$, but the author has observed that it can be removed.)

Applying Theorem 2.1 we get the following multiplicity result for problem $(P_\mu)$

**Theorem 2.2.** (Theorem 2.2 of Ref. [1]) *Assume $p > 1$. Suppose that the functions $f, h : \Omega \times \mathbb{R} \to \mathbb{R}$ satisfy the following growth conditions:*

*there exist $m \ge 1$, $q > 0$, $b > 0$, $\gamma \in L^m(\Omega)$, with $q < \frac{(p-1)N+p}{N-p}$ and $m > \frac{pN}{N+p}$ if $N > p$, $m > 1$ if $N \le p$, and $l \in [0, p-1]$ such that*

    *i) $|h(x,t)| \le b|t|^q + \gamma(x)$ for all $t \in \mathbb{R}$ and for a.e. $x \in \Omega$;*

    *ii) $|f(x,t)| \le b|t|^l + \gamma(x)$ for all $t \in \mathbb{R}$ and for a.e. $x \in \Omega$.*

*Moreover, denote by A the set*

$$\left\{ \xi \in \mathbb{R} : \int_0^\xi h(x,t)dt = \sup_{\tau \in \mathbb{R}} \int_0^\tau h(x,t)dt \ for \ a.e. \ x \in \Omega \right\}$$

*and assume that there exist $s \in [0,p]$, $M > 0$, a sequence $\{\xi_n\}$ in $A \setminus \{0\}$ with*
$\lim_{n \to \infty} |\xi_n| = +\infty$, *a sequence $\{\sigma_n\}$ in $]0,+\infty[$, with $\lim_{n \to \infty} \left| \dfrac{\sigma_n}{\xi_n} \right| \in ]0,+\infty[$, a non degen-*
*erate interval $I \subseteq ]0+\infty[$ and a Lebesgue measurable subset $D$ of $\Omega$ with $m(D) > 0$
such that*

$$\sup_{\sigma \in I \cup (-I)} \int_{\xi_n}^{\xi_n + \sigma(\sigma_n)^{\frac{s}{p}}} h(x,t)dt \leq -M|\xi_n|^s$$

*for all $n \in \mathbb{N}$ and $x \in D$. Then, if $l \leq s\frac{p-1}{p}$, there exists $\mu^* \in ]0,+\infty]$, with $\mu^* = +\infty$
if $l < s\frac{p-1}{p}$, such that, for all $\mu \in ] - \mu^*, \mu^*[$, problem $(P_\mu)$ admits an unbounded
sequence of weak solutions in $W^{1,p}(\Omega)$.*

**Example** A simple example of function $h$ which satisfies the assumptions of
Theorem 2.2 can be otained choosing $h$ defined by

$$\int_0^\xi h(x,t)dt = \begin{cases} -\lambda(x)|\xi|^m g(\ln|\xi|) & \text{if } x \in \Omega \text{ and } \xi \neq 0 \\ 0 & \text{if } x \in \Omega \text{ and } \xi = 0 \end{cases},$$

*where :*
$m \in \left] p, \frac{pN}{N-p} \right[$ *with $p \leq N$,*
$g \in C^2(\mathbb{R})$ *is a nonnegative $T$-periodic function with $g(0) = 0$ and $g''(0) > 0$,*
$\lambda \in L^\infty(\Omega)$ *with $\operatorname{ess\,inf}_\Omega \lambda > 0$.*

A question which arises from the statement of Theorem 2.2 is to check if the set
$A$ can be replaced by the set

$$\left\{ \xi \in \mathbb{R} : \exists \delta > 0 \text{ such that } \int_0^\xi h(\cdot,t)dt = \sup_{\tau \in ]\xi - \delta, \xi + \delta[} \int_0^\tau h(\cdot,t)dt \text{ a.e. in } \Omega \right\}.$$

The next results give a partial answer to this question:

**Theorem 2.3.** (Theorem 2.1 of Ref.[3]). *Let $\lambda \in L^\infty(\Omega)$ with $\operatorname{ess\,inf}_\Omega \lambda > 0$. Sup-
pose $f(x,0) = 0$ and $h(x,t) = -\lambda(x)|t|^{p-2}t$ for a.e. $x \in \Omega$ and all $t \in \mathbb{R}$, with $f$
satisfying the following growth conditions:*

  i) $1 < p \leq N$ *and* $\displaystyle\sup_{t \in \mathbb{R}} \frac{|f(\cdot,t)|}{1 + |t|^q} \in L^\infty(\Omega)$ *where $q > p - 1$ with*
  $q < \frac{(p-1)N+p}{N-p}$ *if $p < N$;*

  ii) $p > N$ *and* $\displaystyle\sup_{|t| \leq r} |f(\cdot,t)| \in L^\infty(\Omega)$ *for all $r > 0$.*

*Assume that there exist two sequences $\{\xi_n\}, \{\xi_n'\} \subseteq [0,+\infty[$ with $\xi_n' > \xi_n$ for all
$n \in \mathbb{N}$ and $\xi_n \to +\infty$ such that*

1) $\displaystyle\int_0^{\xi_n} f(x,t)dt = \sup_{\xi \in [\xi_n, \xi_n']} \int_0^{\xi} f(x,t)dt \quad$ *for a.a.* $x \in \Omega$

*and, further, assume that*

2) $\displaystyle\limsup_{\xi \to +\infty} \frac{\int_\Omega \int_0^{\xi} f(x,t)dtdx}{\xi^p} > 0.$

*Then, for every* $\mu > \displaystyle\frac{\int_\Omega \lambda(x)dx}{p} \liminf_{\xi \to +\infty} \frac{\xi^p}{\int_\Omega \int_0^{\xi} f(x,t)dtdx}$, *problem* $(P_\mu)$ *admits an unbounded sequence* $\{u_n\}$ *of non-negative weak solutions in* $W^{1,p}(\Omega)$.

In the above result, condition 1) says that the primitive of $f(x, \cdot)$ must have an oscillating behavior near $+\infty$. In this case we have the existence of a sequence of arbitrarily large weak solutions. We note that no global maximum of the primitive of $f(x, \cdot)$ is involved there. Slightly modifying the assumptions in Theorem 2.3 we can also obtain the existence of a sequence of non-zero arbitrarily small weak solutions. In particular, in this case, we have to require that the primitive of $f(x, \cdot)$ have an oscillating behavior near to 0 (see condition 1) below). The statement of the result is as follows

**Theorem 2.4.** (Theorem 3.1 of Ref.[3]) *Let* $\lambda \in L^\infty(\Omega)$ *with* ess inf$_\Omega$ $\lambda > 0$. *Suppose* $\sup_{t \in [0,\bar{t}]} |f(\cdot, t)| \in L^\infty(\Omega)$ *for some* $\bar{t} > 0$ *and* $f(x, 0) = 0$, $h(x, t) = -\lambda(x)|t|^{p-2}t$ *for a.e.* $x \in \Omega$ *and all* $t \in \mathbb{R}$. *Assume that there exist two sequences* $\{\xi_n\}, \{\xi_n'\} \subseteq [0, +\infty[$ *with* $\xi_n' > \xi_n$ *for all* $n \in \mathbb{N}$ *and* $\xi_n' \to 0$ *such that*

1) $\displaystyle\int_0^{\xi_n} f(x,t)dt = \sup_{\xi \in [\xi_n, \xi_n']} \int_0^{\xi} f(x,t)dt \quad$ *for a.a.* $x \in \Omega$.

*Moreover, suppose that*

2) $\displaystyle\limsup_{\xi \to 0^+} \frac{\int_\Omega \int_0^{\xi} f(x,t)dtdx}{\xi^p} > 0.$

*Then, for every* $\mu > \displaystyle\frac{\int_\Omega \lambda(x)dx}{p} \liminf_{\xi \to 0^+} \frac{\xi^p}{\int_\Omega \int_0^{\xi} f(x,t)dtdx}$, *problem* $(P_\mu)$ *admits a sequence* $\{u_n\}$ *of non-zero non-negative weak solutions in* $W^{1,p}(\Omega)$ *strongly converging to zero and such that* $\lim_{n \to +\infty} \max u_n = 0$.

Theorems 2.3, 2.4 assure the existence of a sequence of pairwise distinct weak solutions to problem $(P_\mu)$ for the special case in which $h(x, t) = \lambda(x)|t|^{p-2}t$. To prove these results one makes use of a method inspired by a paper of J. Saint-Raymond (see Ref. [12]). Dealing with more general nonlinearity $h$, this method does not work. Nevertheless, when $h$ is of the form $\lambda(x)g(t)$, we can use the following existence result, which allows us to localize the values of the solution, to get, in a straightforward way, a multiplicity theorem of solutions for problem $(P_\mu)$. Its proof comes out by applying Theorem 2.1 of Ref. 10 together with a regularity of solutions argument

**Theorem 2.5.** (Theorem 1 of Ref.[2]) *Let $[a, b]$ be a compact real interval and $g :$ $[a, b] \to \mathbb{R}$ a continuous function satisfying*

$$\max\{G(a), G(b)\} < \max_{\xi \in [a,b]} G(\xi)$$

*where $G$ is a primitive of $g$. Moreover, let $\lambda \in L^\infty(\Omega)$ with ess $\inf_\Omega \lambda > 0$. Then, if $h(x, t) = \lambda(x)g(t)$ for a.a. $x \in \Omega$ and $t \in [a, b]$ and $f$ satisfies $\sup_{t \in [a,b]} |f(\cdot, t)| \in L^q(\Omega)$ for some $q > N$, there exist $\overline{\mu}, \sigma > 0$ such that, for every $\mu \in [-\overline{\mu}, \overline{\mu}]$, there exist a strong solution $u_\mu \in W^{2,2}(\Omega) \cap C^1(\overline{\Omega})$ of problem $(P_\mu)$ satisfying $u_\mu(x) \in ]a, b[$ for all $x \in \Omega$, and*

$$\int_\Omega (|\nabla u_\mu|^2 + |u_\mu|^2)dx \leq \sigma.$$

As said before, we can now derive from Theorem 2.5 the following multiplicity result. We want to stress out the quite simple oscillating behavior required on the nonlinearity $h$.

**Theorem 2.6.** (Theorem 2 of Ref.[2]) *Let $[a_1, b_1], .., [a_n, b_n]$ be $n$ compact pairwise disjoint real intervals, $D = \cup_{i=1}^n [a_i, b_i]$, and $g : D \to \mathbb{R}$ a continuous function satisfying*

$$\max\{G(a_i), G(b_i)\} < \max_{\xi \in [a_i, b_i]} G(\xi)$$

*for all $i = 1, ..., n$, where $G$ is a primitive of $g$. Moreover, let $\lambda \in L^\infty(\Omega)$ with ess $\inf_\Omega \lambda > 0$. Then, if $h(x, t) = \lambda(x)g(t)$ for all $x \in \Omega$, $t \in D$ and $f$ satisfies $\sup_{t \in D} |f(\cdot, t)| \in L^q(\Omega)$ for some $q > N$, there exist $\overline{\mu}, \sigma > 0$ such that, for every $\mu \in [-\overline{\mu}, \overline{\mu}]$, there exist $n$ strong solutions $u_\mu^{(1)}, .., u_\mu^{(n)} \in W^{2,2}(\Omega) \cap C^1(\overline{\Omega})$ of problem $(P_\mu)$ satisfying $u_\mu^{(i)}(x) \in ]a_i, b_i[$ for all $x \in \Omega$, $i = 1, .., n$, and*

$$\max_{i \in \{1, .., n\}} \int_\Omega (|\nabla u_\mu^{(i)}|^2 + |u_\mu^{(i)}|^2)dx \leq \sigma.$$

Note that in the statement of the above result no global maximum of the primitive of $h(x, \cdot)$ is involved.

## References

1. G. Anello, Existence of infinitely many weak solutions for a Neumann problem, *Nonlinear Anal.*, **57** (2004), 199-209;
2. G. Anello, Existence and multiplicity of solutions to a pertubed Neumann problem, *Math. Nach.*, to appear.
3. G.Anello, G. Cordaro, Infinitely many positive solutions for the Neumann problem involving the p-Laplacian, *Colloq. Math.* **97**(2) (2003),221-231.
4. G. Bilotta, Existence of infinitely many solutions for a quasilinear Neumann problem, *Panamer. Math. J.*, **13** (2) 19-36 (2003).

5. F. Cammaroto and A. Chinnì, Infinitely many solutions for a two points boundary value problem, *Far East J. Math. Sci.*, **11** (1) (2003), 41-51 (2003).
6. F. Cammaroto and A. Chinnì and B. Di Bella, Infinitely many solutions for the Dirichlet problem involving the p-Laplacian, *Nonlinear Anal.*, **61** No.1-2 (A) (2005), 41-49.
7. P. Candito, Infinitely many solutions to the Neumann problem for elliptic equations involving the p-Laplacian and with discontinuous nonlinearities, *Proc. Math. Soc. Edinburgh*, **45** (2002), 397-409.
8. F. Faraci and R. Livrea, Infinitely many periodic solutions for a second order nonautonomous system, *Nonlinear Anal.*, **54** No.3(A) (2003), 417-429.
9. S. A. Marano and D.Montreanu, Infinitely many critical points of non-differentiable functions and applications to a Neumann type problem involving the p-Laplacian, *J. Differential Equations*, **182** (2002), 108-120.
10. B. Ricceri, A general variational principle and some of its applications, *J. Comput. Appl. Math.* **113** (2000), 401-410.
11. B. Ricceri, Infinitely many solutions of the Neumann problem for elliptic equaitons involving the p-Laplacian, *Bull. London Math. Soc.* **33** (2001), 331-340
12. J. Saint Raymond, On the multiplicity of the solutions of the equation $-\Delta u = \lambda \cdot f(u)$, *J. Differential Equations* **180** (2002), 65-88.

# INFINITELY MANY SOLUTIONS FOR THE DIRICHLET PROBLEM FOR THE P-LAPLACIAN

F. CAMMAROTO *, A. CHINNÍ , B. DI BELLA

*Department of Mathematics, University of Messina*
*Contrada Papardo, Salita Sperone n. 31*
*98166 - Messina, Italy*
*E-mail: filippo@dipmat.unime.it*

Using a recent variational principle of B. Ricceri, we present some results of existence of infinitely many solutions for the Dirichlet problem involving the p-Laplacian.

**Key words:** Dirichlet problem, p-Laplacian, multiple solutions
**Mathematics Subject Classification:** 35J20, 35J25

## 1. Introduction

The aim of this paper is to present a list of multiplicity results for the following autonomous Dirichlet problem

$$
\begin{cases}
- \Delta_p u = f(u) & \text{in } \Omega \\
\\
u = 0 & \text{on } \partial\Omega
\end{cases}
\tag{$D_{n,p}$}
$$

where $\Omega$ is a bounded open subset of the euclidean space $(\mathbb{R}^n, |\cdot|)$ with boundary of class $C^1$, $p > n$, $\Delta_p u = \text{div}(|\nabla u|^{p-2}\nabla u)$ and $f : \mathbb{R} \to \mathbb{R}$ is a continuous function having a suitable oscillating behaviour. Let us recall that a weak solution of $(D_{n,p})$ is any $u \in W_0^{1,p}(\Omega)$ such that

$$
\int_\Omega |\nabla u(x)|^{p-2}\nabla u(x)\nabla v(x) \, dx - \int_\Omega f(u(x))v(x) \, dx = 0
$$

for each $v \in W_0^{1,p}(\Omega)$.

The existence of infinitely many solutions of the problem $(D_{n,p})$ has been widely studied under sublinearity or superlinearity conditions at 0 and at $+\infty$ of function $f$ (see, for instance, [5]). More rarely, multiplicity of solutions has been investigated when $f$ has an oscillating behaviour (see [6, 7] and [9]).

All the results contained in this paper are obtained making use of a recent general

---

*Because of a surprising coincidence of names within the same Department, we have to point out that the author was born on August 4, 1968.

variational principle obtained by B. Ricceri in [8]. The following result is a direct consequence of Theorem 2.5 of [8].

**Theorem 1.1.** *Let $X$ be a reflexive real Banach space, and let $\Phi$, $\Psi : X \to \mathbb{R}$ be two sequentially weakly lower semicontinuous and Gâteaux differentiable functionals. Assume also that $\Psi$ is (strongly) continuous and satisfies $\lim_{\|x\|\to+\infty} \Psi(x) = +\infty$. For each $r > \inf_X \Psi$, put*

$$\varphi(r) = \inf_{x\in\Psi^{-1}(]-\infty,r[)} \frac{\Phi(x) - \inf_{\overline{(\Psi^{-1}(]-\infty,r[))}_w} \Phi}{r - \Psi(x)},$$

*where $\overline{(\Psi^{-1}(]-\infty,r[))}_w$ is the closure of $\Psi^{-1}(]-\infty,r[)$ in the weak topology. Fixed $\lambda \in \mathbb{R}$, then*

(a) *if $\{r_n\}_{n\in\mathbb{N}}$ is a real sequence with $\lim_{n\to\infty} r_n = +\infty$ such that $\varphi(r_n) < \lambda$, for each $n \in \mathbb{N}$, the following alternative holds: either $\Phi + \lambda\Psi$ has a global minimum, or there exists a sequence $\{x_n\}$ of critical points of $\Phi + \lambda\Psi$ such that $\lim_{n\to\infty} \Psi(x_n) = +\infty$.*

(b) *if $\{s_n\}_{n\in\mathbb{N}}$ is a real sequence with $\lim_{n\to\infty} s_n = (\inf_X \Psi)^+$ such that $\varphi(s_n) < \lambda$, for each $n \in \mathbb{N}$, the following alternative holds: either there exists a global minimum of $\Psi$ which is a local minimum of $\Phi + \lambda\Psi$, or there exists a sequence $\{x_n\}$ of pairwise distinct critical points of $\Phi + \lambda\Psi$, with $\lim_{n\to\infty} \Psi(x_n) = \inf_X \Psi$, which weakly converges to a global minimum of $\Psi$.*

Throughout the sequel, $f : \mathbb{R} \to \mathbb{R}$ is a continuous function such that $f(x) = 0$ for each $x \in ]-\infty,0]$ and $F : \mathbb{R} \to \mathbb{R}$ is the function defined by setting

$$F(\xi) = \int_0^\xi f(t)dt$$

for each $\xi \in \mathbb{R}$.

We shall consider the Sobolev space $W_0^{1,p}(\Omega)$ endowed with the norm

$$\|u\| := \left(\int_\Omega |\nabla u(x)|^p \, dx\right)^{1/p}.$$

We recall that there exists a constant $c > 0$ such that

$$\sup_{x\in\Omega} |u(x)| \le c\|u\| \tag{1}$$

for each $u \in W_0^{1,p}(\Omega)$. Moreover we put $\omega := \dfrac{\pi^{n/2}}{\frac{n}{2}\Gamma(\frac{n}{2})}$ the measure of the n-dimensional unit ball.

## 2. Results

The following result, appeared in [2], guarantees that the problem $(D_{n,p})$ has infinitely many weak solutions that form an unbounded set in $W_0^{1,p}(\Omega)$.

**Theorem 2.1.** *Assume that, for each $\xi \in \mathbb{R}$, $F(\xi) \geq 0$. Moreover suppose that there exist $x_0 \in \Omega$, a positive number $\delta \leq d(x_0, \partial\Omega)$ and four real sequences $\{r_k\}_{k\in\mathbb{N}}$, $\{\gamma_k\}_{k\in\mathbb{N}}$, $\{\varepsilon_k\}_{k\in\mathbb{N}}$, $\{\xi_k\}_{k\in\mathbb{N}}$ with $\lim_{k\to\infty} r_k = +\infty$, $0 < \gamma_k \leq \mathrm{dist}(x_0, \partial\Omega)$, $\varepsilon_k \in {]0, \gamma_k[}$ and $\xi_k \in {]0, +\infty[}$ for all $k \in \mathbb{N}$, such that*

$$(i) \qquad F(\xi_k) = \max_{[0, cr_k^{\frac{1}{p}}]} F \quad \text{for each } k \in \mathbb{N};$$

$$(ii) \qquad \xi_k < (\gamma_k - \varepsilon_k)\left(\frac{r_k}{\omega(\gamma_k^n - \varepsilon_k^n)}\right)^{\frac{1}{p}} \quad \text{for each } k \in \mathbb{N};$$

$$(iii) \qquad F(\xi_k) < \frac{1}{p(|\Omega| - \omega\varepsilon_k^n)}\left[r_k - \frac{\omega\xi_k^p}{(\gamma_k - \varepsilon_k)^p}(\gamma_k^n - \varepsilon_k^n)\right] \quad \text{for each } k \in \mathbb{N};$$

$$(iv) \qquad \limsup_{\xi\to+\infty} \frac{F(\xi)}{\xi^p} > \frac{2^p}{p\delta^p}(2^n - 1).$$

*Then, the problem $(D_{n,p})$ has infinitely many weak solutions that form an unbounded set in $W_0^{1,p}(\Omega)$.*

A possible function that verifies Theorem 2.1 is the following

**Example 2.1.** Let $\Omega = B(0, r)$ the open ball of $\mathbb{R}^2$, $p = 3$ and $F : \mathbb{R} \to \mathbb{R}$ be the function defined by setting

$$F(x) = \begin{cases} 0 & \text{if } x \in {]-\infty, 0[} \\ A(x) & \text{if } x \in [0, e] \\ B_k(x) & \text{if } x \in {]e^{8k-7}, e^{8k-3}]} \\ C_k(x) & \text{if } x \in {]e^{8k-3}, e^{8k+1}]} \end{cases}$$

with $k \in \mathbb{N}$,

$$A(x) = a(-2x^3 + 3e\, x^2)$$

$$B_k(x) = \frac{a}{(e^4 - 1)^3}(2x^3 - 3e^{8k-7}(e^4 + 1)x^2 + 6e^{16k-10}x + e^{24k-13}(e^4 - 3))$$

$$C_k(x) = \frac{a}{(e^4 - 1)^3}(-2e^{12}x^3 + 3e^{8k+9}(e^4 + 1)x^2 - 6e^{16k+10}x + e^{24k+3}(3e^4 - 1))$$

where $a$ is a real number such that

$$\frac{64}{r^3} < a < \frac{8e^{12}}{45\pi^4 r^3} - \frac{64}{15r^3}.$$

This function satisfies all assumptions of Theorem 2.1 taking $x_0 = 0$, $\delta = \dfrac{r}{2}$, $r_k = \dfrac{e^{24k-9}}{2\pi^3 r}$, $\gamma_k = \dfrac{r}{2}$, $\epsilon_k = \dfrac{r}{4}$ and $\xi_k = e^{8k-7}$; in particular, the choice of $a$ makes true hypotheses (iii) and (iv).

It is interesting to note that, in this case, one has

$$a \leq \limsup_{\xi \to +\infty} \frac{F(\xi)}{\xi^3} < +\infty.$$

In the following result, the case $n = 1$ and $p = 2$ is considered. It appeared in 1.

**Theorem 2.2.** *Assume that, for each $\xi \in \mathbb{R}$, $F(\xi) \geq 0$ and that there exist three real sequences $\{r_k\}_{k \in \mathbb{N}}$, $\{\epsilon_k\}_{k \in \mathbb{N}}$, $\{\xi_k\}_{k \in \mathbb{N}}$ with $\lim_{k \to \infty} r_k = +\infty$, $\{\epsilon_k : k \in \mathbb{N}\} \subseteq ]0, \frac{1}{2}[$ and $\{\xi_k : k \in \mathbb{N}\} \subseteq ]0, +\infty[$, such that*

*(i)*      $F(\xi_k) = \max_{[0, \frac{\sqrt{r_k}}{2}]} F$;

*(ii)*      $\xi_k < \sqrt{\dfrac{r_k \epsilon_k}{2}}$ *for each $k \in \mathbb{N}$;*

*(iii)*     $F(\xi_k) < \dfrac{1}{4\epsilon_k}\left( r_k - 2\dfrac{\xi_k^2}{\epsilon_k}\right)$ *for each $k \in \mathbb{N}$;*

*(iv)*     $\limsup_{\xi \to +\infty} \dfrac{F(\xi)}{\xi^2} > 8$.

*Then, the problem*

$$\begin{cases} -u'' = f(u) & \text{in } ]0, 1[ \\ \\ u(0) = u(1) = 0 \end{cases} \qquad (D_{1,2})$$

*has infinitely many classical solutions that form an unbounded set in $W_0^{1,2}(]0, 1[)$.*

An explicit example of function $F$ that fits all the hypotheses of Theorem 2.2 is the following.

**Example 2.2.** Let $a$ a real number such that

$$4 < a < 4e^{\frac{\pi}{2}} - 8 := \sigma.$$

Let $F : \mathbb{R} \to \mathbb{R}$ be the function defined by setting

$$F(x) = \begin{cases} 0 & \text{if } x \in ]-\infty, 0] \\ \\ ax^2 \left( \sin\left( \ln x^2 \right) + 1 \right) & \text{if } x \in ]0, +\infty[ \end{cases}$$

The function $F$ assumes its local maxima in $x = e^{\frac{\pi}{2}+k\pi}$, for each $k \in \mathbf{Z}$ and its local minima in $x = e^{\frac{3}{4}\pi+k\pi}$, for each $k \in \mathbf{Z}$. Moreover it satisfies all the hypotheses of Theorem 2.2.

To justify this assertion we choose, for each $k \in \mathbb{N}$ $(k \geq 0)$, $r_k = 4e^{\frac{3}{2}\pi+2k\pi}$, $\xi_k = e^{\frac{\pi}{2}+k\pi}$ and $\epsilon_k = \frac{1}{4}$. It is easy to prove that (i), (ii), (iii) and (iv) are satisfied: in particular, the choice of $a$ makes true hypotheses (iii) and (iv).

Note that, in this case, one has

$$\limsup_{\xi\to+\infty} \frac{F(\xi)}{\xi^2} = \limsup_{\xi\to+\infty} a\left(\sin\left(\ln x^2\right) + 1\right) = 2a < +\infty$$

and

$$\liminf_{\xi\to+\infty} \frac{F(\xi)}{\xi^2} = \liminf_{\xi\to+\infty} a\left(\sin\left(\ln x^2\right) + 1\right) = 0.$$

With a slight modification on function $F$ (see Example 2.1 of [1]) it is possible to obtain

$$\liminf_{\xi\to+\infty} \frac{F(\xi)}{\xi^2} > 0.$$

The next two results, appeared in [3], are two simpler but less general forms of Theorem 2.1.

**Theorem 2.3.** *Assume that, for each $\xi \in \mathbb{R}$, $F(\xi) \geq 0$. Moreover, suppose that there exist two real sequences $\{a_k\}_{k\in\mathbb{N}}$ and $\{b_k\}_{k\in\mathbb{N}}$ in $]0,+\infty[$ with $a_k < b_k$, $\lim_{k\to\infty} b_k = +\infty$, such that*

*(i)* $\lim_{k\to\infty} \dfrac{b_k}{a_k} = +\infty$;

*(ii)* $\max_{[a_k,b_k]} f \leq 0$ *for all* $k \in \mathbb{N}$;

*(iii)* $\dfrac{2^p}{p(\sup_{x\in\Omega} \mathrm{dist}(x,\partial\Omega))^p}(2^n - 1) < \limsup_{\xi\to+\infty} \dfrac{F(\xi)}{\xi^p} < +\infty.$

*Then, problem $(D_{n,p})$ admits an unbounded sequence of non-negative weak solutions in $W_0^{1,p}(\Omega)$.*

**Theorem 2.4.** *Assume that, for each $\xi \in \mathbb{R}$, $F(\xi) \geq 0$. Moreover, suppose that there exist two real sequences $\{a_k\}_{k\in\mathbb{N}}$ and $\{b_k\}_{k\in\mathbb{N}}$ in $]0,+\infty[$ with $a_k < b_k$, $\lim_{k\to\infty} b_k = 0$, such that*

(j) $\lim\limits_{k\to\infty} \dfrac{b_k}{a_k} = +\infty$;

(jj) $\max\limits_{[a_k,b_k]} f \leq 0$ for all $k \in \mathbb{N}$;

(jjj) $\dfrac{2^p}{p(\sup\limits_{x\in\Omega} \operatorname{dist}(x,\partial\Omega))^p}(2^n - 1) < \limsup\limits_{\xi\to 0^+} \dfrac{F(\xi)}{\xi^p} < +\infty.$

Then, problem $(D_{n,p})$ admits a sequence of nonzero weak solutions which strongly converges to 0 in $W_0^{1,p}(\Omega)$.

**Remark 2.1.** Observe that, in the mere condition:

$$0 < \limsup\limits_{\xi\to +\infty} \frac{F(\xi)}{\xi^p} < +\infty$$

we can apply Theorem 2.3 and 2.4 taking $\Omega$ sufficiently large.

An explicit example of function which satisfies all the assumptions of Theorem 2.3 is the following.

**Example 2.3.** Let $\Omega$ be a bounded open subset of $\mathbb{R}^n$ with boundary of class $C^1$ and $p > n$. Let $f : \mathbb{R} \to \mathbb{R}$ the function defined by setting

$$f(\xi) = \sum_{k=1}^{\infty} \frac{2Lh_k}{k!}\operatorname{dist}(\xi, \mathbb{R} \setminus [k!k, (k+1)!]) \ ,$$

for each $\xi \in \mathbb{R}$, where

$$L > \frac{2^p}{p(\sup\limits_{x\in\Omega} \operatorname{dist}(x,\partial\Omega))^p}(2^n - 1)$$

and

$$h_k = 2(k!)^{p-1}[(k+1)^p - 1]$$

for each $k \in \mathbb{N}$. A more explicit expression of $f$ is

$$f(\xi) = \begin{cases} 0 & \text{if } \xi \in \mathbb{R} \setminus \bigcup_{k\in\mathbb{N}}[k!k, (k+1)!] \\[2mm] \dfrac{2Lh_k}{k!}\min\{\xi - k!k, (k+1)! - \xi\} & \text{if } \xi \in [k!k, (k+1)!] \ , \ k \in \mathbb{N} \end{cases}$$

By choosing, for each $k \in \mathbb{N}$,

$$a_k = k!$$

$$b_k = k!k$$

the hypotheses of Theorem 2.3 are satisfied and one has

$$\limsup_{\xi \to +\infty} \frac{F(\xi)}{\xi^p} = L.$$

In fact

$$\frac{F(a_k)}{a_k^p} = \frac{L}{2(k!)^p} \sum_{i=1}^{k-1} ((i+1)! - i!i)h_i$$

$$= \frac{L}{(k!)^p} \sum_{i=1}^{k-1} [((i+1)!)^p - (i!)^p)] = L\frac{(k!)^p - 1}{(k!)^p}.$$

On the other side, for each $\xi \in [b_k, b_{k+1}]$, one has

$$\frac{F(\xi)}{\xi^p} \le \frac{F(a_{k+1})}{b_k^p} = L\frac{((k+1)!)^p - 1}{(k!)^p k^p}.$$

## 3. Concluding remarks

Now we wish to recall some other results existing in literature concerning the existence of infinitely many solutions for the problem $(D_{n,p})$.

The following result comes directly from a recent theorem obtained by J. Saint-Raymond (Theorem 3.1 of [9]).

**Theorem 3.1.** *Let $f : \mathbb{R} \to \mathbb{R}$ be a continuous function and let $F : \mathbb{R} \to \mathbb{R}$ the function defined by setting*

$$F(\xi) = \int_0^\xi f(t)dt$$

*for each $\xi \in \mathbb{R}$. Assume that*

*(1) there exists $M > 0$ such that, for every $\rho > 0$, there exists $t > 0$ satisfying $F(t) \ge \rho(1 + t^2)$ and $F(s) \ge -MF(t)$ for each $s \in [0, t]$;*
*(2) $\sup \{t \in \mathbb{R} : f(t) < 0\} = +\infty$;*
*(3) $\inf \{t \in \mathbb{R} . f(t) > 0\} < 0$.*

*Then there are unboundedly (infinitely) many solutions of the problem $(D_{1,2})$.*

We wish to emphasize that Theorem 3.1 cannot be applied to the function of the Example 2.2. In fact the hypotheses (1) and (3) of Theorem 3.1 are surely not satisfied; if we consider the function $F$ of Example 2.2 it is easy to observe that $\inf\{t \in \mathbb{R} : f(t) > 0\} = 0$. Moreover, for each $t > 0$, one has

$$F(t) \le a(1 + b)t^2 < 3at^2 < 3\sigma(1 + t^2).$$

This means that, in particular, hypothesis (2) cannot be satisfied when $\rho \geq 3\sigma$.

Other recent results in which infinitely many solutions of the problem $(D_{n,p})$ are assured is contained in [6] and [7].

In [6] Omari and Zanolin obtain the following result

**Theorem 3.2.** (Corollary 1.2 of [6]) *Assume that*

$$\liminf_{\xi \to +\infty} \frac{F(\xi)}{\xi^p} = 0 \quad \text{and} \quad \limsup_{\xi \to +\infty} \frac{F(\xi)}{\xi^p} = +\infty \tag{2}$$

*then problem $(D_{n,p})$ has a sequence $\{u_n\}_{n \in \mathbb{N}}$ of positive solutions in $W_0^{1,p}(\Omega)$ with $\max_{\overline{\Omega}} u_n \to +\infty$.*

In [7] the same authors replace the conditions (2) at $+\infty$ by similar ones at 0, in order to produce arbitrarily small positive solutions of problem $(D_{n,p})$. Namely, the following holds

**Theorem 3.3.** *Assume that*

$$\liminf_{\xi \to 0^+} \frac{F(\xi)}{\xi^p} = 0 \quad \text{and} \quad \limsup_{\xi \to 0^+} \frac{F(\xi)}{\xi^p} = +\infty \tag{3}$$

*then problem $(D_{n,p})$ has a sequence $\{u_n\}_{n \in \mathbb{N}}$ of positive solutions in $W_0^{1,p}(\Omega)$ with $\max_{\overline{\Omega}} u_n$ decreasing to zero and $\frac{1}{p} \int_\Omega |\nabla u_n(x)|^p dx - \int_\Omega F(u_n(x))dx$ increasing to zero.*

We note that in these results it is requested that $\limsup_{\xi \to +\infty} \frac{F(\xi)}{\xi^p} = +\infty$ and $\limsup_{\xi \to 0^+} \frac{F(\xi)}{\xi^p} = +\infty$; these are stronger requests with respect of hypothesis (iii) of Theorem 2.3 and (jjj) of Theorem 2.4. Moreover, in our results, nothing is said about the behaviour of $\liminf_{\xi \to +\infty} \frac{F(\xi)}{\xi^p}$ and $\liminf_{\xi \to 0^+} \frac{F(\xi)}{\xi^p}$. In fact, as we have already observed, the function $F$ of Example 2.1 of [1] doesn't satisfy any of the (2) and the function $F$ of Example 2.1 doesn't satisfy $\limsup_{\xi \to +\infty} \frac{F(\xi)}{\xi^3} = +\infty$.

## Bibliography

1. F. Cammaroto and A. Chinnì, *Infinitely many solutions for a two points boundary value problem. Far East Journal of Mathematical Sciences*, 11(1):41–51, 2003.
2. F. Cammaroto, A. Chinnì and Di Bella B, Infinitely many solutions for the Dirichlet problem via a variational principle of Ricceri. In F. Giannessi and A. Maugeri, editors, *Variational Analysis and Applications*, pages 215–229. Springer, 2005.
3. F. Cammaroto, A. Chinnì and Di Bella B, *Infinitely many solutions for the Dirichlet problem involving the p-Laplacian. Nonlinear Analysis TMA*, 61:41–49, 2005.

4. P. Korman and Y. Li, *Infinitely many solutions at a resonance. Nonlinear Differential Equations, Electron. J. Diff. Eqns.*, Conf. **05**:105–111, 2000.

5. G.B. Li and H.S. Zhou, *Multiple solutions to p-Laplacian problems with asymptotic nonlinearity as $u^{p-1}$ at infinity. J. London Math. Soc.*, **65**, n. 2:123–138, 2002.

6. P. Omari and F. Zanolin, *Infinitely many solutions of a quasilinear elliptic problem with an oscillatory potential. Commun. in Partial Differential Equations*, **21**:721–733, 1996.

7. P. Omari and F. Zanolin, *An elliptic problem with arbitrarily small positive solutions. Nonlinear Differential Equations, Electron. J. Diff. Eqns.*, Conf. **05**:301–308, 2000.

8. B. Ricceri *A general variational principle and some of its applications. J. Comput. Appl. Math.*, **113**:401–410, 2000.

9. J. Saint Raymond, *On the multiplicity of the solutions of the equation $-\Delta u = \lambda f(u)$. J. Differential Equations*, **180**:65–88, 2002.

# INFINITELY MANY SOLUTIONS TO DIRICHLET AND NEUMANN PROBLEMS FOR QUASILINEAR ELLIPTIC SYSTEMS

ANTONIO GIUSEPPE DI FALCO

*Department of Mathematics & Computer Science*
*University of Catania*
*Viale A. Doria, 6*
*95125 Catania, Italy*
*E-mail address: difalco@dmi.unict.it*

Let $\Omega \subset \mathbb{R}^N$ be a bounded open set. We deal with the existence of weak solutions for the following Neumann problem

$$\begin{cases} -\Delta_p u + \lambda(x)|u|^{p-2}u = \alpha(x)f(u,v) \text{ in } \Omega \\ -\Delta_q v + \mu(x)|v|^{q-2}v = \alpha(x)g(u,v) \text{ in } \Omega \\ \frac{\partial u}{\partial \nu} = 0 \text{ on } \partial\Omega \\ \frac{\partial v}{\partial \nu} = 0 \text{ on } \partial\Omega \end{cases}$$

where $\nu$ is the outward unit normal to the boundary $\partial\Omega$ of $\Omega$ We deal with the existence of weak solutions for the following Dirichlet problem

$$\begin{cases} -\Delta_p u = f(u,v) \text{ in } \Omega \\ -\Delta_q v = g(u,v) \text{ in } \Omega \\ u = 0 \text{ on } \partial\Omega \\ v = 0 \text{ on } \partial\Omega \end{cases}$$

The existence of solutions is proved by applying the critical point theorem obtained by B. Ricceri (see [2]).

**Key words:** Gradient system, Dirichlet problem, Neumann problem
**Mathematics Subject Classification:** 35J50, 35J55

## 1. Neumann Problem

Here and in the sequel:

$\Omega \subset \mathbb{R}^N$ is a bounded open set with boundary of class $C^1$;

$N \geq 1$; $p > N$; $q > N$;

$\lambda, \mu \in L^\infty(\Omega)$, such that $\text{essinf}_\Omega \lambda > 0$, $\text{essinf}_\Omega \mu > 0$;

$\alpha \in C^0(\overline{\Omega})$ nonnegative;

$f, g \in C^0(\mathbb{R}^2)$ such that the differential form $f(u,v)du + g(u,v)dv$ be exact.

We are interested in the following problem:

$$(P) \qquad \begin{cases} -\Delta_p u + \lambda(x)|u|^{p-2}u = \alpha(x)f(u,v) \text{ in } \Omega \\ -\Delta_q v + \mu(x)|v|^{q-2}v = \alpha(x)g(u,v) \text{ in } \Omega \\ \frac{\partial u}{\partial \nu} = 0 \text{ on } \partial\Omega \\ \frac{\partial v}{\partial \nu} = 0 \text{ on } \partial\Omega \end{cases}$$

More precisely we are interested in the existence of infinitely many weak solutions to such a problem (see [1]).

Let $G : \mathbb{R}^2 \to \mathbb{R}$ be the differentiable function such that $G_u(u,v) = f(u,v)$, $G_v(u,v) = g(u,v)$, $G(0,0) = 0$ and let $F(x,u,v) = \alpha(x)G(u,v)$,

We first consider the space $W^{1,p}(\Omega)$ with the norm

$$\|u\|_\lambda = \left( \int_\Omega \lambda(x)|u(x)|^p dx + \int_\Omega |\nabla u(x)|^p dx \right)^{\frac{1}{p}}$$

and the space $W^{1,q}(\Omega)$ with the norm

$$\|v\|_\mu = \left( \int_\Omega \mu(x)|v(x)|^q dx + \int_\Omega |\nabla v(x)|^q dx \right)^{\frac{1}{q}}.$$

The norm $\|u\|_\lambda$ is equivalent to the usual one in $W^{1,p}(\Omega)$, defined by

$$\|u\|_{1,p} = \left( \int_\Omega |\nabla u(x)|^p dx + \int_\Omega |u(x)|^p dx \right)^{\frac{1}{p}}.$$

Therefore, $W^{1,p}(\Omega)$ with such norm is compactly embedded in $C^0(\overline{\Omega})$.

Since by hypotheses $p > N$ and $q > N$, $W^{1,p}(\Omega)$ and $W^{1,q}(\Omega)$ are both compactly embedded in $C^0(\overline{\Omega})$. So, if we put

$$c_1 = c(\lambda) = \sup_{u \in W^{1,p}(\Omega) \setminus \{0\}} \frac{\sup_{x \in \Omega} |u(x)|}{\|u\|_\lambda}$$

and

$$c_2 = c(\mu) = \sup_{u \in W^{1,q}(\Omega) \setminus \{0\}} \frac{\sup_{x \in \Omega} |u(x)|}{\|u\|_\mu}$$

then both $c_1$ and $c_2$ are finite. Then we take $X = W^{1,p}(\Omega) \times W^{1,q}(\Omega)$ with the norm $\|(u,v)\|_X = \sqrt{\|u\|_\lambda^2 + \|v\|_\mu^2}$ and $Y = C^0(\overline{\Omega}) \times C^0(\overline{\Omega})$ with the norm $\|(u,v)\|_Y = \sqrt{\|u\|_{C^0(\overline{\Omega})}^2 + \|v\|_{C^0(\overline{\Omega})}^2}$. Of course the space $X$ is compactly embedded in $Y$ and if we put

$$c = \sup_{(u,v) \in X \setminus \{(0,0)\}} \frac{\|(u,v)\|_Y}{\|(u,v)\|_X}$$

we have $c = \max\{c_1, c_2\}$. In order to apply Ricceri's theorem we set

$$\Psi(u,v) = \frac{1}{p}\|u\|_\lambda^p + \frac{1}{q}\|v\|_\mu^q$$

and

$$\Phi(u,v) = -\int_\Omega F(x, u(x), v(x))dx$$

for all $(u,v) \in X$. Since $X$ is compactly embedded in $Y$, the constant $c$ is finite. Moreover the functionals $\Phi$ and $\Psi$ fit the hypotheses of Ricceri's theorem. The critical points of $\Phi + \Psi$ are precisely the weak solutions to Problem (P).

The sets $A(r)$, $B(r)$, $r > 0$, below specified, play an important role in our exposition:

$$A(r)$$

$$= \left\{ (\xi, \eta) \in \mathbb{R}^2 \text{ such that } \frac{1}{pc_1^p} |\xi|^p + \frac{1}{qc_2^q} |\eta|^q \le r \right\}$$

$$B(r) = \left\{ (\xi, \eta) \in \mathbb{R}^2 \text{ such that } \frac{\int_\Omega \lambda(x)dx}{p} |\xi|^p \right.$$

$$\left. + \frac{\int_\Omega \mu(x)dx}{q} |\eta|^q \le r \right\}.$$

The following inclusion holds:

$$B(r) \subseteq A(r).$$

## 1.1. *Results*

**Theorem 1.1.** *Assume that there are $r > 0$ and $\xi_0 \in \mathbb{R}$, $\eta_0 \in \mathbb{R}$ such that*

$$\frac{1}{p} |\xi_0|^p \int_\Omega \lambda(x)dx + \frac{1}{q} |\eta_0|^q \int_\Omega \mu(x)dx < r$$

*and*

$$\max_{A(r)} G(\xi, \eta) = G(\xi_0, \eta_0).$$

*Then Problem (P) admits a weak solution $(u, v)$ satisfying $\Psi(u, v) < r$*

**Theorem 1.2.** *Assume that there are sequences $\{r_n\}$ in $\mathbb{R}^+$ with $\lim_{n \to \infty} r_n = +\infty$, and $\{\xi_n\}$, $\{\eta_n\}$ in $\mathbb{R}$ such that for all $n \in \mathbb{N}$, one has*

$$\frac{1}{p} |\xi_n|^p \int_\Omega \lambda(x)dx + \frac{1}{q} |\eta_n|^q \int_\Omega \mu(x)dx < r_n$$

*and*

$$\max_{(\xi, \eta) \in A(r_n)} G(\xi, \eta) = G(\xi_n, \eta_n).$$

*Finally assume that*

$$\limsup_{(\xi, \eta) \to \infty} \frac{G(\xi, \eta) \int_\Omega \alpha(x)dx}{|\xi|^p \int_\Omega \lambda(x)dx + |\eta|^q \int_\Omega \mu(x)dx}$$

$$> \max \left( \frac{1}{p}, \frac{1}{q} \right).$$

*Then, Problem (P) admits an unbounded sequence of weak solutions in $X$.*

**Theorem 1.3.** *Assume that there are sequences $\{r_n\}$ in $\mathbb{R}^+$ with $\lim_{n\to\infty} r_n = 0$, and $\{\xi_n\}$, $\{\eta_n\}$ in $\mathbb{R}$ such that for all $n \in \mathbb{N}$, one has*

$$\frac{1}{p}|\xi_n|^p \int_\Omega \lambda(x)dx + \frac{1}{q}|\eta_n|^q \int_\Omega \mu(x)dx < r_n$$

*and*

$$\max_{(\xi,\eta)\in A(r_n)} G(\xi,\eta) = G(\xi_n,\eta_n).$$

*Finally assume that*

$$\limsup_{(\xi,\eta)\to(0,0)} \frac{G(\xi,\eta)\int_\Omega \alpha(x)dx}{|\xi|^p \int_\Omega \lambda(x)dx + |\eta|^q \int_\Omega \mu(x)dx}$$

$$> \max\left(\frac{1}{p},\frac{1}{q}\right).$$

*Then, Problem (P) admits a sequence of non-zero weak solutions which strongly converges to $\theta_X$ in $X$.*

## 1.2. Examples

Here is an example of application of theorem 1.2

**Example 1.1.** Let $N = 1$, $p = q = 2$, $\lambda \equiv 1$, $\mu \equiv 1$, $\Omega = ]0,1[$, $f(u,v) = G_u(u,v)$ and $g(u,v) = G_v(u,v)$ where $G : \mathbb{R}^2 \to \mathbb{R}$ is the function defined by setting

$$G(u,v) = \frac{1}{2}\left[(u^2 + v^2)\sin\log(u^2 + v^2 + 1)\right]$$

Then for each $\alpha \in C^0(\overline{\Omega})$ with $\alpha(t) \geq 0$ in $\Omega$ and $\int_0^1 \alpha(t)dt > 1 = m(\Omega)$, the following problem

$$\begin{cases} -u'' + u = \alpha(t)f(u,v) \\ -v'' + v = \alpha(t)g(u,v) \\ u'(0) = u'(1) = 0 \\ v'(0) = v'(1) = 0 \end{cases}$$

admits an unbounded sequence of weak solutions in $X = H^1(\Omega) \times H^1(\Omega)$.

Here is an example of application of theorem 1.3 :

**Example 1.2.** Let $N = 1$, $p = q = 2$, $\lambda \equiv 1$, $\mu \equiv 1$, $\Omega = ]0,1[$, $f(u,v) = G_u(u,v)$ and $g(u,v) = G_v(u,v)$ where $G : \mathbb{R}^2 \to \mathbb{R}$ is the function defined by setting

$$G(\xi,\eta)$$

$$= \begin{cases} \frac{1}{2}(\xi^2 + \eta^2)\cos\log\frac{1}{\xi^2+\eta^2} & \text{if } (\xi,\eta) \neq (0,0) \\ 0 & \text{if } (\xi,\eta) = (0,0) \end{cases}$$

Then for each $\alpha \in C^0(\overline{\Omega})$ with $\alpha(t) \geq 0$ in $\Omega$ and $\int_0^1 \alpha(t)dt > 1 = m(\Omega)$, the following problem

$$\begin{cases} -u'' + u = \alpha(t)f(u, v) \\ -v'' + v = \alpha(t)g(u, v) \\ u'(0) = u'(1) = 0 \\ v'(0) = v'(1) = 0 \end{cases}$$

admits a sequence of nonzero weak solutions which strongly converges to $\theta_X$ in $X = H^1(\Omega) \times H^1(\Omega)$.

## 2. Dirichlet Problem

Here and in the sequel:
$\Omega \subset \mathbb{R}^N$ is a bounded open set with boundary of class $C^1$;
$N \geq 1; p > N; q > N$;
$f, g \in C^0(\mathbb{R}^2)$ such that the differential form $f(u, v)du + g(u, v)dv$ be exact.

We are interested in the following problem:

$(P)$
$$\begin{cases} -\Delta_p u = f(u, v) \text{ in } \Omega \\ -\Delta_q v = g(u, v) \text{ in } \Omega \\ u = 0 \text{ on } \partial\Omega \\ v = 0 \text{ on } \partial\Omega \end{cases}$$

More precisely we are interested in the existence of infinitely many weak solutions to such a problem [3].

Let $G : \mathbb{R}^2 \to \mathbb{R}$ be the differentiable function such that $G_u(u, v) = f(u, v)$, $G_v(u, v) = g(u, v)$, $G(0, 0) = 0$.

We first consider the space $W_0^{1,p}(\Omega)$ with the norm

$$\|u\|_{W_0^{1,p}(\Omega)} = \left( \int_\Omega |\nabla u(x)|^p dx \right)^{\frac{1}{p}}$$

and the space $W_0^{1,q}(\Omega)$ with the norm

$$\|v\|_{W_0^{1,q}(\Omega)} = \left( \int_\Omega |\nabla v(x)|^q dx \right)^{\frac{1}{q}}.$$

Since by hypotheses $p > N$ and $q > N$, $W^{1,p}(\Omega)$ and $W^{1,q}(\Omega)$ are both compactly embedded in $C^0(\overline{\Omega})$. Then we put

$$c_1 = \sup_{u \in W^{1,p}(\Omega)\backslash\{0\}} \frac{\sup_{x \in \Omega} |u(x)|}{\|u\|}$$

that is finite since $W^{1,p}(\Omega)$ is compactly embedded in $C^0(\overline{\Omega})$ and

$$c_2 = \sup_{u \in W^{1,q}(\Omega)\backslash\{0\}} \frac{\sup_{x \in \Omega} |u(x)|}{\|u\|}$$

that is finite since $W^{1,q}(\Omega)$ is compactly embedded in $C^0(\overline{\Omega})$. In order to apply the former theorem we set

$$\Psi(u,v) = \frac{1}{p}\|u\|^p + \frac{1}{q}\|v\|^q$$

and

$$\Phi(u,v) = -\int_\Omega G(u(x),v(x))dx$$

for all $(u,v) \in X$. The critical points of $\Phi + \Psi$ are precisely the weak solutions to Problem (P).

If the following definitions are used

$$\alpha = \frac{1}{pc_1^p}$$

$$\beta = \frac{1}{qc_2^q}$$

and for each $r > 0$

$$A(r) = \{(\xi,\eta) \in \mathbb{R}^2 \text{ such that } \alpha|\xi|^p + \beta|\eta|^q \le r\}$$

$$S(r) = \{(\xi,\eta) \in \mathbb{R}^2 \text{ such that } |\xi|^p + |\eta|^q \le r\}$$

then

$$S\left(\frac{r}{\max(\alpha,\beta)}\right) \subseteq A(r) \subseteq S\left(\frac{r}{\min(\alpha,\beta)}\right)$$

Moreover we put $\omega := \frac{\pi^{n/2}}{\frac{n}{2}\Gamma(\frac{n}{2})}$ the measure of the $n$-dimensional unit ball.

## 2.1. Results

We wish to establish two multiplicity results for Problem (P). Let $D = \sup_{x\in\Omega} d(x,\partial\Omega)$.

**Theorem 2.1.** *Assume that* $\inf_{\mathbb{R}^2} G \ge 0$. *Moreover, suppose that there exist two real sequences* $\{a_n\}$ *and* $\{b_n\}$ *in* $]0,+\infty[$ *with* $a_n < b_n$, $\lim_{n\to\infty} b_n = +\infty$, *such that*

$$\lim_{n\to+\infty} \frac{b_n}{a_n} = +\infty$$

$$\max_{S(a_n)} G = \max_{S(b_n)} G > 0$$

$$\max\left\{\frac{2^p(2^N-1)}{pD^p}, \frac{2^q(2^N-1)}{qD^q}\right\}$$

$$< \limsup_{(\xi,\eta)\to\infty} \frac{G(\xi,\eta)}{|\xi|^p + |\eta|^q} < +\infty$$

*Then Problem (P) admits an unbounded sequence of weak solutions.*

**Theorem 2.2.** *Assume that* $\inf_{\mathbb{R}^2} G \geq 0$. *Moreover, suppose that there exist two real sequences* $\{a_n\}$ *and* $\{b_n\}$ *in* $]0, +\infty[$ *with* $a_n < b_n$, $\lim_{n\to\infty} b_n = 0$, *such that*

$$\lim_{n\to+\infty} \frac{b_n}{a_n} = +\infty$$

$$\max_{S(a_n)} G = \max_{S(b_n)} G > 0$$

$$\max \left\{ \frac{2^p(2^N - 1)}{pD^p}, \frac{2^q(2^N - 1)}{qD^q} \right\}$$

$$< \limsup_{(\xi,\eta)\to(0,0)} \frac{G(\xi,\eta)}{|\xi|^p + |\eta|^q} < +\infty$$

*where* $D = \sup_{x\in\Omega} d(x, \partial\Omega)$. *Then Problem* $(P)$ *admits a sequence of non-zero weak solutions which strongly converges to* $\theta_X$ *in* $X$.

## 2.2. Examples

Let $A$ be a positive number such that

$$A > \max \left\{ \frac{2^p(2^N - 1)}{pD^p}, \frac{2^q(2^N - 1)}{qD^q} \right\}$$

Let $b_0 = 0$. The sequences $\{a_n\}_{n\in\mathbb{N}}$ and $\{b_n\}_{n\in\mathbb{N}}$ with $a_n = (n + 1)!$ and $b_n = (n + 1)(n + 1)!$ satisfy the hypothesis of theorem 1.1 and besides $b_{n-1} < a_n$ for all $n \in \mathbb{N}$.

Moreover the sequences $\{a_n\}_{n\in\mathbb{N}}$ and $\{b_n\}_{n\in\mathbb{N}}$ with $a_n = \frac{1}{(n+1)(n+1)!}$ and $b_n = \frac{1}{(n+1)!}$ satisfy the hipothesis of theorem 1.2 and besides $b_{n+1} < a_n$ for all $n \in \mathbb{N}$. Here is an example of application of theorem 2.1:

**Example 2.1.** Let $b_0 = 0$ and let $\{a_n\}$ and $\{b_n\}$ be two sequences satisfying the hypothesis of theorem 1.1 and such that $b_{n-1} < a_n$ for all $n \in \mathbb{N}$. Let us consider the countable family of pairwise disjoint closed bounded intervals

$$\{[b_{n-1}, a_n]\}_{n\in\mathbb{N}}$$

Then for each $n \in \mathbb{N}$ the function

$$t \to \frac{2\pi t - (a_n + b_{n-1})\pi}{a_n - b_{n-1}}$$

is an homeomorphism between the interval $[b_{n-1}, a_n]$ and the interval $[-\pi, \pi]$. For each $n \in \mathbb{N}$ the function

$$f_n(t) = \frac{1}{2} \left\{ 1 + \cos\left( \frac{2\pi t - (a_n + b_{n-1})\pi}{a_n - b_{n-1}} \right) \right\}$$

satisfies

- $f_n \in C^1([b_{n-1}, a_n])$

- $0 \leq f_n(t) \leq 1$
- $f_n(b_{n-1}) = f_n(a_n) = 0$
- $f_n'(b_{n-1}) = f_n'(a_n) = 0$

For each $n \in \mathbb{N}$ let

$$\alpha_n(t) = \begin{cases} f_n(t) & \text{if } t \in [b_{n-1}, a_n] \\ 0 & \text{if } t \notin [b_{n-1}, a_n] \end{cases}$$

and let

$$\alpha(t) = \sum_{n=1}^{\infty} \alpha_n(t) \text{ for each } t \in \mathbb{R}$$

Let $t_0 \in \mathbb{R}$. If there exist $n \in \mathbb{N}$ such that $t_0 \in [b_{n-1}, a_n]$ then $\alpha(t_0) = \alpha_n(t_0) = f_n(t_0)$. Else, if $t_0 \in \mathbb{R} \setminus \cup_{n=1}^{\infty}[b_{n-1}, a_n]$ then $\alpha_n(t_0) = 0$ for each $n \in \mathbb{N}$ so $\alpha(t_0) = 0$. The function $G : \mathbb{R}^2 \to \mathbb{R}$

$$G(\xi, \eta) = A(|\xi|^p + |\eta|^q)\alpha(|\xi|^p + |\eta|^q)$$

satisfies the hipothesis of theorem 1.1. Infact $G(\xi, \eta) \geq 0$ for each $(\xi, \eta) \in \mathbb{R}^2$ and $G(0,0) = 0$. Moreover $A > 0$, $|\xi|^p + |\eta|^q \geq 0$ and $\alpha(t) \geq 0$ for each $t \in \mathbb{R}$ therefore

$$\inf_{\mathbb{R}^2} G \geq 0$$

From $a_n \leq t \leq b_n$ it follows that $\alpha(t) = 0$, so if $a_n \leq |\xi|^p + |\eta|^q \leq b_n$ then $G(\xi, \eta) = 0$, whence

$$\max_{S(a_n)} G = \max_{S(b_n)} G$$

Finally

$$\limsup_{(\xi, \eta) \to \infty} \frac{G(\xi, \eta)}{|\xi|^p + |\eta|^q} = A$$

Here is an example of application of theorem 2.2:

**Example 2.2.** Let $\{a_n\}$ and $\{b_n\}$ be two sequences satisfying the hypothesis of theorem 1.2 and such that $b_{n+1} < a_n$ for all $n \in \mathbb{N}$. Let us consider the countable family of pairwise disjoint closed bounded intervals

$$\{[b_{n+1}, a_n]\}_{n \in \mathbb{N}}$$

Then for each $n \in \mathbb{N}$ the function

$$t \to \frac{2\pi t - (a_n + b_{n+1})\pi}{a_n - b_{n+1}}$$

is an homeomorphism between the interval $[b_{n+1}, a_n]$ and the interval $[-\pi, \pi]$. For each $n \in \mathbb{N}$ the function

$$f_n(t) = \frac{1}{2}\left\{1 + \cos\left(\frac{2\pi t - (a_n + b_{n+1})\pi}{a_n - b_{n+1}}\right)\right\}$$

satisfies

841

- $f_n \in C^1([b_{n+1}, a_n])$
- $0 \le f_n(t) \le 1$
- $f_n(b_{n+1}) = f_n(a_n) = 0$
- $f'_n(b_{n+1}) = f'_n(a_n) = 0$

For each $n \in \mathbb{N}$ let

$$\alpha_n(t) = \begin{cases} f_n(t) & \text{if } t \in [b_{n+1}, a_n] \\ 0 & \text{if } t \notin [b_{n+1}, a_n] \end{cases}$$

and let

$$\alpha(t) = \sum_{n=1}^{\infty} \alpha_n(t) \text{ for each } t \in \mathbb{R}$$

Let $t_0 \in \mathbb{R}$. If there exist $n \in \mathbb{N}$ such that $t_0 \in [b_{n+1}, a_n]$ then $\alpha(t_0) = \alpha_n(t_0) = f_n(t_0)$. Else, if $t_0 \in \mathbb{R} \setminus \cup_{n=1}^{\infty}[b_{n-1}, a_n]$ then $\alpha_n(t_0) = 0$ for each $n \in \mathbb{N}$ so $\alpha(t_0) = 0$.

It is easy to see that the function $G : \mathbb{R}^2 \to \mathbb{R}$

$$G(\xi, \eta) = A(|\xi|^p + |\eta|^q)\alpha(|\xi|^p + |\eta|^q)$$

satisfies the hypothesis of theorem 1.2.

## References

1. A. G. Di Falco, Infinitely many solutions to the Neumann problem for quasilinear elliptic systems. *Le Matematiche* **58**, 117–130 (2003).
2. B. Ricceri, A general variational principle and some of its applications. *J. Comput. Appl. Math.* **113**, 401–410 (2000).
3. A. G. DI FALCO, Infinitely many solutions to the Dirichlet problem for quasilinear elliptic systems, *Le Matematiche* **60** (2005), 163-179.
4. A. G. DI FALCO, Esistenza di infinite soluzioni per i problemi di Dirichlet e di Neumann relativi a sistemi ellittici quasilineari, *Ph. D. Thesis*, Catania, 2004.

# MULTIPLICITY RESULTS FOR A NEUMANN-TYPE PROBLEM INVOLVING THE P-LAPLACIAN

F. CAMMAROTO\*, A.CHINNÌ , B. DI BELLA[†]

*Department of Mathematics, University of Messina*
*98166 Sant'Agata-Messina, Italy*
*E-mail: filippo@dipmat.unime.it*

The aim of this paper is to provide suitable conditions under which the following Neumann problem:

$$\begin{cases} -\Delta_p u + \alpha(x)|u|^{p-2}u = \alpha(x)f(u) + \lambda g(x,u) & \text{in } \Omega \\ \dfrac{\partial u}{\partial \nu} = 0 & \text{on } \partial\Omega \end{cases}$$

has at least two or even three solutions, for each $\lambda$ nonnegative, small enough.

**Key words:** Neumann problem, p-Laplacian, multiple solutions
**Mathematics Subject Classification:** 35J65

## 1. Introduction

The existence of multiple solutions for this kind of problem has been widely investigated. The results that will be presented here differ from those obtained by various authors because the key assumption, which regard the nonlinearity, is based on a recent and abstract result of Ricceri, that is the following:

**Theorem 1.1.** *(Ref. [3], Theorem 10). Let $X$ be a uniformly convex and separable real Banach space, $p > 1$, $J, \varphi : X \to \mathbb{R}$ two sequentially weakly lower semicontinuous functionals, $J$ also continuous. For every $x \in X$, put*

$$\psi(x) = \frac{1}{p}||x||^p + J(x).$$

*Assume that the functional $\psi$ is coercive and has a strict, not global, local minimum, say $x_0$.*
*Then, for every $r > \psi(x_0)$, there exists $\lambda_r^* > 0$ such that, for each $\lambda \in ]0, \lambda_r^*[$, the functional $\psi + \lambda\varphi$ has at least two local minima lying in $\psi^{-1}(] - \infty, r[)$.*

---

\*Because of a surprising coincidence of names within the same Department, we have to point out that the author was born on August 4, 1968.
[†]Corresponding author.

The aim of this paper is to obtain an application of Theorem 1.1 to the following Neumann problem:

$$
\begin{cases}
-\Delta_p u + \alpha(x)|u|^{p-2}u = \alpha(x)f(u) + \lambda g(x,u) & \text{in } \Omega \\
\dfrac{\partial u}{\partial \nu} = 0 & \text{on } \partial\Omega
\end{cases}
\tag{1}
$$

where $\Delta_p u = \operatorname{div}(|\nabla u|^{p-2}\nabla u)$, $\nu$ is the outward unit normal to $\partial\Omega$ and $\lambda \in ]0,+\infty[$. In the sequel, $\Omega \subset \mathbb{R}^n (n \geq 1)$ is a bounded open set with boundary of class $C^1$, $\alpha \in L^\infty(\Omega)$ with $\operatorname{ess\,inf}_\Omega \alpha > 0$, $p > n$, $f : \mathbb{R} \to \mathbb{R}$ a continuous function and $g : \Omega \times \mathbb{R} \to \mathbb{R}$ a Carathéodory function such that, for any $\rho > 0$

$$
\Omega \ni x \to \sup_{|\xi| \leq \rho} |g(x,\xi)| \in L^1(\Omega)
\tag{2}
$$

In the Banach space $W^{1,p}(\Omega)$ we introduce the norm

$$
\|u\| = \left( \int_\Omega |\nabla u(x)|^p dx + \int_\Omega \alpha(x)|u(x)|^p dx \right)^{\frac{1}{p}}
$$

which is equivalent to the standard one. It is known that $W^{1,p}(\Omega)$ is a separable, reflexive, uniformly convex Banach space. Since $p > n$, $W^{1,p}(\Omega)$ is compactly embedded in $C^0(\bar{\Omega})$ and hence there exists $c > 0$ such that

$$
\|u\|_{C^0(\bar{\Omega})} \leq c\|u\|
$$

for all $u \in W^{1,p}(\Omega)$. Moreover, we denote by $\|\cdot\|_q$ $(q \geq 1)$ the usual norm on the space $L^q(\Omega)$.

Let us define in $W^{1,p}(\Omega)$ the two functionals $\Psi$, $\Phi$ by setting, for each $u \in W^{1,p}(\Omega)$,

$$
\Psi(u) = \frac{1}{p}\|u\|^p - \int_\Omega \alpha(x)F(u(x))dx
$$

and

$$
\Phi(u) = -\int_\Omega G(x,u(x))dx
$$

where $F(\xi) = \int_0^\xi f(t)dt$ and $G(x,\xi) = \int_0^\xi g(x,t)dt$. Let us recall that a weak solution of (1) is any $u \in W^{1,p}(\Omega)$ such that

$$
\int_\Omega |\nabla u(x)|^{p-2}\nabla u(x)\nabla v(x)dx + \int_\Omega \alpha(x)|u(x)|^{p-2}u(x)v(x)dx
$$

$$
- \int_\Omega \alpha(x)f(u(x))v(x)dx - \int_\Omega \lambda g(x,u(x))v(x)dx = 0
$$

for all $v \in W^{1,p}(\Omega)$.

Then, the weak solutions of (1) are precisely the critical points of the energy functional $\Psi + \lambda\Phi$.

## 2. Results

Our first result is the following.

**Theorem 2.1.** *Assume (2) and*

$(i_1)$ $\displaystyle \limsup_{|\xi| \to +\infty} \frac{F(\xi)}{|\xi|^p} < \frac{1}{p}$;

$(i_2)$ *the function $h : \mathbb{R} \to \mathbb{R}$ defined by putting*

$$h(\xi) = \frac{1}{p}|\xi|^p - F(\xi) \quad \forall \xi \in \mathbb{R}$$

*has a strict not global local minimum, say $\xi_0$.*

*Then, for each $r > \Psi(\xi_0)$, there exists $\lambda_r^* > 0$ such that, for each $\lambda \in ]0, \lambda_r^*[$, the problem (1) has at least two weak solutions in $\Psi^{-1}(] - \infty, r[)$.*

*Proof.* We apply Theorem 1.1 taking $X = W^{1,p}(\Omega)$, $J(u) = -\int_\Omega \alpha(x)F(u(x))dx$ and $\psi = \Psi$, $\varphi = \Phi$ defined as above. Let us show that these functionals satisfy the required conditions. First of all, by classical results, the functional $J$ is well defined and sequentially weakly continuous. In such a way, the functional $\Psi(u) = \frac{1}{p}\|u\|^p + J(u)$ is well defined, sequentially weakly lower semicontinuous and Gâteaux differentiable in $W^{1,p}(\Omega)$. We claim that $\Psi$ is coercive. Since $(i_1)$ holds, there exist two constants $a, b \in \mathbb{R}$ such that $0 < a < \frac{1}{p}$ and $F(\xi) \le a|\xi|^p + b$, for all $\xi \in \mathbb{R}$.
Fix $u \in X$. We have

$$\int_\Omega \alpha(x)F(u(x))dx \le a \int_\Omega \alpha(x)|u(x)|^p dx + b\|\alpha\|_1$$

$$\le a\|u\|^p + b\|\alpha\|_1.$$

So,

$$\Psi(u) \ge \left(\frac{1}{p} - a\right)\|u\|^p - b\|\alpha\|_1$$

Consequently, $\lim_{\|u\| \to +\infty} \Psi(u) = +\infty$, as claimed.
Now, we prove that the functional $\Psi$ has strict, not global, local minimum in $X$.
By $(i_2)$ there exists $\delta > 0$ such that

$$h(\xi_0) < h(\xi)$$

for any $\xi$ with $|\xi - \xi_0| < \delta$, and there exists $\xi_1 \in \mathbb{R}$ such that

$$h(\xi_1) < h(\xi_0).$$

Put $w_0(x) = \xi_0$. For each $u \in X$ such that $u \ne w_0$ and $\|u - w_0\| < \frac{\delta}{c}$ we deduce

$$\|u - w_0\|_{C^0(\overline{\Omega})} < \delta.$$

By continuity of $u$, the set

$$\Omega_0 = \{x \in \Omega : u(x) \neq \xi_0\}$$

has positive measure. Then, we get

$$\int_{\Omega_0} \alpha(x) \left( \frac{1}{p}|\xi_0|^p - F(\xi_0) \right) dx < \int_{\Omega_0} \alpha(x) \left( \frac{1}{p}|u(x)|^p - F(u(x)) \right) dx .$$

This implies that

$$\Psi(w_0) < \Psi(u) .$$

Therefore $w_0$ is a strict local minimum for $\Psi$. Moreover, since $h(\xi_1) < h(\xi_0)$ we have

$$\Psi(w_1) < \Psi(w_0) ,$$

where $w_1(x) = \xi_1$ for all $x \in \Omega$. This ensure that $w_0$ is a not global minumum for $\Psi$. Finally the functional $\Phi$ is well defined and sequentially weakly lower semicontinuous as it follows from condition (2).

Hence, all assumptions of Theorem 1.1 are satisfied. For every $r > \Psi(\xi_0)$, there exists $\lambda_r^* > 0$ such that, for each $\lambda \in ]0, \lambda_r^*[$, $\Psi + \lambda\Phi$ has at least two local minima in $\Psi^{-1}(]-\infty, r[)$. $\qquad\square$

An example concerning the previous result is the following.

**Example 2.1.** Let $2 < q < p$ and $a, b \in ]0, +\infty[$ such that $\dfrac{a}{q} - \dfrac{b}{2} > \dfrac{1}{p}$.

Then, the problem

$$\begin{cases} -\Delta_p u + \alpha(x)|u|^{p-2}u = \alpha(x)(a|u|^{q-2}u - bu) + \lambda g(x, u) & \text{in } \Omega \\ \dfrac{\partial u}{\partial \nu} = 0 & \text{on } \partial\Omega \end{cases}$$

has at least two weak solutions for each $\lambda > 0$ sufficiently small.

In this case

$$F(\xi) = \frac{a}{q}|\xi|^q - \frac{b}{2}\xi^2$$

and

$$\limsup_{|\xi| \to +\infty} \frac{F(\xi)}{|\xi|^p} = 0$$

The function

$$h(\xi) = \frac{1}{p}|\xi|^p - \frac{a}{q}|\xi|^q + \frac{b}{2}\xi^2$$

has in 0 a strict not global local minumum.

In the following results we require that the functional $\Psi + \lambda\Phi$ has the Palais-Smale property in order to obtain three solutions of (1) instead of two. We recall that a Gâteaux differentiable functional $S$ on a real Banach space $X$ is said to satisfy the Palais-Smale condition if each sequence $\{x_n\}$ in $X$ such that $\sup_{n\in\mathbb{N}} |S(x_n)| < +\infty$ and $\lim_{n\to+\infty} \|S'(x_n)\|_X = 0$ admits a strongly converging subsequence.

## Theorem 2.2.
*Let assume the same assumptions of Theorem 2.1 and suppose that*

(i$_3$)

$$\limsup_{|\xi|\to+\infty} \frac{\sup_{x\in\Omega} G(x,\xi)}{|\xi|^p} < +\infty .$$

*Then, for each $r > \Psi(\xi_0)$, there exists $\lambda_r^* > 0$ such that, for each $\lambda \in ]0, \lambda_r^*[$, the problem (1) has at least three weak solutions in $\Psi^{-1}(]-\infty, r[)$.*

*Proof.* Like in the proof of Theorem 2.1, we get that the functional $\Psi + \lambda\Phi$ has two distinct local minima.
Let us check the Palais-Smale condition for $\Psi + \lambda\Phi$.
By (i$_3$) there exist $k \in L^1(\Omega)$ and $\sigma > 0$ such that $G(x,\xi) < \sigma|\xi|^p + k(x)$ for all $\xi \in \mathbb{R}$, $x \in \Omega$. Therefore, for suitable constant $C$,

$$\Psi(u) + \lambda\Phi(u) \geq \frac{1}{p}\|u\|^p - a\|u\|^p - b\|\alpha\|_1 - \lambda\left(\frac{C\sigma}{\lambda_1}\|u\|^p + \|k\|_1\right)$$

for all $u \in X$, where

$$\lambda_1 = \inf_{u\in W^{1,p}(\Omega)\setminus\{0\}} \frac{\|u\|^p}{\|u\|_p^p} .$$

Therefore, for $\lambda$ small enough the functional $\Psi + \lambda\Phi$ is coercive.
By Proposition 1 of Ref.[1], the operator $\Psi'$ admits a continuous inverse on $X^*$ and it is easily seen that $\Phi'$ is compact. So, by Example 38.25 of Ref.[4] we deduce that $\Psi + \lambda\Phi$ has the Palais-Smale property.
Since the functional $\Psi + \lambda\Phi$ is $C^1$ in $X$, our conclusion follows by Corollary 1 of Ref.[2] which assures that there exists a third critical point of the functional $\Psi + \lambda\Phi$ which is a solution of the problem (1). □

**Example 2.2.** If we consider the problem of the Example 2.1 with, in addition, $g(x, u) = aq\dfrac{|u|^q}{u}$ , the corrispondenting integral function

$$G(x, \xi) = a|\xi|^q + \beta(x) ,$$

where $\beta$ is a bounded function in $\Omega$, satisfies condition (i$_3$).

In the following result, in order to get another result of three solutions for the Neumann problem, we require again that the functional $\Psi + \lambda\Phi$ has the Palais-Smale property, but it is not necessarily coercive.

**Theorem 2.3.** *In addition to the hypotheses of Theorem 2.1 assume that*

(j$_3$) *there exist $q > p$, $R > 0$ such that*

$$qF(\xi) \le f(\xi)\xi \quad \text{for each } |\xi| \ge R$$

$$\sup_{\xi \in \mathbb{R}}[qG(x, \xi) - g(x, \xi)\xi] \in L^1(\Omega).$$

*Then, for each $r > \Psi(\xi_0)$, there exists $\lambda_r^* > 0$ such that, for each $\lambda \in ]0, \lambda_r^*[$, the problem (1) has at least three weak solutions in $\Psi^{-1}(] - \infty, r[)$.*

*Proof.* Reasoning as in the proof of Theorem 2.2, it is enough to show that the functional $\Psi + \lambda\Phi$ satisfies the Palais-Smale condition. To this end, let $\{u_n\}$ be a sequence in $X$ such that

$$\sup_{n \in \mathbb{N}} |\Psi(u_n) + \lambda\Phi(u_n)| \le M \tag{3}$$

$$\lim_{n \to \infty} ||\Psi'(u_n) + \lambda\Phi'(u_n)||_{X^*} = 0 . \tag{4}$$

From (4) there exists $\nu \in \mathbb{N}$ such that

$$|\Psi'(u_n)(v) + \lambda\Phi'(u_n)(v)| \le ||v||$$

for all $v \in X$ and $n > \nu$. On the other hand, by (j$_3$), there exists $k \in L^1(\Omega)$ such that $qG(x, \xi) - g(x, \xi)\xi \le k(x)$ for each $\xi \in \mathbb{R}$, $x \in \Omega$. Consequently, for each $n > \nu$, we have

$$qM + ||u_n|| \ge q\left(\Psi(u_n) + \lambda\Phi(u_n)\right) - \Psi'(u_n)(u_n) - \lambda\Phi'(u_n)(u_n)$$

$$= \frac{q}{p}||u_n||^p - q\int_\Omega \alpha(x)F(u_n(x))\, dx - q\lambda \int_\Omega G(x, u_n(x))\, dx - ||u_n||^p$$

$$+ \int_\Omega \alpha(x)f(u_n(x))u_n(x)\, dx + \lambda \int_\Omega g(x, u_n(x))u_n(x)\, dx$$

$$= \left(\frac{q}{p} - 1\right) ||u_n||^p - \int_\Omega \alpha(x) \left[qF(u_n(x)) - f(u_n(x))u_n(x)\right] dx$$

$$-\lambda \int_\Omega \left[qG(x, u_n(x)) - g(x, u_n(x))u_n(x)\right] dx$$

$$\geq \left(\frac{q}{p} - 1\right) ||u_n||^p - \max_{|\xi| \leq R} \left(qF(\xi) - f(\xi)\xi\right) ||\alpha||_1 - \lambda ||k||_1$$

and then $\{u_n\}$ is bounded in $X$. So, there exsits a subsequence (still denoted by $\{u_n\}$) weakly convergent to some $u \in X$. Therefore, there exists a constant $c_1$ such that $||u_n - u|| \leq c_1$, for each $n \in \mathbb{N}$. By (4), fixed $\varepsilon > 0$, there exists $\nu \in \mathbb{N}$ such that

$$|\Psi'(u_n)(v) + \lambda\Phi'(u_n)(v)| < \frac{\varepsilon}{c_1}||v|| .$$

Hence,

$$\lim_{n\to\infty} \left[\Psi'(u_n)(u_n - u) + \lambda\Phi'(u_n)(u_n - u)\right] = 0 .$$

Since $W^{1,p}(\Omega)$ is compactly embedded in $C^0(\overline{\Omega})$, $\{u_n\}$ converges to $u$ in $C^0(\overline{\Omega})$. Applying the Lebesgue dominate convergence theorem we obtain

$$\lim_{n\to\infty} \Phi'(u_n)(u_n - u) = 0.$$

So

$$\lim_{n\to\infty} \Psi'(u_n)(u_n - u) = 0 \tag{5}$$

Since $\Psi'(u)$ is a linear and continuous functional on $X$ we have

$$\lim_{n\to\infty} \Psi'(u)(u_n - u) = 0. \tag{6}$$

Finally, we prove that $\{u_n\}$ converges strongly to $u$ in $X$. Observe that, if $x, y \in \mathbb{R}^n$ and $\langle \cdot, \cdot \rangle$ denotes the standart inner product in $\mathbb{R}^n$, there exist two positive constants $c_p$ and $c_p^*$ such that

$$\langle |x|^{p-2}x - |y|^{p-2}y, \ x - y \rangle \geq \begin{cases} c_p|x - y|^p & \text{if } p \geq 2 \\ c_p \dfrac{|x - y|^2}{(|x| + |y|)^{2-p}} & \text{if } p \leq 2 \end{cases}$$

Thus

$$\Psi'(u_n)(u_n - u) - \Psi'(u)(u_n - u)$$

$$\geq \begin{cases} c_p||u_n - u||^p - \displaystyle\int_\Omega \left[\alpha(x)(f(u_n(x)) - f(u(x)))(u_n(x) - u(x))\right] dx & \text{if } p \geq 2 \\ \dfrac{c_p^*}{L^{2-p}}||u_n - u||^2 - \displaystyle\int_\Omega \left[\alpha(x)(f(u_n(x)) - f(u(x)))(u_n(x) - u(x))\right] dx & \text{if } p \leq 2 \end{cases}$$

$$\tag{7}$$

where $L = \max\{\|u\|, \sup_{n \in \mathbb{N}} \|u_n\|\}$. Therefore, our claim follows immediately by (5), (6) and (7). □

**Example 2.3.** As an example of nonlinearity $f$ satisfying $(i_1)$, $(i_2)$ and $(j_3)$ of Theorem 2.3 let $f : \mathbb{R} \to \mathbb{R}$ be defined by setting, for each $t \in \mathbb{R}$,

$$f(t) = \begin{cases} r \left[ \dfrac{|t|^p}{t} \sin \dfrac{1}{|t|^{p-1}} - \dfrac{p-2}{p} t \cos \dfrac{1}{|t|^{p-2}} \right] & \text{if } t \neq 0 \\ 0 & \text{if } t = 0 \end{cases}$$

where $r < 0$, $p > 2$ and $g : \Omega \times \mathbb{R} \to \mathbb{R}$ be defined by putting

$$g(x, \xi) = |\xi|^{p+1} - \beta(x)$$

with $\beta \in L^1(\Omega)$, $\beta \geq 0$. In this case the problem (1) has at least three weak solutions for each $\lambda > 0$ sufficiently small.

Finally, in the last result we prove the existence of positive solutions of our Neumann problem.

**Theorem 2.4.** *Assume (2), $f(0) = g(x,0) = 0$ for all $x \in \Omega$ and*

$(i_1)$ $\limsup\limits_{|\xi| \to +\infty} \dfrac{F(\xi)}{|\xi|^p} < \dfrac{1}{p}$;

$(i_2)$ *the function $h : \mathbb{R} \to \mathbb{R}$ defined by putting*

$$h(\xi) = \frac{1}{p}|\xi|^p - F(\xi) \quad \forall \xi \in \mathbb{R}$$

*has a strict not global local minimum in zero;*

$(k_3)$ $\exists \delta > 0 \; : \; F(\xi) > 0 \quad \forall \xi \in ]0, \delta]$;

$(k_4)$ $\limsup\limits_{\xi \to 0^+} \dfrac{\inf_{x \in \Omega} G(x, \xi)}{\xi^p} = +\infty.$

*Then, for each $r > 0$, there exists $\lambda_r^* > 0$ such that, for each $\lambda \in ]0, \lambda_r^*[$, the problem (1) has at least two nontrivial and non negative weak solutions in $\Psi^{-1}(] - \infty, r[)$.*

*Proof.* Let us truncate nonlinearities as the following

$$f_0(u) := \begin{cases} 0 & \text{if } u < 0 \\ f(u) & \text{if } u \geq 0 \end{cases}$$

$$g_0(x, u) := \begin{cases} 0 & \text{if } u < 0 \\ g(x, u) & \text{if } u \geq 0 \end{cases}$$

and consider the problem

$$
\begin{cases}
-\Delta_p u + \alpha(x)|u|^{p-2}u = \alpha(x)f_0(u) + \lambda g_0(x,u) & \text{in } \Omega \\
\dfrac{\partial u}{\partial \nu} = 0 & \text{on } \partial\Omega
\end{cases}
\tag{8}
$$

Going through the proof of Theorem 2.1, the new problem (8) has at leat two weak solutions, say $u_1$ and $u_2$, which are local minima of the restriction of $\Psi_0 + \lambda\Phi_0$ to the set $\Psi_0^{-1}(]-\infty, r[)$.

First of all, the weak solutions to the problem (8) are non negative and, consequently, they are also solutions for the problem (1). In fact, arguing by contradiction, if we assume that a solution $u_1$ of (8) is negative at a point of $\Omega$, the set $A = \{x \in \Omega : u_1(x) < 0\}$ is nonempty and open. For each $v \in W^{1,p}(\Omega)$ we have

$$
\int_\Omega |\nabla u_1(x)|^{p-2}\nabla u_1(x)\nabla v(x)dx + \int_\Omega \alpha(x)|u_1(x)|^{p-2}u_1(x)v(x)dx
$$

$$
- \int_{\Omega\setminus A} \alpha(x)f_0(u_1(x))v(x)dx - \int_{\Omega\setminus A} \lambda g_0(x, u_1(x))v(x)dx = 0 \ .
$$

Then, for $v = \min\{u_1, 0\}$ we obtain

$$
\int_A |\nabla u_1(x)|^p \, dx + \int_A \alpha(x)|u_1(x)|^p \, dx = 0
$$

that is $u_1 = 0$ a.e. in A, an absurd.

Now, we prove that $u_1$ and $u_2$ are different from zero. It is sufficient to prove that 0 is not a local minimum for $\Psi_0 + \lambda\Phi_0$. Fix $\lambda \in ]0, \lambda_r^*[$, let $T = \dfrac{\|\alpha\|_1}{p\lambda|\Omega|} > 0$. From (k$_4$) we obtain a sequence $\{r_n\} \subseteq ]0, \delta[$ converging to zero, with $\inf_{x\in\Omega} G_0(x, r_n) > Tr_n^p$, for every $n \in \mathbb{N}$. Let $v_n \in W^{1,p}(\Omega)$ be define by putting, for every $n \in \mathbb{N}$, $v_n(x) = r_n$ for each $x \in \Omega$. Since 0 is a strict local minimum for $\Psi_0$, and the sequence $\{v_n\}$ converges to 0 in $W^{1,p}(\Omega)$, $\Psi_0(v_n) > 0$, for $n$ big enough. So, there exists $n^* \in \mathbb{N}$ such that for every $n > n^*$ one has

$$
-\frac{\Phi_0(v_n)}{\Psi_0(v_n)} = \frac{\int_\Omega G_0(x, r_n)\, dx}{\frac{r_n^p\|\alpha\|_1}{p} - F(r_n)\|\alpha\|_1} > \frac{Tr_n^p|\Omega|}{\frac{r_n^p\|\alpha\|_1}{p}} = \frac{1}{\lambda} \ .
$$

Then, definitively, $\Psi_0(v_n) + \lambda\Phi_0(v_n) < 0$, namely 0 can not be a local minimum for such functional. $\qquad\square$

## References

1. G. Bonanno and P. Candito, *Three solutions to a Neumann problem for elliptic equations involving the p-Laplacian*, Arch. Math., **80**, 424–429, 2003.
2. P. Pucci and J. Serrin, *A mountain pass theorem*, J. Differential Equations, **60**, 142–149, 1985.

3. B. Ricceri, *Sublevel sets and global minima of coercive functionals and local minima of their perturbations*, J. Nonlinear Convex Anal., **5**, 157–168, 2004.
4. E. Zeidler, *Nonlinear Functional Analysis and Applications, Vol. III Springer*, 1985.

# MULTIPLE SOLUTIONS TO A CLASS OF ELLIPTIC DIFFERENTIAL EQUATIONS

GIUSEPPE CORDARO

*Department of Mathematics, University of Messina,*
*98166 Sant'Agata-Messina, Italy.*
*E-mail:cordaro@dipmat.unime.it*

We present some of our results concerning the existence of multiple solutions to elliptic differential equations. In particular, we deal with the Dirichlet problem involving the $p$-Laplacian and the periodic solutions to second order Hamiltonian systems. In all of these results, we follow a variational approach. We look for solutions of the considered problem which are in turn local minima for the underlying energy functional.

**Key words:** Weak solution, oscillating behaviour, positive solution, multiple solutions, perturbation problem, periodic solution

**Mathematics Subject Classification:** 34B15, 35J20

## 1. Infinitely many solutions

The first result deals with the following Dirichlet problem

$$\begin{cases} -\Delta_p u = \lambda f(x, u) & \text{in } \Omega \\ u = 0 & \text{on } \partial\Omega \end{cases} \tag{1}$$

$\Omega \subseteq \mathbb{R}^N$ is a bounded open set with sufficiently smooth boundary $\partial\Omega$, $p > 1$, $\Delta_p$ is the $p$-Laplacian operator, that is $\Delta_p u = \text{div}(|\nabla u|^{p-2}\nabla u)$, $\lambda > 0$, $f : \Omega \times \mathbb{R} \to \mathbb{R}$ is a Carathéodory function.

A weak solution of (1) is any $u \in W_0^{1,p}(\Omega) \cap L^\infty(\Omega)$ such that

$$\int_\Omega |\nabla u|^{p-2} \nabla u \nabla v \, dx - \lambda \int_\Omega f(x, u(x)) v(x) \, dx = 0,$$

for each $v \in W_0^{1,p}(\Omega)$.

The existence of infinitely many solutions for problem (1) has been widely investigated. The most classical results on this topic are essentially based on the Ljusternik-Schnirelman theory. In them, the key role is played by the oddness of the nonlinearity. Moreover, in order to check the Palais-Smale condition (or some of its variants), one assumes certain conditions which do not allow an oscillating behaviour of the nonlinearity. Results of this types are for example in [1,4,5].

Multiplicity results for problem (1), when $f(x, \cdot)$ has an oscillating behaviour, are certainly more rare.

Our result concerns the existence of infinitely many small positive solutions for these types of nonlinearities. By positive solution we mean a weak solution of (1) in $W_0^{1,p}(\Omega)$ which is nonnegative almost everywhere in $\Omega$.

**Theorem 1.1.** [2] *Suppose that the function $f$ satisfies the following conditions:*

(i)   *There exists $\bar{t} > 0$ such that*

$$\sup_{t \in [0,\bar{t}]} |f(\cdot, t)| \in L^\infty(\Omega).$$

(ii)   *For every $n \in \mathbb{N}$ there exist $\xi_n, \xi_n' \in \mathbb{R}$, with $0 \leq \xi_n < \xi_n'$ and $\lim_{n \to +\infty} \xi_n' = 0$, such that, for a.e. $x \in \Omega$,*

$$\int_0^{\xi_n} f(x,s)ds = \sup_{t \in [\xi_n, \xi_n']} \int_0^t f(x,s)ds.$$

(iii)   *There exists a non-empty open set $D \subseteq \Omega$, a constant $M \geq 0$ and a sequence $\{t_n\}_{n \in \mathbb{N}} \subset \mathbb{R}_+ \setminus \{0\}$, with $\lim_{n \to +\infty} t_n = 0$, such that*

$$\lim_{n \to +\infty} \frac{\operatorname{ess\,inf}_{x \in D} \int_0^{t_n} f(x,s)ds}{t_n^p} = +\infty,$$

*and*

$$\operatorname*{ess\,inf}_{x \in D} \left( \inf_{t \in [0,t_n]} \int_0^t f(x,s)ds \right) \geq -M \operatorname*{ess\,inf}_{x \in D} \left( \int_0^{t_n} f(x,s)ds \right).$$

*Then, for every $\lambda > 0$, problem $(P_\lambda)$ admits a sequence $\{u_n\}$ of non-zero and non-negative weak solutions strongly convergent to zero and such that $\lim_{n \to +\infty} \max_{\overline{\Omega}} u_n = 0$.*

The proof can be summarized as follows.
*Sketch of the proof.*
Consider a suitable truncation of $f$, which we denote by $g : \Omega \times \mathbb{R} \to \mathbb{R}$, defined as follows

$$g(x,t) = \begin{cases} f(x,\bar{t}) & \text{if } t > \bar{t}; \\ f(x,t) & \text{if } 0 \leq t \leq \bar{t}; \\ 0 & \text{if } t < 0. \end{cases}$$

So, we look for local minima in $W_0^{1,p}(\Omega)$ of the functional

$$\Phi_\lambda(u) = \frac{1}{p\lambda} \int_\Omega |\nabla u|^p dx - \int_\Omega \left( \int_0^{u(x)} g(x,t)dt \right) dx.$$

Such local minima, being in particular critical points for $\Phi_\lambda$, turn to be weak solutions in $W_0^{1,p}(\Omega)$ for problem (1) with $g$ in the place of $f$.

Fix $n \in \mathbb{N}$ and put

$$E_n = \{u \in W_0^{1,p}(\Omega) : 0 \le u(x) \le \xi'_n \text{ a.e. in } \Omega\},$$

there exists $u_n \in E_n$, such that

$$\Phi_\lambda(u_n) = \inf_{E_n} \Phi_\lambda,$$

which is a local minimum for $\Phi_\lambda$. Since $\Phi_\lambda(u_n) < 0$ and there exists a sequence of pairwise distinct $u_n$. Clearly such $u_n$ are solutions also of (1) because $\xi'_n \le \bar{t}$ for $n \in \mathbb{N}$ large enough. $\triangle$

The existence of infinitely many small solutions to (1) has been also studied by Omari and Zanolin[11] who obtained the same conclusion as in Theorem 1 by only supposing that

$$\liminf_{t \to 0^+} \frac{F(t)}{t^p} = 0 \qquad \text{and} \qquad \limsup_{t \to 0^+} \frac{F(t)}{t^p} = +\infty,$$

where $F(t) = \int_0^t f(s)ds$. In the following example the conditions of Omari and Zanolin are not satisfied while Theorem 1.1 can be applied:

**Example 1.1.** Let us consider the following Dirichlet problem

$$\begin{cases} -\Delta u = \lambda f(u) & \text{in } \Omega \\ u = 0 & \text{on } \partial\Omega \end{cases}$$

$\lambda > 0$, $f : \mathbb{R} \to \mathbb{R}$ is defined by

$$f(t) = \begin{cases} 9t^{\frac{1}{2}} \sin(\frac{1}{t^{1/3}}) - 2t^{\frac{1}{6}} \cos(\frac{1}{t^{1/3}}) & \text{if } t > 0; \\ 0 & \text{if } t \le 0. \end{cases}$$

This is an example of nonlinearity, case $p = 2$, such that

$$\liminf_{t \to 0^+} \frac{F(t)}{t^2} = -\infty,$$

and Theorem 1.1 can be applied.

## 2. Three distinct solutions

We now show two multiplicity results concerning the existence of at least three distinct solutions for two different types of problems. Both of them are based on a three critical points theorem obtained by Ricceri[12].

Let us consider the following eigenvalue problem

$$\begin{aligned} \ddot{u} - A(t)u &= \lambda \nabla F(t, u) & \text{a.e in } [0, T] \\ u(T) - u(0) &= \dot{u}(T) - \dot{u}(0) = 0, \end{aligned} \tag{2}$$

with $T > 0$ and $\lambda \ge 0$.

$A : [0, T] \to \mathbb{R}^N \times \mathbb{R}^N$ is a mapping into the space of $N$-order symmetric matrices with $A \in L^\infty([0, T])$ and there exists $\mu > 0$ such that

$$(A(t)x, x) \ge \mu |x|^2,$$

for a.e. $t \in [0, T]$ and each $x \in \mathbb{R}^N$.

$F : [0, T] \times \mathbb{R}^N \to \mathbb{R}$ is such that

$F(\cdot, x)$ is measurable in $t$, for each $x \in \mathbb{R}^N$;

$F(t, \cdot)$ is continuously differentiable in $x$, for almost every $t \in [0, T]$;

and satisfies the following conditions:

$$\max\{|F(t, x)|, |\nabla F(t, x)|\} \le a(|x|)b(t),$$

for some $a \in C(\mathbb{R}_+, \mathbb{R}_+)$ and $b \in L^1(0, T; \mathbb{R}_+)$.

Our result assures the existence of an open interval $\Lambda \subset [0, \infty[$ and real number $\rho > 0$ such that problem (2), for each $\lambda \in \Lambda$, admits at least three distinct solutions whose norms are less than $\rho$. Moreover, we are able to give information about the location of such interval $\Lambda$, since it results that $\Lambda \subseteq [0, \bar{a}]$, where $\bar{a}$ is a positive real number whose dependence from data is given.

We define

$$k(A) = \sup_{u \in H_T^1 \setminus \{0\}} \frac{\|u\|_\infty}{\left( \int_0^T (A(t)u(t), u(t)) dt + \int_0^T |\dot{u}(t)|^2 dt \right)^{\frac{1}{2}}},$$

where $\| \cdot \|_\infty$ denotes the sup-norm in $C^0(0, T; \mathbb{R}^N)$. The result is as follows:

**Theorem 2.1.** [7] *Assume that the following conditions hold:*

(i) *There exist $r > 0$ and $c \in \mathbb{R}^N$ such that*

$$\int_0^T (A(t)c, c) dt > r$$

*and*

$$\frac{\int_0^T F(t, c) dt}{\int_0^T (A(t)c, c) dt} < \frac{1}{r} \int_0^T \inf_{|x| \le k(A)\sqrt{r}} F(t, x) dt.$$

(ii) *Put*

$$\bar{a} = \frac{h r}{2 \left( \int_0^T \inf_{|x| \le k(A)\sqrt{r}} F(t, x) dt - r \frac{\int_0^T F(t, c) dt}{\int_0^T (A(t)c, c) dt} \right)},$$

*with $h > 1$, and assume that there exist $M > 0$ and $\alpha \in L^1(0, T; \mathbb{R}_+)$, with $\|\alpha\|_{L^1} < \frac{1}{2k(A)^2 \bar{a}}$, such that*

$$F(t, x) \ge -\alpha(t)|x|^2$$

*for every $x \in \mathbb{R}^N$, with $|x| > M$, and a.e. $t \in [0, T]$.*

*Then there exist an open interval $\Lambda \subset [0, \bar{a}]$ and a real number $\rho > 0$ such that, for each $\lambda \in \Lambda$, problem $(P_\lambda)$ has at least three distinct solutions whose norms in $H_T^1$ are less than $\rho$.*

*Sketch of the proof.*
Consider the following functionals

$$\Phi(u) = \frac{1}{2}\left(\int_0^T |\dot{u}(t)|^2 dt + \int_0^T (A(t)u(t), u(t)) dt\right)$$

and

$$\Psi(u) = \int_0^T F(t, u(t))\, dt.$$

They are continuously differentiable and weakly lower semicontinuous on $H_T^1$, where

$$H_T^1 = \left\{ u : [0, T] \to \mathbb{R}^N \,\middle|\, \begin{array}{l} u \text{ is absolutely continuous,} \\ u(0) = u(T) \text{ and } \dot{u} \in L^2(0, T; \mathbb{R}^N) \end{array} \right\}$$

is a Hilbert space with norm defined by

$$\|u\| = \left(\int_0^T |u(t)|^2 dt + \int_0^T |\dot{u}(t)|^2 dt\right)^{\frac{1}{2}}.$$

The critical points of functional $\Phi + \lambda\Psi$ are solutions of problem (2).

So, the thesis is a consequence of the following result due to Bonanno[6] which specified the Ricceri's three critical points theorem cited above. $\triangle$

The authors who gave the major contribute to the existence of three solutions for second order Hamiltonian systems are Tang and Wu[13-15]. Another interesting result on this topic has been recently obtained by Faraci[9].

The condition present in all the papers of Tang is as follows:
There exist $r > 0$ and an integer $k \geq 0$ such that

$$-\frac{1}{2}(k+1)^2 w^2 |x|^2 \leq F(t, x) - F(t, 0) \leq -\frac{1}{2}k^2 w^2 |x|^2, \ (T)$$

for each $|x| \leq r$ and a.e. $t \in [0, T]$, where $w = \frac{2\pi}{T}$.

Tang, using condition (T) together with coercive assumption on $F$ or in presence of a sublinear behaviour of the nonlinearity, proved that problem

$$\ddot{u} = \nabla F(t, u) \qquad\qquad \text{a.e. in } [0, T]$$
$$u(T) - u(0) = \dot{u}(T) - \dot{u}(0) = 0$$

admits three solutions.

Problem $(P_\lambda)$, with $\lambda = 1$, has been studied by Faraci when $F(t, x) = b(t)V(x)$, with $b \in L^1(0, T; \mathbb{R}_+) \setminus \{0\}$ and $V \in C^1(\mathbb{R}^N)$.

We also cite the result recently obtained by Barletta and Livrea[3], where the authors deal with problem (2) when the nonlinearity is of the type $b(t)\nabla G(x)$.

Finally, we give the following simple application of Theorem 2.1.

**Example 2.1.** Let us consider the following eigenvalue problem

$$\ddot{u}_i - \sum_j a_{i,j}(t)u_j = \lambda[a(t)(u_i^3 + 3u_i^2) + b(t)]\, e^{u_i}\ i=1,2,\ldots N \text{ a.e. in } [0, T]$$
$$u(T) - u(0) = \dot{u}(T) - \dot{u}(0) = 0$$

858

with $\lambda \geq 0$, $a, b \in L^1(0, T; \mathbb{R}_+) \setminus \{0\}$, and $a_{i,j}(t)$ denoting the $(i,j)$-entry of a $N$-order matrix $A(t)$.

The last result we present concern the following perturbation problem:

$$\begin{cases} -\Delta u = f(x, u) + \lambda g(x, u) & \text{a.e. in } \Omega \\ u = 0 & \text{on } \partial\Omega \end{cases} \qquad (3)$$

where $\Omega \subset \mathbb{R}^N$ is a bounded open set with sufficiently smooth boundary $\partial\Omega$, $\lambda \in \mathbb{R}$, $f, g : \Omega \times \mathbb{R} \to \mathbb{R}$ are Carthéodory functions.

Denote by $\lambda_1$ and $\lambda_2$ respectively the first and second eigenvalue of the following eigenvalue problem

$$\begin{cases} -\Delta u = \lambda u & \text{a.e. in } \Omega \\ u = 0 & \text{on } \partial\Omega. \end{cases}$$

We list our assumptions:

(i) For a.e. $x \in \Omega$ and every $t \in \mathbb{R}$, $f(x, -t) = -f(x, t)$.
(ii) There exist $L, M \in \mathbb{R}$, with $L < \lambda_2$ and $M > 0$, such that

$$-M \leq \frac{f(x,t)}{t} \leq L,$$

for a.e. $x \in \Omega$ and all $t \in \mathbb{R} \setminus \{0\}$.
(iii) One has

$$\limsup_{t \to +\infty} \frac{\operatorname{ess\,sup}_{x \in \Omega} \int_0^t f(x,s)ds}{t^2} < \frac{\lambda_1}{2} < \liminf_{t \to 0^+} \frac{\operatorname{ess\,inf}_{x \in \Omega} \int_0^t f(x,s)ds}{t^2}.$$

(iv) When $N \geq 2$, there exist $a > 0$, $p \geq 0$, with $p < \frac{N+2}{N-2}$ if $N \geq 3$, and $\alpha \in L^{\frac{2N}{N+2}}(\Omega)$, such that, for $(x, t) \in \Omega \times \mathbb{R}$,

$$|g(x,t)| \leq a|t|^p + \alpha(x).$$

In the case $N = 1$, for each $r > 0$, $\sup_{|t| \leq r} |g(\cdot, t)| \in L^1(\Omega)$.
(v) There exist $s \in [0, 2]$, $\beta \in L^q(\Omega)$ and $\gamma \in L^1(\Omega)$, with $q = \frac{2N}{2N-(N-2)s}$ if $N > 2$, $q > 1$ if $N = 2$, $q = 1$ if $N = 1$, such that

$$\left| \int_0^t g(x,s)ds \right| \leq \beta(x)|t|^s + \gamma(x),$$

for all $t \in \mathbb{R}$ and a.e. $x \in \Omega$.
(vi) One of the following conditions holds, for a.e. $x \in \Omega$ and all $t \in \mathbb{R} \setminus \{0\}$,

$$t \int_0^t g(x,s)ds > 0, \quad t \int_0^t g(x,s)ds < 0.$$

Our result is as follows.

**Theorem 2.2.** [8] *Assume that the functions $f$ and $g$ satisfy all the conditions listed above. Then, there exists $\bar{b} > 0$ such that, for every $0 < b \le \bar{b}$, there exist an open interval $\Lambda \subset [-b, b]$ and a positive real number $\rho$ so that, for each $\lambda \in \Lambda$, problem (3) admits three weak solutions in $W_0^{1,2}(\Omega)$ whose norms are less than $\rho$.*

*Sketch of the proof.*

The proof relies on the following Lemma which is a consequence of the above cited three critical points theorem of Ricceri:

**Lemma 2.1.** *Let $X$ be a separable and reflexive real Banach space, $\Phi, \Psi : X \to \mathbb{R}$ be two continuously Gâteaux differentiable functionals. Assume that $\Phi$ is sequentially weakly lower semicontinuous, $\Psi$ is sequentially weakly continuous, and there exists $\bar{b} > 0$ such that, for each $\lambda \in [-\bar{b}, \bar{b}]$, the functional $\Phi + \lambda\Psi$ satisfies the Palais-Smale condition and*

$$\lim_{\|x\| \to +\infty} (\Phi(x) + \lambda\Psi(x)) = +\infty.$$

*Finally, suppose that there exist $x_1, x_2 \in X$ and $r \in \mathbb{R}$ such that*

$$\inf_{x \in X} \Phi(x) < \inf_{x \in \Psi^{-1}(r)} \Phi(x), \tag{4}$$

$$\Phi(x_1) = \Phi(x_2) = \inf_{x \in X} \Phi(x), \tag{5}$$

$$\Psi(x_1) < r < \Psi(x_2). \tag{6}$$

*Then, for every $0 < b \le \bar{b}$, there exist an open interval $\Lambda \subseteq [-b, b]$ and a positive real number $\rho$, such that, for each $\lambda \in \Lambda$, the equation*

$$\Phi'(x) + \lambda\Psi'(x) = 0$$

*admits at least three solutions in $X$ whose norms are less than $\rho$.*

We apply Lemma 2.1 taking $X = W_0^{1,2}(\Omega)$ endowed with the norm of gradient

$$\|u\| = \left( \int_\Omega |\nabla u|^2 dx \right)^{\frac{1}{2}},$$

and setting

$$\Phi(u) = \frac{1}{2}\|u\|^2 - \int_\Omega \left( \int_0^{u(x)} f(x, s)ds \right) dx,$$

$$\Psi(u) = -\int_\Omega \left( \int_0^{u(x)} g(x, t)dt \right) dx,$$

for each $u \in X$. Condition (i), (ii) and (iv) imply that $\Phi$ and $\Psi$ are well-defined and continuously Gâteaux differentiable functionals on $X$. Moreover, for each $\lambda \in \mathbb{R}$, the critical points of $\Phi + \lambda\Psi$ in $X$ are the weak solutions of problem (3). $\triangle$

## Example 2.2.

Let us consider the following problem

$$\begin{cases} -\Delta u = \alpha(x)\sin(u) + \lambda(\beta(x)|u|^q + \gamma(x)) & \text{in } \Omega \\ u = 0 & \text{on } \partial\Omega \end{cases}$$

where $\alpha \in C^1(\overline{\Omega})$, $\beta \in L^\infty(\Omega)$, $\gamma \in L^p(\Omega)$ with $p > \frac{N}{2}$, and $0 < q \le 1$.

Assume that, for each $x \in \overline{\Omega}$, $\lambda_1 < \alpha(x) < \lambda_2$, ess $\inf_\Omega \beta > 0$ and $\gamma$ is nonnegative almost everywhere in $\Omega$. Then, it is easily seen that all hypotheses of Theorem 2.2 are satisfied.

Another example where the odd nonlinearity may not be Lipschitzean.

Let $f : \mathbb{R} \to \mathbb{R}$ be a continuous bounded even function such that $f(0) > \lambda_1$ and $\lim_{t\to+\infty} f(t) < \lambda_1$, $a \in \mathbb{R}$, $\alpha \in L^{\frac{2N}{N+2}}(\Omega)$ be nonnegative almost everywhere in $\Omega$ and $n \in \mathbb{N}$ be odd. Put $M = \sup_{t\in\mathbb{R}} |f(t)|$ and suppose that $M < \lambda_2$. Then, it is easily seen that the following problem

$$\begin{cases} -\Delta u = uf(u) + \lambda((u(u-a)^{n-1})^{\frac{1}{n}} + \alpha(x)) & \text{in } \Omega \\ u = 0 & \text{on } \partial\Omega \end{cases}$$

satisfies all hypotheses of Theorem 2.2.

The paper of Li and Liu[10] is one of the most interesting paper which deal with problem (3).

Li and Liu obtained the existence of multiple solutions to (3) when $f, g \in C(\overline{\Omega} \times \mathbb{R})$ and with no other assumption on the perturbation term $g$.

We list the set of conditions and then we cite below both the results of Li and Liu.

(f1) $f(x, -t) = -f(x, t)$ for all $x \in \Omega$ and $t \in \mathbb{R}$.
(f2) There exist $a > 0$ and $1 < p < (N+2)/(N-2)$ such that

$$|f(x, t)| \le a(1 + |t|^p), \qquad x \in \overline{\Omega}, \ t \in \mathbb{R}$$

if $N \ge 3$; $\lim_{|t|\to+\infty} \ln(|f(x,t)| + 1)|t|^{-2} = 0$ uniformly for $x \in \Omega$ if $N = 2$; and no assumption if $N = 1$.
(f3) There exist constant $M > 0$ and $2 < \mu < \frac{2N}{N-2}$ such that

$$0 < \mu F(x, t) \le tf(x, t), \qquad x \in \overline{\Omega}, \ |t| \ge M,$$

where $F(x, t) = \int_0^t f(x, s)ds$.
(f4) There exists $\delta$ such that $f(x, -t) = -f(x, t)$ for all $x \in \Omega$ and $|t| \le \delta$.
(f5) $\lim_{|t|\to 0} F(x, t)t^{-2} = +\infty$.
(f6) There exists $\delta_1 > 0$ such that $2F(x, t) > tf(x, t)$ for all $x \in \Omega$ and $0 < |t| \le \delta_1$.

**Theorem A** [10]. *Suppose that (f1)-(f3) are satisfied. Then for any $j \in \mathbb{N}$, there exists $\epsilon_j > 0$ such that if $|\epsilon| \le \epsilon_j$ then problem $(P\epsilon)$ possesses at least $j$ distinct solutions corresponding to positive critical values. That is, for such a solution $u$,*

$$\frac{1}{2} \int_\Omega |\nabla u|^2 dx - \int_\Omega F(x,u)dx - \epsilon \int_\Omega G(x,u)dx > 0,$$

*where $G(x,t) = \int_0^t g(x,s)ds$.*

**Theorem B** [10]. *Suppose that (f4)-(f6) are satisfied. Then for any $j \in \mathbb{N}$, there exists $\epsilon_j > 0$ such that if $|\epsilon| \le \epsilon_j$ the problem $(P_\epsilon)$ possesses at least $j$ distinct solutions corresponding to negative critical values. That is, for such a solution $u$,*

$$\frac{1}{2} \int_\Omega |\nabla u|^2 dx - \int_\Omega F(x,u)dx - \epsilon \int_\Omega G(x,u)dx < 0.$$

In comparing our conditions (i)-(vi) with those of Li and Liu, (f1)-(f6), we note that assumption (f3) is incompatible with our hypothesis (iii) while (f5) is stronger than condition (iii). This fact allow us to treat different classes of nonlinearities. Moreover, we stress out the information, given by Theorem 2.2, that, for every $\lambda \in \Lambda$, the three solutions of (3) belong to a ball of $W_0^{1,2}(\Omega)$, centered in the origin, whose radius $\rho$ does not depend on $\lambda$.

## References

1. A.Ambrosetti, H.Brezis and G.Cerami, Combined effect of concave and convex nonlinearities in some elliptic problems, *J.Functional Anal.*, **122** (1994), 519-543.
2. G.Anello and G.Cordaro, Infinitely many arbitrarily small positive solutions for the Dirichlet problem involving the $p$-Laplacian, *Proc. R. Soc. Edinb. Sect. A.*, **132** (2002), 511-519.
3. G. Barletta and R. Livrea, Existence of three solutions for a non autonomous second order system, *Le Matematiche*, 57 (2002), 205-215.
4. T.Bartsch, Infinitely many solutions of a symmetric Dirichlet problem, *Nonlinear Analisys T.M.A.*, **20** (1993), 1205-1216.
5. T.Bartsch and M.Willem, On an elliptic equation with concave and convex nonlinearities, *Proc. Amer. Math. Soc.*, **123** (1995), 3555-3561.
6. G. Bonanno, Some remarks on a three critical points theorem, *Nonlinear Anal.*, **54**, 651-665.
7. G. Cordaro, *Three periodic solutions to an eigenvalue problem for a class of second order Hamiltonian systems*, Abstr. Appl. Anal., **18** (2003), 1037-1045.
8. G.Cordaro, A multiplicity result for a perturbation of a symmetric Dirichlet problem, *Applicable Analysis*, **83** no. 8 (2004), 799-806.
9. F. Faraci, Three periodic solutions for a second order nonautonomous system, *J. Nonlinear Convex Anal.*, **3** (2002), no. 3, 393-399.
10. S. Li and Z.Liu, Perturbations from Symmetric Elliptic Boundary Value Problems, *J. Differential Equations* **185** (2002), 271-280.
11. P.Omari and F.Zanolin, An elliptic problem with arbitrarily small positive solutions, *Electron. J. Differential Equations Conf.*, **5** (2000), 301-308.
12. B.Ricceri, On a three critical points theorem, *Arch. Math. (Basel)*, **75** (2000), 220-226.
13. C.L. Tang, Existence and multiplicity of periodic solutions of nonautonomous second order systems, *Nonlinear Anal.* **32** (1998), 299-304.

14. C.L. Tang, Periodic solutions of nonautonomous second order systems sublinear nonlinearity, *Proc. Amer. Math. Soc.* **126** no.11 (1998), 3263-3270.
15. C.L. Tang, X.P. Wu, Periodic solutions for second order systems with not uniformly coercive potential, *J. Math. Anal. Appl.* **259** no.2 (2001), 386-397.

# ONE CAN HEAR THE SHAPES OF SOME NON-CONVEX DRUMS*

*Dedicated to Professor Mitsuru IKAWA*

WAICHIRO MATSUMOTO

Minoru Murai and Shoji Yotsutani

*Ryukoku University, Faculty of Science and Technology,
Department of Applied Mathematics and Informatics,
Seta,
520-2194 Otsu, JAPAN
E-mail: waichiro@math.ryukoku.ac.jp*

We show the existence of a series of non-convex domains whose shapes are determined by those eigenvalues of Dirichlet Laplacian. Further, we give those concrete shapes.

**Key words:** Isospectrality, area with the minimal curvature energy, elliptic functions and elliptic integrals, explicit solutions of ordinary differential equation

**Mathematics Subject Classification:** 58J53, 49Q10, 34A05, 33E05

## 1. Introduction

In the paper titled "Can one hear the shape of a drum?" [4], M. Kac proposed a problem: Do the shapes of domains coincide each other if these domains have the same eigenvalues of the Laplacian $\triangle = \dfrac{\partial^2}{\partial x^2} + \dfrac{\partial^2}{\partial y^2}$ with the Dirichlet boundary condition? When two domains has the same eigenvalues, we say that these domains are isospectral. Thus, the question is

> *"if two planar domains are isospectral, isometric are these two?"*
>
> *(the uniqueness of the shape of domains)*

On this problem, we have known few results. M. Kac [4] announced that the disk has the uniqueness. S. Marvizi and R. Melrose [6] and R. Melrose [9] gave the convex domains with the uniqueness through the research of the wave equation. K. Watanabe [14] [15] gave the domains with the uniqueness through the research of the heat equation. Watanabe considered the shapes of the domain as a perturbation of a disk and they stay convex.

A counter-exqample was given by C. Gordon, D. Webb and S. Wolpert [1] [2] standing on the Sunada's framework [13]. Their example is non-convex and has the corners.

---

*This work is supported by Ryukoku Univ.

The general condition for the uniqueness has not yet given. S. Zelditch [17] gave a condition but it assures the uniqueness in the domains with analytic boundary. (See also a survey by us [7].)

In this paper, we analyze the Watanabe's domains. They are determined by the Euler-Lagrange equation of the minimizing problem of the curvature energy and include two parameters. We consider the cases for all possible parameters and show that some of the domains are non-convex. Some concrete shapes of Watanabe's domains are given in the last part of this paper as Fig. 1.

Let $\{\lambda_j\}$ be the eigenvalues of $-\triangle$ in a bounded domain $\Omega$ with the Dirichlet boundary condition. The trace of the kernel $k(x, y)$ of the heat equation in $\Omega$ with the Dirichlet boundary condition is determined by the eigenvalues $\{\lambda_j\}$ and, if $\partial\Omega$ is of $C^\infty$ class, it has the asymptotic expansion near $t \sim 0$ :

$$\int_\Omega k(x, x; t)\, dx = \sum_{j=1}^\infty \exp(-\lambda_j t) \sim \frac{1}{4\pi t} \sum_{i=0}^\infty D_i\, t^{i/2}\,, \qquad (1)$$

where

$$D_0 = M = the\ area\ of\ \Omega\,,$$

$$D_1 = -\frac{\sqrt{\pi}}{2}L\,, \quad L = the\ length\ of\ \partial\Omega\,,$$

$$D_2 = \frac{1}{3}\int_{\partial\Omega}\kappa(s)ds\,,$$

$$D_3 = \frac{\sqrt{\pi}}{64}\int_{\partial\Omega}\kappa(s)^2 ds\,, \qquad (2)$$

$$D_4 = \frac{4}{315}\int_{\partial\Omega}\kappa(s)^3 ds\,,$$

$$D_5 = \frac{37\sqrt{\pi}}{8192}\int_{\partial\Omega}\kappa(s)^4 ds - \frac{\sqrt{\pi}}{1024}\int_{\partial\Omega}(\kappa(s)')^2 ds\,,$$

$$e.t.c.$$

(See H. Weyl [16], A. Pleijel [10], H. P. McKean Jr and I. M. Singer [8], G. Louchard [5], K. Stewartson and R. Waechter [12], L. Smith [11], e.t.c.)

**Remark 1.1.** Instead of the Dirichlet boundary condition, we can take the Neumann boundary condition. Further, instead of the heat equation, we can consider, for example, the wave equation.

## 2. Watanabe's Variational Problem and Results

In the case of a disk, announced by M. Kac, it is the solution of the variational problem of minimizing $-D_1$ under the fixed $D_0$. As $\Omega$ is a domain, $\partial\Omega$ has the winding number 1 and then $D_2$ must be $\frac{2\pi}{3}$. Watanabe proposed the variational problem: the minimizing $D_3$, that is, minimizing the curvature energy $\frac{1}{2}\int_{\partial\Omega}\kappa(s)^2 ds$ for given $M$ and $L$ with the relation $4\pi M < L^2$ assuming $\int_{\partial\Omega}\kappa(s)ds = 2\pi$. He showed the following facts [14] [15].

Using Lagrange's multiplier,

$$E(\theta) = \frac{\lambda_0}{2}\int_0^L \theta'(s)^2 ds + \lambda_1 \int_0^L \cos\theta(s)ds + \lambda_2 \int_0^L \sin\theta(s)ds$$

$$+ \frac{\lambda_3}{2}\int_0^L \int_0^s \sin(\theta(s) - \theta(\xi))d\xi ds$$

takes a minimizer without restriction condition. This leads us to the Euler-Lagrange equation:

$$\theta''(s) = \mu_1 \sin\theta(s) + \mu_2 \cos\theta(s) + \mu_3 \int_0^s \cos(\theta(s) - \theta(\xi))d\xi . \tag{3}$$

This is reduced to

$$\kappa(s)''(s) + \frac{1}{2}\kappa(s)^3 + \mu_4\kappa(s) - \mu_3 = 0 . \tag{4}$$

**Remark 2.1.** $\mu_3$ and $\mu_4$ satisfy the relations (6). If $\mu_4$ satisfies the second relation, the first one is obtained under the equation (4).

Integrating this, we obtain

$$K(\kappa(s)) + U(\kappa(s)) = \mu_5 , \tag{5}$$

$$K(\kappa(s)) = \frac{1}{2}(\kappa'(s))^2 , \quad U(\kappa(s)) = \frac{1}{8}\kappa(s)^4 + \frac{\mu_4}{2}\kappa(s)^2 - \mu_3\kappa(s) ,$$

where

$$\mu_3 = \frac{1}{L^2 - 4\pi M}\left(\frac{L}{2}\int_0^L \kappa(s)^3 ds - \pi\int_0^L \kappa(s)^2 ds\right),$$

$$\mu_4 = \frac{1}{L^2 - 4\pi M}\left(M\int_0^L \kappa(s)^3 ds - \frac{L}{2}\int_0^L \kappa(s)^2 ds\right), \tag{6}$$

$$\mu_5 = -105\pi\mu_3 + 36\left(\left(\int_0^L \kappa(s)^2 ds\right)\mu_4 + \frac{3072}{\sqrt{\pi}}D_5\right).$$

**Remark 2.2.** Obviously, the third relation of (6) also obtained from the first relation of (6) and the equation (4).

From now on, we take the start point of $s$ as $\kappa(0) = \max_{0\leq s\leq L}\kappa(s)$. Thus, $\kappa'(0) = 0$ and the equation (4) implies two parameteers $\mu_4$ and $\kappa(0)$.

**Theorem 2.1.** *(K. Wtanabe* [14]*)*
*For every $n \geq 2$, the equation (4) has a $C^\infty$ periodic solution with the prime period*

$L/n$ ( the solution of mode $n$ ), which is symmetric in $[0, L/n]$ at $L/2n$, decreasing in $[0, L/2n]$ and increasing in $[L/2n, L/n]$. Further, it hold that $\kappa(s)(L/2n) = \min_{0 \le s \le L} \kappa(s)$ and $\kappa'(0) = \kappa'(L/2n) = 0$.

**Remark 2.3.** $\kappa(s)$ is in reality analytic.

**Remark 2.4.** K. Watanabe intrinsically considered simple closed curves. There exists a positive number $M_1(n) < L^2/4\pi$ and K. Watanabe's curves rest simple for $M_1 < M < L^2/4\pi$. The general results will be announced in Theorem 2.3.

**Theorem 2.2.** *(K. Watanabe [14] [15])*

If $(40/49)L^2 \le 4\pi M < L^2$ [a], *minimizers are simple curves and take even modes. Further, the domains enclosed by the minimizers have the uniqueness.*

*Further, if $L^2/4\pi$ is sufficiently close to $M$, the minimizer is unique, of mode 2 and the domain enclosed by this curve is convex.*

K. Watanabe conjectured by a numerical try that whenever the solution gives a simple curve, the minimizer is unique and of mode 2. From now on, we allow non-simple curves and negative areas. Thus, we pose the restriction $-L^2 < 4\pi M < L^2$. We obtained the following theorem.

**Theorem 2.3.**

*(1) For each integer $n \ge 2$ , the equation (4) has at least one and at most three solutions of mode $n$ for $-\dfrac{L^2}{4\pi(n-1)} < M < \dfrac{L^2}{4\pi}$. Further, there exist positive number $M_0(n) < \dfrac{L^2}{4\pi(n+1)}$ such that the number of the solutions is one for $-\dfrac{L^2}{4\pi(n-1)} < M < M_0$, two for $M = M_0$ and $\dfrac{L^2}{4\pi(n+1)} \le M < \dfrac{L^2}{4\pi}$ and three for $M_0 < M < \dfrac{L^2}{4\pi(n+1)}$ . Both solutions tend to the circle with the radius $\dfrac{L}{2\pi}$ as $M$ tends to $\dfrac{L^2}{4\pi}$, one regularly and the other singularly. The singular limit of the solution as $M$ tends to $-\dfrac{L^2}{4(n-1)\pi} + 0$ is the negative $(n-1)$-fold circle with the radius $\dfrac{L}{2\pi(n-1)}$. One solution singularly tends to the $(n+1)$-fold circle with the radius $\dfrac{L}{2\pi(n+1)}$ as $M$ tends to $\dfrac{L^2}{4\pi(n+1)} - 0$. In case of $n = 1$, the number of solutions decreases 1 from the case of $n \ge 2$.*

*(2) For $n \ge 2$, there exists a positive number $M_1(n) < L^2/4\pi$ such that the unique solution to (4) gives a simple closed curve for $M_1 < M < \dfrac{L^2}{4\pi}$ . In the case of $n = 1$, no solution gives a closed and simple curve.*

---

[a] K. Watanabe announced $(16/25)M \le 4\pi M < L^2$, but it seems that his proof holds good under this restriction.

*(3) The minimizer of the curvature energy is always unique and of mode 2. Further, there exists a positive number $M_2$ $(M_1(2) < M_2 < \dfrac{L^2}{4\pi})$ and the minimizer gives a convex curve for $M_2 \le M < \dfrac{L^2}{4\pi}$ and a non-convex curve for $M_1 < M < M_2$*

**Remark 2.5.** For the general winding number $\omega$, we can also obtain the similar results on the solutions to the equation (4). Of course, the relations in (6) change depending on $\omega$.

## 3. Representation of the solution

In order to obtain the solutions to the non-local equation (4), first, we regard $\mu_j$'s as the constants independent of the solutions and obtain the formal solutions and secondarily we pick up admissible parameters by the relations of the winding number and the area. We fix the arc-length $L$. Let $n$ be the mode of the solution.

By the equation (5), in the case of mode $n$ $(n \ge 2)$, we have

$$\kappa(s)'(s) = -\frac{1}{2}\sqrt{-\kappa(s)^4 - 4\mu_4\kappa(s)^2 + 8\mu_3\kappa(s) + 8\mu_5} \quad (0 \le s \le L/2n), \quad (7)$$

$$\kappa(s)(0) = \max \kappa(s) = p, \quad \kappa(s)(L/2n) = \min \kappa(s) = q,$$

$$\kappa(s) = \kappa(s)(L/n - s), \quad (L/2n \le s \le L/n).$$

As we orient the curve anti-clockwise, $p$ is positive. We introduce new parameters $p$ and $q$ and use another subsidiary $\delta$, where $p$ and $q$ are those in (5):

$$-\kappa(s)^4 - 4\mu_4\kappa(s)^2 + 8\mu_3\kappa(s) + 8\mu_5 = (p - \kappa(s))(\kappa(s) - q)\{(\kappa(s) + \frac{p+q}{2})^2 + 4\delta\}$$
$$(8)$$

$$(q \le \kappa(s) \le p, \quad p > 0),$$

$$\begin{cases} \mu_4 = -(1/16)(3p^2 + 2pq + 3q^2 - 16\delta), \\ \mu_3 = (1/32)(p+q)\{(p-q)^2 + 16\delta)\}, \\ \mu_5 = -(1/32)pq\{(p+q)^2 + 16\delta)\}. \end{cases} \quad (9)$$

We replace $p$ and $q$ by $P$ and $Q$, and set $\xi = \dfrac{1}{2}\left(\kappa(s) + \dfrac{p+q}{2}\right)$ : we have

$$P = (3p+q)/4, \quad p = (3P - Q)/2, \quad p+q = P+Q,$$
$$Q = (p+3q)/4, \quad q = (-P+3Q)/2, \quad p-q = 2(P-Q)$$

and

$$\frac{d\kappa(s)}{d\xi} = -\sqrt{(P-\xi)(\xi - Q)(\xi^2 + \delta)} . \quad (10)$$

Thus we arrive at

$$s = \int_{\kappa(s)(0)}^{\kappa(s)} \frac{ds}{d\kappa(s)} d\kappa(s) = \int_{(1/2)\{\kappa(s)+(P+Q)/2\}}^{P} \frac{d\xi}{\sqrt{(P-\xi)(\xi-Q)(\xi^2+\delta)}} . \tag{11}$$

Changing the the variable of integral:

$$\begin{cases} \xi = Q + (1/\eta), \\ \eta = \{1/(P-Q)\}\{1 + \sqrt{(P^2+\delta)/(Q^2+\delta)}\tan^2(\varphi/2)\}, \end{cases} \tag{12}$$

(11) becomes

$$s = \frac{1}{\sqrt[4]{(P^2+\delta)(Q^2+\delta)}} \int_0^{\varphi(\kappa(s))} \frac{d\varphi}{\sqrt{1-k^2\sin^2\varphi}} , \tag{13}$$

where

$$\tan^2 \frac{\varphi(\kappa(s))}{2} = \sqrt{\frac{Q^2+\delta}{P^2+\delta}} \frac{p-\kappa(s)}{\kappa(s)-q} , \tag{14}$$

$$k^2 = \frac{1}{2}(1 - \frac{(PQ+\delta)}{\sqrt{(P^2+\delta)(Q^2+\delta)}}) . \tag{15}$$

This means

$$s = \begin{cases} \frac{1}{\sqrt[4]{(P^2+\delta)(Q^2+\delta)}} F(\varphi(\kappa(s)); k) & (0 \le s \le L/4n), \\ \frac{1}{\sqrt[4]{(P^2+\delta)(Q^2+\delta)}} \{2K(k) - F(\pi - \varphi(\kappa(s)); k)\} & (L/4n \le s \le L/2n), \end{cases} \tag{16}$$

where

$$F(\varphi; k) = \int_0^{\varphi} \frac{d\theta}{\sqrt{1-k^2\sin^2\theta}} \quad \text{(the elliptic integral of the first kind)},$$

$$K(k) = F(\pi/2; k) \quad \text{(the complete elliptic integral of the first kind)}.$$

Finally, using the elliptic function, we arrive at

$$\kappa(s) = \frac{2(P-Q)\sqrt{Q^2+\delta}\left(1 + \mathrm{cn}(\hat{K}(k)s; k)\right)}{\sqrt{P^2+\delta} + \sqrt{Q^2+\delta} - \left(\sqrt{P^2+\delta} - \sqrt{Q^2+\delta}\right)\mathrm{cn}(\hat{K}(k)s; k)} - \frac{P-3Q}{2} \tag{17}$$

where $\hat{K}(k) = \dfrac{4n}{L}K(k)$ and $\mathrm{sn}(z; k)$ is Jacobi's elliptic function $i.e.$ the inverse function of $z = \displaystyle\int_0^x \frac{du}{\sqrt{(1-u^2)(1-k^2u^2)}}$ and $\mathrm{cn}(z; k) = \sqrt{1 - \mathrm{sn}^2 z}$ .

the detail proof will be given in the forthcoming papers.

## 4. Restriction Condition and New Parameters

Taking $s = L/4n$ in (11), we obtain the relation:

$$\sqrt[4]{(P^2 + \delta)(Q^2 + \delta)} = \frac{4nK(k)}{L} \equiv \hat{K}(k),\tag{18}$$

which corresponds to the relation that $L$ is the length of the curve, that is,

$$\delta = \frac{1}{2}\{-(P^2 + Q^2) \pm \sqrt{(P^2 - Q^2)^2 + 4\hat{K}(k)^4}\,\}.\tag{19}$$

By this relation, $\delta$ is represented by $P$ and $Q$ that is $p$ and $q$. Let us introduce new parameters $k$ and $h$:

$$\left\{ \begin{array}{l} k^2 = (1/2)\,\{1 - (PQ + \delta)/\sqrt{(P^2 + \delta)(Q^2 + \delta)}\,\}, \\ P - Q = hP, \quad (Q = (1 - h)P). \end{array} \right.\tag{20}$$

Using

$$\Pi(c; k) = \int_0^{\pi/2} \frac{d\varphi}{(1 + c\sin^2\varphi)\sqrt{1 - k^2\sin^2\varphi}} \quad \text{(the elliptic integral of the third kind)}.$$

the relation that the winding number is one corresponds

$$z_n(k, t) \equiv -[(3 - 4k^2)\{t^2 + 16(1 - k^2)\} + \{8k(1 - k^2) + \sqrt{(1 - 2k^2)^2t^2 + 16(1 - k^2)}\,\}t]\times$$
$$[t^2 + \{8k(1 - k^2) + \sqrt{(1 - 2k^2)^2t^2 + 16(1 - k^2)}\,\}t + 16(1 - k^2)]\hat{K}(k)$$
$$- 4(1 - k^2)\{t^2 + 16(1 - k^2)\}\times$$
$$[t^2 + \{8k(1 - k^2) + \sqrt{(1 - 2k^2)^2t^2 + 16(1 - k^2)}\,\}t + 16(1 - k^2)]\times$$
$$\Pi(\frac{k^2\{t^2 - 8k(1 - k^2)t + 16(1 - k^2) - t\sqrt{(1 - 2k^2)^2t^2 + 16(1 - k^2)}\,\}}{(1 - 2k^2)t^2 + \{8k(1 - k^2) + \sqrt{(1 - 2k^2)^2t^2 + 16(1 - k^2)}\,\}t + 16(1 - k^2)(1 - 2k^2)}, k)$$
$$- \frac{\sqrt{2}\pi}{n}\sqrt{(1 - k^2)\{t^2 + 16(1 - k^2)\}}\{-kt^2 + 2\sqrt{(1 - 2k^2)^2t^2 + 16(1 - k^2)}\,\}\times$$
$$\sqrt{(1 - 2k^2)\{t^2 + 16(1 - k^2)\} + \{8k(1 - k^2) + \sqrt{(1 - 2k^2)^2t^2 + 16(1 - k^2)}\,\}t} = 0.$$

$$\tag{21}$$

On the other hand, the relation that the area is $M$ corresponds

$$M_n(k,t) \equiv -\sqrt{2\{t^2 + 16(1-k^2)\}} \times \tag{22}$$

$$\sqrt{(1-2k^2)t^2 + \{8k(1-k^2) + \sqrt{(1-2k^2)^2 t^2 + 16(1-k^2)}\}t + 16(1-k^2)(1-2k^2)} \times \tag{23}$$

$$[\{(1-2k^2)t^4 + 8k(1-k^2)t^3 - 8(1-k^2)(1-2k^2)(2k^2-5)t^2 + 128k(1-k^2)^2 t$$
$$- 128(1-k^2)^2(4k^2-3)$$
$$+ t(t^2 - 8k(1-k^2)t + 16(1-k^2))\sqrt{(1-2k^2)^2 t^2 + 16(1-k^2)}\}\hat{K}(k)$$
$$- 2\{t^2 + 16(1-k^2)\}\{(1-2k^2)t^2 + (8k(1-k^2) + \sqrt{(1-2k^2)^2 t^2 + 16(1-k^2)})t$$
$$+ 16(1-k^2)(1-2k^2)\}E(k)]L^2$$
$$/[16\sqrt{1-k^2}(kt^2 - 2\sqrt{(1-2k^2)^2 t^2 + 16(1-k^2)} \times$$
$$\{(-2k^4 + 2k^2 - 1)t^4 - 8k(1-k^2)(1-2k^2)t^3 + 8(1-k^2)(1-2k^2)(2k^2-3)t^2$$
$$- 128(1-k^2)^2(1-2k^2)t - 128k(1-k^2)^2$$
$$- t((1-4k^2)t^2 - 8k(1-k^2)t + 16(1-k^2)(1-2k^2))\sqrt{(1-2k^2)^2 t^2 + 16(1-k^2)}\} \times$$
$$n\hat{K}(k)^2]$$
$$= M .$$

We consider the curves defined by $z_n(k,t) = 0$. We obtain two curves $C_r(n)$ and $C_s(n)$.

**Lemma 4.1.** *On $C_r(n)$, $M_n(k,t)$ monotonically decreases from $\dfrac{L^2}{4\pi}$ up to $-\dfrac{L^2}{4\pi(n-1)}$ and the curvature energy monotonically increases from $\dfrac{2\pi^2}{L}$ to infinity. On $C_s(n)$, $M_n(k,t)$ and curvature energy first monotonically decrease from $\dfrac{L^2}{4\pi}$ and from infinity respectively up to the positive minimums, and secondarily they monotonically increase up to $\dfrac{L^2}{4\pi(n+1)}$ and up to infinity respectively.*

**Lemma 4.2.**
*(i) For the same area $M$ and $n_1 < n_2$, the curvature energy on $C_r(n_1)$ is less than that on $C_r(n_2)$.*
*(ii) For the same $n$, the curvature energy on $C_r(n)$ with $M = 0$ is less than the minimum of the curvature energy on $C_s(n)$.*

As K. Watanabe showed that $C_r(1)$ is empty and the solution of (4) with the odd mode cannot be the minimizer, the minimizer of the curvature energy is the solution on $C_r(2)$ for each positive $M$.

Some of the shapes of the domains are given in Fig. 1 and the relation between the area and the curvature energy is given in Fig. 2.

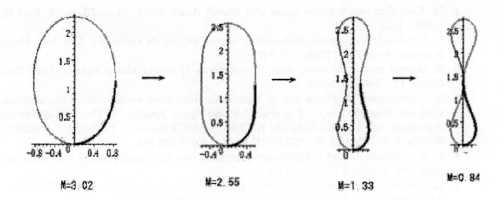

Fig. 1. shape of domain $n = 2$

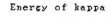

Energy of kappa

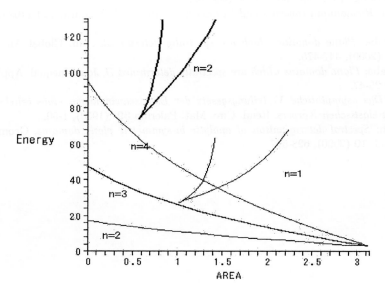

Fig. 2. area-curveture energy

# References

1. C. Gordon, D. Webb and S. Wolpert; *Isospectral plane domains and surfaces via Riemannian orbifolds*, Inv. Math. **110** (1992), 1-22.
2. C. Gordon, D. Webb and S. Wolpert; *One cannot hear the shape of a drum*, Bull. Amer. Math. Soc. **27** (1992), 134-138.

3. P. Greiner; *An asymptotic expansion for the heat equation*, Arch. Rat. Mech. **41** (1971), 163-218.

4. M. Kac; *Can one hear the shape of a dram?*, Amer. Math. Mon. **73**, no. 4, Part II, (1966), 1-23.

5. G. Louchard; *Mouvement Brownien et valeurs propres du Laplacien*, Ann. Inst. Henri Poincaré, Sect. **B 4** (1968), 331-342.

6. S. Marvizi and R. Melrose; *Spectral invariants of convex planar regions*, Jour. Diff. Geom. **17** (1982), 475-502.

7. W. Matsumoto, M. Murai and S. Yotsutani; *What have we learned on the problem: Can one hear the shape of a drum?*, Phase Space Analysis of Partial Differential Equations, vol. II, CENTRO DE RICERCA MATEMATICA ENNIO DE GIORGI, SCUOLA NORMALE SUPERIORE PISA, (2005) 345-361.

8. H. P. McKean, Jr and I. M. Singer; *Curvature and the eigenvalues of the Laplacian*, Jour. Diff. Geom. **1** (1967), 43-69

9. R. Melrose; *Inverse spectral problems*, Proc Cent. Math. Appl., Australian Nat. Univ. **34** (1996), Part I, P. D. F., ed. T. Cranny et al., 137-160.

10. A. Pleijel; *A study of certain Green's functions with applications in the theory of vibrating membranes*, Ark. Mat. **2** (1952-54), 553-569.

11. L. Smith; *The asymptotics of the heat equation for a boundary value problem*, Inv. Math. **63** (1981), 467-493.

12. K. Stewartson and R. T. Waechter; *On hearing the shape of a drum: further results*, Proc. Cambridge Phil. Soc. **69** (1971), 353-363.

13. T. Sunada; *Riemannian covering and isospectral manifolds*, Ann. Math. **121** (1985), 169-186.

14. K. Watanabe; *Plane domains which are spectrally determined*, Ann. Global Anal. Geom. **18** (2000), 447-475.

15. K. Watanabe; *Plane domains which are spectrally determined II*, Jour. Inequal. Appl. **7** (2002), 25-47.

16. H. Weyl; *Das asymptotiche Verteilungsgesetz der Eigenschwingungen eines beliebig gestalteten elastischen Körpers*, Rend. Circ. Mat. Palermo **39** (1950), 1-50.

17. S. Zelditch; *Spectral determination of analytic bi-symmetric plane domains*, Geom. Funct. Anal. **10** (2000), 628-677.

# MULTIPLICITY RESULTS FOR TWO POINTS BOUNDARY VALUE PROBLEMS

A. CHINNÌ, F. CAMMAROTO* , B. Di BELLA

*Department of Mathematics*
*University of Messina*
*98166 Sant'Agata Messina, Italy*
*E-mail: chinni@dipmat.unime.it*

In this paper, we establish the existence of two (or three) solutions for a Dirichlet problem related to a parametric second order ordinary equation. Our main tool is a recent two local minima theorem of B. Ricceri.

**Keywords:** Critical points, three solutions, Dirichlet problem
**Mathematics Subject Classification:** 34B08

## 1. Introduction

In this paper, we consider the following Dirichlet problem

$$\begin{cases} -u'' = \lambda f(u) + \mu g(t,u) & \text{in } [0,1] \\ \\ u(0) = u(1) = 0 \end{cases} \qquad (D_{\lambda,\mu})$$

where $f : \mathbb{R} \to \mathbb{R}$ and $g : [0,1] \times \mathbb{R} \to \mathbb{R}$ are two continuous functions and $\lambda$, $\mu$ are two positive parameters. Precisely, we want to prove the existence of two (or three) solutions for the problem $(D_{\lambda,\mu})$.

In several recent papers (see for example [2], [3], [4]), multiplicity results for Dirichlet boundary value problem have been deduced from a three critical point theorem of Ricceri (6, Theorem 1) making use of the following thechnical lemma on the minimax inequality:

**Proposition 1.1.** *([7], Proposition 3.1.) Let $X$ be a nonempty set, and $\Phi$, $\Gamma$ two real functions on $X$. Assume that there are $\sigma > 0$, $x_0$, $x_1 \in X$, such that*

$$\Phi(x_0) = \Gamma(x_0) = 0, \qquad \Phi(x_1) > \sigma,$$

$$\sup_{x \in \Phi^{-1}(]-\infty,\sigma])} \Gamma(x) < \sigma \frac{\Gamma(x_1)}{\Phi(x_1)}.$$

---

*Because of a surprising coincidence of names within the same Department, we have to point out that the author was born on August 4, 1968.

874

*Then, for each $\rho$ satisfying*

$$\sup_{x\in\Phi^{-1}(]-\infty,\sigma])}\Gamma(x) < \rho < \sigma\,\frac{\Gamma(x_1)}{\Phi(x_1)},$$

*one has*

$$\sup_{\lambda\geq 0}\inf_{x\in X}(\Phi(x)+\lambda(\rho-\Gamma(x))) < \inf_{x\in X}\sup_{\lambda\geq 0}(\Phi(x)+\lambda(\rho-\Gamma(x))).$$

In order to solve our problem we will make use of another recent result due to Ricceri deduced from a purely topological minimax result. We recall it in a convenient form

**Theorem 1.1.** *([1], Theorem 4).Let $X$ be a reflexive real Banach space, $I \subseteq \mathbb{R}$ an interval, and $\Psi : X \times I \to \mathbb{R}$ a function such that $\Psi(x, \cdot)$ is concave in $I$ for all $x \in X$, $\Psi(\cdot, \lambda)$ is continuous, coercive and sequentially weakly lower semicontinuous in $X$ for all $\lambda \in I$. Further, assume that*

$$\sup_{\lambda\in I}\inf_{x\in X}\Psi(x,\lambda) < \inf_{x\in X}\sup_{\lambda\in I}\Psi(x,\lambda).$$

*Then, for each $\alpha > \sup_I \inf_X \Psi$ there exists a non-empty open set $A_\alpha \subseteq I$ with the following property: for every $\lambda \in A_\alpha$ and every sequentially weakly lower semi-continuous functional $H : X \to \mathbb{R}$, there exists $\delta_{\lambda,H} > 0$ such that, for each $\mu \in ]0, \delta_{\lambda,H}[$, the functional $\Psi(\cdot, \lambda) + \mu H(\cdot)$ has at least two local minima lying in the set $\{x \in X : \Psi(x, \lambda) < \alpha\}$.*

The most remarkable feature of Theorem 1.1, compared with Theorem 1 of [6], is its wide applicability to nonlinear differential equations. Before introducing our results, we specify that, in the sequel, $F$ will be the integral function of $f$ defined by putting

$$F(\xi) = \int_0^\xi f(t)\,dt$$

for each $\xi \in \mathbb{R}$. Moreover, we recall that, given two topological spaces $X, Y$, a multifunction $S : X \to 2^Y$ is said to be upper semicontinuous if, for every closed set $C \subseteq Y$, the set

$$S^-(C) = \{x \in X : S(x) \cap C \neq \emptyset\}$$

is closed in $X$. Moreover, we say that $S$ is sequentially upper semicontinuous if, for every pair of sequences $\{x_n\}, \{y_n\}$ in $X$ and $Y$ respectively, with $y_n \in S(x_n)$ for all $n \in \mathbb{N}$ and $\{x_n\}$ converging to some $x_0 \in X$, there exists a subsequence of $\{y_n\}$ converging to some $y_0 \in S(x_0)$.

## 2. Main results

**Theorem 2.1.** *Assume that there exist three positive numbers $c, d, l$ with $l > 2$ and $c < \sqrt{\frac{l}{2}}\, d$ such that*

*(i) $f(\xi) \geq 0$ for each $\xi \in [-c, max\{c, d\}]$;*

*(ii) $\dfrac{F(c)}{c^2} < 2\,\dfrac{l-2}{l^2}\,\dfrac{F(d)}{d^2}$;*

*(iii) $\limsup\limits_{|\xi| \to +\infty} \dfrac{F(\xi)}{\xi^2} \leq 0$.*

*Then, there exist a number $r \in \mathbb{R}$ and a non-degenerate compact interval $C \subseteq [0, +\infty[$ with the following property: for every $\lambda \in C$ and for every continuous function $g : [0, 1] \times \mathbb{R} \to \mathbb{R}$, there exists $\delta_{\lambda, g} > 0$ such that, for each $\mu \in ]0, \delta_{\lambda, g}[$, the problem $(D_{\lambda, \mu})$ has at least two solutions whose norms are less than $r$.*

*Proof.*   We put $X = W_0^{1,2}([0,1])$ endowed with the norm

$$\|u\| = \left( \int_0^1 |u'(t)|^2 \, dt \right)^{\frac{1}{2}} \qquad \forall u \in X$$

and we define the functionals $\Phi$ and $J$ as follows

$$\Phi(u) = \frac{1}{2}\|u\|^2 \quad \text{and} \quad J(u) = -\int_0^1 F(u(t)) \, dt$$

for each $u \in X$. Let $\bar{u} \in X$ defined by

$$\bar{u}(t) = \begin{cases} ldt & t \in [\,0, \frac{1}{l}\,[ \\[2mm] d & t \in [\,\frac{1}{l}, \frac{l-1}{l}\,] \\[2mm] ld(1-t) & t \in ]\,\frac{l-1}{l}, 1\,]. \end{cases} \tag{1}$$

It is easy to verify that

$$\Phi(\bar{u}) = ld^2.$$

Put $\sigma = 2c^2$, taking into account that for every $u \in X$, one has

$$\max_{0 \leq t \leq 1} |u(t)| \leq \frac{1}{2}\|u\| \,,$$

from conditions (i) and (ii) it follows that

$$\sup_{u \in \Phi^{-1}(]-\infty, \sigma])} (-J(u)) \leq \max_{|\xi| \leq c} F(\xi) = F(c) < \sigma \, \frac{-J(\bar{u})}{\Phi(\bar{u})}.$$

At this point, chosen

$$\sup_{x \in \Phi^{-1}(]-\infty, \sigma])} (-J(u)) < \rho < \sigma \frac{-J(\bar{u})}{\Phi(\bar{u})},$$

Proposition 1.1 assures that

$$\sup_{\lambda \geq 0} \inf_{u \in X} \Psi_\rho(u, \lambda) < \inf_{u \in X} \sup_{\lambda \geq 0} \Psi_\rho(u, \lambda)$$

where

$$\Psi_\rho(u, \lambda) = \Phi(u) + \lambda(J(u) + \rho)$$

for each $(u, \lambda) \in X \times [0, +\infty[$. We apply Theorem 1.1 by choosing $\Psi = \Psi_\rho$ and $I = [0, +\infty[$.

Easily, we can observe that $\Psi_\rho(u, \cdot)$ is concave in $I$ for each $u \in X$ while classical arguments provide the sequential weak lower semicontinuity and the continuity of $\Psi_\rho(\cdot, \lambda)$ for each $\lambda \in I$.

Now, we prove the coercivity of $\Psi_\rho(\cdot, \lambda)$ for each $\lambda \in I$. It is obvious that $\Psi_\rho(\cdot, 0)$ is coercive. Fixed $\lambda \in ]0, +\infty[$ and $0 < \epsilon < \dfrac{2}{\lambda}$, condition (iii) implies that there exists a convenient positive number $b_\epsilon$ such that

$$F(\xi) \leq \epsilon \, \xi^2 + b_\epsilon$$

for all $\xi \in \mathbb{R}$. Then, for each $u \in X$, we have that

$$\Psi_\rho(u, \lambda) \geq \frac{1}{2}\|u\|^2 + \lambda \left( -\int_0^1 [\epsilon(u(t))^2 + b_\epsilon] dt \right) + \lambda \rho$$

$$\geq \left( \frac{1}{2} - \lambda \frac{\epsilon}{4} \right) \|u\|^2 - \lambda b_\epsilon + \lambda \rho$$

i.e. $\Psi_\rho(., \lambda)$ is coercive. Fixed $\alpha > \sup_{\lambda \geq 0} \inf_{u \in X} \Psi_\rho(u, \lambda)$, Theorem 1.1 ensures that there exists a non-empty open set $A_\alpha \subseteq I$ with the following property: for every $\lambda \in A_\alpha$ and every continuous function $g : [0, 1] \times \mathbb{R} \to \mathbb{R}$, there exists $\delta_{\lambda, g} > 0$ such that, for each $\mu \in ]0, \delta_{\lambda, g}[$, the functional $E_{\lambda, \mu, g}(u) = \Psi_\rho(u, \lambda) + \mu H_g(u)$ has at least two local minima lying in the set $\{u \in X : \Psi_\rho(u, \lambda) < \alpha\}$, where $H_g$ is the weakly sequentially lower semicontinuous functional defined by

$$H_g(u) = -\int_0^1 \left( \int_0^{u(t)} g(t, \xi) \, d\xi \right) dt$$

for each $u \in X$. These two local minima are critical points of $E_{\lambda, \mu, g}$ and then are weak solutions of the problem $(D_{\lambda, \mu})$.

Fixed $C_\alpha \subseteq A_\alpha$ non-degenerate compact interval, we consider the multifunction $S : C_\alpha \to 2^X$ defined by

$$S(\lambda) = \{u \in X : \Psi_\rho(u, \lambda) \leq \alpha\},$$

for each $\lambda \in C_\alpha$. Because of the sequential weak lower semicontinuity of $\Psi_\rho$ with respect to $u$, $S$ has sequentially weakly compact (and so sequentially weakly closed) values. In particular, the choice of $\alpha$, assures that these values are non-empty. Finally, from affinity of $\Psi_\rho$ with respect to $\lambda$, it is easy to prove that the set $C_\alpha \setminus S^-(u)$ is connected and open in $C_\alpha$ for all $u \in X$. At this point, thanks to Proposition 1 of [6] $S$ is sequentially upper semicontinuous and then, $S(C_\alpha)$ is sequentially weakly compact (and so bounded) in $X$. The conclusion follows taking $r = \sup_{u \in S(C_\alpha)} \|u\|$ and taking into account that, by well known arguments, these solutions are classical solutions. □

The following result allows us to obtain three solutions of $(D_{\lambda,\mu})$ instead of two. We recall that a Gâteaux differentiable functional $S$ on a real Banach space $X$ is said to satisfy the Palais-Smale condition if each sequence $\{x_n\}$ in $X$ such that $\sup_{n \in \mathbb{N}} |S(x_n)| < +\infty$ and $\lim_{n \to +\infty} \|S'(x_n)\|_X = 0$ admits a strongly converging subsequence.

**Theorem 2.2.** *Let assume the same hypotheses of Theorem 2.1. Then, there exists a non-empty open set $A \subseteq [0, +\infty[$ such that, for every $\lambda \in A$ and every continuous function $g : [0, 1] \times \mathbb{R} \to \mathbb{R}$ with*

*(iv)* $\displaystyle \limsup_{|\xi| \to +\infty} \frac{\max_{t \in [0,1]} \int_0^\xi g(t, x)\, dx}{\xi^2} < +\infty,$

*there exists $\delta_{\lambda,g} > 0$ such that, for each $\mu \in ]0, \delta_{\lambda,g}[$, the problem $(D_{\lambda,\mu})$ has at least three solutions.*

*Proof.*

As in the proof of Theorem 2.1, conditions (i), (ii) and (iii) assure the existence of a non-empty open set $A \subseteq [0, +\infty[$ with certain property.

In particular, fixed a continuous function $g : [0, 1] \times \mathbb{R} \to \mathbb{R}$ satisfyng (iv), for each $\lambda \in A$, there exists $\delta_{\lambda,g} > 0$ such that for every $\mu \in ]0, \delta_{\lambda,g}[$ the problem $(D_{\lambda,\mu})$ has at least two solutions which are critical points of the functional

$$E_{\lambda,\mu,g}(u) = \Psi_\rho(u, \lambda) + \mu H_g(u)$$

where $\rho$ is a convenient positive number. In order to obtain a third solution of $(D_{\lambda,\mu})$, we prove the coercivity of $E_{\lambda,\mu,g}$. From (iv) there exist two constants $p$, $q$ with $p > 0$ such that

$$\int_0^\xi g(t, x)\, dx \le p|\xi|^2 + q$$

for all $t \in [0,1]$ and $\xi \in \mathbb{R}$. Then, for each $u \in X$, we have

$$H_g(u) = -\int_0^1 \left( \int_0^{u(t)} g(t,\xi)\, d\xi \right) dt \geq -\frac{p}{4}\|u\|^2 - q.$$

Fix $\lambda \in A$ and $0 < bar\delta < \min\{\delta_{\lambda,g}, \frac{2}{p}\}$. Then, for each $\mu \in ]0, \bar\delta[$, choosen $0 < \epsilon < \frac{2}{\lambda+1}(1-\mu\frac{p}{2})$, condition (iii) implies that there exists $b_\epsilon \in \mathbb{R}$ such that the inequality

$$\Psi_\rho(u,\lambda) \geq \left( \frac{1}{2} - \lambda\,\frac{\epsilon}{4} \right) \|u\|^2 - \lambda b_\epsilon + \lambda\rho$$

holds for each $u \in X$ and so that

$$E_{\lambda,\mu,g}(u) \geq \left( \frac{1}{2} - \lambda\,\frac{\epsilon}{4} - \mu\,\frac{p}{4} \right) \|u\|^2 - \lambda b_\epsilon + \lambda\rho - \mu q$$

for each $u \in X$. The last condition provides the coercivity of $E_{\lambda,\mu,g}$. Standard arguments assure that the functional $\Phi'$ admits a continuous inverse on $X^*$ while $J'$ and $H'_g$ are compact. Then, by Example 38.25 of [8] we deduce that the functional

$$E_{\lambda,\mu,g}(\cdot) = \Phi(\cdot) + \lambda(J(\cdot) + \rho) + \mu H_g(\cdot)$$

has the Palais-Smale property. Finally, by using Corollary 1 of [5] and taking into account that the functional $E_{\lambda,\mu,g}$ is $C^1$ in $X$, there exists a third critical point of $E_{\lambda,\mu,g}$ which is a third solution of problem $(D_{\lambda,\mu})$. □

Let us conclude with an example in which Theorem 2.2 is applied.

**Example 2.1.** Let $f : \mathbb{R} \to \mathbb{R}$ be defined by setting, for each $\xi \in \mathbb{R}$,

$$f(\xi) = \begin{cases} \xi^{2\beta} & \text{if } \xi \leq 1 \\ \xi^\alpha & \text{if } \xi > 1 \end{cases}$$

with $\beta > \frac{3}{2}$ and $0 < \alpha < 1$. Then, there exists a non-empty open set $A \subseteq [0, +\infty[$ with the following property: for every $\lambda \in A$ and every $\gamma \in ]-1, 1]$, there exists $\delta > 0$ such that , for each $\mu \in ]0, \delta[$, the problem

$$\begin{cases} -u'' = \lambda f(u) + \mu t|u|^\gamma & \text{in } [0,1] \\ \\ u(0) = u(1) = 0 \end{cases} \tag{D}$$

has at least two nontrivial solutions.

In this case conditions (i), (ii) and (iii) are satisfied taking $c = \frac{1}{2}$, $d = 1$ and $l \in ]l_1, l_2[$ with $l_1, l_2$ solutions of the equation $x^2 - 4^\beta x + 2 \cdot 4^\beta = 0$, while condition (iv) is satisfied taking $g(t, x) = t|x|^\gamma$, for each $(t, x) \in [0, 1] \times \mathbb{R}$.

Appliyng Theorem 2.1 instead of Theorem 2.2, with $f$ defined as in Example 2.1, we obtain the following conclusion:

there exist a number $r \in \mathbb{R}$ and a non-degenerate compact interval $C \subseteq [0, +\infty[$ with the following property: for every $\lambda \in C$ and every $\gamma > -1$, there exists $\delta > 0$ such that , for each $\mu \in ]0, \delta[$, the problem $(D)$ has at least a nontrivial solution whose norm is less than $r$ .

## References

1. B. Ricceri, Minimax theorems for limits of parametrized functions having at most one local minimum lying in a certain set, *Topology Appl. (to appear)*
2. G. Bonanno, Existence of three solutions for a two point boundary value problem *Appl. Math. Lett.* **13**, 53-57 (2000).
3. P. Candito, Existence of three solutions for a nonautonomous two point boundary value problem *J. Math. Anal. Appl.* **252**, 532-537 (2000).
4. R. Livrea, Existence of three solutions for a quasilinear two point boundary value problem *Arch. Math.* **79**, 288-298 (2002).
5. P. Pucci and J. Serrin, A mountain pass theorem *J. Differential Equations* **60**, 142-149 (1985).
6. B. Ricceri, On a three critical points theorem *Arch. Math.* **75**, 220-226 (2000).
7. B. Ricceri, Existence of three solutions for a class of elliptic eigenvalue problems *Math. Comput. Modelling* **32**, 1485-1494 (2000).
8. E. Zeidler, *Nonlinear Functional Analysis and Applications, Vol. III*, Springer (1985).

# MULTIPLE SOLUTIONS FOR AN ORDINARY SECOND ORDER SYSTEM

R. LIVREA*

*Dipartimento di Patrimonio Architettonico e Urbanistico
Università degli Studi Mediterranea di Reggio Calabria
Salita Melissari - 89100 Reggio Calabria, Italy
E-mail: roberto.livrea@unirc.it*

The aim of this lecture is to establish some multiplicity results for an eigenvalue second order Hamiltonian system. Some critical points arguments are exploited in order to prove the existence of an exactly determined open interval of positive eigenvalues for which the system admits at least three distinct periodic solutions. An existence result of two distinct periodic solutions, when the energy functional related to the Hamiltonian system is not coercive, is then pointed out.

**Key words:** Second order Hamiltonian systems, eigenvalue problem, periodic solutions, critical points, multiple solutions
**Mathematics Subject Classification:** 34B15, 34C25

## 1. Introduction

The aim of this lecture is to point out some multiplicity results for a class of second order Hamiltonian systems of the following type

$$(P) \quad \begin{cases} \ddot{u} = \nabla_u F(t, u) & \text{a.e. in } [0, T] \\ u(T) - u(0) = \dot{u}(T) - \dot{u}(0) = 0. \end{cases}$$

Basic results for problem (P) can be found in the monograph of J. Mawhin and M. Willem [8], while, more recently, many other authors were interested in studying this subject. In particular, H. Brézis and L. Nirenberg in [6], applying a critical points result in the presence of splitting, ensure that (P) has at least three solutions when $F : [0, T] \times \mathbb{R}^N \to \mathbb{R}$ is a smooth function, with $F(t, 0) = 0$, $\nabla_w F(t, 0) = 0$, which, among the other, satisfies

(a) $F(t, w) \to +\infty$ as $|w| \to \infty$ *uniformly in t.*
(b) *For $|w| \leq r$ small, and some integer $k \geq 0$,*

$$-\frac{1}{2}(k+1)^2 \omega^2 |w|^2 \leq F(t, w) \leq -\frac{1}{2}k^2 \omega^2 |w|^2,$$

*where $\omega = 2\pi/T$.*

*Research supported by RdB (ex 60% MIUR) of Reggio Calabria University

882

Subsequently, C.L. Tang and the same author with X.P. Wu, in [9] and [10,11] respectively, relaxing assumption (a), with or without requiring assumption (b), obtained either existence or multiplicity results.

Here we want to stress problem (P) when it takes the following form

$$(\text{P}_\lambda) \quad \begin{cases} \ddot{u} = A(t)u - \lambda b(t)\nabla G(u) & \text{a.e. in } [0,T] \\ u(T) - u(0) = \dot{u}(T) - \dot{u}(0) = 0, \end{cases}$$

where $A$ satisfies the following condition

($\mathcal{A}$)  $A(t) = [a_{ij}(t)]$ is a $N \times N$ symmetric matrix valued function with $a_{ij} \in L^\infty([0,T])$ and there exists a positive constant $\mu$ such that

$$A(t)w \cdot w \geq \mu|w|^2$$

for every $w \in \mathbb{R}^N$ and a.e. in $[0,T]$,

$b : [0,T] \to \mathbb{R}$ is an a.e. nonnegative function in $L^1([0,T]) \setminus \{0\}$ and $G : \mathbb{R}^N \to \mathbb{R}$ is a continuously differentiable function.

A first result for problem (P$_1$) was obtained in [7] where the existence of three periodic solutions is ensured without assuming (b), but still requiring a condition that implies the coercivity of the energy functional related to the Hamiltonian system, in addition to the following:

(c)  There exist $\sigma > 0$ and $u_0 \in \mathbb{R}^N$ such that $|u_0| < \sqrt{\frac{\sigma}{\sum_{i,j=1}^N \|a_{ij}\|_\infty\, T}}$ and

$$G(u_0) = \sup_{|u| \leq \overline{k}\sqrt{\sigma}} G(u),$$

namely $G$ achieves its maximum in the interior of the ball of radius $\overline{k}\sqrt{\sigma}$,

where $\overline{k}$ is the best constant in the embedding $H_T^1 \hookrightarrow C^0([0,T],\mathbb{R}^N)$ and $a_{ij}$ are the entries of the matrix $A$ (see assumptions 1 and 3 of Theorem 2.1 in [7])

Exploiting a critical point theorem due to G. Bonanno [3], a multiple solutions result is obtained in [2] and it can be stated as follows

**Theorem 1.1.** *Assume that $G(0) \geq 0$ and that there exist $d > 0$ and $w_0 \in \mathbb{R}^N$, with $|w_0| > \frac{d}{c\sqrt{\mu T}}$, such that*

(i)  $\frac{\max_{|w| \leq d} G(w)}{d^2} < \frac{1}{c^2 T \sum_{ij} \|a_{ij}\|_\infty} \cdot \frac{G(w_0)}{|w_0|^2}.$

*Put*

$$\lambda^* = \frac{p\, d^2}{\frac{d^2}{T \sum_{ij} \|a_{ij}\|_\infty} \cdot \frac{G(w_0)}{|w_0|^2} - c^2 \max_{|w| \leq d} G(w)}$$

*with $p > \frac{1}{2}$ and suppose that*

(ii)  $\limsup_{|w| \to +\infty} \frac{G(w)}{|w|^2} < \frac{1}{2c^2\lambda^*}.$

*Then, for every function $b \in L^1([0,T]) \setminus \{0\}$ that is a.e. nonnegative, there exist an open interval $\Lambda \subseteq \left[0, \frac{\lambda^*}{\|b\|_1}\right]$ and a positive real number $\rho$ such that for every $\lambda \in \Lambda$ problem $(\mathrm{P}_\lambda)$ admits at least three solutions in $H_T^1$ whose norms are less than $\rho$.*

Here we establish the existence of an exactly determined open interval of positive parameters $\lambda$ for which $(P_\lambda)$ admits at least three (Theorem 2.1) or two (Theorem 2.3) distinct periodic solutions.

Our main tools in order to prove Theorem 2.1 and Theorem 2.3 are this two critical point results due to D. Averna and G. Bonanno [1] and G. Bonanno [4] respectively.

**Theorem A.** [1] *Let $X$ be a reflexive Banach space, $\Phi : X \to \mathbb{R}$ a continuously Gâteaux differentiable, coercive and sequentially weakly lower semicontinuous functional whose Gâteaux derivative admits a continuous inverse on $X^*$, $\Psi : X \to \mathbb{R}$ a continuously Gâteaux differentiable functional whose Gâteaux derivative is compact. Put, for each $r > \inf_X \Phi$,*

$$\varphi_1(r) := \inf_{x \in \Phi^{-1}(]-\infty, r[)} \frac{\Psi(x) - \inf_{\overline{\Phi^{-1}(]-\infty, r[)}^w} \Psi}{r - \Phi(x)},$$

$$\varphi_2(r) := \inf_{x \in \Phi^{-1}(]-\infty, r[)} \sup_{y \in \Phi^{-1}([r, +\infty[)} \frac{\Psi(x) - \Psi(y)}{\Phi(y) - \Phi(x)},$$

*where $\overline{\Phi^{-1}(] - \infty, r[)}^w$ is the closure of $\Phi^{-1}(] - \infty, r[)$ in the weak topology, and assume that*

*(j) there is $r \in \mathbb{R}$, with $\inf_X \Phi < r$, such that:*

$$\varphi_1(r) < \varphi_2(r).$$

*(jj) $\lim_{\|x\| \to +\infty} (\Phi(x) + \lambda\Psi(x)) = +\infty$, for all $\lambda \in \left] \frac{1}{\varphi_2(r)}, \frac{1}{\varphi_1(r)} \right[.$*

*Then, for each $\lambda \in \left] \frac{1}{\varphi_2(r)}, \frac{1}{\varphi_1(r)} \right[$ the equation*

$$\Phi'(x) + \lambda\Psi'(x) = 0 \tag{E}$$

*has at least three solutions in $X$.*

**Theorem B.** [4] *Let $X$ be a reflexive real Banach space, and let $\Phi, \Psi : X \to \mathbb{R}$ be two sequentially weakly lower semicontinuous and Gâteaux differentiable functions. Assume that $\Phi$ is (strongly) continuous and satisfies $\lim_{\|x\| \to +\infty} \Phi(x) = +\infty$. Assume also that there exist two constants $r_1$ and $r_2$, with $\inf_X \Phi < r_1 < r_2$, such that*

*(jj') $\max\{\varphi_1(r_1), \varphi_1(r_2)\} < \varphi_2^*(r_1, r_2),$*

*where $\varphi_1$ is defined as in Theorem A and*

$$\varphi_2^*(r_1, r_2) := \inf_{x \in \Phi^{-1}(]-\infty, r_1[)} \sup_{y \in \Phi^{-1}([r_1, r_2[)} \frac{\Psi(x) - \Psi(y)}{\Phi(y) - \Phi(x)},$$

*Then, for each $\lambda \in \left] \frac{1}{\varphi_2^*(r_1, r_2)}, \min\left\{ \frac{1}{\varphi_1(r_1)}, \frac{1}{\varphi_1(r_2)} \right\} \right[$, the functional $\Phi + \lambda\Psi$ admits at least two critical points which lie in $\Phi^{-1}(]-\infty, r_1[)$ and $\Phi^{-1}([r_1, r_2[)$ respectively.*

## 2. Main results

Recall that $A(t) = [a_{ij}(t)]$ satisfies condition $(\mathcal{A})$. Hence, in particular, there exists a positive constant $\mu$ such that

$$A(t)w \cdot w \geq \mu |w|^2$$

for every $w \in \mathbb{R}^N$ and a.e. in $[0, T]$. If we consider

$$k = \sqrt{\frac{2}{m}} \max\left\{ \sqrt{T}, \frac{1}{\sqrt{T}} \right\}$$

where $m = \min\{1, \mu\}$, we can put

$$L = \frac{1}{k^2 T \sum_{ij} \|a_{ij}\|_\infty}, \qquad R = \frac{L}{1 + L}.$$

Now, we can present our first multiplicity theorem and give a sketch of its proof.

**Theorem 2.1.** [5] *Let $G \in C^1(\mathbb{R}^N, \mathbb{R})$, with $G(0) = 0$. Assume that there exist a positive constant $\gamma$ and a vector $w_0 \in \mathbb{R}^N$ with $\gamma < |w_0|$, such that:*

*(1)* $\quad \dfrac{\max_{|w| \leq \gamma} G(w)}{\gamma^2} < R \dfrac{G(w_0)}{|w_0|^2};$

*(2)* $\quad \limsup_{|w| \to +\infty} \dfrac{G(w)}{|w|^2} < \dfrac{\max_{|w| \leq \gamma} G(w)}{\gamma^2}.$

*Then, for every function $b \in L^1([0, T]) \setminus \{0\}$ that is a.e. nonnegative and for every $\lambda$ belonging to the interval $\left] \frac{1}{2\|b\|_1 k^2} \frac{1}{R} \frac{|w_0|^2}{G(w_0)}, \frac{1}{2\|b\|_1 k^2} \frac{\gamma^2}{\max_{|w| \leq \gamma} G(w)} \right[$, problem $(P_\lambda)$ admits at least three solutions.*

**Proof.** We want to apply Theorem A, with

$$\Phi(u) = \frac{1}{2} \int_0^T |\dot{u}(t)|^2 dt + \int_0^T A(t)u(t) \cdot u(t) dt,$$

$$\Psi(u) = -\int_0^T b(t)G(u(t)) dt$$

for every $u \in H_T^1$. Fix $b \in L^1([0, T]) \setminus \{0\}$ that is a.e. nonnegative. For the sake of simplicity, let us consider only the case $\max_{|w| \leq \gamma} G(w) > 0$ and fix $\lambda \in \left] \frac{1}{2\|b\|_1 k^2} \frac{1}{R} \frac{|w_0|^2}{G(w_0)}, \frac{1}{2\|b\|_1 k^2} \frac{\gamma^2}{\max_{|w| \leq \gamma} G(w)} \right[$. Exploiting (2) and the elliptic condition $(\mathcal{A})$, we can find two positive constants $\alpha$, $\beta$ such that

$$\Phi(u) + \lambda\Psi(u) > \alpha\|u\|^2 - \beta\|b\|_1,$$

for each $u \in H_T^1$. Hence, $\Phi + \lambda \Psi$ is coercive.

After that, putting $r = \gamma^2/(2k^2)$ and using assumption (1), by technical computations one can shows that

$$\varphi_1(r) < \varphi_2(r)$$

and

$$\left] \frac{1}{2\|b\|_1 k^2} \frac{1}{R} \frac{|w_0|^2}{G(w_0)}, \frac{1}{2\|b\|_1 k^2} \frac{\gamma^2}{\max_{|w| \leq \gamma} G(w)} \right[ \subseteq \left] \frac{1}{\varphi_2(r)}, \frac{1}{\varphi_1(r)} \right[ .$$

Hence, all the assumptions of Theorem A are satisfied and the conclusion is obtained once observed that the solutions of equation (E) are functions $u \in C^1([0,T], \mathbb{R}^N)$ such that $\dot{u}$ is absolutely continuous and

$$\ddot{u} = A(t)u - \lambda b(t)\nabla G(u) \quad \text{a.e. in } [0,T]$$
$$u(T) - u(0) = \dot{u}(T) - \dot{u}(0) = 0,$$

that is, $u$ is a solution to our problem $(P_\lambda)$. $\qquad \square$

**Remark 2.1.** We explicitly observe that, when $\max_{|w| \leq \gamma} G(w) = 0$, the interval of parameters for which problem $(P_\lambda)$ admits at least three solutions is $\left] \frac{1}{2\|b\|_1 k^2} \frac{1}{L} \frac{|w_0|^2}{G(w_0)}, +\infty \right[ .$

**Example 2.1.** Let $G : \mathbb{R}^2 \to \mathbb{R}$ be defined by

$$G(x,y) = \frac{(x^2 + y^2)^6}{e^{x^2+y^2}} + x$$

for every $w \equiv (x,y) \in \mathbb{R}^2$. By choosing $\gamma = 1$ and $w_0 \equiv (\sqrt{6}, 0)$ all assumption of Theorem 2.1 are satisfied and so, for every function $b \in L^1([0,1]) \setminus \{0\}$ that is a.e. nonnegative and for every $\lambda \in \left] \frac{1}{\|b\|_1} \frac{7}{100}, \frac{1}{\|b\|_1} \frac{18}{100} \right[ $, the following problem

$$\begin{cases} \ddot{u} = u - \lambda b(t)\nabla G(u) & \text{a.e. in } [0,1] \\ u(1) - u(0) = \dot{u}(1) - \dot{u}(0) = 0 \end{cases}$$

admits at least three non trivial solutions. In fact, it is enough to observe that

$$\frac{\max_{|w| \leq \gamma} G(w)}{\gamma^2} = \frac{1}{e} + 1,$$

$R = \frac{1}{5}$, $G(w_0) = \left( \frac{6}{e} \right)^6 + \sqrt{6}$ and

$$\lim_{|w| \to +\infty} \frac{G(w)}{|w|^2} = 0.$$

**Remark 2.2.** Let $G$ be as in Example 2.1, fix $b \in C^0([0,1], \mathbb{R}^+)$ and $\lambda > 0$. It is easy to see that, if we put $F(t,w) = \frac{1}{2}|w|^2 - \lambda b(t)G(w)$ for every $(t,w) \in [0,1] \times \mathbb{R}^2$, one has that

$$\liminf_{|w| \to 0} \frac{F(t,w)}{|w|^2} = -\infty$$

uniformly with respect to $t$. Therefore, assumption (b) of Brézis-Nirenberg [6] does not hold.

**Example 2.2.** Let $G : \mathbb{R} \to \mathbb{R}$ be defined by

$$G(w) = \begin{cases} e^{e^w} - e & \text{if } w < 2 \\ e^{e^2}(e^2 w + 1 - 2e^2) - e & \text{if } w \geq 2. \end{cases}$$

By choosing $\gamma = 1$ and $w_0 = 2$ we are able to apply Theorem 2.1 and affirm that for every function $b \in L^1([0,1]) \setminus \{0\}$ that is a. e. nonnegative and for every $\lambda \in \left] \frac{1}{\|b\|_1} \frac{19}{1000}, \frac{1}{\|b\|_1} \frac{17}{100} \right[$, the following problem

$$\begin{cases} \ddot{u} = u - \lambda b(t) \dot{G}(u) & \text{a.e. in } [0,1] \\ u(1) - u(0) = \dot{u}(1) - \dot{u}(0) = 0 \end{cases}$$

admits at least three non trivial solutions. In fact, a simple computation shows that

$$\frac{\max_{|w| \leq \gamma} G(w)}{\gamma^2} = e^e - e,$$

$R = \frac{1}{3}$ and $G(w_0) = e^{e^2} - e$ so that assumption (1) holds. Moreover

$$\lim_{|w| \to +\infty} \frac{G(w)}{|w|^2} = 0$$

and assumption (2) is also true.

**Remark 2.3.** By the fact that the function $\overline{\lambda} G$, where $\overline{\lambda} \in \left] \frac{1}{\|b\|_1} \frac{19}{1000}, \frac{1}{\|b\|_1} \frac{17}{100} \right[$ and $G$ is as in Example 2.2, is increasing, assumption (c) is not verified. Hence, Theorem 2.1 of [7] does not apply. Moreover, fixed $b \in C^0([0,1], \mathbb{R}^+)$, if we consider $F(t, w) = \frac{1}{2}|w|^2 - b(t)[\overline{\lambda} G(w)]$ for every $(t, w) \in [0,1] \times \mathbb{R}^2$, it is easy to verify that

$$\liminf_{|w| \to 0} \frac{F(t, w)}{|w|^2} = -\infty$$

uniformly with respect to $t$ and assumption (b) fails too.

As an immediate consequence of Theorem 2.1, we can obtain the following

**Theorem 2.2.** [5] *Let $G$, $\gamma$ and $w_0$ like in Theorem 2.1.*
*Then, for every $b \in L^1([0,T]) \setminus \{0\}$ that is a.e. nonnegative and such that $\|b\|_1 \in \left] \frac{1}{2k^2} \frac{1}{R} \frac{|w_0|^2}{G(w_0)}, \frac{1}{2k^2} \frac{\gamma^2}{\max_{|w| \leq \gamma} G(w)} \right[$, problem $(P_1)$ admits at least three solutions.*

Making use of Theorem B, we can obtain another multiplicity result. In particular, without requiring any coercivity condition, we can assure the existence of an exactly determined open interval of positive parameters $\lambda$ for which $(P_\lambda)$ admits at least two distinct solutions that, in addition, satisfy a stability condition.

**Theorem 2.3.** [5] *Let $G \in C^1(\mathbb{R}^N)$, with $G(0) = 0$. Assume that that there exist two positive constants $\gamma_1$, $\gamma_2$ such that, if we put $l = \min\left\{1, \dfrac{1}{k\left(T\sum_{ij}\|a_{ij}\|_\infty\right)^{1/2}}\right\}$, one has $\gamma_1 < |w_0| < l\gamma_2$. Moreover, let $w_0$ be a vector in $\mathbb{R}^N$ and assume that:*

*(1')*
$$\max\left\{\frac{\max_{|w|\leq\gamma_1} G(w)}{\gamma_1^2}, \frac{\max_{|w|\leq\gamma_2} G(w)}{\gamma_2^2}\right\} < R\frac{G(w_0)}{|w_0|^2}.$$

*Then, for every function $b \in L^1([0,T]) \setminus \{0\}$ that is a.e. nonnegative, if we put*

$$I_\lambda = \left]\frac{1}{2\|b\|_1 k^2}\frac{1}{R}\frac{|w_0|^2}{G(w_0)}, \frac{1}{2\|b\|_1 k^2}\min\left\{\frac{\gamma_1^2}{\max_{|w|\leq\gamma_1} G(w)}, \frac{\gamma_2^2}{\max_{|w|\leq\gamma_2} G(w)}\right\}\right[,$$

*for every $\lambda \in I_\lambda$, problem $(P_\lambda)$ admits at least two solutions $u_{1,\lambda}$ and $u_{2,\lambda}$ such that $\|u_{1,\lambda}\|_{C^0} \leq \gamma_1$ and $\|u_{2,\lambda}\|_{C^0} \leq \gamma_2$.*

**Example 2.3.** Let $G : \mathbb{R}^2 \to \mathbb{R}$ be as follows

$$G(w) = \begin{cases} \frac{|w|^6}{e^{|w|^2}} & \text{if } |w| \leq \sqrt{3} \\ \left(\frac{3}{e}\right)^3 \cos(|w|^2 - 3) & \text{if } \sqrt{3} < |w| \leq \sqrt{3 + \frac{15}{2}\pi} \\ \left(\frac{3}{e}\right)^3 \left[e^{|w|^2 - 3 - \frac{15}{2}\pi} - 1\right] & \text{if } |w| > \sqrt{3 + \frac{15}{2}\pi}. \end{cases}$$

Theorem 2.3 guarantees that for every $b \in L^1([0,1]) \setminus \{0\}$ that is a. e. nonnegative and for every $\lambda \in \left]\frac{3}{\|b\|_1}, \frac{4}{\|b\|_1}\right[$ the following problem

$$(P_\lambda^I) \begin{cases} \ddot{u} = u - \lambda b(t)\nabla G(u) & \text{a.e. in } [0,1] \\ u(1) - u(0) = \dot{u}(1) - \dot{u}(0) = 0 \end{cases}$$

admits at least one non trivial solution $u_\lambda$ such that $\|u_\lambda\|_{C^0} \leq \sqrt{3 + \frac{15}{2}\pi}$. To see this, we can observe that

$$k = \sqrt{2}, \quad \sum_{ij}\|a_{ij}\|_\infty = 2, \quad l = \frac{1}{2}, \quad R = \frac{1}{5}.$$

Hence, if we choose $\gamma_1 = \frac{1}{2}$, $\gamma_2 = \sqrt{3 + \frac{15}{2}\pi}$ and $w_0 \in \mathbb{R}^2$ with $|w_0| = \sqrt{3}$, Theorem 2.3 applies.

**Remark 2.4.** We explicitly observe that in the previous Example 2.3, for every positive $\lambda$, the energy functional related to problem $(P_\lambda^I)$ is not coercive. Hence, we cannot apply the results due to Tang [9] and Tang-Wu [10,11].

### References

1. D. Averna and G. Bonanno, *A three critical point theorem and its applications to the ordinary Dirichlet problem*, Topological Methods Nonlinear Anal. **22** (2003), 93-103.
2. G. Barletta and R. Livrea, *Existence of three periodic solutions for a non autonomous second order system*, Le Matematiche, **57** (2002), no. 2, 205-215.
3. G. Bonanno, *Some remarks on a three critical points theorem*, Nonlinear Anal., **54** (2003), 651-665.

888

4. G. Bonanno, *Multiple critical points theorems without the Palais-Smale condition*, J. Math. Anal. Appl., **299** (2004), 600-614.

5. G. Bonanno and R. Livrea, *Periodic solutions for a class of second-order Hamiltonian systems*, Electron. J. Diff. Eqns. **115**, Vol 2005(2005), 1-13.

6. H. Brézis and L. Nirenberg, *Remarks on finding critical points*, Comm. Pure. Appl. Math., **44** (1991), 939-963.

7. F. Faraci, *Three periodic solutions for a second order nonautonomous system*, J. Nonlinear Convex Anal. **3** (2002), 393-399.

8. J. Mawhin and M. Willem *"Critical Point Theory and Hamiltonian Systems"*, Springer-Verlag, New York 1989.

9. C.L. Tang, *Existence and multiplicity periodic solutions of nonautonomous second order systems*, Nonlinear Anal. **32** (1998), 299-304.

10. C.L. Tang and X.P. Wu, *Periodic solutions for nonautonomous second order systems with sublinear nonlinearity*, Proc. Amer. Math. Soc. **126** (1998), 3263-3270.

11. C.L. Tang and X.P. Wu, *Periodic solutions for second order systems with not uniformly coercive potential*, J. Math. Anal. Appl. **259** (2001), 386-397.

# SOME MULTIPLICITY RESULTS FOR SECOND ORDER NON-AUTONOMOUS SYSTEMS

FRANCESCA FARACI

*Department of Mathematics and Computer Science*
*University of Catania*
*Viale A. Doria 6*
*95125 Catania, Italy*
*E-mail: ffaraci@dmi.unict.it*

In this paper variational methods are employed for establishing the existence of multiple solutions for second order non-autonomous systems. Different kinds of potentials are considered.

**Key words:** Multiple periodic solutions, second order non-autonomous systems, variational methods

**Mathematics Subject Classification:** 34C25, 35A15

## 1. Introduction

In this paper we consider the second order non-autonomous system

$$(S) \quad \begin{cases} \ddot{u} = A(t)u + \nabla_x F(t, u) \quad \text{a.e. in } [0,\text{T}] \\ \\ u(0) - u(T) = \dot{u}(0) - \dot{u}(T) = 0. \end{cases}$$

where $A(t)$ is a $N \times N$ positive definite matrix, $F(t, x) : [0, T] \times \mathbb{R}^N \to \mathbb{R}$ is measurable in $t$ for all $x \in \mathbb{R}^N$ and continuously differentiable in $x$ for a.e. $t \in [0, T]$.

We are concerned with an overview of multiplicity results for system $(S)$ in the following cases:

- $F(t, x) = b(t)V(x)$, with $b \geq 0$;
- $F(t, x)$ is a "changing sign" potential;
- $F(t, x) = b(t)V(x) + \lambda G(t, x)$, with $b > 0$.

The existence of at least three periodic solutions for the problem

$$\begin{cases} \ddot{u} = \nabla_x \phi(t, u) \quad \text{a.e. in } [0,\text{T}] \\ \\ u(0) - u(T) = \dot{u}(0) - \dot{u}(T) = 0 \end{cases}$$

was studied in [2,9,10] and [12]. In the quoted papers the key assumption, firstly introduced by Brezis and Nirenberg is the following: there exist $r > 0$ and an integer

890

$k \geq 0$ such that

$$(BN) \qquad -\frac{1}{2}(k+1)^2 w^2 |x|^2 \leq \phi(t,x) - \phi(t,0) \leq -\frac{1}{2}k^2 w^2 |x|^2$$

for each $|x| \leq r$, a.e. $t \in [0,T]$, where $w = \frac{2\pi}{T}$.

A similar condition was required in [1] and in [11] where the authors proved a multiplicity result for the perturbed system

$$\begin{cases} \ddot{u} = \nabla_x \phi(t,u) + \lambda \psi(t) \quad \text{a.e. in } [0,T] \\ u(0) - u(T) = \dot{u}(0) - \dot{u}(T) = 0 \end{cases}$$

under the following hypothesis : there exist $r > 0$ and an integer $k \geq 0$ such that

$$(BN_\star) \qquad -\mu|x|^2 \leq \phi(t,x) - \phi(t,0) \leq -\nu|x|^2$$

for each $|x| \leq r$, a.e. $t \in [0,T]$, where $\nu > \frac{1}{2}k^2 w^2$, $\mu < \frac{1}{2}(k+1)^2 w^2$ and $w = \frac{2\pi}{T}$.

Our attempt is to provide a new contribution to the subject without assuming conditions $(BN)$ or $(BN_\star)$: using two recent local minima theorems by Ricceri we propose a new set of hypotheses rather different to those of the quoted papers.

## 2. The variational setting

Throughout the sequel $T$ is a positive number, $A : [0,T] \rightarrow \mathbb{R}^{N \times N}$ is a symmetric matrix valued function with bounded coefficients $a_{ij}$ in $[0,T]$ and $\|A\| = \sum_{i,j} \max_{[0,T]} |a_{ij}|$, $F(t,x) : [0,T] \times \mathbb{R}^N \rightarrow \mathbb{R}$ is measurable in $t$ for all $x \in \mathbb{R}^N$ and continuously differentiable in $x$ a.e. in $[0,T]$.

Let suppose that $A$ is positive definite, i.e. there exists a positive constant $\alpha$ such that

$$A(t)x \cdot x \geq \alpha|x|^2 \tag{1}$$

for every $x \in \mathbb{R}^N$ and a.e. in $[0,T]$.

Let us recall that a solution of $(S)$ is a function $u \in C^1([0,T], \mathbb{R}^N)$ with $\dot{u}$ absolutely continuous, such that

$$\begin{cases} \ddot{u}(t) = A(t)u(t) + \nabla_x F(t,u(t)) \quad \text{a.e. in } [0,T] \\ u(0) - u(T) = \dot{u}(0) - \dot{u}(T) = 0. \end{cases}$$

That is, introduced the Sobolev space $H_T^1$ of the functions $u \in L^2([0,T], \mathbb{R}^N)$ having a weak derivative $\dot{u} \in L^2([0,T], \mathbb{R}^N)$ and such that $u(0) = u(T)$, we are looking for functions $u \in H_T^1$ such that

$$\int_0^T \dot{u}(t) \cdot \dot{v}(t)dt + \int_0^T A(t)u(t) \cdot v(t)dt + \int_0^T \nabla_x F(t,(u(t)) \cdot v(t)dt = 0$$

for all $v \in H^1_T$. So, solutions of $(S)$ can be found as critical points in $H^1_T$ of the functional

$$u \to \int_0^T |\dot{u}(t)|^2 dt + \int_0^T A(t)u(t) \cdot u(t) dt + \int_0^T F(t, u(t)) dt.$$

We endow $H^1_T$ with the norm

$$\|u\| = \left( \int_0^T |\dot{u}(t)|^2 dt + \int_0^T A(t)u(t) \cdot u(t) dt \right)^{\frac{1}{2}}$$

that is equivalent to the standard one. Let us recall that $H^1_T$ is compactly embedded in $C^0([0,T], \mathbb{R}^N)$. For the embedding constant

$$c = \sup_{u \in H^1_T \setminus \{0\}} \frac{\|u\|_{C^0}}{\|u\|}$$

the following estimate holds:

$$c \leq \sqrt{\frac{2}{\min\{1, \alpha\}}} \max\{\sqrt{T}, \frac{1}{\sqrt{T}}\}.$$

## 3. Results

### 3.1. *Positive potentials*

We first consider the case when

$$F(t, x) = b(t)V(x)$$

where $b : [0, T] \to \mathbb{R}$ is a non-negative function in $L^1([0,T]) \setminus \{0\}$ (whose norm in $L^1([0,T])$ is denoted by $\|b\|_1$), $V : \mathbb{R}^N \to \mathbb{R}$ is a continuously differentiable function. System $(S)$ reads as follows:

$$(S) \quad \begin{cases} \ddot{u} = A(t)u + b(t)\nabla V(u) & \text{a.e. in } [0,T] \\ u(0) - u(T) = \dot{u}(0) - \dot{u}(T) = 0 \end{cases}$$

In Theorem 2.5 of [7], Ricceri provides a powerful tool for obtaining information about the existence and the localization of a minimum of variational problems. In smooth cases this minimum is a critical point of the functional associated to the problem. Applying this result together with the Mountain Pass Lemma (precisely Corollary 1 of [6]) we get the following theorem:

**Theorem 3.1.** ([3] Theorem 2.1) *Assume*

*(I) there exist $\rho > 0$, and $x_0 \in \mathbb{R}^N$ such that*

*i)*

$$|x_0| < \sqrt{\rho(\|A\|T)^{-1}} \quad \text{and} \quad V(x_0) = \inf_{|x| \leq c\sqrt{\rho}} V(x);$$

*ii) there exists $x_1 \in \mathbb{R}^N$ such that*

$$V(x_0) - V(x_1) > \frac{\|A\|T}{2\|b\|_1}|x_1|^2;$$

*(II)*

$$\liminf_{|x| \to +\infty} \frac{V(x)}{|x|^2} > -\frac{1}{2c^2\|b\|_1}.$$

*Then, system (S) has at least three solutions in $H_T^1$.*

### 3.2. "Sign changing" potentials

In this subsection another application of Theorem 2.5 of [7], allows us to consider more general potentials, namely functions $F$ satisfying

(F) there exist non-negative functions $a, \bar{a} \in C^0(\mathbb{R}_0^+)$ and $b, \bar{b} \in L^1([0,T])$ such that

$$|F(t,x)| \le a(|x|)b(t), \qquad |\nabla_x F(t,x)| \le \bar{a}(|x|)\bar{b}(t).$$

For system

$$(S) \quad \begin{cases} \ddot{u} = A(t)u + \nabla_x F(t,u) & \text{a.e. in } [0,T] \\ u(0) - u(T) = \dot{u}(0) - \dot{u}(T) = 0 \end{cases}$$

we obtain the following result:

**Theorem 3.2.** ([4] Theorem 2) *Assume (F) and*

*(I) there exist $\rho > 0$ and $x_0 \in \mathbb{R}^N$ such that*

   *i)*

$$\max_{[0,c\rho]} \bar{a} < \frac{\rho}{c\|\bar{b}\|_1};$$

   *ii)*

$$\int_0^T F(t,x_0)dt < -\left\{\left(\max_{[0,c\rho]} a\right)\|b\|_1 + \frac{1}{2}|x_0|^2 T\|A\|\right\};$$

*(II) there is $M > 0$ and a non-negative function $k \in L^1([0,T])$ with $\|k\|_1 < \frac{1}{2c^2}$ such that*

$$F(t,x) \ge -k(t)|x|^2 \qquad \text{for } |x| \ge M \text{ and a.e.in } [0,T].$$

*Then, system (S) has at least three solutions in $H_T^1$.*

**Remark 3.1.** Note that when $F(t,x) = b(t)V(x)$ condition $(II)$ of Theorem 3.1 and condition $(II)$ of Theorem 3.2 are equivalent.

## 3.3. *The perturbed case*

The last subsection is devoted to a perturbed nonlinearity of the type

$$F(t, u) = b(t)V(u) + \lambda G(t, u)$$

where $b : [0, T] \to \mathbb{R}$ is a bounded function, with $\operatorname{essinf}_{[0,T]} b > 0$, $V : \mathbb{R}^N \to \mathbb{R}$ is a continuously differentiable function, $\lambda$ is a positive parameter, $G(t, x)$ is measurable in $t$ for all $x \in \mathbb{R}^N$ and continuously differentiable in $x$ for a.e. $t \in [0, T]$ and satisfying $\sup_{|x| \le s} |\nabla_x G(\cdot, x)| \in L^1([0, T])$ for every $s > 0$, $G(\cdot, 0) \in L^1([0, T])$. In this case $A(t) = b(t)A$ where $A$ is a $N \times N$ symmetric matrix, satisfying

$$Ax \cdot x \ge \alpha |x|^2$$

for all $x \in \mathbb{R}^N$, for some positive constant $\alpha$.

Our system reads as follows:

$$(S_\lambda) \quad \begin{cases} \ddot{u} = b(t)(Au + \nabla V(u)) + \lambda \nabla_x G(t, u) & \text{a.e. in [0,T]} \\ u(0) - u(T) = \dot{u}(0) - \dot{u}(T) = 0. \end{cases}$$

The next result is an application of Theorem 6 of [8]. In this paper Ricceri proved that when sufficient information about the set of global minima of some functional are known, then it is possible to estimate from below the number of local minima of suitable perturbations of the same functional.

Before stating our result we introduce in the space $H_T^1$ the functional

$$\Psi(u) = \frac{1}{2}\|u\|^2 + \int_0^T \alpha(t)V(u(t))dt.$$

**Theorem 3.3.** ([5] Theorem 1) *Assume*

*(I)* $\liminf\limits_{|x| \to +\infty} \dfrac{V(x)}{|x|^2} > -\dfrac{\alpha}{2}.$

*(II) Let $H(x) = \frac{1}{2}Ax \cdot x + V(x)$. Assume that the set of global minima of $H$ has at least $k$ connected components in $\mathbb{R}^N$ ($k \ge 2$).*

*Then, for every $r > \|b\|_1 \inf_{\mathbb{R}^N} H$, there exists $\lambda_r > 0$ such that, for every $\lambda \in ]0, \lambda_r[$, problem $(S_\lambda)$ has at least $k$ solutions in $\Psi^{-1}(] - \infty, r[)$.*

## 4. Examples

We first give an example of nonlinearity to which it is possible to apply Theorem 3.1.

**Example 4.1.** Let $N = 2$, $\xi_0 = (\frac{1}{2}, \frac{1}{2})$ and $V$ defined by

$$V(u) = -\cos(|u - \xi_0|^2 + 1).$$

Let $A$ a symmetric matrix satisfying (1) and

$$\|A\| < \frac{\min\{1,\alpha\}}{\max\{T^2,1\}} \left(\sqrt{\frac{3}{2}\pi - 1} - \frac{\sqrt{2}}{2}\right)^2,$$

$b : [0,T] \to \mathbb{R}$ a non-negative function satisfying

$$\|b\|_1 > \|A\|T\frac{(2\pi - 1)^2}{2(1 - cos1)}.$$

With this choice of the data, an application of Theorem 3.1 yields the existence of at least three solutions in $H^1_T$ for system $(S)$.

In the next two examples we are going to apply Theorem 3.2. In the first one we consider a nonlinearity $F(t,x) = b(t)V(x)$ where $b$ is allowed to change sign; in the second one we deal with a nonlinearity where we can not "split" the variables.

**Example 4.2.** Let $[0,T] = [0,\frac{3}{2}\pi]$, $\gamma$ and $r$ two positive constants satisfying

$$\gamma < \frac{\min\{1,\alpha\}}{36\pi}, \qquad (r + \sqrt[4]{\frac{3}{2}\pi})^2 > 1 + \frac{3}{4}\sqrt{\frac{3}{2}\pi}\frac{\|A\|}{\gamma}$$

respectively and $\xi_0 = r\delta_{1i}$, where $\delta_{ij} = 0$ for $i \neq j$, $\delta_{ij} = 1$ for $i = j$.
Define

$$F(t,x) = \gamma(\sin t)|x + \xi_0|^2 \sin |x|^4.$$

Then, system $(S)$ has at least three solutions.

**Example 4.3.** Define

$$F(t,x) = \begin{cases} (t + |x|^3)\sin |x|^4 & \text{if } |x| \leq M \\ (t + M^3) + 3M^2(|x| - M) & \text{if } |x| > M \end{cases}$$

where $M^4 = \frac{\pi}{2} + 2n\pi$, $n$ big enough. Theorem 3.2 applied to the nonlinearity $F$ gives at least three solutions in $H^1_T$ for system $(S)$.

Finally we give an example of application of Theorem 3.3.

**Example 4.4.** Let $A$ be the identity matrix $(\alpha = 1)$, $f \in C^1([0,+\infty[)$ be a periodic function such that $f(0) > b = \inf_{\mathbb{R}} f$, $q \in ]0,1[$ and $p \in [2,+\infty[$. Define $V$ as follows:

$$V(x) = \begin{cases} (f(|x|^{-q}) - b)|x|^p - \frac{|x|^2}{2} & \text{if } x \neq 0 \\ 0 & \text{if } x = 0 \end{cases}$$

Note that $H(x) = (f(|x|^{-q}) - b)|x|^p$. The set of its global minima is $\{x \in \mathbb{R}^N : f(|x|^{-q}) = b\} \cup \{0\}$, which has infinitely many connected components.

Then, with this choice of $V$, for every $k \geq 2$, $(S_\lambda)$ has at least $k$ solutions for $\lambda$ small enough.

**Remark 4.1.** Note that conditions $(BN)$ (in Examples 4.1, 4.2, 4.3) and $(BN_*)$ (in Example 4.4) do not hold.

## References

1. I.Birindelli, Periodic solutions for a class of second order systems with a small forcing term, *Nonlinear Anal.* **27** (1996), 261–270.
2. H.Brezis and L.Nirenberg, Remarks on finding critical points, *Comm. Pure Appl. Math.* **44** (1991), 939-963.
3. F.Faraci, Three periodic solutions for a second order nonautonomous system, *J. Nonlinear Convex Anal.* **3** (2002), 393-399.
4. F.Faraci, Multiple periodic solutions for second order systems with changing sign potential, *J. Math. Anal. Appl.* **319** (2006), 567–578
5. F.Faraci and A.Iannizzotto, A multiplicity theorem for a perturbed second order nonautonomous system, *Proc. Edinb. Math. Soc.* **49** (2006), 267–275.
6. P.Pucci and J.Serrin, A mountain pass theorem, *J. Differential Equations* **60** (1985), 142-149.
7. B.Ricceri, A general variational principle and some of its applications, *J. Comput. Appl. Math.* **113** (2000), 401-410.
8. B.Ricceri, Sublevel sets and global minima of coercive functionals and local minima of their perturbations, *J. Nonlinear Convex Anal.* **5** (2004), 157-168.
9. C.L.Tang, Periodic solutions of nonautonomous second order systems with sublinear nonlinearity, *Proc. Amer. Math. Soc.* **126** (1998), 3263-3270.
10. C.L.Tang, Existence and multiplicity of periodic solutions of nonautonomous second order systems, *Nonlinear Anal.* **32** (1998), 299-304.
11. C.L.Tang, Multiplicity of periodic solutions for second order systems with a small forcing term, *Nonlinear Anal.* **38** (1999), 471-479.
12. C.L.Tang and X.P.Wu, Periodic solutions for second order systems with not uniformly coercive potential, *J. Math. Anal. Appl.* **259** (2001), 386-397.

# A PURELY VECTORIAL CRITICAL POINT THEOREM

BIAGIO RICCERI

*Department of Mathematics, University of Catania*
*Viale A. Doria 6*
*95125 Catania, Italy*
*E-mail: ricceri@dmi.unict.it*

In this paper, we establish a critical point theorem for functionals on vector spaces endowed with a sequential structure in the most general sense.

**Key words:** Vector space, critical point, sequential convergence structure
**Mathematics Subject Classification:** 58E05

The aim of this very short contribution is to establish a most general version of Theorem 2.6 of [1]. In fact, detecting the truly necessary tools in the proof, we formulate the new result in a purely vectorial setting, using also the most general notion of a sequential convergence structure.

Here, we are simply interested in the formulation and proof of this result. We plan to offer some applications in future papers.

We start by giving the relevant definitions.

Let $X$ be a set in a real vector space $E$. We say that $X$ is radial at $x_0 \in X$ if for each $y \in E$ there is some $\delta > 0$ such that $x_0 + \lambda y \in X$ for all $\lambda \in [0, \delta]$. We say that $X$ is algebraically open if it is radial at each of its points.

Let $J : X \to \mathbf{R}$ and let $X$ be radial at $x_0 \in X$. We say that $J$ is differentiable at $x_0$ in algebraic sense if there exists a functional $g : E \to \mathbf{R}$ such that

$$\lim_{\lambda \to 0} \frac{J(x_0 + \lambda y) - J(x_0)}{\lambda} = g(y)$$

for all $y \in E$.

The functional $g$ is the derivative of $J$ at $x_0$ and is denoted by $J'(x_0)$.

We say that $u_0 \in X$ is a local minimum in algebraic sense for $J$ if there exists a set $U \subseteq E$ which is radial at $u_0$, such that $J(u_0) \leq J(x)$ for all $x \in U \cap X$.

In a classical way, one proves

PROPOSITION 1. - *If $X$ is radial at $u_0$ and if $u_0$ is a local minimum for $J$ in algebraic sense, then $J'(u_0) = 0$.*

898

Let $Y$ be a non-empty set. A sequential convergence structure on $Y$ is any pair $(\mathcal{A}, F)$ where $\mathcal{A}$ is a non-empty set of $Y^{\mathbf{N}}$ and $F$ is a multifunction from $\mathcal{A}$ into $Y$, with non-empty values. If $\{x_n\} \in \mathcal{A}$, we say that $\{x_n\}$ is a convergent sequence and that any point belonging to $F(\{x_n\})$ is a limit of $\{x_n\}$.

The set $Y$ is said to be a sequential convergence space if it is considered with a given sequential convergence structure on it.

Let $Y$ be a sequential convergence space and $C$ a subset of $Y$.

$C$ is said to be sequentially compact if every sequence in $C$ admits a convergent subsequence a limit of which belongs to $C$.

$C$ is said to be sequentially closed if for every convergent sequence in $C$ the set of its limits is contained in $C$.

A function $f : C \to \mathbf{R}$ is said to be sequentially lower semicontinuous at $x_0 \in C$ if for every convergent sequence in $C$ of which $x_0$ is a limit, one has

$$f(x_0) \leq \liminf_{n \to \infty} f(x_n) \ .$$

The function $f$ is said to be sequentially lower semicontinuous if it is so at each point of $C$.

In a classical way, one proves

PROPOSITION 2. - *If $C$ is non-empty and sequentially compact and if $f$ is sequentially lower semicontinuous, then $f$ has a global minimum in $C$.*

We now are in a position to state our result:

THEOREM 1. - *Let $X$ be a non-empty algebraically open subset of a real vector space $E$ and let $J : X \to \mathbf{R}$ be a function which is differentiable in algebraic sense at each point of $X$.*
*Assume that there exist a sequential convergence structure on $X$, a sequentially lower semicontinuous function $\Psi : X \to \mathbf{R}$ which is differentiable in algebraic sense at each point of $X$, a decreasing sequence $\{\rho_n\}$ in $]\inf_X \Psi, +\infty[$ converging to $\inf_X \Psi$, a sequence $\{K_n\}$ of sequentially compact subsets of $X$ and a constant $\lambda^*$ such that*
*(a) the set $\Psi^{-1}(] -\infty, \rho_n[)$ is algrebraically open and is contained in $K_n$ for all $n \in \mathbf{N}$ ;*
*(b) $\inf_{x \in \Psi^{-1}(]-\infty, \rho_n[)} \frac{J(x) - \inf_{K_n} J}{\rho_n - \Psi(x)} < \lambda^*$ for all $n \in \mathbf{N}$ ;*
*(c) the function $J + \lambda^* \Psi$ is sequentially lower semicontinuous ;*
*(d) the set*

$$\{x \in X : J'(x) + \lambda^* \Psi'(x) = 0\}$$

*is sequentially closed.*
*Then, there exists a global minimum $x^*$ of $\Psi$ such that $J'(x^*) = 0$.*

PROOF. Fix $n \in \mathbf{N}$. Thanks to (b), there exists some $x_n \in X$ satisfying $\Psi(x_n) < \rho_n$ and

$$J(x_n) - \inf_{K_n} J < \lambda^*(\rho_n - \Psi(x_n)) \ . \tag{1}$$

Note also that, since $\Psi^{-1}(]-\infty, \rho_n[) \subseteq K_n$, one has $\lambda^* > 0$. Due to $(c)$ and to the sequential compactness of $K_n$, thanks to Proposition 2, the restriction of the function $J + \lambda^* \Psi$ to $K_n$ has a global minimum, say $y_n$. We claim that $\Psi(y_n) < \rho_n$. Indeed, arguing by contradiction, assume that $\Psi(y_n) \geq \rho_n$. Then, by (1), we would have

$$J(x_n) + \lambda^* \Psi(x_n) < \lambda^* \rho_n + \inf_{K_n} J \leq \lambda^* \Psi(y_n) + J(y_n)$$

which is absurd. Consequently, since $\Psi^{-1}(]-\infty, \rho_n[)$ is algebraically open, $y_n$ is a local minimum point in algebraic sense for the function $J + \lambda^* \Psi$, and hence, by Proposition 1, we have

$$J'(y_n) + \lambda^* \Psi'(y_n) = 0 . \tag{2}$$

Now, note that the sequence $\{y_n\}$ is contained in $K_1$ which is sequentially compact. Hence, there exists a sub-sequence $\{y_{n_k}\}$ converging to some point $y^* \in K_1$. Note that, by (2) and $(d)$, we have

$$J'(y^*) + \lambda^* \Psi'(y^*) = 0 . \tag{3}$$

Further, since $\Psi$ is sequentially lower semicontinuous, we have

$$\Psi(y^*) \leq \liminf_{k \to \infty} \Psi(y_{n_k}) = \inf_X \Psi .$$

So, $y^*$ is global minimum of $\Psi$ and hence, since $X$ is algebraically open, we have $\Psi'(y^*) = 0$. Now, the conclusion follows from (3). △

## References

1. B. Ricceri. *A general variational principle and some of its applications.* J. Comput. Appl. Math., **113** 401–410, 2000.

# ON MINIMIZATION IN INFINITE DIMENSIONAL BANACH SPACES

E. BRÜNING

*School of Mathematical Sciences*
*University of KwaZulu-Natal*
*Private Bag X54001*
*Durban 4000, South Africa*
*E-mail: bruninge@ukzn.ac.za*

After a brief review of the main strategies to prove the existence of solutions to minimization problems in infinite dimensional reflexive Banach spaces, we point out some shortcomings. We then present a framework in which we can give necessary and sufficient conditions for the existence of minimizers. These necessary and sufficient conditions are then transformed into powerful sufficient conditions. A concrete realization of these sufficient conditions is developed for the case of functionals relevant to the study of the existence of solutions of nonlinear PDE's. We conclude with an overview of several existing applications and an outlook on possible future developments.

**Key words:** Existence of minimizers, necessary and suffcient conditions, systems of quasilinear PDE's
**Mathematics Subject Classification:** 49J45, 49K27, 49J27, 35D05

## 1. Introduction

### 1.1. *Brief review of existing strategies*

Typically minimization problems are solved by applying one or the other version of the so-called "Generalized Weierstraß Theorem" (a weakly sequentially lower semi-continuous function $f$ attains its infimum on a bounded and weakly sequentially closed subset $M$ of a real reflexive Banach space $E$). Various other versions are for instance given in chapter 38 of [20] or in [19]. In particular the results have proven to be extremely powerful for the proof of the existence of solutions of many classes of nonlinear partial differential equations [2,5,4,6−10,14,15,18,20,21,19].

The generalized Weierstraß Theorem in the above form relies on the following assumptions:

(a) The underlying space $E$ of "a priori solutions" is a real reflexive Banach space;
(b) The subset $M$ is

   (i) bounded in $E$ and
   (ii) weakly sequentially closed.

(c) The function $f$ is weakly sequentially lower semi-continuous (w.s.l.s.c) on $E$.

We now comment on the possibility of relaxing these assumptions respectively to implement them. It is fairly well understood what can be done when the underlying space $E$ is a not necessarily reflexive Banach space [14,20].

The condition of boundedness of the subset $M$ can be replaced by various somewhat weaker conditions (see for instance chapter 38 of [20] and our condition (c) formulated later in section 2).

In concrete applications the condition of the subset $M$ being weakly sequentially closed is often rather difficult to check. For instance in the variational approach to semi-linear global elliptic partial differential equations where $M$ is some level surface of a sufficiently smooth function $u : E \to \mathbb{R}$ the realization of this condition turns out to be the major difficulty of the problem (chapter IX of [2] and [19]). Here in our further investigations we will work with a weakened version of this condition (see Condition 8).

Finally we comment on the condition of weak sequential lower semi-continuity for the function $f : E \to \mathbb{R}$ to be minimized. In applications to semi-linear elliptic PDE's this condition usually is ensured by the appropriate choice of the underlying Banach space of a priori solutions ([2,20] and references there.) However in various other applications, for instance to quasilinear PDE's, it is not at all straightforward that it can be expected to hold and/or how to implement it and accordingly there have been considerable efforts to implement weak sequential lower semi-continuity in terms of other more easily accessible conditions on $f$ and its Gâteaux-derivative $f'$. These efforts are summarized in Proposition 4.1.8 of [20].

## 1.2. *Some limitations of existing strategies*

It is very important to realize that in all these existence results weak sequential lower semi-continuity on $E$ plays a fundamental role and that one could not do without it. Nevertheless one is aware that weak sequential lower semi-continuity on $M \subset E$ would be sufficient (see for instance Theorem 1.2 of [19]), but typically no method is known to get weak sequential lower semi-continuity on $M$ without knowing it on all of $E$.

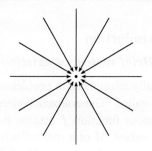

Fig. 1.   Limit along certain directions

In finite dimensional analysis we make effective use of directional limits of functions at certain points and we speak about continuity at a point if all the directional limits exist uniformly and agree. However in problems of infinite dimensional analysis in Banach spaces for the weak topology we typically consider for instance weak sequential lower semi-continuity of a function only on the whole space, not as a property of the function in individual points of the space and typically also no directional restrictions are considered.

The following drawings point into the direction where one can expect that a weakened version of the standard condition of weak sequential lower semi-continuity on the whole space could be sufficient to solve minimization problems. In Figure 2 the set $M$ is a part of a cone in $E$ (it could also be a full cone) while in Figure 3 the set $M$ is a level surface of a (sufficiently smooth) function $g$ on $E$. In both cases weak sequential lower semi-continuity on $E$ is a much more restrictive condition than weak lower semi-continuity on $M$. We are going to develop the necessary concepts and results which allows us to take an intermediate approach. For this the concept of weak lower semi-continuity along directions or along cones of directions is needed, in particular when weak lower semi-continuity of functions at boundary points of the set $M$ has to be investigated.

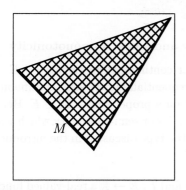

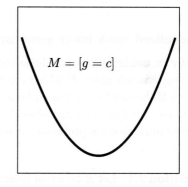

FIGURE 2          FIGURE 3

To formulate such a concept of directional weak lower lower semi-continuity for a function $f$ is straightforward, however one also needs ways to realize respectively to implement it in concrete cases. We will offer a solution in terms of the concept of "weak $K$-monotonicity" for the Gâteaux derivative $f'$ of $f$, for a certain class of subsets $K \subseteq E$. This concept of weak $K$-monotonicity naturally plays a decisive role in solutions of minimization problems.

### 1.3. *Implementing weak sequential lower semi-continuity*

When minimization problems are considered in the variational approach for solving (systems of) nonlinear partial differential equations it is natural to involve the Gâteaux derivative $f'$ of the function $f$ in the hypotheses. In terms of various concepts of monotonicity of $f'$ this is done for instance in Proposition 4.1.8 of [20].

For functionals of the form

$$f(u) \stackrel{\text{def}}{=} \int_\Omega F(x, u(x), Du(x))dx \tag{1}$$

for some bounded domain $\Omega \subset \mathbb{R}^n$ with Lipschitz boundary defined on some Sobolev space $E = W^{1,p}(\Omega)$ the characterization of weak sequential lower semi-continuity has a long history [14,20]. Under certain technical restrictions on $F$ the functional $f$

is w.s.l.s.c. if, and only if, the function $y \mapsto F(x, u, y)$ is convex [14,17,18,1] and this then implies that the Euler equation for $f$, i.e the equation $f'(u) = 0$, is elliptic.

The case of vector-valued functions $u : \Omega \to \mathbb{R}^m, m > 1$ is much more complicated and various concepts of convexity have been suggested and used with some success. Finally a comprehensive characterization of weak sequential lower semi-continuity of the functional (1) in terms of quasi-convexity of the integrand $F$ could be established (see again [14,17,18,1] and references there). Unfortunately these results do not indicate how to check or how to implement quasi-convexity in concrete cases, arising for instance in the variational approach to PDE's. So the question to establish weak sequential lower semi-continuity offers still substantial challenges which we suggest to meet, and we offer an answers in terms of the concept of weak $K$-monotonicity.

## 2. Localized weak lower semi-continuity and weak $K$-monotonicity

Usually in minimization problems we consider continuous functionals $f : E \to \mathbb{R}$ on a real Banach space $E$ which are weakly sequentially lower semi-continuous on $E$, even if the functional $f$ is to be minimized on a proper subset $M \subset E$. Here we introduce a localized concept of weak sequential lower semi-continuity which will be useful in minimization problems for sets $M$ of the type discussed in the introduction (see [13]).

**Definition 2.1.** Let $E$ be a real Banach space and $f : E \to \mathbb{R}$ a real-valued function on $E$. Furthermore, let $K$ be any nonempty subset of $E$ such that $0 \in E$ belongs to the weak sequential closure of $K$, i.e., $0 \in \overline{K}^w$. The functional $f$ is called $K$**-weakly sequentially lower semi-continuous at** $x \in E$ if, and only if, for all infinite sequences $\underline{a} = (a_i)_{i \in \mathbb{N}}$ of elements $a_i \in K$ with $w - \lim_{i \to \infty} a_i = 0$ one has

$$f(x) \le \liminf_{i \to \infty} f(x + a_i). \tag{2}$$

$f$ is called $K$**-weakly sequentially lower semi-continuous on a subset** $M \subset E$ if, and only if, it is $K$-weakly sequentially lower semi-continuous at all points $x \in M$.

Obviously, for $K = E$ this definition gives just the standard definition of weak sequential lower semi-continuity.

Before we proceed we introduce the important concept of weak $K$-monotonicity (see [11]).

**Definition 2.2.** Let $K \subset E$ be a nonempty subset such that $0 \in E$ is in the weak sequential closure of $K$, i.e., $0 \in \overline{K}^w$. For such a subset we define: A mapping $T : E \to E'$ is called **weakly K-monotone at** $u \in E$ if, and only if, for every sequence $\underline{a} = (a_i)_{i \in \mathbb{N}} \subset K$ which converge weakly in $E$ to 0 one has

$$\limsup_{i \to \infty} \langle T(u + a_i), a_i \rangle \ge 0. \tag{3}$$

Note that this condition is actually equivalent to the following

$$\limsup_{i \to \infty} \langle T(u + a_i) - T(u), a_i \rangle \geq 0$$

which is closer in form to the original definition of monotonicity.

Note also that in this definition $K$ can be any subset of $E$ with the property $0 \in \overline{K}^w$. And if we know weak $K$-monotonicity for one such subset $K_0$ we know it for all other subsets with this property which are contained in $K_0$.

There is a natural way to implement $K$-weak sequential lower semi-continuity, for a fairly general class of subsets $K$, in terms of weak $K$-monotonicity of its Gâteaux derivative $f'$.

**Prop 2.1.** Suppose that $f : E \to \mathbb{R}$ is a continuous real-valued function on a real Banach space $E$ which has a hemi-continuous and locally bounded Gâteaux derivative $f'$. Let $K \subset E$ be a subset such that $0 \in \overline{K}^w$ and $t \cdot K \subseteq K$ for all $0 < t \leq 1$.

If $f'$ is weakly $K$-monotone at $x \in E$ then $f$ is $K$-weakly sequentially lower semi-continuous at $x$.

**Proof.** For the indirect proof assume that $f$ is not $K$-weakly sequentially lower semi-continuous at $x$. Then there is an infinite sequence $\underline{a}$ with $a_i \in K$ for all $i \in \mathbb{N}$ such that $w - \lim_{i \to \infty} a_i = 0$ and

$$f(x) > \liminf_{i \to \infty} f(x + a_i).$$

Then, by definition of the $\liminf$ there are $\delta > 0$ and a subsequence (denoted in the same way) such that $f(x + a_i) - f(x) < -2\delta$. For any $\epsilon > 0$ we have, for some $t_i = t_i(\epsilon) \in (0, 1)$,

$$f(x + \epsilon a_i) - f(x) = \epsilon \langle f'(x + \epsilon t_i a_i), a_i \rangle.$$

Since $f'$ is locally bounded,

$$\sup\{|\langle f'(x + s a_i), a_i \rangle| : i \in \mathbb{N}, \ 0 \leq s \leq 1\} = C < \infty.$$

Introduce $\epsilon_0 = \min\{\frac{3}{4}, \frac{\delta}{C}\}$. Then, for every $0 < \epsilon < \epsilon_0$ we find $|f(x + \epsilon a_i) - f(x)| < \delta$ for all $i \in \mathbb{N}$. Since

$$f(x + a_i) - f(x + \epsilon a_i) + f(x + \epsilon a_i) - f(x) < -2\delta$$

we deduce

$$f(x + a_i) - f(x + \epsilon a_i) < -\delta \quad \forall i \in \mathbb{N}, \forall \epsilon \in (0, \epsilon_0).$$

Since $f'$ is hemi-continuous there is $t_i \in (\epsilon, 1)$ such that

$$\int_{\epsilon}^{1} \langle f'(x + t a_i), a_i \rangle dt = (1 - \epsilon)\langle f'(x + t_i a_i), a_i \rangle,$$

and, since

$$f(x + a_i) - f(x + \epsilon a_i) = \int_\epsilon^1 \langle f'(x + t a_i), a_i \rangle dt$$

we get, for all $i \in \mathbb{N}$,

$$\langle f'(x + t_i a_i), t_i a_i \rangle < -t_i \frac{\delta}{1 - \epsilon} < -\frac{\epsilon \delta}{1 - \epsilon}.$$

By assumption on $K$ we know that $b_i = t_i a_i$ also belongs to $K$ and this sequence satisfies $w - \lim_{i \to \infty} b_i = 0$. The above estimate implies

$$\limsup_{i \to \infty} \langle f'(x + b_i), b_i \rangle \leq -\frac{\epsilon \delta}{1 - \epsilon}$$

which contradicts (3). Thus we conclude. □

## Remark 2.3.

(1) The core of the argument in the proof of this proposition is the similar to that in the proof of the fact that condition (P) of Browder and Hess implies weak sequential lower semi-continuity (see [3]).

(2) Later in applications to minimization problems we will present a concrete way to implement weak $K$-monotonicity of $f'$, for certain subsets $K$ of the class considered above.

## 3. Basic minimization

We proceed in two steps: At first we prove a criterium for the existence of a minimum for $(f, M)$ in terms of weak $K$-monotonicity for $f'$. Here the set $K$ depends in a subtle way on the data of the problem. Then we present some ways to estimate the set $K$ so that the necessary and sufficient conditions for the existence of a minimizer can be implemented. In this way some sufficient conditions for the existence of a minimizer can be obtained without assuming or implementing first weak sequential lower semi-continuity of $f$ on $E$.

Let $E$ be a real reflexive Banach space and $f : E \to \mathbb{R}$ a continuous function on $E$ which has a hemi-continuous Gâteaux-derivative $f' : E \to E'$. Then it follows for all $u, v \in E$

$$f(u) - f(v) = \int_0^1 \langle f'(v + t(u - v)), u - v \rangle dt. \tag{4}$$

For a nonempty subset $M$ of $E$ without isolated points introduce

$$I = I(f, M) = \inf\{f(v)|\, v \in M\} \in [-\infty, +\infty) \tag{5}$$

and

$$\mu = \mu(f, M) = \{u \in M|\, f(u) = I = I(f, M)\}. \tag{6}$$

Clearly, if $I$ is finite, $\mu(f, M)$ denotes the set of minimizers for the pair $(f, M)$.

We study the problem of the existence of minimizers for $(f, M)$ under the following additional hypotheses:

(H)

(a) $f' : E \to E'$ is locally bounded;
(b) $f$ is bounded from below on M;
(c) The pre-image under $f_M = f|M$ of a bounded set in $\mathbb{R}$ is a bounded set in the Banach space $E$.

While (b) is a necessary condition for our problem conditions (a) and (c) are not unnatural convenient technical restrictions on the class of functions considered. In particular (a) implies in conjunction with (4) that the function f is locally bounded too.

There are a several ways to realize conditions (b) and (c). The simplest case is to assume that $M$ is bounded in $E$. Then condition (c) is trivially satisfied while condition (b) follows from the fact that $f$ is locally bounded. When $M$ is not bounded in $E$ the standard way to implement conditions (b) and (c) is to assume that $f$ is coercive on $M$, i.e., $f(u) \to \infty$ for $u \in M$ and $\|u\| \to \infty$.

By assumption (b) the number $I = I(f, M)$ is finite and thus there are minimizing sequences for our minimization problem. Assumption (c) now assures that there are minimizing sequences which converge weakly in $E$.

Note that in our basic hypothesis (H) we did not assume any kind of closedness of $M$. Certainly one cannot proceed without such an assumption. Accordingly we assume that the set $S = S(f, M)$ of minimizing sequences $\underline{u} = (u_i)_{i \in \mathbb{N}}$ in $M$ for the minimization problem $(f, M)$ with

$$u = w - \lim_{i \to \infty} u_i \in M \tag{7}$$

is not empty:

$$S = S(f, M) \neq \emptyset. \tag{8}$$

The easiest way to have (8) is to assume that $M$ is weakly sequentially closed. If, for instance, $M$ is a level set of a sufficiently smooth function $g$ on $E$, then one has to prove (8). In many important minimization problems for solving semi-linear (global elliptic) partial differential equations this is actually the major difficulty [2,19].

As the following criterium shows, in terms of weak $K$-monotonicity one can formulate necessary and sufficient conditions for the existence of a minimizer.

**Theorem 3.1.** *Let $E$ be a real reflexive Banach space and $M \subseteq E$ a nonempty subset without isolated points. Assume furthermore that $f : E \to \mathbb{R}$ is a continuous functions with hemi-continuous Gâteaux derivative $f'$ such that hypothesis (H) and condition (8) hold. Then the minimization problem for $(f, M)$ has a solution if, and only if, there is a minimizing sequence $(\underline{u}) = (u_j) \in S$ such that $f'$ is weakly $K_\varepsilon(\underline{u})$-monotone at $u = w - \lim u_j \in M$ for all $0 < \varepsilon \leq \varepsilon_0$ for some $0 < \varepsilon_0 \leq 3/4$ where*

$$K_\varepsilon(\underline{u}) = \{v \in E \mid v = t(u_j - u), \ \varepsilon \leq t \leq 1, \ j = 1, 2, \ldots\} \tag{9}$$

**Proof.** Necessity: Suppose that our minimization problem has a solution $u \in M$. Then there is a sequence $\underline{u} = (u_i)_{i \in \mathbb{N}}$ in $M$ which converges strongly in $E$ to $u$. Now consider any $0 < \varepsilon < \varepsilon_0$ as specified in the condition of the theorem. Take any sequence $\underline{a} = (a_i)_{i \in \mathbb{N}}$ in $K_\varepsilon(\underline{u})$ which converges weakly to $0 \in E$. The elements $a_i$ of such a sequence are of the form $a_i = t_i(u_{j(i)} - u)$ for some subsequence of the sequence $\underline{u}$ and some $t_i \in [\varepsilon, 1]$. It follows

$$\limsup_{i \to \infty} \langle f'(u + a_i), a_i \rangle = \limsup_{i \to \infty} \langle f'(u + t_i(u_{j(i)} - u)), t_i(u_{j(i)} - u) \rangle = 0.$$

Hence $f'$ is weakly $K_\varepsilon(\underline{u})$-monotone at $u$, for any fixed $0 < \varepsilon_0 \leq 3/4$ and any $0 < \varepsilon \leq \varepsilon_0$. Since the sequence $\underline{u}$ belongs to $S$ we conclude that the condition of the theorem is necessary.

Sufficiency: Suppose that there is $\underline{u} \in S$ such that for some $0 < \varepsilon_0 \leq 3/4$ $f'$ is weakly $K_\varepsilon(\underline{u})$-monotone at $u = w - \lim u_j \in M$ for all $0 < \varepsilon \leq \varepsilon_0$. We have to show that our minimization problem has a solution.

Proposition 2.1 implies that $f$ is $K_\varepsilon(\underline{u})$-weakly lower semi-continuous at $u$, for all $\varepsilon \in (0, \varepsilon_0)$; hence (2) holds for all $\underline{a} \in K_\varepsilon(\underline{u})$, in particular for $a_i = u_i - u$ and thus, by (2),

$$f(u) \leq \liminf_{i \to \infty} f(u_i).$$

Since $\underline{u} \in S$, we know

$$\liminf_{i \to \infty} f(u_i) = I$$

and $f(u) \geq I$. We conclude $f(u) = I$ and our minimization problem $(f, M)$ has a minimizer $u$. $\square$

## Remark 3.2.

(a) In its sharp form the condition of this theorem is nearly impossible to check or to implement. Thus Theorem 3.3 is more of theoretical value. Nevertheless it gives precise guidance about what one can expect and what one has to do. This will be shown by looking at various special cases.

(b) Suppose that the Gâteaux-derivative of $f$ is weakly $K$-monotone at $u \in M$. Then clearly we can successfully apply Theorem 3.3 when we can find a sequence $\underline{u} \in S$ with $u = w - \lim_{j \to \infty} u_j$ and some $0 < \varepsilon_o < 1$ such that $K_\varepsilon(\underline{u}) \subseteq K$ for all $0 < \varepsilon \leq \varepsilon_0$. For a suitable choice of $K$ this can happen for all $\underline{u} \in S$, because of the geometry of the set $M$.

(c) Suppose $f'$ is weakly $E$-monotone. Then for any $\underline{u} \in S$ we know that $f'$ is weakly $K_\varepsilon(\underline{u})$-monotone at $u = w - \lim u_j \in M$ for all $0 < \varepsilon \leq \varepsilon_0$, for any $0 < \varepsilon_0 \leq 3/4$. Theorem 3.1 implies that the minimization problem for the pair $(f, M)$ has a solution. Since weak $E$-monotonicity is the same as condition (P) (see section 2) which is known to imply in the context of Theorem 3.3 weak sequential lower semi-continuity of $f$ we have thus a new version of a generalized Weierstraß Theorem.

In Theorem 3.1 we considered a fairly general class of sets $M$ on which a function $f$ was to be minimized. In this general situation we have no à priori knowledge about the possible location of the points of a minimizing sequence for the pair $(f, M)$. If however the set $M$ is restricted in certain ways one can often use some "estimate" for the location of these points and thus gain some information about the "size" and the location of the sets $K_\epsilon(\underline{u})$, $\underline{u} \in S$, used in Theorem 3.3. According to Remark 3.4(b) it suffices to have some estimate for the sets $K_\varepsilon(\underline{u})$ from above, $K_\varepsilon(\underline{u}) \subseteq K$, and to check weak $K$-monotonicity of $f'$ for a successful application of Theorem 3.3.

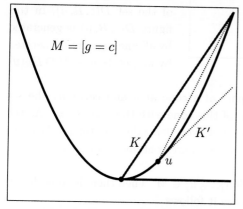

A particularly promising example of this situation is the case of a constrained minimization problem where the set $M$ is the level surface $g^{-1}(c)$ of some smooth function $g : E \to \mathbb{R}$ (see the figure to the left). Whatever $\underline{u} \in S$ and $0 < \epsilon < 1$ is, the set $K_\epsilon(\underline{u})$ will be contained in the cone $K$ drawn with solid lines in the figure on the left. One instance is indicated. Suppose $\underline{u} \in S$ has $u \in M$ as weak limit. Then $K_\epsilon(\underline{u})$ is contained in the cone $K'$ drawn with dotted lines. Certainly $K' \subset K$.

Naturally to left hand side of the graph of $g$ the same considerations apply.

A quantitative version of this idea is given in the next theorem which we have to prepare by some definitions.

For $R > 0$ denote by $B_R$ the open ball in the Banach space $E$ with radius $R$ centered at $u = 0$. For any $0 < \varepsilon \leq 1$ and any $u \in M$ introduce the sets

$$D(\varepsilon, R, u) = \{w \in E \mid w = t(v - u), \ \varepsilon \leq t \leq 1, \ v \in M, \ v - u \in B_R\} \qquad (10)$$

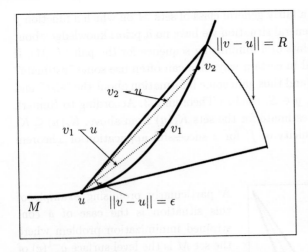

that is the set of line segments in $E$ generated by all difference vectors $u - v$ of points $u$ of $M$ in the neighborhood $B_R(v) = v + B_R$ of the point $v \in M$. In our figure above $D(\varepsilon, R, u)$ is contained in the intersection of the cone $K'$ with the ball $B_R$. The figure to the left gives some details of the set $D(\varepsilon, R, u)$. In this figure $D(\varepsilon, R, u)$ is generated by all line segments generated by all points $v_i \in M \cap B_R(u)$

of which two are shown.

The two blue solid lines indicate a part of a cone at $u$ and certainly the set $D(\varepsilon, R, u)$ is contained in the intersection of this cone with the ball $B_R(u)$. As the point $u$ runs through $M$ this cone changes; however, depending on the set $M$ there might be a cone $C$ in $E$ such that all these cones are contained in $C$, compare the figure above.

Now let $\underline{u} \in S$ be given with $u = w - \lim_{j \to \infty} u_j \in M$. Then there is some $R > 0$ such that $\|u_j - u\| < R$ for all $j \in \mathbb{N}$. Hence it follows

$$K_\varepsilon(\underline{u}) \subseteq D(\varepsilon, R, u).$$

Next we formulate a geometric restriction on the set $M$.

**Definition 3.3.** Suppose that $C \subset E$ is some convex non-empty cone in $E$. We say that $M$ satisfies the *cone condition* for the cone $C$ at a point $v \in M$ if, and only if, for $R > 0$ one has

$$D(\varepsilon, R, v) \subseteq C \cap B_R. \tag{11}$$

If there exists a convex cone $C \subset E$ such that M satisfies the cone condition for this cone at every $v \in M$ we say that $M$ satisfies the cone condition for the cone $C$.

**Remark 3.4.**

(a) If $v \in M$ is an interior point $M$ then $D(\varepsilon, R, v)$ contains an open ball in $E$ and thus allows any direction in $E$.

(b) If however $v \in M$ is not an interior point, in particular when $M$ has no interior points at all as in the case of a level surface $M = h^{-1}(c)$, then $D(\varepsilon, R, v)$ fails to contain an open ball centered at the origin of $E$ and the set of "directions" in $D(\varepsilon, R, v)$ is restricted accordingly.

(c) Condition (3.8) means that the set of directions in $D(\varepsilon, R, v)$ is contained in some cone $C \subset E$ depending on the point $v \in M$. Intuitively this requires, in the case of a level surface, that the "curvature" of $M$ does not vary too much (how much actually is allowed depends on the cone $C$).

The following result on the existence of a solution for minimization problems without assuming weak sequential lower semi-continuity appears to be very attractive for applications. It is a straightforward implication of Theorem 3.3 and the above discussion.

**Theorem 3.5.** *Suppose that the pair $(f, M)$ satisfies conditions (H) and (8). In addition assume:*

*(1) $M$ satisfies the cone - condition for a convex cone $C \subset E$.*
*(2) $f' : E \to E'$ is weakly $C \cap B_R$-monotone for $R > 0$.*

*Then the minimization problem for the pair $(f, M)$ has a solution.*

**Proof.** For $\underline{v} = (v_j) \in S$ we know $v = w - \lim_{j \to \infty} v_j \in M$. Now assumption (1) in conjunction with (10) - (11) yields $K_\varepsilon(\underline{v}) \subset D(\varepsilon, R, v) \subset C \cap B_R$, for a suitable $R > 0$. Thus by assumption (2) and Proposition 4.1, part (4), we know that $f'$ is weakly $K_\varepsilon(\underline{v})$-monotone at $v$. Therefore by Theorem 3.1 we conclude. $\square$

**Remark 3.6.** Note the following fact about assumption (2) which follows from our discussion of the relation of the concept of weak $K$-monotonicity with condition (P) in section 2: The smaller the cone $C$ in assumption (2) is, the more is missing to ensure weak sequential continuity of $f$ and thus the stronger is the statement of Theorem 3.5.

## 4. $K$-monotone and weakly $K$-monotone mappings

It is well known that various concepts of monotonicity for nonlinear operators played a very important role in the solution theory of nonlinear partial differential equations. Some of the main references are [7,8,3,10,20]. In the last section we learned about the importance of weakly $K$-monotone mappings for minimization problems. This section introduces basic properties of $K$-monotone and weakly $K$-monotone mappings and discusses their relation to the main established notions of monotonicity.

Let $E$ be a real Banach space and $E'$ its topological dual with canonical bilinear form $\langle \cdot, \cdot \rangle$. In the following $K$ will always denote a nonempty subset of $E$ whose weak closure $\overline{K}^w$ contains $0 \in E$. Our basic definitions are as follows (see [11]):

**Definition** A map $T : E \to E'$ is called

(a) **$K$-monotone at $v \in E$** iff for all $a \in K$

$$\langle T(v + a) - T(v), a \rangle \geq o$$

(b) **weakly $K$-monotone at $v \in E$** iff

$$\overline{\lim_{i \to \infty}} \langle T(v + a_i) - T(v), a_i \rangle \geq o$$

for all sequences $a_i \epsilon K$ converging weakly (in $E$) to zero: $0 = w - \lim_{i \to \infty} a_i$.

(c-d) **$K$-monotone (weakly $K$-monotone)** iff $T$ is $K$-monotone (weakly $K$-monotone) at every $v \in E$.

Clearly whenever a map $T$ is $K$-monotone (at $v \in E$) then $T$ is also weakly $K$-monotone (at $v \in E$). In the literature various notions of monotonicity have been used quite successfully [7,8,3,9,20]. We comment on those which are related to the notions introduced in the above definition. To this end we recall these definitions.

A map $T : E \to E'$ is called

($a'$) **monotone** iff $\langle T(u) - T(v), u - v \rangle \geq 0$ for all $u, v \in E$;

($b'$) **pseudo-monotone** iff for all sequences $u_i \in E$ converging weakly in $E$ to some $u \in E$ the inequality $\overline{\lim_{j \to \infty}} \langle T(u_j), u_j - u \rangle \leq 0$ implies $\langle T(u), u - v \rangle \leq \underline{\lim_{j \to \infty}} \langle T(u_j), u_j - v \rangle$ for all $v \in E$;

($c'$) $T : E \to E'$ is said to satisfy **condition (P)** iff $\overline{\lim_{i \to \infty}} \langle T(u_i), u_i - u \rangle \geq 0$ for all sequences $u_i \in E$ converging weakly to $u \in E$.

There are some obvious relations to Definition 2.1: Monotonicity is the same as $K$-monotonicity for $K = E$ in the sense of part (c). A map $T : E \to E'$ satisfies condition (P) if and only if it is weakly $E$-monotone in the sense of part (d). Furthermore the following implication holds: A pseudo-monotone map $T : E \to E'$ is weakly $K$-monotone for every subset $K \subseteq E$.

Further relations among these concepts of monotonicity a'), b'), and c') and in particular how to implement them are given for instance in propositions 27.6-27.7 of [20]. This implies that there are many weakly $K$-monotone operators.

We mention some elementary properties of (weakly) $K$-monotone maps. For a proof see [11].

## Prop 4.1.

(1) The set of mappings $T : E \to E'$ which are (weakly) $K$-monotone at $v \in E$ is a convex cone.

(2) If $T : E \to E'$ is weakly K-monotone at $v \in E$ and if $S : E \to E'$ is completely continuous at $v$ then $T + S$ is weakly $K$-monotone at $v \in E$.

(3) Let $T : E \to E'$ be weakly $K$-monotone at $v \in E$ and let $A : E \to E$ be a weakly sequentially continuous linear map with adjoint $A' : E' \to E'$. Then $A' \circ T \circ A$ is weakly $A^{-1}(K)$-monotone at $A^{-1}v$.

(4) If $T : E \to E'$ is weakly K-monotone at $v \in E$ and if $K_0$ is some subset of $K$ with $0 \in \overline{K_0}^\sigma$ then $T$ is weakly $K_0$-monotone.

Elementary examples show in particular

(i) There are mappings $T : E \to E'$ which are not monotone but $K$-monotone for some proper subset of $K$ of $E$.
(ii) There are mappings $T : E \to E'$ which are not $K$-monotone but weakly $K$-monotone (for some proper subset $K$ of $E$).
(iii) There are mappings $T : E \to E'$ which are not pseudo-monotone but weakly $K$-monotone for some proper subset $K \subseteq E$.

**Example 1**: Let $E$ be a real Hilbert space and $A$ and $B$ two non-negative bounded linear operators on $E$. Define a mapping $T : E \to E'$ by

$$\langle T(u), v \rangle = (a, Av) - (u, Bv)$$

where $(\cdot, \cdot)$ denotes the inner product of $E$. For all $u,v \in E$ one has $\langle T(u) - T(v), u - v \rangle = (w, Aw) - (w, Bw)$, $w = u - v$, and thus

(1) $T$ is monotone iff $B \leq A$.
(2) Define $K = \{u \in E | (u, Bu) \leq (u, Au)\}$; if $K \neq \emptyset$, then $T$ is $K$-monotone but not monotone if $K \neq E$.
(3) If $A \geq 0$ and if $B$ is compact, then $T$ is weakly $E$-monotone.

**Example 2**: Let $E$ be a real Hilbert space which is densely contained in some larger Hilbert space $H$. Suppose $J : E \to H$ to be some continuous linear operator and let $A, B$ be two bounded linear operators in $H$. Now define a map $T : E \to E'$ by

$$\langle T(u), v \rangle = (Ju, AJv) - (Ju, BJv)$$
$$= (u, J^*AJv)_E - (u, J^*BJv)_E$$

for all $u, v \in E$. Here $(\cdot, \cdot)$ inner product of the Hilbert space $H$, $(\cdot, \cdot)_E$ inner product of the Hilbert space $E$, $J^*$ adjoint of $J$. Case study as in Example 1; interesting case: $J$ compact embedding; concrete case: $E = H^1(\Omega)$, $H = L^2(\Omega)$.

The following lemma shows that a locally bounded map can fail to be weakly $K$-monotone only in the case of "infinite-dimensional" sets $K$, i.e., sets which are not contained in some finite dimensional subspace. This indicates that the concept of a weakly $K$-monotone map is at the core of infinite dimensional minimization.

**Lemma 4.1.** *Suppose $T : E \to E'$ is locally bounded, i.e, $T$ maps bounded subsets of $E$ onto bounded subsets of $E'$ and suppose that $K$ is a subset of $E$ which is contained in some finite dimensional subspace $F$ of $E$. Then $T$ is weakly $K$-monotone.*

**Proof.** The weak and the strong topology agree on finite dimensional subspaces.□

## 5. Applications

In this section we report on some successful applications of these ideas to the minimization of functionals of the form

$$f(u) = \int_\Omega F(x, u(x), Du(x))dx, \tag{12}$$

$\Omega \subset \mathbb{R}^n$ open with Lipschitz boundary, $u : \Omega \longrightarrow \mathbb{R}$ in a suitable Sobolev(-type) space $E$, relevant for the study of quasi-linear partial differential operators. If $\Omega$ is bounded we speak about a local problem, otherwise it is considered a global problem.

We begin with a simple class of linear problems.

### 5.1. *A linear problems*

Let $\Omega \subset \mathbb{R}^n$, $n \geq 3$, be a bounded domain with sufficiently smooth boundary $\partial\Omega$. Then the Sobolev space $E = H^1(\Omega)$ is compactly and continuously embedded by the identity map into the Lebesgue space $H = L^2(\Omega)$.

Consider the functional $f : E \longrightarrow \mathbb{R}$ of class $\mathcal{C}^1$ defined by

$$f(u) = \frac{1}{2}(Du, ADu) + \frac{1}{2}(u, Bu) \tag{13}$$

where $B$ is a bounded symmetric linear operators in the real Hilbert space $H = L^2(\Omega)$ with inner product $(\cdot, \cdot)$. For the symmetric operator $A = (A_{ij})$, $A_{ij} \in L^\infty(\Omega)$, $i, j = 1, \ldots, n$ in $L^2(\Omega)^n$ we assume:

There exists a proper subspace $L$ of $\mathbb{R}^n$ and a number $c > 0$ such that for almost all $x \in \Omega$

$$(i) \quad \sum_{i,j=1}^n \xi_j A_{ij}(x)\xi_j \geq 0 \quad \forall \, \xi = (\xi_1, \ldots, \xi_n) \in L$$

$$\tag{14}$$

$$(ii) \quad \sum_{i,j=1}^n \eta_i A_{ij}(x)\eta_j \leq -c\sum_{j=1}^n \eta_j^2 \quad \forall \, \eta \in L^\perp$$

Introduce the set

$$C = \{u \in E \mid Du(x) \in L \text{ for almost all } x \in \Omega\}. \tag{15}$$

One shows that $C$ is a weakly closed proper linear subspace of $E$.

The *Fréchet*-derivative $f'$ of $f$ is easily calculated. For $u, w \in E$ one has

$$\langle f'(u + w) - f'(u), w \rangle = (Dw, ADw) + (w, Bw)$$

and therefore by assumption (14) and definition (15) for all $w \in C$

$$\langle f'(u + w) - f'(u), w \rangle \geq (w, Bw)$$

and hence, if $w_j \to 0$ weakly in $E$, $w_j \in C$, it follows

$$\liminf_{j\to\infty} \langle f'(u + w_j) - f'(u), w_j \rangle \geq 0$$

since $B$ is a bounded operator in $H$ and $w_j \to 0$ strongly in $H$. Therefore $f'$ is weakly $C$-monotone at every $u \in E$.

Since $C$ is a proper subspace of $E$, $f'$ is not weakly $E$-monotone, and thus $f$ is not w.s.l.s.c. Nevertheless we have

**Corollary 5.1.** $f$ attains a minimum on each set $M \subset E$ which satisfies the 'cone' condition for $C$ and for which $(f, M)$ satisfies hypotheses (H) and (8).

**Proof.** Theorem 3.5 applies. □

**Remark 5.2.** Note that the functional $f$ of (13) has a derivative $f'$ which is neither monotone nor pseudo-monotone and it also does not satisfy condition (P).

### 5.2. *Quasi-linear problems*

Assume that the integrand $F : \Omega \times \mathbb{R}^{n+1} \longrightarrow \mathbb{R}$ in (12) is a Carathéodory function which has partial derivatives $F_j = \frac{\partial F}{\partial y_j}$ with respect to $y_j$, $j = 0, 1, \ldots, n$ which are Carathéodory functions too and which satisfy natural growth restrictions. Explicitly we state the growth restriction only for $F$.

$$|F(x, y_0, \underline{y})| \le a(x) + b(x)|y_0|^{s_0/k} + a|\underline{y}|^p \tag{16}$$

with $0 \le a \in L^1(\Omega)$, $0 \le b \in L^{k'}(\Omega)$, $a > 0$, $1 < p < n$, $s = \frac{pn}{n-p}$, $1 < k < \infty$, $k'$ the Hölder conjugate exponent to $k$.

Denote $\underline{F} = (F_1, \ldots, F_n)$ and for a proper subspace $L \subset \mathbb{R}^n$ assume the condition of restricted monotonicity: For all $y_0 \in \mathbb{R}$, all $\underline{y} = (y_1, \ldots, y_n) \in \mathbb{R}^n$, almost all $x \in \Omega$, and all $\underline{a} \in L$:

$$[\underline{F}(x, y_0, \underline{y} + \underline{a}) - \underline{F}(x, y_0, \underline{y})] \cdot \underline{a} \ge 0. \tag{17}$$

Then, under growth restrictions for the $F_j$ corresponding to (16) and involved estimates (see [11] for details) one can show that the derivative $f'$ of the functional $f$ of Equation (12) is weakly $C$-monotone where $C$ is defined in (15). Hence the analogue of Corollary 5.1 holds in this case too.

### 5.3. *Global eigen-value problems*

For a domain $\Omega \subset \mathbb{R}^n$ with (piecewise) smooth boundary $\partial\Omega$ but not necessarily bounded we look at the problem of finding a number $\lambda \in \mathbb{R}$ and a function $u : \Omega \longrightarrow \mathbb{R}$ such that (at least in the weak sense)

$$A(u) = \lambda B(u) \tag{18}$$

where $A(\cdot)$ is a second order quasi-linear differential operator in divergence form, i.e., for almost every $x \in \Omega$

$$A(u)(x) = F_0(x, u(x), Du(x)) - \sum_{j=1}^{n} D_j F_j(x, u(x), Du(x)) \tag{19}$$

with certain coefficient functions $F_j$ and where $B(\cdot)$ is of zero order:

$$B(u)(x) = H_0(x, u(x)) \qquad (20)$$

for a.e. $x \in \Omega$ with a certain function $H_0$. The coefficient functions $F_j$ and $H_0$ are assumed to satisfy growth restrictions of the type (16) and accordingly we work in the Sobolev type space

$$E = E^{1,p}(\Omega) = \text{completion of } C_0^\infty(\Omega) \text{ with respect to the norm} \qquad (21)$$
$$u \mapsto ||u|| := ||Du||_p \equiv ||Du||_{L^p(\Omega)}.$$

The strategy is standard. Since our differential operators are in divergence form, Equation (18) is rewritten as

$$f'(u) = \lambda h'(u) \qquad (22)$$

where the functional $f$ is of the form (12) with an integrand $F$ such that the coefficient functions $F_j$ in (19) are given by $F_j = \frac{\partial F}{\partial y_j}$, $j = 0, 1, \ldots, n$ and where the functional $h$ is of the form

$$h(u) = \int_\Omega H(x, u(x)) dx, \qquad u \in E \qquad (23)$$

where $\frac{\partial H}{\partial u}$ equals the function $H_0$ in (20). Then (22) can be established by showing that the theorem of the existence of a Lagrange multiplier applies for the minimizer $u$ for the functional $f$ on a level surface $[h = c]$ of the functional $h$. In the present context the standard strategies and results for proving weak sequential lower semi-continuity of $f$ on $E$ do not apply. Accordingly, under the assumption of (17) for almost all $x \in \Omega$, all $y_0 \in \mathbb{R}$, and all $\underline{y}, \underline{a} \in \mathbb{R}^n$ we show that $f'$ is weakly $E$-monotone and apply Proposition 2.1. The strategy for the proof of weak $E$-monotonicity of $f'$ is similar to that used in the following section for the case of vector fields.

Then, under a suitable assumption of coerciveness of the functional $f$ (expressed in terms of the coefficient functions $F_j$, Theorem 3.1 can be used to prove the existence of a minimizer for the functional $f$ on $M = [h = c]$ for $c > 0$. A major difficulty here is to show that there are weakly convergent minimizing sequences with their weak limit belonging to $M = [h = c]$. A simplified version of this has been presented in [12].

## 6. Non-quasiconvex minimization

Now we address the problem of minimizing functionals of the form (12) for the case of vector fields, i.e., the following type of functionals:

$$f(u) = \int_\Omega F(x, u(x), Du(x)) dx, \qquad (24)$$

where $\Omega \subset \mathbb{R}^n$ is open and has a Lipschitz boundary and $u : \Omega \longrightarrow \mathbb{R}^m$, $m > 1$, $u \in W^{1,p}(\Omega; \mathbb{R}^m)$, relevant for the study of systems of quasi-linear partial differential operators.

First, for the study of boundary value problems, we want to minimize the special functional

$$f(u) = \int_\Omega F(Du) dx \qquad (25)$$

over a set

$$M = \left\{ u \in W^{1,p}(\Omega; \mathbb{R}^m) \mid u = g \text{ on } \partial\Omega \right\}$$

where $1 < p < \infty$, $g : \partial\Omega \longrightarrow \mathbb{R}^m$ and $F : \mathcal{M}^{m \times n} \longrightarrow \mathbb{R}$ are given with $\mathcal{M}^{m \times n}$ denoting the space of real $m \times n$ matrices. We recall the following important result (see [1,16]):

**Theorem 6.1.** *Suppose $F$ is continuously differentiable and satisfies the growth condition*

$$0 \le F(P) \le C(1 + |P|^p) \qquad \forall P \in \mathcal{M}^{m \times n}. \qquad (26)$$

*Then the functional $f$ is lower semi-continous with respect to weak convergence in $W^{1,p}(\Omega; \mathbb{R}^m)$ if, and only if, $F$ is quasi-convex, i.e.,*

$$\int_Q F(P) dx \le \int_Q F(P + Dv) dx \qquad (27)$$

*for each $P \in \mathcal{M}^{m \times n}$ and all $v \in C_c^\infty(Q; \mathbb{R}^m)$ for any open bounded set $Q \subset \Omega$ .*

For the more general case of functionals of the form (24) under similar growth conditions weak lower semi-continuity is characterized by $F(x, u, \cdot)$ being quasi-convex.

In many minimization problems of practical interest, for instance in nonlinear elasticity, thermoplasticity and visco-elasticity the condition of quasi-convexity for the functional in question (for instance of potential energy, of complementary energy, or strain energy) is often not available. In addition, it is not known how to verify the condition of quasi-convexity in concrete examples.

However, from our discussion of the theory of minimization without weak lower semi-continuity, we deduce that one can expect that these functionals can be minimized quite a large class of subsets $M$ of the underlying Banach space $E$, in particular when $M$ is a set without interior points in $E$, or is contained in the intersection of some half-spaces of $E$.

Now we indicate briefly a situation where a functional of the form (24) can be minimized in circumstances where quasi-convexity is not available or not known. When a functional of the form (24) is studied for proving the existence of (weak) solutions of a system of nonlinear PDE's it is natural to assume that the integrand $F(x, u, P)$ is not only a Caratheodory function but has derivatives with respect to $u$ ($F_{,u}$) and $P$ ($F_{,P}$) which are Caratheodory functions too.

In order to study minimizations for the functional (24) some standard growth restrictions have to be imposed. We give here explicitly and in full details only the growth restrictions for integrand $F$, but not for the partial derivatives $F_{,u}$ and

$F_{,P}$. Suppose that there are nonnegative functions $A \in L^1(\Omega)$ and $B \in L^\sigma(\Omega)$, $1 < \sigma < \infty$ and a constant $C$ such that for almost all $x \in \Omega$, all $u \in \mathbb{R}^m$ and all $P \in \mathcal{M}^{m \times n}$ one has

$$|F(x, u, P)| \le A(x) + B(x)|u|^{\frac{s}{\sigma'}} + C|P|^p \qquad (28)$$

where $s = \frac{pn}{n-p}$ is the Sobolev exponent and $2 \le p < n$. For the partial derivatives $F_{,u}$ and $F_{,P}$ we use similar growth restrictions where however in the estimate the exponents for the variables $u$ and $P$ are reduced by 1. (Full details are given in [11]). Then one obtains for the Gâteaux derivative of $f$ at $u \in W^{1,p}(\Omega; \mathbb{R}^m)$

$$f'(u)(v) = \langle F_{,u}(\cdot, u, Du) \cdot v \rangle + \langle F_{,P}(\cdot, u, Du) \cdot Dv \rangle \qquad (29)$$

for all $v \in E = W^{1,p}(\Omega; \mathbb{R}^m)$. Here we use

$$\langle \cdot \rangle = \int_\Omega \cdot \, dx.$$

Now suppose that $C \subset \mathcal{M}^{m \times n}$ is a closed convex cone and define

$$K = \{u \in E| \ Du(x) \in C \text{ for almost all x} \in \Omega\}. \qquad (30)$$

One verifies that the zero element of $E$ is in the weak closure of $K$. Certainly, for $C \ne \mathcal{M}^{m \times n}$ we have $K \ne E$ and thus weak $K$-monotonicity does not imply weak lower semi-continuity.

**Prop 6.1.** Suppose that for all $u \in \mathbb{R}^m$, all $P \in \mathcal{M}^{m \times n}$, almost all $x \in \Omega$, and all $Q \in C$ we have

$$[F_{,P}(x, u, P + Q) - F_{,P}(x, u, P)] \cdot Q \ge 0. \qquad (31)$$

Then, under growth and smoothness restrictions as mentioned above, the Gâteaux derivative $f'$ is weakly $K$-monotone at any $v \in M$.

**Proof.**

Take any sequence $u_i \in K$ which converges weakly to zero and $v \in M$. A first step shows

$$\overline{\lim_{i \to \infty}} \langle f'(v + u_i) - f'(v), u_i \rangle =$$
$$\overline{\lim_{i \to \infty}} \langle [F_{,P}(\cdot, v + u_i, Dv + Du_i) - F_{,P}(\cdot, v + u_i, Dv)]Du_i \rangle. \qquad (32)$$

Evaluate the right hand side:

$$\langle [F_{,P}(\cdot, v + u_i, Dv + Du_i) - F_{,P}(\cdot, v + u_i, Dv)]Du_i \rangle$$
$$= \int_\Omega [F_{,P}(x, v(x) + u_i(x), Dv(x) + Du_i(x)) -$$
$$F_{,P}(x, v(x) + u_i(x), Dv(x))] \cdot Du_i(x)dx.$$

Since $u_i \in K$ we know $Q = Du_i(x) \in C$ for almost all $x \in \Omega$ and thus our assumption (31) of restricted monotonicity implies that the integral is non-negative. We conclude

$$\varlimsup_{i \to \infty} \langle f'(v + u_i) - f'(v), u_i \rangle \geq 0 \tag{33}$$

for any sequence $u_i \subset K$ which converges weakly to zero. Hence $f'$ is weakly $K$-monotone at $v$.

□

The growth and smoothness restriction used above also ensure that our standard assumptions (H) and (8) are satisfied. Thus we can apply Theorem 3.5 to conclude:

**Theorem 6.2.** *Consider the functional $f$ (Equation (24)) on $E = W^{1,p}(\Omega; \mathbb{R}^m)$ and assume the condition (31) of restricted monotonicity. Then $f$ has a minimum on any set $M \subset E$ which satisfies the cone condition for the cone $K$, defined in (30).*

**Remark 6.3.** We give a simple example of how to realize condition (31). Let $F_{,P}$ be of the form

$$F_{,P}(x, u, Q) = A(x, u) + B(x, u)Q$$

where $A$ is a Carathédory function with values in $\mathcal{M}^{m \times n}$ and $B$ a Carathédory function with values in $\mathcal{M}^{m \times m}$, with suitable growth restrictions. Then condition (31) reads

$$[F_{,P}(x, u, P + Q) - F_{,P}(x, u, P)] \cdot Q = \mathrm{Tr}\,(Q^t B(x, u)Q) \geq 0$$

for almost all $x \in \Omega$, all $u \in \mathbb{R}^m$, and all $Q \in C$.

**Remark 6.4.** Though in Theorem 6.2 the set $M$ over which the functional $f$ is minimized and the set $K$ for which the Gâteaux derivative $f'$ is weakly $K$-monotone are related in a complicated way, one can use this result for the solution of various model problems. In praxis however the set $M$ and the functional $f$ are given and we have to determine the cone $K$ such that a) $M$ satisfies the cone condition for this cone and b) $f'$ is weakly $K$-monotone on $M$. At present we are looking into material sciences in the expectation to find there interesting concrete problems where Theorem 6.2 can be applied successfully.

## 7. Conclusions

As our discussion shows one can minimize functionals $f$ of vector fields $u$, for instance of the form (24), over certain sets $M \subset E$, even if the integrand $F$ is not quasi-convex in the variable $Du$ whenever the Gâteaux derivative is weakly $K$-monotone for some subset $K \subset E$. In some detail we have presented the case where $K$ is some cone in the space $E = W^{1,p}(\Omega; \mathbb{R}^m)$ and the set $M$ satisfies the cone condition for this cone. We expect that several important classes of functionals from nonlinear elasticity, thermo-elasticity, visco-elasticity, plasticity etc. could be minimized by this approach based on weak $K$-monotonicity (work in progress).

# References

1. J. M. Ball and F. Murat. $w^{1,p}$-quasiconvexity and variational problems for multiple integrals. *J. Funct. Anal.*, 58:225–253, 1984.
2. Ph. Blanchard and E. Brüning. *Calculus of Variations. A Unified Approach.* Texts and Monographs in Physics. Springer–Verlag, New York Heidelberg Berlin, 1992.
3. F. Browder and P. Hess. Monotone mappings of monotone type. *J. Funct. Anal.*, 11:251–294, 1972.
4. F. E. Browder. Variational boundary value problems for quasilinear elliptic equations. III. *Proc. Nat. Acad. Sci. U.S.A.*, 50:794–798, 1963.
5. F. E. Browder. Variational boundary value problems for quasilinear elliptic equations of arbitrary order. *Proc. Nat. Acad. Sci. U.S.A.*, 50:31–37, 1963.
6. F. E. Browder. Variational methods for non–linear elliptic eigen–value problems. *Bull. Amer. Math. Soc.*, 71:176–183, 1965.
7. F. E. Browder. Existence theorems for nonlinear partial differential equations. In S.–S. Chern and S. Smale, editors, *Global Analysis*, Volume 16 of *Proceedings of Symposia in Pure Mathematics*, pages 1–60, Providence, RI, 1970. American Math. Soc.
8. F. E. Browder. Pseudo–monotone operators and the direct method of the calculus of variations. *Arch. Rational Mech. Anal.*, 38(4):268–277, 1970.
9. F. E. Browder. *Nonlinear operators and nonlinear equations of evolution in Banach spaces*, Volume 18, Part 2 of *Proceedings of Symposia in Pure Mathematics*. American Math. Soc., Providence, RI, 1976.
10. F. E. Browder. Nonlinear functional analysis. In F. E. Browder, editor, *Mathematical Developments arising from Hilbert Problems*, Volume 28 of *Proceedings of Symposia in Pure Mathematics*, pages 68-73, Providence, RI, 1977. American Math. Soc.
11. E. Brüning. Minimization without weak lower semi–continuity. *Applicable Analysis*, 54:91–111, 1994.
12. E. Brüning. Eigenvalue problems for global quasi–linear partial differential operators. *Commun. Appl. Nonlinear Anal.*, 4(3):55–67, 1997.
13. E. Brüning. A note on localized directional weak lower semi–continuity. *Preprint, University of KwaZulu–Natal, submitted*, 2005.
14. B. Dacorogna. *Direct methods in the calculus of variations*, Volume 78 of *Applied Mathematical Sciences*. Springer–Verlag, Berlin Heidelberg New York, 1989.
15. I. Ekeland and R. Temam. *Convex analysis and variational problems*. Monograph. North Holland, Amsterdam Oxford, 1976.
16. L. C. Evans. *Weak convergence methods for nonlinear partial differential equations*, Volume 74 of *CBMS — Regional Conference Series in Mathematics*. American Mathematical Society, Providence, Rhode Island, 1990.
17. C. B. Morrey. Quasi–convexity and semi–continuity of multiple integrals. *Pacific J. Math.*, 2:25–53, 1952.
18. C. B. Morrey. *Multiple integrals in the calculus of variations*. Monograph. Springer, Berlin, 1966.
19. M. Struwe. *Variational Methods. Applications to nonlinear partial differential equations and Hamiltonian systems. 2nd Edition*, Volume 34 of *Ergebnisse der Mathematik und ihrer Grenzgebiete. 3. Folge.* Springer–Verlag, Berlin Heidelberg New York, 1996.
20. E. Zeidler. *Variational methods and optimization.* Volume III of *Nonlinear functional analysis and its applications.* Springer–Verlag, New York Berlin Heidelberg London Paris Tokyo, 1985.
21. E. Zeidler. *Nonlinear monotone operators.* Volume II 1B of *Nonlinear functional analysis and its applications.* Springer–Verlag, New York Berlin Heidelberg London Paris Tokyo, 1990.

# EXISTENCE RESULTS FOR NONLINEAR HEMIVARIATIONAL INEQUALILTIES

P. CANDITO*

*Dipartimento di Informatica, Matematica, Elettronica e Trasporti*
*Facoltà di Ingegneria*
*Università degli Studi Mediterranea di Reggio Calabria*
*Via Graziella (Feo di Vito), 89100 Reggio Calabria, Italy*
*E-mail: pasquale.candito@unirc.it*

This paper deals with existence results for nonlinear hemivariational inequalities. Our approach is based on critical point theory for Motreanu-Panagiotopoulos type functionals [2].

**Keywords:** Hemivariational inequalities, critical points for non-differentiable functionals
**Mathematics Subject Classification:** 49J40, 49J52, 35J85

## 1. Introduction

Let $(X, \|\cdot\|)$ be a Banach space and $(X, \|\cdot\|)$ and let $\varphi : X \to \mathbb{R}$ be a locally Lipschitz continuous function. We consider the following hemivariational inequality
*Find $x \in X$ such that*

$$\varphi^0(x; z) \geq 0 \quad \forall z \in X, \qquad (H)$$

where

$$\varphi^0(x; z) := \limsup_{w \to x,\, t \to 0^+} \frac{\varphi(w + tz) - \varphi(w)}{t}$$

indicates the generalized directional derivative of $\varphi$ at $x$ in the direction $z$. Denoted by $\partial \varphi(x)$, the generalized gradient of $\varphi$ at $x$, that is

$$\partial \varphi(x) := \left\{ x^* \in X^* : \langle x^*, z \rangle \leq \varphi^0(x; z); \forall z \in X \right\}.$$

It is well known that problem $(H)$ becomes

$$0 \in \partial \varphi(x). \qquad (H')$$

By definition a solution of $(H')$ is said a critical point of $\varphi$. For basic definitions we mention [6,7] and [8]; for a complete study of the topics treating here, we refer the

---

*Research supported by RdB (ex 60% MIUR) of Reggio Calabria Unecessity

922

reader to [2] and the reference given therein.

The start point of our research is related to paper [1] where, putting

$$\varphi(x) = \Phi(x) + \Psi(x), \quad x \in X, \qquad (*)$$

with $\Phi, \Psi : X \to \mathbb{R}$ locally Lipschitz continuous, the following result it has been established:

**Theorem A.** *Assume that:*

($i_1$) *Let $X$ be reflexive and let $X = V \oplus E$, where $V$ is finite dimensional. Consequently each $x \in X$ can be uniquely written as $x = \bar{x} + \tilde{x}$, with $\bar{x} \in V$ and $\tilde{x} \in E$;*

($i_2$) *if $x_n \rightharpoonup x$ in $X$ and for some $\{\xi_n^*\} \subseteq X^*$ one has*

$$\xi_n^* \in \partial\Phi(x_n) \; \forall n \in \mathbb{N}, \; \limsup_{n \to +\infty}\langle \xi_n^*, x_n - x \rangle \leq 0$$

*then $\{x_n\}$ possesses a strongly convergent subsequence.*

($i_3$) *Let $X$ be compactly embedded in a Banach space $Y$, with $X$ dense in $Y$.*

($i_4$) *Suppose $\Psi : Y \to \mathbb{R}$ is a globally Lipschitz continuous functional and, moreover,*

($i_5$) $\lim_{\|\bar{x}\| \to +\infty} \Psi|_V(\bar{x}) = -\infty$.

*Then $\varphi$ admits a critical point.*

In other words, owing to Theorem A, the hemivariational inequality
*Find $x \in X$ such that*

$$\Phi^0(x; z) + \Psi^0(x; z) \geq 0 \quad \forall z \in X,$$

*admits at least one solution.*

In our opinion, inside to the framework of critical point theory, Theorem A exhibits at least two interesting features:

- no direct verification of some Palais-Smale like condition is required;
- the assumptions on $\Phi$ involve only the subspace $E$, whereas those on $\Psi$ merely concern the subspace $V$.

However, both ($i_3$) and especially ($i_4$) look rather restrictive. They prevent to use Theorem A in a large number of cases or force to make hypotheses which are not always very natural when studying boundary value problems. To overcome these difficulties we will study problem $(H)$ by a general point of view.

## 2. Preliminary results

We look for critical points on $X$ to functions $f$ that satisfy the structural assumption

($H'_f$) $f(x) := \varphi(x) + \alpha(x)$ *for each $x \in X$, where $\varphi : X \to \mathbb{R}$ is locally Lipschitz continuous while $\alpha : X \to \mathbb{R} \cup \{+\infty\}$ turns out convex, proper, and lower* *semicontinuous. Moreover, $\alpha$ is continuous on any nonempty compact set $A \subseteq X$ such that $\sup_{x \in A} \alpha(x) < +\infty$.*

Put $D_\alpha := \{x \in X : \alpha(x) < +\infty\}$. To simplify notation, always denote by $\partial \alpha(x)$ the subdifferential of $\alpha$ at $x$ in the sense of convex analysis. We say that $x \in D_\alpha$ is a critical point of $f$ when

$$\varphi^0(x; z - x) + \alpha(z) - \alpha(x) \geq 0 \quad \forall z \in X.$$

If $\alpha \equiv 0$, it clearly signifies $0 \in \partial \varphi(x)$, i.e. $x$ is a critical point of $\varphi$. On this direction we apply Theorem 3.3 of [5]. To state it we need some basic definitions which now I am going to recall. Given a real number $c$, write

$$K_c(f) := \{x \in X : f(x) = c, \ x \text{ is a critical point of } f\}.$$

Let $S$ be a nonempty closed subset of $X$. The function $f$ is said to fulfill the Palais-Smale condition at the level $c$ and around the set $S$ provided

$(PS)_{S,c}$*Every sequence $\{x_n\} \subseteq D_\alpha$ with $d(x_n, S) \to 0$, $f(x_n) \to c$, and*

$$\varphi^0(x_n; x - x_n) + \alpha(x) - \alpha(x_n) \geq -\varepsilon_n \|x - x_n\| \quad \forall n \in \mathbb{N}, \ x \in X,$$

*where $\varepsilon_n \to 0^+$, possesses a convergent subsequence.*

When $S = X$ we merely write $(PS)_c$ in place of $(PS)_{X,c}$. Condition $(PS)_c$ coincides with the one exploited by Motreanu-Panagiotopoulos[6]. If $\alpha \equiv 0$ then it reduces to the Palais-Smale condition introduced in [3], which can be easily verified via Proposition 3.1 of [6].

Now, let $Q$ be a compact set in $X$ and let $Q_0$ be a nonempty closed subset of $Q$. Put

$$\Gamma := \{\gamma \in C^0(Q, X) : \gamma|_{Q_0} = id|_{Q_0}\}.$$

We say that the pair $(Q, Q_0)$ links with $S$ provided $Q_0 \cap S = \emptyset$ and for every $\gamma \in \Gamma$ one has $\gamma(Q) \cap S \neq \emptyset$. Now we are in position to state Theorem 3.3 of [5] which is the main tool to investigate problem $(H)$.

**Theorem 2.1.** *Suppose $(Q, Q_0)$ links with $S$, while the function $f$ satisfies the following conditions in addition to $(H'_f)$,*

$(a_1)$ $\sup_{x \in Q} \alpha(x) < +\infty$.

$(a_2)$ $\sup_{x \in Q_0} f(x) \leq \inf_{x \in S} f(x)$.

$(a_3)$ *Setting*

$$c := \inf_{\gamma \in \Gamma} \sup_{x \in Q} f(\gamma(x)) \tag{1}$$

*either $(PS)_c$ or $(PS)_{S,c}$ holds true, according to whether $\inf_{x \in S} f(x) < c$ or $\inf_{x \in S} f(x) = c$.*

*Then $K_c(f) \neq \emptyset$. If, moreover, $\inf_{x \in S} f(x) = c$ then $K_c(f) \cap S \neq \emptyset$.*

924

Several classical results can be reformulated in this framework through Theorem 2.1. For later use, we state here appropriate versions of Saddle Point Theorem (SPT)[9].

**Theorem 2.2.** *Let* $X = V \oplus E$, *where* $V \neq \{0\}$ *is finite dimensional, and let* $f$ *fulfill* $(H'_f)$. *Assume there exists an* $r > 0$ *such that*

$(a_4)$ $\sup_{x \in \overline{B}_r \cap V} \alpha(x) < +\infty,$
$(a_5)$ $\sup_{x \in \partial B_r \cap V} f(x) \leq \inf_{x \in E} f(x),$ *and*
$(a_6)$ *either* $(PS)_c$ *or* $(PS)_{E,c}$ *holds true, according to whether* $\inf_{x \in E} f(x) < c$ *or* $\inf_{x \in E} f(x) = c$, *where* $c$ *is given by (1) written for* $Q := \overline{B}_r \cap V$, $Q_0 := \partial B_r \cap V$.

*Then* $K_c(f) \neq \emptyset$. *If, moreover,* $\inf_{x \in E} f(x) = c$ *then* $K_c(f) \cap E \neq \emptyset$.

As well as of Generalized Mountain Pass Theorem (GMT)[9].

**Theorem 2.3.** *Suppose* $X = V \oplus E$, *where* $V$ *is finite dimensional,* $(H'_f)$ *holds, and there exist an* $r > 0$, $e \in \partial B_1 \cap E$, $\rho \in ]0, r[$ *such that* $(a_1)$–$(a_3)$ *are satisfied for* $Q := (\overline{B}_r \cap V) \oplus [0, re]$, $Q_0$ *the boundary of* $Q$ *relative to* $V \oplus span\{e\}$, $S := \partial B_\rho \cap E$. *Then the conclusion of Theorem 2.1 is true.*

We explicit observe that at this point it is clear that by using (SPT) and (GMT) we can improve Theorem A in two directions:

- $\varphi$ is the sum of two locally Lipschitz continuous functions;
- $\varphi$ is perturbed by a proper, convex and lower semicontinuous function.

However in any case a crucial point in applying these results is to prove the Palais-Smale condition. The first step in obtaining our goal is the following

**Lemma 2.1.** *Assume that:*

$(a_7)$ *If* $\{x_n\} \subseteq D_\alpha$, $x_n \rightharpoonup x$ *in* $X$, *and there exists a* $\{\xi_n^*\} \subseteq X^*$ *fulfilling*

$$\xi_n^* \in \partial\Phi(x_n) \,\forall n \in \mathbb{N}, \quad \limsup_{n \to +\infty} \langle \xi_n^*, x_n - x \rangle \leq 0,$$

*then* $\{x_n\}$ *has a strongly convergent subsequence.*
$(a_8)$ $\limsup_{n \to +\infty} \Psi^0(x_n; x - x_n) \leq 0$ *provided* $\{x_n\} \subseteq D_\alpha$ *and* $x_n \rightharpoonup x$ *in* $X$.

*Then every bounded sequence* $\{x_n\} \subseteq D_\alpha$ *such that*

$$\varphi^0(x_n; z - x_n) + \alpha(z) - \alpha(x_n) \geq -\varepsilon_n \|z - x_n\| \,\forall n \in \mathbb{N}; z \in X,$$

*where* $\varepsilon_n \to 0^+$, *possesses a strongly convergent subsequence.*

**Remark 2.1.** Condition $(a_8)$ holds whenever
$(a'_8)$ $X$ compactly embeds in a Banach space $Y$, while $\Psi$ is defined and locally Lipschitz continuous on the whole $Y$.

**Remark 2.2.** An equivalent formulation of $(a_8)$ is:

$(a_8'')$ *Every sequence $\{x_n\} \subseteq D_\alpha$ weakly converging to some point $x$ in $X$ admits a subsequence $\{x_{k_n}\}$ such that $\lim \sup_{n \to +\infty} \Psi^0(x_{k_n}; x - x_{k_n}) \leq 0$.*

The second step is to ensure that a Palais-Smale sequence to $f$ turns be bounded.

**Lemma 2.2.** *Assume that:*

$(a_9)$ *There exist $c_1, c_2 \in \mathbb{R}$, $\delta > 0$, $\theta : \mathbb{R}_0^+ \to \mathbb{R}$ such that*

$$\liminf_{t \to +\infty} \frac{\theta(t)}{t} > c_1$$

*and to each $x \in D_\alpha$, $x = \bar{x} + \tilde{x}$ with $\|\tilde{x}\| > \delta$, there corresponds a $\bar{\zeta}^* \in \partial \alpha(\bar{x})$ satisfying*

$$\langle \zeta^* + \bar{\zeta}^*, \tilde{x} \rangle \geq \theta(\|\tilde{x}\|) - c_1 \|\tilde{x}\| - c_2 \quad \forall \zeta^* \in \partial\varphi(x).$$

$(a_{10})$ *$|f(x_n)| \to +\infty$ whenever $\{x_n\} \subseteq D_\alpha$, $x_n = \bar{x}_n + \tilde{x}_n \in V \oplus E$, $n \in \mathbb{N}$, with $\|\bar{x}_n\| \to +\infty$ and $\{\tilde{x}_n\}$ bounded.*

*Then every sequence $\{x_n\}$ in $X$ such that $\{f(x_n)\}$ is bounded and (3) holds true is also bounded.*

From now on we give attention to locally Lipschitz case only, that is $\alpha \equiv 0$ and combining together the previous results we obtain

**Theorem 2.4.** *Let $X$, $\varphi$, $Q$, $Q_0$, $S$ be as in Theorem 2.3 (GMT) and let $(*)$, $(a_7)$, $(a_8)$ be satisfied. Suppose that $\inf_{x \in S} f(x) = c$, with $c$ given by (1). Then $K_c(f) \cap S \neq \emptyset$.*

**Theorem 2.5.** *Let $\varphi$ be given by $(*)$. If $(a_5)$ holds true for some $r > 0$, $(a_7)$-$(a_8)$ are satisfied and, moreover,*

$(a_{11})$ *$\Phi(x) \leq \beta(\|\tilde{x}\|) \; \forall x \in X$, $x = \bar{x} + \tilde{x} \in V \oplus E$, $\beta \in C^0(\mathbb{R}_0^+)$,*

$(a_{12})$ *there exists a function $\theta : \mathbb{R}_0^+ \to \mathbb{R}$ fulfilling $\lim_{t \to +\infty} \frac{\theta(t)}{t} = +\infty$ as well as*

$$\langle \xi^*, \tilde{x} \rangle \geq \theta(\|\tilde{x}\|) \quad \forall \xi^* \in \partial\Phi(x), \; x \in X, \; x = \bar{x} + \tilde{x} \text{ with } \bar{x} \in V \; \tilde{x} \in E,$$

$(a_{13})$ *$\lim \sup_{\|\tilde{x}\| \to +\infty} \frac{\Psi^0(\bar{x} + \tilde{x}; -\tilde{x})}{\|\tilde{x}\|} \leq c_3$, $c_3 \in \mathbb{R}$, uniformly in $\bar{x} \in V$,*

$(a_{14})$ *there is a constant $L \geq 0$ such that*

$$|\Psi(\bar{x} + \tilde{x}) - \Psi(\bar{x})| \leq L\|\tilde{x}\| \quad \forall \bar{x} \in V; \tilde{x} \in E,$$

$\lim_{\|\bar{x}\| \to +\infty} \Psi|_V(\bar{x}) = -\infty$.

*Then $f$ possesses at least a critical point.*

A special case of Theorem 2.5 is the next result which contains Theorem A as well as Theorem 2.3 in [1].

**Theorem 2.6.** *Suppose $(a_7)$, $(a_8)$, $(a_{11})$-$(a_{15})$ hold true and*

$(a_{16})$ $\lim_{\|\tilde{x}\|\to+\infty} \frac{\Phi|_{E'}(\tilde{x})}{\|\tilde{x}\|} = +\infty;$

*Then the function $\varphi$ admits a critical point.*

## 3. Results

Let $\Omega$ be a nonempty, bounded, open subset of the real Euclidean $N$-space $(\mathbb{R}^N, |\cdot|)$, $N \geq 3$, having a smooth boundary $\partial\Omega$ and let $H_0^1(\Omega)$ be the closure of $C_0^\infty(\Omega)$ with respect to the norm

$$\|u\| := \left( \int_\Omega |\nabla u(x)|^2 dx \right)^{1/2}.$$

Denote by $2^*$ the critical exponent for the Sobolev embedding $H_0^1(\Omega) \subseteq L^p(\Omega)$. Recall that $2^* = \frac{2N}{N-2}$, if $p \in [1, 2^*]$ then there exists a constant $C_p > 0$ fulfilling

$$\|u\|_{L^p(\Omega)} \leq C_p \|u\|, \quad u \in H_0^1(\Omega)$$

and the embedding is compact whenever $p \in [1, 2^*[$. Now, let $\{\lambda_n\}$ be the sequence of eigenvalues of the operator $-\Delta$ in $H_0^1(\Omega)$, with $0 < \lambda_1 < \lambda_2 \leq \cdots \leq \lambda_n \leq \cdots$, and let $\{\varphi_n\}$ be a corresponding sequence of eigenfunctions normalized as follows:

$$\|\varphi_n\|^2 = 1 = \lambda_n \|\varphi_n\|^2_{L^2(\Omega)}, \quad n \in \mathbb{N};$$

$$\int_\Omega \nabla\varphi_m(x) \cdot \nabla\varphi_n(x)dx = \int_\Omega \varphi_m(x)\varphi_n(x)dx = 0 \quad m \neq n.$$

Given $k \in \mathbb{N}$, $\lambda \in [\lambda_k, \lambda_{k+1}]$, $\Psi : H_0^1(\Omega) \to \mathbb{R}$ locally Lipschitz continuous function. Denote by $(P_{\lambda,\Psi})$ the elliptic hemivariational inequality problem:

*Find $u \in H_0^1(\Omega)$ such that*

$$-\int_\Omega \nabla u(x)\cdot\nabla(v-u)(x)\,dx + \lambda\int_\Omega u(x)(v(x)-u(x))dx \leq \Psi^0(u; v-u) \text{ for all } v \in H_0^1(\Omega)$$

A meaningful special case occurs when

$$\Psi(u) = \int_\Omega J(x, u(x))\,dx, \quad u \in H_0^1(\Omega),$$

where

$$J(x, \xi) = \int_0^\xi -j(x, t)\,dt, \quad (x, \xi) \in \Omega \times \mathbb{R},$$

and the function $j : \Omega \times \mathbb{R} \to \mathbb{R}$ satisfies the conditions

$(j_1)$ *$j$ is measurable with respect to each variable separately,*

$(j_2)$ *there exist $a_1 > 0$, $p \in [1, 2^*]$ such that*

$$|j(x, t)| \leq a_1 \left(1 + |t|^{p-1}\right) \quad \forall (x, t) \in \Omega \times \mathbb{R}.$$

Indeed, under $(j_1)$–$(j_2)$, $J$ turns out well defined, $J(\cdot,\xi)$ is measurable, while $J(x,\cdot)$ is locally Lipschitz continuous. So it makes sense to consider its generalized directional derivative $J_\xi^0$ with respect to the variable $\xi$. Moreover on account of formula (9) at p. 84 [4], any solution $u$ to $(P_{\lambda,\Psi})$ also fulfills the inequality

$$-\int_\Omega \nabla u(x)\cdot\nabla(v-u)(x)\,dx+\lambda\int_\Omega u(x)(v(x)-u(x))\,dx \le \int_\Omega J_x^0(u(x);v(x)-u(x))dx$$

for all $v \in H_0^1(\Omega)$.

If $j \in C^0(\Omega\times\mathbb{R})$ then the function $u \in H_0^1(\Omega)$ turns out a weak solution to the Dirichlet problem

$$-\Delta u = \lambda u + j(x,u) \quad \text{in } \Omega, \qquad u=0 \quad \text{on } \partial\Omega.$$

Throughout the sequel we shall write

$$X := H_0^1(\Omega), \quad V := \text{span}\{\varphi_1,\varphi_2,\ldots,\varphi_k\}, \quad E := V^\perp,$$

and

$$Q(r) := (\overline{B}_r \cap V) \oplus [0, r\varphi_{k+1}] \quad r>0.$$

Each $u \in Q(r)$ can be uniquely written as $u = \bar{u}+t\varphi_{k+1}$, with $\bar{u}\in\overline{B}_r\cap V$, $t\in[0,r]$.

Now we are in position to state the main results

**Theorem 3.1.** *Let $(a_8)$ be satisfied and let $r>0$ be such that*

$(\Psi_1)$ $\Psi(\bar{u}) \le \frac{1}{2}\left(\frac{\lambda}{\lambda_k}-1\right)\|\bar{u}\|^2$ $\forall \bar{u}\in\overline{B}_r\cap V$;

$(\Psi_2)$ $\Psi(u) \le \frac{1}{2}\left(\frac{\lambda}{\lambda_k}-1\right)\|\bar{u}\|^2 - \frac{1}{2}\left(1-\frac{\lambda}{\lambda_{k+1}}\right)t^2$ $\forall u\in Q(r)$.

$(\Psi_3)$ *There exists a* $\rho\in]0,r[$ *fulfilling* $\inf_{\tilde{u}\in\partial B_\rho\cap E}\frac{1}{\rho^2}\Psi(\tilde{u}) \ge -\frac{1}{2}\left(1-\frac{\lambda}{\lambda_{k+1}}\right)$.

*Then $(P_{\lambda,\Psi})$ admits at least one nontrivial solution $u\in E$ such that $\|u\|=\rho$, as well as*

$$\frac{\lambda}{2}\|u\|_{L^2(\Omega)}^2 - \Psi(u) = \frac{\rho^2}{2}.$$

Finally we have

**Theorem 3.2.** *Suppose that the function $j:\Omega\times\mathbb{R}\to\mathbb{R}$ satisfies $(j_1)$ and $(j_2)$ with $a_1\in]0,\lambda_{k+1}-\lambda_k[$, $p=2$. If, moreover, for suitable $a_3\in]0,(\lambda_{k+1}-\lambda_k)/2]$ one has*

$(j_5)$ $-a_3\xi^2 \le J(x,\xi) \le 0$ *in $\Omega\times\mathbb{R}$, where $J$ is given above.*

*Then there exists an $\tilde{u}\in E$ such that*

$$-\int_\Omega \nabla\tilde{u}(x)\cdot\nabla(v-\tilde{u})(x)\,dx+\lambda_k\int_\Omega \tilde{u}(x)(v(x)-\tilde{u}(x))dx \le \int_\Omega J_x^0(\tilde{u}(x);v(x)-\tilde{u}(x))dx$$

*for all $v\in H_0^1(\Omega)$.*

928

## References

1. S. Adly, G. Buttazzo, and M. Théra, *Critical points for nonsmooth energy functions and applications,* Nonlinear Anal. **32** (1998), 711-718.
2. P. Candito. D. Motreanu and S.A. Marano, *Critical points for a class of non-differentiable functions and applications,* Discrete Contin. Dyn. Syst. **13** (2005), no. 1, 175–194.
3. K.-C. Chang, *Variational methods for non-differentiable functionals and their applications to partial differential equations,* J. Math. Anal. Appl. **80** (1981), 102–129.
4. F.H. Clarke, *Optimization and Nonsmooth Analysis,* Classics Appl. Math. **5**, SIAM, Philadelphia, 1990.
5. R. Livrea and S.A. Marano, *Existence and classification of critical points for non-differentiable functions,* Advances in Differential Equations, **9** (2004), 961-978.
6. D. Motreanu and P.D. Panagiotopoulos, *Minimax Theorems and Qualitative Properties of the Solutions of Hemivariational Inequalities,* Nonconvex Optim. Appl. **29**, Kluwer, Dordrecht, 1998.
7. D. Motreanu and V. Radulescu, *Variational and Non-variational Methods in Nonlinear Analysis and Boundary Value Problems,* Nonconvex Optim. Appl. **67**, Kluwer, Dordrecht, 2003.
8. P.D. Panagiotopoulos, *Hemivariational Inequalities. Applications in Mechanics and Engeneering,* Springer-Verlag, Berlin, 1993.
9. P. H. Rabinowitz, *Minimax methods in critical point theory with applications to differential equations,* CBMS Reg. Conf. Ser. in Math. **65**, Amer. Math. Soc., Providence, 1986.

## III.1   Complex analysis and potential theory

Organizers: Massimo Lanza de Cristoforis, Promarz Tamrazov

This session has been devoted to recent advances in complex analysis and potential theory. It includes contributions in fine potential theory, and in particular the extension to finely meromorphic functions of 'contour-solid' estimates which have been known for finely holomorphic functions. It includes results on the connection between different structures relating quasiconformal extensions and reflections and Grunsky coefficient inequalities, and reports results concerning the Moser conjecture. It contains results of singular perturbation for the Poisson equation and results on degenerate elliptic equations on the plane via analytic functions taking values in a suitable Banach algebra. It contains results on dynamical systems associated to Clifford Algebras and various applications of dynamical systems associated to discrete Laplacians.

# MOSER'S CONJECTURE ON GRUNSKY INEQUALITIES AND BEYOND

SAMUEL KRUSHKAL

*Department of Mathematics,*
*Bar-Ilan University, 52900 Ramat-Gan, Israel*
*E:mail: krushkal@macs.biu.ac.il*

The paper relates to the quantitative estimation of quasiconformal extensions and reflections in connection with fundamental Grunsky coefficient inequalities, Fredholm eigenvalues and complex Finsler geometry of Teichmüller spaces. The main subject concerns two conjectures on the Grunsky coefficients of univalent functions and sheds new light on their features. The first conjecture was posed by Moser in 1985. We present the main ideas in the proofs and discuss the related topics.

**Key words:** Grunsky inequalities, Moser conjecture, Teichmüller spaces, univalent functions

**Mathematics Subject Classification:** 30C62

## 1. Grunsky inequalities and related results

### 1.1.

We consider quasiconformal maps (homeomorphisms) $f$ of the Riemann sphere $\widehat{\mathbb{C}} = \mathbb{C} \cup \{\infty\}$ whose **Beltrami coefficients** $\mu_f = \partial_{\bar{z}} f / \partial_z f$ are supported in the unit disk $\Delta = \{z \in \mathbb{C} : |z| < 1\}$, and such that

$$f(z) = z + b_0 + b_1 z^{-1} + \dots$$

on the complementary disk $\Delta^* = \{z \in \widehat{\mathbb{C}} : |z| > 1\}$. The **dilatation** $k(f) = \|\mu\|_\infty$ estimates the deviation of $f$ from conformality.

The conformal maps of the disk $\Delta^*$ into $\widehat{\mathbb{C}} \setminus \{0\}$ with such hydrodynamical normalization form the well-known class $\Sigma$; its subset $\Sigma(k)$ consists of the maps with $k$-quasiconformal extensions to $\widehat{\mathbb{C}}$. Let $\Sigma^0 = \cup_k \Sigma(k)$. The functions from $\Sigma^0$ are uniquely determined by adding a third condition, for example, $f(0) = 0$.

This collection naturally relates to the universal Teichmüller space $\mathbf{T}$ modelled as a bounded domain in the complex Banach space $\mathbf{B}$ of hyperbolically bounded holomorphic functions in $\Delta^*$ with the norm

$$\|\varphi\|_{\mathbf{B}} = \sup_{\Delta^*} (|z|^2 - 1)^2 |\varphi(z)|. \tag{1}$$

The elements of this space are the **Schwarzian derivatives**

$$S_f = (f''/f')' - (f''/f')^2/2$$

of locally univalent holomorphic functions in $\Delta^*$, while the indicated domain $\mathbf{T}$ consists of the Schwarzian derivatives of those functions which are univalent in the whole disk $\Delta^*$ and have quasiconformal exstensions onto the sphere $\widehat{\mathbb{C}}$. Note that $\varphi(z) = O(|z|^{-4})$ as $z \to \infty$.

The fundamental Grunsky theorem states that a holomorphic function in $\Delta^*$ with the above expansion is univalent in this disk if and only if it satisfies the inequality

$$\left| \sum_{m,n=1}^{\infty} \sqrt{mn}\, \alpha_{mn} x_m x_n \right| \le 1, \tag{2}$$

where the **Grunsky coefficients** $\alpha_{mn}(f)$ are determined by

$$\log[(f(z) - f(\zeta))/(z - \zeta)] = - \sum_{m,n=1}^{\infty} \alpha_{mn} z^{-m} \zeta^{-n}, \quad (z,\zeta) \in (\Delta^*)^2,$$

taking the principal branch of the logarithmic function, and $\mathbf{x} = (x_n)$ ranges over the unit sphere $S(l^2)$ of the Hilbert space $l^2$ of sequences with $\|\mathbf{x}\|^2 = \sum_1^{\infty} |x_n|^2$ (cf. [8]).

The global univalence is an important and rather rigid property of holomorphic functions. It has been studied by many authors. There were established some other criteria which in the case of the disk are equivalent to (2).

It was shown by Kühnau [20] that for $f \in \Sigma^0$, the inequality (2) is strengthened as follows:

$$\varkappa(f) := \sup \left\{ \left| \sum_{m,n=1}^{\infty} \sqrt{mn}\, \alpha_{mn} x_m x_n \right| : \mathbf{x} = (x_n) \in S(l^2) \right\} \le k(f), \tag{3}$$

where $k(f)$ beginning from here denotes the minimal dilatation among all quasiconformal extensions of $f$ from $\Delta^*$ onto $\widehat{\mathbb{C}}$ (cf. [20]). The quantity $\varkappa(f)$ is called the **Grunsky constant** of the map $f$.

A crucial point here is that for a generic function $f \in \Sigma^0$, one has in (3) the strict inequality

$$\varkappa(f) < k(f) \tag{4}$$

(see, e.g., [21,22,10]), and it is important to know for which functions the equality in (3) occurs. This relates to various topics.

On the other hand, the important result of Pommerenke [29](Theorem 9.12), reproved by Zhuravlev using another method (see [35,18], pp. 82-84), states that if $f \in \Sigma$ satisfies the inequality (3) for all indicated $\mathbf{x}$, then it admits a $k'$-quasiconformal extension to $\widehat{\mathbb{C}}$ with $k' \ge k$. An explicit $k' = k'(k)$ is given in [23].

A natural and important question posed by various authors (see, e.g., [3,20,25]) is whether the inequalities (3) ensure a $k$-quasiconformal extension of $f \in \Sigma$ with the same $k$.

A characterization of the functions for which the inequality (2) is both necessary and sufficient to belong to $\Sigma(k)$ was given in [11,12], and in [22] more explicitly for functions which transform $|z| = 1$ onto an analytic Jordan curve. To present it, we need some notations. We call the quantity

$$\varkappa(f) = \sup\left\{\left|\sum_{m,n=1}^{\infty} \sqrt{mn}\,\alpha_{mn}x_mx_n\right| : \sum_{1}^{\infty}|x_n|^2 = 1\right\}$$

the *Grunsky constant* of a function $f \in \Sigma$ and denote by $k(f)$ the minimal dilatation among all quasiconformal extensions of $f$ to $\widehat{\mathbb{C}}$.

Let $A_1(\Delta)$ denote the subspace of $L_1(\Delta)$ formed by holomorphic functions in $\Delta$, and let

$$A_1^2 = \{\psi \in A_1(\Delta) : \psi = \varphi^2\};$$

this set consists of integrable holomorphic functions in $\Delta$ having only zeros of even order. Put

$$< \mu, \psi >_\Delta = \int_\Delta \mu\psi\, dm_2 \quad \text{for } \mu \in L_\infty(\Delta),\ \psi \in L_1(\Delta),$$

where $m_2$ is the Lebesgue measure on $\mathbb{C}$.

The indicated result of [11,12] says that *the equality*

$$\varkappa(f) = k(f) \tag{5}$$

*holds if and only if the function $f$ is the restriction to $\Delta^*$ of a quasiconformal homeomorphism $w^{\mu_0}$ of $\widehat{\mathbb{C}}$ whose Beltrami coefficient $\mu_0$ satisfies the equality*

$$\sup | < \mu_0, \varphi >_\Delta | = \|\mu_0\|_\infty, \tag{6}$$

*where the supremum is taken over holomorphic functions $\varphi \in A_1^2$ with $\|\varphi\|_{A_1(\Delta)} = 1$.*

This result reveals a deep connection between the Grunsky coefficients and integrable holomorphic quadratic differentials with zeros of even order. Such differentials play a crucial role in various problems, in particular, in applications to Teichmüller spaces (see, e.g., [11,14]).

It turned out that for a wide class of the boundary curves $f(\partial\Delta)$ the equality (4) is both necessary and sufficient for the existence of the Teichmüller-Kühnau extension of a function $f \in \Sigma$, i.e., with the Beltrami coefficient of the form $\mu_f = k|\varphi|/\varphi$ with $\varphi \in A_1^2$ (cf., e.g., [13,22]).

There is a geometric interpretation of these results in terms of invariant metrics on the universal Teichmüller space (see Section 6).

## 2. Two conjectures

In 1985, at the XI. Österreichischer Mathematikerkongress in Graz, Jürgen Moser had conjectured that the set of the functions satisfying (4) must be rather sparse in $\Sigma^0$ so that any function $f \in \Sigma$ is approximated by functions satisfying (6), uniformly on compact sets in $\Delta^*$.

Moser's conjecture was recently proved in our joint paper with Reiner Kühnau [19], where two different proofs of this conjecure are given.

We also posed in [19] another, still open, conjecture that, in contrast to the above duration, the functions $f \in \Sigma^0$ with $\varkappa(f) < k(f)$, cannot be approximated by $f_n \in \Sigma^0$ with $\varkappa(f_n) = k(f_n)$ (again in topology of uniform convergence on compact sets in $\Delta^*$).

Both conjectures shed new light on the features of Grunsky coefficients.

## 2.1.

The convergence of the maps $f_n$ to $f$ yields that their Schwarzian derivatives $S_{f_n}$ are convergent to $S_f$ uniformly on compact sets in $\Delta^*$. It would be interesting to know the situation when $S_{f_n}$ converge in stronger topology determined by the norm of **B**, and how sparse is the set of points $S_f$ satisfying (4) in the universal Teichmüller space **T**. Nothing is known in this direction.

Another important question for applications is to provide the explicit sufficient conditions for the functions $f \in \Sigma$ ensuring the existence of $k$-quasiconformal extensions with an explicit $k = k(\varkappa(f))$. Certain conditions have been given in [23,16].

## 3. Main results

The first main result, which we wish to present here, confirms Moser's conjecture:

**Theorem 3.1.** [19] *For every function $f \in \Sigma$, there exists a sequence of functions $f_n \in \cup_k \Sigma(k)$ with $\varkappa(f_n) < k(f_n)$ for all $n$, which is uniformly convergent to $f$ on compact sets in $\Delta^*$.*

There are obtained in [19] two alternate proofs of this theorem. Both proofs will be discussed here.

## 3.1.

As for the second conjecture, we can prove its weakened version:

**Theorem 3.2.** [17] *Any sequence of the functions $f_n \in \Sigma^0$ with $\varkappa(f_n) = k(f_n)$ and such that the curves $f_n(|z| = 1)$ are asymptotically conformal, cannot converge locally uniformly in $\Delta^*$ to a function $f \in \Sigma$ with $\varkappa(f) < k(f)$.*

We recall that a Jordan curve $L \subset \mathbb{C}$ is called **asymptotically conformal** if for any pair of points $a, b \in L$,

$$\max_{z \in L(a,b)} \frac{|a - z| + |z - b|}{|a - b|} \to 1 \quad \text{as} \quad |a - b| \to 0,$$

where $L(a, b)$ denotes the subarc of $L$ with the endpoints $a, b$ of smaller diameter. Such curves are quasicircles (i.e., the images of the circle under quasiconformal homeomorphisms of the sphere $\widehat{\mathbb{C}}$) and cannot have corners. In particular, all $C^1$-smooth curves are asymptotically conformal. On the other hand, asymptotically conformal curves can be rather pathological (see, e.g., [30], p. 249).

There are some equivalent conditions of asymptotic conformality. For example, when a curve $L$ is obtained as the image of the unit circle $S^1 = \partial D$ under a map $f \in \Sigma^0$, one of the other characterizing properties is $S_f(z) = o((|z| - 1)^{-2})$ as $|z| \to 1+$.

## 4. Fredholm eigenvalues

The above results can be also expressed in the terms of Fredholm eigenvalues of quasiconformal curves which are intrinsically connected with the Grunsky coefficients. These values, being important for various problems, are determined for smooth bounded Jordan curves $L$ as the eigenvalues of the double-layer potential over $L$:

$$ h(z) = \frac{\rho}{\pi} \int_L h(\zeta) \frac{\partial}{\partial n} \log \frac{1}{|\zeta - z|} ds, $$

with the outer normal $n$.

The least nontrivial Fredholm eigenvalue $\rho_1 = \rho_L$ plays a crucial role. It is defined for any oriented closed Jordan curve $L \subset \widehat{\mathbb{C}}$ by

$$ \frac{1}{\rho_L} = \sup \frac{|\mathcal{D}_G(u) - \mathcal{D}_{G^*}(u)|}{\mathcal{D}_G(u) + \mathcal{D}_{G^*}(u)}, $$

where $G$ and $G^*$ are respectively the interior and exterior of $L$; $\mathcal{D}$ denotes the Dirichlet integral, and the supremum is taken over all functions $u$ continuous on $\widehat{\mathbb{C}}$ and harmonic on $G \cup G^*$. Note that all quantities in the last equality remain invariant under the action of the Möbius group $PSL(2,\mathbb{C})/\pm 1$, so it suffices to consider the maps normalized by (1).

The important theorem of Kühnau and Schiffer states that for any $f \in \Sigma^0$, the quantities $\varkappa(f)$ and $\rho_{f(S^1)}$ are reciprocal (cf. [21,32]).

## 5. Sketch of the proofs of Theorem 3.1

The first proof was obtained by the author and consists of two steps. We first consider quasiconformal deformations of Fuchsian groups and establish the following result.

**Theorem 5.1.** *Let $\Gamma$ be a torsion free finitely generated Fuchsian group of the first kind acting on $\Delta$ (so that the orbit space $\Delta/\Gamma$ is a hyperbolic Riemann surface of finite analytic type). Then every extremal Beltrami differential*

$$ \mu_0 \in \mathrm{Belt}(\Delta, \Gamma)_1 := \{\mu \in L_\infty(\mathbb{C}) : \mu|\Delta^* = 0, \|\mu\| < 1; \ (\mu \circ \gamma)\overline{\gamma'}/\gamma' = \mu \text{ for } \gamma \in \Gamma\} $$

*determines a quasiconformal homeomorphism $f^{\mu_0}$ of $\widehat{\mathbb{C}}$ compatible with the group $\Gamma$ and such that*

$$ \varkappa(f^{\mu_0}) < k(f^{\mu_0}). $$

**Proof.** As is well-known, the Teichmüller space $\mathbf{T}(\Gamma)$ of the group $\Gamma$ is embedded into $\mathbf{T}$, and $\mathbf{T}(\Gamma) = \mathbf{T} \cap \mathbf{B}(\Gamma)$, where $\mathbf{B}(\Gamma)$ is the subspace of $\mathbf{B}$ formed by $\Gamma$-automorphic forms $\varphi$ of the weight $-4$ (quadratic differentials) in $\Delta^*$, i.e., $(\varphi \circ \gamma)(\gamma')^2 = \varphi$ for all $\gamma \in \Gamma$ (see, e.g., [26]).

Denote by $A_1(\Delta, \Gamma)$ the space of holomorphic $\Gamma$-automorphic forms of the weight $-4$ in $\Delta$ integrable over $\Delta/\Gamma$, with the norm $\|\psi\| = \int_{\Delta/\Gamma} |\psi| dm_2$. For every $\mu \in \text{Belt}(\Delta, \Gamma)_1$, we have the pairings

$$< \mu, \psi >_{\Delta/\Gamma} = \int_{\Delta/\Gamma} \mu \psi dm_2, \quad < \mu, \Psi >_\Delta = \int_\Delta \mu \Psi dm_2$$

with $\psi \in L_1(\Delta/\Gamma)$ and $\Psi \in L_1(\Delta)$, respectively.

It is well-known (cf. [4]) that the Poincaré theta-operator

$$\Theta_\Gamma \Psi(z) = \sum_{\gamma \in \Gamma} \Psi(\gamma z) \gamma'(z)^2 : A_1(\Delta) \to A_1(\Delta, \Gamma)$$

acts on $A(\Delta)$ surjectively, and by McMullen's theorem [27] its norm $\|\Theta_\Gamma\| < 1$.

Any extremal Beltrami differential $\mu_0 \in \text{Belt}(\Delta, \Gamma)_1$ is of Teichmüller form $\mu_0 = \|\mu_0\|_\infty |\psi_0|/\psi_0$ with $\psi_0 \in A_1(\Delta, \Gamma)$, $|\psi_0\| = 1$, and for any $\Psi_0$ with $\Theta_\Gamma \Psi_0 = \psi_0$, we have

$$\|\mu_0\|_\infty = \max\{| < \mu_0, \psi >_{\Delta/G} | : \|\psi_0\|_{A_1(\Delta, \Gamma)} = 1\}$$
$$= | < \mu_0, \psi_0 >_{\Delta/G} | = | < \mu_0, \Psi_0 >_\Delta |.$$

Let us now compare the value $\|\mu_0\|_\infty$ with the minimal dilatation in the class $[f^{\mu_0}]_{\partial\Delta}$ of all quasiconformal extensions of $f^{\mu_0}|\partial\Delta$ into the disk $\Delta$, i.e., with the quantity $k(f^{\mu_0})$.

Using the Hamilton-Krushkal-Reich-Strebel theorem that a Beltrami coefficient $\mu$ is extremal in $\text{Belt}(\Delta)_1$ if and only if $\|\mu\|_\infty = \sup\{| < \mu, \Psi >_\Delta | : \|\Psi\|_{A_1(\Delta)} = 1\}$ (see, e.g., [7,9,10,31]), we obtain that

$$| < \mu_0, \Psi >_\Delta | \leq \|\Theta_\Gamma\| \|\mu_0\|_\infty$$

for all functions $\Psi \in A_1(\Delta)$ with $\|\Psi\| = 1$.

This inequality implies that the Beltrami coefficient $\mu_0$ cannot be extremal in the class $[f^{\mu_0}]_{\partial\Delta}$; and, therefore, for any extremal Beltrami coefficient $\nu_0$ in this class, we have

$$\|\nu_0\|_\infty = \sup\{| < \nu_0, \Psi >_\Delta | : \Psi \in A_1(\Delta)\} = k(f^{\mu_0}) < \|\mu_0\|_\infty. \quad (7)$$

Put $\mu_0^* = \mu_0/\|\mu_0\|_\infty$ and consider in the space $\mathbf{T}(\Gamma) \subset \mathbf{T}$ the holomorphic disk

$$\Delta(\mu_0^*) = \{\phi_{\mathbf{T}(\Gamma)}(t\mu_0^*) : t \in \Delta\}, \quad (8)$$

where $\phi_{\mathbf{T}(\Gamma)}$ is the holomorphic projection $\mu \to S_{f^\mu} : \text{Belt}(\Delta, \Gamma)_1 \to \mathbf{T}(\Gamma) \subset \mathbf{B}(\Gamma)$. The same map $\mu \to S_{f^\mu}$ induces the defining projection $\phi_{\mathbf{T}} : \text{Belt}(\Delta)_1 \to \mathbf{T}$.

The tangent vector to the disk (8) in $\mathbf{T}(\Gamma)$ at the origin is $\phi'_{\mathbf{T}(\Gamma)}(0)\mu_0^*$ and its Teichmüller norm is equal to one. On the other hand, the tangent vector to this disk in $\mathbf{T}$ is $\phi'_{\mathbf{T}}(0)\mu_0^*$ and has by (7) the Teichmüller norm $\|\phi'_{\mathbf{T}}(0)\mu_0^*\| = \|\nu_0\|_\infty/\|\mu_0\|_\infty < 1$.

Therefore, we must have $\varkappa(f^{t\mu_0^*}) < k(f^{t\mu_0^*})$ for all $t \in \Delta \setminus \{0\}$.

Otherwise, due to [11], the equality $\varkappa(f^{t\mu_0^*}) = k(f^{t\mu_0^*})$, even for one $t \neq 0$, would imply, due to the result of [11] mentioned in Section 1, that the disk $\Delta(\mu_0^*)$ must be extremal (geodesic) in both Carathéodory and Teichmüller metrics. On the other hand, by [6] and [13], a holomorphic disk $h(\Delta) \subset \mathbf{T}$ is extremal if and only if the tangent vector $h'(0)$ has Teichmüller length 1. This contradiction proves Theorem 5.1.

The second step in the proof of Theorem 3.1 consists of an approximation of univalent functions. Take the set of points

$$E = \{z_{n,m} = e^{\pi m i/2^n}, \ m = 0, 1, \ldots, 2^{n+1} - 1; \ n = 1, 2, \ldots\}$$

on the unit circle and consider the punctured spheres

$$X_n = \widehat{\mathbb{C}} \setminus \{e^{\pi m i/2^n}, \ m = 0, 1, \ldots, 2^{n+1} - 1\}, \quad n = 1, 2, \ldots.$$

Let us normalize their universal holomorphic covering maps $g_n : \Delta \to X_n$ by $g_n(0) = 0$, $g'_n(0) > 0$. The classical Carathéodory theorem on convergence of planar domains to a kernel (which in our case is the disk $\Delta$) implies that the sequence $\{g_n\}$ is convergent locally uniformly on $\Delta$ to the identity map.

To complete the proof of Theorem 3.1, it suffices to consider the functions admitting quasiconformal extensions. Such functions $f^\mu$ are approximated by homeomorphisms $\widetilde{f}^{\mu_n}$ with the Beltrami differentials

$$\mu_n = (\mu_f \circ g_n)\overline{g'_n}/g'_n, \quad n = 1, 2, \ldots.$$

These differentials are compatible with the group $\Gamma_n$ and converge to $\mu$ almost everywhere on $\mathbb{C}$, hence $f^{\mu_n}$ are convergent to $f^\mu$ uniformly in the spherical metric on $\widehat{\mathbb{C}}$.

**The second proof** of Theorem 3.1 was given by R. Kühnau. The idea of this proof is to approximate a given fixed $f \in \Sigma$ by such maps of $\Sigma^0$ which have an extremal quasiconformal extension of Teichmüller type with at least one single zero (near the unit circle) of the corresponding quadratic differential $\varphi_r dz^2$. This relies on the Strebel frame mapping criterion [34] which uniquely determines an extremal quasiconformal extension (of Teichmüller type) of the values of $f(|z| = r)$ to the disk $\{|z| < r\}$. Here $r - 1$ is chosen sufficiently small to have $f(rz)/r$ close to $f(z)$.

The differential $\varphi_r$ is approximated by

$$\varphi^*(z)dz^2 = \varphi_r(z)(z - N)dz^2/(z - P).$$

Then for the Beltrami coefficients $\mu_r$ and $\mu^*$ of the extremal extensions determined by these quadratic differentials, we have the estimate

$$|\mu^*(z) - \mu_r(z)| = k_r \left| \frac{|z - N|}{z - N} \frac{z - P}{|z - P|} - 1 \right|,$$

and the right-hand side can be made arbitrary small.

Since $\varphi^*$ has a simple zero in $\Delta$, we have for the correspnding map $f^*$ the strong inequality $\varkappa(f^*) < k_0(f^*)$.

This proof yields also that *the functions $f_n$ in Theorem 1.1 can be chosen in such a way that the images of the unit circle $\{|z| = 1\}$ are closed analytic Jordan curves.*

## 6. Sketch of the proof of Theorem 3.2

### 6.1.

We now apply completely different methods coming from complex and Finsler geometry of hyperbolic Banach manifolds and especially from metric geometry of the universal Teichmüller space. For readers convenience, we briefly recall the main definitions and results needed in the sequel.

There are certain natural intrinsic complete metrics on the space $\mathbf{T}$, first of all, its **Teichmüller metric** defined by

$$\tau_{\mathbf{T}}(\phi(\mu), \phi(\nu)) = \frac{1}{2} \inf \left\{ \log K\left(w^{\mu_*} \circ \left(w^{\nu_*}\right)^{-1}\right) : \ \mu_* \in \phi(\mu), \nu_* \in \phi(\nu) \right\},$$

where $\phi = \phi_{\mathbf{T}}$ is the defining canonical projection (factorization) from the Banach ball

$$\mathrm{Belt}(\Delta)_1 = \{\mu \in L_\infty(\mathbb{C}) : \ \|\mu\|_\infty < 1 < \ \mu|\Delta^* = 0\}$$

of measurable conformal structures on the disk $\Delta$ onto $\mathbf{T}$. This metric is generated by the **Finsler structure** on the tangent bundle of $\mathbf{T}$ defined by

$$F_{\mathbf{T}}(\phi(\mu), \phi'(\mu)\nu) = \inf \big\{ \big\| \nu_*(1 - |\mu|^2)^{-1} \big\|_\infty :$$
$$\phi'(\mu)\nu_* = \phi'(\mu)\nu; \ \mu \in \mathrm{Belt}(\Delta)_1; \ \nu, \nu_* \in L_\infty(\mathbb{C}) \big\}. \tag{9}$$

Like other complex manifolds, the main invariant metrics on $\mathbf{T}$ are defined as follows.

The **Kobayashi metric** $d_{\mathbf{T}}$ is the largest pseudometric $d$ on $\mathbf{T}$ contracted by holomorphic maps $h : \ \Delta \to \mathbf{T}$ so that for any two poins $\psi_1$, $\psi_2 \in \mathbf{T}$, we have

$$d_{\mathbf{T}}(\psi_1, \psi_2) \leq \inf \{ d_\Delta(0, t) : \ h(0) = \psi_1, \ h(t) = \psi_2 \},$$

where $d_\Delta$ is the hyperbolic Poincaré metric on $\Delta$ of Gaussian curvature $-4$. The **Carathéodory metric** $c_{\mathbf{T}}$ is the least pseudometric on $\mathbf{T}$ with such a property.

The infinitesimal (differential) Kobayashi metric $\mathcal{K}_{\mathbf{T}}(\psi, v)$ is a Finsler metric on the tangent bundle of $\mathbf{T}$ defined by

$$\mathcal{K}_{\mathbf{T}}(\psi, v) = \inf \{ |t| : \ h \in \mathrm{Hol}(\Delta, \mathbf{T}), \ h(0) = \psi, \ dh(0)t = v \}$$
$$= \inf \left\{ \frac{1}{r} : \ r > 0, \ h \in \mathrm{Hol}(\Delta_r, \mathbf{T}), \ h(0) = \psi, \ h'(0) = v \right\}. \tag{10}$$

Here $v$ is a tangent vector at the point $\psi \in \mathbf{T}$, $\Delta_r$ denotes the disk $\{|z| < r\}$, and $\mathrm{Hol}(\Delta, \mathbf{T})$ denotes the collection of holomorphic maps from $\Delta$ into $\mathbf{T}$ (for details we refer, e.g., to [5,7,10]).

Note that the basic equality (5) is equivalent to the equality of the Carathéodory and Teichmüller metrics on the holomorphic disk

$$\mathbb{D}(\mu) := \{\phi_{\mathbf{T}}(t\mu) : t \in \mathbb{C}\} \tag{11}$$

(which gives also that $\mathbb{D}(\mu)$ is a geodesic disk).

## 6.2.

The proof of Theorem 3.2 is obtained along the following line.

We first construct for any $f \in \Sigma$ the complex isotopy $f(z,t) : \Delta^* \times \Delta \to \widehat{\mathbb{C}}$ by

$$f(z,t) = tf(t^{-1}z) = z + b_0 t + b_1 t^2 z^{-1} + b_2 t^3 z^{-2} + \ldots.$$

The Schwarzian derivatives $\varphi(t,z) = S_{f_t}(z)$ of the fiber maps $f_t(z) := f(z,t)$ relate to $S_f$ by

$$S_{f_t}(z) = t^{-2} S_f(t^{-1}z).$$

This equality defines a holomorphic map

$$\Phi_f(t) = \varphi(t,\cdot) : \Delta \to \mathbf{T} \tag{12}$$

so that the image $\Phi_f(\Delta)$ is a holomorphic disk in $\mathbf{T}$ with a single singular point $\Phi_f(0) = \mathbf{0}$ (see, e.g., [11]). Further, the functions

$$k_f(t) := k_{f_t}, \quad \text{and} \quad \varkappa_f(t) := \varkappa_{f_t}$$

are logarithmically subharmonic and continuous on the disk $\Delta$; regarded as radial functions of $r = |t|$, they are strictly monotone increasing and continuous on $[0,1]$ (cf. [14,21,33]).

The essential part of the proof of Theorem 3.2 consists of establishing how to relate $\varkappa_f(t)$ and $k_f(t)$ on $\Delta$. This is derived by examining certain Finsler metrics of generalized Gaussian curvature $\kappa \leq -4$ on the disk $\Phi_f(\Delta)$.

Recall that the **generalized Laplacian** $\Delta u$ of an upper semicontinuous function $u : \Omega \to [-\infty, \infty)$ in a domain $\Omega \subset \mathbb{C}$ is defined by

$$\Delta u(z) = 4 \liminf_{r \to 0} \frac{1}{r^2} \left\{ \frac{1}{2\pi} \int_0^{2\pi} u(z + re^{i\theta})d\theta - u(z) \right\}.$$

It reduces to the usual Laplacian $4\partial\bar\partial$ for $C^2$ functions and has similar properties: a function $u$ is subharmonic in $\Omega$ if and only if $\Delta u(z) \geq 0$; at a point $z_0$ of a local maximum of an upper semicontinuous function $u$ with $u(z) > -\infty$, we have $\Delta u(z_0) \leq 0$. Respectively, the **generalized Gaussian curvature** $\kappa_\lambda$ of a upper semicontinuous Finsler metric $ds = \lambda(t)|dt|$ on a domain $\Omega$ is defined by

$$\kappa_\lambda(t) = -\frac{\Delta \log \lambda(t)}{\lambda(t)^2}, \tag{13}$$

using the generalized Laplacian $\Delta$.

One defines in a similar way also the **sectional holomorphic curvature** of a Finsler metric on complex Banach manifolds $X$ as the supremum of the curvatures

(13) over appropriate collections of holomorphic maps $\Delta \to X$ for a given tangent direction (for details, we refer e.g. to [5]). An important fact is that *the holomorphic curvature of the Kobayashi metric $\lambda_K$ of the universal Teichmüller space* **T** *equals* $-4$ *everywhere* (cf. [1,13]).

We use in an essential way the **maximum principle of Minda**, which states:

**Lemma 6.1.** [28] *If a function $u : \Omega \to [-\infty, +\infty)$ is upper semicontinuous in a domain $\Omega \subset \mathbb{C}$ and its generalized Laplacian satisfies the inequality $\Delta u(z) \geq K u(z)$ with some positive constant $K$ at any point $z \in \Omega$, where $u(z) > -\infty$, then if $\limsup_{z \to \zeta} u(z) \leq 0$ for all $\zeta \in \partial\Omega$, then either $u(z) < 0$ for all $z \in \Omega$ or else $u(z) = 0$ for all $z \in \Omega$.*

This principle is applied to the comparison of the Kobayashi metric $\lambda_K$ on **T** with the metric $\lambda_\varkappa$ generated by Grunsky coefficients of $\alpha_{mn}(S_f)$ of the functions $f \in \Sigma^0$ as the pull-back of hyperbolic metric $\lambda_{\text{hyp}}(\zeta)|d\zeta|$ in $\Delta$ of curvature $-4$, i.e., with the density

$$\lambda_{\text{hyp}}(\zeta) = 1/(1 - |\zeta|^2,$$

via the holomorphic maps (for appropriate tangent vectors $v$ at $\varphi$)

$$h_{\mathbf{x}}(\varphi) = \sum_{m,n=1}^{\infty} \sqrt{mn}\, \alpha_{mn}(\varphi) x_m x_n : \mathbf{T} \to \Delta \tag{14}$$

parametrized by the points $\mathbf{x} = (x_n) \in S(l^2)$. Here $\varphi \in \Phi_f(\Delta)$. Explicitly,

$$\lambda_{h_{\mathbf{x}} \circ \Phi_f}(t) := (h_{\mathbf{x}} \circ \Phi_f)^*(\lambda_0) = \frac{|(h_{\mathbf{x}} \circ \Phi_f)'(t)|}{1 - |h_{\mathbf{x}} \circ \Phi_f(t)|^2}. \tag{15}$$

This allows us to prove the following two lemmas:

**Lemma 6.2.** *If $f \in \Sigma^0$ is such that $\varkappa(f) < k(f)$, then $\varkappa_f(r) < k_f(r)$ for all $0 < r < 1$.*

**Lemma 6.3.** *Let $f \in \Sigma^0$ with $\varkappa(f) = k(f)$ map the unit circle $\partial\Delta$ onto an analytic curve. Then $\varkappa_f(r) = k_f(r)$ on the whole interval $[0,1]$.*

Due to the well-known properties of conformal maps of domains with analytic boundaries, the function $f$ extends to a conformal map $\widehat{f}$ of some disk $\Delta_a^* := \{|z| > a\}$, $a < 1$. This fact is also important for the proofs of both Lemma 6.3 and Theorem 3.2.

The next step consists of approximation:

**Lemma 6.4.** *Every function $f \in \Sigma^0$ with $\varkappa(f) = k(f)$ mapping the unit circle $\partial\Delta$ onto an asymptotically conformal curve is approximated locally uniformly on $\Delta^*$ by functions $f_n \in \Sigma^0$ with $\varkappa(f_n) = k(f_n)$ mapping the circle $\partial\Delta$ onto analytic curves.*

The **proof** of this lemma relies on the frame mapping criterion of Strebel, already applied in the above, as well as on some deep results on extremal quasiconformal extensions of conformal maps over asymptotically conformal curves (cf. [12,22]).

We show that $f$ has (a unique) extremal quasiconformal extension $\widehat{f}^{\mu_0}$ to $\Delta$ which is of Teichmüller-Kühnau type; its Beltrami coefficient $\mu_0 = k_0|\psi_0|/\psi_0$ is determined by quadratic differential $\psi_0$ of the form

$$\psi_0(z) = \frac{1}{\pi} \sum_{m+n=2}^{\infty} \sqrt{mn}\, x_m x_n z^{m+n-2} \quad \text{with } \mathbf{x} = (x_n) \in l^2,$$

which has in $\Delta$ only zeros of even order.

Now, letting $\psi_n(z) = c_n \psi_0(r_n z)$ with $r_n \to 1$ and appropriate $c_n > 0$ so that $\|\psi_n\|_{L_1(\Delta)} = 1$, one obtains the desired approximation which is prescribed by Lemma 6.4.

To complete the proof of Theorem 3.2, one needs to exploit the continuity of the functions $\varkappa(f)$ and $k(f)$ on $\mathbf{T}$.

### 6.3.

Note that the above arguments give in fact somewhat more: the functions $f \in \Sigma$ with $\varkappa(f) < k(f)$ cannot be approximated by functions $f_n \in \Sigma^0$ with $\varkappa(f_n) = k(f_n)$, provided $f_n$ admit quasiconformal extensions $\widetilde{f}_n^\mu$ to $\widehat{\mathbb{C}}$ of Teichmüller-Kühnau type, i.e., with Beltrami coefficients $\mu_n$ in $\Delta$ determined by holomorphic quadratic differentials having only zeros of even order. So the question remains open whether the conjecture is true for $f_n \in \Sigma^0$, for which the supremum in the fundamental equality

$$\sup | < \mu_0, \psi >_\Delta | = \{ \|\mu\|_\infty : \psi \in A_1(\Delta),\ \|\psi\|_1 = 1 \}$$

for exremal Beltrami coefficients $\mu_0$ does not be attained on the distinguished subset $A_1^2$.

## 7. A glimpse at applications of property (5)

The equality (5) naturally arises in certain topics of geometric function theory. It ensures the remarkable features of quasiconformal maps and of curve functionals, for example, of Fredholm eigenvalues. Though by Theorem 3.1 the set of maps obeying this property is sparse, their role is crucial in qualitative estimating.

For example, it allowed us to establish the deep plurisubharmonic features of the basic Teichmüller and Kobayashi metrics on certain Teichmüller spaces (see [14]).

The role of the property (5) is crucial in the questions related to Fredholm eigenvalues and quasiconformal reflections. One of the basic facts in this theory is the well-known inequality of Ahlfors [2] that if a Jordan curve $L$ admits quasiconformal reflection with the dilatation $q < 1$, then its Fredholm eigenvalue $\rho_L$ is estimated by $q \geq 1/\rho_L$. The exact bounds are obtained for various curves by establishing that

942

for these curves we have the equality in the above relation. For details and results, we refer to the surveys [23,15].

Each of the quantities in (5) is equal to the value $\mathbf{g}(S_f, \mathbf{0})$ of the pluricomplex Green function of the space $\mathbf{T}$, which gives rise to other applications.

## References

1. M. Abate and G. Patrizio, *Isometries of the Teichmüller metric*, Ann. Scuola Super. Pisa Cl. Sci.(4) **26** (1998), 437-452.
2. L. Ahlfors, *Remarks on the Neumann-Poincaré equation*, Pacific J. Math. **2** (1952), 271-280.
3. J. Becker, *Conformal mappings with quasiconformal extensions*, Aspects of Contemporary Complex Analysis, D.A. Brannan and J.G. Clunie, eds., Academic Press, London, 1980, pp. 37-77.
4. L. Bers, *Automorphic forms and Poincaré series for infinitely generated Fuchsian groups*, Amer. J. Math. **87** (1965), 196-214.
5. S. Dineen, *The Schwarz Lemma*, Clarendon Press, Oxford, 1989.
6. C.J. Earle, I. Kra and S.L. Krushkal, *Holomorphic motions and Teichmüller spaces*, Trans. Amer. Math. Soc., **944** (1994), 927-948.
7. F.P. Gardiner and N. Lakic, *Quasiconformal Teichmüller Theory*, Amer. Math. Soc., Providence, 2000.
8. H. Grunsky, *Koeffizientenbedingungen für schlicht abbildende meromorphe Funktionen*, Math. Z. **45** (1939), 29-61.
9. R. Hamilton, *Extremal quasiconformal mappings with prescribed boundary values*, Trans. Amer. Math. Soc. **138** (1969), 399-406.
10. S. L. Krushkal, *Quasiconformal Mappings and Riemann Surfaces*, Wiley, New York, 1979.
11. S. L. Krushkal, *Grunsky coefficient inequalities, Carathéodory metric and extremal quasiconformal mappings*, Comment. Math. Helv. **64** (1989), 650-660.
12. S.L. Krushkal *Extension of conformal mappings and hyperbolic metrics*, Siberian Math. J. **30** (1989), 730-744.
13. S.L. Krushkal *On Grunsky conditions, Fredholm eigenvalues and asymptotically conformal curves*, Mitt. Math. Sem. Gissen, **228** (1996), 17-23.
14. S.L. Krushkal *Plurisubharmonic features of the Teichmüller metric*, Publications de L'Institut Mathématique-Beograd, Nouvelle série **75(89)** (2004), 119-138.
15. S. L. Krushkal, *Quasiconformal extensions and reflections*, in: Handbook of Complex Analysis: Geometric Function Theory, Vol. 2, R. Kühnau, ed., Elsevier Science, Amsterdam, 2004, pp. 507-553.
16. S. L. Krushkal, *Schwarzian derivative and complex Finsler metrics*, Contemporary Math. **382** (2005), 243-262.
17. S. L. Krushkal, *Beyond Moser's conjecture on Grunsky inequalities*, Georgian Math. J. **12** (2005), 485-492.
18. S. L. Kruschkal und R. Kühnau, *Quasikonforme Abbildungen - neue Methoden und Anwendungen*, Teubner-Texte zur Math., **54**, Teubner, Leipzig, 1983 (in Russian: Novosibirsk, 1984).
19. S. L. Krushkal and R. Kühnau, *Grunsky inequalities and quasiconformal extension*, Israel J. Math. (2006), to appear.
20. R. Kühnau, *Verzerrungssätze und Koeffizientenbedingungen vom Grunskyschen Typ für quasikonforme Abbildungen*, Math. Nachr. **48** (1971), 77-105.
21. R. Kühnau, *Quasikonforme Fortsetzbarkeit, Fredholmsche Eigenwerte und Grun-*

*skysche Koeffizientenbedingungen*, Ann. Acad. Sci. Fenn. Ser. AI. Math. **7** (1982), 383-391.

22. R. Kühnau, *Wann sind die Grunskyschen Koeffizientenbedingungen hinreichend für Q-quasikonforme Fortsetzbarkeit?*, Comment. Math. Helv. **61** (1986), 290-307.

23. R. Kühnau, *Möglichst konforme Spiegelung an einer Jordankurve*, Jber. Deutsch. Math. Verein. **90** (1988), 90-109.

24. R. Kühnau, *Über die Grunskyschen Koeffizientenbedingungen*, Ann. Univ. Mariae Curie-Skłodowska Lublin, Sect.A, **54** (2000), 53-60.

25. O. Lehto, *Quasiconformal mappings and singular integrals*, Symposia Mathematica (Istituto Naz. di Alta Matematica) **18** (1976), 429-453.

26. O. Lehto, *Univalent functions and Teichmüller spaces*, Springer, New York, 1987.

27. C. T. McMullen, *Amenability, Poincaré series and quasiconformal maps*, Inv. Math. **97** (1989), 95-127.

28. D. Minda, *The strong form of Ahlfors' lemma*, Rocky Mountain J. Math., **17** (1987), 457-461.

29. Chr. Pommerenke, *Univalent Functions*, Vandenhoeck & Ruprecht, Göttingen, 1975.

30. Chr. Pommerenke, *Boundary Behavior of Conformal Maps*, Springer, Berlin, 1992.

31. E. Reich and K. Strebel, *Extremal quasiconformal mappings with given boundary values*, Contributions to Analysis (L. V. Ahlfors et al. eds.), Academic Press, New York and London, 1974, pp. 375-391.

32. M. Schiffer, *Fredholm eigenvalues and Grunsky matrices*, Ann. Polon. Math. **39** (1981), 149-164.

33. Y.L. Shen, *Pull-back operators by quasisymmetric functions and invariant metrics on Teichmüller spaces*, Complex Variables **42** (2000), 289-307.

34. K. Strebel, *On the existence of extremal Teichmueller mappings*, J. Analyse Math. **30** (1976), 464-480.

35. I. V. Zhuravlev, *Univalent functions and Teichmüller spaces*, Inst. of Mathematics, Novosibirsk, preprint, 1979, pp. 1-23 (Russian).

# FINELY MEROMORPHIC FUNCTIONS IN CONTOUR-SOLID PROBLEMS

T. ALIYEV AZEROGLU

*Department of Mathematics*
*Gebze Institute of Technology, Gebze, 41410 Kocaeli, Turkey*
*E-mail: aliyev@gyte.edu.tr*


P.M. TAMRAZOV

*Institute of Mathematics of National Academy of Sciences of Ukraine*
*Tereshshenkivska str., 3, 01601, Kiev, MSP, Ukraine*
*E-mail: tamrazov@gyte.edu.tr*

We establlich contour-solid theorems for finely meromorphic functions taking into account zeros and the multivalence of functions.

**Key words:** Finely holomorphic functions, finely meromorphic functions, contour-solid problems, fine potential theory, finely subharmonic functions

**Mathematics Subject Classification:** 30G12, 31C40

## 1. Introduction

In [1] the purely fine contour-solid theory for finely holomorphic and finely hypoharmonic functions was established. That theory contains refined, strengthened and extended theorems for the mentioned classes of functions in finely open sets of the complex plane with preservable majorants (from the maximal classes of such majorants for the mentioned function classes). On the other hand, in [2-6] certain contour-solid theorems for analytic functions from earlier authors' works were extended onto meromorphic functions and strengthened with taking into account zeros and the multivalence of functions.

In the present work we extend and strengthen some results of [1] related to finely holomorphic functions. The generalization is related to considering finely meromorphic functions instead of finely holomorphic, and strengthening is connected with taking into account zeros and the multivalence of functions.

We need to recall a number of definitions and notations from [1].

Let $\mathfrak{M}$ be the class of all functions $\mu : (0, +\infty) \to [0, +\infty)$ for each of which the set $I^\mu := \{x : \mu(x) > 0\}$ is connected and the restriction of the function $\log \mu(x)$ to $I^\mu$ is concave with respect to $\log x$. Let $\mathfrak{M}^*$ be the class of all $\mu \in \mathfrak{M}$ for which $I^\mu$ is non-empty.

For $\mu \in \mathfrak{M}^*$ let us denote by $x_-^\mu$ and $x_+^\mu$ the left and the right ends of the interval $I^\mu$, respectively. Obviously, $0 \le x_-^\mu \le x_+^\mu \le +\infty$. When $x_-^\mu < x_+^\mu$, the mentioned concavity condition is equivalent to the combination of the following conditions: the function $\log \mu(x)$ is concave with respect to $\log x$ (and therefore continuous) in the interval $(x_-^\mu, x_+^\mu)$ and lower semicontinuous on $I^\mu$. For $\mu \in \mathfrak{M}$ the limits

$$\mu_0 := \lim_{x \to 0} \frac{\log \mu(x)}{\log x}, \quad \mu_\infty := \lim_{x \to +\infty} \frac{\log \mu(x)}{\log x}$$

exist, and we have

$$\mu_0 \ge \mu_\infty, \quad \mu_0 > -\infty, \quad \mu_\infty < +\infty.$$

In particular, if $x_-^\mu > 0$ (analogously, if $x_+^\mu < +\infty$), then $\mu_0 = +\infty$ ($\mu_\infty = -\infty$, respectively). When $\mu_0 < +\infty$, define the integer $m_0$ by the conditions $m_0 - 1 < \mu_0 \le m_0$, and when $\mu_\infty > -\infty$, define the integer $m_\infty$ by the conditions $m_\infty \le \mu_\infty < m_\infty + 1$.

For every fixed $\alpha \in \mathbb{R}$, $\beta \in (0, +\infty)$, the function $\mu(x) := \beta x^\alpha$ belongs to $\mathfrak{M}^*$, and for it we have $\mu_0 = \mu_\infty = \alpha$. If, moreover, $\alpha$ is an integer, then $m_0 = m_\infty = \alpha$.

Let $\widehat{\mathbb{C}}$ be the compact Riemann sphere.

We refer to [7-9] concerning the fine topology and related notions such as thinness, the fine boundary and the fine closure of a set, fine limits of functions, fine superior and fine inferior limits of functions, finely hypoharmonic, finely hyperharmonic, finely subharmonic, finely harmonic, finely meromorphic functions, the generalized harmonic measure, the Green's function for a fine domain and so on.

Let $E \subset \widehat{\mathbb{C}}$. Denote by $\widehat{E}$ the standard closure of $E$ in $\widehat{\mathbb{C}}$, and by $\overline{E}$ the standard closure of a set $E \subset \mathbb{C}$ in $\mathbb{C}$. The set of all points $x \in \widehat{\mathbb{C}}$ in which $E$ is not thin is called *the base of the set* E *in* $\widehat{\mathbb{C}}$ and is denoted by $b(E)$. The set $\tilde{E} := E \cup b(E)$ is called *the fine closure of the set* E *in* $\widehat{\mathbb{C}}$. Clearly, $\tilde{E} \subset \widehat{E}$. Denote by $\widehat{\partial_f} E$ the fine boundary of $E$ in $\widehat{\mathbb{C}}$. Let $\partial_f E := \mathbb{C} \cap \widehat{\partial_f} E$, $(E)_i := E \setminus b(E)$, $(E)_r := E \setminus (E)_i$. Points $x \in (E)_r$ and $x \in (E)_i$ are called *regular* and *irregular* points, respectively, of the set $E$.

For a set $E \subset \widehat{\mathbb{C}}$ let us denote $\widehat{\mathbb{C}} \setminus E =: FE$, and for a set $E \subset \mathbb{C}$, denote also $\mathbb{C} \setminus E =: CE$.

Let $G \subset \widehat{\mathbb{C}}$ be a finely open set, the set $FG$ be non-polar and $z \in G$. A set $Q \subset FG$ will be called *nearly negligible relative to* $G$ if for every finely connected component $T$ of $G$ the set $Q \cap \partial_f T$ contains no compact subset $K$ of the harmonic measure $\omega_z^T(K) > 0$ at some (and therefore at any) point $z \in T$. This requirement is equivalent to the following alternative: either $FG$ is polar (and then $Q$ is also polar), or for every finely connected component $T$ of $G$ both $\partial_f T$ is non-polar and $Q$ is a set of inner harmonic measure zero relative to $T$ and any point $z \in T$.

If a set $E \subset FG$ is such that for every finely connected component $T$ of $G$ it contains no compact subset $K \subset \partial_f T$ of logarithmic capacity $\mathrm{Cap}\, K > 0$, then $E$ is nearly negligible relative to $G$.

In particular, any set $E \subset FG$ of inner logarithmic capacity zero is nearly negligible relative to $G$.

For any finely open set $G \subset \mathbb{C}$ with a non-polar complement $\mathbb{C} \setminus G$, and $w \in \widehat{\mathbb{C}}$, $\zeta \in G$, $w \neq \zeta$ there exists Green's function $g_G(w, \zeta)$.

Let $D$ be a fine domain in $\widehat{\mathbb{C}}$, i. e. a finely open, finely connected set. Then $\check{D} = \widehat{D}$. In particular, then $D$ is finely separable from a point $z \in \widehat{\mathbb{C}}$ if and only if it is separable from $z$ in the standard topology. So in such a situation we may speak of separability not specifying in what sense.

Let $G$ be a finely open set in $\widehat{\mathbb{C}}$, and $z \in \widehat{\partial_f G}$. Given any functions $u : G \to [-\infty, +\infty]$ and $h : G \to \widehat{\mathbb{C}}$ we introduce the following notations for fine superior limits of functions:

$$\text{fine lim sup}_{\zeta \to z, \ \zeta \in G} \ u(\zeta) =: \check{u}_{G,f}(z),$$

$$\text{fine lim sup}_{\zeta \to z, \ \zeta \in G} \ |h(\zeta)| =: \bar{h}_{G,f}(z).$$

Let $G$ be a finely open set in $\mathbb{C}$ and $a \in CG$ be a fixed point. For a function $h : G \to \widehat{\mathbb{C}}$ let us denote:

$$h_{a,G,f} := \begin{cases} \text{fine lim sup}_{\zeta \to a, \ \zeta \in G} \ \frac{\log|f(\zeta)|}{|\log|\zeta - z||} & \text{when} \quad a \in \partial_f G, \\ \\ 0 & \text{when} \quad a \notin \partial_f G. \end{cases}$$

$$h_{\infty,G,f} := \begin{cases} \text{fine lim sup}_{\zeta \to \infty, \ \zeta \in G} \ \frac{\log|f(\zeta)|}{\log|\zeta|} & \text{when} \quad \infty \in \widehat{\partial_f G}, \\ \\ 0 & \text{when} \quad \infty \notin \widehat{\partial_f G}, \end{cases}$$

If the function $f$ is finely meromorphic in $G$ then we denote by $k(f, w)$ the multiplicity of its value $f(w)$ at the point $w \in G$.

Given functions $p : X \to \widehat{\mathbb{C}}$ and $q : X \to \widehat{\mathbb{C}}$ on a set $X \subset \widehat{\mathbb{C}}$ and a finely limit point $w$ for $X$, we use the following notations. If there exist a finite number $l \geq 0$ and a fine neighbourhood $U$ of $w$ for which $|p(z)| \leq l \, |q(z)| \ \forall z \in X \cap U$, then we write

$$p(z) = \text{fine } O(q(z)) \quad (z \to w, z \in X),$$

and if for every $\epsilon > 0$ there exists a fine neighbourhood $U$ of $w$ for which $|p(z)| \leq \epsilon \, |q(z)| \ \forall z \in X \cap U$, then we write

$$p(z) = \text{fine } o(q(z)) \quad (z \to w, z \in X).$$

Let $G \subset \mathbb{C}$ be a finely open set, $h : G \to \widehat{\mathbb{C}}$ be a finely meromorphic function, and $\mu \in \mathfrak{M}$. Consider the following conditions:

$(A, \infty)$ $\quad \infty \in b(CG)$ and for every finely connected component $T$ of $G$ with $\infty \in b(T)$ there holds $h_{\infty,T,f} < +\infty$;

$(B, \infty)$ $\quad \infty \notin b(CG)$, $\mu_\infty > -\infty$ and

$$h(\zeta) = \text{fine } o(|\zeta|^{m_\infty + 1}) \ (\zeta \to \infty, \ \zeta \in G); \tag{1}$$

$(B_0, \infty)$    $\infty \notin b(CG)$,    $\mu_\infty \geq 0$  and (1) is true.

If $z \in \mathbb{C}$ is a fixed point, then we consider also the following conditions:

$(A, z)$    $z \in b(CG)$     and for every finely connected component $T$ of $G$ with $z \in b(T)$ there holds $h_{z,T,f} < +\infty$;

$(B, z)$    $z \notin b(CG)$,    $\mu_0 < +\infty$  and

$$h(\zeta) \;=\; \text{fine } o(|\zeta - z|^{m_0 - 1}) \; (\zeta \to z, \; \zeta \in G); \tag{2}$$

$(B_1, z)$    $z \notin b(CG)$,    $\mu_0 \leq 1$  and (2) is true.

We will use the following commutative rule for possible indefinite expressions: $0 \cdot (\pm\infty) = 0$, $-\infty + \infty = -\infty$.

## 2. Main Results

We prove the following results.

**Theorem 2.1.** *Let $a \in \mathbb{C}$ be a fixed point; $G \subset \mathbb{C} \setminus \{a\}$ be a finely open set with non-polar fine boundary $\partial_f G$; $\mu \in \mathfrak{M}$; $h : G \to \widehat{\mathbb{C}}$ be a finely meromorphic function for which*

$$\bar{h}_{G,f}(z) \leqslant \mu(|z - a|) \qquad \forall z \in \partial_f G \setminus \{a\}.$$

*Denote $z_1 := a$, $z_2 := \infty$ and suppose that for each $s = 1, 2$ (independently from each other) one of the conditions $(A, z_s)$ or $(B, z_s)$ is satisfied. Let $\mathfrak{P}$ be the set of all poles of $h$ in $G$. Then*

$$|h(\zeta)| \exp \left[ -\sum_{p \in \mathfrak{P}} g_G(p, \zeta) \cdot k(h, p) \right] \leqslant$$

$$\mu(|\zeta - a|) \exp \left[ -\sum_{w:h(w)=0} g_G(w, \zeta) \cdot k(h, w) \right] \quad \forall \zeta \in G \setminus \mathfrak{P}. \tag{3}$$

**Theorem 2.2.** *Let $a \in \mathbb{C}$ be a fixed point; $G \subset \mathbb{C} \setminus \{a\}$ be a finely open set with non-polar fine boundary $\partial_f G$; $Q$ be a set contained in $FG$ and containing the points $a$ and $\infty$; $\mu \in \mathfrak{M}$; $h : G \to \widehat{\mathbb{C}}$ be a finely meromorphic function. Denote $z_1 := a$, $z_2 := \infty$ and suppose that for each $s = 1, 2$ (independently from each other) one of the conditions $(A, z_s)$ or $(B, z_s)$ is satisfied. Let $\mathfrak{P}$ be the set of all poles of $h$ in $G$.*

*Denote*

$$\lambda(x) := \log \mu(x) \qquad \forall x > 0,$$

$$u(\zeta) := \log |h(\zeta)| \; -\sum_{p \in \mathfrak{P}} g_G(p, \zeta) \cdot k(h, p) \;\; \forall \zeta \in G.$$

*Suppose that for every finely connected component $T$ of $G$ the following conditions are satisfied: $Q$ is nearly negligible relative to $T$ and*

$$\check{u}_{T,f}(z) < \infty \qquad \forall z \in \partial_f T \setminus \{a\};$$

$$\check{u}_{T,f}(z) \le \lambda(|z-a|) \qquad \forall z \in \partial_f T \backslash Q.$$

*Then (3) holds.*

Notice that in formulations of Theorems 2, 8.1, 8.2 of [1] the following obvious correction should be done: $Q$ must be assumed to be contained in $\widehat{FG}$ instead of $\widehat{\partial_f G}$.

**Theorem 2.3.** *Let $G \subset \mathbb{C}$ be a finely open set; $\mu \in \mathfrak{M}$; $h : \tilde{G} \cap \mathbb{C} \to \widehat{\mathbb{C}}$ be a function finely meromorphic in $G$ and satisfying the condition*

$$|h(z) - h(\zeta)| \le \mu(|z-\zeta|) \quad \forall z, \zeta \in \partial_f G, \ \zeta \ne z \tag{4}$$

*Let $\mathfrak{P}$ be the set of all poles of $h$ in $\tilde{G} \cap \mathbb{C}$. Let one of the conditions $(A, \infty)$ or $(B_0, \infty)$ be satisfied for the restriction of $h$ onto $G$ (instead of $h$). Suppose also that for every finely connected component $T$ of $G$ the restriction of $h$ onto $(\tilde{T} \cap \mathbb{C})\backslash\mathfrak{P}$ is finely continuous.*

*Under these assumptions we have*

$$|h(\zeta) - h(z)| \exp\left[-\sum_{p\in\mathfrak{P}} (g_G(p,\zeta) + g_G(p,z)) \cdot k(h,p)\right] \le \mu(|\zeta-z|)\times$$

$$\exp\left[-\sum_{\substack{w\in G \\ h(w)=h(\zeta)}} g_G(w,z) \cdot k(h,w)\right] \quad \forall z \in (\partial_f G)_r \quad \forall \zeta \in (\tilde{G}\cap\mathbb{C})\backslash\mathfrak{P}, \ \zeta \ne z. \tag{5}$$

**Theorem 2.4.** *Let $G$, $\mu$, $h$ satisfy all assumptions of Theorem 2.3. Additionally suppose that $z_0 \in (CG)_i \cup G$ is a fixed point, $\mu_0 < +\infty$ and*

$$|h(\zeta) - h(z_0)| = \text{ fine } o(|\zeta - z_0|^{m_0 - 1}) \qquad (\zeta \to z_0, \ \zeta \in G\backslash\mathfrak{P}).$$

*Then*

$$|h(\zeta) - h(z_0)| \exp\left[-\sum_{p\in\mathfrak{P}} (g_G(p,\zeta) + g_G(p,z_0)) \cdot k(h,p)\right] \le \mu(|\zeta-z_0|)\times$$

$$\exp\left[-\sum_{\substack{w\in G \\ h(w)=h(\zeta)}} g_G(w,z_0) \cdot k(h,w)\right] \qquad \forall \zeta \in \tilde{G}\backslash\mathfrak{P}, \ \zeta \ne z_0, \infty.$$

**Theorem 2.5.** *Let $G$, $\mu$, $h$ satisfy all assumptions of Theorem 2.3. Additionally suppose that $\mu_0 < 1$. Then*

$$|h(\zeta) - h(z)| \exp\left[-\sum_{p\in\mathfrak{P}} (g_G(p,\zeta) + g_G(p,z)) \cdot k(h,p)\right] \le \mu(|\zeta-z|)\times$$

$$\exp \left[ - \sum_{\substack{w \in G \\ h(w)=h(\zeta)}} g_G(w,z) \cdot k(h,w) \right] \qquad \forall z, \zeta \in (\tilde{G} \cap \mathbb{C}) \backslash \mathfrak{P}, \ \zeta \neq z. \qquad (6)$$

**Theorem 2.6.** *Let* $G$, $\mu$, $h$ *satisfy all assumptions of Theorem 2.3. If* $\infty \in \tilde{G}$, *additionally suppose that* $h$ *can be extended to* $\infty$ *in such a way that the extended function* $\tilde{h}$ *satisfies the following hypothesis: the restriction* $\tilde{h}|_{\tilde{T}}$ *of* $\tilde{h}$ *to* $\tilde{T}$ *is finely continuous (and finite) at* $\infty$ *for every finely connected component* $T$ *of* $G$ *with* $\infty \in \tilde{T}$. *Under these assumptions the inequality (6) is true.*

## 3. Proof of Theorems

We need to recall a number of notations from [1].

Let $L$ be the class of all functions $\lambda : (0, +\infty) \rightarrow [-\infty, +\infty]$ for each of which the set $I_\lambda := \{x : \lambda(x) > -\infty\}$ is connected and the restriction of $\lambda$ to $I_\lambda$ is concave with respect to $\log x$. Let $L^*$ be the class of all $\lambda \in L$ for which $I_\lambda$ is non-empty. For $\lambda \in L^*$ let us denote by $x_\lambda^-$ and $x_\lambda^+$, respectively, the left and the right ends of the interval $I_\lambda$. Obviously, $0 \leq x_\lambda^- \leq x_\lambda^+ \leq +\infty$. When $\lambda(\cdot)$ runs through the classes $L$ or $L^*$, the function $\exp \lambda(\cdot)$ runs through the classes $\mathfrak{M}$ or $\mathfrak{M}^*$, respectively. When $x_\lambda^- < x_\lambda^+$, the mentioned concavity condition is equivalent to the combination of the following conditions: the function $\lambda(x)$ is concave with respect to $\log x$ (and therefore continuous) in the interval $(x_\lambda^-, x_\lambda^+)$ and lower semicontinuous on $I_\lambda$. For $\lambda \in L$ the limits

$$\lambda^0 := \lim_{x \to 0} \frac{\lambda(x)}{\log x}, \qquad \lambda^\infty := \lim_{x \to +\infty} \frac{\lambda(x)}{\log x}$$

exist, and we have

$$\lambda^0 \geq \lambda^\infty, \quad \lambda^0 > -\infty, \quad \lambda^\infty < +\infty.$$

For every fixed $\alpha \in \mathbb{R}$, $\theta \in \mathbb{R}$ the function $\lambda(x) := \alpha \log x + \theta$ belongs to $L^*$, and for it we have $\lambda_0 = \lambda_\infty = \alpha$.

Let $a \in CG$ be a fixed point.

Let $G$ be a finely open set in $\widehat{\mathbb{C}}$ and $a \in CG$ be a fixed point. For any function $u : G \to [-\infty, +\infty]$ let us denote:

$$u_{G,f}^a := \begin{cases} \text{fine lim sup}_{\zeta \to a, \ \zeta \in G} \ \frac{u(\zeta)}{||\log|\zeta - z||} & \text{when} \quad a \in \partial_f G, \\ \\ 0 & \text{when} \quad a \notin \partial_f G. \end{cases}$$

$$u_{G,f}^\infty := \begin{cases} \text{fine lim sup}_{\zeta \to \infty, \ \zeta \in G} \ \frac{u(\zeta)}{\log|\zeta|} & \text{when} \quad \infty \in \widehat{\partial_f G}, \\ \\ 0 & \text{when} \quad \infty \notin \widehat{\partial_f G}. \end{cases}$$

Let $\lambda \in L$. Consider the following conditions:

$(A', \infty)$   $\infty \in b(CG)$ and for every finely connected component $T$ of $G$ with $\infty \in b(T)$ there holds   $u^{\infty}_{T,f} < +\infty$;

$(B', \infty)$   $\infty \notin b(CG)$ and there exist a constant $t \in \mathbb{R}$ and a fine neighbourhood $V$ of $\infty$ for which

$$u(\zeta) \le \lambda(|\zeta - a|) + t \quad \forall \zeta \in G \cap V;$$

$(A', a)$   $a \in b(CG)$ and for every finely connected component $T$ of $G$ with $a \in b(T)$ there holds   $u^a_{T,f} < +\infty$;

$(B', a)$   $a \notin b(CG)$ and there exist a constant $t \in \mathbb{R}$ and a fine neighbourhood $V$ of $a$ for which

$$u(\zeta) \le \lambda(|\zeta - a|) + t \quad \forall \zeta \in G \cap V.$$

The validity of Theorem 2.1 follows from Theorem 2.2.

In $^1$, pp. 355-357] the following statement was actually proved, although not formulated.

**Lemma 3.1.** *Let $a \in \mathbb{C}$ be a fixed point; $G\backslash\{a\}$ be a finely open set; $\mu \in \mathfrak{M}$; $h : \tilde{G} \to \widehat{\mathbb{C}}$ be a finely meromorphic function.*

*Denote $z_1 := a$, $z_2 := \infty$ and suppose that for $G$, $h$, $\mu$ and some $s = 1,2$ one of the conditios $(A, z_s)$ or $(B, z_s)$ is satisfied. Then for $G$, the functions $\log|h(\zeta)|$, $\lambda(x) := \log\mu(x)$ and the same $s$ the corresponding of the conditions $(A', z_s)$ or $(B', z_s)$ is fulfilled, and $z_s$ is finely distincted from the set $\mathfrak{P}$ of all poles of $h$ in $G$ and $h$ is finely holomorphically extendable into $z_s$.*

Notice that the presenting in $^1$ fine holomorphicity of $h$ in the context of the given statement was used only in a fine neighbourhoods of the points $a$ and $\infty$, not in the whole $G$.

**Proof of Theorem 2.2.** Applying Lemma 3.1 to the set $G$ and the function $h$ we obtain, that for $G$, the functions $u(\zeta) := \log|h(\zeta)|$, $\lambda(x) := \log\mu(x)$, and the same $s$ the corresponding of the conditions $(A', z_s)$ or $(B', z_s)$ is fulfilled.

We omit from consideration those connected components of the set $G$ at each point of which the series $\sum_{p \in \mathfrak{P} \cap G(\zeta)} g_G(p, \zeta) \cdot k(h, p)$ diverges. Therefore, without loss of generality, we shal assume that it converges on the set $G\backslash\mathfrak{P}$. Assume now that $Z$ is an arbitrary finite set of zeros of the function $h$ in $G$. We consider in $G$ the function

$$v_Z(\zeta) := u(\zeta) + \sum_{z \in Z} g_G(w, \zeta)k(h, w), \qquad \zeta \in G.$$

The set $Q_Z := Q \cup (FG)_i$ is nearly negligible relative to every finely connected component $T$ of $G$. The function $v_Z(\zeta)$ is finely hypoharmonic in $G$ and

$$\check{u}_{G,f}(z) = (\check{v}_Z)_{G,f}(z) \quad \forall z \in (\partial_f G)\backslash Q_Z.$$

So, the function $v_Z(\zeta)$ fulfils all assumptions of Theorem 8.2 from [1], and we obtain

$$v_Z(\zeta) \leq \lambda(|\zeta - a|)$$

and finally

$$|h(\zeta)| \exp\left[-\sum_{p \in \mathfrak{P}} g_G(p, \zeta) \cdot k(h, p)\right] \leqslant$$

$$\mu(|\zeta - a|) \exp\left[-\sum_{z \in Z} g_G(w, \zeta) \cdot k(h, w)\right] \quad \forall \zeta \in G \setminus \mathfrak{P}.$$

Since $Z$ is an arbitrary finite subset of the set of zeros of $h$, from here we obtain (3). □

**Proof of Theorem 2.3.** Fix an arbitrary point $w \in (\partial_f G)_r$. Let us check that we may apply Theorem 2.2 with $w$ as $a$, and the function

$$\phi(\zeta) := h(\zeta) - h(w) \quad (\zeta \in G)$$

instead of $h(\zeta)$. Indeed from (4) we get (3) with $w$ as $a$, and $\phi$ as $h$. Since for every finely connected component $T$ of $G$ the function $h|_{(\bar{T} \cap \mathbb{C}) \setminus \mathfrak{P}}$ is finely continuous and $w \in (\partial_f G)_r \subset b(CG)$, therefore $h$ satisfies the condition $(A, w)$, and the same is true for $\phi$.

One of the conditions $(A, \infty)$ or $(B_0, \infty)$ is assumed to hold for $h$, and it implies the same condition for $\phi$.

Hence, Theorem 2.2 is applicable in the situation mentioned above. Since $w \in (\partial_f G)_r$, we have $G \neq \mathbb{C} \setminus \{w\}$ and the exceptional case of Theorem 2.1 is impossible in the situation under consideration. Thus we get (3) for $\phi$ instead of $h$, and the estimate of (5) with $w$ as $z$.

Because of the choice of $w$, it gives us (5).

Theorem 2.3 is proved. □

The proofs of Theorems 2.4-2.6 use our Theorem 2.3 and are similar to the proofs of Theorems 4-6 in [1].

### References

1. P.M. Tamrazov, Finely holomorphic and finely subharmonic functions in contour-solid problems. Annales Academia Scientiarum Fennica. Mathematica, Vol. 26, 2001, 325-360.
2. P.M. Tamrazov and T. Aliyev, A contour-solid problem for meromorphic functions taking into account zeros and nonunivalence. Dokl. Akad. Nauk SSSR 288 (1986), No. 2, 304-308 (Russian); English translations in Soviet Math. Dokl. Vol. 33(1986), No. 3, 670-674.

3. T.G. Aliyev and P.M. Tamrazov, *A contour–solid problem for meromorphic functions, taking into account nonunivalence*, Ukrainian Math. Zh. 39 (1987), 683–690 (Russian).

4. T. Aliyev and P. Tamrazov, Contour-solid theorems for meromorphic functions taking multivalence into account. In: Progress in Analysis. Proceedings of 3-d International ISAAC Congress, Vol. I, 469-476. World Scientific, Singapore, 2003.

5. T.G. Aliyev, *Irregular boundary zeros of analytic functions in contour-solid theorems*, Bull. Soc. Sci. Lettres Lodz 49 Ser. Rech. Deform. 28(1999), 45-53.

6. T.G. Aliyev, *Contour-solid theorems*, Preprint 85.84, Math. Inst. Acad. Sci. Ukrainian SSR, Kiev, 1985 (Russian).

7. M. Brelot, On Topologies and Boundaries in Potential Theory. -Lecture Notes in Math. 175, Springer-verlag,1971.

8. B.Fuglede, Finely harmonic Functions.-Lecture Notes in Math. 175, Springer-verlag,1971.

9. B.Fuglede, Finely holomorphic Functions.A survey.- Rev. Roumaine Math. Pures Appl. 33, 1988, 283-295.

# A SINGULAR DOMAIN PERTURBATION PROBLEM FOR THE POISSON EQUATION

MASSIMO LANZA DE CRISTOFORIS

*Dipartimento di Matematica Pura ed Applicata*
*Università di Padova*
*Via Trieste 63*
*35121 Padova, Italy.*
*E-mail: mldc@math.unipd.it*

We consider an approach based on Functional Analysis and Potential Theory to analyze singular domain perturbation problems for boundary value problems for nonhomogeneous elliptic equations. In particuar, we consider the Dirichlet problem for the Poisson equation.

**Key words:** Dirichlet boundary value problem, Poisson equation, singular domain perturbation

**Mathematics Subject Classification:** 35J25, 31B10, 45F15, 47H30

## 1. Introduction

In this paper, we consider an approach based on Functional Analysis and Potential Theory to analyze singular domain perturbation problems for boundary value problems for nonhomogeneous elliptic equations. This paper can be considered as a continuation of [9,10], where we have considered the case of homogeneous equations. For simplicity, we confine our attention to the Poisson equation, and to Dirichlet boundary conditions.

We consider a domain $\mathbb{I}[\phi^o]$ with boundary $\phi^o(\partial \mathbb{B}_n)$, where $\mathbb{B}_n$ denotes the unit ball $\{x \in \mathbb{R}^n : |x| < 1\}$ in $\mathbb{R}^n$ with $n \geq 2$ and $\partial \mathbb{B}_n$ denotes the boundary of $\mathbb{B}_n$, and where $\phi$ is a diffeomorphism of $\partial \mathbb{B}_n$ into $\mathbb{R}^n$ which belongs to the class

$$\mathcal{A}_{\partial \mathbb{B}_n} \equiv \left\{ \phi \in C^1\left(\partial \mathbb{B}_n, \mathbb{R}^n\right) : \phi \text{ is injective, } d\phi(y) \text{ is injective for all } y \in \partial \mathbb{B}_n \right\}.$$

Then we select a point $w$ of $\mathbb{I}[\phi^o]$, and we consider the 'small hole' $\mathbb{I}[w + \epsilon \xi]$, where $\xi \in \mathcal{A}_{\partial \mathbb{B}_n}$, and $\epsilon > 0$ is a small parameter. Then we remove from the domain $\mathbb{I}[\phi^o]$ the closure $\text{cl} \mathbb{I}[w + \epsilon \xi]$ of $\mathbb{I}[w + \epsilon \xi]$, and we consider a Dirichlet problem in the perforated domain $\mathbb{A}[w, \epsilon, \xi, \phi^o] \equiv \mathbb{I}[\phi^o] \setminus (w + \epsilon \text{cl} \mathbb{I}[\xi])$. Namely, we assign the boundary data by means of two functions $g^i$, $g^o$ of $\partial \mathbb{B}_n$ to $\mathbb{R}$, and the data in the interior by means of a function $F$ of a bounded open connected subset $\Omega$ of $\mathbb{R}^n$ containing the closure of $\mathbb{I}[\phi^o]$ to $\mathbb{R}$, and we consider the problem

$$\begin{cases} \Delta u = F & \text{in } \mathbb{A}[w,\epsilon,\xi,\phi^o], \\ u = g^i \circ (w + \epsilon\xi)^{(-1)} & \text{on } w + \epsilon\xi(\partial\mathbb{B}_n), \\ u = g^o \circ \phi^{o(-1)} & \text{on } \phi^o(\partial\mathbb{B}_n), \end{cases} \tag{1}$$

where the exponent $(-1)$ denotes that we are taking the inverse function. Under reasonable conditions on $g^i$, $g^o$, $F$, problem (1) has a unique solution $u = u[w,\epsilon,\xi,\phi^o,g^i,g^o,F]$, and here the question is what happens to the unique solution of problem (1), and to related functionals such as the energy integral $\int_{\mathbb{A}[w,\epsilon,\xi,\phi^o]} |Du(t)|^2 \, dt$ around the 'degenerate' case when $\epsilon = 0$ and the hole degenerates to a point. Such problem is not new and has been investigated with the methods of Asymptotic Analysis, which aims at giving complete asymptotic expansions of the solutions of problem (1) in terms of the parameter $\epsilon$. Here, we mention the work of Kozlov, Maz'ya and Movchan [4], Kühnau [5], Maz'ya, Nazarov and Plamenewskii [15], Ozawa [16], Ward and Keller [20]. In particular, we mention that a complete asymptotic expansion of the solution of (1) in terms of $\epsilon$ for fixed values of $w$, $\xi$, $\phi^o$ can be found in Maz'ya, Nazarov and Plamenewskii [15] by the so-called compound asymptotic expansion method. We also mention the point of view of Dal Maso [1] (see also references therein), which aims at giving continuity results as $\epsilon$ tends to 0 and which has the virtue of being capable of handling very low boundary regularity assumptions on the domain and on the data.

This paper is in the spirit of [7–10], and can be considered a continuation of [9,10], where (1) has been considered with $F = 0$. We think of $\mathbf{q} \equiv (w,\epsilon,\xi,\phi^o,g^i,g^o,F)$ as a point in a suitable Banach space and we try to represent the behaviour of the solution $u[\mathbf{q}]$ and of the corresponding energy integral in terms of real analytic operators depending on $\mathbf{q}$ and of possible singular terms in $\epsilon$. We now briefly indicate our strategy.

We denote by $\Upsilon_n$ the function of $]0,+\infty[$ to $\mathbb{R}$ defined by

$$\Upsilon_n(r) \equiv \begin{cases} \frac{1}{s_n} \log r & \forall r \in ]0,+\infty[, \text{ if } n = 2, \\ \frac{1}{(2-n)s_n} r^{2-n} & \forall r \in ]0,+\infty[, \text{ if } n > 2, \end{cases} \tag{2}$$

where $s_n$ denotes the $(n-1)$ dimensional measure of $\partial\mathbb{B}_n$. We denote by $S_n$ the function of $\mathbb{R}^n \setminus \{0\}$ to $\mathbb{R}$ defined by

$$S_n(\xi) \equiv \Upsilon_n(|\xi|) \qquad \forall \xi \in \mathbb{R}^n \setminus \{0\}.$$

$S_n$ is well-known to be the fundamental solution of the Laplace operator. Then under reasonable conditions on the data, the solution of problem (1) can be written in the form

$$u[w,\epsilon,\xi,\phi^o,g^i,g^o,F] = u[w,\epsilon,\xi,\phi^o,g^i - P[F] \circ (w+\epsilon\xi), g^o - P[F] \circ \phi^o, 0] + P[F],$$

where

$$P[F](t) \equiv \int_\Omega S_n(t-s)F(s)\,ds \qquad \forall t \in \Omega, \tag{3}$$

is the Newtonian potential of $F$ in $\Omega$. Now by [9,10], we can represent $u[w, \epsilon, \xi, \phi^o, g^i, g^o, 0]$ in terms of $\Upsilon_n(\epsilon)$ and in terms of real analytic operators depending on $(w, \epsilon, \xi, \phi^o, g^i, g^o)$ when $\xi$, $\phi^o$, $g^i$, $g^o$ belong to suitable Schauder spaces $C^{m,\alpha}$. Thus what remains to be done here is to choose an appropriate Banach space for $F$ so that $P[F]$, $P[F] \circ \phi$ depend real analytically on $F$, $(\phi, F)$. Now for a large variety of choices of function spaces for $F$, $P[F]$ depends real analytically of $F$, and this is so in particular for the Schauder spaces $C^{m,\alpha}$. Less clear instead is the choice for the function spaces for $F$, $P[F] \circ \phi$ in order that $P[F] \circ \phi$ depends real analytically on $(F, \phi)$ when $\phi$ is in a Schauder space. Then we resort to results on the composition operators of Preciso [17,18], which indicate that a right choice for the spaces for $F$, $P[F]$ is a Romieu class, and thus to the corresponding real analyticity results for $P[F] \circ \phi$ of [6], where a regular perturbation problem for the Poisson equation has been treated. Then we can prove Theorems 2.1, 2.2 for the behaviour of $u[\mathbf{q}]$, and Theorem 2.3 for the behaviour of the corresponding energy integral, which extends the corresponding results for $F = 0$ of [9,10].

## 2. Introduction of the Romieu classes and of the representation Theorem

Let $\Omega$ be a bounded open subset of $\mathbb{R}^n$. For each $m \in \mathbb{N}$, $C^m(\mathrm{cl}\,\Omega)$ denotes the space of real valued functions of class $C^m$ in $\mathrm{cl}\,\Omega$ endowed with the norm of the uniform convergence on all derivatives up to order $m$. Let $\alpha \in ]0,1[$. $C^{m,\alpha}(\mathrm{cl}\,\Omega)$ denotes the subspace of $C^m(\mathrm{cl}\,\Omega)$ of those functions whose $m$-th order derivatives are $\alpha$-Hölder continuous. Then $C^{m,\alpha}(\partial\Omega)$ is defined as the space of traces of functions of $C^{m,\alpha}(\mathrm{cl}\,\Omega)$ on $\partial\Omega$. We think of the Schauder spaces $C^{m,\alpha}(\mathrm{cl}\,\Omega)$, $C^{m,\alpha}(\partial\Omega)$ as endowed with their usual norm. For standard properties of functions in Schauder spaces, we refer the reader to Gilbarg and Trudinger [3] (see also [13, §2, Lem. 3.1, 4.26, Thm. 4.28], Lanza and Rossi [14, §2].) Let $\mathbb{D} \subseteq \mathbb{R}^n$. Then $C^{m,\alpha}(\mathrm{cl}\,\Omega, \mathbb{D})$ denotes $\{f \in (C^{m,\alpha}(\mathrm{cl}\,\Omega))^n : f(\mathrm{cl}\,\Omega) \subseteq \mathbb{D}\}$. Also, we note that $\mathcal{A}_{\partial\mathbb{B}_n}$ is open in $C^1(\partial\mathbb{B}_n, \mathbb{R}^n)$ (cf. e.g., [13, Prop. 4.29], Lanza and Rossi [14, Lem. 2.5].)

We now turn to the Romieu classes. For all bounded open subsets $\Omega$ of $\mathbb{R}^n$ and $\rho > 0$, we set

$$C^0_{\omega,\rho}(\mathrm{cl}\,\Omega) \equiv \left\{ u \in C^\infty(\mathrm{cl}\,\Omega) : \sup_{\beta \in \mathbb{N}^n} \frac{\rho^{|\beta|}}{|\beta|!} \|D^\beta u\|_{C^0(\mathrm{cl}\,\Omega)} < +\infty \right\},$$

and

$$\|u\|_{C^0_{\omega,\rho}(\mathrm{cl}\,\Omega)} \equiv \sup_{\beta \in \mathbb{N}^n} \frac{\rho^{|\beta|}}{|\beta|!} \|D^\beta u\|_{C^0(\mathrm{cl}\,\Omega)} \qquad \forall u \in C^0_{\omega,\rho}(\mathrm{cl}\,\Omega),$$

where $|\beta| \equiv \beta_1 + \cdots + \beta_n$ for all $\beta \equiv (\beta_1, \ldots, \beta_n) \in \mathbb{N}^n$. As is well known, the Romieu class $\left( C^0_{\omega,\rho}(\mathrm{cl}\,\Omega), \| \cdot \|_{C^0_{\omega,\rho}(\mathrm{cl}\,\Omega)} \right)$ is a Banach space.

Then we have the following technical Lemma.

958

**Lemma 2.1.** *Let $m \in \mathbb{N}$, $\alpha \in ]0,1[$, $\rho > 0$. Let $\Omega$ be a bounded open connected subset of $\mathbb{R}^n$. Let $\Omega_2$ be an open connected subset of $\mathbb{R}^n$ of class $C^1$ such that $\mathrm{cl}\Omega_2 \subseteq \Omega$. Then the following statements hold.*

*(i) If $\tilde{\Omega}$ is a bounded open subset of $\mathbb{R}^n$ such that $\mathrm{cl}\tilde{\Omega} \subseteq \Omega_2$, then there exists $\rho_1 \in ]0,\rho]$ such that the map of $C^0_{\omega,\rho}(\mathrm{cl}\Omega)$ to $C^0_{\omega,\rho_1}(\mathrm{cl}\tilde{\Omega})$ which takes $F$ to $P[F_{|\mathrm{cl}\Omega_2}]_{|\mathrm{cl}\tilde{\Omega}}$ is real analytic.*

*(ii) The map of $\left\{ (F,t) \in C^0_{\omega,\rho}(\mathrm{cl}\Omega) \times \mathbb{R}^n : t \in \Omega_2 \right\}$ to $\mathbb{R}$ which takes $(F,t)$ to $P[F_{|\mathrm{cl}\Omega_2}](t)$ is real analytic.*

*(iii) The map of $C^0_{\omega,\rho}(\mathrm{cl}\Omega) \times C^{m,\alpha}(\partial\mathbb{B}_n, \Omega_2)$ to $C^{m,\alpha}(\partial\mathbb{B}_n)$ which takes $(F,\phi)$ to $P[F_{|\mathrm{cl}\Omega_2}] \circ \phi$ is real analytic.*

For a proof of Lemma 2.1, we refer to [[6], Lem. 2.15, and its proof].

Clearly, not all quadruples $(w, \epsilon, \xi, \phi^o)$ correspond to annular domains such as $\mathbb{A}[w, \epsilon, \xi, \phi^o]$ above. In other words, in order to ensure that $\mathbb{A}[w, \epsilon, \xi, \phi^o]$ is actually a domain with a hole and that it has the regularity we need, and that accordingly it is 'admissible' for our analysis, we must impose on $(w, \epsilon, \xi, \phi^o)$ certain restrictions. Thus we introduce the following set of quadruples $(w, \epsilon, \xi, \phi^o)$, which we shall regard as the set of 'admissible' quadruples. For each open connected subset $\Omega$ of $\mathbb{R}^n$, $m \in \mathbb{N}$, $\alpha \in ]0,1[$, we set

$$
\mathcal{E}^{m,\alpha}_\Omega \equiv \left\{ \mathbf{a} \equiv (w, \epsilon, \xi, \phi^o) \in \mathbb{R}^n \times \mathbb{R} \times \left( C^{m,\alpha}\left(\partial\mathbb{B}_n, \mathbb{R}^n\right) \cap \mathcal{A}_{\partial\mathbb{B}_n} \right)^2 : \right.
$$
$$
\left. 0 \in \mathbb{I}[\xi], w + \epsilon\xi(\partial\mathbb{B}_n) \subseteq \mathbb{I}[\phi^o], \mathrm{cl}\,\mathbb{I}[\phi^o] \subseteq \Omega \right\}.
$$

The set $\mathcal{E}^{m,\alpha}_\Omega$ is easily seen to be open in $\mathbb{R}^n \times \mathbb{R} \times \left( C^{m,\alpha}\left(\partial\mathbb{B}_n, \mathbb{R}^n\right) \right)^2$, and a simple topological argument shows that if $(w, \epsilon, \xi, \phi^o) \in \mathcal{E}^{m,\alpha}_\Omega$, then $w + \epsilon\mathrm{cl}\,\mathbb{I}[\xi] \subseteq \mathbb{I}[\phi^o]$. To simplify our notation we shall sometimes write $\mathbf{a}$ instead of $(w, \epsilon, \xi, \phi^o)$. Also, we find convenient to set $\mathbb{A}[w, 0, \xi, \phi^o] \equiv \mathbb{I}[\phi^o] \setminus \{w\}$ for all $(w, 0, \xi, \phi^o)$ in $\mathcal{E}^{m,\alpha}$. As a set of admissible $\mathbf{q}$'s, we will take $\mathcal{E}^{m,\alpha}_\Omega \times (C^{m,\alpha}(\partial\mathbb{B}_n))^2 \times C^0_{\omega,\rho}(\mathrm{cl}\Omega)$.

**Theorem 2.1.** *Let $m \in \mathbb{N} \setminus \{0\}$, $\alpha \in ]0,1[$, $\rho \in ]0,1[$. Let $\Omega$ be a bounded open connected subset of $\mathbb{R}^n$. Let $\mathbf{q}_0 \equiv (w_0, 0, \xi_0, \phi^o_0, g^i_0, g^o_0, F_0) \in \mathcal{E}^{m,\alpha}_\Omega \times (C^{m,\alpha}(\partial\mathbb{B}_n))^2 \times C^0_{\omega,\rho}(\mathrm{cl}\Omega)$. Let $\tilde{\Omega}$ be an open subset of $\mathbb{R}^n$ such that $\mathrm{cl}\tilde{\Omega} \subseteq \mathbb{I}[\phi^o_0] \setminus \{w_0\}$. Then there exist $\tilde{\rho} \in ]0,\rho]$ and an open neighborhood $\mathcal{U}_0$ of $(w_0, 0, \xi_0, \phi^o_0)$ in $\mathcal{E}^{m,\alpha}_\Omega$, and two real analytic operators $V_1$ and $V_2$ of $\mathcal{U}_0$ to $\mathbb{R}$, and an open neighborhood $\mathcal{Q}_0$ of $\mathbf{q}_0$ in $\mathcal{E}^{m,\alpha}_\Omega \times (C^{m,\alpha}(\partial\mathbb{B}_n))^2 \times C^0_{\omega,\rho}(\mathrm{cl}\Omega)$ and two real analytic operators $Q_1, Q_2$ of $\mathcal{Q}_0$ to $C^0_{\omega,\tilde{\rho}}(\mathrm{cl}\tilde{\Omega})$ such that the following conditions hold.*

*(i) $V_2[\mathbf{a}] \neq 0$ if $\mathbf{a} \equiv (w, \epsilon, \xi, \phi^o) \in \mathcal{U}_0$, and $V_1[\mathbf{a}] + V_2[\mathbf{a}]\Upsilon_n(\epsilon) \neq 0$ if $\mathbf{a} \in \mathcal{U}_0$ with $\epsilon > 0$.*

*(ii) $(w, \epsilon, \xi, \phi^o) \in \mathcal{U}_0$ for all $\mathbf{q} \in \mathcal{Q}_0$.*

*(iii) $\mathrm{cl}\tilde{\Omega} \subseteq \mathbb{A}[w, \epsilon, \xi, \phi^o]$ for all $(w, \epsilon, \xi, \phi^o, g^i, g^o, F) \in \mathcal{Q}_0$.*

*(iv)*

$$u[\mathbf{q}](t) = Q_1[\mathbf{q}](t) + \frac{Q_2[\mathbf{q}](t)}{V_1[w, \epsilon, \xi, \phi^o] + V_2[w, \epsilon, \xi, \phi^o] \Upsilon_n(\epsilon)} \qquad \forall t \in \mathrm{cl}\tilde{\Omega}, \qquad (4)$$

for all $\mathbf{q} \equiv (w, \epsilon, \xi, \phi^o, g^i, g^o, F) \in \mathcal{Q}_0$ such that $\epsilon > 0$.

*(v)* For each $(w, 0, \xi, \phi^o, g^i, g^o, F) \in \mathcal{Q}_0$, $Q_1[w, 0, \xi, \phi^o, g^i, g^o, F]$ equals the unique solution $u^o[\phi^o, g^o, F] \in C^{m,\alpha}(\mathrm{cl}\mathbb{I}[\phi^o])$ of problem $\Delta u = F$ in $\mathbb{I}[\phi^o]$, $u = g^o \circ \phi^{o(-1)}$ on $\phi^o(\partial \mathbb{B}_n)$.

**Proof.** Let $\Omega_1$, $\Omega_2$ be open connected subsets of $\mathbb{R}^n$ of class $C^\infty$ such that $\mathrm{cl}\tilde{\Omega} \subseteq \Omega_1 \subseteq \mathrm{cl}\Omega_1 \subseteq \mathbb{I}[\phi_0^o] \setminus \{w_0\} \subseteq \mathrm{cl}\mathbb{I}[\phi_0^o] \subseteq \Omega_2 \subseteq \mathrm{cl}\Omega_2 \subseteq \Omega$ (for the well known existence of $\Omega_1$, $\Omega_2$ cf. e.g., [11, p. 935].) We can clearly choose $\mathcal{U}_0$ sufficiently small so that $\mathrm{cl}\Omega_1 \subseteq \mathbb{A}[w, \epsilon, \xi, \phi^o] \subseteq \mathrm{cl}\mathbb{I}[\phi^o] \subseteq \Omega_2$ for all $(w, \epsilon, \xi, \phi^o) \in \mathcal{U}_0$. Thus if $(w, \epsilon, \xi, \phi^o) \in \mathcal{U}_0$, $\epsilon > 0$ and $(g^i, g^o, F) \in (C^{m,\alpha}(\partial \mathbb{B}_n))^2 \times C^0_{\omega,\rho}(\mathrm{cl}\Omega)$, we have

$$u[w, \epsilon, \xi, \phi^o, g^i, g^o, F] \qquad (5)$$
$$= u[w, \epsilon, \xi, \phi^o, g^i - P[F_{|\mathrm{cl}\Omega_2}] \circ (w + \epsilon\xi), g^o - P[F_{|\mathrm{cl}\Omega_2}] \circ \phi^o, 0] + P[F_{|\mathrm{cl}\Omega_2}]$$

on $\mathrm{cl}\tilde{\Omega}$. Now we introduce the following abbreviations

$$\gamma_0^i \equiv g_0^i - P[F_{0|\mathrm{cl}\Omega_2}](w_0), \qquad \gamma_0^o \equiv g_0^o - P[F_{0|\mathrm{cl}\Omega_2}] \circ \phi_0^o.$$

By [9, Thm. 5.7, §5], there exist two real analytic operators $V_1$ and $V_2$ of an open neighborhood of $(w_0, 0, \xi_0, \phi_0^o)$, which we still denote $\mathcal{U}_0$, to $\mathbb{R}$ such that (i)–(iii) hold and an open neighborhood $\mathcal{U}$ of $\mathbf{p}_0 \equiv (w_0, 0, \xi_0, \phi_0^o, \gamma_0^i, \gamma_0^o)$ with $(w, \epsilon, \xi, \phi^o) \in \mathcal{U}_0$ for all $\mathbf{p} \in \mathcal{U}$, and two real analytic operators $U_1$, $U_2$ of $\mathcal{U}$ to the space

$$C_h^0(\mathrm{cl}\Omega_1) \equiv \{u \in C^0(\mathrm{cl}\Omega_1) \cap C^2(\Omega_1) : \Delta u(t) = 0 \ \forall t \in \Omega_1\}$$

endowed with the norm of the uniform convergence such that

$$u[\mathbf{p}, 0](t) = U_1[\mathbf{p}](t) + \frac{U_2[\mathbf{p}](t)}{V_1[w, \epsilon, \xi, \phi^o] + V_2[w, \epsilon, \xi, \phi^o] \Upsilon_n(\epsilon)} \qquad \forall t \in \mathrm{cl}\Omega_1, \qquad (6)$$

for all $\mathbf{p} \equiv (w, \epsilon, \xi, \phi^o, \gamma^i, \gamma^o) \in \mathcal{U}$ with $\epsilon > 0$. Moreover, $U_1[w, 0, \xi, \phi^o, \gamma^i, \gamma^o]$ coincides with the restriction to $\mathrm{cl}\tilde{\Omega}$ of the unique solution in $C^{m,\alpha}(\mathrm{cl}\mathbb{I}[\phi^o])$ of problem $\Delta u = 0$ in $\mathbb{I}[\phi^o]$, $u = \gamma^o$ on $\phi^o(\partial \mathbb{B}_n)$. By Lemma 2.1 (iii), there exists a neighborhood $\mathcal{Q}_0$ of $\mathbf{q}_0$ such that $(w, \epsilon, \xi, \phi^o, g^i - P[F_{|\mathrm{cl}\Omega_2}] \circ (w + \epsilon\xi), g^o - P[F_{|\mathrm{cl}\Omega_2}] \circ \phi^o) \in \mathcal{U}$ and $(w, \epsilon, \xi, \phi^o) \in \mathcal{U}_0$ for all $\mathbf{q} \in \mathcal{Q}_0$. Let $\rho_1$ be as in Lemma 2.1 (i). Let $Q_1$, $Q_2$ be the operators of $\mathcal{Q}_0$ to $C^0_{\omega,\rho_1}(\mathrm{cl}\tilde{\Omega})$ defined by

$$Q_1[\mathbf{q}] \equiv U_1[w, \epsilon, \xi, \phi^o, g^i - P[F_{|\mathrm{cl}\Omega_2}] \circ (w + \epsilon\xi), g^o - P[F_{|\mathrm{cl}\Omega_2}] \circ \phi^o]_{|\mathrm{cl}\tilde{\Omega}}$$
$$+ P[F_{|\mathrm{cl}\Omega_2}]_{|\mathrm{cl}\tilde{\Omega}}$$

$$Q_2[\mathbf{q}] \equiv U_2[w, \epsilon, \xi, \phi^o, g^i - P[F_{|\mathrm{cl}\Omega_2}] \circ (w + \epsilon\xi), g^o - P[F_{|\mathrm{cl}\Omega_2}] \circ \phi^o]_{|\mathrm{cl}\tilde{\Omega}}$$

for all $\mathbf{q} \in \mathcal{Q}_0$. Possibly shrinking $\rho_1$, one easily verifies that the restriction map is linear and continuous from $C_h^0(\mathrm{cl}\Omega_1)$ to $C^0_{\omega,\rho_1}(\mathrm{cl}\tilde{\Omega})$ (cf. [12, Prop. 8].) Then the properties of $U_1$, $U_2$ and Lemma 2.1 (i), (iii) imply that $Q_1$ and $Q_2$ are real analytic. By (5), (6), we immediately deduce the validity of statements (iv), (v). $\qquad \square$

960

We note that in case $n \geq 3$, the right hand side of the formula in (4) can be continued real analytically in the whole of $\mathcal{Q}_0$, while for $n = 2$, the right hand side of (4) displays a logarithmic behaviour. We also note that by [9, Thm. 6.19], the term $(V_1[w, \epsilon, \xi, \phi^o] + V_2[w, \epsilon, \xi, \phi^o]\Upsilon_n(\epsilon))^{-1}$ in (4) coincides with the opposite of the electrostatic capacity of $w + \epsilon \mathrm{cl}\mathbb{I}[\xi]$ with respect to $\mathbb{I}[\phi^o]$. We now wish to investigate the dependence of $u[\mathbf{q}](t)$ on the variable $(\mathbf{q}, t)$. Then we introduce the sets

$$\mathcal{K}^{m,\alpha} \equiv \big\{(w, \epsilon, \xi, \phi^o, g^i, g^o, F, t)$$

$$\in \mathcal{E}_\Omega^{m,\alpha} \times (C^{m,\alpha}(\partial \mathbb{B}_n))^2 \times C^0_{\omega,\rho}(\mathrm{cl}\Omega) \times \mathbb{R}^n : t \in \mathbb{A}[w, \epsilon, \xi, \phi^o]\big\},$$

and $\mathcal{K}_+^{m,\alpha} \equiv \{(w, \epsilon, \xi, \phi^o, g^i, g^o, F, t) \in \mathcal{K}^{m,\alpha} : \epsilon > 0\}$, and we consider the map $K$ of $\mathcal{K}_+^{m,\alpha}$ to $\mathbb{R}$ defined by

$$K[w, \epsilon, \xi, \phi^o, g^i, g^o, F, t] \equiv u[w, \epsilon, \xi, \phi^o, g^i, g^o, F](t)$$

for all $(w, \epsilon, \xi, \phi^o, g^i, g^o, F, t) \in \mathcal{K}_+^{m,\alpha}$, and we ask whether we can describe the behaviour of $K$ around points with $\epsilon = 0$. To do so, we introduce the following technical statement.

**Proposition 2.1.** *Let $\tilde{\Omega}$ be a bounded open subset of $\mathbb{R}^n$. Let $\rho > 0$. Then the map $\Xi$ of $C^0_{\omega,\rho}(\mathrm{cl}\tilde{\Omega}) \times \tilde{\Omega}$ to $\mathbb{R}$ which takes a pair $(u, t)$ to $u(t)$ is real analytic.*

For a proof of Proposition 2.1, which follows by a restatement of a result of Preciso [18, Prop 11, p. 101], we refer to [12, Cor. 1, App. A].

The map $\Xi$ of Proposition 2.1 is linear in its first variable. Then by combining Theorem 2.1 and Proposition 2.1, we deduce the validity of the following.

**Theorem 2.2.** *Let $m \in \mathbb{N} \setminus \{0\}$, $\alpha \in ]0,1[$, $\rho \in ]0,1[$. Let $\Omega$ be a bounded open connected subset of $\mathbb{R}^n$. Let $\mathbf{k}_0 \equiv (w_0, 0, \xi_0, \phi_0^o, g_0^i, g_0^o, F_0, t_0) \in \mathcal{K}^{m,\alpha}$. Then there exist $\mathcal{U}_0, V_1, V_2$ as in Theorem 2.1 (i), and an open neighborhood $\mathcal{K}$ of $\mathbf{k}_0$ in $\mathcal{K}^{m,\alpha}$, and two real analytic operators $K_1, K_2$ of $\mathcal{K}$ to $\mathbb{R}$ such that $(w, \epsilon, \xi, \phi^o) \in \mathcal{U}_0$ for all $\mathbf{k} \in \mathcal{K}$ and*

$$u[w, \epsilon, \xi, \phi^o, g^i, g^o, F](t) = K_1[\mathbf{k}] + \frac{K_2[\mathbf{k}]}{V_1[w, \epsilon, \xi, \phi^o] + V_2[w, \epsilon, \xi, \phi^o]\Upsilon_n(\epsilon)}, \quad (7)$$

*for all $\mathbf{k} \equiv (w, \epsilon, \xi, \phi^o, g^i, g^o, F, t) \in \mathcal{K}$ such that $\epsilon > 0$.*

We now turn to consider the energy integral, *i.e.*, the integral

$$\int_{\mathbb{A}[w,\epsilon,\xi,\phi^o]} |Du[\mathbf{q}](t)|^2 \, dt.$$

To do so, we need the following two technical statements, which generalize the corresponding statements for regular perturbations of [6].

**Lemma 2.2.** *Let $m \in \mathbb{N} \setminus \{0\}$, $\alpha \in ]0,1[$, $\rho > 0$. Let $\Omega$ be a bounded open connected subset of $\mathbb{R}^n$. Let $(w_0, 0, \xi_0, \phi_0^o, G_0) \in \mathcal{E}_\Omega^{m,\alpha} \times C^0_{\omega,\rho}(\mathrm{cl}\Omega)$. Then there exist an open neighborhood $\mathcal{V}_0$ of $(w_0, 0, \xi_0, \phi_0^o, G_0)$ in $\mathcal{E}_\Omega^{m,\alpha} \times C^0_{\omega,\rho}(\mathrm{cl}\Omega)$ and a real analytic operator*

$\Gamma$ of $\mathcal{V}_0$ to $\mathbb{R}$ such that $\Gamma[w, \epsilon, \xi, \phi^o, G] = \int_{\mathbb{A}[w,\epsilon,\xi,\phi^o]} G \, dt$ for all $(w, \epsilon, \xi, \phi^o, G) \in \mathcal{V}_0$ such that $\epsilon > 0$, and such that $\Gamma[w, 0, \xi, \phi^o, G] = \int_{\mathbb{I}[\phi^o]} G \, dt$ for all $(w, 0, \xi, \phi^o, G) \in \mathcal{V}_0$.

**Proof.** First we note that

$$\int_{\mathbb{A}[\mathbf{a}]} G \, dt = \int_{\mathbb{I}[\phi^o]} G \, dt - \epsilon^n \int_{\mathbb{I}[\xi]} G(\epsilon \tau + w) \, d\tau, \qquad (8)$$

for all $(\mathbf{a}, G) \in \mathcal{E}_\Omega^{m,\alpha} \times C_{\omega,\rho}^0(\mathrm{cl}\Omega)$ such that $\epsilon > 0$. Since $\mathbf{a}_0 \equiv (w_0, 0, \xi_0, \phi_0^o) \in \mathcal{E}_\Omega^{m,\alpha}$, then we have $\{w_0\} \subseteq \mathbb{I}[\phi_0^o]$. Since the map of $\mathbb{R}^n \times \mathbb{R} \times \mathbb{R}^n$ to $\mathbb{R}$ which takes $(w, \epsilon, \tau)$ to $w + \epsilon\tau$ is continuous and maps the compact set $\{w_0\} \times \{0\} \times \mathrm{cl}\mathbb{I}[\xi_0] \equiv A$ to $\Omega$, there exists $\delta > 0$ such that if $(w, \epsilon, \tau) \in \mathbb{R}^{2n+1}$, $|w - w_0| < \delta$, $|\epsilon| < \delta$, and if the Euclidean distance of $\tau$ to $\mathrm{cl}\mathbb{I}[\xi_0]$ is less than $\delta$, then $w + \epsilon\tau \in \Omega$. Possibly shrinking $\delta$ we can also assume that if $(w, \epsilon, \xi, \phi^o)$ belongs to the set

$$\mathcal{V} \equiv \Big\{ (w, \epsilon, \xi, \phi^o) \in \mathbb{R}^{n+1} \times C^{m,\alpha}(\partial \mathbb{B}_n, \mathbb{R}^n)^2 : |w - w_0| < \delta, |\epsilon| < \delta,$$

$$\|\xi - \xi_0\|_{C^{m,\alpha}(\partial \mathbb{B}_n, \mathbb{R}^n)} < \delta, \|\phi^o - \phi_0^o\|_{C^{m,\alpha}(\partial \mathbb{B}_n, \mathbb{R}^n)} < \delta \Big\}$$

then $(w, \epsilon, \xi, \phi^o) \in \mathcal{E}_\Omega^{m,\alpha}$. As is well known, there exists an open connected subset $W$ of $\mathbb{R}^n$ of class $C^\infty$ such that $\mathrm{cl}\mathbb{I}[\xi_0] \subseteq W \subseteq \mathrm{cl}W \subseteq \mathbb{I}[\xi_0] + \mathbb{B}_n(0, \delta/2)$. Obviously, the map of $\mathcal{V}$ to $C_{\omega,\rho}^0(\mathrm{cl}W, \Omega)$ which takes $\mathbf{a}$ to $w + \epsilon\,\mathrm{id}_{\mathrm{cl}W}$, where $\mathrm{id}_{\mathrm{cl}W}$ denotes the identity map in $\mathrm{cl}W$, is real analytic. Moreover, by Preciso [18, Prop. 1.3], there exists $\rho_1 \in ]0, \rho]$ such that the map of $C_{\omega,\rho}^0(\mathrm{cl}\Omega) \times C_{\omega,\rho}^0(\mathrm{cl}W, \Omega)$ to $C_{\omega,\rho_1}^0(\mathrm{cl}W)$ which takes the pair $(G, G_1)$ to the composite function $G \circ G_1$ is real analytic. Then we define $\mathcal{V}_0$ to be the set of elements of $\mathcal{V} \times C_{\omega,\rho}^0(\mathrm{cl}\Omega)$ such that $\mathrm{cl}\mathbb{I}[\xi] \subseteq W$. Then the map of $\mathcal{V}_0$ to $C_{\omega,\rho_1}^0(\mathrm{cl}W)$ which takes $(\mathbf{a}, G)$ to $G \circ (w + \epsilon\,\mathrm{id}_{\mathrm{cl}W})$ is real analytic. By [6, Thm. 2.26], the map of $\mathcal{V}_0$ to $\mathbb{R}$ which takes $(\mathbf{a}, G)$ to $\int_{\mathbb{I}[\phi^o]} G \, dt$ is real analytic, and the map of $\{\mathbf{a} \in \mathcal{E}_\Omega^{m,\alpha} : \mathrm{cl}\mathbb{I}[\xi] \subseteq W\} \times C_{\omega,\rho_1}^0(\mathrm{cl}W)$ to $\mathbb{R}$ which takes $(\mathbf{a}, G)$ to $\int_{\mathbb{I}[\xi]} G \, dt$ are real analytic. Hence, it suffices to define $\Gamma$ as the operator of $\mathcal{V}_0$ to $\mathbb{R}$ which takes $(\mathbf{a}, G)$ to the right hand side of (8). $\qquad \square$

Then we have the following.

**Proposition 2.2.** *Let $m \in \mathbb{N} \setminus \{0\}$, $\alpha \in ]0, 1[$, $\rho > 0$. Let $\Omega$ be a bounded open connected subset of $\mathbb{R}^n$. Let $\Omega_2$ be an open connected subset of $\mathbb{R}^n$ of class $C^1$ such that $\mathrm{cl}\Omega_2 \subseteq \Omega$. Let $(w_0, 0, \xi_0, \phi_0^o, F_0) \in \mathcal{E}_{\Omega_2}^{m,\alpha} \times C_{\omega,\rho}^0(\mathrm{cl}\Omega)$. Then there exist an open neighborhood $\mathcal{V}_1$ of $(w_0, 0, \xi_0, \phi_0^o, F_0)$ in $\mathcal{E}_{\Omega_2}^{m,\alpha} \times C_{\omega,\rho}^0(\mathrm{cl}\Omega)$ and a real analytic operator $\Pi$ of $\mathcal{V}_1$ to $\mathbb{R}$ such that $\Pi[w, \epsilon, \xi, \phi^o, F] = \int_{\mathbb{A}[w,\epsilon,\xi,\phi^o]} |DP[F_{|\mathrm{cl}\Omega_2}](t)|^2 \, dt$ for all $(w, \epsilon, \xi, \phi^o, F) \in \mathcal{V}_1$ such that $\epsilon > 0$, and such that $\Pi[w, 0, \xi, \phi^o, F] = \int_{\mathbb{I}[\phi^o]} |DP[F_{|\mathrm{cl}}\Omega_2](t)|^2 \, dt$ for all $(w, 0, \xi, \phi^o, F) \in \mathcal{V}_1$.*

**Proof.** Let $\Omega_1$ be an open connected subset of $\mathbb{R}^n$ such that $\mathrm{cl}\mathbb{I}[\phi_0^o] \subseteq \Omega_1 \subseteq \mathrm{cl}\Omega_1 \subseteq \Omega_2$. We set $\mathcal{V}_1 \equiv \mathcal{E}_{\Omega_1}^{m,\alpha} \times C_{\omega,\rho}^0(\mathrm{cl}\Omega)$. By [6, Prop. 2.4], there exists

$\rho_1 \in ]0, \rho]$ such that $P[\cdot]_{|\mathrm{cl}\Omega_1}$ is linear and continuous from $C^0_{\omega,\rho}(\mathrm{cl}\Omega_2)$ to $C^0_{\omega,\rho_1}(\mathrm{cl}\Omega_1)$. It can be easily verified that there exists $\rho_2 \in ]0, \rho_1]$ such that the map of $C^0_{\omega,\rho_1}(\mathrm{cl}\Omega_1)$ to $C^0_{\omega,\rho_2}(\mathrm{cl}\Omega_1)$ which takes a function $f$ to $|Df|^2$ is real analytic (cf. *e.g.*, [6, Proof of Prop. 2.25].) Now we take $\Gamma$, $\mathcal{V}_0$ as in Lemma 2.2 relative to $\rho_2$, $\Omega_1$, $G_0 \equiv |DP[F_{0|\mathrm{cl}\Omega_2}]_{|\mathrm{cl}\Omega_1}|^2$. Possibly shrinking $\mathcal{V}_1$, we can assume that $(w, \epsilon, \xi, \phi^o, |DP[F_{|\mathrm{cl}\Omega_2}]_{|\mathrm{cl}\Omega_1}|^2) \in \mathcal{V}_0$ for all $(w, \epsilon, \xi, \phi^o, F) \in \mathcal{V}_1$. Thus it suffices to set $\Pi[w, \epsilon, \xi, \phi^o, F] \equiv \Gamma[w, \epsilon, \xi, \phi^o, |DP[F_{|\mathrm{cl}\Omega_2}]_{|\mathrm{cl}\Omega_1}|^2]$ for all $(w, \epsilon, \xi, \phi^o, F) \in \mathcal{V}_1$. $\qquad \square$

Then we are ready to prove the following.

**Theorem 2.3.** *Let $m \in \mathbb{N} \setminus \{0\}$, $\alpha \in ]0, 1[$, $\rho > 0$. Let $\Omega$ be a bounded open connected subset of $\mathbb{R}^n$. Let $\mathbf{q}_0 \equiv (w_0, 0, \xi_0, \phi_0^o, g_0^i, g_0^o, F_0)$ be an element of $\mathcal{E}_\Omega^{m,\alpha} \times (C^{m,\alpha}(\partial \mathbb{B}_n))^2 \times C^0_{\omega,\rho}(\mathrm{cl}\Omega)$. Then there exist $\mathcal{U}_0$, $V_1$, $V_2$ as in Theorem 2.1 (i), and an open neighborhood $\mathcal{Q}_0$ of $\mathbf{q}_0$ in $\mathcal{E}_\Omega^{m,\alpha} \times (C^{m,\alpha}(\partial \mathbb{B}_n))^2 \times C^0_{\omega,\rho}(\mathrm{cl}\Omega)$ and two real analytic operators $\Phi_1$, $\Phi_2$ of $\mathcal{Q}_0$ to $\mathbb{R}$ such that $(w, \epsilon, \xi, \phi^o) \in \mathcal{U}_0$ for all $\mathbf{q} \in \mathcal{Q}_0$, and*

$$\int_{\mathbb{A}[w,\epsilon,\xi,\phi^o]} |Du[\mathbf{q}](t)|^2 \, dt = \Phi_1[\mathbf{q}] + \frac{\Phi_2[\mathbf{q}]}{V_1[w,\epsilon,\xi,\phi^o] + V_2[w,\epsilon,\xi,\phi^o]\Upsilon_n(\epsilon)}, \qquad (9)$$

*for all $\mathbf{q} \equiv (w, \epsilon, \xi, \phi^o, g^i, g^o, F) \in \mathcal{Q}_0$ such that $\epsilon > 0$. Moreover,*

$$\Phi_1[w, 0, \xi, \phi^o, g^i, g^o, F] = \int_{\mathbb{I}[\phi^o]} |Du^o[\phi^o, g^o, F]|^2 \, dt + \delta_{2,n} \int_{\mathbb{R}^n \setminus \mathrm{cl}\mathbb{I}[\xi]} |Du^i[\xi, g^i]|^2 \, dt \qquad (10)$$

*for all $(w, 0, \xi, \phi^o, g^i, g^o, F) \in \mathcal{Q}_0$, where $u^i[\xi, g^i] \in C^0(\mathbb{R}^n \setminus \mathbb{I}[\xi])$ is the unique harmonic function in $\mathbb{R}^n \setminus \mathrm{cl}\mathbb{I}[\xi]$ which equals $g^i \circ \xi^{(-1)}$ on $\xi(\partial \mathbb{B}_n)$ and which is such that $|u^i[\xi, g^i](t)||t|^{n-2}$ is bounded in $t \in \mathbb{R}^n \setminus \mathbb{I}[\xi]$, and where $\delta_{2,n} = 1$ if $n = 2$, $\delta_{2,n} = 0$ if $n \geq 3$ (see also Theorem 2.1 (v).)*

**Proof.** Let $\Omega_2$ be an open bounded connected subset of $\mathbb{R}^n$ of class $C^\infty$ such that $\mathrm{cl}\mathbb{I}[\phi_0^o] \subseteq \Omega_2 \subseteq \mathrm{cl}\Omega_2 \subseteq \Omega$. Then we have

$$u[\mathbf{q}] = u_1[\mathbf{q}] + P[F_{|\mathrm{cl}\Omega_2}]$$

for all $\mathbf{q} \in \mathcal{E}_{\Omega_2}^{m,\alpha} \times (C^{m,\alpha}(\partial \mathbb{B}_n))^2 \times C^0_{\omega,\rho}(\mathrm{cl}\Omega)$ with $\epsilon > 0$, where

$$u_1[\mathbf{q}] \equiv u[w, \epsilon, \xi, \phi^o, g^i - P[F_{|\mathrm{cl}\Omega_2}] \circ (w + \epsilon\xi), g^o - P[F_{|\mathrm{cl}\Omega_2}] \circ \phi^o, 0].$$

Now we denote by $\nu_\phi$ the exterior unit normal to $\mathbb{I}[\phi]$ for all $\phi \in \mathcal{A}_{\partial \mathbb{B}_n}$. Then the Divergence Theorem implies that

$$\int_{\mathbb{A}[\mathbf{q}]} |Du[\mathbf{q}]|^2 \, dt = \int_{\mathbb{A}[\mathbf{q}]} |Du_1[\mathbf{q}]|^2 \, dt + \int_{\mathbb{A}[\mathbf{q}]} |DP[F_{|\mathrm{cl}\Omega_2}]|^2 \, dt \qquad (11)$$

$$+2 \int_{\partial \mathbb{B}_n} (P[F_{|\mathrm{cl}\Omega_2}] \circ \phi^o) \left\{ \left( \frac{\partial u_1[\mathbf{q}]}{\partial \nu_{\phi^o}} \right) \circ \phi^o \right\} \tilde{\sigma}[\phi^o] \, d\sigma$$

$$-2\epsilon^{n-1} \int_{\partial \mathbb{B}_n} (P[F_{|\mathrm{cl}\Omega_2}] \circ (w + \epsilon\xi)) \left\{ \left( \frac{\partial u_1[\mathbf{q}]}{\partial \nu_{w+\epsilon\xi}} \right) \circ (w + \epsilon\xi) \right\} \tilde{\sigma}[\xi] \, d\sigma$$

for all $\mathbf{q} \in \mathcal{E}_{\Omega_2}^{m,\alpha} \times (C^{m,\alpha}(\partial \mathbb{B}_n))^2 \times C_{\omega,\rho}^0(\mathrm{cl}\Omega)$ with $\epsilon > 0$. Here for each $\phi \in \mathcal{A}_{\partial \mathbb{B}_n}$, $\tilde{\sigma}[\phi]$ is defined by the equality

$$\int_{\phi(\partial \mathbb{B}_n)} \omega(s) \, d\sigma_s = \int_{\partial \mathbb{B}_n} \omega \circ \phi(y) \tilde{\sigma}[\phi](y) \, d\sigma_y \qquad \forall \omega \in L^1(\phi(\partial \mathbb{B}_n)) \, ,$$

and one can easily verify that the map $\tilde{\sigma}[\cdot]$ of $C^{m,\alpha}(\partial \mathbb{B}_n, \mathbb{R}^n) \cap \mathcal{A}_{\partial \mathbb{B}_n}$ to the space $C^{m-1,\alpha}(\partial \mathbb{B}_n)$ which takes $\phi$ to $\tilde{\sigma}[\phi]$ is real analytic (see also [14, p. 166].) By [9, Thm. 6.1], and by Lemma 2.1 (iii), the first integral in the right hand side of (11) admits a representation as in the right hand side of (9), and the corresponding $\Phi_1$ equals $\int_{\mathrm{I}[\phi^o]} |Du^o[\phi^o, g^o - P[F_{|\mathrm{cl}\Omega_2}] \circ \phi^o, 0]|^2 \, dt + \delta_{2,n} \int_{\mathbb{R}^n \setminus \mathrm{cl}\mathrm{I}[\xi]} |Du^i[\xi, g^i - P[F_{|\mathrm{cl}\Omega_2}](w)]|^2 \, dt$ if $\epsilon = 0$. By Proposition 2.2, the second integral in (11) admits a real analytic continuation in the variable $\mathbf{q}$ around $\mathbf{q}_0$, and accordingly it admits a representation as in the right hand side of (9) and the corresponding $\Phi_1$, $\Phi_2$ equal $\Pi$ and $0$, respectively. Thus it suffices to show that the third and fourth integrals in the right hand side of (11) admit a representation as in the right hand side of (9), and to compute its term corresponding to (10). By Lemma 2.1 (iii), both the functions $g^i - P[F_{|\mathrm{cl}\Omega_2}] \circ (w + \epsilon\xi)$ and $g^o - P[F_{|\mathrm{cl}\Omega_2}] \circ \phi^o$ depend real analytically on $\mathbf{q} \in \mathcal{E}_{\Omega_2}^{m,\alpha} \times (C^{m,\alpha}(\partial \mathbb{B}_n))^2 \times C_{\omega,\rho}^0(\mathrm{cl}\Omega)$. Then by [9, Proof of Thm. 6.1], possibly shrinking $\mathcal{Q}_0$, there exist real analytic operators $H_1$, $H_2$, $H_3$, $H_4$ of $\mathcal{Q}_0$ to $C^{m-1,\alpha}(\partial \mathbb{B}_n, \mathbb{R}^n)$ such that

$$\epsilon^{n-1} \left( \frac{\partial u_1[\mathbf{q}]}{\partial \nu_{w+\epsilon\xi}} \right) \circ (w + \epsilon\xi) = H_1[\mathbf{q}] + \frac{H_2[\mathbf{q}]}{V_1[w, \epsilon, \xi, \phi^o] + V_2[w, \epsilon, \xi, \phi^o] \Upsilon_n(\epsilon)} ,$$

$$\left( \frac{\partial u_1[\mathbf{q}]}{\partial \nu_{\phi^o}} \right) \circ \phi^o = H_3[\mathbf{q}] + \frac{H_4[\mathbf{q}]}{V_1[w, \epsilon, \xi, \phi^o] + V_2[w, \epsilon, \xi, \phi^o] \Upsilon_n(\epsilon)} ,$$

for all $\mathbf{q} \equiv (w, \epsilon, \xi, \phi^o, g^i, g^o, F) \in \mathcal{Q}_0$ such that $\epsilon > 0$. Moreover,

$$H_1[w, 0, \xi, \phi^o, g^i, g^o, F] = \delta_{2,n} \left( \frac{\partial u^i[\xi, g^i - P[F_{|\mathrm{cl}\Omega_2}](w)]}{\partial \nu_\xi} \right) \circ \xi, \qquad (12)$$

$$H_3[w, 0, \xi, \phi^o, g^i, g^o, F] = \frac{\partial u^o[\phi^o, g^o - P[F_{|\mathrm{cl}\Omega_2}] \circ \phi^o, 0]}{\partial \nu_{\phi^o}} \circ \phi^o \, .$$

Clearly, if $n = 2$, then $Du^i[\xi, g^i - P[F_{|\mathrm{cl}\Omega_2}](w)] = Du^i[\xi, g^i]$. Since $\tilde{\sigma}[\cdot]$ is real analytic, and the pointwise product in $C^{m-1,\alpha}(\partial \mathbb{B}_n)$ is bilinear and continuous, and the integration on $\partial \mathbb{B}_n$ is linear and continuous from $C^{m-1,\alpha}(\partial \mathbb{B}_n)$ to $\mathbb{R}$, and by Lemma 2.1 (iii), the proof of the existence of $\Phi_1$, $\Phi_2$ for the third and fourth integral in the right hand side of (9) is complete. Finally,

$$\Phi_1[w, 0, \xi, \phi^o, g^i, g^o, F] = \int_{\mathbb{I}[\phi^o]} |Du^o[\phi^o, g^o - P[F_{|cl\Omega_2}] \circ \phi^o, 0]|^2 \, dt \qquad (13)$$

$$+\delta_{2,n} \int_{\mathbb{R}^n \setminus cl\mathbb{I}[\xi]} |Du^i[\xi, g^i]|^2 \, dt + \int_{\mathbb{I}[\phi^o]} |DP[F_{|cl\Omega_2}]|^2 \, dt$$

$$+2 \int_{\partial \mathbb{B}_n} \left( P[F_{|cl\Omega_2}] \circ \phi^o \right) \left\{ \left( \frac{\partial u^o[\phi^o, g^o - P[F_{|cl\Omega_2}] \circ \phi^o, 0]}{\partial \nu_{\phi^o}} \right) \circ \phi^o \right\} \tilde{\sigma}[\phi^o] \, d\sigma$$

$$-2\delta_{2,n} \int_{\partial \mathbb{B}_n} \left( P[F_{|cl\Omega_2}](w) \right) \left\{ \left( \frac{\partial u^i[\xi, g^i]}{\partial \nu_\xi} \right) \circ \xi \right\} \tilde{\sigma}[\xi] \, d\sigma \, .$$

If $n = 2$, then $u^i[\xi, g^i]$ is bounded at infinity, and thus by standard properties of harmonic functions (cf. *e.g.*, Folland [2, p. 149]), one can apply the Divergence Theorem in the exterior of $\mathbb{I}[\xi]$ and deduce that $\int_{\partial \mathbb{B}_n} \left( \frac{\partial u^i[\xi, g^i]}{\partial \nu_\xi} \circ \xi \right) \tilde{\sigma}[\xi] \, d\sigma = \int_{\xi(\partial \mathbb{B}_n)} \frac{\partial u^i[\xi, g^i]}{\partial \nu_\xi} \, d\sigma = 0$. Then again the Divergence Theorem and (13) imply the validity of the formula (10) and thus the proof is complete. □

## References

1. G. Dal Maso, Comportamento asintotico delle soluzioni di problemi di Dirichlet, *Bollettino U.M.I.* (7), **11**-A (1997), 253–277.

2. G.B. Folland, Introduction to partial differential equations, *Princeton University Press*, Princeton, 1976.

3. D. Gilbarg and N.S. Trudinger, Elliptic Partial Differential Equations of Second Order, *Springer Verlag*, Berlin, 1983.

4. V.A. Kozlov, V.G. Maz'ya and A.B. Movchan, Asymptotic analysis of fields in multistructures, *Oxford Mathematical Monographs, Oxford: Clarendon Press*, 1999.

5. R. Kühnau, Die Kapazität dünner Kondensatoren, *Math. Nachr.*, **203**, 125–130, (1999).

6. M. Lanza de Cristoforis, A domain perturbation problem for the Poisson equation, *Complex Variables*, **50**, 851–867, (2005).

7. M. Lanza de Cristoforis, Asymptotic behaviour of the conformal representation of a Jordan domain with a small hole, and relative capacity, in 'Complex Analysis and Dynamical Systems', edited by M. Agranovsky, L. Karp, D. Shoikhet, and L. Zalcman, *Contemporary Mathematics*, **364** (2004), 155-167.

8. M. Lanza de Cristoforis, Asymptotic behaviour of the conformal representation of a Jordan domain with a small hole in Schauder spaces, *Computational methods and function theory*, **2** (2002), 1–27.

9. M. Lanza de Cristoforis, Asymptotic behaviour of the solutions of the Dirichlet problem for the Laplace operator in a domain with a small hole. A functional analytic approach, *submitted*, 2004.

10. M. Lanza de Cristoforis, A singular perturbation Dirichlet boundary value problem for harmonic functions on a domain with a small hole, Proceedings of the 12th International Conference on Finite or Infinite Dimensional Complex Analysis and Applications, Tokyo July 27–31 2004, edited by H. Kazama, M. Morimoto, C. Yang, Kyushu University Press, (2005), 205–212.

11. M. Lanza de Cristoforis, Differentiability properties of an abstract autonomous composition operator, *J. London Math. Soc.* **61**, 923–936 (2000).

12. M. Lanza de Cristoforis, Perturbation problems in potential theory. A functional analytic approach, *(to appear on Journal of Applied Functional Analysis)*, 2004.
13. M. Lanza de Cristoforis, Properties and pathologies of the composition and inversion operators in Schauder spaces, *Acc. Naz. delle Sci. detta dei XL* **15**, 93–109 (1991).
14. M. Lanza de Cristoforis and L. Rossi, Real analytic dependence of simple and double layer potentials upon perturbation of the support and of the density, *Journal of Integral Equations and Applications*, **16** (2004), 137–174.
15. V.G. Mazya, S.A. Nazarov and B.A. Plamenewskii, Asymptotic theory of elliptic boundary value problems in singularly perturbed domains, vol I, II, (translation of the original in German published by Akademie Verlag 1991), *Operator Theory: Advances and Applications*, **111**, **112**, *Birkhäuser Verlag*, Basel, 2000.
16. S. Ozawa, Electrostatic capacity and eigenvalues of the Laplacian, *J. Fac. Sci. Univ. Tokyo*, **30** (1983) , 53–62.
17. L. Preciso, Perturbation analysis of the conformal sewing problem and related problems, *Doctoral Dissertation, University of Padova*, 1998.
18. L. Preciso, Regularity of the composition and of the inversion operator and perturbation analysis of the conformal sewing problem in Romieu type spaces, National Academy of Sciences of Belarus, *Proceedings of the Institute of Mathematics* **5**, 99–104 (2000).
19. G. Prodi and A. Ambrosetti, Analisi non lineare, *Editrice Tecnico Scientifica*, Pisa, 1973.
20. M.J. Ward and J.B. Keller, Strong localized perturbations of eigenvalue problems, *SIAM J. Appl. Math.*, **53** (1993), 770–798.

# ITERATION DYNAMICAL SYSTEM OF DISCRETE LAPLACIANS
# AND
# THE EVOLUTION OF EXTINCT ANIMALS

KAZUO KOSAKA *

*Department of Geosystem Sciences, College of Humanities and Sciences*
*Setagaya-ku Sakurajousui 3-25-40,*
*156 Setagaya,Tokyo, Japan*
*E-mail: kosaka@chs.nihon-u.ac.jp*


OSAMU SUZUKI †

*Department of Computer and System Analysis, College of Humanities and Sciences*
*Setagaya-ku Sakurajousui 3-25-40,*
*156 Setagaya,Tokyo, Japan*
*E-mail: osuzuki@cssa.chs.nihon-u.ac.jp*

The time changes of numbers of families are simulated by the iteration of dynamical systems of the discrete Laplacian and the evolution of extinct animals is discussed. The following results are obtained: (1) The time changes of families for the Cambrian fauna, Paleozonic fauna and Modern fauna are simulated by suitable choices of neighborhoods. (2) The mutation can be realized by the change of neighborhoods. (3) The change of environments can be realized by the change of sources

**Key words:** Simulation of structure changes, dynamical systems, discrete Laplacian
**Mathematics Subject Classification:** 65N06

## 1. Introduction

It is well known that the numbers of families or species of extinct animals diverged at the Cambrian era and the changes of the numbers satisfy the logistic equation [3] and that a large number of families/species were extinguished at the Permian era [5]. In [1,2], we have made computer simulations on the organizations, i.e., the crystalization of waters, the growth of cities and generations of designs by use of the iteration dynamical system of discrete Laplacians.

In this paper we shall make computer simulations of the time changes of numbers of families by use of the dynamical system, choosing neighborhoods and seeds. We notice that the behaviors of orbits depend on the evenness/oddness of the neighborhoods. A neighborhood is called even/odd neighborhood, when the number of the

---

*This work is supported by the natural scientific reserach project of Nihon Univ.,
†Work partially supported by grant 16540122 Kakenhi of the Ministrium of Academy of Japan (2004-5)

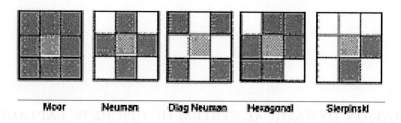

Moor     Neuman     Diag Neuman     Hexagonal     Sierpinski

Fig. 1.   Examples of neighborhoods

cells of the neighborhood is even/odd. Here we consider only even neighborhoods. Our observations are divided into three parts: (1) The increase of the numbers of families, (2) the decrease and extinctions of the families and (3) the mutation and the change of environments. Here we want to make a stress on the fact that the both increase and decrease can be described in terms of the same dynamical systems.

## 2. Iteration dynamical systems of discrete Laplacians

We choose the lattice $\mathbf{L}$ on the real plane. Each lattice point is identified with the corresponding cell $\Delta_p(p = (i, j))$. We consider a function $f$ on $\mathbf{L}$ whose value is taken in $\{0,1\}$. The set of such functions constitute the commutative algebra $\mathbf{F}$ where we calculate sums and products in the mod 2 calculation rule. Next we introduce neighborhoods of a lattice point $p \in \mathbf{L}$. A set of cells which attach the referenced cell is called a neighbourhood of $p$ which is denoted by $U_p$. We list several examples which will be used in this paper:

(1) Moor neighborhood: $U_M(i, j) = U_{\rho,\tau=0,\pm1}\Delta_{i+\rho,j+\tau}$
(2) Neumann neighborhood: $U_N(i, j) = U_{\rho,\tau=0,\pm1}(\Delta_{i+\rho,j} \cup \Delta_{i,j+\rho})$
(3) Diagonal Neumann neighborhood: $U_{dN}(i, j) = \cup_{\rho,\tau=0,\pm1}(\Delta_{i+\rho,j+\rho} \cup \Delta_{i+\rho,j-\rho})$
(4) Hexagonal neighborhood: $U_H(i, j) = U_M(i, j) - (\Delta_{i-1,j+1} \cup \Delta_{i+1,j-1})$
(5) Sierpinski neighborhood: $U_S(i, j) = \Delta_{i+1,j} \cup \Delta_{i,j-1}$

The definition is given uniquely up to orthogonal transformations.

We give several basic notations on the discrete Laplacian:

(1) (**Discrete Laplacian**) Choosing a neighborhood $U_p$, we define the discrete Laplacian:

$$\Delta_{U_p} f(p) = \sum_{q \in U_p - p} (f(q) - f(p)). \tag{1}$$

(2) (**Iteration dyamical system of discrete Laplacian**) Choosing an initial function $f_0 \in \mathbf{F}$, we define the dynamical system defined by the iteration of the Laplacian:

$$\{f_n\}, f_n = \Delta_U f_{n-1}(n \geq 1). \tag{2}$$

(3) (**Source**) We call $p$ a source(or seed) of the dynamical system when $f_n(p) = 1$

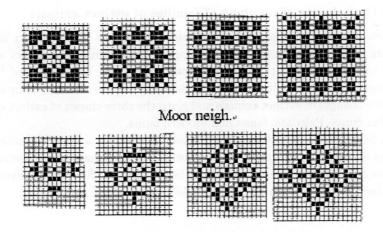

Moor neigh.

Neumann neigh.

Fig. 2.   Examples of iterations

for any $n \in \mathbf{N}$. We regard the sources as boundary conditions.

(4) We give computer simulations in the case of a single source:

Choosing an integer $M$ (the size of the lattice), we consider the subalgebra $\mathbf{F}(M)$:

$$\mathbf{F}(M) = \{f \in \mathbf{F} | f(i,j) = f(i + kM, j + lM), k, l \in \mathbf{Z}\}. \tag{3}$$

Choosing the sublattice $\mathbf{L}'(M) = \{\Delta_{kM,lM} | k, l \in \mathbf{Z}\}$, we put $\mathbf{L}(M) = \mathbf{L}/\mathbf{L}'(M)$. Then, we see that $\Delta_U : \mathbf{F}(M) \mapsto \mathbf{F}(M)$ on $(M)$.

(5) (**Periodicity and stability**): (1) The orbit $\{f_n\}$ is called periodic with a period $T$, if there exists an integer $k$(minimal) and $T$ such that $f_{k+nT} = f_k(n \in \mathbf{N})$. (2)Especially in the case of $T = 1$, the orbit is called stable and $k$ is called stability speed.

(6) (**Observations on the behaviors of dynamical systems**): We can observe the following facts by computer simulations:

(I) In the case where $M$ is even and the source is only one point, any orbit is stable, when the neighborhood is even. The stability speed depends on the symmetry structure of the neighborhood. In fact, the stability speed is $2^{p-1}$ in the case where Moor neigh., Neumann neigh., Diagonal Neumann. neigh., etc.

(II) When neighhood is odd, the orbit $\{f_n\}$ is periodic. The period is one of $2^p$ or $2^{p-1}$, or $2^p - 1$ which depends on the type of neighborhood.

(III) In the odd size $M$, we may ask "Is an orbit periodic ?". We give the table of periods for small odd integers in the case of the Neumann neigh. and one source (by Excel): The upper line is size $M$ and the lower line is the corresponding period.

| 3 | 5 | 7 | 9 | 11 | 13 | 15 | 17 | 19 | 21 | 23 | 25 | 27 | 29 | 31 |
|---|---|---|---|----|----|----|----|-----|-----|------|------|------|-------|----|
| 1 | 5 | 6 | 13 | 30 | 62 | 29 | 30 | 511 | 124 | 2046 | 2045 | 1021 | 16384 | 61 |

## 3. The time change of numbers of families of extinct animals

We review on the time change of the numbers of families of the extinct animals. There are many references on the time change of the numbers of extinct animals and the background extinctions and mass extinctions have been discussed [4,5]. One of the most important results are given by Sepkoski. He has collected many samples of the time changes of extinct animals and given the three classes of extinct animals :Cambrian fauna, Paleozoic fauna and Modern fauna.

In [6], he has given a mathematical model described by a system of differential equations of logistic type and has shown that the three classes can be described by "the three equation model" quite well. Also he has given models for mass extinctions by the changes of growth ratios in the models and tried the fittings.

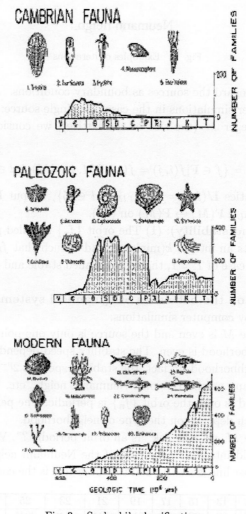

Fig. 3.   Sepkoski's classifications

## 4. Increase of number of families and their computer simulations

We analyze the increase of the number of families in the Sepkoski's table and obtain several laws for the increasing properties. Then we give several computer simulations

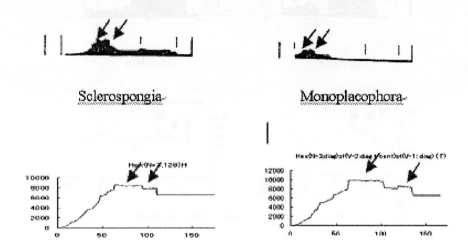

Fig. 4.   Two peaks law in Cambrian fauna.

### Two peaks law for Cambrian fauna

The only convex increasing can be observed and the concave increasing can not be found. Two tops on the hills in the curve of the time change can be observed which is called the "Two peaks law for Cambrian fauna". This law can be well realized in our simulations (see Figure 4).

### Three peaks law for Paleozoic fauna

There are two kinds of increasings, the convex increasing and the concave increasing. Three peaks in the curve of the time change are observed which we call "Three peaks law for Paleozoic fauna (see Figure 5)".

### Two pauses-in-increasing law for Modern fauna

There are two kinds of increasing, one is the convex increasing and the other is the concave increasing. Two pauses can be observed in the process of the increasing (see Figure 6).

## 5. Decrease of number of families and their computer simulations

We analyze the decrease of the number of families and the mass extinction. Next we give several computer simulations.

### The moderate decreasing of families for Cambrian fauna

(1) The one is concave decreasing and the other is convex decreasing. The first

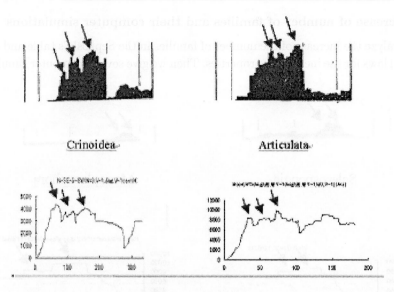

Fig. 5.   Three peaks law in Paleozoic fauna.

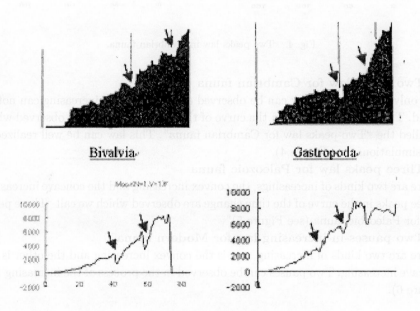

Fig. 6.   Two pauses in increasing law in Modern fauna.

increasing is popular and the second one is very few. Moreover, the decreasing is usually mild. (2) When the decreasing begins and it is concave, we can not observe the mass extinction. Only the back ground extinctions can be seen (see Figure 7).

**The rapid decreasing for Paleozoic fauna**

(1) The both of concave and convex decreasings are observed. The first increasing is

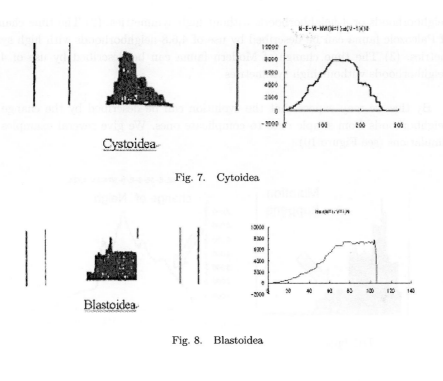

Cystoidea

Fig. 7. Cytoidea

Blastoidea

Fig. 8. Blastoidea

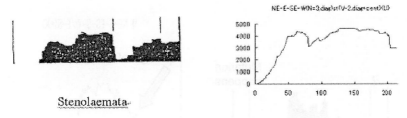

Stenolaemata

Fig. 9. Stenolaemata

popular and usually it is concave and rapid. The second one is very few (see Figure 8).

### The decreasing for Modern fauna

(2) The decreasing can be observed in the early stage whose curve is similar to that of the Cambrian fauna. But the decreasing does not continue to the final extinction (see Figure 9).

## 6. The mutation and the change of environments

**Mutation**: At first we notice that our simulations tell us the following facts:

(1) The time change of Cambrian fauna can be described by use of 2-

neighborhoods or 4-neighborhoods without high symmetries. (2) The time change of Paleozoic fauna can be described by use of 4,6,8-neighborhoods with high symmetries. (3) The time change of Modern fauna can be described by use of 4,6-neighborhoods without high symmetries.

By this we may expect that the evolution can be described by the change of neighborhoods from simple ones to complicate ones. We give several examples of simulations (see Figure 10).

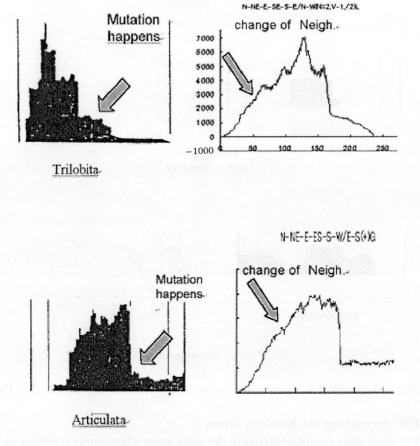

Fig. 10.   Trilobita and Articulata

**Environments**: By the following experiments, we may expect to describe the changes of environments by the changes of sources:

(1) The mass extinctions can be described by the take off of sources. (2) The take off of sources in the early stage in the evolution induces wild behaviors in the change. On the other side the take off of sources in the stable stage does not give

strong changes and keeps the stability in the change.

We give two examples (Figure 11):

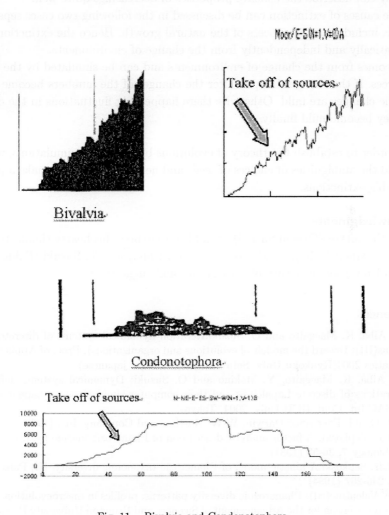

Fig. 11.   Bivalvia and Condonotophora

## 7. Conclusions and discussions

From above simulations, we can obtain the following conclusions :

(1) Discrete Laplacians can describe the increase of number of families well.
(i) They can realize the logistic curves quite well when the numbers of seeds are big.

(ii) They can describe the three peaks law for Paleozoic fauna, the two peaks law for Cambrian fauna and the two pauses law for Modern fauna respectivelly.

(2) Discrete Laplacians can describe the decreases and extinctions quite well:

(i) They can describe the concave properties of decreasings quite well.

(ii) The causes of extinction can be discussed in the following two cases separately:

(a) It is included in the process of the natural growth. Hence the extinction arises automatically and independently from the change of environments.

(b) It comes from the change of environments and can be simulated by the change of sources. If the changes happen after the changes of the numbers become stable, then the changes are mild. Otherwise there happen big fluctuations in the changes and they become mild finally.

In order to establish the theory of evolutions by use of our simulations, we have to avoid the ambiguities of choices of seeds and neighborhoods and make a prophet on the big extinctions.

## Acknowledgments

One of the author (Osamu Suzuki) would like to express his hearty thanks to Profs. K. Endo (Nihon Univ.), T. Murakami (Kyoto Univ.) and T. Suzuki (Fukuoka Education Univ.) for their variable discussions and suggestions.

## References

1. Y. Aiba, K. Maegaito and O. Suzuki:Iteration dynamical systems of discrete Laplacians(III)(Toward the models of evolutions and organizations), Proc. of Applied Mathematics 2004(Ryukoku Univ. Seta Campas),59-64(in japanese)
2. Y. Aiba, K. Maegaito, Y. Makino and O. Suzuki: Dynamical systems defined by iterations of discrete Lapalcians and their computer simulations, to aapear in Proc. ISSAC Int. Conf. (ICU Univ. 2004, Tokyo)
3. S.J. Gould: Ever since Darwin, W.W. Northon and Company, Inc. (1977)
4. J.J.Jr. Sepkoski: A factor analytic description of Phanerozic marine fossil record, Paleobiology 7, 36-53 (1981)
5. J.J.Jr. Sepkoski:A kinetic model of Phanerozic taxonomic diversity. III, Paleobiology 10, 246-267 (1984)
6. J.W.Valentine(ed): Phanerozoic diversity patterns: profiles in macroevolution, American Association for the Advancement of Sicences and Princton University Press, Princton(1986)

# COMMUTATIVE ALGEBRAS OF HYPERCOMPLEX MONOGENIC FUNCTIONS AND SOLUTIONS OF ELLIPTIC TYPE EQUATIONS DEGENERATING ON AN AXIS *

S. PLAKSA

*Institute of Mathematics of the National Academy of Sciences of Ukraine,*
*3 Tereshchenkivska Street,*
*Kiev, Ukraine*
*E-mail: plaksa@imath.kiev.ua*

For investigation of equations with partial derivatives we develop a method analogous to the analytic function method in the complex plane. We have obtained expressions of solutions of elliptic equations degenerating on an axis via components of analytic functions taking values in a commutative associative Banach algebra.
**Key words:** Elliptic equations degenerating on an axis, hypercomplex analytic functions, commutative associative Banach algebra
**Mathematics Subject Classification:** 30G35, 35J99

## 1. Introduction

Analytic function methods in the complex plane for plane potential fields inspire searching of analogous methods for spatial potential solenoid fields. The problem to construct such methods for spatial potential solenoid fields was posed by M.A. Lavrentyev [1], pp. 205, 18]). Besides, even independently of connections with applications in mathematical physics, a variety and an effectiveness of the analytic function methods stimulate developing of analogous methods for equations with partial derivatives.

The theory of monogenic functions in commutative associative Banach algebras presents a fruitful approach to equations with partial derivatives and gives effective methods for studying them. An idea of realization of such an algebraic-analytic approach to equations with partial derivatives consists in a construction of commutative Banach algebras such that monogenic functions taking values in these algebras have components satisfying the given equations.

Such algebras are constructed for the three-dimensional Laplace equation [2,3] and the two-dimensional biharmonic equation [4] and some other equations.

In the paper [5] (see also [6]) we constructed explicitly analytic functions taking values in an infinite-dimensional commutative Banach algebra and proved that components of these functions generate the axial-symmetric potential $\varphi$ and the Stokes

---

*This work is supported partially by the State Program of Ukraine No. 0105U000433

flow function $\psi$ satisfying the following system of equations

$$y \, \frac{\partial \varphi(x,y)}{\partial x} = \frac{\partial \psi(x,y)}{\partial y}, \qquad y \, \frac{\partial \varphi(x,y)}{\partial y} = -\frac{\partial \psi(x,y)}{\partial x} \qquad (1)$$

in a meridian plane $xOy$ of a spatial potential solenoid field symmetric with respect to the axis $Ox$. Thus, for investigation of spatial axial-symmetric potential solenoid fields we developed a new method analogous to the analytic function method in the complex plane. On this way, in the papers [7-9] we developed a functionally-analytic method for effective solving boundary problems for solutions of equations

$$y \, \Delta \varphi(x,y) + \frac{\partial \varphi(x,y)}{\partial y} = 0, \qquad (2)$$

$$y \, \Delta \psi(x,y) - \frac{\partial \psi(x,y)}{\partial y} = 0 \qquad (3)$$

(here $\Delta := \frac{\partial^2}{\partial x^2} + \frac{\partial^2}{\partial y^2}$) which are sequent of the system (1). Furthermore, in the paper [10] for a boundary problem about a streamline of the ideal incompressible fluid along an axial-symmetric body we constructed solutions unknown before.

In this paper we establish a relation between components of hypercomplex analytic functions constructed in [5] and solutions of other elliptic equations degenerating on the axis $Ox$.

## 2. Expressions of solutions of elliptic equations degenerating on an axis via components of hypercomplex analytic functions

### 2.1. *Equations in proper domains*

Let $\mathbf{R}$ be the algebra of real numbers and $\mathbf{C}$ be the algebra of complex numbers. Let $\mathbf{H} := \{a = \sum\limits_{k=1}^{\infty} a_k e_k \ : \ a_k \in \mathbf{R}, \ \sum\limits_{k=1}^{\infty} |a_k| < \infty\}$ be a commutative associative Banach algebra over $\mathbf{R}$ with the norm $\|a\|_{\mathbf{H}} := \sum\limits_{k=1}^{\infty} |a_k|$ and the basis $\{e_k\}_{k=1}^{\infty}$. The multiplication table for elements of basis was offered in the paper [11] and is of the following form

$$e_n e_1 = e_n, \quad e_m e_n = \frac{1}{2} \left( e_{m+n-1} + (-1)^{n-1} e_{m-n+1} \right) \quad \forall \, m \geq n \geq 1.$$

Denote by $\mathbf{H_C} := \mathbf{H} \oplus i\mathbf{H} \equiv \{c = a + ib \ : \ a, b \in \mathbf{H}\}$ a comlexification of the algebra $\mathbf{H}$ such that the norm of element $c := \sum\limits_{k=1}^{\infty} c_k e_k \in \mathbf{H_C}$ is given by means the equality $\|c\|_{\mathbf{H_C}} := \sum\limits_{k=1}^{\infty} |c_k|$.

Consider the Cartesian plane $\mu := \{\zeta = x e_1 + y e_2 \ : \ x, y \in \mathbf{R}\}$ in $\mathbf{H}$.

For a domain $D \subset \mathbf{R}^2$ we consider the domains $D_z := \{z = x + iy \ : \ (x,y) \in D\} \subset \mathbf{C}$ and $D_\zeta := \{\zeta = x e_1 + y e_2 \ : \ (x,y) \in D\} \subset \mu$ which are congruent to the domain $D$.

We shall call $D_z$ a *proper* domain in $\mathbf{C}$, provided that for every $z \in D_z$ with $\mathrm{Im}\, z \neq 0$ the domain $D_z$ contains the segment connecting points $z$ and $\bar{z}$.

In what follows, $(x, y) \in D$ and $z = x + iy$ and $\zeta = xe_1 + ye_2$.

In the paper [5], for every function $F : D_z \to \mathbf{C}$ analytic in a proper domain $D_z$ we have explicitly constructed the expansion in the basis $\{e_k\}_{k=1}^{\infty}$ of the principal extension into the domain $D_\zeta$:

$$\frac{1}{2\pi i} \int_{\gamma} (te_1 - \zeta)^{-1} F(t)\, dt = U_1(x, y)\, e_1 + 2 \sum_{k=2}^{\infty} U_k(x, y)\, e_k. \qquad (4)$$

Here

$$U_k(x, y) := \begin{cases} \frac{1}{2\pi i} \int_{\gamma} \frac{F(t)}{\sqrt{(t-z)(t-\bar{z})}} \left( \frac{\sqrt{(t-z)(t-\bar{z})} - (t-x)}{y} \right)^{k-1} dt\,, & \\ & \text{when } y \neq 0\,, \\ F(x)\,, & \text{when } y = 0 \text{ and } k = 1\,, \\ 0\,, & \text{when } y = 0 \text{ and } k > 1\,, \end{cases} \qquad (5)$$

where $\gamma$ is an arbitrary closed rectifiable Jordan curve in $D_z$ which embraces the segment connecting $z$ and $\bar{z}$, and $\sqrt{(t-z)(t-\bar{z})}$ is that continuous branch of the analytic function $G(t) = \sqrt{(t-z)(t-\bar{z})}$ outside of the segment mentioned above for which $G(t) > 0$ for all $t > \mathrm{Re}\, z$.

We have proved (see [5,6]) that the first and the second components of the principal extension (4) of analytic function $F$ generate the solutions $\varphi$ and $\psi$ of the system (1) in $D$ by the formulas

$$\varphi(x, y) = U_1(x, y)\,, \qquad \psi(x, y) = y\, U_2(x, y)\,. \qquad (6)$$

Moreover, the functions (6) are solutions of Eqs. (2) and (3), respectively.

Generalizing this result, in the following theorem we obtain expressions of solutions of elliptic equations degenerating on the axis $Ox$ via components $U_k$ of hypercomplex analytic function (4).

**Theorem 2.1.** *If $F : D_z \to \mathbf{C}$ is an analytic function in a proper domain $D_z$, then the components $U_k$ of the principal extension (4) of function $F$ into the domain $D_\zeta$ satisfy the equations*

$$y^2 \Delta U_k(x, y) + y \frac{\partial U_k(x, y)}{\partial y} - (k-1)^2 U_k(x, y) = 0\,, \quad k = 1, 2, \dots\,, \qquad (7)$$

*on the set $\{(x, y) \in D : y \neq 0\}$. In addition, the function*

$$\psi_k(x, y) := y^{k-1}\, U_k(x, y) \qquad (8)$$

*is a solution in $D$ of the equation*

$$y\, \Delta \psi_k(x, y) - (2k-3) \frac{\partial \psi_k(x, y)}{\partial y} = 0\,, \quad k = 1, 2, \dots\,. \qquad (9)$$

**Proof.** Using the equalities (8) and (5), we obtain the following integral expression for the function $\psi_k$:

$$\psi_k(x,y) = \begin{cases} \dfrac{1}{2\pi i} \displaystyle\int\limits_{\gamma} \dfrac{F(t)\left(\sqrt{(t-z)(t-\bar{z})} - (t-x)\right)^{k-1}}{\sqrt{(t-z)(t-\bar{z})}} \, dt, & \text{when } y \neq 0, \\[4mm] F(x), & \text{when } y = 0 \text{ and } k = 1, \\[2mm] 0, & \text{when } y = 0 \text{ and } k > 1. \end{cases} \tag{10}$$

Applying the standard reasoning [12, p. 661], we prove that the function (10) has continuous derivatives of arbitrary order in $D$, and for any $(x,y) \in D : y \neq 0$ the following equalities are fulfilled:

$$\frac{\partial^2 \psi_k(x,y)}{\partial x^2} = \frac{1}{2\pi i} \int\limits_{\gamma} \frac{F(t)\left(\sqrt{(t-z)(t-\bar{z})} - (t-x)\right)^{k-1}}{\left(\sqrt{(t-z)(t-\bar{z})}\right)^3}$$

$$\times \left( (k-1)^2 + \frac{3(k-1)(t-x)}{\sqrt{(t-z)(t-\bar{z})}} + \frac{2(t-x)^2 - y^2}{(t-z)(t-\bar{z})} \right) dt,$$

$$\frac{\partial \psi_k(x,y)}{\partial y} = \frac{1}{2\pi i} \int\limits_{\gamma} \frac{y\, F(t)\left(\sqrt{(t-z)(t-\bar{z})} - (t-x)\right)^{k-2}}{(t-z)(t-\bar{z})}$$

$$\times \left( k - 2 + \frac{t-x}{\sqrt{(t-z)(t-\bar{z})}} \right) dt,$$

$$\frac{\partial^2 \psi_k(x,y)}{\partial y^2} = \frac{1}{2\pi i} \int\limits_{\gamma} \frac{F(t)\left(\sqrt{(t-z)(t-\bar{z})} - (t-x)\right)^{k-3}}{\left(\sqrt{(t-z)(t-\bar{z})}\right)^3}$$

$$\times \left( (k-2)\left((k-3)y^2 + (t-x)^2\right) + \frac{(k-3)(t-x)\left(2y^2 - (t-x)^2\right)}{\sqrt{(t-z)(t-\bar{z})}} \right.$$

$$\left. + \frac{(t-x)^2\left(2y^2 - (t-x)^2\right)}{(t-z)(t-\bar{z})} \right) dt.$$

Substituting the obtained partial derivatives of the function (10) into Eq. (9) and taking into account the continuity of them in points of the axis $Ox$, we establish that the equality (9) becomes an identity in $D$.

Finally, substituting the function $\psi_k$ of the form (8) in the equality (9), we establish that the function $U_k$ satisfies Eq. (7) for $(x,y) \in D : y \neq 0$. Theorem is proved.

Equations of the form (7), (9) were studied in many papers (see, for example, [13,14]). At $k = 1$ Eqs. (7) and (9) take the form (2), at $k = 2$ Eq. (9) is turned into Eq. (3).

## 2.2. *Equations outside of proper domains*

Let's establish a relation between hypercomplex analytic functions taking values in the algebra $\mathbf{H_C}$ and solutions of Eqs. (7), (9) in the complement of the closure of a bounded proper domain $D$.

Introduce an element $e_0 \notin \mathbf{H_C}$ which satisfies the following rules of multiplication:

$$e_0 e_1 = e_0, \quad e_0 e_2 = -e_1, \quad e_0 e_{2k+1} = e_0 - 2\sum_{m=1}^{k} e_{2m}, \quad e_0 e_{2k+2} = -e_1 - 2\sum_{m=1}^{k} e_{2m+1}$$

for $k = 1, 2, \ldots$ . We assume that the axioms of associativity and commutativity and distributivity remain valid.

Include the algebra $\mathbf{H_C}$ into a Banach space $\widetilde{\mathbf{H}}_{\mathbf{C}} := \{d = \sum_{k=0}^{\infty} d_k e_k : d_k \in$ $\mathbf{C}, \sum_{k=0}^{\infty} |d_k| < \infty\}$ with the norm $\|d\|_{\widetilde{\mathbf{H}}_{\mathbf{C}}} := \sum_{k=0}^{\infty} |d_k|$. Note that $\widetilde{\mathbf{H}}_{\mathbf{C}}$ is only an extension of the linear space of algebra $\mathbf{H_C}$ and is no algebra because the product $e_0 e_0$ is not definite.

Consider the Cartesian plane $\widetilde{\mu} := \{\widetilde{\zeta} = x e_1 + y e_0 : x, y \in \mathbf{R}\}$ in $\widetilde{\mathbf{H}}_{\mathbf{C}}$ and the domain $D_{\widetilde{\zeta}} := \{\widetilde{\zeta} = x e_1 + y e_0 : (x, y) \in D\}$ in $\widetilde{\mu}$ which is congruent to the domain $D \subset \mathbf{R}^2$. In what follows, $\widetilde{\zeta} = x e_1 + y e_0$.

In the paper [5], for every analytic function $F : \mathbf{C} \setminus \overline{D_z} \to \mathbf{C}$ vanishing in infinity we have explicitly constructed the expansion in the basis $\{e_k\}_{k=1}^{\infty}$ of the principal extension into the domain $\widetilde{\mu} \setminus \overline{D_{\widetilde{\zeta}}}$:

$$-\frac{1}{2\pi i}\int_{\Gamma}(te_1 - \widetilde{\zeta})^{-1} F(t)\,dt = -\frac{e_1}{2\pi i}\int_{\Gamma}\frac{F(t)}{\pm\sqrt{(t-z)(t-\bar{z})}}\frac{\pm\sqrt{(t-z)(t-\bar{z})}-y}{t-x}\,dt$$

$$-\frac{y}{\pi i}\sum_{k=2}^{\infty}(-1)^k e_k \int_{\Gamma}\frac{F(t)}{\pm\sqrt{(t-z)(t-\bar{z})}\,(t-x)}$$

$$\times\left(\frac{\pm\sqrt{(t-z)(t-\bar{z})}-y}{t-x}\right)^{k-1}dt, \qquad \widetilde{\zeta} \in \widetilde{\mu}\setminus\overline{D_{\widetilde{\zeta}}}, \pm y > 0, \qquad (11)$$

where $\Gamma$ is an arbitrary closed bounded rectifiable Jordan curve in $\mathbf{C}$ which bounds a proper domain $D'_z$ such that $\overline{D_z} \subset D'_z$ and $z \in \mathbf{C} \setminus \overline{D'_z}$, and $\sqrt{(t-z)(t-\bar{z})}$ is that continuous branch of the analytic function $G(t) = \sqrt{(t-z)(t-\bar{z})}$ outside of the cut $\{t \in \mathbf{C} : \operatorname{Re} t = x, |\operatorname{Im} t| \geq |y|\}$ for which $G(t) > 0$ for all $t > \operatorname{Re} z$.

The function (11) is an extension of the function $F$ into the domain $\widetilde{\mu} \setminus \overline{D_{\widetilde{\zeta}}}$ in the following sense: if $x_0 \in \mathbf{R}$ and the point $\widetilde{\zeta} \in \widetilde{\mu} \setminus \overline{D_{\widetilde{\zeta}}}$ tends to the point $x_0 e_1 \in \widetilde{\mu} \setminus \overline{D_{\widetilde{\zeta}}}$, then the function (11) converges by coordinates to $F(x_0)e_1$.

In the paper [5] we have shown that functions of the form (11) form an algebra.

In the following theorem solutions of Eqs. (7), (9) in $\mathbf{R}^2 \setminus \overline{D}$ are constructed explicitly via hypercomplex analytic function (11).

**Theorem 2.2.** *If a proper domain $D_z$ is bounded and an analytic function $F$ : $\mathbf{C} \setminus \overline{D_z} \to \mathbf{C}$ have zero of multiplicity at least $k$ in infinity, then for $\widetilde{\zeta} = x e_1 + y e_0$ : $y \neq 0$ the function (11) is expressed in the form*

$$-\frac{1}{2\pi i} \int_{\Gamma} (t e_1 - \widetilde{\zeta})^{-1} F(t) \, dt = -(e_2)^k \sum_{m=1}^{\infty} V_{k,m}(x, y) \, e_m, \tag{12}$$

*where*

$$V_{k,1}(x, y) := \frac{(-1)^{k-1}}{2\pi i \, y^{k-1}} \int_{\Gamma} \frac{F(t) \, (t-x)^{k-1}}{\pm\sqrt{(t-z)(t-\bar{z})}} \, dt, \tag{13}$$

$$V_{k,m}(x, y) := -\frac{(-1)^{m+k-1}}{\pi i \, y^{k-1}} \int_{\Gamma} \frac{F(t) \, (t-x)^{k-1}}{\pm\sqrt{(t-z)(t-\bar{z})}}$$

$$\times \left( \frac{\pm\sqrt{(t-z)(t-\bar{z})} - y}{t-x} \right)^{m-1} dt, \qquad m = 2, 3, \ldots. \tag{14}$$

*Furthermore, the function*

$$U_k(x, y) := \begin{cases} \pm \sum_{j=0}^{[\frac{k-1}{2}]} V_{k,2j+1}(x, y) \sum_{m=j}^{[\frac{k-1}{2}]} 2^{-2m} \, C_{k-1}^{2m} \, C_{2m}^{m-j}, & \text{when } \pm y > 0, \\ \lim_{\eta \to 0} U_k(x, \eta), & \text{when } y = 0, \end{cases} \tag{15}$$

*(here $C_n^m$ are binomial coefficients) satisfies Eq. (7) on the set $\{(x, y) \in \mathbf{R}^2 \setminus \overline{D} : y \neq 0\}$. Moreover, the function (8) is a solution of Eq. (9) in $\mathbf{R}^2 \setminus \overline{D}$, if the function $U_k$ is of the form (15).*

**Proof.** We establish easily the equality (12). Really, performing the multiplication on the right-hand side of the equality (12) and taking account that the function $F$ have zero of multiplicity $k$ in infinity, we obtain the equality (11).

Let $(x, y) \in \mathbf{R}^2 \setminus \overline{D}$ and $y \neq 0$. Taking account the equalities (13) – (15) and the fact that the function $F$ have zero of multiplicity $k$ in infinity, we transform the expression (8):

$$\psi_k(x, y) = \frac{(-1)^{k-1}}{2\pi i} \int_{\Gamma} \frac{F(t)}{\sqrt{(t-z)(t-\bar{z})}} \left( (t-x)^{k-1} \sum_{m=0}^{[\frac{k-1}{2}]} 2^{-2m} \, C_{k-1}^{2m} \, C_{2m}^m \right.$$

$$+ \sum_{j=1}^{[\frac{k-1}{2}]} (t-x)^{k-1-2j} \left( \pm\sqrt{(t-z)(t-\bar{z})} - y \right)^{2j} \sum_{m=j}^{[\frac{k-1}{2}]} 2^{1-2m} \, C_{k-1}^{2m} \, C_{2m}^{m-j} \right) dt$$

$$= \frac{(-1)^{k-1}}{2\pi i} \int_{\Gamma} \frac{F(t)}{\sqrt{(t-z)(t-\bar{z})}} \left( (t-x)^{k-1} \sum_{m=0}^{[\frac{k-1}{2}]} 2^{-2m} \, C_{k-1}^{2m} \, C_{2m}^m \right.$$

$$+ \sum_{j=1}^{j} (t-x)^{k-1-2j} \sum_{p=0}^{j} C_{2j}^{2p} \left( (t-x)^2 + y^2 \right)^p y^{2j-2p} \sum_{m=j}^{[\frac{k-1}{2}]} 2^{1-2m} C_{k-1}^{2m} C_{2m}^{m-j} \right) dt$$

$$= \frac{(-1)^{k-1}}{2\pi i} \int_{\Gamma} \frac{F(t)}{\sqrt{(t-z)(t-\bar z)}} \left( (t-x)^{k-1} \sum_{m=0}^{[\frac{k-1}{2}]} 2^{-2m} C_{k-1}^{2m} C_{2m}^{m} \right.$$

$$+ \sum_{j=1}^{[\frac{k-1}{2}]} \sum_{m=j}^{[\frac{k-1}{2}]} 2^{1-2m} C_{k-1}^{2m} C_{2m}^{m-j} \sum_{p=0}^{j} C_{2j}^{2p} \sum_{q=0}^{p} C_p^q (t-x)^{k-1-2j+2q} y^{2j-2q} \right) dt$$

$$= \frac{(-1)^{k-1}}{2\pi i} \int_{\Gamma} \frac{F(t)}{\sqrt{(t-z)(t-\bar z)}} \left( (t-x)^{k-1} \sum_{m=0}^{[\frac{k-1}{2}]} 2^{-2m} C_{k-1}^{2m} \left( C_{2m}^{m} + 2 \sum_{j=0}^{m-1} C_{2m}^{j} \right) \right.$$

$$+ \sum_{n=1}^{[\frac{k-1}{2}]} y^{2n} (t-x)^{k-1-2n} \sum_{j=n}^{[\frac{k-1}{2}]} \sum_{m=j}^{[\frac{k-1}{2}]} 2^{1-2m} C_{k-1}^{2m} C_{2m}^{m-j} \sum_{p=j-n}^{j} C_{2j}^{2p} C_p^{j-n} \right) dt. \quad (16)$$

Taking account the equalities

$$\sum_{p=j-n}^{j} C_{2j}^{2p} C_p^{j-n} = \frac{2^{2n} j}{2n} C_{j+n-1}^{2n-1},$$

$$\sum_{j=n}^{[\frac{k-1}{2}]} \frac{j}{n} C_{j+n-1}^{2n-1} \sum_{m=j}^{[\frac{k-1}{2}]} 2^{2n-2m} C_{k-1}^{2m} C_{2m}^{m-j}$$

$$= \sum_{p=0}^{[\frac{k-1}{2}]} 2^{-2p} C_{k-1}^{2n+2p} \sum_{q=0}^{p} \frac{n+q}{n} C_{2n+q-1}^{q} C_{2n+2p}^{p-q},$$

$$\sum_{q=0}^{p} \frac{n+q}{n} C_{2n+q-1}^{q} C_{2n+2p}^{p-q} = 2^{2p} C_{n+p}^{p}$$

and the fact that the function $F$ have zero of multiplicity $k$ in infinity, we transform further the expression (16):

$$\psi_k(x, y) = \frac{(-1)^{k-1}}{2\pi i} \int_{\Gamma} \frac{F(t)}{\sqrt{(t-z)(t-\bar z)}} \sum_{n=0}^{[\frac{k-1}{2}]} y^{2n} (t-x)^{k-1-2n} \sum_{j=n}^{[\frac{k-1}{2}]} C_{k-1}^{2j} C_j^n \, dt$$

$$= \frac{(-1)^{k-1}}{2\pi i} \int_{\Gamma} \frac{F(t)}{\sqrt{(t-z)(t-\bar z)}} \sum_{j=0}^{[\frac{k-1}{2}]} C_{k-1}^{2j} (t-x)^{k-1-2j} \sum_{n=0}^{j} C_j^n y^{2n} (t-x)^{2j-2n} \, dt$$

$$= \frac{(-1)^{k-1}}{2\pi i} \int_\Gamma \frac{F(t)}{\sqrt{(t-z)(t-\bar{z})}} \sum_{j=0}^{[\frac{k-1}{2}]} C_{k-1}^{2j} (t-x)^{k-1-2j} \left((t-x)^2 + y^2\right)^j dt$$

$$= \frac{1}{2\pi i} \int_\Gamma \frac{F(t) \left(\sqrt{(t-z)(t-\bar{z})} - (t-x)\right)^{k-1}}{\sqrt{(t-z)(t-\bar{z})}} \, dt \, .$$

Now, the statement of theorem is proved in exactly the same way as Theorem 2.1.

## 3. Integral expressions for solutions of families of elliptic equations degenerating on an axis

Let's generalize integral expressions (5) and (10) for solutions of Eqs. (7) and (9), respectively, in the case where the domain $D$ is more general than one in Theorems 2.1, 2.2.

In what follows, $D$ is a bounded domain in $\mathbf{R}^2$ such that the domain $D_z$ is simply connected and symmetric with respect to the real axis.

For every $z \in D_z$ with $\operatorname{Im} z \neq 0$, we fix an arbitrary Jordan rectifiable curve $\Gamma_{z\bar{z}}$ in $D_z$ which is symmetric with respect to the real axis and connects the points $z$ and $\bar{z}$. In the case where $D_z$ is an unbounded domain with the bounded boundary $\partial D_z$, we assume that the curve $\Gamma_{z\bar{z}}$ crosses the real axis on the interval $(-\infty, \min_{t \in \partial D_z} \operatorname{Re} t)$.

For $z \in D_z$ with $\operatorname{Im} z \neq 0$, let $\sqrt{(t-z)(t-\bar{z})}$ be that continuous branch of the analytic function $G(t) = \sqrt{(t-z)(t-\bar{z})}$ outside of the cut along $\Gamma_{z\bar{z}}$ for which $G(t) > 0$ for all $t > \max_{\tau \in \Gamma_{z\bar{z}}} \operatorname{Re} \tau$.

Now, the following theorem is proved similarly to Theorem 2.1.

**Theorem 3.1.** *If $F : D_z \to \mathbf{C}$ is an analytic function in a simply connected domain $D_z$ symmetric with respect to the real axis, then the function*

$$\psi_k(x,y) := \begin{cases} \dfrac{1}{2\pi i} \int_\gamma \dfrac{F(t)\left(\sqrt{(t-z)(t-\bar{z})} - (t-x)\right)^{k-1}}{\sqrt{(t-z)(t-\bar{z})}} \, dt \, , & \text{when } y \neq 0 \, , \\[2ex] \lim_{\eta \to 0} \psi_k(x, \eta) \, , & \text{when } y = 0 \, , \end{cases}$$

*is a solution of Eq. (9) in the domain $D$ and the function*

$$U_k(x,y) := \frac{1}{2\pi i} \int_\gamma \frac{F(t)}{\sqrt{(t-z)(t-\bar{z})}} \left(\frac{\sqrt{(t-z)(t-\bar{z})} - (t-x)}{y}\right)^{k-1} dt \qquad (17)$$

*is a solution of Eq. (5) on the set $\{(x,y) \in D : y \neq 0\}$; here $\gamma$ is an arbitrary closed Jordan rectifiable curve in $D_z$ which embraces $\Gamma_{z\bar{z}}$.*

Note that in the case where the domain $D$ is unbounded with the bounded boundary $\partial D$, the limit

$$U_k(x,0) := \lim_{\eta \to 0} U_k(x,\eta), \tag{18}$$

exists for all $x < \min\{x \in \mathbf{R} : (x,0) \in \partial D\}$ but, generally speaking, does not exist for $x > \max\{x \in \mathbf{R} : (x,0) \in \partial D\}$. At the same time, if the function $F$ have zero of multiplicity at least $k$ in infinity, then the limit (18) exists for all $(x,0) \in D$.

## Acknowledgments

I am very grateful to Prof. Igor Mel'nichenko, which died a sudden death in 2004, for our long-term scientific collaboration and friendship. I acknowledge gratefully that Theorem 2.1 was proved by me jointly with Igor Mel'nichenko.

## References

1. M. A. Lavrentyev and B. V. Shabat, *Problems of hydrodynamics and theirs mathematical models*, Moskow: Nauka, 1977, 408 p. [in Russian]
2. I. P. Mel'nichenko, On expression of harmonic mappings by monogenic functions, *Ukr. Math. J.*, **27** (1975), no. 5, 606–613.
3. I. P. Mel'nichenko, Algebras of functionally-invariant solutions of the three-dimensional Laplace equation, *Ukr. Math. J.*, **55** (2003), no. 9, 1284–1290.
4. V. F. Kovalev and I. P. Mel'nichenko, Biharmonic functions on a biharmonic plane, *Dokl. Akad. Nauk Ukrain. SSR : ser. A*, (1981), no. 8, 26–28. [in Russian]
5. I. P. Mel'nichenko and S. A. Plaksa, Potential fields with axial symmetry and algebras of monogenic functions of vector variable, III, *Ukr. Math. J.*, **49** (1997), no. 2, 253–268.
6. S. Plaksa, Algebras of hypercomplex monogenic functions and axial-symmetrical potential fields, in: *Proceedings of the Second ISAAC Congress, Fukuoka, August 16–21, 1999*, Netherlands–USA: Kluwer Academic Publishers, **1** (2000), 613–622.
7. S. Plaksa, Boundary properties of axial-symmetrical potential and Stokes flow function, in: *Finite or Infinite Dimensional Complex Analysis: Proceedings of the Seventh International Colloquium. — Lecture Notes in Pure and Applied Mathematics*, New York–Basel: Marcel Dekker Inc., **214** (2000), 443–455.
8. S. Plaksa, Singular and Fredholm integral equations for Dirichlet boundary problems for axial-symmetric potential fields, in: *Factotization, Singular Operators and Related Problems: Proceedings of the Conference in Honour of Professor Georgii Litvinchuk, Funchal, January 28–February 1, 2002*, Netherlands–USA: Kluwer Academic Publishers, 2003, 219–235.
9. S. A. Plaksa, Dirichlet problem for the Stokes flow function in a simply connected domain of the meridian plane, *Ukr. Math. J.*, **55** (2003), no. 2, 197–231.
10. I. P. Mel'nichenko and S. A. Plaksa, Outer boundary problems for the Stokes flow function and steady streamline along axial-symmetric bodies, in: *Complex Analysis and Potential Theory: Proceedings of Ukrainian Mathematical Congress–2001*, Kiev: Institute of Mathematics of the National Academy of Sciences of Ukraine, 2003, 82–91.
11. I. P. Mel'nichenko, On a method of description of potential fields with axial symmetry, in: *Contemporary Questions of Real and Complex Analysis*, Kiev: Institute of Mathematics of Ukrainian Academy of Sciences, 1984, 98–102. [in Russian]
12. G. M. Fikhtengol'ts, *A course of differential and integral calculus*, Moskow: Nauka, 1966, **2**, 800 p. [in Russian]

13. R. P. Gilbert, *Function theoretic methods in partial differential equations*, New York–London: Academic Press, 1969, 311 p.
14. N. Rajabov, Integral equations and boundary problems for certain equations of elliptic type with a singular line, *Dokl. Akad. Nauk Tadzh. SSR : phys.-math. ser.*, (1972), no. 4, 3–12. [in Russian]

# QUATERNIONIC BACKGROUND OF THE PERIODICITY OF PETAL AND SEPAL STRUCTURES IN SOME FRACTALS OF THE FLOWER TYPE

JULIAN LAWRYNOWICZ*

*Institute of Physics, University of Łódź*
*Pomorska 149/153, PL-90-236 Łódź, Poland*
*Institute of Mathematics, Polish Academy of Sciences*
*Łódź Branch, Banacha 22, PL-90-238 Łódź, Poland*
*E-mail: jlawryno@uni.lodz.pl*


STEFANO MARCHIAFAVA

*Departimento di Matematica "Guido Castelnuovo"*
*Universita di Roma I "La Sapienza"*
*Piazzale Aldo Moro, 2, I-00-185 Roma, Italia*
*E-mail: marchiaf@mat.uniroma1.it*


MAŁGORZATA NOWAK-KĘPCZYK

*High School of Business*
*Kolejowa 22, PL-26-600 Radom, Poland*
*E-mail: gosianmk@poczta.onet.pl*

It is well known that starting with real structure, the Cayley–Dickson process gives complex, quaternionic, and octonionic (Cayley) structures related to the Adolf Hurwitz composition formula for dimensions $p = 2$, 4 and 8, respectively, but the procedure fails for $p = 16$ in the sense that the composition formula involves no more a triple of quadratic forms of the same dimension; the other two dimensions are $n = 2^7$. Instead, Lawrynowicz and Suzuki (2001) have considered graded fractal bundles of the flower type related to complex and Pauli structures and, in relation to the iteraton process $p \rightarrow p + 2 \rightarrow p + 4 \rightarrow \ldots$, they have constructed $2^4$-dimensional "bipetals" for $p = 9$ and $2^7$-dimensional "bisepals" for $p = 13$. The objects constructed appear to have an interesting property of periodicity related to the gradating function on the fractal diagonal interpreted as the "pistil" and a family of pairs of segments parallel to the diagonal and equidistant from it, interpreted as the "stamens". The present paper aims at an effective, explicit determination of the periods and expressing them in terms of complex and quaternionic structures, thus showing the quaternionic background of that periodicity. The proof of the Periodicity Theorem is given in the case where the index of the generator of the algebra in question does not exeed the order of the initial algebra. In contrast to earlier results, the fractal bundle flower structure, in particular petals, bipetals, the ovary, ovules, pistils, and stamens are not introduced *ab initio*; they are

*Research of the author is partially supported by the State Committee for Scientific Research (KBN) grant PB1 P03A 001 26 (Sections 1-2 of the paper), and partially by the grant of the University of Łódź no. 505/692 (Sections 3-4).

quoted *a posteriori*, when they are fully motivated.

**Key words:** Clifford algebra, quaternion, billinear form, quadratic form
**Mathematics Subject Classification:** 81R25, 32L25, 53A50, 15A66

## 1. Introductory

Given generators $A_1^1, A_2^1, \ldots, A_{2p-1}^1$ of a Clifford algebra $Cl_{2p-1}(\mathbb{C})$, $p = 2, 3, \ldots$, in particular the generators

$$\sigma_1 = \begin{pmatrix} 0 & 1 \\ 1 & 0 \end{pmatrix}, \quad \sigma_2 = \begin{pmatrix} 0 & -i \\ i & 0 \end{pmatrix}, \quad \sigma_3 = \begin{pmatrix} 1 & 0 \\ 0 & -1 \end{pmatrix}$$

of the Pauli algebra [19,20], consider the sequence

$$A_\alpha^{q+1} = \sigma_3 \otimes A_\alpha^q \equiv \begin{pmatrix} A_\alpha^q & 0 \\ 0 & -A_\alpha^q \end{pmatrix}, \quad \alpha = 1, 2, \ldots, 2p + 2q - 3;$$

$$A_{2p+2q-2}^{q+1} = \sigma_1 \otimes I_{p,q} \equiv \begin{pmatrix} 0 & I_{p,q} \\ I_{p,q} & 0 \end{pmatrix}, \quad A_{2p+2q-1}^{q+1} = -\sigma_2 \otimes I_{p,q} \equiv \begin{pmatrix} 0 & iI_{p,q} \\ -iI_{p,q} & 0 \end{pmatrix},$$

$$\tag{1}$$

of generators of Clifford algebras $Cl_{2p+2q-1}(\mathbb{C})$ $q = 1, 2, \ldots$, and the sequence of corresponding systems of closed squares $Q_q^\alpha$ of diameter 1, centered at the origin of $\mathbb{C}$, where $I_{p,q} = I_{2^{p+q-2}}$, the unit matrix of order $2^{p+q-2}$. It is convenient to start with $q$ always from 1, i.e., to shift $q$ for $\alpha \geq 2p$ correspondingly.

Within a closed square $Q_q^\alpha$ consider its diameter

$$L_\infty = \left[ \frac{1}{2\sqrt{2}}(-1+i); \frac{1}{2\sqrt{2}}(1-i) \right]$$

and two segments, symmetric and equidistant with respect to $L_\infty$:

$$L_h^- = \left[ \frac{1}{2\sqrt{2}}(-1+i-i\varepsilon_q^r); \frac{1}{2\sqrt{2}}(1-\varepsilon_q^r-i) \right],$$

$$L_h^+ = \left[ \frac{1}{2\sqrt{2}}(-1+\varepsilon_q^r+i); \frac{1}{2\sqrt{2}}(1-i+i\varepsilon_q^r) \right],$$

where

$$\varepsilon_q^r = 1/2^h, \quad h = p+q-1-r, \quad r = \begin{cases} 2 & \text{for } \alpha = 1, 2, \ldots, 2p-1; \\ [\frac{1}{2}\alpha] & \text{for } \alpha = 2p, 2p+1, \ldots \end{cases}$$

and [ ] denotes the function "entier". Clearly, dist $(L_h^\pm, L_\infty) = 1/2^{h+2}$.

Consider then the sets: $L_\infty^0$ of points

$$z = \frac{1}{2\sqrt{2}} \frac{m}{2^n}(1-i), \quad m = 0, \pm 1, \ldots, \pm(2^n - 1); \quad n = 0, 1, \ldots, \quad \text{of } L_\infty$$

and $L_h^0$ of points

$$z_-(h) = \frac{1}{2\sqrt{2}} \frac{m}{2^n}(1-i) - \frac{1}{2\sqrt{2}}\varepsilon_q^r \quad \text{and} \quad z_+(h) = \frac{1}{2\sqrt{2}} \frac{m}{2^n}(1-i) - \frac{i}{2\sqrt{2}}\varepsilon_q^r,$$

$$m = 0, \pm 1, \ldots, \pm(2^n - 1); \quad n = 0, 1, \ldots, \quad \text{of } L_h^-,$$

$$z_+^+(h) = \frac{1}{2\sqrt{2}} \frac{m}{2^n} (1 - i) + \frac{1}{2\sqrt{2}} \varepsilon_q^r \quad \text{and} \quad z_-^+(h) = \frac{1}{2\sqrt{2}} \frac{m}{2^n} (1 - i) + \frac{i}{2\sqrt{2}} \varepsilon_q^r,$$

$$m = 0, \pm 1, \ldots, \pm(2^n - 1); \quad n = 0, 1, \ldots, \quad \text{of } L_h^+.$$

Let

$$A_\alpha^q = (a_{\alpha j}^{qk}), \ A_\alpha = (a_{\alpha j}^k), \ j, k = 1, 2, \ldots, 2^{p+q-2}. \tag{2}$$

Let further

$$g_q^\alpha(a_{\alpha j}^{qk}; z) = a_{\alpha j}^{qk} \ \text{if} \ g_q^\alpha(z) = a_{\alpha j}^{qk}; \quad g_q^\alpha(a_{\alpha j}^{qk}; z) = 0 \ \text{if} \ g_q^\alpha(z) \neq a_{\alpha j}^{qk},$$

where $g_q^\alpha$ is the gradating function equal $a_{\alpha j}^{qk}$ on the closed square $Q_{qk}^{\alpha j}$ corresponding to the pair $(j, k)$; we suppose that the original square is divided into $4^{p+q-2}$ squares with sides parallel to the sides of $Q_q^\alpha$ for $\alpha \leq 2p - 1$, and into $4^{p+q-1}$ analogous squares for $\alpha \geq 2p$. We shall call the squares $Q_{qk}^{\alpha j}$ *basic squares* for $Q_q^\alpha$.

Given $z \in L_\infty^0$, consider the sequences

$$g_1^\alpha(z), g_2^\alpha(z), \ldots \quad \text{for } \alpha < 2p, \tag{3}$$

$$\hat{g}_1^\alpha(z_-^1), \hat{g}_2^\alpha(z_-^2), \ldots \quad \text{for } \alpha < 2p, \tag{4}$$

$$\hat{g}_1^\alpha(z_+^1), \hat{g}_2^\alpha(z_+^2), \ldots \quad \text{for } \alpha < 2p, \tag{5}$$

where

$$\hat{g}_q^\alpha(z_-^q) = (g_q^\alpha(z_-^-(h)), g_q^\alpha(z_-^+(h)), \quad \hat{g}_q^\alpha(z_+^q) = (g_q^\alpha(z_+^-(h)), g_q^\alpha(z_+^+(h)), \tag{6}$$

as well as

$$\hat{g}_1^{2r}(z_-^1), \hat{g}_2^{2r}(z_-^2), \ldots \quad \text{for } 2r = \alpha \geq 2p, \tag{7}$$

$$\hat{g}_1^{2r}(z_+^1), \hat{g}_2^{2r}(z_+^2), \ldots \quad \text{for } 2r = \alpha \geq 2p, \tag{8}$$

$$\hat{g}_1^{2r+1}(z_-^1), \hat{g}_2^{2r+1}(z_-^2), \ldots \quad \text{for } 2r + 1 = \alpha > 2p, \tag{9}$$

$$\hat{g}_1^{2r+1}(z_+^1), \hat{g}_2^{2r+1}(z_+^2), \ldots \quad \text{for } 2r + 1 = \alpha > 2p, \tag{10}$$

with the notation (6), and

$$\hat{g}_1^{2r}(z_1^-), \hat{g}_2^{2r}(z_2^-), \ldots \quad \text{for } 2r = \alpha \geq 2p, \tag{11}$$

$$\hat{g}_1^{2r}(z_1^+), \hat{g}_2^{2r}(z_2^+), \ldots \quad \text{for } 2r = \alpha \geq 2p, \tag{12}$$

$$\hat{g}_1^{2r+1}(z_1^-), \hat{g}_2^{2r+1}(z_2^-), \ldots \quad \text{for } 2r + 1 = \alpha > 2p, \tag{13}$$

$$\hat{g}_1^{2r+1}(z_1^+), \hat{g}_2^{2r+1}(z_1^+), \ldots \quad \text{for } 2r + 1 = \alpha > 2p, \tag{14}$$

where

$$\hat{g}_q^\alpha(z_q^-) = (g_q^\alpha(z_-^-(h)),\, g_q^\alpha(z_+^-(h)), \quad \hat{g}_q^\alpha(z_q^+) = (g_q^\alpha(z_-^+(h)),\, g_q^\alpha(z_+^+(h)). \tag{15}$$

Denote by $\mathbf{1}, \mathbf{i}, \mathbf{j}, \mathbf{k}$ the matrices representing the orthogonal unit vectors of the algebra $\mathbb{H}$ of (real) quaternions:

$$\mathbf{1}=\begin{pmatrix} 1 & 0 \\ 0 & 1 \end{pmatrix}, \quad \mathbf{i}=\begin{pmatrix} 0 & i \\ i & 0 \end{pmatrix}=i\sigma_1, \quad \mathbf{j}=\begin{pmatrix} 0 & 1 \\ -1 & 0 \end{pmatrix}=i\sigma_2, \quad \mathbf{k}=\begin{pmatrix} i & 0 \\ 0 & -i \end{pmatrix}=i\sigma_3. \tag{16}$$

Then (1) becomes

$$A_\alpha^{q+1} = \frac{1}{i}\mathbf{k}\otimes A_\alpha^q, \quad \alpha = 1,2,\ldots,2p+2q-3; \tag{17}$$

$$A_{2p+2q-2}^{q+1} = \frac{1}{i}\mathbf{i}\otimes\mathbf{1}^{\otimes(p+q-2)}, \quad A_{2p+2q-1}^{q+1} = -\frac{1}{i}\mathbf{j}\otimes\mathbf{1}^{\otimes(p+q-2)}.$$

## 2. Statement of the periodicity theorem

We have

PERIODICITY THEOREM (quaternionic formulation). (i) *If* $a_{\alpha\lambda}^\lambda \neq 0$ *for* $\lambda = 2^{p+q-2}$, *the sequences* (3) *are periodic of period* 2, *starting from some term. The periods are:*

$$\frac{1}{2}\eta(a_{\alpha\lambda}^\lambda - a_{\alpha 1}^1)\mathbf{1} + \frac{1}{2i}\eta(a_{\alpha\lambda}^\lambda + a_{\alpha 1}^1)\mathbf{k}, \quad \frac{1}{2}\eta(a_{\alpha\lambda}^\lambda + a_{\alpha 1}^1)\mathbf{1} - \frac{1}{2i}\eta(a_{\alpha\lambda}^\lambda - a_{\alpha 1}^1)\mathbf{k} \tag{18}$$

*where* $\eta = 1$ *or* $-1$.

(ii) *If*

$$a_{\alpha\lambda}^\lambda = 0 \quad and \quad a_{\alpha\lambda-1}^{\lambda-1} = a_{\alpha 2}^2 = a_{\alpha 1}^1 = 0, \quad where \quad \lambda = 2^{p+q-2}, \tag{19}$$

*the sequences* (3) *are constant-valued, starting from some term; it amounts at*

$$-\frac{1}{2}\eta a_{\alpha 1}^1\mathbf{1} + \frac{1}{2i}\eta a_{\alpha 1}^1\mathbf{k}, \quad where \quad \eta = 1 \quad or \quad -1. \tag{20}$$

(iii) *If* (19) *holds, the sequences* (4) *are periodic of period* 2, *starting from some term. The periods are:*

$$\frac{1}{2}\eta\left(a_{\alpha\lambda}^{\lambda-1} + a_{\alpha,\lambda-1}^\lambda\right)\mathbf{j} + \frac{1}{2i}\eta\left(a_{\alpha\lambda}^{\lambda-1} - a_{\alpha,\lambda-1}^\lambda\right)\mathbf{i}, \tag{21}$$

$$-\frac{1}{2}\eta\left(a_{\alpha\lambda}^{\lambda-1} + a_{\alpha,\lambda-1}^\lambda\right)\mathbf{j} - \frac{1}{2i}\eta\left(a_{\alpha\lambda}^{\lambda-1} - a_{\alpha,\lambda-1}^\lambda\right)\mathbf{i},$$

*where* $\eta = 1$ *or* $\eta = -1$.

(iv) *If* (19) *holds, the sequences* (5) *are constant-valued, starting from some term; it amounts at*

$$-\frac{1}{2}\eta\left(a_{\alpha 2}^1 + a_{\alpha 1}^2\right)\mathbf{j} - \frac{1}{2i}\eta\left(a_{\alpha 2}^1 - a_{\alpha 1}^2\right)\mathbf{i}, \quad where \quad \eta = 1 \quad or \quad -1. \tag{22}$$

(v) *The sequences (7) and (9) are periodic of period 2, starting from some term. The periods are:*

$$\left(\frac{1}{2}1+\frac{1}{2i}k,\ \frac{1}{2}1+\frac{1}{2i}k\right),\ \left(-\frac{1}{2}1-\frac{1}{2i}k,\ -\frac{1}{2}1-\frac{1}{2i}k\right) \qquad (23)$$

*or*

$$\left(-\frac{1}{2}1-\frac{1}{2i}k,\ -\frac{1}{2}1-\frac{1}{2i}k\right),\ \left(\frac{1}{2}1+\frac{1}{2i}k,\ \frac{1}{2}1+\frac{1}{2i}k\right) \qquad (24)$$

*in the case of (7), and*

$$\left(\frac{1}{2i}1-\frac{1}{2}k,\ -\frac{1}{2i}1+\frac{1}{2}k\right),\ \left(-\frac{1}{2i}1+\frac{1}{2}k,\ \frac{1}{2i}1-\frac{1}{2}k\right) \qquad (25)$$

*or*

$$\left(\frac{1}{2i}1+\frac{1}{2}k,\ -\frac{1}{2i}1-\frac{1}{2}k\right),\ \left(-\frac{1}{2i}1-\frac{1}{2}k,\ \frac{1}{2i}1+\frac{1}{2}k\right) \qquad (26)$$

*in the case of (9).*

(vi) *The sequences (8) and (10) are constant-valued, starting from some term; it amounts at*

$$\left(-\frac{1}{2}1+\frac{1}{2i}k,\ -\frac{1}{2}1+\frac{1}{2i}k\right)\ \text{or}\ \left(\frac{1}{2}1-\frac{1}{2i}k,\ \frac{1}{2}1-\frac{1}{2i}k\right) \qquad (27)$$

*in the case of (8), and*

$$\left(\frac{1}{2i}1-\frac{1}{2}k,\ -\frac{1}{2i}1-\frac{1}{2}k\right)\ \text{or}\ \left(-\frac{1}{2i}1+\frac{1}{2}k,\ -\frac{1}{2i}1-\frac{1}{2}k\right) \qquad (28)$$

*in the case of (10).*

(vii) *The sequences (13) and (14) are periodic of period 2, starting from some term. The periods are:*

$$\left(\frac{1}{2i}1-\frac{1}{2}k,\ -\frac{1}{2}1-\frac{1}{2i}k\right),\ \left(-\frac{1}{2i}1+\frac{1}{2}k,\ -\frac{1}{2}1-\frac{1}{2i}k\right) \qquad (29)$$

*or*

$$\left(-\frac{1}{2i}1+\frac{1}{2}k,\ \frac{1}{2}1+\frac{1}{2i}k\right),\ \left(\frac{1}{2i}1-\frac{1}{2}k,\ \frac{1}{2}1+\frac{1}{2i}k\right) \qquad (30)$$

*in the case of (13), and (30) or (29) in the case of (14) (given $z \in L_\infty^0$, the choices (29) and (30) are mutually correlated ).*

(viii) *The sequences (11) and (12) are periodic of period 2, starting from some term. The periods are:*

$$\left(\frac{1}{2}1+\frac{1}{2i}k,\ -\frac{1}{2}1+\frac{1}{2i}k\right),\ \left(-\frac{1}{2}1-\frac{1}{2i}k,\ -\frac{1}{2}1+\frac{1}{2i}k\right) \qquad (31)$$

*or*

$$\left(-\frac{1}{2}1-\frac{1}{2i}k,\ \frac{1}{2}1-\frac{1}{2i}k\right),\ \left(\frac{1}{2}1+\frac{1}{2i}k,\ \frac{1}{2}1-\frac{1}{2i}k\right). \qquad (32)$$

The Periodicity Theorem is a consequence of Theorems 1-6 [18]. We do not repeat the assertions of these theorems that precise the starting terms of the periodicity – here we have no further contribution in that field. We are dealing only with the assertions that can be reformulated to involve quaternions explicitly. More precisely, we restrict ourselves to proving the assertion (i). Assertions (ii)-(viii) will be proved in the forthcoming papers [8,9].

### 3. Periodicity in the case $\alpha \leq 2p - 1$, period 2

According to Theorem 1 [16], the sequences (3) are periodic of period 2, starting from some term, determined in that theorem, for $a_{\alpha\lambda}^{\lambda} \neq 0$, $\lambda = 2^{p+q-2}$. The periods are:

$$\begin{pmatrix} \eta a_{\alpha\lambda}^{\lambda} & 0 \\ 0 & -\eta a_{\alpha 1}^{1} \end{pmatrix}, \begin{pmatrix} -\eta a_{\alpha\lambda}^{\lambda} & 0 \\ 0 & -\eta a_{\alpha 1}^{1} \end{pmatrix}, \text{ where } \eta = 1 \text{ or } -1. \quad (33)$$

The proof given [16] relies upon three lemmas that analyze two iterations preceding the iteration where periodicity starts.

Alternatively, we may observe that, in order to calculate

$$g_q^{\alpha}(\frac{1}{2\sqrt{2}}\frac{m}{2^n}(1-i)), \quad m = 0, \pm 1, \ldots, \pm(2^n - 1); \quad n = 0, 1, \ldots,$$

we have to look for

$$g_{q-1}^{\alpha}(\frac{1}{2\sqrt{2}}\frac{m}{2^{n-1}}(1-i)), \quad g_{q-2}^{\alpha}(\frac{1}{2\sqrt{2}}\frac{m}{2^{n-2}}(1-i)), \ldots$$

To this end, in the case of $\Sigma_3 = (Q_q^3)$, $p = 2$ and $A_3^1 = \sigma_3$, we consider the table where rows represent the diameter $L_\infty$ in the $q$-th iteration step, and columns represent the configuration related to $s_m = m/2^n$ (in Fig. 1(a) we take $n = 11$); hereafter, for $z \in L_\infty$, where we are also using the notation $s = (1/\sqrt{2})((\text{re}z - \text{im}z) + \frac{1}{2})$. Later on $\Sigma_3$ and, generally, $\Sigma_\alpha$ will be rediscovered as graded Clifford-type fractals [2,6].

The configuration in question consists of one value of the generating function if $z$ belongs to the interior of the square $Q_{qk}^{\alpha j}$ corresponding to a pair $(j, k)$; of two values if $z$ is on the boundary but not in a vertex; finally, of four values if $z$ is in a vertex (of course, it may happen that some of these values coincide). In the case of $\sigma_3$ we have the following possibilities for the two values corresponding to the direction of $L_\infty$:

$$\boxed{\frac{\boxed{1}}{\boxed{1}}}, \boxed{\frac{\boxed{1}}{\boxed{-1}}}, \boxed{\frac{\boxed{-1}}{\boxed{-1}}}, \boxed{\frac{\boxed{-1}}{\boxed{1}}}.$$

As far as the direction perpendicular to $L_\infty$ at $z \in L_\infty$ is concerned, we have three possibilities:

$\boxed{1}\ \boxed{1}$ or $\boxed{-1}\ \boxed{-1}$, denoted in the both cases (in Fig. 1) as $\boxed{/}\ \boxed{/}$

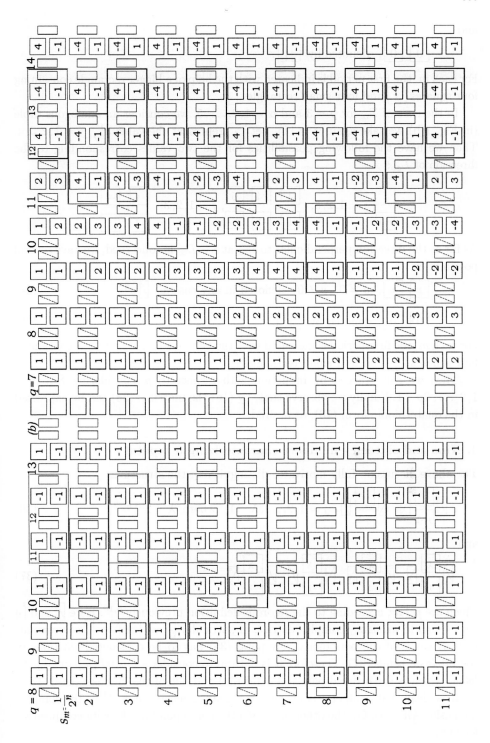

Fig. 1.   Checking the periodicity in construction of a graded Clifford-type fractal: (a) $\Sigma_3$ for $p = 2$, (b) $\Sigma_\alpha$ for $p = 3$, $\alpha = 1, 2, \ldots, 5$; $a^\lambda_{\alpha\lambda} \neq 0$.

and

$$\boxed{0}\ \boxed{0}, \quad \text{denoted by} \ \boxed{}\ \boxed{}.$$

For each $m$, the first period is indicated with the help of a bigger rectangle containing four squares $\boxed{x}$ and four small rectangles $\boxed{}$.

In the case of $\Sigma_\alpha$ for $p = 3$ and $\alpha = 1, 2, \ldots, 5$, we consider an analogous table (in Fig. 1(b) we take again $n = 11$). Let $\boxed{j}$ represent the value $a_{\alpha j}^j$, $j = 1, 2, 3, 4$. Then we have the following possibilities for the two values corresponding to the direction of $L_\infty$:

$$\boxed{\begin{array}{c} j \\ j+1 \end{array}}, \ j = 1, 2, 3; \quad \boxed{\begin{array}{c} 4 \\ 1 \end{array}}, \boxed{\begin{array}{c} 4 \\ -1 \end{array}}, \boxed{\begin{array}{c} -j \\ -j-1 \end{array}}, \ j = 1, 2, 3; \quad \boxed{\begin{array}{c} -4 \\ -1 \end{array}}, \boxed{\begin{array}{c} -4 \\ 1 \end{array}}.$$

As far as the direction perpendicular to $L_\infty$ at $z \in L_\infty$ is concerned, we have several possibilities:

$$\boxed{u}\ \boxed{v}, \ \ u \neq 0, \ v \neq 0, \ \text{denoted in any case as} \ \boxed{/}\ \boxed{/} \tag{34}$$

$$\boxed{0}\ \boxed{0}, \ \text{denoted by} \ \boxed{}\ \boxed{}. \tag{35}$$

For each $m$, the first period is indicated with the help of a bigger rectangle containing four squares $\boxed{x}$ and four small rectangles $\boxed{}$.

Then we can proceed by induction with respect to $p$, considering $\Sigma_\alpha$, $\alpha = 1, 2, \ldots, 2p - 1$. Finally we observe that, by (16) the matrices (33) can be expressed in the quaternionic form (18), as desired.

## 4. Nine and sixteen - numbers characterizing a bipetal

The above discussion fully motivates distinguishing the structures

$$\boxed{/}\ \boxed{\begin{array}{c} j \\ j+1 \end{array}}\ \boxed{\backslash}, \ \boxed{\backslash}\ \boxed{\begin{array}{c} -j \\ -j-1 \end{array}}\ \boxed{/}, \ j = 1, 2, \ldots, \lambda - 1, \ \text{where} \ \lambda = 2^{p+q-2}, \tag{36}$$

and

$$\boxed{/}\ \boxed{\begin{array}{c} \lambda \\ 1 \end{array}}\ \boxed{\backslash}, \ \boxed{/}\ \boxed{\begin{array}{c} \lambda \\ -1 \end{array}}\ \boxed{\backslash}, \ \boxed{\backslash}\ \boxed{\begin{array}{c} -\lambda \\ -1 \end{array}}\ \boxed{/}, \ \boxed{/}\ \boxed{\begin{array}{c} -\lambda \\ -1 \end{array}}\ \boxed{\backslash}, \tag{37}$$

which are *petals*, according to the rigorous definition previously [6,16]. Here $\boxed{/}$ and $\boxed{\backslash}$ correspond to arbitrary complex entries, in particular $\boxed{/} = \boxed{}$ or $\boxed{\backslash} = \boxed{}$ corresponds to 0. Because of the periodicity, any petal (36) has to be considered with one of the *intrinsic petals* (37). Therefore it is natural to consider a *bipetal (a pair of petal labia)* which is the ordered pair of a petal (36) and one of the intrinsic petals (37).

It is clear that a bipetal is an object of real dimension $n = 2^4 = 16$ (complex dimension $\frac{1}{2}n = 8$). On the other hand, via the corresponding Hurwitz composition formula [3]:

$$(x_1^2 + \cdots + x_9^2)(y_1^2 + \cdots + y_{16}^2) \equiv (x \circ_9 y)_1^2 + \cdots + (x \circ_9 y)_{16}^2$$

with $(x \circ_9 y)_j = \sum_{\alpha=1}^{9} \sum_{k=1}^{16} C_j^{\alpha k} x_\alpha y_k, \quad (x_1, \ldots, x_9) \in \mathbb{R}^9, \quad (y_1, \ldots, y_{16}) \in \mathbb{R}^{16},$

it is also characterized by the real dimension $2p - 1 = 9$.

The research including the rest of proof of the Periodicity Theorem, will be continued in the forthcoming papers [8,9]. It is worthwhile to stress the relationship of fractal bundles of algebraic structure with studies of the Hurwitz problem [3], especially on its geometrical aspects introduced and studied in other papers [6,7,10,15,17].

## References

1. J. Cuntz, *Simple $C^*$-algebras generated by isometries*, Comm. Math. Phys. **57** (1977), 173–185.

2. K. J. Falconer, *Fractal Geometry. Mathematical Foundations and Applications*, Wiley, Chichester 1990.

3. A. Hurwitz, *Über die Komposition der quadratischen Formen*, Math. Ann. **88** (1923), 1–25; reprinted in: A. Hurwitz, *Mathematische Werke II*, Birkhäuser Verlag, Basel 1933, 641–666.

4. S. Kakutani, *On equivalence of infinite product measures*, Ann. of Math. **49** (1948), 214–224.

5. J. Kigami, *Analysis on fractals* (Cambridge Tracts in Mathematics 143), Cambridge Univ. Press, Cambridge 2001.

6. Ralitza Kovacheva, J. Ławrynowicz, and Małgorzata Nowak-Kępczyk, *Critical dimension 13 in approximation related to fractals of algebraic structure*, Bull. Soc. Sci. Lettres Łódź **52** Sér. Rech. Déform. **37** (2002) 77–102.

7. J. Ławrynowicz, *Type-changing transformations of pseudo-euclidean Hurwitz pairs, Clifford analysis, and particle lifetimes*, in: Clifford Algebras and Their Applications in Mathematical Physics. Ed. by V. Dietrich, K. Habetha, and G. Jank, Kluwer Academic, Dordrecht 1998, pp. 217–226.

8. —, S. Marchiafava, and Małgorzata Nowak-Kępczyk, *Periodicity theorem for structure fractals in quaternionic formulation*, J. of Geom. Meth. in Modern Phys. **3** (2006), 1167–1197.

9. —, —, —, *Applied periodicity theorem for structure fractals in quaternionic formulation*, in: Applied Complex Approximation and Quaternionic Structures. Ed. by R. Kovacheva, J. Ławrynowicz, and S. Marchiafava, to appear.

10. —, Małgorzata Nowak-Kępczyk, and Osamu Suzuki, *A duality theorem for inoculated graded fractal bundles vs. Cuntz algebras and their central extensions*, in: $C^*$-Algebras in Geometry and Physics, Ed. by B. Bojarski, Birkhäuser Verlag, Basel, to appear.

11. — and J. Rembieliński, *Pseudo-euclidean Hurwitz pairs and generalized Fueter equations*, (a) Inst. of Math. Polish Acad. Sci. Preprint no. 355 (1985), ii+10 pp., (b) in: Clifford Algebras and Their Applications in Mathematical Physics. Ed. by J. S. R. Chisholm and A. K. Common (NATO-ASI Series C: Mathematical and Physical Sciences 183), Reidel, Dordrecht 1986, pp. 39–48.

12. —, —, *Pseudo-euclidean Hurwitz pairs and the Kałuża-Klein theories*, (a) Inst. of Phys. Univ. of Łódź Preprint no. 86-8 (1986), 28 pp., (b) J. Phys. A: Math. Gen. **20** (1987), 5831–5848.

13. —, —, *On the composition of nondegenerate quadratic forms with an arbitrary index*, (a) Inst. of Math. Polish Acad. of Sci. Preprint no. 369 (1986), ii+29 pp., (b) Ann. Fac. Sci. Toulouse Math. (5) **10** (1989), 141-168 [due to a printing error in vol. **10** the whole article was reprinted in vol. **11** (1990), no. 1, of the same journal, pp. 141–168].

996

14. —, and O. Suzuki, *An introduction to pseudotwistors: Spinor solutions vs. harmonic forms and cohomology groups*, Progress in Physics **18** (2000), 393–423.

15. —, —, *An introduction to pseudotwistors: Basic constructions*, in: Quaternionic Structures in Mathematics and Physics. Ed. by S. Marchiafava, P. Piccinni and M. Pontecorvo, World Scientific, Singapore 2001, pp. 241–252.

16. —, —, *Periodicity theorems for graded fractal bundles related to Clifford structures*, Internat. J. of Pure and Appl. Math. **24** (2005), 181–209.

17. —, —, and F. L. Castillo Alvarado, *Basic properties and applications of graded fractal bundles related to Clifford structures*, in: Clifford Algebras and Their Applications in Mathematical Physics, Ed. by P. Anglès, Birkhäuser Verlag, Basel, to appear.

18. P. Lounesto, *Clifford Algebras and Spinors*, (London Math. Soc. Lecture Notes Series 239), Cambridge Univ. Press, 1997; 2nd ed. (vol. 286), ibid. 2001.

19. W. Pauli, *Zur Quantenmechanik des magnetischen Elektrons*, Z. Phys. **42** (1927), 601–623.

20. —, *Contributions mathématiques à la théorie des matrices de Dirac*, Ann. Inst. H. Poincaré **6** (1936), 109–136.

## III.2 Dirac operators in analysis and related topics

Organizers: John Ryan, Irene Sabadini

There were approximately twenty speakers in this section representing much of the cutting edge of research in modern Clifford analysis. All contributions to this section were refereed. Each contribution represents a significant and original paper in the growing and youthful field of Clifford analysis. In S. L. Eriksson's contribution we see a complete introduction to Cauchy integral formulas for $k$-hypermonogenic functions. These functions are solutions to the Dirac equation on upper half space endowed with a metric that is a deformation of the hyperbolic metric. These results should prove to be extremely important in the future development of Clifford analysis and spin geometry in hyperbolic spaces. Further in the contribution of P. Cerejeiras and F. Sommen we see the use of Witt bases in the Clifford algebra setting applied to the factorization of inhomogeneous operators like the heat operator. This should have interesting applications to symplectic geometry. In the contribution of F. Colombo, A. Damiano, I. Sabadini and D.C. Struppa we see further developments in the applications of sheaf theory and computational methods to the theory of functions of several quaternionic and vector variables. Again this is an area that should see significant applications in the study of generalized Riemann surfaces in the context of Dirac operators and spin geometry. Further in A. Perotti's paper there are interesting links made between certain classes of quaternionic regular functions and holomorphic functions. Martin's paper shows that work on Hartogs-Rosenthal Theorems developed earlier in the context of complex and Clifford analysis carry over to first order differential operators with coefficients in a Banach algebra. This has interesting applications to spin geometry and non-commutative differential (spin) geometry. Y. Krasnov's paper deals with the very interesting topic of function theory over unital algebras. In particular, the methods developed here are successfully applied to constructing solutions to the classical Cauchy problem. In Z. Zhang's paper we see integral representations in Clifford analysis with applications to boundary value problems, while in M.E. Luna-Elizarrarás and M. Shapiro's contribution we see a reawakening of an interest in quaternionic linear structures, a field visited many years ago by Teichmüller. E. Lehman's paper studies some properties of particular subsets of the hypermonogenic functions introduced by H. Leutwiler and of the holomorphic Cliffordian functions introduced by G. Laville and I. Ramadanoff. The contribution written by J. Lawrynowicz, K. Nôno and O. Suzuki introduces a fractal method for the renormalization theory of an infinite dimensional Clifford algebra, thus presenting an interesting connection between fractals and Clifford analysis.

Overall these papers together represent a very good insight to many future interesting and worthwhile directions of research in Clifford analysis and its applications. The papers are written with a view of communicating basic ideas in this area of research to non-experts with the intention of inviting wider interest in this young and growing area of research.

# ON SOME RELATIONS BETWEEN REAL, COMPLEX AND QUATERNIONIC LINEAR SPACES

M.E. LUNA-ELIZARRARÁS* and M. SHAPIRO†

*Departamento de Matemáticas E.S.F.M. del I.P.N.*
*07338 México D.F., México*
*E-mail: eluna@esfm.ipn.mx, shapiro@esfm.ipn.mx*

Continuing our previous research we consider here the internal quaternionization of real and complex linear spaces; categories of quaternionic linear spaces, and decompositions of two-sided-quaternionic linear spaces, generated by the analogs of the conjugations of the set of quaternions.

Key words: Quaternionic linear spaces
Mathematics Subject Classification: 46-99, 46A99

## 1. Introduction

**1.1.** This work can be seen as a continuation of our articles [4], [5] and [1] where we studied what happened with the norms of linear operators acting on $\mathbb{R}$-linear spaces and with the norms of their quaternionic extensions acting already on the quaternionizations of the original real spaces. In particular there were considered linear operators acting on some classic function spaces. The term quaternionization here refers to the so called external quaternionization, that is, the original real space is embedded into a wider quaternionic linear space. The latter always exists and in this paper we are concerned more with the existence of the internal quaternionization meaning the possibility to introduce a quaternionic multiplication on the same set. This is considered with enough detail in Section 2.

In Section 3 we treat the categories of real and quaternionic linear spaces where we introduce and use the notion of a quaternionic associated space.

In section 4 we are establishing, for a two-sided-quaternionic linear space, the analogs of the different conjugations on the skew-field of quaternions leading to the corresponding representation of the elements of those quaternionic spaces.

**1.2.** To fix the notation, the quaternionic imaginary units in the paper are $e_1 = i$, $e_2 = j$, $e_3 = k$, which satisfy the known properties: $e_1 e_2 = -e_2 e_1 = e_3$; $e_2 e_3 = $

---

*On leave from Universidad Autónoma Metropolitana-Azcapotzalco. Research partially supported by COFAA-IPN.
†Research partially supported by CONACYT projects as well as by Instituto Politécnico Nacional in the framework of COFAA and CGPI programs.

$-\mathbf{e}_3\mathbf{e}_2 = \mathbf{e}_1$; $\mathbf{e}_3\mathbf{e}_1 = -\mathbf{e}_1\mathbf{e}_3 = \mathbf{e}_2$; $\mathbf{e}_1^2 = \mathbf{e}_2^2 = \mathbf{e}_3^3 = -1$. The real unit 1 is written, sometimes, as $\mathbf{e}_0$, so that given $\alpha \in \mathbb{H}$ we write $\alpha = \displaystyle\sum_{\ell=0}^{3} \alpha_\ell \mathbf{e}_\ell$ with $\{\alpha_0, \alpha_1, \alpha_2, \alpha_3\} \subset \mathbb{R}$ or, when it would be convenient to consider $\mathbb{H}$ as a $\mathbb{C}$-linear space, $\alpha = z_0 + z_1\,\mathbf{e}_2$, with $\{z_0, z_1\} \subset \mathbb{C}$. The quaternionic conjugate to $\alpha$ is $\overline{\alpha} := \alpha_0 - \displaystyle\sum_{\ell=1}^{3} \alpha_\ell \mathbf{e}_\ell$.

## 2. Internal quaternionization of real and complex linear spaces

**2.1.** Let $\mathbb{K}$ be the field $\mathbb{R}$ or $\mathbb{C}$. Given a $\mathbb{K}$-linear space E, then its elements "understand" what does it mean to be multiplied by elements belonging to $\mathbb{K}$, since there is defined a multiplication by scalar

$$\cdot : (r, x) \in \mathbb{K} \times E \mapsto r \cdot x \in E.$$

We are interested in finding out whether E can be converted into an $\mathbb{H}$-linear space for which E is a $\mathbb{K}$-subspace, i.e., whether there exists a function

$$* : (\alpha, x) \in \mathbb{H} \times E \mapsto \alpha * x \in E,$$

which extends the multiplication on E making it an $\mathbb{H}$-linear space denoted by $E_\mathbb{H}$.

**2.2. Definition.** *Whence the above is valid, it is said that E admits to be internally quaternionized and $E_\mathbb{H}$ is called the internal quaternionization of E.*

The next statement refines a result in [6].

**2.3. Proposition.** *If E is an $\mathbb{R}$-linear space, then it admits to be internally quaternionized if and only if there exist automorphisms $u$, $v$ on E such that*

$$u^2 = v^2 = -Id_E \quad \text{and} \quad uv = -vu, \tag{1}$$

*with $Id_E$ the identity automorphism on E.*

Indeed, for E that admits to be internally quaternionized, the automorphisms $u$ and $v$ are defined by:

$$u(x) := \mathbf{e}_1 x; \quad v(x) := \mathbf{e}_2 x.$$

Reciprocally, if there exist automorphisms $u$, $v$ with (1) then set

$$\alpha * x := \left( \sum_{\ell=0}^{3} \alpha_\ell\,\mathbf{e}_\ell \right) * x := \alpha_0\,x + \alpha_1\,u(x) + \alpha_2\,v(x) + \alpha_3\,(u \circ v)(x). \tag{2}$$

In what follows, the product by quaternionic scalars will be written as $\alpha\,x$ instead of $\alpha * x$. As we will need the concept of an anti-linear map, we recall its definition.

## 2.4. Definition.

(1) *Given $\mathbb{C}$-linear spaces $V$ and $W$, an additive map $T : V \to W$ is called anti-linear if $T(\lambda\,x) = \overline{\lambda}\,T(x)$ for all $\lambda \in \mathbb{C}$ and for all $x \in V$.*

(2) *Let $E$ be a left-$\mathbb{H}$-linear space and let $F$ be a right-$\mathbb{H}$-linear space.*

- *Given an additive map $T : E \to F$, it is called anti-linear if for any $\lambda \in \mathbb{H}$ and $x \in E$, $T(\lambda\,x) = T(x)\,\overline{\lambda}$ .*
- *Given an additive map $T : F \to E$, it is called anti-linear if for any $\lambda \in \mathbb{H}$ and $y \in F$, $T(y\,\lambda) = \overline{\lambda}\,T(y)$.*

(3) *If $E$ and $F$ are two-sided-$\mathbb{H}$-linear spaces, an additive map $T : E \to F$ is called two-sided-anti-linear if $T(\lambda\,x\,\mu) = \overline{\mu}\,T(x)\,\overline{\lambda}$ for any $x \in E$ and any $\lambda, \mu \in \mathbb{H}$.*

**2.5. Corollary.** *If $V$ is a $\mathbb{C}$-linear space, then it admits to be internally quaternionized if and only if there exists an anti-linear automorphism $v$ such that $v^2 = -Id_V$.*

Notice that if such a $v$ exists then for a quaternion $\alpha + \beta\,\mathbf{j}$ we set:

$$(\alpha + \beta\,\mathbf{j}) \cdot x := \alpha\,x + \beta\,v(x).$$

**2.6. Example.** *The spaces $\mathbb{C}^{2n}$ are internally quaternionized.*

**Proof.** Write any element of $\mathbb{C}^{2n}$ as $z = (z_1, w_1, z_2, w_2, \ldots, z_n, w_n)$ . Define $v : \mathbb{C}^{2n} \to \mathbb{C}^{2n}$ by

$$v((z_1, w_1, z_2, w_2, \ldots, z_n, w_n)) := (-\overline{w}_1, \overline{z}_1, -\overline{w}_2, \overline{z}_2, \ldots, -\overline{w}_n, \overline{z}_n) .$$

Clearly $v$ satisfies the conditions of Corollary 2.5. □

**2.7.** The following notation and definitions serve for another characterization of $\mathbb{C}$-linear spaces that admit to be internally quaternionized (compare with Chapter I, Subsection 4.1 in [2]).

Given a $\mathbb{C}$-linear space $V$, its associated space $V^\sharp$ is defined as follows: as a set, it coincides with $V$: $V^\sharp = V$ , but the multiplication by complex scalars is defined, for given $\lambda \in \mathbb{C}$ and $v^\sharp \in V^\sharp$, by $\lambda \bullet v^\sharp \in V^\sharp := \overline{\lambda} \cdot v \in V$.

Denote by $\rho_V : V \longrightarrow V^\sharp$ the identity map, hence $\rho_V(v) := v^\sharp$ and $\rho_V(\lambda \cdot v) = \overline{\lambda} \bullet v^\sharp = \overline{\lambda} \bullet \rho_V(v)$.

Observe that $(V^\sharp)^\sharp = V$, that $\rho_V$ is an anti-linear map and that $\rho_V^{-1} = \rho_{V^\sharp}$.

Given a map $A : V \longrightarrow W$, with $V$ and $W$ being $\mathbb{C}$-linear spaces, it induces the associated map

$$A^\sharp : V^\sharp \longrightarrow W^\sharp$$

defined by

$$A^\sharp(v^\sharp) := (A(v))^\sharp.$$

Hence $A^\sharp = \rho_W \circ A \circ \rho_V^{-1} = \rho_W \circ A \circ \rho_{V^\sharp}$. This means, in particular, that, if $A$ is a $\mathbb{C}$-linear (anti-linear) map, then $A^\sharp$ is a $\mathbb{C}$-linear (anti-linear) map also.

**2.8. Definition.** *A* $\mathbb{C}$*-linear map* $c : V \longrightarrow V^\sharp$ *such that*

$$c^\sharp \circ c = -Id_V,$$

*is called an inductor of a quaternionic structure on* $V$.

**2.9.** Since $c^\sharp = \rho_{V^\sharp} \circ c \circ \rho_V^{-1} = \rho_{V^\sharp} \circ c \circ \rho_{V^\sharp}$ we get for an inductor of a quaternionic structure $c$:

$$\rho_{V^\sharp} \circ c \circ \rho_{V^\sharp} \circ c = -Id_V.$$

Defining $\widetilde{c} := \rho_{V^\sharp} \circ c$, we see that $\widetilde{c}$ is an anti-linear automorphism on $V$ such that $(\widetilde{c})^2 = -Id_V$. Hence we arrived at the conditions of Corollary 2.5.

Clearly this reasoning is invertible, that is, if a $\mathbb{C}$-linear space $V$ admits an internal quaternionization and if $J(v) := \mathbf{j}\, v$ then the map $c := \rho_V \circ J$ is an inductor of a quaternionic structure on $V$. Thus we have proved

**2.10. Proposition.** *A* $\mathbb{C}$*-linear space* $V$ *admits an internal quaternionization if and only if it possesses an inductor of a quaternionic structure.*

**2.11. Theorem.** *A real linear space* $E$ *admits to be internally quaternionized if and only if it admits an internal complexification* $E_\mathbb{C}$ *and the latter possesses an inductor of a quaternionic structure.*

**2.12. Example.** Let us show that among all $\mathbb{R}^m$ only those with $m = 4n$ admit to be internally quaternionized. Indeed, if $m = 4n$ then the elements of $\mathbb{R}^{4n}$ are

$$(x_1, y_1, t_1, s_1, x_2, y_2, t_2, s_2, \ldots, x_n, y_n, t_n, s_n);$$

define the automorphisms $u$, $v$ on $\mathbb{R}^{4n}$ by the formulas:

$$u((x_1, y_1, t_1, s_1, \ldots, x_n, y_n, t_n, s_n)) := (-y_1, x_1, -s_1, t_1, \ldots, -y_n, x_n, -s_n, t_n);$$

$$v((x_1, y_1, t_1, s_1, \ldots, x_n, y_n, t_n, s_n)) := (-t_1, s_1, x_1, -y_1, \ldots, -t_n, s_n, x_n, -y_n).$$

It is clear that $u^2 = v^2 = -I_{\mathbb{R}^{4n}}$ and $uv = -vu$. Proposition 2.3. completes this part of the proof.

If $m = 2n + 1$ then it is known that $\mathbb{R}^{2n+1}$ is not even complexifiable. Hence, both $\mathbb{R}^{4n+1}$ and $\mathbb{R}^{4n+3}$ are not internally quaternionizable.

Finally, consider the spaces $\mathbb{R}^{4n+2}$. It is known that $\mathbb{R}^{4n+2} \cong \mathbb{C}^{2n+1}$. Since $\mathbb{C}^{2n+1} \cong \mathbb{C}^{2n} \oplus \mathbb{C}$, by Example 2.6., to prove that $\mathbb{C}^{2n+1}$ is internally quaternionized is equivalent to prove that $\mathbb{C}$ does. So assume that there exists an anti-linear map $\widetilde{c} : \mathbb{C} \longrightarrow \mathbb{C}$ such that $\widetilde{c}^2 = -Id_\mathbb{C}$. Denote by $Z_\mathbb{C}$ the complex conjugation and define $\varphi := \widetilde{c} \circ Z_\mathbb{C}$, which is a $\mathbb{C}$-linear operator, hence there exists $a \in \mathbb{C}$ such that $\varphi(z) = a\, z$ for all $z \in \mathbb{C}$. We have $\widetilde{c} = \varphi \circ Z_\mathbb{C}$ and $\widetilde{c}^2(z) = -z$ for all $z \in \mathbb{C}$. On the other hand $\widetilde{c}^2(z) = (\varphi \circ Z_\mathbb{C})^2(z) = \varphi \circ Z_\mathbb{C}(a\, \overline{z}) = \varphi(\overline{a}\, z) = a\, \overline{a}\, z = |a|^2\, z$, which implies $|a|^2 = -1$.

$\square$

## 3. Categories of quaternionic linear spaces

**3.1.** Given a left-$\mathbb{H}$-linear space E, restricting the multiplication by scalars to $\mathbb{R}$, we obtain that E becomes a real linear space denoted by $\mathrm{E}^{(r)}$. This space will be called the *dequaternionization* of the left-quaternionic linear space E. Similarly if $T : \mathrm{E}_1 \longrightarrow \mathrm{E}_2$ is an $\mathbb{H}$-linear operator, its restriction $T^{(r)} : \mathrm{E}_1^{(r)} \longrightarrow \mathrm{E}_2^{(r)}$ is a real linear operator, called the *dequaternionization of the operator* $T$.

Denote by $\mathfrak{C}(\mathbb{H})$ and $\mathfrak{C}(\mathbb{R})$ the categories whose objects are the left-$\mathbb{H}$-linear spaces and the real linear spaces respectively, and the morphisms between these objects are, respectively, $\mathbb{H}$-linear and $\mathbb{R}$-linear maps. Recall that a covariant functor $F$ from a category $\mathfrak{C}$ to a category $\mathfrak{D}$ is a couple $(F_1, F_2)$, such that $F_1$ assigns to each object $C \in \mathfrak{C}$ an object $F_1(C)$ in $\mathfrak{D}$, and $F_2$ assigns to each morphism $\varphi \in Mor(A, B)$, with $A$ and $B$ objects in $\mathfrak{C}$, a morphism $F_2(\varphi)$ in $Mor(F_1(A), F_1(B))$; besides $F_2(Id_A) = Id_{F_1(A)}$ and $F_2(\varphi \circ \psi) = F_2(\varphi) \circ F_2(\psi)$. Hence, we have the next

**3.2. Proposition.** *The dequaternionizations of left-$\mathbb{H}$-linear spaces and of $\mathbb{H}$-linear operators define a covariant functor from the category of left-$\mathbb{H}$-linear spaces $\mathfrak{C}(\mathbb{H})$ to the category of real-linear spaces $\mathfrak{C}(\mathbb{R})$.*

The proof consists of a rigorous usage of the corresponding definitions.

Any real linear space F (including those that do admit to be internally quaternionized) can be injected into the left-$\mathbb{H}$-linear space $^{\mathbb{H}}\mathrm{F} := \mathbb{H} \otimes_{\mathbb{R}} \mathrm{F}$, which is called the left-$\mathbb{H}$ (external) extension of F. It is known that any $x \in {}^{\mathbb{H}}\mathrm{F}$ is written in a unique way as $x = \sum_{\ell=0}^{3} \mathbf{e}_\ell \, x_\ell$. Let E and F be real linear spaces and let $^{\mathbb{H}}\mathrm{E}$ and $^{\mathbb{H}}\mathrm{F}$ be their respective $\mathbb{H}$-extensions. Given the real linear map $T : \mathrm{E} \longrightarrow \mathrm{F}$, the $\mathbb{H}$-linear map $T^{\mathbb{H}} : {}^{\mathbb{H}}\mathrm{E} \longrightarrow {}^{\mathbb{H}}\mathrm{F}$ defined by $T^{\mathbb{H}}(x) = T^{\mathbb{H}} \left( \sum_{\ell=0}^{3} \mathbf{e}_\ell \, x_\ell \right) := \sum_{\ell=0}^{3} \mathbf{e}_\ell \, T(x_\ell)$ is called the $\mathbb{H}$-extension of $T$. For more details and properties about these extensions see for instance [4], [5] and [1]. The above relations between real and quaternionic linear spaces brings us to the next

**3.3. Proposition.** *The $\mathbb{H}$-extensions of $\mathbb{R}$-linear spaces $\mathrm{F} \mapsto \mathbb{H} \otimes_{\mathbb{R}} \mathrm{F}$ and the $\mathbb{H}$-extension of $\mathbb{R}$-linear operators $T \mapsto T^{\mathbb{H}}$, give rise to a covariant functor between the category $\mathfrak{C}(\mathbb{R})$ of real linear spaces and the category $\mathfrak{C}(\mathbb{H})$ of left-$\mathbb{H}$-linear spaces.*

**3.4.** Recall that the collection of covariant functors from a category $\mathfrak{C}$ to another category $\mathfrak{D}$ forms itself the category $\mathrm{Cov}(\mathfrak{C}, \mathfrak{D})$. Two categories $\mathfrak{C}$ and $\mathfrak{D}$ are said to be equivalent if there exist covariant functors $F : \mathfrak{C} \longrightarrow \mathfrak{D}$ and $G : \mathfrak{D} \longrightarrow \mathfrak{C}$ such that the functors $F \circ G$ and $G \circ F$ are isomorphic to the identity functors in the respective categories $\mathrm{Cov}(\mathfrak{D}, \mathfrak{D})$ and $\mathrm{Cov}(\mathfrak{C}, \mathfrak{C})$. The statement of the complex linear spaces antecedent of the next theorem can be found in Chapter II, problem 66 of [3].

**3.5. Proposition.** *The categories* $\mathfrak{C}(\mathbb{R})$ *and* $\mathfrak{C}(\mathbb{H})$ *are not equivalent.*

**Proof.** Assume that there exist covariant functors $F = (F_1, F_2)$ and $G = (G_1, G_2)$ which realize the equivalence between the given categories. Considering in particular a real linear space $X$ of finite dimension, say $dim_{\mathbb{R}} X = n$. In this case the set of morphisms on $X$, $Mor(X, X)$ is an $\mathbb{R}$-linear space isomorphic to the space of real matrices $n \times n$, $\mathcal{M}_{n \times n}(\mathbb{R})$. Similarly $Mor(F_1(X), F_1(X)) \cong_{\mathbb{C}} \mathcal{M}_{n \times n}(\mathbb{C})$. But both $\mathcal{M}_{n \times n}(\mathbb{R})$ and $\mathcal{M}_{n \times n}(\mathbb{C})$ are semigroups also, and the above functor would give rise to an isomorphism between them, which is impossible. $\square$

**3.6.** Observe that we have considered the category $\mathfrak{C}(\mathbb{H})$ of left-$\mathbb{H}$-linear spaces only. Of course the category $\mathfrak{C}_r(\mathbb{H})$ of right-$\mathbb{H}$-linear spaces can be considered as well. In both categories the morphisms between objects are $\mathbb{H}$-linear operators.

Note here that it is impossible to define $\mathbb{H}$-linear maps between a left-$\mathbb{H}$-linear space and a right-$\mathbb{H}$-linear space. Indeed, if X is a left-$\mathbb{H}$-linear space and Y is a right-$\mathbb{H}$-linear space, suppose that there exists an $\mathbb{H}$-linear map $T : X \longrightarrow Y$. Then, for $\lambda, \mu \in \mathbb{H}$ and $x \in X$, we would have

$$T((\lambda \mu) \cdot x) = T(x) \cdot (\lambda \mu),$$

but on the other hand:

$$T((\lambda \mu) \cdot x) = T(\mu \cdot x) \cdot \lambda = (T(x) \cdot \mu) \cdot \lambda = T(x) \cdot (\mu \lambda),$$

and in general, the right-hand sides of both equations are not the same.

**3.7.** Now we define the quaternionic analogs of the concepts of the associated spaces and operators. Given a left-$\mathbb{H}$-linear space E, we define the right-$\mathbb{H}$-linear space $E^\star$ which, as a set, coincides with E, and where the multiplication by quaternionic scalars is defined as follows: for $\lambda \in \mathbb{H}$ and $x^\star \in E^\star$, $x^\star \bullet \lambda \in E^\star := \overline{\lambda} \cdot x \in E$. Denote by $\rho_E : E \longrightarrow E^\star$ the identity map. Hence it satisfies

$$\rho_E(\overline{\lambda} \cdot x) = x^\star \bullet \lambda = \rho_E(x) \bullet \lambda,$$

that is, it is an $\mathbb{H}$-anti-linear map. The right-$\mathbb{H}$-linear space $E^\star$ will be called the $\mathbb{H}$-associated space of E. In the same fashion all works for a right-$\mathbb{H}$-linear space E. Observe that $(E^\star)^\star = E$.

Given a map $A : E \longrightarrow F$, with E and F being left- or right-$\mathbb{H}$-linear spaces, we define the $\mathbb{H}$-associated operator

$$A^\star : E^\star \longrightarrow F^\star$$

by

$$A^\star(v^\star) := (A(v))^\star.$$

Hence:

$$A^\star = \rho_F \circ A \circ \rho_E^{-1} = \rho_F \circ A \circ \rho_{E^\star} \, ,$$

and, whence $A$ is an $\mathbb{H}$-linear (anti-linear) map, then $A^\star$ is an $\mathbb{H}$-linear (anti-linear) map also. As a consequence of these definitions one gets:

**3.8. Proposition.** *The relation between left and right-$\mathbb{H}$-linear spaces given by $E \mapsto E^\star$ and the relation between linear operators acting on left-$\mathbb{H}$-linear spaces and linear operators acting on right-$\mathbb{H}$-linear spaces given by $A \mapsto A^\star$, define a covariant functor between the categories $\mathfrak{C}(\mathbb{H})$ of left-$\mathbb{H}$-linear spaces and $\mathfrak{C}_r(\mathbb{H})$ of right-$\mathbb{H}$-linear spaces.*

## 4. Decompositions of a two-sided-quaternionic linear space

We begin this section with some notation. Given a two-sided-$\mathbb{H}$-linear space E, and $\lambda, \mu \in \mathbb{H}$, we denote by ${}^\lambda M^\mu$ the operator defined on E by ${}^\lambda M^\mu(x) := \lambda \, x \, \mu$.

Denote by $Z_k : \mathbb{H} \to \mathbb{H}$ the $k$-th-partial-quaternionic conjugation. For instance, given any $\lambda = \sum_{\ell=0}^{3} \lambda_\ell \, \mathbf{e}_\ell \in \mathbb{H}$, $Z_1(\lambda) = Z_1 \left( \sum_{\ell=0}^{3} \lambda_\ell \, \mathbf{e}_\ell \right) = \lambda_0 - \lambda_1 \, \mathbf{e}_1 + \lambda_2 \, \mathbf{e}_2 + \lambda_2 \, \mathbf{e}_2$.

All of them satisfy $Z_k(\lambda \mu) = Z_k(\mu) \, Z_k(\lambda)$. Denoting by $Z_{\mathbb{H}}$ the quaternionic conjugation, then $Z_{\mathbb{H}} = - \, {}^{\mathbf{e}_k} M^{\mathbf{e}_k} \circ Z_k$ and $Z_k = - \, {}^{\mathbf{e}_k} M^{\mathbf{e}_k} \circ Z_{\mathbb{H}}$ for $k \in \{1, 2, 3\}$.

### 4.1. Definition.

(1) *Let* E *be a left-$\mathbb{H}$-linear space and let* F *be a right-$\mathbb{H}$-linear space.*

 - *Given an additive map $T : E \to F$, it is called $Z_k$-anti-linear if for any $\lambda \in \mathbb{H}$ and $x \in E$, $T(\lambda x) = T(x) \, Z_k(\lambda)$.*
 - *Given an additive map $T : F \to E$, it is called $Z_k$-anti-linear if for any $\lambda \in \mathbb{H}$ and $x \in F$, $T(x \lambda) = Z_k(\lambda) \, T(x)$.*

(2) *If* E *and* F *are two-sided-$\mathbb{H}$-linear spaces, an additive map $T : E \to F$ is called (two-sided)-$Z_k$-anti-linear if $T(\lambda x \mu) = Z_k(\mu) \, T(x) \, Z_k(\lambda)$, for any $x \in E$ and $\lambda, \mu \in \mathbb{H}$. If the context is clear, we will only say that $T$ is anti-linear.*

**4.2.** Given a two-sided-$\mathbb{H}$-linear space E, what one could say about anti-linear and $Z_k$-anti-linear involutions on it? Let $h_k : E \to E$, $k \in \{0, 1, 2, 3\}$ be such that for all $k$, $h_k^2 = Id_E$ and

(i) $h_0$ is a two-sided-anti-linear map;
(ii) for $k \in \{1, 2, 3\}$, $h_k$ is a two-sided-$Z_k$-anti-linear map.

**4.3. Lemma.** *All the above maps $h_k$, $k \in \{0, 1, 2, 3\}$ exist simultaneously only.*

**Proof.** Assume there exists $h_0$. Set $h_k := -\,^{\mathbf{e}_k}M^{\mathbf{e}_k} \circ h_0$, hence $h_k^2(x) = (-\,^{\mathbf{e}_k}M^{\mathbf{e}_k} \circ h_0)^2(x) = \,^{\mathbf{e}_k}M^{\mathbf{e}_k} \circ h_0 \circ \,^{\mathbf{e}_k}M^{\mathbf{e}_k} \circ h_0(x) = \,^{\mathbf{e}_k}M^{\mathbf{e}_k}(h_0(\mathbf{e}_k\,h_0\,(x)\,\mathbf{e}_k)) = \,^{\mathbf{e}_k}M^{\mathbf{e}_k}(\mathbf{e}_k\,x\,\mathbf{e}_k) = x$, thus, $h_k^2 = Id_{\mathrm{E}}$.

On the other hand, given $\lambda \in \mathbb{H}$, $h_k(\lambda\,x\,\mu) = -\,^{\mathbf{e}_k}M^{\mathbf{e}_k} \circ h_0(\lambda\,x\,\mu) = -\mathbf{e}_k\,(\overline{\mu}\,h_0(x)\,\overline{\lambda})\,\mathbf{e}_k = (-\mathbf{e}_k\,\overline{\mu}\,\mathbf{e}_k)(-\mathbf{e}_k\,h_0(x)\,\mathbf{e}_k)(-\mathbf{e}_k\,\overline{\lambda}\,\mathbf{e}_k) = Z_k(\mu)\,h_k(x)\,Z_k(\lambda)$, that is, $h_k$ satisfies the property (ii).

Similarly if there exists $h_{k_0}$ with $k_0 \in \{1,2,3\}$, then $h_0 := \,^{\mathbf{e}_{k_0}}M^{\mathbf{e}_{k_0}} \circ h_{k_0}$ and $h_\ell := \,^{\mathbf{e}_\ell\mathbf{e}_{k_0}}M^{\mathbf{e}_\ell\mathbf{e}_{k_0}} \circ h_{k_0}$, for $\ell \in \{1,2,3\} \setminus \{k_0\}$, are the looked-for anti-linear involutions. $\qquad\square$

**4.4.** The skew-field of quaternions $\mathbb{H}$, admits various the decompositions:

$$\mathbb{H} = \mathbb{R} \oplus \mathbf{e}_1\,\mathbb{R} \oplus \mathbf{e}_2\,\mathbb{R} \oplus \mathbf{e}_3\,\mathbb{R}\,, \tag{3}$$

$$\mathbb{H} = \mathbb{R} \oplus Vect\mathbb{H}\,, \tag{4}$$

$$\mathbb{H} = \mathbf{e}_k\,\mathbb{R} \oplus V_k \qquad \text{where} \quad V_k := \{\,x \in \mathbb{H} \mid Z_k(x) = x\,\}\,, \tag{5}$$

which are valid due to the properties of the quaternionic conjugations, in particular, the principal conjugation is a two-sided-anti-linear map on $\mathbb{H}$ and all the elements of $\mathbb{R}$, the center of the non-abelian group $(\mathbb{H}, \cdot)$, are its fixed points.

**4.5. Remark.** If on a two-sided-$\mathbb{H}$-linear space E there exist the maps $h_0, h_1, h_2, h_3$, then the sets $\mathrm{E}_0 := \{\,x \in \mathrm{E} \mid h_0(x) = x\,\}$ and $\mathrm{E}_k := \{\,x \in \mathrm{E} \mid h_k(x) = -x\,\}$, $k \in \{1,2,3\}$ are nonempty, since 0 belongs to all of them. Similarly the $\mathbb{R}$-linear subspace of E

$$C_{\mathrm{E}} := \{\,x \in \mathrm{E} \mid \lambda\,x = x\,\lambda \text{ for any } \lambda \in \mathbb{H}\,\}\,,$$

is nonempty. From now on we assume that $C_{\mathrm{E}} \neq \{0\}$.

Next lemmas provide some relations between $C_{\mathrm{E}}$, $h_0, h_1, h_2, h_3, \mathrm{E}_0, \mathrm{E}_1, \mathrm{E}_2$ and $\mathrm{E}_3$, whenever all the objects exist.

**4.6. Lemma.** $C_{\mathrm{E}}$ *is invariant under* $h_0$. *Moreover* $h_0(C_{\mathrm{E}}) = C_{\mathrm{E}}$.

**Proof.** Given $x \in C_{\mathrm{E}}$ and $\lambda \in \mathbb{H}$, there holds $\lambda\,h_0(x) = h_0(x\,\overline{\lambda}) = h_0(\overline{\lambda}\,x) = h_0(x)\,\lambda$, thus $h_0(x) \in C_{\mathrm{E}}$. Reciprocally, given $x \in C_{\mathrm{E}}$, $x = h_0(h_0(x))$ which means that $x \in h_0(C_{\mathrm{E}})$. $\qquad\square$

**4.7. Lemma.** *For* $k \in \{1,2,3\}$ *it follows:*

$$\mathrm{E}_k = \{\,x \in \mathrm{E} \mid h_0(\mathbf{e}_k\,x) = \mathbf{e}_k\,x\,\}\,. \tag{6}$$

**Proof.** Denote by $X_k$ the set in the right-hand side of (6). Take $x \in \mathrm{E}_k$; then $h_0(\mathbf{e}_k\,x) = -\,^{\mathbf{e}_k}M^{\mathbf{e}_k} \circ h_k(\mathbf{e}_k\,x) = -\mathbf{e}_k\,h_k(\mathbf{e}_k\,x)\,\mathbf{e}_k = -\mathbf{e}_k\,h_k(x)\,\overline{\mathbf{e}}_k\,\mathbf{e}_k = \mathbf{e}_k\,x$, that is, $\mathbf{e}_k\,x \in X_k$. Reciprocally given $x \in X_k$, $h_k(x) = -\,^{\mathbf{e}_k}M^{\mathbf{e}_k} \circ h_0(x) = -\mathbf{e}_k\,h_0(x)\,\mathbf{e}_k = \mathbf{e}_k\,h_0(\mathbf{e}_k\,x) = -x$, hence $x \in \mathrm{E}_k$. $\qquad\square$

**4.8. Lemma.** *For every* $k \in \{1,2,3\}$, *we have that* $\mathrm{E}_k = \mathbf{e}_k\,\mathrm{E}_0$.

**Proof.** If $x \in E_0$ hence $h_k(e_k\,x) = -\,^{e_k}M^{e_k} \circ h_0(e_k\,x) = -e_k\,h_0(x)\bar{e}_k\,e_k = -e_k\,x$, that is, $e_k\,x \in E_k$. Reciprocally given $y \in E_k$, $h_0(e_k\,y) = e_k\,y$, thus $e_k\,y \in E_0$. $\quad\square$

**4.9. Proposition.** *Let* E *be a two-sided-$\mathbb{H}$-linear space on which there exist the maps* $h_k,;\ k \in \{0,1,2,3\}$. *Then* E *admits the decompositions:*

$$E = E_0 \oplus V_0 \quad \text{where} \quad V_0 := \{x \in E \mid h_0(x) = -x\}, \tag{7}$$

$$E = e_k\,E_0 \oplus V_k, \quad \text{where} \quad V_k := \{x \in E \mid h_k(x) = x\}, \quad k \in \{1,2,3\}. \tag{8}$$

**Proof.** It is clear that for $k \in \{0,1,2,3\}$, $e_k\,E_0 \cap V_K = \{0\}$. Given $x \in E$, define

$$x_0 := \frac{1}{2}(x + h_0(x)) \in E_0\ ; \quad x_1 := \frac{1}{2}(x - h_0(x)) \in V_0$$

and for $k \in \{1,2,3\}$ define

$$w_0 := \frac{1}{2}(x - Z_k(x)) \in E_k\ ; \quad w_1 := \frac{1}{2}(x + Z_k(x)) \in V_k\ .$$

Thus $x = x_0 + x_1 \in E_0 \oplus V_0$ and $x = w_0 + w_1 \in E_k \oplus V_k$, $k \in \{1,2,3\}$. $\quad\square$

Note that by Lemma 4.3., E has the decomposition (7) if and only if it has the decomposition (8). One may think that it would be enough to have these decompositions in order to get a decomposition similar to (3). Paradoxically we can prove only that it is true under certain additional assumptions.

**4.10. Lemma.** *If* $E_0 = C_E$, *then* $E_\ell \cap E_k = \{0\}$, *for* $\ell, k \in \{0,1,2,3\}$, $\ell \neq k$.

**Proof.** If $x \in E_0 \cap E_k$, $k \in \{1,2,3\}$, then $h_0(x) = x$ and $h_0(e_k\,x) = e_k\,x = h_0(x)\bar{e}_k = -x\,e_k = -e_k\,x$, which implies $x = 0$. The rest is similar. $\quad\square$

**4.11. Theorem.** *Let* E *be a two-sided-$\mathbb{H}$-linear space on which there exist the maps* $h_k$, $k \in \{0,1,2,3\}$ *and* $E_0 = C_E$. *Then* E *admits the decomposition*

$$E = E_0 \oplus e_1\,E_0 \oplus e_2\,E_0 \oplus e_3\,E_0 \tag{9}$$

*if and only if the maps* $h_k$ *satisfy*

$$\frac{1}{2}\left(\sum_{k=1}^{3} h_k - h_0\right) = Id_E\ . \tag{10}$$

**Proof.** Assume that (10) is valid. Given $x \in E$, define

$$x_0 := \frac{1}{2}(x + h_0(x)) \in E_0\ ;$$

and

$$x_k := -\frac{1}{2}e_k\,(x - h_k(x)), \quad k \in \{1,2,3\}\ .$$

Since $h_0 \circ h_k = - \,^{\mathbf{e}_k} M^{\mathbf{e}_k}$, we have that $h_0(x_k) = -\frac{1}{2}h_0(\mathbf{e}_k\,(x - h_k(x))) = \frac{1}{2}h_0(x - h_k(x))\,\mathbf{e}_k = \frac{1}{2}(h_0(x) - h_0 \circ h_k(x))\,\mathbf{e}_k = \frac{1}{2}(-\mathbf{e}_k\,h_k(x)\,\mathbf{e}_k + \mathbf{e}_k\,x\,\mathbf{e}_k)\,\mathbf{e}_k = -\frac{1}{2}\mathbf{e}_k(x - h_k(x))$, thus $x_k \in \mathrm{E}_0$ and this implies that $\mathbf{e}_k\,x_k \in \mathrm{E}_k = \mathbf{e}_k\,\mathrm{E}_0$.

Finally using (10) we obtain:

$$x_0 + \sum_{k=1}^3 \mathbf{e}_k\,x_k = \frac{1}{2}(x + h_0(x)) + \sum_{k=1}^3 \frac{1}{2}(x - h_k(x))$$
$$= 2x + \frac{1}{2}h_0(x) - \frac{1}{2}\sum_{k=1}^3 h_k(x) = x\,,$$

hence (9) is proved. The inverse implication is obvious. $\qquad\square$

**Remark.** Property (10) is valid for the case $\mathrm{E} = \mathbb{H}$. We don't know if there exist two-sided-$\mathbb{H}$-linear spaces where this property is not true and what could be said in general about the sets $\mathrm{E}_0$ and $C_{\mathrm{E}}$.

# References

1. D. Alpay, M.E. Luna-Elizarrarás, M. Shapiro, *Normes des extensions quaternionique d'opérateurs réels*. Comptes Rendus Acad. Sci. Paris, Ser. I, **340**, 639–643 (2005).
2. R. Delanghe, F. Sommen, V. Souček, *Clifford Algebra and Spinor-Valued Functions: a function theory for the Dirac operator*. Series: Mathematics and its applications, Kluwer Academic Publishers, v. **53**, (1992).
3. A.A. Kirilov, A.D. Gvishiani. *Theorems and problems in functional analysis*. Problem Books in Mathematics, Springer-Verlag, 1982.
4. M.E. Luna-Elizarrarás, M. Shapiro. *Preservation of the norms of linear operators acting on some quaternionic function spaces*. Operator Theory: Advances and Applications, Vol. **157**, 205-220 (2005), Birkhäuser Verlag, Basel, Switzerland.
5. M.E. Luna-Elizarrarás, M. Shapiro. *On some properties of quaternionic inner product spaces*. Proceedings of the XXV International Colloquium on Group Theoretical Methods in Physics, Cocoyoc, México, 2-6 Agosto 2004, IOP Publishing Ltd, Inst. of Phys. Conf. Ser. No. **185** 371–376 (2005).
6. C.S. Sharma *Complex structure on a real Hilbert space and symplectic structure on a complex Hilbert space*. J. Math. Phys., **29**, ♯ 5, 1069–1078 (1988).
7. C.S. Sharma and T.J. Coulson, *Spectral theory for unitary operators on a quaternionic Hilbert space*. J. Math. Phys., **28**, 1941–1946 (1987).

# HOLOMORPHIC FUNCTIONS AND REGULAR QUATERNIONIC FUNCTIONS ON THE HYPERKÄHLER SPACE $\mathbb{H}$

A. PEROTTI*

*Department of Mathematics*
*University of Trento*
*Via Sommarive, 14*
*I-38050 Povo Trento ITALY*
*E-mail: perotti@science.unitn.it*

Let $\mathbb{H}$ be the space of quaternions, with its standard hypercomplex structure. Let $\mathcal{R}(\Omega)$ be the module of $\psi$-regular functions on $\Omega$. For every unitary vector $p$ in $\mathbb{S}^2 \subset \mathbb{H}$, $\mathcal{R}(\Omega)$ contains the space of holomorphic functions w.r.t. the complex structure $J_p$ induced by $p$. We prove the existence, on any bounded domain $\Omega$, of $\psi$-regular functions that are not $J_p$-holomorphic for any $p$. Our starting point is a result of Chen and Li concerning maps between hyperkähler manifolds, where a similar result is obtained for a less restricted class of quaternionic maps. We give a criterion, based on the energy-minimizing property of holomorphic maps, that distinguishes $J_p$-holomorphic functions among $\psi$-regular functions.

**Key words:** Quaternionic regular functions, hypercomplex structure, hyperkähler space
**Mathematics Subject Classification:** Primary 32A30; Secondary 53C26, 30G35

## 1. Introduction

Let $\mathbb{H}$ be the space of quaternions, with its standard hypercomplex structure given by the complex structures $J_1, J_2$ on $T\mathbb{H} \simeq \mathbb{H}$ defined by left multiplication by $i$ and $j$. Let $J_1^*, J_2^*$ be the dual structures on $T^*\mathbb{H}$.

We consider the module $\mathcal{R}(\Omega) = \{f = f_1 + f_2 j \mid \overline{\partial} f_1 = J_2^*(\partial \overline{f_2}) \text{ on } \Omega\}$ of left $\psi$-*regular* functions on $\Omega$. These functions are in a simple correpondence with Fueter left regular functions, since they can be obtained from them by means of a real coordinate reflection in $\mathbb{H}$. They have been studied by many authors (see for instance Sudbery[8], Shapiro and Vasilevski[6] and Nōno[5]). The space $\mathcal{R}(\Omega)$ contains the identity mapping and any holomorphic mapping $(f_1, f_2)$ on $\Omega$ defines a $\psi$-regular function $f = f_1 + f_2 j$. This is no more true if we replace the class of $\psi$-regular functions with that of regular functions. The definition of $\psi$-regularity is also equivalent to that of *q-holomorphicity* given by Joyce[2] in the setting of hypercomplex manifolds.

*Work partially supported by MIUR (Project "Proprietà geometriche delle varietà reali e complesse") and GNSAGA of INDAM.

For every unitary vector $p$ in $\mathbb{S}^2 \subset \mathbb{H}$, $\mathcal{R}(\Omega)$ contains the space $Hol_p(\Omega, \mathbb{H}) = \{f : \Omega \to \mathbb{H} \mid df + pJ_p(df) = 0 \text{ on } \Omega\}$ of holomorphic functions w.r.t. the complex structure $J_p = p_1 J_1 + p_2 J_2 + p_3 J_3$ on $\Omega$ and to the structure induced on $\mathbb{H}$ by left-multiplication by $p$ ($J_p$-holomorphic functions on $\Omega$).

We show that on every domain $\Omega$ there exist $\psi$-regular functions that are not $J_p$-holomorphic for any $p$. A similar result was obtained by Chen and Li[1] for the larger class of $q$-maps between hyperkähler manifolds.

This result is a consequence of a criterion (cf. Theorem 4.1) of $J_p$-holomorphicity, which is obtained using the energy-minimizing property of $\psi$-regular functions (cf. Proposition 4.1) and ideas of Lichnerowicz[4] and Chen and Li[1].

In Sec. 4.4 we give some other applications of the criterion. In particular, we show that if $\Omega$ is connected, then the intersection $Hol_p(\Omega, \mathbb{H}) \cap Hol_{p'}(\Omega, \mathbb{H})$ ($p \neq \pm p'$) contains only affine maps. This result is in accord with what was proved by Sommese[7] about *quaternionic maps* (cf. Sec. 3.2 for definitions).

## 2. Fueter-regular and $\psi$-regular functions

### 2.1. *Notations and definitions*

We identify the space $\mathbb{C}^2$ with the set $\mathbb{H}$ of quaternions by means of the mapping that associates the pair $(z_1, z_2) = (x_0 + ix_1, x_2 + ix_3)$ with the quaternion $q = z_1 + z_2 j = x_0 + ix_1 + jx_2 + kx_3 \in \mathbb{H}$. Let $\Omega$ be a bounded domain in $\mathbb{H} \simeq \mathbb{C}^2$. A quaternionic function $f = f_1 + f_2 j \in C^1(\Omega)$ is *(left) regular* on $\Omega$ (in the sense of Fueter) if

$$\mathcal{D}f = \frac{\partial f}{\partial x_0} + i\frac{\partial f}{\partial x_1} + j\frac{\partial f}{\partial x_2} + k\frac{\partial f}{\partial x_3} = 0 \quad \text{on } \Omega.$$

Given the "structural vector" $\psi = (1, i, j, -k)$, $f$ is called *(left) $\psi$-regular* on $\Omega$ if

$$\mathcal{D}'f = \frac{\partial f}{\partial x_0} + i\frac{\partial f}{\partial x_1} + j\frac{\partial f}{\partial x_2} - k\frac{\partial f}{\partial x_3} = 0 \quad \text{on } \Omega.$$

We recall some properties of regular functions, for which we refer to the papers of Sudbery[8], Shapiro and Vasilevski[6] and Nōno[5]:

(1) $f$ is $\psi$-regular $\Leftrightarrow$ $\dfrac{\partial f_1}{\partial \bar{z}_1} = \dfrac{\partial \bar{f}_2}{\partial z_2}$, $\dfrac{\partial f_1}{\partial \bar{z}_2} = -\dfrac{\partial \bar{f}_2}{\partial z_1}$.

(2) Every holomorphic map $(f_1, f_2)$ on $\Omega$ defines a $\psi$-regular function $f = f_1 + f_2 j$.

(3) The complex components are both holomorphic or both non-holomorphic.

(4) Every regular or $\psi$-regular function is harmonic.

(5) If $\Omega$ is pseudoconvex, every complex harmonic function is the complex component of a $\psi$-regular function on $\Omega$.

(6) The space $\mathcal{R}(\Omega)$ of $\psi$-regular functions on $\Omega$ is a *right* $\mathbb{H}$-module with integral representation formulas.

### 2.2. q-holomorphic functions

A definition equivalent to $\psi$-regularity has been given by Joyce[2] in the setting of hypercomplex manifolds. Joyce introduced the module of *q-holomorphic* functions on a hypercomplex manifold. On this module he defined a (commutative) product. A hypercomplex structure on the manifold $\mathbb{H}$ is given by the complex structures $J_1, J_2$ on $T\mathbb{H} \simeq \mathbb{H}$ defined by left multiplication by $i$ and $j$. Let $J_1^*, J_2^*$ be the dual structures on $T^*\mathbb{H}$. In complex coordinates

$$\begin{cases} J_1^* dz_1 = i\, dz_1, & J_1^* dz_2 = i\, dz_2 \\ J_2^* dz_1 = -d\bar{z}_2, & J_2^* dz_2 = d\bar{z}_1 \\ J_3^* dz_1 = i\, d\bar{z}_2, & J_3^* dz_2 = -i\, d\bar{z}_1 \end{cases}$$

where we make the choice $J_3^* = J_1^* J_2^* \Rightarrow J_3 = -J_1 J_2$.

A function $f$ is $\psi$-regular if and only if $f$ is *q-holomorphic*, i.e.

$$df + iJ_1^*(df) + jJ_2^*(df) + kJ_3^*(df) = 0.$$

In complex components $f = f_1 + f_2 j$, we can rewrite the equations of $\psi$-regularity as

$$\bar{\partial} f_1 = J_2^*(\bar{\partial} \bar{f}_2).$$

## 3. Holomorphic maps

### 3.1. Holomorphic functions w.r.t. a complex structure $J_p$

Let $J_p = p_1 J_1 + p_2 J_2 + p_3 J_3$ be the complex structure on $\mathbb{H}$ defined by a unit imaginary quaternion $p = p_1 i + p_2 j + p_3 k$ in the sphere $\mathbb{S}^2 = \{p \in \mathbb{H} \mid p^2 = -1\}$. It is well-known that every complex structure compatible with the standard hyperkähler structure of $\mathbb{H}$ is of this form. If $f = f^0 + if^1 : \Omega \to \mathbb{C}$ is a $J_p$-*holomorphic* function, i.e. $df^0 = J_p^*(df^1)$ or, equivalently, $df + iJ_p^*(df) = 0$, then $f$ defines a $\psi$-regular function $\tilde{f} = f^0 + pf^1$ on $\Omega$. We can identify $\tilde{f}$ with a holomorphic function

$$\tilde{f} : (\Omega, J_p) \to (\mathbb{C}_p, L_p)$$

where $\mathbb{C}_p = \langle 1, p \rangle$ is a copy of $\mathbb{C}$ in $\mathbb{H}$ and $L_p$ is the complex structure defined on $T^*\mathbb{C}_p \simeq \mathbb{C}_p$ by left multiplication by $p$.

More generally, we can consider the space of holomorphic maps from $(\Omega, J_p)$ to $(\mathbb{H}, L_p)$

$$Hol_p(\Omega, \mathbb{H}) = \{f : \Omega \to \mathbb{H} \mid \bar{\partial}_p f = 0 \text{ on } \Omega\} = Ker\bar{\partial}_p$$

(the $J_p$-*holomorphic maps* on $\Omega$) where $\bar{\partial}_p$ is the Cauchy-Riemann operator w.r.t. the structure $J_p$

$$\bar{\partial}_p = \frac{1}{2}\left(d + pJ_p^* \circ d\right).$$

For any positive orthonormal basis $\{1, p, q, pq\}$ of $\mathbb{H}$ $(p, q \in \mathbb{S}^2)$, the equations of $\psi$-regularity can be rewritten in complex form as

$$\overline{\partial}_p f_1 = J_q^*(\partial_p \overline{f}_2),$$

where $f = (f^0 + pf^1) + (f^2 + pf^3)q = f_1 + f_2 q$. Then every $f \in Hol_p(\Omega, \mathbb{H})$ is a $\psi$-regular function on $\Omega$.

**Remark 3.1.** 1) The *identity* map is in $Hol_i(\Omega, \mathbb{H}) \cap Hol_j(\Omega, \mathbb{H})$, but not in $Hol_k(\Omega, \mathbb{H})$.

2) $Hol_{-p}(\Omega, \mathbb{H}) = Hol_p(\Omega, \mathbb{H})$

3) If $f \in Hol_p(\Omega, \mathbb{H}) \cap Hol_{p'}(\Omega, \mathbb{H})$, with $p \neq \pm p'$, then $f \in Hol_{p''}(\Omega, \mathbb{H})$ for every $p'' = \frac{\alpha p + \beta p'}{\|\alpha p + \beta p'\|}$.

4) $\psi$-regularity distinguishes between holomorphic and anti-holomorphic maps: if $f$ is an *anti-holomorphic* map from $(\Omega, J_p)$ to $(\mathbb{H}, L_p)$, then $f$ can be $\psi$-regular or not. For example, $f = \overline{z}_1 + \overline{z}_2 j \in Hol_j(\Omega, \mathbb{H}) \cap Hol_k(\Omega, \mathbb{H})$ is a $\psi$-regular function induced by the anti-holomorphic map

$$(\overline{z}_1, \overline{z}_2) : (\Omega, J_1) \to (\mathbb{H}, L_i),$$

while $(\overline{z}_1, 0) : (\Omega, J_1) \to (\mathbb{H}, L_i)$ induces the function $g = \overline{z}_1 \notin \mathcal{R}(\Omega)$.

## 3.2. *Quaternionic maps*

A particular class of $J_p$-holomorphic maps is constituted by the *quaternionic maps* on the quaternionic manifold $\Omega$. Sommese[7] defined quaternionic maps between hypercomplex manifolds: a quaternionic map is a map

$$f : (X, J_1, J_2) \to (Y, K_1, K_2)$$

that is holomorphic from $(X, J_1)$ to $(Y, K_1)$ *and* from $(X, J_2)$ to $(Y, K_2)$.

In particular, a quaternionic map

$$f : (\Omega, J_1, J_2) \to (\mathbb{H}, J_1, J_2)$$

is an element of $Hol_i(\Omega, \mathbb{H}) \cap Hol_j(\Omega, \mathbb{H})$ and then a $\psi$-regular function on $\Omega$. Sommese showed that quaternionic maps are affine. They appear for example as transition functions for 4-dimensional *quaternionic manifolds*.

## 4. Non-holomorphic $\psi$-regular maps

A natural question can now be raised: can $\psi$-regular maps always be made holomorphic by rotating the complex structure or do they constitute a new class of harmonic maps? In other words, does the space $\mathcal{R}(\Omega)$ contain the union $\bigcup_{p \in \mathbb{S}^2} Hol_p(\Omega, \mathbb{H})$ properly?

Chen and Li[1] posed and answered the analogous question for the larger class of *q-maps* between hyperkähler manifolds. In their definition, the complex structures of the source and target manifold can rotate *independently*. This implies that also anti-holomorphic maps are q-maps.

## 4.1. *Energy and regularity*

The *energy* (w.r.t. the euclidean metric $g$) of a map $f : \Omega \to \mathbb{C}^2 \simeq \mathbb{H}$, of class $C^1(\overline{\Omega})$, is the integral

$$\mathcal{E}(f) = \frac{1}{2} \int_\Omega \|df\|^2 dV = \frac{1}{2} \int_\Omega \langle g, f^* g \rangle dV = \frac{1}{2} \int_\Omega tr(J_{\mathbb{C}}(f)\overline{J_{\mathbb{C}}(f)}^T) dV,$$

where $J_{\mathbb{C}}(f)$ is the Jacobian matrix of $f$ with respect to the coordinates $\bar{z}_1, z_1, \bar{z}_2, z_2$.

Lichnerowicz[4] proved that holomorphic maps between Kähler manifolds minimize the energy functional in their homotopy classes. Holomorphic maps $f$ smooth on $\overline{\Omega}$ minimize energy in the homotopy class constituted by maps $u$ with $u_{|\partial\Omega} = f_{|\partial\Omega}$ which are homotopic to $f$ relative to $\partial\Omega$.

From the theorem, functions $f \in Hol_p(\Omega, \mathbb{H})$ minimize the energy functional in their homotopy classes (relative to $\partial\Omega$). More generally:

**Proposition 4.1.** *If $f$ is $\psi$-regular on $\Omega$, then it minimizes energy in its homotopy class (relative to $\partial\Omega$).*

**Proof.** We repeat arguments of Lichnerowicz, Chen and Li. Let $i_1 = i, i_2 = j, i_3 = k$ and let

$$\mathcal{K}(f) = \int_\Omega \sum_{\alpha=1}^3 \langle J_\alpha, f^* L_{i_\alpha} \rangle dV, \quad \mathcal{I}(f) = \frac{1}{2} \int_\Omega \|df + \sum_{\alpha=1}^3 L_{i_\alpha} \circ df \circ J_\alpha\|^2 dV.$$

Then $\mathcal{K}(f)$ is a homotopy invariant of $f$ and $\mathcal{I}(f) = 0$ if and only if $f \in \mathcal{R}(\Omega)$. A computation similar to that made by Chen and Li[1] gives

$$\mathcal{E}(f) + \mathcal{K}(f) = \frac{1}{4}\mathcal{I}(f) \geq 0.$$

From this the result follows immediately. $\square$

## 4.2. *A criterion for holomorphicity*

We now come to our main result. Let $f : \Omega \to \mathbb{H}$ be a function of class $C^1(\overline{\Omega})$.

**Theorem 4.1.** *Let $A = (a_{\alpha\beta})$ be the $3 \times 3$ matrix with entries $a_{\alpha\beta} = -\int_\Omega \langle J_\alpha, f^* L_{i_\beta} \rangle dV$. Then*

*(1) $f$ is $\psi$-regular if and only if $\mathcal{E}(f) = tr A$.*
*(2) If $f \in \mathcal{R}(\Omega)$, then $A$ is real, symmetric and*

$$tr A \geq \lambda_1 = \max\{eigenvalues\ of\ A\}.$$

*It follows that $\det(A - (tr A)I_3) \leq 0$.*
*(3) If $f \in \mathcal{R}(\Omega)$, then $f$ belongs to some space $Hol_p(\Omega, \mathbb{H})$ if and only if $\mathcal{E}(f) = tr A = \lambda_1$ or, equivalently, $\det(A - (tr A)I_3) = 0$.*
*(4) If $\mathcal{E}(f) = tr A = \lambda_1$, $X_p = (p_1, p_2, p_3)$ is a unit eigenvector of $A$ relative to the largest eigenvalue $\lambda_1$ if and only if $f \in Hol_p(\Omega, \mathbb{H})$.*

### 4.3. *The existence of non-holomorphic $\psi$-regular maps*

The criterion can be applied to show that on every domain $\Omega$ in $\mathbb{H}$, there exist $\psi$-regular functions that are not holomorphic.

**Example 4.1.** Let $f = z_1 + z_2 + \bar{z}_1 + (z_1 + z_2 + \bar{z}_2)j$. Then $f$ is $\psi$-regular, but not holomorphic, since on the unit ball $B$ in $\mathbb{C}^2$, $f$ has energy $\mathcal{E}(f) = 6$ and the matrix $A$ of the theorem is

$$A = \begin{bmatrix} 2 & 0 & 0 \\ 0 & 2 & 0 \\ 0 & 0 & 2 \end{bmatrix}.$$

Therefore $\mathcal{E}(f) = tr A > 2 = \lambda_1$.

In the preceding example, the Jacobian matrix of the function has even rank, a necessary condition for a holomorphic map. In the case when the rank is odd, the non-holomorphicity follows immediately. For example, $g = z_1 + \bar{z}_1 + \bar{z}_2 j$ is $\psi$-regular (on any $\Omega$) but not $J_p$-holomorphic, for any $p$, since $rk J_{\mathbb{C}}(f)$ is odd.

**Example 4.2.** The linear, $\psi$-regular functions constitute a $\mathbb{H}$-module of dimension 3 over $\mathbb{H}$, generated e.g. by the set $\{z_1 + z_2 j, z_2 + z_1 j, \bar{z}_1 + \bar{z}_2 j\}$. An element

$$f = (z_1 + z_2 j)q_1 + (z_2 + z_1 j)q_2 + (\bar{z}_1 + \bar{z}_2 j)q_3$$

is holomorphic if and only if the coefficients $q_1 = a_1 + a_2 j$, $q_2 = b_1 + b_2 j$, $q_3 = c_1 + c_2 j$ satisfy the $6^{th}$-degree real homogeneous equation

$$\det(A - (tr A)I_3) = 0$$

obtained after integration on $B$. The explicit expression of this equation is given in the Appendix. So "almost all" (linear) $\psi$-regular functions are non-holomorphic.

**Example 4.3.** A positive example (with $p \neq i, j, k$). Let $h = \bar{z}_1 + (z_1 + \bar{z}_2)j$. On the unit ball $h$ has energy 3 and the matrix $A$ is

$$A = \begin{bmatrix} -1 & 0 & 2 \\ 0 & 2 & 0 \\ 2 & 0 & 2 \end{bmatrix}$$

then $\mathcal{E}(h) = tr A$ is equal to the (simple) largest eigenvalue, with unit eigenvector $X = \frac{1}{\sqrt{5}}(1, 0, 2)$. It follows that $h$ is $J_p$-holomorphic with $p = \frac{1}{\sqrt{5}}(i + 2k)$, i.e. it satisfies the equation

$$df + \tfrac{1}{5}(i + 2k)(J_1^* + 2J_3^*)(df) = 0.$$

**Example 4.4.** We give a quadratic example. Let $f = |z_1|^2 - |z_2|^2 + \bar{z}_1 \bar{z}_2 j$. $f$ has energy 2 on $B$ and the matrix $A$ is

$$A = \begin{bmatrix} -2/3 & 0 & 0 \\ 0 & 4/3 & 0 \\ 0 & 0 & 4/3 \end{bmatrix}$$

Then $f$ is $\psi$-regular but not holomorphic w.r.t. any complex structure $J_p$.

### 4.4. *Other applications of the criterion*

1) If $f \in Hol_p(\Omega, \mathbb{H}) \cap Hol_{p'}(\Omega, \mathbb{H})$ for two $\mathbb{R}$-independent $p, p'$, then $X_p, X_{p'}$ are independent eigenvectors relative to $\lambda_1$. Therefore the eigenvalues of the matrix $A$ are $\lambda_1 = \lambda_2 = -\lambda_3$.

If $f \in Hol_p(\Omega, \mathbb{H}) \cap Hol_{p'}(\Omega, \mathbb{H}) \cap Hol_{p''}(\Omega, \mathbb{H})$ for three $\mathbb{R}$-independent $p, p', p''$ then $\lambda_1 = \lambda_2 = \lambda_3 = 0 \Rightarrow A = 0$ and therefore $f$ has energy 0 and $f$ is a (locally) *constant* map.

2) If $\Omega$ is connected, then $Hol_p(\Omega, \mathbb{H}) \cap Hol_{p'}(\Omega, \mathbb{H})$ $(p \neq \pm p')$ contains only *affine* maps (cf. Sommese[7]).

We can assume $p = i$, $p' = j$ since in view of property 3) of Remark 3.1 we can suppose $p$ and $p'$ orthogonal quaternions and then we can rotate the space of imaginary quaternions. Let $f \in Hol_i(\Omega, \mathbb{H}) \cap Hol_j(\Omega, \mathbb{H})$ and $a = \left( \dfrac{\partial f_1}{\partial z_1}, \dfrac{\partial f_2}{\partial z_1} \right)$, $b = \left( \dfrac{\partial f_2}{\partial z_2}, -\dfrac{\overline{\partial f_1}}{\partial z_2} \right)$. Since $f \in Hol_i(\Omega, \mathbb{H})$, the matrix $A$ is obtained after integration on $\Omega$ of the matrix

$$\begin{bmatrix} |a|^2 + |b|^2 & 0 & 0 \\ 0 & 2Re\langle a, b \rangle & -2Im\langle a, b \rangle \\ 0 & -2Im\langle a, b \rangle & -2Re\langle a, b \rangle \end{bmatrix}$$

where $\langle a, b \rangle$ denotes the standard hermitian product of $\mathbb{C}^2$.

Since $f \in Hol_j(\Omega, \mathbb{H})$, we have $\int_\Omega Im\langle a, b \rangle dV = 0$ and $\int_\Omega |a - b|^2 dV = 0$. Therefore $a = b$ on $\Omega$. Then $a$ is holomorphic and anti-holomorphic w.r.t. the standard structure $J_1$. This means that $a$ is constant on $\Omega$ and $f$ is an affine map with linear part of the form

$$(a_1 z_1 - \bar{a}_2 z_2) + (a_2 z_1 + \bar{a}_1 z_2)j$$

i.e. the right multiplication of $q = z_1 + z_2 j$ by the quaternion $a_1 + a_2 j$.

3) We can give a classification of $\psi$-regular functions based on the dimension of the set of complex structures w.r.t. which the function is holomorphic. Let $\Omega$ be connected. Given a function $f \in \mathcal{R}(\Omega)$, we set

$$\mathcal{J}(f) = \{ p \in \mathbb{S}^2 \mid f \in Hol_p(\Omega, \mathbb{H}) \}.$$

The space $\mathcal{R}(\Omega)$ of $\psi$-regular functions is the disjoint union of subsets of functions of the following four types:

(i) $f$ is $J_p$-holomorphic for three $\mathbb{R}$-independent structures
 $\Longrightarrow f$ is a constant and $\mathcal{J}(f) = \mathbb{S}^2$.
(ii) $f$ is $J_p$-holomorphic for exactly two $\mathbb{R}$-independent structures
 $\Longrightarrow f$ is a $\psi$-regular, invertible affine map and $\mathcal{J}(f)$ is an equator $S^1 \subset \mathbb{S}^2$.
(iii) $f$ is $J_p$-holomorphic for exactly one structure $J_p$ (up to sign of $p$) $\Longrightarrow \mathcal{J}(f)$ is a two-point set $S^0$.

(iv) $f$ is $\psi$-regular but not $J_p$-holomorphic w.r.t. any complex structure $\Longrightarrow \mathcal{J}(f) = \emptyset$.

We will return in a subsequent paper to the application of the criterion to the study of $\psi$-*biregular* functions, which are invertible $\psi$-regular functions with $\psi$-regular inverse (see Królikowski and Porter[3] for the class of *biregular* functions). This class contains as a proper subset the invertible holomorphic maps.

## 5. Sketch of proof of Theorem 4.1

If $f \in \mathcal{R}(\Omega)$, then $\mathcal{E}(f) = -\mathcal{K}(f) = tr A$. Let

$$\mathcal{I}_p(f) = \frac{1}{2} \int_\Omega \|df + L_p \circ df \circ J_p\|^2 dV.$$

Then we obtain, as in Chen and Li[1]

$$\mathcal{E}(f) + \int_\Omega \langle J_p, f^* L_p \rangle dV = \frac{1}{4} \mathcal{I}_p(f).$$

If $X_p = (p_1, p_2, p_3)$, then

$$XAX^T = \sum_{\alpha,\beta} p_\alpha p_\beta a_{\alpha\beta} = -\int_\Omega \langle \sum_\alpha p_\alpha J_\alpha, f^* \sum_\beta p_\beta L_{i_\beta} \rangle dV$$

$$= -\int_\Omega \langle J_p, f^* L_p \rangle dV = \mathcal{E}(f) - \frac{1}{4} \mathcal{I}_p(f).$$

Then $tr A = \mathcal{E}(f) = XAX^T + \frac{1}{4}\mathcal{I}_p(f) \geq XAX^T$, with equality if and only if $\mathcal{I}_p(f) = 0$ i.e if and only if $f$ is a $J_p$-holomorphic map.

Let $M_\alpha$ ($\alpha = 1,2,3$) be the matrix associated to $J_\alpha^*$ w.r.t. the basis $\{d\bar{z}_1, dz_1, d\bar{z}_2, dz_2\}$. The entries of the matrix $A$ can be computed by the formula

$$a_{\alpha\beta} = -\int_\Omega \langle J_\alpha, f^* L_{i_\beta} \rangle dV = \frac{1}{2} \int_\Omega tr(\overline{B_\alpha}^T C_\beta) dV$$

where $B_\alpha = M_\alpha J_C(f)^T$ for $\alpha = 1,2$, $B_\alpha = -M_\alpha J_C(f)^T$ for $\alpha = 3$ and $C_\beta = J_C(f)^T M_\beta$ for $\beta = 1,2,3$.

A direct computation shows how from the particular form of the Jacobian matrix of a $\psi$-regular function it follows the symmetry property of $A$.

## Appendix

We give the explicit expression of the $6^{th}$-degree real homogeneous equation satisfied by the complex coefficients of a linear $J_p$-holomorphic $\psi$-regular function.

$\frac{1}{16} det(A - (trA)I_3) = a_1 a_2 b_2 c_1^2 \bar{b}_1 - a_1 a_2 b_1 c_1 c_2 \bar{b}_1 - a_1^2 b_2 c_1 c_2 \bar{b}_1 + a_1^2 b_1 c_2^2 \bar{b}_1 - a_1 c_1^2 \bar{a}_1 \bar{b}_1^2 - a_1 c_1 c_2 \bar{a}_2 \bar{b}_1^2 + a_2^2 b_2 c_1^2 \bar{b}_2 - a_2^2 b_1 c_1 c_2 \bar{b}_2 - a_1 a_2 b_2 c_1 c_2 \bar{b}_2 + a_1 a_2 b_1 c_2^2 \bar{b}_2 - a_2 c_1^2 \bar{a}_1 \bar{b}_1 \bar{b}_2 - a_1 c_1 c_2 \bar{a}_1 \bar{b}_1 \bar{b}_2 - a_2 c_1 c_2 \bar{a}_2 \bar{b}_1 \bar{b}_2 - a_1 c_2^2 \bar{a}_2 \bar{b}_1 \bar{b}_2 - a_2 c_1 c_2 \bar{a}_1 \bar{b}_2^2 - a_2 c_2^2 \bar{a}_2 \bar{b}_2^2 + a_1 a_2 b_1 b_2 c_1 \bar{c}_1 - a_1^2 b_2^2 c_1 \bar{c}_1 - a_1 a_2 b_1^2 c_2 \bar{c}_1 + a_1^2 b_1 b_2 c_2 \bar{c}_1 - 2a_1 b_1 c_1 \bar{a}_1 \bar{b}_1 \bar{c}_1 - a_1 b_2 c_1 \bar{a}_2 \bar{b}_1 \bar{c}_1 -$

$$a_1b_1c_2\bar{a}_2\bar{b}_1\bar{c}_1 - a_2b_1c_1\bar{a}_1\bar{b}_2\bar{c}_1 - 2a_1b_2c_1\bar{a}_1\bar{b}_2\bar{c}_1 + a_1b_1c_2\bar{a}_1\bar{b}_2\bar{c}_1 - 2a_2b_2c_1\bar{a}_2\bar{b}_2\bar{c}_1 +$$
$$a_2b_1c_2\bar{a}_2\bar{b}_2\bar{c}_1 - a_1b_2c_2\bar{a}_2\bar{b}_2\bar{c}_1 + c_1\bar{a}_1\bar{a}_2\bar{b}_1\bar{b}_2\bar{c}_1 + c_2\bar{a}_2^2\bar{b}_1\bar{b}_2\bar{c}_1 - c_1\bar{a}_1^2\bar{b}_2^2\bar{c}_1 - c_2\bar{a}_1\bar{a}_2\bar{b}_2^2\bar{c}_1 -$$
$$a_1b_1^2\bar{a}_1\bar{c}_1^2 - a_1b_1b_2\bar{a}_2\bar{c}_1^2 + b_1\bar{a}_1\bar{a}_2\bar{b}_2\bar{c}_1^2 + b_2\bar{a}_2^2\bar{b}_2\bar{c}_1^2 + a_2^2b_1b_2c_1\bar{c}_2 - a_1a_2b_2^2c_1\bar{c}_2 - a_2^2b_1^2c_2\bar{c}_2 +$$
$$a_1a_2b_1b_2c_2\bar{c}_2 - a_2b_1c_1\bar{a}_1\bar{b}_1\bar{c}_2 + a_1b_2c_1\bar{a}_1\bar{b}_1\bar{c}_2 - 2a_1b_1c_2\bar{a}_1\bar{b}_1\bar{c}_2 + a_2b_2c_1\bar{a}_2\bar{b}_1\bar{c}_2 -$$
$$2a_2b_1c_2\bar{a}_2\bar{b}_1\bar{c}_2 - a_1b_2c_2\bar{a}_2\bar{b}_1\bar{c}_2 - c_1\bar{a}_1\bar{a}_2\bar{b}_1^2\bar{c}_2 - c_2\bar{a}_2^2\bar{b}_1^2\bar{c}_2 - a_2b_2c_1\bar{a}_1\bar{b}_2\bar{c}_2 - a_2b_1c_2\bar{a}_1\bar{b}_2\bar{c}_2 -$$
$$2a_2b_2c_2\bar{a}_2\bar{b}_2\bar{c}_2 + c_1\bar{a}_1^2\bar{b}_1\bar{b}_2\bar{c}_2 + c_2\bar{a}_1\bar{a}_2\bar{b}_1\bar{b}_2\bar{c}_2 - a_2b_1^2\bar{a}_1\bar{c}_1\bar{c}_2 - a_1b_1b_2\bar{a}_1\bar{c}_1\bar{c}_2 -$$
$$a_2b_1b_2\bar{a}_2\bar{c}_1\bar{c}_2 - a_1b_2^2\bar{a}_2\bar{c}_1\bar{c}_2 - b_1\bar{a}_1\bar{a}_2\bar{b}_1\bar{c}_1\bar{c}_2 - b_2\bar{a}_2^2\bar{b}_1\bar{c}_1\bar{c}_2 - b_1\bar{a}_1^2\bar{b}_2\bar{c}_1\bar{c}_2 - b_2\bar{a}_1\bar{a}_2\bar{b}_2\bar{c}_1\bar{c}_2 -$$
$$a_2b_1b_2\bar{a}_1\bar{c}_2^2 - a_2b_2^2\bar{a}_2\bar{c}_2^2 + b_1\bar{a}_1^2\bar{b}_1\bar{c}_2^2 + b_2\bar{a}_1\bar{a}_2\bar{b}_1\bar{c}_2^2 = 0$$

## References

1. J. Chen J. and Li, *J. Differential Geom.* **55** 355–384 (2000).
2. D. Joyce, *Quart. J. Math. Oxford* **49** 129–162 (1998).
3. W. Królikowski and R.M. Porter, *Ann. Polon. Math.* **59** 53–64 (1994).
4. A. Lichnerowicz, *Symp. Math.* **III**, Bologna, 341-402 (1970).
5. K. Nōno, *Bull. Fukuoka Univ. Ed. III* **35** 11–17 (1985).
6. M.V. Shapiro and N.L. Vasilevski, *Complex Variables Theory Appl.* **27** no.1 17–46 (1995).
7. A.J. Sommese, *Math. Ann.* **212** 191–214 (1975).
8. A. Sudbery, *Mat. Proc. Camb. Phil. Soc.* **85** 199–225 (1979).

# A NEW DOLBEAULT COMPLEX IN QUATERNIONIC AND CLIFFORD ANALYSIS

F. COLOMBO and I. SABADINI

*Dipartimento di Matematica*
*Politecnico di Milano*
*Via Bonardi, 9*
*20133 Milano, Italy*
*E-mail: fabcol@mate.polimi.it*
*E-mail: sabadini@mate.polimi.it*

A. DAMIANO and D. C. STRUPPA

*Department of Mathematical Sciences*
*George Mason University*
*Fairfax, VA 22030 USA*
*E-mail: adamiano@gmu.edu*
*E-mail: dstruppa@gmu.edu*

The papers introduces a new complex of differential forms which provides a fine resolution for the sheaf of regular functions in two quaternionic variables and the sheaf of monogenic functions in two vector variables. The paper announces some applications of this complex to the construction and interpretation of sheaves of quaternionic and Clifford hyperfunctions as equivalence classes of such differential forms.

**Key words:** Quaternions, differential forms, hyperfunctions
**Mathematics Subject Classification:** 30G25, 13D25.

## 1. Introduction

In this paper we present some new ideas on complexes of differential forms in the quaternionic and the Clifford settings. In particular we announce some new results and some research directions, which are particularly promising and are giving a new insight in the possibility of constructing a theory of multivariable quaternionic and Clifford valued hyperfunctions.

The study of the Cauchy-Fueter systems for functions of a quaternionic variable dates back to the twenties, when Fueter introduced the system as a way to generalize holomorphicity to the case of functions defined on the space of quaternions. The theory for the case of a single quaternionic variable is quite well known (and the solutions of the Cauchy-Fueter system are known as regular functions), while the case of several quaternionic variables (and correspondingly several Cauchy-Fueter equations) has only recently been understood, and the most up-to-date description is given in [3]. More recently, the field of Clifford analysis has developed as the study of solutions of the Dirac equation (or system) for Clifford valued functions (these functions are usually referred to as monogenic functions). Once again, one finds properties which generalize holomorphicity, and even though the theory is by now very well developed, see for example [1], and [4], most of the results are restricted to the case of a single Dirac equation. The only attempt to extend the study to the case of several Dirac equations is brought forward in [3], but the results are far from being complete.

Given the many formal similarities between the theory of regular functions of quaternionic variables, the theory of monogenic functions, and the classical theory of holomorphic functions, it is no surprise that several attempts have been made towards the creation of a theory of quaternionic and Clifford hyperfunctions to be interpreted as suitable boundary values of regular and monogenic functions. The only known work in the case of regular functions is due to the authors (see the references in [3]), while the most important work for the Clifford case is due to Sommen (see [6] and references therein). The state of the art on this problem is given in the recent monograph [3].

All the existing work, however, only deals with rather simple extensions of the theory of one-dimensional hyperfunctions. In other words, the existing literature always consider solutions to a single differential equation (though of course both the Cauchy-Fueter and the Dirac equation can be thought of as systems). Because of the way in which the Cauchy-Fueter and the Dirac operators are defined, and since the theory of regular functions in one quaternionic variable and the theory of monogenic functions are essentially uni-dimensional theories, the corresponding hyperfunction and microfunction theories always share the flavor of the standard theory of hyperfunctions in one variable.

Until recently, however, very little progress had been done in the attempt to find a way to develop and interpret a hyperfunction theory for boundary values of regular and monogenic functions in several quaternionic or Clifford variables. A significant obstacle to progress in this direction was the lack of an appropriate theory of differential forms which could fulfill the role that the Dolbeault theory plays in the complex case. In the last couple of years, there was finally some advance with the introduction, by F. Sommen, of the notion of megaforms and with the understanding of their connection with our analysis of multidimensional Dirac systems [5].

In this paper we announce several important results on megaforms for regular functions of several quaternionic variables and for multivariable monogenic functions. Specifically, in sections two and three, we show that it is possible to construct a Dolbeault-like complex, which provides a fine resolution for the sheaves of regular and monogenic functions in two variables. In section four we point out the many important consequences of our construction and we discuss how this will lead to a full hyperfunction theory for multivariable regular and monogenic functions. The detailed proofs will appear elsewhere ([2]).

## 2. Clifford megaforms for the two variables Dirac system

Let $\mathbb{R}^m$ be the real Euclidean space: we define $\mathbb{R}_m$ as the real Clifford algebra generated by the basis elements $\{e_1, \ldots, e_m\}$ together with the defining relations

$$e_j e_k + e_k e_j = -2\delta_{jk}.$$

Let $x_1, \ldots, x_m$ denote the standard (commuting) variables, $\underline{x} = \sum_{j=1}^m x_j e_j$ and let $f(\underline{x})$ be an $\mathbb{R}_m$–valued function. One may consider the action of the Dirac operator, or vector derivative,

$$\partial_{\underline{x}} : \ f(\underline{x}) \to \partial_{\underline{x}} f(\underline{x}) = \sum_{j=1}^m e_j \partial_{x_j} f(\underline{x})$$

and the solutions of the homogeneous equation $\partial_{\underline{x}} f(\underline{x}) = 0$, that are called monogenic functions.

Consider now several vector variables

$$\underline{x}_1, \ldots, \underline{x}_n, \qquad \underline{x}_j = \sum_{k=1}^m x_{jk} e_k,$$

and functions $f(\underline{x}_1, \ldots, \underline{x}_n)$ with values in $\mathbb{R}_m$. We then define monogenic functions in $n$ copies of $\mathbb{R}^m$ to be those functions which satisfy the following multivariable Dirac system:

$$\begin{cases} \partial_{\underline{x}_1} f = 0 \\ \quad \cdots \\ \partial_{\underline{x}_n} f = 0 \end{cases} \tag{1}$$

The final piece of notation we want to introduce is the notion of radial algebra.

**Definition 2.1.**
    Let $S$ be a set of objects which we will consider as "abstract vector variables". The radial algebra $R(S)$ is defined to be the associative algebra generated by $S$ over $\mathbb{R}$ with the defining relations

$$[\{x, y\}, z] = xyz + yxz - zxy - zyx = 0, \quad \text{for} \ \ x, y, z \in S. \tag{2}$$

Note that such an algebra could be constructed over any suitable ring, though in this paper we will confine ourselves to $\mathbb{R}$. Let $T(S)$ be the tensor algebra generated by the elements of $S$, and let $I(S)$ be the two-sided ideal generated by the polynomials

$$[\{x, y\}, z].$$

Then

$$R(S) = T(S)/I(S).$$

The notion of radial algebra is particularly useful, because it turns out that Dirac operators in several variables satisfy exactly the relations which define the radial algebra, and nothing more.

This section is devoted to a recap of some of the main results from [5], with special interest for the case of two variables (this means that our monogenic functions are defined on $(\mathbb{R}^m)^2$, and have values in $\mathbb{R}_m$). The ideas contained in this section are the inspiration for section three.

We begin by introducing what we have called megaforms as the analogue of classical differential forms for real coordinates. To define the classical basic differential forms one starts from the operator $d = \sum \partial_{x_j} dx_j$ acting on the algebra generated by $x_1, ..., x_m$, $dx_1, ..., dx_m$ together with the properties

(1) $d(\omega) = d(\sum F_j dg_j) = \sum dF_j \wedge dg_j$,
(2) $d(f\omega) = df \wedge \omega + f d\omega$,
(3) $dx_j \wedge dx_k = -dx_k \wedge dx_j$.

If one replaces $x_1, ..., x_m$ with the generators of a radial algebra, which we will denote in the same way, it is natural to replace the partial derivatives with the vector derivatives $\partial_{x_1}, ..., \partial_{x_n}$, but it is not so clear how to generalize the properties above. Instead, we try to generalize the formula "$d = \sum \partial_{x_j} dx_j$" in the case of vector derivatives $\partial_{x_1}, ..., \partial_{x_n}$, which themselves satisfy the defining relations for a radial algebra. To be able to make use of the radial algebra defining relations, we will construct spaces $F_k$ of forms, and differentials $d^k : F_k \to F_{k+1}$ such that $d^{k+1} d^k = 0$. However, we will recognize that the differentials $d^k$ may consist of two pieces: a "degree one" piece $d_1^k$ of the form $\sum_{j=1}^m D_j^k \partial_j$ and a "degree two" piece $d_2^k$ of the form $\sum_{i,j=1}^m D_{ij}^k \partial_i \partial_j$. The symbols of degree one, $D_j^k$, are the Dirac analogue of the $dx_j$ which are used in the classical de Rham complex, while the symbols of degree two, $D_{ij}^k$, are new symbols which will be necessary to reflect the existence of quadratic syzygies. As such, these symbols will have to satisfy some axioms in order to guarantee $d^{k+1} d^k = 0$. Note also that the symbol $D_j^k$ and the symbols $D_{ij}^k$ may, a priori, depend on $k$, though we will see that the full theory can be developed with only a minimal dependence on $k$.

**Definition 2.2.** Let $x_1, ..., x_n \in S$; the algebra $M(x_1, ..., x_n; S)$ of megaforms in the variables $x_1, ..., x_n$ with coefficients in $R(S)$ is the algebra which is generated over the algebra $R(S)$ by the set of "basic megaforms" $\{D_i^k, D_{j\ell}^k : i, j, \ell = 1, ..., n\}$ together with the identities derived from $d^{k+1} d^k = 0$.

To our purposes, the position of the various symbols $D_i^k$, $D_{j\ell}^k$ in the relations we obtain makes clear that we can omit the top label. It suffices to keep in mind that the relations obtained for $D_{j\ell} = D_{j\ell}^k$ at the $k$–th step cannot be used for $D_{j\ell} = D_{j\ell}^{k+1}$. From now on we will write $D_i$, $D_{j\ell}$, and we will leave it to the reader to distinguish among the different levels (in fact the order in which they appear identifies their level immediately).

We are now ready to discuss directly the case of systems of two Dirac operators.

Let $\mathcal{M}$ be the space of monogenic functions in two variables $x_1$, $x_2$. Let $F_0$ be the space of $\mathcal{C}^\infty$ functions in $x_1$, $x_2$ and let $F_1$ be the space of "1–forms" whose elements are written as $g = \sum_{i=1}^2 D_i g_i$, $g_i \in F_0$, so that if $d^0 = \sum_{i=1}^2 D_i \partial_i$, where

$\partial_i = \partial_{x_i}$, we have an exact sequence

$$0 \to \mathcal{M} \hookrightarrow F_0 \xrightarrow{d^0} F_1.$$

The next step in the construction of the complex consists in defining a space $F_2$ of "2–forms" and a suitable "differential" $d^1 : F_1 \to F_2$ such that $d^1 d^0 = 0$. As we explained above, we postulate that $d^1$ be made of two components $d_1^1$ and $d_2^1$ of degrees, respectively, one and two in the $\partial_i$' s. Thus we assume, for $g \in F_1$,

$$d^1 g = d_1^1 g + d_2^1 g = \sum_{j,k=1}^{2} D_k D_j \partial_k g_j + \sum_{i,j,k=1}^{2} D_{ki} D_j \partial_k \partial_i g_j.$$

The condition $d^1 d^0 = 0$ implies that $\sum_{j,k=1}^{2} D_k D_j \partial_k \partial_j f = 0$, i.e., that for any $k$ and $j$, $D_k D_j = 0$. But then, because of the form of elements in $F_1$, one may assume that $d_1^1 \equiv 0$, and therefore

$$d^1 g = d_2^1 g = \sum_{i,j,k=1}^{2} D_{ki} D_j \partial_k \partial_i g_j.$$

**Prop 2.1.** Let $d^0 = D_1 \partial_1 + D_2 \partial_2$ and $d^1 = D_{11} \partial_1^2 + D_{12} \partial_1 \partial_2 + D_{21} \partial_2 \partial_1 + D_{22} \partial_2^2$. Then $d^1 d^0 = 0$ implies the following:

$$D_{ij} D_i = 0 \quad i,j = 1,2 \qquad D_{ii} D_j + D_{ji} D_i = 0 \quad i,j = 1,2, \; i \neq j. \tag{3}$$

**Remark 2.1.** The relations (3) are the analogues of the complex relations $d\bar{z}_1 \wedge d\bar{z}_2 = -d\bar{z}_2 \wedge d\bar{z}_1$, $d\bar{z}_i \wedge d\bar{z}_i = 0$.

We now study the kernel of the map $d^1 : F_1 \to F_2$ and we note that in [5] we have shown that a 1–form $g$ is $d^1$–closed if and only if its components $g_j$ satisfy the compatibility conditions of the system $d^0 f = g$.

**Prop 2.2.** Let $g = D_1 g_1 + D_2 g_2$ be an element of $F_1$. Then $d^1 g = 0$ if and only if

$$\partial_i^2 g_j - \partial_j \partial_i g_i = 0, \quad i = 1,2 \tag{4}$$

i.e. $d^1 g = 0$ if and only if $(g_1, g_2)$ satisfy the compatibility conditions for the solvability of the system

$$\begin{cases} \partial_1 f = g_1 \\ \partial_2 f = g_2 . \end{cases}$$

We know, from the general theory, that the complex closes with one more linear condition that is the compatibility condition for the solvability of the system

$$\begin{cases} \partial_1^2 g_2 - \partial_2 \partial_1 g_1 = h_{12} \\ \partial_2^2 g_1 - \partial_1 \partial_2 g_2 = h_{21} . \end{cases} \tag{5}$$

Interestingly enough this condition can be derived using megaforms and their closure.

**Prop 2.3.** Let $d^1 = D_{11}\partial_1^2 + D_{12}\partial_1\partial_2 2 + D_{21}\partial_2\partial_1 + D_{22}\partial_2^2$ and let $d^2 = D_1\partial_1 + D_2\partial_2 + D_{11}\partial_1^2 + D_{12}\partial_1\partial_2 2 + D_{21}\partial_2\partial_1 + D_{22}\partial_2^2$ then $d^2 d^1 = 0$ implies for $i,j = 1,2,\ i \neq j$

$$D_{ii}^2 D_j = 0, \quad D_{ii}D_{jj}D_i - D_{ij}D_{ii}D_j = 0, \quad D_{ij}D_{jj}D_i = 0,$$

and

$$D_i D_{ii} D_j = 0, \quad D_2 D_{11} D_2 + D_1 D_{22} D_1 = 0.$$

**Prop 2.4.** Let $h = D_{11}D_2 h_{12} + D_{22}D_1 h_{21}$ be a generic element in $F_2$. Then $d^2 h = 0$ if and only if

$$\partial_1 h_{21} + \partial_2 h_{12} = 0,$$

i.e. $d^2 h = 0$ if and only if $h = (h_{12}, h_{21})$ satisfies the compatibility condition for the system (5).

On the basis of what we know from the syzygies of the Dirac complex in two operators (see for example [3]) we would now expect that the complex of differential forms should naturally close to zero. In fact, in [5] we establish that the differential operator $d^3$ is identically zero on $F_3$ and therefore we have the following result that concludes the description in the case of two variables:

**Theorem 2.1.** *Monogenic functions in two variables can be embedded in the following de Rham–like complex:*

$$0 \to \mathcal{R} \hookrightarrow F_0 \xrightarrow{d^0} F_1 \xrightarrow{d^1} F_2 \xrightarrow{d^2} F_3 \xrightarrow{d^3} 0.$$

**Remark 2.2.** The de Rham–like complex which we have constructed is self dual in the sense that $d^2$ is the transpose of $d^0$. This is not surprising because the same structure occurs in the resolution of the Dirac system in two variables

$$0 \longrightarrow R^{2^m}(-4) \longrightarrow R^{2 \cdot 2^m}(-3) \longrightarrow R^{2 \cdot 2^m}(-1) \xrightarrow{P^t} R^{2^m} \longrightarrow \mathcal{M}_2 \longrightarrow 0,$$

where

$$\mathcal{M}_2 = \frac{R^{2^m}}{P^t R^{2 \cdot 2^m}}.$$

**Remark 2.3.** In [5] we have shown how to apply the same ideas to the case of three Dirac operators. Unfortunately the results are not yet complete because of the lengthy computations, and so, in this paper, we have decided to limit our attention to the case of two operators.

**Remark 2.4.** Note that the sequence we have constructed is in fact a fine resolution of the sheaf of monogenic functions. This is a consequence of the fact that each space $F_j$ can be thought of as a direct sum of a certain number of copies of the sheaf of infinitely differentiable functions, which in itself is a fine sheaf. Another fine resolution, which in fact turns out to be isomorphic to this one, is described in [3] as a consequence of our general theory of syzygies for the Dirac system.

## 3. Quaternionic megaforms for the two dimensional Cauchy-Fueter system

In this section we use the same ideas introduced in the case of Dirac systems to discuss megaforms for the two-dimensional Cauchy-Fueter system. We denote by $\mathbb{H}$ the algebra of quaternions and by $q = x_0 + \mathbf{i}x_1 + \mathbf{j}x_2 + \mathbf{k}x_3$ a quaternion, where $x_\ell \in \mathbb{R}$ for $\ell = 0, ..., 3$. We define the Cauchy-Fueter operator as

$$\frac{\partial}{\partial \bar{q}} = \frac{\partial}{\partial x_0} + \mathbf{i}\frac{\partial}{\partial x_1} + \mathbf{j}\frac{\partial}{\partial x_2} + \mathbf{k}\frac{\partial}{\partial x_3}$$

with obvious meaning of the symbols. Differentiable functions which belong to the kernel of $\partial/\partial \bar{q}$ are called regular functions.

When we consider several quaternionic variables $q_r$, for $r = 1, ..., n$, we can define a Cauchy-Fueter operator for each of them by setting

$$\frac{\partial}{\partial \bar{q}_r} = \frac{\partial}{\partial x_{r0}} + \mathbf{i}\frac{\partial}{\partial x_{r1}} + \mathbf{j}\frac{\partial}{\partial x_{r2}} + \mathbf{k}\frac{\partial}{\partial x_{r3}}$$

where the double index for the variables has obvious meaning. Then, a differentiable function $f : \mathbb{H}^n \to \mathbb{H}$ is said to be regular if it is regular separately in each of its $n$ quaternionic variables. There are substantial, and important, differences between the Dirac case and the Cauchy-Fueter case; nevertheless, as indicated in [3], the case of dimension two still presents more similarities than differences, and so the techniques used in the previous section can be suitably modified for the quaternionic case. Again, as in the Dirac case, we introduce new symbols that will act as "differential forms". We construct the associative algebra generated by the degree one quaternionic megaforms

$$\{\check{D}_i^k, D_i^k \mid i = 1, ..., n, k \in \mathbb{N}\}$$

together with the degree two megaforms

$$\{\check{D}_{ij}^k, D_{ij}^k, \tilde{D}_{ij}^k, D_{ij}^{*k} \mid i, j = 1, ..., n, k \in \mathbb{N}\}.$$

Note that, as remarked before, we will omit the superscript $k$. As coefficients of the quaternionic megaforms we use the quaternionic derivatives $\partial_{\bar{q}_i}$, $i = 1 ... n$ and their conjugates $\partial_{q_i}$. Together they generate an algebra and satisfy the relations

$$\Delta_j \partial_{\bar{q}_i} = \partial_{\bar{q}_i} \Delta_j, \quad i, j = 1, ..., n$$

and

$$\Delta_j \partial_{q_i} = \partial_{q_i} \Delta_j, \quad i, j = 1, ..., n$$

where the Laplacian symbol is defined as $\Delta_j = \partial_{\bar{q}_j}\partial_{q_j} = \partial_{q_j}\partial_{\bar{q}_j}$. The proofs for the results in this section can be found in [2], but we will include a couple of the simplest computations to convey the flavor of this approach. Let $\mathcal{R}$ be the space of regular functions in two variables $q_1$, $q_2$. Let $F_0$ be the space of $C^\infty$ functions in $q_1$, $q_2$ and let $F_1$ be the space of "1-forms" whose elements are written as

$$g = \sum_{i=1}^{2} \check{D}_i \check{g}_i, \quad \check{g}_i \in F_0, \quad i = 1, 2$$

so that if $d^0 = \sum_{i=1}^{2} \check{D}_i \bar{\partial}_i$, where $\bar{\partial}_i = \partial_{\bar{q}_i}$, we have an exact sequence

$$0 \to \mathcal{R} \hookrightarrow F_0 \xrightarrow{d^0} F_1.$$

As before, we want to define a space $F_2$ of "2-forms" and a suitable "differential" $d^1 : F_1 \to F_2$ such that $d^1 d^0 = 0$, and as before we assume that $d^1$ is made of two components $d_1^1$ and $d_2^1$ of degrees, respectively, one and two. Let us define

$$d_1^1 = \check{D}_1 \bar{\partial}_1 + \check{D}_2 \bar{\partial}_2 + D_1 \partial_1 + D_2 \partial_2$$

where $\partial_i$ denotes $\partial/\partial q_i$ and

$$d_2^1 = \sum_{i,j=1}^{2} (\check{D}_{ij} \bar{\partial}_i \bar{\partial}_j + D_{ij} \partial_i \partial_j + \tilde{D}_{ij} \partial_i \bar{\partial}_j + D_{ij}^* \bar{\partial}_i \partial_j).$$

**Remark 3.1.** The condition $d^1 d^0 = 0$ implies that $d_1^1 d^0 = 0$, i.e.

$$\sum_{j,k=1}^{2} (\check{D}_k \check{D}_j \bar{\partial}_k \bar{\partial}_j + D_k \check{D}_j \partial_k \bar{\partial}_j) f = 0.$$

As a consequence, for any $k$ and $j$, $\check{D}_k \check{D}_j = 0$ and $D_k \check{D}_j = 0$. So one can assume that $d_1^1 \equiv 0$, and therefore

$$d^1 g = d_2^1 g.$$

**Prop 3.1.** Let $d^0 = \check{D}_1 \bar{\partial}_1 + \check{D}_2 \bar{\partial}_2$ and

$$d^1 = \sum_{i,j=1}^{2} (\check{D}_{ij} \bar{\partial}_i \bar{\partial}_j + D_{ij} \partial_i \partial_j + \tilde{D}_{ij} \partial_i \bar{\partial}_j + D_{ij}^* \bar{\partial}_i \partial_j).$$

Then $d^1 d^0 = 0$ implies the following:

$$\begin{aligned}
(\tilde{D}_{ii} + D_{ii}^*)\check{D}_i &= 0, \quad i = 1, 2, \\
D_{ji}^* \check{D}_i + (\tilde{D}_{ii} + D_{ii}^*)\check{D}_j &= 0, \quad i, j = 1, 2, \ i \neq j, \\
D_{ij} \check{D}_k &= 0, \quad i, j, k = 1, 2, \\
\check{D}_{ij} \check{D}_k &= 0, \quad i, j, k = 1, 2, \\
\tilde{D}_{ij} \check{D}_k &= 0, \quad i, j, k = 1, 2, \ i \neq j, \\
D_{ij}^* \check{D}_i &= 0, \quad i, j = 1, 2, \ i \neq j.
\end{aligned} \quad (6)$$

**Proof.** The condition $d_2^1 d^0 = 0$ implies that for any $f \in F_0$ it is

$$\sum_{i,j=1}^{2} (\check{D}_{ij}\check{D}_k\bar{\partial}_i\partial_j\bar{\partial}_k + D_{ij}\check{D}_k\partial_i\partial_j\bar{\partial}_k + \tilde{D}_{ij}\check{D}_k\partial_i\bar{\partial}_j\bar{\partial}_k + D_{ij}^*\check{D}_k\bar{\partial}_i\partial_j\bar{\partial}_k)f = 0.$$

By writing explicitly the right hand term, and using the relation $\partial_i\bar{\partial}_i\bar{\partial}_j = \bar{\partial}_j\partial_i\bar{\partial}_i$, i.e. $\Delta_i\bar{\partial}_j = \bar{\partial}_j\Delta_i$, $i,j = 1,2$ where $\Delta_i$ denotes the Laplacian with respect to the $i$-th variable, and grouping the various terms we get the statement. $\square$

We now consider the kernel of the map $d^1 : F_1 \to F_2$. We have that a 1–form $g$ is $d^1$–closed if and only if its components $g_j$ satisfy the compatibility conditions of the system $d^0 f = g$.

**Prop 3.2.** Let $g = \check{D}_1 g_1 + \check{D}_2 g_2$ be an element of $F_1$. Then $d^1 g = 0$ if and only if

$$\bar{\partial}_i\partial_i g_j - \bar{\partial}_j\partial_i g_i = 0, \quad i,j = 1,2,\ i \neq j, \tag{7}$$

i.e., $d^1 g = 0$ if and only if $(g_1, g_2)$ satisfy the compatibility conditions for the solvability of the system

$$\begin{cases} \bar{\partial}_1 f = g_1 \\ \bar{\partial}_2 f = g_2. \end{cases}$$

**Proof.** By the definition of $g$ and $d^1$ we have that $d^1 g = 0$ can be written as

$$\check{D}_{11}\check{D}_1\partial_1^2 g_1 + \check{D}_{12}\check{D}_1\bar{\partial}_1\bar{\partial}_2 g_1 + \check{D}_{21}\check{D}_1\bar{\partial}_2\bar{\partial}_1 g_1 + \check{D}_{22}\check{D}_1\partial_2^2 g_1$$

$$+D_{11}\check{D}_1\partial_1^2 g_1 + D_{12}\check{D}_1\partial_1\partial_2 g_1 + D_{21}\check{D}_1\partial_2\partial_1 g_1 + D_{22}\check{D}_1\partial_2^2 g_1$$

$$+\tilde{D}_{11}\check{D}_2\partial_1^2 g_2 + \tilde{D}_{12}\check{D}_2\partial_1\partial_2 g_2 + \tilde{D}_{21}\check{D}_2\partial_2\partial_1 g_2 + \tilde{D}_{22}\check{D}_2\partial_2^2 g_2$$

$$+D_{11}\check{D}_2\partial_1^2 g_2 + D_{12}\check{D}_2\partial_1\partial_2 g_2 + D_{21}\check{D}_2\partial_2\partial_1 g_2 + D_{22}\check{D}_2\partial_2^2 g_2 = 0.$$

In view of (6) this can be rewritten as

$$D_{11}^*\check{D}_1(\bar{\partial}_1\partial_1 g_1 - \partial_1\bar{\partial}_1 g_1) + D_{22}^*\check{D}_2(\bar{\partial}_2\partial_2 g_2 - \partial_2\bar{\partial}_2 g_2)$$

$$+D_{22}^*\check{D}_1(\bar{\partial}_2\partial_2 g_1 - \bar{\partial}_1\partial_2 g_2) + D_{22}\check{D}_1(\partial_2\bar{\partial}_2 g_1 - \bar{\partial}_1\partial_2 g_2)$$

$$+D_{11}^*\check{D}_2(\bar{\partial}_1\partial_1 g_2 - \bar{\partial}_2\partial_1 g_1) + \tilde{D}_{11}\check{D}_2(\partial_1\bar{\partial}_1 g_2 - \bar{\partial}_2\partial_1 g_1) = 0$$

which completes the proof. $\square$

From the general theory, it is known that the complex closes with one more linear condition that is the compatibility condition for the solvability of the system

$$\begin{cases} \Delta_1 g_2 - \bar{\partial}_2\partial_1 g_1 = h_{12} \\ \Delta_2 g_1 - \bar{\partial}_1\partial_2 g_2 = h_{21}. \end{cases} \tag{8}$$

This condition can be derived by simply using megaforms and their closure.

1028

**Prop 3.3.** Let

$$d^1 = \sum_{i,j=1}^{2} (\check{D}_{ij}\bar{\partial}_i\bar{\partial}_j + D_{ij}\partial_i\partial_j + \tilde{D}_{ij}\partial_i\bar{\partial}_j + D_{ij}^*\bar{\partial}_i\partial_j)$$

and let

$$d^2 = d_1^2 + d_2^2 = \check{D}_1\bar{\partial}_1 + \check{D}_2\bar{\partial}_2 + D_1\partial_1 + D_2\partial_2$$

$$+ \sum_{i,j=1}^{2} (\check{D}_{ij}\bar{\partial}_i\bar{\partial}_j + D_{ij}\partial_i\partial_j + \tilde{D}_{ij}\partial_i\bar{\partial}_j + D_{ij}^*\bar{\partial}_i\partial_j);$$

then $d^2 d^1 = 0$ implies for $i,j = 1,2$, $i \neq j$

$$\check{D}_1 D_{11}^* \check{D}_2 + \check{D}_1 \tilde{D}_{11} \check{D}_2 = 0,$$

$$\check{D}_i D_{jj}^* \check{D}_i + \check{D}_i \tilde{D}_{jj} \check{D}_i = 0,$$

$$(D_1 D_{22}^* \check{D}_1 + D_1 \tilde{D}_{22} \check{D}_1) - (D_2 D_{11}^* \check{D}_2 + D_2 \tilde{D}_{11} \check{D}_2) = 0,$$

$$D_i D_{ii}^* \check{D}_j + D_i \tilde{D}_{ii} \check{D}_j = 0,$$

$$D_2 D_{22}^* \check{D}_1 + D_2 \tilde{D}_{22} \check{D}_1 = 0,$$

$$\tilde{D}_{ii} D_{ii}^* \check{D}_j + \tilde{D}_{ii} \tilde{D}_{ii} \check{D}_j + D_{ii}^* D_{ii}^* \check{D}_j + D_{ii}^* \tilde{D}_{ii} \check{D}_j = 0, \quad i,j=1,2, \quad i \neq j,$$

$$D_{ii} D_{jj}^* \check{D}_i + D_{ii} \tilde{D}_{jj} \check{D}_i - D_{ij} D_{ii}^* \check{D}_j - D_{ij} \tilde{D}_{ii} \check{D}_j = 0, \quad i,j=1,2, \quad i \neq j,$$

$$\tilde{D}_{ii} D_{jj}^* \check{D}_i + \tilde{D}_{ii} \tilde{D}_{jj} \check{D}_i + D_{ii}^* D_{jj}^* \check{D}_i + D_{ii}^* \tilde{D}_{jj} \check{D}_i - D_{ij}^* D_{ii}^* \check{D}_j - D_{ij}^* \tilde{D}_{ii} \check{D}_j = 0, \quad i,j=1,2, \quad i \neq j,$$

$$\check{D}_{ij} D_{kk}^* \check{D}_\ell + \check{D}_{ij} \tilde{D}_{kk} \check{D}_\ell = 0, \quad i,j,k,\ell = 1,2, \quad k \neq \ell,$$

$$D_{ii} D_{ii}^* \check{D}_j + D_{ii} \tilde{D}_{ii} \check{D}_j = 0, \quad i,j,=1,2, \quad i \neq j,$$

$$\tilde{D}_{ij} D_{kk}^* \check{D}_\ell + \tilde{D}_{ij} \tilde{D}_{kk} \check{D}_\ell = 0, \quad i,j,k,\ell = 1,2, \quad k \neq \ell, \quad i \neq j,$$

$$D_{ji}^* D_{ii}^* \check{D}_j + D_{ji}^* \tilde{D}_{ii} \check{D}_j = 0, \quad i,j,=1,2, \quad i \neq j,$$

$$D_{ji} D_{ii}^* \check{D}_j + D_{ji} \tilde{D}_{ii} \check{D}_j = 0, \quad i,j,=1,2, \quad i \neq j.$$

**Prop 3.4.** Let

$$h = (D_{11}^* \check{D}_2 + \tilde{D}_{11} \check{D}_2)h_{12} + (D_{22}^* \check{D}_1 + \tilde{D}_{22} \check{D}_1)h_{21}$$

be a generic element in $F_2$. Then $d^2 h = 0$ if and only if

$$\partial_1 h_{21} + \partial_2 h_{12} = 0$$

i.e. $d^2 h = 0$ if and only if $h = (h_{12}, h_{21})$ satisfies the compatibility condition for the system (8).

The analysis can be pushed one more step, to the closure of the complex, as indicated in the next two propositions from [2].

**Prop 3.5.** Let

$$d^3 = d_1^3 + d_2^3 = \check{D}_1\bar{\partial}_1 + \check{D}_2\bar{\partial}_2 + D_1\partial_1 + D_2\partial_2$$

$$+ \sum_{i,j=1}^{2}(\check{D}_{ij}\bar{\partial}_i\bar{\partial}_j + D_{ij}\partial_i\partial_j + \tilde{D}_{ij}\partial_i\bar{\partial}_j + D_{ij}^*\bar{\partial}_i\partial_j).$$

Then $d^3 d_1^2 = 0$ implies

(1) $\check{D}_i D_1 D_{22}^* \check{D}_1 + \check{D}_i D_1 \tilde{D}_{22}\check{D}_1 = 0, \quad i = 1,2,$

(2) $D_i D_1 D_{22}^* \check{D}_1 + D_i D_1 \tilde{D}_{22}\check{D}_1 = 0, \quad i = 1,2,$

(3) $(\tilde{D}_{ii} D_1 D_{22}^* \check{D}_1 + \tilde{D}_{ii} D_1 \tilde{D}_{22}\check{D}_1) + (D_{ii}^* D_1 D_{22}^* \check{D}_1 + D_{ii}^* D_1 \tilde{D}_{22}\check{D}_1) = 0, \quad i,j = 1,2, i \neq j,$

(4) $(D_{ij} D_1 D_{22}^* \check{D}_1 + D_{ij} D_1 \tilde{D}_{22}\check{D}_1) = 0, \quad i,j = 1,2,$

(5) $(\check{D}_{ij} D_1 D_{22}^* \check{D}_1 + \tilde{D}_{ij} D_1 \tilde{D}_{22}\check{D}_1) = 0, \quad i,j = 1,2,$

(6) $(\tilde{D}_{ij} D_1 D_{22}^* \check{D}_1 + \tilde{D}_{ij} D_1 \tilde{D}_{22}\check{D}_1) = 0, \quad i,j = 1,2, i \neq j,$

(7) $(D_{ij}^* D_1 D_{22}^* \check{D}_1 + D_{ij}^* D_1 \tilde{D}_{22}\check{D}_1) = 0, \quad i,j = 1,2, i \neq j.$

**Prop 3.6.** Let

$$k = (D_1 D_{22}^* \check{D}_1 + D_1 \tilde{D}_{22}\check{D}_1)k_0$$

be a generic element in $F_3$. Then $d^3 k = 0$.

The interest of these results is clarified by the next theorem and remark.

**Theorem 3.1.** *Regular functions in two quaternionic variables can be embedded in the following Dolbeault–like complex:*

$$0 \to \mathcal{R} \hookrightarrow F_0 \xrightarrow{d^0} F_1 \xrightarrow{d^1} F_2 \xrightarrow{d^2} F_3 \xrightarrow{d^3} 0.$$

**Remark 3.2.** If we denote by $\mathcal{C}^\infty$ the sheaf of quaternionic valued infinitely differentiable functions on $\mathbb{H}^2$, we can rewrite the various spaces of megaforms as follows:

$$F_0 := \mathcal{C}^\infty,$$

$$F_1 := \{\textstyle\sum_{i=1}^{2} \check{D}_i \check{g}_i, \quad \check{g}_i \in \mathcal{C}^\infty\},$$

$$F_2 := \{(D_{11}^* \tilde{D}_2 + \tilde{D}_{11}\check{D}_2)h_{12} + (D_{22}^* \check{D}_1 + \tilde{D}_{22}\check{D}_1)h_{21} : h_{12}, h_{21} \in \mathcal{C}^\infty\},$$

$$F_3 := \{(D_1 D_{22}^* \check{D}_1 + D_1 \tilde{D}_{22}\check{D}_1)k_0 : k_0 \in \mathcal{C}^\infty\}.$$

This shows, in particular, that we have in fact constructed a fine resolution for the sheaf of regular functions in two variables. As in the case of monogenic functions, this resolution is isomorphic to the one described in [3].

## 4. Applications of megaforms, and further directions for research

The previous two sections have shown that, at least when we restrict our attention to the case of two operators, it is possible to embed the sheaves $\mathcal{M}$ and $\mathcal{R}$ in suitable fine resolutions whose objects are sheaves of differential forms (actually, what we have called megaforms) with infinitely differentiable coefficients.

These embeddings are very important as they allow us to calculate explicitly the relative cohomology of the sheaves of regular and monogenic functions. This, in turn, will allow the construction and the interpretation of suitable sheaves of hyperfunctions (see [3] for a quick review of related notions, and their interpretation in the quaternionic and Clifford settings).

Let us therefore mention here some of the most important results which we have discussed and fully proved in [2]. To begin with, the usual arguments using fine sheaves prove the following result:

**Theorem 4.1.** *Let $U$ be an open convex set in $\mathbb{H}^2$. Then, for any $j \geq 1$, it is*

$$H^j(U, \mathcal{R}) = 0.$$

Its monogenic version states:

**Theorem 4.2.** *Let $U$ be an open convex set in $\mathbb{R}^{2m}$. Then, for any $j \geq 1$, it is*

$$H^j(U, \mathcal{M}) = 0.$$

These results allow us to introduce suitable spaces of hyperfunctions.

**Definition 4.1.** Let $L$ be an initial variety for the Cauchy-Fueter system in two variables (see [2] for details). The presheaf associated to the pairing $(V, H_L^3(V, \mathcal{R}))$, with $V$ an open set in $L$, is a sheaf which we will call sheaf of $\mathbb{H}$-hyperfunctions. In the sequel we will denote this sheaf by $\mathcal{B}_{\mathbb{H}^2, L}$.

The vanishing properties of the cohomology groups of $\mathcal{R}$ and the definition above imply the following:

**Theorem 4.3.** *Let $U$ be any open convex set in $\mathbb{H}^2$ and let $L$ be a five-dimensional initial variety for the Cauchy-Fueter system. Then $\mathcal{B}_{\mathbb{H}^2, L}(U \cap L) \cong H^2(U \setminus L, \mathcal{R})$.*

Note now that the open set $U \setminus L$ is not convex, and therefore its $\mathcal{R}$ cohomology does not vanish. In fact, our results show that its elements are equivalence classes of two-forms so that we can identify any quaternionic hyperfunction with an equivalence class of two-forms. By recalling the characterization of $F_2$ and $F_1$, we can identify a quaternionic hyperfunction $f$ on $U \cap L$ with an equivalence class of pairs $(f_1, f_2)$ of quaternionic valued, infinitely differentiable functions on $U \setminus L$, such that $\partial_1 f_1 + \partial_2 f_2 = 0$. Two pairs $(f_1, f_2)$ and $(g_1, g_2)$ are equivalent if there exist another pair $(\alpha_1, \alpha_2)$ such that

$$\begin{cases} \Delta_1 \alpha_2 - \bar{\partial}_2 \partial_1 \alpha_1 = f_1 - g_1 \\ \Delta_2 \alpha_1 - \bar{\partial}_1 \partial_2 \alpha_2 = f_2 - g_2. \end{cases}$$

One can proceed in a totally analogous way for the case of monogenic functions.

**Definition 4.2.** Let $L$ be an initial variety for the Dirac system in two variables. The presheaf associated to the pairing $(V, H_L^3(V, \mathcal{M}))$, with $V$ an open set in $L$, is a sheaf which we will call sheaf of Dirac-hyperfunctions. In the sequel we will denote this sheaf by $\mathcal{D}_{\mathbb{R}^{2m}, L}$.

Using the fine resolution we have described for the sheaf of monogenic functions we can deduce the Clifford analogues of all the previous results. The cohomology of $\mathcal{M}$ vanishes on all convex open sets in $\mathbb{R}^{2m}$, and the sheaf of hyperfunctions ends up being isomorphic to an appropriate sheaf of 2-megaforms.

There are still many interesting open problems. For example, one would like to extend these arguments to the case of more than two variables. In [5] we have shown how to construct a theory of megaforms for the Dirac operators. However, the theory we have constructed is not fully complete, and is not sufficient to recuperate all the results announced in this paper. The case of Dirac operators seems to be independent of the number of operators, but we would like to be able to prove this conclusively, so to have a general theory of Dirac-hyperfunctions, irrespective of dimension.

The situation is much more complicated in the case of quaternions. As we have shown, the case of three Cauchy-Fueter operators is quite distinctive, and introduces new phenomena which did not occur in dimension two. We do not have, at this time, an appropriate megaform theory for this case, nor we have a clear idea on how this could be done. On the positive side, we know that no new phenomena emerge when we consider more than three Cauchy-Fueter operators, so that if we are able to understand the case of three operators, we should be able to have a general theory.

Finally, and more ambitiously, we believe it should be possible to construct a general theory of abstract megaforms for general systems of differential equations. Since such a theory would imply some vanishing of cohomologies, we should be prepared to find conditions on a system for it to have an associated theory of megaforms (in other words, we can envision differential operators, whose kernels have non-vanishing cohomology, and for these operators we could not construct appropriate megaforms). The development of such a theory would rely on an appropriate understanding of the syzygies of the operators, and one should be open to the case in which higher order megaforms may be needed.

### Acknowledgements

The authors are grateful to the organizers of ISAAC 2005 for inviting them to speak at the conference. The authors are also grateful to George Mason University for its financial support during various stages of this work.

1032

## References

1. F. Brackx, R. Delanghe, F. Sommen, Clifford Analysis, *Pitman Res. Notes in Math.*, **76**, 1982.
2. F. Colombo, A. Damiano, I. Sabadini, D. C. Struppa, Quaternionic hyperfunctions on 5-dimensional varieties in $\mathbf{H}^2$, *J. Geom. Anal.* **17**, 459 (2007).
3. F. Colombo, I. Sabadini, F. Sommen, D. C. Struppa, Analysis of Dirac systems and computational algebra, *Progress in Mathematical Physics*, Birkhäuser Boston, 2004.
4. J. Gilbert, M. Murray, Clifford algebras and Dirac operators in harmonic analysis, *Cambridge Univ. Press*, **26**, 1990.
5. I. Sabadini, F. Sommen, D.C. Struppa, The Dirac complex on abstract vector variables: megaforms, *Exp. Math.*, **12**, 351 (2003).
6. F. Sommen, , Microfunctions with values in a Clifford algebra II, *Sci. Papers College Arts Sci. Univ. Tokyo*, **36**, 15 (1986).

# SOME INTEGRAL REPRESENTATIONS AND ITS APPLICATIONS IN CLIFFORD ANALYSIS *

ZHANG ZHONGXIANG

School of Mathematics and Statistics,
Wuhan University, Wuhan 430072, P. R. China

E-mail: zhangzx9@sohu.com

In this paper, we mainly study the integral representations for functions $f$ with values in a universal Clifford algebra $C(V_{n,n})$, where $f \in \Lambda(f, \overline{\Omega})$, $\Lambda(f, \overline{\Omega}) = \left\{ f | f \in C^\infty(\overline{\Omega}, C(V_{n,n})), \max_{x \in \overline{\Omega}} \left| D^j f(x) \right| = O(M^j)(j \to +\infty), \text{for some } M, 0 < M < +\infty \right\}$. The integral representations of $T_i f$ are also given. Some properties of $T_i f$ and $\Pi f$ are shown. As applications of the higher order Pompeiu formula, we get the solutions of the Dirichlet problem and the inhomogeneous equations $D^k u = f$.

**Key words:** Universal Clifford algebra, integral representation, $T_i$-operator, $\Pi$−operator
**Mathematics Subject Classification:** 30G35, 45J05

## 1. Introduction and Preliminaries

Integral representation formulas of Cauchy-Pompeiu type expressing complex valued, quaternionic and Clifford algebra valued functions have been well developed in [1−9,12−23,25,28,30] etc. These integral representation formulas serve to solve boundary value problems for partial differential equations. In [2,3], H. Begehr gave the different integral representation formulas for functions with values in a Clifford algebra $C(V_{n,0})$, the integral operators provide particular weak solutions to the inhomogeneous equations $\partial^k \omega = f$, $\triangle^k \omega = g$ and $\partial \triangle^k \omega = h$. In [5,28], the higher order Cauchy-Pompeiu formulas for functions with values in a universal Clifford algebra $C(V_{n,n})$ are obtained. In [18], G.N. Hile gave the detailed properties of the $T$-operator by following the techniques of Vekua. In [15,16], K. Gürlebeck gave many properties of the $\Pi$-operator. In [20], H. Malonek and B. Müller gave some properties of the vectorial integral operator $\overrightarrow{\Pi}$. In [7,21,25], the integral representations related with the Helmholtz operator are given, the weak solutions of the inhomogeneous equations $L^k u = f$ and $L_*^k u = f$, $k \geq 1$, are obtained, where $Lu = Du + uh$ and $L_* u = uD - hu$, $h = \sum_{i=1}^{n} h_i e_i$, $D$ is the Dirac operator. In this paper, we shall continue to study the properties of Cauchy-Pompeiu operator, higher order Cauchy-Pompeiu operator and $\Pi$ operator for $f \in \Lambda(f, \overline{\Omega})$, where $\Lambda(f, \overline{\Omega}) =$

---

* "Project supported by DAAD K. C. Wong Fellowship and SRF for ROCS, SEM.").

$$\left\{ f \mid f \in C^\infty(\overline{\Omega}, C(V_{n,n})), \max_{x \in \overline{\Omega}} \left| D^j f(x) \right| = O(M^j)(j \to +\infty), \text{for some } M, 0 < M < +\infty \right\},$$

the integral representations of $T_i f$ are given, some properties of $T_i f$ and $\Pi f$ are shown. As applications, we get the strong solutions of the Dirichlet problem and the inhomogeneous equations $D^k u = f$ which improve the results in [2,30]. We refer to the preliminaries and notations of the universal Clifford algebra $C(V_{n,n})$ to be found in [5,7,8,28-30].

## 2. Integral Representations

In this section, we shall give the integral representations for $f$ and $T_i f$, $i \geq 1$, $f \in \Lambda(f, \overline{\Omega})$, where

$$\Lambda(f, \overline{\Omega})$$

$$= \left\{ f \mid f \in C^\infty(\overline{\Omega}, C(V_{n,n})), \max_{x \in \overline{\Omega}} \left| D^j f(x) \right| = O(M^j)(j \to +\infty), \text{for some } M, 0 < M < +\infty \right\}.$$

In [5,28] the kernel functions

$$H_j^*(x) = \begin{cases} \dfrac{A_j}{\omega_n} \dfrac{\mathbf{x}^j}{\rho^n(x)}, & n \text{ is odd}; \\[2ex] \dfrac{A_j}{\omega_n} \dfrac{\mathbf{x}^j}{\rho^n(x)}, & 1 \leq j < n, \ n \text{ is even}; \\[2ex] \dfrac{A_{j-1}}{2\omega_n} \log(\mathbf{x}^2), & j = n, \ n \text{ is even}; \\[2ex] \dfrac{A_{n-1}}{2\omega_n} C_{l,0} \mathbf{x}^l \left( \log(\mathbf{x}^2) - 2 \sum_{i=0}^{l-1} \dfrac{C_{i+1,0}}{C_{i,0}} \right), & j = n+l, l > 0, \ n \text{ is even}; \end{cases}$$

$$(1)$$

are constructed for any $j \geq 1$, where $\mathbf{x} = \sum_{k=1}^n x_k e_k$, $\rho(x) = \left( \sum_{k=1}^n x_k^2 \right)^{\frac{1}{2}}$, $\omega_n$ denotes the area of the unit sphere in $\mathcal{R}^n$, and

$$A_j = \frac{1}{2^{\left[\frac{j-1}{2}\right]}\left[\frac{j-1}{2}\right]! \prod\limits_{r=1}^{\left[\frac{j}{2}\right]}(2r-n)}, \quad 1 \leq j < n \text{ if n is even and } j \in N^* \text{if n is odd}, \quad (2)$$

$$C_{j,0} = \begin{cases} 1, & j = 0, \\[2ex] \dfrac{1}{2^{\left[\frac{j}{2}\right]}\left[\frac{j}{2}\right]! \prod\limits_{\mu=0}^{\left[\frac{j-1}{2}\right]}(n+2\mu)}, & j \in \mathbf{N}^* = \mathbf{N}\backslash\{0\}. \end{cases} \quad (3)$$

**Lemma 2.1 (Higher order Cauchy-Pompeiu formula)** (see[28]) *Suppose that $M$ is an $n$–dimensional differentiable compact oriented manifold contained in some open non empty subset $\Omega \subset \mathcal{R}^n$, $f \in C^{(r)}(\Omega, C(V_{n,n}))$, $r \geq k$, moreover $\partial M$ is* *given the induced orientation, for each $j = 1, \cdots, k$, $H_j^*(x)$ is as above. Then, for $z \in \overset{\circ}{M}$*

$$f(z) = \sum_{j=0}^{k-1} (-1)^j \int_{\partial M} H_{j+1}^*(x - z) \mathrm{d}\sigma_x D^j f(x) + (-1)^k \int_M H_k^*(x - z) D^k f(x) \mathrm{d}x^N. \quad (4)$$

In the following, $\Omega$ is supposed to be an open non empty subset of $\mathcal{R}^n$ with a Liapunov boundary $\partial\Omega$. Denote

$$T_i f(z) = (-1)^i \int_\Omega H_i^*(x - z) f(x) \mathrm{d}x^N \quad (5)$$

where $H_i^*(x)$ is denoted as in (1), $i \in N^*$, $f \in L^p(\Omega, C(V_{n,n}))$, $p \geq 1$. The operator $T_1$ is the Pompeiu operator $T$. Especially, we denote $f$ as $T_0 f$.

In [30], it is shown that, if $f \in L^p(\Omega, C(V_{n,n})), p \geq 1$, then $Tf \in C^\alpha(\overline{\Omega}, C(V_{n,n})), \alpha = \dfrac{p - n}{p}$. $T_k f$ provides a particular weak solution to the inhomogeneous equation $D^k \omega = f$(weak) in $\Omega$. In this section, we shall show that, if $f \in \Lambda(f, \overline{\Omega})$, then $T_i f \in C^\infty(\Omega, C(V_{n,n}))$, $i \in \mathbf{N}^*$ and $T_k f$ provides a particular solution to the inhomogeneous equation $D^k \omega = f$ in $\Omega$.

**Theorem 2.1.** *Let $\Omega$ be an open non empty bounded subset of $\mathcal{R}^n$ with a Liapunov boundary $\partial\Omega$, $f \in \Lambda(f, \overline{\Omega})$. Then, for $z \in \Omega$*

$$T_i f(z) = \sum_{j=0}^{\infty} (-1)^{j+i} \int_{\partial\Omega} H_{j+i+1}^*(x - z) \mathrm{d}\sigma_x D^j f(x), \quad i \in \mathbf{N}. \quad (6)$$

**Proof** Step 1. For $f \in \Lambda(f, \overline{\Omega})$, we shall firstly prove

$$T_i f(z) = \sum_{j=0}^{k} (-1)^{j+i} \int_{\partial\Omega} H_{j+i+1}^*(x - z) \mathrm{d}\sigma_x D^j f(x)$$
$$+ (-1)^{i+k+1} \int_\Omega H_{i+k+1}^*(x - z) D^{k+1} f(x) \mathrm{d}x^N, \quad (7)$$

where $i, k \in \mathbf{N}$, $z \in \Omega$. It is obvious that (7) is the direct result of Lemma 2.1 for $i = 0$.

For $i \geq 1$, in view of the properties of the kernel functions of $H_j^*(x - z)$

$$D\left[H_{j+1}^*(x - z)\right] = \left[H_{j+1}^*(x - z)\right] D = H_j^*(x - z), \quad x \in \mathcal{R}_z^n, \text{ for any } j \geq 1. \quad (8)$$

Combining Stokes formulas with (8), we have

$$(-1)^i \int_{\Omega \backslash B(z,\varepsilon)} H_i^*(x - z) f(x) \mathrm{d}x^N = \sum_{j=0}^{k} (-1)^{j+i} \int_{\partial(\Omega \backslash B(z,\varepsilon))} H_{j+i+1}^*(x - z) \mathrm{d}\sigma_x D^j f(x)$$
$$+ (-1)^{i+k+1} \int_{\Omega \backslash B(z,\varepsilon)} H_{i+k+1}^*(x - z) D^{k+1} f(x) \mathrm{d}x^N. \quad (9)$$

For $i \geq 1$ and $j \geq 0$, it is easy to check that,

$$\lim_{\varepsilon \to 0} \int_{\partial B(z,\varepsilon)} H^*_{j+i+1}(x - z)\mathrm{d}\sigma_x D^j f(x) = 0. \tag{10}$$

In view of the weak singularity of the kernel functions and (10), taking limits as $\varepsilon \to 0$ in (9), (7) holds.

Step 2. For $f \in \Lambda(f, \overline{\Omega})$, we shall show that

$$\lim_{k \to \infty} \max_{z \in \overline{\Omega}} \left| \int_{\Omega} H^*_{i+k+1}(x - z)D^{k+1}f(x)\mathrm{d}x^N \right| = 0. \tag{11}$$

Since $f \in \Lambda(f, \overline{\Omega})$, then there exist constants $C_0, M, 0 < C_0, M < +\infty$, and $N \in \mathbf{N}^*$, such that for any $k \geq N$

$$\max_{x \in \overline{\Omega}} \left| D^k f(x) \right| \leq C_0 M^k. \tag{12}$$

Case 1. $n$ is odd. For any $k \geq N$, we have

$$\left| \int_{\Omega} H^*_{i+k+1}(x - z)D^{k+1}f(x)\mathrm{d}x^N \right| \leq 2^n A_{i+k+1} C_0 V(\Omega) M^{k+1} \delta^{i+k+1-n}, \tag{13}$$

where $\delta = \sup_{x_1, x_2 \in \Omega} \rho(x_1 - x_2)$, $V(\Omega)$ denotes the volume of $\Omega$. It is obvious that the series

$$\sum_{k=1}^{\infty} 2^n A_{i+k+1} C_0 V(\Omega) M^{k+1} \delta^{i+k+1-n} \tag{14}$$

converges. Then

$$\lim_{k \to \infty} 2^n A_{i+k+1} C_0 V(\Omega) M^{k+1} \delta^{i+k+1-n} = 0, \tag{15}$$

thus (11) holds.

Case 2. $n$ is even. In view of (1) and (3), it can be similarly proved that (11) holds.

Combining (7) with (11), taking limits $k \to \infty$ in (7), (6) follows.

By Theorem 2.1, we have

**Corollary 2.1.** *Suppose that $f$ is $k$-regular in a domain $U$ in $\mathcal{R}^n$, $\Omega$ is an open non empty bounded subset of $U$ with a Liapunov boundary $\partial\Omega$. Then, for $z \in \Omega$*

$$T_i f(z) = \sum_{j=0}^{k-1} (-1)^{j+i} \int_{\partial\Omega} H^*_{j+i+1}(x - z)\,\mathrm{d}\sigma_x D^j f(x), \quad i \in \mathbf{N}. \tag{16}$$

**Remark 2.1.** For $i = 0$, (16) is exactly the higher order Cauchy integral formula which has been obtained in [5,28]. Analogous higher order Cauchy integral formula can be also found in [2,3,12,23].

**Corollary 2.2.** *Let $\Omega$ be an open non empty bounded subset of $\mathcal{R}^n$ with a Liapunov boundary $\partial\Omega$, $f \in \Lambda(f,\overline{\Omega})$. Then, for $z \in \Omega$*

$$D[T_{i+1}f] = T_i f, \quad i \in \mathbf{N}. \tag{17}$$

**Remark 2.2.** Corollary 2.2 implies that $T_k f$ provides a particular solution to the inhomogeneous equation $D^k \omega = f$ in $\Omega$ for $f \in \Lambda(f,\overline{\Omega})$. Especially, suppose $U$ is a domain in $\mathcal{R}^n$, $\Omega$ is an open non empty bounded subset of $U$ with a Liapunov boundary $\partial\Omega$, $f$ is regular in $U$, then $T_k f$ is $(k+1)$-regular in $\Omega$. This result gives an improved result in [2,30] under the assumption of $f \in \Lambda(f,\overline{\Omega})$.

**Corollary 2.3.** *Let $U$ be a domain in $\mathcal{R}^n$, $\Omega$ be an open non empty bounded subset of $U$ with a Liapunov boundary $\partial\Omega$, $f$ be a solution of equation $Lu = 0$ in $U$, where $Lu = Du + uh$, $h = \sum\limits_{i=1}^{n} h_i e_i, h_i \in \mathcal{R}$ or $h$ be a real (complex) number. Then for $z \in \Omega$*

$$T_i f(z) = \sum_{j=0}^{\infty} (-1)^{j+i} \int_{\partial\Omega} H^*_{j+i+1}(x-z) d\sigma_x D^j f(x), \quad i \in \mathbf{N}. \tag{18}$$

**Proof.** Obviously, if $f$ is a solution of equation $Lu = 0$ in $U$, where $Lu = Du + uh$, $h = \sum\limits_{i=1}^{n} h_i e_i$ or $h$ is a real (complex) number, then $f \in \Lambda(f,\overline{\Omega})$. By Theorem 2.1, the result follows.

**Example 2.1.** Suppose $u_i(x) = \sum\limits_{k=0}^{\infty} \dfrac{(\alpha x_i e_i)^k}{k!} \triangleq e^{\alpha x_i e_i}$, $i = 1, \cdots, n$, where $\alpha$ is a real number. Clearly, $Du_i(x) = \alpha u_i(x)$. Thus for $u_i(x)$, $z \in \Omega$, by Corollary 2.3, (18) holds.

**Example 2.2.** Suppose $h = \sum\limits_{i=1}^{n} h_i e_i, h_i \in \mathcal{R}$. Denote $R = |h| = \sqrt{\sum\limits_{i=1}^{n} h_i^2}$. Obviously, $e^{Rx_i e_i}$ satisfies $Du - Ru = 0$, thus $e^{Rx_i e_i}$ is also a solution of the Helmholtz equation $\triangle u - R^2 u = 0$. Then $e^{Rx_i e_i}(R - h)$ is a solution of equation $Du + uh = 0$. For $e^{Rx_i e_i}(R - h)$, $z \in \Omega$, by Corollary 2.3, (18) holds.

$\Omega$ is supposed to be an open non empty subset of $\mathcal{R}^n$ with a Liapunov boundary $\partial\Omega$. Denote

$$\Pi f(z) = \begin{cases} \displaystyle\int_{\Omega} K(x-z)f(x)\mathrm{d}x^N, & z \in \Omega, \\[2mm] \displaystyle\lim_{\substack{\xi \to z \\ \xi \in \Omega}} \int_{\Omega} K(x-\xi)f(x)\mathrm{d}x^N & z \in \partial\Omega, \end{cases} \tag{19}$$

where

$$K(x) = \frac{1}{\omega_n}\left(\frac{(2-n)e_1}{\rho^n(x)} - \frac{nxe_1x}{\rho^{n+2}(x)}\right), \quad x \in \mathcal{R}_0^n. \tag{20}$$

$f \in H^\alpha(\overline{\Omega}, C(V_{n,n})), 0 < \alpha \leq 1, \Pi f$ is a singular integral to be taken in the Cauchy principal sense. In [30], we have proved the existence and Hölder continuity of $\Pi f$ in $\overline{\Omega}$.

For $u \in H^\alpha(\partial\Omega, C(V_{n,n})), 0 < \alpha \leq 1$, denote

$$(F_{\partial\Omega} u)(x) = \int_{\partial\Omega} H_1^*(y - x) d\sigma_y u(y), \quad x \in \mathcal{R}^n \setminus \partial\Omega. \tag{21}$$

$$(S_{\partial\Omega} u)(x) = \int_{\partial\Omega} H_1^*(y - x) d\sigma_y u(y), \quad x \in \partial\Omega. \tag{22}$$

$$\left(F_{\partial\Omega}^+ u\right)(x) = \begin{cases} (F_{\partial\Omega} u)(x), & x \in \Omega^+, \\ \dfrac{1}{2} u(x) + (S_{\partial\Omega} u)(x) & x \in \partial\Omega. \end{cases} \tag{23}$$

**Lemma 2.1.** *(see [30]) Let $\Omega$ be an open non empty bounded subset of $\mathcal{R}^n$ with a Liapunov boundary $\partial\Omega$, $f \in C^1(\overline{\Omega}, C(V_{n,n}))$, $\Pi f$ is defined as in (19). Then*

$$\Pi f(z) = \left(F_{\partial\Omega}^+(\alpha e_1 \alpha f)\right)(z) + T\left(e_1 D[f]\right)(z) - \frac{2-n}{n} e_1 f(z), \quad z \in \overline{\Omega}, \tag{24}$$

*where $\alpha(x)$ denotes the unit outer normal of $\partial\Omega$.*

**Theorem 2.2.** *Let $\Omega$ be an open non empty bounded subset of $\mathcal{R}^n$ with a Liapunov boundary $\partial\Omega$, $f \in \Lambda(f, \overline{\Omega})$, $\Pi f$ is defined as in (19). Then in $\Omega$*

$$D[\Pi f] = e_1 D[f] + \frac{n-2}{n} D[e_1 f]. \tag{25}$$

**Theorem 2.3.** *Suppose that $f$ is regular in a domain $U$ in $\mathcal{R}^n$, $\Omega$ is an open non empty bounded subset of $U$ with a Liapunov boundary $\partial\Omega$. $\Pi f$ is defined as in (19). Then in $\Omega$*

$$\triangle [\Pi f] = 0, \tag{26}$$

*where $\triangle$ is the Laplace operator.*

## 3. Some applications

In this section, we shall give some applications of the higher order Cauchy-Pompeiu formula. The solutions of Dirichlet problems as well as the inhomogeneous equations $D^k u = f$ are obtained. In the sequel, $K_n$ denotes the unit ball in $\mathcal{R}^n$ ($n \geq 3$), more clearly,

$$K_n = \{x | x = (x_1, x_2, \cdots, x_n) \in \mathcal{R}^n, |x| < 1\}.$$

Denote

$$G(y, x) = \frac{1}{\rho^{n-2}(y-x)} - \frac{1}{|y|^{n-2}\rho^{n-2}\left(\dfrac{y}{|y|^2} - x\right)}, \quad x \in K_n, y \in \overline{K_n}, x \neq y. \tag{27}$$

**Remark 3.1.** $G(y, x)$ has the following properties:
(1) $\triangle_x G(y, x) = 0$, $x \in K_n \setminus \{y\}$.
(2) $G(y, x) = G(x, y)$, $x, y \in K_n$, $x \neq y$.
(3) $G(y, x) = 0$, $y \in \partial K_n$, $x \in K_n$.

**Theorem 3.1.** *Suppose* $f \in C^2(\overline{K_n}, C(V_{n,n}))$, *then for* $x \in K_n$

$$f(x) = \frac{1}{\omega_n} \int_{\partial K_n} \frac{1 - |x|^2}{\rho^n(y - x)} f(y) \, dS_y + \frac{1}{(2 - n)\omega_n} \int_{K_n} G(y, x) \triangle_y f(y) \, dy^N. \tag{28}$$

**Proof.** By Lemma 2.1, for $x \in K_n$, we have

$$\begin{aligned}
f(x) = {} & \frac{1}{\omega_n} \int_{\partial K_n} \frac{\mathbf{y} - \mathbf{x}}{\rho^n(y - x)} d\sigma_y f(y) - \frac{1}{(2 - n)\omega_n} \int_{\partial K_n} \frac{1}{\rho^{n-2}(y - x)} d\sigma_y D[f](y) \\
& + \frac{1}{(2 - n)\omega_n} \int_{K_n} \frac{1}{\rho^{n-2}(y - x)} \triangle_y f(y) \, dy^N.
\end{aligned} \tag{29}$$

By Stokes formula, for $x \in K_n$ and $x \neq 0$, we have

$$\begin{aligned}
0 = {} & \frac{1}{\omega_n} \int_{\partial K_n} \frac{\mathbf{y} - \frac{\mathbf{x}}{|x|^2}}{\rho^n(y - \frac{x}{|x|^2})} d\sigma_y f(y) - \frac{1}{(2 - n)\omega_n} \int_{\partial K_n} \frac{1}{\rho^{n-2}(y - \frac{x}{|x|^2})} d\sigma_y D[f](y) \\
& + \frac{1}{(2 - n)\omega_n} \int_{K_n} \frac{1}{\rho^{n-2}(y - \frac{x}{|x|^2})} \triangle_y f(y) \, dy^N.
\end{aligned} \tag{30}$$

(30) can be rewritten as

$$\begin{aligned}
0 = {} & \frac{1}{\omega_n} \int_{\partial K_n} \frac{|x|^2 \left( \mathbf{y} - \frac{\mathbf{x}}{|x|^2} \right)}{|x|^n \rho^n(y - \frac{x}{|x|^2})} d\sigma_y f(y) - \frac{1}{(2 - n)\omega_n} \int_{\partial K_n} \frac{1}{|x|^{n-2}\rho^{n-2}(y - \frac{x}{|x|^2})} d\sigma_y D[f](y) \\
& + \frac{1}{(2 - n)\omega_n} \int_{K_n} \frac{1}{|x|^{n-2}\rho^{n-2}(y - \frac{x}{|x|^2})} \triangle_y f(y) \, dy^N.
\end{aligned} \tag{31}$$

In view of

$$|x|^k \rho^k (y - \frac{x}{|x|^2}) = |y|^k \rho^k (\frac{y}{|y|^2} - x), \quad k \in \mathbf{N}^*, \tag{32}$$

combining (29), (31) with (32), (28) follows.

For $x = 0$, by Stokes formula and (29), (28) still holds. Thus the result is proved.

**Remark 3.2.** Suppose $f \in C^2(\overline{K_n}, C(V_{n,n}))$, moreover, $f$ is harmonic in $K_n$. Then for $x \in K_n$

$$f(x) = \frac{1}{\omega_n} \int_{\partial K_n} \frac{1 - |x|^2}{\rho^n(y - x)} f(y) \, dS_y. \tag{33}$$

(33) is exactly the Poisson expression of harmonic functions.

**Theorem 3.2.** *The solution of the Dirichlet problem for the Poisson equation in the unit ball $K_n$*

$$\triangle u = f \ in \ K_n, \ u = \gamma \ on \ \partial K_n,$$

*for $f \in \Lambda(f, \overline{K_n})$ and $\gamma \in C(\partial K_n, C(V_{n,n}))$ is uniquely given by*

$$u(x) = \frac{1}{\omega_n} \int_{\partial K_n} \frac{1 - |x|^2}{\rho^n(y - x)} \gamma(y) \, dS_y + \frac{1}{(2-n)\omega_n} \int_{K_n} G(y, x) f(y) \, dy^N. \quad (34)$$

**Proof.** It can be directly proved by Corollary 2.2, Theorem 3.1, Remark 3.1 and Remark 3.2.

**Lemma 3.1.** *(see [29]) If $f$ is $k$-regular in an open neighborhood $\Omega$ of the origin, then in a suitable open ball $\overset{\circ}{B}(0, R) \subset \Omega$*

$$f(x^N) = f(0) + \sum_{p=1}^{\infty} \sum_{j=0}^{k-1} \sum_{(l_1, \cdots, l_{p-j})} C_{j,p-j} x^j V_{l_1, \cdots, l_{p-j}}(x^N) C_{l_1, \cdots, l_{p-j}}, \quad (35)$$

*$C_{j,p-j}$ and $C_{l_1, \cdots, l_{p-j}}$ are constants which are suitably chosen.*

By Lemma 3.1 and Corollary 2.2, we have

**Theorem 3.3.** *The solutions of inhomogeneous equations in the unit ball $K_n$*

$$D^k u = f \ in \ K_n,$$

*for $f \in \Lambda(f, \overline{K_n})$ are given in a suitable open ball $\overset{\circ}{B}(0, R) \subset K_n$ by*

$$u = C_0 + \sum_{p=1}^{\infty} \sum_{j=0}^{k-1} \sum_{(l_1, \cdots, l_{p-j})} C_{j,p-j} x^j V_{l_1, \cdots, l_{p-j}}(x^N) C_{l_1, \cdots, l_{p-j}} + T_k f. \quad (36)$$

*$C_0$, $C_{j,p-j}$ and $C_{l_1, \cdots, l_{p-j}}$ are constants which are suitably chosen.*

**Acknowledgements**

This paper is done while the author is visiting at Free University, Berlin during 2004 and 2005 supported by DAAD K. C. Wong Fellowship. The author would like to thank sincerely Prof. H. Begehr for his helpful suggestions and discussions.

**References**

1. H. Begehr, Iterations of Pompeiu operators. Mem. Diff. Eq. Math. Phys. 12 (1997), 3–21.
2. H. Begehr, Iterated integral operators in Clifford analysis. Journal for Analysis and its Applications 18 (1999), 361–377.
3. H. Begehr, Representation formulas in Clifford analysis. Acoustics, Mechanics,and the Related Topics of Mathematical Analysis. World Scientific Singapore, 2002, 8–13.

4. H. Begehr, Dai D Q, Li X, Integral representation formulas in polydomains. Complex Variables, Theory Appl. 47 (2002), 463–484.
5. H. Begehr, Du Jinyuan, Zhang Zhongxiang, On higher order Cauchy-Pompeiu formula in Clifford analysis and its applications, General Mathematics 11 (2003), 5-26.
6. H. Begehr, Zhang Zhongxiang, Du Jinyuan, On Cauchy-Pompeiu formula for functions with values in a universal Clifford algebra, Acta Mathematica Scientia 23B(1) (2003), 95–103.
7. H. Begehr, Zhang Zhongxiang, Vu Thi Ngoc Ha, Generalized integral representations in Clifford analysis. Complex Var., elliptic Eq. 51 (2006), 745–762.
8. F. Brack, R. Delanghe and F. Sommen F, Clifford Analysis, Research Notes in Mathematics 76. Pitman Books Ltd, London, 1982.
9. R. Delanghe, On regular analytic functions with values in a Clifford algebra. Math. Ann. 185 (1970), 91–111.
10. R. Delanghe, On the singularities of functions with values in a Clifford algebra. Math. Ann. 196 (1972), 293–319.
11. R. Delanghe, Morera's theorem for functions with values in a Clifford algebra, Simon Stevin 43 (1969–1970), 129–140.
12. R. Delanghe, F. Brackx, Hypercomplex function theory and Hilbert modules with reproducing kernel. Proc. London Math. Soc. 37 (1978), 545–576.
13. R. Delanghe, F. Sommen, V. Soucek, Clifford Algebra and Spinor Valued Functions, Kluwer Academic, Dordrecht, 1992.
14. Du Jinyuan, Zhang Zhongxiang, A Cauchy's integral formula for functions with values in a universal Clifford algebra and its applications. Complex Variables, Theory Appl. 47 (2002), 915–928.
15. K. Gürlebeck, On some classes of $\Pi-$operators. Research Notes in Math. 394 (1998), 41–57.
16. K. Gürlebeck, On some operators in Clifford analysis. Contemporary Mathematics. 212 (1998), 95–107.
17. K. Gürlebeck, W. Sprössig, Quaternionic and Clifford Calculus for Physicists and Engineers, Wiley, New York, 1997.
18. G.N. Hile, Hypercomplex function theory applied to partial differential equations. Ph.D. Thesis, Indiana Univ., Bloomington, Indiana, 1972.
19. V. Iftimie, Functions hypercomplex. Bull. Math. Soc. Sci. Math. R. S. Romania. 9(57), 1965, 279–332.
20. H. Malonek and B. Müller, Definition and properties of a hypercomplex singular integral operator. Results in Mathematics 22 (1992), 713–724.
21. E. Obolashvili, Higher order partial differential equations in Clifford analysis. Birkhauser, Boston, Basel, Berlin, 2002.
22. J. Ryan, Basic Clifford analysis, Cubo Math. Educ. 2 (2000) 226-256.
23. J. Ryan, Cauchy-Green type formulae in Clifford analysis, Trans. Amer. Math. Soc. 347 (1995) 1331-1341.
24. I.N.Vekua, Generalized analytic functions, Pergamon Press, Oxford, 1962.
25. Vu, Thi Ngoc Ha, Integral representations in quaternionic analysis related to Helmholtz operator. Complex Var. Theory Appl. 12(48), 2003, 1005–1021.
26. R. I. Yeh, Analysis and applications of holomorphic functions in higher dimensions. Trans. Amer. Math. Soc. 345 (1994), 151–177.
27. R. I. Yeh, Hyperholomorphic functions and second order partial differential equations. Trans. Amer. Math. Soc. 325(1991), 287–318.
28. Zhang, Zhongxiang, A revised higher order Cauchy-Pompeiu formula in Clifford analysis and its application. Proceedings of the workshop on recent trends in Applied

1042

Complex Analysis, held in METU., Ankara, Turkey, June, 2004. eds. H. Begehr, O. Celebi, R.P. Gilbert, special issue of Journal of Applied Functional Analysis, 2007, 269-278.

29. Zhang Zhongxiang, On k-regular functions with values in a universal Clifford algebra. J. Math. Anal. Appl. 315 (2006), 491-505.

30. Zhang Zhongxiang, Some properties of operators in Clifford analysis. Accepted for publication in Complex Var., elliptic Eq.. To appear spring 2007.

# CLIFFORD ALGEBRA APPLIED TO THE HEAT EQUATION

P. CEREJEIRAS

*Departamento de Matemática,*
*Universidade de Aveiro, Portugal*
*E-mail: pceres@mat.ua.pt*

F. SOMMEN

*Clifford Research Group - Department of Mathematical Analysis,*
*Ghent University, Belgium*
*E-mail: fs@cage.UGent.be*

In this paper we show how to factorize the heat operator $\triangle_x - \partial_t$, the heat operator with velocity term $\triangle_x - \partial_t + <v, \nabla_x>$ and the Schroedinger operator $\triangle_x - \partial_{it} + V(x,t)$. We discuss the basic operators of Clifford analysis in the affine and hermitean setting.

**Key words:** In-stationary heat equation, in-stationary Schrödinger equation, convection term

**Mathematics Subject Classification:** 30G35, 35Q55

## 1. Introduction

In recent years Clifford analysis has proven to be a good tool for studying elliptic partial differential equations of mathematical physics, and this by providing the possibility to factorize second order partial operators in terms of appropriate Dirac operators. This is of particular importance to the theory of boundary value problems developed by K. Gürlebeck and W. Sprößig which is based on orthogonal decompositions of the underlying function space where one of the spaces is the subspace of null-solutions of the corresponding Dirac operator.

Factorizations of the Helmholtz operator for particular values of the wave number have been considered by several authors (see, for e.g., Gürlebeck / Sprößig[9] [8], Xu[13], Brackx / Van Acker[4], Brackx / Delanghe / Sommen / Van Acker[3]). For the factorization of the Schroedinger operator we refer to Bernstein[1], Kravchenko[10] [11] [12], and De Schepper / Peña Peña [7]. But all of these cases are limited to stationary problems, i.e. elliptic problems. If one wants to deal with instantionary problems one immediately has one major obstacle to overcome in that the heat operator is not a homogeneous operator and, therefore, a direct factorization is not possible.

In this paper we propose an approach in terms of basis elements of a Witt basis of a Clifford algebra. This idea allows to construct a parabolic Dirac operator which factorizes the heat operator and which only contains partial derivatives.

## 2. Preliminaries

### 2.1. *Clifford Algebras*

Consider the $m$-dimensional vector space $\mathbb{R}^m$ provided with a standard orthonormal basis $\{e_1, \ldots, e_m\}$ and endowed with the following multiplication rules

$$e_i^2 = +1, \; i = 1, \ldots, p,$$
$$e_i^2 = -1, \; i = p+1, \ldots, m$$
$$e_i e_j + e_j e_i = 0, \; i \neq j,$$

where $p + q = m$.

We denote by $\mathbb{R}_{p,q}$ the minimal enlargement of $\mathbb{R}^m$ to a $2^m$-dimensional real associative algebra preserving the above multiplication rules. The basis for $\mathbb{R}_{p,q}$ is generated by $e_0 = 1$ and $e_A = e_{h_1} \ldots e_{h_k}$, where $A = \{h_1, \ldots, h_k\} \subset M = \{1, \ldots, m\}$, for $1 \leq h_1 < \ldots < h_k \leq m$. The particular linear combination of products of basic elements with equal length $k$ is called a $k$−vector and we shall denote by $[x]_k$ the $k$-vector part of $x \in \mathbb{R}_{p,q}$.

In this algebra we define three types of involutions: the *main involution* is denoted as $a \to \tilde{a}$, where we have $\tilde{1} = 1$, $\tilde{e}_j = -e_j, j = 1, \ldots, m$ and $\widetilde{(ab)} = \tilde{a}\tilde{b}$. The *conjugation* $a \to \bar{a}$ is defined by $\bar{1} = 1$, $\bar{e}_j = -e_j$ and $\overline{ab} = \bar{b}\bar{a}$, while the *reversion* $a \to a^*$ is defined by $1^* = 1$, $e_j^* = e_j$ and $(ab)^* = b^* a^*$. We define an inner product in $\mathbb{R}_{p,q}$ as $< a, b >:= [\bar{a}b]_0$, for all $a, b \in \mathbb{R}_{p,q}$.

Given the universal Clifford algebra $\mathbb{R}_{0,m}$, we call *complexified Clifford algebra* $\mathbb{C}_m$ of $\mathbb{R}_{0,m}$ to the tensor product $\mathbb{C} \otimes \mathbb{R}_{0,m}$, that is to say, $\mathbb{C}_m$ is the set of all $a + bi$, such that $a, b \in \mathbb{R}_{0,m}$. In the complexified Clifford algebra we define the *hermitean conjugate* $z \to z^+$ as $(a + ib)^+ = \bar{a} - i\bar{b}$. Finally we define the *hermitean inner product* as $(z, w) := [z^+ w]_0$, for all $z, w \in \mathbb{C}_m$.

### 2.2. *Clifford Analysis*

A $\mathbb{R}_{p,q}$-valued function $f$ from a open domain $\Omega \in \mathbb{R}^m$ will be written as $f = \sum_A f_A e_A$ where the function components $f_A$ are $\mathbb{R}$-valued. We denote by $D$ the (massless) Dirac operator acting on $f$ as

$$Df = \sum_{j=1}^{m} e_j \partial_{x_j} f = \sum_{j=1}^{m} \sum_A e_j e_A \partial_{x_j} f_A, \qquad (1)$$

which verifies $D^2 = \Delta_p - \Delta_q$, where $\Delta_n$ is the $n$-dimensional Laplacian. For more details, see Delanghe/Sommen/Soucek[5] and Brackx/Delanghe/Sommen[2].

We denote by

$$\{a, b\} := ab + ba$$

the *anti-commutator* of $a$ and $b$, while its *commutator* is defined as

$$[a, b] := ab - ba.$$

## 3. A suitable Witt basis for $\mathbb{R}_{n,n}$

We consider the Clifford algebra $\mathbb{R}_{n,n}$ generated by the vector basis elements $e_1, \ldots, e_n, \ \epsilon_1, \ldots, \epsilon_n$ verifying

$$e_j \epsilon_l + \epsilon_l e_j = 0,$$
$$e_l e_j + e_j e_l = -2\delta_{lj},$$
$$\epsilon_l \epsilon_j + \epsilon_j \epsilon_l = +2\delta_{lj}$$

for $j, l = 1, \ldots, n$.

We construct the corresponding Witt basis $\mathfrak{f}_1, \mathfrak{f}_1^+, \ldots, \mathfrak{f}_n, \mathfrak{f}_n^+$ for $\mathbb{R}_{n,n}$ by setting

$$\mathfrak{f}_j = \frac{1}{2}(e_j - \epsilon_j), \ \mathfrak{f}_j^+ = -\frac{1}{2}(e_j + \epsilon_j), \tag{2}$$

for $j = 1, \ldots, n$. We notice that

$$\mathfrak{f}_1^2 = (\mathfrak{f}_1^+)^2 = \ldots = \mathfrak{f}_n^2 = (\mathfrak{f}_n^+)^2 = 0, \tag{3}$$

with anti-commutation relationship

$$\{\mathfrak{f}_i, \mathfrak{f}_j^+\} = \delta_{ij},$$

while for the commutation of the Witt basis elements we have

$$[\mathfrak{f}_i, \mathfrak{f}_j^+] = \frac{1}{2}(-\delta_{ij} e_i e_j + \delta_{ij} \epsilon_i \epsilon_j + (\epsilon_i e_j + \epsilon_j e_i)),$$

for $i, j = 1, \ldots, n$.

Moreover, we assume that these elements behave with respect to the basis elements $e_{n+1}, \ldots, e_{n+m}$ of the Clifford algebra $\mathbb{R}_{0,m}$ as

$$e_{n+j} e_i + e_i e_{n+j} = \epsilon_i e_{n+j} + e_{n+j} \epsilon_i = 0,$$

for $j = 1, \ldots, m$ and $i = 1, \ldots, n$. The commutation with the Witt basis elements is given by

$$[\mathfrak{f}_i, e_{n+l}] = -2e_{n+l} \mathfrak{f}_i, \ \ [\mathfrak{f}_i^+, e_{n+l}] = -2e_{n+l} \mathfrak{f}_i^+,$$

while we have for the anti-commutation that

$$\{\mathfrak{f}_i, e_{n+l}\} = \{\mathfrak{f}_i^+, e_{n+l}\} = 0,$$

for $i = 1, \ldots, n$ and $l = 1, \ldots, m$.

We will now make use of the advantages provided by the Witt basis (2) in order to treat in-stationary equations linked to the heat operator instead of the Laplacian. This procedure will allow us to avoid the fractional derivatives which arise from the factorization of the heat operator.

## 4. Factorization of the Heat equation

We will consider $\mathbb{R}_{1,m+1} = \mathbb{R}_{1,1} \otimes \mathbb{R}_{0,m}$. We use the basis vectors $e, \epsilon$ of $\mathbb{R}_{1,1}$ in order to construct the Witt basis elements $\mathfrak{f}$ and $\mathfrak{f}^+$, which satisfy

$$\mathfrak{f}\mathfrak{f}^+ + \mathfrak{f}^+\mathfrak{f} = 1$$
$$\mathfrak{f}^2 = (\mathfrak{f}^+)^2 = 0$$
$$\mathfrak{f}e_j + e_j\mathfrak{f} = 0$$
$$\mathfrak{f}^+e_j + e_j\mathfrak{f}^+ = 0,$$

for $j = 1, \ldots, m$.

Let us consider the operators

$$D_{x,t}^{\pm} = \sum e_j \partial_{x_j} + \mathfrak{f}\partial_t \pm \mathfrak{f}^+$$

For the square of these operators we have

$$D_{x,t}^{\pm 2} = -\Delta \pm \partial_t,$$

which means that $D_{x,t}^+$ factorizes the forward Heat operator. For more details, see Cerejeiras / Kähler / Sommen[6].

## 5. Factorization of the In-stationary Heat equation with Convection Term

Now we consider the Clifford algebra $\mathbb{R}_{m+1,m+1}$. We construct the appropriated Witt basis $\mathfrak{f}_0, \mathfrak{f}_0^+, \ldots, \mathfrak{f}_m, \mathfrak{f}_m^+$ for $\mathbb{R}_{m+1,m+1}$ by setting

$$\mathfrak{f}_j = \frac{1}{2}(e_j - \epsilon_j), \quad \mathfrak{f}_j^+ = -\frac{1}{2}(e_j + \epsilon_j), \tag{4}$$

for $j = 0, 1, \ldots, m$.

Moreover, we construct the primitive idempotent element

$$I = \mathfrak{f}_1\mathfrak{f}_1^+ \mathfrak{f}_2\mathfrak{f}_2^+ \cdots \mathfrak{f}_m\mathfrak{f}_m^+ \tag{5}$$

which verifies $\mathfrak{f}_i I = 0$, $i = 1, \ldots, m$.

We assume $\tilde{\Omega} \subset \mathbb{R}^m \times \mathbb{R}^+$ to be an open and connected domain with sufficiently smooth boundary $\partial\tilde{\Omega}$. Any element $\underline{x} \in \mathbb{R}^m$ is to be identified with $\underline{x} = x_1 e_1 + \ldots + x_m e_m$, a 1-vector in $\mathbb{R}_{0,m}$.

**Definition 5.1.** Given a $\mathbb{R}_{0,m}$-valued function $u = u(\underline{x}, t)$ on the domain $\tilde{\Omega} \subset \mathbb{R}^m \times \mathbb{R}^+$, we define the parabolic first order operator acting on $u$ as

$$\mathcal{D}_{\underline{x},t} u(\underline{x}, t) = \left( D_{\underline{x}} + \mathfrak{f}_0 \partial_t + \mathfrak{f}_0^+ + \sum_{j=1}^{m} v_j(\underline{x}, t)\mathfrak{f}_j \right) u(\underline{x}, t). \tag{6}$$

where $v_j = v_j(\underline{x}, t)$ represents a differentiable real-valued function on the domain $\tilde{\Omega}$, for $j = 1, \ldots, m$.

**Theorem 5.1.** *The parabolic first order operator (6) factorizes the operator of second order*

$$-\Delta_{\underline{x}} + \partial_t + \sum_{j=1}^{m}\{D_{\underline{x}}, v_j \mathfrak{f}_j\},$$

*where $\{\cdot,\cdot\}$ is the anti-commutator operator.*

**Proof.** Considering the square of the above defined operator (6) we have

$$\mathcal{D}_{\underline{x},t}^2 u(\underline{x},t) = (D_{\underline{x}} + \mathfrak{f}_0 \partial_t + \mathfrak{f}_0^+ + \sum_{j=1}^{m} v_j(\underline{x},t)\mathfrak{f}_j)^2 u(\underline{x},t)$$

$$= (-\Delta_{\underline{x}} + \partial_t + \sum_{j=1}^{m}\{D_{\underline{x}}, v_j \mathfrak{f}_j\})u(\underline{x},t). \qquad \square$$

We now remark that, as $\{\mathfrak{f}_i, \mathfrak{f}_j^+\} = \{\mathfrak{f}_i^+, e_j\} = -\{\mathfrak{f}_i, e_j\} = \delta_{ij}, \ i,j = 0,1,\ldots,m$, we have

$$\{D_{\underline{x}}, v_j \mathfrak{f}_j\}u(\underline{x},t) = \sum_{i=1}^{m}(e_i \mathfrak{f}_j \partial_{x_i}(v_j u) + \mathfrak{f}_j e_i v_j (\partial_{x_i} u))$$

$$= \sum_{i=1}^{m}(e_i \mathfrak{f}_j \partial_{x_i}(v_j u) - (\delta_{ij} + e_i \mathfrak{f}_j)v_j(\partial_{x_i} u))$$

$$= \sum_{i=1}^{m}(e_i \mathfrak{f}_j (\partial_{x_i} v_j)u - \delta_{ij}v_j(\partial_{x_i} u))$$

$$= -v_j \partial_{x_j} u - \mathfrak{f}_j(D_{\underline{x}}v_j)u \qquad (7)$$

Therefore, we have obtained the following result.

**Theorem 5.2.** *The field $\Phi = \varphi I$, where $\varphi$ is a $\mathbb{R}_{1,m+1}$-valued function and $I$ is the idempotent element (5), is a solution of the in-stationary Heat equation with (real-valued) convection term*

$$(\Delta_{\underline{x}} - \partial_t + \sum_{j=1}^{m} v_j \partial_{x_j})\Phi(\underline{x},t) = 0, \qquad (8)$$

*if and only if it is a null solution of the square of the operator (6).*

## 6. A parabolic Dirac operator for the in-stationary Schroedinger equation with a potential term

In this section we aim at obtaining a suitable factorization of the in-stationary Schroedinger operator.

Let us assume that $\mathbb{R}_{n,n+m} = \mathbb{R}_{0,m} \otimes \mathbb{R}_{n,n}$. We remark that, in the case where $m = 1$, $\mathbb{R}_{n,n+1}$ can be identified with the complexified Clifford algebra $\mathbb{C}_{2n}$, since $\mathbb{R}_{0,1}$ is isomorphic to $\mathbb{C}$. Moreover, $\mathbb{C} \otimes \mathbb{R}_{n,n} = \mathbb{C}_{2n}$ corresponds to $\text{End}(\mathbb{C}_{2n})$.

In what follows, let $\tilde{\Omega} \subset \mathbb{R}^m \times \mathbb{R}^+$ denote an open and connected domain with sufficiently smooth boundary $\partial\tilde{\Omega}$.

**Definition 6.1.** Given a complex-valued function $u = u(\underline{x}, t)$ on the domain $\tilde{\Omega} \subset \mathbb{R}^m \times \mathbb{R}^+$, we define the (forward/backward) parabolic Dirac operator $\mathbb{D}^{\pm}_{\underline{x},t}$ acting on $u$ as

$$\mathbb{D}^{\pm}_{\underline{x},t} u(\underline{x},t) = \left(D_{\underline{x}} + \mathfrak{f}_1 \partial_{it} \pm (\mathfrak{f}_1^+ + \mathfrak{f}_2 V(\underline{x},t) + \mathfrak{f}_2^+)\right) u(\underline{x},t), \qquad (9)$$

where the potential $V(\underline{x}, t)$ is a complex-valued function on the domain $\tilde{\Omega}$.

Note that the action $\mathbb{D}^{\pm}_{\underline{x},t} u$ can be identified with a complex vector in $\mathbb{C}^{m+4}$.

**Theorem 6.1.** *The (forward/backward) parabolic Dirac operator $\mathbb{D}^{\pm}_{\underline{x},t}$ with potential term factorizes the (forward/backward) operator of second order*

$$-\Delta_{\underline{x}} \pm \partial_{it} + V \pm \left\{D_{\underline{x}} + \mathfrak{f}_1 \partial_{it}, \mathfrak{f}_2^+ V\right\},$$

*where the potential $V = V(\underline{x}, t)$ is assumed to be a differentiable complex-valued function and $\{\cdot, \cdot\}$ represents the anti-commutator operator.*

**Proof.** The theorem follows directly from

$$(\mathbb{D}^{\pm}_{\underline{x},t})^2 u(\underline{x},t) = \left(D_{\underline{x}} + \mathfrak{f}_1 \partial_{it} \pm (\mathfrak{f}_1^+ + \mathfrak{f}_2 V + \mathfrak{f}_2^+)\right)^2 u(\underline{x},t)$$

$$= \Big( (D_{\underline{x}}^2 + \mathfrak{f}_1(\partial_{it} D_{\underline{x}} - D_{\underline{x}} \partial_{it})) \pm (D_{\underline{x}} \mathfrak{f}_1^+ + \mathfrak{f}_1^+ D_{\underline{x}}) \pm$$
$$(D_{\underline{x}} \mathfrak{f}_2 V + \mathfrak{f}_2 V D_{\underline{x}}) \pm (D_{\underline{x}} \mathfrak{f}_2^+ + \mathfrak{f}_2^+ D_{\underline{x}}) + (\mathfrak{f}_1 \partial_{it})^2 \pm$$
$$(\mathfrak{f}_1 \partial_{it} \mathfrak{f}_1^+ + \mathfrak{f}_1^+ \mathfrak{f}_1 \partial_{it}) \pm (\mathfrak{f}_1 \partial_{it} \mathfrak{f}_2 V + \mathfrak{f}_2 V \mathfrak{f}_1 \partial_{it}) \pm$$
$$(\mathfrak{f}_1 \partial_{it} \mathfrak{f}_2^+ + \mathfrak{f}_2^+ \mathfrak{f}_1 \partial_{it}) + (\mathfrak{f}_1^+)^2 + (\mathfrak{f}_1^+ \mathfrak{f}_2 V + \mathfrak{f}_2 V \mathfrak{f}_1^+) +$$
$$(\mathfrak{f}_1^+ \mathfrak{f}_2^+ + \mathfrak{f}_2^+ \mathfrak{f}_1^+) + (\mathfrak{f}_2 V \mathfrak{f}_2^+ + \mathfrak{f}_2^+ \mathfrak{f}_2 V) + (\mathfrak{f}_2^+)^2 \Big) u(\underline{x},t)$$

$$= \left(-\Delta_{\underline{x}} \pm \{D_{\underline{x}}, \mathfrak{f}_2 V\} \pm \partial_{it} \pm \{\mathfrak{f}_1 \partial_{it}, \mathfrak{f}_2 V\} + V\right) u(\underline{x},t)$$

$$= \left(-\Delta_{\underline{x}} \pm \partial_{it} + V \pm \{D_{\underline{x}} + \mathfrak{f}_1 \partial_{it}, \mathfrak{f}_2 V\}\right) u(\underline{x},t). \qquad \square$$

Moreover, easy calculations show that we have

$$\{D_{\underline{x}} + \mathfrak{f}_1 \partial_{it}, \mathfrak{f}_2 V\} u(\underline{x},t) = (D_{\underline{x}} + \mathfrak{f}_1 \partial_{it})(\mathfrak{f}_2 V u) + \mathfrak{f}_2 V (D_{\underline{x}} + \mathfrak{f}_1 \partial_{it}) u$$

$$= -\mathfrak{f}_2 D_{\underline{x}}(V u) - \mathfrak{f}_2 \mathfrak{f}_1 \partial_{it}(V u) + \mathfrak{f}_2 V D_{\underline{x}} u + \mathfrak{f}_1 \partial_{it} u$$

$$= -\mathfrak{f}_2 ((D_{\underline{x}} + \mathfrak{f}_1 \partial_{it}) V) u(\underline{x},t).$$

**Theorem 6.2.** *The field $\Phi = \varphi \mathfrak{f}_2$, where $\varphi$ is a $\mathbb{C}$-valued function, is a solution of the (backward/forward) in-stationary Schroedinger equation with (complex-valued) potential term $V = V(\underline{x}, t)$, that is,*

$$\left(-\Delta_{\underline{x}} \pm \partial_{it} + V(\underline{x},t)\right) \Phi(\underline{x},t) = 0,$$

*if and only if it is a solution of the square of the (forward/backward) parabolic Dirac operator with potential term $\mathbb{D}^{\pm}_{\underline{x},t} = D_{\underline{x}} + \mathfrak{f}_1 \partial_{it} \pm (\mathfrak{f}_1^+ + \mathfrak{f}_2 V(\underline{x},t) + \mathfrak{f}_2^+)$.*

**Proof.** This result follows immediately from

$$(\mathbb{D}_{\underline{x},t}^{\pm})^2 \Phi(\underline{x},t) = \left(-\Delta_{\underline{x}} \pm \partial_{it} + V(\underline{x},t) \pm \left\{D_{\underline{x}} + \mathfrak{f}_1 \partial_{it}, \mathfrak{f}_2 V(\underline{x},t)\right\}\right) \varphi(\underline{x},t)\mathfrak{f}_2$$
$$= \left(-\Delta_{\underline{x}} \pm \partial_{it} + V(\underline{x},t)\right) \varphi(\underline{x},t)\mathfrak{f}_2$$
$$= \left(-\Delta_{\underline{x}} \pm \partial_{it} + V(\underline{x},t)\right) \Phi(\underline{x},t). \qquad \square$$

### Acknowledgments

The work of the first author was partially supported by grant SFRH / BSAB / 495 / 2005 of the Fundação para a Ciência e a Tecnologia.

The work of the second author was supported by FWO-Krediet aan Navorsers, 1.5.065.04.

### References

1. S. Bernstein, *Factorization of solutions of the Schrödinger equation*, in Sprößig, Wolfgang (ed.) et al., Proceedings of the symposium on analytical and numerical methods in quaternionic and Clifford Analysis, June 5–7, 1996, Seiffen, Germany. Freiberg: TU Bergakademie Freiberg, 1–6.
2. F. Brackx, R. Delanghe and F. Sommen, *Clifford Analysis*, Research Notes in Mathematics **No. 76**, Pitman, London, 1982.
3. F. Brackx, R. Delanghe, F. Sommen and Van Acker, *Reproducing kernels on the unit sphere*, **in:** Pathak, R. S. (**ed.**), Generalized functions and their applications. Proceedings of the international symposium, December 23-26, 1991, Varanasi, India. New-York, Plenum Press, 1-10, 1993.
4. F. Brackx and Van Acker, *Boundary value theory for eigenfunctions of the Dirac operator*, Bull. Soc. Math. Belg., **vol. 45** 2, Ser. B, 113-123 (1993).
5. R. Delanghe, F. Sommen and V. Souček, *Clifford algebras and spinor-valued functions*, Kluwer Academic Publishers, 1992.
6. P. Cerejeiras, U. Kähler and F. Sommen, *Parabolic Dirac operators and the Navier-Stokes equations over time-varying domains*, Math. Meth. Appl. Sci., **28**, 1715-1724 (2005)
7. N. De Schepper and D. Peña Peña, *Factorization of the Schrödinger Operator and the Ricatti Equation in the Clifford Analysis Setting*, **In:** F. Brackx, H. De Schepper (eds.): Liber Amicorum Richard Delanghe: een veelzijdig wiskundige, Academia Press, Gent, 69-84, 2005.
8. K. Gürlebeck and W. Sprößig, *Quaternionic analysis and elliptic boundary value problems*, Akademieverlag, Berlin, 1989.
9. K. Gürlebeck and W. Sprößig, *Quaternionic and Clifford calculus for Engineers and Physicists*, John Wiley &. Sons, Cinchester, 1997.
10. V. G. Kravchenko and V. V., Kravchenko, *Quaternionic factorization of the Schrödinger operator and its applications to some first-order systems of mathematical physics*, J. Phys. A: Math. Gen. **36** (2003), No.44, 11285–11297
11. V. V. Kravchenko, *On the reduction of the multidimensional stationary Schroedinger equation to a first-order equation and its relation to the pseudoanalytic function theory*, J. Phys. A: Math. Gen. **38** (2005), 851–868.
12. V. V. Kravchenko, *On force free magnetic fields. Quaternionic approach*, Math. Meth. Appl. Sci. **28** (2005), No.4, 379–386.

1050

13. Z. Xu, *Helmholtz equations and boundary value problems*, **in:** Partial Differential Equations with Complex Analysis, **eds.** H. Begehr, A. Jeffrey, Pitman Research Notes in Mathematics Series 262, 204-214, 1992.

# CAUCHY-TYPE INTEGRAL FORMULAS FOR
# $k$-HYPERMONOGENIC FUNCTIONS

SIRKKA-LIISA ERIKSSON

*Institute of Mathematics*
*Tampere University of Technology*
*P.O. Box 553*
*FI-33101 Tampere, Finland*
*E-mail: sirkka-liisa.eriksson@tut.fi*

Let $C\ell_n$ be the (universal) Clifford algebra generated by $e_1, ..., e_n$ satisfying $e_i e_j + e_j e_i = -2\delta_{ij}$, $i,j = 1, ..., n$. The modified Dirac operator is introduced by $M_k f = Df + kx_n^{-1}Q'f$, where $'$ is the main involution and $Qf$ is given by the decomposition $f(x) = Pf(x) + Qf(x)e_n$ with $Pf(x)$ and $Qf(x)$ in $C\ell_{n-1}$. A continuously differentiable function $f$ is called $k$-hypermonogenic, if $x_n M_k f(x) = 0$. Note that 0-hypermonogenic are monogenic and $n-1$-hypermonogenic functions are hypermonogenic defined by H. Leutwiler and the author. The function $|x|^{k-n+1} x^{-1}$ is $k$-hypermonogenic. Hypermonogenic functions are related to harmonic functions with respect to the hyperbolic metric $ds^2 = x_n^{2k/(1-n)} \left( dx_0^2 + ... + dx_n^2 \right)$. We present the Cauchy integral formula for $k$-hypermonogenic functions and related results.

**Key words:** Monogenic, hypermonogenic, Dirac operator, hyperbolic metric
**Mathematics Subject Classification:** 30G35, Secondary 30A05, 30F45

## 1. Introduction

In the nineties Heinz Leutwiler started to work with the hyperbolic version of the theory of monogenic functions. The choice of hyperbolic version has the surprising effect that the power function $x^m$ ($m \in \mathbb{N}_0$) becomes a solution of the modified system $Mf = 0$, where $M$ is the hyperbolic counterpart of the Dirac operator $D = \sum_{i=0}^{n} e_i \frac{\partial}{\partial x_i}$. Hence elementary functions can be defined similarly as in classical complex analysis.

The systematic treatment of solution of the system $Mf = 0$, called hypermonogenic functions, after the quaternionic version [10], began with the joint paper [6] and continued with [8,9] and [5].

A Cauchy-type formula for hypermonogenic functions was proved in [5]. A key concept in the proof was $k-$hypermonogenic functions introduced in [3]. In this paper we prove the Cauchy formula for $k$-hypermonogenic functions.

## 2. Hyperbolic harmonic functions

Let $C\ell_n$ be the universal Clifford algebra generated by the elements $e_1, ..., e_n$, satisfying the relation $e_i e_j + e_j e_i = -2\delta_{ij}$, where $\delta_{ij}$ is the usual Kronecker delta. We briefly recall some standard notations. The elements $x = x_0 + x_1 e_1 + ... + x_n e_n$, for $x_0, ..., x_n \in \mathbb{R}$, are called *paravectors*. The set $\mathbb{R}^{n+1}$ is identified with the set of paravectors.

We consider the equation

$$\triangle f - \frac{k}{x_n} \frac{\partial f}{\partial x_n} = 0. \tag{1}$$

on the upper half space $\{x \in \mathbb{R}^{n+1} \mid x_n > 0\}$, denoted by $\mathbb{R}^{n+1}_+, n \geq 2$. This equation is the Laplace-Beltrami equation with respect to the Riemannian metric

$$ds^2 = x_n^{-2k/n-1} \sum_{i=0}^{n} dx_i^2. \tag{2}$$

Real valued solutions of this equation are called $k-hyperbolic$ *harmonic*. The value $k = n - 1$ is specially important. For this reason $(n - 1)$-hyperbolic harmonic functions are called hyperbolic harmonic functions.

Besides equations on the upper half space we also need their models on the unit ball denoted by

$$B_{n+1} = \{x \in \mathbb{R}^{n+1} \mid \|x\| < 1\},$$

where $\|x\| = \sum x_i^2$.

**Proposition 1** ($^{1,12}$) *Let $T$ be the Möbius transformation mapping the unit ball $B_{n+1}$ onto $\mathbb{R}^{n+1}_+$ defined by*

$$T(x) = (x + e_n)(e_n x + 1)^{-1}$$

*If a function $u$ satisfies the equation (1) then the function*

$$v(x) = \|x - e_n\|^{k-n+1} u(T(x))$$

*satisfies the equation*

$$\triangle v + \frac{2k}{1 - \|x\|^2} \sum_{i=0}^{n} x_i \frac{\partial v}{\partial x_i} - \frac{k(1 + k - n)}{1 - \|x\|^2} v = 0 \tag{3}$$

*on the unit ball $B_{n+1}$. Conversely, if the function $v$ satisfies the equation (3) on the unit ball $B_{n+1}$ then the function*

$$u(x) = \|x + e_n\|^{k-n+1} v(T^{-1}(x))$$

*satisfies the equation (1) on $\mathbb{R}^{n+1}_+$, where*

$$T^{-1}(x) = (x - e_n)(-e_n x + 1)^{-1}.$$

We also recall the interplay between the solutions of the equation (3) in the case $k < -1$ and in the case $k > -1$.

**Lemma 2.1.** ([1][p. 27]) *A function u satisfies the equation (3) on the unit ball $B_{n+1}$ in the case $k < -1$ if and only if the function*

$$\bar{u}(x) = \left(1 - \|x\|^2\right)^{-k-1} u(x)$$

*satisfies the equation*

$$\Delta v + \frac{2(k+2)}{1 - \|x\|^2} \sum_{i=0}^{n} x_i \frac{\partial v}{\partial x_i} + \frac{(k+2)(1+n-k)}{1 - \|x\|^2} v = 0$$

*on the unit ball $B_{n+1}$.*

There is also an interesting relation between $k$-hyperbolic functions and the eigen functions of the equation (1) with the value $k = n - 1$.

**Lemma 2.2.** ([12])*A function $f$ is k-hyperbolic harmonic if and only if the function $g(x) = x_n^{-\frac{1-n+k}{2}} f(x)$ is a solution of the equation*

$$\Delta g - \frac{n-1}{x_n} \frac{\partial g}{\partial x_n} + l\frac{g}{x_n^2} = 0, \tag{4}$$

*where*

$$l = \frac{1}{4}\left(n^2 - (k+1)^2\right).$$

Solutions of the equation (4) are easy to transform to the unit ball model as follows.

**Proposition 2.1.** *Let $T$ be the Möbius transformation mapping $B_{n+1}$ onto $\mathbb{R}_+^{n+1}$ defined by*

$$T(x) = (x + e_n)(e_n x + 1)^{-1}.$$

*A function $u$ satisfies the equation (4) on $\mathbb{R}_+^{n+1}$ if and only if the function*

$$v(x) = u(T(x))$$

*satisfies the equation*

$$\Delta v + \frac{2(n-1)}{1 - \|x\|^2} \sum_{i=0}^{n} x_i \frac{\partial v}{\partial x_i} - \frac{4lv}{\left(1 - \|x\|^2\right)^2} = 0 \tag{5}$$

*on the unit ball.*

Möbius transformations in $\mathbb{R}^{n+1}$ may be characterized as follows.

**Definition 2.1 (Vahlen, Ahlfors, Waterman).** *Denote by $GL(2, Cl_n)$ the group of all matrices*

$$\begin{pmatrix} a & b \\ c & d \end{pmatrix}, \qquad a, b, c, d \in Cl_n$$

*so that the mapping*

$$T(x) = (ax + b)(cx + d)^{-1}$$

*is a bijection from* $\mathbb{R}^{n+2} \cup \{\infty\} = \widehat{\mathbb{R}}^{n+2} \to \widehat{\mathbb{R}}^{n+2}$. *We say that the mapping* $T$ *is a Möbius transformation in* $\widehat{\mathbb{R}}^{n+1}$ *induced by the matrix* $\begin{pmatrix} a & b \\ c & d \end{pmatrix}$.

Using Lemma 2.2 we obtain a simple proof for the result stated in [1][Theorem 2.1].

**Theorem 2.1.** *Let* $\Omega$ *be an open set contained in* $\mathbb{R}^{n+1}_+$ *and* $T : \Omega \to \mathbb{R}^{n+1}$ *be the Möbius transformation induced by* $\begin{pmatrix} a & b \\ c & d \end{pmatrix} \in GL(2, Cl_{n-1})$, *where* $-c^{-1}d \notin \Omega$. *If* $f$ *is* $k$-*hyperbolic harmonic on* $T(\Omega)$, *then the function*

$$F(x) = \frac{f(T(x))}{|cx + d|^{n-k-1}}$$

*is* $k$-*hyperbolic harmonic on* $\Omega$.

**Proof** Assume $T : \Omega \to \mathbb{R}^{n+1}$ is the Möbius transformation induced by the matrix

$$\begin{pmatrix} a & b \\ c & d \end{pmatrix} \in GL(2, Cl_{n-1}).$$

If $k = n - 1$, the result holds, since the metric (2) is invariant under $T$. If $k \neq n - 1$, then we have

$$(Tx)_n = \frac{(ad^* - bc^*) x_n}{|cx + d|^2}.$$

Since $ad^* - bc^* \in \mathbb{R}$, applying Lemma 2.2 we infer the result.

We are looking for solutions of the equation (3) depending only on $r$. These solutions satisfy the equation

$$(1 - r^2) r \frac{\partial^2 f}{\partial r^2} + (n - (n - 2k) r^2) \frac{\partial f}{\partial r} - k(1 - n + k) rf = 0. \tag{6}$$

**Lemma 2.3.** *In the case* $k \geq -1$ *the function*

$$\varphi_k(r) = \frac{(1 - r^2)^{1+k}}{\omega_0} \int_{S_n} \frac{d\sigma_0(v)}{\|rv - e_n\|^{n+k+1}}$$

$$= \frac{1}{\omega_0} \int_{S_n} \frac{d\sigma_0(v)}{\|v - re_n\|^{n-k-1}}$$

*is a solution of the equation (6) in the unit ball, where* $S_n$ *is the surface of the unit ball in* $\mathbb{R}^{n+1}$. *If* $k > -1$, *then*

$$\lim_{r \to 1} \varphi_k(r) = \frac{1}{\omega_0} \int_{S_n} \frac{d\sigma_0(v)}{\|v - e_n\|^{n-k-1}} = \frac{2^{k+1}\omega_0'}{\omega_0} \int_0^\infty \frac{t^{n-1}dt}{(1 + t^2)^{\frac{n+k+1}{2}}}$$

$$= \frac{2^{k+1}\omega_0'\Gamma\left(\frac{n}{2}\right)\Gamma\left(\frac{k+1}{2}\right)}{\omega_0\Gamma\left(\frac{n+k+1}{2}\right)}.$$

where $\omega_0'$ is the surface measure of the unit ball in $\mathbb{R}^n$.and $\omega_0$ the surface measure of the unit ball in $\mathbb{R}^{n+1}$. Moreover we have

$$\lim_{r \to 0} \varphi_k(r) = 1.$$

**Proof** Assume that $k > -1$. The first statement follows from ([1][p.23–24]). In order to compute the limits, we denote $2s = n - k - 1$ and $z = re_n = T^{-1}(a)$ for $a = \frac{1+r}{1-r}e_n$. The mapping $T$ is defined as earlier

$$T^{-1}(u) = (u - e_n)(-e_n u + 1)^{-1}.$$

If $v = T^{-1}(u)$, then using the same ideas as in [13][p. 58–59] we compute $1 - \|z\|^2 = \frac{4a_n}{1+\|a\|^2+2a_n}$ and

$$1 - 2z_n v_n + \|z\|^2 = \frac{4\left(a_n^2 + \sum_{i=0}^{n-1} u_i^2\right)}{\left(1 + \|a\|^2 + 2a_n\right)\left(1 + u_0^2 + \dots + u_{n-1}^2\right)}$$

$$\frac{\left(1 - \|z\|^2\right)^s}{\|v - re_n\|^{2s}} = \frac{\left(1 - \|z\|^2\right)^s}{\left(1 - 2z_n v_n + \|z\|^2\right)^s} = \frac{a_n^s \left(1 + u_0^2 + \dots + u_{n-1}^2\right)^s}{\left(a_n^2 + \sum_{i=0}^{n-1} u_i^2\right)^s}.$$

Hence we have

$$\int_{S_n} \frac{d\sigma_0(v)}{\|v - re_n\|^{2s}} = \frac{2^n a_n^s}{\left(1 - \|z\|^2\right)^s} \int_{-\infty}^{\infty} \dots \int_{-\infty}^{\infty} \frac{\left(1 + u_0^2 + \dots + u_{n-1}^2\right)^{s-n}}{\left(a_n^2 + \sum_{i=0}^{n-1} u_i^2\right)^s} du_0 \dots du_{n-1}$$

$$= 2^n \omega_0' \int_0^\infty \frac{t^{n-1}}{(1+t^2)^{\frac{n+k+1}{2}}\left((1+r)^2 + (1-r)^2 t^2\right)^{\frac{n-k-1}{2}}} dt.$$

where $\omega_0'$ is $n$-dimensional surface measure of the unit ball. The statements of the limits follow directly.

**Theorem 2.2.** *In the case $k > -1$ the general solution of the equation (6) is the following*

$$f_k(r) = \varphi_k(r)\left(C \int_r^1 \varphi_k^{-2}(t) \frac{(1 - t^2)^k}{t^n} dt + C_0\right).$$

*When $k < -1$, the general solution of the equation (6) is the following*

$$f_k(r) = \left(1 - \|r\|^2\right)^{k+1} \varphi_{-k-2}(r)\left(C \int_r^1 \varphi_{-k-2}^{-2}(t) \frac{(1 - t^2)^{-k-2}}{t^n} dt + C_0\right).$$

*If $k = -1$, then the general solution of the equation (6) in $B_a = \{x \mid \|x\| < a\}$ ($a < 1$) is the following*

$$f_k(r) = \varphi_k(r)\left(C \int_r^a \varphi_k^{-2}(t) \frac{(1 - t^2)^{-1}}{t^n} dt + C_0\right),$$

**Proof** Denoting $f_k(r) = g(r)\varphi(r)$ we obtain

$$0 = (1-r^2)\, r\varphi \frac{\partial^2 g}{\partial r^2} + 2(1-r^2)\, r \frac{\partial \varphi}{\partial r} \frac{\partial g}{\partial r}$$
$$+ (n - (n-2k)\, r^2)\, \varphi \frac{\partial g}{\partial r}.$$

Setting $\frac{\partial g}{\partial r} = h$, we infer

$$(1-r^2)\, r\varphi \frac{\partial h}{\partial r} + \left(2(1-r^2)\, r \frac{\partial \varphi}{\partial r} + (n - (n-2k)\, r^2)\, \varphi \right) h = 0.$$

Hence we solve

$$\log h = -2 \log \varphi + k \log (1-r^2) - n \log r + C$$

and therefore

$$\frac{\partial g}{\partial r} = h = C \frac{(1-r^2)^k}{\varphi^2 r^n}.$$

Since $\varphi^{-2}(t) \frac{(1-t^2)^k}{t^n} \sim o\left((1-t)^k\right)$, when $r \to 1$, we conclude

$$g(r) = C \int_r^1 \varphi^{-2}(t) \frac{(1-t^2)^k}{t^n}\, dt + C_0$$

in the case $k > -1$. In the case $k < -1$, the result is obtained using Lemma 2.1. The case $k = -1$ is treated similarly.

Boundary behavior of the function $f_k$ is easy to see as follows.

**Lemma 2.4.** *If $r \to 1$, then*

$$f_k(r) = \begin{cases} O(1-r)^{k+1}, & \text{if } k > -1 \\ O(1), & \text{if } k < -1 \end{cases}.$$

*In the both cases,*

$$f(r) = O\left(r^{-n+1}\right)$$

*when $r \to 0$.*

The solution $\varphi_k$ leads to a bounded function $\mathbb{R}_+^{n+1}$, if $k > -1$.

**Theorem 2.3.** *In the case $k > -1$ the function*

$$u_k(x) = \|x + e_n\|^{k-n+1} \varphi_k(|Tx|)$$
$$= \frac{\|x + e_n\|^{k-n+1}}{\omega_0} \int_{S_n} \frac{d\sigma_0(v)}{\|v - |Tx|\, e_n\|^{n-k-1}}$$

*is bounded.*

**Proof** Note that the function $\varphi_k\left(\|x\|\right)$ is a solution of the equation (3) on the unit ball. Using Theorem 2 we see that it is a positive $k$-hyperbolic harmonic on the whole upper half space $\mathbb{R}_+^{n+1}$. Moreover we have

$$\frac{u_k\left(x\right)}{2^n\omega_0'} = \int_0^\infty \frac{t^{n-1}dt}{(1+t^2)^{\frac{n+k+1}{2}}\left(\left(\|x+e_n\|+\|x-e_n\|\right)^2+\left(\|x+e_n\|-\|x-e_n\|\right)^2 t^2\right)^{\frac{n-k-1}{2}}}$$

$$\leq \frac{1}{\|x+e_n\|^{n-k-1}}\int_0^\infty \frac{t^{n-1}dt}{(1+t^2)^{\frac{n+k+1}{2}}} \leq \int_0^\infty \frac{t^{n-1}dt}{(1+t^2)^{\frac{n+k+1}{2}}}.$$

Hence $u_k\left(x\right)$ is a positive bounded $k$-hyperbolic harmonic function.

A fundamental $k$-hyperbolic harmonic functions is given next.

**Theorem 2.4.** *Let $T$ and $T^{-1}$ be the same as in Proposition 2 and $f_k$ as in Theorem 2.2. Then the function*

$$h_k\left(x, e_n\right) = \|x+e_n\|^{k-n+1} f_k\left(T^{-1}\left(x\right)\right)$$

$$= \|x+e_n\|^{k-n+1} f_k\left(\frac{\|x-e_n\|}{\|x+e_n\|}\right)$$

*is $k$–hyperbolic harmonic on $\mathbb{R}_+^{n+1}\backslash\{e_n\}$. Generally the function*

$$h_k\left(x, y\right) = \|x-\hat{y}\|^{k-n+1} f_k\left(\frac{\|x-y\|}{\|x-\hat{y}\|}\right)$$

*is $k$–hyperbolic harmonic on $\mathbb{R}_+^{n+1}\backslash\{y\}$, where*

$$\hat{y} = y_0 + y_1 e_1 + \ldots + y_{n-1}e_{n-1} - y_n e_n$$

$$= y - 2y_n e_n.$$

**Proof** The first part follows from Proposition 2. If $L$ is a Möbius transformation

$$Lx = \frac{x - y_0 - y_1 e_1 - \ldots - y_{n-1}e_{n-1}}{y_n}$$

then $L$ is induced by

$$\begin{pmatrix} 1 & y_0 - y_1 e_1 - \ldots - y_{n-1}e_{n-1} \\ 0 & y_n \end{pmatrix} \in GL\left(2, Cl_{n-1}\right).$$

On the account of Theorem 2.1 the function $h_k\left(x, y\right) = \frac{h(L(x), e_n)}{y_n^{n-k-1}}$ is $k$-hypermonogenic. Since

$$\frac{\|Lx - e_n\|}{\|Lx + e_n\|} = \frac{\|x - y\|}{\|x - \hat{y}\|}$$

and

$$\|Lx + e_n\| = \frac{\|x - \hat{y}\|}{y_n}$$

we obtain

$$h_k(x, y) = \frac{h(L(x), e_n)}{y_n^{n-k-1}}$$

$$= \|x - \hat{y}\|^{k-n+1} f_k\left(\frac{\|x - y\|}{\|x - \hat{y}\|}\right),$$

completing the proof.

## 3. Hypermonogenic functions

We briefly recall some concepts of Clifford algebras needed in this section. The main involution $' : Cl_n \to Cl_n$ is the algebra isomorphism defined by $e_0' = 1$ and $e_i' = -e_i$ for $i = 1, ..., n$. Thus $(ab)' = a'b'$. The involution $\hat{} : Cl_n \to Cl_n$ is defined by $\widehat{e_n} = -e_n$, $\widehat{e_i} = e_i$, for $i = 0, ..., n-1$, and $\widehat{ab} = \widehat{a}\widehat{b}$. It is easy to calculate that for arbitrary $a \in Cl_n$

$$a'e_n = e_n\widehat{a} \qquad \text{and} \qquad e_n a' = \widehat{a}e_n, \tag{7}$$

and for arbitrary $b \in Cl_{n-1}$ (the Clifford algebra generated by $e_1, ..., e_{n-1}$, embedded in $Cl_n$)

$$b'e_n = e_n b \qquad \text{and} \qquad e_n b' = be_n. \tag{8}$$

The antiautomorphism $* : Cl_n \to Cl_n$, called *reversion*, is defined by $e_i^* = e_i$ for $i = 0, ..., n$ and $(ab)^* = b^*a^*$. The conjugation $\bar{a}$ is given by $\bar{a} = (a')^*$.

Any element $a \in Cl_n$ may be uniquely decomposed as $a = b + ce_n$ for $b, c \in Cl_{n-1}$. Using this decomposition we define the mappings $P : Cl_n \to Cl_{n-1}$ and $Q : Cl_n \to Cl_{n-1}$ by $Pa = b$ and $Qa = c$. Note that if $w \in Cl_n$ then

$$Qw = \frac{e_n w' - we_n}{2} = \frac{\widehat{w} - w}{2} e_n,$$

$$Pw = \frac{w - e_n w' e_n}{2} = \frac{w + \widehat{w}}{2}.$$

The following calculation rules are proved in ($^6$[Lemma 2] and $^8$[Lemma 1])

$$P(ab) = (Pa)Pb + (Qa)Q(b'),$$

$$Q(ab) = (Pa)Qb + (Qa)P(b'),$$

$$Q(ab) = aQb + (Qa)b'.$$

We often use the abbreviations $P'a = (Pa)'$ and $Q'a = (Qa)'$.

Let $\Omega$ be an open subset of $\mathbb{R}^{n+1}$. The left Dirac operator in $Cl_n$ is defined by $D_l f = \sum_{i=0}^n e_i \frac{\partial f}{\partial x_i}$ and the right one by $D_r f = \sum_{i=0}^n \frac{\partial f}{\partial x_i} e_i$, for a mapping $f : \Omega \to Cl_n$, whose components are continuously differentiable. The operators $\overline{D_l}$ and $\overline{D_r}$ are defined by $\overline{D_l} f = \sum_{i=0}^n \overline{e_i} \frac{\partial f}{\partial x_i}$ and $\overline{D_r} f = \sum_{i=0}^n \frac{\partial f}{\partial x_i} \overline{e_i}$, respectively. As usual we abbreviate $D_l f = Df$ if there is no confusion possible. The corresponding modified Dirac operators $M_k^l$, $\overline{M}_k^l$ and $M_k^r$, $\overline{M}_k^r$ are introduced for any $k \in \mathbb{R}$ by

$$M_k^l f(x) = D_l f(x) + \frac{k}{x_n} Q' f(x)$$

$$M_k^r f(x) = D_r f(x) + \frac{k}{x_n} Q f(x)$$

and

$$\overline{M}_k^l f(x) = \overline{D}_l f(x) - \frac{k}{x_n} Q' f(x),$$

$$\overline{M}_k^r f(x) = \overline{D}_r f(x) - \frac{k}{x_n} Q f(x),$$

where $f \in C^1(\Omega, C\ell_n)$ and $'$ denotes the main involution. In what follows the operator $M_k^l$ is often denoted by $M_k$ and $M_{n-1}^l$ by $M$.

**Definition 3.1.** Let $\Omega \subset \mathbb{R}^{n+1}$ be an open set. A mapping $f : \Omega \to C\ell_n$ is called left $k$-hypermonogenic, if $f \in C^1(\Omega)$ and $x_n M_k^l f(x) = 0$, for any $x \in \Omega$. The $0$-hypermonogenic functions are the classical monogenic ones. The $n-1$-hypermonogenic functions are called briefly hypermonogenic. The right $k$-hypermonogenic functions are defined similarly.

Hypermonogenic functions were introduced in [6] and developed further in [8,9] and [5], where also references to other related papers are given. Paravector-valued hypermonogenic functions are called $H$-solutions which were introduced by H. Leutwiler in [11].

**Theorem 3.1 ([4]).** If $T : \Omega \to \mathbb{R}^{n+1}$ is the Möbius transformation induced by $\begin{pmatrix} a & b \\ c & d \end{pmatrix} \in GL(2, C\ell_{n-1})$., then

$$F(x) = \frac{(cx+d)^{-1}}{\|cx+d\|^{n-k-1}} f(T(x))$$

is k-hypermonogenic.

**Corollary 3.1.** The function $\frac{x^{-1}}{\|x\|^{n-k-1}}$ is k-hypermonogenic.

Using the simple observations

$$\overline{D} g(r) = g'(r) \overline{D} r$$

and

$$\overline{D}\left(\frac{\|x-y\|}{\|x-\hat{y}\|}\right) = \frac{2y_n \|x-y\|}{\|x-\hat{y}\|} (x-y)^{-1} e_n (x-\hat{y})^{-1}$$

we may compute the kernels for k-hyperharmonic functions.

**Theorem 3.2.** *Set* $\tau(x,y) = \frac{\|x-y\|}{\|x-\hat{y}\|}$. *In the case* $k > -1$ *the function*

$$
\begin{aligned}
p_k(x,y) = {} & (k-n+1)(x-\hat{y})^{-1} h_k(\tau(x,y)) \\
& - \frac{2y_n h_k(\tau(x,y))}{\varphi_k(\tau(x,y))} \tau(x,y)(x-y)^{-1} e_n (x-\hat{y})^{-1} \\
& + \frac{2^{2k+1} y_n^{k+1} x_n^k}{\varphi_k(\tau(x,y))} \frac{(x-y)^{-1}}{\|x-y\|^{n-1}} e_n \frac{(x-\hat{y})^{-1}}{\|x-\hat{y}\|^k}
\end{aligned}
$$

*is $k$-hypermonogenic, when $h_k$ is defined in Theorem 2.4. In the case $k < -1$ the function*

$$
\begin{aligned}
p_k(x,y) = {} & (k-n+1)(x-\hat{y})^{-1} h_k(\tau(x,y)) \\
& - \frac{2y_n h_k(\tau(x,y))}{\varphi_k(\tau(x,y))} \tau(x,y)(x-y)^{-1} e_n (x-\hat{y})^{-1} \\
& - (k+1) \frac{2 h_k(\tau(x,y)) \|x-y\|^2}{4x_n y_n}(x-y)^{-1} e_n (x-\hat{y})^{-1} \\
& - \frac{1}{2x_n \varphi_{-k-2}(\tau(x,y))} \frac{(x-y)^{-1}}{\|x-y\|^{n-1}} e_n \frac{(x-\hat{y})^{-1}}{\|x-\hat{y}\|^{-k-2}}
\end{aligned}
$$

*is $k$-hypermonogenic. In the case $k = -1$, the function $p_k(x,y)$ is defined as above, but only for values $\frac{\|x-y\|}{\|x-\hat{y}\|} < a$ for some strictly positive $a < 1$. Moreover, we have*

$$
p_k(x,y) = O\left( \frac{(x-y)^{-1}}{\|x-y\|^{n-1}} \right)
$$

*when $x \to y$.*

**Proof** We just compute in the case $k \geq -1$ as follows

$$
\begin{aligned}
p_k(x,y) = {} & \overline{D_x} h_k \\
= {} & (k-n+1) \frac{(x-\hat{y})^{-1}}{\|x-\hat{y}\|^{n-1-k}} f_k(\tau(x,y)) + \\
& 2y_n f_k'(\tau(x,y)) \|x-y\| (x-y)^{-1} e_n \frac{(x-\hat{y})^{-1}}{\|x-\hat{y}\|^{n-k}} \\
= {} & (k-n+1) \frac{(x-\hat{y})^{-1}}{\|x-\hat{y}\|^{n-1-k}} f_k(\tau(x,y)) \\
& - \frac{2y_n h_k(\tau(x,y))}{\varphi_k(\tau(x,y))} \tau(x,y)(x-y)^{-1} e_n (x-\hat{y})^{-1} \\
& + \frac{2^{2k+1} y_n^{k+1} x_n^k}{\varphi_k(\tau(x,y))} \frac{(x-y)^{-1}}{\|x-y\|^{n-1}} e_n \frac{(x-\hat{y})^{-1}}{\|x-\hat{y}\|^k}.
\end{aligned}
$$

In the case $k > -1$, the computations are

$$p_k(x, y) = \overline{D_x} h_k = (k - n + 1) \frac{(x - \hat{y})^{-1}}{\|x - \hat{y}\|^{n-1-k}} f_k(\tau(x, y))$$

$$- \frac{2y_n h_k(\tau(x, y))}{\varphi_k(\tau(x, y))} \tau(x, y)(x - y)^{-1} e_n (x - \hat{y})^{-1}$$

$$- (k + 1) \frac{h_k(\tau(x, y))}{x_n} \|x - y\|^2 (x - y)^{-1} e_n (x - \hat{y})^{-1}$$

$$- \|x - \hat{y}\|^k \frac{2y_n (1 - \tau^2(x, y))^{k+1} (1 - \tau^2(x, y))^{-k-2}}{\varphi_{-k-2}(\tau(x, y)) \|x - y\|^{n-1}} (x - y)^{-1} e_n (x - \hat{y})^{-1}$$

$$= (k - n + 1) \frac{(x - \hat{y})^{-1}}{\|x - \hat{y}\|^{n-1-k}} f_k(\tau(x, y))$$

$$- \frac{2y_n h_k(\tau(x, y))}{\varphi_k(\tau(x, y))} \tau(x, y)(x - y)^{-1} e_n (x - \hat{y})^{-1}$$

$$- (k + 1) \frac{h_k(\tau(x, y))}{x_n} \|x - y\|^2 (x - y)^{-1} e_n (x - \hat{y})^{-1}$$

$$- \frac{1}{2x_n \varphi_{-k-2}(\tau(x, y))} \frac{(x - y)^{-1}}{\|x - y\|^{n-1}} e_n \frac{(x - \hat{y})^{-1}}{\|x - \hat{y}\|^{-k-2}}.$$

**Proposition 3.1.** ([5][*Proposition 5*]) *Let* $\Omega$ *be an open subset of* $\mathbb{R}^{n+1}$ *and* $f : \Omega \to Cl_n$ *be a* $C^1(\Omega, Cl_n)$ *function. If* $k \in \mathbb{R}$, *then*

$$M^l_{-k}\left(\frac{f e_n}{x_n^k}\right) = \frac{(M_k f) e_n}{x_n^k},$$

$$M^r_{-k}\left(\frac{e_n f}{x_n^k}\right) = \frac{e_n (M_k f)}{x_n^k}.$$

*Moreover a function* $f : \Omega \to Cl_n$ *is* $k$-*hypermonogenic if and only if the function* $\frac{f e_n}{x_n^k}$ *is* $-k$-*hypermonogenic.*

Let $\Omega$ be an open subset of $\mathbb{R}^{n+1} \setminus \{x_n = 0\}$ and $K$ is an $n + 1$-chain satisfying $\overline{K} \subset \Omega$. Define a paravector valued $n$-form by $d\sigma_k = \sum_{i=0}^{n} \frac{(-1)^i e_i}{x_n^k} dx_0 \wedge \ldots \wedge dx_{i-1} \wedge dx_{i+1} \ldots dx_n$. and a real $n + 1$-form is introduced by

$$dm_k = x_n^{-k} dx_0 \wedge \ldots \wedge dx_n.$$

The equation

$$\int_{\partial K} g \, d\sigma_0 f = \int_K ((D_r g) f + g D_l f) \, dm_0.$$

has two hyperbolic versions.

**Theorem 3.3.** ([5][*Theorem 6, Theorem 10*]) *Let* $\Omega$ *be an open subset of* $\mathbb{R}^{n+1}_+$ *and*

$K$ an $n+1$-chain satisfying $\overline{K} \subset \Omega$. If $f, g \in C^1(\Omega, C\ell_n)$, then

$$\int_{\partial K} g d\sigma_k f = \int_K (M_k^r g)\, f + g M_k^l f - \frac{k}{x_n} P(gf')\, e_n dm_k,$$

$$\int_{\partial K} g d\sigma_0 f = \int_K (M_{-k}^r g)\, f + g M_k^l f + \frac{k}{x_n} Q(gf')\, dm_0.$$

Moreover we have

$$\int_{\partial K} P(g d\sigma_k f) = \int_K P\left((M_k^r g)\, f + g M_k^l f\right) dm_k,$$

$$\int_{\partial K} Q(g d\sigma_0 f) = \int_K Q\left((M_{-k}^r g)\, f + g M_k^l f\right) dm_0.$$

**Corollary 3.2.** *Let $\Omega$ be an open subset of $\mathbb{R}_+^{n+1}$ and $K$ an $n+1$-chain satisfying $\overline{K} \subset \Omega$. If $f$ is left $k$-hypermonogenic and $g$ is right $k$-hypermonogenic, then*

$$\int_{\partial K} P(g d\sigma_k f) = 0.$$

**Corollary 3.3.** *Let $\Omega$ be an open subset of $\mathbb{R}_+^{n+1}$ and $K$ an $n+1$-chain satisfying $\overline{K} \subset \Omega$. If $f$ is left $k$-hypermonogenic and $g$ is right $-k$-hypermonogenic, then*

$$\int_{\partial K} Q(g d\sigma_0 f) = 0.$$

Applying the usual methods we obtain the Borel-Pompeiu formula for $P$- and $Q$-parts and Cauchy integral formulas.

**Theorem 3.4.** *Let $\Omega$ be an open subset of $\mathbb{R}_+^{n+1}$ and $K$ an $n+1$-chain satisfying $\overline{K} \subset \Omega$. If $f \in C^1(\Omega, C\ell_{0,n})$, then*

$$Pf(y) = \frac{1}{2^k \omega_{n+1}} \int_{\partial K} P(p_k(x,y)\, d\sigma_k(x)\, f(x))$$
$$- \frac{1}{2^k \omega_{n+1}} \int_K P(p_k(x,y)\, M_k^l f)\, dm_k$$

*for $y \in K$ and $\omega_{n+1}$ the surface measure of the unit ball .*

**Theorem 3.5.** *Let $\Omega \subset \mathbb{R}_+^{n+1}$ be open and $K$ an $n+1$-chain satisfying $\overline{K} \subset \Omega$. If $f$ is hypermonogenic in $\Omega$ and $y \in K$ , then*

$$Pf(y) = \frac{1}{2^k \omega_{n+1}} \int_{\partial K} P(p_k(x,y)\, d\sigma_k(x)\, f(x))$$
$$= \frac{1}{2^k \omega_{n+1}} \int_{\partial K} p_k(x,y)\, d\sigma_k(x)\, f(x)$$
$$+ \frac{1}{2^k \omega_{n+1}} \int_{\partial K} \widehat{p_k(x,y)} d\widehat{\sigma}_k(x)\, \widehat{f}(x)$$

*where $\omega_{n+1}$ the surface measure of the unit ball in $\mathbb{R}^{n+1}$.*

**Theorem 3.6 (Borel-Pompeiu).** *Let $\Omega$ be an open subset of $\mathbb{R}^{n+1}_+ \cap \{x_n > 0\}$ and $K$ an $n+1$-chain satisfying $\overline{K} \subset \Omega$. If $f \in C^1(\Omega, C\ell_{0,n})$, then*

$$Qf(y) = \frac{1}{2^k \omega_{n+1}} \int_{\partial K} Q\left(p_{-k}(x,y)\, d\sigma_0(x)\, f(x)\right)$$

$$- \frac{1}{2^k \omega_{n+1}} \int_K Q\left(p_{-k}(x,y)\, M_k^l f\right) dm_0$$

*where $\omega_{n+1}$ is the surface measure of the unit ball in $\mathbb{R}^{n+1}$.*

**Theorem 3.7.** *Let $\Omega \subset \mathbb{R}^{n+1}_+$ be open and $K$ an $n+1$-chain satisfying $\overline{K} \subset \Omega$. If $f$ is hypermonogenic in $\Omega$ and $y \in K$, then*

$$Qf(y) = \frac{1}{2^k \omega_{n+1}} \int_{\partial K} Q\left(p_{-k}(x,y)\, d\sigma_0(x)\, f(x)\right)$$

$$= -\frac{1}{2^k \omega_{n+1}} \int_{\partial K} p_{-k}(x,y)\, d\sigma_0(x)\, f(x)\, e_n$$

$$+ \frac{1}{2^k \omega_{n+1}} \int_{\partial K} \widehat{p_{-k}(x,y)} \widehat{d\sigma_0}(x)\, \widehat{f}(x)\, e_n.$$

Combining the previous results we obtain.

**Corollary 3.4.** *Let $\Omega$ be an open subset of $\mathbb{R}^{n+1}_+$ and $K$ an $n+1$-chain satisfying $\overline{K} \subset \Omega$. If $f$ is $k$-hypermonogenic, then*

$$f(x) = \frac{1}{2^k \omega_{n+1}} \left( \int_{\partial K} g(x,y)\, d\sigma_k(x)\, f(x) - \int_{\partial K} w(x,y)\, \widehat{d\sigma_k}(x) \widehat{f}(x) \right),$$

*where*

$$g(x,y) = x_n^k p_{-k}(x,y) + p_k(x,y)$$

*and*

$$w(x,y) = x_n^k \widehat{p_{-k}(x,y)} - \widehat{p_k(x,y)}.$$

Applying Proposition 3.1 we obtain an additional interesting result.

**Proposition 3.2.** *The functions $g$ and $\widehat{w}$ may be written as*

$$g(x,y) = h(x,y)\, e_n + p_k(x,y)$$
$$\widehat{w}(x,y) = h(x,y)\, e_n + p_k(x,y)$$

*where $h(x,y) = -x_n^k p_{-k}(x,y)\, e_n$ and $p_k(x,y)$ are $k$-hypermonogenic function with respect to $x$.*

## References

1. Akin, Ö. and H. Leutwiler, On the invariance of the solutions of the Weinstein equation under Möbius transformations, In Classical and modern potential theory and applications. Proceedings of the NATO advanced research workshop, Chateau de Bonas, France, July 25–31, 1993. NATO ASI Ser., Ser. C, Math. Phys. Sci. 430, Kluwer, Dordrecht, 1994, 19–29.

2. Brackx, R. Delanghe and F. Sommen, *Clifford Analysis*, Pitman, Boston, London, Melbourne, 1982.

3. Eriksson-Bique, S.-L., $k-$hypermonogenic functions, *In Progress in Analysis*, Vol I, World Scientific (2003), 337–348.

4. Eriksson-Bique, S.-L., Möbius transformations and $k$-hypermonogenic functions, Manuscript.

5. Eriksson, S.-L., Integral formulas for hypermonogenic functions, *Bull. Bel. Math. Soc.* **11** (2004), 705–717.

6. Eriksson-Bique, S.-L. and H. Leutwiler, Hypermonogenic functions, In *Clifford Algebras and their Applications in Mathematical Physics*, Vol. 2, Birkhäuser, Boston, 2000, 287–302.

7. Eriksson-Bique, S.-L. and Leutwiler, H., Hyperholomorphic functions, *Computational methods and function theory* Vol **1** (2001), No. 1, 179–192.

8. Eriksson-Bique, S.-L. and Leutwiler, H., Hypermonogenic functions and Möbius transformations, *Advances in Applied Clifford algebras*, Vol **11** (S2), December (2001), 67–76.

9. Eriksson, S.-L. and H. Leutwiler, Hypermonogenic functions and their Cauchy-type theorems. In Advances in analysis and geometry. Trends in Mathematics Birkhäuser, Basel, (2004), 97-112.

10. Hengartner, W. and Leutwiler, H., Hyperholomorphic functions in $\mathbb{R}^3$, *Advances in Applied Clifford algebras* Vol **11** (S1), November (2001), 247–265.

11. H. Leutwiler, Modified Clifford analysis, *Complex Variables* **17** (1992), 153–171.

12. H. Leutwiler, Best constants in the Harnack inequality for the Weinstein equation, *Aequation Math.* 34 (1987), 304–315.

13. Hua, L.K., *Starting with the unit circle, Background to higher analysis,* Springer, New York, 1981.

# DECONSTRUCTING DIRAC OPERATORS
## I: QUANTITATIVE HARTOGS-ROSENTHAL THEOREMS

MIRCEA MARTIN

*Department of Mathematics*
*Baker University*
*Baldwin City, KS 66006, USA*
*E-mail: mircea.martin@bakeru.edu*

The main purpose of our article is to prove a quantitative Hartogs-Rosenthal theorem concerning uniform approximation on compact sets by solutions of elliptic first-order differential operators with coefficients in a Banach algebra. This theorem, as well as some other related results and consequences, points out that certain properties of the standard Euclidean Dirac operators originally established in the settings of Clifford analysis, or single-variable complex analysis, persist under more general circumstances.

**Key words:** Clifford analysis, Dirac operator, quantitative Hartogs-Rosenthal theorems
**Mathematics Subject Classification:** 35E05, 41A30

## 1. Introduction

We start with a quick review of some basic constructions in Clifford analysis. Throughout our article, $\mathfrak{A}_n$ denotes the real Clifford algebra associated with the Euclidean space $\mathbb{R}^n$, $n \geq 2$. Specifically, $\mathfrak{A}_n$ is the unital associative real algebra generated by the standard orthonormal basis $\{e_1, e_2, \ldots, e_n\}$ for $\mathbb{R}^n$, with relations

$$e_i e_j + e_j e_i = -2\delta_{ij} e_0, \qquad 1 \leq i, j \leq n, \tag{1.1}$$

where $e_0$ is the identity of $\mathfrak{A}_n$, and $\delta_{ij}$ equals 1 or 0, according as $i = j$ or $i \neq j$. The set consisting of $e_0$ and all products $e_{i_1} \cdots \cdot e_{i_p}, 1 \leq i_1 < \cdots < i_p \leq n$, yields a basis for $\mathfrak{A}_n$ as a real vector space, and $\mathfrak{A}_n$ has an inner product $(\cdot|\cdot)$ and a norm $|\cdot|$ such that the basis just defined is orthonormal. Further, by regarding $\mathfrak{A}_n$ as an algebra of left multiplication operators acting on the Hilbert space $(\mathfrak{A}_n, (\cdot|\cdot))$ we convert $\mathfrak{A}_n$ into a real $C^*$-algebra. The operator norm on $\mathfrak{A}_n$ is denoted by $\|\cdot\|$.

One identifies any $x = (x_1, \ldots, x_n) \in \mathbb{R}^n$ with $x = x_1 e_1 + \cdots + x_n e_n \in \mathfrak{A}_n$, and gets an embedding of $\mathbb{R}^n$ into $\mathfrak{A}_n$. The norms $|\cdot|$ and $\|\cdot\|$ coincide on $\mathbb{R}^n$.

Let now $\mathfrak{H}$ be a Hilbert left $\mathfrak{A}_n$-module, that is, a Hilbert space upon which the algebra $\mathfrak{A}_n$ acts on the left such that each generator $e_i$ of $\mathfrak{A}_n$, $1 \leq i \leq n$, determines a skew-adjoint unitary operator. The space $C^\infty(\mathbb{R}^n, \mathfrak{H})$ of smooth $\mathfrak{H}$-valued functions on $\mathbb{R}^n$ is a left $\mathfrak{A}_n$-module under pointwise multiplication. Therefore, it makes sense to introduce a differential operator $D_{\text{euc}} : C^\infty(\mathbb{R}^n, \mathfrak{H}) \to C^\infty(\mathbb{R}^n, \mathfrak{H})$ by setting

$$D_{\text{euc}} = e_1 D_1 + e_2 D_2 + \cdots + e_n D_n, \tag{1.2}$$

where $D_k = \partial/\partial x_k$, $1 \leq k \leq n$. Operator $D_{\text{euc}}$ is called the *Euclidean Dirac operator* on $C^\infty(\mathbb{R}^n, \mathfrak{H})$. It is easy to check, based on (1.1), that $D_{\text{euc}}$ is elliptic. Moreover, its fundamental solution is an $\mathfrak{A}_n$-valued function $E$, called the *Euclidean Cauchy kernel* on $\mathbb{R}^n$, and defined as

$$E(x) = \frac{1}{|\mathbb{S}^{n-1}|} \cdot \frac{-x}{|x|^n}, \qquad x \in \mathbb{R}_0^n = \mathbb{R}^n \setminus \{0\}, \tag{1.3}$$

where $|\mathbb{S}^{n-1}|$ is the surface area of the unit sphere $\mathbb{S}^{n-1}$ in $\mathbb{R}^n$, and on the right-hand side $x \in \mathbb{R}^n$ is regarded as an element of $\mathfrak{A}_n$.

The study of Euclidean Dirac operators is part of what is known nowadays as Clifford analysis, a far reaching several variables extension of single-variable complex analysis. Excellent accounts on the subject can be found in Brackx, Delanghe, and Sommen [4], Delanghe, Sommen, and Souček [5], Gilbert and Murray [7], Gürlebeck and Sprössig [8], Mitrea [15], Ryan [16], and Ryan and Sprössig [17].

Our specific goal in this article is to pinpoint the origin of some properties of Euclidean Dirac operators by investigating first-order differential operators with coefficients in a Banach algebra. Though this class is limited, one might expect that the study of such operators will lead to a better understanding of the class of Dirac operators, and perhaps connect their theory with some new basic issues of harmonic and complex analysis in several variables, or of multi-dimensional operator theory.

To begin this program, suppose that $\mathfrak{A}$ is a real unital Banach algebra and $\mathfrak{M}$ is a real Banach left $\mathfrak{A}$-module. In other words, we assume that $\mathfrak{M}$ is a real Banach space and $\mathfrak{A}$ is realized as a subalgebra of $\mathfrak{L}(\mathfrak{M})$, the algebra of all bounded linear operators on $\mathfrak{M}$. We will denote the norms on $\mathfrak{A}$, $\mathfrak{M}$, and $\mathfrak{L}(\mathfrak{M})$ by $\|\cdot\|$.

Next, let $C^\infty(\mathbb{R}^n, \mathfrak{M})$ be the space of smooth $\mathfrak{M}$-valued functions on $\mathbb{R}^n$, $n \geq 2$, that becomes an $\mathfrak{A}$-module by extending the action of $\mathfrak{A}$ to $\mathfrak{M}$-valued functions pointwise. In particular, given an $n$-tuple $A = (a_1, a_2, \ldots, a_n)$ of elements of $\mathfrak{A}$, we introduce a differential operator $D_A$ on $C^\infty(\mathbb{R}^n, \mathfrak{M})$ by setting

$$D_A = a_1 D_1 + a_2 D_2 + \cdots + a_n D_n, \tag{1.4}$$

where, as in (1.1), $D_k = \partial/\partial x_k$, $1 \leq k \leq n$. The symbol of operator $D_A$ is the linear mapping $\sigma(D_A) : \mathbb{R}^n \to \mathfrak{A}$ given by $\sigma(D_A)(\xi) = A(\xi)$, $\xi \in \mathbb{R}^n$, with

$$A(\xi) = \xi_1 a_1 + \xi_2 a_2 + \cdots + \xi_n a_n, \qquad \xi = (\xi_1, \xi_2, \ldots, \xi_n) \in \mathbb{R}^n. \tag{1.5}$$

The operator norm $\|\sigma(D_A)\|$ of the symbol mapping will be denoted by $|A|$, that is,

$$|A| = \sup \|A(\xi)\|, \qquad \xi \in \mathbb{S}^{n-1}. \tag{1.6}$$

We are now in a position to formulate the problem addressed in our article. Suppose that $\Omega \subset \mathbb{R}^n$ is a compact set, and let $C(\Omega, \mathfrak{M})$ be the Banach space of all $\mathfrak{M}$-valued continuous functions on $\Omega$, with the uniform norm given by

$$\|u\|_{\Omega,\infty} = \sup\{\|u(x)\| : x \in \mathbb{R}^n\}, \qquad u \in C(\Omega, \mathfrak{M}). \tag{1.7}$$

Associated with $D_A$ and $\Omega$ we define the solution space $R_A(\Omega, \mathfrak{M})$ as the closure in $C(\Omega, \mathfrak{M})$ of the subspace consisting of restrictions to $\Omega$ of smooth functions

$u_0 \in C^\infty(\mathbb{R}^n, \mathfrak{M})$ such that $D_A u_0 = 0$ on some open neighborhoods of $\Omega$ in $\mathbb{R}^n$. Our purpose is to find estimates of $\text{dist}_{C(\Omega,\mathfrak{M})}[u, R_A(\Omega, \mathfrak{M})]$, the distance in $C(\Omega, \mathfrak{M})$ from a function $u \in C^\infty(\mathbb{R}^n, \mathfrak{M})$ to the solution subspace $R_A(\Omega, \mathfrak{M})$. As a first partial answer to this problem we have the following result.

**Theorem A.** *Suppose that $\Omega \subset \mathbb{R}^n$, $n \geq 2$, is a compact set of volume $|\Omega|$ different from 0, that has a smooth oriented boundary $\Sigma$.*
*(i) If $u \in C^\infty(\mathbb{R}^n, \mathfrak{M})$, then*

$$\text{dist}_{C(\Omega,\mathfrak{M})}[u, R_A(\Omega, \mathfrak{M})] \geq \frac{1}{|A| \cdot |\Sigma|} \cdot \| \int_\Omega D_A u(x) \text{dvol}(x)\|,$$

*where $|A|$ is given by (1.6) and $|\Sigma|$ stands for the surface area of $\Sigma$.*
*(ii) If $u : \mathbb{R}^n \to \mathfrak{M}$ is a linear function, then*

$$\text{dist}_{C(\Omega,\mathfrak{M})}[u, R_A(\Omega, \mathfrak{M})] \geq \frac{|\Omega|}{|A| \cdot |\Sigma|} \cdot \|D_A u\|.$$

In order to state the next results we assume that the differential operator $D_A$ is elliptic, that is, the values $\sigma(D_A)(\xi)$ of the symbol are invertible in $\mathfrak{A}$ for any $\xi \in \mathbb{R}^n_0 = \mathbb{R}^n \setminus \{0\}$. As a consequence one gets that $D_A$ has a fundamental solution $E_A : \mathbb{R}^n_0 \to \mathfrak{A}$ which is smooth, with a singularity at the origin. For more details we refer to Hörmander [10]. If $u \in C^\infty(\mathbb{R}^n, \mathfrak{M})$ is compactly supported, then

$$D_A(E_A * u) = u, \tag{1.8}$$

where $E_A * u$ is given by

$$E_A * u(y) = \int_{\mathbb{R}^n} E_A(y - x)u(x)\text{dvol}(x), \qquad y \in \mathbb{R}^n. \tag{1.9}$$

In addition, $E_A$ is an odd function homogeneous of degree $1 - n$, that is,

$$E_A(tx) = t^{1-n} E_A(x), \qquad t \in (0, \infty), \qquad x \in \mathbb{R}^n_0. \tag{1.10}$$

Actually, the integral transform (1.9) makes sense for $u \in L^\infty(\mathbb{R}^n, \mathfrak{M}) \cap L^1(\mathbb{R}^n, \mathfrak{M})$. The $L^\infty$- and $L^1$-norms of $u$ are subsequently denoted by $\|u\|_\infty$ and $\|u\|_1$.

It will be convenient to express the fundamental solution $E_A$ as

$$E_A(x) = \frac{1}{|\mathbb{S}^{n-1}|} \cdot \Phi_A(x), \qquad x \in \mathbb{R}^n_0,$$

and to introduce the compact set

$$\mathbb{B}^n_A = \{x \in \mathbb{R}^n_0 : \|\Phi_A(x)\| \geq 1\} \cup \{0\}. \tag{1.11}$$

**Theorem B.** *Suppose $D_A$ is an elliptic operator, and $u \in L^\infty(\mathbb{R}^n, \mathfrak{M}) \cap L^1(\mathbb{R}^n, \mathfrak{M})$, $n \geq 2$. Then $E_A * u \in L^\infty(\mathbb{R}^n, \mathfrak{M})$, and*

$$\|E_A * u\|_\infty \leq \frac{n}{|\mathbb{S}^{n-1}|} \cdot |\mathbb{B}^n_A|^{(n-1)/n} \cdot \|u\|_\infty^{(n-1)/n} \cdot \|u\|_1^{1/n}.$$

*where $|\mathbb{B}^n_A|$ equals the volume of the compact set $\mathbb{B}^n_A$.*

On $C(\Omega, \mathfrak{M})$ we also need the $L^1$-norm $\|\cdot\|_{\Omega,1}$, given by

$$\|u\|_{\Omega,1} = \int_\Omega \|u(x)\| \mathrm{dvol}(x), \qquad u \in C(\Omega, \mathfrak{M}). \tag{1.12}$$

Theorem B proves to be quite useful for the proof of the next quantitative Hartogs-Rosenthal theorem, which complements the first distance estimate in Theorem A.

**Theorem C.** *Suppose $D_A$ is an elliptic operator, $u \in C^\infty(\mathbb{R}^n, \mathfrak{M})$, and let $\Omega \subset \mathbb{R}^n$, $n \geq 2$, be a compact set. Then*

$$\mathrm{dist}_{C(\Omega,\mathfrak{M})}[u, R_A(\Omega, \mathfrak{M})] \leq \frac{n}{|\mathbb{S}^{n-1}|} \cdot |\mathbb{B}_A^n|^{(n-1)/n} \cdot \|D_A u\|_{\Omega,\infty}^{(n-1)/n} \cdot \|D_A u\|_{\Omega,1}^{1/n}.$$

The absolute constant $|\mathbb{B}_A^n|$ in Theorem C is no longer the best when $u : \mathbb{R}^n \to \mathfrak{M}$ is a linear function. In this regard we need to consider the dual space $\mathfrak{A}^*$ of $\mathfrak{A}$, and then to introduce, for each $\alpha \in \mathfrak{A}^*$ with $\|\alpha\| = 1$, the compact set $\mathbb{B}_{A,\alpha}^{n,+}$ given by

$$\mathbb{B}_{A,\alpha}^{n,+} = \{x \in \mathbb{R}_0^n : \alpha \circ \Phi_A(x) \geq 1\} \cup \{0\}, \tag{1.13}$$

which is a subset of the set $\mathbb{B}_A^n$. In addition, since the quantities $|\mathbb{B}_{A,\alpha}^{n,+}|$ provide a measure of the size of the $n$-tuple $A$, we set

$$|A|_* = \sup |\mathbb{B}_{A,\alpha}^{n,+}|, \qquad \alpha \in \mathfrak{A}^*, \qquad \|\alpha\| = 1. \tag{1.14}$$

The next result complements the second part of Theorem A.

**Theorem D.** *Suppose $D_A$ is an elliptic operator, and let $\Omega \subset \mathbb{R}^n$, $n \geq 2$, be a compact set. Then, for any linear function $u : \mathbb{R}^n \to \mathfrak{M}$ we have*

$$\mathrm{dist}_{C(\Omega,\mathfrak{M})}[u, R_A(\Omega, \mathfrak{M})] \leq \frac{n}{|\mathbb{S}^{n-1}|} \cdot |A|_*^{(n-1)/n} \cdot \|D_A u\| \cdot |\Omega|^{1/n}.$$

We would like to mention that the problem of uniform approximation on compact sets by solutions of elliptic equations has a rather long history. Early contributions addressing this problem are due to Browder [3]. For the classical Hartogs-Rosenthal theorem on rational approximation in single-variable complex analysis we refer to Hartogs and Rosenthal [9]. Other important contributions in this area related to the four theorems stated above are due to Ahlfors and Beurling [1], Alexander [2], and Gamelin and Khavinson [6]. General qualitative results and additional references are presented in Tarkhanov [18].

The remainder of the article is organized as follows. In Section 2 we will prove Theorem A. Section 3 is concerned with two techical lemmas that eventually are used to prove Theorem B. In Section 4 we complete the proofs of Theorems C and D. Finally, in Section 5 we comment on some consequences of the main results.

## 2. Lower Distance Estimates

This section provides a proof a Theorem A. The setting is the same as in Section 1.

**2.1.** Suppose $A = (a_1, a_2, \ldots, a_n)$ is the $n$-tuple used to define $D_A$ by (1.4). For each $1 \leq i \leq n$ we denote by $dx_i^c$ the $(n-1)$-form on $\mathbb{R}^n$ defined as

$$dx_i^c = dx_1 \wedge \cdots \wedge dx_{i-1} \wedge dx_{i+1} \cdots dx_n,$$

and let $\omega_A$ be the $\mathfrak{A}$-valued form given by

$$\omega_A = \sum_{i=1}^{n} (-1)^{i-1} a_i dx_i^c. \tag{2.1}$$

For any $u \in C^\infty(\mathbb{R}^n, \mathfrak{M})$ we introduce the $\mathfrak{M}$-valued form $\omega_A \cdot u$ and observe that its exterior derivative equals $d(\omega_A \cdot u) = D_A u \cdot dx$, where $dx = dx_1 \wedge \cdots \wedge dx_i \wedge \cdots dx_n$. With $\Omega$ and $\Sigma$ as in Theorm A, we next apply Stokes' theorem and get

$$\int_\Sigma \omega_A \cdot u = \int_\Omega D_A u \cdot dx. \tag{2.2}$$

Both sides of (2.2) can be expressed as integrals of $\mathfrak{M}$-valued functions by taking the surface area measure darea on $\Sigma$, and the volume measure dvol on $\Omega$, respectively. For each $x \in \Sigma$ we let $\nu(x) = (\nu_1(x), \nu_2(x), \ldots, \nu_n(x)) \in \mathbb{R}^n$ be the outer normal unit vector to $\Sigma$ at $x$. Then $(-1)^{i-1} dx_i^c|_x = \nu_i(x) \cdot \text{darea}(x), 1 \leq i \leq n$, whence, by (2.1) and (1.5) we have $\omega_A(x) = \nu_A(x) \cdot \text{darea}(x)$, with

$$\nu_A(x) = A(\nu(x)), \qquad x \in \Sigma. \tag{2.3}$$

Therefore, equation (2.2) amounts to

$$\int_\Sigma \nu_A(x) \cdot u(x) \cdot \text{darea}(x) = \int_\Omega D_A u(x) \cdot \text{dvol}(x). \tag{2.4}$$

**2.2. Proof of Theorem A.** Let $u_0 \in C^\infty(\mathbb{R}^n, \mathfrak{M})$ be such that $D_A u_0 = 0$ on an open neighborhood of $\Omega$. Since $D_A(u - u_0) = D_A u$ on $\Omega$, by using (2.4) for $u - u_0$ we get

$$\int_\Omega D_A u(x) \text{dvol}(x) = \int_\Sigma \nu_A(x)[u(x) - u_0(x)]\text{darea}(x),$$

whence

$$\left\| \int_\Omega D_A u(x) \text{dvol}(x) \right\| \leq \int_\Sigma \|\nu_A(x)\| \cdot \|u(x) - u_0(x)\| \text{darea}(x). \tag{2.5}$$

From (2.3) and (1.6) we have $\|\nu_A(x)\| \leq |A|$, and obviously $\|u(x) - u_0(x)\| \leq \|u - u_0\|_{\Omega,\infty}$ for any $x \in \Sigma$. Therefore, by (2.5) we obtain

$$\left\| \int_\Omega D_A u(x) \text{dvol}(x) \right\| \leq |A| \cdot |\Sigma| \cdot \|u - u_0\|_{\Omega,\infty}. \tag{2.6}$$

The estimate in part $(i)$ of Theorem A follows by taking the infimum in (2.6) with respect to functions $u_0$. The estimate in part $(ii)$ for linear functions is a direct consequence of part $(i)$. The proof is complete.

## 3. Integral Estimates

This section deals with two technical lemmas and the proof of Theorem B. For some other integral estimates of the same kind and their applications we refer to Martin and Szeptycki [14] and Martin [13].

**3.1.** The first lemma is stated in a rather general setting. Suppose $k : M \to [0, \infty]$ is a measurable function defined on a measurable space $M$ equipped with a positive measure $\mu$. For every $t \in (0, \infty)$ one introduces the measurable set

$$X[t] = \{x \in M : k(x) \geq t\}, \tag{3.1}$$

and let $\delta : (0, \infty) \to [0, \infty]$ be the distribution function of $k$ given by

$$\delta(t) = \mu(X[t]), \qquad t \in (0, \infty). \tag{3.2}$$

The specific result we are interested in can be stated as follows.

**Lemma 1.** *Suppose $\kappa > 1$ and $\lambda > 0$ are two constants such that*

$$\delta(t) \leq \lambda t^{-\kappa}, \qquad t \in (0, \infty). \tag{3.3}$$

*Then, for any measurable set $X$ in $M$ we have*

$$\int_X k(x) \mathrm{d}\mu(x) \leq \frac{\kappa}{\kappa - 1} \cdot \lambda^{1/\kappa} \cdot [\mu(X)]^{1-1/\kappa}. \tag{3.4}$$

**Proof.** There is no loss of generality if we assume that $\mu(X)$ is finite and different from 0. With $\kappa$ and $\lambda$ as in (3.2), we let $\tau$ be a positive number such that

$$\mu(X) = \lambda \tau^{-\kappa}, \tag{3.5}$$

and split the integral in (3.4) into

$$\int_X k(x) \mathrm{d}\mu(x) = I_1 + I_2, \tag{3.6}$$

where

$$I_1 = \int_{X \cap X[\tau]} [k(x) - \tau] \mathrm{d}\mu(x), \qquad I_2 = \tau \mu(X \cap X[\tau]) + \int_{X \setminus X[\tau]} k(x) \mathrm{d}\mu(x).$$

Next, from (3.1) we get the following estimate for $I_2$,

$$I_2 \leq \tau \mu(X \cap X[\tau]) + \tau \mu(X \setminus X[\tau]) = \tau \mu(X). \tag{3.7}$$

With regard to $I_1$ we clearly have

$$I_1 \leq \int_{X[\tau]} [k(x) - \tau] \mathrm{d}\mu(x).$$

The last integral can be now expressed as a Stieltjes integral associated with the distribution function $\delta$ defined by (3.2), and then an integration by parts yields

$$\int_{X[\tau]} [k(x) - \tau] \mathrm{d}\mu(x) = \int_\tau^\infty \delta(t) \mathrm{d}t.$$

Using this equation and (3.3) we get

$$\int_{X[\tau]} [k(x) - \tau] d\mu(x) \leq \int_{\tau}^{\infty} \lambda t^{-\kappa} dt = \frac{\lambda}{\kappa - 1} \cdot \tau^{1-\kappa}.$$

Therefore,

$$I_1 \leq \frac{\lambda}{\kappa - 1} \cdot \tau^{1-\kappa}. \tag{3.8}$$

Estimates (3.7) and (3.8) in conjunction with (3.6) result in

$$\int_X k(x) d\mu(x) \leq \tau \mu(X) + \frac{\lambda}{\kappa - 1} \cdot \tau^{1-\kappa}. \tag{3.9}$$

Inequality (3.4) follows by substituting into (3.9) the value of $\tau$ obtained from (3.5), and simplifying the resulting expression. The proof of Lemma 1 is complete.

**3.2.** We are going to refine Lemma 1 by taking $M = \mathbb{R}^n$ with the Lebesque measure, and assuming that $k : \mathbb{R}^n \to [0, \infty]$ is homogeneous of degree $-n/\kappa$, that is,

$$k(tx) = t^{-n/\kappa} k(x), \qquad t \in (0, \infty), \qquad x \in \mathbb{R}^n, \tag{3.10}$$

where $\kappa > 1$ is a given constant. From this condition we easily get that the sets $X[t]$ defined by (3.1) are related as follows, $X[t] = t^{-\kappa/n} X[1], t \in (0, \infty)$, whence,

$$|X[t]| = t^{-\kappa} |X[1]|, \qquad t \in (0, \infty). \tag{3.11}$$

Suppose next that $\varphi$ is a Lebesque measurable function in $L^\infty(\mathbb{R}^n, \mathbb{R}) \cap L^1(\mathbb{R}^n, \mathbb{R})$. The following result estimates the convolution product $k * \varphi$.

**Lemma 2.** *If* $\varphi \in L^\infty(\mathbb{R}^n, \mathbb{R}) \cap L^1(\mathbb{R}^n, \mathbb{R})$, $\varphi \geq 0$, *and* $k$ *satisfies* (3.10), *then*

$$k * \varphi(y) \leq \frac{\kappa}{\kappa - 1} \cdot |X[1]|^{1/\kappa} \cdot \|\varphi\|_\infty^{1/\kappa} \cdot \|\varphi\|_1^{1-1/\kappa}, \qquad y \in \mathbb{R}^n. \tag{3.12}$$

**Proof.** Suppose that $y \in \mathbb{R}^n$ is fixed and let $\mu$ be the measure on $\mathbb{R}^n$ given by

$$d\mu(x) = \varphi(y - x) d\text{vol}(x), \qquad x \in \mathbb{R}^n. \tag{3.13}$$

Obviously $k * \varphi(y) = \int_{\mathbb{R}^n} k(x) d\mu(x)$, $y \in \mathbb{R}^n$, so we can use Lemma 1, and to this end we first show that (3.2) takes place provided

$$\lambda = |X[1]| \cdot \|\varphi\|_\infty. \tag{3.14}$$

Since $\delta(t) = \mu(X[t]) \leq \|\varphi\|_\infty \cdot |X[t]|$, by (3.11) we get $\delta(t) \leq |X[1]| \cdot \|\varphi\|_\infty \cdot t^{-\kappa}$, which amounts to (3.2) with $\lambda$ as in (3.14). In addition, by taking $X = \mathbb{R}^n$ in Lemma 1 we observe that

$$\mu(X) = \|\varphi\|_1. \tag{3.15}$$

By (3.14) and (3.15) estimate (3.4) in Lemma 1 reduces to (3.12). The proof is complete.

1072

**3.3. Proof of Theorem B.** Under the assumptions of Theorem B, we observe that for any $y \in \mathbb{R}^n$ we have

$$\|E_A * u(y)\| \leq \frac{1}{|\mathbb{S}^{n-1}|} \cdot k_A * \varphi(y),$$

where $k_A(\cdot) = \|\Phi_A(\cdot)\|$ and $\varphi(\cdot) = \|u(\cdot)\|$. By (1.10) we deduce that $k_A$ is homogeneous of degree $-n/\kappa$, with $\kappa = n/(n-1)$. On the other hand, the set $X[1]$ associated with $k_A$ equals the set $\mathbb{B}_A^n$ given by (1.11). The complete proof of Theorem B is now a straightforward consequence of Lemma 2.

## 4. Upper Distance Estimates

This section includes proofs of Theorems C and D.

**4.1. Proof of Theorem C.** We need an auxiliary result.

**Lemma 3.** *If* $u \in C^\infty(\mathbb{R}^n, \mathfrak{M})$ *and* $\chi_\Omega$ *is the characteristic function of* $\Omega$, *then*

$$\mathrm{dist}_{C(\Omega,\mathfrak{M})}[u, R_A(\Omega, \mathfrak{M})] \leq \|E_A * (\chi_\Omega D_A u)\|_{\Omega,\infty}. \tag{4.1}$$

**Proof.** Let $\chi_m$, $m \geq 1$, be a sequence of smooth compactly supported functions on $\mathbb{R}^n$ taking values in $[0,1]$, with each $\chi_m$ equal to 1 on an open relatively compact neighborhood $U_m$ of $\Omega$ in $\mathbb{R}^n$, such that $\mathrm{supp}\,\chi_{m+1} \subseteq \mathrm{supp}\,\chi_m$ and

$$\bigcap_{m \geq 1} \mathrm{supp}\,\chi_m = \Omega. \tag{4.2}$$

Suppose $u \in C^\infty(\mathbb{R}^n, \mathfrak{M})$ is given and let $u_m \in C^\infty(\mathbb{R}^n, \mathfrak{M})$ be defined as

$$u_m = u - E_A * (\chi_m D_A u), \qquad m \geq 1.$$

By (1.8) we get $D_A u_m = D_A u - \chi_m D_A u = (1 - \chi_m) D_A u$, whence $D_A u_m = 0$ on $U_m$, that is, $u_m \in R_A(\Omega, \mathfrak{M})$. Consequently,

$$\mathrm{dist}_{C(\Omega,\mathfrak{M})}[u, R_A(\Omega, \mathfrak{M})] \leq \|u - u_m\|_{\Omega,\infty} = \|E_A * (\chi_m D_A u)\|_{\Omega,\infty}.$$

Inequality (4.1) follows from (4.2) and the last estimate by taking the limit. The proof of Lemma 3 is complete.

Theorem C is now a direct consequence of Lemma 3 and Theorem B applied to the function $\chi_\Omega D_A u$.

**4.2. Proof of Theorem D.** If $u : \mathbb{R}^n \to \mathfrak{M}$ is a linear function, then $D_A u$ is a vector in $\mathfrak{M}$, and inequality (4.1) in Lemma 3 implies

$$\mathrm{dist}_{C(\Omega,\mathfrak{M})}[u, R_A(\Omega, \mathfrak{M})] \leq \|E_A * \chi_\Omega\|_{\Omega,\infty} \cdot \|D_A u\|. \tag{4.3}$$

Using the dual space $\mathfrak{A}^*$ of $\mathfrak{A}$ and the function $\Phi_A$ introduced in Section 1 we get

$$\|E_A * \chi_\Omega\|_{\Omega,\infty} \leq \frac{1}{|\mathbb{S}^{n-1}|} \cdot \sup\{\|(\alpha \circ \Phi_A) * \chi_\Omega\|_\infty : \alpha \in \mathfrak{A}^*, \|\alpha\| = 1\}. \tag{4.4}$$

For $\alpha \in \mathfrak{A}^*$ we define the functions $k_{A,\alpha}^{\pm}$ as $k_{A,\alpha}^{\pm}(x) = \max\{\pm\alpha \circ \Phi_A(x), 0\}, x \in \mathbb{R}^n$, and notice that

$$\|(\alpha \circ \Phi_A) * \chi_\Omega\|_\infty \leq \max\{\|k_{A,\alpha}^+ * \chi_\Omega\|_\infty, \|k_{A,\alpha}^- * \chi_\Omega\|_\infty\}. \tag{4.5}$$

On the other, hand $k_{A,\alpha}^- = k_{A,-\alpha}^+$, so from (4.4) and (4.5) we have

$$\|E_A * \chi_\Omega\|_{\Omega,\infty} \leq \frac{1}{|\mathbb{S}^{n-1}|} \cdot \sup\{\|k_{A,\alpha}^+ * \chi_\Omega\|_\infty : \alpha \in \mathfrak{A}^*, \|\alpha\| = 1\}. \tag{4.6}$$

Once more we are in a position to apply Lemma 2, this time to the functions $k = k_{A,\alpha}^+$ and $\varphi = \chi_\Omega$, with $\kappa = n/(n-1)$. The set $X[1]$ associated with $k_{A,\alpha}^+$ turns out to be the set $\mathbb{B}_{A,\alpha}^{n,+}$ defined by (1.13), and from (3.12) we get

$$\|k_{A,\alpha}^+ * \chi_\Omega\|_\infty \leq n \cdot |\mathbb{B}_{A,\alpha}^{n,+}|^{(n-1)/n} \cdot |\Omega|^{1/n}. \tag{4.7}$$

The complete proof of Theorem D follows from (4.3), (4.6), (4.7), and (1.14).

## 5. Concluding Remarks

We continue with some direct consequences of the theorems stated and proved above. First we notice that Theorem C in conjunction with the Stone-Weierstrass theorem obviously leads to the following qualitative Hartogs-Rosenthal theorem:

**Corollary 1.** *Under the assumptions of Theorem C, if $|\Omega| = 0$, then*

$$R_A(\Omega, \mathfrak{M}) = C(\Omega, \mathfrak{M}).$$

Next we combine part $(i)$ in Theorem A with Theorem D and get:

**Corollary 2.** *Suppose that $\Omega \subset \mathbb{R}^n$, $n \geq 2$, is compact with a smooth oriented boundary $\Sigma$. Let $A = (a_1, a_2, \ldots, a_n)$ be an $n$-tuple of elements of a Banach algebra $\mathfrak{A}$, such that $A(\xi) = \xi_1 a_1 + \xi_2 a_2 + \cdots + \xi_n a_n$ is invertible in $\mathfrak{A}$ for any $\xi \in \mathbb{R}_0^n$, and let $|A|$ and $|A|_*$ be the quantities associated with $A$ by (1.6) and (1.14). Then*

$$\frac{|\Omega|^{(n-1)/n}}{|\Sigma|} \leq \frac{n}{|\mathbb{S}^{n-1}|} \cdot |A| \cdot |A|_*^{(n-1)/n}. \tag{5.1}$$

We would like to mention that in the special case when $D_A$ is the Euclidean Dirac operator $D_{\mathrm{euc}}$, so $A = (e_1, e_2, \ldots, e_n)$, all the constants in our previous results can be explicitly computed. With regard to Theorem A, we notice that $|A| = 1$. The compact set $\mathbb{B}_A^n$ defined by (1.11) and used in Theorems B and C turns out to be the unit ball $\mathbb{B}^n$ in $\mathbb{R}^n$. Moreover, based on some results proved in Martin [11,12] one can show that the quantity $|A|_*$ defined by (1.14) and used in Theorem D is actually given by $|A|_* = |\mathbb{B}_\xi^{n,+}|$, with $\xi \in \mathbb{S}^{n-1}$ and $\mathbb{B}_\xi^{n,+} = \{x \in \mathbb{R}^n : |x|^n \leq (x|\xi)\}$, where $(\cdot|\cdot)$ is the inner product on $\mathbb{R}^n$. In particular, one may take $\xi = e_1$, whence

$$|A|_* = \mathrm{vol}\{x = (x_1, x_2, \ldots, x_n) \in \mathbb{R}^n : |x|^n \leq x_1\}. \tag{5.2}$$

1074

The simplest case $n = 2$ deserves a special comment. The compact set in (5.2) is a disk of radius $1/2$, so $|A|_* = \pi/4$, and because $|A| = 1$, Corollary 2 reduces to an inequality proved by Gamelin and Khavinson [GK] in the setting of single-variable complex analysis, that solves the isoperimetric problem in $\mathbb{R}^2$. Specifically, one gets that for any compact subset $\Omega \subset \mathbb{R}^2$ with a smooth oriented boundary $\Sigma$, the area $|\Omega|$ of $\Omega$ and the length $|\Sigma|$ of $\Sigma$ satisfy the inequality

$$|\Omega| \le \frac{1}{4\pi} \cdot |\Sigma|^2. \tag{5.3}$$

## References

1. Ahlfors, L. and Beurling, A., Conformal invariants and function theoretic null sets, *Acta Math.* **83**(1950), 101–129.
2. Alexander, H., Projections of polynomial hulls, *J. Funct. Anal.* **13**(1973), 13–19.
3. Browder, F., Approximation by solutions of partial differential equations, *Amer. J. Math.* **84**(1962), 134–160.
4. Brackx, F., Delanghe, R., and Sommen, F., *Clifford Analysis*, Pitman Research Notes in Mathematics Series, **76**, 1982.
5. Delanghe, R., Sommen, F., and Souček, V., *Clifford Algebra and Spinor-Valued Functions*, Kluwer Academic Publishers, 1992.
6. Gamelin, T. W. and Khavinson, D., The isoperimetric inequality and rational approximation, *Amer. Math. Monthly* **96**(1989), 18–30.
7. Gilbert, J. E. and Murray, M. A. M., *Clifford Algebras and Dirac Operators in Harmonic Analysis*, Cambridge Studies in Advanced Mathematics, **26**, Cambridge University Press, 1991.
8. Gürlebeck, K. and Sprössig, W., *Quaternionic and Clifford Calculus for Physicists and Engineers*, John Wiley & Sons, New York, 1997.
9. Hartogs, F. and Rosenthal, A., Über Folgen analytischer Funktionen, *Math. Ann.* **104**(1931), 606–610.
10. Hörmander, L., *The Analysis of Linear Partial Differential Operators, vol I: Distribution Theory and Fourier Analysis*, Springer-Verlag, Berlin, 1983.
11. Martin, M., Higher-dimensional Ahlfors-Beurling inequalities in Clifford analysis, *Proc. Amer. Math. Soc.* **126**(1998), 2863–2871.
12. Martin, M., Convolution and maximal operator inequalities in Clifford analysis, *Clifford Algebras and Their Applications in Mathematical Physics, Vol. 2: Clifford Analysis*, Progress in Physics **9**, Birkhäuser Verlag, Basel, 2000, 95–113.
13. Martin, M., Self-commutator inequalities in higher dimension, *Proc. Amer. Math. Soc.* **130**(2002), 2971–2983.
14. Martin, M. and Szeptycki, P., Sharp inequalities for convolution operators with homogeneous kernels and applications, *Indiana Univ. Math. J.* **46** (1997), 975–988.
15. Mitrea, M., *Singular Integrals, Hardy Spaces, and Clifford Wavelets*, Lecture Notes in Mathematics, **1575**, Springer-Verlag, Heidelberg, 1994.
16. Ryan, J. (Ed.), *Clifford Algebras in Analysis and Related Topics*, CRC Press, Boca Raton, FL, 1995.
17. Ryan, J. and Sprößig, W. (Eds.), *Clifford Algebras and Their Applications in Mathematical Physics, Volume 2: Clifford Analysis*, Progress in Physics **19**, Birkhäuser, Basel, 2000.
18. Tarkhanov, N. N., *The Cauchy Problem for Solutions of Elliptic Equations*, Akademie Verlag, Berlin, 1995.

# HYPERMONOGENIC AND HOLOMORPHIC CLIFFORDIAN FUNCTIONS

ERIC LEHMAN

*Université de Caen - Basse Normandie*
*E-mail: lehman@math.unicaen.fr*

In Clifford analysis there are two well known generalizations of the concept of holomorphic functions issued from the idea of monogenic functions: the hypermonogenic functions introduced by H. Leutwiler and the holomorphic Cliffordian functions introduced by G. Laville and I. Ramadanoff. When these functions are of type $\overline{D}h$ with $h$ scalar valued, they are called respectively $H$-solutions and $p$-holomorphic Cliffordian functions. The parity of the dimension $d$ of the vector space $\mathbb{R}^{0,d}$ generating the Clifford algebra $\mathbb{R}_{0,d}$ induces quite different behaviours of these functions. We summarize the properties of the analytic Cliffordian monomials and classify the properties of $H$-solutions and of $p$-holomorphic Cliffordian functions with respect to the parity of $d$. We show that the restriction of an $H$-solution on $\mathbb{R} \oplus \mathbb{R}^{0,d}$ to $x_d = 0$ is a $p$-holomorphic Cliffordian function. By an explicit construction of the homogeneous polynomials which are $H$-solutions, we show the converse. Finally, a paravector-valued function of paravectors is an $H$-solution if and only if it is the restriction of a $p$-holomorphic Cliffordian function.

**Key words:** Clifford algebra, Clifford analysis, H-solution, holomorphic Cliffordian function, iterated Laplacian

**Mathematics Subject Classification:** 30G30

## 0. Introduction

Let $d \in \mathbb{N}_0$. We denote by $\mathbb{R}^{0,d}$ the anti-Euclidean real vector space of dimension $d$. Let $e_1, \ldots, e_d$ be an orthonormal basis of $\mathbb{R}^{0,d}$. We denote by $\mathbb{R}_{0,d}$ the Clifford algebra constructed on $\mathbb{R}^{0,d}$ by the rules

$$\forall i \in \{1, \ldots, d\} \quad e_i^2 = -1 \tag{1}$$

$$\forall i, j \in \{1, \ldots, d\} \; i \neq j \Rightarrow e_i e_j = -e_j e_i. \tag{2}$$

For each subset $I$ of $\{1, \ldots, d\}$, if $I = \{i_1, \ldots, i_k\}$ with $i_1 < \ldots < i_k$, we put

$$e_I = e_{i_1 i_2 \ldots i_k} = e_{i_1} e_{i_2} \ldots e_{i_k}.$$

We denote by $e_0$ the real number $1 = e_\phi$. Then $\{e_{i_1 \ldots i_k}; i_1 < \ldots < i_k\}$ is a basis for the real vector space $\Lambda^k \mathbb{R}^{0,d}$. We have $\dim \Lambda^k \mathbb{R}^{0,d} = \binom{d}{k}$,

$$\mathbb{R}_{0,d} = \bigoplus_{k=0}^{d} \Lambda^k \mathbb{R}^{0,d}$$

and $\dim \mathbb{R}_{0,d} = 2^d$. Notice that $\Lambda^0 \mathbb{R}^{0,d} = \mathbb{R}$, $\Lambda^1 \mathbb{R}^{0,d} = \mathbb{R}^{0,d}$ and $\Lambda^d \mathbb{R}^{0,d} = \mathbb{R}e_{12...d}$. The elements of $\mathbb{R}$ are called scalars, those of $\mathbb{R}^{0,d}$ vectors, those of $\mathbb{R}e_{12...d}$ pseudoscalars and the elements of $\mathbb{R} \oplus \mathbb{R}^{0,d}$ are called paravectors. We denote the Clifford conjugation by

$$\bar{e}_0 = e_0, \quad \forall i \in \{1,\ldots,d\} \quad \bar{e}_i = -e_i, \quad \forall a,b \in \mathbb{R}_{0,d} \quad \overline{ab} = \bar{b}\bar{a}.$$

If $a$ is a paravector

$$a = a_0 + \sum_{i=1}^{d} a_i e_i, \quad \text{where } a_0, a_1, \ldots, a_d \in \mathbb{R},$$

we have

$$\bar{a} = a_0 - \sum_{i=1}^{d} a_i e_i$$

and

$$a\bar{a} = \bar{a}a = a_0^2 + a_1^2 + \ldots + a_d^2 \in \mathbb{R}_+.$$

We define $|a|$ by $|a| := \sqrt{a\bar{a}}$. Clifford analysis generalizes the concept of holomorphic function to functions $f$ from $\mathbb{R} \oplus \mathbb{R}^{0,d}$ to $\mathbb{R}_{0,d}$ or more simply as we will do in the following from $\mathbb{R} \oplus \mathbb{R}^{0,d}$ to $\mathbb{R} \oplus \mathbb{R}^{0,d}$ :

$$f : \mathbb{R} \oplus \mathbb{R}^{0,d} \to \mathbb{R} \oplus \mathbb{R}^{0,d}, (x_0, x_1, \ldots, x_d) \longmapsto f(x_0, x_1, \ldots, x_d) = f_0 + f_1 e_1 + \ldots + f_d e_d.$$

One defines the Dirac operator

$$D := e_0 \partial_0 + e_1 \partial_1 + \ldots + e_d \partial_d$$

where $\partial_\alpha := \dfrac{\partial}{\partial x^\alpha}$ for $\alpha \in \{0, 1, \ldots, d\}$. We put

$$\bar{D} := e_0 \partial_0 - e_1 \partial_1 - \ldots - e_d \partial_d.$$

Notice that the Laplacian $\Delta = (\partial_0)^2 + (\partial_1)^2 + \ldots + (\partial_d)^2$ is given by $\Delta = D\bar{D}$ or $\Delta = \bar{D}D$. If $d = 1$, we have the case of complex functions of a complex variable. If we write $z = x_0 + x_1 e_1$, $e_1 = i$, we have

$$\frac{\partial}{\partial z} = \frac{1}{2}\bar{D} \quad \text{and} \quad \frac{\partial}{\partial \bar{z}} = \frac{1}{2}D$$

and $f$ is holomorphic if and only if $Df = 0$ that is in a sense if $f$ "is independent of $\bar{z}$". If $d = 0$, we have the case of real functions of a real variable. In this case there

is no differential characterisation of holomorphy but a function is analytic if it is a (finite or infinite) linear combination of $x^n$. We generalize the monomials $x^n$ by the analytic Cliffordian monomials $(ax)^n a$ where $a$ and $x$ are paravectors.

**Remark 1.** The two-variables functions $(\mathbb{R} \oplus \mathbb{R}^{0,d})^2 \rightarrow \mathbb{R} \oplus \mathbb{R}^{0,d}, (x,y) \longmapsto (y^{-1}x)^n y^{-1}$ are even more interesting since both $x \longmapsto (y^{-1}x)^n y^{-1}$ for fixed $y$ and $y \longmapsto (y^{-1}x)^n y^{-1}$ for fixed $x$ are holomorphic Cliffordian functions.

**Remark 2.** Notice that $(ax)^n a = a(xa)^n = b(bxb)^n b$ for any of the two paravectors $b$ such that $b^2 = a$.

## 1. Properties of the analytic Cliffordian monomials $(ax)^n a$

**Algebraic property.** For any paravectors $a$ and $x$ and any integer $n$, $(ax)^n a$ is a paravector.

*Proof.* Write $axa = (a\,x + \overline{x}\,\overline{a})a - \overline{x}\,\overline{a}\,a$ and notice that both $\overline{a}\,a$ and $a\,x + \overline{x}\,\overline{a}$ are scalars. Thus $axa$ is a paravector as well as $xax$, and thus also $axaxa$. A trivial induction gives the result for $n$ positive. For $n$ negative, $n = -n'$, use

$$(ax)^{-n'}a = (x^{-1}a^{-1})^{n'}a = (x^{-1}a^{-1})^{n'-1}x^{-1}$$

and remember that $x^{-1}$ and $a^{-1}$ are paravectors.

**Scalar potential.** There is a scalar function $h$ of the paravector variable $x$ such that $(ax)^n a = \overline{D}h(x)$. More precisely

$$(ax)^n a = \tfrac{1}{n+1}\overline{D}\{\text{scalar part of } (ax)^{n+1}\} \qquad (3)$$

$$= \tfrac{1}{n+1}\tfrac{1}{2}\overline{D}\{(ax)^{n+1} + (\overline{x}\,\overline{a})^{n+1}\}. \qquad (4)$$

For the proof see [7]. We call the function $h$ a scalar potential of the monomial $(ax)^n a$. In the case of the complex functions of a complex variable, the last formula reduces for $a = 1$ to

$$z^n = \frac{1}{n+1}\frac{\partial}{\partial z}\{z^{n+1} + \overline{z}^{n+1}\}$$

and for $a = e^{i\frac{\pi}{2(n+1)}}$ to

$$iz^n = (az)^n a = \frac{1}{n+1}\frac{\partial}{\partial z}\{(az)^{n+1} + (\overline{a}\,\overline{z})^{n+1}\}.$$

## Symmetry property
**Notation.** Let $\breve{\phantom{a}}$ denote the change of $e_d$ into $-e_d$. If $a$ is a paravector

$$a = a_0 + a_1 e_1 + \ldots + a_{d-1}e_{d-1} + a_d e_d$$

then

$$\breve{a} = a_0 + a_1 e_1 + \ldots + a_{d-1}e_{d-1} - a_d e_d.$$

**Definition.** A paravector $a$ is called short if $a_d = 0$, or equivalently if $\breve{a} = a$.

**Definition.** A paravector valued function $f$ of a paravector variable $x$ is said to be even with respect to $e_d$ if for every paravector $x$

$$(f(x))^{\check{}} = f(\check{x}).$$

$f$ is called odd with respect to $e_d$ if for every paravector $x$ :

$$(f(x))^{\check{}} = -f(\check{x}).$$

**Remark.** Let us write $f$ explicitly as

$$f(x_0, x_1, \ldots, x_d) = f_0(x_0, x_1, \ldots, x_d) + f_1(x_0, x_1, \ldots, x_d)e_1 + \ldots + f_d(x_0, x_1, \ldots, x_d)e_d.$$

Then $f$ is even with respect to $e_d$ if and only if $f_0, f_1, \ldots, f_{d-1}$ are even functions of $x_d$ and $f_d$ is an odd function of $x_d$. The function $f$ is odd with respect to $e_d$ if and only if $f_0, f_1, \ldots, f_{d-1}$ are odd functions of $x_d$ and $f_d$ is a even function of $x_d$.

**Remark.** If $f$ admits a scalar potential $h$, that is a function $h : \mathbb{R} \oplus \mathbb{R}^{0,d} \to \mathbb{R}$ such that $f = \overline{D}h$, then $f$ is even (respectively odd) with respect to $e_d$ if and only if $h$ is even (respectively odd) in the variable $x_d$.

**Proposition.** If $a$ is short then $(ax)^n a$ is even with respect to $e_d$.

*Proof.* Write $x = \xi + x_d e_d$ with $\xi = x_0 + x_1 e_1 + \ldots + x_{d-1}e_{d-1}$ and put $f(x) = (ax)^n a$, where $a$ is short, that is independent of $e_d$. To compute $f(\check{x})$ from $f(x)$ you have to change $x_d$ into $-x_d$. To compute $(f(x))^{\check{}}$ from $f(x)$ you have to change $e_d$ into $-e_d$. But since $\xi + (-x_d)e_d = \xi + x_d(-e_d)$ you get the same result.

**Differential properties.** We define by $\overline{D}_{\text{right}}$ the differential operator such that for any differentiable function $g$ of a paravector variable

$$\overline{D}_{\text{right}}g = \partial_0 g - (\partial_1 g)e_1 - \ldots - (\partial_d g)e_d.$$

A paravector valued function $f$ of a paravector variable has a scalar potential if and only if

$$Df = \overline{D}_{\text{right}}\overline{f} \qquad (*)$$

The analytic Cliffordian monomials $f(x) = (ax)^n a$ satisfy $(*)$ since $f$ has a scalar potential. In the following paragraphs we will see that if $a$ is short then $f$ is an $H$-solution, that is $f$ satisfies the equation

$$x_d Df + (d-1)f_d = 0.$$

For $d$ odd and any paravector $a$, $f$ is a $p$-holomorphic function, that is $f$ satisfies [7]

$$D\Delta^{\frac{d-1}{2}}f = 0.$$

## 2. $H$-solutions

A differentiable function $f$ of a paravector $x$

$$f = \sum_{I \subset \{1,\dots,d\}} f_I e_I = \sum_{J \subset \{1,\dots,d-1\}} f_J e_J + \sum_{K \subset \{1,\dots,d-1\}} f_{K \cup \{d\}} e_{K \cup \{d\}}$$

is called hypermonogenic [5,6] if

$$x_d D f + (d-1) \sum_{K \subset \{1,\dots,d-1\}} f_{K \cup \{d\}} e_K = 0.$$

If $f$ is also paravector valued it is called an $H$-solution and the above equation becomes

$$x_d D f + (d-1) f_d = 0.$$

The characteristic $\psi_f$ of a hypermonogenic function (and thus also of an $H$-solution) is defined by [1,4]

$$\psi_f = \frac{1}{(d-1)!} \left[ (\partial_d)^{d-1} f \right]_{x_d=0}.$$

The properties of the $H$-solutions depend on the parity of $d$.

**First case:** If $d$ is odd, then:

(i) There are even $H$-solutions and there are odd $H$-solutions. The polynomials called $T$-polynomials by Sirka-Liisa Eriksson [2,3] are odd $H$-solutions.

(ii) For all $n \in \mathbb{N}$, the monomials $(ax)^n a$ with $a$ short, generate $\mathbb{R}$-linearly the linear space of homogeneous polynomials of degree $n$ which are $H$-solutions and which are even with respect to $e_d$.

(iii) Every $H$-solution $f$ even with respect to $e_d$ is uniquely determined by $f|_{x_d=0}$.

(iv) Every $H$-solution $f$ odd with respect to $e_d$ is uniquely determined by $\psi_f$.

**Second case:** If $d$ is even, then:

(i) All $H$-solutions are even with respect to $e_d$.

(ii) For all $n \in \mathbb{N}$, the monomials $(ax)^n a$ with $a$ short, generate $\mathbb{R}$-linearly the space of homogeneous polynomials of degree $n$ which are $H$-solutions.

(iii) Any $H$-solution is uniquely determined by $f|_{x_d=0}$ and $\psi_f$.

The statements (i), (iii) and (iv) are equivalent to results stated in [4] and [9]; (ii) follows easily from [7] and from the last theorem of this paper.

## 3. *p*-holomorphic Cliffordian functions

The concepts of Cliffordian holomorphy [8] and Cliffordian analyticity are equivalent [7]. The set of homogeneous polynomials of degree $n$ which are holomorphic Cliffordian is a right $\mathbb{R}_{0,d}$-module generated by the analytic monomials $(ax)^n a$ with $a$ a paravector. Let $f : \mathbb{R} \oplus \mathbb{R}^{0,d} \to \mathbb{R}_{0,d}$.

**First case**: $d$ odd. $f$ is holomorphic Cliffordian if

$$D\Delta^{\frac{d-1}{2}} f = 0.$$

**Second case**: $d$ even. $f$ is holomorphic Cliffordian if there is a holomorphic Cliffordian function of one more variable $F : \mathbb{R} \oplus \mathbb{R}^{0,d+1} \to \mathbb{R}_{0,d+1}$, even with respect to $e_{d+1}$ and such that $F\mid_{x_{d+1}=0} = f$.

**Definition.** A holomorphic Cliffordian function $f$ is called $p$-holomorphic Cliffordian if it arises from a scalar potential $h$, that is $f = \overline{D}h$, or equivalently $Df = \overline{D}_{\text{right}}\overline{f}$. The set $pH_n$ of homogeneous polynomials of degree $n$ which are $p$-holomorphic Cliffordian is an $\mathbb{R}$-linear space generated by the analytic monomials since it is $\mathbb{R}$-linearly generated by the polynomials $P_\alpha$ [8] et [7]. It is of dimension $N_{n,d} = \binom{n+d+1}{d} - \binom{n}{d}$ if $d$ is odd and $N_{n,d} = \binom{n+d+1}{d}$ if $d$ is even. Thus there are $N = N_{n,d}$ paravectors $a_1, a_2, \ldots, a_N$ such that $(a_1 x)^n a_1, \ldots, (a_N x)^n a_N$ is a basis of $pH_n$.

**Remark 1.** The $p$-holomorphic Cliffordian functions are paravector valued, but there are paravector valued holomorphic Cliffordian functions which an not $p$-holomorphic Cliffordian. For example, if $d = 3$ the function $f(x) = x_0^2 + x_1^2 + x_2^2 + x_3^2$ satisfies $D\Delta f = 0$ and is scalar valued and thus paravector valued. If there were a scalar valued function $h$ such that $f = \overline{D}h$, one would have $\partial_1 h = \partial_2 h = \partial_3 h = 0$ and thus $f = \partial_0 h$ would be independent of $x_1, x_2$ and $x_3$. Notice that $f(x)$ belongs to the right-$\mathbb{R}_{0,3}$-module generated by the $(ax)^2 a$ with $a$ paravectors since :

$$2f(x) = -[(1x)^2 1]1 + [(e_1 x)^2 e_1]e_1 + [(e_2 x)^2 e_2]e_2 + [(e_3 x)^2 e_3]e_3.$$

**Remark 2.** All hypermonogenic functions are holomorphic Cliffordian. Let us recall one proof in the case when $d$ is odd : the function $f : \mathbb{R} \oplus \mathbb{R}^d \to \mathbb{R}_{0,d}$ is hypermonogenic if and only if there is a function $h : \mathbb{R} \oplus \mathbb{R}^d \to \mathbb{R}_{0,d-1}$ such that

$$f = \overline{D}h \quad \text{and} \quad x_d \Delta h - (d-1)\partial_d h = 0.$$

Here $\mathbb{R}_{0,d-1}$ is the Clifford algebra constructed on $e_1, e_2, \ldots, e_{d-1}$. Since for any function $F$ we have $\Delta(x_d F) = 2\partial_d F + x_d \Delta F$, we have $0 = \Delta^{\frac{d-1}{2}}[x_d \Delta h - (d-1)\partial_d h] = x_d \Delta^{\frac{d+1}{2}} h = 0$. The identity $\Delta = D\overline{D}$, gives then $\Delta^{\frac{d-1}{2}} f = 0$, which means that $f$ is holomorphic Cliffordian. In the case $d$ even, we notice that $f$ is hypermonogenic if it belongs to the right $\mathbb{R}_{0,d}$ module generated by the analytic monomials $(ax)^n a$ with $a$ short. This module is included in the module generated by all the analytic monomials.

**Remark 3.** All $H$-solutions are $p$-holomorphic Cliffordian functions. In fact, let $h$ be scalar valued in the preceeding remark.

**Remark 4.** If $d$ is odd, then for every holomorphic Cliffordian (respectively $p$-holomorphic Cliffordian) function $f$ there are $d+1$ hypermonogenic functions (resp. $H$-solutions) $g_0, g_1, \ldots, g_d$ such that

$$f = g_0 + g_1 e_d + \Delta(g_2 + g_3 e_d) + \ldots + \Delta^{\frac{d-1}{2}}(g_{d-1} + g_d e_d).$$

**Example.** Let $d = 3$, then

$$|x|^2 = g_0 + g_1 e_d + \Delta(g_2 + g_3 e_d)$$

with $g_0 = \dfrac{1}{2}|x|^2 - x_3 e_3 x$, $g_1 = -x_3^2 e_3$, $g_3 = 0$ and

$$g_2 = \frac{1}{48}(x_0^4 + x_1^4 + x_2^4) - \frac{3}{16}x_3^4 + \frac{1}{8}(x_0^2 + x_1^2 + x_2^2)x_3^2 - \frac{1}{12}x_3 e_3 [x_0(x_0^2 - 3x_3^2) + x_1(x_1^2 - 3x_3^2)e_1 + x_2(x_2^2 - 3x_3^2)e_2].$$

What happens if $d$ is even?

## 4. Restriction and inflating process

**Restriction.**

**Theorem.** The restriction to $x_{d+1} = 0$ of any $H$-solution in $\mathbb{R}_{0,d+1}$ is a $p$-holomorphic Cliffordian function in $\mathbb{R}_{0,d}$.

*Proof.* Let $f$ be an $H$-solution in $\mathbb{R}_{0,d+1}$. It is possible to decompose $f$ as $f = f^+ + f^-$ where $f^+$ is an $H$-solution which is even with respect to $e_{d+1}$ and $f^-$ is an $H$-solution which is odd with respect to $e_{d+1}$ : $f^+(x) = \dfrac{1}{2}f(x) + (f(\breve{x}))\check{}$ and $f^-(x) = \dfrac{1}{2}f(x) - (f(\breve{x}))\check{}$. Since $f^-|_{x_{d+1}} = 0$, we may suppose that $f$ is even with respect to $e_{d+1}$. Then $f$ is in the $\mathbb{R}$-linear space generated by the analytic monomials $(ax)^n a$ where $a$ is short in $\mathbb{R}_{0,d+1}$, that is $a$ is a paravector of $\mathbb{R}_{0,d}$. Thus $(ax)^n a|_{x_{d+1}=0}$ is an analytic monomial of $\mathbb{R}_{0,d}$. These monomials generate the $\mathbb{R}$-linear space of $p$-holomorphic Cliffordian functions, thus $f|_{x_{d+1}=0}$ belongs to that set. $\square$

Let us now investigate the converse.

**Preliminary example.** Let $u$ be a real function of a real variable, analytic on an open interval containing 0. For $|x|$ less than the radius of convergence $R$ we have

$$u(x) = a_0 + a_1 x + a_2 x^2 + \ldots + a_n x^n + \ldots$$

Then there is a unique holomorphic function $f$ that satisfies, in a neighbourghood of 0 in $\mathbb{C}$

$$\begin{cases} f(x+iy)|_{y=0} = u(x) \\ f(\overline{x+iy}) = \overline{f(x+iy)} \end{cases}$$

where $-$ denotes the complex conjugation. This function $f$ is the analytic function such that for $|x+iy| < R$

$$f(x+iy) = a_0 + a_1(x+iy) + \ldots + a_n(x+iy)^n + \ldots$$

where $a_0, a_1, \ldots, a_n$ are the same real numbers as above.

**General problem.** Given a $p$-holomorphic Cliffordian function

$$u : \mathbb{R} \oplus \mathbb{R}^{0,d-1} \to \mathbb{R} \oplus \mathbb{R}^{0,d-1}$$

find the $H$-solutions $f$, even with respect to $e_d$

$$f : \mathbb{R} \oplus \mathbb{R}^{0,d} \to \mathbb{R} \oplus \mathbb{R}^{0,d}$$

such that $f|_{x_d=0} = u$.

**First case:** $d$ odd ; $d = 2m+1$. Inflate from $\mathbb{R}_{0,2m}$ to $\mathbb{R}_{0,2m+1}$.
Let $u$ be $p$-holomorphic Cliffordian in $\mathbb{R}_{0,2m}$. Since it is holomorphic Cliffordian there is a holomorphic Cliffordian function $f$ even with respect to $e_{2m+1}$ such that $f|_{x_{2m+1}=0} = u$. In the case $u = (ax)^n a$ we find the trivial solution $f(z) = (az)^n a$. This shows that the $p$-holomorphic Cliffordian $f$ may be taken as an $H$-solution.

**Remark.** By this process we do not get the $H$-solutions which are odd with respect to $e_{2m+1}$. These odd $H$-solutions are generated by the polynomials $T_\alpha$ where $\alpha = (\alpha_0, \alpha_1, \ldots, \alpha_{2m})$ of characteristic $\psi_{T_\alpha} = x^\alpha := x_0^{\alpha_0} x_1^{\alpha_1} \ldots x_{2m}^{\alpha_m}$, that is

$$\frac{1}{(2m)!} \left[ (\partial_{2m+1})^{2m} T_\alpha(x) \right]_{x_{2m+1}=0} = x_0^{\alpha_0} x_1^{\alpha_1} \ldots x_{2m}^{\alpha_{2m}} e_{2m+1}$$

**Second case:** $d$ even ; $d = 2m$. Inflating from $\mathbb{R}_{0,2m-1}$ to $\mathbb{R}_{0,2m}$.
Notice that for $m = 1$ we have the first step out of $\mathbb{C}$. These steps with $d$ even are more difficult than those with $d$ odd, since the constraint on $u$ is stronger, that is $D\Delta^m u = 0$, and since there is not a unique solution $f$ for a given $p$-holomorphic Cliffordian function $u$. If we write $f = f_0 + f_1 e_1 + \ldots + f_{2m} e_{2m}$, only the polynomial parts of $f_0, \ldots, f_{2m-1}$ of degree less or equal to $2m-1$ and the polynomial part of $f_{2m}$ of degree less or equal to $2m-2$ are unique. These results are a direct consequence of the study in next paragraph.

**Example:** $m = 0$, $u : \mathbb{R} \oplus \mathbb{R} \to \mathbb{R} \oplus \mathbb{R}$, such that $Du = 0$. The solutions are $f(x_0, x_1, x_2) = u(x_0, x_1) + \overline{D}\big(x_2^2 \tau(x_0, x_1, x_2)\big)$ where $\tau$ is any solution of $x_2 \Delta\tau + 3\partial_2\tau = 0$.

## 5. Computation of the homogeneous polynomials which are $H$-solutions

Every $H$-solution $f$ may be written $f = \overline{D}h$, where $h$ is scalar valued and satisfies

$$x_d\Delta h - (d-1)\partial_d h = 0 \qquad (*)$$

Finding the homogeneous polynomials of degree $(n-1)$ in $x_0, x_1, \ldots, x_d$ which are $H$-solutions, is thus equivalent to finding the real homogeneous polynomials $h$ of degree $n$ satisfying $(*)$. Let us write $h$ as :

$$h = A_0 + A_1 x_d + A_2(x_d)^2 + \ldots + A_n(x_d)^n$$

where for $j \in \{0, 1, \ldots, n\}$, $A_j$ is a real homogeneous polynomial of degree $n-j$ in the variable $x_0, x_1, \ldots, x_{d-1}$. The equation $(*)$ is then equivalent to

$$\begin{cases} (d-1)A_1 = 0 \\ \forall j \in \{2, 3, \ldots, n\} \quad \Delta A_{j-2} - j(d-j)A_j = 0. \end{cases}$$

The last relation shows that unless $j = d$, we can compute $A_j$ knowing $A_{j-2}$ by $A_j = \dfrac{1}{j(d-j)}\Delta A_{j-2}$. To compute $h$ explicitly we just have to know $A_0$ and $A_d$ [4].

**First case**: If $d$ is even. From $(d-1)A_1 = 0$ we get $A_1 = 0$ and thus $A_{2g+1} = 0$ for any integer $g$. If we look at solution of degree $n$ greater than or equal to $d$, we have

$$0 = \Delta A_{d-2} = \Delta^2 A_{d-4} = \ldots = \Delta^{\frac{d}{2}} A_0.$$

Thus there is a solution $h$ if and only if $\Delta^{\frac{d}{2}} A_0 = 0$ which means that $\overline{D}A_0$ is $p$-holomorphic Cliffordian. Then we may choose any homogeneous polynomial of degree $n - d$ in $x_0, x_1, \ldots, x_{d-1}$ for $A_d$.

**Second case**: If $d$ is odd. From $(d-1)A_1 = 0$, we get $0 = A_1 = A_3 = \ldots = A_{d-2}$.

We may choose $A_d$ freely. We may also choose $A_0$ freely. If we take any $A_0$ and $A_d = 0$ then $\overline{D}h$ is an $H$-solution even with respect to $e_d$, thus also a $p$-holomorphic Cliffordian function. This shows that $\overline{D}A_0$ is a $p$-holomorphic Cliffordian function, for any choice of $A_0$.
If we make the choice $A_0 = 0$, $A_d = x_0^{\alpha_0} x_1^{\alpha_1} \ldots x_{d-1}^{\alpha_{d-1}}$ we get $\overline{D}h = T_\alpha$, the $T$-polynomial for $\alpha = (\alpha_0, \alpha_1, \ldots, \alpha_{d-1})$.
General case. Writing functions as series we can deduce the following theorem from the above results.

**Theorem.** Given a paravector valued function $f : \mathbb{R} \oplus \mathbb{R}^{0,d-1} \to \mathbb{R} \oplus \mathbb{R}^{0,d-1}$ there is an $H$-solution $g : \mathbb{R} \oplus \mathbb{R}^{0,d} \to \mathbb{R} \oplus \mathbb{R}^{0,d}$ such that $g\,|_{x_d=0} = f$ if and only if $f$ is $p$-holomorphic Cliffordian.

1084

# References

1. J. Cnops: Hurwitz Pairs and Applications of Möbius Transformations, 1994, Habilitation thesis, Univ. Gent.
2. S.L. Eriksson-Bique: Real analytic functions on modified Clifford analysis, Longman, Pitman Res. Notes Math. Sen., 394 (1998), 109–121.
3. S.L. Eriksson-Bique: On Modified Clifford Analysis, Complex Variables, Vol. 45 (2001), 11–33.
4. Th. Hempfling: Beiträge zur Modifizierten Clifford Analysis, Dissertation, Univ. Erlangen-Nürnberg 1997.
5. H. Leutwiler: Modified Clifford analysis, Complex Variables, Vol. 17 (1982), 153-171.
6. H. Leutwiler, P. Zeilinger: On Quaternionic Analysis and its Modifications, Computational Methods and Function Theory, Volume X, N° X (2000), 1–24.
7. G. Laville, E. Lehman: Analytic Cliffordian Functions, Annales Academiae Scientiarum Fermicae Mathematica, Vol. 29 (2004), 251–268.
8. G. Laville, I. Ramadanoff: Holomorphic Cliffordian functions, Adv. Appl. Clifford Algebra 8, N° 2 (1998), 323–340.
9. P. Zeilinger: Beiträge zur Clifford Analysis und deren Modifikation, Dissertation Univ. Erlangen-Nürnberg, 2005.

# A FRACTAL RENORMALIZATION THEORY OF INFINITE DIMENSIONAL CLIFFORD ALGEBRA AND RENORMALIZED DIRAC OPERATOR

JULIAN LAWRYNOWICZ *

*Institute of Physics, University of Łódź*
*Pomorska 149/153, PL-90-236 Łódź, Poland*
*Institute of Mathematics, Polish Academy of Sciences*
*Łódź Branch, Banacha 22, PL-90-238 Łódź, Poland*
*E-mail: jlawryno@uni.lodz.pl*


KIYOHARU NÔNO

*Department of Mathematics, Fukuoka Education University*
*Bunkyou-cho 1,*
*156 Munakata,Fukuoka, Japan*
*E-mail: nouno@fukuoka-edu.ac.jp.*


OSMAU SUZUKI †

*Department of Computer and System Analysis, College of Humanities and Sciences*
*Setagaya-ku Sakurajousui 3-25-40,*
*156 Setagaya,Tokyo, Japan*
*E-mail: osuzuki@cssa.chs. nihon-u.ac.jp*

A fractal method for the renormalization theory of an infinite dimensional Clifford algebra is presented and the renormalized Dirac operator is given. A fractal set $K$ which is defined by four self similar mappings between a rectangle is introduced, which is called a self similar fractal set of Peano type. We show that the Hilbert space $L^2(K, d\mu^D)$ with respect to the Hausdorff measure $\mu^D$ is regarded as the renormalization space of an infinite dimensional Clifford algebra. The concept of degrees is introduced for basis of the Hilbert space and is orthonormalized by different degrees. The obtained space is denoted by $L_*^2(K, d\mu^D)$. The following results are obtained:
(1) An orthonormal basis of $L_*^2(K, d\mu^D)$ is constructed (Theorem I).
(2) Infinite dimensional Clifford algebra $Cl_\Omega(\infty, \mathbf{C})$ is defined by a sequence of inclusions $\Omega = \{\omega_{2p-1}\}, \omega_{2p-1} : Cl(2p-1, \mathbf{C}) \mapsto Cl(2p+1, \mathbf{C})$ and its representation is constructed on $L_*^2(K, d\mu^D)$ (Theorem II).
(3) The derivation structures are introduced on $L_*^2(K, d\mu^D)$ and the Euclidean spaces of arbitrary dimensions can be realized as subspaces of $L_*^2(K, d\mu)$ which preserve the derivation structures (Theorem III).

*This work is partiallly supported by research project of Nihon University (Sections 1-2 of the paper), partially by the State Committee for Scientific Research (KBN) grant PBI P03A 001 26 (Sections 3-4), and partially by the grant of the University of Łódź no. 505/692 (Section 5),
†Work partially supported by grant 16540122 Kakenhi of the Ministrium of Academy of Japan(2004-5)

(4) The Dirac operator is defined for $Cl_\Omega(\infty, \mathbf{C})$ and its renormalization is given. The embedding of the Dirac operator of $Cl_\Omega(\infty, \mathbf{C})$ into $L^2_*(K, d\mu^D)$ is given (Theorem IV).

**Key words:** Clifford algebra, Dirac algebra, fractal, renormalization
**Mathematics Subject Classification:** 81R25, 32L25, 53A50, 15A66

## 1. Introduction

It is well known that we encounter the problems of divergence when we treat infinite dimensional systems. Then we need to obtain finite quantities from the infinity in a rigorous manner. The process is called the renormalization. The typical example of the divergence can be found in the theory of quantum field theory and several kinds of renormalization theory are presented ([1]).

On the other hand, it is well known that fractal structures can be observed in many fields, for examples, the behavior of stock prices in the stock market, the shape of sea shows, amorphus materials and in solutions of Navier-Stokes equation. Also a theory of fractal geometry has been developed in [2]. Recently analysis on fractal sets has been well developed, especially the theory of Laplacians on the Sierpinski's gasket is one of the remarkable ones ([6],[10]).

In this paper we shall propose a fractal method for a construction of a renormalization theory of an infinite dimensional Clifford algebra. The basic idea can be described in the following manner. We choose a fractal set of Peano type. This fractal set $K$ is defined by a system of four contractible self similar mappings $\tau_{i,j} : K_0 \mapsto K_0$ with contraction ration $\lambda_{ij}(i, j = 1, 2)$ between the unit rectangle $K_0$ in a well known manner:

$$K = \cap_{n=1}^\infty K_n; K_n = \cup_{i,j=1}^2 \tau_{i,j}(K_{n-1}). \tag{1}$$

We choose the following subsets

$$\begin{cases} K_{2p+1} = \cup K_{i_p,j_p...,i_1,j_1} \\ K_{i_p,j_p...i_1,j_1} = \tau_{i_p,j_p} \circ ... \circ \tau_{i_1,j_1}(K) \end{cases} \tag{2}$$

and consider the function space which constitute of linear combination of characteristic functions of $K_{i_p,j_p...i_1,j_1}$:

$$\Gamma_c^{(2p+1)}(K_{2p+1}, \mathbf{C}) = \{\sum c_{i_p,j_p..,i_1,j_1} \chi_{i_p,j_p...i_1,j_1}\}. \tag{3}$$

The basic idea of this paper can be stated as follow. We have the following isomorphism:

$$\Phi_{2p+1} : M(2^p, \mathbf{C}) \mapsto \Gamma_c^{(2p+1)}(K_{2p+1}, \mathbf{C}) \tag{4}$$

by $\Phi_{2p+1}((c_{i_p,j_p..,i_1,j_1})) = \sum c_{i_p,j_p..,i_1,j_1} \chi_{i_p,j_p...i_1,j_1}$, where we represent an element of $M(2^p, \mathbf{C})$ as a linear combination of tensor products of 2-sized matrices. Then we can introduce the product structure of functions by $\Phi_{2p+1}(A)\Phi_{2p+1}(B) = \Phi_{2p+1}(AB)$. By this isomorphism, we can realize finite and infinite dimensional Clifford algebras on $L^2(K, d\mu^D)$ and make analysis by use of the results in [9]. Then we can make the desired renormalizations. We want to make a stress on the fact that we can introduce the derivations on the fractal set by this isomorphism and we can give the Dirac operators and their renormalizations.

## 2. Finite and infinite dimensional Clifford algebras

In this section we recall some basic materials on finite dimensional Clifford algebras and introduce infinite dimensional Clifford algebras. Here we treat Clifford algebras over the complex number $\mathbf{C}$ which is denoted by $Cl(2p+1; \mathbf{C})$. The algebra generated by $\{A_j : j = 1, 2, .., 2p+1\}$ is called the Clifford algebra if they satisfy the following commutation relations:

$$A_i A_j + A_j A_i = 2\delta_{ij}\mathbf{1} \quad (i, j = 1, 2, .., 2p+1), \tag{5}$$

where $\mathbf{1}$ is the identity of $M(2^p, \mathbf{C})$. We give some basic properties on the Clifford algebras:

(1) (**The orthonormal property of generators**):The following sequence of the products of generators $\{1, A_{i_1} A_{i_2}, .., A_{i_k}(1 \le i_1 < i_2 < ... < i_k \le 2p+1)\}$ constitute a complete orthonormal basis of the full matrix algebra $M(2^p, \mathbf{C})$. Here we take a metric by

$$(A, B) = \frac{1}{2^p} tr A^* B. \tag{6}$$

(2) (**The basic construction of Clifford algebras**): We always assume that the algebra has a matrix representation. Then we know that there exists a unique irreducible matrix representation of $Cl(2p+1, \mathbf{C})$ on $M(2^p, \mathbf{C})$. We have a standard construction method which is called the basic construction. $Cl(3, \mathbf{C})$ is generated by the Pauli matrices $\{\sigma_1, \sigma_2, \sigma_3\}$. We have $\sigma_i \sigma_j + \sigma_j \sigma_i = 2\delta_{ij}\mathbf{1}(i, j = 1, 2, 3)$. Next we give the construction of Clifford algebras $Cl(2p + 1, \mathbf{C})$ for general $p$ by induction: For the generators $A_j (j = 1, 2, .., 2p - 1)$ of $Cl(2p - 1, \mathbf{C})$, putting

$$\begin{pmatrix} A_j & 0 \\ 0 & -A_j \end{pmatrix} \begin{pmatrix} 0 & I_n \\ I_n & 0 \end{pmatrix} \begin{pmatrix} 0 & iI_n \\ -iI_n & 0 \end{pmatrix} (j = 1, 2, .., 2p - 1), \tag{7}$$

where $n = 2^{p-1}$, we have a system of generators of $Cl(2p + 1, \mathbf{C})$ which is called basic construction with respect to $\sigma_3$. In a similar manner, we can make the basic constructions with respect to $\sigma_1$ and $\sigma_2$ respectively.

$$\begin{pmatrix} 0 & A_j \\ A_j & 0 \end{pmatrix} \begin{pmatrix} 0 & I_n \\ I_n & 0 \end{pmatrix} \begin{pmatrix} I_n & 0 \\ 0 & -I_n \end{pmatrix} (j = 1, 2, .., 2p - 1) \tag{8}$$

and

$$\begin{pmatrix} 0 & iA_j \\ -iA_j & 0 \end{pmatrix} \begin{pmatrix} 0 & I_n \\ I_n & 0 \end{pmatrix} \begin{pmatrix} I_n & 0 \\ 0 & -I_n \end{pmatrix} (j = 1, 2, .., 2p - 1). \tag{9}$$

Next we proceed to the definition of infinite dimensional Clifford algebras. We assume that a sequence of embeddings $\Omega = \{\omega_{2p-1}\} : \omega_{2p-1} : Cl(2p - 1, \mathbf{C}) \to Cl(2p + 1, \mathbf{C})$. Then we can define the infinite dimensional Clifford algebra:

$$Cl_\Omega(\infty, \mathbf{C}) = \underline{lim}_\Omega Cl(2p + 1, \mathbf{C}) \tag{10}$$

Typical examples are given by the basic constructions:Putting $\Omega_i = \{\sigma_i\}(i = 1, 2, 3)$, we define $Cl_{\sigma_i}(\infty, \mathbf{C})(i = 1, 2, 3)$.

## 3. A fractal set of Peano type

In this section we recall some basic facts on fractal sets which are defined by self similar mappings ([2],[3],[4]). Let $(X, d)$ be a compact metric space and let $\tau_j : X \mapsto X(j = 1, 2, .., L)$ be contractible mappings with contraction ratios $\lambda_j(j = 1, 2, .., L)$ : $d(\tau_j(x), \tau_j(y)) = \lambda_j d(x, y)$. We assume that the following separation condition:

$$\tau_i(X^\circ) \cap \tau_j(X^\circ) = \emptyset(i \neq j), \tag{11}$$

where $X^\circ$ denotes the open kernel of $X$. Then we have the fractal set as in (1). We give some basic properties of self similar fractal sets:
(1)$K$ is invariant under the $\{\tau_j\}$:

$$\cup_{j=1}^L \tau_j(K) = K. \tag{12}$$

(2) The Hausdorff dimension $D = dim_H(K)$ is given by the equation:

$$\sum_{j=1}^L \lambda_j^D = 1. \tag{13}$$

(3) The Hausdorff measure is a $\sigma$-additive measure([2]). The measure can be calculated by use of the mass distribution law:

$$\mu^D(K) = 1, \mu^D(K_{j_p, j_{p-1}, .., j_1}) = \lambda_{j_p}^D \lambda_{j_{p-1}}^D ... \lambda_{j_1}^D, \tag{14}$$

where we put (2). The Borel algebra of the Hausdorff measure is generated by $\{K_{j_p, j_{p-1}..j_1}\}$.
Here we make the following definition:

### Definition (A fractal set of Peano type)
A self similar fractal set which is defined by four contraction mappings between the unit rectangle $E_0, \tau_{i,j} : E_0 \mapsto E_0(i; j = 1; 2)$ is called a fractal set of Peano type.

## 4. Renormalization of infinite dimensional Clifford algebras

In this section we introduce a system of orthonormal basis on a fractal set of Peano type and construct a renormalization of an infinite dimensional Clifford algebra. Here we make the algebra by the inductive limit of finite dimensional Clifford algebras. We choose the following matrix:

$$\sigma_0 = \begin{pmatrix} 1 & 1 \\ 1 & 1 \end{pmatrix}. \tag{15}$$

Identifying $\hat{\sigma}_0 : M(2^{p-1}, \mathbf{C}) \mapsto M(2^p, \mathbf{C})$ by $\hat{\sigma}_0(A) = A \otimes \sigma_0$, we define the infinite dimensional algebra:

$$M(2^\infty, \mathbf{C}) = \underline{lim}_{\hat{\sigma}_0} M(2^p, \mathbf{C}). \tag{16}$$

Next we introduce an inner product on $\Gamma_c^{(2p+1)}(K_{2p+1}, \mathbf{C})$ by

$$(\Phi_{2p+1}(A), \Phi_{2p+1}(B))$$
$$= \int_K \sum \lambda_{i_1,j_1}^{-D} \ldots \lambda_{i_p,j_p}^{-D} \overline{c}_{i_p,j_p\ldots i_1,j_1} d_{i_p,j_p\ldots i_1,j_1} |\chi_{i_p,j_p\ldots i_1,j_1}|^2 d\mu^D \quad (17)$$

for $A = (c_{i_p,j_p,..,i_1,j_1})$, $B = (d_{i_p,j_p,..,i_1,j_1}) \in M(2^p, \mathbf{C})$. The space is denoted by $L_c^2(K_{2p+1}, d\mu^D)$. Then we see that $\{\chi_0, \chi_{i_p,j_p\ldots i_1,j_1}\}$ constitute a system of orthonormal basis of the space. We define $\tilde{\sigma}_0 : L_c^2(K_{2p-1}, d\mu^D) \mapsto L_c^2(K_{2p+1}, d\mu^D)$ by the following commutative diagram:

$$\hat{\sigma}_0 : M(2^{p-1}, \mathbf{C}) \mapsto M(2^p, \mathbf{C}) \quad (18)$$
$$\downarrow \Phi \qquad\qquad \downarrow \Phi$$
$$\tilde{\sigma}_0 : L_c^2(K_{2p-1}, d\mu^D) \mapsto L_c^2(K_{2p+1}, d\mu^D) \quad (19)$$

We notice that $\tilde{\sigma}_0$ gives the natural inclusions:

$$\iota_{2p-1} : \Gamma_c^{(2p-1)}(K_{2p-1}, \mathbf{C}) \mapsto \Gamma_c^{(2p+1)}(K_{2p+1}, \mathbf{C}). \quad (20)$$

Next we give a system of orthonormal basis on the space $L^2(K, d\mu^D)$ which is given by the Clifford algebras. From the orthonormality condition (5) of generators of Clifford algebras, we have

$$(\chi_{A_i}, \chi_{A_j}) = \delta_{ij}(i, j = 1, 2, ..., 2p+1) \quad (21)$$

for generators $\{A_j : j = 1, 2, .., 2p+1\}$ of the Clifford algebra $Cl(2p+1, \mathbf{C})$. Putting

$$\chi_{i_p,i_{p-1}\ldots,i_1} = \chi_{\sigma_{i_p} \otimes \sigma_{i_{p-1}} \otimes \ldots \otimes \sigma_{i_1}}, \quad (22)$$

where $\sigma_{i_k}(k = 1, 2, .., p)$ are Pauli matrices, we have a system of orthonormal basis of $L_c^2(K_{2p+1}, d\mu^D)$. Here we introduce a special orthonormalized inner product:

**Definition (Orthonormalized inner product)**
(1) We put

$$L_*^{2p+1} = \{\chi_{i_p,..,i_1} | i_1 \neq 0\} \quad (23)$$

An element $\chi_{i_p,..,i_1}$ of $L_*^{2p+1}$ is called an element of degree $2p+1$.
(2) We introduce an inner product such that

$$L_*^{2p+1} \perp L_*^{2q+1}(p \neq q). \quad (24)$$

(3) We put the following notation:

$$L_*^{(2p+1)} = \bigoplus_{r=0}^{p} L_*^{2r+1}(p = 0, 1, 2, ...). \quad (25)$$

Then we can prove the following theorem:

**Theorem I (Orthonormalized basis generated by Clifford algebras)**
(I)The following elements constitute a system of linear basis of $L_*^2(K, d\mu^D)$:

$$L_*^2(K, d\mu) = \underline{lim}_{\tilde{\sigma}_0} L_*^{(2p+1)}, \tag{26}$$

where the completion is taken with respect to the orthonormalized metric.

(II) Introducing the orthonormalized norm, we have

$$(\chi_{i_k,i_{k-1},..,i_1}, \chi_{j_l,j_{l-1},...,j_1}) = \delta_{kl}\delta_{i_k j_k}\delta_{i_{k-1}j_{k-1}}...\delta_{i_1 j_1}, \tag{27}$$

and we see that they become a complete orthonormal basis of $L_*^2(K, d\mu^D)$. Hence we have the orthonormal expansion of $f \in L_*^2(K, d\mu^D)$:

$$f = a_0\chi_0 + \sum a_{i_k,i_{k-1},...,i_1}\chi_{i_k,i_{k-1},...,i_1}, \tag{28}$$

$$a_0 = \int_K f\overline{\chi_0}d\mu^D, a_{i_k,i_{k-1},...,i_1} = \int_K f\overline{\chi}_{i_k,i_{k-1},...,i_1}d\mu^D \tag{29}$$

Finally we give the definition:

**Definition (The renormalization of the $Cl_\Omega(\infty; \mathbf{C})$)**
(1)The following space is called the renormalization of $Cl_\Omega(\infty; \mathbf{C})$:

$$L_*^2(K, d\mu^D) = \underline{lim}_{\tilde{\sigma}_0} L_*^{(2p+1)} \tag{30}$$

(2) The mapping $\Phi : Cl_\Omega(\infty, \mathbf{C}) \mapsto L_*^2(K, d\mu^D)$ is called the renormalization mapping, where $\Phi = \oplus_{p=0}^\infty \Phi_{2p+1}$.
(3)The Hilbert space $L_*^2(K, d\mu^D)$ endowed with the product structure becomes an algebra which is called a fractal renormalization of an infinite dimensional Clifford algebra.

## 5. Representation of infinite dimensional Clifford algebras on the renormalized space

In this section we make representations of infinite dimensional Clifford algebras on the renormalized space and prove their unitary equivalence. We choose an infinite dimensional Clifford algebra $Cl_\Omega(\infty, \mathbf{C})$. Then we can prove the following theorem:

**Theorem II (Representation of $Cl_\Omega(\infty, \mathbf{C})$ on $L_*^2(K, d\mu^D)$)**
(I)(**Existence Theorem**):There exists an irreducible representation:

$$\hat{\pi} : Cl_\Omega(\infty, \mathbf{C}) \times L_*^2(K, d\mu^D) \mapsto L_*^2(K, d\mu^D) \tag{31}$$

so that the following hold:
(1) Restricting the representation on $Cl(2p + 1, \mathbf{C})$:$\hat{\pi}_{2p+1} = \hat{\pi}_{Cl(2p+1,\mathbf{C})}$, we have
$\hat{\pi}_{2p+1} : Cl(2p + 1, \mathbf{C}) \times L_*^{(2p+1)} \mapsto L_*^{(2p+1)}$. Then we have

$$\hat{\pi}_{2p+1}(A)\chi_X = \chi_{AX}. \tag{32}$$

(2) We have the following commutative diagram:

$$\omega_{2p-1} : Cl(2p-1, \mathbf{C}) \mapsto Cl(2p+1, \mathbf{C}) \tag{33}$$

$$\downarrow \Phi_{2p-1} \qquad \downarrow \Phi_{2p+1}$$

$$\hat{\omega}_{2p-1} : L_*^{(2p-1)} \quad \mapsto \quad L_*^{(2p+1)}. \tag{34}$$

Hence we see that

$$\hat{\pi} = \underline{lim}_\Omega \hat{\pi}_{2p+1} \tag{35}$$

(2) **(Unitary Equivalence Theorem)** The representations associated to a fixed inclusion $\Omega$(see (10))

$$\hat{\pi}^{(i)} : Cl_\Omega(\infty, \mathbf{C}) \times L_*^2(K^{(i)}, d\mu^{D^{(i)}}) \mapsto L_*^2(K^{(i)}, d\mu^{D^{(i)}})(i = 1, 2) \tag{36}$$

are unitary equivalent, if the fractal sets have the same Kakutani invariant:

$$\{\lambda_{i,j}^{(1)D^{(1)}}\} = \{\lambda_{i,j}^{(2)D^{(2)}}\} \tag{37}$$

The equivalence can be proved by use of the representations of the Cuntz algebras. A $C^*$-algebra is called the Cuntz algebra $\mathcal{O}(N)$, if the generators $\{S_j|j = 1, 2, .., N\}$ satisfy the following commutation relations:

$$S_i^* S_i = 1(i = 1, 2, .., N) \text{ and } \sum_{i=1}^{N} S_i S_i^* = 1. \tag{38}$$

The first relation implies the isometry condition on $S_i(i = 1, 2, .., N)$ and the second condition implies the decomposition of the total space into $N$ parts by the projections $P_i = S_i S_i^*(i = 1, 2, .., N)$. Then we have a representation of the Cuntz algebra on a self similar fractal set.

**Proposition** ([7],[8])
(1) Let $K$ be a proper fractal set defined by $\{\sigma_j(j = 1, 2, ..., N)\}$. Then we have the following representation $\pi : \mathcal{O}(N) \times L^2(K, d\mu^D) \mapsto L^2(K, d\mu^D)$:

$$\pi(S_j)f(x) = \begin{cases} \Phi_j^{1/2}(\sigma_j^{-1}(x))f(\sigma_j^{-1}(x)); x \in \sigma_j(K), \\ 0 \qquad x \notin \sigma_j(K)(j = 1, 2, .., N), \end{cases} \tag{39}$$

$$\pi(S_j^*)f(x) = \Phi_j^{-1/2}(x)f(\sigma_j(x))(j = 1, 2, .., N) \tag{40}$$

where $\Phi_j$ is the Radon-Nikodym derivative of $\sigma_j(j = 1, 2, .., N)$.
(2) The representations are unitary equivalent, i.e., there exists a unitary operator $U : L^2(K; d\mu^D) \mapsto L^2(K', d\mu^D)$ such that the following diagram becomes commutative:

$$\pi : L^2(K, d\mu^D) \mapsto L^2(K, d\mu^D) \tag{41}$$

$$\downarrow U \qquad \downarrow U$$

$$\pi' : L^2(K', d\mu^{D'}) \mapsto L^2(K', d\mu^{D'}), \tag{42}$$

if and only if the Kakutani invariant is identical each other.

## 6. The renormalized Dirac operator

In this section we introduce the Dirac operator for an infinite dimensional Clifford algebra and construct its renormalization. We begin with an introduction of the derivations on the renormalization space $L^2_*(K; d\mu^D)$. When a product strucuture, which is called *-product, is given, a linear operator $\delta : L^2_*(K, d\mu^D) \mapsto L^2_*(K, d\mu^D)$ is called derivation, if it satisfies the Leibnitz rule:

$$\delta(f * g) = \delta f * g + f * \delta g. \tag{43}$$

We can introduce derivations associtated to the Clifford algebra. We introduce a product structure which is defined by the following bilinear relations:

$$\begin{cases} \chi_0 * \chi_0 = \chi_0, \\ \chi_0 * \chi_{i_k, i_{k-1}, .., i_1} = \chi_{i_k, i_{k-1}, .., i_1} * \chi_0 = \chi_{i_k, i_{k-1}, .., i_1}, \\ \chi_{i_k, i_{k-1}, .., i_1} * \chi_{j_l, j_{l-1}, .., j_1} = \chi_{i_k, i_{k-1}, .., i_1, j_l, j_{l-1}, .., j_1}. \end{cases} \tag{44}$$

Then we make the following definition:

### Definition (Derivations for Clifford algebras) ([9])
The linear operator $\delta_j (j = 0, 1, 2, 3)$ defined by

$$\begin{cases} (1)\delta_j \chi_0 = 0, \\ (2)\delta_j \chi_{i_k, i_{k-1}, .., i_1} = \sum_{l=1}^k \delta_{j, i_k} \chi_{i_k, i_{k-1}, .., \hat{i}_l ... i_1} \end{cases} \tag{45}$$

become derivations which are called derivations of the Clifford algebras.

Next we shall find Euclidean spaces of arbitrary dimensions and their noncommutative embeddings into $L^2_*(K, d\mu^D)$ which preserve the derivation structure. By this embedding we can discuss relationships between Dirac operators on the Euclidean spaces and that on $L^2_*(K, d\mu^D)$. The key point of the introduction of the derivation is the following theorem:

### Theorem III (The existences of spatial subspaces of $L^2_*(K, d\mu^D)$)
There exists an algebraic isomorphism:$\Psi_{2p+1} : \mathbf{C}[x_1, x_2, .., x_{2^p}] \mapsto L^{(2p+1)}_*$, where $\mathbf{C}[x_1, x_2, .., x_{2^p}]$ is the ring of polynomial of degree $2^p$ and a system of derivations $\{\delta_i (i = 1, 2, .., 2^p)\}$ such that the following commutative diagram holds:

$$\partial_i : \mathbf{C}[x_1, x_2, .., x_{2^p}] \mapsto \mathbf{C}[x_1, x_2, .., x_{2^p}] \tag{46}$$

$$\downarrow \Psi_{2p+1} \qquad \downarrow \Psi_{2p+1}$$

$$\delta_i : L^{(2p+1)}_* \quad \mapsto \quad L^{(2p+1)}_*. \tag{47}$$

The proof can be given by introduction of derivations of general type. We introduce the space of "$k$-monics" and derivations for elements of the $k$-monic space. We consider the algebra which is generated by $k$-monics:

$$\mathcal{K}_k = \{\chi_{I_i^{(k)}}\}, \quad \chi_{I_i^{(k)}} = \chi_{i_1, i_2, .., i_k}. \tag{48}$$

Choosing $I_i^{(k)}$, we have the derivations $\delta_{I_i^{(k)}} : \mathcal{K}_k \mapsto \mathcal{K}_k$ :

$$
\begin{cases}
(1)\delta_{I_i^{(k)}}(\chi_0) = 0, \\
(2)\delta_{I_i^{(k)}}(\chi_{I_{i_k}^{(k)}} * \chi_{I_{i_{k-1}}^{(k)}} * \dots * \chi_{I_{i_1}^{(k)}}) \\
= \sum_{l=1}^{r} \delta_{I_i^{(k)}, I_{i_k}^{(k)}} \chi_{I_{i_r}^{(k)}} * \chi_{I_{i_{r-1}}^{(k)}} * \dots * \check{\chi}_{I_{i_l}^{(k)}} * \dots * \chi_{I_{i_1}^{(k)}},
\end{cases}
\tag{49}
$$

which we call the derivation of order $k$.

Next we proceed to the introduction of the Dirac operator for an infinite dimensional Clifford algebra $Cl_\Omega(\infty, \mathbf{C})$. We take the Dirac operator of $Cl(2p+1, \mathbf{C})$:

$$
\hat{D}_{2p+1} = \sum_{j=1}^{2p+1} A_j \frac{\partial}{\partial x_j},
\tag{50}
$$

where $\{A_j : j = 1, 2, .., 2p+1\}$ are generators of $Cl(2p+1, \mathbf{C})$. We notice that $\hat{D}_{2p+1}^2 = \hat{\Delta}_{2p+1}$, where $\hat{\Delta}_{2p+1}$ is the Laplacian on $\mathbf{R}^{2p+1}$. We can define the Dirac operator $\hat{D}_\Omega(\infty : \mathbf{C})$ of $Cl_\Omega(\infty : \mathbf{C})$ by

$$
\hat{D}_\Omega(\infty : \mathbf{C}) = \underline{lim}_\Omega \hat{D}_{2p+1}.
\tag{51}
$$

Next we define the renormalized Dirac operator. For a system of the generators $\{A_j\}$ of $Cl(2p+1, \mathbf{C})$, putting $\hat{\chi}_{j_p, j_{p-1}, .., j_1} = \Phi(A_j)$, we obtain an element of the $p$-monic space, which is denoted by $\hat{\chi}_{I_j^{(p)}}$. We define the Hilbert space $\mathcal{H}$ by

$$
\mathcal{H} = \mathbf{C}[\hat{\chi}_{I_0^{(0)}}, \hat{\chi}_{I_1^{(1)}}, \dots, \hat{\chi}_{I_j^{(p)}}, \dots] \cap L_*^2(K, d\mu^D),
\tag{52}
$$

we have the following expression for $f \in \mathcal{H}$(see(23)):

$$
f = \sum_{p=0}^{\infty} f_p (f_p \in L_*^{2p+1}).
\tag{53}
$$

We can make the following definition:

**Definition (The renormalized Dirac operator of $Cl_\Omega(\infty; \mathbf{C})$)**
(1) We define the renormalized Dirac operator:$D_\Omega(\infty) : L_*^2(K, d\mu^D) \times \mathcal{H} \mapsto L_*^2(K, d\mu^D) \times \mathcal{H}$ by

$$
\tilde{D}_\Omega(\infty)f = \underline{lim}_\Omega \tilde{D}_{2p+1} f_p, \quad \tilde{D}_{2p+1} f_p = \sum \chi_{I_j^{(p)}} \delta_{I_j^{(p)}} f_p,
\tag{54}
$$

where $\delta_{I_j^{(p)}} (j = 1, 2, .., 2p-1)$ are derivations for $p$-monics.
(2) The renormalized Laplacian of the Dirac operator is given by

$$
\tilde{\Delta}_\Omega(\infty) = \underline{lim}_\Omega \tilde{\Delta}_{2p+1}, \tilde{\Delta}_{2p+1} = \sum \delta_{I_j^{(p)}}^2.
\tag{55}
$$

Then we can prove the following theorem:

**Theorem IV (The renormalized Dirac operator of the infinite dimensional Clifford algebra)**

(1) The Dirac operator is the root of the Laplacian:

$$\tilde{D}_\Omega^2(\infty) = \tilde{\Delta}_\Omega(\infty). \tag{56}$$

(2) For the embedding $\Psi_{2p+1} : \mathbf{C}[x_1; x_2, .., x_{2^p}] \mapsto L_*^{(2p+1)}$, we have the following commutative diagram:

$$\hat{D}_{2p-1} : M(2^{p+1}; \mathbf{C}) \times \mathbf{C}[x_1 x_2, .., x_{2^p}] \mapsto M(2^p; \mathbf{C}) \times \mathbf{C}[x_1, x_2, .., x_{2^p}] \tag{57}$$

$$\downarrow \Phi_{2p+1} \otimes \Psi_{2p+1} \qquad \downarrow \Phi_{2p+1} \otimes \Psi_{2p+1}$$

$$\tilde{D}_{2p+1} : L_*^{(2p+1)} \times \mathcal{H} \qquad \mapsto \qquad L_*^{(2p+1)} \times \mathcal{H}. \tag{58}$$

In a similar manner to (30), we can define the renormalization mapping for Dirac operators.

## Acknowledgement

One of the authors express his hearty thanks to the referee for pointing out non rigorous parts of the manuscript.

## References

1. N.N.Bogolubov and D.V.Shrikov: Introduction to the theory of quantum field theory, John Wiley and Sons, U.S.A.(1976).
2. K.Falconer: Fractal Geometry, Mathematical foundations and Applications,John Wiley and Sons, U.S.A.(1990).
3. J.E.Hutchinson: Fractal and self-similarity, Indiana Univ. Math. J., 30 (1981), 713-747.
4. S and S. Ishimura: Fractal Mathematics, Tokyo-Tosho, Japan(in japanese) (1990).
5. S.Kakutani: On equivalence of infinite product measures Ann. Math., 47 (1948), 214-224.
6. J.Kigami: Analysis on fractal, Cambridge University Press, U.K.(2001).
7. M.Mori, O.Suzuki and Y.Watatani: Representations of Cuntz algebras on fractal sets, in preparation.
8. M.Mori, O.Suzuki and Y.Watatani: A noncommutative differential geometric method to fractal geometry(I) (Representations of Cuntz algebras of Hausdorff type on fractal sets), to appear in the Proc. of Int. ISSAC Conf.(2001).
9. T.Ogata and O.Suzuki: Differential and integral calculus on fractal sets, in preparation.
10. M.Yamaguchi, M.Hata and J.Kigami: Mathematics on fractal, Iwanami Publisher (in Japanese)(1993).

# ANALYTIC FUNCTIONS IN ALGEBRAS

Y. KRASNOV

*Department of Mathematics,*
*Bar-Ilan University,*
*Ramat-Gan 52900, Israel*
*E-mail: krasnov@math.biu.ac.il*

Representations of the solutions to PDE's in algebras as function in symmetry operator indeterminates is presented. Essentially classes of these solutions in terms of formal power series is discussed. We touch also upon applications to the Cauchy problem for many important PDE's in algebras.

**Key words:** A-analytic functions, constant coefficient system, operator indeterminate, series solutions

**Mathematics Subject Classification:** 35E20, 35C10

## 1. Introduction

An *algebra* $\mathbb{A}$ for us will always be a real $n$-dimensional Euclidean vector space that is finitely generated by the orthonormal system of basis vectors $e_1, e_2, \ldots, e_n$ and is equipped with a bilinear map $m : \mathbb{A} \times \mathbb{A} \to \mathbb{A}$ called "multiplication". The symbol $\circ$, unless there is ambiguity, stands for the abbreviation $m(x, y)$ as $x \circ y$ in $\mathbb{A}$. Of course, knowing the tensor form $a_{ij}^m$ of a bilinear map in an orthonormal basis $e_1, e_2, \ldots, e_n$ one can rewrite the multiplication $m(e_i, e_j)$ as follows:

$$e_i \circ e_j = \sum_{m=1}^{n} a_{ij}^m e_m. \tag{1}$$

We use $x = x_1 e_1 + \ldots + x_n e_n$ to denote vector in $\mathbb{R}^n$ as well as element $x \in \mathbb{A}$, and $\partial_i$ to denote $\partial/\partial x_i$, for $i \in \{1, \ldots, n\}$. The symbol $<,>$ stands for the Euclidean scalar product.

Every $\mathbb{A}$-valued function $f(x)$ will be for us always locally real analytic that is represented as:

$$f(x) := e_1 u_1(x_1, \ldots, x_n) + e_2 u_2(x_1, \ldots, x_n) + \ldots + e_n u_n(x_1, \ldots, x_n), \tag{2}$$

where $u_i(x)$ are also real analytic functions, $i = 1, 2, \ldots, n$.

Begin with the *Dirac operator* in $\mathbb{A}$ definition:

$$D := e_1 \partial_1 + \ldots + e_n \partial_n. \tag{3}$$

**Definition 1.1.** A real analytic function $f(x)$ is called an A-**analytic** if $f(x)$ is a solution to the system of partial differential equations

$$D \circ f(x) := \sum_{i,j=1}^{n} e_i \circ e_j \partial_i u_j(x_1, \ldots, x_n) = 0. \tag{4}$$

**Definition 1.2.** Recall a system of PDEs for $x \in \mathbb{R}^n$ is an **analytic PDE** if

(1) it may be written in form of first order (linear) system

$$\mathcal{L}(x, D)u := \{\mathcal{L}_k(x, D)u\}_{k=1}^{n} := \sum_{i,j=1}^{n} a_{ij}^{k}(x)\partial_i u^j(x) = 0, \tag{5}$$

(2) the coefficients $a_{ij}^{k}(x)$ in PDOs $\mathcal{L}_k$ all are entire functions.
(3) $\mathcal{L}(x, D)$ is an *involutive system*, meaning that there exists entire functions $b_{ij}^{k}(x)$ such that the commutators $[\mathcal{L}_i(x, D), \mathcal{L}_j(x, D)]$ fulfill the relation

$$[\mathcal{L}_i(x, D), \mathcal{L}_j(x, D)] = \sum_{k=1}^{n} b_{ij}^{k}(x)\mathcal{L}_k(x, D). \tag{6}$$

Given a system of analytic PDEs in form $L(x, D)u(x) = 0$. One can easily verify that solution to analytic PDE at least locally at point $x_0$ is an $\mathbb{A}_{x_0}$ - analytic function where $\mathbb{A}_{x_0}$ is a local algebra with the multiplication rule $\circ$, such that

$$D_0 \circ u(x) := \mathcal{L}(x_0, D)u(x). \tag{7}$$

The main goal of this article is to show that the technique of the *local algebras bundle* (cf. (7) is a natural language in qualitative theory to analytic PDEs .

## 2. Algebraic approach to function theories

Let $\mathbb{A}$ be a real algebra,(not necessarily commutative and/or associative). The literature on function theory over such algebras has been developed by many authors (see [8], [4], [22], [23]) and contains a range of definitions for analyticity (holomorphicity, monogenicity). Three distinct approaches in these investigations are mentioned.

- The first one (Weierstrass approach) regards functions on $\mathbb{A}$ as their convergent in some sense power series expansions (cf. [11]).
- The second (Cauchy-Riemann) approach concentrated on the solution to the Dirac equation in algebra $\mathbb{A}$ (cf [12],[13] ).
- The third one is based on the function-theoretic properties known for complex analytic functions, such as Cauchy's theorem, residues theory, Cauchy integral formula etc. (cf [19], [9] )

All these methods look like generalization of the *analytic function theory* of complex variables (cf. [22],[23]). We use the term $\mathbb{A}$-*analysis* for such cases (cf. Clifford or quaternionic analysis [4], [22] if algebra $\mathbb{A}$ is embedded into a Clifford algebra).

If $\mathbb{A}$ in the Definition 1.1 is an algebra of complex numbers $\mathbb{C}$ then (4) coincide with the Cauchy-Riemann equations. Thus gives us a good reason to denote the space of $\mathbb{A}$-analytic functions as $Hol(\mathbb{A})$ in a complete agreement with the definition of holomorphic functions in complex analysis (denoted by $Hol(\mathbb{C})$).

**Remark 2.1.** By $\mathbb{A}$-analysis we mean the systematic study of $Hol(\mathbb{A})$.

## 2.1. *Isotopy classes*

In this subsection we will be concerned with *Albert isotopies* [1] of algebras :

**Definition 2.1.** Two $n$-dimensional algebras $\mathbb{A}_1$ and $\mathbb{A}_2$ with multiplication $\circ$ and $\star$ are said to be **isotopic** ($\mathbb{A}_1 \sim \mathbb{A}_2$) if there exist nonsingular linear operators $K, L, M$ such that

$$x \circ y = M(Kx \star Ly), \tag{8}$$

Obviously, if in addition, $K \equiv L \equiv M^{-1}$, then two algebras $\mathbb{A}_1$ and $\mathbb{A}_2$ are isomorphic ($\mathbb{A}_1 \simeq \mathbb{A}_2$) .

**Definition 2.2.** If for a given two algebras $\mathbb{A}_1$ and $\mathbb{A}_2$ there exist a nonsingular linear operators $P, Q$ such that for every $g(x) \in Hol(\mathbb{A}_2)$ the function $f(x) = Pg(Qx)$ belongs to $Hol(\mathbb{A}_1)$ and vice versa, we would say that two function theories are **equivalent** and write $Hol(\mathbb{A}_1) \simeq Hol(\mathbb{A}_2)$.

Using these results we obtain the important

**Theorem 2.1.** *Two function theories are equivalent iff the corresponding algebras are isotopic.*

**Definition 2.3.** If $a_{ij}^k(x_0)$ in (5) for all fixed $x_0 \in \Omega \subset \mathbb{R}^n$ forms a set of an isotopic algebras then we will say that $\mathcal{L}$ is an unique defined type PDO in $\Omega$, else a *mixed type* PDO. An algebra $\mathbb{A}_0$ with multiplication tensor $a_{ij}^k(x_0)$ (see. (1)) is called corresponding algebra for $\mathcal{L}$ in $x_0 \in \Omega$.

Obviously, if the coefficients $a_{ij}^k(x)$ are thought as constants and $b_i^k$ are identically zero then operator $\mathcal{L}$ in (5) coincides with the Dirac operator $D$ defined by (3) and one can obtain the solution to homogeneous equation $\mathcal{L}f = 0$ as a (left) $\mathbb{A}$-analytic function for the operator $D$.

**Theorem 2.2.** *Decomposition an algebras into the isotopy classes in turn is a powerful classification tools for the corresponding PDO.*

**Theorem 2.3.** *Totality of a functions on (in general noncommutative and/or non associative) regular algebras are splitting into the non equivalent classes. These classes unique characterized by its **unital hearts**. If such an heart is in addition an associative algebra then $\mathbb{A}$-analytic function may be expanded into the commutative operator valued power series.*

We will proof above stated theorems in the next sections.

## 3. Evolution equations

In this Section we consider evolution PDE in form

$$Qu(t,x) := \partial_t u(t,x) - P(x, \partial_x) u(t,x) = 0, \qquad (9)$$

where one of the independent variables is distinguished as a time and the other independent variables $x = \{x_1, \ldots, x_n\} \in X$ are spatial.

We consider $P$ to be an operator on the manifold $X \times U$ where dependent variables $u$ are elements of $U$. Let $G$ be a local one parametric group of transformations on $X \times U$. In turn, we consider the first order differential operator of special type as a generator of a local group $G$. The next our step is

**Definition 3.1.** A **local symmetry group** of an equation (9) is one parameter group $g^t$ of transformation $G$, acting on $X \times U$, such that if $u(t,x) \in U$ is an arbitrary smooth enough solution of (9) and $g^t \in G$ then $g^t[u(\tau,x)]$ is also solution of (9) for all small enough $t, \tau > 0$.

S.Lie developed a technique for computing local groups of symmetries. His observations was based on theory of jet bundles and prolongation of the vector fields whenever $u$ is a solution of (9). An explicit formulas one can found in [16].

In general, symmetries of (9) one computes by Lie-Bäcklund (LB) method. Everything necessary for our method one can find using explicit computations due to Baker-Campbell-Hausdorff formula (in sequel: BCH formula). This formula is equivalent to LB method and based on successive commutators calculation:

$$K_i = e^{tP(x,\partial)} x_i e^{-tP(x,\partial)} = \sum_{m \geq 0} \frac{1}{m!} t^m [P(x, \partial_x), x_i]_m. \qquad (10)$$

Here $[a,b]_m = [a, [a,b]_{m-1}]$, $[a,b]_1 = ab - ba$ and $[a,b]_0 = b$.

If all $K_i \in \mathbb{C}[D]$ in (10) are PDEs of finite order then $Q$ in (9) is of **finite type**.

If an evolution equations (9) is of a finite type then the initial value problem $u(0,x) = f(x)$ is well posed. If in addition $e^{\lambda t}$ is a solution, then the explicit solution to (9) may be written in form $u = f(K)[e^{\lambda t}]$.

**Proposition 3.1.** *If $Q$ is of finite type, then symmetry operators $K_i$ defined in (10) and completing with identity operator, forms a commutative, associative, unital subalgebra of an algebra $Sym(Q)$ of all symmetries of $Q$.*

**Proof.** The commutator relations $[K_i, K_j] = 0$ are proper for all $i, j$. $\qquad \square$

## 4. Classification of the first order PDE

Here we start to study classification of PDE theory by treating it in the algebraic terms. We begin by examining the conditions that the Dirac operator in $\mathbb{A}$ (3) is a well defined system of PDE.

## 4.1. *Under- and overdetermined system*

Let $P(D)u(x) = f(x)$ be a system of partial differential equations, where $P(D)$ is a given $k \times l$ matrix of differential operators with constant coefficients, the given $f(x)$ (respectively, unknown $u(x)$) being $k$- ($l$-)tuples of functions or distributions in $x \in \mathbb{R}^m$. Many authors assumes usually that the system is under- (over-) determined, if the rank of $P(\xi)$ (cf. of its transpose $P'(\xi)$) is less than $l$ for all (cf. for some) nonzero $\xi \in \mathbb{R}^m$.

The algebraic formulation of the fact that PDE with the constant coefficient (4) is under- (over-) determined can be roughly described as follows.

**Definition 4.1.** A real $n$-dimensional algebra $\mathbb{A}$ is called left (right) **regular** if there exists $v \in \mathbb{A}$, such that the linear operators $L_v, R_v : \mathbb{R}^n \to \mathbb{R}^n$ defined by $x \to v \circ x$ (resp. $x \to x \circ v$) are both invertible. Otherwise, $\mathbb{A}$ is called a *left (right) singular* algebra. In other words, $\mathbb{A}$ is regular iff $\mathbb{A} \subset \mathbb{A}^2$

Recall an elements $u, v \in \mathbb{A}$ a left (cf. right **annihilator**) if $u \circ x = 0$, $(x \circ v = 0)$ for all $x \in \mathbb{A}$.

**Theorem 4.1.** *The Dirac operator $D$ in algebra $\mathbb{A}$ is under determined iff $\mathbb{A}$ is singular and is over determined iff $\mathbb{A}$ is regular and contains an annihilators.*

**Proof.** For a given Dirac operator $D$ in the corresponding algebra $\mathbb{A}$ define left (right) multiplication operators $L_v, R_v : \mathbb{R}^n \to \mathbb{R}^n$ by $x \to v \circ x$ (resp. $x \to x \circ v$) as in Definition 4.1. If $L_\xi, R_\xi$ are both invertible for some $\xi$ then $D$ is well determined. Conversely, let $L_v$ (respectively, $R_v$) be $k_1 \times l_1$ ($k_2 \times l_2$) matrices of differential operators. Then $D$ is under determined if $k_1 < l_1$ and/or $k_2 > l_2$ and is overdetermined if $k_1 > l_1$ and/or $k_2 < l_2$. The only case $k_1 = l_1 = k_2 = l_2 = n$ stands for the regular algebras $\mathbb{A}$ without annihilators and therefore, for a well determined Dirac operator $D$ in $\mathbb{R}^n$. $\qquad\qquad\square$

In Definition 2.3 we consider some properties of $\mathcal{L}(x, D)$ to be of the same type in a given open set $x \in \Omega$. This properties was given in terms of the existence of one common isotopy relations between the set of the corresponding algebras. If there exists a point on the boundary $x_0 \in \partial\Omega$ and such that an algebra $\mathbb{A}_0$ with multiplication tensor (1) $a(x_0)_{ij}^k$ is not isotopic to any local algebra $\mathbb{A}_1$ with multiplication tensor $a(x_1)_{ij}^k$ for $x_1 \in \Omega$, we will say that the PDO $\mathcal{L}(x, D)$ is of *degenerate type* in $\overline{\Omega}$.

## 4.2. *Elliptic type PDE*

An ellipticity of the Dirac operator in the regular algebra $\mathbb{A}$ one can reformulate as property of $\mathbb{A}$ to be a *division* algebra.

**Definition 4.2.** An algebra $\mathbb{A}$ is a **division algebra** iff the both operations of left and right multiplications by any nonzero element are invertible.

**Proposition 4.1.** *The well determined Dirac operator $D$ in the necessary regular (by Theorem 4.1) algebra $\mathbb{A}$ is elliptic iff $\mathbb{A}$ is a division algebra.*

The proof is given in [7].

## 5. Power series expansions

As was proven in [19] every regular algebra is isotopic to unital. Assume that $e_0, e_1, \ldots, e_n$ forms a basis in **unital** algebra $\mathbb{A}$ and $e_0$ is it two sided unit element. In order to construct an $\mathbb{A}$ - analytic function theory the following $\mathbb{A}$ - analytic variable are used:

$$z_m = x_m e_0 - x_0 e_m, \quad m = 1, 2, \ldots, n. \tag{11}$$

In turn, $Dz_k = 0$ for all $k = 1, 2, \ldots, n$ where $D = e_0 \partial_{x_0} + e_1 \partial_{x_1} + \ldots + e_n \partial_{x_n}$ is the Dirac operator in the algebra $\mathbb{A}$.

Denote a canonical spherical homogeneous polynomial solution of the Dirac equation in $\mathbb{A}$ by formula:

$$V_0(x) = e_0, \quad V_m(x) = z_m, \quad m = 1, 2, \ldots, n, \tag{12}$$

$$V_\mu(x) := V_{m_1 \ldots, m_k}(x) = \sum_{\pi(m_1, \ldots, m_k)} z_{m_1}(z_{m_2}(\cdots(z_{m_{k-1}} z_{m_k})\cdots)), \tag{13}$$

where the sum runs over all distinguishable permutations of $m_1, \ldots, m_k$.

**Proposition 5.1.** *(cf. [4]) The polynomials $V_\mu(x)$ of order $k$ for $m_i \in \{1, 2, \ldots, n\}$ and multi indexes $\mu = \{m_1, \ldots, m_k\}$ are both left and right $\mathbb{A}$-analytic. Any homogeneous of order $k$ $\mathbb{A}$-analytic polynomial $p_k(x)$ have a Taylor like expansion:*

$$p_k(x) = \sum_{m_1 + \ldots + m_k = k} \frac{V_{m_1, \ldots, m_k}(x)}{m_1! \cdots m_k!} \partial_{x_{m_1}} \ldots \partial_{x_{m_k}} p_k(x) \tag{14}$$

*where the sum runs over all possible combinations of $m_1, \ldots, m_k$ of $k$ elements out of $1, 2, \ldots, n$ and repetition being allowed.*

**Proof.** (cf. [4], Theorem 11.2.3,5) Clearly, for $\mu = m_1, \ldots, m_k$

$$x_0 D V_\mu(x) = \sum_{\pi(m_1, \ldots, m_k)} \sum_{i=0}^{n} x_0 e_i \partial_{x_i}(z_{m_1}(z_{m_2} \cdots z_{m_k})\ldots) =$$

$$x_0 \sum_{i=1}^{n} \sum_{\pi(\mu)} (e_{m_i} z_{m_1} \cdots z_{m_{i-1}} z_{m_{i+1}} \cdots z_{m_k} - z_{m_1} \cdots z_{m_{i-1}} e_{m_i} z_{m_{i+1}} \cdots z_{m_k})$$

$$= \sum_{i=1}^{n} \sum_{\pi(\mu)} (z_{m_i}(z_{m_1} \cdots z_{m_{i-1}}(z_{m_{i+1}} \cdots z_{m_k}) \ldots) - z_{m_1} \cdots z_{m_k}) = 0.$$

For homogeneous of order $k$ $\mathbb{A}$-analytic polynomial $p_k(x)$ in (14) the Euler's formula result:

$$kp_k(x) = x_0 \partial_{x_0} p_k(x) + \sum_{i=1}^{n} x_i \partial_{x_i} p_k(x) = \sum_{i=1}^{n} z_i \partial_{x_i} p_k(x)$$

Obviously, $q_{k-1}(x) := \partial_{x_i} p_k(x)$ is homogeneous of order $k-1$ $\mathbb{A}$-analytic polynomial and one can use an induction. $\qquad\square$

**Theorem 5.1.** *(cf.* [4]*)* $\mathbb{A}$*-analytic in an open neighborhood of the origin function* $f(x)$ *can be developed into a normally convergent series of spherical homogeneous polynomials*

$$f(x) = \sum_{k=0}^{\infty} \sum_{m_1+\ldots+m_k=k} \frac{V_{m_1,\ldots,m_k}(x)}{m_1! \cdots m_k!} \partial_{x_{m_1}} \cdots \partial_{x_{m_k}} f(0)$$

The proof is similar to the method described in ([4], Theorem 11.3.4) and is generalization of results [4] to the general unital associative algebra if the series by spherical harmonics are convergent.

**Theorem 5.2.** *The polynomials* $V_\mu(x)$ *introduced in (11)-(13) play an analogous role as the powers of the complex variable* $z = x + iy$ *in the theory of complex variables.*

## 5.1. *Symmetries*

Let $\mathbb{A}$ be a unital associative algebra and $D$ is the Dirac operator in $A = (\mathbb{R}^{n+1}, \circ)$ with unit $e_0$. Then $D = \partial_t e_0 + \partial_{x_1} e_1 + \ldots + \partial_{x_n} e_n$ is an evolution operator.

**Theorem 5.3.** *The first order PDO* $K_i \in \mathbb{C}[x, \partial_x]$, $i = 1, 2, \ldots, n$ *builded in (10) are pairwise commute symmetry operators for* $D$ *and such that any polynomial solution to Dirac equation* $Du = 0$, $u(0, x) = P_0(x)e_0 + P_1(x)e_1 + \ldots + P_n(x)e_n$ *may be represented as*

$$u(t, x) = P_0(K)[e_0] + P_1(K)[e_1] + \ldots + P_n(K)[e_n] \tag{15}$$

*with real polynomials* $P_i(x)$, $i = 0, 1, \ldots, n$.

The proof was given in [7].

Below we give some examples of above stated theorem:

1102

### 5.1.1. *Complex analysis*

In complex analysis symmetry operator $K$ constructed by formula (10) as a symmetry of the Cauchy-Riemann operator $D = \partial_x + i\partial_y$ is operator multiplication $K[u] = (x - iy)u$. The solution to Cauchy-Riemann equations $(\partial_x + i\partial_y)u(x, y) = 0$ with initial data $u(0, y) = P_1(y) + P_2(y)i$ is obviously $u(x, y) = P_1(x-iy)[1] + P_2(x-iy)[i]$.

### 5.1.2. *Quaternion analysis*

From the point of view of quaternion analysis (cf. [4]) the entire smooth enough (differentiable in neighborhood of origin) solution of the Dirac equation can be represented as a convergent series of quaternion harmonics defined in $X \subset \mathbb{R}^4$. They are the only homogeneous polynomial solutions of degree $m$ to the Dirac equation and

$$Y^m(q) = \sum_{|\alpha|=m} c_\alpha x^\alpha, \quad m = 0, 1, \ldots, \quad D \circ Y^m(q) = 0, \tag{16}$$

where $q = x_1 + x_2i + x_3j + x_4k$, $\alpha$ is multi index, $|\alpha| = \alpha_1 + \alpha_2 + \alpha_3 + \alpha_4$, $x^\alpha = x_1^{\alpha_1} x_2^{\alpha_2} x_3^{\alpha_3} x_4^{\alpha_4}$, and $c_\alpha$ are the quaternion valued constants.

**Theorem 5.4.** [12] *The quaternion harmonics fulfill the relation*

$$2m(m+1)Y^m(q) = \sum_{i=1}^{4} K_i[\partial_{x_i} Y^m(q)], \tag{17}$$

*where $K_1, \ldots, K_4$ are the generators of the "conformal group" in quaternion space*

$$K_1 = (x_1^2 - x_2^2 - x_3^2 - x_4^2)\partial_{x_1} + 2x_1x_2\partial_{x_2} + 2x_1x_3\partial_{x_3} + 2x_1x_4\partial_{x_4} + 2x_1 + q;$$
$$K_2 = (x_2^2 - x_1^2 - x_3^2 - x_4^2)\partial_{x_2} + 2x_2x_1\partial_{x_1} + 2x_2x_3\partial_{x_3} + 2x_2x_4\partial_{x_4} + 2x_2 - iq;$$
$$K_3 = (x_3^2 - x_2^2 - x_1^2 - x_4^2)\partial_{x_3} + 2x_3x_2\partial_{x_2} + 2x_3x_1\partial_{x_1} + 2x_3x_4\partial_{x_4} + 2x_3 - jq;$$
$$K_4 = (x_4^2 - x_2^2 - x_3^2 - x_1^2)\partial_{x_4} + 2x_4x_2\partial_{x_2} + 2x_4x_3\partial_{x_3} + 2x_4x_1\partial_{x_1} + 2x_4 - kq.$$

**Theorem 5.5.** *Homogeneous polynomial $Y^m(q)$ is a quaternion harmonics iff*

$$Y^m(q) := P_{m0}(K)[1] + P_{m1}(K)[i] + P_{m2}(K)[j] + P_{m3}(K)[k], \tag{18}$$

*where $P_{mi}$ are homogeneous of the same order $m$ real polynomials.*

Proof of above stated theorems is given in [7].

### 5.1.3. *Clifford analysis*

Define the Clifford algebra $Cl_{0,n} \in Alg(\mathbb{R}^{2^n})$ as associative unital algebra freely generated by $\mathbb{R}^n$ with usual inner product $< x, y >$ modulo the relation $x^2 = -||x||^2$ for all $x \in \mathbb{R}^n$. Equivalently, the Clifford algebra $Cl_{0,n}$ is generated by the orthonormal basis $e_0, e_1, \ldots, e_n$ in $\mathbb{R}^{n+1}$, and all theirs permutations. Here $e_0$ is an unit element and $e_i$ satisfies the relationships $e_ie_j + e_je_i = -2 < e_i, e_j > e_0$ for $1 \leq j \leq n$. In fact more details on Clifford algebras can be found in [4], [22].

Below we present some results from [13].

There are exactly four classes of 1-symmetry operators for the Dirac operator $D$ in $Cl_{0,n}$, namely:

- the generators of the translation group in $\mathbb{R}^{n+1}$

$$\partial_{x_k}, \quad k = 0, 1, \ldots, n; \tag{19}$$

- the dilatations

$$R_0 = x_0 \partial_{x_0} + x_1 \partial_{x_1} + \ldots + x_n \partial_{x_n} + \frac{n}{2}; \tag{20}$$

- the generators of the rotation group

$$J_{ij} = -J_{ji} = x_j \partial_{x_i} - x_i \partial_{x_j} + \frac{1}{2} e_{ij}, \quad i, j = 1, 2, \ldots, n, \quad i \neq j$$

$$J_{i0} = -J_{0i} = x_0 \partial_{x_i} - x_i \partial_{x_0} + \frac{1}{2} e_i, \quad i = 1, 2, \ldots, n, \tag{21}$$

- and the generators of the "conformal group"

$$K_i = \sum_{s=0}^{n} 2x_i x_s \partial_{x_s} - x\bar{x}\partial_{x_i} + (n+1)x_i - \bar{x}e_i, \tag{22}$$

for $i = 0, 1, \ldots, n$. Here $x = x_0 + x_1 e_1 + \ldots + x_n e_n$ and $\bar{x}$ are conjugate in the sense of Clifford valued functions.

Using these basic 1-symmetries we can construct the Clifford-valued operator indeterminates $K - A$ in the space $Hol(Cl_{0,n})$ as operator action similar to multiplication on $x - a$. Namely, let $a = a_1 e_1 + \ldots + a_n e_n$ and $\bar{a}$ be conjugate in the sense of Clifford algebra. Define $A = A_0 + A_1 e_1 + \ldots + A_n e_n$ and $A_i$ for $i = 0, 1, \ldots, n$ where

$$A_i = 2 \sum_{j \neq i, j=0}^{m} a_j J_{ji} - 2a_i \sum_{j=0}^{m} a_j \partial_{x_j} + 2a_i R_0 + \bar{a}a \partial_{x_i}.$$

**Theorem 5.6.** *All $Cl_{0,n}$-analytic polynomial functions $f(x)$ can be represented in neighborhood of a given point $a$ in form*

$$u(x) = U_0(K_0 - A_0, \ldots, K_n - A_n)[1] + \ldots + U_i(K_0 - A_0, \ldots, K_n - A_n)[e_i],$$

*where $U_i(x)$, $i = 0, 1, 2, \ldots, 2^n$, are real homogeneous polynomials being factorized by the relation $x_0^2 + x_1^2 + \ldots + x_n^2 = 0$.*

**Proof.** is analogous [13] to the proof of Theorem 5.5. □

**Theorem 5.7.** *The Clifford valued analytic functions have an unique power series expansion in pairwise commutative operator independents $K = \{K_0, \ldots K_n\}$.*

1104

## Conclusions

- Every first order PDO with constant coefficient is the *Dirac operator* in the corresponding algebra.
- Solution to Dirac equation in isotopic algebras forms an equivalent function theories.
- The A-analysis in the regular algebras is equivalent to the canonical function theory on their unital hearts.

## References

1. A.A. Albert, Non-associative algebras *Ann. of Math.*, **43**, 1942, 685-707.
2. G.W. Bluman, J.D. Cole, *Similarity methods for differential equations.* Applied Mathematical Sciences, Vol. 13. Springer-Verlag, New York, 1974. 332 pp.
3. G.W. Bluman, S. Kumei, *Symmetries and differential equations.* Applied Mathematical Sciences, 81. Springer-Verlag, New York, 1989. 412 pp.
4. F. Brackx, R. Delanghe, F. Sommen, *Clifford Analysis,* Pitman Research Notes in Math; **76**, 1982, 308pp.
5. M.J. Craddock, A.H. Dooley, *Symmetry group methods for heat kernels.* J. Math. Phys. 42 (2001), no. 1, 390–418.
6. L. Ehrenpreis, *A fundamental principle for systems of linear equations with constant coefficients.* in Proc. Intern. Symp. Linear Spaces, Jerusalem, 1960, pp. 161-174
7. S.D. Eidelman, Y. Krasnov, Operator method for solution of PDEs based on their symmetries. Oper. Theory Adv. Appl., 157, Birkhäuser, Basel, 2005, pp. 107–137,
8. R.P. Gilbert, J.L. Buchanan, *First Order Elliptic Systems,* Mathematics in Science and Engineering; **163**, *Academic Press*, 1983, 281pp.
9. I.J. Good, *A simple generalization of analytic function theory.* Exposition. Math, **6**, no. 4, 1988, 289–311.
10. L. Hormander, *The analysis of linear partial differential operators II*, Springer Verlag, Berlin, 1983.
11. P.W. Ketchum, *Analytic functions of hypercomplex variables,* Trans.Amer.Mat.Soc., **30**, # 4, pp 641-667, 1928.
12. Y. Krasnov, *Symmetries of Cauchy-Riemann-Fueter equation,* Complex Variables, vol. **41**, pp 279-292, 2000.
13. Y. Krasnov, *The structure of monogenic functions,* Clifford Algebras and their Applications in Mathematical Physics, vol 2, **Progr. Phys., 19,** *Birkhauser, Boston* pp. 247-272 ,2000.
14. B. Malgrange, *Sur les systémes différentiels à coefficients constants.* (French) 1963 Les Équations aux Dérivées Partielles (Paris, 1962) pp. 113–122
15. W. Miller, *Symmetry and Separation of Variables,* Encyclopedia of Mathematics and its Applications, Addison-Wesley, **4**, 1977, 285pp.
16. P.J. Olver, *Applications of Lie Groups to Differential Equations.* Graduate Texts in Mathematics 107, Springer, New York, 1993.
17. P.J. Olver, *Symmetry and explicit solutions of partial differential equations.* Appl. Numer. Math. 10 (1992), no. 3-4, 307–324.
18. P.J. Olver, V.V. Sokolov, *Integrable Evolution Equations on Associative Algebras,* Commun. Math. Phys. 193, (1998), 245 – 268
19. P.S. Pedersen, *Cauchy's integral theorem on a finitely generated, real, commutative, and associative algebra.* Adv. Math. 131 (1997), no. 2, 344–356.
20. P.S. Pedersen, *Basis for power series solutions to systems of linear, constant coefficient partial differential equations.* Adv. Math. 141 (1999), no. 1, pp. 155–166.
21. S.P. Smith, *Polynomial solutions to constant coefficient differential equations.* Trans. Amer. Math. Soc. 329, (1992), no. 2, 551–569.

22. F. Sommen, N. Van Acker, *Monogenic differential operators,* Results in Math. Vol. **22**, 1992, pp. 781-798.
23. I. Vekua, *Generalized analytic functions.* London. Pergamon, 1962.

### III.4   Complex and functional analysis methods in pdes

Organizers: Heinrich Begehr, Dao-Qing Dai, Alexandre Soldatov

Complex and functional analytic methods in partial differential equations is a session which was organized at all the ISAAC congresses so far. The topics in this session are divers. They vary from orthogonal polynomials on the unit circle, a subject related to the Riemann jump problem, boundary value problems for the Cauchy-Riemann equation in a quarter plane, the Riemann-Hilbert boundary value problem for meta-analytic functions in the unit disc, some boundary value problems for the inhomogeneous poly-analytic-harmonic equation, the differential operator of which is the product of the poly-analytic and the poly-harmonic operators, a system of equations of crystal optics, eigenvalues and eigenfunctions of the curl and Stokes operators in three dimensions, to over determined systems of first order singular differential equations, and nonlinear ordinary differential equations in algebras, and to Abel's ordinary differential equation of second kind.

The activity of this special interest group is also reflected in the at the beginning of 2007 published proceedings of a workshop on Recent Trends in Applied Complex Analysis, June 1-5, 2004, at METU, Ankara. They have appeared in the Journal of Applied Functional Analysis as issues 1 to 3 of volume 2 under the title Snapshots in Applied Complex Analysis and were edited by H. Begehr, A.O. Celebi, and R.P. Gilbert.

# ON DISTRIBUTION OF ZEROS AND ASYMPTOTICS OF SOME RELATED QUANTITIES FOR ORTHOGONAL POLYNOMIALS ON THE UNIT CIRCLE *

ZHIHUA DU and JINYUAN DU [†]

*Department of Mathematics, Wuhan University*
*Wuhan 430072, P.R. China*
*E-mail: zhdu80@126.com, jydu@whu.edu.cn*

In this paper, we consider orthogonal polynomials with respect to strictly positive and analytic weights on the unit circle. The asymptotic distribution of their zeros in some complex domains is given and the asymptotics of some related quantities for them, such as recurrence coefficients, Toeplitz determinants, reciprocal polynomials and Christoffel functions are obtained by Riemann-Hilbert approach.

**Key words:** Orthogonal polynomials, distribution of zeros, asymptotics, Riemann-Hilbert approach

**Mathematics Subject Classification:** 42C05, 30E25, 30E10

## 1. Introduction

A few years ago, P. Deift and X. Zhou introduced successfully a new approach so called Riemman-Hilbert approach to study asymptotic for orthogonal polynomials [1]. A large number of investigations for asymptotic problems based on such approach have been published [1-12]. A sufficiently extensive overview can be found in the book by Deift [6]. By using Riemman-Hilbert approach, A.B.J. Kuijlaars and his collaborators yielded an excellent work about the asymptotics for orthogonal polynomials on $[-1,1]$ (see [8,10]). Recently, in terms of the same approach, the authors established strong asymptotics for orthogonal polynomials with respect to strictly positive weights on the unit circle in the complex plane [17]. However, we find that such approach is also powerful to discuss the asymptotic distribution of zeros of orthogonal polynomails in some complex domains and the asymptotics of some related quantities for orthogonal polynomials, such as recurrence coefficients, Toeplitz determinants, reciprocal polynomials and Christoffel functions. This is just our main consideration in the present paper. In order to do so, in the sequent section, we first need a brief review of the theory of orthogonal polynomials on the unit

---

*This work is supported by NNSF of China.

[†]Corresponding author. The revision of this work is done while the corresponding author is visiting Free University Berlin in summer term 2005 on basis of the State Scholarship Fund Award of China. The authors are very grateful to Professor H. Begehr for his helpful support.

circle which can be found in [13,18,19,21–23]. In Section 3, we restate partial results obtained in [17] which will be used as the fundamentals of the present paper.

## 2. Preliminaries

Let $w$ be a nonnegative function defined on the unit circle $\Gamma = \{z : |z| = 1\}$ in the complex plane $C$ and be integrable with

$$\frac{1}{2\pi i} \int_\Gamma w(\tau) \frac{d\tau}{\tau} > 0, \tag{1}$$

which is called a weight function on the unit circle. For integrable functions $f$ and $g$ defined on the unit circle $\Gamma$, we set up the inner product

$$(f, g)_w = \frac{1}{2\pi i} \int_\Gamma w(\tau) f(\tau) \overline{g(\tau)} \frac{d\tau}{\tau}, \tag{2}$$

then, obviously an inner product space $\mathbf{L}_{w,\Gamma}$ follows with the corresponding norm

$$\|f\|_w = \left[ \frac{1}{2\pi i} \int_\Gamma w(\tau) |f(\tau)|^2 \frac{d\tau}{\tau} \right]^{\frac{1}{2}}. \tag{3}$$

It is well known that, by using the Gram–Schmidt orthogonalization process to the system of linear independent polynomials $\{1, z, \cdots, z^n, \cdots\}$, we may get the unique system of monic orthogonal polynomials with respect to the weight $w$ on the unit circle [24]

$$p_0(z) = 1, p_1(z), p_2(z), \cdots, p_n(z), \cdots \tag{4}$$

with the relationship

$$\begin{cases} (p_j, p_k) = 0, & j \neq k, \ j, k = 0, 1, \cdots, \\ \|p_n\|_w = \left[ \dfrac{1}{2\pi i} \displaystyle\int_\Gamma w(\tau) |p_n(\tau)|^2 \dfrac{d\tau}{\tau} \right]^{\frac{1}{2}} = \dfrac{1}{\lambda_n}, & n = 0, 1, 2, \cdots. \end{cases} \tag{5}$$

Then the system of normal orthogonal polynomials on the unit circle is

$$\pi_n(z) = \lambda_n p_n(z), \quad n = 0, 1, \cdots. \tag{6}$$

Let

$$c_n = \frac{1}{2\pi i} \int_\Gamma w(\tau) \overline{\tau}^{n+1} d\tau, \quad n = 0, \pm 1, \pm 2, \cdots \tag{7}$$

and

$$T_n = \begin{vmatrix} c_0 & c_{-1} & c_{-2} & \cdots & c_{-n} \\ c_1 & c_0 & c_{-1} & \cdots & c_{-n+1} \\ \cdots & \cdots & \cdots & \cdots & \cdots \\ c_{n-1} & c_{n-2} & c_{n-3} & \cdots & c_{-1} \\ c_n & c_{n-1} & c_{n-2} & \cdots & c_0 \end{vmatrix}, \quad n = 0, 1, 2, \cdots, \tag{8}$$

which are usually called Toeplitz determinants. By the uniqueness of orthogonal polynomials, one may easily find

$$\pi_0(z) = T_0^{-\frac{1}{2}} \tag{9}$$

and

$$\pi_n(z) = (T_{n-1}T_n)^{-\frac{1}{2}} \begin{vmatrix} c_0 & c_{-1} & c_{-2} & \cdots & c_{-n} \\ c_1 & c_0 & c_{-1} & \cdots & c_{-n+1} \\ \cdots & \cdots & \cdots & \cdots & \cdots \\ c_{n-1} & c_{n-2} & c_{n-3} & \cdots & c_{-1} \\ 1 & z & z^2 & \cdots & z^n \end{vmatrix}, \quad n = 1, 2, \cdots . \tag{10}$$

So

$$\lambda_n = \sqrt{\frac{T_{n-1}}{T_n}}. \tag{11}$$

Define the reciprocal polynomial of $p_n(z)$ as

$$p_n^*(z) = z^n \overline{p_n(\bar{z}^{-1})}. \tag{12}$$

Introducing reproducing kernels

$$S_n(\xi, z) = \sum_{k=0}^{n-1} \overline{p_k(\xi)} p_k(z) \tag{13}$$

which are sometimes called Szegő kernels, Szegő proved the recurrence formula [13]

$$p_{n+1}(z) = z p_n(z) - \overline{\alpha_n} p_n^*(z) \quad \text{with} \quad \alpha_n = -\overline{p_{n+1}(0)} \tag{14}$$

which are well called Verblunsky coefficients (also reflection coefficients or Schur parameters) [20,21], and the Christoffel-Darboux-Szegő formula

$$S_n(\xi, z) = \frac{\overline{\pi_n^*(\xi)} \pi_n^*(z) - \overline{\pi_n(\xi)} \pi_n(z)}{1 - \bar{\xi} z}. \tag{15}$$

The Christoffel functions associated with $w$ for orthogonal polynomials on the unit circle $\Gamma$ are defined as

$$\mu_n(w, z) = \inf_{P \in \Pi_{n-1}, P(z)=1} \left\{ \frac{1}{2\pi i} \int_\Gamma w(\tau) |P(\tau)|^2 \frac{d\tau}{\tau} \right\}, \tag{16}$$

where $\Pi_{n-1}$ denotes the class of polynomials of degrees at most $n-1$. By using the Cauchy-Schwarz inequality, we can easily prove

$$\mu_n^{-1}(w, z) = S_n(z, z). \tag{17}$$

The Christoffel functions play an important role in many problems for orthogonal polynomials [5,18,19,23]. We will give asymptotics for them in $\mathcal{C} \setminus \Gamma$ in Section 5.

## 3. Notations and lemma

In what follows we need to use some notations introduced in [14–17]. Now let us concisely recall them.

Let $L$ be a set of smooth closed curves $L_1, L_2, \cdots, L_n$ in the complex plane, non-intersecting each other and each $L_j$ oriented positively. Therefore $L$ is also oriented positively, denoted by $L = \sum_{j=1}^{n} L_j$. $L_j$'s divide the extended complex plane into a finite number of regions. The region containing the point at infinity is put into $S^-$, the regions neighboring to it is put into $S^+$, and so on. Then the extended complex plane is divided into two open sets $S^+$ and $S^-$.

**Remark 1.** Certainly, we may also put the point at infinity in $S^+$ at the very beginning and then proceed as above, then $S^+$ and $S^-$ are interchanged in position and all the $L_j$'s and hence $L$ are oppositely oriented.

Let $\varphi$ be defined on $L$ and Hölder continuous, denoted as $\varphi \in H(L)$. Now we introduce the Cauchy singular integral operator

$$\mathbf{C}[\varphi](z) = \begin{cases} \dfrac{1}{2\pi i} \displaystyle\int_L \dfrac{\varphi(\tau)}{\tau - z} d\tau, & z \notin L, \\[3mm] \dfrac{1}{\pi i} \displaystyle\int_L \dfrac{\varphi(\tau)}{\tau - t} d\tau, & z = t \in L, \end{cases} \tag{18}$$

and the projection operators

$$\mathbf{C}^{\pm}[\varphi](z) = \begin{cases} \mathbf{C}[\varphi](z), & z \in S^{\pm}, \\[2mm] \pm \dfrac{1}{2}\varphi(t) + \dfrac{1}{2}\mathbf{C}[\varphi](t), & z = t \in L. \end{cases} \tag{19}$$

Let $A^+H(A^-H)$ denote the class of functions analytic in $S^+(S^-)$ and $\in H(\overline{S^+})$ ($\in H(\overline{S^-})$). We know that [16]

$$\begin{cases} \mathbf{C}^{\pm}[\varphi] \in A^{\pm}H, \\[2mm] \mathbf{C}[\varphi](\infty) = \lim_{z \to \infty} \mathbf{C}[\varphi](z) = 0. \end{cases} \tag{20}$$

A function $F(z)$ is said to be sectionally holomorphic with $L$ as its jump curve, if it is holomorphic in $S^+$ and $S^-$, probably possesses a pole at infinity and has the finite boundary values $F^+(t)$ and $F^-(t)$ when $z$ tends to any point $t \in L$ from $S^+$ and $S^-$ respectively. Sometimes it is convenient that we introduce two projection functions

$$F^{\pm}(z) = \begin{cases} F(z), & z \in S^{\pm}, \\[2mm] F^{\pm}(t), & z = t \in L. \end{cases} \tag{21}$$

Obviously, $\mathbf{C}[\varphi]|_{S^+ \cup S^-}$ is sectionally holomorphic with $L$ as its jump curve.

From now on, we assume that $w$ is a strictly positive and analytic weight on the unit circle $\Gamma$, more precisely, $w$ is a non-vanishing analytic function in an annulus $\Omega_r = \{z : r < |z| < r^{-1}\}$ with $0 < r < 1$ and strictly positive on the unit circle $\Gamma$. For convenience, we denote $w \in PA(\Omega_r)$ for this situation.

**Example 1.** Let $\Gamma$ be the unit circle oriented counter–clockwisely. Under the preceding assumption, we know that $w \in H(L)$ is a positive function. Obviously, the Szegö function

$$D(z) = \begin{cases} \exp\left\{\mathbf{C}[\log w](z)\right\}, & z \in S^+ = \{z : |z| < 1\}, \\ \exp\left\{-\mathbf{C}[\log w](z)\right\}, & z \in S^- = \{z : |z| > 1\} \end{cases} \tag{22}$$

is sectionally holomorphic with $\Gamma$ as its jump curve and $D(\infty) = 1$.

Let

$$\mathbf{\Phi}(z) = \begin{pmatrix} \Phi_{1,1}(z) & \Phi_{1,2}(z) \\ \Phi_{2,1}(z) & \Phi_{2,2}(z) \end{pmatrix} \tag{23}$$

be a $(2 \times 2)$ matrix-valued function defined on a set $\Omega$ in the complex plane, where each element $\Phi_{j,k}$ is a function defined on $\Omega$. Whenever a property such as continuity, analyticity, etc., is ascribed to $\mathbf{\Phi}$ it is clear that in fact all the component functions $\Phi_{j,k}$'s possess the cited property. So the meanings of $\mathbf{\Phi} \in H(L)$, $\mathbf{\Phi}$ be sectionally holomorphic with $L$ as its jump curve, etc. all are obvious. In particular,

$$\lim_{z \to \infty} \mathbf{\Phi}(z) = \mathbf{a} =: \begin{pmatrix} a_{1,1} & a_{1,2} \\ a_{2,1} & a_{2,2} \end{pmatrix} \tag{24}$$

means

$$\lim_{z \to \infty} \Phi_{j,k}(z) = a_{j,k}, \quad j, k = 1, 2, \tag{25}$$

where $\mathbf{a}$ is a complex valued matrix. We would rather treat (24) as the convergence in the sense of the norm

$$\|\mathbf{\Phi}(z)\| = \sum_{j,k=1}^{2} |\Phi_{j,k}(z)|. \tag{26}$$

Sometimes we use also the following norm

$$\|\mathbf{\Phi}\|_\Omega = \sum_{j,k=1}^{2} \|\Phi_{j,k}\|_\Omega \quad \text{with} \quad \|\Phi_{j,k}\|_\Omega = \sup\left\{|\Phi_{j,k}(z)|, \ z \in \Omega\right\}. \tag{27}$$

With the above preliminaries, following [17], we have

**Lemma 1.** *Let $w$ is a strictly positive analytic weight on the unit circle $\Gamma$ and analytic in $\Omega_r = \{z : r < |z| < r^{-1}\}\, (r > 0)$, $p_n(z)$ are the monic orthogonal* polynomials associated with $w$ and $p_n^*(z)$ are the reciprocal polynomials of $p_n(z)$. Write $L_\sharp = \Gamma_{r_\sharp} + \Gamma_{r_\sharp^{-1}}$ with

$$\Gamma_{r_\sharp} = \{z : |z| = r_\sharp\} \quad \text{and} \quad \Gamma_{r_\sharp^{-1}} = \{z : |z| = r_\sharp^{-1}\} \tag{28}$$

where $r < r_\sharp < 1$, then there exists a sequence $\{\mathbf{R}^{(n)}(z)\}$ of $(2 \times 2)$ matrices, which are sectionally holomorphic matrix-valued functions with $L_\sharp$ as its jump curve fulfilling

$$\left\| \mathbf{R}^{(n)} - I \right\|_{C \backslash L_\sharp} \leq \exp\{C_1 - C_2 n\} \tag{29}$$

for sufficiently large $n$, and

$$\mathbf{R}^{(n)}(\infty) = \lim_{z \to \infty} \mathbf{R}^{(n)}(z) = I \tag{30}$$

for all $n$, such that

$$\begin{cases} p_n(z) = \dfrac{z^n}{D(z)} R_{1,1}^{(n)}(z), \\[2mm] \lambda_{n-1}^2 p_{n-1}^*(z) = -\dfrac{z^n}{D(z)} R_{1,2}^{(n)}(z), \end{cases} \quad \text{for } z \in S_{\sharp 2}^+ = \{z : r_\sharp^{-1} < |z| \leq \infty\}, \tag{31}$$

$$\begin{cases} p_n(z) = -\dfrac{1}{D(z)} R_{2,1}^{(n)}(z), \\[2mm] \lambda_{n-1}^2 p_{n-1}^*(z) = \dfrac{1}{D(z)} R_{2,2}^{(n)}(z), \end{cases} \quad \text{for } z \in S_{\sharp 1}^- = \{z : |z| < r_\sharp\}, \tag{32}$$

$$\begin{cases} p_n(z) = \dfrac{z^n D(z)}{w(z)} R_{1,1}^{(n)}(z) - \dfrac{1}{D(z)} R_{2,1}^{(n)}(z), \\[2mm] \lambda_{n-1}^2 p_{n-1}^*(z) = -\dfrac{z^n D(z)}{w(z)} R_{1,2}^{(n)}(z) + \dfrac{1}{D(z)} R_{2,2}^{(n)}(z), \end{cases} \quad z \in S_{\sharp 1}^+ = \{z : r_\sharp < |z| < 1\}, \tag{33}$$

where $C_1$, $C_2$ are positive constants independent of $n$ and only dependent on $r_\sharp$, $I$ denotes the $(2 \times 2)$ identity matrix, and $\lambda_{n-1}$, $D(z)$ are respectively given in (5) and (22).

Moreover, we need the fact (also see [17])

$$\lim_{n \to \infty} \lambda_n = \left[ \Theta(w) \right]^{-\frac{1}{2}} \tag{34}$$

with

$$\Theta(w) = \exp\left\{ \mathbf{C}[\log w](0) \right\} = \exp\left\{ \frac{1}{2\pi} \int_0^{2\pi} \log\left[ w\left(e^{i\theta}\right) \right] d\theta \right\} > 0. \tag{35}$$

## 4. Distribution of zeros for orthogonal polynomials

From Lemma 1, we can obtain the asymptotic distribution of zeros for orthogonal polynomials with respect to $w$ on the unit circle in some domains of the complex plane. This is just the main results in the present section. In order to do so, let

$$\rho = \inf_{0<r<1}\{r : w \in PA(\Omega_r)\}. \tag{36}$$

Then, we have

**Theorem 1.** *Let $w$ be a strictly positive and analytic weight on the unit circle $\Gamma$, $\rho$ be defined as above and $p_n$ be the monic orthogonal polynomials associated with $w$, then, for any $\rho' > \rho$, there exists a natural number $N(\rho')$ such that $p_n(z) \neq 0$ in $K_{\rho'} = \{z : |z| \geq \rho'\}$ for any $n > N(\rho')$.*

**Proof.** Take

$$r_\sharp = \begin{cases} \dfrac{\rho+\rho'}{2}, & \text{if } \rho' \leq 1, \\[3mm] \max\left\{\dfrac{1+\rho}{2}, \dfrac{1+\rho'}{2\rho'}\right\}, & \text{if } \rho' > 1, \end{cases} \tag{37}$$

obviously,

$$\begin{cases} \rho < r_\sharp < \rho' < 1, & \text{if } \rho' \leq 1, \\[3mm] \rho < r_\sharp < 1, \quad r_\sharp^{-1} < \rho', & \text{if } \rho' > 1. \end{cases} \tag{38}$$

Fixing $\rho'$, by Lemma 1, (29), (30), (31) and (33), we have

$$\left|\frac{p_n(z)D(z)}{z^n} - 1\right|_{z\in S_{\sharp 2}^+} \leq \exp\{C_1 - C_2 n\} \quad \text{for sufficiently large } n, \tag{39}$$

and

$$\left|\frac{p_n(z)w(z)}{z^n D(z)} - 1\right|_{z\in S_{\sharp 1}^+} \leq \left[1+\left\|\frac{w}{D^2}\right\|_{z\in S_{\sharp 1}^+}\right]\exp\{C_1-C_2 n\} \quad \text{for sufficiently large } n, \tag{40}$$

where the positive constants $C_1$ and $C_2$ are given in (29) which are only related to $r_\sharp$, $S_{\sharp 1}^+, S_{\sharp 2}^+$ have the same meaning as in (33) and (31), and $D(z)$ is given by (22).

Let $N(r_\sharp) = [C_1/C_2] + 1$ with $[C_1/C_2]$ is the integral part of $C_1/C_2$, since $r_\sharp$ is related to $\rho'$ by (37), we may also say it $N(\rho')$. Thus, for any $n > N(\rho')$, by (38), (39) and (40), we immediately know that $p_n(z) \neq 0$ in $K_{\rho'} = \{z : |z| \geq \rho'\}$.

**Remark 2.** In other words, Theorem 1 shows that, if $w$ is a strictly positive and analytic weight on the unit circle $\Gamma$, then the zeros of the orthogonal polynomials $p_n(z)$ associated with $w$ are laid in the closed disc $D_\rho = \{z : |z| \leq \rho\}$ in the sence of asymptotic distribution. Under the same assumptions, the same assertion in Theorem 1 is also obtained by A. Martínez-Finkelshtein, K.T.- R. McLaughlin and E.B. Saff by using an analogous treatment in their recent paper [20]. However, although they obtain more information for zeros of orthogonal polynomials under their assumptions, the proof here seems more simple.

## 5. Asymptotics of related quantities for orthogonal polynomials

One and the same, as consequences of Lemma 1, in this section we respectively give aymptotics of $\alpha_n$, $T_n$, $p_n^*(z)$ and $\mu_n(w, z)$ introduced in Section 2 as follows.

**Theorem 2.** (Recurrence Coefficients) *Let $w$ is a strictly positive and analytic weight on the unit circle $\Gamma$ and $\alpha_n$ be given as in* (14), *then*

$$\lim_{n \to \infty} \alpha_n = 0. \tag{41}$$

**Proof.** Note that $\alpha_n = -\overline{p_{n+1}(0)}$, take $z = 0$, by (32), then

$$|\alpha_n| \leq \frac{1}{D(0)} \exp\{C_1 - C_2(n+1)\} \tag{42}$$

for sufficiently large $n$, where the positive constants $C_1$ and $C_2$ are given in (29), $D(0) = \Theta(w)$ is given in (22) and (35). Thus, (41) follows.

**Theorem 3.** (Teoplitz Determinants) *Let $w$ is a strictly positive and analytic weight on the unit circle $\Gamma$ and $T_n$ be given in* (8), *then*

$$\lim_{n \to \infty} \sqrt[n]{T_n} = \Theta(w) \tag{43}$$

*where $\Theta(w)$ is the same as in* (35).

**Proof.** For any positive sequence $\{a_n\}$, we know that, if $\lim_{n \to \infty} a_n = a$ then $\lim_{n \to \infty} \sqrt[n]{a_1 \cdots a_n} = a$. Noting this fact and quoting (11) and (34), we see that (43) is obvious.

**Theorem 4.** (Reciprocal Polynomials) *Let $w$ is a strictly positive and analytic weight on the unit circle $\Gamma$ and $p_n^*(z)$ be given in* (12), *then*

$$\lim_{n \to \infty} \frac{D(z)p_n^*(z)}{z^{n+1}} = 0 \tag{44}$$

*holds uniformly for $|z| \geq R > 1$ and*

$$\lim_{n \to \infty} D^+(z)p_n^*(z) = \Theta(w) \tag{45}$$

*holds uniformly for $|z| \leq 1$, in which $\Theta(w)$, $D(z)$ have the preceding meaning and $D^+(z)$ is explained by* (19) *and* (22).

**Proof.** By Lemma 1, quoting (29), we rewrite (31), (32) and (33) in the following forms respectively, for sufficiently large $n$,

$$\left| \frac{\lambda_n^2 p_n^*(z) D(z)}{z^{n+1}} \right|_{|z| \geq R > 1} \leq \exp\{C_1 - C_2(n+1)\}, \tag{46}$$

$$\left|\lambda_n^2 p_n^*(z)D(z)-1\right|_{|z|\le r<1}\le\exp\left\{C_1-C_2(n+1)\right\},\tag{47}$$

and

$$\left|\lambda_n^2 p_n^*(z)D(z)-1\right|_{r\le|z|<1}\le\left[1+\left\|\frac{D^2}{w}\right\|_{r\le|z|<1}\right]\exp\left\{C_1-C_2(n+1)\right\}.\tag{48}$$

Note that (34), (44) follow immediately from (46) while (45) follows from (47) and (48). When $|t|=1$, one needs to let $z$ tend to $t$ in (48). Thus $D(z)$ is attained at its boundary value $D^+(t)$.

**Theorem 5.** (Christoffel Functions) *Let $w$ is a strictly positive and analytic weight on the unit circle $\Gamma$ and $\mu_n(w,z)$ be given as in (16), then*

$$\lim_{n\to\infty}\frac{\mu_n^{-1}(w,z)}{|z|^{2n+2}}=\frac{1}{\Theta(w)(|z|^2-1)|zD(z)|^2}\tag{49}$$

*holds uniformly for $|z|\ge R>1$ and*

$$\lim_{n\to\infty}\mu_n^{-1}(w,z)=\frac{1}{\Theta(w)(1-|z|^2)|D(z)|^2}\tag{50}$$

*holds uniformly for $|z|\le r<1$, where $\Theta(w)$ and $D(z)$ are respectively given in (35) and (22).*

**Proof.** Noting (6) and using Lemma 1, for orthogonal polynomials $\pi_n(z)$ with respect to $w$, by (31) and (32), we have

$$\begin{cases}\pi_n(z)=\dfrac{\lambda_n z^n}{D(z)}R_{1,1}^{(n)}(z),\\ \pi_{n-1}^*(z)=-\dfrac{z^n}{\lambda_{n-1}D(z)}R_{1,2}^{(n)}(z),\end{cases}\quad z\in\{z:|z|\ge R>1,z=\infty\},\tag{51}$$

and

$$\begin{cases}\pi_n(z)=-\dfrac{\lambda_n}{D(z)}R_{2,1}^{(n)}(z),\\ \pi_{n-1}^*(z)=\dfrac{1}{\lambda_{n-1}D(z)}R_{2,2}^{(n)}(z),\end{cases}\quad z\in\{z:|z|\le r<1\}.\tag{52}$$

Quoting (15), (17) and (34), then, we can easily see that, (49) follows from (51), (29) and (30) while (50) follows from (52) and (29).

**Remark 3.** (50) is due to Szegö [13]. By using the Riemann-Hilbert approach, we reobtain it here and simultaneously obtain asymptotics (49), which is a result in [25].

### References

1. P. Deift, X. Zhou, A steepest descent method for oscillatory Riemann-Hilbert problem, asymptotics for the MKdV equation, *Ann. Math.*, **137**(1993), 295–368.
2. P. Deift, X. Zhou, Asymptotics for the Painlevé II equation, *Comm. Pure Appl. Math.*, 48(1995), 277–337.

1118

3. P. Deift, T. Kriecherbauer, K.T.R. McLaughlin, S. Venakides, X. Zhou, Uniform asymptotics for polynomials orthogonal with respect to varying exponential weights and applications to universality questions in random matrix theory, *Comm. Pure Appl. Math.*, **52**(1999), 1335–1425.

4. P. Deift, T. Kriecherbauer, K.T.R. McLaughlin, S. Venakides, X. Zhou, Strong asymptotics of orthogonal polynomials with respect to exponetial weights, *Commu. Pure Appl. Math.*, **52**(1999), 1491–1552.

5. D.S. Lubinsky, Asymptotics of orthogonal polynomials: some old, some new, some identities, *Acta Appl. Math.*, **61**(2000), 207–256.

6. P. Deift, *Orthogonal Polynomials and Matrices: a Riemann-Hilbert Approach*, Amer. Math. Soc., Providence, RI, 2000.

7. P. Deift, T. Kriecherbauer, K.T.R. McLaughlin, S. Venakides, X. Zhou, A Riemann-Hilbert approach to asymptotic questions for orthogonal polynomials, *J. Comp. Appl. Math.*, **133**(2001), 47–63.

8. A.B.J. Kuijlaars, Riemann-Hilbert analysis for orthogonal polynomials, *Lecture Notes in Mathematics* (E. Koelink & W. Van Assche eds.), **1817**, Springer, Berlin, 2003, 167–210.

9. A.I. Aptekarev, Sharp constants for rational approximations of analytic functions, *Sb. Math.*, **193**(2002), 1–72.

10. A.B.J. Kuijlaars, K.T.R. McLaughlin, W. Van Assche, M. Vanlessen, The Riemann-Hilbert approach to strong asymptotics for orthogonal polynomials on $[-1, 1]$, *Adv. Math.*, **188**(2004), 337–398.

11. J. Baik, P. Deift, K. Johansson, On the distribution of the length of the longest increasing subsequence of random permutations, *J. Amer. Math. Soc.*, **12**(4)(1999), 1119–1178.

12. J. Baik, Riemann-Hilbert problems for last passage percolation, *Contemp. Math.* (K.McLaughlin & X.Zhou eds.), **326**(2003), 1–21.

13. G. Szegö, *Orthogonal Polynomials*, Colloquium Publications, **23**, Amer. Math. Soc., New York, 1939.

14. N.I. Muskhelishvili, *Singular Integral Equations*, 2nd. ed., Dover, New York, 1992.

15. F.D. Gakhov, *Boundary Value Problems*, Dover, New York, 1990.

16. J. Lu, *Boundary Value Problems for Analytic Functions*, World Sci., Singapore, 1993.

17. Z. Du, J. Du, Riemann-Hilbert approach to strong asymptotics for orthogonal polynomials on the unit circle, *Chinese Annual Maths.*, **27**A(5) (2206), 701–718.

18. P. Nevai, Géza Freud, Orthogonal polynomials and Christoffel functions. A case study, *J. Approx. Theory*, **48**(1986), 3–167.

19. G. Freud, *Orthogonal Polynomials*, Akadémiai Kiadó/Pergamon, Budapest, 1971.

20. A. Martínez-Finkelshtein, K.T.R. McLaughlin and E.B. Saff, Szegö orthogonal polynomials with respect to an analytic weight: canonical representation, strong asymptotics, *Constr. Apprxo.*, **24** (2006), 319–363.

21. B. Simon, *Orthogonal Polynomials on the Unit Circle*, 1: Classical Theory, 2: Spectral Theory, AMS colloquium series, Amer. Math. Soc., Providence, R.I., 2005.

22. Ya.L. Geronimus, *Polynomials Orthogonal on a Circle and Interval*, Pergamon Press Oxford, 1960.

23. P. Nevai, *Orthogonal Polynomials*, Memoirs Amer. Math. Soc., **213**, Amer. Math. Soc., Providence, R.I., 1979.

24. W. Rudin, *Functional Analysis, 2nd. ed.*, McGraw-Hill, New York, 1991.

25. A. Máté, P. Nevai, V. Totik, Szegö extremum problem on the unit circle, *Ann. Mat.*, **134** (1991), 433–453.

# REPRESENTATION OF PSEUDOANALYTIC FUNCTIONS IN THE SPACE

P. BERGLEZ

*Department of Mathematics,*
*Graz University of Technology,*
*Graz, Austria*
*E-mail: berglez@tugraz.at*

We consider a generalized Vekua equation in biquaternionic formalism where the Cauchy-Riemann operator is replaced by the differential operator $D$ of Dirac. For particular classes we construct differential operators of higher order which give a relation between the monogenic functions as solutions of $Dw = 0$ and the generalized pseudoanalytic functions as solutions of the generalized Vekua equation. This is done by considering a corresponding differential equation of second order. Using generating functions in the sense of L. Bers we can give further representations of such functions and we can obtain related pseudoanalytic functions of the second kind as solutions of another differential equation of first order.

**Key words:** Pseudoanalytic functions, quaternions, monogenic functions, differential operators

**Mathematics Subject Classification:** 30G20, 30G35, 35C05

## 1. Introduction

We denote by $\mathbb{H}(\mathbb{C})$ the algebra of complex quaternions (so called biquaternions) defined by

$$\mathbb{H}(\mathbb{C}) = \{a \mid a = a_0 + a_1 i_1 + a_2 i_2 + a_3 i_3\}$$

where the $i_k$ are the standard basic quaternions with $i_3 = i_1 i_2$, $a_k \in \mathbb{C}$, and the complex imaginary unit $i$ commutes with the $i_k, k = 1, 2, 3$. The quaternionic conjugation is given by $\bar{a} := a_0 - a_1 i_1 - a_2 i_2 - a_3 i_3$. Let $\mathcal{S}$ denote the subset of zero divisors from $\mathbb{H}(\mathbb{C})$.

Generalizing the Cauchy–Riemann operator $\partial_{\bar{z}} = \frac{1}{2}(\partial_x + i\partial_y)$ we introduce the Dirac operator

$$D := \sum_{k=1}^{3} i_k \partial_k \quad \text{where} \quad \partial_k = \frac{\partial}{\partial x_k}$$

(see e.g. K. Gürlebeck and W. Sprössig[2]). For functions $f : \mathbb{R}^3 \to \mathbb{H}(\mathbb{C})$ the set $\ker D$ defines the class of regular functions with respect to $D$. Since the algebra of quaternions is non-commutative we have to distinguish between $f \in \text{Ker} D$ such that

$Df = 0$ or $0 = fD$. Such functions are called left monogenic or right monogenic respectively.

H.R. Malonek[5,6] extended the concept of the so called $(F, G)$–derivative in the sense of L. Bers to quaternionic-valued functions to study a spatial version of pseudoanalytic functions. He proved that these pseudoanalytic functions in the space obey a generalized Vekua equation.

The generalized pseudoanalytic functions considered here are solutions of the generalized Vekua equation

$$D^* w = \frac{m}{x_1} \bar{w}, \; m \in \mathbb{N},$$

with $D^* = -i_1 D$ for which we give a general representation theorem using differential operators. Thus we establish a relation between the monogenic functions and the pseudoanalytic functions in the space. Furthermore relations between the solutions of such differential equations with different parameters $m$ are proved.

Finally we give generating pairs for such classes of generalized pseudoanalytic functions and we represent generalized pseudoanalytic functions of the second kind also.

For example such pseudoanalytic functions in the space are of interest in treating the Dirac equation with a vectorial electromagnetic potential (see e.g. V.V. Kravchenko et al.[4]) or the multidimensional stationary Schrödinger equation (see V.V. Kravchenko[3]).

## 2. Differential operators for the solutions

We first consider the generalized Vekua equation

$$D^* w = \varphi(x_1) \bar{w} \tag{1}$$

and look for a solution of the form

$$w = \sum_{k=0}^{m} A_k(x_1) \left( g \overline{D^{*k}} \right) + \sum_{k=0}^{n} B_k(x_1) \left( D^{*k} \bar{g} \right) \tag{2}$$

with $m, n \in \mathbb{N}$ where $Dg = 0$ i.e. the function $g$ is assumed to be left monogenic and $\overline{D^*} = \partial_{x_1} + i_3 \partial_{x_2} - i_2 \partial_{x_3}$ is used. Doing so we are led to the relation

$$n = m - 1,$$

and the following conditions

$$\begin{aligned}
&A'_m(x_1) = 0, \\
&A'_k(x_1) = \varphi(x_1) B_k(x_1), \;\; k = m - 1, \dots, 0, \\
&B_{m-1}(x_1) = \varphi(x_1) A_m(x_1), \\
&B_{k-1}(x_1) = -B'_k(x_1) + \varphi(x_1) A_k(x_1), \;\; k = m - 1, \dots, 1, \\
&0 = -B'_0(x_1) + \varphi(x_1) A_0(x_1).
\end{aligned} \tag{3}$$

The conditions (3) form a system of $2m + 2$ ordinary differential equations for the $2m + 1$ functions $A_k, k = 0, \ldots, m, B_k, k = 0, \ldots, m - 1$, which is overdetermined in general. This means that in the general case it is not possible to obtain solutions of this generalized Vekua equation in the form (2). But as we will see in the following there are factors $\varphi(x_1)$ in (1) such that the system (3) can be satisfied. We can assume that there will be only a few classes of differential equations of such type for which a representation of the solutions using differential operators acting on monogenic functions exists. Indeed it is a disadvantage of the method presented here that the differential equations which can be solved by this method are very particular. On the other hand we have the advantage that the application of such differential operators is a method consisting of a finite number of processes which can be carried out for a wide range of monogenic functions in an explicit and easy way.

For further investigations we will not consider the system (3) but we will turn over to a differential equation of second order to obtain a general representation theorem for the solutions of such a generalized Vekua equation.

## 3. A differential equation of second order

Consider the generalized Vekua equation

$$D^* w = \frac{m}{x_1} \bar{w}, \ m \in \mathbb{N}. \tag{4}$$

We apply the operator $\overline{D^*}$ from the right to this equation and get the differential equation of second order

$$D^* w \overline{D^*} + \frac{1}{x_1} D^* w - \frac{m^2}{x_1^2} w = 0. \tag{5}$$

For this differential equation we will deduce the general solution now. First let us consider the differential equation

$$D^* w \overline{D^*} - \frac{1}{x_1} D^* w - \frac{m^2 - 1}{x_1^2} w = 0 \tag{6}$$

which is associated to Eq. (5) in the following manner. We denote by $\Omega$ a suitable domain in $\mathbb{R}^3$ not containing the plane $x_1 = 0$ and by $\mathcal{L}_m(\Omega)$ and $\mathcal{L}_{m-1}(\Omega)$ the set of solutions of the differential equations (5) and (6) respectively defined in $\Omega$. For $w \in \mathcal{L}_m(\Omega)$ we have

$$u = Rw := \frac{x_1^2}{m^2} D^* w \in \mathcal{L}_{m-1}(\Omega) \tag{7}$$

and for $w \in \mathcal{L}_{m-1}(\Omega)$ we have

$$v = Sw := w \overline{D^*} - \frac{1}{x_1} w \in \mathcal{L}_m(\Omega) \tag{8}$$

and furthermore

$$S \circ R = \mathrm{id}_{\mathcal{L}_m(\Omega)}, \quad R \circ S = \mathrm{id}_{\mathcal{L}_{m-1}(\Omega)}$$

1122

holds. Thus we have the following

**Lemma 3.1.** *The mapping*

$$R : \mathcal{L}_m(\Omega) \ \to \ \mathcal{L}_{m-1}(\Omega)$$

*with R from (7) is an isomorphism. Its inverse $R^{-1}$ is given by*

$$R^{-1} = S.$$

Now let $\mathcal{L}_{m-k}(\Omega)$ denote the set of solutions of the differential equation

$$D^* w_{m-k} \overline{D^*} + \frac{1-2k}{x_1} D^* w_{m-k} - \frac{m^2 - k^2}{x_1^2} w_{m-k} = 0 \, , k = 0, 1, 2, \dots \quad (9)$$

With $k = 0$ we get Eq. (5). By iterated application of the mapping $R_{m-k}$ : $\mathcal{L}_{m-k}(\Omega) \ \to \ \mathcal{L}_{m-k-1}(\Omega)$ with $R_{m-k}u := \dfrac{x_1^2}{m^2 - k^2} D^* u$, which in general is an isomorphism too, the solutions of these differential equations are related.

With $k = m$ in (9) we get the equation

$$D^* w_0 \overline{D^*} + \frac{1-2m}{x_1} D^* w_0 = 0.$$

Let the operator $I$ generate the antiderivative from a biquaternionic-valued function that means

$$I(f) = F \quad \text{with} \quad D^* F = f.$$

With this we have

$$w_0 = I(x_1^{2m-1} \, \hat{h}) + g \quad (10)$$

where the functions $\hat{h}$ and $g$ obey the conditions $\hat{h}\overline{D^*} = 0$ and $D^* g = 0$ respectively. By

$$w \equiv w_m = S_{m-1} \circ \dots \circ S_0 \, w_0 \quad (11)$$

with $S_k : \mathcal{L}_{k-1}(\Omega) \ \to \ \mathcal{L}_k(\Omega)$, $S_k u := u\overline{D^*} - \frac{2m-2k-1}{x_1} u$ we get the solutions of Eq. (5). With $w_0$ from (10) and the substitution $\hat{h} = D^{*\, 2m} h$, where $h\overline{D^*} = 0$ holds, we can transform the representation (11) for $w$ into the form

$$w = \sum_{j=0}^{m} \frac{(-1)^{m-j}(2m-1-j)!}{j!(m-j)!x_1^{m-j}} (g\overline{D^*}^j) + \sum_{j=0}^{m-1} \frac{(-1)^{m-1-j}(2m-1-j)!}{j!(m-1-j)!x_1^{m-j}} (D^{*\, j} h).$$

$$(12)$$

Now we have the following

**Theorem 3.1.**

(1) *For each solution of (5) there exist two functions g left monogenic in $\Omega$ (i.e. $D^* g = 0$) and h right antimonogenic in $\Omega$ (i.e. $h\overline{D^*} = 0$) respectively such that the representation (12) holds.*

(2) *For each function $g$ left monogenic in $\Omega$ and $h$ right antimonogenic in $\Omega$ the expression in (12) represents a solution of (5) in $\Omega$.*

(3) *For a solution $w$ of (5) in the form (12) the functions*

$$\left(g\overline{D^*}^{2m}\right) = \frac{m}{x_1^{2m+1}}\left(P^m(x_1\,w)\right) \quad \text{with} \quad Pw = x_1^2(w\overline{D^*})$$

*and*

$$\left(D^{*\,2m}h\right) = \frac{1}{mx_1^{2m+1}}\left(Q^{m+1}w\right) \quad \text{with} \quad Qw = x_1^2(D^*\,w)$$

*are determined uniquely.*

Between the solutions of Eq. (5) with different parameters $m$ there exist interesting relations. Let $\mathcal{F}_m(\Omega)$ be the set of solutions of (5)

$$D^*\,w\,\overline{D^*} + \frac{1}{x_1}\,D^*\,w - \frac{m}{x_1^2}\,w = 0$$

and $\mathcal{F}_{m\pm1}(\Omega)$ the set of solutions of the differential equation

$$D^*\,w\,\overline{D^*} + \frac{1}{x_1}\,D^*\,w - \frac{(m\pm1)^2}{x_1^2}\,w = 0$$

For $w \in \mathcal{F}_m(\Omega)$ we have

$$v_1 = w\,\overline{D^*} - D^*\,w \in \mathcal{F}_m(\Omega),$$

$$v_2 = w\,\overline{D^*} + \frac{m-1}{m}D^*\,w + \frac{2m-1}{x_1}\,w \in \mathcal{F}_{m-1}(\Omega),$$

$$v_3 = w\,\overline{D^*} + \frac{m+1}{m}D^*\,w - \frac{2m+1}{x_1}\,w \in \mathcal{F}_{m+1}(\Omega).$$

## 4. Representation of pseudoanalytic functions in the space

Now we can deduce a representation for the solutions of Eq. (4). From the preceeding section we see that among the solutions of (5) we find the solutions of the generalized Vekua equation (4). Inserting the solution $w$ of (5) according to (12) into Eq. (4) we are led to the condition

$$g - m\bar{h} = 0.$$

After the substitution $g = mf$, $f$ left monogenic, we have the following representation for the solutions of Eq. (4)

$$w = \sum_{j=0}^{m} \frac{(-1)^{m-j}(2m-1-j)!}{j!(m-j)!x_1^{m-j}}\left[m(f\overline{D^*}^j) - (m-j)(D^{*\,j}\bar{f})\right], \tag{13}$$

and we can prove the following

**Theorem 4.1.**

*(1) For each function $f$ left monogenic in $\Omega \subset \mathbb{R}$ (i.e. $Df = 0$) the function $w$ according to (13) is a solution of the generalized Vekua equation (4)*

$$D^* w = \frac{m}{x_1} \bar{w}, \ m \in \mathbb{N}.$$

*(2) For each solution $w$ of equation (4) defined in $\Omega$ there exists a function $f$ monogenic in $\Omega$ such that (13) holds.*

*(3) The function $f$ in (13) is not determined uniquely by $w$. Only the expression $(f\overline{D^{*2m}})$ is determined uniquely by*

$$(f\overline{D^{*2m}}) = \frac{1}{x_1^m} \left[ (x_1^m \, w) \, \overline{D^*}^m \right].$$

There exist connections between the solutions of the generalized Vekua equation (4) with different parameters. Let

$$\mathcal{V}_m(\Omega) = \left\{ w \, : \, D^* w = \frac{m}{x_1} \bar{w} \right\}$$

be se set of the solutions of Eq. (4) defined in $\Omega$. Then we have

$$u_1 = w\overline{D^*} + \frac{m+1}{x_1}\bar{w} - \frac{2m+1}{x_1}w \ \in \ \mathcal{V}_{m+1}(\Omega),$$

$$u_2 = w\overline{D^*} + \frac{m-1}{x_1}\bar{w} + \frac{2m-1}{x_1}w \ \in \ \mathcal{V}_{m-1}(\Omega),$$

$$u_3 = w\overline{D^*} - \frac{m}{x_1}\bar{w} \ \in \ \mathcal{V}_{-m}(\Omega).$$

With the last relation solutions may be obtained for Eq. (4) even in the case when $m$ is a negative integer.

## 5. Generating pairs

H.R. Malonek[6] extended the concept of generating functions introduced by L. Bers[1] for the representation of pseudoanalytic functions in the space. Let $\mathbb{H}_k$ denote the set of reduced complex quaternions which have the form $a = a_0 + a_k i_k$, $a_0, a_k \in \mathbb{C}, k = 1, 2$. Two pairs of functions $\mathcal{H}_1 = (F, G), \mathcal{H}_2 = (M, N)$ with $F, G \in \mathbb{H}_1, M, N \in \mathbb{H}_2, F, G, M, N \notin S$, are called generating pairs for the solutions of (4) if $G\bar{F} - \bar{G}F \neq 0, N\bar{M} - \bar{N}M \neq 0$ in $\Omega$ and the products $FG, FN, GM, GN$ are solutions of the generalized Vekua equation (4). For $m = 1$ such generating pairs are given by

$$F = 1 \quad M = x_1^{-m}(x_1 - (2m-1)x_3 i_2),$$
$$G = i_1 \quad N = x_1^{-m} i_2.$$

With these functions the solutions of Eq. (4) can be represented in the form

$$w = F(M\phi + N\psi) + G(M\mu + N\nu)$$

where $\phi, \psi, \mu, \nu$ are suitable complex valued functions, which are determined uniquely by the solution $w$.

## 6. Pseudoanalytic functions of the second kind in the space

Generalizing the concept of L. Bers the corresponding spatial version of a pseudo-analytic function of second kind $\omega$ is defined as

$$\omega(x) = \phi(x) + \mu(x)\, i_1 + \psi(x)\, i_2 + \nu(x)\, i_3$$

(cf. H.R. Malonek[5]). In the case of our particular pairs of generating functions (again for $m = 1$) it obeys the differential equation

$$(D^* \omega)\,(x_1 + 1 - (2m - 1)x_3 i_2) + J_2(\overline{D^* \omega})\,(x_1 - 1 - (2m - 1)x_3 i_2) = 0 \qquad (14)$$

where $J_2$ denotes the linear mapping $J_2 : \mathbb{H}(\mathbb{C}) \to \mathbb{H}(\mathbb{C})$ with $J_2(i_1) = i_1$ and $J_2(i_2) = -i_2$.

A solution can be obtained by

$$\omega = x_1^{m-1}\,(w_0 + i_1 w_1 + i_2(x_1\, w_2 + (2m - 1)x_3\, w_0) + i_3(x_1\, w_3 + (2m - 1)x_3\, w_1))$$

$$(15)$$

where

$$w = w_0 + w_1\, i_1 + w_2\, i_2 + w_3\, i_3$$

is the corresponding pseudoanalytic function of the first kind. Using (15) with $w$ in the form (13) we finally get a representation of the solutions of the differential equation (14) using differential operators.

## References

1. L. Bers, *Theory of Pseudo–Analytic Functions*, New York University, 1953.
2. K. Gürlebeck, W. Sprössig, *Quaternionic and Clifford Calculus for Physicists and Engineers*, John Wiley, Chichester, 1997.
3. V.V. Kravchenko, *On the reduction of the multidimensional stationary Schrödinger equation to a first-order equation and its relation to the pseudoanalytic function theory*, J. Phys. A **38** (2005), 851–868.
4. V.V. Kravchenko, H.R. Malonek, G. Santana, *Biquaternionic integral representations for massive Dirac spinors in a magnetic field and generalized biquaternionic differentiability*, Math. Methods Appl. Sci. **19** (1996), 1415–1431.
5. H.R. Malonek, *Remarks on a spatial version of the pseudoanalytic functions in the sense of Bers*, W. Sprössig, K Gürlebeck (eds.), Analytical and Numerical Methods in Quaternionic and Clifford Analysis, 113–119, Proc. Symposium, Seiffen, 1996.
6. H.R. Malonek, *Generalizing the (F,G)-derivative in the sense of Bers*, in: V. Dietrich, K. Habetha, G. Jank (eds.), Clifford Algebras and their Application in Mathematical Physics, 247–257, Kluwer, Dordrecht, 1998.
7. I.N. Vekua, *Verallgemeinerte analytische Funktionen*, Akademie Verlag, Berlin, 1963.

# HILBERT BOUNDARY VALUE PROBLEM FOR A CLASS OF METAANALYTIC FUNCTIONS ON THE UNIT CIRCUMFERENCE*

YUFENG WANG

*School of Mathematics and Statistics,*
*Wuhan University,*
*Wuhan 430072, P.R.China*
*E-mail: wh_yfwang@163.com*

In this article, the Hilbert boundary value problem for a class of metaanalytic functions with different factors on the unit circumference is discussed in sixteen cases. Using the appropriate decomposition of metaanalytic function, the explicit expression of solutions and the condition of solvability are obtained by transforming the problem into two independent Hilbert boundary value problems for analytic functions.

**Key words:** Hilbert boundary value problem, metaanalytic functions, analytic functions
**Mathematics Subject Classification:** 30E25, 30G30, 35J25

## 1. Introduction

Suppose $G \subset C$ is an open set. Introduce the differential operator

$$\mathbf{M} = \frac{\partial^2}{\partial \bar{z}^2} - 2\lambda \frac{\partial}{\partial \bar{z}} + \lambda^2 \equiv \left( \frac{\partial}{\partial \bar{z}} - \lambda \right)^2 \tag{1}$$

where $\frac{\partial}{\partial \bar{z}} = \frac{1}{2}(\frac{\partial}{\partial x} + i \frac{\partial}{\partial y})$ is the Cauchy-Riemann operator and $\lambda$ a complex constant. If $f \in C^2(G)$ satisfies the generalized Cauchy-Riemann equation $\mathbf{M}f = 0$, then $f$ is called an $\mathbf{M}$-analytic function on $G$, or simply, a metaanalytic function. The class of such functions is denoted by $M(G)$. Clearly, $M(G) = H_2(G)$ while $\lambda = 0$. $H_2(G)$ is the set of bianalytic functions in $G$, so metaanalytic function is a generalization of polyanalytic function[1]. In addition, $U \in M(G)$ may be uniquely expressed as[1]

$$U(z) = V^*(z) \exp\{\lambda \bar{z}\} \quad \text{with} \quad V^* \in H_2(G). \tag{2}$$

Thus $U \in M(G)$ may also be uniquely written as

$$U(z) = V(z) \exp\{\lambda \bar{z} + \bar{\lambda} z\} \quad \text{with} \quad V \in H_2(G). \tag{3}$$

As is well-known, the classical Hilbert boundary value problem (BVP) of analytic function are solved by two methods[2-4]. Various kinds of BVPs for polyanalytic

---
*Project supported by Tianyuan Fund of Mathematics(10626039) and NNSF of China(10471107) and RFDP of Higher Education(20060486001)

function or generalized polyanalytic functions have been widely investigated in recent time [5-12]. Besides, BVPs for different classes of metaanalytic functions are also studied. For instance, the expression of solution and the condition of solvability of Haseman type BVPs for a class of metaanalytic functions on the unit circumference are obtained by B. F. Fatulaev[10]. We have also considered Riemann BVP of another class of metaanalytic functions[12]. In general, the way to solve BVPs for metaanalytic function is to change the problem into equivalent BVPs of polyanalytic function or analytic function by using the appropriate decomposition of metaanalytic functions.

In this article, Hilbert BVP for a class of metaanalytic functions with different factors on the unit circumference is discussed in details. Using the decomposition (3) of metaanalytic functions, the expression of solution and the condition of solvability are obtained by transforming the problem into two equivalent and independent Hilbert BVPs of analytic functions. Finally, we assume $\lambda \neq 0$ in the following.

## 2. Hilbert BVP for Metaanalytic Function

In this section, our problem is to find a function $U \in M(D) \cap C^1(\overline{D})$, satisfying the following Hilbert type boundary condition

$$\text{Re} \left\{ [a_k(t) + ib_k(t)] \left[ \frac{\partial^k U}{\partial \bar{z}^k} \right]^+ (t) \right\} = c_k(t), \ t \in \partial D, \ k = 1, 2, \tag{4}$$

where $D$ is the unit disc and $\partial D$ its boundary, $a_k(t), b_k(t), c_k \in H(\partial D)$ $(k = 1, 2)$ are real-valued functions and $a_k^2(t) + b_k^2(t) = 1$ $(k = 1, 2)$. This problem is simply called MH problem.

Let

$$U(z) = e^{\lambda \bar{z} + \bar{\lambda} z} [\varphi_1(z) + \bar{z} \varphi_2(z)], \tag{5}$$

where $\varphi_k$ $(k = 1, 2)$ are analytic in $D$ and continuous up to the boundary $\partial D$. Substituting (5) into the boundary condition (4), one immediately gets

$$\begin{cases} \text{Re} \left\{ t^{-1} [a_1(t) + i\, b_1(t)] [t\varphi_1^+(t) + \varphi_2^+(t)] \right\} = e^{-(\lambda \bar{t} + \bar{\lambda} t)} c_1(t), & t \in \partial D, \\ \text{Re} \left\{ t^{-1} [a_2(t) + i\, b_2(t)] [\lambda t\varphi_1^+(t) + (\lambda + t)\varphi_2^+(t)] \right\} = e^{-(\lambda \bar{t} + \bar{\lambda} t)} c_2(t), & t \in \partial D, \end{cases} \tag{6}$$

which are two independent boundary conditions for H problems of analytic functions. Let

$$\kappa_j = \frac{1}{2\pi} \left[ \arg \left\{ t [a_j(t) - i\, b_j(t)] \right\} \right]_{\partial D} = \frac{1}{2\pi} \left[ \arg \left\{ a_j(t) - i\, b_j(t) \right\} \right]_{\partial D} + 1, \ j = 1, 2, \tag{7}$$

$$X_j(z) = z^{\kappa_j} Y_j(z), \quad Y_j(z) = \exp \left\{ \Gamma_j(z) \right\}, \ j = 1, 2, \tag{8}$$

where

$$\Gamma_j(z) = \frac{1}{\pi} \int_{\partial D} \frac{\Theta_j(\tau)}{\tau - z} d\tau - \frac{1}{2\pi} \int_{\partial D} \frac{\Theta_j(\tau)}{\tau} d\tau + \frac{\pi i}{2}, \ j = 1, 2 \tag{9}$$

with

$$\Theta_j(\tau) = \arg\left\{\tau^{-\kappa_j+1}\left[a_j(\tau) - i\,b_j(\tau)\right]\right\}. \tag{10}$$

Introduce the symbol for the class of the symmetric Laurent polynomials[11]

$$S\Pi_k = \begin{cases} \left\{\pi_k(z) = \displaystyle\sum_{j=-k}^{k} c_j z^j : c_j = \bar{c}_{-j} \text{ for } j = 0,1,2,\cdots,k\right\}, & k \geq 0, \\ \{0\}, & k < 0. \end{cases} \tag{11}$$

Now we discuss the solution of MH problem (4) in details.

**Case 1:** $\kappa_1 > 1$, $\kappa_2 > 1$. The solutions of the H problems (6) for analytic functions may be written as[11]

$$\begin{cases} z\varphi_1(z) + \varphi_2(z) = \dfrac{X_1(z)}{2\pi i}\left[\displaystyle\int_{\partial D} \dfrac{e^{-(\bar{\lambda}\tau+\lambda\bar{\tau})}c_1(\tau)}{[a_1(\tau) + i\,b_1(\tau)]X_1^+(\tau)}\dfrac{\tau+z}{\tau-z}\mathrm{d}\tau + 2\pi i\,q_{\kappa_1}^1\right] \\[4mm] \lambda z\varphi_1(z) + (\lambda+z)\varphi_2(z) = \dfrac{X_2(z)}{2\pi i}\left[\displaystyle\int_{\partial D} \dfrac{e^{-(\bar{\lambda}\tau+\lambda\bar{\tau})}c_2(\tau)}{[a_2(\tau) + i\,b_2(\tau)]X_2^+(\tau)}\dfrac{\tau+z}{\tau-z}\mathrm{d}\tau + 2\pi i\,q_{\kappa_2}^2\right] \end{cases} \tag{12}$$

with $q_{\kappa_j}^j \in S\Pi_{\kappa_j}$ $(j = 1, 2)$. Setting

$$I_j(z) = \int_{\partial D} \frac{\Lambda_j(\tau)}{X_j^+(\tau)}\frac{\tau+z}{\tau-z}\mathrm{d}\tau \text{ with } \Lambda_j(\tau) = \frac{e^{-(\bar{\lambda}\tau+\lambda\bar{\tau})}c_j(\tau)}{a_j(\tau) + i\,b_j(\tau)}$$

for $j = 1, 2$, one has

$$\begin{cases} \varphi_1(z) = \dfrac{1}{z^2}\left\{\dfrac{(\lambda+z)X_1(z)}{2\pi i}\left[I_1(z) + 2\pi i\,q_{\kappa_1}^1(z)\right] - \dfrac{X_2(z)}{2\pi i}\left[I_2(z) + 2\pi i\,q_{\kappa_2}^2(z)\right]\right\}, \\[4mm] \varphi_2(z) = \dfrac{1}{z}\left\{\dfrac{X_2(z)}{2\pi i}\left[I_2(z) + 2\pi i\,q_{\kappa_2}^2(z)\right] - \dfrac{\lambda X_1(z)}{2\pi i}\left[I_1(z) + 2\pi i\,q_{\kappa_1}^1(z)\right]\right\}. \end{cases} \tag{13}$$

Let

$$q_{\kappa_j}^j(z) = \sum_{\ell=-\kappa_j}^{\kappa_j} a_{j,\ell}z^\ell \in S\Pi_{\kappa_j}, \quad j = 1, 2. \tag{14}$$

Then $\varphi_j(z)$ $(j = 1, 2)$ given in (2.10) is analytic in the unit disc $D$ if and only if

$$\begin{cases} a_{2,-\kappa_2} = \dfrac{\lambda}{\mu}\,a_{1,-\kappa_1}, \\[4mm] a_{2,-\kappa_2+1} = \dfrac{1 + \lambda(\nu_1 - \nu_2)}{\mu}\,a_{1,-\kappa_1} + \dfrac{\lambda}{\mu}\,a_{1,-\kappa_1+1}, \end{cases} \tag{15}$$

where

$$\mu = \frac{Y_2(0)}{Y_1(0)} = \exp\left\{\frac{1}{2\pi}\int_{\partial D}\frac{\Theta_2(\tau) - \Theta_1(\tau)}{\tau}d\tau\right\} = e^{i\phi}, \tag{16}$$

$$\nu_j = \frac{Y_j'(0)}{Y_j(0)} = \frac{1}{\pi}\int_{\partial D}\frac{\Theta_j(\tau)}{\tau^2}d\tau, \quad j = 1,2 \tag{17}$$

with

$$\phi = \frac{1}{2\pi}\int_0^{2\pi}\left[\Theta_2\left(e^{i\theta}\right) - \Theta_1\left(e^{i\theta}\right)\right]d\theta \in \mathcal{R}. \tag{18}$$

**Case 2:** $\kappa_1 = 1$, $\kappa_2 > 1$. Similarly, $\varphi_j$ $(j = 1,2)$ given by (13) is analytic in $D$ if and only if

$$\begin{cases} a_{2,-\kappa_2} = \dfrac{\lambda}{\mu}a_{1,-1}, \\[3mm] a_{2,-\kappa_2+1} = \dfrac{\lambda}{2\pi\mu i}\displaystyle\int_{\partial D}\dfrac{\Lambda_1(\tau)}{X_1^+(\tau)}d\tau + \dfrac{1+\lambda(\nu_1-\nu_2)}{\mu}a_{1,-1} + \dfrac{\lambda}{\mu}a_{1,0}. \end{cases} \tag{19}$$

**Case 3:** $\kappa_1 = 0$, $\kappa_2 > 1$. $\varphi_j$ $(j = 1,2)$ given in (13) is analytic in $D$ if and only if

$$\begin{cases} a_{2,-\kappa_2} = \dfrac{\lambda}{\mu}\left[\dfrac{1}{2\pi i}\displaystyle\int_{\partial D}\dfrac{\Lambda_1(\tau)}{X_1^+(\tau)}d\tau + a_{1,0}\right], \\[3mm] a_{2,-\kappa_2+1} = \dfrac{1+\lambda(\nu_1-\nu_2)}{\mu}\left[\dfrac{1}{2\pi i}\displaystyle\int_{\partial D}\dfrac{\Lambda_1(\tau)}{X_1^+(\tau)}d\tau + a_{1,0}\right] + \dfrac{\lambda}{\mu\pi i}\displaystyle\int_{\partial D}\dfrac{\Lambda_1(\tau)}{X_1^+(\tau)}\dfrac{d\tau}{\tau}, \end{cases} \tag{20}$$

where $a_{1,0}$ is a free real constant.

**Case 4:** $\kappa_1 < 0$, $\kappa_2 > 1$. If and only if the condition[11]

$$\int_{\partial D}\frac{\Lambda_1(\tau)}{X_1^+(\tau)}\frac{d\tau}{\tau^{k-1}} = 0, \quad k = 1,\cdots,-\kappa_1 \tag{21}$$

is satisfied, we get from (2.3)

$$\begin{cases} \varphi_1(z) = \dfrac{1}{z^2}\left\{\dfrac{(\lambda+z)Y_1(z)}{\pi i}\displaystyle\int_{\partial D}\dfrac{\Lambda_1(\tau)}{Y_1^+(\tau)}\dfrac{\tau d\tau}{\tau-z} - \dfrac{X_2(z)}{2\pi i}\left[I_2(z) + 2\pi i\, q_{\kappa_2}^2(z)\right]\right\}, \\[3mm] \varphi_2(z) = \dfrac{1}{z}\left\{\dfrac{X_2(z)}{2\pi i}\left[I_2(z) + 2\pi i\, q_{\kappa_2}^2(z)\right] - \dfrac{\lambda Y_1(z)}{\pi i}\displaystyle\int_{\partial D}\dfrac{\Lambda_1(\tau)}{Y_1^+(\tau)}\dfrac{\tau d\tau}{\tau-z}\right\}. \end{cases} \tag{22}$$

Similarly, when

$$\begin{cases} a_{2,-\kappa_2} = \dfrac{\lambda}{\pi\mu i}\displaystyle\int_{\partial D}\dfrac{\Lambda_1(\tau)}{Y_1^+(\tau)}d\tau, \\[3mm] a_{2,-\kappa_2+1} = \dfrac{1+\lambda(\nu_1-\nu_2)}{\pi\mu i}\displaystyle\int_{\partial D}\dfrac{\Lambda_1(\tau)}{Y_1^+(\tau)}d\tau + \dfrac{\lambda}{\pi\mu i}\displaystyle\int_{\partial D}\dfrac{\Lambda_1(\tau)}{Y_1^+(\tau)}\dfrac{d\tau}{\tau} \end{cases} \tag{23}$$

is fulfilled, $\varphi_j$ $(j = 1,2)$ given in (2.19) is analytic in $D$.

**Case 5:** $\kappa_1 > 1$, $\kappa_2 = 1$. When

$$
\begin{cases}
a_{2,-1} = \dfrac{\lambda}{\mu} a_{1,-\kappa_1}, \\[2mm]
a_{2,0} = \dfrac{1 + \lambda(\nu_1 - \nu_2)}{\mu} a_{1,-\kappa_1} + \dfrac{\lambda}{\mu} a_{1,-\kappa_1+1} - \dfrac{1}{2\pi i} \displaystyle\int_{\partial D} \dfrac{\Lambda_2(\tau)}{X_2^+(\tau)} d\tau \in \mathcal{R},
\end{cases}
\tag{24}
$$

$\varphi_j$ $(j = 1, 2)$ given in (2.10) is analytic in $D$.

**Case 6:** $\kappa_1 = 1$, $\kappa_2 = 1$. When

$$
\begin{cases}
a_{2,-1} = \dfrac{\lambda}{\mu} a_{1,-1}, \\[2mm]
a_{2,0} = \dfrac{1 + \lambda(\nu_1 - \nu_2)}{\mu} a_{1,-1} + \dfrac{\lambda}{\mu} \left[ a_{1,0} + \dfrac{1}{2\pi i} \displaystyle\int_{\partial D} \dfrac{\Lambda_1(\tau)}{X_1^+(\tau)} d\tau \right] - \dfrac{1}{2\pi i} \displaystyle\int_{\partial D} \dfrac{\Lambda_2(\tau)}{X_2^+(\tau)} d\tau,
\end{cases}
\tag{25}
$$

$\varphi_j$ $(j = 1, 2)$ given in (13) is analytic in $D$.

**Remark 2.1.** When $\lambda = \frac{1}{\nu_2 - \nu_1}$ $(\nu_1 \neq \nu_2)$, the second formula in (25) is equivalent to

$$
a_{2,0} = \frac{1}{\mu(\nu_2 - \nu_1)} \left[ a_{1,0} + \frac{1}{2\pi i} \int_{\partial D} \frac{\Lambda_1(\tau)}{X_1^+(\tau)} d\tau \right] - \frac{1}{2\pi i} \int_{\partial D} \frac{\Lambda_2(\tau)}{X_2^+(\tau)} d\tau.
\tag{26}
$$

Under this case, if $\text{Im}\{(\nu_2 - \nu_1)\mu\} = 0$, say $\phi + \psi \equiv 0 \pmod{\pi}$ with $\phi$ defined by (18) and

$$
\psi = \arctan \left\{ \frac{\int_0^{2\pi} [\Theta_2(e^{i\theta}) - \Theta_1(e^{i\theta})] \cos\theta d\theta}{\int_0^{2\pi} [\Theta_2(e^{i\theta}) - \Theta_1(e^{i\theta})] \sin\theta d\theta} \right\}
$$

when $\int_0^{2\pi} [\Theta_2(e^{i\theta}) - \Theta_1(e^{i\theta})] \sin\theta d\theta \neq 0$, then (26) is equivalent to

$$
\begin{cases}
a_{2,0} = \dfrac{1}{\mu(\nu_2 - \nu_1)} a_{1,0}, \\[3mm]
\dfrac{1}{\mu(\nu_2 - \nu_1)} \displaystyle\int_{\partial D} \dfrac{\Lambda_1(\tau)}{X_1^+(\tau)} d\tau - \displaystyle\int_{\partial D} \dfrac{\Lambda_2(\tau)}{X_2^+(\tau)} d\tau = 0,
\end{cases}
$$

in which the second formula is a real condition of solvability.

**Case 7:** $\kappa_1 = 0$, $\kappa_2 = 1$. One gets $\varphi_j$ $(j = 1, 2)$ given in (13) and

$$
\begin{cases}
a_{2,-1} = \dfrac{\lambda}{\mu} \left[ a_{1,0} + \dfrac{1}{2\pi i} \displaystyle\int_{\partial D} \dfrac{\Lambda_1(\tau)}{X_1^+(\tau)} d\tau \right], \\[3mm]
a_{2,0} = \dfrac{1 + \lambda(\nu_1 - \nu_2)}{\mu} \left[ a_{1,0} + \dfrac{1}{2\pi i} \displaystyle\int_{\partial D} \dfrac{\Lambda_1(\tau)}{X_1^+(\tau)} d\tau \right] \\[3mm]
\qquad + \dfrac{\lambda}{\pi \mu i} \displaystyle\int_{\partial D} \dfrac{\Lambda_1(\tau)}{X_1^+(\tau)} \dfrac{d\tau}{\tau} - \dfrac{1}{2\pi i} \displaystyle\int_{\partial D} \dfrac{\Lambda_2(\tau)}{X_2^+(\tau)} d\tau.
\end{cases}
\tag{27}
$$

1132

**Remark 2.2.** If $\mathrm{Im}\{\frac{1+\lambda(\nu_1-\nu_2)}{\mu}\} = 0$, the second formula in (27) produces a real condition of solvability

$$\mathrm{Im}\left\{\frac{\lambda}{\pi\mu i}\int_{\partial D}\frac{\Lambda_1(\tau)}{X_1^+(\tau)}\frac{\mathrm{d}\tau}{\tau}\right\} - \frac{1+\lambda(\nu_1-\nu_2)}{2\pi\mu}\int_{\partial D}\frac{\Lambda_1(\tau)}{X_1^+(\tau)}\mathrm{d}\tau + \frac{1}{2\pi}\int_{\partial D}\frac{\Lambda_2(\tau)}{X_2^+(\tau)}\mathrm{d}\tau = 0.$$

**Case 8:** $\kappa_1 < 0$, $\kappa_2 = 1$. When the condition of solvability (21) is satisfied, one has $\varphi_j$ $(j = 1, 2)$ given in (22) and

$$\begin{cases} a_{2,-1} = \dfrac{\lambda}{\pi\mu i}\displaystyle\int_{\partial D}\frac{\Lambda_1(\tau)}{Y_1^+(\tau)}\mathrm{d}\tau, \\ a_{2,0} = \dfrac{1+\lambda(\nu_1-\nu_2)}{\pi\mu i}\displaystyle\int_{\partial D}\frac{\Lambda_1(\tau)}{Y_1^+(\tau)}\mathrm{d}\tau + \dfrac{\lambda}{\pi\mu i}\displaystyle\int_{\partial D}\frac{\Lambda_1(\tau)}{Y_1^+(\tau)}\frac{\mathrm{d}\tau}{\tau} - \dfrac{1}{2\pi i}\displaystyle\int_{\partial D}\frac{\Lambda_2(\tau)}{X_2^+(\tau)}\mathrm{d}\tau. \end{cases} \tag{28}$$

**Remark 2.3.** The second formula in (28) gives a real condition of solvability

$$\mathrm{Im}\left\{\frac{1+\lambda(\nu_1-\nu_2)}{\pi\mu i}\int_{\partial D}\frac{\Lambda_1(\tau)}{Y_1^+(\tau)}\mathrm{d}\tau + \frac{\lambda}{\pi\mu i}\int_{\partial D}\frac{\Lambda_1(\tau)}{Y_1^+(\tau)}\frac{\mathrm{d}\tau}{\tau}\right\} + \frac{1}{2\pi}\int_{\partial D}\frac{\Lambda_2(\tau)}{X_2^+(\tau)}\mathrm{d}\tau = 0.$$

**Case 9:** $\kappa_1 > 1$, $\kappa_2 = 0$. One gets $\varphi_j$ $(j = 1, 2)$ given in (13) with

$$\begin{cases} a_{2,0} = \dfrac{\lambda}{\mu}a_{1,-\kappa_1} - \dfrac{1}{2\pi i}\displaystyle\int_{\partial D}\frac{\Lambda_2(\tau)}{X_2^+(\tau)}\mathrm{d}\tau \in \mathcal{R}, \\ \dfrac{1}{\pi i}\displaystyle\int_{\partial D}\frac{\Lambda_2(\tau)}{X_2^+(\tau)}\frac{\mathrm{d}\tau}{\tau} = \dfrac{1+\lambda(\nu_1-\nu_2)}{\mu}a_{1,-\kappa_1} + \dfrac{\lambda}{\mu}a_{1,-\kappa_1+1}. \end{cases} \tag{29}$$

**Case 10:** $\kappa_1 = 1$, $\kappa_2 = 0$. One has $\varphi_j$ $(j = 1, 2)$ given in (13) with

$$\begin{cases} a_{2,0} = \dfrac{\lambda}{\mu}a_{1,-1} - \dfrac{1}{2\pi i}\displaystyle\int_{\partial D}\frac{\Lambda_2(\tau)}{Y_2^+(\tau)}\mathrm{d}\tau, \\ \dfrac{1}{\pi i}\displaystyle\int_{\partial D}\frac{\Lambda_2(\tau)}{Y_2^+(\tau)}\frac{\mathrm{d}\tau}{\tau} = \dfrac{1+\lambda(\nu_1-\nu_2)}{\mu}a_{1,-1} + \dfrac{\lambda}{\mu}\left[a_{1,0} + \dfrac{1}{2\pi i}\displaystyle\int_{\partial D}\frac{\Lambda_2(\tau)}{Y_2^+(\tau)}\mathrm{d}\tau\right]. \end{cases} \tag{30}$$

**Remark 2.4.** If $\lambda = \frac{1}{\nu_2-\nu_1}$ $(\nu_1 \neq \nu_2)$, then the second formula in (30) also produces a real condition of solvability

$$\mathrm{Im}\left\{\frac{\mu}{\lambda\pi i}\int_{\partial D}\frac{\Lambda_2(\tau)}{Y_2^+(\tau)}\frac{\mathrm{d}\tau}{\tau}\right\} + \frac{1}{2\pi}\int_{\partial D}\frac{\Lambda_2(\tau)}{Y_2^+(\tau)}\mathrm{d}\tau = 0.$$

**Case 11:** $\kappa_1 = 0$, $\kappa_2 = 0$. When

$$\begin{cases} \dfrac{\lambda}{\mu}\left[\dfrac{1}{2\pi i}\displaystyle\int_{\partial D}\frac{\Lambda_1(\tau)}{Y_1^+(\tau)}\mathrm{d}\tau + a_{1,0}\right] = \dfrac{1}{2\pi i}\displaystyle\int_{\partial D}\frac{\Lambda_2(\tau)}{Y_2^+(\tau)}\mathrm{d}\tau + a_{2,0}, \\ \dfrac{1}{\pi i}\displaystyle\int_{\partial D}\frac{\Lambda_2(\tau)}{Y_2^+(\tau)}\frac{\mathrm{d}\tau}{\tau} = \dfrac{1+\lambda(\nu_1-\nu_2)}{\mu}\left[\dfrac{1}{2\pi i}\displaystyle\int_{\partial D}\frac{\Lambda_1(\tau)}{Y_1^+(\tau)}\mathrm{d}\tau + a_{1,0}\right] \\ \qquad\qquad + \dfrac{\lambda}{\mu\pi i}\displaystyle\int_{\partial D}\frac{\Lambda_1(\tau)}{Y_1^+(\tau)}\frac{\mathrm{d}\tau}{\tau} \end{cases} \tag{31}$$

is satisfied, $\varphi_j$ $(j = 1, 2)$ given in (13) is analytic.

**Remark 2.5.** When $1 + \lambda(\nu_1 - \nu_2) \neq 0$, (31) produces two real conditions

$$
\begin{cases}
\operatorname{Im}\left\{\dfrac{\mu}{[1 + \lambda(\nu_1 - \nu_2)]\pi i}\left[\displaystyle\int_{\partial D}\dfrac{\Lambda_2(\tau)}{Y_2^+(\tau)}\dfrac{\mathrm{d}\tau}{\tau} - \dfrac{\lambda}{\mu}\int_{\partial D}\dfrac{\Lambda_1(\tau)}{Y_1^+(\tau)}\dfrac{\mathrm{d}\tau}{\tau}\right]\right\} + \dfrac{1}{2\pi}\displaystyle\int_{\partial D}\dfrac{\Lambda_1(\tau)}{Y_1^+(\tau)}\mathrm{d}\tau \\
= 0, \\
\operatorname{Im}\left\{\dfrac{\lambda}{\mu}\left[\dfrac{1}{2\pi i}\displaystyle\int_{\partial D}\dfrac{\Lambda_1(\tau)}{Y_1^+(\tau)}\mathrm{d}\tau + a_{1,0}\right]\right\} + \dfrac{1}{2\pi}\displaystyle\int_{\partial D}\dfrac{\Lambda_2(\tau)}{Y_2^+(\tau)}\mathrm{d}\tau = 0
\end{cases}
$$

with

$$
a_{1,0} = \operatorname{Re}\left\{\dfrac{\mu}{[1 + \lambda(\nu_1 - \nu_2)]\pi i}\left[\int_{\partial D}\dfrac{\Lambda_2(\tau)}{Y_2^+(\tau)}\dfrac{\mathrm{d}\tau}{\tau} - \dfrac{\lambda}{\mu}\int_{\partial D}\dfrac{\Lambda_1(\tau)}{Y_1^+(\tau)}\dfrac{\mathrm{d}\tau}{\tau}\right]\right\}.
$$

When $\lambda = \frac{1}{\nu_2 - \nu_1}$ $(\nu_1 \neq \nu_2)$, the second formula in (31) is the complex condition

$$
\int_{\partial D}\dfrac{\Lambda_2(\tau)}{Y_2^+(\tau)}\dfrac{\mathrm{d}\tau}{\tau} = \dfrac{1}{\mu(\nu_2 - \nu_1)}\int_{\partial D}\dfrac{\Lambda_1(\tau)}{Y_1^+(\tau)}\dfrac{\mathrm{d}\tau}{\tau}.
$$

Under this case, if $\operatorname{Im}(\lambda/\mu) = 0$, the first formula in (31) also leads to another real condition of solvability

$$
\dfrac{\lambda}{\mu}\int_{\partial D}\dfrac{\Lambda_1(\tau)}{Y_1^+(\tau)}\mathrm{d}\tau - \int_{\partial D}\dfrac{\Lambda_2(\tau)}{Y_2^+(\tau)}\mathrm{d}\tau = 0;
$$

if $\operatorname{Im}(\lambda/\mu) \neq 0$, the first formula in (31) does not give a real condition.

**Case 12:** $\kappa_1 < 0$, $\kappa_2 = 0$. If and only if the condition of solvability (21) is satisfied, one has $\varphi_j$ $(j = 1, 2)$ given in (22) with

$$
\begin{cases}
a_{2,0} = \dfrac{\lambda}{\mu\pi i}\displaystyle\int_{\partial D}\dfrac{\Lambda_1(\tau)}{Y_1^+(\tau)}\mathrm{d}\tau - \dfrac{1}{2\pi i}\displaystyle\int_{\partial D}\dfrac{\Lambda_2(\tau)}{X_2^+(\tau)}\mathrm{d}\tau, \\
\displaystyle\int_{\partial D}\dfrac{\Lambda_2(\tau)}{X_2^+(\tau)}\dfrac{\mathrm{d}\tau}{\tau} = \dfrac{1 + \lambda(\nu_1 - \nu_2)}{\mu}\displaystyle\int_{\partial D}\dfrac{\Lambda_1(\tau)}{Y_1^+(\tau)}\mathrm{d}\tau + \dfrac{\lambda}{\mu}\displaystyle\int_{\partial D}\dfrac{\Lambda_1(\tau)}{Y_1^+(\tau)}\dfrac{\mathrm{d}\tau}{\tau}.
\end{cases}
\tag{32}
$$

**Remark 2.6.** The first formula in (32) produces a real condition of solvability

$$
\operatorname{Im}\left\{\dfrac{\lambda}{\mu\pi i}\int_{\partial D}\dfrac{\Lambda_1(\tau)}{Y_1^+(\tau)}\mathrm{d}\tau\right\} + \dfrac{1}{2\pi}\int_{\partial D}\dfrac{\Lambda_2(\tau)}{X_2^+(\tau)}\mathrm{d}\tau = 0
$$

and the second formula in (32) is a complex condition of solvability.

**Case 13:** $\kappa_1 > 1$, $\kappa_2 < 0$. If and only if the condition[11]

$$
\int_{\partial D}\dfrac{\Lambda_2(\tau)}{X_2^+(\tau)}\dfrac{\mathrm{d}\tau}{\tau^{k-1}} = 0, \quad k = 1, \cdots, -\kappa_2,
\tag{33}
$$

is satisfied, we get from (6)

$$
\begin{cases}
\varphi_1(z) = \dfrac{1}{z^2}\left\{\dfrac{(\lambda + z)X_1(z)}{2\pi i}\left[I_1(z) + 2\pi i\, q_{\kappa_1}^1(z)\right] - \dfrac{Y_2(z)}{\pi i}\displaystyle\int_{\partial D}\dfrac{\Lambda_2(\tau)}{Y_2^+(\tau)}\dfrac{\tau\mathrm{d}\tau}{\tau - z}\right\}, \\
\varphi_2(z) = \dfrac{1}{z}\left\{\dfrac{Y_2(z)}{\pi i}\displaystyle\int_{\partial D}\dfrac{\Lambda_2(\tau)}{Y_2^+(\tau)}\dfrac{\tau\mathrm{d}\tau}{\tau - z} - \dfrac{\lambda X_1(z)}{2\pi i}\left[I_1(z) + 2\pi i\, q_{\kappa_1}^1(z)\right]\right\}.
\end{cases}
\tag{34}
$$

1134

Clearly, when

$$
\begin{cases}
a_{1,-\kappa_1} = \dfrac{\mu}{\pi\,\lambda\,i} \displaystyle\int_{\partial D} \dfrac{\Lambda_2(\tau)}{Y_2^+(\tau)}\mathrm{d}\tau, \\[3mm]
a_{1,-\kappa_1+1} = \dfrac{\mu}{\pi\,\lambda\,i} \displaystyle\int_{\partial D} \dfrac{\Lambda_2(\tau)}{Y_2^+(\tau)}\dfrac{\mathrm{d}\tau}{\tau} - \dfrac{[1+\lambda(\nu_1-\nu_2)]\mu}{\pi\,\lambda^2\,i}\displaystyle\int_{\partial D}\dfrac{\Lambda_2(\tau)}{Y_2^+(\tau)}\mathrm{d}\tau,
\end{cases}
\tag{35}
$$

$\varphi_j(z)\,(j=1,2)$ given in (34) is analytic in $D$.

**Case 14:** $\kappa_1 = 1$, $\kappa_2 < 0$. If and only if (33) is fulfilled, one gets $\varphi_j(z)\,(j=1,2)$ given in (34) with

$$
\begin{cases}
a_{1,-1} = \dfrac{\mu}{\pi\,\lambda\,i}\displaystyle\int_{\partial D}\dfrac{\Lambda_2(\tau)}{Y_2^+(\tau)}\mathrm{d}\tau, \\[3mm]
a_{1,0} = \dfrac{\mu}{\pi\,\lambda\,i}\displaystyle\int_{\partial D}\dfrac{\Lambda_2(\tau)}{Y_2^+(\tau)}\dfrac{\mathrm{d}\tau}{\tau} - \dfrac{[1+\lambda(\nu_1-\nu_2)]\mu}{\pi\,\lambda^2\,i}\displaystyle\int_{\partial D}\dfrac{\Lambda_2(\tau)}{Y_2^+(\tau)}\mathrm{d}\tau \\[3mm]
\qquad - \dfrac{1}{2\pi i}\displaystyle\int_{\partial D}\dfrac{\Lambda_1(\tau)}{X_1^+(\tau)}\mathrm{d}\tau.
\end{cases}
\tag{36}
$$

**Remark 2.7.** The second formula in (36) produces a real condition of solvability

$$
\mathrm{Im}\left\{\dfrac{\mu}{\pi\,\lambda\,i}\int_{\partial D}\dfrac{\Lambda_2(\tau)}{Y_2^+(\tau)}\dfrac{\mathrm{d}\tau}{\tau} - \dfrac{[1+\lambda(\nu_1-\nu_2)]\mu}{\pi\,\lambda^2\,i}\int_{\partial D}\dfrac{\Lambda_2(\tau)}{Y_2^+(\tau)}\mathrm{d}\tau\right\} + \dfrac{1}{2\pi}\int_{\partial D}\dfrac{\Lambda_1(\tau)}{X_1^+(\tau)}\mathrm{d}\tau = 0.
$$

**Case 15:** $\kappa_1 = 0$, $\kappa_2 < 0$. If and only if condition (33) is fulfilled, one gets $\varphi_j(z)\,(j=1,2)$ given in (34) with

$$
\begin{cases}
a_{1,0} = \dfrac{\mu}{\pi\,\lambda\,i}\displaystyle\int_{\partial D}\dfrac{\Lambda_2(\tau)}{Y_2^+(\tau)}\mathrm{d}\tau - \dfrac{1}{2\pi i}\displaystyle\int_{\partial D}\dfrac{\Lambda_1(\tau)}{Y_1^+(\tau)}\mathrm{d}\tau, \\[3mm]
\dfrac{1+\lambda(\nu_1-\nu_2)}{2\pi\,\mu\,i}\left[\displaystyle\int_{\partial D}\dfrac{\Lambda_1(\tau)}{Y_1^+(\tau)}\mathrm{d}\tau + 2\pi i\,a_{1,0}\right] = \dfrac{1}{\pi i}\displaystyle\int_{\partial D}\dfrac{\Lambda_2(\tau)}{Y_2^+(\tau)}\dfrac{\mathrm{d}\tau}{\tau} \\[3mm]
\qquad - \dfrac{\lambda}{\pi\,\mu\,i}\displaystyle\int_{\partial D}\dfrac{\Lambda_1(\tau)}{X_1^+(\tau)}\dfrac{\mathrm{d}\tau}{\tau}.
\end{cases}
\tag{37}
$$

**Remark 2.8.** It is easy to see that (37) also produces three real conditions of solvability.

**Case 16:** $\kappa_1 < 0$, $\kappa_2 < 0$. If and only if the conditions of solvability (21) and (34) are satisfied, we have

$$
\begin{cases}
\varphi_1(z) = \dfrac{1}{z^2}\left\{\dfrac{(\lambda+z)Y_1(z)}{\pi i}\displaystyle\int_{\partial D}\dfrac{\Lambda_1(\tau)}{Y_1^+(\tau)}\dfrac{\tau\mathrm{d}\tau}{\tau-z} - \dfrac{Y_2(z)}{\pi i}\displaystyle\int_{\partial D}\dfrac{\Lambda_2(\tau)}{Y_2^+(\tau)}\dfrac{\tau\mathrm{d}\tau}{\tau-z}\right\}, \\[3mm]
\varphi_2(z) = \dfrac{1}{z}\left\{\dfrac{Y_2(z)}{\pi i}\displaystyle\int_{\partial D}\dfrac{\Lambda_2(\tau)}{Y_2^+(\tau)}\dfrac{\tau\mathrm{d}\tau}{\tau-z} - \dfrac{\lambda Y_1(z)}{\pi i}\displaystyle\int_{\partial D}\dfrac{\Lambda_1(\tau)}{Y_1^+(\tau)}\dfrac{\tau\mathrm{d}\tau}{\tau-z}\right\}
\end{cases}
\tag{38}
$$

and

$$\begin{cases} \dfrac{\lambda}{\mu} \displaystyle\int_{\partial D} \dfrac{\Lambda_1(\tau)}{Y_1^+(\tau)}\mathrm{d}\tau = \int_{\partial D} \dfrac{\Lambda_2(\tau)}{Y_2^+(\tau)}\mathrm{d}\tau, \\[4mm] \dfrac{1+\lambda(\nu_1-\nu_2)}{\mu} \displaystyle\int_{\partial D} \dfrac{\Lambda_1(\tau)}{Y_1^+(\tau)}\mathrm{d}\tau + \dfrac{\lambda}{\mu}\int_{\partial D} \dfrac{\Lambda_1(\tau)}{Y_1^+(\tau)}\dfrac{\mathrm{d}\tau}{\tau} = \int_{\partial D} \dfrac{\Lambda_2(\tau)}{Y_2^+(\tau)}\dfrac{\mathrm{d}\tau}{\tau}. \end{cases} \tag{39}$$

Clearly, (39) is also the condition of solvability compatible with (21) and (33).
To sum up the above discussion, we obtain the following result.

**Theorem 2.1.** *For MH problem (4), sixteen cases arise.*

*(1) When $\kappa_1 > 1$ and $\kappa_2 > 1$, MH problem (4) is solvable and its solution may be written as (5), where $\varphi_j$ ($j = 1, 2$) is given in (13) with the coefficients of the terms in $q_{\kappa_j}^j \in S\Pi_{\kappa_j}$ ($j = 1, 2$) satisfying the relation (15).*

*(2) When $\kappa_1 = 1$ and $\kappa_2 > 1$, MH problem (4) is solvable and its solution may be written as (5), where $\varphi_j$ ($j = 1, 2$) is given in (13) with the coefficients of the terms in $q_{\kappa_j}^j$ ($j = 1, 2$) satisfying the relation (19).*

*(3) When $\kappa_1 = 0$ and $\kappa_2 > 1$, MH problem (4) is solvable and its solution may be written as (5), where $\varphi_j$ ($j = 1, 2$) is given in (13) with $a_{1,0} \in \mathbb{R}$ and two coefficients of the terms in $q_{\kappa_2}^2$ satisfying the relation (20).*

*(4) When $\kappa_1 < 0$ and $\kappa_2 > 1$, MH problem (4) is solvable if and only if the condition (21) is satisfied, and its solution ay be written as (5), where $\varphi_j$ ($j = 1, 2$) is given in (22) ith two of the coefficients of the terms in $q_{\kappa_2}^2$ given by (23).*

*(5) When $\kappa_1 > 1$ and $\kappa_2 = 1$, MH problem (4) is solvable and its solution may be written as (5), where $\varphi_j$ ($j = 1, 2$) is given in (13) with the coefficients of the terms in $q_{\kappa_j}^j$ ($j = 1, 2$) satisfying (24).*

*(6) When $\kappa_1 = 1$ and $\kappa_2 = 1$, MH problem (4) is solvable and its solution may be written as (5), where $\varphi_j$ ($j = 1, 2$) is given in (13) with the relation (25).*

*(7) When $\kappa_1 = 0$ and $\kappa_2 = 1$, MH problem (4) is solvable and its solution may be written as (5), where $\varphi_j$ ($j = 1, 2$) is given in (13) with the relation (27).*

*(8) When $\kappa_1 = 0$ and $\kappa_2 = 1$, MH problem (4) is solvable if and only if the condition of solvability (21) is satisfied, and its solution may be written as (5), where $\varphi_j$ ($j = 1, 2$) is given in (22) with the relation (28).*

*(9) When $\kappa_1 > 1$ and $\kappa_2 = 0$, MH problem (4) is solvable and its solution may be written as (5), where $\varphi_j$ ($j = 1, 2$) is given in (13) with $a_{2,0} \in \mathbb{R}$ and two coefficients of the terms in $q_{\kappa_1}^1$ satisfying the relation (29).*

*(10) When $\kappa_1 = 1$ and $\kappa_2 = 0$, MH problem (4) is solvable and its solution may be written as (5), where $\varphi_j$ ($j = 1, 2$) is given in (13) with the relation (30).*

*(11) When $\kappa_1 = 0$ and $\kappa_2 = 0$, MH problem (4) is solvable and its solution may be written as (5), where $\varphi_j$ ($j = 1, 2$) is given in (13) with the relation (31).*

*(12) When $\kappa_1 < 0$ and $\kappa_2 = 0$, MH problem (4) is solvable if and only if the condition (21) is satisfied, and its solution may be written as (5), where $\varphi_j$ ($j = 1, 2$) is given in (13) with the relation (32).*

*(13)* When $\kappa_1 > 1$ and $\kappa_2 < 0$, MH problem (4) is solvable if and only if the condition (33) is satisfied, and its solution may be written as (5), where $\varphi_j$ $(j = 1, 2)$ is given in (34) with two coefficients of the terms in $q^1_{\kappa_1}$ given in (35).

*(14)* When $\kappa_1 = 1$ and $\kappa_2 < 0$, MH problem (4) is solvable if and only if the condition (33) is satisfied, and its solution may be written as (5), where $\varphi_j$ $(j = 1, 2)$ is given in (34) with the relation (36).

*(15)* When $\kappa_1 = 1$ and $\kappa_2 < 0$, MH problem (4) is solvable if and only if the condition (33) is satisfied, and its solution may be written as (5), where $\varphi_j$ $(j = 1, 2)$ is given in (34) with the relation (38).

*(16)* When $\kappa_1 < 0$ and $\kappa_2 < 0$, MH problem (4) is solvable if and only if the conditions (21) (33) and (39) are satisfied, and its solution may be written as (5) with $\varphi_j$ $(j = 1, 2)$ given in (38).

## References

1. M. B. Balk, *Polyanalytic Functions*, Akademie Verlag, Berlin, 1991.
2. Jianke Lu, *Boundary Value Problems for Analytic Functions*, World Scientific Publication, Singapore, 1993.
3. N. I. Muskhelishvili, *Singular Integral Equations*, 2nd ed., Noordhoff, Groningen, 1968.
4. A. S. Mshimba, On Hilbert boundary value problem for holomorphic function in the Sobolev spaces, *Appl. Anal.*, 30(1988), 87-99.
5. H. Begehr and G. N. Hile, A hierarchy of integral operators, *Rochy Mountains J. Math.*, 27(1997), 669-706.
6. A. S. Mshimba, The generalized Riemann-Hilbert boundary value problem for nonhomogeneous polyanalytic differential equation of order $n$ in the Sobolev space $W_{n,p}(D)$, *ZAA*, 18(3)(1999), 611-624.
7. Jinyuan Du and Yufeng Wang, On boundary value problem of polyanalytic function on the real axis, *Complex Variables, Theory Appl.*, 48(6)(2003), 527-542.
8. Yufeng Wang and Jinyuan Du, On Riemann boundary value problem of polyanalytic function on the real axis, *Acta. Math. Sci.*, 24B(4)(2004), 663-671.
9. A. S. Mshimba, A mixed boundary value problem for polyanalytic function of order $n$ in the Sobolev space $W_{n,p}(D)$, *Complex Variables, Theory Appl.*, 47(12)(2002), 1107-1114.
10. B. F. Fatulaev, The main Haseman type boundary value problem for metaanalytic function in the case of circular domains, *Mathematical Modelling and Analysis*, 6(1)(2001), 68-76.
11. Yufeng Wang and Jinyuan Du, Hilbert boundary value problem of polyanalytic function on the unit circumference, *Complex Variables, Elliptic Eqs.* 51 (2006), 923–943.
12. Jinyuan Du and Yufeng Wang, Riemann boundary value problem of polyanalytic function and metaanalytic function on the closed curve, *Complex Variables, Theory Appl.*, 50(7-11)(2005), 521-533.

# FOUR BOUNDARY VALUE PROBLEMS FOR THE
# CAUCHY-RIEMANN EQUATION IN A QUARTER PLANE

S.A. ABDYMANAPOV[1], H. BEGEHR[2], G. HARUTYUNYAN[3], A.B. TUNGATAROV[1]

[1] *Gumilyev Eurasian National University*
*5 Munaitpassov Str.*
*Astana, Kazakhstan 010008*
*E-mail:tun-mat@list.ru*

[2] *Freie Universität Berlin*
*Institut für Mathematik*
*Arnimallee 3, 14195 Berlin, Germany*
*E-mail: begehr@math.fu-berlin.de*

[3] *Yerevan State University*
*E-mail: gohar@ysu.am*

The Schwarz, Dirichlet, Neumann and a Robin boundary value problems are investigated in the upper right quater plane of the complex plane for the inhomogeneous Cauchy-Riemann equation.

**Key words:** Schwarz, Dirichlet, Neumann, Robin boundary value problem, inhomogeneous Cauchy-Riemann equation, quarter plane
**Mathematics Subject Classification:** 30E25, 35C15, 35F15

## 1. Introduction

Basic for the following considerations is the complex form of the Gauss divergence theorem in two real variables [2]

$$\frac{1}{2\pi i} \int_{\partial \mathbb{Q}_{1R}} w(z)dz = \frac{1}{\pi} \int_{\mathbb{Q}_{1R}} w_{\bar{z}}(z)dxdy \tag{1}$$

for functions $w \in C^1(\mathbb{Q}_{1R}; \mathbb{C}) \cap C(\overline{\mathbb{Q}_{1R}}; \mathbb{C})$ where $\mathbb{Q}_{1R}$ is the special domain

$$\mathbb{Q}_{1R} = \{z = x + iy, |z| < R, \ 0 < x, 0 < y\}.$$

These function are representable by the Cauchy-Pompeiu formula

$$w(z) = \frac{1}{2\pi i} \int_{\partial \mathbb{Q}_{1R}} w(\zeta)\frac{d\zeta}{\zeta - z} - \frac{1}{\pi} \int_{\mathbb{Q}_{1R}} w_{\bar{\zeta}}(\zeta)\frac{d\xi d\eta}{\zeta - z}. \tag{2}$$

By a limiting process a Cauchy-Pompeiu representation follows for proper functions in the right upper quarter plane

$$\mathbb{Q}_1 = \{z \in \mathbb{C} : \ 0 < \operatorname{Re} z, 0 < \operatorname{Im} z\}.$$

**Theorem 1** *Any* $w \in C^1(\mathbb{Q}_1; \mathbb{C}) \cap C(\bar{\mathbb{Q}}_1; \mathbb{C})$ *for which for some* $0 < \delta$ *the function* $(1 + r)^\delta M(r, w)$ *with*

$$M(r, w) = \max\{|w(z)| : |z| = r, 0 \le \text{ Re } z, 0 \le \text{ Im } z\}$$

*is bounded for* $1 \le r$ *and* $w_{\bar{z}} \in L_1(\mathbb{Q}_1; \mathbb{C})$ *is representable as*

$$w(z) = \frac{1}{2\pi i} \int\limits_0^{+\infty} w(t) \frac{dt}{t - z} - \frac{1}{2\pi i} \int\limits_0^{+\infty} w(it) \frac{dt}{t + iz} - \frac{1}{\pi} \int\limits_{\mathbb{Q}_1} w_{\bar{\zeta}}(\zeta) \frac{d\xi d\eta}{\zeta - z}. \tag{3}$$

As in the case of the unit disc [4,6,7] and for the upper half plane [10] the basic boundary value problems can be studied by modifying (3). Cauchy-Pompeiu representations are also available for other domains [11]. They can also be used for solving boundary value problems. Higher order equations are studied in [5,9]. For another modification of the Cauchy-Pompeiu formula see [3].

## 2. Schwarz problem

Besides (2) the Cauchy-Pompeiu formula covers a second part stating for $z \notin \overline{\mathbb{Q}_{1R}}$

$$\frac{1}{2\pi i} \int\limits_{\partial \mathbb{Q}_{1R}} w(\zeta) \frac{d\zeta}{\zeta - z} - \frac{1}{\pi} \int\limits_{\mathbb{Q}_{1R}} w_{\bar{\zeta}}(\zeta) \frac{d\xi d\eta}{\zeta - z} = 0 \tag{2'}$$

resulting in

$$\frac{1}{2\pi i} \int\limits_{\partial \mathbb{Q}_1} w(\zeta) \frac{d\zeta}{\zeta - z} - \frac{1}{\pi} \int\limits_{\mathbb{Q}_1} w_{\bar{\zeta}}(\zeta) \frac{d\xi d\eta}{\zeta - z} = 0 \tag{3'}$$

for $z \in \mathbb{C} \backslash \bar{\mathbb{Q}}_1$. Taking (3') evaluated for the three symmetric points $\bar{z}, -z, -\bar{z}$ of $z \in \mathbb{Q}_1$ and combining them properly with (3), see [1] the following result is attained.
**Theorem 2** *Any* $w \in C^1(\mathbb{Q}_1; \mathbb{C}) \cap C(\bar{\mathbb{Q}}_1; \mathbb{C})$ *for which for some* $0 < \delta$ *the function* $(1 + r)^\delta M(r, w)$ *is bounded and* $w_{\bar{\zeta}} \in L_1(\mathbb{Q}_1; \mathbb{C})$ *is representable as*

$$w(z) = \frac{2}{\pi i} \int\limits_0^{+\infty} \text{Re } w(t) \frac{z}{t^2 - z^2} dt - \frac{2}{\pi i} \int\limits_0^{+\infty} \text{Im } w(it) \frac{z}{t^2 + z^2} dt$$

$$- \frac{2}{\pi} \int\limits_{\mathbb{Q}_1} \left[ w_{\bar{\zeta}}(\zeta) \frac{z}{\zeta^2 - z^2} - \overline{w_{\bar{\zeta}}(\zeta)} \frac{z}{\bar{\zeta}^2 - z^2} \right] d\xi d\eta. \tag{4}$$

This formula provides the solution to the Schwarz problem.
**Schwarz problem** Let $f \in L_1(\mathbb{Q}_1; \mathbb{C}), \gamma_1, \gamma_2 \in C(\mathbb{R}^+; \mathbb{R})$ be bounded on $\mathbb{R}^+ = [0, +\infty)$. Find a solution of

$$w_{\bar{z}} = f \text{ in } \mathbb{Q}_1$$

satisfying

$$\text{Re } w = \gamma_1 \text{ on } 0 < x, y = 0,$$

$$\text{Im} w = \gamma_2 \text{ on } 0 < y, x = 0.$$

**Theorem 3** *The Schwarz problem is uniquely weakly solvable. The solution is*

$$w(z) = \frac{2}{\pi} \int\limits_0^{+\infty} \gamma_1(t) \frac{z}{t^2 - z^2} dt - \frac{2}{\pi i} \int\limits_0^{+\infty} \gamma_2(t) \frac{z}{t^2 + z^2} dt$$

$$- \frac{2}{\pi} \int\limits_{\mathbb{Q}_1} \left[ \frac{z f(\zeta)}{\zeta^2 - z^2} - \frac{z \overline{f(\zeta)}}{\bar{\zeta}^2 - z^2} \right] d\xi d\eta. \tag{5}$$

For verifying (5) to satisfy the boundary conditions the Poisson kernel

$$P(z,t) = \frac{1}{\pi} \frac{\text{Im } z}{|t - z|^2}, \ 0 < \text{Im } z, \ t \in \mathbb{R},$$

for the upper half plane [9] is used. Similarly

$$P(z,it) = \frac{1}{\pi} \frac{\text{Re } z}{|it - z|^2}, \ 0 < \text{Re } z, \ t \in \mathbb{R},$$

serves as the Poisson kernel for the right half plane.

**Remark** $\lim_{z \to 0} \text{Re} w(z)$ exists if and only if $\gamma_1$ is continuous in 0 and $\gamma_1(0) = 0$. Then $\lim_{z \to 0} \text{Re} w(z) = 0$. Similarly $\lim_{z \to 0} \text{Im} w(z) = 0$ if and only if $\gamma_2$ is continuous in 0 and $\gamma_2(0) = 0$. Then $\lim_{z \to 0} w(z) = 0$. Otherwise, if $\gamma_1, \gamma_2 \in C([0, +\infty); \mathbb{R})$ then $\lim_{z \to 0} \text{Re} w(z) = 1/2\gamma_1(0), \lim_{z \to 0} \text{Im} w(z) = 1/2\gamma_2(0)$.

## 3. Dirichlet problem

The Cauchy-Pompeiu representation formula (3) is a candidate for the solution to the Dirichlet problem. However this problem is known to be overdetermined for the Cauchy-Riemann equation. It is overdetermined as the Schwarz data already determine the solution to the (inhomogeneous) Cauchy-Riemann equation. The solvability conditions consist of (3′) applied for $\bar{z}, -z, -\bar{z}$ where $z \in \mathbb{Q}_1$.

**Dirichlet problem** Let $f \in L_1(\mathbb{Q}_1; \mathbb{C}), \gamma_1, \gamma_2 \in C([0, +\infty); \mathbb{C})$ such that $(1 + t)^\delta \gamma_1(t), (1 + t)^\delta \gamma_2(t)$ are bounded for some $0 < \delta$ and satisfying the compatibility condition $\gamma_1(0) = \gamma_2(0) = 0$. Find a function $w$ satisfying

$$w_{\bar{z}} = f \text{ in } \mathbb{Q}_1, \ w = \gamma_1 \text{ for } 0 \le x, y = 0, \ w = \gamma_2 \text{ for } x = 0, \ 0 \le y.$$

**Theorem 4** *This Dirichlet problem is uniquely solvable in the weak sense if and only if for any $z \notin \bar{\mathbb{Q}}_1$*

$$\frac{1}{2\pi i} \int\limits_0^{+\infty} \gamma_1(t) \frac{dt}{t - z} - \frac{1}{2\pi i} \int\limits_0^{+\infty} \gamma_2(t) \frac{dt}{t + iz} - \frac{1}{\pi} \int\limits_{\mathbb{Q}_1} f(\zeta) \frac{d\xi d\eta}{\zeta - z} = 0. \tag{6}$$

*This unique solution is*

$$w(z) = \frac{1}{2\pi i} \int\limits_0^{+\infty} \gamma_1(t)\frac{dt}{t-z} - \frac{1}{2\pi i} \int\limits_0^{+\infty} \gamma_2(t)\frac{dt}{t+iz} - \frac{1}{\pi} \int\limits_{Q_1} f(\zeta)\frac{d\xi d\eta}{\zeta - z}. \qquad (7)$$

For verifying (7) to satisfy the boundary conditions, (6) applied to $\bar{z}, -z, -\bar{z}$ for $z \in Q_1$ is combined with (7) in proper ways, see [8], leading to

$$w(z) = \frac{2}{\pi} \int\limits_0^{+\infty} \gamma_1(t)\frac{y}{|t-z|^2}\frac{t^2+|z|^2}{|t+z|^2}dt + \frac{2i}{\pi} \int\limits_0^{+\infty} \gamma_2(t)\frac{y}{|t-iz|^2}\frac{t^2-|z|^2}{|t-iz|^2}dt$$

$$- \frac{2}{\pi} \int\limits_{Q_1} f(\zeta)\left(\frac{z}{\zeta^2-z^2} - \frac{\bar{z}}{\zeta^2-\bar{z}^2}\right)d\xi d\eta. \qquad (8)$$

and

$$w(z) = \frac{2}{\pi i} \int\limits_0^{+\infty} \gamma_1(t)\frac{x}{|t-z|^2}\frac{t^2-|z|^2}{|t+z|^2}dt + \frac{2}{\pi i} \int\limits_0^{+\infty} \gamma_2(t)\frac{x}{|t+iz|^2}\frac{t^2+|z|^2}{|t-iz|^2}dt$$

$$- \frac{2}{\pi} \int\limits_{Q_1} f(\zeta)\left(\frac{z}{\zeta^2-z^2} + \frac{\bar{z}}{\zeta^2-\bar{z}^2}\right)d\xi d\eta. \qquad (9)$$

Using again the properties of the Poisson kernels the boundary conditions are seen to be satisfied. The Dirichlet problem serves to solve the Neumann boundary value problem.

## 4. Neumann problem

The Neumann boundary condition consists in prescribing the normal derivatives of the functions considered on the boundary of the domain. In the case of the quarter plane $Q_1$ it is appropriate to take the inner normal direction derivative because this is the $x$-direction on the $y$-axis and the $y$-direction on the $x$-axis. With $z = x + iy, \bar{z} = x - iy$ they are

$$\partial_y = i(\partial_z - \partial_{\bar{z}}) \text{ on } y = 0, \quad \partial_x = \partial_z + \partial_{\bar{z}} \text{ on } x = 0.$$

**Neumann problem** Let $f \in L_{p,2}(Q_1; \mathbb{C}) \cap C^\alpha(\overline{Q_1}; \mathbb{C})$ for $2 < p, 0 < \alpha < 1, \gamma_1, \gamma_2 \in C(\mathbb{R}^+; \mathbb{C})$ such that $(1+t)^\delta\gamma_1(t), (1+t)^\delta\gamma_2(t), (1+t)^\delta f(t), (1+t)^\delta f(it)$ are bounded for some $0 < \delta, c \in \mathbb{C}$. Find $w \in C^1(\overline{Q_1}; \mathbb{C})$ satisfying

$$w_{\bar{z}} = f \text{ in } Q_1, \partial_y w = \gamma_1 \text{ for } 0 < x, y = 0, \partial_x w = \gamma_2 \text{ for } 0 < y, x = 0, w(0) = c.$$

**Theorem 5** *This Neumann problem is uniquely solvable in the weak sense if and* *only if for any* $z \notin \overline{Q_1}$

$$\frac{1}{2\pi} \int_0^{+\infty} [\gamma_1(t) + if(t)] \frac{dt}{t-z} + \frac{1}{2\pi i} \int_0^{+\infty} [\gamma_2(t) - f(it)] \frac{dt}{t+iz} + \frac{1}{\pi} \int_{Q_1} f(\zeta) \frac{d\xi d\eta}{(\zeta-z)^2} = 0$$

(10)

*holds. The solution is*

$$w(z) = c + \frac{1}{2\pi} \int_0^{+\infty} [\gamma_1(t) + if(t)] \log \left| \frac{t^2 - z^2}{t^2} \right|^2 dt$$

$$+ \frac{1}{2\pi} \int_0^{+\infty} [\gamma_2(t) - f(it)] \log \left| \frac{t^2 + z^2}{t^2} \right|^2 dt - \frac{z}{\pi} \int_{Q_1} \frac{f(\zeta)}{\zeta} \frac{d\xi d\eta}{\zeta - z}.$$

(11)

For a proof, see [8], the analytic function

$$\varphi = w - Tf, \quad Tf(z) = -\frac{1}{\pi} \int_{Q_1} f(\zeta) \frac{d\xi d\eta}{\zeta - z}$$

is satisfying

$$\varphi' = -i\partial_y(w - Tf) = -i\gamma_1 - (\Pi f)^+ + f \text{ for } 0 < x, y = 0$$

and

$$\varphi' = \partial_x(w - Tf) = \gamma_2 - (\Pi f)^+ - f \text{ for } 0 < y, x = 0,$$

where the Calderon-Zygmund type operator

$$\Pi f(z) = -\frac{1}{\pi} \int_{Q_1} f(\zeta) \frac{d\xi d\eta}{(\zeta - z)^2}$$

is Hölder-continuous in $\bar{Q}_1$ such that $(1 + |z|)^{(p+2)/p} \Pi f(z)$ is bounded in $\bar{Q}_1$.

Theorem 4 applied to this Dirichlet problem for the analytic function $\varphi'$ leads to the condition (10) and to the solution

$$\varphi'(z) = -\frac{1}{2\pi i} \int_0^{+\infty} [i\gamma_1(t) - f(t)] \frac{dt}{t-z} - \frac{1}{2\pi i} \int_0^{+\infty} [\gamma_2(t) - f(it)] \frac{dt}{t+iz}.$$

Integration gives the solution (11).

Dirichlet and Neumann problems are particular cases of the Robin boundary value problem being a linear combination of the latter two.

1142

## 5. Particular Robin problem

Instead of investigating the general linear combination of the Dirichlet and Neumann conditions here a special case is considered. For the unit disc [7] other conditions were considered.

**A Robin problem** Let $f \in L_{p,2}(\mathbb{Q}_1; \mathbb{C}) \cap C^\alpha(\bar{\mathbb{Q}}_1; \mathbb{C})$ for $2 < p, 0 < \alpha < 1, \gamma_1, \gamma_2 \in C(\mathbb{R}^+; \mathbb{C})$ such that for some $0 < \delta$ the functions $(1+t)^\delta \gamma_1(t), (1+t)^\delta \gamma_2(t), (1+t)^\delta f(t), (1+t)^\delta f(it)$ are bounded on $\mathbb{R}^+, c \in \mathbb{C}$. Find $w \in C^1(\bar{\mathbb{Q}}_1; \mathbb{C})$ satisfying

$$w_{\bar{z}} = f \text{ in } \mathbb{Q}_1, w(0) = c,$$
$$w - i\partial_y w = \gamma_1 \text{ for } 0 < x, y = 0,$$
$$w + \partial_x w = \gamma_2 \text{ for } 0 < y, x = 0.$$

**Theorem 6** *This particular Robin problem is uniquely solvable in the weak sense if and only if for $z \notin \bar{\mathbb{Q}}_1$*

$$\frac{1}{2\pi i} \int_0^{+\infty} \int_0^t \left[ \gamma_1(\tau) + \frac{3}{2} f(\tau) \right] e^{\tau - t} d\tau \frac{dt}{t - z}$$

$$- \frac{i}{2\pi i} \int_0^{+\infty} \int_0^t \left[ \gamma_2(\tau) - \frac{1}{2} f(i\tau) \right] e^{i(\tau - t)} d\tau \frac{i dt}{it - z} = 0. \qquad (12)$$

*The solution is*

$$w(z) = [c - Tf(0)] e^{-z} + \frac{1}{2\pi i} \int_0^{+\infty} \int_0^t \left[ \gamma_1(\tau) + \frac{3}{2} f(\tau) \right] e^{\tau - t} d\tau \frac{dt}{t - z}$$

$$- \frac{1}{2\pi i} \int_0^{+\infty} \int_0^t \left[ \gamma_2(\tau) - \frac{1}{2} f(i\tau) \right] e^{i(\tau - t)} i dt \frac{i dt}{it - z} \qquad (13)$$

$$- \int_0^z T(f + f_\zeta)(\zeta) e^{\zeta - z} d\zeta + \int_0^z \frac{1}{2\pi i} \int_{\partial \mathbb{Q}_1} f(\tilde{\zeta}) \frac{d\tilde{\zeta}}{\tilde{\zeta} - \zeta} e^{\zeta - z} d\zeta + Tf(z).$$

**Proof** Representing $w = \varphi + Tf$ with analytic $\varphi$ in $\mathbb{Q}_1$ and observing

$$w - i\partial_y w = w + w_z - w_{\bar{z}} = \varphi + \varphi' + Tf + (\Pi f)^+ - f \text{ for } 0 < x, y = 0,$$

$$w + \partial_x w = w + w_z + w_{\bar{z}} = \varphi + \varphi' + Tf + (\Pi f)^+ + f \text{ for } 0 < y, x = 0$$

the analytic function $\varphi$ is seen to have to satisfy the boundary conditions

$$\varphi' + \varphi = \gamma_1 + f - Tf - (\Pi f)^+ \text{for } 0 < x, y = 0,$$

$$\varphi' + \varphi = \gamma_2 - f - Tf - (\Pi f)^+ \text{for } 0 < y, x = 0.$$

Accordiong to [12], p. 63 the boundary values of $\Pi f$ in $\mathbb{Q}_1$ are

$$(\Pi f)^+(t) = -\frac{1}{2}f(t) - \frac{1}{2\pi i} \oint_0^{+\infty} f(\tau)\frac{d\tau}{\tau - it}$$

$$+\frac{1}{2\pi i}\int_0^{+\infty} f(i\tau)\frac{id\tau}{i\tau - t} + Tf_\zeta(t) \text{ for } 0 < t, \qquad (14)$$

$$(\Pi f)^+(it) = -\frac{1}{2}f(t) - \frac{1}{2\pi i}\int_0^{+\infty} f(\tau)\frac{d\tau}{\tau - it}$$

$$+\frac{1}{2\pi i}\oint_0^{+\infty} f(i\tau)\frac{d\tau}{\tau - t} + Tf_\zeta(it) \text{ for } 0 < t, \qquad (14')$$

moreover $\varphi'(t) = d_t\varphi(t)$, $\varphi'(it) = -id_t\varphi(it)$. Hence,

$$d_t\varphi(t) + \varphi(t) = \hat{\gamma}_1(t), \ d_t\varphi(it) + i\varphi(it) = \hat{\gamma}_2(t) \text{ for } 0 < t$$

with

$$\hat{\gamma}_1(t) = \gamma_1(t) + \frac{3}{2}f(t) - T(f + f_\zeta)(t)$$

$$+\frac{1}{2\pi i}\oint_0^{+\infty} f(s)\frac{ds}{s - t} - \frac{1}{2\pi i}\int_0^{+\infty} f(is)\frac{ids}{is - t},$$

$$\hat{\gamma}_2(t) = i\gamma_2(t) - \frac{i}{2}f(it) - iT(f + f_\zeta)(it)$$

$$+\frac{1}{2pi}\int_0^{+\infty} f(s)\frac{ds}{s - it} - \frac{1}{2\pi i}\oint_0^{+\infty} f(is)\frac{ds}{s - t}.$$

These are ordinary differential equations on $\mathbb{R}^+$. Their solutions are

$$\varphi(t) = \varphi(0)e^{-t} + \int_0^t \hat{\gamma}_1(\tau)e^{\tau - t}d\tau = \tilde{\gamma}_1(t) \text{ for } 0 \le t,$$

$$\varphi(it) = \varphi(0)e^{-it} + \int_0^t \hat{\gamma}_2(\tau)e^{i(\tau - t)}d\tau = \tilde{\gamma}_2(t) \text{ for } 0 \le t.$$

These equations are Dirichlet data for the analytic function $\varphi$ in $\mathbb{Q}_1$. This Dirichlet problem is solvable according to Theorem 4 if and only if for all $z \notin \bar{\mathbb{Q}}_1$

$$\frac{1}{2\pi i}\int_0^{+\infty} \tilde{\gamma}_1(t)\frac{dt}{t - z} - \frac{1}{2\pi i}\int_0^{+\infty} \tilde{\gamma}_2(t)\frac{idt}{it - z} = 0. \qquad (15)$$

1144

The solution is for $z \in \mathbb{Q}_1$

$$\varphi(z) = \frac{1}{2\pi i} \int\limits_0^{+\infty} \tilde{\gamma}_1(t) \frac{dt}{t-z} - \frac{1}{2\pi i} \int\limits_0^{+\infty} \tilde{\gamma}_2(t) \frac{idt}{it-z}. \tag{16}$$

From the Cauchy formula for $\mathbb{Q}_1$ follows

$$\frac{1}{2\pi i} \int\limits_0^{+\infty} e^{-t} \frac{dt}{t-z} - \frac{1}{2\pi i} \int\limits_0^{+\infty} e^{-it} \frac{idt}{it-z} = \frac{1}{2\pi i} \int\limits_{\partial \mathbb{Q}_1} e^{-\zeta} \frac{d\zeta}{\zeta-z} = \begin{cases} e^z, z \in \mathbb{Q}_1, \\ 0, z \notin \bar{\mathbb{Q}}_1, \end{cases}$$

$$-\frac{1}{2\pi i} \int\limits_0^{+\infty} \int\limits_0^t T(f+f_\zeta)(\tau) e^{\tau-t} d\tau \frac{dt}{t-z} + \frac{1}{2\pi i} \int\limits_0^{+\infty} \int\limits_0^t T(f+f_\zeta)(i\tau) e^{i(\tau-t)} id\tau \frac{idt}{it-z}$$

$$= -\frac{1}{\pi} \int\limits_{\mathbb{Q}_1} [f(\zeta) + f_\zeta(\zeta)] \frac{1}{2\pi i} \int\limits_{\partial \mathbb{Q}_1} \int\limits_0^{\hat{\zeta}} \frac{e^{\tilde{\zeta}-\hat{\zeta}}}{\tilde{\zeta}-\zeta} d\tilde{\zeta} \frac{d\hat{\zeta}}{\hat{\zeta}-z} d\xi d\eta$$

$$= \begin{cases} -\int\limits_0^z T(f+f_\zeta)(\zeta) e^{\zeta-z} d\zeta, & z \in \mathbb{Q}_1, \\ 0, & z \notin \bar{\mathbb{Q}}_1, \end{cases}$$

$$\frac{1}{2\pi i} \int\limits_0^{+\infty} \int\limits_0^t \frac{1}{2\pi i} \oint\limits_0^{+\infty} f(s) \frac{ds}{s-\tau} e^{\tau-t} d\tau \frac{dt}{t-z}$$

$$-\frac{1}{2\pi i} \int\limits_0^{+\infty} \int\limits_0^t \frac{1}{2\pi i} \int\limits_0^{+\infty} f(s) \frac{ds}{s-i\tau} e^{i(\tau-t)} id\tau \frac{idt}{it-z}$$

$$= \frac{1}{2\pi i} \oint\limits_0^{+\infty} \frac{f(s)}{2\pi i} \int\limits_{\partial \mathbb{Q}_1} \int\limits_0^{\hat{\zeta}} \frac{e^{\zeta-\hat{\zeta}}}{s-\zeta} d\zeta \frac{d\hat{\zeta}}{\hat{\zeta}-z} ds = \begin{cases} \int\limits_0^z \frac{1}{2\pi i} \oint\limits_0^{+\infty} f(s) \frac{ds}{s-\zeta} e^{\zeta-z} d\zeta, z \in \mathbb{Q}_1, \\ 0, z \notin \bar{\mathbb{Q}}_1, \end{cases}$$

$$-\frac{1}{2\pi i} \int\limits_0^{+\infty} \int\limits_0^t \frac{1}{2\pi i} \int\limits_0^{+\infty} f(is) \frac{ids}{is-\tau} e^{\tau-t} d\tau \frac{dt}{t-z}$$

$$+\frac{1}{2\pi i} \int\limits_0^{+\infty} \int\limits_0^t \frac{1}{2\pi i} \oint_0^{+\infty} f(is) \frac{ds}{s-\tau} e^{i(\tau-t)} id\tau \frac{idt}{it-z}$$

$$= -\frac{1}{2\pi i} \oint\limits_0^{+\infty} \frac{f(is)}{2\pi i} \int\limits_{\partial \mathbb{Q}_1} \int\limits_0^{\hat{\zeta}} \frac{e^{\zeta-\hat{\zeta}}}{is-\zeta} d\zeta \frac{d\hat{\zeta}}{\hat{\zeta}-z} ids$$

$$= \begin{cases} -\int\limits_0^z \frac{1}{2\pi i} \oint_0^{+\infty} f(is) \frac{ids}{is-\zeta} e^{\zeta-z} d\zeta, z \in \mathbb{Q}_1, \\ 0, z \notin \bar{\mathbb{Q}}_1. \end{cases}$$

Thus

$$\varphi(z) = \varphi(0)e^{-z} + \frac{1}{2\pi i} \int\limits_0^{+\infty} \int\limits_0^t \left[\gamma_1(\tau) + \frac{3}{2}f(\tau)\right] e^{\tau - t} \frac{dt}{t - z}$$

$$- \frac{1}{2\pi i} \int\limits_0^{+\infty} \int\limits_0^t \left[\gamma_2(\tau) - \frac{1}{2}f(i\tau)\right] e^{i(\tau - t)} i d\tau \frac{idt}{it - z} \tag{17}$$

$$- \int\limits_0^z e^{\zeta - z} T(f + f_\zeta)(\zeta) d\zeta + \int\limits_0^t e^{\zeta - z} \frac{1}{2\pi i} \int\limits_{\partial \mathbb{Q}_1} f(\tilde{\zeta}) \frac{d\tilde{\zeta}}{\tilde{\zeta} - \zeta} d\zeta$$

if and only if (12) is satisfied.

Obviously (12) has also to be satisfied for $z = 0$ as is seen from (17). Also (17) implies (13).

In order to verify (13) as the solution to the particular Robin problem under the condition (12) $w$ is immediately seen to satisfy the differential equation in $\mathbb{Q}_1$. For checking the boundary conditions consider

$$w(z) + w_z(z) = \frac{1}{2\pi i} \int\limits_0^{+\infty} \int\limits_0^t \left[\gamma_1(\tau) + \frac{3}{2}f(\tau)\right] e^{\tau - t} d\tau \left[\frac{1}{t - z} + \frac{1}{(t - z)^2}\right] dt$$

$$- \frac{1}{2\pi i} \int\limits_0^{+\infty} \int\limits_0^t \left[\gamma_2(\tau) - \frac{1}{2}f(\tau)\right] e^{i(\tau - t)} i dt \left[\frac{1}{it - z} + \frac{1}{(it - z)^2}\right] i dt$$

$$- T f_\zeta(z) + \frac{1}{2\pi i} \int\limits_{\partial \mathbb{Q}_1} f(\zeta) \frac{d\zeta}{\zeta - z} + \Pi f(z).$$

Integration by parts leads to

$$w(z) + w_z(z) = \frac{1}{2\pi i} \int\limits_0^{+\infty} \left[\gamma_1(t) + \frac{3}{2}f(t)\right] \frac{dt}{t - z} - \frac{1}{2\pi i} \int\limits_0^{+\infty} \left[\gamma_2(t) - \frac{1}{2}f(it)\right] \frac{idt}{it - z}$$

$$- T f_\zeta(z) + \frac{1}{2\pi i} \int\limits_{\partial \mathbb{Q}_1} f(\zeta) \frac{d\zeta}{\zeta - z} + \Pi f(z).$$

Letting $z$ tend to the boundary and using (14) and (14′), respectively shows for

$0 < t$

$$w(t) + w_z(t) - w_{\bar{z}}(t) = \lim_{z \to t} \left[ \frac{1}{2\pi i} \int_0^{+\infty} \left[ \gamma_1(\tau) + \frac{3}{2} f(\tau) \right] \frac{d\tau}{\tau - z} \right.$$

$$\left. - \frac{1}{2\pi i} \int_0^{+\infty} \left[ \gamma_2(\tau) - \frac{1}{2} f(i\tau) \right] \frac{id\tau}{i\tau - z} \right] - \frac{3}{2} f(t),$$

$$w(it) + w_z(it) - w_{\bar{z}}(it) = \lim_{z \to it} \left[ \frac{1}{2\pi i} \int_0^{+\infty} \left[ \gamma_1(\tau) + \frac{3}{2} f(\tau) \right] \frac{d\tau}{\tau - z} \right.$$

$$\left. - \frac{1}{2\pi i} \int_0^{+\infty} \left[ \gamma_2(\tau) - \frac{1}{2} f(i\tau) \right] \frac{id\tau}{i\tau - z} \right] + \frac{1}{2} f(it).$$

Using the solvability condition (12) for $\bar{z}, -z, -\bar{z}$ where $z \in \mathbb{Q}_1$ as in the case of the Dirichlet problem verifies the boundary conditions.

The methods used here can, obviously, immediately be applied to the general Robin problem of the form

$$w_{\bar{z}} = f \text{ in } \mathbb{Q}, \ w(0) = c,$$
$$\alpha_1 w - i\partial_y w = \gamma_1 \text{ for } 0 \le x, y = 0,$$
$$\alpha_2 w + \partial_x w = \gamma_2 \text{ for } 0 \le y, x = 0$$

with some continuous coefficients $\alpha_1, \alpha_2$ being bounded and nonnegative.

## References

1. S.A. Abdymanapov, H. Begehr, A.B. Tungatarov: Some Schwarz problems in a quarter plane, Preprint, FU Berlin, 2005.

2. H. Begehr, Complex analytic methods for partial differential equations. An introductory text, World-Sci., Singapore, 1994.

3. H. Begehr, Orthogonal decompositions of the function space $L_2(\bar{D}; \mathbb{C})$. J. Reine Angew. Math. 549 (2002), 191-219.

4. H. Begehr, Boundary value problems in complex analysis, I, II. Bol. Asoc. Mat. Venezolana 12 (2005), 65-85; 217-250.

5. Begehr, H., Schmersau, D.: The Schwarz problem for polyanalytic functions. ZAA 24(2005), 341-351.

6. H. Begehr, Basic boundary value problems in complex analysis, Snapshot in Appl. Complex Anal. Proc. Workshop Recent Trends in Appl. Complex Analysis, METU, Ankara, 2004, eds. H. Begehr et al., J. Appl. Funct. Anal. 2 (2007), 57-71.

7. H. Begehr, G. Harutyunyan, Robin boundary value problem for the Cauchy-Riemann operator, Complex Var., Theory Appl. 50 (2005), 1125-1136.

8. H. Begehr, G. Harutyunyan, Complex boundary value problems in a quarter plane. Complex Anal. Appl., Proc. 13th Intern. Conf. on Finite or Infinite Dimensional Complex Analysis and Appl., Shantou, 2005, eds. Y. Wang et al., World Sci., New Jersey, 2006, 1-10.

9. H. Begehr, J.C. Vanegas, Iterated Neumann problem for the higher order Poisson equation, Math. Nachr. 279 (2006), 38–57.

10. E. Gaertner, Basic complex boundary value problems in the upper half plane. Ph.D. thesis, FU Berlin, 2006; http://www.diss.fu-berlin.de/2006/320.

11. A. Krausz, Intergraldarstellungen mit Greenschen Funktionen höherer Ordnung in Gebieten und Polygebieten, Dissertation, FU Berlin, 2005; http://www.diss.fu-berlin.de/2005/128/ .

12. I.N. Vekua, Generalized analytic functions, Pergamon, Oxford, 1962.

# MIXED BOUNDARY VALUE PROBLEM FOR INHOMOGENEOUS POLY-ANALYTIC-HARMONIC EQUATION

AJAY KUMAR

*Department of Mathematics*
*University of Delhi*
*Delhi-110007, India*
*E-mail: ajaykr@bol.net.in*

RAVI PRAKASH

*Department of Mathematics*
*Rajdhani College (University of Delhi)*
*Raja Garden, New Delhi-110015, India*
*E-mail: rprakash_rachit@yahoo.com*

The mixed boundary value problem arising from Schwarz and Dirichlet boundary conditions for the polyharmonic and polyanalytic inhomogeneous equations is studied.

**Key words:** Polyanalytic functions, polyharmonic functions, Schwarz-Dirichlet problem, Cauchy-Pompeiu representation, Green function
**Mathematics Subject Classification:** 30E25, 30G20, 31A10, 35J40, 35J55

## 1. Introduction

An investigation of boundary value problems for complex partial differential equations needs a study of equation of the type $\partial_z^m \partial_{\bar{z}}^n w = f$. Two particular cases namely, the polyanalytic equation $\partial_{\bar{z}}^n w = f$ and the polyharmonic equation $(\partial_{\bar{z}} \partial_z)^n w = f$ have already been studied in [2-15]. Of course this classical equation is very well known and studied e.g. see [16,17]. A variety of boundary value problems arising from the Schwarz, the Dirichlet and the Neumann problem for the inhomogeneous polyanalytic equation have been investigated in [3-7,13,14]. The Neumann problem and Dirichlet problem for $n$-Poisson equation have been studied in [8-10,12,15]. For more literature on these boundary conditions see [11].

In the following section we study the mixed Dirichlet and Schwarz problem for the equation of the type $\partial_z^m \partial_{\bar{z}}^n w = f$ and give the explicit solution of the problem. For the inhomogeneous polyanalytic equation, the Schwarz problem is solved in [8]. The solution is given as follows:

**Theorem 1.** The Schwarz problem for the inhomogeneous polyanalytic equation in the unit disc $\mathbb{D}$, $\partial_{\bar{z}}^m w = f$ in $\mathbb{D}$, Re $\partial_{\bar{z}}^\mu w = \gamma_\mu$ on $\partial \mathbb{D}$, Im $\partial_{\bar{z}}^\mu w(0) = b_\mu$, is uniquely solvable for $f \in L_1(\mathbb{D}, \mathbb{C})$, $\gamma_\mu \in C(\partial \mathbb{D}, \mathbb{R})$, $b_\mu \in \mathbb{R}$, $0 \le \mu \le m-1$. The solution is given by

$$w(z) = i \sum_{s=0}^{m-1} \frac{b_s}{s!}(z+\bar{z})^s + \sum_{s=0}^{m-1} \frac{(-1)^s}{2\pi i s!} \int_{|\zeta|=1} \gamma_s(\zeta) \frac{\zeta+z}{\zeta-z}(\zeta-z+\overline{\zeta-z})^s \frac{d\zeta}{\zeta}$$

$$+ \frac{(-1)^m}{2\pi(m-1)!} \int_{|\zeta|<1} \left( \frac{f(\zeta)}{\zeta} \frac{\zeta+z}{\zeta-z} + \frac{\overline{f(\zeta)}}{\bar{\zeta}} \frac{1+z\bar{\zeta}}{1-z\bar{\zeta}} \right) (\zeta - z + \overline{\zeta - z})^{m-1} d\xi d\eta \qquad (1)$$

The solution of the Dirichlet problem for Poisson equation

$$w_{z\bar{z}} = f \text{ in } D, w = \beta_0 \text{ on } \partial\mathbb{D}$$

for a regular domain $D$ of the complex plane $\mathbb{C}$ is given by

$$w(z) = \frac{1}{2\pi i} \int_{\partial D} H_1^0(z,\zeta)\beta_0(\zeta) \frac{d\zeta}{\zeta} - \frac{1}{\pi} \int_D G_1(z,\zeta)f(\zeta)d\xi d\eta \qquad (2)$$

where $H_1^0(z,\zeta) = g_1(z,\zeta) = \dfrac{1}{1-z\bar{\zeta}} + \dfrac{1}{1-\bar{z}\zeta}$ and $G_1(z,\zeta)$ is the Green function for the Laplacian i.e.

$$G_1(z,\zeta) = \log \left| \frac{1-z\bar{\zeta}}{\zeta-z} \right|^2$$

see [1,2,15]. Explicit solution of $\partial_{z\bar{z}}^n w = f$ in $\mathbb{D}$, $\partial_{z\bar{z}}^\mu w = \gamma_\mu$ on $\partial\mathbb{D}$, $0 \le \mu \le n-1$ for $n = 2, 3, 4, 5$ are obtained in [12,15], whereas the solution for arbitrary $n$ through a recursive function is obtained in [15].

## 2. Dirichlet-Schwarz problem

We begin with a generalisation of the Poisson kernel $g_1(z,\zeta)$ considered in [2].

**Definition 1** *For* $k \in \mathbb{N}$, $r \in \mathbb{N}_0 = \mathbb{N} \cup \{0\}$ *and* $z$, $\zeta \in \bar{\mathbb{D}}$ *with* $z\bar{\zeta} \ne 1$, *let*

$$H_k^r(z,\zeta) = \frac{1}{k^r} + \sum_{m=1}^{\infty} \frac{1}{(k+m)^r}((z\bar{\zeta})^m + (\bar{z}\zeta)^m) \qquad (3)$$

**Lemma 1** *For* $p, q, r \in \mathbb{N}_0$, $k \in \mathbb{N}$, $|z| < 1$, *if* $A_{k,p,q}^r(z) = \dfrac{1}{\pi} \displaystyle\int_{|\zeta|<1} \zeta^p \bar{\zeta}^q H_k^r(z,\zeta)d\xi d\eta$

*then* $A_{k,p,q}^r(z)$ *is equal to* $\dfrac{(u(z,\bar{z}))^{|p-q|}}{(k+|p-q|)^r(q+1)}$, *where* $u(z,\bar{z})$ *is equal* $z$ *or* $\bar{z}$ *according-ing as* $p \ge q$ *or* $p < q$ *respectively.*

**Proof** Using (3) above, $A_{k,p,q}^r(z)$ can be written as $\frac{1}{k^r}X_{p,q}$

$+ \sum_{m=1}^{\infty} \frac{1}{(k+m)^r} [\bar{z}^m X_{p+m,q} + z^m X_{p,q+m}]$ where $X_{r,s}$ is given by $\frac{1}{\pi} \int\limits_{|\varsigma|<1} \varsigma^r \bar{\varsigma}^s d\xi d\eta$.

Using the Gauss theorem, it follows that $X_{r,s}$ is equal $\frac{1}{s+1}$ if $r = s$, otherwise 0. Thus, we obtain the required value. $\qquad\square$

**Lemma 2** *For* $p,q,r \in \mathbb{N}_0$, $k \in \mathbb{N}$, $|z|, |\tilde{\varsigma}| < 1$, *if* $C_{k,p,q}^r(z,\tilde{\varsigma}) = \frac{1}{\pi} \int\limits_{|\varsigma|<1} \varsigma^p \bar{\varsigma}^q H_k^r$

$(\varsigma,\tilde{\varsigma}) \frac{d\xi d\eta}{\varsigma - z}$ *then* $C_{k,p,q}^r(z,\tilde{\varsigma})$ *can be expressed as*

$$\frac{z^p}{q+1} \left[ \sum_{j=1}^{\infty} \frac{\bar{\tilde{\varsigma}}^{q+j} z^{j-1}}{(k+q+j)^r} + \sum_{j=1}^{\infty} \frac{j\tilde{\varsigma}^j \bar{z}^{q+j+1}}{(k+j)^r(q+1+j)} - \bar{z}^{q+1} H_k^r(z,\tilde{\varsigma}) \right]$$

$$+ \sum_{j=0}^{p-1} z^{p-1-j} A_{k,j,q}^r(\tilde{\varsigma}), \qquad (4)$$

*where* $A_{k,j,q}^r$ *are as in Lemma 1 above.*

**Proof** Writing $\frac{\varsigma^p}{\varsigma - z} = \sum_{j=0}^{p-1} \varsigma^{p-j-1} z^j + \frac{z^p}{\varsigma - z}$, we can express $C_{k,p,q}^r(z,\tilde{\varsigma})$ in the form

$$\sum_{j=0}^{p-1} z^{p-1-j} A_{k,j,q}^r(\tilde{\varsigma}) + z^p B_{k,q}^r(z,\tilde{\varsigma})$$

where $B_{k,q}^r(z,\tilde{\varsigma}) = \frac{1}{\pi} \int\limits_{|\varsigma|<1} \bar{\varsigma}^q H_k^r(\varsigma,\tilde{\varsigma}) \frac{d\xi d\eta}{\varsigma - z}$.

The last boundary integral can be written as

$$\frac{1}{q+1} \left[ \frac{1}{\pi} \int\limits_{|\varsigma|<1} \frac{\partial}{\partial\bar{\varsigma}} (\bar{\varsigma}^{q+1} H_k^r(\varsigma,\tilde{\varsigma})) \frac{d\xi d\eta}{\varsigma - z} \right.$$

$$\left. - \frac{1}{\pi} \int\limits_{|\varsigma|<1} \frac{\partial}{\partial\bar{\varsigma}} \left( \sum_{j=1}^{\infty} \frac{j\tilde{\varsigma}^j \bar{\varsigma}^{q+j+1}}{(k+j)^r(q+j+1)} \right) \frac{d\xi d\eta}{\varsigma - z} \right] \qquad (5)$$

Using the Cauchy Pompeiu formula [1] the last expression in (5) reduces to

$$\frac{1}{q+1} \left[ \frac{1}{2\pi i} \int\limits_{|\varsigma|=1} \bar{\varsigma}^{q+1} H_k^r(\varsigma,\tilde{\varsigma}) \frac{d\varsigma}{\varsigma - z} - \bar{z}^{q+1} H_k^r(z,\tilde{\varsigma}) \right.$$

$$\left. - \frac{1}{2\pi i} \int\limits_{|\varsigma|=1} \sum_{j=1}^{\infty} \frac{j\tilde{\varsigma}^j \bar{\varsigma}^{q+j+1}}{(k+j)^r(q+j+1)} \frac{d\varsigma}{\varsigma - z} + \sum_{j=1}^{\infty} \frac{j\tilde{\varsigma}^j \bar{z}^{q+j+1}}{(k+j)^r(q+j+1)} \right]$$

But $\dfrac{1}{2\pi i}\displaystyle\int_{|\zeta|=1}\bar\zeta^{q+1}H_k^r(\zeta,\tilde\zeta)\dfrac{d\zeta}{\zeta-z}=\sum_{j=0}^{\infty}\dfrac{\bar{\tilde\zeta}^{q+j+1}z^j}{(k+q+j)^r}$, so we obtain (3). $\qquad\square$

**Lemma 3** *For $p,q,r\in\mathbb{N}_0$, $k\in\mathbb{N}$, $|z|,|\tilde\zeta|<1$,*

$$D_{k,p,q}^r(z,\tilde\zeta)=\frac{1}{\pi}\int_{|\zeta|<1}\zeta^p\bar\zeta^q H_k^r(\zeta,\tilde\zeta)\frac{1+z\bar\zeta}{1-z\bar\zeta}\frac{d\xi d\eta}{\bar\zeta}$$

$$=\alpha_{p,q}(\tilde\zeta)+2z\sum_{l=0}^{\infty}z^l A_{k,p,q+l}^r(\tilde\zeta)$$

*where*

$$\alpha_{p,q}(\tilde\zeta)=\begin{cases}A_{k,p,q-1}^r(\tilde\zeta) & \text{if } q\geq 1\\[2mm]\dfrac{\tilde\zeta^p}{(k+p+1)^r} & \text{if } q=0\end{cases}$$

**Proof** We first write $D_{k,p,q}^r(z,\tilde\zeta)$ as

$$\alpha_{p,q}(\tilde\zeta)+2z\sum_{l=0}^{\infty}z^l A_{k,p,q+l}^r(\tilde\zeta)$$

where

$$\alpha_{p,q}(\tilde\zeta)=\frac{1}{\pi}\int_{|\zeta|<1}\zeta^p\bar\zeta^{q-1}H_k^r(\zeta,\tilde\zeta)d\xi d\eta$$

If $q\geq 1$, then

$$\alpha_{p,q}(\tilde\zeta)=A_{k,p,q-1}^r(\tilde\zeta)$$

For $q=0$

$$\alpha_{p,q}(\tilde\zeta)=\frac{1}{\pi}\int_{|\zeta|<1}\zeta^p H_k^r(\zeta,\tilde\zeta)\frac{d\xi d\eta}{\bar\zeta}$$

$$=\frac{1}{k^r}Y_p+\sum_{m=1}^{\infty}\frac{1}{(k+m)^r}[\tilde\zeta^m X_{p,m-1}+\bar{\tilde\zeta}^m Y_{p+m}]$$

where $X_{p,r}$ being as in proof of Lemma 6.2.1 and $Y_p=\dfrac{1}{\pi}\displaystyle\int_{|\zeta|<1}\zeta^p\dfrac{d\xi d\eta}{\zeta}$.

It is easy to conclude that $Y_p=0$.

Thus, $\alpha_{p,0}(\tilde\zeta)=\dfrac{\tilde\zeta^p}{(k+p+1)^r}$. $\qquad\square$

For $p,q,r\in\mathbb{N}_0$, we denote $\dfrac{r!(-z-\bar z)^{r-p-q}}{p!q!(r-p-q)!}$ by $M(p,q,r;z)$ and the subset

$\{(p,q) : p+q \le r\}$ of $\mathbb{N}_0 \times \mathbb{N}_0$ by $T(r)$. Using Lemma 2 and multinomials, we obtain the following:

**Corollary 4** For $r \in \mathbb{N}_0$, $m, k \in \mathbb{N}, |z|, |\tilde\zeta| < 1$,

$$\text{(i)} \quad E_{k,m}^r(z,\tilde\zeta) = \frac{1}{\pi} \int\limits_{|\varsigma|<1} (\zeta - z + \overline{\zeta - z})^{m-1} H_k^r(\varsigma,\tilde\zeta) \frac{\varsigma+z}{\varsigma(\varsigma-z)} d\xi d\eta$$

$$= \sum_{(p,q)\in T(m-1)} M(p,q,m-1;z)[2C_{k,p,q}^r(z,\tilde\zeta) - C_{k,p,q}^r(0,\tilde\zeta)]$$

$$\text{(ii)} \quad F_k^r(z,\tilde\zeta) = \frac{1}{\pi} \int\limits_{|\varsigma|<1} (\zeta - z + \overline{\zeta - z})^{m-1} H_k^r(\varsigma,\tilde\zeta) \frac{1+z\tilde\zeta}{\overline{\zeta}(1-z\overline{\zeta})} d\xi d\eta$$

$$= \sum_{(p,q)\in T(m-1)} M(p,q,m-1;z) D_{k,p,q}^r(z,\tilde\zeta)$$

**Lemma 5** For $p,q \in \mathbb{N}_0$, $|z|, |\tilde\zeta| < 1$,

$$N_{p,q}(z,\tilde\zeta) = \frac{1}{\pi} \int\limits_{|\varsigma|<1} \varsigma^p \bar\varsigma^q G_1(\varsigma,\tilde\zeta) \frac{d\xi d\eta}{\varsigma - z} = \sum_{j=0}^{p-1} z^j L_{p-j-1,q}(\tilde\zeta) + z^p M_q(z,\tilde\zeta)$$

*where*

$$L_{p,q}(z,\tilde\zeta) = \frac{u^{\alpha-\beta}}{(\alpha+1)(\beta+1)}[1 - |\tilde\zeta|^{2\beta+2}] \tag{6}$$

*where $u = \bar{\tilde\zeta}$ if $p > q$ and $\tilde\zeta$ if $p \le q$, $\alpha = \max\{p,q\}, \beta = \min\{p,q\}$ and*

$$M_q(z,\tilde\zeta) = \frac{1}{q+1}\left[ \log \frac{(1-z\bar{\tilde\zeta})}{|z-\tilde\zeta|^2} - \bar z^{q+1} \left\{ G_1(z,\tilde\zeta) + \sum_{j=1}^{\infty} \frac{(\bar{\tilde\zeta}z)^j}{q+j+1} \right\} \right.$$

$$\left. - \sum_{j=0}^{q} \frac{\bar{\tilde\zeta}^{q-j}\bar z^{j+1}}{j+1} \right]$$

**Proof** As in Lemma 2, we write $N_{p,q}(z,\tilde\zeta)$ as

$$\sum_{j=0}^{p-1} z^j L_{p-j-1,q}(\tilde\zeta) + z^p M_q(z,\tilde\zeta)$$

where

$$L_{p,q}(\tilde\zeta) = \frac{1}{\pi} \int\limits_{|\varsigma|<1} \varsigma^p \bar\varsigma^q G_1(\varsigma,\tilde\zeta) d\xi d\eta$$

and

$$M_q(z, \tilde{\zeta}) = \frac{1}{\pi} \int\limits_{|\varsigma|<1} \bar{\varsigma}^q G_1(\varsigma, \tilde{\zeta}) \frac{d\xi d\eta}{\varsigma - z}$$

We can write

$$L_{p,q}(\tilde{\zeta}) = \frac{1}{q+1} \left[ \frac{1}{2\pi i} \int\limits_{|\varsigma|=1} \varsigma^p \bar{\varsigma}^{q+1} G_1(\varsigma, \tilde{\zeta}) d\varsigma + \frac{\tilde{\zeta}}{\pi} \int\limits_{|\varsigma|<1} \varsigma^p \bar{\varsigma}^{q+1} \frac{d\xi d\eta}{1 - \bar{\tilde{\zeta}} \tilde{\zeta}} \right.$$

$$\left. + \frac{1}{\pi} \int\limits_{|\varsigma|<1} \varsigma^p \bar{\varsigma}^{q+1} \frac{d\xi d\eta}{\bar{\varsigma} - \bar{\tilde{\zeta}}} \right]$$

The first integral on the right side is 0, as $G_1(\varsigma, \tilde{\zeta}) = 0$ for $|\varsigma| = 1$.
  By the Gauss theorem.

$$\frac{1}{\pi} \int\limits_{|\varsigma|<1} \varsigma^p \bar{\varsigma}^{q+1} \frac{d\xi d\eta}{1 - \bar{\tilde{\zeta}} \tilde{\zeta}} = -\frac{1}{p+1} \frac{1}{2\pi i} \int\limits_{|\varsigma|=1} \frac{\bar{\varsigma}^{q-p}}{1 - \bar{\tilde{\zeta}} \tilde{\zeta}} d\bar{\varsigma}$$

For $q \geq p$, this integral is 0 and for $q < p$, it is given by $\tilde{\zeta}^{p-q-1}$.
  Next, by the Cauchy Pompeiu formula

$$\frac{1}{\pi} \int\limits_{|\varsigma|<1} \varsigma^p \bar{\varsigma}^{q+1} \frac{d\xi d\eta}{\bar{\varsigma} - \bar{\tilde{\zeta}}} = -\frac{1}{p+1} \frac{1}{2\pi i} \int\limits_{|\varsigma|=1} \varsigma^{p+1} \bar{\varsigma}^{q+1} \frac{d\varsigma}{\bar{\varsigma} - \bar{\tilde{\zeta}}} - \tilde{\zeta}^{p+1} \bar{\tilde{\zeta}}^{q+1}$$

$$= \begin{cases} \dfrac{1}{p+1}(\bar{\tilde{\zeta}}^{q-p} - \tilde{\zeta}^{p+1} \bar{\tilde{\zeta}}^{q+1}) & \text{if } q \geq p \\[2mm] -\dfrac{1}{p+1} \tilde{\zeta}^{p+1} \bar{\tilde{\zeta}}^{q+1} & \text{if } q < p \end{cases}$$

Therefore,

$$L_{p,q}(z, \tilde{\zeta}) = \begin{cases} \dfrac{1}{q+1} \dfrac{1}{p+1}(\bar{\tilde{\zeta}}^{q-p} - \tilde{\zeta}^{p+1} \bar{\tilde{\zeta}}^{q+1}) & \text{if } q \geq p \\[2mm] \dfrac{1}{q+1} \dfrac{1}{p+1}(\tilde{\zeta}^{p-q} - \tilde{\zeta}^{p+1} \bar{\tilde{\zeta}}^{q+1}) & \text{if } p > q \end{cases}$$

$$= \frac{u^{\alpha-\beta}}{(\alpha+1)(\beta+1)}[1 - |\tilde{\zeta}|^{2\beta+2}]$$

where $\alpha = \max\{p, q\}$, $\beta = \min\{p, q\}$ and $u = \tilde{\zeta}$ if $p > q$ and $\bar{\tilde{\zeta}}$ if $p \leq q$.
  Next, by using the Cauchy Pompeiu formula, we can write

$$M_q(z, \tilde{\zeta}) = \frac{1}{q+1} \left[ \frac{1}{2\pi i} \int\limits_{|\varsigma|=1} \bar{\varsigma}^{q+1} G_1(\varsigma, \tilde{\zeta}) \frac{d\varsigma}{\varsigma - z} - \bar{z}^{q+1} G_1(z, \tilde{\zeta}) \right]$$

$$+\frac{\tilde{\zeta}}{\pi}\int\limits_{|\varsigma|<1}\bar{\zeta}^{q+1}\frac{1}{1-\bar{\zeta}\zeta}\frac{d\xi d\eta}{\zeta-z}+\frac{1}{\pi}\int\limits_{|\varsigma|<1}\bar{\zeta}^{q+1}\frac{1}{\bar{\zeta}-\bar{\tilde{\zeta}}}\frac{d\xi d\eta}{\zeta-z}$$

As $G_1(\varsigma,\tilde{\varsigma})=0$ for $|\varsigma|=1$, the first boundary integral equals zero.

Next,

$$\frac{\tilde{\varsigma}}{\pi}\int\limits_{|\varsigma|<1}\frac{\bar{\zeta}^{q+1}}{1-\bar{\zeta}\zeta}\frac{d\xi d\eta}{\zeta-z}=\sum_{j=1}^{\infty}\tilde{\zeta}^j\frac{1}{\pi}\int\limits_{|\varsigma|<1}\bar{\zeta}^{q+j}\frac{d\xi d\eta}{\zeta-z}$$

As for $k\geq 1$, $\frac{1}{\pi}\int\limits_{|\varsigma|<1}\bar{\zeta}^k\frac{d\xi d\eta}{\zeta-z}=-\frac{\bar{z}^{k+1}}{k+1}$,

we get

$$\frac{\tilde{\varsigma}}{\pi}\int\limits_{|\varsigma|<1}\frac{\bar{\zeta}^{q+1}}{1-\bar{\zeta}\zeta}\frac{d\xi d\eta}{\zeta-z}=-\sum_{j=1}^{\infty}\tilde{\zeta}^j\frac{\bar{z}^{q+j+1}}{q+j+1}$$

Using

$$\frac{\bar{\varsigma}^{q+1}}{\bar{\zeta}-\bar{\tilde{\zeta}}}=\sum_{j=0}^{q}\bar{\zeta}^{q-j}\tilde{\zeta}^j+\frac{\bar{\tilde{\zeta}}^{q+1}}{\bar{\zeta}-\bar{\tilde{\zeta}}},$$

we can write

$$\frac{1}{\pi}\int\limits_{|\varsigma|<1}\frac{\bar{\zeta}^{q+1}}{\bar{\zeta}-\bar{\tilde{\zeta}}}\frac{d\xi d\eta}{\zeta-z}=\sum_{j=0}^{q}\bar{\tilde{\zeta}}^{q-j}\frac{1}{\pi}\int\limits_{|\varsigma|<1}\bar{\zeta}^j\frac{d\xi d\eta}{\zeta-z}+\frac{\bar{\tilde{\zeta}}^{q+1}}{\pi}\int\limits_{|\varsigma|<1}\frac{d\xi d\eta}{(\bar{\zeta}-\bar{\tilde{\zeta}})(\zeta-z)}$$

$$=-\sum_{j=0}^{q}\frac{\bar{\tilde{\zeta}}^{q-j}\bar{z}^{j+1}}{j+1}+\bar{\tilde{\zeta}}^{j+1}I$$

where

$$I=\frac{1}{\pi}\int\limits_{|\varsigma|<1}\frac{d\xi d\eta}{(\bar{\zeta}-\bar{\tilde{\zeta}})(\zeta-z)}$$

It is easy to show that

$$I=\log\frac{(1-z\bar{\tilde{\zeta}})}{|z-\tilde{\zeta}|^2}$$

Thus,

$$M_q(z,\tilde{\varsigma})=\frac{1}{q+1}\left[\log\frac{(1-z\bar{\tilde{\zeta}})}{|z-\tilde{\zeta}|^2}-\bar{z}^{q+1}G_1(z,\tilde{\varsigma})\right.$$

$$\left.-\bar{z}^{q+1}\sum_{j=1}^{\infty}\frac{(\tilde{\zeta}\bar{z})^j}{q+j+1}-\sum_{j=0}^{q}\frac{\bar{\tilde{\zeta}}^{q-j}\bar{z}^{j+1}}{j+1}\right]$$

$\square$

**Lemma 6** *For* $p, q \in \mathbb{N}_0$, $|z|, |\tilde{\zeta}| < 1$,

$$Q_{p,q}(z, \tilde{\zeta}) = \frac{1}{\pi} \int\limits_{|\zeta|<1} \zeta^p \bar{\zeta}^q G_1(\zeta, \tilde{\zeta}) \frac{1 + z\bar{\zeta}}{(1 - z\bar{\zeta})\zeta} d\xi d\eta$$

$$= u(\tilde{\zeta}) + 2z \sum_{j=0}^{\infty} z^j L_{p,q+j}(\tilde{\zeta})$$

*with*

$$u(\tilde{\zeta}) = \begin{cases} \overline{M_p(0, \tilde{\zeta})} & \text{if } q = 0 \\ L_{p,q-1}(\tilde{\zeta}) & \text{if } q > 0 \end{cases}$$

*where* $M_p$ *and* $L_{p,q}$ *are as in Lemma 5.*

**Proof** Write

$$\frac{1 + z\bar{\zeta}}{(1 - z\bar{\zeta})\bar{\zeta}} = \frac{1}{\bar{\zeta}} + 2z \sum_{j=0}^{\infty} z^j \bar{\zeta}^j$$

and observe that for $q = 0$, we have

$$Q_{p,0}(z, \zeta) = \frac{1}{\pi} \int\limits_{|\zeta|<1} \zeta^p G_1(\zeta, \tilde{\zeta}) \frac{d\xi d\eta}{\bar{\zeta}} + 2z \left[ \sum_{j=0}^{\infty} \frac{z^j}{\pi} \int\limits_{|\zeta|<1} \zeta^p \bar{\zeta}^j G_1(\zeta, \tilde{\zeta}) d\xi d\eta \right]$$

$$= \overline{M_p(0, \tilde{\zeta})} + 2z \sum_{j=0}^{\infty} z^j L_{p,j}(\tilde{\zeta})$$

and for $q > 0$

$$Q_{p,q}(z, \tilde{\zeta}) = L_{p,q-1}(\tilde{\zeta}) + 2z \sum_{j=0}^{\infty} z^j L_{p,q+j}(\tilde{\zeta}) \qquad \square$$

The following Lemma can be proved as the Lemma 5.

**Lemma 7** *If* $m \in \mathbb{N}$, $|z|, |\tilde{\zeta}| < 1$, *then*

(i) $R_{1,m}(z, \tilde{\zeta}) = \dfrac{1}{\pi} \int\limits_{|\zeta|<1} (\zeta - z + \bar{\zeta} - \bar{z})^{m-1} G_1(\zeta, \tilde{\zeta}) \dfrac{\zeta + z}{\zeta(\zeta - z)} d\xi d\eta$

$$\sum_{(i,j) \in T(m-1)} M(i, j, m-1; z)[2N_{i,j}(z, \tilde{\zeta}) - N_{i,j}(0, \tilde{\zeta})]$$

(ii) $S_{1,m}(z, \tilde{\zeta}) = \dfrac{1}{\pi} \int\limits_{|\zeta|<1} (\zeta - z + \bar{\zeta} - \bar{z})^{m-1} G_1(\zeta, \tilde{\zeta}) \dfrac{1 + z\bar{\zeta}}{1 - z\bar{\zeta}} \dfrac{d\xi d\eta}{\zeta} =$

$$\sum_{(i,j) \in T(m-1)} M(i, j, m-1; z) Q_{i,j}(z, \tilde{\zeta})$$

**Lemma 8** *If* $r \in \mathbb{N}_0$, $k, m, s \in \mathbb{N}$, $|z|, |\tilde{\zeta}| < 1$, *then*

(i) $\quad T_{k,m,s}^r(z,\tilde\zeta) \;=\; \dfrac{1}{\pi}\displaystyle\int\limits_{|\zeta|<1}(1-|\zeta|^{2s})(\zeta-z+\overline{\zeta-z})^{m-1}H_k^r(\zeta,\tilde\zeta)\dfrac{\zeta+z}{\zeta(\zeta-z)}d\xi d\eta \;=\;$

$$\sum_{(p,q)\in T(m-1)} M(p,q,m-1;z)[2(C_{k,p,q}^r - C_{k,p+s,q+s}^r)(z,\tilde\zeta)$$

$$- (C_{k,p,q}^r - C_{k,p+s,q+s}^r)(0,\tilde\zeta)]$$

(ii) $\quad U_{k,m,s}^r(z,\tilde\zeta) = \dfrac{1}{\pi}\displaystyle\int\limits_{|\zeta|<1}(1-|\zeta|^{2s})(\zeta-z+\tilde\zeta-\bar z)^{m-1}H_k^r(\zeta,\tilde\zeta)\dfrac{1+z\bar\zeta}{(1-z\bar\zeta)\zeta}d\xi d\eta$

$$= \sum_{(p,q)\in T(m-1)} M(p,q,m-1;z)(E_{k,p,q}^r - E_{k,p+s,q+s}^r)(z,\tilde\zeta)$$

(iii) $\quad R_{2,m}(z,\tilde\zeta) = \dfrac{1}{\pi}\displaystyle\int\limits_{|\zeta|<1}(\zeta-z+\tilde\zeta-\bar z)^{m-1}G_2(z,\tilde\zeta)\dfrac{\zeta+z}{\zeta(\zeta-z)}d\xi d\eta$

$$= \sum_{(p,q)\in T(m-1)} M(p,q,m-1;z)[(\tilde\zeta\bar{\tilde\zeta}N_{p,q}-\tilde\zeta N_{p,q+1} \quad -\bar{\tilde\zeta}N_{p+1,q}+N_{p+1,\,q+1}]$$

$$+ (1-|\tilde\zeta|^2)\{2(I_{p+1,q+1}-I_{p,q})(z)+(I_{p,q}-I_{p+1,q+1})(0)\}]$$

*where*

$$I_{p,q}(z) = \begin{cases} \dfrac{1}{q+1}(z^{p-q-1}-z^p\bar z^{q+1}) & \text{if } p \geq q+1 \\[2mm] -\dfrac{1}{q+1}z^p\bar z^{q+1} & \text{if } p < q+1 \end{cases}$$

(iv) $\quad S_{2,m}(z,\tilde\zeta) = \dfrac{1}{\pi}\displaystyle\int\limits_{|\zeta|<1}(\zeta-z+\overline{\zeta-z})^{m-1}G_2(\zeta,\tilde\zeta)\dfrac{1+z\bar\zeta}{1-z\bar\zeta}\dfrac{d\xi d\eta}{\bar\zeta}$

$$= \sum_{(p,q)\in T(m-1)} M(p,q,m-1;z)[|\tilde\zeta|^2 Q_{p,q} - \tilde\zeta Q_{p,q+1} -\bar{\tilde\zeta}Q_{p+1,q}$$

$$- Q_{p+1,q+1}\,(z,\tilde\zeta)] + (1-|\tilde\zeta|^2)(zW_{p+1,q+1}-zW_{p,q}+W_{p+1,q}-W_{p,q-1})(z)]$$

*where*

$$W_{p,q}(z) = \begin{cases} \dfrac{1}{p+1}z^{p-q} & \text{if } p \geq q \\[2mm] 0 & \text{if } p < q \end{cases}$$

**Proof** (i) and (ii) involve similar computations as in Lemma 2 and 3. For (iii), we note that

$$G_2(\zeta,\tilde\zeta) = (\zeta\bar{\tilde\zeta}-\zeta\tilde\zeta-\zeta\bar{\tilde\zeta}+\zeta\tilde\zeta)G_1(\zeta,\tilde\zeta)+(1-|\tilde\zeta|^2)(\zeta\tilde\zeta-1)$$

Thus,

$$R_{2,m}(z,\tilde\zeta) = \sum_{(p,q)\in T(m-1)} M(p,q,m-1;z)[|\tilde\zeta|^2 N_{p,q}-\tilde\zeta N_{p,q+1}$$

$$-\bar{\tilde\zeta}N_{p+1,q}+N_{p+1,q+1})+ (1-|\tilde\zeta|^2)\{2(I_{p+1,q+1}-I_{p,q})(z)$$

$$+(I_{p,q}-I_{p+1,q+1})(0)\}]$$

where

$$I_{p,q}(z) = \frac{1}{\pi} \int\limits_{|\zeta|<1} \zeta^p \bar\zeta^q \frac{d\xi d\eta}{\zeta - z}$$

$$= \begin{cases} \dfrac{1}{q+1}(z^{p-q-1} - z^p \bar z^{q+1}) & \text{if } p \geq q+1 \\[2mm] -\dfrac{1}{q+1} z^p \bar z^{q+1} & \text{if } p < q+1 \end{cases}$$

As in (iii), we can write

$$S_{2,m}(z,\tilde\zeta) = \sum_{(p,q)\in T(m-1)} M(p,q,m-1;z)[(|\tilde\zeta|^2 Q_{p,q} - \tilde\zeta Q_{p,q+1} - \bar{\tilde\zeta} Q_{p+1,q}$$

$$+ Q_{p+1,q+1})(z,\tilde\zeta) + (1 - |\tilde\zeta|^2)(z W_{p+1,q+1} - z W_{p,q} + W_{p+1,q} - W_{p,q-1})(z)]$$

where for $p \geq 0 \; q \geq -1$

$$W_{p,q}(z) = \frac{1}{\pi} \int\limits_{|\zeta|<1} \zeta^p \bar\zeta^q \frac{d\xi d\eta}{1 - z\bar\zeta}$$

For $q = -1$

$$W_{p,-1}(z) = \frac{z^{p+1}}{p+1}$$

and for $q \geq 0$

$$W_{p,q}(z) = \begin{cases} 0 & \text{if } p < q \\[2mm] \dfrac{1}{p+1} z^{p-q} & \text{if } p \geq q \end{cases}$$

Thus, $W_{p,q}(z) = \dfrac{1}{p+1} z^{p-q}$ if $p \geq q$ and $0$ if $p < q$ $\qquad\qquad\square$

**Theorem 1** *The Dirichlet-Schwarz problem for the polyharmonic and polyanalytic inhomogeneous equation*

$$\partial_{\bar z}^{m+j} \partial_z^j w = f \text{ in } \mathbb{D}, \qquad \mathrm{Re}\,(\partial_{\bar z}^\mu w) = \gamma_\mu \tag{7}$$

$$\partial_{\bar z z}^{j-1}(\partial_{\bar z}^m w) = \beta_{j-1}, \qquad \mathrm{Im}\,(\partial_{\bar z}^\mu w(0)) = a_\mu$$

*is uniquely solvable for $f \in L_1(\mathbb{D};\mathbb{C})$, $\gamma_\mu \in C(\partial\mathbb{D};\mathbb{R})$, $\beta_j \in C(\partial\mathbb{D};\mathbb{C})$, $a_\mu \in \mathbb{R}$ for $j = 1,2; \; 0 \leq \mu \leq m-1$. The solution for $j = 1,2$ is given by*

$$w(z) = i \sum_{s=0}^{m-1} \frac{b_s}{s!}(z+\bar z)^s + \sum_{s=0}^{m-1} \frac{(-1)^s}{2\pi i s!} \int\limits_{|\zeta|=1} \gamma_s(\zeta)(\zeta - z + \bar\zeta - \bar z)^{m-1} \frac{\zeta + z}{\zeta(\zeta - z)} d\zeta$$

$$+\frac{(-1)^m}{2(m-1)!}\left[\frac{1}{2\pi i}\int\limits_{|\zeta|=1}\beta_0(\zeta)E^0_{1,m}+\overline{\beta_0(\zeta)}\ F^0_{1,m})(z,\zeta)\frac{d\zeta}{\zeta}\right.$$

$$-\frac{1}{\pi}\int\limits_{|\zeta|<1}f(\zeta)R_{j,m}+\overline{f(\zeta)}S_{j,m}](z,\zeta)d\xi d\eta$$

$$-r(j)\left[\frac{1}{2\pi i}\int\limits_{|\zeta|=1}\beta_1(\zeta)T^1_{1,m,1}+\overline{\beta_1(\zeta)}U^1_{1,m,1}](z,\zeta)\frac{d\zeta}{\zeta}\right.$$

$$\left.+\frac{1}{\pi}\int\limits_{|\zeta|<1}(|\zeta|^2-1)\{f(\zeta)T^1_{1,m,1}+\overline{f(\zeta)}U^1_{1,m,1}]\ (z,\zeta)d\xi d\eta\right] \tag{8}$$

*where $r(1)=0$ and $r(2)=1$.*

**Proof** We rewrite the problem (7) as the system

$$\partial^m_{\bar z}w=\omega\text{ in }\mathbb{D},\ \mathrm{Re}(\partial^\mu_{\bar z}w)=\gamma_\mu\text{ on }\partial\mathbb{D} \tag{9}$$

$$\mathrm{Im}(\partial^\mu_{\bar z}w(0))=a_\mu,\ 0\le\mu\le m-1$$

$$\partial^j_{\bar z z}\omega=f\text{ in }\mathbb{D},\ \partial^j_{\bar z z}\omega=\beta_{j-1}\text{ on }\partial\mathbb{D},\ j=1,2 \tag{10}$$

Using Theorem 1, the solution of (9) is given by

$$w(z)=i\sum_{s=0}^{m-1}\frac{b_s}{s!}(z+\bar z)^s+\sum_{s=0}^{m-1}\frac{(-1)^s}{2\pi!s!}\int\limits_{|\zeta|=1}\gamma_s(\zeta)\frac{\zeta+z}{\zeta-z}(\zeta-z+\overline{\zeta-z})^s\frac{d\zeta}{\zeta}$$

$$+\frac{(-1)^m}{2\pi(m-1)!}\int\limits_{|\zeta|<1}\left(\frac{\omega(\zeta)}{\zeta}\frac{\zeta+z}{\zeta-z}+\frac{\overline{\omega(\zeta)}}{\bar\zeta}\frac{1+z\bar\zeta}{1-z\bar\zeta}\right)(\zeta-z+\overline{\zeta-z})^{m-1}d\xi d\eta \tag{11}$$

For $j=1$, the solution of (10) is given by

$$w(z)=\frac{1}{2\pi i}\int\limits_{|\zeta|=1}\beta_0(\zeta)H^0_1(z,\zeta)\frac{d\zeta}{\zeta}-\frac{1}{\pi}\int\limits_{|\zeta|<1}G_1(z,\zeta)f(\zeta)d\xi d\eta \tag{12}$$

and for $j=2$, the solution of (10) is given by

$$w(z)=\frac{1}{2\pi i}\int\limits_{|\zeta|=1}\beta_0(\zeta)H^0_1(z,\zeta)-(1-|\zeta|^2)H^1_1(z,\zeta)\beta_1(\zeta))\frac{d\zeta}{\zeta}$$

$$-\frac{1}{\pi}\int\limits_{|\zeta|<1}f(\zeta)[G_2(z,\zeta)+(1-|z|^2)(|\zeta|^2-1)H^1_1(z,\zeta)]d\xi d\eta \tag{13}$$

To obtain the required solution, it is sufficient to obtain the area integral in (11).

Iterating the integrals from (12), the area integral of (11) can be expressed as

$$= \frac{1}{\pi} \int\limits_{|\tilde{\zeta}|<1} \left[ \frac{1}{2\pi i} \int\limits_{|\zeta|=1} H_1^0(\zeta,\tilde{\zeta})\beta_0(\tilde{\zeta})\frac{d\tilde{\zeta}}{\tilde{\zeta}} - \frac{1}{\pi} \int\limits_{|\tilde{\zeta}|<1} G_1(\zeta,\tilde{\zeta})f(\tilde{\zeta})d\tilde{\xi}d\tilde{\eta} \right]$$

$$(\zeta - z + \bar{\zeta} - \bar{z})^{m-1}\frac{\zeta+z}{\zeta(\zeta-z)}d\xi d\eta - \frac{1}{\pi} \int\limits_{|\tilde{\zeta}|<1} \left[ \frac{1}{2\pi i} \int\limits_{|\zeta|=1} H_1^0(\zeta,\tilde{\zeta})\overline{\beta_0(\tilde{\zeta})}\frac{\overline{d\tilde{\zeta}}}{\overline{\tilde{\zeta}}} \right.$$

$$\left. + \frac{1}{\pi} \int\limits_{|\tilde{\zeta}|<1} G_1(\zeta,\tilde{\zeta})\overline{f(\tilde{\zeta})}d\tilde{\xi}d\tilde{\eta} \right] (\zeta - z + \bar{\zeta} - \bar{z})^{m-1}\frac{1+z\bar{\zeta}}{(1-z\zeta)\bar{\zeta}}d\xi d\eta$$

Usual complex analysis results and Lemmas 4 and 7 enable us to apply Fubini's theorem. So the above expression can be written as

$$\frac{1}{\pi} \int\limits_{|\zeta|=1} [\beta_0(\zeta)E_{1,m}^0 + \overline{\beta_0(\zeta)}F_{1,m}^0](z,\zeta)\frac{d\zeta}{\zeta}$$

$$- \frac{1}{\pi} \int\limits_{|\zeta|<1} [f(\zeta)R_{1,m} + \overline{f(\zeta)}S_{1,m}](z,\zeta)d\xi d\eta$$

Next, substituting the value of $\omega$ from (13) in the area integral of (11), we get it as

$$\frac{1}{\pi} \int\limits_{|\tilde{\zeta}|<1} \left[ \frac{1}{2\pi i} \int\limits_{|\zeta|=1} [\beta_0(\tilde{\zeta})H_1^0(\zeta,\tilde{\zeta}) - (1-|\zeta|^2)H_1^1(\zeta,\tilde{\zeta})\beta_1(\tilde{\zeta})]\frac{d\tilde{\zeta}}{\tilde{\zeta}} \right]$$

$$- \frac{1}{\pi} \int\limits_{|\tilde{\zeta}|<1} f(\zeta)[G_2(\zeta,\tilde{\zeta}) + (1-|\zeta|^2)(|\tilde{\zeta}|^2-1)H_1^1(\zeta,\tilde{\zeta})d\tilde{\xi}d\tilde{\eta}]$$

$$(\zeta - z + \bar{\zeta} - \bar{z})^{m-1}\frac{\zeta+z}{\zeta(\zeta-z)}d\xi d\eta$$

$$- \frac{1}{\pi} \int\limits_{|\zeta|<1} \left[ \frac{1}{2\pi i} \int\limits_{|\zeta|=1} \overline{\beta_0(\tilde{\zeta})}H_1^0(\zeta,\tilde{\zeta}) - (1-|\zeta|^2)H_1^1(\zeta,\tilde{\zeta})\overline{\beta_1(\tilde{\zeta})} \right] \frac{\overline{d\tilde{\zeta}}}{\overline{\tilde{\zeta}}}$$

$$+ \frac{1}{\pi} \int\limits_{|\tilde{\zeta}|<1} \overline{f(\zeta)}[G_2(\zeta,\tilde{\zeta}) + (1-|\zeta|^2)(|\tilde{\zeta}|^2-1)H_1^1(\zeta,\tilde{\zeta})]d\tilde{\xi}d\tilde{\eta}]$$

$$(\zeta - z + \bar{\zeta} - \bar{z})^{m-1}\frac{1+z\bar{\zeta}}{\bar{\zeta}(1-z\zeta)}d\xi d\eta$$

$$= \frac{1}{2\pi i} \int\limits_{|\zeta|=1} [\beta_0(\zeta)E^0_{1,m} + \overline{\beta_0(\zeta)}F^0_{1,m} - (\beta_1(\zeta)T^1_{1,m,1} + \overline{\beta_1(\zeta)}U^1_{1,m,1})](z,\zeta)\frac{d\zeta}{\zeta}$$

$$-\frac{1}{\pi} \int\limits_{|\zeta|<1} f(\zeta)[\{R_{2,m} + (|\zeta|^2 - 1)T^1_{1,m,1}\}$$

$$+\overline{f(\zeta)}\{S_{2,m} + (|\zeta|^2 - 1)U^1_{1,m,1}\}](z,\zeta)d\xi d\eta \qquad \square$$

The solution of (7) given in (8) is indeed a solution can be verified using differentiability of the operators $T_{m,n}$ [4]. $\qquad \square$

We can explicitly write the solution of problem (7) for $j = 3, 4, 5$ using explicit solutions in [15] for the Dirichlet problem on $n$-Poisson equation but this has been avoided due to lack of space.

## References

1. H. Begehr, Complex analytic methods for partial differential equations. An introductory text. World Scientific, Singapore, 1994.
2. H. Begehr, Orthogonal decomposition of the function space $L_2(\mathbb{D},\mathbb{C})$. *J. Reine Angew. Math.* 549 (2002), 191–219.
3. H. Begehr, Boundary value problems in complex analysis, I, I, Bol. Asoc. Mat. Venezolana, 12 (2005), 65–85; 217–250.
4. H. Begehr, G.N. Hile, A hierarchy of integral operators. *Rocky Mountain J. Math.* 27 (1997), 669–706.
5. H. Begehr, A. Kumar, Boundary value problems for the inhomogeneous polyanalytic equation I. *Analysis* 25 (2005), 55–71.
6. H. Begehr, A. Kumar, Boundary value problems for the inhomogeneous polyanalytic equation II. *Analysis* (2007), to appear.
7. H. Begehr, A. Kumar, Boundary value problems for bi-polyanalytic functions. Applicable Analysis 85 (2006), 1045–1077.
8. H. Begehr, D. Schmersau, The Schwarz problem for polyanalytic functions. ZAA 24 (2005), 341–351.
9. H. Begehr, C.J. Vanegas, Iterated Neumann problem for the higher order Poisson equation. *Math. Nachr.* 279 (2006), 38–57.
10. H. Begehr, T.N.H. Vu, Z.X. Zhang, Polyharmonic Dirichlet Problems. Proc. Steklov Inst. Math. 255 (2006), 13–34.
11. W. Haack, W. Wendland, Lectures on partial and Pfaffian differential equations. Pergamon Press, Oxford, 1972.
12. A. Kumar, R. Prakash, Boundary value problems for the Poisson equation and bi-analytic functions. *Complex Variables Theory Appl.* 50 (2005), 597–609.
13. A. Kumar, R. Prakash, Neumann and mixed boundary value problems. *Journal of Applied Functional Analysis* (2007), to appear.
14. A. Kumar, R. Prakash, Mixed boundary value problems for the inhomogeneous polyanalytic equation. (Preprint)
15. A. Kumar, R. Praksah, Dirichlet problem for inhomogeneous polyharmonic equation. (preprint)
16. M. Niolescu, Les functions polyharmonic, Hermann, Paris, 1936.
17. I.N. Vekua, New methods for solving elliptic equations, North Holland, Amsterdam, John Wiley, New York, 1967.

# INITIAL VALUE PROBLEM FOR A SYSTEM OF EQUATIONS OF CRYSTAL OPTICS

N.A. ZHURA

*Lebedev Physical Institute of Russian Academy of Sciences*
*Leninskiy pr. 53, Moscow, 119991, Russian Federation*
*E-mail: nzhura@sci.lebedev.ru*

An explicit formula for a solution of the initial value problem for a system of equations of crystal optics is obtained. For a particular class of nonnegative biquadratic ternary forms a reduction to canonical forms is obtained.

**Key words:** Initial value problem, system of equations of crystal optics, nonnegative biquadratic ternary form
**Mathematics Subject Classification:** 35J25, 35J55, 45F15

## 1. Introduction

Let $a$, $b \in \mathbb{R}^{3 \times 3}$ be arbitrary symmetric positive definite matrices. We consider the system

$$\frac{\partial u_1}{\partial t} = \operatorname{rot} b u_2, \quad \operatorname{div} u_1 = 0, \quad \frac{\partial u_2}{\partial t} = -\operatorname{rot} a u_1, \quad \operatorname{div} u_2 = 0 \qquad (1)$$

with respect to vector- valued functions $u_1(x,t)$ and $u_2(x,t)$ in the domain $\{x \in \mathbb{R}^3, t > 0\}$. It becomes an ordinary system of crystal optics in the case $a = \varepsilon^{-1}$, $b = \mu^{-1}$, where $\varepsilon$ and $\mu$ are so called electric and magnetic susceptibility. If $\mu$ is a scalar this system was considered in [1]. In the literature in particular of physics the notation $u_1 = D$, $u_2 = B$, where $D = \varepsilon E$, $B = \mu H$ is widely known . The number of equations in (1) is greater than the number of unknown functions and so the system is overdetermined [2,3]. As it was noted in [2] this system is a significant example of such one.

Setting $U = (u_1, u_2)$ we can rewrite (1) in the form

$$A_0 \frac{\partial U}{\partial t} = A_1 \frac{\partial U}{\partial x_1} + A_2 \frac{\partial U}{\partial x_2} + A_3 \frac{\partial U}{\partial x_3}, \qquad (2)$$

with the correspondence matrixes $A_j \in \mathbb{R}^{8 \times 6}$. The rank of the matrix symbol

$$A(\lambda, \xi) = A_0 \lambda - A_1 \xi_1 - A_2 \xi_2 - A_3 \xi_3 \qquad (3)$$

of this system is equal to 6. Besides all minors of sixth order coincide, provided they are not identically equal to zero. The common value $h(\lambda, \xi)$ of these minors we call a characteristic form of the system (2).

The system (2) of such type is called hyperbolic, if for every $\xi \neq 0$ all roots of the characteristic equation

$$h(\lambda, \xi) = 0 \tag{4}$$

with respect to $\lambda$ are real. If additionally these roots are mutually different the system is called strictly hyperbolic. We will prove later that system (1) is hyperbolic, but not strictly hyperbolic. The above definitions are analogous to the classical case[4].

In this paper we give a solution of the initial value problem in an explicit form for the system (1). To obtain this result we describe the reduction to canonical type of one class of biquadratic ternary forms [5] (see also [6,7]). As is known from a D. Hilbert result [5] an arbitrary nonnegative biquadratic ternary form can be represented as a sum of three quarters of irreducible quadratic forms. However a constructive form of that result apparently is absent (see for example a commentary in [6]). In this paper for a particular class of such forms we obtain the correspondence representation as a sum of two quarters of quadratic forms. With the help of this result we implicitly construct a solution of the initial value problem for (1). Very interesting results were obtained in [8,9](see also commentary about one of them [10]). At the end of the paper we consider the structure of a normal surface [4] and some generalizations [11,12] of (1).

## 2. Characteristic form and geometry of a normal surface

Let us introduce the skew-symmetric matrix

$$[\xi] = \begin{pmatrix} 0 & -\xi_3 & \xi_2 \\ \xi_3 & 0 & -\xi_1 \\ -\xi_2 & \xi_1 & 0 \end{pmatrix}, \quad \xi \in \mathbb{R}^3.$$

**Lemma 1.** *The characteristic form of (1) can be represented in the form*

$$h(\lambda, \xi) \equiv \lambda^4 + p(\xi)\lambda^2 + q(\xi), \tag{5}$$

$$p(\xi) = \operatorname{tr}([\xi]b[\xi]a), \quad q(\xi) = (a^*\xi, \xi)(b^*\xi, \xi). \tag{6}$$

Here $\operatorname{tr} g$ means the trace of the matrix $g$ and $g^*$ is an adjoint matrix. Note that $g^* = (\det g)g^{-1}$ under the assumption $\det g \neq 0$.

It is seen from (6) that all roots of (4) are equal to $\pm\lambda_\pm(\xi)$ (all combinations of signs are possible), where

$$\lambda_\pm^2(\xi) = (-p(\xi) \pm \sqrt{d(\xi)})/2, \quad d(\xi) = p^2(\xi) - 4q(\xi), \tag{7}$$

are roots of biquadratic equation (5). The roots are real if the quadratic form $-p(\xi)$ is positively defined and the discriminant $d(\xi)$ is nonnegative for all $\xi \neq 0$. The biquadratic ternary form $d(\xi)$ will play an important role.

Let us put

$$s = a^{-1/2}ba^{-1/2}, \tag{8}$$

and consider the orthogonal matrix $e$ reducing the symmetric positively defined matrix $s$ to a diagonal form:

$$e^{-1}se = \nu, \quad \nu = \text{diag}(\nu_1, \nu_2, \nu_3). \tag{9}$$

If all eigenvalues of $s$ are mutually distinct then we can choose a permutation $\{i, k, j\}$ of indices $\{1, 2, 3\}$ such that

$$\nu_i < \nu_k < \nu_j. \tag{10}$$

If there exist only two distinct eigenvalues then we set

$$\nu_i = \nu_k < \nu_j, \quad \nu_i < \nu_k = \nu_j. \tag{11}$$

**Theorem 1.** (a) *Under assumption (10) the form $d(\xi)$ in the new variables*

$$\zeta = e^{-1}(a^*)^{1/2}\xi, \tag{12}$$

*admits the representation*

$$d(\zeta) = P^2(\zeta) + Q^2(\zeta) \tag{13}$$

*with the quadratic forms*

$$P(\zeta) = (\nu_k - \nu_j)\zeta_i^2 + (\nu_k - \nu_i)\zeta_j^2 + (\nu_j - \nu_i)\zeta_k^2, \quad Q(\zeta) = 2\sqrt{(\nu_j - \nu_i)(\nu_j - \nu_k)}\zeta_i\zeta_k. \tag{14}$$

(b) *Let the first or second condition in (11) be satisfied. Then the equality*

$$d(\zeta) = R^2(\zeta) \tag{15}$$

*is valid with the corresponding quadratic form*

$$R(\zeta) = (\nu_j - \nu_i)(\zeta_i^2 + \zeta_k^2) \quad \text{or} \quad R(\zeta) = (\nu_j - \nu_i)(\zeta_j^2 + \zeta_k^2). \tag{16}$$

**Corollary 1.** *Let the condition of Theorem 10 and inequalities (10) be satisfied. Then the form $d(\xi)$ can be represented as a product of nonnegative quadratic forms*

$$d(\zeta) = d_+(\zeta)d_-(\zeta), \tag{17}$$

*where*

$$d_\pm(\zeta) = (\nu_j - \nu_i)\zeta_k^2 + [(\sqrt{\nu_j - \nu_k})\zeta_i \pm (\sqrt{\nu_k - \nu_i})\zeta_j]^2.$$

In order to deduce this result we observe that $(\nu_k - \nu_i)(\nu_k - \nu_j) < 0$ and take (14) into account.

**Corollary 2.** *Under assumption (10) a set of zeroes of form (13) coincides with a union of lines*

$$\{\zeta_k = 0, (\sqrt{\nu_j - \nu_k})\zeta_i + (\sqrt{\nu_k - \nu_i})\zeta_j = 0\},$$

$$\{\zeta_k = 0, (\sqrt{\nu_j - \nu_k})\zeta_i - (\sqrt{\nu_k - \nu_i})\zeta_j = 0\}. \tag{18}$$

*If only one of the inequalities (11) is fulfilled then this set coincides with one of the lines*

$$\{\zeta_i = \zeta_k = 0\}, \quad \{\zeta_j = \zeta_k = 0\}. \tag{19}$$

**Corollary 3.** *The eigenvalues of the matrices $s$ and $t = ba^{-1}$ coincide, but their eigenvectors do not coincide. In particular the main result of [9] is valid if and only if the matrixes $a$ and $b$ commute.*

The proof of the theorem is a consequence of the following propositions.

**Lemma 2.** *The quadratic form $-p(\xi)$ is positively definite and*

$$-p(\xi) = (c_0^* \xi, \xi), \quad c_0^* = (a+b)^* - a^* - b^*. \tag{20}$$

The equality in (20) may be verified directly. The proof of the positive definiteness of that form is based on the Silvester criterion.

**Lemma 3.** *The matrices $s$, $s^*$ and $c^* = (a^*)^{-1/2} c_0^* (a^*)^{-1/2}$ are mutually commutative. In particular*

$$e^{-1} c^* e = m, \quad e^{-1} s^* e = n. \tag{21}$$

*We have also the following relations*

$$m_k = s_0 - \nu_k, \quad n_k = \nu_i \nu_j,$$

*between their eigenvalues where $s_0 = \operatorname{tr} s$ and $\{i, k, j\}$ is a permutation (10) of indices $\{1, 2, 3\}$.*

**Lemma 4.** *Under assumption (10) the equality*

$$d(\zeta) = (m\zeta, \zeta) - 4(\zeta, \zeta)(n\zeta, \zeta)$$

*is valid for the variables (12) and the relations*

$$m_k 2 - 4 n_k = (\nu_j - \nu_i)2, \quad m_i m_j - 2(n_i + n_j) = (\nu_k - \nu_i)(\nu_k - \nu_j),$$

*hold.*

Let us consider a particular case, when $a$ and $b$ do not commute but they have a common principal axes. This case admits some simplification. Namely the matrix $a$ and $b$ written in the form

$$a = \begin{pmatrix} a_1 & 0 & 0 \\ 0 & a_2 & 0 \\ 0 & 0 & a_3 \end{pmatrix}, \quad b = \begin{pmatrix} b_1 & 0 & 0 \\ 0 & b_2 & b_4 \\ 0 & b_4 & b_3 \end{pmatrix},$$

are block-diagonal with respect to a partition of the set $\{1, 2, 3\}$ in two subsets. That properties are also valid for the matrices $c^*$, $s$, $s^*$, and $e$. In particularly their eigenvalues can be easily found. For example, if $s_i = b_i/a_i$, $i = 1, 2, 3$, and $g = b_4/\sqrt{a_2 a_3}$, then the eigenvalue of $s$ is $\nu_2 = s_2$ and the roots of the quadratic equation

$$\nu^2 - \nu(s_1 + s_2) + s_1 s_2 - g^2 = 0.$$

In particular all coefficients of the forms $P, Q, R$ have an explicit form. They can be expressed by means of elements of the matrices $a$ and $b$. In this case the form $d(\xi)$ with respect to one of the variables has only an even degree and so one can not use the transformation (12).

At last if $a$ and $b$ are commutative, then $e$ is the unit matrix in (12). Moreover the matrices $s$ and $t$ coincide and $d(\xi) \equiv 0$. This fact essentially simplifies the proof of the theorem in this case. The details are omitted.

Passing to affine coordinates $n$, $\xi = \lambda n$ in (4), we get the algebraic surface $\sigma$ of fourth order

$$\sigma : \quad 1 + p(n) + q(n) = 0, \tag{22}$$

which are decomposed in two pieces $\sigma_\pm$ with the equations

$$\sigma_\pm : \quad -p(n) \pm \sqrt{d(n)} = 2. \tag{23}$$

By transformation (13) these surfaces are reduced to canonical type, which is analogous the classical case $\mu = 1$ [1]. Moreover $\varepsilon_k^{-1}$ in [1] must be replaced by $\nu_k$, $k = 1, 2, 3$. In particular $\sigma_\pm$ have four or two common points, which are singular for the surface $\sigma$. That points are located on one or two lines with equation (18) ,(19) respectively. Other geometrical properties of the surface (22) in the variables $\zeta$ are omitted, because they are similar to the classical case $\mu = 1$.

## 3. Initial value problem

Let us consider the Cauchy problem

$$u_k(0, x) = g_k(x), \quad k = 1, 2, \tag{24}$$

for the system (1). The vector-valued functions $g_k$ here are taken from the Schwartz class $S(\mathbb{R}^3)$ i.e these functions together with their derivatives of arbitrary order are decrease at infinity faster than every power of $(x_1^2 + x_2^2 + x_3^2)^{-1}$. Let us consider the Fourier transform

$$\hat{u}_k(t, \xi) = \int_{\mathbb{R}^3} u_k(t, x) e^{i\xi x} dx \tag{25}$$

of the unknown functions $u_k(t, x)$, $k = 1, 2$. The problem (1),(24) is equivalent to the initial value problem for the system of ordinary differential equations with parameter

$$\frac{d\hat{u}_1(t, \xi)}{dt} = i[\xi] b \hat{u}_2(t, \xi), \quad \xi \hat{u}_1(t, \xi) = 0,$$

$$\frac{d\hat{u}_2(t, \xi)}{dt} = -i[\xi] a \hat{u}_1(t, \xi), \quad \xi \hat{u}_2(t, \xi) = 0, \tag{26}$$

where the initial conditions

$$\hat{u}_k(0, \xi) = \hat{g}_k(\xi), \quad k = 1, 2, \tag{27}$$

also belong to $S(\mathbb{R}^3)$.

**Lemma 5.** *Let a vector-valued functions $g_k(x)$, $k = 1, 2$ belong to $S$ and $\mathrm{div} g_k(x) = 0$, $k = 1, 2$. Then problem (26), (27) has a unique classical solution*

$$\hat{u}_1(t, \xi) = \hat{g}_1(\xi) \cos t\lambda_+(\xi) + i[\xi]b\hat{g}_2(\xi)\frac{\sin t\lambda_-(\xi)}{\lambda_-(\xi)},$$

$$\hat{u}_2(t, \xi) = \hat{g}_2(\xi) \cos t\lambda_-(\xi) - i[\xi]a\hat{g}_1(\xi)\frac{\sin t\lambda_+(\xi)}{\lambda_+(\xi)},$$

*where the functions $\lambda_\pm(\xi)$ are defined by (7).*

Let $\Gamma = \sigma_+(x, t) \cup \sigma_-(x, t)$ be the compact piecewise smooth surface which is obtained from $\sigma_\pm$ in (23) by translation of the origin to a point $x$ and dilation $t > 0$ along each axes . Let $\delta_\Gamma$ be a distribution which is defined by the formula

$$\langle \delta_\Gamma, \varphi \rangle = \int_\Gamma \varphi(\eta) d\sigma_\eta,$$

where $d\sigma$ is the element of surface measure.

Then for the Fourier transform $\left(\delta_{\sigma_\pm(x,t)}\right)^\wedge$ the relation

$$\left(\frac{\delta_{\sigma_\pm(x,t)}}{|\sigma_\pm(x,t)|}\right)^\wedge = \frac{\sin t\lambda_\pm(\xi)}{t\lambda_\pm(\xi)},$$

is valid. In this formula the functions $\lambda_\pm(\xi)$ are defined as above and $|\sigma_\pm(x,t)|$ is a measure of $\sigma_\pm$. Let us set now

$$(M_\pm\varphi)(t, x) = \frac{1}{|\sigma_\pm|}\int_{\sigma_\pm} \varphi(\eta)d\sigma_\eta.$$

**Theorem 2.** *Let vector-valued functions $g_k(x)$ belong to the Schwartz space $S$ and $\mathrm{div} g_k(x) = 0$, $k = 1, 2$. Then the initial value problem (1), (24) has a unique classical solution which is described by the formula*

$$u_1(t, x) = \frac{\partial}{\partial t}\left(t(M_+g_1)(t, x)\right) + t(M_-\mathrm{rot}\, bg_2)(t, x),$$

$$(28)$$

$$u_2(t, x) = -t(M_+\mathrm{rot}\, ag_1)(t, x) + \frac{\partial}{\partial t}(t(M_-g_2)(t, x).$$

**Corollary 1.** *The existence and uniqueness theorem for problem (1),(24)in $C^\infty(\mathbb{R}^3) \cap L2(\mathbb{R}^3)$ hold.*

**Corollary 2** (Huygens principle). *The solution $u(t, x)$ of (1) at a point $(t, x)$ is defined by the values of the initial function $g(x)$ in some neighborhood of the surface $\sigma(x, t)$.*

## 4. Generalizations of crystal optics system

At first consider system (1) with lower order terms

$$\frac{\partial u_1}{\partial t} = \mathrm{rot}\, bu_2 - \gamma u_1, \quad \frac{\partial u_2}{\partial t} = -\mathrm{rot}\, au_1 + \gamma u_2, \qquad (29)$$

where the constant $\gamma > 0$. Accordingly to Section 2 we have the following conjecture.

*The system (29) has a characteristic polynomial of type (5), where $\lambda$ must be replaced by $\lambda + i\gamma$. In particular this system is dissipative hyperbolic. A solution of this system is equal to a solution of problem (1),(24) times $e^{-\gamma t}$. The value*

$$\int_{\mathbb{R}^3} [(u_1, au_1) + (u_2, bu_2)]dx$$

*for each $t > 0$ is equal to that value for $t = 0$ times $e^{-2\gamma t}$.*

A next generalizations of a crystal optics system of equation is analogous [11]

$$\frac{\partial u_1}{\partial t} = \operatorname{rot} bu_2 - \beta \operatorname{grad} \operatorname{div} u_2, \quad \frac{\partial u_2}{\partial t} = -\operatorname{rot} bu_1 + \beta \operatorname{grad} \operatorname{div} u_1,$$

where constant $\beta > 0$.

This system is a generalization of a system from [9], however problem (24) for it have some speciality. Namely , if $u_k = v_k + w_k$, $\operatorname{div} v_k = 0$, $\operatorname{rot} w_k = 0$, $k = 1, 2$, and if $w = w_1 + iw_2$ then a Schrödinger equation without potential $iw_t = -\Delta w$ occur. However, the solenoidal vector-valued functions $v_1$ and $v_2$ will be connected with potential vectors $w_k$ because

$$\frac{\partial v_1}{\partial t} = \operatorname{rot} bv_2 + \operatorname{rot} bw_2, \quad \operatorname{div} v_1 = 0, \quad \frac{\partial v_2}{\partial t} = -\operatorname{rot} av_1 + \operatorname{rot} aw_1, \quad \operatorname{div} v_2 = 0.$$

The initial value problem for this system is also studied and its solution is obtained in explicit form.

### References

1. Courant, R. *Partial differential equatins*, New York -London,1962.
2. Petrovskiy, I.G., *Selected works. Systems of partial differential equations. Algebraic geometry.* Moscow, Nauka, 1986 (in Russian).
3. Spencer, D.C. *Overdetermined systems of linear partial differential equations*, Bull. Amer. Math. Soc. 75, No 2, (1969), 179–239.
4. Bers, L., John, F., Schechter, M. *Partial differential equations*, Intersciense Publishers, New York-London-Sydney, 1964.
5. Hilbert, D. *Ueber die Darstellung definierter Formen als Summe von Formenquadraten*, Math. Ann. 32, (1888), 342–350.
6. Hilbert, D. *Selected works*, Volume 1, 1998.
7. Fedorov, F.I. *Optics of anisotropic media.* Minsk, 1958; second edition, Moscow, 2004 (Russian).
8. John, F. *Algebraic conditions for hyperbolicity of systems of partial differential equations*, Comm. Pure Appl. Math., XXXI (1978), 89–106.
9. John, F. *Addendum to: Algebraic conditions for hyperbolicity of systems of partial differential equations*, Comm. Pure Appl. Math., XXXI (1978), 787–793.
10. Garding, L. *Hyperbolic differential operators.* In Perspectives in Mathematics. Anniversary of Oberwolfach 1984, Birkhauser Verlag. Basel, 215–247.
11. Zhura, N.A. *Initial value problem for a system equations of crystal optic.* Intern. Conf. Diff. Eq. Dynamical Systems. Book of abstract, Suzdal, 1–6 July 2002, 72–73 (Russian).
12. Zhura, N.A., Oraevskii, A.N. *Initial value problem for one hyperbolic system with constant coefficients.* Doklady RAS, 396, No 5, (2004), 590–594 (Russian).

# ON REPRESENTATION OF SOLUTIONS OF SECOND ORDER ELLIPTIC SYSTEMS IN THE PLANE

A.P. SOLDATOV

*Belgorod State University*
*Pobeda str. 85, Belgorod, 308015, Russian Federation*
*E-mail: soldatov@bsu.edu.ru*

The analogue of the Privalov theorem is established for solutions of elliptic systems of second order in Lipshitz domains. The corresponding result is also obtained with respect to the weighted Hölder spaces. The notion of conjugate functions to solutions of elliptic systems is introduced and representation formulas for them are received. For so called strengthen elliptic systems the class of degenerate solutions is described.

**Key words**: Dirichlet problem, elliptic systems, Hölder spaces, Lipshitz domains, conjugate functions, strong and strengthen elliptic systems
**Mathematics Subject Classification**: 35J25, 35J55

## 1. Harmonic functions

Let $C^\mu(E) = C^{0,\mu}(E)$, $0 < \mu \leq 1$, be the ordinary Hölder space of functions $\varphi(z), z \in E \subseteq \mathbb{C}$, which is Banach with respect to the norm

$$|\varphi|_\mu = |\varphi|_0 + [\varphi]_\mu; \quad |\varphi|_0 = \sup_{z \in E} |\varphi(z)|, \quad [\varphi]_\mu = \sup_{z,z' \in E, z \neq z'} \frac{|\varphi(z) - \varphi(z')|}{|z - z'|^\mu}. \quad (1.1)$$

In the case $\mu = 1$ it is called the Lipschitz space and denoted by $C^{0,1}(E)$. If the set $E$ is not closed a function $\varphi \in C^\mu(E)$ is extended to the function $\tilde{\varphi} \in C^\mu(\overline{E})$ and $[\varphi]_{\mu,E} = [\tilde{\varphi}]_{\mu,\overline{E}}$. Besides if the set $E$ is bounded then (1.1) is equivalent to the norm

$$|\varphi| = |\varphi(c)| + [\varphi]_\mu, \quad c \in E.$$

Let a function $u(z) = u(x,y)$ be harmonic in the unit disk $B = \{|z| < 1\}$ and belong to $C^\mu(\overline{B})$, $0 < \mu < 1$. The known Privalov theorem [1] asserts that the analytic function $\phi(z)$ for which

$$u = \operatorname{Re}\phi, \quad (1.2)$$

belongs to the same class and the estimate $[\phi]_\mu \leq M|u|_\mu$ is valid where the constant $M > 0$ depends only on $\mu$. We extend this result to the case when $B$ is an arbitrary Lipschitz domain but $u$ is a solution of a second order elliptic system with constant and only leading coefficient.

The work was supported by the program of Russian Universities (project No. UR 04.01.486)

Recall the definition of the mentioned domains. Following Stein [2] the set

$$\{z = x + iy \mid f(x) < y\}, \tag{1.3}$$

where $f \in C^{0,1}(\mathbb{R})$, is called by a special Lipschitz domain. We attribute the domains to the same type which result by motion from (1.3). In the general case a finite domain $D$ is Lipschitz if there exist open sets $V_1, \ldots, V_n$ and special Lipschitz domains $D_1, \ldots, D_n$ such that

$$\partial D \subseteq V_1 \cup \ldots \cup V_n, \quad D \cap V_i = D_i \cap V_i, \ i = 1, \ldots, n. \tag{1.4}$$

To illustrate our approach we consider first the classic case of harmonic functions. Our approach is based on the following property of Hölder spaces.

**Lemma 1.** *Let a function $\varphi(x, y)$ be bounded and continuously differential in a Lipschitz domain $D$, and the following estimate*

$$\left| \frac{\partial \varphi}{\partial x}(z) \right| + \left| \frac{\partial \varphi}{\partial y}(z) \right| \leq C[\rho(z, \partial D)]^{\mu - 1}, \quad 0 < \mu \leq 1, \tag{1.5}$$

*hold where $\rho(z, \partial D)$ denotes the distance from a point $z \in D$ to the boundary $\partial D$. Then $\varphi \in C^{0,\mu}(\overline{D})$ and*

$$[\varphi]_{\mu, D} \leq MC, \tag{1.6}$$

*where the constant $M > 0$ depends only on $\mu$ and $D$.*

Apparently this result is known but for completeness we produce its proof below.
**Proof.** Let us first assume that $D$ is the special Lipschitz domain $D$ of the form (1.4). Let $\theta(z)$ be the angle between the vector $z \in \mathbb{C}$ and the $y$–axis and thus $\theta(z) = \arctan(|x|/y)$, $z = x + iy$. Let $K(\alpha)$, $0 < \alpha < \pi/2$, be a sector $\{z \mid \theta(z) < \alpha\}$. Note that the distance from $z \in K$ to its boundary is defined by the equality

$$\rho[z, \partial K(\alpha)] = |z| \sin[\alpha - \theta(z)]. \tag{1.7}$$

Let the sector $K_a$ be received by parallel translation of $K$ to the vertex $a = a_1 + i a_2 \in \Gamma = \partial D$. Verify that $K_a(\alpha) \subseteq D$ under assumption

$$[f]_1 < \cot \alpha. \tag{1.8}$$

In fact let

$$\tan \theta(z - a) = \frac{|x - a_1|}{y - a_2} < \tan \alpha,$$

or $y > f(a_1) + |x - a_1| \cot \alpha$. As $|f(x) - f(a_1)| \leq [f]_1 |x - a_1|$, we have $y > f(x) + (\cot \alpha - [f]_1)|x - a_1|$. Taking (1.8) into account we receive $y > f(x)$.

Let us set $0 < \alpha < \alpha^0$, where $\alpha^0$ satisfies (1.8) and put for brevity $K_a = K_a(\alpha)$, $K_a^0 = K_a(\alpha^0)$. Then by virtue of (1.7) for $z \in K_a$ we have

$$\rho(z, \partial D) \geq \rho(z, \partial K_a^0) \geq |z - a| \sin(\alpha^0 - \alpha).$$

So the estimate (1.5) in $K_a$ transforms into

$$\left| \frac{\partial \varphi}{\partial x}(z) \right| + \left| \frac{\partial \varphi}{\partial y}(z) \right| \leq M_1 C |z - a|^{\mu - 1}, \quad z \in K_a. \tag{1.9}$$

Let us consider the ray $L = \{a + te \mid t > 0\} \subseteq K_a$ with the end $a$ and a unit vector $e$. For the function $\psi(t) = \varphi(a + te)$ on this ray (1.9) takes the form $|\psi'(t)| \leq M_1 C t^{\mu-1}$, so

$$|\psi(t) - \psi(t')| \leq M_1 C \int_t^{t'} s^{\mu-1} ds \leq \mu^{-1} M_1 C (t' - t)^\mu, \quad t \leq t'.$$

Thus we have the estimate

$$|\varphi(z) - \varphi(z')| \leq M_2 C |z - z'|^\mu \tag{1.10}$$

for every $z, z' \in L \subseteq K_a$, which does not depend on $a \in \Gamma$ and the ray $L$.

If the points $z, z' \in D$ satisfy the condition

$$y \leq y', \quad \theta(z' - z) < \alpha, \tag{1.11}$$

then these points lie on the ray $L \subseteq K_a$, where $a$ is a point of intersection of the straight line through $z$ and $z'$ with $\partial D$. So (1.10) holds for these points.

In the opposite case $y \leq y', \theta(z' - z) \geq \alpha$ let us consider the triangle with the vertices $z, z'$ and $z'' = x' + it$, $t > f(x')$. Let $\theta, \theta'$ and $\theta''$ be the angles in these vertices. Let us choose the value $t$ such that $\theta'' = \alpha$. Then the pairs of points $z, z''$ and $z', z''$ satisfy (1.11) and therefore

$$|\varphi(z) - \varphi(z')| \leq |\varphi(z) - \varphi(z'')| + |\varphi(z') - \varphi(z'')| \leq M_2 C(|z - z''|^\mu + |z' - z''|^\mu).$$

On the other hand from the sine theorem it follows that

$$\frac{|z - z''|}{|z - z'|} = \frac{\sin \theta'}{\sin \alpha} \leq \frac{1}{\sin \alpha}, \quad \frac{|z' - z''|}{|z - z'|} = \frac{\sin \theta}{\sin \alpha} \leq \frac{1}{\sin \alpha},$$

and hence $|z - z''|^\mu + |z' - z''|^\mu \leq M_3 |z - z'|^\mu$. Combined with the previous inequality we receive (1.10) with a constant $M_4$, and (1.5) is proved in the considered case.

Suppose that the function $\varphi$ satisfies (1.5) only in a disk $|z - z_0| < r$ with center $z_0 \in \partial D$. Then analogues reasonings show that there exists $0 < r_0 < r$ such that

$$[\varphi]_{\mu, G} \leq M_0 C, \quad G = D \cap \{|z - z_0| < r_0\}. \tag{1.12}$$

Let us turn to a general Lipschitz domain $D$. By virtue of (1.4) there exists $r > 0$ such that for every point $z_0 \in \partial D$ the set $D \cap \{|z - z_0| < r\}$ is contained in $V_i$ for some $i$. So we can suppose that (1.12) holds for every point $z_0 \in \partial D$. Let us consider the compact $K = \{z \in D, \rho(z, \partial D) \geq r_0\}$. The estimate (1.4) shows that the partial derivatives of $\varphi$ do not exceed $C r_0^{\mu-1}$ by module on $K$ and therefore $[\varphi]_{\mu, K} \leq M_1 C$. Together with (1.12) this estimate proves (1.5).

With the help of the lemma it is easy to establish the following generalization of Privalov theorem.

**Theorem 1.** *Let a function $u \in C^\mu(\overline{D})$, $0 < \mu < 1$, be harmonic in a simply connected Lipschitz domain $D$. Then the analytic function $\phi$ in (1.2) belongs to the same class and*

$$[\phi]_\mu \leq M[u]_\mu, \tag{1.13}$$

*where the constant $M > 0$ depends only on $\mu$ and $D$.*

**Proof.** If a disk $B = \{|z - z_0| < r\} \subseteq D$ then the estimate

$$\left|\frac{\partial u}{\partial x}(z_0)\right| + \left|\frac{\partial u}{\partial y}(z_0)\right| \leq M r^{\mu-1} [u]_{\mu,B} \tag{1.14}$$

is valid, where the constant $M > 0$ does not depend on $B$. In fact without loss of generality we can assume $u(z_0) = 0$. Putting $z_0 = 0$ and $\tilde{u}(z) = u(rz)$, we can also content ourself with the case $r = 1$. So $B$ is the unit circle and $u(0) = 0$, and therefore $|u|_0 \leq [u]_\mu$. In this case (1.14) follows from Poisson formula.

Obviously we can set $r = \rho(z_0, \partial D)$ in (1.14). By virtue of the relation

$$\phi' = \frac{\partial u}{\partial x} - i\frac{\partial u}{\partial y}, \tag{1.15}$$

which is inverse to (1.2) we can substitute $u$ for $\phi$ in the left side of this estimate. On the basis of Lemma 1 applying to $\phi$ we receive (1.13).

## 2. Elliptic systems

Let us consider the elliptic system

$$A_{11}\frac{\partial^2 u}{\partial x^2} + (A_{12} + A_{21})\frac{\partial^2 u}{\partial x \partial y} + A_{22}\frac{\partial^2 u}{\partial y^2} = 0 \tag{2.1}$$

with constant coefficients $A_{ij} \in \mathbb{R}^{l \times l}$ for an unknown vector-valued function $u = (u_1, \ldots, u_l) \in C^2$. The condition of ellipticity means that $\det A_{22} \neq 0$ and the characteristic polynomial

$$\chi(z) = \det P(z), \quad P(z) = A_{11} + (A_{12} + A_{21})z + A_{22}z^2, \tag{2.2}$$

has no real roots.

Recall the representation formula [3] of a general solution of the system (2.1) in a simply connected domain $D$. Let us introduce the block- matrix

$$A_* = \begin{pmatrix} 0 & 1 \\ -A_{22}^{-1}A_{11} & -A_{22}^{-1}(A_{12} + A_{21}) \end{pmatrix},$$

its spectrum $\sigma(A_*)$ consists of roots of the characteristic polynomial (2.2). Then (2.1) can be written in the form

$$\frac{\partial \nabla u}{\partial y} = A_* \frac{\partial \nabla u}{\partial x}, \quad \nabla u = \left(\frac{\partial u}{\partial x}, \frac{\partial u}{\partial y}\right).$$

Let the matrix $B_*$ reduce $A_*$ to the Jordan form

$$B_*^{-1} A_* B_* = J_*, \quad J_* = \text{diag}(J, \overline{J}), \tag{2.3}$$

where the spectrum of the matrix $J \in \mathbb{C}^{l \times l}$ lies in the upper half plane. As the matrix $A_*$ is real the matrix $B_*$ can be chosen in block form

$$B_* = \begin{pmatrix} B & \overline{B} \\ BJ & \overline{BJ} \end{pmatrix}. \tag{2.4}$$

So the $l$-vector-valued function $\psi$ from the equality

$$B_* \psi_* = 2\nabla u, \quad \psi_* = (\psi, \overline{\psi}),$$  (2.5)

satisfies the system

$$\frac{\partial \psi}{\partial y} - J \frac{\partial \psi}{\partial x} = 0.$$  (2.6)

Solutions of this system are said to be Douglis analytic functions or shortly $J$-analytic functions. For the case of a Toeplitz matrix $J$ the system was firstly investigated by Douglis [4] in the frame of so called hypercomplex numbers. Later first order elliptic systems studied by many authors ([5-10]). It is easy to show [3] that a Douglis analytic function $\psi(z)$ in the neighborhood of each point $z_0 \in D$ expands in a generalized power series

$$\psi(z) = \sum_{k=0}^{\infty} \frac{1}{k!} (z - z_0)_J^k \psi^{(k)}(z_0), \quad \psi^{(k)} = \frac{\partial^k \psi}{\partial x^k},$$

where $z_J$, $z = x + iy$, denotes the matrix $x1 + yJ$. Besides, if this function is continuous up to the boundary $\partial D$, then the generalized Cauchy formula

$$\psi(z) = \frac{1}{2\pi i} \int_{\partial D} dt_J (t - z)_J^{-1} \psi(t), \quad z \in D,$$  (2.7)

holds, where the matrix differential $dt_J$ has the same sense and the contour $\partial D$ is oriented positively with respect to $D$. The function $\psi$ can be also expressed through usual analytic functions. Let the matrix $J$ in (2.3) be written in the form $J = \text{diag}(J_1, \ldots, J_n)$, $J_k \in C^{l_k \times l_k}$, where $\sigma(J_k) = \nu_k$, $\text{Im}\,\nu_k > 0$, and therefore $(J_k - \nu_k)^{l_k} = 0$. Then the system (2.6) decompose into

$$\frac{\partial \psi_k}{\partial y} - J_k \frac{\partial \psi_k}{\partial x} = 0, \quad \psi = (\psi_1, \ldots, \psi_n).$$

In these notations $J$-analytic vector-valued functions $\psi = (\psi_1, \ldots, \psi_n)$ can be uniquely represented by the formula

$$\psi = \Lambda \tilde{\psi}, \quad (\Lambda \tilde{\psi})_k (x + iy) = \sum_{r=0}^{l-1} \frac{y^r}{r!} (J_k - \nu_k)^r \tilde{\psi}_k^{(r)} (x + \nu_k y),$$  (2.8)

where the $l_k$-vector $\tilde{\psi}_k$ is analytic in the domain $D_k = \{x + \nu_k y \mid z = x + iy \in D\}$. The inverse formula is

$$\tilde{\psi}_k (x + \nu_k y) = \sum_{r=0}^{l-1} \frac{(-y)^r}{r!} (J_k - \nu_k)^r \psi_k^{(r)} (x + iy), \quad k = 1, \ldots, n.$$

Let us turn to the relation (2.5). After integrating it takes the form

$$u = \text{Re}\, B\phi,$$  (2.9)

where $\phi$ is the Douglis analytic function

$$\phi(z) = \int_{z_0}^{z} dt_J \psi(t) + \phi(z_0).$$

Accordingly the inverse relation is

$$\phi' = B_1 \frac{\partial u}{\partial x} + B_2 \frac{\partial u}{\partial y}, \quad 2B_*^{-1} = \begin{pmatrix} B_1 & B_2 \\ \overline{B_1} & \overline{B_2} \end{pmatrix}. \tag{2.10}$$

The representation (2.9) of a general solution of the elliptic system (2.1) was firstly received by Soldatov [11] and Yeh [12]. Note that (2.9), (2.10) are analogous to (1.2), (1.15) with respect to solutions of (2.1). Substitution the formula (2.8) for corresponding functions $\phi$ and $\tilde{\phi}$ into (2.9) gives the known Bitsadze representation [13] through the analytic function $\tilde{\phi}$.

Recall that the elliptic system (2.1) is called weakly connected [13] if the matrix $B$ in (2.4) is invertible. This condition is equivalent [14] to invertibility of the matrix

$$\int_{\mathbb{R}} P^{-1}(t)dt \in \mathbb{R}^{l \times l}, \tag{2.11}$$

where $P$ is defined by (2.2).

**Theorem 2.** *Let the elliptic system (2.1) be weakly connected and a function $u \in C^\mu(\overline{D})$, $0 < \mu < 1$, be its solution in a simply connected Lipschitz domain $D$. Then the function $\phi$ in (2.9) belongs to the same class and (1.13) holds with a constant $M$ depending only on $\mu$ and $D$.*

**Proof.** The Proof is identical with the one of Theorem 1 under the assumption that the following fact is true. If $u \in C^\mu(\overline{B})$ is a solution of (2.1) in the unit disk $B = \{|z| < 1\}$, then the inequality

$$\left| \frac{\partial u}{\partial x}(0) \right| + \left| \frac{\partial u}{\partial y}(0) \right| \le M[u]_\mu \tag{2.12}$$

is valid. Substituting $B$ for the circle $\{|z| < 1/2\}$, we can assume $u \in C^1(\overline{B})$. Then taking (2.10) into account the function $\phi$ in (2.9) belongs to the same class and (2.12) reduces to

$$|\phi'(0)| \le M[\operatorname{Re} B\phi]_\mu, \tag{2.13}$$

where the constant $M > 0$ does not depend on the Douglis analytic function $\phi \in C^\mu(\overline{B})$.

Let $X$ be the class of these functions with the additional condition $\phi(0) = 0$. It follows from the Cauchy formula (2.7) that

$$|\phi'(0)| \le M[\phi]_\mu$$

and (2.13) reduces to the estimate

$$[\phi]_\mu \le M[\operatorname{Re} B\phi]_\mu \quad \phi \in X. \tag{2.14}$$

As it was shown in [15] the Riemann-Hilbert problem

$$\operatorname{Re} B\phi|_{\partial B} = f$$

is Fredholm in the class $X$, i.e. the operator $(R\phi)(t) = \operatorname{Re} B\phi(t)$, $t \in \partial B$, is Fredholm $X \to C^\mu(\partial B)$. So its image $Y = R(X)$ is a closed subspace of $C^\mu(\partial B)$ of finite co-dimension but its kernel $X_0 = \ker R$ is a finite dimensional subspace of $X$. Thus there exists a bounded operator $R^{(-1)} : Y \to X$ such that $RR^{(-1)}f = f, f \in Y$, and hence

$$X = X_0 \oplus X_1, \quad X_1 = R^{(-1)}(Y). \tag{2.15}$$

By virtue of the boundedness of $R^{(-1)}$ the norms $[\phi]_\mu$ and $[\operatorname{Re} B\phi]_\mu$ are equivalent on $X_1$. Obviously the same is true on the finite dimensional space $X_0$. Taking (2.15) into account (2.14) follows. That completes the proof.

## 3. Estimates in weighted Hölder spaces

Let $D$ be a simply connected domain with piecewise smooth boundary $\partial D$ and one sided tangents at its angular points are different. In particular $D$ is a Lipschitz domain. Let $0 \in \partial D$ and a solution $u$ of the weakly connected system (2.1) belongs to $C_{loc}^\mu(\overline{D}\backslash 0)$. The last implies that $u$ satisfies a Hölder condition in $D_\varepsilon = D \cap \{|z| > \varepsilon\}$ for each $\varepsilon > 0$. Then on basis of Theorem 2 the Douglis analytic function in (2.9) belongs to the same class. We are interested in the question whether if $u(z) = O(|z|^\lambda)$ as $z \to 0$ it is true for $\phi$. More exactly we describe the power behavior of functions with the help of weighted Hölder spaces. Recall [15] that the Banach space $C_\lambda^\mu = C_\lambda^\mu(D,0), \lambda \in \mathbb{R}$, consists of all functions $\varphi(z)$ on $D \backslash 0$ with finite norm

$$|\varphi| = ||z|^{-\lambda}\varphi(z)|_0 + [|z|^{\mu-\lambda}\varphi(z)]_\mu. \tag{3.1}$$

In notations (1.1) we can give another definition of this space.

**Lemma 2.** Let the integer $m$ be such that $D_k = \{z' \mid 2^{-k}z' \in D, 1/2 < |z'| < 2\} \neq \emptyset, k \geq m$ and $D_k = \emptyset, k < m - 1$. Then (3.1) is equivalent to the norm

$$|\varphi|' = \sup_{k \geq m} 2^{k\lambda}|\varphi(2^{-k}z')|_{\mu,D_k}. \tag{3.2}$$

**Proof.** Obviously

$$||z|^{-\lambda}\varphi|_0 \leq |2|^{|\lambda|}|\varphi|'.$$

Let $|z_1| \leq |z_2|$ and $z_j = 2^{-k}z'_j$, where $k$ is defined by condition $1 \leq |z'_1| < 2$. Then there are two cases, when $|z'_2| \leq 1/2$ and $|z'_2| > 1/2$. In the first case

$$A = \frac{||z_1|^{\mu-\lambda}\varphi(z_1) - |z_2|^{\mu-\lambda}\varphi(z_2)|}{|z_1 - z_2|^\mu} \leq \left\{\frac{|z'_1|^{\mu-\lambda} + |z'_2|^{\mu-\lambda}}{|z'_1 - z'_2|^\mu}\right\}|\varphi|'$$

and in the second case

$$A \leq \left\{|z'_1|^{\mu-\lambda} + \frac{||z'_1|^{\mu-\lambda} - |z'_2|^{\mu-\lambda}|}{|z'_1 - z'_2|^\mu}\right\}|\varphi|'.$$

In both cases the figured expressions are bounded by a constant depending only on $\mu$ and $\lambda$. According to (3.1), (3.2) these estimates yield $|\varphi| \leq M|\varphi|'$.

To receive the inverse inequality let us denote $\psi(z) = |z|^{\mu-\lambda}\varphi(z)$. Then $|\varphi(z)| \leq |z|^\lambda[\psi]_\mu \leq |z|^\lambda|\varphi|$ and hence

$$2^{k\lambda}|\varphi(2^{-k}z')| \leq |z'|^\lambda|\varphi|, \quad 1/2 < |z'| < 2.$$

Further we have

$$2^{k\lambda}\frac{|\varphi(2^{-k}z_1') - \varphi(2^{-k}z_2')|}{|z_1' - z_2'|^\mu} = \frac{|z_1'|^{\mu-\lambda}\psi(z_1) - |z_2'|^{\mu-\lambda}\psi(z_2)}{|z_1 - z_2|^\mu}$$

$$\leq \left\{|z_1'|^{\mu-\lambda} + \frac{||z_1'|^{\lambda-\mu} - |z_2'|^{\lambda-\mu}|}{|z_1' - z_2'|^\mu}|z_2'|^\mu\right\}|\varphi|.$$

Together with the previous inequality we receive the required estimate $|\varphi|' \leq M|\varphi|$.

By definition the space $C_{(\lambda)}^\mu = C_{(\lambda)}^\mu(D, 0), 0 < \lambda < 1$, consists of functions $\varphi \in C(\overline{D})$ such that $\varphi(z) - \varphi(0) \in C_\lambda^\mu$. It is convenient to set $C_{(\lambda)}^\mu = C_\lambda^\mu$ for $\lambda \leq 0$.

**Lemma 3.** *The space $C_{(\lambda)}^\mu$, $\lambda \neq 0$, can be defined by the equivalent norm*

$$|\varphi|' = |\varphi(c)| + [\varphi]_{\mu,(\lambda)}, \quad [\varphi]_{\mu,(\lambda)} = \sup_{k\geq m} 2^{k\lambda}[\varphi(2^{-k}z')]_{\mu,D_k}, \tag{3.3}$$

*where $c \in D$ is a fixed point.*

**Proof.** Let us first consider the case $\lambda < 0$. Let us choose points $z_i' \in D_i \cap D_{i+1}$, $i = m, m+1, \ldots$. By virtue of Lemma 1 the equality

$$|\varphi| = \sup_{k\geq m} 2^{k\lambda}\{|\varphi(2^{-k}z_k')| + [\varphi(2^{-k}z')]_{\mu,D_k}\}$$

defines an equivalent norm in $C_\lambda^\mu$. Without loss of generality we can set $c = 2^{-m}z_m'$ in (3.3). Then

$$|\varphi(2^{-k}z_k')| \leq |\varphi(c)| + \sum_{i=m}^{k-1}|\varphi(2^{-k}z_{i+1}') - \varphi(2^{-k}z_i')|.$$

As

$$|\varphi(2^{-k}z_{i+1}') - \varphi(2^{-k}z_i')| \leq 2^{-i\lambda}[\varphi]_{\mu,(\lambda)}|z_{i+1}' - z_i'|^\mu \leq 4\,2^{-i\lambda}[\varphi]_{\mu,(\lambda)},$$

it follows that

$$|\varphi(2^{-k}z_k')| \leq |\varphi(c)| + 4(2^{-\lambda} - 1)^{-1}2^{-k\lambda}[\varphi]_{\mu,(\lambda)}$$

and thus $|\varphi| \leq M|\varphi|'$. The inverse inequality is obvious.

The case $0 < \lambda < 1$ can be considered analogously. Without loss of generality we can set $c = 0$ in (3.3). On basis of Lemma 2 the equality

$$|\varphi| = |\varphi(0)| + \sup_{k\geq m} 2^{k\lambda}\{|\tilde{\varphi}(2^{-k}z_k')| + [\tilde{\varphi}(2^{-k}z')]_{\mu,D_k}\}, \quad \tilde{\varphi}(z) = \varphi(z) - \varphi(0),$$

defines an equivalent norm in $C_{(\lambda)}^\mu$. So

$$|\tilde{\varphi}(2^{-k}z_k')| \leq \sum_{i=k}^\infty |\varphi(2^{-k}z_{i+1}') - \varphi(2^{-k}z_i')| \leq 4(1 - 2^{-\lambda})^{-1}2^{-k\lambda}[\varphi]_{\mu,(\lambda)},$$

and therefore $|\varphi| \leq M|\varphi|'$.

**Theorem 3.** *Let $D$ be a simply connected domain with piecewise smooth boundary $\partial D$ and with different one-sided tangents at its angular points. Let $0 \in \partial D$ and a function $u \in C_{(\lambda)}^\mu(\overline{D}, 0)$, where $0 < \mu < 1, \lambda < 1, \lambda \neq 0$, satisfy the weakly connected* *elliptic system* (2.1) *in D. Then the function* $\phi$ *in* (2.9) *belongs to the same class and the estimate*

$$[\phi]_{\mu,(\lambda)} \le M[u]_{\mu,(\lambda)}, \tag{3.4}$$

*holds with a constant M depending only on* $\mu, \lambda$ *and D.*

**Proof.** Let us set

$$D_0 = D \cap \{|z| < \varepsilon\}, \quad D_1 = D \cap \{|z| > \varepsilon/2\},$$

where $\varepsilon > 0$ is sufficiently small. Then $D_1$ is a Lipschitz domain and on basis of Theorem 2 the function $\phi \in C^\mu(\overline{D}_1)$. So without loss of generality we can assume $D = D_0$. Then $\partial D$ is formed by an arc $L$ of the circumference $|z| = \varepsilon$ and two smooth arcs $\Gamma_1, \Gamma_2$ with the common end $z = 0$. In polar coordinates $r = |z|, \theta = \arg z$ these arcs are described by equations $\theta = h_j(r)$, where

$$h_j(r) \in C[0,\varepsilon] \cap C^1(0,\varepsilon], \quad \lim_{r \to 0} r h'_j(r) = 0, \quad j = 1,2. \tag{3.5}$$

Assuming $h_1(0) \ne h_2(0)$ modulo $2\pi$, we can suppose that $h_1(r) < h_2(r)$ for all $0 \le r \le \varepsilon$ and $D$ is defined by the inequalities $h_1(r) < \theta < h_2(r), 0 < r < \varepsilon$. In particular the domains $D_k = \{z' \mid 2^{-k}z' \in D, 1/2 < |z'| < 2\}$ in (3.1) are described by the inequalities $h_1(2^{-k}r) < \theta < h_2(2^{-k}r), 1/2 < r < 2$. By virtue of (3.5) the functions $h_j(2^{-k}r) \to h_j(0)$ as $k \to \infty$ in $C^1[1/2,2]$. Hence the estimate (1.13) in Theorem 2 with respect to $D_k$ holds uniformly by $k \ge m$. In particular

$$[\phi(2^{-k}z')]_{\mu,D_k} \le M[u(2^{-k}z')]_{\mu,D_k},$$

where the constant $M > 0$ does not depend on $k$. On basis of Lemma 3 it follows that the function $\phi \in C^\mu_{(\lambda)}(\overline{D}, 0)$ and the estimate (3.4) is valid.

## 4. Conjugate functions

Let $u(z)$ be a solution of the elliptic system (2.1). Consider the function $v(z)$ which is defined by the relation

$$\frac{\partial v}{\partial x} = -\left(A_{21}\frac{\partial u}{\partial x} + A_{22}\frac{\partial u}{\partial y}\right), \quad \frac{\partial v}{\partial y} = A_{11}\frac{\partial u}{\partial x} + A_{12}\frac{\partial u}{\partial y}. \tag{4.1}$$

The existence of this function follows from (2.1) written in the form

$$\frac{\partial}{\partial x}\left(A_{11}\frac{\partial u}{\partial x} + A_{12}\frac{\partial u}{\partial y}\right) + \frac{\partial}{\partial y}\left(A_{21}\frac{\partial u}{\partial x} + A_{22}\frac{\partial u}{\partial y}\right) = 0.$$

The function $v$ is said to be conjugate to the solution $u$. It is defined with accuracy of a constant vector $\xi \in \mathbb{R}^l$. This function is closely connected with the second boundary value problem

$$\sum_{i,j=1}^{2} A_{ij} n_i \frac{\partial u}{\partial x_j}\bigg|_{\partial D} = g, \tag{4.2}$$

where $x_1 = x$, $x_2 = y$ and $n = (n_1, n_2)$ denotes the external normal to the boundary. In fact by virtue of (4.1) this boundary condition we can write in the form

$$v'_s = g, \tag{4.3}$$

where $(\ )'_s$ means the tangent differentiation.

As it is seen from (4.1) theorems 2 and 3 are valid with respect to the conjugate function $v$. This fact also follows from the representation of this function

$$v = \xi + \operatorname{Re} C\phi, \quad C = -(A_{21}B + A_{22}BJ), \tag{4.4}$$

where the Douglis analytic function $\phi$ is given in (2.9).

The proof of this representation is not complicated. On the basis of (2.6), (2.9) the equation (2.1) gives the identity

$$\operatorname{Re}\left[A_{11}B + (A_{12} + A_{21})BJ + A_{22}BJ^2\right]\phi'' = 0.$$

It follows that $A_{11}B + (A_{12} + A_{21})BJ + A_{22}BJ^2 = 0$ or

$$C = -(A_{21}B + A_{22}BJ), \quad CJ = A_{11}B + A_{12}BJ. \tag{4.5}$$

Thus we can rewrite the relations (4.1) in the form

$$\frac{\partial v}{\partial x} = \operatorname{Re} C\phi', \quad \frac{\partial v}{\partial y} = \operatorname{Re} CJ\phi'.$$

So the partial derivatives of $v - \operatorname{Re} C\phi$ are equal to 0 that proves (4.4).

It follows from (4.3), (4.4) that the second boundary value problem (4.2) reduces to the Riemann-Hilbert problem

$$\operatorname{Re} C\phi|_{\partial D} = f, \quad f'_s = g,$$

with a constant matrix coefficient $C$. Hence the condition $\det C \neq 0$ is necessary [15] for this problem to be Fredholm. The equality $\det C = 0$ is closely connected with the case of a constant conjugate function $v$, when the right-hand side of (4.1) is identically equal to 0. In this case the solution $u$ of the system (2.1) is called degenerate.

In such a way the degenerate solutions are defined by the over-determined first order system

$$A_{i1}\frac{\partial u}{\partial x} + A_{i2}\frac{\partial u}{\partial y} = 0, \quad i = 1, 2. \tag{4.6}$$

It is obvious that the polynomials of first degree $u(x) = \xi_0 + \xi_1 x + \xi_2 y$, where $A_{i1}\xi_1 + A_{i2}\xi_2 = 0$, give the simplest example of degenerate solutions.

It is convenient below to consider numerical $l \times l$−matrices as linear operators in $\mathbb{C}^l$. Let us put

$$P = A_{11}^{-1}A_{12}, \quad Q = A_{22}^{-1}A_{21}, \quad X = \operatorname{Ker}(1 - PQ) \cap \operatorname{Ker}(1 - QP), \tag{4.7}$$

and introduce the subspace $Y$ of vectors $\eta \in \mathbb{C}^l$ such that

$$\operatorname{Re} CJ^k\eta = 0, \quad k = 0, 1, 2. \tag{4.8}$$

The class of all degenerate solutions can be described with the help of this space. Namely the solution $u$ is degenerate if and only if all partial derivatives of the second order of the function (4.4) are equal to 0, that is equivalent to the condition $\eta = \phi''(z) \in Y$ for all $z$.

**Lemma 4.** *The space $Y$ is invariant with respect to the operator $J$ and described by the equivalent conditions*

$$\operatorname{Re} B\eta \in X, \quad \operatorname{Re} C\eta = 0. \tag{4.9}$$

**Proof.** Make sure first that the space defined by (4.9) is invariant with respect to $J$. Let $\eta$ satisfy (4.9). It is obviously that the operators $P$ and $Q$ are invariant and mutually inverse on $X$. Besides we can rewrite (4.5) in the form

$$C = -A_{22}(QB + BJ), \quad CJ = -A_{11}(B + PBJ). \tag{4.10}$$

So (4.8) becomes

$$\operatorname{Re} BJ\eta = -Q\operatorname{Re} B\eta \in X,$$

$$\operatorname{Re} CJ\eta = A_{11}P\operatorname{Re}(B\eta + PBJ)\eta = A_{11}P\operatorname{Re}(QB\eta + BJ\eta) = 0$$

and (4.9) is really invariant with respect to $J$.

In particular (4.9) implies (4.8). Conversely let $\eta$ satisfy (4.9). It follows from (4.10) that

$$CJ^2 = A_{11}BJ - A_{11}P(A_{22}^{-1}C + QB)J = A_{11}(1 - PQ)BJ - A_{11}PA22^{-1}CJ.$$

Let us set $x = \operatorname{Re} B\eta, y = \operatorname{Re} BJ\eta$ for brevity and substitute (4.10) and the last expression into (4.8). Then we receive the relations $Qx + y = 0$, $x + Py = 0$, $(1 - PQ)y = 0$, that imply $x = \operatorname{Re} B\eta \in X$.

**Theorem 4.** *If $X = 0$, then every degenerate solution of the elliptic system is a polynomial of first order. Otherwise the class of degenerate solutions is infinitely dimensional. The case $X = 0$ is provided by $\det C \neq 0$. In general case the dimension of $X$ is even.*

**Proof.** By virtue of Lemma 4 the dimension of $Y$ coincides with $\dim X$. So if $X = 0$, then $\phi''(z) \in Y$ is equivalent $\phi'' = 0$.

If $X \neq 0$, then the class of all Douglis analytic functions $\phi$, such that $\phi''(z) \in Y$ is invariant with respect to $J$ and therefore it contains all functions of the form

$$\phi(z) = (z - z_0)_J^{-k}\eta, \quad \eta \in Y, k = 1, 2 \ldots, \tag{4.11}$$

where the fixed point $z_0 = x_0 + iy_0$ lies outside $D$. So this class is infinitely dimensional.

Suppose further by contradiction that $\det C \neq 0$ but $\dim X > 0$. Let us consider the system (2.1) in the upper half-plane $\operatorname{Im} z > 0$. Then the functions (4.11) for $\operatorname{Im} z_0 < 0$ define degenerate solutions $u = \operatorname{Re} B\phi$ of this system. In particular, $\operatorname{Re} C\phi'(x) = 0, x \in \mathbb{R}$, where $\mathbb{R}$ is the real axis of the complex plane $\mathbb{C}$. According to (2.8) we can write $\phi = \Lambda\tilde{\phi}$, $\tilde{\phi} = (\tilde{\phi}_1, \ldots, \tilde{\phi}_n)$, where the $l_k$-vector valued function $\tilde{\phi}_k$

is analytic on $\mathbb{C} \setminus \{x_0 + \nu_k y_0\}$ and has the same degree $-k$ at infinity as $\phi$. Formula (2.8) also shows that $\phi(x) = \tilde{\phi}(x)$, $x \in \mathbb{R}$. Hence $\operatorname{Re} C\tilde{\phi}'(x) = 0, x \in \mathbb{R}$, and in particular the function $C\tilde{\phi}'(z)$ is analytically extended to the lower half-plane. As $\tilde{\phi}'(z) \to 0$ at $\infty$ and under the assumption $\det C \neq 0$, it follows $\tilde{\phi}' = 0$. Hence $\phi' = 0$, that contradicts the choice of $\phi$.

Recall that $P$ and $Q$ from Lemma 4 act on $X$ as mutually inverse operators. We assert that these operators have no real eigenvalues. Really assuming the contrary let $P\xi = \mu\xi$ and $Q\xi = \mu^{-1}\xi$ for $\mu \in \mathbb{R}$ and $\xi \in X$ not equal to 0. Then

$$\sum_{i,j=1}^{2} (A_{ij} t_i t_j)\xi = [A_{11}(t_1 + Pt_2)t_1 + A_{22}(Qt_1 + t_2)t_2]\xi = 0, \quad t_1 + \mu t_2 = 0.$$

That contradicts to the ellipticity condition

$$\det \left( \sum_{i,j=1}^{2} A_{ij} t_i t_j \right) \neq 0, \quad |t_1| + |t_2| > 0, \tag{4.12}$$

of the system (2.1). Thus the operators $P$ and $Q$ acting on $X$ have no real eigenvalues and hence $\dim X$ is even.

## 5. Strengthen elliptic systems

Due to Vishik[16] the system (2.1) is strongly elliptic if

$$\left( \left( \sum_{i,j=1}^{2} A_{ij} t_i t_j \right) \xi, \xi \right) > 0 \tag{5.1}$$

for all $t_j \in \mathbb{R}$, $|t_1| + |t_2| \neq 0$ and nonzero vectors $\xi \in \mathbb{R}^l$. Here and below $(\,,\,)$ denotes the inner product in $\mathbb{R}^l$. In particular the condition (2.11) for the system to be weakly connected is fulfilled. A narrower class of elliptic systems is defined [14] by the conditions

$$A_{ji}^{\top} = A_{ij}, \quad \sum_{i,j=1}^{2} (A_{ij}\xi_j, \xi_i) \geq 0, \quad \xi_i \in \mathbb{R}^l, \tag{5.2}$$

i.e the block matrix

$$A = \begin{pmatrix} A_{11} & A_{12} \\ A_{21} & A_{22} \end{pmatrix} \tag{5.3}$$

is positive semidefinite. These systems are said to be strengthened elliptic. Note that the ellipticity condition (4.12) for these systems is equivalent to the following property. If $A\xi = 0, \xi = (\xi_1, \xi_2)$, where $t_1\xi_1 + t_2\xi_2 = 0, |t_1| + |t_2| \neq 0$, then $\xi_1 = \xi_2 = 0$.

**Theorem 5.** *Let the system (2.1) be strengthened elliptic. Then $X = 0$ is equivalent to $\det C \neq 0$ and is provided by*

$$\operatorname{rang} A \geq 2l - 1. \tag{5.4}$$

**Proof.** On basis of Theorem 4 $\det C \neq 0$ implies $X \neq 0$. To prove that $\det C = 0$ implies $\dim X > 0$ we suppose by contradiction that $\det C = 0$ but $X = 0$. Let $C\eta = 0$ for a nonzero $\eta \in \mathbb{C}^l$ and set

$$\tilde{\phi}(z) = (z - z_0)^{-1}\eta, \quad \operatorname{Im} z_0 < 0, \tag{5.5}$$

Let the Douglis analytic function $\phi$ be connected with $\tilde{\phi}$ by (2.8) i.e $\phi = \Lambda\tilde{\phi}$. As $\phi(x) = \tilde{\phi}(x)$, $x \in \mathbb{R}$, we have the equality

$$C\phi(x) = 0, \quad x \in \mathbb{R}. \tag{5.6}$$

In particular the conjugate function $v = \operatorname{Re} C\phi$, to the solution $u = \operatorname{Re} B\phi$ of (2.1) is equal to 0 on the boundary $\mathbb{R} = \partial D$ of the half-plane $D = \{\operatorname{Im} z > 0\}$.

The Green formula applied to the scalar product of (2.1) with $u$ gives the equality

$$\int_D \sum_{i,j=1}^2 \left(A_{ij}\frac{\partial u}{\partial x_i}, \frac{\partial u}{\partial x_j}\right) dx_1 dx_2 = -\int_{\mathbb{R}} \left(A_{21}\frac{\partial u}{\partial x} + A_{22}\frac{\partial u}{\partial y}, u\right) dx,$$

where $x_1 = x, x_2 = y$. By virtue of (5.6)

$$A_{21}\frac{\partial u}{\partial x} + A_{22}\frac{\partial u}{\partial y} = -\frac{\partial v}{\partial x} = 0$$

on the boundary $\mathbb{R} = \partial D$, so we have

$$\int_D \sum_{i,j=1}^2 \left(A_{ij}\frac{\partial u}{\partial x_i}, \frac{\partial u}{\partial x_j}\right) dx_1 dx_2 = 0.$$

Together with (5.2) the relations (4.4) follow. Thus the solution $u$ is degenerate and by virtue of Theorem 4 it is a polynomial of the first degree. In particular $\operatorname{Re} B\phi'' = 0$ on the half-plane $D$ and therefore the function $\phi'$ is constant. But this fact contradicts to (5.5).

Let us turn to the second assertion of the theorem. It follows from (4.7), (5.3) that

$$A = \begin{pmatrix} A_{11} & 0 \\ 0 & A_{22} \end{pmatrix}\begin{pmatrix} 1 & P \\ Q & 1 \end{pmatrix}$$

and therefore $\xi \in X$ implies $A\tilde{\xi} = 0$, $\tilde{\xi} = (\xi, Q\xi)$. Hence $\dim X \leq \dim(\operatorname{Ker} A) = 2l - \operatorname{rang} A$. Taking (5.4) into account we have the inequality $\dim X \leq 1$. But on basis of Theorem 4 the dimension of $X$ is even, and this dimension has to be equal to 0.

The assertion of Theorem 4 on the evenness of $\dim X$ we can complete in the following way.

**Lemma 5.** *For every even number $s$ between 0 and $l$ there exists a strengthened elliptic system such that $\dim X = s$.*

**Proof.** Let the matrices $P_0, Q_0 = P_0^{-1} \in \mathbb{R}^{s \times s}$ be orthogonal and have no real eigenvalues. Let us set $\mathbb{R}^l = \mathbb{R}^s \times \mathbb{R}^{l-s}$ and according to this representation introduce

1184

the matrices $P = \operatorname{diag}(P_0, 0)$ and $Q = \operatorname{diag}(Q_0, 0)$. Then $P^\top = O$, $PQ = \operatorname{diag}(1, 0)$ and therefore

$$|P(\xi_1, \xi_2)| = |(\xi_1, Q\xi_2)| \le |\xi_1||\xi_2|, \quad \xi_j \in \mathbb{R}^l,$$

$$|P(\xi, \xi)| = |(\xi, Q\xi)| < |\xi|^2, \quad \xi \ne 0. \tag{5.7}$$

It is also clear that the space $X$ in (4.7) coincides with $\mathbb{R}^s \times 0$ for these matrices.

Let us consider the system (2.1) with the coefficients $A_{11} = A_{22} = 1$, $A_{12} = P$, $A_{21} = Q$ and make sure that it is strengthened elliptic. By virtue of (5.7) we have for this system

$$(\xi_1 + P\xi_2, \xi_1) + (Q\xi_1 + \xi_2, \xi_2) = |\xi_1|^2 + |\xi_2|^2 + 2(Q\xi_1, \xi_2) \ge 0,$$

i.e. condition (3.2) is fulfilled. Besides the left hand side of this expression is positive for $\xi_j = t_j \xi$ and therefore the considered system is strengthened elliptic.

### References

1. Mushelishvili, N.I. *Singular integral equations.* Noordhoff, Groningen, 1953.
2. Stein, E.M. *Singular integrals and differentiability properties of functions.* Princeton Univ. Press, Princeton, N. J., 1970.
3. Soldatov, A.P., *Hyperanalytic functions and their applications.* J. Math. Sciences, 2004, V. 15, P.142–199.
4. Douglis, A. *A function theoretic approach to elliptic systems of equations in two variables.* Comm. Pure Appl. Math. 1953, 6, 259–289.
5. Gilbert, R.P., Buchanan, J.L. *First order elliptic systems.* New York, Acad. Press, 1983.
6. Begehr, H., Wen, G.-C. *Boundary value problem for elliptic equations and systems.* Pitman Monographs and Surveys in Pure and Appl. Math., 46, Pitman, Harlow, 1990.
7. Wendland, W. *Elliptic systems in the plane.* Pitman, London etc., 1979.
8. Gilbert R.P., Hile G.N. *Generalized hypercomplex function theory.* Trans. Amer. Math. Soc. 195 (1974), 1–29.
9. Hile, G.N. *Elliptic systems in the plane with lower order terms and constant coefficients.* Comm. Pure Appl. Math. 3(10), (1978), 949–977.
10. Begehr H., Gilbert, R.P., *Pseudohyperanalytic functions,* Complex Variables, Theory Appl. 9 (1988), 343–357.
11. Soldatov, A.P., *Higher order elliptic systems,* Differentcial'nye Uravneniya, 25 (1989), 136–144; Differential Equations, 25 (1989) (English transl).
12. Yeh, R.Z. *Hyperholomorphic functions and higher order partial differentials equations in the plane.* Pacific Journ. of Mathem, 142, 2, (1990), 379–399.
13. Bitsadze, A.V. *Boundary value problems for elliptic equations of second order.* North Holland Publ. Co., Amsterdam, 1968.
14. Soldatov, A.P., *On the first and second boundary value problems for elliptic systems on the plane,* Differentcial'nye Uravneniya, 39, (2003), 674-686;. Differential Equations, 39, (2003) (English transl.).
15. Soldatov, A.P., *On Dirichlet problem for elliptic systems on the plane,* 2005, to appear.
16. Vishik, M.I. *On strong elliptic systems of differencial equations,* Matematicheskiy Sbornik, 29, (1951), No 3, 615–676 (Russian).

# QUANTITATIVE TRANSFER OF SMALLNESS FOR SOLUTIONS OF ELLIPTIC EQUATIONS WITH ANALYTIC COEFFICIENTS AND THEIR GRADIENTS

E. MALINNIKOVA

*Department of Mathematics*
*Norwegian University of Science and Technology*
*7491, Trondheim, Norway*
*E-mail: eugenia@math.ntnu.no*

This article starts with a short survey of the results on the quantitative transfer of smallness. We prove a new version of the three sphere theorem for solutions to Cauchy-Riemann system. Then we use an earlier result on propagation of smallness for real analytic functions [1] to derive new estimates for solutions of elliptic equations with analytic coefficients and their gradients.

**Key words:** Hadamard three-circle theorem, generalized Cauchy-Riemann system, harmonic functions, transfer of smallness, first order elliptic systems
**Mathematics Subject Classification:** 35J45, 31B35

## 1. Preliminaries

### 1.1. *Three sphere theorems*

Our starting point is the classical Hadamard three-circle theorem.
*Let $f(z)$ be a complex function defined and analytic on $\{z \in \mathbb{C} : r_1 \leq |z| \leq r_2\}$. And let $\|f\|_r = \max_{|z|=r} |f(z)|$, $\|f\|_{r,2} = \int_{|z|=r} |f(z)|^2$. Then*

$$\|f\|_r \leq \|f\|_{r_1}^{\alpha} \|f\|_{r_2}^{1-\alpha}, \quad \text{and} \quad \|f\|_{r,2} \leq \|f\|_{r_1,2}^{\alpha} \|f\|_{r_2,2}^{1-\alpha}, \quad \text{where } r = r_1^{\alpha} r_2^{1-\alpha}.$$

We will refer to this inequalities as $L^{\infty}$- and $L^2$-versions of three-circle theorem. This theorem has many generalizations. We will discuss those concerning solutions to second order elliptic equations and their gradients.

First versions of the three-sphere theorem for harmonic functions in $\mathbb{R}^n$ and more generally for solutions of second order elliptic equations appeared in 1960's in the works of E.M. Landis[2], S. Agmon[3], E. Solomentsev[4], and Yu.K. Gerasimov[5].

A fresh look at this problem is due to J. Korevaar and J.L.H. Meyers[6]. They proved that the $L^2$-version of the three-sphere theorem is valid for harmonic functions in $\mathbb{R}^n$ with the same $\alpha$ that appears in Hadamard's theorem. Then R. Brummelhuis[7] pushed their technique to prove the $L^2$-version of the three-sphere theorem for solutions to elliptic equations.

Our first result, presented in Sec.2, is the extension of the three-sphere inequality for solutions of a class of generalized Cauchy-Riemann systems. The main feature here is that functions are defined in a **spherical shell** exactly as it is in the classical three circle theorem. All results on solutions to elliptic equations sited above require functions to be defined on the whole ball. Clearly, this condition is crucial since there exists a function harmonic on a spherical shell that equals zero on one of the spheres that bounds the shell.

## 1.2. *Transfer of smallness from small sets*

In the second part of the article, Sec.3, we will be interested in more delicate results concerning transfer of smallness. Before explaining the history of the question and formulating the results, we need

**Definition 1.1.** Given a subset $E$ of the unit ball $B \subset \mathbb{R}^n$ and a class of (vector)functions $\mathcal{A} \subset C(B)$, we say that $\mathcal{A}$ admits transfer of smallness from $E$ if for any $u \in \mathcal{A}$

$$|u(x)| \leq C\|u\|_B^{1-\alpha}\|u\|_E^\alpha, \tag{1}$$

where $\alpha = \alpha(x, E) > 0$ is independent of $u$ and bounded away from 0 on compact subsets of $B$.

Clearly, our last equation follows from the three spheres theorem if $E$ contains a (small) ball. However we would think of $E$ being a very "thin" set.

This question for harmonic functions and subsets of analytic hypersurfaces of positive surface measure was addressed in the work by M.A. and M.M. Lavrentiev [8]. Two other series of works on transfer of smallness when $E$ is a set of positive Lebesgue measure are due to N. Nadirashvili [9–11] and S. Vessella [12,13].

## 1.3. *Transfer of smallness for real analytic functions*

If we look at the known results on transfer of smallness for *harmonic* functions, we see that what was used is mainly real analyticity. In this subsection we will describe transfer of smallness for real analytic functions of several real variables [1].

We denote by $A(B_R)$ the set of real analytic functions $f$ on $B_R$ that admit analytic continuation $\tilde{f}$ to the closed complex ball $B_{R,\mathbb{C}} = \{z \in \mathbb{C}^N : |z| \leq R\}$ such that $|\tilde{f}| \leq 1$ on $B_{R,\mathbb{C}}$. Clearly $A(B_R)$ admits transfer of smallness form $E \subset \mathbb{R}^N$ if and only if $E$ is not a **pluri-polar** subset of $\mathbb{C}^N$. However this condition is not easy to check or compare to other geometric quantities related to $E$.

Let $E$ be a compact subset of $B_r$, we want to obtain an estimate

$$\sup_{B_r} |f| \leq \sup_K |f|^\alpha, \tag{2}$$

for any $f \in A(B_R)$.

In order to get some quantitative versions of (2), we will consider sets of Hausdorff dimension larger than $N - 1$. We use the fact that the Hausdorff dimension coincides with the capacitary dimension and consider sets of positive Riesz

$(N - 1 + \delta)$-capacity. We remind, that the Riesz $s$-capacity of a set $A \subset \mathbb{R}^N$ is defined by

$$C_s(A) = \sup\{I_s(\mu)^{-1} : \mu(A) = 1\},$$

here the supremum is taken over all Radon measures $\mu$ compactly supported on $A$, and

$$I_s(\mu) = \int \int |x - y|^{-s} d\mu(x) d\mu(y).$$

We fix $\delta > 0$ for the rest of the article.

**Lemma 1.1.** *Let $K \subset B_r$ be a compact set of positive Riesz $(N - 1 + \delta)$-capacity. Then*

$$\sup_{B_r} |f| \leq \sup_K |f|^\alpha,$$

*for any $f \in A(B_R)$, where $\alpha = \alpha(N, \delta, r^{1-N-\delta} C_{N-1+\delta}(K))$.*

This lemma was obtained in order to prove transfer of smallness for solutions of GCR systems from subsets of a hyperplane of dimension greater than $N - 2$. In the present article we will use it to prove new estimates for harmonic functions, solutions to elliptic equations and their gradients.

## 2. Three-spheres for solutions of the Cauchy-Riemann systems

### 2.1. *Formulation of the result*

Let $A$ be a first-order linear differential operator with constant coefficients defined on vector-functions, $A = \sum_{j=1}^n A_j \partial_j$, where $A_j$ are constant matrices. The system of equations

$$(AF =) \sum_{j=1}^n A_j \frac{\partial F}{\partial x_j} = 0 \tag{3}$$

is called a generalized Cauchy–Riemann (GCR) system if for any solution $F$ to this system, $F : \mathbb{R}^n \to \mathbb{R}^m$, $F = (f_1, ..., f_m)$, all its components $f_1, f_2, ..., f_m$ are harmonic. (We suppose that $F$ is well-defined and satisfies (3) in an open subset of $\mathbb{R}^n$.) Standard examples of solutions to generalized Cauchy–Riemann systems include gradients of harmonic functions, harmonic differential forms, Clifford monogenic functions.

**Definition 2.1.** A generalized Cauchy–Riemann system (3) is called *normal* if there exists a first order differential operator $B = \sum_{j=1}^N B_j \frac{\partial}{\partial x_j}$ with constant coefficients such that the following conditions are satisfied:

(1) $BA = \Delta$ (this identity means that

$$B_j A_j = I_M, \ j = 1, ..., N, \text{ and } B_j A_k + B_k A_j = 0, \ 1 \leq j < k \leq N,$$

where $I_M$ is the identity $M \times M$ matrix.)

(2) $B_j A_k$ is a skew-symmetric matrix for any $j \neq k$.

All Cauchy–Riemann systems we mentioned above are normal. Further examples and a description of rotationally invariant normal GCR systems can be found in the article by K.M. Davis, E.J. Gilbert, and R.A. Kunze [14].

We consider a vector valued function $F$ and by $|F|$ we mean the usual norm in Euclidean space. Then by $\|F\|_R$ we denote the $L^2$-norm of the function $|F|$ over the sphere $\{|x| = R\}$, i.e. $\|F\|_R^2 = \int_{\mathbb{S}} |F(Rx')|^2 ds(x')$. We use the standard notation $r = |x| = \sqrt{x_1^2 + ... x_n^2}$ and $x' = x/r \in \mathbb{S}^{n-1}$.

**Theorem 2.1.** *Let (3) be a normal GCR system. Suppose that $F$ is defined and satisfies (3) in the spherical shell $R = \{x \in \mathbb{R}^n : r_1 \leq |x| \leq r_2\}$. Then*

$$\|F\|_r \leq \|F\|_{r_1}^{\alpha} \|F\|_{r_2}^{1-\alpha}, \tag{4}$$

*where $r_1 < r < r_2$ and $\alpha = \ln(r_2/r) / \ln(r_2/r_1)$, i.e. $r = r_1^{\alpha} r_2^{1-\alpha}$.*

## 2.2. *Proof of Theorem 2.1*

All the components of $F$ are harmonic in $R$. Hence we can represent $F$ as a sum of spherical harmonics,

$$F = \sum_{q=0}^{\infty} \sum_{s=1}^{n_q} \left( C_{sq}^+ + C_{sq}^- r^{-2q-n+2} \right) Y_{sq} = \sum_{q=0}^{\infty} F_q^+ + F_q^-,$$

where $C_{sq}^+$ and $C_{sq}^-$ are constant vectors and $Y_{1q}, ..., Y_{n_q q}$ are spherical harmonics of order $q$ that form an orthonormal basis. The $L^2$-norm of $F$ over a sphere centered in the origin is easy to calculate. We have

$$\|F\|_r = \sum_{q=0}^{\infty} \sum_{s=1}^{n_q} \left( C_{sq}^+ + C_{sq}^- r^{-2q-n+2} \right) \left( C_{sq}^+ + C_{sq}^- r^{-2q-n+2} \right)^T.$$

In order to prove that this function is logarithmically convex with respect to $r$ it suffices to show that

$$\sum_{s=1}^{n_q} C_{sq}^+ \left( C_{sq}^- \right)^T = 0, \tag{5}$$

then $\|F\|_r$ is a sum of powers of $r$ with positive coefficients.

By the hypothesis of the theorem $F$ is a solution to $AF = 0$, and this system is normal, so we can find matrices $B_1, ..., B_n$ such that

$$B_j A_j = I_m, \quad B_j A_k + B_k A_j = 0, \text{ and } B_j A_k = -(B_j A_k)^T, \quad j \neq k.$$

Multiplying (3) by $x_1 B_1 + ... x_n B_n$ from the left, we get

$$\sum_{i=1}^{n} \sum_{j=1}^{n} x_i B_i A_j \frac{\partial F}{\partial x_j} = 0.$$

Then, taking the products with $i \neq j$ to the right-hand side, we obtain

$$r\frac{\partial F}{\partial r} = \sum_{j=1}^{n} x_j \frac{\partial F}{\partial x_j} = -\sum_{i \neq j} x_i B_i A_j \frac{\partial F}{\partial x_j}.$$

It is clear that both sides of the last identity are obtained by applying some operators of order $-1$ homogeneous with respect to $r$ to $F$. Thus the same identity is valid for any homogeneous part of $F$. In particular, for any $q$,

$$qF_q^+ = r\frac{\partial F_q^+}{\partial r} = -\sum_{i \neq j} x_i B_i A_j \frac{\partial F_q^+}{\partial x_j} \tag{6}$$

and

$$qF_q^- = r\frac{\partial F_q^-}{\partial r} = -\sum_{i \neq j} x_i B_i A_j \frac{\partial F_q^-}{\partial x_j}. \tag{7}$$

Our aim is to prove (5). We fix $q$ and in what follows we omit index $q$ writing $Y_s = Y_{sq}$ and $C_s^{\pm} = C_{sq}^{\pm}$. now, we multiply (6) by $Y_t$ and integrate it over the spherical shell $R$ to get

$$qC_t^+ = a \sum_s \int_R \sum_{i \neq j} x_i B_i A_j C_s^+ \frac{\partial Y_s}{\partial x_j} Y_t,$$

where $a = a(n, r_1, r_2)$. The last identity means that $\mathcal{C}^+ = \{C_s^+\}_{s=1}^{n_q} \in \mathbb{R}^{m \times n_q}$ is an eigenvector corresponding to the eigenvalue $q$ of the following matrix $\mathcal{M}$ (we enumerate the elements of $\mathcal{M}$ using four indices):

$$\mathcal{M}(ks, lt) = a \int_R \sum_{i \neq j} x_i B_i A_j(k, l) \frac{\partial Y_s}{\partial x_j} Y_t.$$

Next, we multiply (7) by $r^{2q+n-2} Y_t$ and integrate it over the spherical shell $R$,

$$-(q+n-2)C_t^- = a \sum_s \int_R \sum_{i \neq j} x_i B_i A_j C_s^- \frac{\partial r^{-2q-n+2} Y_s}{\partial x_j} r^{2q+n-2} Y_t$$

$$= a \sum_s \int_R \sum_{i \neq j} x_i B_i A_j C_s^- \frac{\partial Y_s}{\partial x_j} Y_t + \tilde{q} a \sum_s \int_R \sum_{i \neq j} x_i B_i A_j C_s^- Y_s Y_t$$

$$= a \sum_s \int_R \sum_{i \neq j} x_i B_i A_j C_s^- \frac{\partial Y_s}{\partial x_j} Y_t + \tilde{q} a \sum_s \int_R \sum_{i < j} \frac{x_i x_j}{r^2} (B_i A_j + B_j A_i) C_s^- Y_s Y_t,$$

where $\tilde{q} = 2q + n - 2$. The last term is equal to $0$ since $B_i A_j + B_j A_i = 0$ for $i < j$. Hence $\mathcal{C}^- = \{C_s^-\}_{s=1}^{n_q}$ is an eigenvector of $\mathcal{M}$ corresponding to eigenvalue $-(q+n-2)$.

To finish the proof we shall show that $\mathcal{M}$ is self-adjoint. Then $\mathcal{C}^+$ and $\mathcal{C}^-$ are orthogonal and (5) holds. In fact,

$$
\begin{aligned}
\mathcal{M}(ks, lt) &= a \int_R \sum_{i \neq j} x_i B_i A_j(k, l) \frac{\partial Y_s}{\partial x_j} Y_t \\
&= -a \int_R \sum_{i \neq j} x_i B_i A_j(k, l) Y_s \frac{\partial Y_t}{\partial x_j} + a \int_{\partial R} \sum_{i \neq j} x_i B_i A_j(k, l) Y_s Y_t x_j \\
&= a \int_R \sum_{i \neq j} x_i B_i A_j(l, k) \frac{\partial Y_t}{\partial x_j} Y_s + a \int_{\partial R} \sum_{i < j} x_i x_j (B_i A_j + B_j A_i)(k, l) Y_s Y_t.
\end{aligned}
$$

The last identity follows from $B_i A_j(k, l) = -B_i A_j(l, k)$. Now the first term in the last expression is just $\mathcal{M}(il, ks)$ and the second one equals zero as $B_i A_j + B_j A_i = 0$. This completes the proof of the theorem.

### 2.3. Concluding remarks

The inequality (4) is exactly the same as for analytic functions. Using standard technique [6,15], one can obtain some $L^\infty$ estimates from it. This theorem is a an extension of the earlier result of the author [15], where the sphere theorem was proved for harmonic differential forms. We do not know if the theorem is valid for any GCR system.

## 3. Transfer of smallness from sets of measure zero

### 3.1. Harmonic functions

We now want to combine some ideas from the work of M.A. and M.M. Lavrentiev [8] with Lemma 1.1. It provides us with new examples of sets which allow the transfer of smallness for harmonic functions but not for real analytic. We will not try to formulate most general results and will restrict ourselves to simple hypersurfaces.

**Theorem 3.1.** *Suppose that $\bar{B}(a, r)$ is a subset of the unit ball $B$, $E$ is a compact subset of the sphere $S(a, r)$, and $C_{n-2+\delta}(E) = C > 0$. Then for any harmonic function $u$ on $B$*

$$
|u(x)| \leq C \left( \max_E |u| \right)^\alpha \left( \max_B |u| \right)^{1-\alpha}, \tag{8}
$$

*where $\alpha = \alpha(x, n, \delta, C, \text{dist}(B(a, r), \partial B))$ is bounded away from 0 on compact subsets of $B$.*

**Proof.** Using the subadditivity of the Riesz capacity, we can reduce the situation to set $E$ that is contained in a ball $\tilde{B}$ such that $5\tilde{B} \subset B$. Further we may assume that there exists a diffeomorphism $g : 3\tilde{B} \to B$ that maps $3\tilde{B} \cap S(a, r)$ to the intersection of a hyperplane that passes through the center of the ball $B$ with this ball. Since $C_{n-2+\delta}(g(E)) \geq A C_{n-2+\delta}(E)$, we may apply the result on transfer of smallness for real analytic functions to the function $u \circ g^{-1}(x)$ restricted to the hyperplane containing $g(3\tilde{B} \cap S(a,r))$. Then we get the estimate (8) for $x$ on the (relatively) open subset of the sphere $S(a,r)$. Taking a chain of small balls, we then extend the estimate to the whole sphere. At last, applying the maximum principle and three spheres inequality, we have (8) for any $x \in B$ with $\alpha(x)$ bounded away from zero on compact subsets of $B$. □

**Remark 3.1.** If $B(a,r) \not\subset B$, then $S(a,r) \cap B$ is not a uniqueness set for harmonic functions. One can take a linear combination of two fundamental solutions to the Laplace equation with singularities that are symmetric with respect to $S(a,r)$ and does not lie in $B$, that vanishes on $S(a,r)$ and is harmonic on $B$.

### 3.2. *Solutions of elliptic equations with analytic coefficients and their gradients*

The result of the previous subsection can be easily extended to solutions of second order elliptic equations with analytic coefficients and analytic closed hypersurfaces compactly contained by the unit ball.

We will here give an estimate that involves the gradient of the solution. Let $L$ be a second order uniformly elliptic differential operator with analytic coefficients.

**Theorem 3.2.** *Suppose that $E$ is a compact subset of a hyperplane, $E \subset B_{1/2}$, and $C_{n-2+\delta}(E) = C > 0$. Then for any solution $u$ of $Lu = 0$ in $B$*

$$|u(x) - u(0)| \le C \left(\max_E |\nabla u|\right)^\alpha \left(\max_B |u|\right)^{1-\alpha}, \tag{9}$$

*where $\alpha = \alpha(x, n, L, \delta, C)$ is bounded away from $0$ on compact subsets of $B$.*

**Proof.** Let $S$ be the hyperplane that contains $E$ and $P = S \cap B_{1/2}$. We first note that Lemma 1.1 implies

$$\max_P |\nabla u| \le C \left(\max_E |\nabla u|\right)^\alpha \left(\max_{B_{3/4}} |\nabla u|\right)^{1-\alpha} \le C \left(\max_E |\nabla u|\right)^\alpha \left(\max_B |u|\right)^{1-\alpha}.$$

We have used the standard elliptic estimate for the gradient as well. Now for $x, y \in P$ we have

$$|u(x) - u(y)| \le |x - y| \max_P |\nabla u|.$$

If we fix $y \in P$, we get

$$\max_{x \in P} |u(x - u(y)| \le C \left(\max_E |\nabla u|\right)^\alpha \left(\max_B |u|\right)^{1-\alpha}.$$

Finally, applying a quantitative version of the Cauchy uniqueness theorem [16] to $u - u(y)$ we obtain (9). □

1192

### 3.3. *Open problems*

- When do harmonic functions admit transfer of smallness from $E$?
  We have seen that not only the size of $E$ plays a role here but also the geometry of the set.
- When do harmonic vector fields in $\mathbb{R}^N$ admit transfer of smallness?
  When $N = 2$ harmonic vector fields can be identified with analytic functions and the complete answer is known, $\text{cap}(E) > 0$.
  In higher dimensions we showed that if $E$ is a subset of a hyperplane (analytic surface) and $\dim E > N - 2$, then we have transfer of smallness. Is it true for any set of dimension bigger than $N - 2$?
- Transfer of smallness for solutions to elliptic equations with non-analytic coefficients.

### Acknowledgments

This article is an extended version of the talk given at the 5th ISAAC Congress in Catania, Italy, July 2005. It is my pleasure to thank the organizers of the Congress, special thanks go to Heinrich Begehr, who was the leader of the section "Complex and functional analytic methods in PDE".

This work was supported by the Research Council of Norway, project no. 160192/V30, "PDE and Harmonic Analysis"

### References

1. E. Malinnikova, Propagation of smallness for solutions of generalized Cauchy-Riemann systems, *Proc. Edinb. Math. Soc. (2)* **47** (2004) no.1, 191–204.
2. E. M. Landis, Some problems of the qualitative theory of second order elliptic equations, *Russian Math. Surveys* **18** (1963), 1–62.
3. S. Agmon, Unicité et convexité dans problémes différentiels, *Sém. Math. Sup.* **13** (1965), Les Presses, Univ. de Montéal, Montéal, 1965.
4. E. Solomentsev, A three-sphere theorem for harmonic functions, *Armjan. SSR Dokl.* **42** (1966), 274–278.
5. Ju. Gerasimov, The three spheres theorem for a certain class of elliptic equations of higher order and a refinement of this theorem for a linear elliptic equation of the second order, *Mat. Sb. (N.S.)* **71 (113)** (1966), 563–585.
6. J. Korevaar and J.L.H. Meyers, Logarithmic convexity for supremum norms of harmonic functions, *Bull. London Math. Soc.* **26** (1994), 353–362.
7. R. Brummelhuis, Three-spheres theorem for second order elliptic equations, *J. Anal. Math.* **65** (1995), 179–206.
8. M. A. Lavrentiev and M. M. Lavrentiev, On some inequalities for space harmonic functions, *Ann. Acad. Sci. Fenn. Ser. A I Math.* **2** (1976), 303–306.
9. N. S. Nadirasvili, A generalization of Hadamrd's three circles theorem, *Vestnik Moscow. Univ. Ser. I Mat. Meh.* **31** (1976) no. 3, 39–42.
10. N. S. Nadirashvili, Estimation of the solutions of elliptic equations with analytic coefficients which are bounded on some set, *Vestnik Moscow. Univ. Ser. I Mat. Mekh.* **2** (1979), 42–46, 102.

11. N. S. Nadirashvili, Uniqueness and stability of continuation from a set to the domain of solution of an elliptic equation, *Mat. Zametki* **40** (1986), 218–225, 287.

12. S. Vessella, A continuous dependence result in the analytic continuation problem, *Forum Math.* 11 (1999), 695–703.

13. S. Vessella, Quantitative continuation from a measurable set of solutions of elliptic equations, *Proc. Roy. Soc. Edinburgh Sect. A* **130** (2000), 909–923.

14. K.M. Davis, E.J. Gilbert and R.A. Kunze, Elliptic differential operators in harmonic analysis. I. Generalized Cauchy-Riemann systems, *Amer. J. Math.* **113** (1991), 75–116.

15. E. Malinnikova, The theorem on three spheres for harmonic differential forms,. Complex analysis operators, and related topics, *Oper. Theory Adv. Appl.* **113** (2000), 213–220.

16. F.-H. Lin, Nodal sets of solutions of elliptic and parabolic equations, *Comm. Pure Appl. Math.* **44** (1991), 287–308.

# THE SOLUTION OF SPECTRAL PROBLEMS FOR THE CURL AND STOKES OPERATORS WITH PERIODIC BOUNDARY CONDITIONS AND SOME CLASSES OF EXPLICIT SOLUTIONS OF NAVIER-STOKES EQUATIONS

R.S. SAKS*

*450077, Russia, Ufa.*
*112 Chernyshevskogo Street*
*E-mail: saks@ic.bashedu.ru*

The relations between eigenvalues and eigenfunctions of the curl operator and the Stokes operator (with periodic boundary condition) are indicated. The eigenvalues of the curl operator are square roots of the eigenvalues of Stokes operator if the parameter $\nu = 1$. The multiplicity of every nonzero eigenvalue is finite but the multiplicity of the zero eigenvalue of curl is infinite. The eigenfunctions of these operators are calculated explicitly. The functional vector space $\mathbf{L}_2(Q, 2\pi)$ may be decomposed into the direct sum of eigensubspaces of the curl operator. The *Curl* and *Stokes* systems of differential equations with complex parameter $\lambda$ are solved. For any complex number $\lambda$ the conditions of solvability and exact spaces for solutions of Eq.(8) and Eq.(12) are indicated. These results are useful in studying the Cauchy problem for the Navier-Stokes system with initial and periodic boundary conditions. We reduce this problem to the Cauchy problem for a system of first order (ordinary) differential equations in Hilbert space, which may be solved by the method of Galerkin approximation. Local solutions of these finite systems exist but we are interested in the existence of global solutions and their bifurcation. We are going to perform computing experiments. Some families of explicit solutions of the nonlinear Navier-Stokes system are found.

**Key words:** Spectral problems, curl operator, Stokes operator, Navier-Stokes equation, periodic boundary conditions

**Mathematics Subject Classification:** 35Q30

## 1. Operator *curl*

The *curl* operator is one of the most important first-order differential operator, which appears in various physical models, for example in the Maxwell equations of electricity and magnetism [21]. In fluid and plasma physics *curl* appears to measure the velocity of various flows [15,18]. The eigenfunctions of curl are called force-free fields or free-decay fields [7,8,12,20] in astrophysics; in the theory of fusion plasmas they are called Taylor states [15,22] and in hydrodynamics – Beltrami fields [1,6,9,10,20] also.

*Work supported by Presidium of RAS program No. 17 and by grants 03-07-90077 and 05-01-00515 of the Russian Bureau Foundation Research.

Let us touch some investigations in which a mathematical theory of the boundary value problems for *curl* is studied and, in particular, the spectral problem

$$curl\,u = \lambda u \quad \text{in} \quad G \subset \mathbb{R}^3 \tag{1}$$

with the boundary condition

$$n \cdot u = 0 \quad \text{on} \quad \partial G, \tag{2}$$

from a rigorous mathematical point of view. Here $u$ is a vector-function, $G$ is 3-dimensional bounded domain with a smooth or piece-wise smooth boundary $\partial G$, and $n$ is the unit normal vector on $\partial G$.

S. Chandrasekhar and P.S. Kendall [8] have given an original method of deriving the general solution of Eq. (1) with constant $\lambda$. By this method D. Montgomery, L. Turner and G. Vahala [18] constructed the family of (periodic along the axle) eigenfunctions of the curl operator in an unbounded cylinder.

B.R. Vainberg and V.V. Grushin have proved the Fredholm solvability for every uniformly nonelliptic operator (and, in particular, $d + *$) on a closed manifold.

The diverse boundary value problems for Eq. (1) with constant $\lambda$ (and more general) was studied in R.Saks works [13,14,17] and conditions of (Fredholm) solvability were given. See also [19] where boundary value problems for the system of Maxwell equations in the case of steady-state processes was considered and [25] where the eigenfunctions of curl operator in the ball was calculated explicitly.

P. Berhin [16], Y. Giga and Z. Yoshida [22], R. Picard[23] studied selfadjoint realization of curl and more general operator $*d$.

R. Saks and C.J. Vanegas [31] considered three methods of studying one special boundary value problem for Eq. (1) with constant $\lambda$ in the ball and in work [30] they proposed one method of deriving the eigenfunctions of curl in a bounded cylinder.

In the case $G = Q$ where $Q$ is a cube with an edge $(0, 2\pi)$ we consider a periodic boundary condition instead of (2). The eigenfunctions of the operator *curl* with minimal eigenvalue are known and used in some papers. For example, V.I. Arnold used the eigenfunctions of the operator *curl* with an eigenvalue equal to one in the class of $2\pi$-periodic vector-functions in papers [2–5].

The solution of spectral problem for *curl* operator on periodic vector-functions was given in R.Saks and Ju.Polyakov work [26] (see also [29]). The following theorems have been proved.

**Theorem 1.1.** *The eigenvalues of the curl operator in the class of $2\pi$-periodic vector-functions are equal to $\pm|\mathbf{k}| = \pm\sqrt{k_1^2 + k_2^2 + k_3^2}$ for all $\mathbf{k} = (k_1, k_2, k_3)$ from $\mathbb{Z}^3$.*
*The multiplicity of the non-zero eigenvalue $|\mathbf{k}_0|$ (and $-|\mathbf{k}_0|$ too) is equal to the number of points $\mathbf{k} \in \mathbb{Z}^3$ lying on the sphere of radius $|\mathbf{k}_0|$.*
*The multiplicity of $\lambda = 0$ is infinite.*

**Theorem 1.2.** *For every $\mathbf{k} \in \mathbb{Z}^3$ we can associate two vector functions $\mathbf{c}_{\mathbf{k}}^+ e^{i\mathbf{k}\cdot\mathbf{x}}$ and $\mathbf{c}_{\mathbf{k}}^- e^{i\mathbf{k}\cdot\mathbf{x}}$, which are the eigenfunctions of the curl operator with eigenvalues*

$\lambda = |\mathbf{k}|$ and $\lambda = -|\mathbf{k}|$. The vector functions $\mathbf{k}e^{i\mathbf{k}\cdot\mathbf{x}}$ are eigenfunctions of the curl with eigenvalues $\lambda = 0$. With $\mathbf{k} = 0$ we associate the basic eigenvector-functions $\mathbf{e}_1$, $\mathbf{e}_2$, $\mathbf{e}_3$.

The vectors $\mathbf{c}_{\mathbf{k}}^+$ and $\mathbf{c}_{\mathbf{k}}^-$ are given by the formulas

$$\mathbf{c}_{\mathbf{k}}^{\pm} = \frac{\sqrt{2}}{2} \begin{pmatrix} \pm k_3|k_3|^{-1} \\ i \\ 0 \end{pmatrix}, \quad \text{if } k_1 = k_2 = 0,$$

$$\mathbf{c}_{\mathbf{k}}^{\pm} = \frac{\sqrt{2}}{2|\mathbf{k}||\mathbf{k}'|} \begin{pmatrix} \pm \lambda k_2 + i k_1 k_3 \\ \mp \lambda k_1 + i k_2 k_3 \\ -i(k_1^2 + k_2^2) \end{pmatrix}, \quad \text{if } |\mathbf{k}'|^2 \equiv k_1^2 + k_2^2 \neq 0.$$

Let us consider the vector function $\mathbf{f}(\mathbf{x}) \in \mathbf{L}_2(Q, 2\pi)$ and its Fourier decomposition

$$\mathbf{f}(\mathbf{x}) = \sum_{|\mathbf{k}|^2=0}^{\infty} \mathbf{f}_{\mathbf{k}} e^{i\mathbf{k}\cdot\mathbf{x}}, \tag{3}$$

where

$$\mathbf{f}_{\mathbf{k}} = \frac{1}{(2\pi)^3} \int_Q \mathbf{f}(\mathbf{x}) e^{-i\mathbf{k}\cdot\mathbf{x}} d\mathbf{x}. \tag{4}$$

This series is convergent in norm of the space $\mathbf{L}_2(Q, 2\pi)$. As

$$\mathbf{f}_{\mathbf{k}} = \alpha_{\mathbf{k}} \dot{\mathbf{k}} + \beta_{\mathbf{k}}^+ \mathbf{c}_{\mathbf{k}}^+ + \beta_{\mathbf{k}}^- \mathbf{c}_{\mathbf{k}}^-,$$

if $|\mathbf{k}| > 0$, and $\mathbf{f}_0 = \sum_{j=1}^{3} f_0^j \mathbf{e}_j$, where

$$\alpha_{\mathbf{k}} = (\mathbf{f}_{\mathbf{k}}, \dot{\mathbf{k}}), \quad \dot{\mathbf{k}} = \frac{\mathbf{k}}{|\mathbf{k}|}, \quad \beta_{\mathbf{k}}^{\pm} = (\mathbf{f}_{\mathbf{k}}, \mathbf{c}_{\mathbf{k}}^{\pm}), \tag{5}$$

we have also the following decomposition

$$\mathbf{f}(\mathbf{x}) = \sum_{j=1}^{3} f_0^j \mathbf{e}_j + \sum_{|\mathbf{k}|^2=1}^{\infty} (\alpha_{\mathbf{k}} \dot{\mathbf{k}} + \beta_{\mathbf{k}}^+ \mathbf{c}_{\mathbf{k}}^+ + \beta_{\mathbf{k}}^- \mathbf{c}_{\mathbf{k}}^-) e^{i\mathbf{k}\cdot\mathbf{x}}. \tag{6}$$

This series will be denoted as modified Fourier decomposition.

**Theorem 1.3.** *For every vector function* $\mathbf{f}(\mathbf{x}) \in \mathbf{L}_2(Q, 2\pi)$ *there is a modified Fourier decomposition (6) (in basis of the curl eigenvector-functions) and*

$$\| \mathbf{f}(\mathbf{x}) \|^2 = \sum_{j=1}^{3} |f_0^j|^2 + \sum_{|\mathbf{k}|^2=1}^{\infty} (|\alpha_{\mathbf{k}}|^2 + |\beta_{\mathbf{k}}^+|^2 + |\beta_{\mathbf{k}}^-|^2). \tag{7}$$

The following six complex eigenvector-functions

$$\begin{pmatrix} 0 \\ \mp i \\ 1 \end{pmatrix} e^{\pm i x_1}, \quad \begin{pmatrix} \pm i \\ 0 \\ 1 \end{pmatrix} e^{\pm i x_2}, \quad \begin{pmatrix} \mp i \\ 1 \\ 0 \end{pmatrix} e^{\pm i x_3}$$

correspond to the eigenvalue $\lambda = 1$. One can form six real eigenvector-functions from them.

The values of multiplicities from the following table are calculated by computer program of my student Yu. N. Polyakov.

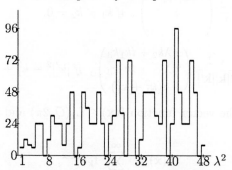

*The diagram of multiplicities*

*The multiplicity of eigenvalue $\pm\lambda$ is equal to the number of points of integer lattice $\mathbb{Z}^3$ lying on the sphere of radius $\lambda = |\mathbf{k}_0|$.*

| $\lambda^2$ | 1 | 2 | 3 | 4 | 5 | 6 | 7 | 8 | 9 | 10 | 11 | 12 |
|---|---|---|---|---|---|---|---|---|---|---|---|---|
| Multiplicity of $\lambda$ | 6 | 12 | 8 | 6 | 24 | 24 | 0 | 12 | 30 | 24 | 24 | 8 |
| $\lambda^2$ | 13 | 14 | 15 | 16 | 17 | 18 | 19 | 20 | 21 | 22 | 23 | 24 |
| Multiplicity of $\lambda$ | 24 | 48 | 0 | 6 | 48 | 36 | 24 | 24 | 48 | 24 | 0 | 24 |
| $\lambda^2$ | 25 | 26 | 27 | 28 | 29 | 30 | 31 | 32 | 33 | 34 | 35 | 36 |
| Multiplicity of $\lambda$ | 30 | 72 | 32 | 0 | 72 | 48 | 0 | 12 | 48 | 48 | 48 | 30 |
| $\lambda^2$ | 37 | 38 | 39 | 40 | 41 | 42 | 43 | 44 | 45 | 46 | 47 | 48 |
| Multiplicity of $\lambda$ | 24 | 72 | 0 | 24 | 96 | 48 | 24 | 24 | 72 | 48 | 0 | 8 |

There are a computer program calculating Fourier coefficients $\mathbf{f_k}$ and modified Fourier coefficients $\alpha_{\mathbf{k}}$ and $\beta_{\mathbf{k}}^{\pm}$ of the given function $\mathbf{f(x)}$ and a program calculating scalar products of vectors $\dot{\mathbf{k}}$, $\mathbf{c_k^+}$ and $\mathbf{c_k^-}$ for different $\mathbf{k}$. These programs were written by students M.Yusupov and S.Rabtsevich.

## 2. Solution of an equation

$$\mathrm{curl}\,\mathbf{u} + \lambda\mathbf{u} = \mathbf{f} \tag{8}$$

in the space $\mathbf{L_2}(Q, 2\pi)$ for all $\lambda \in \mathbb{C}$.

**Theorem 2.1.** *If $\lambda$ is not equal to an eigenvalue of the curl operator in the class of $2\pi$-periodic vector functions (that is $\lambda \neq \pm|\mathbf{k}|$ for all $\mathbf{k}$ from $\mathbb{Z}^3$) than the equation (8) has a unique solution $\mathbf{u} = \mathbf{G}_\lambda\mathbf{f}$ for all vector function $\mathbf{f} \in \mathbf{L_2}(Q, 2\pi)$. This solution is given by formula*

$$\mathbf{u} = \mathbf{G}_\lambda\mathbf{f} \equiv \sum_{j=1}^{3} \frac{f_0^j}{\lambda}\mathbf{e_j} + \sum_{|\mathbf{k}|^2=1}^{\infty} \left( \frac{\alpha_{\mathbf{k}}}{\lambda}\dot{\mathbf{k}} + \frac{\beta_{\mathbf{k}}^+}{\lambda+|\mathbf{k}|}\mathbf{c_k^+} + \frac{\beta_{\mathbf{k}}^-}{\lambda-|\mathbf{k}|}\mathbf{c_k^-} \right)e^{i\mathbf{k}\cdot\mathbf{x}}. \tag{9}$$

*It belongs to the space* $\mathbf{L}_2(Q, 2\pi)$ *and has an estimate*

$$\|\mathbf{G}_\lambda \mathbf{f}\| \leq \frac{1}{\rho_\lambda} \|\mathbf{f}\|,$$

*where* $\rho_\lambda$ *is the distance from* $\lambda$ *to the points of the spectrum of curl.*

If $curl\mathbf{f} = 0$, $\mathbf{G}_\lambda \mathbf{f} = \lambda^{-1}\mathbf{f}$, so there are some $\mathbf{f} \in \mathbf{L}_2(Q, 2\pi)$ such that the solution $\mathbf{G}_\lambda \mathbf{f}$ do not belong to the Sobolev space $\mathbf{W}_2^\alpha(Q, 2\pi)$ for every $\alpha > 0$. Let us consider the Hilbert space $\mathbf{L}_{2,\ div}(Q, 2\pi)$, the subspace of $\mathbf{L}_2(Q, 2\pi)$ which is defined by the conditions: vector function $\mathbf{f} \in \mathbf{L}_2(Q, 2\pi)$ and function $div\mathbf{f} \in \mathbf{L}_2(Q, 2\pi)$ also. There are inclusions

$$\mathbf{W}_2^1(Q, 2\pi) \subset \mathbf{L}_{2,\ div}(Q, 2\pi) \subset \mathbf{L}_2(Q, 2\pi).$$

If $\mathbf{f}$ belongs to $\mathbf{L}_{2,\mathbf{div}}(Q, 2\pi)$ then $\mathbf{G}_\lambda \mathbf{f}$ belongs to the space $\mathbf{W}_2^1(Q, 2\pi)$ and if $\mathbf{u} \in \mathbf{W}_2^1(Q, 2\pi)$ than the vector function $curl\mathbf{u} + \lambda\mathbf{u}$ belongs to $\mathbf{L}_{2,\ div}(Q, 2\pi)$. So we have a corollary.

**Theorem 2.2.** *If* $\lambda$ *is not equal to an eigenvalue of the curl operator in a class of* $2\pi$*-periodic vector functions then the operator* $curl + \lambda I$ *is a homeomorphism of the Hilbert spaces* $\mathbf{W}_2^1(Q, 2\pi)$ *in* $\mathbf{L}_{2,\ div}(Q, 2\pi)$.

Let $\lambda = 0$ in (8), then this equation is solvable for given function $\mathbf{f}$ if and only if its modified Fourier coefficients $f_0^j$ and $\alpha_\mathbf{k}$ are equal to zero. These conditions are equivalent to the differential equation

$$div\mathbf{f}(\mathbf{x}) = 0$$

and an integral condition

$$\mathbf{f}_0 \equiv \frac{1}{(2\pi)^3} \int_Q \mathbf{f}(\mathbf{x}) \, d\mathbf{x} = 0.$$

If these conditions are satisfied a general solution of the $curl\mathbf{u} = \mathbf{f}$ equation is given by row

$$\mathbf{u} = \sum_{j=1}^{3} a_j \mathbf{e}_j + \sum_{|\mathbf{k}|^2=1}^{\infty} (b_\mathbf{k}\mathbf{k} + \frac{\beta_\mathbf{k}^+}{|\mathbf{k}|} \mathbf{c}_\mathbf{k}^+ - \frac{\beta_\mathbf{k}^-}{|\mathbf{k}|} \mathbf{c}_\mathbf{k}^-)e^{i\mathbf{k}\cdot\mathbf{x}},$$

where the constants $a_j$ and $b_\mathbf{k}$ are arbitrary.

Thus the equation: $curl\mathbf{u} = \mathbf{f}$ is solvable in sense of Hausdorf theory.

Let $\lambda = |\mathbf{k}_0|$ in (8), where $\mathbf{k}_0 \in \mathbb{Z}^3 \setminus \{0\}$. Then this equation is solvable for given function $\mathbf{f}$ if and only if its modified Fourier coefficients $\beta_\mathbf{k}^- = 0$ are equal to zero for all $\mathbf{k}$ such that $|\mathbf{k}| = |\mathbf{k}_0|$. That is

$$\int_Q (f(\mathbf{x}), \mathbf{c}_\mathbf{k}^-)e^{-i\mathbf{k}\mathbf{x}} \, d\mathbf{x} = 0 \ \forall \mathbf{k} : |\mathbf{k}| = |\mathbf{k}_0|.$$

The number of these conditions is finite and equal to $\varkappa(\mathbf{k}_0)$. If these conditions are satisfied a general solution of the $curl\,\mathbf{u} + |\mathbf{k}_0|\mathbf{u} = \mathbf{f}$ equation is given by row

$$\mathbf{u} = \sum_{j=1}^{3} \frac{f_0^j}{|\mathbf{k}_0|}\,\mathbf{e}_j + \sum_{|\mathbf{k}|^2=1}^{\infty}\left(\frac{\alpha_\mathbf{k}}{|\mathbf{k}_0|}\mathbf{k} + \frac{\beta_\mathbf{k}^+}{|\mathbf{k}_0|+|\mathbf{k}|}\mathbf{c}_\mathbf{k}^+\right)e^{i\mathbf{k}\cdot\mathbf{x}}$$

$$+ \sum_{|\mathbf{k}|\neq|\mathbf{k}_0|}^{\infty}\frac{\beta_\mathbf{k}^-}{|\mathbf{k}_0|-|\mathbf{k}|}\mathbf{c}_\mathbf{k}^- e^{i\mathbf{k}\cdot\mathbf{x}} + \sum_{|\mathbf{k}|=|\mathbf{k}_0|} b_\mathbf{k}\cdot\mathbf{c}_\mathbf{k}^- e^{i\mathbf{k}\cdot\mathbf{x}},$$

where the constants $b_\mathbf{k}$ are arbitrary. Thus this equation is solvable in sense of Fredholm theory and its index is equal to zero.

## 3. Stokes operator

Eigenfunctions $(\mathbf{u}_n, p_n)$ and eigenvalues $\mu_n$ of the Stokes operator are defined by the equations

$$-\nu\Delta\mathbf{u}_n + \nabla p_n = \mu_n\mathbf{u}_n, \quad \operatorname{div}\mathbf{u}_n = 0.$$

From these equations it follows: $\Delta p_n = 0$ in $Q$. But periodic harmonic functions are constant. Thus $\nabla p_n = 0$ and we can consider eigenfunctions $\mathbf{u}_n$ instead of $(\mathbf{u}_n, p_n)$.

**Theorem 3.1.** *Eigenvalues of Stokes operator with periodic boundary conditions are equal to $\nu|\mathbf{k}|^2$, where $\mathbf{k} \in \mathbb{Z}^3$. The multiplicity of zero eigenvalue is three, the multiplicity of every non zero eigenvalue $\nu|\mathbf{k}_0|^2$ is equal to $2\varkappa(|\mathbf{k}_0|)$.*

**Theorem 3.2.** *The vectors $\mathbf{e}_1$, $\mathbf{e}_2$, $\mathbf{e}_3$ are eigenfunctions of the Stokes operator with zero eigenvalue, and vectors $\mathbf{u}_\mathbf{k}^+ = \mathbf{c}_\mathbf{k}^+ e^{i\mathbf{k}\mathbf{x}}$ and $\mathbf{u}_\mathbf{k}^- = \mathbf{c}_\mathbf{k}^- e^{i\mathbf{k}\mathbf{x}}$ are its eigenfunctions with non zero eigenvalue $\nu|\mathbf{k}_0|^2$, if $|\mathbf{k}| = |\mathbf{k}_0|$. They are also the eigenfunctions of the curl operator with zero and $\pm|\mathbf{k}_0|$ eigenvalues.*

**Theorem 3.3.** *Each vector function $\mathbf{f}(\mathbf{x})$ of $\mathbf{L}_2(Q, 2\pi)$, which satisfy the $\operatorname{div}\mathbf{f} = 0$ equation, has a modified Fourier decomposition*

$$\mathbf{f}(\mathbf{x}) = \sum_{j=1}^{3} f_0^j\mathbf{e}_j + \sum_{|\mathbf{k}|^2=1}^{\infty}(\beta_\mathbf{k}^+\mathbf{c}_\mathbf{k}^+ + \beta_\mathbf{k}^-\mathbf{c}_\mathbf{k}^-)e^{i\mathbf{k}\cdot\mathbf{x}} \tag{10}$$

*(on the basis of the Stokes eigenvector-functions) and*

$$\|\mathbf{f}(\mathbf{x})\|^2 = \sum_{j=1}^{3}|f_0^j|^2 + \sum_{|\mathbf{k}|^2=1}^{\infty}(|\beta_\mathbf{k}^+|^2 + |\beta_\mathbf{k}^-|^2). \tag{11}$$

## 4. Solution of Stokes equations

Consider

$$-\nu(\Delta \mathbf{v} + \lambda^2 \mathbf{v}) + \nabla p = \mathbf{f}, \quad div\mathbf{v} = 0, \tag{12}$$

in the space $\mathbf{L}_2(Q, 2\pi)$ for all $\lambda \in \mathbb{C}$.

**Theorem 4.1.** *If $\nu\lambda^2$ is not equal to an eigenvalue of the Stokes operator in the class of $2\pi$-periodic vector functions (that is $\lambda \neq |\mathbf{k}|^2$ for all $\mathbf{k}$ from $\mathbb{Z}^3$) the equations (12) have a unique solution $(\mathbf{v}, p)$ of the form $(\mathbf{S}_\lambda \mathbf{f}, P\mathbf{f})$ for all vector function $\mathbf{f} \in \mathbf{L}_2(Q, 2\pi)$, where the operators $\mathbf{S}_\lambda \mathbf{f}$ and $P\mathbf{f}$ are given by formulas*

$$\mathbf{v} = \mathbf{S}_\lambda \mathbf{f} \equiv -\sum_{j=1}^{3} \frac{f_0^j}{\nu\lambda^2} \mathbf{e_j} + \nu^{-1} \sum_{|\mathbf{k}|^2=1}^{\infty} \left( \frac{\beta_\mathbf{k}^+}{|\mathbf{k}|^2 - \lambda^2} \mathbf{c}_\mathbf{k}^+ + \frac{\beta_\mathbf{k}^-}{|\mathbf{k}|^2 - \lambda^2} \mathbf{c}_\mathbf{k}^- \right) e^{i\mathbf{k}\cdot\mathbf{x}}, \tag{13}$$

$$P\mathbf{f} = -i \sum_{|\mathbf{k}|^2=1}^{\infty} \frac{\alpha_\mathbf{k}}{|\mathbf{k}|} e^{i\mathbf{k}\mathbf{x}}.$$

*The vector $\mathbf{S}_\lambda \mathbf{f}$ belongs to the Sobolev space $\mathbf{W}_2^2(Q, 2\pi)$, a subspace of $\mathbf{L}_2(Q, 2\pi)$, and its norms in these spaces have the estimates*

$$\|\mathbf{S}_\lambda \mathbf{f}\| \leq \frac{1}{\nu\rho_\lambda} \|\mathbf{f}\|, \quad \|\mathbf{S}_\lambda \mathbf{f}\|_2 \leq \frac{c_\lambda}{\nu} \|\mathbf{f}\|,$$

*where $\rho_\lambda$ is a distance from $\lambda^2$ to points of spectra of the Stokes operator with $\nu = 1$, $c_\lambda$ is constant. The function $P\mathbf{f} \in W_2^1(Q, 2\pi)$, a subspace of $L_2(Q, 2\pi)$, and has the properties*

$$\|P\mathbf{f}\| \leq \|\mathbf{f}\|, \quad \|\nabla P\mathbf{f}\| \leq \|\mathbf{f}\|, \quad P0 = 0.$$

*A general solution of the Stokes equations has the form $(\mathbf{S}_\lambda \mathbf{f}, P\mathbf{f} + q)$, where the constant $q$ is arbitrary.*

**Theorem 4.2.** *If $\lambda = 0$ the equations (12) have a solution $(\mathbf{v}, p)$ of the form $(\mathbf{S}_0 \mathbf{f}, P\mathbf{f})$ for all vector functions $\mathbf{f} \in \mathbf{L}_2(Q, 2\pi)$, such that*

$$\mathbf{f}_0 \equiv \frac{1}{(2\pi)^3} \int_Q \mathbf{f}(\mathbf{x}) \, d\mathbf{x} = 0;$$

*the operators $\mathbf{S}_0 \mathbf{f}$ and $P\mathbf{f}$ are given by the formulas*

$$\mathbf{S}_0 \mathbf{f} \equiv \nu^{-1} \sum_{|\mathbf{k}|^2=1}^{\infty} \left( \frac{\beta_\mathbf{k}^+}{|\mathbf{k}|^2} \mathbf{c}_\mathbf{k}^+ + \frac{\beta_\mathbf{k}^-}{|\mathbf{k}|^2} \mathbf{c}_\mathbf{k}^- \right) e^{i\mathbf{k}\cdot\mathbf{x}}, \tag{14}$$

$$P\mathbf{f} = -i \sum_{|\mathbf{k}|^2=1}^{\infty} \frac{\alpha_\mathbf{k}}{|\mathbf{k}|} e^{i\mathbf{k}\mathbf{x}}.$$

The vector $\mathbf{S}_0\mathbf{f}$ belongs to the Sobolev space $\mathbf{W}_2^2(Q, 2\pi) \subset \mathbf{L}_2(Q, 2\pi)$ and its norms in these spaces have the following estimates

$$\|\mathbf{S}_0\mathbf{f}\| \le \frac{1}{\nu}\|\mathbf{f}\|, \quad \|\mathbf{S}_0\mathbf{f}\|_2 \le \frac{c_0}{\nu}\|\mathbf{f}\|,$$

where $c_0$ is a constant. The function $Pf \in W_2^1(Q, 2\pi) \subset L_2(Q, 2\pi)$ and has the properties

$$\|Pf\| \le \|\mathbf{f}\|, \quad \|\nabla Pf\| \le \|\mathbf{f}\|, \quad P0 = 0.$$

The general solution of Stokes equations has the form

$$\left(\mathbf{S}_0\mathbf{f} + \sum_{j=1}^{3} a_j \mathbf{e}_j, Pf + q\right),$$

where the constants $a_j$ and $q$ are arbitrary.

**Theorem 4.3.** *If $\nu\lambda^2$ is equal to an eigenvalue of the Stokes operator in the class of $2\pi$-periodic vector functions (for example, $\lambda$ is $|\mathbf{k}_0|$ for some $\mathbf{k}_0$ from $\mathbb{Z}^3\backslash 0$) the equations (12) have a solution $(\mathbf{v}, p)$ of the form $(\mathbf{S}_{|\mathbf{k}_0|}\mathbf{f}, Pf)$ if and only if a vector function $\mathbf{f} \in \mathbf{L}_2(\mathbf{Q}, 2\pi)$ satisfy the integral conditions*

$$\int_Q (f(\mathbf{x}), c_{\mathbf{k}}^{\pm})e^{-i\mathbf{k}\mathbf{x}}\,d\mathbf{x} = 0 \ \forall \mathbf{k}: |\mathbf{k}| = |\mathbf{k}_0|.$$

*The operators $\mathbf{S}_{|\mathbf{k}_0|}\mathbf{f}$ and $Pf$ are given by the formulas*

$$\mathbf{v} = \mathbf{S}_{|\mathbf{k}_0|}\mathbf{f} \equiv -\sum_{j=1}^{3}\frac{f_0^j}{\nu|\mathbf{k}_0|^2}\mathbf{e}_j + \nu^{-1}\sum_{|\mathbf{k}|^2 \ne |\mathbf{k}_0|^2}^{\infty}\left(\frac{\beta_{\mathbf{k}}^+}{|\mathbf{k}|^2 - |\mathbf{k}_0|^2}c_{\mathbf{k}}^+ + \frac{\beta_{\mathbf{k}}^-}{|\mathbf{k}|^2 - |\mathbf{k}_0|^2}c_{\mathbf{k}}^-\right)e^{i\mathbf{k}\cdot\mathbf{x}},$$

$$\tag{15}$$

$$Pf = -i\sum_{|\mathbf{k}|^2=1}^{\infty}\frac{\alpha_{\mathbf{k}}}{|\mathbf{k}|}e^{i\mathbf{k}\mathbf{x}}.$$

*The vector $\mathbf{S}_{|\mathbf{k}_0|}\mathbf{f}$ belongs to the Sobolev space $\mathbf{W}_2^2(Q, 2\pi)$ and has two estimates*

$$\|\mathbf{S}_{|\mathbf{k}_0|}\mathbf{f}\| \le \frac{1}{\nu\rho_{|\mathbf{k}_0|}}\|\mathbf{f}\|, \quad \|\mathbf{S}_{|\mathbf{k}_0|}\mathbf{f}\|_2 \le \frac{c_{|\mathbf{k}_0|}}{\nu}\|\mathbf{f}\|,$$

*where $\rho_{|\mathbf{k}_0|}$ is the distance from $|\mathbf{k}_0|^2$ to the other points of the spectrum of the Stokes operator with $\nu = 1$, $c_{|\mathbf{k}_0|}$ is a constant. The function $Pf \in W_2^1(Q, 2\pi)$ and has the properties*

$$\|Pf\| \le \|\mathbf{f}\|, \quad \|\nabla Pf\| \le \|\mathbf{f}\|, \quad P0 = 0.$$

*A general solution of the Stokes equations has the form*

$$\left(\mathbf{S}_{|\mathbf{k}_0|}\mathbf{f} + \sum_{|\mathbf{k}|^2=|\mathbf{k}_0|^2}(b_{\mathbf{k}}^+ c_{\mathbf{k}}^+ + b_{\mathbf{k}}^- c_{\mathbf{k}}^-)e^{i\mathbf{k}\mathbf{x}}, Pf + q\right),$$

*where constants $b_{\mathbf{k}}^{\pm}$ and $q$ are arbitrary.*

These results are published in Proceeding of Saint-Petersburg Steklov's Mathematical Institute (2004, vol. 318)[29].

## 5. Application

The Cauchy problem for the Navier-Stokes system

$$\partial_t \mathbf{v} - \nu \boldsymbol{\Delta} \mathbf{v} + (\mathbf{v}, \nabla)\mathbf{v} + \nabla p = \mathbf{f}, \quad \operatorname{div} \mathbf{v} = 0, \tag{16}$$

with periodic boundary conditions and initial conditions $\mathbf{v}|_{t=0} = \mathbf{g}(\mathbf{x})$.

Let us consider the modified Fourier decompositions of the given vectors $\mathbf{f}(\mathbf{x}, t)$, $\mathbf{g}(\mathbf{x})$ and the unknown vector $\mathbf{v}(\mathbf{x}, t)$,

$$\mathbf{f}(\mathbf{x}, t) = \mathbf{f}_0(t) + \sum_{|k|^2=1}^{\infty} (\alpha_{\mathbf{k}}(t)\dot{\mathbf{k}} + \beta_{\mathbf{k}}^+(t)\mathbf{c}_{\mathbf{k}}^+ + \beta_{\mathbf{k}}^-(t)\mathbf{c}_{\mathbf{k}}^-)e^{i\mathbf{k}\cdot\mathbf{x}}, \tag{17}$$

$$\mathbf{g}(\mathbf{x}) = \mathbf{g}_0(t) + \sum_{|k|^2=1}^{\infty} (g_{\mathbf{k}}^+\mathbf{c}_{\mathbf{k}}^+ + g_{\mathbf{k}}^-\mathbf{c}_{\mathbf{k}}^-)e^{i\mathbf{k}\cdot\mathbf{x}}, \tag{18}$$

$$\mathbf{v}(\mathbf{x}, t) = \mathbf{v}_0(t) + \sum_{|k|^2=1}^{\infty} (w_{\mathbf{k}}^+(t)\mathbf{c}_{\mathbf{k}}^+ + w_{\mathbf{k}}^-(t)\mathbf{c}_{\mathbf{k}}^-)e^{i\mathbf{k}\cdot\mathbf{x}}; \tag{19}$$

and Fourier decomposition of function $p(\mathbf{x}, t)$,

$$p(x, t) = p_0(t) + \sum_{|k|^2=1}^{\infty} p_{\mathbf{k}}(t)e^{i\mathbf{k}\cdot\mathbf{x}}. \tag{20}$$

Using these decompositions we reduce the Cauchy problem for the Navier-Stokes system to the Cauchy problem for the infinite system of first order (ordinary) differential equations:

$$\frac{dv_0^j(t)}{dt} = f_0^j(t), \quad v_0^j(0) = g_0^j, \quad j = 1, 2, 3, \tag{21}$$

$$\frac{dw_{\mathbf{k}}^+(t)}{dt} + \left(\nu|\mathbf{k}|^2 + i(\mathbf{v}_0(t), \mathbf{k})\right)w_{\mathbf{k}}^+ + \tag{22}$$

$$i\sum_{m} \left(w_{\mathbf{k}-\mathbf{m}}^+(\mathbf{c}_{\mathbf{k}-\mathbf{m}}^+, \mathbf{m}) + w_{\mathbf{k}-\mathbf{m}}^-(\mathbf{c}_{\mathbf{k}-\mathbf{m}}^-, \mathbf{m})\right) \cdot \left(w_{\mathbf{m}}^+(\mathbf{c}_{\mathbf{m}}^+, \mathbf{c}_{\mathbf{k}}^+) + w_{\mathbf{m}}^-(\mathbf{c}_{\mathbf{m}}^-, \mathbf{c}_{\mathbf{k}}^+)\right) = \beta_{\mathbf{k}}^+(t),$$

$$\frac{dw_{\mathbf{k}}^-(t)}{dt} + \left(\nu|\mathbf{k}|^2 + i(\mathbf{v}_0(t), \mathbf{k})\right)w_{\mathbf{k}}^- + \tag{23}$$

$$i\sum_{m} \left(w_{\mathbf{k}-\mathbf{m}}^+(\mathbf{c}_{\mathbf{k}-\mathbf{m}}^+, \mathbf{m}) + w_{\mathbf{k}-\mathbf{m}}^-(\mathbf{c}_{\mathbf{k}-\mathbf{m}}^-, \mathbf{m})\right) \cdot \left(w_{\mathbf{m}}^+(\mathbf{c}_{\mathbf{m}}^+, \mathbf{c}_{\mathbf{k}}^-) + w_{\mathbf{m}}^-(\mathbf{c}_{\mathbf{m}}^-, \mathbf{c}_{\mathbf{k}}^-)\right) = \beta_{\mathbf{k}}^-(t),$$

$$w_{\mathbf{k}}^{\pm}(0) = g_{\mathbf{k}}^{\pm},$$

with vectors $\mathbf{m}, \mathbf{k} \in \mathbb{Z}^3 \backslash 0$.

It may be solved by the method of Galerkin approximations let $n$ be a fixed natural number, we shall consider finite parts of the series (17), (18), (19), (20) which contain their terms with $|\mathbf{k}|^2 \leq n$ and finite subsystems of the systems (22), (23) which contain unknowns $w_{\mathbf{k}}^{\pm}$, $w_{\mathbf{m}}^{\pm}$, $w_{\mathbf{k}-\mathbf{m}}^{\pm}$ with $|\mathbf{m}|^2$, $|\mathbf{k}|^2$ and $|\mathbf{k} - \mathbf{m}|^2 \leq n$.

This subsystem will be denoted by $GAS_n$. There is a computer program calculating their coefficients.

The system $GAS_2$ contain 39 complex equations with 39 complex unknowns $v_0^j$ and $w_{\mathbf{k}}^{\pm}$, where $j = 1, 2, 3$ and $1 \leq |\mathbf{k}|^2 \leq 2$. The equations (21) are the first three equations of this system $GAS_2$ and the following equation has the form

$$\frac{d}{dt} w_{1(-1,0,0)}^+ = (-\nu + iv_0^1) w_{1(-1,0,0)}^+$$
$$- \big( w_{2(0,-1,0)}^+ (0.603553i) w_{10(-1,1,0)}^+ + w_{2(0,-1,0)}^+ (0.103553i) w_{10(-1,1,0)}^-$$
$$+ w_{3(0,0,-1)}^+ (0.603553i) w_{9(-1,0,1)}^+ + w_{3(0,0,-1)}^+ (-0.103553i) w_{9(-1,0,1)}^-$$
$$+ w_{4(0,0,1)}^+ (-0.603553i) w_{8(-1,0,-1)}^+ + w_{4(0,0,1)}^+ (0.103553i) w_{8(-1,0,-1)}^-$$
$$+ w_{5(0,1,0)}^+ (-0.603553i) w_{7(-1,-1,0)}^+ + w_{5(0,1,0)}^+ (-0.103553i) w_{7(-1,-1,0)}^-$$
$$+ w_{7(-1,-1,0)}^+ (0.25i) w_{5(0,1,0)}^+ + w_{7(-1,-1,0)}^+ (0.25i) w_{5(0,1,0)}^-$$
$$+ w_{8(-1,0,-1)}^+ (0.25i) w_{4(0,0,1)}^+ + w_{8(-1,0,-1)}^+ (0.25i) w_{4(0,0,1)}^-$$
$$+ w_{9(-1,0,1)}^+ (-0.25i) w_{3(0,0,-1)}^+ + w_{9(-1,0,1)}^+ (-0.25i) w_{3(0,0,-1)}^-$$
$$+ w_{10(-1,1,0)}^+ (-0.25i) w_{2(0,-1,0)}^+ + w_{10(-1,1,0)}^+ (-0.25i) w_{2(0,-1,0)}^-$$
$$+ w_{2(0,-1,0)}^- (-0.603553i) w_{10(-1,1,0)}^+ + w_{2(0,-1,0)}^- (-0.103553i) w_{10(-1,1,0)}^-$$
$$+ w_{3(0,0,-1)}^- (-0.603553i) w_{9(-1,0,1)}^+ + w_{3(0,0,-1)}^- (0.103553i) w_{9(-1,0,1)}^-$$
$$+ w_{4(0,0,1)}^- (0.603553i) w_{8(-1,0,-1)}^+ + w_{4(0,0,1)}^- (-0.103553i) w_{8(-1,0,-1)}^-$$
$$+ w_{5(0,1,0)}^- (0.603553i) w_{7(-1,-1,0)}^+ + w_{5(0,1,0)}^- (0.103553i) w_{7(-1,-1,0)}^-$$
$$+ w_{7(-1,-1,0)}^- (-0.25i) w_{5(0,1,0)}^+ + w_{7(-1,-1,0)}^- (-0.25i) w_{5(0,1,0)}^-$$
$$+ w_{8(-1,0,-1)}^- (0.25i) w_{4(0,0,1)}^+ + w_{8(-1,0,-1)}^- (0.25i) w_{4(0,0,1)}^-$$
$$+ w_{9(-1,0,1)}^- (-0.25i) w_{3(0,0,-1)}^+ + w_{9(-1,0,1)}^- (-0.25i) w_{3(0,0,-1)}^-$$
$$+ w_{10(-1,1,0)}^- (0.25i) w_{2(0,-1,0)}^+ + w_{10(-1,1,0)}^- (0.25i) w_{2(0,-1,0)}^- \big) + \beta_{1(-1,0,0)}^+(t).$$

Here we used the double numeration of known and unknown functions; for example $w_{1(-1,0,0)}^+$ means that element $w_{(-1,0,0)}^+$ is a first in this numeration.

Unfortunately we have not enough place to write another equation of this system.

Local solutions of the systems $GAS_n$ exist but we are interested by the existence of global solutions and their bifurcation. In case $n = 2$ we are going to preform computing experiment.

Studying the structure of the systems $GAS_n$ we get some families of explicit solutions $(\mathbf{v}, p)$ of the Navier-Stokes system (16) with $\mathbf{f} = 0$.

The first family of complex solutions $(\mathbf{v}_{\mathbf{k}}^{\pm}, p_{\mathbf{k}}^{\pm})$, where

$$\mathbf{v}_{\mathbf{k}}^{\pm} = a_{\mathbf{k}}^{\pm} \mathbf{c}_{\mathbf{k}}^{\pm} \exp(i\mathbf{k} \cdot \mathbf{x} - \nu |\mathbf{k}|^2 t), \quad p_{\mathbf{k}}^{\pm} = p_{\mathbf{k}}^{\pm}(t),$$

depends on arbitrary complex constants $a_{\mathbf{k}}^{\pm}$ and functions $p_{\mathbf{k}}^{\pm}(t)$ ( for which $\mathrm{grad}\, p(t) = 0$); remember that $(\mathbf{c}_{\mathbf{k}}^{\pm}, \mathbf{k}) = 0$ for all vector $\mathbf{k} \in \mathbb{Z}^3 \backslash 0$.

**Important remarks**

a) The real and imaginary parts of every solution $(\mathbf{v}_{\mathbf{k}}^{\pm}, p_{\mathbf{k}}^{\pm})$ are also solutions of this nonlinear system (16) (with $\mathbf{f} = 0$).

b) For every proportional vectors $\mathbf{k}$ and $\mathbf{k}'$ (i.e., $\mathbf{k} = l\mathbf{k}'$ with $l \in \mathbb{Z}\backslash 0$) the sum of these solutions $(\mathbf{v}_{\mathbf{k}}^{\pm}, p_{\mathbf{k}}^{\pm})$ and $(\mathbf{v}_{\mathbf{k}'}^{\pm}, p_{\mathbf{k}'}^{\pm})$ is a solution of the homogenous system (16) too.

2nd family of complex solutions $(\mathbf{w}_{\mathbf{k}}^{\pm}, p_{\mathbf{k}}^{\pm})$, where

$$\mathbf{w}_{\mathbf{k}}^{\pm} = \mathbf{g}_0 + a_{\mathbf{k}}^{\pm}\mathbf{c}_{\mathbf{k}}^{\pm}\exp(i\mathbf{k}\cdot\mathbf{x} - \nu|\mathbf{k}|^2 t - i(\mathbf{g}_0, \mathbf{k})t), \quad p_{\mathbf{k}}^{\pm} = p_{\mathbf{k}}^{\pm}(t),$$

depend on arbitral complex constants $a_{\mathbf{k}}^{\pm}$, vector $\mathbf{g}_0$ and functions $p_{\mathbf{k}}^{\pm}(t)$.

**Remark.** If $\mathbf{g}_0 = i\mu\mathbf{k}$ and $\mu > \nu$ than these functions $\mathbf{w}_{\mathbf{k}}^{\pm}$ grow in time.

3rd family of real solutions $(\mathbf{v}_{\mathbf{b}}, p_{\mathbf{b}})$, where

$$\mathbf{v}_{\mathbf{b}} = e^{-\nu t}(b_2 \sin x_2 + b_3 \sin x_3, \quad b_1 \sin x_1 + b_3 \cos x_3, \quad b_1 \cos x_1 - b_2 \cos x_2),$$

$$p_{\mathbf{b}} = e^{-2\nu t}(-b_1 b_2 \cos x_1 \cos x_2 + b_1 b_3 \sin x_1 \cos x_3 + b_2 b_3 \sin x_2 \sin x_3),$$

depend on real constants $b_1, b_2, b_3$ which are arbitrary.

**References**

1. E. Beltrami, *Considerazioni idrodinamiche*, Rend Ist. Lomabardo No. 22, (1889), 121–130.

2. V.I. Arnold, *Hopf asymptotic invariant and its applications*, Izbrannoe - 60 (1974), no. 28, Moscow, FAZIS, 1997, 215–236.

3. V.I. Arnold, *Sur la géométrie différentielle des groups de Lie de dimension infinie et ses applications à l'hydrodinamique des fluides parfaits*, Ann. Inst. Fourier (Grenoble)**16** (1966), no. 1, 319–361.

4. V.I. Arnold, E.I. Korkina, *Growth of the magnetic field in 3-dimensional stationary stream of incompressible fluid*, Vestnik MGU. Seria 1. Math. Mech., (1983), no. 3, 43–46.

5. V.I. Arnold, *Some remarks about antidinamo theorem*, Vestnik MGU. Seria 1. Math. Mech., (1982), no. 6, 50–57.

6. V.I. Arnold, *Sur la topologie des écoulements stationnaires des fluides parfaits.* C. R. Acad. Sci. Paris 261 (1965), 17–20.

7. S. Chandrasekhar, *On force-free magnetic fields.*, National Academy of Sciences, 42, No. 1, (1956).

8. S. Chandrasekhar, P.S. Kendall, *On force-free magnetic fields.* Astrophys. J. 126 (1957), 457–460.

9. M. Hénon, *Sur la topologie des lignes de courant dans un case particulier*, C. R. Acad. Sci. Paris **262** (1966), 312–314.

10. O.A. Ladyzhenskaya, *The mathematical theory of viscous incompressible flow.* Gordon and Breach. New York, 1969.

11. B.R. Vainberg, V.V. Grushin, *On uniformly nonelliptic problems.* I. Math. Sbornik 72 (1967), 126–154.

12. D.A. Jette, *Force-free magnetic fields in resistive magnetohydrostatics*, J. Math. Anal. Appl., 29 (1970), 109–122.

13. R.S. Saks, *On boundary value problems for the system* $\mathrm{rot} u + \lambda u = h$. Soviet. Math. Dokl. 12 (1971), 1240–1244.

1206

14. R.S. Saks, *On boundary value problems for the system* $rotu + \lambda u = h$. Diff. Equat. 8 (1972), 126–133.

15. J.B. Taylor, *Relaxation of toroidal plasma and generation of reversed magnetic fields.* Phys. Rev. Lett. 33 (1974), 1139–1141.

16. P.E. Berhin, *Selfadjoint boundary value problem for the system* $*du + \lambda u = f$. Soviet. Math. Dokl. 222 (1975), 15–17.

17. R.S. Saks, *On boundary value problems of Noether type for some classes of weakly elliptic systems of differential equations.* In: Math. Analysis and Related Problems of Mathematics, Proc. of Inst. Math. Sibirian Depart. Akad. of Sci. USSR, Novosibirsk, Nauka, Moscow, 1978, 237–253.

18. D. Montgomery, L. Turner, G. Vahala, *Tree-dimensional magnetohydrodynamic turbulece in cylindrical geometry.* Phys. Fluids 21(5) (1978), 757–764.

19. R.S. Saks, *Normally solvable and Noeterian boundary value problems for the system of Maxwell equations in the case of steady-state processes.* Soviet. Math. Dokl. 28 (1983), 391–395.

20. R. Kress, *On constant-alpha force-free fields in a torus,* Journal of Engineering Mathematics, 20 (1986), 323–344 .

21. V.S. Vladimirov, *Equations of mathematical physics.* 5th edition, Moscow, Nauka, 1988.

22. Y. Giga, Z. Yoshida, *Remark on spectra of operator rot.* Math. Z. 204 (1990), 235–245.

23. R. Picard, *On selfadjoint realization of curl and some of its applications.* Preprint Math-AN-02-96, Techn. Univ. Dresden, 1996.

24. D. Elton, D.G. Vassiliev, *The Dirac equation without spinors.* University of Sussex at Brighton, Research Report No: 98/02.(1998).

25. R.S. Saks, *Spectrum of the operator curl in a ball under the condition not passing through the boundary and eigen frequencies of oscillations of an elastic spherical volume fixed at the boundary.* Complex Analysis, Diff. Equations and Related Topics, IV. Applicable Mathematics (Proc. of Intern. Conference Ufa, Inst. of Math. RAS), 2000, 61–69.

26. R.S. Saks, Ju.N. Polykov, *On Spectrum of the operator curl in the class of periodic vector functions.* In "Methods of Functional Analysis and Theory of Functions in Various Problems of Math. Physics, II" (Proc. of Intern. Workshop, Ufa, Bashkir St. University, 2000, red. R. Saks), Ufa, Math. Inst. RAS, 2002, 175–194.

27. R. Saks, *On Spectrum of the operator curl.* In "Progress in Analysis", vol. 2 (Proceeding of 3-rd ISAAC Congress, Freie Univers., Berlin 2001, eds. H. Begehr et al.), Wold Sci., Singapore, 2003, 811–817.

28. A. Mahalov, B. Nicolaenko, *Global regularity of the 3D Navier-Stokes equations with uniformly large initial vorticity.* Russian Math. Surveys (transl. of Uspekhi Mat. Nauk), 58 (2003), No. 2, 32–64.

29. R.S. Saks, *The solution of a spectral problem for the curl and Stokes operators with periodic boundary condition.* In "Boundary Value Problems of Math. Physics and Some Problems of Functions Theory. 35" (Zapiski nauchnykh seminarov POMI, redact. S. Repin and G. Seregin) 318 (2004), 246–276.

30. R.S. Saks, C.J. Vanegas, *Solution of a spectral problem for the curl operator on a cylinder* (Colloquium on Differ. Equations and Applications, Maracaibo, Venezuela 2003). In electronic Journal of Differential Equations, Conference 13 (2005), 95–99.

31. R.S. Saks, C.J. Vanegas, *Solution of one boundary value problem for* $curl + \lambda I$ *operator in the ball.* Complex Variables, Theory Appl. 50, No.7-11, (2005), 837–850.

# ABOUT ONE CLASS OF LINEAR FIRST ORDER OVERDETERMINED SYSTEMS WITH INTERIOR SINGULAR AND SUPER-SINGULAR MANIFOLDS

NUSRAT RAJABOV

*Presidium of the Academy of Sciences*
*of the Republic of Tajikistan, TSNU*
*E-mail: nusrat38@mail.ru*

In this paper we find the solution of the first order linear overdetermined system of partial differential equations with interior singular and super-singular manifolds, in depence of the order of singularity and of the sign of the coefficients. We investigate the behavior of the obtained solution in the neighborhood of singular and super-singular manifolds. The obtained integral representations of the manifold of solutions make use of the correct formulation of the linear conjugate problems and their investigation.

**Key-words:** Overdetermined system of pdes, singular and super-singular coefficients, singular and super-singular manifold, integral representation of manifold of solutions
**Mathematics Subject Classification:** 35N10

Let $D_0$ denote the rectangle $D_0 = \{(x,y) : -c < x < c, -d < y < d\}$; $D_1 = \{(x,y) : -c < x < 0, 0 < y < d\}$, $D_2 = \{(x,y) : 0 < x < c, 0 < y < d\}$, $D_3 = \{(x,y) : -c < x < 0, -d < y < 0\}$, $D_4 = \{(x,y) : 0 < x < c, -d < y < 0\}$, $\Gamma_1 = \{-c < x < c, y = 0\}$, $\Gamma_2 = \{x = 0, -d < y < d\}$, $D = D_0 \backslash (\Gamma_1 \cup \Gamma_2)$, $\Gamma_1^1 = \{-c < x < 0, y = 0\}$, $\Gamma_1^2 = \{0 < x < c, y = 0\}$, $\Gamma_2^1 = \{x = 0, -d < y < 0\}$, $\Gamma_2^2 = \{x = 0, 0 < y < d\}$. In the domain $D$ we shall consider the following first order overdetermined systems

$$\frac{\partial U}{\partial x} + \frac{a(x,y)}{|x|^\alpha} U = \frac{f_1(x,y)}{|x|^\alpha}, \quad \frac{\partial U}{\partial y} + \frac{b(x,y)}{r^\beta} U = \frac{f_2(x,y)}{r^\beta}, \tag{1}$$

where $r^2 = x^2 + y^2$, $a(x,y), b(x,y), f_j(x,y), j = 1,2$ is a given functions in $D$. Moreover these functions in $\Gamma_1^j, \Gamma_2^j, j = 1,2$, may have first order singularities, $\alpha = constant > 0, \beta = constant > 0$.

First order linear overdetermined systems of type (1) in domains $D_1, D_1 \cup D_2$ are investigated in [1-3]. In case, when $\alpha = 0, \beta = 0$, functions $a(x,y), b(x,y), f_j(x,y), j = 1,2$, sufficiently smooth, it follows from the theory of total differentials, that any solution of the system (1) from the class $C'(D) \cap C(\bar{D})$ contains one arbitrary constant.

In [1] the linear first order overdetermined system with two singular lines and singular point

$$\frac{\partial U}{\partial x} + \frac{a(x,y)}{x^\alpha} U = \frac{f_1(x,y)}{x^\alpha}, \quad \frac{\partial U}{\partial y} + \frac{b(x,y)}{y^\beta} U = \frac{f_2(x,y)}{y^\beta}, \tag{2}$$

$$\frac{\partial U}{\partial x} + \frac{xa(x,y)}{r^\alpha} U = \frac{f_1(x,y)}{r^\alpha}, \quad \frac{\partial U}{\partial y} + \frac{yb(x,y)}{r^\beta} U = \frac{f_2(x,y)}{r^\beta}, \tag{3}$$

was investigated in the domain $D_2$ where $\alpha = constant > 0, \beta = constant > 0$.

In [2,3] the system (1) was investigated in the domain $D_1 \cup D_2$.

In this work on the basis of a method proposed in [1-3], system (1) is investigated in the domain $D$, containing the lines $\Gamma_1$ and $\Gamma_2$ inside. It is proved that the solution of the system (1) essentially is influenced by the numbers $\alpha, \beta(\alpha < 1, \beta < 1; \alpha = 1, \beta = 1; \alpha > 1, \beta > 1)$ and the value $a(\pm 0, \pm 0), a(\pm 0, \mp 0), b(\pm 0, \pm 0), b(\pm 0, \mp 0)$. The system (1) is investigated in all possible cases. The general solution of the system (1) contains four constants and in any possible case is found in explizit form.

**Case 1.** Let in system (1) $\alpha = 1, \beta = 2, b(x,y) = b_1(x,y)y, b_1(x,y) \in C'_x(D)$ and on $\Gamma_2$ it may have a first kind singularity. In this case the following confirmations hold.

**Theorem 1.** Let in (1) $\alpha = 1, \beta = 2, b(x,y) = yb_1(x,y), a(x,y), f_1(x,y) \in C'_y(D), b_1(x,y), f_2(x,y) \in C'_x(D)$. The functions $a(x,y), b_1(x,y), f_j(x,y), j = 1,2$, on $\Gamma^j_1, \Gamma^j_2, j = 1,2$, may have first order singularities, and in $D$ satisfy the conditions

$$\frac{\partial}{\partial y}\left(\frac{a(x,y)}{|x|}\right) = \frac{\partial}{\partial x}\left(\frac{yb_1(x,y)}{r^2}\right),$$

$$\frac{\partial}{\partial y}\left(\frac{f_1(x,y)}{|x|}\right) + \frac{yb_1(x,y)}{r^2}\frac{f_1(x,y)}{|x|} = \frac{\partial}{\partial x}\left(\frac{f_2(x,y)}{r^2}\right) + \frac{a(x,y)}{|x|}\frac{f_2(x,y)}{r^2}. \tag{4}$$

*Besides let the functions* $a(x,y), b_1(x,y)$ *correspondingly in the neighborhood of* $\Gamma_1$ *and of the origin satisfy the conditions*

$$|a(x,y) - a(\pm 0, \pm 0)| \le H_1|x|^{\gamma_1}, \gamma_1 > 0, \tag{A}$$

$$|b(x,y) - b(\pm 0, \pm 0)| \le H_2 r^{\gamma_2}, \gamma_2 > 0, \tag{B}$$

$$|a(x,y) - a(\pm 0, \mp 0)| \le H_3|x|^{\gamma_3}, \gamma_3 > 0 \tag{C}$$

$$|b(x,y) - b(\pm 0, \mp 0)| \le H_4 r^{\gamma_4}, \gamma_4 > 0, \tag{D}$$

*and* $a(\ddagger 0, \pm 0) > 0, a(= 0, \pm 0) < 0, b_1(\pm 0, \ddagger 0) > 1, b_1(\mp 0, = 0) < -1$.

*Let the limit functions*

$$f_2^1(y) = \lim_{x \to -0}(|x|^{-a(-0,+0)}f_2(x,y)), \tag{$f_2^1$}$$

$$f_2^2(y) = \lim_{x \to +0}(x^{a(+0,+0)}f_2(x,y)), \tag{$f_2^2$}$$

$$f_2^3(y) = \lim_{x \to -0} (|x|^{-a(-0,-0)} f_2(x,y)), \qquad (f_2^3)$$

$$f_2^4(y) = \lim_{x \to +0} (x^{a(+0,-0)} f_2(x,y)), \qquad (f_2^4)$$

$f_2^j(y) \in C(\bar{\Gamma}_2), 1 \le j \le 4$, *exist.*

*Then any solution of the system (1) in class $C'(D)$ are representable in the form*

$$U(x,y) = \begin{cases} K_{1,1}^{-;+}[c_1, f_1(x,y), f_2^1(y)], & when \quad (x,y) \in D_1, \\ K_{1,1}^{+;+}[c_2, f_1(x,y), f_2^2(y)], & when \quad (x,y) \in D_2, \\ K_{1,1}^{-;-}[c_3, f_1(x,y), f_2^3(y)], & when \quad (x,y) \in D_3, \\ K_{1,1}^{+;-}[c_4, f_1(x,y), f_2^4(y)], & when \quad (x,y) \in D_4, \end{cases}$$

$$\equiv K_1^1[c_1, c_2, c_3, c_4, f_1(x,y), f_2^1(y), f_2^2(y), f_2^3(y), f_2^4(y)], \qquad (5)$$

*where $c_j$ $(1 \le j \le 4)$ are arbitrary constants.*

In the integral representation (5), the integral operators $K_{1,1}^{-;+}, K_{1,1}^{+;+}, K_{1,1}^{-;-}, K_{1,1}^{+;-}$ are determined by the equalities

$$K_{1,1}^{-;+}[c_1, f_1(x,y), f_2^1(y)] \equiv \exp[W_{a,\alpha}^{-;+}(x,y)]|x|^{a(-0,+0)}$$

$$\times [y^{-b_1(-0,+0)} \exp[W_{b_1}^{-,+}(0,y)](c_1 + \int_0^y s^{b_1(-0,+0)-2} \exp(-W_{b_1}^{-,+}(0,s)) f_2^1(s) ds)$$

$$- \int_x^0 |t|^{-a(-0,+0)} \exp(-W_{a,\alpha}^{-;+}(t,y)) \frac{f_1(t,y)}{|t|} dt],$$

$$K_{1,1}^{+;+}[c_2, f_1(x,y), f_2^2(y)] \equiv \exp[-W_{a,\alpha}^{+;+}(x,y)]x^{-a(+0,+0)}$$

$$\times \{y^{-b_1(+0,+0)-2} \exp(W_{b_1}^{+,+}(0,y))(c_2 + \int_0^y s^{b_1(+0,+0)-2} \exp(W_{b_1}^{+,+}(0,s)) f_2^2(s) ds)$$

$$+ \int_0^x \exp(W_{a,\alpha}^{+;+}(t,y)) t^{a(+0,+0)-1} f_1(t,y) dt\},$$

$$K_{1,1}^{-;-}[c_3, f_1(x,y), f_2^3(y)] \equiv |x|^{a(-0,-0)} \exp[W_{a,\alpha}^{-;-}(x,y)]$$

$$\times \{|y|^{b_1(-0,-0)} \exp(W_{b_1}^{-,-}(-0,y))(c_3 - \int_y^0 |s|^{-b_1(-0,-0)-2} \exp(-W_{b_1}^{-,-}(-0,s)) f_2^3(s) ds)$$

$$- \int_x^0 |t|^{-a(-0,-0)-1} \exp(-W_{a,\alpha}^{-,-}(t,y)) f_1(t,y) dt\},$$

$$K_{1,1}^{+,-}[c_4, f_1(x,y), f_2^4(y)] \equiv x^{-a(+0,-0)} \exp(-W_{a,\alpha}^{+,-}(x,y))$$

$$\times \{|y|^{b_1(+0,-0)} \exp(W_{b_1}^{+,-}(0,y))(c_4 - \int_y^0 |s|^{-b_1(+0,-0)-2} \exp(-W_{b_1}^{+,-}(+0,s)) f_2^4(s) ds)$$

$$+ \int_0^x t^{a(+0,-0)-1} \exp(W_{a,\alpha}^{+,-}(t,y)) f_1(t,y) dt\}.$$

**Remark 1.** The solution of type (5) satisfies the properties

$$P_{1,1}^{-,+}(u) \equiv (|x|^{-a(-0,+0)} y^{b_1(-0,+0)} U(x,y))_{\substack{x=-0 \\ y=+0}} = c_1,$$

$$P_{1,1}^{+,+}(u) \equiv (x^{a(+0,+0)} y^{b_1(+0,+0)} U(x,y))_{\substack{x=+0 \\ y=+0}} = c_2,$$

$$P_{1,1}^{-,-}(u) \equiv (|x|^{-a(-0,-0)} |y|^{-b_1(-0,-0)} U(x,y))_{\substack{x=-0 \\ y=-0}} = c_3,$$

$$P_{1,1}^{+,-}(u) \equiv (x^{a(+0,-0)} |y|^{-b_1(+0,-0)} U(x,y))_{\substack{x=+0 \\ y=-0}} = c_4.$$

In particular, if $a(x,y) \in C(\bar{D}), b_1(x,y) \in C(\bar{D})$, then $a(-0,-0) = a(-0,+0) = a(+0,-0) = a(+0,+0) = a(0,0), b_1(-0,-0) = b_1(-0,+0) = b_1(+0,-0) = b_1(+0,+0) = b_1(0,0)$, and in this case from Theorem 1 the next result follows

**Theorem 2.** *Let in system (1)* $\alpha = 1, \beta = 2, b(x,y) = yb_1(x,y)$, *the functions* $a(x,y), b_1(x,y), f_j(x,y), j = 1,2$, *satisfy the conditions of Theorem 1, except the conditions* $a(\mp 0, \pm 0) > 0, a(= 0, \pm 0) < 0, b_1(\pm 0, \mp 0) > 1, b_1(\mp 0, = 0) < -1$. *Besides let* $a(x,y) \in C(\bar{D}), b(x,y) \in C(\bar{D})$, *that is* $a(\pm 0, \pm 0) = a(0,0) < 0, b_1(\pm 0, \pm 0) = b_1(0,0) > 1, a(\pm 0, \mp 0) = a(0,0) < 0, b_1(\mp 0, \pm 0) = b_1(0,0) > 1, f_1(x,y) \in C(\bar{D})$ *and as* $x \to \pm 0$ *vanish, and its asymptotic behavior be determined by*

$$f_1(x,y) = o(|x|^{\delta_1}), \quad \delta_1 > |a(0,0)|, \quad \text{at} \quad x \to \pm 0.$$

*Then any solution of the system (1) in the class* $C'(D)$ *is represented in the form (5) if* $a(\pm 0, \mp 0) = a(\pm 0, \pm 0) = a(0,0), b_1(\mp 0, \pm 0) = b_1(\pm 0, \pm 0) = b_1(0,0) (a(0,0) < 0, b(0,0) > 1), c_j (1 \le j \le 4)$ *are arbitrary constants.*

**Remark 2.** In particular, if $a(x,y) = 0, b_1(x,y) = 0$ in $D$, $\alpha = 1, \beta = 2$, from Theorem 1 another result follows.

**Theorem 3.** *Let in system (1)* $\alpha = 1, \beta = 2, a(x,y) = b(x,y) = 0, f_1(x,y) \in C_y'(D), f_2(x,y) \in C_x'(D)$ *and satisfy condition (4) at* $a = b = 0$, *let the limit functions* $f_2^j(y), 1 \le j \le 4$, *exist and let the functions* $f_2^j(y) \in C(\bar{\Gamma}_2)$ *vanish when* $y \to \pm 0$ *and its asymptotic behavior be determined by*

$$f_2^j(y) = o(|y|^{\delta_2}), \quad \delta_2 > 1, \quad \text{at} \quad y \to \pm 0.$$

Let the function $f_1(x,y) \in C(\bar{D})$ vanish when $x \to \pm 0$ vanish and its asymptotic behavior be determined by

$$f_1(x,y) = o(|x|^{\delta_3}), \ \delta_3 > 0.$$

Then any solution of the system (1) in class $C'(D)$ are representable in the form

$$U(x,y) = \begin{cases} c_1 + \int\limits_0^y s^{-2} f_2^1(s)ds - \int\limits_x^0 |t|^{-1} f_1(t,y)dt, & when \ (x,y) \in D_1, \\ c_2 + \int\limits_0^y s^{-2} f_2^2(s)ds + \int\limits_0^x t^{-1} f_1(t,y)dt, & when \ (x,y) \in D_2, \\ c_3 - \int\limits_y^0 s^{-2} f_2^3(s)ds - \int\limits_x^0 |t|^{-1} f_1(t,y)dt, & when \ (x,y) \in D_3, \\ c_4 - \int\limits_y^0 s^{-2} f_2^4(s)ds + \int\limits_0^x t^{-1} f_1(t,y)dt, & when \ (x,y) \in D_4, \end{cases} \tag{6}$$

where $c_j \ (1 \le j \le 4)$ are arbitrary constants.

**Remark 3.** The solution of type (6) has the characteristics $U(-0,+0) = c_1, U(+0,+0) = c_2, U(-0,-0) = c_3, U(+0,-0) = c_4$.

**Remark 4.** Under the conditions in Theorem 1, the solution of type (5) tends to infinity as $x \to \pm 0$ and $y \to \pm 0$ and its asymptotic behavior is determined by

$$U(x,y) = O(|x|^{a(=0,\pm 0)}) \quad at \quad x \to -0, \ y \ne 0,$$

$$U(x,y) = O(x^{-a(\ddagger 0, \pm 0)}) \quad at \quad x \to +0, \ y \ne 0,$$

and

$$U(x,y) = O(y^{-b_1(\mp 0, \ddagger 0)}) \quad at \quad y \to +0, \ x \ne 0,$$

$$U(x,y) = O(|y|^{-b_1(\mp 0, =0)}) \quad at \quad y \to -0, \ x \ne 0.$$

**Case 2.** Case, when $\alpha = 1, \beta = 2, b(x,y) = yb_1(x,y), a(\ddagger 0, \pm 0) < 0, a(= 0, \pm 0) > 0, b_1(\pm 0, \ddagger 0) < 1, b_1(\mp 0, = 0) > -1$.

In this case the following statement holds.

**Theorem 4.** Let in system (1) $\alpha = 1, \beta = 2, b(x,y) = yb_1(x,y), a(x,y), f_1(x,y) \in C_y'(D), b_1(x,y), f_2(x,y) \in C_x'(D)$, the functions $a(x,y), b_1(x,y), f_j(x,y), j = 1, 2$, on $\Gamma_1^j, \Gamma_2^j, j = 1, 2$ may have first order singularities, and in $D$ satisfy the conditions (4). Besides let the functions $a(x,y), b_1(x,y)$ correspondingly in the neighborhood of $\Gamma_1$ and of the origin satisfy the conditions $(A), (B), (C), (D)$ and $a(\ddagger 0, \pm 0) < 0, a(= 0, \pm 0) > 0, b_1(\pm 0, \ddagger 0) < 1, b_1(\mp 0, = 0) > -1$. Let the limiting functions $(f_2^1), (f_2^2), (f_2^3), (f_2^4)$ exist. Let the functions $f_2^j(y) \in C(\bar{\Gamma}_2), 1 \le j \le 4$, vanish as $y \to \pm 0$ vanish and its asymptotic behavior be determined by

$$f_2^1(y) = o(y^{\gamma_5}), \ \gamma_5 > 1 - b_1(-0, +0), \ at \ y \to +0,$$

$$f_2^2(y) = o(y^{\gamma_6}), \ \gamma_6 > 1 - b_1(+0, +0), \ at \ y \to +0,$$

1212

$$f_2^3(y) = o(|y|^{\gamma_7}),\ \gamma_7 > 1 + b_1(-0, -0),\ at\ y \to -0,$$

$$f_2^4(y) = o(|y|^{\gamma_8}),\ \gamma_8 > 1 + b_1(+0, -0),\ at\ y \to -0.$$

*Let the function $f_1(x,y)$ vanish as $x \to \pm 0$ and its asymptotic behavior be determined by*

$$f_1(x, y) = o(|x|^{\gamma_9}),\ \gamma_9 > a(= 0, \pm 0),\ at\ x \to -0,$$

$$f_1(x, y) = o(x^{\gamma_{10}}),\ \gamma_{10} > |a(\ddagger 0, \pm 0)|,\ at\ x \to +0.$$

*Then any solution of the system (1) from the class $C'(D)$ are representable in the form (5), where $c_j\ (1 \le j \le 4)$ are arbitrary constants.*

**Remark 5.** Statements analogous to Theorems 1, 2 are obtained in the cases
1) $a(-0, -0) < 0, a(+0, -0) < 0, a(-0, +0) < 0, a(+0, +0) < 0, b_1(\pm 0, \pm 0) > 1,$ $b_1(\pm 0, \mp 0) > 1;$
2) $a(-0, -0) < 0, a(+0, -0) < 0, a(-0, +0) < 0, a(+0, +0) < 0, b_1(\pm 0, \pm 0) < 1,$ $b_1(\pm 0, \mp 0) < 1;$
3) $a(-0, -0) > 0, a(+0, -0) > 0, a(-0, +0) > 0, a(+0, +0) > 0, b_1(\pm 0, \pm 0) > 1,$ $b_1(\pm 0, \mp 0) > 1;$
4) $a(-0, -0) > 0, a(+0, -0) > 0, a(-0, +0) > 0, a(+0, +0) > 0, b_1(\pm 0, \pm 0) < 1,$ $b_1(\pm 0, \mp 0) < 1,$
and in the other possible cases.

**Remark 6.** Under the conditions in Theorem 4, the solution of type (5) vanishes as $x \to \pm 0, y \to \pm 0$ and its asymptotic behavior is determined by

$$U(x, y) = o(|x|^{a(=0,\pm 0)})\quad at\quad x \to -0,\ y \ne 0,$$

$$U(x, y) = o(x^{|a(\ddagger 0,\pm 0)|})\quad at\quad x \to +0,\ y \ne 0,$$

$$U(x, y) = o(y^{-b_1(\mp 0,\ddagger 0)})\quad at\quad y \to +0,\ x \ne 0,\ b(\mp 0, \ddagger 0) < 0,$$

$$U(x, y) = o(|y|^{-b_1(\mp 0,\mp 0)})\quad at\quad y \to -0,\ x \ne 0,\ b(\mp 0, -0) < 0.$$

For the integral representations obtained in Theorem 1 let us give a possible application and investigate the following linear conjugate problem.

**Problem $R_1$.** Find the solution of system (1) for $\alpha = 1, \beta = 2, b(x, y) = yb_1(x, y)$ in the class $C'(D)$ satisfying of the linear conjugate conditions

$$\left.\begin{array}{l} A_{11}P_{1,1}^{-,+}(u) + A_{12}P_{1,1}^{+,+}(u) + A_{13}P_{1,1}^{-,-}(u) + A_{14}P_{1,1}^{+,-}(u) = B_1,\\ A_{21}P_{1,1}^{-,+}(u) + A_{22}P_{1,1}^{+,+}(u) + A_{23}P_{1,1}^{-,-}(u) + A_{24}P_{1,1}^{+,-}(u) = B_2,\\ A_{31}P_{1,1}^{-,+}(u) + A_{32}P_{1,1}^{+,+}(u) + A_{33}P_{1,1}^{-,-}(u) + A_{34}P_{1,1}^{+,-}(u) = B_3,\\ A_{41}P_{1,1}^{-,+}(u) + A_{42}P_{1,1}^{+,+}(u) + A_{43}P_{1,1}^{-,-}(u) + A_{44}P_{1,1}^{+,-}(u) = B_4, \end{array}\right\}\quad (R_1)$$

where $A_{ks}\ (1 \le k, s \le 4)\ B_j\ (1 \le j \le 4)$ are given constants.
**Solution of problem $R_1$.** Let the coefficients of system (1) satisfy all conditions of Theorem 1. Using the integral representation (5), the properties indicated in

Remark 1 and the conditions $(R_1)$, for the definition of the constants $c_j$ $(1 \le j \le 4)$, we obtain the following algebraic system

$$\sum_{k=1}^{4} c_k A_{jk} = B_j, \quad 1 \le j \le 4. \tag{7}$$

If $\det \|A_{jk}\| \neq 0$, then system (7) has a unique solution. From system (7) defining the constants $c_j$ $(1 \le j \le 4)$ and substituting them in the integral representation (5), we find the solution of Problem $R_1$.

**Theorem 5.** *Let in system (1) $\alpha = 1, \beta = 2, b(x,y) = yb_1(x,y)$, the functions $a(x,y), b_1(x,y), f_j(x,y), j = 1,2$, satisfy the conditions in Theorem 1, and in conditions $(R_1) \det \|A_{jk}\| \neq 0$. Then Problem $(R_1)$ has a unique solution, which is given by formula (5), in which $c_j, 1 \le j \le 4$, are determined from the algebraic system (7).*

**Case 3.** Let in system (1) $\alpha > 1, \beta = 2, b(x,y) = yb_1(x,y)$ and in $\Gamma_2$, $b_1(x,y)$ may have first order singularities. In this case the following confirmations hold.

**Theorem 6.** *Let in system (1) $\alpha > 1, \beta = 2, b(x,y) = yb_1(x,y), a(x,y), f_1(x,y) \in C_y'(D), b_1(x,y), f_2(x,y) \in C_x'(D)$. The functions $a(x,y), b_1(x,y), f_j(x,y), j = 1,2$, may have first order singularities on $\Gamma_1^j, \Gamma_2^j, j = 1,2$, and in $D$ satisfy the following conditions*

$$\frac{\partial}{\partial y}\left(\frac{a(x,y)}{|x|^\alpha}\right) = \frac{\partial}{\partial x}\left(\frac{yb_1(x,y)}{r^2}\right),$$

$$\frac{\partial}{\partial y}\left(\frac{f_1(x,y)}{|x|^\alpha}\right) + \frac{yb_1(x,y)}{r^2}\frac{f_1(x,y)}{|x|^\alpha}\frac{\partial}{\partial x}\left(\frac{f_2(x,y)}{r^2}\right) + \frac{a(x,y)}{|x|^\alpha}\frac{f_2(x,y)}{r^2}.$$

*Besides let the functions $a(x,y), b_1(x,y)$ correspondingly in the neighborhood of $\Gamma_1$ and of the origin satisfy conditions $(A), (C)$ for $\gamma_1 > \alpha - 1, \gamma_3 > \alpha - 1$, and condition $(B), (D)$ for $\gamma_2 > 0, \gamma_4 > \alpha - 1$, and $a(= 0, \pm 0) < 0, a(\ddag 0, \pm 0) > 0, b_1(\pm 0, \ddag 0) > 1, b_1(\mp 0, = 0) < -1$.*

*Let the limiting functions*

$$F_2^1(y) = \lim_{x \to -0}[\exp(a(-0, +0)\omega_\alpha(|x|))f_2(x,y)], \tag{$F_2^1$}$$

$$F_2^2(y) = \lim_{x \to +0}[\exp(-a(+0, +0)\omega_\alpha(x))f_2(x,y)], \tag{$F_2^2$}$$

$$F_2^3(y) = \lim_{x \to -0}[\exp(a(-0, -0)\omega_\alpha(|x|))f_2(x,y)], \tag{$F_2^3$}$$

$$F_2^4(y) = \lim_{x \to +0}[\exp(-a(+0, -0)\omega_\alpha(x))f_2(x,y)], \tag{$F_2^4$}$$

$F_2^j(y) \in C(\bar{\Gamma}_2), 1 \le j \le 4$, exist.

*Then any solution of the system (1) in class $C'(D)$ is representable in the form*

$$U(x,y) = \begin{cases} K_{\alpha,1}^{-;+}[c_1, f_1(x,y), F_2^1(y)], & when \quad (x,y) \in D_1, \\ K_{\alpha,1}^{+;+}[c_2, f_1(x,y), F_2^2(y)], & when \quad (x,y) \in D_2, \\ K_{\alpha,1}^{-;-}[c_3, f_1(x,y), F_2^3(y)], & when \quad (x,y) \in D_3, \\ K_{\alpha,1}^{+;-}[c_4, f_1(x,y), F_2^4(y)], & when \quad (x,y) \in D_4, \end{cases}$$

$$\equiv K_\alpha^1[c_1, c_2, c_3, c_4, f_1(x,y), F_2^1(y), F_2^2(y), F_2^3(y), F_2^4(y)], \tag{8}$$

*where $c_j$ ($1 \leq j \leq 4$) are arbitrary constants.*

In integral representation (8), the integral operators $K_{\alpha,1}^{-;+}, K_{\alpha,1}^{+;+}, K_{\alpha,1}^{-;-}, K_{\alpha,1}^{+;-}$ are determined by

$$K_{\alpha,1}^{-;+}[c_1, f_1(x,y), F_2^1(y)] \equiv \exp[W_{\alpha,a}^{-;+}(x,y) - a(-0, +0)\omega_\alpha(|x|)]$$

$$\times \{y^{-b_1(-0,+0)} \exp[W_{b_1}^{-,+}(-0,y)](c_1 + \int_0^y s^{b_1(-0,+0)-2} \exp[-W_{b_1}^{-,+}(0,s)]F_2^1(s)ds)$$

$$- \int_x^0 \exp[a(-0, +0)\omega_\alpha(|t|) - W_\alpha^{-,+}(t,y)]|t|^{-\alpha}f_1(t,y)dt\},$$

$$K_{\alpha,1}^{+;+}[c_2, f_1(x,y), F_2^2(y)] \equiv \exp[-W_{\alpha,a}^{+;+}(x,y) + a(+0, +0)\omega_\alpha(x)]$$

$$\times \{y^{-b_1(+0,+0)} \exp[W_{b_1}^{+,+}(x,y)](c_2 + \int_0^y s^{b_1(+0,+0)-2} \exp[-W_{b_1}^{+,+}(+0,s)]F_2^2(s)ds)$$

$$+ \int_0^x \exp[W_{\alpha,a}^{+,+}(t,y)) - a(+0, +0)\omega_\alpha(t)]t^{-\alpha}f_1(t,y)dt\},$$

$$K_{\alpha,1}^{-;-}[c_3, f_1(x,y), F_2^3(y)] \equiv \exp[W_{\alpha,a}^{-;-}(x,y) - a(-0, -0)\omega_\alpha(|x|)]$$

$$\times \{|y|^{b_1(-0,-0)} \exp[W_{b_1}^{-,-}(-0,y)](c_3 - \int_y^0 |s|^{-b_1(-0,-0)-2} \exp[-W_{b_1}^{-,-}(-0,s)]F_2^3(s)ds)$$

$$- \int_x^0 \exp[a(-0, -0)\omega_\alpha(|t|) - W_{a,\alpha}^{-,-}(t,y)]|t|^{-\alpha}f_1(t,y)dt\},$$

$$K_{\alpha,1}^{+;-}[c_4, f_1(x,y), F_2^4(y)] \equiv \exp[a(+0, -0)\omega_\alpha(x) - W_{a,\alpha}^{+;-}(x,y)]$$

$$\times \{|y|^{b_1(+0,-0)} \exp[W_{b_1}^{-,-}(-0,y)](c_4 - \int_y^0 |s|^{-b_1(+0,-0)-2} \exp(-W_{b_1}^{+,-}(0,s)]F_2^4(s)ds)$$

$$+ \int_0^x \exp[W_{\alpha,a}^{+,-}(t,y) - a(+0,-0)\omega_\alpha(t)]t^{-\alpha}f_1(t,y)dt\},$$

$$W_{\alpha,a}^{=,\pm}(x,y) = \int_x^0 \frac{a(t,y) - a(=0,\pm 0)}{|t|^\alpha} \, dt,$$

$$W_{\alpha,a}^{\ddagger,\pm}(x,y) = \int_0^x \frac{a(t,y) - a(\ddagger 0, \pm 0)}{t^\alpha} \, dt,$$

$$W_{b_1}^{\mp,\ddagger}(x,y) = \int_0^y \frac{(b_1(x,s) - b_1(\mp 0, \ddagger 0))s}{x^2 + s^2} \, ds,$$

$$W_{b_1}^{\pm,=}(x,y) = \int_y^0 \frac{(b_1(x,s) - b_1(\pm 0, = 0))s}{x^2 + s^2} \, ds, \omega_\alpha(x) = [(\alpha-1)x^{\alpha-1}]^{-1}.$$

**Remark 7.** The solution of type (8) satisfy the properties

$$P_{\alpha,1}^{-,+}(u) \equiv (\exp(a(-0,+0)\omega_\alpha(|x|))y^{b_1(-0,+0)}u)_{\substack{x=-0 \\ y=+0}} = c_1,$$

$$P_{\alpha,1}^{+,+}(u) \equiv (\exp(-a(+0,+0)\omega_\alpha(x))y^{b_1(+0,+0)}u)_{\substack{x=+0 \\ y=+0}} = c_2,$$

$$P_{\alpha,1}^{-,-}(u) \equiv (\exp(a(-0,-0)\omega_\alpha(|x|))|y|^{-b_1(-0,-0)}U(x,y))_{\substack{x=-0 \\ y=-0}} = c_3,$$

$$P_{\alpha,1}^{+,-}(u) \equiv (\exp(-a(+0,-0)\omega_\alpha(x))|y|^{-b_1(+0,-0)}U(x,y))_{\substack{x=+0 \\ y=-0}} = c_4.$$

**Remark 8.** Under the conditions of Theorem 6, the solution of type (8) tends to infinity as $x \to \pm 0$ and $y \to +0$, and vanihes as $y \to -0$ and its asymptotic behavior is determined by

$$U(x,y) = O[\exp(-a(=0,\pm 0)\omega_\alpha(|x|))] \, at \, x \to -0, \, y \neq 0,$$

$$U(x,y) = O[\exp(a(\ddagger 0, \pm 0)\omega_\alpha(x))] \, at \, x \to +0, \, y \neq 0,$$

$$U(x,y) = O(y^{-b_1(\mp 0, \ddagger 0)}) \, at \, x \neq 0, \, y \to +0,$$

$$U(x,y) = o(|y|^{+b_1(\mp 0, = 0)}) \, at \, x \neq 0, \, y \to -0.$$

In particular, if $a(x,y) \in C(\bar{D}), b_1(x,y) \in C(\bar{D})$, then $a(\pm 0, \mp 0) = a(\pm 0, \pm 0) = a(0,0), b(\pm 0, \mp 0) = b(\pm 0, \pm 0) = b(0,0)$. In this case, from Theorem 6, follows

**Theorem 7.** *Let in system (1) $\alpha > 1, \beta = 2, b(x,y) = yb_1(x,y)$, the functions $a(x,y), b_1(x,y), f_j(x,y), j = 1,2$, satisfy the conditions of Theorem 6, except the conditions $a(= 0, \pm 0) < 0, a(\ddagger 0, \pm 0) > 1, b_1(\mp 0, \ddagger 0) > 1, b_1(\mp 0, = 0) < -1$. Besides let $a(x,y) \in C(\bar{D}), b(x,y) \in C(\bar{D})$, that is $a(\pm 0, \mp 0) = a(\pm 0, \pm 0) = a(0,0) < 0, b_1(\pm 0, \mp 0) = b_1(\pm 0, \pm 0) = b_1(0,0) > 1, f_1(x,y) \in C(\bar{D})$ and vanish as $x \to +0$, and its asymptotic behavior be determined by*

$$f_1(x,y) = o[\exp(a(0,0)\omega_\alpha(x))t^{\delta_4}], \ \delta_4 > \alpha \ at \ x \to +0.$$

*Let the function $F_2^3(y)$ vanish as $y \to -0$ and its asymptotic behavior be determined by*

$$F_2^3(y) = o(|y|^{\delta_5}), \quad \delta_5 > 1 + b_1(-0,-0).$$

*Then any solution of the system (1) in the class $C'(D)$ is represented in the form (8) as $a(\pm 0, \mp 0) = a(\pm 0, \pm 0) = a(0,0), b_1(\pm 0, \mp 0) = b_1(\pm 0, \pm 0) = b_1(0,0)$.*

**Remark 9.** Under the conditions of Theorem 7, the solution of type (8) tends to infinity as $x \to -0$ and $y \to +0$ vanishes as $x \to +0$ and $y \to -0$ and its asymptotic behavior is determined by

$$U(x,y) = O(\exp(|a(0,0)|\omega_\alpha(|x|))) \ at \ x \to -0,$$

$$U(x,y) = o[\exp(-|a(0,0)|\omega_\alpha(x))] \ at \ x \to +0,$$

$$U(x,y) = O[|y|^{-b_1(0,0)}] \ at \ y \to +0,$$

$$U(x,y) = o[|y|^{b_1(0,0)}] \ at \ y \to -0.$$

**Case 4.** Let in system (1) $\alpha > 1, \beta = 2, b(x,y) = yb_1(x,y)$ and on $\Gamma_2$, $b_1(x,y)$ have first order singularities. Besides let $a(\mp 0, \pm 0) > 0, a(\ddagger 0, \pm 0) < 0, b_1(\pm 0, \ddagger 0) < 1, b_1(\mp 0, = 0) > -1$. In this case we have the following statement.

**Theorem 8.** *Let in system (1) $\alpha > 1, \beta = 2, b(x,y) = yb_1(x,y)$, the functions $a(x,y), b_1(x,y)$, $f_j(x,y), j = 1,2$ satisfy all conditions from Theorem 6, except the conditions $a(= 0, \pm 0) < 0, a(\ddagger 0, \pm 0) > 0, b_1(\mp 0, \ddagger 0) > 1, b_1(\mp 0, = 0) < -1$. Let $a(= 0, \pm 0) > 0, a(\ddagger 0, \pm 0) < 0, b_1(\pm 0, \ddagger 0) < 1, b_1(\mp 0, = 0) > -1$. Let the functions $F_2^j(y), 1 \le j \le 4$, vanish as $y \to \pm 0$ and its asymptotic behavior be determined by*

$$F_2^j(y) = o(|y|^{\delta_j}), \quad \delta_j > 1 - b_1(\pm 0, \ddagger 0) \ j = 6,7, \ as \ y \to +0,$$

$$F_2^j(y) = O(|y|^{\delta_j}), \quad \delta_j > 1 - b_1(\pm 0, = 0) \ j = 8,9, \ as \ y \to -0.$$

*Let the function $f_1(x,y)$ vanish as $x \to \pm 0$ and its asymptotic behavior be determined by*

$$f_1(x,y) = o(|x|^{\delta_{10}} \exp[-a(= 0, \mp 0)\omega_\alpha(|t|)]), \delta_{10} > \alpha - 1 \ at \ x \to -0,$$

$$f_1(x,y) = o[\exp(a(\ddagger 0, \pm 0)\omega_\alpha(x))x^{\delta_{11}}], \delta_{11} > \alpha - 1 \ at \ x \to +0.$$

*Then any solution of the system (1) in the class $C'(D)$ is represented in the form (8), where $c_j$ $(1 \leq j \leq 4)$ are arbitrary constants.*

**Remark 10.** Under the conditions of Theorem 8, the solution of type (8) as $x \to \pm 0, y \to \pm 0$ $(b_1(\pm 0, \ddagger 0) < 0, b_1(\mp 0, = 0) > 0)$ vanishes and its behavior asymtotic is determined by

$$U(x,y) = o[\exp(a(\ddagger 0, \pm 0)\omega_\alpha(x))] \text{ at } x \to -0, \ y \neq 0,$$

$$U(x,y) = o[\exp(-a(= 0, \pm 0)\omega_\alpha(x))] \text{ at } x \to +0, \ y \neq \pm 0,$$

$$U(x,y) = o[y^{|b_1(\pm 0, \ddagger 0)|}] \text{ at } b_1(\pm 0, \pm 0) < 0, \ y \to +0, \ x \neq \pm 0$$

and

$$U(x,y) = o[|y|^{+b_1(\pm 0, = 0)}] \text{ at } b_1(\mp 0, = 0) > 0, \ y \to -0, \ x \neq \pm 0.$$

For the integral representations obtained in Theorem 6 let us give a possible application and investigate the following linear conjugate problem.

**Problem $R_2$.** Find the solution of system (1) for $\alpha > 1, \beta = 2, b(x,y) = yb_1(x,y)$ in the class $C'(D)$ satisfying of the linear congugate conditions

$$\left. \begin{array}{l} B_{11}P_{\alpha,1}^{-;+}(u) + B_{12}P_{\alpha,1}^{+;+}(u) + B_{13}P_{\alpha,1}^{-;-}(u) + B_{14}P_{\alpha,1}^{+;-}(u) = D_1, \\ B_{21}P_{\alpha,1}^{-;+}(u) + B_{22}P_{\alpha,1}^{+;+}(u) + B_{23}P_{\alpha,1}^{-;-}(u) + B_{24}P_{\alpha,1}^{+;-}(u) = D_2, \\ B_{31}P_{\alpha,1}^{-;+}(u) + B_{32}P_{\alpha,1}^{+;+}(u) + B_{33}P_{\alpha,1}^{-;-}(u) + B_{34}P_{\alpha,1}^{+;-}(u) = D_3, \\ B_{41}P_{\alpha,1}^{-;+}(u) + B_{42}P_{\alpha,1}^{+;+}(u) + B_{43}P_{\alpha,1}^{-;-}(u) + B_{44}P_{\alpha,1}^{+;-}(u) = D_4, \end{array} \right\} \qquad (R_2)$$

where $B_{ks}$ $(1 \leq k, s \leq 4)$, $D_j$ $(1 \leq j \leq 4)$ are given constants.

**Remark 11.** The system (1) is investigated also in the case $\alpha > 1, \beta = 2, \beta > 2$, depending on the signs of the numbers $a(\pm 0, \pm 0), b(\pm 0, \pm 0), a(\mp 0, \pm 0), b(\mp 0, \pm 0)$.

## References

1. Rajabov N., An Introduction to the theory of partial differential equations with super-singular coefficients, Dushanbe, 1992, TSNU Publ. (Russian); Tehran University, Tehran, 1997 (English).

2. Rajabov N., Mirzoev N. To the theory of one class of linear first order overdetermined systems with interior singular line and singular point. Bulletin National University (Scientific J.), Mathematical Series, N 1, Dushanbe, 2004, 81–101.

3. Rajabov N., Introduction to ordinary differential equations with singular and super-singular coefficients, Dushanbe, 1998.

4. Rajabov N., To the theory of a class of linear first order overdetermined systems with singular and super-singular manifolds. Proc. Intern. Sci. Conf. "Partial Differential Equations and Related Analysis and Informatic Problems", 16-19 November, 2004, I, Tashkent, 2004, 124–127.

Then any solution of the system (7) in the class $C^1(\overline{\Gamma})$ is represented in the form (8), where $c_j (1 \le j \le 4)$ are arbitrary constants.

Remark 10. Under the conditions of Theorem 8, the solution of type (8) as $x \to +0, y \to +0$ $\big(\alpha (+0,+0) < 0, b(+0,+0) > 0\big)$ vanishes and its behavior asymptotic is determined by

$$U(x,y) = o[\exp[a(+0,+0)\omega_1(x)]] \text{ as } x \to +0, y \ne +0,$$

$$U(x,y) = o[\exp[-c(+0,+0)\omega_2(x)]] \text{ as } x \to +0, y \ne +0,$$

$$U(x,y) = o[y^{b(+0,+0)}]\, c(h,(+0,+0)) > 0, y \to +0, x \ne +0$$

and

$$U(x,y) = o[y^{(+b(+0,+0))}]\, ch(+0,+0) > 0, y \to +0, x \ne +0.$$

For the integral representations obtained in Theorem 6 let us give a possible application and investigate the following linear conjugate problem

Problem $B_{14}$. Find the solution of system (1) for $\alpha > 1, b = -2, b(x,y) + yb(x,y)$ in the class $C^1(\overline{\Gamma})$ satisfying of the linear conjugate conditions:

$$(N) \quad \begin{cases} B_{11}P_{a_1}^{-1}(u) + B_{12}P_{a_2}^{-1}(u) + B_{13}P_{a_3}^{-1}(u) + B_{14}P_{a_4}^{-1}(u) = D_1, \\ B_{21}P_{a_1}^{-1}(u) + B_{22}P_{a_2}^{-1}(u) + B_{23}P_{a_3}^{-1}(u) + B_{24}P_{a_4}^{-1}(u) = D_2, \\ B_{31}P_{a_1}^{-1}(u) + B_{32}P_{a_2}^{-1}(u) + B_{33}P_{a_3}^{-1}(u) + B_{34}P_{a_4}^{-1}(u) = D_3, \\ B_{41}P_{a_1}^{-1}(u) + B_{42}P_{a_2}^{-1}(u) + B_{43}P_{a_3}^{-1}(u) + B_{44}P_{a_4}^{-1}(u) = D_4, \end{cases}$$

where $B_{jk} (1 \le k, n \le 4), D_j (1 \le j \le 4)$ are given constants.

Remark 11. The system (1) is investigated also in the case $\alpha < 1, b = -2, b = -2$, depending on the signs of the numbers $a(+0,+0), b(+0,+0), c(+0,+0), d(+0,+0)$.

References

1. Rajabov N. An introduction to the theory of partial differential equations with super-singular coefficients. Dushanbe 1992, TSNU Publ. (Russian). Tehran University, Tehran, 1375 (English).

2. Rajabov N., Mirzoev S. To the theory of one class of linear first order overdetermined systems with likerise singular line and singular point. Bulletin National University (Scientific J.), Mathematical Series, N 1, Dushanbe, 2004, 81-101.

3. Rajabov N. Introduction to ordinary differential equations with singular and super-singular coefficients, Dushanbe, 1998

4. Rajabov N. To the theory of a class of linear first order overdetermined systems with singular and super-singular manifolds. Proc. Intern. Sci. Conf. "Partial Differential Equations and Related Analysis and Information Problems", 16-19 November, 2004, 1. Tashkent, 2004, 124-127.

# ON THE CONSTRUCTION OF THE GENERAL SOLUTIONS OF THE CLASSES OF ABEL'S EQUATIONS OF THE SECOND KIND

DIMITRIOS E. PANAYOTOUNAKOS

*Division of Mechanics of Applied Mathematical and Physical Sciences*
*National Technical University of Athens, NTUA*
*5 Heroes of Polytechniou Avenue, Zographou, GR 157 73, Athens, Hellas, Greece*

Combining both mathematical techniques included in Refs. [3,5,6] we construct the general solution for the Abel equation of the second kind of the normal form $yy'_x - y = F(x)$. Since there exist admissible functional transformations that can reduce the general Abel equation of the second kind; the Emden-Fowler equation, and the generalized Emden-Fowler equation to the Abel equation of the second kind of normal form, the developed construction concerns also the general solutions of these classes of equations.

**Key words:** Abel equation of second kind, Enden-Fourier equation
**Mathematics Subject Classification:** 34A34

## 1. Introduction

Several basic particular nonlinear ordinary differential equations (ODEs) of the second and higher-order in mathematical physics and nonlinear mechanics are reduced to equivalent equations of the Abel normal form $yy'_x - y = f(x)$ by means of various admissible functional transformations [1]. These equivalent equations do not admit exact analytic solutions in terms of known (tabulated) functions, since only very special cases of the above type of Abel equation can be solved in parametric form [2,3]. In 1992 a mathematical technique was developed [4] that leads to the construction of the general solution of the above Abel equation, if $n$ distinct particular solutions $y_k(x)$ $(k = 1, ..., n)$ are known. This technique presupposes concrete forms of the second member $F(x)$ and the obtained results are tabulated in [3]. On the other hand, another mathematical technique was recently developed [5,6], that permits us to construct exact analytic solutions of an Abel equation of the second kind. In this paper, combining both previous techniques, a successful attempt is made to present a mathematical construction leading to the general solution of the above Abel equation. Since there are admissible functions transformations that reduce the general Abel equation of the second kind; the Emden-Fowler equation, and the generalized Emden-Fowler equation to the Abel equation of the second kind of the normal form, the developed construction concerns also the general solutions of these classes of equations.

## 2. Some results on the class of Abel's equations of the second kind

Before we address the issue of the possibility of construction of the general solution of the Abel equation of the second kind of the normal form, we will provide two mathematical results given by Julia in 1933 (see Ref. [2, p. 27]) and by Alexeeva, Zaitsev and Shvets in 1992 (see Ref. [4]) concerning the general solutions of an Abel nonlinear ODE of the second kind.

*1st result: [2, p. 27]*

According to this result, if the variable coefficients of the equation

$$[g_1(x) y + g_0(x)] y_x' = f_2(x) y^2 + f_1(x) y + f_0(x) \tag{1}$$

satisfy the functional relation

$$g_0 \left(2f_2 + g_{1_x}'\right) = g_1 \left(f_1 + g_{0_x}'\right), \quad g_1 \neq 0, \tag{2}$$

then, there exists its general solution given by the formula

$$\frac{g_1 y^2 + 2g_0 y}{g_1 J} = 2 \int \frac{f_0}{g_1 J} dx + C, \tag{3}$$

where $C$ is an integration constant and $J$ the integrating factor

$$J(x) = \exp \left( \int \frac{2f_2}{g_1} dx \right).$$

*2nd result: [4; 3, pp. 12-13]*

Consider the Abel equation of the second kind of the normal form

$$yy_x' - y = F(x), \tag{4}$$

whose general solution can be represented in the special form

$$\prod_{k=1}^{n} |y - y_k|^{m_k} = C. \tag{5}$$

In the above notation $y_x' = dy/dx$, $y_{xx}'' = d^2y/dx^2 \ldots$ is used for the total derivatives, and the particular solutions $y_k(x)$ correspond to $C = 0$ (if $m_k > 0$) and $C = \infty$ (if $m_k < 0$), where $C$ is the integration constant. The logarithmization of (5), followed by the differentiation of the resulting expression and rearrangement, leads to the equation

$$\sum_{j=1}^{n} \left[ m_j \left(y_x' - (y_j)_x'\right) \prod_{\substack{k=1 \\ k \neq j}}^{n} (y - y_k) \right] \equiv y_x' \sum_{s=1}^{n-1} \Phi_s y^s + \sum_{s=1}^{n-1} \Psi_s y^s = 0. \tag{6}$$

We require that equation (6) be the Abel equation (4). To this end, we set $\Psi_v = -\Phi_v$, $\Psi_{v-1} = -F(x) \Phi_v$ and equate the other $\Phi_1$ and $\Psi_1$ with zero. Selecting different values $v = 1, 2, \ldots, n-1$, we obtain $n-1$ systems of differential-algebraic equations; only one of the systems, corresponding to $m_k \neq 0$ for all $k = 1, \ldots, n$ and $y_i \neq y_j$ for $i \neq j$, leads to a non-degenerate solution of the form (5). Consider now the Abel equation (4) corresponding to the simplest solutions of the form (5) in more detail.

$1^{st}$ Case ($n = 2$). The above system of differential-algebraic equations is in the following form:

$$\begin{aligned} m_1 + m_2 &= M, \\ m_1 y_2 + m_2 y_1 &= 0, \\ m_1 y'_{1_x} + m_2 y'_{2_x} &= M, \\ m_1 y'_{1_x} y_2 + m_2 y_1 y'_{2_x} &= -MF(x), \end{aligned} \qquad (7)$$

where $M$ is an arbitrary constant. It follows from the second and third equations (after integration) that

$$y_1 = \frac{m_1}{m_1^2 - m_2^2}(Mx + N), \qquad y_2 = -\frac{m_2}{m_1^2 - m_2^2}(Mx + N), \qquad (8)$$

where $N$ is a new arbitrary constant. Introducing the new constants

$$A = \frac{m_1 m_2 (m_1 + m_2)}{(m_1 - m_2)^2}M, \qquad B = \frac{m_1 m_2 (m_1 + m_2)}{(m_1 - m_2)^2}N, \qquad (9)$$

we find from Eqs. (7) that

$$F(x) = Ax + B, \qquad (10)$$

which means that for $n = 2$ the right hand side of the Abel equation (4) is a linear function of $x$.

The particular solutions $y_1$, $y_2$ and the corresponding exponents $m_1$, $m_2$ in the general integral (5) are defined as follows:

$$\begin{aligned} y_1 &= \frac{1+\sqrt{4A+1}}{2A}(Ax + B), \qquad y_2 = -\frac{1+\sqrt{4A+1}}{2A+1+\sqrt{4A+1}}(Ax + B), \\ m_1 &= 2A + 1 + \sqrt{4A+1}, \qquad m_2 = 2A. \end{aligned} \qquad (11)$$

$2^{nd}$ Case ($n = 3$). In this case Eq. (6) leads to the Abel equation (4) with the right-hand side

$$F(x) = -\frac{2}{9}x + A + Bx^{-1/2}, \qquad (12)$$

while the particular solutions are

$$y_s = \frac{2}{3}x + \frac{2}{3}\lambda_s x^{1/2} + \frac{3B}{\lambda_s}, \qquad m_s = \frac{2A}{3(2\lambda_s^2 - 3A)}, \qquad (13)$$

where the $\lambda_s$ are roots of the cubic equation

$$\lambda^3 - \frac{9}{2}A\lambda - \frac{9}{2}B = 0, \qquad s = 1, 2, 3. \qquad (14)$$

The cases $n = 4$, $n > 4$ are also examined in [3].

## 3. A new construction concerning exact analytic solutions of the Abel equation of the second kind of the normal form

Consider the Abel equation of the second kind of the normal form

$$yy'_x - y = F(x),\tag{15}$$

where $F(x)$ is an arbitrary smooth function of the variable $x$. We will present a new construction concerning exact analytic solutions of the Eq. (15) ([5,6]).

We introduce the admissible functional transformation

$$y(x) = f_1(x)\,n(\xi), \quad \xi = \xi(x)\tag{16}$$

which reduces (15) to the form

$$f_1^2\xi'_x nn'_\xi + f_1 f'_{1_x} n^2 - f_1 n = F.\tag{17}$$

Here, $f_1(x)$, $\xi(x)$ and $n(\xi)$ are to be determined. Introducing a new function $g(x)$ one rewrites (17) as

$$\left(f_1^2\xi'_x n + g\right) n'_\xi - 2F = \left(-f_1^2\xi'_x n + g\right) n'_\xi - 2f_1 f'_{1_x} n^2 + 2f_1 n.\tag{18}$$

The last equation splits into the following two equations

$$\left(f_1^2\xi'_x n + g\right) n'_\xi - 2F = G(x),\tag{19}$$

$$\left(-f_1^2\xi'_x n + g\right) n'_\xi - 2f_1 f'_{1_x} n^2 + 2f_1 n = G(x).\tag{20}$$

Here, $g(x)$ and $G(x)$ are also arbitrary smooth functions to be determined. We develop now the following steps.

**Step 1:** We apply Julia's construction (Sec. 2) on the Abel equation (19) and we obtain, after integration, the following results:

$$g = f_1^2\xi'_x;\tag{21}$$

$$n^2 + 2n - 8\int\frac{G+2F}{f_1^2}dx + C = 0,\tag{22}$$

where $C$ is an integration constant.

**Step 2:** Similarly, we apply Julia's construction on the Abel equation (20) and obtain the following results:

$$f_1 = \frac{(x+2\lambda)}{2};\quad n^2 - 2n + \frac{8}{(x+2\lambda)^4}\int(x+2\lambda)^2\,G(x)\,dx - \frac{\overset{*}{C}}{(x+2\lambda)^4} = 0,\tag{23}$$

where $\lambda$ is a parameter and $\overset{*}{C}$ is a new constant of integration.

The problem under consideration demands that from the already introduced three constants of integration $\lambda$, $C$ and $\overset{*}{C}$ only the one exists. If $\lambda = 0$, then $f_1(x) = x/2$, and supposing that the original Abel equation (15) obeys to the condition for $x = 0$, $y(0) = y_0 \neq 0$, by (16) we admit the result $y_0 = 0 \cdot n(\xi_0) = 0$. This is unacceptable and thus one concludes that $\lambda \neq 0$ and $C = \overset{*}{C} = 0$.

Summarizing, the obtained results are the following:

$$f_1(x) = \frac{x + 2\lambda}{2} \ , \quad \frac{g(x)}{\xi'_x(x)} = f_1^2 \ , \tag{24}$$

$$n^2 + 2n - 8 \int \frac{G + 2F}{(x + 2\lambda)^2} dx = 0 \ , \quad n^2 - 2n + \frac{8}{(x + 2\lambda)^4} \int (x + 2\lambda)^2 \, G dx = 0 \ , \tag{25}$$

where $g(x)$, $\xi(x)$ and $G(x)$ are to be determined.

**Step 3:** In what follows we shall try to construct an exact analytic solution of the problem under consideration, that is of the original Abel equation (15). Supposing, without loss of generality, real roots of the two last equations (25), solving them for $n$ and equating the results we deduce the unique relation

$$\sqrt{1 + 8 \int \frac{G}{\omega} dx + 16 \int \frac{F}{\omega} dx} = 2 + \sqrt{1 - \frac{8}{\omega^2} \int \omega G dx} \ , \tag{26}$$

where

$$\omega = (x + 2\lambda)^2 = (2 f_1)^2 \ . \tag{27}$$

We rewrite (26) in the form

$$\sqrt{\omega^2 - M(x)} = \omega \sqrt{1 + N(x)} - 2\omega \tag{28}$$

with

$$M(x) = 8 \int \omega G dx \ , \quad N(x) = 8 \int \frac{G}{\omega} dx + 16 \int \frac{F}{\omega} dx \ , \tag{29}$$

and, since

$$M'_x = 8\omega G, \quad N'_x = 8\frac{G}{\omega} + 16\frac{F}{\omega},$$

squaring and differentiating (28), one extracts the equation

$$6\omega\omega'_x + M'_x + 2(1 + N)\omega\omega'_x + \omega^2 N'_x - 8\sqrt{1 + N}\omega\omega'_x - 2\omega^2 \frac{N'_x}{\sqrt{1 + N}} = 0 \ . \tag{30}$$

Finally, introducing $M'_x$ and $N'_x$ by way of (29) into the last relation we perform the following cubic equation for $(1 + N)^{1/2}$:

$$(1 + N)^{3/2} - 4(1 + N) + \left[3 + \frac{4(G + F)}{x + 2\lambda}\right](1 + N)^{1/2} - 4\frac{G + 2F}{x + 2\lambda} = 0 \ . \tag{31}$$

The substitution

$$(1 + N)^{1/2} = \bar{N} + \frac{4}{3} \ , \tag{32}$$

transforms (31) to the Cardan form:

$$\bar{N}^3 + p\bar{N} + q = 0 \ , \tag{33}$$

1224

where

$$p = -\frac{a^2}{3} + b, \quad q = 2\left(\frac{a}{3}\right)^3 - \frac{ab}{3} + c, \quad a = -4,$$
$$b = 3 + \frac{4(G+F)}{x+2\lambda}, \quad c = -\frac{4(G+2F)}{x+2\lambda} \tag{34}$$

($F$ is the second member of the original Abel equation (15); $G$ is a subsidiary function to be determined).

It is well known that the solution of the cubic equation (33) can be expressed in analytic form, depending on the sign of the discriminant

$$Q = \left(\frac{p}{3}\right)^3 + \left(\frac{q}{2}\right)^2. \tag{35}$$

When $Q < 0$, (33) possesses three real roots, whereas when $Q > 0$ there exist a single real root and a complex conjugate pair of roots. The case $Q = 0$ furnishes three real roots the two of which are equal. Based on these observations and noting that only real roots are interested, we list the specific forms of the roots of (33) as follows:

Case 1: $Q < 0$ $(p < 0)$

$$N(x) = \left(\bar{N}(x) + \frac{4}{3}\right)^2 - 1 ;$$
$$\bar{N}_1(x) = 2\sqrt{-\frac{p}{3}}\cos\frac{a}{3}, \quad \bar{N}_2(x) = 2\sqrt{-\frac{p}{3}}\cos\frac{a-\pi}{3}, \quad \bar{N}_3(x) = -2\sqrt{-\frac{p}{3}}\cos\frac{a+\pi}{3} ;$$
$$\cos a = -\frac{q}{2\sqrt{-\left(\frac{p}{3}\right)^3}} ; \quad 0 < a < \pi, \quad p,q \text{ as in (34)} \tag{36}$$

Case 2: $Q > 0$

$$N(x) = \left(\bar{N}(x) + \frac{4}{3}\right)^2 - 1 ;$$
$$\bar{N}_1(x) = \sqrt[3]{-\frac{q}{2} + \sqrt{Q}} + \sqrt[3]{-\frac{q}{2} - \sqrt{Q}} ,$$
$$\bar{N}_{2,3}(x) = -\frac{K+\Lambda}{2} \pm i\frac{K-\Lambda}{2}\sqrt{3}, \quad K = \sqrt[3]{-\frac{q}{2} + \sqrt{Q}}, \quad \Lambda = \sqrt[3]{-\frac{q}{2} - \sqrt{Q}}, \quad i = \sqrt{-1}$$
$$p,q \text{ as in (34).} \tag{37}$$

Case 3: $Q = 0$

$$N(x) = \left(\bar{N}(x) + \frac{4}{3}\right)^2 - 1 ; \quad \bar{N}_1(x) = 2\sqrt[3]{-\frac{q}{2}} ,$$
$$\bar{N}_2(x) = \bar{N}_3(x) = -\sqrt[3]{-\frac{q}{2}} , \quad p,q \text{ as in (34).} \tag{38}$$

By now the problem can be solved according to the following procedure.

a) Combination of the second of Eqs. (29) together with the substitution (32) results

$$N(x) = 8\int\frac{G+2F}{\omega} = \left[N(x) + \frac{4}{3}\right]^2 - 1 . \tag{39}$$

Furthermore, Eq. (28) together with the second of (25) perform the expressions

$$\sqrt{1 - \frac{M}{\omega^2}} = \sqrt{1+N} - 2 , \quad (n-1)^2 = 1 - \frac{M}{\omega^2} , \tag{40}$$

by means of which we extract the equation

$$\sqrt{1+8\int\frac{G+2F}{\omega}dx}=2+(n-1)\ .\qquad(41)$$

Thus, combining (39) and (41) one obtains the following solution for $n\,(x)$:

$$n\,(x)=\bar{N}\,(x)+\frac{1}{3}=[N\,(x)+1]^{1/2}-1\ ,\qquad(42)$$

where $\bar{N}\,(x)$ as in Eqs. (36) to (38). Since $N\,(x)$ has been already evaluated in (39), the subsidiary function $M\,(x)$ provided in (29) can be also evaluated.

Summarizing, we perform the following solutions:

$$\begin{array}{c}n\,(x)=\bar{N}+\frac{1}{3}=\sqrt{1+N}-1\ ,\\ N\,(x)=8\int\frac{G+2F}{(x+2\lambda)^2}dx=\left(\bar{N}+\frac{4}{3}\right)^2-1\ ,\\ \frac{M(x)}{(x+2\lambda)^4}=\frac{8}{(x+2\lambda)^4}\int(x+2\lambda)^2\,Gdx=-\left(\bar{N}+\frac{1}{3}\right)\left(\bar{N}-\frac{5}{3}\right)\ ,\\ \bar{N}\,(x)\quad\text{as in Eqs. (36) to (38) .}\end{array}\qquad(43)$$

Note that both $N\,(x)$ and $M\,(x)$ denote integrals and they are expressed in terms of the unknown subsidiary function $G\,(x)$, which is to be determined.

b) It is easy now to show that the already constructed solution $n\,(x)$ in Eq. (42) verifies all the transformed Abel equations (17) and (19), (20) if and only if the derivative $\bar{N}'_x$ of the function $\bar{N}\,(x)$ is given by the differentiation of the second of (43). In other words, the set

$$n=\bar{N}+\frac{1}{3}\ ,\qquad\bar{N}'_x=\frac{4\,(G+2F)}{(x+2\lambda)^2\,(\bar{N}+4/3)}\qquad(44)$$

verifies all the before mentioned Abel equations.

Introducing (44) together with the expression for $f_1\,(x)$ given in (23) into the above Abel equations, by setting simultaneously $\xi'_x=1\big[\xi=x;g=f_1^2\xi'_x=(x+2\lambda)^2/4\big]$, we extract the cubic equations (31). For example Eq. (20) furnishes

$$\left(\sqrt{1+N}-2\right)\frac{2\,(G+2F)}{x+2\lambda}$$
$$=-\left(\sqrt{1+N}-1\right)^2\sqrt{1+N}+2\left(\sqrt{1+N}-1\right)\sqrt{1+N}-\frac{2G}{x+2\lambda}\sqrt{1+N}\ ,$$

while Eq. (17) furnishes too

$$\left(\sqrt{1+N}-1\right)\frac{4\,(G+2F)}{(x+2)\sqrt{1+N}}=-\left(\sqrt{1+N}-1\right)^2+2\left(\sqrt{1+N}-1\right)+\frac{4F}{x+2}\ ,$$

that coincide with the cubic equation (31).

Summarizing we note that the set of equations

$$\begin{array}{c}y\,(x)=\frac{1}{2}\,(x+2\lambda)\left[\bar{N}\,(x)+\frac{1}{3}\right],\quad y'_x=\frac{1}{2}\left[\bar{N}\,(x)+\frac{1}{3}\right]+\frac{1}{2}\,(x+2\lambda)\,\bar{N}'_x,\\ \bar{N}'_x=\frac{4(G+2F)}{(x+2\lambda)^2[\bar{N}(x)+\frac{4}{3}]};\quad\bar{N}\,(x)=\text{as in Eqs. (36)-(38).}\end{array}\qquad(45)$$

including the second member $F(x)$ and the subsidiary function $G(x)$ (which will be determined), constitutes an exact solution of the original Abel equation (15). We note also that in each of the above cases a, b, c (Eqs. (36)-(38)) the Abel equation (15) performs three or two particular solutions including the subsidiary function $G(x)$.

In what follows combining the second result of Sec. 2 together with the already constructed solution (45) we will try to extract the general solution of the Abel equation (15).

## 4. Construction of the general solution for the Abel equation (15)

Consider case 3 $(Q = 0)$ where the Abel equation performs two particular solutions (Eqs. (38)), that is

$$y_1(x) = \tfrac{1}{2}(x + 2\lambda)\left[\bar{N}_1(x) + \tfrac{1}{3}\right], \quad y_2(x) = \tfrac{1}{2}(x + 2\lambda)\left[\bar{N}_2(x) + \tfrac{1}{3}\right],$$
$$\bar{N}_1(x) = -2\bar{N}_2(x). \tag{46}$$

The whole problem is focused on the determination of the above mentioned subsidiary function $G(x)$, as well as of the constants $m_1$, $m_2$ and $M$ being introduced in Eqs. (7). Based on (46) and (45) Eqs. (7) can be rewritten as follows:

$$m_1 + m_2 = M,$$
$$m_1(x + 2\lambda)\left(\bar{N}_2 + \tfrac{1}{3}\right) + m_2(x + 2\lambda)\left(\bar{N}_1 + \tfrac{1}{3}\right) = 0,$$
$$m_1\left[\left(\bar{N}_1 + \tfrac{1}{3}\right) + (x + 2\lambda)\bar{N}'_{1_x}\right] + m_2\left[\left(\bar{N}_2 + \tfrac{1}{3}\right) + (x + 2\lambda)\bar{N}'_{2_x}\right] = 2M, \tag{47}$$
$$m_1\left[\left(\bar{N}_1 + \tfrac{1}{3}\right) + (x + 2\lambda)\bar{N}'_{1_x}\right](x + 2\lambda)\left(\bar{N}_2 + \tfrac{1}{3}\right)$$
$$+ m_2\left[\left(\bar{N}_2 + \tfrac{1}{3}\right) + (x + 2\lambda)\bar{N}'_{2_x}\right](x + 2\lambda)\left(\bar{N}_1 + \tfrac{1}{3}\right) = -4MF.$$

Solving the first and the second of these equations for $m_2(x + 2\lambda)(\bar{N}_1 + 1/3)$, and $m_1\left[(\bar{N}_1 + 1/3) + (x + 2\lambda)\bar{N}'_{1_x}\right]$ respectively and introducing the results into the third of them, we lead to the following nonlinear algebraic equation

$$\left[\left(\bar{N}_2 + \tfrac{1}{3}\right) + (x + 2\lambda)\bar{N}'_{2_x}\right]\left(\bar{N}_2 + \tfrac{1}{3}\right) = \frac{2M\left[2F + (x + 2\lambda)\left(\bar{N}_2 + \tfrac{1}{3}\right)\right]}{(m_1 + m_2)(x + 2\lambda)}, \tag{48}$$

where

$$\bar{N}'_{2_x} = \frac{4(G + 2F)}{(x + 2\lambda)^2\left(\bar{N}_2 + 4/3\right)}, \quad \bar{N}_2(x) = \text{as in Eqs. (38)}.$$

This last equation performs the subsidiary function $G(x)$ in terms of the second member $F(x)$ of the Abel equation (15), as well as of the parameters $m_1$, $m_2$ and $M$, which are to be determined.

On the other hand, by the second and third of Eqs. (7), equivalently by the second and third of (47), we derive the results

$$y_1 = \frac{m_1}{m_1^2 - m_2^2}(Mx + N), \quad y_2 = -\frac{m_2}{m_1^2 - m_2^2}(Mx + N)$$

or

$$\frac{1}{2}(x+2\lambda)\left(\bar{N}_1+\frac{1}{3}\right) = \frac{m_1}{m_1^2-m_2^2}(Mx+N),$$

$$\frac{1}{2}(x+2\lambda)\left(\bar{N}_2+\frac{1}{3}\right) = -\frac{m_2}{m_1^2-m_2^2}(Mx+N),$$

(49)

where $N$ is a new integration constant. But since $\bar{N}_1(x) = -2\bar{N}_2(x)$ (Eqs. (38)), we conclude that

$$\bar{N}_2(x) = -\frac{m_1}{m_1^2-m_2^2}\frac{(Mx+N)}{x+2\lambda}+\frac{1}{6} = -\frac{2m_2}{m_1^2-m_2^2}\frac{(Mx+N)}{x+2\lambda}-\frac{1}{3},$$

(50)

by means of which one extracts:

$$M=1, \qquad N=2\lambda, \qquad \frac{m_1-2m_2}{m_1^2-m_2^2}=\frac{1}{2}.$$

(51)

By the system of equations

$$2m_1-4m_2 = m_1^2-m_2^2, \qquad m_1+m_2 = M = 1,$$

we evaluate the parameters $m_1$ and $m_2$, that is $m_1 = 3/4$ and $m_2 = 1/4$.

Summarizing, the general solution of the Abel equation of the second kind (15) in case when $Q = 0$ (Eqs. (38)) can be written as follows:

$$\left|y-\tfrac{1}{2}(x+2\lambda)\left[\bar{N}_1(x)+\tfrac{1}{3}\right]\right|^{\frac{3}{4}}\left|y-\tfrac{1}{2}(x+2\lambda)\left[\bar{N}_2(x)+\tfrac{1}{3}\right]\right|^{\frac{1}{4}} = C;$$
$$\bar{N}_1(x),\ \bar{N}_2(x) = \text{as in Eqs. (38)};$$
$$\left\{\left(\bar{N}_2+\tfrac{1}{3}\right)+\frac{4[G(x)+2F(x)]}{(x+2\lambda)[\bar{N}_2(x)+\frac{4}{3}]}\right\}\left[\bar{N}_2(x)+\tfrac{1}{3}\right] = \frac{4F}{x+2\lambda}+2\left(\bar{N}_2+\tfrac{1}{3}\right);$$
$$\lambda = \text{free parameter}, \qquad C = \text{integration constant}.$$

(52)

## References

1. D. E. Panayotounakos and N. Sotiropoulos, Exact analytic solutions of unsolvable classes of first – and second-order nonlinear ODEs (Part II: Emden-Fowler and relative equations), *Applied Math. Letters*, 18 (2005), 367-374.

2. E. Kamke, *Differentialgleichungen, Lösungsmethoden und Lösungen*, Vol. 1, B. G. Teubner, Stuttgard (1977).

3. A. D. Polyanin and V. F. Zaitsev, *Handbook of Exact Solutions for Ordinary Differential Equations*, $2^{nd}$ Ed., Chapman and Hall / CRC, Boca Raton, London, New York, Washington, D. C., 2002.

4. T. A. Alexeeva, V. F. Zaitsev and T. B. Shvets, On discrete symmetries of the Abel equation of the $2^{nd}$ kind (in Russian), *Appl. Mech. and Mathematics, MIPT*, Moscow, (1992), 4-11.

5. D. E. Panayotounakos, Exact analytic solutions of unsolvable classes of first and second order nonlinear ODEs (Part I: Abel's equations), *Applied Math. Letters*, 18 (2005), 155-162.

6. D. E. Panayotounakos, N. B. Sotiropoulos, A. B. Sotiropoulou and N. D. Panayotounakou, Exact analytic solutions of nonlinear boundary value problems in fluid mechanics (Blasius equations), *J. Math. Physics, JMP*, 46 (2005), 033101-033026.

# QUALITATIVE THEORY OF NONLINEAR ODE's IN ALGEBRAS

Y. KRASNOV

*Department of Mathematics,*
*Bar-Ilan University,*
*Ramat-Gan 52900, Israel*
*E-mail: krasnov@math.biu.ac.il*

S. ZUR

*Department of Mathematics,*
*Bar-Ilan University,*
*Ramat-Gan 52900, Israel*
*E-mail: zur@math.biu.ac.il*

We establish the basic qualitative properties of the solutions to nonlinear ODE's in commutative algebras via arrangement of their Pierce numbers. We touch also the connections between homogeneous nonlinear operators of order $m$ in two dimensional space and agreeable $m$-nary algebras. We study the influence of the Pierce numbers allocation on these non associative algebras in the same fashion as spectral analysis of a matrices serves the qualitative properties of linear operators. As an application we mention stability theory of the nonlinear (quadratic) ODE's.

**Key words:** Differential equation in algebra, ray solutions, Peirce numbers
**Mathematics Subject Classification:** 34C41, 47H60, 37E30

## 1. Introduction

The main objective of this paper is to study qualitative nature of the solutions to system of the **homogeneous polynomial ODEs** of degree $m$ in $\mathbb{R}^n$

$$\frac{dx(t)}{dt} = P_m(x(t)), \quad x(t) \in \mathbb{R}^n, \quad P_m(\lambda x) = \lambda^m P_m(x). \tag{1}$$

with emphasis on the *"spectral properties"* of their solutions.

In 1960, L. Markus (see [10]) for $m = 2$ proposed investigate (1) as the quadratic **ODE in algebras.** Namely, he suggested to attach to the homogeneous system (1) of order $m = 2$ a certain binary algebra and to reformulate (1) as the Riccati equation $\dot{x} = x^2$. Later C.Coleman (see [6], [10], [13], [7]) showed that there is a natural generalization of Markus' method for an arbitrary $m$ in (1) by attaching to each system (1) the commutative but not necessarily associative $m$-ary algebra with $\omega$ as the $m$-linear commutative multiplication. That enable to rewrite (1) in form

$$\frac{dx}{dt} = x^2 := \omega(\underbrace{x, x, \ldots, x}_{m \ times}).$$

In turn, a linear system may be thinking as an ODE in a unary algebra also.

The phase-portrait of system (1) may be particulary composed of a finite number of an **integral ray solutions** curves which are straight lines and which become coincide geometrically with the direction of the $\lambda \neq 0$ **idempotents** in algebra **A**:

**Definition 1.1.** For real $\lambda \in \mathbb{R}$, a $\lambda$-idempotent of **A** is a nonzero element $c \in \mathbf{A}$ such that $c^2 = \lambda c$ or, equivalently, $\omega(c, \ldots, c) = \lambda c$.

By an *integral ray* one means a ray, flow out from/to the origin, such that the solution $x(t) = exp(\lambda t)c$ to the equation (1) starting from the $\lambda$-idempotent $c$ remains lie on that ray all the time.

In two dimensional case the integral ray solutions divide the plane into a finite number of sectors of three possible kinds: elliptic, hyperbolic and parabolic.

In 1950, G. E. Shilov (see [11]) propose to investigate the solution of (1) in terms of the *behavior of the trajectories* in a neighborhood of each integral rays. He gives criteria (in two dimension only) for the *specially membership* of an integral ray (hyperbolic, parabolic of first and second types) as a function of the qualitative behavior of the solutions to (1) near the ray itself. G.E. Shilov find necessary and sufficient conditions for the integral rays under consideration to belong to one or another of the above types.

We generalize the aforesaid G. E. Shilov conditions on the integral ray in pure algebraic form, as the *"spectral properties"* of the **Peirce numbers** in algebra **A**.

## 2. Algebraic background

A projective line $\mathbb{RP}^1$ passing trough a $\lambda$-idempotent of **A** is an affine invariant [14] (i.e., all points which lie on a line collinear to the $\lambda$-idempotent remain on that line after affine transformation). In particular, the $\lambda$-eigenvector in a unary algebra with (matrix) multiplication $\omega(v) = Av$ is such an invariant. Situation in the $m$-ary algebras is slightly different. Instead of a $\lambda$-idempotent, up to affine equivalence in $2k$-ary algebras, we will use the only idempotents and nilpotents. Finally, in a $m$-ary algebra with odd $m = 2k + 1$, a negative idempotent, i.e. the (-1)-idempotent is not equivalent up to dilatations to the idempotent. Therefore, the *negative idempotent* notion require to be considered in the $2k + 1$-ary algebras.

Usually the term *idempotent* is used instead of 1-idempotent, *negative idempotent* is used instead of $(-1)$-idempotent and *nilpotent* is used instead of 0-idempotent.

Denote by $\mathcal{A}_m(\mathbb{C}^n)$ the *complexification* of the real $m$-ary algebra $\mathcal{A}_m(\mathbb{R}^n)$. Let $e$ be a (possible complex) $\lambda$-idempotent in $\mathbf{A} \in \mathcal{A}_m(\mathbb{C}^n)$ with $\lambda \in \mathbb{C}$. Denote by $L_e$ the *"multiplication by $e$"* linear operator:

$$L_e[a] := \omega(e, e, \ldots, e, a), \qquad a, e \in \mathbf{A} \tag{2}$$

Using the Euler's homogeneous function theorem, we can write explicitly the matrix of the linear operator $L_e : x \to L_e[x]$ in the following form [8]:

$$L_e := \frac{1}{m} J(e), \quad \text{where} \quad J(x) = \frac{DP_m(x)}{Dx}. \tag{3}$$

Here $J(x)$ is an Jacoby matrix of the mapping $x \to P_m(x)$ defined in (1).

Clearly, $\lambda$ is one of the eigenvalues of the matrix $L_e$. Denote by $\lambda_1, \ldots \lambda_{n-1}$ the remaining eigenvalues of $L_e$ in $\mathcal{A}_m(\mathbb{C}^n)$. This gives rise to the following important

**Definition 2.1.** Let $e \in \mathbf{A}$ be a $\lambda$-idempotent ($\lambda \neq 0$) in $\mathbf{A} \in \mathcal{A}_m(\mathbb{C}^n)$. Then $\lambda$ is an eigenvalue of the linear operator $L_e$ in $\mathcal{A}_m(\mathbb{C}^n)$. The ratio of the remaining eigenvalues $\lambda_1, \ldots \lambda_{n-1}$ of $L_e$ to $\lambda$ will be called the **Peirce numbers** $\mu_1, \ldots \mu_{n-1}$ associated to the $\lambda$-idempotent $e$. Namely $\mu_i = \frac{\lambda_i}{\lambda}$, $i = 1, \ldots, n-1$.

The Peirce numbers may not be chosen always arbitrarily. For example, in the two dimensional algebras, there exists syzygy between these numbers:

**Theorem 2.1.** [9] *Let $\mathbf{A}$ be the two dimensional complex commutative m-ary algebra. Then for every $\lambda$-idempotent $e \in \mathbf{A}$ the only one Peirce number $\mu$ is correctly defined. Moreover, if there exist exactly $m + 1$ (possible complex) pairwise different $\lambda$-idempotents $e_0, e_1, \ldots, e_m$ equipped with the Peirce numbers $\mu_0, \mu_1, \ldots, \mu_m$ correspondingly, then the following syzygy holds:*

$$\sum_{i=0}^{m} \frac{1}{1 - m\mu_i} = 1. \tag{4}$$

We prove the theorem 2.1 in the next section.

## 3. Planar homogeneous ODE's

Consider a system of two differential equations

$$x' = P_m(x, y), \quad y' = Q_m(x, y), \tag{5}$$

where $P_m$ and $Q_m$ are homogeneous polynomials of degree $m$ in variables $x$ and $y$. The common situation for a phase-portrait of the non degenerate system (5) is the following. There is a finite number of possible complex integral rays solution curves which are the (complex) straight lines $y = zx$. The complex numbers $z$ are determined from the characteristic equation

$$R(z) := \frac{Q_m(1, z)}{P_m(1, z)} = z. \tag{6}$$

The only real solutions to (6) divide the plane into a finite number of sectors of three kinds: elliptic, hyperbolic and parabolic. If (6) does not admit any real solution, then the only singular point is either a *center* or *focus*. Consider the only case with the simple roots of the equation (6).

**Theorem 3.1.** [9] *Let $z_i$ be a simple root of (6). Then a projective line $\mathbb{RP}^1$ passing trough $\{1, z_i\}$ represent a $\lambda$-idempotent $e_i$ in attached to the (5) m-ary algebra $\mathbf{A} \in \mathcal{A}_m(\mathbb{C}^n)$. The associated to $e_i$ Peirce number $\mu_i$ may be calculated as follows:*

$$\mu_i = \frac{R'(z_i)}{m}. \tag{7}$$

**Proof.** The first assertion in theorem is trivial. Let $\lambda$-idempotent $e_i = (1, z_i)$. That means $z = z_i$ is the root of the system:

$$mP_m(1, z) \equiv \partial_x P_m(1, z) + z \partial_y P_m(1, z) = \lambda m,$$
$$mQ_m(1, z) \equiv \partial_x Q_m(1, z) + z \partial_y Q_m(1, z) = \lambda m z. \tag{8}$$

Let $L_{e_i}$ be the matrix of a multiplication by $e_i = (1, z_i)$, defined in (3). Using (8) we can rewrite the Jacoby matrix $J(e_i)$ in the form:

$$J(e_i) := \begin{pmatrix} m\lambda - z_i \partial_y P_m(1, z_i) & \partial_y P_m(1, z_i) \\ m\lambda z_i - z_i \partial_y Q_m(1, z_i)) & \partial_y Q_m(1, z_i) \end{pmatrix} \tag{9}$$

Then the characteristic polynomial $p_2(t) := \det(L_{e_i} - tI)$ of a matrix $L_{e_i}$ will have two eigenvalues: $t = \lambda$ and

$$t = \frac{\partial_y Q_m(1, z_i) - z_i \partial_y P_m(1, z_i)}{m} = \frac{\lambda}{m} R'(z_i); \tag{10}$$

where $R(z)$ was defined in (6).

The theorem is proven. $\qquad\square$

## 3.1. *Proof of the Theorem 2.1*

**Proof.** The syzygy between Peirce numbers (4) one can put into effect using the Residual Theorem for the rational function $F(z)$ of complex variable $z$:

$$F(z) = \frac{1}{z - R(z)} \tag{11}$$

Without lost of generality, we can choose basis such, that there are no root of the equation $z = R(z)$ at infinity. Namely, suppose $R(z)$ is bounded at infinity. Then

$$\lim_{z \to \infty} zF(z) = 1; \qquad \operatorname*{res}_{z=\infty} F(z) = 1.$$

The assertion of Theorem 2.1 about existence of exactly $m + 1$ pairwise different $\lambda$-idempotents lead to the convention, that all $m + 1$ poles of $F(z)$ are simple. Thus

$$\sum_{i=0}^{m} \operatorname*{res}_{z=z_i} F(z) := \sum_{i=0}^{m} \frac{1}{1 - R'(z_i)} = \operatorname*{res}_{z=\infty} F(z) := 1.$$

Using Theorem 3.1 the proof is finished. $\qquad\square$

## 3.2. *Integral rays schema*

In order to distinguish the global qualitative behavior of the solutions to (5) in the neighborhood of the integral ray $\mathbb{R}[e]$ with the Peirce number $\mu$ define the following three intervals: $p_1 = \{\mu : \mu > 1/m\}$, $p_2 = \{\mu : 0 < \mu < 1/m\}$, $h = \{\mu : \mu < 0\}$ and two **exceptional** sets: $e_0 = \{\mu : \mu = 0\}$, $e_1 = \{\mu : \mu = 1/m\}$. Then

**Theorem 3.2.** *(see* [11]*) The trajectories of (5) starting at the point sufficient closest to the integral ray $\mathbb{R}[e]$ with the Peirce number $\mu$:*

*(1) touch the integral ray at the origin if $\mu \in p_1$,*
*(2) tends to the integral ray at the infinity but do not touch it if $\mu \in p_2$,*
*(3) touch the integral ray at the infinity if $\mu \in h$,*
*(4) yield an exceptional manner of the contact to the integral ray if $\mu \in \{e_0, e_1\}$.*

**Proof.** In [11] theorem was proven in terms of values of function $R'(z)$. Using Theorem 3.1 we only reformulate it in terms of the Peirce numbers. We add the graphical explanations of the G.E. Shilov results in Figure 1.

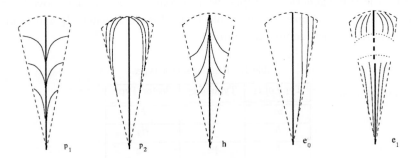

Fig. 1.   Rays types

$\square$

The phase portrait at the origin looks like a gluing together the local phase portraits near any two consecutive rays. In this way, one can determine the six possible non exceptional sectorial types (see [1], [2]) shown on Figure 2.

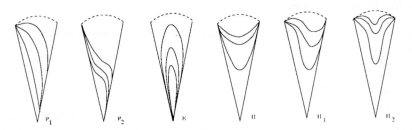

Fig. 2.   Regular sector types

The integral rays divide the phase portrait in the neighborhood of the origin into subregions which are called *sectors*. Since the integral rays separate each sector from the next one, they are called *separatrices*. Each sector is filled by paths whose qualitative behavior is identical at the origin and at infinity. The sign of the angular velocity of each point in a certain sector is fixed. Note that two consecutive fixed directions create two symmetric sectors up to inversion. Since $P$ and $Q$ are homogeneous these two sectors have the same qualitative dynamics. It is known (see [5], [1]) that all isolated equilibrium states of the planar dynamical system (1) have neighborhoods which are constructed from areas of three different types of flow behavior. Namely:

1. *Parabolic sector*, (briefly, type $P$-sector) where all paths lead to, or away from, the equilibrium state,
2. *Hyperbolic sector*, (briefly, type $H$-sector) where all paths sweep past the equilibrium state.
3. *Elliptic sector*, (briefly, type $E$-sector) where all paths originate and end at the equilibrium state.

Based on the three generic ray types, we are going to extend the above list to six *regular sector types* as specified in Table 1 and Figure 2, where $l_1$ and $l_2$ are two consecutive rays which delimit a sector.

Table 1.  Regular sector types

| Type of $l_1$ | Type of $l_2$ | Sector Type |
|:---:|:---:|:---:|
| $p_1$ | $p_1$ | $E$ |
| $h$ | $h$ | $H_1$ |
| $h$ | $p_2$ | $H_2$ |
| $p_2$ | $p_2$ | $H_3$ |
| $p_1$ | $p_2$ | $P_1$ |
| $h$ | $p_1$ | $P_2$ |

**Definition 3.1.** The sectors, which are composed by the two consecutive rays and such that at least one of the ray is an exceptional would be called an *exceptional type $E_0$, $H_0$ and $P_0$ sectors.*

In fact, no rigorous topological behavior (except of a type) may be defined in exceptional types of sectors.

## 4. Combinatorial Schemes

It is well known (see [1], [3], [5]), that a phase portrait of a planar homogeneous polynomial ODE's is constrained by pasting together sectors of concrete types, following the combinatorial scheme. Two neighboring rays forms a type of sector. Conversely, sector of a given type is coupled by pair of a rays of the consensual types. Suppose, enumeration of the sectors in the phase portrait is chosen counterclockwise and cyclic. The same is true for the rays forms the sectors.

**Definition 4.1.** The cyclic ordered sequence of the rays types (see Theorem 3.2) yield the *ray scheme* of the planar system (5).

Knowing the *ray scheme* one can uniquely rearrange it and ordering by pairs. Thus, one can construct (see Table 1) the ordered set of sectors.

**Definition 4.2.** The cyclic ordered sequence of the types of the sectors, obtained from the rays scheme, yield the *sector scheme* of the planar system (5).

This tool will be used to explore the phase portrait of the planar ODE as well as to understanding some properties of the associated to the planar ODE algebra. In particulary, the components of the sector schemes are responsible for the Bendixon formula for the topological degree of the origin (see [5], [4]).

Note, that the notion of the **combinatorial scheme** in the contents of the quadratic ODEs first was appear in [10]. We concretize Definition 4.1 and 4.2 in formal automata theory language as follows:

Denote by $\Sigma_r$ an *alphabet* with lower-case letters $\Sigma_r = \{h, p_1, p_2, e_0, e_1\}$ and by $\Sigma_s$ an alphabet with capital letters $\Sigma_s = \{E, H_1, H_2, H_3, P_1, P_2, E_0, H_0, P_0\}$. Here $E_0$, $H_0$, $P_0$ stand for the common families of an exceptional type sectors. The word composition in both the two alphabets respect some special rules (*grammar* $\mathbb{G}$). The presence of the main syzygy (4) is responsible for the arrangement the letters in a word $\omega \in \Sigma_r$ correctly in the grammar $\mathbb{G}(\Sigma_r)$. Accordance of the sector scheme to the ray scheme is the best possibility for checking that a word $\omega \in \Sigma_s$ is allowed in the grammar $\mathbb{G}(\Sigma_s)$.

Define a string $w_1^r$ as a shuffle of $k \leq m+1$ letters $p_1$, $p_2$, $h$, $e_0$, $e_1$ and $w_2^s$ as a shuffle of $k \leq m+1$ letters $E$, $H_1$, $H_2$, $H_3$, $P_1$, $P_2$ and $E_0$, $H_0$, $P_0$ in alphabets $\Sigma_r$ and $\Sigma_s$ respectively. We will use the cyclic ordered words of letters in both two alphabets only.

**Definition 4.3.** We will say that a word $w_i^r$ is *allowed* in the grammar $\mathbb{G}(\Sigma_r)$ and write $w_i^r \in \mathbb{G}(\Sigma_r)$ iff $w_i^r$ can be formed by interspersing the characters from $\Sigma_r$ in a way that preserves the main syzygy (4).

**Definition 4.4.** We will say that word $w_i^s$ is *allowed* in the grammar $\mathbb{G}(\Sigma_s)$ and write $w_i^r \in \mathbb{G}(\Sigma_s)$ iff $w_i^s$ can be build using rule in Table 1 from the word $w_i^r \in \mathbb{G}(\Sigma_r)$ allowed in the grammar $\mathbb{G}(\Sigma_r)$.

Denote by $w^{r,s}$ henceforth as the only *allowed word* in the grammar $\mathbb{G}(\Sigma_{r,s})$.

Next, rewrite two types of combinatorial schemes formally:

1. Redefine a ***ray scheme*** as an allowed *word* $w_i^r$ in grammar $\mathbb{G}(\Sigma_r)$ which is the cyclic ordering sequence of consecutive types of the system's fixed directions (ray types).

2. Redefine a ***sector scheme*** as an allowed *word* $w_i^s$ in grammar $\mathbb{G}(\Sigma_s)$ represented by the sequence of consecutive sector types up to cyclic ordering.

**Remark 4.1.** Once the allowed combination of letters in $w_1^s$ is chosen, the combinatorial scheme is fixed, whereas different and thus dissimilar sectorial combinatorial schemes $w_2^s$ may be chosen even for equivalent up to order ray schemes.

We stress that the words are *cyclic ordered*, because we will be treating them as arrangement of the sectorial schemes with due regard to ensure that the Peirce numbers fulfill (4).

## 5. Similarities

**Definition 5.1.** Recall a planar system $x' = P(x,y)$, $y' = Q(x,y)$ similar to a homogeneous system (5) if its phase-portrait is **topologically equivalent**. As it was described above they both have the only one singular point and the same (finite) numbers of the same type sectors that are equivalently alternating.

**Theorem 5.1.** *Two planar homogeneous systems (5) are similar reciprocally iff they have the same sector scheme.*

The proof of this theorem repeat almost in exactly the same words sentences as were used directly from the Definition 5.1.

**Definition 5.2.** Following [12] we say that system $x' = P(x,y)$, $y' = Q(x,y)$ is **weakly structurally stable** if for *"sufficiently small"* perturbations $p$ and $q$, the system $x' = P(x,y) + p(x,y)$, $y' = Q(x,y) + q(x,y)$ is similar to a homogeneous system.

The smallness of $p$ and $q$ and their derivatives can be characterized by a finite number of constants determined by the right-hand sides of (5).

The following theorem consist of sufficient conditions for system (5) to be weakly structurally stable.

**Theorem 5.2.** *The system $x' = P(x,y)$, $y' = Q(x,y)$ is weakly structurally stable if it is similar to the homogeneous system with no exceptional Peirce numbers.*

**Proof.** The only bifurcation of the spectrum of the multiplication by an idempotent operator is an eigenvalue doubling. Therefore, the small perturbation of the homogeneous system lead out from the class of similarity if there is at least one exceptional Peirce number. □

## 6. Concluding remarks

In the previous sections we investigate the disposition of the invariant rays of the system (5), among which we note that $Q_m(x,y)$ and $P_m(x,y)$ are homogeneous functions of the same order of homogeneity and do not vanish simultaneously at no point other than the origin.

**Definition 6.1.** Recall the $\lambda$-idempotents and theirs Peirce numbers the **spectral characteristics** of the multiplication operator $\omega(x)$ in (1).

Thus, at least from point of view the qualitative theory of a homogeneous ODEs, the Definition 6.1 fulfill a role in a generalization of the well known spectral theory of the linear operators. The eigenvectors notion interchange with a $\lambda$-idempotent. But the eigenvalues notion (in sense of the linear spectral theory) is unaffected even for a quadratic operators. Assuming now a Peirce numbers in the associated to the nonlinear homogeneous operator $m$-ary algebra be substitute of an eigenvalues. Thus the basic idea of a spectral characteristics technique covering the qualitative theory is as follows:

(1) the integral rays in phase portrait of (1) indicate the locus of an eigenvectors,
(2) the Peirce numbers are responsible for a local phase portrait near the rays,
(3) syzygy between Peirce numbers fixed the possible types of a phase portrait.

Table 2. Combinatorial schemes in binary algebras without nilpotents

| Length | Type $h$ | Type $p_1$ | Type $p_2$ | Type $e_0$ | Type $e_1$ | Sector scheme |
|---|---|---|---|---|---|---|
| | 3 | 0 | 0 | 0 | 0 | $H_1 H_1 H_1$ |
| | 2 | 1 | 0 | 0 | 0 | $H_1 P_2 P_2$ |
| | 1 | 1 | 1 | 0 | 0 | $H_2 P_1 P_2$ |
| | 0 | 1 | 2 | 0 | 0 | $H_3 P_1 P_1$ |
| 3 | 0 | 2 | 1 | 0 | 0 | $E P_1 P_1$ |
| | 1 | 1 | 0 | 1 | 0 | $H_0 P_0 P_0$ |
| | 0 | 1 | 1 | 1 | 0 | $H_0 P_0 P_0$ |
| | 0 | 1 | 0 | 2 | 0 | $H_0 P_0 P_0$ |
| | 1 | 0 | 0 | 0 | 1 | $H_0 P_0$ |
| 2 | 0 | 1 | 0 | 0 | 1 | $E_0 P_0$ |
| | 0 | 0 | 1 | 0 | 1 | $H_0 P_0$ |
| | 0 | 0 | 0 | 1 | 1 | $H_0 P_0$ |
| | 1 | 0 | 0 | 0 | 0 | $H_1$ |
| | 0 | 1 | 0 | 0 | 0 | $E$ |
| 1 | 0 | 0 | 1 | 0 | 0 | $H_3$ |
| | 0 | 0 | 0 | 1 | 0 | $H_0$ |
| | 0 | 0 | 0 | 0 | 1 | $E_0$ or $H_0$ |

We can see from Table 2, for example, that the sector schemes $HHH$, $HPP$ and $EPP$ are available in the binary two dimensional algebras but $EEE$ not. If an algebras contain a nilpotent, there exist rays consisting only of singular points.

We do not covering the various cases of behavior of the phase portraits, depending on whether or not there exists the nilpotents in an algebra. Some of the generic cases mentioned in Table 2 are present on Fig. 3.

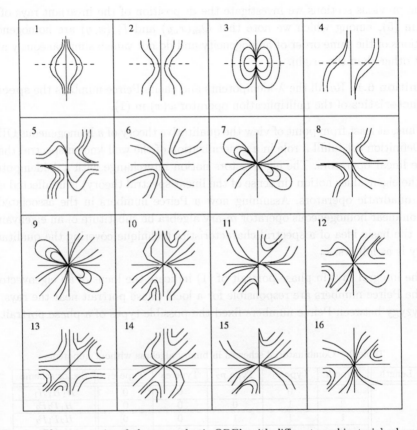

Fig. 3.    Phase portraits of planar quadratic ODE's with different combinatorial schemes

# References

1. A. Andronov; E. Leontovic; I. Gordon; A. Maier, Qualitative theory of second-order dynamic systems, Translated from the Russian, Halsted Press [A division of John Wiley & Sons], New York-Toronto, Ont., (1973).

2. D. V. Anosov and V. I. Arnold, Dynamical Systems I , Encyclopaedia of Mathematical Sciences 1, Springer- Verlag, Berlin, Heidelberg, New York, (1988).

3. J. Artes, R. Kooij, J. Llibre, Structurally stable quadratic vector fields, Mem. Amer. Math. Soc. 134 (1998).

4. Z. Balanov and Y. Krasnov, Complex structures in real algebras. I. Two-dimensional commutative case. Comm. Algebra 31 (2003), no. 9, 4571–4609.

5. I. Bendixson, Sur les courbes définies par des équations différentielles, Acta Math. 24 (1901), 1-88.

6. W. A. Coppel, A survey of quadratic systems, J. Differential Equations 2 (1966), 293-304.

7. M. K. Kinyon, A. A. Sagle, Quadratic dynamical systems and algebras. J. Differential Equations 117 no. 1 (1995), 67-126.

8. Y. Krasnov, Commutative algebras in Clifford analysis. Progress in analysis, Vol. I, II (Berlin, 2001), 349–359, World Sci. Publishing, River Edge, NJ, (2003).

9. M. F. Atiyah, R. Bott, A Lefschetz fixed point formula for elliptic complexes: II. Applications, Ann. of Math. (2) **88** (1968), 451–491.

10. L. Markus, Quadratic differential equations and non-associative algebras, Contributions to the theory of nonlinear oscillations , Vol. V (L. Cesari, J .P. LaSalle, and S. Lefschetz, eds.), Princeton Univ. Press, Princeton (1960), 185-213. J. Math. Pures Appl. 8 (1960), 187-216.

11. G. E. Shilov, Integral curves of a homogeneous equation of the first order. (Russian) Uspehi Matem. Nauk (N.S.) 5, (1950), no. 5(39), 193–203.

12. K. S. Sibirsky, Introduction to the algebraic theory of invariants of differential equations. Translated from the Russian. Nonlinear Science: Theory and Applications. Manchester University Press, Manchester, (1988).

13. S. Walcher, Algebras and differential equations, Hadronic Press, Palm Harbor, FL, (1991)

14. S. Walcher, On algebras of rank three, Comm. Algebra 27, no. 7 (1999), 3401-3438.

### III.5    Complex analytic methods in the applied sciences

Organizers: Vladimir V. Mityushev, Sergei V. Rogosin

The main attention at the session was paid to analytic type results in Complex Analysis especially those which have applications in Mathematical Physics, Mechanics, Biology, etc. Two articles are devoted to developing methods of complex analysis. It is shown that the Weierstrass $\mathcal{P}$-function is not topologically elementary. This function is highly used at the study of composite materials with periodic structure. Monotone Operators' Principle is applied to nonlinear singular integral equations.

Other contributions are connected with special applications. Namely, the effective conductivity of two-dimensional composite materials with a finite number of circular reinforcements in a cell is investigated for structures with different symmetrical arrangements: square and pseudofractal. It is shown that in average symmetrical arrangements possess the minimal effective conductivity with respect to perturbed arrangements if the conductivity of reinforcements is larger than the conductivity of the matrix.

On a biological example two methods of depicting dislocations are given: by Delone triangulations as pentagon-heptagon pairs and as vortices in the complex plane. Some aspects of development of biological tissues in animals and plants can be described in a 2D space. The osteon growth and phyllotaxis are perhaps the best known examples.

An analytic solution of the Ornstein-Zernike equation for a multicomponent fluid with a multi-Screened Coulomb Plus Power Series closure is found. It offers sufficient flexibility and opens access to systems with any smooth, realistic isotropic potentials where the potentials can be fitted by the multi-SCPPS closure under the mean-spherical approximation.

# THE WEIERSTRASS $\mathcal{P}$–FUNCTION IS NOT TOPOLOGICALLY ELEMENTARY *

V.V. MITYUSHEV

*Department of Mathematics, Krakow Pedagogical Academy,*
*ul. Podchorazych 2, 30-084 Krakow, Poland*
*E-mail: mityu@ipgp.jussieu.fr*

S.V. ROGOSIN

*Mechanical-Mathematical Faculty, Belarusian State University,*
*Fr.Skaryny ave, 4, 220050, Minsk, Belarus*
*E-mail: rogosin@bsu.by*

We prove that the Weierstrass $\mathcal{P}$–function cannot be homeomorphically transformed to any elementary function.

**Key words:** Weierstraß $\mathcal{P}$-function, topologically elemtary functions, Arnold problem
**Mathematics Subject Classification:** 33E05

## 1. Introduction

One can hear first from the secondary school that it is impossible to express roots of the algebraic equations of degree 5 or higher in terms of the coefficients using only arithmetic operations and radicals. This assertion were proved by Ruffini in 1799 with minor gaps (see historical notes [3]) and it is known in our days as the Abel (- Ruffini) theorem.

In 1963 Vladimir Igorevich Arnold gave special course *Abel's theorem* for pupils of the College of the Moscow State University. Latter V. M. Alekseev prepared the book [2] according to this course. In 1963-1964 V.I. Arnold had proved that equation $x^5 + ax + 1 = 0$ cannot be solved in wider sense, namely the roots of this equation cannot be presented as a topologically elementary function $x(a)$. In 1963 V.I. Arnold had stated the following question. Whether the elliptic integral and the Weierstrass $\mathcal{P}$–function are not topologically elementary? He also proposed a plan of the long proof of this conjecture which is not yet realized. One can find a discussion devoted to this question and to many other interesting facts in [2-8].

In what follows elementary functions of complex variable are those which can be obtained from the basic elementary functions (exp and polynomials) by using finite

---

*The work of the second author is partially supported by Belarusian Fund for Fundamental Scientific Research.

number of arithmetic operations and compositions. Therefore in the complex case we exclude from the whole collection of elementary functions single-valued branches of logarithmic function, powers with non-integer exponents etc. (cf. e.g. [9]).

We recall that the elliptic integral is defined as follows [1]

$$u(w) = \int_w^\infty \frac{dt}{\sqrt{4t^3 - g_2 t - g_3}}, \tag{1}$$

where $g_2$ and $g_3$ are given constants. The Weierstrass $\mathcal{P}$-function with the periods $\omega_1, \omega_2$ ($Im\,\omega_2/\omega_1 > 0$) can be defined as the series

$$\mathcal{P}(z) = \frac{1}{z^2} + \sum_{m^2+n^2 \neq 0} \left( \frac{1}{(z - m\omega_1 - n\omega_2)^2} - \frac{1}{(m\omega_1 + n\omega_2)^2} \right). \tag{2}$$

It satisfies the differential equation

$$dz = -\frac{d\mathcal{P}(z)}{\sqrt{4[\mathcal{P}(z)]^3 - g_2\mathcal{P}(z) - g_3}}, \tag{3}$$

where $g_2$ and $g_3$ are related with the periods $\omega_1$ and $\omega_2$. Comparing (1) and (3) one can see that the elliptic integral $u(w)$ is the inverse function to $\mathcal{P}(z)$.

## 2. Main result

Let $\mathbb{C}_j$ ($j = 1, 2, 3, 4$) be four copies of the complex plane $\mathbb{C}$.

**Definition 2.1.** A function $\tilde{f}\colon \mathbb{C}_3 \to \mathbb{C}_4$ is called the topologically elementary if there exist an elementary function $f(z)$ and homeomorphisms $h$, $k$ of the complex plane such that the following diagram is commutative

$$
\begin{array}{ccc}
 & f & \\
\mathbb{C}_1 & \longrightarrow & \mathbb{C}_2 \\
h \downarrow & & \downarrow k \\
 & \tilde{f} & \\
\mathbb{C}_3 & \longrightarrow & \mathbb{C}_4
\end{array}
\tag{4}
$$

In the present note we propose a simple proof of the following conjecture.

**Conjecture 2.1.** *(V. I. Arnold) The Weierstrass $\mathcal{P}$-function is not topologically elementary.*

The idea is based on possibility to construct such a curve $\gamma_1 \in \mathbb{C}_1$ going to infinity that the restriction of the diagram (4) cannot be commutative for any elementary function $f(z)$ and any homeomorphisms $h$, $k$ of the complex plane.

In order to distinguish the behavior of the curves at infinity we introduce the following definition.

**Definition 2.2.** Let a continuous curve $\gamma$ be defined by the parametrization $x \mapsto g(x)$, where $0 \leq x < 1$, $g(x) \in \mathbb{C}$. One says that $\gamma$ goes by infinity if for any $R > 0$ there exist points of $\gamma$ which do not lie in the disk $|z| < R$.

In particular, this definition includes the case when $\gamma$ tends to infinity.

Note that instead of all elementary functions one can consider only exponent, and polynomials. For the exponent and polynomials there are unbounded domains along which these functions tends to $\infty$ when $z \to \infty$. It is true also for compositions, sums, differences and products of exponential functions and polynomials. Really, these functions are either polynomials or entire functions of order not less than 1 having at most finite number of zeros. Polynomials tends to infinity uniformly in the whole complex plane. From the Fragmen-Lindelöf theorem (see e.g. [1]) follows that for the above said entire functions of finite order $\rho$ there are sectors of angle less than $\frac{\pi}{\rho}$ along which they have to tend to infinity. In the case of infinite order it is sufficient to show an existence of $\gamma_1$ for the function $\exp \circ \exp$. Analogous consideration is valid also for (transcendental) meromorphic functions and for rational functions $\frac{P}{Q}$, $deg\,P > deg\,Q$,. In the case of rational functions $\frac{P}{Q}$, $deg\,P \le deg\,Q$, we have uniformly $\lim_{|z|\to\infty} \frac{P(z)}{Q(z)} = a \in \mathbb{C}$. The discussion of this case we present at the end of the paper.

Thus we will consider all those elementary functions $f(z)$ mapping infinity to infinity. We assume that the diagram (4) is commutative. It is possible to find the curves $\gamma_j \in \mathbb{C}_j$ $(j = 1, 2, 3, 4)$ having the following properties. The curve $\gamma_4$ with the end–points at $z_1$ and at $\infty$ lays on $\mathbb{C}_4$ and is such that $\gamma_1 = f^{-1}(\gamma_2)$ goes by infinity on $\mathbb{C}_1$, $\gamma_2 = k^{-1}(\gamma_4)$ goes by infinity on $\mathbb{C}_2$. Here $f^{-1}(\gamma_2)$ means the full pre–image of $\gamma_2$. The curve $\gamma_3 = h(\gamma_1)$ goes by infinity on $\mathbb{C}_3$. The curve $\gamma_1$ goes by infinity on $\mathbb{C}_1$ again due to the Fragmen-Lindelöf theorem. The mapping $k \circ f = \mathcal{P} \circ h \colon \gamma_1 \to \gamma_4$ homeomorphically transforms the curve $\gamma_1$ onto the curve $\gamma_4 := k \circ f(\gamma_1) \subset \mathbb{C}_4$.

Consider now the curve $\gamma_3 := h(\gamma_1)$. By assumption the mapping $\mathcal{P} \circ h \colon \gamma_1 \to \gamma_4$ coincides with $k \circ f$, i.e., it have to map the curve $\gamma_1$ onto $\gamma_4$ homeomorphically. This yields $\mathcal{P}[h(t)]$, $t \in \gamma_1$, goes by infinity iff $\mathcal{P}(\zeta)$, $t \in \gamma_3$, goes by infinity, since $h \colon \mathbb{C}_1 \to \mathbb{C}_3$ is a homeomorphism. We note that only the curve $\gamma_4$ tends to infinity and $\gamma_1, \gamma_2, \gamma_3$ go by infinity.

Consider a doubly periodic family $\Pi_{(l_1,l_2)}$, $(l_1, l_2) \in \mathbb{Z}^2$, of parallelograms on $\mathbb{C}_3$ such that the Weierstrass $\mathcal{P}$-function is univalent in the interior of each parallelogram. It follows from the properties of $\mathcal{P}$-function that such a family exists, cover the whole complex plane and on one of four sides of any parallelogram there is a pole of $\mathcal{P}$-function (see [1]). Denote this side by $\Gamma_{(l_1,l_2)}^{(0)}$. Due to periodicity of the Weierstrass function we can note that its values on the remaining sides of the parallelograms $\Gamma_{(l_1,l_2)}^{(j)}$, $j = 1, 2, 3$, are connected to each other in such way that the sets

$$L^{(j)} := \mathcal{P}(\Gamma_{(l_1,l_2)}^{(j)}) \tag{5}$$

do not depend on indexes $l_1, l_2$. Moreover, $L^{(j)}$ are bounded sets on the complex plane $\mathbb{C}_4$.

Let us examine the behavior of the curve $\gamma_3$ and its image $\mathcal{P}(\gamma_3)$. If $\gamma_3$ crosses at least one pole of $\mathcal{P}$-function then $\mathcal{P}(\gamma_3)$ cannot coincide with $\gamma_4$. Hence we already have a contradiction. If not, we take an infinite sequence $\{\Pi_m\}_{m=1}^{\infty}$ of the above

1246

family of parallelograms for which $\gamma_3$ intersects their sides out of the poles of $\mathcal{P}$-function. We choose further an infinite subset $\{\Pi_{m_k}\}_{k=1}^{\infty}$ of the parallelograms from $\{\Pi_m\}_{m=1}^{\infty}$ for which the curve $\gamma_3$ crosses at least one of the sides $\Gamma_{m_k}^{(j)}$, $j = 1, 2, 3$. Choosing at least one crossing point $\zeta_k^{(j)}$ on $\gamma_3 \cap \Gamma_{m_k}^{(j)}$ we order them in accordance with the orientation on $\gamma_3$. The obtained sequence $(\zeta_{k_p}^{(j)})_{p=1}^{\infty}$ tends to infinity.

It follows from the commutativity of the diagram (4) that $\gamma_4 \equiv \mathcal{P}(\gamma_3)$ tends to infinity. However, $\lim_{p\to\infty} \mathcal{P}(\zeta_{k_p}^{(j)})$ could not be equal to $\infty$. This contradiction completes the proof.

In the remaining case of rational functions $f = \frac{P}{Q}$, $\deg P \leq \deg Q$, we have

$$\lim_{t\to\infty, t\in\gamma_1} f(t) = a \in \mathbb{C}$$

for any unbounded curve $\gamma_1$. Hence $\lim_{t\to\infty, t\in\gamma_1}(k \circ f)(t) = b \in \mathbb{C}$. The same argument as before show that the function $\mathcal{P} \circ h$ do not satisfy the relation

$$\lim_{t\to\infty, t\in\gamma_1}(\mathcal{P} \circ h)(t) = b.$$

Therefore the diagram (4) cannot be commutative in this case too.

We are grateful to V. I. Arnold for explanations concerning the problem of topological non–elementary of functions and for the literature.

## References

1. N. I. Akhiezer, Elements of the Theory of Elliptic Functions, Moscow 1970; AMS Translations of Mathematical Monographs, v. 79, Rhode Island, 1990.
2. V. B. Alekseev, Abel's Theorem in Problems and Solutions, MCNMO, Moskwa, 2001; Kluwer Academic Pub, 2004.
3. R. G. Ayoub, Paolo Ruffini's Contributions to the Quintic, Archive for History of Exact Sciences 23, 1980, 253-277.
4. V. I. Arnold, Chto takoe matematika, MCCME, Moskva, 2002, 20-27 (in Russian).
5. V. I. Arnold, Problems to the Seminar, 2003-2004, Cahiers du CEREMADE, Universite Paris-Dauphine, N 0416, 1 Mars 2004, 17-18.
6. V. I. Arnold, Zadachi Seminara 2003-2004, MCCME, Moscow, 2005, 23-27 (in Russian).
7. V. I. Arnold, Abel's theory and modern mathematics, Springer book "Stockholm Intelligencer", European Mathematical Congress at Stokholm, 2004, 6-7.
8. V. I. Arnold, Arnold's Problems, Springer and Phasis, 2005, 168-170.
9. W. Rudin, Real and Complex Analysis. McGraw-Hill, New York (1966).

# ON APPLICATION OF THE MONOTONE OPERATOR METHOD TO SOLVABILITY OF NONLINEAR SINGULAR INTEGRAL EQUATIONS*

S.V. ROGOSIN

*Mechanical-Mathematical Faculty, Belarusian State University,*
*Nezavisimosti ave, 4, 220050, Minsk, Belarus*
*E-mail: rogosin@bsu.by*

A variant of the monotone operator method is applied to the study of solvability of nonlinear singular integral equations of Hammerstein type in Orlicz spaces.

**Key words:** Nonlinear singular integral equation, monotone operator, Minty-Browder-Vainberg theorem, Orlicz spaces, superposition operator
**Mathematics Subject Classification:** 45G05, 47H05

## 1. Introduction

We consider here solvability of nonlinear singular integral equations of the following type

$$x(t) + \frac{\lambda}{\pi i} \int_{\gamma} \frac{k(t,s)}{s-t} f(s, x(s)) ds = g(t), \ t \in \gamma, \tag{1}$$

where $\gamma$ is a simple closed smooth curve on the complex plane $\mathbb{C}$, $\lambda \in \mathbb{R} \setminus \{0\}$ is a real parameter, under different assumptions on given functions $k : \gamma \times \gamma \to \mathbb{C}$, $g : \gamma \to \mathbb{C}$, $f : \gamma \times \mathbb{C} \to \mathbb{C}$. Applicability of a variant of the monotone operator method to equation (1) in Orlicz spaces is studied.

Fundamental Monotonicity Principle worked out by Minty-Browder-Vainberg (see, [5,6,16,20]) was generalized recently in different directions (e.g. [4]). Based on this principle method is effectively applied (see [3,10,11,21]) to the study of nonlinear singular integral equations starting from pioneering paper by Gusejnov & Mukhtarov [9]. The main role in these investigations played the properties of nonlinear composition operator in different functional spaces as described, e.g., in the monographs [2,17]. At the study of nonlinear integral equations in the Lebesgue-type spaces one have to take into account the classical Krasnosel'skij result (see e.g. [15]) which states that nonlinearity $f(s,u)$ has power growth in $u$ whenever the corresponding superposition operator acts in certain Lebesgue spaces. Avoiding power-type behavior of

---

*This work is partially supported by Belarusian Fund for Fundamental Scientific Research (projects F05-036, F06R-106)

$f(s, u)$ gives us possibility to consider the nonlinear integral equations (1) in more general spaces of summable functions, in particular in Orlicz spaces. Since in general (see [14,18]) Orlicz spaces are non-reflexive it leads to necessity to apply theorems on monotone operators for non-reflexive Banach spaces (e.g. [4,20]).

A general scheme of the study of nonlinear singular integral equation (1) in the (non-reflexive) Orlicz spaces is given in [22]. In the present article these results are corrected and completed by the concrete assumptions on the given functions $k, f$ and $g$ under which the above scheme guarantee an existence (and uniqueness) of the solution to (1) in Orlicz spaces.

## 2. Notation and Auxiliary Results

We present here the necessary definitions and auxiliary results connecting with application of the Monotone Operators Principle following [12,15] and theory of Orlicz spaces as presented in [14,18].

Let $\mathcal{X}, \mathcal{X}'$ be a pair of dual complex Banach ideal spaces, $\mathcal{X} \subseteq \mathcal{X}'$. We denote in what follows by $\langle f, x \rangle$ the value of a functional $f \in \mathcal{X}'$ on an element $x \in \mathcal{X}$. Let $\Phi$ be a (nonlinear) operator acting from $\mathcal{X}$ to $\mathcal{X}'$. The operator $\Phi$ is called *monotone* (see [20]) if

$$\operatorname{Re} \langle \Phi x - \Phi y, x - y \rangle \geq 0, \quad \text{for all } x, y \in \mathcal{X}, \tag{2}$$

and *strictly monotone* if the equality in (2) holds only for $x = y$. The operator $\Phi$ satisfies *Rothe condition on a ball* $B(0, R) \subset \mathcal{X}$, if

$$\operatorname{Re} \langle \Phi x, x \rangle \geq 0, \quad \text{for all } x \in \mathcal{X}, \; \|x\| = R. \tag{3}$$

The operator $\Phi$ is called *hemicontinuous* on $\mathcal{X}$ if for any $x, h \in \mathcal{X}$ and for each sequence $(t_n)$ of positive numbers, $t_n \to +0$, the sequence $\Phi(x + t_n h)$ is weakly converging to $\Phi(x)$. It follows from [20] that any hemiconinuous on $\mathcal{X}$ operator $\Phi$ does satisfy the relation

$$\lim_{t \to +0} \langle \Phi(x + th), y \rangle = \langle \Phi(x), y \rangle \geq 0, \quad \text{for all } y \in \mathcal{X}. \tag{4}$$

The main role in the below consideration plays the following "local" variant of theorem on monotone operators ([22]).

**Theorem 2.1.** *Let $\mathcal{X}, \mathcal{X}'$ be a pair of dual complex Banach ideal spaces, $\mathcal{X} \subseteq \mathcal{X}'$, such that $\mathcal{X} = \Gamma$, where*

$$\Gamma = (\mathcal{X}')^0 \equiv \{ f \in \mathcal{X}' : |\langle f, x \rangle| \leq 1, \; \text{for all } y \in \mathcal{X} \}.$$

*Let $\Phi$ be a hemicontinuous and (strictly) monotone operator acting from $\mathcal{X}$ to $\mathcal{X}'$. If $\Phi$ satisfies the Rothe condition on the ball $B(0, R) \subset \mathcal{X}$ then equation*

$$\Phi x = 0 \tag{5}$$

*has at least one (unique) solution $x_* \in B(0, R)$.*

*The same conclusion is true also for inhomogeneous nonlinear operator equation*

$$\Phi x = h$$

*for each $h \in \mathcal{X}' \cap \Phi\left(B(0, R)\right)$.*

For $\mathcal{X}$ being a reflexive Banach space theorem 2.1 coincides with the complex variant of the classical Browder-Minty theorem (e.g., Theorem 18.1 [20]). In general case its statement follows from the $\Gamma$-weak (sequential) compactness of the unit ball of $\mathcal{X}$ (see, e.g., p. 113 [15]). In particular (see, e.g. pp. 75, 105, 325[18]), each dual pair of Orlicz spaces $\mathcal{L}_M \subset \mathcal{L}_{M'}$ satisfies the condition

$$\left(\mathcal{L}_{M'}^0\right)^* = \mathcal{L}_M.$$

In this connection we have to note that the space $\mathcal{L}_M$ is reflexive iff its generating N-function $M$ does satisfy $\triangle_2$- and $\nabla_2$-conditions simultaneously (see p. 113[18]).

Let us recall a number of definitions and results from the theory of Orlicz spaces. A convex function $M : \mathbb{R} \to \mathbb{R}_+ \cup \{+\infty\}$ satisfying conditions

$$M(-t) = M(t), \ M(0) = 0, \ \lim_{t \to \infty} M(t) = +\infty \tag{6}$$

is called the *Young function*. The Young function is called *N-function* (nice-function) if additionally

$$\lim_{t \to \infty} \frac{M(t)}{t} = 0, \ \lim_{t \to \infty} \frac{M(t)}{t} = +\infty \tag{7}$$

hold. For each N-function $M$ its dual N-function $M'$ can be introduced via Fenchel relation

$$M'(u) = \sup_{v \in \mathbb{R}} \{uv - M(v)\}. \tag{8}$$

Usually a pair of dual N-function is normalizing by the condition

$$M(1) + M'(1) = 1. \tag{9}$$

Its allows us, in particular, to carry out a comparison with the case of Lebesgue spaces which corresponds to the following choice of the N-functions

$$M(t) = \frac{1}{p}|t|^p, \ M'(t) = \frac{1}{q}|t|^q, \ \frac{1}{p} + \frac{1}{q} = 1. \tag{10}$$

Let $(\Omega, \Sigma, \mu)$ be (real or complex) measure space. An Orlicz class $\widetilde{\mathcal{L}}_M$ generated by an N-function $M$ is a set of all $\mu$-measurable functions $f : \Omega \to \mathbb{R}(\mathbb{C})$ for which the following condition

$$\int_\Omega M(|f|)d\mu < +\infty \tag{11}$$

holds. An Orlicz space $\mathcal{L}_M$ generated by an N-function $M$ is a family of all $\mu$-measurable functions $f : \Omega \to \mathbb{R}(\mathbb{C})$ such that $\alpha f \in \widetilde{\mathcal{L}}_M$ for at least one $\alpha > 0$. This space becomes a normed space endowing the Luxemburg norm (see p. $56^{18}$)

$$\|f\|_{\mathcal{L}_M} \equiv \inf \left\{ k > 0 : \int_\Omega M \left( \frac{|f|}{k} \right) d\mu \leq M(1) \right\}. \tag{12}$$

The normed spaces $L_M$ endowing by the linear functionals' norm (called Orlicz-Amemiya norm, see p. $59^{18}$)

$$\|f\|_{L_M} \equiv \sup \left\{ k > 0 : \int_\Omega |fg| d\mu : \int_\Omega M'(|g|) \, d\mu \leq 1 \right\} \tag{13}$$

are introduces too. It is known (see, e.g. p. $69^{18}$) that Orlicz-Amemiya norm can be represented by using N-function $M$ directly:

$$\|f\|_{L_M} \equiv \inf \left\{ k > 0 : \frac{1}{k} \left( 1 + \int_\Omega M(k|f|) \, d\mu \right) \right\}. \tag{14}$$

If equivalent functions are identified in $\mathcal{L}_M$, $L_M$ then these spaces become Banach ones. The above introduce norms are equivalent (see, e.g. Ch. 2, $\S 9^{18}$), but their values coincides quite seldom (see p. $126^{18}$).

For analysis of the linear structure of (real) spaces $L_M$ Morse and Transue have introduced the space

$$\mathcal{M}_M \equiv \left\{ f \in L_M : kf \in \widetilde{\mathcal{L}}_M, \, \forall k > 0 \right\}. \tag{15}$$

If $\mu$ is $\sigma$-finite measure and both dual functions $M, M'$ are N-functions, then (p. $105^{18}$) $(\mathcal{M}_M)^* = L_{M'}$, i.e. for each $x^* \in (\mathcal{M}_M)^*$ there exists a unique element $g \in L_{M'}$ such that

$$x^*(f) = \langle x^*, f \rangle = \int_\Omega fg d\mu.$$

If $M$ satisfies $\triangle_2$-condition then $(\mathcal{M}_M)^* = (L_M)^* = L_{M'}$.

In the case of complex Orlicz spaces the (real) linear continuous functionals can be taken in the following form (see p. $265^{12}$)

$$y^*(f) = \operatorname{Re} \langle y^*, f \rangle = \operatorname{Re} \int_\Omega f \bar{g} d\mu. \tag{16}$$

At last we have to mention embedding-like result for Orlicz spaces (p. $155^{18}$): let $M_1, M_2$ be a pair of (not necessary dual) Young functions. It is said that the function $M_2$ majorizes $M_1$ (in symbols $M_1 \prec M_2$ or $M_2 \succ M_1$) if there exists $b > 0$ such that

$$M_1(x) \leq M_2(bx), \, \forall x \geq x_0 \geq 0$$

$(x_0 = 0$ if $\mu(\Omega) = +\infty)$. If $M_1 \prec M_2$ then

$$L_{M_1} \subseteq L_{M_2}.$$

An inverse is true if the measure $\mu$ is non-atomic one and the functions $M_i$ are continuous and strictly increasing.

## 3. Singular integral operator

Nonlinear singular integral equation (1) can be rewritten in the following operator form

$$(I - \lambda L_k f)\, x = g, \tag{17}$$

where $I$ is the identity operator, $L_k$ is linear singular integral operator generated by the kernel-function $k : \gamma \times \gamma \to \mathbb{C}$:

$$(L_k y)\,(t) \equiv \frac{1}{\pi i} \int\limits_\gamma \frac{k(t,\tau)}{\tau - t} y(\tau) d\tau, \tag{18}$$

in particular, $L_1 = S$ is the classical linear singular integral equation with Cauchy kernel, and $f$ is the nonlinear superposition operator (or Nemytskii' operator) generated by the function $f : \gamma \times \mathbb{C} \to \mathbb{C}$:

$$(f z)\,(t) \equiv f(t, z(t)). \tag{19}$$

**Lemma 3.1.** *Let $\gamma$ be a simple closed Lyapunov curve and let function $k$ be satisfied Hölder condition in both variables, i.e. for certain $C > 0$, $0 < \nu_1, \nu_2 \le 1$ the following inequality holds:*

$$|k(t_1, \tau_1) - k(t_2, \tau_2)| \le C\left(|t_1 - t_2|^{\nu_1} + |\tau_1 - \tau_2|^{\nu_2}\right), \quad \forall t_1, t_2, \tau_1, \tau_2 \in \gamma. \tag{20}$$

*Then the linear singular operator $L_k$ is bounded on each Orlicz space $L_M$ generated by an N-function $M$.*

**Proof.** The integral in (18) can be rewritten in the following form

$$(L_k y)\,(t) = \frac{k(t,t)}{\pi i} \int\limits_\gamma \frac{y(\tau)}{\tau - t} d\tau + \frac{1}{\pi i} \int\limits_\gamma \frac{k(t,\tau) - k(t,t)}{\tau - t} y(\tau) d\tau.$$

Boundedness of the singular integral operator with Cauchy kernel $S$ in Orlicz spaces follows from the same argument as used for the corresponding proof in Lebesgue spaces (see, e.g., [13]). At first one can established the boundedness in $L_M$ of the Hilbert transform

$$(H y)\,(t) \equiv \frac{1}{2\pi} \int\limits_0^{2\pi} y(e^{i\sigma}) \cot \frac{\sigma - s}{2} d\sigma, \quad t = e^{is},$$

which is related to the operator $S$ on the unit circle $\mathbb{T}$ via formula

$$S y = i H y + C_0,$$

where $C_0$ is a real constant. Next it can be shown that if $\gamma$ is a simple closed Lyapunov curve then the kernel of the operator

$$\frac{1}{\pi i} \int\limits_{\gamma} \left[ \frac{1}{\tau - t} - \frac{\alpha'}{\alpha(\tau) - \alpha(t)} \right] y(\tau) d\tau,$$

where $\alpha(t)$ is the boundary function of the conformal mapping of the domain $\text{int}\,\gamma$ onto unit disc $\mathbb{U}$ can be represented in the form

$$\frac{k_0(t, \tau)}{|\tau - t|^{\nu}}, \quad 0 < \nu < 1,$$

with $k_0$ being bounded on $\gamma \times \gamma$ and continuous except possibly the diagonal $\{(t, \tau) \in \gamma \times \gamma : t = \tau\}$. An analogous representation takes place to the kernel

$$\frac{k(t, \tau) - k(t, t)}{\tau - t}$$

because of condition (20).

The final conclusion follows from [8].                           □

**Lemma 3.2.** *Let $\gamma$ be a simple closed Lyapunov curve and let function $k$ be satisfied Hölder condition in both variables, $k(t, t) \neq 0$, $\forall t \in \gamma$. Let the following relations hold*

$$k(t, \tau) = \frac{1}{\pi i} \int\limits_{\gamma} \frac{k(s, \tau)}{s - t} ds, \quad \text{for all} \quad t, \tau \in \gamma, \tag{21}$$

$$k(t, \tau) = \frac{1}{\pi i} \int\limits_{\gamma} \frac{k(t, \sigma)}{\sigma - \tau} ds, \quad \text{for all} \quad t, \tau \in \gamma. \tag{22}$$

*Then the linear singular operator $\mathbf{L}_k$ has a bounded inverse in $\mathcal{L}_2$ represented in the form*

$$\left( \mathbf{L}_k^{-1} u \right)(t) = \frac{1}{\pi i} \int\limits_{\gamma} \frac{k(t, \tau)}{k(t, t) k(\tau, \tau)} \frac{u(\tau)}{\tau - t} d\tau. \tag{23}$$

*The operator $\mathbf{L}_k^{-1}$ is defined and bounded on each Orlicz space $L_M$ generated by an N-function $M$.*

**Proof.** Conditions (21), (22) are necessary and sufficient conditions (see p. 41[7]) for the function $k(t, \tau)$ to be analytically continued in both variables onto domain $D^+ \times D^+ \equiv \text{int}\,\Gamma \times \text{int}\,\Gamma$. In this case an existence of the inverse operator in $\mathcal{L}_2$ and its representation in the form (23) is proved in [19].

It follows from the assumptions of Lemma that the function $k_1(t, \tau) \equiv \frac{k(t, \tau)}{k(t, t) k(\tau, \tau)}$ satisfies the Hölder condition in both variables. Hence the operator $\mathbf{L}_k^{-1}$ can be extended on each Orlicz space $L_M$. Its boundedness on $L_M$ is due to Lemma 3.1.□

**Lemma 3.3.** *Let the kernel function $k : \gamma \times \gamma \to \mathbb{C}$ satisfies conditions (20)–(22). Then for the inverse operator $\mathbf{L}_k^{-1}$ defined by (23) the following inequalities*

$$\operatorname{Re}\langle \mathbf{L}_k^{-1}u, u\rangle \geq \delta_-\langle u, u\rangle, \quad u \in \mathbf{L}_M, \qquad (24)$$

$$\operatorname{Re}\langle \mathbf{L}_k^{-1}u, u\rangle \leq \delta_+\langle u, u\rangle, \quad u \in \mathbf{L}_M, \qquad (25)$$

*are valid, where $\delta_-, \delta_+$ are certain constants satisfying relations*

$$-\left\|\mathbf{L}_k^{-1}\right\| \leq \delta_- \leq \delta_+ \leq \left\|\mathbf{L}_k^{-1}\right\|.$$

**Proof.** It follows from Lemma 3.2 that under conditions (20)–(22) the operator $\mathbf{L}_k^{-1}$ can be represented in the form

$$\left(\mathbf{L}_k^{-1}u\right)(t) = \frac{1}{\pi i} \int_\gamma \frac{m(t, \tau)}{\tau - t} u(\tau) d\tau, \qquad (26)$$

where $m : \gamma \times \gamma \to \mathbb{C}$ satisfies Hölder condition in both variables. By Lemma 3.1 we have the boundedness of the operator $\mathbf{L}_k^{-1}$ in each Orlicz space $L_M$ generated by an N-function $M$. Then the validity of (24)–(25) follows from the definition of the operator norm and the pairing on $L_M \times L_{M'}$. $\qquad\square$

**Lemma 3.4.** *Let $\gamma$ be a simple closed Lyapunov curve. Let $M$, $M'$ be a pair of mutually dual N-functions, and let $L_M \subset \mathcal{L}_2 \subset L_{M'}$.*

*Then the characteristic singular integral operator $\mathbf{S}$ satisfies the relation*

$$\langle \mathbf{S}x, \overline{x}\rangle = 0, \quad \forall x \in \mathbf{L}_M. \qquad (27)$$

**Proof.** Lemma 3.1 gives that for each pair of elements $x \in \mathbf{L}_M$, $z \in \mathbf{L}_{M'}$ the expression $\langle \mathbf{S}x, z\rangle$ is well-defined. Then by applying Poincare-Bertrand formula (see [7]) we obtain

$$\langle \mathbf{S}x, z\rangle = \int_\gamma \overline{z(t)} dt \frac{1}{\pi i} \int_\gamma \frac{x(\tau)}{\tau - t} d\tau = -\int_\gamma x(\tau) d\tau \frac{1}{\pi i} \int_\gamma \frac{\overline{z(t)}}{t - \tau} dt = -\langle \mathbf{S}\overline{z}, \overline{x}\rangle.$$

In particular, this relation holds for $z = \overline{x} \in \mathbf{L}_M \subset \mathbf{L}_{M'}$, i.e.

$$\langle \mathbf{S}x, \overline{x}\rangle = -\langle \mathbf{S}x, \overline{x}\rangle, \quad \forall x \in \mathbf{L}_M.$$

It gives formula (27). $\qquad\square$

**Remark 3.1.** Relation (27) remains true also for each operator $\mathbf{L}_k$ generated by a function $k(t, \tau)$ satisfying the Hölder condition in both variables and also the symmetry condition

$$k(t, \tau) \equiv k(\tau, t). \qquad (28)$$

**Remark 3.2.** For each function $x \in \mathbf{L}_M$ taking either only real values or only purely imaginary values the relation (27) can be rewritten in the form

$$\langle \mathbf{S}x, x\rangle = 0. \qquad (29)$$

**Remark 3.3.** The relation (29) and even weaker relation

$$\operatorname{Re} \langle Sx, x \rangle = 0. \tag{30}$$

is not valid in general for any function $x \in L_M$.

The conclusions of the Remarks 3.1, 3.2 can be checked directly. Let us gives a consideration led to the conclusion of the Remark 3.3. If $x \in L_M$ admits an analytic continuation into domain $D^+ \equiv \operatorname{int} \gamma$ (i.e. $x \equiv \varphi^+$), then $(S\varphi^+)(t) = \varphi^+(t)$. Thus

$$\langle S\varphi^+, \varphi^+ \rangle = \int_\gamma \left| \varphi^+(t) \right|^2 dt.$$

The last integral does not vanish. In the case of contour $\gamma$ being the unit circle $\mathbb{T}$ the above conclusion can be performed by using Fourier series expansion.

It follows from Remark 3.3 that in contrast to the results on nonlinear singular equation presented in [3,10] we need in our situation certain "monotonicity" assumptions either on operator $\mathbf{f}$, or on $\mathbf{L}_k$.

## 4. Superposition operator

The following Lemmas is simply restatement of Lemma 2 and 3 [22]. The proof of the first Lemma follows from the general properties of the superposition operators (see e.g. [2]). The second one is due to the definition of the norm in Orlicz spaces and general properties of convex functions (see e.g. [14]).

**Lemma 4.1.** *Let $\mathcal{X}$, $\mathcal{X}'$ be a pair of dual Banach spaces. Let nonlinear function $f(s, u)$ be a Caratheodory function generating superposition operator $\mathbf{f}$ acting from $\mathcal{X}$ into $\mathcal{X}'$. Then the operator $\mathbf{f}$ is hemicontinuous on $\mathcal{X}$.*

**Lemma 4.2.** *Let $L_M \subset L_{M'}$ be a pair of dual Orlicz spaces generated by an $N$-functions $M$, $M'$. The superposition operator $\mathbf{f}$ generated by a nonlinear function $f(t, u)$ is well-defined on the ball $B(0, R)$ of the space $L_M$ and acting from $L_M$ into $L_{M'}$ iff the function $f(t, u)$ satisfies the following inequality*

$$M'[af(t, u)] \le bM(R^{-1}u) + c(t), \tag{31}$$

*where $a, b$ are real constants and $c(\cdot)$ is an integrable function.*

## 5. Solvability of nonlinear singular integral equation

In this section we apply Theorem 2.1 to a nonlinear operator equation of type (5) related to the nonlinear singular integral equation (1). Let us suppose that the kernel function $k$ satisfies the conditions of Lemma 3.2. Then the operator equation (17) can be rewritten in the equivalent form

$$\Phi x = h, \tag{32}$$

where
$$\Phi \equiv \mathbf{L}_k^{-1} - \lambda \mathbf{f}, \quad h \equiv \mathbf{L}_k^{-1} g. \tag{33}$$

**Lemma 5.1.** *Let the kernel function $k : \gamma \times \gamma \to \mathbb{C}$ satisfies conditions (20)–(22). Let the function $f(t, u)$ be a Caratheodory function satisfying the relation (31).*

*If the nonlinear superposition operator $\mathbf{f}$ satisfy the following inequalities*

$$\operatorname{Re} \langle \mathbf{f}u - \mathbf{f}v, u - v \rangle \geq a_- \langle u - v, u - v \rangle, \quad \forall u, v \in L_M, \tag{34}$$

*where $a_-$ is a real constant, then for all negative real $\lambda < 0$ such that*

$$\delta_- - \lambda a_- \geq 0 \ (> 0) \tag{35}$$

*the operator $\Phi = \mathbf{L}_k^{-1} - \lambda \mathbf{f}$ is monotone (strictly monotone) as an operator acting from $L_M$ to $L_{M'}$, and for all positive real $\lambda > 0$ such that*

$$\lambda a_- - \delta_+ \geq 0 \ (> 0) \tag{36}$$

*the operator $-\Phi = \lambda \mathbf{f} - \mathbf{L}_k^{-1}$ is monotone (strictly monotone) as an operator acting from $L_M$ to $L_{M'}$.*

The proof of Lemma follows immediately from the general form of the real linear functionals on (complex) Orlicz spaces.

It is not hard to give sufficient condition for inequalities (34) in terms of nonlinear function $f(t, u)$ (see e.g. [22]), namely

$$a_- = \inf_{t \in \gamma} \inf_{u, v \in \mathbb{C}, u \neq v} \operatorname{Re} \frac{f(t, u) - f(t, v)}{u - v}. \tag{37}$$

**Lemma 5.2.** *Let the kernel function $k : \gamma \times \gamma \to \mathbb{C}$ satisfies conditions (20)–(22). Let the function $f(t, u)$ be a Caratheodory function satisfying the relation (31).*

*Let the following inequality*

$$\operatorname{Re} f(t, u)u \geq M(u) - d(t), \quad \forall t \in \gamma, \ \forall u \in \mathbb{C}, \tag{38}$$

*holds for certain N-function $M$ and integrable function $d(\cdot)$.*

*If $\delta_- \geq 0$, then there exists sufficiently large $R_0 > 0$ such that for any $\lambda < 0$ and any $R = \|u\|_{L_M} > R_0$ the operator $\Phi = \mathbf{L}_k^{-1} - \lambda \mathbf{f}$ satisfies the Rothe condition on the ball $B(0, R) \subset L_M$.*

*If $\delta_+ \leq 0$, then there exists sufficiently large $R_0 > 0$ such that for any $\lambda > 0$ and any $R = \|u\|_{L_M} > R_0$ the operator $-\Phi = \lambda \mathbf{f} - \mathbf{L}_k^{-1}$ satisfies the Rothe condition on the ball $B(0, R) \subset L_M$.*

**Proof.** Show first the validity of the Rothe condition for the operator $\Phi$ for any real $\lambda < 0$.

It follows from (24), (38) that

$$\operatorname{Re} \langle \Phi u, u \rangle \geq \delta_- \|u\|_{\mathcal{L}_2}^2 + (-\lambda) \left( \operatorname{Re} \int_\gamma f(t, u) u \, dt - \|d\|_{\mathcal{L}} \right) \geq \delta_- \|u\|_{\mathcal{L}_M}^2$$

$$+ (-\lambda) \left( (R - \varepsilon) \int_\gamma M \left( (R - \varepsilon)^{-1} u(t) \right) dt - \|d\|_{\mathcal{L}} \right) \geq \delta_- R^2 - \lambda (R - \varepsilon - \|d\|_{\mathcal{L}}).$$

The last expression is nonnegative for each sufficiently large $R$. It gives the fulfilment of the first assertion of Lemma for the balls in Luxemburg norm. The same is true also for balls in Orlicz-Amemiya norm due to their equivalence.

In the same manner we obtain in the case $\lambda > 0$ that

$$\operatorname{Re}\langle -\Phi u, u \rangle \geq \lambda(R - \varepsilon - \|d\|_{\mathcal{L}}) - \delta_+ \|u\|_{\mathcal{L}_2}^2 \geq \lambda(R - \varepsilon - \|d\|_{\mathcal{L}}).$$

It completes the proof. $\qquad\qquad\qquad\qquad\qquad\qquad\qquad\qquad\qquad\qquad\qquad\qquad\qquad\square$

Combining the results of Lemmas 3.2, 5.1, 5.2 and applying Theorem 2.1 either to the operator $\Phi = \mathbf{L}_k^{-1} - \lambda \mathbf{f}$ or to the operator $-\Phi = \lambda \mathbf{f} - \mathbf{L}_k^{-1}$ we obtain the main theorem of the paper

**Theorem 5.1.** *Let the kernel function* $k : \gamma \times \gamma \to \mathbb{C}$ *satisfies conditions (20)–(22). Let the function* $f(t, u)$ *be a Caratheodory function satisfying the relations (31) and (38).*

*If for* $\lambda < 0$ *the inequality (35) is satisfied with certain* $\delta_- \geq 0$ *(for* $\lambda > 0$ *the inequality (36) is satisfied with certain* $\delta_+ \leq 0$*) then the corresponding nonlinear singular (1) has solution for each* $g \in \mathcal{L}_{M'} \cap \Phi(B(0, R))$*. This solution is unique if (35) is a strict inequality ((36) is a strict inequality).*

### References

1. J. Appell, Upper estimates for superposition operators and some applications, *Ann. Acad. Sci. Fenn., Ser. A I.* **8**, 149-159 (1983).
2. J. Appell and P. P. Zabreiko, *Nonlinear Superposition Operators.* Cambridge University Press, Cambridge (1990).
3. S. N. Askhabov, Singular integral equations with monotone nonlinearity in complex Lebesgue spaces, *Z. Anal. Anwend.* **11**, No.1, 77-84 (1992).
4. I. A. Bakhtin, *The method of monotone approximations in the theory of nonlinear equations.* Voronezhskij Gos. Ped. Inst., Voronezh (1989).
5. H. Brezis and F. E. Browder, Maximal monotone operators in nonreflexive Banach spaces and nonlinear equations of Hammerstein type, *Bull. Amer. Math. Soc.* **81**, 82-88 (1975).
6. F. E. Browder, Singular nonlinear equations of Hammerstein type, *Lecture Notes Math.* **446**, 75-95 (1975).
7. F. D. Gakhov, *Boundary value problems.* Nauka, Moscow, 3rd ed. (1977) (in Russian).
8. I. Gohberg and N. Krupnik, *One-dimensional linear singular integral equations. Vol. II: General theory and applications.* Birkäuser Verlag, Basel, 1992.
9. A. I.. Gusejnov and Kh. Sh. Mukhtarov, Application of the monotone operators method to a class of singular integral equations, *Dokl. AN Azerb. SSR.* **35**, 3-6 (1979) (in Russian).
10. A. I.. Gusejnov and Kh. Sh. Mukhtarov, *Introduction to the Theory of Nonlinear Singular Equations*, Nauka, Moscow (1980) (in Russian).
11. Deng Yaohua, Singular integral equations in Orlicz spaces. I. *Acta Math. Sci.* **3**, 71-83 (1983).
12. L. V. Kantorovich and G. P. Akilov, *Functional Analysis.* 3rd ed., Nauka, Moscow (1984) (in Russian).

13. B. V. Khvedelidze, The method of Cauchy-type integrals in the discontinuous boundary value problems of the theory of holomorphic functions of a complex variable. *J. Sov. Math.* **7**, 309-414 (1977).

14. M. A. Krasnosel'skij and Ya. B. Rutitskij, *Convex functions and Orlicz spaces*. Fizmatgiz, Moscow (1958) (in Russian).

15. M. A. Krasnosel'skij and P. P. Zabreiko, *Geometrical methods of nonlinear analysis*. Springer-Verlag, Berlin etc. (1984).

16. G. Minty, On a "monotonicity" method for the solution of nonlinear integral equations in Banach spaces, *Proc. Nat. Acad. Sci. USA*. **50**, 1038-1041 (1963).

17. T. Runst and W. Sickel, *Sobolev spaces of fractional order, Nemytskij operators and nonlinear partial differential equations*. de Gruyter, Berlin (1996).

18. M. M. Rao and Z. D. Ren, *Theory of Orlicz spaces*. Marcel Dekker, New York (1991).

19. S. G. Samko, On closed form solvability of singular integral equations. *Sov. Math. Dokl.* **10**, 1445-1448 (1969).

20. M. M. Vainberg, *Variational method and method of monotone operators in the theory of nonlinear equations*. John Wiley & Sons, Jerusalem-London (1973).

21. L. von Wolfersdorf, Some recent developments in the theory of nonlinear singular integral equations, *Z. Anal. Anwend.* **6**, 83-92 (1987).

22. P. P. Zabreiko and S. V. Rogosin, On solvability and unique solvability of nonlinear singular integral equations in Orlisz spaces, *Dokl. AN Belarusi.* **36**, No. 5, 398-402 (1992) (in Russian).

# ANALYTICAL AND NUMERICAL RESULTS FOR THE EFFECTIVE CONDUCTIVITY OF 2D COMPOSITE MATERIALS WITH RANDOM POSITION OF CIRCULAR REINFORCEMENTS *

E.V. PESETSKAYA and T. FIEDLER

*Centre for Mechanical Technology and Automation,*
*Campus Universitrio de Santiago,*
*3810-193 Aveiro, Portugal*
*E-mail: kate@ua.pt*

A. ÖCHSNER and J. GRÁCIO

*Centre for Mechanical Technology and Automation,*
*Department of Mechanical Engineering, University of Aveiro,*
*Campus Universitrio de Santiago,*
*3810-193 Aveiro, Portugal*
*E-mail: aoechsner@mec.ua.pt*

S.V. ROGOSIN

*Mechanical-Mathematical Faculty, Belarusian State University,*
*Fr.Skaryny ave, 4,*
*220050, Minsk, Belarus*
*E-mail: rogosin@bsu.by*

The effective conductivity of two-dimensional composite materials with a finite number of circular reinforcements in a cell is investigated for structures with different symmetrical arrangements: square and pseudofractal. Also structures with small random perturbation of the reinforcements are considered, and it is shown that in average symmetrical arrangements possess the minimal effective conductivity with respect to perturbed arrangements if the conductivity of reinforcements is larger than the conductivity of the matrix.

**Key words:** Composite materials, boundary value problems, method of functional equations, R-linear conjugation problem, complex potentials

**Mathematics Subject Classification:** 30E25, 35B27, 74Q05

## 1. Introduction

The effective conductivity of two-dimensional composite materials with a finite number of circular disjoint reinforcements is investigated. Using homogenisation concepts [2], it is possible to represent composite materials as a structure consisting of

---

*This work is supported by TEMA (c.c. 3.66.5 - UI 481/98)

periodically located cells occupied by reinforcements. These reinforcements can generate different arrangements, e.g. square or pseudofractal. 2D composite materials with such structures are mathematically considered as multiply connected domains. Physical-mathematical fields are described in terms of the solution of a boundary value problem of the potential theory for multiply connected domains. Macroparameters of a structure are explicitly defined as functionals on the solutions of functional equations for an appropriate boundary value problem. The macroparameters contain all necessary information about the structure. This approach allows to construct model composite materials which are close to real reinforcement arrangements.

In this paper, a boundary value problem is analytically and numerically solved for structures with a square arrangement. For structures with a pseudofractal arrangement, an analytical description of the effective conductivity is derived. Independently obtained analytical and numerical results are compared, and an excellent correlation is found. To get numerical results for the effective conductivity, the finite element method is used [9]. An analytical solution of the boundary value problem is found by the method of functional equations [5].

A random perturbation of the reinforcements in a cell is considered for both arrangements. The reinforcements move independently of each other with restriction that they lie in a prescribed part of the original cell and do not cross or touch each other. It is shown that in the case when the conductivity of reinforcements is larger than the conductivity of the matrix, in average a small perturbation increases the effective conductivity of a composite material as a whole.

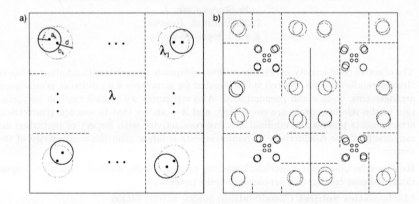

Fig. 1.   a) square arrangement; b) fractal arrangement.

## 2. The boundary value problem and the effective conductivity

A lattice $L$ which is defined by the two fundamental translation vectors 1 and $i$ ($i^2 = -1$) in the plane $\mathbb{C} \cong \mathbb{R}^2$ of the complex variable $z = x + iy$ is considered. According to the homogenisation concepts to describe properties of a composite material as a whole, it is enough to do it for one of its cell which is called the fundamental. The fundamental cell $Q_{(0,0)}$ is the square $\{z = t_1 + \imath t_2 \in \mathbb{C} : -\frac{1}{2} < t_p < \frac{1}{2}, p = 1, 2\}$. Let $\mathcal{E} := \bigcup_{m_1,m_2} \{m_1 + \imath m_2\}$ be the set of the lattice points, where $m_1, m_2 \in \mathbf{Z}$. The cells corresponding to the points of the lattice $L$ are denoted as $Q_{(m_1,m_2)} = Q_{(0,0)} + m_1 + \imath m_2 := \{z \in \mathbb{C} : z - m_1 - \imath m_2 \in Q_{(0,0)}\}$. First, the situation when $N$ mutually disjoint disks of equal radii $D_k := \{z \in \mathbb{C} : |z - a_k| < r\}$ $(k = 1, 2, \ldots, N)$ are located inside of fundamental cell $Q_{(0,0)}$, and periodically repeated in all cells $Q_{(m_1,m_2)}$ is considered. The boundary of the corresponding reinforcements is denoted by $T_k := \{z \in \mathbb{C} : |z - a_k| = r\}$ $(k = 1, 2, \ldots, N)$, and considered the connected domain $D_0 := Q_{(0,0)} \backslash (\bigcup_{k=1}^{N} D_k \cup T_k)$ obtained by removing of the reinforcements from the fundamental cell $Q_{(0,0)}$.

The problem is to define the effective conductivity of the doubly periodic composite material with the matrix $\mathcal{D}_{per} = \bigcup_{m_1,m_2} ((D_0 \cup \partial Q_{(0,0)}) + m_1 + \imath m_2)$ and reinforcements $\mathcal{D}_r = \bigcup_{m_1,m_2} \bigcup_{k=1}^{N} (D_k + m_1 + \imath m_2)$ occupied by materials of conductivities $\lambda$ and $\lambda_1 > 0$, respectively. This problem is equivalent to the determination of the potential of the corresponding field, i.e. to find a function $u(z)$ satisfying the Laplace equation in each component of the composite material:

$$\nabla^2 u = 0, \quad z \in \mathcal{D}_{per} \bigcup \mathcal{D}_r, \tag{1}$$

and conjugation conditions:

$$u^+(t) = u^-(t), \quad \lambda \frac{\partial u^+}{\partial n}(t) = \lambda_1 \frac{\partial u^-}{\partial n}(t), \tag{2}$$

$t \in \bigcup_{m_1,m_2} T_k$, $k = 1, 2, \ldots, N$, where $\frac{\partial}{\partial n}$ is the outward normal derivative and $u^+(t) := \lim_{z \to t, z \in D_0} u(z)$, $u^-(t) := \lim_{z \to t, z \in D_k} u(z)$, $t \in \bigcup_{m_1,m_2} T_k$, $k = 1, 2, \ldots, N$. The equalities of Eq. (2) form so-called ideal contact conditions on the boundary between the reinforcements and the matrix.

The quasi periodicity conditions are also imposed on $u(z)$, where $u(z)$ has a constant jump in the direction of each fundamental translation vector: $u(z + 1) = u(z) + 1$, $u(z + \imath) = u(z)$.

Furthermore, the complex potentials $\varphi(z)$, $\varphi_k(z)$ which are analytic in $D_0$ and $D_k$ and continuously differentiable in the closures of $D_0$ and $D_k$, respectively, are introduced. The harmonic potential $u$ and complex analytic potentials $\varphi$ and $\varphi_k$ are related by the equalities

$$u(z) = \begin{cases} \Re(\varphi(z) + z), \ z \in D_0, \\ \frac{2\lambda}{\lambda + \lambda_1} \Re\varphi_k(z), \ z \in D_k, \end{cases} \tag{3}$$

$k = 1, 2, \ldots, N$.

Conditions Eq. (2) can be rewritten in terms of the complex potentials [11]

$$\varphi(t) = \varphi_k(t) - \rho \ \overline{\varphi_k(t)} - t, \quad |t - a_k| = r, \tag{4}$$

$k = 1, 2, \ldots, N$, where $\rho = \frac{\lambda_1 - \lambda}{\lambda_1 + \lambda}$ is the contrast parameter which express the difference between the conductivities of both materials. The problem of Eq. (4) is a particular case of the so-called $\mathbb{R}$-linear conjugation problem [5].

To determine the flux $\nabla u(x, y)$, it is necessary to obtain the derivatives of the complex potentials:

$$
\begin{cases}
\psi(z) := \frac{\partial \varphi}{\partial z} = \frac{\partial u}{\partial x} - i\frac{\partial u}{\partial y}, & z \in D_0 \\[2mm]
\psi_k(z) := \frac{\partial \varphi_k}{\partial z} = \frac{\lambda_1 + \lambda}{2\lambda}\left(\frac{\partial u}{\partial x} - i\frac{\partial u}{\partial y}\right), & z \in D_k.
\end{cases}
\tag{5}
$$

Differentiating Eq. (4), the following problem is obtained

$$\psi(t) = \psi_k(t) + \rho \left(\frac{r}{t - a_k}\right)^2 \overline{\psi_k(t)} - 1, \quad |t - a_k| = r, \quad k = 1, 2, \ldots, N, \tag{6}$$

which can be solved by the method of functional equations [5]. In this problem, $N$ contours $T_k$ and $N$ conjugation conditions on each contour $T_k$ are considered. It is necessary to find $N + 1$ functions $\psi, \psi_1, \psi_2, \ldots, \psi_N$, i.e. satisfying $N$ conditions, so it is necessary to complete the system. The main idea is on the base of functions $\psi, \psi_1, \psi_2, \ldots, \psi_N$ to define a new function $\Phi$ (see formula Eq. (7) below) such that $\Phi$ is analytic on $Q_{(0,0)}$ and $\bigcup\limits_{k=1}^{N} D_k$, has a zero jump along each $T_k$ $(k = 1, 2, \ldots, N)$, and is doubly periodic on $\mathbb{C}$. Applying Liouville's theorem for doubly periodic functions, it can be concluded that $\Phi(z) = c$ (actually, $c = 0$ [1]). This gives an additional condition on $\psi, \psi_1, \psi_2, \ldots, \psi_N$.

For a description of the realisation of this idea, the Banach space $C_k$ of the functions continuous on $T_k$ with the norm $\|\psi_k\| := \max\limits_{T_k} |\psi_k(t)|$, $k = 1, 2, \ldots, N$, and the closed subspace $C_k^+ \subset C_k$ of functions $T_k$ which admit an analytic continuation into $D_k$ are considered. It is also introduced the Banach space $C^+$ consisting of the functions $\Psi(t) := \psi_k(t) \in C_k^+$ for all $k = 1, 2, \ldots, N$ with the norm $\|\Psi\| := \max\limits_{k} \|\psi_k\|$ (i.e., $\Psi(t)$ is defined in all disks $T_k$).

Furthermore, an auxiliary function is introduced by the following formula:

$$
\Phi(z) =
\begin{cases}
\psi_m(z) - \rho \sum\limits_{k=1}^{N} \sum\limits_{m_1, m_2} {}^{*}(W_{m_1, m_2, k}\psi_k)(z) - 1, & |z - a_m| \leq r, \ m = 1, 2, \ldots, N, \\[4mm]
\psi(z) - \rho \sum\limits_{k=1}^{N} \sum\limits_{m_1, m_2} (W_{m_1, m_2, k}\psi_k)(z), & z \in D_0,
\end{cases}
\tag{7}
$$

where

$$(W_{m_1, m_2, k}\psi_k)(z) = \left(\frac{r}{z - a_k - m_1 - im_2}\right)^2 \overline{\psi_k\left(\frac{r^2}{t - a_k - m_1 - im_2} + a_k\right)},$$

and

$$\sum_{k=1}^{N}\sum_{m_1,m_2} {}^*W_{m_1,m_2,k} := \sum_{k\neq m}\sum_{m_1,m_2} W_{m_1,m_2,k} + \sum_{m_1,m_2} {}' W_{m_1,m_2,m}.$$

In the sum $\sum_{m_1,m_2}'$ the integer numbers $m_1, m_2$ range from $-\infty$ to $+\infty$ except $m_1^2 + m_2^2 = 0$. Since $\Phi(z) = 0$, the following system of linear functional equations is obtained

$$\psi_m(z) = \rho \sum_{k=1}^{N}\sum_{m_1,m_2} {}^*(W_{m_1,m_2,k}\psi_k)(z) + 1, \ |z - a_m| \leq r, \ m = 1, 2, \ldots, N,$$

with respect to $\psi_m \in C_m^+$. This system can be considered as an equation for the function $\Psi(z)$ in the space $C^+$

$$\Psi(z) = \rho \sum_{k=1}^{N}\sum_{m_1,m_2} {}^*(W_{m_1,m_2,k}\Psi)(z) + 1, \ z \in \bigcup_{m=1}^{N} (D_m \cup T_m), \qquad (8)$$

where $\Psi(z) = \psi_m(z)$ in $|z - a_m| \leq r, \ m = 1, 2, \ldots, N$. The functional Eq. (8) has a unique solution in $C^+$ [1,5].

The function $\psi_m(z)$ can be determined as an analytic function in variable $r^2$ in the disk $V$ of the radius $r_0^2$, where $r_0^2$ is the minimum radius of the disks with the given set of the centres $a_k \ (k = 1, 2, \ldots, N)$ for which at least two disks are touching [1]. Thus, it is possible to express $\psi_m(z)$ in the form of series in $r^2$:

$$\psi_m(z) = \psi_m^{(0)}(z) + r^2\psi_m^{(1)}(z) + r^4\psi_m^{(2)}(z) + \ldots, \qquad (9)$$

where each term $\psi_m^{(s)}(z)$ is expanded into the Taylor series

$$\psi_m^{(s)}(z) = \sum_{l=0}^{\infty} \psi_{lm}^{(s)}(z - a_m)^l. \qquad (10)$$

The method of functional equations allows to find a simple recursive algorithm which defines the flux $\nabla u(x,y)$ in terms of Eisenstein's functions $E_l(z) = \sum_{m_1,m_2,m_1+\imath m_2 \neq 0} (z - m_1 - \imath m_2)^{-l}$ of order $l$, modified Eisenstein's functions $\sigma_l = E_l(z) - z^{-l}$ of order $l$, properties of which are studied in [8] in detail, and the contrast parameter $\varrho$ [1]:

$$\psi_m^{(0)}(z) = 1,$$

$$\psi_m^{(1)}(z) = \rho\Big[\sum_{k\neq m}^N \overline{\psi_{0k}^{(0)}}E_2(z-a_k) + \overline{\psi_{0m}^{(0)}}\sigma_2(z-a_m)\Big],$$

$$\psi_m^{(2)}(z) = \rho\Big[\sum_{k\neq m}^N \overline{\psi_{0k}^{(1)}}E_2(z-a_k) + \overline{\psi_{0m}^{(1)}}\sigma_2(z-a_m)$$

$$+ \sum_{k\neq m}^N \overline{\psi_{1k}^{(0)}}E_3(z-a_k) + \overline{\psi_{1m}^{(0)}}\sigma_3(z-a_m)\Big],$$

$$\ldots$$ (11)

$$\psi_m^{(p+1)}(z) = \rho\Big[\sum_{k\neq m}^N \overline{\psi_{pk}^{(0)}}E_{p+2}(z-a_k) + \overline{\psi_{pm}^{(0)}}\sigma_{p+2}(z-a_m)$$

$$+ \sum_{k\neq m}^N \overline{\psi_{p-1,k}^{(1)}}E_{p+1}(z-a_k) + \overline{\psi_{p-1,m}^{(1)}}\sigma_{p+1}(z-a_m) + \ldots +$$

$$+ \sum_{k\neq m}^N \overline{\psi_{0k}^{(p)}}E_2(z-a_k) + \overline{\psi_{0m}^{(p)}}\sigma_2(z-a_m)\Big].$$

The effective conductivity of an isotropic composite material with $N$ identical reinforcements is obtained in the form of series in the concentration $\nu = N\pi r^2$ and given by the equality [1]

$$\lambda_e = \lambda + 2\lambda\rho\nu\sum_{p=0}^{s}A_p\nu^p + o(\nu^{s+1}),$$ (12)

where the coefficients

$$A_p = \frac{1}{\pi^p N^{p+1}}\sum_{m=1}^N \psi_m^{(p)}(a_m), \quad p = 0, 1, 2, \ldots,$$ (13)

are represented as [10]

$$A_0 = 1,$$

$$A_1 = \frac{\rho}{\pi N^2} X_2,$$

$$A_2 = \frac{\rho^2}{\pi^2 N^3} X_{22},$$

$$A_3 = \frac{1}{\pi^3 N^4} [-2!\rho^2 X_{33} + \rho^3 X_{222}],$$

$$A_4 = \frac{1}{\pi^4 N^5} [3!\rho^2 X_{44} - 2!\rho^3 [X_{332} + X_{233}] + \rho^4 X_{2222}],$$

$$A_5 = \frac{1}{\pi^5 N^6} [-4!\rho^2 X_{55} + 3!\rho^3 [X_{442} + X_{343} + X_{244}] - 2!\rho^4 [X_{3322} + X_{2332} + X_{2233}],$$

$$+ \rho^5 X_{22222}],$$

$$A_6 = \frac{1}{\pi^6 N^7} (5!\rho^2 X_{66} - 4!\rho^3 [X_{255} + X_{354} + X_{453} + X_{552}] + 3!\rho^4 [X_{2244} + X_{2343} +$$

$$+ X_{2442} + X_{3432} + X_{4422}] + 4\rho^4 X_{3333} - 2!\rho^5 [X_{22233} + X_{22332} + X_{23322} + X_{33222}]$$

$$+ \rho^6 X_{222222}) \ldots$$

$$\text{(14)}$$

with $X_{p_1, \ldots, p_M}$ defined as

$$X_{p_1} := \sum_{m=1}^{N} \sum_{k \neq m} E_{p_1}(a_m - a_k) + \sigma_{p_1}(0),$$

$$X_{p_1, p_2} := \sum_{m=1}^{N} \left( \sum_{k \neq m} \sum_{k_1 \neq k} E_{p_1}(a_m - a_k) \overline{E_{p_2}(a_k - a_{k_1})} \right.$$

$$\left. + \sigma_{p_2}(0) \sum_{k \neq m} E_{p_1}(a_m - a_k) + \sigma_{p_1}(0)\sigma_{p_2}(0) \right)$$

and so on.

Numerical investigations of the effective conductivity of composites with square arrangement (cf. Figure 1a) were carried out using the finite element method. The basic idea of the finite element method is the decomposition of a domain with a complicated geometry into geometrically simple elements, such that the governing differential equation (e.g. describing a temperature field) can be approximately solved for these finite elements. The single element solutions are then assembled to obtain the complete system solution using given boundary conditions. The assembly process uses appropriate balance equations at the nodes which are used to define the elements and serve also as connection points between the elements. In the scope of the finite element approach, three different geometries have been considered, namely square arrangement of reinforcements for $\nu = 0.1, 0.3, 0.5$. Results obtained from both approaches are presented in Figure 2, and a good correlation of the results is found.

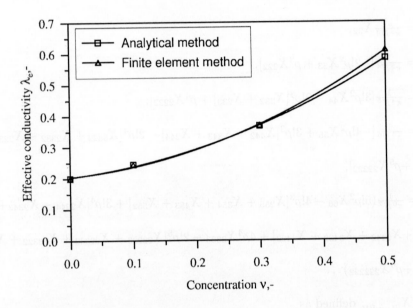

Fig. 2.   Effective conductivity $\lambda_e$ in dependence on the concentration $\nu$.

An analytical formula for the effective conductivity of composite materials with a fractal arrangement has the form of Eq. (12) [12], where coefficients $A_s, s = 1, 2, \ldots$, are defined as follows [6]

$$A_s = \frac{1}{16^{s+1} \pi^s \left( \sum\limits_{g=1}^{G} \frac{1}{9^{g-1}} \right)^{s+1}} \sum_{g=1}^{G} \sum_{h=1}^{4} \sum_{k=1}^{4} \frac{1}{9^{g-1}} \psi_{ghk}^{(s)}(a_{ghk}). \tag{15}$$

For a fractal structure, it is considered that the cell $Q_{(0,0)}$ is divided by four equal parts which are numbered in the counter-clock wise direction (cf. Figure 1b). Thus, $g$ denotes the generation of disk set, $G$ is the number of the generations, $h$ is the number of the part of the cell $Q_{(0,0)}$, $k$ is the number of the disk at a generation on the part of $Q_{(0,0)}$. The radius of a disk of the $g$ generation $D_{ghk} := \{z \in \mathbb{C} : |z - a_{ghk}| < r_g\}$ $(g = 1, \ldots, G, \ h, k = 1, \ldots, 4)$ is defined by the following relation $r_g = \frac{1}{3^{g-1}} r_1$, where $r_1$ is the radius of a disk of the first generation. The concentration is defined as

$$\nu = \sum_{g=1}^{G} \nu_g = 16 \pi r_1^2 \sum_{g=1}^{G} \frac{1}{9^{g-1}}.$$

For coefficients $A_s$, $s = 0, 1, 2, \ldots$, the functions $\psi^{(s)}_{ghk}(z)$ are calculated by the recursive algorithm of Eq. (11).

## 3. Effect of perturbation

First, the case of $N$ circular reinforcements of equal radii in the cell $Q_{(0,0)}$ is considered. Let $b_1, b_2, \ldots, b_N$ be the centres of these reinforcements which form a symmetric array in $Q_{(0,0)}$. $N$ identically distributed random variables $d_1, d_2, \ldots, d_N$ with uniform distribution inside the disk $|z| < d$ are considered, i.e. they have the probability density

$$f_k(z) := \begin{cases} \frac{1}{\pi d^2}, & |z| < d, \\ 0, & |z| \geq d. \end{cases}$$

Let $a_1, a_2, \ldots, a_N$ be the random variables defined by formula $a_k = d_k + b_k$, $k = \overline{1, N}$, where the variables $b_k$ possess the constant values $b_k$. In the following, the values of the above introduced random variables $a_k$ are denoted by $a_k$. Thus, the disks $D_k = \{z : |z - b_k| < d\}$ are randomly distributed inside the unit periodic cell $Q_{(0,0)}$. It is assumed that $d + r < \frac{1}{\sqrt{N}}$. This means that each disk lies in the prescribed part of the original square cell and does not cross or touch the boundary of corresponding part. In the following, the dependence of the effective conductivity upon the random positions of the reinforcements in the symmetric array is investigated.

For the sake of simplicity, each coefficient $\mathbf{A}_p$ in the representation $\lambda_e$ is decomposed as follows

$$\mathbf{A}_p = A_p^{(0)} + \widetilde{\mathbf{A}}_p, \tag{16}$$

where $A_p^{(0)}$ corresponds to the deterministic periodic structure ($d = 0$) and $\widetilde{\mathbf{A}}_p$ is the correction due to randomness of the positions of $a_k$. The expected value derived from the last equality is

$$\langle \mathbf{A}_p \rangle = A_p^{(0)} + \langle \widetilde{\mathbf{A}}_p \rangle. \tag{17}$$

Since $\lambda_e$ is linear in $A_p$, the same decomposition for $\lambda_e$: $\lambda_e = \lambda_e^{(0)} + \widetilde{\lambda}_e$ is obtained. Calculating the expected value from the latter relation, one gets

$$\langle \lambda_e \rangle = \lambda_e^{(0)} + \langle \widetilde{\lambda}_e \rangle. \tag{18}$$

The part $\lambda_e^{(0)}$ corresponds to the deterministic symmetric structure. The case of a single reinforcement in a square periodicity cell has been calculated in [3,4]. The expected value of $\langle \widetilde{\lambda}_e \rangle$ is defined by the formula:

$$\langle \widetilde{\lambda}_e \rangle = 2\rho\nu[\langle \widetilde{\mathbf{A}}_0 \rangle + \langle \widetilde{\mathbf{A}}_1 \rangle \nu + \langle \widetilde{\mathbf{A}}_2 \rangle \nu^2 + \ldots]. \tag{19}$$

Table 1. Effect of a small perturbation on the effective conductivity ($\lambda_1 = 190$, $\lambda = 0.2$).

| Concentration $\nu$ | Symmetric structure | Perturbed structure |
|---|---|---|
| square arrangement | | |
| 0.1 | 0.24425 | 0.24444 |
| 0.2 | 0.29892 | 0.29998 |
| 0.3 | 0.3679 | 0.37139 |
| 0.4 | 0.45822 | 0.46783 |
| 0.5 | 0.58292 | 0.60713 |
| fractal arrangement | | |
| 0.1 | 0.2438 | 0.24401 |
| 0.2 | 0.29678 | 0.29788 |
| 0.3 | 0.36198 | 0.36504 |
| 0.4 | 0.44426 | 0.45118 |
| 0.5 | 0.55145 | 0.56542 |

The coefficients $\langle \widetilde{\mathbf{A}_p} \rangle$ in Eq. (19) are defined as

$$\langle \widetilde{\mathbf{A}_p} \rangle := \int_{[Q_\iota(0,0)]^N} \widetilde{\mathbf{A}_p}(a_1, ..., a_N) \prod_{k=1}^{N} f_k(a_k) \, (d\sigma)^N, \tag{20}$$

where $p = 0, 1, 2, \ldots$. Here, the integration is performed with respect to the complex-valued variables $a_k = x + \iota y$ ($k = 1, 2, ..., N$) which are the values of random variables $a_k$.

It is shown in [7] that the expected values of the random correction of the coefficients $\mathbf{A}_p$ in the representation of Eq. (19) are non-negative

$$\langle \widetilde{\mathbf{A}_p} \rangle \geq 0, \qquad p = 0, 1, 2, 3, \ldots \tag{21}$$

for the contrast parameter $\rho > 0$.

For composite materials with fractal structure, identically distributed random variables $a_{ghk}$ have uniform distributions inside the disk $|z| < d_g$, i.e. the probability density has a form

$$f_{ghk}(z) := \begin{cases} \frac{1}{\pi d_g^2}, & |z| < d_g, \\ 0, & |z| \geq d_g. \end{cases}$$

In this case, equalities Eq. (16)-(20) hold [6]. It is also shown in [6] that the expected values of $\mathbf{A}_p$ are non-negative in the case when the contrast parameter $\rho > 0$.

It follows from Eq. (18), (19), (21) that in average a symmetric structure with circular reinforcements has a lower effective conductivity than a nonsymmetric structure which consists of reinforcements distributed in an arbitrary way. Thus, in average the effective conductivity $\lambda_e$ attains its minimum when $d = 0$.

## 4. Conclusions

The effective conductivity of 2D composite materials with different topology is investigated. Obtained values for the effective conductivity as a function of the concentration $\nu$ for composites with a square arrangement are compared with the values independently obtained by the numerical finite element method. An excellent correlation is shown in Figure 2.

Small random perturbations of the reinforcement position is also considered. It has been proven for different arrangements that for $\rho > 0$ (i.e. the reinforcements conductivity is more than the matrix conductivity) and disjoint reinforcements in average a small perturbation increases the effective conductivity.

## References

1. L. Berlyand and V. V. Mityushev, *J. Stat. Phys.* **102**(N112), 115-145 (2001).
2. V. V. Jikov, S. M. Kozlov and O. A. Olejnik, *Homogenization of Differential Operators and Integral Functionals* Springer Verlag, Berlin (1994).
3. R. C. McPhedran, *Proc. Roy. Soc. Lond. Ser. A* **408**, 31-43 (1986).
4. V. V. Mityushev, *ZAMM* **77**, 115-120 (1997).
5. V. V. Mityushev and S. V. Rogosin, *Constructive Methods for Linear and Nonlinear Boundary Value Problems for Analytic Functions. Theory and Application* Chapman & Hall/CRC, Boca Raton (2000).
6. E. V. Pesetskaya, *Proc. Inst. Math.* **12**, 117-122 (2004).
7. E. V. Pesetskaya, *Applicable Analysis*, 84, 843–865 (2005)..
8. A. Weil, *Elliptic Functions According to Eisenstein and Kronecker* Springer Verlag, Berlin (1976).
9. O. C. Zienkiewicz and R. L.Taylor, Finite Element Method, **1**, *Fifth ed., Butterworth-Heinemann, Oxford*, (2000).
10. Szczepkowski J., Malevich A. E. and Mityushev V., *Quart J. Appl. Math. Mech.* **56**, 617-628 (2003).
11. Mityushev V., *Complex Variables* **50**, N 7-10, 621-630 (2005).
12. Adler P. M. and Mityushev V., *Proc. Inst. Math.* **9**, 7-15 (2001).

# STRAINS IN TISSUE DEVELOPMENT: A VORTEX DESCRIPTION

R. WOJNAR

*IPPT Polska Akademia Nauk*
*ul. Świętokrzyska 21*
*00-049 Warszawa, Poland*
*E-mail: rwojnar@ippt.gov.pl*

On a biological example two methods of depicting dislocations are given: by Delone triangulations as pentagon-heptagon pairs and as the vortices in complex plane.

Some aspects of development of biological tissues in animals and plants can be described in a 2D space. The osteon growth and phyllotaxis are perhaps the best known examples. A phyllotaxial spiral growth of meristem by dislocations of primordia is discussed. It covers two types of phyllotaxial growth: one described by the Fibonacci sequence, and the second, more rare in nature described by the Lucas sequence. In this approach the dislocations can be interpreted as the vortices, and appear as poles of the meromorphic function $w = \frac{i}{2\pi} \sum_{k=1}^{n} \frac{\kappa_k}{z - z_{B_k}}$.

**Key words:** Growth, living tissue, dislocations, pentagons-heptagons
**Mathematics Subject Classification:** 92C15, 76B47, 65E05

## 1. Introduction

### 1.1. *Solid state and biological structures*

According to Alexander Braun (1831), Carl Schimper (1831), brothers Louis and Auguste Bravais (1837), Simon Schwendener (1878, 1883), Georgii Wulff (1908) and Frederic Thomas Lewis (1931, 1949) the crystallization and growing of living tissue proceed along a similar pattern, cf. e.g. [1].

In the present paper we regard that the dislocations gliding and climbing is basis for such similarity. In 2D packing, both in the crystals and the biological quasicrystals, the growth is realized by motion of pentagons and heptagons (5-7) among crystalline hexagons (6).

D'Arcy Thompson [2] emphasized the deep correlation between mathematical statements, physical laws and fundamental phenomena of organic growth of biological structures.

At the end of 'On Growth and Form' we read:

I know that in the study of material things number, order, and position are the threefold clue to exact knowledge.

Remarkable symmetry characterizes crystals and organic forms because both are subdued primarily to the topological laws of close - packing by the Descartes-Euler theorem. In an analogous manner as in propagation of defects during crystallization, the growing of tissue stress leads to buckling and undulation.

Recently, Karl J. Niklas and Brian J. Enquist [3], cf. also [4], presented empirical scaling relationships involving the rates of plant growth in species ranging from unicellular algae to large trees. Their analysis reveals growth scales among plants similar to those among animals, and further underscores the growing realization that the same scaling rules may apply to both animals and plants, and for much the same reasons.

## 1.2. *Vortices in nature*

The vortices occur at different levels of the nature organization, from subatomic dimensions to cosmic scale. Perhaps René Descartes was the first who observed this fact and postulated that the sun and planets are swept along in vortices of subtle ethereal matter in a constant motion.

The dislocation pairs of penta- and heptagons in 2D hexagonal lattice regarded as a continuous medium can also be interpreted as vortices.

## 1.3. *Mathematics of structure*

Primaries are units of a structure, such as atoms in crystal lattice, osteons in compact bone, primordia of leaves in meristem of plant. These units can be counted and in this meaning the natural numbers are fundamental for our knowledge, cf. e.g. Kronecker's dictum [5].

Important physical phenomena connected with change of phase, such as increase of crystal grains or growth of biological tissues are accomplished on 2D surfaces, interfaces or boundaries.

Thus it is natural to describe such phenomena using a complex plane in which topological defects are introduced. These defects are known as dislocations. They appear when plastic deformations described by the Burgers integral formula are observed.

## 1.4. *Organization of the paper*

As an example of 2D structure the growth process of the plant meristem is considered. The role of pairs of topological defects (pentagon-heptagon pairs) or pairs of vortices in the process is stressed.

The problem of growth of 2D structures is considered as:

(i) a development of spirals during tissue growth, and

(ii) a development of vortices.

In the case (i) a number theory is used, and in the case (ii) complex number theory is employed.

## 2. Dislocations in 2D

### 2.1. *2D elasticity*

Deformation of 2D body - monolayer of rigid centers of forces - is a state of plane stress. This can be seen as follows. The thermodynamic relation determines the stress from the free energy $F$

$$\sigma_{ij} = \left( \frac{\partial F}{\partial u_{ij}} \right)_T \qquad i, j = 1, 2 \tag{1}$$

where $T$ denotes the temperature and

$$u_{ij} = \frac{1}{2} \left( \frac{\partial u_i}{\partial x_j} + \frac{\partial u_j}{\partial x_i} \right)$$

is the strain tensor with $u_i$ being the displacement vector.

The expression for $F$ of an isotropic body in 2D elasticity is taken similar as in 3D theory

$$F = \frac{1}{2} \lambda u_{ii} u_{jj} + \mu u_{ij} u_{ij}$$

where $\lambda$ and $\mu$ are the Lamé coefficients. Decomposing a strain into a pure shear and hydrostatic compression, the last expression is replaced by

$$F = \frac{1}{2} \lambda \left( u_{kk} \right)^2 + \mu \left[ \left( u_{ij} - \frac{1}{2} u_{kk} \delta_{ij} \right) \left( u_{ij} - \frac{1}{2} u_{ll} \delta_{ij} \right) + \frac{1}{4} u_{kk} u_{ll} 2 \right]$$

(where $\delta_{ii} = 2$) or

$$F = \mu \left( u_{ij} - \frac{1}{2} \lambda u_{kk} \delta_{ij} \right)^2 + \frac{1}{2} K \left( u_{kk} \right)^2 \tag{2}$$

Here $K$ and $\mu$ are respectively the bulk modulus and the shear modulus. $K$ is related to the Lamé coefficients by

$$K = \lambda + \mu \tag{3}$$

The moduli $\mu$ and $K$ are always positive $\mu > 0, K > 0$.

Using (1) we find Hooke's law: the stress in terms of the strain tensor for an isotropic body

$$\sigma_{ij} = 2\mu \left( u_{ij} - \frac{1}{2} u_{kk} \delta_{ij} \right) + K u_{kk} \delta_{ij} \tag{4}$$

The converse formula expresses strain in terms of stress tensor

$$u_{ij} = \frac{1}{2\mu} \sigma_{ij} + \left( \frac{1}{4K} - \frac{1}{4\mu} \right) \sigma_{kk} \delta_{ij} \tag{5}$$

Young's modulus $E$ and Poisson's ratio $\nu$ are defined as

$$E = \frac{4K\mu}{K + \mu} \quad \text{and} \quad \nu = \frac{K - \mu}{K + \mu}$$

Hence

$$K = \frac{1 + \nu}{1 - \nu}\mu \quad \text{and} \quad 2\mu = \frac{E}{1 + \nu} \tag{6}$$

and

$$u_{ij} = \frac{1 + \nu}{E}\left(\sigma_{ij} - \frac{\nu}{1 + \nu}\sigma_{kk}\delta_{ij}\right) \qquad i, j = 1, 2 \tag{7}$$

This is Hooke's law written with coefficients $E$ and $\nu$.

## 2.2. Burgers vector and distorsion tensor

A dislocation deformation of a crystal regarded as a continuous medium has the property: after a passage round any closed contour which encircles the dislocation, the elastic displacement vector **u** receives a finite increment **b** which is equal to crystal lattice vector [6]

$$\oint du_i = \oint \frac{\partial u_i}{\partial x_j} dx_j = -b_i \qquad i, j = 1, 2 \tag{8}$$

After introducing the distortion tensor

$$w_{ji} = \frac{\partial u_i}{\partial x_j} \tag{9}$$

condition (8) becomes

$$\oint w_{ji} dx_j = -b_i \tag{10}$$

According to Stokes' theorem the contour integral can be transformed into one over a surface $S$ spanning this contour. For 2D problem the surface $S$ is perpendicular to $x_3$ axis and the theorem reads

$$\oint w_{ji} dx_j = \int_S e_{3pq}\frac{\partial w_{qi}}{\partial x_p} dS$$

where $e_{abc}$ is the antisymmetric unit tensor, or

$$\oint w_{ji} dx_j = \int_S T_{12i} dS$$

Here $T_{12i} = \partial w_{2i}/\partial x_1 - \partial w_{1i}/\partial x_2$ is tensor of dislocation density in 2D.

## 2.3. Dislocations in 2D

Structural units are denoted as centers.

Voronoi polygons (V) precisely describe contacts of centers and Delone triangulations, dual to V, describe a distribution of the centers.

Interaction of centers is revealed by flips (exchange of contacts).

Single flip in a pure hexagonal packing causes creation of two pentagons (5) and two heptagons (7). Global compensation is guaranteed in the planar close-packing by the Descartes-Euler theorem. Most often the 2D dislocation is composed by a pair: pentagon and heptagon, shortly 5-7. 2D dislocations correspond to edge dislocations in 3D.

The dislocation pair 5-7 (if 5 is up) goes in the same direction as the shear goes, and in the opposite direction if 5 is down.

In present case only pentagons (5), hexagons (6) and heptagons (7) are observed.

Disclination 5 may be interpreted as a vortex of orientations (well seen in colour scale) and 7 an opposite vortex. Reduction of disclinations to vortices only is, however, a simplification.

The 5 is 1/12 Gaussian curvature of the sphere, while the 7 is a saddle with negative Gaussian curvature of 1/12 Lobachevski plane curvature.

In 1868, from his microscopic study of plant meristems, the botanist Wilhem Hofmeister proposed that a new primordium always forms in the least crowded spot along the meristem ring, at the periphery of shoot apical meristem (SAM).

The emerging primordia areas are represented in our model penta-, hexa- and heptagons. Neighbouring pentagon and heptagon create edge dislocation (5-7) among hexagons. Motion of such edge dislocation (exchange of contacts) is called 5-7 flip.

## 3. Vortices on the plane: plane incompressible flow

Let $\mathbf{v} = (v_x, v_y)$, where $v_x = v_x(x, y, t)$ and $v_y = v_y(x, y, t)$ be the velocity of fluid depending on the Cartesian coordinates $x, y$. The fluid continuity equation $\operatorname{div} \mathbf{v} = 0$ is satisfied if

$$v_x = \frac{\partial \Psi}{\partial y} \qquad v_y = -\frac{\partial \Psi}{\partial x}$$

where $\Psi = \Psi(x, y, t)$ is a function, known as a flow or stream function. This means that the motion of fluid is described by equations

$$\dot{x} = \frac{\partial \Psi}{\partial y} \qquad \dot{y} = -\frac{\partial \Psi}{\partial x}$$

in which the role of Hamiltonian is played by the flow function $\Psi$.

In the stationary case the functions $\Psi(x, y)$ and $\mathbf{v}(x, y)$ can be written in terms of complex potentials. Introduce the complex variable $z = x + iy$ and denote $z_B = x_B + iy_B$. Then the function $\Omega(z) = \Phi(x, y) + i\Psi(x, y)$ analytic in $0 < |z - z_B| < r_0$ is called the complex potential corresponding to the velocity $\mathbf{v}(x, y)$. Here $\Phi(x, y)$ is harmonically conjugated to $\Psi(x, y)$, i.e., the Cauchy–Riemann equations hold $\Phi_x = \Psi_y$, $\Phi_y = -\Psi_x$. Then $\mathbf{v}(x, y)$ can be identified with complex function $\mathbf{v}(x, y) \equiv v_x + iv_y = \overline{\Omega'(z)}$, where the bar means the complex conjugation.

Especially, if

$$\Psi = -\frac{\kappa}{2\pi} \ln \frac{r}{r_0} \quad \text{where} \quad r^2 = (x - x_B)^2 + (y - y_B)^2 \quad \text{and} \quad r_0 = \text{const.}$$

the velocity field is

$$\mathbf{v} = \frac{\kappa}{2\pi}\left(-\frac{y-y_B}{r^2}, \frac{x-x_B}{r^2}\right) \tag{11}$$

and flow produces the vortex of intensity $\kappa$ localized at the point $(x_B, y_B)$.

The corresponding complex potential in this case has the form

$$\Omega_B(z) = -\frac{\kappa}{2\pi}i\ln(z-z_B) = \frac{\kappa}{2\pi}(\arg(z-z_B) - i\ln|z-z_B|), \tag{12}$$

where $\arg(z-z_B)$ is the argument of the complex number $z-z_B$. We cut the complex plane along the ray connecting the points $z = z_B$, $z = -\infty + iy_B$ and choose a branch of the argument in such a way that $-\pi < \arg(z-z_B) < \pi$.

For instance, $\arg(z-z_B) = \arctan\frac{y-y_B}{x-x_B}$ for $x - x_B > 0$. Here the definition of the complex logarithm is used

$$\ln(z-z_B) = \ln|z-z_B| + i\arg(z-z_B). \tag{13}$$

The velocity field is the potential one

$$\mathbf{v} = \nabla\Phi, \tag{14}$$

where

$$\Phi(x,y) = Re\Omega_B(z) = \frac{\kappa}{2\pi}\arg(z-z_B). \tag{15}$$

The vortex intensity is given by the velocity circulation

$$\kappa = \oint(v_x dx + v_y dy) \tag{16}$$

along an arbitrary closed path that encircles the point $(x_B, y_B)$ one time counterclockwise.

To show this, let the integration path be a circle centered at $(x_B, y_B)$. Then unit normal is $\mathbf{n} = [(x-x_B)/r, (y-y_B)/r]$ and unit tangent $\mathbf{t} = [-(y-y_B)/r, (x-x_B)/r]$. Moreover, in circular coordinates $r, \phi$, the circle arc element is $ds = rd\phi$. Then, if components of $\mathbf{v}$ are given by (11)

$$v_x dx + v_y dy = \frac{\kappa}{2\pi}\left(-\frac{y-y_B}{r^2}, \frac{x-x_B}{r^2}\right)\cdot\left(-\frac{y-y_B}{r}, \frac{x-x_B}{r}\right)rd\phi = \frac{\kappa}{2\pi}d\phi$$

and integration over the round angle, from 0 to $2\pi$, gives (16).

In the general case of few singularities we have the complex potential

$$\Omega(z) = \frac{-i}{2\pi}\sum_{k=1}^{n}\kappa_k\ln(z-z_{B_k}).$$

The velocity is represented in the complex form

$$\mathbf{v} = \frac{i}{2\pi}\sum_{k=1}^{n}\frac{\kappa_k}{z-z_{B_k}}. \tag{17}$$

## 4. Biological structures

One of the most striking aspects of symmetry in plants is in phyllotaxis - the arrangement of leaves on a stem or of flowers in the inflorescenses. It is an inter-disciplinary study mathematics, botanics and crystallography among others.The phyllotaxis should be properly studied at the shoot apical meristem (SAM). It is at the meristemic apex that the organs of shoot originate such as primordia of leaves, buds or flowers. From the position of the growing tip of plant, one might visualize the SAM as being propelled by a jet of tissue behind it as it advances.

The arrangement of leaf primordia at the SAM may be modelled as a cen-tric array of points in te plane. The points are arrayed along generative spiral or, equivalently, at the intersections of a set of secondary spirals of opposite chirality, parastichies. Position of each primordium is identified in polar coordinates by its radial distance $r$ from the center and its angle $\phi$ along the generative spiral.

Thus a clear and specific relationship is observed between the primordial phyl-lotactic pattern at the SAM and the phyllotaxis of the mature shoot: the relationship can be formulated as a transformation (mapping) of the primordial pattern into the mature pattern and the phyllotactic symmetry of the plant is inherent in its growth processes.

During the growth of SAM, most frequently as result of stretching, the wall of primordia move proportionally to their age. Then the Fibonacci spirals develop. However, sometimes (4% of cases) certain irregularity appears at beginning and a different stretching is developed; then we have the Lucas spirals.

The boths, the Fibonacci and Lucas spirals are built below by common relation-ship. Namely, in polar coordinates $r$ and $\varphi$ the $n$-th primordium has the position

$$r_n = \sqrt{n} \qquad \varphi_n = C \tag{18}$$

where for the Fibonacci case

$$C = C_F = 360^0 u \approx 222.5^0$$

or, what is equivalent for structural form

$$C = C'_F = \frac{360^0}{2 + u} \approx 137.5^0$$

and for the more rare (only 4% of cases) Lucas case

$$C = C_L = \frac{360^0}{3 + u} \approx 99.5^0$$

with

$$u = \frac{\sqrt{5} - 1}{2}$$

The corresponding structures are shown in Figs. 1-4.

## 5. Ring shaped grain boundary

In Fig.1 (and 3) the emerging primordia areas are shown by the Delone triangulation. Such areas will pass through exchange of contacts (5-7 dislocation glide - 5-7 flip) sequentially establishing the Fibonacci differencess 1,2,3,5,8,... (or respectively the Lucas differences 1,3,4,7, 11,...) up to highest number reached by a plant.

The pentagons are indicated by black dots, while heptagons are shown by white ones. The corresponding representation by vortices in complex plane, as the arguments of function of type (17), is represented in Figs.2 and 4. Ring of pentagons and heptagons with strictly equilibrated Fibonacci structure is clearly seen at the periphery of Fig.1 where we observe two complete circles of 5-7 pairs and two outer demonstrated partially.

It exchanges a single contact (during the appearance of next primordium) to next Fibonacci difference by the most gentle - Fibonacci structured - 5-7 flip.

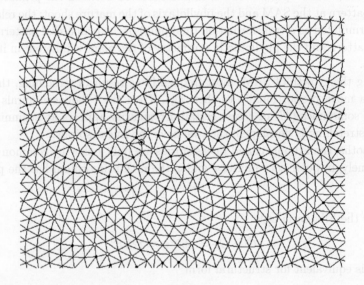

Fig. 1. Delone triangulation of primordia in meristem ordered according to the Fibonacci sequence. The big circle shown in the center indicates a point in which a virtual primordium appers. This virtual primordium has 5 neighbours.

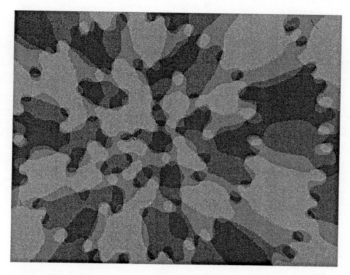

Fig. 2. The phase argument picture corresponding to Fig. 1. The 5-7 pairs are represented as dipoles of vortices.

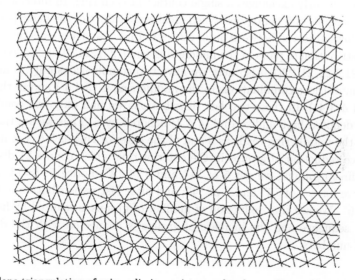

Fig. 3. Delone triangulation of primordia in meristem ordered according to the Lucas sequence.

In Fig. 1 and 2 we see the same rings of dislocations (overlapping in the center) in different representations. The dislocations appear in those rings either separately (1) or in pairs (2). The arrangement of the pairs in the interior ring is: 2|11122|1 (the vertical bars end the full rotation). We observe only two combinations: combination (2 1) denoted further by 3 and combination (221) denoted by 4. So the structure of this ring is : (3 4).

Near the center of SAM surface the complicated overlapping of such rings is seen

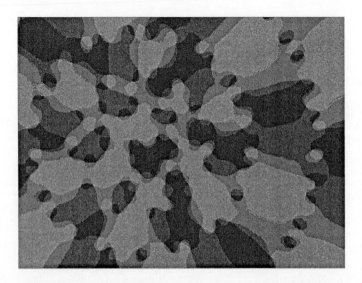

Fig. 4. The phase argument picture corresponding to Fig. 3.

but 5-7 flip similarly exchanges a single contact at each ring. In interior of the first circles it is surplus of 6 pentagons which creates curvature of the hemisphere. The 6 radial shadows observed between outer circles in Figs. 2 and 4 reveals this curvature.

The distribution of pairs in the middle ring is: 2|12212212|1 and in exterior ring: 22|1, 21, 221, 221, 21, 22|1. Also here we have two combinations only: 3 and 4. The distribution of those combinations in the middle ring is (3 4 4 ) and in the exterior ring: ( 3 4 4 3 4 ). In these two last distributions we find only two combinations: first (4 3) denoted by 5 and second (4 4 3) denoted by 6. Thus we see that the full period on the middle ring reads as 6 and of the exterior ring as 5 6. Let us consider for instance reduction of groups in the exterior ring which counts 21 dislocations; after the first grouping we get 5 of "1" and 8 of "2"; notice 5+8=13; after next grouping we have 2 of "3" and 3 of "4"; and after last redistribution it is one "5" and one "6". We see that reduction of number of groups is realized by hierarchical Fibonacci sequence and the increase of groups is connected with the decrease of Fibonacci numbers.

## 6. Final remarks

The unique opportunity to study biological systems is provided by phenomenon of spiral phyllotaxis - leaf primordia packing with Fibonacci (rarely with Lucas) differences between nearest neighbours. As they grow, older primordia are displaced radially away from the center of the circular meristem. The newest primordium initiates in the least crowded space at the edge of the meristem. The growth process is accomplished in an exceptional order. Phyllotaxis compromises local interactions giving rise to long range order and assures the best way of optimal close packing.

We propose natural extension of the Hofmeister rule in phyllotaxis as edge dislocation motion: close-packing of virtual areas of soon emerging primordia is approaching final Fibonacci differences through 5-7 flips from previous lower Fibonacci differences. In sequel, those virtual areas will be for short simply called "primordia". The actual, already established primordia, are called "real primordia". For example, the contact on the parastichy spiral with difference 5 (between 8 and 13 spirals) changes after 5-7 flip to the contact on the spiral with difference 21, cf. Fig.1. Ring of pentagonal and heptagonal promordia with strictly equilibrated Fibonacci structure appears at the periphery of SAM. It exchanges a single contact (during the emerging of next primordium) to next Fibonacci difference by the most gentle - fully Fibonacci structured - 5-7 flip. If a new primordium emerges from the meristem only one flip is performed in every ring - so there is so much flips as rings. As the effect of flip, ring rotates but distribution of dislocations in it never changes. Morphogenesis in plants is precisely controlled plastic deformation of cells, tissues and organs brought about by the growth processes. We observe at example of SAM development that:

(i) Hofmeister rule follows from addition of flips on consecutive dislocation rings;

(ii) Emerging of a new initial (central) virtual primordium area is associated with exactly one flip in each ring;

(iii) The higher Fibonacci (or Lucas) numbers are reached assuring the best geometrical packing.

## Acknowledgements

I am indebted to Professor V. V. Mityushev for corrections and remarks concerning this paper. The paper was partially supported by grant 4T07A 00327 of the Ministry of Scientific Research and Information Technology (MNiI), Poland.

## References

1. R. O. Erickson, Phyllotactic symmetry in plant growth, in: *Symmetry in Plants*, eds. R.V. Jean, D. Barabé, pp. xvii-xxvi, World Scientific, Singapore 1998.
2. D'Arcy W. Thompson ,*On Growth and Form*, Cambridge Univ. Press, 1917, 1942.
3. K. J. Niklas and B. J. Enquist, *Proc. Natl. Acad. Sci. USA (PNAS)* **98**, 2922 (2001).
4. J. Damuth, *PNAS* **98**, 2113 (2001).
5. H. Weber, *Jahresberichte der Deutschen Mathematiker-Vereingung* **2**, 5 (1891/92); also: *Math. Annalen* **43**, 1 (1893).
6. L. D. Landau and E. M. Lifshitz, *Theory of Elasticity*, Pergamon Press, Oxford 1970.

# AN ANALYTIC SOLUTION OF THE ORNSTEIN-ZERNIKE EQUATION FOR A FLUID AND ITS APPLICATION

M. YASUTOMI

*Department of Physics and Earth Sciences,*
*College of Science, University of the Ryukyus,*
*Nishihara-Cho, Okinawa 903-0213, Japan*
*E-mail:g800002@lab.u-ryukyu.ac.jp*

The physical properties of a fluid are determined by the interaction potentials between constituent particles. Recently, we have found an analytic solution of the Ornstein-Zernike equation for a multicomponent fluid with a *multi-Screened Coulomb Plus Power Series (multi-SCPPS)* closure:

$$c_{ij}(r) = \sum_n e^{-z_n r} \sum_{\tau=-1}^{L_n} K_{ij}^{(n,\tau)} z_n^{\tau+1} r^\tau, \quad \text{for } r \geq \sigma_{ij},$$

where $c_{ij}(r)$ are the direct correlation functions, $z_n$ and $K_{ij}^{(n,\tau)}$ are constants, $L_n$ are arbitrary integers, $r$ is the distance and $\sigma_{ij}$ is the distance of closest approach of two particles $i$ and $j$. The multi-SCPPS closure includes the multi-Yukawa closure as a special case. Our analytic solution offers sufficient flexibility and opens access to systems with any smooth, realistic isotropic potentials where the potentials can be fitted by the multi-SCPPS closure under the mean-spherical approximation (MSA). We will overview our recent analyses and apply the multi-SCPPS-MSA method to a fluid of Lennard-Jones particles as a demonstration.

**Key words:** Ornstein-Zernike equation, multi-screened coulomb plus power series, fluid Lennard-Jones particles **Mathematics Subject Classification:** 76B02, 82D20

## 1. Introduction

It is well known that the solution of the Ornstein-Zernike (OZ) equation with a closure is a very useful tool for the study of both simple and complex fluids. There are a number of problems, ranging from engineering applications to chemistry, biology, pharmaceutical industries and various medical disciplines, which can be formulated as closures of the OZ equation.

Analytic solutions of the OZ equation presented so far are restricted to those for the following three types of closures: $g_{ij}(r) = 0$ for $r < \sigma_{ij}$ and

(1) $c_{ij}(r) = 0$ (neutral hard-sphere),[1-5]
(2) $c_{ij}(r) = \Lambda_{ij}\delta(r - \sigma_{ij}^*)$ (sticky hard-sphere),[6-9] or
(3) $c_{ij}(r) = \sum_n K_{ij}^{(n)} e^{-z_n r}/r$ (multi-Yukawa tail)[10-14]

for $r > \sigma_{ij}$, where $g_{ij}(r)$ are the radial distribution functions, $c_{ij}(r)$ are the direct correlation functions, $\sigma_{ij}$ is the distance of closest approach of two particles $i$ and $j$, $\delta(r)$ is the Dirac delta function, $\Lambda_{ij}$, $z_n$ and $K_{ij}^{(n)}$ are constants, $\sigma_{ij}^* = \sigma_{ij} + 0^+$ ($0^+$: positive infinitesimal) and $r$ is the distance.

Recently, we have found an analytic solution of the OZ equation for a multi-component fluid with a sticky hard-sphere Yukawa closure:[15]

$$c_{ij}(r) = \Lambda_{ij}\delta(r - \sigma_{ij}^*) + K_{ij}e^{-zr}/r$$

for $r > \sigma_{ij}$ where $K_{ij}$ and $z$ are constants. Further, we have found that with a *multi-Screened Coulomb Plus Power Series (multi-SCPPS)* closure:[16,17–20]

$$c_{ij}(r) = \sum_{n=1}^{N} e^{-z_n r} \sum_{\tau=-1}^{L_n} K_{ij}^{(n,\tau)} z_n^{\tau+1} r^\tau \quad \text{for} \quad r \geq \sigma_{ij},$$

where $K_{ij}^{(n,\tau)}$ are constants, $N$ and $L_n$ are arbitrary integers. The multi-SCPPS closure includes the multi-Yukawa closure as a special case. The analytic solution offers sufficient flexibility and opens access to systems with any smooth, realistic isotropic potentials where the potentials can be fitted by the multi-SCPPS closure under the mean-spherical approximation (MSA). We will outline our recent analyses and apply the multi-SCPPS-MSA method to a fluid of Lennard-Jones (LJ) particles as a demonstration.[17–20]

## 2. An analytic solution of the OZ equation

In this section we summarise our recent works on the multi-SCPPS-MSA method introducing some new notations. Thermodynamic quantities of a fluid mixture will be derived from the total correlation function $h_{ij}(r)$ or the radial distribution function $g_{ij}(r) = 1 + h_{ij}(r)$ between constituent particles $i$ and $j$. These can be obtained by solving the OZ equation for the mixture written as

$$h_{ij}(r) = c_{ij}(r) + \sum_l \rho_l \int c_{il}(x)h_{lj}(|\mathbf{r} - \mathbf{x}|)d\mathbf{x}, \tag{1}$$

where $\rho_l$ represents the density of the component $l$ in the mixture.

Let us consider here the interaction potential for the mixture of hard spheres of diameter $\sigma_i$ with multi-SCPPS tails outside the core given by

$$\phi_{ij}(r) = \begin{cases} \infty & \text{for } r < \sigma_{ij}, \\ -k_\mathrm{B}T \sum_{n=1}^{N} e^{-z_n r} \sum_{\tau=-1}^{L_n} K_{ij}^{(n,\tau)} z_n^{\tau+1} r^\tau & \text{for } r \geq \sigma_{ij}, \end{cases} \tag{2}$$

where $\sigma_{ij} = (\sigma_i + \sigma_j)/2$, $k_\mathrm{B}$ is the Boltzmann constant and $T$ is a temperature.

In the mean spherical approximation (MSA) the closure for the OZ equation is given by the following two sets of equations:

$$h_{ij}(r) = -1 \quad \text{for} \quad r < \sigma_{ij}, \tag{3}$$

$$c_{ij}(r) = -\frac{\phi_{ij}(r)}{k_\mathrm{B}T} \quad \text{for} \quad r \geq \sigma_{ij}. \tag{4}$$

where Eq. (3) reflects the impenetrability to the hard cores and Eq. (4) accounts for the influence of the interaction potential outside the cores.

We will use Baxter's factorized version of the OZ equation (1) for mixtures as given by Blum and Høye:[11]

$$2\pi r h_{ij}(r) = -Q'_{ij}(r) + 2\pi \sum_l \rho_l \int_{\lambda_{jl}}^{\infty} (r-t) h_{il}(|r-t|) Q_{lj}(t) dt, \qquad (5)$$

$$2\pi r c_{ij}(r) = -Q'_{ij}(r) + \sum_l \rho_l \int_{\lambda_{lj}}^{\infty} Q_{jl}(t) Q'_{il}(r+t) dt, \qquad (6)$$

where $\lambda_{jl} = (\sigma_j - \sigma_l)/2$ and $Q_{ij}(r)$ are the factor correlation functions, $Q'_{ij}(r)$ being their derivatives.

The functions $Q_{ij}(r)$ must have the following form:

$$Q_{ij}(r) = Q^0_{ij}(r) + 2\pi \sum_{n=1}^{N} e^{-z_n r} \sum_{\tau=-1}^{L_n} K^{(n,\tau)}_{n_1,n_2} \mathcal{D}^{(\tau)}_{ij}(z_n) z_n^\tau r^{\tau+1}, \qquad (7)$$

where $K^{(n,\tau)}_{n_1 n_2}$ are normalization constants,

$$Q^0_{ij}(r) = \frac{1}{2} A_j (r^2 - \sigma_{ij}^2) + B_j(r - \sigma_{ij})$$
$$+ 2\pi \sum_{n=1}^{N} \sum_{k=0}^{L_n+1} z_n^{k-1} \breve{C}^{(k)}_{ij}(z_n) k! \sum_{\xi=0}^{k} \frac{r^{k-\xi} e^{-z_n r} - \sigma_{ij}^{k-\xi} e^{-z_n \sigma_{ij}}}{z_n^\xi (k-\xi)!} \qquad (8)$$

for $\lambda_{ji} \le r < \sigma_{ij}$ and $Q^0_{ij}(r) = 0$ otherwise. The coefficients $A_j$, $B_j$ and $\breve{C}^{(k)}_{ij}(z_n)$ in Eq. (8) are given in Appendix A of Ref. 20.

The two sets of Eqs. (5) and (6) can be led, respectively, to the following two sets of algebraic equations:

$$F^{(m)}_{1ij}(z_k) = \breve{A}^{(m)}_{ij}(z_k) A_j + \breve{B}^{(m)}_{ij}(z_k) B_j + \sum_{n=1}^{N} \sum_{\tau=0}^{L_n+1} \sum_l \breve{C}^{(m,\tau)}_{ilj}(z_k, z_n) \breve{C}^{(\tau)}_{lj}(z_n)$$
$$+ \sum_{n=1}^{N} \sum_{\tau=-1}^{L_n} \sum_l \breve{D}^{(m,\tau)}_{ilj}(z_k, z_n) \mathcal{D}^{(\tau)}_{lj}(z_n) K^{(n,\tau)}_{n_1 n_2} - e^{\mathcal{G}^{(m+1)}_{ij}(z_k)} = 0, \qquad (9)$$

and

$$F^{(m)}_{2ij}(z_n) = -\frac{K^{(n,m-1)}_{ij}}{K^{(n,m-1)}_{n_1 n_2}} + \sum_l \mathcal{D}^{(m-1)}_{il}(z_n) \left[\delta_{jl} - \rho_l \tilde{Q}^{(0)}_{jl}(iz_n)\right]$$
$$- (m+1) \sum_l \mathcal{D}^{(m)}_{il}(z_n) \frac{K^{(n,m)}_{n_1 n_2}}{K^{(n,m-1)}_{n_1 n_2}} \left[\delta_{jl} + z_n \rho_l \tilde{Q}^{(1)}_{jl}(iz_n) - \rho_l \tilde{Q}^{(0)}_{jl}(iz_n)\right]$$
$$+ \sum_{\tau=m+1}^{L_n} \sum_l \mathcal{D}^{(\tau)}_{il}(z_n) \frac{K^{(n,\tau)}_{n_1 n_2}}{K^{(n,m-1)}_{n_1 n_2}} \frac{z_n^\tau}{z_n^m} \rho_l$$
$$\times \left[(\tau+1) C^\tau_{\tau-m} \tilde{Q}^{(\tau-m)}_{jl}(iz_n) - C^{\tau+1}_{\tau+1-m} z_n \tilde{Q}^{(\tau+1-m)}_{jl}(iz_n)\right] = 0, \qquad (10)$$

where $m = 0, 1, 2, \cdots$, and $L_n + 1$, and the coefficients $\breve{A}_{ij}^{(m)}(z_k)$, $\breve{B}_{ij}^{(m)}(z_k)$, $\breve{C}_{ilj}^{(m,\tau)}(z_k, z_n)$, $\breve{D}_{ilj}^{(m,\tau)}(z_k, z_n)$ and functions $\widetilde{Q}_{lj}^{(m)}(is)$ are given in Appendix B of Ref. 20.

All quantities presented above are determined by the two sets of unknowns $\left\{ \mathcal{D}_{il}^{(\tau)}(z_n) \right\}$ and $\left\{ \mathcal{G}_{ij}^{(\tau)}(z_k) \right\}$. These two sets of unknowns are obtained by solving the two sets of algebraic Eqs. (9) and (10). Those equations are solved using the Newton-Raphson technique, with the partial derivatives evaluated analytically. This iterative technique provides a fast and reliable convergence to the final solution. Since $\mathcal{G}_{ij}^{(m)}(z_k) = \mathcal{G}_{ji}^{(m)}(z_k)$ for all $m$ and $k$, one can reduce the number of Eqs. (9) by $[I(I-1)/2 \cdot \sum_{n=1}^{N} (L_n+2)]$ where $I$ is the number of species of particles in the fluid mixture. However, a simplification of the system is possible only through the use of a symbolic manipulation program and leads to a very considerable complication of the remaining equations. In the process, the symmetric structure of the equations (which facilitates their evaluation) is likely to be lost and their convergence properties may be altered. In the meantime, Eqs. (10) show that when one of $K_{ij}^{(n,m-1)}$ is zero an equation $F_{2ij}^{(m)}(z_n) = 0$ becomes equivalently zero. It makes one of the unknowns $\mathcal{D}_{il}^{(m)}(z_n)$ zero. Therefore, unknowns $\mathcal{D}_{il}^{(m)}(z_n)$ are reduced by the number $N_0$ of parameters $K_{ij}^{(n,m)}$ with a zero value. Thus, the system is finally reduced to $[2I^2 \sum_{n=1}^{N} (L_n + 2) - N_0]$ equations.

In most cases, only a few iterations are needed to reach the final solution (see Ref. 20 for more details). The correctness of the present algorithm was verified by the method which will be presented in the next Sec. 3.2.

## 3. Thermodynamic properties

### 3.1. Static structure factors

The Fourier-transformed OZ equation is given by

$$\sum_l [\delta_{il} - \rho_l \widetilde{c}_{il}(k)] \widetilde{h}_{lj}(k) = \widetilde{c}_{ij}(k), \tag{11}$$

where

$$\widetilde{h}_{ij}(k) \equiv \int d\mathbf{r}\, e^{i\mathbf{k}\cdot\mathbf{r}} h_{ij}(r) = \frac{4\pi}{k} \int_0^\infty dr\, r h_{ij}(r) \sin(kr) \tag{12}$$

and

$$\widetilde{c}_{ij}(k) \equiv \int d\mathbf{r}\, e^{i\mathbf{k}\cdot\mathbf{r}} c_{ij}(r) = \frac{4\pi}{k} \int_0^\infty dr\, r c_{ij}(r) \sin(kr). \tag{13}$$

In the absence of infinitely ranged correlations the Fourier-transformed direct correlation functions $\widetilde{c}_{ij}(k)$ can be factorized and given by

$$\widetilde{c}_{ij}(k) = \widetilde{Q}_{ij}^{(0)}(k) + \widetilde{Q}_{ji}^{(0)}(-k) - \sum_l \rho_l \widetilde{Q}_{il}^{(0)}(k) \widetilde{Q}_{jl}^{(0)}(-k). \tag{14}$$

The Fourier-transformed total correlation functions $\tilde{h}_{ij}(k)$ are obtained by solving the linear Eq. (11) with Eq. (14).

The partial and total structure factors $S_{ij}(k)$ and $S(k)$ are given by

$$S_{ij}(k) = \delta_{ij} + \rho(c_i c_j)^{1/2} \tilde{h}_{ij}(k) \tag{15}$$

and

$$S(k) = 1 + \rho \sum_{ij} c_i c_j \tilde{h}_{ij}(k). \tag{16}$$

where $c_j = \rho_j/\rho$ ($\rho$ is the total number density).

## 3.2. Thermodynamic quantities

From the correlation parameters, thermodynamic properties can be calculated directly. Configurational energy $U^c$ per molecule is given by

$$\frac{U^c}{\mathcal{N}} = -k_B T \sum_{ij} c_i \sum_{n=1}^{N} \sum_{\tau=-1}^{L_n} K_{ij}^{(n,\tau)} z_n \gamma_{ij}^{(\tau+2)}(z_n), \tag{17}$$

where $\mathcal{N}$ is the total number of molecules in the system and

$$\gamma_{ij}^{(m)}(z_k) = \frac{2\pi\rho_j}{z_k^{3-m} e^{z_k \sigma_{ij}}} e^{\mathcal{G}_{ij}^{(m)}(z_k)}, \tag{18}$$

or

$$s^{2-m}\gamma_{il}^{(m)}(s) = 2\pi\rho_l \tilde{g}_{il}^{(m)}(s) = 2\pi\rho_l \int_{\sigma_{il}}^{\infty} dx \, e^{-sx} x^m g_{il}(x). \tag{19}$$

The virial pressure, $p^V$, is obtained through

$$\frac{p^V}{\rho k_B T} = 1 + \frac{2\pi\rho}{3} \sum_{ij} c_i c_j \sigma_{ij}^3 g_{ij}(\sigma_{ij}) + J, \tag{20}$$

where

$$J = \frac{1}{3} \sum_{ij} c_i \sum_{n=1}^{N} \sum_{\tau=-1}^{L_n} K_{ij}^{(n,\tau)} z_n \left[ \tau \gamma_{ij}^{(\tau+2)}(z_n) - \gamma_{ij}^{(\tau+3)}(z_n) \right]. \tag{21}$$

We have already obtained $\gamma_{ij}^{(\tau)}(z_n)$ for $\tau \leq L_n + 2$ by solving Eqs. (9) and (10) with (18), while the computation of $\gamma_{ij}^{(L_n+3)}(z_n)$ is presented in Appendix C of Ref. 20.

As mentioned in the previous section, the correctness of the MSA solutions yielded by this multi-SCPPS algorithm can be verified by comparing the values of $\gamma_{ij}^{(L_n+3)}(z_n)$ calculated by the method presented in Appendix C of Ref. 20 with those calculated by direct numerical integraion using relations (19).

In most applications the equation of state obtained by differentiating the energy equation is more accurate than those obtained from either the pressure equation or the compressibility.[21] According to the method of Høye and Stell[22] the incremental energy pressure $\Delta p^{\mathrm{E}}$ is found from

$$\frac{\Delta p^{\mathrm{E}}}{\rho k_{\mathrm{B}}T} = \frac{\pi}{3}\rho \sum_{ij} c_i c_j \sigma_{ij}^3 \left([g_{ij}(\sigma_{ij})]^2 - [g_{ij}^0(\sigma_{ij})]^2\right) + J. \tag{22}$$

The incremental Helmholtz free energy is yielded by

$$\frac{\Delta A}{\mathcal{N}k_{\mathrm{B}}T} = \frac{U^{\mathrm{c}}}{\mathcal{N}k_{\mathrm{B}}T} - \frac{\Delta p^{\mathrm{E}}}{\rho k_{\mathrm{B}}T} + \frac{1}{2}(\chi^{-1} - \chi_0^{-1}). \tag{23}$$

Inverse isotermal compressibility is given by

$$\chi^{-1} = \frac{1}{k_{\mathrm{B}}T}\left(\frac{\partial p}{\partial \rho}\right)_T = 1 - \rho \sum_{ij} c_i c_j \tilde{c}_{ij}(0) = \sum_j c_j \left(\frac{A_j}{2\pi}\right)^2 \tag{24}$$

and the corresponding hard-sphere quantity is

$$\chi_0^{-1} = \sum_j c_j \left(\frac{A_j^0}{2\pi}\right)^2. \tag{25}$$

The contact values of the distribution functions are given by

$$2\pi\sigma_{ij}g_{ij}(\sigma_{ij}) = A_j\sigma_{ij} + B_j - 2\pi \sum_{n=1}^{N} e^{-z_n\sigma_{ij}} \sum_{k=0}^{L_n+1} (z_n\sigma_{ij})^k \check{C}_{ij}^{(k)}(z_n). \tag{26}$$

The corresponding hard-sphere quantity is

$$2\pi\sigma_{ij}g_{ij}^0(\sigma_{ij}) = A_j^0\sigma_{ij} + B_j^0, \tag{27}$$

where

$$A_j^0 = \frac{2\pi}{\Delta}\left(1 + \frac{\pi\zeta_2}{2\Delta}\sigma_j\right), \quad B_j^0 = -\frac{\zeta_2}{2}\left(\frac{\pi}{\Delta}\right)^2\sigma_j^2. \tag{28}$$

Here, $\zeta_k = \sum_l \rho_l \sigma_l^k$ and $\Delta = 1 - \pi\zeta_3/6$.

## 4. An application to a Lennard-Jones fluid mixture

In this section we demonstrate the power of the multi-SCPPS-MSA algorithm for a binary-component LJ system.

### 4.1. *Fitting procedure of the SCPPS closure to the Lennard-Jones potentials*

Let us consider here a fluid mixture with LJ interactions which are given by

$$\phi_{ij}^{\mathrm{LJ}}(r) = 4\epsilon_{ij}\left[\left(\frac{\sigma_{ij}}{r}\right)^{12} - \left(\frac{\sigma_{ij}}{r}\right)^6\right]. \tag{29}$$

For any values of LJ parameters $\epsilon_{ij}$ and $\sigma_{ij}$, the potentials $\phi_{ij}^{\mathrm{LJ}}(r)$ can be reproduced with arbitrary accuracy by the single-SCPPS closure, all of whose parameters are obtained from only one $z$ and $(L_n + 2)$ of $a_\tau$ defined by

$$z = z_{n_{ij}} \sigma_{ij}, \qquad a_\tau = -\frac{k_B T}{\epsilon_{ij}} \frac{K_{ij}^{(n_{ij}, \tau)}}{\sigma_{ij}} z^{\tau+1}. \qquad (30)$$

The fitting parameters $z$ and $a_\tau$ are adjusted, using the Levenberg-Marquardt technique,[23] to fit the single-SCPPS potential $\phi(x)$ to an LJ potential $\phi^{\mathrm{LJ}}(x)$ as follows:

$$\phi^{\mathrm{LJ}}(x) \equiv 4\left(\frac{1}{x^{12}} - \frac{1}{x^6}\right) = \phi(x) \equiv \sum_{\tau=-1}^{L} a_\tau x^\tau e^{-zx}. \qquad (31)$$

As a measure of the error for the values of the fitting parameters, the following norm of the deviations is used:

$$\Omega_\phi = \sqrt{\frac{1}{K} \sum_{k=1}^{K} [\phi(x_k) - \phi^{\mathrm{LJ}}(x_k)]^2}. \qquad (32)$$

All fitting points $x_k$ are spaced with an even step $\Delta s$ along the curve of $\phi^{\mathrm{LJ}}(x)$ as follows:

$$x_{k+1} - x_k = \frac{\Delta s}{\sqrt{1 + \phi'^{\mathrm{LJ}}(x_k)^2}}. \qquad (33)$$

Under the deviation norm $\Omega_\phi = 10^{-4}$ with $K = 200$ fitting points between $1 \leq x \leq 5$, the fitting parameters obtained are shown in Ref. 20.

This work provides numerical results for a specific binary mixture. The mixture has model parameters chosen to resemble an argon-xenon mixture: $\sigma_2/\sigma_1 = 1.167$, $\epsilon_2/\epsilon_1 = 1.919$, $\epsilon_{12} = \sqrt{\epsilon_1 \epsilon_2}$ and $c_1 = c_2$.

For the binary LJ mixture, the symbols $n_{ij}$ in Eq. (30) can be expressed as $n_{11} = 1$, $n_{12} = n_{21} = 2$ and $n_{22} = 3$. In this case, the nonzero parameters $z_n$ ($n = 1, 2$ and 3), $K_{11}^{(1,\tau)}$, $K_{12}^{(2,\tau)} = K_{21}^{(2,\tau)}$ and $K_{22}^{(3,\tau)}$ are given by Eqs. (30). The rest of the parameters $K_{ij}^{(n,\tau)}$ are all zero and we can see from Eq. (10) that $F_{211}^{(\tau)}(z_3)$, $F_{212}^{(\tau)}(z_3)$, $F_{221}^{(\tau)}(z_1)$ and $F_{222}^{(\tau)}(z_1)$ are equivalently zero and $\mathcal{D}_{11}^{(\tau)}(z_3) = \mathcal{D}_{12}^{(\tau)}(z_3) = \mathcal{D}_{21}^{(\tau)}(z_1) = \mathcal{D}_{22}^{(\tau)}(z_1) = 0$ for all $\tau$. Thus, the normalization constants $K_{n_1 n_2}^{(n,\tau)}$ in Eq. (10) are set as $K_{n_1 n_2}^{(1,\tau)} = K_{11}^{(1,\tau)}$, $K_{n_1 n_2}^{(2,\tau)} = K_{12}^{(2,\tau)}$ and $K_{n_1 n_2}^{(3,\tau)} = K_{22}^{(3,\tau)}$.

### 4.2. Thermodynamic properties of model mixture

The isotherms of energy pressure $p^E$ are shown in Fig. 1 by solid curves.[20] The isotherm at critical temperature $T_{\mathrm{crit}}^*$ $(= k_B T_{\mathrm{crit}}/\epsilon_{11}) = 2.1322$ shown by curve A has a critical point $c$ where $(\partial p^E/\partial \eta^{-1})_T = (\partial^2 p^E/\partial(\eta^{-1})^2)_T = 0$. At temperatures above $T_{\mathrm{crit}}^*$ the isotherms are single-valued decreasing functions of $\eta^{-1}$, but below $T_{\mathrm{crit}}^*$ the isotherms exhibit so-called van der Waals loops (curve B at $T_B^* = 1.8182$ and curve C at $T_C^* = 1.6667$): at low and high densities $p^E$ decreases with $\eta^{-1}$, but in the intermediate-density range $p^E$ has both a maximum and a minimum, separated by a region where $p^E$ is an increasing function of $\eta^{-1}$. This region corresponds to thermodynamically unstable states, because the isothermal compressibility $(\partial p^E / \partial \eta^{-1})_T$ is positive. The unstable states can be eliminated by replacing the loops by horizontal segments (line $B_1$-$B_2$ for curve B and line $C_1$-$C_2$ for curve C) through the Maxwell equal-area construction in the $p^E$-$\eta^{-1}$ plane.[24,25] The dashed curve shows a spinodal curve which is the locus of points where $(\partial p^E / \partial \eta^{-1})_T = 0$. The dotted curve is a binodal curve which shows the locus of coexistence points satisfying the Maxwell construction. All thermodynamic states falling between the spinodal and binodal curves are metastable, but can be reached experimentally if sufficient care is taken.[26] Below the temperature $T_C^*$ the stable branch of the isotherms in the high-density range disappears as shown by curves D, E and F at $T_D^* = 1.6129$, $T_E^* = 1.$ and $T_F^* = 0.5$, respectively.

## 5. Summary

We have overviewed our recent works on solving the Blum and Høye MSA equations for multicomponent fluid mixtures interacting through multi-SCPPS potentials.[16-20] Analytic expressions are presented for the calculation of all MSA thermodynamic properties. The only numerical part of the algorithm is the solution of a system of nonlinear equations. This algorithm is sufficiently fast and is reliable in that it converges to the physical solution, if it exists. It offers sufficient flexibility and opens access to systems with any smooth, realistic isotropic potentials where the potentials can be fitted by the multi-SCPPS closure. The quasianalytic multi-SCPPS-MSA algorithm can treat realistic fluid mixtures with computational convenience.

The present approximate method can be further extended: a cluster expansion method, such as the exponential approximation (EXP),[27,28] can be used to improve the relatively poor MSA radial distribution functions. A generalized mean spherical approximation (GMSA)[10,29] type of approach or self-consistent Ornstein Zernike approximation (SCOZA) may further increase the accuracy of the MSA.[22,30-33]

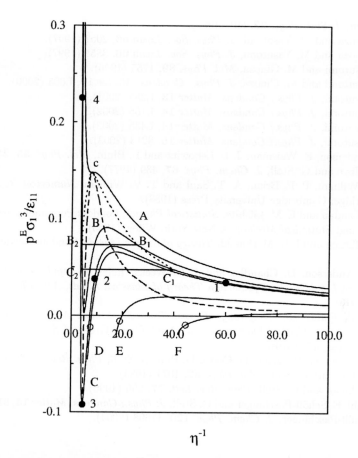

Fig. 1. Isotherms of energy pressure $p^E$ for model mixture. Solid curves labeled A, B, C, D, E and F show the isotherms for $T^*_{\text{crit}} = 2.1322$, $T^*_B = 1.8182$, $T^*_C = 1.6667$, $T^*_D = 1.6129$, $T^*_E = 1$ and $T^*_F = 0.5$, respectively. The dashed curve is a spinodal, $c$ is the critical point, while the dotted curve is a binodal. Horizontal lines $B_1$-$B_2$ and $C_1$-$C_2$ are Maxwell constructions for isotherms B and C, respectively. Open circles on the isotherms D, E and F are the bounrdary states beyond which distribution functions $g_{ij}(r)$ diverge at infinite $r$ [20].

## References

1. D. Henderson, F. F. Abraham and J. A. Barker, *Mol. Phys.* **31**, 1291 (1976).
2. E. Waisman, D. Henderson and J. L. Lebowitz, *Mol. Phys.* **32**, 1373 (1976).
3. N. E. Thompson, D. J. Isbister, R. J. Bearman and B. C. Freasier, *Mol. Phys.* **39**, 27 (1980).
4. D. Henderson, J. L. Lebowitz, L. Blum and E. Waisman, *Mol. Phys.* **39**, 47 (1980).
5. M. Plischke and D. Henderson, *Proc. R. Soc.* A **404**, 323 (1986).
6. R. J. Baxter, *J. Chem. Phys.* **49**, 2770 (1968).
7. J. W. Perram and E. R. Smith, *Chem. Phys. Lett.* **35**, 138 (1975).
8. B. Barboy and R. Tenne, *Chem. Phys.* **38**, 369 (1979).
9. M. Ginoza and M. Yasutomi, *Mol. Phys.* **87**, 593 (1996).
10. E. Waisman, *Mol. Phys.* **25**, 45 (1973).
11. L. Blum and J. S. Høye, *J. Stat. Phys.* **19**, 317 (1978).

12. L. Blum, *J. Stat. Phys.* **22**, 661 (1980).
13. M. Ginoza and M. Yasutomi, *J. Phys. Soc. Japan* **66**, 2057 (1997).
14. M. Ginoza and M. Yasutomi, *J. Phys. Soc. Japan* **66**, 3854 (1997).
15. M. Yasutomi and M. Ginoza, *Mol. Phys.* **89**, 1755 (1996).
16. M. Yasutomi and M. Ginoza, *J. Phys.: Condens. Matter* **12**, L605 (2000).
17. M. Yasutomi, *J. Phys.: Condens. Matter* **13**, L255 (2001).
18. M. Yasutomi, *J. Phys.: Condens. Matter* **14**, L165 (2002).
19. M. Yasutomi, *J. Phys.: Condens. Matter* **14**, L435 (2002).
20. M. Yasutomi, *J. Phys.: Condens. Matter* **15**, 8213 (2003).
21. D. Henderson, E. Waisman, J. L. Lebowitz and L. Blum, *Mol. Phys.* **35**, 241 (1978).
22. J. S. Høye and G. Stell, *J. Chem. Phys.* **67**, 439 (1977).
23. H. P. William, P. F. Brian, A. T. Saul and T. V. William, *Numerical Recipes in C*, Cambridge: Cambridge University Press (1988).
24. L. D. Landau and E. M. Lifshitz, *Statistical Physics*, London: Pergamon Press, (1958).
25. K. Huang, *Statistical Mechanics*, New York: Wiley, (1963).
26. J. P. Hansen and I. R. McDonald, *Theory of Simple Liquids*, London: Academic Press, (1976).
27. H. C. Anderson, D. Chandler and J. D. Weeks, *Roles of Repulsive and Attractive Forces in Liquids: The Equilibrium Theory of Classical Fluids*, Adv. Chem. Phys. **34**, 105 (1976).
28. G. Stell, *Fluids with Long-range Forces: Toward a Simple Analytic Theory*, Phase Transition and Critical Phenomena **15**, Domb C and Green M S, eds., London: Academic Press, 47 (1976).
29. C. Caccamo, G. Pellicane, D. Costa, D. Pini, G. Stell, *Phys. Rev. E* **60**, 5533 (1999).
30. J. S. Høye and G. Stell, *Mol. Phys.* **52**, 1071 (1984).
31. R. Dickman and G. Stell, *Phys. Rev. Lett.* **77**, 996 (1996).
32. G. Kahl, E. Schöll-Paschinger and G. Stell, *J. Phys.: Condens. Matter* **14**, 9153 (2002).
33. E. Schöll-Paschinger, *J. Chem. Phys.* **120**, 11698 (2004).

**III.6** Value Distribution Theory and Related Topics

Organizers: P.C. Hu, P. Li, C.C. Yang

# FINITE FOURIER TRANSFORMS AND THE ZEROS OF THE RIEMANN ξ-FUNCTION II

GEORGE CSORDAS

*Department of Mathematics*
*University of Hawaii, Honolulu, HI 96822*
*E-mail: george@math.hawaii.edu*

CHUNG-CHUN YANG

*Department of Mathematics,*
*Hong Kong University of Science & Technology*
*Clear Water Bay, Kowloon, Hong Kong*
*E-mail: mayang@ust.hk*

This is a continuation of the authors' investigation of the distribution of zeros of entire functions represented by Fourier transforms. Applications include the Riemann ξ-function and the Fourier transforms of kernels related to the Jacobi theta function.

**Key words:** Zeros of Fourier transforms, Jacobi theta function, Riemann ξ-function, Riemann hypothesis

**Mathematics Subject Classification:** Primary 30C10; Secondary 30C15

## 1. Introduction

In[12] we pointed out that today, there are no known necessary *and* sufficient conditions that a "nice" kernel must satisfy in order that its Fourier transform have only real zeros. It is this fundamental issue that motivates us to consider questions and results dealing with the distribution of zeros of real entire functions represented by Fourier transforms. Specifically, for suitably "nice" kernels $K(t)$ (cf. Definition 1.1), we consider real entire functions $F(x)$ and $F_R(x)$, $R > 0$, which can be represented by the Fourier cosine transform of $K(t)$ and the finite Fourier cosine transform of $K(t)$, respectively:

$$F(x) := \int_0^\infty K(t)\cos(xt)\,dt \quad \text{and} \quad F_R(x) := \int_0^R K(t)\cos(xt)\,dt. \quad (1.1)$$

The primary purpose of this paper is to investigate the distribution of zeros of finite Fourier transforms. For some classical results pertaining to the distribution of zeros of finite transforms, we refer to [3] ([20] Part I Problem #147, Part III, Problems #149, and #205, Part V, Problems #164, #170–175) and ([19] pp. 166–197). Analogous problems involving the (infinite) Fourier transform of $K(t)$ seem considerably more difficult and elusive [1,18] or ([19] pp. 265–277).

In order to expedite our presentation (and for the sake of simplicity), it will be convenient to introduce the following definitions.

**Definition 1.1.** A function $K : \mathbb{R} \longrightarrow \mathbb{R}$ is termed an *admissible kernel*, if $K(t)$ satisfies (i) $K(t) \in C^{\infty}(\mathbb{R})$ and (ii) $K^{(n)}(t) = O\left(\exp\left(-|t|^{2+\alpha}\right)\right)$ as $t \longrightarrow \infty$ for some $\alpha > 0$ and $n = 0, 1, 2, \ldots$.

Let $S(\tau)$ denote the closed strip of width $2\tau$, $\tau \geq 0$, in the complex plane, $\mathbb{C}$, symmetric about the real axis:

$$S(\tau) = \{z \in \mathbb{C} \mid |\mathrm{Im}(z)| \leq \tau\}. \tag{1.2}$$

**Definition 1.2.** Let $0 \leq \tau < \infty$. We say that a real entire function $f$ belongs to the class $\mathfrak{S}(\tau)$, if $f$ is of the form

$$f(z) = Ce^{-az^2 + bz} z^m \prod_{k=1}^{\infty} (1 - z/z_k) e^{z/z_k}, \tag{1.3}$$

where $a \geq 0$, $z_k \in S(\tau) \setminus \{0\}$, and $\sum_{k=1}^{\infty} 1/|z_k|^2 < \infty$ and where the zeros $\{z_k\}_{k=1}^{\infty}$ of $f$ are counted according to multiplicity and are arranged so that $0 < |z_1| \leq |z_2| \leq \cdots$. We allow functions in $\mathfrak{S}(\tau)$ to have only finitely many zeros by letting, as usual, $z_k = \infty$ and $0 = 1/z_k$, $k \geq k_0$, so that the canonical product in (1.3) is a finite product. If $f \in \mathfrak{S}(\tau)$, for some $\tau \geq 0$, and if $f$ has only real zeros (i.e., if $\tau = 0$), then $f$ is said to belong to the *Laguerre-Pólya class*, and we write $f \in \mathcal{L}\text{-}\mathcal{P}$. In addition, we write $f \in \mathcal{L}\text{-}\mathcal{P}^*$, if $f = pg$, where $g \in \mathcal{L}\text{-}\mathcal{P}$ and $p$ is a real polynomial. Thus, $f \in \mathcal{L}\text{-}\mathcal{P}^*$ if and only if $f \in \mathfrak{S}(\tau)$, for some $\tau \geq 0$, and $f$ has at most finitely many non-real zeros.

The relevance of the class $\mathcal{L}\text{-}\mathcal{P}^*$ in connection with the distribution of zeros of $F_R(x)$ (cf. (1.1)) stems from the fact that by virtue of the Pólya-Wiman Theorem [5] and [14], for each $R > 0$, there exits a positive integer $p_0$, where $p_0$ depends on $R$, such that $F_R^{(p)}(x) \in \mathcal{L}\text{-}\mathcal{P}$ for all $p \geq p_0$. Unfortunately, there is no known result which describes the relationship between relation between $p_0$ and $R$.

The following elementary lemma highlights some of the differences that exist between the distribution of zeros of finite and infinite Fourier cosine transforms of admissible kernels.

**Lemma 1.3** ([12] *Lemma 3.2*) *Let $K(t)$ be an admissible kernel. Let $F(x)$ and $F_R(x)$ $(R > 0)$ denote the real entire functions defined in (1.1). Then the following assertions are valid.*

(a) *If $K'(0) \neq 0$, then the entire function $F(x)$ has an infinite number of non-real zeros and at most a finite number of real zeros.*

(b) *If $K(R) \neq 0$, then the entire function of exponential type, $F_R(x)$, has an infinite number of real zeros and at most a finite number of non-real zeros; that is, with the notation introduced above, $F_R(x) \in \mathcal{L}\text{-}\mathcal{P}^*$.*

There is little known about the connection between the distribution of zeros of $F(x)$ and $F_n(x)$, even in the important special case when $K(t)$ is the the Jacobi theta function, $\Phi(t)$, (see Section 2 below). In the sequel we will investigate the following general problems for kernels $K(t)$ related to $\Phi(t)$.

**Problem 1.4** Let $F(x)$ and $F_n(x)$ $(R > 0)$ denote the real entire functions defined in (1.1), where $K(t)$ is an admissible kernel. Suppose that $F(x) \in \mathcal{L}\text{-}\mathcal{P}$. What additional assumptions on the kernel $K(t)$ will imply that $F_n(x) \in \mathcal{L}\text{-}\mathcal{P}$, for all $R$ sufficiently large?

**Problem 1.5** Let $F(x)$ and $F_R(x)$ $(R > 0)$ denote the real entire functions defined in (1.1). Suppose that $F(x) \in \mathfrak{S}(1)$ (cf. Definition 1.2), so that the zeros of the real entire function $F(x)$ all lie in the horizontal strip $S(1)$ (cf. (1.2)). When, under what conditions on $K(t)$, is it true that $F_R(x) \in \mathfrak{S}(1)$ for all $R$ sufficiently large?

We remark that for kernels, $K(t)$, which are not necessarily admissible, even if $F(x) \in \mathcal{L}\text{-}\mathcal{P}$, it need not follow that $F_R(x) \in \mathcal{L}\text{-}\mathcal{P}$ for all $R > 0$. Consider, for example, $K(t) = e^{-t^2}$. Then, it is well known that $\int_0^\infty e^{-t^2} \cos(xt)\, dt = \frac{\sqrt{\pi}}{2} e^{-x^2/4} \in$ $\mathcal{L}\text{-}\mathcal{P}$. However, it is not true that the entire function $F_R(x) = \int_0^R e^{-t^2} \cos(xt)\, dt$ has only real zeros for all $R > 0$. For the sake of brevity, we will only mention here that when $R = 2.35$, then in the vicinity of $x = 5.4$, $F_R(x)$ has a positive local minimum and whence when $R = 2.35$, $F_R(x)$ has a non-real zero (approximately $z = 5.4 \cdots + i\, 0.31 \ldots$). We are unaware of results which provide, even in this ostensibly elementary situation, a complete description of the distribution of zeros of $F_R(x)$, for all $R > 0$. An argument similar to the one used to prove Lemma 1.3 (b), reveals that for each $R > 0$, $F_R(x)$, has an infinite number of real zeros and at most a finite number of non-real zeros.

We conclude this section with two additional, concrete, problems involving admissible kernels.

**Problem 1.6** For each positive integer $n$, set $K_n(t) := \exp(-t^{2n})$. Let

$$F(x; n) := \int_0^\infty K_n(t) \cos(xt)\, dt \quad \text{and} \quad F_R(x; n) := \int_0^R K_n(t) \cos(xt)\, dt.$$

Then the Taylor series expansion of the even entire function, $F(x; n)$, is given by

$$F(x; n) = \frac{1}{2n} \sum_{k=0}^{\infty} (-1)^k \frac{\Gamma\left(\frac{2k+1}{2n}\right)}{\Gamma(2k+1)} x^{2k}, \tag{1.4}$$

where $\Gamma(x)$ denotes the gamma function. Using the theory of multiplier sequences, Pólya ([19] p. 237) has shown that $F(x; n) \in \mathcal{L}\text{-}\mathcal{P}$ for $n = 1, 2, 3, \ldots$. Thus, the problem is to determine the values of $R > 0$, such that the finite Fourier transform $F_R(x; n) \in \mathcal{L}\text{-}\mathcal{P}$, for $n = 1, 2, 3, \ldots$.

**Remark 1.7** The explicit expressions for the Taylor coefficients of $F(x; n)$ in terms of the gamma function (cf. (1.4)) suggests that the theory of mutiplier sequences may provide the tools for investigating Problem 1.6.

The next problem we propose here is of particular interest to us, since it involves an admissible kernel whose (double exponential) decay order is similar to that of the Jacobi theta function (cf. (2.2) and Theorem 2.1).

**Problem 1.8** Fix $a > 0$ and set $K_a(t) := \exp(-a\cosh(t))$. Let

$$F(x; a) := \int_0^\infty K_a(t)\cos(xt)\,dt \quad \text{and} \quad F_R(x; a) := \int_0^R K_a(t)\cos(xt)\,dt.$$

In ([19] p. 272), Pólya used a second order differential equation (an idea due to A. Hurwitz and developed by E. Hille) to prove that $F(x; a) \in \mathcal{L}\text{-}\mathcal{P}$ for any $a > 0$. Now the problem is to determine the values of $R > 0$, such that the finite Fourier transform $F_R(x; a) \in \mathcal{L}\text{-}\mathcal{P}$, for $a > 0$.

## 2. The finite Fourier transform of $\Phi(t)$

The specific entire functions we study, in the sequel, are intimately connected with the Riemann $\xi$-function. We recall that the Riemann $\xi$-function admits the Fourier integral representation (see, for example [17], ([19] pp. 278–308) or [6]

$$H(x) := \frac{1}{8}\xi\left(\frac{x}{2}\right) = \int_0^\infty \Phi(t)\cos(xt)\,dt, \tag{2.1}$$

where

$$\Phi(t) := \sum_{n=1}^\infty a_n(t) \tag{2.2}$$

(called the Jacobi theta function) and where, for $n = 1, 2, 3 \ldots$,

$$a_n(t) := \pi n^2(2\pi n^2 e^{4t} - 3)\exp\left(5t - \pi n^2 e^{4t}\right) \quad (t \in \mathbb{R}). \tag{2.3}$$

The Riemann Hypothesis is equivalent to the statement that all the zeros of $H(x)$ are real (cf. [21] p. 255). We also recall that $H(x)$ is an entire function of order one ([21] p. 16) of maximal type (cf.,[7] Appendix A). Thus, with the above nomenclature the Riemann Hypothesis is true if and only if $H \in \mathcal{L}\text{-}\mathcal{P}$. It is also known ([13] p. 7) or ([15] p. 25) that all the zeros of $H$ lie in the *interior* of the strip $S(1)$, so that $H(x) \in \mathfrak{S}(\tau)$, with $\tau = 1$. Moreover, it is known that $H(x)$ (cf. (2.1)) has an infinite number of real zeros ([21] p. 256).

We begin here with a brief review of some of the known properties of $\Phi(t)$.

**Theorem 2.1** ([6] Theorem A) *The function $\Phi(t)$ of (2.2) satisfies the following properties. (i) For each $n \geq 1$, $a_n(t) > 0$ for all $t \geq 0$, so that $\Phi(t) > 0$ for all $t \geq 0$; (ii) $\Phi(z)$ is analytic in the strip $-\pi/8 < \operatorname{Im} z < \pi/8$; (iii) $\Phi(t)$ is an even function, so that $\Phi^{(2m+1)}(0) = 0$ $(m = 0, 1, \ldots)$; (iv) $\Phi'(t) < 0$ for all $t > 0$; (v) For any $\varepsilon > 0$,*

$$\lim_{t \to \infty} \Phi^{(n)}(t)\exp\left[(\pi - \varepsilon)e^{4t}\right] = 0 \quad \text{for each} \quad n = 0, 1, \ldots.$$

Thus, it is clear from Theorem 2.1 that $\Phi(t)$ is an admissible kernel. Among the convexity properties of $\Phi(t)$, we also mention that $\log \Phi(\sqrt{t})$ is strictly concave for

$t > 0$ ($^9$ Theorem 2.1) and $\Phi(\sqrt{t})$ is strictly convex for $t > 0$ ($^2$ Theorem 2.12). This latter property implies that the (infinite) Fourier cosine transform of $\Phi(\sqrt{t})$ is strictly positive on the real axis ($^2$ Corollary 2.13).

The convexity properties of $\Phi(t)$, summarized above, can be used to prove that some finite Fourier cosine transforms of $\Phi(t)$ have only real zeros. Indeed, $\Phi(t) > 0$ and $\Phi'(t) < 0$ for all $t > 0$ by Theorem 2.1 and a detailed ("hard") analysis of $\Phi''(t)$ shows that $\Phi''(t) < 0$ on the interval $I = [0, 0.11]$ ($^{10}$ Lemma 3.5). Therefore, it follows from a classical theorem ($^{20}$ Part V, Problem #173) that

$$H_R(x) := \int_0^R \Phi(t) \cos(xt)\, dt \in \mathcal{L}\text{-}\mathcal{P} \quad \text{for all} \quad 0 < R \le 0.11, \tag{2.4}$$

see ($^{10}$ Lemma 3.5) for the details. We remark, parenthetically, that *if $H_R(x) \in \mathcal{L}\text{-}\mathcal{P}$ for all sufficiently large $R$*, then an easy argument, using the properties of functions in the Laguerre-Pólya class ($\mathcal{L}\text{-}\mathcal{P}$) would establish the validity of the Riemann Hypothesis. Thus, we are led to the following intriguing question.

**Problem 2.2** For what values of $R > 0.11$, if any, does $H_R(x)$ possess some non-real zeros?

We know that the zeros of $H(x)$ (cf. (2.1)) lie in the *interior* of the horizontal strip $S(1)$. Thus, in the present setting, a concrete version of the general theoretical question raised in Section 1 (cf. Problem 1.5) is the following problem.

**Problem 2.3** For all $R$ sufficiently large, do the zeros of $H_R(x)$ (cf. (2.4)) all lie in the horizontal strip $S(1)$?

**Conjecture 2.4**($^{12}$ Conjecture 3.7) *For each fixed $\lambda \in \mathbb{R}$, $|\lambda| \ge 1$, and for all $R$ sufficiently large, the entire function*

$$H_R(x; \lambda) := \int_0^R \Phi(t) \cos(xt) \cosh(\lambda t)\, dt \in \mathcal{L}\text{-}\mathcal{P}. \tag{2.5}$$

## 3. Analysis of the transforms $H(x)$, $H_R(x)$ and $H_R(x; \lambda)$

In this section our goal is to investigate the transforms $H(x)$ (cf. (2.1)), $H_R(x)$ (cf. (2.4)) and $H_R(x; \lambda)$(cf. (2.5)) and to provide some results pertaining to the above problems. To begin with, we remark that since the zeros of $H(x)$ lie in the interior of the strip $S(1)$, it follows from the Hermite-Biehler Theorem ($^{11}$ Proposition 3.1), ($^{16}$ p. 314) and ($^1$ Theorem 9a) hat for each fixed $\lambda \in \mathbb{R}$, $|\lambda| \ge 1$,

$$H(x; \lambda) := \int_0^\infty \Phi(t) \cos(xt) \cosh(\lambda t)\, dt \in \mathcal{L}\text{-}\mathcal{P}. \tag{3.1}$$

In order to elucidate the relationship between $H_R(x)$ and $H_R(x; \lambda)$, we recall that the (infinite-order) differential operator $\cos(\lambda D)$, $D := d/dx$, acting on entire functions, $f(x)$, may be defined by (cf.$^4$)

$$2\cos(\lambda D)f(x) := \left(e^{i\lambda D} + e^{-i\lambda D}\right) f(x) = f(x + i\lambda) + f(x - i\lambda). \tag{3.2}$$

In reference to Problem 2.3 and Conjecture 2.4, we have the following result.

**Theorem 3.1** *For each $R > 0$, there exists a real number $\lambda_0 := \lambda_0(R)$ such that the finite Fourier transform $H_R(x; \lambda) \in \mathcal{L}\text{-}\mathcal{P}$ for all $\lambda \geq \lambda_0$.*

**Proof** Fix $R > 0$ and note that $\cos(\lambda D) H_R(x) = H_R(x; \lambda)$ (cf. (3.2)). Since for any $R > 0$, $\Phi(R) \neq 0$, by Lemma 1.3 (b), $H_R(x)$ has at most a finite number of non-real zeros $z_1, \cdots, z_n$. If $H_R(x)$ has only real zeros, then it follows from a generalization of the classical Hermite-Poulain theorem ([19] p. 142) that $H_R(x; \lambda) \in \mathcal{L}\text{-}\mathcal{P}$ for all $\lambda \in \mathbb{R}$ and therefore in this case the assertion of the theorem is valid. Otherwise, set $\lambda_0 := \max(|z_1|, \cdots, |z_n|)$. Now it is known ([1] Theorem 8) that if $f(x) \in \mathfrak{S}(\tau)$, then the zeros of $\cos(\lambda D) f(x)$ lie in the strip $S(\tau_\lambda)$, where $\tau_\lambda := \sqrt{\max(\tau^2 - \lambda^2, 0)}$. In particular, if $\lambda \geq \tau$, then $f(x)$ has only real zeros. Thus, if $\lambda \geq \lambda_0$, then $H_R(x; \lambda) \in \mathcal{L}\text{-}\mathcal{P}$ and the proof of the theorem is complete. $\square$

It follows from Theorem 3.1 that *if* Problem 2.3 has an affirmative answer, then Conjecture 2.4 is valid.

Problem 2.2 appears to be a very deep question, which is not surprising in light of its intimate connection with the Riemann Hypothesis. Nevertheless, we attempt to provide here some results which shed light on various aspects of the problem. To this end, we define the following kernel related to $\Phi(t)$ (cf. (2.2)):

$$\Psi(t) := \frac{e^t}{8} \sum_{n=1}^{\infty} \exp(-\pi n^2 e^{4t}) \quad t \geq 0. \tag{3.3}$$

Our next lemma summarizes some of the properties of $\Psi(t)$.

**Lemma 3.2** *Consider the kernels $\Phi(t)$ (cf. (2.2)) and $\Psi(t)$ (cf. (3.3)). Then*

- (a) $\Phi(t) = \Psi''(t) - \Psi(t)$;
- (b) $\Psi'(0) = -\frac{1}{16}$;
- (c) $\Psi'(t) < 0$ for $t \geq 0$;
- (d) $\Psi''(t) > 0$ for $t \geq 0$;
- (e) $\lim_{t \to \infty} \Psi(t) = \lim_{t \to \infty} \Psi'(t) = 0$.

Parts (a) and (b) of Lemma 3.2, follow from the known ([17] p. 10) or ([19] p. 285) properties of the related function.

$$\omega(t) := \frac{1}{4} e^t + 4\Psi(t). \tag{3.4}$$

Since $\Phi(t) = \frac{1}{4}(\omega''(t) - \omega(t))$ and since $\omega(t)$ is an even function, we infer that (a) and (b) hold. Parts (c) and (d) may be established by elementary considerations. Alternatively, part (d) is a consequence of Lemma 2.1 (i) and part (a). Since

$$0 \leq \frac{e^t}{8} \sum_{n=1}^{\infty} \exp(-\pi n^2 e^{4t}) \leq \frac{e^t}{8} \left( \frac{\exp(-\pi e^{4t})}{1 - \exp(-\pi e^{4t})} \right) \to 0 \text{ as } t \to \infty,$$

$\lim_{t \to \infty} \Psi(t) = 0$. A similar argument shows that $\lim_{t \to \infty} \Psi'(t) = 0$.

**Remarks 3.3** While the functions $\omega(t)$ and $\Psi(t)$ are closely related, we choose to work with $\Psi(t)$, instead of $\omega(t)$, because $\omega(t)$ is not integrable on $[0, \infty)$. With the aid of parts (a), (b) and (e) of Lemma 3.2 and two integrations by parts, we obtain the following representation of $H(x)$:

$$H(x) = \int_0^\infty \Phi(t) \cos(xt)\,dt = \frac{1}{16} - (1 + x^2) \int_0^\infty \Psi(t) \cos(xt)\,dt. \qquad (3.5)$$

We remark that the proof of Lemma 1.3 shows (see[12] Lemma 3.2) that it remains valid under less restrictive assumptions. Thus, since $\Psi(t)$ is not an even function, it follows that the (infinite) Fourier cosine transform of $\Psi(t)$ has at most a finite number of real zeros. In fact, it has no real zeros! Indeed, by virtue of the convexity property of $\Psi(t)$ (see Lemma 3.2 (d)), we are able to establish the following positivity result.

**Theorem 3.4** *For all $x \in \mathbb{R}$,*

$$\int_0^\infty \Psi(t) \cos(xt)\,dt > 0. \qquad (3.6)$$

**Proof** Another appeal to Lemma 3.2 and two integrations by parts yield, for $x \neq 0$,

$$x^2 \int_0^\infty \Psi(t) \cos(xt)\,dt = \frac{1}{16} - \int_0^\infty \Psi''(t) \cos(xt)\,dt > \frac{1}{16} - \int_0^\infty \Psi''(t)\,dt \geq \frac{1}{16} - \frac{1}{16} = 0.$$

$\square$

The truncation of the integrals in (3.5) yields an interesting relation which once again highlights the possible advantages of working with $\Psi(t)$ in connection with the analysis of Problem 2.2.

**Proposition 3.5** *Fix $R > 0$. Then*

$$H_R(x) = \int_0^R \Phi(t) \cos(xt)\,dt = f(x; R) + \frac{1}{16} - (1 + x^2) \int_0^R \Psi(t) \cos(xt)\,dt, \qquad (3.7)$$

*where $\Psi(t)$ is defined by (3.3) and*

$$f(x; R) := \Psi'(R) \cos(xR) + \Psi(R)\, x\, \sin(xR) \in \mathcal{L}\text{-}\mathcal{P}. \qquad (3.8)$$

*Moreover, for each $R > 0$, the zeros of $f(x; R)$ are simple.*

**Proof** By Lemma 3.2 (a), we have $\Phi(t) = \Psi''(t) - \Psi(t)$. Then two integrations by parts yield (3.7). In order to prove that $f(x; R) \in \mathcal{L}\text{-}\mathcal{P}$, set $\mu := \frac{|\Psi'(R)|}{\Psi(R)}$ and observe that $\Psi(R) > 0$ for any $R > 0$. Since $\exp\left(\frac{-x^2 R}{2\mu}\right) \sin(xR) \in \mathcal{L}\text{-}\mathcal{P}$ and the Laguerre-Pólya class is closed under differentiation, the entire function

$$-\Psi(R)\frac{\mu}{R}\frac{d}{dx}\exp\left(\frac{-x^2 R}{2\mu}\right)\sin(xR) = \exp\left(\frac{-x^2 R}{2\mu}\right)[\Psi(R)x\sin(xR) - |\Psi'(R)|\cos(xR)] \qquad (3.9)$$

belongs to the Laguerre-Pólya class. But $-|\Psi'(R)| = \Psi'(R)$ (see Lemma 3.2 (c)) and whence we conclude that $f(x; R) \in \mathcal{L}\text{-}\mathcal{P}$. Moreover, it follows from (3.9) that for each fixed $R > 0$, the zeros of $f(x; R)$ are simple. $\square$

**Acknowledgement** This research was partially supported by UGC grants of Hong Kong, Project No.: 604103.

## References

1. N. G. de Bruïjn, The roots of trigonometric integrals, Duke Math. J. bf 17 (1950), 197–226.
2. G. Csordas, Convexity and the Riemann$\xi$-function, Glas. Mat. Ser. III **33(53)** (1998), 37–50.
3. M.L. Cartwright, The zeros of certain integral functions, Quart. J. Math. **1** (1930), 38–59.
4. T. Craven and G. Csordas, Differential operators of infinite order and the distribution of zeros of entire functions, J. Math. Anal. Appl. **186** (1994), 799–820.
5. T. Craven, G. Csordas and W. Smith, The zeros of derivatives of entire functions and the Pólya-Wiman conjecture, Ann. of Math. **125** (1987), 405–431.
6. G. Csordas, T.S. Norfolk and R.S. Varga, The Riemann hypothesis and the Turán inequalities, Trans. Amer. Math. Soc. **296** (1986), 521–541.
7. G. Csordas, T.S. Norfolk and R.S. Varga, A lower bound for the de Bruijn-Newman constant $\Lambda$, Numer. Math. Soc. **52** (1988), 483–497.
8. G. Csordas, W. Smith and R.S. Varga, Level sets of real entire functions and the Laguerre inequalities, Analysis **12** (1992), 377–402.
9. G. Csordas and R.S. Varga, Moment inequalities and the Riemann hypothesis, Constr. Approx. **4** (1988), 175–198.
10. G. Csordas and R.S. Varga, Necessary and sufficient conditions and the Riemann hypothesis, Adv. in Appl. Math. **11** (1990), 328–357.
11. G. Csordas and R.S. Varga, Fourier transforms and the Hermite-Biehler Theorem, Proc. Amer. Math. Soc. **107** (1989), 645–652.
12. G. Csordas and C.C. Yang, Finite Fourier transforms and the zeros of the Riemann $\xi$-function, J. Math. Anal. Appl. **314** (2006), 109–125.
13. A. Ivić, The Riemann Zeta-Function. The Theory of the Riemann Zeta-Function with Applications, John Wiley & Sons, Inc., New York 1985 .
14. H. Ki and Y.-O. Kim, On the number of nonreal zeros of real entire functions and the Fourier-Pólya conjecture, Duke Math. J. **104** (2000), 45–73.
15. A.A. Karatsuba and S.M. Voronin, The Riemann Zeta-Function, Translated from the Russian by Neal Koblitz. Walter de Gruyter & Co., Berlin 1992.
16. B.Ja. Levin, Distribution of Zeros of Entire Functions, Transl. Math. Mono. **5**, Amer. Math. Soc., Providence, RI 1964; revised ed. 1980.
17. G. Pólya, Über die algebraisch–funktionentheoritischen Untersuchungen von J.L.W.V. Jensen, Kgl. Danske Vid. Sel. Math.–Fys. Medd. **7** (1927), 3–33.
18. G. Pólya, Über trigonometrische Integrale mit nur reellen Nullstellen, J. Reine Angew. Math. **158** (1927), 6–18.
19. G. Pólya, Collected Papers, Vol. II Location of Zeros, (R P. Boas, ed.), MIT Press, Cambridge, MA, 1974.
20. G. Pólya and Szegö, Problems and Theorems in Analysis, I,II, Springer-Verlag, New York, 1976.
21. E.C. Titchmarsh, The Theory of the Riemann Zeta-Function, (2nd ed., D. R. Heath-Brown, ed.) The Clarendon Press, Oxford University Press.

# CONDITION POUR LES ZÉROS DE LA FONCTION HOLOMORPHE, BORNÉE ET SECOND PROBLÉM DE COUSIN DANS LE BIDISQUE-UNITÉ DE $\mathbb{C}^2$

KAZUKO KATÔ

*Université Ryukoku*

*2-407. Takehanakinomoto-cho, Yamashina-ku, Kyoto, Japon*

*E-mail:kato@math.ryukoku.ac.jp*

On cherche la condition nécéssaire et surffisante pour les zèros de la fonction holomorphe et bornée dans le bidisque-unité de $\mathbb{C}^2$.

**Key words:** Zeros of holomorphic functions, 2nd Cousin problem, unit bidisc
**Mathematics Subject Classification:** 32A60

## 1. Introduction

A propos de l'ensemble de zèros de la fonction holomorphe et bornée, nous avons beacoup d'articles dans la boule de l'espace $\mathbb{C}^n$ de $n$ variables complexes.

D'autre part, nous en n'avons pas beacoup dans le polydisque de $\mathbb{C}^n$, et nous avons dans le cas spéciale, les articles dus à W. Rudin,[1] (1967), E.L. Stout[2] (1968), K.Katô[3] (1997),[4] (2000).

Dans cet article, nous allons rechercher la condition nécessaire et suffisante pour les zèros de la fonction holomorphe, bornée dans le bidisque-unité de $\mathbb{C}^2$.

Or, dans le théorie de la fonction d'une variable complexe, on a le théorème de Blaschke:

"*Soit $f(z)$ une fonction holomorphe de $z$, vérifiant $|f(z)| < 1$, dans le cercle-unité $C(1)$ sur le plan de $z$, et soit $\{z_i\}$, $(i \in \mathbb{N})$, l'ensemble de zèros de $f(z) = 0$, dans $C(1)$, on a alors,*

$$\sum_{i=1}^{\infty}(1 - |z_i|) < \infty, \quad (Condition\ de\ Blaschke).$$

*Et en plus, soit $\{z_i\}$, $(i \in \mathbb{N})$ une suite de points dans $C(1)$ sur le plan de $z$, vérifiant la condition de Blaschke, on a alors une fonction $B(z)$, holomorphe dans $C(1)$, telle que $B(z_i) = 0$, $(i \in \mathbb{N})$,*

$$B(z) = z^n \cdot \prod_{i=1}^{\infty} \left( \frac{z - z_i}{1 - \overline{z_i} \cdot z} \right) e^{\sqrt{-1}\theta_i}, \quad (produit\ de\ Blaschke),$$

*où $|B(z)| < 1$, dans $C(1)$ et $|B(z)| = 1$, sur $\{|z| = 1\}$.*"

Dans cet article, nous considérons les problèmes suivants:

**Problème 1.** Soit $f(x, y)$ une fonction holomorphe et bornée dans le bidisque - unité $D(1)$ de $\mathbb{C}^2$, et soit $\Sigma$ la surface analytique définie par $f(x, y) = 0$.

Nous allons chercher la condition nécessaire pour $\Sigma$.

**Problème 2.** Soit $\Sigma$ une surface analytique vérifiant la condition nécessaire dans $D(1)$. Nous allons construire la fonction $f(x, y)$ holomorphe et bornée dans $D(1)$, telle que $\Sigma$ soit définie par $f(x, y) = 0$, (solution du second problèm de Cousin).

Nous considérons le problem 1 dans les sections 2, 3, et le problème 2 dans les sections 4, 5, 6.

## 2. Condition nécéssaire pour $\Sigma$

Soit $D(1) = [C_x(1), C_y(1)]$ le bidisque-unité de l'espace de deux variables complexes $x, y$, où

$$C_x(1) = \{x \mid |x| < 1\}, \ C_y(1) = \{y \mid |y| < 1\}.$$

Soit $f(x, y)$ une fonction holomorphe dans $D(1)$, telle que

$$\log|f(x, y)| < M, \ f(0, 0) = 1, \ (M \in \mathbb{R}^+),$$

et soit $\Sigma$ la surface analytique définie par $f(x, y) = 0$ dans $D(1)$. Soit $x \in C_x(1)$ un point fixé, on désigne par $\{y_j(x)\}$, $(j \in \mathbb{N})$, la section de $\Sigma$ par $x$. On a alors d'après la formule de Jensen,

$$\sum_{j=1}^{\infty}(1 - |y_j(x)|) \leq M - \log|f(x, 0)|.$$

Considérons la moyenne $m(R)$ sur $C_x(R) = \{x \mid |x| < R\}$, $(0 < R < 1)$. Puisque $\log|f(x, 0)|$ est une fonction subharmonique de $x$ dans $C_x(1)$, on a

$$m(R)\sum_{j=1}^{\infty}\left(1 - \left|y_j\left(re^{\sqrt{-1}\theta}\right)\right|\right) = \frac{1}{\pi R^2}\int_0^{2\pi}\int_0^R \sum_{j=1}^{\infty}\left(1 - \left|y_j\left(re^{\sqrt{-1}\theta}\right)\right|\right) r \, dr \, d\theta$$

$$\leq M - \frac{1}{\pi R^2}\int_0^{2\pi}\int_0^R \log\left|f\left(re^{\sqrt{-1}\theta}, 0\right)\right| r \, dr \, d\theta$$

$$\leq M - \log|f(0, 0)|.$$

D' après l'hypothèse, on a

$$m(R)\sum_{j=1}^{\infty}(1 - |y_j(x)|) \leq M, \ \left(x = re^{\sqrt{-1}\theta}\right). \tag{1}$$

En plus, soit $y \in C_y(1)$ un point fixé et soit $\{x_j(y)\}$, $(j \in \mathbb{N})$, la section de $\Sigma$ par $y$. On a alors

$$\sum_{j=1}^{\infty}(1 - |x_j(y)|) \leq M - \log|f(0, y)|.$$

On pose $\Gamma(R'', \ R') = \{y \mid R'' < |y| < R'\}, \ (0 < R'' < R' < 1)$.
Considérons la moyenne $m(R'', \ R')$ sur $\Gamma(R'', \ R')$, et on a

$$m(R'', \ R') \sum_{j=1}^{\infty} \left(1 - \left|x_j\left(re^{\sqrt{-1}\theta}\right)\right|\right)$$

$$= \frac{1}{\pi(R'^2 - R''^2)} \int_0^{2\pi} \int_{R''}^{R'} \sum_{j=1}^{\infty} \left(1 - \left|x_j\left(re^{\sqrt{-1}\theta}\right)\right|\right) r \ dr \ d\theta$$

$$\leq M - \log|f(0, \ 0)|,$$

on a donc

$$m(R'', \ R') \sum_{j=1}^{\infty} (1 - |x_j(y)|) \leq M, \ \left(y = re^{\sqrt{-1}\theta}\right). \tag{2}$$

Ensuite, nous allons représenter (1), (2) par les formes plus simples.
Soit $\{R_i\}$ une suite de nombres $(R_i \in \mathbb{R}^+)$, telle que

$$0 < R_i < R_{i+1} < 1; \ R_i \to 1, \ (i \to \infty); \ \sum_{i=1}^{\infty} d_i < \infty, \ (d_i = 1 - R_i, \ i \in \mathbb{N}).$$

On désigne par $n^i(x)$ le nombre de $\{y_j(x)\}$ dans $C_y(R_i) = \{y \mid |y| < R_i\}$, pour $x \in C_x(R)$, on a alors par (1)

$$m(R) \sum_{i=1}^{\infty} d_i \left(n^i(x) - n^{i-1}(x)\right) \leq M, \ \left(n^0(x) = 0\right), \tag{3}$$

et plus, on note simplement $n_x^i = m(R) \cdot n^i(x)$, et on désigne par $\sigma_x[C_x(R), \ C_y(R_i)]$ l'aire de la projection (surface de Riemann) sur $C_x(R)$ de $\Sigma \cap [C_x(R), \ C_y(R_i)]$. On a alors,

$$n_x^i = m(R) \cdot n^i(x) = \frac{1}{\pi R^2} \sigma_x[C_x(R), \ C_y(R_i)],$$

ce qui signifie la moyenne de nombres des feuilles de la projection sur $C_x(R)$ de $\Sigma \cap [C_x(R), \ C_y(R_i)]$, et on a par (3)

$$\sum_{i=1}^{\infty} d_i \left(n_x^i - n_x^{i-1}\right) \leq M, \ \left(n_x^0 = 0\right). \tag{4}$$

En plus, soit $n^i(y)$ le nombre de $\{x_j(y)\}$ dans $C_x(R_i)$ pour $y \in \Gamma(R'', \ R')$, et on considère la moyenne $m(R'', \ R')$ sur $\Gamma(R'', \ R')$, on a alors,

$$m(R'', \ R') \sum_{i=1}^{\infty} d_i \left(n^i(y) - n^{i-1}(y)\right) \leq M, \ \left(n^0(y) = 0\right). \tag{5}$$

En notant simplement $n_y^i = m(R'', \ R')n^i(y)$, on a

$$\sum_{i=1}^{\infty} d_i \left(n_y^i - n_y^{i-1}\right) \leq M, \ \left(n_y^0 = 0\right), \tag{6}$$

où $n_y^i$ signifie la moyenne de nombres des feuilles de la projection sur $\Gamma(R', R'')$ de $\Sigma \cap [C_x(R_i), \Gamma(R'', R')]$.

En particulier, pour $y = 0$, on a $n_0^i = n^i(0)$, et

$$\sum_{i=1}^{\infty} d_i \left( n_0^i - n_0^{i-1} \right) \leq M, \quad \left( n_0^0 = 0 \right). \tag{7}$$

## 3. Déformation de la condition

Pour construire la solution associée à $\Sigma$, nous allons déformer la condition (4).

Or, grâce à K. Oka,[5] on a le théorème:

"*Soit* $D = [C_x(R), C_y(R')]$ *le bidisque de* $\mathbb{C}^2$, $(R < 1, R' < 1)$, *et soit* $\Sigma$ *une surface analytique dans* $D$.

*Soit* $\Delta^0, \Delta^1$ *la portion dans* $D$ *respectivement comme suivante:*

$$\Delta^0 = [C_x(R), \Gamma(R'_0, R')]; \ \Gamma(R'_0, R') = \{y \mid R'_0 < |y| < R'\}, \ (0 < R'_0 < R'),$$
$$\Delta^1 = [C_x(R_0), C_y(R')], \ (0 < R_0 < R).$$

*On désigne par* $\sigma^0, \sigma^1$ *l'aire de* $\Sigma$ *respectivement dans* $\Delta^0, \Delta^1$. *Soit* $\sigma$ *l'aire de* $\Sigma$ *dans* $D'$ $(D' \Subset D)$, *on a alors,*

$$\sigma_x \leq C(\sigma^0 + \sigma^1),$$

*où* $\sigma_x$ *est l'aire de la projection sur la plan de* $x$ *de* $\Sigma \cap D'$, *et* $C$ *est une constante* $(C \in \mathbb{R}^+)$ *indépendant de* $\Sigma$." Soit $(\delta) = \{x \mid |x - x_0| < \delta\}$, $(|x_0| \leq R - \delta)$, on

désigne par $m(\delta)$ la moyenne sur $(\delta)$. D'après le théorème de K.Oka, on a par (3), (5),

$$m(\delta) \cdot \sum_{i=1}^{\infty} d_i(n^i(x) - n^{i-1}(x)) \leq M, \quad \left( x = x_0 + re^{\sqrt{-1}\theta} \right).$$

En notant à nouveau, $n_x^i = m(\delta) \cdot n^i(x)$, on a

$$\sum_{i=1}^{\infty} d_i(n_x^i - n_x^{i-1}) \leq M, \quad (n_x^0 = 0). \tag{8}$$

On a donc le théorème suivant: **Théorème 1.** "*Soit* $f(x, y)$ *une fonction holomorphe dans le bidisque-unité* $D(1)$ *de* $\mathbb{C}^2$, *telle que*

$$\log |f(x, y)| < M, \ f(0, 0) = 1, \ (M \in \mathbb{R}^+)$$

*Soient* $C_x(R_i) = \{x \mid |x| < R_i\}$, $C_y(R_i) = \{y \mid |y| < R_i\}$, $(0 < R_i < 1)$.

$(\delta) = \{x \mid |x - x_0| \leq \delta\}$; $(\delta) \subset C_x(1)$,

$\Gamma(R'', R') = \{y \mid R'' \leq |y| \leq R'\}$; $\Gamma(R'', R') \subset C_y(1)$, $(0 \leq R'' \leq R' < 1)$.

*On désigne par* $n_x^i, n_y^i$ *la moyenne de nombres respectivement, de* $\{y_j(x)\}$ *dans* $C_y(R_i)$ *pour* $x \in (\delta)$, *de* $\{x_j(y)\}$ *dans* $C_x(R_i)$ *pour* $y \in \Gamma(R'', R')$.

*On a alors une suite $\{R_i\}$, vérifiant $0 < R_i < R_{i+1} < 1$; $R_i \to 1$, $(i \to \infty)$,* $\sum_{i=1}^{\infty} d_i < \infty$, $(d_i = 1 - R_i)$, $(i \in \mathbb{N}$, $R_i \in \mathbb{R}^+)$, *telle que*

$$\begin{cases} \displaystyle\sum_{i=1}^{\infty} d_i \left( n_x^i - n_x^{i-1} \right) < M, \\ \displaystyle\sum_{i=1}^{\infty} d_i \left( n_y^i - n_y^{i-1} \right) < M, \end{cases} \tag{9}$$

*pour tous $(\delta)$, $\Gamma(R'', R')$, et pour $n_0^i$ où $n_x^0 = 0$, $n_y^0 = 0$, $n_0^0 = 0$."*

Nous appelons les rélations (9) *"la condition de Blaschke."*

## 4. Construction de la solution

Soit $\Sigma$ une donnée de Cousin (surface analytique) vérifiant la condition de Blaschke dans $D(1)$.

Soit $\{R_i\}$ une suite de nombres $(R_i \in \mathbb{R}^+)$, telle que

$$0 < R_i < R_{i+1} < 1; \ R_i \to 1, \ (i \to \infty); \ \sum_{i=1}^{\infty} d_i < \infty, \ (d_i = 1 - R_i),$$

et que, pour tout $\delta$ $(0 < \delta < 1)$, et pour tous $R''$, $R'$ $(0 < R'' < R' < 1)$, les rélations (9) soient vérifies.

Soit $D_i = [C_x(R_i), \ C_y(R_i)]$, $(i \in \mathbb{N})$, et on suppose que $\Sigma$ soit réguliere sur la forntière de $D_i$, $(i \in \mathbb{N})$.

Soit $\alpha^\lambda$ la projection sur $C_x(R_i)$ de $\Sigma \cap [C_x(R_i)\{|y| = R_\lambda\}]$, $(1 \leq \lambda \leq i)$, et on partage $\alpha^\lambda$ en quelques courbes $\{\alpha_k^\lambda\}$, $(1 \leq k \leq k_\lambda, \ 1 \leq \lambda \leq i)$, qui sont connexes sur la surface de Riemann. $\alpha_k^\lambda$ est représentée par

$$\alpha_k^\lambda = \{x \in C_x(R_i) \mid |y_k(x)| = R_\lambda\}, \ (1 \leq k \leq k_\lambda, \ 1 \leq \lambda \leq i).$$

Et on partage $C_x(R_i)$ par $\bigcup_{\lambda=1}^{i} \left\{\bigcup_k^{k_\lambda} \alpha_k^\lambda\right\}$ en quelques portions $\{\Delta_h^i\}$, $(1 \leq h \leq h_i)$, de telle façon que, à tout $x \in \Delta_h^i$, ils correspondent le même nombre de $\{y_j(x)\}$ dans $C_y(R_i)$, qu'on notera par $n_h^i = n^i(x)$.

Soit $\{x_\nu(0)\} = \Sigma \cap \{y = 0\}$, $(\nu \in \mathbb{N})$, on désigne par $\nu_i$ le nombre de $x_\nu(0)$ dans $C_x(R_i)$, et on pose

$$(\gamma_\nu) = \{x \mid |x - x_\nu(0)| < \gamma_\nu\}, \ (\gamma_\nu) \subset C_x(1),$$
$$(\gamma_\nu^0) = \{x \mid |x - x_\nu(0)| < \gamma_\nu^0\}, \ (\gamma_\nu^0 < \gamma_\nu),$$

où $(\gamma_\nu) \cap (\gamma_{\nu'}) = \emptyset$, $(x_\nu(0) \neq x_{\nu'}(0))$, $\sum_{\nu=1}^{\infty} \gamma_\nu < \infty$.

On peut supposer que $\{x_\nu(0)\} \notin \left\{\bigcup_{\lambda=1}^{i} \alpha^\lambda\right\}$, $(1 \leq \nu \leq \nu_i)$, et que $\{x_\nu(0)\}$ ne contienne pas de point multiple.

On pose

$$C_x^0(R_i) = C_x(R_i) \setminus \left\{\bigcup_{\nu=1}^{\nu_i}(\gamma_\nu^0)\right\} \cap C_x(R_i),$$
$$\Delta_h^{i0}(x, \ y) = \Delta_h^i \setminus \left\{\bigcup_{\nu=1}^{\nu_i}(\gamma_\nu^0)\right\} \cap \Delta_h^i, \ (1 \leq h \leq h_i).$$

Considérons la fonction dans $[\Delta_h^{i0}, \; C_y(R_i)]$:

$$\varphi_h^{i0}(x, \; y) = \prod_{\lambda=1}^{i} \left[ \prod_{j>n^{\lambda-1}(x)}^{n^\lambda(x)} \left( \frac{y - y_j(x)}{y - C_\lambda y_j(x)} \right) e^{-P_{\lambda j}(x, \; y)} \right] \qquad (10)$$

avec $C_\lambda = 1 + cd_\lambda$,

$$P_{\lambda j}(x, \; y) = -\log \frac{1}{C_\lambda} - \sum_{\mu=1}^{N_\lambda} \frac{1}{\mu} \left( \frac{y}{C_\lambda y_j(x)} \right)^\mu, \; (1 \le \lambda \le i). \qquad (11)$$

$\varphi_h^{i0}(x, \; y)$ est une fonction holomorphe dans $[\Delta_h^{i0}, \; C_y(R_i)]$, vérifiant $\varphi_h^{i0}(x, \; 0) = 1$, dans $\Delta_h^{i0}$. Et puis, on considère l'intégrale de Cousin:

$$I_k^\lambda(x, \; y) = \frac{-1}{2\pi\sqrt{-1}} \int_{\alpha_k^\lambda} \mathrm{Log} \left( \frac{y - C_{\lambda+1} y_k(t)}{y - C_\lambda y_k(t)} \right) e^{-P_{\lambda k}(t, \; y) + P_{\lambda+1, k}(t, \; y)} \cdot \frac{dt}{t - x},$$

où $1 \le k \le k_\lambda$, $1 \le \lambda \le i$, $C_{i+1} = 1$, $P_{i+1, \, k}(t, \; y) = 0$.
Et on pose dans $[C_x^0(R_i), \; C_y(R_i)]$,

$$f_0^i(x, \; y) = \varphi_h^{i0}(x, \; y) \cdot \exp \left\{ \sum_{\lambda=1}^{i} \sum_{k=1}^{k_\lambda} I_k^\lambda(x, \; y) \right\}, \qquad (12)$$

$$\text{pour } x \in \Delta_h^{i0}, \; (1 \le h \le h_i), \; (i \in \mathbb{N}).$$

$f_0^i(x, \; y)$ est une solution dans $[C_x^0(R_i), \; C_y(R_i)]$ vérifiant $f_0^i(x, \; 0) = 1$, $(i \in \mathbb{N})$.
On considère la fonction dans $[(\gamma_\nu), \; C_y(R_i)]$:

$$\varphi_\nu^i(x, \; y) = \left[ \left( \frac{y - y_\nu(x)}{y - C_1(y_\nu(x) + d_1)} \right) \cdot e^{-P_{i\nu}^0(x, \; y)} \right]$$

$$\times \prod_{\lambda=1}^{i} \left[ \prod_{j\neq\nu}^{n^\lambda(x) - n^{\lambda-1}(x)} \left( \frac{y - y_j(x)}{y - C_\lambda y_j(x)} \right) \cdot e^{-P_{\lambda j}(x, \; y)} \right] \qquad (13)$$

où $y_\nu(x_\nu(0)) = 0$,

$$P_{i\nu}^0(x, \; y) = \log C_1(y_\nu(x) + d_1) - \sum_{\mu=1}^{N_1} \frac{1}{\mu} \left( \frac{y}{C_1(y_\nu(x) + d_1)} \right)^\mu, \; (1 \le \nu \le \nu_i).$$

Et on pose dans $[(\gamma_\nu), \; C_y(R_i)]$,

$$f_\nu^i(x, \; y) = \varphi_\nu^i(x, \; y) \cdot \exp \left\{ \sum_{\lambda=1}^{i} \sum_{k}^{k_\lambda} I_k^\lambda(x, y) \right\}, \; (1 \le \nu \le \nu_i), \; (i \in \mathbb{N}). \qquad (14)$$

Et puis, on considère l'intégrale de Cousin:

$$J_\nu^i(x, \; y) = \frac{-1}{2\pi\sqrt{-1}} \int_{<\gamma_\nu'>} \mathrm{Log} \left( \frac{y - C_1(y_\nu(t) + R_1)}{y - C_1 y_\nu(t)} \right) e^{-P_{1\nu}(t, \; y) + P_{1\nu}^0(t, \; y)} \cdot \frac{dt}{t - x} \qquad (15)$$

où $< \gamma_\nu' > = \{|x - x_\nu(0)| = \gamma_\nu'\}$, $(\gamma_\nu^0 < \gamma_\nu' < \gamma_\nu)$, $(1 \le \nu \le \nu_i)$.

Et on pose dans $D_i = [C_x(R_i),\ C_y(R_i)]$:

$$f_i(x,\ y) = \begin{cases} f_0^i(x,\ y)\cdot\exp\left\{\displaystyle\sum_{\nu=1}^{\nu_i} J_\nu^i(x,\ y)\right\},\ x\in C_x^0(R_i),\\[2mm] f_\nu^i(x,\ y)\cdot\exp\left\{\displaystyle\sum_{\nu=1}^{\nu_i} J_\nu^i(x,\ y)\right\},\ x\in(\overline{\gamma_i^0}),\ (1\le\gamma\le\gamma_i), \end{cases} \tag{16}$$

$f_i(x,\ y)$ est une solution dans $D_i$, vérifiant $f_i(x,\ y)\not\equiv 0,\ (i\in\mathbb{N})$.

## 5. Estimation de la solution $f_i(x,\ y)$

De la même manière que nous avons calculé dans les articles,[3] [4], nous avons dans $D_i' = [C_x(R_i'),\ C_y(R_i')],\ (R_i' = R_i - \delta,\ \delta\le\gamma_\nu - \gamma_\nu^0)$,

$$m(\delta)\log|\varphi_h^{i0}(x,y)| \le \sum_{\lambda=1}^{i} cd_\lambda(n_x^\lambda - n_x^{\lambda-1}),$$

$$m(\delta)\log|\varphi_\nu^i(x,y)| \le \sum_{\lambda=1}^{i} cd_\lambda(n_x^\lambda - n_x^{\lambda-1}),\ (1\le\nu\le\nu_i).$$

Et puis, on a pour $t\in\alpha_k^\lambda,\ y\in C_y(R_i')$ et pour $N_\lambda$ assez grand dans (11), (13),

$$\left|\mathrm{Log}\left(\frac{y - C_{\lambda+1}y_k(t)}{y - C_\lambda y_k(t)}\right)\cdot e^{-P_{\lambda k}(t,\ y)+P_{\lambda+1,k}(t,\ y)}\right| \le cd_\lambda^{N_\lambda},$$

on a alors pour $\delta\le R' - R''$,

$$m(\delta)\cdot\sum_{k}^{k_\lambda}|I_k^\lambda(x,\ y)| \le C\cdot\frac{d_\lambda^{N_\lambda}}{\delta}\sum_{k}^{k_\lambda}\int_{\alpha_k^\lambda}|dt| \le cd_\lambda\cdot n_y^\lambda,\ (1\le\lambda\le i),$$

$$m(\delta)\cdot\sum_{\lambda=1}^{i}\cdot\sum_{k}^{k_\lambda-k_{\lambda-1}}|I_k^\lambda(x,\ y)| \le \sum_{\lambda=1}^{i}cd_\lambda(n_y^\lambda - n_y^{\lambda-1}) + \sum_{\lambda=1}^{i}cd_\lambda(d_{\lambda-1} - d_\lambda)n_y^{\lambda-1}.$$

On a donc dans $D_i'$,

$$\log|f_i(x,\ y)|$$
$$\le \sum_{\lambda=1}^{i}cd_\lambda\cdot\left\{(n_x^\lambda - n_x^{\lambda-1}) + (n_y^\lambda - n_y^{\lambda-1})\right\} + \sum_{\lambda=1}^{i}cd_\lambda(n_0^\lambda - n_0^{\lambda-1}) \tag{17}$$

et par suite, on a

$$\log|f_i(x,\ y)| \le C\cdot M. \tag{18}$$

## 6. Solution globale dans $D(1)$

Soit $\{D_i\}$ une suite de bidisques dans $D(1)$, telle que $D_i \Subset D_{i+1},\ D_i \to D(1)$, $(i\to\infty)$, et soit $f_i(x,\ y)$ la solution dans $D_i'$, vérifiant les rélations (17), (18).

On pose dans $D'_i$, $(i \in \mathbb{N})$,

$$F(x, y) = f_i(x, \ y) \cdot \prod_{\mu=1}^{\infty} \frac{f_{i+\mu}(x, \ y)}{f_{i+\mu-1}(x, \ y)}$$

On a alors dans $D'_i$,

$$\frac{f_{i+\mu}(x, \ y)}{f_{i+\mu-1}(x, \ y)} = \left[ \prod_{n^{i+\mu-1}(x)}^{n^{i+\mu}(x)} \left( \frac{y - y_j(x)}{y - c_{i+\mu} \cdot y_j(x)} \right) e^{-P_{i+\mu, \ j}(x, \ y)} \right]$$

$$\times \exp \left\{ \sum_{\nu=1}^{\nu_{i+\mu}} J_\nu^{i+\mu}(x, \ y) - \sum_{\nu=1}^{\nu_{i+\mu-1}} J_\nu^{i+\mu-1}(x, \ y) \right\}$$

$$\times \exp \left\{ \sum_{k=1}^{k_{i+\mu}} I_k^{i+\mu}(x, \ y) - \sum_{k=1}^{k_{i+\mu-1}} I_k^{i+\mu-1}(x, \ y) \right\}, \ (\mu \in \mathbb{N}). \quad (19)$$

Donc, on a d'après l'hypothèse dans $D(1)$, par (17), (19)

$$\log |F(x, \ y)| \leq \sum_{\lambda=1}^{\infty} cd_\lambda \left\{ (n_x^\lambda - n_x^{\lambda-1}) + (n_y^\lambda - n_y^{\lambda-1}) + (n_0^\lambda - n_0^{\lambda-1}) \right\}$$

$$\leq C \cdot M. \quad (20)$$

Nous avons en fin le théorème: **Théorème 2.** *"Soit $\Sigma$ une surface analytique vérifiant la condition de Blaschke dans le bidisque-unité $D(1)$ de $\mathbb{C}^2$. On a alors une solution bornée du second probléme de Cousin pour $\Sigma$ dans $D(1)$."*

## References

1. W.Rudin, Zero-sets in polydiscs. Bull. Amer. Math. Soc. 73(1967).
2. E.L.Stout, The second Cousin problem with bounded data. Pac. J. Math. 26(1968).
3. K.Katô, Construction de la solution bornée du second problème de Cousin dans le bidisque. Finite or infinite dimensional Complex Analysis. The Inst. of math. of Peking univ.(1997)
4. K.Katô, Généralisation du produit de Blaschke dans le bidisque-unité. Lecture Note in pure and applied Math. vol.214, Marcel Decher, 6(2000)
5. K.Oka, Sur les fonction analytiques de plusieurs variables. X-une mode nouvelle engendrant les domaines pseudoconvexes, Jap. J. Math.,32(1962).

# CRITERIA FOR BIHOLOMORPHIC CONVEX MAPPINGS ON $p$-BALL IN $C^n$ *

MING-SHENG LIU

*School of Mathematical Sciences*
*South China Normal University,*
*Guangzhou 510631, Guangdong, P. R. China*
*Email:liumsh65@163.com*

In this paper, we shall summarize the necessary or sufficient conditions for biholomorphic convex mappings on Reinhardt domain $B_p^n$ in $C^n$. Some new results are given. From these, we may construct some concrete biholomorphic convex mappings on $B_p^n$.

**Key words:** Locally biholomorphic mapping, biholomorphic convex mapping, Reinhardt domain

**Mathematics Subject Classification:** 32H05, 30C45

## 1. Introduction and preliminaries

The analytic functions of one complex variable which map the unit disk onto starlike or convex domains have been extensively studied. These functions are easily characterized by simple analytic or geometric conditions. In moving to higher dimensions, several difficulties arise. Some are predictable, some are somewhat surprising. In the following we first give some definitions and notations.

Let $C^n$ denote the space of $n$-complex variables $z = (z_1, z_2, \cdots, z_n)$ with the Euclidean inner product

$$\langle z, w \rangle = \sum_{j=1}^{n} z_j \overline{w}_j,$$

where $w = (w_1, w_2, \cdots, w_n) \in C^n$, and the Euclidean norm $\|z\| = \sqrt{\langle z, z \rangle}$.

Assume $p > 1$, we introduce the $p$-norm of $C^n$ such that $\|z\|_p = (\sum_{j=1}^{n} |z_j|^p)^{1/p}$, and let Reinhardt domain $B_p^n = \{z = (z_1, \cdots, z_n) \in C^n : \|z\|_p < 1\}$. In particular, let $\Delta = B_p^1 = \{z \in C : |z| < 1\}$ be the unit disc in $C$. It is evident that $\|z\| = \|z\|_2$.

Let $H(B_p^n)$ be the class of all mappings $f(z) = (f_1(z), \cdots, f_n(z)) : B_p^n \to C^n$, which are holomorphic mappings on $B_p^n$ in $C^n$. The matrix representation of the

---

*This work is supported by the National Natural Science Foundation of China(No.10471048)

1312

first Fréchet derivative of $f \in H(B_p^n)$ is

$$Df(z) = \left(\frac{\partial f_j(z)}{\partial z_k}\right)_{1 \le j,k \le n}.$$

The second Fréchet derivative of a mapping $f \in H(B_p^n)$ is a symmetric bilinear operator $D^2 f(z)(\cdot,\cdot)$ on $C^n \times C^n$. The matrix representation of $D^2 f(z)(b,\cdot)$ is

$$D^2 f(z)(b,\cdot) = \left(\sum_{l=1}^{n} \frac{\partial^2 f_j(z)}{\partial z_k \partial z_l} b_l\right)_{1 \le j,k \le n},$$

where $f(z) = (f_1(z), \cdots, f_n(z)), b = (b_1, \cdots, b_n) \in C^n$.

A mapping $f \in H(B_p^n)$ is said to be local biholomorphic in $B_p^n$ if the first Fréchet derivative $Df(z)$ is nonsingular at each point in $B_p^n$ or, equivalently, if $f$ has a local inverse at each point $z \in B_p^n$.

If $f \in H(B_p^n)$, then for every $k = 0, 1, \cdots$, there is a bounded symmetric $k-$linear operator $D^k f(0) : C^n \times C^n \times \cdots \times C^n \to C^n$ such that

$$f(z) = \sum_{k=0}^{\infty} \frac{D^k f(0)}{k!}(z^k)$$

for $z \in B_p^n$, where $D^0 f(0)(z^0) = f(0), D^k f(0)(z^k) = D^k f(0)(z, z, \cdots, z)$.

Let $N(B_p^n)$ denote the class of all local biholomorphic mappings $f : B_p^n \to C^n$ such that $f(0) = 0, Df(0) = I$, where $I$ is the unit matrix of $n \times n$. If $f \in N(B_p^n)$ is a biholomorphic mapping on $B_p^n$ and $f(B_p^n)$ is a convex domain in $C^n$, then we say that $f$ is a biholomorphic convex mapping on $B_p^n$. The class of all biholomorphic convex mappings on $B_p^n$ with $f(0) = 0, Df(0) = I$ is denoted by $K(B_p^n)$. In particular, we let $K = K(B_p^1), H(\Delta) = N(B_p^1)$. Then $H(\Delta) = \{f : \Delta \to C$ are analytic and $f'(z) \ne 0$ for $z \in \Delta$ with $f(0) = 0, f'(0) = 1\}$, and

$$f \in K \Longleftrightarrow f \in H(\Delta) \text{ and } \operatorname{Re}\left\{\frac{zf''(z)}{f'(z)} + 1\right\} > 0(\forall z \in \Delta).$$

It is difficult to construct concrete biholomorphic convex mappings on some domains in $C^n$, even if on the unit ball $B_2^n$. Until a few years ago, we only know a few concrete examples about the convex mappings on $B_2^n$. In 1995, Roper-Suffridge[10] proved that: If $f \in K$ and $F(z_1, z_2) = (f(z_1), \sqrt{f'(z_1)}z_2)$, then $F \in K(B_2^2)$. From this, we may construct a lot of concrete examples about the convex mappings on $B_2^2$. However, its proof is very complex. They also gave two concrete examples of convex mappings on $B_2^2$ in.[11] Recently, we generalized Roper-Suffridge operator to Banach spaces in.[6,15] But according to Theorem 3 in section 2.2, none of these concrete examples belong to $K(B_p^n)(p > 2)$. In the present paper, we shall summarize some necessary or sufficient conditions for a mapping $f(z)$ to be a biholomorphic convex mappings on the domain $B_p^n$ in $C^n$. Some new sufficient conditions are also given. From these, we construct some concrete biholomorphic convex mappings on Reinhardt domain $B_p^n$.

## 2. Criteria for biholomorphic convex mappings on $B_p^n$

### 2.1. *The necessary and sufficient conditions*

In 1973 Kikuchi[5] obtained a necessary and sufficient condition as follows.

**Theorem 1.**(Kikuchi[5]) Let $f \in N(B_2^n)$. Then $f \in K(B_2^n)$ if and only if for any $z = (z_1, \cdots, z_n) \in B_2^n \setminus \{0\}$ and $b = (b_1, \cdots, b_n) \in C^n \setminus \{0\}$ with $\mathrm{Re}\langle b, z \rangle = 0$, such that

$$\| b \|^2 - \mathrm{Re}\langle Df(z)^{-1} D^2 f(z)(b,b), z \rangle \geq 0.$$

In 1993 S.Gong, S.K.Wang and Q.H.Yu[1,2] obtained an equivalent condition via a different method. In 1998 Y.C. Zhu[13] generalized the above results from $B_2^n$ to $B_p^n$.

**Theorem 2.**(Y.C.Zhu[13]) Suppose that $p \geq 2, u(z) = \sum\limits_{j=1}^{n} | z_j |^p$ and $f \in N(B_p^n)$. Then $f \in K(B_p^n)$ if and only if for any $z = (z_1, \cdots, z_n) \in B_p^n \setminus \{0\}$ and $b = (b_1, \cdots, b_n) \in C^n \setminus \{0\}$ such that $\mathrm{Re}\langle b, \frac{\partial u}{\partial \bar{z}} \rangle = 0$, we have

$$J_f(z,b) = \mathrm{Re}\left\{ \frac{p^2}{4} \sum_{k=1}^{n} | z_k |^{p-2} | b_k |^2 + \frac{p}{2}(\frac{p}{2} - 1) \sum_{k=1}^{n} \frac{| z_k |^p}{z_k^2} b_k^2 \right.$$
$$\left. - \left\langle Df(z)^{-1} D^2 f(z)(b,b), \frac{\partial u}{\partial \bar{z}} \right\rangle \right\} \geq 0,$$

where $\frac{\partial u}{\partial \bar{z}} = (\frac{\partial u}{\partial \bar{z}_1}, \cdots, \frac{\partial u}{\partial \bar{z}_n})$.

### 2.2. *The necessary conditions*

In 1997 T.S.Liu and W.J.Zhang obtained a decomposition theorem of biholomorphic convex mappings on $B_p^n$ as follows.

**Theorem 3.**(T.S.Liu and W.J.Zhang[8]) Suppose that $p > 2$ and $k$ is a positive integer such that $k < p \leq k+1$. If $f \in K(B_p^n)$, then

$$f(z) = \begin{pmatrix} z_1 \\ z_2 \\ \vdots \\ z_n \end{pmatrix} + \begin{pmatrix} a_{12} z_1^2 \\ a_{22} z_2^2 \\ \vdots \\ a_{n2} z_n^2 \end{pmatrix} + \cdots + \begin{pmatrix} a_{1k} z_1^k \\ a_{2k} z_2^k \\ \vdots \\ a_{nk} z_n^k \end{pmatrix} + O(\| z \|^{k+1}),$$

where $| a_{ij} | \leq 1, 1 \leq i \leq n, 2 \leq j \leq k$.

### 2.3. *The sufficient conditions*

In 2003 Y.C.Zhu[14] obtained two sufficient conditions for biholomorphic convex mappings on $B_p^n$. In 2004, we have generalized the two sufficient conditions and obtained the following theorem.

**Theorem 4.**(M.S.Liu and Y.C.Zhu[7]) Suppose that $n \geq 2, p \geq 2, q = p/(p-1)$ and $f_j : \Delta \to C$ are analytic on $\Delta$ with $f_j(0) = f_j'(0) = 0(j = 1, 2, \cdots, n-1)$,

$p_j(\zeta) \in H(\Delta)$ satisfy the conditions $|\zeta p_j''(\zeta)| \leq |p_j'(\zeta)|, (\zeta \in \Delta, j = 1, 2, \cdots, n)$. Let

$$f(z) = (p_1(z_1) + f_1(z_n), p_2(z_2) + f_2(z_n), \cdots, p_{n-1}(z_{n-1}) + f_{n-1}(z_n), p_n(z_n)).$$

If for any $z = (z_1, \cdots, z_n) \in B_p^n$, we have

$$(1 - |z_n|^p)^{1/q} \Big( \sum_{j=1}^{n-1} \Big| \frac{f_j''(z_n)}{p_j'(z_j)} \Big|^p \Big)^{1/p} + (1 - |z_n|^p)^{1/q} \Big( \sum_{j=1}^{n-1} \Big| \frac{f_j'(z_n)}{p_j'(z_j)} \Big|^p \Big| \frac{p_n''(z_n)}{p_n'(z_n)} \Big|^p \Big)^{1/p}$$
$$\leq \Big( 1 - \Big| \frac{z_n p_n''(z_n)}{p_n'(z_n)} \Big| \Big) |z_n|^{p-2},$$

then $f \in K(B_p^n)$.

**Remark 1.** Setting $p_j(\xi) = \xi(j = 1, 2, \cdots, n)$ in Theorem 4, we get Theorem 2 of.[14] Setting $f_j(\zeta) \equiv 0(j = 1, 2, \cdots, n-1)$ in Theorem 4, we get Theorem 4 of.[14] Setting $f_j(\zeta) \equiv 0(j = 1, 2, \cdots, n-1)$ and $n = 2$ in Theorem 5, we get Theorem 4.1 in.[9]

**Theorem 5.** Suppose that $n \geq 2, p \geq 2$, and $f_j : \Delta \to C$ are analytic on $\Delta$ with $f_j(0) = f_j'(0) = 0(j = 1, 2, \cdots, n-1)$, $p_j(\zeta) \in H(\Delta)$ satisfy the conditions $|\zeta p_j''(\zeta)| \leq |p_j'(\zeta)|, (\zeta \in \Delta, j = 1, 2, \cdots, n)$. Let

$$f(z) = (p_1(z_1) + f_1(z_n), p_2(z_2) + f_2(z_n), \cdots, p_{n-1}(z_{n-1}) + f_{n-1}(z_n), p_n(z_n)).$$

If for any $z = (z_1, \cdots, z_n) \in B_p^n \setminus \{0\}$, we have

$$\sum_{j=1}^{n-1} \Big[ \Big| \frac{f_j''(z_n)}{p_j'(z_j)} \Big| + \Big| \frac{f_j'(z_n)}{p_j'(z_j)} \Big| \Big| \frac{p_n''(z_n)}{p_n'(z_n)} \Big| \Big] \leq \Big( 1 - \Big| \frac{z_n p_n''(z_n)}{p_n'(z_n)} \Big| \Big) |z_n|^{p-2},$$

then $f \in K(B_p^n)$.

**Proof.** By direct computing the Fréchet derivatives of $f(z)$, we obtain

$$Df(z) = \begin{pmatrix} p_1'(z_1) & 0 & \cdots & 0 & f_1'(z_n) \\ 0 & p_2'(z_2) & \cdots & 0 & f_2'(z_n) \\ \cdots & \cdots & \cdots & \cdots & \cdots \\ 0 & 0 & \cdots & p_{n-1}'(z_{n-1}) & f_{n-1}'(z_n) \\ 0 & 0 & \cdots & 0 & p_n'(z_n) \end{pmatrix},$$

$$Df(z)^{-1} = \begin{pmatrix} \frac{1}{p_1'(z_1)} & 0 & \cdots & 0 & -\frac{f_1'(z_n)}{p_1'(z_1)p_n'(z_n)} \\ 0 & \frac{1}{p_2'(z_2)} & \cdots & 0 & -\frac{f_2'(z_n)}{p_2'(z_2)p_n'(z_n)} \\ \cdots & \cdots & \cdots & \cdots & \cdots \\ 0 & 0 & \cdots & \frac{1}{p_{n-1}'(z_{n-1})} & -\frac{f_{n-1}'(z_n)}{p_{n-1}'(z_{n-1})p_n'(z_n)} \\ 0 & 0 & \cdots & 0 & \frac{1}{p_n'(z_n)} \end{pmatrix},$$

$$D^2 f(z)(b,b) = \begin{pmatrix} p_1''(z_1)b_1 & 0 & \cdots & 0 & f_1''(z_n)b_n \\ 0 & p_2''(z_2)b_2 & \cdots & 0 & f_2''(z_n)b_n \\ \cdots & \cdots & \cdots & \cdots & \cdots \\ 0 & 0 & \cdots p_{n-1}''(z_{n-1})b_{n-1} & f_{n-1}''(z_n)b_n \\ 0 & 0 & \cdots & 0 & p_n''(z_n)b_n \end{pmatrix} \begin{pmatrix} b_1 \\ b_2 \\ \cdots \\ b_{n-1} \\ b_n \end{pmatrix}$$

$$= \begin{pmatrix} p_1''(z_1)b_1^2 + f_1''(z_n)b_n^2 \\ p_2''(z_2)b_2^2 + f_2''(z_n)b_n^2 \\ \cdots \\ p_{n-1}''(z_{n-1})b_{n-1}^2 + f_{n-1}''(z_n)b_n^2 \\ p_n''(z_n)b_n^2 \end{pmatrix}.$$

Taking $z = (z_1, \cdots, z_n) \in B_p^n \setminus \{0\}, b = (b_1, \cdots, b_n) \in C^n \setminus \{0\}$ such that $\mathrm{Re}\langle b, \frac{\partial u}{\partial \bar{z}} \rangle = 0$, we obtain

$$\frac{2}{p} J_f(z,b) \geq \sum_{j=1}^{n} |b_j|^2 |z_j|^{p-2} - \frac{2}{p} \mathrm{Re}\langle Df(z)^{-1} D^2 f(z)(b,b), \frac{\partial u}{\partial \bar{z}} \rangle$$

$$= \sum_{j=1}^{n} |b_j|^2 |z_j|^{p-2}$$

$$- \mathrm{Re}\Big\{ \sum_{j=1}^{n-1} [\frac{p_j''(z_j)}{p_j'(z_j)} b_j^2 + \frac{f_j''(z_n)}{p_j'(z_j)} b_n^2 - \frac{f_j'(z_n)p_n''(z_n)b_n^2}{p_j'(z_j)p_n'(z_n)}] \frac{|z_j|^p}{z_j} + \frac{p_n''(z_n)}{p_n'(z_n)} b_n^2 \frac{|z_n|^p}{z_n} \Big\}$$

$$\geq \sum_{j=1}^{n-1} |b_j|^2 |z_j|^{p-2} \left[ 1 - \left| \frac{z_j p_j''(z_j)}{p_j'(z_j)} \right| \right] + |b_n|^2 \left[ \left( 1 - \left| \frac{z_n p_n''(z_n)}{p_n'(z_n)} \right| \right) |z_n|^{p-2} \right.$$

$$\left. - \sum_{j=1}^{n-1} \left| \frac{f_j''(z_n)}{p_j'(z_j)} \right| |z_j|^{p-1} - \sum_{j=1}^{n-1} \left| \frac{f_j'(z_n)}{p_j'(z_j)} \right| \left| \frac{p_n''(z_n)}{p_n'(z_n)} \right| |z_j|^{p-1} \right].$$

Hence we conclude from the above inequalities and the hypothesis of Theorem 5 that

$$\frac{2}{p} J_f(z,b) \geq |b_n|^2 \left[ (1 - \left| \frac{z_n p_n''(z_n)}{p_n'(z_n)} \right|) |z_n|^{p-2} - \sum_{j=1}^{n-1} \left[ \left| \frac{f_j''(z_n)}{p_j'(z_j)} \right| + \left| \frac{f_j'(z_n)}{p_j'(z_j)} \right| \left| \frac{p_n''(z_n)}{p_n'(z_n)} \right| \right] \right] \geq 0$$

for all $z = (z_1, \cdots, z_n) \in B_p^n \setminus \{0\}, b = (b_1, \cdots, b_n) \in C^n \setminus \{0\}$ with $\mathrm{Re}\langle b, \frac{\partial u}{\partial \bar{z}} \rangle = 0$.

By Theorem 2, we obtain that $f \in K(B_p^n)$, and the proof is complete.

**Example 1.** Suppose that $p \geq 2, 0 < |\lambda| < 1$ and $k$ is a positive integer such that $k < p \leq k+1$, let

$$f(z) = \Big( z_1 + a_1' z_1^2 + a_1 z_n^{k+1} + b_1 z_n^{k+2}, \cdots, z_{n-1} + a_{n-1}' z_{n-1}^2 + a_{n-1} z_n^{k+1} + b_{n-1} z_n^{k+2},$$

$$\frac{e^{\lambda z_n} - 1}{\lambda} \Big),$$

and $c = \max\{|a_j'| : j = 1, 2, \cdots, n-1\}$. If $c \leq \frac{1}{4}$ and

$$\sum_{j=1}^{n-1} \left[ (k+1)(k+|\lambda|)|a_j| + (k+2)(k+1+|\lambda|)|b_j| \right] \leq (1-2c)(1-|\lambda|),$$

then $f(z) \in K(B_p^n)$.

$$f_j'(\xi) = (k+1)a_j\xi^k + (k+2)b_j\xi^{k+1}, \quad f_j''(\xi) = (k+1)ka_j\xi^{k-1} + (k+2)(k+1)b_j\xi^k,$$

so it follows from $|a_j'| \le c \le \frac{1}{4}$ that

$$|z_jp_j''(z_j)| = 2|a_j'||z_j| \le 1 - 2|a_j'||z_j| \le |p_j'(z_j)|, j = 1, 2, \cdots, n-1,$$

and

$$\left|\frac{z_np_n''(z_n)}{p_n'(z_n)}\right| = |\lambda||z_n| < |\lambda| \le 1.$$

Since $c = \max\{|a_j'| : j = 1, 2, \cdots, n-1\} \le \frac{1}{4}$, then we have $|p_j'(z_j)| \ge 1 - 2c > 0$. Hence according to the hypothesis, we have that

$$\sum_{j=1}^{n-1}\left[\left|\frac{f_j''(z_n)}{p_j'(z_j)}\right| + \left|\frac{f_j'(z_n)}{p_j'(z_j)}\right|\left|\frac{p_n''(z_n)}{p_n'(z_n)}\right|\right] \le \frac{1}{1-2c}\sum_{j=1}^{n-1}\left[(k+1)|ka_jz_n^{k-1} + (k+2)b_jz_n^k|\right.$$

$$\left. +|\lambda||(k+1)a_jz_n^k + (k+2)b_jz_n^{k+1}|\right]$$

$$\le \frac{1}{1-2c}\sum_{j=1}^{n-1}\left[(k+1)(k+|\lambda|)|a_j|\right.$$

$$\left. +(k+2)(k+1+|\lambda|)|b_j|\right]|z_n|^{k-1}$$

$$\le (1-|\lambda|)|z_n|^{p-2}\cdot|z_n|^{k+1-p}$$

$$\le (1-|\frac{z_np_n''(z_n)}{p_n'(z_n)}|)|z_n|^{p-2}.$$

for any $z = (z_1, \cdots, z_n) \in B_p^n \setminus \{0\}$.

By Theorem 5, we obtain that $f \in K(B_p^n)$, and the proof is complete.

**Example 2.** Suppose that $p \ge 2$ and $k$ is a positive integer such that $k < p \le k+1$, let

$$f(z) = (z_1 + a_1'z_1^2 + a_1z_n^{k+1} + b_1z_n^{k+2}, \cdots, z_{n-1} + a_{n-1}'z_{n-1}^2 + a_{n-1}z_n^{k+1} + b_{n-1}z_n^{k+2}, z_n),$$

and $c = \max\{|a_j'| : j = 1, 2, \cdots, n-1\}$. If $c \le \frac{1}{4}$ and

$$\sum_{j=1}^{n-1}\left[k|a_j| + (k+2)|b_j|\right] \le \frac{1-2c}{k+1},$$

then $f(z) \in K(B_p^n)$.

**Example 3.** Suppose that $n \ge 2$, $p \ge 2$ and $k$ is a positive integer such that $k < p \le k+1$, let

$$f(z) = (z_1 + a_1'z_1^2 + a_1z_n^{k+1} + b_1z_n^{k+2}, \cdots, z_{n-1} + a_{n-1}'z_{n-1}^2 + a_{n-1}z_n^{k+1} + b_{n-1}z_n^{k+2}, \frac{z_n}{1-a_n}),$$

and $c = \max\{|a_j'| : j = 1, 2, \cdots, n-1\}$. If $c \le \frac{1}{4}$, $|a_n| < \frac{1}{3}$ and

$$\sum_{j=1}^{n-1}\left[(k+1)(k+\frac{2|a_n|}{1-|a_n|})|a_j| + (k+2)(k+1+\frac{2|a_n|}{1-|a_n|})|b_j|\right] \le (1-2c)\frac{1-3|a_n|}{1-|a_n|},$$

then $f(z) \in K(B_p^n)$.

**Theorem 6.**(M.S.Liu and Y.C.Zhu[7]) Suppose that $n \geq 2, p \geq 2$ and $k$ is a positive integer such that $k < p \leq k+1$. Let

$$f(z) = (f_1(z_1, z_2, \cdots, z_n), f_2(z_2), \cdots, f_n(z_n)),$$

where $z = (z_1, z_2, \cdots, z_n) \in B_p^n, f_j \in H(\Delta)(j = 2, 3, \cdots, n)$, and $f_1(z_1, \cdots, z_n) : B_p^n \to C$ is holomorphic with $f_1(0, 0, \cdots, 0) = 0, \frac{\partial f_1}{\partial z_1}(0, 0, \cdots, 0) = 1$. If $f$ satisfies the following conditions

$$\frac{\partial f_1}{\partial z_1} \cdot \prod_{j=2}^n f_j'(z_j) \neq 0, \quad \sum_{j=1}^n |z_1 \frac{\partial^2 f_1}{\partial z_1 \partial z_j}| \leq |\frac{\partial f_1}{\partial z_1}|,$$

$$|z_1|^{p-1} \left| \frac{\frac{\partial f_1}{\partial z_j} \cdot \frac{f_j''(z_j)}{f_j'(z_j)}}{\frac{\partial f_1}{\partial z_1}} \right| + |z_1|^{p-1} \sum_{l=1}^n \left| \frac{\frac{\partial^2 f_1}{\partial z_j \partial z_l}}{\frac{\partial f_1}{\partial z_1}} \right| \leq \left( 1 - \left| \frac{z_j f_j''(z_j)}{f_j'(z_j)} \right| \right) \cdot |z_j|^{p-2}, j = 2, \cdots, n$$

for all $z = (z_1, \cdots, z_n) \in B_p^n$, then $f \in K(B_p^n)$.

In fact, Theorem 6 is equivalent to the following theorem.

**Theorem 6'.** Suppose that $n \geq 2, p \geq 2$ and $k$ is a positive integer such that $k < p \leq k+1$. Let

$$f(z) = (f_1(z_1, z_2, \cdots, z_n), f_2(z_2), \cdots, f_n(z_n)),$$

where $z = (z_1, z_2, \cdots, z_n) \in B_p^n, f_j \in H(\Delta)(j = 2, 3, \cdots, n)$, and $f_1(z_1, \cdots, z_n) : B_p^n \to C$ is holomorphic with $f_1(0, 0, \cdots, 0) = 0, \frac{\partial f_1}{\partial z_1}(0, 0, \cdots, 0) = 1$. If $f$ satisfies the following conditions

$$\frac{\partial f_1}{\partial z_1} \neq 0, \quad \sum_{j=1}^n |z_1 \frac{\partial^2 f_1}{\partial z_1 \partial z_j}| \leq |\frac{\partial f_1}{\partial z_1}|,$$

$$\left| \frac{\partial f_1}{\partial z_j} \cdot \frac{f_j''(z_j)}{f_j'(z_j)} \right| + \sum_{l=1}^n \left| \frac{\partial^2 f_1}{\partial z_j \partial z_l} \right| \leq |\frac{\partial f_1}{\partial z_1}| \left( 1 - \left| \frac{z_j f_j''(z_j)}{f_j'(z_j)} \right| \right) \cdot |z_j|^{p-2}, j = 2, \cdots, n$$

for all $z = (z_1, \cdots, z_n) \in B_p^n$, then $f \in K(B_p^n)$.

**Example 4.** Suppose that $n \geq 2, p \geq 2$ and $k$ is a positive integer such that $k < p \leq k+1$, and

$$f(z) = \left( z_1 + a_1 z_1 z_n^{k+1}, \frac{z_2}{1 - a_2 z_2}, \cdots, \frac{z_n}{1 - a_n z_n} \right).$$

If $|a_j| \leq \frac{1}{3}(j = 2, 3, \cdots, n)$ and

$$|a_1| \leq \min \left\{ \frac{1}{k+2}, \frac{1 - 3|a_n|}{(k+1)[k+1 - (k-1)|a_n|] + 1 - 3|a_n|} \right\},$$

then $f(z) \in K(B_p^n)$.

**Example 5.** Suppose that $n \geq 2, p \geq 2$ and $k$ is a positive integer such that $k < p \leq k+1$, and let

$$f(z) = (z_1 + \frac{a_1}{k+2} z_1^{k+2} + a_2 z_1 z_2^{k+1} + a_3 z_1 z_3^{k+1} + \cdots + a_n z_1 z_n^{k+1}, z_2, \cdots, z_n).$$

If $\sum_{j=1}^n |a_j| \leq \frac{1}{(k+1)^2+1}$, then $f(z) \in K(B_p^n)$.

In 2003, H. Hamada and G. Kohr obtained a sufficient condition for biholomorphic convex mappings on $B_2^n$ as follows.

**Theorem 7.**(H. Hamada and G. Kohr[4]) Suppose that $p \geq 2$ and $f \in N(B_p^n)$. If for any $b = (b_1, \cdots, b_n) \in C^n$ and $z = (z_1, \cdots, z_n) \in B_2^n$ with $\|b\| = 1$, we have

$$\|Df(z)^{-1}D^2f(z)(b,b)\| \leq 1,$$

then $f \in K(B_2^n)$.

We have generalized Theorem 7 to the case of $B_p^n$ as follows.

**Theorem 8.**(M.S.Liu and Y.C.Zhu[7]) Suppose that $p \geq 2$ and $f \in N(B_p^n)$. If for any $b = (b_1, \cdots, b_n) \in C^n$ and $z = (z_1, \cdots, z_n) \in B_p^n$ such that $\operatorname{Re}\langle b, \frac{\partial u}{\partial z}\rangle = 0$, we have

$$\|Df(z)^{-1}D^2f(z)(b,b)\|_p \leq \sum_{j=1}^{n} |b_j|^2 |z_j|^{p-2}, \tag{1}$$

then $f \in K(B_p^n)$.

**Example 6.** Suppose that $n \geq 2, p \geq 2$ and $k$ is a positive integer such that $k < p \leq k+1$, and let

$$f(z) = (z_1 + a_1 z_2^{k+1}, z_2 + a_2 z_1^{k+1}, z_3, \cdots, z_n).$$

If $a_1$ and $a_2$ satisfy the following inequality

$$\left\{ |a_1|^p [1 + (k+1)|a_2|]^p + |a_2|^p [1 + (k+1)|a_1|]^p \right\}^{1/p} \leq \frac{1 - (k+1)^2 |a_1||a_2|}{k(k+1)},$$

then $f(z) \in K(B_p^n)$.

**Corollary 1.**(M.S.Liu and Y.C.Zhu[7]) Suppose that $p \geq 2$ and $f \in H(B_p^n)$ with $f(0) = 0, Df(0) = I$. If $f$ satisfies $\|Df(z) - I\|_p \leq c < 1$ for each $z \in B_p^n$, and

$$\|D^2f(z)(b,b)\|_p \leq (1 - c) \sum_{j=1}^{n} |b_j|^2 |z_j|^{p-2},$$

for all $b = (b_1, \cdots, b_n) \in C^n$ and $z = (z_1, \cdots, z_n) \in B_p^n$ with $\operatorname{Re}\langle b, \frac{\partial u}{\partial z}\rangle = 0$. Then $f \in K(B_p^n)$.

In 1999, K.A.Roper and T.J.Suffridge obtained a sufficient condition for biholomorphic convex mappings on $B_2^n$ as follows.

**Theorem 9.**(K.A.Roper and T.J.Suffridge[3,11]) Suppose that $f \in H(B_2^n)$ with $f(0) = 0, Df(0) = I$, and $\sum_{k=2}^{\infty} \frac{k^2 \|D^k f(0)\|_2}{k!} \leq 1$. Then $f \in K(B_2^n)$.

**Theorem 10.**(M.S.Liu and Y.C.Zhu[7]) Suppose that $p \geq 2, f \in H(B_p^n)$ with $f(0) = 0, Df(0) = I$, and $\sum_{k=2}^{\infty} \frac{k^2 \|D^k f(0)\|_p}{k!} \leq 1$. If for any $b = (b_1, \cdots, b_n) \in C^n$ and $z = (z_1, \cdots, z_n) \in B_p^n$ such that $\operatorname{Re}\langle b, \frac{\partial u}{\partial z}\rangle = 0$, we have

$$\|D^2f(z)(b,b)\|_p \leq \left(1 - \sum_{k=2}^{\infty} \frac{k}{k!} \|D^k f(0)\|_p\right) \sum_{j=1}^{n} |b_j|^2 |z_j|^{p-2}, \tag{2}$$

then $f \in K(B_p^n)$.

**Remark 2.** When $p = 2$, by a straightforward calculation, it is obvious that the inequality (2) holds, hence set $p = 2$ in Theorem 10, we get Theorem 9.[11]

**Theorem 11.** Suppose that $p \geq 2, f \in H(B_p^n)$ with $f(0) = 0, Df(0) = I$, and $\sum_{k=2}^{\infty} \frac{k\|D^k f(0)\|_p}{k!} < 1$. If for any $b = (b_1, \cdots, b_n) \in C^n$ and $z = (z_1, \cdots, z_n) \in B_p^n$ such that $\text{Re}\langle b, \frac{\partial u}{\partial \bar{z}} \rangle = 0$, we have

$$\|D^2 f(z)(b, b)\|_p \leq \Big(1 - \sum_{k=2}^{\infty} \frac{k}{k!}\|D^k f(0)\|_p\Big) \sum_{j=1}^{n} |b_j|^2 |z_j|^{p-2},$$

then $f \in K(B_p^n)$.

**Proof.** Since $f \in H(B_p^n)$ and $f(0) = 0, Df(0) = I$, then we have

$$f(z) = z + \sum_{k=2}^{\infty} \frac{D^k f(0)}{k!}(z^k),$$

$$Df(z) = I + \sum_{k=2}^{\infty} \frac{kD^k f(0)}{k!}(z^{k-1}, \cdot).$$

for any $z \in B_p^n$. Notice that the norm of symmetric $k-$ linear operator $A_k : \prod_{j=1}^{k} C^n \to C^n$ is

$$\|A_k\|_p = \sup_{\|z^{(j)}\|_p=1, 1\leq j\leq k} \|A_k(z^{(1)}, \cdots, z^{(k)})\|_p,$$

we obtain

$$\|Df(z) - I\|_p \leq \sum_{k=2}^{\infty} \frac{k\|D^k f(0)\|_p}{k!} = c < 1$$

for any $z \in B_p^n$.

According to the hypothesis of Theorem 11, for any $b = (b_1, \cdots, b_n) \in C^n$ and $z = (z_1, \cdots, z_n) \in B_p^n$ such that $\text{Re}\langle b, \frac{\partial u}{\partial \bar{z}} \rangle = 0$, we have

$$\|D^2 f(z)(b, b)\|_p \leq \Big(1 - \sum_{k=2}^{\infty} \frac{k}{k!}\|D^k f(0)\|_p\Big) \sum_{j=1}^{n} |b_j|^2 |z_j|^{p-2}.$$

It follows from Corollary 1 that $f \in K(B_p^n)$ and the proof is complete.

### References

1. Sheng Gong, Convex and starlike mappings in several complex variables, *Since Press/Kluwer Academic Publishers*, (1998).
2. Sheng Gong, Shikun Wang and Quhuang Yu, Biholomorphic convex mappings of ball in $C^n$, *Pacific J. Math.*, **161**(2)(1993), 287-306.
3. I. Graham, G. Kohr, Geometric Function Theory in One and Higher Dimensions, *Dekker, New York*, (2003).
4. H. Hamada and G. Kohr, Simple criterions for strongly starlikeness and starlikeness of certain order, *Math. Nachr.*, **254-255**(2003), 165-171.

1320

5. K. Kikuchi, Starlike and convex mappings in several complex variables, *Pacific Jour of Math.*, **44**(1973), 569-580

6. Mingsheng Liu and Yucan Zhu, On the Generalized Roper-Suffridge Extension Operator in Banach Spaces, *International J. Math. Math. Sci.*, **2005**(2005), 1171-1187.

7. Mingsheng Liu and Yucan Zhu, Some Sufficient Conditions for Biholomorphic Convex Mappings On $B_p^n$, *J. Math. Anal. Appl.*, 316 (2006), 210–228.

8. Taishun Liu and Wenjun Zhang, Homogeneous expansions of normalized biholomorphic convex mappings over $B^p$, *Science in China(Series A)*, **40**(8)(1997), 799-806.

9. Jerry R. Muir,Jr. and T.J.Suffridge, Construction of convex mappings of $p$–balls in $C^2$, *Comput. Methods Funct. Theory*, 4(2004)1, 21-34

10. K. A. Roper and T. J. Suffridge, Convex mappings on the unit ball of $C^n$, *J. Anal. Math.*, **65**(1995), 333-347.

11. K. A. Roper and T. J. Suffridge, Convexity properties of holomorphic mappings in $C^n$, *Tran. Amer. Math. Soc.*, **351**(5)(1999), 1803-1833.

12. A. E. Taylor , D. C. Lay, Introduction to Functional Analysis, *New York: John Wiley Sons Inc*, (1980).

13. Yucan Zhu, Biholomorphic convex mappings on $B_p$, *Chinese J. Contemp. Math.*, **19**(3) (1998),271-276.

14. Yucan Zhu, Biholomorphic convex mappings on $B_p^n$(in Chinese), *Chin. Ann. of Math.*, **24A**(3)(2003), 269-278.

15. Yucan Zhu and Mingsheng Liu, The Generalized Roper-Suffridge Extension Operator in Banach Spaces (II), *J. Math. Anal. Appl.*, **303**(2005), 530-544.

**III.7 Geometric Theory of Real and Complex Functions**

Organizers: G. Barsegian

# RADIAL CLUSTER SET OF A BOUNDED HOLOMORPHIC MAP IN THE UNIT BALL OF $\mathbb{C}^n$

TOSHIO MATSUSHIMA

*Ishikawa National College of Technology*
*Kita-Chuhjoh, Tsubata, Ishikawa 929-0392*
*Japan*
*e-mail: matsush@ishikawa-nct.ac.jp*

The aim of this notes is to see the construction of bounded holomorphic maps with wild boundary behavior in the unit ball of $\mathbb{C}^n$. We construct a bounded holomorphic map in the unit ball of $\mathbb{C}^n$ whose cluster set along a radius contains a ball at given points on the boundary. For the details, see [3].

**Key words:** Bounded holomorphic map, unit ball of $\mathbb{C}^n$, cluster set along a radius
**Mathematics Subject Classification:** 32A40, 32H99

## 1. Introduction

Fatou's theorem states that every bounded holomorphic function in the unit disk has non-tangential limits at almost all points of the unit circle. A higher dimensional version of this theorem is also available. This theorem is the image of mildness of boundary behavior of the bounded holomorphic functions, which is one of the backgrounds of the inner function conjecture [4]. The fact that the existence of the inner functions was proved against the conjecture causes a question, which asks us the existence of a bounded holomorphic function that has wild boundary behavior at given points. The author's answer to this question is the following:

**Theorem 1.1** *Let* $\{\zeta_k\}_{k=1}^m$ *be an arbitrary discrete subset of the boundary of the unit ball of* $\mathbb{C}^n$, *where* $1 \leq m \leq +\infty, n \geq 1$ *and* $\zeta_k \neq \zeta_l$ *if* $k \neq l$. *Then there exists a bounded holomorphic function* $f(z)$ *in the unit ball of* $\mathbb{C}^n$ *whose radial cluster set at* $\zeta_k$

$$\bigcap_{T<1} \overline{\{f(t\zeta_k) : T < t < 1\}}$$

*includes a closed disk of positive radius for all* $k$.

To extend this result, we discuss a bounded holomorphic map defined in the unit ball of $\mathbb{C}^n$. We first construct a bounded holomorphic map whose cluster set along a radius at a boundary point is a ball (Proposition 2.1). Then, we obtain the bounded holomorphic map whose boundary behavior is wild at every point of an arbitrary sequence in the boundary of the unit ball of $\mathbb{C}^n$ (Theorem 3.1 and Theorem 3.2).

Throughout this notes, we use the following notation:

$$\langle z, w \rangle = \sum_{k=1}^{n} z_k \overline{w_k}, \ |z| = \sqrt{\langle z, z \rangle} \text{ for } z = (z_1, \ldots, z_n), \ w = (w_1, \ldots, w_n) \in \mathbb{C}^n;$$
$$B_n = \{z \in \mathbb{C}^n : |z| < 1\}, \ \partial B_n = \{z \in \mathbb{C}^n : |z| = 1\}.$$

$H(B_n)$ denotes the set of all holomorphic functions on $B_n$. Moreover we define

$$\|f\| = \sup_{z \in B_n} |f(z)|$$

for a function $f$ in $B_n$. We write $H^\infty(B_n)$ for the set $\{f \in H(B_n) : \|f\| < +\infty\}$. $H^\infty(B_n, \mathbb{C}^g)$ denotes the set of all maps from $B_n$ to $\mathbb{C}^g$ whose components are all in $H^\infty(B_n)$.

We introduce the radial cluster set of a map $F$ defined in $B_n$ at $\zeta \in \partial B_n$:

$$C_r(F, \zeta) = \bigcap_{T < 1} \overline{\{F(t\zeta) : T < t < 1\}}.$$

For simplicity, we construct the maps in $H^\infty(B_n, \mathbb{C}^2)$. See [3] for the other cases.

## 2. Construction of the basic map

**Proposition 2.1** *For any $\zeta \in \partial B_n$ and $n \geq 1$, there exists $F \in H^\infty(B_n, \mathbb{C}^2)$ such that $C_r(F, \zeta)$ is a ball of $\mathbb{C}^2$.*

*Sketch of proof.* Define $l(z) = -\log(1 - \langle z, \zeta \rangle)$ for $z \in B_n$. We easily notice that we can choose the argument of $1 - \langle z, \zeta \rangle$ as

$$-\frac{\pi}{2} < \arg(1 - \langle z, \zeta \rangle) < \frac{\pi}{2}.$$

Thus $l(z)$ is a single-valued holomorphic function in $B_n$. Let $M$ be an arbitrary positive real number, and let $K$, $L_1$, $L_2$, and $L_3$ are positive real numbers which are linearly independent over $\mathbb{Z}$. We define the following functions in $B_n$:

$$x_\zeta^1(z) = \frac{M}{C} \sin^2(2\pi K l(z)) \sin(2\pi L_1 l(z)) \sin(2\pi L_2 l(z)) \sin(2\pi L_3 l(z)),$$

$$y_\zeta^1(z) = \frac{M}{C} \sin^2(2\pi K l(z)) \sin(2\pi L_1 l(z)) \sin(2\pi L_2 l(z)) \cos(2\pi L_3 l(z)),$$

$$x_\zeta^2(z) = \frac{M}{C} \sin^2(2\pi K l(z)) \sin(2\pi L_1 l(z)) \cos(2\pi L_2 l(z)),$$

$$y_\zeta^2(z) = \frac{M}{C} \sin^2(2\pi K l(z)) \cos(2\pi L_1 l(z)),$$

where $C = \exp\left(\pi^2(K + L_1 + L_2 + L_3)\right)$. It is trivial that these functions are all bounded holomorphic in $B_n$. Furthermore, we define

$$f^1(z) = x_\zeta^1(z) + i y_\zeta^1(z) \quad \text{and} \quad f^2(z) = x_\zeta^2(z) + i y_\zeta^2(z),$$

and set

$$F(z) = \left(f^1(z), f^2(z)\right).$$

Then $F \in H^\infty(B_n, \mathbb{C}^2)$ and $\|F\| = \sup_{z \in B_n} |F(z)| < M$. Notice that $z = t\zeta$ for $0 \leq t < 1$ if $z$ is in the radius of $B_n$ terminating at $\zeta$. Then, it is clear that the image of this radius by $F$ in $\mathbb{C}^2$ is naturally identified with

$$D = \left\{ \left( x^1(T), y^1(T), x^2(T), y^2(T) \right) : T \geq 0 \right\} \subset \mathbb{R}^4,$$

where

$$x^1(T) = \frac{M}{C} \sin^2(2\pi K T) \sin(2\pi L_1 T) \sin(2\pi L_2 T) \sin(2\pi L_3 T),$$

$$y^1(T) = \frac{M}{C} \sin^2(2\pi K T) \sin(2\pi L_1 T) \sin(2\pi L_2 T) \cos(2\pi L_3 T),$$

$$x^2(T) = \frac{M}{C} \sin^2(2\pi K T) \sin(2\pi L_1 T) \cos(2\pi L_2 T),$$

$$y^2(T) = \frac{M}{C} \sin^2(2\pi K T) \cos(2\pi L_1 T).$$

Since $K$, $L_1$, $L_2$, and $L_3$ are linearly independent over $\mathbb{Z}$, $D$ is a dense subset of the ball

$$\left\{ w \in \mathbb{C}^2 : |w| \leq \frac{M}{C} \right\}.$$

Therefore, we conclude that

$$C_r(F, \zeta) = \left\{ w \in \mathbb{C}^2 : |w| \leq \frac{M}{C} \right\}.$$

For the details, see [3]. $\qquad\qquad\qquad\qquad\qquad\qquad\qquad\qquad\qquad\qquad\qquad\qquad$ □

## 3. Main results

**Theorem 3.1** *Let $\{\zeta_k\}_{k=1}^m$ be a finite subset of $\partial B_n$. Then there exists $F \in H^\infty(B_n, \mathbb{C}^2)$ such that $C_r(F, \zeta_k)$ is a ball of $\mathbb{C}^2$ for all $k$.*

*Sketch of proof.* Let $M$ be an arbitrary positive real number. For $z \in B_n$, we define

$$f_{\zeta_k}^1(z) = x_{\zeta_k}^1(z) + i y_{\zeta_k}^1(z) \quad \text{and} \quad f_{\zeta_k}^2(z) = x_{\zeta_k}^2(z) + i y_{\zeta_k}^2(z) \quad \text{for} \quad 1 \leq k \leq m.$$

Note that $f_{\zeta_k}^1$ and $f_{\zeta_k}^2$ are the functions $f^1$ and $f^2$ in the proof of Proposition 2.1 respectively if we substitute $\zeta_k$ for $\zeta$. Moreover, we define the map

$$F(z) = \left( \frac{1}{m} \sum_{k=1}^m f_{\zeta_k}^1(z), \frac{1}{m} \sum_{k=1}^m f_{\zeta_k}^2(z) \right).$$

Then $F \in H^\infty(B_n, \mathbb{C}^2)$ and $\|F\| < M$. Decompose $F(t\zeta_j)$ for $1 \leq j \leq m$ into

$$F(t\zeta_j) = \left( \frac{1}{m} f_{\zeta_j}^1(t\zeta_j), \frac{1}{m} f_{\zeta_j}^2(t\zeta_j) \right) + \left( \frac{1}{m} \sum_{k \neq j} f_{\zeta_k}^1(t\zeta_j), \frac{1}{m} \sum_{k \neq j} f_{\zeta_k}^2(t\zeta_j) \right).$$

Since $f_{\zeta_k}^1$ and $f_{\zeta_k}^2$ are continuous at $\zeta_j$ if $k \neq j$, we derive that

$$C_r\left(F, \zeta_j\right) = C_r\left(\left(\frac{1}{m}f_{\zeta_j}^1, \frac{1}{m}f_{\zeta_j}^2\right), \zeta_j\right) + \left(\frac{1}{m}\sum_{k \neq j} f_{\zeta_k}^1(\zeta_j), \frac{1}{m}\sum_{k \neq j} f_{\zeta_k}^2(\zeta_j)\right),$$

which is a ball of $\mathbb{C}^2$. For the details, see [3].                    $\square$

**Theorem 3.2** *Let* $\{\zeta_k\}_{k=1}^{\infty}$ *be an arbitrary sequence of* $\partial B_n$. *Then there exists* $F \in H^{\infty}(B_n, \mathbb{C}^2)$ *such that* $C_r(F, \zeta_k)$ *contains a ball of* $\mathbb{C}^2$ *for all* $k$.

*Sketch of proof.* Let $\{a_k\}_{k=1}^{\infty}$ be a sequence of positive real numbers satisfying $\sum_{k=1}^{\infty} a_k = 1$, and let $M$ be an arbitrary positive real number. We define

$$F_p(z) = \left(\sum_{k=1}^{p} a_k f_{\zeta_k}^1(z), \sum_{k=1}^{p} a_k f_{\zeta_k}^2(z)\right),$$

and set

$$F(z) = \left(\sum_{k=1}^{\infty} a_k f_{\zeta_k}^1(z), \sum_{k=1}^{\infty} a_k f_{\zeta_k}^2(z)\right),$$

where $f_{\zeta_k}^1$ and $f_{\zeta_k}^2$ are as in the proof of Theorem 3.1. We easily notice that the components of $F(z)$ absolutely and uniformly converge to the bounded holomorphic functions in $B_n$. Then, it is clear that $F \in H^{\infty}(B_n, \mathbb{C}^2)$ and $\|F\| < M$. Moreover, we see that

$$C_r\left(F_p, \zeta_j\right) = C_r\left(\left(a_j f_{\zeta_j}^1, a_j f_{\zeta_j}^2\right), \zeta_j\right) + \left(\sum_{k \neq j} a_k f_{\zeta_k}^1(\zeta_j), \sum_{k \neq j} a_k f_{\zeta_k}^2(\zeta_j)\right) \quad \text{for every } j \geq 1,$$

which is a ball of $\mathbb{C}^2$. Since $F_p$ uniformly converges to $F$ as $p \to +\infty$, we can choose some integer $N$ for any $\delta > 0$ such that the distance between the centers of two balls $C_r(F_N, \zeta_j)$ and $C_r(F_p, \zeta_j)$ is less than $\delta$ if $p \geq N$. This shows that $C_r(F_p, \zeta_j)$ contains some ball of $\mathbb{C}^2$ if $p \geq N$. Hence, $C_r(F, \zeta_j)$ contains this ball. For the details, see [3].                    $\square$

**Remark.** In [2] the author shows the existence of a bounded holomorphic map that is defined on some kind of domain of $\mathbb{C}^n$ whose cluster sets along segments are "big" at given boundary points.

## References

1. T. Matsushima, Bounded holomorphic function with some boundary behavior in the unit ball of $\mathbb{C}^n$, Kodai Math. J., **24** (2001) 305-312.
2. T. Matsushima, Bounded holomorphic functions and maps with some boundary behavior, J. Math. Anal. Appl., **285** (2003) 691-707.
3. T. Matsushima, Bounded holomorphic function with some radial cluster set in the unit ball of $\mathbb{C}^n$, in preparation.
4. W. Rudin, Function theory in the unit ball of $\mathbb{C}^n$, Berlin-Heidelberg-New York, Springer, 1980.

## IV.2 Mathematical and computational aspects of kinetic models

Organizer: Armando Majorana

The session was devoted to mathematical models described by kinetic equations, and to recent techniques for their numerical solutions. Several different topics have been treated, ranging from semiconductor modeling to biological applications. The session took place in two different days, hosting a total of eight speakers, who gave interesting talks on different aspects of the thematic. Four papers have been submitted for publication in the proceedings of the conference. We report here a brief description of the other four talks. Dr. Groppi gave a review on kinetic model of chemical reaction in multispecies mixtures. Analytical and numerical results were presented. Dr. Mascali presented a hydrodynamical model for charge transport in semiconductor devices, based on the moment closure of an underlying kinetic model, obtained by the maximum entropy technique. Prof. Majorana spoke about finite difference schemes for the numerical solution of Boltzmann-Poisson system for electron transport in semiconductor devices. A comparison between the deterministic scheme and detailed Monte Carlo simulation have been presented. The work, in collaboration with J.A.Carrillo, I.Gamba, and C.W.Shu, will appear on Journal of Computational Physics. Dr. Dietmar Oeltz presented some recent results on diffusion limit of some kinetic equations with relaxation collision kernels.

# EFFICIENCY CONSIDERATIONS FOR THE
# BOLTZMANN-POISSON SYSTEM FOR SEMICONDUCTORS*

A. DOMAINGO

*Institute of Theoretical and Computational Physics,*
*Graz University of Technology,*
*Petersgasse 16,*
*8010 Graz, Austria*
*E-mail: domaingo@itp.tu-graz.ac.at*


A. MAJORANA

*Dipartimento di Matematica e Informatica,*
*Università di Catania,*
*Città Universitaria, viale Andrea Doria 6,*
*95125 Catania, Italy*
*E-mail: majorana@dmi.unict.it*

The simulation of semiconductor devices has become a standard tool in the construction process. A very detailed description of the transport processes in semiconductors is provided by kinetic transport models. However, the numerical approaches to this kind of equation which have been presented so far are not competitive with respect to the computation time to other methods. In this paper, we investigate approaches to speed up numerical methods like box methods and WENO schemes for solving the Boltzmann transport equation.

**Key words:** Boltzmann-Poisson system, semiconductor, transport process
**Mathematics Subject Classification: 35Q35**

## 1. Introduction

The numerical simulation of carrier transport in semiconductors has been established in the development process of semiconductor devices as a cheap and efficient method. However, as the scale integration approaches gate lengths around 50 nm the standard drift-diffusion equations[1], which are the nowadays work horse in commercial simulators, provide no longer a sufficient description. One approach to deal with this problem is to introduce quantum corrections to the transport equations[1,2]. Another possibility is to describe the transport by more sophisticated semiclassical kinetic transport equations. Among these, the Boltzmann transport equation (BTE) is a widely used model[3]. Of course, this integro-differential integration is compu-

*This work has been supported by the European community program IHP under the contract number HPRN-CT-2002-00282 on behalf of the CNR.

tationally more demanding than the simple drift diffusion equations. Nevertheless, many different solutions have been developed to solve the BTE.

The first method applied to solve the BTE directly has been the stochastic Monte Carlo (MC) method[4]. The clear advantage of this method is that it can be interpreted very easily in physical terms. However, the drawback is the presence of statistical noise. Noise-free alternatives are deterministic solution methods.

Starting with a paper of Fatemi and Odeh[5], where they have presented an up-wind finite difference scheme, several fully discretized deterministic methods have been introduced. A box method, which allows a more simple evaluation of the scattering operator has been published by Majorana and Pidatella[6]. Later, high order shock capturing algorithms have been applied by Carrillo et al.[7] to cope with the strong carrier density gradients due to doping. A combination of finite element and finite difference methods has been presented in Ref. 8. All mentioned methods have one point in common. Due the their fully discretized nature, they are naturally slower than solvers for drift diffusion like equations. Therefore, methods must be found, which can make new kinetic methods more competitive with respect to computation time. To this end, the concept of suitable weight functions has been introduced in Ref. 9.

In this paper, we present several approaches by which an acceleration of the numerical schemes can be achieved. In Sec. 2, we introduce in short the necessary equations to model the charge transport in devices. A short review, how the model equations can be obtained from these equations is given in Sec. 3. In Sec. 4, the numerical performance of several different weight functions according to the idea presented in Ref. 9 is investigated. Finally, the effect of introducing non-uniform computational grids is considered in Sec. 5.

## 2. The Boltzmann-Poisson system

The Boltzmann transport (BTE) equation allows a semiclassical description of the electron transport in semiconductors. It governs the time evolution of the electron distribution function (DF) $f(\boldsymbol{k}, \boldsymbol{x}, t)$ where $\boldsymbol{k}$ denotes the quasi-momentum, $\boldsymbol{x}$ is the position and $t$ time. For planar devices, i.e. the doping varies only along a certain direction $z$, made of semiconductors with a spherically symmetric, non-parabolic band structure

$$\gamma(\varepsilon) := \varepsilon(1 + \alpha\varepsilon) = \frac{\hbar^2}{2m^*}k^2. \tag{1}$$

the BTE can be written as

$$\frac{\partial f}{\partial t} + \mu v \frac{\partial f}{\partial z} - eE\left[\mu v \frac{\partial f}{\partial \varepsilon} + \frac{1-\mu^2}{\sqrt{2m^*\gamma}}\frac{\partial f}{\partial \mu}\right] = Q[f]. \tag{2}$$

Here, $m^*$ denotes the electron effective mass, $\alpha$ the non-parabolicity factor, $e$ the positive elementary charge, $E$ the space dependent electric field and the cosine of the polar angle is introduced by the one-to-one mapping

$$\boldsymbol{k} \iff (\varepsilon, \mu, \varphi), \quad \mu := (\boldsymbol{k}/k) \cdot \boldsymbol{e}_z. \tag{3}$$

Thus, the new DF is $f(\varepsilon, \mu, z, t)$ due to symmetry considerations. Finally, the modulus of the group velocity $v$ is given by

$$v(\varepsilon) = \frac{1}{\gamma'(\varepsilon)} \sqrt{\frac{2\gamma(\varepsilon)}{m^*}} \tag{4}$$

and the collision term $Q[f]$ accounts for the change of the DF due to various scattering mechanisms. By applying the low density approximation, the collision term is given by

$$Q[f] = \iint \mathcal{D}(\varepsilon')[S(\varepsilon', \varepsilon)f(\varepsilon', \mu', z, t) - S(\varepsilon, \varepsilon')f(\varepsilon, \mu, z, t)]\mathrm{d}\varepsilon'\mathrm{d}\mu', \tag{5}$$

where the density of states

$$\mathcal{D}(\varepsilon) = \frac{1}{2\pi^2} \left(\frac{2m^*}{\hbar^2}\right)^{3/2} \gamma'(\varepsilon)\sqrt{\gamma(\varepsilon)} \tag{6}$$

has been introduced as the Jacobian of the transformation (3). The isotropic scattering rate $S(\varepsilon, \varepsilon')$ accounts for elastic scattering of electrons with acoustic phonons and inelastic scattering with optical phonons of energy $\hbar\omega_{\mathrm{op}}$. Both sorts of phonon are assumed to be in equilibrium characterized by the lattice temperature $T_{\mathrm{L}}$. Explicitly, the scattering rate reads

$$S(\varepsilon, \varepsilon') = K_{\mathrm{ac}}\delta(\varepsilon'-\varepsilon) + K_{\mathrm{op}}\left[(n_{\mathrm{op}}+1)\,\delta(\varepsilon'-\varepsilon+\hbar\omega_{\mathrm{op}}) + n_{\mathrm{op}}\,\delta(\varepsilon'-\varepsilon-\hbar\omega_{\mathrm{op}})\right], \tag{7}$$

where $K_{\mathrm{ac}}$ and $K_{\mathrm{op}}$ are constants and the phonon occupation number $n_{\mathrm{op}}$ is given by the Bose-Einstein distribution: $n_{\mathrm{op}} = [\exp(\hbar\omega_{\mathrm{op}}/k_{\mathrm{B}}T) - 1]^{-1}$.

Since electron transport in the semiconductor causes space charge regions, the electric field $E$ and potential $\Phi$ must be self consistently coupled with the BTE via the Poisson equation:

$$-\frac{\mathrm{d}E}{\mathrm{d}z} = \frac{\mathrm{d}^2\Phi}{\mathrm{d}z^2} = \frac{e}{\epsilon}[n(z, t) - n_{\mathrm{D}}(z)]. \tag{8}$$

In this equation, the semiconductors permittivity is labelled by $\epsilon$, $n_{\mathrm{D}}$ is the fixed profile of donors and the electron density $n$ can be obtained as the zeroth moment from the DF by

$$n(z, t) = \frac{1}{2} \iint \mathcal{D}(\varepsilon)f(\varepsilon, \mu, z, t)\mathrm{d}\varepsilon\mathrm{d}\mu. \tag{9}$$

## 3. Discretized equations

The first step in discretizing the Boltzmann-Poisson system is to choose an appropriate computational domain for the independent variables. For the angle variable $\mu$ and the position $z$ this is straightforward, because these quantities have well defined physical bounds. The model band structure defined by Eq. (1) has only a lower bound, contrary to real band structures which always have a certain band width. However, it can be assumed even for high fields that the DF is almost vanishing for energies $\varepsilon > \varepsilon_{\mathrm{max}}$.

In the $(\varepsilon, \mu)$-plane a non-uniform computational grid $\mathcal{G} = \{(\varepsilon_\nu, \mu_\lambda)\}$ with $1 \le \nu \le N_\nu$ and $1 \le \lambda \le N_\lambda$ is introduced. Around each of the grid points cells $\mathcal{Z}_{\nu\lambda} = [\varepsilon_{\nu-1/2}, \varepsilon_{\nu+1/2}] \times [\mu_{\lambda-1/2}, \mu_{\lambda+1/2}]$ are constructed, where $\varepsilon_{\nu\pm1/2} = \varepsilon_\nu \pm \Delta\varepsilon_\nu/2$ and $\mu_{\lambda\pm1/2} = \mu_\lambda \pm \Delta\mu_\lambda/2$. These cells must not overlap and fill the entire computational domain:

$$\overset{\circ}{\mathcal{Z}}_{\nu\lambda} \cap \overset{\circ}{\mathcal{Z}}_{n\ell} = \emptyset \quad \forall (\nu, \lambda) \ne (n, \ell) \qquad \wedge \qquad \bigcup_{\nu\lambda} \mathcal{Z}_{\nu\lambda} = [0, \varepsilon_{\max}] \times [-1, 1]. \quad (10)$$

Additionally, with regard to the collision with optical phonons, the following requirements are stated:

$$\hbar\omega_{\text{op}}/\Delta\varepsilon_\nu = s_\nu \in \mathbb{N}, \qquad s_\nu \ge s_{\nu+1}, \quad (11)$$

which will be further discussed when discretizing the collision operator.

The BTE, Eq. (2), can be multiplied by the density of states to be written in conservative form, which is much more convenient for the following discretization:

$$\frac{\partial(\mathcal{D}f)}{\partial t} + \frac{\partial}{\partial z}(\mu v \mathcal{D}f) - eE\left\{\frac{\partial}{\partial\varepsilon}(\mu v \mathcal{D}f) + \frac{\partial}{\partial\mu}\left(\frac{1-\mu^2}{\sqrt{2m^*\gamma}}\mathcal{D}f\right)\right\} = \mathcal{D}Q[f] \quad (12)$$

with

$$\mathcal{D}Q[f] = \int \left\{ K_{\text{ac}}\mathcal{D}^2 \left[f(\varepsilon, \mu') - f(\varepsilon, \mu)\right] + n_{\text{op}}^+ K_{\text{op}}\mathcal{D}\left[\mathcal{D}^+ f(\varepsilon^+, \mu') - \mathcal{D}^- f(\varepsilon, \mu)\right] \right.$$
$$\left. + n_{\text{op}} K_{\text{op}}\mathcal{D}\left[\mathcal{D}^- f(\varepsilon^-, \mu') - \mathcal{D}^+ f(\varepsilon, \mu)\right] \right\} d\mu'. \quad (13)$$

Here, the abbreviations $\mathcal{D}^\pm = \mathcal{D}(\varepsilon \pm \hbar\omega_{\text{op}})$ and $n_{\text{op}}^+ = n_{\text{op}} + 1$ have been introduced and the independent variables $z$ and $t$ have been omitted for the DF.

The procedure to obtain discretized model equations is an enhancement of the method presented in Ref. 9 for non-uniform grids. In a first step, the product of the DF and the density of states is replaced by the product of a known weight function $w(\varepsilon)$ and a new unknown $g(\varepsilon, \mu, z, t)$:

$$\mathcal{D}(\varepsilon)f(\varepsilon, \mu, z, t) = w(\varepsilon)g(\varepsilon, \mu, z, t). \quad (14)$$

This ansatz is inserted into the BTE and the resulting equation is then integrated over each of the cells $\mathcal{Z}_{\nu\lambda}$ with respect to $\varepsilon$ and $\mu$ under the assumption

$$g(\varepsilon, \mu, z, t) \approx g_{\nu\lambda}(z, t) \quad \forall (\varepsilon, \mu) \in \mathcal{Z}_{\nu\lambda}. \quad (15)$$

The resulting model equations are given by

$$\Delta\mu_\lambda \langle w \rangle_\nu \frac{\partial g_{\nu\lambda}}{\partial t} + \mu_\lambda \Delta\mu_\lambda \langle vw \rangle_\nu \frac{\partial g_{\nu\lambda}}{\partial z} - eE\mu_\lambda \Delta\mu_\lambda \left[v(\varepsilon)w(\varepsilon)\tilde{g}_\lambda(\varepsilon)\right]_{\varepsilon_{\nu-1/2}}^{\varepsilon_{\nu+1/2}}$$
$$- eE\langle w\tilde{\gamma} \rangle_\nu \left[(1-\mu^2)\hat{g}_\nu(\mu)\right]_{\mu_{\lambda-1/2}}^{\mu_{\lambda+1/2}} = Q_{\nu\lambda}[g] \quad (16)$$

with the definitions $\langle \cdot \rangle_\nu = \int_{\varepsilon_{\nu-1/2}}^{\varepsilon_{\nu+1/2}} \cdot \, d\varepsilon$ and $\tilde{\gamma}(\varepsilon) = [2m^*\gamma(\varepsilon)]^{-1/2}$. To evaluate the values of the functions $\tilde{g}_\lambda(\varepsilon) = g(\varepsilon, \mu_\lambda)$ and $\hat{g}_\nu(\mu) = g(\varepsilon_\nu, \mu)$, a MinMod slope limiter[10] is applied. By denoting $\tilde{g}_\lambda(\varepsilon_{\nu\pm1/2})$ by $g_{(\nu\pm1/2)\lambda}$, the contribution of the first term in square brackets in (16) can be expressed as

$$a_{(\nu+1/2)\lambda}g_{(\nu+1/2)\lambda} - a_{(\nu-1/2)\lambda}g_{(\nu-1/2)\lambda} \tag{17}$$

with $a_{(\nu\pm1/2)\lambda} = -eE\mu_\lambda\Delta\mu_\lambda v(\varepsilon_{\nu\pm1/2})w(\varepsilon_{\nu\pm1/2})$. The slope limiter approximates $g_{(\nu+1/2)\lambda}$ by

$$g_{(\nu+1/2)\lambda} \approx \begin{cases} g_{\nu\lambda} + \frac{\Delta\varepsilon_\nu}{2}g'_{\nu\lambda} & \text{if } a > 0 \\ g_{(\nu+1)\lambda} - \frac{\Delta\varepsilon_{\nu+1}}{2}g'_{(\nu+1)\lambda} & \text{if } a < 0 \end{cases}, \tag{18}$$

where the slopes $g'_{\nu\lambda}$ read

$$g'_{\nu\lambda} = \begin{cases} \min\left(|G'_{\nu\lambda}|,\left|G'_{(\nu+1)\lambda}\right|\right)\text{sgn}\left(G'_{\nu\lambda}\right) & \text{if } G'_{\nu\lambda}G'_{(\nu+1)\lambda} > 0 \\ 0 & \text{else} \end{cases}, \tag{19}$$

with

$$G'_{\nu\lambda} = \frac{2(g_{\nu\lambda} - g_{(\nu-1)\lambda})}{\Delta\varepsilon_\nu + \Delta\varepsilon_{\nu-1}}. \tag{20}$$

An analogous scheme applies to the terms containing $\hat{g}_\nu(\mu)$.

By applying the steps above, the BTE of integro-differential type has been reduced to a system of partial differential equations. To deal with the derivate with respect to $z$ in Eq. (16), the third order WENO scheme for non uniform grids presented in Ref. 8 is applied, which results in a system of ordinary differential equations, which is solved by using a forward Euler scheme. The time-step used in this scheme is bounded by a suitable CFL condition to guarantee stability. In the present case, this CFL condition reads

$$\Delta t \leq \frac{\text{CFL}}{\max_{\nu\lambda}\left[\left|\frac{\mu_\lambda\langle vw\rangle_\nu}{\langle w\rangle_\nu\Delta z_i}\right| + \left|\frac{eE\mu_\lambda vw}{\langle w\rangle_\nu}\right| + \left|\frac{eE\langle w\tilde{\gamma}\rangle_\nu(1-\mu^2)}{\langle w\rangle_\nu\Delta\mu_\lambda}\right|\right]} \tag{21}$$

with a suitably chosen number CFL $\leq 1$.

The discretization of the collision term is quite straightforward. However, the non-uniform grid introduces additional effort compared to the case with an uniform grid. Acoustic phonons do not cause the electrons to change their energy during the collision. Therefore, elastic scattering couples only cells $Z_{\tilde{\nu}\lambda}$ which belong to the same energy group $\tilde{\nu}$. For inelastic collisions, the electron energy is increased or decreased by the optical phonon energy $\hbar\omega_{\text{op}}$. In the case of an uniform grid, where the energy cell-width is chosen so that $\Delta\varepsilon = s\hbar\omega_{\text{op}}, s \in \mathbb{N}$, an energy cell with index $\nu$ couples with exactly two other cells $(\nu + s)$ and $(\nu - s)$. This is no longer the case for a non-uniform grid. However, due to the condition (11), a cell with energy index $\nu$ always couples with a certain number of cells $n_\nu^+$ of higher energies and $n_\nu^-$ of lower energies. The discretized collision operator can therefore be written as

$$Q_{\nu\lambda}[g] = 2\pi K_{\text{ac}}\Delta\mu_\lambda\langle w\mathcal{D}\rangle_\nu\left[\sum_{\lambda'}(\Delta\mu_{\lambda'}g_{\nu\lambda'}) - 2g_{\nu\lambda}\right]$$
$$+ 2\pi K_{\text{op}}\Delta\mu_\lambda n_{\text{op}}\left[\sum_{\nu',\lambda'}(\Delta\mu_{\lambda'}\langle w\mathcal{D}^+\rangle_{\nu'}g_{\nu'\lambda'}) - 2\langle w\mathcal{D}^+\rangle_\nu g_{\nu\lambda}\right] \tag{22}$$
$$+ 2\pi K_{\text{op}}\Delta\mu_\lambda n_{\text{op}}^+\left[\sum_{\nu'',\lambda'}(\Delta\mu_{\lambda'}\langle w^+\mathcal{D}\rangle_\nu g_{\nu''\lambda'}) - 2\langle w\mathcal{D}^-\rangle_\nu g_{\nu\lambda}\right].$$

Concerning the summations of the energy cells, the following conditions must hold:

$$\nu' \in \left\{ n \mid (\varepsilon_{n-1/2}, \varepsilon_{n+1/2}) \cap (\varepsilon_{n-1/2} - \hbar\omega_{\mathrm{op}}, \varepsilon_{n+1/2} - \hbar\omega_{\mathrm{op}}) \neq \emptyset \right\}, \tag{23a}$$

$$\nu'' \in \left\{ n \mid (\varepsilon_{n-1/2}, \varepsilon_{n+1/2}) \cap (\varepsilon_{n-1/2} + \hbar\omega_{\mathrm{op}}, \varepsilon_{n+1/2} + \hbar\omega_{\mathrm{op}}) \neq \emptyset \right\}. \tag{23b}$$

It should be noted that due to the condition (11) the summation index $\nu''$ for upscattering may take on at most one value.

## 4. Admissible weight functions

So far, no statement concerning the weight function $w(\varepsilon)$ has been made. Of course, this choice is not a trivial task, because several aspects must be considered for this problem. First the weight function should guarantee a stable and numerically efficient scheme. And second one would like to make an interpretation on physical grounds why a chosen weight function might be a good choice. The simplest admissible weight function

$$w_{\mathrm{u}}(\varepsilon) \equiv 1 \tag{24}$$

is also the one with the clearest physical interpretation. The unknown function $g(\varepsilon, \mu.z, t)$ in this case is exactly the particle density (i.e. the density of states times the DF). By choosing the weight function

$$w_{\mathcal{D}}(\varepsilon) = \mathcal{D}(\varepsilon), \tag{25}$$

one returns again to the original formulation and obtains the electron distribution as unknown. So far, the defined weight functions do not contain any hint to the structure of the solution. By applying the weight function

$$w_{\mathrm{eu}}(\varepsilon) = \exp\left[ \frac{\min(\varepsilon, \varepsilon_{\mathrm{cut}})}{k_{\mathrm{B}}T} \right], \tag{26}$$

some information about the possible solution is included into the weight function. Here, $\varepsilon_{\mathrm{cut}}$ is a suitably chosen cut-off energy to avoid instabilities in the numerical scheme. Considering the model equation (16), the equation must be divided by $\langle w \rangle_{\nu}$. If the cut-off energy is too high, this term becomes very small and instabilities might occur due to rounding errors. Considering that the well known solution of the BTE in low density approximation without external fields is the Boltzmann distribution

$$f_{\mathrm{B}}(\varepsilon) = \exp\left( -\frac{\varepsilon - \mu_{\mathrm{c}}}{k_{\mathrm{B}}T} \right), \tag{27}$$

where $\mu_{\mathrm{c}}$ is the chemical potential, this weight function can be interpreted in the following way. In the case without external field the exact solution for the unknown $g(\varepsilon, \mu, z, t)$ by applying the weight function $w_{\mathrm{eu}}$ would be the energy independent term $\exp[\mu_{\mathrm{c}}/(k_{\mathrm{B}}T)]$, which is the smoothest of all possible solutions and therefore suited very well for a numerical approximation. In the case that external fields are present, the chemical potential will also depend on the energy $\varepsilon$. However, if this variation is not too strong, an efficient numerical treatment is still possible. Finally, one might also think of a combined weight function

$$w_{e\mathcal{D}}(\varepsilon) = \mathcal{D}(\varepsilon) \exp\left[\frac{\min(\varepsilon, \varepsilon_{\mathrm{cut}})}{k_{\mathrm{B}}T}\right]. \tag{28}$$

The different weight functions are tested by considering a bulk silicon sample. For the detailed physical parameters refer to Ref. 8. In particular, the dependence of the average velocity on the applied external electric field is investigated. The energy grid is fixed for these simulations, i.e. $\Delta\varepsilon = \hbar\omega_{\mathrm{op}}/s = const.$ The top-left plot in Fig. 1 shows the error of the average velocity in dependence of the applied electric field of a high resolution run with $s = 12$ with respect to a more sophisticated method[8]. Here, the error for any quantity $u$ with respect to a reference $u_{\mathrm{ref}}$ is calculated by

$$r_u = \left|\frac{u - u_{\mathrm{ref}}}{u_{\mathrm{ref}}}\right|. \tag{29}$$

The other plots in Fig. 1 show the convergence behavior of the average velocity in dependence of the grid refinement parameter $s$, where the case with $s = 12$ has been taken as reference. As expected, one observes a monotone convergence, except for the weight functions $w_u$ and $w_\mathcal{D}$ at fields around $E = 500$ kV/cm. The reason for this irregular convergence behavior is not yet clarified and can not be explained from the model equations. However, it must be remarked that from the numerical point of view the order of the error is around $10^{-3}$ which is at least as good as for lower fields. By considering that the typical electric fields are of the magnitude of 100 kV/cm in not too high integrated devices, i.e. with channel lengths around 100 nm[6], the convergence behavior at that fields must be the criterion to choose a suitable weight function. Therefore, the superior weight functions are $w_u$ and $w_\mathcal{D}$. By considering the comparison with the more sophisticated method, too, the function $w_u$ is identified as the optimal weight function.

## 5. Grid layout

To examine the effects of the grid with respect to $\varepsilon$ and $z$ a Schottky barrier diode (SBD) is considered. This device is characterized by a planar device layout and strong carrier density gradients. The doping profile consists of a small region with high doping $n^+$ and a longer region with low doping $n$. The high doping region is equipped with an ohmic contact while there is a Schottky contact at the end of the low doping region. In the present case, we consider the diode in forward bias regime. In this situation, the Schottky contact can be modeled by a perfect absorbing contact (cf. Ref. 11). The high doping concentration is given by $n^+ = 10^{17}$ cm$^{-3}$ over a length of $L^+ = 0.3$ $\mu$m the low doping region is characterized by $n = 10^{16}$ cm$^{-3}$ and $L = 0.8$ $\mu$m.

As already pointed out in the previous section, the equilibrium distribution function shows the strongest variation with respect to energy at low energies. Therefore,

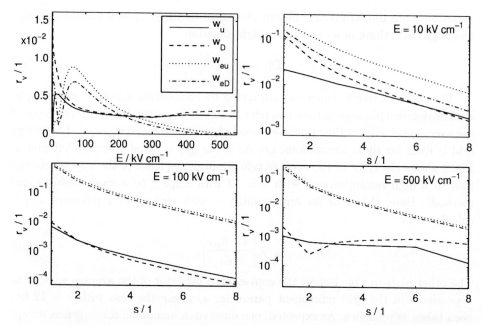

Fig. 1. The top-left plot shows the relative error $r_v$ of the average velocity in comparison to a more sophisticated model. The other plots show the relative error in dependence of the grid refinement factor $s$ for different electric field strengths.

the idea is to set up a computational energy grid with a small cell-width at low energies and a bigger cell-width at high energies even in the case with an applied electric field. Thus, the energy steps can be defined by

$$\Delta\varepsilon_\nu = \begin{cases} \Delta\varepsilon_< & \text{for } \nu \leq \nu^*, \\ \Delta\varepsilon_> & \text{for } \nu > \nu^*, \end{cases} \tag{30}$$

with an appropriate index $\nu^*$ and $\Delta\varepsilon_< < \Delta\varepsilon_>$. The maximum energy for the discretization is given by $\varepsilon_{\max} = 10\hbar\omega_{op}$. The reference result is obtained with a constant energy resolution of $s = 12$ and 120 energy groups (12:120). It is compared to a low resolution test with constant energy resolution, $s = 4$ and 40 energy groups (4:40). The test for the non-uniform energy grid is performed with 36 small energy cells with $s_< = \hbar\omega_{op}/\Delta\varepsilon_< = 12$ and 28 large cells with $s_> = 4$ (12:36,4:28). The comparison in the top-right plot of Fig. 2 shows that the most problematic zone is the depletion region next to the Schottky contact. Here the (4:40) discretization leads to an unacceptable error, while the (12:36,4:28) method still behaves fairly well. Concerning the consumed CPU time, the (12:36,4:28) method saves 35% compared to the (12:120) reference.

As it can be clearly seen from the carrier density profile in the top-left of Fig. 2, there is a strong gradient next to the Schottky contact, while in other regions there is an almost constant density. Therefore, a gain in speed could be obtained by using a coarser grid with respect to $z$ in regions with almost constant density without

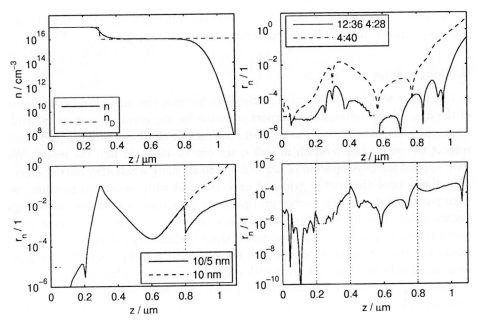

Fig. 2. Electron density in a SBD (top-left) and relative errors with respect to a high resolution run for variation of $\varepsilon$-grid (top-right), $z$-grid (bottom-left) and both of them (bottom-right).

a severe loss of accuracy. This comparison is shown in the bottom-left of Fig. 2. The derivative with respect to $z$ has been approximated by a third order WENO scheme for non-uniform grids. Fur further implementation details refer to Ref. 8. For the reference data a uniform grid with $\Delta z = 5$ nm is used. This is compared with a uniform grid width a cell width of 10 nm and a non-uniform grid with cells of widths 10 nm and 5 nm respectively. The border between the two zones of different resolution was fixed at $z_{\mathrm{B}} = 0.8$ $\mu$m. The comparison shows that the error is of about the same magnitude as for the test with the non-uniform energy grid. At the $n^+ - n$ interface we observe a pronounced peak which is due to the steep carrier density gradient. However, concerning the relative error at the Schottky barrier, the scheme with the non-uniform grid performs a factor of two magnitudes better than the scheme with the coarse grid. Compared to the high resolution run there is a reduction of computation time of 35%.

Finally, a combination of both grid refinements is investigated. The grid with respect to energy is splitted into 72 cells with $s_< = 12$ and 16 cells with $s_> = 4$. This higher resolution compared to the (12:36,4:28)-test accounts for the too large error next to the Schottky contact. For the non-uniform grid with respect to $z$, an additional high resolution area with $\Delta z = 5$ nm is introduced in a region of 20 nm length around the $n^+ - n$ junction. The relative error of this test is presented in the bottom-right of Fig. 2. The time saving is 64% with respect to the full high resolution run. Compared to the tests where only one grid has been refined, the error of this test is much smaller. A very interesting aspect is that a refinement of the grid in real space at the $n^+ - n$ junction influences the error next to the Schottky contact, but hardly the error between these two regions.

## 6. Conclusions

In this paper we have investigated methods to increase the numerical performance of the discretized Boltzmann transport equation for the carrier transport in semi-conductors. We have shown that without considerable effort non-uniform grids with respect to energy and position in real space can be included into the model. We have applied the developed numerical scheme to simulate a Schottky barrier diode. By refining once the energy grid and once the grid with respect to position, we have pointed out that the energy grid plays an important role in regions of high electric fields. The discretization in real space must be very fine in regions of steep electron density gradients. By considering these preliminary calculations, we have finally tested a combination of both grid refinements. Up to 64% of CPU time have been saved with less than 1% error with respect to a high resolution run.

Moreover, we have investigated the influence of different weight functions on the accuracy. The results suggest the following empirical rule. The simpler and better interpretable in physical terms a weight function is, the more accurate it turns out to be. In this sense it is advisable to use either a constant weight function or the density of states. Using a constant weight function, i.e., evaluating the density of states times the distribution function (DF) as unknown function, has the advantage that a singularity near the center of the Brillouin zone of the force term vanishes. When using the density of states, the DF is the unknown and, therefore, this function is bounded due to the probabilistic nature of the distribution function.

## References

1. A. Jüngel, *Quasi-hydrodynamic Semiconductor Equations* (Birkhäuser Verlag, Basel, 2001)
2. M. Ancona and G. Iafrate, *Phys. Rev. B*, **39**, 9536 (1989)
3. D. K. Ferry, *Semiconductors* (Maxwell Macmillian International Editions, New York, 1991)
4. C. Jacoboni and P. Lugli, *The Monte Carlo Method for Semiconductor Device Simulation* (Springer, Wien, 1989)
5. E. Fatemi and F. Odeh, *J. Comp. Phys.*, **108**, 209 (1993)
6. A. Majorana and R. M. Pidatella, *J. Comp. Phys.*, **174**, 649 (2001)
7. J. A. Carrillo, I. M. Gamba, A. Majorana and C-W. Shu, *J. Comp. Phys.*, **184**, 498 (2003)
8. A. Domaingo, M. Galler and F. Schürrer, *COMPEL*, **in press**, (2005)
9. M. Galler and A. Majorana, *Transport Theory and Statistical Physics*, **in press**, (2005)
10. R. J. LeVeque, *Numerical Methods for Conservation Laws* (Birkhäuser Verlag, Basel, 1992)
11. A. Domaingo and F. Schürrer, *Journal of Computational Electronics*, **in press**, (2005)

# KINETIC THEORY APPLICATIONS IN EVAPORATION/CONDENSATION FLOWS OF POLYATOMIC GASES

A. FREZZOTTI

*Dipartimento di Matematica del Politecnico di Milano*
*Piazza Leonardo da Vinci 32 - 20133 Milano, Italy*
*E-mail: aldo.frezzotti@polimi.it*

The evaporation and condensation of a polyatomic vapor in contact with its condensed phase has received much less attention than the monatomic case. In this paper we investigate the structure of the Knudsen layer formed in the steady evaporation and condensation of a vapor whose molecules behave as rigid rotators. The vapor motion is obtained by the numerical solution of the Boltzmann equation by the Direct Simulation Monte Carlo (DSMC) method. The obtained results are also compared with the solutions of a simplified kinetic BGK-like model equation. The relationships between the problem parameters are determined in numerical form for evaporation and condensation flows. It is shown that the present results are in good agreement with previous moment method investigations of evaporation flows.

**Key words:** Knudsen layer, evaporation, condensation of vapor, Boltzmann equation, Monte Carlo method, polyatomic gases
**Mathematics Subject Classification:** 82C40, 65C05,

## 1. Introduction

In many situations, both of theoretical and practical interest, it is required to model the transfer of mass, momentum and energy between the vapor phase and its condensed phase (liquid or solid) coexisting in the same flowfield. Although hydrodynamic equations (Euler or Navier-Stokes) generally provide an adequate description of flow properties far from the vapor-liquid or vapor-solid interface, it is well known that there exists a narrow region (Knudsen layer) where the mechanical interaction between the vapor and the condensed phase is to be treated by the microscopic approach of the kinetic theory of gases[1]. The Knudsen layer formed during evaporation or condensation processes is only a few mean free paths wide, but the macroscopic quantities appearing in hydrodynamic equations may suffer strong jumps across this kinetic region. The necessity of describing the structure of the Knudsen layer and providing jump relationships to be used as boundary conditions for a hydrodynamic treatment of multi-phase flows has triggered the production of a huge number of papers where the problem has been extensively investigated[2]. However, most of the research activity has been concentrated on studying monatomic substances whereas

evaporation or condensation of polyatomic vapors has not attracted the same attention, in spite of their importance for many applications. It is worth mentioning that
the state of the art is not the same for applications of kinetic theory to evaporation
and condensation flows. Actually, a small number of investigations of the steady
evaporation of a polyatomic vapor is already present in the literature.

A paper by Cercignani[3] is the first reference which covers the whole range of admissible downstream Mach number values and provides relatively simple jump formulas
which can be used to match the Knudsen layer to a hydrodynamic region which
might exist in the flowfield. In Ref. [3], the steady one-dimensional evaporation of a
polyatomic gas possessing $j$ internal indistinguishable degrees of freedom is studied
by a BGK-like kinetic model equation[4,5] solved by an extension of the method of
moments used by Ytrehus[6] for a monatomic gas. A more accurate solution method
has been proposed by Soga[7] to study *weak* evaporation and condensation of a polyatomic gas by the linearized version of a similar kinetic model equation. Further
approximate methods have been proposed by Skovorodko[8] and Zharov et al.[9] to
correct some inaccuracies which were noted in Cercignani's results when the downstream Mach number approached the sonic limit. With the exception of the study
of a linearized model equation described in Ref. [7], investigations quoted above are
based on strongly simplified solution methods in order to obtain closed form jump
relationships. The accuracy of these approximate methods does not seem to have
been assessed yet. No systematic kinetic theory study of condensation flows seems
to be present in the literature which only covers the linear regime[7] and very few
flow conditions in the strong condensation regime[10] for very specific applications.
According to the analysis described above, the present paper aims at presenting a
summary of recent results about the numerical study of the Knudsen layer structure
in the steady evaporation and condensation of a polyatomic vapor. A more complete
account is given in Refs. [11,12].

The vapor motion is assumed to be described by the Bolzmann equation for
a gas whose molecule behave as classical rigid rotators that collide according to
Borgnakke-Larsen phenomenological model[13]. The Boltzmann equation has been
solved numerically by the Direct Simulation Monte Carlo (DSMC) method[14]. In
order to increase confidence in the numerical results, a few solutions has also been
obtained from the finite difference numerical solution of a model kinetic equation[15].
The paper is organized as follows: section 2 is devoted to the formulation of the
problem and to the description of the mathematical models, whereas section 3
describe and discuss the numerical results for evaporation and condensation flows.

## 2. Problem statement and basic equations

We consider the steady one-dimensional flow of a dilute polyatomic vapor evaporating from or condensing onto an infinite planar surface (interface) kept at constant
and uniform temperature $T_w$. The surface, located at $x = 0$, separates the condensed phase, which occupies the half-space $x < 0$, from the vapor phase flowing

in the half-space $x > 0$. The coordinate $x$ spans the direction normal to the surface in a reference frame at rest with respect to the interface. The validity of the results obtained from the one-dimensional flow geometry also extends to problems in which the vapor-liquid/solid interface is not planar but its curvature radius is much larger than the reference mean free path. In this case, the kinetic region can be locally treated as one-dimensional. The problem treatment presented here is limited to the temperature range where vibrational excitation can be neglected. Moreover, since the gap between quantized rotational energy levels is supposed to be much smaller than $\kappa T_w$, vapor molecules are supposed to behave as classical rigid rotators of *average* diameter $a$ and mass $m$ with $j = 2$ (linear molecule) or $j = 3$ (non-linear molecule) rotational degrees of freedom. Accordingly, the vapor motion in the positive half-space is assumed to be governed by the steady form of the following one-dimensional kinetic equation[1]:

$$\frac{\partial f}{\partial t} + v_x \frac{\partial f}{\partial x} = I(f, f) \tag{1}$$

where $f(\boldsymbol{v}, \epsilon | x, t)$ is the distribution function of molecular velocity $\boldsymbol{v}$ and rotational energy $\epsilon$ at location $x$ and time $t$. The term $I(f, f)$, whose specific forms are given below, represents the collisional rate of change of $f$. It is to be noted that rotational degrees of freedom are described by the rotational energy $\epsilon$, not by the angular momentum vector $\boldsymbol{l}$ since no preferential alignment of molecular spinning motion is possible in the present problem.

The flowfield structure in the condensed phase has not been calculated. Molecular exchange processes between the vapor and condensed phase across the interface have been described through the following boundary condition for $f$ at $x = 0$:

$$v_x f(\boldsymbol{v}, \epsilon | 0, t) = v_x f_e(\boldsymbol{v}, \epsilon) + \int_{v_{1x}<0} K(\boldsymbol{v}, \epsilon | \boldsymbol{v}_1, \epsilon_1) |v_{1x}| f(\boldsymbol{v}_1, \epsilon_1 | 0, t) \, d\boldsymbol{v}_1 \, d\epsilon_1, \quad v_x > 0 \tag{2}$$

The distribution function of evaporating molecules, $f_e$, is assumed to be a half-range Maxwellian:

$$f_e(\boldsymbol{v}, \epsilon) = \alpha \frac{N_w}{(2\pi RT_w)^{3/2}} \exp\left(-\frac{\boldsymbol{v}^2}{2RT_w}\right) \frac{\epsilon^{j/2-1}}{\Gamma(j/2)(\kappa T_w)^{j/2}} \exp\left(-\frac{\epsilon}{\kappa T_w}\right), \quad v_x > 0 \tag{3}$$

The velocities of vapor molecules re-emitted into the gas phase after interacting with the interface are obtained by the following scattering kernel $K$ which describes a pure diffusive re-emission[1]:

$$K(\boldsymbol{v}, \epsilon | \boldsymbol{v}_1, \epsilon_1) = (1 - \alpha) \frac{1}{2\pi (RT_w)^2} \exp\left(-\frac{\boldsymbol{v}^2}{2RT_w}\right) \frac{\epsilon^{j/2-1}}{\Gamma(j/2)(\kappa T_w)^{j/2}} \exp\left(-\frac{\epsilon}{\kappa T_w}\right) \tag{4}$$

In Eqs. (3,4) $N_w$ denotes the saturated vapor density at temperature $T_w$, $\alpha$ is the evaporation/condensation coefficient, $\kappa$ is the Boltzmann constant, $R$ denotes the ratio $\kappa/m$ and $\Gamma$ is the complete Gamma function. It is further assumed that the far downstream vapor state is described by the equilibrium Maxwellian distribution function:

$$f_\infty(\boldsymbol{v}, \epsilon) = \frac{N_\infty}{(2\pi R T_\infty)^{3/2}} \exp\left[-\frac{(\boldsymbol{v} - \boldsymbol{u}_\infty)^2}{2R T_\infty}\right] \frac{\epsilon^{j/2-1}}{\Gamma(j/2)(\kappa T_\infty)^{j/2}} \exp\left(-\frac{\epsilon}{\kappa T_\infty}\right) \quad (5)$$

in which the bulk velocity $\boldsymbol{u}_\infty$ has both a component $u_{\infty x}\hat{\boldsymbol{x}}$, normal to the interface, and a component $u_{\infty y}\hat{\boldsymbol{y}}$ along the unit vector $\hat{\boldsymbol{y}}$, parallel to the interface. Normalizing $x$ to the reference mean free path $\lambda_w = \frac{1}{\sqrt{2}\pi N_w a^2}$, $\boldsymbol{v}$ to $\sqrt{R T_w}$ and $\epsilon$ to $\kappa T_w$ immediately shows that the problem solutions, for fixed gas properties, depend on the following four parameters: $\frac{N_\infty}{N_w}$, $\frac{T_\infty}{T_w}$, $M_{\infty x} = \frac{u_{\infty x}}{\sqrt{\gamma R T_\infty}}$, $M_{\infty y} = \frac{u_{\infty y}}{\sqrt{\gamma R T_\infty}}$, $\gamma$ being the specific heat ratio $\frac{c_p}{c_v} = \frac{5+j}{3+j}$. The parameters listed above cannot be arbitrarily specified since a number of relationships have to be satisfied for the existence of a steady flow. The number and the nature of such relationships are different for evaporation and condensation flows. In the case of a monatomic gas ($j = 0$), the general properties of steady evaporation/condensation flows can be summarized as follows[2] :

- Steady evaporation

  (1) Only *subsonic* steady evaporation flows are possible, $0 \le M_{\infty x} < 1$.
  (2) If purely diffusive reemission at $x = 0$ is present, then the vapor mean velocity has to be normal to the interface ($M_{\infty y} = 0$).
  (3) Two out of the three remaining parameters depend on the third one. Therefore, for example, one may write

$$\frac{N_\infty}{N_w} = \frac{N_\infty}{N_w}(M_\infty) \qquad \frac{T_\infty}{T_w} = \frac{T_\infty}{T_w}(M_\infty) \qquad (6)$$

- Steady condensation

  (1) When $\alpha = 1$, steady subsonic condensation ($|M_{\infty x}| < 1$) is possible only if the upstream pressure ratio $p_\infty/p_w$ satisfies the relationship

$$p_\infty/p_w = F_{sub}(M_{\infty x}, M_{\infty y}, T_\infty/T_w) \qquad (7)$$

  Approximate values of the function $F_{sub}(M_{\infty x}, M_{\infty y}, T_\infty/T_w)$ have been determined by numerical solution of BGK model equation [2] and DSMC simulations [16].

  (2) When $\alpha = 1$, steady supersonic condensation ($|M_{\infty x}| > 1$) is possible if the upstream pressure ratio $p_\infty/p_w$ satisfies the inequality

$$p_\infty/p_w \ge F_{sup}(M_{\infty x}, M_{\infty y}, T_\infty/T_w) \qquad (8)$$

  where the points on the limit surface

$$p_\infty/p_w = F_{sup}(M_{\infty x}, M_{\infty y}, T_\infty/T_w)$$

  are connected to those on the subsonic surface

$$p_\infty/p_w = F_{sub}(M_{\infty x}, M_{\infty y}, T_\infty/T_w)$$

  by the Rankine-Hugoniot shock relationships.

The important relationships (6,7,8) give the jumps of macroscopic flow properties across the Knudsen layer and provide boundary conditions for the hydrodynamic description of the flowfield outside the Knudsen layers separating the regions occupied by the vapor from those occupied by the condensed phase. Although there is no reason to doubt that the general features outlined above would still hold for a polyatomic gas, the particular form of the jump formulas is affected by internal degrees of freedom.

## 2.1. *Collision models*

Two different expressions of the collision term $I(f, f)$ have been used to study the problem described above. The first and simpler expression is based on Holway's model[15] which approximates the collision term by the following BGK-like expression[4]:

$$\nu_{el}(n, T_t)\,(\Phi_{el} - f) + \nu_{in}(n, T_t)\,(\Phi_{in} - f) \tag{9}$$

where the "frozen" local equilibrium and local equilibrium Maxwellian distribution functions, $\Phi_{el}(\boldsymbol{v}, \epsilon)$ and $\Phi_{in}(\boldsymbol{v}, \epsilon)$, are defined as follows:

$$\Phi_{el}(\boldsymbol{v}, \epsilon) = \frac{n(\epsilon|x, t)}{[2\pi R T_t(x, t)]^{3/2}} \exp\left\{-\frac{[\boldsymbol{v} - \boldsymbol{u}(x, t)]^2}{2R T_t(x, t)}\right\} \tag{10}$$

$$\Phi_{in}(\boldsymbol{v}, \epsilon) = \frac{N(x, t)}{[2\pi R T(x, t)]^{3/2}} \exp\left\{-\frac{[\boldsymbol{v} - \boldsymbol{u}(x, t)]^2}{2R T(x, t)}\right\} \frac{\epsilon^{j/2-1} \exp\left[-\frac{\epsilon}{\kappa T(x,t)}\right]}{\Gamma(j/2)[\kappa T(x, t)]^{j/2}} \tag{11}$$

The first term describes elastic collisions, whereas the second one models inelastic collisions. The macroscopic fields associated with $\Phi_{el}(\boldsymbol{v}, \epsilon)$ and $\Phi_{in}(\boldsymbol{v}, \epsilon)$ are defined as:

$$n(\epsilon|x, t) = \int f(\boldsymbol{v}, \epsilon|x, t)\, d^3v, \quad N(x, t) = \int n(\epsilon|x, t)\, d\epsilon \tag{12}$$

$$\boldsymbol{u}(x, t) = \frac{1}{N(x, t)} \int \boldsymbol{v} f(\boldsymbol{v}, \epsilon|x, t)\, d^3v\, d\epsilon = u_x(x, t)\hat{\boldsymbol{x}} \tag{13}$$

$$T_t(x, t) = \frac{1}{3RN(x, t)} \int [\boldsymbol{v} - \boldsymbol{u}(x, t)]^2 f(\boldsymbol{v}, \epsilon|x, t)\, d^3v\, d\epsilon \tag{14}$$

$$T_r(x, t) = \frac{2}{j\kappa N(x, t)} \int \epsilon f(\boldsymbol{v}, \epsilon|x, t)\, d^3v\, d\epsilon, \quad T(x, t) = \frac{3T_t(x, t) + jT_r(x, t)}{3 + j} \tag{15}$$

The quantity $n(\epsilon|x, t)$ is the number density of molecules having internal energy $\epsilon$, $N(x, t)$ is the total number density, $\boldsymbol{u}(x, t)$ is the mean velocity of the gas, whereas $T_t(x, t)$, $T_r(x, t)$ and $T(x, t)$ are the translational, internal and overall temperature, respectively. The elastic and inelastic collision frequencies, $\nu_{el}(n, T_t)$ and $\nu_{in}(n, T_t)$ have been computed from the shear viscosity $\mu_{hs}$ of the hard sphere gas as:

$$\nu_{el}(N, T_t) = (1 - z)\nu_{tot}(N, T_t), \quad \nu_{in}(N, T_t) = z\nu_{tot}(N, T_t) \tag{16}$$

$$\nu_{tot}(N, T_t) = \frac{N\kappa T_t}{\mu_{hs}(T_t)}, \quad \mu_{hs}(T_t) = \frac{5}{16} \frac{\sqrt{\pi m \kappa T_t}}{\pi a^2} \tag{17}$$

In Eqs. (16) $1 - z$ and $z$ are the fractions of elastic and inelastic number of collisions. The model parameter $z$, which tunes the strength of translational-rotational coupling, has been set equal to a constant, although it could be a function of the temperature.

The numerical solution of Eq. (9) can be greatly simplified by a transformation first considered by Chu[17]. Let us define the reduced distribution functions $F(v_x|x,t)$, $G(v_x|x,t)$, $H(v_x|x,t)$ and $K(v_x|x,t)$ as

$$\begin{pmatrix} F(v_x|x,t) \\ G(v_x|x,t) \\ H(v_x|x,t) \\ K(v_x|x,t) \end{pmatrix} = \int \begin{pmatrix} 1 \\ v_y \\ (v_y^2 + v_z^2) \\ \epsilon \end{pmatrix} f(\boldsymbol{v}, \epsilon|x,t)\, dv_y\, dv_z\, d\epsilon \qquad (18)$$

It can be easily verified that the reduced distribution functions defined above obey the following system of kinetic equations:

$$\frac{\partial}{\partial t}\begin{pmatrix} F \\ G \\ H \\ K \end{pmatrix} + v_x \frac{\partial}{\partial x}\begin{pmatrix} F \\ G \\ H \\ K \end{pmatrix} = \nu_{el}\begin{pmatrix} \hat{F}_{el} - F \\ \hat{G}_{el} - G \\ \hat{H}_{el} - H \\ \hat{K}_{el} - K \end{pmatrix} + \nu_{in}\begin{pmatrix} \hat{F}_{in} - F \\ \hat{G}_{in} - G \\ \hat{H}_{in} - H \\ \hat{K}_{in} - K \end{pmatrix} \qquad (19)$$

where

$$\begin{pmatrix} \hat{F}_{el}(v_x|x,t) \\ \hat{G}_{el}(v_x|x,t) \\ \hat{H}_{el}(v_x|x,t) \\ \hat{K}_{el}(v_x|x,t) \end{pmatrix} = \begin{pmatrix} 1 \\ u_y(x,t) \\ u_y^2(x,t) + 2RT_t(x,t) \\ \frac{j\kappa T_i}{2} \end{pmatrix} \frac{N(x,t)}{\sqrt{2\pi RT_t}} \exp\left\{-\frac{(v_x - u_x)^2}{2RT_t}\right\} \qquad (20)$$

$$\begin{pmatrix} \hat{F}_{in}(v_x|x,t) \\ \hat{G}_{in}(v_x|x,t) \\ \hat{H}_{in}(v_x|x,t) \\ \hat{K}_{in}(v_x|x,t) \end{pmatrix} = \begin{pmatrix} 1 \\ u_y(x,t) \\ u_y^2(x,t) + 2RT(x,t) \\ \frac{j\kappa T(x,t)}{2} \end{pmatrix} \frac{N(x,t)}{\sqrt{2\pi RT}} \exp\left\{-\frac{(v_x - u_x)^2}{2RT}\right\} \qquad (21)$$

Boundary conditions for the reduced distribution functions at $x = 0$ are easily derived from Eqs. (2-4).

Although the interest in the numerical solutions of a kinetic model equation is justified by the necessity of assessing the quality of previous studies, a further comparison with a more sophisticated collision model seems in order. A particle scheme has then been adopted to solve the Boltzmann equation for a polyatomic gas in which binary collisions are described by Borgnakke-Larsen phenomenological model[14,13,18]. Accordingly, Eq. (1) now takes the following form

$$\frac{\partial f}{\partial t} + v_x \frac{\partial f}{\partial x} = \int [f(\boldsymbol{v}_1^*, \epsilon_1^*|x)f(\boldsymbol{v}^*, \epsilon^*|x) - f(\boldsymbol{v}_1, \epsilon_1|x)f(\boldsymbol{v}, \epsilon|x)]\, Q\epsilon_1^\mu\, d^3v_1\, d\epsilon_1 \qquad (22)$$

where $Q$ is a function related to the differential collision cross-section $\sigma(E, \hat{\boldsymbol{k}} \circ \boldsymbol{v}_r, \epsilon, \epsilon \to \epsilon^*, \epsilon_1^*)$ which depends on the total energy $E = \frac{1}{4}mv_r^2 + \epsilon + \epsilon_1$ in the center of mass reference frame, on the orientation of the impact unit vector $\hat{\boldsymbol{k}}$ with respect to the relative velocity $v_r = v_1 - v$ as well as on the internal energy states $\epsilon, \epsilon_1, \epsilon^*, \epsilon_1^*$ before and after a collision. The exponent $\mu$ in Eq. (22) takes the values 0 for $j = 2$ and $\frac{1}{2}$ for $j = 3$. In the particular form of the Borgnakke-Larsen model adopted here, $\sigma$ has been written as the product of the hard sphere collision cross-section times a transition probability $\theta(\epsilon, \epsilon_1, \epsilon^*, \epsilon_1^*)$ in the form:

$$\theta(\epsilon, \epsilon_1, \epsilon^*, \epsilon_1^*) = (1 - z)\theta_{el}(\epsilon, \epsilon_1, \epsilon^*, \epsilon_1^*) + z\theta_{in}(\epsilon, \epsilon_1, \epsilon^*, \epsilon_1^*) \tag{23}$$

where $\theta_{el}$ and $\theta_{in}$ are the elastic and maximally inelastic [18] contributions to $\theta$. Again, the parameter $z$ represent the fraction of inelastic collisions in which energy is transferred from translational to rotational degrees of freedom of colliding molecules.

## 3. Description and discussion of numerical results

Steady solutions of the kinetic equations described in the previous section have been obtained as the long time limit of unsteady solutions in the finite spatial domain

$$D = \{x \in \mathcal{R} : 0 < x < L\}$$

bounded by the condensed phase at $x = 0$ and by an imaginary plane located at position $x = L$, far enough from the interface to avoid distortions in the flowfield. Numerical solutions of Holway's kinetic model have been obtained by the simple finite difference scheme described in Ref. [19], whereas the Boltzmann equation has been solved by a variant of Direct Simulation Monte Carlo (DSMC) method proposed by Koura[20]. The application of both techniques to the present problem is straightforward. However, it is worth mentioning that the boundary condition at infinity given by Eq. (5) requires some attention since not all the parameters of the upstream equilibrium state can be freely specified in a steady subsonic solution, as shown by Eqs. (6,7). Actually, the relationship among the problem parameters is the main outcome of the calculations and it is not known beforehand. Therefore, the distribution function of molecules entering the computational domain from the boundary at $x = L$ is not completely specified. The problem, common to both numerical methods, can be solved by adopting the inflow/outflow boundary conditions proposed in Ref. [21] for evaporation and in Ref. [22] for condensation. Such boundary conditions automatically produce the correct downstream/upstream Maxwellian equilibrium state during the system time evolution.

In most of the numerical calculations presented here the parameter $z$ has been set equal to 0.3. The chosen value is not specific for a particular substance and it simply appears to be reasonable for low temperature flows of diatomic gases like nitrogen[23]. It should be observed that the particular choice of $z$ does not appear to be crucial to obtain Eqs. (6,7) since the strength of translational-rotational coupling does affect the local structure of the Knudsen layer but it seems to have little or no effects on the jump relationships[7]. In all cases presented here the evaporation coefficient has been set equal to 1.0 to follow the choice made in previous investigations.

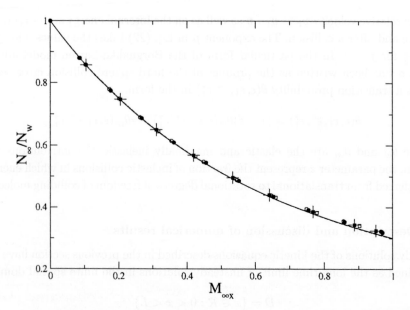

Fig. 1. Normalized downstream density as a function of $M_\infty$. **Moment method**, Ref. [3]: solid line, $j = 2$; dashed line, $j = 3$. **DSMC solution of Boltzmann equation**: ○, $j = 2$, $z = 0.3$; □, $j = 3$, $z = 0.3$. **Numerical solution of Holway's model kinetic equation**: ●, $j = 2$, $z = 0.3$; ■, $j = 3$, $z = 0.3$; +, $j = 3$, $z = 1.0$.

### 3.1. *Evaporation*

The numerical calculation for evaporation flows were mainly aimed at obtaining numerical values of relationships (6) which allow expressing the jumps suffered by macroscopic quantities across the Knudsen layer as functions of the downstream Mach number only. The downstream Mach number $M_{\infty x}$ has been varied form zero to about 0.97. Obtaining solutions with almost sonic downstream state requires a considerable amount of computing time because of the increasing Knudsen layer thickness[6] and the slower rate of convergence to the steady state. The highest value of $M_{\infty x}$ considered here is close enough to the sonic limit to judge about the quality of previous studies.

As shown in Fig. 1, the density ratio $N_\infty/N_w$ is almost unaffected by internal degrees of freedom. The values obtained from the numerical solution of the Boltzmann equation are very close to those obtained from Holway's model kinetic equation and both data sets compare well with the curve computed from the moment method calculation[3].

The effects of internal degrees of freedom is particularly evident in the temperature ratio $T_\infty/T_w$ which turns out to be a monotonically increasing function of $j$, as shown in Fig. 2. Again, Holway's model kinetic equation and Boltzmann equation predictions almost coincide and the results do not appear to be affected by the selected value for $z$. The present results deviates from the moment method predictions which overestimate the temperature ratio. It is to be noted that the sonic values

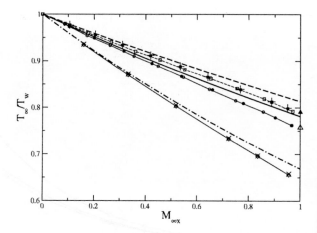

Fig. 2. Normalized downstream temperature as a function of $M_\infty$. **Moment method**, Ref. [3]: dotted-dashed line, monatomic gas, $j = 0$; solid line, $j = 2$; dashed line, $j = 3$. **DSMC solution of Boltzmann equation**: $\diamond$, monatomic gas, $j = 0$ (from Ref. [21]); $\circ$, $j = 2$, $z = 0.3$; $\square$, $j = 3$, $z = 0.3$. **Numerical solution of Holway's model kinetic equation**: $\bullet$, $j = 2$, $z = 0.3$; $\blacksquare$, $j = 3$, $z = 0.3$; $+$, $j = 3$, $z = 1.0$; $\times$, monatomic gas, $j = 0$, (BGK model results from Ref. [21]). **Sonic limit values** Ref. [8]: $\triangle$, $j = 2$; $\blacktriangle$, $j = 3$.

given by the simple method proposed in Ref. [8] seem to provide a better approximation. It also worth stressing that the strength of translational rotational coupling does not affect density and temperature jumps. Actually, when $z$ is changed to its maximum value ($z = 1.0$), the new density and temperature curves (represented by the symbol $+$) superpose to the data obtained from setting $z = 0.3$.

### 3.2. *Condensation*

The analysis of condensation is intrinsically more complicated than evaporation because, as Eqs. (7,8) show, the number of parameters that can be freely specified is higher. In steady subsonic condensation, upstream temperature and velocity can be both arbitrarily chosen whereas the pressure (or density) is given by $F_{sub}$. In steady supersonic condensation, all macroscopic parameters of the upstream equilibrium state can be arbitrarily chosen, provided the pressure ratio $p_\infty/p_w$ satisfies the inequality (8). However, it should be noted that studying subsonic condensation is sufficient for a complete characterization, since $F_{sub}$ determines the limit surface $F_{sup}$ as well. The effect of internal degrees of freedom on $F_{sub}$ is shown in Fig. 3 which presents numerical values of the pressure ratio $p_\infty/p_w$ as a function of $-M_{\infty x}$ and $j$ for condensation flows having $T_\infty/T_w = 1.0$ and $M_{\infty y} = 0.0$. The results reported in Fig. 3 have been obtained by the numerical solution of the Boltzmann equation. The coupling parameter $z$ has been set equal to 0.3. A small number of exploratory calculations indicated that changing $z$ has negligible effects on the pressure ratio. An extensive numerical analysis[11] revealed that Holway's kinetic model predictions are very close to the Boltzmann equation results for condensation

1348

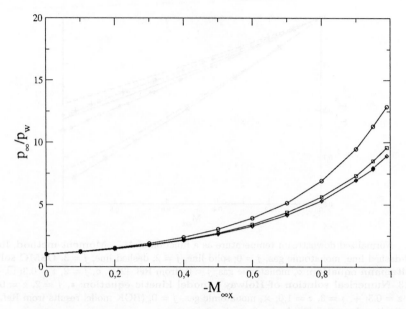

Fig. 3. **Subsonic condensation** - $T_\infty/T_w = 1.0$, $M_{\infty y} = 0.0$. Upstream pressure ratio $p_\infty/p_w$ as a function of $M_{\infty x}$ and $j$. ○, $j = 0$; □, $j = 2$; ◇, $j = 3$.

flows, too. The curves reported in Fig. 3 show that internal degrees of freedom do affect $F_{sub}$, but the change is small at small Mach numbers whereas the deviation from the monatomic gas curve is more evident in the vicinity of sonic upstream conditions where blind application of the monatomic gas theory would certainly give incorrect results.

# References

1. C. Cercignani, *The Boltzmann Equation and Its Applications*, Springer-Verlag, Berlin, 1988.
2. Y. Sone, *TTSP* **29**(3-5), 227 (2000).
3. C. Cercignani, *Rarefied gas dynamics*, Proceedings of the 12th International Symposium on Rarefied Gas dynamics, Vol. 1, 305, AIAA, N.Y, 1981.
4. P. L. Bhatnagar, E. P. Gross, and M. Krook, *Phys. Rev.* **94**, 511 (1954).
5. T.F. Morse, *Phys. Fluids* **7**, 159 (1964).
6. T. Ytrehus, *Rarefied gas dynamics*, Proceedings of the 10th International Symposium on Rarefied Gas dynamics, Vol. 2, 1197, AIAA, N.Y., 1977.
7. T. Soga, *Phys. Fluids* **28**, 1280 (1985).
8. , P.A. Skovorodko, *AIP Conference Proceedings* **585**, 588 (2001).
9. N.A. Zharov, I.A. Kudnetsova and A.A. Yushkanov,*High Temperature* **36**, 109 (1998).
10. S. Fujikawa, T. Yano, K. Kobajashi, K. Iwanami and M. Ichijo, *Exp. Fluids*, **37**, 80 (2004).
11. A. Frezzotti and T.Ytrehus,"Kinetic Theory Study of Steady Condensation of a Polyatomic Gas", to appear on *Phys. Fluids* (2005).
12. A. Frezzotti, "A Numerical Investigation of the Steady Evaporation of a Polyatomic Gas", submitted to the *European Journal of Mechanics/B Fluids* (2005).

13. C. Borgnakke, P.S. Larsen, *J. Comput. Phys.* **18**, 405 (1975).

14. Bird, G. A. *Molecular Gas Dynamics and the Direct Simulation of Gas Flows*, Clarendon Press, Oxford, 1994.

15. L.H. Holway, *Phys. Fluids* **9**, 1658 (1966).

16. M.N. Kogan, A.A. Abramov, *Rarefied Gas Dynamics*, Proceeding of the 17th International Symposium on Rarefied Gas Dynamics, 1278, VCH Verlag, Weinheim, 1991.

17. C. K. Chu, *Phys. Fluids* **8**, 12 (1965).

18. I. Kuščer, *Physica A* **158**, 784 (1989).

19. A. Frezzotti, *Rarefied Gas Dynamics*, Proceedings of the 17th International Symposium on Rarefied Gas Dynamics, 1243, VCH Verlag, Weinheim, 1991.

20. K. Koura, *Phys. Fluids* **29**, 3509 (1986).

21. A. Frezzotti, in *Rarefied Gas Dynamics*, Proceedings of the 20th International Symposium on Rarefied Gas Dynamics, 837, Peking University Press, Beijing, 1997.

22. R. Meland, A. Frezzotti, T. Ytrehus, and B. Hafskjold, *Phys. Fluids* **16**, 223 (2004).

23. I. J. Wysong and D.C. Wadsworth, *Phys. Fluids* **10** 2983 (1998).

# MATHS AGAINST CANCER*

## F. PAPPALARDO, S. MOTTA

*Dept. of Mathematics & Computer Science and Faculty of Pharmacy*
*University of Catania*
*V.le A. Doria, 6*
*95125 Catania, Italy*
*E-mail: {francesco,motta}@dmi.unict.it*

## P.-L. LOLLINI

*Sezione di Cancerologia, Dipartimento di Patologia Sperimentale*
*and Centro Interdipartimentale di Ricerche sul Cancro "Giorgio Prodi"*
*University of Bologna*
*Viale Filopanti 22, I-40126 Bologna, Italy*
*E-mail: pierluigi.lollini@unibo.it*

## E. MASTRIANI

*Dept. of Mathematics*
*University of Catania*
*V.le A. Doria, 6*
*95125 Catania, Italy*
*E-mail: mastriani@dmi.unict.it*

The risk of cancer can be already written in our genoma. Nowadays this risk can be discovered before the cancer appears. A biologist group of the University of Bologna has found a vaccine for mammary carcinoma (Triplex Vaccine). The vaccine has been tested on genetic modified mice (HER-2/neu mice) which develop the mammary carcinoma. Each *in vivo* experiment takes at least **one year** and roughly 20 mice. In vivo experiments show that the vaccine is successful **but only if** the mice are treated for all its lifetime (chronic schedule). We have been requested to built up a model of the vaccine action in order to find a *lighter* vaccination schedule.

**Key words:** Model for vaccine action
**Mathematics Subject Classification:** 92C50

*F.P. and S.M. acknowledge partial support from University of Catania research grant and MIUR (PRIN 2004: *Problemi matematici delle teorie cinetiche*). This work has been done while F.P. is research fellow of the Faculty of Pharmacy of University of Catania. P.-L.L. acknowledges financial support from the University of Bologna, the Department of Experimental Pathology ("Pallotti" fund) and MIUR

## 1. Introduction

The Immune System (IS) is a complex adaptive system of cells and molecules, distributed in the vertebrates body, that provides a basic defense against pathogenic organisms

The IS performs two major tasks:

- it **distinguishes** between *self* and *nonself*, which are respectively elements of the body and foreign elements. The success of the IS depends on its ability to distinguish between harmful nonself, and everything else.
- it **prevents** harmful nonself to enter into the body and/or **destroys them**.

Recognizing *harmful non-self* is a hard problem for IS. It has to distinguish roughly $10^{16}$ nonself patterns from $10^6$ self patterns in a highly distributed environment. Recognition occurs in a changing living environments and the body resources are scarce with respect to external variety.

The Immune System has a twin solution, namely the **Innate Immunity** and the **Adaptive Immunity**. The **innate immune system** provides the non specific immunity of the body. It works with a two layer strategy. It prevents the pathogen to enter the body setting up a barrier. This barrier which is both physical (skin) and physical-chemical (pH, temperature, etc) create inappropriate living conditions for foreign organism. Inside the body the innate Immune System primarily consists of the endocytic and phagocytics systems, which involve roaming scavenger cells, such as phagocytes, that ingest extracellular molecules and materials, clearing the system of both debris and pathogens.

The **acquired immune response** is the most sophisticated and involves a host of cells, chemicals and molecules. It is called **acquired** because it is responsible for the immunity that is adaptively acquired during the lifetime of the organism. Acquired immune response is mainly driven by T-cell and B-cell. Those cells are derived from stems cell and selected in such a way that they do not recognize as harmful the body cells (self cells). When an external pathogen is recognized as harmful the IS produce two specific response: the **cellular response** produced by T-cells which *change their internal state* and specifically kill that pathogen; **humoral response** produced by B-cells previously stimulate by Th cells. A fraction of the B-cells become **plasma cells** and produce a large number of specific **antibodies** able to degenerate the pathogen. Degenerated pathogens are then ingested and decomposed by **macrophages**. Both cellular and humoral response are triggered by adjuvants (molecules) like cytokine and interleukines. Both T and B cell keep memory of the encountered pathogen as stimulated cells lives longer than unstimulated ones.

Many approaches have been proposed to describe the competition between the adaptive Immune System cells against pathogens. These can be divided into three major class according to their scale of representation: macroscale, mesoscale and cellular scale.

- Continuous models describe the evolution of entities densities at macroscopic scale using a dynamical systems representation.

- Kinetic models describe the evolution of distribution function of entities at cellular scale.
- Discrete models describe individual entities and their interactions.

Cancer cells are body cells (*self*) so they are not usually recognized as harmful. Immune Response to cancer cells can however be induced by drugs. The **Cancer Immuno-Therapy** tries to reduce solid cancer using the Immune Response. This therapy has not been very successful as established solid cancer reduce the Immune System efficiency. Cancer Immuno-prevention tries to prevent the solid tumor formation by setting an Immune Response at the very early stage of tumor formation. Immuno-prevention looks to be effective also in reducing the growth of metastasis.

## 2. Models for cancer - immune system competition

Immune competition involves several interacting populations each one characterized by microscopic internal states which may differ from one population to another. The evolution of this collective behavior can be described by the so called **Generalized Kinetic Cellular Theory** [1] which provides a statistical description of large populations undergoing kinetic type of interactions. The substantial difference with respect to the equation of kinetic theory is that the state of the cells is defined not only by mechanical variables but also by an *internal biological microscopic state* related to the typical activities of the cells of a certain population. The description of the state of the system can be given by the normalized density distribution function:

$$f_i = f_i(t, \mathbf{x}, \mathbf{v}, \mathbf{u}) = \frac{1}{\mu_0} N_i(t, \mathbf{x}, \mathbf{v}, \mathbf{u})$$

where $\mu_0$ is a positive constant and subscript $i$ refer to the different entities. Marginal density over the mechanical state is

$$f_i^m(t, \mathbf{x}, \mathbf{v}) = \int_{D_u} f_i(t, \mathbf{x}, \mathbf{v}, \mathbf{u}) d\mathbf{u}$$

Marginal density over the biological state is

$$f_i^b(t, \mathbf{u}) = \int_{D_x \otimes D_v} f_i(t, \mathbf{x}, \mathbf{v}, \mathbf{u}) dxdv$$

Bellomo and coworkers [2] considered three entities (immune cells, cancer cells and environmental cells) and described the evolution of the cells' populations using the system of equations

$$\frac{\partial}{\partial u} f_i(t, u) + \frac{\partial}{\partial u} \left[ f_i(t, u) \sum_{k=1}^{3} \int_0^\infty \varphi_{ik}(u, w) f_k(t, w) dw \right]$$

$$= f_i(t, u) \sum_{k=1}^{3} \int_0^\infty \psi_{ik}(u, w) f_k(t, w) dw$$

where conservative actions are modeled by the function $\varphi = \varphi(u, w)$ and non-conservative actions are modeled by the function $\psi(v, w)\delta(v - u)$. Their analysis is very rich and interesting. They show that even a simplified model can give insight of the biological problems using the appropriate mathematical tools. However referring to *modeling therapeutical actions* the authors point out that "*even if one can prove well posedness, carry out asymptotic analysis and so one, this may be pure mathematical results which unfortunately does not always correspond to realistic conditions of a therapeutical procedure*".

This framework looks very promising but it is still too naive to be able to describe realistic biological experiments. Here we point out some of the main reasons:

(1) the *recognition phase* is not taken into account;
(2) individuals difference (*phenotype*) should be inserted in the framework in order to fit experimental data;
(3) as the interaction rules between entities are biologically poorly known and knowledge is rapidly changing, one needs to find an easy way to adjust the models;
(4) inserting all the relevant quantities which play a role into a biological experiment will lead to an untreatable set of kinetic equations;
(5) binary encounters hypothesis may not be appropriate because adjuvants (cytokine and interleukines) do play an important role in cells interaction.

In spite of these difficulties kinetic models can suggest the approach in which the problem can be treated computationally.

Discrete models are based on automata models such as lattice gas cellular automata. They can be seen as a discrete Lattice Boltzmann method with discrete states. These models have some advantages:

- can incorporate a lot of biological details;
- components and processes of interest are described in biological terms;
- can be easily adapted to new biological discoveries;
- allow easy modification of the complexity of the problem without introducing new difficulties in computing a solution.
- allow different investigations using the same model, i.e. easily integrate hypothesis and compare results;
- are very suitable for *in silico experiments*.

Unfortunately, they also have all the well known disadvantages of numerical solutions.

The most biological accurate discrete model was proposed by Celada and Seiden [3]. Entities are represented by their relevant mechanical and biological properties (biological states, finite lifetime, receptors, etc.). Receptors/ligands are represented in the *shape space* [5] with a bitstring of length $l$. Recognition phase, i.e. *affinity*, between cells is measured as function of the Hamming distance on the bitstring space. *Note that* recognition is not linked to the *still mostly unknown* receptor-ligand interaction probability. The state of each entity probabilistically evolve according *known* biological and interaction rules. When entities and rules are fixed the model can be implemented on a Lattice and the evolution followed on discrete time steps. Each time step represents roughly 8 hours. Entities can move from one lattice site to another (*diffusion* or *chemotaxis effect*); It is included external environment representing *thymus* and *bone marrow* which respectively create new virgin T cells and B cells.

## 3. IS vs cancer competition: the vaccine effect

Here we outline a discrete model [8] which describe the effect of the **Triplex** [4] immunoprevention vaccine. This method can be referred as a *Lattice Boltzmann method* applied to biological problems. As this approach is able to reproduce *in vivo* experiments on HER-2/neu mice, it can help in finding a more appropriate mathematical formulation of the immune system - cancer competition. **Triplex** is a three component vaccine. It contains

- the specific antigen (viz. p185, the product of HER-2/neu)
- a nonspecific signals like allogeneic histocompatibility antigens
- interleukin 12 molecules which play the role of adjuvants.

The vaccine cell that we will model is similar to the latter one and consists of a HER-2/neu transgenic mammary carcinoma cell allogeneic with respect to the host and transduced with IL-12 genes.

The biological entities (cells and molecules) we consider in the model and their half life (in timesteps) are listed in the following table.

| entity | type | half life |
|---|---|---|
| B Cells | cell | 3.33 days |
| Antibody Secreting Plasma Cells | cells | 10 |
| T-helper lymphocytes | cells | 10 |
| T-cytotoxic lymphocytes | cells | 10 |
| Macrophages | cells | 3.33 days |
| Dendritic Cells | cells | 10 |
| Interleukin-2 | molecules | 5 |
| Immunoglobulins | molecules | 5 |
| Danger Signal | molecules | 5 |
| Major Histocompatibility Complex Class I | proteins | - |
| Major Histocompatibility Complex Class II | proteins | - |
| Tumor Associated Antigens | molecules | 3 |
| Immunocomplexes | molecules | 100 |
| Cancer Cells | cells | 1095 |
| Vaccine Cells | cells | 5 |
| Natural Killer Cells | cells | 10 |
| Interleukin-12 | molecules | 5 |

Cells play different roles and have different internal states as listed in the table.

| Cell | Active | Re-sting | Intern | PresI | PresII | Duplica | Bound-ToAb |
|------|--------|----------|--------|-------|--------|---------|------------|
| B    | •      | •        | •      |       | •      | •       |            |
| TH   | •      | •        |        |       |        | •       |            |
| TC   | •      | •        |        |       |        | •       |            |
| DC   | •      | •        | •      | •     | •      |         |            |
| MP   | •      | •        | •      |       | •      |         |            |
| CC   |        |          |        | •     |        | •       | •          |
| VC   |        |          |        | •     |        |         | •          |
| NK   | •      |          |        |       |        |         |            |

The following table shows interactions between entities. An interaction will change, according to a given probability, the internal state of the two interacting entities. This reproduces the effect of the collision integral in the Boltzmann equation.

| Entity 1 / Entity 2 | B | Ag | Ab | TH | TC | MP | DC | IC | VC | CC | NK |
|---------------------|---|----|----|----|----|----|----|----|----|----|----|
| B  |   | •  |    | •  |    |    |    |    |    |    |    |
| Ag | • |    | •  |    |    | •  | •  |    |    |    |    |
| Ab |   | •  |    |    |    |    |    |    | •  | •  | •  |
| TH | • |    |    |    |    | •  | •  |    |    |    |    |
| TC |   |    |    |    |    |    |    |    | •  | •  |    |
| MP |   | •  |    | •  |    |    |    | •  |    |    |    |
| DC |   | •  |    | •  |    |    |    |    |    |    |    |
| IC |   |    |    |    |    | •  |    |    |    |    |    |
| VC |   |    | •  |    | •  |    |    |    |    |    | •  |
| CC |   |    | •  |    | •  |    |    |    |    |    | •  |
| NK |   |    | •  |    |    |    |    |    | •  | •  |    |

## 4. Results

Experiments on HER-2/neu mice have been carried out using different vaccination schedule. Each experiment uses 20 mice which has been treated for one year with the same vaccine schedule. The vaccination schedules used in experiments are named: *Early, Late, Very Late, Chronic*. An experiment with *untreated* mice was also performed in order to have a reference point. The experiment measures the time in which a solid tumor is formed on the mice. The goal of the vaccine is to keep the mouse free from solid tumor for all its lifetime. Figure 1 shows *in vivo* experiments [6].

Changing the random generator seed one gets a different random number sequence. This mimic different phenotypes by changing, for instance, the *repertoire* of the mouse. We randomly choose 100 random seed and get a population of 100

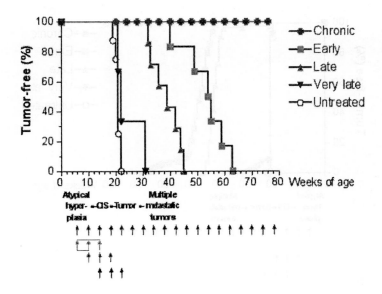

Fig. 1.  Tumor-free survival curves of HER-2/neu transgenic mice receiving the Triplex vaccine according to different protocols. Each arrow at the bottom of the graph represents one cycle of vaccination. The sequence of neoplastic progression in Untreated mice is outlined under the x axis; CIS, carcinoma in situ.

different mice For each vaccine schedule we treat all *virtual mice* with the vaccine and we measure the time in which the solid tumor, defined as a threshold of cancer cells on the lattice, is formed. Figure 2 shows obtained results.

A vaccination schedule is the sequence of vaccine's injections for the time length of the experiment (one year); Chronic Vaccination schedule uses 59 vaccine injections in one year. We search for a better therapy by a genetic algorithm which uses the discrete model shown before as *fitness function*. For a single mouse it is easy to find therapy which prevent solid tumor formation with only 29 vaccine injections. This solution is not biologically acceptable as one should identify *a priori* the *phenotype*. A therapy which is good for a mouse will, in general, be good only for ~ 30% of the sample. We then decided to search a therapy using a new strategy. We randomly select 8 mice of the sample and search for a therapy which will be acceptable for all of them; we find a therapy with 35 vaccine injections; this therapy is able to prevent the formation of solid tumor in 88% mice of the sample [7] A therapy for a single mouse takes about ~ 1 min. on a PC. Searching a better therapy on 8 mice, ~ 48 hrs on 32 node of a parallel computer.

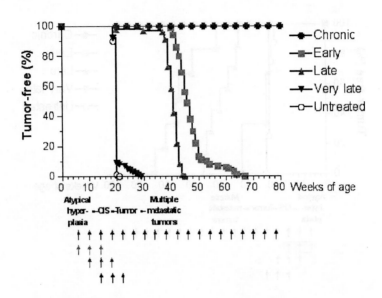

Fig. 2. Tumor-free survival curves of virtual mice receiving the Triplex vaccine according to different protocols. Each arrow at the bottom of the graph represents one cycle of vaccination. The sequence of neoplastic progression in Untreated mice is outlined under the x axis; CIS, carcinoma in situ.

## 5. Conclusions

We presented a *discrete* Lattice Boltzmann method for modeling the Immune System - Cancer competition stimulated by an immunoprevention vaccine. We have been able to reproduce data from *in vivo* experiments on HER-2/neu mice; We use the model to search for a better vaccination schedule and we find a safe schedule using only 60% of the number of vaccine injections used in the experiments. The biology group is planning the experiment to confirm the model. We hope that this work can help in finding a detailed mathematical approach able to describe biological experiments.

## References

1. N. Bellomo and G. Forni, *Math.Comp.Model.* **20**, 107 (1994).
2. N. Bellomo, E. De Angelis and L. Preziosi, *Journal of Theoretical Medicine* **5**, 111 (2003).
3. F. Celada, and P. Seiden, P., *Immunol. Today*, **13**, 52 (1992).
4. C. De Giovanni, G. Nicoletti, L. Landuzzi, A. Astolfi, S. Croci, A. Comes, S. Ferrini, R. Meazza, M. Iezzi, E. Di Carlo, P. Musiani, F. Cavallo, P. Nanni, and P.-L. Lollini, *Cancer Res.*, **64:11**, 4001 (2004).
5. J.D. Farmer, N.H. Packard and A.S. Perelson, *Physica D*, **22**, 187 (1986).
6. P.-L. Lollini, C. De Giovanni, T. Pannellini, F. Cavallo, G. Forni and P. Nanni, *Future*

*Oncology*, **1:1**, 57 (2005).

7. P.-L. Lollini, S. Motta and F. Pappalardo, *submitted to Bioinformatics*, (2005).

8. F. Pappalardo, P.-L. Lollini, F. Castiglione and S. Motta, *Bioinformatics*, **21:12**, 2891 (2005).

# IV.4 Inverse Problems, Theory and Numerical Methods

Organizers: M. Klibanov, M. Yamamoto

# LIPSCHITZ STABILITY IN AN INVERSE HYPERBOLIC
# PROBLEM BY BOUNDARY OBSERVATIONS

M. BELLASSOUED

*Faculté des Sciences de Bizerte,*
*Dép. des Mathematiques,*
*7021 Jarzouna Bizerte, Tunisia,*
*E-mail:mourad.bellassoued@fsb.rnu.tn*


M. YAMAMOTO

*Department of Mathematical Sciences,*
*The University of Tokyo, 3-8-1 Komaba,*
*Meguro, Tokyo 153, Japan,*
*E-mail:myama@ms.u-tokyo.ac.jp*

Under a weak regularity assumption and a geometrical condition on the metric, we consider a a multidimensional hyperbolic inverse problem of determining a wave speed with a single measurement of Neumann data on a suitable subboundary and we prove a Lipschitz stability estimate in $L^2$ space for solution to the inverse problem.

**Key words:** Lipschitz stability, inverse hyperbolic problem
**Mathematics Subject Classification:** 35R30, 35B35

## 1. Introduction

In this paper we will consider an inverse problem of determining a function in an anisotropic speed of propagation for a second-order hyperbolic equation from data of the solution on a subboundary part over a time interval. In a bounded domain $\Omega \subset \mathbb{R}^n$ with sufficiently smooth boundary $\Gamma = \partial\Omega$, we consider a second-order differential operator

$$A(x, D_x)u = -\sum_{i,j=1}^{n} \frac{\partial}{\partial x_i}\left(\gamma_{ij}(x)\frac{\partial u}{\partial x_j}\right) \tag{1}$$

with real-valued coefficients $\gamma_{ij} = \gamma_{ji}$ of class $C^3(\mathbb{R}^n)$ such that the uniform ellipticity condition is satisfied, that is,

$$\sum_{i,j=1}^{n} \gamma_{ij}(x)\xi_i\xi_j \geq \gamma_0 \sum_{i=1}^{n} \xi_i^2, \quad x \in \mathbb{R}^n, \tag{2}$$

for some positive constant $\gamma_0 > 0$.

For fixed positive constant $c_0 > 0$, we denote by $\mathcal{C}$ the set of functions $c(x)$ of class

$C^1(\mathbb{R}^n)$ which satisfy $c_0^{-1} \le c(x) \le c_0, \ \forall x \in \mathbb{R}^n$.

We consider the second-order hyperbolic problem with Dirichlet boundary condition

$$\begin{aligned}
\partial_t^2 u + c^2(x) A(x, D_x) u &= 0 && \text{in } Q = \Omega \times [0,T], \\
u(x,0) = a(x), \quad \partial_t u(x,0) &= b(x) && \text{in } \Omega, \\
u(x,t) &= g_0(x,t) && \text{on } \Sigma = \Gamma \times [0,T].
\end{aligned} \tag{3}$$

Here we assume that $a \in H^1(\Omega)$; $b \in L^2(\Omega)$.

Let $\Gamma_0 \subset \Gamma$ is a part of the boundary $\Gamma = \partial \Omega$ which is given suitably. In this paper, we consider the following inverse problem: determine the coefficient $c(x)$, $x \in \Omega$ in (3) from the extra data $\partial_\nu u_{|\Sigma_0}$, where $\Sigma_0 = \Gamma_0 \times [0,T]$, $a$, $b$ and $T > 0$ are fixed, $\partial_\nu u = \displaystyle\sum_{i,j=1}^n c(x) \gamma_{ij}(x) (\partial_{x_j} u) \nu_i$ is the conormal derivative, and $\nu = (\nu_1, ..., \nu_n)$ is the outward normal on $\Gamma$.

Proofs of uniqueness theorems of multidimensional inverse problems for differential equations are based on the following two points; the Bukhgeim-Klibanov [3] method and Carleman estimates near the boundary for boundary value problems. See also Klibanov [11]. We remark the first method is an application of Carleman estimate to inverse problems and effective for various problems of determining coefficients in the equations where a Carleman estimate holds. Since the Carleman estimate depends essentially on the type of differential equation and the shape of the domain, many serious difficulties arise in particular for hyperbolic operators with variables coefficients.

Stability estimates play a special role in the theory of inverse problems of mathematical physics that are ill-posed in the classical sense. As for ill-posed problems, see Isakov [8] and Lavrent'ev [14], for example. Stability rates determine the choice of regularization parameters and the rate at which solutions of regularized problems converge to an exact solution.

When the extra data is required only on subboundary part $\Gamma \subset \partial \Omega$, for the isotropic case Isakov & Yamamoto [9] prove the uniqueness and Lipschitz stability in multidimensional inverse problems of determining source terms or potential (under some conditions on $\Gamma \subseteq \partial \Omega$ and $T > 0$). In Bellassoued [1], similar uniqueness and the Lipschitz stability are proved by means of data on the whole boundary over a time interval. In this paper, we will prove the Lipschitz stability by means of boundary data on a subboundary. As for the Lipschitz stability for inverse problems, see Klibanov and Yamamoto [13]. Also we refer to Khaidarov [10], Klibanov and Timonov [12].

We will study Carleman estimates near the boundary for the basic second order hyperbolic operator $\partial_t^2 + c^2(x) A(x, D)$ with variable coefficients. This increases the technical difficulty of the proofs substantially. Moreover, even for isotropic wave operators (i.e. $A(x, D) = -\Delta$) and without boundary conditions, we have to impose additional conditions on the speed of propagation $c(x)$ since otherwise pseudo-convexity fails and non-uniqueness may occur (see Isakov [8]). Triggiani & Yao [15] considers a general second-order hyperbolic equation defined in a bounded domain, and under checkable conditions on the principal part, they proved Carleman type estimates where some weighted $H^1$-norm of solution is dominated by the $L^2(\Sigma)$-norm of the boundary traces $\partial_\nu u$ and $u_t$. See also Imanuvilov [4].

## 1.1. Standing Assumptions

Throughout this section we use the following notations: $a(x, \xi) = \sum\limits_{i,j=1}^{n} a_{ij}(x)\xi_i\xi_j$.

Given two symbols $p$ and $q$ we define their Poisson bracket as

$$\{p, q\}(x, \xi) = \frac{\partial p}{\partial \xi} \cdot \frac{\partial q}{\partial x} - \frac{\partial p}{\partial x} \cdot \frac{\partial q}{\partial \xi} = \sum_{j=1}^{n} \left( \frac{\partial p}{\partial \xi_j} \frac{\partial q}{\partial x_j} - \frac{\partial p}{\partial x_j} \frac{\partial q}{\partial \xi_j} \right). \tag{4}$$

**Assumption H.1-H.2:** We assume that there exists a non-negative function $\vartheta : \overline{\Omega} \longrightarrow \mathbb{R}$ of class $C^2$ such that:

(H.1) $\qquad \{a, \{a, \vartheta\}\}(x, \xi) \geq 8\varrho\,|A(x)\xi|^2, \quad \forall \xi \in \mathbb{R}^n,\ x \in \overline{\Omega},$

(H.2) $\qquad \{\{a, \vartheta\}, a(x, \nabla\vartheta)\}(x, \xi) \geq 8\varrho\,|\nabla_g\vartheta|^2, \quad \forall \xi \in \mathbb{R}^n,\ x \in \overline{\Omega}.$

Here $(A(x)\xi)_j = \sum\limits_{i=1}^{n} a_{ij}(x)\xi_i$ and $\varrho > 0$ and we set

$$h(x) = \nabla_g \vartheta(x) \equiv \sum_{i=1}^{n} \left( \sum_{j=1}^{n} a_{ij}(x) \frac{\partial \vartheta}{\partial x_j} \right). \tag{5}$$

**Assumption H.3:** Let a subboundary $\Gamma_0$ satisfy $\{x \in \Gamma;\ h(x) \cdot \nu(x) \geq 0\} \subset \Gamma_0$.

Define $T_0 = 2 \left( \dfrac{\max \vartheta(x)}{\varrho} \right)^{\frac{1}{2}}$. We remark that the time $T_0$ depends on $\vartheta$ and $\varrho$, and therefore on the fixed reference speed $c(x)$.

Now we can state our main result.

**Theorem 1.** *Assume (H.1), (H.2), (H.3) and let the observation time $T > T_0$. Let $u$ be the solution of (3) such that $u \in W^{3,\infty}(\Omega \times (0, T))$ and $|A(x, D_x)a(x)| \geq \epsilon_0 > 0$ almost everywhere on $\overline{\Omega}$ with some constant $\epsilon_0 > 0$. Let $u_*$ be a solution of (3) with unknown $c_*$. Then there exists a constant $C > 0$ such that*

$$\left\| c^2 - c_*^2 \right\|_{L^2(\Omega)} \leq C \left\| \partial_t(\partial_\nu u - \partial_\nu u_*) \right\|_{L^2(\Sigma_0)}. \tag{6}$$

**Corollary 1.** *Assume (H.1), (H.2), (H.3) and let the observation time $T > T_0$. Let $u$ be the solution of (3) such that $u \in W^{3,\infty}(\Omega \times (0, T))$ and $|A(x, D_x)a(x)| \geq \epsilon_0 > 0$ almost everywhere on $\overline{\Omega}$ with some constant $\epsilon_0 > 0$. Let $u_*$ be a solution of (3) with unknown $c_*$. Then $\partial_\nu u = \partial_\nu u_*$, $(x, t) \in \Sigma_0 = \Gamma_0 \times [0, T]$ implies $c(x) = c_*(x)$ for $x \in \Omega$.*

This paper is composed of four sections. In section 2, we state a Carleman estimate. In section 3 we show our main result concerning the Lipschitz e stimate in our inverse problem and we discuss the uniqueness source problem for the hyperbolic operator with variable coefficients.

## 2. Carleman inequalities for hyperbolic equation

In this section we present the Carleman estimate which is the starting point of the proof of Theorem 1. In order to formulate our Carleman type estimate, we introduce some notations. Let $\vartheta : \overline{\Omega} \longrightarrow R$ be a strictly convex function with respect to the Riemannian metric $g$, provided by assumptions (H.1)-(H.2). For the observation time $T > T_0$, let us define

$$\psi(x,t) = \vartheta(x) - \gamma \left( t - \frac{T}{2} \right)^2, \tag{7}$$

and we fix $\delta > 0$ and $\gamma > 0$ such that

$$\varrho T^2 > 4 \max_{x \in \Omega} \vartheta(x) + 4\delta, \quad \text{and} \quad \gamma T^2 > 4 \max_{x \in \Omega} \vartheta(x) + 4\delta, \quad 0 < \gamma < \varrho, \tag{8}$$

where $\varrho$ is given by (H.1)-(H.2). Then $\psi(x,t)$ verifies the following properties:

(i) $\psi(x,0) < -\delta$ and $\psi(x,T) < -\delta$ for all $x \in \Omega$.
(ii) There exists $\eta \in \left(0, \frac{T}{2}\right)$ such that: $\min \psi(x,t) \geq -\frac{\delta}{2}, \quad x \in \Omega, \quad \eta \leq t \leq T - \eta$.
(iii) $\psi \left( x, \frac{T}{2} \right) = \vartheta(x) \geq 0$ for all $x \in \Omega$.

We define the pseudo-convex function $\varphi : \Omega \times R \longrightarrow R$ by $\varphi(x,t) = e^{\beta \psi(x,t)}$, where $\beta > 0$ is a large parameter selected in the following.

Now we would like to consider the following second-order hyperbolic operator

$$P(x,D) = \partial_t^2 + c^2(x)A(x,D_x), \tag{9}$$

where $A(x,D_x)$ is the elliptic operator of the second order defined by (1). For $\rho \in \left(0, \frac{T}{2}\right)$ we set

$$Q_\rho = \Omega \times [\rho, T - \rho] \subset Q. \tag{10}$$

Finally we introduce the notation: $\nabla v(x,t) = (\nabla_x v, \partial_t v)$. The following inequality of Carleman type holds:

**Theorem 2.** (Bellassoued-Yamamoto[2]) Assume that (H.1), (H.2), (H.3) hold and $T > T_0$. Then there exist a constant $C > 0$ independent of $\lambda$ and a parameter $\lambda_*$ such that for all $\lambda \geq \lambda_*$ the following Carleman estimate holds:

$$\lambda \int_{Q_\rho} e^{2\lambda\varphi} \left( |\nabla v|^2 + \lambda^2 |v|^2 \right) dx dt \leq C \int_Q e^{2\lambda\varphi} |P(x,D)v|^2 dx dt$$

$$+ C\lambda \int_{\Sigma_0} e^{2\lambda\varphi} \left( |\partial_\nu v|^2 \right) ds dt + \lambda \int_{Q \setminus Q_\rho} e^{2\lambda\varphi} \left( |\nabla v|^2 + \lambda^2 |v|^2 \right) dx dt, \tag{11}$$

whenever $v \in H^1(Q)$, $v_{|\Sigma} = 0$ and the right hand side is finite.

## 3. Proof of the main result

This section is devoted to the proof of Theorem 1. The idea of the proof is based on a usual method by a Carleman estimate.

### 3.1. Preliminaries

We need the following preliminaries, which are essentially known although most of them are not explicitly stated in the literatures. In the proof of the main result, we shall make use of the following Green formula:

$$\int_\Omega A(x, D_x)u \cdot v dx = \int_\Omega \langle \nabla_g u, \nabla_g v \rangle_g \, dx - \int_\Gamma v \partial_\nu u d\sigma, \quad v \in C^1(\overline{\Omega}). \tag{12}$$

Let $F \in L^2(Q)$. Let $\phi$ be a given solution of the second-order hyperbolic equation

$$\begin{aligned} \partial_t^2 \phi + c^2(x)A(x, D_x)\phi &= F(x, t), \quad (x, t) \in Q, \\ \phi(x, t) &= 0, \quad (x, t) \in \Sigma = \Gamma \times [0, T], \end{aligned} \tag{13}$$

within the following class: $\phi \in L^2(0, T, H^1(\Omega)) \cap H^1(0, T, L^2(\Omega))$. Then the following identity holds true for each $t_1, t_2 \in [0, T]$:

$$\int_\Omega c^{-2} |\partial_t \phi(t_1)|^2 + |\nabla_g \phi(t_1)|_g^2 \, dx - \int_\Omega c^{-2} |\partial_t \phi(t_2)|^2 + |\nabla_g \phi(t_2)|_g^2 \, dx$$
$$= \int_{t_1}^{t_2} \int_\Omega c^{-2} F(x, t)\partial_t \phi dx dt. \tag{14}$$

In fact, to prove (14) we multiply both sides (13) by $c^{-2}(x)\partial_t \phi$ and integrate over $[t_1, t_2] \times \Omega$, using the Green formula (12).

### 3.2. Linearized inverse problem

First of all, without loss of generality, we may assume in (14) $t = T/2 := t^*$ as an initial moment. This is not essential, because the change of independent variables $t \longrightarrow t - T/2$ transforms $t = T/2$ to $t = 0$.

Next we consider the difference $w = u - u_*$ which satisfies

$$\begin{aligned} \partial_t^2 w + c^2(x)A(x, D)w &= f(x)g(x, t) \text{ in } Q = \Omega \times [0, T], \\ w(x, t^*) = \partial_t w(x, t^*) &= 0 \quad \text{in } \Omega, \\ w(x, t) &= 0 \quad \text{on } \Sigma = \Gamma \times [0, T], \end{aligned} \tag{15}$$

where $f$ and $g$ are given by

$$f(x) = c^2(x) - c_*^2(x) \quad \text{and} \quad g(x, t) = -A(x, D_x)u_*(x, t). \tag{16}$$

In this subsection we discuss a linearized inverse problem of determining $f$ from $\partial_\nu w_{|\Sigma_0}$.

It is well known (e.g., Bellassoued[1], Bugheim-Klibanov[3], Imanuvilov-Yamamoto[5-7], Isakov[8], Isakov-Yamamoto[9]) that the uniqueness problem in determining coefficients can be reduced to the study of uniqueness in inverse source problem (15). More precisely, assuming that $c(x)$ and $g(x,t)$ are given functions, our inverse problem is the identification of $f(x)$. For the linearized inverse problem, we need the following lemmas, which are simple consequences of (14).

**Lemma 3.1.** *Let us consider $F \in L^2(Q)$, $\phi_1 \in L^2(\Omega)$. Let $\phi$ be a given solution of the second-order hyperbolic system*

$$\begin{aligned}
\partial_t^2 \phi + c^2(x) A(x, D)\phi &= F(x,t) \ \ in \ Q = \Omega \times [0, T], \\
\phi(x, t^*) = 0, \quad \partial_t \phi(x, t^*) &= \phi_1 \ \ in \ \Omega, \\
\phi(x, t) &= 0 \hspace{2.3cm} on \ \Sigma = \Gamma \times [0, T]
\end{aligned} \tag{17}$$

*within the following class $\phi \in H^{1,1}(Q) = L^2(0, T; H^1(\Omega)) \cap H^1(0, T; L^2(\Omega))$. Then the following estimate holds true:*

$$\|\phi_1\|_{L^2(\Omega)}^2 \le C \left\{ \|\phi\|_{H^1(Q_\rho)}^2 + \int_{Q_\rho} |F(x,t)\partial_t \phi(x,t)| \, dx dt \right\} \tag{18}$$

*for some positive constant $C > 0$.*

**Proof .** Applying the energy identity (14) to the solution $\phi$ of (17) with $t_1 = \frac{T}{2}$ and $t_2 = t$, where $t \in (\rho, T - \rho)$, we obtain

$$\int_\Omega |\phi_1|^2 \, dx \le C \int_\Omega (|\partial_t \phi(t)|^2 + |\nabla_x \phi(t)|^2) dx + \int_{Q_\rho} |F(x,t)\partial_t \phi| \, dx dt. \tag{19}$$

Thus we obtain (18), after (19) is integrated over $[\rho, T - \rho]$. $\qquad\square$

**Lemma 3.2.** *Let $w \in H^2(Q)$ be the solution of (15). Assume that*

$$g, g_t \in L^2(0, T; L^\infty(\Omega)). \tag{20}$$

*Then there exists a constant $C > 0$ such that*

$$\int_Q \left( |\partial_t w|^2 + |\nabla_x w|^2 \right) dx dt \le C \|f\|_{L^2(\Omega)}^2, \tag{21}$$

*and moreover*

$$\int_Q \left( |\partial_t^2 w|^2 + |\Delta w|^2 + |\nabla_x \partial_t w|^2 \right) dx dt \le C \|f\|_{L^2(\Omega)}^2. \tag{22}$$

**Proof .** Apply the energy identity (14) to the solution $w$ of (15) with $t_1 = T/2 = t^*$ and $t_2 = t$ where $0 < t < T$. Since $\partial_\nu w = 0$ in $\Sigma$, for any $\varepsilon > 0$, we obtain

$$\int_\Omega \left( |\partial_t w(t)|^2 + |\nabla_x w(t)|^2 \right) dx \le C_\varepsilon \int_Q |f(x)g(x,t)|^2 \, dx dt + \varepsilon \int_Q |\partial_t w|^2 \, dx dt. \tag{23}$$

By (20), the first integral term in $Q$ on the right-hand side of estimate (23) satisfies the following estimate:

$$\int_Q |f(x)g(x,t)|^2 \, dxdt \le C \int_\Omega |f(x)|^2 \left( \int_0^T \|g(\cdot,t)\|_{L^\infty(\Omega)}^2 \, dt \right) dx \le C \|f\|_{L^2(\Omega)}^2 \,.$$
$$(24)$$

Next, after (23) is integrated over $[0,T]$, we insert (24) into it and choose a small $\varepsilon$, so that we obtain (21).

In order to prove (22), set $v = \partial_t w$. Then we have

$$\begin{aligned}
\partial_t^2 v + c^2(x) A(x, D_x) v &= f(x)g_t(x,t) & &\text{in } \Omega \times [0,T], \\
v(x, t^*) = 0, \quad \partial_t v(x, t^*) &= f(x)g(x, t^*) & &\text{in } \Omega, \\
v(x,t) &= 0 & &\text{on } \Sigma = \Gamma \times [0,T].
\end{aligned}$$
$$(25)$$

Using (14) we obtain

$$\int_\Omega \left( |\partial_t v(t)|^2 + |\nabla_x v(t)|^2 \right) dx \le C \int_\Omega |f(x)g(x, t^*)|^2 \, dx$$
$$+ C_\varepsilon \int_Q |f(x)g_t(x,t)|^2 \, dxdt + \varepsilon \int_Q |\partial_t v|^2 \, dxdt.$$
$$(26)$$

Similarly to (24), by (20) we have

$$\int_Q |f(x)g_t(x,t)|^2 \, dxdt \le C \int_\Omega |f(x)|^2 \left( \int_0^T \|g_t(\cdot,t)\|_{L^\infty(\Omega)}^2 \, dt \right) dx \le C \|f\|_{L^2(\Omega)}^2$$
$$(27)$$

and

$$\int_\Omega |f(x)g_t(x, t^*)|^2 \, dx \le C \|f\|_{L^2(\Omega)}^2 \,.$$
$$(28)$$

Next we integrate (26) over $[0,T]$ and we insert (27) and (28) into it and choose a small $\varepsilon$, so that we obtain (22). $\qquad\square$

The above preparations now allow us to begin the proof of Theorem 1.

### 3.3. *Proof of Theorem 1*

We use the global Carleman estimate (11) and an idea inspired by Imanuvilov & Yamamoto's works [5–7]. Henceforth we assume that (H.1) and (H.2) hold true. Let $\varphi(x,t)$ be the pseudo-convex function defined in section 2. Then we have $\varphi(x,t) = e^{\beta \psi(x,t)} =: \rho(x)\sigma(t)$, where $\rho(x) \ge 1$ and $\sigma(t) \le 1$ are defined by

$$\rho(x) = e^{\beta \vartheta(x)} \ge 1, \ \forall x \in \Omega \quad \text{and} \quad \sigma(t) = e^{-\beta \gamma \left( t - \frac{T}{2} \right)^2} = e^{-\beta \gamma (t - t^*)^2} \le 1, \ \forall t \in [0, T].$$
$$(29)$$

Let $v$ be the solution to

$$\begin{aligned}
\partial_t^2 v + c^2(x) A(x, D_x) v &= f(x)g_t(x,t) & &\text{in } \Omega \times [0,T], \\
v(x, t^*) = 0, \quad \partial_t v(x, t^*) &= f(x)g(x, t^*) & &\text{in } \Omega, \\
v(x,t) &= 0 & &\text{on } \Sigma = \Gamma \times [0,T],
\end{aligned}$$
$$(30)$$

where $f(x) = c^2(x) - c_*^2(x)$ and $g(x,t) = -A(x, D_x)u_*(x,t)$.
We apply Theorem 2 to obtain

$$\lambda \int_{Q_\rho} e^{2\lambda\varphi} \left( |\nabla v|^2 + \lambda^2 |v|^2 \right) dxdt \leq C \int_Q e^{2\lambda\varphi} |f(x)g_t(t,x)|^2 dxdt$$

$$+\lambda \int_{\Sigma_0} e^{2\lambda\varphi} \left( |\partial_\nu v|^2 \right) dxdt + \lambda \int_{Q\setminus Q_\rho} e^{2\lambda\varphi} \left( |\nabla v|^2 + \lambda^2 |v|^2 \right) dxdt \qquad (31)$$

provided $\lambda > 0$ is large enough.

**Lemma 3.3.** *Let $v$ be the solution of (30). Then there exist constants $C > 0$ and $\kappa < 1$ such that for all $\lambda > 0$ large enough, there exists $C_\lambda > 0$ such that the following estimate holds true:*

$$\lambda \int_{Q_\rho} e^{2\lambda\varphi} \left( |\nabla v|^2 + \lambda^2 |v|^2 \right) dxdt \leq C \left[ \int_Q e^{2\lambda\varphi} |f(x)g_t(t,x)|^2 dxdt \right.$$

$$\left. +Ce^{2\kappa\lambda} \|f\|_{L^2(\Omega)}^2 \right] + C_\lambda \|\partial_\nu v\|_{L^2(\Sigma_0)}^2 . \qquad (32)$$

**Proof .** It follows from (22) and the condition (i) of $\psi$ (section 2) that we can choose sufficiently small $\rho > 0$ such that

$$\lambda \int_{Q\setminus Q_\rho} e^{2\lambda\varphi} \left( |\nabla v|^2 + \lambda^2 |v|^2 \right) dxdt \leq Ce^{2\kappa\lambda} \|f\|_{L^2(\Omega)}^2 , \qquad (33)$$

where $\kappa < 1$ and $C > 0$ are generic constants.
Inserting (33) to (31) we obtain

$$\lambda \int_{Q_\rho} e^{2\lambda\varphi} \left( |\nabla v|^2 + \lambda^2 |v|^2 \right) dxdt \leq C \int_Q e^{2\lambda\varphi} |f(x)g_t(t,x)|^2 dxdt$$

$$+\lambda \int_{\Sigma_0} e^{2\lambda\varphi} \left( |\partial_\nu v|^2 \right) dxdt + e^{2\lambda\kappa} \|f\|_{L^2(\Omega)}^2 . \qquad (34)$$

This completes the proof of (32). $\qquad\square$

**Lemma 3.4.** *Let $v$ be the solution of (30). Then there exists a constant $C > 0$ such that for all $\lambda > 0$ large enough there exists $C_\lambda > 0$ such that the following estimate holds true:*

$$\|e^{\lambda\varphi}f\|_{L^2(\Omega)}^2 \leq C \left[ \lambda \int_{Q_\rho} e^{2\lambda\varphi} \left( \lambda^2 |v|^2 + |\nabla v|^2 \right) dxdt + \int_Q |f(x)g_t(x,t)|^2 e^{2\lambda\varphi} dxdt \right]$$

$$+ C_\lambda \|\partial_\nu v\|_{L^2(\Sigma_0)}^2 . \qquad (35)$$

**Proof .** Let $\phi = e^{\lambda\varphi}v$. Then we have

$$P(x,D)\phi = P(x,D)(e^{\lambda\varphi}v) = e^{\lambda\varphi} \left[ e^{-\lambda\varphi} P(x,D)e^{\lambda\varphi}v \right] = e^{\lambda\varphi} \left[ P(x, D - i\lambda\nabla\varphi)v \right].$$

Now decompose the conjugated operator $P(x, D - i\lambda\nabla\varphi)$ as follows

$$P(x, D - i\lambda\nabla\varphi(z)) = P_s(z, D, \lambda) + \lambda P_a(z, D), \quad z = (x,t), \qquad (36)$$

where

$$P_s(z, D, \lambda) = P(x, D) - \lambda^2 p(x, \nabla\varphi) = P(x, D) - \lambda^2 \left( \varphi_t^2 - \sum_{i,j=1}^{n} a_{ij}(x) \frac{\partial\varphi}{\partial x_i} \frac{\partial\varphi}{\partial x_j} \right)$$

$$P_a(z, D) = \left( \varphi_t \partial_t + \partial_t \varphi_t - \sum_{i,j=1}^{n} \frac{\partial\varphi}{\partial x_i} a_{ij}(x) \frac{\partial}{\partial x_j} - \sum_{i,j=1}^{n} \frac{\partial}{\partial x_i} \left( a_{ij}(x) \frac{\partial\varphi}{\partial x_j} \right) \right)$$

(37)

is respectively the symmetric part and the skew-symmetric part of $P(x, D - i\lambda\nabla\varphi)$. Therefore by (30) we obtain

$$\partial_t^2 \phi + c^2(x) A(x, D_x)\phi = f(x)g_t(x, t)e^{\lambda\varphi} + e^{\lambda\varphi} K(z, D, \lambda)v \text{ in } \Omega \times [0, T],$$
$$\phi(x, t^*) = 0, \quad \partial_t\phi(x, t^*) = f(x)g(x, t^*)e^{\lambda\rho(x)} \qquad \text{in } \Omega, \qquad (38)$$
$$\phi(x, t) = 0 \qquad\qquad\qquad\qquad\qquad \text{in } \Sigma = \Gamma \times [0, T],$$

where $K(z, D, \lambda)v = -\lambda^2 p(x, \nabla\varphi)v + \lambda P_a(x, D)v$. Next using (18) with $\phi = e^{\lambda\varphi}v$, we obtain that

$$\|e^{\lambda\rho} f\|_{L^2(\Omega)}^2 \leq C \left[ \int_{Q_\rho} e^{2\lambda\varphi} \left( \lambda^2 |v|^2 + |\nabla v|^2 \right) dxdt + \int_{Q_\rho} e^{2\lambda\varphi} |f(x)g_t(x, t)|^2 dxdt \right.$$

$$\left. + \int_{Q_\rho} e^{\lambda\varphi} |K(x, D, \lambda)v.\partial_t\phi| dxdt \right] + C_\lambda \|\partial_\nu v\|_{L^2(\Sigma_0)}^2 . \qquad (39)$$

Since

$$e^{\lambda\varphi} K(x, D, \lambda)v.\partial_t\phi = e^{2\lambda\varphi} \left[ -\lambda^2 p(x, \nabla\varphi)v + \lambda P_a(x, D)v \right] \left[ \lambda\varphi_t v + \partial_t v \right], \qquad (40)$$

we obtain by the Schwarz inequality

$$\int_{Q_\rho} e^{\lambda\varphi} |K(x, D, \lambda)v.\partial_t\phi| dxdt \leq C\lambda \int_{Q_\rho} e^{2\lambda\varphi} \left( \lambda^2 |v|^2 + |\nabla v|^2 \right) dxdt. \qquad (41)$$

Inserting (41) into the right-hand side of (39), we obtain (35). $\qquad\qquad\square$

We will now complete the proof of Theorem 1. Substituting (32) into the right-hand side of (35), we have

$$\|e^{\lambda\rho} f\|_{L^2(\Omega)}^2 \leq C \left\{ \int_Q e^{2\lambda\varphi} |f(x)g_t(x, t)|^2 dxdt + e^{2\lambda\kappa} \|f\|_{L^2(\Omega)}^2 \right\} + C_\lambda \|\partial_\nu v\|_{L^2(\Sigma_0)}^2 . \qquad (42)$$

Now we return to the first integral term in $Q$ on the right-hand side of (42)

$$\int_Q e^{2\lambda\varphi} |f(x)g_t(x, t)|^2 dxdt \leq \int_\Omega e^{2\lambda\rho(x)} |f(x)|^2 \left( \int_0^T e^{-2\lambda(\rho-\varphi)} \|g_t(t, .)\|_{L^\infty(\Omega)}^2 dt \right) dx.$$

On the other hand, by the Lebesgue theorem, we obtain

$$\int_0^T e^{-2\lambda(\rho-\varphi)} \|g_t(t, .)\|_{L^\infty(\Omega)}^2 dt = \int_0^T e^{-2\lambda\rho(x)(1-\sigma(t))} \|g_t(t, .)\|_{L^\infty(\Omega)}^2 dt$$

1372

$$\leq \int_0^T e^{-2\lambda(1-\sigma(t))} \|g_t(t,.)\|_{L^\infty}^2 \, dt = o(1), \tag{43}$$

as $\lambda \to \infty$. By (43) and (42), we obtain

$$\left\|e^{\lambda\rho}f\right\|_{L^2(\Omega)}^2 \leq o(1)\left\|e^{\lambda\rho}f\right\|_{L^2}^2 + Ce^{2\lambda\kappa}\|f\|_{L^2(\Omega)}^2 + C_\lambda\|\partial_\nu v\|_{L^2(\Sigma_0)}^2. \tag{44}$$

Here we note that the first term of the right-hand side of (44) can be absorbed into the left hand side if we take large $\lambda > 0$. On the other hand, since $\rho(x) \geq 1$ for all $x \in \Omega$ and $\kappa < 1$, we have for $\lambda$ sufficiently large

$$\|f\|_{L^2(\Omega)}^2 \leq C\|\partial_t(\partial_\nu u - \partial_\nu u_*)\|_{L^2(\Sigma_0)}^2. \tag{45}$$

The proof of Theorem 1 is complete.

**Acknowledgements.** The second named author was partly supported by Grant 15340027 from the Japan Society for the Promotion of Science and Grant 17654019 from the Ministry of Education, Cultures, Sports and Technology.

### References

1. M.Bellassoued: *Uniqueness and stability in determining the speed of propagation of second-order hyperbolic equation with variable coefficients*, Appl.Anal. 83, 983–1014 (2004).
2. M.Bellassoued-M.Yamamoto: *Determination of a coefficient in the wave equation with a single measurement* , appear in SMF Seminar and Congress.
3. A.L. Bugheim-M.V.Klibanov: *Global uniqueness of class of multidimentional inverse problems*, Soviet Math. Dokl. 24, 244-247 (1981).
4. O. Yu. Imanuvilov: *On Carleman estimates for hyperbolic equations*, Asymptotic Analysis 32, 185 – 220 (2002).
5. O.Yu.Imanuvilov-M.Yamamoto: *Global uniqueness and stability in determining coefficients of wave equations*, Comm. Part. Diff. Eq. 26, 1409-1425 (2001).
6. O.Yu.Imanuvilov-M.Yamamoto: *Global Lipshitz stability in an inverse hyperbolic problem by interior observations*, Inverse Problems 17, 717–728 (2001).
7. O.Yu.Imanuvilov-M.Yamamoto: *Determination of a coefficient in an acoustic equation with single measurement*, Inverse Problems 19, 157-171 (2003).
8. V.Isakov: *Inverse Problems for Partial Differential Equations*, Springer-Verlag, Berlin, 1998.
9. V.Isakov-M.Yamamoto: *Carleman estimate with the Neumann boundary condition and its applications to the observability inequality and inverse hyperbolic problems*, Comptemporary Mathematics 268, 191-225 (2000).
10. A.Khaidarov: *On stability estimates in multidimensional inverse problems for differential equation*, Soviet Math. Dokl. 38, 614-617 (1989).
11. M.V.Klibanov: *Inverse problems and Carleman estimates*, Inverse Problems 8, 575-596, (1992).
12. M.V. Klibanov - A.Timonov: *Carleman Estimates for Coefficient Inverse Problems and Numerical Applications*, VSP, Utrecht, 2004.
13. M.V. Klibanov - M. Yamamoto: *Lipschitz stability of an inverse problem for an accoustic equation*, Preprint No. 2004-03. Technical Report Series of The Department of Mathematics of The University of North Carolina at Charlotte. Also available at http://www.math.uncc.edu/preprint/2004/index.php/ and at http://kyokan.ms.u-tokyo.ac.jp/users/preprint/preprint2004.html.

14. M.M.Lavrent'ev: *Some Improperly Posed Problems of Mathematical Physics*, Springer-Verlag, Berlin, 1967.

15. R.Triggiani and P.F.Yao: *Carleman estimates with no lower-order terms for general Riemann wave equation. global uniqueness and observability in one shot*, Appl. Math. Optim. 46, 331-375 (2002).

# NUMERICAL CAUCHY PROBLEMS FOR THE LAPLACE EQUATION

T. MATSUURA

*Department of Mechanical Engineering, Faculty of Engineering,*
*Gunma University,Kiryu, Gunma 376-8515, Japan*
*E-mail: matsuura@me.gunma-u.ac.jp*

S. SAITOH

*Department of Mathematics, Faculty of Engineering,*
*Gunma University,Kiryu, Gunma 376-8515, Japan*
*E-mail: ssaitoh@math.sci.gunma-u.ac.jp*

M. YAMAMOTO

*Department of Mathematical Sciences,The University of Tokyo,*
*Komaba, Meguro, Tokyo 153-8914, Japan*
*E-mail: myama@ms.cc.u-tokyo.ac.jp*

We give a new algorithm constructing harmonic functions from data on a part of a boundary. Our approach is based on a general concept and we can apply our methodology to many problems, but here we numerically deal with an ill-posed Cauchy problem which appears in many applications. On some part $\Gamma$ of a boundary, for suitably given functions $f$ and $g$, we look for a constructing formula of an approximate harmonic function $u$ satisfying $u = f$ and $\frac{\partial}{\partial \nu} u = g$, the outer normal derivative. Our method is based on the Dirichlet principle by combinations with generalized inverses, Tikhonov's regularization and the theory of reproducing kernels.

**Key words:** Laplace equation, Dirichlet principle, Cauchy problem, approximation of functions, reproducing kernel, Tikhonov regularization

**Mathematics Subject Classification:** 30C40,31A05,31B05,35J05,45B05

## 1. Introduction

We shall propose a new algorithm for constructing approximate solutions for the Laplace equation

$$\Delta u = 0 \tag{1}$$

in a domain $D \subset \mathbf{R}^n$ satisfying the boundary conditions

$$u = f \quad \text{on} \quad \Gamma \tag{2}$$

and

$$\frac{\partial}{\partial \nu} u = g \quad \text{on} \quad \Gamma \tag{3}$$

where $\frac{\partial}{\partial \nu}$ denotes the outer normal derivative, $\Gamma \subset \partial D$. Henceforth $H^s$ denotes the Sobolev space $H^s(\mathbf{R}^n)$. For functions $f$ and $g$, they must satisfy strong conditions, in order that equations (1), (2) and (3) hold true. However, in our method, for any given functions $f, g$ in $L_2(\Gamma)$, we can consider the best approximation $u$ for equations (1), (2) and (3).

For numerics for the Cauchy problem for the Laplace equation, there have been very many papers and we refer to some of them and the readers can consult also the references therein [1,3-12,16-18,24]. Moreover as a substantial source book on inverse problems, we refer for example to Isakov ([10]).

For convenience, we introduce in $H^m = H^m(\mathbf{R}^n)$ with $m \in \mathbf{N}$ an equivalent norm:

$$\|F\|_{H^m}^2$$

$$= \sum_{\nu=0}^{m} {}_mC_\nu \sum_{r_1+r_2+\cdots+r_n=\nu, \ r_1,r_2,\ldots,r_n \geq 0} \frac{\nu!}{r_1!r_2!\cdots r_n!} \int_{\mathbf{R}^n} \left( \frac{\partial^\nu F(x)}{\partial x_1^{r_1} \partial x_2^{r_2} \cdots \partial x_n^{r_n}} \right)^2 dx. \tag{4}$$

Then the Hilbert space $H^m$ admits the reproducing kernel

$$K(x,y) = \frac{1}{(2\pi)^n} \int_{\mathbf{R}^n} \frac{1}{(1+|\xi|^2)^m} e^{i(x-y)\cdot\xi} d\xi, \tag{5}$$

as we see easily by using Fourier's transform (cf. [19], page 58). Note that the $H^s$ admitting the reproducing kernel (5) for $m = s$ can be defined for any positive number $s(s > n/2)$ in terms of Fourier integrals $\hat{F}$ of $F$

$$\hat{F}(\xi) = \frac{1}{(2\pi)^{n/2}} \int_{\mathbf{R}^n} e^{-i\xi \cdot x} F(x) dx$$

as follows:

$$\|F\|_{H^s}^2 = \int_{\mathbf{R}^n} |\hat{F}(\xi)|^2 (1+|\xi|^2)^s d\xi.$$

For our problem, we shall consider the extremal problem based on our recent methods in [2,13-15,20-23]: for fixed $\lambda > 0$ and for any $f, g \in L_2(\Gamma)$,

$$\inf_{F \in H^s} \left\{ \lambda \|F\|_{H^s}^2 + \|\Delta F\|_{L_2(\mathbf{R}^n)}^2 + \|F - f\|_{L_2(\Gamma)}^2 + \|\frac{\partial}{\partial \nu} F - g\|_{L_2(\Gamma)}^2 \right\}. \tag{6}$$

In order to simplify the problem, we consider here as follows:

(1) As a function space, we use $H^s$; in this case the space admits the reproducing kernel (5), which is extremely simple.

(2) For the integral of $\Delta F$, we shall consider it in the whole space, not in the domain $D$. Then the extremal problem will become simple.

For these simplifications, we see that the reproducing kernel $K_\lambda(x, y)$ of the Hilbert space $H_{K_\lambda}$ with the norm square

$$\lambda \|F\|_{H^s}^2 + \|\Delta F\|_{L_2(\mathbf{R}^n)}^2 \tag{7}$$

is given by

$$K_\lambda(x, y) = \frac{1}{(2\pi)^n} \int_{\mathbf{R}^n} \frac{e^{ip\cdot(x-y)}}{\lambda(|p|^2 + 1)^s + |p|^4} dp \tag{8}$$

([2]).

## 2. New Algorithm

Our strategy is first to represent the extremal function $F_{s,\lambda,g}^*(x)$ in (6) in an explicit form and second to consider the limit of this extremal function as $\lambda \to 0$. We can formulate these procedures and discuss theoretically by means of a Fredholm integral equation of the second kind. Here we propose a new numerical approach for the present problem.

In order to consider the boundary value problems in (2) and (3), we shall consider it in (6) as follows:

For any fixed points $\{x_j\}_{j=1}^N$ on the boundary $\Gamma$ and for any given values $\{A_j\}_{j=1}^N$ and $\{B_j\}_{j=1}^N$, we consider the extremal problem, for any fixed $\{\lambda_j\}_{j=1}^N (\lambda_j > 0)$ and $\{\mu_j\}_{j=1}^N (\mu_j > 0)$

$$\inf_{F \in H^s} \left\{ \lambda \|F\|_{H^s}^2 + \|\Delta F\|_{L_2(\mathbf{R}^n)}^2 + \sum_{j=1}^N \lambda_j |F(x_j) - A_j|^2 + \sum_{j=1}^N \mu_j \left| \frac{\partial}{\partial \nu} F(x_j) - B_j \right|^2 \right\}; \tag{9}$$

that is, we approximate the integral in (6) by weighted sums. Then, we wish to look for the reproducing kernel $K_{\lambda_j,\mu_j}(x, y)$ of the Hilbert space $H_{K_{\lambda_j,\mu_j}}$ with the norm

$$\left( \lambda \|F\|_{H^s}^2 + \|\Delta F\|_{L_2(\mathbf{R}^n)}^2 + \sum_{j=1}^N \lambda_j |F(x_j)|^2 + \sum_{j=1}^N \mu_j \left| \frac{\partial}{\partial \nu} F(x_j) \right|^2 \right)^{\frac{1}{2}}. \tag{10}$$

We can represent the reproducing kernel $K_{\lambda_j,\mu_j}(x, y)$ in terms of $K_\lambda(x, y)$ directly. However, for many points $\{x_j\}_{j=1}^N$ on the boundary $\Gamma$, the size of a matrix is very large so that this direct representation will not be effective ([19], pp. 81-83).

In order to overcome this difficulty, we shall propose a new approach.

The reproducing kernel $K^{(1,0)}(x, y)$ of the Hilbert space with the norm

$$\left( \lambda \|F\|_{H^s}^2 + \|\Delta F\|_{L_2(\mathbf{R}^n)}^2 + \sum_{j=1}^1 \lambda_j |F(x_j)|^2 \right)^{\frac{1}{2}} \tag{11}$$

is given by

$$K^{(1,0)}(x,y) = K_\lambda(x,y) - \frac{\lambda_1 K_\lambda(x,x_1)K_\lambda(x_1,y)}{1+\lambda_1 K_\lambda(x_1,x_1)}. \tag{12}$$

Next, the reproducing kernel $K^{(1,1)}(x,y)$ of the Hilbert space with the norm

$$\left(\lambda\|F\|_{H^s}^2 + \|\Delta F\|_{L_2(\mathbf{R}^n)}^2 + \sum_{j=1}^{1}\lambda_j|F(x_j)|^2 + \sum_{j=1}^{1}\mu_j|\frac{\partial}{\partial\nu}F(x_j)|^2\right)^{\frac{1}{2}} \tag{13}$$

is given by

$$K^{(1,1)}(x,y) = K^{(1,0)}(x,y) - \frac{\mu_1\partial_\nu K^{(1,0)}(x,x_1)\partial_\nu K^{(1,0)}(x_1,y)}{1+\mu_1\partial_\nu\partial_\nu K^{(1,0)}(x_1,x_1)}. \tag{14}$$

For two points $x_1, x_2$, the reproducing kernel $K^{(2,1)}(x,y)$ of the Hilbert space with the norm

$$\left(\lambda\|F\|_{H^s}^2 + \|\Delta F\|_{L_2(\mathbf{R}^n)}^2 + \sum_{j=1}^{2}\lambda_j|F(x_j)|^2 + \sum_{j=1}^{1}\mu_j|\frac{\partial}{\partial\nu}F(x_j)|^2\right)^{\frac{1}{2}} \tag{15}$$

is given by

$$K^{(2,1)}(x,y) = K^{(1,1)}(x,y) - \frac{\lambda_2 K^{(1,1)}(x,x_2)K^{(1,1)}(x_2,y)}{1+\lambda_2 K^{(1,1)}(x_2,x_2)} \tag{16}$$

in terms of the reproducing kernel $K^{(1,1)}(x,y)$.

In the same way, for $N$ points $x_1, x_2, \cdots, x_N$, the reproducing kernel $K^{(N,N)}(x,y)$ of the Hilbert space with the norm (9) is given by

$$K^{(N,N)}(x,y) = K^{(N,N-1)}(x,y) - \frac{\mu_N\partial_\nu K^{(N,N-1)}(x,x_N)\partial_\nu K^{(N,N-1)}(x_N,y)}{1+\mu_N\partial_\nu\partial_\nu K^{(N,N-1)}(x_N,x_N)}. \tag{17}$$

In this way we can proceed as $K^{(2,1)}(x,y) \to K^{(2,2)}(x,y) \to K^{(3,2)}(x,y)$ to construct $K^{(N,N)}$, $N \in \mathbf{N}$. When we have the $N$-th kernel $K^{(N,N)}(x,y)$, it is the reproducing kernel of the Hilbert space with the norm (10) and the extremal function in the minimum problem in (9) is given by

$$F_{s,\lambda,x_j,\lambda_j,\mu_j,A_j,B_j}^*(x) = \sum_{j=1}^{N}A_j\lambda_j K^{(N,N)}(x,x_j) + \sum_{j=1}^{N}B_j\mu_j\partial_\nu K^{(N,N)}(x,x_j). \tag{18}$$

## 3. Error Estimate

In representation (18), when the data $A_j$ and $B_j$ contain noises, we can derive the error estimation and the proof will be given in our forthcoming paper.

**Theorem 1.** In (18), we obtain the estimate

$$|F^*_{s,\lambda,x_j,\lambda_j,\mu_j,A_j,B_j}(x)|$$

$$\leq \frac{1}{\sqrt{\lambda}}\left(\sum_{j=1}^{N}(\lambda_j|A_j|^2 + \mu_j|B_j|^2)\right)^{1/2}\sqrt{C_{n,s}}, \qquad (19)$$

where

$$C_{n,s} = \frac{\Gamma(s - n/2)}{(4\pi)^{n/2}\Gamma(s)}.$$

## 4. Numerical Experiments

In the following experiments, we take

$$n = 2, s = 2, \lambda = 10^{-8}$$

in (18) and we change values of $\lambda$ (in Figures 1,2), and we see that $\lambda = 10^{-8}$ can be regarded to be an optimal choice in order to obtain good approximations (18) in reconstructing the following two exact solutions.

$$u(x_1, x_2) = e^{x_1}\cos x_2, \quad \cos x_1 \sinh x_2, \quad -1 < x_1, x_2 < 1.$$

We try three types $\{\lambda_j\}$ and $\{\mu_j\}$ for each model.

On the square $[-1, +1] \times [-1, +1]$ of the $x$-plane, we take uniformly distributing grid points with mesh size 0.05 ( in total $41 \times 41 = 1681$ points ) and the similar points on the $y$-plane. Then we calculate the kernel (8) for the

$$1681 \times 1681 = 2825761$$

points in $\mathbf{R}^2 \times \mathbf{R}^2$.

For this purpose, note that the kernel (8) depends only on the function $|x - y|$ and for practical calculations we may assume that $y = 0$ on $\mathbf{R}^2$.

We write

$$x = (x_1, x_2) = (R\cos\theta, R\sin\theta)$$

and

$$p = (p_1, p_2) = (r\cos\Theta, r\sin\Theta).$$

Then,

$$p \cdot x = Rr\cos(\Theta - \theta)$$

and the dominator is only the function in $r$, and so we can replace $\theta = 0$ for our numerical reconstruction.

Therefore, by setting $x = (x_1, 0)$, when we calculate the kernel for $(z_1, z_2)$ by integrating (8) in $p_1$ and $p_2$, by setting $x_1$ by $x_1^2 = z_1^2 + z_2^2$ we obtain the value of $K_\lambda((z_1, z_2), (0, 0))$.

The kernel (8) is symmetric with respect to the origin and it is sufficient to calculate the kernel (8) on the 1/8 plane $\{x_2 \leq x_1, 0 \leq x_1, x_2\}$. Here, for $x, y \in [-2, +2] \times [-2, +2]$, we need the data on

$$\{x_2 \leq x_1, 0 \leq x_1, x_2 \leq 4\},$$

because we have to consider $x - y$.

When we calculate the kernel (8) by integrating in $p_1$ and $p_2$, the numerator of (8) is $e^{ip_1 x_1}$ and it is independent of $p_2$ and so we integrate (8) first by $p_2$ and we obtain the analytic expression for (8)

$$K_\lambda((x_1, 0), (0, 0)) = \frac{1}{(2\pi)^2} \int_{\mathbf{R}^2} \frac{e^{p_1 x_1}}{\lambda(p_1^2 + p_2^2 + 1)^2 + (p_1^2 + p_2^2)^2} dp_1 dp_2$$

$$= \frac{1}{8\sqrt{\lambda}\pi} \int_{\mathbf{R}} \left[ \frac{1+\lambda}{p_1^4 + \lambda(1+p_1^2)^2} \right]^{\frac{1}{4}} \cos(p_1 x_1) \cdot 2\sin\left[\frac{1}{2}\arctan\left[\frac{\sqrt{\lambda}}{(1+\lambda)p_1^2 + \lambda}\right]\right] dp_1.$$

We use Mathematica to calculate this integral by the Gauss-Kronrod method. We assume 6 significant digits.

## On the boundary conditions

We consider the domain as $[-0.95, +0.95] \times [-0.95, +0.95]$, because for the calculation of $K^{(i,i)}(x, y)$ ( $i = 1, 2, \cdots N$), we need the normal derivative and we have data on $[-1, +1] \times [-1, +1]$. For the derivatives we consider the central difference. At the corner we consider the derivative for the bisector direction.

As the boundary points we take 77 points (in other words $N = 77$ in (17)) as follows:

We take the coordinates as $x = (x_1, x_2)$ and $x_1 = -0.95 \sim +0.95, x_2 = 0.95; x_1 = 0.95, x_2 = -0.95 \sim +0.95$ with mesh size 0.05.

Our numerical tests show that the final results are almost same for any sequences of the 77 points. This fact will mean that our numerical method on the basis of the iteration given by (12)-(17) is appropriate and robust. In our Figures, we calculated the kernels for the iteration from the first point (-0.95,+0.95) to the last point(+0.95,-0.95)( the midway point (+0.95,+0.95)) for the 77 points.

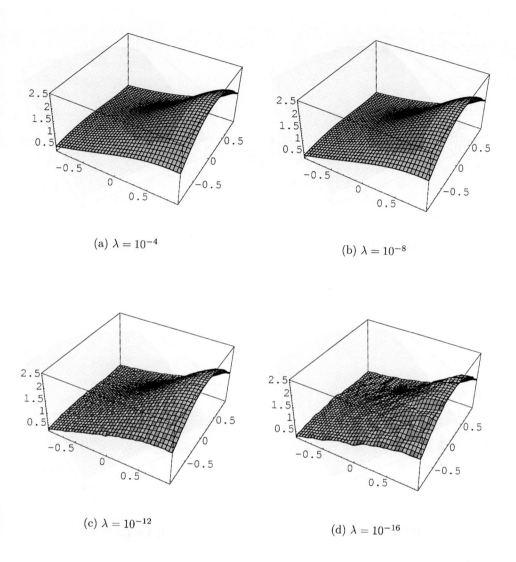

(a) $\lambda = 10^{-4}$

(b) $\lambda = 10^{-8}$

(c) $\lambda = 10^{-12}$

(d) $\lambda = 10^{-16}$

Fig. 1.   $u(x_1, x_2) = e^{x_1} \cos x_2$ and $\lambda_j = \mu_j = 10^3$

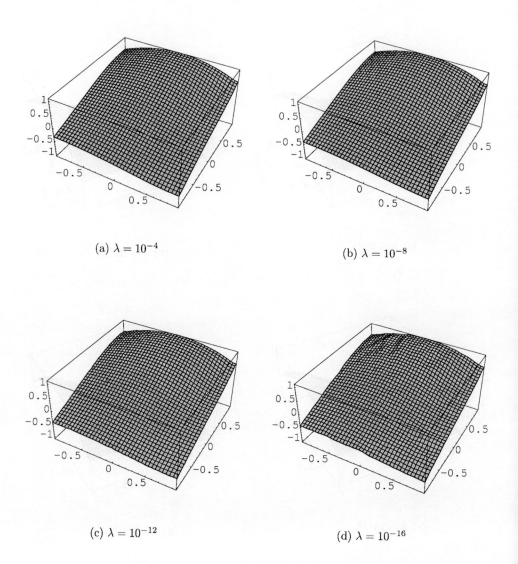

(a) $\lambda = 10^{-4}$

(b) $\lambda = 10^{-8}$

(c) $\lambda = 10^{-12}$

(d) $\lambda = 10^{-16}$

Fig. 2. $u(x_1, x_2) = \cos x_1 \sinh x_2$ and $\lambda_j = \mu_j = 10^3$

## Acknowledgements

T. Matsuura is supported in part by the Gunma University Foundation for the Promotion of Science and Engineering. S.Saitoh is supported in part by the Grant-in-Aid for Scientific Research (C)(2)(No.16540137) from the Japan Society for the Promotion Science. Saitoh and Matsuura are partially supported by the Mitsubishi Foundation, the 36th, Natural Sciences, No.20 (2005-2006). M.Yamamoto is supported in part by grant 15340027 from the Japan Society for the Promotion of Science and grant 17654019 from the Ministry of Education, Cultures, Sports and Technology.

## References

1. D. D. Ang, D. N. Thanh and V. V. Thanh, *Regularized solutions of a Cauchy problem for the Laplace equation in an irregular strip*, J. Integral Equations Appl. **5** 429(1993).
2. M. Asaduzzaman, T. Matsuura, and S. Saitoh, *Constructions of approximate solutions for linear differential equations by reproducing kernels and inverse problems*, Advances in Analysis, Proceedings of the 4th International ISAAC Congress, World Scientific , **30**, 335(2005).
3. F. Berntsson and L. Elden, *Numerical solution of a Cauchy problem for the Laplace equation.* Special issue to celebrate Pierre Sabatier's 65th birthday (Montpellier, 2000), Inverse Problems **17** 839 (2001).
4. L. Bourgeois, *A mixed formulation of quasi-reversibility to solve the Cauchy problem for Laplace's equation*, Inverse Problems **21** 1087 (2005).
5. J. R. Cannon and P. DuChateau, *Approximating the solution to the Cauchy problem for Laplace's equation*, SIAM J. Numer. Anal. **14** 473 (1977).
6. J. Cheng, Y. C. Hon, T. Wei and M. Yamamoto, *Numerical computation of a Cauchy problem for Laplace's equation*, ZAMM Z. Angew. Math. Mech. **81** 665 (2001).
7. N.H. Dinh and M. H. Pham, *Stability results for the Cauchy problem for the Laplace equation in a strip*, Inverse Problems **19** 833 (2003).
8. D. N. Hao, N. Dinh and D. Lesnic, *The Cauchy problem for Laplace's equation via the conjugate gradient method*, IMA J. Appl. Math. **65** 199 (2000).
9. Y. C. Hon and T. Wei, *Backus-Gilbert algorithm for the Cauchy problem of the Laplace equation*, Inverse Problems **17** 261 (2001).
10. V. Isakov, *Inverse Problems for Parital Differential Equation*, Springer, (1998).
11. M. V. Klibanov and F. Santosa, *A computational quasi-reversibility method for Cauchy problems for Laplace's equation*, SIAM J. Appl. Math. **51** 1653 (1991).
12. J. Y. Lee and J. R.. Yoon, *A numerical method for Cauchy problem using singular value decomposition*, Second Japan-Korea Joint Seminar on Inverse Problems and Related Topics (Seoul, 2001). Commun. Korean Math. Soc. **16** 487 (2001).
13. T. Matsuura, S. Saitoh and D.D. Trong, *Numerical solutions of the Poisson equation*, Applicable Analysis, **83** 1037 (2004).
14. T. Matsuura, S. Saitoh and D.D. Trong, *Approximate and analytical inversion formulas in heat conduction on multidimensional spaces*, J. of Inverse and Ill-posed Problems **13** 1 (2005).
15. T. Matsuura and S. Saitoh, *Analytical and numerical solutions of linear ordinary differential equations with constant coefficients*, Journal of Analysis and Applications, **3** 1 (2005).
16. T. Ohe and K. Ohnaka, *Uniqueness and convergence of numerical solution of the Cauchy problem for the Laplace equation by a charge simulation method*, Japan J.

Indust. Appl. Math. **21** 339 (2004).

17. K. Onishi and Q. Wang, *Numerical solutions of the Cauchy problem in potential and elastostatics*, Inverse problems and related topics (Kobe, 1998), 115–132, Chapman & Hall/CRC Res. Notes Math., **419**, Chapman & Hall/CRC, Boca Raton, FL, 2000.

18. H. J. Reinhardt, H. Han and N. H. Dinh, *Stability and regularization of a discrete approximation to the Cauchy problem for Laplace's equation*. SIAM J. Numer. Anal. **36** 890 (1999),(electronic).

19. S. Saitoh, *Integral Transforms, Reproducing Kernels and their Applications*, Pitman Res. Notes in Math. Series **369**(1997), Addison Wesley Longman Ltd, UK.

20. S. Saitoh, *Approximate Real Inversion Formulas of the Gaussian Convolution*, Applicable Analysis, **83** 727 (2004).

21. S. Saitoh, *Applications of Reproducing Kernels to Best Approximations, Tikhonov Regularizations and Inverse Problems*, Advances in Analysis, Proceedings of the 4th International ISAAC Congress, World Scientific , **39**, 439–446 (2005).

22. S. Saitoh, *Best approximation, Tikhonov regularization and reproducing kernels*, Kodai. Math. J. **28** 59 (2005).

23. S. Saitoh, *Tikhonov regularization and the theory of reproducing kernels*, Proceedings of the 12th ICFIDCAA (to appear).

24. T. Wei, Y. C. Hon and J. Cheng, *Computation for multidimensional Cauchy problem*, SIAM J. Control Optim. **42** , no. 2, 381 (2003) (electronic).

# CONDITIONAL STABILITY IN RECONSTRUCTION OF INITIAL TEMPERATURES

MASAHIRO YAMAMOTO

*Department of Mathematical Sciences,*
*The University of Tokyo, 3-8-1 Komaba,*
*Meguro, Tokyo 153-8914, Japan,*
*E-mail:myama@ms.u-tokyo.ac.jp*


JUN ZOU

*Department of Mathematics,*
*The Chinese University of Hong Kong,*
*Shatin, N.T., Hong Kong, China*
*E-mail:zou@math.cuhk.edu.hk*

In this paper, we shall present a conditional stability estimate of logarithmic rate in the reconstruction of initial temperatures.

**Key words:** Carleman estimate, conditional stability, backward heat problem
**Mathematics Subject Classification:** 35R30, 35B35, 35K45, 35K05

## 1. Conditional stability in ill-posed problems

Our main target is the reconstruction of initial values from restricted data in a heat equation. This problem is important from various industrial purposes, for example, effective monitoring techniques for working processes. On the other hand, from the mathematical viewpoint, our problem is ill-posed in the sense of Hadamard, that is, small errors in data may cause huge deviations in determining solutions. However this kind of instability can be restored frequently by taking into consideration practically acceptable a priori assumptions for the solutions. As one typical case, we can refer to the backward heat problem with a priori boundedness of solutions, where we would like to determine an initial value from the temperature distribution at a positive moment. The backward heat problem with a priori boundedness, can be stable, and in this case, a priori boundedness can be interpreted as the melting point of the material under consideration.

Before presenting our results, we will state the significance of the conditional stability from a general point of view: the conditional stability estimates can be stated as follows:

$$\|\text{deviations of solutions}\|$$
$$\leq \text{constant} \times \|\text{errors in data}\| : \quad \text{the Lipschitz type,}$$
$$\leq \text{constant} \times \|\text{errors in data}\|^{\alpha} : \text{Hölder type with } \alpha \in (0,1),$$
$$\leq \text{constant} \times \frac{1}{\log \|\text{errors in data}\|} : \quad \text{log-type,}$$

provided that deviations in solutions are assumed to be in $\mathcal{A}$, which is called an admissible set. Usually an admissible set is defined as a bounded set in a function space. Conditional stability of the logarithmic rate is much worse than the Lipschitz stability.

The conditional stability is interesting not only in view of the mathemtical theory, but also for giving guidelines to numerical computations of practical inverse problems. For example, the conditional stability rates give optimal (or quasi-optimal) choices of the Tikhonov regularizing parameters and discretization sizes (see Cheng, Yamamoto and Zou [2], Theorem 5 in Klibanov [4] and Section 2.5 in Klibanov and Timonov [5]).

The conditional stability rates essentially depend on the choice of $\mathcal{A}$, and other choices may improve the rates drastically. Even if we can presently know only the logarithmic type of stability, the proved stability does not necessarily mean the real badness in the numerical computations, because the proved estimate does not imply the logarithmic rate for really chosen data for the numerics (e.g., numerical tests in Yamamoto and Zou [13]). Moreover even if we can prove the conditional stability of the Lipschitz type, it does not necessarily mean the goodness in the numerical computations, because the constants in conditional stability estimate which we prove, may be tremendously large in comparison with the error level of data.

Thus the conditional stability is a very useful guideline for the numerical computation (e.g., [2]), while we should not regard the proved conditional stability as the absolute. In that sense, the stability which we have established is not the best possible and might admits possibilities for future improvements.

As for the ill-posed heat problems, there have been many works on the conditinal stability, but in several important cases, the conditional stability is not known. We aimed at filling such gaps in one typical case, and for the readability we do not state our result with full generality.

## 2. Initial temperature reconstruction

Consider the heat equation

$$
\begin{cases}
y_t = \Delta y & \text{in } \Omega \times (0, T), \\[2mm]
y = 0 & \text{on } \partial\Omega,
\end{cases}
\tag{1}
$$

where $\Omega \in \mathbb{R}^n$ is a bounded domain with smooth boundary $\partial\Omega$. Moreover let $\omega \subset \Omega$ be an arbitrary subdomain.

For a given fixed $\delta > 0$, our inverse problem is to reconstruct $y(\cdot, 0)$, when the measurement data of $y$ in $\omega \times (\delta, T)$ is available.

For any fixed $\varepsilon > 0$ and $M > 0$, we introduce an admissible set by

$$\mathcal{A} = \left\{ a \in H^{2\varepsilon}(\Omega); \ \|a\|_{H^{2\varepsilon}(\Omega)} \leq M \right\}. \tag{2}$$

We have the following stability estimate

**Theorem 1.** *If $y(\cdot, 0) \in \mathcal{A}$, then there exists $\kappa = \kappa(M, \varepsilon) \in (0, 1)$ such that*

$$\|y(\cdot, 0)\|_{L^2(\Omega)} \leq C(M, \varepsilon) \left( -\log \|y\|_{L^2(\omega \times (\delta, T))} \right)^{-\kappa}. \tag{3}$$

As for similar results, see Klibanov [4], Saitoh and Yamamoto [11], Xu and Yamamoto [12]. In particular, in [4], the author establishes a conditional stability estimate for a parabolic equation $y_t = A(x, t)y$ by lateral Cauchy data, where the coefficients are dependent both on $x$ and $t$, and an elliptic operator $A(x, t)$ of the second order, is not necessarily symmetric. The stability in [4] holds for a parabolic inequality but ours is only for a parabolic equation. Moreover the result in [4] is applied to a similar numerical method to the quasi-reversibilty. In the stability by [4], one can take $\kappa = 1$ but more smoothness for unknown initial values is necessary (i.e., $\varepsilon = 1/2$ in (2)).

**Proof:** Without loss of generality, we may assume that $\|y\|_{L^2(\omega \times (\delta, T))}$ is sufficiently small.

*First step; Application of a Carleman estimate.* We first show for any fixed $\kappa_1 \in (0, 1)$ and $\theta \in (\delta, T)$, we can choose a constant $C = C(M, \kappa_1) > 0$ such that

$$\|y(\cdot, \theta)\|_{L^2(\Omega)} \leq C \|y\|_{L^2(\omega \times (\delta, T))}^{\kappa_1}. \tag{4}$$

Since $\|y\|_{L^2(\omega \times (\delta, T))}$ is sufficiently small, we can assume that $\|y(\cdot, \theta)\|_{L^2(\Omega)} < 1$.

To prove (4), we apply the $H^{-1}$ Carleman estimates in [3] to obtain for arbitrarily fixed $\theta_1, \theta_2$ with $0 < \theta_1 < \theta < \theta_2 < T$,

$$\|y\|_{L^2(\theta_1, \theta_2; H^1(\Omega))} \leq C_M \|y\|_{L^2(\omega \times (\delta, T))}. \tag{5}$$

On the other hand, multiplying equation (1) by any $v \in H_0^1(\Omega)$, we derive

$$\int_\Omega y_t \, v dx = -\int_\Omega \nabla y \cdot \nabla v dx, \quad \forall v \in H_0^1(\Omega),$$

and this gives

$$\|y_t\|_{L^2(\theta_1, \theta_2; H^{-1}(\Omega))} \leq C \|y\|_{L^2(\theta_1, \theta_2; H^1(\Omega))}$$
$$\leq C_M \|y\|_{L^2(\omega \times (\delta, T))}.$$

With (5), we have

$$\|y\|_{H^1(\theta_1, \theta_2; H^{-1}(\Omega))} \leq C_M \|y\|_{L^2(\omega \times (\delta, T))}.$$

Therefore, by the Sobolev embedding, we have

$$\|y\|_{C([\theta_1, \theta_2]; H^{-1}(\Omega))} \leq C_M \|y\|_{L^2(\omega \times (\delta, T))}. \tag{6}$$

Furthermore, by the semigroup theory (e.g., Pazy [10]), we have $y(t) = y(\cdot, t) = e^{-tA}y(\cdot, 0)$, where $A$ is an operator in $L^2(\Omega)$ defined by $A = -\Delta$ and $\mathcal{D}(A) = H_0^1(\Omega) \cap H^2(\Omega)$. Then we see that for any $\gamma > 0$, there exists a constant $C_\gamma > 0$ such that

$$\|A^\gamma e^{-tA}\| \le C_\gamma t^{-\gamma},$$
$$\|a\|_{H^{2\gamma}(\Omega)} \le C_\gamma \|A^\gamma a\|_{L^2(\Omega)} \tag{7}$$

(e.g., [10]). Hence

$$\|y\|_{C([\theta_1,\theta_2];H^{2\gamma}(\Omega))}$$
$$\le C_\gamma \|A^\gamma e^{-tA}y(\cdot, 0)\|_{C([\theta_1,\theta_2];L^2(\Omega))} \tag{8}$$
$$\le \frac{C_\gamma}{\theta_1^\gamma}\|y(\cdot, 0)\|_{L^2(\Omega)} \le \frac{C_\gamma M}{\theta_1^\gamma}.$$

Therefore by the interpolation theory (e.g., Proposition 2.3 (p.19) and Theorem 12.4 (p.73) in Lions and Magenes [8]), (6) and (8) yield

$$\|y\|_{C([\theta_1,\theta_2];L^2(\Omega))}$$
$$\le \|y\|_{C([\theta_1,\theta_2];H^{-1}(\Omega))}^{\frac{2\gamma}{2\gamma+1}}\|y\|_{C([\theta_1,\theta_2];H^{2\gamma}(\Omega))}^{\frac{1}{2\gamma+1}}$$
$$\le \left(\frac{C_\gamma M}{\theta_1^\gamma}\right)^{\frac{1}{2\gamma+1}}\|y\|_{L^2(\omega\times(\delta,T))}^{\frac{2\gamma}{2\gamma+1}},$$

which completes the proof of (4).

*Second step; Logarithmic convexity.* We next show the following logarithmic convexity inequality (e.g., [1,9]):

$$\|y(\cdot, t)\|_{L^2(\Omega)} \le \|y(\cdot, 0)\|_{L^2(\Omega)}^{1-\frac{t}{\theta}}\|y(\cdot, \theta)\|_{L^2(\Omega)}^{\frac{t}{\theta}}. \tag{9}$$

If this is true, then we square the both sides and integrate over $t \in (0, \theta)$ to obtain using the a priori $L^2$ boundedness of the initial value $y(\cdot, 0)$ and estimate (4) that

$$\|y\|_{L^2(0,\theta;L^2(\Omega))}$$
$$\le C\left(-\log \|y(\cdot, \theta)\|_{L^2(\Omega)}\right)^{-\frac{1}{2}} \tag{10}$$
$$\le C\left(-\log \|y\|_{L^2(\omega\times(\delta,T))}\right)^{-\frac{1}{2}}.$$

Here we note at the last inequality that $\|y\|_{L^2(\omega\times(\delta,T))}$ is sufficiently small. To prove (9), we consider

$$V(t) = \|y(\cdot, t)\|_{L^2(\Omega)}^2.$$

It is easy to see

$$V'(t) = 2\int_\Omega y\, y_t dx$$
$$= 2\int_\Omega y\, \Delta y dx = -2\int_\Omega \nabla y \cdot \nabla y dx.$$

Further differentiating, we have

$$V''(t) = -4\int_\Omega \nabla y \cdot \nabla y_t dx$$
$$= 4\int_\Omega y_t\, \Delta y dx = 4\int_\Omega y_t(\cdot, t)^2 dx.$$

Using the above formulae for $V'(t)$ and $V''(t)$ and the Cauchy-Schwarz inequality, we have

$$V'(t)^2 - V''(t)V(t)$$
$$= \left(2\int_\Omega yy_t dx\right)^2 - 4\int_\Omega y_t^2 dx \int_\Omega y^2 dx \le 0,$$

this yields

$$(\log V(t))'' = \frac{V''(t)V(t) - V'(t)^2}{V(t)^2} \ge 0.$$

Therefore we know that $\log V(t)$ is convex, which leads to

$$\log V(t) \le \left(1 - \frac{t}{\theta}\right)\log V(0) + \frac{t}{\theta}\log V(\theta),$$

or

$$V(t) \le V(0)^{1-\frac{t}{\theta}} V(\theta)^{\frac{t}{\theta}},$$

from which (9) follows.

*Third step.* We have

$$y_t(\cdot, t) = -Ae^{-tA}y(\cdot, 0) = -A^{1-\varepsilon}e^{-tA}A^\varepsilon y(\cdot, 0),$$

and

$$\|y_t(\cdot, t)\|_{L^2(\Omega)} \le Ct^{\varepsilon-1}\|A^\varepsilon y(\cdot, 0)\|_{L^2(\Omega)}.$$

Now for any $1 < p < 1/(1-\varepsilon)$, using (7) and $y(\cdot, 0) \in \mathcal{A}$, we obtain that

$$\int_0^\theta \|y_t(\cdot, t)\|_{L^2(\Omega)}^p dt$$
$$\le C\int_0^\theta t^{p(\varepsilon-1)}dt \|A^\varepsilon y(\cdot, 0)\|_{L^2(\Omega)}^p$$
$$\le C\|y(\cdot, 0)\|_{H^{2\varepsilon}(\Omega)}^p \le C(M).$$

This proves

$$\|y\|_{W^{1,p}(0,\theta;L^2(\Omega))} \le C(M). \tag{11}$$

We note that we can choose $p$ such that $1 < p \le 2$. On the other hand, we know from (10) and $\|\eta\|_{L^p(0,T)} \le C\|\eta\|_{L^2(0,T)}$ for $p \le 2$ that

$$\|y\|_{L^p(0,\theta;L^2(\Omega))} \le C(M)\left(-\log\|y\|_{L^2(\omega\times(\delta,T))}\right)^{-\frac{1}{2}}. \tag{12}$$

Using this, (11), (12) and the Sobolev interpolation, we derive for $0 < s < 1$,

$$\|y\|_{W^{1-s,p}(0,\theta;L^2(\Omega))}$$
$$\le C(M)\left(-\log\|y\|_{L^2(\omega\times(\delta,T))}\right)^{-\frac{s}{2}}.$$

We can choose $s \in (0,1)$ such that $(1-s)p > 1$. Therefore the space $W^{1-s,p}(0,\theta;L^2(\Omega))$ is continuously embedded into $C([0,\theta];L^2(\Omega))$, so that this leads to

$$\|y\|_{C([0,\theta];L^2(\Omega))} \le C\|y\|_{W^{1-s,p}(0,\theta;L^2(\Omega))}$$
$$\le C(M)\left(-\log\|y\|_{L^2(\omega\times(\delta,T))}\right)^{-\frac{s}{2}}.$$

Thus the proof of the theorem is complete. ☐

**Remark** For conditional stability for the backward heat problem, some Carleman-type estimates are possible tools and see e.g., Lees and Protter [6], M.M. Lavrent'ev, V.G. Romanov and S.P. Shishat·skiĭ[7, §2 of Chapter IV].

**Acknowledgements:** Masahiro Yamamoto was partly supported by Grant 15340027 from the Japan Society for the Promotion of Science and Grant 17654019 from the Ministry of Education, Cultures, Sports and Technology. Jun Zou was fully supported by Hong Kong RGC grants (Project No. 404105 and Project No. 403403).

# References

1. K. A. Ames and B. Straughan, *Non-standard and Improperly Posed Problems*, Academic Press, San Diego 1997.
2. J. Cheng, M. Yamamoto and J. Zou, *A priori strategy in discretization of the Tikhonov regularization*, in Progress in Analysis, World Scientific, New Jersey 2003, 1405–1412.
3. O. Yu. Imanuvilov and M. Yamamoto, *Carleman estimates for a parabolic equation in a Sobolev space of negative order and its applications*, Lecture Notes in Pure Appl. Math. 218, 2001, 113–137.
4. M. V. Klibanov, *Estimates of initial conditions of parabolic equations and inequalities via the lateral Cauchy data*, submitted for publication, available in the pdf format online at http://rene.ma.utexas.edu/mp_arc/index-05.html#end, preprint number 05-378.
5. M. V. Klibanov and A. Timonov, *Carleman Estimates for Coefficient Inverse Problems and Numerical Applications*, VSP, Utrecht 2004.
6. M. Lees and M. H. Protter, *Unique continuation for parabolic differential equations and inequalities*, Duke Math. J. 28, 1961, 369–382.
7. M. M. Lavrent'ev, V. G. Romanov and S. P. Shishat·skiĭ, *Ill-posed Problems of Mathematical Physics and Analysis*, American Mathematical Society, Providence, RI 1986.
8. J. L. Lions and E. Magenes, *Non-homogeneous Boundary Value Problems and Applications*, Vol. I, Springer-Verlag, Berlin 1972.
9. L. E. Payne, *Improperly Posed Problems in Partial Differential Equations*, SIAM, Philadelphia 1975.
10. A. Pazy, *Semigroups of Linear Operators and Applications to Partial Differential Equations*, Springer-Verlag, Berlin 1983.
11. S. Saitoh and M. Yamamoto, it Stability of Lipschitz type in determination of initial heat distribution, J. Inequal. Appl. 1, 1997, 73–83.
12. D. Xu and M. Yamamoto, *Stability estimates in state-estimation for a heat process*, in Proceedings of the Second ISAAC Congress, Vol. 1 (Fukuoka, 1999): Int. Soc. Anal. Appl. Comput. 7, 193–198, Kluwer Acad. Publ., Dordrecht 2000.
13. M. Yamamoto and J. Zou, *Simultaneous reconstruction of the initial temperature and heat radiative coefficient*, Inverse Problems 17, 2001, 1181–1202.

# IV.6   Mathematical Biology and Medicine

Organizers: R.P. Gilbert, A. Wirgin. Y. Xu

# INVERSE PROBLEM FOR WAVE PROPAGATION IN A PERTURBED LAYERED HALF-SPACE WITH A BUMP

R.P. GILBERT, N. ZHANG and N. ZEEV

*Department of Mathematical Sciences*
*University of Delaware*
*Newark DE 19716, USA*
*E-mail: gilbert@math.udel.edu*

Y. Xu

*Department of Mathematics*
*University of Louisville*
*Louisville, KY 40292*
*E-mail: ysxu0001@louisville.edu*

We investigate a problem of potential interest to geophysicists. Namely, we attempt to find the shape of a magnum chamber lying in the earth crust below a volcano. This is posed as an inverse acoustics problem where we use scattered seismic waves to determine the location and form of the chamber. A simplification is to consider the finite crust to be over a semi-finite magma region. An earlier work of ours discussed the inverse problem associated with the Perkeris model [3], whereas the present work represents the crust with volcano as a perturbation of this model. In order to reformulate this inverse problem in integral equation form we first construct the Green's function for this problem. A numerical example is presented for the case of the bump being a half circle.

**Key words:** Canellous bone, Biot model, wave propagation, poroelastic medium
**Mathematics Subject Classification:** 76B25, 35Q51

## 1. Introduction

In this paper, we study the undetermined object scattering problem in a perturbed, layered half-space with a bump. Without loss of generality, we may assume the function describing the bump is $x_2 = g(x_1)$, $-0.1 < x_1 < 0.1$. The magnum chamber is represented as a wave penetrable object located in the crust, i.e. an object with different refraction index from the other part of the earth's crust, i.e. the half space . The crust is represented by $R_b^2 = R_+^2 \cup B = \{(x_1, x_2) \in R^2 | x_2 \geq 0\} \cup B$, where $B = \{(x_1, x_2) | -0.1 < x_1 < 0.1, x_2 = g(x_1)\}$ represents the volcano. The index $n(\mathbf{x})$ is defined on $R_b^2$ with the following properties:

Let

$$n_0(x_2) = \begin{cases} 1, & for \ h < x_2 < \infty \\ n_0, & \text{otherwise} \end{cases} \tag{1}$$

where $n_0 > 1$ is a constant. We assume that the inhomogeneity is contained in the crust minus the volcano $\Omega \in R_h^2 = \{\mathbf{x} \in R^2 | 0 < x_2 < h\}$ and that $\partial\Omega$ is a $C^2$ boundary whose outward-pointing normal vector is denoted by $\nu$. This corresponds to the investigation of geological objects located in the earth crust.

$$n(\mathbf{x}) = n_0(x_2), \text{ for } \mathbf{x} \notin \Omega. \tag{2}$$

We consider the following acoustic problem: given a point source or a plane, incident wave $u^i$ satisfying

$$\triangle u^i + k^2 n_0^2(x_2)u^i = f(\mathbf{x}) , \ in \ R_b^2, \tag{3}$$

find the total field $u = u^s + u^i$ such that

$$\triangle u^s + k^2 n^2(\mathbf{x})u^s = 0 , \ in \ R_b^2, \tag{4}$$

$$u(\mathbf{x}) = u^s(\mathbf{x}) + u^i(\mathbf{x}) = 0 , \ when \ x_2 = \begin{cases} g(x_1), \ for \ -0.1 < x_2 < 0.1 \\ 0, \qquad\qquad otherwise \end{cases} \tag{5}$$

Let $\frac{\partial u_-}{\partial\nu}(x_1, x_2)$ and $u_-(x_1, x_2)$ be the limits of $\frac{\partial u}{\partial\nu}(x_1, x_2)$ and $u(x_1, x_2)$ as $(x_1, x_2)$ approaches $\partial\Omega$ from the interior of $\Omega$, and $\frac{\partial u_+}{\partial\nu}(x_1, x_2)$ and $u_+(x_1, x_2)$ be the limits of $\frac{\partial u}{\partial\nu}(x_1, x_2)$ and $u(x_1, x_2)$ as $(x_1, x_2)$ approaches $\partial\Omega$ from the exterior of $\Omega$, respectively. We assume for some constants $\rho_0$ and $\rho$, $u$ satisfies interface conditions

$$\rho_0 u^+(\mathbf{x}) = \rho u^-(\mathbf{x}), \quad \frac{\partial u^+(\mathbf{x})}{\partial\nu} = \frac{\partial u^-(\mathbf{x})}{\partial\nu}, \ on \ \partial\Omega. \tag{6}$$

Moreover, $u^s$ satisfies also the out-going radiation condition; that is, no wave is coming in from infinity except the prescribed incident wave plane wave. Here $\triangle = \frac{\partial^2}{\partial x_1^2} + \frac{\partial^2}{\partial x_2^2}$, $u^i$, $u^s$ and $u$ are the incident, scattered and total field of the time harmonic waves.

## 2. Green's function and its far-field behavior

In this section we outline the construction of the Green's function for the time-harmonic waves in a layered half-space with the bump $g(x_1)$ and obtain its free-wave and guided-wave far-field behavior. A function $G(\ \cdot\ ; x_1^0, x_2^0)$ is called the outgoing Green's function from the source at $\mathbf{x}^0 = (x_1^0, x_2^0)$ for the time-harmonic wave in a layered half-space, if $G(x_1, x_2; x_1^0, x_2^0)$ satisfies

$$\triangle G + k^2 n_0^2(x_2)G = -\delta(|\mathbf{x} - \mathbf{x}^0|) , \ in \ R_b^2 \tag{7}$$

From the paper [2], we know that the Green's function for the layered half-space can be expressed as

$$G(x_1, x_2; x_1^0, x_2^0) = G_f(x_1, x_2; x_1^0, x_2^0) + G_e(x_1, x_2; x_1^0, x_2^0) + G_g(x_1, x_2; x_1^0, x_2^0)$$

$$= \frac{1}{2\pi} \int_{\Gamma} \frac{\phi^+(\xi, x_{2>})\phi^-(\xi, x_{2<})e^{i\xi|x_1 - x_1^0|}}{e^{i\tau h}[\tau_0 \cos(\tau_0 h) - i\tau \sin(\tau_0 h)]} d\xi$$

$$+ \frac{1}{2\pi} \int_{\Gamma_0} \frac{\phi^+(\xi, x_{2>})\phi^-(\xi, x_{2<})e^{i\xi|x_1 - x_1^0|}}{e^{i\tau h}[\tau_0 \cos(\tau_0 h) - i\tau \sin(\tau_0 h)]} d\xi$$

$$+ \sum_{n=1}^{N} \frac{\phi^+(\xi_n, x_{2>})\phi^-(\xi_n, x_{2<})e^{i\xi_n|x_1 - x_1^0|}}{\frac{d}{d\xi}W(\phi^+, \phi^-)\Big|_{\xi = \xi_n}}, \tag{8}$$

where the functions $\phi^+$ and $\phi^-$ are of the form

$$\phi^+(\xi, x_2) = \begin{cases} e^{i\tau x_2}, & h < x_2 < \infty \\ B_1 e^{i\tau_0 x_2} + B_2 e^{-i\tau_0 x_2}, & 0 < x_2 < h, \end{cases}$$

$$\phi^-(\xi, x_2) = \begin{cases} \sin(\tau_0 x_2), & 0 < x_2 < h \\ C_1 \cos(\tau x_2) + C_2 \sin(\tau x_2), & h < x_2 < \infty, \end{cases}$$

$$B_1 = \frac{1}{2}\left(1 + \frac{\tau}{\tau_0}\right)e^{i(\tau - \tau_0)h},$$

$$B_2 = \frac{1}{2}\left(1 - \frac{\tau}{\tau_0}\right)e^{i(\tau + \tau_0)h},$$

$$C_1 = \cos(\tau h)\sin(\tau_0 h) - \frac{\tau_0}{\tau}\sin(\tau h)\cos(\tau_0 h),$$

$$C_2 = \sin(\tau h)\sin(\tau_0 h) + \frac{\tau_0}{\tau}\cos(\tau h)\cos(\tau_0 h).$$

$$x_{2>} = \max\{x_2, x_2^0\}, \quad x_{2<} = \min\{x_2, x_2^0\},$$

$$W(\phi^+, \phi^-) = \phi^+ \frac{d}{dx_2}\phi^- - \phi^- \frac{d}{dx_2}\phi^+,$$

$$\tau = \sqrt{k^2 - \xi^2}$$

$$\tau_0 = \sqrt{k^2 n_0^2 - \xi^2}$$

and $\xi_n$ are the positive poles that are larger than k.

From [1], assume D is a finite or infinite domain with a smooth boundary curve C. If s is the length parameter along C, we define a positive twice continuously differentiable function $v(s)$ on C and shift every boundary point $t(s)$ in the direction of the interior normal of D by an amount

$$\delta = \epsilon v(s)$$

where $\epsilon > 0$ is a smallness parameter. For sufficiently small value of $\epsilon$, the points obtained will form a new curve $C^*$ and bounds a new domain $D^*$. Let $G^*$ be the Green's function for $D^*$, the difference $d(x_1, x_2; x_1^0, x_2^0)$ between the two Green's functions is

$$d(x_1, x_2; x_1^0, x_2^0) = G^*(x_1, x_2; x_1^0, x_2^0) - G(x_1, x_2; x_1^0, x_2^0) \qquad (9)$$

which is harmonic and regular in $D^*$ since the singularities of the two Green's functions cancel each other. On the boundary $C^*$ of $D^*$, $d(x_1, x_2; x_1^0, x_2^0)$ has the boundary value $-G(x_1, x_2; x_1^0, x_2^0)$ since by definition $G^*$ vanishes on $C^*$. Hence, we can represent $d(x_1, x_2; x_1^0, x_2^0)$ in terms of its boundary values, applied to the domain $D^*$, we obtain [1]

$$d(x_1, x_2; x_1^0, x_2^0) = -\oint_{C^*} G(t_1, t_2; x_1^0, x_2^0) \frac{\partial G^*(t_1, t_2; x_1, x_2)}{\partial \nu_t} \delta ds, \qquad (10)$$

where $\nu_t$ is the normal direction in terms of $(t_1, t_2)$.

For fixed points in $D^*$, we have the uniform estimate

$$G(x_1, x_2; x_1^0, x_2^0) - G^*(x_1, x_2; x_1^0, x_2^0) = O(\epsilon),$$

$$\left| \text{grad}[G(x_1, x_2; x_1^0, x_2^0) - G^*(x_1, x_2; x_1^0, x_2^0)] \right| = O(\epsilon). \qquad (11)$$

Since $G^*$ vanishes on $C^*$, $G(x_1, x_2; x_1^0, x_2^0)$ is obviously of the order $\epsilon$ on this curve and if we replace in (10) $G^*$ by G, we only generate an error of order $\epsilon^2$. Thus, we have

$$G^*(x_1, x_2; x_1^0, x_2^0) - G(x_1, x_2; x_1^0, x_2^0)$$
$$= -\int\!\!\int_{D-D^*} \text{grad}G(t_1, t_2; x_1^0, x_2^0) \cdot \text{grad}G(t_1, t_2; x_1, x_2) \delta ds \qquad (12)$$

This formula is derived under the assumption that $\delta$ is everywhere positive on C. It may be extended to the case in which $\delta = \epsilon v(s)$ where $v(s)$ is twice continuously differentiable but which is no longer restricted in sign. Furthermore, this formula holds for bounded domains. It can be proved that it also holds for infinite domains where $\nu_s$ has compact support.

Therefore, in our case, we can write the Green's function $G_b$ for the layered, half-space with a bump in terms of the Green's function G for the layered, half-space as follows:

$$G_b(x_1 x_2; x_1^0, x_2^0) = G(x_1, x_2; x_1^0, x_2^0) - \int_C \frac{\partial G(x_1, x_2; t_1, t_2)}{\partial \nu_t} \frac{\partial G(t_1, t_2; x_1^0, x_2^0)}{\partial \nu_t} g(t_1) dt_1$$

where $g(t_1)$ is the bump function.

By straightforward calculations, we can obtain $G_b(x_1 x_2; x_1^0, x_2^0)$ as follows.

For $0 < x_2 < x_2^0$,

$$\phi^+(\xi, x_{2>}) = B_1 e^{i\tau_0 x_2^0} + B_2 e^{-i\tau_0 x_2^0},$$

$$\phi^-(\xi, x_{2<}) = \sin(\tau_0 x_2),$$

$$G_g(x_1, x_2; t_1, t_2) = i \sum_{n=1}^{N} \frac{\left(B_1 e^{i\tau_0 t_2} + B_2 e^{-i\tau_0 t_2}\right) \sin(\tau_0 x_2) e^{i\xi_n |x_1 - t_1|}}{\frac{\xi(i + h\tau_n) e^{i\tau_n h}}{\tau_{0n}^2 \tau_n} (k^2 n_0^2 - k^2) \sin(\tau_{0n} h)}$$

$$\frac{\partial G_g(x_1, x_2; t_1, t_2)}{\partial t_2} = \sum_{n=1}^{N} \frac{\tau_0 \left(B_2 e^{-i\tau_0 t_2} - B_1 e^{i\tau_0 t_2}\right) \sin(\tau_0 x_2) e^{i\xi_n |x_1 - t_1|}}{\frac{\xi(i + h\tau_n) e^{i\tau_n h}}{\tau_{0n}^2 \tau_n} (k^2 n_0^2 - k^2) \sin(\tau_{0n} h)}$$

$$\left.\frac{\partial G_g(x_1, x_2; t_1, t_2)}{\partial t_2}\right|_{t_2=0} = \sum_{n=1}^{N} \frac{\tau_0 \left(B_2 - B_1\right) \sin(\tau_0 x_2) e^{i\xi_n |x_1 - t_1|}}{\frac{\xi(i + h\tau_n) e^{i\tau_n h}}{\tau_{0n}^2 \tau_n} (k^2 n_0^2 - k^2) \sin(\tau_{0n} h)}$$

$$\left.\frac{\partial G_g(t_1, t_2; x_1^0, x_2^0)}{\partial t_2}\right|_{t_2=0} = \sum_{n=1}^{N} \frac{i\tau_0 \left(B_1 e^{i\tau_0 x_2^0} + B_2 e^{-i\tau_0 x_2^0}\right) e^{i\xi_n |t_1 - x_1^0|}}{\frac{\xi(i + h\tau_n) e^{i\tau_n h}}{\tau_{0n}^2 \tau_n} (k^2 n_0^2 - k^2) \sin(\tau_{0n} h)}$$

$$G_f(x_1, x_2; t_1, t_2) = \frac{1}{2\pi} \int_{-\infty}^{\infty}$$

$$\times \frac{\left(B_1 e^{i\sqrt{\zeta^2 + k^2(n_0^2 - 1)} t_2} + B_2 e^{-i\sqrt{\zeta^2 + k^2(n_0^2 - 1)} t_2}\right) \sin(\sqrt{\zeta^2 + k^2(n_0^2 - 1)} x_2)}{\sqrt{\zeta^2 + k^2(n_0^2 - 1)} \cos(\sqrt{\zeta^2 + k^2(n_0^2 - 1)} h) - i\zeta \sin(\sqrt{\zeta^2 + k^2(n_0^2 - 1)} h)}$$

$$\times \frac{-\zeta e^{i\sqrt{k^2 - \zeta^2} |x_1 - t_1|}}{e^{i\zeta h} \sqrt{k^2 - \zeta^2}} d\zeta$$

$$\frac{\partial G_f(x_1, x_2; t_1, t_2)}{\partial t_2} = \frac{1}{2\pi} \int_{-\infty}^{\infty}$$

$$\times \frac{\left(B_1 e^{i\sqrt{\zeta^2 + k^2(n_0^2 - 1)} t_2} - B_2 e^{-i\sqrt{\zeta^2 + k^2(n_0^2 - 1)} t_2}\right) \sin(\sqrt{\zeta^2 + k^2(n_0^2 - 1)} x_2)}{\cos(\sqrt{\zeta^2 + k^2(n_0^2 - 1)} h) - i\zeta \sin(\sqrt{\zeta^2 + k^2(n_0^2 - 1)} h)}$$

$$\times \frac{-i\zeta \sqrt{\zeta^2 + k^2(n_0^2 - 1)} e^{i\sqrt{k^2 - \zeta^2} |x_1 - t_1|}}{e^{i\zeta h} \sqrt{\zeta^2 + k^2(n_0^2 - 1)} \sqrt{k^2 - \zeta^2}} d\zeta$$

$$\left.\frac{\partial G_f(x_1, x_2; t_1, t_2)}{\partial t_2}\right|_{t_2=0}$$

$$= \frac{1}{2\pi} \int_{-\infty}^{\infty} \frac{(B_2 - B_1) \sin(\sqrt{\zeta^2 + k^2(n_0^2 - 1)} x_2)}{\sqrt{\zeta^2 + k^2(n_0^2 - 1)} \cos(\sqrt{\zeta^2 + k^2(n_0^2 - 1)} h) - i\zeta \sin(\sqrt{\zeta^2 + k^2(n_0^2 - 1)} h)}$$

$$\times \frac{i\zeta \sqrt{\zeta^2 + k^2(n_0^2 - 1)} e^{i\sqrt{k^2 - \zeta^2} |x_1 - t_1|}}{e^{i\zeta h} \sqrt{k^2 - \zeta^2}} d\zeta$$

$$\frac{\partial G_f(t_1, t_2; x_1^0, x_2^0)}{\partial t_2}\bigg|_{t_2=0} = \frac{1}{2\pi}\int_{-\infty}^{\infty}$$

$$\times \frac{\left(B_1 e^{i\sqrt{\zeta^2 + k^2(n_0^2 - 1)}x_2^0} + B_2 e^{-i\sqrt{\zeta^2 + k^2(n_0^2 - 1)}x_2^0}\right)}{\sqrt{\zeta^2 + k^2(n_0^2 - 1)}\cos(\sqrt{\zeta^2 + k^2(n_0^2 - 1)}h) - i\zeta\sin(\sqrt{\zeta^2 + k^2(n_0^2 - 1)}h)}$$

$$\times \frac{-\zeta\sqrt{\zeta^2 + k^2(n_0^2 - 1)}e^{i\sqrt{k^2 - \zeta^2}|t_1 - x_1^0|}}{e^{i\zeta h}\sqrt{k^2 - \zeta^2}}d\zeta$$

$$G_e(x_1, x_2; t_1, t_2) = \frac{1}{2\pi}\int_{-\infty}^{\infty}\frac{\left(B_1 e^{i\zeta t_2} + B_2 e^{-i\zeta t_2}\right)\sin(\zeta x_2)}{\zeta\cos(\zeta h) + \sqrt{k^2(n_0^2 - 1) + \zeta^2}\sin(\zeta h)}$$

$$\times \frac{\zeta e^{i\sqrt{k^2 n_0^2 + \zeta^2}|x_1 - t_1|}}{e^{-\sqrt{k^2(n_0^2 - 1) + \zeta^2}}\sqrt{\zeta^2 + k^2 n_0^2}}d\zeta$$

$$\frac{\partial G_e(x_1, x_2; t_1, t_2)}{\partial t_2} = \frac{1}{2\pi}\int_{-\infty}^{\infty}\frac{\left(B_1 e^{i\zeta t_2} - B_2 e^{-i\zeta t_2}\right)\sin(\zeta x_2)}{\zeta\cos(\zeta h) + \sqrt{k^2(n_0^2 - 1) + \zeta^2}\sin(\zeta h)}$$

$$\times \frac{i\zeta^2 e^{i\sqrt{k^2 n_0^2 + \zeta^2}|x_1 - t_1|}}{e^{-\sqrt{k^2(n_0^2 - 1) + \zeta^2}}\sqrt{\zeta^2 + k^2 n_0^2}}d\zeta$$

$$\frac{\partial G_e(x_1, x_2; t_1, t_2)}{\partial t_2}\bigg|_{t_2=0} = \frac{1}{2\pi}\int_{-\infty}^{\infty}\frac{(B_1 - B_2)\sin(\zeta x_2)}{\zeta\cos(\zeta h) + \sqrt{k^2(n_0^2 - 1) + \zeta^2}\sin(\zeta h)}$$

$$\times \frac{i\zeta^2 e^{i\sqrt{k^2 n_0^2 + \zeta^2}|x_1 - t_1|}}{e^{-\sqrt{k^2(n_0^2 - 1) + \zeta^2}}\sqrt{\zeta^2 + k^2 n_0^2}}d\zeta$$

$$\frac{\partial G_e(t_1, t_2; x_1^0, x_2^0)}{\partial t_2}\bigg|_{t_2=0} = \frac{1}{2\pi}\int_{-\infty}^{\infty}\frac{\left(B_1 e^{i\zeta x_2^0} + B_2 e^{-i\zeta x_2^0}\right)\sin(\zeta x_2)}{\zeta\cos(\zeta h) + \sqrt{k^2(n_0^2 - 1) + \zeta^2}\sin(\zeta h)}$$

$$\times \frac{\zeta^2 e^{i\sqrt{k^2 n_0^2 + \zeta^2}|t_1 - x_1^0|}}{e^{-\sqrt{k^2(n_0^2 - 1) + \zeta^2}}\sqrt{\zeta^2 + k^2 n_0^2}}d\zeta$$

Plugging the above equations into equation (13) will give us $G_b(x_1 x_2; x_1^0, x_2^0)$. Similarly, for $x_2^0 < x_2 < h$, we have

$$\phi^+(\xi, x_{2>}) = B_1 e^{i\tau_0 x_2} + B_2 e^{-i\tau_0 x_2},$$

$$\phi^-(\xi, x_{2<}) = \sin(\tau_0 x_2^0),$$

$$\frac{\partial G_g(x_1, x_2; t_1, t_2)}{\partial t_2}\bigg|_{t_2=0} = \sum_{n=1}^{N} \frac{i\tau_0 \left( B_1 e^{i\tau_0 x_2} - B_2 e^{-i\tau_0 x_2} \right) e^{i\xi_n |x_1 - t_1|}}{\frac{\xi(i + h\tau_n) e^{i\tau_n h}}{\tau_{0n}^2 \tau_n} (k^2 n_0^2 - k^2) \sin(\tau_{0n} h)}$$

$$\frac{\partial G_g(t_1, t_2; x_1^0, x_2^0)}{\partial t_2}\bigg|_{t_2=0} = \sum_{n=1}^{N} \frac{\tau_0 \left( B_2 - B_1 \right) \sin(\tau_0 x_2^0) e^{i\xi_n |t_1 - x_1^0|}}{\frac{\xi(i + h\tau_n) e^{i\tau_n h}}{\tau_{0n}^2 \tau_n} (k^2 n_0^2 - k^2) \sin(\tau_{0n} h)}$$

$$\frac{\partial G_f(x_1, x_2; t_1, t_2)}{\partial t_2}\bigg|_{t_2=0} = \frac{1}{2\pi} \int_{-\infty}^{\infty} \frac{B_1 e^{i\sqrt{\zeta^2 + k^2(n_0^2 - 1)} x_2} + B_2 e^{-i\sqrt{\zeta^2 + k^2(n_0^2 - 1)} x_2}}{\tau_0 \cos(\tau_0 h) - i\tau \sin(\tau_0 h)}$$

$$\times \frac{-\zeta \sqrt{\zeta^2 + k^2(n_0^2 - 1)} e^{i\sqrt{k^2 - \zeta^2} |x_1 - t_1|}}{e^{i\tau h} \sqrt{k^2 - \zeta^2}} d\zeta$$

$$\frac{\partial G_f(t_1, t_2; x_1^0, x_2^0)}{\partial t_2}\bigg|_{t_2=0} = \frac{1}{2\pi} \int_{-\infty}^{\infty} \frac{(B_2 - B_1) \sin(\sqrt{\zeta^2 + k^2(n_0^2 - 1)} x_2^0)}{\tau_0 \cos(\tau_0 h) - i\tau \sin(\tau_0 h)}$$

$$\times \frac{i\zeta \sqrt{\zeta^2 + k^2(n_0^2 - 1)} e^{i\sqrt{k^2 - \zeta^2} |t_1 - x_1^0|}}{e^{i\tau h} \sqrt{k^2 - \zeta^2}} d\zeta$$

$$\frac{\partial G_e(x_1, x_2; t_1, t_2)}{\partial t_2}\bigg|_{t_2=0} = \frac{1}{2\pi} \int_{-\infty}^{\infty} \frac{B_1 e^{i\zeta x_2} + B_2 e^{-i\zeta x_2}}{\zeta \cos(\zeta h) + \sqrt{k^2(n_0^2 - 1) + \zeta^2} \sin(\zeta h)}$$

$$\times \frac{\zeta^2 e^{i\sqrt{k^2 n_0^2 + \zeta^2} |x_1 - t_1|}}{e^{-\sqrt{k^2(n_0^2 - 1) + \zeta^2} h} \sqrt{\zeta^2 + k^2 n_0^2}} d\zeta$$

$$\frac{\partial G_e(t_1, t_2; x_1^0, x_2^0)}{\partial t_2}\bigg|_{t_2=0} = \frac{1}{2\pi} \int_{-\infty}^{\infty} \frac{(B_1 - B_2) \sin(\zeta x_2^0)}{\zeta \cos(\zeta h) + \sqrt{k^2(n_0^2 - 1) + \zeta^2} \sin(\zeta h)}$$

$$\times \frac{i\zeta^2 e^{i\sqrt{k^2 n_0^2 + \zeta^2} |t_1 - x_1^0|}}{e^{-\sqrt{k^2(n_0^2 - 1) + \zeta^2} h} \sqrt{\zeta^2 + k^2 n_0^2}} d\zeta$$

For $h < x_2 < \infty$, we have

$$\phi^+(\xi, x_{2>}) = e^{i\tau x_2},$$

$$\phi^-(\xi, x_{2<}) = \sin(\tau_0 x_2^0),$$

$$\frac{\partial G_g(x_1, x_2; t_1, t_2)}{\partial t_2}\bigg|_{t_2=0} = \sum_{n=1}^{N} \frac{i\tau_0 e^{i\tau x_2} e^{i\xi_n |x_1 - t_1|}}{\frac{\xi(i + h\tau_n) e^{i\tau_n h}}{\tau_{0n}^2 \tau_n} (k^2 n_0^2 - k^2) \sin(\tau_{0n} h)}$$

$$\frac{\partial G_g(t_1, t_2; x_1^0, x_2^0)}{\partial t_2}\bigg|_{t_2=0} = \sum_{n=1}^{N} \frac{-\tau \sin(\tau_0 x_2^0) e^{i\xi_n |t_1 - x_1^0|}}{\frac{\xi(i + h\tau_n) e^{i\tau_n h}}{\tau_{0n}^2 \tau_n} (k^2 n_0^2 - k^2) \sin(\tau_{0n} h)}$$

$$\frac{\partial G_f(x_1,x_2;t_1,t_2)}{\partial t_2}\bigg|_{t_2=0} = \frac{1}{2\pi}\int_{-\infty}^{\infty}$$

$$\times \frac{e^{i\zeta x_2}}{\sqrt{\zeta^2+k^2(n_0^2-1)}\cos(\sqrt{\zeta^2+k^2(n_0^2-1)}h) - i\zeta\sin(\sqrt{\zeta^2+k^2(n_0^2-1)}h)}$$

$$\times \frac{-\zeta\sqrt{\zeta^2+k^2(n_0^2-1)}e^{i\sqrt{k^2-\zeta^2}|x_1-t_1|}}{e^{i\zeta h}\sqrt{k^2-\zeta^2}}d\zeta$$

$$\frac{\partial G_f(t_1,t_2;x_1^0,x_2^0)}{\partial t_2}\bigg|_{t_2=0} = \frac{1}{2\pi}\int_{-\infty}^{\infty}$$

$$\frac{\sin(\sqrt{\zeta^2+k^2(n_0^2-1)}x_2^0)}{\sqrt{\zeta^2+k^2(n_0^2-1)}\cos(\sqrt{\zeta^2+k^2(n_0^2-1)}h) - i\zeta\sin(\sqrt{\zeta^2+k^2(n_0^2-1)}h)}$$

$$\times \frac{-i\zeta^2 e^{i\sqrt{k^2-\zeta^2}|t_1-x_1^0|}}{e^{i\zeta h}\sqrt{k^2-\zeta^2}}d\zeta$$

$$\frac{\partial G_e(x_1,x_2;t_1,t_2)}{\partial t_2}\bigg|_{t_2=0} = \frac{1}{2\pi}\int_{-\infty}^{\infty}\frac{e^{-i\sqrt{\zeta^2+k^2(n_0^2-1)}x_2}}{\zeta\cos(\zeta h)+\sqrt{k^2(n_0^2-1)+\zeta^2}\sin(\zeta h)}$$

$$\times \frac{\zeta^2 e^{i\sqrt{k^2n_0^2+\zeta^2}|x_1-t_1|}}{e^{-\sqrt{k^2(n_0^2-1)+\zeta^2}h}\sqrt{\zeta^2+k^2n_0^2}}d\zeta$$

$$\frac{\partial G_e(t_1,t_2;x_1^0,x_2^0)}{\partial t_2}\bigg|_{t_2=0} = \frac{1}{2\pi}\int_{-\infty}^{\infty}\frac{\sin(\zeta x_2^0)\sqrt{k^2(n_0^2-1)+\zeta^2}}{\zeta\cos(\zeta h)+\sqrt{k^2(n_0^2-1)+\zeta^2}\sin(\zeta h)}$$

$$\times \frac{-\zeta e^{i\sqrt{k^2n_0^2+\zeta^2}|t_1-x_1^0|}}{e^{-\sqrt{k^2(n_0^2-1)+\zeta^2}h}\sqrt{\zeta^2+k^2n_0^2}}d\zeta$$

Furthermore, for $x_2 > h$ and $r = |x|$ large,

$$G_g(x_1,x_2;x_1^0,x_2^0) = \sum_{n=1}^{N}\frac{ie^{i\tau x_2}\sin(\tau_0 x_2^0)e^{i\xi_n|x_1-x_1^0|}}{\frac{\xi(i+h\tau_n)e^{i\tau_n h}}{\tau_{0n}^2\tau_n}(k^2n_0^2-k^2)\sin(\tau_{0n}h)} = O\left(e^{i\tau x_2}\right)$$

$$G_f(x_1,x_2;x_1^0,x_2^0) = \frac{e^{ikr}}{\sqrt{2\pi ir}}\left[G_f^{\infty}(\theta;x_1^0,x_2^0)\right] + O\left(\frac{1}{r^{3/2}}\right)$$

$$= \frac{e^{ikr}}{\sqrt{2\pi ir}}\left[\frac{\Phi_f(\zeta_+)}{\sqrt{\Theta''(\zeta_+)}} + \frac{\Phi_f(\zeta_-)}{\sqrt{\Theta''(\zeta_-)}}\right] + O\left(\frac{1}{r^{3/2}}\right), \tag{13}$$

$$G_e(x_1, x_2; x_1^0, x_2^0) = O\left(\frac{1}{r^{3/2}},\right).$$ (14)

where

$$\Theta_f(\zeta) = \sqrt{k^2 - \zeta^2}|\cos\theta| + \zeta\sin\theta,$$

$$\Phi_f(\zeta) = \frac{-\zeta}{e^{i\zeta h}\sqrt{k^2 - \zeta^2}}$$

$$\times \frac{\sin(\sqrt{\zeta^2 + k^2(n_0^2 - 1)}x_2^0)}{\sqrt{\zeta^2 + k^2(n_0^2 - 1)}\cos(\sqrt{\zeta^2 + k^2(n_0^2 - 1)}h) - i\zeta\sin(\sqrt{\zeta^2 + k^2(n_0^2 - 1)}h)}$$

and

$$\zeta_\pm = \pm k\sin\theta.$$

$$\left.\frac{\partial G(x_1, x_2; t_1, t_2)}{\partial t_2}\right|_{t_2=0} = O\left(e^{i\tau x_2}\right) + \frac{e^{ikr}}{\sqrt{2\pi i r}}\left[\frac{\Phi_f^*(\zeta_+)}{\sqrt{\Theta''(\zeta_+)}} + \frac{\Phi_f^*(\zeta_-)}{\sqrt{\Theta''(\zeta_-)}}\right] + O\left(\frac{1}{r^{3/2}}\right),$$

where

$$\Phi_f^*(\zeta) = \frac{-\zeta}{e^{i\zeta h}\sqrt{k^2 - \zeta^2}}$$

$$\times \frac{\sqrt{\zeta^2 + k^2(n_0^2 - 1)}\cos(\sqrt{\zeta^2 + k^2(n_0^2 - 1)}t_2)}{\sqrt{\zeta^2 + k^2(n_0^2 - 1)}\cos(\sqrt{\zeta^2 + k^2(n_0^2 - 1)}h) - i\zeta\sin(\sqrt{\zeta^2 + k^2(n_0^2 - 1)}h)}.$$

Hence we can obtain the far-field behavior of $G_b(x_1 x_2; x_1^0, x_2^0)$.

$$G_b(x_1 x_2; x_1^0, x_2^0) = G_f(x_1, x_2; x_1^0, x_2^0) + G_e(x_1, x_2; x_1^0, x_2^0) + G_g(x_1, x_2; x_1^0, x_2^0)$$

$$- \int_{-0.1}^{0.1} \left.\frac{\partial G(x_1, x_2; t_1, t_2)}{\partial t_2}\right|_{t_2=0} \left.\frac{\partial G(t_1, t_2; x_1^0, x_2^0)}{\partial t_2}\right|_{t_2=0} g(t_1)dt_1$$

$$G_b(x_1 x_2; x_1^0, x_2^0) = O\left(e^{i\tau x_2}\right) + \frac{e^{ikr}}{\sqrt{2\pi i r}}\left[\frac{\Phi_f(\zeta_+)}{\sqrt{\Theta''(\zeta_+)}} + \frac{\Phi_f(\zeta_-)}{\sqrt{\Theta''(\zeta_-)}}\right] + O\left(\frac{1}{r^{3/2}}\right)$$

$$- \int_{-0.1}^{0.1} \left[O\left(e^{i\tau_1 x_2}\right) + \frac{e^{ikr}}{\sqrt{2\pi i r}}\left[\frac{\Phi_f^*(\zeta_+)}{\sqrt{\Theta''(\zeta_+)}} + \frac{\Phi_f^*(\zeta_-)}{\sqrt{\Theta''(\zeta_-)}}\right] + O\left(\frac{1}{r^{3/2}}\right)\right]$$

$$\times \left.\frac{\partial G(t_1, t_2; x_1^0, x_2^0)}{\partial t_2}\right|_{t_2=0} g(t_1)dt_1.$$

## 3. Scattered wave in perturbed layered half-space with a bump

Now we consider the case where there exists a given inhomogeneity $\Omega$ in the layer $0 < x_2 < h$. We assume that the incident wave is from a point source located at $(x_1^0, x_2^0)$ below $\Omega$; i.e., $x_2^0 < \min\{x_2 | x \in \Omega\}$. The propagating solution

$$u(x) = \begin{cases} u_1(x), & if\ x \in \tilde{R}_h^2 \setminus \Omega, \\ u_2(x), & if\ x \in R_b^2 \setminus \tilde{R}_h^2, \\ u_3(x), & if\ x \in \Omega \end{cases} \tag{15}$$

satisfies

$$\Delta u_1 + n_0^2 u_1 = -\delta\left(x_1 - x_1^0\right)\delta\left(x_2 - x_2^0\right),\ in\ \tilde{R}_h^2$$

$$\Delta u_2 + u_2 = 0,\ in\ R_b^2 \setminus \tilde{R}_h^2,$$

$$\Delta u_3 + n(x)^2 u_3 = -\delta\left(x_1 - x_1^0\right)\delta\left(x_2 - x_2^0\right),\ in\ \Omega$$

If we denote by

$$m\left(\mathbf{x}\right) = k^2[n(\mathbf{x}) - n_0(x_2)],\ \text{for } \mathbf{x} \in R_b^2, \tag{16}$$

then we may rewrite the above three equations in the form

$$\Delta u + k^2 n_0^2 u = -m(\mathbf{x})u - \delta\left(x_1 - x_1^0\right)\delta\left(x_2 - x_2^0\right),\ a.e.\text{in } R_+^2. \tag{17}$$

**Theorem 3.1** If $(u, \phi)$ satisfies the direct scattering problem, then $(u, \phi)$ satisfies the following integral equations. Conversely, if $(u, \phi) \in C(\Omega) \times C(\partial\Omega)$ is a solution of the integral equations, then $(u, \phi)$ is a solution of the direct scattering problem.(See [2] for the proof and analysis) $\square$

$$u\left(x_1, x_2\right) + \int_\Omega G_b\left(\xi_1, \xi_2; x_1, x_2\right) m(\xi_1, \xi_2) u\left(\xi_1, \xi_2\right) d\xi_1\, d\xi_2$$

$$-\int_{\partial\Omega} \phi\left(\xi_1, \xi_2\right) \frac{\partial G_b}{\partial \nu}\left(\xi_1, \xi_2; x_1, x_2\right) ds \tag{18}$$

$$= G_b\left(x_1^0, x_2^0; x_1, x_2\right),\ \left(x_1, x_2\right) \in \Omega.$$

$$\phi\left(x_1, x_2\right) + \frac{2\left(\rho_0 - \rho\right)}{\rho_0 + \rho} \int_{\partial\Omega} \phi\left(\xi_1, \xi_2\right) \frac{\partial G_b}{\partial \nu}\left(\xi_1, \xi_2; x_1, x_2\right) ds$$

$$-\frac{2\left(\rho_0 - \rho\right)}{\rho_0 + \rho} \int_\Omega G_b\left(\xi_1, \xi_2; x_1, x_2\right) m(\xi_1, \xi_2) u\left(\xi_1, \xi_2\right) d\xi_1\, d\xi_2 \tag{19}$$

$$= -\frac{2\left(\rho_0 - \rho\right)}{\rho_0 + \rho} G_b\left(x_1^0, x_2^0; x_1, x_2\right),\ \left(x_1, x_2\right) \in \partial\Omega.$$

where

$$\phi\left(\xi_1, \xi_2\right) = u_+\left(\xi_1, \xi_2\right) - u_-\left(\xi_1, \xi_2\right). \tag{20}$$

**Theorem 3.2** If $M := max\{m(\xi_1, \xi_2)\}$ and $|\rho_0 - \rho|$ are small enough, then the system of integral equations (20) and (21) have a unique solution. Similar to the proof of Theorem 3.2 of Ref. [2]. $\square$

For given $\Omega$, $m(x)$, $\rho_0$, $\rho$ and source point $(x_1^0, x_2^0)$, we can determine $\phi(x_1, x_2)$ on $\Omega$ and on $\partial\Omega$ by equations (19) and (20). In the case that $\rho_0 = \rho$, we have

$$\phi(x_1, x_2) = 0, \ (x_1, x_2) \in \partial\Omega, \tag{21}$$

and the system of integral equations reduces to a single integral equation

$$u(x_1, x_2) + \int_\Omega G_b(\xi_1, \xi_2; x_1, x_2)\, m(\xi_1, \xi_2) u(\xi_1, \xi_2)\, d\xi_1\, d\xi_2 = G_b\left(x_1^0, x_2^0; x_1, x_2\right). \tag{22}$$

If $M$ is small, we can use the following algorithm to approximate $u(x_1, x_2)$: Let

$$u_0\left(\xi_1, \xi_2; x_1^0, x_2^0\right) = G_b\left(\xi_1, \xi_2; x_1, x_2\right), \ (x_1, x_2) \in \Omega$$

and for $n = 1, 2, 3, \cdots$, let

$$u_{n+1}(x_1, x_2) = G_b\left(x_1^0, x_2^0; x_1, x_2\right)$$

$$- \int_\Omega G_b(\xi_1, \xi_2; x_1, x_2)\, m(\xi_1, \xi_2) u_n\left(\xi_1, \xi_2; x_1^0, x_2^0\right)\, d\xi_1\, d\xi_2. \tag{23}$$

## 4. Inverse scattering problem and a uniqueness theorem

In this section we assume that $\rho = \rho_0$. Let

$$\Gamma = \{(x_1, x_2) \in R_+^2 | x_2 = x_2^0 = \text{constant}\}, \tag{24}$$

and

$$\Gamma_s = \{(x_1^s, x_2^s) \in R_h^2 | x_2^s = x_2^{s0} = \text{constant}\}. \tag{25}$$

We assume that both $\Gamma$ and $\Gamma_s$ are 'below' the layer $R_h^2$; i.e., $\max\{x_2 | x \in \Omega\} < h$. The inverse problem we consider is the following:

Given $u^s(x, x^s)$ for $x \in \Gamma$ and $x^s \in \Gamma_s$, determine $n(x)$. Using the proof in [2], we have the following uniqueness theorem.

**Theorem 4.1** Assume that $\rho = \rho_0$. Let $n_1, n_2$ be two indices of refraction with $n_1(x) = n_2(x) = n_0(x_2)$ for all $x \notin \Omega$ where $\Omega \subset R_h^2$, such that $n_1 - n_0$, $n_2 - n_0 \in C^2(R_h^2)$. Let $u_1(x, x_s)$, $u_2(x, x_s)$ be the corresponding scattered waves from an acoustic source $x_s$. If $u_1(x, x_s) = u_2(x, x_s)$ for all $x \in \Gamma$ and $x_s \in \Gamma_s$, then $n_1 = n_2$. $\square$

Now we present a numerical example for the inverse problem. We use a regularized Born approximation method to reconstruct the unknown inhomogeneity. Assume that $M$ is small, we have $u(x, z) \simeq G_b(x_{1s}, x_{2s}; x_1, x_2)$ for $(x_1, x_2) \in \Omega$. Therefore, the scattered field operator

$$F(mG_b)(x, z) := \int_\Omega G_b(\xi_1, \xi_2; x_1, x_2)\, \widehat{k}^2 G_b(\xi_1, \xi_2; x_{1s}, x_{2s})\, d\xi_1\, d\xi_2,$$

$$for\ (x_1, x_2) \in \Gamma$$

is the approximation of $F(mu)(x_1, x_2)$ and

$$F(mG_b)(x_1, x_2) \simeq G_b(x_{1s}, x_{2s}; x_1, x_2) - u(x_1, x_2) =: u_*^s(x_1, x_2),\ for\ (x_1, x_2) \in \Gamma.$$
$$(26)$$

Note that $F$ is a linear operator of $m$. Discretizing (4.8) we obtain an ill-conditioned linear system $\mathbf{F}m = u_*^s$. The regularized Born approximation gives the system

$$(\epsilon I + \mathbf{F}^*\mathbf{F})m = \mathbf{F}^*u^*.$$

## 5. Numerical Analysis

In our numerical example, we use the following parameters.
the height of earth crust h=4,
the source point position (-0.5, 0.001),
the height of the bump (volcano) 0.2,
the measured data are from along a line x2= .5: 0.05 : 5
We use iteration (23) until $|u_{n+1} - u_n| < 10^{-12}$. The minimization problem is solved using the regularized Born approximation with $\epsilon = 10^{-20}$.
The first graph is the original index function, and the reconstruction is shown in the second one. The contour of the real part of the index is shown in the third and the reconstruction is shown in the fourth.

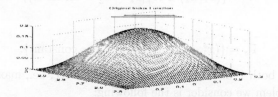

Fig. 1.  Original index function

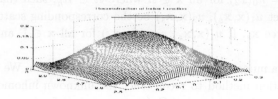

Fig. 2.  Reconstruction of index function

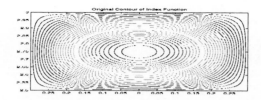

Fig. 3.  Original index function

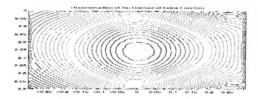

Fig. 4.  Reconstruction of index function

# References

1. Stefan Bergman, M. Schiffer. *Kernel Functions and Elliptic Differential Equations in Mathematical Physics*, Acaademic Press, New York, (1953).
2. Robert Gilbert, Klaus Hackl, Yongzhi Xu.: *Inverse Problem for Wave Propagation in a Perturbed Layered Half-Space*, Math. Comp. Modelling 45 (2007), 21-33.
3. F. B. Jensen, W. A. Kuperman, M. B. Porter and H. Schmidt: *Computational Ocean Acoustics* AIP Series in Mdern Acoustics and Signal Processing, New York (1994).

# A TIME DOMAIN METHOD TO MODEL VISCOELASTIC WAVE PROPAGATION IN LONG CORTICAL BONES*

J.-P. GROBY, E. OGAM, A. WIRGIN and Z.E.A. FELLAH

*LMA/CNRS*
*31 Chemin Joseph Aiguier,*
*13402, Marseille cedex 20, France*
*E-mail: groby@lma.cnrs-mrs.fr*


C. TSOGKA

*Department of Mathematics, University of Chicago,*
*5734 University Avenue,*
*Chicago, IL 60637, USA*
*E-mail: tsogka@math.uchicago.edu*

Bone characterisation, especially for osteoporosis diagnosis, is still a relevant problem due to the complexity of bone tissue. Non-ionizing techniques based on acoustic wave propagation have mostly been developed using ultrasonic waves considering bones to be elastic media [2] or poroelastic media [6], while other authors investigate and develop vibroacoustic methods, [3,5,4]. Bones are composed of complex porous media at the microscopic scale. This means that at macroscopic scales, we deal with homogeneized properties, and the latter have to be determined. A first approximation of long bones consists in considering the cortical part to be filled with an isotropic, linear, viscoelastic medium, and neglecting the trabecular part. The configuration in the sagittal plane takes the form of a circular viscoelastic ring cylinder, placed in a softer viscoelastic medium (in the sense that its shear velocity is very low compared with the shear velocity of the cortical part). By applying the methods proposed in [18,1], we show how dissipation can be taken into acount in the time domain to model viscoelastic wave propagation. Computational results are provided for the solution of the direct problem, which, in subsequent work, will be fed to an appropriate algorithm to solve the inverse problem (for retrieving the material and geometrical parameters of the bone).

**Key words:** Long cortical bone, viscoelastic wave propagation, mathematical modeling
**Mathematics Subject Classification:** 76B25, 35Q51

## 1. Introduction

Bone afflictions, such as osteoporosis (OS), osteosarcoma..., were not prevalent in ancient times because human beings did not live long enough for them to be of particular notice. Actually, this assertion is controversial [7], because the number of human deaths due to bone afflictions is not known due to the lack of diagnostic means in the remote past. Presently, an estimated 10 million persons, mostly white,

*This research was supported by Action CNRS/Etats-Unis 2005 no. 3321

postmenopausal women, suffer from osteporosis in the USA [8].

OS is due to an imbalance in resorption and formation of new bone. The remodeling process continues after maturity is attained, but at a slower pace in that bone formation steadily lags behind bone removal. In men, this begins at age $\approx 40$ and the rate of slowing down is linear (with age). In women, the rate of slowing down is nonlinear, with an acceleration of bone mineral density (BMD) loss of $2 - 3\%$ annually in the $5 - 10$ years following the menopause. The mass of bone tissue present at an instant of adult life is the difference between the amount accumulated at maturity, i.e. the peak bone mass (PBM), and the amount lost with aging.

OS is a systemic skeletal disease characterised by low bone density and microarchitectural deterioration of bone tissue. The consequences of this are fragile bones and fractures, mostly of the hip (which reduce considerably the possibilities of movement and which are painful), vertebrae (which can even make it impossible to stand) and distal radius bones (which make it difficult to carry out the ordinary tasks of everyday life). It is also necessary to develop bone characterisation methods for early diagnosis of OS and to monitor the evolution of this affliction.

A review of the sites (calcaneum, hip, long bones like radius...) and techniques (Dual X-ray Absorptiometry, Computed Tomography appealing to X-rays, Quantitative Ultrasound, Vibration ...) used to make a diagnosis of OS, suggests that a good choice of the probe radiation is acoustic/elastic waves. Subsequently, one is faced with the choice of: i) the most appropriate mathematical/physical model for predicting vibroacoustic phenomena associated with bones in both in vitro and in vivo contexts, ii) the most suitable model for inverting either simulated or real data relative to vibroacoustic phenomena produced by bones for the purpose of giving an accurate measure of the strengh and structural integrity of human bones, iii) the nature of the measured quantity and the sensitivity of this data to the parameters of the models.

Another choice has to be made concerning the target site which is dictated by the physical properties of bone. There exist two forms of bone: cortical and cancellous. Cortical bones are predominant in the appendicular portion and cancellous bone in the axial portion, notably the spine. Cortical bone occupies the exterior portion of the long (appendicular) bones, and is arranged as bundles of osteons packed tightly together to resist bending forces. Cancellous bone is lighter and more porous (up to $90\%$ as compared to $3 - 5\%$ in cortical bone) and structured to resist compressive forces. Bone is also a highly heterogneous porous anisotropic media at both micro and macro scale. As we deal with wavelengths large compared to the size of bundles of osteon, homogenized models involving homogenized parameters are appropriate. Our study is focused on cortical bones, particularly those of the appendicular part of the human radius.

## 2. Basic ingredients of the model

Fig.1 depicts a) longitudinal b) and sagittal cut of a human radius. The media surrounding long bones, constituted by muscles, skin and flesh, are usually considered as a fluid, whose properties are close to those of water [10,11], because the shear wave velocity therein is very low compared to the longitudinal wave velocity [12].

Fig. 1.  X-ray computed tomography of a human radius[2]

Here, we neglect the variations of the soft tissue properties and we consider that the medium $M^0$, occupying the host domain $\Omega_0$, composed of the surrounding domain and the medullar cavity, in figure 1 is viscoelastic (i.e. dissipative) such that $\rho^0 = 1030 kg.m^{-3}$, $\left(c_S^0\right)_R = 200 m.s^{-1}$, $Q_S^0 = 100$, $\left(c_P^0\right)_R = 1500 m.s^{-1}$, $Q_P^0 = 150$, values which are representative of muscles, except the shear wave velocity $\left(c_S^0\right)_R$ whose value $(200 m.s^{-1})$ is a somewhat too large but still acceptable compared with the value of the longitudinal wave velocity, [12]. In the sagittal plane, the media filling the bone can be considered to be istropic. Assuming that the bone is solicited in its central portion (indicated by "measurement region" in figure 1) by a incident P wave radiated by a line source located in the surrounding medium, and that the excited portion of the bone is sufficiently long for the latter to be assimilated with an infinitely-long cylinder, both the geometry of the configuration, and the polarization and geometry of the solicitation, enable us to treat a 2-D problem involving elastic(viscoelastic) waves. Effectively, in figure 1: a) we can distinguish from the center towards the periphery: the medullar cavity, the spongy bone and the cortical bone. The spongy bone can be neglected since its thickness is very small. The porosity of cortical bone $(3-5\%)$ can be accounted for by considering medium $M^1$ filling $\Omega_1$, figure 2 to be a homogenized viscoelastic solid such that $\rho^1 = 1850 kg.m^{-3}$, $\left(c_S^1\right)_R = 1800 m.s^{-1}$, $Q_S^1 = 30$, $\left(c_P^1\right)_R = 3050 m.s^{-1}$, $Q_P^0 = 50$. Most of the aforementioned parameters are taken from [2,12] except for the quality factor (in [2], the authors consider 3-D elastic, non-lossy isotropic or anisotropic media) and are fairly representative of cortical bone and tissue in the radius site. Particular attention must be paid to the dissipation in cortical bone (which is relatively important in our calculation), since this parameter is of great interest in osteporosis diagnosis. Effectively, an increase of the structural disorder and of porosity, can be interpreted as leading to an increase of the dissipation in cortical bones.

The final configuration is represented figure 2. We assume that the shape of the radius is circular in the sagittal plane. The circular cylinder radius is in contact across interface $\Gamma_{\text{ext}}$ (at $r_{ext} = 6mm$) with the surrounding medium. The same type of medium is assumed to fill the medullary cavity and to be in contact with the cortical portion across $\Gamma_{int}$ (at $r_{int} = 2.56mm$). We assume that on both interfaces $\Gamma_{ext}$ and $\Gamma_{int}$, the displacement and normal stress are continuous. Media $M^0$ and $M^1$ are assumed to be isotropic, linear, viscoelastic, macroscopically-homogeneous, and initially stress-free. The problem is to model P-SV wave propagation in a viscoelastic medium in the time domain.

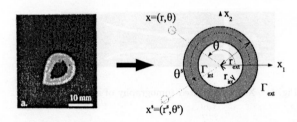

Fig. 2.  Cross-section view of a humaun radius (right) and of the simplified configuration (left).

## 3. Governing equations in viscoelastic solid media

In a viscoelastic medium $\Omega \in \mathbb{R}^2$ the (Hooke-Cauchy) constitutive relation between the stress tensor $\boldsymbol{\sigma}(\omega) = \boldsymbol{\sigma}(\mathbf{x}, \omega)$ and the strain tensor $\boldsymbol{\epsilon}(\omega) = \boldsymbol{\epsilon}(\mathbf{x}, \omega)$, related to the displacement $\mathbf{u}(\mathbf{x}, \omega)$, under the small deformation hypothesis by $\epsilon_{ij} = \frac{1}{2}(u_{i,j} + u_{j,i})$, takes the following form in the frequency domain:

$$\sigma_{ij}(\omega) = \lambda(\mathbf{x}, \omega)\epsilon_{mm}(\omega)\delta_{ij} + \mu(\mathbf{x}, \omega)\epsilon_{ij}(\omega) \tag{1}$$

wherein the Einstein sommation convention is implicit, $\lambda(\mathbf{x}, \omega)$ and $\mu(\mathbf{x}, \omega)$ are the Lamé coefficients, which are complex and frequency-dependent. The dissipative aspect of a linear viscoelastic material in P-SV wave propagation is often described by two quality factors, one relative to the P (pressure) waves and the other to SV (vertically polarized shear) waves. To be consistent, the bulk modulus $\Pi(\mathbf{x}, \omega) = \lambda(\mathbf{x}, \omega) + 2\mu(\mathbf{x}, \omega)$ is often preferred to $\lambda(\mathbf{x}, \omega)$, [13]. The expression for the two quality factors, relative to the P wave (denoted by index P) and SV wave (denoted by index S) $Q_P(\omega) = Q_P(\mathbf{x}, \omega)$ and $Q_S(\omega) = Q_S(\mathbf{x}, \omega)$ are then:

$$Q_P(\omega) = \frac{\Re(\Pi(\mathbf{x}, \omega))}{\Im(\Pi(\mathbf{x}, \omega))}, \quad Q_S(\omega) = \frac{\Re(\mu(\mathbf{x}, \omega))}{\Im(\mu(\mathbf{x}, \omega))} \tag{2}$$

The constitutive relation then becomes

$$\sigma_{ij}(\omega) = (\Pi(\mathbf{x}, \omega) - 2\mu(\mathbf{x}, \omega))\,\epsilon_{mm}(\omega)\delta_{ij} + \mu(\mathbf{x}, \omega)\epsilon_{ij}(\omega) \tag{3}$$

In the time domain, this relation is expressed in terms of a convolution operator, denoted by $\star_t$

$$\sigma_{ij}(t) = (\Pi(\mathbf{x}, t) - 2\mu(\mathbf{x}, t)) \star_t \epsilon_{mm}(t)\delta_{ij} + \mu(\mathbf{x}, t) \star_t \epsilon_{ij}(t) \tag{4}$$

The discretisation of this equation requires saving in memory the whole history of the solution at all points of the computational domain and is thus very cumbersome and expensive. To overcome this inconvenience, we approximate the two viscoelastic modulii $\Pi(\mathbf{x}, \omega)$ and $\mu(\mathbf{x}, \omega)$ by rational functions in the frequency domain, as proposed in [15,17,18,16,14]. Here we follow the approach exposed in [1], which is mostly based on [18], to get a final system close to the one studied in [14].

The approximated modulii, $\mu_L(\mathbf{x}, \omega)$ and $\Pi_L(\mathbf{x}, \omega)$ of $\mu(\mathbf{x}, \omega)$ and $\Pi(\mathbf{x}, \omega)$, are then:

$$\mu_L(\mathbf{x}, \omega) = \mu_R(\mathbf{x}) \left( 1 + \sum_{l=1}^{L} \frac{i\omega y_l^S(\mathbf{x})}{i\omega - \omega_l} \right) ; \; \Pi_L(\mathbf{x}, \omega) = \Pi_R(\mathbf{x}) \left( 1 + \sum_{l=1}^{L} \frac{i\omega y_l^P(\mathbf{x})}{i\omega - \omega_l} \right) \quad (5)$$

where $\mu_R$ (respectively $\Pi_R$) is the relaxed rigidity ($\lim_{\omega \to 0} \mu(\mathbf{x}, \omega)$) (respectively the relaxed bulk modulus), $y_l^S(\mathbf{x})$ ($y_l^P(\mathbf{x})$), $l \in [1, ..., L]$, are weights associated with shear waves (respectively longitudinal waves), and $\omega_l^S$ ($\omega_l^P$), $l \in [1, ..., L]$, are the relaxation frequences. Once these relaxation frequencies are determined –equidistant on a logarithmic scale in $\left[ \frac{\omega_{max}}{100}, \omega_{max} \right]$, $\omega_{max}$ being the maximal frequency of the solicitation spectrum, [18]– the weights are determined by solving an overdetermined system obtained by equaling the required functions of the frequency $Q_S^{-1}(\mathbf{x}, \omega)$ (respectively $Q_P^{-1}(\mathbf{x}, \omega)$) with its approximate value $\frac{\Im(\mu_L(\mathbf{x}, \omega))}{\Re(\mu_L(\mathbf{x}, \omega))}$ (respectively $\frac{\Im(\Pi_L(\mathbf{x}, \omega))}{\Re(\Pi_L(\mathbf{x}, \omega))}$) at $2L + 1$ frequencies [15].

**Remark:** The linear properties of the considered medium, and the mode conversion between P and SV waves, suggest the use of the same relaxation frequencies and the same development order for the approximations of the two modulii; this is why the relaxation frequencies $\omega_l$, $\forall l \in [1, ..., L]$ do not depend on $\mathbf{x}$, but the relaxed modulii $\Pi_R(\mathbf{x})$ and $\mu_R(\mathbf{x})$) and weights ($y_l^S(\mathbf{x})$ and $y_l^P(\mathbf{x})$, $\forall l \in [1, ..., L]$) do depend on $\mathbf{x}$.

The introduction of this approximation into Eq.(3) leads to

$$\sigma_{ij}(\omega) = (\Pi_R(\mathbf{x}) - 2\mu_R(\mathbf{x})) \, \epsilon_{mm}(\omega)\delta_{ij} + \mu_R \epsilon_{ij}(\omega) +$$
$$\sum_{l=1}^{L} \frac{i\omega}{i\omega - \omega_l} \left( (\Pi_R y_l^P(\mathbf{x}) - 2\mu_R(\mathbf{x}) y_l^S(\mathbf{x})) \, \epsilon_{mm}(\omega)\delta_{ij} + \mu_R y_l^S(\mathbf{x})\epsilon_{ij}(\omega) \right) \quad (6)$$

We then introduce $L$ new tensors, $\zeta_l(\omega) = \zeta(\mathbf{x}, \omega)$ such that

$$\zeta^l(\omega) = \frac{i\omega}{i\omega - \omega_l} \times$$
$$\left( (\Pi_R(\mathbf{x}) y_l^P(\mathbf{x}) - 2\mu_R(\mathbf{x}) y_l^S(\mathbf{x})) \, \epsilon_{mm}(\omega)\delta_{ij} + \mu_R(\mathbf{x}) y_l^S(\mathbf{x})\epsilon_{ij}(\omega) \right) \quad (7)$$

Finally, multiplying Eq.(6) by $-i\omega$ and going back to the time domain, Eqs.(6) and (7) become a system of coupled first-order-in-time partial differential equations:

$$\begin{cases} \dfrac{\partial \sigma_{ij}(t)}{\partial t} - \displaystyle\sum_l^L \dfrac{\partial \zeta_{ij}(t)}{\partial t} = & \text{in } \Omega \times [0,T] \\[2mm] (\Pi_R(\mathbf{x}) - 2\mu_R(\mathbf{x}))\, d_{mm}(t)\delta_{ij} + \mu_R(\mathbf{x})d_{ij}(t) & \\[2mm] \dfrac{\partial \zeta_{ij}(t)}{\partial t} + \omega_l \zeta_{ij}(t) = & \text{in } \Omega \times [0,T] \\[2mm] (\Pi_R(\mathbf{x})y_l^P(\mathbf{x}) - 2\mu_R(\mathbf{x})y_l^S(\mathbf{x}))\, d_{mm}(t)\delta_{ij} + \mu_R(\mathbf{x})y_l^S(\mathbf{x})d_{ij}(t) & \end{cases} \qquad (8)$$

wherein $\mathbf{d}(t) = \mathbf{d}(\mathbf{x},t)$ is the deformation rate tensor defined such that $\mathbf{d}(t) = \dfrac{\partial \boldsymbol{\epsilon}(t)}{\partial t} = \dfrac{1}{2}\left(\nabla \mathbf{v}(t) + \nabla^T \mathbf{v}(t)\right)$, $\mathbf{v}(t) = \mathbf{v}(\mathbf{x},t) = \partial_t \mathbf{u}(t)$ being the particle velocity.

## 4. Outline of the mathematical procedure for solving the problem

We now consider a procedure for solving the problem of 2D P-SV wave motion, in response to the solicitation $\mathbf{f}(\mathbf{x},t)$, in the temporal interval $[0,T]$ and the spatial domain $\Omega$, occupied by an isotropic, linear, viscoelastic medium $M$ characterised by the density $\rho(\mathbf{x})$, the relaxed modulii $\Pi_R(\mathbf{x})$ and $\mu_R(\mathbf{x})$, and the quality factors $Q_P(\omega)$ and $Q_S(\omega)$. Dissipation of biological media (soft tissue and cortical bone) can be represented, to a first approximation, and in a given frequency range, by a constant-in-frequency quality factor $Q = Q(\omega)$.

**Remark:** A constant quality factor with respect to the frequency does not lead to an attenuation that is independent of frequency; rather, the attenuation $\alpha(\omega)$ is defined by the ratio of the imaginary part to the real part of the wavenumber of the considered wave and medium. Thus, the aforemention procedure is:

- obtain the weight functions $y_l^S(\mathbf{x})$ from $Q_S^{-1} = \dfrac{\Im\left(\mu_L(\mathbf{x},\omega)\right)}{\Re\left(\mu_L(\mathbf{x},\omega)\right)}$ and $y_l^P(\mathbf{x})$ from $Q_P^{-1} = \dfrac{\Im\left(\Pi_L(\mathbf{x},\omega)\right)}{\Re\left(\Pi_L(\mathbf{x},\omega)\right)}$ via Eq.(5),with $\mathbf{x} \in \Omega$, $\omega \in \left[\dfrac{\omega\max}{100}, \omega\max\right]$

- solve the coupled system of three first-order-in-time partial differential equations, resulting from Eq.( 8) coupled with the conservation of the momentum relation:

$$\begin{cases} \rho \dfrac{\partial \mathbf{v}}{\partial t} = \nabla \cdot \boldsymbol{\sigma} & \text{in } \Omega \times [0,T] \\[3mm] \mathbf{A}\dfrac{\partial \boldsymbol{\sigma}}{\partial t} - \displaystyle\sum_l^L \mathbf{A}\dfrac{\partial \boldsymbol{\zeta}}{\partial t} = \mathbf{d} & \text{in } \Omega \times [0,T] \\[3mm] \mathbf{B}\dfrac{\partial \boldsymbol{\zeta}}{\partial t} + \omega_l \mathbf{B}\boldsymbol{\zeta} = \mathbf{d} & \text{in } \Omega \times [0,T] \end{cases} \qquad (9)$$

for the three unknown tensors $\mathbf{v}(\mathbf{x},t)$, $\boldsymbol{\sigma}(\mathbf{x},t)$ and $\boldsymbol{\zeta}(\mathbf{x},t)$.

## 5. Numerical procedure

The preceding governing equations are those of linear elastodynamics for a heterogeneous and isotropic medium. The domain $\Omega$ is conveniently considered to be rectanglar consisting of subregions (from the center towards the periphery): a homogeneous material with low shear wave velocity (modelling the marrow in the medicullar cavity), a hard homogeneous material (modelling the cortical bone), and the same homogeneous material as in the medullar cavity modelling here the homogenized soft tissues. To discretize the problem, we devide the domain $\Omega$ into $2D$ mesh of identical square elements. The material properties (density, velocity and memory data) are discretized by picewise constant functions (one value per element of the mesh). This means that in principle macroscopically heterogeneous materials could be taken into account. Similarly, the fact that the shape of cortical bone is circular is of no qualitative consequence concerning the numerical effort, since values of the material parameters have to be assigned to all cells, whether they are occupied by soft tissue, marrow or cortical bone. Of course, the shape of the cortical bone, gives rise to discrete interfaces $\Gamma_{ext}$ and $\Gamma_{int}$ described by piecewise linear functions, so that this shape entails smaller cells and an increased numerical effort. As is testified by the material in the previous sections, rather than deal with the single second- order-in-time vectorial partial differential or integrodifferential equation in an unbounded domain ($\mathbb{R}^2$ in 2D problems), as is common in studies of wave propagation problems, we chose to solve a mixed first-order-in-time system of equations in a bounded sub-domain of $\mathbb{R}^2$. Working with the mixed first-order formulation instead of the second-order wave equation presents the main advantage that it enables us to model wave propagation in infinite domains.

For the space discretization of the problem we use the finite elements proposed in [20] for the stress tensor and piecewise linear discontinuous functions for the velocity. For the time discretization we rely on a centered second order finite diffrence scheme for the time discretisation. The finite elements are compatible with mass-lumping, which leads to explicit time discretisation shemes. To model wave propagation in infinite domains (the case of interest herein, since the host domain is assumed to be infinite extent) we use the Perfectly Matched Layer (PML) [19]. More details on the numerical method can be found in [22,21,23].

In the example presented in this paper the computational domain was a $36mm \times 40mm$ square discretised by a grid of $400 \times 500$ nodes. This domain was surrounded by a PML layer 50 nodes thick.

## 6. Validation of the method

We validated the time domain method on the canonical example of diffraction of an incident cylindrical P wave by a circular cylinder as depicted in figure 3. The viscoelastic modulii are given by Kjartansson's formula [24]:

$$\mu(\mathbf{x},\omega) = |\mu_R(\mathbf{x})| \left( \frac{-i\omega}{\omega_R} \right)^{\frac{2}{\pi} \arctan(Q_S^{-1}(\mathbf{x}))} \quad ; \quad \Pi(\mathbf{x},\omega) = |\Pi_R(\mathbf{x})| \left( \frac{-i\omega}{\omega_R} \right)^{\frac{2}{\pi} \arctan(Q_P^{-1}(\mathbf{x}))}$$

(10)

We compute the velocity at three points denoted by $R_j$, $j = 1, 2, 3$ such that $\mathbf{x}_j = (j\pi/2, 1mm)$. Medium $M^0$ filling $\Omega_0$ is caracterised by $\rho^0 = 1300 kg.m^{-3}$, $(c_P^0)_R = 1500 m.s^{-1}$, $Q_P^0 = 150$, $(c_S^0)_R = 500 m.s^{-1}$, $Q_S^0 = 100$, and medium $M^1$ filling $\Omega_1$ is described by $\rho^1 = 1800 kg.m^{-3}$, $(c_P^1)_R = 3050 m.s^{-1}$, $Q_P^1 = 50$, $(c_S^1)_R = 1800 m.s^{-1}$, $Q_S^1 = 30$. The external radius is $0.5mm$ and the solicitation is delivered by a line source, located at $\mathbf{x}^s = (3\pi/2, 2mm)$, radiating a Ricker-like P wave, such that $\nu_0 = 1MHz$.

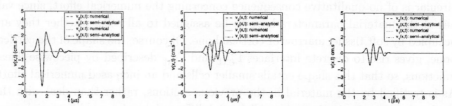

Fig. 3. Comparison of the two components (i.e. along the $\mathbf{x}_2$ axis and $\mathbf{x}_1$ axis) of the particle velocity calculated with the numerical procedure and the semi-analytical one. The results on the left corresponds to observation point $R_1$, in the center to point $R_2$, and on the right to point $R_3$.

## 7. Results

The aim of this work was first to perform calculations for a configuration as close as possible to the real configuration (i.e., one with a relatively low velocity in the host medium, and which takes into account dissipation in both the cortical bone and in the host and central regions).

Fig.4 depicts the two components of the velocity (i.e. along $\mathbf{x}_1$ and $\mathbf{x}_2$). The line source, located at $\mathbf{x}^s = (3\pi/2, 20mm)$, radiates a P wave which takes the form of a Ricker wavelet such that, $\nu_0 = 100kHz$. The components along the $x_1$ axis at $\mathbf{x} = (\pi/2, 10mm)$ and $\mathbf{x} = (3\pi/2, 10mm)$ are nil.

Figure 5 constitute snapshots of the modulus of the total particle velocity at time $t = 16\mu s$, $t = 28\mu s$, $t = 36\mu s$, and $t = 60\mu s$.

## 8. Discussion

The time histories in Fig.4 show: (i) that the location of the largest amplitudes, at which it is approriate to acquire data in order to solve the inverse problem, is in the lower portion of the domain between $[\pi, 0]$. This is as expected, considering the source location. In particular, points around $\frac{5\pi}{4}$ seem to be those for which both the $x_1$ and $x_2$ components of velocity have the highest amplitudes. Nevertheless, records in the upper portion of the domain do not seem to suffer from the large attenuation in bones. Solving inverse problems by taking into account the whole history of the total particle velocity all around the bone could be useful and be carried out in pratice for further in-vitro or in-vivo experiments.

The snapshots in Fig.5 show that the waves diffracted by bones are composed of both P and SV components. In particular, these results show that shear waves are important for large times. This means that shear wave should not be neglected in muscles, skin... These results could be confirmed by the use of elastography techniques (employed heretofore for breast cancer evaluation) which suppose that shear waves are not too attenuated in the medium (i.e., the breast). These results seem to show that a non- negligible part of the information, certainly related to shear waves in the bone (phenomena due to mode excitation because of the high contrasts between shear wave velocity in bone and in the surrounding medium), is neglected when one considers that the surrounding medium is a simple fluid. This is particularly true when the solicitation is low-frequency in nature.

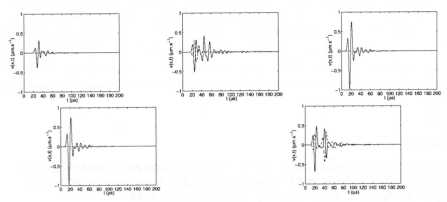

Fig. 4. Time record of the total particle velocity at various locations: on top from the left to the right $\mathbf{x} = (\pi/2, 10mm)$, $\mathbf{x} = (\pi, 10mm)$ and $\mathbf{x} = (3\pi/2, 10mm)$ and on the bottom, on the left $\mathbf{x} = (3\pi/4, 10mm)$ and on the right $\mathbf{x} = (5\pi/4, 10mm)$. The solid curves correspond to the component along the $x_2$ axis, and the dashed curves to the component along the $x_1$ axis.

Fig. 5. Snapshot of the total particle velocity at $t = 16\mu s$, $t = 28\mu s$, $t = 36\mu s$ and $t = 60\mu s$.

## 9. Conclusion

The traditional, and even quite recent choice, concerning bone caracterisation by ultrasound, has been to consider ether the surrounding medium as a fluid, or the surrounding medium plus the medium filling the bone as two fluids. Nowadays, these models, which were of great interest in understanding the phenomena that have been observed, and which have led to more or less reliable inversion algorithms, seem to be incomplete, in the sense that: i) bone is an anelastic medium, i.e., to the very least, porous and/or viscoelastic in its appendicular part, ii) muscles, marrow and flesh are viscoelastic media. We show that by taking into account the shear waves, (the velocity of which considered herein is certainly too large, i.e. $(c_S^1)_R = 200 m.s^{-1}$), with an excitation at $\nu_0 = 100 kHz$), induces a noticeable modification of the type of diffracted waves, essentially for large times. After a certain time, the field diffracted by the circular bone is essentially composed of shear waves. This indicates a strong dependence on shear wave velocity in such a configuration.

We show that the amplitudes of the diffracted waves are of the same order for any angle around the circular cylinder supposed to model the bone. This indicates that we could use the whole information recorded around the circular cylinder to carry out the inversion. The problem, which at present is still largely unsolved, is to determine the most appropriate descriptor of osteoporosis, such that it is sensitive to variations in the state of the bone, and can be extracted from the simplest mathematical model by means of the least amount, and most easily-measured type, of data.

## References

1. J.P. Groby, C. Tsogka, and A. Wirgin, *Soil.Dyn&Earth.Engrg*, **25:487-504**,2005.
2. E. Bossy, M. Talmant and P. Laugier, *J.Acout.Soc.Am*, **115:2315-2324**, 2004.
3. E. Ogam, and A. Wirgin, *Proc. of ISMA2004*, **2437-2446**,
4. J.L. Buchanan, R. Gilbert, and K. Khashanah, *J. Compt.Acoust.*, **12:99-126**, 2004.
5. R. Gilbert, A. Wirgin, and Y. Xu, *Appl.Math.Computation*, **142:561-573**, 2003.
6. Z.E.A. Fellah, J.-Y. Chapelon, S. Berger, W. Lauriks, and C. Depollier, *J.Acout.Soc.Am*, **115:61-73**, 2004.
7. J.M. McEwan, S. Mays, and G.M. Blake, *Calcified tissue Intl.*, 2003 .
8. R.C. Mellors, *Cornell medical website*, 2004.
9. D. Taylor and T.C. Lee, *J. Biomech.*, **31:1177-1180**, 1998.
10. L. Le Marrec, Thése de doctorat, Universit de la Mediterrane, 2004.
11. P. Lasaygue, E. Ouedraogo, J-P Lefebvre, M. Gindre, M. Talmant and P. Laugier, *Phys.Med.&Biol.*, **5:2633-2649**, 2005.
12. E.L. Masen, H.J. Sathoff and J.A. Zagzebski, *J.Acoust.Soc.Am.*, **74:1346-1355**, 1983.
13. J.O.A. Robertsson, J.O. Blanch and W.W. Symes, *Geophys.*, **9:1444-1456**, 1994.
14. E. Bécache, A. Ezziani and P. Joly, *Proc. WAVE'S 2003*, **916-921**, 2003.
15. H. Emmerich and M. Korn *Geophys.*, **52:1252-1264**, 1987.
16. J. Carcione, D. Losloff and R. Kosloff, *Geophys.J.R.Astr.Soc.*, **93:393-407**, 1988.
17. S.M. Day and J.B. Minster, *Geophys.J.R.Astr.Soc.*, **78:105-118**, 1984.
18. J.-P. Groby and C. Tsogka, *J.Comput.Acoust.*, in press.
19. F. Collino and C. Tsogka, *Geophys.*, **66:294-305**, 2001.

20. E. Bcache, P. Joly and C. Tsogka, *SIAM J.Num.Anal.*, **39:2109-2132**, 2002.
21. C. Tsogka, PhD thesis, University Paris Dauphine, 2000.
22. A. Ezziani, PhD thesis, University Paris Dauphine, 2004.
23. J.-P. Groby, PhD thesis, University Aix-Marseille *II*, 2005.
24. E. Kjartansson, *J.Geophys.Res.*, **84:4737-4748**, 1979.

# EIGENMODE ANALYSIS OF THE CORTICAL OSSEOUS TISSUe-MARROW COUPLED SYSTEM*

E. OGAM, Z.E.A FELLAH, J.-P GROBY, A. WIRGIN

*CNRS Laboratoire de Mécanique et d'Acoustique,*
*31 chemin Joseph Aiguier,*
*13402 Marseille cedex 20*

Osseous tissue contains innumerable cavities filled with various fluids such as blood, synovial fluid and bone marrow. The potential importance of the fluid occupying the medullary canal (the bone marrow) on wave propagation in the bone has generally been ignored. In this study, we address this issue by comparing the vibrational response, to point-like (both harmonic and transient) excitation, of a hollow elastic cylinder representing hard compact cortical bone in a dry state (air filled), with the response to the same excitation of the same bone in which the medullary canal is filled with marrow (found largely in the shafts of the long bones that surround the marrow cavities), represented by a fluid. The fluid is assumed to be inviscid and incompressible, and its motion, as well as that of the cortical solid is modeled in the framework of small displacements. Firstly, the Timoshenko beam equations (flexural two mode theory) are solved using fourth order finite differences to obtain the phase velocities which are compared to the experimental ones. An eigenmode analysis of the coupled system is then developed using 3D finite element modeling. The fluid-structure interaction is materialized on a vibroacoustic test-rig from which vibration response data, constituting the input to inversion algorithms, is retrieved.

**Key words:** Cortical osseous tissue-marow coupled system, eigenmode analysis, wave propagation
**Mathematics Subject Classification:** 76B25, 35Q51

## 1. Introduction

Using the vibration response of bone to predict its mechanical properties in vitro can provide clinically-valuable information for assessing fracture risk due to the weakening of bone by disease. Most of the research on characterization of human bone, using ultrasonic waves [1,3] or vibrations [7], has been devoted to osteoporosis. Osteoporosis is characterized by low bone mass and microarchitectural deterioration of bone tissue leading to enhanced bone fragility and a consequent increase in fracture risk.

The vibrational method can only be successful if there is a thorough understanding of the fundamental interactions between the bone and its surrounding structure and the fluid. The coupled motion of the fluid (marrow) and the flexible structure

---

*Research supported by action CNRS/ETATS-UNIS 2005 *N*°3321

(bone) is a problem which occurs in various engineering applications.

Some important questions to be answered, are: what is the influence of the fluid on the eigenmodes of the structure, and what happens when one principal mode of the fluid is near a principal mode of the structure?

The potential importance of the fluid occupying the medullary canal (the bone marrow in vivo) to wave propagation has generally been ignored. The marrow composition is variable, but a distinction is commonly drawn between yellow, or fatty, marrow, which is composed primarily of adipocytes, and red marrow, which is composed primarily of fibroblastic/reticular cells and is active in hematopoiesis [11]. Marrow composition varies with anatomical site, age, race, and disease state [9,12]. With aging, the proportion of fat in the marrow increases [9,12]. The age-related increase in fatty marrow is associated with decreased osteogenic potential, and has been implicated in age-related bone loss [11].

First, the one dimensional Timoshenko beam model equations for flexural vibration of the hollow cylinder, mimicking the *in vacuo* bone, is solved using Fourth order finite difference. This simple model constitutes a means of retrieving, Young's modulus, shear modulus, Poisson ratio and the wall thickness of the cylinder (incarnated in the shear correction factor and the cross sectional area moment of inertia) from the measured vibration modes. Then, the domain equations and boundary conditions to be used in solving the vibration problem of the fluid-filled elastic cylinder, mimicking the cortical bone containing marrow are derived. The equations are then solved in three dimensions using finite element modelling. Finally, the computed results, in phase velocities form, are compared against those obtained from a resonance experiment on our vibroacoustic test rig.

## 2. Simple 1D flexural vibration model

In 1921, Timoshenko included the effects of both shear and rotational inertia and obtained results in agreement with the exact theory[15] for the flexural vibration of beams.

The two modes of deformation with respect to the total lateral deflection, $y$, and the bending slope, $\varphi$, are given by two coupled equations [4,5,8]

$$
\begin{aligned}
GA\kappa \left( \frac{\partial \varphi(x,t)}{\partial x} - \frac{\partial^2 y(x,t)}{\partial x^2} \right) + \rho A \frac{\partial^2 y(x,t)}{\partial t^2} &= q(x,t) \\
GA\kappa \left( \frac{\partial y(x,t)}{\partial x} - \varphi(x,t) \right) + EI \frac{\partial^2 \varphi(x,t)}{\partial x^2} &= \rho I \frac{\partial^2 \varphi(x,t)}{\partial t^2}
\end{aligned}
\tag{1}
$$

where $q(x,t)$ is the external force, $\rho$ is the density, $E$ is Young's modulus of elasticity and $I$ is the cross sectional area moment of inertia about an axis normal to $x$ and $y$ passing through the center of the cross-sectional area (for a filled circular area of radius $R$, $I = \frac{\pi R^4}{4}$, for a hollow cylinder with external and internal radii, $R_1$ and $R_0$ respectively $I = \frac{\pi(R_1^4 - R_0^4)}{64}$). $A$ is the cross-sectional area of the beam.

The values of the shear correction factor, $\kappa$, for several cross-sections, have been derived by several authors [2,6]. Cowper's coefficients are based on a static elasticity solution. Hutchinson gave modified shear correction factors for the solid and hollow, circular cylinders [6]. We have used his values in all the computations that follow. The shear factors for the hollow and solid cylinder developed by the three authors are resumed in Table(1) for reference sake.

Table 1. The shear factors $\kappa$ for circular cylinders as given by three authors. The ratio of the inner radius $a$ to the outer radius $b$ is given by $m = \frac{a}{b}$.

| model | solid circular cylinder | hollow cylinder |
|---|---|---|
| Cowper | $\frac{6(1+\nu)}{7+6\nu}$ | $\frac{6(1+\nu)(1+m^2)^2}{(7+6\nu)(1+m^2)^2+(20+12\nu)m^2}$ |
| Hutchinson | $\frac{6(1+\nu)^2}{7+12\nu+4\nu^2}$ | $\frac{6(a^2+b^2)^2(1+\nu)^2}{7a^4+34a^2b^2+7b^4+\nu(12a^4+48a^2b^2+12b^4)+\nu^2(4a^4+16a^2b^2+4b^4)}$ |
| Timoshenko | $\frac{(6+12\nu+6\nu^2)}{(7+12\nu+4\nu^2)}$ | - |

## 2.1. Numeric resolution of the TBT for a finite length cylinder using the finite difference method

In order to keep the complexities of engineering problems intact, approximate numerical solutions, based on, for example, the finite difference or finite element methods, are required.

We employ the finite difference method (FDM) [13] to solve the one-dimensional equation for the axially-symmetric motion of an elastic rod of circular cross section. The FDM uses the approximation of the derivative of the partial differential equations governing the wave propagation.

The equations in Eqn.(1) can be reduced to a single one by differentiating the second equation with respect to $x$, and solving for $\frac{\partial \varphi}{\partial x}$ in the first equation, whereby one finds

$$\frac{\partial \varphi(x,t)}{\partial x} = \frac{\partial^2 y(x,t)}{\partial x^2} - \frac{\rho}{G\kappa}\frac{\partial^2 \varphi(x,t)}{\partial t^2}. \qquad (2)$$

Differentiating Eqn. (2) gives expressions for $\frac{\partial^3 \varphi(x,t)}{\partial x^3}$, $\frac{\partial^3 \varphi(x,t)}{\partial x \partial t^2}$ and substituting these in the second equation in (1), one obtains

$$\frac{EI}{\rho A}\frac{\partial^4 y(x,t)}{\partial x^4} - \frac{I}{A}\left(1+\frac{E}{G\kappa}\right)\frac{\partial^4 y(x,t)}{\partial x^2 \partial t^2} + \frac{\partial^2 y(x,t)}{\partial t^2} + \frac{\rho I}{GA\kappa}\frac{\partial^4 y(x,t)}{\partial t^4} = 0. \qquad (3)$$

If we seek a time-harmonic solution of the form

$$y(x,t) = Y(x)e^{-j\omega t}, \qquad (4)$$

wherein $j = \sqrt{-1}$, then substituting Eqn. (4) into Eqn. (3) yields the following fourth order linear ODE for $Y(x)$

$$\frac{EI}{\rho A}\frac{d^4Y(x)}{dx^4} + \frac{\omega^2 I}{A}\left(1 + \frac{E}{G\kappa}\right)\frac{d^2Y(x)}{dx^2} - \omega^2 Y(x) + \frac{\omega^4 \rho I}{GA\kappa}Y(x) = 0. \qquad (5)$$

Replacing the derivatives by appropriate fourth-order central-difference approximations, enables the Timoshenko beam equation to be approximated by

$$\begin{aligned}(-\alpha y_{n-3} + 12\alpha y_{n-2} - 39\alpha y_{n-4} + 56\alpha y_n - 39\alpha y_{n+1} - \alpha y_{n+3}) + \\ \omega^2(-\beta y_{n-2} + 16\beta y_{n-1} - (30\beta + 1)y_n + 16\beta y_{n+1} - \beta y_{n+2}) + \omega^4 \gamma y_n = 0\end{aligned} \qquad (6)$$

wherein $i = 0, 1, 2, ...,$ $\alpha = \frac{EI}{\rho A}\left(\frac{1}{6h^4}\right)$, $\beta = \frac{I}{A}\left(1 + \frac{E}{G\kappa}\right)\left(\frac{1}{12h^2}\right)$ and $\gamma = \left(\frac{\rho I}{GA\kappa}\right)$

In Eqn. (6), the points for which the values of $i = 0$ ($y_{-1}$, $y_{-2}$, $y_{-3}$ ) and $i = n$ ($y_{n+1}$, $y_{n+2}$, $y_{n+3}$ ), are called fictive points. They are not situated on the cylinder. Their values are computed using the boundary conditions and the first- and second-order FDM approximations.

### 2.1.1. Boundary conditions

### 2.1.2. free-free cylinder

The boundary conditions associated with the flexural vibrations of a free-free cylinder in the $xy$-plane ($0 \leq x \leq L$) at $x = 0$ and $x = L$ ( wherein L is the length of the cylinder) are: $\frac{\partial^2 y}{\partial x^2} = 0$ and $\frac{\partial^3 y}{\partial x^3} = 0$.

The first-order theory of central differences gives

$$\begin{aligned}y_i^{(2)} = \frac{\partial^2 y}{\partial x^2} &= \left(\frac{1}{2h}\right)(y_{i+1} - 2y_i + y_{i-1}) \\ y_i^{(3)} = \frac{\partial^3 y}{\partial x^3} &= \left(\frac{1}{2h^3}\right)(-y_{i-2} + 2y_{i-1} - 2y_{i+1} + y_{i+2})\end{aligned} \qquad (7)$$

Using first-order theory, we find, at cylinder position $x = L$, $i = n$,

$$(y_{n+1} - y_{n-1}) = 2(y_n - y_{n-1}), \quad (y_{n+2} - y_{n-2}) = 4(y_n - y_{n-1}) \qquad (8)$$

At position, $x = 0$, using first-order finite differences and the boundary conditions, we obtain $y_{-2} = 4(y_0 - y_1) + y_2$.

The first-order equations that satisfy these conditions at $x = 0$ should also satisfy the second-order equations for $u_0^{(3)} = 0$, $i = 0$: $y_{-3} = 6y_0 - 6y_1 + y_3$,.

The aforementioned equations should also satisfy $y_n^{(3)} = 0$ at beam positions, $x = L$ and $x = 0$ for the second-order theory.

At $x = L$, ($i = n$): $y_{n+3} = 6y_n - 6y_{n-1} + y_{n-3}$

The equations obtained after substitution of the boundary conditions into Equation (6) are written in Table (2) for different positions (stations) on the cylinder.

The critical frequency or cutoff frequency is $\omega_c = \sqrt{\frac{GA\kappa}{\rho I}}$.

## 2.2. Solving the TBM FDM matrix equation

Equation (6) can be rewritten in matrix form as

Table 2.  Entries of the left hand side ( LHS ) matrix for the free-free Timoshenko beam formed from the station positions $(St_i)$.

| $St_i$ | LHS |
|---|---|
| 0 | $20\alpha y_0 - 42\alpha y_1 + 24\alpha y_2 - 2\alpha y_3 + \omega^2\left(-(2\beta+1)y_0 + 4\beta y_1 - 2\beta y_2\right) + \omega^4\gamma y_0$ |
| 1 | $-19\alpha y_0 + 48\alpha y_1 - 40\alpha y_2 + 12\alpha y_3 - \alpha y_4 + \omega^2\left(14\beta y_0 - (29\beta+1)y_1 + 16\beta y_2 - \beta y_3\right) + \omega^4\gamma y_1$ |
| 2 | $10\alpha y_0 - 38\alpha y_1 + 56\alpha y_2 - 39\alpha y_3 + 12\alpha y_4 - \alpha y_5$ $\omega^2\left(-\beta y_0 + 16\beta y_1 - (30\beta+1)y_2 + 16\beta y_3 - \beta y_4\right) + \omega^4\gamma y_2$ |
| 3 to n-4 | $(-\alpha y_{n-3} + 12\alpha y_{n-2} - 39\alpha y_{n-4} + 56\alpha y_n - 39\alpha y_{n+1} - \alpha y_{n+3}) +$ $\omega^2\left(-\beta y_{n-2} + 16\beta y_{n-1} - (30\beta+1)y_n + 16\beta y_{n+1} - y_{n+2}\right) + \omega^4\gamma y_n$ |
| n-3 | $(-2\alpha y_{n-6} + 24\alpha y_{n-5} - 42\alpha y_{n-4} + 20\alpha y_{n-3}) +$ $\omega^2\left(-2\beta y_{n-5} + 4\beta y_{n-4} - (2\beta+1)y_{n-3}\right) + \omega^4\gamma y_{n-3}$ |
| n-2 | $(-2\alpha y_{n-5} + 24\alpha y_{n-4} - 42\alpha y_{n-3} + 20\alpha y_{n-2}) +$ $\omega^2\left(-2\beta y_{n-4} + 4\beta y_{n-3} - (2\beta+1)y_{n-2}\right) + \omega^4\gamma y_{n-2}$ |
| n-1 | $(-2\alpha y_{n-4} + 24\alpha y_{n-3} - 42\alpha y_{n-2} + 20\alpha y_{n-1}) +$ $\omega^2\left(-2\beta y_{n-3} + 4\beta y_{n-2} - (2\beta+1)y_{n-1}\right) + \omega^4\gamma y_{n-1}$ |
| n | $(-2\alpha y_{n-3} + 24\alpha y_{n-2} - 42\alpha y_{n-1} + 20\alpha y_n) +$ $\omega^2\left(-2\beta y_{n-2} + 4\beta y_{n-1} - (2\beta+1)y_n\right) + \omega^4\gamma y_n$ |

$$[C]\{y\} + \omega^2[B]\{y\} + \omega^4[A]\{y\} = 0 \tag{9}$$

Use of the identity $\Omega = \omega^2$ gives the frequency equation. It is a quadratic matrix equation whose solutions are given by

$$[\Omega] = \frac{-[B] \pm \sqrt{[B]^2 - 4[A][C]}}{2}[A]^{-1} \tag{10}$$

In expression (10) it is necessary to calculate the matrix

$$[X] = \sqrt{[B][B] - 4[A][C]} = [M]^{\frac{1}{2}}, \tag{11}$$

$[X]$ is the principal square root of matrix $[M]$, i.e., $[X][X] = [M]$. Matrix $[X]$ is the unique square root for which every eigenvalue has nonnegative real part. If $[M]$ has any eigenvalues with negative real parts then a complex result is produced. If $[M]$ is singular, then $[M]$ may not have a square root. The values of $\Omega$ obtained in this case are complex.

$$\Omega_k = \nu_k + j\varpi_k \tag{12}$$

wherein, $\nu_k$ is called the damping factor and $\varpi_k$ the natural frequency of mode ($k$). Other related and commonly-used terms are the damping ratio and resonant frequency.

In the presence of losses, the damping of the displacement field amplitude is usually accounted-for through a complex frequency (Eqn. 12), whose imaginary part provides the temporal decay rate. The two solutions of Eq. (10) correspond to the two flexural modes of propagation having two different phase velocities.

### 2.3. *Comparison with other theories*

The results from the TBM FDM for hollow cylinders are compared in Table (3) with those published by Hutchinson [6], and So and Leissa [14] ( S&L). S&L calculated their values using a 3D Rayleigh-Ritz method. The results are computed for different ratios of interior ($D_0$) to outer diameters ($D_1$), $\frac{D_0}{D_1}$. For easy comparison with the published results, the frequency values are given in the reduced form $\omega D_1 \sqrt{\frac{\rho}{G}}$. The length-to-diameter ratio is $\frac{L}{D_1} = 0.5$ and the Poisson ratio is chosen to be $\nu = 0.3$. The eigenfrequencies are in close agreement with the target values from the published results for free vibration.

Table 3.   Comparison between the resonance frequencies computed using the TBM FDM method with those from Hutchinson, and So and Leissa (S&L). S and A refer to the symmetric and antisymmetric modes.

| | | $\frac{D_0}{D_1}$ 0.1 | 0.1 | 0.5 | 0.5 |
|---|---|---|---|---|
| | Mode Number | S | A | S | A |
| S&L | 1 | 0.1651 | 0.3990 | 0.1776 | 0.4096 |
| Hutchinson | | 0.1651 | 0.3989 | 0.1776 | 0.4098 |
| Cowper | | 0.1649 | 0.3972 | 0.1771 | 0.4064 |
| TBM FDM | | 0.1657 | 0.3926 | 0.1771 | 0.4037 |

### 2.4. *Fluid-structure interaction (FSI)*

It has been reported that the marrow of the cancellous bone in the calcanea is responsible for: i) a significant reduction of the ultrasound (US) velocity, and ii) increased attenuation, attenuation slope and backscatter, compared to the water-saturated state in the calcanea [10]. In the present study, we develop a model and experiment to study the influence of the filling fluid on the resonance modes of vibration.

Three coupling mechanisms determine FSI of the hollow cylinder filled with an inviscid fluid. Friction coupling is due to shear stresses resisting relative axial motion between the fluid and the cylinder wall. Poisson coupling is due to normal stresses acting at this same interface. For example, an increase in fluid pressure causes an increase in pipe hoop stress and hence a change in axial wall stress. Two types of modes exist, namely modes related to solid vibrations and coupled fluid-solid modes.

The fluid is assumed herein to be inviscid and incompressible, and the structure equations are derived for small displacements. The simulation is carried out using 3D finite element modeling. The equations for each subdomain and the boundary conditions are given in the following sections.

### 2.5. The mathematical model for the acoustic analysis in the fluid subdomain

The acoustics in the fluid subdomain is described by the wave equation

$$\nabla.(-\frac{1}{\rho_f}\nabla p + \mathbf{q}) + \frac{1}{\rho_f c_f^2}\frac{\partial^2 p}{\partial t^2} = 0 \tag{13}$$

wherein $p$ is the pressure, $c_f$ the sound velocity, $\rho_f$ the fluid density and $\mathbf{q}$ a source term.

For a time-harmonic wave of frequency $f$, the pressure variation in time $(t)$ is given by : $p = p_0 \exp(-j\omega t)$, with $\omega = 2\pi f$.

In this case, the wave equation for acoustic waves reduces to the well-known Helmholtz equation:

$$\nabla.(-\frac{1}{\rho_f}\nabla p + \mathbf{q}) + \frac{\omega^2 p}{\rho_f c_f^2} = 0, \tag{14}$$

The *eigenvalue* equation for the pressure is

$$\nabla.(-\frac{1}{\rho_f}\nabla p) - \frac{\lambda}{\rho_f c_f^2}p = 0 \tag{15}$$

the eigenvalue, $\lambda$ and the the eigenfrequency, $f$ are related by, $\lambda = (2\pi f)^2$.

### 2.6. The solid subdomain

The wall dynamics are described under the assumption of small elastic deformations.

The symmetric strain tensor $\varepsilon$ consists of both normal and shear strain components and the stress in a material is described by the symmetric stress tensor :

$$\varepsilon = \begin{bmatrix} \varepsilon_x & \varepsilon_{xy} & \varepsilon_{xz} \\ \varepsilon_{xy} & \varepsilon_y & \varepsilon_{yz} \\ \varepsilon_{xz} & \varepsilon_{yz} & \varepsilon_z \end{bmatrix} \quad \sigma = \begin{bmatrix} \sigma_x & \tau_{xy} & \tau_{xz} \\ \tau_{yx} & \sigma_y & \tau_{yz} \\ \tau_{xz} & \tau_{yz} & \sigma_z \end{bmatrix} \tag{16}$$

wherein $\sigma$ is the stress tensor, consisting of three normal stresses $(\sigma_x, \sigma_y, \sigma_z)$ and six, or if symmetry is invoked, three shear stresses $(\tau_{xy}, \tau_{yz}, \tau_{xz})$. The linear stress-strain relationship is $\sigma = D\varepsilon$. $D$ [16] is the $6 \times 6$ elasticity matrix.

Cauchy's first equation of motion is: $div\sigma + \mathbf{F} = \rho\ddot{\mathbf{u}}$, wherein $\mathbf{u}$ denotes the displacement field, and $\ddot{\mathbf{u}} = \frac{d^2\mathbf{u}}{dt^2}$, $\mathbf{F}$ the body force.

Applying the product rule, $div\sigma.\eta = div(\sigma\eta) - \sigma : grad\eta$, leads to the principal of virtual work in the spatial description (weak formulation)

$$f(\mathbf{u}, \delta\mathbf{u}) = \int_\Omega (\sigma : \delta e - (\mathbf{F} - \rho\ddot{\mathbf{u}}).\delta\mathbf{u}).\delta V - \int_{\partial\Omega} \bar{t}\delta\mathbf{u}ds = 0 \tag{17}$$

wherein $\delta e$ is the virtual strain, $\sigma : \delta e$, the virtual stress, $\bar{\mathbf{t}} = \sigma \mathbf{n}$ the traction on the boundary surface $\partial \Omega$, $\mathbf{n}$, the unit exterior vector normal to the boundary surface, and $dV$ the material volume element.

The initial conditions are:

$$\int_\Omega \mathbf{u}(\mathbf{x}, t)|_{t=0} . \delta \mathbf{u} dV = \int_\Omega \mathbf{u}_0(\mathbf{X}) . \delta \mathbf{u} dV, \tag{18}$$

$$\int_\Omega \dot{\mathbf{u}}(\mathbf{x}, t)|_{t=0} . \delta \mathbf{u} dV = \int_\Omega \dot{\mathbf{u}}_0(\mathbf{X}) . \delta \mathbf{u} dV, \tag{19}$$

wherein $\dot{\mathbf{u}} = \frac{d\mathbf{u}}{dt}$, $\mathbf{x}$ and $\mathbf{X}$ are the the current and referential positions respectively. Eqn. (17) is typically called a *variation equation* and forms the basis of the finite element method.

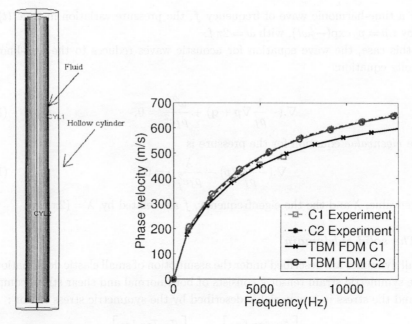

Fig. 1. (Left) The 3D fluid-filled cylinder geometry. (Right) The phase velocity varies with the thickness of the cylinder wall (*air filled*).

## 2.6.1. *Boundary Conditions*

The sound-hard boundary condition for a rigid wall is that the normal acceleration vanishes. For a moving wall, such as the one in the interior of the hollow cylinder, the condition is

$$\mathbf{n}.\nabla p = -\rho_f a_n, \tag{20}$$

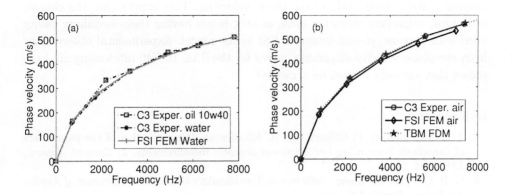

Fig. 2. Phase velocity dispersion curves, comparison of the theory (FSI FEM) against experiment, for (a) water and oil filled (b) air filled.

wherein $a_n$ is the normal acceleration of the wall. As the wall of the hollow cylinder undergoes harmonic motion, its acceleration is proportional to its displacement

$$\mathbf{n}.\nabla p = \rho_f \lambda \mathbf{u}. \tag{21}$$

The FSI equations are implemented using finite elements in the FEMLAB software package [16]. The problem geometry is shown on Fig. (1).

## 3. The numerical simulation and experimental validation of FSI

Three hollow cylinders, $23cm$ long, with outer diameter $1.5cm$, were used in this study. They are denoted C1, C2 and C3. They have walls of thickness $5.5mm$, $8.75mm$ and $7mm$ respectively. To study the influence of the wall thickness, the results (theoretical and experimental) are presented in the form of phase velocity dispersion curves for C1 and C2 (Fig. (1)). The phase velocity $v_\phi = \frac{\omega}{k}$ ($k$ is the wavenumber) is the rate at which the phase of the wave propagates in space. Material dispersion is a frequency-dependent response of a material to waves. Waveguide dispersion (exists within a finite waveguide and depends upon the wavelength and size of the waveguide). FSI was studied using C3, filling it with different fluids (air, water, SAE 10W40 lubricating oil). Simulations of the FSI experiment were carried out for the two inviscid fluids, air and water. The phase velocities for the fluid-filled cylinder are lower than that for the air-filled one (Fig. 2). The TBM FDM compares well with the experiment for the air-filled case showing that air can be considered as a very light fluid so that its influence on the vibration modes is negligible.

## 4. Conclusion

It has been shown both theoretically (using FSI) and experimentally that fluid loading the column of the hollow cylinder results in a decrease in the resonance frequency and consequently of the phase velocities. This implies that the characterization, employing vibration data, of long bones having their medullary cavity filled with marrow, should integrate FSI in the model. Experimental observations from the phase velocity dispersion curves for the SAE 10W40 lubricating oil, have shown that viscosity cannot be neglected.

## References

1. J. L Buchanan, R. P. Gilbert, and K. Khashanan. Determination of the parameters of cancellous bone using low frequency acoustic measurements. *J. Comput. Acoust.*, 12:99–126, 2004.

2. G. R. Cowper. The shear coefficient in Timoshenko's beam theory. *Journal of Applied Mechanics*, 33:335–340, 1966.

3. ZEA Fellah, JY Chapelon, S. Berger, W. Lauriks, and C. Depollier. Ultrasonic wave propagation in human cancellous bone: Application of Biot theory. *J. Acoust. Soc. Am.*, 116:61–73, 2004.

4. Karl F. Graff. *Wave Motion in Elastic Solids*. Dover Publications, Inc., New York, 1991.

5. Seon M. Han, Haym Benaroya, and Timothy Wei. Dynamics of transversely vibrating beams using four engineering theories. *Journal of Sound and Vibration*, 225(5):935–988, 1999.

6. J. R Hutchinson. Shear coefficient for timoshenko beam theory. *J. Applied Mechanics*, 68:87–92, 2001.

7. C M Langton and C F Njeh. *The Physical Measurement of Bone*. IoP Publishing, Bristol, UK, 2003.

8. M.R Maheri and R.D. Adams. On the flexural vibration of timoshenko beams, and the applicability of the analysis to a sandwich configuration. *Journal of Sound and Vibration*, 209(3):419–442, 1998.

9. P. Meunier, J. Aron, C. Eduoard, and G. Vignon. Osteoporosis and the replacement of cell populations of the marrow by adipose tissue: A quantitative study of 84 iliac crest biopsies. *Clin Orthop*, 80:147–154, 1971.

10. Patrick H. F. Nicholson and Mary L. Bouxsein. Bone marrow influences quantitative ultrasound measurements in human cancellous bone. *Ultrasound in Medicine & Biology*, 28(3):369–375, 2002.

11. PG Robey and P. Bianco. Cellular mechanisms of bone loss. In Rosen CJ, Glowacki J, and Bilezikian JP, editors, *The aging skeleton.*, volume 10, pages 145–157, San Diego, CA, 1999. Academic Press.

12. CM Schnitzler and J. Mesquita. Bone marrow composition and bone microarchitecture and turnover in blacks and whites. *J Bone Min Res*, 13:1300–1307, 1998.

13. G. D. Smith. *Numerical Solution of Partial Differential Equations: Finite Difference Methods*. Oxford Univ. Press, third edition, 1985.

14. Jinyoung So and A. W. Leissa. Free vibrations of thick hollow circular cylinders from three-dimensional analysis. *ASME J. Vibration and Acoustics*, 119:89–95, 1997.

15. Stephen Timoshenko. On the correction for shear of the differential equation for transverse vibrations of prismatic bars. *Philisophical Magazine*, 41:744–746, 1921.

16. Femlab V3.1. *Structural Mechanics Module User's Guide*. Comsol AB, 2004.

# AN INVERSE PROBLEM FOR THE FREE BOUNDARY MODEL
# OF DUCTAL CARCINOMA IN SITU

Y.S. XU

*Department of Mathematics*
*University of Louisville*
*Louisville, KY 40292, USA*
*E-mail: ysxu0001@louisville.edu*

In an earlier paper, we developed a free boundary model to describe the growth of ductal carcinoma in situ (DCIS). Assuming that we know the coefficients of the model, we analyzed the growth tendency of DCIS. The analysis and computation of the problem show interesting results that are similar to the patterns found in DCIS. In this paper we consider an inverse problem of determining the coefficient function of the free boundary model of DCIS from information obtained by incisional biopsy. Assuming that from incisional biopsy we observe the tumor pattern and some information of its growing rate, which is corresponding to the solution of the free boundary problem and some information of its derivative at a time, we consider the inverse problem of finding the potential function of the free boundary problem. We show that mathematically we can determine the coefficient of the model from incisional biopsy information, hence we may develop mathematical methods to diagnose growth tendency of DCIS from incisional biopsy information.

**Key words:** Cancer model, reaction-diffusion equation, free boundary problem, inverse problem

**Mathematics Subject Classification:** 35K60

## 1. Introduction

Ductal carcinoma in situ (DCIS) refers to a specific diagnosis of cancer that is isolated within the breast duct, and has not spread to other parts of the breast. There are two categories of DCIS: non-comedo and comedo. The non-comedo type DCIS tends to be less aggressive than the comedo types of DCIS. The most common non-comedo types of DCIS are: (1) Solid DCIS: cancer cells completely fill the affected breast ducts. (2) Cribiform DCIS: cancer cells do not completely fill the affected breast ducts; there are gaps between the cells. (3) Papillary DCIS: the cancer cells arrange themselves in a fern-like pattern within the affected breast ducts.

In [18] [19] we modify a model proposed by Byrne and Chaplain [3], (also see Friedman and Reitich [11]), for the growth of a tumour consisting of live cells (nonnecrotic tumour) to describe the homogeneous growth inside a cylinder, a model mimicking the growth of a ductal carcinoma. The model is in the form of a free boundary problem. (Recently Franks, Byrne et al. [7], [8] [9] have developed models for DCIS at different stages. For publications on the growth of avascular tumours, see, for examples, Greenspan [12] [13], Byrne [2], Chaplain and Sleeman [4], Please et al. [14], Ward and King [16], [17], Franks and King [6], Preziosi [10].)

Use this model, we reproduced some commonly observed morphologies in DCIS. We found that for different choices of the parameters, the model has solutions that mimik some typical patterns of DCIS, including the three types of non-comedo DCIS: solid DCIS, Cribiform DCIS, and Papillary DCIS. The analysis also shows that there may be two other kinds of patterns that resemble the non-growing tumour. (See Figure 1.)

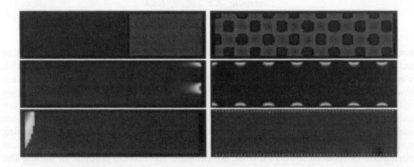

Fig. 1.   Computational results of DCIS patterns

These interesting results show that the free boundary problem model of DCIS reflects, to some extent, the growth process of DCIS, and hence it may be applied to study clinical data of DCIS. A possible process is that we obtain information of a particular DCIS from a kind of biopsy or imaging technique, then we use the clinical information to determine the coefficients (usually functions of time and space variables). After we know the coefficients of the model, we can use the model to predict the growth tendence of DCIS, which will provide a new method to diagnose growth tendency of DCIS from biopsy data. This problem of determining unknown coefficient functions of a partial differential equation is classified as an inverse problem of the free boundary problem. To the best knowledge of the author, there is little research done for the kind of inverse problems presented here. Therefore, it is also significant mathematically.

## 2. Free boundary problem model of DCIS

One typical type of DCIS is that cancer cells completely fill the affected breast duct (solid type). We describe the solid type of DCIS by an one dimensional model. We assume the tumour to be within the interval $[-s(t)/2, s(t)/2]$ at each time $t$; the growing boundary of the tumour is given by $x = -s(t)/2$ and $x = s(t)/2$, where

$x = s(t)$ is an unknown function. As tumour growth strongly depends upon the availability of nutrients its diffusion through the growing material is introduced in the description of model. We model the tumour growth using dimensionless nutrient concentration $\sigma(x, t)$ which satisfies a reaction-diffusion equation. In the one-dimensional case, the model is simplified to [18]:

$$c\frac{\partial \sigma}{\partial t} = \frac{\partial^2 \sigma}{\partial x^2} - \lambda(x)\sigma, \quad \text{in } B(t), t > 0. \tag{1}$$

$$\sigma(-s(t)/2, t) = \sigma_1, \quad t > 0, \tag{2}$$

$$\sigma(s(t)/2, t) = \sigma_1, \quad t > 0, \tag{3}$$

$$\sigma(x, 0) = \sigma_0(x), \quad \text{in } B(0); \tag{4}$$

$$\frac{\partial s}{\partial t} = \mu \int_{-s(t)/2}^{s(t)/2} (\sigma - \tilde{\sigma})dx, \quad s(0) = s_0, \tag{5}$$

where

$$B(t) = \{x| - s(t)/2 < x < s(t)/2\}, \quad B(0) = (-s_0/2, s_0/2).$$

Here $\lambda(x)\sigma(x, t)$ is the nutrient consumption rate at the location $x$ at time $t$. $c = T_{diffusion}/T_{growth}$ is the ratio of the nutrient diffusion time scale to the tumour growth time scale. Typically $c << 1$. (cf. [1] and [12].) $c, \mu, \sigma_1, s_0$ and $\tilde{\sigma}$ are known constants. $\lambda(x) \geq 0$ and $\sigma_0(x)$, the initial data, are given. The free boundary problem is to determine $(\sigma(x, t), s(t))$ for given $\lambda(x), \sigma_0(x), c, \mu, \sigma_1$, and $s_0$.

In general, $\lambda(x)$ is not a continuous function. To define the solution of (1)-(5) properly, we adopt the following notations for spaces involving time. Let $H^i(0, 1)$ and $H^0(0, 1) = L^2(0, 1)$ denote the standard Sobolev spaces with the standard norms $\| \cdot \|_i$, $H_0^i(0, 1)$ be the subspace of $H^i(0, 1)$ with zero boundary data. Let $X$ be a Banach space, $L(0, T; X)$ the space of all weakly (Lebesgue) measurable functions from [0,T] into $X$, and $C(0, T; X)$ the space of all continuous functions from [0,T] into $X$. Define

$$L(0, T; X) = \{v : [0, T] \to X \mid \|v(.)\|_X \text{ is measurable for all } v \in X\},$$

$$L^p(0, T; X) = \{v : v \in L(0, T; X), |v|_p \equiv \left(\int_0^T \|v(t)\|_X^p dt\right)^{1/p} < \infty\}, \quad 1 \leq p < \infty,$$

$$W^{i,p}(0, T; X) = \{v : v \in L^p(0, T; X), \frac{dv}{dt}, ..., \frac{d^i v}{dt^i} \in L^p(0, T; X)\}, \quad 1 \leq p < \infty,$$

$$H^i(0, T; X) = W^{i,2}(0, T; X)$$

$$L^\infty(0, T; X) = \{v : v \in L(0, T; X), |v|_\infty := ess \sup_{0 \leq t \leq T} \|v(t)\|_X < \infty\}.$$

The initial and boundary value problem of parabolic equation has been well studied. The following result is a special case of that in [5] Ch.7, replacing $a(x,t)$ by $\nu(t)/c$.

Let

$$Au = -\frac{\nu(t)}{c}u_{xx}, \quad F(u) = b(x,t)u_x + \frac{\lambda_1(x,t)}{c}u. \tag{6}$$

where $c > 0$ is a constant, $\lambda_1 \in L^\infty([0,1] \times [0,T])$.

Under the assumption $\nu(t) \geq \delta_1 > 0$ for $0 \leq t \leq T$, the partial differential operator $\frac{\partial}{\partial t} + A$ is uniformly parabolic for all $(x,t) \in [0,1] \times [0,T]$.

Consider the initial and boundary value problem of parabolic equation

$$u_t + Au + F(u) = f \text{ in } [0,1] \times [0,T], \tag{7}$$

$$u = 0 \text{ on } \{0,1\} \times [0,T] \tag{8}$$

$$u = g \text{ on } [0,1] \times \{t = 0\}, \tag{9}$$

where $f : [0,1] \times [0,T] \to \mathbf{R}$, $g : [0,1] \to \mathbf{R}$ are given such that $f \in L^2([0,1] \times [0,T])$, $g \in L^2(0,1)$. $u : [0,1] \times [0,T] \to \mathbf{R}$ is the unknown function $u = u(x,t)$.

Denote the time-dependent bilinear form

$$B[u,v;t] := \int_0^1 \left[ \frac{\nu(t)}{c}u_x v_x + b(\cdot,t)u_x v + \lambda_1(\cdot,t)uv \right] dx \tag{10}$$

for $u,v \in H_0^1(0,1)$ and a.e. $0 \leq t \leq T$.

We say a function $u \in L^2(0,T;H_0^1(0,1))$ with $u' \in L^2(0,T;H^{-1}(0,1))$ is a weak solution of the parabolic initial boundary value problem (7)-(9) if $(u',v) + B(u,v;t) = (f,v)$ for each $v \in H_0^1(0,1)$ and a.e. $0 < t < T$, and $u(0) = g$.

Using the arguments in [5] Ch.7, we obtain the following lemma.

**Lemma 2.1:** (1) Assume that $b, \lambda_1 \in L^\infty([0,1] \times [0,T])$, $f \in L^2([0,1] \times [0,T])$ and $g \in L^2(0,1)$. There exists a unique weak solution of (7)-(9).

(2) If $f \in L^2(0,T;L^2(0,1))$ and $g \in H_0^1(0,1)$. Suppose also $u \in L^2(0,T;H_0^1(0,1))$ with $u' \in L^2(0,T;H^{-1}(0,1))$ is a weak solution of the parabolic initial boundary value problem (7)-(9). Then $u \in L^2(0,T;H^2(0,1)) \cap L^\infty(0,T;H_0^1(0,1))$, $u' \in L^2(0,T;L^2(0,1))$.

By Sobolev imbedding theorem, if $u \in L^2(0,T;H^2(0,1))$, with $u' \in L^2(0,T;L^2(0,1))$, then $u \in C(0,T;H^1(0,1))$.

Now we come back to the free boundary problem. In addition we assume that $\lambda \in L^\infty(0,\infty)$ such that $\lambda \geq 0$. In view of the maximum principle and condition (5), we have

**Lemma 2.2** Assume that $\sigma_0(x)$ is continuous, and $\bar\sigma = \max_{0<x<s(0)}\{\sigma_0(x),\sigma_1\}$. If $\sigma(x,t)$ satisfies (1)-(5), then

$$0 < \sigma(x,t) < \bar\sigma, \text{ if } -\frac{s(t)}{2} < x < \frac{s(t)}{2}, \, t > 0, \tag{11}$$

$$-\mu\tilde{\sigma}s(t) \leq s'(t) \leq \mu(\bar{\sigma} - \tilde{\sigma})s(t), \text{ if } t > 0. \tag{12}$$

$$0 < s_0 e^{-\mu\tilde{\sigma}t} \leq s(t) \leq s_0 e^{\mu(\bar{\sigma}-\tilde{\sigma})t}, \text{ if } 0 < t < T. \tag{13}$$

We change variable $x \in [-s(t)/2, s(t)/2]$ to $\xi \in [0,1]$ by setting $x = s(t)\xi - s(t)/2$. Let

$$v(x,t) = \sigma\left(s(t)x - \frac{s(t)}{2}, t\right) - \sigma_1, \ 0 \leq x \leq 1, \ t \geq 0.$$

Then

$$\sigma(x,t) = v\left(\frac{2x - s(t)}{2s(t)}, t\right) + \sigma_1, \ \sigma_x = \frac{1}{s(t)}v_x, \ \sigma_t = -v_x\frac{xs'(t)}{s^2(t)} + v_t,$$

$$\sigma_{xx} = \frac{1}{s^2(t)}v_{xx}.$$

From (5) we have

$$s'(t) = \mu \int_{-\frac{s(t)}{2}}^{\frac{s(t)}{2}} [\sigma(x,t) - \tilde{\sigma}]\,dx = \mu \int_0^1 \left[\sigma\left(s(t)\xi - \frac{s(t)}{2}, t\right) - \tilde{\sigma}\right] s(t)d\xi,$$

$$s(t) = s_0 e^{\mu \int_0^t \int_0^1 [v(\xi,\tau) + \sigma_1 - \tilde{\sigma}]d\xi d\tau}, \tag{14}$$

where $s_0 = s(0)$ is known.

The free boundary problem is equivalent to the initial-boundary problem of nonlinear parabolic equation

$$\frac{\partial v}{\partial t} - \frac{1}{s^2(t)}\frac{\partial^2}{\partial x^2}v - \frac{xs'(t)}{s^2(t)}\frac{\partial v}{\partial x} + \lambda(x)v = -\lambda(x)\sigma_1, \ 0 < x < 1, t > 0. \tag{15}$$

$$v(0,t) = 0, \tag{16}$$

$$v(1,t) = 0, \tag{17}$$

$$v(x,0) = \sigma_0(x) - \sigma_1, \ 0 < x < 1, \tag{18}$$

where

$$s(t) = s_0 e^{\mu \int_0^t \int_0^1 [v(\xi,\tau) - \sigma_1 - \tilde{\sigma}]d\xi d\tau}. \tag{19}$$

An iteration algorithm has been developed in [20], in a way inspired by [15], for free boundary problem of diffusion equation with integral condition. The method can be applied to the problem discussed here. We consider the iterative scheme as follows:

$$\frac{\partial v^{n+1}}{\partial t} - \frac{1}{s_n^2(t)}\frac{\partial^2 v^{n+1}}{\partial x^2} - \frac{xs_n'(t)}{s_n^2(t)}\frac{\partial v^{n+1}}{\partial x} + \lambda v^{n+1} = -\lambda(x)\sigma_1, \tag{20}$$

$$0 < x < 1, 0 < t < T.$$

1434

$$v^{n+1}(0,t) = 0, \ 0 < t < T \tag{21}$$

$$v^{n+1}(1,t) = 0, \ 0 < t < T \tag{22}$$

$$v^{n+1}(x,0) = \sigma_0(x) - \sigma_1, \ 0 < x < 1, \tag{23}$$

$$s_{n+1}(t) = s(0)e^{\mu \int_0^t \int_0^1 [v^{n+1}(\xi,\tau)+\sigma_1-\tilde{\sigma}_1]d\xi d\tau}, \ s_0 = s(0), \ 0 < t < T \tag{24}$$

for $n = 0, 1, 2, \dots$. The apriori estimates of $v$ and $s$ imply that $a = 1/s^2 \geq \delta > 0$ for some constant $\delta$ such that $a$, $b = -\frac{xs_n'(t)}{s_n^2(t)}$, $c = \lambda$ and $f = -\lambda(x)\sigma_1$ are in $L^\infty[0,1]$, and $\sigma_0(x) - \sigma_1$ are in $H_0^1(0,1)$. Hence, the initial-boundary value problem (20)-(24) has a unique solution for each $n$ by Lemma 2.1. Moreover, $\{v_n\}$, $\{s_n\}$ are bounded in the apropriate spaces. Using Sobolev embedding theorem we can conclude they have unique limit $\{v,s\}$ which is the solution of the free boundary problem. For details please see [20].

## 3. Determine potential function from terminal data

There are a number of inverse problems motivated by cancer diagnose. Here we consider the case that clinical data is obtained by one incisional biopsy. That is, when noticing possible breast cancer, one opts to do an incisional biopsy to find out the DCIS pattern along with the changing rate at the moment. Since no information is available before, we can assume no initial data given, instead we assume that two conditions are given at terminal time $t = T$. The problem is as follows: find $\lambda(x)$ such that

$$c\frac{\partial\sigma}{\partial t} = \frac{\partial^2\sigma}{\partial x^2} - \lambda(x)\sigma, \ -\frac{s(t)}{2} < x < \frac{s(t)}{2}, \ 0 < t < T, \tag{25}$$

$$\sigma(-\frac{s(t)}{2},t) = \sigma_1, \ 0 < t < T, \tag{26}$$

$$\sigma(\frac{s(t)}{2},t) = \sigma_1, \ 0 < t < T, \tag{27}$$

$$\sigma(x,T) = \sigma_T(x), \ -\frac{s(T)}{2} < x < \frac{s(T)}{2}, \tag{28}$$

$$\sigma_t(x,T) = \eta_T(x), \ -\frac{s(T)}{2} < x < \frac{s(T)}{2}, \tag{29}$$

$$\frac{\partial s}{\partial t} = \mu \int_{-\frac{s(t)}{2}}^{\frac{s(t)}{2}} (\sigma(x,t) - \tilde{\sigma})dx, \ 0 < t < T. \tag{30}$$

where $c$ and $\sigma_1$ are known constants, $\sigma_T, \eta_T \in L^\infty\left(-\frac{s(T)}{2}, \frac{s(T)}{2}\right)$ are given, and $\sigma_T$ has at most countable many zeros. $s(t)$ and $\sigma(x,t)$ are the unknown free boundary and unknown function. We call the problem (25)-(30) the IP 1(a).

By change of variables, (25)-(30) is equivalent to finding $\lambda(x)$ such that

$$\frac{\partial v}{\partial t} - \frac{1}{s^2(t)}\frac{\partial^2}{\partial x^2}v - \frac{xs'(t)}{s^2(t)}\frac{\partial v}{\partial x} + \lambda(x)v = 0, \ 0 < x < 1, \ 0 < t < T, \qquad (31)$$

$$v(0,t) = v_1, \ 0 < t < T, \qquad (32)$$

$$v(1,t) = v_1, \ 0 < t < T, \qquad (33)$$

$$v(x,T) = v_T(x), \ 0 < x < 1, \qquad (34)$$

$$v_t(x,T) = \eta_T(x), \ 0 < x < 1, \qquad (35)$$

$$s(t) = s_T e^{-\mu \int_t^T \int_0^1 [v(\xi,\tau)+\sigma_1-\bar\sigma]d\xi d\tau}. \qquad (36)$$

**Theorem 3.1:** The inverse problem (31)-(36) has at most one solution.
**Proof:** Let $\phi(x,t)$ satisfy

$$-c\frac{\partial \phi}{\partial t} - \frac{1}{s^2(t)}\frac{\partial^2}{\partial x^2}\phi + \frac{xs'(t)}{s^2(t)}\frac{\partial \phi}{\partial x} = 0, \ 0 < x < 1, \ 0 < t < T, \qquad (37)$$

$$\phi(0,t) = 0, \ 0 < t < T, \qquad (38)$$

$$\phi(1,t) = 0, \ 0 < t < T, \qquad (39)$$

Let $(v_1, \lambda_1, s_1)$ and $(v_2, \lambda_2, s_2)$ be two solutions of the inverse problem (32)-(37). Let $v = v_1 - v_2$. Then

$$-cv_t - \frac{1}{s_1^2}v_{xx} - \frac{xs_1'}{s_1^2}v_x + \lambda_1 v_1 - \lambda_2 v_2 + \left(\frac{1}{s_2^2} - \frac{1}{s_2^2}\right)v_{2,xx} - x\left(\frac{s_2'}{s_2^2} - \frac{s_1'}{s_1^2}\right)v_{2,x} = 0, \quad (40)$$

$$v(0,t) = 0, \ v(1,t) = 0, \qquad (41)$$

$$v(x,T) = 0, \ v_t(x,T) = 0. \qquad (42)$$

Integrating the difference of $\phi$ multiplying (40) and $v$ multiplying (37), we have

$$0 = \int_0^1 \left[c\left(v_t\phi + v\phi_t\right) - \frac{1}{s_1^2}\left(v_{xx}\phi - v\phi_{xx}\right) - \frac{cs_1'}{s_1^2}\left(v_x\phi + v\phi_x\right) + \left(\lambda_1 v_1 - \lambda_2 v_2\right)\phi\right] dx$$

$$+ \int_0^1 \left[\left(\frac{1}{s_2^2} - \frac{1}{s_1^2}\right)v_{2,xx}\phi + x\left(\frac{s_2'}{s_2^2} - \frac{s_1'}{s_1^2}\right)v_{2,x}\phi\right] dx$$

$$= \int_0^1 \left[c\left(v_t\phi + v\phi_t\right) - \frac{cs_1'}{s_1^2}\left(v\phi\right) + \left(\lambda_1 v_1 - \lambda_2 v_2\right)\phi\right] dx$$

$$+ \int_0^1 \left[ \left( \frac{1}{s_2^2} - \frac{1}{s_1^2} \right) v_{2,xx}\phi + x \left( \frac{s_2'}{s_2^2} - \frac{s_1'}{s_1^2} \right) v_{2,x}\phi \right] dx.$$

When $t = T$, $v_1 = v_2$ and $v_{1,t} = v_{2,t}$. Hence $s_1(T) = s_2(T)$, and

$$\frac{1}{s_2^2} - \frac{1}{s_1^2} = \frac{s_1^2 - s_2^2}{s_1^2 s_2^2} = \frac{s_1 + s_2}{s_1^2 s_2^2}(s_1 - s_2) = 0,$$

$$\frac{s_2'}{s_2^2} - \frac{s_1'}{s_1^2} = S_T \left( \frac{1}{s_2} - \frac{1}{s_1} \right) = \frac{S_T}{s_2 s_1}(s_1 - s_2) = 0,$$

where

$$S_T := \frac{s_2'}{s_2} = \mu \int_0^1 (v - \tilde{\sigma})d\xi = \frac{s_1'}{s_1}.$$

Therefore,

$$\int_0^1 (\lambda_1(x) - \lambda_2(x))v_1(x,T)\phi(x,T)dx = 0, \tag{43}$$

for all $\phi$ satisfying (37)-(39).

Now we show that the set of all solutions of (37)-(39) $\{\phi(x)\}$ is dense in $L^2[0,1]$. For any $\phi_T \in C_0[0,1]$, we consider the terminal-boundary value problem (37)-(39), and

$$\phi(x,T) = \phi_T(x), \ 0 \le x \le 1. \tag{44}$$

Let $\tau = T - t$. The problem (37)-(39) and (44) is equivalent to the initial-boundary value problem

$$c\frac{\partial \phi}{\partial \tau} - \frac{1}{s^2(T-\tau)}\frac{\partial^2}{\partial x^2}\phi + \frac{xs'(T-\tau)}{s^2(T-\tau)}\frac{\partial \phi}{\partial x} = 0, \ 0 < x < 1, \ 0 < \tau < T, \tag{45}$$

$$\phi(0, T-\tau) = 0, \ 0 < \tau < T, \tag{46}$$

$$\phi(1, T-\tau) = 0, \ 0 < \tau < T, \tag{47}$$

$$\phi(x,0) = \phi_T(x), \ 0 \le x \le 1, \tag{48}$$

which has a unique solution by Lemma 2.1.

Since $C_0[0.1]$ is dense in $L^2[0,1]$, the solution set $\{\phi(x)\}$ is dense in $L^2[0,1]$. Moreover, $v_1 \ge 0$ and $v_1(x,T)$ has at most countable many zeros. therefore we obtain $\lambda_1 = \lambda_2$ a.e. in $[0,1]$. $\square$

Now we present an algorithm for construction of $\lambda(x)$.
Let $\psi(x)$ be the solution of the eiganvalue problem

$$\psi''(x) + \zeta^2\psi(x) = 0, \ -\frac{s(T)}{2} < x < \frac{s(T)}{2}, \tag{49}$$

$$\psi(-\frac{s(T)}{2}) = 0, \quad \psi(\frac{s(T)}{2}) = 0, \tag{50}$$

which has eigenvalues and eigenfunctions

$$\zeta_n = (2n+1)\frac{\pi}{s(T)}, \quad \psi_n(x) = \cos\left(\frac{(2n+1)\pi}{s(T)}x\right), \quad n = 0, 1, 2, \ldots$$

From

$$\int_{-s(T)/2}^{s(T)/2} \left[c\frac{\partial\sigma}{\partial t} - \frac{\partial\sigma}{\partial x^2} + \lambda(x)\sigma\right]\psi_n dx = 0,$$

when $t = T$ we have

$$\int_{-s(T)/2}^{s(T)/2} \left[c\eta_T(x) + \zeta_n^2\sigma_T(x) + \lambda(x)\sigma_T(x)\right]\psi_n(x)dx$$

$$+ \left[\psi_n'(s(T)/2) - \psi_n'(s(T)/2)\right]\sigma_1 = 0, \quad \text{for } n = 0, 1, 2, \ldots \tag{51}$$

Hence,

$$\int_{-s(T)/2}^{s(T)/2} \lambda(x)\sigma_T(x)\psi_n(x)dx = \alpha_n, \tag{52}$$

where

$$\alpha_n := \int_{-s(T)/2}^{s(T)/2} \left[c\eta_T(x) + \zeta_n^2\sigma_T(x)\right]\psi_n(x)dx + \left[\psi_n'(s(T)/2) - \psi_n'(s(T)/2)\right]\sigma_1$$

is known. We recover $\lambda(x)$ from

$$\lambda(x) = \frac{1}{\sigma_T(x)} \sum_{n=0}^{\infty} \alpha_n\psi_n(x), \quad \text{a.e.in } [-\frac{s(T)}{2}, \frac{s(T)}{2}].$$

## 4. Discussion

We presented mathematical analysis for an inverse problem of using a free boundary problem model to diagnose the growth tendency of DCIS. The problem is formulated as an inverse problem of free boundary problem of partial differential equation. This problem has not been studied before. A better understanding of this problem may enable scientists to develop mathematical methods to support the diagnosing process of DCIS.

Our analysis shows that if we have information of the tumor pattern and some information related to its growing rate at a time, which may be obtained through an incisional biopsy, then we can use a regularized minimization method to determine the coefficient function of the free boundary problem model of DCIS, hence to estimate the growth tendency of the tumor.

Since this is a rather simple model that neglects to account for some important factors such as tumor cell density, etc., further study is neccessary. Nevertheless, this result shows a way to use mathematical model in the diagnosing process of the growth tendency of DCIS. Therefore, it is worth to further study.

## References

1. Adam, J. A., A simplified mathematical model of tumour growth, *Math. Biosciences*, 81, 229-244, 1986.
2. Byrne, H.M., The effect of time delays on the dynamics of avascular tumour growth, *Math. Biosciences*, 144, 83-117, 1997.
3. Byrne, H.M. and M.A.J. Chaplain, Growth of nonnecrotic tumours in the presence and absence of inhibitors, *Math. Biosciences*, 130, 151-181, 1995.
4. Chaplain, M.A.J. and Sleeman, B.D., Modelling the growth of solid tumours and incorporating a method for their classification using nonlinear elasticity theory, *J. Math. Biol.* 31, 431-473. 1992.
5. Evans, L., *Partial differential equations*, American Mathematical Society, 1998.
6. Franks, S.J., King, J.K., Interaction between a uniformly proliferating tumour and its surroundings: uniform material properties, *Math. Med. Biol.* 20, 47-89, 2003.
7. Franks, S. J., H. M. Byrne, et. al, Mathematical modelling of comedo ductal carcinoma in situ of the breast, *Mathematical Medicine and Biology*, 20, 277-308, (2003).
8. Franks, S. J., H. M. Byrne, et. al, Modelling the early growth of ductal carcinoma in situ of the breast, *J. Math. Biol.* 47, 424-452, (2003).
9. Franks, S. J., H. M. Byrne, et. al, Biological inferences from a mathematical model of comedo ductal carcinoma in situ of the breast, *Journal of Theoretical Biology*, to appear. (2004)
10. Preziosi, L. (Ed.), *Cancer Modelling and Simulation*, Taylor and Francis Ltd, Chapman, Hall/CRC, London, Boca Raton, FL, 2003.
11. Friedman, A. and F. Reitich, Analysis of a mathematical model for the growth of tumours, *J. Math. Biol.* 38, 262-284, 1999.
12. Greenspan, H. P., Models for the growth of a solid tumour by diffusion, *Studies in Appl. Math.*, 52, 317-340, 1972.
13. Greenspan, H. P., On the growth and stability of cell cultures and solid tumours, *J. Theor. Biol.*, 56, 229-242, 1976.
14. Please, C.P., Pettet, G.J., McElwain, D.L.S., Avascular tumour dynamics and necrosis, *Math. Mod. Methods Appl. Sci.* 9, 569-579, 1999.
15. Taylor, M, *Partial Differential Equations*, III, Springer,
16. Ward, J.P., King, J.R., Mathematical modelling of avasculartumour growth. *IMA J. Math. Appl. Med. Biol.* 14, 39-69, 1997.
17. Ward, J.P., King, J.R., Mathematical modelling of avasculartumour growth II: Modelling growth saturation. *IMA J. Math. Appl. Med. Biol.* 16, 171-211, 1999.
18. Xu, Y., A free boundary problem model of ductal carcinoma in situ, *Discrete and Continous dynamical systems - Series B*, Vol. 4, No. 1 (2004), 337-348.
19. Xu, Y., A Mathematical Model of Ductal Carcinoma in Situ and its Characteristic Stationary Solutions, *Advances in Analysis*, edited by H. Begehr, etc, World Scientific, 2005.
20. Xu, Y., A free boundary problem of diffusion equation with integral condition, to appear in *Applicable Analysis*, (2004).

# COMPARING MATHEMATICAL MODELS OF THE HUMAN LIVER BASED ON BSP TEST

L. ČELECHOVSKÁ - KOZÁKOVÁ*

*Mathematical Institute of the Silesian University in Opava,*
*Na Rybníčku 1,*
*746 01 Opava, Czech Republic*
*E-mail: Lenka.Kozakova@math.slu.cz*

The BSP dynamical test is employed for quantitative assesing of the liver function. The process of extraction of BSP is described via ODE's systems. Parameters characterizing these systems are determined by some modification of Bellman's quasilinearization method.

**Key words:** Model of human liver, bromsalphtalein dynmical test, system of ODE
**Mathematical Subjects Classification:** 93A30

## 1. The physiological problem and mathematical models

The BSP (Bromsulphtalein) dynamical test is employed for quantitative assesing of the liver function. BSP is a hepatotrophic matter, which is injected into the blood. The liver is the only organ in the body which takes BSP and secrets it directly into the bile. We can represent this process by a three compartment model:

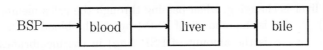

The extraction of BSP can be described by systems of ordinary differential equations.

A simple model of the process describing the extraction of BSP in these individual compartments (the blood, the liver and the bile) can be given by a system of linear ordinary differential equations

$$x'(t) = -a_1 x + a_2 y,$$
$$y'(t) = a_1 x - (a_2 + a_3)y,$$
$$z'(t) = a_3 y, \tag{1}$$

---

*Work partially supported by contracts 201/03/1153 from the Grant Agency of the Czech Republic and MSM4781305904 from the Czech Ministry of Education.

where
$x(t)$         is the amount of BSP $(mg)$ in the blood at the time $t$,
$y(t)$         is the amount of BSP $(mg)$ in the liver at the time $t$,
$z(t)$         is the amount of BSP $(mg)$ in the bile at the time $t$,
$a_1, a_2, a_3$     are the transfer rate constants $(min^{-1})$.

The system (1) we will denote SLM, i.e. simple linear model, which was done in
[12].

Suppose that some quantity $I > 0$ $(mg)$ of BSP is injected into the blood at once. This leads to the initial condition

$$x(0) = I, \; y(0) = z(0) = 0. \tag{2}$$

The hepatotrophic matter is cumulated in the liver. This organ is able to take in only a limited amount of BSP, i.e. the liver has some capacity $K > 0$. In this case the process of extraction can be described by the simple nonlinear system SNM of ordinary differential equations

$$x'(t) = -b_1 x(K - y),$$
$$y'(t) = b_1 x(K - y) - b_2 y,$$
$$z'(t) = b_2 y, \tag{3}$$

with the initial condition (2).

BSP is "working" inside the hepatic cells. Suppose that the rate of transfer from blood to the liver is changing, when passing through the cell's membrane. Denote
$X(t)$         is the amount of BSP $(mg)$ in the blood at the time $t$,
$Y(t)$         is the amount of BSP $(mg)$ in the membranes of hepatic cells at the time $t$,
$Z(t)$         is the amount of BSP $(mg)$ inside the cells at the time $t$,
$V(t)$         is the amount of BSP $(mg)$ in the bile at the time $t$,
$d_1, d_2, d_3, d_4$     are the transfer rate constants $(min^{-1})$.

Now we represent the extraction of BSP by the four compartment model:

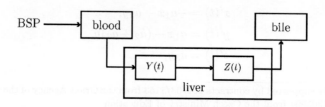

This situation can be described by the system of linear diferential equations LM

$$X'(t) = -d_1 X + d_2 Y,$$
$$Y'(t) = d_1 X - (d_2 + d_3)Y,$$
$$Z'(t) = d_3 Y - d_4 Z,$$
$$V'(t) = d_4 Z \tag{4}$$

with initial condition

$$X(0) = I, \ Y(0) = Z(0) = V(0) = 0. \tag{5}$$

If we consider the capacity of the liver, then we can describe the extraction from the respective components of the model by the system of nonlinear diferential equations NM

$$X'(t) = -c_1 X(K_1 - Y),$$
$$Y'(t) = c_1 X(K_1 - Y) - c_2 Y(K_2 - Z),$$
$$Z'(t) = c_2 Y(K_2 - Z) - c_3 Z,$$
$$V'(t) = c_3 Z, \tag{6}$$

where $K_1$ is the capacity of the cell's membranes, $K_2$ denotes the capacity of the interior of the cells, and $c_1, c_2, c_3$ are the rates transfer constants. The system NM was proposed in [9].

All the parameters characterizing the systems are unknown.

## 2. Clinical data

In order to determine the unknown parameters (the capacities and the rate transfer constants) we employ the measured data. In the first table there are presented measurements of the decay of BSP from the blood:

| Time | $t_i$ [min] | 0 | 5 | 11 | 16 | 20 | 30 | 45 |
|------|------------|-----|-----|-----|----|----|----|------|
| BSP | $r_i$ [mg] | 280 | 221 | 152 | 94 | 47 | 25 | 13.5 |

The values of the amount of BSP in the bile are contained in the second table:

| Time | $s_j$ [min] | 0 | 15 | 30 | 45 | 60 | 75 | 90 |
|------|------------|------|------|------|------|------|------|----|
| BSP | $e_j$ [mg] | 0 | 2.8 | 12.8 | 25.6 | 38.9 | 51.8 | 64 |
| Time | $s_j$ [min] | 105 | 120 | 135 | 150 | | | |
| BSP | $e_j$ [mg] | 75.3 | 85.7 | 95.1 | 103.7 | | | |

Note that the second set of measurements we get only in the case that a drain was inserted into the channel connecting the liver with the gall-bladder.

For numerical computations we use cubic splines interpolating these data.

## 3. Modified quasilinearization method for the inverse problem

We use a modification of the quasilinearization method for determining the unknown parameters characterizing the mathematical models describing the BSP kinetics in the human liver. The algorithm of the modified quasilinearization method consist of the steps displayed below.

ALGORITHM

Step 1.      Choose the initial approximation $\alpha^{(1)}$ and put $k = 1$.

Step 2.      Compute the solution $\mathbf{x}^{(k)}(t)$ of the system

$$\dot{\mathbf{x}}(t) = \mathbf{g}(\mathbf{x}),$$

with the initial condition

$$\mathbf{x}(0) = (c_1, \ldots, c_n, \alpha_1^{(k)}, \ldots, \alpha_N^{(k)}).$$

Step 3.      Evaluate the solution $\mathbf{y}^{(k+1)}(t)$ of the linearized equation in the form

$$\mathbf{y}^{(k+1)}(t) = \mathbf{p}^{(k+1)}(t) + \sum_{j=1}^{N} \beta_j \mathbf{h}^{(j,k+1)}(t).$$

Step 4.      Determine the minimum $\beta^*$ of the function $S$ (penalty function)

$$S(\mathbf{y}^{(k+1)}) = S_{k+1}(\beta).$$

Step 5.      Choose $\zeta_k > 0$, i.e. maximum distance between the parameter $\alpha^{(k+1)}$ and $\alpha^{(k)}$.

Step 6.      (a) If the inequality $\|\alpha^{(k+1)} - \alpha^{(k)}\| \leq \zeta_k$ is fulfilled for $\alpha^{(k+1)} = \beta^*$, go to Step 2.

               (b) If the inequality $\|\alpha^{(k+1)} - \alpha^{(k)}\| \leq \zeta_k$ is not true for $\alpha^{(k+1)} = \beta^*$, then suitably change the value $\alpha^{(k+1)}$. The possibility of this change is based on the proposition in Lemma 3.1.

Step 7.      Repeat Steps 2, 3, 4, 5, 6(a), respectively 6(b) until the condition

$$0 \leq S(\mathbf{x}^{(k)}) - S(\mathbf{x}^{(k+1)}) < \varepsilon$$

is satisfied for a given $\varepsilon > 0$.

Step 8.      If $S(\mathbf{x}^{(k+1)}) > S(\mathbf{x}^{(k)})$ then go back to Step 1 and start the algorithm with a better choice $\alpha^{(1)}$.

Now we describe the method in more detail.

Let $Q \subset R^n$ be closed convex set of the variables $x = (x_1, \ldots, x_n)^\top$ and $D \subset R^N$ be closed convex set of the parameters $\alpha = (\alpha_1, \ldots, \alpha_N)^\top$. Let $f: Q \times D \to R^n$ have continuous bounded partial derivatives up to the second order. Consider a nonlinear autonomous system of ordinary differential equations with the initial condition

$$\dot{x} = f(x, \alpha), \quad x(0) = c. \tag{7}$$

The aim is to find the unknown parameters $\alpha$ such that the solution of the initial problem (7) fits in some sense to the measured data, respectively to the continuous function which approximates these data. In order to avoid considering two different types of vectors we will suppose that the vector $\alpha$ satisfies the differential equation

$$\dot{\alpha} = 0$$

with the initial condition

$$\alpha(0) = \beta.$$

Define a new vector $\mathbf{x}$ by

$$\mathbf{x} = (x, \alpha)^\top = (x_1, \ldots, x_n, \alpha_1, \ldots, \alpha_N)^\top \in R^{n+N},$$

and a vector $\mathbf{c}$ (corresponding to the initial condition) by

$$\mathbf{c} = (c, \beta)^\top = (c_1, \ldots, c_n, \beta_1, \ldots, \beta_N)^\top \in R^{n+N}.$$

The vector $\mathbf{x}(t)$ satisfies the nonlinear differential equation

$$\dot{\mathbf{x}} = \mathbf{g}(\mathbf{x}), \tag{8}$$

where $\mathbf{g}(\mathbf{x}) = (f(x, \alpha), \underbrace{0, \ldots, 0}_{N})^\top$, with the initial condition

$$\mathbf{x}(0) = \mathbf{c}. \tag{9}$$

Let $\mathbf{x}^{(k)}(t)$ ($k$–th approximation) be a solution to (8) on the interval $[0, T]$ with the initial condition (9) for $\beta_1 = \alpha_1^{(k)}, \ldots, \beta_N = \alpha_N^{(k)}$, i.e.

$$\mathbf{x}^{(k)}(0) = (c_1, \ldots, c_n, \alpha_1^{(k)}, \ldots, \alpha_N^{(k)})^\top$$

(the $k$–th approximation of the solution).

The deviation between this solution and the functions $r(t) = (r_1(t), \ldots, r_n(t))^\top$, $e(t)$ approximating the experimental (measured) values has the form

$$S(\mathbf{x}^{(k)}) = \sum_{l=1}^{n} (\int_0^T (x_l^{(k)}(t) - r_l(t))^2 \, dt +$$

$$+ \int_0^T ((\gamma + \sum_{l=1}^{n} \gamma_l x_l^{(k)}(t)) - e(t))^2 \, dt, \tag{10}$$

where $\gamma, \gamma_l$ are real constants.

We would like to find a new vector of parameters $\beta = \alpha^{(k+1)}$ so that

$$S(\mathbf{x}^{(k+1)}) < S(\mathbf{x}^{(k)}). \tag{11}$$

The dependence $\mathbf{x}^{(k)}(t)$ on the parameters $\beta$ ($\beta = \alpha^{(k)}$) is not clear, therefore we approximate $\mathbf{x}^{(k)}(t)$ by the solution $\mathbf{y}^{(k+1)}(t)$ of linearized system

$$\dot{\mathbf{y}}(t) = \mathbf{g}(\mathbf{x}^{(k)}(t)) + \mathbf{J}(\mathbf{x}^{(k)}(t))(\mathbf{y}(t) - \mathbf{x}^{(k)}(t)), \tag{12}$$

where $\mathbf{J}(\mathbf{x})$ is the Jacobian matrix of $\mathbf{g}(\mathbf{x})$ with elements

$$\mathbf{J}_{ij} = \frac{\partial g_i}{\partial x_j}$$

in the $i$-th row and $j$-th column, $i, j = 1, \ldots, n + N$.

The equation (12) represents a linear system of $n + N$ differential equations and its general solution $\mathbf{y}(t)$ with

$$y_l(0) = c_l, \quad l = 1, \ldots, n \tag{13}$$

can be expressed in the form

$$\mathbf{y}(t) = \mathbf{y}^{(k+1)}(t) = \mathbf{p}^{(k+1)}(t) + \sum_{j=1}^{N} \beta_j \mathbf{h}^{(j,k+1)}(t). \tag{14}$$

The function $\mathbf{p}^{(k+1)}(t)$ is the (particular) solution of the nonhomogeneous equation

$$\dot{\mathbf{p}}(t) = \mathbf{g}(\mathbf{x}^{(k)}(t)) + \mathbf{J}(\mathbf{x}^{(k)}(t))(\mathbf{p}(t) - \mathbf{x}^{(k)}(t)) \tag{15}$$

which fulfils the initial condition

$$\mathbf{p}(0) = (c_1, \ldots, c_n, 0, \ldots, 0)^{\mathsf{T}}.$$

The $(n + N)$-column vectors $\mathbf{h}^{(j,k+1)}(t)$, $j = 1, \ldots, N$, are solutions of the homogeneous system

$$\dot{\mathbf{h}}^{(j,k+1)}(t) = \mathbf{J}(\mathbf{x}^{(k)}(t))\mathbf{h}^{(j,k+1)}(t) \tag{16}$$

with

$$h_i^{(j,k+1)}(0) = \begin{cases} 0, & \text{for } i \neq j + n \\ 1, & \text{for } i = j + n, \ i = 1, \ldots, n + N. \end{cases} \tag{17}$$

The equality (14) immediately implies that the dependence of $\mathbf{y}^{(k+1)}(t)$ on the parameters $\beta_j$, $j = 1, \ldots, N$ is linear. The parameters $\beta_j$, $j = 1, \ldots, N$ are free and they can be used for minimizing the function

$$S(\mathbf{y}^{(k+1)}) = S_{k+1}(\beta) = S_{k+1}(\beta_1, \ldots, \beta_N) =$$

$$\sum_{l=1}^{n} \left( \int_0^T (y_l^{(k+1)}(t) - r_l(t))^2 \, dt \right) + \int_0^T \left( \gamma + \sum_{l=1}^{n} \gamma_l y_l^{(k+1)}(t) - e(t) \right)^2 dt.$$

It is easy to see that the functional $S_{k+1}(\beta)$ is a strictly convex function with a unique point of minimum $\beta^*$. Put

$$\alpha^{(k+1)} = \beta^* = (\beta_1^*, \ldots, \beta_N^*)^{\mathsf{T}}$$

if

$$\|\alpha^{(k+1)} - \alpha^{(k)}\| \leq \zeta_k, \tag{18}$$

for arbitrary small $\zeta_k > 0$, and $S(\mathbf{x}^{(k+1)}) < S(\mathbf{x}^{(k)})$, where $\mathbf{x}^{(k+1)}(t)$ is the solution of the equation (8) wtih the initial condition

$$\mathbf{x}^{(k+1)}(0) = (c, \alpha^{(k+1)})^\top.$$

Then we can repeat the whole process of evaluation until the condition

$$0 \leq S(\mathbf{x}^{(k)}) - S(\mathbf{x}^{(k+1)}) < \varepsilon, \tag{19}$$

where $\varepsilon$ is fixed positive real number, is satisfied.

If $S(\mathbf{x}^{(k+1)}) > S(\mathbf{x}^{(k)})$, we do not repeat the whole process of computation, but we have to start with a better choice of the initial approximation $\alpha_1$.

If the equality

$$S(\mathbf{x}^{(k+1)}) = S(\mathbf{x}^{(k)}),$$

holds, we get the required values of parameters $\alpha = \alpha^{(k)}$. Noting that the deviation is not altered we finish our computation.

If the inequality (11) is fulfiled, but

$$\|\alpha^{(k+1)} - \alpha^{(k)}\| \geq \zeta_k,$$

we have to modify the value of the parameter $\alpha^{(k+1)}$. The modification is based on the following lemma.

**Lemma 3.1.** *Let $\alpha^{(k)}$ be fixed for given $k$.*
*Then for arbitrary $\zeta_k > 0$ there is a parameter $\alpha^{(k+1)} \in M_k$, where*

$$M_k := \{\beta | \beta \in D, S_{k+1}(\beta) \leq S(\mathbf{x}^{(k)})\}$$

*is a convex set, such that*

$$\|\alpha^{(k+1)} - \alpha^{(k)}\| \leq \zeta_k.$$

We are able to choose such $\zeta_k$ that the sequence $\{\zeta_k\}_{k=1}^\infty$ is decreasing and $\lim_{k \to \infty} \zeta_k = 0$. In addition, we can construct this sequence in such a way that

$$\sum_{k=1}^\infty \zeta_k < \infty.$$

**Theorem 3.1.** *Let $\alpha^{(k)} \in D$, for every $k = 1, 2, \ldots$, where $D \subset R^N$ is closed convex subset. Let the sequence of the parameters $\{\alpha^{(k)}\}_{k=1}^\infty$ satisfy the inequality (18)*

$$\|\alpha^{(k+1)} - \alpha^{(k)}\| \leq \zeta_k,$$

*for every $k$. Let the sequence $\zeta_k$ be convergent and decreasing such that*

$$\sum_{k=1}^\infty \zeta_k = \zeta < \infty.$$

*Then $\{\alpha^{(k)}\}_{k=1}^\infty$ is a Cauchy sequence.*

## 4. Numerical results

In this paper we give the results for the stated data in section 2. We used in turn the models SLM, SNM, LM, NM.

(i) Simple linear model SLM. We start the calculation with

$$\alpha^{(1)} = (a_1^{(1)}, a_2^{(1)}, a_3^{(1)})^\top = (0.107, 0.004, 0.01)^\top, \text{ and } S(\mathbf{x}^{(1)}) = 859\,473.$$

After 6 steps the condition 19 is fulfilled $(S(\mathbf{x}^{(6)}) = S(\mathbf{x}^{(7)}))$, we recieve

$$\alpha^{(6)} = (0.108602, 0.0072054, 0.00340841)^\top, \text{ and } S(\mathbf{x}^{(6)}) = 2\,591, 36.$$

(ii) Linear model LM. We begin with

$$\alpha^{(1)} = (d_1^{(1)}, d_2^{(1)}, d_3^{(1)}, d_4^{(1)})^\top = (0.109825, 0.0079841, 0.1, 0.00343389)^\top.$$

For this parameters we have $S(\mathbf{x}^{(1)}) = 5\,468, 77$. Since $S(\mathbf{x}^{(7)}) = S(\mathbf{x}^{(8)})$ we get $S(\mathbf{x}^{(7)}) = 360, 92$ and

$$\alpha^{(7)} = (0.136284, 0.0585446, 0.112485, 0.00367095).$$

(iii) Simple nonlinear model SNM. The initial choice is $\varepsilon = 0.7$,

$$\alpha^{(1)} = (K^{(1)}, b_1^{(1)}, b_2^{(1)})^\top = (220, 0.0009, 0.005)^\top, \text{ and } S(\mathbf{x}^{(1)}) = 80\,677.$$

After 1594 steps we get $S(\mathbf{x}^{(1594)}) = 1\,517, 171$.

$$\alpha^{(1594)} = (K^{(1594)}, b_1^{(1594)}, b_2^{(1594)})^\top = (379.4, 0.000396646, 0.00329456)^\top.$$

(iv) Nonlinear model NM. Initial aproximation is

$$\alpha^{(1)} = (K_1^{(1)}, K_2^{(1)}, c_1^{(1)}, c_2^{(1)})^\top$$
$$= (20, 300, 0.0088, 0.008, 0.0034)^\top, \text{ and } S(\mathbf{x}^{(1)}) = 8\,439, 18.$$

After 3500 steps of enumeration we get $S(\mathbf{x}^{(3500)}) = 1\,183, 03$

$$\alpha^{(3500)} = (K_1^{(3500)}, K_2^{(3500)}, c_1^{(3500)}, c_2^{(3500)})^\top$$
$$= (133.557, 156.716, 0.00124054, 0.00250425, 0.00494137)^\top.$$

The condition (19) is fulfilled for $\varepsilon = 0.012$.

For illustration we give the graphs for simple nonlinear model. The first one corresponds to the initial aproximation $\alpha^{(1)}$, the second one to $\alpha^{(1594)}$.

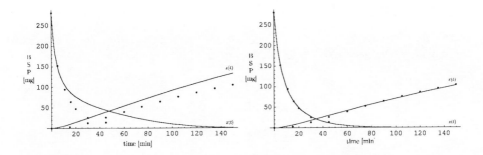

## 5. Conclusions

For the final choice of the best model we keep at disposal only the penalty function (10), respectively its discrete form. This function depends on the clinical data, namely on the initial condition, which is different for all patients.

In the most cases it is sufficient to calculate discrete form of the penalty function $S(\mathbf{x})$. In this case we need less steps of enumeration and the value of deviation is smaller too. The number of steps depends mainly on the initial choice $\alpha^{(1)}$.

The best model of liver tissue describing its biological structure is the nonlinear model NM. However, this model is sensitive to the initial parameters approximation.

## References

1. Abdullayev, U. G., *Numerical Method of Solving Inverse Problems for Non-linear Differential Equations*, Comput. Maths Math. Phys., Vol. 33, No. 8, pp. 1043-1057, (1993).
2. Abdullayev, U. G., *Quasilinearization and Inverse Problems of Nonlinear dynamics*, J. of Optimization Theory and Applications, Vol. 85, No. 3, pp. 509-526, (1995).
3. Bellman, R., Roth, R., *Quasilinearization and the Identification Problem*, World Scientific Publishing Co Pte Ltd, Singapore, (1983).
4. Čelechovská, L., *A simple mathematical model of the human liver*, Appl. Math.,Vol. 49, No.3, pp. 227-246, (2004).
5. Čelechovská, L., *Mathematical Models of BSP Kinetics in the Human Liver*, 3rd International Conference APLIMAT 2004, Bratislava.
6. Gantmakher, F.R., *Teorija Matric*, Gostekhteorizdat, Moscow, (1953).
7. Giannakis, G. B., Serpedin, E., *A bibliography on nonlinear system identification*, Signal Processing 81, pp. 533-580, (2001).
8. Hartman, P., *Ordinary Differential Equations*, Baltimore, (1973).
9. Hrnčíř, E. *Personal notes*, unpublish, (1985).
10. Kalaba, R., Spingarn, K., *Control, Identification, and Input Optimization*, Plenum Press, New York and London, (1982).
11. Sánchez, D., A., *Ordinary Differential Equations and Stability Theory*, W. H. Freeman and Company, San Francisco and London, (1968).
12. Watt, J.M., Young, A., *An attempt to simulate the liver on a computer*, Comput. J. 5, pp. 221-227, (1962).

## 5. Conclusions

For the final choice of the best model we keep at disposal only the penalty function (10), respectively its discrete form. This function depends on the clipped data, namely on the initial condition, which is different for all periods.

In the most case it is sufficient to calculate discrete form of the penalty function $S(x)$. In this case we need less steps of enumeration and the value of deviation is smaller too. The number of steps depends mainly on the initial choice of $c_i^0$.

The best model of liver tissue describing its biological structure is the nonlinear model NM. However, this model is sensitive to the initial parameter's approximation.

## References

1. Abdulla, U. G., Numerical Method of Solving Inverse Problems for Nonlinear Differential Equations, Comput. Maths. Math. Phys., Vol. 33, No. 8, pp. 1033-1087 (1993).
2. Abdulla, U. G., Own Iterations and Inverse Problems of Nonlinear dynamics, J. of Optimization Theory and Applications, Vol. 85, No. 3, pp. 509-526 (1995).
3. Bellman, R., Roth, R., Quasilinearization and the Identification Problem, World Scientific Publishing Co Pte Ltd, Singapore, (1983)
4. Celechovska, L., A simple mathematical model of the human liver, Appl. Math. Vol. 49, No.3, pp. 227-246, (2004).
5. Celechovska, L., Mathematical Models of ICG Kinetics in the Human Liver, 3rd International Conference APLIMAT 2004, Bratislava.
6. Ozmutska, P.R., Teorie Abrye Geolekij vzralat, Moscow, (1975).
7. Ozmutska, G. R., Serpedin, E., A bibliography on nonlinear system identification, Signal Processing 81, pp. 533-580, (2001).
8. Hartman, P., Ordinary Differential Equations, Baltimore, (1973).
9. Herout, P., Personal notes, unpublished, (1985).
10. Kubina, R., Stipsteru, L., Control, Identifications and Inverse Optimization, Plenum Press, New York and London, (1983).
11. Sanchez, D. A., Ordinary Differential Equations and Stability Theory, W.H. Freeman and Company, San Francisco and London, (1968).
12. Wah, J.M., Young, A., An attempt to simulate the liver on a computer, Comput. J. 8, pp. 221-225, (1982).

# IMPLEMENTATION OF ADAPTIVE RANDOMIZATIONS FOR CLINICAL TRIALS

E.R. MILLER

*Interactive Clinical Technologies Incorporated (ICTI)*
*1040 Stony Hill Rd., Suite 200, Yardley, PA 19067*
*E-mail: Eva.Miller@ICTI-Global.com*

Adaptive randomization is used to support the statistical methodology of a clinical trial by maintaining balance among treatments for the analysis population and subgroups (or strata). With this technology individual subject response information is maximized, especially important for studies involving rare conditions, terminally ill patients, and for studies terminated early.

In determining the random adaptive algorithm the clinical trial statistician considers the statistical methodology, or model statement, to be used for the primary efficacy variable. Prognostic factors are incorporated into the random adaptive algorithm in a parallel design to best support the planned statistical analyses. Theoretical frameworks are either frequentist or Bayesian.

The process of developing the random adaptive module of the automated randomization system is delineated including re-sequencing, re-randomization and reporting of randomization transactions. Implementation of an adaptive randomization is relatively easy so that efficiencies of design to achieve balances which support the statistical analysis and reporting accountability are well worth the implementation effort.

Interactive Clinical Technologies Incorporated (ICTI) has been implementing adaptive randomization designs in clinical trials since 1999 and has participated in over 75 adaptive randomization studies of varying complexity.

**Key words:** Clinical trial, randomization, random adaptive algorithm
**Mathematics Subject Classification:** 60K37

## 1. Introduction

Baseline adaptive randomization is a technique that is especially appropriate for clinical trials involving small sample sizes, rare conditions, medical devices, surgical procedures, and many strata or subgroups. The randomization must be structured to support the analysis, and stratification alone is of limited utility especially if many strata result in too few subjects per stratum [4] (p. 84) and [7]. Consideration is given to the hierarchy of balancing decisions and the weights of prognostic factors. Hierarchical ordering of prognostic factors and weights of factors may be derived from previous clinical trials or preliminary data.

Regulatory authorities in the US are beginning to consider the benefits of utilization of probabilistic Baseline Adaptive Randomization techniques because they may do better than simple randomization, especially for small trials[2]. Balancing on covariates tends to decrease variance of estimates of the covariates and improves efficiency of the treatment effect estimate.

A modified Pocock Simon design will be shown that includes some element of chance in every randomization decision. A method especially appropriate if blinding in not feasible, e.g. in a surgical trial or medical device trial is the Adaptive Biased Urn Randomization described by [5]. Here an application of the Adaptive Biased Urn Randomization for a rare medical condition will be shown.

## 2. The Development of the Random Adaptive Algorithm

The principal investigator and clinical trial statistician determine the primary efficacy endpoint, the appropriate study population and the analytical model. In determining the random adaptive algorithm to best support the statistical analysis, the clinical trial statistician considers the statistical methodology, or model statement, to be used for the primary efficacy variable. Prognostic factors are incorporated into the random adaptive algorithm in a parallel design to the planned statistical analyses. The randomization design must support the planned statistical analysis[1].

An adaptive randomization algorithm which follows either a classical or Bayesian theoretical framework is clearly delineated by the client statistician. At this time within the body of statistical knowledge there is not a set of rules governing how to determine the single definitive algorithm appropriate for a particular clinical trial. The client statistician develops the algorithm including what prognostic factors are important and how they are to be handled (possibly hierarchically or by weighting) based upon clinical judgment, previous studies, or historical values reported in clinical literature. Then simulation testing is begun both by the client statistician and ICTI. Simulation testing is used to:

- Address the robustness of the random adaptive algorithm to conditions not under the researcher's control, especially the order in which the subjects enroll [3]. The random adaptive algorithm must accommodate this variation. Re-sequencing is a technique for reordering seeded data in a simulation data set and checking the resultant balances while holding the random adaptive algorithm and random numbers constant.
- Verify the degree to which chance alone will alter the overall balance and the balance for strata or subgroups. This may be accomplished by using multiple randomization schedules against the same simulation data set. Some authors refer to this as re-randomization.
- Test the contributions of the prognostic factors, their hierarchical structure and/or their weighting and how these impact balance.
- Validate the implementation within the Interactive Voice Response or Electronic Data Capture system by verifying that randomization counts are

1451

being computed correctly, verifying that all factor boundary values are correct, that definitions stated in the algorithm are appropriately programmed, and that information from the randomization module about each subject is being appropriately captured.

The clinical trial statistician may decide that too much variation is occurring based upon re-randomization or re-sequencing results during black box testing. Areas for possible reconsideration are:

(1) If a boundary is placed on $|q_1 - q_2| \leq b$, where $q_1$ = the number of subjects assigned to treatment 1, $q2$ = the number of subjects assigned to treatment 2, and $b$ = boundary value, as the boundary value increases, more treatment assignments will be based upon later steps in the hierarchy.
(2) Hierarchical steps in the adaptive randomization algorithm may be changed.
(3) Prognostic factors may be added or eliminated from the adaptive randomization algorithm.
(4) If weights are used among prognostic factors, these may be altered.
(5) To assure that every treatment determination is based upon an element of randomization, weighted-coin decisions are stated with a maximum of 0.90 in any treatment arm; these weights may be adjusted.

The adaptive randomization algorithm is accepted by the client statistician and clinical team if it shows a high degree of convergence of balance within each level of each prognostic factor.

Randomization reports are designed to facilitate the decision-making process by showing resultant balances overall and for subgroups or strata. Also, individual listings of subjects randomized (in simulations and later in the live study) show the stratification factors of the subjects, the randomization counts upon which the treatment assignment was made, the random number, the probabilities, and the logical decision making rules (in footnotes).

The Randomization Logic (balanced coin or weighted coin and weighted in what direction), Randomization Logic Probability Outcome (statements showing the treatment assignment based upon the random number value) and Randomization Probabilities (decision in relation to the random number value) on the reports allow for a complete audit trail for each patient randomized and an accurate assessment of the robustness of the random adaptive algorithm.

After a clinical trial is completed (or stopped) it is possible to hypothetically re-sequence the subjects and run through the random adaptive algorithm used in the Interactive Voice Response or Electronic Data Capture System to verify the specificity of the overall treatment assignment balance among treatment groups, the balance within hierarchical steps in the random adaptive algorithm, and the balances for subgroups or strata.

## 3. Simple Case Using Two Stratifying Factors

Pocock and Simon, classical theoretical statisticians, developed a general procedure for treatment assignment which concentrates on minimizing imbalances in the distribution of subjects to treatment groups within the levels of each prognostic factor [3]. The method involves determining the amount of imbalance for each of the prognostic factors, hypothetically assigning the next subject to each treatment group and then assigning that subject to the treatment group which will minimize the treatment assignment imbalances. The original statement of the minimization method was deterministic in that random number values were only used in tie-breaking situations. Modified Pocock-Simon methodology is recommended here, in which every randomization decision is dependent on a random number and the maximum probability of treatment assignment possible for any specific treatment group assignment is .90. This is an example of a probabilistic baseline adaptive randomization.

This example of a Pocock and Simon's range method was designed to support a clinical study with two stratifying factors: Factor 1 (with four levels) and Factor 2 (with two levels). The study had three treatment groups: active drug at two dose levels (referred to as T1 and T2) and a comparator, T3. The sample size for this study is approximately 1000 subjects. There were more than 25 investigational sites, but site was not a stratifying factor. Sampling was without replacement.

Assignment of subjects to treatments in this study applied a weighted coin approach with a maximum probability component of 0.80. The treatment allocation resulting in the smaller overall measure of imbalance is performed with probability p=0.80. This allocation scheme focuses on achieving an approximate balance with respect to the predefined factors, while also including a random component that tends to balance arms generally with respect to any other prognostic factors.

A spreadsheet illustrating an individual treatment assignment is shown on Figure 1. The user would input subject identifiers, randomization counts corresponding to the subject's Factor levels, and the random number. The spreadsheet automatically computes the **D** or **Diff** pre-randomization and **Diff** for hypothetical assignment to each treatment group. The user then checks the appropriate logical decision making rule and assigns the treatment based upon the decision making rule with the associated random number. The spreadsheet serves to help users follow the logic within the protocol specific adaptive randomization algorithm, and may be used to check randomization assignments derived from the automated system which is written in TSQL and functions on an SQL platform.

## 4. Case Using the Adaptive Biased Urn Randomization in Small Strata When Blinding is Impossible

Urn randomization was introduced by Wei [8] [9], a Bayesian statistical theorist. In urn randomization the assignment of probabilities is adapted to the degree of imbalance in relation to the number of patients already entered into the trial. Schoeten [5] describes an even more flexible urn model than that described by Wei. Schoeten's model, called biased urn randomization, is especially useful because it is appropriate for small sample sizes and when blinding is not feasible.

Schoeten's method of biased urn randomization is randomization without replacement, where:

s = number of balls (chances) for each treatment groups at study start,
x = extra balls (chances) for each of the treatment group not selected after each randomization
g = number of treatment groups in the clinical trial

The integers s and x must be chosen such that there never can be a negative number of balls in the urn. Also, care must be taken in the selection of s and x such that it is not possible to predict the next patient treatment assignment, given previous treatment assignments.

In our example, for three treatment groups, $g = 3$ with $s = 0$ and $x = 1$. Schoeten's formula of

$$p_i = \frac{\text{the number of balls of one treatment type}}{\text{the total number of balls in the urn after N subjects have been randomized}}$$

$$= \frac{s + (N - n_i)x - n_i}{gs + N(g-1)x - N}$$

reduces to:

$$\frac{N - 2n_i}{N}$$

in the example that we have implemented.

A crucial point is that the total number of balls in the urn after N drawings is independent of the balls that have been drawn so unpredictability is maintained.

Our real-world example clinical-trial was for a rare medical condition; the clinical trial had three treatment groups representing different dose levels of the active study drug. Given the rare medical condition, the total anticipated sample size was small (under 50 subjects). The design called for one continuous factor (Factor 1) with two levels to be considered. The design is particularly interesting because it required attenuation of a second continuous factor (Factor 2) for clinical considerations: only treatments A and B were administered to patients in the highest level of Factor 2, and only treatments B and C were administered to patients in the lowest level of Factor 2. Balancing was not required for Factor 2; it was used to accomplish the desired clinical dose regulated attenuation. Balance within this randomization design was achieved overall and for Factor 1.

The algorithm is stated as follows:

Let s = 0 = the number of chances to get each treatment at study start (balls in the urn for each treatment)

Let x = 1 = the number of extra chances added to the urn after each randomization for each treatment type not selected

| If the number of patients in the study within the same Factor 1 stratum as the patient to be randomized > 0, Find the probability of assigning Treatment A: | Find the probability of assigning Treatment B where there are patients within the same Factor 1 stratum. | Find the probability of assigning Treatment C where there are patients within the same Factor 1 stratum. |
|---|---|---|
| A = Number of treatments minus 1 | A = Number of treatments minus 1 | A = Number of treatments minus 1 |
| B = A multiplied by x | B = A multiplied by x | B = A multiplied by x |
| C = B minus 1 | C = B minus 1 | C = B minus 1 |
| D = Number of treatments multiplied by s | D = Number of treatments multiplied by s | D = Number of treatments multiplied by s |
| E = D divided by the number of patients in the study within the same Factor 1 stratum as the patient to be randomized | E = D divided by the number of patients in the study within the same Factor 1 stratum as the patient to be randomized | E = D divided by the number of patients in the study within the same Factor 1 stratum as the patient to be randomized |
| F = C plus E | F = C plus E | F = C plus E |
| G = x plus 1 | G = x plus 1 | G = x plus 1 |
| H = G divided by F | H = G divided by F | H = G divided by F |
| I = Number of patients assigned to Treatment A in the study (within the same Factor 1 stratum as the patient to be randomized) divided by the number of patients in the study (within the same Factor 1 stratum as the patient to be randomized) | I = Number of patients assigned to Treatment B in the study (within the same Factor 1 stratum as the patient to be randomized) divided by the number of patients in the study (within the same Factor 1 stratum as the patient to be randomized) | I = Number of patients assigned to Treatment C in the study (within the same Factor 1 stratum as the patient to be randomized) divided by the number of patients in the study (within the same Factor 1 stratum as the patient to be randomized) |
| J = 1 divided by number of treatments | J = 1 divided by number of treatments | J = 1 divided by number of treatments |
| K = J minus I | K = J minus I | K = J minus I |
| L = K multiplied by H | L = K multiplied by H | L = K multiplied by H |
| $P_a$ = J plus L | $P_a$ = J plus L | $P_a$ = J plus L |
| If $P_a$ < 0, set $P_a$ = 0.1 | If $P_a$ < 0, set $P_a$ = 0.1 | If $P_a$ < 0, set $P_a$ = 0.1 |
| $P_a$ = probability of assigning treatment A | $P_a$ = probability of assigning treatment A | $P_a$ = probability of assigning treatment A |

To perform appropriate clinical attenuation of treatment groups:
After determining $P_a$, $P_b$ and $P_c$ then:
If Factor 2 stratum is low then determine $P_b'$ and $P_a'$

$$P_b' = P_b + (P_b * (P_a/(P_b + P_c))), \quad P_a' = 0.$$

If Factor 2 stratum is high then determine $P_a'$ and $P_c'$

$$P_a' = P_a + (P_a * (P_c/(P_a + P_b))), \quad P_c' = 0.$$

To standardize the final notation before programming:
Whether Factor 2 stratum is low, middle, or high:

$$P_a'' = (P_a/(P_a + P_b + P_c)),$$

$$P_b'' = (P_b/(P_a + P_b + P_c)),$$

$$P_c'' = (P_c/(P_a + P_b + P_c)),$$

$$(P_a'' + P_b'' + P_c'') = 1.$$

An excerpt from a spreadsheet illustrating an individual treatment assignment is shown on Figure 2. As the user inputs the randomization counts for subjects already in the study (or simulation) with the same Factor 1 level, the spreadsheet automatically computes the probability boundaries: $P_a^{''}, P_b^{''}, and P_c^{''}$ for the three levels of Factor 2 The user then checks the appropriate logical decision making rule and assigns the treatment based upon the decision making rule with the associated random number.

A summary report for a simulation using this random adaptive algorithm is shown in Figure 3. The example reflects 36 randomization transactions and balance achieved for the study overall and for each level of Factor 1. Factor 2 is a controlling factor related to clinical requirements for assignment of clinically appropriate dose groups (Treatment arms A and B for Factor 2 High level, all treatment arms for Factor 2 middle level and Treatment arms B and C for Factor 2 Low level). Subjects will not necessarily present with balanced levels of each factor. In this example, Factor 1 is balanced with 18 subjects in each of the two strata. For Factor 2: Low level: 1 patient, Middle level: 27 patients, and High level: 8 patients. If an individual clinical trial requires certain numbers of patients per baseline factor level, the system can utilize ceilings or caps to stop enrollment of patients with certain baseline characteristics beyond the sample frame caps.

An excerpt from a randomization report listing each randomization is in Figure 4. This report would serve as an audit trail of each randomization decision in a live clinical trial since it is possible to trace the randomization decision given the subject ID, factor values and levels, random number, randomization counts, Pa, Pb, and Pc boundary values for each treatment group. The logical decision making probability statements which are random adaptive algorithm specific are in the footnote of each page of the report.

## 5. Conclusions

Baseline adaptive randomization approaches are flexible, easy to customize for specific clinical trial statistical designs and easy to implement within Interactive Voice Response or Electronic Data Capture systems. They are especially valuable in clinical trials of small and moderate sample size and studies in which subgroup analyses are planned. It is very important to do thorough simulation testing on a random adaptive algorithm before implementation because there is no one definitive algorithm for each clinical trial protocol. A random adaptive algorithm for which simulation testing resulted in a high degree of convergence regardless of the random number schedule, the order of patients enrolling in the study, or the background characteristics of the simulated subjects can be used with confidence. The added effort to perform baseline adaptive randomization is frequently small in comparison to other efforts involved in designing and administering a clinical trial and the design

| Subject ID: | 45051 | | Assigned Treatment: | | | | T1 |
|---|---|---|---|---|---|---|---|
| Site Number: | Client sim | | Next sequential number from rand list: 18 | | | | Random Number: 0.1092 |
| | | | | | | | Max.Diff |

| | T1 | T2 | T3 | \|T1-T2\| | \|T1-T3\| | \|T2-T3\| | Max.Diff |
|---|---|---|---|---|---|---|---|
| Factor $1_{(i)}$ | 0 | 0 | 1 | 0 | 1 | 1 | 1 |
| Factor $2_{(i)}$ | 3 | 3 | 3 | 0 | 0 | 0 | 0 |
| Total | | | | | $Diff_{pre-rand}$ | | 1 |

Hypothetical values if subject is assigned to treatment arm T1:

| | T1 | T2 | T3 | \|T1-T2\| | \|T1-T3\| | \|T2-T3\| | Max.Diff |
|---|---|---|---|---|---|---|---|
| Factor $1_{(i)}$ | 1 | 0 | 1 | 1 | 0 | 1 | 1 |
| Factor $2_{(i)}$ | 4 | 3 | 3 | 1 | 1 | 0 | 1 |
| Total, | | | | | $Diff_{T1}$ | | 2 |

Hypothetical values if subject is assigned to treatment arm T2:

| | T1 | T2 | T3 | \|T1-T2\| | \|T1-T3\| | \|T2-T3\| | Max.Diff |
|---|---|---|---|---|---|---|---|
| Factor $1_{(i)}$ | 0 | 1 | 1 | 1 | 1 | 0 | 1 |
| Factor $2_{(i)}$ | 3 | 4 | 3 | 1 | 0 | 1 | 1 |
| Total | | | | | $Diff_{T2}$ | | 2 |

Hypothetical values if subject is assigned to treatment arm T3:

| | T1 | T2 | T3 | \|T1-T2\| | \|T1-T3\| | \|T2-T3\| | Max.Diff |
|---|---|---|---|---|---|---|---|
| Factor $1_{(i)}$ | 0 | 0 | 2 | 0 | 2 | 2 | 2 |
| Factor $2_{(i)}$ | 3 | 3 | 4 | 0 | 1 | 1 | 1 |
| Total | | | | | $Diff_{T3}$ | | 3 |

Fig. 1. Pocock Range Method for Two Factors, Three treatment groups: T1, T2, and T3

efficiencies gained.

| Subject ID: | 111112049 | | |
| Site Number: | 22 | | |

s= 0
x= 1

**Treatment**

| | A | B | C |
|---|---|---|---|
| Factor 2: Low level | | | |
| Factor 1(i) | | 1 | 8 |
| Factor 2: Middle level | | | |
| Factor 1(i) | 9 | 6 | 7 |
| Factor 2: High level | | | |
| Factor 1(i) | 9 | 8 | |

**Treatment**

| | A | B | C |
|---|---|---|---|
| Factor 2: Middle level | | | |
| A= | 2 | 2 | 2 |
| B= | 2 | 2 | 2 |
| C= | 1 | 1 | 1 |
| D= | 0 | 0 | 0 |
| E= | 0 | 0 | 0 |
| F= | 1 | 1 | 1 |
| G= | 2 | 2 | 2 |
| H= | 2 | 2 | 2 |
| I= | 0.375 | 0.3125 | 0.3125 |
| J= | 0.3333 | 0.3333 | 0.3333 |
| K= | -0.0417 | 0.0208 | 0.0208 |
| L= | -0.0833 | 0.0417 | 0.0417 |
| $P_{(T)}$ = | 0.2500 | 0.3750 | 0.3750 |
| neg. corrected $P_{(T)}$ = | 0.25 | 0.375 | 0.375 |
| $P''_{(T)}$ = | 0.25 | 0.375 | 0.375 |

Random Number: 0.3789     Assigned Treatment B

Logical Decision Rules for Factor 2 Middle Level

| | |
|---|---|
| $0.0001 \leq$ random number $< 0.2500$ | A |
| $0.2500 \leq$ random number $< 0.6250$ | B |
| $0.6250 \leq$ random number $< 1.00000$ | C |

Fig. 2.   Attenuated Example with Two Factors, Factor 1 and Factor 2 Three Treatment Groups Representing Different Levels of Active Drug

**RandCounts Simulation**

**Summary of Counts by Stratification Factor**

Factor 1: Stratum 1

| Treatment Arm | Overall | | Factor 2 | |
|---|---|---|---|---|
| | | Low | Middle | High |
| A | 6 | 0 | 5 | 1 |
| B | 6 | 0 | 4 | 2 |
| C | 6 | 1 | 5 | 0 |
| Total Stratum 1 | 18 | 1 | 14 | 3 |

Factor 1: Stratum 2

| Treatment Arm | Overall | Factor 2 | | |
|---|---|---|---|---|
| | | Low | Middle | High |
| A | 6 | 0 | 3 | 3 |
| B | 7 | 0 | 5 | 2 |
| C | 5 | 0 | 5 | 0 |
| Total Stratum 2 | 18 | 0 | 13 | 5 |

| Total A | Total B | Total C | Grand Total |
|---|---|---|---|
| 12 | 13 | 11 | 36 |

Fig. 3. Adaptive Randomization Using Urn Method Attenuated Example with Two Factors, Factor 1 and Factor 2 Three Treatment Groups Represent Different Levels of Active Drug

**Summary of Simulation Number: 2**

| | | | | | | | Count Values for Subject's Factor 1 Stratum Before Rand | | | | | | | | | |
|---|---|---|---|---|---|---|---|---|---|---|---|---|---|---|---|---|
| Subj. ID | Tx | Random Number | Factor 1 (value) | Factor 1 Stratum | Factor 2 (value) | Factor 2 Stratum | TxArm A middle | TxArm A high | TxArm B low | TxArm B mid. | TxArm B high | TxArm C low | TxArm C middle | $P''_a$ | $P''_b$ | $P''_c$ |
| 1040 | C | 0.9462 | 6 | 1 | 47 | middle | 12 | 1 | 0 | 10 | 2 | 4 | 10 | ≤0.3333 | >0.3333 & ≤0.7149 | >0.7149 |
| 2061 | B | 0.6368 | 10 | 2 | 62 | high | 12 | 11 | 0 | 14 | 7 | 1 | 15 | ≤0.4375 | >0.4375 | 0 |
| 2062 | A | 0.1063 | 14 | 2 | 56 | middle | 12 | 11 | 0 | 14 | 8 | 1 | 15 | ≤0.2459 | >0.2459 & ≤0.5246 | >0.5246 |

| Randomization-Logic-Probability-Tx-Assignment Key | | | |
|---|---|---|---|
| Factor 2 | Treatment A | Treatment B | Treatment C |
| Low | $P''_a = 0$ | Random number ≤ $P''_b$ | Random number > $P''_b$ |
| Middle | Random number ≤ $P''_a$ | If Random number > $P''_a$ and Random number ≤ ($P''_a$+$P''_b$) | Random number > ($P''_a$+$P''_b$) |
| High | Random number < $P''_a$ | Random number > $P''_a$ | $P''_c = 0$ |

Fig. 4. Adaptive Randomization Using Urn Method Attenuated Example with Two Factors, Factor 1 and Factor 2 Three Treatment Groups Represent Different Levels of Active Drug

# References

1. Fisher, R.A. (1956). Statistical Methods, Experimental Design and Scientific Inference, J.H. Bennet, ed. (Oxford University, Oxford).
2. Hung, J. (2005). A regulatory perspective on adaptive randomization. Presented: 41st Annual DIA Annual Meeting, Washington DC, June 27, 2005.
3. Pocock, S.J. and Simon, R.(1975). Sequential treatment assignment with balancing for prognostic factors in the controlled clinical trial, Biometrics 31, 103-115.
4. Pocock, S.J. (1983). Clinical Trials, A Practical Approach, Chapter 5. Chichester: Wiley.
5. Schouten, H.J.A. (1995). Adaptive biased urn randomization in small strata when-blinding is impossible, Biometrics 51, 1529-1535.
6. Rosenberger and Lachin (2002). Randomization in Clinical Trials.Wiley.
7. Therneau, T. (1993). How many stratification factors are too many to use in a randomization plan? Controlled Clinical Trials 14: 98-108.
8. Wei, L.J. (1977). A class of designs for sequential clinical trials. Journal of the American Statistical Association 72: 382-386.
9. Wei, L.J. (1978). An application of an urn model to the design of sequential controlled clinical trials. Journal of the American Statistical Association 73: 559-563.
10. Wei, L.J. (1988). Properties of the urn randomization in clinical trials. Controlled Clinical Trials 9: 345-364.

# List of Session Organizers

| | | | |
|---|---|---|---|
| Alpay, D. | 123 | Reissig, M. | 481 |
| | | Ricceri, B. | 803 |
| Barsegian, G. | 1321 | Rogosin, S.V. | 1241 |
| Begehr, H. | 1107 | Ryan, J. | 997 |
| Berlinet, A. | 123 | | |
| Burenkov, V.I. | 67 | Sabadini, I. | 997 |
| | | Samko, St. | 91 |
| Dai, D.-Q. | 1107 | Saitoh, S. | 123, 189 |
| | | Soldatov, A. | 1107 |
| Gilbert, R.P. | 1391 | | |
| Grudski, S. | 325 | Tamrazov, P. | 929 |
| | | Toft, J. | 421 |
| Hochmuth, R. | 397 | Tuan, V. | 189 |
| Holschneider, M. | 397 | | |
| Hu, P.C. | 1293 | Vasilevski, N. | 325 |
| | | | |
| Jacob, N. | 439 | Wirgin, A. | 1391 |
| | | Wirth, J. | 481 |
| Kilbas, A. | 189 | Wong, M.W. | 421 |
| Klibanov, M. | 1361 | | |
| Krutitskii, P. | 615 | Xiao, Y. | 439 |
| | | Xu, Y. | 1391 |
| Lanza de Cristoforis, M. | 929 | | |
| Li, P. | 1293 | Yamamoto, M. | 1361 |
| | | Yang, C.C. | 1293 |
| Majorana, A. | 1327 | | |
| Mityushev, V.V. | 1241 | Zayed, A.I. | 189 |
| | | | |
| Nicolosi, F. | 657 | | |

# List of Authors

| | | | | |
|---|---|---|---|---|
| Abdymanapov, S.A. | 1137 | Colombini, F. | | 591 |
| Akhmetov, D.R. | 737 | Colombo, F. | | 1019 |
| Alexandrov, O. | 627 | Cordaro. G. | | 853 |
| Aliyev Azeroglu, T. | 945 | Cordero, E. | | 415 |
| Andriulli, F.P. | 407 | Csordas, G. | | 1295 |
| Anello, G. | 815 | | | |
| Anile, A.M. | 13 | Damiano, A. | | 1019 |
| Antontsev, S.N. | 681 | D'Asero, S. | | 771 |
| Aoyagi, M. | 259, 279 | De Mari, F. | | 415 |
| Ayache, A. | 441 | Di Bella, B. | 823, 843, | 873 |
| | | Di Falco, A.G. | | 833 |
| Bartolo, R. | 763 | Diening, L. | | 101 |
| Bauer, W. | 327 | Domaingo, A. | | 1329 |
| Begehr, H. | xv, 1137 | Du, J.Y. | | 1109 |
| Bellassoued, M. | 1363 | Du, Z.H. | | 1109 |
| Benaissa, A. | 561 | | | |
| Berglez, P. | 1119 | Eisner, J. | | 691 |
| Berlinet, A. | 153 | Eriksson, S.-L. | | 1051 |
| Besov, O.V. | 55 | Estrade, A. | | 441 |
| Biroli, M. | 711 | | | |
| Bonafede, S. | 747 | Fan, X.-L. | | 93 |
| Bonami, A. | 441 | Faraci, F. | | 889 |
| Boscain, U. | 701 | Fellah, Z.E.A. | 1407, | 1419 |
| Brüning, E. | 901 | Fiedler, T. | | 1259 |
| Bučhovska, A. | 399 | Fiores, J.L. | | 763 |
| Buescu, J. | 175 | Frezzotti, A. | | 1339 |
| Burenkov, V.I. | xi | Fujita, K. | | 125 |
| | | Fujiwara, D. | | 429 |
| Cammaroto, F. | 823, 843, 873 | Futamura, T. | | 107 |
| Candela, A.M. | 753, 763 | | | |
| Candito, P. | 921 | Gilbert, R.P. | | 1393 |
| Canu, St. | 163 | Girardie, M. | | 721 |
| Castro, L.P. | 349 | Gorenflo, R. | | 451 |
| Cataldo, V. | 771 | Gourdin, D. | | 543 |
| Čelechovská-Kozáková, L. | 1439 | Grácio, J. | | 1259 |
| Cerejeiras, P. | 1043 | Groby, J.-P. | 1407, | 1419 |
| Chang, K.S. | 221 | | | |
| Chen, S. | 39 | Harjulehto, P. | | 107 |
| Chinnì, A. | 823, 843, 873 | Harutyunyan, G. | | 1137 |
| Cho, D.H. | 211 | Hasanov, A. | | 521 |
| Cianci, P. | 785 | Häströ, P. | 101, | 107 |
| Ciraolo, G. | 627 | Hihnala, M. | | 637 |
| | | Hsiao, G.C. | | 23 |

| | | |
|---|---|---|
| Ichinose, W. | 423 | |
| Isakov, V. | 3 | |
| Itou, H. | 135 | |
| | | |
| Jerbashian, A.M. | 289 | |
| | | |
| Kamoun, H. | 543 | |
| Karlovich, Yu.I. | 339 | |
| Katayama, S. | 581 | |
| Katô, K. | 1303 | |
| Kempfle, S. | 301 | |
| Khalifa, O.B. | 543 | |
| Kilbas, A.A. | 201, 311 | |
| Kim, B.S. | 221 | |
| Koroleva, A.A. | 201 | |
| Kosaka, K. | 967 | |
| Krasnov, Y. | 1095, 1229 | |
| Kristály, A. | 805 | |
| Krüger, K. | 301 | |
| Krushkal, S. | 931 | |
| Krutitskii, P.A. | 647 | |
| Kryvko, A. | 483, 511 | |
| Kubo, H. | 581 | |
| Kučera, M. | 617, 691 | |
| Kucherenko, V.V. | 483, 511 | |
| Kumano-Go, N. | 429 | |
| Kumar, A. | 1149 | |
| | | |
| Lanza de Cristoforis, M. | xiii, 955 | |
| Lavrentiev, M.M. | 737 | |
| Lawrynowicz, J. | 987, 1085 | |
| Lehman, E. | 1075 | |
| Liu, M.-S. | 1311 | |
| Livrea, R. | 881 | |
| Lollini, P.-L. | 1351 | |
| Luna-Elizarrarás, M.E. | 999 | |
| Lytvynov, E. | 467 | |
| | | |
| Mainardi, F. | 451 | |
| Majorana, A. | 1329 | |
| Makhmudov, O.I. | 69 | |
| Malinnikova, E. | 1185 | |
| Marchiafava, St. | 987 | |
| Martin, M. | 1065 | |

| | | |
|---|---|---|
| Mary, S. | 163 | |
| Mastriani, E. | 1351 | |
| Mataloni, S. | 721 | |
| Matsumoto, W. | 863 | |
| Matsushima, T. | 1323 | |
| Matsuura, T. | 191, 1375 | |
| Matsuyama, T. | 553 | |
| Matzeu, M. | 721 | |
| Miller, E.R. | 1449 | |
| Mityushev, V.V. | 1243 | |
| Miyatake, S. | 603 | |
| Mizuta, Y. | 107 | |
| Mochizuki, K. | 533 | |
| Motta, S. | 1351 | |
| | | |
| Nicolosi, F. | 747, 771, 785 | |
| Niyozov, I.E. | 69 | |
| Nolasco, A.P. | 349 | |
| Nôno, K. | 1085 | |
| Nowak-Kępczyk, M. | 987 | |
| | | |
| Öchsner, A. | 1259 | |
| Ogam, E. | 1407, 1419 | |
| Ong, Ch.S. | 163 | |
| | | |
| Paixão, A.C. | 175 | |
| Panayotounakos, D.E. | 1219 | |
| Pappalardo, F. | 1351 | |
| Perotti, A. | 1009 | |
| Pesetskaya, E.V. | 1259 | |
| Pessoa, L.V. | 339 | |
| Plaksa, S. | 977 | |
| Polidoro, S. | 701, 729 | |
| Prakash, R. | 1149 | |
| | | |
| Rabinovich, V. | 359 | |
| Ragusa, M.A. | 729 | |
| Rajabov, N. | 1207 | |
| Rappoport, J.M. | 269 | |
| Recke, L. | 691 | |
| Ricceri, B. | 897 | |
| Rivero, M. | 311 | |
| Rogosin, S.V. | 1243, 1247, 1259 | |
| Romano, V. | 13 | |

| | | | |
|---|---|---|---|
| Sabadini, I. | 1019 | Tungatarov, A.B. | 1137 |
| Saitoh, S. | 135, 191, 1375 | Uesaka, H. | 571 |
| Saks, R.S. | 1195 | | |
| Salakhitdinov, M.S. | 521 | Vaillant, J. | 493 |
| Salvatore, A. | 753 | Vecchi, G. | 407 |
| Samko, N. | 369 | Vivaldi, M.A. | 659 |
| Saneva, K. | 399 | Vivoli, A. | 451 |
| Schäfer, I. | 301 | | |
| Seikkala, S. | 637 | Wang, J. | 251 |
| Seung, B.I. | 211 | Wang, Y.F. | 1127 |
| Shapiro, M. | 999 | Watanabe, S. | 241, 279 |
| Shimomura, T. | 107 | Wendland, W.L. | 23 |
| Shmarev, S.I. | 681 | Wirgin, A. | 1407, 1419 |
| Soldatov, A. P. | 1171 | Wojnar, R. | 1271 |
| Sommen, F. | 1043 | | |
| Song, T.S. | 221 | Xu, Y. | 1393, 1429 |
| Spigler, R. | 737 | | |
| Struppa, D.C. | 1019 | Yakubovich, S.B. | 231 |
| Suquet, Ch. | 143 | Yamamoto, M. | 1363, 1375, 1385 |
| Suzuki, O. | 967, 1085 | Yang, C.C. | 1295 |
| | | Yasutomi, M. | 1283 |
| Tabacco, A. | 407, 415 | Yoo, I. | 221 |
| Taglialatella, G. | 591 | | |
| Tamrazov, P.M. | 945 | Zabroda, O.N. | 377 |
| Tarsia, A. | 669 | Zeev, N. | 1393 |
| Tchou, N.A. | 711 | Zhang, N. | 1393 |
| Trenogin, V.A. | 83 | Zhang, Z.X. | 1033 |
| Trombetta, G. | 775 | Zhura, N.A. | 1163 |
| Trujillo, J.J. | 311 | Zorboska, N. | 387 |
| Tsogka, C. | 1407 | Zou, J. | 1385 |
| Tsuji, K. | 259 | Žubrinić, D. | 793 |
| | | Zur, S. | 1229 |

Sabadini, I. 1019
Saitoh, S. 185, 191, 1375
Saka, H.S. 1195
Sakhliddinov, M.S. 521
Salvatore, A. 763
Sanko, N. 309
Saneva, K. 383
Schafer, I. 301
Seikkala, S. 637
Seung, B.L. 711
Shapiro, M. 659
Shimomura, T. 107
Shmarev, S.I. 681
Soldatov, A. P. 1171
Sombrea, F. 1013
Song, T.S. 231
Spigler, R. 737
Struppa, D.C. 1019
Suquet, Ch. 147
Suzuki, O. 987, 1055
Tabacco, A. 107, 115
Tagliafiello, G. 291
Tamrazov, P.M. 915
Tarata, A. 309
Tatton, N.A. 711
Trenogin, V.A. 81
Trombetta, G. 779
Trujillo, J.J. 311
Tsogka, C. 1407
Tsuji, R. 260

Tangatarov, A.B. 1137
Uesaka, H. 571
Vaillant, J. 493
Vecchi, G. 107
Vivaldi, M.A. 659
Vivoli, A. 451
Wang, J. 251
Wang, Y.F. 1137
Watanabe, S. 241, 979
Wendland, W.L. 28
Wrgin, A. 1407, 1419
Wojnar, B. 1271
Xu, Y. 1393, 1429
Yakubovich, S.B 251
Yamamoto, M 1363, 1375, 1385
Yang, C.C. 1293
Yasutomi, M. 1283
Yoo, J. 231
Zabreiko, O.N 377
Zeev, N. 1393
Zhang, N. 1393
Zhang, Z.X. 1053
Zhura, N.A. 1163
Zobaska, N 387
Zou, J. 1385
Zabinnie, D 793
Zin, S. 1920